Electronics Engineer's Reference Book

Fifth Edition

Electronics Engineer's Reference Book

Fifth Edition

Edited by
F F Mazda
DFH, MPhil, CEng, MIEE, DMS, MBIM

With specialist contributors

Butterworths
London · Boston · Durban · Singapore
Sydney · Toronto · Wellington

First published 1983

British Library Cataloguing in Publication Data

Electronics engineer's reference book—5th ed.
1. Electronics
I. Mazda, F.
621.381 TK7870

ISBN 0-408-00589-0

Typeset by Mid-County Press Ltd. 2a Merivale Road, London SW15 2NW
Printed and bound by Robert Hartnoll Ltd., Bodmin, Cornwall

Preface

Many new developments have occurred, in Electronics, since the publication of the 4th edition of the *Electronics Engineer's Reference Book*. Every editor also has his own idea on what a reference book should contain. On taking over the editorship of the 5th edition of the *Electronics Engineer's Reference Book*, I have therefore extensively rearranged and revised the book.

The 5th edition has been set by modern methods to enable rapid corrections, and so minimise the delays, which can so often occur, between the time the manuscripts are written and the book is published. The page dimensions have also been increased, to make the book of a more conventional size.

In selecting the contents of the book I have placed emphasis on the latest technologies, and on maintaining a balance between the various aspects of electronic engineering. Many chapters from the 4th edition, on older technologies such as valves, have been eliminated or reduced in size. Some of the chapters on broadcasting have also been reduced. Several application chapters from the 4th edition have been deleted, and some of the more relevant material which they contained, on components or design techniques, have been distributed into other chapters, so that they can be more readily accessed.

Thirty-two new chapters have been added to the 5th edition of the Reference Book. Many of these are on relatively new topics such as semiconductor memories, microprocessors, fibre-optic communications and office communication. Other chapters, such as on attenuators, filters and telephone systems, have been added to give a wider, and more balanced coverage of topics which are useful to electronics engineers.

The 5th edition has been arranged into five parts. The first part contains details on mathematical and electrical techniques, which are required in design analysis, but have often been forgotten, from college days, by the practising engineer.

Part two contains information on physical phenomena, such as electricity, light and radiation. The third part, along with the fourth, represents the core of the book. It contains chapters on basic electronic components and materials, which are the building blocks of any electronic system. The information presented covers a wide spectrum of components, from the humble resistor to the glamorous memory and microprocessor.

Part four has chapters on electronic circuit design and on instrumentation. A range of design techniques are covered from linear to digital circuits, and from signal power levels to those in the megawatt region. The fifth part of the book contains chapters on topics, such as radar and computers, which form well recognised application areas of electronics.

I have greatly enjoyed planning and editing the 5th edition of the *Electronics Engineer's Reference Book*. I am grateful to Neil Warnock-Smith of Butterworths for his support and advice during the preparation of the book, and to the sub-editors who have worked tirelessly, in the background, on the manuscripts and proofs. In particular I am grateful to the many, many authors for producing first-class technical contributions, and for working to a very tight publication schedule. I can honestly say that without them this book would not have been written!

FFM
Sawbridgeworth
March 1983

Contents

Digital transmission . Basic digital techniques . Time division multiplexing . Higher-order multiplexing . Digital transmission . Coding of other types of information . Submarine telecommunications systems . Submarine telephone cables . Performance of cable . Submerged repeaters . Submerged equalisers . Terminal power feed . Termination translation apparatus . Recent developments . Future development

57 Fibre-optic Communication
Introduction . Practical optical fibres . Optical-fibre characteristics . Light sources . Photodetectors . Device packages . Coupling optical fibres . System design considerations . Applications

58 Videotape Recording
Introduction . Tape transport . Quadruplex machines . Helical machines . Modulation . Colour under . Sound tracks . Control track . Record . Reply . Time base correction . Drop out compensation . Velocity errors . Capstan and head drum servos . Tape editing . Domestic systems . Slow and stop motion . Composite recorders . Digital recording

59 Radar Systems
Introduction . Primary radar . Secondary surveillance radar (SSR)

60 Computers and Their Application
Introduction . Types of computers . Generations of digital computers . Digital computer systems . Central processor unit . Memory . Instruction set . CPU enhancements . Input/output . Peripherals . Terminals . High-speed printers and plotters . Direct input . Disc storage . Magnetic tape . Digital and analogue input/output . Data communications . Software

61 Office Communications
Basic principles of xerography . Extension to a dynamic copier . Xerographic materials . Raster scanning . Image input scanners . Output printers . Facsimilie . Word processors . Local area networks

62 Medical Electronics
Introduction . Diagnosis . Electronic instruments . Safety aspects . Lasers . New developments in imaging

Acknowledgements

The production of this reference book would have been impossible without the good will, help and cooperation of the electronics industry, the users of electronic equipment and members of the educational profession. Bare acknowledgements are very inadequate but the editor wishes to thank the following firms and organisations which so readily made available information and illustrations and permitted members of their specialist staffs to write contributions:

AMP of Great Britain Ltd
British Aircraft Corporation
British Broadcasting Corporation
British Railways Board, Research and Development Division
British Telecom International
Cambridge Scientific Instruments Ltd
Clinical Research Centre
Digital Equipment Co. Ltd
Duracell Batteries Ltd
English Electric Valve Company Ltd
Ever Ready (Special Batteries) Ltd
GEC-Marconi Electronics Ltd
Gould Advance Ltd
Greater London Council, Department of Planning & Transportation
Heriot-Watt University, Edinburgh
Hewlett-Packard Ltd
International Rectifier Ltd
ITT Components Group
Lucas Electrical Company Ltd
Marconi Company Ltd
Meteorological Office
Mostek Corporation
Motorola Ltd
Mullard Ltd
North East Thames Regional Health Authority
N.V. Philips Gloeilampenfabrieken
Plessey Company Ltd
Polytechnic of North London
Post Office
Rank Xerox Ltd
RCA
Rediffusion Ltd
Rutherford and Appleton Laboratories, Science Research Council
SAFT (UK) Ltd
SEEQ Corporation
Standard Telephones and Cables Ltd
Texas Instruments Incorporated
University of Aston
University of Cambridge
University of Leeds
University of Manchester Institute of Science and Technology
University of Oxford
University of Sussex
University of Technology, Loughborough
Xerox Corporation

Acknowledgement is made to the Director of the International Radio Consultative Committee (C.C.I.R.) for permission to use information and reproduce diagrams and curves from the C.C.I.R. Documents.

Extracts from British Standard publications are reproduced by permission of the British Standards Institution.

List of Contributors

S Amos, BSc(Hons), CEng, MIEE
Freelance Technical Editor and Author

J Barron, BA, MA(Cantab)
Cambridge University

G L Bibby, BSc, CEng, MIEE
University of Leeds

W Bowry
Ever Ready (Special Batteries) Ltd

P A Bradley, BSc, MSc, CEng, MIEE
Rutherford and Appleton Science and Engineering Research Council

D Bulgin, PhD
Cambridge Instruments Ltd

M Burchall, CEng, FIERE
CELAB Ltd

J Camarata
Electronics Division, Xerox Corporation

P M Chalmers, BEng, AMIEE
English Electric Valve Co. Ltd

A Clark, DipEE, CEng, MIEE
Electrical Engineering Consultant

G T Clayworth, BSc
High Power Klystron Department, English Electric Valve Co. Ltd

H W Cole
Marconi Radar Systems Ltd

A P O Collis, BA, CEng, MIEE
Broadcast Tubes Division, English Electric Valve Co. Ltd

P Cottam, BSc
STC Components

J A Dawson, CChem, MRSC
Rank Xerox Ltd

J P Duggan
Mullard Ltd

G W A Dummer, MBE, CEng, FIEE, FIEEE, FIERE
Electronics Consultant

Duracell Batteries Ltd

J L Eaton, BSc, MIEE, CEng
Consultant (Broadcasting)

M D Edwards, BSc, MSc, PhD
UMIST

S M Edwardson, CEng, FIEE
BBC Research Department

M Ewing
SAFT (UK) Ltd

M E Fabian, BSc, DMS
ITT Components Group UK

R P Gabriel, BSc, CEng, FIEE, MIERE, FIEEE
Rediffusion Ltd

C L S Gilford, MSc, PhD, FInstP, MIEE, FIOA
Independent Acoustic Consultant

I Graham, BSc, CEng, MIEE
Hewlett-Packard Ltd

V J Green, BSc
Standard Telephones and Cables PLC

J E Harry, BSc(Eng), PhD
Loughborough University of Technology

P Hawker
Independent Broadcasting Authority

D R Heath, BSc, PhD
Rank Xerox Ltd

E W Herold, BSc, MSc, DSc, FIEE

D W Hill, MSc, PhD, FInstP, FIEE
North East Thames Regional Health Authority

P J Howard, BSc, CEng, MIEE
Transmission Products Division, Standard Telephones and Cables PLC

A K Jefferis, BSc, CEng, FIEE
Satellite Systems Division, British Telecom International

F Jones, BSEE
MOSTEK Corporation

C Kindell
Amp of Great Britain Ltd

T Kingham
Amp of Great Britain Ltd

P R Knott, MA, MInst, AI Ceram
Microwave Division, Marconi Electronic Devices Ltd

J A Lane, DSc, CEng, FIEE
Consultant-Directorate of Radio Technology, Home Office

R F G Linford, BSc, CEng, MIEE
Rank Xerox Ltd

R Lomax, MSc, CEng, MIEE, MInstP
Advanced Engineering Development, ITT Relay Division

S Lowe
British Broadcasting Corporation

P G Lund
Oxford University

R C Marshall, MA, CEng, FIEE
Rank Xerox Ltd

F Mazda, DFH, MPhil, CEng, MIEE, DMS, MBIM
Rank Xerox Ltd

D Meyerhofer, BEng, PhD
David Sarnoff Research Center, RCA

A D Monk, MA, CEng, MIEE
Marconi Research Centre, GEC Research Laboratories

R Mosedale
MITEL Corporation

P Musson, BA
Digital Equipment Co. Ltd

T Oswald, BSc, MIEE
Submarines Systems Division, Standard Telephones and Cables PLC

H J M Otten
N V Philips Gloeilampenfabrieken

S C Pascall, BSc, PhD, CEng, MIEE
Satellite Systems Division, British Telecom International

A Pope, BSc
Fairchild Camera and Instrument (UK) Ltd

H Remfry, CEng, MIEE, MBIM, MIIM
ITTE Component Group

J Riley
Amp of Great Britain Ltd

I G Robertson, BSc
Digital Equipment Co. Ltd

M J Rose, BSc(Eng)
Mullard Ltd

R Saxby, BSc
Motorola Ltd

M G Say, PhD, MSc, CEng, ACGI, DIC, FIEE, FRSE
Heriot-Watt University

M J B Scanlon, BSc, ARCS
Marconi Research Laboratories

A Shewan, BSc
Motorola Ltd

J A Smith, CEng, MIM
Cambridge Instruments Ltd

S F Smith, BSc(Eng), CEng, MIEE
Switching Main Exchange Products Division, Standard Telephones and Cables PLC

C R Spicer, AMIEE
BBC Designs Department

K R Sturley, BSc, PhD, FIEE, FIEEE
Telecommunications Consultant

M Trowbridge, MA
Fairchild Camera and Instrument (UK) Ltd

W E Turk, BSc, CEng, FIEE, FRTS, MInstM

L W Turner, CEng, FIEE, FRTS
Consultant Engineer

J A Van Raalte, BSc, MSc, PhD, MAPS, FSID, SMIEEE
David Sarnoff Research Center, RCA

W D C Walker, BSc, CChem, MRSC
Ever Ready (Special Batteries) Ltd

D B Waters, BSc(Eng)
Transmission Products Division, Standard Telephones and Cables PLC

F J Weaver, BSc, MIEE
English Electric Valve Co. Ltd

R C Whitehead, CEng, MIEE
Polytechnic of North London

S Young, BSEE, MBA
SEEQ Corporation

Part 1

Techniques

1 Trigonometric Functions and General Formulae

J Barron BA, MA (Cantab)
Lecturer, University of Cambridge

Contents

1.1 Mathematical signs and symbols

Sign, symbol	Quantity
$=$	equal to
$\neq$ $\neq$	not equal to
$\equiv$	identically equal to
Δ	corresponds to
$\approx$	approximately equal to
$\rightarrow$	approaches
$\simeq$	asymptotically equal to
$\sim$	proportional to
∞	infinity
$<$	smaller than
$>$	larger than
$\leq$ $\leqslant$ $\leqq$	smaller than or equal to
$\geq$ $\geqslant$ $\geqq$	larger than or equal to
$\ll$	much smaller than
$\gg$	much larger than
$+$	plus
$-$	minus
$.$ $\times$	multiplied by
$\frac{a}{b}$ a/b	a divided by b
$\lvert a\rvert$	magnitude of a
a^n	a raised to the power n
$a^{1/2}$ $\sqrt{a}$	square root of a
$a^{1/n}$ $\sqrt[n]{a}$	nth root of a
$\bar{a}$ $\langle a\rangle$	mean value of a
$p!$	factorial p, $1\times 2\times 3\times\ldots\times p$
$\binom{n}{p}$	binomial coefficient, $\frac{n(n-1)\ldots(n-p+1)}{1\times 2\times 3\times\ldots\times p}$
Σ	sum
Π	product
$f(x)$	function f of the variable x
$[+(x)]_a^b$	$f(b)-f(a)$
$\lim\limits_{x\to a} f(x)$; $\lim_{x\to a} f(x)$	the limit to which $f(x)$ tends as x approaches a
Δx	delta x = finite increment of x
δx	delta x = variation of x
$\frac{df}{dx}$; df/dx; $f'(x)$	differential coefficient of $f(x)$ with respect to x
$\frac{d^nf}{dx^n}$; $f^{(n)}(x)$	differential coefficient of order n of $f(x)$
$\frac{\partial f(x,y,\ldots)}{\partial x}$; $\left(\frac{\partial f}{\partial x}\right)_{y\ldots}$	partial differential coefficient of $f(x, y, \ldots)$ with respect to x, when $y, \ldots$ are held constant
df	the total differential of f
$\int f(x)\,dx$	indefinite integral of $f(x)$ with respect to x
$\int_a^b f(x)\,dx$	definite integral of $f(x)$ from $x=a$ to $x=b$
e	base of natural logarithms
e^x; $\exp x$	e raised to the power x
$\log_a x$	logarithm to the base a of x
$\lg x$; $\log x$; $\log_{10} x$	common (Briggsian) logarithm of x
$\mathrm{lb}\, x$; $\log_2 x$	binary logarithm of x
$\sin x$	sine of x
$\cos x$	cosine of x
$\tan x$; $\mathrm{tg}\, x$	tangent of x
$\cot x$; $\mathrm{ctg}\, x$	cotangent of x
$\sec x$	secant of x
$\mathrm{cosec}\, x$	cosecant of x
$\arcsin x$	arc sine of x
$\arccos x$	arc cosine of x
$\arctan x$, $\mathrm{arctg}\, x$	arc tangent of x
$\mathrm{arccot}\, x$, $\mathrm{arcctg}\, x$	arc cotangent of x
$\mathrm{arcsec}\, x$	arc secant of x
$\mathrm{arccosec}\, x$	arc cosecant of x
$\sinh x$	hyperbolic sine of x
$\cosh x$	hyperbolic cosine of x
$\tanh x$	hyperbolic tangent of x
$\coth x$	hyperbolic cotangent of x
$\mathrm{sech}\, x$	hyperbolic secant of x
$\mathrm{cosech}\, x$	hyperbolic cosecant of x
$\mathrm{arcsinh}\, x$	inverse hyperbolic sine of x
$\mathrm{arcosh}\, x$	inverse hyperbolic cosine of x
$\mathrm{arctanh}\, x$	inverse hyperbolic tangent of x
$\mathrm{arcoth}\, x$	inverse hyperbolic cotangent of x
$\mathrm{arsech}\, x$	inverse hyperbolic secant of x
$\mathrm{arcosech}\, x$	inverse hyperbolic cosecant of x
i, j	imaginary unity, $i^2=-1$
$\mathrm{Re}\, z$	real part of z
$\mathrm{Im}\, z$	imaginary part of z
$\lvert z\rvert$	modulus of z
$\arg z$	argument of z
z^*	conjugate of z, complex conjugate of z
$\tilde{A}$, A', A^t	transpose of matrix A
A^*	complex conjugate matrix of matrix A
A^+	Hermitian conjugate matrix of matrix A
$\mathbf{A}$, $\mathbf{a}$	vector
$\lvert\mathbf{A}\rvert$, A	magnitude of vector
$\mathbf{A}\cdot\mathbf{B}$	scalar product
$\mathbf{A}\times\mathbf{B}$, $\mathbf{A}\wedge\mathbf{B}$	vector product
∇	differential vector operator
$\nabla\varphi$, grad φ	gradient of φ
$\nabla\cdot A$, div $\mathbf{A}$	divergence of $\mathbf{A}$
$\nabla\times\mathbf{A}$, $\nabla\wedge\mathbf{A}$, curl $\mathbf{A}$, rot $\mathbf{A}$	curl of $\mathbf{A}$
$\nabla^2\varphi$, $\Delta\varphi$	Laplacian of φ

1.2 Trigonometric formulae

$$\sin^2 A+\cos^2 A=\sin A\,\mathrm{cosec}\, A=1$$

$$\sin A=\frac{\cos A}{\cot A}=\frac{1}{\mathrm{cosec}\, A}=(1-\cos^2 A)^{1/2}$$

$$\cos A=\frac{\sin A}{\tan A}=\frac{1}{\sec A}=(1-\sin^2 A)^{1/2}$$

$$\tan A=\frac{\sin A}{\cos A}=\frac{1}{\cot A}$$

$$1+\tan^2 A=\sec^2 A$$

$$1+\cot^2 A=\mathrm{cosec}^2 A$$

$$1-\sin A=\mathrm{coversin}\, A$$

$$1-\cos A=\mathrm{versin}\, A$$

$$\tan\tfrac{1}{2}\theta=t;\quad \sin\theta=2t/(1+t^2);\quad \cos\theta=(1-t^2)/(1+t^2)$$

$\cot A = 1/\tan A$

$\sec A = 1/\cos A$

$\operatorname{cosec} A = 1/\sin A$

$\cos(A \pm B) = \cos A \cos B \mp \sin A \sin B$

$\sin(A \pm B) = \sin A \cos B \pm \cos A \sin B$

$$\tan(A \pm B) = \frac{\tan A \pm \tan B}{1 \mp \tan A \tan B}$$

$$\cot(A \pm B) = \frac{\cot A \cot B \mp 1}{\cot B \pm \cot A}$$

$\sin A \pm \sin B = 2 \sin \tfrac{1}{2}(A \pm B) \cos \tfrac{1}{2}(A \mp B)$

$\cos A + \cos B = 2 \cos \tfrac{1}{2}(A + B) \cos \tfrac{1}{2}(A - B)$

$\cos A - \cos B = 2 \sin \tfrac{1}{2}(A + B) \sin \tfrac{1}{2}(B - A)$

$$\tan A \pm \tan B = \frac{\sin(A \pm B)}{\cos A \cos B}$$

$$\cot A \pm \cot B = \frac{\sin(B \pm A)}{\sin A \sin B}$$

$\sin 2A = 2 \sin A \cos A$

$\cos 2A = \cos^2 A - \sin^2 A = 2\cos^2 A - 1 = 1 - 2\sin^2 A$

$\cos^2 A - \sin^2 B = \cos(A + B)\cos(A - B)$

$\tan 2A = 2 \tan A/(1 - \tan^2 A)$

$$\sin \tfrac{1}{2}A = \left(\frac{1 - \cos A}{2}\right)^{1/2}$$

$$\cos \tfrac{1}{2}A = \pm\left(\frac{1 + \cos A}{2}\right)^{1/2}$$

$$\tan \tfrac{1}{2}A = \frac{\sin A}{1 + \cos A}$$

$\sin^2 A = \tfrac{1}{2}(1 - \cos 2A)$

$\cos^2 A = \tfrac{1}{2}(1 + \cos 2A)$

$$\tan^2 A = \frac{1 - \cos 2A}{1 + \cos 2A}$$

$$\tan \tfrac{1}{2}(A \pm B) = \frac{\sin A \pm \sin B}{\cos A + \cos B}$$

$$\cot \tfrac{1}{2}(A \pm B) = \frac{\sin A \pm \sin B}{\cos B - \cos A}$$

1.3 Trigonometric values

Angle	*0°*	*30°*	*45°*	*60°*	*90°*	*180°*	*270°*	*360°*
Radians	0	$\pi/6$	$\pi/4$	$\pi/3$	$\pi/2$	π	$3\pi/2$	2π
Sine	0	$\frac{1}{2}$	$\frac{1}{2}\sqrt{2}$	$\frac{1}{2}\sqrt{3}$	1	0	-1	0
Cosine	1	$\frac{1}{2}\sqrt{3}$	$\frac{1}{2}\sqrt{2}$	$\frac{1}{2}$	0	-1	0	1
Tangent	0	$\frac{1}{3}\sqrt{3}$	1	$\sqrt{3}$	∞	0	∞	0

1.4 Approximations for small angles

$\sin\theta = \theta - \theta^3/6$; $\cos\theta = 1 - \theta^2/2$; $\tan\theta = \theta + \theta^3/3$; ($\theta$ in radians)

1.5 Solution of triangles

$$\frac{\sin A}{a} = \frac{\sin B}{b} = \frac{\sin C}{c} \qquad \cos A = \frac{b^2 + c^2 - a^2}{2bc}$$

$$\cos B = \frac{c^2 + a^2 - b^2}{2ca} \qquad \cos C = \frac{a^2 + b^2 - c^2}{2ab}$$

where A, B, C and a, b, c are shown in *Figure 1.1*. If $s = \tfrac{1}{2}(a + b + c)$,

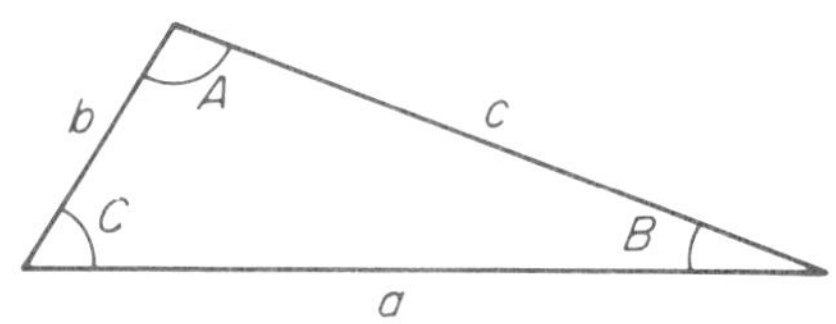

Figure 1.1 Triangle

$$\sin\frac{A}{2} = \sqrt{\frac{(s-b)(s-c)}{bc}} \qquad \sin\frac{B}{2} = \sqrt{\frac{(s-c)(s-a)}{ca}}$$

$$\sin\frac{C}{2} = \sqrt{\frac{(s-a)(s-b)}{ab}}$$

$$\cos\frac{A}{2} = \sqrt{\frac{s(s-a)}{bc}} \qquad \cos\frac{B}{2} = \sqrt{\frac{s(s-b)}{ca}}$$

$$\cos\frac{C}{2} = \sqrt{\frac{s(s-c)}{ab}}$$

$$\tan\frac{A}{2} = \sqrt{\frac{(s-b)(s-c)}{s(s-a)}} \qquad \tan\frac{B}{2} = \sqrt{\frac{(s-c)(s-a)}{s(s-b)}}$$

$$\tan\frac{C}{2} = \sqrt{\frac{(s-a)(s-b)}{s(s-c)}}$$

1.6 Spherical triangle

$$\frac{\sin A}{\sin a} = \frac{\sin B}{\sin b} = \frac{\sin C}{\sin c}$$

$\cos a = \cos b \cos c + \sin b \sin c \cos A$

$\cos b = \cos c \cos a + \sin c \sin a \cos B$

$\cos c = \cos a \cos b + \sin a \sin b \cos C$

where A, B, C and a, b, c are now as in *Figure 1.2.*

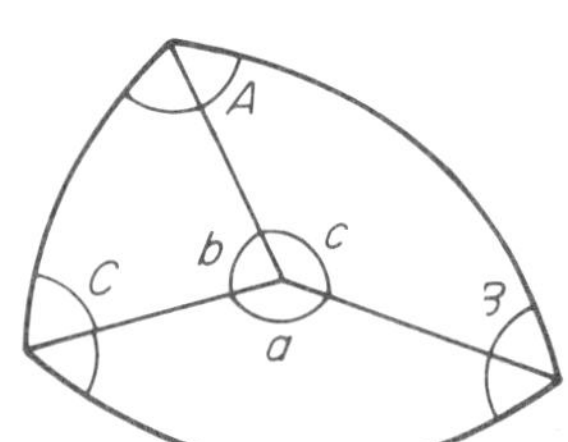

Figure 1.2 Spherical triangle

1.7 Exponential form

$$\sin\theta = \frac{e^{i\theta} - e^{-i\theta}}{2i} \qquad \cos\theta = \frac{e^{i\theta} + e^{-i\theta}}{2}$$

$$e^{i\theta} = \cos\theta + i\sin\theta \qquad e^{-i\theta} = \cos\theta - i\sin\theta$$

1.8 De Moivre's theorem

$$(\cos A + i\sin A)(\cos B + i\sin B)$$
$$= \cos(A+B) + i\sin(A+B)$$

1.9 Euler's relation

$$(\cos\theta + i\sin\theta)^n = \cos n\theta + i\sin n\theta = e^{in\theta}$$

1.10 Hyperbolic functions

$$\sinh x = (e^x - e^{-x})/2 \qquad \cosh x = (e^x + e^{-x})/2$$

$$\tanh x = \sinh x/\cosh x$$

Relations between hyperbolic functions can be obtained from the corresponding relations between trigonometric functions by reversing the sign of any term containing the product or implied product of two sines, e.g.:

$$\cosh^2 A - \sinh^2 A = 1$$

$$\cosh 2A = 2\cosh^2 A - 1 = 1 + 2\sinh^2 A$$
$$= \cosh^2 A + \sinh^2 A$$

$$\cosh(A \pm B) = \cosh A \cosh B \pm \sinh A \sinh B$$

$$\sinh(A \pm B) = \sinh A \cosh B \pm \cosh A \sinh B$$

$$e^x = \cosh x + \sinh x \qquad e^{-x} = \cosh x - \sinh x$$

1.11 Complex variable

If $z = x + iy$, where x and y are real variables, z is a complex variable and is a function of x and y. z may be represented graphically in an Argand diagram (*Figure 1.3*).

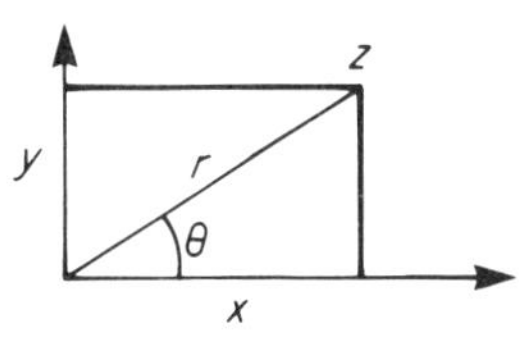

Figure 1.3 Argand diagram

Polar form:

$$z = x + iy = |z|\, e^{i\theta} = |z|(\cos\theta + i\sin\theta)$$

$$x = r\cos\theta \qquad y = r\sin\theta$$

where $r = |z|$.

Complex arithmetic:

$$z_1 = x_1 + iy_1 \qquad z_2 = x_2 + iy_2$$

$$z_1 \pm z_2 = (x_1 \pm x_2) + i(y_1 \pm y_2)$$

$$z_1 . z_2 = (x_1x_2 - y_1y_2) + i(x_1y_2 + x_2y_1)$$

Conjugate:

$$z^* = x - iy \qquad z . z^* = x^2 + y^2 = |z|^2$$

Function: another complex variable $w = u + iv$ may be related functionally to z by

$$w = u + iv = f(x + iy) = f(z)$$

which implies

$$u = u(x, y) \qquad v = v(x, y)$$

e.g.,

$$\cosh z = \cosh(x + iy) = \cosh x \cosh iy + \sinh x \sinh iy$$
$$= \cosh x \cos y + i\sinh x \sin y$$

$$u = \cosh x \cos y \qquad v = \sinh x \sin y$$

1.12 Cauchy–Riemann equations

If $u(x, y)$ and $v(x, y)$ are continuously differentiable with respect to x and y,

$$\frac{\partial u}{\partial x} = \frac{\partial v}{\partial y} \qquad \frac{\partial u}{\partial y} = -\frac{\partial v}{\partial x}$$

$w = f(z)$ is continuously differentiable with respect to z and its derivative is

$$f'(z) = \frac{\partial u}{\partial x} + i\frac{\partial v}{\partial x} = \frac{\partial v}{\partial y} - i\frac{\partial u}{\partial y} = \frac{1}{i}\left(\frac{\partial u}{\partial y} + i\frac{\partial v}{\partial y}\right)$$

It is also easy to show that $\nabla^2 u = \nabla^2 v = 0$. Since the transformation from z to w is conformal, the curves u = constant and v = constant intersect each other at right angles, so that one set may be used as equipotentials and the other as field lines in a vector field.

1.13 Cauchy's theorem

If $f(z)$ is analytic everywhere inside a region bounded by C and a is a point within C

$$f(a) = \frac{1}{2\pi i}\int_C \frac{f(z)}{z - a}\,dz$$

This formula gives the value of a function at a point in the interior of a closed curve in terms of the values on that curve.

1.14 Zeros, poles and residues

If $f(z)$ vanishes at the point z_0 the Taylor series for z in the region of z_0 has its first two terms zero, and perhaps others also: $f(z)$ may then be written

$$f(z) = (z - z_0)^n g(z)$$

where $g(z_0) \neq 0$. Then $f(z)$ has a *zero* of order n at z_0. The reciprocal

$$q(z) = 1/f(z) = h(z)/(z - z_0)^n$$

where $h(z) = 1/g(z) \neq 0$ at z_0. $q(z)$ becomes infinite at $z = z_0$ and is said to have a *pole* of order n at z_0. $q(z)$ may be expanded in the form

$q(z)=c_{-n}(z-z_0)^n+\ldots+c_{-1}(z-z_0)^{-1}+c_0+\ldots$

where c_{-1} is the *residue* of $q(z)$ at $z=z_0$. From Cauchy's theorem, it may be shown that if a function $f(z)$ is analytic throughout a region enclosed by a curve C except at a finite number of poles, the integral of the function around C has a value of $2\pi i$ times the sum of the residues of the function at its poles within C. This fact can be used to evaluate many definite integrals whose indefinite form cannot be found.

1.15 Some standard forms

$$\int_0^{2\pi} e^{\cos\theta}\cos(n\theta-\sin\theta)\,d\theta=2\pi/n!$$

$$\int_0^{\infty}\frac{x^{a-1}}{1+x}dx=\pi\,\text{cosec}\,a\pi$$

$$\int_0^{\infty}\frac{\sin\theta}{\theta}d\theta=\frac{\pi}{2}$$

$$\int_0^{\infty}x\exp(-h^2x^2)dx=\frac{1}{2h^2}$$

$$\int_0^{\infty}\frac{x^{a-1}}{1-x}dx=\pi\cot a\pi$$

$$\int_0^{\infty}\exp(-h^2x^2)\,dx=\frac{\sqrt{\pi}}{2h}$$

$$\int_0^{\infty}x^2\exp(-h^2x^2)\,dx=\frac{\sqrt{\pi}}{4h^3}$$

1.16 Coordinate systems

The basic system is the rectangular Cartesian system (x, y, z) to which all other systems are referred. Two other commonly used systems are as follows.

1.16.1 Cylindrical coordinates

Coordinates of point P are (x,y,z) or (r,θ,z) (see *Figure 1.4*), where

$x=r\cos\theta \qquad y=r\sin\theta \qquad z=z$

In these coordinates the volume element is $r\,dr\,d\theta\,dz$.

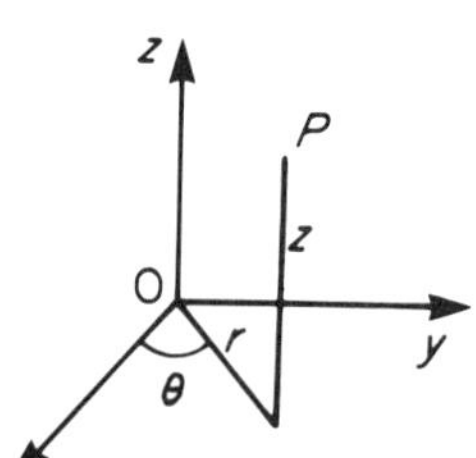

Figure 1.4 Cylindrical coordinates

1.16.2 Spherical polar coordinates

Coordinates of point P are (x,y,z) or (r,θ,φ) (see *Figure 1.5*), where

$x=r\sin\theta\cos\phi \qquad y=r\sin\theta\sin\phi \qquad z=r\cos\theta$

In these coordinates the volume element is $r^2\sin\theta\,dr\,d\theta\,d\phi$.

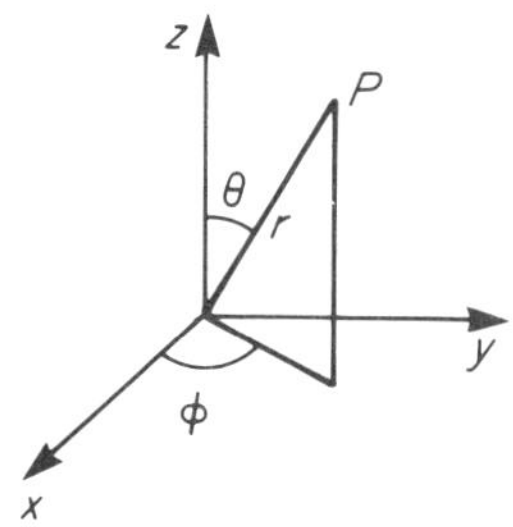

Figure 1.5 Spherical polar coordinates

1.17 Transformation of integrals

$$\iiint f(x,y,z)\,dx\,dy\,dz=\iiint \varphi(u,v,w)|J|\,du\,dv\,dw$$

where

$$J=\begin{vmatrix}\frac{\partial x}{\partial u} & \frac{\partial y}{\partial u} & \frac{\partial z}{\partial u}\\ \frac{\partial x}{\partial v} & \frac{\partial y}{\partial v} & \frac{\partial z}{\partial v}\\ \frac{\partial x}{\partial w} & \frac{\partial y}{\partial w} & \frac{\partial z}{\partial w}\end{vmatrix}=\frac{\partial(x,y,z)}{\partial(u,v,w)}$$

is the Jacobian of the transformation of coordinates. For Cartesian to cylindrical coordinates, $J=r$, and for Cartesian to spherical polars, it is $r^2\sin\theta$.

1.18 Laplace's equation

The equation satisfied by the scalar potential from which a vector field may be derived by taking the gradient is Laplace's equation, written as:

$$\nabla^2\phi=\frac{\partial^2\phi}{\partial x^2}+\frac{\partial^2\phi}{\partial y^2}+\frac{\partial^2\phi}{\partial z^2}=0$$

In cylindrical coordinates:

$$\nabla^2\phi=\frac{1}{r}\frac{\partial}{\partial r}\left(r\frac{\partial\phi}{\partial r}\right)+\frac{1}{r^2}\frac{\partial^2\phi}{\partial\theta^2}+\frac{\partial^2\phi}{\partial z^2}$$

In spherical polars:

$$\nabla^2\phi=\frac{1}{r^2}\frac{\partial}{\partial r}\left(r^2\frac{\partial\phi}{\partial r}\right)+\frac{1}{r^2\sin\theta}\frac{\partial}{\partial\theta}\left(\sin\theta\frac{\partial\phi}{\partial\theta}\right)+\frac{1}{r^2\sin^2\theta}\frac{\partial^2\phi}{\partial\phi^2}$$

The equation is solved by setting

$\phi=U(u)V(v)W(w)$

in the appropriate form of the equation, separating the variables and solving separately for the three functions, where (u, v, w) is the coordinate system in use.

In Cartesian coordinates, typically the functions are trigonometric, hyperbolic and exponential; in cylindrical coordinates the function of z is exponential, that of θ trigonometric and that of r is a Bessel function. In spherical polars, typically the function of r is a power of r, that of φ is trigonometric, and that of θ is a Legendre function of $\cos\theta$.

1.19 Solution of equations

1.19.1 Quadratic equation

$$ax^2+bx+c=0$$

$$x=-\frac{b}{2a}\pm\frac{\sqrt{b^2-4ac}}{2a}$$

In practical calculations if $b^2>4ac$, so that the roots are real and unequal, calculate the root of larger modulus first, using the same sign for both terms in the formula, then use the fact that $x_1x_2=c/a$ where x_1 and x_2 are the roots. This avoids the severe cancellation of significant digits which may otherwise occur in calculating the smaller root.

For polynomials other than quadratics, and for other functions, several methods of successive approximation are available.

1.19.2 Bisection method

By trial find x_0 and x_1 such that $f(x_0)$ and $f(x_1)$ have opposite signs (see *Figure 1.6*). Set $x_2=(x_0+x_1)/2$ and calculate $f(x_2)$. If

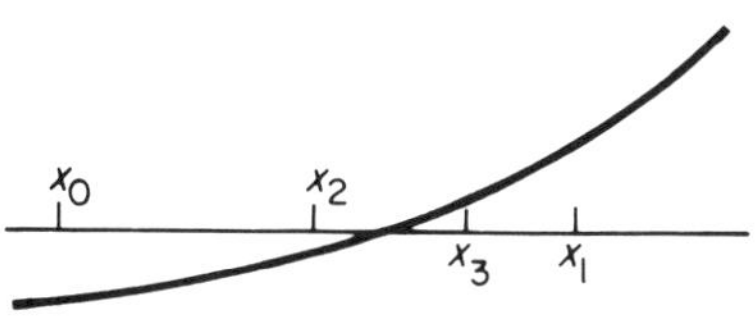

Figure 1.6 Bisection method

$f(x_0)f(x_2)$ is positive, the root lies in the interval (x_1, x_2); if negative in the interval (x_0, x_2); and if zero, x_2 is the root. Continue if necessary using the new interval.

1.19.3 Regula Falsi

By trial, find x_0 and x_1 as for the bisection method; these two values define two points $(x_0, f(x_0))$ and $(x_1, f(x_1))$. The straight line joining these two points cuts the x-axis at the point (see *Figure 1.7*)

$$x_2=\frac{x_0f(x_1)-x_1f(x_0)}{f(x_1)-f(x_0)}$$

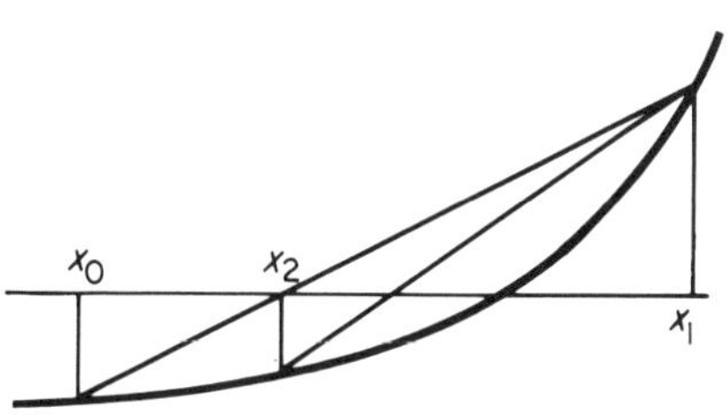

Figure 1.7 Regula Falsi

Evaluate $f(x_2)$ and repeat the process for whichever of the intervals (x_0, x_2) or (x_1, x_2) contains the root. This method can be accelerated by halving at each step the function value at the retained end of the interval, as shown in *Figure 1.8*.

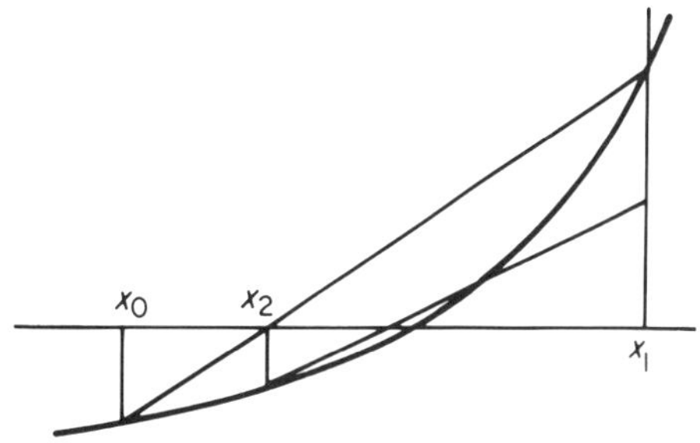

Figure 1.8 Accelerated method

1.19.4 Fixed-point iteration

Arrange the equation in the form

$$x=f(x)$$

Choose an initial value of x by trial, and calculate repetitively

$$x_{k+1}=f(x_k)$$

This process will not always converge.

1.19.5 Newton's method

Calculate repetitively (*Figure 1.9*)

$$x_{k+1}=x_k-f(x_k)/f'(x_k)$$

This method will converge unless: (a) x_k is near a point of inflexion of the function; or (b) x_k is near a local minimum; or (c) the root is

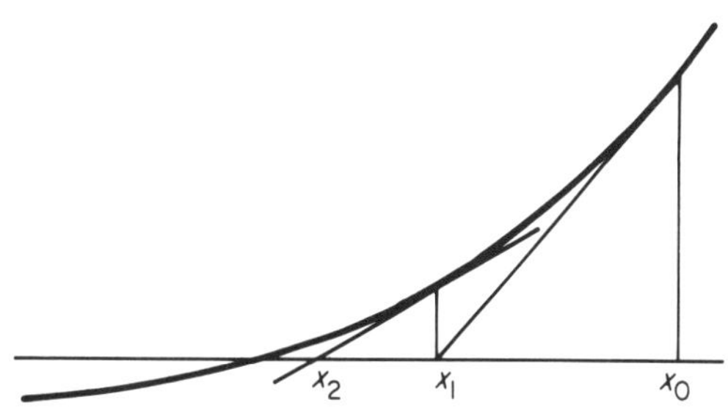

Figure 1.9 Newton's method

multiple. If one of these cases arises, most of the trouble can be overcome by checking at each stage that

$$f(x_{k+1})<f(x_k)$$

and if not, halving the preceding value of $|x_{k+1}-x_k|$.

1.20 Method of least squares

To obtain the best fit between a straight line $ax+by=1$ and several points $(x_1, y_1), (x_2, y_2), \ldots, (x_n, y_n)$ found by observation, the coefficients a and b are to be chosen so that the sum of the squares of the errors

$e_i = ax_i + by_i - 1$

is a minimum. To do this, first write the set of inconsistent equations

$$ax_1 + by_1 - 1 = 0$$
$$ax_2 + by_2 - 1 = 0$$
$$\vdots$$
$$ax_n + by_n - 1 = 0$$

Multiply each equation by the value of x it contains, and add, obtaining

$$a \sum_{i=1}^{n} x_i^2 + b \sum_{i=1}^{n} x_i y_i - \sum_{i=1}^{n} x_i = 0$$

Similarly multiply by y and add, obtaining

$$a \sum_{i=1}^{n} x_i y_i + b \sum_{i=1}^{n} y_i^2 - \sum_{i=1}^{n} y_i = 0$$

Lastly, solve these two equations for a and b, which will be the required values giving the least squares fit.

1.21 Relation between decibels, current and voltage ratio, and power ratio

$$\text{dB} = 10 \log \frac{P_1}{P_2} = 20 \log \frac{V_1}{V_2} = 20 \log \frac{I_1}{I_2}$$

dB	I_1/I_2 or V_1/V_2	I_2/I_1 or V_2/V_1	P_1/P_2	P_2/P_1
0.1	1.012	0.989	1.023	0.977
0.2	1.023	0.977	1.047	0.955
0.3	1.035	0.966	1.072	0.933
0.4	1.047	0.955	1.096	0.912
0.5	1.059	0.944	1.122	0.891
0.6	1.072	0.933	1.148	0.871
0.7	1.084	0.923	1.175	0.851
0.8	1.096	0.912	1.202	0.832
0.9	1.109	0.902	1.230	0.813
1.0	1.122	0.891	1.259	0.794
1.1	1.135	0.881	1.288	0.776
1.2	1.148	0.871	1.318	0.759
1.3	1.162	0.861	1.349	0.741
1.4	1.175	0.851	1.380	0.724
1.5	1.188	0.841	1.413	0.708
1.6	1.202	0.832	1.445	0.692
1.7	1.216	0.822	1.479	0.676
1.8	1.230	0.813	1.514	0.661
1.9	1.245	0.804	1.549	0.645
2.0	1.259	0.794	1.585	0.631
2.5	1.334	0.750	1.778	0.562
3.0	1.413	0.708	1.995	0.501
3.5	1.496	0.668	2.24	0.447
4.0	1.585	0.631	2.51	0.398
4.5	1.679	0.596	2.82	0.355
5.0	1.778	0.562	3.16	0.316
5.5	1.884	0.531	3.55	0.282
6.0	1.995	0.501	3.98	0.251
6.5	2.11	0.473	4.47	0.224
7.0	2.24	0.447	5.01	0.200
7.5	2.37	0.422	5.62	0.178
8.0	2.51	0.398	6.31	0.158
8.5	2.66	0.376	7.08	0.141
9.0	2.82	0.355	7.94	0.126
9.5	2.98	0.335	8.91	0.112
10.0	3.16	0.316	10.00	0.100
10.5	3.35	0.298	11.2	0.089 1
11.0	3.55	0.282	12.6	0.079 4
15.0	5.62	0.178	31.6	0.031 6
15.5	5.96	0.168	35.5	0.028 2
16.0	6.31	0.158	39.8	0.025 1
16.5	6.68	0.150	44.7	0.022 4
17.0	7.08	0.141	50.1	0.020 0
17.5	7.50	0.133	56.2	0.017 8
18.0	7.94	0.126	63.1	0.015 8
18.5	8.41	0.119	70.8	0.014 1
19.0	8.91	0.112	79.4	0.012 6
19.5	9.44	0.106	89.1	0.011 2
20.0	10.00	0.100 0	100	0.010 0
20.5	10.59	0.094 4	112	0.008 91
21.0	11.22	0.089 1	126	0.007 94
21.5	11.88	0.084 1	141	0.007 08
22.0	12.59	0.079 4	158	0.006 31
22.5	13.34	0.075 0	178	0.005 62
23.0	14.13	0.070 8	200	0.005 01
23.5	14.96	0.066 8	224	0.004 47
24.0	15.85	0.063 1	251	0.003 98
24.5	16.79	0.059 6	282	0.003 55
25.0	17.78	0.056 2	316	0.003 16
25.5	18.84	0.053 1	355	0.002 82
26.0	19.95	0.050 1	398	0.002 51
26.5	21.1	0.047 3	447	0.002 24
27.0	22.4	0.044 7	501	0.002 00
27.5	23.7	0.042 2	562	0.001 78
28.0	25.1	0.039 8	631	0.001 58
28.5	26.6	0.037 6	708	0.001 41
29.0	28.2	0.035 5	794	0.001 26
29.5	29.8	0.033 5	891	0.001 12
30.0	31.6	0.031 6	1 000	0.001 00
31.0	35.5	0.028 2	1 260	7.94×10^{-4}
32.0	39.8	0.025 1	1 580	6.31×10^{-4}
33.0	44.7	0.022 4	2 000	5.01×10^{-4}
34.0	50.1	0.020 0	2 510	3.98×10^{-4}
35.0	56.2	0.017 8		3.16×10^{-4}
36.0	63.1	0.015 8	3 980	2.51×10^{-4}
37.0	70.8	0.014 1	5 010	2.00×10^{-4}

2 Calculus

J Barron BA, MA (Cantab)
Lecturer, University of Cambridge

Contents

2.1 Derivative

$$f'(x) = \lim_{\delta x \to 0} \frac{f(x+\delta x) - f(x)}{\delta x}$$

If u and v are functions of x,

$$(uv)' = u'v + uv'$$

$$\left(\frac{u}{v}\right)' = \frac{u'v - uv'}{v^2}$$

$$(uv)^{(n)} = u^{(n)}v + nu^{(n-1)}v^{(1)} + \ldots + {}^nC_p u^{(n-p)}v^{(p)} + \ldots + uv^{(n)}$$

where

$${}^nC_p = \frac{n!}{p!\,(n-p)!}$$

If $z = f(x)$ and $y = g(z)$, then

$$\frac{\mathrm{d}y}{\mathrm{d}x} = \frac{\mathrm{d}y}{\mathrm{d}z}\frac{\mathrm{d}z}{\mathrm{d}x}$$

2.2 Maxima and minima

$f(x)$ has a stationary point wherever $f'(x) = 0$: the point is a maximum, minimum or point of inflexion according as $f''(x) <, >$ or $= 0$.

$f(x, y)$ has a stationary point wherever

$$\frac{\partial f}{\partial x} = \frac{\partial f}{\partial y} = 0$$

Let (a, b) be such a point, and let

$$\frac{\partial^2 f}{\partial x^2} = A, \qquad \frac{\partial^2 f}{\partial x\,\partial y} = H \qquad \frac{\partial^2 f}{\partial y^2} = B$$

all at that point, then:

If $H^2 - AB > 0$, $f(x, y)$ has a saddle point at (a, b).
If $H^2 - AB < 0$ and if $A < 0$, $f(x, y)$ has a maximum at (a, b), but if $A > 0$, $f(x, y)$ has a minimum at (a, b).
If $H^2 = AB$, higher derivatives need to be considered.

2.3 Integral

$$\int_a^b f(x)\,\mathrm{d}x = \lim_{N\to\infty} \sum_{n=0}^{N-1} f\left(a + \frac{n(b-a)}{N}\right)\left(\frac{b-a}{N}\right)$$

$$= \lim_{N\to\infty} \sum_{n=1}^{N} f(a + (n-1)\,\delta x)\,\delta x$$

where $\delta x = (b-a)/N$.

If u and v are functions of x, then

$$\int uv'\,\mathrm{d}x = uv - \int u'v\,\mathrm{d}x \quad \text{(integration by parts)}$$

2.4 Derivatives and integrals

y	$\frac{\mathrm{d}y}{\mathrm{d}x}$	$\int y\,\mathrm{d}x$
x^n	nx^{n-1}	$x^{n+1}/(n+1)$
$1/x$	$-1/x^2$	$\ln(x)$
e^{ax}	$a\,\mathrm{e}^{ax}$	e^{ax}/a
$\ln(x)$	$1/x$	$x[\ln(x) - 1]$
$\log_a x$	$\frac{1}{x}\log_a \mathrm{e}$	$x \log_a\left(\frac{x}{\mathrm{e}}\right)$
$\sin ax$	$a\cos ax$	$-\frac{1}{a}\cos ax$
$\cos ax$	$-a\sin ax$	$\frac{1}{a}\sin ax$
$\tan ax$	$a\sec^2 ax$	$-\frac{1}{a}\ln(\cos ax)$
$\cot ax$	$-a\,\mathrm{cosec}^2\, ax$	$\frac{1}{a}\ln(\sin ax)$
$\sec ax$	$a\tan ax\sec ax$	$\frac{1}{a}\ln(\sec ax + \tan ax)$
$\mathrm{cosec}\, ax$	$-a\cot ax\,\mathrm{cosec}\, ax$	$\frac{1}{a}\ln(\mathrm{cosec}\, ax - \cot ax)$

y	$\dfrac{dy}{dx}$	$\int y\,dx$
$\sin^{-1}(x/a)$	$1/(a^2-x^2)^{1/2}$	$x\sin^{-1}(x/a)+(a^2-x^2)^{1/2}$
$\cos^{-1}(x/a)$	$-1/(a^2-x^2)^{1/2}$	$x\cos^{-1}(x/a)-(a^2-x^2)^{1/2}$
$\tan^{-1}(x/a)$	$a/(a^2+x^2)$	$x\tan^{-1}(x/a)-\frac{1}{2}a\ln(a^2+x^2)$
$\cot^{-1}(x/a)$	$-a/(a^2+x^2)$	$x\cot^{-1}(x/a)+\frac{1}{2}a\ln(a^2+x^2)$
$\sec^{-1}(x/a)$	$a(x^2-a^2)^{-1/2}/x$	$x\sec^{-1}(x/a)-a\ln[x+(x^2-a^2)^{1/2}]$
$\mathrm{cosec}^{-1}(x/a)$	$-a(x^2-a^2)^{-1/2}/x$	$x\,\mathrm{cosec}^{-1}(x/a)+a\ln[x+(x^2-a^2)^{1/2}]$
$\sinh ax$	$a\cosh ax$	$\dfrac{1}{a}\cosh ax$
$\coth ax$	$a\sinh ax$	$\dfrac{1}{a}\sinh ax$
$\tanh ax$	$a\,\mathrm{sech}^2\,ax$	$\dfrac{1}{a}\ln(\cosh ax)$
$\coth ax$	$-a\,\mathrm{cosech}^2\,ax$	$\dfrac{1}{a}\ln(\sinh ax)$
$\mathrm{sech}\,ax$	$-a\tanh ax\,\mathrm{sech}\,ax$	$\dfrac{2}{a}\tan^{-1}(e^{ax})$
$\mathrm{cosech}\,ax$	$-a\coth ax\,\mathrm{cosech}\,ax$	$\dfrac{1}{a}\ln\left(\tanh\dfrac{ax}{2}\right)$
$\sinh^{-1}(x/a)$	$(x^2+a^2)^{-1/2}$	$x\sinh^{-1}(x/a)-(x^2+a^2)^{1/2}$
$\cosh^{-1}(x/a)$	$(x^2-a^2)^{-1/2}$	$x\cosh^{-1}(x/a)-(x^2-a^2)^{1/2}$
$\tanh^{-1}(x/a)$	$a(a^2-x^2)^{-1}$	$x\tanh^{-1}(x/a)+\frac{1}{2}a\ln(a^2-x^2)$
$\coth^{-1}(x/a)$	$-a(x^2-a^2)^{-1}$	$x\coth^{-1}(x/a)+\frac{1}{2}a\ln(x^2-a^2)$
$\mathrm{sech}^{-1}(x/a)$	$-a(a^2-x^2)^{-1/2}/x$	$x\,\mathrm{sech}^{-1}(x/a)+a\sin^{-1}(x/a)$
$\mathrm{cosech}^{-1}(x/a)$	$-a(x^2+a^2)^{-1/2}/x$	$x\,\mathrm{cosech}^{-1}(x/a)+a\sinh^{-1}(x/a)$
$(x^2\pm a^2)^{1/2}$		$\frac{1}{2}x(x^2\pm a^2)^{1/2}\pm\frac{1}{2}a^2\sinh^{-1}(x/a)$
$(a^2-x^2)^{1/2}$		$\frac{1}{2}x(a^2-x^2)^{1/2}+\frac{1}{2}a^2\sin^{-1}(x/a)$
$(x^2\pm a^2)^p x$		$\frac{1}{2}(x^2\pm a^2)^{p+1}/(p+1)$ $(p\neq-1)$ $\frac{1}{2}\ln(x^2\pm a^2)$ $(p=-1)$
$(a^2-x^2)^p x$		$-\frac{1}{2}(a^2-x^2)^{p+1}/(p+1)$ $(p\neq-1)$ $-\frac{1}{2}\ln(a^2-x^2)$ $(p=-1)$
$x(ax^2+b)^p$		$(ax^2+b)^{p+1}/2a(p+1)$ $(p\neq-1)$ $[\ln(ax^2+b)]/2a$ $(p=-1)$
$(2ax-x^2)^{-1/2}$		$\cos^{-1}\left(\dfrac{a-x}{a}\right)$
$(a^2\sin^2x+b^2\cos^2x)^{-1}$		$\dfrac{1}{ab}\tan^{-1}\left(\dfrac{a}{b}\tan x\right)$
$(a^2\sin^2x-b^2\cos^2x)^{-1}$		$-\dfrac{1}{ab}\tanh^{-1}\left(\dfrac{a}{b}\tan x\right)$
$c^{ax}\sin bx$		$e^{ax}\dfrac{a\sin bx-b\cos bx}{a^2+b^2}$
$e^{ax}\cos bx$		$e^{ax}\dfrac{(a\cos bx+b\sin bx)}{a^2+b^2}$

y	$\int y\,dx$	
$\sin mx \sin nx$	$\frac{1}{2}\frac{\sin(m-n)x}{m-n} - \frac{1}{2}\frac{\sin(m+n)x}{m+n}$	$(m \neq n)$
	$\frac{1}{2}\left(x - \frac{\sin 2mx}{2m}\right)$	$(m = n)$
$\sin mx \cos nx$	$-\frac{1}{2}\frac{\cos(m+n)x}{m+n} - \frac{1}{2}\frac{\cos(m-n)x}{m-n}$	$(m \neq n)$
	$-\frac{1}{2}\frac{\cos 2mx}{2m}$	$(m = n)$
$\cos mx \cos nx$	$\frac{1}{2}\frac{\sin(m+n)x}{m+n} + \frac{1}{2}\frac{\sin(m-n)x}{m-n}$	$(m \neq n)$
	$\frac{1}{2}\left(x + \frac{\sin 2mx}{2m}\right)$	$(m = n)$

2.5 Standard substitutions

Integral a function of	*Substitute*
$a^2 - x^2$	$x = a\sin\theta$ or $x = a\cos\theta$
$a^2 + x^2$	$x = a\tan\theta$ or $x = a\sinh\theta$
$x^2 - a^2$	$x = a\sec\theta$ or $x = a\cosh\theta$

2.6 Reduction formulae

$$\int \sin^m x\,dx = -\frac{1}{m}\sin^{m-1}x\cos x + \frac{m-1}{m}\int \sin^{m-2}x\,dx$$

$$\int \cos^m x\,dx = \frac{1}{m}\cos^{m-1}x\sin x + \frac{m-1}{m}\int \cos^{m-2}x\,dx$$

$$\int \sin^m x\cos^n x\,dx = \frac{\sin^{m+1}x\cos^{n-1}x}{m+n} + \frac{n-1}{m+n}\int \sin^m x\cos^{n-2}x\,dx$$

If the integrand is a rational function of $\sin x$ and/or $\cos x$, substitute $t = \tan\frac{1}{2}x$, then

$$\sin x = \frac{1}{1+t^2}, \qquad \cos x = \frac{1-t^2}{1+t^2}, \qquad dx = \frac{2dt}{1+t^2}$$

2.7 Numerical integration

2.7.1 Trapezoidal rule (*Figure 2.1*)

$$\int_{x1}^{x2} y\,dx = \tfrac{1}{2}h(y_1 + y_2) + O(h^3)$$

2.7.2 Simpson's rule (*Figure 2.1*)

$$\int_{x1}^{x2} y\,dx = 2h(y_1 + 4y_2 + y_3)/6 + O(h^5)$$

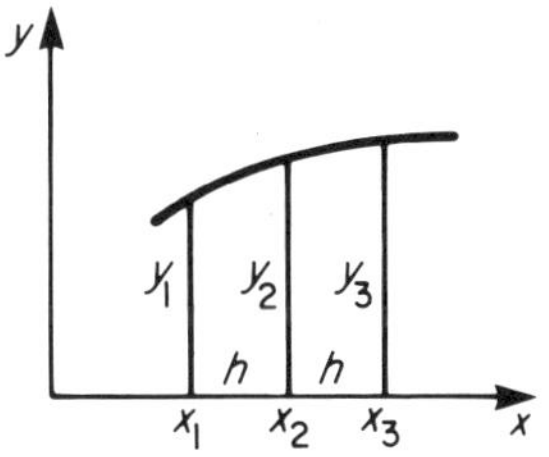

Figure 2.1 Numerical integration

2.7.3 Change of variable in double integral

$$\iint f(x, y)\,dx\,dy = \iint F(u, v)|J|\,du\,dv$$

where

$$J = \frac{\partial(x, y)}{\partial(u, v)} = \begin{vmatrix} \frac{\partial x}{\partial u} & \frac{\partial x}{\partial v} \\ \frac{\partial y}{\partial u} & \frac{\partial y}{\partial v} \end{vmatrix} = \begin{vmatrix} \frac{\partial x}{\partial u} & \frac{\partial y}{\partial u} \\ \frac{\partial x}{\partial v} & \frac{\partial y}{\partial v} \end{vmatrix}$$

is the Jacobian of the transformation.

2.7.4 Differential mean value theorem

$$\frac{f(x+h) - f(x)}{h} = f'(x + \theta h) \qquad 0 < \theta < 1$$

2.7.5 Integral mean value theorem

$$\int_a^b f(x)g(x)\,dx = g(a + \theta h)\int_a^b f(x)\,dx$$

$h = b - a,\ 0 < \theta < 1$

2.8 Vector calculus

Let $s(x, y, z)$ be a scalar function of position and let

$\mathbf{v}(x, y, z) = \mathbf{i}v_x(x, y, z) + \mathbf{j}v_y(x, y, z) + \mathbf{k}v_z(x, y, z)$

be a vector function of position. Define

$$\nabla = \mathbf{i}\frac{\partial}{\partial x} + \mathbf{j}\frac{\partial}{\partial y} + \mathbf{k}\frac{\partial}{\partial z}$$

so that

$$\nabla.\nabla = \nabla^2 = \frac{\partial^2}{\partial x^2} + \frac{\partial^2}{\partial y^2} + \frac{\partial^2}{\partial z^2}$$

then

$$\text{grad}\, s = \nabla s = \mathbf{i}\frac{\partial s}{\partial x} + \mathbf{j}\frac{\partial s}{\partial y} + \mathbf{k}\frac{\partial s}{\partial z}$$

$$\text{div}\,\mathbf{v} = \nabla.\mathbf{v} = \frac{\partial v_x}{\partial x} + \frac{\partial v_y}{\partial y} + \frac{\partial v_z}{\partial z}$$

$$\text{curl}\,\mathbf{v} = \nabla \times \mathbf{v} = \mathbf{i}\left(\frac{\partial v_z}{\partial y} - \frac{\partial v_y}{\partial z}\right) + \mathbf{j}\left(\frac{\partial v_x}{\partial z} - \frac{\partial v_z}{\partial x}\right) + \mathbf{k}\left(\frac{\partial v_y}{\partial x} - \frac{\partial v_x}{\partial y}\right)$$

The following identities are then true:

$\text{div}(s\mathbf{v}) = s\,\mathit{div}\,\mathbf{v} + (\text{grad}\, s).\mathbf{v}$

$\text{curl}\,(s\mathbf{v}) = s\,\text{curl}\,\mathbf{v} + (\text{grad}\,s) \times \mathbf{v}$

$\text{div}\,(\mathbf{u} \times \mathbf{v}) = \mathbf{v}\,.\,\text{curl}\,\mathbf{u} - \mathbf{u}\,.\,\text{curl}\,\mathbf{v}$

$\text{curl}\,(\mathbf{u} \times \mathbf{v}) = \mathbf{u}\,\text{div}\,\mathbf{v} - \mathbf{v}\,\text{div}\,\mathbf{u} + (\mathbf{v}\,.\,\nabla)\mathbf{u} - (\mathbf{u}\,.\,\nabla)\mathbf{v}$

$\text{div}\,\text{grad}\,s = \nabla^2 s$

$\text{div}\,\text{curl}\,\mathbf{v} = 0$

$\text{curl}\,\text{grad}\,s = 0$

$\text{curl}\,\text{curl}\,\mathbf{v} = \text{grad}\,(\text{div}\,\mathbf{v}) - \nabla^2\mathbf{v}$

where ∇^2 operates on each component of $\mathbf{v}$.

$\mathbf{v} x\,\text{curl}\,\mathbf{v} + (\mathbf{v}\,.\,\nabla)\mathbf{v} = \text{grad}\,\tfrac{1}{2}\mathbf{v}^2$

Potentials:

If $\text{curl}\,\mathbf{v} = 0$, $\mathbf{v} = \text{grad}\,\varphi$ where φ is a scalar potential.
If $\text{div}\,\mathbf{v} = 0$, $\mathbf{v} = \text{curl}\,\mathbf{A}$ where $\mathbf{A}$ is a vector potential.

3 Series and Transforms

J Barron BA, MA (Cantab)
Lecturer, University of Cambridge

Contents

3.1 Arithmetic series

Sum of n terms,

$$S_n = a + (a+d) + (a+2d) + \ldots + [a+(n-1)d]$$
$$= n[2a+(n-1)d]/2$$
$$= n(a+l)/2$$

3.2 Geometric series

Sum of n terms,

$$S_n = a + ar + ar^2 + \ldots + ar^{n-1} = a(1-r^n)/(1-r)$$

$(|r|<1)$

$$S_\infty = a/(1-r)$$

3.3 Binomial series

$$(1+x)^p = 1 + px + \frac{p(p-1)}{2!}x^2 + \frac{p(p-1)(p-2)}{3!}x^3 + \ldots$$

If p is a positive integer the series terminates with the term in x^p and is valid for all x; otherwise the series does not terminate, and is valid only for $-1<x<1$.

3.4 Taylor's series

Infinite form

$$f(x+h) = f(x) + hf'(x) + \frac{h^2}{2!}f''(x) + \ldots + \frac{h^n}{n!}f^{(n)}(x) + \ldots$$

Finite form

$$f(x+h) = f(x) + hf'(x) + \frac{h^2}{2!}f''(x) + \ldots + \frac{h^n}{n!}f^{(n)}(x) + \frac{h^{n+1}}{(n+1)!}f^{(n+1)}(x+\lambda h)$$

where $0 \leqslant \lambda \leqslant 1$.

3.5 Maclaurin's series

$$f(x) = f(0) + xf'(0) + \frac{x^2}{2!}f''(0) + \ldots + \frac{x^n}{n!}f^{(n)}(0) + \ldots$$

Neither of these series is necessarily convergent, but both usually are for appropriate ranges of values of h and of x respectively.

3.6 Laurent's series

If a function $f(z)$ of a complex variable is analytic on and everywhere between two concentric circles centre a, then at any point in this region

$$f(z) = a_0 + a_1(z-a) + \ldots + b_1/(z-a) + b_2/(z-a)^2 + \ldots$$

This series is often applicable when Taylor's series is not.

3.7 Power series for real variables

	Math	*Comp*
$e^x = 1 + x + \frac{x^2}{2!} + \ldots$	all x	$\lvert x\rvert \leqslant 1$
$\ln(1+x) = x - \frac{x^2}{2} + \frac{x^3}{3} - \frac{x^4}{4} + \ldots$		$-1 < x \leqslant 1$
$\sin x = x - \frac{x^3}{3!} + \frac{x^5}{5!} - \frac{x^7}{7!} + \ldots$	all x	$\lvert x\rvert \leqslant 1$
$\cos x = 1 - \frac{x^2}{2!} + \frac{x^4}{4!} - \frac{x^6}{6!} + \ldots$	all x	$\lvert x\rvert \leqslant 1$
$\tan x = x + \frac{x^3}{3} + \frac{2x^5}{15} + \frac{17x^7}{315} + \ldots$		$\lvert x\rvert < \frac{\pi}{2}$
$\arctan x = x - \frac{x^3}{3} + \frac{x^5}{5} - \frac{x^7}{7} + \ldots$		$\lvert x\rvert \leqslant 1$
$\sinh x = x + \frac{x^3}{3!} + \frac{x^5}{5!} + \frac{x^7}{7!} + \ldots$	all x	$\lvert x\rvert \leqslant 1$
$\cosh x = 1 + \frac{x^2}{2!} + \frac{x^4}{4!} + \frac{x^6}{6!} + \ldots$	all x	$\lvert x\rvert \leqslant 1$

The column headed 'Math' contains the range of values of the variable x for which the series is convergent in the pure mathematical sense. In some cases a different range of values is given in the column headed 'Comp', to reduce the rounding errors which arise when computers are used.

3.8 Integer series

$$\sum_{n=1}^{N} n = 1+2+3+4+\ldots+N = N(N+1)/2$$

$$\sum_{n=1}^{N} n^2 = 1^2+2^2+3^2+4^2+\ldots+N^2 = N(N+1)(2N+1)/6$$

$$\sum_{n=1}^{N} n^3 = 1^3+2^3+3^3+4^3+\ldots+N^3 = N^2(N+1)^2/4$$

$$\sum_{n=1}^{\infty} \frac{(-1)^{n+1}}{n} = 1 - \frac{1}{2} + \frac{1}{3} - \frac{1}{4} + \ldots = \ln(2) \qquad \text{(see } \ln(1+x)\text{)}$$

$$\sum_{n=1}^{\infty} \frac{(-1)^{n+1}}{2n-1} = 1 - \frac{1}{3} + \frac{1}{5} - \frac{1}{7} + \ldots = \frac{\pi}{4} \qquad \text{(see } \arctan x\text{)}$$

$$\sum_{n=1}^{\infty} \frac{1}{n^2} = 1 + \frac{1}{4} + \frac{1}{9} + \frac{1}{16} + \ldots = \frac{\pi^2}{6}$$

$$\sum_{n=1}^{N} n(n+1)(n+2)\ldots(n+r)$$
$$= 1.2.3\ldots + 2.3.4\ldots + 3.4.5\ldots + \ldots + N(N+1)(N+2)\ldots(N+r)$$
$$= \frac{N(N+1)(N+2)\ldots(N+r+1)}{r+2}$$

3.9 Fourier series

$$f(\theta) = \tfrac{1}{2}a_0 + \sum_{n=1}^{\infty} (a_n \cos n\theta + b_n \sin n\theta)$$

with

$$a_n = \frac{1}{\pi}\int_0^{2\pi} f(\Theta)\cos n\Theta \, d\Theta$$

$$b_n = \frac{1}{\pi}\int_0^{2\pi} f(\Theta)\sin n\Theta \, d\Theta$$

or

$$f(\Theta) = \sum_{n=-\infty}^{\infty} c_n e^{jn\Theta}$$

with

$$c_n = \frac{1}{2\pi}\int_0^{2\pi} f(\Theta) e^{-jn\Theta} \, d\Theta = \begin{cases} \frac{1}{2}(a_n + jb_n) & n<0 \\ \frac{1}{2}(a_n - jb_n) & n>0 \end{cases}$$

3.10 Rectified sine wave

Half wave (*Figure 3.1*)

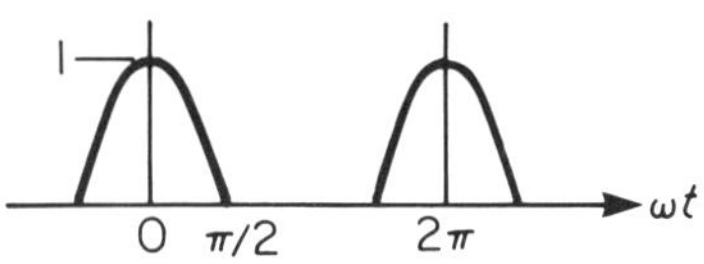

Figure 3.1 Half wave

$$f(\omega t) = \frac{1}{\pi} + \frac{1}{2}\cos \omega t + \frac{2}{\pi}\sum_{n=1}^{\infty}(-1)^{n+1}\frac{\cos 2n\omega t}{4n^2 - 1}$$

p-phase (*Figure 3.2*)

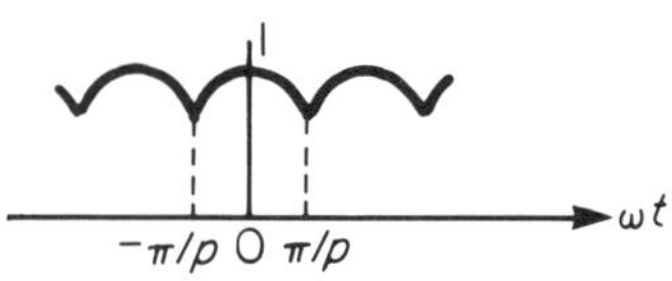

Figure 3.2 *p*-phase

$$f(\omega t) = \frac{\sin(\pi/p)}{\pi/p} + \frac{2p}{\pi}\sin\left(\frac{\pi}{p}\right)\sum_{n=1}^{\infty}(-1)^{n+1}\frac{\cos np\omega t}{p^2 n^2 - 1}$$

3.11 Square wave (*Figure 3.3*)

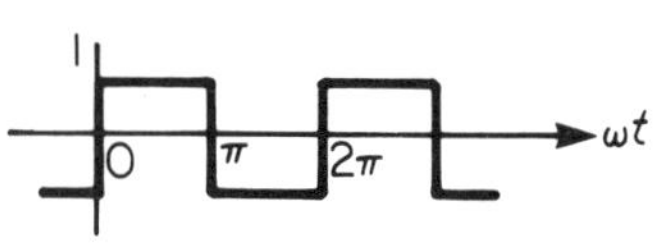

Figure 3.3 Square wave

$$f(\omega t) = \frac{4}{\pi}\sum_{n=1}^{\infty}\frac{\sin(2n-1)\omega t}{(2n-1)}$$

3.12 Triangular wave (*Figure 3.4*)

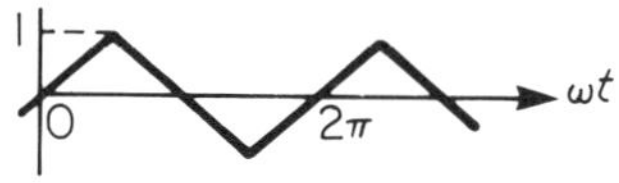

3.4 Triangular wave

$$f(\omega t) = \frac{8}{\pi^2}\sum_{n=1}^{\infty}(-1)^{n+1}\frac{\sin(2n-1)\omega t}{(2n-1)^2}$$

3.13 Sawtooth wave (*Figure 3.5*)

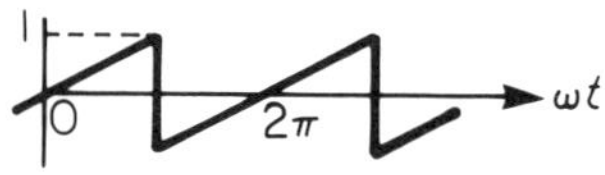

3.5 Sawtooth wave

$$f(\omega t) = \frac{2}{\pi}\sum_{n=1}^{\infty}(-1)^{n+1}\frac{\sin n\omega t}{n}$$

3.14 Pulse wave (*Figure 3.6*)

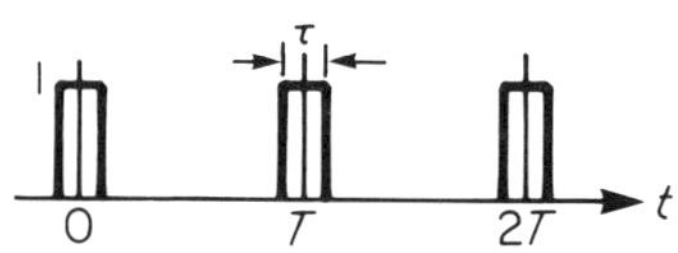

3.6 Pulse wave

$$f(t) = \frac{\tau}{T} + \frac{2\tau}{T}\sum_{n=1}^{\infty}\frac{\sin(n\omega\tau/T)}{n\pi\tau/T}\cos\left(\frac{2n\pi t}{T}\right)$$

3.15 Fourier transforms

Among other applications, these are used for converting from the time domain to the frequency domain.

Basic formulae:

$$\int_{-\infty}^{\infty} U(f)\exp(j2\pi ft)\,df = u(t) \leftrightarrows U(f) = \int_{-\infty}^{\infty} u(t)\exp(-j2\pi ft)\,dt$$

Change of sign and complex conjugates:

$u(-t) \leftrightarrows U(-f)$, $\quad u^*(t) \leftrightarrows U^*(-f)$

Time and frequency shifts (τ and φ constant):

$u(t-\tau) \leftrightarrows U(f)\exp(-j2\pi f\tau)\exp(j2\pi\varphi t)u(t) \leftrightarrows U(f-\varphi)$

Scaling (T constant):

$u(t/T) \leftrightarrows TU(fT)$

Products and convolutions:

$u(t) \dagger v(t) \leftrightarrows U(f)V(f)$, $\quad u(t)v(t) \leftrightarrows U(f) \dagger V(f)$

Differentiation:

$u'(t) \leftrightarrows j2\pi f U(f), \qquad -j2\pi t u(t) \leftrightarrows U'(f)$

$\partial u(t,\alpha)/\partial\alpha \leftrightarrows \partial(U-f,\alpha)/\partial\alpha$

Integration ($U(0)=0$, a and b real constants):

$$\int_{-\infty}^{t} u(\tau)\,d\tau \leftrightarrows U(f)/j2\pi f$$

$$\int_{a}^{b} v(t,\alpha)\,d\alpha \leftrightarrows \int_{a}^{b} V(f,\alpha)\,d\alpha$$

Interchange of functions:

$U(t) \leftrightarrows u(-f)$

Dirac delta functions:

$\delta(t) \leftrightarrows 1, \qquad \exp(j2\pi f_0 t) \leftrightarrows \delta(f-f_0)$

Rect(t) (unit length, unit amplitude pulse, centred on $t=0$):

$\mathrm{rect}(t) \leftrightarrows \sin \pi f/\pi f$

Gaussian distribution:

$\exp(-\pi t^2) \leftrightarrows \exp(-\pi f^2)$

Repeated and impulse (delta function) sampled waveforms:

$$\sum_{-\infty}^{\infty} u(t-nT) \leftrightarrows (1/T)U(f) \sum_{-\infty}^{\infty} \delta(f-n/T)$$

$$u(t) \sum_{-\infty}^{\infty} \delta(t-nT) \leftrightarrows (1/T) \sum_{-\infty}^{\infty} U(f-n/T)$$

Parseval's lemma:

$$\int_{-\infty}^{\infty} u(t)v^*(t)\,dt = \int_{-\infty}^{\infty} U(f)V^*(f)\,df$$

$$\int_{-\infty}^{\infty} |u(t)|^2\,dt = \int_{-\infty}^{\infty} |U(f)|^2\,df$$

3.16 Laplace transforms

$$\bar{x}_s = \int_0^{\infty} x(t)\,e^{-st}\,dt$$

Function	*Transform*	*Remarks*	
$e^{-\alpha t}$	$\dfrac{1}{s+\alpha}$		
$\sin \omega t$	$\dfrac{\omega}{s^2+\omega^2}$		
$\cos \omega t$	$\dfrac{s}{s^2+\omega^2}$		
$\sinh \omega t$	$\dfrac{\omega}{s^2-\omega^2}$		
$\cosh \omega t$	$\dfrac{s}{s^2-\omega^2}$		
t^n	$n!/s^{n+1}$		
1	$1/s$		
$H(t-\tau)$	$\dfrac{1}{s}e^{-s\tau}$	Heaviside step function	0
$x(t-\tau)H(t-\tau)$	$e^{-s\tau}\bar{x}(s)$	Shift in t	0
$\delta(t-\tau)$	$e^{-s\tau}$	Dirac delta function	0
$e^{-\alpha t}x(t)$	$\bar{x}(s+\alpha)$	Shift in s	
$e^{-\alpha t}\sin \omega t$	$\dfrac{\omega}{(s+\alpha)^2+\omega^2}$		
$e^{-\alpha t}\cos \omega t$	$\dfrac{(s+\alpha)}{(s+\alpha)^2+\omega^2}$		
$tx(t)$	$-\dfrac{d\bar{x}(s)}{ds}$		
$\dfrac{dx(t)}{dt}=x'(t)$	$s\bar{x}(s)-x(0)$		
$\dfrac{d^2x(t)}{dt^2}=x''(t)$	$s^2\bar{x}(s)-sx(0)-x'(0)$		
$\dfrac{d^nx(t)}{dx^n}=x^{(n)}(t)$	$s^n\bar{x}(s)-s^{n-1}x(0)-s^{n-2}x'(0)\ldots -sx^{(n-2)}(0)-x^{(n-1)}(0)$		

Convolution integral

$$\int_0^t x_1(\sigma)x_2(t-\sigma)\,d\sigma \rightarrow \bar{x}_1(s)\bar{x}_2(s)$$

4 Matrices and Determinants

J Barron BA, MA (Cantab)
Lecturer, University of Cambridge

Contents

4.1 Linear simultaneous equations

The set of equations

$$a_{11}x_1+a_{12}x_2+\ldots+a_{1n}x_n=b_1$$
$$a_{21}x_1+a_{22}x_2+\ldots+a_{2n}x_n=b_2$$
$$\ldots$$
$$a_{n1}x_1+a_{n2}x_2+\ldots+a_{nn}x_n=b_n$$

may be written symbolically

$$\mathbf{Ax}=\mathbf{b}$$

in which $\mathbf{A}$ is the *matrix* of the coefficients a_{ij}, and $\mathbf{x}$ and $\mathbf{b}$ are the *column matrices* (or vectors) $(x_1\ldots x_n)$, and $(b_1\ldots b_n)$. In this case the matrix $\mathbf{A}$ is square $(n\times n)$. The equations can be solved unless two or more of them are not independent, in which case

$$\det\mathbf{A}=|\mathbf{A}|=0$$

and there then exist non-zero solutions x_i only if $\mathbf{b}=0$. If $\det\mathbf{A}\neq 0$, there exist non-zero solutions only if $\mathbf{b}\neq 0$. When $\det\mathbf{A}=0$, $\mathbf{A}$ is *singular.*

4.2 Matrix arithmetic

If $\mathbf{A}$ and $\mathbf{B}$ are both matrices of m rows and n columns they are *conformable*, and

$$\mathbf{A}\pm\mathbf{B}=\mathbf{C}\quad\text{where } C_{ij}=A_{ij}\pm B_{ij}.$$

4.2.1 Product

If $\mathbf{A}$ is an $m\times n$ matrix and $\mathbf{B}$ an $n\times l$, the product $\mathbf{AB}$ is defined by

$$(\mathbf{AB})_{ij}=\sum_{k=1}^{n}(\mathbf{A})_{ik}(\mathbf{B})_{kj}$$

In this case, if $l\neq m$, the product $\mathbf{BA}$ will not exist.

4.2.2 Transpose

The transpose of $\mathbf{A}$ is written $\mathbf{A}'$ or $\mathbf{A}^t$ and is the matrix whose rows are the columns of $\mathbf{A}$, i.e.

$$(\mathbf{A}^t)_{ij}=(\mathbf{A})_{ji}$$

A square matrix may be equal to its transpose, and it is then said to be *symmetrical.* If the product $\mathbf{AB}$ exists, then

$$(\mathbf{AB})^t=\mathbf{B}^t\mathbf{A}^t$$

4.2.3 Adjoint

The *adjoint* of a square matrix $\mathbf{A}$ is defined as $\mathbf{B}$, where

$$(\mathbf{B})_{ij}=(\mathbf{A})_{ji}$$

and A_{ji} is the *cofactor* of a_{ji} in $\det\mathbf{A}$.

4.2.4 Inverse

If $\mathbf{A}$ is non-singular, the *inverse* $\mathbf{A}^{-1}$ is given by

$$\mathbf{A}^{-1}=\operatorname{adj}\mathbf{A}/\det\mathbf{A}\quad\text{and}\quad\mathbf{A}^{-1}\mathbf{A}=\mathbf{AA}^{-1}=\mathbf{I}$$

the *unit* matrix.

$$(\mathbf{AB})^{-1}=\mathbf{B}^{-1}\mathbf{A}^{-1}$$

if both inverses exist. The original equations $\mathbf{Ax}=\mathbf{b}$ have the solutions $\mathbf{x}=\mathbf{A}^{-1}\mathbf{b}$ if the inverse exists.

4.2.5 Orthogonality

A matrix $\mathbf{A}$ is orthogonal if $\mathbf{AA}^t=\mathbf{I}$. If $\mathbf{A}$ is the matrix of a coordinate transformation $\mathbf{X}=\mathbf{AY}$ from variables y_i to variables x_i, then if $\mathbf{A}$ is orthogonal $\mathbf{X}^t\mathbf{X}=\mathbf{Y}^t\mathbf{Y}$, or

$$\sum_{i=1}^{n}x_i^2=\sum_{i=1}^{n}y_i^2$$

4.3 Eigenvalues and eigenvectors

The equation

$$\mathbf{Ax}=\lambda\mathbf{x}$$

where $\mathbf{A}$ is a square matrix, $\mathbf{x}$ a column vector and λ a number (in general complex) has at most n solutions $(\mathbf{x},\lambda)$. The values of λ are *eigenvalues* and those of $\mathbf{x}$ *eigenvectors* of the matrix $\mathbf{A}$. The relation may be written

$$(\mathbf{A}-\lambda\mathbf{I})\mathbf{x}=0$$

so that if $\mathbf{x}\neq 0$, the equation $\mathbf{A}-\lambda\mathbf{I}=0$ gives the eigenvalues. If $\mathbf{A}$ is symmetric and real, the eigenvalues are real. If $\mathbf{A}$ is symmetric, the eigenvectors are orthogonal. If $\mathbf{A}$ is not symmetric, the eigenvalues are complex and the eigenvectors are not orthogonal.

4.4 Coordinate transformation

Suppose $\mathbf{x}$ and $\mathbf{y}$ are two vectors related by the equation

$$\mathbf{y}=\mathbf{Ax}$$

when their components are expressed in one orthogonal system, and that a second orthogonal system has unit vectors $\mathbf{u}_1,\mathbf{u}_2,\ldots,\mathbf{u}_n$ expressed in the first system. The components of $\mathbf{x}$ and $\mathbf{y}$ expressed in the new system will be $\mathbf{x}'$ and $\mathbf{y}'$, where

$$\mathbf{x}'=\mathbf{U}^t\mathbf{x},\qquad\mathbf{y}'=\mathbf{U}^t\mathbf{y}$$

and $\mathbf{U}^t$ is the orthogonal matrix whose rows are the unit vectors $\mathbf{u}_1^t$, $\mathbf{u}_2^t$, etc. Then

$$\mathbf{y}'=\mathbf{U}^t\mathbf{y}=\mathbf{U}^t\mathbf{Ax}=\mathbf{U}^t\mathbf{Ax}=\mathbf{U}^t\mathbf{AUx}'$$

or

$$\mathbf{y}'=\mathbf{A}'\mathbf{x}'$$

where

$$\mathbf{A}'=\mathbf{U}^t\mathbf{AU}$$

Matrices $\mathbf{A}$ and $\mathbf{A}'$ are *congruent.*

4.5 Determinants

The determinant

$$D=\begin{vmatrix}a_{11} & a_{12} & \ldots & a_{1n}\\ a_{21} & a_{22} & \ldots & a_{2n}\\ \vdots & \vdots & \vdots & \vdots\\ a_{n1} & a_{n2} & \ldots & a_{nn}\end{vmatrix}$$

is defined as follows. The first suffix in a_{rs} refers to the row, the second to the column which contains a_{rs}. Denote by M_{rs} the determinant left by deleting the rth row and sth column from D, then

$$D=\sum_{k=1}^{n}(-1)^{k+1}a_{1k}M_{1k}$$

gives the value of D in terms of determinants of order $n-1$, hence by repeated application, of the determinant in terms of the elements a_{rs}.

4.6 Properties of determinants

If the rows of $|a_{rs}|$ are identical with the columns of $|b_{sr}|$, $a_{rs}=b_{sr}$ and

$$|a_{rs}|=|b_{sr}|$$

that is, the *transposed* determinant is equal to the original.

If two rows or two columns are interchanged, the numerical value of the determinant is unaltered, but the sign will be changed if the permutation of rows or columns is odd.

If two rows or two columns are identical, the determinant is zero.

If each element of one row or one column is multiplied by k, so is the value of the determinant.

If any row or column is zero, so is the determinant.

If each element of the pth row or column of the determinant c_{rs} is equal to the sum of the elements of the same row or column in determinants a_{rs} and b_{rs}, then

$$|c_{rs}|=|a_{rs}|+|b_{rs}|$$

The addition of any multiple of one row (or column) to another row (or column) does not alter the value of the determinant.

4.6.1 Minor

If row p and column q are deleted from $|a_{rs}|$, the remaining determinant M_{pq} is called the *minor* of a_{pq}.

4.6.2 Cofactor

The *cofactor* of a_{pq} is the minor of a_{pq} prefixed by the sign which the product $M_{pq}a_{pq}$ would have in the expansion of the determinant, and is denoted by A_{pq}:

$$A_{pq}=(-1)^{p+q}M_{pq}$$

A determinant a_{ij} in which $a_{ij}=a_{ji}$ for all i and j is called *symmetric*, whilst if $a_{ij}=-a_{ji}$ for all i and j, the determinant is *skew-symmetric*. It follows that $a_{ii}=0$ for all i in a skew-symmetric determinant.

4.7 Numerical solution of linear equations

Evaluation of a determinant by direct expansion in terms of elements and cofactors is disastrously slow, and other methods are available, usually programmed on any existing computer system.

4.7.1 Reduction of determinant or matrix to upper triangular or to diagonal form

The system of equations may be written

$$\begin{bmatrix} a_{11} & a_{12} & \cdots & a_{1n} \\ a_{21} & a_{22} & \cdots & a_{2n} \\ \vdots & \vdots & \vdots & \vdots \\ a_{n1} & a_{n2} & \cdots & a_{nn} \end{bmatrix} \begin{matrix} \\ x_1 \\ \vdots \\ \end{matrix} \begin{bmatrix} x_1 \\ x_2 \\ \vdots \\ x_n \end{bmatrix} = \begin{bmatrix} b_1 \\ b_2 \\ \vdots \\ b_n \end{bmatrix}$$

The variable x_1 is eliminated from the last $n-1$ equations by adding a multiple $-a_{i1}/a_{11}$ of the first row to the ith, obtaining

$$\begin{bmatrix} a_{11} & a_{12} & \cdots & a_{1n} \\ 0 & a'_{22} & ,,, & a'_{2n} \\ \vdots & \vdots & \cdots & \vdots \\ 0 & 0 & \cdots & a''_{nn} \end{bmatrix} \begin{matrix} \\ x_1 \\ \vdots \\ \end{matrix} \begin{bmatrix} x_1 \\ x_2 \\ \vdots \\ x_n \end{bmatrix} = \begin{bmatrix} b_1 \\ b'_1 \\ \vdots \\ b''_n \end{bmatrix}$$

where primes indicate altered coefficients. This process may be continued by eliminating x_2 from rows 3 to n, and so on. Eventually the form will become

$$\begin{bmatrix} a_{11} & a_{12} & \cdots & a_{1n} \\ 0 & a'_{22} & \cdots & a'_{2n} \\ \vdots & \vdots & \cdots & \vdots \\ 0 & 0 & \cdots & a''_{nn} \end{bmatrix} \begin{matrix} \\ x_1 \\ \vdots \\ \end{matrix} \begin{bmatrix} x_1 \\ x_2 \\ \vdots \\ x_n \end{bmatrix} = \begin{bmatrix} b_1 \\ b'_2 \\ \vdots \\ b''_n \end{bmatrix}$$

x_n can now be found from the nth equation, substituted in the $(n-1)$th to obtain x_{n-1} and so on.

Alternatively the process may be applied to the system of equations in the form

$$\mathbf{Ax}=\mathbf{Ib}$$

where $\mathbf{I}$ is the unit matrix, and the same operations carried out upon $\mathbf{I}$ as upon $\mathbf{A}$. If the process is continued after reaching the upper triangular form, the matrix $\mathbf{A}$ can eventually be reduced to diagonal form. Finally, each equation is divided by the corresponding diagonal element of $\mathbf{A}$, thus reducing $\mathbf{A}$ to the unit matrix. The system is now in the form

$$\mathbf{Ix}=\mathbf{Bb}$$

and evidently $\mathbf{B}=\mathbf{A}^{-1}$. The total number of operations required is $O(n^3)$.

5 Electric Circuit Theory

P G Lund
Department of Engineering Science,
Oxford University

Contents

5.1 Types of source

If we were to measure the terminal voltage of a source as an increasing current was drawn we should find the relationships shown in *Figure 5.1(a)* where A is a line of constant voltage V_0 obtained when the generator is perfect and without internal impedance, and where B shows the practical case in which there is internal impedance Z and $V = V_0 - IZ$. The corresponding graphs for a constant current generator are shown in *Figure 5.1(b)* where A shows the perfect case and B the imperfect case. Apart from the lack of familiarity the difficulty with a constant current generator is that the perfect case calls for an infinite impedance in any equivalent circuit and a practical case requires an impedance which is large but not infinite. This can best be seen by considering the equations which represent line B in *Figure 5.1(b)*, i.e. $I = I_0 - VY$ or $I = I_0 - V/Z$, where Y is the admittance of an element and equals $1/Z$ where Z is the impedance.

The equivalent circuits for the two types of generator are usually shown as in *Figures 5.2(a)* and *(b)*.

In practice the constant current generator is a useful aid towards the understanding of many transistors in which, crudely but often sufficiently accurately, the output is a current constant over a range of loads.

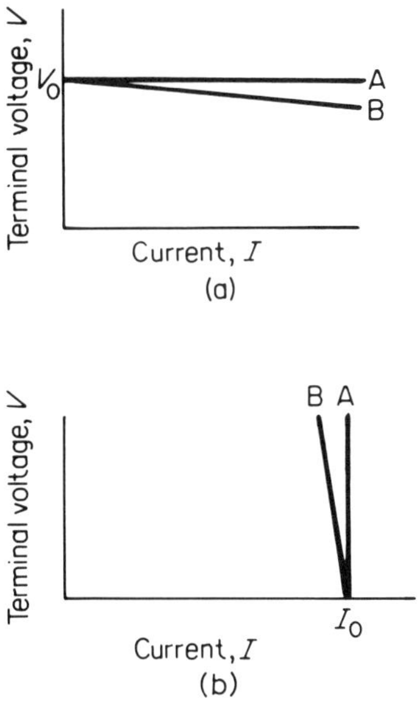

Figure 5.1 Types of sources: (a) constant voltage; (b) constant current

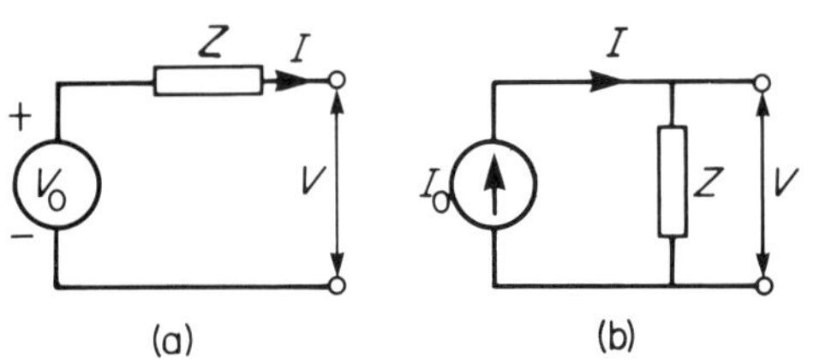

Figure 5.2 Equivalent circuit: (a) voltage source; (b) current source

5.2 Alternating current theory

Because it is possible to analyse all periodic functions of time into series of sinusoids and because of the mathematical properties of sinusoids we always consider currents and voltages varying in such a way that

$$\left.\begin{aligned} i &= I\sin\omega t = I\sin 2\pi f t \\ v &= V\sin\omega t = V\sin 2\pi f t \end{aligned}\right\} \tag{5.1}$$

or

$$\left.\begin{aligned} i &= I\cos\omega t = I\sin 2\pi f t \\ v &= V\cos\omega t = V\cos 2\pi f t \end{aligned}\right\} \tag{5.2}$$

where

i, v = instantaneous values of current, voltage
I, V = maximum or peak values of current, voltage
ω = angular frequency (in radians)
f = frequency (in cycles/s or hertz).

An alternative approach is to make use of complex number ideas based on de Moivre's theorem which states that

$$e^{j\theta} = \cos\theta + j\sin\theta \tag{5.3}$$

where $j = \sqrt{-1}$, so that we can write

$$v = V\,\mathrm{Re}\,(e^{j\omega t}) \tag{5.4}$$

for $V\cos\omega t$, or

$$v = V\,\mathrm{Im}\,(e^{j\omega t}) \tag{5.5}$$

for $V\sin\omega t$, where Re and Im stand for the real and imaginary parts of $e^{j\omega t}$. The convenient mathematical properties mentioned above are the simple forms of the differentials:

$$dv/dt = V\omega\cos\omega t \tag{5.6}$$

if $v = V\sin\omega t$, and

$$dv/dt = \mathrm{Re}\ \text{or}\ \mathrm{Im}\,(j\omega V e^{j\omega t}) \tag{5.7}$$

dependent on whether v was assumed to be the real or the imaginary part of $Ve^{j\omega t}$.

Integration produces similar expressions:

$$\int v\,dt = -(V/\omega)\cos\omega t \tag{5.8}$$

for $v = V\sin\omega t$, and

$$\int v\,dt = \mathrm{Re}\ \text{or}\ \mathrm{Im}\left(\frac{V}{j\omega}e^{j\omega t}\right) = \mathrm{Re}\ \text{or}\ \mathrm{Im}\left(-\frac{jV}{\omega}e^{j\omega t}\right) \tag{5.9}$$

The nature of these variations is shown in *Figure 5.3*. Other properties of sinusoids and various trigonometrical relations will be developed in later sections.

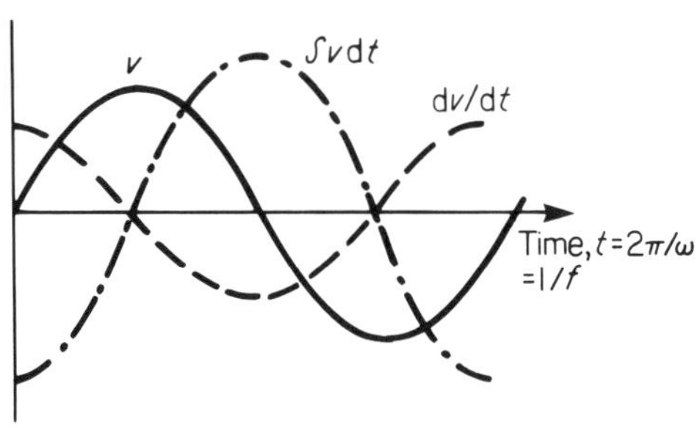

Figure 5.3 Sinusoidal variation

5.3 Resistance, inductance, capacitance, and related quantities

Resistance, inductance and capacitance are the three main elements that either absorb or store electrical energy. They constitute the passive as opposed to the active parts of a circuit such as voltage and current sources. The relationship between voltage and current for each is of fundamental importance.

5.3.1 Resistance

Ohm's law states that

$$v=Ri \tag{5.10}$$

The unit of resistance is the ohm (symbol Ω).

5.3.2 Inductance

$$v=L\frac{di}{dt} \tag{5.11}$$

The unit of inductance is the henry (symbol H).

This equation is sometimes found with a minus sign, but then the 'v' is voltage induced in the inductor which has to be overcome by an applied voltage. Because our purpose is to study circuits we think in terms of the applied voltage.

5.3.3 Capacitance

The basic relation for a capacitor is

$$q=vC \tag{5.12}$$

where q is charge in coulombs (symbol C) and C is capacitance in farads (symbol F). Differentiation leads to the more usual expressions

$$\frac{dq}{dt}=i=C\frac{dv}{dt} \tag{5.13}$$

or

$$v=\frac{1}{C}\int i\,dt \tag{5.14}$$

In some circumstances it is convenient to use the reciprocal quantities, the main occasion being when many elements are in parallel, as in *Figure 5.4*. If we introduce the conductance G (equal to $1/R$) then the conductance of the whole circuit becomes $G=G_1+G_2+G_3$ which is easier than solving

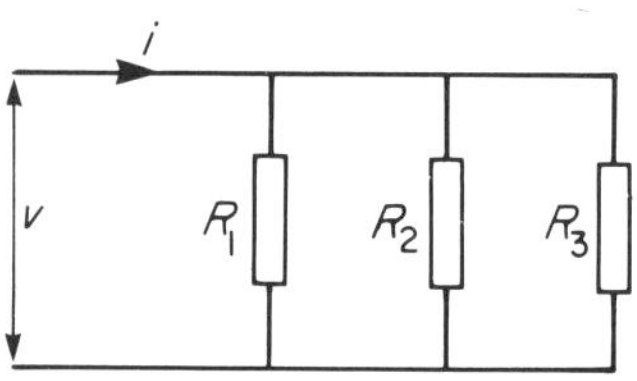

Figure 5.4 Parallel circuit to show advantage of using conductances

$$\frac{1}{R}=\frac{1}{R_1}+\frac{1}{R_2}+\frac{1}{R_3}$$

for R. Of course Ohm's law now takes the form $i=Gv$.

The unit of conductance is the siemen (symbol S). The phrase 'reciprocal ohms' (symbol ℧) is sometimes used.

Reciprocals for inductance and capacitance are not used in the same way but related quantities will be introduced when discussing a.c. circuits. Similarly the idea of impedance and its reciprocal, admittance, will be introduced later.

5.4 A.C. analysis of electric circuits

5.4.1 Mathematical approach

Consider a circuit such as that shown in *Figure 5.5* in which the elements are connected in series. The applied voltage must be

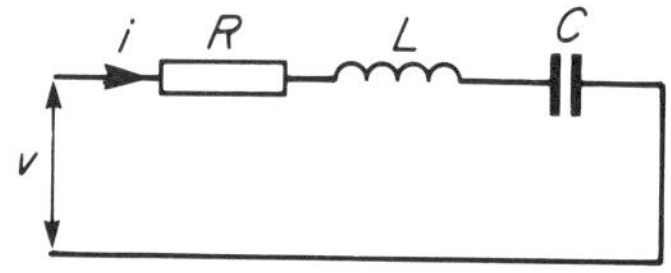

Figure 5.5 Alternating current series circuit

$$v=Ri+L\frac{di}{dt}+\frac{1}{C}\int i\,dt \tag{5.15}$$

If it is assumed that v and i vary sinusoidally i can be expressed as $I\,e^{j\omega t}$ and Equation (5.15) becomes

$$\begin{aligned} v &= \left(RI+j\omega LI+\frac{1}{j\omega C}I\right)e^{j\omega t} \\ &= \left(R+j\omega L-\frac{j}{\omega C}\right)I\,e^{j\omega t} \\ &= \left[R+j\left(\omega L-\frac{1}{\omega C}\right)\right]I\,e^{j\omega t} \end{aligned} \tag{5.16}$$

The expression $R+j(\omega L-1/\omega C)$ can be written

$$\sqrt{R^2+\left(\omega L-\frac{1}{\omega C}\right)^2}\;e^{j\varphi}$$

where

$$\varphi=\tan^{-1}\left(\frac{\omega L-1/\omega C}{R}\right)$$

The quantity $\sqrt{R^2+(\omega L-1/\omega C)^2}$ is known as the impedance, Z. Equation (5.16) can now be written

$$v=Z\,e^{j\varphi}I\,e^{j\omega t}=ZI\,e^{j(\omega t+\varphi)} \tag{5.17}$$

The product ZI is the magnitude V of this applied voltage and the angle φ is termed the phase angle.

5.4.2 Approach using phasor diagrams

Sinusoidal variations can be understood in terms of the projection onto a straight line of a point that moves round a circle, or of a radial line that sweeps round a circle as shown in *Figure 5.6*. If $\theta=\omega t$ then the lengths OA, OB represent $V\sin\omega t$ and $V\cos\omega t$ respectively. The line OP represents V and is known as a phasor. Such lines used to be called vectors but this is now regarded as misleading because vectors, as understood in

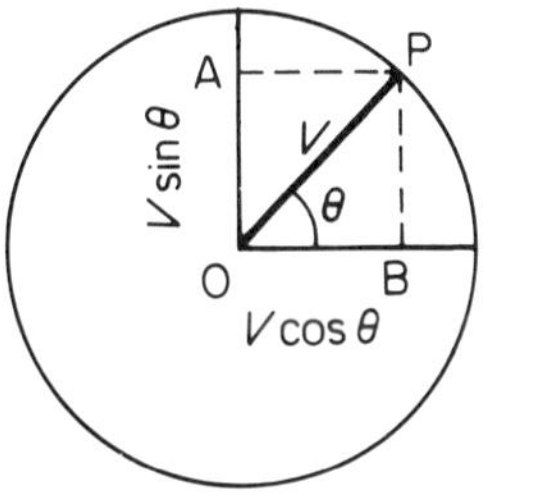

Figure 5.6 Phasor representation of sinusoidal variation

mechanics, represent *space*-dependent quantities such as force and momentum but the electrical quantities which we represent in a similar graphical manner are *time* dependent.

Figure 5.7 shows the phasor diagram for the circuit of *Figure 5.5*.

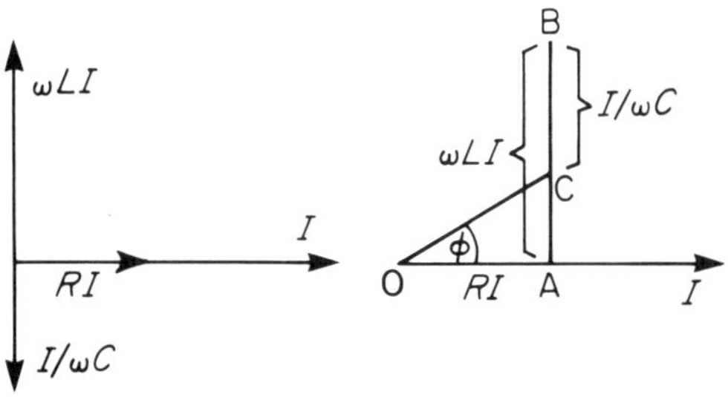

Figure 5.7 Phasor diagram for the series circuit of *Figure 5.5*

It will be realised that the triangle OAC is a graphical method of displaying and calculating the quantities Z and φ introduced in the mathematical approach. If $i=I\sin\omega t$ we can write $v=ZI\sin(\omega t+\varphi)$.

5.4.3 Summary

For a.c. circuits in which the frequency is ω the impedance to current flow of a simple series combination of R, L and C is denoted by $Z=\sqrt{R^2+(\omega L-1/\omega C)^2}$ so that, confining our attention to magnitudes, $V=IZ$ or $I=V/Z$. There also exists a phase shift because the voltage and current will not, in general, be in phase, i.e. their maxima, minima, and zero values will not occur at the same instant of time.

Figure 5.8 shows how voltage and current vary with time. In this case the current lags behind the voltage or the voltage leads

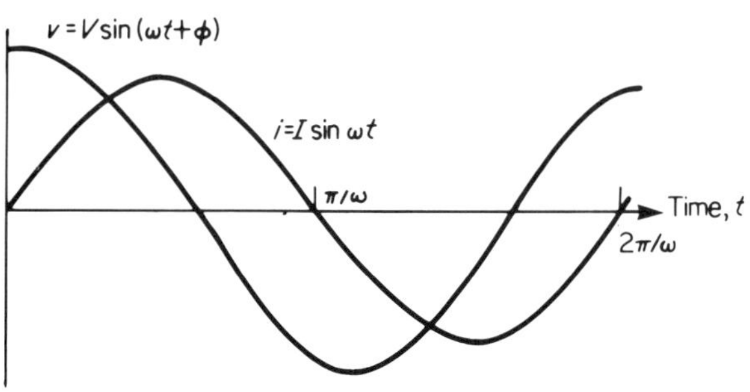

Figure 5.8 Voltage and current in a circuit in which current lags behind voltage

the current. If the effect of the capacitor were to exceed that of the inductor (i.e. if $I/\omega C>\omega LI$) the current would lead the voltage.

It should be appreciated that although diagrams could be drawn to scale and the answer for, say, the impedance could be measured, it is much more common to sketch the diagrams and use trigonometry to find the answer.

5.5 Impedance, reactance, admittance and susceptance

The impedance Z is best regarded as the quantity used in a.c. circuits analogous to resistance in d.c. circuits. In general it is a complex quantity consisting of two parts: the real part is resistance and represents energy dissipated in the circuit; while the imaginary part represents the energy stored and ultimately returned to the circuit. In this case, for an LCR series circuit,

$$Z=R+\mathrm{j}\left(\omega L-\frac{1}{\omega C}\right)$$

The terms representing the energy storage elements are $(\omega L-1/\omega C)$ and we introduce the term reactance, X, to represent the energy storage element. Consequently $Z=R+\mathrm{j}X$. The units of Z and X must be the same as those of R, that is ohms. The inverse of impedance Z is the admittance Y and that of reactance X is susceptance B. As the reciprocal of resistance R is conductance G we can write $Y=G+\mathrm{j}B$.

5.6 Technique for a.c. circuits

Drawing on the example of the series LCR circuits we can form general rules. Considering a frequency ω we write $\mathrm{j}\omega L$ for every inductance, $1/\mathrm{j}\omega C$ for every capacitance and proceed as for a d.c. circuit containing only resistances. For example, consider the circuit in *Figure 5.9*, having written in $\mathrm{j}\omega L$ and $1/\mathrm{j}\omega C$ as necessary:

$$Z=\frac{R_2(R_1+\mathrm{j}\omega L)}{R_2+R_1+\mathrm{j}\omega L}+\frac{1}{\mathrm{j}\omega C_1}+\frac{R_3/\mathrm{j}\omega C_2}{R_3+1/\mathrm{j}\omega C_2}$$

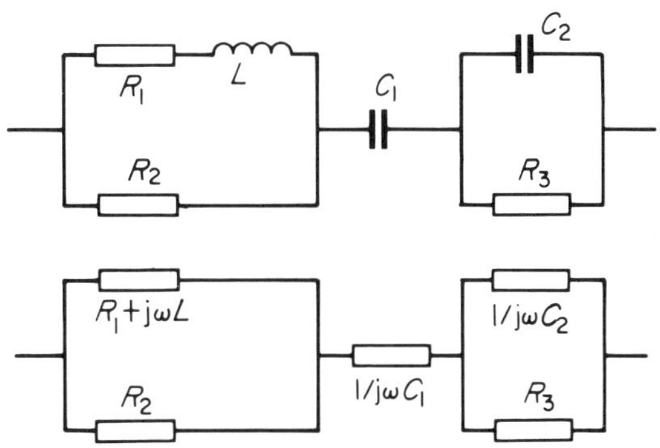

Figure 5.9 Circuit for analysis and first step in solution

This will reduce to the form $R+\mathrm{j}X$ enabling the relationship between overall voltage and current to be established. It is possible to achieve the same result using phasor diagrams, but when combinations of series and parallel elements are needed the work becomes tricky. The principles are that parts in series carry the same current whereas those in parallel experience the same voltage.

5.7 Average and r.m.s. values

So far we have dealt with the instantaneous values of voltage v and current i and the peak or maximum V and I. The true average of any quantity which is varying sinusoidally is zero. However, circumstances exist, such as full wave rectification, where we have waveforms like that shown in *Figure 5.10*. If the original voltage waveform was $v = V\sin\omega t$ the average of the fully rectified wave is $V_{av} = (2/\pi)V$.

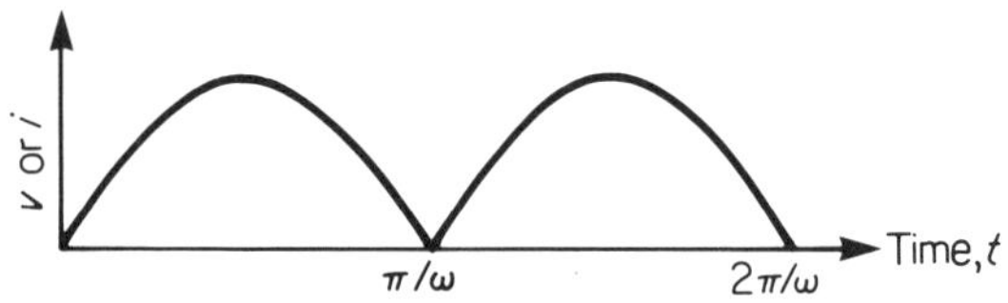

Figure 5.10 Full wave rectification

Another form of average is that associated with the current's ability to deliver power to a resistor R. If the current waveform is $i = I\sin\omega t$ then the power delivered is $i^2R = (I\sin\omega t)^2R$ averaged over a cycle. We introduce I_{rms} to represent this average such that $I_{rms} = (1/\sqrt{2})I$. The subscript stands for root mean square (r.m.s.) and those are the operations which, in the order square, mean and root, have been performed on the current.

Similarly we have $V_{rms} = (1/\sqrt{2})V$. It is the r.m.s. value of an a.c. quantity that is usually quoted and, unless stated to the contrary, it is to be assumed that a given voltage or current is the r.m.s. value.

5.8 Power, power factor

Instantaneously the power delivered to a circuit is given by $w = vi$ where w is the power and v and i the instantaneous voltage and current. The average power delivered is the mean of vi; therefore

$$W = \frac{\omega}{2\pi}\int_0^{2\pi/\omega} vi\,dt$$

If $v = V\sin\omega t$ and $i = I\sin(\omega t + \varphi)$

$$W = \frac{\omega}{2\pi}\int_0^{2\pi/\omega} VI\sin\omega t\sin(\omega t + \varphi)\,dt = \tfrac{1}{2}VI\cos\varphi \tag{5.18}$$

In this expression V and I are the peak values and φ is the phase angle. Introducing r.m.s. values we write

$$W = V_{rms}I_{rms}\cos\varphi \tag{5.19}$$

The quantity $\cos\varphi$ is known as the power factor and φ is sometimes known as the power factor angle. This form has the advantage that, for a circuit in which $\varphi = 0$, such as a pure resistance, the expression for power is simply $V_{rms}I_{rms}$ as would be expected from simple d.c. considerations.

Figure 5.11 shows how power varies through a cycle for different power factors. When current and voltage have the same polarity, either positive or negative, power is supplied to the circuit, and is shown shaded. When current and voltage have opposite polarities power is recovered from the circuit and is shown cross-hatched. The shaded areas represent power and it will be seen that power is always positive, i.e. being absorbed by the resistor in case I. In case II the shaded and cross-hatched areas are equal, indicating that energy is alternately stored and recovered with no net consumption. Case III is intermediate and it will be noticed that the shaded areas exceed the cross-hatched ones indicating a net consumption of power.

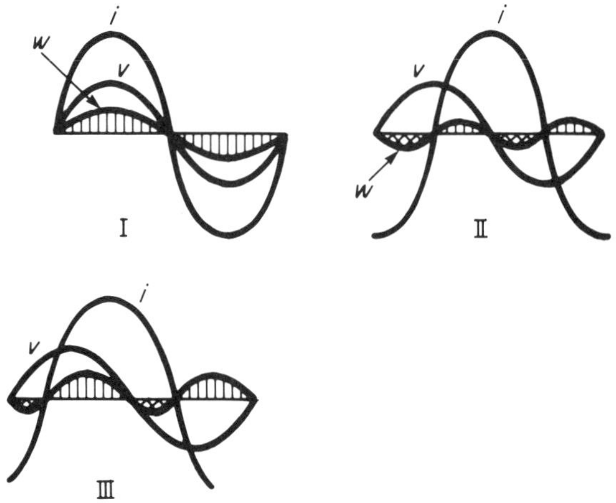

Figure 5.11 Power transfer in a.c. circuit. Case I: load entirely resistive, v and i in phase, maximum power w. Case II: load entirely inductive, i lags v by 90°, no net power. Case III load partially inductive, i lags v by less than 90°, some net power

5.9 Network laws and theorems: Kirchhoff's laws

The basis of systematic analysis of any circuit other than the most elementary is the two laws attributed to Kirchhoff. The first is the current law which states that the sum of all currents flowing into a node must be zero so in *Figure 5.12(a)*

$$i_1 + i_2 + i_3 + i_4 = 0$$

The second is the voltage law which states that there is no net change of voltage round a closed loop so in *Figure 5.12(b)* $V = iR_1 + iR_2$ when V is the rise of voltage from A to B and $iR_1 + iR_2$ is the fall of voltage from B through C and D to A.

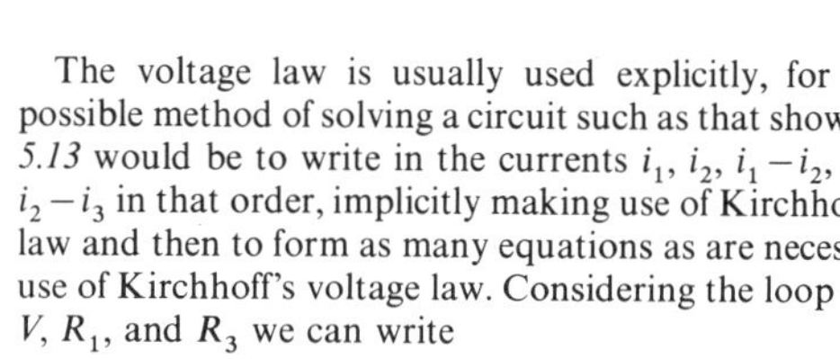

Figure 5.12 Kirchhoff's laws: (a) current law; (b) voltage law

The voltage law is usually used explicitly, for example a possible method of solving a circuit such as that shown in *Figure 5.13* would be to write in the currents i_1, i_2, $i_1 - i_2$, i_3 and $i_1 - i_2 - i_3$ in that order, implicitly making use of Kirchhoff's current law and then to form as many equations as are necessary by the use of Kirchhoff's voltage law. Considering the loop comprising V, R_1, and R_3 we can write

$$V = i_1R_1 + i_2R_3$$

Considering R_2, R_3 and R_4

$$0 = (i_1 - i_2)R_2 + i_3R_4 - i_2R_3$$

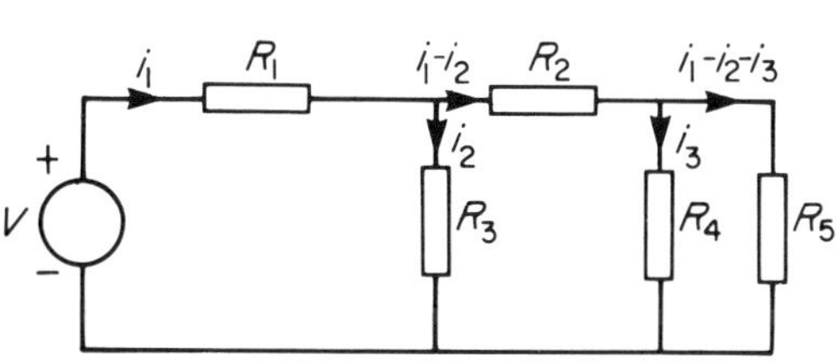

Figure 5.13 Illustration of Kirchhoff's laws

Considering R_4 and R_5

$$0=(i_1-i_2-i_3)R_5-i_3R_4$$

There are alternatives such as that obtained by considering V, R_1, R_2 and R_5 which would give

$$V=i_1R_1+(i_1-i_2)R_2+(i_1-i_2-i_3)R_5$$

However, this is not independent of the previous set of three equations as can be seen by adding these equations.

5.9.1 Loop or mesh currents

A device that is often helpful is to consider the current flowing round a loop rather than that actually in a wire. *Figure 5.14* shows the method. The current in any part is the sum of the loop currents which flow in the adjacent loops. If we apply the idea to the circuit of *Figure 5.13* we will consider loop currents i_4, i_5, i_6 as shown in *Figure 5.15*. In this case $i_1=i_4$, $i_2=i_4-i_5$, $i_3=i_5-i_6$.

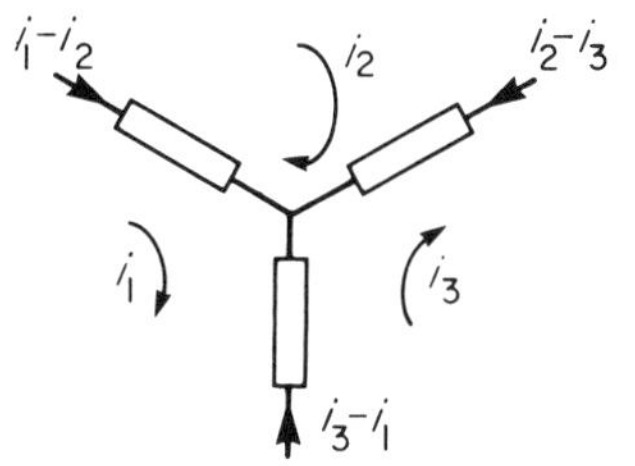

Figure 5.14 Loop currents

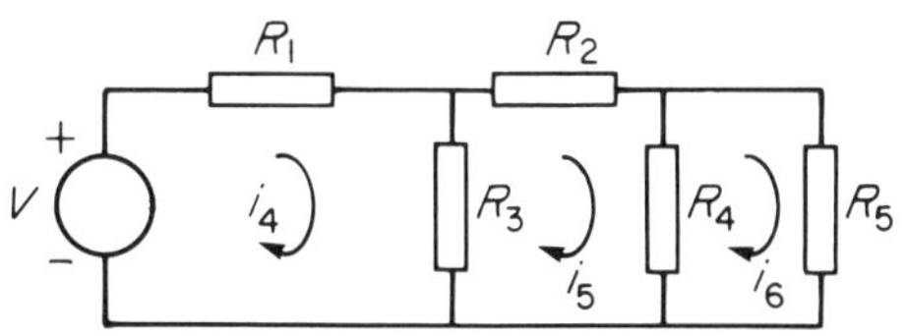

Figure 5.15 Illustration of loop currents

Writing the voltage loop equation we have the following equations. Considering loop i_1

$$V=i_4R_1+(i_4-i_5)R_3$$

Considering loop i_2

$$0=i_5R_2+(i_5-i_6)R_4+(i_5-i_4)R_3$$

Considering loop i_3

$$0=i_6R_5+(i_6-i_5)R_4$$

These can be arranged in a systematic manner which makes checking easy and leads to a matrix solution:

$$\begin{aligned}V&=(R_1+R_3)i_4 && -R_3i_5 &&\\ 0&=-R_3i_4+(R_2+R_3+R_4)i_5 && -R_4i_6 &&\\ 0&= \quad -R_4i_5 && +(R_4+R_5)i_6 &&\end{aligned}$$

The expression is symmetrical about the leading diagonal and the terms of the leading diagonal are the resistances as one traverses the appropriate loop.

5.9.2 Superposition

If a circuit is linear and contains more than one source the principle of superposition may be useful. Linearity means that currents are proportional to voltages. This means that diodes etc., which do not conduct equally for both directions of applied voltage, and devices (such as incandescent light bulbs) where the resistance changes and so the current depends on the voltage to some power other than one, are excluded. Consider a circuit such as that shown in *Figure 5.16(a)*. Supposing we require the current I in R_2, the principle of superposition tells us that it is the sum of

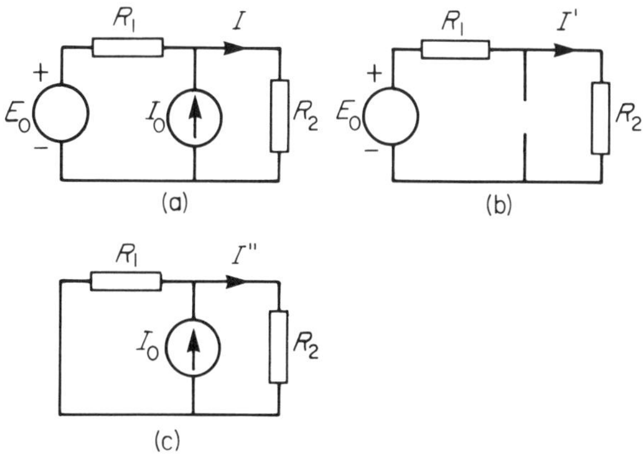

Figure 5.16 Illustration of superposition: (a) complete circuit; (b) current source deactivated or suppressed; (c) voltage source deactivated or suppressed

the currents I' and I'' where I' is the result of deactivating the current source and calculating the current due to the voltage source and I'' is the result of deactivating the voltage source and calculating the current due to the current source. A deactivated voltage source offers no impedance to the flow of current and so becomes a short circuit, whereas a deactivated current source offers an infinite impedance to the flow of current and so becomes an open circuit. Therefore *Figures 5.16(b)* and *(c)* show the two constituent parts and we can see that

$$I'=\frac{E_0}{R_1+R_2} \qquad I''=I_0\left(\frac{R_1}{R_1+R_2}\right)$$

and so

$$I=I'+I''=\frac{E_0+I_0R_1}{R_1+R_2}$$

5.9.3 Star–delta and delta–star transformations

These transformations may be useful in simplifying a circuit and are illustrated in *Figures 5.17* and *5.18*:

$$Y_{AB}=\frac{Y_A Y_B}{Y_A+Y_B+Y_C} \tag{5.20}$$

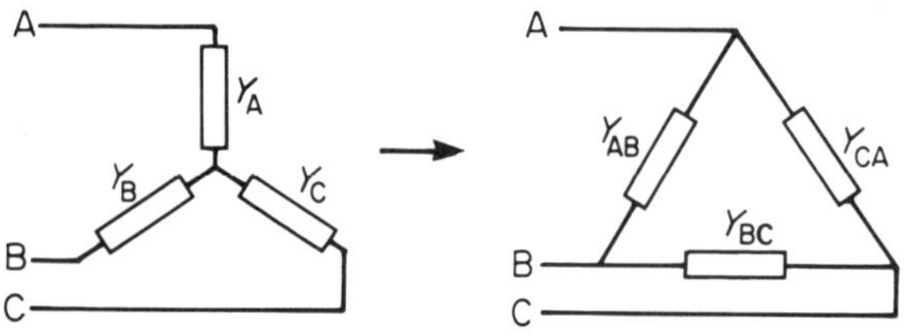

Figure 5.17 Star–delta transformation

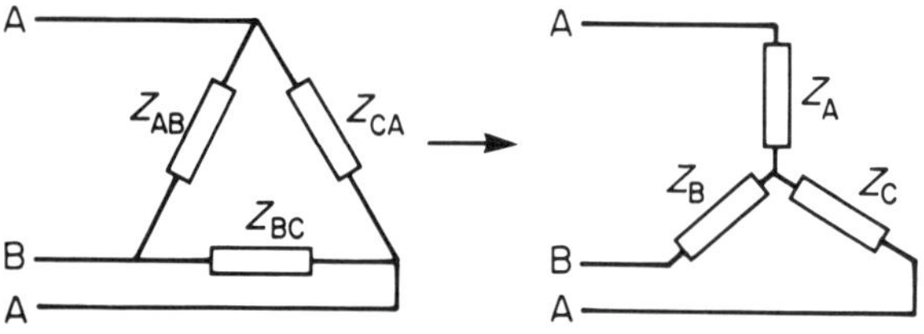

Figure 5.18 Delta–star transformation

Similarly for Y_{BC} and Y_{CA}.

$$Z_A = \frac{Z_{AB}Z_{CA}}{Z_{AB}+Z_{BC}+Z_{CA}} \tag{5.21}$$

Similarly for Z_B and Z_C.

5.10 Thevenin's theorem and Norton's theorem

Acceptance of these theorems enables one to analyse circuits consisting of one or more 'black boxes'. The so-called 'black box' approach encourages one to regard a piece of equipment in terms of what is observable at its terminals without regard to what is going on inside. Thevenin's and Norton's theorems give the quantities which must be used in any analysis. Previous sections have introduced current generators alongside voltage generators and the basic difference between Thevenin and Norton is that Thevenin expresses the circuit in terms of an equivalent voltage generator whereas Norton gives an equivalent current generator. Thevenin's theorem states that any two-terminal linear network can be represented by an ideal voltage source V_T in series with an impedance Z_T. The value of V_T is the voltage observed between the terminals when on open circuit and the value of Z_T is the impedance measured between the terminals with independent sources of voltage and current deactivated. The implications of deactivation were explained in Section 5.9.2. There is an alternative approach to the calculation of Z_T; it is that Z_T can be expressed as $Z_T = V_{oc}/I_{sc}$ when $V_{oc} = V_T$ and I_{sc} is the current which would flow in a short circuit placed across the terminals. Norton's theorem states that any two-terminal linear network can be represented by an ideal current generator I_N in parallel with an impedance. The value of the impedance is the same as that introduced with the Thevenin equivalent circuit but now it is in parallel with the source. The value of I_N equals I_{sc}, namely the short circuit current.

5.11 Resonance, 'Q' factor

If a circuit containing inductance and capacitance is subjected to a voltage of constant amplitude but varying frequency the current will have a maximum or minimum value for a particular frequency. Alternatively if the frequency is held fixed and the value of one of the components varied the same maximum or minimum will be observed. This phenomenon is known as resonance. To develop the important relations we will consider a simple circuit as shown in *Figure 5.19(a)* containing capacitance

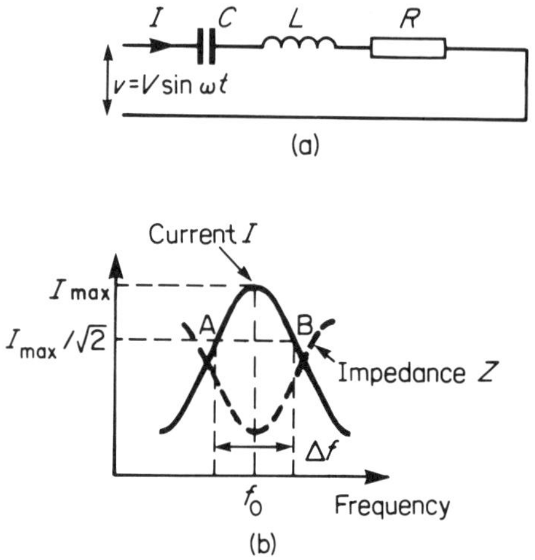

Figure 5.19 Resonance: (a) series circuit; (b) resonance curves

C, inductance L and resistance R. Using the techniques already developed for a.c. circuit analysis we write

$$v = \left(R + j\omega L + \frac{1}{j\omega C}\right) I\, e^{j\omega t} \tag{5.22}$$

where the instantaneous current i is given by $I\,e^{j\omega t}$. The instantaneous voltage v can be similarly represented by

$$v = V\, e^{j(\omega t+\varphi)} \tag{5.23}$$

where φ is the angle by which V leads I. It follows that V, the peak value or modulus of the voltage, is related to I, the peak value or modulus of the current, by the relationship

$$V = \sqrt{R^2 + (\omega L - 1/\omega C)^2}\, I \tag{5.24}$$

and the phase angle φ is given by

$$\tan \varphi = \frac{\omega L - 1/\omega C}{R} \tag{5.25}$$

If we consider the usual elementary situation in which the circuit is supplied with a voltage of constant amplitude but varying frequency then the current I is given by

$$I = \frac{V}{\sqrt{R^2 + (\omega L - 1/\omega C)^2}} \tag{5.26}$$

and a possible shape is sketched in *Figure 5.19(b)*.

The current has a maximum value at a frequency f_0 $(=\omega_0/2\pi)$ known as the resonant frequency. Its value is given by putting $\omega_0^2 = 1/LC$ in Equation (5.24). At this frequency it should be noticed that the phase angle φ is zero. If $\omega > \omega_0$ the phase angle is positive, that is, the voltage leads the current as is typical of an inductance; and if $\omega < \omega_0$ the phase angle is negative as is found with a capacitance. *Figure 5.19(b)* also shows the manner in which the impedance Z varies.

Resonance circuits, because of their frequency-dependent characteristics, are used for tuning purposes; that is, to pick out signals of one particular frequency from a range of signals, e.g. the tuning stages of a radio receiver. In these cases, of course, one of the circuit elements—normally the capacitor C—is varied until the resonant frequency equals that of the desired signal. For good selectivity the peak of the resonance curve needs to be sharp. The measure of the sharpness is the resonant frequency f_0 divided by the width Δf at some particular level. The level that is usually chosen is that at which the current has a value $1/\sqrt{2}$ of its maximum value. Bearing in mind that the power developed by a current in a resistor is I^2R it will be realised that if the current is $1/\sqrt{2}$ of its maximum then the power will be 1/2 of the maximum. These points are therefore known as the half power points (A and B in *Figure 5.19(b)*).

The symbol Q, defined as $f_0/\Delta f$ or $\omega_0/\Delta\omega$, is used for the sharpness as it measures the quality of the circuit when used as a tuning device. If we return to the simple series circuit shown in *Figure 5.19(a)* and if we draw the full phasor diagram as in *Figure 5.20* it will be seen that at resonance the voltages across both the

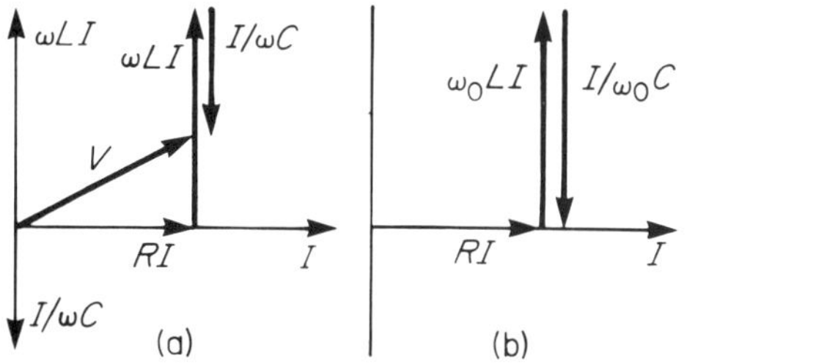

Figure 5.20 Phasor diagram for the circuit of *Figure 5.19(a)*: (a) circuit not in resonance; (b) circuit in resonance

inductor and the capacitor are equal. This leads to an alternative definition of Q, based on voltage magnification:

$$\frac{\text{Voltage across inductor}}{\text{Total voltage}}=\frac{\omega_0 LI}{RI}=\frac{\omega_0 L}{R} \tag{5.27}$$

$$\frac{\text{Voltage across capacitor}}{\text{Total voltage}}=\frac{I}{\omega_0 CRI}=\frac{1}{\omega_0 CR} \tag{5.28}$$

So we can define Q as $\omega_0 L/R$ or $1/\omega_0 CR$.

One should point out that, although voltage has been magnified, there has been no magnification of power. The voltages across both the inductor and capacitor are in quadrature with the current so no power is available. Looked at from the energy point of view, relatively large amounts of energy are stored in the inductor and the capacitor but they are always returned to the source.

This section has, so far, only considered series circuits; the results for a parallel circuit are similar except that whereas the series circuit has a minimum impedance at resonance the parallel one has a maximum. The analysis is complicated by the fact that a true parallel circuit (R, L and C in parallel) is not a correct representation of a real inductor connected in parallel with a real capacitor. It will be realised that to obtain a high Q in a series circuit it is necessary to have a small R. A real good quality capacitor has negligible resistance but a real inductor, made of a coil of wire, must possess some resistance. However, it can be shown that the practical situation of a real capacitor in parallel with a real inductor complete with parasitic resistance (*Figure 5.21(a)*) is equivalent to a true parallel circuit (*Figure 5.21(b)*) provided that Q is large and we put $C=C^1$, $L=L^1$ and $R=Q^2r$. In other words, the small parasitic 'r' in series is equivalent to a large R in parallel.

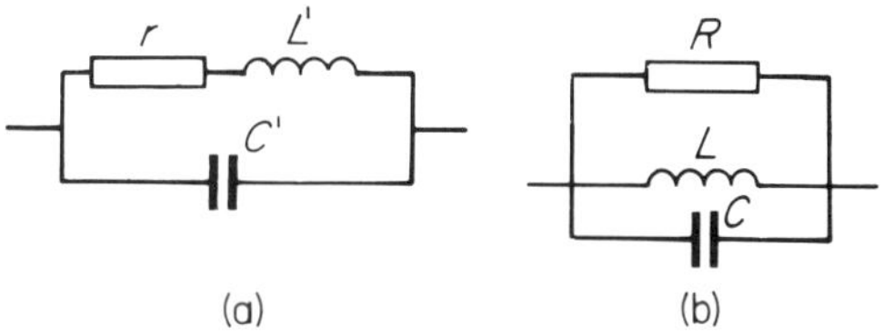

Figure 5.21 To show equivalence of real parallel circuit (a) with circuit (b) which is easier to analyse

5.12 Mutual inductance

This is the phenomenon whereby a changing current in one circuit produces a voltage in another. It is usually explained by appealing to the ideas of lines of magnetic flux which, when they change, produce an electromotive force in a circuit. Consider two coils as shown in *Figure 5.22*. If current in coil 1 is changing at a

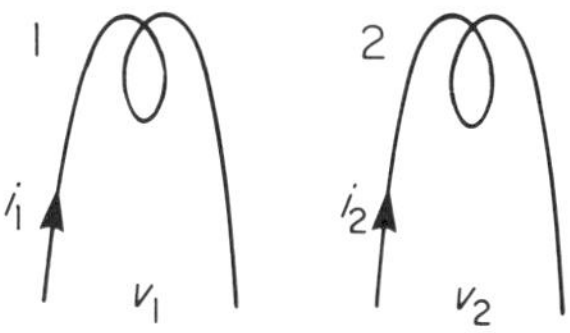

Figure 5.22 Illustration of mutual inductance (air cored)

rate $\mathrm{d}i_1/\mathrm{d}t$ then a voltage proportional to $\mathrm{d}i_1/\mathrm{d}t$ will be induced in coil 2. The coefficient of proportionality is known as the coefficient of mutual inductance M so that

$$v_2=M\,\mathrm{d}i_1/\mathrm{d}t \tag{5.29}$$

and

$$v_1=M\,\mathrm{d}i_2/\mathrm{d}t \tag{5.30}$$

The fact that the same value of M is used in both equations will probably be intuitively obvious; it can be proved by considering the energy stored in the coupled circuits when first one and then the other current is switched on. For two coils such as those shown in *Figure 5.22* the value of M will be much smaller than it would be if the coils were closer together to reduce leakage and if they were wound on an iron former. Much of the flux emanating from, say, coil 1 will not go through coil 2 and there is said to be a lot of leakage; this could be reduced by bringing the coils together. If, however, we were to wind the coils on an iron former the value of M would be much greater because more flux is produced due to ferromagnetism.

When analysing a circuit containing mutual inductance it is often helpful to redraw the mutual inductance adding generators as shown in *Figure 5.23*. The directions shown for positive current flow have the merit of symmetry but if one is considering

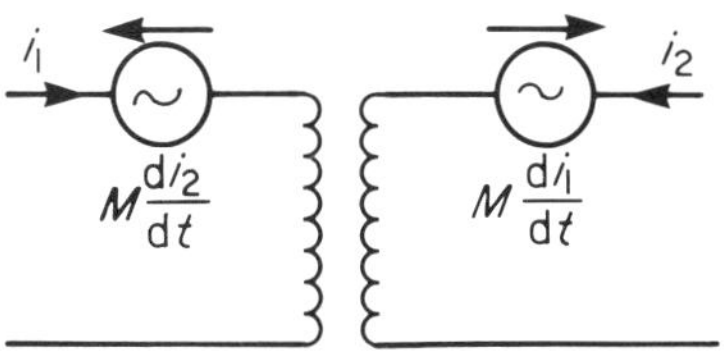

Figure 5.23 Circuit representation of mutual inductance

a transformer, the most commonly occurring example of mutual inductance, it is normal to think of an input current i, as shown, and of an output current, which would be $-i_2$ in *Figure 5.23*.

In line with the introduction to L and C and the analysis of a.c. circuits which has been presented earlier it should be pointed out that, if, as is often the case, we are dealing with sinusoidally varying quantitives, then $\mathrm{d}i/\mathrm{d}t$ becomes '$\mathrm{j}\omega i$'.

5.13 Differential equations and Laplace transforms

Equations containing integral and differential expressions are known as differential equations and many books have been devoted to their solution. To illustrate the process we will consider in more detail the equation developed in Section 5.4. The basic equation is (see Equation (5.15))

$$v = Ri + L\frac{\mathrm{d}i}{\mathrm{d}t} + \frac{1}{C}\int i\,\mathrm{d}t$$

An alternative form which may appear simpler, because it contains only differential coefficients and not a mixture of differential coefficients and integrals, is obtained by differentiating throughout and rearranging to give

$$L\frac{\mathrm{d}^2 i}{\mathrm{d}t^2} + R\frac{\mathrm{d}i}{\mathrm{d}t} + \frac{i}{C} = \frac{\mathrm{d}v}{\mathrm{d}t} \tag{5.31}$$

There are two parts to the solution of such an equation and the complete solution is the sum of both. The term on the right-hand side ($\mathrm{d}v/\mathrm{d}t$) is known as the forcing or driving function. In this case it is the voltage which is applied to the circuit even though in this analysis it is its rate of change that is used. Without a driving function nothing would happen unless, say, there was an initial charge on the capacitor in which case a solution is required to describe what happens as a result of this initial charge. The two parts of the solution referred to above are the complimentary function and the particular integral.

(*1*) *The complimentary function* is the solution to the equation with the independent variable, which is usually written on the right-hand side, put equal to zero. The complimentary function describes the result of any initial charges or currents. In a stable system it decays to zero and represents the transient behaviour.

(*2*) *The particular integral* This requires the inclusion of the independent variable, the driving voltage in the case we are considering. The particular integral is any solution to the full equation resulting from the inclusion of the forcing function. In our case it cannot be found until the amplitude and frequency of the driving voltage are known.

Traditionally, for this type of equation, the method of finding these two solutions goes as follows. For the complimentary function we let $i = A\,\mathrm{e}^{mt}$ and substitute into Equation (5.31) which produces values for m. For each m there will be a different value of A; these arbitrary constants depend on the initial conditions. The solution for the particular integral calls for an element of guesswork and intuition. If the driving function is sinusoidal it is a sensible guess that the response will also be sinusoidal and of the same frequency. It represents the steady-state, or long-term, solution.

All the guesswork can be taken out of the solution of these differential equations by the use of the Laplace transform. The effect of this is to transform the differential (and integral) expressions into straightforward algebraic equations. Once this has been done to the complete equation the transform of the dependent variable, in our case the current i, is expressed in terms of the other quantities and the use of the inverse transform produces the solution we require.

The Laplace transform of a function of time $f(t)$ is denoted by $F(s)$ and is defined as

$$F(s) = \int_0^\infty f(t)\,\mathrm{e}^{-st}\,\mathrm{d}t$$

The inverse transform is

$$f(t) = \frac{1}{2\pi\mathrm{j}}\int_{\sigma-\mathrm{j}\infty}^{\sigma+\mathrm{j}\infty} F(s)\,\mathrm{e}^{st}\,\mathrm{d}s$$

However, tables are frequently used rather than these formulae.

Use of the Laplace transform method quickly leads to the realisation that the equations can easily be set up by considering the impedance of an inductor to be sL and that of a capacitor to be $1/sC$. This is satisfactory as long as there are no initial currents or voltages respectively. The term generalised impedance is used for impedances expressed in this manner. The advantage of the method is that any type of input can be handled equally easily as long as the transform exists. In addition to sinusoids the other inputs that are encountered are step functions, repeated step functions or square waves and ramp functions. For sinusoids replace s with $\mathrm{j}\omega$ and we arrive back at the equation with which we started.

5.14 Transients and time constants

As the word transient implies, any phenomenon to which it is applied is short-lived and the time constant is a measure of its duration. A simple circuit that is often used to illustrate these ideas is that of a capacitor connected to a source of voltage through a switch and a resistor as shown in *Figure 5.24*(*a*).

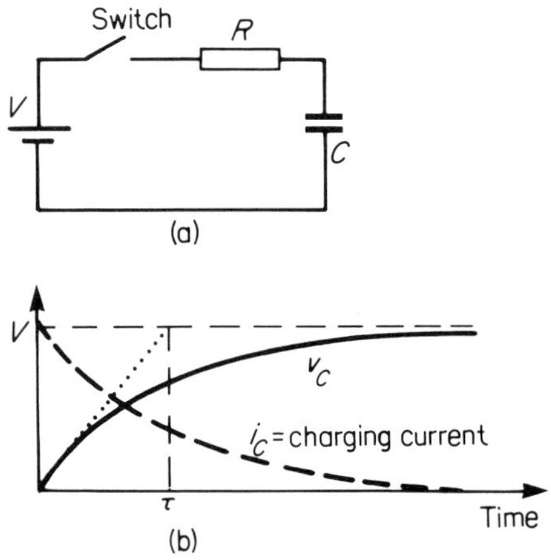

Figure 5.24 Illustration of transient phenomena: (a) circuit; (b) voltage and current graphs

If the capacitor is initially uncharged it can be shown that the voltage v_C across the capacitor at time t after closing the switch is given by

$$v_C = V(1 - \mathrm{e}^{-t/RC}) \tag{5.32}$$

The current in the circuit decays according to the expression

$$i=(V/R)e^{-t/RC} \quad (5.33)$$

The time constant τ can be regarded in two ways. Firstly it is defined, for this circuit, as being RC which means that it is the time it takes for the voltage to rise to $1-1/e=0.632$ of its final value. The second approach is to calculate how long it would take for v_C to reach V if the initial rate of change of v_C were to be maintained. The dotted line in *Figure 5.24(b)* is drawn tangential to the curve of v_C at the origin and illustrates this approach to the time constant.

If a circuit were to contain an inductor L instead of the capacitor the time constant would be L/R, and the current in a circuit containing an inductor and a resistor in series would be

$$i=\frac{V}{R}(1-e^{-Rt/L}) \quad (5.34)$$

The final current must be V/R because there will be no voltage across the inductor in the ultimate steady state. If one is faced with a circuit containing both inductance and capacitance the results are more complicated and, in the absence of resistance, there will be no steady long-term solution. The solution can best be approached by setting up the differential equation and using Laplace transforms to solve it. The closing of a switch to apply a constant voltage is to apply a 'step function' and this can be handled easily using Laplace transforms because the transform of a step is $1/s$ times the size of the step.

5.14.1 Pulses and square waves

The response of a circuit to a pulse is often easily understood by considering the response to two step functions of opposite signs and one delayed relative to the other, as shown in *Figure 5.25(a)*. The shape depends on the relative sizes of the time constant τ and the delay interval T. *Figure 5.25(c)* shows the situation when $\tau \simeq T$.

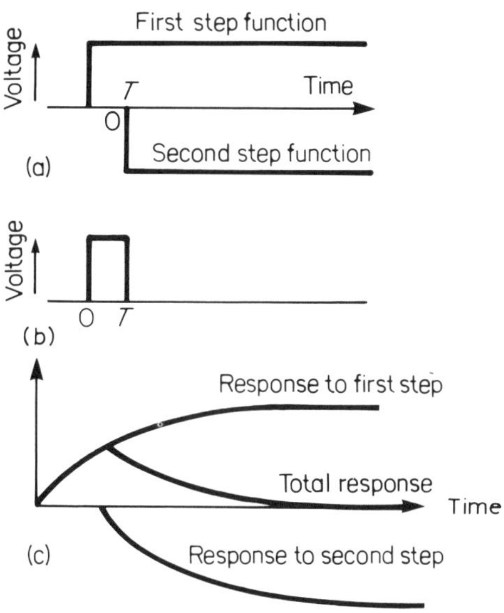

Figure 5.25 Effect of a pulse: (a) two-step representation; (b) pulse; (c) circuit response

If $\tau \ll T$ then the response would be much closer to the square wave input. If $\tau \gg T$ the output voltage would hardly change because the second and negative pulse would come so soon after the first and positive one.

Square waves are a series of such pulses and the response can be built up accordingly. The time gap between pulses and its relation to the time constant is the important quantity.

5.15 Three-phase circuits

Alternating current has many advantages over direct current when considering generation and transmission; no commutator is needed in the generator and the voltage can easily be changed to the level appropriate for economical transmission. However, single-phase a.c. systems are not as efficient as d.c. when considering the quantity of material required to transmit a given amount of energy. A further drawback to single-phase a.c. motors is that they are not inherently self-starting. A three-phase system overcomes both these disadvantages. Imagine a generator with three coils (R (red), Y (yellow) and B (blue)) on the stator and a magnet turning inside. The magnet could be a permanent one but is more likely to be an electromagnet fed from a d.c. source. If the pole faces are shaped so that the distribution of flux is sinusoidal then sinusoidal waves of voltage will be induced in the coils (see *Figure 5.26*).

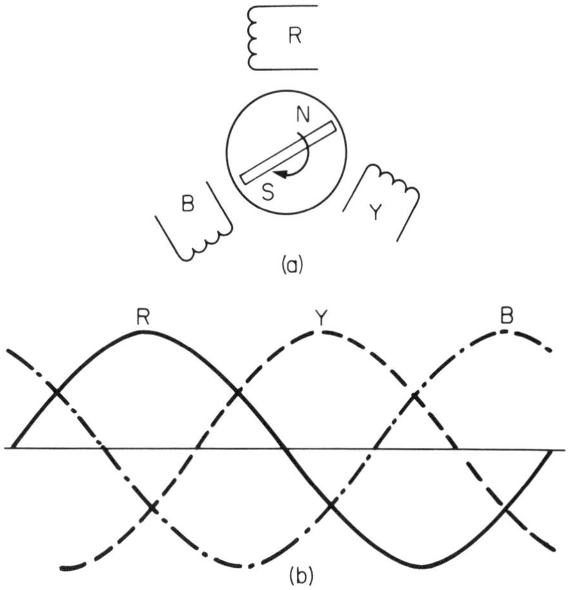

Figure 5.26 Elementary three-phase generator (a) and waveforms (b)

5.15.1 Star or Y connection

It is not necessary to carry all six wires from the three coils to the load. The easiest way to connect them together is to join one end of each coil to a common point and call that point the star or neutral point. This is known as a star connection as shown in *Figure 5.27*.

There are two ways in which the voltage can be expressed, namely the phase value or the line value. The three-phase values V_R, V_Y and V_B are equal in magnitude (V_ϕ) but differ in phase by 120°; similarly the line values V_{RB}, V_{BY} and V_{YR} are also equal in magnitude (V_L), but differ in phase by 120°. The relationship between V_L and V_ϕ is

$$V_L=\sqrt{3}\,V_\phi \quad (5.35)$$

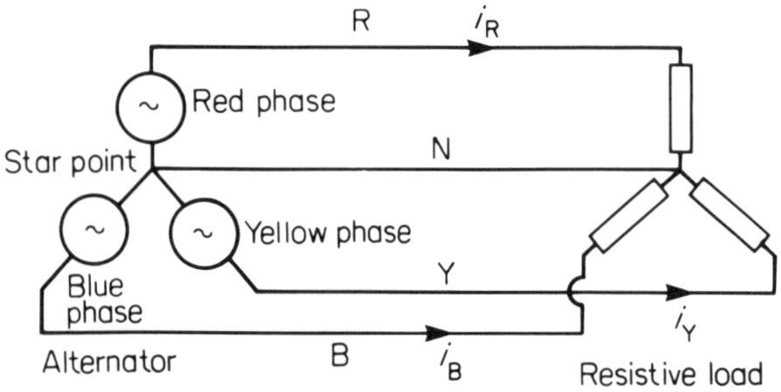

Figure 5.27 Star or Y connected alternator and load

If we consider the red phase to be the reference one we can express the driving voltages as follows:

$$v_R = V \sin \omega t \tag{5.36}$$

$$v_Y = V \sin(\omega t + 120°) \tag{5.37}$$

$$v_B = V \sin(\omega t + 240°) \tag{5.38}$$

If the impedances are of magnitude Z and produce a phase shift θ then the currents are

$$i_R = (V/Z) \sin(\omega t - \theta) \tag{5.39}$$

$$i_Y = (V/Z) \sin(\omega t + 120° - \theta) \tag{5.40}$$

$$i_B = (V/Z) \sin(\omega t + 240° - \theta) \tag{5.41}$$

In the neutral wire the current is the sum of these values and can easily be shown to be equal to zero, since the loads are equal (balanced).

If the loads are not balanced and there is no neutral conductor there will be a voltage between the star point of the supply and that of the load. In any analysis one introduces an unknown such as V_N to represent the drop between the star point of the source and the neutral point of the load. The voltages across the loads are written in terms of v_R, v_Y and v_B as defined above and of V_N remembering that this is a phasor quantity. The value of V_N is found from the equation which expresses the fact that the sum of the three line currents equals zero.

It can be shown that the total power consumed is

$$3V_\phi I_\phi \cos\theta = \sqrt{3}\, V_L I_L \cos\theta \tag{5.42}$$

5.15.2 Delta or mesh connection

The three phases of *Figure 5.26* could be connected together as shown in *Figure 5.28*. In this case $V_L = V_\phi$ and it can be shown that in magnitude $I_L = \sqrt{3}\, I_\phi$. Methods of analysis are similar and it is often helpful to use the delta–star transformation described in Section 5.9.3 to find the equivalent star-connected load. In general, as far as the supply is concerned, one does not know (or care) if it is star or delta connected and one is free to choose whichever is more convenient, which is usually star.

The result for the power consumed is the same for delta as for star, namely $\sqrt{3}\, V_L I_L \cos\theta$.

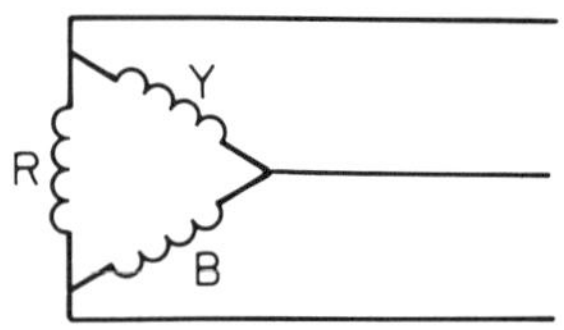

Figure 5.28 Delta or mesh connection

5.16 The decibel

When using the bel (symbol B), or more frequently the decibel (symbol dB) as a unit in measuring such electrical quantities as voltage and current it must be remembered that it is the power that the voltage or current could produce that is being considered. This power is proportional to the square of the voltage or current, consequently the expression in decibels is

$$20 \log_{10}(V/V_{ref}) \tag{5.43}$$

and similarly for current ratios.

These expressions assume the existence of a reference value of voltage V_{ref} or current I_{ref} and also that both the voltage or current being measured and the reference values are developing power in the same resistor. In electrical work, these have been chosen as 1 mW and 600 Ω which implies that $V_{ref} = \sqrt{0.6} = 0.775$. Information to this effect is often found in small figures somewhere on the dial of a decibel meter. If the same meter or one with the same value of V_{ref} is used for measuring both the input voltage (V_1) and the output voltage (V_2), the value of (V_2/V_1) dB is obtained by subtracting V_1 expressed in dB from V_2 expressed in dB. The following explains why:

$$(V_2 \text{ in dB}) - (V_1 \text{ in dB})$$
$$= 20 \log_{10}(V_2/V_{ref}) - 20 \log_{10}(V_1/V_{ref})$$
$$= 20 \log_{10}\left(\frac{V_2}{V_{ref}} \cdot \frac{V_{ref}}{V_1}\right)$$
$$= V_2/V_1 \text{ in dB}$$

It will be seen that V_{ref} cancels.

5.17 Frequency response and Bode diagrams

This is a most important description of the behaviour of a circuit; in the case of a voltage amplifier, for example, it is an expression of the manner in which the voltage gain varies with frequency. The information is often given graphically in the form of a Bode diagram, or occasionally in the form of an Argand diagram. This latter can be regarded either as a polar plot of the magnitude and phase relation or as a complex number plot.

There are two parts to a Bode diagram, one is a plot of the amplitude (in decibels) against frequency and the other is a plot of the phase shift against frequency. In both cases the frequency is plotted on a logarithmic scale. The transfer function (i.e. the expression for output in terms of input) can often be broken down into simple terms whose Bode diagrams can be quickly sketched. As the complete transfer function is the *product* of several such terms the complete Bode diagram is the *sum* of the Bode diagrams of the individual terms.

For example, consider the circuit in *Figure 5.29*, assuming sinusoidal conditions, the transfer function is:

$$\frac{e_2}{e_1}(j\omega) = \frac{Z_2}{Z_1 + Z_2}$$

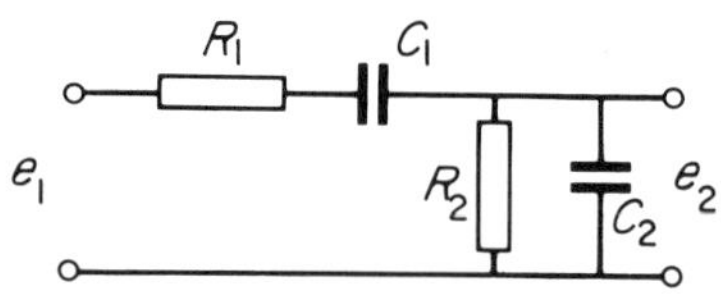

Figure 5.29 Circuit used to illustrate Bode diagrams

where

$$Z_2 = \frac{R_2}{1 + j\omega C_2 R_2} \quad \text{and} \quad Z_1 = R_1 + \frac{1}{j\omega C_1}$$

Analysis shows that, provided $C_1R_2 \ll C_1R_1 + C_2R_2$,

$$\frac{e_2}{e_1}(j\omega) = A\frac{j\omega C_1 R_1}{(1 + j\omega C_1 R_1)(1 + j\omega C_2 R_2)}$$

where A is a numerical factor independent of frequency. Alternatively, this can be expressed as

$$\frac{e_2}{e_1}(j\omega) = A\frac{j\omega/\omega_1}{(1 + j\omega/\omega_1)(1 + j\omega/\omega_2)}$$

where $\omega_1 = 1/R_1C_1$ and $\omega_2 = 1/R_2C_2$. If $R_1C_1 \gg R_2C_2$ then $\omega_1 \ll \omega_2$.

To draw the Bode diagram for the amplitude of this transfer function we express e_2/e_1 in decibels as follows:

$$\left(\frac{e_2}{e_1}\right)\text{dB} = 20\log_{10} A + 20\log_{10}\left(\frac{\omega/\omega_1}{\sqrt{1 + (\omega/\omega_1)^2}}\right)$$
$$- 20\log_{10}\sqrt{1 + (\omega/\omega_2)^2}$$

The Bode diagrams of the three terms are shown in *Figure 5.30*.

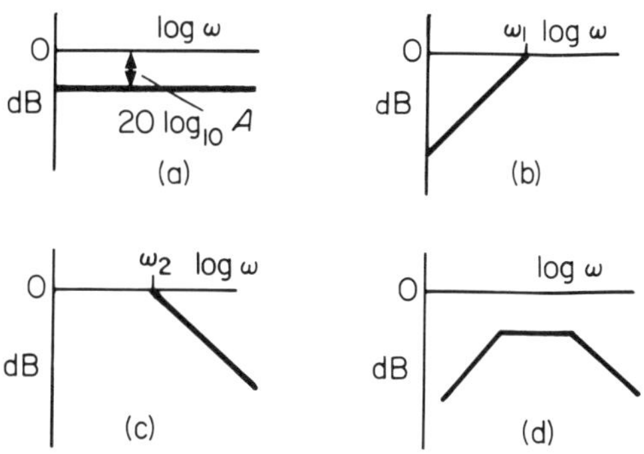

Figure 5.30 Bode diagrams for amplitude of transfer function of circuit shown in *Figure 5.29*

Part (a) shows $20\log_{10} A$ as a negative number because A will be less than 1. Part (b) shows

$$20\log_{10}\left(\frac{\omega/\omega_1}{\sqrt{1 + (\omega/\omega_1)^2}}\right)$$

which is best regarded as

$$20\log_{10}(\omega/\omega_1) - 20\log_{10}\sqrt{1 + (\omega/\omega_1)^2}$$

Part (c) shows

$$-20\log_{10}\sqrt{1 + (\omega/\omega_2)^2}$$

Part (d) shows the whole transfer function made by adding up the constituent parts. The sloping lines rise or fall at a rate of 20 dB for every decade, a decade being a tenfold change of frequency.

The other part of the Bode diagram showing phase is built up in a similar manner as shown in *Figure 5.31*. The first term

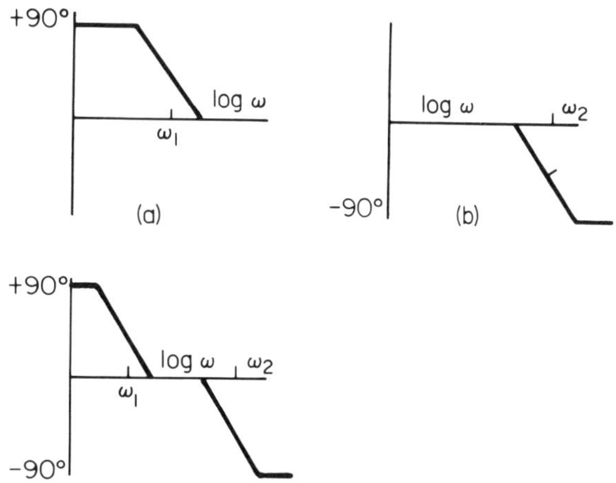

Figure 5.31 Bode diagram for phase shift of transfer function of circuit shown in *Figure 5.29*

introduces no phase shift, the second one a shift of $90° - \tan^{-1}(\omega/\omega_1)$ (*Figure 5.31(a)*) and the third a shift of $-\tan^{-1}(\omega/\omega_2)$.

In *Figures 5.30* and *5.31* the lines shown are the asymptotes to which the true curves approach.

6 Statistics

F F Mazda DFH, MPhil, CEng, MIEE, MBIM
Rank Xerox Ltd

Contents

6.1 Introduction

Data are available in vast quantities in all branches of electronic engineering. This chapter presents the more commonly used techniques for presenting and manipulating data to obtain meaningful results.

6.2 Data presentation

Probably the most common method used to present engineering data is by tables and graphs. For impact, or to convey information quickly, pictograms and bar charts may be used. Pie charts are useful in showing the different proportions of a unit.

A strata graph shows how the total is split amongst its constituents. For example, if a voltage is applied across four parallel circuits, then the total current curve may be as in *Figure 6.1*. This shows that the total current is made up of currents in the four parallel circuits, which vary in different ways with the applied voltage.

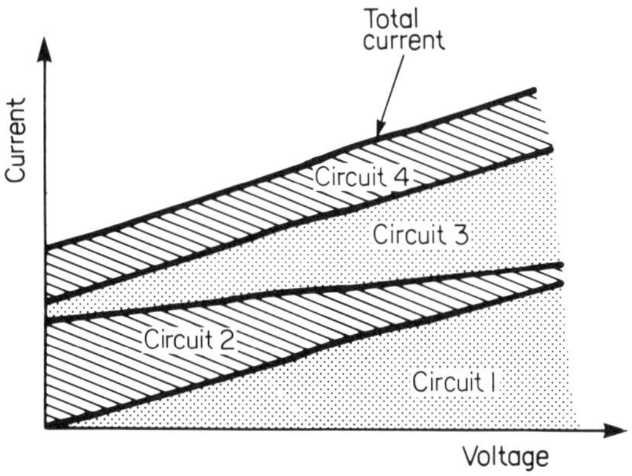

Figure 6.1 Illustration of a strata graph

Logarithmic or ratio graphs are used when one is more interested in the change in the ratios of numbers rather than their absolute value. In the logarithmic graph, equal ratios represent equal distances.

Frequency distributions are conveniently represented by a histogram as in *Figure 6.2*. This shows the voltage across a batch of diodes. Most of the batch had voltage drops in the region 0.65 to 0.75 V, the next largest group being 0.55 to 0.65 volts. In a histogram, the areas of the rectangles represent the frequencies in the different groups. Ogives, illustrated in *Figure 6.3*, show the cumulative frequency occurrences above or below a given value. From this curve it is possible to read off the total number of devices having a voltage greater than or less than a specific value.

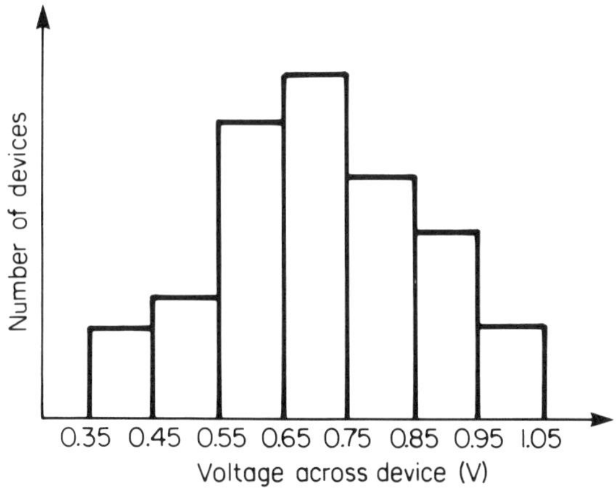

Figure 6.2 An histogram

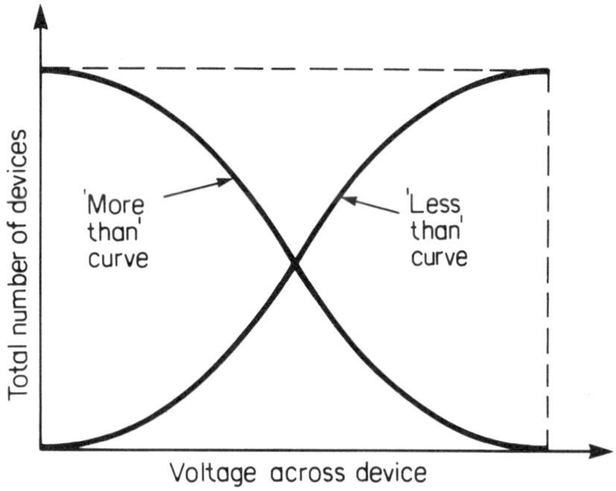

Figure 6.3 Illustration of ogives

6.3 Averages

6.3.1 Arithmetic mean

The arithmetic mean of n numbers $x_1, x_2, x_3, \ldots, x_n$ is given by

$$\bar{x} = \frac{x_1 + x_2 + x_3 + \cdots + x_n}{n}$$

or

$$\bar{x} = \frac{\sum_{r=1}^{n} x_r}{n} \tag{6.1}$$

The arithmetic mean is easy to calculate and it takes into account all the figures. Its disadvantages are that it is influenced unduly by extreme values and the final result may not be a whole number, which can be absurd at times, e.g. a mean of $2\frac{1}{2}$ men.

6.3.2 Median and mode

Median or 'middle one' is found by placing all the figures in order and choosing the one in the middle, or if there are an even number of items, the mean of the two central numbers. It is a useful technique for finding the average of items which cannot be expressed in figures, e.g. shades of a colour. It is also not influenced by extreme values. However, the median is not representative of all the figures.

The mode is the most 'fashionable' item, that is, the one which appears the most frequently.

6.3.3 Geometric mean

The geometric mean of n numbers $x_1, x_2, x_3, \ldots, x_n$ is given by

$$x_g = \sqrt[n]{(x_1 \times x_2 \times x_3 \times \ldots \times x_n)} \tag{6.2}$$

This technique is used to find the average of quantities which follow a geometric progression or exponential law, such as rates of changes. Its advantage is that it takes into account all the numbers, but is not unduly influenced by extreme values.

6.3.4 Harmonic mean

The harmonic mean of n numbers $x_1, x_2, x_3, \ldots, x_n$ is given by

$$x_h = \frac{n}{\sum_{r=1}^{n} (1/x_r)} \tag{6.3}$$

This averaging method is used when dealing with rates or speeds or prices. As a rule when dealing with items such as A per B, if the figures are for equal As then use the harmonic mean but if they are for equal Bs use the arithmetic mean. So if a plane flies over three equal distances at speeds of 5 m/s, 10 m/s and 15 m/s the mean speed is given by the harmonic mean as

$$\frac{3}{\frac{1}{5}+\frac{1}{10}+\frac{1}{15}} = 8.18 \text{ m/s}$$

If, however, the plane were to fly for three equal times, of say, 20 seconds at speeds of 5 m/s, 10 m/s and 15 m/s, then the mean speed would be given by the arithmetic mean as $(5+10+15)/3 = 10$ m/s.

6.4 Dispersion from the average

6.4.1 Range and quartiles

The average represents the central figure of a series of numbers or items. It does not give any indication of the spread of the figures, in the series, from the average. Therefore, in *Figure 6.4*, both curves, A and B, have the same average but B has a wider deviation from the average than curve A.

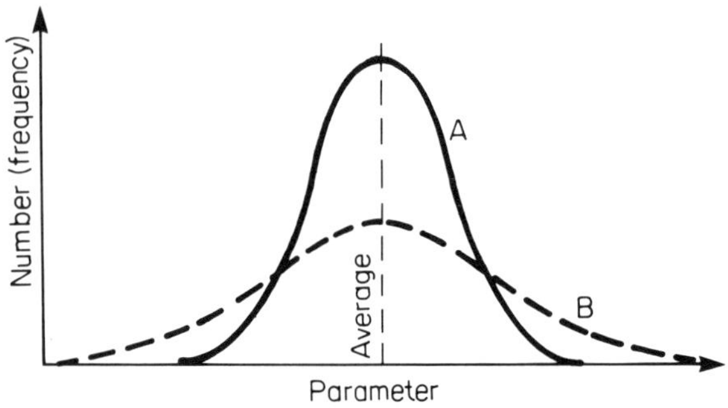

Figure 6.4 Illustration of deviation from the average

There are several ways of stating by how much the individual numbers, in the series, differ from the average. The range is the difference between the smallest and largest values. The series can also be divided into four quartiles and the dispersion stated as the interquartile range, which is the difference between the first and third quartile numbers, or the quartile deviation which is half this value.

The quartile deviation is easy to use and is not influenced by extreme values. However, it gives no indication of distribution between quartiles and covers only half the values in a series.

6.4.2 Mean deviation

This is found by taking the mean of the differences between each individual number in the series and the arithmetic mean, or median, of the series. Negative signs are ignored.

For a series of n numbers $x_1, x_2, x_3, \ldots, x_n$ having an arithmetic mean of $\bar{x}$ the mean deviation of the series is given by

$$\frac{\sum_{r=1}^{n} |x_r - \bar{x}|}{n} \tag{6.4}$$

The mean deviation takes into account all the items in the series. But it is not very suitable since it ignores signs.

6.4.3 Standard deviation

This is the most common measure of dispersion. For this the arithmetic mean must be used and not the median. It is calculated by squaring deviations from the mean, so eliminating their sign, adding the numbers together and then taking their mean and then the square root of the mean. Therefore, for the series in Section 6.4.2 the standard deviation is given by

$$\sigma = \left(\frac{\sum_{r=1}^{n} (x_r - \bar{x})^2}{n} \right)^{1/2} \tag{6.5}$$

The unit of the standard deviation is that of the original series. So if the series consists of the heights of a group of children in metres, then the mean and standard deviation are in metres. To compare two series having different units, such as the height of children and their weights, the coefficient of variation is used, which is unitless:

$$\text{coefficient of variation} = \frac{\sigma}{\bar{x}} \times 100 \tag{6.6}$$

6.5 Skewness

The distribution shown in *Figure 6.4* is symmetrical since the mean, median and mode all coincide. *Figure 6.5* shows a skewed distribution. It has positive skewness although if it bulges the other way, the skewness is said to be negative.

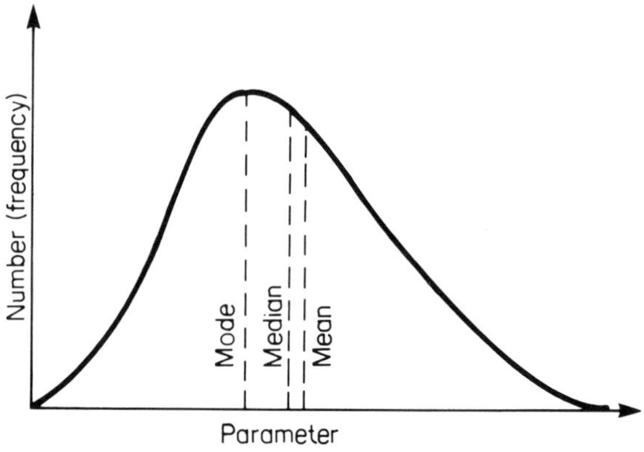

Figure 6.5 Illustration of skewness

There are several mathematical ways for expressing skewness. They all give a measure of the deviation between the mean, median and mode and they are usually stated in relative terms, for ease of comparison between series of different units. The Pearson coefficient of skewness is given by

$$P_k = \frac{\text{mean} - \text{mode}}{\text{standard deviation}} \tag{6.7}$$

Since the mode is sometimes difficult to measure this can also be stated as

$$P_k = \frac{3\ (\text{mean} - \text{median})}{\text{standard deviation}} \tag{6.8}$$

6.6 Combinations and permutations

6.6.1 Combinations

Combinations are the number of ways in which a proportion can be chosen from a group. Therefore the number of ways in which two letters can be chosen from a group of four letters A, B, C, D is equal to 6, i.e. AB, AC, AD, BC, BD, CD. This is written as

$$^4C_2 = 6$$

The factorial expansion is frequently used in combination calculations where

$$n! = n \times (n-1) \times (n-2) \times \cdots \times 3 \times 2 \times 1$$

Using this the number of combinations of n items from a group of n is given by

$$^nC_r = \frac{n!}{r!\,(n-r)!} \tag{6.9}$$

6.6.2 Permutations

Combinations do not indicate any sequencing. When sequencing within each combination is involved the result is known as a permutation. Therefore the number of permutations of two letters out of four letters A, B, C, D is 12, i.e. AB, BA, AC, CA, AD, DA, BC, CB, BD, DB, CD, DC. The number of permutations of r items from a group of n is given by

$$^nP_r = \frac{n!}{(n-r)!} \tag{6.10}$$

6.7 Regression and correlation

6.7.1 Regression

Regression is a method for establishing a mathematical relationship between two variables. Several equations may be used to establish this relationship, the most common being that of a straight line. *Figure 6.6* shows the plot of seven readings. This is called a scatter diagram. The points can be seen to lie approximately on the straight line AB.

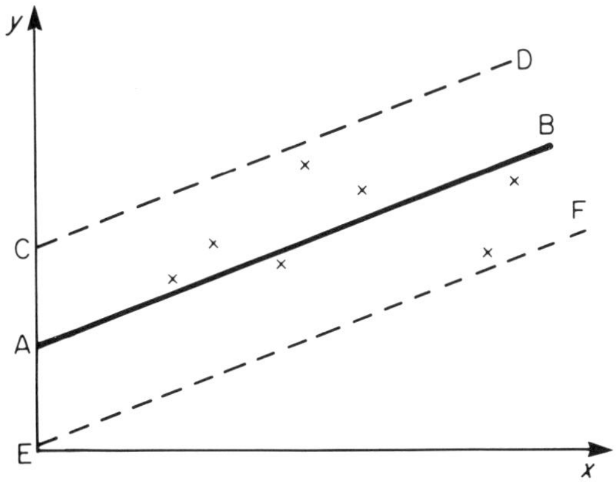

Figure 6.6 A scatter diagram

The equation of a straight line is given by

$$y = mx + c \tag{6.11}$$

where x is the independent variable, y the dependent variable, m is the slope of the line and c its interception on the y-axis. c is negative if the line intercepts the y-axis on its negative part and m is negative if the line slopes the other way to that shown in *Figure 6.6*.

The best straight line to fit a set of points is found by the method of least squares as

$$m = \frac{\sum xy - (\sum x \sum y)/n}{\sum x^2 - (\sum x)^2/n} \tag{6.12}$$

and

$$c = \frac{\sum x \sum xy - \sum y \sum x^2}{(\sum x)^2 - n \sum x^2} \tag{6.13}$$

where n is the number of points. The line passes through the mean values of x and y, i.e. $\bar{x}$ and $\bar{y}$.

6.7.2 Correlation

Correlation is a technique for establishing the strength of the relationship between variables. In *Figure 6.6* the individual figures are scattered on either side of a straight line and although one can approximate them by a straight line it may be required to establish if there is correlation between the x- and y-readings.

Several correlation coefficients exist. The product moment correlation coefficient (r) is given by

$$r = \frac{\sum (x-\bar{x})(y-\bar{y})}{n\sigma_x\sigma_y} \tag{6.14}$$

or

$$r = \frac{\sum (x-\bar{x})(y-\bar{y})}{\left[\sum (x-\bar{x})^2 \sum (y-\bar{y})^2\right]^{1/2}} \tag{6.15}$$

The value of r varies from $+1$, when all the points lie on a straight line and y increases with x, to -1, when all the points lie on a straight line but y decreases with x. When $r=0$ the points are widely scattered and there is said to be no correlation between x and y.

The standard error of estimation in r is given by

$$S_y = \sigma_y(1 - r^2)^{1/2} \tag{6.16}$$

In about 95% of cases, the actual values will lie between plus or minus twice the standard error of estimated values given by the regression equation. This is shown by lines CD and EF in *Figure 6.6*. Almost all the values will be within plus or minus three times the standard error of estimated values.

It should be noted that σ_y is the variability of the y-values, whereas S_y is a measure of the variability of the y-values as they differ from the regression which exists between x and y. If there is no regression then $r=0$ and $\sigma_y = S_y$.

It is often necessary to draw conclusions from the order in which items are ranked. For example, two judges may rank contestants in a beauty contest and we need to know if there is any correlation between their rankings. This may be done by using the Rank correlation coefficient (R) given by

$$R = 1 - \frac{6\sum d^2}{n^3 - n} \tag{6.17}$$

where d is the difference between the two ranks for each item and n is the number of items. The value of R will vary from $+1$ when the two ranks are identical to -1 when they are exactly reversed.

6.8 Probability

If an event A occurs n times out of a total of m cases then the probability of occurrence is stated to be

$$P(\mathrm{A}) = n/m \tag{6.18}$$

Probability varies between 0 and 1. If $P(\mathrm{A})$ is the probability of occurrence then $1-P(\mathrm{A})$ is the probability that event A will not occur and it can be written as $P(\bar{\mathrm{A}})$.

If A and B are two events then the probability that either may occur is given by

$$P(\mathrm{A\ or\ B}) = P(\mathrm{A}) + P(\mathrm{B}) - P(\mathrm{A\ and\ B}) \tag{6.19}$$

A special case of this probability law is when events are mutually exclusive, i.e. the occurrence of one event prevents the other from happening. Then

$$P(\mathrm{A\ or\ B}) = P(\mathrm{A}) + P(\mathrm{B}) \tag{6.20}$$

If A and B are two events then the probability that they may occur together is given by

$$P(\mathrm{A\ and\ B}) = P(\mathrm{A}) \times P(\mathrm{B|A}) \tag{6.21}$$

or

$$P(\mathrm{A\ and\ B}) = P(\mathrm{B}) \times P(\mathrm{A|B}) \tag{6.22}$$

$P(\mathrm{B|A})$ is the probability that event B will occur assuming that event A has already occurred and $P(\mathrm{A|B})$ is the probability that event A will occur assuming that event B has already occurred. A special case of this probability law is when A and B are independent events, i.e. the occurrence of one event has no influence on the probability of the other event occurring. Then

$$P(\mathrm{A\ and\ B}) = P(\mathrm{A}) \times P(\mathrm{B}) \tag{6.23}$$

Bayes' theorem on probability may be stated as

$$P(\mathrm{A|B}) = \frac{P(\mathrm{A})P(\mathrm{B|A})}{P(\mathrm{A})P(\mathrm{B|A}) + P(\bar{\mathrm{A}})P(\mathrm{B}|\bar{\mathrm{A}})} \tag{6.24}$$

As an example of the use of Bayes' theorem suppose that a company discovers that 80% of those who bought its product in a year had been on the company's training course. 30% of those who bought a competitir's product had also been on the same training course. During that year the company had 20% of the market. The company wishes to know what percentage of buyers actually went on its training course, in order to discover the effectiveness of this course.

If B denotes that a person bought the company's product and T that he went on the training course then the problem is to find $P(\mathrm{B|T})$. From the data $P(\mathrm{B}) = 0.2$, $P(\bar{\mathrm{B}}) = 0.8$, $P(\mathrm{T|B}) = 0.8$, $P(\mathrm{T}|\bar{\mathrm{B}}) = 0.3$. Then from Equation (6.24)

$$P(\mathrm{B|T}) = \frac{0.2 \times 0.8}{0.2 \times 0.8 + 0.8 \times 0.3} = 0.4$$

6.9 Probability distributions

There are several mathematical formulae with well-defined characteristics and these are known as probability distributions. If a problem can be made to fit one of these distributions then its solution is simplified. Distributions can be discrete when the characteristic can only take certain specific values, such as 0, 1, 2, etc., or they can be continuous where the characteristic can take any value.

6.9.1 Binomial distribution

The binomial probability distribution is given by

$$(p+q)^n = q^n + {}^nC_1 pq^{n-1} + {}^nC_2 p^2 q^{n-2} + \cdots + {}^nC_x p^x q^{n-x} + \cdots + p^n \tag{6.25}$$

where p is the probability of an event occurring, $q\ (=1-p)$ is the probability of an event not occurring and n is the number of selections.

The probability of an event occurring m successive times is given by the binomial distribution as

$$p(m) = {}^nC_m p^m q^{n-m} \tag{6.26}$$

The binomial distribution is used for discrete events and is applicable if the probability of occurrence p of an event is constant on each trial. The mean of the distribution $B(M)$ and the standard deviation $B(S)$ are given by

$$B(M) = np \tag{6.27}$$

$$B(S) = (npq)^{1/2} \tag{6.28}$$

6.9.2 Poisson distribution

The Poisson distribution is used for discrete events and, like the binomial distribution, it applies to mutually independent events. It is used in cases where p and q cannot both be defined. For example, one can state the number of goals which were scored in a football match, but not the goals which were not scored.

The Poisson distribution may be considered to be the limiting case of the binomial when n is large and p is small. The probability of an event occurring m successive times is given by the Poisson distribution as

$$p(m) = (np)^m \frac{e^{-np}}{m!} \tag{6.29}$$

The mean $P(M)$ and standard deviation $P(S)$ of the Poisson distribution are given by

$$P(M) = np \tag{6.30}$$

$$P(S) = (np)^{1/2} \tag{6.31}$$

Poisson probability calculations can be done by the use of probability charts as shown in *Figure 6.7*. This shows the probability that an event will occur at least m times when the mean (or expected) value np is known.

6.9.3 Normal distribution

The normal distribution represents continuous events and is shown plotted in *Figure 6.8*. The x-axis gives the event and the y-

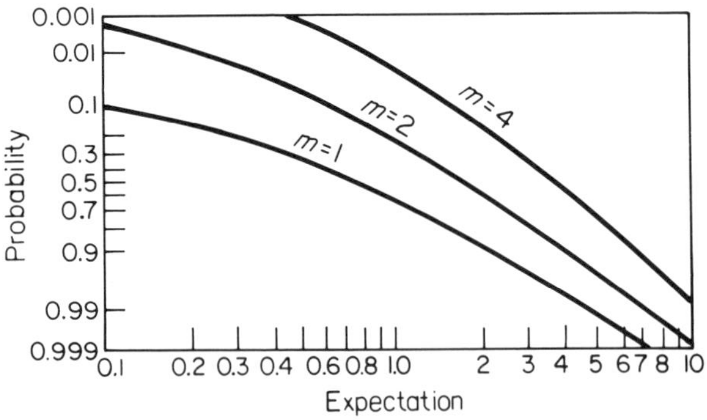

Figure 6.7 Poisson probability paper

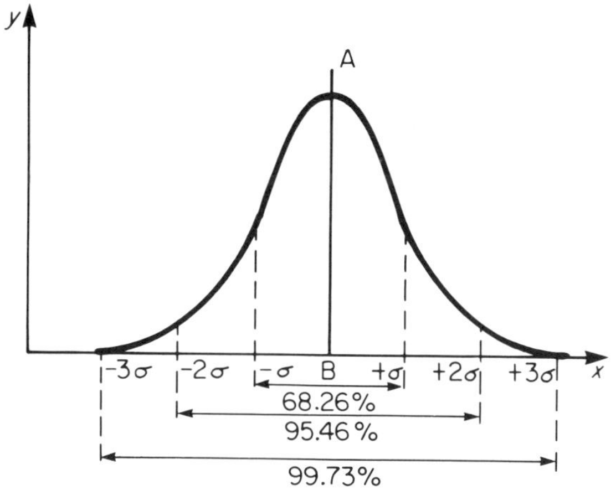

Figure 6.8 The normal curve

axis the probability of the event occurring. The curve shows that most of the events occur close to the mean value and this is usually the case in nature. The equation of the normal curve is given by

$$y=\frac{1}{\sigma(2\pi)^{1/2}}e^{-(x-\bar{x})^2/2\sigma^2} \tag{6.32}$$

where $\bar{x}$ is the mean of the values making up the curve and σ is their standard deviation.

Different distributions will have varying mean and standard deviations but if they are distributed normally then their curves will all follow Equation (6.32). These distributions can all be normalised to a standard form by moving the origin of their normal curve to their mean value, shown as B in *Figure 6.8*. The deviation from the mean is now represented on a new scale of units given by

$$\omega=\frac{x-\bar{x}}{\sigma} \tag{6.33}$$

The equation for the standardised normal curve now becomes

$$y=\frac{1}{(2\pi)^{1/2}}e^{-\omega^2/2} \tag{6.34}$$

The total area under the standardised normal curve is unity and the area between any two values of ω is the probability of an item from the distribution falling between these values. The normal curve extends infinitely in either direction but 68.26% of its values (area) fall between $\pm\sigma$, 95.46% between $\pm 2\sigma$, 99.73% between $\pm 3\sigma$ and 99.994% between $\pm 4\sigma$.

Table 6.1 gives the area under the normal curve for different values of ω. Since the normal curve is symmetrical the area from $+\omega$ to $+\infty$ is the same as from $-\omega$ to $-\infty$. As an example of the use of this table, suppose that 5000 street lamps have been installed in a city and that the lamps have a mean life of 1000 hours with a standard deviation of 100 hours. How many lamps will fail in the first 800 hours? From Equation (6.33)

$$\omega=(800-1000)/100=-2$$

Ignoring the negative sign, Table 6.1 gives the probability of lamps not failing as 0.977 so that the probability of failure is $1-0.977$ or 0.023. Therefore 5000×0.023 or 115 lamps are expected to fail after 800 hours.

Table 6.1 Area under the normal curve from $-\infty$ to ω

ω	0.00	0.02	0.04	0.06	0.08
0.0	0.500	0.508	0.516	0.524	0.532
0.1	0.540	0.548	0.556	0.564	0.571
0.2	0.579	0.587	0.595	0.603	0.610
0.3	0.618	0.626	0.633	0.640	0.648
0.4	0.655	0.663	0.670	0.677	0.684
0.5	0.692	0.700	0.705	0.712	0.719
0.6	0.726	0.732	0.739	0.745	0.752
0.7	0.758	0.764	0.770	0.776	0.782
0.8	0.788	0.794	0.800	0.805	0.811
0.9	0.816	0.821	0.826	0.832	0.837
1.0	0.841	0.846	0.851	0.855	0.860
1.1	0.864	0.869	0.873	0.877	0.881
1.2	0.885	0.889	0.893	0.896	0.900
1.3	0.903	0.907	0.910	0.913	0.916
1.4	0.919	0.922	0.925	0.928	0.931
1.5	0.933	0.936	0.938	0.941	0.943
1.6	0.945	0.947	0.950	0.952	0.954
1.7	0.955	0.957	0.959	0.961	0.963
1.8	0.964	0.966	0.967	0.969	0.970
1.9	0.971	0.973	0.974	0.975	0.976
2.0	0.977	0.978	0.979	0.980	0.981
2.1	0.982	0.983	0.984	0.985	0.985
2.2	0.986	0.987	0.988	0.988	0.989
2.3	0.989	0.990	0.990	0.991	0.991
2.4	0.992	0.992	0.993	0.993	0.993
2.5	0.994	0.994	0.995	0.995	0.995
2.6	0.995	0.996	0.996	0.996	0.996
2.7	0.997	0.997	0.997	0.997	0.997
2.8	0.997	0.998	0.998	0.998	0.998
2.9	0.998	0.998	0.998	0.998	0.999
3.0	0.999	0.999	0.999	0.999	0.999

6.9.4 Exponential distribution

The exponential probability distribution is a continuous distribution and is shown in *Figure 6.9*. It has the equation

$$y=\frac{1}{x}e^{-x/\bar{x}} \tag{6.35}$$

where $\bar{x}$ is the mean of the distribution. Whereas in the normal distribution the mean value divides the population in half, for the exponential distribution 36.8% of the population is above the average and 63.2% below the average. *Table 6.2* shows the area under the exponential curve for different values of the ratio $K=x/\bar{x}$, this area being shown shaded in *Figure 6.9*.

As an example suppose that the time between failures of a piece

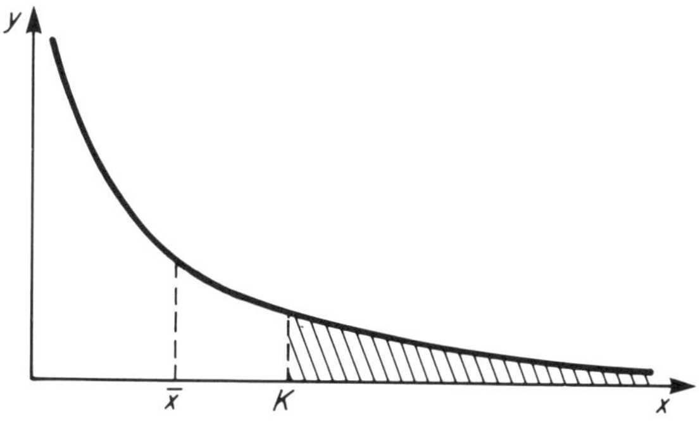

Figure 6.9 The exponential curve

Table 6.2 Area under the exponential curve from K to $+\infty$

K	0.00	0.02	0.04	0.06	0.08
0.0	1.000	0.980	0.961	0.942	0.923
0.1	0.905	0.886	0.869	0.852	0.835
0.2	0.819	0.803	0.787	0.771	0.776
0.3	0.741	0.726	0.712	0.698	0.684
0.4	0.670	0.657	0.644	0.631	0.619
0.5	0.607	0.595	0.583	0.571	0.560
0.6	0.549	0.538	0.527	0.517	0.507
0.7	0.497	0.487	0.477	0.468	0.458
0.8	0.449	0.440	0.432	0.423	0.415
0.9	0.407	0.399	0.391	0.383	0.375

of equipment is found to vary exponentially. If results indicate that the mean time between failures is 1000 hours, then what is the probability that the equipment will work for 700 hours or more without a failure? Calculating K as $700/1000=0.7$ then from *Table 6.2* the area beyond 0.7 is 0.497 which is the probability that the equipment will still be working after 700 hours.

6.9.5 Weibull distribution

This is a continuous probability distribution and its equation is given by

$$y=\alpha\beta(x-\gamma)^{\beta-1}\,e^{-\alpha(x-\gamma)^{\beta}} \tag{6.36}$$

where α is called the scale factor, β the shape factor and γ the location factor.

The shape of the Weibull curve varies depending on the value of its factors. β is the most important, as shown in *Figure 6.10*, and

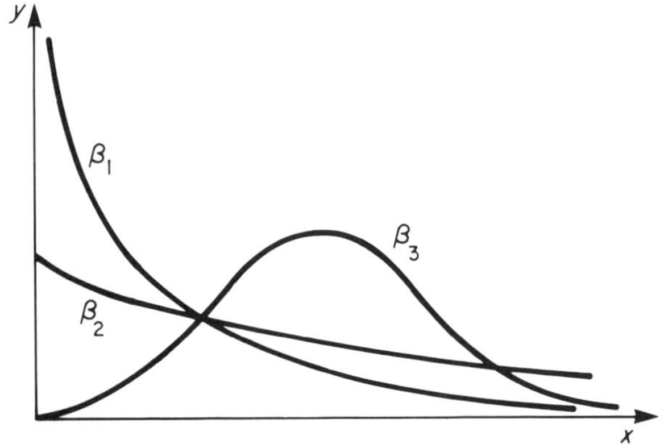

Figure 6.10 Weibull curves ($\alpha=1$)

the Weibull curve varies from an exponential ($\beta=1.0$) to a normal distribution ($\beta=3.5$). In practice β varies from about $\frac{1}{3}$ to 5. Because the Weibull distribution can be made to fit a variety of different sets of data, it is popularly used for probability distributions.

Analytical calculations using the Weibull distribution are cumbersome. Usually predictions are made using Weibull probability paper. The data are plotted on this paper and the probability predictions read from the graph.

6.10 Sampling

A sample consists of a relatively small number of items drawn from a much larger population. This sample is analysed for certain attributes and it is then assumed that these attributes apply to the total population, within a certain tolerance of error.

Sampling is usually associated with the normal probability distribution and, based on this distribution, the errors which arise due to sampling can be estimated. Suppose a sample of n_s items is taken from a population of n_p items which are distributed normally. If the sample is found to have a mean of μ_s with a standard deviation of σ_s then the mean μ_p of the population can be estimated to be within a certain tolerance of μ_s. It is given by

$$\mu_p=\mu_s\pm\frac{\gamma\sigma_s}{n_s^{1/2}} \tag{6.37}$$

γ is found from the normal curve depending on the level of confidence we need in specifying μ_p. For $\gamma=1$ this level is 68.26%; for $\gamma=2$ it is 95.46% and for $\gamma=3$ it is 99.73%.

The standard error of mean σ_e is often defined as

$$\sigma_e=\frac{\sigma_s}{n_s^{1/2}} \tag{6.38}$$

so Equation (6.37) can be rewritten as

$$\mu_p=\mu_s\pm\gamma\sigma_e \tag{6.39}$$

As an example suppose that a sample of 100 items, selected at random from a much larger population, gives their mean weight as 20 kg with a standard deviation of 100 g. The standard error of the mean is therefore $100/(100)^{1/2}=10$ g and one can say with 99.73% confidence that the mean value of the population lies between $20\pm3\times0.01$ or 20.03 kg and 19.97 kg.

If in a sample of n_s items the probability of occurrence of a particular attribute is p_s, then the standard error of probability p_e is defined as

$$p_e=\left(\frac{p_s q_s}{n_s}\right)^{1/2} \tag{6.40}$$

where $q_s=1-p_s$.

The probability of occurrence of the attribute in the population is then given by

$$p_p=p_s\pm\gamma p_e \tag{6.41}$$

where γ is again chosen to cover a certain confidence level.

As an example suppose a sample of 500 items shows that 50 are defective. Then the probability of occurrence of the defect in the sample is $50/500=0.1$. The standard error of probability is $(0.1\times0.9/500)^{1/2}$ or 0.0134. Therefore we can state with 95.46% confidence that the population from which the sample was drawn has a defect probability of $0.1\pm2\times0.0134$, i.e. 0.0732 to 0.1268; or we can state with 99.73% confidence that this value will lie between $0.1\pm3\times0.0134$, i.e. 0.0598 to 0.1402.

If two samples have been taken from the same population and these give standard deviations of σ_{s1} and σ_{s2} for sample sizes of n_{s1} and n_{s2} then Equation (6.38) can be modified to give the standard error of the difference between means as

$$\sigma_{de}=\left(\frac{\sigma_{s1}^2}{n_{s1}}+\frac{\sigma_{s2}^2}{n_{s2}}\right)^{1/2} \tag{6.42}$$

Similarly Equation (6.40) can be modified to give the standard error of the difference between probabilities of two samples from the same population as

$$p_{de}=\left(\frac{p_{s1}q_{s1}}{n_{s1}}+\frac{p_{s2}q_{s2}}{n_{s2}}\right)^{1/2} \tag{6.43}$$

6.11 Tests of significance

In taking samples we often obtain results which deviate from the expected. Tests of signifidance are then used to determine if this deviation is real or if it could have arisen due to sampling error.

6.11.1 Hypothesis testing

In this system a hypothesis is set up and is then tested at a given confidence level. For example, suppose a coin is tossed 100 times and it comes up heads 60 times. Is the coin biased or is it likely that this falls within a reasonable sampling error? The hypothesis is set up that the coin is not biased. Therefore one would expect that the probability of heads is 0.5, i.e. $p_s = 0.5$. The probability of tails, q_s, is also 0.5. Using Equation (6.40) the standard error of probability is given by $p_e = (0.5 \times 0.5/100)^{1/2}$ or 0.05. Therefore from Equation (6.41) the population probability at the 95.45% confidence level of getting heads is $0.5 + 2 \times 0.05 = 0.6$. Therefore it is highly likely that the coin is not biased and the results are due to sampling error.

The results of any significance test are not conclusive. For example, is 95.45% too high a confidence level to require? The higher the confidence level the greater the risk of rejecting a true hypothesis, and the lower the level the greater the risk of accepting a false hypothesis.

Suppose now that a sample of 100 items of production shows that five are defective. A second sample of 100 items is taken from the same production a few months later and gives two defectives. Does this show that the production quality is improving? Using Equation (6.3) the standard error of the difference between probabilities is given by $(0.05 \times 0.95/100 + 0.02 \times 0.98/100)^{1/2} = 0.0259$. This is less than twice the difference between the two probabilities, i.e. $0.05 - 0.02 = 0.03$, therefore the difference is very likely to have arisen due to sampling error and it does not necessarily indicate an improvement in quality.

6.11.2 Chi-square test

This is written as χ^2. If O is an observed result and E is the expected result then

$$\chi^2 = \sum \frac{(O - E)^2}{E} \tag{6.44}$$

The χ^2 distribution is given by tables such as *Table 6.3*, from which the probability can be determined. The number of degrees of freedom is the number of classes whose frequency can be assigned independently. If the data are presented in the form of a table having V vertical columns and H horizontal rows then the degrees of freedom are usually found as $(V-1)(H-1)$.

Returning to the earlier example, suppose a coin is tossed 100 times and it comes up heads 60 times and tails 40 times. Is the coin biased? The expected values for heads and tails are 50 each so that

$$\chi^2 = \frac{(60-50)^2}{50} + \frac{(40-50)^2}{50} = 4$$

The number of degrees of freedom is one since once we have fixed the frequency for heads that for tails is defined. Therefore entering Table 6.3 with one degree of freedom the probability level for $\chi^2 = 4$ is seen to be above 2.5%, i.e. there is a strong probability that the difference in the two results arose by chance and the coin is not biased.

As a further example suppose that over a 24-hour period the average number of accidents which occur in a factory is seen to be as in *Table 6.4*. Does this indicate that most of the accidents occur during the late night and early morning periods? Applying the χ^2 tests the expected value, if there was no difference between the time periods, would be the mean of the number of accidents, i.e. 5.

Table 6.3 The chi-square distribution

Degrees of freedom	*Probability level*				
	0.100	0.050	0.025	0.010	0.005
1	2.71	3.84	5.02	6.63	7.88
2	4.61	5.99	7.38	9.21	10.60
3	6.25	7.81	9.35	11.34	12.84
4	7.78	9.49	11.14	13.28	14.86
5	9.24	11.07	12.83	15.09	16.75
6	10.64	12.59	14.45	16.81	18.55
7	12.02	14.07	16.01	18.48	20.28
8	13.36	15.51	17.53	20.09	21.96
9	14.68	16.92	19.02	21.67	23.59
10	15.99	18.31	20.48	23.21	25.19
12	18.55	21.03	23.34	26.22	28.30
14	21.06	23.68	26.12	29.14	31.32
16	23.54	26.30	28.85	32.00	34.27
18	25.99	28.87	31.53	34.81	37.16
20	28.41	31.41	34.17	37.57	40.00
30	40.26	43.77	46.98	50.89	53.67
40	51.81	55.76	59.34	63.69	66.77

Table 6.4 Frequency distribution of accidents in a factory during 24 hours

Time (24 hour clock)	*Number of accidents*
0–6	9
6–12	3
12–18	2
18–24	6

Therefore from Equation (6.44)

$$\chi^2 = \frac{(9-5)^2}{5} + \frac{(3-5)^2}{5} + \frac{(2-5)^2}{5} + \frac{(6-5)^2}{5}$$
$$= 6$$

There are three degrees of freedom, therefore from *Table 6.3* the probability of occurance of the result shown in *Table 6.4* is seen to be greater than 10%. The conclusion would be that although there is a trend, as yet there are not enough data to show if this trend is significant or not. For example, if the number of accidents were each three times as large, i.e. 27, 9, 6, 18 respectively, then χ^2 would be calculated as 20.67 and from *Table 6.3* it is seen that the results are highly significant since there is a very low probability, less than $\frac{1}{2}$%, that it can arise by chance.

6.11.3 Significance of correlation

The significance of the product moment correlation coefficient of Equations (6.14) or (6.15) can be tested at any confidence level by means of the standard error of estimation given by Equation (6.16). An alternative method is to use the Student t test of significance. This is given by

$$t = \frac{r(n-2)^{1/2}}{(1-r^2)^{1/2}} \tag{6.45}$$

where r is the correlation coefficient and n the number of items. Tables are then used, similar to *Table 6.3*, which give the probability level for $(n-2)$ degrees of freedom.

The Student t for the rank correlation coefficient is given by

$$t=R[(n-2)/(1-R^2)]^{1/2} \qquad (6.46)$$

and the same Student t tables are used to check the significance of R.

Further reading

BESTERFIELD, D. H., *Quality Control*, Prentice Hall (1979)
CAPLEN, R. H., *A Practical Approach to Quality Control*, Business Books (1982)
CHALK, G. O. and STICK, A. W., *Statistics for the Engineer*, Butterworths (1975)
DAVID, H. A., *Order Statistics*, Wiley (1981)
DUNN, R. A. and RAMSING, K. D., *Management Science, a Practical Approach to Decision Making*, Macmillan (1981)
FITZSIMMONS, J. A., *Service Operations Management*, McGraw-Hill (1982)
GRANT, E. L. and LEAVENWORTH, R. S., *Statistical Quality Control*, McGraw-Hill (1980)
HAHN, W. C., *Modern Statistical Methods*, Butterworths (1979)
LYONS, S., *Handbook of Industrial Mathematics*, Cambridge University Press (1978)

Part 2

Physical Phenomena

7 Quantities and Units

L W Turner CEng, FIEE, FRTS
Consultant Engineer

Contents

7.1 International unit system

The International System of Units (SI) is the modern form of the metric system agreed at an international conference in 1960. It has been adopted by the International Standards Organisation (ISO) and the International Electrotechnical Commission (IEC) and its use is recommended wherever the metric system is applied. It is now being adopted throughout most of the world and is likely to remain the primary world system of units of measurement for a very long time. The indications are that SI units will supersede the units of existing metric systems and all systems based on Imperial units.

SI units and the rules for their application are contained in *ISO Resolution* R1000 (1969, updated 1973) and an informatory document *SI-Le Systeme International d'Unités*, published by the Bureau International des Poids et Mesures (BIPM). An abridged version of the former is given in British Standards Institution (BSI) publication PD 5686 *The use of SI Units* (1969, updated 1973) and BS 3763 *International System (SI) Units*; BSI (1964) incorporates information from the BIPM document.

The adoption of SI presents less of a problem to the electronics engineer and the electrical engineer than to those concerned with other engineering disciplines as all the practical electrical units were long ago incorporated in the metre-kilogram-second (MKS) unit system and these remain unaffected in SI.

The SI was developed from the metric system as a fully coherent set of units for science, technology and engineering. A coherent system has the property that corresponding equations between quantities and between numerical values have exactly the same form, because the relations between units do not involve numerical conversion factors. In constructing a coherent unit system, the starting point is the selection and definition of a minimum set of independent 'base' units. From these, 'derived' units are obtained by forming products or quotients in various combinations, again without numerical factors. Thus the base units of length (metre), time (second) and mass (kilogram) yield the SI units of velocity (metre/second), force (kilogram-metre/second-squared) and so on. As a result there is, for any given physical quantity, only one SI unit with no alternatives and with no numerical conversion factors. A single SI unit (joule = kilogram metre-squared/second-squared) serves for energy of any kind, whether it be kinetic, potential, thermal, electrical, chemical..., thus unifying the usage in all branches of science and technology.

The SI has seven base units, and two supplementary units of angle. Certain important derived units have special names and can themselves be employed in combination to form alternative names for further derivations.

Each physical quantity has a quantity-symbol (e.g., *m* for mass) that represents it in equations, and a unit-symbol (e.g., kg for kilogram) to indicate its SI unit of measure.

7.1.1 Base units

Definitions of the seven base units have been laid down in the following terms. The quantity-symbol is given in italics, the unit-symbol (and its abbreviation) in roman type.

Length: *l*; metre (m). The length equal to 1 650 763.73 wavelengths in vacuum of the radiation corresponding to the transition between the levels $2p_{10}$ and $5d_5$ of the krypton-86 atom.

Mass: *m*; kilogram (kg). The mass of the international prototype kilogram (a block of platinum preserved at the International Bureau of Weights and Measures at Sèvres).

Time: *t*; second (s). The duration of 9 192 631 770 periods of the radiation corresponding to the transition between the two hyperfine levels of the ground state of the caesium-133 atom.

Electric current: *i*; ampere (A). The current which, maintained in two straight parallel conductors of infinite length, of negligible circular cross-section and 1 m apart in vacuum, produces a force equal to 2×10^{-7} newton per metre of length.

Thermodynamic temperature: *T*; kelvin (K). The fraction 1/273.16 of the thermodynamic (absolute) temperature of the triple point of water.

Luminous intensity: *I*; candela (cd). The luminous intensity in the perpendicular direction of a surface of 1/600 000 m^2 of a black body at the temperature of freezing platinum under a pressure of 101 325 newtons per square metre.

Amount of substance: *Q*; mole (mol). The amount of substance of a system which contains as many elementary entities as there are atoms in 0.012 kg of carbon-12. The elementary entity must be specified and may be an atom, a molecule, an ion, an electron, etc., or a specified group of such entities.

7.1.2 Supplementary angular units

Plane angle: α, β...; radian (rad). The plane angle between two radii of a circle which cut off on the circumference an arc of length equal to the radius.

Solid angle: Ω; steradian (sr). The solid angle which, having its vertex at the centre of a sphere, cuts off an area of the surface of the sphere equal to a square having sides equal to the radius.

Force: The base SI unit of electric current is in terms of force in newtons (N). A force of 1 N is that which endows unit mass (1 kg) with unit acceleration (1 m/s^2). The newton is thus not only a coherent unit; it is also devoid of any association with gravitational effects.

7.1.3 Temperature

The base SI unit of thermodynamic temperature is referred to a point of 'absolute zero' at which bodies possess zero thermal energy. For practical convenience two points on the Kelvin temperature scale, namely 273.15 K and 373.15 K, are used to define the Celsius (or Centigrade) scale (0°C and 100°C). Thus in terms of temperature *intervals*, 1 K = 1°C; but in terms of temperature *levels*, a Celsius temperature θ corresponds to a Kelvin temperature (θ + 273.15) K.

7.1.4 Derived units

Nine of the more important SI derived units with their definitions are given

Quantity	*Unit name*	*Unit symbol*
Force	newton	N
Energy	joule	J
Power	watt	W
Electric charge	coulomb	C
Electrical potential difference and EMF	volt	V
Electric resistance	ohm	Ω
Electric capacitance	farad	F
Electric inductance	henry	H
Magnetic flux	weber	Wb

Newton That force which gives to a mass of 1 kilogram an acceleration of 1 metre per second squared.

Joule The work done when the point of application of 1 newton is displaced a distance of 1 metre in the direction of the force.

Watt The power which gives rise to the production of energy at the rate of 1 joule per second.

Coulomb The quantity of electricity transported in 1 second by a current of 1 ampere.
Volt The difference of electric potential between two points of a conducting wire carrying a constant current of 1 ampere, when the power dissipated between these points is equal to 1 watt.
Ohm The electric resistance between two points of a conductor when a constant difference of potential of 1 volt, applied between these two points, produces in this conductor a current of 1 ampere, this conductor not being the source of any electromotive force.
Farad The capacitance of a capacitor between the plates of which there appears a difference of potential of 1 volt when it is charged by a quantity of electricity equal to 1 coulomb.
Henry The inductance of a closed circuit in which an electromotive force of 1 volt is produced when the electric current in the circuit varies uniformly at a rate of 1 ampere per second.
Weber The magnet flux which, linking a circuit of one turn, produces in it an electromotive force of 1 volt as it is reduced to zero at a uniform rate in 1 second.

Some of the simpler derived units are expressed in terms of the seven basic and two supplementary units directly. Examples are listed in *Table 7.1.*

Table 7.1 Directly derived units

Quantity	*Unit name*	*Unit symbol*
Area	square metre	m^2
Volume	cubic metre	m^3
Mass density	kilogram per cubic metre	kg/m^3
Linear velocity	metre per second	m/s
Linear acceleration	metre per second squared	m/s^2
Angular velocity	radian per second	rad/s
Angular acceleration	radian per second squared	rad/s^2
Force	kilogram metre per second squared	$kg\ m/s^2$
Magnetic field strength	ampere per metre	A/m
Concentration	mole per cubic metre	mol/m^3
Luminance	candela per square metre	cd/m^2

Units in common use, particularly those for which a statement in base units would be lengthy or complicated, have been given special shortened names (see *Table 7.2*). Those that are named from scientists and engineers have an initial capital letter: all others are in small letters.

Table 7.2 Named derived units

Quantity	*Unit name*	*Unit symbol*	*Derivation*
Force	newton	N	$kg\ m/s^2$
Pressure	pascal	Pa	N/m^2
Power	watt	W	J/s
Energy	joule	J	N m, W s
Electric charge	coulomb	C	A s
Electric flux	coulomb	C	A s
Magnetic flux	weber	Wb	V s
Magnetic flux density	tesla	T	Wb/m^2
Electric potential	volt	V	J/C, W/A
Resistance	ohm	Ω	V/A
Conductance	siemens	S	A/V
Capacitance	farad	F	A s/V, C/V
Inductance	henry	H	V s/A, Wb/A
Luminous flux	lumen	lm	cd sr
Illuminance	lux	lx	lm/m^2
Frequency	hertz	Hz	l/s

The named derived units are used to form further derivations. Examples are given in *Table 7.3.*

Table 7.3 Further derived units

Quantity	*Unit name*	*Unit symbol*
Torque	newton metre	N m
Dynamic viscosity	pascal second	Pa s
Surface tension	newton per metre	N/m
Power density	watt per square metre	W/m^2
Energy density	joule per cubic metre	J/m^3
Heat capacity	joule per kelvin	J/K
Specific heat capacity	joule per kilogram kelvin	J/(kg K)
Thermal conductivity	watt per metre kelvin	W/(m K)
Electric field strength	volt per metre	V/m
Magnetic field strength	ampere per metre	A/m
Electric flux density	coulomb per square metre	C/m^2
Current density	ampere per square metre	A/m^2
Resistivity	ohm metre	Ω m
Permittivity	farad per metre	F/m
Permeability	henry per metre	H/m

Names of SI units and the corresponding EMU and ESU CGS units are given in *Table 7.4.*

Table 7.4 Unit names

Quantity	*Symbol*	*SI*	*EMU & ESU*
Length	l	metre (m)	centimetre (cm)
Time	t	second (s)	second
Mass	m	kilogram (kg)	gram (g)
Force	F	newton (N)	dyne (dyn)
Frequency	f, ν	hertz (Hz)	hertz
Energy	E, W	joule (J)	erg (erg)
Power	P	watt (W)	erg/second (erg/s)
Pressure	p	newton/metre^2 (N/m^2)	dyne/centimetre^2 (dyn/cm^2)
Electric charge	Q	coulomb (C)	coulomb
Electric potential	V	volt (V)	volt
Electric current	I	ampere (A)	ampere
Magnetic flux	Φ	weber (Wb)	maxwell (Mx)
Magnetic induction	B	tesla (T)	gauss (G)
Magnetic field strength	H	ampere turn/metre (At/m)	oersted (Oe)
Magneto-motive force	F_m	ampere turn (At)	gilbert (Gb)
Resistance	R	ohm (Ω)	ohm
Inductance	L	henry (H)	henry
Conductance	G	mho (Ω^{-1}) (siemens)	mho
Capacitance	C	farad (F)	farad

7.1.5 Gravitational and absolute systems

There may be some difficulty in understanding the difference between SI and the Metric Technical System of units which has been used principally in Europe. The main difference is that while mass is expressed in kg in both systems, weight (representing a force) is expressed as kgf, a gravitational unit, in the MKSA system and as N in SI. An absolute unit of force differs from a gravitational unit of force because it induces unit acceleration in a unit mass whereas a gravitational unit imparts gravitational acceleration to a unit mass.

A comparison of the more commonly known systems and SI is shown in *Table 7.5*.

Table 7.5 Commonly used units of measurement

	SI (absolute)	*FPS (gravitational)*	*FPS (absolute)*	*cgs (absolute)*	*Metric technical units (gravitational)*
Length	metre (m)	ft	ft	cm	metre
Force	newton (N)	lbf	poundal (pdl)	dyne	kgf
Mass	kg	lb or slug	lb	gram	kg
Time	s	s	s	s	s
Temperature	°C K	°F	°F °R	°C K	°C K
Energy: mech.	joule*	ft lbf	ft pdl	dyn cm = erg	kgf m
Energy: heat	joule*	Btu	Btu	calorie	kcal
Power: mech.	watt	hp	hp	erg/s	metric hp
Power: elec.	watt	watt	watt	erg/s	watt
Electric current	amp	amp	amp	amp	amp
Pressure	N/m^2	lbf/ft^2	pdl/ft^2	dyn/cm^2	kgf/cm^2

* 1 joule = 1 newton metre or 1 watt second.

7.1.6 Expressing magnitudes of SI units

To express magnitudes of a unit, decimal multiples and submultiples are formed using the prefixes shown in *Table 7.6*. This method of expressing magnitudes ensures complete adherence to a decimal system.

Table 7.6 The internationally agreed multiples and submultiples

Factor by which the unit is multiplied		*Prefix*	*Symbol*	*Common everyday examples*
One million million (billion)	10^{12}	tera	T	
One thousand million	10^9	giga	G	gigahertz (GHz)
One million	10^6	mega	M	megawatt (MW)
One thousand	10^3	kilo	k	kilometre (km)
One hundred	10^2	hecto*	h	
Ten	10^1	deca*	da	decagram (dag)
UNITY	1			
One tenth	10^{-1}	deci*	d	decimetre (dm)
One hundredth	10^{-2}	centi*	c	centimetre (cm)
One thousandth	10^{-3}	milli	m	milligram (mg)
One millionth	10^{-6}	micro	μ	microsecond (μs)
One thousand millionth	10^{-9}	nano	n	nanosecond (ns)
One million millionth	10^{-12}	pico	p	picofarad (pF)
One thousand million millionth	10^{-15}	femto	f	
One million million millionth	10^{-18}	atto	a	

* To be avoided wherever possible.

7.1.7 Auxiliary units

Certain auxiliary units may be adopted where they have application in special fields. Some are acceptable on a temporary basis, pending a more widespread adoption of the SI system. *Table 7.7* lists some of these.

Table 7.7 Auxiliary units

Quantity	*Unit symbol*	*SI equivalent*
Day	d	86 400 s
Hour	h	3600 s
Minute (time)	min	60 s
Degree (angle)	°	$\pi/180$ rad
Minute (angle)	′	$\pi/10\,800$ rad
Second (angle)	″	$\pi/648\,000$ rad
Are	a	$1\ dam^2 = 10^2\ m^2$
Hectare	ha	$1\ hm^2 = 10^4\ m^2$
Barn	b	$100\ fm^2 = 10^{-28}\ m^2$
Standard atmosphere	atm	101 325 Pa
Bar	bar	$0.1\ MPa = 10^5\ Pa$
Litre	l	$1\ dm^3 = 10^{-3}\ m^3$
Tonne	t	$10^3\ kg = 1\ Mg$
Atomic mass unit	u	$1.660\,53 \times 10^{-27}$ kg
Angström	Å	$0.1\ nm = 10^{-10}\ m$
Electron-volt	eV	$1.602\,19 \times 10^{-19}$ J
Curie	Ci	$3.7 \times 10^{10}\ s^{-1}$
Röntgen	R	2.58×10^{-4} C/kg

7.1.8 Nuclear engineering

It has been the practice to use special units with their individual names for evaluating and comparing results. These units are usually formed by multiplying a unit from the cgs or SI system by a number which matches a value derived from the result of some natural phenomenon. The adoption of SI both nationally and internationally has created the opportunity to examine the practice of using special units in the nuclear industry, with the object of eliminating as many as possible and using the pure system instead.

As an aid to this, ISO draft Recommendations 838 and 839 have been published, giving a list of quantities with special names, the SI unit and the alternative cgs unit. It is expected that as SI is increasingly adopted and absorbed, those units based on cgs will go out of use. The values of these special units illustrate the fact that a change from them to SI would not be as revolutionary as might be supposed. Examples of these values together with the SI units which replace them are shown in *Table 7.8*.

Table 7.8 Nuclear engineering

Special unit			*SI replacement*
Name		*Value*	
Angström	(Å)	10^{-10} m	m
Barn	(b)	$10^{-28}\ m^2$	m^2
Curie	(Ci)	$3.7 \times 10^{10}\ s^{-1}$	s^{-1}
Electron-volt	(eV)	$(1.602\,189\,2 \pm .000\,004\,6) \times 10^{-19}$ J	J
Röntgen	(R)	2.58×10^{-4} C/kg	C/kg

7.2 Universal constants in SI units

Table 7.9 Universal constants

The digits in parentheses following each quoted value represent the standard deviation error in the final digits of the quoted value as computed on the criterion of internal consistency. The unified scale of atomic weights is used throughout ($^{12}C=12$). C= coulomb; G = gauss; Hz = hertz; J = joule; N = newton; T = tesla; u = unified nuclidic mass unit; W = watt; Wb = weber. For result multiply the numerical value by the SI unit.

Constant	*Symbol*	*Numerical value*	*SI unit*
Speed of light in vacuum	c	2.997 925(1)	10^{8} m s^{1}
Gravitational constant	G	6.670(5)*	10^{-11} N m^2 kg^2
Elementary charge	e	1.602 10(2)	10^{-19} C
Avogadro constant	N_A	6.022 52(9)	10^{26} $kmol^{-1}$
Mass unit	u	1.660 43(2)	10^{-27} kg
Electron rest mass	m_e	9.109 08(13)	10^{-31} kg
		5.485 97 (3)	10^{-4} u
Proton rest mass	m_p	1.672 52(3)	10^{-27} kg
		1.007 276 63(8)	u
Neutron rest mass	m_n	1.674 82(3)	10^{-27} kg
		1.008 665 4(4)	u
Faraday constant	F	9.648 70(5)	10^{4} C mol^{-1}
Planck constant	h	6.625 59(16)	10^{-34} J s
	$h/2\pi$	1.054 494(25)	10^{-34} J s
Fine-structure constant	α	7.297 20(3)	10^{-3}
	$1/\alpha$	137.038 8(6)	
Charge-to-mass ratio for electron	e/m_e	1.758 796(6)	10^{11} C kg^{-1}
Quantum of magnetic flux	hc/e	4.135 56(4)	10^{-11} Wb
Rydberg constant	R_∞	1.097 373 1(1)	10^{7} m^{-1}
Bohr radius	a_0	5.291 67(2)	10^{-11} m
Compton wavelength of electron	h/m_ec	2.426 21(2)	10^{-12} m
	$\lambda C/2\pi$	3.861 44(3)	10^{-13} m
Electron radius	$e^2/m_ec^2=r_e$	2.817 77(4)	10^{-15} m
Thomson cross section	$8\pi r_e^2/3$	6.651 6(2)	10^{-29} m^2
Compton wavelength of proton	$\lambda_{C,p}$	1.321 398(13)	10^{-15} m
	$\lambda_{C,p}/2\pi$	2.103 07(2)	10^{-16} m
Gyromagnetic ratio of proton	γ	2.675 192(7)	10^{8} rad s^{-1} T^{-1}
	$\gamma/2\pi$	4.257 70(1)	10^{7} Hz T^{-1}
(uncorrected for diamagnetism of H_2O)	γ'	2.675 123(7)	10^{8} rad s^{-1} T^{-1}
	$\gamma'/2\pi$	4.257 59(1)	10^{7} Hz T^{-1}
Bohr magneton	μ_B	9.273 2(2)	10^{-24} J T^{-1}
Nuclear magneton	μ_N	5.050 50(13)	10^{-27} J T^{-1}
Proton magnetic moment	μ_p	1.410 49(4)	10^{-26} J T^{-1}
	μ_p/μ_N	2.792 76(2)	
(uncorrected for diamagnetism in H_2O sample)	μ_p'/μ_N	2.792 68(2)	
Gas constant	R_0	8.314 34(35)	J K^{-1} mol^{-1}
Boltzmann constant	k	1.380 54(6)	10^{-23} J K^{-1}
First radiation constant ($2\pi hc^2$)	c_1	3.741 50(9)	10^{-16} W m^{-2}
Second radiation constant (hc/k)	c_2	1.438 79(6)	10^{-2} m K
Stefan–Boltzmann constant	σ	5.669 7(10)	10^{-8} W m^{-2} K^{-4}

* The universal gravitational constant is not, and cannot in our present state of knowledge, be expressed in terms of other fundamental constants. The value given here is a direct determination by P. R. Heyl and P. Chrzanowski, *J. Res. Natl. Bur. Std.* (*U.S.*) 29, 1 (1942).

The above values are extracts from *Review of Modern Physics* Vol. 37 No. 4 October 1965 published by the American Institute of Physics.

7.3 Metric to Imperial conversion factors

Table 7.10 Conversion factors

SI units	*British units*
SPACE AND TIME	
Length:	
1 µm (micron)	$=39.37\times10^{-6}$ in
1 mm	=0.039 370 1 in
1 cm	=0.393 701 in
1 m	=3.280 84 ft

Table 7.10 *continued*

SI units	*British units*
SPACE AND TIME	
Length:	
1 m	=1.093 61 yd
1 km	=0.621 371 mile
Area:	
1 mm^2	$=1.550\times10^{-3}$ in^2
1 cm^2	=0.155 0 in^2
1 m^2	=10.763 9 ft^2
1 m^2	=1.195 99 yd^2
1 ha	=2.471 05 acre
Volume:	
1 mm^3	$=61.023 7\times10^{-6}$ in^3
1 cm^3	$=61.023 7\times10^{-3}$ in^3
1 m^3	=35.314 7 ft^3
1 m^3	=1.307 95 yd^3
Capacity:	
10^6 m^3	$=219.969\times10^6$ gal
1 m^3	=219.969 gal
1 litre (l)	=0.219 969 gal
	=1.759 80 pint
Capacity flow:	
10^3 m^3/s	$=791.9\times10^6$ gal/h
1 m^3/s	$=13.20\times10^3$ gal/min
1 litre/s	=13.20 gal/min
1 m^3/kW h	=219.969 gal/kW h
1 m^3/s	=35.314 7 ft^3/s (cusecs)
1 litre/s	$=0.588 58\times10^{-3}$ ft^3/min (cfm)
Velocity:	
1 m/s	=3.280 84 ft/s=2.236 94 mile/h
1 km/h	=0.621 371 mile/h
Acceleration:	
1 m/s^2	=3.280 84 ft/s^2
MECHANICS	
Mass:	
1 g	=0.035 274 oz
1 kg	=2.204 62 lb
1 t	=0.984 207 ton=19.684 1 cwt
Mass flow:	
1 kg/s	=2.204 62 lb/s=7.936 64 klb/h
Mass density:	
1 kg/m^3	=0.062 428 lb/ft^3
1 kg/litre	=10.022 119 lb/gal
Mass per unit length:	
1 kg/m	=0.671 969 lb/ft=2.015 91 lb/yd
Mass per unit area:	
1 kg/m^2	=0.204 816 lb/ft^2
Specific volume:	
1 m^3/kg	=16.018 5 ft^3/lb
1 litre/tonne	=0.223 495 gal/ton
Momentum:	
1 kg m/s	=7.233 01 lb ft/s
Angular momentum:	
1 kg m^2/s	=23.730 4 lb ft^2/s
Moment of inertia:	
1 kg m^2	=23.730 4 lb ft^2
Force:	
1 N	=0.224 809 lbf
Weight (force) per unit length:	
1 N/m	=0.068 521 lbf/ft
	=0.205 566 lbf/yd
Moment of force (or torque):	
1 N m	=0.737 562 lbf ft

Table 7.10 *continued*

SI units	*British units*
MECHANICS	
Weight (force) per unit area:	
1 N/m^2	$=0.020\,885$ lbf/ft^2
Pressure:	
1 N/m^2	$=1.450\,38\times10^{-4}$ lbf/in^2
1 bar	$=14.503\,8$ lbf/in^2
1 bar	=0.986 923 atmosphere
1 mbar	=0.401 463 in H_2O
	=0.029 53 in Hg
Stress:	
1 N/mm^2	$=6.474\,90\times10^{-2}$ $tonf/in^2$
1 MN/m^2	$=6.474\,90\times10^{-2}$ $tonf/in^2$
1 hbar	$=0.647\,490$ $tonf/in^2$
Second moment of area:	
1 cm^4	$=0.024\,025$ in^4
Section modulus:	
1 m^3	$=61\,023.7$ in^3
1 cm^3	$=0.061\,023\,7$ in^3
Kinematic viscosity:	
1 m^2/s	$=10.762\,75$ $ft^2/s=10^6$ cSt
1 cSt	$=0.038\,75$ ft^2/h
Energy, work:	
1 J	=0.737 562 ft lbf
1 MJ	=0.372 5 hph
1 MJ	=0.277 78 kW h
Power:	
1 W	=0.737 562 ft lbf/s
1 kW	=1.341 hp=737.562 ft lbf/s
Fluid mass:	
(Ordinary) 1 kg/s	=2.204 62 lb/s=793 6.64 lb/h
(Velocity) 1 kg/m^2 s	$=0.204\,815$ lb/ft^2s
HEAT	
Temperature:	
(Interval) 1 K	=9/5 deg R (Rankine)
1°C	=9/5 deg F
(Coefficient) $1°R^{-1}$	$=1$ deg $F^{-1}=5/9$ deg C
$1°C^{-1}$	$=5/9$ deg F^{-1}
Quantity of heat:	
1 J	$=9.478\,17\times10^{-4}$ Btu
1 J	=0.238 846 cal
1 kJ	=947.817 Btu
1 GJ	$=947.817\times10^3$ Btu
1 kJ	=526.565 CHU
1 GJ	$=526.565\times10^3$ CHU
1 GJ	=9.478 17 therm
Heat flow rate:	
1 W(J/s)	=3.412 14 Btu/h
1 W/m^2	$=0.316\,998$ Btu/ft^2 h
Thermal conductivity:	
1 W/m °C	$=6.933\,47$ Btu in/ft^2 h °F
Coefficient and heat transfer:	
1 W/m^2 °C	$=0.176\,110$ Btu/ft^2 h °F
Heat capacity:	
1 J/°C	$=0.526\,57\times10^{-3}$ Btu/°R
Specific heat capacity:	
1 J/g °C	=0.238 846 Btu/lb °F
1 kJ/kg °C	=0.238 846 Btu/lb °F
Entropy:	
1 J/K	$=0.526\,57\times10^{-3}$ Btu/°R
Specific entropy:	
1 J/kg °C	$=0.238\,846\times10^{-3}$ Btu/lb °F
1 J/kg K	$=0.238\,846\times10^{-3}$ Btu/lb °R

Table 7.10 *continued*

SI units	*British units*
HEAT	
Specific energy/specific latent heat:	
1 J/g	=0.429 923 Btu/lb
1 J/kg	$=0.429\,923\times10^{-3}$ Btu/lb
Calorific value:	
1 kJ/kg	=0.429 923 Btu/lb
1 kJ/kg	=0.773 861 4 CHU/lb
1 J/m^3	$=0.026\,839\,2\times10^{-3}$ Btu/ft^3
1 kJ/m^3	$=0.026\,839\,2$ Btu/ft^3
1 kJ/litre	=4.308 86 Btu/gal
1 kJ/kg	=0.009 630 2 therm/ton
ELECTRICITY	
Permeability:	
1 H/m	$=10^7/4\pi\ \mu_0$
Magnetic flux density:	
1 tesla	$=10^4$ gauss $=1$ Wb/m^2
Conductivity:	
1 mho	=1 reciprocal ohm
1 siemens	=1 reciprocal ohm
Electric stress:	
1 kV/mm	=25.4 kV/in
1 kV/m	=0.025 4 kV/in

7.4 Symbols and abbreviations

Table 7.11 Quantities and units of periodic and related phenomena (based on ISO Recommendation R31)

Symbol	*Quantity*
T	periodic time
τ, (T)	time constant of an exponentially varying quantity
f, v	frequency
η	rotational frequency
ω	angular frequency
λ	wavelength
σ $(\tilde{v})$	wavenumber
k	circular wavenumber
$\log_e (A_1/A_2)$	natural logarithm of the ratio of two amplitudes
$10 \log_{10} (P_1/P_2)$	ten times the common logarithm of the ratio of two powers
δ	damping coefficient
Λ	logarithmic decrement
α	attenuation coefficient
β	phase coefficient
γ	propagation coefficient

Table 7.12 Symbols for quantities and units of electricity and magnetism (based on ISO Recommendation R31)

Symbol	*Quantity*
I	electric current
Q	electric charge, quantity of electricity
ρ	volume density of charge, charge density (Q/V)
σ	surface density of charge (Q/A)
E, (K)	electric field strength

Table 7.12 *continued*

Symbol	*Quantity*
V, (φ)	electric potential
U, (V)	potential difference, tension
E	electromotive force
D	displacement (rationalised displacement)
D'	non-rationalised displacement
ψ	electric flux, flux of displacement (flux of rationalised displacement)
ψ'	flux of non-rationalised displacement
C	capacitance
ε	permittivity
ε_0	permittivity of vacuum
ε'	non-rationalised permittivity
ε'_0	non-rationalised permittivity of vacuum
ε_r	relative permittivity
χ_e	electric susceptibility
χ'_e	non-rationalised electric susceptibility
P	electric polarisation
p, (p_e)	electric dipole moment
J, (S)	current density
A, (α)	linear current density
H	magnetic field strength
H'	non-rationalised magnetic field strength
U_m	magnetic potential difference
F, (F_m)	magnetomotive force
B	magnetic flux density, magnetic induction
Φ	magnetic flux
A	magnetic vector potential
L	self-inductance
M, (L)	mutual inductance
k, (x)	coupling coefficient
σ	leakage coefficient
μ	permeability
μ_0	permeability of vacuum
μ'	non-rationalised permeability
μ'_0	non-rationalised permeability of vacuum
μ_r	relative permeability
k, (χ_m)	magnetic susceptibility
k', (χ'_m)	non-rationalised magnetic susceptibility
m	electromagnetic moment (magnetic moment)
H_i, (M)	magnetisation
J, (B_i)	magnetic polarisation
J'	non-rationalised magnetic polarisation
w	electromagnetic energy density
S	Poynting vector
c	velocity of propagation of electromagnetic waves *in vacuo*
R	resistance (to direct current)
G	conductance (to direct current)
ρ	resistivity
γ, σ	conductivity
R, R_m	reluctance
Λ, (P)	permeance
N	number of turns in winding
m	number of phases
p	number of pairs of poles
φ	phase displacement
Z	impedance (complex impedance)
$[Z]$	modulus of impedance (impedance)
X	reactance
R	resistance
Q	quality factor
Y	admittance (complex admittance)
$[Y]$	modulus of admittance (admittance)
B	susceptance

Table 7.12 *continued*

Symbol	*Quantity*
G	conductance
P	active power
S, (P_s)	apparent power
Q, (P_q)	reactive power

Table 7.13 Symbols for quantities and units of acoustics (based on ISO Recommendation R31)

Symbol	*Quantity*
T	period, periodic time
f, v	frequency, frequency interval
ω	angular frequency, circular frequency
λ	wavelength
k	circular wavenumber
ρ	density (mass density)
P_s	static pressure
p	(instantaneous) sound pressure
ε, (x)	(instantaneous) sound particle displacement
u, v	(instantaneous) sound particle velocity
a	(instantaneous) sound particle acceleration
q, U	(instantaneous) volume velocity
c	velocity of sound
E	sound energy density
P, (N, W)	sound energy flux, sound power
I, J	sound intensity
Z_s, (W)	specific acoustic impedance
Z_a, (Z)	acoustic impedance
Z_m, (w)	mechanical impedance
L_p, (L_N, L_w)	sound power level
L_p, (L)	sound pressure level
δ	damping coefficient
Λ	logarithmic decrement
α	attenuation coefficient
β	phase coefficient
γ	propagation coefficient
δ	dissipation coefficient
r, τ	reflection coefficient
γ	transmission coefficient
α, (α_a)	acoustic absorption coefficient
R	sound reduction index; sound transmission loss
A	equivalent absorption area of a surface or object
T	reverberation time
L_N, (Λ)	loudness level
N	loudness

Table 7.14 Some technical abbreviations and symbols

Quantity	*Abbreviation*	*Symbol*
Alternating current	a.c.	
Ampere	A or amp	
Amplification factor		μ
Amplitude modulation	a.m.	
Angular velocity		ω
Audio frequency	a.f.	

Table 7.14 *continued*

Quantity	*Abbreviation*	*Symbol*
Automatic frequency control	a.f.c.	
Automatic gain control	a.g.c.	
Bandwidth		Δf
Beat frequency oscillator	b.f.o.	
British thermal unit	Btu	
Cathode-ray oscilloscope	c.r.o.	
Cathode-ray tube	c.r.t.	
Centigrade	c	
Centi-	C	
Centimetre	cm	
Square centimetre	cm^2 or sq cm	
Cubic centimetre	cm^3 or cu cm or c.c.	
Centimetre-gram-second	c.g.s.	
Continuous wave	c.w.	
Coulomb	C	
Deci-	d	
Decibel	dB	
Direct current	d.c.	
Direction finding	d.f.	
Double sideband	d.s.b.	
Efficiency		η
Equivalent isotropic radiated power	e.i.r.p.	
Electromagnetic unit	e.m.u.	
Electromotive force instantaneous value	e.m.f.	E or V, e or v
Electron-volt	eV	
Electrostatic unit	e.s.u.	
Fahrenheit	F	
Farad	F	
Frequency	freq.	f
Frequency modulation	f.m.	
Gauss	G	
Giga-	G	
Gram	g	
Henry	H	
Hertz	Hz	
High frequency	h.f.	
Independent sideband	i.s.b.	
Inductance-capacitance		L-C
Intermediate frequency	i.f.	
Kelvin	K	
Kilo-	k	
Knot	kn	
Length		l
Local oscillator	l.o.	
Logarithm, common		log or $\log_{10}$
Logarithm, natural		ln or $\log_e$
Low frequency	l.f.	
Low tension	l.t.	
Magnetomotive force	m.m.f.	F or M
Mass		m
Medium frequency	m.f.	
Mega-	M	
Metre	m	
Metre-kilogram-second	m.k.s.	
Micro-	μ	
Micromicro-	p	
Micron		μ
Milli-	m	
Modulated continuous wave	m.c.w.	
Nano-	n	
Neper	N	
Noise factor		N
Ohm		Ω
Peak to peak	p–p	
Phase modulation	p.m.	
Pico-	p	
Plan-position indication	PPI	
Potential difference	p.d.	V
Power factor	p.f.	
Pulse repetition frequency	p.r.f.	
Radian	rad	
Radio frequency	r.f.	
Radio telephony	R/T	
Root mean square	r.m.s.	
Short-wave	s.w.	
Single sideband	s.s.b.	
Signal frequency	s.f.	
Standing wave ratio	s.w.r.	
Super-high frequency	s.h.f.	
Susceptance		B
Travelling-wave tube	t.w.t.	
Ultra-high frequency	u.h.f.	
Very high frequency	v.h.f.	
Very low frequency	v.l.f.	
Volt	V	
Voltage standing wave ratio	v.s.w.r.	
Watt	W	
Weber	Wb	
Wireless telegraphy	W/T	

Table 7.15 Greek alphabet and symbols

Name	*Symbol*		*Quantities used for*
alpha	A	α	angles, coefficients, area
beta	B	β	angles, coefficients
gamma	Γ	γ	specific gravity
delta	Δ	δ	density, increment, finite difference operator
epsilon	E	ε	Napierian logarithm, linear strain, permittivity, error, small quantity
zeta	Z	ζ	coordinates, coefficients, impedance (capital)
eta	H	η	magnetic field strength, efficiency
theta	Θ	θ	angular displacement, time
iota	I	ι	inertia
kappa	K	κ	bulk modulus, magnetic susceptibility
lambda	Λ	λ	permeance, conductivity, wavelength
mu	M	μ	bending moment, coefficient of friction, permeability

Table 7.15 *continued*

Name	*Symbol*		*Quantities used for*
nu	N	ν	kinematic viscosity, frequency, reluctivity
xi	Ξ	ξ	output coefficient
omicron	O	o	
pi	Π	π	circumference ÷ diameter
rho	P	ρ	specific resistance
sigma	Σ	σ	summation (capital), radar cross section, standard deviation
tau	T	τ	time constant, pulse length
upsilon	Y	u	
phi	Φ	φ	flux, phase
chi	X	χ	reactance (capital)
psi	Ψ	ψ	angles
omega	Ω	ω	angular velocity, ohms

References

1 COHEN, E. R. and TAYLOR, B. N., *Journal of Physical and Chemical Reference Data*, vol. 2, 663, (1973).
2 'Recommended values of physical constants', CODATA (1973).
3 McGLASHAN, M. L., *Physiochemical quantities and units*, London: The Royal Institute of Chemistry, (1971).

8 Electricity

M G Say PhD, MSc, CEng, ACGI, DIC, FIEE, FRSE
Professor Emeritus of Electrical Engineering,
Heriot-Watt University, Edinburgh

Contents

8.1 Introduction

Most of the observed electrical phenomena are explicable in terms of electric *charge* at rest, in motion and in acceleration. Static charges give rise to an *electric field* of force; charges in motion carry an electric field accompanied by a *magnetic field* of force; charges in acceleration develop a further field of *radiation*.

Modern physics has established the existence of elemental charges and their responsibility for observed phenomena. Modern physics is complex: it is customary to explain phenomena of engineering interest at a level adequate for a clear and reliable concept, based on the electrical nature of matter.

8.2 Molecules, atoms and electrons

Material substances, whether solid, liquid or gaseous, are conceived as composed of very large numbers of *molecules*. A molecule is the smallest portion of any substance which cannot be further subdivided without losing its characteristic material properties. In all states of matter molecules are in a state of rapid continuous motion. In a *solid* the molecules are relatively closely 'packed' and the molecules, although rapidly moving, maintain a fixed mean position. Attractive forces between molecules account for the tendency of the solid to retain its shape. In a *liquid* the molecules are less closely packed and there is a weaker cohesion between them, so that they can wander about with some freedom within the liquid, which consequently takes up the shape of the vessel in which it is contained. The molecules in a *gas* are still more mobile, and are relatively far apart. The cohesive force is very small, and the gas is enabled freely to contract and expand. The usual effect of heat is to increase the intensity and speed of molecular activity so that 'collisions' between molecules occur more often; the average spaces between the molecules increase, so that the substance attempts to expand, producing internal pressure if the expansion is resisted.

Molecules are capable of further subdivision, but the resulting particles, called *atoms*, no longer have the same properties as the molecules from which they came. An atom is the smallest portion of matter that can enter into chemical combination or be chemically separated, but it cannot generally maintain a separate existence except in the few special cases where a single atom forms a molecule. A molecule may consist of one, two or more (sometimes many more) atoms of various kinds. A substance whose molecules are composed entirely of atoms of the same kind is called an *element*. Where atoms of two or more kinds are present, the molecule is that of a chemical *compound*. At present 102 atoms are recognised, from combinations of which every conceivable substance is made. As the simplest example, the atom of hydrogen has a mass of 1.63×10^{-27} kg and a molecule (H_2), containing two atoms, has twice this mass. In one gram of hydrogen there are about 3×10^{23} molecules with an order of size between 1 and 0.1 nm.

Electrons, as small particles of negative electricity having apparently almost negligible mass, were discovered by J. J. Thomson, on a basis of much previous work by many investigators, notably Crookes. The discovery brought to light two important facts: (1) that atoms, the units of which all matter is made, are themselves complex structures, and (2) that electricity is atomic in nature. The atoms of all substances are constructed from particles. Those of engineering interest are: *electrons*, *protons* and *neutrinos*. Modern physics concerns itself also with *positrons*, *mesons*, *neutrons* and many more. An *electron* is a minute particle of negative electricity which, when dissociated from the atom (as it can be) indicates a purely electrical, nearly mass-less nature. From whatever atom they are derived, all electrons are similar. The electron charge is $e = 1.6 \times 10^{-19}$ C, so that 1 C $= 6.3 \times 10^{18}$ electron charges. The apparent rest mass of an electron is 1/1850 of that of a hydrogen atom, amounting to $m = 9 \times 10^{-28}$ g. The meaning to be attached to the 'size' of an electron (a figure of the order of 10^{-13} cm) is vague. A *proton* is electrically the opposite of an electron, having an equal charge, but positive. Further, protons are associated with a mass the same as that of the hydrogen nucleus. A *neutron* is a chargeless mass, the same as that of the proton.

8.3 Atomic structure

The mass of an atom is almost entirely concentrated in a nucleus of protons and neutrons. The simplest atom, of hydrogen, comprises a nucleus with a single proton, together with one associated electron occupying a region formerly called the K-shell. Helium has a nucleus of two protons and two neutrons, with two electrons in the K-shell. In these cases, as in all normal atoms, the sum of the electron charges is numerically equal to the sum of the proton charges, and the atom is electrically balanced. The neon atom has a nucleus with 10 protons and 10 neutrons, with its 10 electrons in the K- and L-shells.

The *atomic weight* A is the total number of protons and neutrons in the nucleus. If there are Z protons there will be $A - Z$ neutrons: Z is the *atomic number*. The nuclear structure is not known, and the forces that keep the protons together against their mutual repulsion are conjectural.

A nucleus of atomic weight A and atomic number Z has a charge of $+Ze$ and is normally surrounded by Z electrons each of charge $-e$. Thus copper has 29 protons and 35 neutrons ($A = 64$, $Z = 29$) in its nucleus, electrically neutralised by 29 electrons in an enveloping cloud. The atomic numbers of the known elements range from 1 for hydrogen to 102 for nobelium, and from time to time the list is extended. This multiplicity can be simplified: within the natural sequence of elements there can be distinguished groups with similar chemical and physical properties (see *Table 8.1*). These are the ***halogens*** (F 9, Cl 17, Br 35, I 53); the *alkali metals* (Li 3, Na 11, K 19, Rb 37, Cs 55); the *copper* group (Cu 29, Ag 47, Au 79); the *alkaline earths* (Be 4, Mg 12, Ca 20, Sr 38, Ba 56, Ra 88); the *chromium* group (Cr 24, Mo 42, W 74, U 92); and the *rare gases* (He 2, Ne 10, A 18, Kr 36, Xe 54, Rn 86). In the foregoing the brackets contain the chemical symbols of the elements concerned followed by their atomic numbers. The difference between the atomic numbers of two adjacent elements within a group is always 8, 18 or 32. Now these three bear to one another a simple arithmetical relation: $8 = 2 \times 2 \times 2$, $18 = 2 \times 3 \times 3$ and $32 = 2 \times 4 \times 4$. Arrangement of the elements in order in a periodic table beginning with an alkali metal and ending with a rare gas shows a remarkable repetition of basic similarities. The periods are I, 1–2; II, 3–10; III, 11–18; IV, 19–36; V, 37–54; VI, 55–86; VII, 87–?.

An element is often found to be a mixture of atoms with the same chemical property but different atomic weights (*isotopes*). Again, because of the convertibility of mass and energy, the mass of an atom depends on the energy locked up in its compacted nucleus. Thus small divergences are found in the atomic weights which, on simple grounds, would be expected to form integral multiples of the atomic weight of hydrogen. The atomic weight of oxygen is arbitrarily taken as 16.0, so that the mass of the proton is 1.007 6 and that of the hydrogen atom is 1.008 1.

Atoms may be in various energy states. Thus the atoms in the filament of an incandescent lamp may emit light when excited, e.g., by the passage of an electric heating current, but will not do so when the heater current is switched off. Now heat energy is the kinetic energy of the atoms of the heated body. The more vigorous impact of atoms may not always shift the atom as a whole, but may shift an electron from one orbit to another of higher energy level within the atom. This position is not normally stable, and the electron gives up its momentarily-acquired

Table 8.1 Elements

Period	*Atomic number*	*Name*	*Symbol*	*Atomic weight*
I	1	Hydrogen	H	1.008
	2	Helium	He	4.002
II	3	Lithium	Li	6.94
	4	Beryllium	Be	9.02
	5	Boron	B	10.82
	6	Carbon	C	12.00
	7	Nitrogen	N	14.008
	8	Oxygen	O	16.00
	9	Fluorine	F	19.00
	10	Neon	Ne	20.18
III	11	Sodium	Na	22.99
	12	Magnesium	Mg	24.32
	13	Aluminium	Al	26.97
	14	Silicon	Si	28.06
	15	Phosphorus	P	31.02
	16	Sulphur	S	32.06
	17	Chlorine	Cl	35.46
	18	Argon	A	39.94
IV	19	Potassium	K	39.09
	20	Calcium	Ca	40.08
	21	Scandium	Sc	45.10
	22	Titanium	Ti	47.90
	23	Vanadium	V	50.95
	24	Chromium	Cr	52.01
	25	Manganese	Mn	54.93
	26	Iron	Fe	55.84
	27	Cobalt	Co	58.94
	28	Nickel	Ni	58.69
	29	Copper	Cu	63.57
	30	Zinc	Zn	65.38
	31	Gallium	Ga	69.72
	32	Germanium	Ge	72.60
	33	Arsenic	As	74.91
	34	Selenium	Se	78.96
	35	Bromine	Br	79.91
	36	Krypton	Kr	83.70
V	37	Rubidium	Rb	85.44
	38	Strontium	Sr	87.63
	39	Yttrium	Y	88.92
	40	Zirconium	Zr	91.22
	41	Niobium	Nb	92.91
	42	Molybdenum	Mo	96.00
	43	Technetium	Tc	99.00
	44	Ruthenium	Ru	101.7
	45	Rhodium	Rh	102.9
	46	Palladium	Pd	106.7
	47	Silver	Ag	107.9
	48	Cadmium	Cd	112.4
	49	Indium	In	114.8
	50	Tin	Sn	118.0
	51	Antimony	Sb	121.8
	52	Tellurium	Te	127.6
	53	Iodine	I	126.9
	54	Xenon	Xe	131.3
VI	55	Caesium	Cs	132.9
	56	Barium	Ba	137.4
	57	Lanthanum	La	138.9
	58	Cerium	Ce	140.1
	59	Praseodymium	Pr	140.9
	60	Neodymium	Nd	144.3

Table 8.1 *continued*

Period	*Atomic number*	*Name*	*Symbol*	*Atomic weight*
VI	61	Promethium	Pm	147
	62	Samarium	Sm	150.4
	63	Europium	Eu	152.0
	64	Gadolinium	Gd	157.3
	65	Terbium	Tb	159.2
	66	Dysprosium	Dy	162.5
	67	Holmium	Ho	163.5
	68	Erbium	Er	167.6
	69	Thulium	Tm	169.4
	70	Ytterbium	Yb	173.0
	71	Lutecium	Lu	175.0
	72	Hafnium	Hf	178.6
	73	Tantalum	Ta	181.4
	74	Tungsten	W	184.0
	75	Rhenium	Re	186.3
	76	Osmium	Os	191.5
	77	Iridium	Ir	193.1
	78	Platinum	Pt	195.2
	79	Gold	Au	197.2
	80	Mercury	Hg	200.6
	81	Thallium	Tl	204.4
	82	Lead	Pb	207.2
	83	Bismuth	Bi	209.0
	84	Polonium	Po	210
	85	Astatine	At	211
	86	Radon	Rn	222
VII	87	Francium	Fr	223
	88	Radium	Ra	226.0
	89	Actinium	Ac	227
	90	Thorium	Th	232.1
	91	Protoactinium	Pa	234
	92	Uranium	U	238.1
	93	Neptunium	Np	239
	94	Plutonium	Pu	242
	95	Americium	Am	243
	96	Curium	Cm	243
	97	Berkelium	Bk	245
	98	Californium	Cf	246
	99	Einsteinium	Es	247
	100	Fermium	Fm	256
	101	Mendelevium	Md	256
	102	Nobelium	No	—

potential energy by falling back into its original level, releasing the energy as a definite amount of light, the *light-quantum* or *photon*.

Among the electrons of an atom those of the outside peripheral shell are unique in that, on account of all the electron charges on the shells between them and the nucleus, they are the most loosely bound and most easily *removable*. In a variety of ways it is possible so to excite an atom that one of the outer electrons is torn away, leaving the atom *ionised* or converted for the time into an *ion* with an effective positive charge due to the unbalanced electrical state it has acquired. Ionisation may occur due to impact by other fast-moving particles, by irradiation with rays of suitable wavelength and by the application of intense electric fields.

The three 'structures' of *Figure 8.1* are based on the former 'planetary' concept, now modified in favour of a more complex

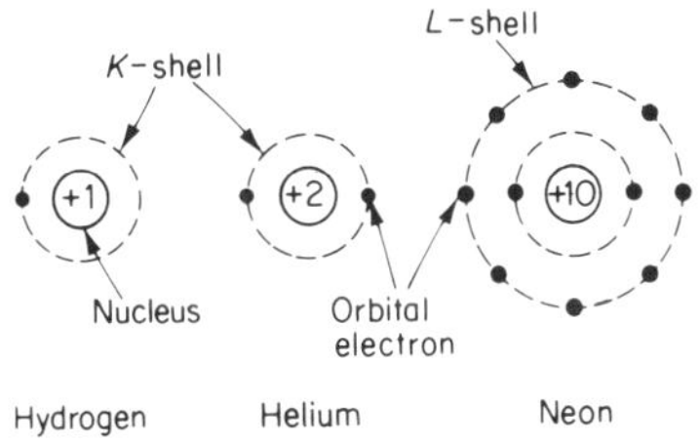

Figure 8.1 Atomic structure. The nuclei are marked with their positive charges in terms of total electron charge. The term 'orbital' is becoming obsolete. Electron: mass $m=9\times10^{28}$ g, charge $e=-1.6\times10^{-19}$ C. Proton: mass $=1.63\times10^{-24}$ g, charge $=+1.6\times10^{-19}$ C. Neutron = mass as for proton; no charge

idea derived from consideration of wave mechanics. It is still true that, apart from its mass, the chemical and physical properties of an atom are given to it by the arrangement of the electron 'cloud' surrounding the nucleus.

8.4 Wave mechanics

The fundamental laws of optics can be explained without regard to the nature of light as an electromagnetic wave phenomenon, and photo-electricity emphasises its nature as a stream or ray of corpuscles. The phenomena of diffraction or interference can only be explained on the wave concept. *Wave mechanics* correlates the two apparently conflicting ideas into a wider concept of 'waves of matter'. Electrons, atoms and even molecules participate in this duality, in that their effects appear sometimes as corpuscular, sometimes as of a wave nature. Streams of electrons behave in a corpuscular fashion in photo-emission, but in certain circumstances show the diffraction effects familiar in wave action. Considerations of particle mechanics led de Broglie to write several theoretic papers (1922–6) on the parallelism between the dynamics of a particle and geometrical optics, and suggested that it was necessary to admit that classical dynamics could not interpret phenomena involving energy quanta. Wave mechanics was established by Schrödinger in 1926 on de Broglie's conceptions.

When electrons interact with matter they exhibit wave properties: in the free state they act like particles. Light has a similar duality, as already noted. The hypothesis of de Broglie is that a particle of mass m and velocity u has wave properties with a wavelength $\lambda=h/mu$, where h is the Planck constant, $h=6.626\times10^{-34}$ J s. The mass m is relativistically affected by the velocity.

When electron waves are associated with an atom, only certain fixed-energy states are possible. The electron can be raised from one state to another if it is provided, by some external stimulus such as a photon, with the necessary energy-difference Δw in the form of an electromagnetic wave of wavelength $\lambda=hc/\Delta w$, where c is the velocity of free-space radiation (3×10^8 m/s). Similarly, if an electron falls from a state of higher to one of lower energy, it emits energy Δw as radiation. When electrons are raised in energy level, the atom is *excited*, but not ionised.

8.5 Electrons in atoms

Consider the hydrogen atom. Its single electron is not located at a fixed point, but can be anywhere in a region near the nucleus with some probability. The particular region is a kind of shell or cloud, of radius depending on the electron's energy state.

With a nucleus of atomic number Z, the Z electrons can have several possible configurations. There is a certain radial pattern of electron probability cloud distribution (or shell pattern). Each electron state gives rise to a cloud pattern, characterised by a definite energy level, and described by the series of quantum numbers n, l, m_l and m_s. The number n ($=1, 2, 3\ldots$) is a measure of the energy level; l ($=0, 1, 2\ldots$) is concerned with angular momentum; m_l is a measure of the component of angular momentum in the direction of an applied magnetic field; and m_s arises from the electron spin. It is customary to condense the nomenclature so that electron states corresponding to $l=0, 1, 2$ and 3 are described by the letters s, p, d and f and a numerical prefix gives the value of n. Thus boron has 2 electrons at level 1 with $l=0$, two at level 2 with $l=0$, and one at level 3 with $l=1$: this information is conveyed by the description $(1s)^2(2s)^2(2p)^1$.

The energy of an atom as a whole can vary according to the electron arrangement. The most stable state is that of minimum energy, and states of higher energy content are *excited*. By Pauli's *exclusion principle* the maximum possible number of electrons in states $1, 2, 3, 4\ldots n$ are $2, 8, 18, 32, \ldots, 2n^2$ respectively. Thus only 2 electrons can occupy the 1s state (or K-shell) and the remainder must, even for the normal minimum-energy condition, occupy other states. Hydrogen and helium, the first two elements, have respectively 1 and 2 electrons in the 1-quantum (K) shell; the next, lithium, has its third electron in the 2-quantum (L) shell. The passage from lithium to neon (*Figure 8.1*) results in the filling up of this shell to its full complement of 8 electrons. During the process, the electrons first enter the 2s subgroup, then fill the 2p subgroup until it has 6 electrons, the maximum allowable by the exclusion principle (see *Table 8.2*).

Table 8.2 Typical atomic structures

Element and atomic number		*Principal and secondary quantum numbers*									
		1s	*2s*	*2p*	*3s*	*3p*	*3d*	*4s*	*4p*	*4d*	*4f*
H	1	1									
He	2	2									
Li	3	2	1								
C	6	2	2	2							
N	7	2	2	3							
Ne	10	2	2	6							
Na	11	2	2	6	1						
Al	13	2	2	6	2	1					
Si	14	2	2	6	2	2					
Cl	17	2	2	6	2	5					
A	18	2	2	6	2	6					
K	19	2	2	6	2	6		1			
Mn	25	2	2	6	2	6	5	2			
Fe	26	2	2	6	2	6	6	2			
Co	27	2	2	6	2	6	7	2			
Ni	28	2	2	6	2	6	8	2			
Cu	29	2	2	6	2	6	10	1			
Ge	32	2	2	6	2	6	10	2	2		
Se	34	2	2	6		6	10	2	4		
Kr	36	2	2	6		6	10	2	6		
		1	2	3	*4s*	*4p*	*4d*	*4f*	*5s*	*5p*	
Rb	37	2	8	18	2	6			1		
Xe	54	2	8	18	2	6	10		2	6	

Very briefly, the effect of the electron-shell filling is as follows. Elements in the same chemical family have the same number of electrons in the subshell that is incompletely filled. The rare gases (He, Ne, A, Kr, Xe) have no uncompleted shells. Alkali metals (e.g., Na) have shells containing a single electron. The alkaline earths have two electrons in uncompleted shells. The good conductors (Ag, Cu, Au) have a single electron in the uppermost quantum state. An irregularity in the ordered sequence of filling (which holds consistently from H to A) begins at potassium (K) and continues to Ni, becoming again regular with Cu, and beginning a new irregularity with Rb.

8.6 Energy levels

The electron of a hydrogen atom, normally at level 1, can be raised to level 2 by endowing it with a particular quantity of energy most readily expressed as 10.2 eV. (1 eV = 1 electronvolt = 1.6×10^{-19} J is the energy acquired by a free electron falling through a potential difference of 1 V, which accelerates it and gives it kinetic energy). 10.2 V is the *first excitation potential* for the hydrogen atom. If the electron is given an energy of 13.6 eV it is freed from the atom, and 13.6 V is the *ionisation potential*. Other atoms have different potentials in accordance with their atomic arrangement.

8.7 Electrons in metals

An approximation to the behaviour of metals assumes that the atoms lose their valency electrons, which are free to wander in the ionic lattice of the material to form what is called an electron gas. The sharp energy-levels of the free atom are broadened into wide bands by the proximity of others. The potential within the metal is assumed to be smoothed out, and there is a sharp rise of potential at the surface that prevents the electrons from escaping: there is a potential-energy step at the surface that the electrons cannot normally overcome: it is of the order of 10 eV. If this is called W, then the energy of an electron wandering within the metal is $-W+\frac{1}{2}mu^2$.

The electrons are regarded as undergoing continual collisions on account of the thermal vibrations of the lattice, and on Fermi–Dirac statistical theory it is justifiable to treat the energy states (which are in accordance with Pauli's principle) as forming an energy-continuum. At very low temperatures the ordinary classical theory would suggest that electron energies spread over an almost zero range, but the exclusion principle makes this impossible and even at absolute zero of temperature the energies form a continuum, and physical properties will depend on how the electrons are distributed over the upper levels of this energy range.

8.8 Conductivity

The interaction of free electrons with the thermal vibrations of the ionic lattice (called 'collisions' for brevity) causes them to 'rebound' with a velocity of random direction but small compared with their average velocities as particles of an electron gas. Just as a difference of electric potential causes a drift in the general motion, so a difference of temperature between two parts of a metal carries energy from the hot region to the cold, accounting for thermal conduction and for its association with electrical conductivity. The free-electron theory, however, is inadequate to explain the dependence of conductivity on crystal axes in the metal.

At absolute zero of temperature (0 K = −273°C) the atoms cease to vibrate, and free electrons can pass through the lattice with little hindrance. At temperatures over the range 0.3–10 K (and usually round about 5 K) the resistance of certain metals, e.g., Zn, Al, Sn, Hg and Cu, becomes substantially zero. This phenomenon, known as *superconductivity*, has not been satisfactorily explained.

Superconductivity is destroyed by moderate magnetic fields. It can also be destroyed if the current is large enough to produce at the surface the same critical value of magnetic field. It follows that during the superconductivity phase the current must be almost purely superficial, with a depth of penetration of the order of 10 μm.

8.9 Electron emission

A metal may be regarded as a potential 'well' of depth $-V$ relative to its surface, so that an electron in the lowest energy state has (at absolute zero temperature) the energy $W = Ve$ (of the order 10 eV): other electrons occupy levels up to a height ε^* (5–8 eV) from the bottom of the 'well'. Before an electron can escape from the surface it must be endowed with an energy not less than $\varphi = W - \varepsilon^*$, called the *work function*.

Emission occurs by *surface irradiation* (e.g., with light) of frequency v if the energy quantum hv of the radiation is at least equal to φ. The threshold of photo-electric emission is therefore with radiation at a frequency not less than $v = \varphi/h$.

Emission takes place at *high temperatures* if, put simply, the kinetic energy of an electron normal to the surface is great enough to jump the potential step W. This leads to an expression for the emission current i in terms of temperature T, a constant A and the thermionic work-function φ:

$$i = AT^2 \exp(-\varphi/kT)$$

Electron emission is also the result of the application of a *high electric-field intensity* (of the order 1–10 GV/m) to a metal surface; also when the surface is bombarded with electrons or ions of sufficient kinetic energy, giving the effect of *secondary* emission.

8.10 Electrons in crystals

When atoms are brought together to form a crystal, their individual sharp and well-defined energy levels merge into energy *bands*. These bands may overlap, or there may be gaps in the energy levels available, depending on the lattice spacing and interatomic bonding. Conduction can take place only by electron migration into an empty or partly filled band: filled bands are not available. If an electron acquires a small amount of energy from the externally applied electric field, and can move into an available empty level, it can then contribute to the conduction process.

8.11 Insulators

In this case the 'distance' (or energy increase Δw in electronvolts) is too large for moderate electric applied fields to endow electrons with sufficient energy, so the material remains an insulator. High temperatures, however, may result in sufficient thermal agitation to permit electrons to 'jump the gap'.

8.12 Semiconductors

Intrinsic semiconductors (i.e., materials between the good conductors and the good insulators) have a small spacing of about 1 eV between their permitted bands, which affords a low conductivity, strongly dependent on temperature and of the order of one-millionth that of a conductor.

Impurity semiconductors have their low conductivity provided by the presence of minute quantities of foreign atoms (e.g., 1 in 10^8) or by deformations in the crystal structure. The impurities 'donate' electrons of energy-level that can be raised into a conduction band (n-type); or they can attract an electron from a filled band to leave a 'hole', or electron deficiency, the movement of which corresponds to the movement of a positive charge (p-type).

8.13 Magnetism

Modern magnetic theory is very complex, with ramifications in several branches of physics. Magnetic phenomena are associated with moving charges. Electrons, considered as particles, are assumed to possess an axial spin, which gives them the effect of a minute current-turn or of a small permanent magnet, called a Bohr *magneton*. The gyroscopic effect of electron spin develops a precession when a magnetic field is applied. If the precession effect exceeds the spin effect, the external applied magnetic field produces less magnetisation than it would in free space, and the material of which the electron is a constituent part is *diamagnetic*. If the spin effect exceeds that due to precession, the material is *paramagnetic*. The spin effect may, in certain cases, be very large, and high magnetisations are produced by an external field: such materials are *ferromagnetic*.

An iron atom has, in the $n=4$ shell (N), electrons that give it conductive properties. The K, L and N shells have equal numbers of electrons possessing opposite spin-directions, so cancelling. But shell M contains 9 electrons spinning in one direction and 5 in the other, leaving 4 net magnetons. Cobalt has 3, and nickel 2. In a solid metal, further cancellation occurs and the average number of unbalanced magnetons is: Fe, 2.2; Co, 1.7; Ni, 0.6.

In an iron crystal the magnetic axes of the atoms are aligned, unless upset by excessive thermal agitation. (At 770°C for Fe, the Curie point, the directions become random and ferromagnetism is lost.) A single Fe crystal magnetises most easily along a cube edge of the structure. It does not exhibit spontaneous magnetisation like a permanent magnet, however, because a crystal is divided into a large number of *domains* in which the various magnetic directions of the atoms form closed paths. But if a crystal is exposed to an external applied magnetic field, (i) the electron spin axes remain initially unchanged, but those domains having axes in the favourable direction grow at the expense of the others (domain-wall displacement); and (ii) for higher field intensities the spin axes orientate into the direction of the applied field.

If wall movement makes a domain acquire more internal energy, then the movement will relax again when the external field is removed. But if wall-movement results in loss of energy, the movement is non-reversible—i.e., it needs external force to reverse it. This accounts for hysteresis and remanence phenomena.

The closed-circuit self-magnetisation of a domain gives it a mechanical strain. When the magnetisation directions of individual domains are changed by an external field, the strain directions alter too, so that an assembly of domains will tend to lengthen or shorten. Thus readjustments in the crystal lattice occur, with deformations (e.g. 20 parts in 10^6) in one direction. This is the phenomenon of *magnetostriction*.

The practical art of magnetics consists in control of magnetic properties by alloying, heat-treatment and mechanical working to produce variants of crystal structure and consequent magnetic characteristics.

8.14 Simplified electrical theories

In the following paragraphs, a discussion of electrical phenomena is given in terms adequate for the purpose of simple explanation.

Consider two charged bodies separated in air (*Figure 8.2*).

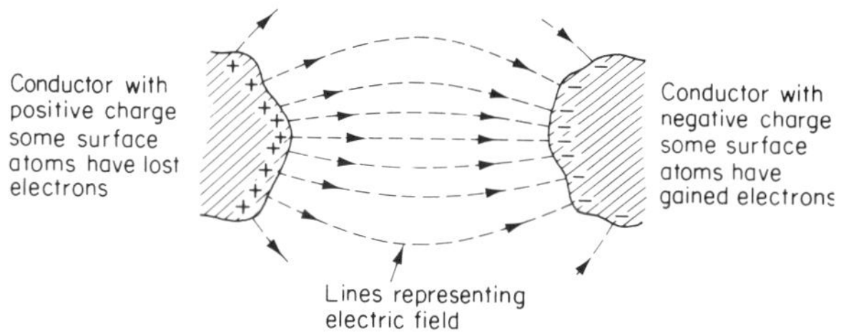

Figure 8.2 Charged conductors and their electric field

Work must have been done in a physical sense to produce on one an excess and on the other a deficiency of electrons, so that the system is a repository of potential energy. (The work done in separating charges is measured by the product of the charges separated and the difference of electrical potential that results.) Observation of the system shows certain effects of interest: (1) there is a difference of electric potential between the bodies depending on the amount of charge and the geometry of the system; (2) there is a mechanical force of attraction between the bodies. These effects are deemed to be manifestations of the *electric field* between the bodies, described as a special state of space and depicted by *lines of force* which express in a pictorial way the strength and direction of the force effects. The lines stretch between positive and negative elements of charge through the medium (in this case, air) which separates the two charged bodies. The electric field is only a concept—for the lines have no real existence—used to calculate various effects produced when charges are separated by any method which results in excess and deficiency states of atoms by electron transfer. Electrons and protons, or electrons and positively ionised atoms, attract each other, and the stability of the atom may be considered due to the balance of these attractions and dynamic forces such as electron spin. Electrons are repelled by electrons and protons by protons, these forces being summarised in the rules, formulated experimentally long before our present knowledge of atomic structure, that 'like charges repel and unlike charges attract one another'.

8.14.1 Conductors and insulators

In substances called *conductors*, the outer-shell electrons can be more or less freely interchanged between atoms. In copper, for example, the molecules are held together comparatively rigidly in the form of a 'lattice'—which gives the piece of copper its permanent shape—through the interstices of which outer electrons from the atoms can be interchanged within the confines of the surface of the piece, producing a random movement of free electrons called an 'electron atmosphere'. Such electrons are responsible for the phenomenon of electrical conductivity.

In other substances called *insulators* all the electrons are more or less firmly bound to their parent atoms so that little or no relative interchange of electron charges is possible. There is no marked line of demarcation between conductors and insulators, but the copper-group metals in the order silver, copper, gold, are outstanding in the series of conductors.

8.14.2 Conduction

Conduction is the name given to the movement of electrons, or ions, or both, giving rise to the phenomena described by the term *electric current*. The effects of a current include a redistribution of charges, heating of conductors, chemical changes in liquid solutions, magnetic effects, and many subsidiary phenomena.

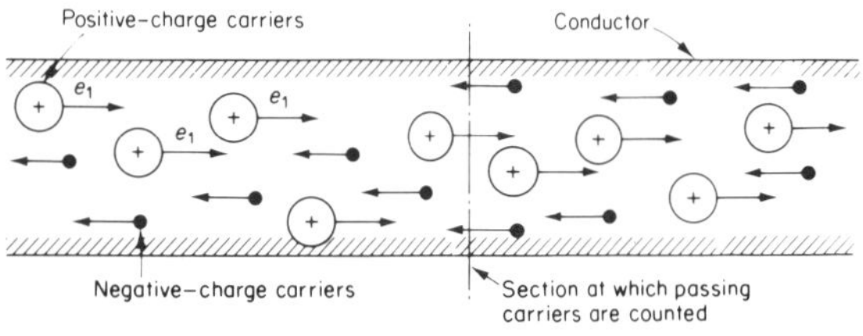

Figure 8.3 Electric current as the result of moving charges

If at some point on a conductor (*Figure 8.3*), n_1 carriers of electric charge (they can be water-drops, ions, dust particles, etc.) each with a positive charge e_1 arrive per second, and n_2 carriers (such as electrons) each with a negative charge e_2 arrive in the opposite direction per second, the total rate of passing of charge is $n_1e_1 + n_2e_2$, which is the charge per second or *current*. A study of conduction concerns the kind of carriers and their behaviour under given conditions. Since an electric field exerts mechanical forces on charges, the application of an electric field (i.e. a potential difference) between two points on a conductor will cause the movement of charges to occur, i.e., a current to flow, so long as the electric field is maintained.

The discontinuous particle nature of current flow is an observable factor. The current carried by a number of electricity carriers will vary slightly from instant to instant with the number of carriers passing a given point in a conductor. Since the electron charge is 1.6×10^{-19} C, and the passage of one coulomb per second (a rate of flow of one *ampere*) corresponds to $10^{19}/1.6 = 6.3 \times 10^{18}$ electron charges per second, it follows that the discontinuity will be observed only when the flow comprises the very rapid movement of a few electrons. This may happen in gaseous conductors, but in metallic conductors the flow is the very slow drift (measurable in mm/s) of an immense number of electrons.

A current may be the result of a two-way movement of positive and negative particles. Conventionally the direction of current flow is taken as the same as that of the positive charges and against that of the negative ones.

8.14.3 Conduction in metallic conductors

Reference has been made above to the 'electron atmosphere' of electrons in random motion within a lattice of comparatively rigid molecular structure in the case of copper, which is typical of the class of good metallic conductors. The random electronic motion, which intensifies with rise in temperature, merges into an average shift of charge of almost (but not quite) zero continuously (*Figure 8.4*). When an electric field is applied along the length of a conductor (as by maintaining a potential difference across its ends), the electrons have a *drift* towards the positive end superimposed upon their random digressions. The drift is slow, but such great numbers of electrons may be involved that very large currents, entirely due to electron drift, can be produced by this means. In their passage the electrons are impeded by the molecular lattice, the collisions producing heat and the opposition called *resistance*. The conventional direction of current flow is actually opposite to that of the drift of charge, which is exclusively electronic.

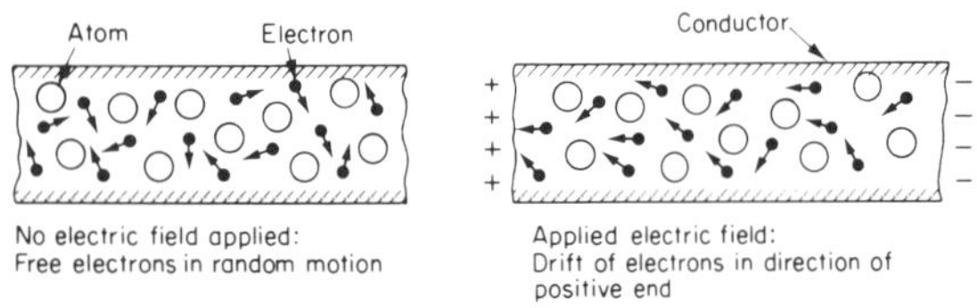

Figure 8.4 Electronic conduction in metals

8.14.4 Conduction in liquids

Liquids are classified according to whether they are *non-electrolytes* (non-conducting) or *electrolytes* (conducting). In the former the substances in solution break up into electrically balanced groups, whereas in the latter the substances form ions, each a part of a single molecule with either a positive or a negative charge. Thus common salt, NaCl, in a weak aqueous solution breaks up into sodium and chlorine ions. The sodium ion Na^+ is a sodium atom less one electron, the chlorine ion Cl^- is a chlorine atom with one electron more than normal. The ions attach themselves to groups of water molecules. When an electric field is applied the sets of ions move in opposite directions, and since they are much more massive than electrons the conductivity produced is markedly inferior to that in metals. Chemical actions take place in the liquid and at the electrodes when current passes. Faraday's electrolysis law states that the mass of an ion deposited at an electrode by electrolyte action is proportional to the quantity of electricity which passes and to the *chemical equivalent* of the ion.

8.14.5 Conduction in gases

Gaseous conduction is strongly affected by the pressure of the gas. At pressures corresponding to a few centimetres of mercury gauge, conduction takes place by the movement of positive and negative ions. Some degree of ionisation is always present due to stray radiations (light, etc.). The electrons produced attach themselves to gas atoms and the sets of positive and negative ions drift in opposite directions. At very low gas pressures the electrons produced by ionisation have a much longer free path before they collide with a molecule, and so have scope to attain high velocities. Their motional energy may be enough to *shock-ionise* neutral atoms, resulting in a great enrichment of the electron stream and an increased current flow. The current may build up to high values if the effect becomes cumulative, and eventually conduction may be effected through a *spark* or *arc*.

8.14.6 Conduction in vacuum

This may be considered as purely electronic, in that any electrons present (there can be no molecular *matter* present if the vacuum is perfect) are moved in accordance with the forces exerted on them by an applied electric field. The number of electrons is always small, and although high speeds may be reached the currents conducted in vacuum tubes are generally measurable only in milli- or micro-amperes.

8.14.7 Vacuum and gas-filled tubes

Some of the effects described above are illustrated in *Figure 8.5*. At the bottom is an electrode, the *cathode*, from the surface of which electrons are emitted, generally by heating the cathode material. At the top is a second electrode, the *anode*, and an electric field is established between anode and cathode, which are enclosed in a vessel which contains a low-pressure inert gas. The electric field causes electrons emitted from the cathode to move upwards. In their passage to the anode these electrons will encounter gas molecules. If conditions are suitable, the gas atoms are ionised, becoming in effect positive charges associated with the nuclear mass. Thereafter the current is increased by the detached electrons moving upwards and by the positive ions

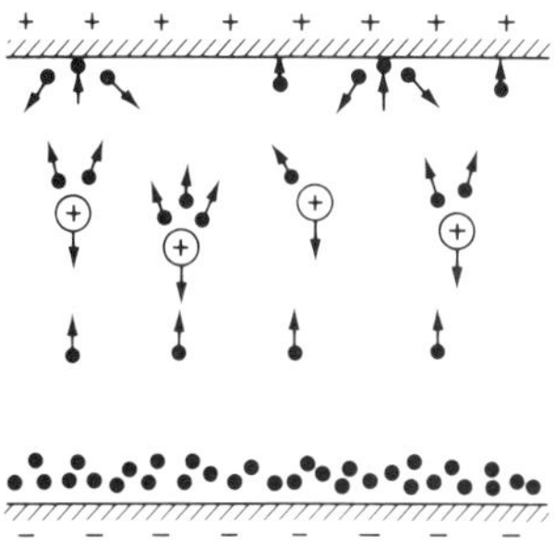

Figure 8.5 Electrical conduction in gases at low pressure

moving more slowly downwards. In certain devices (such as the mercury-arc rectifier) the impact of ions on the cathode surface maintains its emission. The impact of electrons on the anode may be energetic enough to cause the *secondary emission* of electrons from the anode surface. If the gas molecules are excluded and a vacuum established, the conduction becomes purely electronic.

8.14.8 Convection currents

Charges may be moved by mechanical means, on discs, endless belts, water-drops, dust or mist particles. A common example is the electron beam between anode and screen in the cathode-ray oscilloscope. Such a motion of charges, independent of an electric field, is termed a *convection* current.

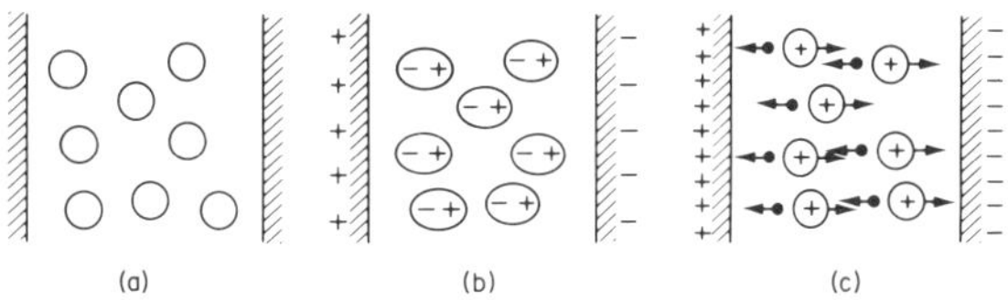

Figure 8.6 Polarisation, displacement and breakdown in a dielectric material: (a) no electric field; atoms unstrained; (b) electric field applied; polarisation; (c) intensified electric field; atoms ionised

8.14.9 Displacement and polarisation currents

If an electric field is applied to a perfect insulator, whether solid, liquid or gaseous, the electric field affects the atoms by producing a kind of 'stretching' or 'rotation' which displaces the electrical centres of negative and positive in opposite directions. This polarisation of the dielectric insulating material may be considered as taking place in the manner indicated in *Figure 8.6.* Before the electric field is applied, in (*a*), the atoms of the insulator are neutral and unstrained; (*b*) as the potential difference is raised the electric field exerts opposite mechanical forces on the negative and positive charges and the atoms become more and more highly strained. On the left face the atoms will all present their negative charges at the surface: on the right face, their positive charges. These surface polarisations are such as to account for the effect known as *permittivity.* The small displacement of the electric charges is an electron shift, i.e., a *displacement current* flows while the polarisation is being established. *Figure 8.6*(*c*) shows that under conditions of excessive electric field atomic disruption or ionisation may occur, converting the insulator material into a conductor, resulting in *breakdown.*

9

Light

D R Heath BSc, PhD
Ranx Xerox Ltd

Contents

9.1 Introduction

In recent years the growth of the field of opto-electronics has required the engineer to furnish himself with a knowledge of the nature of optical radiation and its interaction with matter. The increase in the importance of measurements of optical energy has also necessitated an introduction into the somewhat bewildering array of terminologies used in the hitherto specialist fields of radiometry and photometry.

9.2 The optical spectrum

Light is electromagnetic radiant energy and makes up part of the electromagnetic spectrum. The term *optical spectrum* is used to described the *light* portion of the electromagnetic spectrum and embraces not only the visible spectrum (that detectable by the eye) but also the important regions in optoelectronics of the ultraviolet and infrared.

The electromagnetic spectrum, classified into broad categories according to wavelength and frequency, is given in *Figure 10.1*, Chapter 10. It is observed that on this scale the optical spectrum forms only a very narrow region of the complete electromagnetic spectrum. *Figure 9.1* is an expanded diagram showing more detail of the ultraviolet, visible and infrared regions. By convention, optical radiation is generally specified according to its wavelength. The wavelength can be determined from a specific electromagnetic frequency from the equation:

$$\lambda = c/f \qquad (9.1)$$

where λ is the wavelength (m), f is the frequency (Hz) and c is the speed of light in a vacuum ($\sim 2.99 \times 10^8$ m s^{-1}). The preferred unit of length for specifying a particular wavelength in the visible spectrum is the nanometre (nm). Other units are also in common use, namely the angström (Å) and the micrometre or micron. The relation of these units is as follows:

1 nanometre (nm) $= 10^{-9}$ metre
1 angström (Å) $= 10^{-10}$ metre
1 micron (μm) $= 10^{-6}$ metre

The micron tends to be used for describing wavelengths in the infrared region and the angström for the ultraviolet and visible regions.

The wavenumber (cm^{-1}) is the reciprocal of the wavelength measured in centimetres, i.e. $1/\lambda$ (cm) = wavenumber (cm^{-1}).

9.3 Basic concepts of optical radiation

In describing the measurement of light and its interaction with matter, three complementary properties of electromagnetic radiation need to be invoked: ray, wave and quantum. At microwave and longer wavelengths it is generally true that radiant energy exhibits primarily wave properties while at the shorter wavelengths, X-ray and shorter, radiant energy primarily exhibits ray and quantum properties. In the region of the optical spectrum, ray, wave, and quantum properties will have their importance to varying degrees.

9.4 Radiometry and photometry

Radiometry is the science and technology of the measurement of radiation from all wavelengths within the optical spectrum. The basic unit of power in radiometry is the watt (W).

Photometry is concerned only with the measurement of light detected by the eye, i.e. that radiation which falls between the wavelengths 380 nm and 750 nm. The basic unit of power in photometry is the lumen (lm).

In radiometric measurements the ideal detector is one which has a flat response with wavelength whereas in photometry the ideal detector has a spectral response which approximates to that of the average human eye. To obtain consistent measurement techniques the response of the average human eye was established by the Commission Internationale de l'Eclairage (CIE) in 1924. The response known as the photopic eye response is shown in *Figure 9.2* and is observed to peak in the green/yellow part of the visible spectrum at 555 nm. The curve indicates that it takes approximately ten times as many units of blue light as green light to produce the same visibility effect on the average human eye.

The broken curve in *Figure 9.2* with a peak at 507 nm is termed the scotopic eye response. The existence of the two responses arises out of the fact that the eye's spectral response shifts at very low light levels. The retina of the human eye has two types of optical receptors, cones and rods. Cones are mainly responsible for colour vision and are highly concentrated in a 0.3 mm diameter spot, called the fovea, at the centre of the field of vision. Rods are not present in the fovea but have a very high density in the peripheral regions of the retina. They do not give rise to colour response but at low light levels are significantly more sensitive than cones. At normal levels of illumination (photopic response) the eye's response is determined by the cones in the retina whilst at very low light levels the retina's rod receptors take over and cause a shift in the response curve to the scotopic response.

In normal circumstances photometric measurements are based on the CIE photopic response and all photometric instruments must have sensors which match this response. At the peak wavelength of 555 nm of the photopic response one watt of radiant power is defined as the equivalent of 680 lumens of luminous power.

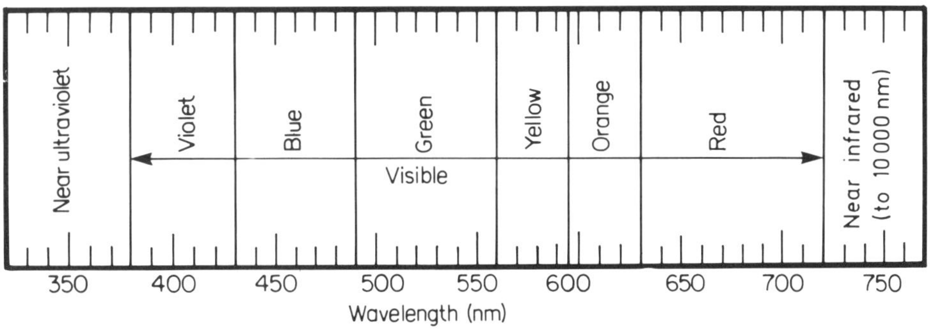

Figure 9.1 The visible spectrum

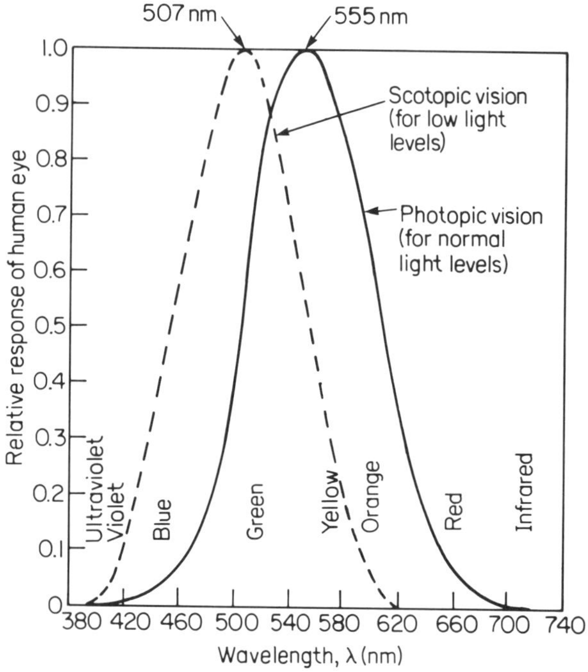

Figure 9.2 The photopic and scotopic eye responses

In order to convert a radiometric power measurement into photometric units both the spectral response of the eye and the spectral output of the light source must be taken into account. The conversion is then achieved by multiplying the energy radiated at each wavelength by the relative lumen/watt factor at that wavelength and summing the results. Note that in the ultraviolet and infrared portions of the optical spectrum although one may have high output in terms of watts the photometric value in lumens is zero due to lack of eye response in those ranges. However, it should be said that many observers can see the 900 nm radiation from a GaAs laser or the 1.06 μm radiation from a Nd:YAG laser since in this instance the intensity can be sufficiently high to elicit a visual response. Viewing of these sources in practice is not to be recommended for safety reasons and the moderately high energy densities at the eye which are involved.

9.5 Units of measurement

There are many possible measurements for characterising the output of a light source. The principles employed in defining radiometric and photopic measurement terms are very similar. The terms employed have the adjective radiant for a radiometric measurement and luminous for a photometric measurement. The subscript e is used to indicate a radiometric symbol and the subscript v for a photometric symbol. A physical visualisation of the terms to be defined is given in *Figure 9.3*. *Figure 9.4* illustrates the concept of solid angle required in the visualisation of *Figure 9.3*.

9.5.1 Radiometric terms and units

Radiant flux or **radiant power**, Φ_e The time rate of flow of radiant energy emitted from a light source. Expressed in J s^{-1} or W.

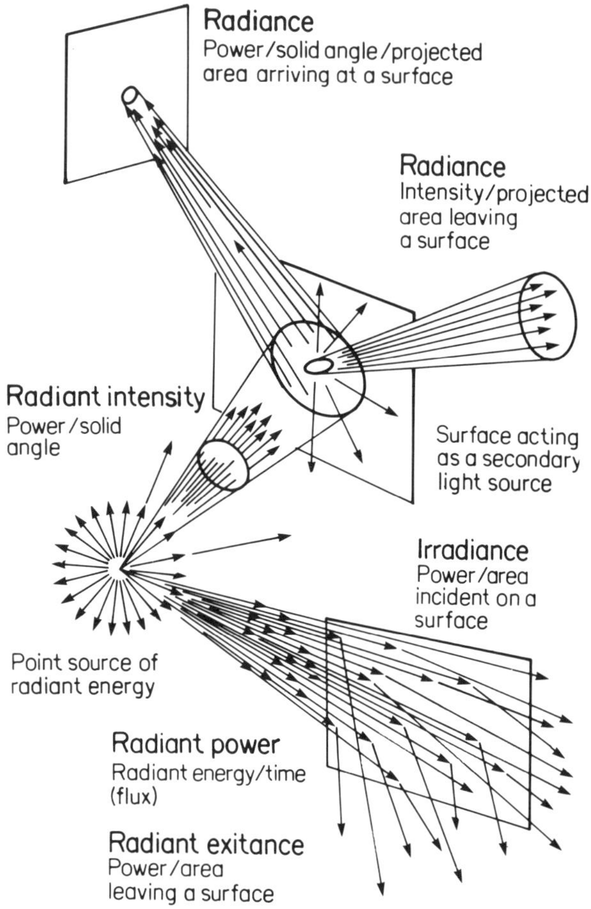

Figure 9.3 A visualisation of radiometric terms (from reference 1)

Irradiance, E_e The radiant flux density incident on a surface. Usually expressed in W cm^{-2}.
Radiant intensity, I_e The radiant flux per unit solid angle travelling in a given direction. Expressed in W sr^{-1}.
Radiant exitance, M_e The total radiant flux divided by the surface area of the source. Expressed in W cm^{-2}.
Radiance, L_e The radiant intensity per unit area, leaving, passing through, or arriving at a surface in a given direction. The surface area is the projected area as seen from the specified direction. Expressed in W cm^{-2} sr^{-1}.

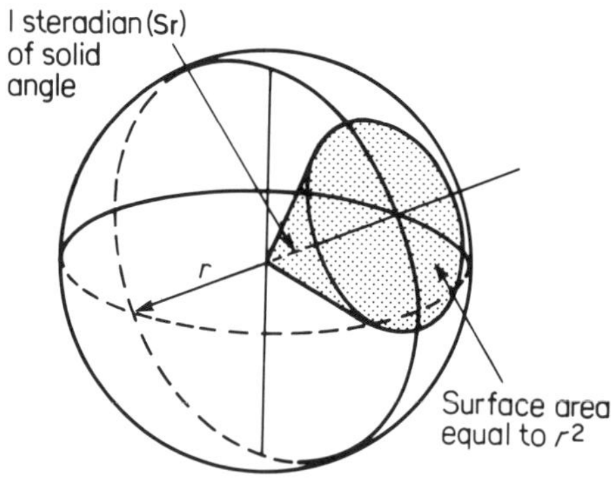

Figure 9.4 Diagram illustrating the steradian (from reference 1)

9.5.2 Photometric terms and units

The equivalent photometric terminologies to the radiometric ones defined above are as follows:

Luminous flux or **power**, Φ_v The time rate of flow of luminous energy emitted from a light source. Expressed in lm.

Illuminance or **illumination**, E_v The density of luminous power incident on a surface. Expressed in lm cm^{-2}. Note the following:

1 lm cm^{-2} = 1 phot
1 lm m^{-2} = 1 lux
1 lm ft^{-2} = 1 footcandle

Luminous intensity, I_v The luminous flux per unit solid angle, travelling in a given direction. Expressed in lm sr^{-1}. Note that 1 lm sr^{-1} = 1 cd.

Luminous exitance, M_v The total luminous flux divided by the surface area of the source. Expressed in lm cm^{-2}.

Luminance, L_v The luminous intensity per unit area, leaving, passing through or arriving at a surface in a given direction. The surface area is the projected area as seen from the specified direction. Expressed in lm cm^{-2} sr^{-1} or cd cm^{-2}.

Mathematically if the area of an emitter has a diameter or diagonal dimension greater than 0.1 of the distance of the detector it can be considered as an area source. Luminance is also called the photometric **brightness**, and is a widely used quantity. In *Figure 9.5* the projected area of the source, A_p varies directly as the cosine of θ, i.e. is a maximum at 0° or normal to the surface and minimum at 90°. Thus

$$A_p = A_s \cos\theta \tag{9.2}$$

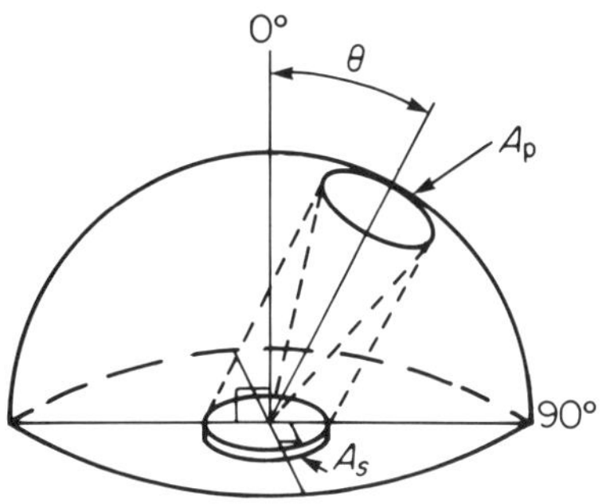

Figure 9.5 Diagram illustrating the projected area

Luminance is then the ratio of the luminous intensity (I_v) to the projected area of the source (A_p):

$$\text{luminance} = \frac{\text{luminous intensity}}{\text{projected area}} = \frac{I_v}{A_p}$$

$$= \frac{I_v}{A_s \cos\theta} \text{ lm sr}^{-1} \text{ per unit area}$$

since one lm sr^{-1} = one cd, depending on the units used for the area we have

1 cd cm^{-2} = 1 stilb
$1/\pi$ cd cm^{-2} = 1 lambert
$1/\pi$ cd ft^{-2} = 1 footlambert

Table 9.1 provides a summary of the radiometric and photometric terms with their symbols and units.

Table 9.1 Radiometric and photometric terms

Quantity	*Symbol*	*Unit(s)*
Radiant flux	Φ_e	W
Luminous flux	Φ_v	lm
Irradiance	E_e	W cm^{-2}
Illuminance	E_v	lm cm^{-2} = phot lm m^{-2} = lux lm ft^{-2} = footcandle
Radiant intensity	I_e	W sr^{-1}
Luminous intensity	I_v	lm sr^{-1} = cd
Radiant exitance	M_e	W cm^{-2}
Luminous exitance	M_v	lm cm^{-2}
Radiance	L_e	W cm^{-2} sr^{-1}
Luminance	L_v	lm cm^{-2} sr^{-1}
(Photometric brightness)		cd cm^{-2} = stilb $1/\pi$ cd cm^{-2} = lambert $1/\pi$ cd ft^{-2} = footlambert

Some typical values of natural scene illumination expressed in units of lm m^{-2} and footcandles are given in *Table 9.2*. *Table 9.3* gives some approximate values of luminance for various sources.

Table 9.2 Approximate levels of natural scene illumination (from reference 1)

	Footcandles	*lm m^{-2}*
Direct sunlight	$1.0–1.3 \times 10^4$	$1.0–1.3 \times 10^5$
Full daylight	$1–2 \times 10^3$	$1–2 \times 10^4$
Overcast day	10^2	10^3
Very dark day	10	10^2
Twilight	1	10
Deep twilight	10^{-1}	1
Full moon	10^{-2}	10^{-1}
Quarter moon	10^{-3}	10^{-2}
Starlight	10^{-4}	10^{-3}
Overcast starlight	10^{-5}	10^{-4}

Table 9.3 Approximate levels of luminance for various sources (from reference 1)

	Footlamberts	*cd m^{-2}*
Atomic fission bomb (0.1 ms after firing, 90 ft diameter ball)	6×10^{11}	2×10^{12}
Lightning flash	2×10^{10}	6.8×10^{10}
Carbon arc (positive crater)	4.7×10^6	1.6×10^7
Tungsten filament lamp (gas-filled, 16 lm W^{-1})	2.6×10^5	8.9×10^5
Sun (as observed from the earth's surface at meridian)	4.7×10^3	1.6×10^4
Clear blue sky	2300	7900
Fluorescent lamp (T-12 bulb, cool white, 430 mA medium loading)	2000	6850
Moon (as observed from earth's surface)	730	2500

9.6 Practical measurements

A wide variety of commercial instruments is available for carrying out optical radiation measurements.

The radiometer is an instrument which will normally employ a photodiode, phototube, photomultiplier or photoconductive cell as its detector. Each of these detectors has a sensitivity which varies with wavelength. It is therefore necessary for the instrument to be calibrated over the full range of wavelengths for which it is to be used. For measurement of monochromatic radiation the instrument reading is simply taken and multiplied by the appropriate factor in the detector sensitivity at the given wavelength. A result in units of power or energy is thereby obtained.

For the characterisation of broadband light sources, where the output is varying with wavelength, it is necessary to measure the source in narrow band increments of wavelength. This can be achieved by using a set of calibrated interference filters.

The spectroradiometer is specifically designed for broadband measurements and has a monochromator in front of the detector which performs the function of isolating all the wavelengths of interest. These can be scanned over the detector on a continuous basis as opposed to the discrete intervals afforded by filters.

The photometer is designed to make photometric measurements of sources. It usually consists of a photoconductive cell, silicon photodiode or photomultiplier with a filter incorporated to correct the total system response to that of the standard photopic eye response curve.

Thermopiles, bolometers and pyrometers generate signals which can be related to the incident power as a result of a change in temperature which is caused by absorption of the radiant energy. They have an advantage that their response as a function of wavelength is almost flat (constant with wavelength), but are limited to measurement of relatively high intensity sources and normally at wavelengths greater than 1 μm.

Calibration of most optical measuring instruments is carried out using tungsten lamp standards and calibrated thermopiles. The calibration accuracy of these lamp standards varies from approximately ±8% of absolute in the ultraviolet to ±5% of absolute in the visible and near infrared. Measurement systems calibrated with these standards will generally have accuracies of 8 to 10% of absolute. It is important to realise that the accuracy of optical measurements is rather poor compared to other spheres of physics. To obtain an accuracy of 5% in a measurement is very difficult; a good practitioner will be doing well to keep his errors to between 10 and 20%.

9.7 Interaction of light with matter

Light may interact with matter by being reflected, refracted, absorbed or transmitted. Two or more of these are usually involved.

9.7.1 Reflection

Some of the light impinging on any surface is reflected away from the surface. The reflectance varies according to the properties of the surface and the wavelength of the impinging radiation.

Regular or *specular reflection* is reflection in accordance with the laws of reflection with no diffusion (surface is smooth compared to the wavelength of the impinging radiation).

Diffuse reflection is diffusion by reflection in which on the microscopic scale there is no regular reflection (surface is rough when compared to the wavelength of the impinging radiation).

Reflectance (ρ) is the ratio of the reflected radiant or luminous flux to the incident flux.

Reflection (optical) density (D) is the logarithm to the base ten of the reciprocal of the reflectance.

$$D(\lambda)=\log_{10}\left(\frac{1}{\rho(\lambda)}\right) \tag{9.3}$$

where $\rho(\lambda)$ is the spectral reflectance.

9.7.2 Absorption

When a beam of light is propagated in a material medium its speed is less than its speed in a vacuum and its intensity gradually decreases as it progresses through the medium. The speed of light in a material medium varies with the wavelength and this variation is known as *dispersion*. When a beam traverses a medium some of the light is scattered and some is absorbed. If the absorption is true absorption the light energy is converted into heat. All media show some absorption—some absorb all wavelengths more or less equally, others show selective absorption in that they absorb some wavelengths very much more strongly than others. The phenomena of scattering, dispersion and absorption are intimately connected.

Absorption coefficient

Lambert's law of absorption states that equal paths in the same absorbing medium absorb equal fractions of the light that enters them. If in traversing a path of length $\mathrm{d}x$ the intensity is reduced from I to $I-\mathrm{d}I$ then Lambert's law states that $\mathrm{d}I/I$ is the same for all elementary paths of length $\mathrm{d}x$. Thus

$$\frac{\mathrm{d}I}{I}=-K\,\mathrm{d}x$$

where K is a constant known as the absorption coefficient. Therefore $\log I=-Kx+C$ where C is a constant. If $I=I_0$ at $x=0$, $C=\log I_0$ and so

$$I=I_0\,\mathrm{e}^{-Kx} \tag{9.4}$$

Note that in considering a medium of thickness x, I_0 is not the intensity of incident light due to there being some reflection at the first surface. Similarly I is not the emergent intensity owing to reflection at the second surface. By measuring the emergent intensity for two different thicknesses the losses due to reflection may be eliminated.

9.7.3 Polarisation

For an explanation of polarisation of light we need to invoke the wave concept and the fact that light waves are of a transverse nature possessing transverse vibrations which have both an electric and magnetic character. *Figures 9.6* and *9.7* set out to illustrate the meaning of unpolarised and linearly polarised light.

In *Figure 9.6* a wave is propagating in the x direction with the vibrations in a single plane. Any light which by some cause possesses this property is said to be linearly polarised. Ordinary light, such as that received from the sun or incandescent lamps, is

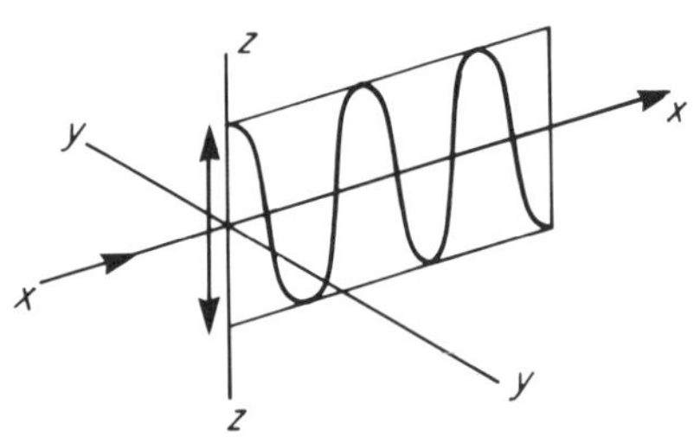

Figure 9.6 Linearly polarised light

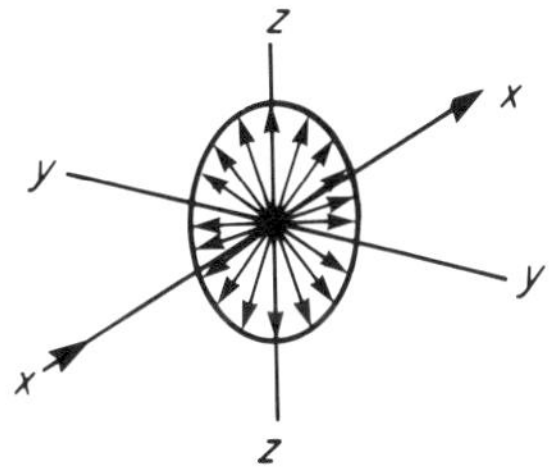

Figure 9.7 Unpolarised light

unpolarised and in this case the arrangement of vibrations is in all possible directions perpendicular to the direction of travel, as in *Figure 9.7*.

There are numerous methods for producing linearly polarised light, those most widely known being birefringence or double refraction, reflection, scattering and dichroism. Double reflection occurs in certain types of natural crystal such as calcite and quartz and will divide a beam of unpolarised light into two separate polarised beams of equal intensity. By eliminating one of the polarised beams a very efficient linear polariser can be made.

Dichroic polarisers make up the great majority of commercially produced synthetic polarisers. They exhibit dichroism, the property of absorbing light to different extents depending on the polarisation form of the incident beam.

The light emerging from a linear polariser can be given a 'twist' so that the vibrations are no longer confined to a single plane but instead form a helix. This is achieved by inserting a sheet of double-refracting material into the polarised beam which divides the beam into two beams of equal intensity but with slightly different speeds, one beam being slightly retarded. The light is said to be circularly polarised.

The application and uses of polarised light are very considerable—liquid crystal displays, control of light intensity, blocking and prevention of specular glare light, measuring optical rotation, measuring propagation of stress and strain are some notable ones.

References

1 ZAHA, M. A., 'Shedding some needed light on optical measurements', *Electronics*, 6 Nov., 91–6 (1972)

Further reading

CLAYTON, R. K., *Light and Living Matter*, Vol. 2, *The Biological Part*, McGraw-Hill, New York (1971)

GRUM, F. and BECHENER, R. J., *Optical Radiation Measurements*, Vol. 1, *Radiometry*, Academic Press, London (1979).

JENKINS, F. A. and WHITE, F. E., *Fundamentals of Optics*, 3rd edn, McGraw-Hill, New York (1957).

KEYS, R. J., *Optical and infrared detectors*, Springer Verlag (1980)

LAND, E. H., 'Some aspects on the development of sheet polarisers', *J. Opt. Soc. Am.*, **41**, 957 (1951)

LERMAN, S., *Radiant energy and the eye*, Macmillan (1980)

LONGHURST, R. S., *Geometrical and Physical Optics*, Longman, Green & Co., London (1957)

MAYER-ARENDT, J. R., *Introduction to Classical and Modern Optics*, Prentice Hall, Englewood Cliffs, NJ (1972)

RCA Electro-Optics Handbook, RCA Commercial Engineering, Harrison, NJ (1974)

WALSH, J. W. T., *Photometry*, Dover, New York (1965)

10 Radiation

L W Turner CEng, FIEE, FRTS
Consultant Engineer

Contents

10.1 Electromagnetic radiation

Light and heat were for centuries the only known kinds of radiation. Today it is known that light and heat radiation form only a very small part of an enormous range of radiations extending from the longest radio waves to the shortest gamma-rays and known as the electromagnetic spectrum, *Figure 10.1*. The wavelength of the radiations extends from about 100 kilometres to fractions of micrometres. The visible light radiations are near the centre of the spectrum. All other radiations are invisible to the human eye.

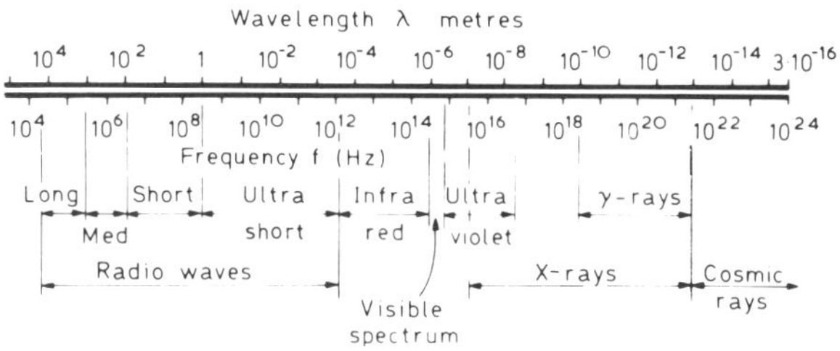

Figure 10.1 Electromagnetic wave spectrum

The researches into electromagnetic radiation can be traced back to 1680, to Newton's theory of the composition of white light. Newton showed that white light is made up from rays of different colours. A prism refracts these rays to varying degrees according to their wavelengths and spreads them out. The result is the visible spectrum of light.

In 1800, William Herschel, during research into the heating effects of the visible spectrum, discovered that the maximum heating was not within the visible spectrum but just beyond the red range. Herschel concluded that in addition to visible rays the sun emits certain invisible ones. These he called infrared rays.

The next year, the German physicist Ritter made a further discovery. He took a sheet of paper freshly coated with silver chloride and placed it on top of a visible spectrum produced from sunlight falling through a prism. After a while he examined the paper in bright light. It was blackened, and it was most blackened just beyond the violet range of the spectrum. These invisible rays Ritter called ultraviolet rays.

The next step was taken in 1805, when Thomas Young demonstrated that light consists of waves, a theory which the frenchman Fresnel soon proved conclusively. Fresnel showed that the waves vibrated transversely, either in many planes, or in one plane, when the waves were said to be plane-polarised. The plane containing both the direction of propagation and the direction of the electric vibrations is called the plane of polarisation.

In 1831 Faraday showed that when a beam of light was passed through a glass block to which a magnetic field was applied in the same direction as the direction of polarisation the plane of polarisation could be rotated. Moreover, when the magnetic field was increased, the angle of rotation also increased. The close relationship between light, magnetism and electricity was thus demonstrated for the first time.

In 1864 James Clerk Maxwell formulated his theory of electromagnetic waves and laid the foundation of the wave theory of the electromagnetic spectrum as it is known today.

The *fundamental Maxwell theory* includes two basic laws and the displacement-current hypothesis:

Ampere's law. The summation of the magnetic force H round a closed path is proportional to the total current flowing across the surface bounded by the path:

F = line-integral of $H.\mathrm{d}l = I$

Displacement current. The symbol I above includes polarisation and displacement currents as well as conduction currents.

Faraday's law. The summation of the electric force E round a closed path is proportional to the rate-of-change of the magnetic flux Φ across the surface bounded by the path:

e = line-integral of $E.\mathrm{d}l = -\mathrm{d}\Phi/\mathrm{d}t$

The magnetic flux is circuital, and representable by 'closed loops' in a 'magnetic circuit'. The electric flux may be circuital, or it may spring from charges. The total flux leaving or entering a charge Q is Q coulombs.

A metallic circuit is not essential for the development of an e.m.f. in accordance with Faraday's law. The voltage-gradient E exists in the space surrounding a changing magnetic flux. The conductor is needed when the e.m.f. is to produce conduction currents. Again, the existence of a magnetic field does not necessarily imply an associated conduction current: it may be the result of a displacement current.

Maxwell deduced from these laws (based on the work of Faraday) the existence of electromagnetic waves in free space and in material media. Waves in free space are classified in accordance with their frequency f and their wavelength λ, these being related to the free-space propagation velocity $c \simeq 3 \times 10^8$ m/s by the expression $c = f\lambda$. Radiant energy of wavelength between 0.4 and 0.8 μm (frequencies between 750 and 375 GHz) is appreciated by the eye as *light* of various colours over the visible spectrum between violet (the shorter wavelength) and red (the longer). Waves shorter than the visible are the *ultraviolet*, which may excite visible fluorescence in appropriate materials. *X-rays* are shorter still. At the longer-wave end of the visible spectrum is *infrared* radiation, felt as *heat*. The range of wavelengths of a few millimetres upward is utilised in *radio* communication.

In 1886 Heinrich Hertz verified Maxwell's theory. At that time Wimshurst machines were used to generate high voltages. A Leyden jar served to store the charge which could be discharged through a spark gap. Hertz connected a copper spiral in series with the Leyden jar; this spiral acted as a radiator of electromagnetic waves. A second spiral was placed a small distance from the first; this also was connected to a Leyden jar and a spark gap. When the wheel of the Wimshurst machine was turned sparks jumped across both gaps. The secondary sparks were caused by electromagnetic waves radiated from the first spiral and received by the second. These waves were what are today called *radio* waves. This experiment was the first of a series by which Hertz established the validity of Maxwell's theory.

In 1895, the German Roentgen found by chance that one of his discharge tubes had a strange effect on a chemical substance which happened to lie nearby: the substance emitted light. It even fluoresced when screened by a thick book. This meant that the tube emitted some kind of radiation. Roentgen called these unknown rays X-rays.

A year later the French physicist Henri Becquerel made a further discovery. He placed a photographic plate, wrapped in black paper, under a compound of uranium. He left it there over night. The plate, when developed, was blackened where the uranium had been. Becquerel had found that there exist minerals which give off invisible rays of some kind.

Later, research by Pierre and Marie Curie showed that many substances had this effect: radioactivity had been discovered. When this radiation was analysed it was found to consist of charged particles, later called alpha- and beta-rays by Rutherford. These particles were readily stopped by thin sheets of paper or metal.

In 1900, Villard discovered another radiation, much more

penetrating and able to pass even through a thick steel plate. This component proved to consist of electromagnetic waves which Rutherford called gamma-rays. They were the last additions to the electromagnetic spectrum as is known today.

Waves are classified according to their uses and the methods of their generation, as well as to their frequencies and wavelengths. Radio waves are divided into various bands: very low frequency (v.l.f.) below 30 kHz, low frequency (l.f.) 30–300 kHz, medium frequency (m.f.) 300–3000 kHz, high frequency (h.f.) 3–30 MHz, very high frequency (v.h.f.) 30–300 MHz, ultra high frequency (u.h.f.) 300–3000 MHz, super high frequency (s.h.f.) 3000–30 000 MHz, extra high frequency (e.h.f.) 30 000–300 000 MHz. Waves of frequencies higher than 100 MHz are generally called microwaves. The microwave band overlaps the infrared band. Actually, *all* wave bands merge imperceptibly into each other; there is never a clear-cut division. Next is the narrow band of visible light. These visible rays are followed by the ultraviolet rays and the X-rays. Again, all these bands merge into each other. Finally, the gamma-rays. They are actually part of the X-ray family and have similar characteristics excepting that of origin.

A convenient unit for the measurement of wavelengths shorter than radio waves is the micrometre (μm) which is 10^{-6} m. The micrometre is equal to the micron (μ), a term still used but now deprecated. Also deprecated is the term mμ, being 10^{-3} micron. For measurement of still shorter wavelengths, the nanometre (10^{-9} m) is used. The angström unit (Å), which is 10^{-10} m, is commonly used in optical physics. These units of wavelength are compared in *Table 10.1*.

Table 10.1 Comparison of units of length

		Å	*nm*	*μm*	*mm*	*cm*	*m*
Å	=	1	10^{-1}	10^{-4}	10^{-7}	10^{-8}	10^{-10}
nm	=	10	1	10^{-3}	10^{-6}	10^{-7}	10^{-9}
μm	=	10^4	10^3	1	10^{-3}	10^{-4}	10^{-6}
mm	=	10^7	10^6	10^3	1	10^{-1}	10^{-3}
cm	=	10^8	10^7	10^4	10	1	10^{-2}
m	=	10^{10}	10^9	10^6	10^3	10^2	1

Electromagnetic waves are generated by moving charges such as free electrons or oscillating atoms. Orbital electrons (see Chapter 8) radiate when they move from one orbit to another, and only certain orbits are permissible. Oscillating nuclei radiate gamma-rays.

The frequency of an electromagnetic radiation is given by the expression:

$$f=\frac{E}{h}$$

where E is the energy and h is Planck's constant ($h \simeq 6.6 \times 10^{-34}$ J s).

The identity of electromagnetic radiations has been established on the following grounds:

(1) The velocity of each *in vacuo* is constant.
(2) They all experience reflection, refraction, dispersion, diffraction, interference and polarisation.
(3) The mode of transmission is by transverse wave action.
(4) All electromagnetic radiation is emitted or absorbed in bursts or packets called quanta (or photons in the case of light).

In connection with (4), Planck established that the energy of each quantum varies directly with the frequency of the radiation (see the above expression).

Modern physics now accepts the concept of the dual nature of electromagnetic radiation, viz. that it has wave-like properties but at the same time it is emitted and absorbed in quanta.

Polarisation. An electromagnetic radiation possesses two fields at right angles to each other as viewed in the direction of the oncoming waves. These are the electric field and the magnetic field. The direction of either of these is known as the polarisation of the field, but the term is more usually related to the electric field, and this is at the same angle as the radiating source. For example, in the case of radio waves, a horizontally positioned receiving dipole will not respond efficiently to waves which are vertically polarised.

The same phenomenon occurs with light radiation which is normally unpolarised, i.e. it is vibrating in all transverse planes. A sheet of polaroid allows light to pass through in one plane, due to the molecular structure of the material, and the resulting plane-polarised light is absorbed if a second sheet of polaroid is set at right-angles to the first.

The 'optical window'. It is important to realise that our knowledge of the universe around us depends upon incoming electromagnetic radiation. However, from the entire spectrum of such radiation, only two bands effectively reach the earth's surface:

(1) the visible light spectrum, together with a relatively narrow band of the adjacent ultraviolet and infrared ranges;
(2) a narrow band of radio waves in the 1 cm to about 10 m band.

Thus the gamma-rays, X-rays, most of the ultraviolet and infrared rays, together with the longer ranges of radio waves fail to reach the earth's surface from outer space. This is mainly due to absorption in the ionosphere and atmosphere. (See Chapter 11.)

The applications of electromagnetic radiations range over an enormous field. Radio waves are used for telecommunication, sound and television broadcasting, navigation, radar, space exploration, industry, research, etc. Infrared rays have many applications including security systems, fire detection, dark photography, industry, medical therapy. Ultraviolet rays, of which the sun is the chief source, have wide industrial and medical applications. X-rays and gamma-rays have become the everyday tools of the doctor, the scientist and in industry. And the narrow band of visible rays not only enables the world around us to be seen but together with ultraviolet radiation makes possible the process of photosynthesis by which plants build up and store the compounds of all our food. Thus the laws governing electromagnetic radiation are relevant to life itself. Many of these applications are described in more detail in later sections.

10.2 Nuclear radiation

There are three main types of radiation that can originate in a nucleus: *alpha*, *beta*, and *gamma* radiation.

Alpha radiation. An alpha particle has a charge of two positive units and a mass of four units. It is thus equivalent to a helium nucleus, and is the heaviest of the particles emitted by radioactive isotopes. Alpha (α) particles are emitted mostly by heavy nuclei and can possess only discrete amounts of energy, i.e. they give a line energy spectrum. The probability of collision between particles increases with the size of the particles. Thus the rate of ionisation in a medium traversed by particles emitted from radioactive isotopes, and hence the rate of loss of energy of the particles, also increases with the size of the particles. Consequently the penetrating power of the large alpha particles is relatively poor.

Beta radiation. Beta particles can be considered as very fast electrons. They are thus much smaller than alpha particles and therefore have greater penetrating powers. Beta (β) radiation will be absorbed in about 100 inches of air or half an inch of Perspex.

Unlike α-particles, β-particles emitted in a nuclear process have a continuous energy spectrum, i.e. β-particles can possess any amount of energy up to a maximum determined by the energy equivalent to the change in mass involved in the nuclear reaction. This has been explained by postulating the existence of the *neutrino*, a particle having no charge and negligible mass. According to this theory, the energy is shared between the β-particle and the neutrino in proportions that may vary, thus giving rise to a continuous energy spectrum.

Gamma radiation. Gamma (γ) radiation is electromagnetic in nature and has, therefore, no charge or mass. Its wavelength is much shorter than that of light or radio waves, and is similar to that of *X-rays*. The distinction between γ-rays and X-rays is that γ-rays are produced within the nucleus while X-rays are produced by the transition of an electron from an outer to an inner orbit.

γ-radiation has well-defined amounts of energy—that is, it occupies very narrow bands of the energy spectrum—since it results from transitions between energy levels within the nucleus. Characteristic X-rays of all but the very lightest of elements also possess well-defined amounts of energy.

γ-radiation has very great penetrating powers. Significant amounts are able to pass through lead bricks 50 mm thick; γ-photons possessing 1 MeV of energy will lose less than 1% of their energy in traversing half a mile of air.

A rough comparison of the penetrating powers of α-, β-, and γ-radiation is given in *Figure 10.2*.

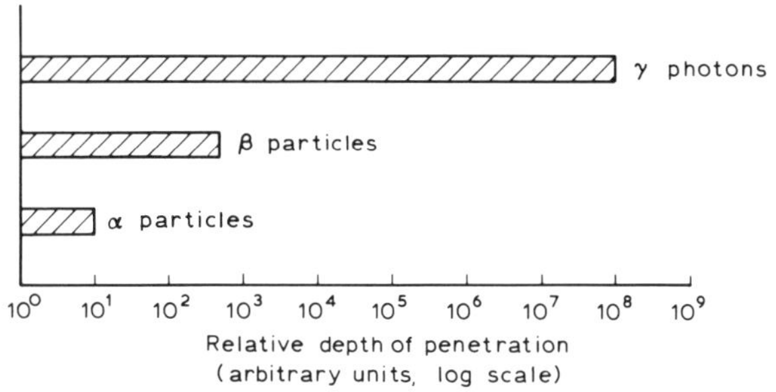

Figure 10.2 Rough comparison of penetrating powers of α-, β- and γ-radiation

10.2.1 Neutrons

Neutrons were discovered in 1932 as a result of bombarding light elements (for example, beryllium and boron) with α-particles. For laboratory purposes, this is still a convenient method of production, but the most useful and intense source is the nuclear reactor in which the neutrons are produced as a by-product of the fission of fissile materials such as uranium-235.

Free neutrons are unstable, and decay to give a proton and a low-energy β-particle. Neutrons when they are produced may have a wide range of energy, from the several millions of electron-volts of *fast neutrons* to the fractions of electron-volts of *thermal neutrons*.

Neutrons lose energy by elastic collision. An *elastic collision* is one in which the incident particles rebound—or are scattered—without the nucleus that is struck having been excited or broken up. An *inelastic collision* is one in which the struck nucleus is excited, or broken up, or captures the incoming particle. For neutrons, the loss of energy in an elastic collision is greater with light nuclei; for example, a 1 MeV neutron loses 28% of its energy in collision with a carbon atom, but only 2% in collision with lead.

By successive collisions, the energy of neutrons is reduced to that of the thermal agitation of the nucleus (that is, some 0.025 eV at 20°C) and the neutrons are then captured.

The consequence of the capture of a neutron may be a new nuclide, which may possibly be radioactive. That is, in fact, the main method of producing radioisotopes. Because they are uncharged, neutrons do not cause direct ionisation, and may travel large distances in materials having a high atomic number. The most efficient materials for shielding against neutron emission are those having light nuclei; as indicated above, these reduce the energy of neutrons much more rapidly than heavier materials. Examples of efficient shielding materials are water, the hydrocarbons and graphite.

10.2.2 Fission

Fission is the splitting of a heavy nucleus into two approximately equal fragments known as fission products. Fission is accompanied by the emission of several neutrons and the release of energy. It can be spontaneous or caused by the impact of a neutron, a fast particle or a photon:

$$^{235}_{92}U + ^{1}_{0}n \rightarrow ^{93}_{38}Sr + ^{140}_{54}Xe + ^{1}_{0}n + ^{1}_{0}n + ^{1}_{0}n$$

The total number of sub-atomic particles is unchanged:

$$235 + 1 = 93 + 140 + 1 + 1 + 1$$

(other combinations of particles are possible)

The number of protons is unchanged.

Figure 10.3 is a diagramatical representation of the foregoing.

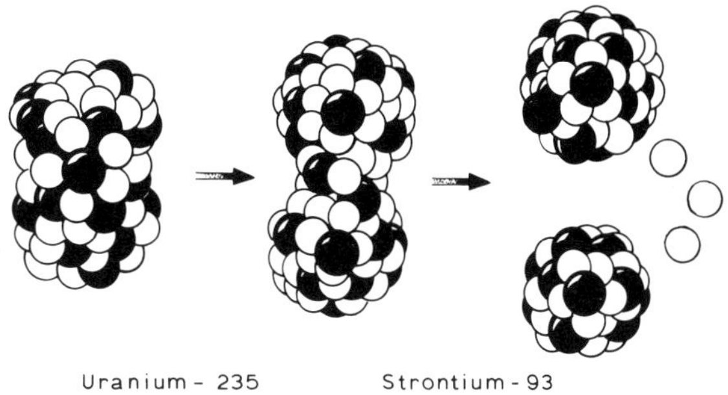

Figure 10.3 Diagrammatic representation of the fission of uranium-235

Other modes of nuclear disintegration

Mention must be made of two other methods of nuclear disintegration:

(1) emission of positively charged electrons or *positrons* (β^+);
(2) *electron capture*.

Positrons interact rapidly with electrons after ejection from the nucleus. The two electrical charges cancel each other, and the energy is released in a form of γ-radiation known as *annihilation radiation*.

In the process of electron capture, the energy of an unstable nucleus is dissipated by the capture into the nucleus of an inner orbital electron. The process is always accompanied by the emission of the characteristic X-rays of the atom produced by electron capture. For example, germanium-71, which decays in this manner, emits gallium X-rays.

10.2.3 Radioactive decay

Radioactive isotopes are giving off energy continuously and if the law of the conservation of energy is to be obeyed this radioactive decay cannot go on indefinitely. The nucleus of the radioactive atom undergoes a change when a particle is emitted and forms a new and often non-radioactive product. The rate at which this nuclear reaction takes place decreases with time in such a way that the time necessary to halve the reaction rate is constant for a given isotope and is known as its half-life. The half-life period can be as short as a fraction of a microsecond or as long as ten thousand million years.

Radioactive decay can be illustrated by considering a radioactive form of bismuth, ${}^{210}_{83}Bi$, which has a half-life of five days. If the number of particles emitted by a sample in one minute is recorded, then after five days, two minutes would be required for the same number to be emitted. After ten days, four minutes would be required, and so on. The amount of the radioactive bismuth, ${}^{210}_{83}Bi$, in the sample will diminish as the emission proceeds. The bismuth nuclei lose electrons as β-particles, and the radioactive bismuth is converted to polonium, ${}^{210}_{84}Po$.

In this particular case the product is itself radioactive. It emits α-particles and has a half-life of 138 days. The product of its disintegration is lead, ${}^{206}_{82}Pb$, which is not radioactive.

The disintegration of the radioactive bismuth can be represented as follows:

$$ {}^{210}_{83}Bi \xrightarrow[5\ \text{days}]{\beta} {}^{210}_{84}Po \xrightarrow[138\ \text{days}]{\alpha} {}^{206}_{82}Pb\ \text{(stable)} $$

10.2.4 Units

It is necessary to have units to define the quantity of radioactivity and its physical nature. The unit of quantity is the curie (Ci). This was originally defined as the quantity of radioactive material producing the same disintegration rate as one gram of pure radium.

The definition of quantity must be couched in different terms in modern times to include the many artificially produced radioisotopes. The curie is now defined as the quantity of radioisotope required to produce 3.7×10^{10} disintegrations per second. Quantity measurements made in the laboratory with small sources are often expressed in terms of disintegrations per second (d.p.s.). What is actually recorded by the detector is expressed in counts per second (c.p.s.). The weights of material associated with this activity can vary greatly. For example, 1 curie of iodine-131 weighs 8 micrograms, whereas 1 curie of uranium-238 weighs 2.7 tons.

The unit of energy is the electron-volt (eV). This is the kinetic energy acquired by an electron when accelerated through a potential difference of one volt. The electron-volt is equivalent to 1.6×10^{-19} joule. With α-, β-, and γ-radiation, it is usual to use thousands of electron-volts (keV) or millions of electron-volts (MeV).

Further reading

FOSTER, K. and ANDERSON, R., *Electro-magnetic Theory*, Vols. 1, 2, Butterworths, Sevenoaks (1970)

YARWOOD, J., *Atomic and Nuclear Physics*, University Tutorial Press (1973)

11 The Ionosphere and The Troposphere

P A Bradley BSc, MSc, CEng, MIEE
Principal Scientific Officer,
Rutherford and Appleton
Science and Engineering Research Council
(Sections 11.1–11.6)

J A Lane DSc, CEng, FIEE
Consultant-Directorate of Radio Technology,
Home Office. Formerly Science and Engineering
Research Council,
Rutherford and Appleton Laboratory

Contents

11.1 The ionosphere

The ionosphere is an electrified region of the Earth's atmosphere situated at heights of from about fifty kilometres to several thousand kilometres. It consists of ions and free electrons produced by the ionising influences of solar radiation and of incident energetic solar and cosmic particles. The ionosphere is subject to marked geographic and temporal variations. It has a profound effect on the characteristics of radio waves propagated within or through it. By means of wave refraction, reflection or scattering it permits transmission over paths that would not otherwise be possible, but at the same time it screens some regions that could be illuminated in its absence (see *Figure 11.1*)

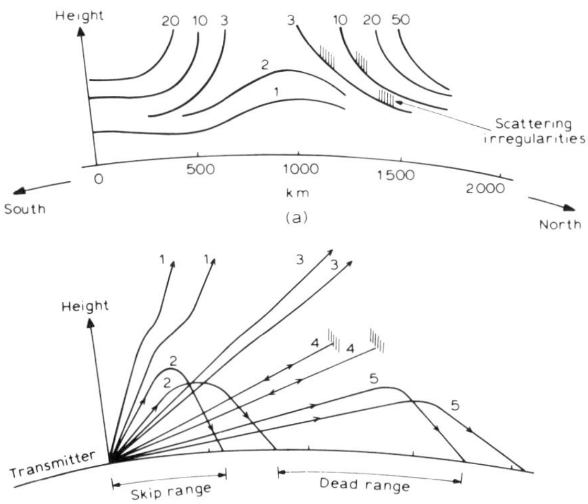

Figure 11.1 High-frequency propagation paths via the ionosphere at high latitudes: (a) Sample distribution of electron density (arbitrary units) in northern hemisphere high-latitude ionosphere (adapted from Buchau[1]); (b) Raypaths for signals of constant frequency launched with different elevation angles

The ability of the ionosphere to refract, reflect or scatter rays depends on their frequency and elevation angle. Ionospheric refraction is reduced at the higher frequencies and for the higher elevation angles, so that provided the frequency is sufficiently great rays 1 in *Figure 11.1* escape whereas rays 2 are reflected back to ground. Rays 3 escape because they traverse the ionosphere at latitudes where the electron density is low (the Muldrew trough[2]). Irregularities in the F-region are responsible for the direct backscattering of rays 4. The low-elevation rays 5 are reflected to ground because of the increased ionisation at the higher latitudes. Note that for this ionosphere and frequency there are two ground zones which cannot be illuminated.

The ionosphere is of considerable importance in the engineering of radio communication systems because:

(*a*) It provides the means of establishing various communication paths, calling for system-design criteria based on a knowledge of ionospheric morphology.

(*b*) It requires specific engineering technologies to derive experimental probing facilities to assess its characteristics, both for communication-systems planning and management, and for scientific investigations.

(*c*) It permits the remote monitoring by sophisticated techniques of certain distant natural and man-made phenomena occurring on the ground, in the air and in space.

11.2 Formation of the ionosphere and its morphology

There is widespread interest in the characteristics of the ionosphere by scientists all over the world. Several excellent general survey books have been published describing the principal known features[3,4] and other more specialised books concerned with aeronomy, and including the ionosphere and magnetosphere, are of great value to the research worker.[5–7] Several journals in the English language are devoted entirely, or to a major extent, to papers describing investigations into the state of the ionosphere and of radio propagation in the ionosphere.

The formation of the ionosphere is a complicated process involving the ionising influences of solar radiation and solar and cosmic particles on an atmosphere of complex structure. The rates of ion and free-electron production depend on the flux density of the incident radiation or particles, as well as on the ionisation efficiency, which is a function of the ionising wavelength (or particle energies) and the chemical composition of the atmosphere. There are two heights where electron production by the ionisation of molecular nitrogen and atomic and molecular oxygen is a maximum. One occurs at about 100 km and is due to incident X-rays with wavelengths less than about 10 nm and to ultraviolet radiation with wavelengths near 100 nm; the other is at about 170 km and is produced by radiation of wavelengths 20–80 nm.

Countering this production, the free electrons tend to recombine with the positive ions and to attach themselves to neutral molecules to form negative ions. Electrons can also leave a given volume by diffusion or by drifting away under the influences of temperature and pressure gradients, gravitational forces or electric fields set up by the movement of other ionisation. The electron density at a given height is given from the so-called *continuity equation* in terms of the balance between the effects of production and loss.

Night-time electron densities are generally lower than in the daytime because the rates of production are reduced. *Figure 11.2* gives examples of a night-time and a daytime height distribution

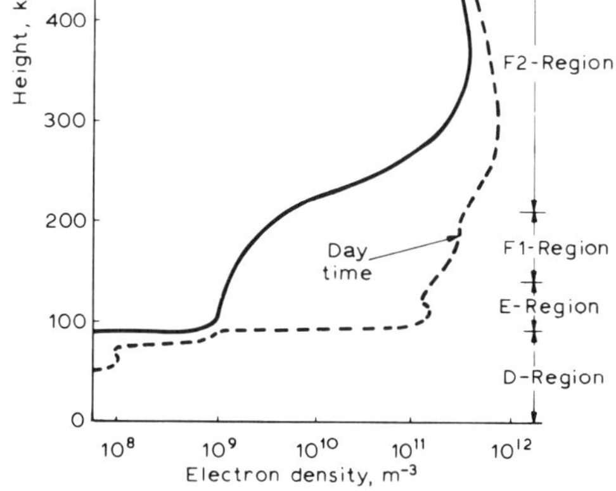

Figure 11.2 Sample night-time and daytime height distributions of electron density at mid-latitudes in summer

of electron density. The ionisation is continuous over a wide height range, but there are certain height regions with particular characteristics, and these are known, following E. V. Appleton, by the letters D, E and F. The E-region is the most regular ionospheric region, exhibiting a systematic dependence of maximum electron density on solar-zenith angle, leading to predictable diurnal, seasonal and geographical variations. There is also a predictable dependence of its electron density on the changes in solar radiation which accompany the long-term fluctuations in the state of the sun. Maximum E-region electron density is approximately proportional to sunspot number, which varies over a cycle of roughly 11 years.

In the daytime the F-region splits into two, with the lower part known as the F1-region and the upper part as the F2-region. This splitting arises because the principal loss mechanism is an ion–atom interchange process followed by dissociative recombination, the former process controlling the loss rates in the F2-region and the latter in the F1-region. Although maximum production is in the F1-region, maximum electron density results in the F2-region, where the loss rates are lower. The maximum electron density of the F1-region closely follows that of the E-region, but there are significant and less predictable changes in its height. The maximum electron density and height of the F2-region are subject to large changes which have important consequences to radio wave propagation. Some of these changes are systematic but there are also major day to day variations. It seems likely that the F2-region is controlled mainly by ionisation transport to different heights along the lines of force of the Earth's magnetic field under the influence of thermospheric winds at high and middle latitudes[8] and by electric fields at low latitudes.[9] These effects, taken in conjunction with the known variations in atmospheric composition, can largely explain characteristics of the F2-region which have in the past been regarded as anomalous by comparison with the E-region—namely, diurnal changes in the maximum of electron density in polar regions in the seasons of complete darkness, maximum electron densities at some middle-latitude locations at times displaced a few hours from local noon with greater electron density in the winter than the summer, and at low latitudes longitude variations linked more to the magnetic equator than to the geographic equator, with a minimum of electron density at the magnetic equator and maxima to the north and south where the magnetic dip is about 30°. At all latitudes electron densities in the F2-region, like those in the E- and F1-regions, increase with increase of sunspot number. The electron densities at heights above the maximum of the F2-region are controlled mainly by diffusion processes.

The D-region shows great variability and fine structure, and is the least well understood part of the ionosphere. The only ionising radiations that can penetrate the upper regions and contribute to its production are hard X-rays with wavelengths less than about 2 nm and Lyman-α radiation at 121.6 nm. Chemical reactions responsible for its formation principally involve nitric oxide and other minor atmospheric constituents.

The D-region is mainly responsible for the absorption of radio waves because of the high electron-collision frequencies at such altitudes (see below). While the electron densities in the upper part of the D-region appear linked to those in the E-region, leading to systematic latitudinal, temporal and solar-cycle variations in absorption, there are also appreciable irregular day to day absorption changes. At middle latitudes anomalously high absorption is experienced on some days in the winter. This is related to warmings of the stratosphere and is probably associated with changes in D-region composition. In the lower D-region at heights below about 70 km the ionisation is produced principally by energetic cosmic rays, uniformly incident at all times of day. Since the free electrons thereby generated tend to collide and become attached to molecules to form negative ions by night, but are detached by solar radiation in the daytime, the lower D-region ionisation, like that in the upper D-region, is much greater by day than night. In contrast, however, electron densities in the lower D-region, being related to the incidence of cosmic rays, are reduced with increase in the number of sunspots. Additional D-region ionisation is produced at high latitudes by incoming particles, directed along the lines of force of the Earth's magnetic field. Energetic electrons, probably originating from the Sun, produce characteristic auroral absorption events over a narrow band of latitudes about 10° wide, associated with the visual auroral regions.[10]

From time to time disturbances occur on the Sun known as solar flares. These are regions of intense light, accompanied by increases in the solar far ultraviolet and soft X-ray radiation. Solar flares are most common at times of high sunspot number. The excess radiation leads to sudden ionospheric disturbances (SID's), which are rapid and large increases in ionospheric absorption occurring simultaneously over the whole sunlit hemisphere. These persist for from a few minutes to several hours giving the phenomena of short-wave fadeouts (SWF's), first explained by Dellinger. Accompanied by solar flares are eruptions from the Sun of energetic protons and electrons. These travel as a column of plasma, and depending on the position of the flare on the Sun's disc and on the trajectory of the Earth, they sometimes impinge on the ionosphere. Then, the protons, which are delayed in transit from fifteen minutes to several hours, produce a major enhancement of the D-region ionisation in polar regions that can persist for several days. This gives the phenomenon of polar cap absorption (PCA) with complete suppression of h.f. signals over the whole of both polar regions.[11] Slower particles, with transit times of from 20–40 hours, produce ionospheric storms. These storms, which result principally from movements in ionisation, take the form of depressions in the maximum electron density of the F2-region.[12] They can last for several days at a time with effects which are progressively different in detail at different latitudes. Since the sun rotates with a period of about 28 days, sweeping out a column of particles into space when it is disturbed, there is a tendency for ionospheric storms to recur after this time interval.

Additional ionisation is sometimes found in thin layers, 2 km or less thick, embedded in the E-region at heights between 90 and 120 km. This has an irregular and patchy structure, a maximum electron density which is much greater than that of the normal E-region, and is known as sporadic-E (or Es) ionisation because of its intermittent occurrence. It consists of patches up to 2000 km in extent, composed of large numbers of individual irregularities each less than 1 km in size. Sporadic-E tends to be opaque to the lower h.f. waves and partially reflecting at the higher frequencies. It results from a number of separate causes and may be classified into different types,[13] each with characteristic occurrence and other statistics. In temperate latitudes sporadic-E arises principally from wind shear, close to the magnetic equator it is produced by plasma instabilities and at high latitudes it is mainly due to incident energetic particles. It is most common at low latitudes where it is essentially a daytime phenomenon.

Irregularities also develop in the D-region due to turbulence and wind shears and other irregularities are produced in the F-region. The F-region irregularities can exist simultaneously over a wide range of heights, either below or above the height of maximum electron density and are referred to as spread-F irregularities. They are found at all latitudes but are particularly common at low latitudes in the evenings where their occurrence is related to rapid changes in the height of the F-region.[14]

11.3 Ionospheric effects on radio signals

A radio wave is specified in terms of five parameters: its amplitude, phase, direction of propagation, polarisation and frequency. The principal effects of the ionosphere in modifying these parameters are considered as follows.

11.3.1 Refraction

The change in direction of propagation resulting from the traverse of a thin slab of constant ionisation is given approximately by Bouger's law in terms of the refractive index and the angle of incidence. A more exact specification including the effects of the Earth's magnetic field is given by the Haselgrove equation solution.[15] The refractive index is determined from the Appleton–Hartree equations of the magnetoionic theory[16,17] as a function of the electron density and electron-collision frequency, together with the strength and direction of the Earth's magnetic field, the wave direction and the wave frequency. The dependence on frequency leads to wave dispersion of modulated signals. Since the ionosphere is a doubly-refracting medium it can transmit two waves with different polarisations (see below). The refractive indices appropriate to the two waves differ. Refraction is reduced at the greater wave frequencies, and at v.h.f. and higher frequencies it is given approximately as a function of the ratio of the wave and plasma frequencies, where the plasma frequency is defined in terms of a universal constant and the square root of the electron density.[16] *Table 11.1* lists the magnitude of the refraction and of other propagation parameters for signals at a frequency of 100 MHz which traverse the whole ionosphere.

11.3.2 Change in phase-path length

The phase-path length is given approximately as the integral of the refractive index with respect to the ray-path length. Ignoring spatial gradients, the change in phase-path length introduced by passage through the ionosphere to the ground of signals at v.h.f. and higher frequencies from a spacecraft is proportional to the total-electron content. This is the number of electrons in a vertical column of unit cross section.

11.3.3 Group delay

The group and phase velocities of a wave differ because the ionosphere is a dispersive medium. The ionosphere reduces the group velocity and introduces a group delay which for transionospheric signals at v.h.f. and higher frequencies, like the phase-path change, is proportional to the total-electron content.

11.3.4 Polarisation

Radio waves that propagate in the ionosphere are called characteristic waves. There are always two characteristic waves known as the ordinary wave and the extraordinary wave; under certain restricted conditions a third wave known as the Z-wave can also exist.[16] In general the ordinary and extraordinary waves are elliptically polarised. The polarisation ellipses have the same axial ratio, orientations in space that are related such that under many conditions they are approximately orthogonal, and electric vectors which rotate in opposite directions.[16] The polarisation ellipses are less elongated the greater the wave frequency. Any wave launched into the ionosphere is split into characteristic ordinary and extraordinary wave components of appropriate power. At m.f. and above these components may be regarded as travelling independently through the ionosphere with polarisations which remain related, but continuously change to match the changing ionospheric conditions. The phase paths of the ordinary and extraordinary wave components differ, so that in the case of transionospheric signals when the components have comparable amplitudes, the plane of polarisation of their resultant slowly rotates. This effect is known as Faraday rotation.

11.3.5 Absorption

Absorption arises from inelastic collisions between the free electrons, oscillating under the influence of the incident radio wave, and the neutral and ionised constituents of the atmosphere. The absorption experienced in a thin slab of ionosphere is given by the Appleton–Hartree equations[16] and under many conditions is proportional to the product of electron density and collision frequency, inversely proportional to the refractive index and inversely proportional to the square of the wave frequency. The absorption is referred to as non-deviative or deviative depending on whether it occurs where the refractive index is close to unity. Normal absorption is principally a daytime phenomenon. At frequencies below 5 MHz it is sometimes so great as to completely suppress effective propagation. The absorptions of the ordinary and extraordinary waves differ, and in the range 1.5–10 MHz the extraordinary wave absorption is significantly greater.

11.3.6 Amplitude fading

If the ionosphere were unchanging the signal amplitude over a fixed path would be constant. In practice, however, fading arises as a consequence of variations in propagation path, brought about by movements or fluctuations in ionisation. The principal causes of fading are:

(*a*) Variations in absorption.
(*b*) Movements of irregularities producing focusing and defocusing.
(*c*) Changes of path length among component signals propagated via multiple paths.
(*d*) Changes of polarisation, such as for example due to Faraday rotation.

These various causes lead to different depths of fading and a range of fading rates. The slowest fades are usually those due to

Table 11.1 Effect of one-way traverse of typical mid-latitude ionosphere at 100 MHz on signals with elevation angle above 60 degrees[22]

Effect	*Day*	*Night*	*Frequency dependence, f*
Total electron content	5×10^{13} cm^{-2}	5×10^{12} cm^{-2}	
Faraday rotation	15 rotations	1.5 rotations	f^{-2}
Group delay	12.5 μs	1.2 μs	f^{-2}
Change in phase-path length	5.2 km	0.5 km	f^{-2}
Phase change	7500 radians	750 radians	f^{-2}
Phase stability (peak-to-peak)	±150 radians	±15 radians	f^{-1}
Frequency stability (r.m.s.)	±0.04 Hz	±0.004 Hz	f^{-1}
Absorption	0.1 dB	0.01 dB	f^{-2}
Refraction	$\leqslant 1°$	—	f^{-2}

absorption changes which have a period of about 10 minutes. The deepest and most rapid fading occurs from the beating between two signal components of comparable amplitude propagated along different paths. A regularly reflected signal together with a signal scattered from spread-F irregularities can give rise to so-called *flutter* fading, with fading rates of about 10 Hz. A good general survey of fading effects, including a discussion of fading statistics, has been produced.[18] On operational communication circuits fading may be combated by space diversity or polarisation-diversity receiving systems and by the simultaneous use of multiple-frequency transmissions (frequency diversity).

11.3.7 Frequency deviations

Amplitude fading is accompanied by associated fluctuations in group path and phase path, giving rise to time and frequency-dispersed signals. When either the transmitter or receiver is moving, or there are systematic ionospheric movements, the received signal is also Doppler-frequency shifted. Signals propagated simultaneously via different ionospheric paths are usually received with differing frequency shifts. Frequency shifts for reflections from the regular layers are usually less than 1 Hz, but shifts of up to 20–30 Hz have been reported for scatter-mode signals at low latitudes.[19]

11.3.8 Reflection, scattering and ducting

The combined effect of refraction through a number of successive slabs of ionisation can lead to ray reflection. This may take place over a narrow height range as at l.f. or rays may be refracted over an appreciable distance in the ionosphere as at h.f. Weak incoherent scattering of energy occurs from random thermal fluctuations in electron density, and more efficient aspect-sensitive scattering from ionospheric irregularities gives rise to direct backscattered and forward-scatter signals. Ducting of signals to great distances can take place at heights of reduced ionisation between the E- and F-regions, leading in some cases to round-the-world echoes.[20] Ducting can also occur within regions of field-aligned irregularities above the maximum of the F-region.

11.3.9 Scintillation

Ionospheric irregularities act as a phase-changing screen on transionospheric signals from sources such as Earth satellites or radiostars. This screen gives rise to diffraction effects with amplitude, phase and angle-of-arrival scintillations.[21]

11.4 Communication and monitoring systems relying on ionospheric propagation

Ionospheric propagation is exploited for a wide range of purposes, the choice of system and the operating frequency being largely determined by the type and quantities of data to be transmitted, the path length and its geographical position.

11.4.1 Communication systems

Radio communication at very low frequencies (v.l.f.) is limited by the available bandwidth, but since ionospheric attenuation is very low, near world-wide coverage can be achieved. Unfortunately the radiation of energy is difficult at such frequencies and complex transmitting antenna systems, coupled with large transmitter powers, are needed to overcome the high received background noise from atmospherics—the electromagnetic radiation produced by lightning discharges. Because of the stability of propagation, v.l.f. systems are used for the transmission of standard time signals and for c.w. navigation systems which rely on direction-finding techniques, or on phase comparisons between spaced transmissions as in the Omega system (10–14 kHz).[23] At low frequencies (l.f.) increased propagation losses limit area coverage, but simpler antenna systems are adequate and lower transmitter powers can be employed because of the reduced atmospheric noise. Low frequency systems are used for communication by on–off keying and frequency-shift keying. Propagation conditions are more stable than at higher frequencies because the ionosphere is less deeply penetrated. Low frequency signals involving ionospheric propagation are also used for communication with submarines below the surface of the sea, with receivers below the ground and with space vehicles not within line-of-sight of the transmitter. Other l.f. systems[24] relying principally on the ground wave, which are sometimes detrimentally influenced by the sky wave at night, include the Decca c.w. navigation system (70–130 kHz), the Loran C pulse navigation system (100 kHz) and long-wave broadcasting.

At medium frequencies (m.f.) daytime absorption is so high as to completely suppress the sky wave. Some use is made of the sky wave at night-time for broadcasting, but generally medium frequencies are employed for ground-wave services. Despite the advent of reliable multichannel satellite and cable systems, high frequencies continue to be used predominantly for broadcasting, fixed, and mobile point-to-point communications, via the ionosphere—there are still tens of thousands of such circuits.

Very high frequency (v.h.f.) communication relying on ionospheric scatter propagation between ground-based terminals is possible. Two-way error-correcting systems with scattering from intermittent meteor trains can be used at frequencies of 30–40 MHz over ranges of 500–1500 km.[25] Bursts of high-speed data of about 1 s duration with duty cycles of the order of 5% can be achieved, using transmitter powers of about 1 kW. Meteor-burst systems find favour in certain military applications because they are difficult to intercept, since the scattering is usually confined to 5–10 degrees from the great-circle path. Forward-scatter communication systems at frequencies of 30–60 MHz, also operating over ranges of about 1000 km, rely on coherent scattering from field-aligned irregularities in the D-region at a height of about 85 km.[26] They are used principally at low and high latitudes. Signal intensities are somewhat variable, depending on the incidence of irregularities. During magnetically disturbed conditions signals are enhanced at high latitudes, but are little affected at low latitudes. Directional transmitting and receiving antennas with intersecting beams are required. Particular attention has to be paid to avoiding interference from signal components scattered from sporadic-E ionisation or irregularities in the F-region. Special frequency-modulation techniques involving time division multiplex are used to combat Doppler effects. Systems with 16 channels, automatic error correction and operating at 100 words per minute, now exist.

11.4.2 Monitoring systems

High frequency (h.f.) signals propagated obliquely via the ionosphere and scattered at the ground back along the reverse path may be exploited to give information on the characteristics of the scattering region.[27] Increased scatter results from mountains, from cities and from certain sea waves. Signals backscattered from the sea are enhanced when the signal wavelength is twice the component of the sea wavelength along the direction of incidence of the signal, since round-trip signals reflected from successive sea-wave crests then arrive in phase to give coherent addition. The Doppler shift of backscatter returns from sea waves due to the sea motion usually exceeds that imposed on the received signals by the ionosphere, so that Doppler filtering enables the land- and sea-scattered signals to be examined separately. This permits studies of distant land–sea boundaries,[28] and since the wavelength of a sea wave is directly related to its velocity, provides a means of synoptic monitoring of distant ocean waves. Doppler filtering can also resolve signals reflected or scattered back along ionosphere paths from aircraft, rockets, rocket trails or ships.

Studies of the Doppler shift of stable frequency, h.f., c.w. signals propagated via the ionosphere between ground-based terminals provide important information about infrasonic waves in the F-region originating from nuclear explosions,[29] earthquakes,[30] severe thunderstorms[31] and air currents in mountainous regions.

High-altitude nuclear explosions lead to other effects which may be detected by radio means.[32] An immediate wideband electromagnetic pulse is produced which can be monitored throughout the world at v.l.f. and h.f. Also, enhanced D-region ionisation, lasting for several days over a wide geographical area, produces an identifiable change in the received phase of long-distance v.l.f. ionospheric signals. Other more localised effects which can be detected include the generation of irregularities in the F-region.

Atmospherics may be monitored at v.l.f. out to distances of several thousand kilometres and by recording simultaneously the arrival azimuths at spaced receivers the locations of thunderstorms may be determined and their movements tracked as an aid to meteorological warning services.

11.5 Ionospheric probing techniques

There are a wide variety of methods of sounding the state of the ionosphere involving single- and multiple-station ground-based equipments, rocket-borne and satellite probes (see *Table 11.2*). A comprehensive survey of the different techniques has been produced by a Working Group of the International Union of Radio Science (URSI).[33] Some of the techniques involve complex analysis procedures and require elaborate and expensive equipment and antenna systems (*Figure 11.3*), others need only a single radio receiver.

The swept-frequency ground-based *ionosonde* consisting of a co-located transmitter and receiver was developed for the earliest of ionospheric measurements and is still the most widely used probing instrument. The transmitter and receiver frequencies are swept synchronously over the range from about 0.5–1 MHz to perhaps 20 MHz depending on ionospheric conditions, and short pulses typically of duration 100 μs with a repetition rate of 50/s are transmitted. Calibrated film records of the received echoes give the group path and its frequency dependence.

In practice these records require expert interpretation because: (*a*) multiple echoes occur, corresponding to more than one traverse between ionosphere and ground (so-called *multiple-hop* modes) or when partially reflecting sporadic-E or F-region irregularities are present; (*b*) the ordinary and extraordinary waves are sometimes reflected from appreciably different heights; and (*c*) oblique reflections occur when the ionosphere is tilted. Internationally agreed procedures for scaling ionosonde records (ionograms) have been produced.[34]

Table 11.2 Principal ionospheric probing techniques

Height range	*Technique*	*Parameters monitored*	*Site*
Above 100 km	vertical-incidence sounding	up to height of maximum ionisation–electron density	ground
	topside sounding	from height of satellite to height of maximum ionisation–electron density; electron and ion temperatures; ionic composition	satellite
	incoherent scatter	up to few thousand km—electron density; electron temperature; ion temperature; ionic composition; collision frequencies; drifts of ions and electrons	ground
	Faraday rotation and differential Doppler	total-electron content	satellite-ground
	in-situ probes	wide range of parameters	satellite
	c.w. oblique incidence	solar flare effects; irregularities; travelling disturbances; radio aurorae	ground
	pulse oblique incidence	oblique modes by ground backscatter and oblique sounding; meteors; radio aurorae; irregularities and their drifts	ground
	whistlers	out to few Earth radii–electron density; ion temperature; ionic composition	ground or satellite
Below 100 km	vertical-incidence sounding	absorption	ground
	riometer	absorption	ground
	c.w. and pulse oblique incidence	electron density and collision frequency	ground
	wave fields	electron density and collision frequency	rocket-ground
	in-situ probes	electron density and collision frequency; ion density; composition of neutral atmosphere	rocket
	cross-modulation	electron density and collision frequency	ground
	partial reflection	electron density and collision frequency	ground
	lidar	neutral air density; atmospheric aerosols; minor constituents	ground

Figure 11.3 Part of the array of dipoles of the radar antenna at the Jicamarca Radar Observatory, Lima, Peru (US National Bureau of Standards, Cover photo, 1 Feb. 1963, Vol. 139, *Science*, © 1963 by the American Association for the Advancement of Science)

Since reflection takes place from a height where the sounder frequency is equal to the ionospheric plasma frequency, and since the group path can be related to that height provided the electron densities at all lower heights are known, the data from a full frequency sweep can be used to give the true-height distribution of electron density in the E- and F-regions up to the height of maximum electron density of the F2-region. The conversion of group path to true height requires assumptions regarding missing data below the lowest height from which echoes are received and over regions where the electron density does not increase monotonically with height. The subject of true-height analysis is complex and controversial.[35] Commercially manufactured pulse sounders use transmitter powers of about 1 kW. Sounders with powers of around 100 kW and a lower frequency limit of a few kHz have been operated successfully in areas free from m.f. broadcast interference, to study the night-time E-region—this has a maximum plasma frequency of around 0.5 MHz.

Pulse-compression systems[36] and c.w. chirp sounders[37] offer the possibility of improved signal/noise ratios and echo resolution. In the chirp sounder system, originally developed for use at oblique incidence, the transmitter and receiver frequencies are swept synchronously so that the finite echo transit time leads to a frequency modulation of the receiver i.f. signals. These signals are then spectrum-analysed to produce conventional ionograms. Receiver bandwidths of only a few tens of hertz are needed so that transmitter powers of a few watts are adequate. Other ionosondes have been produced and used operationally which record data digitally on magnetic tape.[34] An ionosonde has been successfully flown in an aircraft to investigate geographical changes in electron density at high latitudes.[38] Over 100 ground-based ionosondes throughout the world make regular soundings each hour of each day; data are published at monthly intervals.[39]

Since 1962 swept-frequency ionosondes have been operated in satellites orbiting the earth at altitudes of around 1000 km. These are known as *topside sounders* and they give the distributions of electron density from the satellite height down to the peak of the F2-region. They also yield other plasma-resonance information, together with data on electron and ion temperatures and ionic composition. Fixed-frequency topside sounders are used to study the spatial characteristics of spread-F irregularities and other features with fine structure.

Many different monitoring probes are mounted in satellites orbiting the earth at altitudes above 100 km, to give direct measurements of a range of ionospheric characteristics. These include r.f. impedance, capacitance and upper-hybrid resonance probes for local electron density, modified Langmuir probes for electron temperature, retarding potential analysers and sampling mass spectrometers for ion density, quadrupole and monopole mass spectrometers for ion and neutral-gas analysis and retarding potential analysers for ion temperature measurements.

Ground measurements of satellite beacon signals permit studies of total-electron content, either from the differential Doppler frequency between two harmonically related h.f./v.h.f. signals,[40] or from the Faraday rotation of a single v.h.f. transmission.[41] Beacons on geostationary satellites are valuable for investigations of temporal variations. The scintillation of satellite signals at v.h.f. and u.h.f. gives information on the incidence of ionospheric irregularities, their heights and sizes.[42]

A powerful tool for ionospheric investigations up to heights of several thousand kilometres is the vertical-incidence incoherent-scatter radar. The technique makes use of the very weak scattering from random thermal fluctuations in electron density which exist in a plasma in quasi-equilibrium. Several important parameters of the plasma affect the scattering such that each of these can be determined separately. The power, frequency spectrum and polarisation of the scattered signals are measured and used to give the height distributions of electron density, electron temperature, ionic composition, ion-neutral atmosphere and ion-ion collision frequencies, and the mean plasma-drift velocity. Tristatic receiving systems enable the vertical and horizontal components of the mean plasma drift to be determined. Radars operate at frequencies of 40–1300 MHz using either pulse or c.w. transmissions. Transmitter peak powers of the order of 1 MW, complex antenna arrays (*Figure 11.3*), and sophisticated data processing procedures are needed. Ground clutter limits the lowest heights that can be investigated to around 100 km.

Electron densities, ion temperatures and ionic composition out to several Earth radii may be studied using naturally occurring whistlers originating in lightning discharges. These are dispersed audio-frequency trains of energy, ducted through the ionosphere and then propagated backwards and forwards along the Earth's magnetic-field lines to conjugate points in the opposite hemisphere. Whistler dispersions may be observed either at the ground or in satellites.[43]

Continuous wave and pulsed signals, transmitted and received at ground-based terminals, may be used in a variety of ways to study irregularities or fluctuations in ionisation. Cross-correlation analyses of the amplitudes on three spaced receivers, of pulsed signals of fixed frequency reflected from the ionosphere at near vertical incidence, give the direction and velocity of the horizontal component of drift.[44] The heights, patch sizes and incidence of F-region irregularities responsible for oblique-path forward-scatter propagation at frequencies around 50 MHz may be investigated by means of highly-directional antennas and from signal transit times.[14] Measurements of the Doppler frequency variations of signals from stable c.w. transmitters may be used to study: (*a*) ionisation enhancements in the E- and F-regions associated with solar flares; (*b*) travelling ionospheric disturbances;[45] and (*c*) the frequency-dispersion component of the ionospheric channel-scattering function.[46]

Sporadic-E and F-region irregularities associated with visual aurorae may be examined by pulsed-radar techniques over a wide range of frequencies from about 6 MHz to 3000 MHz. They may also be investigated using c.w. bistatic systems in which the transmitter and receiver are separated by several hundred kilometres. Since the irregularities are known to be elongated and aligned along the direction of the Earth's magnetic field and since at the higher frequencies efficient scattering can only occur under restricted conditions, the scattering centres may readily be located. Using low-power v.h.f. beacon transmitters, this technique has proved very popular with radio amateurs. Pulsed meteor radars incorporating Doppler measurements indicate the properties and movements of meteor trains.[25]

Two other oblique-path techniques, giving information on the regular ionospheric regions, are high-frequency ground backscatter sounding and variable-frequency oblique sounding. The former uses a co-located transmitter and receiver, and record interpretation generally involves identifying the skip distance (see *Figure 11.1*) where the signal returns are enhanced because of ray convergence. It is important to use antennas with azimuthal beamwidths of only a few degrees to minimise the ground area illuminated. Long linear antenna arrays with beam slewing, and circularly-disposed banks of log-periodic antennas with monopulsing are used. Oblique-incidence sounders are adaptions of vertical-incidence ionosondes with the transmitter and receiver controlled from stable synchronised sources. Atlases of characteristic records obtained from the two types of sounder under different ionospheric conditions have been produced. Mean models of the ionosphere over the sounding paths may be deduced by matching measured data with ray-tracing results.[47]

So far, no mention has been made of the height region below about 100 km. As already noted, the D-region is characterised by a complex structure and high collision frequencies which lead to large daytime absorption of h.f. and m.f. waves. This absorption may be measured using fixed-frequency vertical-incidence pulses,[48] or by monitoring c.w. transmissions at ranges of 200–500 km, where there is no ground-wave component and the dominant signals are reflected from the E-region by day and the sporadic-E layer by night. There is then little change in the raypaths from day to night so that, assuming night absorption can be neglected, daytime reductions in amplitude are a measure of the prevailing absorption. Multifrequency absorption data give information on the height distributions of electron density.[49] Auroral absorption is often too great to be measured in such ways, but special instruments known as riometers can be used.[50] These operate at a frequency around 30 MHz and record changes in the incident cosmic noise at the ground caused by ionospheric absorption.

D-region electron densities and collision frequencies may be inferred from oblique or vertical-path measurements of signal amplitude, phase, group-path delay and polarisation at frequencies of 10 Hz to 100 kHz, with atmospherics as the signal sources at the lower frequencies. Vertically radiated signals in the frequency range 1.5–6 MHz suffer weak partial reflections from heights of 75–90 km. Measurements of the reflection coefficients of both the ordinary and extraordinary waves, which can be of the order of 10^{-5}, enable electron density and collision-frequency data to be deduced.[51] Pulsed signals with high transmitter antenna systems and very sensitive receivers are needed. As well as aerial systems and very sensitive receivers are needed. As well as *in-situ* probes in rockets, there are a wide range of other schemes for determining electron density and collision frequency, involving the study of wave-fields radiated between the ground and a rocket. These use combinations of frequencies in the v.l.f.–v.h.f. range and include the measurement of differential-Doppler frequency, absorption, differential phase, propagation time and Faraday rotation.

Theory shows that signals propagated via the ionosphere can become cross-modulated by high-power interfering signals which heat the plasma electrons through which the wanted signals pass. This heating causes the electron-collision frequency, and therefore the amplitude of the wanted signal to fluctuate at the modulation frequency of the interfering transmitter. Investigations of this phenomenon (known as the Luxembourg effect after the first identified interfering transmitter) usually employ vertically transmitted and received wanted pulses, modulated by a distant disturbing transmitter radiating synchronised pulses at half the repetition rate. Changes in signal amplitude and phase between successive pulses are measured, and by altering the relative phase of the two transmitters, the height at which the cross-modulation occurs can be varied. Such data enable the height distributions of electron density in the D-region to be determined.[52]

Using a Laser radar (Lidar), the intensity of the light backscattered by the atmospheric constituents at heights above 50 km gives the height distributions of neutral-air density and the temporal and spatial statistics of high-altitude atmospheric aerosols. Minor atmospheric constituents may be detected with tunable dye lasers from their atomic and molecular-resonance scattering.

11.6 Propagation prediction procedures

Long-term predictions based on monthly median ionospheric data are required for the circuit planning of v.l.f.–h.f. ground-based systems. Estimates of raypath launch and arrival angles are needed for antenna design, and of the relationship between transmitter power and received field strength at a range of frequencies, so that the necessary size of transmitter and its frequency coverage can be determined. Since there are appreciable day to day changes in the electron densities in the F2-region, in principle short-term predictions based on ionospheric probing measurements or on correlations with geophysical indices should be of great value for real-time frequency management. In practice, however, aside from the technical problems of devising schemes of adequate accuracy: (*a*) not all systems are frequency agile (e.g. broadcasting); (*b*) effective schemes may require two-way transmissions; and (*c*) only assigned frequencies may be used. An alternative to short-term predictions is real-time channel sounding; certain procedures involve a combination of the two techniques.

11.6.1 Long-term predictions

The first requirement of any long-term prediction is a model of the ionosphere. At v.l.f. waves propagate between the Earth and the lower boundary of the ionosphere at heights of 70 km by day and 90 km by night as if in a two-surface waveguide. Very low frequency field-strength predictions are based on a full-wave theory that includes diffraction and surface-wave propagation. For paths beyond 1000 km range only three or fewer waveguide modes need be considered. A general equation gives field strength as a function of range, frequency, ground-electrical properties and the ionospheric reflection height and reflection coefficients.[53] Unfortunately the reflection coefficients vary in a complex way with electron density and collision frequency, the direction and strength of the Earth's magnetic field, wave frequency and angle of incidence, so that in the absence of accurate D-region electron-density data, estimates are liable to appreciable error. At l.f. propagation is more conveniently described by wave-hop theory in terms of component waves with different numbers of hops. As at v.l.f. reflection occurs at the base of the ionosphere and the accuracy of the field-strength prediction is largely determined by uncertainties in ionospheric models and reflection coefficients.

Medium-frequency signals penetrate the lower ionosphere and are usually reflected from heights of 85–100 km, except over distances of less than 500 km by night when reflection may be from the F-region. Large absorptions occur near the height of reflection and daytime signals are very weak. It is now realised that because of the uncertainties in ionospheric models, signal-strength predictions are best based on empirical equations fitted to measured signal-strength data for other oblique paths.

Prediction schemes for h.f. tend to be complicated because they must assess the active modes and elevation angles; these vary markedly with ionospheric conditions and transmitter frequency. Equations are available for the raypaths at oblique incidence through ionospheric models composed of separate layers, each with a parabolic distribution of electron density with height.[54] They are employed in one internationally used prediction scheme,[55] with the parameters of the parabolas determined from numerical prediction maps of the vertical-incidence ionospheric characteristics, as given by data from the world network of ionosondes[56] (see *Figure 11.4*). Calculations over a fixed path for

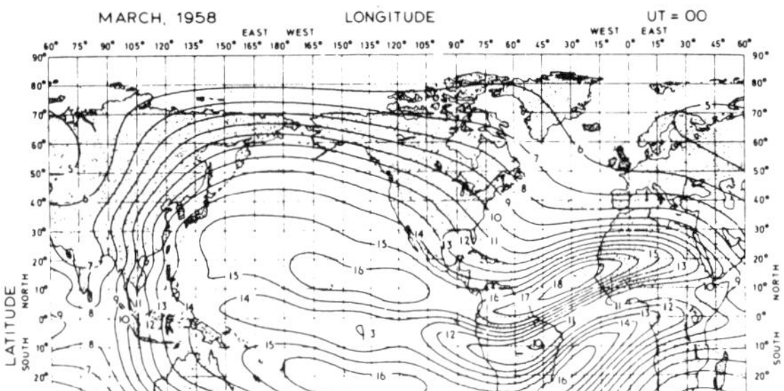

Figure 11.4 Predicted median, foF2, MHz for 00h, UT in March 1958 (Reproduced by permission of the Institute for Telecommunication Sciences, Boulder, USA)

a range of frequencies indicate the largest frequency (the basic m.u.f. or maximum usable frequency) that propagates via a given mode. Assuming some statistical law for the day to day variability of the parameters of the model they also give the *availability*, which is the fraction of days that the mode can exist. Received signal strengths are then determined in terms of the transmitter power and a number of transmission loss and gain factors. These include transmitting and receiving antenna gain, spatial attenuation, ray convergence gain, absorption, intermediate-path ground-reflection losses and polarisation-coupling losses. Predictions may be further extended by including estimates of the day to day variability in signal intensity. Calculations are prohibitively lengthy without computing aids and a number of computerised prediction schemes have been produced. By means of estimates of background noise intensity, and from the known required signal/noise ratio, the mode reliability may also be determined. This is the fraction of the days that the signals are received with adequate strength. For some systems involving fast data transmission, predictions of the probability of multipath, with two or more modes of specified comparable amplitude with propagation delays differing by less than some defined limit, are also useful and can be made.

11.6.2 Short-term predictions and real-time channel sounding

Some limited success has been achieved in the short-term prediction of the ionospheric characteristics used to give the parameters of the ionospheric models needed for h.f. performance assessment. Schemes are based either on spatial or temporal extrapolation of near real-time data or on correlations with magnetic activity indices. Regression statistics have been produced for the change in the maximum plasma frequency of the F2-region (foF2) with local magnetic activity index *K*, and other work is concerned with producing joint correlations with *K* and with solar flux.

In principle at h.f. the most reliable, although costly, way of ensuring satisfactory propagation over a given path and of optimising the choice of transmission frequency involves using an oblique-incidence sounder over the actual path; in practice, however, sounder systems are difficult to deploy operationally, require expert interpretation of their data, lead to appreciable spectrum pollution, and give much redundant information. Some schemes involve low-power channel monitoring of the phase-path stability on each authorised frequency, to ensure that at all times the best available is used. Real-time sounding on one path can aid performance predictions for another. Examples include ray tracing through mean ionospheric models simulated from measured backscatter or oblique-incidence soundings. Many engineers operating established radio cricuits prefer, for frequency management, to rely on past experience, rather than to use predictions. This is not so readily possible for mobile applications. Real-time sounding schemes involving ground transmissions on a range of frequencies to an aircraft, but only single-frequency transmission in the reverse direction, have proved successful.[57]

11.7 The troposphere

The influence of the lower atmosphere, or troposphere, on the propagation of radio waves is important in several respects. At all frequencies above about 30 MHz refraction and scattering, caused by local changes in atmospheric structure, become significant—especially in propagation beyond the normal horizon. In addition, at frequencies above about 5 GHz, absorption in oxygen and water vapour in the atmosphere is important at certain frequencies corresponding to molecular absorption lines. An understanding of the basic characteristics of these effects is thus essential in the planning of very high frequency communication systems. The main features of tropospheric propagation are summarised from a practical point of view as follows.

There are two general problems: firstly, the influence of the troposphere on the *reliability* of a communication link. Here attention is concentrated on the weak signals which can be received for a large percentage of the time, say 99.99%. Secondly, it is necessary to consider the problem of *interference* caused by abnormal propagation and unusually strong, unwanted signals of the same frequency as the wanted transmission.

In both these aspects of propagation, the radio refractive index of the troposphere plays a dominant role. This parameter depends on the pressure, temperature and humidity of the atmosphere. Its vertical gradient and local fluctuations about the mean value determine the mode of propagation in many important practical situations. Hence the interest in the subject of radio meteorology, which seeks to relate tropospheric structure and radio-wave propagation. In most ground-to-ground systems the height range 0–2 km above the Earth's surface is the important region, but in some aspects of Earth—space transmission, the meteorological structure at greater heights is also significant.

11.7.1 Historical background

Although some experiments on ultra-short-wave techniques were carried out by Hertz and others more than sixty years ago, it was only after about 1930 that any systematic investigation of tropospheric propagation commenced. For a long time it was widely believed that at frequencies above about 30 MHz transmission beyond the geometric horizon would be impossible. However, this view was disputed by Marconi as early as 1932. He demonstrated that, even with relatively low transmitter powers, reception over distances several times the optical range was possible. Nevertheless, theoreticians continued for several years to concentrate on studies of diffraction of ultra-short waves around the Earth's surface. However, their results were found to over-estimate the rate of attenuation beyond the horizon. To correct for this disparity, the effect of refraction was allowed for by assuming a process of diffraction around an Earth with an effective radius of 4/3 times the actual value. In addition, some experimental work began on the effect of irregular terrain and the diffraction caused by buildings and other obstacles.

However, it was only with the development of centimetric radar in the early years of World War II that the limitations of earlier concepts of tropospheric propagation were widely recognised. For several years attention was concentrated on the role of unusually strong refraction in the surface layers, especially over water, and the phenomenon of trapped propagation in a *duct*. It was shown experimentally and theoretically that in this mode the rate of attenuation beyond the horizon was relatively small. Furthermore, for a given height of duct or surface layer having a very large, negative, vertical gradient of refractive index, there was a critical wavelength above which trapping did not occur; a situation analogous to that in waveguide transmission.

Again however it became apparent that further work was required to explain experimental observations. The increasing use of v.h.f., and later u.h.f., for television and radio communication emphasised the need for a more comprehensive approach on beyond-the-horizon transmission. The importance of refractive-index variations at heights of the order of a kilometre began to be recognised and studies of the correlation between the height-variation of refractive index and field strength began in several laboratories.

With the development of more powerful transmitters and antennae of very high gain it proved possible to establish communication well beyond the horizon even in a 'well-mixed' atmosphere with no surface ducts or large irregularities in the height-variation of refractive index. To explain this result, the concept of *tropospheric scatter* was proposed. The trans-horizon field was assumed to be due to incoherent scattering from the random, irregular fluctuations in refractive index produced and maintained by turbulent motion. This procedure has dominated much of the experimental and theoretical work of recent years and it certainly explains some characteristics of troposphere propagation. However, it is inadequate in several respects. It is now known that some degree of stratification in the troposphere is more frequent than was hitherto assumed. The possibility of reflection from a relatively small number of layers or *sheets* of large vertical gradient must be considered, especially at v.h.f. At u.h.f. and s.h.f. strong scattering from a 'patchy' atmosphere, with local regions of large variance in refractive index filling only a fraction of the common volume of the antenna beams, is probably the mechanism which exists for much of the time.

The increasing emphasis on microwaves for terrestrial and space systems has recently focused attention on the effects of precipitation on tropospheric propagation. While absorption in atmospheric gases is not a serious practical problem below 40 GHz, the attenuation in rain and wet snow can impair the performance of links at frequencies of about 10 GHz and above. Moreover, scattering from precipitation may prove to be a significant factor in causing interference between space and terrestrial systems sharing the same frequency. The importance of interference-free sites for Earth stations in satellite links has also stimulated work on the shielding effect of hills and mountains. In addition, the use of large antennae of high gain in space systems requires a knowledge of refraction effects (especially at low angles of elevation), phase variations over the wavefront, and the associated effects of scintillation fading and gain degradation. Particularly at the higher microwave frequencies, thermal noise radiated by absorbing regions of the troposphere (rain, clouds, etc.) may be significant in space communication. Much of the current research is therefore being directed towards a better understanding of the spatial structure of precipitation.

11.8 Survey of propagation modes in the troposphere

Figure 11.5 illustrates qualitatively the variation of received power with distance in a homogeneous atmosphere at frequencies above about 30 MHz. For antenna heights of a wavelength or more, the propagation mode in the free-space range is a *space wave* consisting of a direct and a ground-reflected

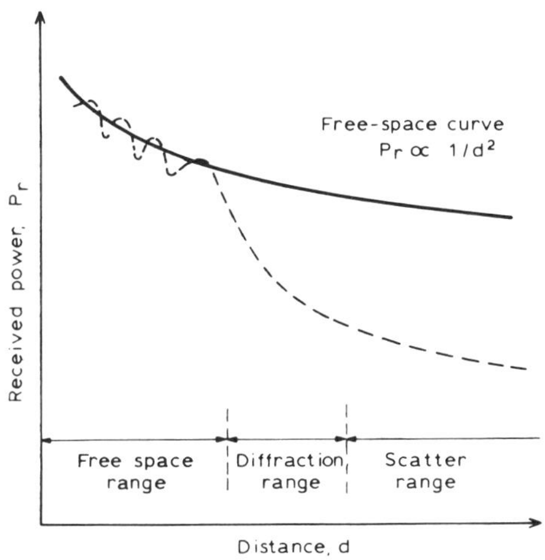

Figure 11.5 Tropospheric attenuation as a function of distance in an homogeneous atmosphere. Direct and ground-reflected rays interfere in the free-space range; obstacle-diffraction effects predominate in the diffraction range; and refractive-index variations are important in the scatter range

ray. For small grazing angles the reflected wave has a phase change of nearly 180° at the Earth's surface, but imperfect reflection reduces the amplitude below that of the direct ray. As the path length increases, the signal strength exhibits successive maxima and minima. The most distant maximum will occur where the path difference is $\lambda/2$, where λ is the wavelength.

The range over which the space-wave mode is dominant can be determined geometrically allowing for refraction effects. For this purpose we can assume that the refractive index, n, decreases linearly by about 40 parts in 10^6 (i.e. 40 N units) in the first kilometre. This is the equivalent to increasing the actual radius of the Earth by a factor of 4/3 and drawing the ray paths as straight lines. The horizon distance d, from an antenna at height h above an Earth of effective radius a is

$$d=(2ah)^{1/2} \tag{11.1}$$

For two antennae 100 m above ground the total range is about 82 km, 15% above the geometric value.

Beyond the free-space range, diffraction around the Earth's surface and its major irregularities in terrain is the dominant mode, with field strengths decreasing with increasing frequency and being typically of the order of 40 dB below the free space value at 100 km at v.h.f. for practical antenna heights. As the distance increases, the effect of reflection or scattering from the troposphere increases and the rate of attenuation with distance decreases. In an actual inhomogeneous atmosphere the height-variation of n is the dominant factor in the *scatter zone* as illustrated in *Figure 11.6*. However, in practice the situation is rarely as simple as that indicated by these simple models.

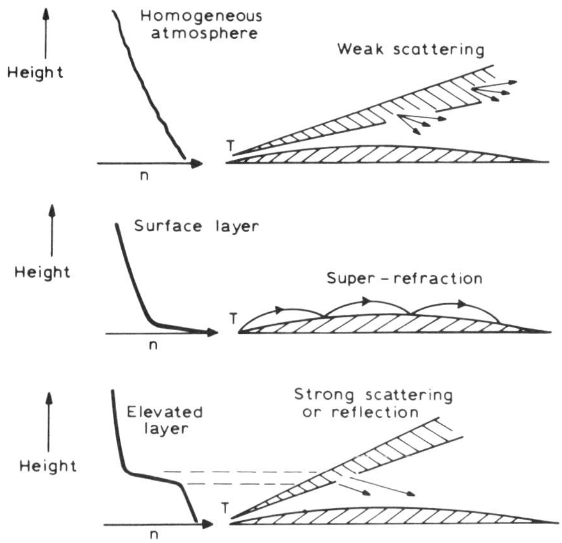

Figure 11.6 Tropospheric propagation modes and height-variation index, n

At frequencies above about 40 GHz, absorption in atmospheric gases becomes increasingly important. This factor may determine system design and the extent to which co-channel sharing is possible; for example, between terrestrial and space communication services. There are strong absorption lines due to oxygen at 60 and 119 GHz, with values of attenuation, at sea level, of the order of 15 and 2 dB km^{-1} respectively. At 183 GHz, a water vapour line has an attenuation of about 35 dB km^{-1}. Between these lines there are 'windows' of relatively low attenuation; e.g. around 35, 90 and 140 GHz. These are the preferred bands for future exploitation of the millimetric spectrum; for example, for short-range communication systems or radar. Further details are given in Report 719 of the CCIR.

11.8.1 Ground-wave terrestrial propagation

When the most distant maximum in *Figure 11.5* occurs at a distance small compared with the optical range, it is often permissible to assume the Earth flat and perfectly reflecting, particularly at the low-frequency end of the v.h.f. range. The space-wave field E at a distance d is then given by

$$E=(90W^{1/2}h_t h_r)/\lambda d^2 \qquad (11.2)$$

where W is the power radiated from a $\lambda/2$ dipole, and h_t and h_r are the heights of the transmitting and receiving antennae respectively.

The effects of irregular terrain are complex. There is some evidence that, for short, line-of-sight links, a small degree of surface roughness increases the field strength by eliminating the destructive interference between the direct and ground-reflected rays. Increasing the terrain irregularity then reduces the field strength, particularly at the higher frequencies, as a result of shadowing, scattering and absorption by hills, buildings and trees. However, in the particular case of a single, obstructing ridge visible from both terminals it is sometimes possible to receive field-strengths greater than those over level terrain at the same distance. This is the so-called *obstacle gain.*

In designing microwave radio-relay links for line-of-sight operation it is customary to so locate the terminals that, even with unfavourable conditions in the vertical gradient of refractive index (with *sub-refraction* effects decreasing the effective radius of the Earth), the direct ray is well clear of any obstacle. However, in addition to multipath fading caused by a ground-reflected ray, it is possible for line-of-sight microwave links to suffer fading caused by multi-path propagation via strong scattering or abnormal refraction in an elevated layer in the lower troposphere. This situation may lead to a significant reduction in usable bandwidth and to distortion, but the use of spaced antennae (space diversity) or different frequencies (frequency diversity) can reduce these effects. Even in the absence of well-defined layers, scintillation-type fading may occasionally occur at frequencies of the order of 30–40 GHz on links more than say 10 km long.

As a guide to the order of magnitude of multipath fading, Report 338 of the CCIR gives the following values for a frequency of 4 GHz, for the worst month of the year, for average rolling terrain in northwest Europe:

0.01% of time;	path length 20 km; 11 dB or more below free space
	path length 40 km; 23 dB or more below free space

Over very flat, moist ground the values will be greater and, for a given path length, will tend to increase with frequency. But at about 10 GHz and above, the effect of precipitation will generally dominate system reliability.

The magnitude of attenuation in rain can be estimated theoretically and the reliability of microwave links can then be forecast from a knowledge of rainfall statistics. But the divergence between theory and experiment is often considerable. This is partly due to the variation which can occur in the drop-size distribution for a given rainfall rate. In addition, many difficulties remain in estimating the intensity and spatial characteristics of rainfall for a link. This is an important practical problem in relation to the possible use of *route diversity* to minimise the effects of absorption fading. Experimental results show that for very high reliability (i.e. for all rainfall rates less than say 50–100 mm/h in temperate climates) any terrestrial link operating at frequencies much above 30 GHz must not exceed say 10 km in length. It is possible, however, to design a system with an alternative route so that by switching between the two links the worst effects of localised, very heavy rain can be avoided. The magnitude of attenuation in rain is shown in *Figure 11.7(a)* and the principle of route diversity is illustrated in *Figure 11.7(b)*. For temperate climates, the diversity gain (i.e. the difference between the attenuation, in dB, exceeded for a specified, small percentage of time on a single link and that exceeded simultaneously on two parallel links) varies as follows:

(a) it tends to decrease as the path length increases from 12 km, for a given percentage of time and for a given lateral path separation;
(b) is generally greater for a spacing of 8 km than for 4 km, though an increase to 12 km does not provide further improvement;
(c) is not strongly dependent on frequency in the range 20–40 GHz for a given geometry.

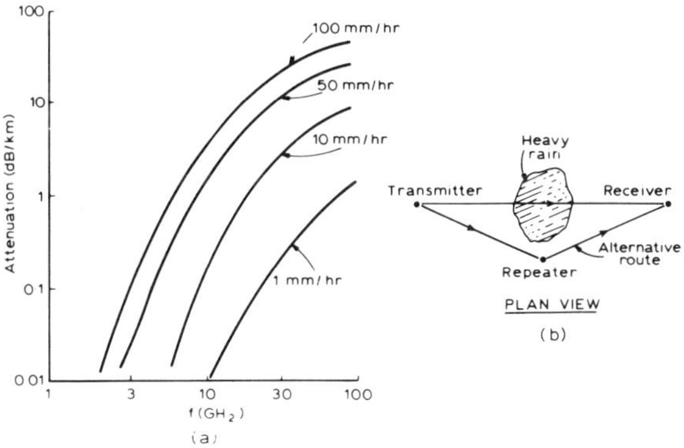

Figure 11.7 (a) Attenuation in rain. (b) The application of route diversity to minimise effects of fading

The main problem in the v.h.f. and u.h.f. broadcasting and mobile services (apart from prediction of interference) is to estimate the effect of irregularities in terrain and of varying antenna height on the received signal. The site location is of fundamental importance. Prediction of received signal strengths, on a statistical basis, has been made using a parameter, Δh, which characterises terrain roughness (see CCIR Recommendation 370 and Reports 239 and 567). However, there is considerable path-to-path variability, even for similar Δh values. Especially in urban areas, screening and multipath propagation due to buildings are important. Moreover, in such conditions—and especially for low heights of receiving antenna—depolarization effects can impair performance of orthogonally-polarized systems sharing a common frequency. At the higher u.h.f. frequencies, attenuation due to vegetation (e.g. thick belts of trees) is beginning to be significant.

11.8.2 Beyond-the-horizon propagation

Although propagation by surface or elevated layers (see *Figure 11.6*) cannot generally be utilised for practical communication circuits, these features remain important as factors in co-channel interference. Considerable theoretical work, using waveguide mode theory, has been carried out on duct propagation and the results are in qualitative agreement with experiment. Detailed comparisons are difficult because of the lack of knowledge of refractive index structure over the whole path, a factor common to all beyond-the-horizon experiments. Nevertheless, the theoretical predictions of the maximum wavelength trapped in a duct are in general agreement with practical experience. These values are as follows:

λ (max) in cm	*Duct height in m*
1	5
10	25
100	110

Normal surface ducts are such that complete trapping occurs only at centimetric wavelengths. Partial trapping may occur for the shorter metric wavelengths. Over land the effects of irregular terrain and of thermal convection (at least during the day) tend to inhibit duct formation. For a ray leaving the transmitter horizontally, the vertical gradient of refractive index must be steeper than -157 parts in 10^6 per kilometre.

Even when super-refractive conditions are absent, there remains considerable variability in the characteristics of the received signal. This variability is conveniently expressed in terms of the transmission loss, which is defined as $10\log(P_t/P_r)$, where P_t and P_r are the transmitted and received powers respectively. In scatter propagation, both slow and rapid variations of field strength are observed. Slow fading is the result of large-scale changes in refractive conditions in the atmosphere and the hourly median values below the long-term median are distributed approximately log-normally with a standard deviation which generally lies between 4 and 8 decibels, depending on the climate. The largest variations of transmission loss are often seen on paths for which the receiver is located just beyond the diffraction region, while at extreme ranges the variations are less. The slow fading is not strongly dependent on the radio frequency. The rapid fading has a frequency of a few fades per minute at lower frequencies and a few hertz at u.h.f. The superposition of a number of variable incoherent components would give a signal whose amplitude was Rayleigh-distributed. This is found to be the case when the distribution is analysed over periods of up to five minutes. If other types of signal form a significant part of that received, there is a modification of this distribution. Sudden, deep and rapid fading has been noted when a frontal disturbance passes over a link. In addition, reflections from aircraft can give pronounced rapid fading.

The long-term median transmission loss relative to the free-space value increases approximately as the first power of the frequency up to about 3 GHz. Also, for most temperate climates, monthly median transmission losses tend to be higher in winter than in summer, but the difference diminishes as the distance increases. In equatorial climates, the annual and diurnal variations are generally small. The prediction of transmission loss, for various frequencies, path lengths, antenna heights, etc., is an important practical problem. An example of the kind of data required is given in *Figure 11.8*.

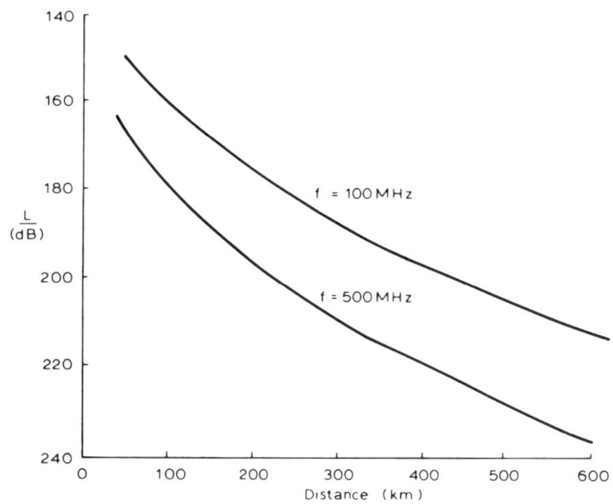

Figure 11.8 Median transmission loss, L, between isotropic antennas in a temperate climate and over an average rolling terrain. The height of the transmitting antenna is 40 m, and the height of the receiving antenna is 10 m

At frequencies above 10 GHz, the heavy rain occurring for small percentages of the time causes an additional loss due to absorption, but the accompanying scatter from the rain may partly offset the effect of absorption.

11.8.3 Physical basis of tropospheric scatter propagation

Much effort has been devoted to explaining the fluctuating trans-horizon field in terms of scattering theory based on statistical models of turbulent motion. The essential physical feature of this approach is an atmosphere consisting of irregular *blobs* in random motion which in turn produce fluctuations of refractivity about a stationary mean value. Using this concept, some success has been achieved in explaining the approximate magnitude of the scattered field but several points of difficulty remain. There is now increasing evidence, from refractometer and radar probing of the troposphere, that some stratification of the troposphere is relatively frequent.

By postulating layers of varying thickness, horizontal area and surface roughness, and of varying lifetime it is possible in principle to interpret many of the features of tropospheric propagation. Indeed, some experimental results (e.g. the small distance-dependence of v.h.f. fields at times of anomalous propagation) can be explained by calculating the reflection coefficient of model layers of constant height and with an idealised height-variation of refractive index such as half-period sinusoidal, exponential, etc. The correlation between field strength and layer height has also been examined and some results can be explained qualitatively in terms of *double-hop* reflection from extended layers. Progress in ray-tracing techniques has also been made. Nevertheless, the problems of calculating the field strength variations on particular links remain formidable, and for many practical purposes statistical and empirical techniques for predicting link performance remain the only solution.

Other problems related to fine structure are space and frequency diversity. On a v.h.f. scatter link with antennae spaced normal to the direction of propagation, the correlation coefficient may well fall to say 0.5 for spacings of 5–30λ in conditions giving fairly rapid fading. Again, however, varying meteorological factors play a dominant role. In frequency diversity, a separation of say 3 or 4 MHz may ensure useful diversity operation in many cases, but occasionally much larger separations are required. The irregular structure of the troposphere is also a cause of gain degradation. This is the decrease in actual antenna gain below the ideal free-space value. Several aspects of the irregular refractive-index structure contribute to this effect and its magnitude depends somewhat on the time interval over which the gain measurement is made. Generally, the decrease is only significant for gains exceeding about 50 dB.

11.9 Tropospheric effects in space communications

In space communication, with an Earth station as one terminal, several problems arise due to refraction, absorption and scattering effects, especially at microwave frequencies. For low angles of elevation of the Earth station beam, it is often necessary to evaluate the refraction produced by the troposphere, i.e. to determine the error in observed location of a satellite. The major part of the bending occurs in the first two kilometres above ground and some statistical correlation exists between the magnitude of the effect and the refractive index at the surface. For high-precision navigation systems and very narrow beams it is often necessary to evaluate the variability of refraction effects from measured values of the refractive index as a function of height. A related phenomenon important in tracking systems is the phase distortion in the wave-front due to refractive index fluctuations, a feature closely linked with gain degradation. This phase distortion also affects the stability of frequencies transmitted through the troposphere.

Absorption in clear air may affect the choice of frequencies, above about 40 GHz, to minimise co-channel interference. *Figure 11.9* shows the zenith attenuation from sea level for an average clear atmosphere as a function of frequency. It illustrates the 'window' regions mentioned in the survey of propagation modes.

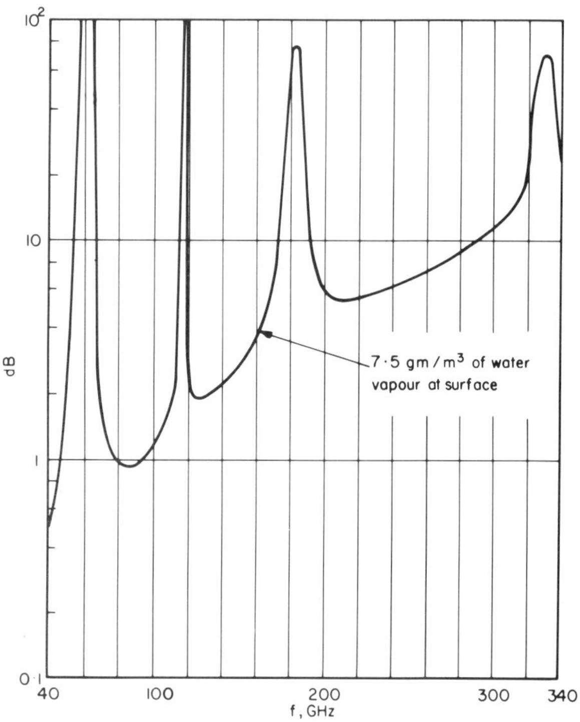

Figure 11.9 Zenith attenuation (dB) in clear air

From an altitude of 4 km, the values would be about one-third of those shown. This indicates the potential application of frequencies above 40 GHz for communication on paths located above the lower layers of the troposphere.

Clouds produce an additional loss which depends on their liquid–water content. Layer-type cloud (stratocumulus) will not cause additional attenuation of more than about 2 dB, even at 140 GHz. On the other hand, cumulonimbus will generally add several decibels to the total attenuation, the exact value depending on frequency and cloud thickness.

Absorption in precipitation (see *Figure 11.7(a)*) has already been mentioned in relation to terrestrial systems. Water drops attenuate microwaves both by scattering and by absorption. If the wavelength is appreciably greater than the drop-size then the attenuation is caused almost entirely by absorption. For rigorous calculations of absorption it is necessary to specify a drop-size distribution; but this, in practice, is highly variable and consequently an appreciable scatter about the theoretical value is found in experimental measurements. Moreover, statistical information on the vertical distribution of rain is very limited. This makes prediction of the reliability of space links difficult and emphasises the value of measured data. Some results obtained, using the Sun as an extraterrestrial source, are shown in *Figure 11.10*. In addition, scatter from rain (being approximately isotropic) may cause appreciable interference on co-channel terrestrial and space systems even when the beams from the two systems are not directed towards each other on a great-circle path. This aspect of 'scatter propagation' requires further study, particularly in relation to the screening effect of local hills on other modes of propagation.

Because precipitation (and to a smaller extent the atmospheric gases) absorb microwaves, they also radiate thermal-type noise. It is often convenient to specify this in terms of an *equivalent black-body temperature* or simply *noise temperature* for an antenna pointing in a given direction. With radiometers and low-

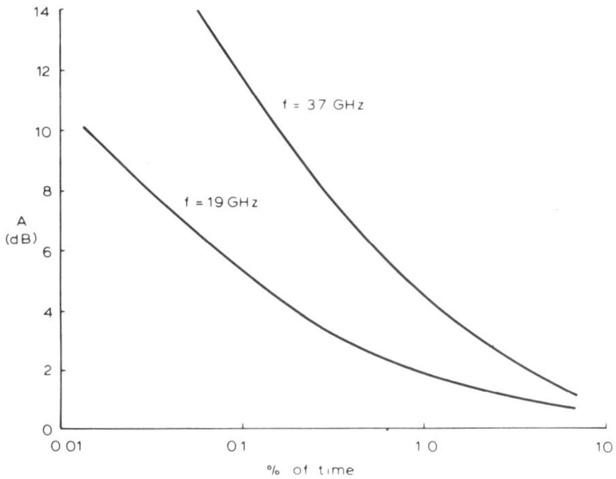

Figure 11.10 Measured probability distribution of attenuation *A* on earth-space path at 19 GHz and 37 GHz (southeast England: Elevation angles 5° to 40°: data from solar-tracking radiometers)

noise receivers it is now possible to measure this tropospheric noise and assess its importance as a factor in limiting the performance of a microwave Earth–space link. For a complete solution, it is necessary to consider not only direct radiation into the main beam but also ground-reflected radiation, and emission from the ground itself, arriving at the receiver via side and back lobes. From the meteorological point of view, radiometer soundings (from the ground, aircraft, balloons or satellites) can provide useful information on tropospheric and stratospheric structure. Absorption in precipitation becomes severe at frequencies above about 30 GHz and scintillation effects also increase in importance in the millimetre range. However, for space links in or near the vertical direction the system reliability may be sufficient for practical application even at wavelengths as low as 3–4 mm. Moreover, spaced receivers in a site-diversity system can be used to minimise the effects of heavy rain.

In recent years, extensive studies of propagation effects (attenuation, scintillation, etc.) have been carried out by direct measurements using satellite transmissions. Special emphasis has been given to frequencies between 10 and 30 GHz, in view of the effect of precipitation on attenuation and system noise. Details are given in Report 564 of the CCIR. For typical elevation angles of 30–45°, the total attenuation exceeded for 0.01% of the time has values of the following order:

f=12 GHz; 5 dB in temperate climates
(e.g. northwest Europe)
20 dB in tropical climates (e.g. Malaysia)
10 dB in East Coast, USA (Maryland)
f=20 GHz; 10 dB (northwest Europe)

Site-diversity experiments using satellite transmissions show that site spacings of the order of 5–10 km can give a useful improvement in reliability. However, the improvement may depend on the site geometry and on topographical effects. At frequencies above say 15 GHz, the advantage of site diversity may be quite small if the sites are so chosen that heavy rain in, for example, frontal systems tends to affect both sites simultaneously.

Frequency re-use is envisaged, in space telecommunication systems, by means of orthogonal polarisation. But this technique is restricted by depolarisation due to rain and ice clouds and—to a lesser extent—by the system antennae. Experimental data on polarisation distortion, obtained in satellite experiments, are given in CCIR Report 564.

11.10 Techniques for studying tropospheric structure

The importance of a knowledge of the structure of the troposphere in studies of propagation is clearly evident in the above sections. The small-scale variations in refractive index and in the intensity of precipitation are two important examples. They form part of the general topic of *tropospheric probing*.

Much useful information on the height-variation of refractive index can be obtained from the radio-sondes carried on free balloons and used in world-wide studies of meteorological structure. However, for many radio applications these devices do not provide sufficient detail. To obtain this detail instruments called refractometers have been developed, mainly for use in aircraft, on captive balloons or on tall masts. They generally make use of a microwave cavity for measuring changes in a resonance frequency, which in turn is related to the refractive index of the enclosed air. Such refractometers are robust, rapid-response instruments which have been widely used as research tools, though they have yet to be developed in a form suitable for widespread, routine use.

High-power, centimetric radar is also a valuable technique. By its use it is possible to detect layers or other regions of strong scatter in the troposphere, and to study their location and structure. Joint radar-refractometer soundings have proved of special interest in confirming that the radar does indeed detect irregularities in clear-air structure. The application of radar in precipitation studies is, of course, a well-known and widely used technique in meteorology; although to obtain the detail and precision necessary for radio applications requires careful refinements in technique.

Optical radar (lidar) and acoustic radar have also been used to probe the troposphere, although the information they provide is only indirectly related to radio refractive index.

The millimetre and sub-millimetre spectrum, as yet not exploited to any significant degree for communications, is nevertheless a fruitful region for tropospheric probing. In particular, the presence of several absorption lines (in water vapour, oxygen and minor constituents such as ozone) makes it possible to study the concentration and spatial distribution of these media. Near the ground, direct transmission experiments are feasible; for example, to study the average water-vapour concentration along a particular path. In addition, it is possible to design radiometers for use on the ground, in an aircraft or in a satellite, which will provide data on the spatial distribution of absorbing atmospheric constituents by measurement of the emission noise they radiate. This topic of *remote probing* is one exciting considerable current interest in both radio and meteorology.

References

1 BUCHAU, J., 'Instantaneous versus averaged ionosphere', *Air Force Surveys in Geophysics No. 241* (*Air Force Systems Command, United States Air Force*), **1** (1972)
2 MULDREW, D. B., 'F-layer ionisation troughs deduced from Alouette data', *J. Geophys. Res.*, **70**, 2635 (1965)
3 DAVIES, K., *Ionospheric Radio Propagation*, Monograph 80, National Bureau of Standards, Washington (1965)
4 RATCLIFFE, J. A., *Sun, Earth and Radio—an Introduction to the Ionosphere and Magnetosphere*, Weidenfeld and Nicolson, London (1970)
5 RISHBETH, H. and GARRIOTT, O. K., *Introduction to Ionospheric Physics*, Academic Press, London (1969)
6 DAVIES, K., *Ionospheric Radio Waves*, Blaisdell, Waltham, Mass. (1969)
7 RATCLIFFE, J. A., *An Introduction to the Ionosphere and Magnetosphere*, Cambridge University Press, Cambridge (1972)
8 RISHBETH, H., 'Thermospheric winds and the F-region: a review', *J. Atmosph. Terr. Phys.*, **34**, 1 (1972)

9 DUNCAN, R. A., 'The equatorial F-region of the ionosphere', *J. Atmosph. Terr. Phys.*, **18**, 89 (1960)
10 HARTZ, T. R., 'The general pattern of auroral particle precipitation and its implications for high latitude communication systems', *Ionospheric Radio Communications*, ed. K. Folkestad, Plenum, New York, 9 (1968)
11 BAILEY, D. K., 'Polar cap absorption', *Planet. Space Sci.*, **12**, 495 (1964)
12 MATSUSHITA, S., 'Geomagnetic disturbances and storms', *Physics of Geomagnetic Phenomena*, ed. Matsushita, S. and Campbell, W. H., Academic Press, London, 793 (1967)
13 SMITH, E. K. and MATSUSHITA, S., *Ionospheric Sporadic-E*, Macmillan, New York (1962)
14 COHEN, R. and BOWLES, K. L., 'On the nature of equatorial spread-F', *J. Geophys. Res.*, **66**, 1081 (1961)
15 HASELGROVE, J., 'Ray theory and a new method for ray tracing', *Report on Conference on Physics of Ionosphere*, Phys. Soc. London, 355 (1954)
16 RATCLIFFE, J. A., *The Magnetoionic Theory*, Cambridge University Press, Cambridge (1959)
17 BUDDEN, K., *Radio Waves in the Ionosphere*, Cambridge University Press, Cambridge (1961)
18 C.C.I.R. REPORT 266-5. 'Ionosphere propagation characteristics pertinent to terrestrial radio communication systems design', *Documents of 15th Plenary Assembly, Geneva*, ITU (1982)
19 NIELSON, D. L., 'The importance of horizontal F-region drifts to transequatorial VHF propagation', *Scatter Propagation of Radio Waves*, ed. Thrane, E., *AGARD Conference Proceedings No. 37*, NATO, Neuilly-sur-Seine, France (1968)
20 FENWICK, F. B. and VILLARD, O. G., 'A test of the importance of ionosphere reflections in long distance and around-the-world high frequency propagation', *J. Geophys. Res.*, **68**, 5659 (1963)
21 RATCLIFFE, J. A., 'Some aspects of diffraction theory and their application to the ionosphere', *Reports on Progress in Physics*, Phys. Soc., London, **19**, 188 (1956)
22 C.C.I.R. REPORT 263–5, 'Ionospheric effects upon earth-space propagation', *Documents of 15th Plenary Assembly, Geneva*, ITU (1982)
23 PIERCE, J. A., 'OMEGA', *IEEE Trans. Aer. and Elect. Syst.*, **1**, 206 (1965)
24 STRINGER, F. S., 'Hyperbolic radionavigation systems', *Wireless World*, **75**, 353 (1969)
25 SUGAR, G. R., 'Radio propagation by reflection from meteor trails', *Proc. IEEE*, **52**, 116 (1964)
26 BAILEY, D. K., BATEMAN, R. and KIRBY, R. C., 'Radio transmission at VHF by scattering and other processes in the lower ionosphere', *Proc. IRE*, **43**, 1181 (1955)
27 CROFT, T. A., 'Skywave backscatter: a means for observing our environment at great distances', *Rev. Geophys. and Space Physics*, **10**, 73 (1972)
28 BLAIR, J. C., MELANSON, L. L. and TVETEN, L. H., 'HF ionospheric radar ground scatter map showing land-sea boundaries by a spectral separation technique', *Electronics Letters*, **5**, 75 (1969)
29 BAKER, D. M. and DAVIES, K., 'Waves in the ionosphere produced by nuclear explosions', *J. Geophys. Res.*, **73**, 448 (1968)
30 DAVIES, K. and BAKER, D. M., 'Ionospheric effects observed around the time of the Alaskan earthquake of March 28 1964', *J. Geophys. Res.*, **70**, 2251 (1965)
31 BAKER, D. M. and DAVIES, K., 'F2-region acoustic waves from severe weather', *J. Atmosph. Terr. Phys.*, **31**, 1345 (1969)
32 PIERCE, E. T., 'Nuclear explosion phenomena and their bearing on radio detection of the explosions', *Proc. IEEE*, **53**, 1944 (1965)
33 SMITH, E. K., 'Electromagnetic probing of the upper atmosphere', ed. U.R.S.I. Working Group, *J. Atmosph. Terr. Phys.*, **32**, 457 (1970)
34 PIGGOTT, W. R. and RAWER, K., *U.R.S.I. Handbook of Ionogram Interpretation and Reduction*, 2nd edn, Rep. UAG-23, Dept. of Commerce, Boulder, USA (1972)
35 BEYNON, W. J. G., 'Special issue on analysis of ionograms for electron density profiles', ed. U.R.S.I. Working Group, *Radio Science*, **2**, 1119 (1967)
36 COLL, D. C. and STOREY, J. R., 'Ionospheric sounding using coded pulse signals', *Radio Science*, **68D**, 1155 (1964)
37 FENWICK, R. B. and BARRY, G. H., 'Sweep frequency oblique ionospheric sounding at medium frequencies', *IEEE Trans. Broadcasting*, **12**, 25 (1966)
38 WHALEN, J. A., BUCHAU, J. and WAGNER, R. A., 'Airborne ionospheric and optical measurements of noontime aurora', *J. Atmosph. Terr. Phys.*, **33**, 661 (1971)
39 Ionospheric Data—Series FA—published monthly for National Geophysical and Solar-Terrestrial Data Centre, Boulder, USA
40 GARRIOTT, G. K. and NICHOL, A. W., 'Ionospheric information deduced from the Doppler shifts of harmonic frequencies from earth satellites', *J. Atmosph. Terr. Phys.*, **22**, 50 (1965)
41 ROSS, W. J., 'Second-order effects in high frequency transionospheric propagation', *J. Geophys. Res.*, **70**, 597 (1965)
42 AARONS, J., 'Total-electron content and scintillation studies of the ionosphere', ed. *AGARDograph 166*, NATO, Neuilly-sur-Seine, France (1973)
43 HELLIWELL, R. A., *Whistlers and Related Ionospheric Phenomena*, Stanford University Press, Stanford, California (1965)
44 MITRA, S. N., 'A radio method of measuring winds in the ionosphere', *Proc. IEE*, **46**, Pt. III, 441 (1949)
45 MUNRO, G. H., 'Travelling disturbances in the ionosphere', *Proc. Roy. Soc.*, **202A**, 208 (1950)
46 BELLO, P. A., 'Some techniques for instantaneous real-time measurements of multipath and Doppler spread', *IEEE Trans. Comm. Tech.*, **13**, 285 (1965)
47 CROFT, T. A., 'Special issue on ray tracing', ed. *Radio Science*, **3**, 1 (1968)
48 APPLETON, E. and PIGGOTT, W. R., 'Ionospheric absorption measurements during a sunspot cycle', *J. Atmosph. Terr. Phys.*, **5**, 141 (1954)
49 BEYNON, W. J. G. and RANGASWAMY, S., 'Model electron density profiles for the lower ionosphere', *J. Atmosph. Terr. Phys.*, **31**, 891 (1969)
50 HARGREAVES, J. K., 'Auroral absorption of H. F. radio waves in the ionosphere—a review of results from the first decade of riometry', *Proc. I.E.E.E.*, **57**, 1348 (1969)
51 BELROSE, J. S. and BURKE, M. J., 'Study of the lower ionosphere using partial reflections', *J. Geophys. Res.*, **69**, 2799 (1964)
52 FEJER, J. A., 'The interaction of pulsed radio waves in the ionosphere', *J. Atmosph. Terr. Phys.*, **7**, 322 (1955)
53 C.C.I.R. REPORT 265–5, 'Sky-wave propagation and circuit performance at frequencies between about 30 kHz and 500 kHz', *Documents of 15th Plenary Assembly, Geneva*, ITU (1982)
54 APPLETON, E. V. and BEYNON, W. J. G., 'The application of ionospheric data to radio communications', *Proc. Phys. Soc.*, **52**, 518 (1940); and **59**, 58 (1947)
55 C.C.I.R. REPORT 252–2, 'C.C.I.R. interim method for estimating sky-wave field strength and transmission loss at frequencies between the approximate limits of 2 and 30 MHz', *Documents of 12th Plenary Assembly, New Delhi*, ITU, Geneva (1970)
56 C.C.I.R. REPORT 340–4, 'C.C.I.R. atlas of ionospheric characteristics', *Documents of 15th Plenary Assembly, Geneva*, ITU (1970)
57 STEVENS, E. E., 'The CHEC sounding system', *Ionospheric Radio Communications*, ed. K. Kolkestad, Plenum, New York, 359 (1968)
58 BEAN, B. R. and DUTTON, E. J., *Radio Meteorology, Monograph 92*, US Government Printing Office, Washington (1966)
59 CASTEL, F. Du., *Tropospheric Radiowave Propagation Beyond the Horizon*, Pergamon Press, Oxford (1966)
60 C.C.I.R. 15th PLENARY ASSEMBLY, Vol. V, *Propagation in Non-Ionized Media*, International Telecommunications Union, Geneva (1982)
61 HALL, M. P. M., *Effects of the Troposphere on Radio Communication*, Peter Peregrinus (for I.E.E.), London (1979)
62 *I.E.E. Conference Publications in Propagation*; No. 48, London (1968); No. 98, London (1973); No. 169, London (1978); No. 195, York (1981); Norwich (1983)
63 SAXTON, J. A. (Ed.) *Advances in Radio Research*, Academic Press, London (1964)
64 U.R.S.I. Commission F, *Colloquium Proceedings*; La Baule, France (CNET, Paris, 1977). Also, at Lennoxville, Canada (Proceedings edited by University of Bradford, England, 1980)

Part 3

Materials and Components

12 Resistive Materials and Components

G W A Dummer MBE, CEng, FIEE, FIEEE, FIERE
Electronics Consultant
(Sections 12.1–12.3 and 12.6)

M J Rose BSc (Eng)
Production Officer,
Marketing Communications Group,
Mullard Ltd

Contents

12.1 Basic laws of resistivity

Ohm's law states that: the ratio of the potential difference E between the ends of a conductor to the current I flowing in it is a constant, provided the physical conditions of the conductor are unaltered. Ohm defined the resistance, R, of the given conductor as the ratio E/I.

The practical units in which the law is expressed are resistance in ohms, potential difference in volts and current in amperes. Until 1948 the standard of resistance was the International Ohm, the resistance offered to an unvarying electric current by a column of mercury of mass 14.452 1 g, of constant cross-section and 106.300 cm in length, maintained at the temperature of melting ice. Resistors made in this way by the five main national standardising laboratories of the world attain equality to within a few parts in one hundred thousand.

The Absolute Ohm, as from the 1 January 1948, replaced the International Ohm. One Absolute Ohm = 0.999 51 International Ohm. The Absolute Ohm has been determined, in England, by rotation of a conductor under specified conditions (Lorenz method).

Electromagnetic theory shows that resistance has the same dimensions as the product of inductance and frequency, and the Lorenz method makes use of this fact by rotating a conductor to cut the entire flux produced by coils of accurately known inductance. The current supplying the inductance coils passes through the standard resistor being calibrated, and the e.m.f. produced by the rotating conductor is arranged, by adjustment of known speed, to balance the IR drop across the resistor. The method is accurate to a few parts in 100 000.

If M is total flux when the current is I amperes, the change in flux linkage per revolution is MI. If the conductor makes n rev/s, the e.m.f. induced in the circuit by rotation is MIn. The potential drop across the resistor is IR; at balance

$$MIn = IR$$

and so

$$R = Mn$$

M is known, by calculation from the geometry and spacing of the known coils, and n is obtained by accurate frequency measurement.

The resistance R of a given conductor is proportional to its length l and inversely proportional to its area of cross-section a. Thus R varies as l/a and $R = \rho(l/a)$ where ρ is a constant of the material known as its *specific resistance* or *resistivity*. When l and a are unity, $R = \rho$; thus the resistivity of a given material may be defined as the resistance of *unit cube* of the material. In the SI system ρ may be expressed in *ohms per metre cube* or simply *ohm-metres*.

It will be seen that resistivity may be expressed in many terms, of which the following are those in general use:

(*a*) Microhms per centimetre cube ($\mu\Omega$-cm)
(*b*) Ohms square mil per foot (Ω-sq mil/foot)
(*c*) Ohms circular mil per foot (Ω-cir mil/foot).

In connection with the above terms, there are three points which should be noted:

(i) That a mil is an expression used to denote one thousandth of an inch (0.001 in) and should not be confused with millimetres.
(ii) Microhm cm is sometimes expressed as microhms/cm^3.
(iii) Resistivity must always be qualified by reference to the temperature at which it is measured, usually 20°C.

12.2 Resistance materials

12.2.1 Mass and volume resistivities

The units of mass resistivity and volume resistivity are interrelated through the density. For copper wires this is stated in the IACS (International Annealed Copper Standard) as 8.89 gram/cm^3 at 20°C. The volume resistivity, ρ, the mass resistivity, δ, and density, d, are related in the formula

$$\delta = \rho d \tag{12.1}$$

The IACS in various units of mass and volume resistivity, all at 20°C, is:

0.153 28	ohm-gram/metre^2
875.20	ohm-pound/mile^2
0.017 241	ohm-millimetre^2/metre
1.724 1	microhm-centimetre
0.678 79	microhm-inch
10.371	ohm-circular mil/foot

12.2.2 Temperature coefficient of resistance

At a temperature of 20°C, the coefficient of variation of resistance with temperature of standard annealed copper—measured between two potential points rigidly fixed to the wire—the metal being allowed to expand freely, is given as 0.003 93 = 1/254.45 per degree Celsius. The temperature coefficient of resistance of a copper wire of constant mass and conductivity of 100% at 20°C is therefore 0.003 93 per degree Celsius.

For other conductivities of copper, the 20°C temperature of resistance is given by multiplying the decimal number expressing the per cent conductivity by 0.003 93.

The conductivity of copper may be calculated from the coefficient of resistance. The temperature coefficient of resistance, α, for different initial Celsius temperatures and different conductivities may be calculated from the formula,

$$\alpha_{t_1} = \frac{1}{\dfrac{1}{\eta(0.003\,93)} + (t_1 - 20)} \tag{12.2}$$

in which

η = the conductivity expressed decimally
α_{t_1} = temperature coefficient at t_1

The temperature coefficient of resistivity is generally taken to be 0.006 8 microhm-cm per °C. Expressed in other values, it is

0.000 597	ohm-gram/metre^2
3.31	ohm-pound/mile^2
0.000 681	microhm-centimetre
0.002 68	microhm-inch
0.040 9	ohm-circular mil/foot

The Fahrenheit equivalent for these constants may be found by dividing them by 1.8. The change of resistivity per degree Fahrenheit is

0.001 49 microhm-inch

12.2.3 Resistivities of common metals and alloys

For a rough comparison of resistivities *Figure 12.1* shows a chart of the more common materials.

12.2.4 Temperature coefficients and resistivities for metals and alloys

These are given for many elements, metals and alloys in *Table 12.1.*

Table 12.1 Resistivities and temperature coefficients of metals and alloys

	Resistivity			*Temperature coefficient*		*Authority*
Material	*Composition*	*Temperature (°C)*	*Resistivity (ohm-m × 10⁻⁸)*	*Temperature (°C)*	*Temperature coefficient of resistance per °C*	
Advance* (See Constantan)						
Ni 30, Cr 5, Fe 65				20	+0.000 72	
Alumel*						
Ni 94, Mn 2.5, Fe 4.5, Al 2, Si 1				0	0.001 2	
Aluminum				0	0.001 2	
				18	+0.003 9	
				25	0.003 4	
				100	0.004	
				500	0.005	
annealed, highest purity				0–100	0.004 45	
Aluminum-bronze						
Cu 97, Al 3					0.001 02	
Cu 90, Al 10					0.003 20	
Cu 6, Al 94					0.003 80	
Antimony				20	0.003 6	
Argentan*						
Cu 61.6, Ni 15.8, Zn 22.6				0–160	0.000 387	
Arsenic				20	0.004 2	
Bismuth				20	0.004	
				0–100	0.004 46	
Brass				20	0.002	
Cu 66, Zn 34				15	0.002	
Cu 60, Zn 40				15	0.001	
Brightray B				20–500	+ +0.000 14	
Brightray C				20–500	+0.000 079	
Brightray F				20–500	+0.000 25	
Brightray H				20–500	+0.000 084	
Brightray S				20–500	+0.000 061	
Bronze						
Cu 88, Sn 12				20	0.000 5	
Cadmium				20	0.003 8	
drawn				0–100	0.004 24	
annealed, pure				0	0.004 2	
Carbon		0	3500		−0.005	
		500	2700			
		100	2100			
		2000	1100			
		2500	900			
Chromax*						
Ni 30, Cr 20, Fe 50				20–500	0.000 31	
Chromel A*						
Ni 80, Cr 20				20–500	0.000 13	
Chromel C*						
Ni 60, Cr 16, Fe 24				20–500	0.000 17	
Chromel D*						
Ni 30, Cr 20, Fe 50				20–500	0.000 32	
Climax				20	+0.000 7	
Cobalt				0	0.003 3	
				0–100	0.006 58	
Constantan	Cn 60, Ni 40	20	49	12	0.000 008	Bureau of Standards
		−200	42.4	25	0.000 002	Nicolai
		−150	43.0	100	0.000 033	Nicolai
		−100	43.5	200	0.000 02	Nicolai
		−50	43.9			Nicolai
		0	44.1			Nicolai
		+100	44.6			Nicolai
		400	44.8			Nicolai

Table 12.1 *continued*

	Resistivity			*Temperature coefficient*		*Authority*
Material	*Composition*	*Temperature (°C)*	*Resistivity (ohm-m × 10^{-8})*	*Temperature (°C)*	*Temperature coefficient of resistance per °C*	
Copper, commercial:		20	1.724 1*	20	0.003 93	Bureau of Standards
annealed		20	1.77	20	0.003 82	Bureau of Standards
hard drawn		20	1.692	100	0.003 8	Wolff, Delinger, 1910
pure, annealed		−258.6	0.014	400	0.004 2	Nicolai
		−206.6	0.163	1000	0.006 2	Nicolai
		−150	0.567			Nicolai
		−100	0.904			Nicolai
		+100	2.28			Northrup, 1914
		200	2.96			Northrup, 1914
		500	5.08			Northrup, 1914
		1000	9.42			Northrup, 1914
electrolytic				0	0.004 1	
pure, annealed				0–100	0.004 33	
Copper–manganese	Mn 0.98	0	4.83			Munker, 1912
	Mn 1.49	0	6.66			Munker, 1912
	Mn 4.2	20	17.9			Sebast & Gray, 1916
	Mn 7.4	20	19.7			Sebast & Gray, 1916
	Mn 15	20	50			Klein, 1924
Copper–manganese						
Cu 96.5, Mn 3.5					0.000 22	
Cu 95, Mn 5					0.000 026	
Cu 70, Mn 30					0.000 04	
Copper–manganese–iron	Cu 91, Mn 7.1, Fe 1.9	0	20	0	0.000 12	Blood
	Cu 70.6, Mn 23.2, Fe 6.2	0	77	0	0.000 022	Blood
Copper–manganese–nickel	Cu 73, Mn 24, Ni 3	0	48	0	−0.000 03	Feussner, Lindeck
Copper–nickel						
Cu 60, Ni 40				0	±0.000 02	
Eureka*		0	47	0	+0.000 05	Drysdale, 1907
Evanohm*						
Cr 20, Al 2.5, Cu 2.5, Ni Bal.				−50 to +100	±0.000 02	
Excello				20	0.000 16	
Ferry*				20–100	±0.000 02	
German silver	Cu 60.16, Zn 25.37	−200	27.9			Dewar, Fleming
	Ni 14.03, Fe 0.3	−100	29.3			
	Co and Mn trace	+100	33.1			
German-silver						
Ni 18%				20	0.000 04	
Cu 60, Zn 25, Ni 15				0	0.000 36	
Gold				20	0.003 4	
				100	0.003 5	
				500	0.003 5	
				1000	0.004 9	
Gold–copper–silver						
Au 58.3, Cu 26.5, Ag 15.2				0	0.000 574	
Au 66.5, Cu 15.4, Ag 18.1				0	0.000 529	
Au 7.4, Cu 78.3, Ag 14.3				0	0.001 830	
Gold–silver						
Au 90, Ag 10				0	0.001 2	
Au 67, Ag 33				0	0.000 65	
Graphite‡		0	800			
		500	830			
		1000	870			
		2000	1000			
		2500	1100			

Material	Resistivity: Composition	Resistivity: Temperature (°C)	Resistivity: Resistivity (ohm-m × 10^{-8})	Temperature coefficient: Temperature (°C)	Temperature coefficient: Temperature coefficient of resistance per °C	Authority
Indium				0	0.004 7	
Iridium				0–100	0.004 11	
Iron				20	0.005 0	
				0	0.006 2	
				25	0.005 2	
				100	0.006 8	
				500	0.014 7	
				1000	0.005	
Karma*						
Ni 80, Cr 20				20	0.000 16	
Lead				18	0.004 3	
pure				0–100	0.004 22	
Lithium				0	0.004 7	
				230	0.002 7	
Lohm*						
Ni 6, Cu 94				20–100	0.000 71	
Magnesium				20	0.004	
				0	0.003 8	
				25	0.005 0	
				100	0.004 5	
				500	0.003 6	
				600	0.010 0	
Mancoloy*				200	0.000 5	
Manganese–copper	Mn 30, Cu 70	0	100	0	0.000 040	Feussner, Lindeck
Manganese–nickel						
Mn 2, Ni 98				20–100	0.004 5	
Manganin	Cu 84, Mn 12, Ni 4	20	44	12	0.000 006	Bureau of Standards
		22.5	45	25	0.000 000	Kimura, Sakamaki
		−200	37.8	100	−0.000 042	Nicolai
		−100	38.5	250	−0.000 052	Nicolai
		−50	38.7	475	0.000 000	Nicolai
		0	38.8	500	+0.000 11	Nicolai
		100	38.9			Nicolai
		400	38.3			Nicolai
Mercury				20	0.000 89	
				0	0.000 88	
Midohm*				20–100	0.000 18	
Ni 23, Cu 77				40	0.000 000	
Minalpha*				25	+0.003 3	
Molybdenum				100	0.003 4	
				1000	0.004 8	
Monel-metal*				20	0.002	
Nichrome*				20–500	0.000 17	
Ni 61, Cr 14, Fe 24						
Nichrome V*						
Ni 80, Cr 20				20–500	0.000 13	
Nickel				20	0.006	
				0	0.006	
				25	0.004 3	
				100	0.004 3	
				500	0.003	
				1000	0.003 7	
pure, annealed				0–100	0.006 75	
Nickel–chromium	Ni 80, Cr 20	20	110			Bureau of Standards

Table 12.1 *continued*

Material	*Resistivity*			*Temperature coefficient*		*Authority*
	Composition	*Temperature (°C)*	*Resistivity (ohm-m × 10^{-8})*	*Temperature (°C)*	*Temperature coefficient of resistance per °C*	
Nickel		20	7.8			Bureau of Standards
pure		−182.5	1.44			Fleming, 1900
		−78.2	4.31			Fleming, 1900
		0	6.93			Fleming, 1900
		94.9	11.1			Nicolai, 1907
		400	60.2			
Nickel–copper–zinc	Ni 12.84, Cu 30.59 Zn 6.57 by volume	0	20.3			Matthiessen
Ohmax*				20–500	0.000 066	
Palladium				20	0.003 3	
pure				0–100	0.003 77	
pure				0	0.003 5	
Phosphor–bronze				0	0.004–0.003	
Palladium–gold						
Pd 50, Au 50				0–100	0.000 36	
Palladium–silver						
Pd 60, Ag 40				0–100	0.000 04	
Palladium		20	11			Bureau of Standards
		−183	2.78			Dewar, Fleming
		−78	7.17			Dewar, Fleming
		0	10.21			Dewar, Fleming
		98.5	13.79			Dewar, Fleming
Palladium–copper	Pd 72, Cu 28	20	47			Johansson, Linde
Palladium–gold	Pd 50, Au 50	20	27.5			Sedstrom, Wise
Palladium–silver	Pd 60, Ag 40	20	42			Sedstrom & Svensson
Platinum				20	0.003	
				0	0.003 7	
				0–100	0.003 92	
Platinum–iridium						
Pt 90, Ir 10				0	0.001 2	
Pt 80, Ir 20				0	0.000 8	
Platinum–rhodium						
Pt 90, Rh 10				0	0.001 3	
Platinum–silver						
Pt 33, Ag 67				0	0.000 24	
Platinum–gold						
Pt 40, Au 60				20	0.000 6	
Pt 20, Au 80				20	0.002 5	
Platinum–copper				20	0.000 3	
Pt 75, Cu 25				0	0.005 5	
Potassium liquid				100	0.004 2	
Platinum		20	10			Bureau of Standards
		−203.1	2.44			Dewar, Fleming
		−97.5	6.87			Dewar, Fleming
		0	10.96			Dewar, Fleming
		+100	14.85			Dewar, Fleming
		400	26			Nicolai
		−265	0.10			Nernst
		−253	0.15			Nernst
		−233	0.54			Nernst
		−153	4.18			Nernst
		−73	7.82			Nernst
		0	11.05			Nernst
		+100	14.1			Pirrani
		200	17.9			Pirrani
		400	25.4			Pirrani
		800	40.3			Pirrani

Material	*Resistivity*			*Temperature coefficient*		*Authority*
	Composition	*Temperature (°C)*	*Resistivity (ohm-m × 10⁻⁸)*	*Temperature (°C)*	*Temperature coefficient of resistance per °C*	
		1000	47.0			Pirrani
		1200	52.7			Pirrani
		1400	58.0			Pirrani
		1600	63.0			Pirrani
Platinum–gold	Au 60, Pt 40	20	42.0			Johansson, Linde
	Au 80, Pt 20	20	25.0			Johansson, Linde
Platinum–iridium	Pt 90, Ir 10	0	24			Barnes
	Pt 80, Ir 20	0	31			Barnes
	Pt 65, Ir 35	20	36			Geibel, Carter and Nemilow
Platinum–rhodium	Pt 90, Rh 10	−200	14.40			Dewar, Fleming
		−100	18.05			Dewar, Fleming
		0	21.14			Dewar, Fleming
		+100	24.2			Dewar, Fleming
	Pt 80, Rh 20	20	20			Acken, Nemilow, Voronow and Carter
Platinum–silver	Pt 67, Ag 33	0	24.2			Kurnakow and Nemilow
	Pt 55, Ag 45	20	61			
Platinum–copper	Pt 75, Cu 25	20	92			Sedstron
Rheotan	Cu 53.28, Ni 25.31 Zn 16.80, Fe 4.46 Mn 0.37	0	53	0	0.000 4	Feussner, Lindeck
Rose metal	Bi 49, Pb 28	0	64	0	0.002	
	Sn 23			0	0.006	
Rhodium				0–100	0.004 43	
Rubidium				0	0.006	
Silchrome* Si Cr, Fe				20	0.000 025	
Silicon bronze				0	0.003 8–0.002 3	
				20	0.003 8	
				25	0.003	
				100	0.003 6	
				500	0.004 4	
				0–100	0.004 1	
				9	+0.004 4	
				120	0.003 3	
Silver		−200	0.357			Nicolai
		−100	0.916			Nicolai
		0	1.506			Nicolai
		+100	2.15			Northrup
		+750	6.65			Northrup
pure, annealed						
Sodium liquid						
Steel						
aluminum	Al 5, C 0.2	20	65			Portevin, 1909
	Al 15, C 0.9	20	88			Portevin, 1909
chromium	Cr 13, C 0.7	20	60			Portevin, 1909
	Cr 40, C 0.8	20	71			Portevin, 1909
Invar	35% Ni	20	81			Bureau of Standards
manganese		20	70			Bureau of Standards
nickel	Ni 10, C 0.1	20	29			
	Ni 25, C 0.1	20	39			
	Ni 80, C 0.1	20	82			Portevin, 1909
		20	18			
Siemens–Martin	Si 2.5%	20	45			Bureau of Standards
silicon	Si 4%	20	62			

Table 12.1 *continued*

Material	Resistivity: Composition	Resistivity: Temperature (°C)	Resistivity: Resistivity (ohm-m × 10^{-8})	Temperature coefficient: Temperature (°C)	Temperature coefficient: Temperature coefficient of resistance per °C	Authority
tempered glass-hard			45.7			Stronhal, Barnes
tempered yellow			27			Stronhal, Barnes
tempered blue			20.5			Stronhal, Barnes
tempered soft			15.9			Stronhal, Barnes
titanium	Ti 2.5, C 0.15	20	16			Portevin, 1909
tungsten	W 5, C 0.2	20	20			Portevin, 1909
	W 20, C 0.2	20	24			Portevin, 1909
vanadium	V 5, C 1.1	20	121			Portevin, 1909
Steel						
Invar						
Ni 36, C 0.2				0	0.002	
piano wire				0	0.003 2	
Siemens-Martin				20	0.003	
silicon						
Si 4				20	0.000 8	
tempered glass-hard				0	0.001 6	
tempered blue				0	0.003 3	
Tantalum						
				20	0.003 1	
				0–100	0.003 47	
Thalium				0	0.004	
Therlo*				20	0.000 01	
Thorium				20–1800	0.002 1	
Tin				20	0.004 2	
Tungsten				18	0.004 5	
				500	0.005 7	
				1000	0.008 9	
pure, annealed				0–100	0.004 65	
Tellurium‡		19.6	200 000			Matthiessen
Tin–bismuth	Sn 90, Bi 9.5	12	16			
	Sn 2, Bi 98	0	244			
Vacrom*				0–500	0.000 06	
Ni 80, Cr 20						
Wood's metal*				0	0.002	
Zinc				20	0.003 7	
				0	0.004	
				0–100	0.004 15	

* Trade name.
† N. B. Polycrystalline; the resistivity of a single-crystal in the plane of the hexagonal network is about 60×10^{-8} ohm-m at 20°C.
‡ Resistivity is greatly dependent on purity of the specimen.

12.3 Fixed resistors

12.3.1 Colour codes for discrete fixed resistors

Bands or rings of colour are usually placed round the resistor, as shown in *Figure 12.2.* The colours of the first three rings determine the total value of the resistance—the first ring ***A*** determines the first digit, the second ring ***B*** determines the second digit, and the third ring ***C*** determines the number of noughts. In some resistors the body is colour coded for the first digit, the tip for the second digit and a colour dot on the body indicates the number of noughts. The colour code used universally is given in *Table 12.2.*

A fourth ring ***D***, or end colour, is usually added to denote the tolerance on value, e.g. a gold ring indicates a 5% tolerance, a silver ring a 10% tolerance and the absence of a fourth colour indicates a 20% tolerance. For other tolerance values the normal colour code may be used, e.g. a brown ring would indicate a 1% tolerance. Cracked-carbon resistors and precision wirewound resistors are often required to an accuracy of three digits and then a number of cyphers. The colour coding system is not easily applied to these types of resistor, and so they are not normally colour coded by some manufacturers.

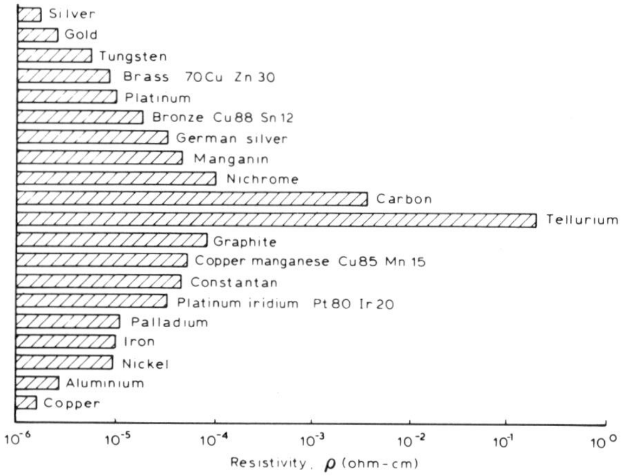

Figure 12.1 Resistivities of common metals and alloys

12.3.2 General characteristics of discrete fixed resistors

A summary of electrical characteristics is given in *Table 12.3.*

Fixed resistors are generally placed in one of two categories—high stability or general purpose. High-stability types include the pyrolytic or cracked-carbon resistors, the wirewound resistors and the metal or metal oxide-film resistors, all of which are capable of providing stable resistance to within 1 or 2%. General purpose types are usually of carbon composition and are cheaper and usually smaller. They are not so stable and resistance variations on load may be from 5 to 20%.

In choosing a resistor for a particular application, a knowledge of some of the following characteristics will be needed.

12.3.2.1 Size

In general, carbon resistors dissipate less power than wirewound resistors of the same resistance value; they are also smaller. The maximum resistance of a wirewound resistor is limited by the length of wire of a given material and the diameter that can be wound upon the available former length. Subminiature cracked-carbon high-stability resistors are made that are comparable in size with the carbon-composition type. Metal-oxide-film resistors are usually larger for high resistance values because of their lower ohms/square.

12.3.2.2 Power-handling capacity

Composition resistors are commonly available for dissipating up to about 2 W, but rarely over 5 W (except in special resistors of low ohmic values), although sintered types can dissipate high powers. Cracked-carbon resistors are available up to 2 W at normal temperatures. All these ratings have to be reduced when the resistors are used at high ambient temperatures. Small metal-film resistors are made up to 2-W dissipation and small oxide-film resistors, up to 6-W dissipation. Wirewound resistors are invariably used when higher powers are to be dissipated, and some vitreous-enamelled wirewound types will handle powers as great as 300–400 W. It is important to remember that the temperatures reached by the resistors when dissipating these wattages can be very high—of the order of several hundred degrees Celsius. Large oxide-film and metal-film resistors are made to dissipate several hundred watts.

There is a 'critical value' of resistance for each wattage rating given by the equation,

$$R = V^2/W$$

In pulse operation (particularly when the duty cycle is low) only the mean power is effective in raising the internal temperature of a resistor. As the power is supplied in short pulses, very high peak ratings are possible, but the mean power should not exceed the continuous rating wattage. Peak pulse voltages for high-stability (cracked-carbon) resistors should be limited to twice the normal rated d.c. voltage; otherwise the limit set by internal sparking or external corona might be exceeded, whereas for general-purpose carbon-composition resistors, the maximum peak pulse voltage should be no greater than the maximum continuous rating.

12.3.2.3 Stability

Stability and accuracy are often confused. Stability is the change in resistance under shelf life or working conditions; accuracy is the tolerance to which the value of the resistor is made or selected. For general purposes, the carbon-composition resistor has been used for many years and is therefore known to have an acceptable long-term stability for domestic and many commercial purposes. Changes in resistance under normal working conditions may be of the order of 5%, but in more severe conditions, such as those encountered in military services, changes of up to 25% may occur. It is found that changes caused by high temperature (due either to ambient or self-generated heat) result in a permanent increase in

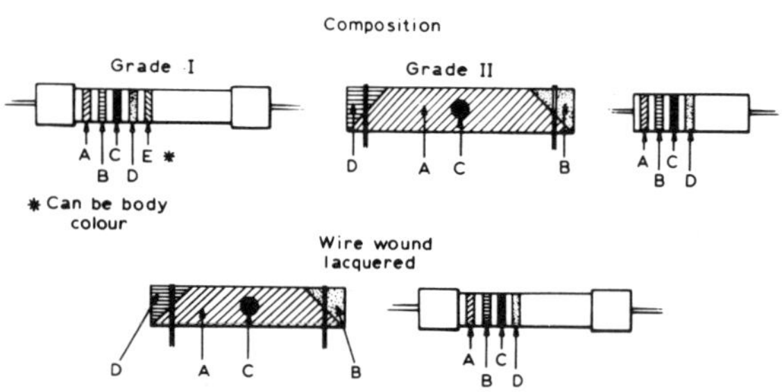

Figure 12.2 British colour coding for fixed resistors

Table 12.2 The standard colour code

Colour	*A 1st digit*	*B 2nd digit*	*C Multiplying factor*	*D Selection tolerance*	*E* Stability (Comp. only)*
Brown	1	1	10	±1%	—
Red	2	2	100	±2%	—
Orange	3	3	1000	—	—
Yellow	4	4	10 000	—	—
Green	5	5	100 000	—	—
Blue	6	6	—	—	—
Violet	7	7	—	—	—
Grey	8	8	—	—	—
White	9	9			
Black	—	0	1	—	—
Gold	—	—	0.1	±5%	—
Silver	—	—	0.01	±10%	—
None	—	—	—	±20%	—
Salmon pink	—	—	—	—	Grade I (High Stability)

* Can be body colour.

Table 12.3 Summary of the electrical characteristics of fixed resistors

Resistor type	*Overall stability (after climatic tests) (%)*	*Mfg. accuracy (%)*	*Best selection accuracy (%)*	*Max. noise (μV/V)*	*Temperature coefficient (ppm/°C)*	*Voltage coefficient (%/V)*	*Max. resistor temp. (°C)*
Moulded carbon composition (insulated and uninsulated)	25	20	5	2.0 (for low values) to 6.0 (for high values)	±1200 ±1200	1 M−0.025 1 M−0.05	115
Carbon composition film type (insulated)	25	20	5	2.0 (for low values) to 6.0 (for high values)	±1200	1 M−0.025 1 M−0.05	115
Cracked carbon (insulated and uninsulated)	2	1	0.25	0.03 (for low values) to 0.5 (for high values)	−200 (for low values) to −1000 (for high values)	0.005	150
Wirewound (general-purpose type)	1	1	0.1	None	±200	None	320
Wirewound (precision type)	0.01 (if hermetically sealed)	0.01	0.01	None	(Ni/Cr)+70 (Cu/Ni)+20	None	70
Metal film (Ni/Cr)	1	1	0.2	Up to 0.3	+220	None	150
Oxide film (Sn/Sb)	2	1	0.5	Up to 0.5	−500 to +500		300

resistance value, whereas exposure to high humidity increases the resistance, but the effect is largely reversible.

The stability of wirewound and cracked-carbon resistors is much higher—of the order of 1 to 2%. Metal-film resistors are comparable in stability to the wirewound types. This stability is dependent mainly on the protection afforded to the resistive element by sealing. Even under the severe conditions encountered in the services, changes are not usually more than 2%. Oxide films have a stability rather better than that of the cracked-carbon film. The highest stability with lowest temperature coefficient is still obtained with wirewound precision resistors, although some of the evaporated metal-film resistors approach this but do not yet equal it.

12.3.2.4 Accuracy (or tolerance)

Carbon-composition resistors are made to approximate target values and then selected to various values after manufacture. Selection tolerances are set up and resistors sorted to ±5%, ±10% and ±20% of the nominal batch value. Carbon-composition resistors cannot be regarded as accurate to better than 5%, because of the lack of precise control in their composition and because of a tendency to drift in value. Pyrolytic or cracked-carbon resistors are usually accurate to 1 or 2% but can be manufactured to about 0.1% if necessary. Wirewound resistors are accurate to 0.25% and can be manufactured to 0.05% or even 0.01% if desired.

If resistors are used to the lowest manufacturing accuracy, the cost will be the minimum. If a greater accuracy is required, it can be provided up to the limits given in *Table 12.3*. It follows, however, that the resistors will be more expensive, not only by reason of the work involved in selection but by the possibility that the resistors not within the required accuracy may be less readily saleable.

12.3.2.5 Maximum operating temperature

Carbon-composition resistors are seriously affected by ambient temperatures over 100°C, mainly by changes in the structure of the binder used in the resistor mixture. The maximum recommended surface temperature is about 110°C to 115°C. This is the total working temperature produced by the power dissipated inside the resistance, the heat from associated components, and the ambient temperature in which the resistor is operating. Cracked-carbon resistors can be operated up to a maximum surface temperature of 150°C under the same conditions, metal films up to 175°C, and oxide films up to 200°C to 250°C, approximately. Some special metal and metal-oxide-film power resistors can operate at 500°C to 600°C when no limiting protective coating is applied.

Wirewound resistors are generally lacquered or vitreous enamelled for protection of the windings. For both types the safe upper limit is set by the protective coating. For lacquered types the maximum recommended temperature is 130°C (some will work up to 450°C).

Free circulation of air should be allowed, and the ends of tubular resistors should not be placed flat against the chassis. If the resistors are badly mounted, or if several resistors are placed together, derating is necessary.

12.3.2.6 Maximum operating voltage

The maximum operating voltage is determined mainly by the physical shape of the resistor and by the resistance value (which determines the maximum current through the resistor and therefore the voltage for a given wattage), that is, the 'critical value' referred to previously.

Commercial ratings at room temperature are some 25% to 50% higher than military ratings, and reference should always be made to the resistor manufacturer for his maximum voltage rating.

12.3.2.7 Frequency range

On a.c. carbon-composition resistors (up to about 10 kΩ in value) behave as pure resistors up to frequencies of several MHz. At higher frequencies the self-capacitance of the resistor becomes predominant, and the impedance falls. The inductance of carbon-composition resistors does not usually cause trouble below 100 MHz (except in special cases such as in attenuator resistors). Cracked-carbon resistors specially manufactured with little or no spiral grinding can be operated at frequencies of many hundreds of MHz, but methods of mounting and connection become important at these frequencies. Other film-type resistors such as metal film and metal-oxide film are also suitable for use at high frequencies, and the effect of spiralling of the film is relatively unimportant below 50 MHz.

For wirewound resistors, the inductance of single-layer windings becomes appreciable, and *Ayrton–Perry* or *back-to-back* windings are often used for so-called non-inductive resistors. At high frequencies the capacitive rather than the inductive effect limits the frequency of operation. For example, the reactance of a typical resistor of 6 kΩ with an Ayrton–Perry winding becomes capacitance at 3 MHz.

In all measurements on resistors at high frequencies, the method of mounting the resistor is important. The direct end-to-end capacitance of the resistor and the capacitance of the two leads to the resistor body are included in the total capacitance being measured, and the resistor should therefore be mounted as near as possible as it is to be mounted in use. Ideally the mounting fixtures should be standardised for comparison measurements.

To summarise, for a resistor to be suitable for operation at high frequencies it should meet the following general requirements:

(*a*) Its dimensions should be as small as possible.
(*b*) It should be low in value.
(*c*) It should be of the film type.
(*d*) A long thin resistor has a better frequency characteristic than a short fat one.
(*e*) All connections to the resistor should be made as short as possible.
(*f*) There should be no sudden geometrical discontinuity along its length.

12.3.2.8 Noise

Carbon-composition resistors generate noise of two types: thermal agitation, or *Johnson* noise, which is common to all resistive impedances, and *current* noise, which is caused by internal changes in the resistor when current is flowing through it. The latter is peculiar to the carbon-composition resistor and other non-metallic films and does not occur in good quality wire-wound resistors. Cracked-carbon resistors generate noise in a similar fashion to the carbon-composition types but at a very much lower level. For low values of resistance (where the film is thick) the noise is difficult to measure. Metal-film and metal-oxide-film resistors generate noise at a very low level indeed.

Measurements have shown that for carbon-composition resistors, current noise increases linearly with current up to about 15 μA. With greater currents the noise curve approximates to a parabola.

12.3.2.9 Temperature coefficient of resistance

A resistor measured at 70°C will have a different resistance value from that at 20°C; the change in value at differing temperatures can be calculated from the temperature coefficient for each class of resistor. Approximate maximum values are given in *Table 12.3.*

The large values for carbon-composition types are partly due to non-cyclic changes, which tend to mask temperature coefficient effects.

12.3.2.10 Voltage coefficient

When a voltage is applied across a carbon resistor, there is an immediate change in resistance, usually a decrease. The change, which is not strictly proportional to the voltage, is usually measured at values not less than 100 kΩ. In carbon-composition resistors, the change in resistance value due to the applied voltage is usually within 0.02%/V d.c. With carcked-carbon resistors, particularly the larger sizes, the effect is negligible for low values of resistance, certainly being less than 0.001%. On the higher values it can rise to 0.002%, and on very small resistors, where the stress is clearly much greater, the maximum values may approach 0.005%/V. The voltage coefficient is frequently quoted at too high a figure because of the difficulty in separating it from effects due to temperature coefficient. Wirewound resistors do not show this effect, provided they are free from leakage between turns. Metal-film resistors have voltage-coefficients from 0.000 1%/V to 0.003%/V depending on wattage, whereas metal-oxide-film resistors approximate from 0.000 1%/V to 0.000 5%/V.

12.3.2.11 Solderability

There is a change in the value of carbon-composition resistors, and to a smaller extent in cracked-carbon resistors, when they are soldered into equipment. This change, which is due to overheating, can be quite serious in miniature constructions if the connecting leads are short, and permanent changes of up to 25% may be caused. If the soldered joint is made 12 mm away from the resistor, there is usually no excessive overheating.

12.3.2.12 Shelf life

There is a change in the resistance of most types of resistor during storage. During one year the resistance of a carbon-composition resistor may change by 5%, while a cracked-carbon or wirewound resistor may change by only 0.5%. Metal-film resistors change by as little as 0.1% or less, oxide-film resistors change by less than 0.5%.

12.3.2.13 Load life (or working life)

Resistors are also tested for their change in resistance after 1000 h at a temperature of 70°C. Under these conditions the resistance of cracked-carbon resistors may vary from 0.1% (for low values) to 3% (for high values). Wirewound and oxide-film resistors do not change in value by more than 1%, but carbon-composition resistors may change by as much as 15%.

12.4 Thermistors

The name thermistor is an acronym for THERMally-sensitive resISTOR. This name is applied to resistors made from semiconductor materials whose resistance value depends on the temperature of the material. This temperature is determined by both the ambient temperature in which the thermistor is operating and the temperature rise caused by the power dissipation within the thermistor itself. The thermistors first developed had a negative temperature coefficient of resistance, that is the resistance value decreases as the temperature increases. This is in contrast to the behaviour of most metals and therefore of wirewound and metal-film resistors whose resistance increases with an increase of temperature. Subsequently thermistors with a positive temperature coefficient of resistance were developed, with the advantage over normal resistors of having a larger temperature coefficient. Thermistors are used in applications as temperature sensors or as stabilising elements to compensate the effects of temperature changes in a circuit.

12.4.1 Thermistor materials

Negative temperature coefficient (n.t.c.) thermistors are manufactured from the oxides of such materials as iron, chromium, manganese, cobalt and nickel. In the pure state, these oxides have a high resistivity. They can be changed into semiconductor materials, however, by the addition of small quantities of a metal with a different valency. Examples of the materials used in the manufacture of commercially available n.t.c. thermistors are given below.

One group of n.t.c. thermistors is manufactured from ferric oxide Fe_2O_3 in which a small number of trivalent ferric Fe^{3+} ions are replaced by tetravalent titanium Ti^{4+} ions. The titanium ions are compensated by an equal number of bivalent ferrous Fe^{2+} ions to maintain the electrical neutrality of the material. At low temperatures, the extra electrons of the Fe^{2+} ions are situated on iron ions next to the Ti^{4+} ions. At higher temperatures, however, the electrons are loosened from these sites and become free charge carriers. Hence the conductivity of the material increases; that is, the resistance decreases with an increase of temperature. Because electrons act as the charge carriers in the material, this oxide mixture is an n-type semiconductor material.

Another group of n.t.c. thermistors is made from nickel oxide NiO or cobalt oxide CoO. A small number of bivalent nickel Ni^{2+} ions or cobalt Co^{2+} ions are replaced by univalent lithium Li^{1+} ions, and these are compensated by an equal number of trivalent nickel Ni^{3+} or cobalt Co^{3+} ions. At low temperatures, the deficiencies of electrons (holes) of the Ni^{3+} or Co^{3+} ions are situated on the Li^{1+} ions, but at higher temperatures the holes are free to move through the crystal as charge carriers. Again there will be a decrease in resistance as the temperature is increased. These oxide mixtures are p-type semiconductor materials.

Stabilising oxides may be added to both n-type and p-type materials to obtain better reproducibility and stability of the characteristics of the manufactured thermistor. The choice of material depends on the required resistance value and temperature coefficient for the particular thermistor.

Positive temperature coefficient (p.t.c.) thermistors are made from barium titanate $BaTiO_3$ or a solid solution of barium titanate and strontium titanate, $BaTiO_3$ and $SrTiO_3$. A semiconductor material is formed by substituting ions of higher valency for either the barium or the titanium ions. Trivalent ions such as lanthanum La^{3+} or bismuth Bi^{3+} are used to replace a small number of bivalent barium Ba^{2+} ions, or pentavalent ions such as antimony Sb^{5+} or niobium Nb^{5+} are used to replace a small number of titanium Ti^{4+} ions. When the barium titanate mixture is prepared in the absence of oxygen, an n-type semiconductor material with a low negative temperature coefficient is obtained. A positive temperature coefficient over a particular temperature range is obtained by heating the material in an oxygen atmosphere. As the mixture cools, oxygen atoms penetrate along the crystal boundaries. The absorbed oxygen atoms attract electrons from a thin layer of the semiconductor crystals along the boundaries. An electrical potential barrier is formed in this zone, consisting of a negative space charge on both sides of which a positive space charge is formed by the now uncompensated added ions. These barriers cause an increase in resistance with an increase in temperature to give the positive coefficient of resistance for the thermistor.

Another material used for p.t.c. thermistors is silicon. In extrinsic silicon, the conductivity is determined by the density and mobility of the charge carriers. The conductivity σ is given by

$$\sigma = Ne\mu \tag{12.3}$$

where N is the density of the charge carriers (electrons or holes depending on whether the silicon is n-type or p-type) and is determined by the doping level; e is the magnitude of the electronic charge; and μ is the mobility of the charge carrier. Thus the variation of conductivity with temperature will be determined by the variation with temperature of the charge carrier density and mobility. The density will increase with temperature because of the thermally generated electron-hole pairs. The mobility of both electrons and holes decreases with increasing temperature because of the increased effectiveness of lattice scattering. At the lower temperatures, the effect of the mobility is greater than that of the thermal generation, and so the conductivity falls with increasing temperature (a positive temperature coefficient is obtained). At higher temperatures, however, the increase in carrier density through thermal generation predominates, and the conductivity increases. The variation of conductivity over a temperature range −50°C to +250°C with doping level as parameter has the form shown in *Figure 12.3*. It can be seen that by choosing a suitable doping level, the required positive temperature coefficient can be obtained over various temperature ranges. Because the minimum value of conductivity occurs approximately 20–30°C higher for n-type material than for p-type material, thermistors made from extrinsic silicon generally use n-type material.

It should be remembered that the semiconductor materials

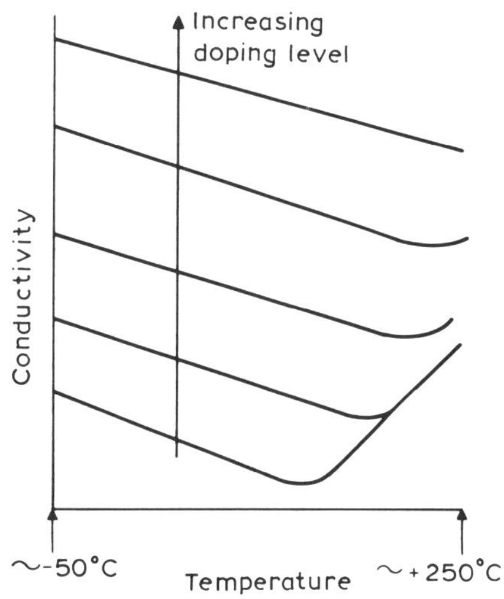

Figure 12.3 Variation of conductivity with temperature for extrinsic silicon with doping level as parameter

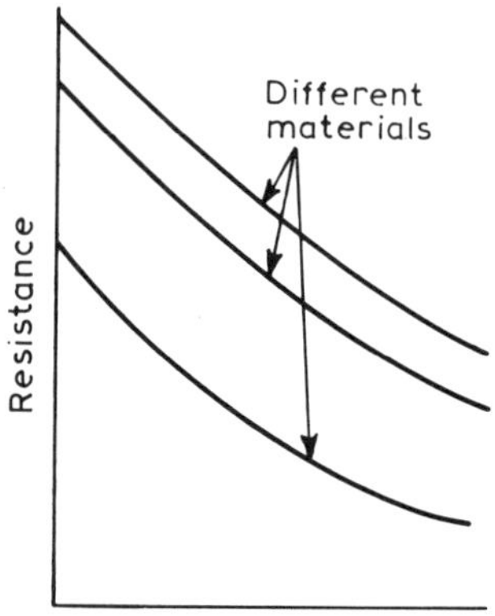

Figure 12.4 Resistance/temperature characteristic for ntc thermistor

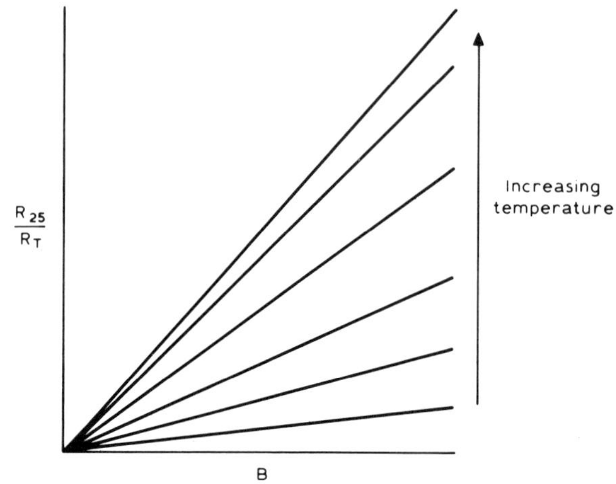

Figure 12.5 Ratio of R_{25}/R_T plotted against B value with temperature as parameter

from which thermistors are manufactured are polycrystalline and not monocrystalline like those used for transistors and diodes.

12.4.2 Electrical characteristics of n.t.c. thermistors

The variation of resistance with temperature for various types of n.t.c. thermistor has the form shown in *Figure 12.4*. The relationship between the resistance and temperature can be expressed as

$$R_T = A\,e^{B/T} \tag{12.4}$$

where R_T is the resistance in ohms at an absolute temperature T in kelvin; e is the base of natural logarithms (2.718); and A and B are constants.

The value of B for a particular thermistor material can be found by measuring the resistance at two values of absolute temperature T_1 and T_2 and using equation (12.4):

$$B = 2.303\left(\frac{T_1T_2}{T_2-T_1}\right)(\log_{10} R_{T_1} - \log_{10} R_{T_2}) \tag{12.5}$$

When B is calculated from equation (12.5), it is found that in practice it is not a true constant but slight deviations occur at high temperatures. More exact expressions for the variation of resistance with temperature have been suggested to replace equation (12.4). These include $R_T = AT^C e^{B/T}$ where C is a small positive or negative constant which may sometimes be zero, or $R_T = A\,e^{B/(T+K)}$, where K is a constant.

The value of B for practical thermistor materials lies between 2000 and 5500; the unit is the kelvin.

Equation (12.5) can be rearranged in terms of R_{T_1} to give

$$\log_{10} R_{T_1} = \log_{10} R_{T_2} + B\left(\frac{T_2-T_1}{T_1T_2}\right)\log_{10} e \tag{12.6}$$

If the resistance of the thermistor at temperature T_2 is known, and the value of B for the thermistor material is known, the resistance value at any temperature in the working range can be calculated from equation (12.6).

Curves relating the ratio of R_T, the resistance at temperature T, and R_{25}, the resistance at 25°C which is taken as a *standard* value for comparing different types of thermistor, to the value of B with temperature as parameter are sometimes given in published data. From these curves, shown in *Figure 12.5*, and the known value of R_{25} and B for the thermistor type given in the data, the value of R_T can be determined.

A temperature coefficient of resistance can be derived for the thermistor material. This coefficient α is obtained by differentiating equation (12.4) with respect to temperature:

$$\alpha = \frac{1}{R}\frac{dR}{dT} = -\frac{B}{T^2} \tag{12.7}$$

At 25°C, the value of α for practical thermistor materials lies, typically, between 2.5 and 7%. It can be seen from equation (12.7) that the temperature coefficient varies inversely as the square of the absolute temperature.

The voltage/current characteristic for an n.t.c. thermistor is shown in *Figure 12.6*. The characteristic relates the current

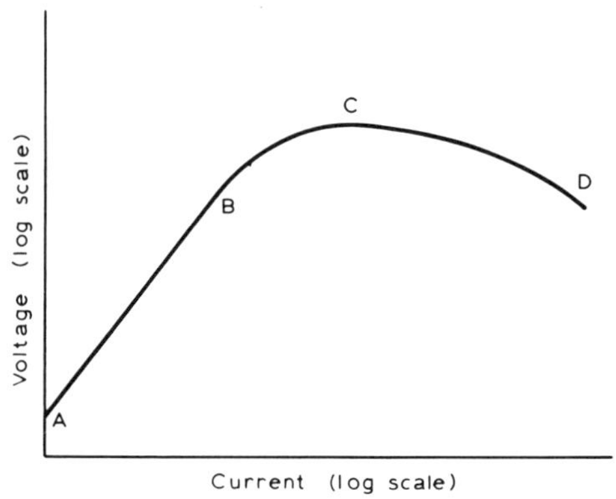

Figure 12.6 Voltage/current characteristic for ntc thermistor

through the thermistor with the voltage drop across it after thermal equilibrium has been established in a constant ambient temperature. This static characteristic is plotted on logarithmic scales.

Over the low-current part of the characteristic, part A–B, the power input to the thermistor is too low to cause any rise in temperature by internal heating. The thermistor therefore acts as a linear resistor. Above point B, however, the current causes sufficient power dissipation within the thermistor to produce a rise in temperature and therefore a fall in resistance. The resistance value is therefore lower than would be expected for a linear resistor. As the current is increased further, the power dissipation within the thermistor causes a progressively larger fall in resistance, so that at some current value the voltage drop across the thermistor reaches a maximum, point C. Above this current value, the fall in resistance caused by the internal heating is large enough to give the thermistor a negative incremental resistance, part C–D.

The temperature corresponding to the maximum voltage across the thermistor can be calculated for a particular thermistor material. If it is assumed that the temperature is constant throughout the body of the thermistor, and that the heat transfer is proportional to the temperature difference between the thermistor and its surroundings (which is true for low temperature differences), then equation (12.4) can be rewritten in terms of natural logarithms as

$$\log_e R_T = \log_e A + \frac{B}{T} \qquad (12.8)$$

At thermal equilibrium, the electrical power input to the thermistor is equal to the heat dissipated; that is

$$VI = D(T - T_{amb}) \qquad (12.9)$$

where V is the voltage drop across the thermistor, I is the current through it, T is the temperature of the thermistor body, T_{amb} is the ambient temperature, and D is the dissipation constant (the power required for unit temperature rise). The dissipation constant may also be represented by the symbol δ.

Substituting $R_T = V/I$ in equation (12.8), taking logarithms of equation (12.9) and adding gives

$$\log_e V = \frac{1}{2}\log_e AD + \frac{1}{2}\log_e (T - T_{amb}) + \frac{B}{2T} \qquad (12.10)$$

At the maximum voltage point, equation (12.10) will be a maximum. This gives

$$T_{V(max)} = \frac{B}{2} \pm \left(\frac{B^2}{4} - BT_{amb}\right)^{1/2} \qquad (12.11)$$

The value of $T_{V(max)}$ corresponding to the maximum voltage across the thermistor is

$$T_{V(max)} = \frac{B}{2} - \left(\frac{B^2}{4} - BT_{amb}\right)^{1/2} \qquad (12.12)$$

It can be seen from equation (12.11) that a solution is possible only if $B > 4T_{amb}$. Also, the temperature corresponding to the maximum voltage is determined by the **B** value only and not by the resistance value. For practical thermistor materials, $T_{V(max)}$ lies between 45°C and 85°C.

In many applications, it is necessary to know the time taken for the thermistor to reach equilibrium. Assuming again that the temperature is constant throughout the body of the thermistor, the cooling in a time dt is given by

$$-H\,dT = D(T - T_{amb})\,dt \qquad (12.13)$$

where H is the thermal capacity of the thermistor in joules/°C. When the thermistor cools from temperature T_1 to T_2, equation (12.13) gives

$$(T_1 - T_{amb}) = (T_2 - T_{amb})\,e^{-t/\tau} \qquad (12.14)$$

where τ is the thermal time-constant of the thermistor and is equal to H/D.

In practice, the temperature is not constant throughout the body of the thermistor, the surface cooling more rapidly than the interior. The time-constant quoted in published data is defined as the time required by the thermistor to change by 63.2% of the total change between the initial and final body temperatures when subjected to an instantaneous temperature change under zero power conditions. Cooling curves showing the rise in resistance with time when the electrical input is removed from the thermistor are often given in published data.

12.4.3 Electrical characteristics of p.t.c. thermistors

The variation of resistance with temperature for a p.t.c. thermistor is more complex than that for an n.t.c. thermistor shown in *Figure 12.4*. Because the temperature coefficient of resistance is positive only over part of the temperature range, a typical resistance/temperature characteristic for a p.t.c. thermistor may have the form shown in *Figure 12.7*. Over parts

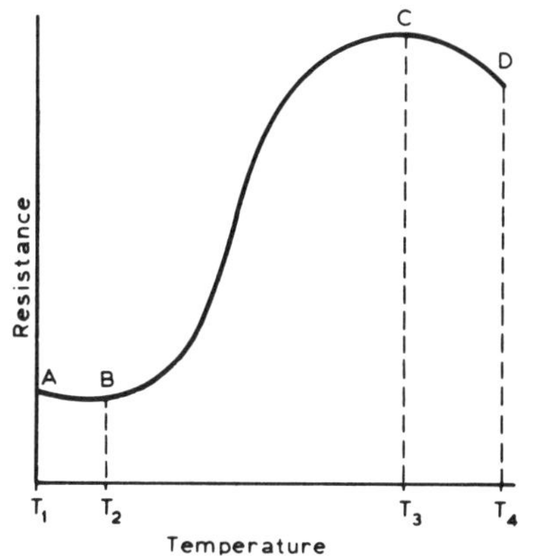

Figure 12.7 Resistance/temperature characteristic for ptc thermistor

A–B (temperature T_1 to T_2) and C–D (temperature T_3 to T_4) the coefficient is negative; it is only over part B–C, corresponding to the temperature range T_2 to T_3, that the required positive coefficient is obtained. Over this range, the relationship between resistance and temperature can be expressed (approximately) as

$$R_T = A + C\,e^{BT} \qquad (12.15)$$

where R_T is the resistance in ohms at an absolute temperature T in kelvin, e is the base of natural logarithms, A, B and C are constants, and T is restricted in value to $T_2 < T < T_3$.

Equation (12.15) can be differentiated to give the temperature coefficient of resistance α:

$$\alpha = \frac{1}{R}\frac{dR}{dT} = \frac{BC\,e^{BT}}{A + C\,e^{BT}}$$

In practice, unfortunately, the resistance/temperature characteristic can seldom be described by such a simple relationship as that of equation (12.15). Unlike n.t.c. thermistors where equation (12.4) provides a reasonable approximation to the practical behaviour, any attempt to modify equation (12.15) to a more accurate form results in complicated expressions. For this reason, graphical methods are often used for design calculations. A quantity called the switch temperature T_s is often quoted in published data. This temperature is the one at which a p.t.c. thermistor begins to have a usable positive temperature coefficient of resistance. T_s is defined as the higher of two temperatures at which the resistance of the thermistor is twice its minimum value.

The voltage/current characteristic for a p.t.c. thermistor is shown in *Figures 12.8* and *12.9*. In both figures the voltage and current axes are interchanged with respect to those of the characteristic for an n.t.c. thermistor shown in *Figure 12.6*. The characteristic shown in *Figure 12.8* is plotted on linear scales, and characteristics for different ambient temperatures are included to show the effect on the thermistor. When the voltage across the thermistor is low, the power dissipation within the thermistor is insufficient to heat it above the ambient temperature. The

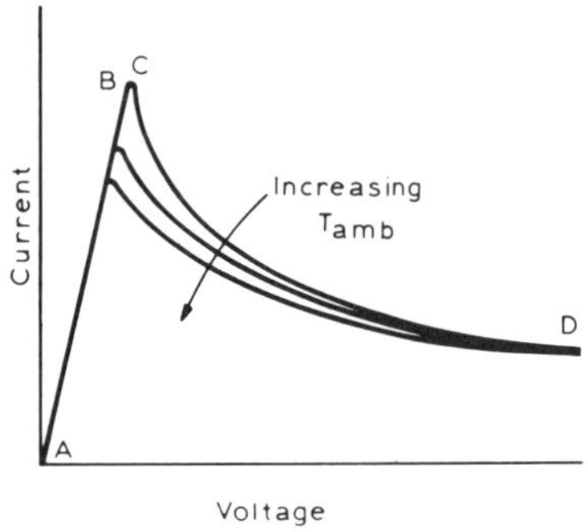

Figure 12.8 Voltage/current characteristic for ptc thermistor with ambient temperature as parameter (linear scales)

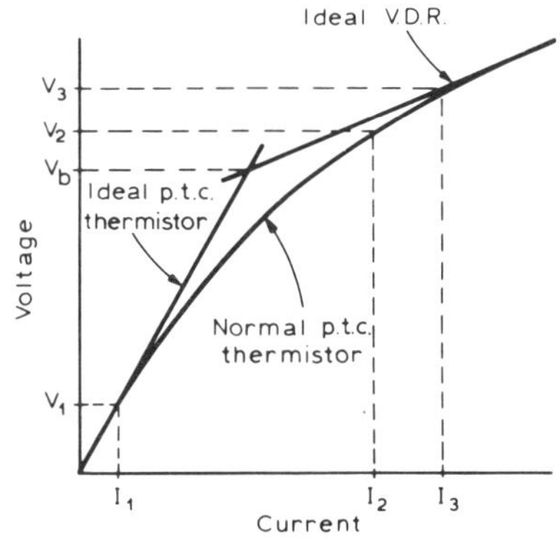

Figure 12.10 Voltage/current characteristic of ptc thermistor compared with *ideal* components of equivalent circuit

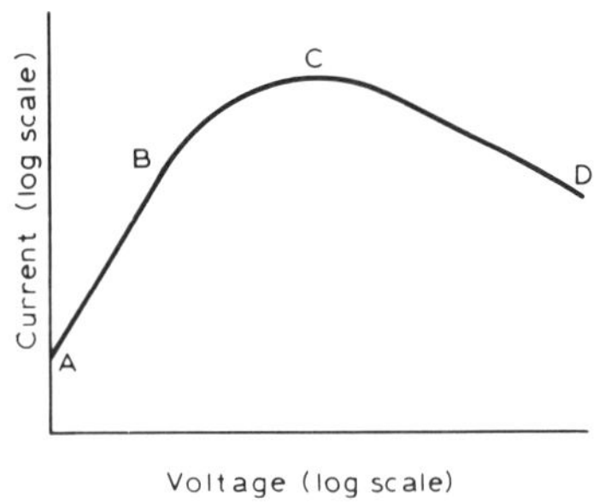

Figure 12.9 Voltage/current characteristic for ptc thermistor (log scales)

thermistor behaves as a linear resistor, part A–B. When the voltage is increased, the power dissipation causes the thermistor temperature to rise above the switch temperature T_s, point C. The resistance of the thermistor rises, and the current falls. Any further increase in voltage results in a progressive fall in current, part C–D. The characteristic for a constant ambient temperature plotted on logarithmic scales is shown in *Figure 12.9*.

At higher voltages, p.t.c. thermistors show a voltage dependency, the resistance value being determined by the voltage across the thermistor as well as by the temperature. The behaviour of the thermistor under these conditions can be represented by an equivalent circuit consisting of an ideal p.t.c. thermistor (no voltage dependency) in parallel with an ideal voltage dependent resistor (the voltage and current being related by the expression: $V \propto I^{\beta}$). The voltage/current characteristics of the components of this equivalent circuit are shown in *Figure 12.10* compared with the characteristic of a normal p.t.c. thermistor. The *normal* characteristic is measured at constant ambient temperature under pulse conditions to avoid self-heating of the thermistor. At low voltages, the characteristic of the normal p.t.c. thermistor coincides with the characteristic of the ideal thermistor. At the higher voltages where the voltage dependency becomes effective, the characteristic of the normal p.t.c. thermistor coincides with the characteristic of the ideal thermistor. At the higher voltages where the voltage dependency becomes effective, the characteristic of the normal thermistor coincides with that of the voltage dependent resistor. The point of intersection of the two 'ideal' characteristics, where the currents through the two components of the equivalent circuit are equal, defines the balance voltage V_b. The value of balance voltage for a specified ambient temperature is given in published data. The voltage dependency of the thermistor β can be calculated from the expression

$$\beta = \frac{\log V_3 - \log V_2}{\log(I_3R - V_3) - \log(I_2R - V_2)} \quad (12.16)$$

where V_2 is a pulse voltage greater than V_b, V_3 is a pulse voltage greater than V_2; I_2 and I_3 are the currents corresponding to V_2 and V_3 respectively; and R is the initial slope of the characteristic, given by V_1/I_1. As the value of β depends on temperature, when quoted in published data the value is qualified by the relevant temperature.

As with n.t.c. thermistors, a thermal time-constant τ is given in published data for p.t.c. thermistors, defined in the same way as previously described. Cooling curves showing the fall in resistance with time after the electrical input is removed from the thermistor are also given.

12.4.4 Choice of thermistor for application

When n.t.c. and p.t.c. thermistors are used in circuits, the designer must consider certain requirements before he can select the correct thermistor for his purpose. These requirements include the resistance value, temperature coefficient of resistance and temperature range; the power dissipation required of the thermistor; and the thermal time-constant.

Most of these factors have been discussed in the previous sections describing the characteristics of n.t.c. and p.t.c. thermistors. In the published data a resistance value, usually at 25°C, is given, supplemented by resistance values at other temperatures, a temperature coefficient of resistance (α or B value), and resistance/temperature curves for various ambient temperatures. Sometimes an operating temperature range or a maximum operating temperature is given. For p.t.c. thermistors, the switch temperature is given. The voltage/current characteristic of the thermistor sometimes has resistance and power axes superimposed, as shown in *Figure 12.11*. The balance voltage V_b and voltage dependency β at a specified temperature are also given for p.t.c. thermistors. A maximum power dissipation may be specified, or a maximum voltage or current. The dissipation factor δ is often given in two forms: for still air assuming cooling by natural convection and radiation, and when mounted on a heatsink to increase the cooling. A thermal time-constant τ is given, supplemented by cooling curves.

From these data, the designer is able to choose a suitable thermistor. For some applications, however, a single thermistor may not meet the requirements, and it is possible to combine a thermistor with a series/parallel combination of linear resistors

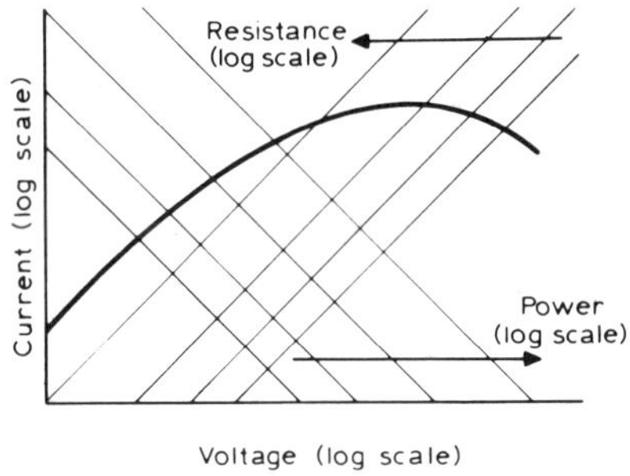

Figure 12.11 Voltage/current characteristic of thermistor with resistance and power axes superimposed

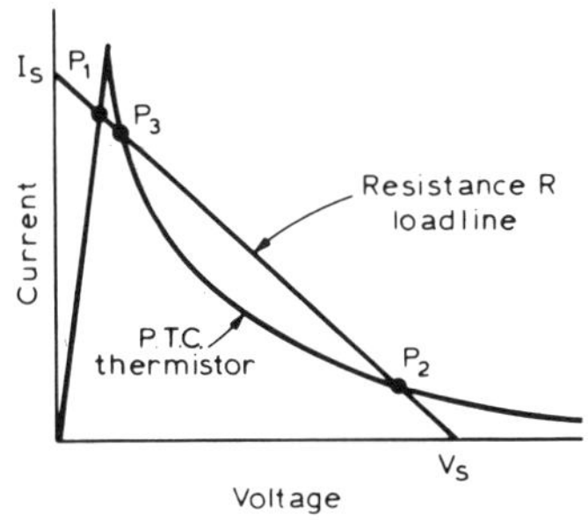

Figure 12.12 Voltage/current characteristic of ptc thermistor with resistive loadlines superimposed

so that the overall characteristic corresponds to that required. For p.t.c. thermistors, it should be noted that with a series resistance three working points are possible. In *Figure 12.12*, the voltage/current characteristic of a p.t.c. thermistor plotted on linear scales is shown with a resistive loadline superimposed. The resistive loadline intersects the voltage axis at the supply voltage V_s, and the current axis at I_s where I_s is given by V_s/R. Of the three working points given by the intersection of the characteristic and loadline, points P_1 and P_2 are stable, while P_3 is an unstable working point. When the supply voltage is first applied, equilibrium will be established at working point P_1 at which the load current is relatively high. Working point P_2 can only be reached if the supply voltage increases, the ambient temperature increases, or the resistance in series with the thermistor decreases. In each case, the displacement of the loadline or the change in the thermistor characteristic with temperature must be large enough to allow the peak of the thermistor characteristic to lie under the loadline.

12.4.5 Manufacture of thermistors

Thermistors are manufactured by sintering the various oxide mixtures previously described in Section 12.4.1. The generally used shapes for unencapsulated thermistors are rods, discs, plates, and beads. Some types are encapsulated to protect the thermistor element, either in plastic or glass.

The oxide mixture used is finely ground and a plastic binder material such as polyvinyl acetate added. Rod thermistors are formed by extrusion through dies of the required diameter, and cutting to length; discs and plates are formed by pressing. The formed element is heated to drive out the binder material, and then sintered at a temperature between 1000°C and 1350°C. Electrical contacts are generally made with silver, gold, or platinum paste which is cured before the lead-out wires are attached. Typical dimensions of rod-type thermistors are diameters of 1–6 mm, and lengths of 5–50 mm. The diameters of disc-type thermistors can be from 1.5 to 15 mm, with thicknesses between 1 and 6 mm. Plate thermistors have a rectangular outline of a comparable size to disc thermistors. The element of this type of thermistor is generally painted to provide some protection and to colour-code the device.

Bead thermistors are made by placing small masses of a slurry formed from the oxide mixture on two closely-spaced parallel platinum wires. Surface tension draws the masses into a bead shape. The beads are allowed to dry, and then gently heated until strong enough to be sintered. Sintering shrinks the beads firmly on to the platinum wires so that good electrical contact is made. The wires can then be cut to form the lead-out wires for the thermistor, and the element enamelled for protection. A typical diameter for bead thermistors is 1 mm.

Thermistors are encapsulated to provide protection when used in dirty or corrosive atmospheres or liquids. Small glass envelopes are used for bead thermistors. An alternative form of encapsulation uses plastic. This enables the thermistor element to be in good thermal contact with the environment and so have a short response time, and at the same time be electrically insulated.

Typical thermistor shapes and encapsulations are shown in *Figure 12.13* and the circuit symbol for a thermistor is shown in *Figure 12.14*. The $t°$ indicates temperature dependency, an n.t.c. thermistor being indicated by $-t°$ and a p.t.c. thermistor by $+t°$. The older symbols for n.t.c. and p.t.c. thermistors which may still be encountered are shown in *Figure 12.15*.

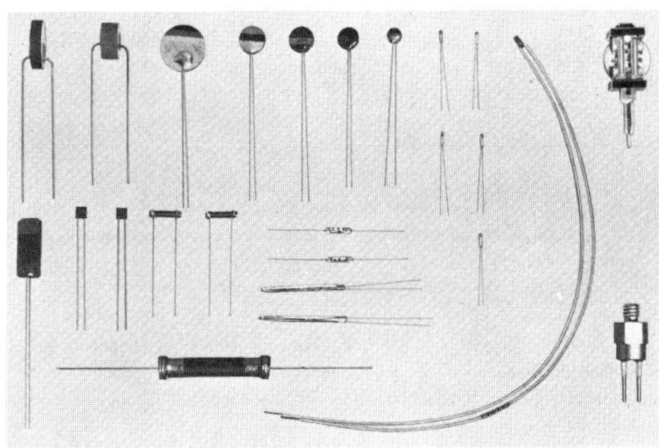

Figure 12.13 Typical thermistors (courtesy Mullard Limited)

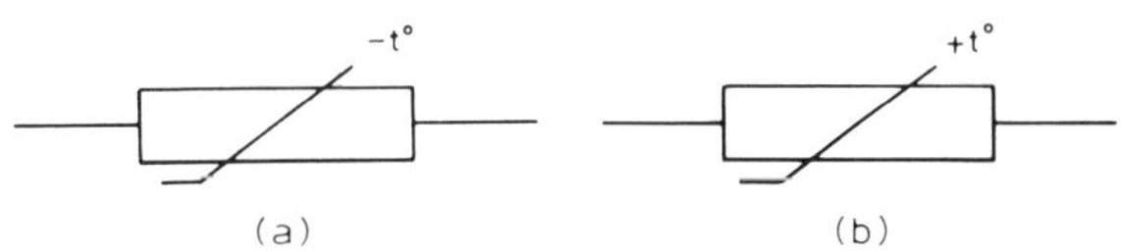

Figure 12.14 Circuit symbol for (a) ntc thermistor and (b) ptc thermistor

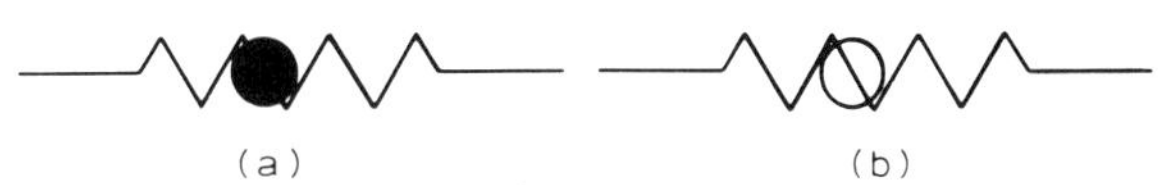

Figure 12.15 Old circuit symbols for (a) ntc thermistor (b) ptc thermistor

12.4.6 Applications of thermistors

It is convenient to classify the applications of thermistors into four groups:

(*a*) Applications where the resistance is determined by the ambient temperature.
(*b*) Applications where the resistance is determined by the power dissipated within the thermistor.
(*c*) Applications making use of the positive or negative temperature coefficient.
(*d*) Applications using the thermal inertia of the thermistor.

The choice between an n.t.c. or p.t.c. thermistor will be determined mainly by the application.

Both n.t.c. and p.t.c. thermistors can be used in temperature measurement and control applications. The thermistor is operated at a low power level so that self-heating is avoided. A bridge circuit with the thermistor forming one arm is generally used, and the out-of-balance current produced by the change of resistance with temperature is used as an indication of temperature or to operate a control circuit. Liquid and gas flows can be measured by the loss of heat from a thermistor heated internally to above ambient temperature. The change in resistance as a result of the different rates of cooling as the flow conducts heat away from the thermistor can be used to measure the rate of flow. Similarly, the different rates of cooling of a thermistor in liquid and air can be used as a level indicator in liquid storage tanks.

Both n.t.c. and p.t.c. thermistors can be used in protection circuits. An n.t.c. thermistor can be connected in series with a switch so that switching surges can be limited to a safe value. When the switch is closed, the initial high resistance of the thermistor will limit the current. The fall in resistance as the thermistor heats up will produce a gradual increase in current to the working value. Current limit circuits can be constructed with p.t.c. thermistors. At the normal working current, the resistance of the thermistor is low. With an overload current, however, the increase in resistance can be used to limit the current directly if the thermistor can withstand the increased dissipation, or the increase in resistance can operate additional protection circuits.

Another group of applications uses the temperature coefficient of the thermistor to compensate temperature changes caused by other elements in a circuit.

The change in resistance as a thermistor heats up can be used in delay circuits. Connecting an n.t.c. thermistor in series with, for example, a relay can delay the energising of the relay until the resistance of the thermistor has fallen sufficiently to allow the operating current to flow. Similarly the increase in resistance of a p.t.c. thermistor can be used to de-energise a relay after an initial energising.

12.5 Voltage-sensitive resistors

These are formed by dry-pressing silicon carbide with a ceramic binder into discs or rods and firing at about 1200°C. The ends of the rods, or the sides of the discs, are sprayed with metal (usually brass) to which connections are soldered. They are often known as voltage dependent resistors, the current through the resistor being given by

$$I = KE^n$$

where K is a constant equal to the current in amperes at $E = 1$ V; and n is a constant dependent on voltage, varying between 3 and 7 for common mixes. It is usually between 4.0 and 5.0.

12.5.1 Applications of voltage-sensitive resistors

Suppressing voltage surges and quenching contact sparks

The resistor is connected across inductive loads and prevents voltage surges. The space required by the resistor is small and the current normally passing through it is quite small. A similar application is in avoiding sparking of relay contacts although the current permissible is limited to about 0.2 A per unit.

12.5.1.1 Protection of smoothing capacitors

In the anode circuit of a valve a smoothing capacitor is often connected from the anode to earth to prevent coupling. After switching on it requires some time for the cathode of the valve to reach a high enough temperature to allow the correct anode current to flow and during this time the capacitor is subjected to full input voltage and it must be rated accordingly. A much 'lighter' type of capacitor can be used if it is shunted by a voltage-sensitive resistor which will limit the voltage across the capacitor.

12.5.1.2 Voltage stabilisation

A resistor is used either directly across a varying power source or in a bridge circuit with linear resistors, but in both cases the loss of energy is somewhat excessive.

Other uses suggested include lightning arrestors, shunting of rectifiers, arc furnaces, thyratrons, armatures, etc.

12.6 Variable resistors (potentiometers)

12.6.1 Types of variable resistor

There are two general classes of variable resistor: general purpose and precision. The general-purpose resistors may be sub-divided into wirewound and carbon-composition types. The precision resistors, which are always wirewound, usually follow linear, sine–cosine or other mathematical laws. Linearities as high as 0.01% (for linear) and 0.1% (for sine–cosine and other laws) are obtainable. The general-purpose types usually follow a linear law, but some follow a logarithmic law. They have overall resistance tolerances of 10% for the wirewound types (although much closer tolerances can be obtained) and 20% for the carbon-composition types.

The metal-film and the high-quality molded-track types can also be considered to be in the precision category. Linearities of 0.5% in the molded type and 0.1% in metal-film types (with the aid of trimming) are obtainable.

12.6.2 Electrical characteristics of variable resistors

A summary of the electrical characteristics of most types is given in *Table 12.4*.

12.6.2.1 Resistance value

For precision-variable resistors the upper limit of resistance value is about 100 kΩ; above this the element size may exceed 150 mm in diameter. General-purpose types are made in values up to 500 kΩ (wirewound) and 5 MΩ (carbon). The lower limit is about 1 Ω for wirewound resistors and about 10 Ω for carbon-composition types.

12.6.2.2 Resistance law

The resistance law is the law relating the change of resistance to the movement of the wiper, and it may be linear, logarithmic, log-

Table 12.4 Summary of electrical characteristics of variable resistors

	Mfg. tolerance (%)	*Selection tolerance (%)*	*Overall stability (after climatic tests) (%)*	*Linearity (%)*	*Resolution (in degrees)*	*Life (number of sweeps)*
Carbon composition. Coated-track types	±20	±20	±20	±15 (at 50% rotation)	(Stepless)	20 000 minimum. Max. depends on construction
Carbon construction. Moulded solid-track types	±20	±20	±5	±15 (at 50% rotation)	(Stepless)	20 000 minimum. Max. depends on construction
Cermet (ceramic/metal)	±20	±20	±20	±15 (at 50% rotation)	(Stepless)	20 000 minimum.
Conductive plastic	±20	±20	±10	±15 (at 50% rotation)	(Stepless)	Up to 25 million
Rotary wirewound general-purpose types	±10	±10	±2	1.0	1.0	20 000 minimum. Max. depends on construction
Wirewound precision linear types	±5	Not applicable	High if sealed	Average 0.5 (can be 0.01)	Average 0.1 (depends on size, wire, etc.)	50 000 minimum. Max. several millions
Wirewound precision toroidal types	±5	Not applicable	High if sealed	Average 0.1 (can be higher)	Average (depends on size, wire, etc.)	50 000 minimum. Max. several millions
Wirewound precision helical types	±5	Not applicable	High if sealed	Average 0.25 (can be higher)	0.01 (for 10-turn pot.) (depends on wire, etc.)	50 000 minimum. Max. several millions
Sine–cosine potentiometer. Card-wound types	±5	Not applicable	High if sealed	Average 0.5 (can be higher)	Varies with slider position	50 000 minimum. Max. several millions

log, sine–cosine, secant and the like, depending on the requirement for which the variable resistor is designed.

12.6.2.3 Linearity

There is often confusion between the terms *linearity*, *resolution*, *discrimination* and *accuracy* in discussing variable resistors. An ideal linear variable resistor has a constant resistance change for each equal increment in angular rotation (or linear movement) of the slider. In practice, this relationship is never achieved, and the linearity, or linear accuracy, is the amount by which the actual resistance at any point on the winding varies from the expected straight line of a 'resistance *vs* rotation' graph in a rotary variable resistor or 'resistance *vs* movement' graph in a linear variable resistor. For example, a 1 kΩ variable resistor held to a linearity of ±0.1% would not vary more than 1 Ω on either side of the line of zero error.

The terms *resolution* and *discrimination* are synonymous. Resolution, or discrimination, is the resistance per turn of resistance wire and is thus a function of the number of turns on the variable resistor. For example, a resistor of 100 Ω containing 100 turns of wire has a resolution of 1 Ω. Resolution may be defined more accurately as *resistance resolution*, *voltage resolution* or *angular resolution*. Resistance resolution is the resistance per turn; voltage resolution is the voltage per turn; and angular resolution is the minimum change in slider angle necessary to produce a change in resistance. In general, the resistance resolution is one-half of the linearity. For example, if the linearity (linear accuracy) of a 1 kΩ resistor is to be held to within 0.1%, the resistance resolution should be 0.05% or less, and the winding should have at least 2000 turns. The word *accuracy* unqualified, has no meaning in defining a variable resistor.

12.6.2.4 Stability

Stability concerns the change of resistance with time, or under severe climatic conditions, as well as the behaviour under normal load conditions. For general-purpose carbon-composition variable resistors the stability tolerance is 25% and for general-purpose wirewound resistors it is 2%. The stability of the precision types of wirewound variable resistor is much higher, as these are usually sealed to exclude moisture and dust.

12.6.2.5 Minimum effective resistance

All variable resistors have some method of *ending off* the resistance element so that the slider goes into a *dead* position at each end, although it may rotate a few degrees more. There is a small jump in resistance, known as the *hop-off* resistance, as the slider touches the element. For general-purpose wirewound types, this should be less than 3% of the nominal resistance, and for carbon composition types it should be less than 5%.

12.6.2.6 Effective angle of rotation

The *dead* positions mentioned in the previous paragraph are known as the *hop-off* angles. For military use the hop-off angle must not exceed 10% of the total angular rotation at either end for general-purpose wirewound resistors and 30% for carbon-composition types. The effective angle of rotation is 360° less the sum of the hop-off angles at the ends and the space allowed for terminations.

12.6.2.7 Life under given conditions

Service specifications require that both wirewound and carbon-composition types should withstand 10 000 sweeps at 30 cycles/min with full-load current through the resistance element, totalling 20 000 cycles. After the test the change in resistance should not be more than 2% for the wirewound types and 5% for the composition types. Precision-variable resistors designed for long life—for example, with low brush pressure and carrying little current—have much longer lives—up to two million sweeps and sometimes up to ten million.

12.6.2.8 Performing under various climatic conditions

The most frequent causes of failure in variable resistors are corrosion of the metal parts and swelling and distortion of plastic parts such as track moldings, cases and the like, due to moisture penetration. To combat these problems, the variable resistor should embody metal parts made from non-corroding metals, which may be difficult to fabricate, or be sealed in a container with a rotating seal for the spindle. The wattage rating is sometimes lowered slightly because of the sealing, but the life of the component is increased by many times. Some present types are made with solid-molded carbon-composition tracks and bases that resist the effects of humidity.

12.6.2.9 Performance under vibration

In variable resistors difficulties may be experienced due to open circuit or intermittent contact if the slider vibrates off the track or due to change of resistance if the slider moves along the track. In general, the second is much more serious, particularly if the vibration occurs sideways to the potentiometer. Resonant frequencies vary between 100 and 300 Hz for the small 45-mm-diameter wirewound potentiometer and at amplitudes of 5–10 g. The shaft-length and knob-weight also affect the resonant frequency. Reduction in shaft length to 6 mm may raise the resonant frequency to 1000 Hz or more, and the knob should be as light and as small as possible.

12.6.2.10 Noise

Electrical noise in carbon-composition variable resistors is usually due to poor or intermittent contact between the slider and the track. Variations of pressure, or the presence of dust or metal particles, cause changes in contact resistance, resulting in noise. Sealing or at least dust-proofing is necessary to avoid trouble due to dust contact variation. In wirewound variable resistors there are several types of noise contact resistance or constriction resistance noise, loading noise, resolution noise and vibrational noise due to slip-rings (if these are used).

Further reading

BROWN, J. A., 'Metal oxide film resistors', *Electronic Equipment News*, 10 (August 1970)

CHURCH, H. F., 'The long-term stability of fixed resistors', *IRE Proc. on Compt Parts*, 31 (1961)

DUMMER, G. W. A., *Fixed Resistors*, 2nd edn, Pitman, London (1967)

DUMMER, G. W. A., *Variable Resistors*, 2nd edn, Pitman, London (1963)

DUMMER, G. W. A., *Materials for Conductive and Resistive Functions*, Hayden Book Co., New York (1970)

DUMMER, G. W. A. and BURKETT, R. H. W., 'Recent developments in fixed and variable resistors', *Proc. IEE, Part B, No. 21, Supplement* (1962)

KARP, H. R., 'Trimmers take a turn for the better', *Electronics*, 79 (17 Jan. 1972)

NEALE, L., 'Manufacturing and testing precision potentiometers', *Electronics & Power*, 497 (27 June 1974)

PRICE, J. R., 'Improvements in the design of precision wirewound resistors', *Electronic Engineering*, 10 (July 1964)

THOMAS PEART, L., 'Picking the right potentiometer', *Electronics*, 78 (14 June 1965)

RAGAN, R., 'Power rating calculations for variable resistors', *Electronics*, 129 (19 July 1973)

WELLAND, C. L., *Resistance and Resistors*, McGraw-Hill, New York (1960)

13

Dielectric Materials and Components

G W A Dummer MBE, CEng, FIEE, FIEEE, FIERE
Electronics Consultant

Contents

13.1 Characteristics of dielectric materials

13.1.1 General characteristics

Dielectric materials used for radio and electronic capacitors can be grouped into the following five main classes:

(1) Mica, glass, low-loss ceramic, etc.: used for capacitors from a few pF to a few hundred pF.
(2) High-permittivity ceramic: used for capacitors from a few hundred pF to a few tens of thousands of pF.
(3) Paper and metallised paper: used for capacitors from a few thousand pF up to some μF.
(4) Electrolytic (oxide film): used for capacitors from a few μF to many μF.
(5) Dielectrics such as polystyrene, polythene, polythylene terephthalate, polycarbonate, etc.: range of use from a few hundred pF to many μF.

Many factors affect the dielectric properties of a material when it is used in a capacitor; among them being the permittivity, power factor, leakage current, dielectric absorption, dielectric strength, operating temperature, etc.

13.1.2 Summary of properties of capacitor dielectrics

A table of the main characteristics of some dielectric materials used in capacitors is given in *Table 13.1.*

13.1.3 Permittivity (dielectric constant)

The permittivity, dielectric constant or specific inductive capacity of any material used as a dielectric is equal to the ratio of the capacitance of a capacitor using the material as a dielectric, to the capacitance of the same capacitor using vacuum as a dielectric. The permittivity of dry air is approximately equal to one. A capacitor with solid or liquid dielectric of higher permittivity (ε) than air or vacuum can therefore store ε times as much energy for equal voltage applied across the capacitor plates. A few typical figures for capacitor dielectrics are:

	Permittivity (ε)
Vacuum	1.0
Dry air	1.000 59
Polythene, polystyrene, etc.	2.0 to 3.0
Impregnated paper	4.0 to 6.0
Glass and mica	4.0 to 7.0
Ceramic (magnesium titanate, etc.)	Up to 20
Ceramic (titania)	80 to 100
Ceramic (high-ε)* (or high-K)	1000 upwards

Dielectrics can be classified in two main groups—polar and non-polar materials. Polar materials have a permanent unbalance in the electric charges within the molecular structure. The dipoles within the structure consist of molecules whose ends are oppositely charged. These dipoles therefore tend to align themselves in the presence of an alternating electric field (if the frequency is not too high). The resultant oscillation causes a large loss at certain frequencies and at certain temperatures.

13.1.4 Losses in dielectrics

Losses occur due to current leakage,dielectric absorption, etc., depending on the frequency of operation. For a good non-polar dielectric the curve relating loss with frequency takes the approximate shape given in *Figure 13.1(a)*. For a polar material the loss-frequency curve may be shown approximately as in *Figure 13.1(b)*.

The variation of permittivity with frequency is negligible so long as the loss is low. Increased losses occur when the process of alignment cannot be completed, owing to molecular collisions, and in these regions there is a fall in permittivity. Viscous drag in the molecular structure limits the frequency at which full alignment can be carried out. If the applied frequency is comparable with the limiting frequency losses still become high.

Equivalent circuits showing series and parallel loss resistance can be given, but are greatly dependent on the system of measurement at any particular frequency. The important criterion is the ratio:

$$\frac{\text{power wasted per cycle}}{\text{power stored per cycle}}$$

This is the power factor of the material and for good dielectrics it is independent of frequency.

13.1.5 Absorption

If a capacitor were completely free from dielectric absorption the initial charging or polarisation current when connected to a d.c. supply would be

$$I=(V/R)e^{-t/CR}$$

where

I = current flowing after a time t
V = applied voltage
R = capacitor series resistance
C = capacitance
e = base of Napierian logs (2.718)

and the polarisation current would die off asymptotically to zero. If R is small, this takes place in a very short time and the capacitor is completely charged. In all solid-dielectric capacitors it is found that, after a fully-charged capacitor is momentarily discharged and left open-circuited for some time, a new charge accumulates

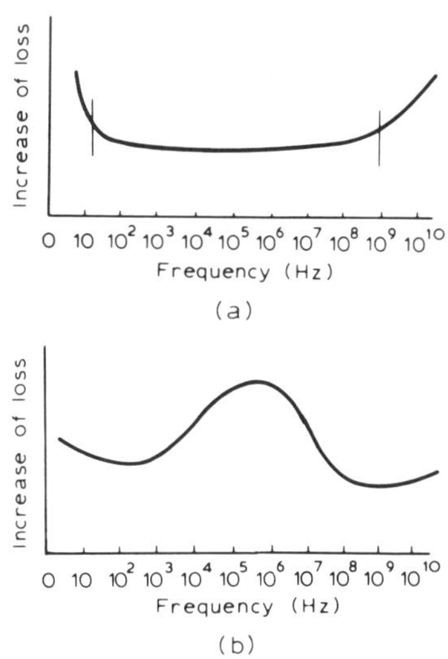

Figure 13.1 Loss/frequency curve for (a) non-polar dielectric (b) polar dielectric

* The high permittivity in high-ε ceramic capacitors comes from the fact that the electric charges in the molecular structure of the material are very loosely bound and can move almost freely under the polarising voltage, resulting in high total capacitance.

Table 13.1 Properties of some capacitor dielectric materials

Material	Loss (at room temperature): Limiting frequency of operation, Approx. min.	Loss (at room temperature): Limiting frequency of operation, Approx. max.	Power factor (Loss angle tan δ) at 1 kHz	Permittivity (over operating frequency range)	Dielectric strength (Volts per mil) (Breakdown)	Temperature limits (°C): Approx. min.	Temperature limits (°C): Approx. max.	Remarks (and some (registered) Trade Names)
1	2	3	4	5	6	7	8	9
Kraft paper (capacitor tissue)	Poor below 100 Hz (can be d.c.)	1 MHz	0.01 to 0.03	4.5	500 to 1000 (depends on impregnant)	No limit but impregnant freezes	85 to 100	Capacitor properties depend greatly on impregnant
with mineral oil	—	—	0.003 5	2.23	—	−55	105	Effect of various impregnants
with castor oil	—	—	0.007	4.7	—	−25	65	
with silicone oil	—	—	0.003 5	2.6	—	−60	125	
with polyisobutylene	—	—	0.003	2.2	—	−55	125	
with chloronaphthalene	—	—	0.005	5.2	—	−20	55	
with diphenyl	—	—	0.003	4.9	—	−55	85	
Mica (Ruby)	Precision work 100 Hz (can be d.c.)	10 000 MHz	0.000 5	7.0	1000	No limit	+200	Properties vary according to source of origin
Ceramic (Low-permittivity types) Magnesium silicate	100 Hz*	10 000 MHz	0.001	5.4 to 7.0	200 to 300	No limit	+150 to 200	Steatite, Temperadex, Frequentex, etc.
(Medium-permittivity types) Titania Rutile (TiO_2)	500 Hz*	5000 MHz	0.001	70 to 90	100 to 150	No limit	+120	Faradex, Condensa, (Has large temperature coefficient, approx. −750 ppm/°C)†
(High-permittivity types) Titanate	1000 Hz*	1000 MHz	0.01	Approx. 1000 to over 7000	100	−100	+120	Characteristics dependent on temperature. Sharp max. in permittivity at 120°C. Permittivity depends on volts. Below 120°C hysteresis occurs and loss dependent on volts. Other high-permittivity materials exist with different peaks

Table 13.1—*contd.*

Material	*Loss (at room temperature)* — *Limiting frequency of operation* — *Approx. min.*	*Approx. max.*	*Power factor (Loss angle tan δ) at 1 kHz*	*Permittivity (over operating frequency range)*	*Dielectric strength (Volts per mil) (Breakdown)*	*Temperature limits (°C)* — *Approx. min.*	*Approx. max.*	*Remarks (and some (registered) Trade Names)*
1	*2*	*3*	*4*	*5*	*6*	*7*	*8*	*9*
Glass (Soft lead–soda)	200 Hz	10 000 MHz	0.001	6.5 to 6.8	<500	No limit	+200	
Glass (Hard, borosilicate)	100 Hz	10 000 MHz	0.001	4.0	<500	No limit	+200	
Quartz (Fused)	100 Hz	>10 000 MHz	0.000 2	3.8	1000	No limit	+300	
Plastics								
Polyethylene	d.c.	10 000 MHz	0.000 2	2.3	1000	Becomes brittle −40	+70	Polythene, Alkathene, Telcothene
Polystyrene film	d.c.	10 000 MHz	0.000 3	2.3	1000	−40	+65	Distrene, Lustron, Styron, Trolitul
Polycarbonate	d.c.	10 000 MHz	0.000 1	2.8	1000	−40	+70	
Polymonochlorotrifluoro-ethylene (PCTFE) film	d.c.	10 000 MHz	0.01 to 0.05	2.3 to 2.8	3000 to 5000	−195	+180	Kel-F, Hostaflon
Polyethylene terephthalate film	Poor between, say, 0.1 to 10 MHz		0.01 at l.f. 0.001 at h.f.	3.1	4000 to 5000	−40	+130	Terylene or Melinex. Could replace paper at l.f.
Paraffin wax	d.c.	10 000 MHz	0.000 1	2.2	1000	No limit but hardens	60	Capacitor impregnant, coil potting, etc.
Liquid insulants								
Paraffin	d.c.	10 000 MHz	0.001	2.2	1000	No limit but solidifies	50	Ozokerite (slightly plastic). Used for coil potting
Transformer oil	d.c.	1 MHz	0.001	2.4	1000	−40	120	
Silicone oil	d.c.	10 000 MHz	0.000 1	2.8	1000	−50	250	
Nitro-benzene	(see remarks)	10 000 MHz	0.000 5	40	50 to 1000	Freezes at +3 then low permittivity	150	Note high permittivity. Cannot be used on d.c. because of leakage. Has large negative temperature coefficient. Has specialised use as capacitor dielectric

* Can be d.c. but slow capacitance change possible.

† ppm/°C = parts per million per degree Celsius.

within the capacitor, because some of the original charge has been 'absorbed' by the dielectric. This produces the effect known as dielectric absorption. A time lag is thus introduced in the rate of charging and of discharging the capacitor which reduces the capacitance as the frequency is increased and also causes unwanted time delays in pulse circuits.

3.1.6 Leakage currents and time constants of capacitors

Losses due to leakage currents when a capacitor is being used on d.c. prevent indefinite storage capacity being realised, and the charge acquired will leak away once the source is removed. The time in which the charge leaks away to 1/e, or 36.8%, of its initial value is given by RC, where R is the leakage resistance and C is the capacitance. If R is measured in megohms and C in microfarads the time constant is in seconds. This can also be expressed as megohm-microfarads or as ohm-farads. Some typical time constants for various dielectrics used in capacitors are

Polystyrene	several days
Impregnated paper	several hours
Tantalum-pellet electrolytic capacitors	one or two hours
High-permittivity capacitors (ceramic)	several minutes
Plain-foil electrolytic capacitors	several seconds

It should be borne in mind that below capacitance values of about 0.1 μF, the time constant is generally determined by the structure, leakage paths, etc., of the capacitor assembly itself rather than the dielectric material. Leakage current increases with increase of temperature (roughly exponentially). In good dielectrics at room temperature it is too small to measure, but at higher temperatures the current may become appreciable, even in good dielectrics.

13.1.7 Insulation resistance

The insulation resistance of a dielectric material may be measured in terms of surface resistivity in ohms or megohms, or as volume resistivity in ohm-centimetres. A method of measurement is given in BS 9070:1970.

13.1.8 Dielectric strength

The ultimate dielectric strength of a material is determined by the voltage at which it breaks down. The stress in kilovolts per metre (or volts per mm), at which this occurs depends on the thickness of the material, the temperature, the frequency and the waveform of the testing voltage, and the method of application, etc., and therefore comparisons between different materials should ideally be made on specimens equal in thickness and under identical conditions of measurement. The ultimate dielectric strength is measured by applying increasing voltage through electrodes to a specimen with recessed surfaces (to ensure that the region of maximum stress shall be as uniform as possible). Preparation of the specimens is important and their previous histories should be known.

The dielectric strength of a material is always reduced when it is operated at high temperatures or if moisture is present. Few materials are completely homogenous and breakdown may take the form of current leak along certain small paths through the material; these become heated and cause rapid deterioration, or flashover along the surface and permanent carbonisation of the surface of organic materials. Inorganic materials such as glass, ceramic and mica are usually resistant to this form of breakdown. The time for which the voltage is applied is important; most dielectrics will withstand a much higher voltage for brief periods. With increasing frequency the dielectric strength is reduced, particularly at radio frequencies, depending on the power factor, etc., of the material.

13.1.9 Effect of frequency

At very low frequencies, also at very high frequencies, there is an increase of loss which sets a limit to the practical use of a capacitor with any given dielectric. At very low frequencies various forms of leakage in the dielectric material have time to become apparent, such as d.c. leakage currents and long time-constant effects, which have no effect at high frequencies. At very high frequencies some of the processes contributing to dielectric polarisation do not have time to become effective and therefore cause loss. These losses might be simply and approximately represented as in *Figure 13.2*.

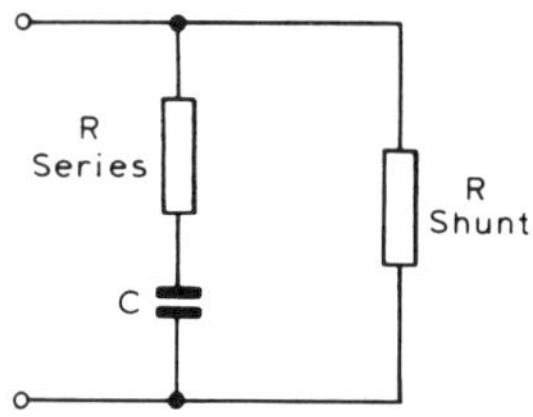

Figure 13.2 Losses in a capacitor

At very low frequencies the circuit is entirely resistive, all the current passing through the shunt resistance (d.c. leakage resistance, etc.). At very high frequencies the current passes through the capacitance C but all the volts are dropped across the series resistance and again the circuit is lossy. This series resistance may be due to the resistance of the capacitor leads, the silvering (in the case of silvered mica or ceramic capacitors), contact resistances, etc., in the capacitor assembly itself. These limit the upper frequency independently of the dielectric material used. Similarly, leakage across the case containing the dielectric may limit the lower frequency so that not all the useful range of the dielectric itself may be realised.

The chart in *Figure 13.3* shows the approximate usable frequency ranges for capacitors with various dielectric materials. The construction of the capacitor assembly will affect the frequency coverage to some extent, so that the chart should be regarded as a guide only.

13.1.10 The impedance of a capacitor

The current (amperes) in a capacitor when an alternating voltage is applied is given by

$$I = 2\pi f C V$$

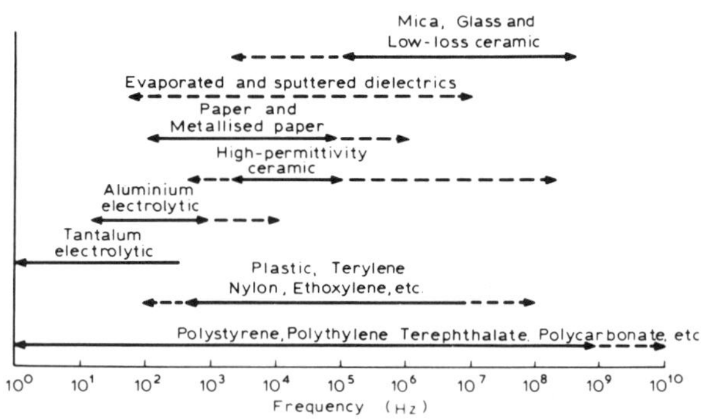

Figure 13.3 Frequency coverage of different classes of capacitor

where

C = capacitance in farads
V = voltage in volts
f = frequency in hertz

and the reactance (ohms) is given by

$X_c = -1/2\pi fC$

An ideal capacitor would have entirely negative reactance but the losses, described previously, due to dielectric, case and leads, preclude this. In addition, inductance is also present in varying amounts and therefore as the frequency is increased the inductive or positive reactance increases, and above a critical frequency the capacitor will behave as an inductor. At the resonant frequency the impedance of the capacitor is controlled by its effective resistance, which in turn is made up of the losses described. Every capacitor will resonate at some given frequency (depending on its construction) and, having inductance and resistance, will exhibit a complex impedance, capacitive in one range of frequencies, resistive in another and inductive in still another.

13.1.11 Insulation resistance of capacitors (or insulance)

The insulation resistance of the assembled capacitor is important in circuit use. The insulation resistance of any capacitor will be lowered in the presence of high humidity (unless it is sealed), and will be reduced when operated in high ambient temperatures (whether sealed or not).

For perfectly-sealed capacitors used under conditions of high humidity there should be no deterioration, but for imperfectly-sealed capacitors the drop in insulation resistance will be roughly inversely proportional to the effectiveness of the sealing. Unsealed capacitors will show a large and rapid drop in insulation resistance under these conditions.

Under high-temperature conditions the fall in insulation resistance for most capacitors is given approximately by the formula

$R_T = R_t/e^{K(T-t)}$

where

R_T = insulation resistance at high temperature T
R_t = insulation resistance at low temperature t
e = base of Napierian logs (2.718)
K = a constant, as described below.

For both impregnated-paper and metallised-paper capacitors, K is taken as 0.1. For mineral jelly impregnation the insulation resistance drops approximately to half its value for every 7°C rise in temperature corresponding to this value of K, and for oil impregnation it drops by approximately half for every 10°C rise in temperature.

For mica-dielectric capacitors K is taken as 0.05. For ceramic-dielectric capacitors as normally used the fall is not so steep and no correction is usually needed.

13.2 General characteristics of discrete fixed capacitors

13.2.1 Summary of electrical characteristics

Capacitors are generally divided into classes according to their dielectric, e.g. paper, mica, ceramic, etc. It is useful to a designer to know the chief characteristics of these classes of capacitor and the main characteristics are briefly outlined in the following paragraphs. It is important to remember that capacitance is never constant, except under certain fixed conditions. It changes with temperature, frequency and age, and the capacitance value marked on the capacitor strictly applies only at room temperature and at low frequencies. A brief summary of their electrical characteristics is given in *Table 13.2*.

13.2.2 Impregnated-paper capacitors

These are general-purpose paper-dielectric capacitors, made by rolling paper as insulation between metal foils and filling with an impregnant. They are relatively inexpensive, have a high capacitance-to-volume ratio and are capable of working at reasonably high voltages, but their power factor is comparatively high and the selection tolerances are fairly wide. The maximum permissible d.c. working voltage of any impregnated-paper capacitor is dependent on the ambient temperature, and the life of the capacitor is approximately inversely proportional to the fifth power of the operating voltage up to 85°C. Irrespective of the d.c. working voltage, the maximum a.c. working voltage of a normal impregnated-paper capacitor with solid or semi-solid impregnant is about 300 V r.m.s. at 50 Hz for the tubular type and 600 V r.m.s. for the rectangular type containing two capacitor units in series. If higher a.c. working voltages are required, specially designed capacitors should be used. Smaller types are made for transistor and integrated circuit use.

The insulation resistance of impregnated-paper capacitors is high at room temperature—of the order of 2000 to 5000 Ω-F (depending on the paper and impregnant)—but falls rapidly as the ambient temperature is increased. Rectangular-cased types of 8 μF capacitance may fall to a few tens of megohms at a temperature of 100°C, while the tubular types of 0.1 μF capacitance (having a higher initial insulation resistance) may fall to a few hundred megohms at the same temperature. The fall in insulation resistance tends to be inversely proportional to the capacitance from one microfarad upwards, depending upon leakage over the case. The temperature coefficient of impregnated-paper capacitors varies from +100 to +200 parts per million per degree Celsius (ppm/°C). The power factor is about 0.005 to 0.01 at 1 Hz and tends to increase with increase of frequency. The capacitance stability under normal operating conditions is about 0.5% to 5%. The inductance of the tubular types is approximately 0.015 μH per 25 mm length of capacitor (including lead lengths).

13.2.3 Metallised-paper capacitors

These obviate voids between paper insulation and the metal, and were introduced in the late 1940s. In this type of capacitor, one side of the paper is metallised before rolling. The main characteristics are small size and self-healing action under voltage stress. If the paper is punctured the metallising quickly evaporates in the area of the puncture and prevents a shortcircuit. The maximum voltage at which the self-healing action will occur without deterioration of the capacitor properties is termed the test voltage. It is about 1.5 times the working voltage and should never be applied for more than one minute at a time. The maximum voltage which may be applied instantaneously without destroying the capacitor is termed the spark voltage. It is approximately 1.75 times the working voltage and should never be applied for more than a few seconds, as continuous sparking will rapidly destroy the capacitor.

13.2.4 Mica-dielectric capacitors

The main characteristics of this type of capacitor are low power factor, high voltage operation and excellent long-term stability at room temperature. The stability of the silvered-mica-plate type is

Table 13.2 Summary of the electrical characteristics of fixed capacitors

Capacitor type	*Capacitor stability (after climatic tests) (%)*	*Normal manu-facturing tolerance on capacitance (%)*	*Best manu-facturing accuracy (%)*	*Permit-tivity (ε)*	*Power factor (at 1 kHz)*	*Temperature coefficient (ppm/°C)*	*Maximum capacitor temperature for long life (°C)*
Impregnated-paper (rect. metal-cased and tubular capactiros)	5	±20	5	approx. 5	0.005 to 0.01	+100 to 200	100
Metallised-paper tubular capacitors	5	±25	5	5	0.005 to 0.01	+150 to 200	Normal 85 Special 125
Moulded stacked-mica capacitors	2	±20 or ±10	5	4 to 7	0.001 to 0.005	±200	Up to 120 Depends on casing
Moulded metallised-mica capacitors	1	±10	±2	4 to 7	0.001 to 0.005	±60	Up to 120 Depends on casing
Glass-dielectric (capacitors)	1	±10	(±1 down to 10 pF)	approx. 8	0.001	+150	200
Glaze-dielectric capacitors	1	±10	1	5 to 10	0.001	+120	150
Ceramic-tubular, normal-ε capacitors	1	±10	±1	6 to 15 80 to 90	0.001	+100 −30 −750 (according to mix)	150
Ceramic-tubular, high-ε capacitors	20	±20	5 (low values) 1 (high values)	1500 and 3000 (may be higher)	0.01 to 0.02 (Varies with tem-perature, etc.)	−1500, varies. Non-linear	100
Polystyrene-film, tubular and rectangular, capacitors	>1	±20	5	2.3	0.000 5	−150	60–70
Polyethylene terephthalate (Melinex) capacitors	5	±20	5	2 to 5, depending on frequency	0.01 at 1 kHz varies with tempera-ture and frequency	Varies with temperature	130
Electrolytic, (normal) capacitors	10	−20 to +100	10	—	0.02 to 0.05	+1000 to 2000 approx.	70–85
Electrolytic, (tantalum-pellet)	5	±20	10	—	0.05	+100 to 200	125
Electrolytic, tantalum-foil capacitors	10	±10	5	—	0.05	+500	85
Precision-type air-dielectric capacitors	±0.01	—	0.01	1	0.000 01	+10	20

about 1% under normal conditions of use and that of the stacked-mica-plate type about 2%. Precision mica capacitors, used as sub-standards, can be adjusted to better than 0.01% for values over 1 μF. They are invariably sealed in cases to prevent moisture, etc., from affecting the stability. Capacitors of this type have remained constant in capacitance within ±0.2 pF on a value of 10 000 pF over a test period of 10 000 h at room temperature. The temperature coefficient is low, between ±100 ppm/°C, but varies according to the source, treatment, etc., of the mica. The silvered-plate capacitor has a better temperature coefficient than the stacked-plate type. Both types, especially the stacked-plate capacitor, show slight non-cyclic capacitance shifts during temperature cycling and, in most of the types available at present, the temperature/capacitance curve is not entirely linear. There is also a wide spread of mean temperature coefficients between different specimens, even of the same batch.

The power factor of mica is approximately 0.000 3 at 1 Hz, but can be as low as 0.000 05 when specially selected and very dry. The permittivity is about 7. The current-carrying capacity of the silvered plate imposes a limit to radio frequency and pulse loading and the silvered-plate capacitor is therefore less suitable for heavy current work than the stacked-plate type, although the latter is less stable and cannot be made to such a close selection tolerance as the silvered-plate capacitor.

13.2.5 Ceramic-dielectric capacitors

Ceramic-dielectric capacitors (ceramic capacitors) are made in three main classes—low permittivity low-loss types, medium-permittivity temperature-compensating types and high-permittivity types.

The low-permittivity low-loss types are generally made of steatite or similar material. Steatite has a permittivity of approximately 8 and other materials may give permittivities between 6 and 15. Their performance at high frequencies, from about 50 Hz upwards, is excellent. The power factor is reasonably low (0.001), approaching that of mica. The temperature coefficient is between +80 and +120 ppm/°C and the capacitors are normally very cyclic in behaviour. The temperature coefficients vary less between different batches than for capacitors of any other dielectric except glass and vacuum. The capacitance stability in normal use is about 1% excluding temperature variations. They operate at comparatively high voltages, 500 V or so (depending on size), over a temperature range from about +150°C down to extremely low temperatures.

The second class, of medium permittivity (ε about 90), are used mainly as temperature-compensating capacitors in tuned circuits and have negative temperature coefficients of the order of −600 to −800 ppm/°C. They are all based on titania or its derivatives. The power factor is again low and may be less than 0.000 3 at radio frequencies. Other temperature coefficients can be obtained by using different mixtures.

The high-permittivity ceramic capacitors provide a very high capacitance in a compact unit. The capacitance and the power factor, however, change widely with temperature, the changes being neither linear nor very cyclic for either property. Capacitors using the $\varepsilon = 1200$ material, for instance, have a high capacitance peak (Curie point) at about 110°C, which is two or three times the value at room temperature, with another much smaller one at about −10°C. The power factor is a minimum around 20°C to 40°C and is in general around 2%. High permittivity materials with other permittivities have peaks at other temperatures. In general, the higher the permittivity, the more temperature-sensitive is the capacitor. In addition to changes with temperature, the capacitance is also reduced under d.c. voltage stress, especially at the peak points: at room temperature a reduction in capacitance of 10 to 20% will occur, but up to 50% can be expected at the Curie points. The d.c. working voltage is rather lower than the low-permittivity ceramic type. The capacitors are subject to hysteresis and accordingly are suitable for working with only very small a.c. voltages. They are used mainly as r.f. bypass capacitors, but can also be used for interstage coupling, provided the capacitance is large enough under all conditions of operation. The properties of high-permittivity capacitors, therefore, vary so much with temperature, voltage stress, etc., that no general electrical characteristics can be given.

13.2.6 Glass-dielectric capacitors

These capacitors are formed of very thin glass sheets (approximately 12 μm thick) which are extruded as foil. The sheets are interleaved with aluminium foil and fused together to form a solid block. Their most important characteristics are the high working voltages obtainable and their small size compared with encased mica capacitors.

Glass-dielectric capacitors (glass capacitors) have a positive temperature coefficient of about 150 ppm/°C, and their capacitance stability and Q are remarkably constant. The processes involved in the manufacture of glass can be accurately controlled, ensuring a product of constant quality, whereas mica, which is a natural product, may vary in quality. As the case of a glass capacitor is made of the same material as the dielectric, the Q maintains its value at low capacitances, while the low-inductance direct connections to the plates maintain the Q at high capacitances.

These capacitors are capable of continuous operation at high temperatures and can be operated up to 200°C. They are also used as high-voltage capacitors in transmitters.

13.2.7 Glaze- or vitreous-enamel-dielectric capacitors

Glaze- or vitreous-enamel-dielectric capacitors (glaze- or vitreous-enamel capacitors) are formed by spraying a vitreous lacquer on metal plates which are stacked and fired at a temperature high enough to *vitrify* the glaze. Capacitors made in this way have excellent r.f. characteristics, exceedingly low loss and can be operated at high temperatures, 150°C to 200°C. As they are *vitrified* into a monolithic block they are capable of withstanding high humidity conditions and can also operate over a wide temperature range. The total change of capacitance over a temperature range of −55°C to +200°C is of the order of 5%. The temperature coefficient is about +120 ppm/°C and the cyclic or retrace characteristics are excellent. As in the glass capacitor the encasing material is the same as the dielectric material and therefore all corona at high voltages is within the dielectric. They are extremely robust and the electrical characteristics cannot normally change unless the capacitor is physically broken.

13.2.8 Plastic-dielectric capacitors

In plastic-dielectric capacitors the dielectric consists of thin films of synthetic polymer material. The chief characteristic of plastic-film capacitors is their very high insulation resistance at room temperature. The main synthetic polymer films used as capacitor dielectrics are.

13.2.8.1 Polyethylene terephthalate

This is a tough polymer with high tensile strength, free from pinholes and with good insulating properties over a reasonably wide temperature range. This is known under a variety of trade names such as Melinex (ICI), Mylor (Du Pont), Hostaphon (Germany) and Terphane (France). It is commercially available in thin films of 3.5 μm in thickness.

13.2.8.2 Polycarbonate

This is a polyester of carbonic acid and bisphenols. It combines in good physical properties with a lower loss (or dissipation factor) than polyethylene terephthalate. It has a temperature characteristic nearer to zero and is available in film form down to 2 μm in thickness.

Table 13.3 Film material properties

	Polyethylene terephthalate	*Polycarbonate*	*Polystyrene*	*Polypropylene*
Permittivity (at 1 kHz)	3.2	2.8	2.5	2.25
Dissipation factor (at 1 kHz)	0.004	0.001	0.000 2	0.000 5
Dielectric strength (V/μm)	304	184	200	204
Insulance (Ω-F)	1×10^5	3×10^5	1×10^6	1×10^5

13.2.8.3 Polystyrene

This is a hydrocarbon material and has a lower permittivity than the previous two dielectrics. It has a better dissipation factor but its tensile strength for winding is much lower and 8 μm is the lowest film thickness available. It is not normally used in metallised capacitors unless heavily derated.

13.2.8.4 Polypropylene

This is a low-price material and has the lowest dissipation factor of the four films discussed. It is not commercially available in films less than 8 μm thick and it has a lower permittivity and its use is therefore limited.

A comparison of the electrical characteristics of the four film materials at 20°C is given in *Table 13.3*.

13.2.9 Electrolytic capacitors

The most notable characteristic of these capacitors is the large capacitance obtainable in a given volume, especially if the working voltage is low. Electrolytic capacitors are used for smoothing and bypassing low frequencies, but they can also be used for high-energy-pulse storage applications, such as photo-flash and pulsed circuits. The electrical properties change widely under different conditions of use and some indication of these is given below.

13.2.8.5 Capacitance

There is a slight increase (about 10%) when the temperature is raised from 20°C to 70°C; a gradual decrease as the temperature is reduced to −30°C, and a very rapid decrease at lower temperatures. The capacitance also decreases slightly as the applied frequency is increased from 50 Hz giving a 10% reduction at 10 000 Hz.

13.2.8.6 Power factor

At 50 Hz and room temperature, the power factor is from 0.02 to 0.05. There is a slight increase at +70°C and a large increase at −30°C. A large increase also takes place as the frequency is increased and the power factor becomes about 0.5 at 10 000 Hz.

13.2.8.7 Leakage current

This is normally considered instead of insulation resistance, which is very low in this type of capacitor. The leakage current varies directly with temperature, having quite a low value at −30°C, but at +70°C it is about ten times the value at room temperature. In addition, the leakage current increases with the applied load, being very high when the load voltage is first applied, but it falls rapidly and after about a minute tends to reach a stable value.

13.2.8.8 Impedance

There is a gradual increase in impedance as the temperature is reduced, until at −30°C it is about twice the impedance at room temperature, while at still lower temperatures a much more rapid increase occurs. At temperatures above normal there are only slight variations. The impedance falls rapidly with increase of frequency and at 10 000 Hz is of the order of 2 Ω for a 16 μF capacitor.

The normal type of electrolytic capacitor is made using plain foils of aluminium, but considerably increased capacitance can be obtained by using etched foils or sprayed gauze foils to increase the surface area. Electrolytic capacitors need to be reformed periodically if they are stored for a considerable time. Reforming is carried out by applying the working voltage through a resistor of approximately 1000 Ω for one hour.

Tantalum-pellet electrolytic capacitors do not need reforming and have an expected shelf life of more than ten years. They have the advantage of even greater capacitance in a small volume and the leakage current is extremely small—of the order of a few microamperes—enabling them to be used in circuits such as multivibrators. They have lower voltage ratings, however, and some types are expensive, but they are capable of operating over a temperature range from −55°C to +125°C with negligible change in capacitance.

Tantalum-foil electrolytic capacitors are also extremely small in size and have a low leakage current. They can operate at higher voltages than the tantalum-pellet types, but cannot operate over as wide a temperature range. The power factor varies considerably with temperature, also with voltage rating.

13.2.8.9 Air-dielectric capacitors

Air-dielectric capacitors are used mainly as laboratory standards of capacitance for measurement purposes. With precision construction and use of suitable materials, they can have a permanence of value of 0.01% over a number of years for large capacitance values.

13.2.8.10 Vacuum and gas-filled capacitors

Vacuum capacitors are used mainly as high-voltage capacitors in airborne radio transmitting equipment and as blocking and decoupling capacitors in large industrial and transmitter equipments. They are made in values up to 500 pF for voltages up to 12 000 V peak. Gas-filled types are used for very high voltages—of the order of 250 000 V. Clean dry nitrogen may be used at pressures up to 10^5 kg/m^2. They are specially designed for each requirement.

13.3 Variable capacitors

Variable capacitors may be grouped into five general classes: precision types, general-purpose types, transmitter types, trimmers and special types such as phase shifters.

Table 13.4 Summary of electrical characteristics of variable capacitors

Type	*Capacitance law*	*Approximate capacitance swing (pF)*	*Power factor*	*Approximate operating voltage (V d.c.)*	*Temperature coefficient (ppm/°C)*
Single unit, precision	SLC	from 100 to 1500	0.000 01	1000	+10 (best)
Multi-gang, precision	SLF	320	0.000 05	500	+20 (−15 with compensating vane)
Single unit, general-purpose	SLC and exp.	15–100 and 350–550	0.001	750	+120
Multi-gang, general-purpose	SLC and exp.	15–100 and 350–550	0.001	750	+120
Multi-gang miniature, general-purpose	SLC	300–350	0.001	500	+150
Trimmers, air-dielectric, vane-type	SLC	Range from 2 to 150	0.001	500 to 1250	+50 to +120
Concentric trimmers, air-dielectric	SLC	6.0 and 27	0.007	300	+300
Mica compression trimmers	Non-linear	Range from 8 to 650 (rotary type to 3000)	0.001	350–500 —rotary type 3000 (up to 1000 pF)	Poor
Ceramic trimmers	Approx. SLC	from 5 to 100	0.002 to 0.005	1000	+100 and −750
Plastic concentric trimmers	SLC	3.0	0.001	350	Depends on plastic

13.3.1 Precision types

Precision types have been developed for many years mainly as laboratory sub-standards of capacitance in bridge and resonant circuits and numerous measuring instruments have been designed around them. Various laws are available and capacitances up to 5000 pF can be obtained in one swing. Capacitance tolerances are of the order of one part in ten thousand and long term stabilities under controlled conditions of 0.02% over as many years as possible.

General-purpose radio types are used as tuning capacitors in broadcast receivers. They have developed from large single capacitors to compact four- or five-gang units which can have a standard capacitance tolerance of within 1% or 1 pF to a stated law. The power factor of a modern air-dielectric variable capacitor, at 1 MHz, varies between 0.03% (at the minimum capacitance setting) and 0.6% (at the maximum capacitance setting). They are available in many laws, e.g. straight-line frequency, straight-line wavelength, straight-line percentage frequency, so that they can be used in test equipment and receivers of many types. The normal capacitance swing of this type of capacitor is about 400–500 pF, but components can be obtained in capacitance swings (in ranges) from 10 to about 600 pF.

13.3.2 Transmitter types

Transmitter types are basically similar in design but have wider spacing between vanes to allow safe operation at much higher voltages. The capacitance swing usually ranges up to about 1000 pF. The edges of the vanes are rounded and polished to avoid flashover, and special attention is paid to shape and mounting for high-voltage operation. The most common laws are square-law capacitance (or linear frequency) and straight-line capacitance. Special split-stator constructions are also used for push–pull circuits. Oil filling increases the capacitance and working voltage from two to five times, depending on the dielectric constant of the oil used. Compressed gas variable capacitors use nitrogen under pressures up to 1.5×10^6 kg/m^2 for broadcast transmitters. Bellows-type variable capacitors are made in the USA, variable from 10 to 60 pF, which are only 3 in (7.62 cm) in diameter and 5 in (12.7 cm) long. These can operate at 20 000 r.f. peak volts and 10 A maximum r.m.s. current at 20 MHz.

13.3.3 Trimmer capacitors

Trimmer capacitors are used mainly for coil trimming at intermediate and radio frequencies. They can be classed into four main groups: air-spaced rotary types, compression types (usually mica), ceramic-dielectric rotary types, and tubular types. The capacitance range for rotary air-spaced types covers from about 2 to about 100 pF. This is effected in several swings, each covering a much smaller range −1.5 to 10, 2 to 30, 5 to 100 pF, etc. Compression types cover a much wider total range from 1.5 to 2000 pF, again in stages. Ceramic-dielectric capacitance ranges are usually smaller, from 1.0 to 3.5 pF, ranging up to 5 to 50 pF, depending on the temperature coefficient required. Ceramic-dielectric trimmer capacitors are made from materials with negative temperature coefficients usually about −300 ppm/°C, and −750 ppm/°C. They are also made with zero temperature coefficient, and with positive temperature coefficients, usually +100 ppm/°C. The power factor of this type of capacitor is about 0.1 to 0.5%, depending on the temperature coefficient. Tubular types are commonly used for the fine adjustment of small capacitance values.

13.3.4 Special types

Special types of variable capacitor are sometimes required, such as differential, split-stator, or phase-shifting capacitors. Phase-shifting capacitors are used mainly in radar systems for the accurate measurement of time intervals, and in high-speed sweep-scanning circuits. In addition, specialised multi-gang capacitors designed for transmitter/receivers are also produced.

13.3.5 Summary of electrical characteristics

A summary of main electrical characteristics of variable capacitors is given in *Table 13.4*.

Further reading

APPS, L. T. and PLASKETT, J. A., 'Capacitors and fixed resistors—a continuing evolution', *Electronics & Power (GB)*, **22**, No. 7, p. 429 (July 1976)

ATKINSON, A. D., 'Fixed and variable capacitors—past, present and future', *New Electron. (GB)*, **12**, No. 6, p. 102 (20 March 1979)

BELL, R. W., 'Tuning devices. Variable capacitors', *Electronic Equipment (GB)*, **17**, No. 9, p. 69 (September 1978)

CAMPBELL, D. S., 'Electrolytic capacitors', *The Radio & Electronic Engineer*, **41**, No. 1 (Jan. 1971)

'CAPACITORS—A COMPREHENSIVE EDN REPORT', *Electronic Design News, USA*, p. 139 (May 1966)

CAPACITORS (Supplement), *Electronic Weekly*, p. 14 (21 March 1973)

DUMMER, G. W. A., *Variable Capacitors and Trimmers*, Pitman, London (1957)

DUMMER, G. W. A., *Fixed Capacitors*, 2nd edn, Pitman, London (1964) (Contains comprehensive bibliography up to 1962)

DUMMER, G. W. A. and NORDENBERG, H. M., *Fixed and Variable Capacitors*, McGraw-Hill, New York (1960)

EVANS, D., 'Fixed ceramic capacitors', *New Electron. (GB)*, **8**, No. 9, p. 86 (29 April 1975)

GIRLING, D. S., 'Quality control in capacitor-production and testing', *The Radio & Electronic Engineer*, **40**, No. 4, p. 173 (Oct. 1970)

'MINIATURE CAPACITOR PROGRESS', *Electron*, p. 15 (29 June 1972)

PAMPLIN, B. F., 'Capacitor selection—facts and figures', *Electronic Equipment News*, p. 44 (Dec. 1969)

'RESISTORS AND CAPACITORS', *Electronics Weekly*, p. 17 (14 June 1978)

VON HIPPEL, A., *Dielectrics & Waves*, Chapman & Hall, London (1954)

VON HIPPEL, A., *Dielectric Materials & Application*, Chapman & Hall, London (1954)

14 Magnetic Materials

P R Knott MA, MInstP, AI Ceram
Section Leader, Materials Studies,
Microwave Division,
Marconi Electronic Devices Ltd

Contents

14.1 Basic magnetic properties

14.1.1 The origins of magnetism

Permanent magnets, in the shape of lodestones, were known to have been in use around 2000 BC for direction finding. The dipolar nature of a magnet was apparent to the ancient Greeks, with its north seeking or N pole and S pole. They also knew that pieces of iron are attracted by a magnet. Gilbert, in the late 16th century, summarised the forces of interaction between two magnets in the famous law 'like poles repel, unlike poles attract'. It is convenient to think of a *magnetic field* existing wherever magnetic effects are experienced, whose direction at any point is given by the N–S axis of a small permanent magnet freely suspended at that point. This is the basis of the familiar compass-needle method of field plotting. Lines of force for a bar magnet are shown in *Figure 14.1*.

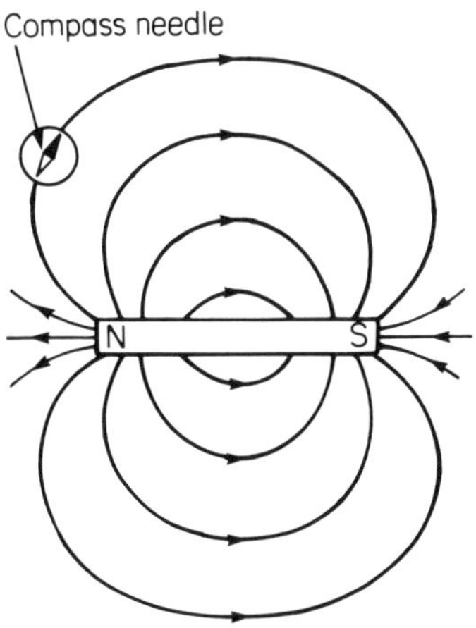

Figure 14.1 Lines of force for a bar magnet (After Duffin[2], reproduced by permission of the publisher)

In the early 1820s Ampère showed that two coils carrying electric currents, or one such coil and a magnet, exerted forces and couples on each other. At separations which are large compared with the dimensions of coils or magnet, these effects are similar, and analogous to the force and couple exerted by one electric dipole on another. The corollary is that the fields of a small permanent dipole and a small current loop should be identical, which is confirmed experimentally. Lines of force can be plotted for a solenoid (*Figure 14.2*); the similarity in patterns outside the bar magnet and solenoid is obvious. Thus an *electric current* is a source of magnetic field.

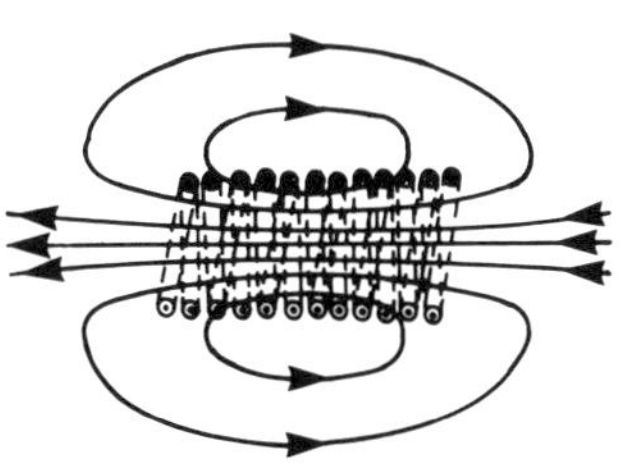

Figure 14.2 Lines of force for a solenoid (After Duffin[2], reproduced by permission of the publisher)

By analogy with electric dipoles, magnetic dipole moments m_1, m_2 may be defined for pairs of small coplanar dipoles 1, 2 of either type, such that the force and couple each exerts on the other are *always* proportional to the product m_1m_2. The effective moment of either dipole is found to vary with orientation; thus m is a vector quantity, denoted by $\mathbf{m}$. Experiments with various current loops show that $\mathbf{m}$ is proportional to the current I and loop area A. For a small single turn loop, by definition

$$\mathbf{m}=I\mathbf{A} \tag{14.1}$$

$\mathbf{m}$ and $\mathbf{A}$ are now *axial* vectors (*Figure 14.3*), the positive direction of each being defined by the sense of current flow. $\mathbf{m}$ is in units of ampere metre^2 (A m^2).

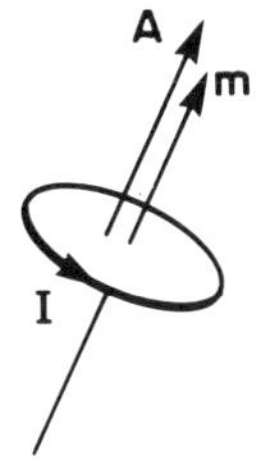

Figure 14.3 Current loop dipole (After Duffin[2], reproduced by permission of the publisher)

Ampère subsequently proposed that the origin of all magnetism in materials lay in *small circulating currents* associated with each atom. They would each give rise to a magnetic dipole moment, the vector sum of which over all atoms would be the bulk magnetic moment. This explains why isolated magnetic poles are never observed. Even on the atomic scale only dipoles exist, which are due to electric currents.

In modern atomic theory the elementary currents of Ampère's theory are replaced by the closed orbits of negatively charged electrons about the atomic nuclei. However, the electron is now known to have an intrinsic 'spin' moment. Thus an atom may have a resultant moment due to both *orbital* and *spin* contributions.

14.1.2 Magnetic fields *in vacuo*

Using a test current loop dipole of known $\mathbf{m}$, the field quantity $\mathbf{B}$, called the magnetic flux density, may be defined at a point as follows. The *direction* of $\mathbf{B}$ is the equilibrium direction of $\mathbf{m}$; the *magnitude* of $\mathbf{B}$ is the couple per unit moment required to keep the test dipole at 90° to $\mathbf{B}$. In general if $\mathbf{m}$ makes an angle θ with $\mathbf{B}$ (*Figure 14.4*) then

$$\mathbf{C}=mB\sin\theta \quad\text{or}\quad \mathbf{C}=\mathbf{m}\times\mathbf{B} \tag{14.2}$$

The unit of $\mathbf{B}$ is the tesla (T) defined in SI units as kg s^{-2} A^{-1} or equivalently as N A^{-1} m^{-1}. While equation (14.2) holds for a permanent magnetic dipole, it is then a definition of $\mathbf{m}$, not $\mathbf{B}$, since $\mathbf{m}$ is independently defined only for a current loop.

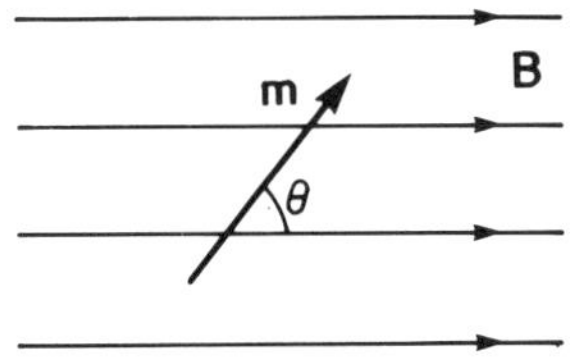

Figure 14.4 Dipole moment **m** in flux density **B** (After Duffin[2], reproduced by permission of the publisher)

The *total* flux of **B** passing through a surface S is defined as the sum of all the products of the normal components of **B** and elements of area d**S**:

$$\varphi=\iint_S \mathbf{B}.\mathrm{d}\mathbf{S} \tag{14.3}$$

φ is in units of tesla metre^2 (T m^2), also called the weber (Wb), and **B** is sometimes given the unit Wb m^{-2}. By convention, B lines of force are said to cross a unit area in a region of flux density **B** so that φ represents the total number of lines crossing S.

By analogy with electromotive force (e.m.f.) which is the line integral of the electric field **E** round a closed circuit C, the magnetomotive force (m.m.f.) is defined as follows:

$$\mathscr{M}=\oint_C \frac{\mathbf{B}.\mathrm{d}\mathbf{s}}{\mu_0} \tag{14.4}$$

where d**s** is an element of length and μ_0 is called the permeability of free space. In SI units $\mu_0=4\pi\times10^{-7}$ N.A^{-1} by definition, so the unit for $\mathscr{M}$ is the ampere. The choice of $\mathbf{B}/\mu_0$ rather than **B** is for convenience as explained in Section 14.1.3. Now in an electrostatic field

$$\oint \mathbf{E}.\mathrm{d}\mathbf{s}=0$$

At large distances the flux density due to a magnetic dipole takes the *same* form as the electric field caused by an electric dipole, so that for a circuit which is far from any dipole

$$\oint_C \mathbf{B}.\mathrm{d}\mathbf{s}=0 \tag{14.5}$$

Figure 14.5 shows the closed lines of force outside a current-carrying wire. They are called a *vortex* of **B**. Evidently any line

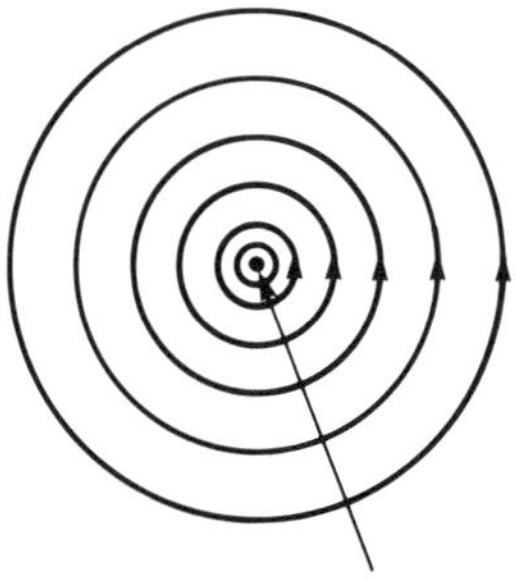

Figure 14.5 Lines of force outside a wire (After Duffin[2], reproduced by permission of the publisher)

integral of **B** around the wire cannot vanish. In fact a closed path C encircling a current I once contributes just I to the m.m.f., i.e.

$$\oint_C \mathbf{B}.\mathrm{d}\mathbf{s}=\mu_0 I \tag{14.6}$$

This is Ampère's law. For N circuits the right-hand side of equation (14.6) becomes $\mu_0 NI$. Clearly in general $\mathscr{M}$ is neither path-independent nor single-valued.

The total flux of **B** through any closed surface S containing or intersecting magnetic dipoles is given by the *magnetic* analogue of Gauss' law. Just as in electrostatics

$$\oiint_S \mathbf{E}.\mathrm{d}\mathbf{S}=0$$

because any number of dipoles contribute zero net charge, so here

$$\oiint_S \mathbf{B}.\mathrm{d}\mathbf{S}=0 \tag{14.7}$$

This is Gauss' law. Equations (14.6) and (14.7) are the general laws summarising the properties of steady magnetic fields in a vacuum.

14.1.3 Magnetisable media

A circuit in a vacuum in which a current I is maintained has a certain flux density $\mathbf{B}_0$ crossing it. The introduction of a magnetisable medium *changes* $\mathbf{B}_0$ to some new value $\mathbf{B}'$, where in general both $\mathbf{B}_0$ and $\mathbf{B}'$ vary from point to point in magnitude and direction. Ampère's theory provides a model in which the effect of the medium is found by replacing it with suitable distributions of *surface* and *volume* currents. Specifically, when the vacuum round the circuit is *completely* filled with the medium, the *mean* flux density $\bar{B}_0$ changes by a factor μ_r which is defined as the *relative permeability* of the medium, i.e.

$$\mu_r=\bar{B}'/\bar{B}_0 \tag{14.8}$$

Evidently the same result would ensue if the vacuum remained and instead I changed to $\mu_r I$. According to the model the increased flux density is due to currents $(\mu_r-1)I$ flowing in the surface next to I. Hereafter the currents producing the fields are called *conduction currents* I_c and those replacing the material *Amperean currents* I_A.

The magnetisation at a point is the magnetic moment per unit volume **M** at that point, given by the dipole moment d**m** of an elementary volume dτ surrounding the point, i.e.

$$\mathbf{M}=\mathrm{d}\mathbf{m}/\mathrm{d}\tau \tag{14.9}$$

where **M** is in units of ampere metre^{-1} (A m^{-1}) and dτ is assumed small enough for **M** to be constant. Let the element have length dl parallel to **M**, and cross-sectional area dS, i.e. $\mathrm{d}\tau=\mathrm{d}l\,\mathrm{d}S$. If it is replaced by loops of total *surface* current dI_A, then by equation (14.1),

$$\mathrm{d}\mathbf{m}=\mathbf{M}\,\mathrm{d}l\,\mathrm{d}S=\mathrm{d}I_A\,\mathrm{d}\mathbf{S}=J_{SA}\,\mathrm{d}l\,\mathrm{d}\mathbf{S}$$

where J_{SA} is the surface current density, i.e.

$$J_{SA}=|\mathbf{M}| \tag{14.10}$$

This result holds for a macroscopic piece of material provided **M** is uniform throughout. Adjacent 'interior' Amperean currents cancel out as regards magnetic effect and only those on the outer surface contribute. $J_{SA}=M\cos\theta$ at a general point where the tangent to the surface is at angle θ to the direction of **M**. A *volume* current distribution J_{VA} need only be invoked when **M** is not uniform, i.e. the interior currents do not balance. This may happen if the medium itself carries conduction currents, or it is *non-linear* (μ_r is a function of **B**).

Since there are now separate contributions to **B** from conduction and Amperean currents, Ampère's law becomes

$$\oint_C \mathbf{B}.\mathrm{d}\mathbf{s}=\mu_0(I_c+I_A) \tag{14.11}$$

where C is a circuit linking conduction currents totalling I_c and magnetised material equivalent to Amperean currents I_A. Let d**s** be a line element of C cutting an elementary volume of the material at some arbitrary angle to **M**. As before, $\mathrm{d}I_A=J_{SA}\,\mathrm{d}l=\mathbf{M}.\mathrm{d}\mathbf{s}$. So equation (14.11) may be rearranged to give

$$\oint_C (\mathbf{B}/\mu_0-\mathbf{M}).\mathrm{d}\mathbf{s}=I_c$$

and the magnetic field strength **H**, in ampere metre^{-1} (A m^{-1}), is *defined* by

$$\mathbf{H} = \mathbf{B}/\mu_0 - \mathbf{M} \tag{14.12}$$

Ampère's law, previously defined *in vacuo* by equation (14.6) can now be generalised:

$$\oint_C \mathbf{H} \cdot d\mathbf{s} = I_c \tag{14.13}$$

Likewise the m.m.f. ($\mathscr{M}$) round a closed path simply equals the total conduction current linked. The utility of **H** lies in the fact that it is generated *solely* by conduction currents and is therefore unaffected by the presence of a medium.

The magnetic susceptibility χ_m is defined at a point by the ratio of magnetic moment per unit volume to magnetic field strength:

$$\mathbf{M} = \chi_m \mathbf{H} \tag{14.14}$$

χ_m is dimensionless and is a *scalar* for an isotropic homogeneous medium, otherwise a tensor, and is independent of **H** *only* for linear media. From equations (14.12) and (14.14),

$$\mathbf{B} = \mu_0(1 + \chi_m)\mathbf{H}$$

so

$$\mu_r = 1 + \chi_m \tag{14.15}$$

$$\mathbf{B} = \mu_0 \mu_r \mathbf{H} \tag{14.16}$$

which now define the relative permeability μ_r *at a point*.

The notion of *magnetic poles* is familiar. It is simple in essence because of the similarity with electric charges. Much of electrostatic theory can be used by analogy to account for magnetic phenomena. While it is often useful in calculations involving permanent magnet materials, the magnetic pole model has several drawbacks:

(a) *Isolated* poles do not exist, whereas atomic magnetic moments can be satisfactorily related to angular momenta via current loops.
(b) While magnetic moment is a measurable quantity, pole location and strength are *ill-defined*.
(c) There is no analogue of charge conduction.
(d) With other than *permanent* magnet materials, problems are generally more difficult to solve using the pole concept.
(e) *Practical* permanent dipoles do not exist. The properties of any magnet are affected by stray fields, including the proximity of other magnetic material. By contrast the moment of a current loop can be maintained constant.

This section has dealt with media filling the whole space in which magnetic fields exist, i.e. effectively *infinite* media. The presence of boundaries may affect **B**, **H** and **M** considerably. In particular demagnetising effects have important practical consequences for ferromagnetic materials. This is considered in Sections 14.2.2 and 14.4.1.

Two excellent though contrasting treatments of the fundamentals of magnetism are given by Bleaney and Bleaney[1] and Duffin[2].

14.2 Ferromagnetic media

Ferromagnetic substances are always solids and usually metals. χ_m is positive, large (often $\gg 1$), non-linear and dependent on previous history (showing hysteresis). Most materials in this class can exhibit a *finite and perhaps a large moment when* $\mathbf{H} = 0$. There is an abrupt change of properties at a characteristic temperature known as the Curie temperature (T_C). Above this, behaviour is simpler, the material becomes *paramagnetic*, i.e. χ_m is still positive but small ($\ll 1$), and it obeys the *Curie–Weiss law*. At absolute temperature T

$$\chi_m = \frac{C}{T - T_p} \tag{14.17}$$

where C is the Curie constant and T_p is the paramagnetic Curie temperature. T_p and T_C usually differ by some 10 or 20°C.

14.2.1 Hysteresis properties

Many of the characteristic properties of a ferromagnetic material are displayed in curves of B (or M) against H. For ease of interpretation the sample used is preferably a toroid wound with primary and secondary coils. **B**, **H** and **M** are uniform and parallel, and there are no demagnetising effects (see Section 14.2.2). A current I in the primary, of n turns per unit length, generates a field $H = nI$ (by Ampère's law) and B is derived from flux linking the secondary. If the material is initially unmagnetised, then increasing H from zero gives an S-shape *magnetisation curve* OABC in *Figure 14.6*, of which the portion OA is reversible. Beyond A, reduction of H from some point B causes the path BB′ to be followed. If H is continuously reversed

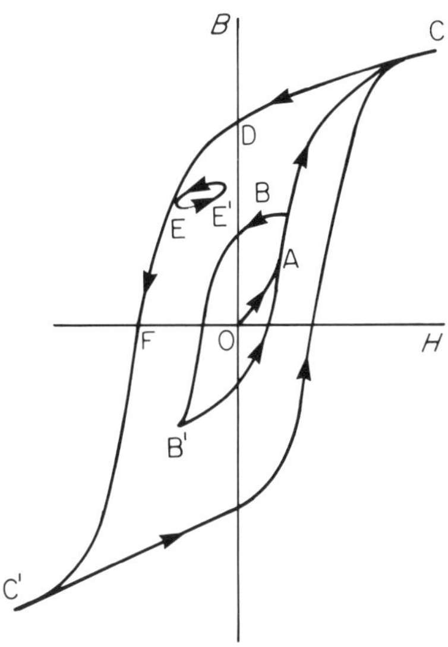

Figure 14.6 Typical hysteresis loops (after Duffin[2], reproduced by permission of the publisher)

through amplitude BB′, a *minor hysteresis loop* is described. The biggest loop CC′, called simply *the hysteresis loop*, is obtained when the amplitude of H is large enough to cause saturation of the magnetisation. Beyond the points CC′, B increases as H simply, as equation (14.12) predicts. If at any point E small reversals of H are made, *subsidiary loops* like EE′ are followed. The *demagnetisation curve* DEF gives the basic information required for permanent magnet design. The intercept OD on the B axis (i.e. $H = 0$) is called the *remanence* B_r, and the intercept on the $-H$ axis where $B = 0$ is the *coercivity* H_{cB}. The *relative permeability* $\mu_r = B/\mu_0 H$ evidently goes through a maximum near point B on the magnetisation curve as H increases from zero, then gradually falls towards unity. The *incremental permeability* μ_{inc}, often of greater practical importance, is defined for small ranges of B and H by $\mu_{inc} = \delta B/\mu_0 \delta H$. The *initial permeability* μ_i clearly equals μ_{inc} for small excursions of H about the origin. The *recoil permeability* μ_{rec} is the slope of the chord joining the tips of a subsidiary loop such as EE′; this is also of relevance in permanent magnet design. A graph of M against H shows similar features but as M saturates the extremities of the major loop are of zero slope. The intercepts here are the *remanent magnetisation* M_r ($\mu_0 M_r = B_r$) and *intrinsic coercivity* (or coercive field) H_{cM}. Since numerically $H_{cM} = M(B=0)$ and $H_{cM} = B/\mu_0$ ($M=0$), $H_{cM} \neq H_{cB}$ in general but is arbitrarily related. Finally, application of a reverse field equal to H_{cB} does not permanently demagnetise the sample as when H is returned to zero, a *recoil line* parallel to EE′ is followed and the sample is left in some remanent state $M_r' < M_r$.

14.2.2 Microscopic theory

Ferromagnetism can be accounted for via the theory of paramagnetism, since the two are clearly intimately related. *Paramagnetism* is exhibited when the atoms or ions in a material possess a resultant moment $\mathbf{m}_p$ which is the vector sum of all the electron moments. While free atoms often have net moments, atoms in combination and ions usually do not. These substances possess *weak* moments in an applied field which disappear when the field is removed. Typically $\chi_m \sim +10^{-3}$ at room temperature. According to Curie's law

$$\chi_m = C/T \tag{14.18}$$

which holds quite well for gases, solutions and some solids. Many liquids and solids obey the modified Curie–Weiss law of equation (14.17). It is assumed that a molecular field $\lambda\mathbf{M}$ is contributed by interactions between atoms. Replacing $\mathbf{H}$ by $\mathbf{H} + \lambda\mathbf{M}$ in equation (14.14) gives to first order

$$\chi_m = \frac{C}{T}(1 + \lambda\chi_m) \quad \text{or} \quad \chi_m = \frac{C}{T - \lambda C} \tag{14.19}$$

which is the Curie–Weiss law with $T_p = \lambda C$. λ is called the *Weiss constant*. Weiss suggested that in ferromagnetic materials neighbouring atomic moments interact strongly, resulting in parallel alignment over considerable regions. T_C measurements confirm that λ is very large, of order 10^4. According to quantum mechanics *electron spins* are almost entirely responsible for ferromagnetism, the large forces between them being called *exchange interactions*.

That ferromagnetic materials can be demagnetised at all by external fields is due to the existence of *domains*. These are small regions, generally 10^{-6} to 10^{-2} cm^3 in size, each spontaneously magnetised to saturation in a definite crystallographic direction, adjacent domains having different directions. They are separated by *Bloch walls* of order 5 nm thick wherein the spin directions change progressively (*Figure 14.7*). They are formed to lower the energy stored in internal and external (flux leakage) fields. Subdivision into a number of domains takes place with broadly opposing magnetisations, with small closure domains at the surfaces (*Figure 14.8*). The process does not go on indefinitely because Bloch walls have a certain energy arising from *magnetocrystalline anisotropy*. In a crystalline material there are preferred directions of magnetisation dictated by the crystal symmetry. The *anisotropy energy* W_A is that required to rotate the magneti-

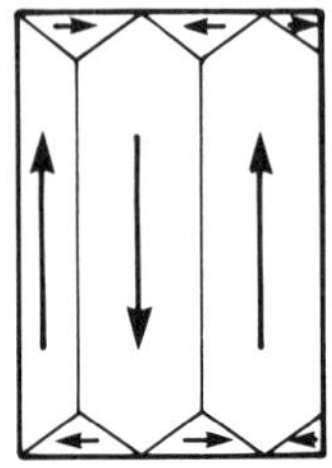

Figure 14.8 Typical domain closure pattern (After Duffin[2], reproduced by permission of the publisher)

sation from an 'easy' to a 'hard' direction. W_A is thought to be due both to the electric field of the ionic charges and spin–orbit coupling whereby the atomic spin direction is coupled to the lattice via the orbit. The wall contribution to W_A arises since inevitably most of the spins in the wall will not be parallel to the easy direction. Domain size is thus largely determined by the balance of field and wall anisotropy energies. The shape of the magnetisation curve (*Figure 14.6*) is now explained by reference to *Figure 14.9*. In unmagnetised material, domains are quasi-randomly orientated and there is no net moment (a). At small **H**, more favourably magnetised domains grow at the expense of others (b), so in the initial part OA of the curve magnetisation proceeds via small reversible displacements. The steeper part AB corresponds to larger irreversible displacements (c) where wall movements are hindered by crystal structure defects. At still higher **H**, between B and C, magnetisation increases by rotation of domains from easy directions towards the field (d). Finally at C, all domain walls are swept away, and alignment with **H** is virtually complete (e).

The irreversible wall movements at moderate **H** can be detected audibly by voltages induced in loudspeaker coils wound on a specimen, as a wall jumps across a defect (Barkhausen effect). Domain patterns may be observed on the surface of a material by applying colloidal ferromagnetic powder (Bitter patterns).

14.2.3 Energy losses in ferromagnetic materials

Two sources of loss of major practical importance are possible when the material is subjected to *alternating* magnetic fields. These are: (a) hysteresis and (b) eddy current losses.

(a) During a cycle, domain rotation is opposed by anisotropy effects, the energy to overcome this being derived from the magnetising field and appearing ultimately as heat. It can be shown that the energy lost per cycle or the net work done by the field (in J m^{-3} per cycle) is

$$W_h = \oint \mathbf{H}.\mathrm{d}\mathbf{B} \tag{14.20}$$

which is just *the area of the BH loop*. Note that this is essentially a d.c. loop. With alternating magnetisation, non-hysteresis losses result in a phase angle between B and H which in turn modifies

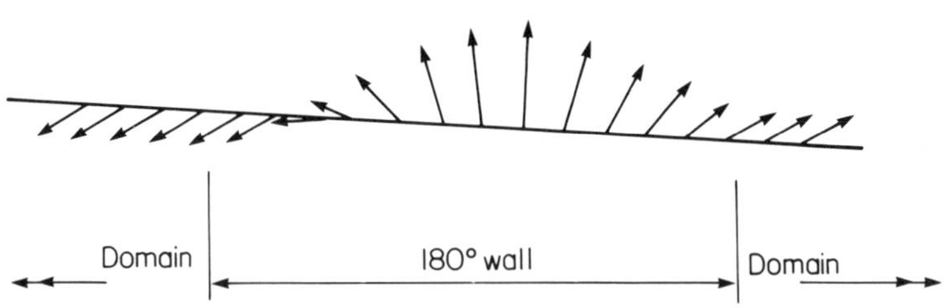

Figure 14.7 Change of spin orientation in a Bloch wall (after Snelling[3], reproduced by permission)

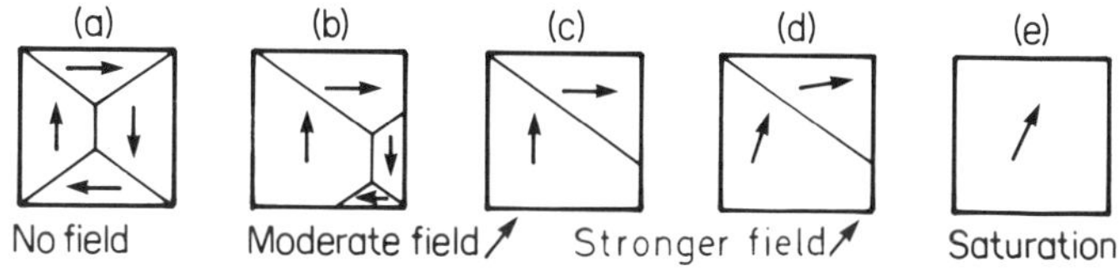

Figure 14.9 Effect of an external field on domains (After Duffin[2], reproduced by permission of the publisher)

loop shape. By making various assumptions about loop properties, or appealing to experiment, power series for $B(H)$ and $W_h(H)$ may be derived which converge for low amplitudes of H. The first term for W_h in one such approach, due to Peterson, is given by

$$W_h \simeq \tfrac{8}{3}\mu_0 a_1 \hat{H}^3 \qquad (14.21)$$

where a_1 is a coefficient with unit m A^{-1} and $\hat{H}$ is the maximum field strength. These aspects of hysteresis loss are discussed by Snelling[4].

(b) An alternating flux in a conductive medium induces *eddy currents* which also result in energy loss. The magnitude of the loss depends partly on medium shape and size, and is reduced by subdivision into electrically insulated regions, e.g. laminations or grains. At low frequencies f where eddy current inductive effects are negligible, and for a peak flux density $\hat{B}$ and bulk resistivity ρ, the power loss per unit volume (in W m^{-3}) in a lamination of thickness d is given by

$$P_E = \frac{(\pi \hat{B} f d)^2}{6\rho} \qquad (14.22)$$

where $\hat{B}$ is assumed to lie in the plane of the lamination. Equation (14.22) is quoted by Snelling[5] and eddy currents are treated in detail by Brailsford[7].

14.2.4 Boundaries and demagnetising factors

At the boundary of a medium of relative permeability μ_r, **B** and **H** cannot both be continuous. Specifically:

(a) for a boundary parallel to **H**, **B** is discontinuous;
(b) for a boundary normal to **B**, **H** is discontinuous.

It is useful to interpret discontinuities in **H** as due to an opposing specimen field $\mathbf{H}_d$. μ_r is so large for ferromagnetic materials that the net internal **H** may oppose **B** and **M**. $\mathbf{H}_d$ is called an *internal demagnetising field*. In case (b) as **B** is continuous, $\mathbf{H}_d = \mathbf{M}$. For a specimen of arbitrary shape, **M** and $\mathbf{H}_d$ vary from point to point in a complex way. A ferromagnetic ellipsoid is the only shape, which when placed in a *uniform* field $\mathbf{H}_a$ has the same magnetisation and demagnetising field at every point. In this case $\mathbf{H}_d = [N_d]\mathbf{M}$, where $[N_d]$ is a *demagnetisation tensor*. If the ellipsoid is orientated with its axes along rectangular coordinate axes x, y, z then $[N_d]$ is diagonalised and the *demagnetising factors* N_x, N_y, N_z along the respective axes are related very simply by

$$N_x + N_y + N_z = 1 \qquad (14.23)$$

It is still difficult to calculate individual demagnetisation factors except in limiting cases by appealing to symmetry. For instance:

(b) sphere: $N_x = N_y = N_z = 1/3$
(b) Thin plate in yz plane: $N_x \simeq 1$, $N_y = N_z \simeq 0$
(c) Long rod, axis in x direction, circular section: $N_x \simeq 0$, $N_y = N_z \simeq 1/2$

Demagnetising effects are of fundamental importance in *any circuit containing air gaps*. Demagnetising factors need to be known in calculating the working point of a permanent magnet, for instance, or the condition for gyromagnetic resonance of a microwave ferrite.

14.2.5 Soft magnetic materials

Soft magnetic materials have small or moderate remanence, low coercivity and hysteresis loss, and high permeability. Ease of magnetisation and demagnetisation implies *freely moving* domain walls. To achieve this a low anisotropy is required, and materials are usually annealed to remove strain and local atomic disorder, which inhibit wall motion. Materials of technological importance are all *iron alloys* (apart from ferrites). Iron itself is often used in electromagnets, including relays and valves, but is sensitive to light element impurities such as C, S, N and O. *Silicon–iron* is extensively used in transformers and large electromagnetic machinery generally. It shows lower eddy current and hysteresis losses, and higher permeabilities than iron. Commercial 3.2% SiFe sheet used for making transformer laminations has enhanced magnetic properties in the rolling direction. Hot-rolled strip is cold reduced in several passes to its final thickness. During subsequent annealing at about 1100°C, uniformly sized crystallites are formed which are highly orientated in the rolling direction. SiFe alloys, like many other soft magnetic materials are *magnetostrictive*, and magnetic losses are increased (reduced) by external compressive (tensile) stresses. Sheets of 3.2% SiFe have a glassy coating which provides interlaminar insulation and also puts the material under tensile stress, partly overcoming the compressive stresses that occur in transformer construction. Higher permeability and lower losses are achieved in *nickel–iron* alloys (~40–80% Ni). They are extensively used for magnetic screening, but they saturate at lower fields than SiFe and are relatively expensive, limiting their use otherwise to small transformers and chokes. High saturation magnetisations are achieved in *iron–cobalt* alloys which may contain low levels of vanadium and nickel to improve malleability and mechanical strength. They are used in aircraft generators where a large power/weight ratio is critical. There is much interest in the recently developed *amorphous ribbon magnetic alloys* of general formula $T_{80}M_{20}$, where T is one or more transition metals (Fe, Co, Ni, Cr, Mn) and M is one or more metalloids (B, C, Si, P). Ribbons up to ~60 μm thick and ~2 cm wide are simply and economically prepared using the *drum quenching technique* (*Figure 14.10*). The melt is forced through an orifice by gas pressure and impinges on two rotating drums. The extremely rapid cooling (~10^6 degrees Celsius per second) ensures that the disordered state of the melt is retained. The ribbons are mechanically *very strong* and do not degrade with mechanical handling. They have low coercivities and fairly high saturation magnetisations. The d.c. magnetic properties compare quite well with those of NiFe alloys, so the new materials are potential replacements. Domain size is large, leading to increased a.c. losses, but is not necessarily inherent to these materials. The origin of ferromagnetism in amorphous substances is not well understood at present. *Soft ferrites*, which though not actually ferromagnetic, behave like soft magnetic alloys in many respects. These are described in Section 14.3.2.

Properties of representative soft ferromagnetic materials are given in *Table 14.1*. A recent review of new magnetic materials, including amorphous alloys has been made by Overshott[8]. Soft magnetic alloys in general are discussed by Brailsford[9].

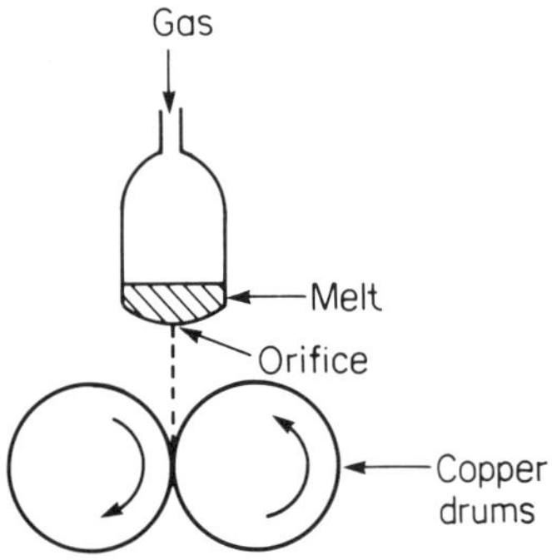

Figure 14.10 Drum quenching technique for manufacture of amorphous ribbon alloys (after Overshott[8], reproduced by permission)

14.2.6 Hard magnetic materials

Hard magnetic materials usually have high coercivity and remanence, with relatively low permeability. Magnetisation and demagnetisation are now deliberately made difficult and there are various ways of introducing structural defects in order to *inhibit* domain wall motion. Some materials may be prepared containing effectively single particle domains in which there are few or no Bloch walls. Magnetisation can only occur by domain rotation requiring high fields. Suitable alloy and ferrite materials, their manufacture and properties, are described in Section 14.4. *Hard ferrites*, like the soft materials, are not strictly ferromagnetic.

14.3 Ferrites

Ferrites are mixed metal oxide ceramics, black or dark grey in colour, and very hard and brittle. They have high electrical resistivities, typically in the range 10–10^{12} Ωcm (*q.v.* values for soft magnetic alloys in *Table 14.1*). Their strong magnetic properties are due to *ferrimagnetism* in which magnetic ions (principally Fe^{3+}) on different crysrallographic sites have spin moments coupled antiparallel via oxygen ions by so-called *superexchange interactions*. Opposing sets of moments or magnetic sublattices are formed, which are not equal in magnitude, resulting in a large magnetisation. Like ferromagnetic materials, ferrites have a well-defined magnetic transition temperature called the Curie temperature (or sometimes the Néel temperature).

14.3.1 Crystal structures

The three technically important classes are: (a) spinels; (b) garnets; and (c) hexaferrites. Spinels and garnets are magnetically soft, while the hexaferrites are hard.

The general formula of an unsubstituted spinel is MFe_2O_4 where M is one or more divalent cations such as Mg, Mn, Co, Ni, Cu, or Zn. The metal ions occupy two types of interstitial site (called A and B) in a cubic close-packed array of oxygen ions. B sublattice moments usually dominate as there are twice as many B as A sites. The site preference of a cation depends on its size and electronic structure, and the nature of the other cation(s) present. The generalised distribution for a cation M with no exclusive preference is:

$(M_\alpha Fe_{1-\alpha})$	$[M_{1-\alpha}Fe_{1+\alpha}]O_4$	(partially inverted spinel)
A sites	B sites	
spin up	spin down	

If M is non-magnetic, the net moment is due to an imbalance of $(1+\alpha)-(1-\alpha)$ or $2\alpha Fe^{3+}$. A practical example is magnesium ferrite, $(Mg_{0.1}Fe_{0.9})[Mg_{0.9}Fe_{1.1}]O_4$ ($\alpha=0.1$). Zinc ferrite, $(Zn)[Fe_2]O_4$ ($\alpha=1$), is a *normal* spinel and is not ferrimagnetic as all the Fe^{3+} are on B sites. By contrast nickel ferrite, $(Fe)[NiFe]O_4$ ($\alpha=0$), is *inverted* and ferrimagnetism arises only because the Ni ions are coupled magnetically. Clearly no net moment arises from equal numbers of Fe^{3+} ions in the two sublattices.

The general formula of an unsubstituted garnet is $R_3Fe_5O_{12}$. R is trivalent yttrium (Y) or a rare earth (or again a mixture of ions). The structure is also cubic but here the cations occupy three types of interstitial site, called c, a, and d. With one or two important exceptions, each cation in a garnet is exclusively confined to one type of site, mainly because now the preferences of ionic species on three types of site have to be mutually compatible. The generalised distribution is

$\{R_3\}$	$[Fe_2]$	$(Fe_3)O_{12}$	(rare earth iron garnet)
c sites	a sites	d sites	
spin up	spin up	spin down	

The moment of the d sublattice dominates that of the a sublattice in the ratio 3:2. Y^{3+} is not magnetic but rare earth ions in the c sublattice couple antiparallel with the d sublattice, which interactions predominate at low temperatures. The most important unsubstituted materials are yttrium iron garnet (YIG), $Y_3Fe_5O_{12}$, and mixed yttrium gadolinium iron garnets (YGdIG), $Y_{3-a}Gd_aFe_5O_{12}$. Barium hexaferrite ($BaFe_{12}O_{19}$) is by far the most important of a number of complex ferrites with *hexagonal* as opposed to cubic symmetry. The structure is a close-packed array of the larger O^{2-} and Ba^{2+} ions with the smaller Fe^{3+} in

Table 14.1 Properties of typical soft magnetic alloys (After Duffin[2] and Brailsford[6])

Material and approximate composition	μ_i	μ_{max}	H_c (*A* m^{-1})	B_{sat} (*T*)	ρ ($\mu\Omega$ *cm*)
Iron, motor grade (99.6% Fe)	250	5×10^3	80	2.1	14
Silicon–iron, cold reduced, grain orientated (3.2% Si, 96.8% Fe)	2×10^3	7×10^4	8	2.0	48
Rhometal (36% Ni, 64% Fe)	1.8×10^3	7×10^3	12	0.9	85
Supermalloy (79% Ni, 15% Fe, 5% Mo, 1% Mn)	10^5	10^6	0.2	0.8	65
Permendur (49% Fe, 49% Co, 2% V)	800	5×10^3	180	2.4	28

Table 14.2 Main properties with typical values of NiZn and MnZn ferrites (After Snelling[3])

Property	*Symbol or expression*	*Measurement conditions*	*Range of values* NiZn	MnZn	*Units*
Initial permeability	μ_i	$\hat{B}\to 0$, $f<10$ kHz	10–2000	300–10^4	—
Saturation flux density	B_{sat}	$\hat{H}<10$ A mm^{-1}	0.1–0.45	0.3–0.55	T
Temperature factor	$\frac{\mu_2-\mu_1}{\mu_2\mu_1(T_2-T_1)}$	$T_1=25°C$, $T_2=55°C$, $\hat{B}\to 0$, $f<10$ kHz	−10 to +40	−1.0 to +3.0	$°C^{-1}\times 10^{-6}$
Disaccommodation factor	$\frac{\mu_1-\mu_2}{\mu_1^2\log(t_2/t_1)}$	$\hat{B}\to 0$, $f<10$ kHz	—	1–10	10^{-6}
Residual loss factor	$\frac{\tan\delta_r}{\mu}$	$\hat{B}\to 0$	—	1–100	10^{-6}
		$f<100$ kHz, $f>100$ kHz	20–2000	—	
Hysteresis loss coefficient	$\frac{\tan\delta_h}{\mu\hat{B}}$	$f=10$ kHz, $\hat{B}=1$–3 mT	2–4000	0.1–2.0	$mT^{-1}\times 10^{-6}$
Resistivity	ρ	d.c.	10^5–10^9	10–2000	Ω cm

no less than five different types of site. Per formula unit, eight Fe^{3+} are 'spin up' and four are 'spin down'. The symmetry bestows a *high uniaxial anisotropy* the material being magnetically hard with the easy direction of magnetisation along the hexagonal axis. $BaFe_{12}O_{19}$ and the very similar strontium compound are considered with other permanent magnet materials in Section 14.4.

14.3.2 Soft ferrites

The term 'soft ferrites' covers those spinel materials used broadly in inductor and transformer applications at frequencies up to about 300 MHz. Among all ferrites they represent the largest application category in terms of tonnes produced per annum.

Soft ferrites are *manganese–zinc* (MnZn) and *nickel–zinc* (NiZn). Substitution of Zn^{2+} (non-magnetic) for Mn^{2+} or Ni^{2+} (magnetic) is on A sites only which initially increases the net moment, and so the magnetisation, and reduces the anisotropy energy. The major practical effect is to increase the initial permeability μ_i and reduce the Curie temperature T_C.

Representative values of the main properties of NiZn and MnZn ferrites are given in *Table 14.2*. A comprehensive review of soft ferrites and their applications has been made by Snelling[3]

14.3.3 Microwave ferrites

These are used basically in the frequency range 1–40 GHz. An electromagnetic wave of frequency ω, propagating through a ferrite in the direction of an external biasing d.c. magnetic field H_a, has *two circularly polarised* components whose behaviour is described by the complex permeability $\mu_\pm' - j\mu_\pm''$ (*Figure 14.11*). The electron spin moments precess about H_a in sympathy with the positive sense component. Precession is sustained by energy coupled from the microwave field. *Gyromagnetic resonance* (peak in μ_+' occurs when

$$\omega=\gamma H_0 \text{ (Larmor precession)} \tag{14.24}$$

where H_0 is the *internal* field for resonance, i.e. H_a corrected for shape demagnetising effects and the effective anisotropy field. γ is the gyromagnetic ratio (~0.035 MHz A^{-1} m). Little energy is absorbed for precession in the anti-Larmor sense (μ_-',μ_-''), i.e. for signal propagation in the reverse direction, or a reversed H_a.

These non-reciprocal effects give rise to the isolation and phase shift necessary for the operation of most microwave devices, wherein the material is housed in a suitable waveguide, coaxial or stripline transmission line. The dispersive and dissipative character of μ is exploited in different ways, depending on where the ferrite is biased, and details of device geometry and construction.

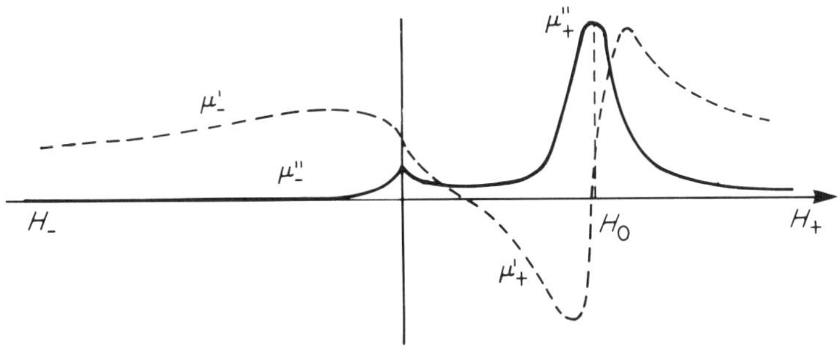

Figure 14.11 Real and imaginary parts of complex permeability versus d.c. magnetic field at fixed frequency

Table 14.3 Typical ranges of properties for microwave ferrite systems

Type	*General formula*	M_s ($kA\ m^{-1}$)	T_C (°C)	ΔH ($kA\ m^{-1}$)	ε'	$\tan\delta_\varepsilon$ ($\times 10^{-4}$)
Garnets						
YAlFe	$Y_3Fe_{5-y}Al_yO_{12}$	140–16	280–95	4.5–2.5	15.5–14	<2
YGdFe	$Y_{3-a}Gd_aFe_5O_{12}$	140–55	280	4.5–19	15.5	<2
Spinels						
NiAlFe	$NiFe_{2-y}Al_yO_4$	250–80	580–400	24–40	13–11	<10
NiZnFe	$Ni_{1-x}Zn_xFe_2O_4$	250–400	580–375	24–14	13–13.5	<10
MgAlFe	$MgFe_{2-y}Al_yO_4$	190–70	275–120	30–12	12.5–11	<5
LiTiFe	$Li_{0.5(1+y)}Fe_{2.5-1.5y}Ti_yO_4$	295–80	620–300	40–24	15.5–18	<5
LiZnFe	$Li_{0.5(1-x)}Fe_{2.5-0.5x}Zn_xO_4$	295–385	620–420	40–16	15.5–15	<5

The important properties of microwave ferrites are:

(a) Magnetisation: saturation magnetisation, M_s; temperature coefficient, $\frac{1}{M_s}\frac{dM_s}{dT}$

(b) Magnetic losses: resonance linewidth, ΔH; spinwave linewidth, ΔH_K

(c) Dielectric properties: dielectric constant, ε'; loss tangent, $\tan\delta_\varepsilon = \varepsilon''/\varepsilon'$

(d) *BH* loop properties: remanence ratio, B_r/B_{sat}; coercivity, H_{cB}

Both garnet and spinel types are widely used as microwave ferrite. Substitution of several different ions can be made to get the best combination of properties.

Major substituents in *garnets* are aluminium (Al^{3+}) and gadolinium (Gd^{3+}). Small quantities of ions such as indium (In^{3+}), holmium (Ho^{3+}), dysprosium (Dy^{3+}) and manganese (Mn^{3+}) are included to control specific properties.

In *spinels*, major substituents in *nickel* (Ni^{2+}) and *magnesium* (Mg^{2+}) ferrites are aluminium and zinc (Zn^{2+}); in *lithium* (Li^+) ferrites they are titanium (Ti^{4+}) and zinc. Again, small amounts of cobalt (Co^{2+}) and manganese control specific properties.

Typical ranges of properties for microwave ferrite systems are given in *Table 14.3*. The effects of individual ions on ferrite properties are summarised in *Table 14.4*. Von Aulock[10] has edited a sourcebook of these materials, while a concise review with emphasis on more recent advances has been made by Dionne[11].

14.3.4 Ferrites for memory application

Small toroids in MgMn, CuMn or LiNi spinel materials have been used in very large numbers as elements for rapid access data storage. High B_r/B_{sat} ratios are needed, with fairly high values of H_{cB} to increase switching speeds.

Garnets are used in the recently developed ***magnetic bubble** memories*. If a thin single crystal plate (~5 μm thick) with orthogonal easy direction is subjected to an increasing magnetic field $\boldsymbol{H}_a$ also normal to the plate, the antiparallel domains shrink until, over a narrow range of $\boldsymbol{H}_a$, small stable cylindrical domains ('magnetic bubbles') are left (*Figure 14.12*), each of which carries one bit of information. Typical bubble diameters are 2–5 μm. To ensure their stability and mobility, the plate must be free from imperfections, with faces as flat and parallel as possible. A promising technique is to grow a magnetic film by *liquid phase epitaxy* on a non-magnetic single crystal substrate. A suitable film composition is the garnet $Y_{2.9}La_{0.1}Fe_{3.8}Ga_{1.2}O_{12}$ using a substrate of $Gd_3Ga_5O_{12}$ (gadolinium gallium garnet or GGG). This technique has been reviewed by Pistorius *et al.*[12]

14.3.5 Manufacture of ferrites

Most ferrites are prepared as polycrystalline materials by ceramic processing. Metal oxides are reacted chemically in a *powder preparation stage* to form a suitably reactive ferrite powder of

Table 14.4 Effect of substituent ions on the properties of microwave ferrites

Property	Ion: Al^{3+} Site: *a, d*	Gd^{3+} *c*	In^{3+} *a*	Ho^{3+}, Dy^{3+} *c*	Mn^{3+} *a*	Al^{3+} *A, B*	Zn^{2+} *A*	Ti^{4+} *B*	Co^{2+} *B*
	Garnets					*Spinels*			
M_s	↓	↓	↑	→	↗	↓	↑	↓	→
T_C	↓	→	↓	→	↘	↓	↓	↓	→
$\frac{1}{M_s}\frac{dM_s}{dT}$ (20°C)	↗	↓	↑	→	↗	↗	↑	↗	→
ΔH	↘	↑	↓	↑	↘	→	↘	→	↘
ΔH_K	→	↗	↘	↑	→	→	↘	→	↑
$\tan\delta_\varepsilon$	→	→	↘	→	↘	→	→	↘	↘
B_r/B_{sat}	↘	→	↘	→	↑	↘	↘	→	↘

Key: ↑ = up, ↓ = down, ↗, ↘ = marginal, → = unchanged

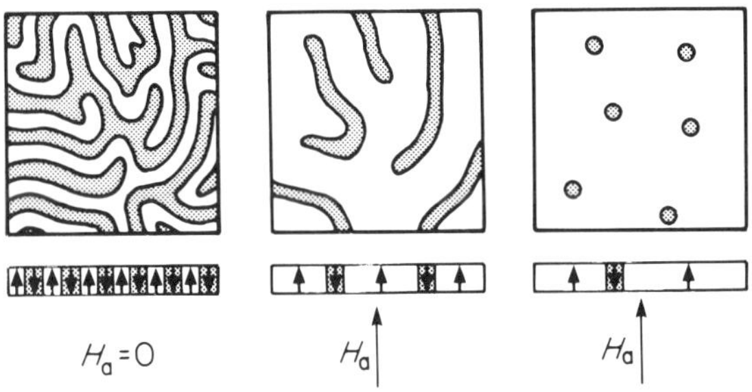

Figure 14.12 Domains in a thin plate of magnetic material with orthogonal easy direction (after Pistorius *et al.*[12], reproduced by permission)

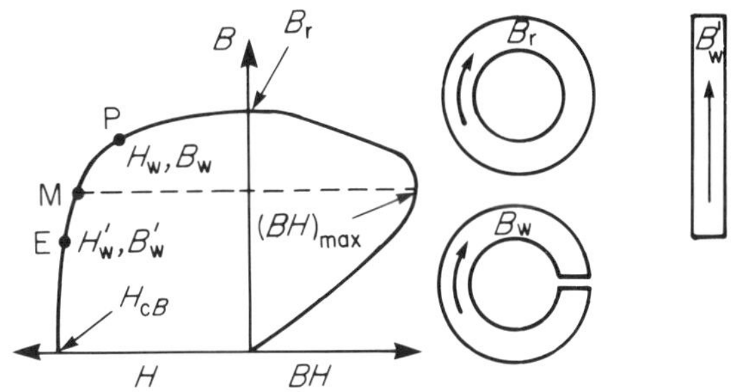

Figure 14.13 Permanent magnet design (after Brailsford[6], reproduced by permission)

correct composition. For instance, YIG is formed by reacting yttrium and iron oxides as follows:

$$3Y_2O_3 + 5Fe_2O_3 \rightarrow 2Y_3Fe_5O_{12}$$

Densification is achieved by green-forming the powder and sintering to make a dense solid with the right mechanical, electrical and magnetic properties.

To prepare the powder suitable grades of oxides (or carbonates, citrates, etc.) are weighed out in the correct amounts and wet ground in a ball mill. Other types of mill, e.g. attritors or colloid mills, are used or sometimes a dry grinding process such as air cyclone milling. The main object is to achieve *very intimate mixing*. After drying the material is usually *presintered* (calcined) at between 800 and 1300°C, below the final firing temperature, converting the fine oxide mixture to a coarse ferrite powder. This reduces shrinkage of the final shape and improves homogeneity and batch repeatability. A *more intensive milling* is usually needed to reduce the reacted oxide to a particle size that is ceramically workable. *Chemical solution methods*, e.g. hydroxide coprecipitation and solution freeze-drying, are used to make extremely reactive materials, where mixing occurs on a molecular scale. These are often difficult to control, and are less flexible than oxide milling processes.

The powder is green-formed as required, the material then having some 40–60% of the final density. The usual methods are die pressing, isostatic pressing or extrusion. Powder binders and lubricants are often necessary, and sometimes plasticisers as well. These are usually organic, and are added before the powder is granulated to improve flow properties. Mouldings are sintered to the final density by firing at between 1000 and 1500°C, depending on composition. Heating up is gradual, to eliminate smoothly any moisture and organic additives. *Microwave ferrites* are usually fired in pure oxygen to achieve the high densities required, whereas *MnZn ferrites* are fired, and more particularly cooled, in atmospheres low in oxygen, otherwise free Fe_2O_3 appears in the material, causing cracking and poor magnetic properties. Peak firing temperature and duration determine the final density, grain size and residual porosity. One special technique is *hot pressing* which is used to make very dense fine grain materials used in some high peak power microwave applications.

Ferrite manufacturing processes are considered in some detail by Snelling[13]. Some of the newer techniques are reviewed by Dionne[11].

14.4 Permanent magnets

14.4.1 Magnetic circuits

If a closed ring of permanent magnet material is magnetised to saturation and the magnetising coil is removed, the internal flux density is the remanence B_r. A small air gap of length g in the ring sets up a demagnetising field $\boldsymbol{H}_w$ causing the internal B to fall to some value B_w at P on the demagnetisation curve (*Figure 14.13*). P is called the *working point*. If the gap field and flux density are H_g and B_g, the pole face area is A_g, and the magnet has length l and cross section A then:

$$\varphi = B_w A = B_g A_g \quad \text{(assuming no flux leakage)}$$
$$H_w l = -H_g g \quad \text{(by Ampères law)}$$

confirming that H_w and H_g are opposed. Hence

$$B_w H_w l A = -B_g H_g g A_g \tag{14.25}$$

i.e. for a given flux density B_g in gap volume gA_g, the magnet volume is a minimum when B_wH_w is a maximum. So point M corresponding to $(BH)_{max}$ on the BH–B curve is the most efficient to work at. A large B_g implies a large $(BH)_{max}$ requiring both a high B_r and high $\boldsymbol{H}_{cB}$. In fact flux leakage affects both B_g and B_w, and is seldom negligible. It certainly has to be taken into account for a bar magnet: as the ratio l/A decreases demagnetising effects increase and the working point E(H_w', B_w') may fall below the *knee* of the BH curve (not necessarily coincident with M). An *external* demagnetising field returning H_w' to zero causes $\boldsymbol{B}$ to follow a recoil line to some remanent point $B_r' < B_r$. Evidently recoil from near P results in little loss in flux density. Finally *any* element of a magnetic circuit of length l, cross section A and permeability μ_r has a *reluctance* (in A Wb^{-1}) defined by

$$R_m = \frac{1}{\mu_0 \mu_r A} \tag{14.26}$$

Magnetic reluctance is analogous to electrical resistance and reluctances in series and in parallel may be combined using the same laws.

14.4.2 Ferrous alloys

Martensitic steels contain 0.5–1.0% carbon and were the first man-made permanent magnets. Properties are improved by adding Cr, Co and W. After alloying they are quench-hardened in oil or water; magnetic properties are stabilised by annealing at 100°C. The relatively high $\boldsymbol{H}_{cB}$ is due to internal strains which impede domain wall motion.

Dispersion hardened alloys are the major group of alloys. They contain basically Al, Ni, Fe and Co and are cast from a melt at 1100–1200°C and annealed at 550–650°C. *Dispersion hardening* occurs during cooling when small ferromagnetic particles of FeCo precipitate in a non-ferromagnetic NiAl matrix. Controlled cooling is vital for correct dispersion of the FeCo particles, which are probably essentially single domains to account for the high values of $\boldsymbol{H}_{cB}$. Materials of this type are called Alnico. *Anisotropic* materials (e.g. Alcomax, Ticonal) are made by cooling Alnico from the melt in an applied field ($\gtrsim 24$ kA m^{-1}). Recrystallisation of both phases along the $\boldsymbol{H}$ direction occurs for favourably

Table 14.5 Properties of typical hard magnetic materials (After Duffin[2], Brailsford[6] and Overshott[8])

Material and approximate composition	H_{cB} ($kA\ m^{-1}$)	B_r (T)	$(BH)_{max}$ ($kJ\ m^{-3}$)	
Carbon steel (1% C, 99% Fe)	4	0.95	1.4	I
Chromium steel (5% Cr, 0.7% C, 94.3% Fe)	5	0.94	2.4	I
Alnico (10% Al, 17% Ni, 12% Co, 6% Cu, 55% Fe)	40	0.80	13.6	I
Alcomax III (8% Al, 13.5% Ni, 24% Co, 3% Cu, 1% Nb, 50.5% Fe)	56	1.3	46.5	A
Columax (as Alcomax III)	60	1.35	62	AC
Samarium cobalt 1–5 ($SmCo_5$)	600	0.78	130	AC
Samarium cobalt 2–17 (Sm_2Co_{17})	400	1.2	300	AC
Feroba I ($BaFe_{12}O_{19}$)	140	0.21	8	I
Magnadur II ($BaFe_{12}O_{19}$)	135	0.38	24	AC

Key: I, isotropic; A, anisotropic; C, crystal orientated.

orientated grains. H_{cB} is increased, B_r is roughly doubled and $(BH)_{max}$ tripled in the principle direction and correspondingly reduced at 90°. Further improvement is made by *differential heat extraction* in the field direction during cooling, and columnar crystals grow with aligned cube edge crystal axes (e.g. Columax, Ticonal GX).

These materials are relatively costly and vulnerable to demagnetisation, obviating their use where large external fields or low working points are involved. Typically they are used in high field applications, e.g. loudspeakers, chucks, clamps, magnetrons, and generally in professional applications requiring high stability of B_r and H_{cB} and/or high environmental temperatures.

Ferrous alloys are discussed by Brailsford[14].

14.4.3 Non-ferrous alloys

Samarium cobalt is the most important of various rare earth cobalt alloys having *extremely high* H_{cB} and $(BH)_{max}$, and high B_r values. It has a linear BH (demagnetisation) characteristic and magnetisation and demagnetisation are both difficult. Manufacture involves fairly complex powder metallurgical techniques. Raw materials and processing are expensive, but as $(BH)_{max}$ is so high $SmCo_5$ is often more cost effective than Alcomax, and the very low working points achievable make for remarkable savings in weight and space. Rare earth cobalt materials in general are the subject of intense research and development effort at present (see, e.g. reference 15).

14.4.4 Hard ferrites

The $BaFe_{12}O_{19}$ structure was described in Section 14.3.1. Powder preparation of the important isotropic (barium) and anisotropic (barium and strontium) compounds is conventional, but must ensure grain sizes under 1 μm in the final ceramic to achieve the *very high* H_{cB} values of which they are capable. B_r by contrast is moderately high. *Isotropic* $BaFe_{12}O_{19}$, also formed and fired in conventional ways, has a linear demagnetisation curve and is used in general applications. The *anisotropic* materials are made by dry or wet pressing the powder in a field of order 500 kA m^{-1}. The crystallites are thus partially aligned, and remarkably alignment is improved on sintering. B_r is roughly doubled and $(BH)_{max}$ tripled in the principle direction (cf. Alnico and Alcomax). The BH curve shows a pronounced knee for anistrotropic $BaFe_{12}O_{19}$ precluding its use in severe demagnetising environments and/or at low temperatures where the working point may move below the knee. It is best suited to high working point *static* applications, e.g. loudspeakers, transducers. By contrast the anisotropic $SrFe_{12}O_{19}$ is particularly resistant to demagnetisation. It has a linear BH characteristic, and its H_{cB} is about 50% higher, though B_r and $(BH)_{max}$ are slightly lower, compared with anisotropic $BaFe_{12}O_{19}$. The Sr material finds use in *dynamic* applications with large external demagnetising fields, e.g. d.c. motor stators, torque drives.

Ba and Sr hexaferrites represent some 80% of all world permanent magnet production. Raw materials and processing are fairly cheap and the good B_r/H_{cB} characteristics make flat and compact assemblies possible. The extremely low eddy current losses are of value in r.f. work and some types of motor. One limitation is that the temperature variations of B_r and H_{cB} are an order higher than in Alnico/Alcomax.

A comprehensive review of hard ferrites has been made by Van den Broek and Stuijts[16]. Properties of representative permanent magnet materials are given in *Table 14.5*.

References

1 BLEANEY, B. I. and BLEANEY, B., *Electricity and Magnetism*, 3rd Edn, Oxford University Press, Oxford (1976)
2 DUFFIN, W. J., *Electricity and Magnetism*, 3rd Edn, McGraw-Hill (1980)
3 SNELLING, E. C., *Soft Ferrites*, Iliffe (1969)
4 SNELLING, E. C., *Soft Ferrites*, p. 23, Iliffe (1969)
5 SNELLING, E. C., *Soft Ferrites*, p. 27, Iliffe (1969)
6 BRAILSFORD, F., *Physical Principles of Magnetism*, van Nostrand (1966)
7 BRAILSFORD, F., *Physical Principles of Magnetism*, p. 227, van Nostrand (1966)
8 OVERSHOTT, K. J., *Electronics and Power*, November/December, 768 (1976)

9 BRAILSFORD, F., *Physical Principles of Magnetism*, p. 226, van Nostrand (1966)

10 VON AULOCK, ed., *Handbook of Microwave Ferrite Materials*, Academic Press (1965)

11 DIONNE, G. F., *Proc. IEEE*, **63**, 777 (1975)

12 PISTORIUS, J. A., ROBERTSON, J. M. and STACY, W. T., *Philips Tech. Rev.*, **35**, 1 (1975)

13 SNELLING, E. C., *Soft Ferrites*, p. 5, Iliffe (1969)

14 BRAILSFORD, F., *Physical Principles of Magnetism*, p. 243, van Nostrand (1966)

15 *Goldschmidt Informiert*, **48**, Th. Goldschmidt AG, 1 (1979)

16 VAN DEN BROEK, C. A. M. and STUIJTS, A. L., *Philips Tech. Rev.*, **37**, 157 (1977)

15 Inductors and Transformers

H C Remfry CEng, MIEE, MBIM, MIIM
Senior Engineer,
ITTE Component Group

Contents

15.1 General characteristics of transformers

The performance of transformer components may be considered in terms of their impedance characteristics and their equivalent circuit, which allows the analysis of transient and transfer functions. The ideal transformer has infinite turns and zero losses. It can therefore represent perfect coupling between input and output circuit or primary to secondary windings. The equivalent circuit includes the parameters shown in *Figure 15.1*.

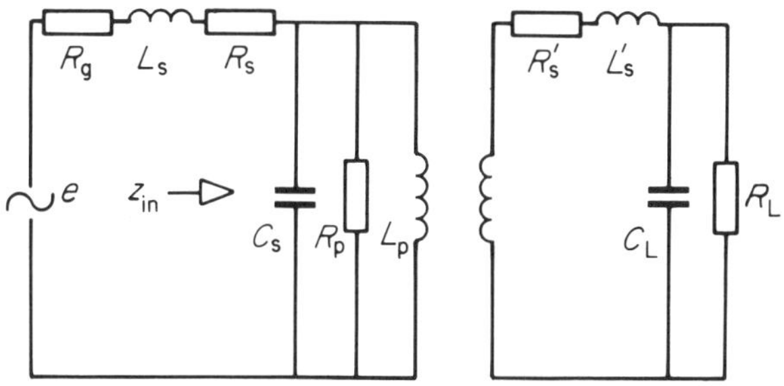

Figure 15.1 Transformer equivalent circuit

The frequency response dependent parameters at low frequencies are the resistive and shunt inductive elements which will attenuate the output voltage as the reactance of the shunt inductance decreases with reducing frequency. At midband frequency the resistance and the reactance of the leakage inductance define the output characteristic and at high frequency the series inductance and shunt capacitance define a stop band. The input impedance Z_{in} is affected by resonance points in the frequency range and these occur due to shunt inductance and self-capacitance acting in current resonance, i.e. apparent infinite impedance, and the self-capacitance with the leakage reactance acting in voltage resonance, i.e. apparent zero impedance. The shunt resonance defines a change in input impedance from inductive to capacitive reactance. Other resonance points occur due to the distributed nature of the reactances, but these will normally lie outside the operating band.

For a step function input the delay parameters are leakage inductance and self-capacitance. Depending on their magnitude either one or the other will be dominant, that is L/R or CR will delay the rise-time or the fall-time and overswing may be present depending on the degree of damping. The equivalent circuit shows discrete component parameters which allow analysis by the use of linear differential equations, the physical model however resembles a transmission line with distributed parameters where non-linear differential equations prevent satisfactory solution[1].

15.2 Transformer design practice

Design practice involves using a series of trials to achieve convergence to a desired performance. A summary of the steps involved is:

(a) Analysis of performance to define parameter specification.
(b) Selection of available material to satisfy parameter specification.
(c) From the defined parameter specification determine the approximate core excitation, that is minimum inductance or Q, or maximum magnetising current or core loss. The number of turns N and conductor size may then be provisionally allocated.
(d) With starting value of N turns and the determined ratio of additional windings, the allocation of conductor size, turns and insulation for subsequent windings may be made with reference to the rating or current capacity.
(e) The geometry of the coil will allow calculation of mean turn, conductor length, winding depth, weight and resistance.
(f) From the estimated dimensions and parameters found in (e), the resistive drop and regulation can be calculated, together with estimates of the leakage inductance, the winding and interwinding capacitance, the voltage gradients, dissipation and thermal gradient.
(g) The mathematical model can then be checked against the performance requirement with corrective action taken on deviations.

15.3 Transformer materials

The principal winding material is round insulated copper conductor which may be solid, stranded or bunched. Variations are round insulated aluminium over a limited size range; insulated rectangular copper or aluminium which may be solid or stranded; sheet or foil in anoidised aluminium or film coated copper.

Published round wire tables are available from manufacturers, e.g. BICC Ltd, and the corresponding British Standards, e.g. BS 4520, give quality assurance and qualifying limit performance data.

Insulation coverings are normally synthetic enamels, but variations include single, double or multiple coverings in paper, glass, rayon or silk, where increased spacings to reduce capacitance or arduous duty requirements have to be met.

15.4 Transformer design parameters

15.4.1 Windings

Using the perimeter length of the tube or spool faces on which the windings is to be wound as a base turn dimension (bt), calculation of the mean turn (MT) of each winding can be made for *Figure 15.2* as,

$$\mathrm{MT}_1(\mathrm{W}_1) = \pi d_1 + \mathrm{bt}$$

$$\mathrm{MT}_2(\mathrm{W}_2) = 2\pi(d_1 + t_1 + \tfrac{1}{2}d_2) + \mathrm{bt}$$

$$\mathrm{MT}_N(\mathrm{W}_N) = 2\pi(d_1 + t_1 + d_2 + t_2 + \tfrac{1}{2}d_N) + \mathrm{bt}$$

where d_1, d_2 and d_N are winding depths (heights) and t_1, t_2 are thicknesses of insulation. The MT value may then be used to calculate the active material weight W and the resistance R as follows:

$$W = \mathrm{MT} \times N \times a \times S_d$$

$$R = \rho l/a = \rho N \mathrm{MT}/a$$

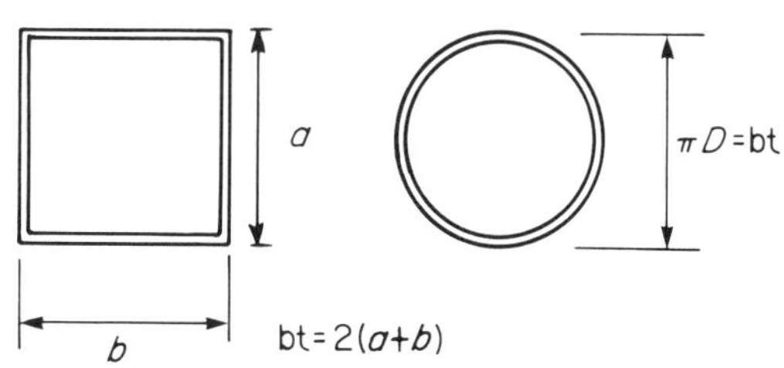

Figure 15.2 Base turn dimension

where N is the number of winding turns, a is the conductor area, S_d is the wire density (9.6 g cm^{-3} for Cu) and ρ is the resistance per unit length (1.73 $\mu\Omega$ mm^{-1} at 20°C for Cu).

For small coils an average mean turn of all the windings will give small error, since the error cancels on the sum of all the windings, the inner windings being overstated, the outer windings understated.

With larger coils it is wise to restrict the error and to compute the individual winding dimensions as above.

The sum of the winding depths, $d_1+d_2+d_N$, must be less than the total winding space d, by a winding space factor which is dependent on the winding practices and skills employed. Variations in factors are great particularly for toroidal, transposed, random or orthocyclic windings.

15.4.2 Capacitance

Capacitive elements are present at low value and consequently become dominant only at high frequency or fast wavefronts. They have significance in terms of voltage stress, r.f. interference, crosstalk and common-mode coupling between circuits.

The capacitive elements affect and may override the inductive voltage distribution interval to the winding, *Figure 15.3* shows distributed capacitance elements present in windings.

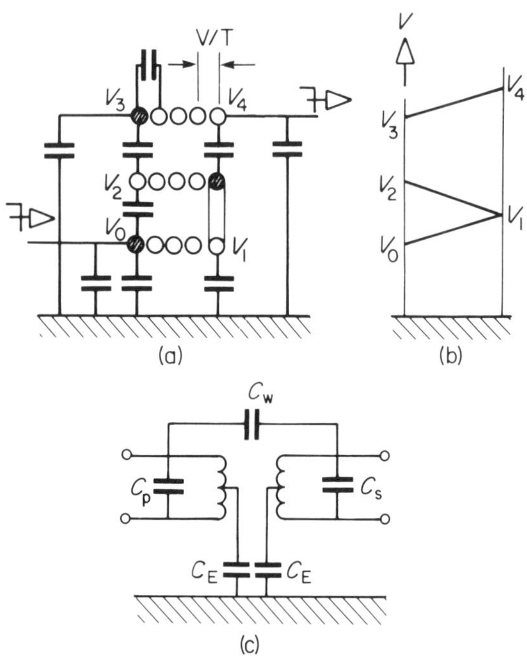

Figure 15.3 Winding capacitance: (a) distributed capacitance in a two-winding coil; (b) voltage gradient; (c) capacitive impedance balance

Considering the first winding layer, the layer capacitance C_l is given approximately by

$$C_l=0.008\,854A\varepsilon/t\ \text{pF}$$

where A is the layer area (in mm^2), t is the dielectric thickness (in mm) and ε is the dielectric constant ($\sim$3–5).

The winding capacitance C_s is given approximately by

$$C_s=\frac{4C_l}{3N}\left(1-\frac{1}{N}\right)\text{pF}$$

where N is the number of layers in the winding. Computation of capacitance value is approximate and measured values may need to be established. For reference see MacFadyen[2], Snelling[3], *Mullard Technical Handbook*[4], and Siemens[8].

The interwinding capacitance C_W may usually be measured directly on *LCR* bridges whilst the self-capacitance requires a Q bridge to determine two values of resonating C for two resonant frequency points:

$$\frac{(f_1)^2}{(f_2)^2}=\frac{C_1}{C_2}$$

and

$$C_s=\frac{(f_1)^2}{(f_2)^2}C_1-C_2$$

Alternatively a dip meter measurement for minimum current point can be made to determine the self-resonant frequency assuming the winding inductance is constant.

The interwinding capacitance C_W is interactive with the core materials and particularly so for ferrite which has a high dielectric constant.

15.4.3 Leakage inductance

The series leakage inductance is calculated in the following general form with reference to *Figure 15.4*. It is dependent on the winding arrangement and it should be noted that the leakage inductance is not frequency dependent, the system acting as an air-cored inductor.

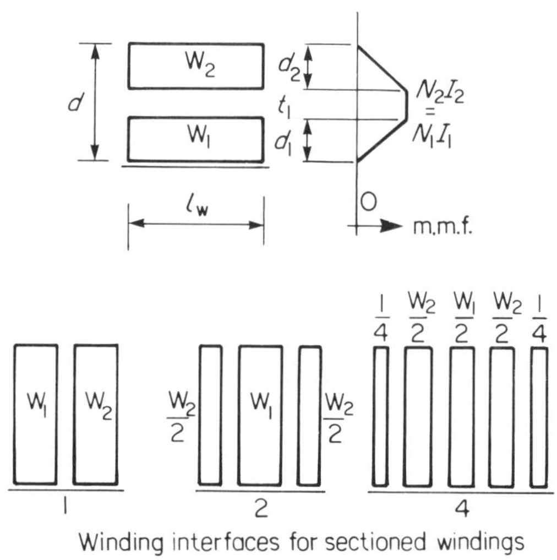

Figure 15.4 Winding leakage inductance

$$L_s=4\pi\times10^{-4}N_1^2\frac{\text{MT}}{l_w}\left(\frac{d_1+d_2}{3}+t_1\right)\mu\text{H}$$

where MT is the average of two windings, l_w is the length of winding traverse, and MT, l_w, d_1, d_2 and t_1 are in mm.

The core excitation for transformers may be considered as a shunt inductance and resistance in parallel with the load, and corresponds to the core magnetisation and loss currents. Values depend on the working flux density B, the frequency and the material used in the core. The definition of self-inductance L is the induced e.m.f. E per unit rate of change of current I where $E=-L\mathrm{d}I/\mathrm{d}t$.

The energy required to establish a current I in an inductance L can be regarded as stored in the magnetic field, so the stored energy

$$W=\int_0^I Li\,\mathrm{d}t$$

and if L is constant the total stored energy is $\frac{1}{2}LI^2$.

The flux density B is related to E by $\mathrm{d}\varphi/\mathrm{d}t$ and $Et=BNA\times10^{-8}$ where A is a uniform core section area in cm^2, N is the number of turns and t the time in seconds.

The m.m.f. H is equal to the work done in carrying a unit pole N times around the current in the winding where

$$H=\frac{4\pi NI}{10l}$$

The permeability μ is the ratio B/H, giving

$$\frac{E}{I}=Z=R+jX=j\omega\frac{4\pi N^2A}{10^9 l_e}\mu$$

and

$$L=\frac{4\pi N^2A\mu}{10^9 l_e}$$

where l_e is the magnetic path length in cm and $\omega=2\pi f$.

15.4.4 Losses

The core loss resistance R may be expressed as

$$R=\frac{4\pi N^2Ak}{10^9 l_e}$$

where k is an experimentally determined factor for the material and flux condition. The permeability μ identifies with values of flux density B. Hence at low flux density μ remains constant[5] and is known as initial permeability μ_i or μ_0. When B has the greatest rate of change with H the permeability is μ_{max}. At operating B_{max} it is called the amplitude or apparent permeability μ_a and for a composite permeability, such as a core interrupted by an air gap space, it is the effective permeability μ_e. Incremental permeability $\Delta\mu$ refers to the shape of a cyclic magnetisation superimposed onto a static magnetisation. For a specific size of core the permeability may be expressed as an inductance factor A_L, usually in nanohenry per turn. Permeability coefficients are evaluated for temperature change, frequency and time.

The loss resistance k is frequency and flux dependent and consists of hysteresis and eddy current loss together with stray or residual loss. It has the form $W=W_h+W_e+C$ where at low induction W_h varies as B^3f, or at high induction as $NB^{1.6}f$, where N is the Steinmetz factor. W_e varies as B^2f^2 or at high induction as

$$\frac{\pi tB^2}{(\mu\rho)^{1/2}}f^{3/2}$$

C represents an energy loss of unknown origin, t is the thickness and ρ the resistivity[5].

The effect of eddy current circulation is such as to oppose the field producing them and to prevent the creation of a uniform field within the lamination or core thickness. The effective permeability is therefore dependent on this skin effect, and at very high frequency 95% of the true permeability may be lost. Welsby[6] defines a critical depth of penetration equal to half the lamination thickness occurring at critical frequency f_c where

$$f_c=\frac{Q}{\pi^2f^2\mu_0}\times10^9\quad\text{for } Q>6$$

The eddy current skin effect is present also in the winding conductor where the eddy currents oppose the flow of current in the centre of the conductor and effectively increase the resistance of the conductor. A similar action is produced by the proximity effect of the alternating magnetic fields of nearby conductors. To limit this increase in losses the use of Litz, bunched or transposed, conductors is necessary, usually at high frequency or in larger windings with high non-active winding space.

15.4.5 Heating

Losses produced within the core and windings create heat and raise the temperature of the materials. Because of the uneven source of heating the structure has thermal gradients recognised as hot spot, and average and surface temperature. With cooling by natural convection the heat is transferred to the surrounding ambient and a steady state temperature will be achieved under constant load conditions, and after a time period dependent on the size and shape of the component. Assessment of thermal gradient for air natural cooling can be made only with extreme simplification. Representation of the heat flow may be made by considering an equivalent electrical circuit as in *Figure 15.5*, but experimental data obtained from physical models are usually necessary to validate a final design.

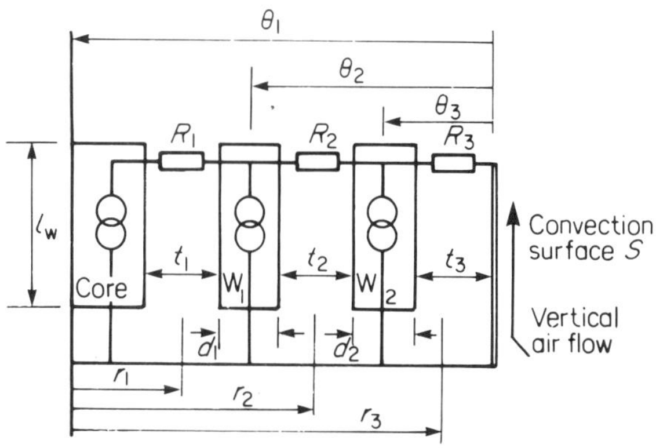

Thermal conductivity ($10^3\ \mu\mathrm{W\ mm^{-1}\ °C^{-1}}$)	
Copper	385
Iron	13
Aluminium	220
Ferrite	3.5
Filled epoxy	0.4
Plastic nylon	0.25
Paper	0.1

Figure 15.5 Radial heat flow gradient

The schematic[3], represents radial heat flow from heat generators (shown as current sources) passing through low conductivity insulating material (shown as resistors) towards the convection surface where the heat is transferred to the ambient air. The surface areas of the low conduction barriers are calculated from the winding length l_w and a mean turn circumference for the barrier of $2\pi r$. Using known values of thermal conductivity a calculation of temperature gradients may be made. For example, using the following values from the outline design for a full-wave centre-tap rectifier and capacitive-input (FWCT–CIF) mains transformer:

$l_w=38.8$ mm

$r_1=14.29$ mm	$r_2=18.29$ mm	$r_3=25.29$ mm
$t_1=1.0$ mm	$t_2=0.25$ mm	$t_3=0.25$ mm
$P_1=1.87$ W	$P_2=3.74$ W	$P_3=3.61$ W

(assuming round limb). Now, the heat flow through R_3 is given by

$$\frac{P_1+P_2+P_3}{2\pi r_3 l_w}=\frac{1.87+3.74+3.61}{6.28\times 25.29\times 38.8}=0.0015 \text{ W mm}^{-2}$$

The thermal resistance of R_3 is given by t_3/λ, where $\lambda=200\ \mu$W mm^{-1} °C^{-1} for tape. Hence

$t_3/\lambda=0.25/200=0.001\,25$°C mm^2 μW^{-1}

Thus the temperature drop θ_3 across R_3 is given by

$$\theta_3=0.001\,25\times 1500=1.87\text{°C} \qquad (15.1)$$

The heat flow through R_2 is

$$\frac{P_1+P_2}{2\pi r_2 l_w}=\frac{5.61}{4457}=0.001\,26 \text{ W mm}^{-2}$$

The thermal resistance of R_2 is t_2/λ and so

$t_2/\lambda=0.25/200=0.001\,25$°C mm^2 μW^{-1}

Thus the temperature drop $(\theta_3-\theta_2)$ across R_2 is

$$\theta_3-\theta_2=0.001\,25\times 1260=1.57\text{°C} \qquad (15.2)$$

Performing the same calculation for R_1 we have

$P_1/2\pi r_1 l_w=0.000\,501\,9$ W mm^{-2}
$t_1/\lambda=1.0/250=0.004$°C mm^2 μW^{-1}

where λ for plastic is 250 μW mm^{-1} °C^{-1}. Thus the temperature drop $(\theta_1-\theta_2)$ across R_1 is

$$\theta_1-\theta_2=0.004\times 502=2.0\text{°C} \qquad (15.3)$$

We can now solve equations (15.1), (15.2) and (15.3) to give

$\theta_3=1.87$°C $\quad \theta_2=3.44$°C $\quad \theta_1=5.44$°C

The temperatures calculated are the temperatures above that of the convection surface S.

The average temperature rise of the winding is conveniently measured by the change in resistance whilst carrying a constant load current. After a period of time $3T$, where T is the time constant of the transformer, a steady state temperature will be reached. The percentage change in resistance R_0 (initial) to R_T (steady state) will give the temperature rise of the winding using the temperature coefficient of the resistivity of copper of 0.393°C/% change.

The resistance measurement is the average resistance of the winding and includes increments of resistance higher than average (hotter) and also of lower resistance (cooler). For coils with height greater than a few centimetres the surface temperatures at the base and at the top of the coil will be different. For coils of 50 cm height the temperature difference becomes extreme. The radial temperature gradient is also distorted at various levels of coil height and for large units the use of cooling ducts becomes necessary to prevent excessive temperature values. The presence and type of varnish, also the degree of penetration and retention have a marked effect on reducing thermal gradient and significant improvements result in the use of liquid cooling.

Safety requirements involve limiting the temperature rise under fault conditions, which may be a continuously applied short circuit of the load. A limit may also apply to rated load so that a dissipation curve may need to be followed closely. These are referred to in BS 3535.

15.4.6 Vector presentation

The input impedance of a transformer connected load may be represented by the magnitude and vector of the constituent parameter losses. The method is used to determine the accuracy of ratio and phase errors in precision measuring current and voltage transformers. It is also useful to consider the performance of audio transformers at a given frequency (*Figure 15.6*). Using the equivalence of a unity ratio transformer the effects on the load of the series and shunt parameters are illustrated in the vector diagrams. The parameters have been treated as partially distributed. If the secondary winding leakage and resistance are transferred to the primary winding values and lumped to them, there is a variation in performance:

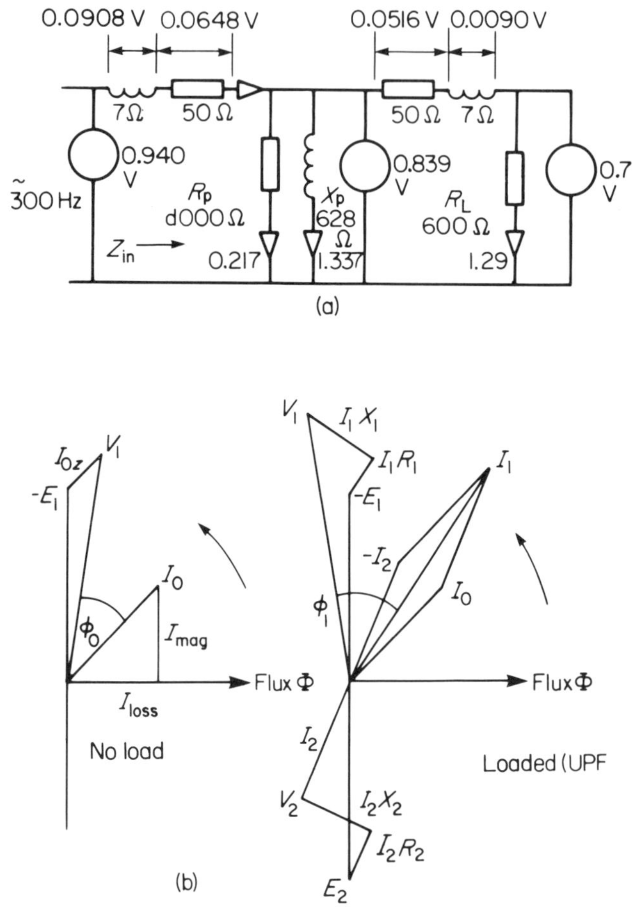

Figure 15.6 (a) Equivalent circuit and (b) vector diagram of an audio transformer

	Lumped parameter	*Partial distributed parameter*
$Z_{in}=E_{in}/I_{in}$	469 Ω	467 Ω
Insertion loss $20\log(E_0/E_{in})$	−1.35 dB	−1.68 dB
Insertion loss $20\log\left(\frac{Z_{in}+R_L}{Z_{in}-R_L}\right)$	−12.23 dB	−18.09 dB

Care is therefore necessary in the allocation of parameters to an equivalent circuit and recognition that the distribution may not be entirely defined. The equivalence of equivalent circuits is given by MacFadyen[2]. Vector representations of loads with lead and lag current in the secondary winding are illustrated in *Figure 15.7*.

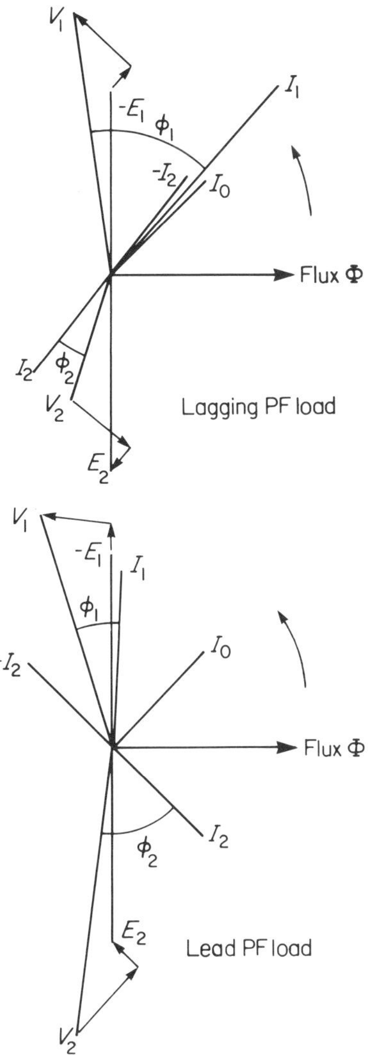

Figure 15.7 Parameter vector

15.5 Inductors

The ability to store energy in the form of electromagnetic charge is a characteristic of inductors and reactors and complements the electrostatic charge energy storage present in capacitance. The inductor has low impedance at zero frequency and high impedance at high frequency. The capacitor has high impedance at zero frequency, low impedance at high frequency. The characteristic equations for the voltage across and the current through these passive components are:

Inductance $\quad v=-L\dfrac{\mathrm{d}i}{\mathrm{d}t} \qquad i=\dfrac{1}{L}\displaystyle\int_0^t v\,\mathrm{d}t$

Capacitance $\quad v=\dfrac{1}{C}\displaystyle\int_0^t i\,\mathrm{d}t \qquad i=C\dfrac{\mathrm{d}v}{\mathrm{d}t}$

The inductor is normally used in series mode and is current driven, whilst the capacitor is used in parallel or shunt mode and is voltage fed. They have linear characteristics but have limit value at overcurrent (saturation) or overvoltage (breakdown).

The inductor dissipation losses are due to the winding conductor resistance and to the loss generated when under magnetic excitation, and a relationship $Q=2\pi fL/R$ may be used to define the inductor performance. It also expresses the loss angle or power factor presented to the circuit.

The use of air gaps in the magnetic path extends the parametric range of inductive components and enables highly stable linear components to be constructed. Provision of moveable core material within the air gap space allows adjustment of inductance. Inductor sizes vary from very small air cored to very large protection reactors where they may be embedded in concrete to withstand electromagnetic forces produced by fault currents.

Their application in electronic circuits however is mainly in suppression, timming and smoothing duties. Here the core requires a large flux range and the silicon–iron materials are usually most suitable to accommodate the dual excitation of high d.c. ampere turns and the superimposed smaller a.c. excitation. In multi-phase rectifier circuits a centre-tapped reactor is often used to solve commutation problems and limit the large circulating current that would otherwise flow between the conduction phases. For power applications the design may be limited by thermal rating due to either winding loss or to the core loss present at high a.c. excitation or high frequency.

Inductors required in frequency selective circuits and particularly in carrier telephony networks are characterised by the need for high stability performance at very low signal amplitude. The construction normally requires ferrite material in pot core form with high shielding from cross-talk coupling. The inductors will be provided with an adjustment to permit setting the value accurately. To assist in the selection of core size it is possible to calculate and plot a range of inductance and loss values against frequency in the form of iso-Q curves. A line joining equal Q-value points provides a Q-contour from which inductance, frequency and turns may be selected. (See *Figure 15.8*.)

At high frequency the filter inductor is designed to have maximum Q-factor at a given frequency. At lower frequencies requiring high inductance value, which in turn requires smaller gap dimension and high μ_1, the variability encountered may require the use of a lower Q-factor with less sharp cut-off frequency. The need to work at low induction requires correspondingly high turns giving increased winding resistance and at the higher frequencies, increased a.c. resistance. The presence of the distributed capacitance within the winding produces dielectric losses which, because of the small signal amplitude, are no longer negligible and contribute to the terminal impedance.

The choice of conductor size becomes complex, as for instance the use of stranded 0.071 mm diameter conductors with higher eddy current loss may give lower overall loss than stranded 0.040 mm conductor which may have higher capacitance and dielectric loss. The performance is therefore dependent on careful manufacture and model validation, before design release, although the use of computer programs is offered by some of the manufacturers of the core material.

Powdered nickel and iron dust cores are currently used for inductors and smoothing chokes at high frequency. The powdered nature of the core material, held together by a binding agent, provides a distributed air gap with low electromagnetic interference field. In normal form the cores are toroidal and are available with toleranced A_L value, stabilised and with stated temperature coefficients. A wide range of values may be produced usually with single layer windings using wire size suited to the current rating. The saturation flux density for this material is twice that of ferrite. The use of nickel core material in lamination form giving high permeability and low hysteresis loss is common in telecommunication components at speech frequency and particularly where superimposed direct current is present—the high inductance and low loss requirement being necessary to

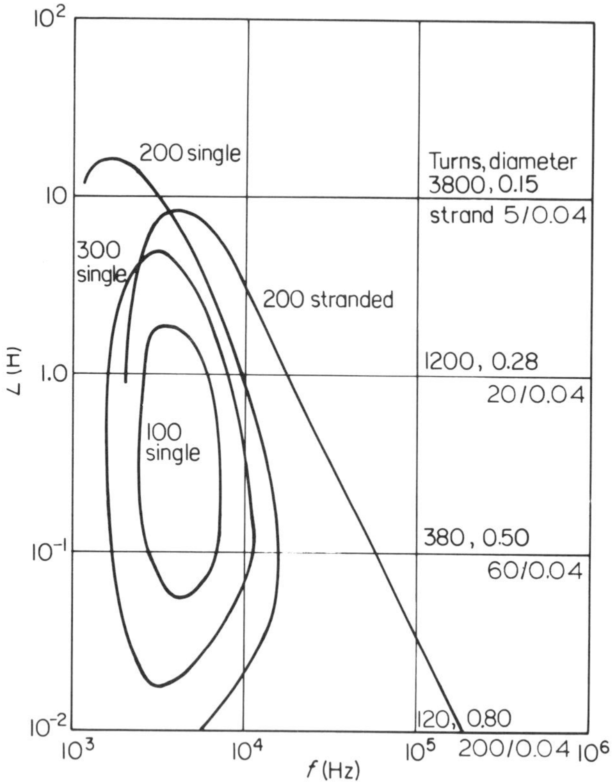

Figure 15.8 Typical full winding Q-factor: 45 mm pot core, μ=100 (from *Mullard Technical Handbook*[8])

reduce attenuation of low frequency signals.

A wide range of nickel alloy materials are available with permeability figures of 260000 being achieved. The effect of orientation and cold reduction in the preparation of nickel strip allows square loop materials with the ratio of remenance flux density to saturation flux density greater than 95%, and which will operate in free running oscillators and have ideal characteristics for saturating reactors or transductors.

15.6 Design examples

Design examples are presented in the form of working notes and establish the scenario, but they need further work to complete the design. The parameters involved in the examples are described briefly in the text and covered by data presented in the illustrations. Magnetisation data have not been included and it is necessary to refer to manufacturers' handbooks for further information.

15.6.1 50 Hz rectifier mains transformer

It is desired to provide a smoothed direct current output of 40 V at 1 A full load and a current of 0.1 A light load, using a full wave centre-tap rectifier and capacitive input filter with values (see *Figure 15.9*): $R_{FL}=40\,\Omega$, $R_{LL}=400\,\Omega$, $N=2$ (number of rectifiers), R_s = source resistance. Using $2\pi fCR_L$ of 0.3 and $R_s/R_L=10\%$ gives $C=24\,\mu$F, $E_0=0.57E_{pk}$, E_{ripple} is almost 100%, $T=CR=$ 0.96 ms. With $2\pi fCR_L$ of 100 and $R_s/R_L=10\%$ then $C=8000\,\mu$F, $E_0=0.76E_{pk}$, $E_{ripple}=1.3\%$, $CR=320$ ms. The first has too low a conversion factor and little smoothing, the second gives high

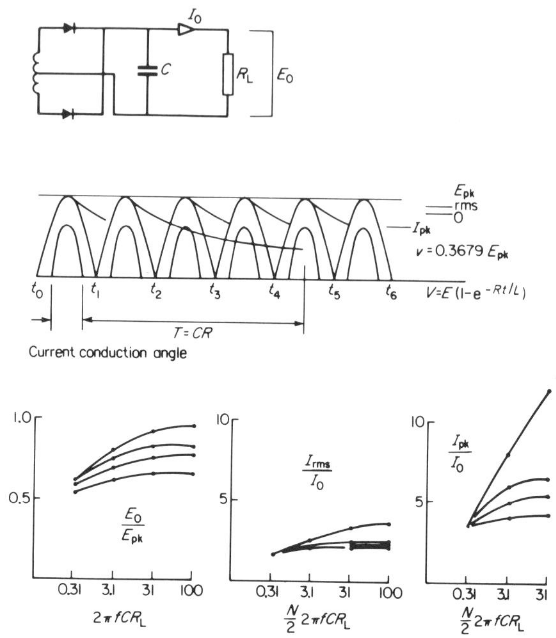

Figure 15.9 Full wave rectification (from O. H. Schade, *Proc. IRE*, July 1943))

conversion but excessive peak currents. Checking alternative $2\pi fCR_L$ values we have (in obvious units):

$2\pi fCR_L$	C	E_0	E_{pk}	E_{rms}	E_{ripple}	CR	I_{rms}	I_{pk}
13	1035	0.75	53.3	37.7	10	41.4	2.2	6.4
28	2230	0.76	53.3	37.7	5	89.2	2.3	6.4
150	12000	0.76	53.3	37.7	1	48.0	2.3	6.4

Satisfactory utilisation occurs at $2\pi fCR_L=30$.

Figure 15.9 for the circuit shows the approximate current conduction for the sine wave period, the relative discharge time constant to T=3 periods or 30×10^{-3} seconds from a 50 Hz supply frequency, e.g. 1.5 cycles. The time constant of 39.2 ms corresponds to a ripple percentage of E_0 of approximately 5%.

The secondary rating for this unit is then 76 V centre-tap at 1.6 A = 60.8 volt-ampere/phase.

Referring to the table of lamination sizes (*Table 15.1*) the pattern 196 in square stock configuration has a turns per volt value of 4.05 at a working flux density of 1.5 T at 50 Hz. Allowing for a winding regulation of 10%, $V_{secondary}=83.6$ V at no load.

The design would proceed as follows:

Primary turns, N_1	240 V × turns/V = 972
Secondary turns, N_2	83.6 V × turns/V = 338 tap at 169
Winding area, A_N	4.65 cm^2 say space factor (SF) 0.7 Allow $\frac{1}{3}$ winding space for primary and $\frac{2}{3}$ space for secondary
Conductor for N_1	$972\times(\frac{1}{3}\times4.65\times0.7)$ turns/cm^2 = 895 turns/cm^2 bare diameter 0.280 mm from wire table has 896 turns/cm^3, grade 2 covering

Table 15.1 Core parameters

No. (8)	Plan L (cm)	Plan W (cm)	A_e (cm^3)	l_e (cm)	Wt. of Fe (kg)	l_w (mm)	d (mm)	bt (mm)	A_N (mm^2)	Wt. of Cu (kg)	A_L (3)	A_R (4)	$LI/R^{1/2}$ (5)	Rating (VA) (2)	Turns/V (1)	Comment
35	57.2	47.6	3.26	11.5	0.296	25.4	7.5	86.5	190	0.075	3572	23.1	0.031	14	9.2	DIN 60 near
29	76.2	63.5	5.8	15.3	0.695	33.8	10.2	117	345	0.209	4776	15.6	0.069	45	5.1	DIN 75 near
196	85.7	71.4	7.34	17.2	0.985	38.8	12	135	465	0.382	5377	11.8	0.101	75	4.05	DIN 84 near
248	133	111	17.8	26.7	3.72	63.5	19.1	203	1210	1.46	8400	7.1	0.344	250	1.65	DIN 135 near
0405	23.8	33.3	0.58	7.45	0.032	12.7	5.5	44.5	70.5	0.018	981	32.9		1	49.5	Single HWR C core
1008	57.2	54	1.98	12.4	0.177	33.3	9.9	79.5	330	0.154	2012	11.4	0.05	5	14.9	Double HWR C core
5014	95.3	88.9	5.25	20.6	0.827	57.3	19	118	1089	0.867	3211	5.1	0.25	75	5.35	Double HWR C core
9024	146	127	13.05	30.2	3.12	79.5	31	184	2465	2.89	5440	3.6	190	300	2.19	Double HWR C core
6/3	28.6	19	0.206	7.48	0.012	56.5	4.5	23	191	0.027	347	8.18	—	0.4	129	HWT ring DEF5193
10/6	47.6	31.8	0.715	12.47	0.068	96.7	7.7	38.8	559	0.142	722	4.41	—	8	37	HWT L=OD, W=ID
12/6	57.2	38.1	0.824	14.96	0.099	116	9.3	42.1	810	0.257	744	3.1	—	16	32.1	HWT L=OD, W=ID
18/9	85.7	57.1	1.93	22.45	0.332	182	14.5	59.2	1990	0.935	1083	1.84	—	150	13.7	HWT L=OD, W=ID
C118	18	11	0.43	2.6	0.006	6	3.05	27.3	16	0.002	2084	93.9			4.3^7	IEC 2 slot BS 4061
RM10	24.7	18.7	0.842	4.13	0.020	11	4.2	39.8	46.2	0.010	2569	44.4	0.002		8.4^7	DIN 41980
K45	45	29.2	3.62	6.7	0.109	16.84	6.7	72.5	113	0.041	6808	31.8	0.014		36.2^7	BS 4061 Range 1
EC35	34.5	34.6	0.865	10.3	0.040	28.8	7.5	51.8	216	0.018	1432	21.7	0.003		8.6^7	
EC70	70	69	2.83	14.1	0.180	41.4	11.3	62.8	468	0.214	2529	6.6	0.026		28.3^7	

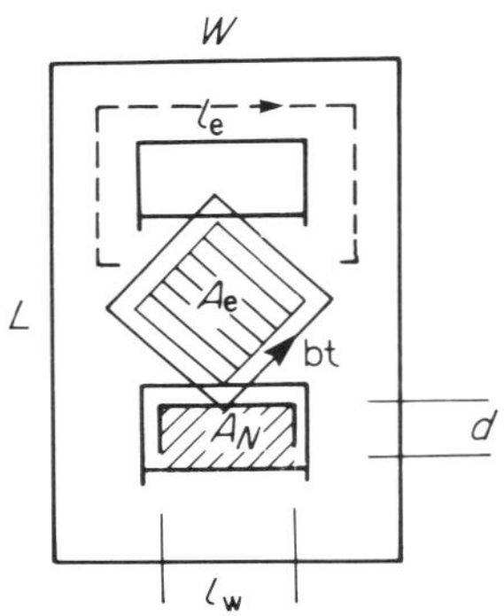

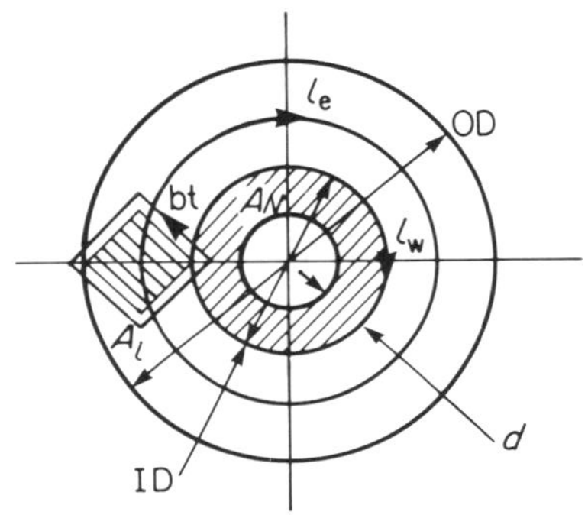

1 Turns/V for $B_{max} = 1.5$ T, 50 H_2
2 Load rating, normal ambient for turns/V
3 $L = N^2 A_L$ in nH at $\mu_R = 1000$
4 $R = N^2 A_R$ in $\mu\Omega$/20°C
5 $LI/R^{1/2}$ constant, incremental excitation at 10 mT 50 H_2 with IDC magnetisation
6 Stacking factor 0.9—square/near-square section
7 $E \times t$ in μV s for $N = 1$, $B = 100$ mT
8 Pattern number or designation.

Conductor for N_2 — $338/(\frac{2}{3}\times 4.65\times 0.7)$ turns/cm^2 = 156 turns/cm^2
bare diameter 0.750 mm from wire table has 144.4 turns/cm^2, grade 2 covering

Winding 1 (primary) — Turns per layer = 116, layer length (traverse) = 38.8 mm
Layers = 8.4; depth, $d_1 = 3.0$ mm
$MT_1 = \pi d_1 + bt = 3\pi + 135 = 144.4$ mm
$a_1 = \pi(0.280/2)^2 = 0.061\,58$ mm^2
$= 144.4 \times 972 \times 0.061\,58 \times 9.6/1000$
$= 86.2$ g

$$R_1 = \frac{\rho l}{a_1} = \frac{1.73\times 10^{-6}\times 144.4\times 972}{0.061\,58}$$
$= 39.4\ \Omega$

Winding 2 (secondary) — Turns per layer = 46, layer length (traverse) = 38.8 mm
Layers = 7.4; depth, $d_2 = 6.7$ mm
$MT_2 = 2\pi(d_{\bar{1}} + t_1 + \frac{1}{2}d_2) + bt$
$= 2\pi(3.0 + 0.25 + 3.35) + 135$
$= 176.4$ mm
$Wt_2 = MT_2 \times N_2 \times a_2 \times S_d$

$= 176.4 \times 338 \times 0.441\,8 \times 9.6/100$
$= 252.8$ g

$$R_2 = \frac{\rho l}{a_2} = \frac{17.3\times 10^{-6}\times 176.4\times 338}{0.441\,8}$$

$= 2.33\ \Omega$

Winding depth — $d_1 + t_1 + d_2 + t_2 = 3 + 0.25 + 6.7 + 0.25 = 10.2$ mm

Calculating losses: — $d = 12$ mm, space factor (SF) = 85%

$R_1 = 39.4\ \Omega/20°C = 47.67\ \Omega/75°C$ reference temperature at full load
$I_1 = 60.8 \times 1.10/240 = 0.28$ A, assuming 90% efficiency
$R_2 = 2.33\ \Omega/20°C = 2.82\ \Omega/75°C = 1.41\ \Omega$/half

Then:

Winding 1 (primary) — 240 V, $I_1 = 0.28$ A, $R_1 = 47.67\ \Omega$, $(I^2R) = 3.74$ W, $(IR)_1 = 5.56\%$

Winding 2 (secondary) — 41.8 V at no load, $I_2 = 1.6$ A, $R_2 = 1.41\ \Omega$, $(I^2R)_2 = 3.61$ W, $(IR)_2 = 5.39\%$
Hence $(I^2R)_{total} = 7.35$ W, $(IR)_{total} = 10.95\%$

Regulation — $41.8 \times 0.89 = 37.2$ V at full load

Core loss at 1.5 T, 50 Hz and 0.985 kg core weight

Material — 800/0.50, 6 w/Kg, 19 va/Kg = 5.91 W 78 mA
M6/0.35, 1.9 W/Kg, 17 va/Kg = 1.87 W 70 MA

from Manufacturers' data

Total loss = copper + iron = 7.35 + 5.91 = 13.26 W, $\eta = 88.1\%$, $\theta_r = 74°C$ or for grain-oriented M6 core: total loss = 7.35 + 1.87 = 9.22 W, $\eta = 86.8\%$, $\theta_r = 57°C$

Temperature rise θ_r is taken from the average conductor temperature rise shown for pattern 196 in *Figure 15.10*. For ambient up to 45°C a permissible temperature rise of $\theta_r = 55°C$ would be acceptable, the design is therefore marginal. It should be noted that the dissipation rate at this steady state temperature is flat, at approximately 6°C/W.

The light load condition of 0.1 A and $R_{LL} = 400\ \Omega$ modifies the conversion to $2\pi f C R_{LL}$ at 300 with reduction in ripple and a rise in E_0. The choice of capacitor value is of course related to available standard values.

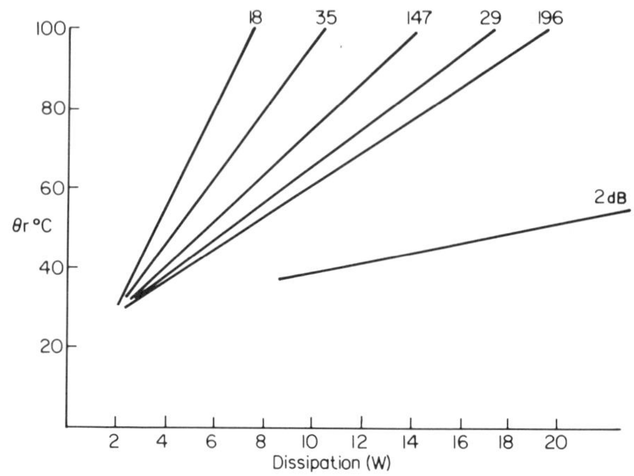

Figure 15.10

Insulations of polyurathane wire, nylon bobbin and polyester tape have continuous temperature category class A to BS 2757 which permits continuous operation at 110°C and therefore allows a hot spot temperature of +10°C above average. For hostile ambients above 45°C an uprating in insulation class would be necessary or an increase in transformer frame size.

15.6.2 50 Hz safety isolating transformer

Rating	240 V to 17.8 V, 4.9 V A
Heating	(a) Temperature rise to be limited to 85°C at FL × 1.1 V in (b) Temperature rise to be limited to 135°C at SC × 1.1 V in
Regulation	Less than 100%, $V_2 > 0.5E_2$
No Load	Current less than 10% of rated output, <2 ma Core loss less than 10% of rated output, <0.49 W
Isolation	4 kV and 6 mm creepage

A selected core would be pattern 35 in square stack configuration from *Table 15.1*. The core area A_c is 3.26 cm^2. After trial calculations a flux density of 660 mT is chosen giving a turns per volt of 20.83. To achieve the specified isolation a two-section spool is chosen giving $l_w = 12$ mm, $d = 7.5$ mm, $A_N = 90$ mm^2 per section, with a separating centre cheek of thickness 1 mm.

The design proceeds as follows:

Winding 1 (primary) — Volts 240 V: turns, $N_1 = 240 \times 20.83 = 5000$
Winding 2 (secondary) — Volts 35.6 V: turns, $N_2 = 35.6 \times 20.83 = 740$

The available winding space per section is 90 mm^2 or with 80% SF = 0.72 cm^2

Conductor for N_1 — 5000/0.72 = 6944 turns/cm^2, diameter = 0.100, grade 2 (from wire table)
Conductor for N_2 — 740/0.72 = 1028 turns/cm^2, diameter = 0.250, grade 2 (from wire table)

Winding 1 (primary) — Turns per layer = 1.2 cm × 77.5 turns/cm = 93

Layers $= 5000/93 = 54$; depth $d_1 = 54 \times 0.129 = 7$ mm
$MT_1 = \pi d_1 + bt = 7\pi + 86.5 = 108.5$ mm
$W_1 = MT_1 \times N_1 \times a_1 \times S_d$
$= 108.5 \times 5000 \times 0.05^2\pi \times 9.6 \times 10^{-6}$
$= 40.9$ g

$$R_1 = \frac{\rho l}{a_1} = \frac{17.3 \times 108.5 \times 5000}{0.05^2\pi}$$

$= 1195\ \Omega$

Winding 2 (secondary) Turns per layer $= 1.2$ cm $\times$ 33.2 turns/cm $= 39$
Layers $= 740/39 = 19$; depth, $d_2 = 19 \times 0.301 = 5.72$ mm
$MT_2 = \pi d_2 + bt = 5.72\pi + 86.5 = 104.5$ mm
$W_2 = MT_2 \times N_2 \times a_2 \times S_d$
$= 104.5 \times 740 \times 0.125^2\pi \times 9.6 \times 10^{-6}$
$= 36.4$ g

$$R_2 = \frac{\rho l}{a_2} = \frac{17.3 \times 10^{-6} \times 104.5 \times 740}{0.125^2\pi}$$

$= 27.25\ \Omega$

Under full load condition the losses are calculated as follows:

$I_1 = 5.8$ V A/240 V $= 0.024$ A,
$I_2 = 4.9$ V A/17.8 V $= 0.275$ A allowing for losses
$R_1 = 1195\ \Omega/20°C = 1446\ \Omega/75°C$
$R_2 = 27.25\ \Omega/20°C = 33\ \Omega/75°C$

Core loss for 800/50 material at 660 mT, 50 Hz, 0.296 kg. From manufacturers' data 1.5 W/kg $= 0.44$ W 1.5 V A/kg $= 0.44$ V A $= 1.85$ mA

$V_1 = 240$ V, $I_1 = 0.024$ A, $R_1 = 1446\ \Omega$,
$(I^2R)_1 = 0.833$ W, $(IR)_1 = 14.46\%$
$V_2 = 35.6$ V, $I_2 = 0.275$ A, $R_2 = 33\ \Omega$,
$(I^2R)_2 = 2.49$ W, $(IR)_2 = 25.49\%$
Loss $= W_1 + W_2 + Fe = 0.833 + 2.49 + 0.44 = 3.76$ W, $\eta = 56.6\%$, $\theta_r = 47°C$ (*Figure 15.5*)
Regulation $14.46 + 25.4 = 39.95\%$ $V_2 = 21.36$ V (17.8 V)

Correction is required to $N_2 = \frac{17.8}{0.6} \times 20.83 = 618$ turns,

30.4 g, 22.75 Ω

Revised V_2 values:

$V_2 = 29.7$ V, $I_2 = 0.275$ A, $R_2 = 27.5\ \Omega$,
$(I^2R)_2 = 2.08$ W, $(IR)_2 = 25.46\%$
Loss $= 0.833 + 2.08 + 0.44 = 3.35$ W, $\eta = 54.4\%$, $\theta_r = 43°C$
Regulation $= 14.46 + 25.46 = 39.92\%$ $V_2 = 17.82$ V

The FL temperature rise is within the specified 85°C and will remain so at $1.1V_1$.

Short circuit load condition entails a possible winding temperature of 135°C, the winding resistance should therefore be corrected to this temperature:

$R_1 = 1195\ \Omega/20°C = 1735\ \Omega/135°C$
$R_2 = 22.75\ \Omega/20°C = 33\ \Omega/135°C$

With the load shorted the output voltage $V_2 = 0$, the transfer voltage E_2 is therefore developed across the internal winding impedance, mainly R. The condition approximates to:

$V_1 = 264$ V, $I_1 = 0.091$ A, $R_1 = 1735\ \Omega$,
$(I^2R)_1 = 14.46$ W, $(IR)_1 = 60\%$
$V_2 = 29.7$ V, $I_2 = 0.36$ A, $R_2 = 33\ \Omega$,
$(I^2R)_2 = 4.28$ W, $(IR)_2 = 40\%$
Loss $= 14.46 + 4.28 + 0.44 = 19.18$ W, $\eta = 0\%$
Regulation $= 60 + 40 = 100\%$ $V_2 = 0$

θ_r at 19.18 W has an indicated dissipation rate of 10°C/W giving a temperature of 192°C (see *Figure 15.11*). The cooling is modified by the heat distribution and with heat flow into the core and to W_2, a value of near 135°C would be found on a physical model.

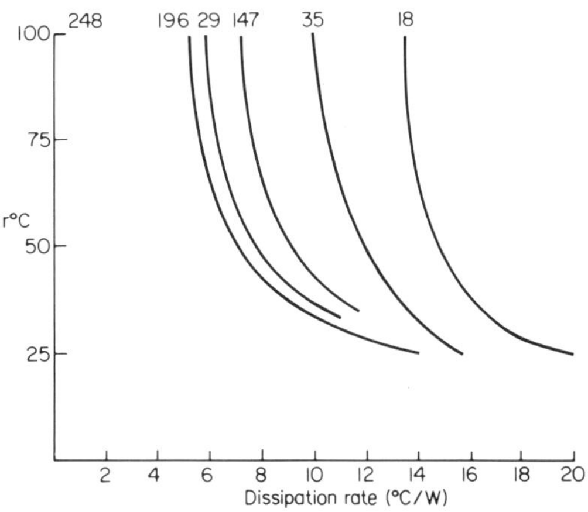

Figure 15.11

References

1 CHENG, D. K., *Analysis of Linear Systems*, Addison-Wesley, New York, 1977
2 MACFADYEN, R. A., *Small Transformers and Inductors*, Chapman and Hall, London (1954)
3 SNELLING, E. C., *Soft Ferrites*, Iliffe Books, London (1969)
4 *Mullard Technical Handbook*, Book Three, Part 3, 'Vinkor inductor cores'
5 *Encyclopaedia Brittanica*, 'Magnetism', p. 594 (1968)
6 WELSBY, W. G., *The Theory and Design of Inductance Coils*, 2nd edn, Macdonald & Co., London (1960)
7 HECK, C., *Magnetic Materials and their Application*, Butterworths, Sevenoaks (1974)
8 SIEMENS, *Data Book*, 'Soft-magnetic siferrit'
9 MATTIN, R., 'The evolution of the toroidal transformer', *Electronic Engineering*, (June 1982)

16 Relays

R W Lomax MSc, CEng, MIEE, MInstP
Principal Engineer,
Advanced Engineering Development,
ITT Relay Division

Contents

16.1 Relay characteristics

Most electromechanical relays are operated electromagnetically. Some can operate thermally or electrostatically, but these are unusual. Electromagnetic relays have a low reluctance, high flux density magnetic circuit. In a gap in this circuit is an armature providing mechanical movement. By passing current through an electric coil on the magnetic circuit, magnetic flux is generated around the circuit. The armature moves to close the gap and operates electric contacts via an insulated actuator.

Consider the change-over contact spring set in *Figure 16.1*. With the relay unenergised the lever contact presses on the break contact. When the relay is energised, the armature moves against a restraining back-stop force and the actuator moves through a pre-travel to touch the lever spring. With increased energisation upon overcoming the break contact force, the break contact is opened. The actuator then moves the lever spring through the contact gap. The lever contact touches the make contact and the lever spring is pushed through an overtravel by the actuator to provide a make contact force.

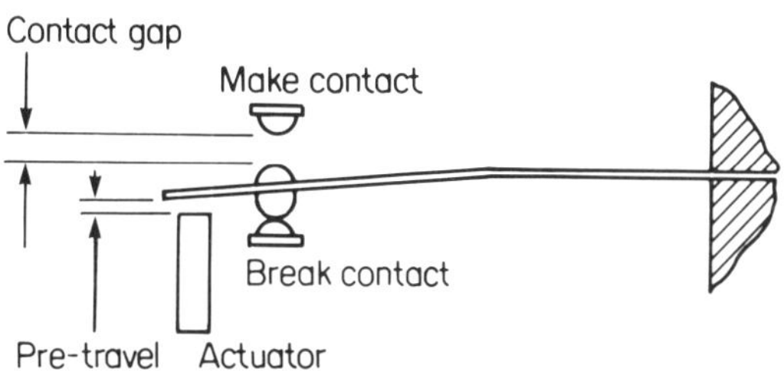

Figure 16.1 The change-over contact

Once operated, coil current change has no effect until it is made so low that it cannot hold the relay operated. Then the armature falls right back to its backstop.

As the armature has mass, external forces may move it, causing transient contact opening. The likelihood of this can be reduced by reducing armature mass, by increasing armature restraining forces and by increasing forces holding contacts together. The relay operate time is largely determined by the coil rise time. Increasing the applied coil voltage reduces the operate time, but increases the coil dissipation. The timing of a change-over contact is shown in *Figures 16.2(a)* and *(b)*. *Figure 16.3* shows that the coil voltage range, which varies with ambient temperature, is limited. However, the operate time is reduced by applying a higher coil voltage along with a series resistance.

Upon open circuiting the coil the relay release time is short, but can cause a surge voltage across the coil. Suppression elements can be added, as described in Section 16.7. With a diode across the coil, the release time is increased by perhaps five times.

Upon operation the contact may bounce. Bounce on contact opening is small, but not so on contact closing. The armature has kinetic energy. Its dissipation can cause contacts to bounce, reducing their lifetime. Contact bounce is reduced by buffer springs and air damping.

16.2 Relay constructions

6.2.1 Dust protected relays

Most relays in use are dust protected. They are inside a case, which is open to the air at gaps around tags and join lines. This is the most important feature of these relays. Any fumes generated

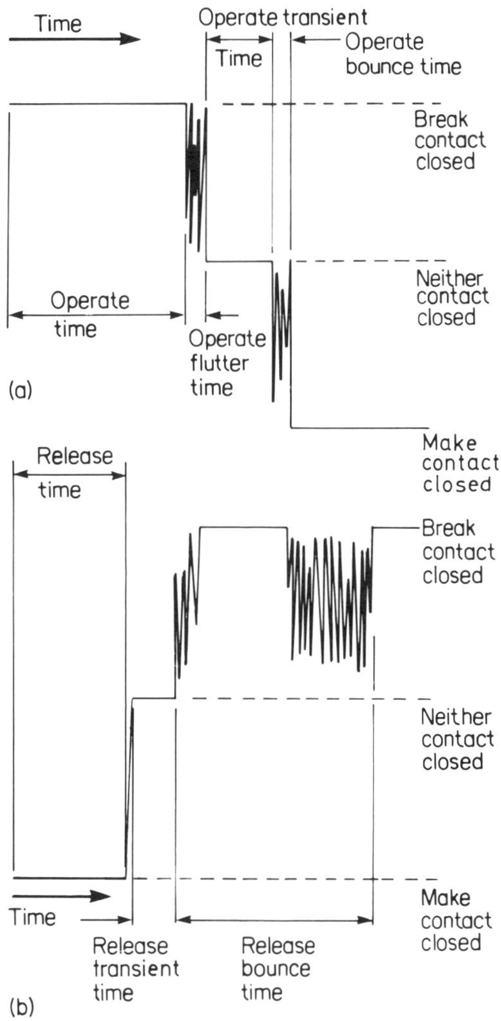

Figure 16.2 Change-over contact timing: (a) operate; (b) release

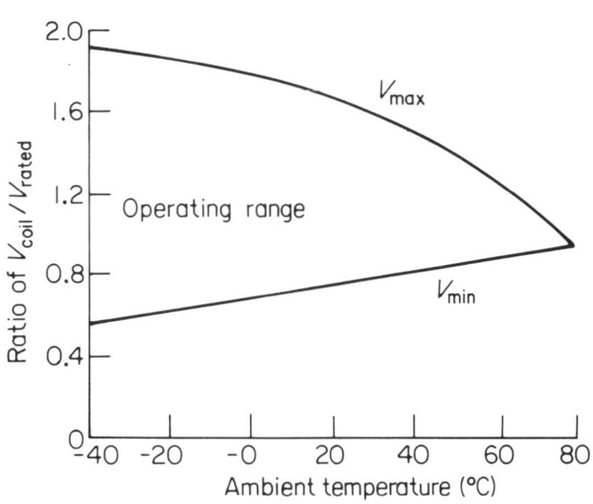

Figure 16.3 Coil voltage operating range

inside the case can escape and do not contaminate the contacts. Oxygen and other gases, which may combine with the contact during operation, are replaced from outside the case. The atmosphere inside the case remains essentially the same as the air outside.

Impurities on surfaces are likely to remain oxidised and produce little electric leakage. However, gases and vapours from the air may enter the case to affect the contacts. This can limit the relay's application.

Versions of dust protected relays are available for low, medium and high level contact loads, with a variety of contact actions, mixed in many combinations, and with numerous mounting methods.

16.2.2 Hermetically sealed relays

Hermetically sealed relays are primarily for military use. They are usually for low and medium level contact loads. They have low contact resistance and are quite sensitive. They have high isolation between circuits and operate over a wide temperature range.

Typical of these relays are crystal can sized electromagnetic relays. They have balanced armatures and can withstand external forces. The termination wires pass through the base, as cores of glass to metal seals. Inside the relay mounted on the wires, close to the base, are the contact assemblies. The magnetic circuit and coil are held above in a strong frame. A metal cover over the assembly is sealed to the edge of the base. There is often an exhaust hole in the base. The relay is vacuum baked and back filled with dry nitrogen after which the exhaust hole is sealed. The relay has a low leak rate, which is what is usually meant by the term hermetically sealed.

16.2.3 Sealed contact unit relays

In these types of relays the actual contact is sealed, whereas the electromagnetic actuating mechanism is not. The moving parts are within the sealed contact units. The contact action within one contact unit is simple, either a single make action or a single change-over action. More complex action relays are made with several contact units operated by one electromagnetic mechanism.

Three types of relays are described in this class, reed relays, mercury wetted relays and diaphragm relays.

In the reed relay the sealed contact unit is the reed contact. A single make action version is shown in *Figure 16.4*. Two long thin

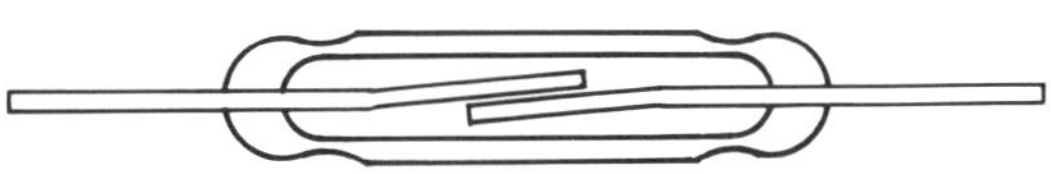

Figure 16.4 Single make reed contact unit

nickel–iron rods, flattened at their ends, looking like reeds, are sealed one at each end of a glass tube. Their flattened ends forming the contacts overlap, separated by a gap. The contact areas are covered with the contact material. The glass tube is filled with special gases.

By passing an electric current through a surrounding coil, a magnetic flux is generated through the reed contact. The flux passes through the contact gap, so attracting one reed blade onto the other. The deflection of the blades produces the required release force.

The reed contact has a relatively high magnetic reluctance. Its blades are long and thin since short and thick blades would require large release forces. In practice, for a reasonable coil power, this limits the contact force. This is compensated by optimising the contact material and the filling gases.

The mercury wetted contact unit is long lived, of stable low resistance and exhibits no electrical bounce. Its contacts are wetted by a film of mercury which is at the upper end of a blade, which moves under the action of magnetic flux excited through the contact unit. The contacts are sealed in a glass tube filled with high pressure hydrogen. A single change-over mercury wetted contact unit is shown in *Figure 16.5*. A relay is completed by a surrounding electric coil and an outer magnetic sheild.

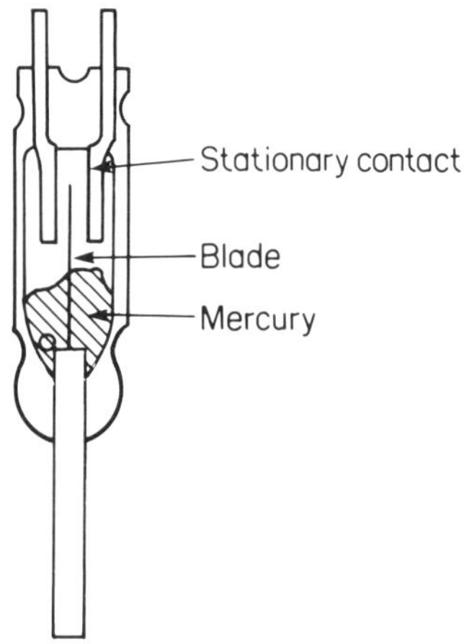

Figure 16.5 Single change-over mercury wetted contact unit

A change-over contact unit may be magnetically biased to remain on one contact, the break contact, when the relay is unenergised. Upon opening a contact the mercury film in between is stretched until it breaks rapidly. A small globule of mercury falls away during this opening, encouraging fresh mercury to rise by capillary action up the blade from a pool at the lower end of the tube. The contact unit is position-sensitive; it cannot be up-ended. The rapid breaking of the connection may produce large surge voltages. Suitable suppression circuit elements are necessary across the contacts, as described in Section 16.7.

In diaphragm relays the sealed contact unit is the diaphragm switch. A single make diaphragm switch is shown in *Figure 16.6* It is designed around a glass to metal seal, of which the core forms the stationary contact. In front of this is suspended a thin nickel–iron diaphragm, which is the moving contact. When magnetic flux is excited through the switch from the core, the diaphragm is attracted onto it, closing the switch. The switch is of low magnetic reluctance giving a large magentic flux. This produces a high contact force for relatively low coil power.

The coil is assembled on the switch core, around which the magnetic circuit is completed by a cover, end cap and tube. There are two types of switch, the single make and the single change-over. Multiples of these switches are built into relays. They are small and are used for printed circuit board mounting.

The contacts are plated with hard gold giving low contact resistances over long lifetimes, on low and medium level loads. *Figure 16.7* shows a section through a single make diaphragm relay.

16.2.4 Solid state relays

At present the relay market is dominated by electromechanical relays (e.m.r.'s). These have some disadvantages such as contact

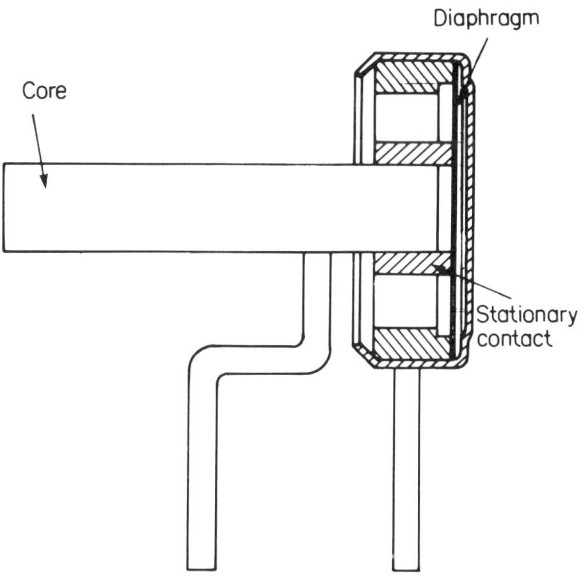

Figure 16.6 Single make diaphragm switch

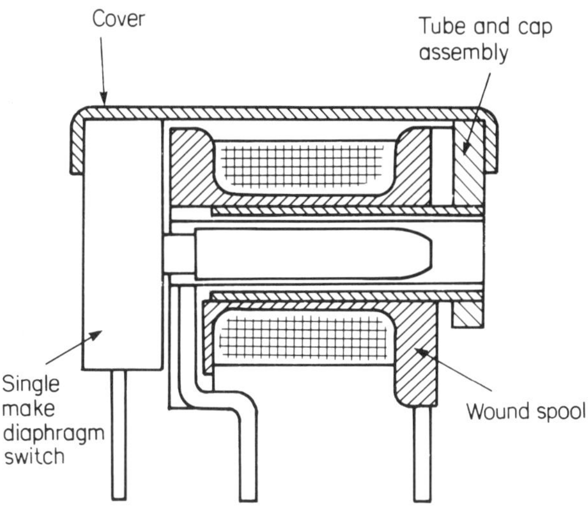

Figure 16.7 Single make diaphragm relay

erosion, arcing, limited operating speed and susceptibility to external forces.

Solid state relays (s.s.r.'s) can overcome these disadvantages. They are of three types: firstly the hybrid e.m.r., which has a solid state input and a mechanically switched output; secondly the more common hybrid s.s.r., which has an electromechanical input and a solid state switched output; and thirdly the true s.s.r., which contains no moving parts.

The hybrid relays are usually cheaper than true s.s.r.'s. The hybrid e.m.r. can provide cheap multipole outputs. The hybrid s.s.r. provides good input isolation with the solid state switching removing the adverse contact effects and is usually used for a.c. loads. It can incorporate trigger circuits, which switch the output when the load voltage is near zero, so that inrush currents and radio frequency interference are reduced.

S.s.r.'s are reliable with a long lifetime. They give arc-free switching, are less susceptible to external forces, and they give noiseless operation. However, they have certain disadvantages. Except for single pole relays of output greater than 10 A, they may be more expensive than e.m.r.'s, multipole outputs being proportionally more expensive. They have no physical break in the output circuit, and a relay with full four terminal isolation is complex. They also run hot, requiring a heatsink for outputs greater than about 5 A and the heatsink may be large.

16.3 Comparison of relay types

Table 16.1 shows a comparison of different relay types with regard to parameters of interest. The larger the number of asterisks the better is the relay for that particular parameter.

The reliability of a relay could be said to be the probability of making a connection through a contact. A dust protected relay is probably the poorest in this respect, as it can be affected by the surrounding atmosphere, although within its contact load range it is good. A reed relay has better reliability as it has a sealed contact. A hermetically sealed relay is even better, as it generally has a greater contact force than a reed relay. However, a mercury wetted relay is best because of its wetted contact. A diaphragm relay is as good, because it has a very high contact force.

The resistance of a relay to external forces, which cause it to be accelerated, can be considered. A mercury wetted relay is probably the poorest, as it is somewhat fragile. A dust protected relay is generally not good, as it often does not have a balanced armature. A reed relay is a little better. A hermetically sealed relay is good, as it usually has a balanced armature. The diaphragm relay is the best from this point of view, as it has a very high contact force coupled with an armature of very low mass.

A dust protected relay can probably deal more easily with the large contact loads, because it is open to the atmosphere and this

Table 16.1 Comparison of relay types

	Dust protected relay	*Hermeticaily selaed relay*	*Reed relay*	*Mercury wetted relay*	*Diaphragm relay*
Reliability	*	***	**	****	****
Life	**	*	**	****	***
Resistance to external forces	**	***	**	*	****
Maximum contact load	****	***	***	*	**
Sealing	unsealed	****	***	***	**
Position sensitive	no	no	no	yes	no
Price	****	**	****	*	**
Multiple contact actions	***	**	*	*	*

can be used to advantage, to reduce the probability of contact welding. Hermetically sealed relays and reed relays are good in this respect. A diaphragm relay may be a little more likely to weld on large contact loads because of its high contact force in a protected atmosphere.

Considering relay sealing, an hermetically sealed relay offers an assured level of sealing with a leak rate below a given limit. Often a reed relay or a mercury wetted relay is practically as well sealed.

The mercury wetted relay suffers because it is position sensitive. To a first approximation other relays are not position sensitive.

Dust protected relays are usually cheap, when the number of contact actions available is considered. Reed relays are also cheap. The complexity of a mercury wetted relay is usually reflected in its price.

Dust protected relays usually offer the largest range of contact actions on a relay. Reed relays, mercury wetted relays and diaphragm relays only offer multiple contact actions by incorporating a multiple of contact units within a relay.

16.4 Electric contact phenomena

All relays except solid state relays rely on the switching of electric contacts. Contact phenomena are complex. All surfaces in the world are covered with a film of deposits, which may chemically bond to them. Metal surfaces are covered with films about 10 nanometres thick. If they were not, i.e. if they were atomically clean, when touched together they would cold weld. An electric contact formed between metal surfaces needs surface films, so that it comes apart.

A voltage across a closed contact causes dielectric breakdown of the surface film. This happens at thin points, where contact metal spikes protrude into the film. This explains why contact wipe is desirable, for then spikes dig deeper into the film. Some breakdown products react with the contact metal. With an electropositive contact metal, its atoms dope the surface film to produce an electric connection through it: the more such connections the lower is the contact resistance. With an electronegative metal anodisation occurs, sealing the dielectric breakdown points. This explains why contact metals like gold and silver are used rather than aluminium and titanium.

Electronegative gases in the atmosphere may permeate the surface film and react with dielectric breakdown products, producing anodisation and preventing electric connection. This explains why some relays are sealed and are outgassed before sealing.

The greater the force between two contact surfaces, the larger is the number of metal spikes protruding into the surface film and the more are the electric connections through it, producing a low contact resistance.

The doping of the surface films enhances their adhesion to the positively charged metal contact. On opening the films may be torn from the negatively charged surface, causing erosion. So the making of an electric contact cannot be divorced from its erosion.

The conditions to make an electric connection are the same as to maintain an electric current after establishing a connection. This explains why a 'wetting' current is often desired. Each parallel connection through the surface film appears to conduct well or not at all. Therefore conduction is noisy, with a large high frequency component. The smaller the total contact current the less reliable is the connection and the larger is its percentage noise. The wider the frequency response of any measuring device, the greater appears the peak contact resistance. Hence the measured contact resistance depends entirely on the method and equipment used.

The making of an electric connection of two metal surfaces and their electric resistance spot welding are just two degrees of the same process. Good electrical contact operation lies between too high a contact resistance developing and the contacts welding up.

When the electric contact is open, but closing, so the surfaces are near, with a high open circuit voltage between them, a large electric field exists. This field exists in the surface films, because they are not metals. Where the film on the negative contact has been doped in a particular way electron emission can be excited into the contact gap. This can initiate an arc. An arc can be similarly produced during opening of a contact, particularly if an inductively generated surge voltage appears across it. To reduce arcing, the initial electron emission needs reducing. Some contact materials dope the surface films in a way less likely to produce arcing. Impurities in the contact can enhance arcing.

16.5 Electric contact materials

Consideration of contact phenomena suggests that certain contact materials are better for particular applications. No contact gives good reliable performance for all loads. On low level contact loads pure electropositive metal contacts like gold are required, perhaps working in protected atmospheres, as obtained in sealed relays and with high contact forces. However, with high contact loads there is a high probability of welding. To compensate for this, material may be added in the contact to enhance any surface film. This could be nickel in gold or cadmium oxide in silver. An amount of oxygen in the surrounding atmosphere may be beneficial, so unsealed relays may be better for high contact loads. A silver or silver alloy contact with a gold flash has often been suggested as a universal contact. This has given problems and was forbidden by the German Post Office in 1971. It appears that, if silver sulphide grows on silver, it is heavily doped with silver and so conducts. With an additional gold flash, silver sulphide grows through pinholes in the gold to cover the contact surface. The gold then acts as a bottleneck between the underlying silver contact and the silver sulphide. The excess silver in the silver sulphide is reduced, which reduces the electric conduction through the surface.

Table 16.2 lists different types of contact materials, the properties each possess, their resistance to atmospheric corrosion and their application. The gold alloys contain several per cent of another transition metal, such as nickel or cobalt. The silver alloys have about 10 per cent or more of another metal or a metal oxide such as nickel or copper or cadmium oxide. The platinum alloys have about 10 per cent of another metal such as ruthenium.

Palladium or platinum in an electrical contact can give rise to the phenomenon of 'brown powder' occurring on the contact. This is because palladium or platinum naturally polymerises gases or vapours in the atmosphere, which are adhering to the contact surface. However, the growth of the resulting polymer film on the contact is self-limiting. The polymer film is usually so thin that it is unseen, until any contact wipe breaks it up and shows it as a 'brown powder'. A further polymer film develops, where the original film was wiped off, and in this way 'brown powder' continues to be produced.

16.6 Relay reliability

From the results of endurance tests on relays, it is found that coil failures due to wire breakages and poor solder joints are far fewer than contact failures. The failure rate of an electric contact determines the failure rate of the relay.

From the contact phenomena given in Section 16.4 it can be said that the probability of making an electric connection at a contact, through a film on it, depends upon:

Table 16.2 Types of contact material

Material type	*Properties*	*Atmospheric corrosion*	*Load level*
Fine gold	tends to cold weld	excellent corrosion resistance	low
Gold alloys	greater hardness; greater wear resistance than fine gold	good corrision resistance; may oxidise	low to medium
Silver	most popular contact material; tends to weld; relatively high erosion	tends to form sulphide in sulphurous atmosphere	medium to high
Silver alloys	higher hardness; resistant to arcing; reduced tendency to weld	tends to form sulphide in sulphurous atmosphere; may oxidise	high
Palladium	Resistant to arcing	corrosion resistance; tends for form 'brown powder'	high
Palladium–silver alloys	hard; resistant to arcing; popular in telecommunications	essentially corrosion resistant	medium
Platinum alloys	very low erosion; extreme demands in high reliability applications	good corrosion resistance; tends to form 'brown powder'	high

(a) the open circuit voltage;
(b) the inverse of the film thickness of the contact;
(c) the electropositivity of the contact metal;
(d) the inverse of the partial pressure of electronegative gas in the atmosphere;
(e) the power dissipated at the contact, i.e. a function of the current;
(f) the inverse of the heat capacity of the contact;
(g) the thermal resistance of the contact to its surroundings;
(h) the ambient temperature of its surroundings.

A large number of such parallel connections are made, so the contact resistance is inversely proportional to the connection probability. However, the probability of welding a contact is directly proportional to this connection probability. Additionally, contact erosion depends on this probability. The connection probability for a given design in a constant environment at a low level load is small; contact resistance may be high and the consequent failure rate, say λ_r, may be high. This is region 1 shown in the diagram of contact failure rate against contact load (*Figure 16.8*).

If just the contact load is increased to a large value, the probability of contact welding and erosion may be large. The consequent failure rate, say λ_w, may be high. This is region 2 of *Figure 16.8*. At a contact load between the low level and large value the sum of the two above failure rates, say λ_b, is least and then contact life is greatest. This is region 3, sometimes called the self-cleaning region.

The contact design in terms of shape, material and its operating environment determines the value of the minimum failure rate and at what load it occurs. Consequently it is difficult to generalise on relay reliability due to the many different relay designs available. However, *Figure 16.9* and *Table 16.3* between them give in orders of magntitude the areas of operation and the corresponding failure rates of contacts of various types of materials, in different general purpose relays. The values given can only be taken as rule of thumb figures. One of the reasons for giving this information is to emphasise that different types of contact have different levels of performance. There is no 'universal' contact.

For all contacts there are suggested lower limits for contact voltages and currents. The values given in the table are generally

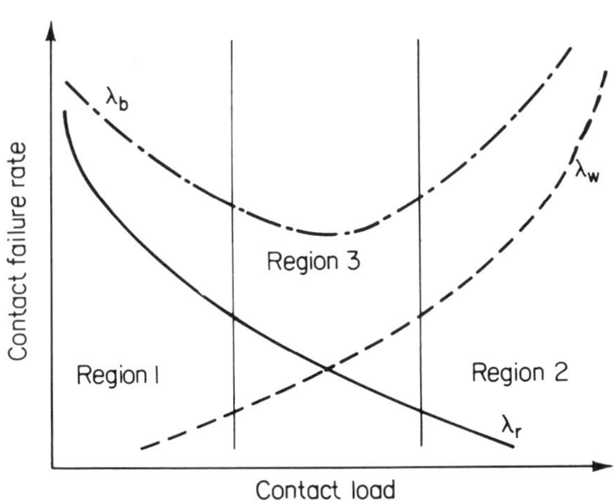

Figure 16.8 Variation of contact failure rate with load

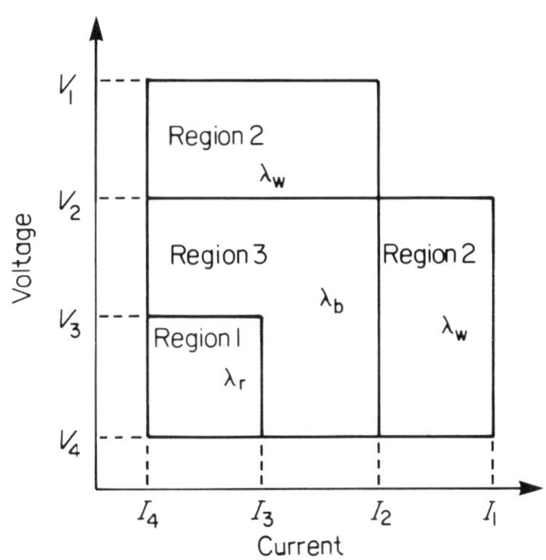

Figure 16.9 Approximate areas of contact operation

Table 16.3 Approximate current and voltage limits and failure rates for various contacts in general purpose relays

Relay type	*Contact material*	*Approximate voltages (V)*				*Approximate currents (A)*				*Contact failure rates ($\times 10^{-9}$)*		
		V_1	V_2	V_3	V_4	I_1	I_2	I_3	I_4	λ_w	λ_b	λ_r
Dust protected	gold alloy	100	10	1	0.1	0.5	0.1	0.01	0.001	30	8	10
Dust protected	silver	250	50	10	3	5	1	0.2	0.05	250	10	10
Dust protected	palladium–silver	100	50	15	5	2	1	0.1	0.01	100	30	10
Dust protected	silver alloy	250	100	30	10	10	1	0.5	0.1	1000	200	300
Sealed	gold alloy	100	10	1	0.01	0.5	0.1	0.01	0.001	100	2	5
Sealed	silver	100	30	3	1	3	1	0.1	0.005	200	5	5
Sealed	silver alloy	100	50	10	3	5	1	0.3	0.05	1000	200	200
Sealed	mercury	200	20	1	0.01	0.5	0.3	0.1	0.001	300	0.5	1

for d.c. resistive loads. The a.c. capability of contacts is greater. The effect of different types of loads on the contact is considered in the next section.

16.7 Switched loads and suppression elements

Electrical loads switched by the relay contacts can be placed in the following groups:

(a) *Resistive load* This is the more easily understood. The load current and voltage should be within that recommended for the contact. This may be approximately as given in *Table 16.3.*

(b) *Capacitive d.c. load* On closing contacts to a capacitive load there is a large initial current. This may cause contact welding.

(c) *Tungsten lamp load* The cold resistance of a tungsten filament lamp is about one tenth of its hot resistance. This produces a large initial current.

(d) *Inductive d.c. load* This is a difficult load to switch. A large transient voltage is generated on breaking the load current. This voltage is of opposite polarity to the supply voltage. It may damage some devices, which are sensitive to the polarity of the voltage applied. It may break down insulation, and it may initiate an arc between the opening contacts which can radiate radio frequency interference (r.f.i.) and increase contact erosion.

A number of circuit elements may be incorporated to suppress these effects. Some such circuit elements are given in *Table 16.4.*

(e) *Inductive a.c. load* This may be a motor or transformer or solenoid. On closing this can cause an initial current of about five times the steady current value. A large transient voltage may also be generated on breaking the load current. This produces adverse effects similar to those described in (d). Two suppression circuits are given in *Table 16.5.*

16.8 Glossary of useful relay terms

All-or-nothing relay A relay intended to be energised either higher than that at which it operates or lower than that at which it releases.

Bistable relay A relay able to stay in either its operated or unoperated condition after the coil energisation has been removed.

Bounce time Time between which a contact first closes and finally closes during an operation (see *Figure 16.2*).

Break contact A contact, which is open when the relay is operated and closed when the relay is released.

Break-before-make contact A change-over contact which on relay operation opens a common connection from a break contact before closing to a make contact. This is the usual form of change-over contact.

Bridging time The time in which both contact circuits of a make-before-break change-over contact are closed.

Change-over contact Contact which on relay operation opens a common connection from a break contact and closes it to a make contact. This can be either a break-before-make contact or a make-before-break contact.

Contact circuit resistance The resistance of a closed contact, when measured by a four pole method at the appropriate relay tags with a given applied voltage and current.

Contact force The force between two contact points when closed.

Contact gap The gap between two contact points when open.

Dielectric test See 'voltage proof test'.

Drop-out See 'release'.

Drop-out energisation See 'release energisation'.

Drop-out time See 'release time'.

Duty cycle The ratio of the coil energisation time to the total period of one cycle of operation, often given as a percentage. Only of interest if a number of similar cycles of operation are performed.

Electromechanical relay A relay in which the intended operation arises through movement produced by energisation.

Energisation Current or voltage or power applied to the relay input, usually its coil.

Hold energisation The energisation at a given temperature, which will hold the relay operated, when energisation is reduced from a larger value.

Table 16.4 Suppression circuit elements for use in switching inductive d.c. loads

Suppression elements	*Advantages*	*Disadvantages*	*Comments*
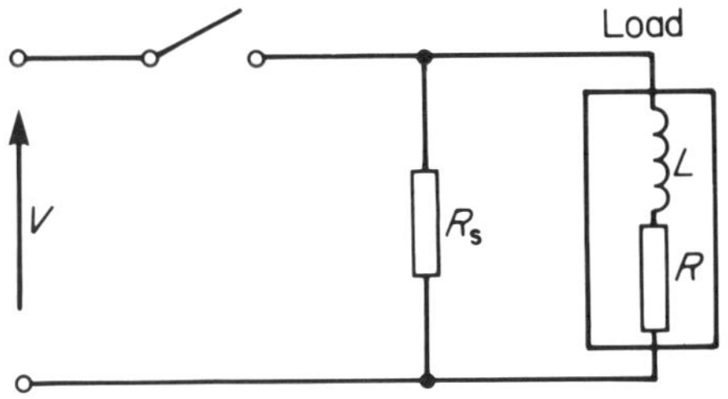	suppresses surge voltage; low cost; reduces contact erosion	extra current drain	if load is a relay coil it lengthens release time
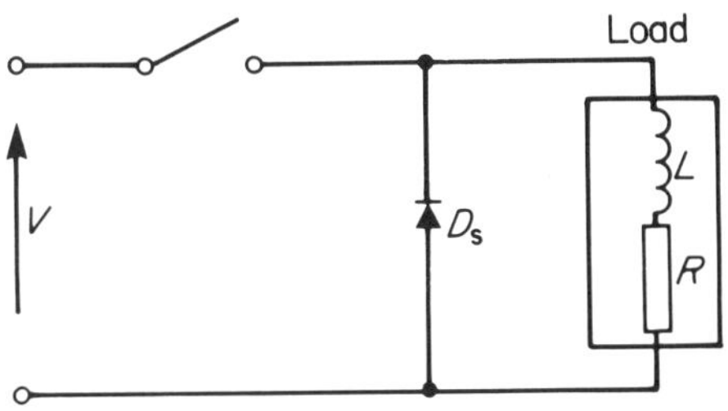	limits surge voltage to 1 V; supresses contact erosion and r.f.i.	greatly lengthens relay release time if load is relay coil	
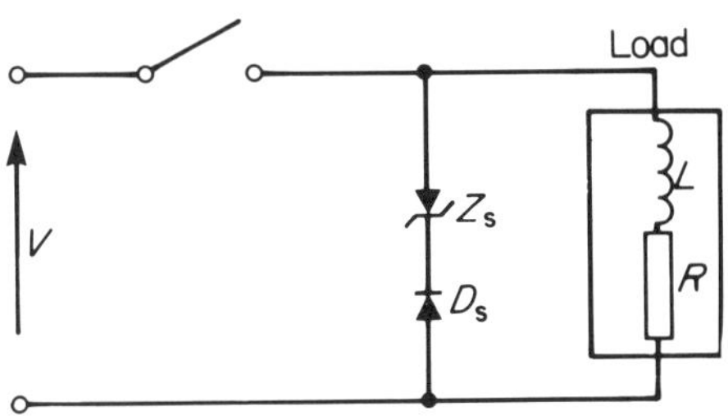	reduces surge voltage; suppresses contact erosion	moderately lengthens relay release time if load is relay coil	costly
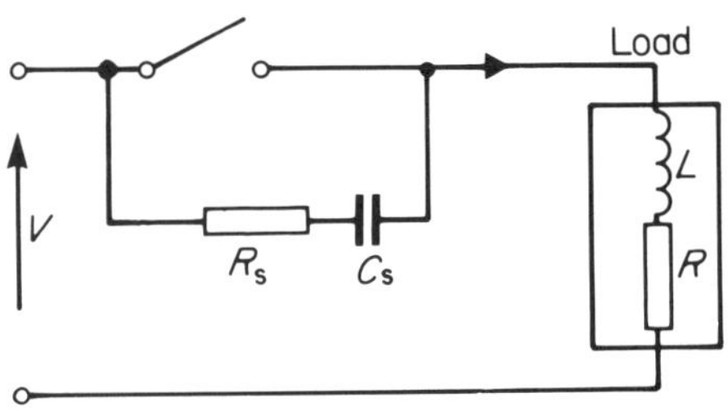	good contact protection	large surge voltage makes this poor for use with solid state devices	C_s (in μF)$=I_L$ (in A) R_s must prevent capacitance current welding contacts
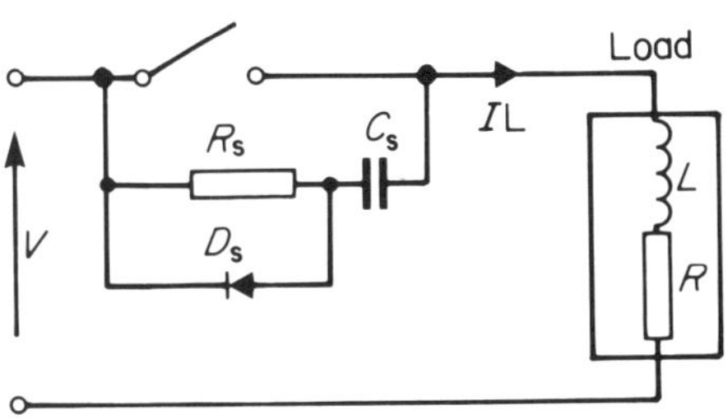	good comprehensive suppression	costly	peak forward current of $D_s=I_L$ $R_s \geqslant R$

Hybrid electromechanical relay A relay in which electromechanical and solid state devices combine to give a mechanically switched output.

Hybrid solid state relay A relay in which electromechanical and solid state devices combine to give a solid state switched output.

Make contact A contact which is closed when the relay is operated and open when the relay is released.

Make-before-break contact A change-over contact which on relay operation closes a common connection to a make contact before opening from a break contact.

Table 16.5 Suppression circuit elements for use in switching inductive a.c. loads

Suppression elements	*Advantages*	*Disadvantages*	*Comments*
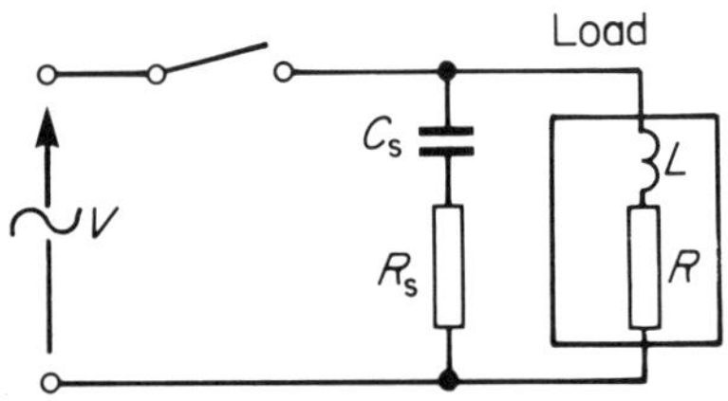	suppresses surge voltage; reduces contact erosion	extra current drain	$R_s \simeq R$ $C_s \simeq L/R^2$
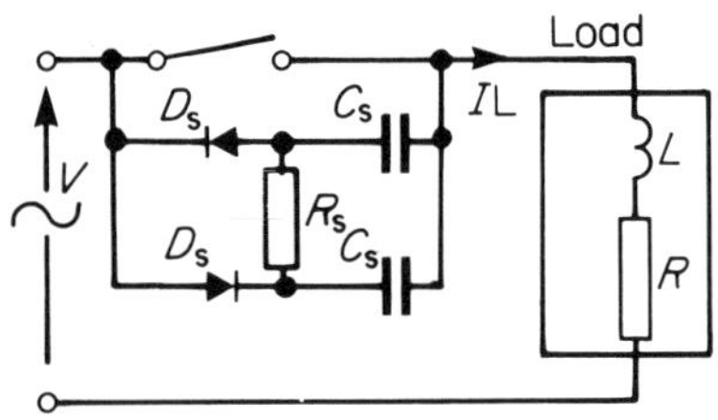	good comprehensive suppression	costly	$R_s \simeq 100$ kΩ peak forward current of $D_s = I_L$ $C_s \simeq L/R^2$

Monostable relay A relay which operates upon energisation and releases after energisation is removed.

Non-operate energisation The energisation at a given temperature, which will not operate the relay.

Operate energisation The energisation at a given temperature, which will operate the relay.

Operate time Time between the application of a given coil energisation and the first closing or opening of a given contact. The operate time is greatly dependent on the value of coil energisation (see *Figure 16.2*).

Pick-up See 'operate'.

Pick-up energisation See 'operate energisation'.

Pick-up time See 'operate time'.

Polarised relay A relay in which the change of condition depends on the sign of the applied energisation.

Pull-in See 'operate'.

Pull-in energisation See 'operate energisation'.

Pull-in time See 'operate time'.

Release energisation The energisation at a given temperature at which the relay will release, when energisation is reduced from a larger value.

Release time The time between the removal of a given coil energisation and the first opening or closing of a given contact. As usually defined this does not include bounce time. The release time is dependent on the value of coil energisation (see *Figure 16.2*).

Solid state relay A relay whose output switching is achieved using solid state devices and without the use of any moving parts.

Thermoelectric electromotive force The electromotive force (e.m.f.) generated by a closed contact of an energised relay at an elevated temperature.

Transit time The time during which both contact circuits of a change-over contact are open (see *Figure 16.2*).

Transfer time See 'transit time'.

Further reading

ANDREIEV, N., 'Power relays: solid state vs. electromechanical', *Control Engineering*, **20** (1), 46–9 (1973)

BEDDOE, S., 'High-performance miniature relays: a design review', *Design Electronics*, **8**, December, 38 (1970)

CAPP, A. O., 'Electrical contact considerations', *Electronics World*, April, 49 (1967)

DUMMER, G. W. A., 'Failure rates, long term changes and failure mechanisms of electronic components: part 1: failure rates and the influence of environments', *Electronic Components*, **5** (10), 835–84 (1964)

GAYFORD, M. L., *Modern Relay Techniques: STC Monograph No. 1*, Newnes–Butterworths, Sevenoaks (1969)

GROSSMAN, M., 'Focus on reed relays', *Electronic Design*, **20** (14), 50–9 (1972)

GROSSMAN, M., 'Focus on relays', *Electronic Design*, No. 26, December, 119 (1978)

HYDE, N., 'Electromechanical relays: part 3: the contact problem', *Electronic Components*, **5** (8), 665–9 (1964)

HYDE, N., 'Electromechanical relays: part 4: reliability testing', *Electronic Components*, **5** (10), 865–72 (1964)

JORDAN, J. S., 'Supressing relay coil transients', *Electronic Industries*, **24** (4), 73–4 (1965)

KODA, A. J., 'Mercury-wetted relays', *Electronics World*, April, 56 (1967)

KOSCO, J. C., 'Fundamentals of electrical contact materials and selection factors', *Insulation/Circuits*, **17** (8), 27–34 (1971)

LOMAX, R. W., 'The design and development of sealed contact change over diaphragm relay', *Proc. 20th NARM Conf., Stillwater, Oklahoma, USA, 18–19 April* (1972)

LYONS, R. E., 'Relays: electromechanical vs. solid state', *Mech. Engng*, **94** (10), 43–6 (1972)

ROSENBERG, R. L., 'Reed relays', *Electronics World*, April, 41 (1967)

ROSINE, L. L., 'Miniature relays', *Electrotechnology*, **84** (1), 47–55 (1969)

ROVNYAK, R. M., 'Arc, surge and noise suppression', *Electronics World*, May, 46 (1967)

WOODHEAD, H. S., 'The diaphragm relay', *Electronic Components*, **9** (3), 290–4 (1968)

17

Piezoelectric Materials and Components

M B J Scanlan BSc, ARCS
Group Chief,
Applied Physics Group Marconi
Research Laboratories

Contents

17.1 Introduction

Piezoelectricity, discovered by the brothers Curie in 1880, takes its name from the Greek piezein, to press, but piezoelectric materials will in general react to any mechanical stress by producing electric charge: in the converse effect, an electric field results in mechanical strain. As a simple introduction, compare small cubes of crystal quartz (a piezoelectric material) and fused quartz (non-piezoelectric). If the former is mechanically stressed, in an appropriate way, electric charge will appear on its faces: there will be some elastic change in dimensions, but the amount of this change will depend on the electric boundary conditions, i.e. whether electrodes on the faces of the cube are open—or shortcircuited. Conversely, if a voltage be applied to these electrodes of the unstressed cube, the cube will not only be charged electrically but strained mechanically: as in the first case, the amount of charge stored electrically will depend on whether the cube is clamped or free mechanically. In contrast, the cube of fused quartz behaves purely elastically under mechanical stress, and as a simple dielectric in an electric field: there is no coupling between the two effects.

As in this example, piezoelectric materials are crystalline in structure and anisotropic in many properties (e.g. thermal conductivity, thermal expansion, dielectric constant) while the non-piezoelectrics are often amorphous and isotropic. It is also worth noting that the same crystalline structure which results in piezoelectricity results also in optical activity[1] (the linear electro-optic, or Pockels effect is possible with all piezoelectric materials, and only with them) and in pyroelectricity, the generation of electricity by heat (every pyroelectric crystal is piezoelectric, and every ferroelectric crystal, but not every piezoelectric crystal, is pyroelectric. Ferroelectricity, an important sub-class of piezoelectricity, will be further discussed below).

Piezoelectricity is a much more common phenomenon than is generally realised. Cady[2] reports that over 1000 piezoelectric crystals have been identified; this is perhaps not so surprising when it is remembered that only 11 out of the 32 crystal classes cannot, because of their symmetrical crystal structure, have any piezoelectric members.

A caution should be given here about notation, both as regards piezoelectricity itself and on crystal structure. Most modern texts follow the IRE *Standards on Piezoelectric Crystals*, dated 1949 and later,[3] but the first edition of Cady[4] (1946) does not, of course, and the second edition[2] (1964) only outlines the standards rather briefly in an appendix. Hueter and Bolt's[5] *Sonics* use Voight's notation, dating from 1890, and explains that conversion into other notations can be made as explained in the standards.

On crystal structure, Cady[2] gives a cross reference between the various crystal notations, which will be useful to those not familiar with the subject.

17.2 The crystalline basis of piezoelectricity

Piezoelectricity is essentially a phenomenon of the crystalline state and the two subjects are dealt with together in Kittel,[6] Cady,[2] Berlincourt *et al.*,[7] Mason[8] and Nye.[9]

Crystals are classified into seven systems, and 32 classes. Of these, eleven have a centre of symmetry, and cannot be piezoelectric: the 21 classes lacking a centre of symmetry all have piezoelectric members. It can therefore be stated as a necessary condition for piezoelectricity that the crystal lacks a centre of symmetry. Only if this is the case will there be a shift of the electrical centre of gravity when the crystal is stressed, and a resultant generation of charge on the crystal faces.

An important sub-class of piezoelectrics is that of the ferroelectrics, which have two or more stable asymmetric states, between which they can be switched by a sufficiently strong electric field. Ferroelectrics have a domain structure, a Curie temperature and show hysteresis; hence their name, by analogy with the ferromagnetics. Cady[2] deals at some length with Rochelle salt, one of the earliest ferroelectrics, and Kittel[6] has a chapter on barium titanate, which has the perovskite structure typical of one family of ferroelectrics.

Another important family (lithium niobate, lithium tantalate, etc.) is of considerable importance for electro-optic purposes as well as (especially lithium niobate) piezoelectricity.

Ferroelectricity can be exploited not only in single crystal material, but also in polycrystalline form, and this gives the commercially important class of piezoelectric ceramics. If the polycrystalline material is *poled*, i.e. subjected to a strong electric field as it is cooled through its Curie point, the domains in the crystals are aligned in the direction of the field. One can in this way have a piezoelectric material whose size, shape and piezoelectric qualities can, within limits, be made to order. These commercially available materials are often solid solutions of lead zirconate and lead titanate with various additives to modify the behaviour. Such a range is marketed under the general name of PZT, with distinguishing numbers, by the Clevite Corporation USA, and under the general name PXE, again with a number, by Mullard in the UK. A good account of these materials (and of piezoelectric materials in general) is given in Berlincourt *et al.*[8] The Mullard materials are listed in a booklet,[10] together with some application notes.

Finally, it is worth noting that some piezoelectric materials, notably cadmium sulphide, CdS, and zinc oxide, ZnO, can be evaporated or sputtered on to a suitable substrate, on which the deposit is polycrystalline, but oriented.

These materials, 'grown' to a thickness of a half wavelength, can be used as electroacoustic transducers to launch elastic waves into the substrate at much higher frequencies than can be achieved by the alternative (low frequency) technique of bonding a halfwave thickness of a crystal transducer (quartz, lithium niobate, ceramic) to the substrate.

17.3 Piezoelectric constants

In a dielectric medium which is elastic but not piezoelectric (e.g. fused quartz, glass), independent relations exist giving its behaviour in an electric field E, and under a mechanical stress T. These are

$$D = \varepsilon E \qquad (17.1)$$

$$S = sT \qquad (17.2)$$

where D is the electric displacement, ε the dielectric constant, S the strain in the material and s the elastic compliance (the reciprocal of Young's modulus, if the stress is tension or compression).

In a piezoelectric medium, the strain or the displacement depend linearly on both the stress and the field, and the equations for D and S become

$$D = dT + \varepsilon^T E \qquad (17.3)$$

$$S = s^E T + dE \qquad (17.4)$$

where d is a piezoelectric constant characteristic of the material, and the superscripts T and E denote that ε and s are to be measured at constant stress and constant electric field respectively.

In words, d may be defined as the displacement (charge per unit area) per unit applied stress, the electric field being constant or as the strain per unit applied field, the stress being constant. From the first definition, d is the piezoelectric charge constant.

The piezoelectric voltage constant g can be introduced by an equivalent pair of equations

$$E = -g^T + D/\varepsilon^T \qquad (17.5)$$

$$S = s^D T + gD \qquad (17.6)$$

By substituting for D from equation (17.3) into equation (17.5),

$$d = g\varepsilon^T \qquad (17.7)$$

Another very important piezoelectric quantity is the electromechanical coupling coefficient, k, the significance of which is seen as follows. The mechanical energy stored in the piezoelectric medium is, from equation (17.4),

$$U_M = \tfrac{1}{2}ST = \tfrac{1}{2}s^E T^2 + \tfrac{1}{2}dET \qquad (17.8)$$

and the electrical energy stored is, from equation (17.3),

$$U_E = \tfrac{1}{2}DE = \tfrac{1}{2}dTE + \tfrac{1}{2}\varepsilon^T E^2$$

U_M contains a mechanical term ($\tfrac{1}{2}s^E T^2$) and a mixed term ($\tfrac{1}{2}dET$): similarly U_E contains an electrical term ($\tfrac{1}{2}\varepsilon^T E^2$) and the same mixed term. By analogy with a transformer, in which the coupling factor is the ratio of the mutual energy to the square root of the product of the primary and secondary energies, k is the ratio of the mixed term to the square root of the products of electrical and mechanical terms, i.e.

$$k = \frac{\tfrac{1}{2}dET}{\sqrt{\tfrac{1}{4}\varepsilon^T s^E E^2 T^2}}$$

or

$$k^2 = \frac{d^2}{\varepsilon^T s^E} \qquad (17.9)$$

At low frequency, i.e. below the frequency of any mechanical resonance of the material, k^2 is a measure of how much of the energy supplied in one form (electrical or mechanical) is stored in the other form: it is not an efficiency since the energy applied is all stored in one form or the other.

17.4 Piezoelectric notations

For the sake of simplicity, the discussion so far of the piezoelectric constants has implied that there is only one piezoelectric axis, and that the stress or electric field is applied along this axis and the resulting strain or charge is also measured in this direction. This situation is appropriate, for instance, to a cylinder of a ceramic piezoelectric material, poled along its axis, and with the stress or field also along this axis. It is not appropriate to a single crystal piezoelectric, with its anisotropic properties, to which stress is applied in an arbitrary direction. To deal with this situation, a more complicated notation must be set up (IRE Standards, 1949).[3]

Figure 17.1 shows a right-handed orthogonal set of axes X, Y, Z, together with a small cube with faces *1*, *2*, *3*, perpendicular to X, Y and Z respectively. Any stress applied to the cube of material can be resolved into forces *1*, *2*, *3*, along the axes, or shears *4*, *5*, *6*, acting about the axes. Any electric field can be resolved into its components along the axes, and any charge resulting from the field or from piezoelectric stress will appear on the faces *1*, *2*, *3*, as shown. As for stress, there are six ways in which a strain can appear, i.e. tension or compression along the three axes, or shear about the axes.

Hence for the general case, there are six permitivities ε_{ij} (i and j, 1–3), twenty-one elastic compliances s_{ij} (i and j, 1–6) and eighteen piezoelectric constants d_{ij} (i, 1–3; j, 1–6). Fortunately, most

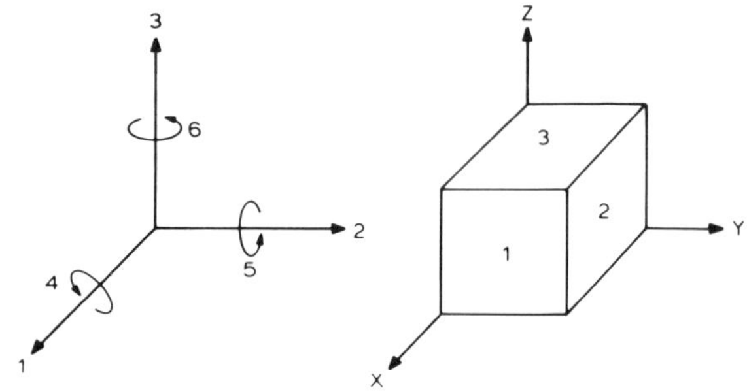

Figure 17.1 Diagram of axes (1, 2, 3), shears (4, 5, 6) and surfaces (1, 2, 3) for piezoelectric notation

crystals are symmetrical enough that the number of independent piezoelectric coefficients is greatly reduced: quartz for instance has only two, d_{11} connecting a field along the X axis with strain in the same direction, and d_{14}, connecting a field along an axis normal to Z and a shear strain in the plane normal to the field.

Berlincourt *et al.*,[8] and Nye[9] give elasto-optic matrices for all 32 of the crystal classes. These are 9×9 matrices, symmetrical about the diagonal, so giving the possible 45 elasto-optic constants listed in the previous paragraph. Such matrices are also given in the IRE Standards (1958).[3]

17.5 Applications of piezoelectric materials

17.5.1 Frequency stabilisation

The use of quartz as a frequency standard is still the most important application of piezoelectricity technically and commercially. This use of quartz relies on its high Q as a mechanical resonator: Mason[8] shows curves of Q against frequency which imply that the internal Q (i.e. that due to internal frictional losses, and excluding mounting losses and air losses) of an AT-cut shear mode crystal can be over 10^7 at 1 MHz. Contributory factors are its high quality as a dielectric, its low dielectric constant, and the relative ease of cutting and polishing. The processes involved in quartz technology are described in some detail by Cady.[2] It is because of the increasing difficulty of mining sufficient high quality natural quartz that the hydrothermal growth of quartz has increased in the last 10 or 15 years. The synthetic quartz can be made with at least as high a Q as natural quartz.

The most important cuts of the crystal for frequency standards at frequencies of 1 MHz upwards are the AT and BT cuts, shown in *Figure 17.2*. A more comprehensive diagram, showing the

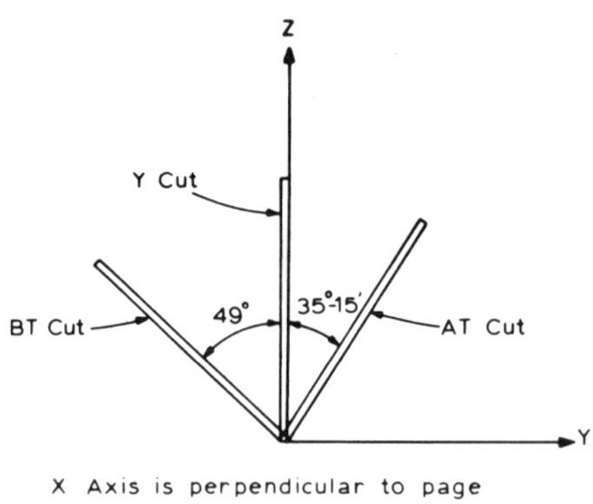

Figure 17.2 Diagram of crystal axes in quartz, illustrating some of the more important crystal cuts

position of a large number of the more important cuts, is in Mason:[8] *AT* and *BT* cuts are *rotated Y-cuts*, i.e. they are obtained by rotating the *Y* plane (perpendicular to *Y*) about the *X* axis, as shown in the diagram. These cuts, which vibrate in a shear mode, are important because they give a zero temperature coefficient of frequency, as shown in *Figure 17.3* (*BT* cut) and *17.4* (*AT* cut).

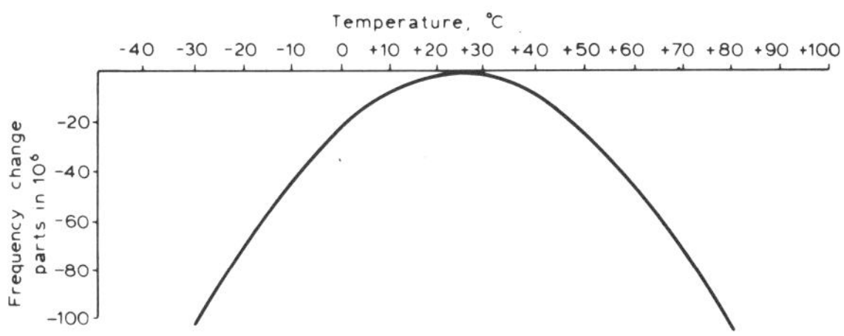

Figure 17.3 The variation of frequency with temperature for a BT cut quartz crystal

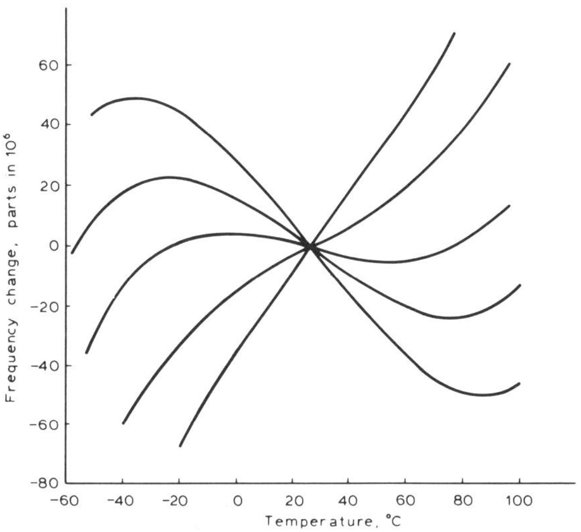

Figure 17.4 Typical family of frequency variation with temperature of AT cut quartz crystals. The curves shown cover an angular range of about 20 minutes of arc

The *BT* temperature coefficient curve is parabolic and is of zero slope at only one temperature: this temperature is a function of the exact angle, and changes by about 1.5°C per minute of arc. *BT* cuts are preferred to *AT* at frequencies above 10 MHz, since they are more stable against changes in drive level and load capacitance.

The *AT* cut, on the other hand, can be chosen to give two temperatures of zero coefficient, and a wide temperature range (the middle curve of *Figure 17.4*) over which the coefficient is very small. The five curves in *Figure 17.4* are spread over about 20 minutes of arc. For best performance as a frequency standard, contoured (i.e. convex on one or both surfaces) ***AT*** cut plates are used, often operating at a harmonic frequency (e.g. a 5th overtone crystal, abbreviated to 5th *OT*). The crystal, despite its low temperature coefficient, will be in a temperature controlled overn, or, for the highest quality, a double oven. A double oven will reduce temperature fluctuations by a factor of 10^4 or more, i.e. the temperature of the crystal will not vary by more than a few millidegrees.

The frequency of a newly made quartz crystal oscillator will drift quite rapidly for a few months, a process known as ageing.[11] After a few months at constant temperature the ageing rate will be about 1 part in 10^{10} per month or less and the frequency will continue to drift at this, or a very slowly decreasing rate, indefinitely. If conditions are disturbed, e.g. the oscillator or its oven is switched off for a few hours, ageing will recommence at near the initial rate. Hence a quartz frequency standard should never be switched off, nor run at a power output above a few microwatts. The specification of a commercial quartz frequency standard,[12] using a 5th *OT AT* crystal, would include a frequency stability better than 15 parts in 10^{12} at constant voltage and temperature measured over 1 second and an ageing rate of less than 1 part in 10^9 per month after three months. These are of course maximum values, and do not contradict the lower ageing rate quoted above.

Because they are not absolute standards (i.e. they require calibration) and because of ageing, quartz crystals have been superseded as frequency/time standards by atomic clocks (rubidium, caesium, hydrogen). However, the short term (1 second) stability of the best quartz crystals is better than that of the atomic standards (except hydrogen) which therefore consist of an atomic clock, used to control and correct the ageing of a good quartz oscillator. It is from the quartz oscillator that the output frequencies are delivered.

Quartz crystal oscillators of high stability are well reviewed in Gerber and Sykes,[13] in an issue of *Proceedings IEEE* devoted to frequency stability: the issue therefore also reviews atomic frequency standards.

Quartz crystals can be made to any frequency from 1 kHz to 750 kHz, and from 1.5 MHz to 200 MHz, the latter being high harmonic overtone crystals. The frequencies between 750 kHz and 1.5 MHz are difficult to cover as strong secondary resonances occur. There is a tendency to use a higher frequency crystal counted down rather than a very low frequency crystal, which tends to be larger and less stable than is desirable. The units may be packaged in glass envelopes, or in a solder-sealed or a cold-welded metal can: this container may also contain a heater and temperature sensor. *Figure 17.5* shows the interior of a microminiature crystal oscillator (Marconi Type F 3187), providing frequencies in the range 6–25 MHz over the temperature range −55°C to +90°C to an accuracy of ±50 parts per million.

Figure 17.5 A micro-miniature quartz crystal oscillator mounted on a TO5 header (courtesy The Marconi Company Ltd)

A new family of quartz crystal cuts, known variously as *SC* (stress compensated), *TTC* (thermal transient compensated) or *TS* (thermal stress) cuts,[25,26] has recently become available. These are doubly rotated cuts, as compared with the simple single rotations shown in *Figure 17.2*, and are more difficult to orient and cut accurately: however, the performance is significantly better than that of the earlier cuts. Particularly important features, especially for frequency standards or in severe environments, are the very low temperature coefficient, over a wide temperature range, low thermal transient effects, excellent short-term stability, low ageing (<1 part in 10^9 per day after three months) and good stability under acceleration and vibration (~ 1 part in 10^9 per g of acceleration).

17.5.2 Crystal filters

Quartz crystals can also be used, and for the same reasons, in very precise filters, used for defining a passband very accurately, and for rejecting adjacent channel interference. *Figure 17.6* shows a very narrow bandpass filter, with a passband of about 50 Hz at a centre frequency of 100 kHz: the 60 dB bandwidth is about 300 Hz. This filter might be used for telegraphy. *Figure 17.7*, on the other hand, shows the passband of a filter for a reasonable quality speech circuit. The passband is 5 kHz wide at 3 dB, and less than 11 kHz at 60 dB. Such filters can be very useful in conditions of severe congestion, or for s.s.b. working.

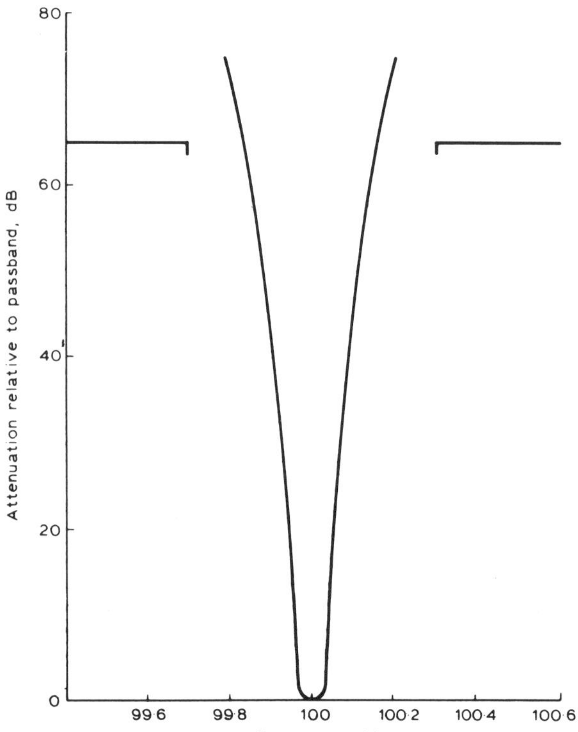

Figure 17.6 Response curve of a very narrow passband crystal filter

A more recent development is the monolithic crystal filter, i.e. a single quartz plate on which a number of mechanically-coupled resonators are printed, the whole assembly forming a multiple element filter.

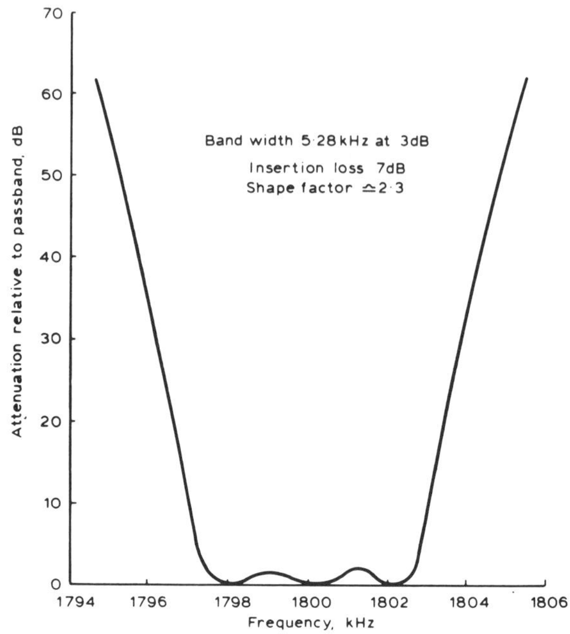

Figure 17.7 Response curve of a multielement narrow passband crystal filter

Filters using surface wave techniques will be discussed later, under that heading.

17.5.3 Bulk delay lines[13,14]

Piezoelectric transducers (quartz, lithium niobate, ceramics) are used to launch ultrasonic waves into liquid (water, mercury) or solid (fused quartz, glass, sapphire, spinel) media, in which the velocity of propagation is about 10^5 times slower than that of radio waves in free space. Shear waves are often used, because the velocity is lower and because, with the correct polarisation of the shear axis, the wave is reflected from a boundary as a pure shear wave. The transducers are either ground to thickness (half a wavelength at the required frequency) and then bonded to the medium, or evaporated or sputtered (cadmium sulphide, zinc oxide) to the required thickness. Whereas ground transducers are limited to, say, 100 MHz by their fragility at this frequency, evaporated or sputtered transducers can be deposited to resonate at frequencies up to 10 GHz. Long delays at such frequencies are, however, precluded by high losses in the available media,[15] and by the fact that the transducers can launch only longitudinal waves. Mode conversion from longitudinal to shear mode can be used for intermediate frequencies and delay times.

Bulk delay lines are used in colour television receivers ($\sim 62\ \mu s$ at the colour subcarrier frequency),[10] in MTI (moving target indication) for the removal of radar clutter from the display, in vertical aperture correction for television cameras, in APEGs (artificial permanent echo generators) for radar performance monitoring, and for a variety of other purposes. Mercury and quartz delay lines were used as data memories in early computers, but are now obsolete in this application. Most bulk wave delay lines are dispersive, but dispersive lines can also be made for pulse compression and other purposes (see *Figure 17.8* and Section 17.5.5).

Acoustic waves in bulk media can also be used for the deflection and modulation of light.[16]

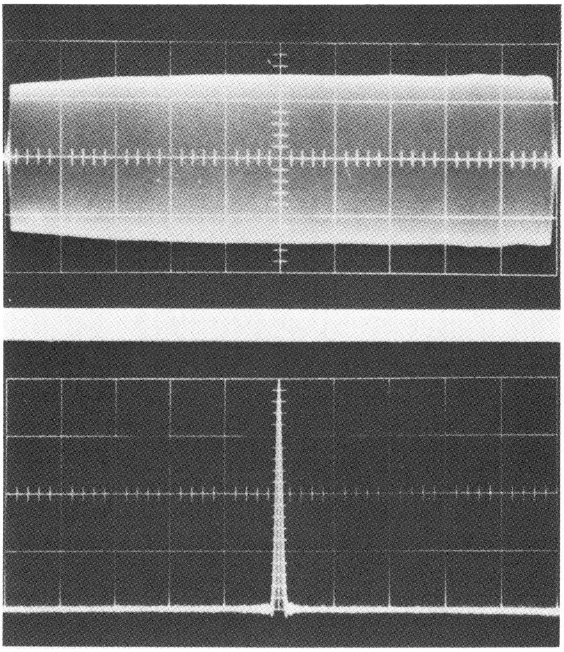

Figure 17.8 Pulse compression from a dispersive ultrasonic delay line. Input (top) is a 5 μs pulse, swept over 25 MHz: the compressed pulse (below) is 52 ns wide, i.e. the compression ratio is about 100 to 1 (courtesy WS Mortley, The Marconi Company Ltd)

17.5.4 Acoustic amplifiers

Some well-known semiconductor materials (gallium arsenide, GaAs, cadmium sulphide, CdS, etc.) are also piezoelectric. It was discovered by Hutson *et al.*,[17] that if electrons were made to travel by means of a d.c. electric field at the same velocity as an acoustic wave in the material, there could be an interaction between the wave and the electrons, resulting in amplification. Since the acoustic wavelength is small, the gain per unit length of material could be high. These devices do not seem to have fulfilled their early promise, doubtless because of their very low efficiency, which makes it difficult to operate them in a c.w. mode.

More recently, surface wave amplifiers have been made, using a single semiconducting piezoelectric material, or two materials, one piezoelectric (e.g. lithium niobate) the other semiconducting (e.g. silicon), in close proximity.[18] It is too early to say whether these amplifiers will be more successful than the bulk wave versions.

17.5.5 Surface acoustic waves

By means of a suitable transducer geometry, a shear or longitudinal wave in a bulk medium may be transferred to the surface of another medium: alternatively, if interdigital transducers (i.e. arrays of fingers one half wavelength apart) are laid down on a piezoelectric substrate and driven electrically, a surface wave is again excited.[19] The wave excited in these cases is the Rayleigh wave, travelling as a surface deformation which dies away very rapidly below the surface of the medium. It can be shown fairly simply that the velocity is a little lower than that of a shear wave in the medium, and that the wave is non-dispersive. Moreover, it travels on the surface and so is accessible as required. The velocity, and therefore the wavelength, is about 10^5 times lower than that of the radio wave in free space, giving the possibility, at least, of truly microelectronic circuits at radio frequencies.

This technology of surface acoustic waves (SAW) is now maturing rapidly; early work,[20] and a more recent review of applications[26] covers such topics as signal processors, code generators, filters, stabilised oscillators, TV filters, convolvers, correlators, etc. As examples, *Figure 17.8* shows a chirp or pulse compression in which an f.m. pulse 5 μs long, swept over 25 MHz is compressed to a pulse 50 ns long. (This result was actually achieved with a bulk wave device, but SAW techniques are now almost universal in this application.) Another relatively simple example is that of a Bragg cell, in which an acoustic wave in SAW[27] (or in a bulk material[28]) is used as a light deflector. Since the deflection angle is proportional to the acoustic frequency, the system can be used as a spectrum analyser which will simultaneously measure the frequencies of all signals present over a bandwidth of 1 GHz or more. The principles of Bragg cells in SAW and bulk wave versions are shown in *Figures 17.9* and *17.10*.

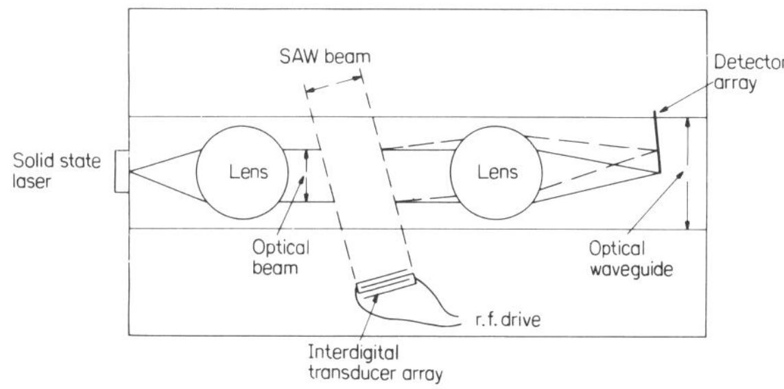

Figure 17.9 A SAW Bragg cell (courtesy of The Marconi Company Ltd)

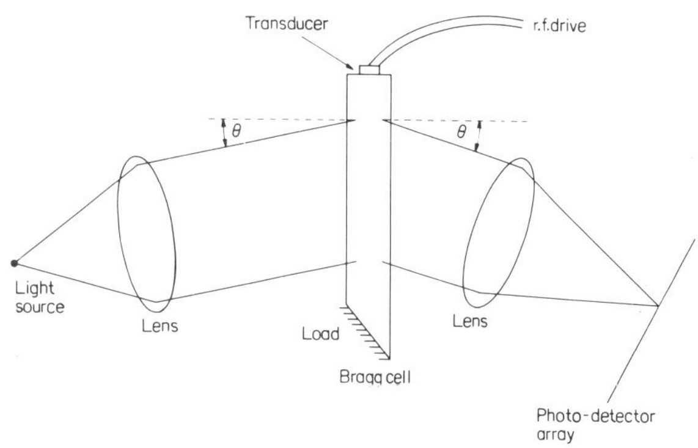

Figure 17.10 A bulk wave Bragg cell (courtesy of The Marconi Company Ltd)

Bragg cells are an example of a new class of acousto-optic devices in which a SAW (or bulk wave) is used to interact with a light beam, often from a laser. Further and more complex examples are given in the reference already quoted.[26]

17.5.6 Piezoelectric plastic films

Considerable interest has been shown in the applications of piezoelectric plastic films, especially polyvinylidene fluoride, PVF_2: films of this material, after polarisation, are strongly piezoelectric (5 to 10 times stronger than quartz) and pyroelectric.

Its structure and applications have been reviewed;[29,30] the applications include microphones, hydrophones (the material couples well to water), pressure switches and pyroelectric radiation detectors: there may also be useful medical applications.

17.5.7 Miscellaneous piezoelectric applications

Besides the major areas dealt with above, piezoelectric materials (usually ceramic materials because of their relatively low cost, ease of fabrication and high coupling coefficient) are used in a wide variety of other applications. These include piezoelectric ('ceramic') pickups for record players[10] (bi-morphs or multimorphs), gas ignition systems, echo sounders, ultrasonic cleaners and soldering irons and ultrasonic drills. (As an alternative to piezoelectric operation, such systems which rely on transmitting high ultrasonic powers, i.e. drills, cleaners and soldering irons, may be driven magnetostrictively.)

Useful accounts of these miscellaneous and high-power applications may be found in books by Crawford[23] and Hueter and Bolt.[5]

17.6 Data on piezoelectric materials

Because of their complexity and volume, it is impossible to give here numerical data on even the most important piezoelectric materials: instead, sources of such data are quoted.

Berlincourt *et al.*,[7] gave a review of the sources (p. 171) and tables of the more important materials (Tables II–VI inclusive pp. 180–4 covering single crystal materials, and Tables VII and VIII (pp. 195 and 202) on ceramic materials). The Mullard Booklet[16] covers the PXE ceramics (pp. 14–15). *The American Institute of Physics Handbook*[24] lists the piezoelectric strain constants of some 50 or 60 materials (Section 9–97), and the temperature coefficients of a few of these.

References

1 KAMINOW, I. P. and TURNER, E. H., 'Electro-optic light modulator', *Proc. IEEE*, **54**, 1374 (1966)

2 CADY, W. G., *Piezoelectricity*, Dover edition, Dover Publications (1964)

3 IRE Standards on Piezoelectric Crystals (1949), *Proc. IRE*, **37**, 1378 (1949)
IRE Standards on Piezoelectric Crystals (1957), *Proc. IRE*, **45**, 354 (1957)
IRE Standards on Piezoelectric Crystals (1958), *Proc. IRE*, **46**, 765 (1958)
IRE Standards on Piezoelectric Crystals (1961), *Proc. IRE*, **49**, 1162 (1961)

4 CADY, W. G., *Piezoelectricity*, 1st edn, McGraw-Hill (1946)

5 HUETER, T. F. and BOLT, R. H., *Sonics*, John Wiley (1955)

6 KITTEL, C., *Introduction to Solid State Physics*, Chapman and Hall (1953)

7 BERLINCOURT, D. A., CURRAN, D. R. and JAFFE, H., 'Piezoelectric and piezomagnetic materials', in *Physical Acoustics*, Vol. 1, Part A (ed. Mason, W. P.), Academic Press (1964)

8 MASON, W. P., *Piezoelectric Crystals and Their Applications to Ultrasonics*, Van Nostrand (1950)

9 NYE, J. F., *The Physical Properties of Crystals*, Oxford U.P. (1957)

10 VAN RANDERAAT, J., ed., *Piezoelectric Ceramics*, Mullard Ltd. (1968)

11 GERBER, E. A. and SYKES, R. A., 'State of the art—quartz crystal units and oscillators', *Proc. IEEE*, **54**, 103 (1966)

12 MARCONI TYPE F.3160, taken from *Catalogue of Quartz Crystal Oscillators, Ovens and Filters*, Marconi Specialised Components Division

13 BROCKELSBY, C. F., PALFREEMAN, J. S. and GIBSON, R. W., *Ultrasonic Delay Lines*, Iliffe Books Ltd (1963)
EVELETH, J. E., 'A survey of ultrasonic delay lines below 100 MHz', *Proc. IEEE*, **53**, 1406

14 MAY, J. E., 'Guided wave ultrasonic delay lines'; and Mason, W. P., 'Multiple reflection ultrasonic delay lines' in *Physical Acoustics*, Vol. 1, Part A (ed. Mason, W. P.), Academic Press (1964)

15 KING, D. G., 'Ultrasonic delay lines for frequencies above 100 MHz', *Marconi Review*, **34**, 314 (1971)

16 GORDON, E. L., 'A review of acoustic optical deflection and modulation devices', *Proc. IEEE*, **54**, 1391

17 HUTSON, A. R., MCFEE, J. H. and WHITE, D. L., *Phys. Rev. Lett.*, **1**, 237 (1961)

18 LAKIN, K. M. and SHAW, H. J., 'Surface wave delay line amplifiers', *IEEE Trans.*, **MTT 17**, 912 (Nov. 1969)

19 STERN, E., 'Microsound components, circuits and applications', *IEEE Trans.*, **MTT 17**, 912, p. 835 (Nov. 1969)

20 *IEEE Trans.*, **MTT 17** (Nov. 1969)
IEEE Trans., **MTT 21** (April 1973)
MAINES, J. D. and PAIGE, E. G. S., 'Surface-acoustic-wave components, devices and applications', *IEE Reviews, Proc. IEE*, **120**, No. 10R, 1078 (1973)

21 KLAUDER, J. R. *et al.*, 'The theory and design of chirp radars', *Bell Syst. Tech. J.*, **39**, 745 (July 1960)

22 MAINES, J. D. and PAIGE, E. G. S., 'Surface-acoustic-wave components, devices and applications', *IEE Reviews, Proc. IEE*, **120**, No. 10R, 1078 (1973)

23 CRAWFORD, A. R., *Ultrasonics for Industry*, Iliffe (1969)
CRAWFORD, A. H., ed., *High Power Ultrasonics—International Conference Proceedings*, IPC Press (1972)

25 *American Institute of Physics Handbook*, 2nd edn, McGraw-Hill (1963)

25 BURGSEN, R. and WILSON, R. L., 'Performance results of an oscillator using the SC-cut crystal', *Proc. 33rd Annual Symp. on Frequency Control*, p. 406 (1979)

26 *IEEE Trans.*, **MTT 29** (5 May 1981)
IEEE Trans., **SU 28** (3 May 1981)

27 ANDERSON, D. B., 'Integrated optical spectrum analyzer: an imminent chip', *IEEE Spectrum*, 22 (Dec. 1978)

28 CARTER, R. W. and WILLATS, T. F., 'The design and performance of a bulk wave Bragg cell spectrum analyzer', *Marconi Rev.*, **44**, 57 (1981)

29 GALLANTRAE, H. R. and QUILLIAM, R. M., 'Polarized polyvinylidene fluoride—its application to pyroelectric and piezoelectric devices', *Marconi Rev.*, **39**, 189 (1976)

30 SUSSNER, H., 'The piezoelectric polymer PVF_2 and its applications', *Proc. IEEE Ultrasonics Symp.*, 491 (1979)

31 POINTON, A. J., 'Pizeoelectric devices', *Proc. IEE*, **129**, Pt. A., No. 5, (July 1982)

18 Connectors

Written by the Product Managers of AMP of Great Britain Ltd, including:

C Kindell
T Kingham
J Riley

Contents

18.1 Connector housings

Connector housings are of different shapes, sizes and form, being able to satisfy requirements for a range of applications and industries—commercial, professional, domestic and military. In a commercial low cost connector the insulator material can be nylon which can be used in a temperature range of −40°C to +105°C and is also available with a flame retardant additive.

The connector can be wire to wire using crimp snap-in contacts, wire to printed circuit board and also printed circuit board to printed circuit board. In the automotive industry it is necessary to have waterproof connectors to prevent any ingress of water thrown up by moving wheels and the velocity of the vehicle through rain. Wire seals are placed at the wire entry point in the connector. These seals, made from neoprene, grip the insulation of the wire very tightly thus preventing any water ingress. When the two halves of the connector are mated it is necessary to have facial seal thus precluding any water ingress between these two parts and preventing any capillary action of the water.

In the professional and military fields the housing material needs to be very stable and to counteract any attack by fluids. One type of material in this form is diallyl phthalate. The primary advantages of diallyl phthalate are exceptional dimensional stability, excellent resistance to heat, acids, alkalies and solvents, low water absorption and good dielectric strength. This combination of outstanding properties makes diallyl phthalate the best choice of plastics for high quality connectors.

Most connectors which are available in diallyl phthalate are also available in phenolic. While phenolic does not have outstanding resistance to acids, alkalies and solvents it nevertheless has many characteristics which make it a good choice for connector housings. Among these characteristics are excellent dimensional stability, good dielectric strength and heat resistance. In addition there are a number of fillers which can be added to phenolic to obtain certain desired properties.

18.2 Connector contacts

The contacts that are used with the majority of connectors are made from brass or phosphor bronze with a variety of platings from tin through to gold. The most common type of brass used in the manufacture of contacts is cartridge brass which has a composition of 70% copper and 30% zinc. This brass possesses good spring properties and strength, has excellent forming qualities and is a reasonably good conductor. Phosphor bronze alloys are deoxidised with phosphorus and contain from about 1 to 10% tin. These alloys are primarily used when a metal is needed with mechanical properties superior to those of brass and where the slightly reduced conductivity is of little consequence.

One extremely important use of phosphor bronze is in locations where the terminal may be exposed to ammonia. Ammonia environments causes stress corrosion cracking in cartridge brass terminals. On the other hand, phosphor bronze terminals are approximately 250 times more resistant to this type of failure. Associated with the materials used in the manufacture of the contacts is a variety of platings. Plating is a thin layer of metal applied to the contact by electrodeposition.

Corrosion is perhaps the most serious problem encountered in contacts and the plating used is designed to eliminate or reduce corrosion. Corrosion can spread uniformly over the surface of the contact covering it with a low conductivity layer, with the thickness of this layer being dependent upon environmental conditions, length of exposure and the type of metal being used. Brass contacts that are unplated and have been in service for a period of time have a reddish brown appearance rather than the bright yellow colour of cartridge brass. This reddish colour is a tarnish film caused by oxidation of the metal. Although this film may not impair conductivity at higher voltages, it does at the very least, destroy the appearance of the contact. To eliminate the problem of this tarnish film the contact is usually tin plated. Although oxides form on tin they are the same colour as the tin and the appearance remains the same. In addition tin is relatively soft, and if it is to be used as a contact plating, most of the oxides will be removed during mating and unmating of the contact.

Tin is the least expensive of the platings and is used primarily for corrosion protection and appearance on contacts which operate at a fairly high voltage. Another important feature of tin is that it facilitates soldering. Gold plating is always used on contacts which operate in low voltage level circuitry and corrosive environments. The presence of films caused by the combination of sulphur or oxygen with most metals can cause open circuit conditions in low voltage equipment. Since gold will not combine with sulphur or oxygen there is no possibility of these tarnish films forming.

18.3 Connector shapes and sizes

A family of connectors would need to include a range of rectangular and circular connectors associated with a variety of size-16 contacts including signal and power contacts, fibre optic contacts and subminiature coaxial contacts. The connectors shown in *Figure 18.1* are rectangular for in-line, panel mounting or rack and panel applications, and circular either for environmentally sealed application or for the commercial portion of MIL-C-5015 connector areas. The housings are available in nylon, diallyl phthalate or phenolic with, in certain areas, a metal shell. The contacts are available in brass or phosphor bronze with tin or gold plating finishes. Associated are a variety of accessories such as strain reliefs, pin hoods, jackscrews, guide pins and pin headers to give greater versatility to the connector ranges.

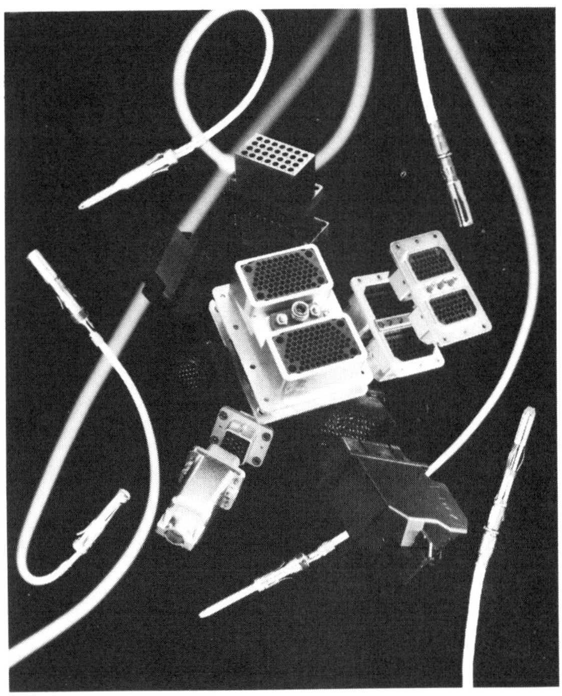

Figure 18.1 Multi-way pin and socket connectors (courtesy Amp of Great Britain Ltd)

18.4 Connector terminations

Crimping has long been recognised as an electrically and mechanically sound technique for terminating wires. Since crimping is a strictly mechanical process, it is relatively easy to automate. Because of this automation capability, crimping has become the accepted terminating technique in many industries.

Terminals or contacts designed for speed crimping in automatic or semi-automatic machines are often significantly different from those designed for handtool assembly, although most machine-crimpable terminals and contacts can also be applied with handtools.

Those designed for hand application cannot normally be used for automatic machinery. The selection of a crimping method is determined by a combination of five factors: (1) access of wire; (2) wire size; (3) production quantity; (4) power availability; and (5) terminal or contact design.

There are, of course, other factors which must be considered, e.g. if the finished leads are liable to be roughly handled, as in the appliance or automotive industries, the conductor insulation and the terminal or contact will have to be larger than electrically necessary in order to withstand misuse.

The user must remember the importance of maintaining the proper combination of wire terminal and tool, only then can the optimum crimp geometry and depth be obtained. In this respect it is best to follow the manufacturer's recommendations since most terminals have been designed for a specific crimp form. The effects of crimp depth are shown in *Figure 18.2*.

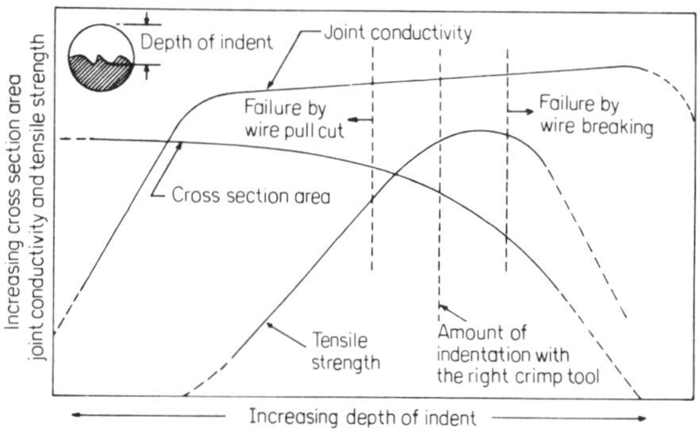

Figure 18.2 Effects of crimp depth

Tensile strength and electrical conductivity increase in proportion to the crimp depth. When the deformation is too great, tensile strength and conductivity suffer because of the reduced cross-sectional area. There is an optimum crimped depth for tensile strength and another for conductivity and, in general, these peaks do not coincide. Thus a design compromise is required to achieve the best combination of properties. However, the use of improper tooling can void the entire design.

Merely selecting the proper wire terminal and tool combination is not enough. The wire must be stripped to the recommended dimension, without nicking, and inserted into the terminal to the correct depth before crimping. A properly crimped terminal is shown in *Figure 18.3*.

It is possible to determine the relative quality of a crimp joint by measuring crimp depth in accordance with manufacturers' suggestions. Tensile strength also provides a relative indication of mechanical quality of crimped connections. This factor has been utilised as a 'user control test' in British and other international specifications.

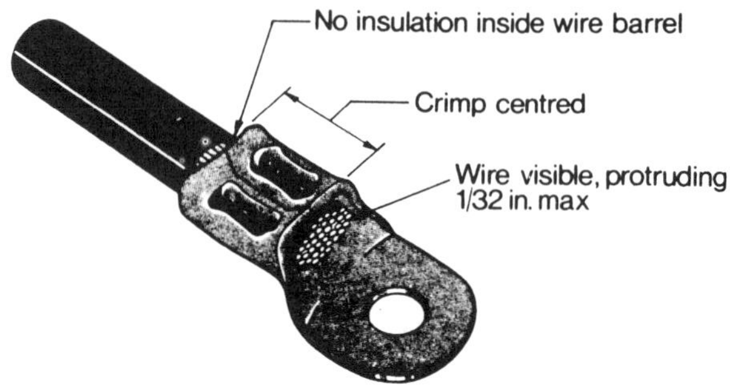

Figure 18.3 A correctly crimped un-insulated terminal

Crimped terminations do not require pre-soldering, and are designed to be used with untinned wires. Soldering may damage the crimp, burn the wire and produce a bad joint. Also the wire could stiffen from wicked solder, and break off later because of vibration. Soldering can effect the characteristics designed into the crimp and seriously affect the performance.

Many terminal designs have some means of supporting the wire insulation. These features are divided into two main categories: (a) insulation gripping; and (b) insulation supporting.

Insulation gripping terminals prevent wire flexing at the termination point and deter movement of the insulation. This feature improves the tensile strength of the crimp for applications where severe vibration is present.

Insulation supporting types of terminals have insulation that extends beyond the crimping barrel and over the wires own insulation. This only provides support and does not grip the insulation in a permanent manner. The latest feature for this type of termination is the funnel entry type. This, as its name implies, has a funnel form on the inside of the insulation sleeve, which aids in the correct placing of the stripped wire in the barrel of the terminal. It has long been possible to snag strands of wire on the edge of the wire barrel while putting the stripped wire into the terminal. This, depending on the number of snagged strands, could impair the electrical characteristics and cause 'hot spots' or under crimping of the terminal.

All these problems are resolved with the introduction of funnel entry, as the wire is able to go straight into the wire barrel without damage. Minimal operator skills, increased production rates and added benefits are possible.

Finally there is the introduction of insulation displacement or slotted beam termination, which has enjoyed wide use in the telecommunication and data systems industries. With the advent of connectors or terminal blocks designed for mass termination, this concept has spread and it is rapidly finding use in many other applications (see Section 18.6).

Although the appearance and materials in insulated displacement connectors vary the design of the slotted areas is basically the same in each.

Insulated wire fits loosely into the wider portion of the V-shaped slot and, as the wire is pushed deeper into the terminal, the narrowing slot displaces the insulation and restricts the conductor as in *Figure 18.4*. Additional downward movement of the insertion tool forces the conductor into the slot where electrical contact is made.

Insulation displacement has recently been applied to terminal blocks. One side has the insulation displacement contact whilst the other side would have the conventional screw terminal accepting ring tongues or bare wire. It would accept a wide range of wire sizes or stranded wire 0.3 mm^2 to 2 mm^2.

Wire termination is accomplished by two simple screwdriver type handtools with different insertion depths. The first is used to

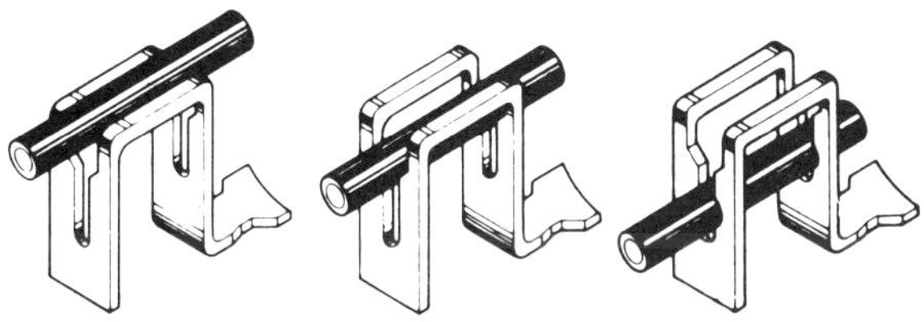

Figure 18.4 Insulation displacement contacts

insert a single wire into the terminal, the other for inserting the second wire.

18.5 Tooling

Crimping tools or machines should be selected after a thorough analysis, as with any other production system.

Generally, the following rates can be achieved with various types of tooling:

manual tools	100–175 per hour
power tools	150–300 per hour
semi-automatic machines	100–4000 per hour
lead making machine	up to 11 000 per hour

As these figures show, manual tools are intended for repair and maintenance, while powered tools and machines are designed for production applications. To the manufacturer whose output in wire terminations is relatively small, automated tooling is not necessary. While the economics of automated wire terminations may vary with different applications, an output of more than one million terminations a year can be taken as a guide for considering semi-automatic tooling.

The basics of a good crimp connection are the same whether the tool to be used is a simple plier type or a fully automatic lead making machine.

The basic type of crimping tool is the simple plier type. It is used for repair, or where very few crimps are to be made. These are similar in construction to ordinary pliers except that the jaws are specifically machined to form a crimp. Most of these tools are dependent upon the operator to complete the crimp properly by closing the pliers until the jaws bottom together. Many of the tools may be used for several functions such as wire stripping, cutting, and crimping a wide range of terminal sizes and types. Tools of this type are in wide use.

Other more sophisticated tooling is available, such as cycle controlled tools. This type normally contains a ratchet mechanism which prevents the tool opening before the crimp has been properly completed. This ratchet action produces a controlled uniform crimp every time regardless of operator skill. However, operator fatigue is normally a limiting factor in production with any manual tool.

Powered handtools, either pneumatic or electronic controlled, can be semi-portable or bench mounted. When larger production quantities of terminations are required, the need for this form of tooling is essential. They not only yield high rates of output at low installation cost but also give high standards of quality and are repeatable throughout the longest production run.

These tools offer the opportunity for the introduction of tape mounted products. A variety of tape mounted terminations are available in either reel or boxed form. Advanced tooling, with interchangeable die sets, gives a fast changeover with minimum downtime. During the crimping cycle the machine will automatically break the tape bonds and free the crimp product for easy extraction, at the same time indexing the termination into position for the next crimp operation.

18.6 Mass termination connectors

Mass termination is a method of manufacturing harnesses by taking wires directly to a connector and eliminating the steps of wire stripping, crimping and contact insertion into housings. It employs a connection technique known as insulation displacement, an idea developed many years ago for the telecommunications industry.

An unstripped insulated wire is forced into a slot which is narrower than the conductor diameter as in *Figure 18.3.* Insulation is displaced from the conductor. The sides of the slot deflect like a spring member and bear against the wire with a residual force that maintains high contact pressure during the life of the termination.

This is the basic principle, but it must be realised that each slot is carefully designed to accept dimensional changes without reducing contact forces. This is accomplished by designing enough deflection into the slot to compensate for creep, stress relaxation and differential thermal expansion.

The force required to push the wire into the slot is approximately 25 times less than for a conventional crimp and it is this factor, in conjunction with the facilitity of not having to strip the wire, that makes insulation displacement readily acceptable for mass termination, by taking several wires to the connector and terminating them simultaneously.

A typical system would employ a pre-loaded connector, with the receptacle having dual slots offering four regions of contact to the wire. The exit of the wire from the connector is at 90° to the mating pin, and can have a maximum current rating of up to 7.5 A.

The average tensile strength of the displacement connection when pulled along the axis of the wire is 70% of the tensile strength of the wire and 20% when pulled on axis parallel to the mating pin. Therefore, plastic strain ears are moulded into the connector to increase the wire removal force in this direction.

The different systems have been developed to accept a wide range of wire including 28 to 22 AWG (0.08 to 0.33 mm^2) wire and 26 to 18 AWG (0.13 to 0.82 mm^2) wire. The connectors are colour coded for each wire gauge since the dimensional difference of the slot width cannot be readily identified.

The pin headers for most systems are available for vertical and right angle applications in flat style for economical wire to post applications, polarised for correct mating and alighnment of housings and polarised with friction lock for applications in a vibration environment.

18.6.1 Types of tooling

To obtain all the benefits for harness manufacture, a full range of tooling is available, from simple 'T' handle tools to cable makers.

The 'T' handle tool would be used only for maintanance and repair. For discrete wires a self-indexing handtool, either manually or air operated, would be used for intermediate volumes.

For terminating ribbon cable, there are small bench presses for relatively low volumes of harnesses, and electric bench presses for higher production needs.

However, it is the innovation of the harness board tool and the cable maker that offers highest production savings. The harness board tool allows connectors to be mass terminated directly onto a harness board. The equipment consists of three parts: power tool, applicator and board mounted comb fixtures. The wires are routed on the harness board and placed through the appropriate comb fingers. The power tool and applicator assembly is placed on the combs to cut and insert the wires into the connectors. After binding with cable ties the harness can be removed from the board.

The cable maker, either double end or single end, will accept up to 20 wires which can be pulled from drums or reels on an appropriate rack. The individual wires can all be the same length, or variable, with a single connector on one end of the cable and multiple connectors on the other end.

In general, a complete cycle would take approximately 15 to 20 seconds according to how many connectors are being loaded. However, three double-ended cables, six-way at each end, able to be produced on the machine in one cycle, would be using the machine to its maximum capacity and the overall time would be expected to be longer.

A comparison can be made between an automatic cut, strip and terminating machine and the cable maker mentioned earlier. This comparison is on 100 000, six-way connectors:

Standard method

Cut, strip and terminate	3400/hour		176 hours
Manually insert contacts	900/hour		666 hours
		Total	842 hours

New method (mass termination)

Cable maker: assume conservative figure of two single-ended cables every 20 seconds			
	360 cables/hour	Total	278 hours

It can clearly be seen that labour savings of 67% are not unrealistic which must be the major benefit from using mass termination techniques. Other such benefits include no strip control, no crimp control, reduced wiring errors, no contact damage, reduced tooling wear.

18.6.2 Ribbon cable connectors

Connectors for 0.050″ pitch ribbon cable can also be considered as mass termination types. The basic four types of ribbon cable are: extruded, bonded, laminated, and woven, with extruded offering the best pitch tolerance and 'tearability'. The connectors, normally loaded with gold-plated contacts, are available in a standard number of ways up to 64, these being 10, 14, 16, 20, 26, 34, 40, 44, 50, 60 and 64.

There are various types of connectors used:

(1) Receptacle connectors, 0.100 grid for plugging to a header.
(2) Card edge connectors, 0.100 pitch to connect to the edge of a PCB.
(3) Pin connectors, to mate with receptacle connectors and offer ribbon to ribbon facility.
(4) Transition connectors, for soldering direct to PCB.
(5) DIL plug, 0.100 × 0.300 grid for either soldering to PCB or connecting to a DIP header.

The normal rating for these type of connectors is 1 ampere with an operating temperature range of −55°C to 105°C and a dielectric withstanding voltage of 500 V r.m.s.

18.7 Fibre optics connectors

When joining fibres light losses will occur in four ways:

(1) Surface finish. The ends of the fibres must be square and smooth and this is usually accomplished by polishing the cut ends.
(2) End separation. Ideally the fibre ends should touch but this could cause damage and so they are normally held between 0.001″ and 0.005″ apart.
(3) Axial misalignment. This causes the highest loss and must be controlled to within 50% of the smaller fibre diameter.
(4) Angular misalignment. The ends of the fibres should be parallel to within 2%.

Any connector system must therefore hold the fibre ends to within these limits, and several different variations have been developed:

(a) *Tube method.* This method uses a metal jack and plug which are usually held together by a threaded coupling. The fit of the plug into the jack provides the primary alignment and guides the fibre in the jack into a tapered alignment hole in the plug. The depth of engagement must be accurately controlled to ensure correct end separation. These connectors are normally made from turned metal parts and have to be produced to close tolerances (*Figure 18.5*).

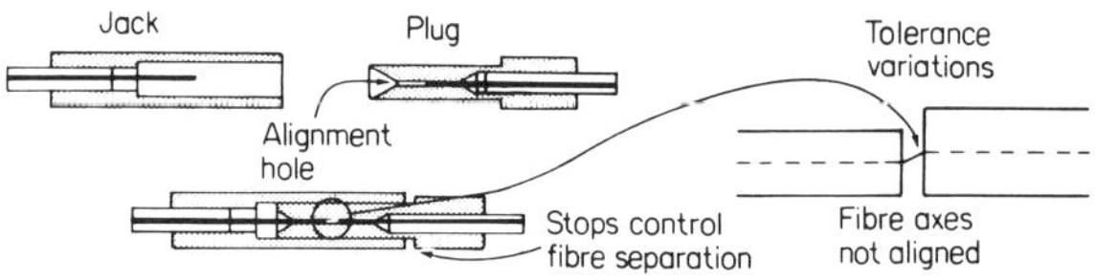

Figure 18.5 Fibre optic tube alignment connector

(b) *Straight sleeve method.* A precision sleeve is used to mate two plugs which are often designed similar to the SMA coaxial connectors and the sleeve aligns the fibres. These connectors are made from very tightly toleranced metal turned parts and due to the design, concentricity needs to be very good (*Figure 18.6*).

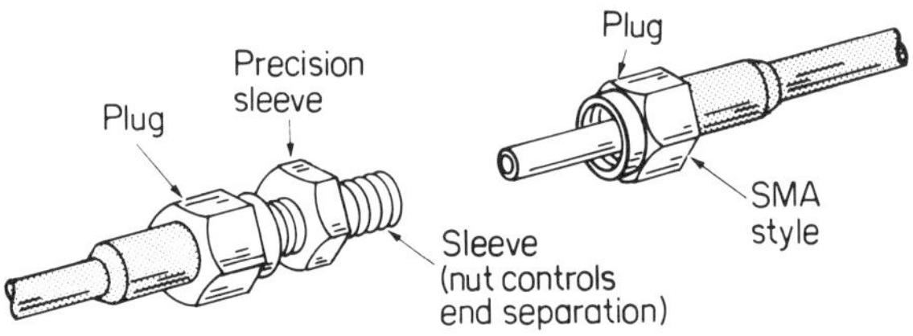

Figure 18.6 Fibre optic straight sleeve connector

(c) *Double eccentric method.* Here the fibres are mounted within two eccentrics which are then mated. The eccentrics are then rotated to bring the fibre axes into very close alignment and locked. This produces a very good coupling with much looser manufacturing tolerances but the adjustment can be cumbersome and must usually be done with some test equipment to measure maximum adjustment (*Figure 18.7*).

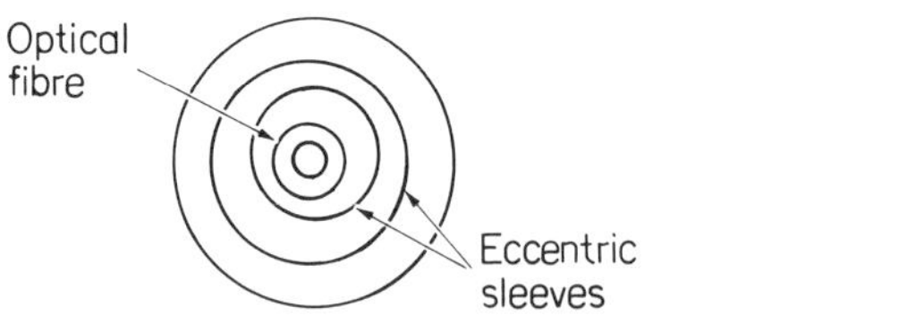

Figure 18.7 Fibre optic double eccentric connector

(d) *Three-rod method.* Three rods can be placed together such that their centre space is the size of the fibre to be joined. The rods, all of equal diameter, compress and centre the fibres radially and usually have some compliancy to absorb fibre variations. With this design it is important that the two mating parts overlap to allow both members to compress each fibre. The individual parts in this design can be moulded plastic but need to be well toleranced (*Figure 18.8*).

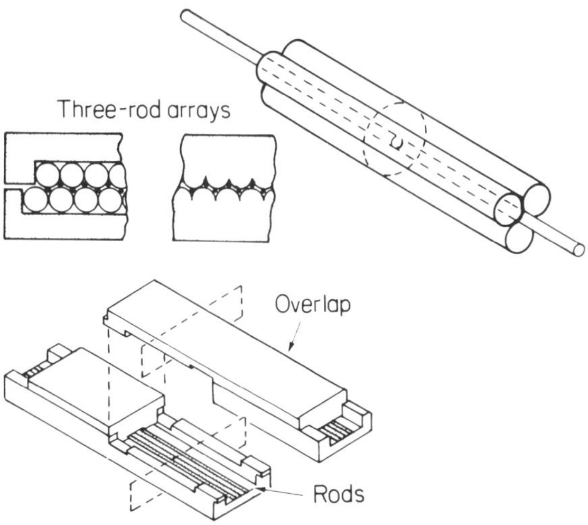

Figure 18.8 Fibre optic three-rod connector

(e) *Four-pin method.* Four pins can be used to centre a fibre and the pins are held in a ferrule. This method is sometimes used with the straight sleeve design when the pins are used to centre the fibre and the sleeve is used to align the mating halves. These parts are normally turned metal and held to tight tolerances (*Figure 18.9*).

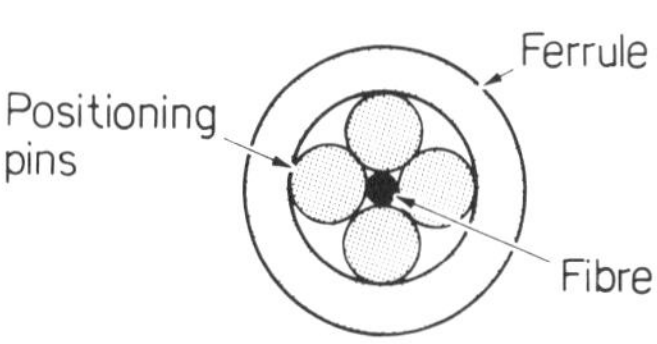

Figure 18.9 Fibre optic four-pin method of connection

(f) *Resilient ferrule.* This method utilises a ferrule and a splice bushing. The front of the ferrule is tapered to match a similar taper in the bush and the two parts are compressed together with a screw-on cap which forces the two tapers together. This moves the fibre in the ferrule on centre and provides a sealed interface between the two parts preventing foreign matter from entering the optical interface. The compression feature accommodates differences in fibre sizes and enables manufacturing tolerances to be considerably relaxed. The parts are plastic mouldings and typically produce a connector loss of less than 2 dB per through way at very low cost (*Figure 18.10*).

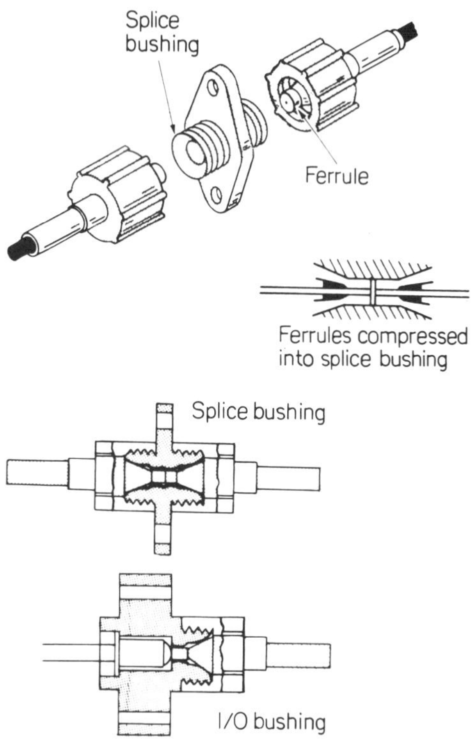

Figure 18.10 Fibre optic resilient ferrule alignment mechanism

18.8 Radio frequency connectors

Radio frequency connectors are used for terminating radio frequency transmission lines which have to be run in coaxial cables. These cables are available in sizes ranging from less than 3 mm diameter for low power applications of around 50 watts, to over 76.2 mm diameter for powers of 100 000 watts. In addition to power handling capabilities cables are also available for high frequency applications, high and low temperature applications, severe environmental applications and many other specialised uses which all require mating connectors.

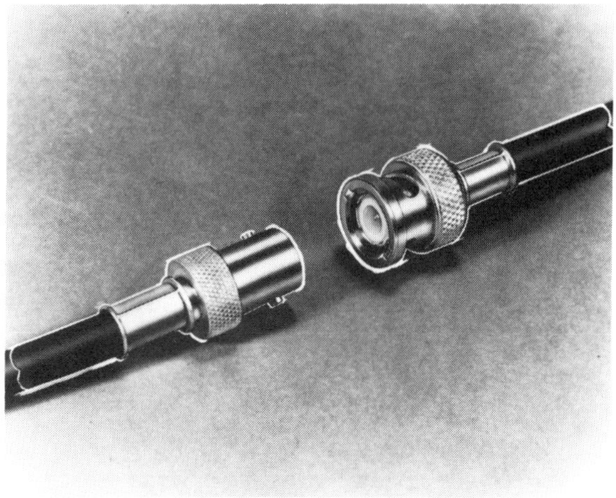

Figure 18.11 A typical r.f. connector (courtesy Amp of Great Britain Ltd)

Some of the more popular ranges are grouped in approximate cable diameter size with their operating frequency ranges as follows:

3 mm diameter	SMA 0–12.4 GHz, SMB 0–1 GHz, SMC 0–1 GHz
5 mm diameter	BNC 0–4 GHz, TNC 0–11 GHz, min. u.h.f. 0–2 GHz
7 mm diameter	N 0–11 GHz, C 0–11 GHz, u.h.f.

The design and construction of the range of connectors is very similar throughout as they all have to terminate a centre conductor and a woven copper braid screen. The variations are in size and materials. For example, to meet the MIL-C-39012 specification it is necessary to use high quality materials such as brass, silver plated for the shells, and teflon for the dielectric with gold-plated copper centre contacts. This is mainly due to the requirement for a temperature range of $-65°C$ to $+165°C$. There are however, three distinct types of termination: (a) soldering and clamping; (b) crimping; and (c) soldering and crimping.

(a) With the soldering and clamping type of design the centre contact is soldered to the centre conductor and the flexible braid is then clamped to the shell of the connector by a series of tapered washers and nuts. The biggest advantage of this type of connector is that it is field repairable and replaceable without the use of special tools. The disadvantages are the possibility of a cold solder joint through underheating or melting the dielectric by overheating. Any solder which gets onto the outside of the centre contact must be removed otherwise the connector will not mate properly. It is easy to assemble the connector wrongly due to the large number of parts involved.

These connectors are used in large numbers by the military.

(b) In the crimping design the centre contact is crimped to the centre conductor and the flexible braid is then crimped between the connector shell and a ferrule. There are versions which require two separate crimps, normally to meet the MIL specification, and versions which can have both crimps made together. The advantages of this method of termination are speed and reliability together with improved electrical performance. Testing has shown that the SWR of a crimped connector is lower than the soldered and clamped version. The crimp is always repeatable and does not rely upon operator skill. The disadvantages of this design are that a special crimp tool is required and the connectors are not field repairable.

(c) With the soldering and crimping type of design the centre contact is soldered to the centre connector and the flexible braid is crimped. Obviously all the advantages and disadvantages of the previous two methods are involved with this design.

Further reading

EVANS, C. J., 'Connector finishes: tin in place of gold', *IEE Trans. Comps, Hybrid & Manf. Technol.*, **CHMT-3**, No. 2 (June 1980)

KINDELL, C. 'Ribon cable review', *Electronic Production* (Nov./Dec. 1980)

PEEL, M., 'Material for contact integrity', *New Electronics* (Jan. 1980)

McDERMOTT, J., 'Hardware and interconnect devices', *EDN* (July 1980)

TANAKA, T., 'Connectors for low-level electronic circuitry', *Electronic Engineering* (Feb. 1981)

McDERMOTT, J., 'Flat cable and its connector systems', *EDN* (Jan. 1981)

CLARK, R., 'The critical role of connectors in modern system design', *Electronics & Power* (Sept. 1981)

MILNER, J. 'LDCs for IDCs', *New Electronics* (Jan. 1982)

ROELOFS, J. A. M. and SVED, A., 'Insulation displacement connections', *Electronic Components and Applications*, **4**, No. 2 (Feb. 1982)

McDERMOTT, J., 'Flat cable and connectors', *EDN* (Aug. 1982)

SAVAGE, J. and WALTON, A., 'The UK connector scene—a review', *Electronic Production* (Sept. 1982)

19 Printed Circuits

G W A Dummer MBE, CEng, FIEE, FIEEE, FIERE
Electronics Consultant

Contents

19.1 Introduction

Printed circuits, consisting of a pattern of wiring on an insulating base, have become the standard, almost universal, method of construction of practically all electronic systems from domestic appliances to aircraft, space vehicles, computers, etc.

The history of printed circuit making shows that many techniques have ceased to become economical, leaving only two basic methods:

(1) the subtractive process where a metal foil, usually copper, is etched from the surface of an insulating substrate to define a required pattern;
(2) the additive process where metal deposition is used to make the pattern on the surface of an insulating substrate.

There are mixtures of these two methods, one of which is the so-called semi-additive method, where a very thin foil (about 5 μm thick), is plated up to produce the requisite pattern, again on an insulating substrate.

19.2 Design and layout factors

Printed circuit boards may contain single-sided or double-sided patterns or sometimes multiple layers; it is simpler and cheaper to use single-sided or double-sided wherever possible. A complex electronic circuit is first broken down into subunits and an initial layout of point-to-point wiring in two dimensions is drawn out. Crossover wires are reduced to a minimum and it is often convenient to make templates of the components so that 'trial and error' methods can be used to finalise the best point-to-point pattern. Some components can be placed to bridge the wires, but care should be taken with respect to mutual coupling and capacitative effects. A typical layout for the printed wiring side of a board is shown in *Figure 19.1*, whilst the component side is shown in *Figure 19.2*.

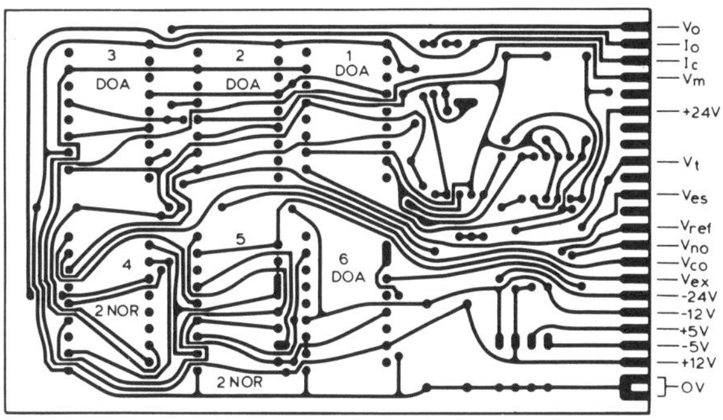

Figure 19.1 Printed wiring board layout

19.3 Artwork and design

This is the term for the original drawing of the pattern of tracks and holes. Black tapes of various widths are available to make a master circuit pattern; the tapes may be placed on transparent materials in order to register one side with the other. Lands and various configurations such as DIL chip patterns can be used to produce an exact copy of the required pattern of tracks, lands and holes.

Coloured pens are used to make complex patterns to register on each side of the board. Filters are used in the photographic reproduction of the master to eliminate one particular colour and thus build up a pattern for each side of the board.

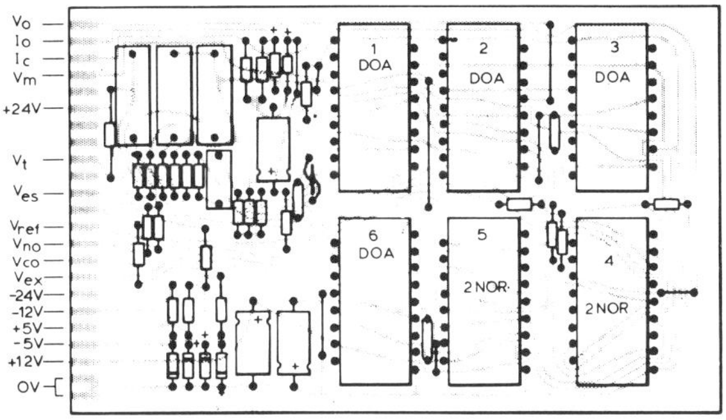

Figure 19.2 Component layout in printed wiring board

The artwork can also be made by automatic methods such as computer aided design (CAD). The most recent method is to take a drawing on graph paper or some form of printed grid, apply a probe from a computer to each coordinate point, press a control and record the exact position of each track, land or hole in a digital manner. The computer then makes a tape or cassette which can be used to photo-plot the results and make a positive or negative of the pattern. A tape can also be produced for use in the NC (numerical control) drilling operation.

This method of digitising the circuit uses an inbuilt datum so that the original drawing need not be exact. Similarly, the NC drilling machine can produce an exact pattern from an inbuilt datum point. The so-called 'interactive system of design' uses the techniques as above but also allows an operator to produce a pattern without any artwork. A computer has a visual display unit (VDU) on which a series of patterns can be displayed. The operator sets up a pattern which includes tracks, holes, lands, etc. The routing of the tracks is then done by the computer but when it is unable to prevent a crossover the operator intervenes to correct its action. With effort from the machine and the operator, a pattern results which can be made up as a trial. Re-routing of tracks can be made as needed to ensure a final performance as expected.

19.3.1 Photographic techniques

The positives and negatives and diazos carrying the wiring pattern are usually produced by specialised companies. Stable base materials (polyester) are used as dimensional stability is extremely important in circuit making because the photographic image is carried through every operation. In the case of double-sided boards, the registration of the two sides must be accurately controlled and this usually means that the photographic materials need to be stored in temperature and humidity controlled conditions.

19.4 Materials for printed wiring

The almost universal material for 'ordinary' printed circuit boards is epoxy resin impregnated glass fibre laminates. Paper laminates are still used in less arduous environments, e.g. hi-fi equipment and domestic appliances,

In recent years, the use of flame retardant materials has become almost universal because of the danger of fire in electronic and electrical apparatus. Exceptions to the use of these materials are irradiated polyethylene and polysulphone for high frequency applications.

The copper foils are usually between 25 and 75 μm (0.001″ and 0.003″) in thickness, although some 5 μm and 12 μm foils are becoming of increasing importance as, because of the overplating of copper and tin lead, less metal needs to be etched away and therefore costs are lower. Additionally, finer lines and spaces can be produced on the thinner foils.

19.5 Mechanical processing

19.5.1 Drilling, routing, blanking and punching

Blanking and punching are used only for low cost paper-based materials, although some glass fibre mat-based materials can be punched. Punching is the process commonly used for low cost components. Drilling is done by two methods: (1) Stylus drilling, where a template is used to locate the stylus and the drill moves upwards to make the necessary holes, usually in a stack of from three to five boards. (2) Numerical control (NC) drilling, which uses a multi-headed machine with a moving XY table on which the boards are located by means of dowels. The drilling operation is controlled either by a 'joystick' and an optical system or, more usually, by means of a computer punched tape or tape cassette. Such machines may have the ability to pick up drills and routers and thus alter the size of the holes to be drilled automatically.

In a routing operation single spindle machines are used to make profiles, polarising slots and panel cutting, but later techniques use the NC drilling machine as a router and all operations are controlled either by punched tapes or tape cassettes.

19.5.2 Forming the wiring pattern

Resists used are of two forms: (1) dry resist, where the resist material is in the form of a thin film with an overlay of plastic film to protect it during processing; and (2) photopolymers, liquids which can be applied to the surface of the boards to make them photosensitive. Both types are exposed to u.v. light through a photographic positive or negative and developed to produce the final image which is etch resistant.

Silk screens may be used for the wiring pattern if it is sufficiently bold, but legends and solder resists are usually produced by silk screen processing.

Etching away the unwanted copper from the patterned board can be achieved by a variety of methods and etchants. Common ones are ferric chloride, cupric chloride or a regenerative system of sulphuric acid and hydrogen peroxide. When plated-through-hole processes are used, and these are almost universal for professional and military applications, ammoniacal etchants are used because it is necessary to etch the tin lead overplate.

19.6 Plated through holes

In the plated-through-hole process, the inside of the holes is made electrically conductive by a chemical sensitiser. The plated-through-hole process consists of:

(1) drilling the laminate;
(2) electroless copper deposition;
(3) electro-copper deposition;
(4) tin lead electrodeposition;
(5) imaging (to define the wiring and hole patterns);
(6) etching;
(7) legend and/or solder resist printing by silk screen.

This ensures that every conductor and each hole is completely covered with about 40 μm thickness of metal to make sound interconnections between the two sides of the board.

19.7 Multilayer printed wiring boards

Multilayer boards are used where the high density of the wiring pattern and the number of components, often integrated circuits with multiple leads, require a large surface area which is not obtainable by either single- or double-sided designs. Most common types are four or six layers, although 32 layers have been made. A typical arrangement is shown in *Figure 19.3*, which shows plated-through holes and a component mounted on a four-layer board. Patterning the inner layers is done by assembl-

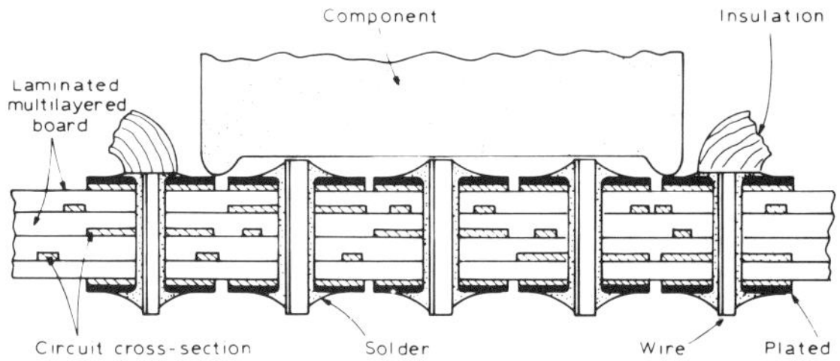

Figure 19.3 Multilayer printed wiring board

ing a stack of inner layers with 'pre-pregs', or partially cured boards, and laminating these together to make a composite board. Holes are drilled from top to bottom and intersecting the pattern as needed in the inner layers. Patterning top and bottom outer layers and then processing by the plated-through-hole process is done as previously described.

This type of board may contain 'via' holes, which may be buried in the inner layers. This provides for a number of internal interconnections while leaving the outer surface free for mounting components.

19.8 Flexible printed wiring

Flexible printed circuits have many uses, particularly for interconnecting other parts of the equipment, such as connectors, large components and rigid boards. The most commonly used material for this is copper-clad polyimide film which has exceptional heat resistant properties, making it resistant to soldering. Copper-clad polyester may be used for printed cabling, etc., where soldering is not necessary.

19.9 Flexi/rigid boards

A combination of rigid and flexible printed circuit boards eliminates some of the interconnections which are nearly always necessary for assembling equipments. In principle, a rigid board has inner layers extended to a flexible film, so that a U-shape can be formed. Such a combination board may be removed from an equipment without electrical disconnection. This makes maintenance easier as the equipment may be operating during removal of a panel.

19.10 Mass soldering systems

Mass soldering systems can be divided into two categories:

(1) making a solderable finish to the board; and
(2) soldering the various components to the board.

The first poses problems during manufacture and, in general, three methods are used to ensure solderability:

(a) reflow in hot oil;
(b) reflow by infrared heat;
(c) solder levelling.

The objective of these processes is to convert the electro-deposited tin lead on the board into a solderable alloy.

The process of mass soldering components into printed circuit boards is done by a process in which the board with its mounted components is first passed over a fluxing wave and then through a solder wave, which ensures that solder penetrates every joint and hole to make reliable connections.

19.11 Components for printed circuits

Modern techniques employ multiple-leaded components, commonly microelectronic DIL packages with either linear or digital circuits, but there is an increasing trend to larger packages, with as many as 98 terminations.

Smaller components such as decoupling capacitors, diodes, resistors, etc., may be mounted adjacent to these packages and present no special problems.

The introduction of chip carriers for microelectronic devices, which are surface mounted, calls for a change of design with lands or interconnection points on 10 mm or 12 mm pitch. This technique enables silicon devices of various types to be mounted in a carrier to form a subsystem, the chip carrier then being bonded to the printed circuit by soldering. Such carriers may have as many as 156 terminations.

19.12 Automatic assembly techniques

The chips may be mounted on a tape which is indexed like a ciné film. They may be mounted automatically and other smaller components may be fitted into holes when fed from bandoliered tapes.

The pressure on printed circuit designers and makers, is to ensure that the fixing holes are closely controlled, in order that automatic assembly is possible.

Later versions of automatic assembly are in the form of robots where sensitive fingers and sensing devices ensure accurate disposition of components on the board.

Further reading

COOMBS, C. F., *Printed Circuits Handbook*, McGraw-Hill (1967)
DRAPER, C. R., *The Production of Printed Circuits and Electronics Assemblies*, R. Draper Ltd, London (1969)
JOWETT, C. E., *Electronic Engineering Processes*, Business Books Ltd, London (1972)
LEONIDA, G., *Handbook of Printed Circuit Design, Manufacture, Components and Assembly*, Electrochemical Publications Ltd, Ayr
Proceedings of INTERNEPCON, Kiver Communications Ltd, Surbiton (annually)
SCARLETT, J. S., *Printed Circuit Boards for Microelectronics*, Electrochemical Publications Ltd, Ayr

20 Power Sources

F F Mazda DFH, MPhil, CEng, MIEE, DMS, MBIM
Rank Xerox Ltd
(Sections 20.1, 20.2.2, 20.2.6, 20.3.1, 20.3.3, 20.5, 20.6)

W Bowry
Ever Ready (Special Batteries) Ltd
(Section 20.2.1)

Duracell Batteries Ltd
(Sections 20.2.3, 20.2.4)

M Ewing
SAFT (UK) Ltd
(Section 20.2.5)

W D C Walker BSc, CChem, MRSC
Applications Manager, Ever Ready (Special Batteries) Ltd
(Sections 20.3.2, 20.4)

Contents

20.1 Cell characteristics

Power sources or batteries are made from several cells which are connected together to give a higher voltage. There are two types of cells:

(i) Primary cell, which is used once, until it is discharged, and then thrown away. Examples of primary cells described in this chapter are zinc–carbon, zinc chloride, alkaline manganese, mercuric oxide, zinc–lithium and zinc–air.
(ii) Secondary cell, which needs to be charged, after it is made, before use. Once discharged the cell can be re-charged and used again. Examples of secondary cells are lead–acid, nickel–cadmium and zinc–air.

Several parameters are important in the choice of a cell for a given application. These parameters are usually determined by the electrochemistry of the material used within a cell.

Open circuit voltage This is the voltage at the terminals of the cell. It depends on several factors such as the history of the cell (*Figure 20.1*) and the amount of energy which it has supplied (*Figure 20.2*). Cells are usually designed to have a relatively flat discharge curve until the cell is almost totally discharged, and then the terminal voltage decays rapidly.

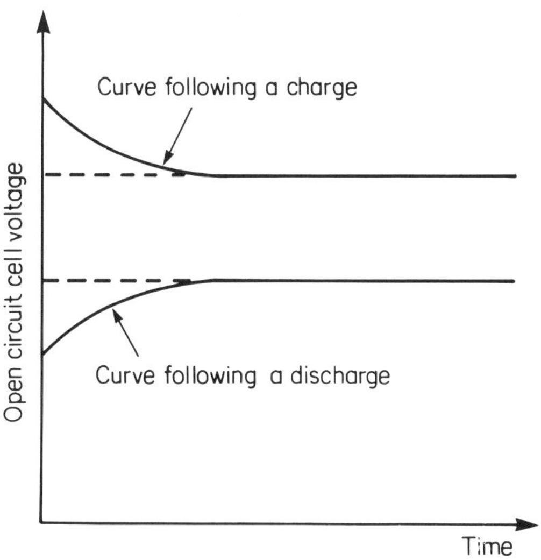

Figure 20.1 Effect of cell history on open circuit cell voltage

Cell capacity It is the amount of energy, usually stated in ampere hours, which the cell can provide without its terminal voltage falling below a given value. This is also specified as its '*C*' rate, which is the rate at which a fully charged cell would be discharged in one hour. So a cell with a capacity of 5 ampere hours has a '*C*' rate of 5 amperes.

Depth of discharge This is the percentage of the cell's capacity by which it has been discharged. So a 50 ampere hours cell which has been discharged by 10 ampere hours has a depth of discharge of 20%.

Charge retention or shelf life This is a measure of the ability of the cell to maintain its charge when stored. It is affected by the state of charge of the cell and the storage temperature.

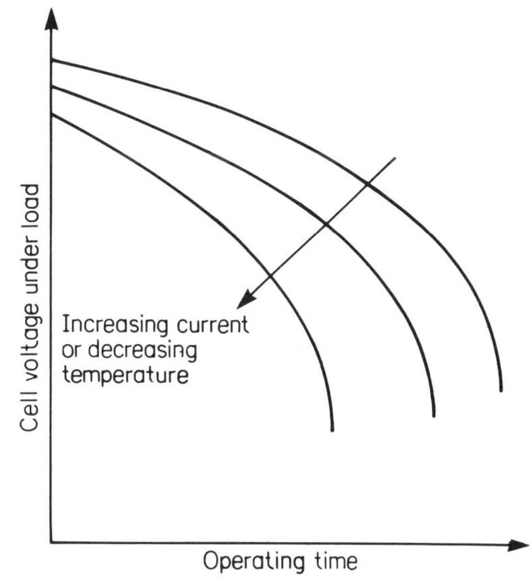

Figure 20.2 Effect of load or temperature on the cell voltage, at constant temperature

Charge rate For a secondary cell this is the charging current expressed as a function of the cell's capacity.

Cycle life This is the number of charge–discharge cycles which a secondary cell can go through before failure occurs.

Charge acceptance This is the ability of a secondary cell to accept energy. It is measured as the proportion of the charge input which the cell can give out again, without its voltage falling below a specified value.

Charge voltage It is the voltage developed across a secondary cell when it is under charge. This voltage can be up to 50% higher than the rated discharge voltage of the cell. It increases with the charge rate and at low temperatures.

20.2 Primary cells

20.2.1 The Leclanché (zinc carbon) cell

The Leclanché (zinc carbon) cell is available in two distinct forms: the round cell and the layer cell. The former (*Figure 20.3*) is marketed both as a single unit or as a multicell battery, whilst the latter (*Figure 20.4*) is sold only in a multicell (layer stack) battery.

20.2.1.1 Theory of operation

The initial open circuit voltage or e.m.f. of a Leclanché cell is the difference between the electrode potentials of a zinc electrode and a manganese dioxide electrode immersed in a solution of battery electrolyte. For a fresh cell it is usually near 1.6 V falling to 1.4–1.2 V for a fully discharged cell.

The internal resistance of a cell is a very complicated property and depends principally on the construction. It is commonly calculated from a measurement of the e.m.f. and the short circuit current read instantaneously on an ammeter whose resistance including leads does not exceed 0.01 Ω:

$$r=\frac{E}{C}-0.01 \tag{20.1}$$

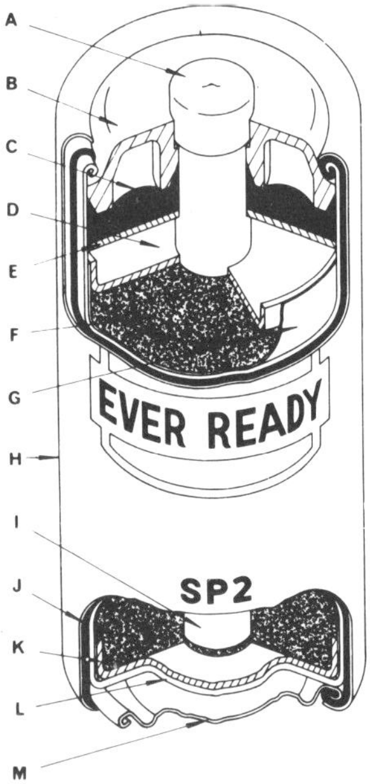

Figure 20.3 A typical round cell. (A) Metal top cap. This is provided with a pointed pip to secure the best possible electrical contact between cells. (B) Plastic top cover. This closes the cell and centralises the positive terminal. (C) Soft bitumen sub-seal. A soft bitumen compound is applied to seal the cell. (D) Top washer. This functions as a spacer and is situated between the depolariser mix and the top collar. (E) Top collar. This centralises the carbon rod and supports the bitumen sub-seal. (F) Positive electrode. This is made from thoroughly mixed, high quality materials. It contains manganese dioxide to act as the electrode material and carbon black or graphite for conductivity. Ammonium chloride and zinc chloride are other necessary ingredients. (G) Paper lining. This is an absorbent paper, impregnated with electrolyte, which acts as a separator. (H) Metal jacket. This is crimped on to the outside of the cell and carries the printed design. This jacket resists bulging, breakage and leakage and holds all components firmly together. (I) Carbon rod. The positive pole is a rod made of highly conductive carbon. It functions as a current collector and remains unaltered by the reactions occurring within the cell. (J) Paper tube. This is made from three layers of paper bonded together by a waterproof adhesive. (K) Bottom washer. This separates the depolariser from the zinc cup. (L) Zinc cup. This consists of zinc metal, extruded to form a seamless cup. It holds all the other constituents making the article clean, compact and easily portable. The cup is also the anode and when the cell is discharged, part of the cup is consumed to produce electrical energy. (M) Metal bottom cover. This is made of tin plate and is in contact with the bottom of the zinc cup. This gives an improved negative contact in the torch or other equipment, and seals off the cell to increase its leakage resistance

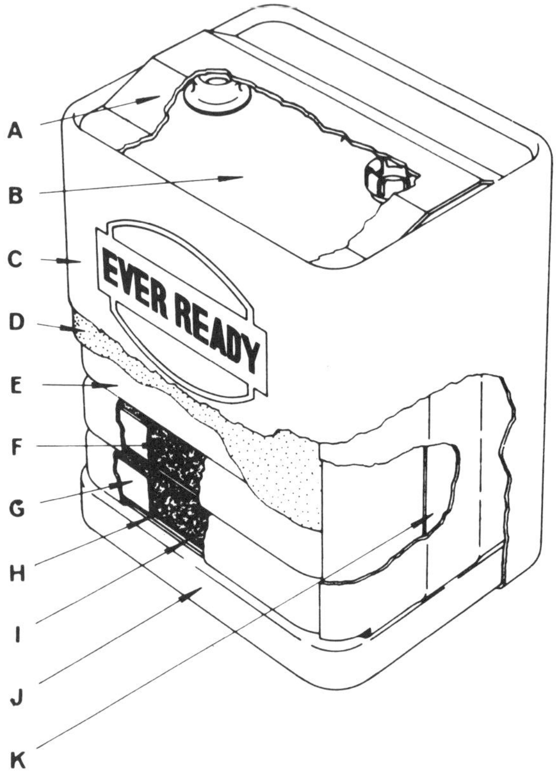

Figure 20.4 The layer cell battery. (A) Protector card. This protects the terminals and is torn away before use. (B) Top plate. This plastic plate carries the snap fastener connectors and closes the top of the battery. (C) Metal jacket. This is crimped on to the outside of the battery and carries the printed design. This jacket helps to resist bulging, breakage and leakage and holds all components firmly together. (D) Wax coating. This seals any capillary passages between cells and the atmosphere, so preventing the loss of moisture. (E) Plastic cell container. This plastic band holds together all the components of a single cell. (F) Positive electrode. This is a flat cake containing a mixture of manganese dioxide as the electrode material and carbon black or graphite for conductivity. Ammonium chloride and zinc chloride are other necessary ingredients. (G) Paper tray. This acts as a separator between the mix cake and the zinc electrode. (H) Carbon coated zinc electrode. Known as a duplex electrode, this is a zinc plate to which is adhered a thin layer of highly conductive carbon which is impervious to electrolyte. (I) Electrolyte impregnated paper. This contains the electrolyte and is an additional separator between the mix cake and the zinc. (J) Bottom plate. This plastic plate closes the bottom of the battery. (K) Conducting strip. This makes contact with the negative zinc plate at the base of the stack and is connected to the negative socket at the other end

where
r = internal resistance (in Ω)
E = e.m.f. (in V)
C = short circuit current (in A)

The value obtained is only approximate because it is impossible to obtain a truly instantaneous reading of the current. More accurate measurements can be made with an oscilloscope. The voltage can be measured within a few microseconds of shunting the cell with a known resistance. The internal resistance can then be calculated using the formula:

$$r=\frac{E-V}{V}R \qquad (20.2)$$

where
r = internal resistance (in Ω)
E = e.m.f. (in V)
R = shunt resistance (in Ω)
V = instantaneous voltage with shunt (in V)

Note the internal resistance of a cell will vary according to the load placed on it, the higher the shunt resistance the higher the internal resistance.

In the Leclanché cell the actual reactions which occur when the cell is discharged through an external circuit are very complicated. *Figure 20.5* represents chemically what happens on discharge.

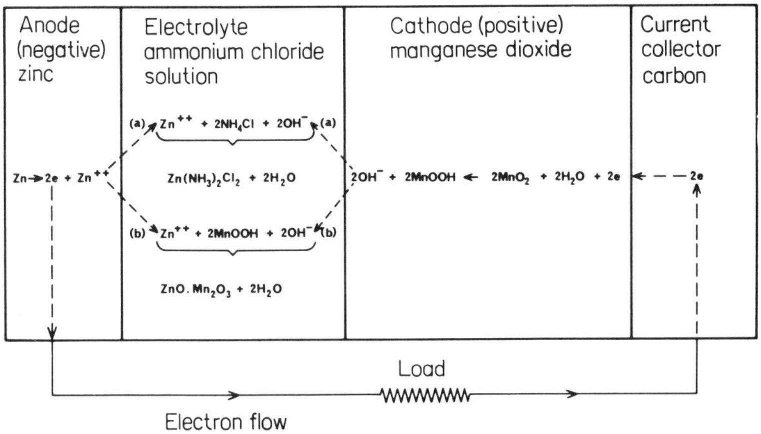

Figure 20.5 Chemical reactions in the Leclanché cell

At the surface of the negative electrode, zinc atoms ionise. The zinc ion goes into solution while electrons travel to the positive electrode through the external load and the carbon rod. In the positive electrode the manganese is reduced from a four valent state in MnO_2 to a trivalent state in MnOOH.

The hydroxyl ions, OH, produced in the cathode reaction and the zinc ions produced in the anode reaction, participate in one of the alternative reactions (a) or (b) (*Figure 20.5*). If reaction (a) takes place ammonium chloride from the electrolyte also takes part. Zinc diammine chloride, $Zn(NH_3)_2Cl_2$, and water are the products. If reactions (b) occurs the trivalent compound of manganese, MnOOH, reacts with zinc and hydroxyl ions to form hetaerolite, $ZnO.Mn_2O_3$, and water. Analyses of discharged cells show that both reactions can occur side by side in one cell because both zinc diammine chloride and hetaerolite are found, but reaction (a) usually predominates. The total water produced by the two reactions is equal to the water used in the cathode process. Consequently, the cell neither gains nor loses water as a result of normal discharge.

If the cell is discharged beyond its useful life, zinc oxychloride, $ZnCl_2.4Zn(OH)_2$, may be formed as a white precipitate. Water is consumed in its formation.

20.2.1.2 Performance and use

The quantity of electricity which can be obtained from a Leclanché dry cell depends on a number of factors, of which the more important are: (1) the physical size of the cell; (2) the rate at which the cell is discharged; (3) the daily duty period; (4) the end point voltage; (5) the method of construction and skill of the manufacturer; (6) the temperature; and (7) the age of the cell.

For cells of the same grade, structure and electrochemical system, the quantity of electricity realised is dependent on cell size. *Figure 20.6* shows the effect of the end point voltage on the service life of an R14PP when discharged through 6.8 Ω for 1 hour per day. If the end point voltage is reduced from 1.2 to 1.1 V an increase of 125% in service life is obtained.

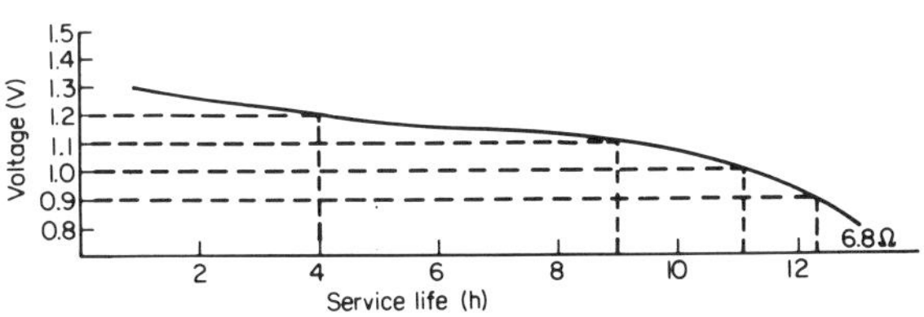

Figure 20.6 Service life curve; discharge period 1 h per day

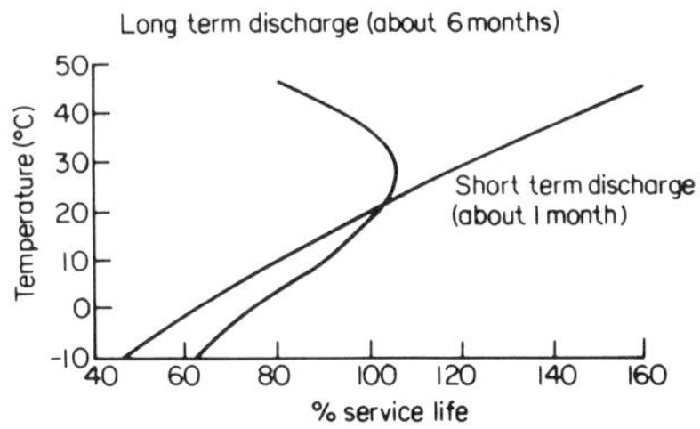

Figure 20.7 Temperature effect

Figure 20.7 shows the rather complicated effect of temperature on the discharge life of a PP9 battery. If the discharge is completed over a fairly short period (about 4 weeks), then raising the temperature from that of IEC test (20°C), may give significant increases in service life. If full discharge takes much longer, the deleterious effects of high temperature storage may be greater than the benefits due to increased efficiency of the discharge reactions. At −20°C the Leclanché cell is virtually inoperative and at −10°C little useful service will be obtained except at low current drains. It is advantageous to keep battery operated equipment in a warm environment before use in subzero temperatures.

Figure 20.8 shows the percentage fall in service life for a PP7 battery when discharged through 450 Ω for 4 hours per day to a 5.4 V end point, after increasing periods of storage at 20°C.

The Leclanché cell is designed to store satisfactorily at 20°C but storage at lower temperatures is beneficial, i.e. −10°C to

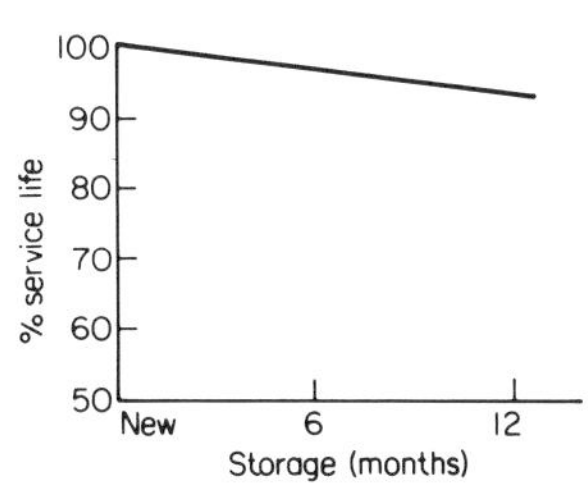

Figure 20.8 The storage effect for a PP7 battery

+10°C, provided the cells are enclosed in sealed containers which should be retained to protect them from condensation when warmed to ambient temperature.

20.2.2 Zinc chloride cell

The zinc chloride cell is similar in construction to a Leclanché cell except that the electrolyte consists of zinc chloride only. In a Leclanché cell some ammonium chloride is added to the zinc chloride to prevent oxygen from the air reacting with the zinc chloride and reducing the shelf life. However, ammonium chloride also reduces electrochemical action and the cell's capacity.

Zinc chloride cells have to be effectively sealed to prevent air reacting with the electrolyte. They have good leak resistance and can operate at temperatures below freezing.

20.2.3 Mercuric oxide cell

20.2.3.1 Theory of operation

In the mercury cell (*Figure 20.9*), the anode is the element zinc (Zn) and the cathode is the compound mercuric oxide (HgO). As current is drawn from the cell the anode is oxidised (that is, the zinc anode attracts the oxygen from the cathode), becoming the compound zinc oxide (ZnO), and the cathode is reduced (or de-

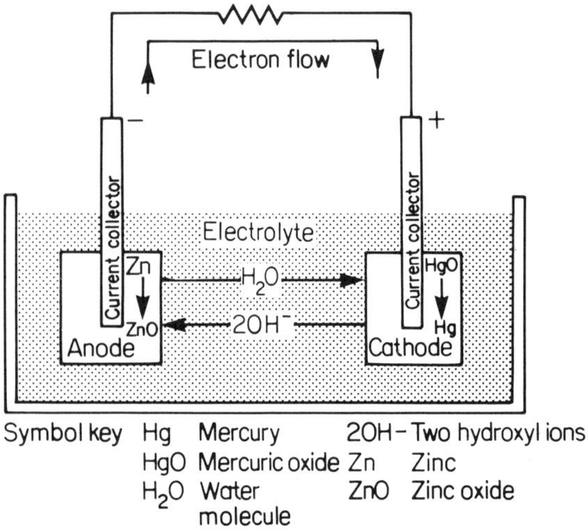

Figure 20.9 Basic reactions in the mercury cell

oxidised), giving up its oxygen to become pure mercury (Hg). Mercury is quite low on the electromotive scale, lying immediately above silver, platnium and gold, and for this reason mercuric oxide will readily yield its oxygen to the zinc.

Each electrode is attached to a current collector, and both are immersed in an electrolyte. When the current collectors of the cell are connected, the strong affinity of the zinc in the anode for the oxygen that is weakly bound to mercury in the cathode is 'felt' over the circuit, and the mercuric oxide begins to break down and give up oxygen. In the reaction that results, the water component of the electrolyte also plays a critical role.

The two electrons are provided by the current collector of the cathode, and attach themselves to the mercury ion, replacing the electrons it donated to its oxygen partner. Thus neutralised, the mercury separates from the oxygen, leaving:

$$O^{2-} + (H^+)_2O^{2-}$$

Or, two oxygen and hydrogen ions which can be written as:

$$2(H^+O^{2-})$$

Or, cancelling a *positive* and *negative*:

$$2HO^-$$

The hydroxyl ion is usually written as:

$$OH^-$$

The end products of the reaction at the cathode are mercury, plus a pair of hydroxyl ions that are released into the electrolyte. The electrolyte *already* has a plentiful supply of these. Consequently there is a 'pressure' exerted across the electrolyte (the 'domino effect') that makes a corresponding pair of ions available at the anode, to react with the zinc. If we write the two ions separately, we can express the 'confrontation' that now occurs at the anode as follows:

$$Zn + \begin{matrix} OH^- \\ OH^- \end{matrix}$$

Given its very strong affinity for oxygen, the zinc will now snatch away one of the two oxygen atoms, and like mercury, it has two outer electrons for doing this. But the oxygen already *has* two electrons. So, to combine with the zinc it must release them. Left behind will be two hydrogen atoms and one oxygen atom, which can then form a simple water molecule. We can now describe the reaction at the anode in simple terms:

$$Zn + 2OH^- \rightarrow ZnO + H_2O + 2e^- \text{ (the two released electrons)}$$

We can do the same for the events at the cathode:

$$HgO + H_2O + 2e^- \text{ ('borrowed' from the anode)} \rightarrow Hg + 2OH^-$$

Everything balances. The water molecule used at the cathode is replaced by one produced at the anode. And the two electrons 'borrowed' at the cathode are restored by the anode. The end result is a steady flow of electrons (electricity) from anode to cathode as a consequence of the sophisticated and dynamic electrochemical processes within the power cell. In the simplest of all terms, then:

$$Zn + HgO \rightarrow ZnO + Hg$$

20.2.3.2 Construction

The effective voltage range is typically 1.3 V down to 1.0 V per cell depending on load and temperature. The mercury system appears in two variants—one with a well defined no-load voltage between 1.35 V and 1.36 V, the second with a voltage which can vary between 1.36 V and 1.55 V. The voltage difference is only important during the first 5–10% of discharge, and only becomes a design consideration when maximum voltage stability is required.

Mercury cells are produced in both cylindrical and button types. Electrochemically both are identical and differ only in can design and internal arrangement. The anode is formed from high purity amalgamated zinc. The cathode is a compressed mercuric oxide/manganese dioxide graphite mixture separated from the anode by an ion permeable barrier. The electrolyte is a solution of an alkali metal hydroxide whose ions act as carriers for the chemical reactions in the cell, but is not part of the reaction. In operation this combination produces metallic mercury which does not inhibit the current flow within the cell. The inside of the cell top is electrochemically compatible with the zinc anode. The cell can is made from nickel plated steel and does not take part in the chemical reaction. *Figure 20.10* illustrates a mercury button cell.

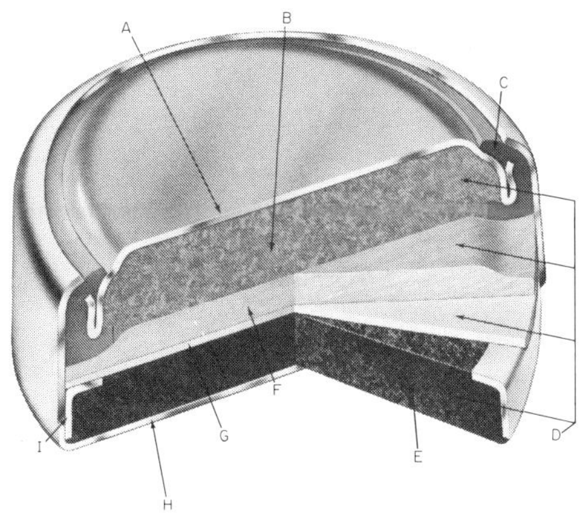

Figure 20.10 Mercury button cell. A, Cell top (negative terminal). Single type. Steel coated with copper on inside and with nickel and gold externally. B, Anode. Powdered zinc (amalgamated), together with gelled electrolyte. C, Nylon grommet. Coated with sealant to ensure freedom from leakage. D, Electrolyte. Alkaline solution. In anode, cathode and separators. E, Cathode. Mercuric oxide with graphite. Highly compacted. F, Absorbent separator. Felted fabric (cotton or synthetic). Prevents direct contact between anode and cathode. Holds electrolyte. G, Barrier separator. Membrane permeable to electrolyte but not to dissolved cathode components. H, Cell can (positive terminal). Nickel, or steel coated on both sides with nickel. I, Sleeve. Nickel-coated steel. Supports grommet pressure. Also aids in consolidating cathode

20.2.3.3 Characteristics

Voltage stability The uniform voltage of the mercury cell is due to the efficient nature of the cathode. Over a long period of discharge a regulation within 1 per cent can be sustained, and for short term operation higher stability can be achieved. This flat voltage characteristic may be utilised as a reference source or in other applications where voltage stability is essential.

Capacity The ability of the mercury system to withstand both continuous and intermittent discharge with relatively constant ampere hour output allows the capacity rating to be specified. Rest periods as for ordinary cells are not required.

Shelf life The capacity retention of the mercury system is excellent in storage and the voltage characteristic is not affected. Mercury cells can be stored for periods of up to 3 years, according to type. It is not, however, good practice to store cells for unnecessarily long periods.

Mechanical strength The mercury cell can withstand severe vibration, shock and acceleration forces.

Vacuum and pressure High vacuum has no detectable effect and maximum permitted pressure depends on the size of the cell.

Corrosion Mercury cells normally have an excellent resistance to corrosive atmospheres and high relative humidity conditions.

Leakage resistance The use of nickel plated steel cans and precision moulded seals are part of meticulous design aimed at eliminating leakage under the most adverse conditions.

Temperature range The mercury system provides a stable voltage over the temperature range −30 to +70°C. Special cells based on a wound anode construction are capable of efficient operations at temperatures 15°C below what is possible in standard cells. Although successful performance of some cells above 120°C for short periods has been reported it is recommended that +70°C should not be exceeded.

20.2.4 Alkaline manganese cells

20.2.4.1 Theory of operation

The basic electrode reactions in the alkaline manganese cell are similar to those occurring in the zinc–mercuric oxide cell. The electrochemically active components of the cell are a zinc metal anode, a strong alkaline electrolyte (KOH) and a manganese dioxide (MnO_2) cathode.

The reaction at the zinc anode is an oxidation process as previously described for the zinc–mercuric oxide cell and this can be simply described by the following chemical equation:

$$Zn + 2OH^- \rightarrow ZnO + H_2O + 2e^-$$

The electrode potential for this reaction is approximately −1.35 V (with reference to a standard hydrogen electrode) and remains relatively constant during discharge of the cell.

At the manganese dioxide cathode, the reaction is more complicated and the reduction process occurs in two steps corresponding to $MnO_2 \rightarrow MnO_{1.5}$ (or Mn_2O_3) $\rightarrow MnO$. During this reduction process, the manganese oxide chemical stoichiometry changes in a continuous fashion, i.e. a homogeneous reaction in contrast to the heterogeneous reaction for mercuric oxide being reduced to mercury metal which does not involve intermediate oxide compositions other than HgO. The consequence of this homogeneous reaction is a change of cathode electrode potential during the discharge life.

Initially, the cathode electrode potential is approximately +0.23 V (with reference to a standard hydrogen electrode) and this falls steadily over the first step of discharge to about −0.5 V. The second step occurs at a more constant potential, around −0.5 V, and then falls sharply in the composition region $MnO_{1.2}$ to $MnO_{1.1}$ to below −0.9 V.

The actual chemical species present during the two step reduction process have been described as:

$$MnO_2 \rightarrow MnOOH \rightarrow \underset{\text{(amorphous)}}{Mn(OH)_2}$$

This implies incorporation of part of the electrolyte into the cathode material during discharge but in simple terms the cathode reaction (first step) can be described by the following chemical equation:

$$2MnO_2 + H_2O + 2e^- \rightarrow 2OH^- + 2MnO_{1.5} \text{ (or } Mn_2O_3\text{)}$$

and the resultant overall cell reaction as:

$$Zn + 2MnO_2 \rightarrow ZnO + 2MnO_{1.5} \text{ (or } Mn_2O_3\text{)}$$

Combination of the two electrode potentials gives an approximate open circuit voltage of 1.58 V.

Most of the useful life of the cell is given by the first step of the manganese dioxide discharge and by the time this is completed, the cell voltage will have fallen in a fairly continuous manner to 0.85 V. This is in contrast to the relatively constant voltage discharge of the zinc–mercuric oxide cell, which occurs until all of the active components have been used, when the voltage falls sharply to a low value. The rate of fall of voltage during discharge is also dependent on the discharge load and will increase with increasing load (i.e. higher current) due to other factors such as electrode polarisation, as is the case for all cell types.

20.2.4.2 Construction

To achieve performance which is far superior to ordinary zinc–carbon cells in most applications, alkaline manganese cells require higher quality and more expensive materials, and a more sophisticated construction. Here the anode, which does not have to double as part of the cell structure, is formed of zinc powder. The zinc particles are of carefully controlled size, shape, and purity, and are amalgamated (i.e. combined with mercury) to supress gassing and to maximise performance at all discharge rates.

The alkaline electrolyte which gives this cell its popular designation is a solution of potassium hydroxide (KOH), which is highly conductive. The electrolyte is diffused throughout the powdered zinc and in intimate contact with its granules, insuring that the anode material is almost completely oxidised by the time the cell's stored energy is exhausted.

Similar design and engineering refinements apply to the cathode of the alkaline manganese cell. The basic cathode material is what is known as *electrolytic manganese dioxide*. This material is produced synthetically by electrolysis. Derived this way it is a much purer oxide than that found in the natural ore, and also has a greater oxygen content per unit volume. This additional oxygen in the cathode material provides increased reactivity and so significantly extends the capacity of the cell.

The supply of reactive oxygen is still further increased because the cathode material is highly compressed, forcing more of it into the available space than it would hold ordinarily. As in the anode, some electrolyte is absorbed into the cathode material during manufacture, assuring good contact with electrolyte throughout the complete cell system. The use of a more conductive electrolyte, and the higher quality of both anode and cathode materials, results in a system that has considerably more usable electrochemical energy stored within it than could be possibly contained in a zinc–carbon cell of a comparable size.

In structural design (*Figure 20.11*) the typical alkaline manganese cell is in many ways the exact opposite of the zinc–carbon configuration. Here the anode is on the inside and the cathode is on the outside, instead of the reverse. The alkaline cell has a central, nail-like anode collector inserted from the bottom, instead of a carbon-rod cathode collector extending down from the top.

In the alkaline manganese system, the entire cell is enclosed in a steel case, which provides a considerably stronger and more secure container than the zinc can of zinc–carbon cells. Just inside the steel case, and in intimate contact with it, is the cathode material. With this arrangement the can becomes the current collector for the cathode, with its positive terminal formed by a protrusion at the top of the case. Lining the thick, cylindrical cathode is a sleeve of absorbent material, which acts as a separator between cathode and anode, as in the zinc–carbon cell. Within the sleeve is the zinc anode. Anode, cathode and separator are all infused with electrolyte for maximum capacity and conductivity.

At the core of the cell, in direct contact with the anode, is the 'nail'. This forms the current collector for the anode and is welded to a cap at the bottom of the can to form the negative terminal of the cell. Because of this somewhat 'reverse' arrangement of design in the alkaline cell, the top terminal is positive just as it is in the zinc–carbon cell (because of the carbon rod), and the bottom terminal is negative (provided by the bottom of the can in the zinc–carbon system).

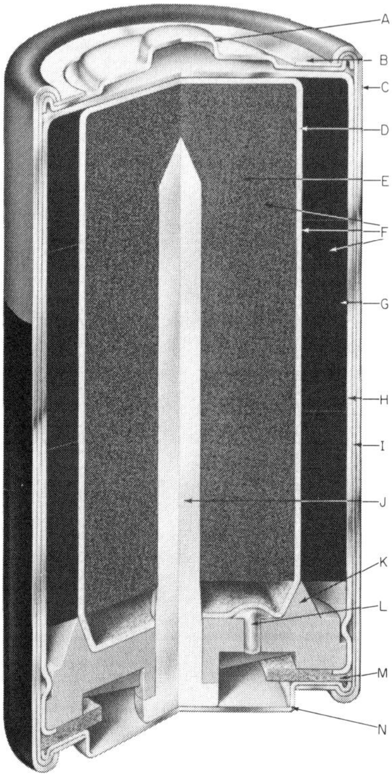

Figure 20.11 A, Cathode cap (positive terminal). Protrusion in contact with steel cell case. B, Insulating washer. C, Outer steel jacket. Lithographed. D, Separator. A sleeve of porous, synthetic fibre impregnated with electrolyte. E, Anode. Powered zinc, highly amalgamated and compacted. F, Electrolyte. Potassium hydroxide (KOH) solution absorbed into separator, anode material and cathode material. G, Cathode. Compressed mix of electrolytic manganese dioxide (MnO_2) and graphite, introduced either by extrusion or by insertion in the form of pre-formed, tight-fitting cylindrical rings. H, Cathode collector. Steel cell case. I, Plastic sleeve. Separates steel case from outer steel jacket. J, Anode collector. Metal 'nail'. K, Plastic grommet. Forms a structural, insulating seal for cell. L, Vent. Wax-sealed hole in plastic grommet (releases gases if they build up and prevents cell rupture). M, Insulator. Separates (and insulates) steel cell case (positive) from the end cap (negative). N, Anode cap (negative terminal). Protrusion in contact with collector 'nail'

20.2.4.3 Characteristics

Voltage range The operating voltage range is 1.3 to 0.8 V per cell under most conditions of load and temperature. Maximum open circuit voltage is typically 1.56 V per cell. The recommended end voltage for single cell operation at room temperature is 0.8 V increasing to 1 V per cell when six or more series cells are used.

Load currents The alkaline battery excels on continuous heavy loads. There is no distinct upper load limit, and the system is typically capable of supplying intermittent loads up to 2 A at room temperature.

Leakproofness Alkaline cells are leakproof under all normal conditions. The following situations should be avoided whenever possible: (a) cell insertion with the wrong polarity; (b) external short circuit; (c) reverse drive of series cells; (d) charging.

Environmental The alkaline system operates efficiently between $-30°C$ and $+70°C$ subject to load and duty cycle regimes. High relative humidity creates no particular problems, and the battery is tolerant to both high pressure and vacuum.

Storage life Alkaline cells may be stored for prolonged periods at room temperature without significant losses in capacity.

Typical capacity retention exceeds 85% after $2\frac{1}{2}$ years storage at 20°C. Short term exposure to temperatures above 45°C is permissible. Long term storage at elevated temperatures causes a progressive deterioration in both capacity and high rate properties and should therefore be avoided whenever possible.

20.2.5 Lithium cells

20.2.5.1 Cell construction

Apart from being the lightest metal known, lithium has the highest electrode potential of any metallic element (3.045 V) with a theoretical electrochemical equivalent capacity of 3860 A h kg^{-1}.

Careful matching with compatible electropositive products has made it possible to develop a number of different lithium couples, although the following are emerging as favourites for the future:

lithium manganese dioxide	3 V
lithium thionyl chloride	3.5 V
lithium carbon monofluoride	2.8 V
lithium sulphur dioxide	2.85 V
lithium iodine	2.8 V
lithium silver chromate	3.1/2.6 V
lithium copper oxide	1.5 V
lithium iron sulphide	1.5 V
lithium iron disulphide	1.5 V
lithium lead bismuthate	1.5 V
lithium bismuth trioxide	1.5 V

Lithium is highly reactive with water and must be processed under strictly controlled moisture free conditions. Whilst it is maleable and easy to manipulate special techniques have had to be developed for it to be precision worked.

Lithium cells feature a high purity lithium anode, compatible cathode and either non-aqueous organic or inorganic electrolytes. The cells are usually hermetically sealed to prevent leakage of electrolyte, cathode materials and byproducts precipitated during discharge by the mainstream electrochemical reaction. Hermetic sealing also inhibits moisture ingress and ensures that cells will function safely even if they are subjected to mechanical and electrical abuse.

The cross sectional arrangements of the three most popular cell constructions are illustrated in *Figure 20.12*.

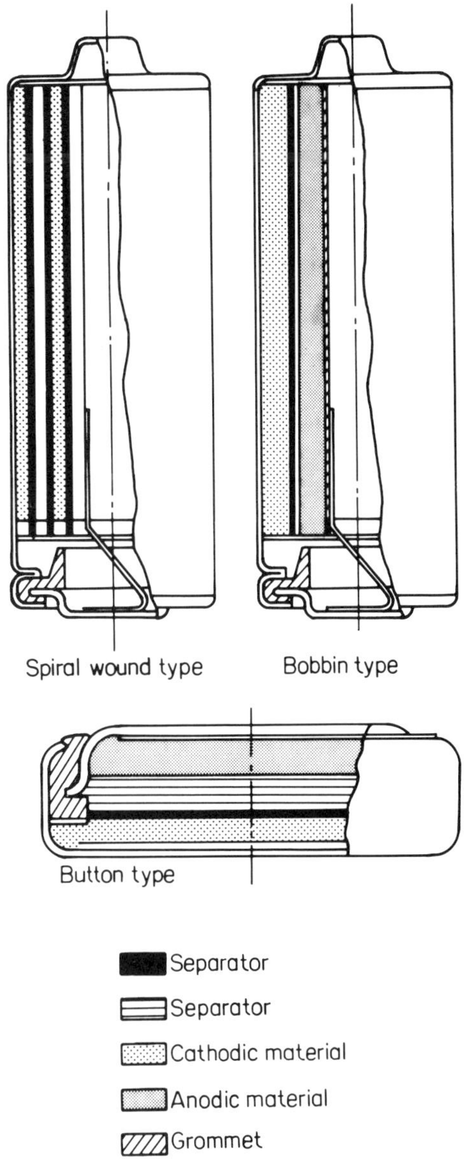

Figure 20.12 Various cell configurations used in lithium primary batteries (Courtesy of SAFT (UK) Ltd)

20.2.5.2 Characteristics of lithium cells

Voltage Unlike zinc–carbon and alkaline manganese cells, whose voltage falls progressively during discharge, all lithium cells exhibit stable on-load voltage characteristics. For this reason it is possible to closely categorise lithium cells in two groups by their voltages. The open circuit and nominal operating voltages of selected lithium couples are listed in *Table 20.1*.

For much of their life the voltages of lithium cells remain relatively flat, falling sharply only near end of life. In those cells with open circuit voltage significantly higher than nominal on-load voltage the initial high voltage can either be accepted or artificially surpressed by 'burn-in', the use of zener diodes or voltage regulators.

In certain applications such as cardiac stimulators it is important to know when the battery needs to be replaced. The Li/Ag_2CrO_4 cells have a second voltage plateau which fulfills the role of battery status indicator, giving advance notice when replacement must be carried out, nominally within a year.

Operating temperature range The use of non-aqueous inorganic and organic electrolytes gives lithium cells their ability to function over extremely wide temperature ranges.

Lithium sulphur dioxide and lithium thionyl chloride cells will function efficiently down to −55°C, whilst lithium copper–oxide cells will, in their encapsulated form, operate quite satisfactorily up to 150°C.

The $Li/SOCl_2$ system is capable of discharging well over 50% of its rated capacity at −55°C under specific load conditions.

Current drain In general terms the ability of a lithium cell to sustain continuous discharge is influenced by cathode resistivity, electrolyte and cell construction.

Liquid cathode spiral wound cells with their high electrode surface areas are able to deliver proportionately greater currents than their bobbin cell counterparts. Drain capability is also proportionate to internal impedance.

Table 20.1 Voltage of lithium couples

Battery couple	*Open circuit voltage (V)*	*Nominal operating voltage (V)*
Lithium copper oxide (Li/CuO)	2.4	1.5
Lithium copper sulphide (Li/Cu_2S)	2.1	1.5
Lithium iron sulphide (Li/FeS_2)	1.8	1.5
Lithium lead copper sulphide (Li/PbCuS)	2.2	1.5
Lithium bismuth trioxide (Li/Bi_2O_3)	2.04	1.5
Lithium lead bismuthate ($Li/Bi_2Pb_2O_5$)	1.8	1.5
Lithium carbon monofluoride (Li/CF_2)	3.1	2.8
Lithium manganese dioxide (Li/MnO_2)	3.7	3.0
Lithium silver chromate (Li/Ag_2CrO_4)	3.45	3.1 first plateau 2.6 second plateau
Lithium sulphur dioxide (Li/SO_2)	3.0	2.85
Lithium thionyl chloride ($Li/SOCl_2$)	3.65	3.5
Lithium iodine (Li/I_2)	3.1	2.8

Shelf life Lithium batteries will be adopted for much the same reasons as those which persuaded industry to use alkaline manganese in preference to traditional zinc–carbon batteries. Selected systems have a proven shelf life of up to 10 years with over 80% retained capacity. Real time desert and accelerated storage discharge testing at elevated temperatures supports this claim.

The excellent shelf life of lithium batteries can be explained by the fact that the lithium anode is passivated during storage.

Load response time This characteristic is particularly important in high drain devices. After long periods of storage at elevated temperature batteries may have to respond quickly on activation and to achieve peak output levels within a matter of a few milliseconds. In the main solid anode/solid cathode systems do not suffer from the voltage delay phenomenon. This is not the case with solid anode, liquid cathode, inorganic electrolyte systems such as $LiSO_2$ and $Li/SOCl_2$. Reformulation of electrolytes and the use of additives although still not fully developed is already alleviating this problem.

Reliability The nature of the manufacturing processes which have had to be specially developed, the care with which all cell components are selected, the hospitalised production facilities all contribute to ensure that cells are less susceptible to failure than any of their predecessors.

High reliability is epitomised by the Li/I_2 and Li/Ag_2CrO_4 batteries used by the medical industry. In over 6 years no reports of Li/Ag_2CrO_4 cells failing prematurely have been received. Reliability level greater than 0.7×10^{-7} and increasing.

Safety This is the single most important parameter likely to influence the rate at which lithium cell technology is adopted. The $LiSO_2$ system features a vent mechanism which opens when the cell is mechanically or electrically abused. The safety vent is designed to exhaust liquid SO_2 at 90°C corresponding to a pressure greater than 31 bar (450 psi).

$LiSOCl_2$ cells on the other hand are non-pressurised. No significant amounts of gas are generated even under abuse conditions consequently no requirement for safety vent. Solid cathode solid and liquid electrolyte systems tend to have a relatively high internal impedance and under a short circuit operational mode the current delivered is kept within safe limits.

The Li/I_2 couple, featuring a poly 2-vinyl pyridine cathode which forms its own separator, is unpressurised. Even under abuse discharge conditions the cell is not prone to swelling.

The greatest safety problems lie with multicell battery packs if individual cells particularly partially discharged ones are inadvertently subjected to a high degree of force charging. The likelihood of battery failure can be eliminated by incorporating diode protection consistent with design and reliability specifications.

Comparison *Table 20.2* compares the characteristics of some 3 V lithium cells.

20.2.5.3 Theory of operation

Lithium sulphur dioxide cells (Li/SO_2) These cells are available in spiral form only. Lithium foil, polypropylene separator and teflon bonded carbon cathode pressed on to a support grid are rolled together to give the required active surface area. Sulphur dioxide under pressure acts as the cathode/depolarising agent.

The overall discharge mechanism is as follows:

$$2Li + 2SO_2 \rightarrow Li_2S_2O_4 \text{ (lithium dithionite)}$$

Li/SO_2 cells feature safety vent. Under force charging or short circuit conditions cell integrity is maintained as SO_2 cathode pressure kept below critical levels.

Lithium thionyl chloride cells ($Li/SOCl_2$) Considered by many to be the natural successor to the Li/SO_2 couple the spiral wound version would have highest theoretical energy density and drain capability of any commercially produceable lithium battery.

In place of pressurised SO_2, the cell relies upon $SOCl_2$ which doubles as depolarising agent and cathode.

Several discharge mechanisms have been proposed although it is now considered by most experts that the discharge reaction is adequately described by the equation:

$$4Li + 2SOCl_2 \rightarrow S + SO_2 + 4LiCl$$

Under normal discharge negligible amounts of gas are generated. Maximum pressure developed under abuse consistent with Raoult's law predicted to be 3.8 bar (55 psi).

Lithium manganese dioxide cells (Li/MnO_2) This was the first solid anode/solid cathode 3 V lithium system to be manufactured commercially. The use of inexpensive materials makes it especially suitable for powering low drain, long life, high volume applications.

Unlike the Li/SO_2 and $Li/SOCl_2$ cells this couple uses an organic lithium perchlorate dioxolane based electrolyte. Overall discharge reaction can be summarised as follows:

$$Li + MnO_2 \rightarrow LiMnO_2$$

Lithium copper oxide cells (Li/CuO) One of the first lithium couples to be released onto the market, it was originally

Table 20.2 Comparison of 3 volt lithium cells

Couple	*$Li/SOCl_2$ AA size*	*Li/MnO_2 AA size*	*Li/SO_2 AA size*	*Li/I_2 button*	*Li/Ag_2CrO_4 button*
Open circuit voltage	3.65 V	3.7 V	3.0 V	3.1 V	3.45 V
Nominal operating voltage	3.5 V	3.0 V	2.85 V	2.8 V	3.1 V 2.6 V
Theoretical capacity (A h)	2.0	1.5	1.4	—	—
Energy density (W h dm^{-3})	800	550	480	400–700	600–700
Capacity loss (% per year)	2–3	2–3	2—3	1	1
Operating temperature range	−55/+71°C	−20/+60°C	−50/+70°C	−40/+60°C	−40/+70°C
Discharge plateau	flat	flat	flat	inner fall	flat
Toxicity	high	mean	high	mean	mean

developed as a replacement for the traditional zinc–carbon and alkaline manganese systems. It is available in bobbin configurations only at present although work on spiral wound cells is advancing.

Cells consist of an annulus of lithium, non-woven separator and copper oxide based cathode hermetically sealed in a steel can.

The overall discharge reaction is as follows:

$$2Li + CuO \rightarrow Li_2O + Cu$$

Lithium silver chromate cells (Li/Ag_2CrO_4) These cells were developed specifically for high integrity medical applications.

The cells in production incorporate the most sophisticated and advanced button cell technology. Cells comprise of a lithium anode, silver chromate cathode, lithium perchlorate electrolyte sealed in stainless steel or titanium laser welded cap and can assembly.

The overall discharge reaction is:

$$2Li + Ag_2CrO_4 \rightarrow Li_2CrO_4 + 2Ag$$

Lithium iodine cells (Li/I_2) As the cell uses solid state chemistry it is suitable only for low drain devices because of current rate limitations.

The cathode consists of iodine and a complex organic conductive charge transfer complex (CTC) which in contact with lithium produces lithium iodine. Continuous discharge promotes precipitation of lithium iodine increasing internal impedance and diminishing terminal voltage.

The active reaction materials are hermetically sealed in laser welded cases complete with tinned nickel ceramic corrosion resistant terminals to prevent diffusion of moisture and gas into the battery, and leakage.

The overall discharge mechanism is as follows:

$$2Li + CTC\ nI_2 \rightarrow CTC\ (n-1)I_2 + 2LiI$$

The volume of cell products within the cell remain unchanged during discharge. Solid electrolyte with restricted ionic conductivity prevents leakage.

Lithium lead bismuthate cells ($Li/Bi_2Pb_2O_5$) These cells, only available as buttons, are less susceptible to leakage and store better at higher temperatures than comparable size AgO/Zn cells. Volumetric energy densities are also superior.

The anode consists of a tablet of lithium pressed into a steel cap matched with a pressed powder lead bismuthate cathode separated by separator material. The separator fulfills two functions: it doubles as a physical barrier preventing internal short circuiting without restricting ionic conductivity and a reservoir holding surplus electrolyte to avoid 'dry out'.

The overall discharge mechanism is as follows:

$$10Li + Bi_2Pb_2O_5 \rightarrow 5Li_2O + 2Bi + 5Pb$$

20.2.6 Zinc–air cell

20.2.6.1 Construction

The cell uses zinc as the anode and oxygen from the air as the cathode material. Since the space usually taken by the cathode oxidising agent can now be allocated to the anode this cell has a high energy to weight ratio.

The key to the zinc–air cell is the construction, which lets air into the cell without allowing any of the electrolyte to leak out. *Figure 20.13* shows the construction of a button cell. The anode is the zinc top and the potassium hydroxide electrolyte is contained within this. An insulating gasket separates the anode and cathode. The air cathode arrangement, shown in *Figure 20.13(b)* is made up of a system of separators and is only about 0.5 mm thick. The oxygen from the air combines with the hydroxide from the electrolyte, under the catalytic action of the carbon, to form water. The metallic mesh gives mechanical support and carries the current, whilst the PTFE film allows air to enter the cell but prevents electrolyte from escaping.

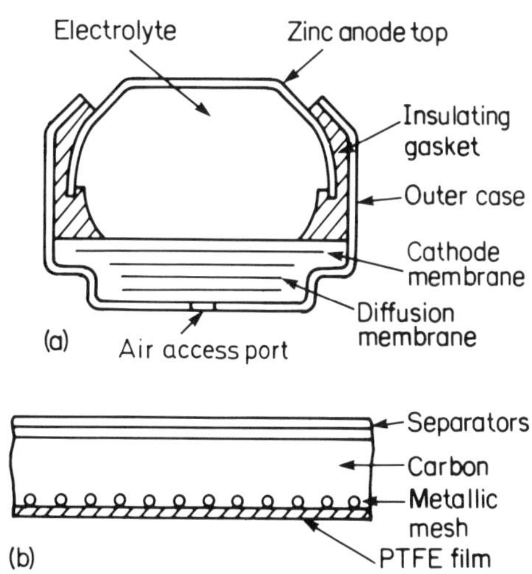

Figure 20.13 Zinc–air cells: (a) button cell; (b) cathode arrangement of button cell

20.2.6.2 Theory of operation

During operation the zinc reacts with the electrolyte leaving electrons on the anode, which give it a negative charge. The chemical reaction is:

$$Zn + 2OH \rightarrow ZnO + H_2O + 2e^-$$

Oxygen from the air combines with the water in the electrolyte, taking electrons from the cathode and replacing the hydroxyl ions lost at the anode. Electrons removed from the cathode leave it positively charged. The reaction is:

$$\tfrac{1}{2}(O_2) + H_2O + 2e^- \rightarrow 2OH^-$$

The current rating of the cell is determined by the rate of air flow and by the cathode surface area. The ampere hour rating depends on the weight of the zinc anode. The cell will be exhausted when all the zinc material is used up.

20.2.6.3 Characteristics

The zinc–air cell has a no load voltage of 1.4 V, with a higher energy to weight ratio, and a higher current than the alkaline or mercuric oxide cells. The zinc–air cell is used in applications which need to work at high currents for long periods.

The internal resistance of the zinc–air cell is low and this resistance is primarily determined by the oxygen diffusion rate. The cell can operate over a wide temperature range of −40°C to +60°C.

The shelf life of a zinc–air cell is very good, the average loss of capacity during storage being about 2% per year. This is because one of the cell's reactants is oxygen, and this can be excluded during storage by airtight wrapping around the cell.

The zinc–air cell is inherently safe in operation since any gasses developed can escape via the air diffusion path. The short circuit current of the cell is limited by the rate at which the cell can absorb oxygen.

The main problem with the zinc–air cell is that it is affected by atmospheric conditions. The water vapour pressure in the cell is equivalent to 55% relative humidity at 20°C, so on wet days the cell will gain moisture and on dry days it will lose moisture. This affects the aqueous electrolyte, which is usually 30% potassium hydroxide, so the cell can fail if operated for long periods at extremes of atmospheric conditions. The carbon dioxide from the air also reacts with the electrolyte to form potassium carbonate, and this increases the internal resistance of the cell.

20.3 Secondary cells

20.3.1 Lead–acid cells

20.3.1.1 Construction

Although the lead–acid cell was developed by Gaston Plante in 1860 it was not until much later, when it was adopted by the automobile industry, that it gained in popularity. The open type of construction used in automobiles is not suitable for use in electronic equipment since the cell must be mounted upright, there is risk of spillage, and the cell needs frequent topping up.

For electronic applications a gelled electrolyte cell is used. This is sealed, does not need topping up with electrolyte and it can be mounted in any position. In this type of construction both plates of the cell are made from lead in the form of a grid. The positive plate is filled with lead dioxide and the negative plate with spongy lead. The plates are formed into thin metal sheets and are interleaved with layers of porous fibreglass separators and wound into a cylinder. This is sealed in a chemically stable polypropylene case and then put into a metal case for strength.

To achieve a low internal resistance, low polarisation and long life the material of the metal plates is kept close to the surface by using thin plates and a spiral wound construction. The separator electrically separates the positive and negative plates and holds the electrolyte, distributing it over the working surface.

The electrolyte is a dilute solution of sulphuric acid. The quantity of electrolyte is such that it is retained by the plates and separators and none of it is free to leak. This gives the cell good gas diffusion and oxygen re-combination, and avoids large gas pressures during overcharging. A safety vent is provided in the cell case to allow for the escape of gas.

20.3.1.2 Theory of operation

During the discharge period the spongy lead in the negative plate, and the lead dioxide in the positive plate, react with the sulphuric acid to give lead sulphate crystals and water. The lead sulphate crystals grow on the lead dioxide and if they are excessive the operation of the cell will slow down. Charging reverses the process. The equations are as follows:

$$PbO_2 + Pb + 2H_2SO_4 \rightleftarrows 2PbSO_4 + 2H_2O$$

During overcharge the charging current electrolyses the water in the electrolyte and forms oxygen and hydrogen. This can lead to a build up of gas pressure.

20.3.1.3 Characteristics

Voltage The nominal cell voltage is 2.1 V. The voltage droops during discharge due to loss in the internal resistance of the cell.

Current capacity The lead–acid cell can withstand high charge and discharge rates, and is specially good on pulsed operation where it can deliver large currents for a short time. During rest periods acid diffuses from the separator back to the working areas of the plates, and allows a greater working capacity. There is a drop in capacity at high discharge currents because of insuffficient ion diffusion caused by the depletion of electrolyte near the active material of the plates. This effect is called concentration polarisation.

Temperature range The cell operates over a temperature range of −60°C to +60°C although the optimum operating temperature is +20°C. The capacity and terminal voltage of the cell decrease at low temperatures due to a reduction of the ionic diffusion rate.

Shelf life The shelf life of the cell is reduced by internal electrochemical discharge, which is worse at high temperatures. The loss of charge per day varies from 0.01% for a nearly discharged cell at 0°C to 2% for a fully charged cell at 45°C.

Chemically the effects are similar during self-discharge as that which occurs during normal operation. However, the lead sulphate crystals are now large and completely surround the active plate material. This is known as sulphation and if it is allowed to continue it can prevent the cell from accepting charge. To avoid sulphation the lead–acid cell should be stored at low temperature and re-charged at periodic intervals.

20.3.2 Nickel–cadmium cells

20.3.2.1 Construction

There are two types of construction: the cylindrical cell and the button cell. Before we consider the two types in turn we should

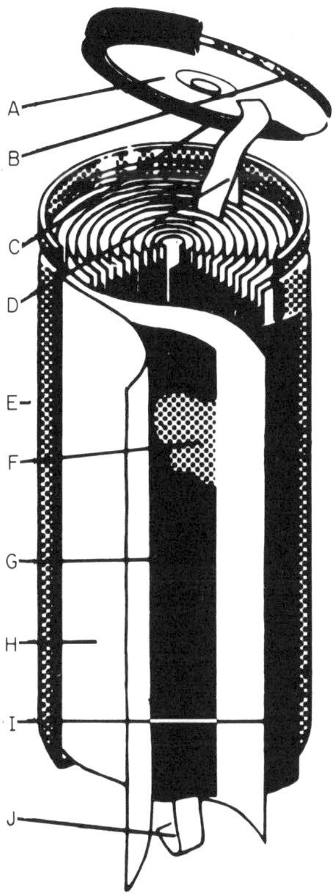

Figure 20.14 Typical cylindrical nickel–cadmium cell construction. A, Resealing safety vent. B, Nickel plated steel top plate (positive). C, Nylon sealing grommet. D, Positive connectors. E, Nickel plated steel can (negative). F, Support. G, Sintered negative electrode. H, Separator. I, Sintered positive electrode, J, Negative connector

summarise a few important differences. Cylindrical cells can be made larger, they have pressure-release safety vents and can give very high discharge currents; button cells are by comparison more compact, give lower currents, but have the ability to hold charge for longer periods. *Figures 20.14* and *20.15* illustrate typical constructions.

Cylindrical cells are generally made to comply with the standards BS 5932:1980 and IEC 285. *Table 20.3* shows some sizes available. The dimensions given include plastic insulating sleeves. Three configurations are specified, in relation to solder tags, as shown in *Figure 20.16* and HB is the most commonly used. It is strongly recommended to solder the cells into the circuits especially as high currents are often required, or perhaps the reliability of electrical contact after years on standby charging. Solder tags are generally fitted at no extra cost. Soldering directly to cell cases or top caps is likely to damage the plastic insulating materials inside the cell with the possibility of internal short circuits at the time or during subsequent service.

Cells or batteries can be encapsulated in resin. This is often required, for example, for mechanical protection for rugged applications, or electrical and thermal insulation for 'intrinsically safe' or similar uses. It is important that steps are taken to avoid sealing up the safety vent on the cells. Various techniques are employed, such as drilling small holes down to voids round the vents, or using a foam with interconnecting pores as the encapsulant. Any external resistors introduced for intrinsic safety reasons can also be housed conveniently within the encapsulation.

20.3.2.2 Theory of operation

We must consider the functions of the four basic components of cells: the positive electrode, the negative electrode, the separator and the electrolyte. During charging the chemical compositions of the two electrodes are progressively changed and during discharge the process is reversed, the alkaline electrolyte allowing the transport of charge through the porous separator.

The 'active materials' in the uncharged electrodes consist of nickel hydroxide in the positive and cadmium hydroxide in the negative and these react in the following way during the passage of the charging and then discharging currents:

$$2Ni(OH)_2 + Cd(OH)_2 \rightleftarrows Cd + 2NiOOH + 2H_2O$$

discharged charged

In order to understand how the maintenance-free, completely sealed system is achieved we must consider what happens on overcharge and in particular the oxygen recombination reaction, which enables the necessarily evolved oxygen gas to be continuously absorbed and reused inside the cell in accordance with the chemical equations:

$$O_2 + 2H_2O + 2Cd \rightarrow 2Cd(OH)_2$$

Thus during overcharge, parts of the negative electrode are being continuously charged to metallic cadmium and discharged back again to cadmium hydroxide.

The oxygen is given off at the overcharged positive nickel electrode and passes through the fine porous separator and is very quickly absorbed at the cadmium negative electrode. To speed this reaction within the cell the two electrodes are mounted in close proximity to each other, insulated only by the thin layer of separator. The cell is designed so that the negative electrode is electrically bigger than the positive electrode to avoid its becoming fully charged and therefore it cannot be overcharged and cannot evolve hydrogen. To absorb the energy of overcharge at least one of these two gases must be produced and as hydrogen cannot be recombined and oxygen easily can, a convenient and reliable cell system presents itself.

20.3.2.3 Characteristics of cylindrical cells

Capacity The ampere hour capacity of a cell is defined as the product of the discharge current and the time in hours for which it can be drawn. It is somewhat dependent on the actual rate of discharge and it is common commercial practice to give the figure

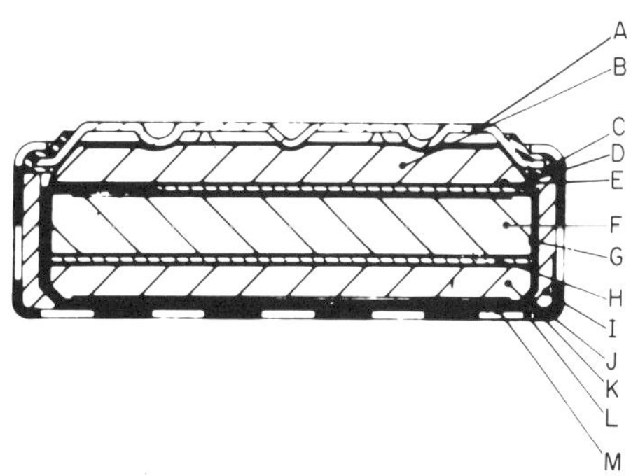

Figure 20.15 Typical button nickel–cadmium cell construction. A, Cell lid. B, Negative electrode (A). C, Grommet. D, Negative connecting strip. E, Negative electrode wrappers (2 off). F, Positive electrode. G, Positive electrode wrapper. H, Separator. I, Insulating sleeve. J, Insulating cup. K, Negative electrode (B). L, Cell case. M, Positive connecting strip

Table 20.3 Nickel–cadmium cylindrical sealed cells

IEC designation	*Size*	*Nominal capacity (A h)*	*Max. dimensions (mm)* A	*Max. dimensions (mm)* B	*Max. dimensions (mm)* C	*Approx. wt (g)*	*Charge rate for 16 h (mA)*
KR/11/45	AAA	0.18	43.0	10.5	44.5	10.0	18
KR/15/18	$\frac{1}{3}$AA	0.11	16.1	14.1	17.0	8.0	12
	$\frac{1}{2}$AA	0.24	27.1	14.3	28.1	14.0	24
KR/16/29	$\frac{1}{2}$A	0.45	27.1	16.7	28.1	19.0	45
KR/15/51	AA	0.50	49.3	14.3	50.3	25.0	50
KR/17/51	super AA	0.60	49.0	15.6	50.0	30.0	60
KR/23/43	RR	1.40	41.0	22.6	42.6	50.0	140
KR/27/50	C	2.20	46.4	26.0	49.0	70.0	220
KR/35/44	$\frac{1}{2}$D	2.60	42.6	32.5	43.7	100.0	260
KR/35/62	D	4.00	58.0	32.5	61.3	140.0	400
KR/35/62	D	4.50	58.4	33.8	61.0	150.0	450
KR/35/92	F	7.00	89.7	33.8	91.0	225.0	700
KR/44/91	super F	10.00	89.9	41.5	91.0	345.0	1000

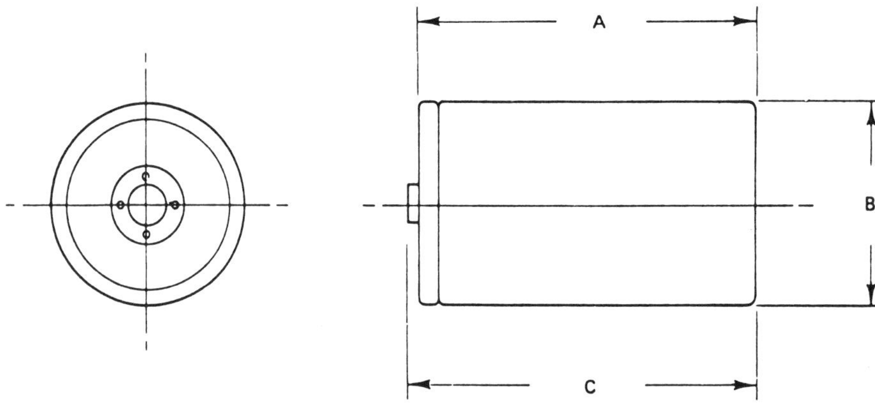

Figure 20.16 Solder tag styles

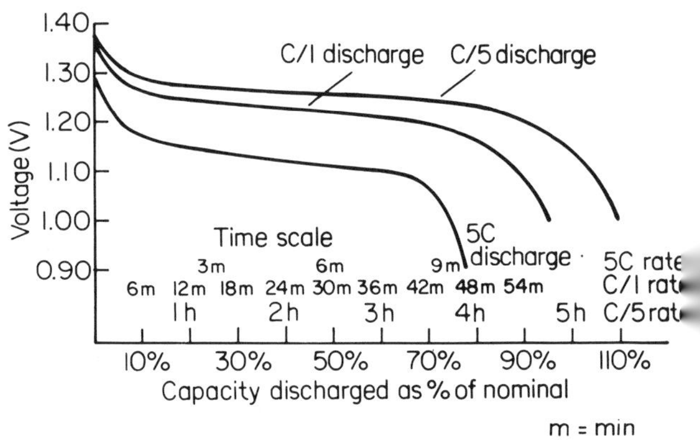

Figure 20.17 Discharge voltage curves of nickel–cadmium cells at 20°C

at the 5 hour rate, the end of discharge being taken as 1 volt per cell. It is a convenient general practice to use common terminology for currents for all sizes of cell or battery. We take the current expected to discharge the cell to 1.0 V in 1 hour as the '*C*' rate. It is obviously equal arithmetically to the ampere hour capacity of the cell. Other currents of charge and discharge are given as multiples and sub-multiples of this. Thus the *C*/5 rate will discharge in 5 hours and the 5*C* rate in 12 minutes. *Figure 20.17* illustrates this by showing the voltage–time discharge curves for *C*/5, *C*/1 and 5*C* rates and it will be seen that the capacity obtained decreases somewhat with increasing current. It will also be seen from this figure that the voltage curve is very flat for a large part of the time. This is a consequence of the low and stable internal resistance and these cylindrical cells are particularly useful when large discharge currents are required. In many applications differing discharge currents are drawn in sequence and to a very good approximation the net capacity taken can be obtained by adding up all the individual current × time pulses. All these cylindrical cells can be charged indefinitely at the *C*/10 rate regardless of their initial state of charge. It should be noted, however, that to allow for a charge efficiency factor of about 1.4, a 'flat' or discharged cell will need at least 14 hours on charge at this '10 hour' rate to attain full capacity. By way of an example, consider a cylindrical nickel–cadmium cell of the 'AA' or 'penlight' size. It has a capacity of about 500 mA h, can be left permanently on charge at 50 mA and its internal resistance is such that it can deliver 10 A for 30 s, 5 A for 3 min, 500 mA for 1 h or 50 mA for 10 h. For many applications it can replace a penlight R6 zinc–carbon or alkaline manganese battery. It can be completely recharged or left continuously on charge at 50 mA.

Nickel–cadmium 'secondary' cells can deliver their full capacity in one continuous discharge without disadvantage as no time is needed for 'depolarising'. Because of this it is very difficult to give a meaningful comparison of the ampere hour capacity per cycle of nickel–cadmium cells and their primary cell equivalents. As the latter give their optimum performance only with intermittent use, and the former can be recharged many times, most manufacturers quote expected cycle lives of 500 to 1000 recharges, at normal temperatures and conditions.

Safety These cylindrical cells are fitted with a resealing vent that relieves any excess internal pressure caused by abuse. It opens, typically, at 14 atm and closes at 12 atm. Abuse conditions could be overcharging at too high a current, or reverse charging. Batteries should not be disposed of by incineration as this may cause a rapid pressure rise. To facilitate the internal chemical reactions only a small amount of electrolyte is present and the cells are said to be 'starved' of this component. Thus on venting only a small amount of oxygen gas should be released.

Temperature Nickel–cadmium batteries give their optimum performance at about 25°C and *Figure 20.18* shows the deviations to be expected at other temperatures. Note that there

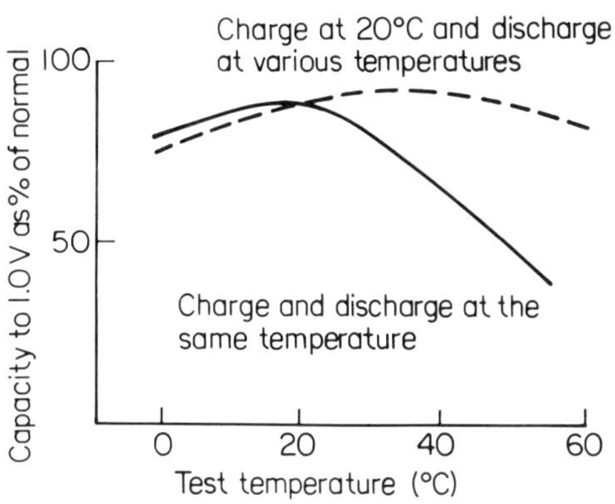

Figure 20.18 Capacity variation with temperature of nickel–cadmium cells

are two curves. The broken curve gives the changes in capacity, in this case after charging at the $C/10$ rate for 14 hours at a 'room temperature' of 20°C and discharging, after conditioning at the test temperature. The full curve is obtained when one charges and discharges at the test temperatures. It can be inferred from this graph that the efficiency of the charging process improves at lower temperatures but can fall off markedly at elevated temperatures.

Shelf life Batteries are by nature chemical entities and are thus affected by temperature, and an important temperature effect is the influence on the ability of the cell to hold charge during storage. All secondary batteries gradually lose their charge and the nickel–cadmium cell is no exception although button cells are better in this respect than cylindrical cells. *Figure 20.19* shows the big effect that temperature has on this property.

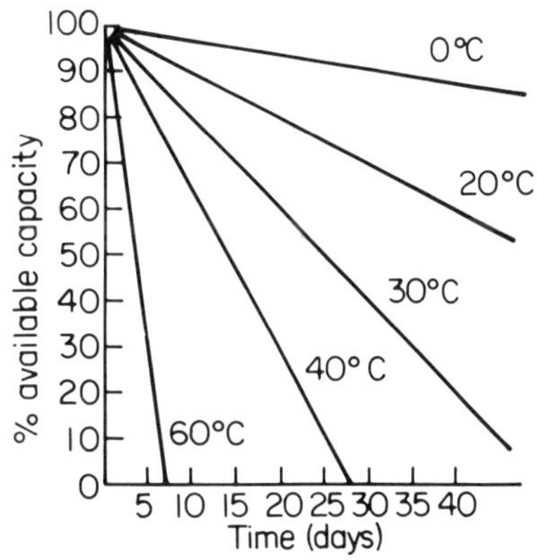

Figure 20.19 Charge retention as a function of temperature for cylindrical nickel–cadmium cells

Effects of temperature on charge characteristics When on continuous charge, once the battery has achieved its fully charged state all the energy (i.e. charging current × battery voltage = power) is converted into heat and the temperature will rise above ambient to a degree largely dictated by the amount of heat insulation and ventilation round the battery. An undesirable effect of this is the amount of deterioration caused in the various plastics components of the cells, and in recent years cells with more stable separators have become available (see below).

Whilst the discharge voltages are altered very little by changes in temperature, the on-charge voltage of a cell is markedly influenced as shown in *Figure 20.20*. It will be seen that there is a

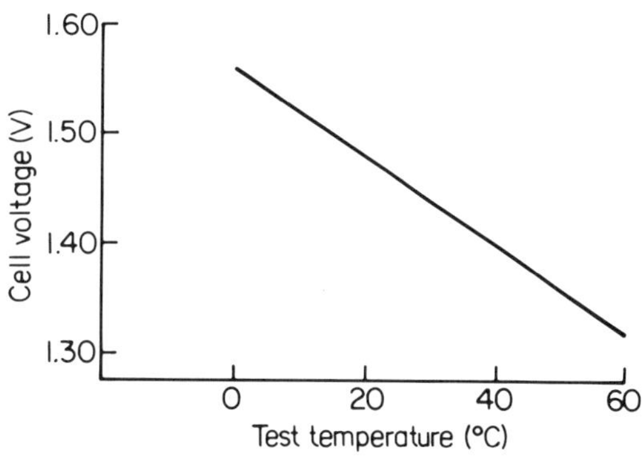

Figure 20.20 Charge–voltage variation with temperature

decrease with an increase in temperature and the slope of the line is -4 mV °C^{-1}. It is largely due to the influence of this phenomenon that sealed nickel–cadmium batteries are almost invariably charged from a constant current source. Control of charge current by cell voltage, or parallel charging, can lead to unstable conditions when the battery temperature rises with a falling voltage and more current passes with consequent further temperature rise and so on. It is possible for batteries, or parts of batteries in parallel systems to be completely destroyed by this sequence of events. The process is called 'thermal runaway'. A clear distinction must be made in this respect between 'open' or 'vented' cells and the 'sealed' cells being considered here. This unstable condition is much more likely to arise with the sealed cells and hence the recommendation for constant current charging.

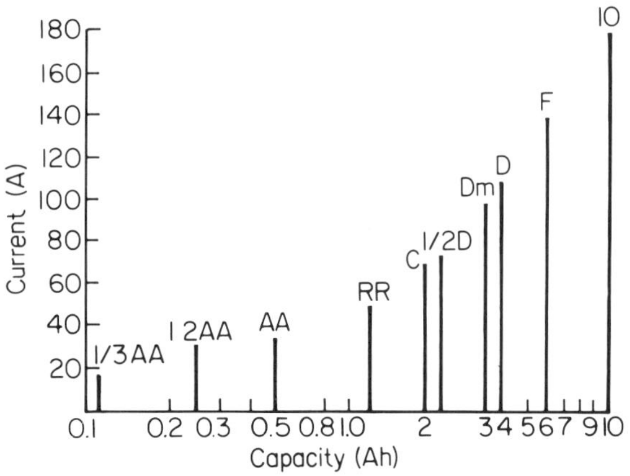

Figure 20.21 Short circuit currents for various cylindrical cells

Internal resistance The very low internal resistances of these cells is best illustrated by the chart (*Figure 20.21*) of short circuit currents of various cylindrical cells. These currents do not harm the cells for short periods but are not normally used as it would be a very inefficient use of the cells as little external work is being done.

20.3.2.4 Characteristics of button cells

Closely related as they are to the cylindrical cells, these sealed nickel–cadmium button cells have features which make them especially suitable for certain applications. Their capacities range from 50 mA h to 600 mA h and details of some sizes available are given in *Table 20.4*. Some cells have electrodes constructed on the so-called mass plate systems and these are characterised by their having a low rate of self-discharge. *Figure 20.22* illustrates this for a commercially available range (see also *Figure 20.19*).

Various combinations of electrodes are used. Some cells have one positive and one negative electrode (ZA in *Table 20.4*); one positive and two negative (DA types) and two positive and two negative (VA types). As will be seen, extra electrodes lower the internal resistance and that has a marked influence on the maximum pulse and maximum discharge currents which can be drawn (see *Tables 20.4* and *20.5* and *Figure 20.23*). It will also be seen from *Figure 20.23* that extra electrodes mean that the capacity given is less dependent on the discharge current.

The trickle-charge current is less than that possible for cylindricals and should be limited to $C/100$. Thus for the 60 mA h size the maximum continuous current is 0.6 mA. The charge efficiency is very good and even this small current will keep the battery fully charged and the capacity can be expected to be still over 60% of nominal after four years overcharge at normal temperatures.

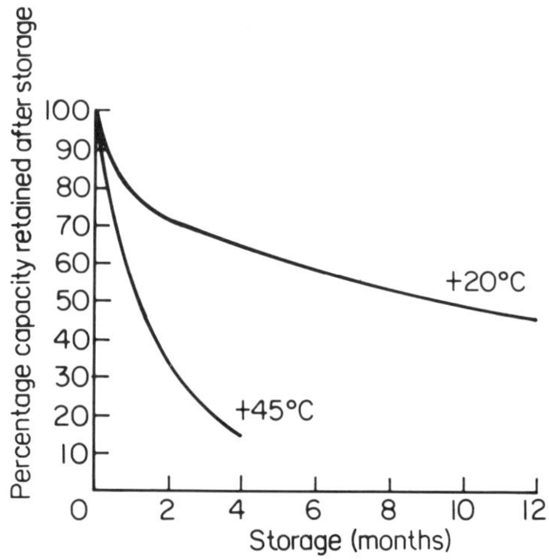

Figure 20.22 Charge retention of mass plate nickel–cadmium button cells

Table 20.5 Maximum discharge currents of some button nickel–cadmium cells

Reference	*Maximum continuous current (mA)*	*Minimum 2 seconds pulse (A)*
NCB 6 ZA	75	0.7
NCB 11ZA	165	1.5
NCB 15ZA	250	2.0
NCB 25ZA	250	2.0
NCB 25DA	500	3.0
NCB 60ZA	600	4.0
NCB 60VA	1500	8.0

As with cylindricals, cell temperature affects the capacity available and this is shown in *Figure 20.24*. The optimum performance is obtained at normal temperatures and again the comparison is made between the two and three electrode types.

20.3.3 Silver–zinc cells

The silver–zinc cell is a relatively new development. It is made in a sealed button form as in *Figure 20.25*. The positive electrode

Table 20.4 Nickel–cadmium button cells

Reference	*Capacity (mA h)*	*Voltage (V)*	*Maximum diameter (mm)*	*Maximum thickness (mm)*	*Approx. weight (g)*	*C/10 charge rate (mA)*	*Internal resistance (mΩ)*
NCB 6 ZA	60	1.2	16	6.1	4	6	280
NCB 11 ZA	110	1.2	23	4.5	6	11	140
NCB 15 ZA	150	1.2	25	5.5	9	15	120
NCB 25 ZA	250	1.2	25	9.0	13	25	100
NCB 25 DA	250	1.2	25	9.0	13.5	25	70
NCB 60 ZA	600	1.2	35	10.0	30.0	60	70
NCB 60 VA	600	1.2	35	10.0	30.5	60	30

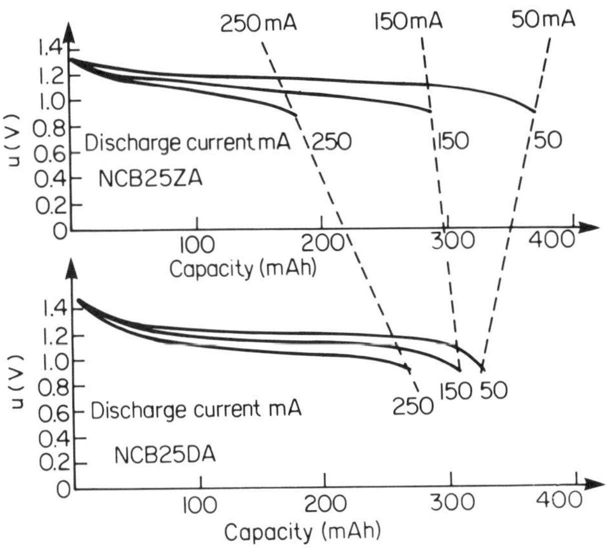

Figure 20.23 Comparison of some button nickel–cadmium cells

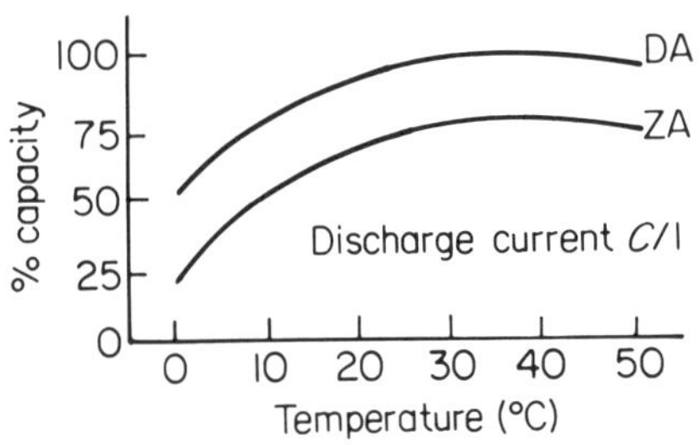

Figure 20.24 Capacity variation with temperature for two button nickel–cadmium cells

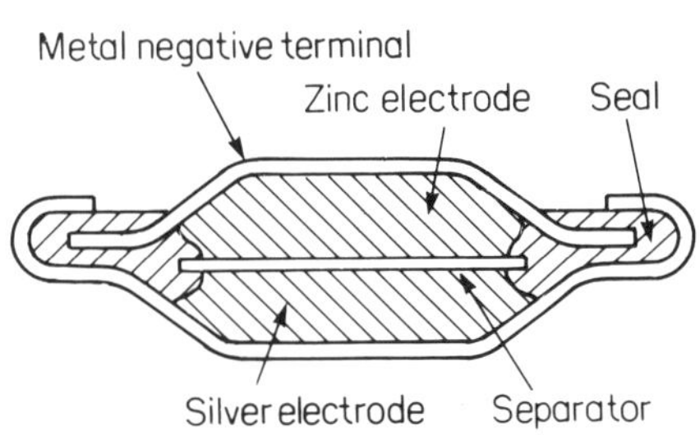

Figure 20.25 The silver–zinc cell

consists of a silver mesh which is coated with silver oxide. The negative electrode uses a perforated metal plate, made from silver or silver coated copper, which is then covered with zinc oxide. The electrodes are saturated with potassium hydroxide which acts as the electrolyte.

The separator consists of cellulose material. It should not become hydrated by the electrolyte as this will cause it to expand and press against the zinc anode. The cellulose material can also oxidise and degenerate. Several new types of material are being developed for the separator.

Silver–zinc cells have more than double the capacity of nickel–cadmium cells of the same size. Their terminal voltage is 1.5 V and the discharge curve is almost flat. During charging the terminal voltage rises to 1.85 V when the cell reaches 95% of its charge and this serves as a control mechanism for the charging circuit.

The silver–zinc cell has a charge efficiency of 95%, and this is useful when it is being charged from low energy sources such as solar cells. The leakage current of silver–zinc cells is also low. The disadvantage of the cell is that it is relatively expensive, primarily due to the cost of the silver, and currently available cells have a low charge–discharge cycle life of about 100 cycles.

20.4 Battery chargers

20.4.1 Charging from d.c. sources

20.4.1.1 Dry cells

This is the simplest form of charging and can be merely leads from a bank of dry cells, a technique used sometimes for small cells for toy applications, when connection to the mains supply might be undesirable for both safety and cost reasons. The method is only suitable for small batteries of low capacity cells and uses dry cells in series/parallel connection having about twice the voltage of the rechargeable battery. The internal resistance of the dry cells controls the rate of charge.

20.4.1.2 Vehicle batteries

Charging from vehicle electrics is a convenient method and this is often accomplished via the cigar-lighter socket. Hand lamps and small transceivers are sometimes charged in this way. Care should be taken that the equipment is not taking current when the vehicle is not in use as this may drain the main battery. Again the battery being charged should have about half the voltage of the supply, but here a resistance is essential in series with the battery.

20.4.1.3 Solar power chargers

A single solar cell can produce a voltage of up to about 0.5 V and will deliver a current dependent on the Sun's intensity (the 'insolation'). Thus, for a given insolation, they can be regarded as constant current devices.

However, the current delivered varies during the day and the maximum in clear weather would be expected at noon. Some of the time these currents could be greater than that needed or recommended for the battery being charged. Under these circumstances it is advisable to control the charge current with a simple constant current circuit, eliminating the high current peaks and ensuring more accurate and evenly spread charging.

20.4.2 Charging from a.c. sources

20.4.2.1 Transformer-less circuits

In its simplest form this can be based on the a.c. reactance of a non-electrolytic capacitor, and a diode to rectify. A circuit is given in *Figure 20.26*. Note that all current carrying parts of this type of charger must be completely insulated as there is no isolation from the mains and could otherwise be dangerous.

Capacitor charging circuits are usually found in appliances using either button cells or small cylindricals as their motive force; e.g. electric razors, toothbrushes, etc., which can have the completely isolated circuitry.

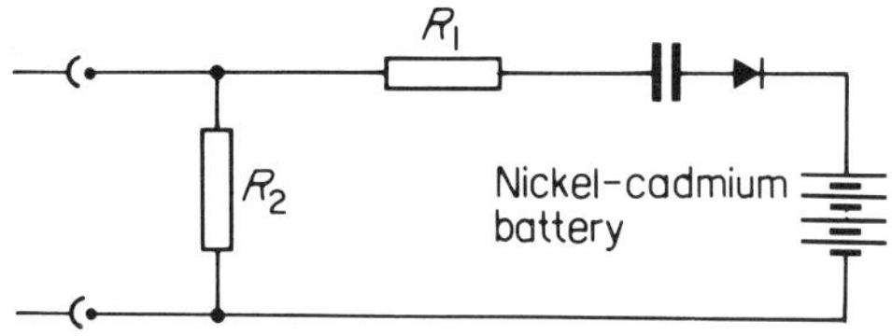

Figure 20.26 A transformer-less circuit for charging small batteries

Table 20.6 Fast charge currents and times for fully discharged nickel–cadmium batteries

$C/8$ charge; normal charge 12 hours; maximum charge indefinite
$C/4$ charge; normal charge 5 hours; maximum charge 6 hours
$C/2$ charge; normal charge $2\frac{1}{4}$ hours; maximum charge $2\frac{1}{2}$ hours
$C/1$ charge; normal charge 1 hour; maximum charge $1\frac{1}{4}$ hours
$2C$ charge; normal charge 27 mins; maximum charge 30 mins
$4C$ charge; normal charge 12 mins; maximum charge 12 mins
$8C$ charge; normal charge 5 mins; maximum charge 5 mins

20.4.2.2 Transformer circuits

A step-down transformer, diode and resistance are the bare essentials of a simple safe charger (see *Figure 20.27*). It is common practice to choose a transformer with a secondary r.m.s. voltage approximately twice that of the battery to be charged. This is to produce a reasonably constant current supply. If all measurements are made using r.m.s. values then calculation of the resistance R for a charge current I is given by Ohm's law:

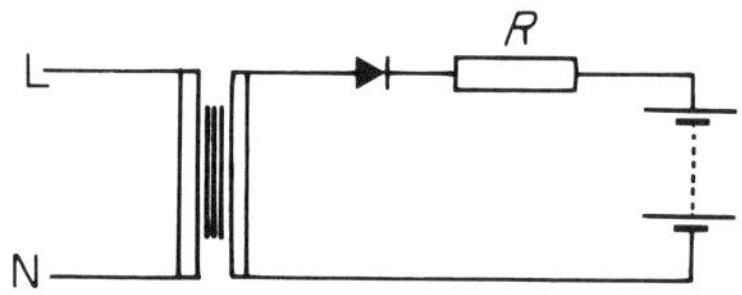

Figure 20.27 A simple charging circuit

$$R = \frac{V_s - V_b}{I}$$

where V_s is the transformer secondary voltage and V_b is the battery on-charge voltage (1.4 × No. of cells).

Improvements may be made by adding a smoothing capacitor and rectifying to full wave with a bridge rectifier.

A transistorised circuit gives a more stable constant current and its simplest form is given in *Figure 20.28*.

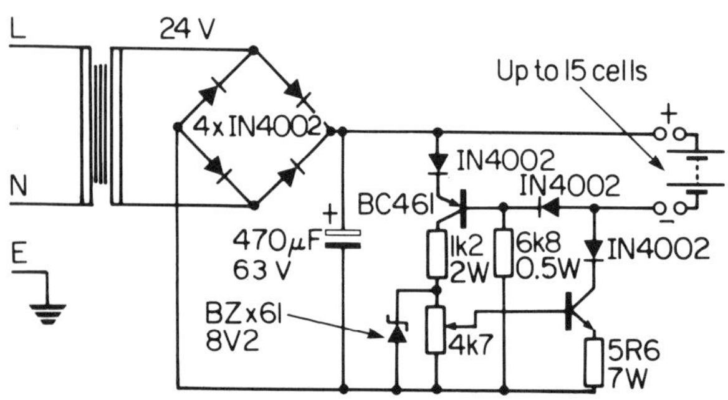

Figure 20.28 A transistorised charger

20.4.3 Fast charging

An easy way to control fast charging is first to ensure the battery is completely discharged either as a consequence of its service routine or by incorporating a discharge circuit into the charger. Charge can then be put back by timing an accurately known current (±5% or greater) and the times given in *Table 20.6* are a guide.

The fast charging can also be controlled by sensing of the voltage rise associated with the attainment of full charge. This technique can replace about 80% or more of the nominal capacity in a few minutes. It is not necessary to discharge the battery first nor know the state of charge. A circuit is shown in *Figure 20.29* for charging a 12 V, 0.5 A h battery to about 80% of nominal in about 12 min. The current peaks at about 15°C.

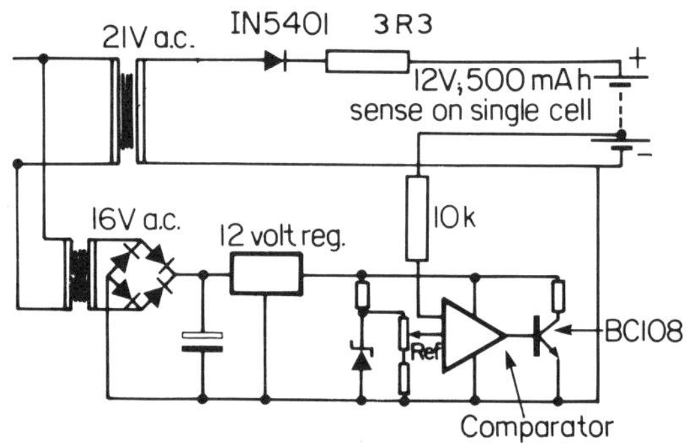

Figure 20.29 A fast-charge circuit

The voltage is sensed by the comparator IC and a usual voltage cut-off is 1.54 V per cell. Inclusion of a negative temperature thermistor in the potential divider network would enable charging to be carried out between 0°C and 45°C. The variation in cut-off must be adjusted to -4 mV $°C^{-1}$ per cell, reaching 1.62 V per cell at 0°C. Often a timer is added to the circuit as an additional precaution.

20.5 Battery selection

The first choice which faces an equipment designer is whether to use primary or secondary batteries. If the application requires a low initial cost, or the life of the equipment is short, then primary batteries are the obvious choice. Secondary batteries can prove cheaper in the long run as they can be re-charged and re-used, but they require a charger, so the initial cost of the battery system is higher.

Table 20.7 compares the parameters of a few types of primary cells. Zinc–carbon is cheap and readily available. It is primarily used in application having a light, intermittent duty cycle. The battery has a low shelf life and a drooping discharge curve.

For heavy duty applications, needing continuous operation at high currents, alkaline manganese dioxide batteries are preferred to zinc–carbon. They have 50% to 100% more energy for the same

Table 20.7 Comparison of primary cells. In the comparative data 1=highest or best, 4=lowest or worst

Parameter	*Carbon–zinc*	*Zinc chloride*	*Alkaline manganese dioxide*	*Mercuric oxide*	*Lithium*	*Zinc–air*
Nominal cell voltage (V)	1.5	1.5	1.5	1.35 or 1.40	3.0	1.4
Energy output (W h kg^{-1})	40	90	60	100	300	200
Shelf life at 20°C (years)	1.5	2	3	3	5	5
Operating temperature range (°C)	+5 to +60	−10 to +60	−10 to +60	−20 to +100	−40 to +80	−40 to +60
Flatness of discharge curve (comparative)	3	3	3	2	1	1
Cost (comparative)	4	3	3	2	1	1

Table 20.8 Comparison of secondary cells. In the comparative data 1=highest or best, 4=lowest or worst

Parameter	*Lead–acid*	*Nickel–cadmium*	*Silver–zinc*
Nominal cell voltage (V)	2.1	1.2	1.5
Energy output (W h kg^{-1})	20	30	110
Shelf life at 20°C (years)	1.0	0.2	0.3
Cycle life (number of operations)	500	2000	100
Operating temperature range (°C)	−60 to +60	−40 to +60	−20 to +80
Flatness of discharge curve (comparative)	2	1	1
Cost (comparative)	3	2	1

weight, and a longer shelf life, but they are more expensive at currents below about 200 mA.

The mercuric oxide battery has a flat discharge curve and a high energy to weight ratio. It is often used in voltage reference applications. Lithium batteries are used in special application which require a long shelf life, or a wide operating temperature range, or a very high energy density.

Table 20.8 summarises the parameters of a few secondary batteries. Lead–acid cells are in wide general use as they cost about one half as much as nickel–cadmium cells, this price difference being more marked at higher current ratings. Nickel–cadmium batteries are used in applications requiring a flat discharge characteristic, or where many charge–discharge cycles are involved. They have a lower weight and volume than lead–acid batteries, and below about 0.5 ampere hour ratings they are competitive on price.

20.6 Fuel cells

The fuel cell converts chemical energy from the oxidation of a fuel into electrical energy. *Figure 20.30* shows the operation of a hydrogen–oxygen fuel cell. The cell reaction is as follows:

at the anode

$$H_2 \rightarrow 2H^+ + 2e^-$$

at the cathode

$$O_2 + 4H^+ + 4e^- \rightarrow 2H_2O$$

overall

$$2H_2 + O_2 \rightarrow 2H_2O$$

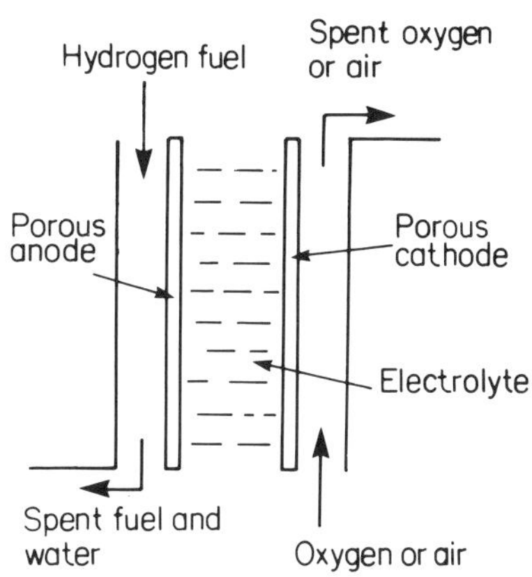

Figure 20.30 Operation of a hydrogen–oxygen fuel cell

This reaction gives a difference of potential between the electrodes of the cell, causing a flow of current in a load connected between them.

The fuel cell works isothermally. Its thermal efficiency is high, of the order of 60% to 80%. In low temperature cells the electrodes are made of finely divided platinum, in the form of wire screens, and the electrolyte is potassium hydroxide.

Figure 20.31 shows the characteristic of a hydrogen–oxygen cell. The power density, measured in milliwatts per square centimetre is greater when pure oxygen is used since it is a better oxidant than air.

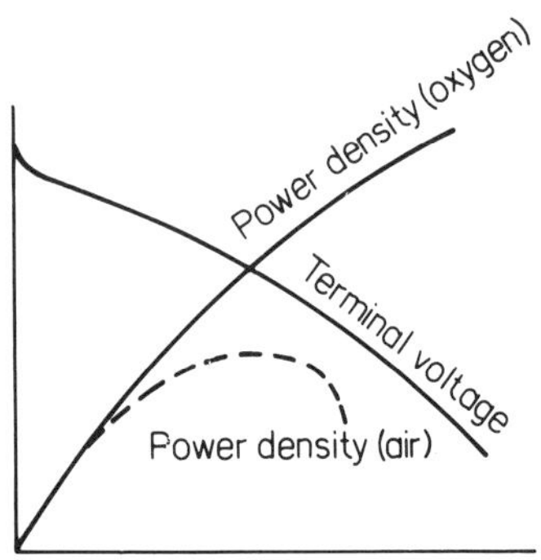

Figure 20.31 Characteristics of a hydrogen–oxygen fuel cell

Many other types of fuel cells have been developed. Menthanol is a good fuel since it has a low carbon content, it is cheap, and it can easily be handled and stored. The chemical reactions of the cell are as follows:

at the anode

$$CH_3OH + H_2O \rightarrow CO_2 + 6H^+ + 6e^-$$

at the cathode

$$O_2 + 4H^+ + 4e^- \rightarrow H_2O$$

Liquid electrolytes present several problems in fuel cells. The walls must be porous to hold the solution, and the chemical reaction occurs in the capillaries of the walls where there is contact between the gas, solution and solid. It is difficult and expensive to make these walls.

With a solid electrolyte the wall surface is no longer critical, and the cell can be run at higher temperatures, giving larger cell voltages. The power to weight ratio is also improved since a very thin layer of electrolyte can be used. The key to future fuel cells lies in the ability to obtain suitable solid electrolytes. Zirconia doped with lime is used as the electrolyte in some cells. This has high ion conductivity at 1000°C but it falls off at lower temperatures.

Acknowledgements

Sections 20.1, 20.2.2, 20.2.6, 20.3.1, 20.3.3, 20.5 and 20.6 are based on material published in *Discrete Electronic Components* by F F Mazda, published by Cambridge University Press, 1981. Berec (Ever Ready) Ltd are thanked for granting permission to reproduce extracts from *Modern Portable Electricity* and *Battery Data Book*.

Further reading

BATTLES, J. E., SMAGA, J. A. and MYLES, K. M., 'Materials requirements for high performance secondary batteries', *Metallurgical Trans. A*, **9A**, February (1978)

BEAUSSART, L., 'Thermal batteries and their applications', *Communications International*, February (1978)

BERKOVITCH, I., 'What prospects for photovoltaic generators?', *Electronics & Power*, August (1980)

BRODD, R. J., KOZAWA, A. and KORDESCH, K. V., 'Primary batteries 1951–1976', *J. Electrochem. Soc.*, July (1978)

BUTTERFIELD, P. N., 'The development, theory and use of nickel–cadmium batteries', *Radio Communication*, May (1978)

DONNELLY, W., 'Battery technology—uses and abuses', *Electronics Industry*, May (1982)

HARRISON, A. I., 'Lead–acid standby-power batteries in telecommunication', *Electronics & Power*, July/August (1981)

HOLMES, L., 'Lithium primary batteries—an expanding technology, *Electronics & Power*, August (1980)

JAY, M. A., 'Lithium/thionyl—chloride cells', *Electronics & Power*, July/August (1981)

KDRDESCH, K. V., '25 years of fuel cell development (1951–1976)', *J. Electrochem. Soc.*, March (1978)

KNUTSEN, J. E., 'A new generation of battery systems', *New Electronics*, July 14 (1981)

KORDESCH, K. V. and TOMANTSCHGER, K., 'The physics teacher', January (1981)

KUWANO, Y., 'Amorphous silicon solar cell seen as power source of future', *JEE*, June (1982)

McDERMOTT, J., 'Manganese-dioxide, lithium power sources sub for high-priced silver-oxide cells', *EDN*, February 3 (1982)

MEEDS, P., 'Lithium sulphur dioxide batteries for long-term memory back up', *New Electronics*, July 13 (1982)

MORRIS, R. J., 'Lithium cells', *New Electronics*, July 14 (1981)

NABESHIMA, T., 'Sealed nickel–cadmium cells at their limits', *New Electronics*, June 10 (1980)

RUETSCHI, P., 'Review on the lead–acid battery science and technology', *J. Power Sources*, **2**, 3–24 (1977/78)

SALKIND, A. J., FERRELL, D. T. and HEDGES, A. J., 'Secondary batteries: 1952–1977', *J. Electrochem. Soc.*, August (1978)

SHIRLAND, F. A. and RAI-CHOUDHURY, P., 'Materials for low cost solar cells', *REP. Prog. Phys.*, **41** (1978)

SPENCER, E. W., 'Lithium batteries: new technology and new problems', *Prof. Safety*, January (1981)

STIRRUP, B. N., 'Charge regimes for nickel–cadmium and lead acid stationary batteries', *Electronics & Power*, July/August (1982)

SUBBARAO, E. C., 'Fuel cell as a direct energy conversion device', *Trans. of the Saest*, **11**, No. 4 (1976)

UMEO, Y., 'Batteries play major role in miniaturization of electronics equipment', *JEE*, November (1981)

WALKER, D. and GILHAM, D., 'Rechargeable nickel–cadmium cells, their construction, reliability and use', *Electron. Eng.*, September (1979)

21 Semiconductor Diodes

M J Rose BSc (Eng)
Production Officer,
Marketing Communications Group,
Mullard Ltd

Contents

21.1 p-n junctions

If part of a single crystal of silicon or germanium is formed into p-type material, and part formed into n-type material, the abrupt interface between the two types of material is called a pn junction. As soon as the junction is formed, majority carriers will diffuse across it. Initially both types of material are electrically neutral. Holes will diffuse from the p-type material into the n-type material, and electrons from the n-type material into the p-type. Thus the p-type material is losing holes and gaining electrons, and so acquires a negative charge. The n-type material is losing electrons and gaining holes, and so acquires a positive charge. The negative charge on the p-type material prevents further electrons crossing the junction, and the positive charge on the n-type material prevents further holes crossing the junction. This space charge creates an internal potential barrier across the junction, shown diagrammatically in *Figure 21.1* by the battery in broken lines.

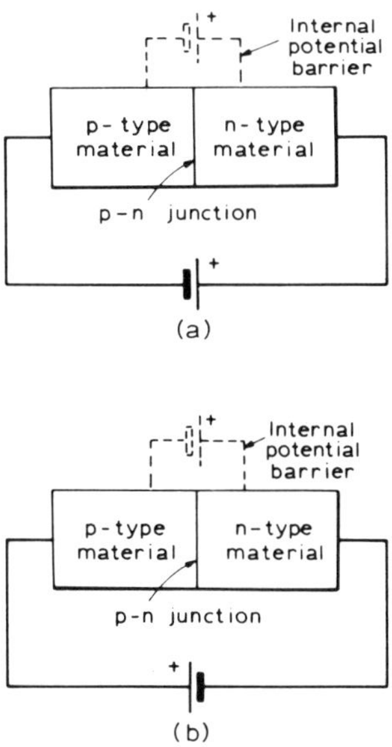

Figure 21.1 p-n junction: (a) reverse bias; (b) forward bias

Because of this internal potential barrier, only electrons and holes with high kinetic energy can cross the junction or remain in the junction region. There are, therefore, only a few majority carriers in the region and a depletion layer is formed around the junction. The potential barrier, however, will assist minority carriers to cross the junction.

If an external battery is connected across the pn junction with the positive terminal connected to the n-type material, *Figure 21.1(a)*, the junction is reverse-biased. The external battery adds to the internal potential barrier and prevents even the few majority carriers in the depletion layer with high energy from crossing the junction. The battery voltage helps minority carriers to cross, but because of the small number only a very small reverse current flows.

Because the depletion layer contains few charge carriers, it is effectively an insulator separating the conducting p-type and n-type regions. A parallel-plate capacitor is therefore formed. The width of the depletion layer is proportional to the reverse voltage, and so the capacitance can be varied by varying the reverse bias across the junction. Constructive use of this effect is made in variable-capacitance (varactor) diodes. On the other hand, the capacitance can limit the performance of switching diodes at high frequencies.

An equivalent circuit for a reverse-biased pn junction is shown in *Figure 21.2*. The depletion layer is represented by the voltage-variable capacitor C_d in parallel with the voltage-variable resistor R_j. The material outside the depletion layer has a resistance, and this bulk resistance is represented by R_b in series with the junction impedance.

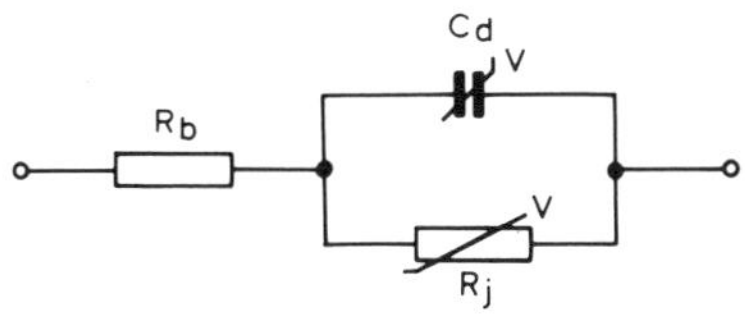

Figure 21.2 Equivalent circuit of reverse-biased p-n junction

As the reverse voltage is increased, the reverse current remains reasonably constant until a voltage is reached at which breakdown occurs. This breakdown is caused by an increase in the number of charge carriers available so that a large, and possibly destructive, current flows.

The increase in carriers may be caused by the avalanche effect or the zener effect. Briefly, avalanche breakdown is caused by ionisation of the semiconductor material to produce the charge carriers, while in zener breakdown electrons are torn away from their atoms to become charge carriers. Constructive use is made of the two breakdown mechanisms in avalanche diodes and zener diodes, and both mechanisms will be discussed in more detail later when describing these two devices.

If the external battery is connected across the junction with the positive terminal connected to the p-type material, *Figure 21.3(b)*, the junction is forward-biased. The battery voltage opposes the internal potential barrier, and as soon as the voltage exceeds this potential, majority carriers flow across the junction. A small increase in forward voltage therefore causes a large increase in forward current.

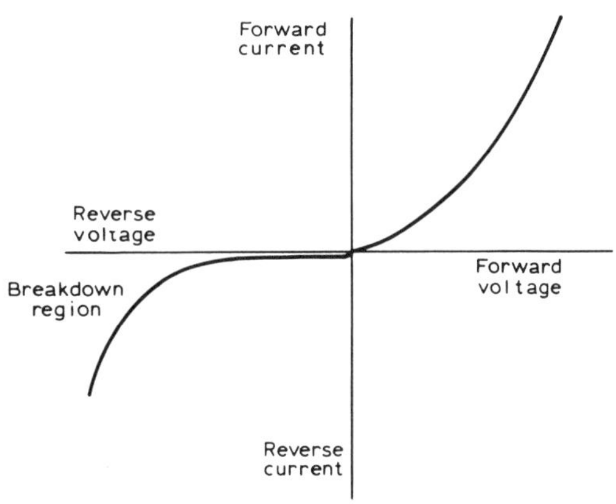

Figure 21.3 Voltage/current characteristic of junction diode

The asymmetrical current flow across a pn junction enables it to be used as a rectifying element. Thus a semiconductor device containing a single pn junction can be used as a diode. The general shape of the voltage/current characteristic of a semiconductor junction diode is shown in *Figure 21.3*.

Another cause of breakdown in a pn junction, besides excessive reverse voltage, is too high a junction temperature. Even with a forward-biased junction, a semiconductor diode will have a small resistance. Current flowing through the device will therefore generate heat, and will raise the temperature of the junction. Above a certain temperature the thermal agitation of the atoms in the crystal will destroy the junction. Sufficient thermal energy is transferred to the valence electrons that a large number of free electrons and holes are produced. These 'minority' carriers swamp the action of the majority carriers, and so prevent normal current flow taking place. Thus a maximum permissible junction temperature is specified for all semiconductor devices to preserve operation.

All semiconductor junction diodes contain a single pn junction, but the properties and structure of the junction can be exploited differently to enable different types of diode to be obtained. For example, besides the normal rectifier diode, other types such as high-speed switching diodes, voltage regulator diodes and variable-capacitance diodes are available.

21.2 Small-signal diodes

Diodes which are used for demodulation and switching applications where only small amounts of power are involved can be conveniently grouped together as small-signal diodes. As such diodes are generally wired directly into circuits in the same way as resistors and capacitors, the usual construction is a glass or plastic envelope with axial connecting wires. The circuit symbol for small-signal diodes is shown in *Figure 21.4*.

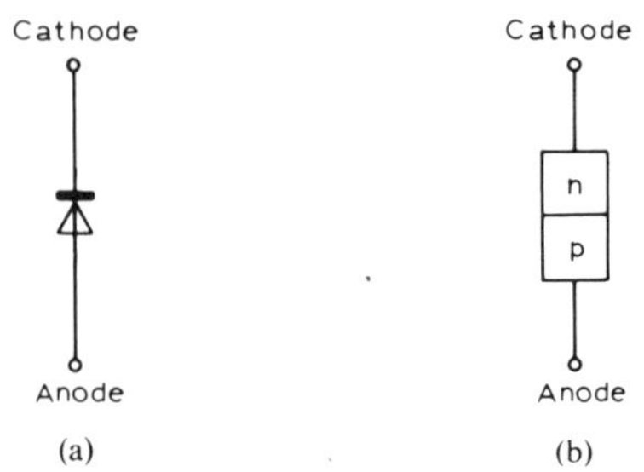

Figure 21.4 Circuit symbol for diode (a) related to schematic structure (b)

The first small-signal junction diodes were introduced in the mid-1950s, and were manufactured from germanium by the alloy-junction process. A small pellet of indium was placed on a slice of n-type germanium, and the assembly heated until some of the indium dissolved into the germanium to form p-type germanium. In this way a pn junction was formed. During the late 1950s silicon started to replace germanium as the generally-used material for semiconductor devices, and silicon diodes were introduced with junctions made with aluminium, first by alloying and later by diffusion. Silicon diodes had the advantage of withstanding higher junction temperatures and having lower reverse currents than germanium diodes. Typical values of the maximum junction temperature of germanium and silicon devices are 75°C and 200°C respectively. The reverse current for a germanium diode at a reverse voltage of 10 V at room temperature is typically 5 μA, while that for a silicon diode at the same voltage is typically 0.02 μA. On the other hand, the forward voltage before conduction occurs in a germanium diode is lower than that for a silicon diode, typically 200 mV instead of 600 mV. Germanium junction diodes are therefore still used today in those applications where the lower forward voltage is advantageous. The junctions, however, are made by diffusion rather than the alloying process.

The introduction of the planar technology in 1960 enabled planar and planar-epitaxial diodes to be manufactured. A slice of n-type monocrystalline silicon is coated with silicon oxide. Windows are cut in this oxide, and boron diffused into the silicon through the windows to form regions of p-type silicon. The silicon oxide is etched away, and the slice cut up so that the individual diode elements containing a pn junction are obtained. Connecting leads to the p-type and n-type regions are attached, and the assembly encapsulated. This method of manufacture allows many thousands diodes to be manufactured at the same time—in fact, mass-production techniques can be applied.

Another advantage is that the diffusion through windows etched in the oxide allows precise control to be exercised over the geometry of the junction.

In the planar-epitaxial diode, the pn junction is formed in an epitaxial layer. This is a layer of silicon grown on the silicon slice (which now forms a substrate for the device) in which the silicon atoms take the same crystal structure as that of the underlying substrate. Thus the near-perfect crystal structure of the substrate (a slice from a single crystal) is transferred into the epitaxial layer. The layer is grown by depositing silicon from a vapour, and impurities can be added to the vapour to produce p-type and n-type epitaxial layers as required. This layer can be of high-resistivity silicon to give a high breakdown voltage for the diode. After the deposition of the epitaxial layer, the method of manufacture is the same as for planar diodes.

Two methods of construction are used for small-signal diodes. In one, the diode element is mounted on a metal backing plate with the same thermal coefficient of expansion as the silicon diode element, for example, molybdenum. Contact with the other side of the diode element is made by a C-spring, as shown in *Figure 21.5(a)*. The connecting wires are attached to the backing plate and C-spring, and the whole assembly encapsulated in glass. The second method of construction produces the 'whiskerless diode'. The diode element is mounted between two metal backing plates, and the whole assembly is held rigidly by the glass envelope, as shown in *Figure 21.5(b)*. Because no spring is used in the whiskerless diode, this type has a greater reliability when used in applications where the diode is subjected to vibration.

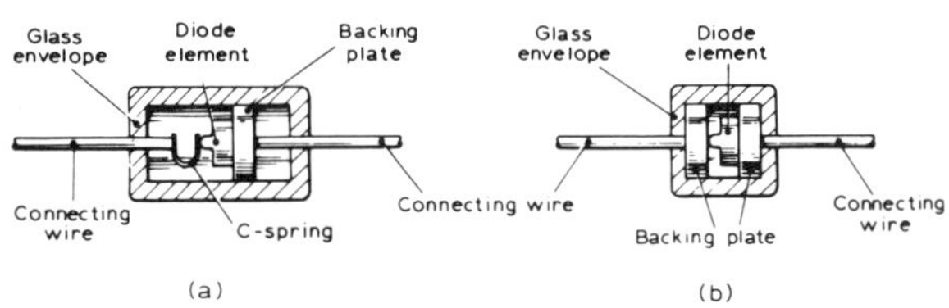

Figure 21.5 Construction of small-signal diodes: (a) spring-contact: (b) whiskerless

Typical values of the forward currents that can be conducted by germanium junction diodes lie between 30 mA and 300 mA. For silicon diodes, typical forward currents lie between 100 mA and 1 A.

When diodes are used for switching, particularly at high frequencies, an important factor is the recovery time. If a diode is switched from the conducting to the non-conducting state, the current does not change instantaneously to the small reverse current but a larger reverse current spike is produced. This is caused by charge storage. When the diode is conducting, majority carriers cross the junction. Those which do not combine with majority carriers of the opposite polarity become, in effect, minority carriers. Thus a hole from the p-type region crossing the junction and not combining with an electron in the n-type region becomes a minority carrier in the n-type region. When the bias across the diode is reversed, these minority carriers recross the junction, and a large reverse current will flow until they have diffused or recombined to allow the depletion layer to be formed. Thus the reverse recovery time will depend on the lifetime of these minority carriers.

Another limitation on the performance of switching diodes is the capacitance associated with the depletion layer. At high frequencies, this capacitance may make the reverse impedance of the diode unacceptably low. To maintain the high-frequency performance, the capacitance can be lowered by decreasing the area of the junction (by such manufacturing techniques as mesa etching described in the transistor section). The small junction area will restrict the value of forward current that can be conducted by the diode.

Gold doping of silicon diodes reduces the lifetime of the minority carriers. By using this technique and using junctions of small area, silicon diodes capable of switching waveforms with rise times of 5 ns can be manufactured. Diodes capable of switching faster waveforms can be made by using gallium arsenide instead of silicon as the semiconductor material.

The circuit designer requires certain data on small-signal diodes to allow him to choose the right type for his needs. For the 'general-purpose' types, he must know the maximum reverse voltage the diode will withstand, the maximum forward current it will conduct continuously and for a short period, the forward voltage drop across the diode, and the reverse current. For switching diodes, additional information on the recovery time is required. This is usually given as a time under specified load and waveform conditions. An alternative method of specifying the switching characteristic is by the recovered charge. This is defined as the area under the reverse current spike, the shape of the spike being determined by the test circuit component values.

21.3 Rectifier diodes

It is convenient to classify rectifier diodes into two groups: the low-power and the high-power types. The difference between the two groups is principally the cooling requirements, and this determines the type of encapsulation used. The circuit symbol used for both types of rectifier diode is the same as that for the small-signal diode, shown in *Figure 21.4*.

The amount of power that can be handled by a semiconductor diode is limited by the junction temperature which must not exceed a certain value if rectification is to be maintained. By 'scaling-up' small-signal diodes to provide a larger junction area, higher currents can be conducted. Such diodes can conduct average forward currents of, typically, 1 A and withstand surge currents of up to 30 A for short periods. With a reverse voltage rating of, typically, 600 V these diodes can be used for such applications as low-current power supplies from the a.c. mains, low-power inverters, and as rectifiers in the deflection circuits of television receivers. Because additional cooling is not required, the diodes can be encapsulated in plastic, and fitted with axial connecting wires for wiring into circuits.

Low-power rectifier diodes used in television deflection circuits and inverters are operating at relatively high frequencies. The diodes should therefore have short reverse recovery times, as with the switching diodes discussed previously. The information is either given as the recovered charge when switched under specified conditions or as the reverse recovery time.

High-power rectifier diodes were developed during the 1960s for use in rectifier systems operating from the mains supply. The aim was to replace existing systems using thermionic valves or mercury-arc rectifiers. Such diodes are required to conduct high currents and withstand large reverse voltages. Cooling, therefore, is an important requirement.

The current-carrying capacity can be increased by increasing the area of the junction and mounting the diode on a heatsink to increase the cooling and so keep the junction temperature below the limiting value. The reverse voltage can be increased by bevelling the edges of the diode element, as shown in *Figure 21.6*.

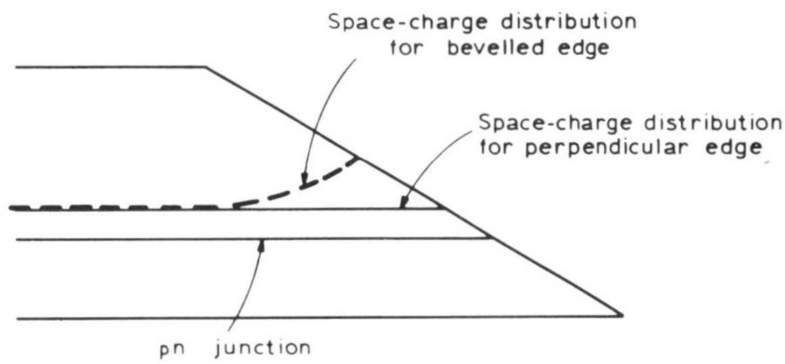

Figure 21.6 Effect of bevelled edges in high-power rectifier diode

When a large reverse voltage is applied across a diode, breakdown usually occurs across the junction at the edge of the element rather than in the centre. If the edges are bevelled, the space charge across the reverse-biased junction is deflected, as shown by the broken lines in *Figure 21.6*. The electric field produced by a large reverse voltage follows the deflection of the space charge, and so the distance over which breakdown may occur at the edge is increased. Thus the value of the breakdown voltage will be increased. Diodes can also be connected in parallel to increase the current-carrying capacity above that of the individual diodes. Similarly, diodes can be connected in series to reduce the reverse voltage across individual diodes.

Despite such improvements to the diodes and methods of connection, considerable difficulties were experienced with early semiconductor-diode rectifier systems, particularly through the effects of voltage transients. Although these transients are of very short duration, they still contained sufficient energy to destroy the diodes. These difficulties were considerably eased by the introduction of the avalanche diode.

The failure of non-avalanche diodes with large reverse voltage transients was caused by local breakdown of the junction. Small irregularities in the junction were inevitable with the alloying method of manufacture used for rectifier diodes at that time. Although these irregularities were of no consequence at lower voltages, when the diodes were subjected to the reverse transients present on the mains, the irregularities produced 'hot spots'. The high local temperature would cause that part of the junction at the 'hot spot' to break down, more current would then flow through that part of the junction raising the temperature further, until finally complete breakdown of the diode would occur.

The avalanche diode uses a diffused junction, and great care is taken during manufacture to ensure that the junction is uniform. When subjected to large reverse voltages, the whole of the junction breaks down at the same time so that the large reverse current is conducted evenly over the junction area. No hot spots are formed therefore. This bulk breakdown is shown by the reverse characteristic for an avalanche diode, *Figure 21.7*. Instead of the more gradual reverse breakdown of the non-avalanche diode, shown in *Figure 21.3*, the avalanche diode has a sharply-defined breakdown voltage. When the reverse voltage falls below the avalanche breakdown value, the diode recovers and resumes its normal blocking action with only the small reverse leakage current flowing.

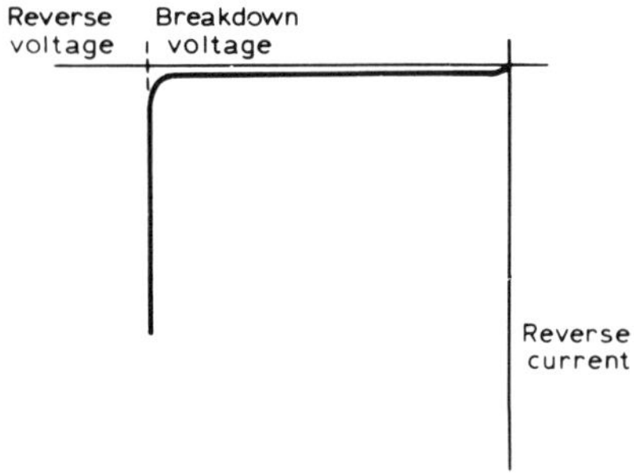

Figure 21.7 Reverse characteristic of avalance diode

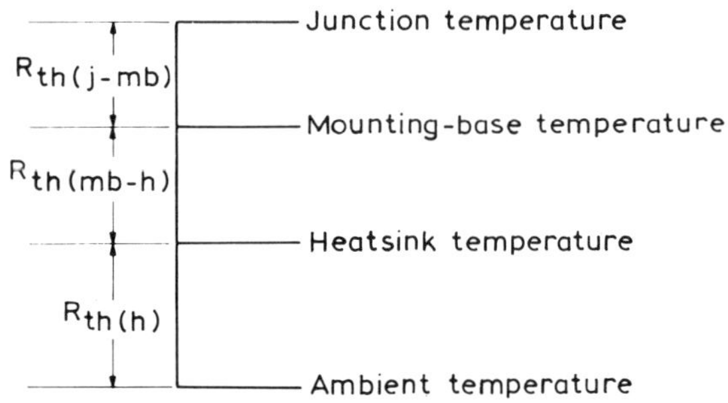

Figure 21.8 Temperatures and thermal resistances in diode mounted on heatsink

The mechanism of avalanche breakdown is essentially ionisation. The voltage across the device is large enough to give the electron minority carriers sufficient velocity to dislodge other electrons on impact with atoms of the crystal lattice. These new free electrons are accelerated sufficiently to dislodge further electrons on impact. There is therefore a sudden build-up of electron–hole pairs through successive impacts to provide a large number of charge carriers, and a large current flows.

The average power that can be absorbed safely by an avalanche diode after the breakdown is limited to some tens of watts. Since the reverse breakdown voltage can be greater than 1 kV, the average current that can flow across the junction is only, at the most, some tens of milliamps. It should be remembered, however, that the duration of the reverse voltage transients is very short so that the pulse power the diode can withstand is considerable. Powers of up to some tens of kilowatts can be withstood for surges with durations of, typically, 10 μs.

Avalanche diodes are also used in inductive circuits where rapid switching can produce large voltage transients. Such applications include thyristor inverters and high-frequency power supplies.

Stress has been laid in the discussion of rectifier diodes on the cooling requirements. Because high-power diodes need to be mounted on heatsinks, a stud-mounting construction is used. Normally the cathode of the diode is connected to the stud for screwing onto the heatsink. 'Reverse-polarity' devices with the anode connected to the stud are also available.

The limiting temperature for operation, the maximum junction temperature of the diode, is quoted by the manufacturer in the published data but cannot be measured directly by the user. Additional information must therefore be given in the data, and this is in the form of thermal resistance, R_{th}. This quantity is analogous to electrical resistance, and represents the opposition to the flow of heat from the device junction. It is expressed in temperature change with power, the usual units being °C/W. Like electrical resistance, thermal resistances can be considered connected in series. Thus a chain of thermal resistance exists from the diode junction to the mounting base. $R_{th(j\text{-}mb)}$, between the mounting base and the heatsink (a contact resistance at a specified torque), $R_{th(mb\text{-}h)}$, and between the heatsink and the ambient temperature. $R_{th(h)}$. The relationship between the thermal resistances and temperatures in the chain is shown in *Figure 21.8*.

Of the various quantities shown, the junction temperature and $R_{th(j\text{-}mb)}$ are fixed, thus fixing the mounting base temperature. The contact thermal resistance is also fixed if the specified torque is applied to the stud mounting of the diode. Therefore the heatsink temperature will be known. The ambient temperature will either be known or a limiting value known, so that the designer has to choose the appropriate thermal resistance for the heatsink to suit his application. The device manufacturer will generally publish data which relate the total power dissipation of the diode (expressed in terms of the average forward current at various form factors) to the thermal resistance and ambient temperature. With avalanche diodes, allowance must be made for the avalanche power; and for diodes that are switched at high frequencies, allowance must be made for the power loss caused by reverse recovery effects.

The circuit designer, in addition to the normal data for a diode (average forward current, reverse current, and forward voltage drop), requires for high-power rectifier diodes information on the transient performance. A crest working reverse voltage V_{RWM} is given which corresponds (with a suitable safety factor) to the peak voltage of the supply on which the diode is operating. A repetitive peak reverse voltage V_{RRM} is given, corresponding to the peak value of transients occurring every cycle, and a non-repetitive peak reverse voltage V_{RSM} corresponding to transients that occur only occasionally. Similar ratings are given for forward current; a repetitive peak forward current I_{FRM}, and a non-repetitive peak forward current I_{FSM}. These current ratings specify the maximum current that the diode can conduct before breakdown, and it may be necessary to provide protective devices such as high-speed fuses to prevent damage to the diode from high-fault currents.

Additional information is given for avalanche rectifier diodes; the avalanche breakdown voltage and the average, repetitive, and non-repetitive avalanche power.

21.4 Zener diodes

The name 'zener diode' is applied to three groups of devices: voltage regulator diodes, voltage reference diodes, and surge suppressor diodes. Strictly speaking, 'zener' is not the correct description because some of the devices use avalanche breakdown rather than zener breakdown for their operation. However, the name zener seems now firmly established among electronic engineers. The circuit symbol for a zener diode is shown in *Figure 21.9*.

Figure 21.9 Circuit symbol for zener diode

The three groups of device operate on the same principle: the sudden breakdown of the junction at a specific reverse voltage, as shown by the characteristic of an avalanche diode in *Figure 21.7*. Because the characteristic after breakdown is nearly parallel to the current axis, the voltage after breakdown remains reasonably constant. The zener diode can therefore be used as a voltage stabilising device. If the voltage falls below the breakdown value, the diode will recover provided the maximum permissible junction temperature has not been exceeded.

In diodes manufactured from semiconductor material with high doping levels, the depletion layer formed when the pn junction is reverse-biased is very narrow. Even with a small voltage across the diode, since practically all the voltage is placed across the junction, the depletion layer is subjected to a very high electric field. Breakdown can occur through internal field emission, the covalent bonds between electrons and atoms being broken to provide a large number of charge carriers. This type of breakdown is called zener breakdown.

Reference to avalanche breakdown has already been made in the description of avalanche diodes. In diodes manufactured from materials with lower doping levels, avalanche breakdown will occur at a lower voltage than that required for zener breakdown. As an approximate guide for practical zener diodes, those with breakdown voltages below 2.5 V use zener breakdown, while those with breakdown voltages above 7 V use avalanche breakdown. For voltages between 2.5 V and 7 V, both effects are used.

It is possible, therefore, to manufacture a range of zener diodes with different breakdown voltages. The value of the breakdown voltage is determined principally by the doping level of the semiconductor material from which the diode is made. The range of voltages obtained is, typically, from 3 V to 75 V. Diodes operating at breakdown voltages below 10 V usually have the junctions manufactured by alloying; those operating at higher voltages have diffused junctions.

In practice, the reverse voltage/current characteristic may differ from that shown in *Figure 21.7*. For diodes with low breakdown voltages, the characteristic does not have such a sharply-defined knee and there is a more gradual transition into the vertical part. Also, the voltage after breakdown, the zener voltage, may not be constant but may increase slightly with the increase in current. The change in voltage is slight, but it may be significant in certain applications.

The basic characteristic of a zener diode is the dynamic characteristic of zener voltage plotted against pulsed zener current. Current pulses are used so that the junction temperature is not changed and heating effects are therefore eliminated. The slope of this characteristic is the dynamic resistance, which is specified in published data at different zener currents to allow for changes in slope. The breakdown voltage and zener voltage are affected slightly by temperature, and so a temperature coefficient is specified (also at different zener currents). This coefficient is negative in diodes with breakdown voltages below 5 V, and positive in diodes operating at higher voltages.

The static characteristic of a zener diode shows the variation of zener voltage with d.c. zener current. This characteristic takes into account the change in junction temperature through internal dissipation. The change in zener voltage with direct current depends on the dynamic resistance, the rise in junction temperature through internal dissipation and the thermal resistance between junction and ambient, and the temperature coeffcient. The static characteristic can be used to determine the operating point at any zener current.

Because of tolerances during the manufacturing process, there is also a tolerance on the value of breakdown voltage and zener voltage. This tolerance is generally ±5%, but special selections can be made to obtain zener voltages with tighter tolerances at specified currents.

When a zener diode is conducting, there is a steady-state dissipation caused by the zener current. In addition, the diode may be subjected to pulses through transients on the voltage supply rail it is stabilising. The sum of the steady-state and pulse power must not exceed the total dissipation permissible for the diode. However, the pulse power can be relatively high if the pulse duration is short.

A voltage regulator diode used in a simple stabilising circuit is shown in *Figure 21.10*. The diode is initially biased beyond the

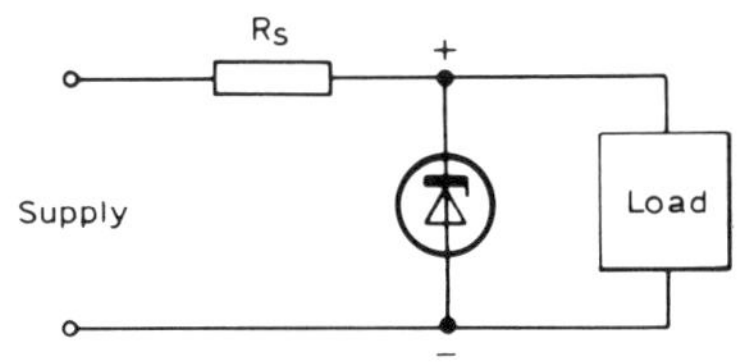

Figure 21.10 Simple stabiliser circuit using voltage regulator diode

breakdown voltage, and then clamps the supply rail at a voltage near the breakdown value. The series resistor R_s should have as high a value as possible to give good stability, but the limit values are determined by the diode. The upper limit of R_s is set by the need to maintain the voltage across the diode greater than the breakdown value when the load is drawing maximum current. The lower limit is set by the need to keep the current through the diode below the value for maximum permissible dissipation, when the load current is a minimum.

A similar circuit is used with surge suppressor diodes to protect circuits against the effects of voltage transients. In this application, however, the diode does not conduct unless a surge is present. The diode then acts as a shunt element to absorb the energy of the surge. Provided the maximum junction temperature is not exceeded, the diode will recover when the surge energy has been dissipated.

The voltage approximately 10% less than the breakdown value is called the stand-off voltage. For a given supply, the suppressor diode is chosen so that the stand-off voltage is equal to or greater than the maximum supply voltage under normal operating conditions. Only the reverse leakage current flows through the diode, and so the power dissipated will be negligibly small. After breakdown by the voltage transient the diode reverse current rises rapidly, the maximum permissible value being the non-repetitive reverse current I_{RSM} which is specified for a particular pulse duration. The voltage also increases to a value called the clamping voltage $V_{(CL)R}$ which is generally less than twice the stand-off voltage. The dissipation at this point is the product of $V_{(CL)R}$ and I_{RSM}, and the diode must be chosen to withstand this dissipation.

The range of stand-off voltages is, typically, 5.6 V to 62 V. Pulse powers up to 2.5 kW with a duration of 1 m s can be dissipated by the diode without requiring a heatsink. The power rating, however, is a non-repetitive rating. In applications where the voltage transients are repetitive at a definite frequency, for example, switching transients occurring every cycle, a heatsink must be used. The published data give information on the repetitive peak reverse power that can be dissipated for various surge durations and frequencies, and from this it can be verified that the diode is suitable for repetitive operation. The thermal resistance for the heatsink can then be calculated, and a suitable heatsink chosen.

It was stated earlier that the temperature coefficient of zener diodes with breakdown voltages below 5 V was negative, while that of diodes operating at higher voltages was positive. Diodes with temperature coefficients of opposite polarity can therefore be matched to provide a combination with a very low overall coefficient. Such combinations are used to form voltage reference diodes. These provide a simple means of obtaining a voltage standard for portable measuring equipment.

In a voltage reference diode, two zener diodes are connected back-to-back as shown in *Figure 21.11*. The diodes are manufactured on separate chips but are encapsulated in the same envelope. One diode with a zener voltage of about 5.5 V and a positive temperature coefficient operates in the reverse direction, diode D_1. The other diode, D_2, operates in the forward direction and has a negative temperature coefficient. As the voltage across the terminals is increased, D_1 breaks down. Because D_2 is conducting in the forward direction, the voltage across it is small. The main component of the reference voltage is therefore the zener voltage of D_1 which varies with current. The current through the diodes must therefore be closely controlled. To maintain a low and constant temperature gradient along the length of the reference diode, it is necessary to mount it in a temperature-controlled environment. A hole in a block of copper or aluminium is generally suitable, although for greater stability in the more accurate applications a temperature-stabilising circuit may be necessary. This form of construction for a voltage reference diode provides a voltage standard between 5.8 V and 6.8 V.

Figure 21.11 Circuit symbol for voltage reference diode

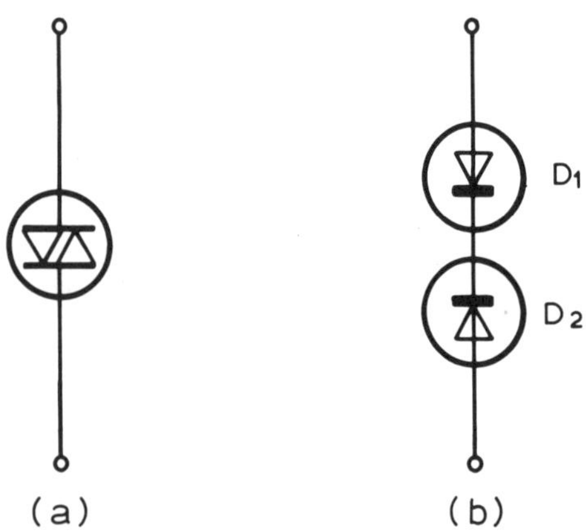

Figure 21.12 Diac: (a) circuit symbol; (b) equivalent circuit

21.5 Diac

The diac is a three-layer device. The circuit symbol for a diac is shown in *Figure 21.12(a)*. It can be considered as two opposed diodes connected in series, as shown in *Figure 21.12(b)*. If a voltage is applied across the diac, say in the forward direction of diode D_1, the current through the devices will be the small reverse leakage current of diode D_2. As the voltage is increased, there will be only a slight increase in current until the breakover voltage V_{BO} for diode D_2 is reached. The current increases rapidly and the voltage falls to the working value V_W. In this condition, the diac is conducting. When the applied voltage is removed, the diac returns to its non-conducting state. If the voltage across the diac is in the forward direction of diode D_2, a similar action occurs with the current direction reversed. The voltage/current characteristic of the diac is shown in *Figure 21.13*.

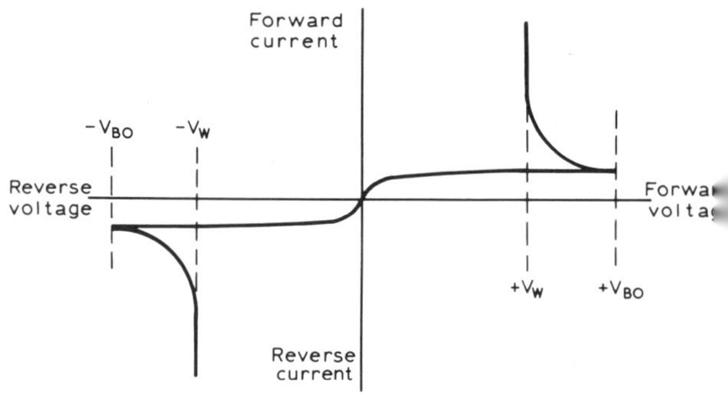

Figure 21.13 Voltage/current characteristic of diac

21.6 Symbols for main electrical parameters of semiconductor diodes

C_d	diode capacitance (reverse bias)
I_F	continuous (d.c.) forward current
$I_{F(AV)}$	average forward current
I_{FRM}	repetitive peak forward current
I_{FSM}	non-repetitive (surge) peak forward current
I_O	average output current
I_{OSM}	non-repetitive (surge) output current
I_R	continuous reverse leakage current
I_{RRM}	repetitive peak reverse current
I_{RSM}	non-repetitive peak reverse current
I_Z	voltage regulator (zener) diode continuous (d.c.) operating current
I_{ZM}	voltage regulator (zener) diode peak current
P_{tot}	total power dissipated within device
Q_s	recovered (stored) charge
$R_{th(h)}$	thermal resistance of heatsink
$R_{th(i)}$	contact thermal resistance
$R_{th(j\text{-}amb)}$	thermal resistance, junction to ambient
$R_{th(j\text{-}mb)}$	thermal resistance, junction to mounting base
r_Z	voltage regulator (zener) diode differential (dynamic) resistance
T_{amb}	ambient temperature
T_{jmax}	maximum permissible junction temperature
t_p	pulse duration
t_{rr}	reverse recovery time
$V_{(CL)R}$	surge suppressor diode clamping voltage

V_F	continuous (d.c.) forward voltage
V_{IRM}	repetitive peak input voltage
$V_{I(RMS)}$	r.m.s. input voltage
V_{ISM}	non-repetitive (surge) input voltage
V_{IWM}	crest working input voltage
V_O	average output voltage
V_R	d.c. reverse voltage
V_{RM}	peak reverse voltage
V_{RRM}	repetitive peak reverse voltage
V_{RSM}	non-repetitive (surge) peak reverse voltage
V_{RWM}	crest working reverse voltage
V_Z	voltage regulator (zener) diode operating voltage

22 Transistors

M J Rose BSc (Eng)
Production Officer,
Marketing Communications Group,
Mullard Ltd

Contents

22.1 Point-contact transistor

The point-contact transistor was invented in 1948 by John Bardeen and Walter H. Brattain, members of a team at Bell Telephone Laboratories working under the direction of William Schockley. 'Invented' may not be the right word: 'discovered' may be more correct, and Shockley himself has used the words 'creative failure' in connection with the work that led to the device.

The work at Bell Telephone Laboratories immediately after World War II was directed towards producing a solid-state amplifier using semiconductor materials. Shockley had predicted from theoretical work that the resistance of a piece of semiconductor material should change when it was subjected to an electric field normal to the current flow. Experiments, however, did not confirm this. The prediction was, in fact, true but it was not until 1963 that it was verified by the invention of the insulated-gate field-effect transistor. The failure to observe the expected change in resistance was ascribed by Bardeen in 1947 to some surface states neutralising the effect of the applied electric field. Further experiments were devised to investigate the surface states of a semiconductor material.

Perhaps the most important function of the point-contact transistor was its demonstration that a practical amplifying device could be made using the flow of charges through a semiconductor material. It provided a verification of the theoretical work on solid-state physics that had preceded it. Much of the material technology that was to be further developed over the next two decades to enable the many new types of solid-state device to be manufactured was developed from work done for the point-contact transistor on the purification of germanium, the growing of single crystals, doping, and preparing small slices of semiconductor material. In many different ways, the point-contact transistor prepared the way for the success of the junction transistor.

22.2 Junction transistor

The junction transistor was invented by William Shockley of Bell Telephone Laboratories in 1949. The original transistor consisted of a piece of a single crystal of germanium containing two n-type regions separated by a p-type region. The two n-type regions were called the emitter and collector, and the p-type region was called the base. The two pn junctions were biased in the same way as the emitter and collector wires in the point-contact transistor had been.

22.2.1 Transistor action

A schematic cross-section of an npn junction transistor is shown in *Figure 22.1*. The n-type emitter and collector regions are more heavily doped than the p-type base region. The base region is also very narrow compared with the other two regions. The biasing of

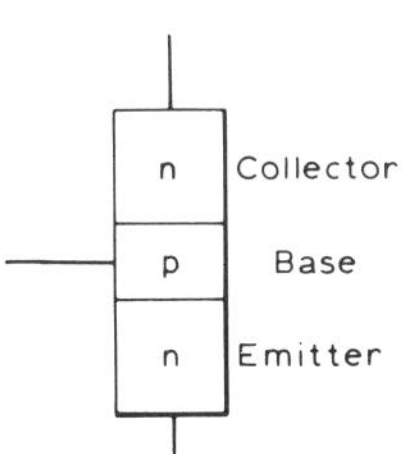

Figure 22.1 Structure of npn transistor

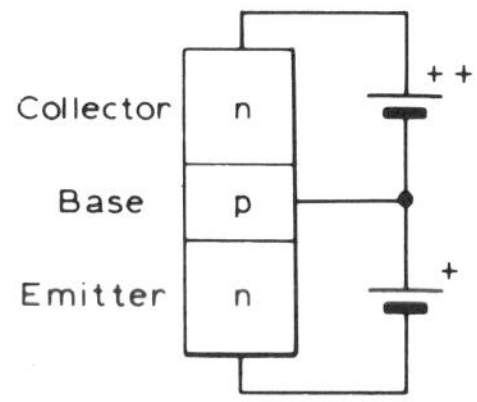

Figure 22.2 Biasing of npn transistor

the transistor is shown in *Figure 22.2*. The base–emitter junction is forward-biased, while the collector–base junction is reverse-biased. The collector is considerably more positive than the base.

Because the base–emitter junction is forward-biased, majority carriers flow across the junction. As the base is only lightly doped, few holes cross into the emitter from the base. Most of the current is therefore provided by electrons flowing from the emitter into the base. The collector–base junction is reverse-biased, so only a small minority carrier current flows: a small number of holes from the collector into the base, and a smaller number of electrons from the base into the collector. However, because the base is narrow and the collector considerably more positive than the base, the large number of electrons crossing the base–emitter junction diffuse across the base to the edge of the collector depletion layer from which they are swept into the collector. A few of these electrons combine with holes in the base, but most cross to the collector to give rise to the collector current. The combination of electrons and holes in the base, and the holes crossing from the base to the emitter, produce the small base current.

The structure described is called an npn transistor from the arrangement of the p and n regions. A second arrangement is possible, the pnp transistor, as shown in *Figure 22.3*. This type of transistor has p-type emitter and collector regions separated by a narrow n-type base region. Again, the base–emitter junction is forward-biased, and the collector–base junction reverse-biased.

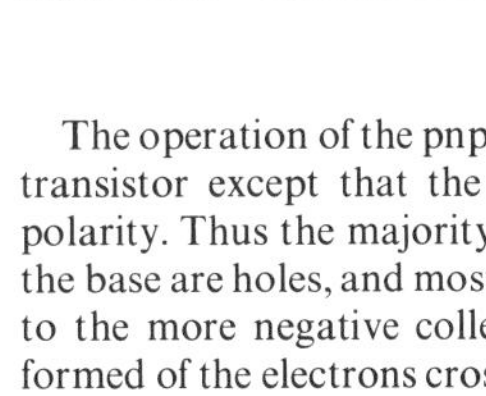

Figure 22.3 Structure and biasing of pnp transistor

The operation of the pnp transistor is similar to that of the npn transistor except that the charge carriers are of the opposite polarity. Thus the majority carriers crossing from the emitter to the base are holes, and most of these cross the narrow base region to the more negative collector. Again, a small base current is formed of the electrons crossing from the base to the emitter, and of the combination of holes and electrons in the base region.

Circuit symbols for the npn and pnp transistors are shown in *Figure 22.4*. The direction of the arrow on the emitter shows the direction of conventional current flow. Also shown are the polarities of the bias supplies V_{BE} and V_{CE} for the two types of transistor, and the directions of the currents.

Thus one terminal of the device must be common to both input and output circuits. It is therefore possible to connect the transistor in three configurations, as shown in *Figure 22.5*. These configurations are called: common emitter, (*a*), because the

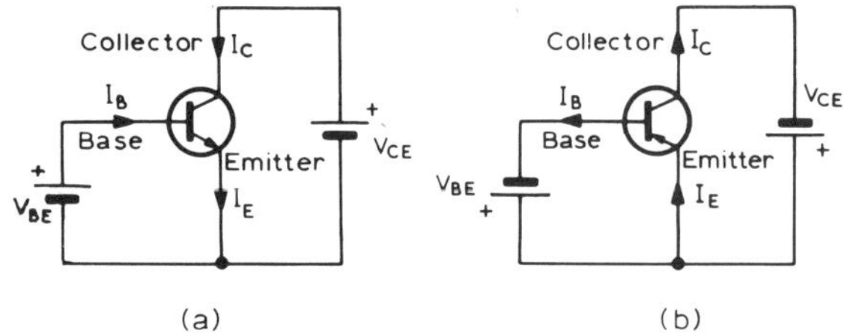

Figure 22.4 Circuit symbols, bias supplies, and current flow in: (a) npn transistor; (b) pnp transistor

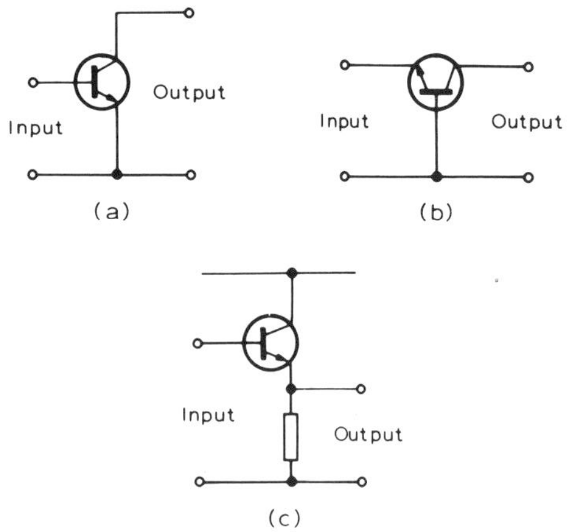

Figure 22.5 Transistor configurations: (a) common emitter; (b) common base; (c) common collector

emitter is common to both input and output circuits; common base, (*b*); and common collector, (*c*). In all the configurations, the currents in the transistor are related by the equation

$$I_E = I_B + I_C \qquad (22.1)$$

In the common-emitter configuration, the input is the base–emitter circuit, and the output is the collector–emitter circuit. The input current is the base current I_B and the output current is the collector current I_C. The ratio of the two currents I_C/I_B is the current gain β, or more precisely, the static common-emitter forward current transfer ratio h_{FE}. Because the base current is small in comparison with the collector current, the value of h_{FE} is large. Thus a small change in the base current produces a large change in the collector current; in other words, the transistor forms a current amplifier.

As the base–emitter junction is forward-biased, a small change in the base-to-emitter voltage V_{BE} will produce a large change in the base current, and a larger change in the collector current. The collector–base junction is reverse-biased and so has a high impedance. A high-value load resistance can therefore be connected in the collector circuit, and the collector current will develop a high voltage across it. Thus the small change in the base voltage produces a larger change in collector voltage, and a high voltage gain is obtained. A power gain between input and output circuits is therefore obtained.

In the common-base configuration, the input is the base–emitter circuit and the output the collector–base circuit. The input current is the emitter current I_E, and the output current is the collector current I_C. The current gain α, or more precisely the static common-base forward current transfer ratio h_{FB}, is I_C/I_E and therefore a little less than unity. However, because of the low impedance of the forward-biased base–emitter junction and the high impedance of the reverse-biased collector–base junction, a high voltage gain is obtained. Thus a power gain between input and output is obtained.

The common-collector configuration is also called the emitter follower by analogy to the thermionic valve cathode follower circuit. The characteristics of this configuration differ somewhat from the other two. The input current is the base current I_B, and the output current is the emitter current I_E. The current gain, the static common-collector forward current transfer ratio h_{FC}, is high but the voltage gain is a little less than unity. The greatest difference between this configuration and the other two, and the feature that provides its main application in practical circuits, is that the emitter follower has a high input impedance. Thus this configuration will respond to voltage input signals.

The main characteristics of the three circuit configurations are summarised below.

	Common emitter	*Common base*	*Common collector*
Current gain	high	nearly 1	high
Voltage gain	high	high	nearly 1
Power gain	high	medium	low
Input impedance	medium	low	high
Output impedance	medium	high	low

It should be noted that the common terminal is common to a.c. and not necessarily to d.c. The term 'grounded' may sometimes be encountered instead of common. 'Grounded' is deprecated as it implies the common terminal is connected to earth, and this may not necessarily be so.

The previous discussion has neglected the leakage currents in the transistor. A small current flows across the reverse-biased collector–base junction caused by minority carriers. For small junction transistors at room temperature, this current is a few nanoamps in silicon devices and a few microamps in germanium devices. It is therefore considerably smaller than the normal collector current.

In the common-base configuration, the leakage current in the output circuit is I_{CBO}. (The value of I_{CBO} is quoted in published data with the emitter open-circuit.) In the common-emitter configuration, however, an increased leakage current is present through transistor action on I_{CBO} causing a current $h_{FE} \times I_{CBO}$ to flow in the emitter circuit (the base being open-circuit). Thus the total leakage current in the output circuit for the common-emitter configuration, I_{CEO}, is given by:

$$I_{CEO} = I_{CBO}(1 + h_{FE}) \qquad (22.2)$$

and this current is considerably higher than that for the common-base configuration. The leakage current increases with temperature, and is undesirable in amplifier circuits as it limits the amount of useful power that can be derived from the transistor. One of the advantages of silicon transistors that enabled them to replace germanium transistors for many applications is a lower leakage current.

The currents I_{CBO} and I_{CEO} are described in published data as cut-off currents, being the currents flowing when the transistor is cut off with the emitter or base open-circuit.

22.2.2 Characteristics and ratings

Because the transistor is a three-terminal device, there are a considerable number of characteristics relating the various currents and voltages. The data published by the transistor manufacturer contain a large number of characteristics, and the most common ones will be described. The shape of these characteristics depends to some extent on the configuration in which the transistor is operated, and the polarities of the currents and voltages depend on the type, npn or pnp. In addition, the manufacturer lists certain current and voltage values which must not be exceeded if the transistor is to remain undamaged.

The input characteristic relates the input current to the input voltage with the output voltage remaining constant. For the common-emitter configuration, the input current is the base current I_B, the input voltage is the base-to-emitter voltage V_{BE}, and the output voltage to be held constant is the collector-to-emitter voltage V_{CE}. The form of the input characteristic, as shown in *Figure 22.6*, is that of a forward-biased diode as would be expected. The reciprocal of the slope at any point gives the input impedance of the transistor when operating at that point. The voltage at which significant current starts to flow depends on the material from which the transistor is manufactured, approximately 200 mV for germanium devices and 600 mV for silicon devices.

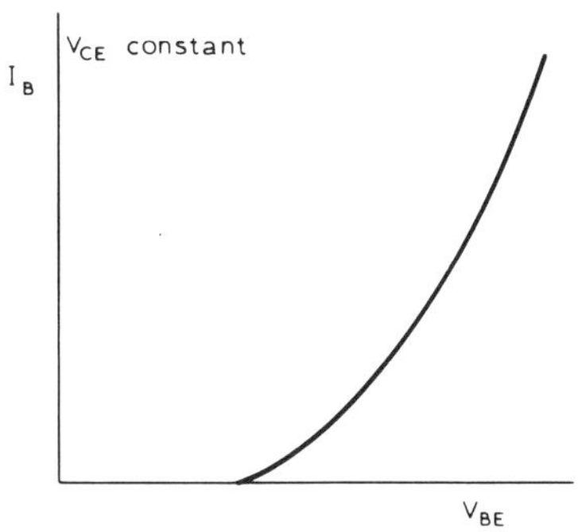

Figure 22.6 Input characteristic of transistor (common emitter)

A similar characteristic is obtained for the common-base configuration. The input current is the emitter current I_E, the input voltage is again the base-to-emitter voltage V_{BE}, and the output voltage to be maintained constant is now the collector-to-base voltage V_{CB}. The input impedance is lower for this configuration.

The output characteristic for the common-emitter configuration relates the collector current I_C to the collector-to-emitter voltage V_{CE}. Two forms of the characteristic are usually given in published data: with base current I_B as parameter, and with base-to-emitter voltage V_{BE} as parameter. Both forms of the characteristic have a similar shape; a sharply-rising curve with a knee, followed by a straight line tending to become nearly parallel to the V_{CE} axis. For low values of V_{CE} below the knee of the characteristic, the collector is not efficient at collecting the charge carriers crossing from the emitter through the base. When V_{CE} exceeds the knee value, the number of carriers collected rises to a 'saturation' value and is nearly independent of the collector-to-emitter voltage. There is, in practice, a slight increase in collector current as V_{CE} increases which is shown by the characteristic not being truly parallel to the V_{CE} axis.

The output characteristic with base current as parameter is shown in *Figure 22.7*. The value of collector current depends on the value of base current, as would be expected through transistor action. However, a small collector current flows even with zero base current. This is the leakage current I_{CEO} already discussed. The output characteristic with base-to-emitter voltage as parameter is shown in *Figure 22.8*. In both forms of the characteristic, the reciprocal of the slope at any point gives the output impedance of the transistor operating at that point. This is high, as would be expected, since the collector–base junction is reverse-biased.

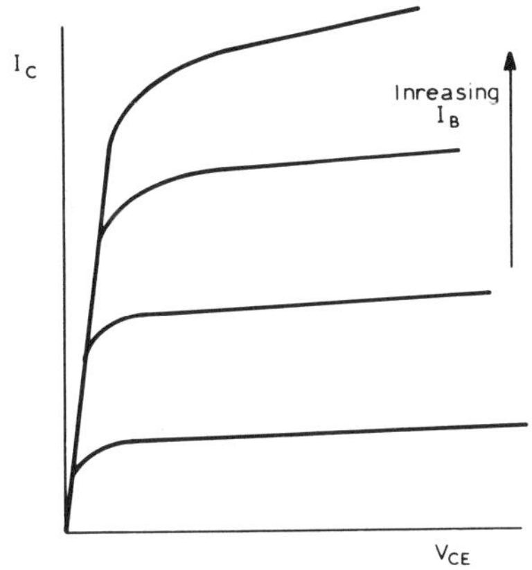

Figure 22.7 Output characteristic of transistor (common emitter) with base current as parameter

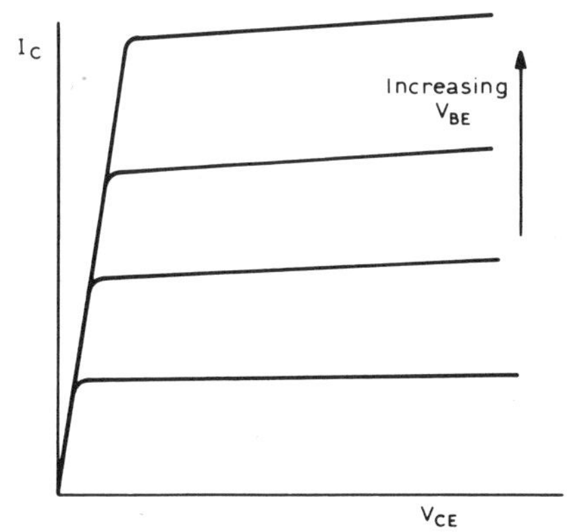

Figure 22.8 Output characteristic of transistor (common emitter) with base-to-emitter voltage as parameter

Similar curves are obtained for the output characteristic in the common-base configuration. The emitter current is the parameter for this configuration, and the value of output current is slightly less than that of the emitter current since the current gain is just below unity. The curves are more nearly parallel to the V_{CB} axis showing the higher output impedance obtained with this configuration.

The mutual characteristic, shown in *Figure 22.9*, relates the collector current I_C to the base-to-emitter voltage V_{BE} with the

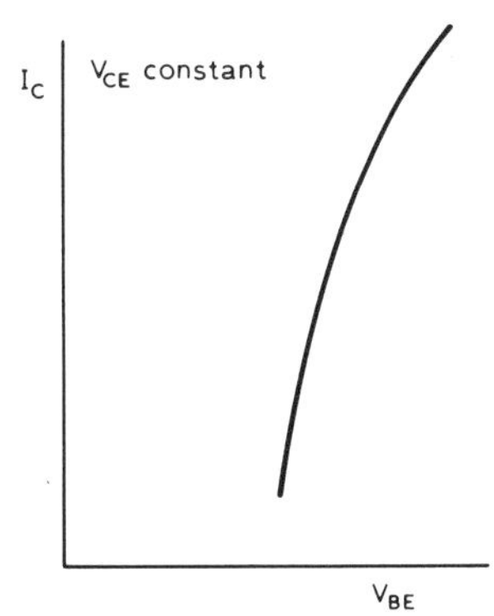

Figure 22.9 Mutual characteristic of transistor

collector-to-emitter voltage constant. (The name mutual characteristic is derived by analogy with thermionic valves where the relationship between anode current and grid voltage is called the mutual conductance.) The characteristic is not linear because of the non-linear relationship between V_{BE} and I_B shown in the input characteristic.

The transfer characteristic relates the input and output currents at constant output voltage. For the common-emitter configuration, the input current is the base current I_B, the output current is the collector current I_C, and the output voltage to be maintained constant is the collector-to-emitter voltage V_{CE}. The characteristic is plotted on logarithmic scales in published data, as shown in *Figure 22.10*, and is a straight line. The slope of the

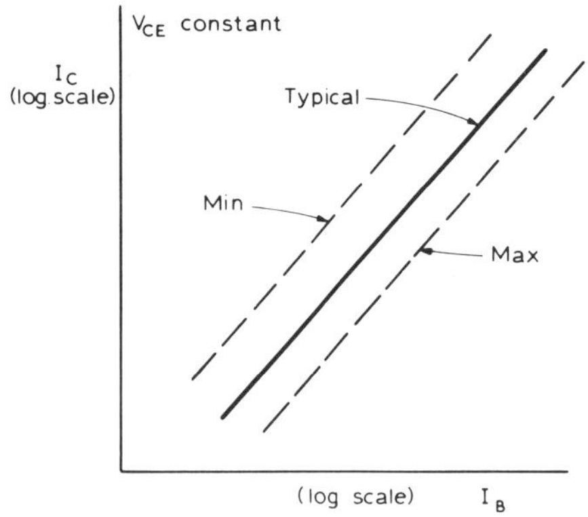

Figure 22.10 Transfer characteristic of transistor (common emitter)

line at any point gives the value of static forward current transfer ratio h_{FE} for that value of collector current. The value of h_{FE} varies with collector current, and so any value quoted should always be qualified with the relevant collector current and voltage. Curves of h_{FE} plotted against collector current are sometimes given in published data. Because of tolerances in the manufacturing processes, there will be a tolerance on the value of h_{FE}. For this reason, three figures are given in published data: a maximum, a minimum, and a typical value. The maximum and minimum values of collector current corresponding to this spread in the value of h_{FE} are shown by the curves in broken lines in *Figure 22.10*.

The transfer characteristic for the common-base configuration relates emitter current I_E to collector current I_C with the collector-to-base voltage V_{CB} maintained constant. The slope of the characteristic at any point is the value of h_{FB} at that collector current and voltage.

Transistor characteristics change with temperature, and the characteristics just discussed are therefore plotted at a constant temperature, usually 25°C. Curves of the variation of base current and h_{FE} with temperature are sometimes given in published data. The maximum operating temperature for the transistor is set by the junction temperature at which the lightly-doped base becomes intrinsic and transistor action cannot take place.

The maximum power dissipation of a transistor P_{tot} is determined by the maximum permissible junction temperature $T_{j,max}$ and the ambient temperature in which the transistor operates T_{amb}. Thus:

$$P_{tot} = \frac{T_{j,max} - T_{amb}}{\Sigma R_{th}} \qquad (22.3)$$

where ΣR_{th} is the total thermal resistance.

The quantity *thermal resistance* represents the opposition to the flow of heat from the transistor. It is expressed as change of temperature with power, and the usual units are °C/W.

For low-power and medium-power transistors, the thermal resistance is specified as that between the junction and ambient, $R_{th(j-amb)}$, or that between the junction and case, $R_{th(j-case)}$. If a cooling clip is used to increase the radiating area of the transistor, a thermal resistance is given in the data for the clip, $R_{th(case-amb)}$. Thus the term ΣR_{th} in equation (22.3) is either $R_{th(j-amb)}$ or $(R_{th(j-case)} + R_{th(case-amb)})$. The value of $T_{j,max}$ and P_{tot} are given in the published data, the value of T_{amb} is known or a limit known, and ΣR_{th} is known. Thus equation (22.3) can be used to ensure that the cooling of the transistor (as shown by ΣR_{th}) is adequate for the total dissipation, or to calculate the maximum ambient temperature in which the transistor can operate.

For high-power transistors, the thermal resistance between the junction and the mounting base, $R_{th(j-mb)}$, is given. A heatsink to increase considerably the cooling area of the transistor is generally required, and a lead washer used to ensure good thermal contact between the transistor and heatsink. If the transistor case has to be insulated from the heatsink, a mica washer is used as well. For these washers, a thermal resistance $R_{th(mb-h)}$ is given. The heatsink itself has a thermal resistance $R_{th(h)}$. Thus a chain of thermal resistance exists between the junction of the transistor and ambient, as shown for a diode in *Figure 21.8*. From a given total dissipation and maximum junction temperature, equation (22.3) can be used to calculate the required ΣR_{th} for a particular ambient temperature, and from ΣR_{th} the thermal resistance of the heatsink calculated. A suitable heatsink can then be chosen.

The value of P_{tot} calculated from thermal considerations must be equal to the total electrical power associated with the transistor. Thus:

$$P_{tot} = V_{CE}I_C + V_{BE}I_B \qquad (22.4)$$

for the common-emitter configuration. Since $V_{CE} > V_{BE}$, and $I_C \gg I_B$, equation (22.4) can be simplified to:

$$P_{tot} = V_{CE}I_C \qquad (22.5)$$

The maximum power dissipation can be plotted on the output characteristic, as shown in *Figure 22.11*. Also shown on *Figure*

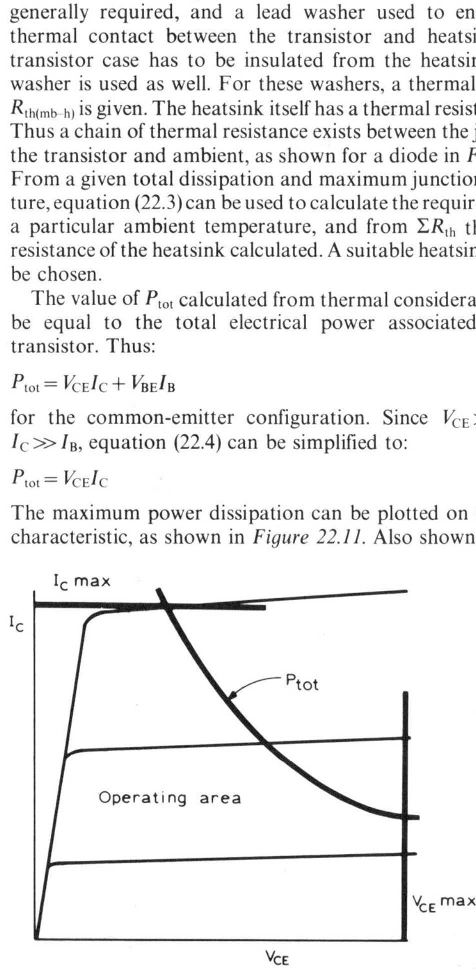

Figure 22.11 Operating area of transistor: limits of collector voltage and current and power dissipation plotted on output characteristic (common emitter)

22.11 are the two other limits for safe operation of the transistor: the maximum collector current I_C max, and the maximum collector-to-emitter voltage V_{CE} max. These three quantities define the permissible operating area, and the static working point of the transistor must be chosen to lie within this area.

Because of the temperature-dependence of transistor characteristics, the power dissipation must be limited at high temperatures. The form of the curve for maximum permissible power dissipation with temperature given in published data is shown in *Figure 22.12*. Up to a certain temperature, the value depending on the type of transistor, the power dissipation can be maintained at the maximum value. Above this temperature, however, the power must be limited (derated) to maintain safe operation.

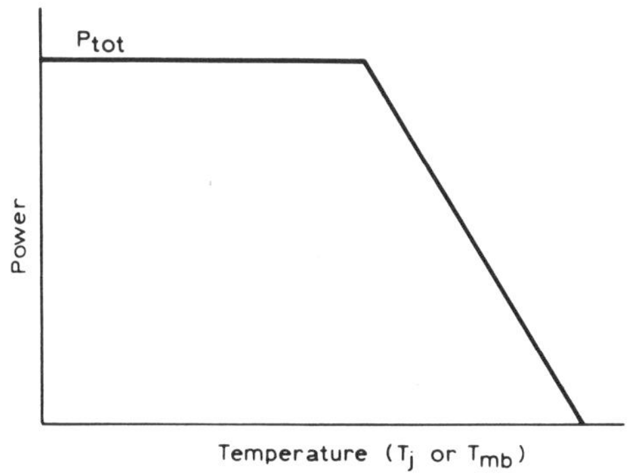

Figure 22.12 Variation of maximum permissible power dissipation with temperature

The limit to collector voltage V_{CE} max shown in *Figure 22.11* is set by the reverse voltage across the collector–base junction. If this voltage is too high, avalanche breakdown occurs, as shown in *Figure 22.13*. The breakdown can occur with either forward or reverse base drive, and the effect of base drive on the breakdown is shown in *Figure 22.13*. The 'limit' curve occurs when the emitter

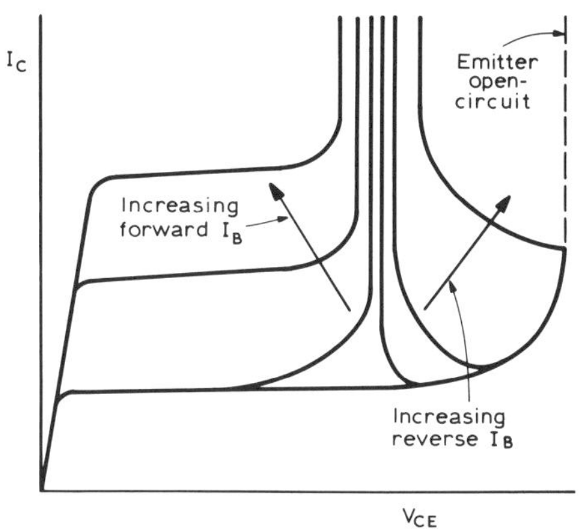

Figure 22.13 Output characteristic of transistor (common emitter) expanded to show avalanche breakdown region

is open-circuit. The maximum collector voltage at which the transistor can operate is therefore chosen to prevent this type of breakdown.

With power transistors, where the area of the junctions is made large to enable large currents to be conducted, a second form of breakdown can occur. Under certain voltage and current conditions, the emitter current tends to concentrate in one area of the emitter–base junction. This effect is called current contraction, and it can cause an increase in the local junction temperature sufficient to cause a 'hot spot', leading to local breakdown of the junction and the consequent destruction of the transistor. This form of breakdown is called *second breakdown*, and can occur with both forward and reverse base drive. With forward base drive, the breakdown usually occurs at the periphery of the emitter; with reverse base drive where the transistor is being switched off after conducting, the emitter periphery is switched off first and the current is concentrated at the centre of the emitter so that breakdown occurs there. The effect of second breakdown is to add a further limit to the operating area of the transistor. The three limits shown on the output characteristic in *Figure 22.11* are shown by the full lines in *Figure 22.14*, which is plotted on logarithmic scales. The maximum collector current and voltage,

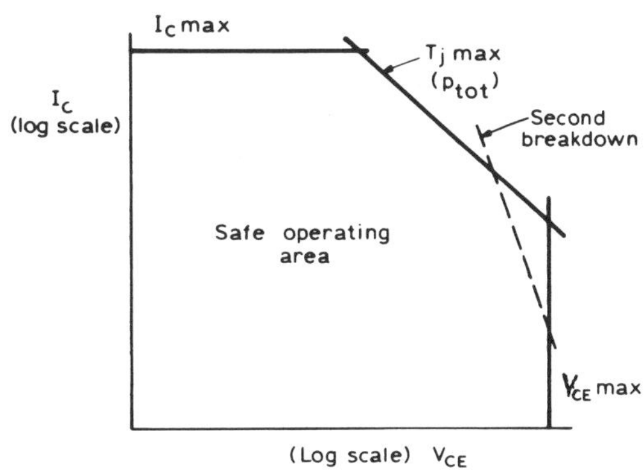

Figure 22.14 Safe operation area of power transistor showing limit imposed by second breakdown

I_C max and V_{CE} max, define two of the boundaries as before. The thermal limit, $T_{j,max}$ which is equivalent to P_{tot}, is a straight line intersecting I_C max and V_{CE} max. The fourth boundary, shown in broken lines, is the limit set by second breakdown. This safe operating area can be extended for pulse operation. Families of curves for various duty cycles and pulse durations are published in the data for power transistors.

When the transistor is used as a switch, it is usually operated with a large base drive to ensure that it is reliably switched on or off. In the on state when a large collector current flows, a high-value collector load resistor is used so that nearly all the supply voltage is dropped across this resistor. The voltage across the transistor, V_{CE} in the common-emitter configuration, is then less than the base-to-emitter voltage V_{BE}. Under these conditions, the collector–base junction becomes forward-biased instead of reverse-biased, and the transistor is said to be bottomed or driven into saturation. The saturation voltage, $V_{CE(sat)}$, varies with collector current and temperature, and curves of $V_{CE(sat)}$ are given in published data.

When the collector–base junction becomes forward-biased, carriers are injected into the more lightly doped side of the junction from the more heavily doped side. (The relative doping levels of the two sides of the junction depend on the type of transistor.) When the transistor is switched off, the collector current does not fall immediately but is delayed while these carriers are removed through the collector. This effect is known as carrier storage, and the delay is known as the storage time. (The effect is similar to minority carrier storage in reverse-biased semiconductor diodes.) Carrier storage can limit the switching frequency of a transistor, and information on switching times is given in published data.

The capacitance of the transistor junctions can affect the operation of the transistor, particularly at high frequencies.

Curves showing the variation of collector–base and emitter–base capacitance with collector-to-base and emitter-to-base voltages respectively for a particular frequency are given in published data.

The preceding discussion on transistor characteristics and ratings can be conveniently summarised in the form of the information presented on a typical data sheet. The information is divided into four groups: ratings, thermal characteristics, electrical characteristics, and characteristic curves.

The ratings are limiting values which must not be exceeded during operation. They are listed below:

- V_{CBO} max — maximum collector-to-base voltage with emitter open-circuit
- V_{CES} max — maximum collector-to-emitter voltage with base short-circuited to emitter
- V_{CEO} max — maximum collector-to-emitter voltage with base open-circuit
- I_C max — maximum continuous collector current
- I_{CM} max — maximum peak collector current
- I_{EM} max — maximum peak emitter current
- I_{BM} max — maximum peak base current
- P_{tot} max — maximum total power dissipation within device, specified at a temperature for which the quoted value is applicable
- T_{stg} — storage temperature
- $T_{j,max}$ — maximum permissible junction temperature

The thermal characteristics give the thermal resistance for the transistor in one of the following forms.

- $R_{th(j-amb)}$ — thermal resistance, junction to ambient
- $R_{th(j-case)}$ — thermal resistance, junction to case
- $R_{th(j-mb)}$ — thermal resistance, junction to mounting base (to allow for heatsink mounting)

The electrical characteristics give information on the static performance of the transistor under specified voltages and currents at a specified ambient temperature. This information supplements that from the characteristic curves. The curves and values given for a particular transistor depend on the type and its range of applications. Typical information could include: input and output characteristics, mutual characteristic, transfer characteristic, variation of forward current transfer ratio with collector current, variation of base current with temperature, variation of collector and emitter capacitance with voltage, variation of $V_{CE(sat)}$ with collector current and temperature, and safe operating areas with respect to voltage, current, and power ratings.

In addition to the static characteristics, the published data include *dynamic* characteristics as discussed in the following section.

22.2.3 a.c. characteristics and equivalent circuits

To enable the behaviour of a transistor under a.c. conditions to be analysed, an equivalent circuit is required. From the earliest days of transistors, much theoretical work has been done to devise an exact equivalent circuit that will reproduce accurately the performance of a transistor in practice. Some of the circuits that have been suggested are complex. Three of the simpler circuits used extensively are described as follows. All use constant voltage or current generators in conjunction with resistors and capacitors. They are applicable to small signals only; other equivalent circuits are needed for large-signal analysis.

The T equivalent circuit is shown in *Figure 22.15* for the common-emitter configuration. Similar circuits are applicable for the other two configurations. Three resistances are connected in a T configuration: r_b representing the resistance of the base

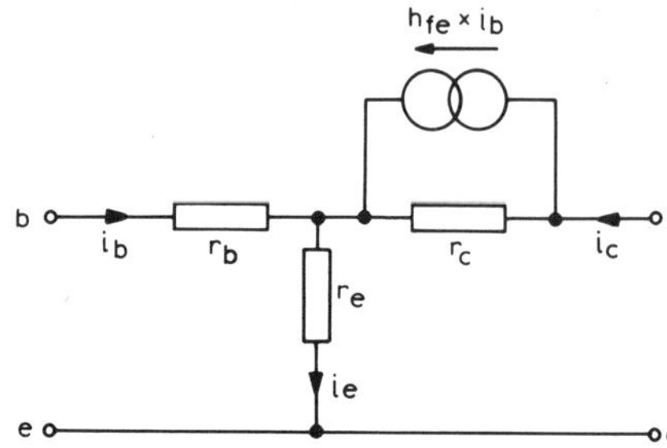

Figure 22.15 T equivalent circuit for common-emitter configuration

region, r_e representing the forward-biased emitter junction and r_c representing the reverse-biased collector junction. The current generator provides a current h_{fe} times the base current to represent the current gain of the transistor. Expressions relating the input and output voltages and currents can be derived in terms of the three resistances and the value of h_{fe}.

The T circuit of *Figure 22.15* applies to low and medium frequencies only. At high frequencies, the capacitances of the transistor junctions must be taken into account. Capacitances can be added to the T circuit but the expressions then become complicated. It is better to use another equivalent circuit.

The hybrid-π equivalent circuit shown in *Figure 22.16*, again for the common-emitter configuration, can be used for any frequency. The gain of the transistor is represented by the

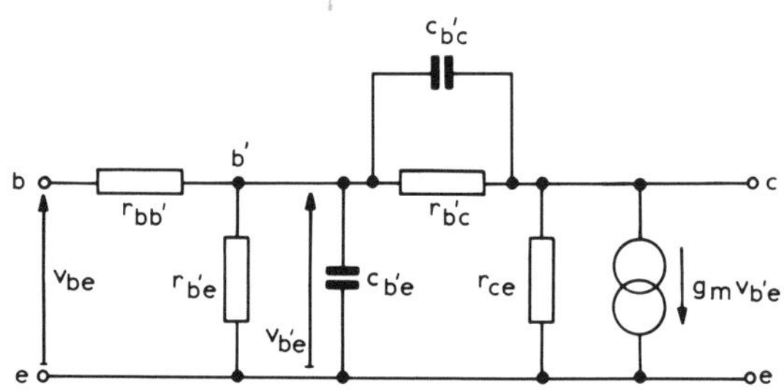

Figure 22.16 Hybrid-π equivalent circuit for common-emitter configuration

forward conductance g_m. In this circuit, a distinction is made between the base connection b and the base–emitter junction b′. Thus the resistance $r_{bb'}$ represents the bulk resistance of the base region. Resistance $r_{b'e}$ is the resistance of the forward-conducting base–emitter junction, while $c_{b'e}$ represents the emitter capacitance. Similarly, the collector–base junction is represented by resistance $r_{b'c}$ and capacitance $c_{b'c}$. Resistance r_{ce} is the output resistance of the transistor. Again, expressions relating input and output voltages and currents in terms of the resistances, capacitances, and g_m can be derived.

A third equivalent circuit, called the hybrid parameter network, is shown in *Figure 22.17*. This time the equivalent circuit does not represent the physical structure of the transistor, but four parameters can be derived which accurately describe the small-signal performance of the transistor at any frequency. These parameters are the *h* parameters quoted in published data.

The circuit contains both a current and a voltage generator. The current generator produces a current h_f times the input current, representing the forward gain of the transistor. The

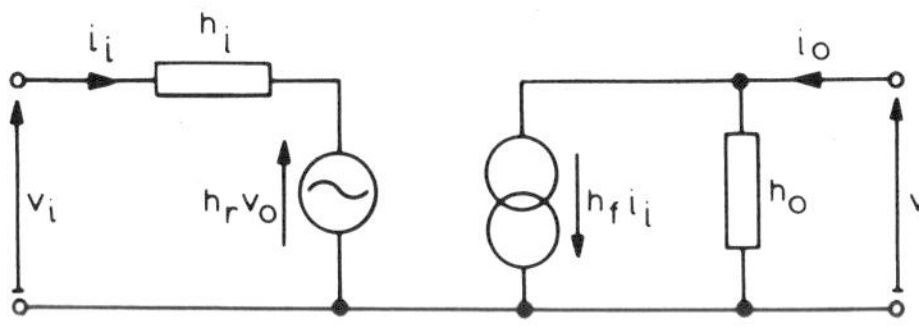

Figure 22.17 Hybrid parameter network

voltage generator produces a voltage h_r times the output voltage, representing the feedback effect of the transistor. The basic equations for the network are:

$$v_i = h_i i_i + h_r v_o \qquad (22.6)$$

$$i_o = h_f i_i + h_o v_o \qquad (22.7)$$

If the output terminals of the network are short-circuited, $v_o = 0$. Thus from equation (22.6)

$$v_i = h_i i_i \quad \text{or} \quad h_i = \frac{v_i}{i_i} \qquad (22.8)$$

that is, h_i is the input impedance of the transistor. Similarly from equation (22.7):

$$i_o = h_f i_i \quad \text{or} \quad h_f = \frac{i_o}{i_i} \qquad (22.9)$$

that is, h_f is the current gain or forward current transfer ratio of the transistor.

If the input terminals are open-circuit, $i_i = 0$. From equation (22.6):

$$h_r = \frac{v_i}{v_o} \qquad (22.10)$$

which is the voltage feedback, ratio. Similarly from equation (22.7):

$$h_o = \frac{i_o}{v_o} \qquad (22.11)$$

which is the output admittance of the transistor.

Subscripts are added to the h parameters to indicate the transistor configuration: b for common base, c for common collector and e for common emitter. Thus h_{fe} is the forward current transfer ratio in the common-emitter configuration. It should be noted that the value of the h parameter will be different for the three configurations, for example:

$$h_{fe} \neq h_{fc} \neq h_{fb}$$

Because the h parameters apply to small signals, they are the modulus values of the ratios of small changes. For the common-emitter configuration, therefore, the parameters are:

h_{ie} input impedance $|\partial V_{BE}/\partial I_B|_{V_{CE}}$

h_{oe} output admittance $|\partial I_C/\partial V_{CE}|_{I_B}$

h_{fe} forward current transfer ratio $|\partial I_C/\partial I_B|_{V_{CE}}$

h_{re} voltage feedback ratio $|\partial V_{BE}/\partial V_{CE}|_{I_B}$

The values will change with changes in I_B and V_{CE}. Curves of the h parameters are given in published data, plotted against collector current I_C for convenience in use. The h parameters are easily measured for actual transistors, and this gives them considerable advantage over other systems of parameters that have been suggested.

The performance of the transistor at high frequencies will be affected by the capacitance of the junctions. The value of h_{fe} will fall with increasing frequency, and so will limit the operation of the transistor. Various frequency values have been suggested as 'figures of merit' for the high-frequency performance. One figure suggested was f_{hfb} or f_{hfe} (originally f_α or f_β). This frequency was defined as that at which the current gain had fallen to 0.7 times its low-frequency value. Another frequency suggested for the common-emitter configuration was f_1, the frequency at which the current gain had fallen to unity. The figure generally quoted in published data is f_T. This is the common-emitter gain-bandwith product, the product of the frequency at which the value of h_{fe} falls off at a rate of 6 dB per octave and the value of h_{fe} at this frequency. For efficient high-frequency performance, the value of f_T should be as high as possible.

The value of f_T varies with collector current, and any value quoted should be qualified by the relevant collector current and voltage. Curves of the variation of f_T with collector current are given in published data.

22.2.4 Alloyed-junction transistors

The first transistors commercially available, in the early 1950s, were made by the germanium alloy-junction process. In this, the pn junctions were formed by alloying germanium with another metal. For pnp transistors, two pellets of indium, which would form the emitter and collector connections in the completed transistor, were placed on opposite sides of a slice of n-type monocrystalline germanium. The germanium slice, which formed the base, was approximately 50 μm thick. The whole assembly was heated in a jig to a temperature of about 500°C, and some of the indium dissolved into the germanium to form on cooling p-type germanium. In this way, two pn junctions were formed, as shown in the simplified cross-section of *Figure 22.18*. Connecting wires were attached to the indium pellets for the emitter and collector leads, and a nickel tab attached to the base to provide the base connection and to mount the transistor on a header. The transistor was then encapsulated in a glass or metal envelope. NPN transistors were made by this process using p-type germanium for the base, and lead-antimony pellets instead of indium.

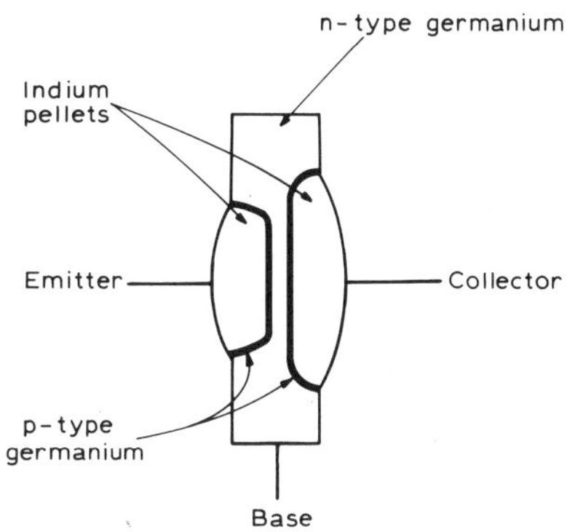

Figure 22.18 Simplified cross-section of pnp alloy-junction transistor

The performance of an alloy-junction transistor is limited by the base width. In particular, the cut-off frequency is inversely proportional to the base width. If the assembly is heated for a longer time to allow more indium to dissolve so that the pn junction penetrates further into the germanium slice, there is a risk of the junctions joining and no transistor action occurring. Because of the difficulty of controlling the base width to an accuracy greater than that of the thickness of the original slice, the base width was kept to approximately 10 μm. This gave the early alloy-junction transistor a cut-off frequency of about 1 MHz, although refinements to the manufacturing process later allowed this to be raised to 5 or 10 MHz.

During the 1950s, other types of transistor were developed to overcome the limitations of the alloy-junction type. One of these was the alloy-diffused transistor. In this, two pellets which formed the emitter and base were alloyed onto the same side of a slice of p-type germanium which formed the collector. The base pellet contained n-type impurities, while the emitter pellet contained both p-type and n-type impurities. On heating the assembly to 800°C alloying took place, but in addition diffusion of the n-type impurities into the p-type germanium ahead of the alloy front occurred. On cooling, a thin n-type layer linking both pellets was formed, with a p-type layer under the emitter pellet only. In this way, a pnp transistor with a narrow base layer was formed. Because of the narrow base layer, the cut-off frequency of this type of transistor was about 600 to 800 MHz. The alloy-diffused transistor was used extensively for high-frequency applications until it was superseded by the silicon planar epitaxial transistor.

Another type of transistor originally developed with germanium in 1956 was the mesa transistor. This transistor was developed to improve the switching characteristics and high-frequency performance of the then existing transistors. The principle was the etching of the edges of the transistor structure to decrease the junction areas and so reduce the capacitances. The resulting shape was a mesa or plateau, as shown in *Figure 22.19*.

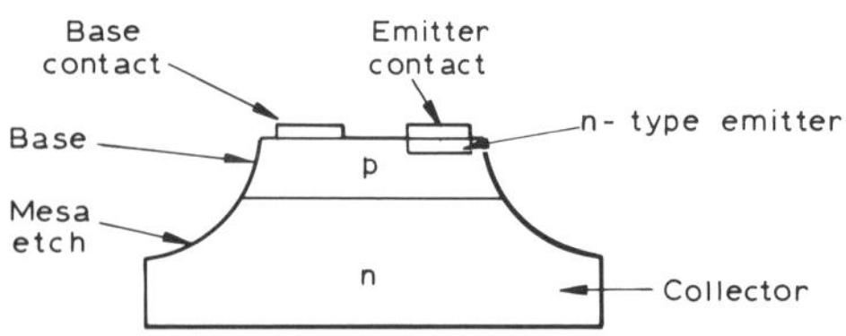

Figure 22.19 Cross-section of mesa transistor

The original germanium mesa transistors were able to switch waveforms with rise times of 1 μs, and had a high-frequency performance comparable to that of the alloy-diffused type.

The principle of mesa etching is used extensively today in many types of transitor. It allows the edges of the transistor to be controlled, in particular the edges of the collector–base junction to be clearly defined. Mesa etching is also a preliminary process in passivation, the 'sealing' of the edges of the transistor to prevent contamination of the junction edges and the consequent change in characteristics during service.

The introduction of silicon during the late 1950s gave the transistor manufacturer a new material with considerable advantages over germanium. In particular, silicon transistors will withstand a higher junction temperature and have lower leakage currents than germanium transistors.

The alloying techniques developed for germanium were applied to silicon. Silicon pnp transistors were manufactured from a slice of n-type silicon (which formed the base) into which emitter and collector pellets were alloyed in the same way as for germanium alloy-junction transistors. The material used for the alloying pellets was aluminium.

22.2.5 Diffused transistors

It was soon realised that the impurity regions in silicon transistors could be diffused into the slice from material deposited on the surface from a vapour, and that this process had considerable advantages over the alloying process. In particular, the greater control possible over the process made it easier to manufacture devices with characteristics superior to those of alloyed transistors.

A silicon npn transistor could be made by two diffusions into an n-type slice which would form the collector of the completed transistor. The first diffusion, forming the base, used p-type impurities such as boron or gallium, and covered the whole surface of the slice. The second diffusion formed the emitter, diffusing n-type impurities such as phosphorus or arsenic into the already diffused base region. Electrical connections to the base were made by alloying rectifying contacts through the emitter to the base. A refinement of this manufacturing process is used today for silicon high-power transistors.

The discovery that thermally-grown silicon oxide on the surface of the slice could form a barrier to diffusion, and so could be used to define the impurity regions, led to the breakthrough in transistor manufacture—the planar process.

22.2.6 Planar transistor

The planar process revolutionised transistor manufacture. For the first time in the manufacture of electronic devices, true mass-production techniques could be applied. The process allowed closer control over the geometry of the device, thus improving the performance. During the 1950s the transistor had first been a novelty, then regarded as an equivalent of the thermionic valve. With the introduction of the planar process in 1960, the transistor established itself as a device with a performance superior to that of the valve in 'general-purpose' electronics. Operation at frequencies up to the microwave region became possible, as did power transistors operating at radio frequencies. And it was the planar process that made possible the most important present-day semiconductor device—the integrated circuit.

The principle of the planar process is the diffusion of impurities into areas of a silicon slice defined by windows in a covering oxide layer. The various stages in the manufacture of a silicon planar npn transistor are shown in *Figure 22.20*.

A slice of n-type monocrystalline silicon (which will form the collector of the completed transistor) is heated to approximately 1100°C in a stream of wet oxygen. The temperature is controlled to better than ±1 deg C, and an even layer of silicon dioxide 0.5 μm thick is grown, *Figure 22.20(a)*. The slice is spun-coated with a photoresist, an organic material which polymerises when exposed to ultraviolet light, producing a layer about 1 μm thick. The photoresist is dried by baking. A mask defining the base area is placed on the slice and exposed. On development of the photoresist, the unexposed area is removed (giving access to the oxide layer), while the exposed area remains and is hardened by further baking to resist the chemical etch. This etch removes the uncovered oxide to define the diffusion area. The remaining photoresist is dissolved away, leaving the slice ready for the base diffusion, *Figure 22.20(b)*.

Boron is used to form the p-type silicon in the base region. The slice is passed through a diffusion tube in a furnace. A gas stream is passed over the slice containing a mixture of oxygen with boron tribromide BBr_3 or diboranne B_2H_6. The boron compounds decompose, and a boron-rich glass is formed on the surface of the slice. From this, boron is diffused into the silicon through the open base window. The glass is then removed by a chemical etch, and the slice heated in an oxygen stream in a second furnace. This drives the boron further into the slice, and grows a sealing oxide layer over the surface. The depth to which the boron penetrates is determined by careful control of the furnace temperature and the time for which the slice is heated. The resulting structure is shown in *Figure 22.20(c)*.

The next stage in the planar process is the emitter diffusion.

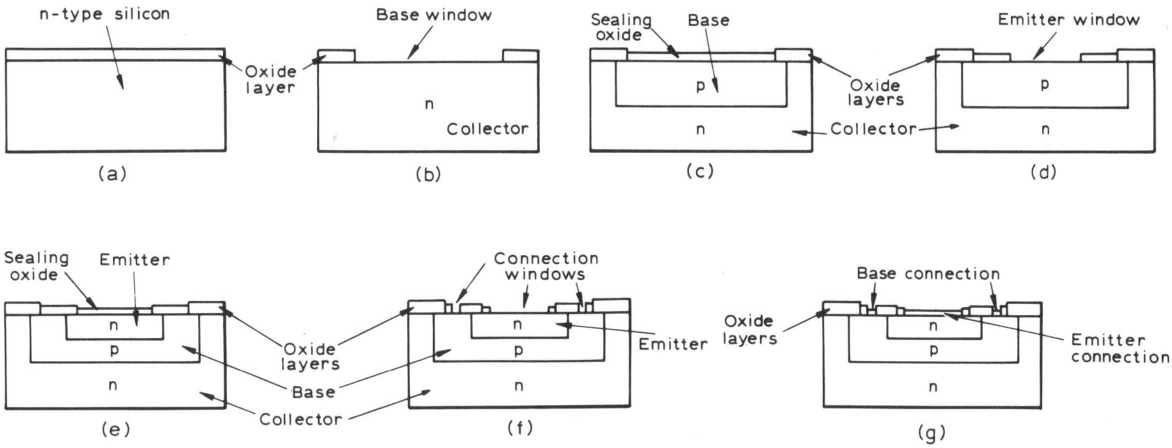

(h)

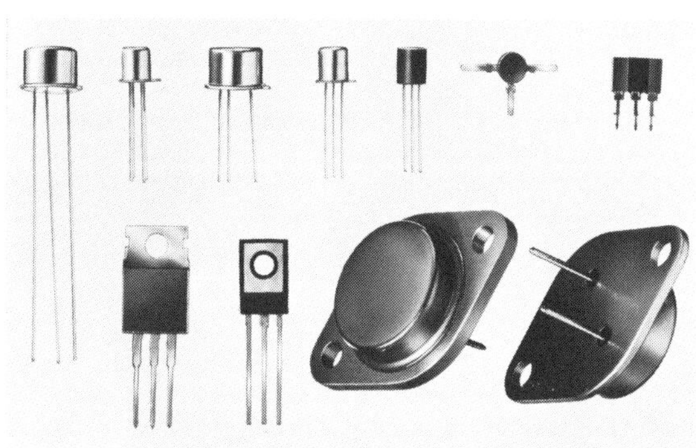
(i)

Figure 22.20 Stages in manufacture of silicon npn planar transistor: (a) oxide layer grown; (b) base window opened; (c) base diffused, sealing oxide grown; (d) emitter window opened; (e) emitter diffused, sealing oxide grown; (f) connection windows opened; (g) connection pads formed; (h) typical diffusion furnace for transistor manufacture (photo—Mullard Limited); (i) typical transistor encapsulations: top row, left to right—TO-5, TO-72, TO-39, TO-18, TO-92, 'T pack', 'Lockfit'; bottom row, left to right—TO-220, TO-126, TO-3 'thick base', TO-3 'thin base' (courtesy Mullard Limited)

The slice is coated with photoresist as before, and a second mask used to define the diffusion area. Exposure, development, and etching are the same as for the base diffusion just described. The structure before emitter diffusion is shown in *Figure 22.20(d)*.

Phosphorus is used to form the n-type emitter region. The gas stream in the diffusion furnace this time contains oxygen and phosphorus oxychloride $POCl_3$, phosphorus tribromide PBr_3, or phosphine PH_3. A phosphorus-rich glass is formed, from which phosphorus is diffused to form n-type silicon in the already-diffused p-type base region. A second furnace drives in the phosphorus to a controlled depth and forms a sealing oxide layer, as shown in *Figure 22.20(e)*.

The final stage in this part of the manufacturing process is forming the connections to the base and emitter regions. A third mask is used to define the base and emitter contacts, the photographic and etching processes being the same as those previously described, *Figure 22.20(f)*. The slice is coated with aluminium by evaporation onto the surface to form a layer about 1 μm thick. A fourth mask, a reversal of the third, is then used to define the areas where the aluminium is to be etched away leaving only contact pads for electrical connections, *Figure 22.20(g)*.

The process so far has been described for clarity as though only one transistor were being manufactured. In practice, many thousand devices are manufactured at the same time. The silicon slices used at present are 5 cm or 7.5 cm in diameter, containing tens of thousands of transistors. More than 100 slices are processed at the same time in a diffusion furnace. Even in the early days of the planar process when silicon slices 2.5 cm in diameter containing 2000 transistors were used, the process represented a considerable increase in production capacity over the alloyed-junction process, with a consequent drop in unit cost.

A typical diffusion furnace is shown in *Figure 22.20(h)*. This is a dual triple-bank furnace, and is shown being loaded with batches of silicon slices for diffusion.

Diffusion processes similar to those described for npn transistors can be used to manufacture pnp transistors. A p-type silicon slice is used, and phosphorus diffused to form the n-type base regions. When the emitter regions are formed, however, there is a tendency for the boron to concentrate in the growing oxide layer rather than the n-type silicon base, and special diffusion techniques have to be used.

From the description of the diffusion processes it will be clear that the accuracy of the masks used to define the diffusion areas is of paramount importance. Computer-aided design techniques are used to design the masks defining the diffusion areas on the transistor chip, the base, emitter, or contact areas of a single transistor. A computer-controlled light unit is used to expose areas on a photographic plate from which the actual diffusion masks will be made. A 10:1 reduction of this plate is made on a second photographic plate which is called the recticle. A step-and-repeat camera is then used to reproduce the recticle with a 10:1 reduction in a predetermined array to form the master, and from this master working copies are made for use in the diffusion process. The accuracy of positioning the image of the recticle on the master is 0.25 μm. To avoid defects in the masks, all the manufacturing and photographic processes take place in 'clean air' areas under carefully-controlled humidity and dust levels.

Careful alignment of the masks with the slices is essential if the geometry of the transistor is to be accurately controlled. The alignment can be done by an operator using a microscope, or in the latest equipment it is done automatically. The exposure of the mask onto the slice can be done with the mask and slice clamped together, or in some equipment by projecting an image of the mask onto the slice. The accuracy of masking is 1.0 μm in the manual system and 0.5 μm when done automatically. The photographic processes preceding the diffusion must take place in 'clean air' conditions.

The manufacturing processes after the transistor element has been formed are the same for both npn and pnp transistors. All the transistors in the slice are individually tested. This is done by means of probes that can move along a row of transistors on the slice testing each one, locate the edge of the slice, step to the next row, and move along this row testing these transistors. Any transistors that do not reach the required specification are marked automatically so that they can be rejected at a later stage. The slice is divided into individual transistors by scribing with a diamond stylus, and cracking the slice into individual chips. It is at this stage that the faulty transistors are rejected. The remaining transistors are prepared for encapsulation.

The oxide layer is removed from the collector side of the chip which is then bonded to a gold-plated header. The bonding action occurs through the eutectic reaction of gold and silicon at 400°C. This contact forms the collector connection. Aluminium or gold wires 25 μm in diameter are used to connect the base and emitter contact pads to the lead-out wires in the header. Thicker wires can be used if the current rating of the transistor requires it. The final stage of assembly is encapsulation, either in a hermetically-sealed case or in moulded plastic, depending on the application of the transistor. Typical transistor encapsulations are shown in *Figure 22.20(i)*.

The planar transistor has advantages over the alloyed-junction type besides the lower cost through mass-production manufacture. During each diffusion process, an oxide layer is grown over the edges of the junction which is not disturbed during the subsequent diffusion and assembly processes. Thus the collector–base junction, once it is formed and sealed by the oxide layer, cannot be easily contaminated by the emitter diffusion, testing and encapsulation, or during the lifetime of the transistor in service. The charge effects that occur at the bare surfaces of semiconductor devices are minimised, giving planar transistors greater reliability and stable characteristics. In addition, diffusion is a process which can be accurately controlled, and so the spacing between the three regions of the transistor can be held to a tolerance of less than 0.1 μm. Transistors with narrow base widths can be manufactured to allow operation at high frequencies, well above 1 GHz. A disadvantage of the planar transistor occurs through the resistivity of the collector. For high breakdown voltages, the resistivity should be high. On the other hand, for a high collector current the resistivity should be low. These two conflicting requirements would mean that a compromise value must be chosen for practical transistors. The conflict can be resolved, however, by the planar epitaxial process.

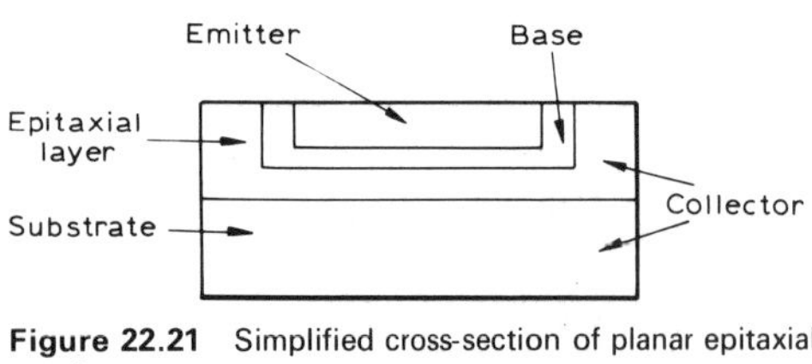

Figure 22.21 Simplified cross-section of planar epitaxial transistor

22.2.7 Planar epitaxial transistor

The planar epitaxial transistor, introduced in 1962, has the same structure as the planar transistor described in the previous section, but the transistor element is formed in an epitaxial layer. This layer is of high-resistivity material grown on a substrate of low resistivity, and so the conflicting requirements for the collector material can be met.

A polished slice of monocrystalline silicon, highly doped and therefore of low resistivity, forms the substrate. The slice can be p-type or n-type according to the type of transistor, and a typical resistivity figure is 1×10^{-5} Ω m. The epitaxial layer is grown on the substrate by vapour deposition in a radio-frequency heated reactor, the substrate being held at a temperature between 1000°C and 1200°C. The silicon vapour is formed by the decomposition of a silicon compound such as silicon tetrachloride $SiCl_4$ with hydrogen, and impurities can be added to the vapour to give the layer the required resistivity. For n^+ substrates, the impurities which can be used include phosphorus, arsenic, and antimony. Of these materials, arsenic and antimony are preferred because they have low diffusion constants. For p^+ substrates, the usual p-type impurities aluminium and gallium cannot be used because their diffusion constants are too high and the impurities would tend to migrate from the epitaxial layer into the substrate during the manufacture of the transistor. Boron is therefore used.

The silicon atoms in the epitaxial layer will take up the same relative positions as those in the substrate. Thus the near-perfect crystal lattice of the monocrystalline silicon substrate is reproduced in the epitaxial layer. The thickness of the layer is between 10 μm and 12 μm, and a typical resistivity value is 1×10^{-2} Ω m. Thus the bulk of the collector of the transistor is formed by the low-resistivity substrate.

The formation of the transistor element in the epitaxial layer follows the same stages as those for the planar transistor described previously. A simplified cross-section of the completed transistor element is shown in *Figure 22.21*.

22.2.8 Special forms of junction transistor

The transistor manufacturer today has a variety of techniques and materials at his disposal. Special geometries for large power handling or radio-frequency operation have been developed so the operating range of the transistor has been extended. In addition, further diffusion, mesa etching, and choice of doping levels enable transistors to be manufactured with special characteristics to meet particular requirements.

Germanium power transistors were made during the early 1950s by 'scaling-up' small-signal alloy-junction transistors. The area of the junctions was increased, and the collector pellet was bonded to the case to ensure a low thermal resistance. Such transistors could dissipate 10 W but showed a rapid fall-off in gain at currents above 1 A. In the late 1950s, the indium emitter was doped with gallium to increase the emitter doping and so improve the high-current gain.

The first silicon power transistors were introduced in the late 1950s, and used diffusion techniques. Base and emitter regions were successively diffused into one side of a slice of n-type silicon, and the electrical connection to the base was made by alloying rectifying contacts through the emitter. This type of transistor showed good gain up to a current of 5 A. Refinements to the manufacturing process during the 1960s led to the present-day diffused power transistor capable of handling currents of up to 30 A and powers up to 150 W. Two manufacturing processes are used for this type of power transistor, the single-diffused and the triple-diffused processes.

The single-diffused or hometaxial process uses a simultaneous diffusion on opposite sides of a homogenous base wafer, forming heavily doped collector and emitter regions. The emitter is mesa etched to allow electrical connection to be made with the base. This type of transistor reduces the risk of hot spots through the use of a homogenous base, the wide base gives good second-breakdown properties, and the heavily-doped collector provides low electrical and thermal resistance.

Triple-diffused power transistors are manufactured by diffusing base and emitter regions on one side of a collector wafer. The third diffusion forms a heavily-doped diffused collector on the other side. This type of transistor has a high voltage rating, often able to withstand voltages of 1 kV or more.

The planar epitaxial process enables further improvements to be made to power transistors. At high current densities, current contraction can occur. (This is the cause of second breakdown, as explained under Section 22.2.2.) The edge of the emitter becomes more forward-biased than the centre, so that the current concentrates along the periphery of the emitter. It is therefore necessary to design base–emitter structures that differ from the annular or pear-drop geometries of small-signal transistors, and scaling-up can no longer be done. An emitter with a long periphery is required. Two structures that have been used successfully are the *star* and *snowflake*, the names being descriptive of the shape of the emitter. These structures could not have been produced in practical transistors without the planar technique of diffusion through a shape in the oxide layer.

More complex base–emitter structures can be produced to combine the large emitter area and long periphery required for high-power handling with the narrow spacing required for high-frequency operation. Geometries have been developed to enable power transistors to operate at radio frequencies. One such geometry is the interdigitated structure where fingers of the base interleave fingers of the emitter. Another is the overlay structure where a large number of separate emitter stripes are interconnected by metalising in a common base region. In effect, a large number of separate high-frequency transistors are connected in parallel to conduct a large current. Transistors using these structures can operate high in the radio frequencies, with typical powers of 175 W at 75 MHz and 5 W at 4 GHz.

Another structure used for power transistors is the epitaxial base or mexa structure. A lightly-doped epitaxial layer is grown on a heavily doped collector, and a single diffusion used to form the emitter in the epitaxial base layer. The resulting structure is mesa etched. Mexa transistors are rugged, and have a low collector resistance.

Power transistors are usually encapsulated in metal cases allowing mounting on a heatsink. In recent years, however, there has been a move towards plastic encapsulations. This has considerably decreased the cost of encapsulating the transistor without affecting the performance. A metal plate is incorporated in the plastic envelope to ensure good thermal contact between the transistor element and a heatsink.

A power transistor used as the output transistor in an amplifier generally requires a driver transistor to provide sufficient input power. If both transistors are mounted on heatsinks, a considerable amount of the volume of the amplifier will be occupied by these two transistors. A further development enables space to be saved by combining the driver and output transistors on the same silicon chip in one encapsulation. This construction is the Darlington power transistor, which can have a current gain of up to 2000 and power outputs up to 150 W.

The circuit diagram of a Darlington transistor is shown in *Figure 22.22*. The two transistors and base–emitter resistors are formed on one chip by successive diffusions using the epitaxial-base process. A diode can also be formed across the collector and emitter terminals for protection if required. The current gains of the two transistors are controlled during manufacture so that the overall gain varies linearly over a range of collector current. This linearity of gain is combined with smaller spreads than would occur with discrete transistors connected in the same circuit. These two advantages of the Darlington transistor are combined with a disadvantage, the high value of $V_{CE(sat)}$.

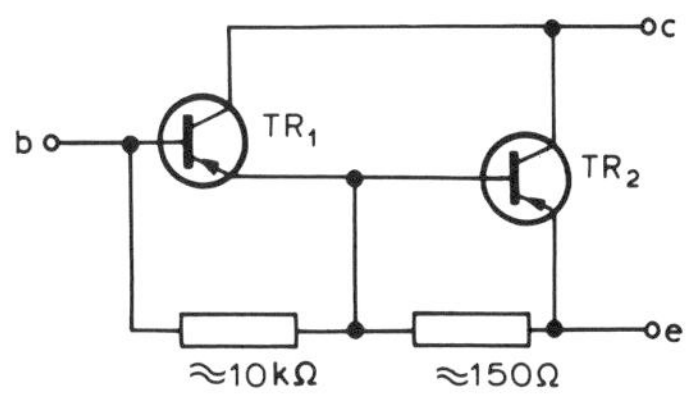

Figure 22.22 Circuit diagram of Darlington power transistor

Transistors for high-frequency operation or fast switching must have narrow spacings between the emitter, base, and collector. Two geometries are generally used: the ring base and the stripe base. The ring-base structure is 'scaled-down' from the annular structure used for low-frequency transistors. The stripe-base structure, which is generally preferred for the higher-frequency operation, is shown in *Figure 22.23*. Many of these structures can be connected in parallel to increase the current-carrying capacity, forming the interdigitated structure already described for r.f. power transistors. The internal capacitances of the transistor, and the stray capacitances of the mounting and case, must be kept as low as possible to prevent restriction of the upper frequency limit. A planar epitaxial manufacturing process must be used to keep the collector resistance low. The doping level is chosen to suit the operating frequency and voltage.

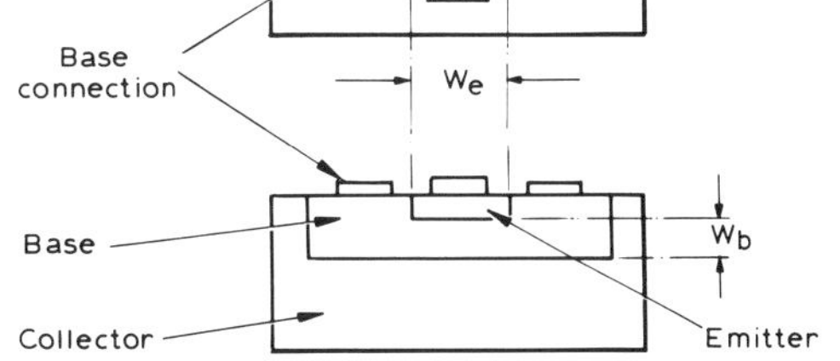

Figure 22.23 Stripe-base structure for high-frequency transistor

In the stripe-base structure, two dimensions are critical for the upper frequency limit. These are the emitter stripe width (W_e in *Figure 22.23*) and the base width W_b. For present-day transistors operating up to the microwave region, the emitter width can be as low as 1 μm and the base width 0.1 μm.

22.2.9 Unijunction transistor

As the name implies, a unijunction transistor contains only one junction although it is a three-terminal device. The junction is formed by alloying p-type impurity at a point along the length of a short bar-shaped n-type silicon slice. This p-type region is called the emitter. Non-rectifying contacts are made at the ends of the bar to form the base 1 and base 2 connections. The structure of a unijunction transistor is shown in *Figure 22.24(a)*, and the circuit symbol in *Figure 22.24(b)*.

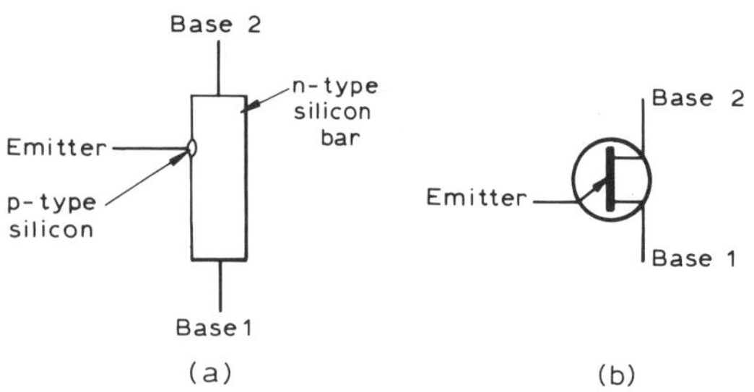

Figure 22.24 Unijunction transistor: (a) simplified structure; (b) circuit symbol

The resistance between the base 1 and base 2 connections will be that of the silicon bar. This is shown on the equivalent circuit in *Figure 22.25* as R_{BB}, and has a typical value between 4 kΩ and 12 kΩ. A positive voltage is applied across the base, the base 2

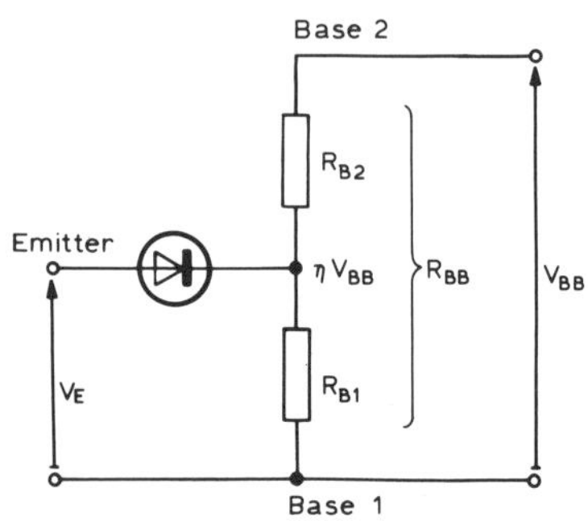

Figure 22.25 Equivalent circuit of unijunction transistor

contact being connected to the positive terminal. The base acts as a voltage divider, and a proportion of the positive voltage is applied to the emitter junction. The value of this voltage depends on the position of the emitter along the base, and is related to the voltage across the base, V_{BB}, by the intrinsic stand-off ratio η. The value of η is determined by the relative values of R_{B1} and R_{B2}, and is generally between 0.4 and 0.8.

The emitter pn junction is represented in the equivalent circuit by the diode. When the emitter voltage V_E is zero, the diode is reverse-biased by the voltage ηV_{BB}. Only the small reverse current flows. If the emitter voltage is gradually increased, a value is reached where the diode becomes forward-biased and starts to conduct. Holes are injected from the emitter into the base, and are attracted to the base 1 contact. The injection of these holes reduces the value of R_{B1} so that more current flows from the emitter to base 1, reducing the value of R_{B1} further. The unijunction transistor therefore acts as a voltage-triggered switch, changing from a high 'off' resistance to a low 'on' resistance at a voltage determined by the base voltage and the value of η.

The voltage/current characteristic for a unijunction transistor is shown in *Figure 22.26*. It can be seen that after the device has

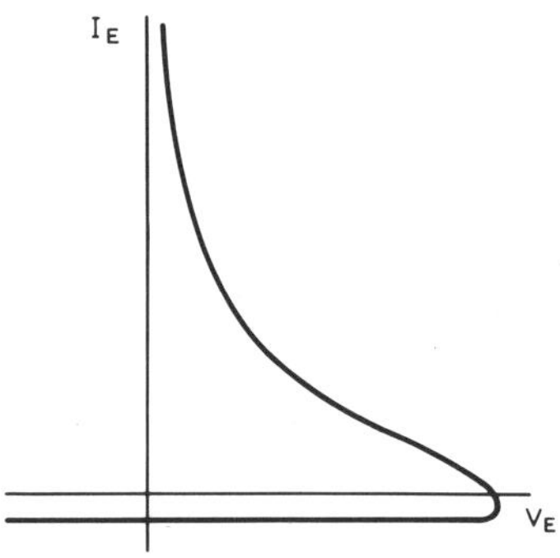

Figure 22.26 Voltage/current characteristic of unijunction transistor

been triggered, there is a negative-resistance region on the characteristic. This enables the unijunction transistor to be used in oscillator circuits as well as in simple trigger circuits.

22.3 Field-effect transistors

A field-effect transistor consists essentially of a current-carrying channel formed of semiconductor material whose conductivity is controlled by an externally-applied voltage. The current is carried by one type of charge carrier only; electrons in channels formed of n-type semiconductor material, holes in channels of p-type material. The field-effect transistor is therefore sometimes called a unipolar transistor. This is to distinguish it from the junction transistor whose operation depends on both types of charge carrier, and which is therefore called a bipolar transistor.

There are two types of field-effect transistor (FET): the junction FET (JFET) and the insulated gate FET (IGFET). The most commonly used insulated-gate FET is the metal oxide semiconductor FET (MOSFET). The shorter name is MOST. The operation of each type of FET is described separately.

The theoretical operation of the FET was described by William Shockley in 1952. It was not until 1963, however, that practical devices were generally available. The delay was a result of the manufacturing techniques not being sufficiently advanced until that date, the planar process being essential for the manufacture of FET's. This was an example of how the theoretical work on solid-state devices in the early days of the transistor was often ahead of the device technology.

22.3.1 Operation of junction FET

The schematic structure of a junction FET is shown in *Figure 22.27*. The device shown is an n-channel FET, formed from a bar-shaped slice of n-type monocrystalline silicon into which two p-

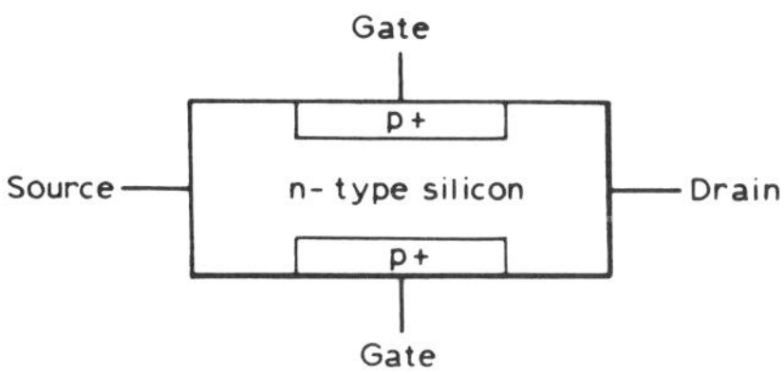

Figure 22.27 Schematic structure of junction field-effect transistor

type regions are diffused. Connections are made to the ends of the channel, the source and the drain, and to the p-regions, the gate.

If a voltage is applied across the channel so that the drain is positive with respect to the source, as shown in *Figure 22.28(a)*, electrons flow through the channel from the source to the drain producing the drain current I_D. The magnitude of the drain

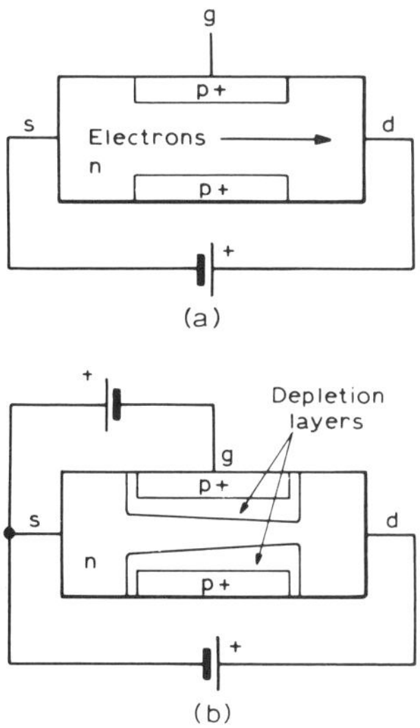

Figure 22.28 Operation of n-channel field-effect transistor: (a) bias to produce drain current; (b) depletion layers produced by gate voltage and drain current voltage drop

current is determined by the conductivity of the channel and the drain-to-source voltage V_{DS}. When a negative voltage is applied to the gate, it will be reverse-biased. A depletion layer is formed about the pn junction between the gate and channel, as shown in *Figure 22.28(b)*. Because the gate is more heavily doped than the channel, the depletion layer extends into the channel rather than into the gate. The depletion layer is virtually devoid of charge carriers and so has a high resistance, and therefore the channel thickness over which the current can flow is decreased. This decrease in channel thickness decreases the conductivity, and therefore decreases the magnitude of the drain current. Because the width of the depletion layer depends on the reverse bias on the gate, the drain current can be controlled by the gate voltage. This method of operation is analogous to that of the triode thermionic valve where the magnitude of the anode current is controlled by the voltage on the grid.

If the gate voltage is sufficiently negative, the depletion layer will extend across the whole of the channel. The channel is then said to be 'pinched off'. The value of gate voltage at which this occurs is called the pinch-off voltage, V_P.

22.3.2 Junction FET characteristics

When the gate voltage is zero, that is the gate is connected directly to the source, the flow of drain current causes a linear drop in voltage along the length of the channel. The gate-to-channel voltage will therefore form a small reverse bias at the source end of the gate, and a larger reverse bias at the drain end of the gate. The depletion layers therefore have the wedge shape shown in *Figure 22.28(b)*. As the drain-to-source voltage is increased, the drain current will increase linearly. But the gate-to-channel voltage will also be increased so that the depletion layer will penetrate further into the channel. A point will be reached where the decrease in channel thickness has a greater effect than the increase in drain-to-source voltage. The characteristic of drain current I_D plotted against drain-to-source voltage V_{DS} will therefore form a knee. As V_{DS} is increased further, a value will be reached above which virtually no further increase in drain current will occur. This is because the depletion layers extend across the whole of the channel at the drain end of the gate and the channel is pinched off. As the value of V_{DS} is increased above the pinch-off value $V_{DS(P)}$, the point along the channel at which the depletion layers meet moves nearer the source. The drain current is maintained by electrons being swept through the depletion layer in the same way minority charge carriers are swept from the base region in junction transistors. Eventually as V_{DS} is increased further, the gate-to-channel breakdown value will be reached.

The characteristic just described is for zero gate voltage. If the gate-to-source voltage V_{GS} has a small constant negative value, the depletion layers will extend into the channel even with no drain current flowing. Thus the knee of the characteristic will occur at a lower drain current, and the voltage at which the channel is pinched off occurs at a lower value of V_{DS}. A higher negative value of V_{GS} will decrease the knee and pinch-off value further. A family of characteristics with gate-to-source voltage V_{GS} as parameter is obtained, as shown in *Figure 22.29*. Also shown is the locus of the pinch-off value $V_{DS(P)}$, shown by the curve in broken lines.

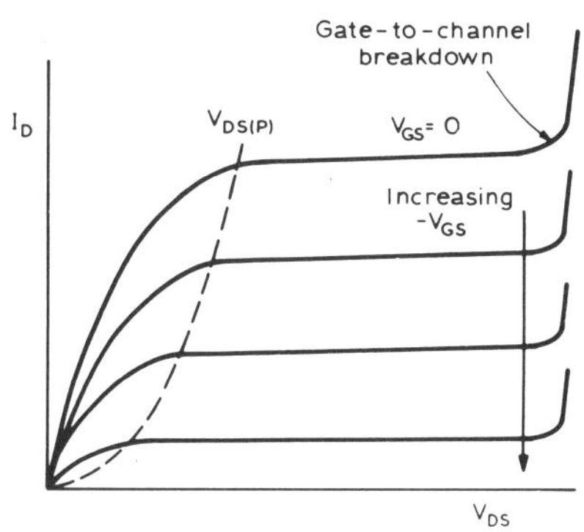

Figure 22.29 Drain voltage/current characteristics for junction FET with gate voltage as parameter

The part of the characteristic where V_{DS} is greater than the pinch-off value is called the pinch-off or saturation region. This is the normal operating region for the FET. With a constant value of source-to-drain voltage, the drain current can be varied by the gate-to-source voltage. The characteristic of drain current against gate voltage with drain-to-source voltage as parameter is shown in *Figure 22.30*. The slope of this transfer characteristic is the transconductance of the FET, g_m. The value of g_m is given by:

$$g_m = |\partial I_D/\partial V_{GS}|_{V_{DS}}$$

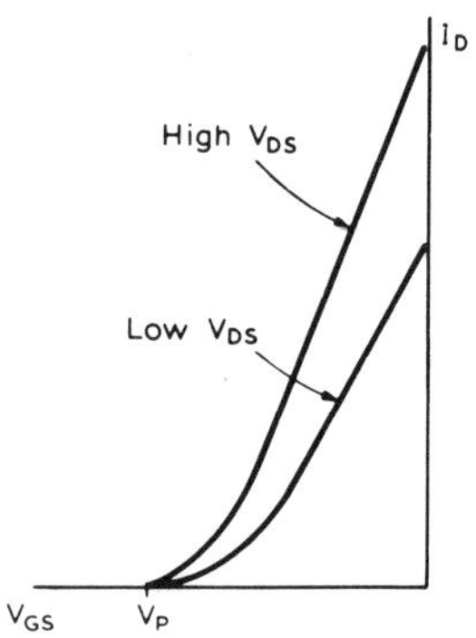

Figure 22.30 Transfer characteristic of junction FET showing effect of drain voltage

The transconductance is used in the equivalent circuit, as described later in the section.

The FET is operated in practical circuits in the common-source configuration; that is, the source is common to both input and output circuits. The input circuit is formed by the gate and source, and the output circuit by the drain and source. Since the gate is reverse-biased, the input impedance of the FET is very high. This is in contrast with the usual configuration (common emitter) of the junction transistor, and so the FET provides the circuit designer with a useful complement to the junction transistor.

The reverse-biased gate also means that the gate current I_G flowing under normal operation is the small reverse current. The total power dissipated by the FET is:

$$P_{tot} = V_{DS}I_D + V_{GS}I_G \tag{22.12}$$

Since $V_{DS} > V_{GS}$ and $I_D \gg I_G$, equation (22.12) can be rewritten as:

$$P_{tot} = V_{DS}I_D \tag{22.13}$$

The maximum power can be plotted on the V_{DS}/I_D characteristic to form one boundary of the operating area. Another boundary is set by the need to operate in the pinch-off region. The other two boundaries of the operating area are set by the maximum value of V_{DS} to prevent breakdown between gate and channel, and the maximum drain current I_D max that can be drawn from the device. These limits are shown in *Figure 22.31*.

Other limiting values are listed in the published data as electrical ratings. These include: the maximum drain-to-gate voltage V_{DGO} max, the maximum gate-to-source voltage V_{GSO} max, and the maximum gate current I_G max.

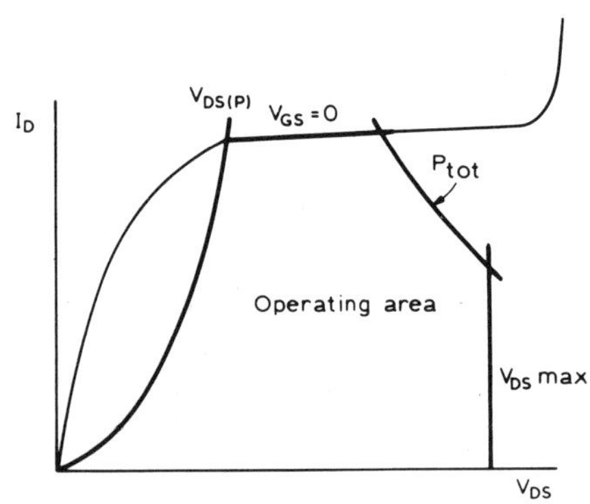

Figure 22.31 Operating area of junction FET

As with the junction transistor, the characteristics of the junction FET are temperature-dependent although to a lesser extent. Curves of the variation with temperature of drain current I_D and the gate cut-off current I_{GSS} (gate current with zero drain-to-source voltage) are given in published data.

An equivalent circuit for the junction FET in the common-source configuration under small-signal conditions is shown in *Figure 22.32*. The gain of the FET is represented by the current

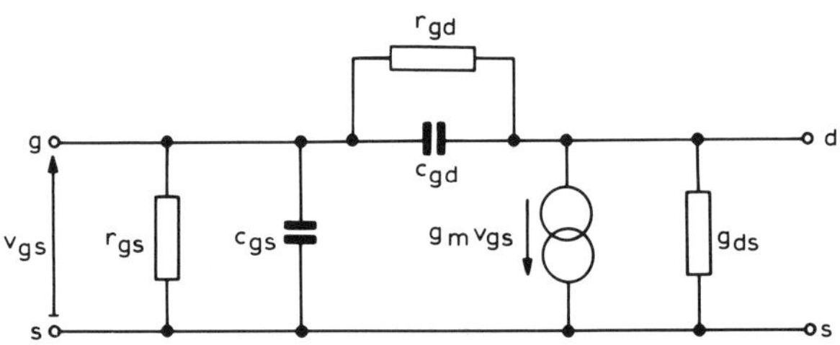

Figure 22.32 Small-signal equivalent circuit for FET

generator $g_m V_{gs}$, where g_m is the transconductance of the FET and V_{gs} is the gate-to-source voltage. The input resistance and capacitance are r_{gs}, the gate-to-source resistance which is high because of the reverse-biased gate, and c_{gs}, the gate-to-channel capacitance formed by the depletion layer between the source and the pinch-off point. The value of r_{gs} is approximately 10^{11} Ω at room temperature. Capacitance c_{gd} is the capacitance of the depletion layer beyond the pinch-off point to the drain. Both c_{gs} and c_{gd} are voltage-dependent. Resistance r_{gd} is the gate-to-drain resistance which has a similar value to r_{gs}. The output conductance g_{ds} is the drain-to-source conductance.

22.3.3 Practical junction FET's

The structure of a practical n-channel junction FET is shown in the cross-section and plan of *Figure 22.33*. An interdigitated structure is used, with interleaved fingers of drain and source between which are the channels. The gate regions are formed on top of the channels.

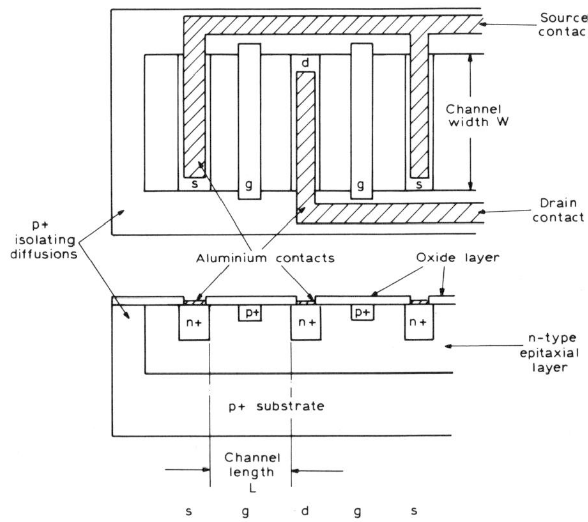

Figure 22.33 Structure of practical n-channel junction FET

An n-type epitaxial layer is grown on a heavily-doped p-type substrate. Using the planar process, p^+ regions are diffused into the lightly-doped epitaxial layer to define the transistor area on the chip by isolating diffusions. A second diffusion process is used to form the p^+ upper gate regions. The individual upper gates are interconnected by the isolating diffusions, which also connect them to the substrate which forms the lower part of the gate. Connection to the gate can therefore be made by one contact at the edge of the transistor area rather than to the individual gates. The final diffusion forms the n^+ regions, the sources and drains. Aluminium fingers are formed on the source and drain areas by vacuum deposition, and the fingers are interconnected to the source and drain contacts on the chip. The chip is mounted on a header and encapsulated in the normal way.

The interdigitated structure is used to provide a favourable channel width-to-length ratio. For a low 'on' resistance, the channel width W should be large but the channel length L should be short. On the other hand, the gate capacitance is proportional to the channel area $W \times L$. Therefore the product WL should be as low as possible while the ratio W/L should be as high as possible. These requirements are met by making the channel length short, a few micrometres, but keeping the width to a practicable value by forming many channels in parallel interconnected in an interdigitated structure.

The n-channel junction FET has been described because this is the device most used in practice at present. P-channel FET's can be manufactured by growing a p-type epitaxial layer on an n-type substrate. N-type regions are then diffused into the epitaxial layer to form the upper gate regions, followed by a p^+ diffusion to form the sources and drains. The current in a p-channel junction FET is carried by holes, and the gate voltage and drain-to-source voltage have the opposite polarities from those used in the n-channel device.

The circuit symbols for n-channel and p-channel junction FET's are shown in *Figure 22.34*. The difference in the two

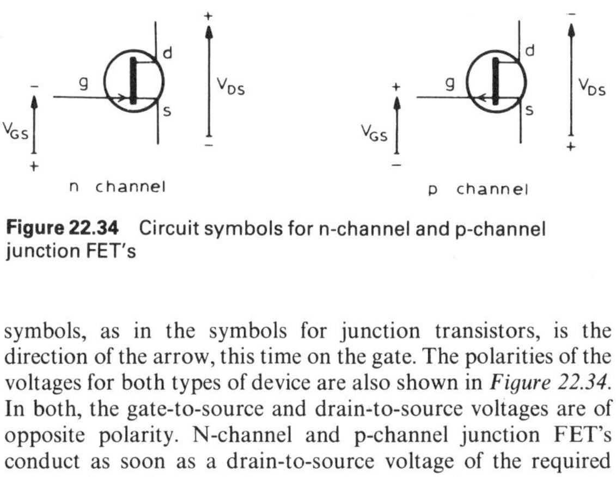

Figure 22.34 Circuit symbols for n-channel and p-channel junction FET's

symbols, as in the symbols for junction transistors, is the direction of the arrow, this time on the gate. The polarities of the voltages for both types of device are also shown in *Figure 22.34*. In both, the gate-to-source and drain-to-source voltages are of opposite polarity. N-channel and p-channel junction FET's conduct as soon as a drain-to-source voltage of the required polarity is applied, and the gate voltage is used to decrease the drain current. They are therefore known as depletion or normally-on devices.

22.3.4 Operation of insulated-gate FET

The structure of the insulated-gate FET differs from that of the junction FET just described in two respects: no separate current-carrying channel is built in, and the construction of the gate is different. A current-carrying channel is formed by the accumulation of charges beneath the gate electrode, as described later. The gate itself is not a diffused region but a thin layer of metal insulated from the rest of the FET by a layer usually of oxide. Thus the structure of the insulated-gate FET is successive layers of metal, oxide, and semiconductor material—giving the transistor its alternative name of MOSFET, or the shorter form, the MOS transistor or MOST. Because the semiconductor material used at present is invariably silicon, the initials MOS are often interpreted as metal–oxide–silicon.

A simplified cross-section of an n-channel MOSFET is shown in *Figure 22.35*. A p-type substrate is used, with heavily-doped n^+

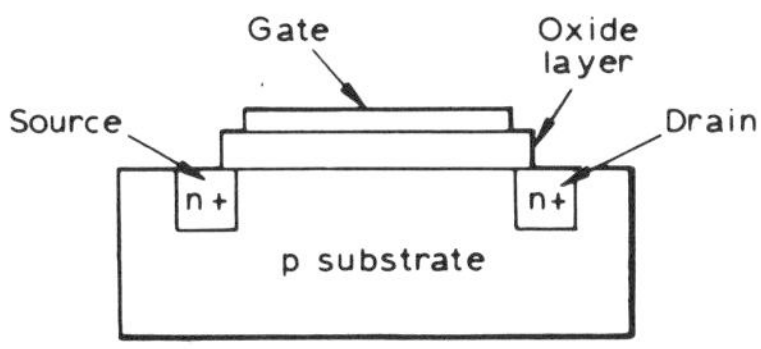

Figure 22.35 Simplified cross-section of n-channel MOSFET

regions forming the source and drain. A thin layer of silicon dioxide is grown on the surface of the substrate between the source and the drain. Aluminium is deposited on top of the oxide layer to form the gate electrode. This gives a form of MOSFET called the aluminium-gate MOSFET. A later development uses polysilicon as the metal for the gate, to give the silicon-gate MOSFET.

With no voltage applied to the gate, only the depletion layers about the pn^+ junctions are present, as shown in *Figure 22.36(a)*. Even if a large positive voltage with respect to the source is applied to the drain, no current will flow.

If a small positive voltage with respect to the source is applied to the gate, free holes in the substrate are repelled from the surface and a depletion layer is formed beneath the gate to link with the two existing layers, *Figure 22.36(b)*. If the gate voltage is increased further, free electrons from the source will be attracted to the region under the gate oxide. In this way, a layer of electrons will be formed in the surface of the p-type substrate. This layer is so thin that it is treated as a sheet of charges. The layer is called an inversion layer, and is shown in *Figure 22.36(c)*. The gate-to-source voltage at which the inversion layer is just formed is called the threshold voltage V_T. The electrons in the inversion layer can be used as charge carriers by applying a positive drain-to-source voltage. A drain current will flow whose magnitude is determined by the drain-to-source voltage and the conductivity of the channel which depends on the charge per unit area in the channel. This charge density is proportional to the voltage difference between the gate and the channel. The flow of drain current will produce a voltage drop along the channel, leading to a reduction in voltage between gate and channel towards the drain. The voltage distribution in the channel is then as indicated in *Figure 22.36(d)* in which the thickness of the depletion layer between channel and substrate is seen to increase towards the drain.

As the drain-to-source voltage is increased, the drain current increases but the voltage drop also increases, thus decreasing the channel conductivity. This decrease of conductivity amounts to a constriction of the inversion layer so that the characteristic of drain current I_D against drain-to-source voltage V_{DS} will form a knee, as in the characteristic for the junction FET. When $V_{DS} = V_{GS}$, the voltage between the gate and the drain end of the channel will be zero, giving zero charge density. This is shown in *Figure 22.36(e)*. This is the pinched-off condition, the value of V_{DS} being $V_{DS(P)}$. Any further increase in V_{DS} merely causes the drain depletion layer to expand, and the end of the channel to move towards the source, *Figure 22.36(f)*. The drain current will be maintained by electrons being swept through the depletion layer

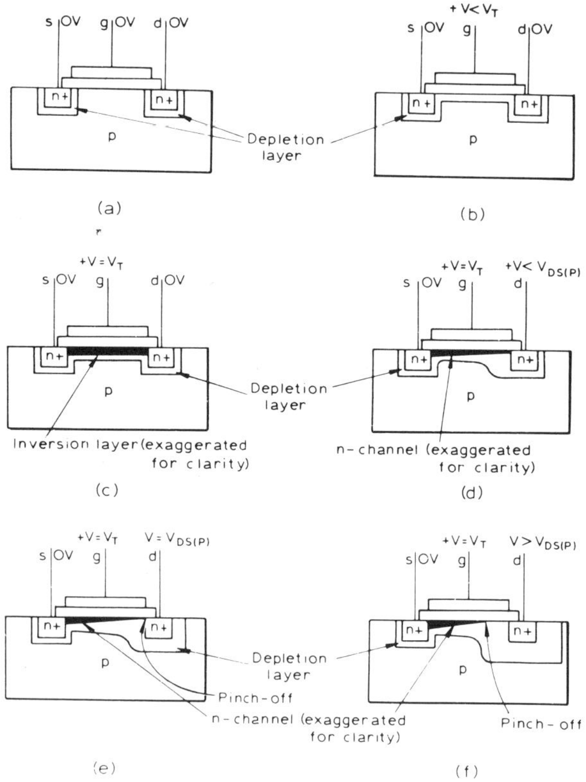

Figure 22.36 Operation of n-channel MOSFET: (a) zero gate voltage; (b) small positive gate voltage; (c) threshold gate voltage, inversion layer formed; (d) drain current flowing, drain voltage < $V_{DS(P)}$; (e) channel pinched off, drain voltage = $V_{DS(P)}$; (f) channel pinched off, drain voltage > $V_{DS(P)}$

as from the base region of a junction transistor, but there will be no significant increase in the drain current with increase in V_{DS}.

The channel conductivity can be varied by the gate voltage, and so a family of characteristics of drain current I_D plotted against drain-to-source voltage V_{DS} with gate-to-source voltage as parameter will be obtained, as shown in *Figure 22.37*. These characteristics are similar to those for the junction FET shown in *Figure 22.29*, and the MOSFET, like the junction FET, is operated in the pinch-off region. The transfer characteristic of drain current plotted against gate-to-source voltage is shown in *Figure 22.38*.

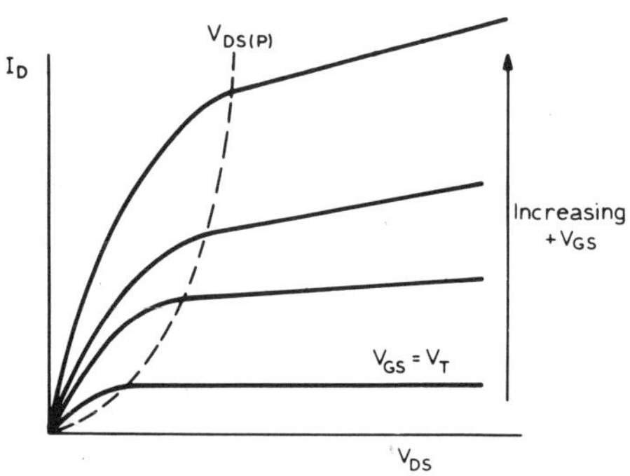

Figure 22.37 Drain voltage/current characteristics for n-channel enhancement MOSFET

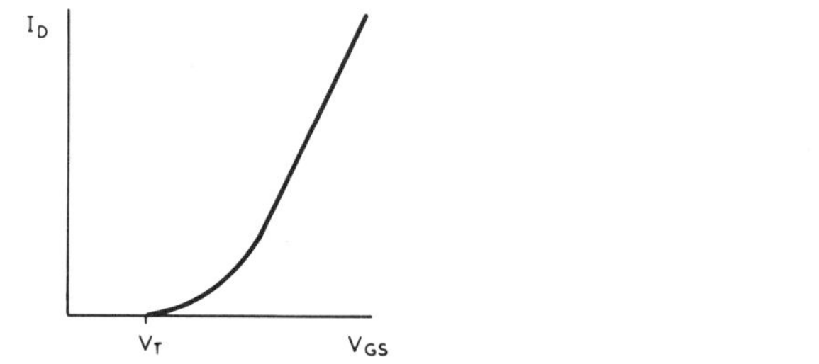

Figure 22.38 Transfer characteristic for n-channel enhancement MOSFET

The MOSFET just described is called n-channel because the charge carriers are electrons. A p-channel MOSFET can also be constructed. This uses an n-type substrate with p^+ source and drain regions. The inversion layer is formed by applying a negative voltage (with respect to the source) to the gate. The free holes from the source are attracted to the gate region to form the inversion layer. If a negative voltage with respect to the source is applied to the drain, the holes in the inversion layer will act as charge carriers and a drain current will flow. The characteristics of the p-channel MOSFET are similar to those of the n-channel except that the polarities of the gate-to-source and drain-to-source voltages are negative instead of positive.

The n-channel and p-channel MOSFET's described so far are ideal in the sense that a small gate voltage of the correct polarity will form the conducting inversion layer between the source and the drain. In practice, naturally-occurring positive charges in the gate oxide, and the effect of the gate material work function, will result in an n-type channel being formed even at zero gate voltage, *Figure 22.39*. Thus the voltage at which the device will start to conduct (V_T) will be negative, as shown in *Figure 22.40*.

The p-channel MOSFET will not conduct, and the n-channel MOSFET will not stop conducting, until V_G is more negative

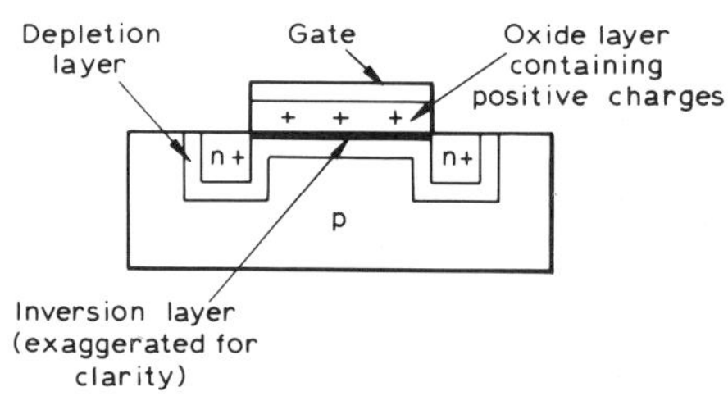

Figure 22.39 Simplified cross-section of n-channel depletion MOSFET

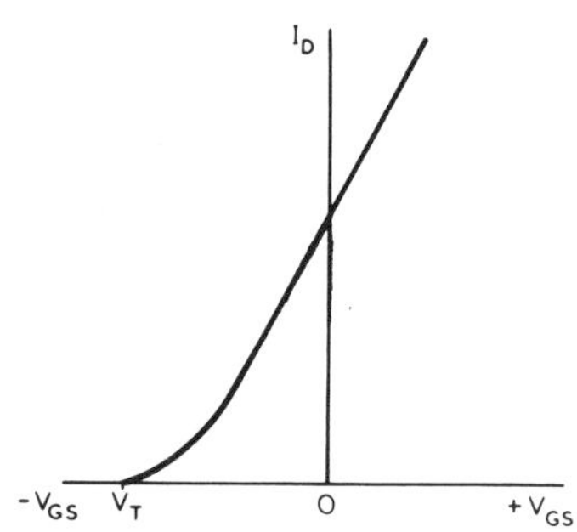

Figure 22.40 Transfer characteristic of n-channel depletion MOSFET

than V_T. The p-channel MOSFET is called an enhancement device as it does not conduct at zero gate voltage, whereas the n-channel MOSFET is a depletion device since, like the junction FET, it conducts at zero voltage. Unlike the junction FET, however, it can be used with the channel either depleted or enhanced.

With very careful processing, the threshold voltage can be controlled so that enhancement n-channel MOSFET's and depletion p-channel MOSFET's can also be made.

The four possible types of insulated-gate FET are shown with the circuit symbols in *Figure 22.41*. The difference in the symbols

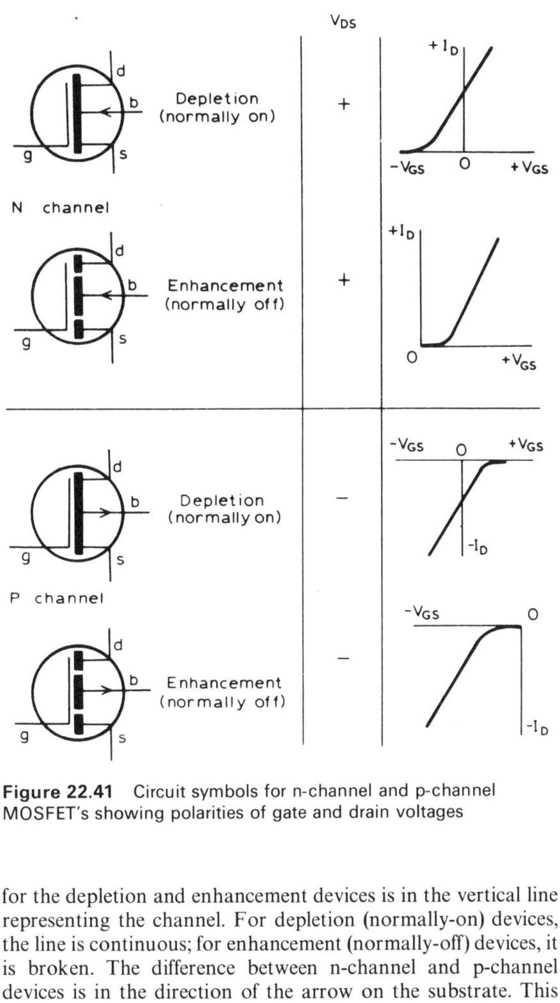

Figure 22.41 Circuit symbols for n-channel and p-channel MOSFET's showing polarities of gate and drain voltages

for the depletion and enhancement devices is in the vertical line representing the channel. For depletion (normally-on) devices, the line is continuous; for enhancement (normally-off) devices, it is broken. The difference between n-channel and p-channel devices is in the direction of the arrow on the substrate. This terminal, marked b (for bulk), is generally connected to the source, often inside the transistor encapsulation. Also shown in *Figure 22.41* are the polarities of the gate and drain voltages.

The equivalent circuit for the junction FET shown in *Figure 22.32* also applies to the MOSFET. The main difference in the circuit between the two types of device is that for the MOSFET the values of r_{gs} and r_{gd} at room temperature are increased to 10^{14} Ω.

22.3.5 Practical MOSFET's

As with the junction FET discussed previously, the structure of a practical MOSFET is considerably more complex than the simple cross-sections used so far. Special structures are necessary for both enhancement and depletion MOSFET's.

The gate electrode in enhancement (normally-off) devices must slightly overlap the source and drain regions to ensure the inversion layer connects the two regions. This overlap can lead to relatively high parasitic capacitances unless the oxide layer underneath the overlapping parts of the gate is thicker than that above the channel. A structure like that shown in *Figure 22.42* is obtained, the thickness of the oxide under the overlapping gate being approximately 1 μm. The extensions to the source and

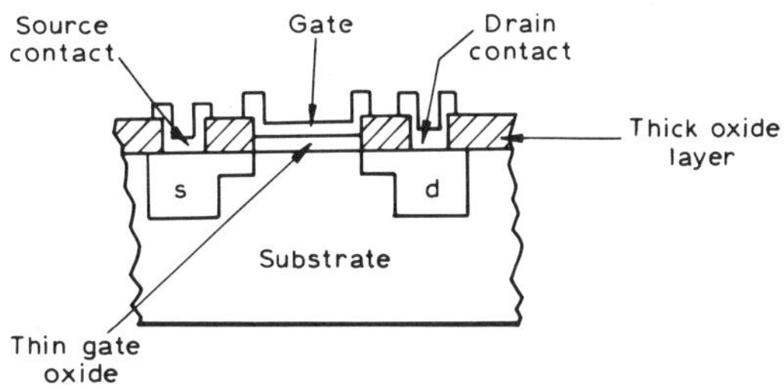

Figure 22.42 Cross-section of practical enhancement MOSFET

drain regions are diffused from the doped thick oxide layer in which a window is etched for the gate oxide before the diffusion. The thin gate oxide, about 0.12 μm thick, is grown in this window and so matches exactly with the channel. The tolerances on the alignment of the gate are then much less critical.

As the operation of the depletion (normally-on) MOSFET depends on the charges contained in the oxide layer, the structure must allow the current-carrying channel to be formed between the source and drain without allowing parasitic channels to be formed. A 'closed' structure is used, the drain area being completely surrounded by the gate area, which in turn is surrounded by the source. To obtain the required high width-to-length ratio, the channel length is made short and the channel width large, but the structure is 'folded' so that it can be accommodated on the transistor chip. This structure combined with a thin oxide layer has low parasitic capacitances and a low 'on' resistance. The feedback capacitance, c_{gd} in the equivalent circuit of *Figure 22.32*, is also low, allowing operation at high frequencies.

The feedback capacitance of the simple MOSFET is too high for operation at u.h.f., approximately 0.5 pF. One method of lowering the capacitance using the same manufacturing technique, to approximately 0.02 pF, is by adding a second gate to the device, to give the structure shown in *Figure 22.43(a)*. The circuit of this dual-insulated-gate FET is shown in *Figure 22.43(b)*. If

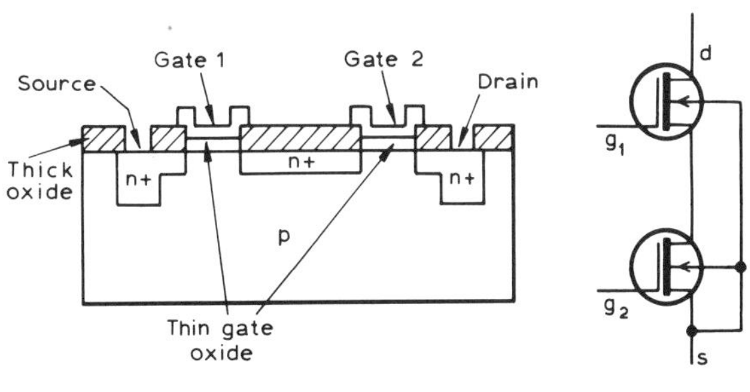

Figure 22.43 Dual-insulated-gate FET: (a) cross-section; (b) circuit diagram

gate 2 is earthed, the feedback between the drain and gate 1 is extremely small because of the screening effect of gate 2. In addition, the output conductance of this device is very low because the drain load resistance of the lower MOSFET is the transconductance of the upper MOSFET in *Figure 22.43(b)*. Thus modulation of the drain-to-source voltage of the lower MOSFET is small. The dual-gate FET can be used at lower frequencies as well. The second gate is then available for use as an a.g.c. input, for example, in the controlled stages of receivers.

Although the substrate is often connected to the source internally, in some MOSFET's it is available as a fourth terminal. The drain current can be controlled by the substrate. The transfer characteristics of drain current plotted against gate-to-source voltage are shown in *Figure 22.44* with substrate-to-source voltage V_{BS} as parameter. This type of MOSFET therefore has two

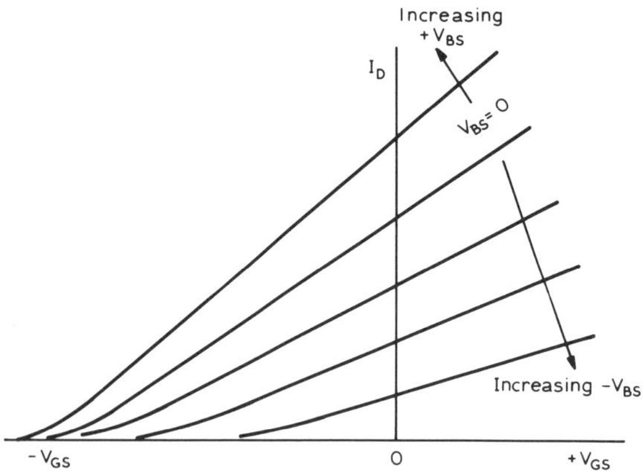

Figure 22.44 Transfer characteristics of n-channel depletion MOSFET with substrate-to-source voltage as parameter

control electrodes, allowing it to be used in such applications as, for example, mixer stages. However, when the substrate is used in this way, it is acting as a junction gate. The input resistance between substrate and source is therefore lower than that between the insulated gate and source.

The ratings and electrical characteristics of the insulated-gate FET given in published data are similar to those of the junction FET discussed previously. When the substrate is available as a separate terminal, additional ratings are given between the substrate and the other regions of the device.

22.3.6 VMOS

The term VMOS stands for vertical MOS and refers to a form of the MOSFET structure that enables higher current flows to be obtained than with the type of MOSFET previously described. Thus power field-effect transistors can be manufactured as well as the longer-established junction types.

In the MOSFET structure shown in *Figure 22.36*, the current flow is 'horizontal', that is parallel to the surface of the chip in which the source, drain, and gate are formed. The maximum current is limited by the dimensions of the current-carrying channel (the inversion layer), in particular by the depth of the channel. In the VMOS structure the current flow is 'vertical', that is across the chip as in junction devices such as power diodes, transistors, and thyristors. Thus a higher current flow is possible and, in addition, each of the VMOS structures can be connected in parallel with others in the chip during manufacture to give a device containing many transistors in parallel. The completed device thus forms a power MOSFET.

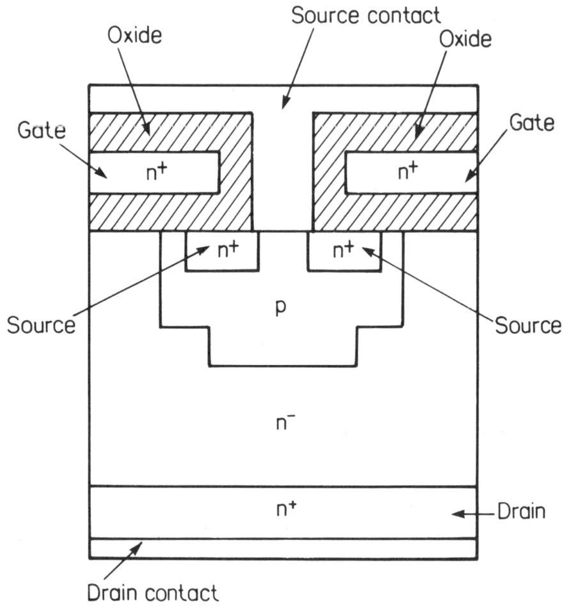

Figure 22.45 Simplified cross-section of a VMOS transistor

A simplified cross-section of the VMOS structure is shown in *Figure 22.45*. The drain is formed by the n^+ substrate on which an n-type epitaxial layer is grown. A p-type region is formed in the surface of this layer, and an n^+ zone which forms the source is diffused into the p-type material. The structure is covered with a layer of silicon dioxide, the thin gate oxide, on top of which the n^+ polysilicon gate is formed. A second oxide layer isolates the gate from the source contact.

The VMOS structure just described can be considered as a 'cell' contained within the gate. In a practical VMOS transistor, the gate forms a lattice within which many cells are situated, connected in parallel by the source and drain contacts which extend over the whole of the upper and lower surfaces of the chip. The connection to the gate is made at the edge of the chip.

The VMOS transistor formed is an n-channel enhancement type. The transistor will not conduct, whatever the drain-to-source voltage, until the gate-to-source voltage exceeds the threshold value. As with the junction FET and MOSFET described previously, the drain voltage/current characteristic consists of a family of curves with gate voltage as parameter, as in *Figure 22.37*.

Typical values for the maximum drain current of VMOS transistors lie between 10 and 20 A, while the maximum drain-to-source voltage can be typically up to 200 V. Typical power dissipation can be about 75 W. Because the VMOS transistor can be switched with a gate voltage of a few volts, and because it has fast turn-on and turn-off times, the transistor can be switched directly by logic circuits and microcomputers. Thus the amplifiers required to interface logic circuits and junction power transistors are not required.

22.3.7 Comparison between junction transistors and FET's

The field-effect transistor can offer the circuit designer certain advantages over the junction transistor. Some of these advantages have already been mentioned. The high input resistance of the FET offers the designer a useful complement to the junction transistor whose input impedance, even in the common-collector configuration, is not as high as that previously obtainable with thermionic valves. As the FET only uses one current carrier, minority carrier storage effects do not occur to limit switching times. The FET can therefore be used for faster switching applications. The FET is also less affected by radiation since carrier lifetimes do not play an important part in its operation. The FET is also less affected by temperature than the junction transistor.

A disadvantage of the FET occurs through the thin gate oxide layer in insulated-gate devices. Electrostatic charges may accumulate on the gate lead during handling, and these can produce a large electric field in the oxide layer. This field may be large enough to puncture the layer. Once the device is connected in a circuit this danger is removed. To protect the insulated-gate FET during transport and handling, the leads are short-circuited to prevent the build-up of electrostatic charges. Another method of protection is to incorporate a diode structure from gate to substrate as a shunt protective element.

22.4 Symbols for main electrical parameters of transistors

Symbol	Meaning
C_{rs}	feedback capacitance in field-effect transistor
C_{Tc}	capacitance of collector depletion layer
C_{Te}	capacitance of emitter depletion layer
f_{hfb}	frequency at which the common-base forward current transfer ratio has fallen to 0.7 × low-frequency value
f_{hfe}	frequency at which the common-emitter forward current transfer ratio has fallen to 0.7 × low-frequency value
f_T	transition frequency (common-emitter gain-bandwidth product)
f_1	frequency at which common-emitter forward current transfer ratio has fallen to 1
G	gain
g_m	transconductance of field-effect transistor
h_{fb}, h_{fc}, h_{fe}	small-signal forward current transfer ratio for transistor configuration indicated by second subscript, output voltage held constant
h_{FB}, h_{FC}, h_{FE}	static forward current transfer ratio for transistor configuration indicated by second subscript, output voltage held constant
h_{ib}, h_{ic}, h_{ie}	small-signal input impedance for transistor configuration indicated by second subscript, output short-circuited to alternating current
h_{ob}, h_{oc}, h_{oe}	small-signal output impedance for transistor configuration indicated by second subscript, input open-circuit to alternating current
h_{rb}, h_{rc}, h_{re}	small-signal reverse voltage transfer ratio (voltage feedback ratio) for transistor configuration indicated by second subscript, output voltage held constant
i_b, i_c, i_e	instantaneous value of varying component of base, collector, emitter, current
i_B, i_C, i_E	instantaneous value of total base, collector, emitter, current
I_b, I_c, I_e	r.m.s. value of varying component of base, collector, emitter, current
I_{bm}, I_{cm}, I_{em}	peak value of varying component of base, collector, emitter, current
I_B, I_C, I_E	continuous (d.c.) base, collector, emitter, current
$I_{B(AV)}$, $I_{C(AV)}$, $I_{E(AV)}$	average value of base, collector, emitter, current
I_{BEX}, I_{CEX}	base, collector, cut-off current in specified circuit
I_{BM}, I_{CM}, I_{EM}	peak value of total base, collector, emitter, current
I_{CBO}	collector cut-off current, emitter open-circuit
I_{CBS}, I_{CES}	collector cut-off current, emitter short-circuited to base
I_{CBX}	collector current with both junctions reverse biased with respect to base
I_{CEO}	collector cut-off current, base open-circuit
I_D	drain current
I_{DM}	peak drain current
I_{EBO}	emitter cut-off current, collector open-circuit
I_G	gate current
I_{GM}	peak gate current
I_{GSS}	gate cut-off current
N	noise figure
P_{tot}	total power dissipated within device
$r_{ds(off)}$	drain-to-source off resistance
$r_{ds(on)}$	drain-to-source on resistance
$R_{th(j-amb)}$	thermal resistance, junction to ambient
$R_{th(j-case)}$	thermal resistance, junction to case
$R_{th(j-mb)}$	thermal resistance, junction to mounting base
T_{amb}	ambient temperature
T_{jmax}	maximum permissible junction temperature
T_{mb}	mounting base temperature
T_{stg}	storage temperature
t_{off}	turn-off time
t_{on}	turn-on time
V_{BE}	base-to-emitter d.c. voltage
$V_{BE(sat)}$	base-to-emitter saturation voltage
V_{CB}	collector-to-base d.c. voltage
V_{CBO}	collector-to-base voltage, emitter open-circuit
V_{CC}	collector d.c. supply voltage
V_{CE}	collector-to-emitter d.c. voltage
V_{CEK}	collector knee voltage
V_{CEO}	collector-to-emitter voltage, base open-circuit
$V_{CE(sat)}$	collector-to-emitter saturation voltage
V_{DB}	drain-to-substrate voltage
V_{DG}	drain-to-gate voltage
V_{DGM}	peak drain-to-gate voltage
V_{DS}	drain-to-source voltage
V_{DSM}	peak drain-to-source voltage
V_{EB}	emitter-to-base d.c. voltage
V_{EBO}	emitter-to-base voltage, collector open-circuit
V_{GB}	gate-to-substrate voltage
V_{GBM}	peak gate-to-substrate voltage
V_{GS}	gate-to-source voltage
V_{GSM}	peak gate-to-source voltage
$V_{GS(P)}$	gate-to-source cut-off voltage
V_{GSO}	gate-to-source voltage, drain open-circuit
V_P	pinch-off voltage in field-effect transistor
V_{SB}	source-to-substrate voltage
V_T	threshold voltage in field-effect transistor
η	intrinsic stand-off ratio in unijunction transistor

23 Thyristors and Triacs

M J Rose BSc (Eng)
Production Officer,
Marketing Communications Group,
Mullard Ltd

Contents

23.1 Switching diode

The switching diode consists of a four-layer pnpn structure with connections to the outer p and n regions, as shown in *Figure 23.1(a)*. The outer p region forms the anode, while the outer n region forms the cathode.

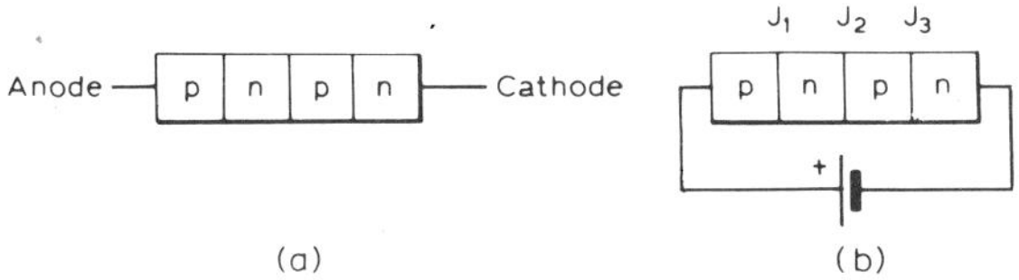

Figure 23.1 Switching diode: (a) structure; (b) biasing

If a voltage is applied across the diode with the positive terminal connected to the anode, as shown in *Figure 23.1(b)*, junctions J_1 and J_3 are forward-biased while junction J_2 is reverse-biased. Because the voltage drops across the forward-biased junctions are small, most of the applied voltage appears across the reverse-biased junction. The current through the diode will be the small reverse current of junction J_2. As the voltage across the diode is increased, the reverse current of junction J_2 will increase only slightly. Thus the forward current through the diode will be small, as shown in the forward-blocking part of the voltage/current characteristic in *Figure 23.2*. The impedance of the diode in this 'off' state is therefore high.

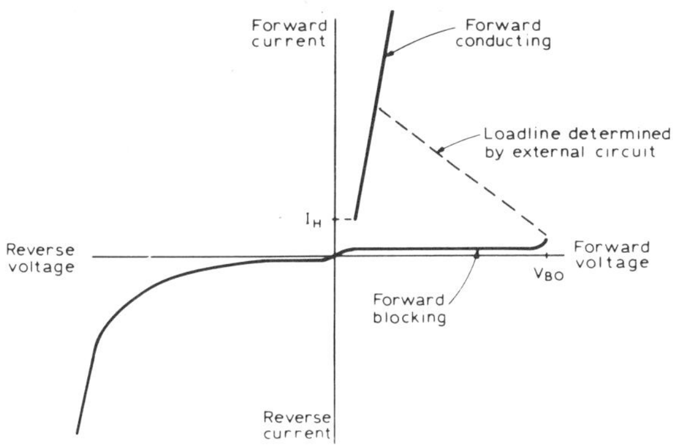

Figure 23.2 Voltage/current characteristic of switching diode

As the voltage is increased further, the reverse voltage across junction J_2 will reach the breakdown value. Avalanche breakdown occurs, and carriers are injected into the diode. The forward current increases along a loadline determined by the external circuit, and the voltage/current characteristic changes to the forward conducting part shown in *Figure 23.2*. The switching diode now acts as a normal forward-biased semiconductor diode, with a low 'on' resistance and with the normal forward voltage/current characteristic.

The voltage at which the change from forward blocking to forward conduction occurs is called the breakover voltage V_{BO}. The switching diode will remain in the conducting state as long as the current exceeds the holding value. The diode is switched off by reducing the current to below the holding value I_H.

The reverse voltage/current characteristic of the switching diode is similar to that of a two-layer semiconductor diode.

A disadvantage of the switching diode occurs if a triggering pulse is applied to the anode or cathode to change the device to the conducting state when the voltage across it is below the breakover value. The triggering circuit is directly connected to the load circuit, which may be a high-power circuit. The switching power may also be high. These disadvantages can be overcome by switching the device with a suitable voltage applied to one of the intermediate layers in the structure. The switching diode is then changed to a thyristor.

23.2 Thyristor

The thyristor is a three-terminal pnpn device. As in the switching diode just described, the outer p and n regions form the anode and cathode respectively, and the third terminal, the gate, is connected to the intermediate p region. By analogy to the switching diode, the original name for this structure was the reverse blocking triode thyristor. As it is the most commonly-used device in the thyristor family, it has generally become known simply as the thyristor.

It is convenient to regard the thyristor as a controlled rectifier diode. Like a rectifier diode, the thyristor will conduct a load current in one direction only, but the current can flow only when the thyristor has been 'triggered'. It is this property of being rapidly switched from the non-conducting to the conducting state that enables the thyristor to be used for the control of electrical power.

The thyristor was developed during the 1960s in parallel with the silicon rectifier power diode. Both devices were intended for use in power engineering. The similarity in application led to the thyristor being called the silicon controlled rectifier (SCR), or the controlled silicon rectifier which gives the device its circuit reference CSR. The circuit symbol for a thyristor is shown in *Figure 23.3*.

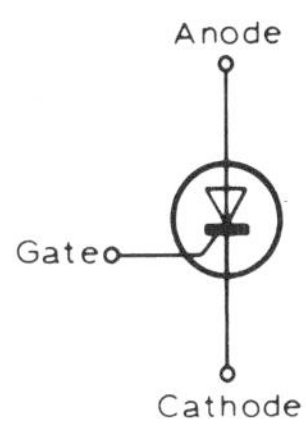

Figure 23.3 Circuit symbol for thyristor

23.2.1 Operation of thyristor

The schematic structure of a thyristor is shown in *Figure 23.4*. If a voltage is applied so that the anode is positive with respect to the cathode, and no voltage is applied to the gate, the operation will be similar to that described for the switching diode. Junctions J_1 and J_3 are forward-biased, junction J_2 is reverse-biased, and the thyristor will remain non-conducting until the voltage across it exceeds the forward breakover value and avalanche breakdown of junction J_2 occurs. The usual method of triggering the thyristor, however, is to apply a small positive voltage to the gate while the anode is positive with respect to the cathode.

The four-layer structure may be considered as forming two interconnected transistors, as shown in *Figure 23.5*. The cathode, gate and intermediate n region form an npn transistor, the cathode acting as the emitter, the gate as the base, and the n region as the collector. Junction J_3 is forward-biased and junction J_2 is reverse-biased as required for normal transistor operation. The anode, n region, and gate form a pnp transistor

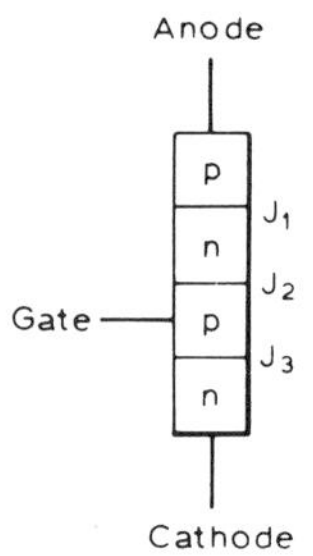

Figure 23.4 Schematic structure of thyristor

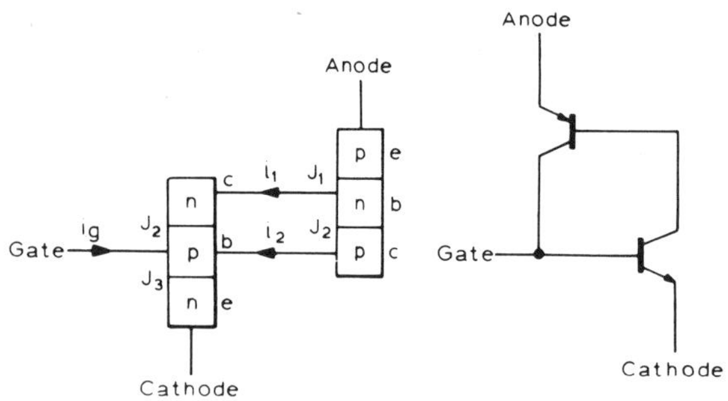

Figure 23.5 Two-transistor analogue of thyristor

with the anode as emitter, the n region as base, and the gate as the collector. Junction J_1 is forward-biased and junction J_2 is reverse-biased so normal transistor action can occur.

With only the anode-to-cathode voltage applied, the current through the thyristor is the small reverse current flowing across junction J_2. The thyristor is in the non-conducting forward blocking state. When the small positive voltage is applied to the gate, electrons flow from the cathode to the gate. This cathode–gate current forms the emitter–base current of the npn transistor, and by normal transistor action some of the electrons cross junction J_2 to enter the n collector region. Because the gate region is thicker than the base region of a transistor, the current gain will be less than unity.

The electron flow across junction J_2 causes the depletion layer to become narrower. The proportion of the anode-to-cathode voltage across the depletion layer is reduced, and the forward voltage across junctions J_1 and J_3 is increased. The flow of holes from the anode to the n-region (emitter–base current of the pnp transistor) is increased by this increase in forward bias across junction J_1. By normal transistor action, some of these holes cross junction J_2 to the gate (collector) region. Again, because the n region is thicker than the base in a transistor, the current gain is less than unity. The flow of holes across junction J_2 reduces further the width of the depletion layer. Consequently the proportion of the anode-to-cathode voltage across the depletion layer is reduced further, the forward voltage across junctions J_1 and J_3 is increased further, and the electron flow in the npn transistor is increased further.

This cumulative action, once initiated by the application of the gate voltage, continues until the depletion layer at junction J_2 collapses. The anode-to-cathode impedance becomes very small, and a large current flows through the thyristor. This current flow is self-sustaining, the thyristor is now in the forward conducting state, and the gate voltage can be removed. The value of current at which the thyristor changes from the non-conducting to the conducting state is called the latching current, and this occurs when the product of the current gains of the two transistors reaches unity.

The input current for the npn transistor is the gate current i_g. The collector current for this transistor is i_1, as shown in *Figure 23.5*, and this is the base current for the pnp transistor. The collector current of the pnp transistor i_2 is fed back to the base of the npn transistor. If the current gain of the npn transistor is h_{FE1}, then:

$$i_1 = h_{FE1}(i_g - i_2) \qquad (23.1)$$

If the gain of the pnp transistor is h_{FE2}, then:

$$i_2 = h_{FE2} i_1 \qquad (23.2)$$

The load current through the thyristor I_L is the sum of i_1 and i_2 and can be found from equations (23.1) and (23.2) as:

$$I_L = \frac{h_{FE1} i_g (1 + h_{FE2})}{1 - h_{FE1} h_{FE2}} \qquad (23.3)$$

The load current becomes infinite, that is the value will depend on the load resistance outside the thyristor rather than on the thyristor itself, when the denominator in equation (23.3) is zero; that is:

$$h_{FE1} h_{FE2} = 1 \qquad (23.4)$$

This is the condition for the thyristor to turn on and remain in the forward conducting state.

The thyristor is turned off by reducing the current through the device to below a value called the holding current. This is often done by making the anode of the thyristor negative with respect to the cathode.

The holding current is the point of extinction of conduction in the thyristor. It is located at that part of the thyristor element which has the highest sensitivity. At this point, the transistor gains are higher than in the rest of the element, and conduction is easier. When the thyristor is triggered by the gate trigger pulse, conduction will start near the gate in a region which is not necessarily the most sensitive. The holding current in this region must be attained before the trigger pulse can be removed, and this value of holding current is the latching current for the thyristor.

The static characteristic of the thyristor is shown in *Figure 23.6*. With a reverse voltage applied, the voltage/current charac-

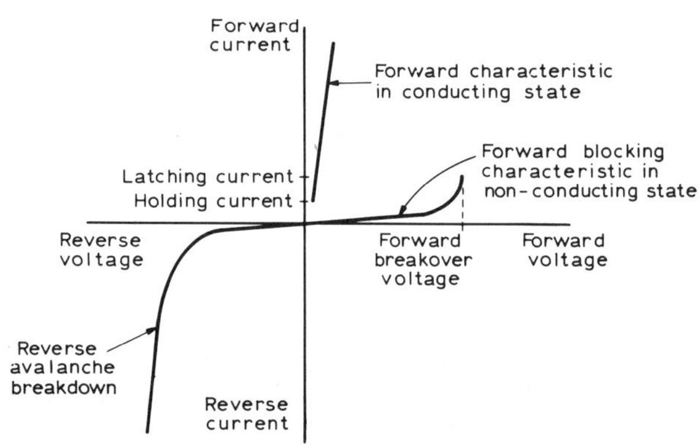

Figure 23.6 Static characteristic of thyristor

teristic is similar to that of any reverse-biased semiconductor diode. A small reverse current flows as the reverse voltage is increased until the reverse avalanche region is reached. The reverse current then increases rapidly as the thyristor breaks down. In the forward direction if the thyristor is not triggered, a small forward leakage current flows until the forward avalanche region is reached. This is the non-conducting forward blocking state. In the avalanche region, the leakage current increases until, at the forward breakover voltage V_{BO}, the thyristor switches rapidly to the conducting state as previously described for the switching diode. If the thyristor is triggered by a positive voltage applied to the gate when the anode-to-cathode voltage is less than V_{BO}, the thyristor again switches rapidly to the conducting state. In the conducting state, the thyristor behaves as a forward-biased semiconductor diode.

When the thyristor is triggered by a gate pulse, the current increases along a loadline determined by the external circuit. The change from the non-conducting to the conducting state for resistive and inductive loads is shown in *Figure 23.7*. With a resistive load, the turn-on may go directly from a high forward voltage to a high forward current, and the holding and latching current levels will only be reached when the current through the thyristor is being reduced during turn-off. With a highly inductive load, the initial turn-on current may be below the holding and latching levels. The forward blocking voltage across the thyristor is transferred to the inductance which produces the current increase shown in *Figure 23.7(b)*.

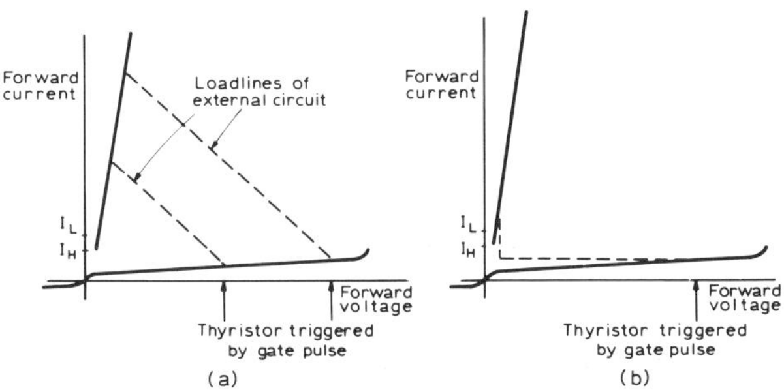

Figure 23.7 Turn-on of gate-triggered thyristor showing increase of current (broken lines): (a) resistive load; (b) inductive load

Also shown on the static characteristic is the holding current. If the load current through the thyristor falls below this value, the thyristor rapidly switches off to the non-conducting state.

The trigger pulse applied to the gate must remain until the current through the thyristor exceeds the latching value I_L. If the load current increases very slowly, for example because the load circuit is highly inductive, and the trigger voltage is removed before the latching value has been reached, the thyristor will stop conducting when the pulse ends. In practice, thyristors are triggered by trains of pulses rather than a single pulse.

23.2.2 Thyristor ratings and characteristics

It is convenient to group the ratings of thyristors into four: anode-to-cathode voltage ratings, current ratings, temperature ratings, and the gate ratings.

Because thyristors are used to control electrical power, or to control electrical machinery where safety is important, it is essential that the thyristors are always under the control of the gate signals and spurious triggering cannot occur. Thyristors often operate directly from the mains supply and are therefore subject to mains-borne voltage transients. Since the magnitude of these transients may be difficult to predict accurately, it may be necessary to provide filter circuits to ensure the thyristor is not triggered by voltages exceeding the forward breakover value.

As with semiconductor power diodes, three reverse voltage ratings are specified for a thyristor: continuous, repetitive, and non-repetitive values. Because thyristors may have to remain non-conducting in the forward direction, three off-state forward voltage ratings are also given. There are therefore six blocking voltage ratings specified in published data. These are listed below.

- V_{RWM} crest working reverse voltage, corresponding to the peak negative value (often with a safety factor) of the sinusoidal mains supply voltage
- V_{RRM} repetitive peak reverse voltage, corresponding to the peak negative value of transients occurring every cycle of the mains supply voltage
- V_{RSM} non-repetitive (surge) peak reverse voltage, corresponding to the peak negative value of transients occurring irregularly in the mains supply voltage
- V_{DWM} crest working off-state voltage, corresponding to the peak positive value (often with a safety factor) of the sinusoidal mains supply voltage
- V_{DRM} repetitive peak off-state voltage, corresponding to the peak positive value of transients occurring every cycle of the mains supply voltage
- V_{DSM} non-repetitive (surge) peak off-state voltage, corresponding to the peak positive value of transients occurring irregularly in the mains supply voltage

The repetitive and non-repetitive ratings are determined partly by the voltage limit and partly by the transient energy that the thyristor can safely absorb. The forward voltage limit is set by the need not to exceed the breakover voltage V_{BO}; the reverse voltage limit ensures that the thyristor is not driven into reverse avalanche breakdown. The transient energy that can be absorbed by the thyristor is limited (as usual) by the maximum permissible junction temperature. Therefore the duration of the repetitive and non-repetitive transients must be short compared with one half-cycle of the mains supply, and sometimes a maximum duration is specified for the transient.

The relationship between the voltage ratings and a typical mains supply waveform is shown diagrammatically in *Figure 23.8*. For clarity, the crest working reverse and off-state voltages have been shown equal to the peak sinusoidal component of the supply. It is considered good practice, however, in applications where thyristors operate directly from the mains supply to choose a device in which these two ratings are twice the peak sinusoidal supply voltage.

Figure 23.8 Diagrammatic relationship between thyristor voltage ratings and typical mains supply waveform with transients

Two other voltage ratings are usually given in published data. These are the on-state voltage V_T, the forward voltage drop across the thyristor when conducting a specified current, and the dV/dt rating. This second rating specifies the maximum rate of rise of the forward off-state anode-to-cathode voltage that will not trigger the thyristor into conduction. Because of the capacitance of the depletion layer associated with the reverse-biased junction, a rapidly-changing voltage across the non-conducting thyristor will induce a current across this junction. The magnitude of this current is:

$$i = C_d \frac{dV}{dt} \tag{23.5}$$

where C_d is the capacitance of the depletion layer. If this current exceeds the latching value, the thyristor will be turned on. This spurious triggering is prevented in practice either by using thyristors with high dV/dt ratings, or by connecting suppression components across the thyristor such as R and C networks to slow down the rate of rise of the applied voltage.

As with the voltage ratings, the current that can be conducted by a thyristor is specified by the continuous, repetitive peak, and non-repetitive peak values. The continuous current is also given as d.c., average, and r.m.s. values. The current ratings are listed below.

I_T	continuous on-state current at specified temperature
$I_{T(AV)}$	average on-state current at specified temperature
$I_{T(RMS)}$	r.m.s. on-state current
I_{TRM}	repetitive peak on-state current
I_{TSM}	non-repetitive peak on-state current, specified as the peak of a half-sinewave of 10 ms duration or as the amplitude of a square pulse of 5 ms duration, both at the maximum junction temperature before the surge current occurs

As well as the I_{TSM} rating, a curve of the maximum r.m.s. surge current plotted against the surge duration is given in the published data.

Two other current ratings are given in published data, the I^2t and the di/dt ratings. The I^2t rating, or surge current capability, is required for selecting fuses to protect the thyristor against excessive current being drawn from the device if a short-circuit occurs in the load circuit. The fuse has an I^2t rating, and this should be equal to or less than that of the thyristor to provide adequate protection. The di/dt rating is the maximum rate of rise of current when the thyristor is turned on which will not cause unequal current distribution and therefore hot spots in the junctions. The rate of rise of current will be determined by the load circuit. If it is greater than the permitted maximum, the value can be limited by connecting a small inductance in series with the thyristor.

The power that can be dissipated within a thyristor (as with all semiconductor devices) is limited by the maximum permissible junction temperature. It is usual to mount the thyristor on a heatsink to increase the cooling area. The heatsink is selected by calculating the required thermal resistance. The quantity *thermal resistance* is a measure of the opposition to the flow of heat, and is measured as the change in temperature with power. The usual units are °C W. For any semiconductor device:

$$P_{tot} = \frac{T_{j,max} - T_{amb}}{\Sigma R_{th}} \tag{23.6}$$

where $T_{j,max}$ is the maximum permissible junction temperature, T_{amb} is the ambient temperature, and ΣR_{th} is the total thermal resistance between the thyristor junction and ambient. The chain of thermal resistance making up ΣR_{th} is the same as that for a diode and is given in *Figure 21.8*. $R_{th(j-mb)}$, the thermal resistance between the thyristor junction and the mounting base; $R_{th(mb-h)}$, the contact thermal resistance between the thyristor and heatsink at a specified torque; and $R_{th(h)}$, the thermal resistance of the heatsink. Of these quantities, $R_{th(j-mb)}$ and $R_{th(mb-h)}$ are given in published data, and $T_{j,max}$ is also given. The value of P_{tot} can be derived from curves in the data. Thus ΣR_{th} can be calculated from equation (23.6), and from this the value of $R_{th(h)}$ derived.

Curves are published in the data enabling the heatsink to be selected without the need for the calculation just described. These

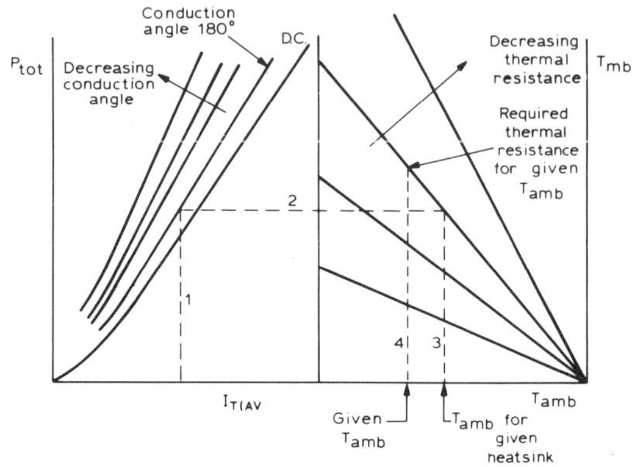

Figure 23.9 Selection of heatsink for thyristor: average current and power with conduction angle as parameter related to mounting-base temperature and ambient temperature with heatsink thermal resistance as parameter

curves are shown in *Figure 23.9*. The left-hand graph relates the power dissipation of the thyristor P_{tot} and the average current $I_{T(AV)}$, with the conduction angle as parameter. The conduction angle is the part of the positive half-cycle of the supply for which the thyristor is conducting. The thyristor is used to control electrical power by variation of the conduction angle. The right-hand graph relates the mounting-base temperature of the thyristor T_{mb} to the ambient temperature T_{amb}, with the thermal resistance of the heatsink as parameter.

For a particular application, the average current and the conduction angle are known. The vertical broken line (1) on the left-hand graph of *Figure 23.9* can therefore be plotted. A horizontal line (2) from the point of intersection of the average current line with the conduction angle curve is drawn to the right-hand graph. If the thermal resistance of the heatsink is known, the intersection of the horizontal line with the thermal resistance curve will give the corresponding ambient temperature, vertical line (3). If a limit to the ambient temperature is specified, the required thermal resistance of the heatsink can be found, as shown by the broken line (4) on the right-hand graph.

For some low-power thyristors, the thermal resistance between junction and ambient, $R_{th(j-amb)}$, is given in published data. From the maximum permissible junction temperature and ambient temperature, the total dissipation without a heatsink can be found.

The gate ratings of the thyristor ensure that all devices of the same type number are reliably triggered over the complete operating temperature range without exceeding the permissible gate dissipation. The principal gate ratings are listed below.

$P_{G(AV)}$	average gate power, averaged over any 20 ms period
P_{GM}	maximum gate power
V_{GD}	maximum continuous gate voltage that will not initiate turn-on
V_{GT}	minimum instantaneous gate voltage to initiate turn-on
I_{GT}	minimum instantaneous gate current to initiate turn-on

A gate characteristic is often given in published data. The form of this characteristic is shown in *Figure 23.10*. The two curves in full lines define the limits between which the gate ratings of all thyristors of the same type number will lie. The curve in broken lines defines the maximum permissible gate dissipation. The minimum voltage and minimum current to initiate turn-on are also shown. These five limit curves define the area of certain triggering for the whole of the operating temperature range of the

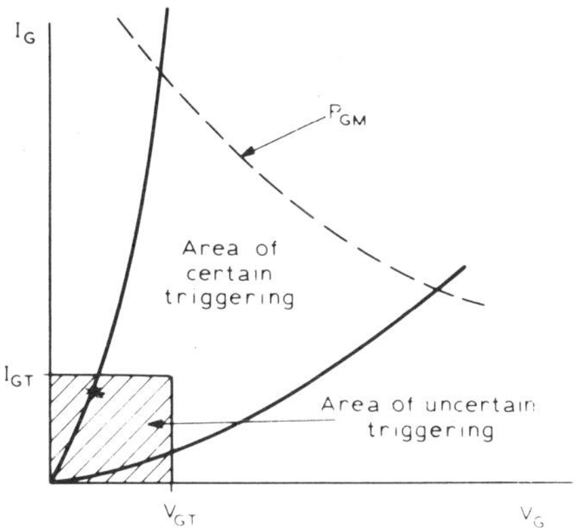

Figure 23.10 Gate characteristic showing area of certain triggering

thyristor. The minimum voltage and current are temperature-dependent, and the area of uncertain triggering will decrease as the junction temperature increases.

When the trigger pulse is applied to the gate of a non-conducting thyristor which has a positive anode voltage with respect to the cathode, there is a short delay, then the thyristor switches rapidly to the conducting state. The switching characteristic showing the rise of thyristor current and fall of anode-to-cathode voltage is given in *Figure 23.11*. The rise time of the

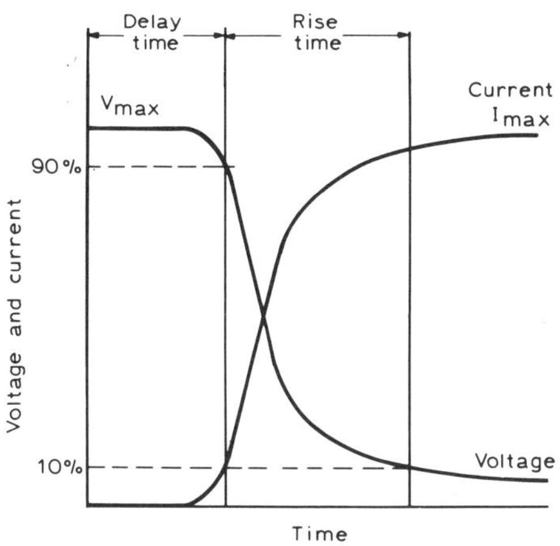

Figure 23.11 Switching characteristic of thyristor

anode current is defined as the time taken for the voltage to fall from the 90% to the 10% level. The sum of the delay time t_d and the rise time t_r is the turn-on time of the thyristor t_{gt}.

The shape of the trigger pulse appreciably affects the turn-on time of the thyristor, in particular the rate of rise of the applied gate current. For most switching applications in which inverter-grade thyristors are used, a rise time for the gate current of 1 μs is satisfactory. Typical values for this type of thyristor with a trigger-pulse rise time of 1 μs are delay times between 0.5 μs and 2.0 μs, and rise times between 0.2 μs and 1.2 μs.

The thyristor is turned off by reducing the current through the device to below the holding value I_H. The value of I_H will vary with the type of thyristor, but typical values are between 10 mA and 200 mA. In practical circuits, the thyristor is often reverse-biased to ensure that it is reliably turned off. The reverse bias must be applied for a minimum time called the turn-off time t_q. For fast-turn-off thyristors, the value of t_q will be between 2.4 μs and 25 μs. For power-control thyristors, the value of I_q will be between 20 μs and 150 μs depending on the type. The value of turn-off time increases with increase in junction temperature.

With fast-turn-off thyristors, as with semiconductor diodes, the turn-off time is limited by minority-carrier storage effects. A majority carrier from one region crossing into a region of the opposite polarity becomes a minority carrier in that region. If the bias across the junction is reversed, the minority carriers must recross the junction or combine with carriers of the opposite polarity before the depletion layer at the junction can be established. Thus the current flowing immediately after the reverse bias is applied may be high as the carriers recross the junction. This current must decay before reverse blocking of the thyristor is established. For fast switching it is necessary to reduce the lifetime of the minority carriers. The technique of gold doping the silicon material used with diodes can be applied to thyristors to give faster switching times.

The first thyristors introduced in the early 1960s were capable of handling currents up to approximately 100 A and withstanding voltages up to 300 V. By the middle of the decade, the current-handling and voltage capability had increased so that thyristors could be operated directly from the 440 V three-phase mains supply. Typical present-day thyristors are able to handle currents up to 1000 A and withstand voltages up to 2.5 kV. Devices with dV/dt ratings up to 1000 V/μs are available, enabling safe operation to be achieved with a minimum of protective components.

23.2.3 Manufacture of thyristors

Two manufacturing processes are used for thyristors: the diffused-alloyed and the all-diffused processes. The diffused-alloyed process is the older, and was used for the first thyristors. The all-diffused process has considerable advantages over the other, and is rapidly replacing it.

The starting material for the diffused-alloyed process is a slice of n-type monocrystalline silicon which will form the intermediate n layer in the completed thyristor, *Figure 23.12(a)*. The doping level of the silicon is chosen to suit the breakdown voltage required of the thyristor. A lightly-doped slice is used for high-voltage devices; a higher-doped slice for the low-voltage devices. The thickness of the slice is a compromise between conflicting requirements. The depletion layer across the anode-to-n-layer junction (J_1 in *Figure 23.4*) with reverse bias extends into the n layer. The layer must be thick enough to prevent the depletion layer extending across the whole layer, otherwise 'punch through' will occur and the thyristor will break down at a lower voltage. If the layer is too thick, a higher value of V_T is obtained, and therefore the forward losses will be higher. A typical thickness for a 1000 V thyristor is 275 μm, with a typical resistivity of 0.3 Ω m.

A diffusion process is used to form p-type regions on both sides of the n-type slice, thus forming the gate and anode of the thyristor, as shown in *Figure 23.12(b)*. The p-type impurity used for the diffusion is gallium or aluminium, and layers approximately 75 μm deep are diffused into the slice. The pnp wafer formed after the diffusion is cut into a round slice whose diameter is determined by the required current-carrying capability. For the highest current devices, the diameter may be 2.5 cm or more. The anode contact is made to one of the p layers using an aluminium solder to bond a molybdenum or tungsten disc to the anode, as shown in *Figure 23.12(c)*. Because of the heavy currents that have to be conducted and the consequent large amount of heat developed, the different temperature coefficients of expansion for the silicon and the copper used for

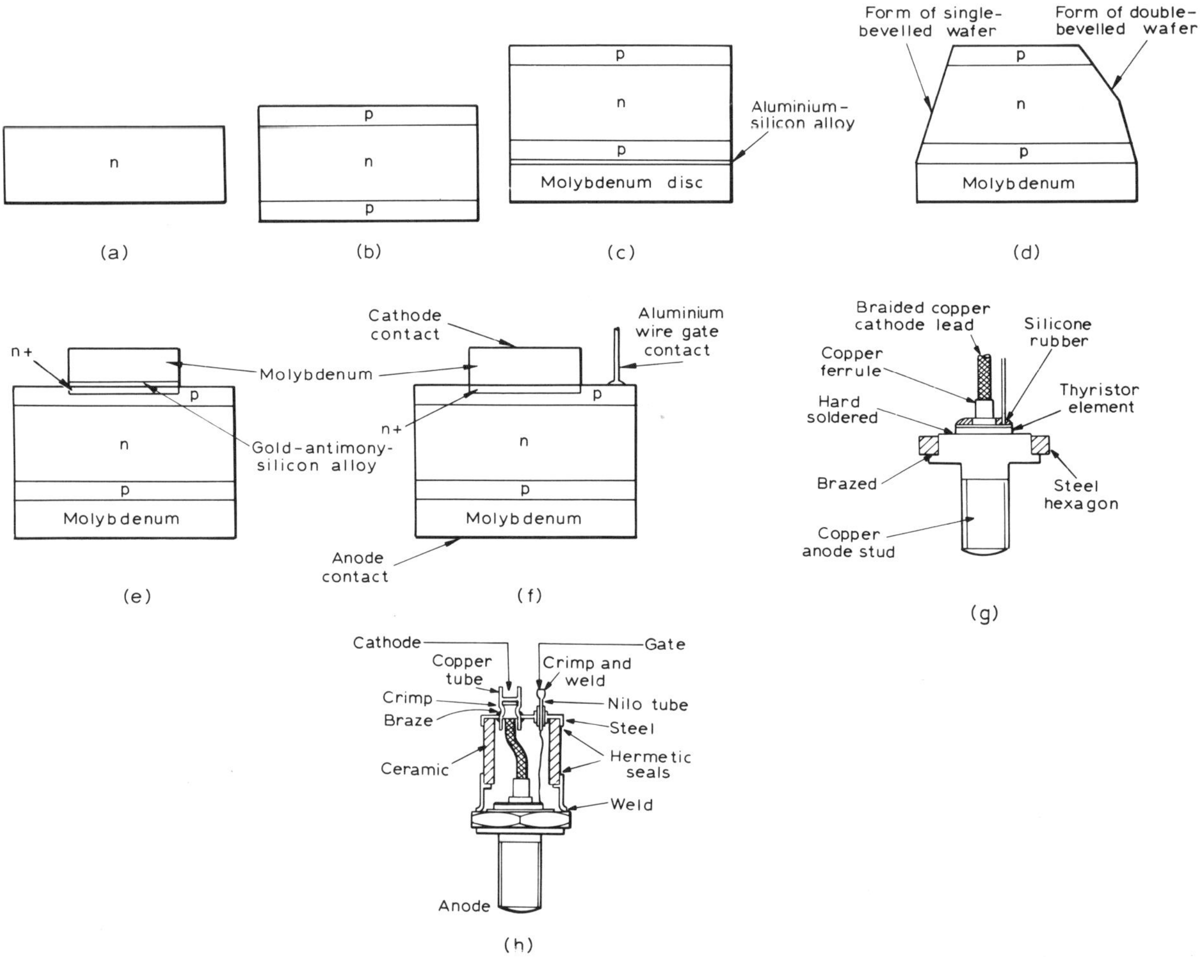

Figure 23.12 Diffused-alloyed manufacturing process for thyristor: (a) starting n-type slice; (b) diffused pnp wafer; (c) anode molybdenum disc attached; (d) edges of wafer bevelled (showing single and double bevel); (e) cathode and molybdenum disc attached; (f) gate contact fitted; (g) mounted on header; (h) encapsulation

connections would lead to stresses within the device, and the possible fracture of the thyristor element. Molybdenum and tungsten have coefficients of expansion comparable with that of silicon. The edges of the pnp wafer may be bevelled to increase the depletion width at the surface and to ensure that any avalanche breakdown occurs in the bulk of the material rather than at the edge, *Figure 23.12(d)*.

The final stage in the manufacture of the thyristor element is the alloying of the cathode. A foil of gold–antimony alloy and a molybdenum backing disc are alloyed in the p-type gate layer, *Figure 23.12(e)*. The gold–antimony foil produces an n^+ region in the p layer, while the molybdenum disc matches the coefficient of expansion of the silicon to that of the copper connecting wire. The depth to which the gold–antimony is alloyed controls the base width of the npn transistor of the two-transistor analogue (*Figure 23.5*) and hence the gain. Too wide a base will also increase the value of the holding current. The gate connection to the thyristor element is made with aluminium wire to ensure a non-rectifying contact. The wire is welded ultrasonically to the exposed p layer, *Figure 23.12(f)*. The element is etched to clean the exposed junctions J_1 and J_2 which provide the high voltage blocking characteristics of the thyristor.

The molybdenum anode contact is hard-soldered to a copper header which forms the anode connection of the device. This is usually in the form of a threaded stud to allow mounting on a heatsink, *Figure 23.12(g)*. The cathode connection is formed by a copper lead of suitable diameter for the currents to be conducted. The lead is soldered to the cathode molybdenum disc. The device is encapsulated in a ceramic and metal top-cap which is welded to the header, as shown in *Figure 23.12(h)*. The gate and cathode connections for the completed thyristor are made by tabs or by flying leads.

The accepted convention for rectifiers is that devices with cathode studs are 'normals', and those with anode studs are 'reverses'. Under this convention, thyristors manufactured in the way just described are 'reverses'. Thyristors with cathode studs are feasible, and some devices are manufactured, but they are not generally available. This is because of manufacturing problems. It can be seen from *Figure 23.12(f)* that if the cathode molybdenum disc is soldered to the header, it would be more difficult to connect the gate lead to its external terminal.

An alternative method of encapsulation can be used for low-power thyristors. This uses a moulded plastic package which is considerably cheaper than the type of encapsulation just described. The thyristor element is mounted on a metal plate which is incorporated in the plastic package so that contact can

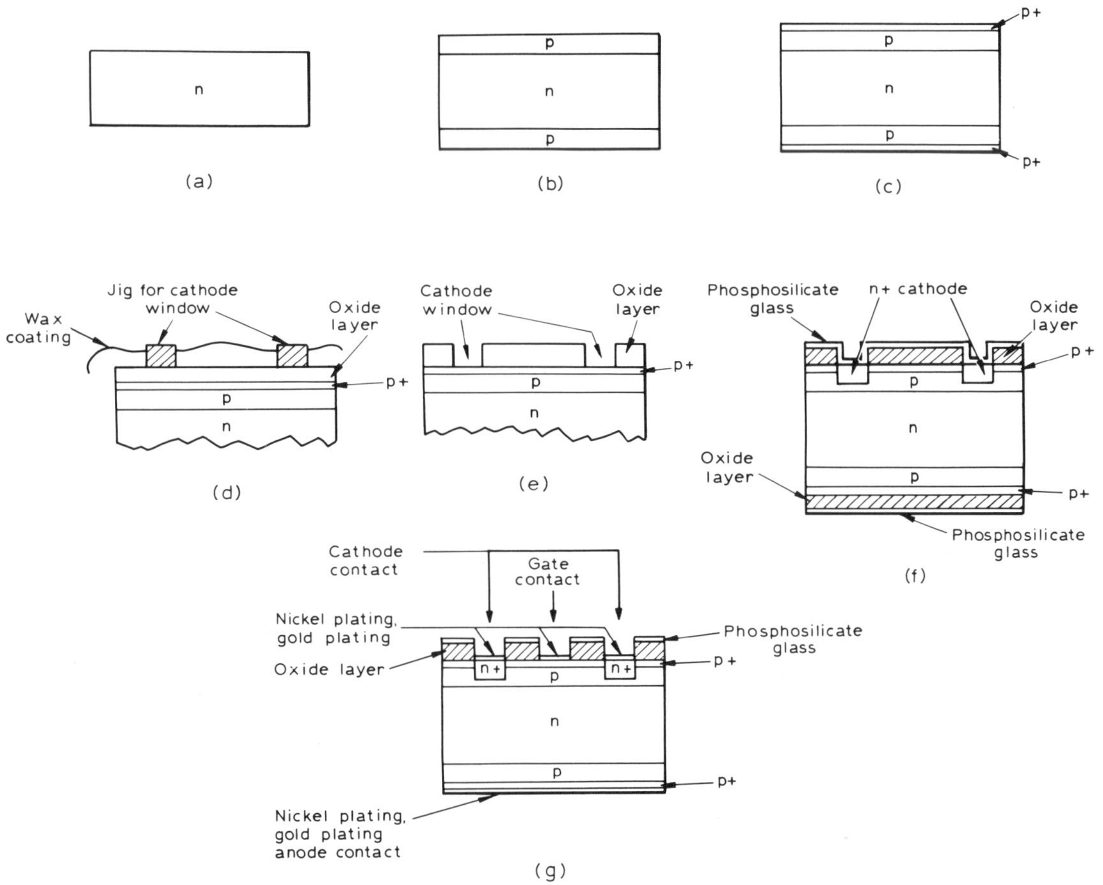

Figure 23.13 All-diffused manufacturing process for thyristor: (a) starting n-type slice; (b) diffused pnp wafer; (c) p^+ diffusion; (d) jig and wax coating to define cathode window; (e) cathode window opened; (f) cathode diffused; (g) contact windows opened and contact pads formed

be made to a heatsink. Thyristors used in such applications as the line output stages of television receivers may use 'transistor' encapsulations such as the TO-39 and TO-66 packages.

In the all-diffused manufacturing process, the starting material is again n-type silicon of similar thickness and doping level to that used in the diffused-alloyed process, *Figure 23.13(a)*. Again, the first stage of manufacture is a p diffusion using aluminium or gallium as the impurity, forming a pnp wafer as shown in *Figure 23.13(b)*. From this stage, however, the two processes differ. Boron is diffused into both sides of the wafer to form a thin p^+ layer, as shown in *Figure 23.13(c)*. This layer compensates for the migration of phosphorus into the p layers that occurs during the contact-forming stage later in the process. The layer also imposes a control on the value of the holding current.

The next stage in the process is the diffusion of the cathode region. For the higher-current devices where one element is made from the pnp wafer, the wafer is cut to the required size. For the smaller-diameter low-current devices, many thyristor elements (say, up to 100) are made on one slice. It is therefore more economical to diffuse the cathodes for all the elements at the same time, and then separate the individual thyristor elements.

A layer of silicon dioxide is grown over the surface of the wafer by heating it in a furnace at 1200°C in a stream of wet oxygen. The thickness of the layer is approximately 1 μm. A window is formed in the oxide through which the cathode is diffused. Two techniques are used to form the window. The simpler one uses a jig where the window is to be, and coats the wafer with wax, *Figure 23.13(d)*. The jig is removed, and a chemical etch used to remove the oxide which is not covered by wax. The other method uses the photographic techniques associated with the planar process. The surface of the oxide is spun-coated with a layer of photoresist about 1 μm thick. The photoresist is then dried by baking. A mask defining the cathode window is placed over the wafer and exposed. On development of the photoresist the unexposed area is removed, giving access to the oxide layer, while the exposed area remains and is hardened by further baking. A chemical etch is used to remove the exposed oxide. The remaining photoresist is then dissolved away, leaving the structure shown in *Figure 23.13(e)*. The usual structure for thyristors manufactured by this process is a central gate with an annular cathode.

Phosphorus is used to form the n-type cathode region. The wafer is heated in a diffusion furnace to a temperature of 1050°C in a stream of phosphorus pentoxide vapour P_2O_5. A phosphosilicate glass is formed, about 0.5 μm thick, from which phosphorus is driven into the p^+ layer by heating in a second furnace at 1250°C. The phosphosilicate glass is not removed but used as the masking for the formation of the contacts in the next stage of manufacture. The resulting structure is shown in *Figure 23.13(f)*.

The top surface of the wafer is sprayed with wax except for

those areas where windows are to be cut in the glass to form the gate and cathode contacts. The exposed glass in these regions, and that on the whole bottom surface of the wafer, is etched away. The wafer is nickel plated, followed by gold plating. The nickel plating is effected by chemical displacement, and therefore nickel is only deposited on clean silicon. The contacts are formed only in the windows and on the bottom of the wafer, *Figure 23.13(g)*. If many thyristor elements have been formed on one wafer, it is at this stage that the wafer is divided into individual elements. The edges of the element may be bevelled to increase the depletion width at the surface to ensure that breakdown takes place in the bulk of the material rather than at the edge.

The mounting and encapsulation of the element are similar to those for the diffused-alloyed types. Molybdenum backing discs are soldered to the anode and cathode to reduce thermal stress caused by the different coefficients of expansion of the silicon element and the copper connections. The element is mounted on a copper header which forms the mounting stud, and encapsulated in a ceramic and metal top cap. Alternatively, a moulded plastic package is used.

In the manufacturing process for all-diffused thyristors, especially for the higher-current types, the yield of devices is greater than with the diffused-alloyed process. In addition, the all-diffused types can have more complex cathode structures which improve the value of forward breakover voltage, the temperature stability, and the dynamic performance. It is possible, for example, to include small p-type regions linking with the p gate layer in the diffused n-type cathode. These small p regions are called p shorts, and the structure formed is called shorted emitter. The p shorts provide a means of charging the junction capacitance of the gate-to-intermediate-n-layer (J_2 in *Figure 23.4*) without injection of carriers from the cathode. This allows the thyristor to withstand high rates of rise of the off-state voltage, giving the device a high d*V*/d*t* rating. This technique is not possible with diffused-alloyed thyristors, and so the all-diffused types have higher d*V*/d*t* ratings.

23.3 Triac

The triac, or bidirectional triode thyristor, is a device which can be used to conduct or block current in either direction. It can therefore be regarded as two thyristors connected in the inverse-parallel configuration, but with a common gate electrode. Unlike the thyristor, however, the triac can be triggered with either a positive or a negative gate pulse.

The circuit symbol for a triac is shown in *Figure 23.14*. The load connections of the device are called main terminals, MT 1 and MT 2. A simplified cross-section of the triac structure is shown in *Figure 23.15*. Because the current can be conducted in either direction, MT 1 and MT 2 are connected to both p and n regions. Similarly, because the triac can be triggered by both positive and negative pulses, the gate is connected to both p and n regions. The actual structure is more complex than the cross-section indicates, and is better shown by the 'exploded' view of *Figure 23.16*. However, for explaining the operation of the device, the simplified cross-section is adequate.

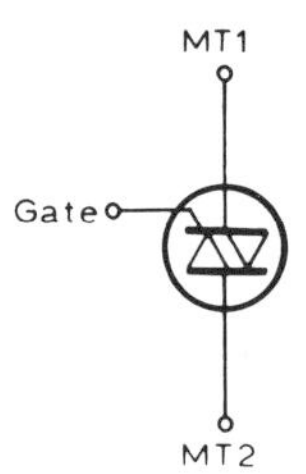

Figure 23.14 Circuit symbol for triac

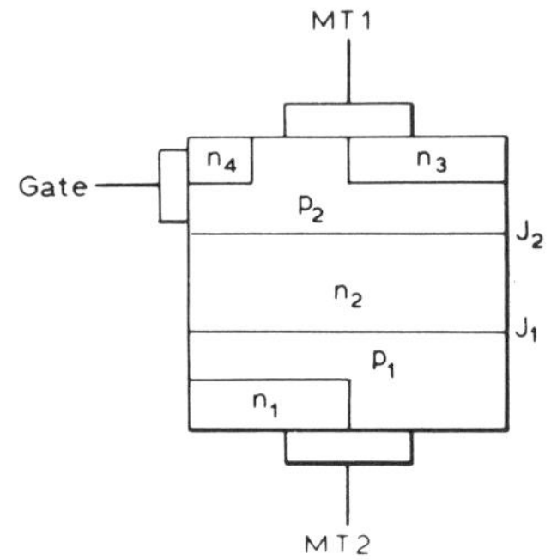

Figure 23.15 Simplified cross-section of triac structure

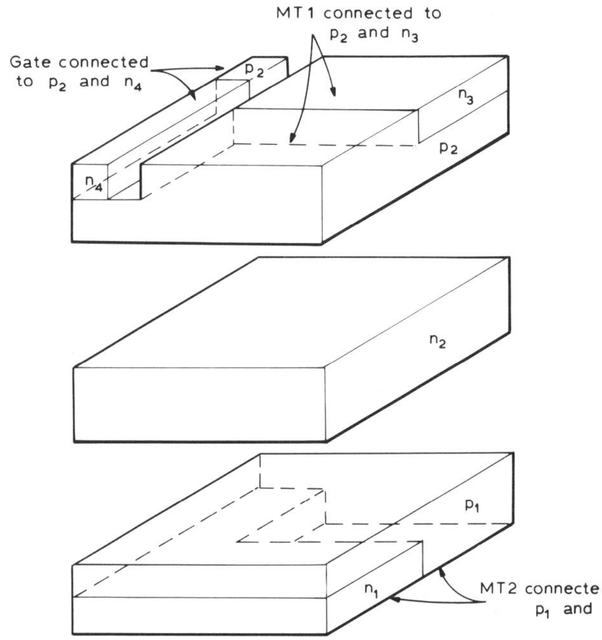

Figure 23.16 Exploded view of triac structure

There are four possible modes of operation with a triac.

MT 2 positive with respect to MT 1, gate pulse positive
MT 2 positive with respect to MT 1, gate pulse negative
MT1 positive with respect to MT2, gate pulse positive
MT 1 positive with respect to MT 2, gate pulse negative

Operation with MT 2 positive with respect to MT 1 is called 'first-quadrant operation'; operation with MT 1 positive with respect to MT 2 is called 'third-quadrant operation'. The quadrants refer to those in which the static characteristics appear when plotted on a coordinate axis system. The operation of the triac in all four modes can be explained by the two-transistor analogue used previously for the thyristor.

In first-quadrant operation, MT 1 acts as the cathode while MT 2 acts as the anode. Junction J_1 (*Figure 23.15*) is forward-biased, and junction J_2 is reverse-biased.

With a positive gate pulse, a forward bias is applied across the gate p-region and cathode junction, the p_2n_3 junction. Electrons flow from n_3, through p_2, and are collected by n_2. These three regions form an npn transistor. The flow of electrons decreases the reverse bias across junction J_2, thereby increasing the forward bias across junction J_1. The hole current in the pnp transistor formed by p_1, n_2, and p_2 increases, decreasing further the reverse bias across junction J_2 and so increasing the electron flow. A cumulative action as in the thyristor occurs, and so the triac is turned on.

With a negative gate pulse, a forward bias is applied across the p_2n_4 junction, causing electrons to flow from n_4 to n_2. The npn transistor for this mode is formed by n_4, p_2, and n_2; the pnp transistor is p_1, n_2, p_2, as with positive triggering. The same cumulative action as before occurs to turn on the triac.

In third-quadrant operation, MT 2 acts as the cathode while MT 1 acts as the anode. Junction J_1 is reverse-biased, and junction J_2 is forward-biased.

With a positive gate pulse, a forward bias is applied across the p_2n_3 junction causing electrons to flow across the forward-biased junction J_2 to n_2. The electron flow causes holes to flow from p_2, through n_2, to be collected by p_1, forming a pnp transistor. The hole current causes electrons to flow in the npn transistor n_1, p_1, n_2. Cumulative action occurs between these two transistors to turn on the triac.

With a negative gate pulse, forward bias is applied across the p_2n_4 junction causing electrons to flow to n_2. This electron flow causes holes to flow in the pnp transistor p_2, n_2, p_1; and electrons to flow in the npn transistor n_1, p_1, n_2. The same cumulative action as with positive triggering occurs to turn on the triac.

The static characteristic of the triac is shown in *Figure 23.17*.

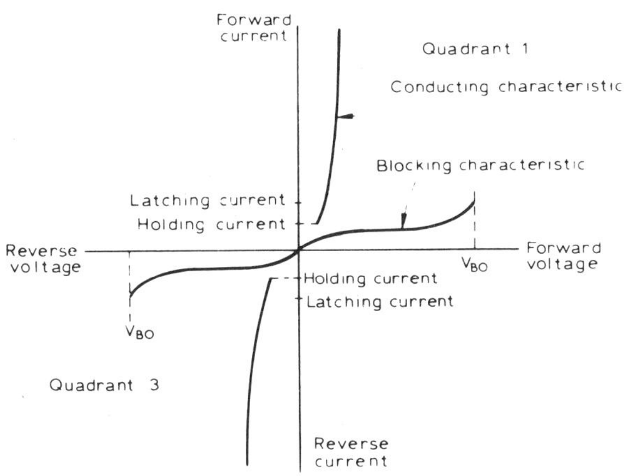

Figure 23.17 Static characteristic of triac

As would be expected, it consists of two thyristor forward voltage/current characteristics combined, the 'forward' current being that when the voltage at MT 2 is positive with respect to MT 1. The characteristics and ratings of the triac are similar to those of the thyristor already described, except that the triac does not have any reverse voltage ratings (a reverse voltage in one quadrant becomes the forward voltage of the opposite quadrant).

The triac is used in power-control applications to replace an inverse-parallel pair of thyristors, that is as a fully-controlled a.c. controller. The triac offers the advantages over the thyristor pair of easier mechanical mounting on the heatsink as only one device is used, and of a simpler triggering system.

23.4 Triggering of thyristors and triacs

Thyristors and triacs can be triggered by a single pulse, a train of pulses or a steady d.c. voltage on the gate. Pulse triggering is most frequently used; single-pulse triggering is used in specific applications and low-cost systems, and d.c. triggering only in cases of difficulty in reaching latching current. A thyristor or triac with the anode positive with respect to the cathode, and with adequate gate drive, will turn on within 10 μs. Conduction will continue irrespective of load current for as long as the gate drive is present. Conduction will continue after the gate drive ceases only if the load current has reached the latching level.

Pulse triggering is preferred to d.c. triggering for general use because the gate dissipation is lower and the trigger system is simpler. Both pulse and d.c. trigger systems must be synchronised to the mains supply, and both use a pulse generator operating through a variable delay to provide the control over trigger angle required for phase control. Whereas in a pulse trigger system the output from the pulse generator can be passed to the thyristor gate through an isolating transformer, a d.c. trigger system requires either rectification of the pulse transformer output or gating of a further supply by the pulse transformer to provide the d.c. gate drive. The more elaborate d.c. gate drive is therefore used only in applications where it is essential, for example with highly inductive loads where difficulty is experienced with latching.

The usual pulse triggering system provides a train of pulses starting at the trigger angle and continuing throughout the half-cycle of the supply. The duration of each pulse is at least 10 μs, generally 20 μs, and the repetition frequency is typically 5 kHz. With inductive loads, a flywheel diode or series *RC* network connected across the load can assist latching.

23.5 Silicon controlled switch

The silicon controlled switch (s.c.s.) is a four-layer device with all four layers accessible. Because it is a four-terminal device, it is also known as the tetrode thyristor. The s.c.s. differs from the thyristor in that not only is it turned on by a gate signal, but it can also be turned off by a gate signal. The circuit symbol for an s.c.s. is shown in *Figure 23.18*.

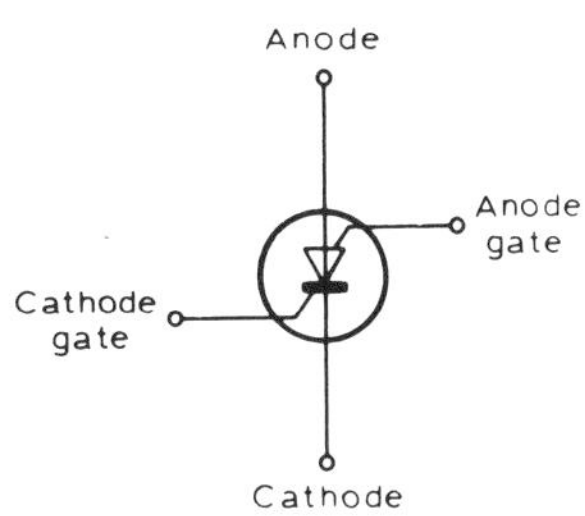

Figure 23.18 Circuit symbol for silicon controlled switch

The schematic structure of an s.c.s. is shown in *Figure 23.19(a)*. The outer p and n regions form the anode and cathode respectively, the intermediate p region is called the cathode gate, and the intermediate n region is called the anode gate. With a voltage applied so that the anode is positive with respect to the cathode, junctions J_1 and J_3 are forward-biased, and junction J_2 is reverse-biased. Because the current through the s.c.s. is the small reverse leakage current of junction J_2, the device is non-conducting.

The operation of the s.c.s. (as with that of the thyristor and triac) can be described in terms of two interconnected transistors.

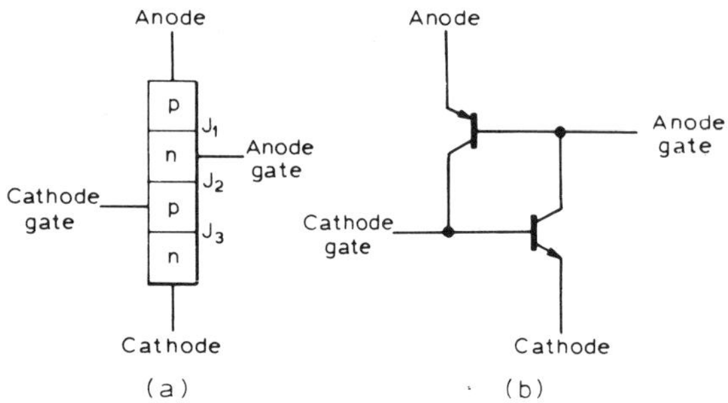

Figure 23.19 Silicon controlled switch: (a) schematic structure; (b) two-transistor analogue

The two-transistor analogue for the s.c.s. is shown in *Figure 23.19(b)*. This differs from that for the thyristor shown in *Figure 23.5* in that the bases of both transistors are accessible. Thus the cumulative action between the two transistors that causes turn-on can be initiated by a pulse on either base: a positive pulse on the cathode gate to start current flowing in the npn transistor, or a negative pulse on the anode gate to start current flowing in the pnp transistor. When the current through the s.c.s. exceeds the latching value (when junction J_2 changes to forward bias), the device remains in the conducting state when the gate pulse is removed.

The s.c.s. (like the thyristor and triac) is turned off by reducing the current through the device to below the holding value. This can be done by applying a negative pulse to the anode, a positive pulse to the anode gate (applying reverse base drive to the pnp transistor), or a negative pulse to the cathode gate (reverse base drive to the npn transistor).

In practice, when the s.c.s. is used as a fast-operating switch, the load is usually connected in the anode-gate circuit, as shown in *Figure 23.20(a)*. The device is then turned off by a negative voltage

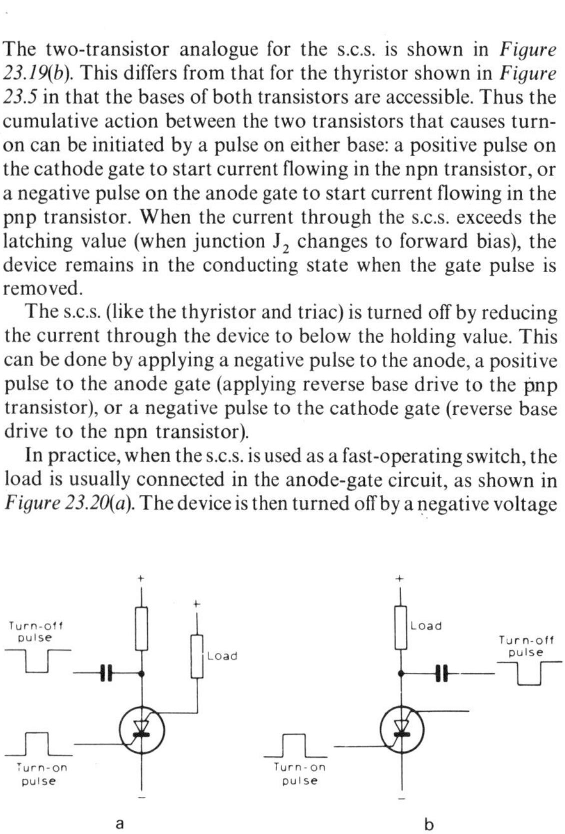

Figure 23.20 Circuit configurations for silicon controlled switch: (a) load in anode-gate circuit; (b) load in anode circuit

pulse to the anode which reduces the current to below the holding value for a time greater than the turn-off time. A typical value for the turn-off time t_q or t_{off} is 5 μs, although the value depends on the junction temperature and external cathode-gate resistance. To ensure that the s.c.s. is reliably turned off and to prevent spurious triggering, it is usual to maintain the anode gate and cathode gate at definite voltages rather than let them float. A positive voltage is applied to the anode gate so that the anode-to-anode-gate junction is reverse-biased. A negative voltage is applied to the cathode gate so that the cathode-to-cathode-gate junction is reverse-biased.

The positive trigger pulse to the cathode gate overcomes this reverse bias to turn on the s.c.s., the positive voltage on the anode gate assisting conduction in the npn transistor of the two-transistor analogue. A typical turn-on time for an s.c.s. is between 0.25 μs and 1.5 μs, depending on the external cathode-gate resistance. The turn-on time is constant over a considerable temperature range. Curves of variation of turn-on and turn-off times with temperature and cathode-gate resistance are often given in published data.

The published data for an s.c.s. in this configuration are given in the form of data for the complementary transistors of the two-transistor analogue. The forward on-state voltage and holding current are specified for the combined device. Characteristic curves showing the variation of the main voltages and currents are also given. For example, the variation of anode current with anode-to-cathode voltage and cathode-gate voltage, and the variation of anode-gate current with cathode-gate voltage.

The s.c.s. can also be used with the load connected in the anode circuit, as shown in *Figure 23.20(b)*. The device now operates as a low-power thyristor. The anode gate is either not connected, or connected to the supply through a high-value resistor to maintain the correct operating voltage on the anode gate. If negative trigger pulses are to be used, then the cathode gate is connected to the cathode line through a high-value resistor, and the negative trigger pulses are applied to the anode gate. The s.c.s. in this configuration is used in sensing networks, being triggered to operate a relay or lamp when the voltage on the gate exceeds a predetermined value. The data for this application are presented in the usual form for a thyristor. The three reverse and off-state voltages are listed, together with the currents and such characteristics as dV/dt, di/dt, gate characteristics, and holding current.

23.6 Gate turn-off switch

The gate turn-off (GTO) switch is switched on by applying a positive pulse to the gate and is switched off by applying a negative pulse to the gate. The GTO combines the high blocking voltage and high overcurrent capability of the thyristor with the ease of gate drive and fast switching of the transistor.

The GTO is a four-layer three-terminal device with a schematic structure the same as that of the thyristor, as shown in *Figure 23.4*. The circuit symbol for the GTO is shown in *Figure 23.21*.

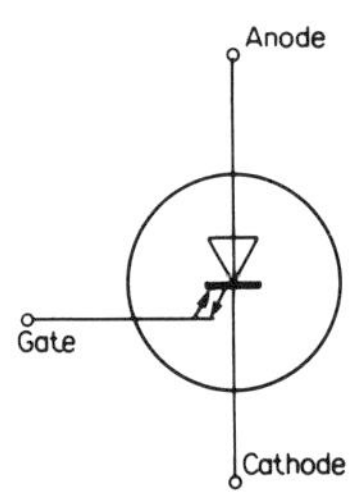

Figure 23.21 Circuit symbol for gate turn-off switch

As with the other four-layer devices described previously, the operation of the GTO can be considered in terms of the two-transistor analogue of *Figure 23.5*. When the anode of the GTO is positive with respect to the cathode, and positive gate drive is applied to the base of the npn transistor, this transistor is turned on. The collector of the npn transistor, which is the base of the pnp transistor, goes low and the pnp transistor is turned on. The collector current of the pnp transistor flows into the base of the npn transistor to reinforce the initial base drive. Cumulative action occurs, and the GTO will be switched on and remain on when the gate drive is removed, when the sum of the transistor currents exceeds the latching current I_L of the GTO. This condition is represented by the product of the transistor gains exceeding unity.

The negative base drive applied to the GTO to switch it off is effective because the gain of the npn transistor is larger than that of the pnp transistor. In a practical device, the gain of that part of the GTO forming the npn transistor is maximised by careful control of the diffusion profiles of the layers. The gain of the part of the GTO forming the pnp transistor is minimised by making the base wide, controlling the carrier lifetimes by gold doping, and by controlled short-circuiting of the emitter region. This control of the structure of the GTO is possible only with modern manufacturing techniques such as ion implantation and neutron doping as well as the close control now possible over the various stages in the manufacturing process.

In practical circuits, the GTO is switched on by injecting a positive current into the gate for the specified turn-on time. The GTO is switched off by drawing current from the gate, achieved in practical circuits by applying a negative voltage pulse of −5 to −10 V between the gate and the cathode.

The published data for the GTO are the same in many respects as those for the thyristor. It is only in the switching characteristics that the data resemble a transistor rather than a thyristor. Typical values of turn-off time for a GTO are less than 0.5 μs.

23.7 Symbols for main electrical parameters of thyristors

Symbol	Meaning
di/dt	rate of rise of on-state current after triggering
dV/dt	maximum rate of rise of off-state voltage which will not trigger any device
I_D	continuous (d.c.) off-state current
I_{DM}	peak off-state current
I_{FG}	forward gate current
I_{FGM}	peak forward gate current
I_{GaM}	peak forward anode-gate current
I_{GaT}	minimum anode-gate current to initiate turn-on
I_{GkM}	peak forward cathode-gate current
I_{GkT}	minimum cathode-gate current to initiate turn-on
I_{GT}	minimum instantaneous gate current to initiate turn-on
I_{GQ}	gate turn-off current
I_H	holding current
I_L	latching current
I_{RG}	reverse gate current
I_{RGM}	peak reverse gate current
I_T	continuous (d.c.) on-state current
$I_{T(AV)}$	average on-state current
$I_{T(ov)}$	overload mean on-state current
$I_{T(RMS)}$	r.m.s. on-state current
I_{TRM}	repetitive peak on-state current
I_{TSM}	non-repetitive peak on-state current
$P_{G(AV)}$	average gate power
P_{GM}	peak gate power
$R_{th(h)}$	thermal resistance of heatsink
$R_{th(i)}$	contact thermal resistance at specified torque
$R_{th(j-amb)}$	thermal resistance, junction to ambient
$R_{th(j-h)}$	thermal resistance, junction to heatsink
$R_{th(j-mb)}$	thermal resistance, junction to mounting base
$R_{th(mb-h)}$	thermal resistance, mounting base to heatsink
T_{amb}	ambient temperature
$T_{j,max}$	maximum permissible junction temperature
T_{mb}	mounting-base temperature
T_{stg}	storage temperature
t_d	delay time
t_{gt}	gate-controlled turn-on time
t_{off}	circuit-commutated turn-off time
t_{on}	turn-on time for s.c.s.
t_q	circuit-commutated turn-off time
t_r	rise time
V_{AK}	forward on-state voltage of s.c.s.
V_{BO}	breakover voltage
V_D	continuous off-state voltage
V_{DRM}	repetitive peak off-state voltage
V_{DSM}	non-repetitive peak off-state voltage
V_{DWM}	crest working off-state voltage
V_{FG}	forward gate voltage
V_{FGM}	peak forward gate voltage
V_{GaM}	peak reverse anode-gate-to-anode voltage
V_{GaT}	minimum anode-gate voltage that will initiate turn-on
V_{GD}	maximum continuous gate voltage which will not initiate turn-on
V_{GkM}	peak reverse cathode-gate-to-cathode voltage
V_{GkT}	minimum cathode-gate voltage to initiate turn-on
V_{GT}	minimum instantaneous trigger voltage to initiate turn-on
V_{RG}	reverse gate voltage
V_{RGM}	peak reverse gate voltage
V_{RRM}	repetitive peak reverse voltage
V_{RSM}	non-repetitive peak reverse voltage
V_{RWM}	crest working reverse voltage
V_T	continuous (d.c.) on-state voltage

24 Microwave Semiconductor Devices

I T Graham BSc, CEng, MIEE
Hewlett-Packard Ltd

Contents

24.1 Introduction

The requirements of high frequency performance impose the greatest possible demands on semiconductor technology. The substrate material must be of the greatest purity so that there will be no imperfections in the active structure. Also, the material itself may need to be a more difficult substance to work with, such as gallium arsenide, because the high mobility of its charge carrier means improved performance at maximum frequencies. The required complexity of the device makes it difficult to define and special techniques such as ion implantation are required. Repeatability of characteristics becomes difficult and production yields are relatively low.

In this section on microwave semiconductors each device is described in terms of its construction and characteristics and the various types are mentioned. The most important parameters are defined and applications examples are given.

24.2 Schottky diodes

Standard pn diodes are limited at higher frequencies by the capacitance provided by the lifetime of the minority carrier. This means that such diodes will not switch on and off quickly compared to the frequency of the signal they are trying to process, resulting in reduced performance in, for example, mixing and detecting in a radio receiver. Schottky diodes overcome this problem by employing a metal–semiconductor barrier at the active junction so that majority carriers only participate in the diode action. It should be noted that Schottky diodes may also be termed 'hot carrier' diodes, 'hot electron' diodes and 'Schottky barrier' diodes.

24.2.1 Schottky diode types

Typical Schottky diodes belong to one of the three types: passivated, hybrid or mesh.

24.2.1.1 Passivated diodes

Figure 24.1 shows a typical passivated diode. The n-type silicon and a metal, such as NiCr (nickel chromium) form the Schottky junction. The surface of the semiconductor is passivated

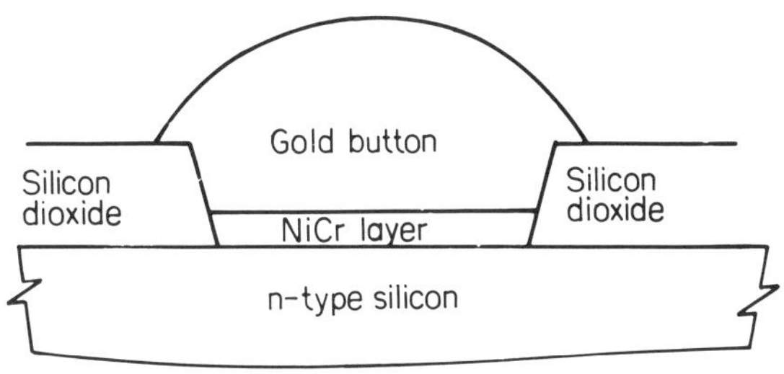

Figure 24.1 Passivated Schottky diode cross-section

(protected) by a layer of silicon dioxide against outside contaminations. Sometimes a layer of silicon nitride is used in addition to the oxide. The thick gold layer (called gold button due to its shape) connects to the outside world as one terminal of the diode, the other terminal being the semiconductor itself.

Usually passivated diodes have small Schottky junction areas, therefore, they have low junction capacitance and are suitable for operation up to 18 GHz. However, under small reverse bias voltages (about 5 V), large electric fields occur where the Schottky junction meets the silicon dioxide, causing voltage breakdown.

24.2.1.2 Hybrid diodes

The hybrid Schottky diode has a higher reverse breakdown voltage than the passivated diode. In addition to the Schottky junction, a pn junction, which does not affect the Schottky junction, is located at the oxide–Schottky interface as in *Figure 24.2*. In this case 'hybrid' indicates the presence of both Schottky

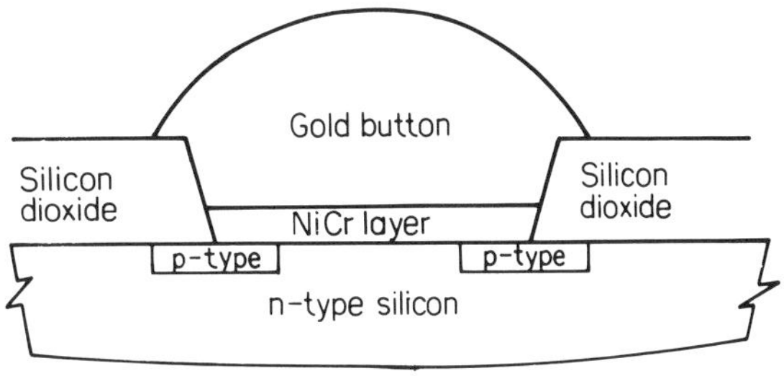

Figure 24.2 Hybrid Schottky diode cross-section

and pn junctions in one diode. This arrangement does not allow high electric fields to build up at the interface, and a reverse breakdown voltage as high as 70 V is achieved. The penalty is high junction capacitance, limiting operating frequency to 4 GHz.

24.2.1.3 Mesh diodes

A mesh diode has an unpassivated Schottky junction. *Figure 24.3* is the top view of a mesh diode chip. Each chip has about 80 pads, each of which is a diode itself. The pads are defined on the

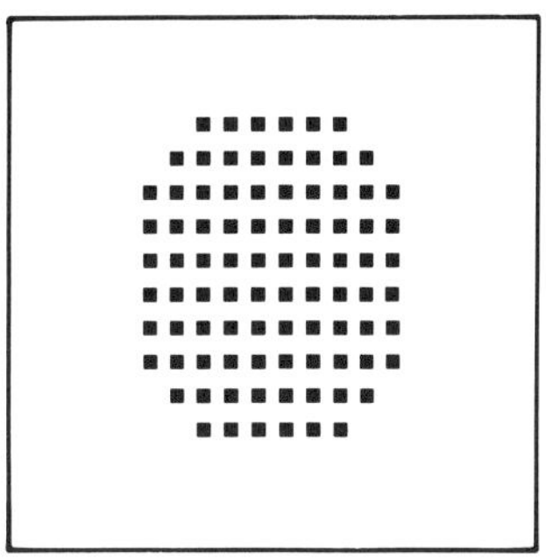

Figure 24.3 Top view of a mesh diode chip

semiconductor chip by a mesh mask. The distance from the centre of one pad to another is about 50 μm (two-thousandths of an inch).

Figure 24.4 shows the cross-section of a single pad on a mesh diode. During assembly the diode chip is put into a package and a metal whisker is placed on the chip. A randomly achieved contact of the whisker with one of the pads is indicated on an oscilloscope. This assembly technique does not lend itself to high volume production.

The mesh diodes usually have medium size Schottky junction areas, and the resulting average junction capacitances limit operation frequencies to about 4 GHz. The breakdown voltage is about 30 V, and the diodes have low flicker ($1/f$) noise.

The Schottky junction in the mesh diode is between the nickel and n-type silicon (*Figure 24.4*). The metal whisker is for diode lead connection only and does not contribute to the formation of

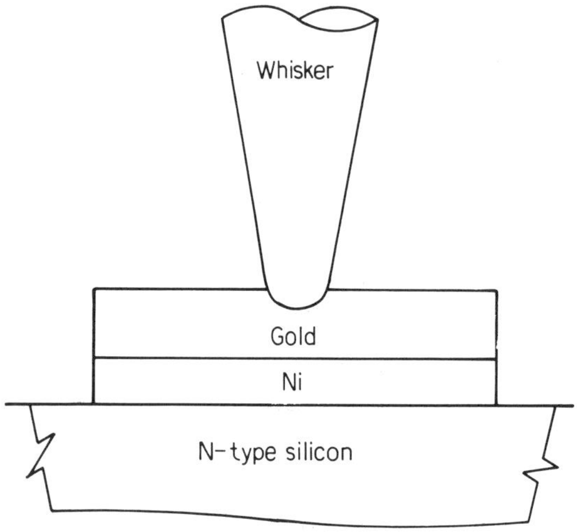

Figure 24.4 Mesh diode cross-section

the junction. So the diode is completely reliable and uniform in its performance. In contrast the junction of a point contact diode is formed by the contact of the metal whisker and the semiconductor.

N-type silicon and nickel chromium (or nickel) are not the only combination of semiconductor and metal which can be used to make a Schottky junction. An alternative semiconductor is gallium arsenide (GaAs) and suitable metals include molybdenum, tungsten, aluminium, titanium and platinum.

Silicon Schottky diodes are capable of operations up to 40 GHz. Gallium arsenide Schottky diodes, due to the higher electron mobility in this semiconductor, are made to cover applications up to 100 GHz and above.

24.2.2 Applications and performance parameters

The three main areas of applications for Schottky diodes are switching (including clipping and sampling), mixing and detecting.

24.2.2.1 Switching

The most important characteristic for a diode used in any form of switching application is speed, and the Schottky diode is unsurpassed in this respect. However, several other parameters are also important and are given in *Table 24.1.* The forward voltage at 1 mA is also known as turn-on voltage. Schottky diodes made with different metal–semiconductor combinations offer different forward voltages.

Table 24.1 Diode performance comparison

Parameters	*Diode*		
	Schottky	*Silicon pn*	*Germanium pn*
Forward voltage drop (mV at 1 mA)	150–450	600	300
Reverse leakage (nA)	<100	<100	>1000
Breakdown voltage (V)	5–70	~5	>1000
Carrier lifetime (ps)	<100	500	2000

The minority carrier lifetime is applicable to the pn junction diode only because its operation involves both majority and minority carriers. The Schottky diode has only majority carriers although their parasitic reactances give rise to an effect similar to minority carrier lifetime. The 'lifetime' is an indication of the switching speed of the diodes; the lower the lifetime, the faster the switching speed.

24.2.2.2 Mixing

Any non-linear element will mix but the Schottky diode is particularly effective for this because of its nearly square law characteristics and low noise performance. The parameters concerned with mixing are noise figure, input admittance, and intermediate frequency noise and impedance.

Noise figure is defined as the ratio of the input signal to noise ratio to the output signal to noise ratio: the lower the noise figure of the mixer diode, the better the performance of the mixer under conditions of weak r.f. signal input. A typical noise figure for a mixer diode is 6 to 7 dB.

The noise figure of the mixer diode is dependent on the local oscillator power (P_{LO}). There is a range of local oscillator power over which the diode exhibits the best noise figure. The noise figure is also dependent on the frequency of the input.

The input admittance of a mixer diode indicates how the diode can be matched to a certain r.f. circuit for best performance. It is plotted on a Smith chart. Since the input admittance varies with local oscillator power and d.c. bias these conditions are always indicated. Often a single standing wave ratio (SWR) is quoted for the test frequency, showing how well the diode matches a fixed tuned test system.

Intermediate frequency (i.f.) noise is also known as flicker noise. Its amplitude increases with decreasing frequency. The mixer diode noise has two components. One is broadband noise which is constant in amplitude at all frequencies and the other is i.f. noise. At high frequencies the former dominates and at low frequencies the latter. Intermediate frequency noise is measured and expressed as noise temperature ratio (T_N) in dB. The higher the number, the noisier the diode. Schottky diodes have low i.f. noise. Of the three types of Schottky diodes the mesh diodes have the lowest i.f. noise, and the passivated diodes the highest. In contrast germanium and silicon point contact diodes have i.f. noise which is 20–30 dB worse.

The intermediate frequencies of some mixer applications are low. For example, in the Doppler radar the i.f. is the audio range (d.c. to 10 kHz) and the Schottky diode, with its low i.f. noise, is the logical choice as a Doppler radar mixer diode.

Similar to the r.f. admittance the i.f. impedance helps the designer put the diode into a suitable i.f. circuit for best performance. It is often plotted against the local oscillator power.

24.2.2.3 Detecting

A Schottky diode can be used as an amplitude modulated (AM) detector or for detecting the presence of any signal. Again, the Schottky diode is particularly useful in this application because of its excellent characteristics of tangential sensitivity, voltage sensitivity, video resistance and i.f. noise.

The tangential sensitivity (TS) of a detector diode describes the performance of the diode under low signal level conditions. It is a subjective measurement using an oscilloscope. The TS is the input signal level which produces a detected signal which is just above the noise floor. At this point the signal to noise ratio is 8 dB. The tangential sensitivity is measured as an input power level in dBm: the smaller the value, the more sensitive the detector diode. For example, the TS for the 5082-2755 detector diode is

-55 dBm for a 2 MHz video bandwidth, at 10 GHz and 20 μA bias. A small amount of d.c. bias usually improves the TS and most detectors are run with some bias.

The voltage sensitivity (γ) is a measure of the output voltage of the detector diode for a given input power. It is measured in mV μW^{-1}. The value of γ is useful to the circuit designer in determining how much voltage he can obtain at the video output of the diode, allowing him to design a video amplifier of the right gain.

The video resistance is simply the output resistance of the diode at video frequencies. It helps the circuit designer load the diode video output properly.

If the frequency of the video signal is low (below 1 MHz) i.f. noise, as in the case of the mixer diode, becomes the main source of detector diode noise. The low i.f. noise of the Schottky diode is therefore a definite advantage over that of the point contact diode.

24.3 PIN diodes

The most important feature of the PIN diode is its basic property of being an almost pure resistor at r.f. frequencies, whose resistance value can be varied from about $10^4\ \Omega$ to less than 1 Ω by the control current flowing through it. Most diodes exhibit this characteristic to some degree, but the PIN diode is specially optimised in design to achieve a relatively wide resistance range, good linearity, low distortion, and low current drive. The characteristics of the PIN diode make it suitable for use in switches, attenuators, modulators, limiters, phase shifters and other signal control circuits.

The PIN diode is a silicon semiconductor consisting of a layer of intrinsic material sandwiched between regions of highly doped p and n material, hence the abbreviation. Under reverse bias, the I-layer is depleted of mobile charges and the PIN diode appears essentially as a capacitor. When forward bias is applied across the PIN diode positive charge (holes) from the p region and negative charge (electrons) from the n region are injected into the I-layer, therefore increasing its conductivity and lowering its resistance.

The conductance (G) of the PIN diode is proportional to the stored charge (Q) in the I-layer, which in turn is proportional to the d.c. bias current. Resistance (R) is therefore inversely proportional to the d.c. bias current.

An important parameter of the PIN diode is the lifetime (τ), which essentially defines the length of time required for the stored charge to deplete by recombination once the forward bias is removed. There is a frequency, f_0, which is related to τ by the expression,

$$f_0 = 1/2\pi\tau \qquad (24.1)$$

used for determining the low frequency limit of the PIN diode.

Since charge depletes for times greater than τ, a signal with cycle period greater than τ will experience varying conductance. This defines a lower frequency limit, given by equation (24.1), for linear performance of a diode. For r.f. signal frequencies below f_0, the PIN diode rectifies the signal much like an ordinary pn diode, and considerable output distortion occurs. At f_0, there is some rectification with resulting distortion. At frequencies above f_0, less and less rectification occurs as charge storage due to r.f. current (I_{rf}) diminishes according to the relationship:

$$Q = \frac{I_{rf}}{2\pi f} \qquad (24.2)$$

Since I_{rf} is dependent on the power absorbed by the diode, distortion is also dependent on the r.f. signal level, r.f. frequency, and d.c. bias current. For most applications the minimum frequency of operation should be about $10f_0$ or

$$f_{min} = \frac{10}{2\pi\tau} = \frac{1.59}{\tau} \qquad (24.3)$$

This restriction is not important in switching applications, where the diode is normally biased either completely off or completely on. In those states, since most of the power is either reflected or transmitted, the effect of r.f. current on the total charge is small and distortion is not a problem.

At frequencies much higher than f_0, the PIN diode with forward bias behaves essentially as a pure resistor. The resistance of the PIN diode is related to the bias current and characteristics of the diode as follows:

$$R = W^2/\mu I\tau \qquad (24.4)$$

where W is the I-layer thickness, μ the combined electron and hole mobilities, τ the lifetime, and I the bias current.

The resistance is inversely proportional to both lifetime and bias current. However, since long lifetime is usually associated with thick I-layer and resistance is proportional to the square of the I-layer thickness, a diode with long lifetime has higher resistance. Exceptions to this case are increases in lifetime by other means without widening the I-layer.

The voltage dependency (or independency) of capacitance is related to the dielectric relaxation frequency, f_D. The undepleted portion of the I-layer has the resistance (R) shunting its capacitance (C). f_D is the frequency where the capacitive reactance is equal to the resistance, i.e.:

$$1/2\pi f_D C = R \qquad (24.5)$$

For an undepleted layer of thickness (W) and area (A)

$$R = \rho W/A \qquad (24.6)$$

$$C = \varepsilon A/W \qquad (24.7)$$

Thus, solving for the dielectric relaxation frequency, gives

$$f_D = 1/2\pi\rho\varepsilon \qquad (24.8)$$

where ρ and ε are, respectively, the resistivity and permittivity of the I-layer silicon, with ε being approximately 10^{-12} F cm^{-1}. At frequencies much higher than f_D, since the reactance of the undepleted layer is much less than its resistance, total capacitance is independent of bias. At frequencies much lower than f_D, the resistance of the undepleted layer shorts out the capacitance, and sufficient reverse bias is required to deplete the I-layer of charge.

24.3.1 PIN diode construction

Most PIN diodes are derived from the four basic chip structures illustrated in *Figure 24.5.*

24.3.1.1 Epitaxial mesa PIN

Because of its small I-layer (both in thickness and area) the epitaxial mesa structure has little charge storage. Therefore it is capable of fast switching. Also because of its thin I-layer, relatively low current drive is required for a given resistance. The mesa structure results in low junction capacitance, and thus enables good frequency response through the microwave bands.

24.3.1.2 Epitaxial planar PIN

The epitaxial planar structure (compared to the mesa) has a somewhat larger junction area, and therefore results in a larger junction capacitance. Its large junction area and thin I-layer also translates into the lowest resistance, requiring very low current drive. This PIN is suited for use in switching applications in the v.h.f./u.h.f. range.

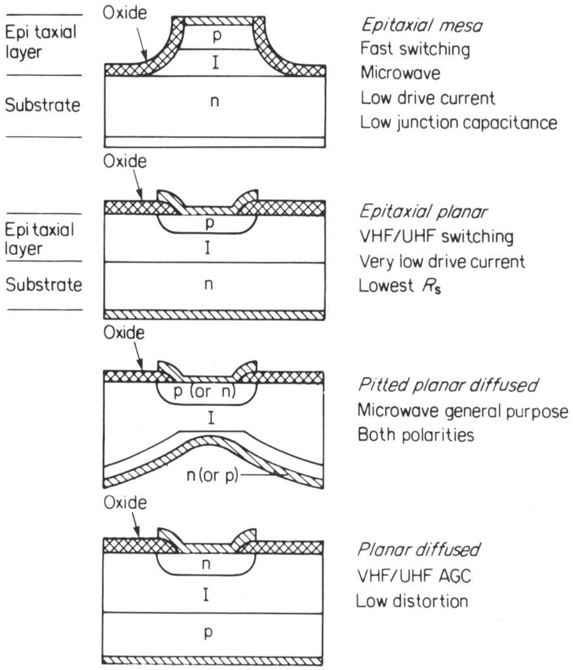

Figure 24.5 PIN chip structures

24.3.1.3 Planar diffused pitted NIP and PIN

The planar diffused structure with a pit etched on the reverse side has low junction capacitance, and therefore is suited for switching and general purpose use in the microwave frequencies. Both polarities are available, i.e. NIP and PIN.

24.3.1.4 Planar diffused NIP

The planar diffused structure has a thick I-layer and is associated with long lifetime diodes capable of performing at low frequencies with low distortion. These diodes are most suited for automatic gain control (AGC) applications in the v.h.f./u.h.f. range.

24.3.2 Applications and diode parameters

PIN diodes are used principally for the control of r.f. and microwave signals from frequencies below 1 MHz to 20 GHz. Applications include switching, attenuating, modulating, limiting and phase shifting. Certain diode requirements are common to all these control functions, while others are more important in a particular type of usage.

24.3.2.1 Switching

Because of its high r.f. resistance when unbiased and its low r.f. resistance when biased with a d.c. current of only 10 mA or less a PIN diode is an ideal circuit element for use in r.f. switches. They may be used as single diode switches in either series or parallel configurations, or for more demanding requirements in combinations using two or more diodes.

The series switch is ON when the diode is forward biased, and it is OFF when the diode is zero or reverse biased. The opposite is true for the shunt switch where zero or reverse bias turns the switch ON, and forward bias turns the switch OFF. For a good approximation, the PIN diode in a switch is essentially a resistor in the forward-biased state and a capacitor in the reverse-biased state.

The loss of signal atrributed to the diode when the switch is ON (transmission state) is insertion loss. The insertion loss is primarily determined in a series switch by the forward-biased resistance of the diode and in a shunt switch both by the diode capacitance and signal frequency. In either case it is the diode impedance in relation to the source and load impedance (generally 50 Ω). For low insertion loss, low resistance is needed in a series switch. Low capacitance (particularly at high frequencies) is needed in a shunt switch.

Isolation is the measure of r.f. leakage between the input and output when the switch is OFF. For high isolation, low capacitance (especially at high frequencies) is required in a series switch. Low resistance is required in a shunt switch.

Reverse recovery time is a measure of switching time, and is dependent on the forward and reverse bias applied. With forward bias current, I_F, charge is stored in the I-layer. When reverse biased, reverse current, I_R, will flow for a short period of time, known as delay time, t_d. When sufficient number of carriers have been removed, the current begins to decrease. The time required for the reverse current to decrease from 90% to 10% is called the transition time, t_t. The sum, $t_d + t_t$, is the reverse recovery time, which is a good indication of the time it takes to switch the diode from ON to OFF.

Standing wave ratio (SWR) which is a measure of the r.f. impedance match, is particularly important in high frequency applications. Since the SWR of most package styles depends on the mounting arrangement, it is only specified for diodes in 50 Ω stripline packages.

The r.f. power (CW or pulse), that can be handled safely by a diode switch is limited by two factors: the breakdown voltage of the diode, and thermal considerations, which involve maximum diode junction temperature and the thermal resistance of the diode and packaging. Other factors affecting power handling capability are ambient temperature, frequency, attenuation level (diode resistance) pulse width, and pulse duty cycle.

When the maximum isolation requirements are greater than that which can be achieved with a single diode switch, multiple diode arrangements can be used. Two diodes connected closely in series or shunt will only increase the isolation by 6 dB over that of a single diode. If n diodes are spaced a quarter wavelength apart, the resulting isolation will be n times that of a single diode plus 6 dB. Where $\lambda/4$ spacing is impractical, isolation can be increased by using a series–shunt combination. This arrangement will give the isolation of a series diode and that of a shunt diode plus 6 dB. The power handling capability is not improved in a pair of diodes spaced $\lambda/4$ apart. The lead diode absorbs a much larger percentage of the incident power than in the case of a closely spaced pair of diodes.

24.3.2.2 Phase shifters

In a phase shifting circuit the PIN diode serves basically as a switch to transfer a transmission line from one electrical length to another, therefore resulting in a phase shift of the transmitted signal.

In each circuit, forward bias applied at one of the bias ports will produce a phase shift in the r.f. output signal. Because of its fast switching speed, low ON and high OFF impedance, the PIN diode is well suited for phase shifter applications.

24.3.2.3 Attenuators

Whereas a switch is used only in its maximum ON or OFF state, an attenuator is operated throughout its dynamic range (or resistance range in the case of a diode attenuator). Although the basic series or shunt diode switch can be used as an attenuator, it cannot offer constant input and output impedance, which is required for good source and load matching in most attenuator applications.

The resistive line attenuator is distributed in structure with a large number of diodes. The diodes in the centre of the structure vary more in impedance than the ones near the ports, resulting in constant impedance at the ports. This structure is capable of large attenuation range, but is not useful at low frequencies because of large size (*Figure 24.6*). The π, T, and bridged-T attenuators are

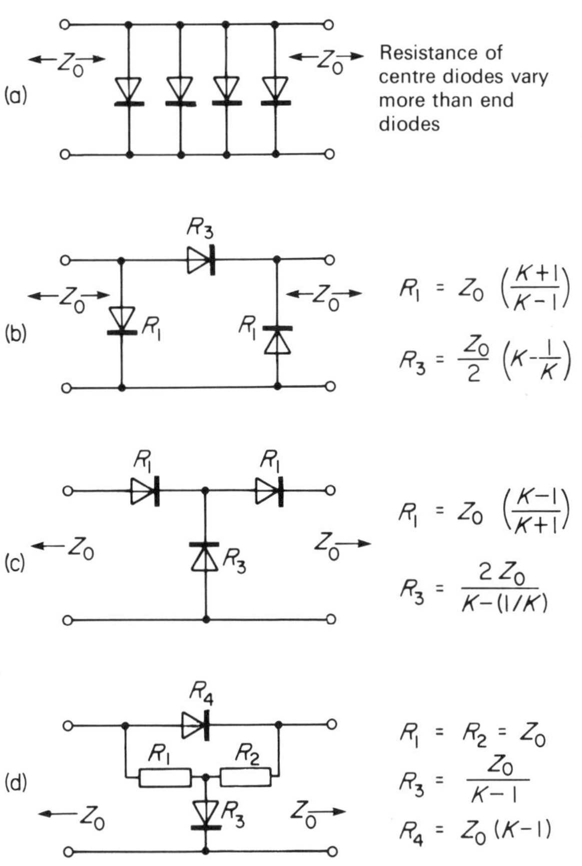

Figure 24.6 Constant impedance absorption attenuators: (a) resistive line; (b) π; (c) T; (d) bridged-T

more compact in structure. Both the π and T configurations use three diodes each, while the bridged-T circuit contains two diodes and two resistors as the basic circuit elements. For a particular value of attenuation, K, the design equations determine the values of resistance to which the series and shunt diodes must be biased in order for each attenuator to achieve constant impedance match.

While certain switching parameters are applicable for attenuators, other diode and circuit considerations are more important for attenuator performance. The range from maximum attenuation (isolation) to minimum attenuation (insertion loss) defines the dynamic range of an attenuator. The theoretical dynamic range achievable by the π, T, and bridged-T attenuators are respectively 52, 40, and 34 dB. The linearity of attenuation versus bias is important in most attenuator applications, and is directly attributable to the basic linearity of resistance versus control current of the diode. In many applications, particularly where more than one diode is driven by a single bias supply, power drain is of prime concern. The resistance requirements of good PIN diodes are achieved with low bias currents. Diode and circuit parasitics limit high frequency performance. Proper choice of circuit components and careful circuit layout will help to minimise limitations and sustain performance at high frequencies. Distortion due to rectification limits low frequency performance. The two principal types of distortion in PIN attenuators are intermodulation and crossmodulation.

24.3.2.4 Modulators

A PIN diode constant impedance attenuator can be used to modulate an r.f. signal by applying the modulation signal to the bias port. For performance with minimum distortion the carrier frequency should be much greater than f_0 given by equation (24.1) and the modulation frequency much less than f_0. For most applications a compromise in these requirements may be necessary.

24.3.2.5 Limiters

Many microwave systems contain sensitive amplifiers, mixers, and detectors, etc., which are subject to burnout by inadvertent high level transient signals. Protection for these systems can be provided by a limiter. A PIN diode limiter is essentially an attenuator that uses self-bias rather than externally applied bias.

A basic passive limiter circuit and its performance characteristics are shown in *Figure 24.7*. When the r.f. input

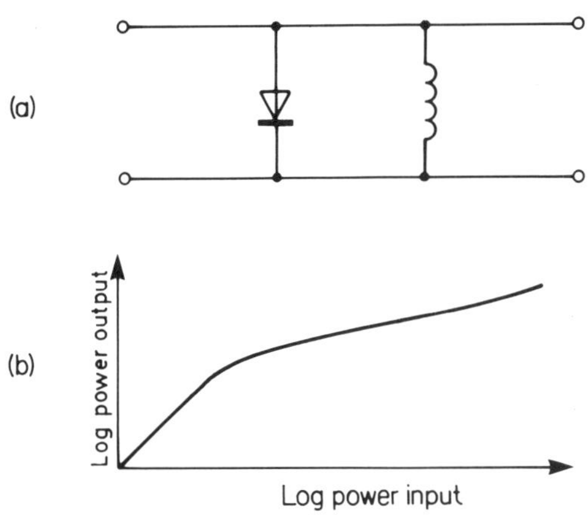

Figure 24.7 Passive limiter circuit and performance characteristics: (a) circuit; (b) characteristic

signal is below the threshold level, the diode is very high in resistance, and the ouptut very closely follows the input signal. Above the threshold level, the r.f. is rectified and the diode resistance changes to a lower state. As a result, much of the r.f. is reflected, allowing only a small, almost constant output with increase in input. For this circuit to be efficient, it is essential that the PIN diode has fast switching time (low lifetime). A PIN with thin I-layer will help rectification efficiency and result in low resistance. Another diode requirement is good heat transfer characteristics.

In a quasi-active limiter, self-rectified current is not needed. Part of the incoming r.f. signal is detected by a Schottky detector diode and the rectified current used to forward bias the PIN limiter diode. This type of circuit enhances the turn-on capability of the limiter. Thus the PIN I-layer thickness may be increased and power handling capability improved.

24.4 Step recovery diode

The step recovery diode is a charge-controlled switch. A forward bias stores charge, a reverse bias depletes this stored charge, and when fully depleted the diode ceases to conduct current. The action of turning off, or ceasing current conduction, takes place so fast that the diode can be used to produce an impulse. If this is done cyclically, a train of impulses is produced. A periodic series of impulses in the time domain converts to an infinitude of frequencies (all multiples of the basic exciting frequency) in the frequency domain. If these impulses are used to excite a resonant circuit, much of the total power in the spectrum can be concentrated into a single frequency. Thus input power at one frequency can be converted to output power at a higher frequency.

Figure 24.8 is a representation of step recovery diode structure. The 'intrinsic' region is actually very lightly doped n-type silicon. It is in this layer that charge is stored.

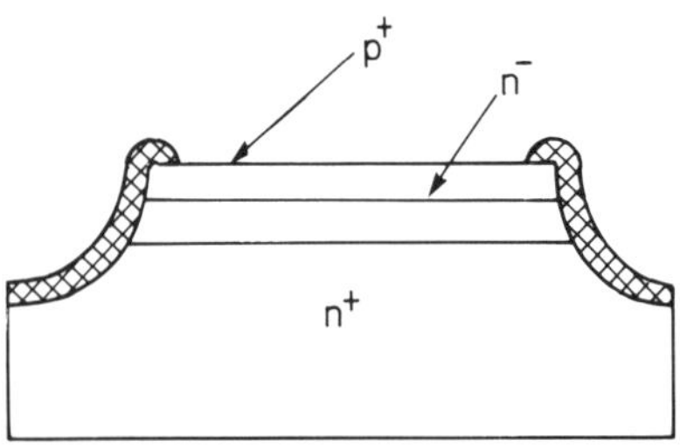

Figure 24.8 Step recovery diode cross-section

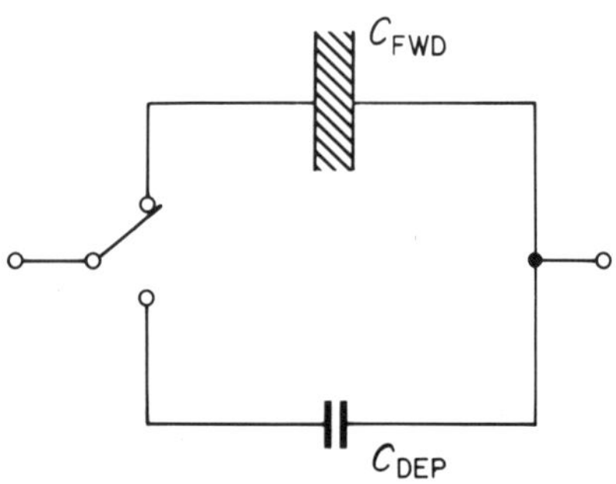

Figure 24.9 Equivalent circuit of step recovery diode

Figure 24.9 is a simple equivalent circuit consisting of an extremely fast switch and two capacitors. When in forward conduction, and until the stored charge is totally depleted by reverse conduction, the switch is connected to the large capacitor, C_{FWD}. When the charge is depleted the switch changes instantaneously (60–400 ps) to the other capacitor, C_{DEP}.

A simple circuit, shown in *Figure 24.10* is the 'impulse generator'. This circuit will help to explain the step recovery diode action. As the signal source overcomes the reverse bias set by the d.c., the diode becomes forward biased by the input source. It then conducts in the forward direction and charge is stored in the intrinsic layer much as it is in a PIN diode. Under reverse bias, conduction continues until all the charge is depleted. (Recall from

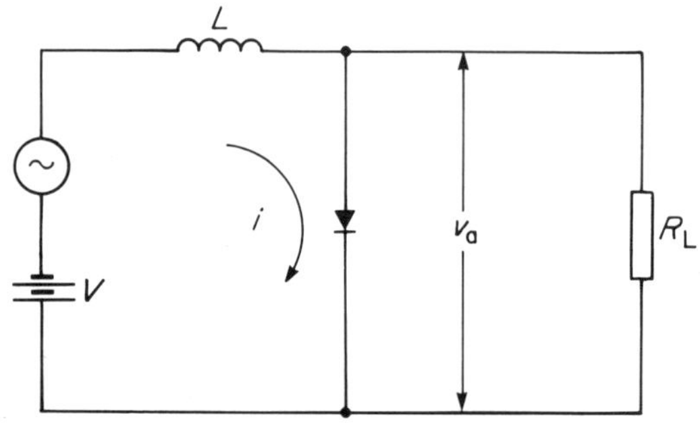

Figure 24.10 Impulse generator circuit

basic electronics that it is impossible to change the current through an inductor instantaneously without producing a voltage impulse.) So we now have a circuit whose output is a series of impulses occurring once each cycle, as shown in *Figure 24.11*. A series of impulses contains a spectrum of frequencies. For practical reasons we cannot produce perfect impulses, so the power contained in the higher harmonics drops off.

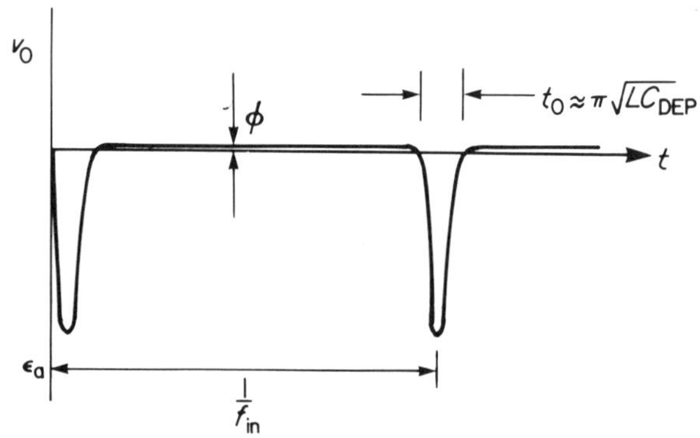

Figure 24.11 Output of impulse generator

24.4.1 Applications and parameters

Applications can be approached in two ways: (1) the way the diode is used; and (2) the equipment in which it is used. The step recovery diode can be made to produce very sharp and narrow pulses, and these in turn contain a virtual infinitude of harmonics of the exciting frequency.

A circuit which exploits the step recovery diode's production of a multitude of frequency components is called a *comb generator*. Comb generators are used in measurement equipment such as spectrum analysers to produce locking signals.

Another type of circuit picks out a single harmonic and optimises the power output around that harmonic. This circuit is called a *multiplier*. The end result of a multiplier is output power at some multiple ($2f_i$, $3f_i$, etc.) of the input frequency. The efficiency of the conversion is high enough to make this a very practical scheme for multiplying up from a readily available transistor oscillator at 600 MHz to get a 2.4 GHz signal ($\times 4$). Multipliers are used as local oscillators, low power transmitters, or transmitter drivers in radar, telemetry, telecommunications, and instrumentation.

Two specifications that determine the *total power output* in any given multiplier mode are *maximum junction temperature* and *thermal resistance*. It may also be necessary to know the *efficiency* of conversion. Efficiency depends heavily on the design of the multiplier, so we do not specify it.

Two other specifications affect the power output by determining the maximum energy in the impulse. The *reverse voltage breakdown* limit, V_{BR}, limits the pulse height and therefore the energy in the pulse. The *reverse bias capacitance*, C_V, does two things. First, it determines the energy which can be put into the pulse, and secondly it sets the impedance level of the impulse circuit.

24.5 Silicon bipolar transistor

Bipolar transistors are commonly used in amplifiers up to 6 GHz and in oscillators up to 12 GHz. The techniques used in bipolar circuit design are well formulated. A bipolar transistor is made of semiconductor materials such as germanium or silicon. Microwave bipolar transistors are almost exclusively silicon.

Figure 24.12 is a simplified model of a bipolar transistor for gain behaviour at microwave frequencies. There are four

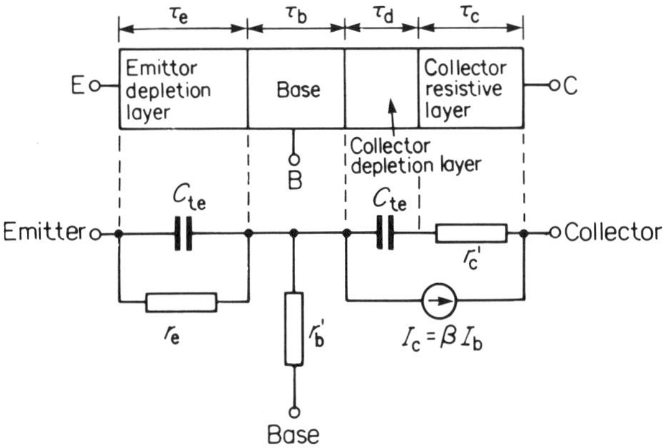

Figure 24.12 Microwave transistor model

important time constants involved. They are:

τ_e = emitter charging time
τ_b = base transit time
τ_d = collector depletion layer transit time
τ_c = collector charging time

The low frequency limit of the exciting signal is set by *minority carrier lifetime*, τ, and the ability to form an effective impulse at the higher frequencies is determined by the *transition time*, t_r. As with PIN diodes, the minority carrier lifetime is the time required for 63% of the charge carriers to recombine. Minority carrier lifetime sets the lower input frequency limit because as the frequency gets lower and lower, more and more of the charge is dissipated by recombination during a cycle. We do not want the charge to dissipate by recombination because that reduces the energy in the impulse. The input frequency should be higher than a number related to τ by:

$$f_{in} \geq \frac{10}{2\pi\tau} \tag{24.9}$$

The emitter charging time is caused by the parallel combination of the base–emitter space charge resistance r_e and base–emitter transition capacitance C_{te}. The base transit time is the time taken by a carrier to cross the neutral base region mainly by diffusion. The collector depletion layer transit time is the time taken by a carrier to cross the base–collector depletion region by electric field and by diffusion. The collector charging time is due to the RC combination of the collector resistance R_c and base–collector transition capacitance C_{tc}.

24.5.1 Microwave transistor construction

Figure 24.13 is the top view of a simplified bipolar transistor. The patterns are the metal contacts which connect the outside circuit to the working parts of the transistor. Each length of the pattern is

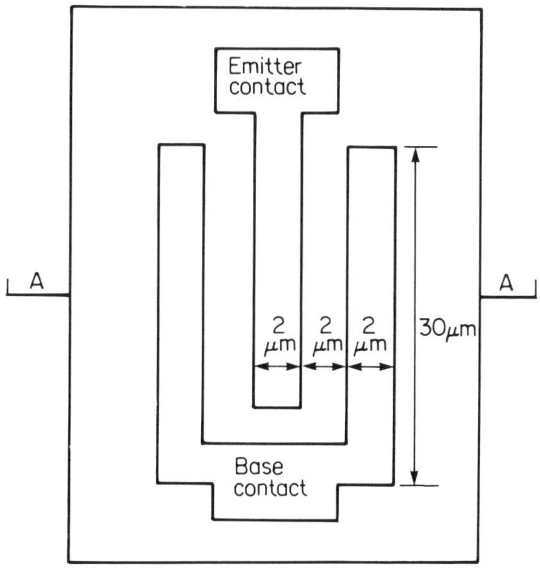

Figure 24.13 Top view of a typical microwave bipolar transistor

called a finger. An actual transistor may have more fingers. The collector is the back of the transistor, underneath the emitter and base.

Figure 24.14 is the cross-section of *Figure 24.13* along A–A. In both figures the relevant dimensions are indicated. A microwave transistor is physically very small. The active surface of some low noise transistors is less than 0.025 mm square.

In *Figure 24.14* the functional parts of the bipolar transistor are outlined by the dotted box. This npn transistor has an n-type

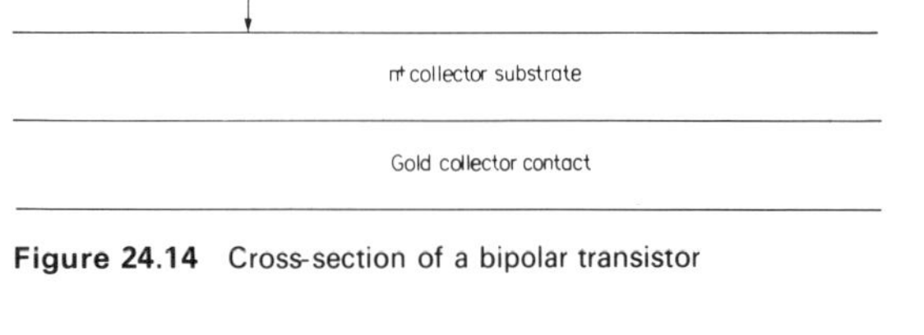

Figure 24.14 Cross-section of a bipolar transistor

emitter, p-type base and an n-type collector. The rest of the device serves the purpose of connecting the transistor to the outside circuit, with a minimum of unwanted resistance and reactance.

24.5.1.1 The emitter and metal contact

The emitter is ion implanted n-type silicon material. The connection to the outside is by the titanium–platinum–gold (Ti–Pt–Au) metallic sandwich. The titanium provides a low ohmic resistance contact and good adhesion to the silicon. The gold carries the electric current well due to its low resistivity; and it does not suffer metal migration problems under high temperature and high current density. The platinum layer prevents the gold from interacting with the titanium and silicon and is known as the barrier layer.

24.5.1.2 The base and base insert

The base is ion implanted p-type silicon. It is connected via the extrinsic base, the base insert and the same metal system to the outside. The extrinsic base does not contribute to transistor action, but rather adds base resistance. The base insert is heavily ion implanted to form a low resistance path to the metal contact.

24.5.1.3 The collector

The collector is a grown epitaxial layer. It is connected to the collector metal contact via the n^+ collector substrate.

24.5.1.4 The local oxide

The local oxide is an oxide layer obtained by oxidising the silicon locally, as opposed to deposited oxide. It is a thick oxide layer onto which the emitter and base bonding pads are located. The thickness minimises the bonding pad to collector capacitance: the smaller the capacitance, the better the microwave performance of the transistor. In the making of a transistor, photomasking is used to define the areas of the silicon to be worked on. For best results the mask should be in intimate contact with the surface. Local oxide is made so that it is flush with the rest of the surface, while deposited oxide creates steps on the surface. So the local oxide allows better masking, leading to finer geometry and precision in the end product.

24.5.1.5 Channel stop

Under conditions of elevated temperature and a reverse biased base–collector junction, a phenomenon known as inversion (or channeling) could occur. The surface of the n-type epitaxial layer may become a p-type layer, joining with the base and causing an undesirable extension of the base. A heavily n-doped surface prevents this from happening. This is the channel stop.

24.5.1.6 Emitter ballasting

Emitter ballasting is useful in providing thermal stability to microwave transistors which operate at collector currents of 30 mA or higher.

A power transistor has many emitter fingers and acts like a large transistor made up of many small transistors. At high current levels the emitters get hot. One will always be the hottest, therefore the lowest resistance. Without emitter ballasting the hottest emitter will let more current through it and this thermal runaway effect will eventually destroy this emitter, thus the whole transistor. With emitter ballasting each emitter finger has a feedback resistor of its own, and each emitter is prevented from thermal runaway. The same transistor with emitter ballasting can be operated at a higher current than the version without.

The advantages of emitter ballasting are better thermal stability and higher output capability due to higher allowable current. The penalty is lower gain and higher noise figure.

24.5.2 Radio frequency parameters

The three most useful r.f. parameters for a bipolar transistor are noise gain, power output and the scattering parameters (*S*-parameters).

24.5.1.7 Noise gain

The minimum noise figure of the transistor is specified when the transistor is d.c. biased and its input and output circuits tuned to achieve minimum noise from the device.

Three noise parameters predict the noise performance of a transistor under input matching conditions other than that for minimum noise figure. The set of three quantities are:

Γ_0 optimal input reflection coefficient
R_n noise resistance
F_{min} minimum noise figure

24.5.1.8 Gain

The maximum available gain is power gain achieved at the given frequency when the input and output circuits are tuned for maximum gain and it is unconditionally stable.

Associated gain is the gain achieved when the transistor is tuned and biased for the minimum noise figure. Associated gain is always lower than the maximum available gain and both are frequency and d.c. bias dependent.

24.5.1.9 Power output at 1 dB compression

This is the power handling parameter, and the output powe power achievable for 1 dB compression in input/output power ratio (*Figure 24.15*).

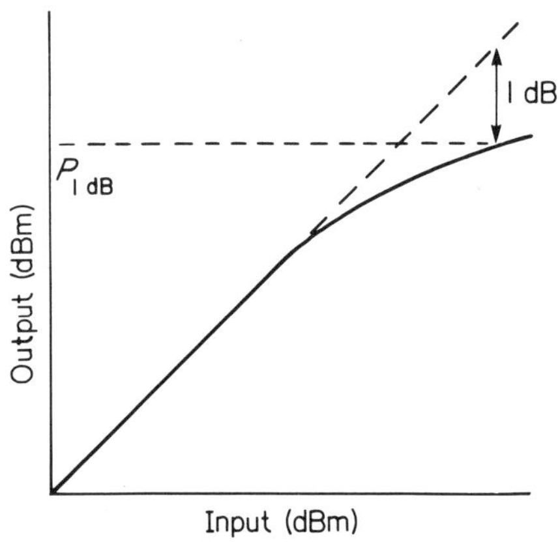

Figure 24.15 Output compression

24.5.1.10 Scattering parameters

S- parameters consist of four vectors (magnitude and angle): S_{11}, S_{21}, S_{12} and S_{22}. They are input reflection coefficient (S_{11}), forward transmission gain (S_{21}), reverse transmission isolation (S_{12}) and output reflection coefficient (S_{22}). They are almost always measured in a 50 Ω system. They are dependent on the d.c. bias conditions and the signal power of the transistor. For the bipolar transistors the *S*-parameters are usually given for small input power (e.g. −20 dBm) and at the d.c. bias condition at which they are expected to perform best.

24.5.3 Applications

As for transistors at lower frequencies, microwave silicon bipolar transistors may be used for amplification up to 4 GHz and are typically configured in a stripline or microstrip circuit. Design is complex and often accomplished with the aid of a computer and most transistor data sheets contain a wealth of data to achieve effective matching characteristics to minimise noise and maximise gain over the desired frequency band.

Computer aided design is used for microwave transistor design to such an extent that the characteristics of many popular devices are contained within powerful proprietary software packages.

Bipolar microwave transistors may be used to make fundamental oscillators up to 6 GHz. They may also be used in applications such as mixing, logic switching, gating, signal detection and sampling.

24.6 Gallium arsenide field-effect transistor

Microwave field-effect transistors are made with gallium arsenide (GaAs) material with a reverse-biased Schottky junction as the gate. Due to the metal-to-semiconductor gate it is often called a metal–semiconductor field-effect transistor (MESFET). The higher electron mobility in gallium arsenide in comparison to that in silicon and the unipolar FET principle make the GaAs MESFET (or GaAs FET) a superior transistor to either silicon bipolar transistors or silicon FET's at high microwave frequencies.

Figure 24.16 is a top view of the GaAs FET chip. The long thin structure is the gate, and the source and drain are labelled

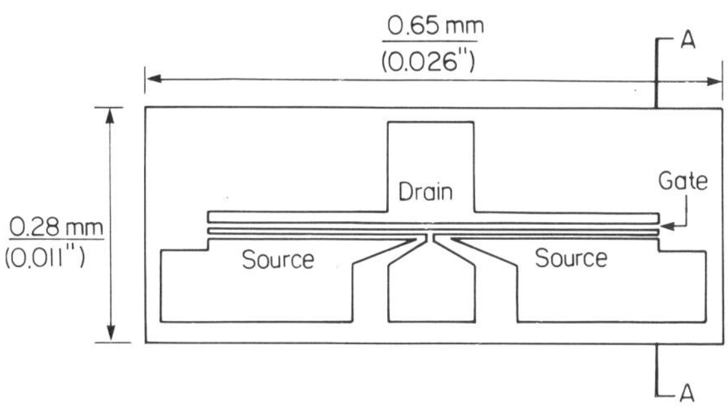

Figure 24.16 GaAs FET chip (top view)

accordingly. By convention, the long dimension of the gate is called its width and the short dimension its length. The cross section of the FET at section AA of *Figure 24.16* is shown in *Figure 24.17*. It gives an indication of the minute device dimensions necessary for good high frequency performance.

24.6.1 Device construction

Figure 24.18 shows the construction of the GaAs FET. The functional layer of the FET is the semiconducting epitaxial layer. It is grown on top of the building block, the semi-insulating substrate. The former is precisely doped for its semiconducting properties and the latter made as pure as possible to act as an

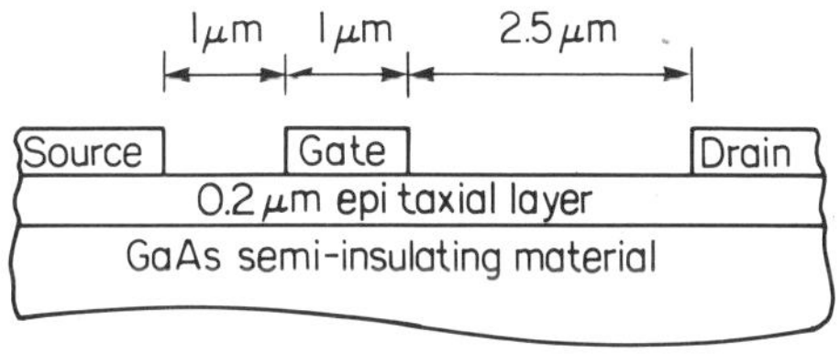

Figure 24.17 GaAs FET cross section view

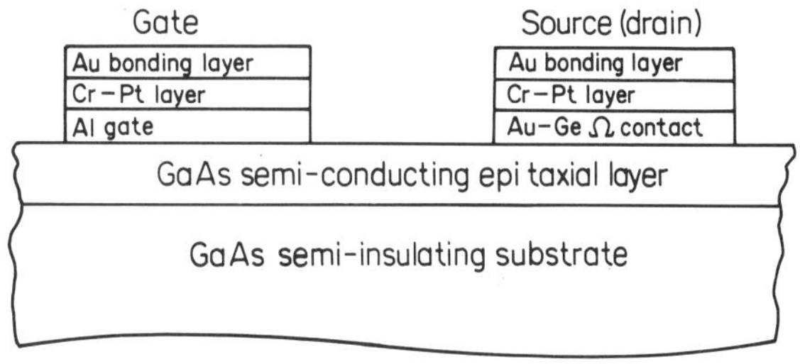

Figure 24.18 GaAs FET construction

insulator. They are both of GaAs material so that one can be grown on top of the other with a perfect crystal interface.

The gate of the FET is the aluminium (Al)–GaAs Schottky barrier, with a gold (Au) layer for wire bonding separated by a chrome (Cr)–platinum (Pt) interface which prevents the migration between gold and aluminium. The chrome is for layer adhesion and the platinum is the actual migration barrier.

The source and drain of the FET are of the same construction. They have a gold–germanium (Ge) ohmic contact with the same gold bonding layer and the chrome–platinum interface. The source, gate and drain surface areas of the FET are covered with a dielectric layer (not shown in *Figure 24.18*) for mechanical scratch protection.

The performance of the FET is controlled largely by the material from which it is made and by the process with which it is manufactured. The long term performance stability of the FET is believed to be related to the quality of the semi-insulating substrate and semiconducting epitaxial layer. It is also dependent on the degree of perfection of the interface between the semi-insulating substrate and the semiconducting epitaxial layer, and the interface of the ohmic contact with the semiconducting epitaxial layer.

24.6.2 Parameters

The d.c. parameters are defined in the same way as low frequency FET's and may be measured in the same way. These are:

I_{DSs} saturation drain current
V_{GSp} pinch-off voltage
g_m transconductance

The important r.f. specifications for an FET are gain, noise figure, power output and *S*-parameters, and these are defined in the same way as for microwave bipolar transistors.

24.6.3 Applications

The microwave GaAs FET is used for a variety of applications, such as in the design of amplifiers, oscillators, mixers, i.f. amplifiers and detectors. In addition, the FET also performs the switching functions in signal processing.

The microwave FET can also be used to replace other

microwave devices such as tunnel diodes and travelling wave tubes and mixer diodes. It offers improved performance, reduced circuit complexity and simplified power supply requirements.

24.7 Tuning varactor diodes

Tuning varactor diodes are pn junction devices which, when reverse biased, exhibit a capacitance which varies inversely with applied reverse bias voltage. The phenomenon occurs because a reverse bias widens the depletion layer so decreasing the capacitance of the device. Tuning varactors may be used in a variety of applications where voltage variable capacitance is required such as tuning tank circuits, AFC loops and filters. It should be noted that signal levels are small so that no signal contribution to bias is evident and the device is never used in the forward-bias configuration. Varactor diodes can be made from both silicon and gallium arsenide.

The capacitance–voltage characteristic is shown for a typical diode in *Figure 24.19*. The curve is of a log/linear type and we define:

$$\text{Tuning ratio} = \frac{C_1 V_1}{C_2 V_2} \tag{24.10}$$

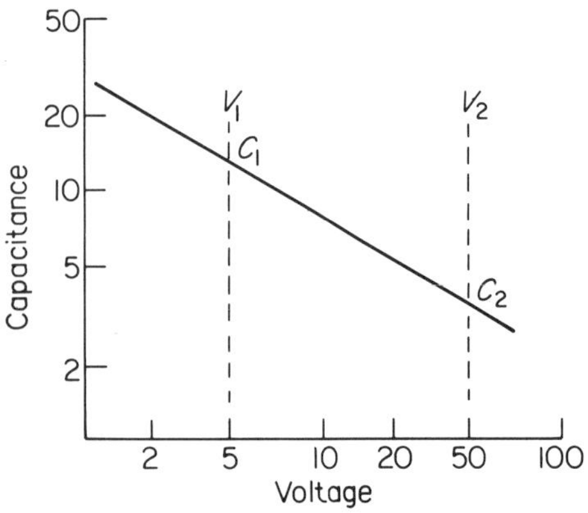

Figure 24.19 Capacitance–voltage for a varactor diode

The tuning ratio determines the magnitude of capacitance change available. It should be noted that the total capacitance includes fixed elements due to package characteristics and this will affect the tuning ratio since, in some cases the package may contribute up to 50% of the total capacitance.

When the space charge across the n region of the diode is totally depleted the varactor action ceases and the capacitance will charge no further with increasing reverse bias. This is called punch-through. Normally this punch-through voltage is less than or equal to the breakdown voltage and the parameter is specified for the designer.

Q-factor is one of the most important characteristics of the varactor diode so that losses in its circuit are kept as low as possible. For high-*Q* the reactances of the device must be kept as low as possible. Obviously the capacitance of the device contributes to the reactance so the *Q*-factor will increase as the reverse bias is increased and the capacitance falls.

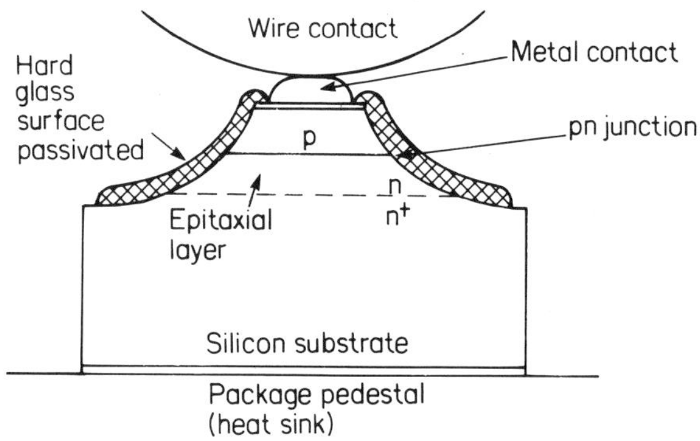

Figure 24.20 Varactor diode construction

For high-*Q* the series resistance must be kept as low as possible and resistance can come from the chip itself, bonding contact and package contacts.

Figure 24.20 shows a commonly used mesa construction of the varactor diode.

Further reading

BARRERA, J. S., 'GaAs field effect-transistors', *Microwave Journal*, February, 28–36 (1976)

WEARDEN, T., 'Varactor diodes', *Electron. Product Des.*, December, 81, 82 (1980)

CASTERLINE, E. T. and BENJAMIN, J. A., 'Trends in microwave power transistors', *Solid State Technol.*, April, 51–56 (1975)

BERSON, B., 'Semiconductors prove fruitful for microwave power devices', *Electronics*, Jan. 22, 83–90 (1976)

DAVIES, R., NEWTON, B. H. and SUMMERS, J. G., 'The TRAPATT oscillator', *Philips Technical Review*, **40**, No. 4 (1982)

25

Light Detectors

M J Rose BSc (Eng)
Production Officer,
Marketing Communications Group,
Mullard Ltd

Contents

25.1 Introduction

Radiation falling on a semiconductor material can produce electron–hole pairs in the material which can be used as charge carriers. Thus the conductivity of the illuminated material is increased considerably. This is the photoconductive effect, and it can be used in various solid-state devices to detect visible and infrared radiation.

If the electron–hole pairs are generated in or near a pn junction, the electrons and holes will be separated by the built-in electric field of the junction. An open-circuit voltage or a short-circuit current will be generated. This photovoltaic effect can be used in photodiodes to produce both voltage and current in an external circuit. A special form of photodiode used considerably in practice is the solar cell. Another device using the photovoltaic effect is the phototransistor. In the phototransistor, the photovoltaic current generated at the collector–base diode is amplified by the transistor action of the emitter.

In certain materials, the energy absorbed from the radiation is sufficient not only for the creation of electron-hole pairs but gives the freed electrons enough energy to be emitted from the material. This is the photoemissive effect which is used in such devices as photoemissive tubes, photomultipliers, and image intensifiers.

Finally, the absorption of radiation increases the temperature of the material, and various detectors have been devised using this thermal effect. One type of detector used at present is based on the pyroelectric effect, the change of electrical polarisation with temperature, observed with certain complex crystals.

25.2 Photoconductive effect

When discussing optoelectronic effects, it is convenient to regard light not as an electromagnetic radiation but rather as a stream of elementary particles. These particles are called photons. Each photon contains a certain amount of energy called a quantum. The energy of a single photon is given by the simple relationship:

$$E = hv \tag{25.1}$$

where E is the energy of the photon, v is the frequency of the radiation, and h is Planck's constant (6.62×10^{-34} J s).

When the radiation is absorbed by a material, the energy of the electrons in the material is raised by the photon energy. However, the electrons in the material can have only certain energy levels, so the radiation is absorbed only when the photon energy can raise the electron from one permissible energy level to another. In semiconductor materials, the two energy levels or bands of interest are the valence band where the electrons are essentially bound to the parent atoms, and the conduction band where the electrons are free and so can be used as charge carriers. The difference between these two energy levels is called the band-gap energy, and is the minimum energy that will generate free carriers. Photons of energy less than the band-gap energy will not be absorbed by the material.

The permissible energy levels in a semiconductor material can be represented by an energy-band diagram, as shown in *Figure 25.1*.

If the energies of photons in the visible and infrared regions are calculated using equation (25.1), it is found that the energies equal or exceed the band-gap energies of many semiconductor materials. Therefore in such materials, illumination with visible or infrared radiation lowers considerably the resistance of the material compared to the 'dark' resistance. This is the operating principle of all solid-state photoconductive detectors. These detectors are called quantum detectors since they depend for their operation on the quantum nature of the radiation.

It is customary to describe radiation in terms of wavelength rather than frequency. Since wavelength and frequency are

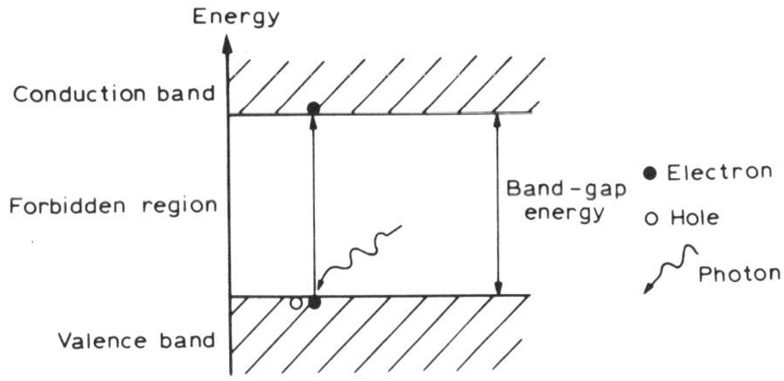

Figure 25.1 Permissible energy levels for electron in semiconductor material, generation of electron-hole pair

related by the speed of light c:

$$v \times \lambda = c \tag{25.2}$$

where λ is the wavelength, equation (25.1) can be rewritten as

$$E = \frac{hc}{\lambda} \tag{25.3}$$

If the minimum band-gap energy for a particular semiconductor material is represented by E_g, equation (25.3) can be rearranged in terms of a critical wavelength λ_c corresponding to this minimum energy:

$$\lambda_c = \frac{hc}{E_g} \tag{25.4}$$

This equation can then be used to determine the wavelength of photons whose energy corresponds to the minimum band-gap energy. Since the energy of the photon is inversely proportional to wavelength, λ_c represents the longest wavelength at which photons contain sufficient energy to produce free carriers within the material. Radiation with wavelengths greater than λ_c does not produce free carriers; with a wavelength shorter than λ_c, the radiation is absorbed and free carriers are produced.

For a fixed total energy of incident radiation, the theoretical relationship between the number of free electrons produced and the wavelengths of the incident radiation is shown in *Figure 25.2*.

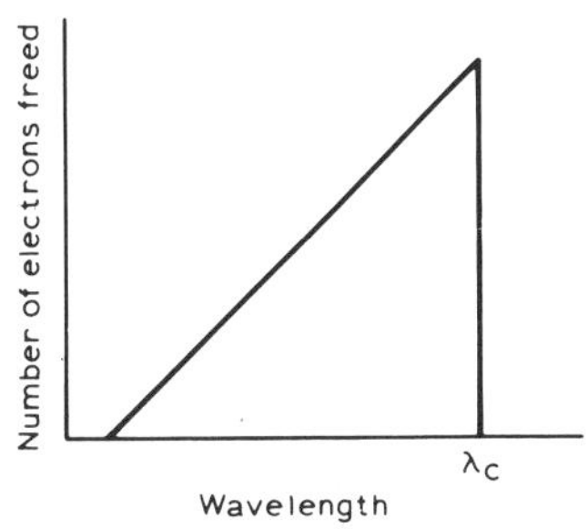

Figure 25.2 Theoretical relationship between number of free electrons produced by incident radiation and wavelength

The sharp cut-off corresponds to λ_c. For wavelengths shorter than λ_c, each photon produces one electron-hole pair. Since the energy of the photons decreases with increasing wavelength, for a constant total energy incident on the material, the number of photons present increases with wavelength. Thus the number of free electrons produced increases, and the triangular characteristic of *Figure 25.2* is obtained.

The higher energy of the photons with shorter wavelengths may excite the electron sufficiently to not only free it from the parent atom, but also to exceed the work function of the material and so allow it to escape from the surface. Obviously such electrons cannot be used as charge carriers within the material, and so a second limiting wavelength is set for the photons λ_m, corresponding to the energy which will cause the electron to be emitted from the surface. An equation similar to Equation (25.4) can be used to relate λ_m to this energy E_t:

$$\lambda_m = \frac{hc}{E_t} \tag{25.5}$$

The wavelengths λ_c and λ_m represent the limiting wavelengths of the incident radiation for use with a particular material. The theoretical relationship between the number of usable charge carriers produced and the wavelength of the radiation is shown in *Figure 25.3*. This curve is called the spectral response of the material. The values of λ_c and λ_m can be calculated from equations (25.4) and (25.5) using the known values of E_g and E_t for the material. For example, the values of λ_c for silicon and germanium are calculated as 1.13 μm and 1.77 μm respectively. The range of wavelengths for visible light is from approximately 0.40 μm (violet) to 0.70 μm (red). Therefore both silicon and germanium can be used in devices operating with visible and near-infrared radiation.

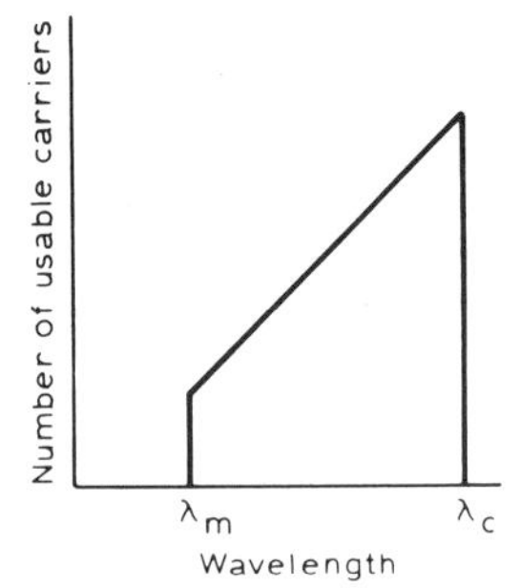

Figure 25.3 Theoretical relationship between number of usable charge carriers produced by incident radiation and wavelength (spectral response)

A practical spectral response is shown in *Figure 25.4* with the theoretical curve in broken lines for comparison. The cut-off wavelengths are the same for both curves since they are determined by the semiconductor material itself. The peak of the practical response, however, is shifted to a shorter wavelength.

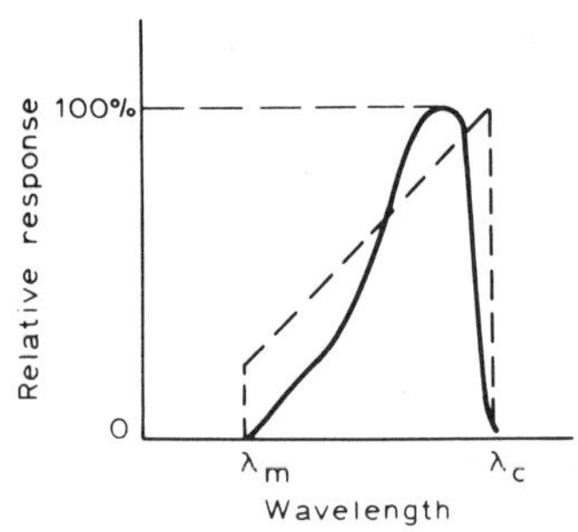

Figure 25.4 Practical spectral response (full line) compared with theoretical response (broken lines)

The shape of the practical curve differs from the theoretical one because the energy-band diagram from which the theoretical curve is derived is an oversimplification of what occurs in practice, and because of surface effects in the semiconductor material. These effects combine to produce the shape of the practical response curve of *Figure 25.4*.

The peak of the response curve can be controlled to some extent by doping the semiconductor material. Adding an impurity to the material creates an intermediate state for the electron between the valence and conduction states. This intermediate state is shown as the trapping level in the energy-band diagram of *Figure 25.5*. Thus an electron can be freed by a photon whose energy is less than the band-gap energy, but sufficient to raise the electron to the trapping level. A second photon can then impart further energy to the electron to transfer it from the trapping level to the conduction band. This two-stage freeing of an electron makes a hole available as a charge carrier before the electron is available. This energy effect combines with the effects mentioned previously to change further the theoretical response curve.

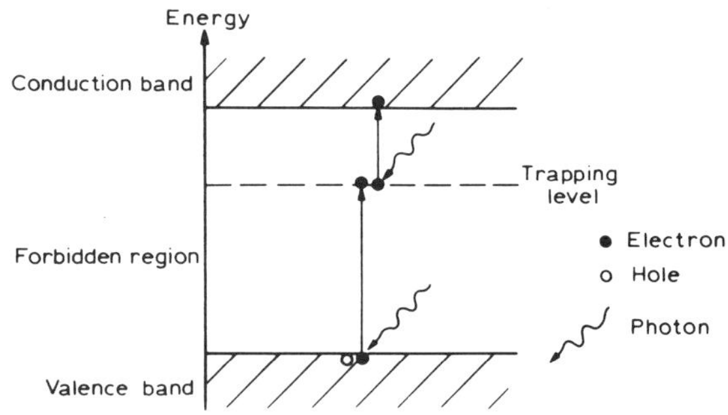

Figure 25.5 Permissible energy levels for electron in doped semiconductor material generation of electron-hole pair in two stages

If a polycrystalline semiconductor material is used, the peak of the practical response curve is flattened so that a wider peak response than that shown in *Figure 25.4* is obtained. Polycrystalline material, however, contains many traps for the generated charge carriers so that a device manufactured from such material will have a slower response time than one using monocrystalline material.

It can be seen from the preceding discussion that by choosing a particular semiconductor material, and by choosing a suitable doping level, a practical device can be constructed to respond to a particular range of wavelengths.

For radiation of a particular wavelength or fixed range of wavelengths, the greater the energy the higher is the number of photons incident on the irradiated material, and hence the higher the number of charge carriers generated within the material. For visible radiation (light), the quantity of light falling on to a given surface is called the *illumination*. The unit of illumination is the lux, defined as the illumination produced when 1 lumen of luminous flux falls on an area of 1 square metre. Thus the fall in resistance of the irradiated semiconductor material is proportional to the illumination of the incident light. For infrared radiation, it is usual to measure the radiation in terms of the power incident on an area. In this case, the fall in resistance of the irradiated material will be proportional to the radiation power per unit area.

25.3 Photoconductive detectors

The performance of a photoconductive detector is limited by noise. The noise in the detector itself is produced by the thermally-generated carriers in the semiconductor material. At very low levels of incident radiation, the thermally-generated carriers may swamp the photo-generated carriers. The effect of the thermally-generated carriers can be minimised by cooling the detector, if necessary to liquid-nitrogen or liquid-helium temperatures. There is, however, a theoretical limit to the performance of the detector set by the thermal background radiation. This radiation is noisy in character, and gives rise to a limiting noise in the detector.

Another technique to improve the performance when the signal produced by the incident radiation is near the noise level is to chop the radiation at a particular frequency before it is incident on the detector. The a.c. signal from the detector produced by the chopped radiation can be extracted from the background noise by an amplifier tuned to the chopping frequency. For the best performance, the chopping frequency should be high and the bandwidth of the system narrow. In practice, the frequency and bandwidth are determined by the application. The amplifier can also be used to produce the constant bias current or bias voltage required for the detector element.

Certain characteristics of the detector are used as 'figures of merit' so that the performance of different types of detector can be compared. The function of a detector is to convert the incident radiation to an electrical signal. Thus a basic property defining the performance of a detector is the ratio of electrical output, expressed as a voltage, to the radiation input, expressed as the incident energy. This ratio is called the responsivity, and is expressed in VW^{-1}. For detectors that use a constant bias voltage, a current responsivity is used, expressed as AW^{-1}.

Various noise figures can be quoted for the detector. The noise of the detector itself is usually expressed as the r.m.s. value of the electrical output measured at a bandwidth of 1 Hz under specified test conditions. Although a low noise enables smaller radiation levels to be detected, it can make the design of suitable amplifiers difficult because the amplifier noise must be low. It is therefore often important to know the ratio of detector noise to the value of Johnson noise in a resistor at room temperature equal in magnitude to the detector. This ratio may be quoted as a noise factor. A signal-to-noise ratio can be measured, but this will apply only for the test conditions under which it was obtained. A more useful quantity, which is often used as a figure of merit for a detector, is the noise equivalent power or NEP. This is the amount of energy that will give a signal equal to the noise in a bandwidth of 1 Hz. It is, in general, a function of the wavelength and the frequency of the measurement. The NEP is equal to the noise per unit bandwidth divided by the responsivity, and is given by the expression:

$$\mathrm{NEP} = \frac{W}{(\Delta f)^{1/2} V_s/V_n} \tag{25.6}$$

where W is the radiation power incident on the detector (r.m.s. value in watts), V_s is the signal voltage across the detector terminals, V_n is the noise voltage across the detector terminals, and Δf is the bandwidth of the measuring amplifier in Hz. The units of NEP are W $Hz^{-1/2}$.

Both responsivity and NEP vary with the size and the shape of the active element of the detector. It is found in practice, and supported by theory, that for similarly-made detectors the NEP is often proportional to the square root of the area of the active element. A quantity called the area normalised detectivity or D^* is commonly used, which is related to NEP by the expression:

$$D^* = \frac{A^{1/2}}{\mathrm{NEP}} \tag{25.7}$$

where A is the area of the active element of the detector in cm^2. The units of D^* are cm $Hz^{1/2}$ W^{-1}.

When responsivity, NEP, and D^* are quoted in published data, certain figures are given in brackets after the quantity. Typical examples of this are responsivity (5.3 μm, 800, 1) and D^* (5.3 μm, 800, 1). These figures refer to the test conditions under which the value was measured. The figure 5.3 μm refers to the wavelength of the monochromatic radiation incident on the detector, 800 to the modulation (chopping) frequency in Hz of the radiation, and 1 to the electronic bandwidth in Hz. An alternative form of the test conditions is given in terms of black-body radiation. An example of this is D^* (500 K, 800, 1). The 500 K refers to the temperature of the black body from which the incident radiation was obtained. The other figures refer to the modulation frequency and bandwidth as before. Details of the other test conditions under which the quantities were measured are also given in published data. These conditions include the distance between the detector element and the source of the radiation, the aperture through which the radiation reaches the element, the operating temperature of the element, details of the chopper system producing the modulation of the radiation, and the bias conditions of the element. The electrical output signal from the detector is amplified by an amplifier tuned to the modulation frequency, which is typically 800 Hz, with a bandwidth of typically 50 Hz.

Both signal and noise vary with the bias current through the element, and so the responsivity, NEP, and detectivity also vary with bias current. At high bias currents the noise increases more rapidly than the signal, and therefore the signal-to-noise ratio has a peak value at some current. The form of the variation of responsivity, noise, and detectivity with bias current for a typical detector is shown in *Figure 25.6*. An optimum value of bias current for the detector can be chosen from these curves. Variations between detectors of the same type may sometimes occur, and so for highly-sensitive applications a fine adjustment of the bias current may be necessary to obtain the optimum performance from the detector. Curves similar to those of *Figure 25.6*, or of the quantities plotted separately, are sometimes given in published data. Similar information may be given in a different form; for example, as the variation of signal-to-noise ratio with bias current.

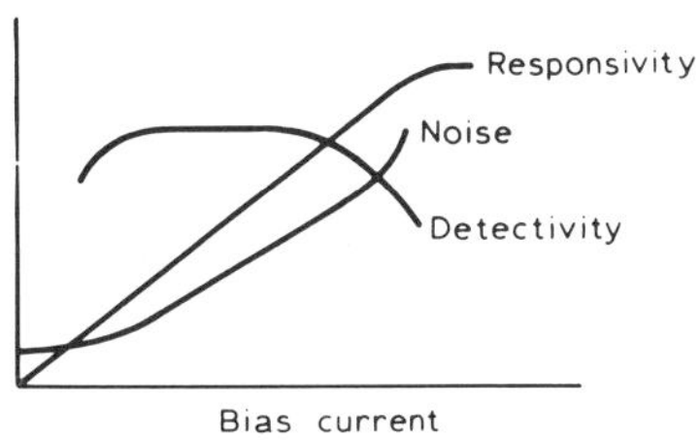

Figure 25.6 Variation of responsivity, detectivity and noise with bias current in infrared photoconductive detector

The values of responsivity, NEP, and D^* will vary with the wavelength of the incident radiation through the variations of signal and noise with wavelength. The variation of D^* can be used as an indication of the spectral performance of the various types of detector, as shown in *Figure 25.7*. The range of wavelengths over which the various semiconductor materials operate can be seen, together with the relative sensitivities. Also shown in *Figure 25.7* is the theoretical limit to operation set by the thermal background radiation. This radiation, as previously explained, limits the attainable detectivity to what is called the background limited value. For wavelengths up to approximately 10 μm, the

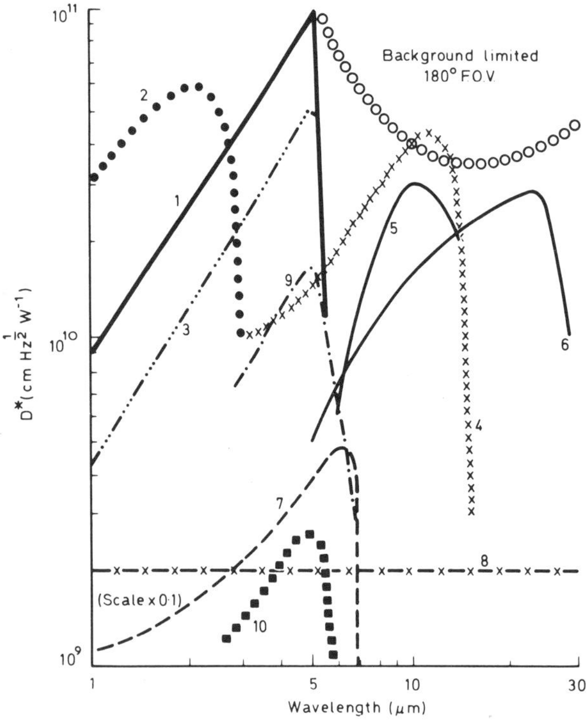

Figure 25.7 Variation of D^* with wavelength: curve 1 indium antimonide, 77 K, 60° FOV; curve 2 lead sulphide, 300 K, 180° FOV; curve 3 indium antimonide, 77 K, 180° FOV; curve 4 cadmium mercury telluride, 77 K, 60° FOV; curve 5 mercury-doped germanium, 35 K, 60° FOV; curve 6 copper-doped germanium, 4.2 K, 60° FOV; curve 7 indium antimonide, 300 K (scale × 0.1); curve 8 TGS pyroelectric detector; curve 9 cadmium mercury telluride, 193 K; curve 10 cadmium mercury telluride, 295 K

background limited value falls with increasing wavelength, and so imposes a maximum value on D^*.

One curve in *Figure 25.7* is shown operating above the theoretical limit. This is because the detector has a cooled aperture restricting the field of view and hence limiting the amount of background radiation received. Because the field of view (FOV) affects the performance, the FOV is stated as well as the temperature for the curves.

The material most suitable to detect radiation of a given wavelength is usually one with the peak spectral response equal to or slightly longer than the given wavelength. Detectors sensitive to wavelengths much longer than the given wavelength require more cooling to reduce the internal thermal noise. It has been shown previously that silicon and germanium are sensitive to radiation of wavelengths up to 1.13 μm and 1.77 μm respectively. These values lie within the infrared region, but silicon and germanium detectors are generally used for visible radiation. These detectors, in the form of photodiodes and phototransistors, are discussed separately. The material used to detect visible radiation is cadmium sulphide (CdS). The most commonly used materials for the detection of infrared radiation are lead sulphide (PbS), for radiation from the visible to that with a wavelength of 3.0 μm; indium antimonide (InSb), visible to 5.6 μm when cooled, visible to 7.0 μm at room temperature; mercury-doped germanium (Ge:Hg), from 2.0 μm to 13 μm; cadmium mercury telluride (CdHgTe), often called CMT, from below 3 μm to 15 μm; and copper-doped germanium (Ge:Cu), from 2.0 μm to 25 μm.

The choice of material for an application may often be influenced by the cooling required. Although detectors requiring cooling to liquid-nitrogen or liquid-helium temperatures are acceptable in scientific work, there are considerable disadvantages in their use for industrial applications. Thus in industrial applications, a less-sensitive detector that can operate at room temperature or with a thermoelectric cooler may be used in preference to a more-sensitive or cheaper detector requiring greater cooling.

25.3.1 Photoconductive detectors for visible radiation

The semiconductor material used in photoconductive detectors for visible light is cadmium sulphide (CdS). The spectral response for this material is shown in *Figure 25.8*. The detector is a two-

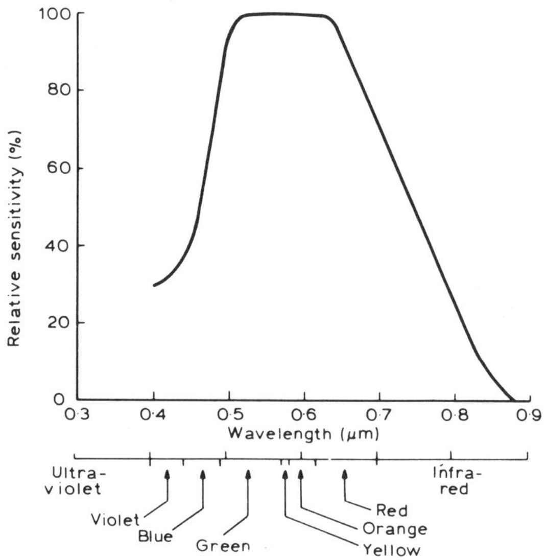

Figure 25.8 Spectral response for cadmium sulphide

terminal device, and is used as a light-dependent resistor. The circuit symbol for a photoconductive detector is shown in *Figure 25.9*, and consists of a resistance symbol with incident arrows representing the radiation.

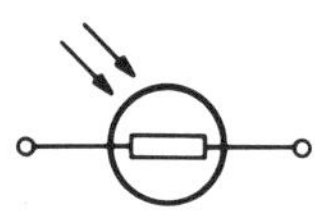

Figure 25.9 Circuit symbol for photoconductive detector (light-dependent resistor)

The active element of the detector is usually made by sintering cadmium sulphide powder into a ceramic-like tablet. Metal electrodes are deposited on to the surface of the tablet to form non-rectifying contacts. The material used has a low resistance compared with that of the cadmium sulphide. An interdigitated structure is used for the contacts, fingers of the contacts connected to one terminal of the detector interleaving fingers connected to the other terminal. The resistance and voltage ratings of the detector are determined by the contact structure, and so detectors with different ratings can be made using the same cadmium sulphide tablet but with different electrode structures. For example, a detector with a small number of widely-spaced contact fingers has a higher resistance and voltage rating than a detector with many closely-spaced contact fingers. The tablet is encapsulated in a glass or transparent plastic envelope. The shape of the encapsulation depends on the direction of the incident light, whether an end-viewing or side-viewing device is required, and the method of termination. Some encapsulations are specially designed for mounting the detector directly on to a printed-wiring board. The spacing of the lead-out wires of the device is made compatible with the standard 2.54 mm grid for printed-wiring boards.

A second and more recent manufacturing process for the sensitive element uses individual grains of cadmium sulphide, the so-called monograin process. The grains, about 40 μm in diameter, are inserted into a thin insulating sheet of a synthetic material so that each grain projects both sides of the material. Each grain is insulated from its neighbours by the sheet, but electrical connection is made by evaporating gold contact pads on to both sides of the sheet. Terminal wires are connected to these contact pads. The complete element is then encapsulated in a transparent plastic. This method of construction enables smaller detectors to be manufactured.

The electrical characteristics of cadmium sulphide detectors depend on several factors. Some of these may be considered as 'direct' factors: the illumination, the wavelength of the incident radiation, the temperature, and the device voltage and current. Other factors, however, affect the operation of the device: the time it has been kept in darkness or the time it has been operating in a circuit, and the operation of the device in the previous 24 hours. The characteristics in the published data are therefore given for specified operating conditions, and the characteristics may vary for different operating conditions. Typical conditions specified in published data could be as follows: operation under d.c. conditions, at the start of life, in an ambient temperature of 25°C, and with an illumination colour temperature of 2700 K. 'Colour temperature' is the temperature to which a black body must be heated to give a similar colour sensation to the source being considered. Sometimes a preconditioning is specified in the data, consisting of an illumination of the detector at a specified level for a specified time before the measurements given in the data are made.

The characteristics of most interest to the circuit designer are the fall of resistance with illumination and the device response time. Values of the 'dark' and 'illuminated' resistance are given under further conditions besides the general ones already stated. For the dark resistance, which is very high, the value is specified with a d.c. voltage applied across the device in series with a high resistance, and after the device has been in darkness for a specified time. Often two values of dark resistance are given, one after a short time in the dark such as 20 s, and the other after a longer time such as 30 min. The difference between these two values of dark resistance can be considerable, the value after the longer time in the dark being at least ten times greater than the value after the shorter time. Two values are also given for the illuminated resistance. An initial value is given under a specified illumination (usually 50 lux) and an applied voltage immediately after removing the device from storage in darkness for 16 hours. (This time allows the chemical changes in the cadmium sulphide to reach equilibrium.) A second value is given after the device has been illuminated for a time such as 15 min. The difference between the two values of illuminated resistance is small. A curve of resistance plotted against illumination is generally given in published data, plotted on logarithmic scales to give a straight line as shown in *Figure 25.10*. Over an illumination range of 1 to 10 000 lux, the resistance value varies typically by three decades. A 'typical' resistance value at 50 lux is often quoted so that different types of detectors can be compared.

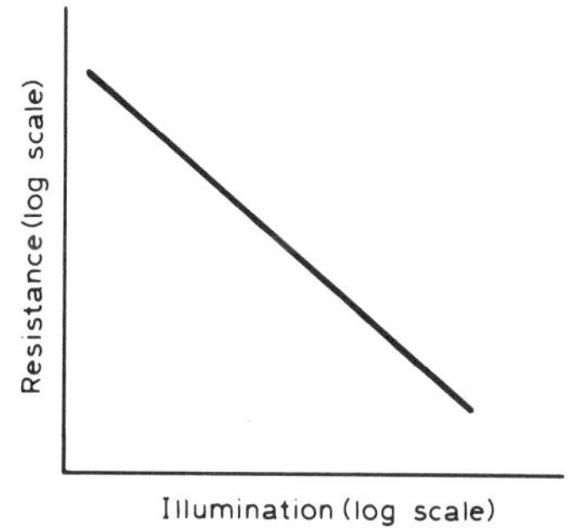

Figure 25.10 Resistance plotted against illumination for cadmium sulphide photoconductive detector

Resistance rise and decay times are specified as the time taken to reach a given resistance value within the range of the device. The resistance decay time is specified as the time for the resistance to fall to a given value from the instant of switching on a given illumination (usually 50 lux) after the device has been in darkness for 16 hours. The resistance rise time is the time for the resistance to reach a given value from the instant of switching off an illumination of 50 lux after the device has been illuminated for a specified time. Curves of the rise and decay times with illumination as parameter are sometimes given in published data. The form of such curves is shown in *Figure 25.11*. Current rise

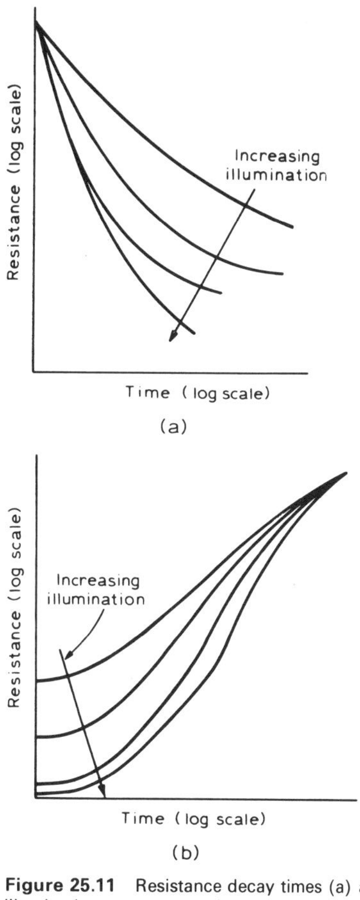

Figure 25.11 Resistance decay times (a) and rise times (b) with illumination as parameter for cadmium sulphide photoconductive detector

and decay times may sometimes be specified. These are the times for the current to rise to 90% of the maximum value, or fall to 10% of the maximum value, after the switching off or on of a given illumination, with a constant voltage across the device. The conditions under which the quoted values are valid are the same as for the resistance times.

Typical values for the rise time of photoconductive detectors lie between 75 ms and 2 s. Typical values of the decay time lie between 25 ms and 3 s.

Ratings given in the data include the maximum device voltage and current, given as d.c. and repetitive peak values, and the maximum device power dissipation. As the power is dependent on the ambient temperature, a power derating curve may sometimes be given. A maximum illumination is also specified.

The material used to encapsulate the active element may affect the spectral response of the detector, causing it to differ from the response for cadmium sulphide shown in *Figure 25.8.* In these cases, a spectral response for the detector may be given in the published data.

When a cadmium sulphide detector is used in a practical circuit, it can be used either as an 'on/off' device or some intermediate resistance value can be used as a trigger level. In the on/off type of operation, the device is used to detect the presence or absence of a light source. The change of resistance between the illuminated and dark values can be used to operate a relay to initiate further action. Applications of this type include alarm systems operated by the interruption of a beam of light directed on to the detector, or counting systems where objects, say on a conveyor belt, interrupt a light beam to produce a series of pulses which operates a counter. In the second type of application, the detector is used to measure the light level. The resistance of the device corresponding to a predetermined light level is used as a threshold value to trigger another circuit. An example of this type of application is a twilight switching circuit. When the daylight has faded to a given level, the corresponding resistance of the detector causes another circuit to switch on the required lights.

25.3.2 Photoconductive detectors for infrared radiation

The operating principle of photoconductive detectors for use with infrared radiation is the same as that of detectors for visible radiation just described. Charge carriers are generated in the illuminated material, and the consequent fall in resistance provides a measure for the incident radiation. Various semiconductor materials can be used to detect infrared radiation, enabling detectors to be chosen to suit particular wavelengths. Some detectors can be used at room temperature, while others must be cooled to low temperatures, for example with liquid nitrogen or liquid helium, to minimise the effects of thermally-generated carriers (thermal noise).

Infrared photoconductive detectors are usually operated with a constant bias current through the active element so that the change in resistance when illuminated appears as a voltage change. For the more sensitive detectors, this voltage can be amplified to produce a larger output signal. Some detectors are operated with a constant bias voltage across the element. For these, the change in resistance produces a change in current.

The infrared detectors that are operated at low temperatures inside some form of cooling vessel require a window through which the incident radiation reaches the active element. The transmission properties of the window may affect the spectral response, causing it to differ from that of the material of the active element. The spectral response of the detector may therefore be given in published data. A time-constant is also given. This is based on the response of the detector to a sudden application of the incident radiation. The time-constant is defined as the time taken from the application of the radiation for the output from the detector to fall to 63% of the peak value. Typical values of time-constant lie between 0.1 μs and 350 μs. Other characteristics that may be given in published data include the variations of responsivity, D^*, and noise with modulation frequency and the operating temperature of the detector.

The ratings given in the data are similar to those already discussed for photoconductive detectors used with visible radiation. They include the maximum detector power, maximum bias current, and the maximum operating and storage temperatures.

The construction of infrared photoconductive detectors depends on the material used for the active element, in particular whether or not it requires cooling. Detectors for operation at room temperature can be encapsulated in a metal envelope with a viewing window, or the element can be deposited on a flat plate. Detectors that require cooling can be deposited on a dewar vessel or arranged for mounting on to a cooling vessel. For detectors that use thermoelectric cooling, the active element is mounted on the cooler in a suitable encapsulation.

Lead sulphide detectors are of two types, depending on the method used to form the active element. The element is deposited as a film either by evaporation or by a chemical reaction. The evaporated-film type can be operated at room temperature or cooled; the chemically-deposited type is generally operated at room temperature only.

A typical construction used for the evaporated-film type is to form the active element on an inner surface of a dewar vessel, as shown in *Figure 25.12.* A metal housing can be used to protect the dewar vessel if the detector is to be used at room temperature only, while for operation below room temperature the dewar vessel can be filled with a suitable coolant. The operating

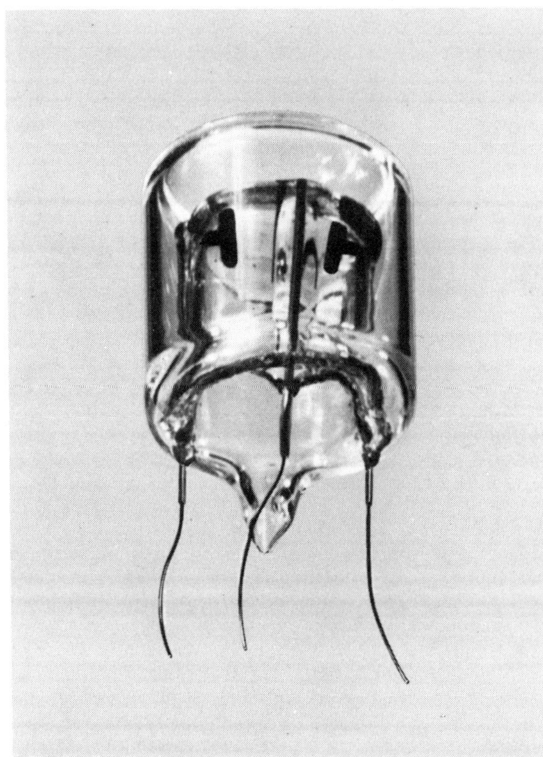

Figure 25.12 Lead sulphide photoconductive detector. The lead sulphide film is deposited on an inner surface of the dewar vessel (photo: Mullard Limited)

temperature range of this type of lead sulphide detector is from 293 K to 173 K (+20°C to −100°C), while the responsivity has a peak value at 240 K and D^* has a peak value at 220 K. A typical value of responsivity (500 K, 800, 1) is 2.0×10^3 V W^{-1}, and typical values of D^* (2.0 μm, 800, 1) are 4.0×10^{10} cm $Hz^{1/2}$ W^{-1} at 293 K and 2.0×10^{11} cm $Hz^{1/2}$ W^{-1} at 230 K. The range of wavelengths over which this detector can be used is from the visible region to 3.0 μm with a peak response at 2.3 μm. The time-constant is typically 100 μs.

Because the chemically-deposited lead sulphide detector is generally operated at room temperature, the construction does not have to allow for cooling. Typical constructions used for this detector are encapsulation in a small metal envelope such as the TO-5 outline, or as a 'flat pack' in which the element is deposited on an insulating substrate. Both types of construction can incorporate a filter to modify the spectral response. This type of detector is operated with a constant bias voltage, and has a typical current responsivity (2.0 μm, 800, 1) of 200 mA W^{-1}. A typical value of D^* (2.0 μm, 800, 1) is 1.0×10^{10} cm $Hz^{1/2}$ W^{-1}. The range of wavelengths is from the visible to 2.8 μm, but this can be restricted by using a germanium window to 1.5 μm to 2.8 μm. The time-constant is typically 250 μs.

Both types of lead sulphide detector have a high resistance compared to detectors using other materials. A typical resistance for the evaporated-film type is 1.5 MΩ, and for the chemically-deposited type 200 kΩ.

Indium antimonide detectors can be operated at room temperature, or cooled by liquid nitrogen to 77 K at which temperature a detectivity near the background limited value is obtained. The photoconductive material is a doped single crystal which is cut into thin slices from which the active element is cut. At room temperature the resistivity of indium antimonide is low, giving detectors with resistances of about 5Ω/square. The most useful forms of element are therefore long strips or 'labyrinths' made from a number of long strips laid in parallel but connected electrically in series. The effect of temperature on performance is considerable at room temperature, and so a good heatsink is required for the element. Typical constructions for room-temperature indium antimonide detectors mount the element either in a copper block or in a flat-pack encapsulation which can be provided with a viewing window (usually sapphire) if it is necessary to protect the element from dirty or corrosive atmospheres. A typical value of responsivity (6.0 μm, 800, 1) is between 1.0 V W^{-1} and 3.5 V W^{-1}, and a typical value of D^* (6.0 μm, 800, 1) is 1.5×10^8 to 3.0×10^8 cm $Hz^{1/2}$ W^{-1}. The range of wavelengths is from the visible region to 7.0 μm, and the time-constant is typically 0.1 μs.

The elements for cooled indium antimonide detectors are made in a similar way to the room-temperature types, but are mounted on the inner surface of a glass dewar vessel. The radiation is transmitted to the element through a sapphire window. The element is cooled with liquid nitrogen either by filling the dewar vessel with the liquid or by using a miniature Joule–Thomson cooler. Cooling increases the resistance of the element to about 2 kΩ/square, and a wide range of shapes and sizes can be made for the element. Arrays as small as 0.1 mm square are possible. Because the detectors of this type are 'background limited', the use of a cooled aperture improves the detectivity. Typical values of responsivity (5.3 μm, 800, 1) and D^* (5.3 μm, 800, 1) are 1.2×10^4 V W^{-1} and 5.0×10^{10} cm $Hz^{1/2}$ W^{-1} respectively. With a restricted field of view, the value of D^* (5.3 μm, 800, 1) can be increased to 1×10^{11} cm $Hz^{1/2}$ W^{-1}. The range of wavelengths is from the visible to 5.6 μm, and the time-constant is typically 2 μs to 5 μs.

The cross-section of a typical construction for a cooled indium antimonide detector is shown in *Figure 25.13*.

Cadmium mercury telluride is an alloy semiconductor material which forms a mixed crystal of cadmium telluride and mercury

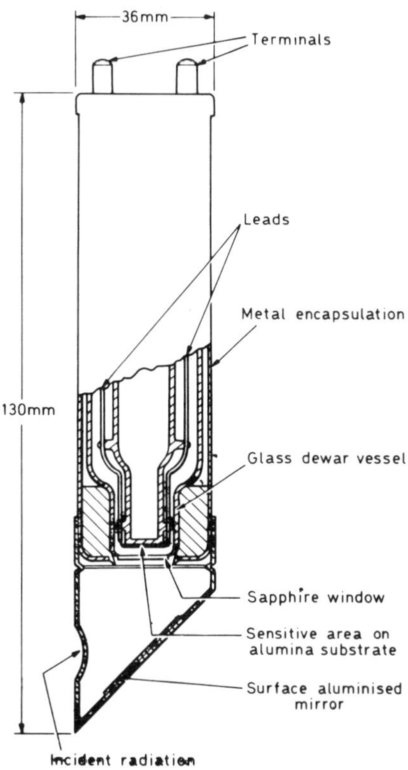

Figure 25.13 Construction of cooled indium antimonide photoconductive detector for operation at 77 K

telluride. The peak spectral response of the material can be varied by the relative proportions of cadmium and mercury telluride in the crystal from 9.5 μm to 15 μm. Cadmium mercury telluride (CMT) detectors can be used to cover a range of wavelengths from approximately 3 μm to 15 μm. Detector elements are made in a similar way to those of indium antimonide just described, being cut from thin slices of a single crystal. CMT detectors can be operated at room temperature, with thermoelectric coolers at temperatures down to approximately 200 K, or with liquid nitrogen at 77 K. For operation with liquid nitrogen, a construction for the detector similar to that for the indium antimonide type shown in *Figure 25.13* can be used, except that the window is made from silicon with an anti-reflective coating ('bloomed' silicon). This window material has a peak transmission between 9 μm and 11 μm. Typical values of D^* (λ_p, 800, 1), λ_p being the peak spectral response, are greater than 10^{10} cm $Hz^{1/2}$ W^{-1}. The typical time-constant is less than 1 μs.

In doped-germanium detectors, the radiation is absorbed by the electrons in the added impurity. This leads to a lower absorption coefficient, and hence the need for a thicker element. It is also essential to cool the element so that the electrons are initially in the impurity centres ready to be excited. The germanium must be extremely pure apart from the added impurity. Mercury and copper are the most widely used impurities.

Mercury-doped germanium detectors are used for wavelengths from 2.0 μm to 13 μm with a peak response at approximately 10 μm. The element requires cooling to 35 K, and this is achieved by using liquid helium either in bulk or in a Joule–Thomson cooler. The cross-section of a typical cryostat for a mercury-doped germanium detector is shown in *Figure 25.14*. The upper tank contains liquid nitrogen, and carries a radiation

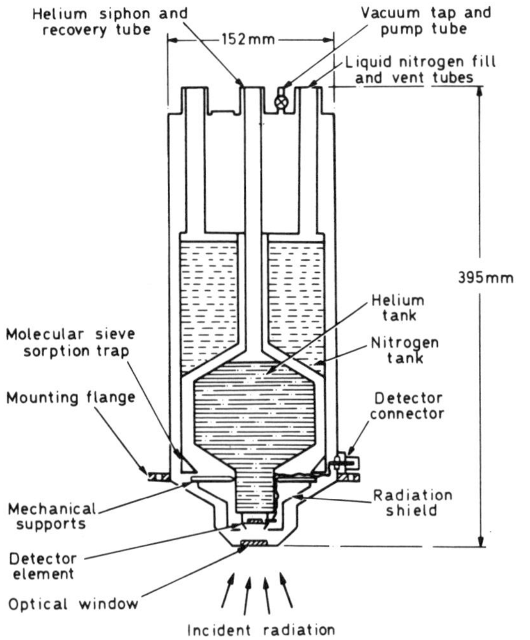

Figure 25.14 Construction of cryostat for doped-germanium photoconductive detectors operating at liquid-helium temperatures

shield which partly surrounds the lower liquid-helium tank. This minimises evaporation of the helium, and also forms a shield for the detector element. A vacuum is maintained inside the cryostat, and the quality of the vacuum is maintained by a molecular sieve trap incorporated in the radiation shield. The cooled window through which the radiation is transmitted to the element is made of bloomed germanium. The resistance of the element is typically between 10 kΩ and 60 kΩ. A typical value of D^* (10 μm, 800, 1) is 1.3×10^{10} cm $Hz^{1/2}$ W^{-1}, and the time-constant is typically less than 1 μs.

Copper-doped germanium detectors are used for wavelengths between 2.0 μm and 25 μm with a peak response at 15 μm. This type of detector requires cooling to 4.2 K, and this is achieved with liquid helium in a cryostat similar to that shown in *Figure 25.14*. The resistance of the element is typically between 2.5 kΩ and 240 kΩ. A typical value of D^* (15 μm, 800, 1) is 10×10^{10} cm $Hz^{1/2}$ W^{-1}, and the time-constant is typically 1 μs.

Photoconductive detectors have been developed for specialised applications of infrared radiation with wavelengths extending beyond 50 μm into the sub-millimetric and microwave regions. A typical detector for these wavelengths uses the impurity level in indium antimonide, and operates at a temperature below 2 K. A complex cooling system is required for such a detector, and the cross-section of a typical cryostat is shown in *Figure 25.15*. Two glass dewar vessels are used: an inner vessel containing liquid helium which is pumped to reduce the pressure inside the vessel and hence lower the boiling point of the helium, and an outer vessel containing liquid nitrogen to minimise the evaporation of the helium. An operating temperature of 1.6 K can be obtained in this way. The detector operates in a magnetic field produced by a superconducting solenoid immersed in the liquid helium. The incident radiation is directed onto the detector element by a light pipe fitted with a polythene window. In a typical detector of this type, the diameter of the window would be approximately 2 cm, and the length of the light pipe between the window and the element approximately 60 cm. This type of detector would operate at

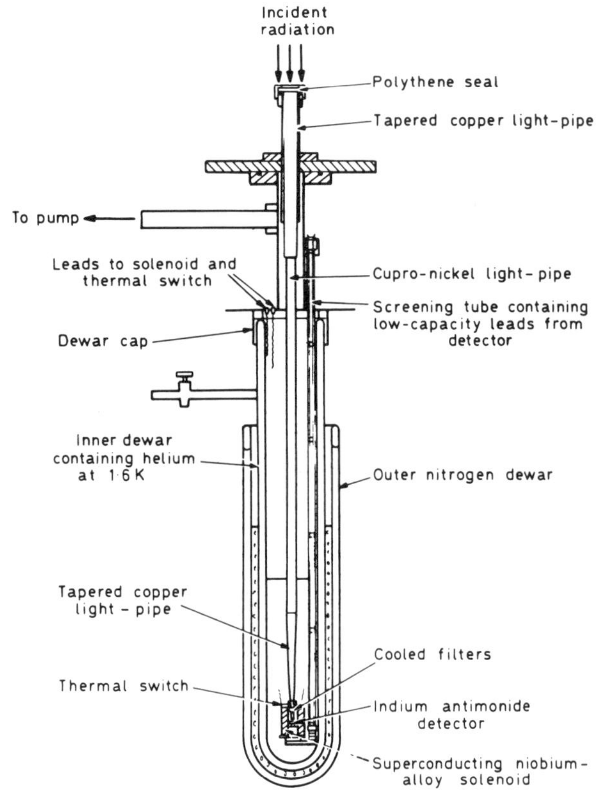

Figure 25.15 Construction of cryostat for indium antimonide photoconductive detector for sub-millimetric radiation operating at 1.6 K

wavelengths, say, between 0.1 mm and 10 mm with a peak response at, typically, 1 mm. A typical value of responsivity (1 mm, 800, 1) is 2×10^3 V W^{-1}, and a typical value of D^* (1 mm, 800, 1) is 1×10^{12} cm $Hz^{1/2}$ W^{-1}. The time-constant is typically 1 μs.

The range of applications of infrared photoconductive detectors is very wide, extending from simple systems using uncooled detectors to give warning of flame-failure in boilers, to complex and sophisticated systems with detectors operating at very low temperatures which are used for physical research or very precise measurements. The detection of the black-body radiation from an object can be used for such applications as detecting overheating in mechanical and electrical systems, intruder detection in security areas, temperature measurement without physical contact with the object, and thermal imaging. The varying absorption of infrared radiation by different materials can be used in such applications as chemical analysis by infrared spectroscopy or leak detection in closed systems.

25.3.3 Other infrared detectors

To complete this brief survey of infrared detectors, two other types are described that use photoelectronic effects other than photoconductivity.

A recently-introduced material for use in infrared detectors is lead tin telluride, known as LTT. The detector uses the photovoltaic effect, a junction formed in the lead tin telluride alloy being used to separate the electron–hole pairs generated by the incident radiation. The photovoltaic effect is discussed more fully in connection with photodiodes; for the operation of the infrared detector it is sufficient to say that the separation of the charge carriers results in an open-circuit voltage or a short-circuit current. Thus an output signal can be produced by the detector without the need for an external bias voltage or current.

The peak spectral response of the detector occurs at, typically, 11 μm, enabling a range of wavelengths from approximately 8 μm to 14 μm to be covered. The detector is operated at a temperature of 77 K. Typical values of responsivity (λ_p, 800, 1) and D^* (λ_p, 800, 1) are 150 V W^{-1} and 8×10^9 cm $Hz^{1/2}$ W^{-1} respectively. The time-constant is typically less than 0.1 μs.

The second type of infrared detector uses the pyroelectric effect. This effect occurs in certain crystals with complex structures in which there is an inbuilt electrical polarisation which is a function of temperature. At temperatures above the Curie point, the pyroelectric properties disappear, but below the Curie point changes in temperature result in changes in the degree of polarisation. This change can be observed as an electrical signal if electrodes are placed on opposite faces of a thin slice of the material to form a capacitor. When the polarisation changes, the charge induced on the electrodes can flow as a current through a comparatively low impedance external circuit, or produce a voltage across the slice if the impedance of the external circuit is comparatively high. The detector produces an electrical signal only when the temperature changes.

Many crystals exhibit the pyroelectric effect, but the one most commonly used for infrared detectors at present is triglycine sulphate $(NH_2CH_2COOH)_3H_2SO_4$, known as TGS. The incident radiation on the active element of the detector produces an increase in temperature by the absorption of energy, and hence a change in the polarisation occurs. Changes in the level of the incident radiation produce an electrical signal from the detector.

The spectral response of TGS detectors, as with other thermal detectors, is wide. It extends from 1 μm, below which incident energy is not absorbed, to the millimetric region. Filters can be used to define the range of wavelengths required for particular applications. TGS detectors generally operate at low modulation frequencies (approximately 10 Hz) but can operate at higher frequencies. This is because although the responsivity falls with increasing frequency (responsivity is inversely proportional to frequency), the noise also falls with increasing frequency. The value of NEP therefore rises only slightly up to frequencies of, say, 10 kHz. The detectivity is constant over the spectral range, D^* (500 K, 10, 1) being typically 2×10^9 cm $Hz^{1/2}$ W^{-1}. A typical value for NEP is 2×10^{-10} W $Hz^{-1/2}$.

25.4 Photovoltaic devices

25.4.1 Photodiode

Visible or infrared radiation incident on a semiconductor material will generate electron–hole pairs by the normal photoconductive action. If these charge carriers are generated near a pn junction, the electric field of the depletion layer at the junction will separate the electrons and holes. This is the normal action of a pn junction which acts on charge carriers irrespective of how they are produced. However, the separation of the carriers gives rise to a short-circuit current or an open-circuit voltage, and this effect is called the photovoltaic effect. It can be used in such devices as photodiodes and phototransistors.

In a normal pn junction with no external bias applied, an equilibrium state is reached with an internal potential barrier across the junction. This potential barrier produces a depletion layer and prevents majority carriers crossing the junction. Minority carriers, however, can still cross the junction, and this gives rise to the small reverse leakage current of the junction diode. An external reverse voltage adds to the internal potential barrier, but the reverse current remains substantially constant because of the limited number of minority carriers available until avalanche breakdown occurs. An external forward voltage overcomes the internal potential barrier, and causes a large majority carrier current to flow. A photodiode differs from a small-signal junction diode only in that visible or infrared radiation is allowed to fall on the diode element instead of being excluded. If there is no illumination of the diode element, the photodiode acts as a normal small-signal junction diode, and has a voltage/current characteristic like that shown in *Figure 25.16(a)*.

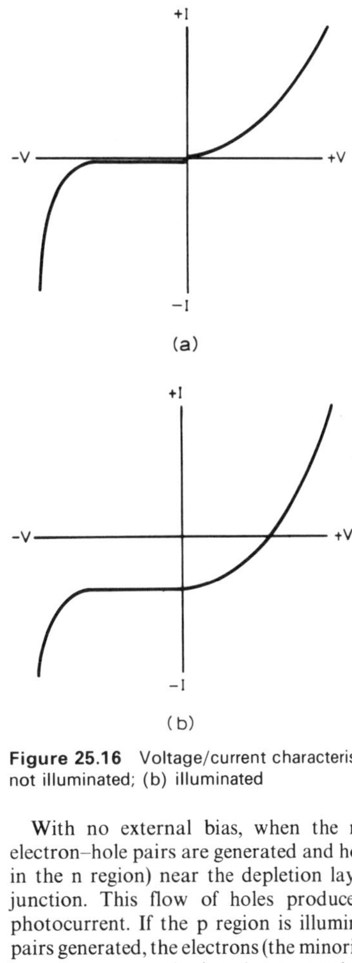

Figure 25.16 Voltage/current characteristic for photodiode: (a) not illuminated; (b) illuminated

With no external bias, when the n region is illuminated electron–hole pairs are generated and holes (the minority carrier in the n region) near the depletion layer are swept across the junction. This flow of holes produces a current called the photocurrent. If the p region is illuminated and electron–hole pairs generated, the electrons (the minority carrier in the p region) are swept across the junction to produce the photocurrent. In practical photodiodes, both sides of the junction are illuminated simultaneously, and the electron and hole photocurrents add together. The effect of the photocurrent on the voltage/current characteristic is to cause a displacement, as shown in *Figure 25.16(b)*. The normally small reverse leakage current is augmented by the photocurrent.

The magnitude of the photocurrent depends on the number of charge carriers generated, and therefore on the illumination of the diode element. Thus a series of characteristics is obtained for various levels of illumination. When forward current flows through the diode, the photocurrent is swamped. Thus the part of the characteristic of interest is that where reverse current flows (quadrants 3 and 4).

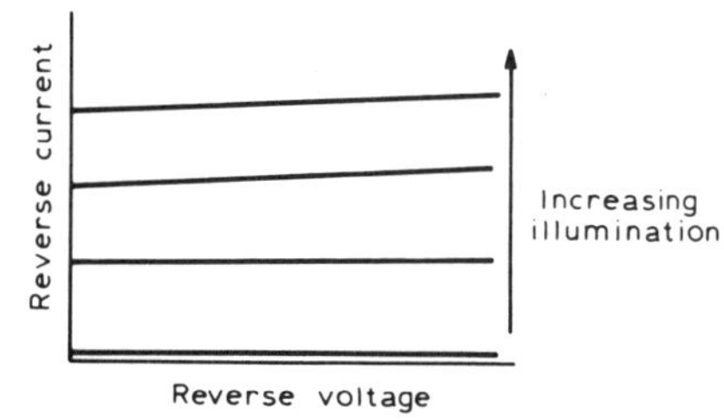

Figure 25.17 Voltage/current characteristic with illumination as parameter for photodiode operating with reverse bias

With a reverse voltage across the diode (quadrant 3), the relationship between the reverse voltage and current with illumination as parameter is shown in *Figure 25.17*. (It is conventional to invert both axes of the characteristic to give the form shown in *Figure 25.17*.) A typical value of the 'dark' current for germanium photodiodes is 6 μA, and for the light current with an illumination of 1600 lux a typical value is 150 μA. For silicon devices, typical values of dark current lie between 0.01 μA and 0.1 μA, and of light current with an illumination of 2000 lux between 250 μA and 300 μA. Both the dark and light currents are temperature-dependent, and information on the variation with temperature is given in published data.

When operated with a reverse bias, the photodiode is sometimes described as being in the photoconductive mode. Although this term is deprecated since the operating principle is strictly speaking the photovoltaic effect, it is descriptive of the action of the photodiode in quadrant 3. The current in the diode is proportional to the number of carriers generated by the incident radiation. The term also provides a convenient distinction from the operation of the diode in quadrant 4.

With no bias voltage across an illuminated photodiode, a reverse current flows, the sum of the small minority-carrier thermal current and the photocurrent. As an increasing forward bias is applied, the magnitude of the reverse current decreases as the majority-carrier forward current increases. Eventually, the magnitude of the majority-carrier current equals that of the photocurrent, and no current flows through the diode. If the forward bias is increased further, a forward current flows as in a normal junction diode. Thus the limiting values in quadrant 4 are the current at zero bias, and the forward voltage at which the current is zero. These values represent the short-circuit current and open-circuit voltage which can be obtained from the photodiode, as shown in *Figure 25.18*. (Again, it is conventional sometimes to invert this characteristic, but this time only about the voltage axis.) With a suitable load resistance connected across the photodiode, it is possible to extract power from the diode.

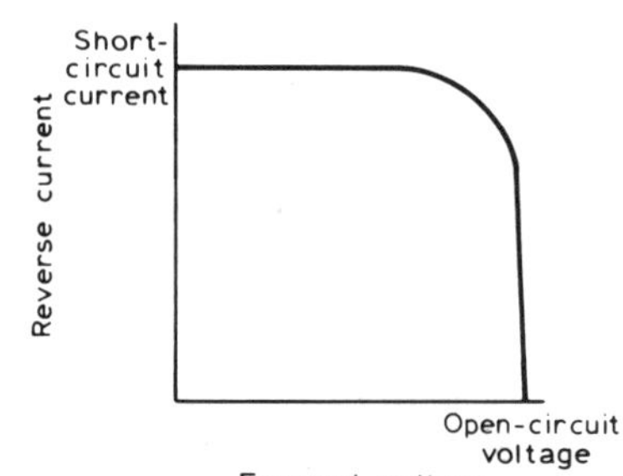

Figure 25.18 Voltage/current characteristic for photodiode with forward bias showing short-circuit current and open-circuit voltage values

This is the principle of the solar cell which is described in Section 25.4.4.

Because the magnitude of the reverse current depends on the illumination, the relationship between the forward voltage and the current with illumination as parameter in quadrant 4 has the form shown in *Figure 25.19*. An alternative method of presenting

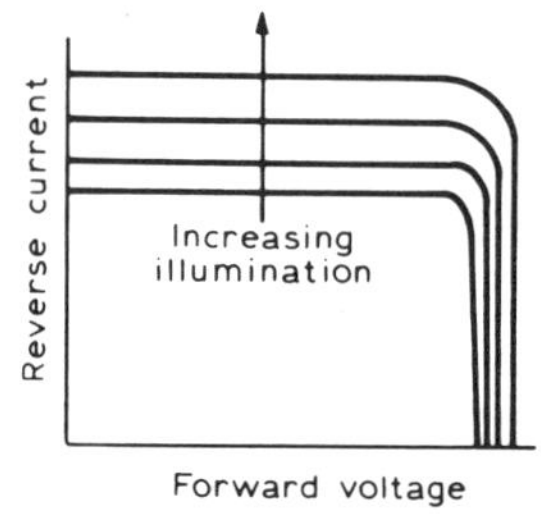

Figure 25.19 Voltage/current characteristic with illumination as parameter for photodiode with forward bias

this information, sometimes used in published data, is a curve of the light current plotted against illumination. This is plotted on logarithmic scales to give a straight line, as shown in *Figure 25.20*. A sensitivity figure may also be given, relating light current to illumination (sensitivity in μA/lux) or to the energy of the incident monochromatic radiation (sensitivity in μA/μW). Information on the variation of both light and dark currents with temperature is also given in published data.

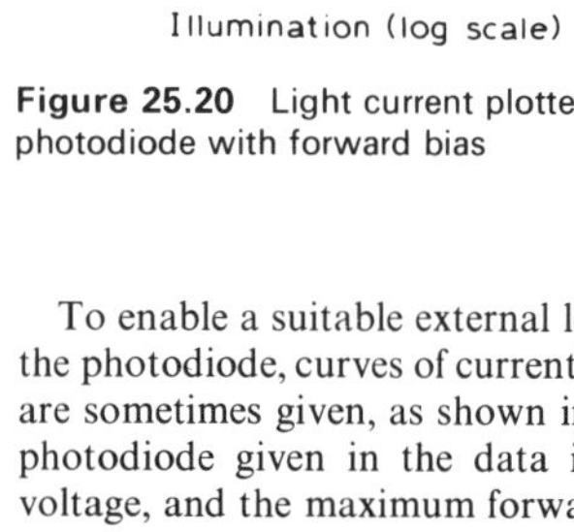

Figure 25.20 Light current plotted against illumination for photodiode with forward bias

To enable a suitable external load resistance to be chosen for the photodiode, curves of current plotted against load resistance are sometimes given, as shown in *Figure 25.21*. Ratings for the photodiode given in the data include the maximum reverse voltage, and the maximum forward and reverse currents. From these ratings and the load resistance curve, a suitable external circuit to be operated by the photodiode for the particular application can be designed.

In many optical applications, the radiation incident on the photosensitive device is modulated or is chopped by some mechanical system. The maximum frequency to which the photodiode responds is determined by three factors: diffusion of carriers, transit time in the depletion layer, and the capacitance of the depletion layer. Carriers generated in the semiconductor material away from the depletion layer must diffuse to the edge of the layer before they can be used to form the photocurrent, and this can cause considerable delay. To minimise the time of

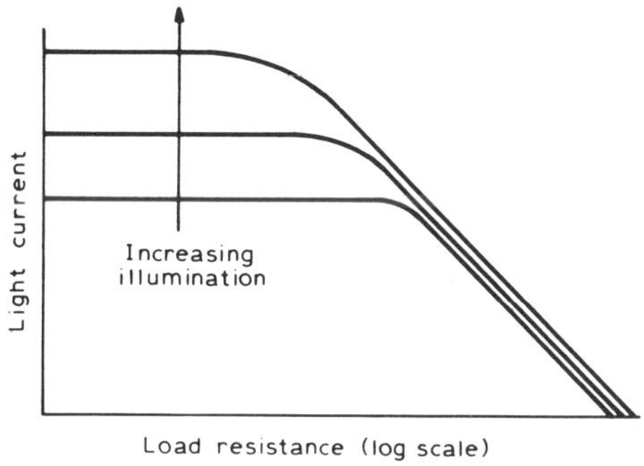

Figure 25.21 Light current plotted against load resistance with illumination as parameter for photodiode with forward bias

diffusion, the junction should be formed close to the illuminated surface. The maximum amount of light is absorbed when the depletion layer is wide (with a large reverse voltage), but the layer cannot be too wide otherwise the transit time of the carriers will be too long. The depletion layer cannot be too thin, however, as the capacitance would be too high and so limit the high-frequency response of the diode. With careful design of the photodiode and choice of a suitable reverse voltage, operation up to at least 1 GHz is possible, enabling the photodiode to be used with rapidly pulsating radiation. Information on the relative response with modulation frequency, or information on the switching time expressed as the light-current rise and fall times in response to the sudden application and removal of the incident radiation, is given in published data. Typical values of rise and fall times for photodiodes operating up to 1 GHz are approximately 0.5 ns. For high-speed applications, say up to a frequency of a few hundred megahertz, a typical value of rise time is 2 ns.

The spectral response for photodiodes generally has a peak about 0.8 μm, as shown in *Figure 25.22*. This value is determined

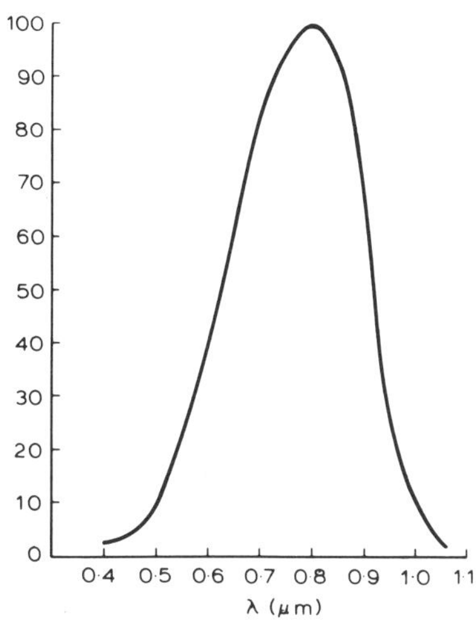

Figure 25.22 Spectral response for silicon photodiode

by the sealing oxide grown over the diode element during the planar manufacturing process. Photodiodes can therefore be used with visible and near-infrared radiation.

Various constructions can be used for photodiodes. In the cheapest form, the diode element is unencapsulated. A moulded plastic encapsulation or small metal envelope such as the TO-5 outline may also be used, with a window to allow the radiation to fall on to the diode element. A glass lens may sometimes be fitted to the device. The circuit symbol for a photodiode is shown in *Figure 25.23*.

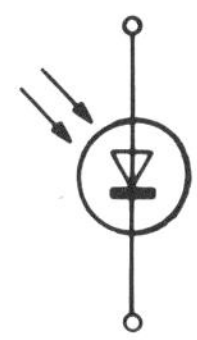

Figure 25.23 Circuit symbol for photodiode

25.4.2 Phototransistor

If a photodiode is formed by the collector–base diode of a transistor, the photocurrent flowing across the collector–base junction will be amplified by normal transistor action. A transistor operating in this way is called a phototransistor.

An npn phototransistor is shown in *Figure 25.24*. The base connection is open-circuit, that is the base is floating and the transistor is operating as a two-terminal device. The normal bias

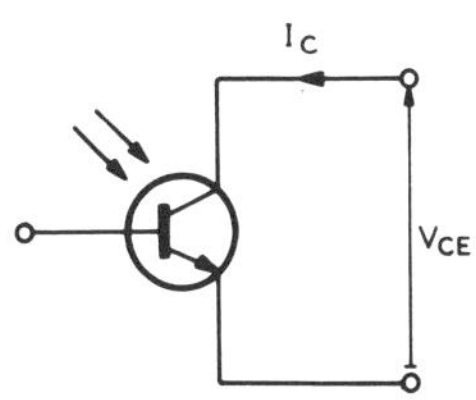

Figure 25.24 NPN phototransistor

conditions are still maintained, however. With no illumination of the transistor, the current in the collector–emitter circuit is the common-emitter leakage current I_{CEO}. When the collector–base diode is illuminated, electron–hole pairs are generated and the minority-carrier photocurrent flows across the reverse-biased collector–base junction. Electrons flow out of the base, and holes flow into the base. Thus the forward bias across the base–emitter junction is increased, which in turn increases the electron flow from the emitter, across the base, into the collector. The collector current is therefore the sum of the electron photocurrent and the electron current from the emitter which is h_{FE} times the photocurrent. In other words, the photocurrent of the diode has been amplified by the current gain of the phototransistor. Thus the phototransistor is used in applications where greater sensitivity is required than can be obtained with a photodiode. In some applications, the base is not left open-circuit but a relatively high value resistor is connected between the base and emitter. This does not affect the operation of the phototransistor just described, but gives improved thermal stability and improves the light-to-dark ratio.

The output characteristic of the phototransistor relates the light collector current $I_{CE(L)}$ to the collector-to-emitter voltage V_{CE} with illumination as parameter. It is similar to the output characteristic of a junction transistor in the common-emitter configuration with illumination rather than base current as parameter. The form of the characteristic is shown in *Figure 25.25*.

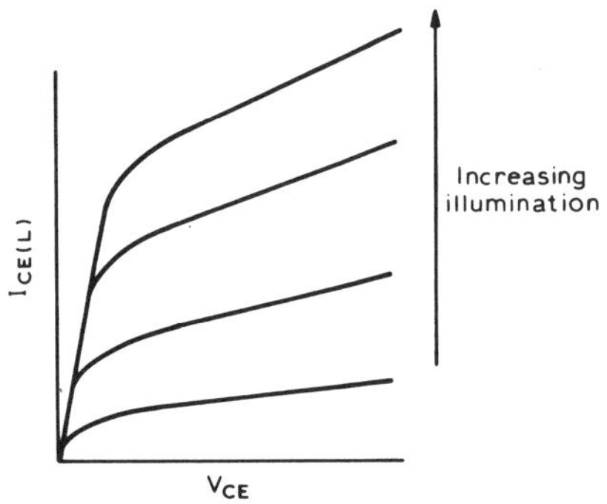

Figure 25.25 Output characteristic for phototransistor with illumination as parameter

Other characteristics included in published data are the variation of light and dark collector currents with temperature, and the variation of collector current with illumination. Information on switching times is given, and sometimes in addition the variation of rise and fall times with emitter current. The spectral response curve is also given. As with the photodiode, the response has a peak at about 0.8 μm determined by the sealing oxide grown over the transistor element during the planar manufacturing process. The ratings for the phototransistor given in published data are similar to those for the junction transistor. They include the maximum collector-to-base and collector-to-emitter voltages, the maximum emitter-to-collector or emitter-to-base voltages, the maximum collector current, and the maximum power dissipation. A thermal resistance is given so that a suitable heatsink can be chosen if required for the application.

The usual construction for a phototransistor is a small metal envelope such as the TO-18 outline with a plane window or a lens.

Photodiodes and phototransistors are used in applications where an electrical signal corresponding to incident visible or infrared radiation is required. An example of this is a paper-tape reader where the photodevice responds to light passing through the holes in the tape to produce an electrical signal based on the pattern of holes in the tape. Other applications include photographic light meters, flame-failure alarm systems for boilers, and detectors for modulated laser beams.

25.4.3 Light-activated silicon controlled switch

A four-layer device such as a silicon controlled switch (s.c.s.) can be used as a photosensitive device. It can be used as a light-controlled switch, the s.c.s. being turned on when illuminated to operate another circuit. Two modes of operation are possible, as shown in *Figure 25.26*: with either the anode gate or cathode gate not connected, and a resistor connected to the other gate. Thus if the anode gate is left floating, a resistor is connected between the cathode gate and the cathode. If the cathode gate is left floating, the resistor is connected between the anode gate and the anode.

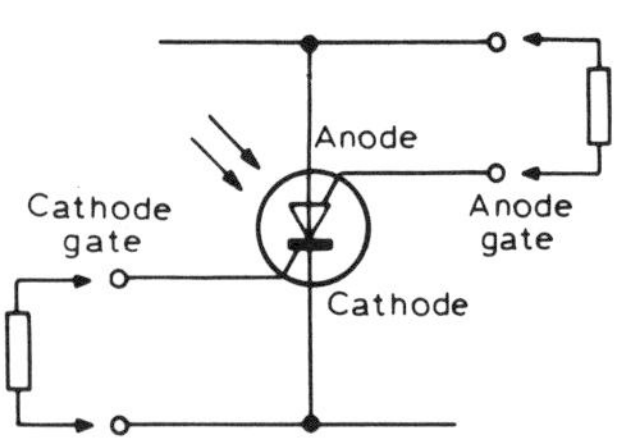

Figure 25.26 Light-activated silicon controlled switch showing alternative modes of operation

The operation of the light-activated s.c.s. can be described in terms of the two-transistor analogue which is shown, with the external resistors in broken lines, in *Figure 25.27*. When the anode gate lead is not connected, *Figure 25.27(a)*, the base of the pnp transistor is left floating, and this device acts as a phototransistor. When the s.c.s. has a positive anode-to-cathode voltage applied, and is not illuminated, only the small leakage current flows. When the device is illuminated, the photocurrent in the pnp transistor flows through the resistor between the cathode gate and cathode. A voltage is developed across this resistor so that the base of the npn transistor becomes positive with respect to its emitter, and so this transistor starts to conduct. (The development of the voltage drop across the resistor is equivalent to placing a small positive voltage on the cathode gate, which is one of the ways in which an s.c.s. can be triggered.) Cumulative action between the two transistors occurs, and the s.c.s. is turned on. If the cathode gate is not connected *Figure 25.27(b)*, the base of the npn transistor is floating and this device acts as the phototransistor. The photocurrent develops a voltage across the resistor between the anode gate and anode, so that the anode gate becomes negative with respect to the anode, and the s.c.s. is turned on. The light-activated s.c.s. is turned off, as with the normal s.c.s., by reducing the current through the device to below the holding value.

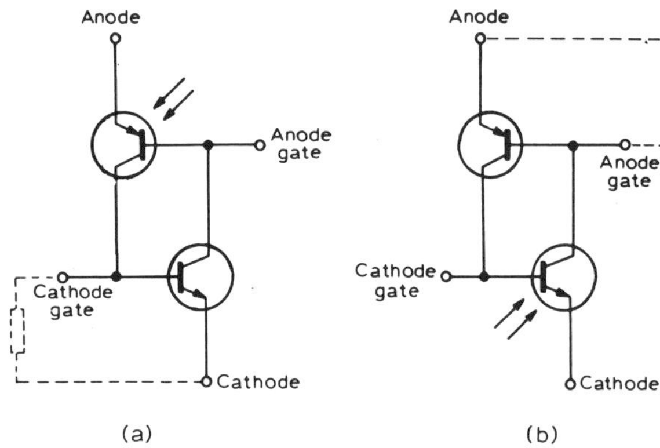

Figure 25.27 Two-transistor analogue for light-activated silicon controlled switch: (a) pnp phototransistor; (b) npn phototransistor

The characteristics and ratings for a light-activated s.c.s. given in published data are similar to those for a normal s.c.s. Additional information is given on the variation of cathode-gate current with illumination, and the spectral response. Two illumination levels are specified: a minimum value $E_{on,min}$ which will trigger all devices, and a maximum non-triggering value $E_{off,max}$. These values are determined by the leakage current of the device.

The usual encapsulation for the light-activated s.c.s. is a small metal envelope such as the TO-72 outline with an end-viewing window. The device is generally used in applications as a relay driver. The relay is connected in the anode circuit, and is energised when the device is turned on by the incident radiation.

25.4.4 Solar cell

The solar cell is a form of photodiode which is optimised for operation from the sun's radiation. The operating principle is the same as that of the photodiode, the generation of electron-hole pairs by the incident radiation, and the separation of these charge carriers by the electric field of the depletion layer at a pn junction. The flow of minority carriers across the junction produces a short-circuit current or open-circuit voltage so that power can be extracted from the device by a suitable load resistance. The surface area of the solar cell is made as large as possible so that the maximum amount of radiation is incident on the device. The pn junction is formed near the surface to minimise carrier recombination.

The most commonly used materials from which practical solar cells are manufactured are silicon and gallium arsenide. The cell can be constructed as a thin p-type layer on an n-type substrate, or as a thin n layer on a p substrate. Both regions of the cell are heavily doped. Non-rectifying contacts must be made to both regions. The contact of the substrate can be made on the back of the device, but the contact to the front layer must be made in such a way that the minimum surface area is obscured. Narrow contact fingers can be deposited on the front surface, as shown in the schematic structure of *Figure 25.28(a)*, or the material of the front layer can be taken round the sides of the cell to the back of the substrate and the contact made there, *Figure 25.28(b)*.

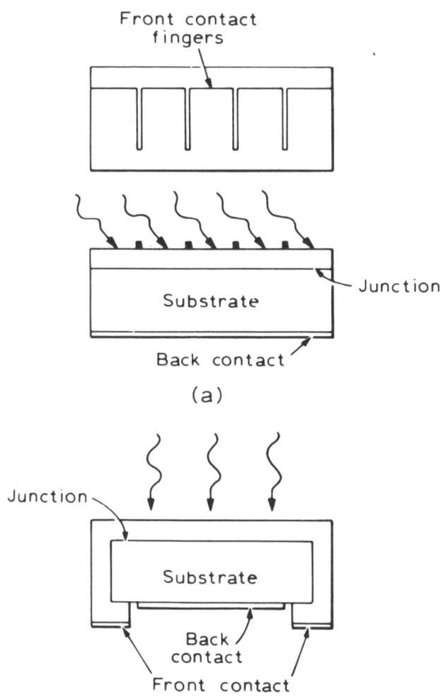

Figure 25.28 Schematic structure of solar cell: (a) contact on front surface; (b) contact on back of cell

The relationship between the voltage and photocurrent for a solar cell is shown in *Figure 25.29* with the maximum-power rectangle superimposed. This rectangle represents the maximum amount of power that can be extracted from the cell, the product $V_{mp} \times I_{mp}$. With a suitable load resistance, this power can be 80%

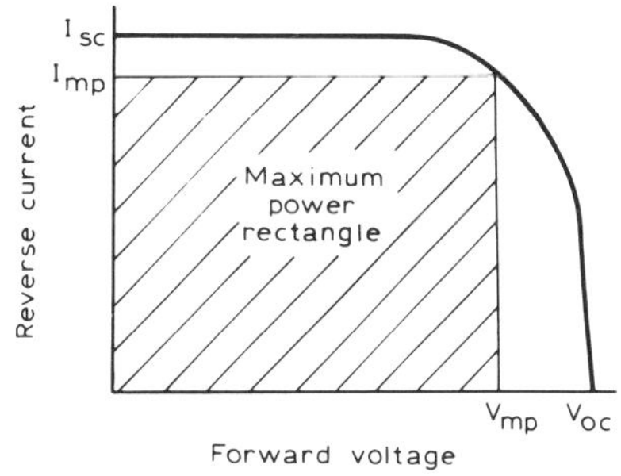

Figure 25.29 Voltage/current characteristic for solar cell with maximum power rectangle superimposed

of the product $V_{oc} \times I_{sc}$, where V_{oc} is the open-circuit voltage and I_{sc} is the short-circuit current. Typical values of the open-circuit voltage and short-circuit current are 0.5 V and 0.1 A respectively.

Solar cells can be connected in series to provide a higher voltage than can be obtained from a single cell, and connected in parallel to provide a higher current. The main application of solar cells is for the power supplies of space vehicles and communication satellites. Large arrays of cells are used, connected in series/parallel, to provide the required values of voltage and current to drive the electronic circuitry in these vehicles. The conversion efficiency of such batteries is between 10% and 15%. Another application of solar cells is for terrestial power supplies. In areas where extended periods of sunlight can be relied on, arrays of solar cells can be used as a power source.

25.5 Symbols for principal parameters of photoelectronic devices

Symbols for the main electrical and optical parameters of photoelectronic devices, other than those common to small-signal diodes and transistors, are listed below.

D^*	area-normalised detectivity of photoconductive detector
E	illumination
$I_{CE(D)}$	dark collector current of phototransistor
$I_{CE(L)}$	light collector current of phototransistor
I_{cell}	current through active element of photoconductive detector
I_{SC}	short-circuit current of solar cell
I_{RD}	dark current of photodiode
$I_{R(SC)}$	short-circuit current of photodiode
NEP	noise equivalent power
P_{cell}	power dissipation of active element of photoconductive detector
S, S_R	sensitivity of photodiode and phototransistor, change of current with incident radiation
T_{tablet}	operating temperature of active element of photoconductive detector
V_{cell}	voltage across active element of photoconductive detector
V_{OC}	open-circuit voltage of solar cell
λ, λ_p	peak spectral response

26 Light Emitters

J A van Raalte
BSc, MSc, PhD, MAPS, FSID, SMIEEE
Head, Videodisc Recording and Playback Research,
David Sarnoff Research Centre,
RCA, Princeton, NJ, USA
(Sections 26.1–26.3)

M E Fabian BSc, DMS
Applications Manager, Electron Devices,
ITT Components Group, UK
(Sections 26.4–8)

Contents

26.1 Light-emitting diodes

26.1.1 Mechanism

Light is emitted from electroluminescent pn junctions as a result of radiative recombination of electrons and holes whose concentrations exceed those statistically permitted at thermal equilibrium. Excess carrier concentrations are obtained in a forward-biased pn junction through minority carrier injection: the lowering of the potential barrier of the junction under forward bias allows conduction band electrons from the n side and valence band holes from the p side to diffuse across the junction. These injected carriers significantly increase the minority carrier concentrations and recombine with the oppositely charged majority carriers. This recombination, which tends to restore the equilibrium carrier densities, can result in the emission of photons, i.e. light, from the junction. The recombination process, hence the amount as well as the wavelength of the light generated, is a strong function of the physical and electrical properties of the material.

Both energy and momentum must be conserved when an electron and a hole recombine to emit a photon. Since the photon has considerable energy but very small momentum ($h\nu/c$), simple recombination only occurs in direct bandgap materials, that is, where the conduction band minimum and valence band maximum both lie at the zero momentum position. This condition was assumed to be a prerequisite for efficient electroluminescence until about 1964 when Grimmeiss and Scholz[1] demonstrated reasonably efficient electroluminescence in the indirect bandgap material GaP.

In an indirect bandgap material where the valence band maximum and conduction band minimum lie at different values of momentum, recombination can only occur when a third momentum-conserving particle is involved; phonons (i.e. lattice vibrations) serve this purpose. The probability of an electron–hole recombination involving both a photon and phonon is considerably smaller than the simpler process involving only a photon in a direct bandgap material. This is clearly illustrated by the differences in recombination coefficients (R) for various materials: this coefficient (R), which relates the radiative recombination rate (r) to the excess minority (Δn) and majority (p) carrier concentrations

$$r = R(\Delta n)p \tag{26.1}$$

is of the order of 10^{-14} to 10^{-15} cm^3 s^{-1} for indirect bandgap materials (such as Si, Ge, and GaP) while it is of the order of 10^{-10} to 10^{-11} cm^3 s^{-1} for direct bandgap materials (such as GaAs, GaSb, InAs, or InSb).

Light-emitting diodes (LED's) can, in principle, be made from any semiconducting compound containing a pn junction, and having a sufficiently wide bandgap. Despite considerable effort on other materials to achieve efficient luminescence, currently only III–V compounds are of practical interest for visible LED's.

26.1.2 Doping

Donor and acceptor impurities determine the magnitude and type of conductivity and also play a major role in the radiative recombination processes. In direct bandgap III–V compounds, donor impurities are usually chosen from group VI of the periodic table (Te, Se, S), while acceptor impurities usually belong to group II (Zn, Cd, Mg). Sometimes group IV impurities (Ge, Si, Sn) are used either as acceptors or donors depending on which of the available lattice sites (III or V) they occupy.

Optimum impurity concentrations are best determined experimentally. High doping levels, i.e. large majority carrier concentrations, are desirable (1) to lower the bulk resistivity, and consequently to minimise heating and voltage drops at high current densities; and also (2) because the recombination probability, which is directly proportional to the carrier concentration, should be as large as possible. The doping concentrations are practically limited, however, by the formation of precipitates and other crystallographic imperfections which introduce competing non-radiative recombination centres and therefore reduce the electroluminescence efficiency. Donor concentrations of 10^{17} to 10^{18} cm^{-3} and acceptor concentrations of 10^{17} to 10^{19} cm^{-3} are typical.

Although the luminescence efficiencies in n- and p-type materials at optimum doping concentrations are essentially equal, the light emitted from the p-region normally dominates for several reasons: (1) electron injection into the p-region is favoured over hole injection into the n-region because of the high electron to hole mobility ratio in most III–V compounds; (2) the Fermi level is slightly higher than the intrinsic energy gap in n-type material, while it is lower in p-type material; the resulting heterojunction effect further favours electron injection into the p-region over hole injection into the n-region; (3) light generated in the n-region is usually of shorter wavelength (higher energy) than that generated in the p-region. Therefore n-generated light is strongly absorbed in the p-region while p-generated light passes through the n-region with reduced absorption losses. This latter fact is usually taken into account in the design of efficient LED's.

In indirect bandgap materials, e.g. GaP, recombination across the gap requires the participation of a momentum-conserving phonon and is therefore inefficient. More efficient recombination can occur when a charged carrier is first trapped at a neutral impurity centre and then used to attract the oppositely charged carrier. Momentum conservation in this case is more easily satisfied because the carrier trapped at a neutral impurity centre is highly localised in space and consequently has a wide range of crystal momentum.

Only a limited number of impurities have been found which enhance the recombination in GaP. Near bandgap ($\sim$2.23 eV) green emission is increased by nitrogen substitution for P in GaP, but competing non-radiative recombination processes limit the internal quantum efficiency to about 1%. Red luminescence ($\sim$1.79 eV) is improved by incorporation of Zn and oxygen centres. The internal quantum efficiency in GaP:Zn-O is higher (10–20%) than for green luminescence, but the emission saturates at relatively low current densities ($\lesssim$10 A cm^{-2}) due to the limited concentration of Zn–O centres ($<10^{17}$ cm^{-3}). Green luminescence in GaP:N does not readily saturate since nitrogen concentrations of 10^{19} cm^{-3} are practical.

26.1.3 Quantum efficiency and brightness

Since most of the applications of LED's involve an observer, the response of the human eye at the emitted wavelength is of primary importance. The eye sensitivity for normal (photopic) vision extends from about 400 nm to 700 nm. It peaks in the green (555 nm) at 680 lm W^{-1} and falls off towards the red and blue regions of the spectrum (*Figure 26.1*). The brightness, and to a large extent the visibility, of an LED is proportional to the product of its external quantum efficiency and the sensitivity of the eye at the emitted wavelength.

The external quantum efficiency of an LED is the ratio of emitted photons to number of electrons passing through the diode, and it is typically 0.1–7% at room temperature. The external quantum efficiency is always less than the internal quantum efficiency ($\gtrsim$50% under optimum conditions), because all the light generated cannot exit from the diode.

The internal quantum efficiency is highly dependent on the perfection of the material near the pn junction. Various defects, contaminants, or dislocations reduce the internal quantum

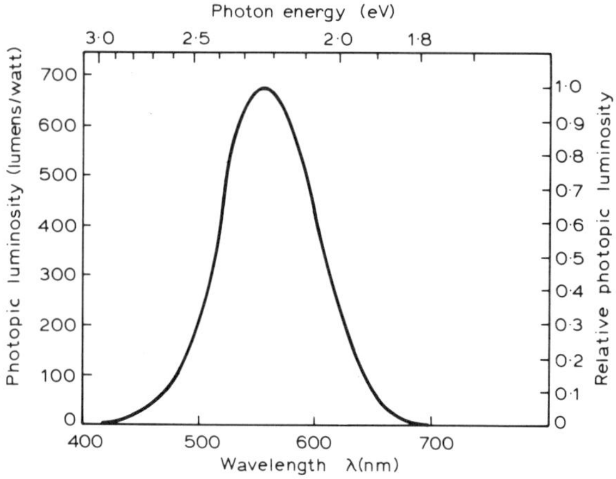

Figure 26.1 Photopic luminosity (normal vision) as a function of incident wavelength

efficiency by producing deep recombination centres, which lead to long wavelength radiation, or by enhancing non-radiative recombination. It is thought that the superiority of GaP:N diodes grown by liquid phase epitaxy (LPE) compared to those grown by vapour-phase epitaxy (VPE) may be due to the lower density of Ga vacancies in LPE material.

Poor quality substrates are a major cause of defects. Additional imperfections can be introduced during the growth of the epitaxial layer, especially when the lattice mismatch is relatively large. These latter imperfections can be reduced by grading the composition of the film during growth from that of the substrate to the composition desired at the junction. Defects are also sometimes introduced during the diffusion process used to form the pn junction.

The high index of refraction of III–V compounds (typically ~3.5) leads to additional light losses. Much of the generated light suffers total internal reflection. This increases the optical path length inside the diode thereby increasing internal optical absorption which is particularly high for near bandgap emission. Thus the external quantum efficiency can be 50 to 100 times smaller than the internal efficiency.

Practically, these losses can be reduced by increasing the transmissivity of the surface. Antireflection coatings can be applied, the diode can be shaped in the form of a hemisphere (but this is very costly), or a hemispherical epoxy or acrylic lens can be used to increase the efficiency by a factor of 2–3 typically. The internal absorption is reduced at longer emission wavelengths, which are obtained by incorporating deeper acceptors (e.g. Si instead of Zn in GaAs), or in indirect bandgap materials, where the emission is much below the energy gap. In indirect gap materials, the reduced internal efficiency (10% to 20% for GaP:Zn,O as compared to ~50% for GaAs:Zn) sets an upper bound for the external efficiency of the device.

The brightness (*B*) of an LED in nits (cd m^{-2}) is given by

$$B = \frac{3940\eta_{ext}LJ}{\lambda}\left(\frac{A_j}{A_s}\right) \tag{26.2}$$

where

η_{ext} = external quantum efficiency
L = luminous efficiency of the eye (*Figure 26.1*) (in lm W^{-1})
J = junction current density (in A cm^{-2})
A_j/A_s = ratio of junction area to observed emitting surface
λ = emission wavelength (in μm)

Brightnesses in excess of 3500 nits at 10 A cm^{-2} are readily achieved in commercial LED's. By way of comparison, the brightness of a TV kinescope is of the order of 300 nits; that of the surface of a frosted light bulb as much as 30 000 nits.

26.1.4 Ternary compounds

In view of the sharp peak of the eye sensitivity in the green region of the spectrum, it is very desirable to obtain yellow or green luminescence. The upper limit for the emission energy is roughly equal to the energy gap. Most direct bandgap III–V binary compounds have energy gaps less than 1.72 eV (720 nm) and therefore can luminesce only in the infrared. Other semiconductors, with energy gaps closer to 2.23 eV (555 nm) are therefore of great interest.

Ternary alloy systems, composed of narrow direct gap and wider indirect gap materials (e.g. $GaAs_{1-x}P_x$, $Al_xGa_{1-x}As$, $In_{1-x}Ga_xP$, $In_{1-x}Al_xP$) provide a monotonically increasing direct energy gap as the relative concentration of the indirect gap material is increased up to a critical composition x_c where the energy gap becomes indirect. This is shown in *Figure 26.2* for the $GaAs_{1-x}P_x$ system.

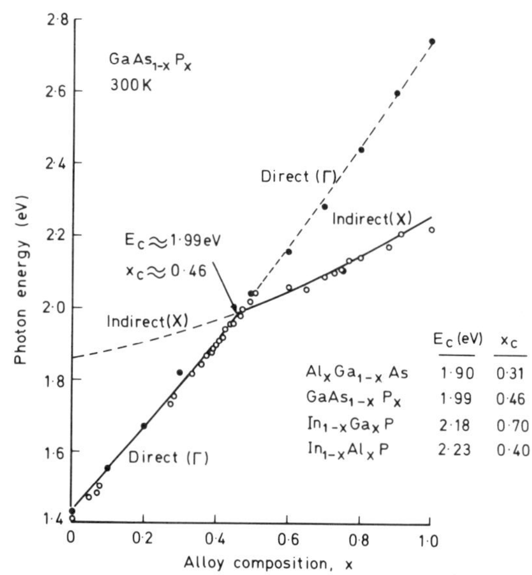

Figure 26.2 Direct (Γ) and indirect (×) conduction band minima for $GaAs_{1-x}P_x$ as a function of alloy composition, *x*. Closed data points are from electroreflectance measurements; open points are from electroluminescence spectra (Reproduced with permission from Smith[17])

The radiation recombination rate decreases sharply near the critical composition because an increasing fraction of electrons is transferred to the indirect band which typically has 50 to 100 times as many available energy states as the direct band.

From a knowledge of the energy bands, the external quantum efficiency and luminous efficiency at different wavelengths can be predicted.[2] This is illustrated in *Figure 26.3* for the ternary compounds of major importance. The alloy systems $In_{1-x}Ga_xP$ and $In_{1-x}Al_xP$ have potentially higher luminous efficiencies than either $GaAs_{1-x}P_x$ or $Al_xGa_{1-x}As$, but other aspects—such as the ease of preparation of the alloy, its defects, the extent of lattice mismatch—are also important in determining the choice of materials.

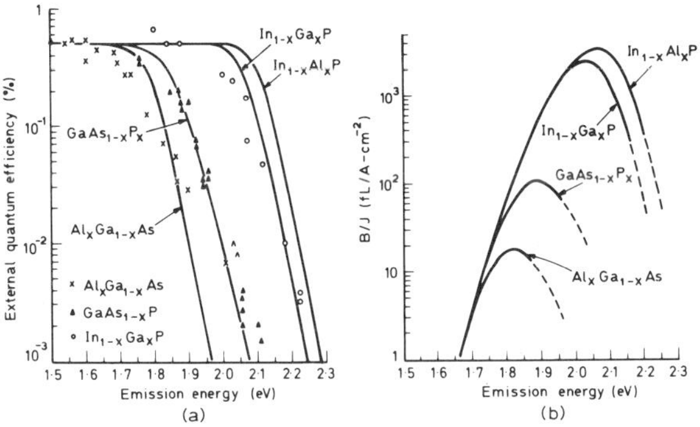

Figure 26.3 (a) Calculated external quantum efficiency for $Al_xGa_{1-x}As$, $GaAs_{1-x}P_x$, $In_{1-x}Ga_xP$ and $In_{1-x}Al_xP$ LED's as a function of emitted photon energy. (b) Calculated brightness values for the same alloys (Reproduced with permission from Smith[17])

26.2 LED synthesis

LED's are generally fabricated from epitaxial material deposited on single-crystal GaAs or GaP substrates. The choice of substrates depends on the lattice constant and thermal expansion coefficient of the epitaxial layer. The lattice constant of ternary alloys depends on the composition. A good match between substrate and epitaxial film can be obtained for $Al_xGa_{1-x}As$, a moderate match for $GaAs_{1-x}P_x$, and relatively poor match for $Al_xIn_{1-x}P$ and $In_{1-x}Ga_xP$.

Two techniques are widely used for the growth of the epitaxial films: liquid phase epitaxy (LPE) and vapour phase epitaxy (VPE). Each technique has advantages and disadvantages; VPE is used for $GaAs_{1-x}P_x$, while LPE produces the best GaP and $Al_xGa_{1-x}As$ devices at this time.

In liquid phase epitaxy (LPE), originally developed by Nelson,[3] the layer, for instance GaP, is grown from a Ga melt to which are added polycrystalline GaP and desired dopants such as Zn (p type) or Te (n type). The melt is positioned at one end of a graphite boat and a polished substrate at the other end (*Figure 26.4*). The boat is heated to about 1060°C in hydrogen and tipped

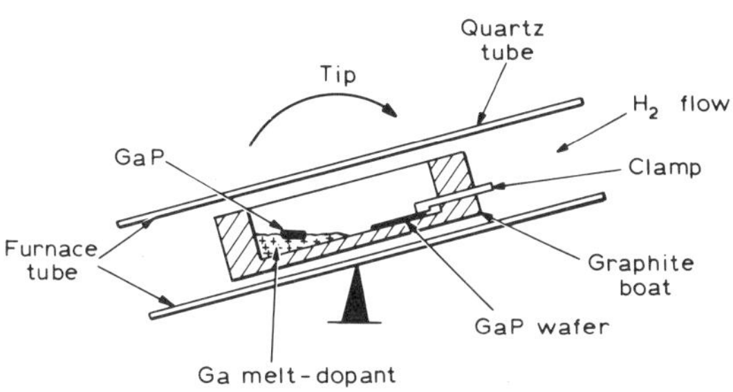

Figure 26.4 Liquid phase epitaxy method used for growing GaP layers on a GaP seed near 1000°C (Reproduced with permission from Smith[17])

so that the melt covers the substrate; upon cooling, an epitaxial GaP layer grows on the substrate. Similar techniques are used for the growth of other compounds.

Variations of the original LPE technique have been developed to improve the quality of the films, ease of fabrication, or to extend the process to large-scale production. A dipping technique,[4] in which the substrate is immersed for a specified time into the melt, provides constant-temperature growth, which is advantageous for the growth of ternary or more complex compounds. A multiple bin technique[5] was used to facilitate sequential deposition of several epitaxial layers. Also, a new technique which uses vapour doping of the melt[6] (e.g. introduction of Zn vapour to grow a p-layer on the previously grown n-layer) has produced excellent results in the fabrication of GaP:N green LED's.

VPE is widely used in the commercial preparation of red $GaAs_{1-x}P_x$ LED's. A typical VPE system, as originally described by Tietjen and Amick,[7] is shown in *Figure 26.5*. Here the Ga is

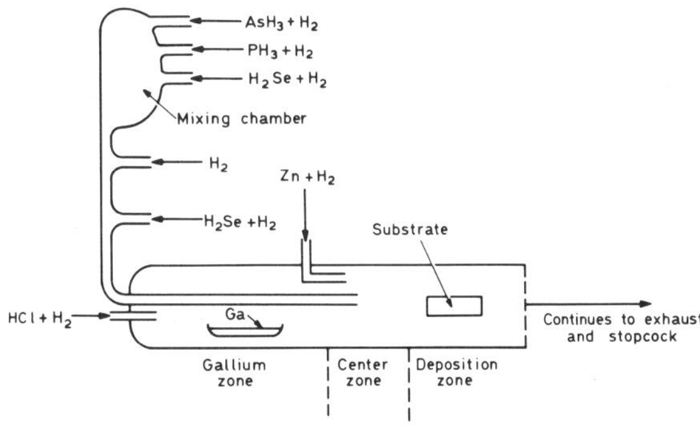

Figure 26.5 Schematic diagram of typical vapour-phase epitaxy growth apparatus used for the deposition of $GaAs_{1-x}P_x$ layers on GaAs substrates (Reproduced with permission from Smith[17])

transported by flowing HCl gas over the molten metal. Arsenic and phosphorus are obtained from the thermal decomposition of arsine and phosphine, respectively. Hydrogen is normally used as the carrier gas.

As the gases pass over the substrate, an epitaxial layer of $GaAs_{1-x}P_x$ forms whose composition is determined by the composition of the gases; these can be controlled accurately over wide ranges with precision flowmeters and valves. Varying the gas composition during growth provides the ability to slowly grade the composition of the epitaxial layer, thereby minimising the lattice mismatch and resultant strains.

Doping is also provided by introducing suitable gases. H_2Se is normally used to incorporate Se donors, and $(CH_3)_2Te$ and SiH_3 have sometimes been used to obtain Te or Si doping, respectively. Acceptor impurities can also be incorporated at the appropriate moment, e.g., by adding Zn vapour and H_2, but more often a postgrowth diffusion of Zn metal is used to form the p-layer. This latter approach has yielded the highest efficiencies to date for $GaAs_{1-x}P_x$ red LED's.

In addition to the original horizontal VPE systems, much larger vertical reactors have been developed for mass production. These systems process many wafers simultaneously (~300 cm^2 of substrates at a time), usually rotating the substrates during growth to improve uniformity. The availability of multiwafer growth equipment and high quality GaAs substrates, coupled with the fact that the epitaxial layers require no postgrowth polishing, have contributed significantly to the commercialisation and reductions in cost of red LED's.

VPE is also used for preparation of GaN[8] and $In_{1-x}Ga_xP$[9] but is less suited for $Al_xGa_{1-x}As$ because of the reactivity of Al-containing gases.

To date, LED's made from III–V materials GaP, GaAs, $Al_xGa_{1-x}As$, $GaAs_{1-x}P_x$, $In_{1-x}Ga_xP$ are of greatest practical

Table 26.1 Performance characteristics of typical LED's

Material	*Commercially available*	*Colour*	*λ (nm)*	η_{ext} *(%)*
GaP:Zn,O	yes	red	690	3–15*
$Al_{0.3}Ga_{0.7}As$	no	red	675	1.3
$GaAs_{0.6}P_{0.4}$	yes	red	660	0.5
$In_{0.42}Ga_{0.58}P$	no	amber	617	0.1
$GaAs_{0.25}P_{0.75}$:N	yes	amber	610	0.04
GaP:N	yes	green	550	0.05–0.7*
SiC	yes	yellow	590	0.003
GaAs:Si		IR	950	10–30

* Range between commercially available and best laboratory results.

and commercial significance. *Table 26.1* shows some of the performance characteristics of these diodes. $Al_xGa_{1-x}P$ diodes have also been reported,[10] with poor ($\sim 10^{-5}$) quantum efficiency. $In_{1-x}Al_xP$, which is ultimately believed to have good potential for high-brightness LED's, is difficult to synthesise and has not yet been made to luminesce.

26.3 Other LED materials

SiC diodes have been studied extensively in the past because they are, in principle, capable of emitting light throughout the whole visible spectrum. Although SiC LED's have been sold commercially by the General Electric Co., the luminous efficiencies to date have been disappointing.

Of some interest also is the use of up-conversion phosphors. These phosphors convert two or more infrared photons into a single higher energy visible photon.[13] When an efficient infrared LED (e.g. GaAs:Si with $\eta_{ext} = 10\%$ @ 930–970 nm) is coated with a thin layer of up-converting phosphor, red, green, or blue light can be obtained. However, since the intensity of the emitted light varies as the second (or higher) power of the infrared light intensity, the LED must be driven very hard in order to achieve useful overall efficiencies.

Up-converting phosphors with high efficiencies (30–50%)[14] have been prepared. However, in practice, the phosphor layers must be thin in order to minimise self-absorption. Consequently, only a small portion of the emitted infrared light is absorbed in the phosphor and the rest is wasted.

There is also renewed interest in II–VI electroluminescent diodes. Previous work was disappointing due to the inability to fabricate pn junctions in the wide bandgap II–VI materials of interest. More recently, luminescence in ZnSe (0.1% power conversion efficiency @ 590 nm)[11] and ZnS:Mn (0.1% power conversion efficiency @ 580 nm)[12] has been reported (see also Section 27.2). Here the excess carriers are generated in a thin high field region or at a metal–semiconductor barrier and consequently these materials operate at higher voltages (10 to 100 V).

26.4 Lasers

The word laser is an acronym for *L*ight *A*mplification by the *S*timulated *E*mission of *R*adiation. The advent of the laser has enabled visible light and infrared radiation sources to be produced with a spectral purity and stability as good as, or better than can be achieved, in the radio spectrum, e.g. a He–Ne laser can be made with a linewidth of a few kHz or less at a carrier frequency of about 5×10^{14} Hz. Such light sources are said to have a high temporal or longitudinal coherence. The meaning of this is illustrated by the fact that the light from the source mentioned above could still interfere with another beam split off from the same source with a path difference between the two beams of up to 300 km. The corresponding figure for the best conventional light source is only 10 cm.

Another property of laser beams is a high degree of transverse coherence, i.e. if one takes any section perpendicular to a laser beam, then there is a very well defined relationship between the phase of all points on the plane at all times, no matter how far from the source (this is only exactly true in free space). With normal light sources the phase relationship between different points on the source is completely random for distances greater than at most a few μm. This high transverse coherence is responsible for the very slow spreading of laser beams.

The unique properties of the laser are due to the process of stimulated emission, first postulated by Einstein in 1917 in order to explain Planck's radiation law. The idea of stimulated emission is that an atom in an excited state can be 'stimulated' to emit radiation by placing it in a radiation field of the same frequency, as it will normally emit in going to one of the lower excited states or to the ground state. It can be seen that such a process can result in amplification, if the process of stimulated emission from atoms in the upper state is more probable than that of absorption by atoms in the lower state. Einstein showed that in general, these processes have equal probabilities for a single atom in the radiation field. Thus in order to get net amplification we must have more atoms in the upper state than in the lower state. This is an inversion of the normal energy distribution, and thus we speak of population inversion being necessary to achieve amplification.

It can also be shown that the light produced by the stimulated emission process is in phase with the stimulating light, i.e. it is coherent with it. This process is the basis of both the laser and the maser. However, at optical frequencies as distinct from microwaves, spontaneous emission, i.e. the random de-excitation of upper state atoms, is a highly probable process which makes the achievement of inversion difficult. All things being equal the probability of spontaneous emission is inversely proportional to the wavelength cubed, thus inversion is much more difficult to achieve in the ultraviolet than the infrared, and extremely difficult in the X-ray region. Of course spontaneous emission is the dominant process in ordinary light sources and is responsible for the poor coherence properties mentioned above since the excited atoms emit in a quite independent and random manner, resulting in poor transverse coherence, and the emission is in the form of very short pulses, resulting in poor longitudinal coherence.

26.4.1 Methods of achieving population inversion

In order to achieve population inversion, we must have a means of excitation and a system of atoms or molecules having energy levels with favourable properties—e.g., the upper laser level should have a long lifetime and the lower level a much shorter one and the means of excitation should excite only the upper level. Different types of lasers are distinguished by the use of different modes of excitation. In gas lasers the primary mode of excitation is electron impact in a gaseous discharge; in solid state (crystalline and glass) and liquid lasers optical excitation is usual, whereas in semiconductor lasers the passage of a high current density through a highly doped, forward biased diode of a suitable material is the preferred mode of excitation. In most cases the excitation process is a multi-step one. For instance in many gas lasers a mixture of gases is used, one of the gases is chosen to have a metastable excited state with a high probability of excitation, whose energy coincides with that of the upper laser level, so that de-excitation by collision with a ground state atom or molecule of the other species is highly probable. Thus helium in the He–Ne laser and N_2 in the 'CO_2' laser (which usually uses a mixture of

CO_2, He and N_2) both have suitable metastable levels for the excitation of Ne atoms and CO_2 molecules respectively.

In solid-state lasers, the light absorbed excites an ion to a high level which then decays very rapidly by a non-radiative process to the upper laser level which usually has a lifetime of at least several hundred microseconds, thus making inversion relatively easy to achieve on a transient basis, i.e. pulsed excitation by a flashtube. In fact solid state laser materials are rather special phosphors, chosen for the narrow linewidth of their emission. Similarly de-excitation of the lower laser level is not usually a simple radiative process. In gas lasers, more often than not, collisions with the walls of the discharge tube (He–Ne) or with another species of atom or molecule (He in the CO_2 laser) are important in de-excitation. In some solid state lasers (so-called four level materials such as neodymium doped materials) de-excitation is achieved by radiationless transfer of energy to the atoms of the lattice in which the Nd^{3+} is embedded. In the case of 'three level' materials such as ruby, the lower laser level is the ground state of the Cr^{3+} ions in an Al_2O_3 crystalline matrix. Thus 'de-excitation' is automatic, but inversion is difficult to achieve since more than half of all the Cr^{3+} ions must be excited in order to achieve inversion.

26.4.2 Oscillators and modes

If we simply have a medium in which population inversion is maintained, then it will act as a tuned amplifier resonant at a frequency corresponding to the transition concerned, with 3 dB points determined by the linewidth of spontaneous emission. If the overall amplification is very high, then amplification of spontaneous emission or super-radiance can depopulate the inversion very rapidly, with some line narrowing.

To make an oscillator, we require positive feedback; this is provided by a pair of mirrors placed at each end of the amplifier as shown in *Figure 26.6*. Usually one of these is as near 100% reflecting as possible, and the other one has a transmission typically of a few per cent or so. At longer infrared wavelengths a hole in the output mirror may be used to couple power out. To obtain oscillation GR_1R_2 must be greater than unity (G is the gain of the medium; R_1,R_2 are the mirror reflectivities). This arrangement is similar to a microwave resonator, except that there are no reflecing walls between the mirrors, and is therefore spoken of as an 'open' resonator.

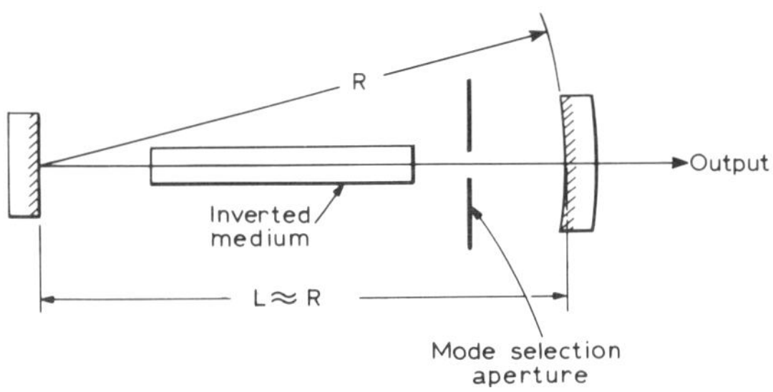

Figure 26.6 Laser with hemispherical resonator incorporating transverse mode selection

Since the distance between the mirrors is typically more than 10^5 wavelengths, and the width of the medium is usually at least 10^3 wavelengths, then it can be seen that many more modes of oscillation are possible than in a microwave resonator. The behaviour of such modes has been described by Fox and Li and others.[15]

The modes in an open resonator of this type are what is known as transverse electromagnetic or t.e.m. waves. They are specified by three subscript mode numbers p, l, q (cylindrical symmetry) or m, n, q (rectangular). The last subscript in each case is the longitudinal mode number, i.e. a very large number. The other two refer to the transverse directions, and in practice here the low order modes (<5) are usually dominant.

Transverse mode patterns corresponding to some of the low order rectangular modes are shown in *Figure 26.7*. The 0,0q mode is of most practical importance since it has the highest degree of transverse coherence and the lowest beam divergence. The radial intensity distribution of energy in any cross section is given by a gaussian function. The propagation of such 'gaussian' beams has been treated in some detail by Kogelnik.[16]

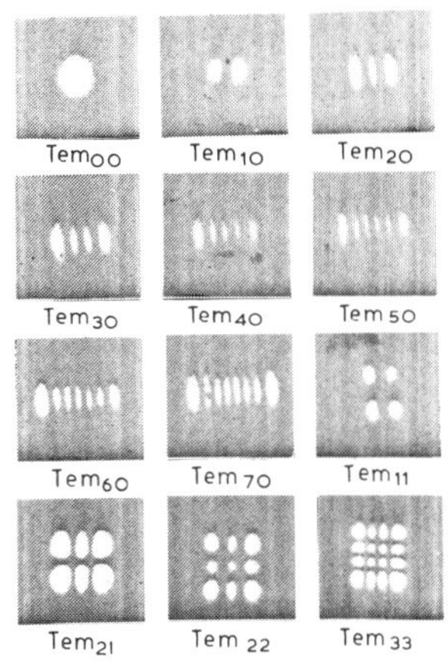

Figure 26.7 Mode patterns of a gas laser oscillator (rectangular symmetry)

Schemes for selecting a particular mode, usually the zero order, depend on the fact that the diffraction losses of different modes propagating in a resonator are different, and by using the hemispherical resonator configuration shown in *Figure 26.6*, the absolute magnitudes of the losses for higher order modes make it easy to select the zero order mode by the use of a suitable aperture as shown in the figure. The longitudinal mode behaviour is determined by the width of the gain curve of the inverted medium and by the frequency response of the resonator. The simplest form of resonator, using a pair of relatively broadband reflecting multilayer dielectric mirrors has very many resonances, separated by a frequency $\Delta f = c/2L$ where c is the velocity of light and L is the optical path length of the resonator.

In rare cases, such as the low pressure CO_2 laser, it is relatively easy to ensure single frequency operation since the linewidth of the gain curve is typically less than $c/2L$. In most types of laser there are very many resonances within the gain curve, and thus many longitudinal modes oscillate, unless a more complex resonator, having only one principle resonance within the width of the gain curve, is used. The subject of mode selection schemes has been reviewed by Smith.[17]

When many longitudinal modes oscillate simultaneously the phases are usually random. By inserting in the cavity a loss modulator or a phase modulator, tuned to the mode spacing Δf, the sidebands so produced cause the phases of all the modes to be locked together, resulting in a pulsed output with a p.r.f. Δf and a pulse length $T = 1/\Delta f_{osc}$ where Δf_{osc} is the width of the gain curve.

Typically $\Delta f \geqslant 100$ MHz, $T \leqslant 1$ nanosecond. Mode locking has also been reviewed in some detail by Smith.[18]

Another pulsed mode of laser operation of great practical importance is the so called 'Q switched' or 'giant pulse' mode. In this case extra loss is introduced into the cavity, by means of a Pockels effect or acousto-optic modulator, while the inversion is built up to a much higher value than required to sustain oscillation in the low loss case. The extra loss is then switched out rapidly, resulting in the stored energy being discharged in a single very short pulse (typically 10–100 ns).

A third pulsed mode operation is 'spiking', which occurs spontaneously in pulse excited solid-state lasers. The output in this case consists of a random train of submicrosecond pulses of duration comparable with the pumping pulse. This type of operation is reviewed by Roess.[19]

26.5 Gas lasers

As mentioned above, these are usually excited by passing a discharge through a gas or, more often, a gas mixture. Only the most common types are described here. These are the helium–neon laser, the argon ion laser and the CO_2 laser.

The helium–neon laser usually contains a mixture of about 5 parts helium with one part neon at a total pressure of a few torr. The discharge tube has a bore of 1–2 mm and a length of 20–100 cm, depending on the power output, usually in the range 0.5–50 mW. The current is usually ~10 mA with voltages up to a few kV. The gain of the medium is very low and losses must be kept to a minimum. Either Brewster angle windows (resulting in plane polarised output) are used with external mirrors, or integral mirrors sealed to the discharge tube are used to ensure low loss. Partly since mirror alignment is critical the sealed mirror method of construction is more difficult to engineer satisfactorily. Although the usual output wavelength is ~633 nm in the red, outputs at 1.15 μm or 3.39 μm can be obtained by using suitable mirrors. Typical beam divergence in the visible is ~1 mrad.

In argon lasers the transitions involved are between excited levels of singly ionised argon ions. Consequently high current densities (>100 A cm^{-2}) are required to achieve threshold, and plasma tubes of BeO are usually used to minimise heating, erosion and gas clean-up. Brewster angle windows of fused silica are usually used with external mirrors. Outputs of up to 15 W, on a number of lines in the blue and green regions of the spectrum simultaneously, are available commercially. Principle wavelengths are ~488 nm and ~515 nm. Plasma tube-lengths of 10–100 cm, with diameters 1–2 mm are used, and more gain is available than with He–Ne. However, the efficiency, ⩽0.1%, is very low. Typical beam divergence ⩽1 mrad.

The CO_2 laser, which gives outputs at a number of wavelengths between 9.6 μm and 10.6 μm, is the most efficient of all the gas lasers (up to 30%). It is the most powerful (CW powers up to 30 kW or so) and the most varied in construction since there are many different versions operating at pressures from a few torr up to atmospheric, with pulsed or CW excitation, and even incorporating Q switching. The CO_2 laser is one of the few gas lasers in which the lifetime of the upper level is great enough to make Q switching worthwhile.

The most common form is still the low pressure flowing gas (He:N_2:CO_2::8:1:1 typical) system using a plasma tube ~2.5 cm in diameter and up to several metres in length, depending on the power output required. Integral mirror or Brewster angle plus external mirror configurations are used. In the simplest form output power may be coupled out using a hole in one mirror, in conjunction with a suitable window. For longer plasma tubes, a folded configuration is used (*Figure 26.8*). A review of CO_2 laser systems has been given by De Maria.[20]

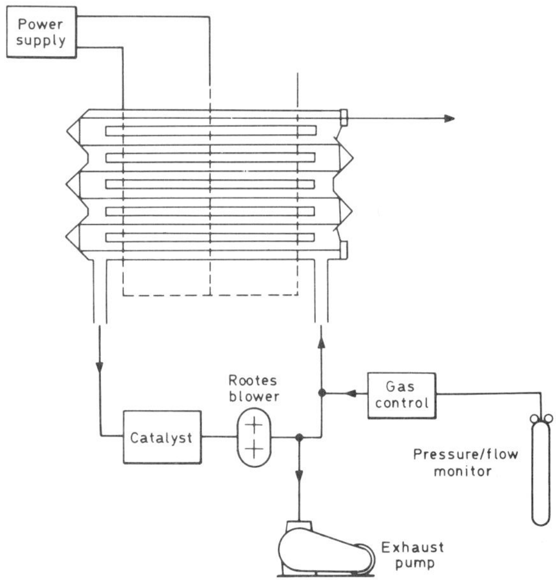

Figure 26.8 Folded CO_2 laser with gas re-circulation (Courtesy Marconi-Elliott Flight Automation Ltd)

26.5.1 Uses of gas lasers

The selection of a laser for a specific application is based primarily on its output power and wavelength. The most common gas laser is the helium–neon laser where its red emission is commonly used for alignment and distance measurement, inspection, code reading, communication and recently videodisc reading.

The argon laser with a higher power blue and green emission is used, together with the helium–neon laser in laser light shows. However, probably more important is its use in industry and medicine for surgical and heating techniques. It is also used for its greater power in communication.

The CO_2 laser with its very high power (up to 30 kW) is known mainly for its use in cutting and welding. However, the penetration of its 10.6 μm line in poor weather makes it of interest to developers of thermal imaging systems working at that wavelength. The multiplicity of lines in its spectrum offer a wide choice for use in aerial pollution monitoring 'Lidar' systems.

26.6 Liquid lasers

These use a solution of a fluorescent molecule, optically pumped either by another laser or by a flash tube, in a configuration similar to that used for solid state lasers. The most common form of liquid laser is the dye laser, so called because the fluorescent molecule is a suitable dye molecule, e.g. rhodamine 6G, in an organic solvent. Usually a substance is added to suppress the triplet excited state which would otherwise be formed, leading to unacceptable losses at the laser wavelength. Often a flowing dye solution is used to remove molecules in the triplet state from the lasing region. Dye laser outputs are usually broad band (~10 nm) unless a grating is used in the cavity. Narrow band tunable outputs can be obtained in this way over most of the visible and near infrared spectrum, using a number of dyes. Argon lasers are usually used to pump CW dye lasers. Outputs up to a few watts have been obtained at some wavelengths as the excitation can be quite efficient (~20%).

26.6.1 Uses of liquid lasers

The wide range of dyes available and the associated wavelengths makes the dye laser a flexible analysis tool. It is being used extensively in research laboratories to study many processes with specific energy absorption, matched by the laser wavelength. The best known of these is probably the analysis of automobile exhaust emission.

26.7 Solid-state lasers

The principle solid-state lasers are the ruby laser, which is a three level laser, using chromium doped aluminium oxide, as mentioned above, and the four level lasers using transitions in the Nd^{3+} ions in various host materials—chiefly various glasses and yttrium aluminium garnet, known as YAG. The ruby laser emits at ~694 nm, the Nd lasers at ~1.06 μm. All lasers in these categories are optically pumped either using a xenon filled flash tube, or a krypton filled arc or a tungsten–iodine lamp in the case of CW Nd–YAG lasers.

The most usual configuration is to use the laser rod in the form of a right cylinder, placed along one focal line of an elliptical cylinder, with the pumping lamp along the other focal line. In 'normal' pulsed operation outputs of up to ~100 J (ruby) and 1000 J (Nd glass) can be attained from a single oscillator in pulses up to a few ms in duration. Q switched pulses of a few joules or more can readily be obtained, even higher values using additional rods as amplifiers. The wavelength spread from Nd glass lasers may be up to ~10 nm in the absence of mode selection. CW Nd–YAG systems have outputs from ~1 W using tungsten–iodine lamps for pumping up to ~1 kW using a cascaded multiple rod configuration pumped by krypton arc lamps. Efficiencies of ~1% are attainable. Repetitively Q switched operation at p.r.f.'s of up to a few kHz is possible with continuously pumped Nd–YAG systems using acousto-optic Q switching. They are often used in lasers for resistor trimming.

26.7.1 Uses of solid-state lasers

Solid-state lasers are more compact than gas or liquid designs and have found uses in portable industrial equipment. The two major types, ruby and neodymium–YAG, are used in industrial welding and laser trimming systems, particularly for the construction of hybrid film circuit products. A major use of Neodymium–YAG lasers is in portable military range-finding equipment. The 100 kW, 20 ns pulses from a binocular sized Tx/Rx system allow ranging up to many kilometres.

26.8 Semiconductor lasers

These lasers use the fact that in the so called 'direct gap' semiconductors such as GaAs, InAs, etc. (and also in some ternary compounds such as $Ga_xAl_{1-x}As$ which exhibit 'direct gap' behaviour over certain ranges of x), carrier recombination between electrons and holes occurs predominantly by the emission of radiation in the near infrared spectrum (800–3000 nm). By constructing a pn junction in which the p- and n-regions are very highly (degenerately) doped, population inversion between electron and hole concentration occurs in the junction region. Since the gain can be very high, up to 30 cm^{-1} or more, laser action can be achieved with cavity lengths of less than 1 mm merely by cleaving the crystal along faces perpendicular to the junction plane, and relying on the relatively high Fresnel reflectivity at the surfaces of the high refractive index material to give sufficient positive feedback. Such cleaved surfaces act as near optically perfect mirrors.

In common with most other semiconductors the diode laser has the potential to be manufactured in high volume and at relatively low cost compared with other types of lasers. Its current injection drive requirements are compatible with conventional semiconductor circuitry and it does not require high voltage supplies or flash tubes for operation. In fact with direct current injection and a divergent output beam the laser diode has more in common with the LED than other laser types.

Three major laser diode designs are currently available. A simple pn laser is only suitable for low duty cycle, short pulse operation at room temperature, since the threshold current density is of the order of 10^5 A cm^{-2} and heating of the junction causes the threshold current to increase rapidly (~100 ns) to the point at which oscillation can no longer occur. More recently more complex structures involving up to five layers of different doping and composition have been devised. The inner layers usually comprise $Ga_xAl_{1-x}As$ where x is near to unity, the object being to achieve confinement of the radiation, as in a dielectric waveguide, and also localisation of carrier recombination. $Ga_xAl_{1-x}As$ has a lower refractive index than GaAs, forming the waveguide.

Figure 26.9 shows schematically the construction and characteristics of the three types of gallium arsenide/gallium

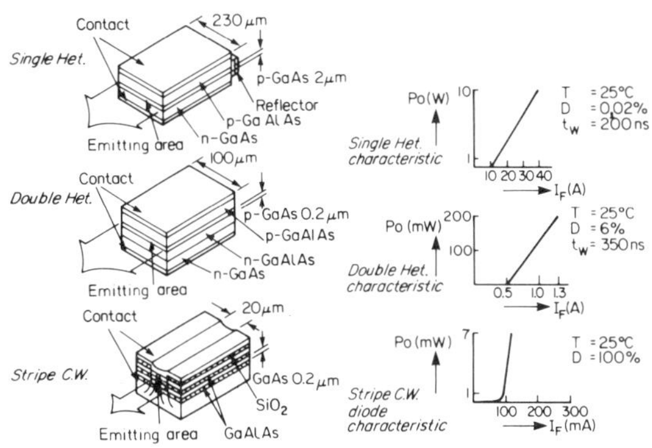

Figure 26.9 Construction and characteristics of the three main diode designs

aluminium arsenide laser currently available for operation[22] up to and above 25°C. It is important to note the relative values of operating current, radiated output and duty factor. Below the lasing threshold the diode outputs a low level of spontaneous emission similar to a light emitting diode. The output from diode lasers diverges due to diffraction and normally requires collimating optics. The emission wavelength depends on the level of aluminium in the active region but is commonly in the near infrared (800–910 nm).

26.8.1 The single heterostructure laser

The single heterostructure diode was developed first and, although requiring high levels of pulse drive current, radiates high power pulses that make it uniquely suitable for particular applications.

A typical example of a single heterostructure diode emits 10 W peak optical power with a peak drive current of 40 A. In order to maintain the laser chip dissipation at an acceptable level the duty cycle is limited to 0.1%. Thus the mean input power is compatible with semiconductor circuits. *Figure 26.10* is an example of a relatively simple circuit capable of supplying 40 A pulses with nanosecond risetimes and pulse widths below 10 ns.

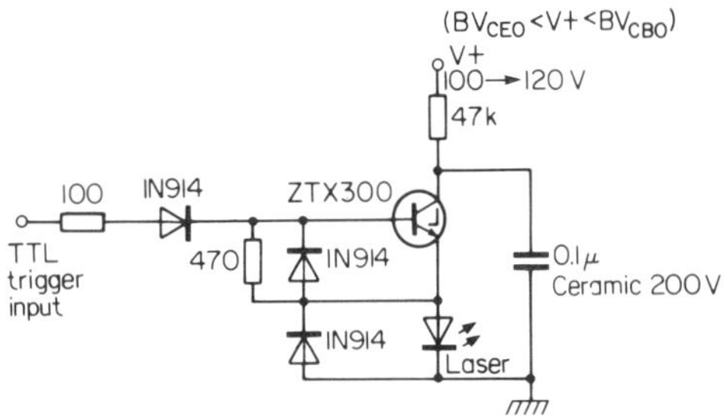

Figure 26.10 Simple 40 A laser drive circuit

Even higher optical powers (1000 W) are available from multichip stacks and arrays.[23] As the array incorporates more laser chips it offers an increasing source size to the collimating optics and degraded performance. Attaching rectangular fibres directly to the chips allows the fibre outputs to be combined into a smaller emitting source than would otherwise be possible. *Figure 26.11* is a schematic of such an array developed by ITT UK, which still transmits 80% of the chip output.

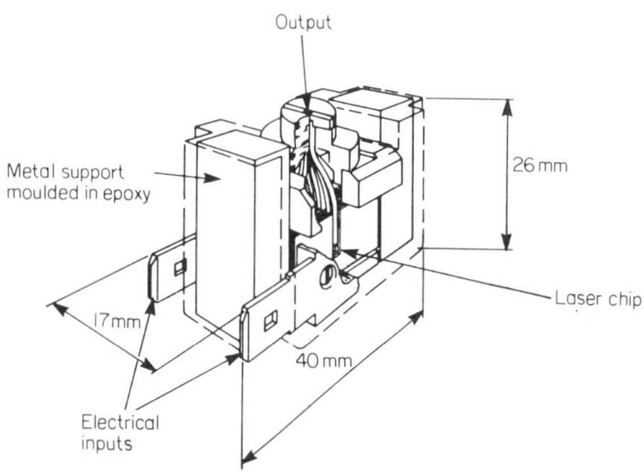

Figure 26.11 High power fibre coupled laser array

The major uses of this design of laser are in military equipment. They include ranging, proximity fusing, missile guidance and weapon simulation. In all cases it is the high optical power in each pulse that provides the operational benefits.

26.8.2 The double heterostructure laser

Improved carrier and optical confinement in the double heterostructure laser chip results in improved efficiency heat sinking and lower operating current requirements. A typical chip requires only 1–2 A drive and the dissipation limit allows operation at duty cycles up to 20% while emitting 200 mW peak power. It is obvious that this performance is significantly different to that of the single heterostructure laser, as are its main uses. Digital (pulse frequency modulation) free space communication requires high pulse powers with a variable duty cycle.[24] The 2 MHz laser modulation capability allows a high quality data or audio link to be constructed with a range well in excess of 1 km and battery operated.

The high mean power of the double heterostructure diode makes it a suitable source for covert illuminators. These are used as auxiliary illumination for the latest designs of night vision equipment.

26.8.3 The continuous wave laser

The application of a stripe contact to the double heterostructure diode further improves confinement and efficiency so that lasing occurs at currents below 200 mA. Some of the latest designs of stripe geometry diodes have operated at currents as low as 20 mA.

Two major features resulted from this design improvement: (1) the ability to sustain CW operation due to reduced consumption; and (2) the reduction of the size of the lasing facet.

The first allowed the removal of the bandwidth limitation inherent in all pulse diodes.[25] There is a delay between an applied current step from zero and the initiation of the optical output in LED's and laser diodes. It is directly related to the recombination time. To remove this delay the laser must be biased near its lasing threshold, and lasing at a low level. Then the addition of a modulating current produces the optical output with no further delay. The stripe geometry diode is the only design capable of sustaining the dissipation associated with the required bias and thus allowing CW and high bandwidth operation.

The second feature resulted in the laser source being compatible with the dimensions of the core of the glass optical fibres designed for optical communication. The 10 μm × 1 μm emission area allows a high launch efficiency of radiation into the 50 μm core of present telecommunication optical fibres.

26.8.4 Telecommunications lasers

It is as the source for optical fibre telecommunications that the main expansion of manufacture will occur. British Telecom has led the world in its support of such development. Current CW lasers emit in the wavelength range 800–880 nm, which is governed by the gallium aluminium arsenide system. However, the attenuation and dispersion of optical fibres reduces markedly in the wavelength region 1.3 μm–1.5 μm compared with 850 nm. Thus current developments in telecommunication lasers are targeted towards producing sources emitting at these wavelengths.

To achieve this the basic stripe geometry double heterostructure chip construction is retained but a quaternary system of gallium indium arsenide phosphide layers is used.[26,27] *Figure 26.12* shows schematically the inverted rib waveguide design. This type of laser will be used at 140 Mbit s^{-1} together with monomode optical fibres of only 8 μm core to project 1920 telephone calls over 25 km without repeaters. Besides the emitted wavelength the operational characteristics of these devices are very similar to those that emit at 850 nm with the exception of an increased temperature coefficient.

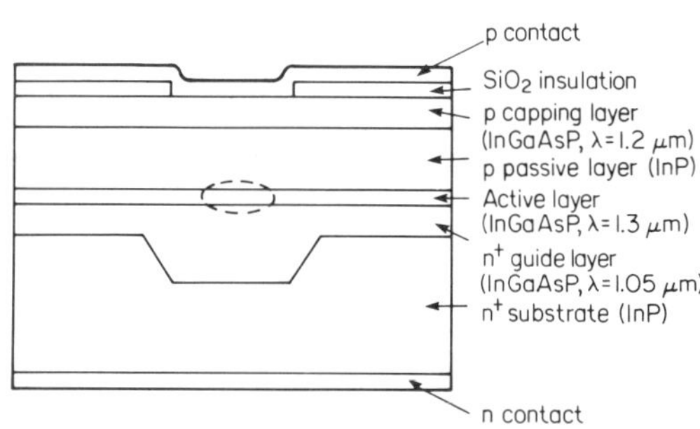

Figure 26.12 Inverted rib waveguide laser design with 1.3 μm wavelength (Courtesy of ITT UK)

26.8.5 Industrial lasers

Two other major applications for the CW laser are in office copier/facsimile machines and video/audio disc replay machines. In the first a short wavelength is required to match the spectral sensitivity of the photosensitive transfer drum. In the second a short wavelength is required to allow a small diffraction limited spot and realise the maximum information density on the disc; this can approach 1 μm at visible wavelengths.

For both these systems the 633 nm line emission from a He–Ne laser has been used but gallium aluminium arsenide diode lasers emitting at 750 nm have been successfully tried. These of course result in more compact, rugged, lower consumption equipment.

26.8.6 Edge-emitting LED

A closely related device to the diode laser, also used in lower grade fibre optic systems, is the edge-emitting LED.[28] Once described as a 'frustrated' laser, it combines small emitting source size and relatively lower output power with ease of driving. *Figure 26.13* shows one method of constructing the diode; in this case by truncating the stripe contact so that the cavity losses

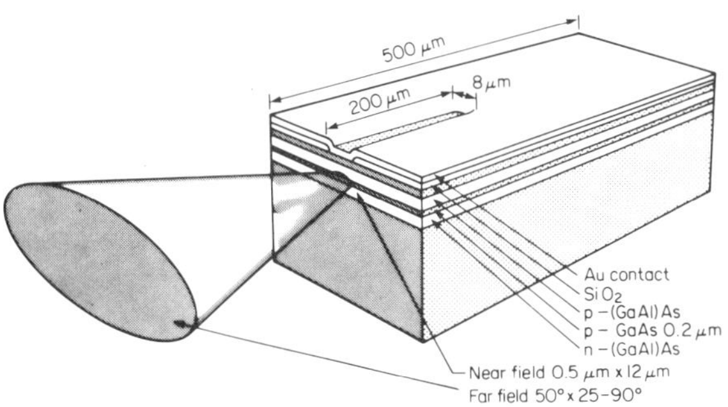

Figure 26.13 High power edge-emitting LED for optical communications (Courtesy of ITT UK)

always exceed the optical gain. The ELED emits 1 mW at 100 mA drive current with a risetime of 6 ns. Such a device can launch up to 1 mW optical power into a communications fibre. Other methods of frustrating lasing action are to de-optimise the active region thickness or doping. Too thick or too lightly doped a region produces a low output but linear device. Too thin or too heavily doped an active region produces a high speed, high power device.

26.8.7 Laser diode life

No description of semiconductor lasers is complete without reference to the life expectancy of the various devices.[29] It is the gradual degradation of performance with time to a specified limit that defines operating life, not a time to catastrophic failure.

Life testing is carried out under conditions representative of those used in operational systems. Thus the single and double heterostructure pulse lasers are run under constant pulse conditions and the fall of output power with time is monitored. Under the conditions given in *Figure 26.9* both device types lose half their initial output power in typically 10 000 hours. In operation optical feedback is employed with CW lasers to maintain the output level constant against characteristic changes with temperature or time. In life testing therefore it is drive current changes at constant output that are quoted. Rapid improvements in reliability and degradation rates have been made in the last two years.

26.8.8 Laser detectors

It is the nature of the use of semiconductor lasers that they are close coupled in systems with suitable detectors. In the 800 to 940 nm spectral region the silicon photodiode is used exclusively. In two forms, PIN (p, intrinsic, n) and APD (avalanche photodiode) it offers bandwidth in excess of 1 GHz with high sensitivity.

However, at the longer wavelengths now being used for telecommunications, ternary (GaInAs) and quaternary (GaInAsP) materials are used. By varying the ratio of the In and P, sensitivity can be maximised at any wavelength from 1.2 μm to 1.6 μm. These materials are made by the same epitaxial growth techniques as the laser diodes.

26.8.9 Safety

The optical output power levels of existing and proposed diode lasers and high power stacks are insufficient to pose a hazard to any part of the human body but the eyes. Authoritative studies and regulations, such as BS 4803 have been published covering hazards presented by all laser types.[30] They are designated class 3B under British and IEC Standards and careful study of the recommended working practices will ensure maximum safety.

References

1. GRIMMEISS, H. G. and SCHOLZ, H., *Phys. Lett.*, **8**, 233 (1964).
2. ARCHER, R. J., 'Light emitting diodes in III–V alloys', Paper 66, *Electrochem. Soc. Mtg, Los Angeles, California, Spring*, 1970. Also ARCHER, R. J., *J. Electr. Mater.*, **1**, 128 (1972).
3. NELSON, H., *RCA Rev.*, **24**, 603 (1963).
4. WOODALL, J. M., RUPPRECHT, H. and REUTER, W., *J. Electrochem. Soc.*, **116**, 899 (1969).
5. NELSON, H., *U.S. Patent* 3 565 702 (23 February 1971).
6. LADANY, I. and KRESSEL, H., *RCA Rev.*, **33**, 517 (1972).
7. TIETJEN, J. J. and AMICK, J. A., *J. Electrochem. Soc.*, **113**, 724 (1966).
8. MARUSKA, H. P. and TIETJEN, J. J., *Appl. Phys. Lett.*, **15**, 327 (1969).
9. NUESE, C. J., RICHMAN, D. and CLOUGH, R. B., *Metall. Trans.*, **2**, 789 (1971).
10. KRESSEL, H. and LADANY, I., *J. Appl. Phys.*, **39**, 5339 (1968).
11. PAUK, Y. S., GEESNER, C. R. and SHIN, B. K., *Phys. Lett.*, **21**, 567 (1972).
12. VECHT, A., WERRING, N. J. and SMITH, P. J. F., 'High efficiency DC electroluminescence in ZnS(Mn,Cu)', *J. Phys. D. Appl. Phys.*, **1**, 134 (1968). Also VECHT, A. and WERRING, N. J., 'Direct current electroluminescence in ZnS', *J. Phys. D: Appl. Phys.*, **3**, 105 (1970).
13. AUZEL, F. E., 'Materials and devices using double-pumped phosphors with energy transfer', *Proc. IEEE*, **61**, 758 (1973).
14. WITTKE, J. P., LADANY, I. and YOCOM, P. N., *J. Appl. Phys.*, **43**, 597 (1972).
15. FOX, A. G. and LI, T., 'Resonant modes in a laser interferometer', *Bell Syst. Tech. J.*, **40**, 453 (1961).
16. KOGELNIK, N. and LI, T., 'Laser beams and resonators', *Appl. Optics*, **5**, 1550 (1966).
17. SMITH, P. W., 'Mode selection in lasers', *Proc. IEEE*, **60**, 422 (1972).
18. SMITH, P. W., 'Mode locking of lasers', *Proc. IEEE*, **58**, 1342 (1970).
19. ROESS, D., *Lasers, Light Amplifiers and Oscillators*, Academic Press, New York (1969).
20. DE MARIA, A. J., 'Review of C.W. high power CO_2 lasers', *Proc. IEEE*, **61**, 731 (1973).
21. HARRY, J. E., *Industrial Lasers and Their Applications*, McGraw-Hill, New York (1974).
22. 'Using semiconductor lasers', *ITT Application Note*.

23. 'High power semiconductor laser arrays', *ITT Application Note.*
24. 'Laser diodes for communication', *ITT Application Note.*
25. 'Modulation of semiconductor laser diodes', *ITT Application Note.*
26. KIRKBY, P. A., 'Semiconductor laser sources for optical communication', *Radio Electron. Engr*, **31**, No. 7/8, 362–76 (1981).
27. TURLEY, S. E. H., HENSHALL, G. D., GREENE, P. D., KNIGHT, V. P., MOULE, D. M. and WHEELER, S. A., 'Properties of inverted rib waveguide, lasers operating at 1.3 μm wavelength', *STL/ITT Application Note.*
28. DAVIES, I. G. A. and GOODWIN, A. R., 'Reliable sources for fibre optic communication—the high power edge emitting LED', *STL/ITT Application Note.*
29. GOODWIN, A. R. and PLUMB, R. G., 'Reliable sources for fibre optic communication—The 20 μm oxide insulated stripe laser', *STL/ITT Application Note.*
30. 'Safety aspects of laser diodes', *ITT Application Note.*

Bibliography

ALLEN, L. and JONES, D. G. C., *Principles of Gas Lasers*, Butterworth, Sevenoaks (1967).

ARECCHI, F. T. and SCHULTZ-DUBOIS, E. O. (Eds), *Laser Handbook*, Vols 1, 2, North Holland, Amsterdam (1972).

BLOOM, A. L., *Gas Lasers*, Wiley, New York (1968).

ELECCION, M., 'The family of lasers: survey', *IEEE Spectrum*, **9**, 326 (1972).

FABIAN, M. E., *Semiconductor Lasers, A Users Handbook*, Electrochemical Publications (1981).

GOOCH, C. H. (Ed.), *Gallium Arsenide Lasers*, Wiley–Interscience, New York (1969).

GOODWIN, A. R. and SELWAY, P., 'Heterostructure injection lasers', *Elect. Commun.*, **47**, 1.49 (1972).

HARVEY, N. F., *Coherent Light*, Wiley, New York (1970).

RIECK, H. *Semiconductor Lasers*, MacDonald, London (1968).

SANDBANK, C. P., (Ed.), *Optical Fibre Communication Systems*, Wiley, New York (1980).

WAGNER, W. G. and LENGYEL, B. A., 'Evolution of the giant pulse in a laser', *J. Appl. Phys.*, **34**, 2040 (1963).

27 Displays

J A van Raalte BSc, MSc, PhD, MAPS, FSID, SMIEEE
Head, Videodisc Recording and Playback Research, David Sarnoff Research Centre, RCA
(Sections 27.1–27.5)

D Meyerhofer BEng, PhD
Member, Technical Staff, David Sarnoff Research Centre, RCA
(Sections 27.5–27.8)

Contents

27.1 Light emitting diode displays

Although the cost of LEDs has come down considerably during the last decade, largely as a result of batch fabrication techniques, the cost of a single diode is still excessive for high resolution ($\gtrsim 10^5$ elements) displays. Nevertheless an experimental television display[1] using LEDs has been demonstrated. LEDs are most widely used for numeric display applications where only a few digits need to be displayed, e.g. calculators or readouts for digital equipment.

The size and geometry of the pn junction, and the fabrication process used depend on the intended application. Small seven-segment displays are often made in a single step on a common substrate, but the substrate cost makes this approach impractical for larger-sized displays; consequently, larger numeric LED displays are generally assembled from seven individual diodes. A seven segment display is shown in *Figure 27.1*. From the seven segments, all the numerals and certain letters such as A, H, P and U can be formed.

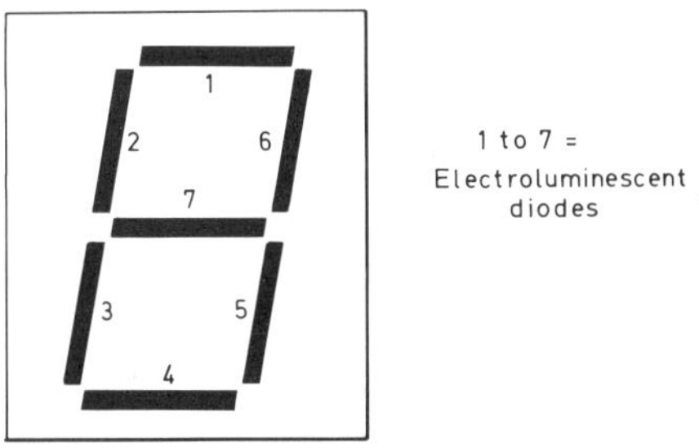

Figure 27.1 Seven-segment character display using electroluminescent diodes

In monolithic displays the contact electrodes can be evaporated or photolithographically deposited on the surface parallel to the pn junction plane (*Figure 27.2(a)*). Light emission is predominantly from this surface. The uppermost layer is kept thin (<5 μm) to minimise self-absorption, especially in direct bandgap materials; it cannot be too thin, however, since current spreading to the whole junction area is desirable. In GaP displays the transparency of the substrate material leads to optical cross-coupling between segments; a special structure (*Figure 27.2(b)*) in which the material is etched away in between the segments and coated with absorbing or reflecting films, was developed to minimise this optical coupling.

By slicing the pn junction into bars (*Figure 27.2(c)*) after electroding, very high line brightnesses from the edge of the junction can be obtained. Electrical connection to the small bar segments and subsequent assembly are more complicated.

Other structures have been designed which use external means (cylindrical lenses, reflecting cavities, diffusing covers—*Figure 27.2(d)*) to enlarge the area from which light is emitted; these techniques are conservative of diode area but require a hybrid process for assembly and have reduced surface brightness.

The reliability, long life and compatibility with low voltage integrated driving circuitry make LEDs very attractive for small display applications. However, their power consumption of typically 5–10 milliwatts/small digit is still considered high for many battery-operated products, e.g. portable calculators, electronic watches.

Additional information about materials, fabrication and applications for LEDs can be found in several excellent review articles.[2,3]

27.2 Electroluminescence—Destriau effect

Some phosphors emit light when a sufficiently high electric field is applied across them. This phenomenon, known as electroluminescence (EL), was first observed by Destriau[4] upon application of a changing electric field. Light is generated by recombination of electrons and holes whereby their excess energy

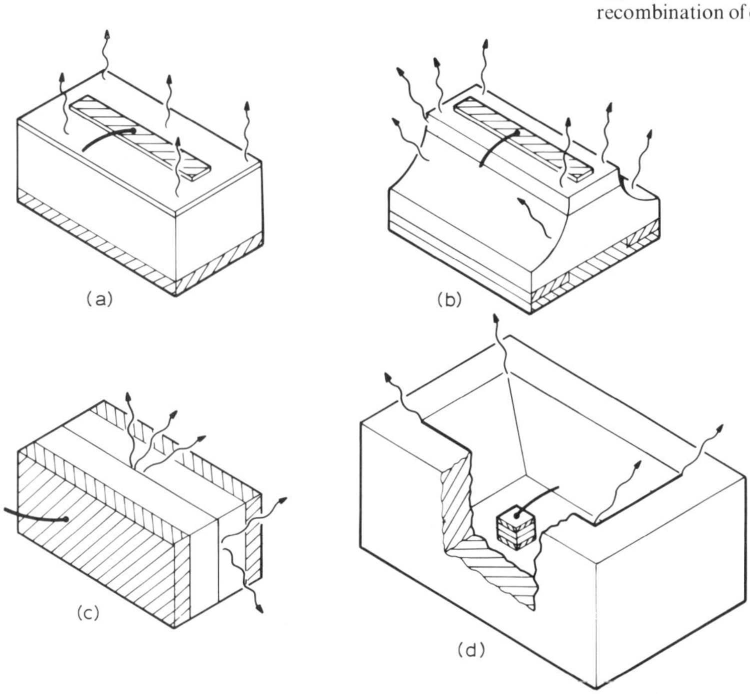

Figure 27.2 LED geometries used for the fabrication of seven-segment numeric displays: (a) and (b) surface emitters; (c) edge emitter; (d) cavity emitter (courtesy American Vacuum Society)

is transferred to an emitted photon. Donors and acceptors play a major role in facilitating this radiative recombination and determining the spectral characteristics of the emitted light, i.e. the increment by which the photon energy is less than the energy gap of the phosphor material.

The mechanism by which electrons and holes are generated in the EL phosphor is not fully understood. The applied voltage is believed to be concentrated near a barrier or thin high-impedance region in the phosphor layer; the electric field strength is thereby locally increased to the point where carrier injection by field emission and possibly avalanche multiplication can occur.

27.2.1 A.C. electroluminescence

Most extensively studied are a.c.-EL phosphors, usually zinc sulphides or zinc sulphoselenides doped with copper or chlorine to produce yellow, green or blue emission; red EL phosphors have been made, but their brightness and efficiency are usually much lower.

In practice, a thin ($\sim$10–50 μm) layer of EL phosphor is sandwiched between two electrodes, at least one of which is transparent and through which the display is viewed. Both rigid (glass, ceramic) and flexible (plastic) substrates can be used. An a.c. excitation voltage, typically 50–5000 Hz, of several hundred volts is applied to the electrodes.

An area of early interest for a.c.-EL was the combination of large-area photoconductor layers and EL phosphors for image intensification or X-ray image storage.[5] However, saturation and trapping effects in the photoconductors and the availability of better alternatives prevented commercialisation of these devices.

Originally a.c.-EL was also believed to be of importance for large-area illumination and flat-panel television. However, in both applications luminous efficiency, average brightness and display life are of prime importance. In particular, for the television application where each display element is excited only during a small fraction of each frame time, high peak brightnesses are required in order to obtain useful average brightnesses.

Despite much effort to develop improved a.c.-EL phosphors, some fundamental materials problems remain. The brightness of EL displays increases with increasing excitation voltage and frequency. On the other hand, both life and luminous efficiency tend to decrease in the same direction. The half-life of most phosphors, i.e. the time over which the brightness drops to one-half of its initial value, varies inversely with frequency of excitation implying a constant number of cycles. Luminous efficiencies in yellow-green phosphors of 1–5 lm W^{-1} (0.1–1%) can be achieved at 35–350 nits brightness with a typical half-life of 1000–3000 hours[6] at 400 Hz excitation. The most important commercial applications are for low power, reliable night lights and instrument panel lighting (there is usually no catastrophic failure mechanism).

The technical feasibility of an a.c.-EL (ZnS, Se: Cu, Br) television panel has also been demonstrated;[7] however, the contrast ratio and brightness, which decrease with increasing number of scan lines due to the decreasing duty factor ($\sim$35 nits and a contrast of 10:1 for an 80-line panel versus $\sim$12 nits for a 225-line panel) are not adequate for commercial television displays. Also power consumption is high due to the relatively large stray capacitance of the thin panel, and the high voltages and short pulses used to excite it.

27.2.2 D.C. electroluminescence

More recently, interest in d.c.-EL materials has increased with the discovery that high brightness (several hundred nits), good life ($>$1000 h), and modest luminous efficiency (0.1% $\approx$0.5 lm W^{-1}) can be achieved in ZnS doped with Mn and coated with copper sulphide. The light emitted from these phosphors originates from a thin (3–5 μm) high-impedance layer near the anode; the rest of the phosphor layer is optically inactive but acts as a resistive protection layer. The high-impedance layer has to be created by a 'forming' process (high voltage and high current) and presumably results from the diffusion of Cu ions out of it.

Of particular interest has been the finding that average brightness and luminous efficiency remain high under low duty-cycle pulse excitation: for instance, an average brightness of 270 nits was achieved using 4 μs-wide 250-volt pulses at 0.5% duty cycle in a panel originally formed at 60 volts. This finding is of special significance for grey-scale displays, such as television where the duty cycle is low due to the large number of scanning lines. Thus an off-the-air 330-mm-diagonal flat d.c.-EL television display has been developed[8] with 224 × 224 elements, producing a 34-nit display with 20:1 contrast and a total power consumption of 150 watts. The addressing circuitry is still complex, however, and a commercial product would additionally require the development of other phosphors which can be used to obtain efficient red, green and blue emission.

27.2.3 Thin film electroluminescence

The combination of long life, stable operation and high brightness under low duty cycle excitation has also been demonstrated in thin film EL panels.[9] These panels typically consist of a sputtered or electron-beam evaporated thin film EL phosphor (Mn-doped ZnS, about 0.5 μm thick) sandwiched between two insulating dielectric layers (about 0.2 μm thick), e.g. Y_2O_3 or Si_3N_4. The dielectric layers block d.c. current flow and provide the hot carriers necessary for efficient electroluminescence; the panels therefore must be addressed by high voltage pulses of alternating polarity. With this type of addressing circuitry high average brightness (200 nits) has been achieved in high resolution, line-at-a-time addressed alphanumeric or grey-scale TV displays where the duty cycle of excitation is 0.2–1%.[10] By modifying the EL panel materials a high resolution storage panel has also been demonstrated.[11] The addressing circuitry for these panels is still quite complex but recent progress in integration of high voltage drivers[12] suggests that these problems will be solved. Presently the only bright, efficient thin film EL material emits orange-yellow light; efficient materials emitting the primary colours have not yet been developed.

27.3 Plasma displays

27.3.1 Operation

When a sufficiently high voltage is applied across two electrodes in a low pressure gas, a breakdown of the insulating properties of the gas is observed.[13] The sudden increase in conductivity is caused by impact ionisation of gas molecules by electrons that have been accelerated to sufficiently high energies by the applied electric field. Each electron leaving the cathode produces an electron avalanche travelling towards the anode and a positive ion avalanche travelling towards the cathode. When the probability of this ion stream regenerating an electron at the cathode becomes unity, then the Townsend breakdown condition is satisfied and the discharge becomes self-supporting. The flow of electrons also excites the gas molecules through collision and this energy is subsequently given off as light emission. This emitted light, which is characteristic of the ambient gas, is the basis for a variety of plasma displays.

Plasma displays normally consist of a single glass envelope filled with a few torr of gas, typically neon for optimum luminous efficiency (orange light at 0.1–1 lm W^{-1}); other gas additives are

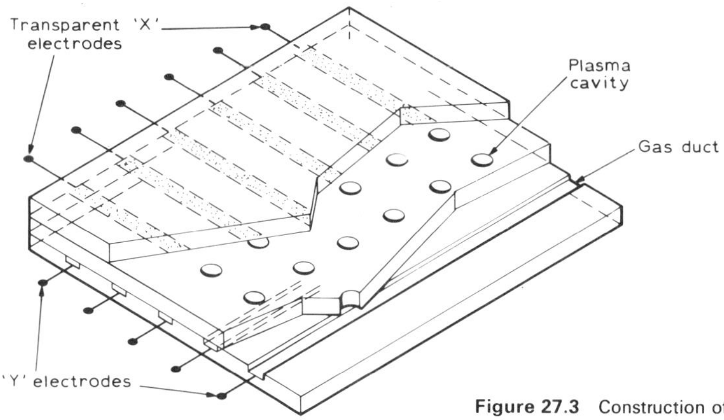

Figure 27.3 Construction of simple plasma matrix display

sometimes used, in particular Hg, to reduce sputtering of the electrodes or to obtain UV emission which can be used to excite phosphors (in order to change the colour of the display). Some internal structure is often used to isolate the display elements and to confine the discharge (*Figure 27.3*).

The extension of plasma displays to other colours can, in principle, be achieved by incorporating fluorescent phosphors which are excited by UV radiation from the discharge[14] or by using carriers from the discharge to bombard and excite a low-voltage cathodoluminescent phosphor directly,[15] or by using electrons extracted from the plasma, and subsequently accelerated to high voltages, to excite more or less standard cathodoluminescent phosphors.[16] Suitable cathodoluminescent colour primaries, either for low voltage applications (i.e. phosphors with negligible 'dead-layers' to provide adequate luminous efficiency) or for high voltage applications (i.e. phosphors resistant to inevitable ion bombardment) are not readily available however.

The high voltage needed to fire a plasma cell (100–200 volts) is a disadvantage in many applications, especially the alphanumeric display field (calculators, digital clocks, digital meters) where competition from other technologies (LED, liquid crystals) is strong. Nevertheless, small and moderate-size plasma displays are being used because of their inherent low fabrication cost, pleasant bright appearance and reliability.

27.3.2 Matrix (multi-element) plasma displays

In larger-sized or high resolution displays, plasmas have many advantages. Since the current through a plasma, and therefore the light output is negligible until breakdown occurs, there is a built-in threshold which makes plasma displays ideal for matrix (coincidence or half-select) addressing. The display elements of a matrix display are arranged at the intersection of a set of orthogonal 'X' and 'Y' electrodes (*Figure 27.3*), so that many elements ($n \times m$) can be addressed with relatively few ($n+m$) electrode connections. Of particular interest would be a flat television display which requires about 2.5×10^5 elements (500×500), and therefore can be addressed with 1000 (500 + 500) wire connections. This basic concept was demonstrated in a 4000 and later 10,000-element TV display.[17] Also extension of the basic d.c. plasma panel to colour has been demonstrated.[18] However, the addressing circuitry required was complex and display uniformity, as in most sampled grey-scale matrix displays, unacceptable for television.

27.3.3 Burroughs SELF-SCAN®

A conceptually simple multiplexing technique has been developed by Holz[19] which drastically reduces the number of external connections required in multi-element plasma displays. The Burroughs SELF-SCAN® panel, shown in *Figure 27.4*, consists of two functionally separate parts: (1) in the back of the panel, a gas discharge is shifted linearly under the influence of external clock pulses and used to address the display elements; (2) the front of the panel contains the video-modulated display elements; these are connected to the scanning cells by means of small holes in the common electrode which are used for 'glow priming' (addressing).

The operation of the scanning part is as follows: the cathodes of a row of plasma elements are connected alternately to one of three common electrodes (*Figure 27.4*). A three-phase sinusoidal voltage whose amplitude is just below threshold is applied to these three electrodes—consequently, none of the elements will be turned 'on'. However, if a particular element, say the first, is externally turned 'on', then metastable ions from this discharge will diffuse to the adjacent element thereby lowering its threshold for firing. This element therefore will turn 'on' when the voltage across it next reaches its maximum; in the meantime the voltage across the previous element is dropping and it is extinguished. In this manner the discharge is 'stepped' along by three elements for every full cycle of the sinusoidal clock voltage. In the Burroughs SELF-SCAN panel the first element of a row is used to start the discharge and the frequency of the clock voltage used to

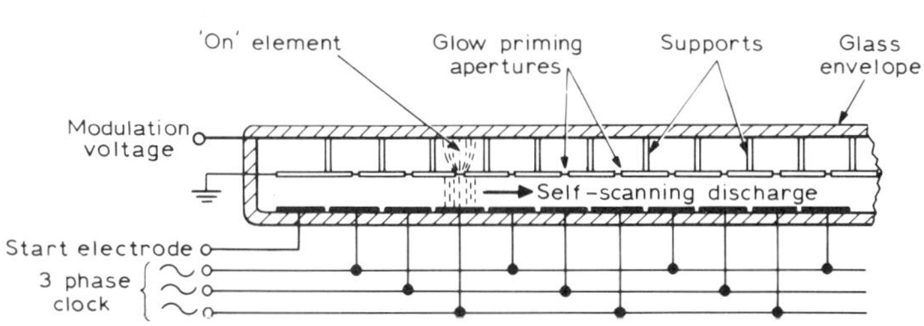

Figure 27.4 Principle of operation of Burroughs SELF-SCAN® plasma display

determine the rate at which it propagates down the row.

This linearly scanning discharge is then used to sequentially address ('glow-prime') the display elements which are interconnected to the scanning elements by means of a small hole in the central common electrode. The diffusion of carriers through these holes from the scanning discharge in the back is sufficient to remove the threshold in the brightness *vs.* voltage characteristic of the display element. Thus by modulating a common display anode, the brightness of the element in front of the scanning discharge can be determined: a single line of 256 elements can be addressed time-sequentially with less than 10 external leads. By scanning at a sufficiently high frequency (60 Hz) a steady, nonflickering display results.

Burroughs SELF-SCAN® plasma panels with 77×222 elements have been used to produce a portion of an off-the-air television display[20,21] which demonstrated the advantages of self-scan addressing (fewer external leads, simplified and lower-voltage circuitry, better grey-scale uniformity) for this type of analog display applications.

27.3.4 Owens-Illinois DIGIVUE®

A very different type of multielement plasma display was developed by Slottow and Bitzer[22] and is now marketed by Owens-Illinois under the trademark DIGIVUE®. The construction of this plasma panel is illustrated in *Figure 27.5*:

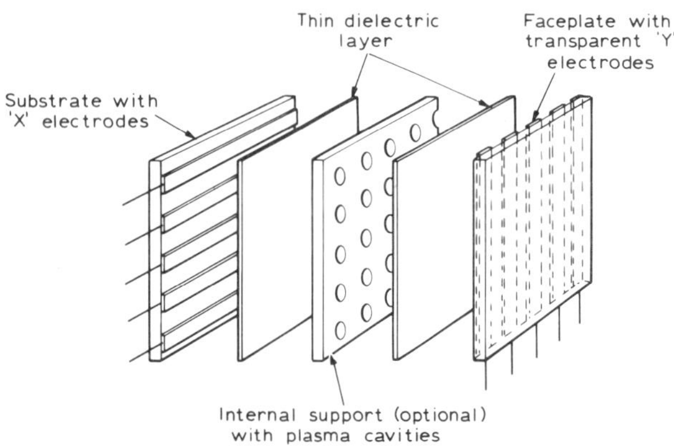

Figure 27.5 Construction of Owens-Illinois DIGIVUE® plasma display

individual plasma elements are located at the intersection of orthogonal X-Y electrodes but these electrode strips are insulated from the gas discharge by means of a thin insulating layer—thus no d.c. current can flow through the panel.

When a sinusoidal voltage is applied to all electrodes, the voltage will divide capacitively across the insulating layers and the plasma element; if the voltage across the plasma is insufficient to cause breakdown, then the element remains 'off'. However, if any element is separately triggered 'on', then the current flow through the plasma will deposit an electric charge on the insulating layer covering the electrodes; this reduces the voltage across the plasma element causing it to extinguish itself. When the polarity of the applied voltage next is reversed, this charge on the insulator produces an increased voltage across the display element thus allowing it to fire again and the charge is then transferred to the opposite electrode. In this manner, any element that has been triggered 'on' continues to fire at every half cycle of the applied voltage while all other elements remain 'off'. Therefore the DIGIVUE® is uniquely suited for alphanumeric (on/off) display applications requiring long-term memory or slow updating.

Although the light from a single-plasma element is emitted in a very short time (10^{-7} s to 10^{-6} s), the average brightness of the panel can be quite high ($\gtrsim 170$ nits) with excitation frequencies of the order of 10–100 kHz. A DIGIVUE® display with more than 10^6 elements (1024×1024) has been demonstrated; simple construction techniques and optimised gas mixtures promise moderate cost and good reliability ($>10{,}000$ h).

The storage plasma panel is undoubtedly best suited for alphanumeric and graphics display applications where frame-storage (memory) was desirable. It has been shown possible to display grey-scale images on such a plasma panel using time modulation—that is, varying the fraction of a frame time during which each element is left 'on'.[23] Thus, in principle, off-the-air television pictures can be displayed at the expense of an analog-to-digital converter and additional storage circuitry; this approach may become economically feasible with future advances in large-scale integrated circuitry. A major advantage of this approach is that the display promises to be extremely uniform and relatively bright.

27.4 Large-area TV displays

The concept of a large-area TV display continues to fascinate and challenge researchers around the world. The bulk and weight of a glass envelope that can withstand the tremendous forces due to atmospheric pressure present a practical size limit for the cathode ray tube or kinescope. On the other hand the requirement in large-area consumer television displays for grey-scale, high resolution, full colour capability, high brightness and uniformity, and acceptable luminous efficiency has so far prevented the other previously discussed display technologies from reaching this market place. Since luminous efficiency and colour appear to be the two most difficult requirements a number of approaches have been investigated recently that would use available cathodoluminescent phosphors in a large, flat-faced, evacuated display. The tremendous atmospheric pressure (e.g. $\pm 75{,}000$ N on a 125 cm diameter display) requires that an internal support structure be incorporated in the panel to keep it from collapsing. The problem then is to develop a panel with an internal support structure that allows the generation, modulation and acceleration of electrons to high energies in order to excite the standard colour phosphors.

A number of different approaches have been considered recently. In one instance the discharge in a type of plasma panel is used as the source of electrons which are modulated and subsequently accelerated to several kilovolts.[16] Normally the gas pressure and electrode spacings in a plasma panel are chosen to minimise the breakdown voltage in order to simplify the addressing circuitry. At low gas pressures and small electrode spacing the breakdown strength of the plasma rises steeply.[13] Thus it is possible to matrix address a portion of the panel with moderately low voltages to establish a localised plasma, then to extract electrons from the discharge, modulate them and accelerate them towards a close-spaced phosphor anode.[16] Indeed by using a plasma discharge as the source of electrons for several picture elements or even one or more horizontal lines, and by utilising the self-scan concepts the addressing circuitry for such a panel can become quite simple.

Another approach has been investigated which operates in a moderately good vacuum, e.g. 10^{-4}–10^{-5} torr.[24] At these low pressures the probability of electron collisions with gas atoms is exceedingly small and a self-sustaining plasma discharge cannot readily be established. In other words there are insufficient gas collisions to produce an electron avalanche and to provide adequate ion flow back to the cathode to regenerate the primary electrons. The necessary electron multiplication is obtained in this panel by incorporating electron multipliers in the form of dynodes into the supporting walls of the panel. The supporting walls, or vanes, run vertically from the cathode plate to the phosphor-coated anode plate. In addition to the dynodes other electrodes are incorporated in the vanes to extract electrons from the self-sustaining discharge, to focus, modulate, accelerate the electrons and to deflect them to the appropriate phosphor stripe of a colour display.

Although there are a large number of electrodes in this panel, many are electrically connected in parallel. Moreover the addressing requires moderate voltages only and the ion bombardment of the phosphors is minimised because of the low pressures, thus promising long life. The stabilisation of the gas pressure in this panel may be difficult, however.

Yet another large-area display panel has been studied which also uses cathodoluminescent phosphors as the source of bright, efficient primary colours.[25] Here also vertical internal support vanes are used to keep the evacuated panel from collapsing under atmospheric pressure. The modulatable area source of electrons is provided by a number of 'electron guides'. These guides use periodic electrostatic focusing to confine and steer a low-voltage electron beam over large distances. This type of periodic confinement of electrons is also used in travelling wave tubes and is the electron equivalent of an optical fibre. The electron guides in this panel have been provided with a number of discrete extraction points where the electron beam can be pushed out of the guide and accelerated towards the phosphor screen. A number of systems' options exist; normally there is one extraction point for each individual, or pair of horizontal display lines.

27.5 Passive electro-optic displays

Passive electro-optic devices modify light that is generated elsewhere. The parameter modulated may be the optical path length or the absorption, reflection or scattering parameters, or a combination of these. If the modulated quantity is wavelength dependent, then the display will be coloured. The display generally consists of a thin sheet of the material sandwiched between two planes of electrodes. The electrode configuration is such that, by applying a voltage to certain of the leads, specified areas of the display are modulated. A typical commercial display is shown in *Figure 27.6*. When only the phase of the transmitted or reflected light is modulated, then a combination of polarisers is required to convert phase changes to amplitude changes which make the display visible.

The properties of interest of an electro-optic material, in addition to the kind and amount of light modulation, are the driving voltage and power required, whether they can be excited by a.c. or d.c., the speed of response, and the life (expressed in operating life or number of addressing cycles).

Figure 27.6 Liquid crystal display

To address a simple display like that shown in *Figure 27.6*, it is practical to have a separate connection leading to each segment of one electrode plane with a common connection on the other plane. However, for more complicated displays, it is necessary to reduce the number of leads by using a grid or matrix system.[26] On one plane, the segments are interconnected in horizontal rows, on the other plane, in vertical columns, so that each segment is uniquely addressed by a combination of one horizontal and one vertical lead. For this scheme to work, it is required that the electro-optic effect have a sharp threshold, fast rise time and long decay time, so that the application of two (or three) times the threshold voltage produces a large optical change, even if that voltage is only applied for a small part of the time (low duty cycle).

Presently, the most widely used passive electro-optic materials are liquid crystals. They exhibit a considerable variety of modulation effects. Other electro-optic effects that may become important are electrophoresis and electrochromism. The properties of these materials are compared in *Table 27.1*.

27.6 Liquid crystals

Liquid crystals[27] are useful in display applications because of their very large electro-optic effects.[28] The materials themselves exist in a phase which is different from both the conventional

Table 27.1

Effect	*Liquid crystals*		*Electrophoresis*	*Electrochromism*
	Hydrodynamic	*Field effect*		
Operation in transmission	T	T		
Operation in reflection	R	R	R	R
Viewing angle	narrow	medium	no restrictions	no restrictions
Operating voltage (V)	15–40	5–10	50–100	1–2
Power consumption (mW cm^{-2}) (continuous operation)	0.1	0.002	0.4	
Energy consumption/cycle (mJ cm^{-2})				2–100
Switching speed (s)	0.05	0.2	1	0.01

liquid and solid phases, and exhibit properties intermediate between the two. Materials having this special phase are found in the group of organic materials with moderately large rod-like molecules. In recent years, numerous compounds and mixtures have been discovered which are liquid crystalline at room temperature and over a temperature range of as much as 100 K. This accounts for the increased practical interest in display applications.

The various material phases are distinguished by their macroscopic symmetry properties. The conventional liquid is isotropic and has no unique symmetry elements while even the lowest order crystals have complete translational symmetry. The intermediate liquid crystals retain the translational freedom of the liquid in which the molecules are free to move in varying degrees. They also exhibit long range ordering which adds certain macroscopic symmetry elements, either translational or orientational. There are presently five known liquid crystalline phases—nematic, cholesteric and three smectic phases. It is primarily the first two which are used in practical electro-optic devices.

A nematic liquid crystal is one where all the molecules align themselves approximately parallel to a unique axis while retaining the complete translational freedom. This symmetry axis is defined locally and may vary in direction in different parts of a volume of liquid. Its direction may be described by a vector (of arbitrary sense) which is called the director (*Figure 27.7(a)*). In the absence of external forces, the lowest energy state of a nematic is that with a uniform orientation of the director in the entire sample.

A cholesteric liquid crystal is a modification of a nematic one. In the undisturbed state, it may be described as a nematic liquid crystal which has been twisted about an axis (the helical axis) lying perpendicular to the orientation of the director at all times (*Figure 27.7(b)*).

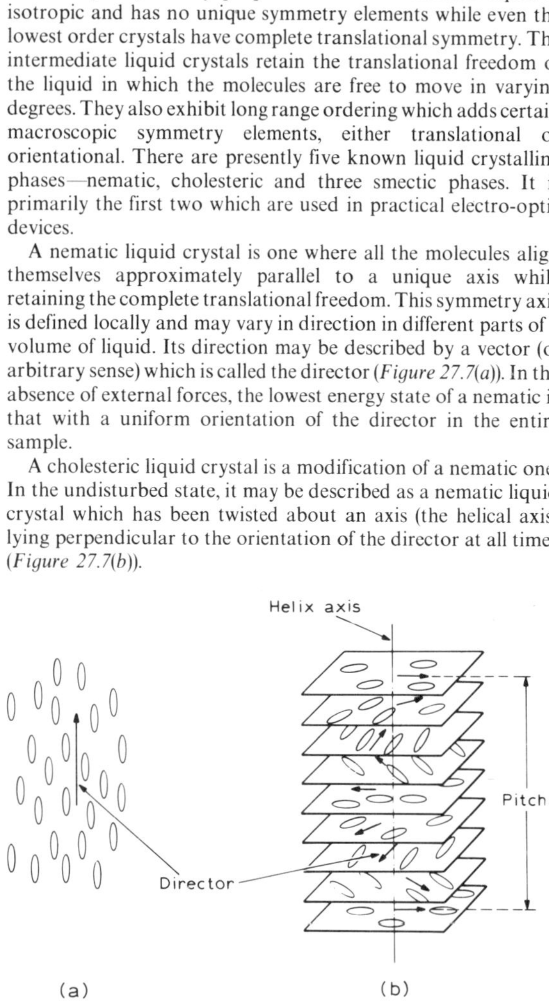

Figure 27.7 Schematic arrangement of molecules in liquid crystals: (a) nematic; (b) cholesteric. The arrow represents the director

The optical properties of these materials are derived from their symmetry. A nematic liquid crystal is optically uniaxial, so that, as in a uniaxial crystal, two different indices of refraction apply for light propagation with the electric vector parallel and perpendicular to the axis. The same is true of the dielectric constant and the conductivity which are tensors of the same order as the index of refraction. The dielectric constant anisotropy may be positive (larger for the fields parallel to the axis than for the perpendicular direction) or negative depending on the polarisation properties of the molecules. The other anisotropies have only been observed to be positive.

The optical properties of the cholesteric fluid are more complicated. It is uniaxial with a screw-type symmetry. The axis is at right angles to the axis in the nematic (or 'untwisted' cholesteric). Light propagating parallel to the helical axis experiences optical rotation, reflection and change of polarisation, depending on the ratio of the optical wavelength to the pitch of the helix.[29] In the extreme case of very small ratio, the polarisation ellipse of the light simply rotates about the axis in the same way that the director does. Light propagating at right angles to the helical axis encounters a sinusoidal variation of index along the axis with the period equal to half the pitch. This is a phase grating which causes diffraction of the light.

The electro-optic properties of a liquid crystal are very different from those of a crystalline solid. This is what causes these materials to be of great interest in display applications. In a solid, the electric field produces small changes in location of ions and in atomic polarisation and correspondingly small changes in the refractive index. In liquids, the electric field can only polarise individual molecules and the effects are even smaller. Liquid crystals respond to an electric field like solids since the molecules remain aligned by long range forces, but now it is possible to rotate the optical axis by a large amount because of the freedom of individual molecules to rearrange themselves while still maintaining the alignment. Consequently, the optical changes are much larger than those in solids and simple liquids. In addition, it is possible to destroy the long range order by changing from a uniform 'single crystal' state to a randomised polycrystalline state which exhibits strong light scattering. This change is reversible and therefore qualifies as an electro-optical effect.

The electro-optic effects can be divided into two different types with respect to the driving force. In one, the electric field acts on the dielectric properties of the molecules and produces reorientation in a uniform manner. The liquid remains static except during the reorientation. In the other, the current is the driving force and the liquid crystal must be conducting. The current is carried by ions, and the moving ions produce movement of the liquid above the electrohydrodynamic threshold.[30] The moving liquid, in turn, causes shear forces and displacements which are accompanied by optical changes. Electrohydrodynamic motion is only created by d.c. and low-frequency a.c. driving voltages. At frequencies above the dielectric relaxation frequency ($f_d = \sigma/2\pi\varepsilon$, where σ is the conductivity and ε the dielectric constant), the current becomes primarily capacitive and only field effects can be observed.

27.6.1 Practical considerations

Liquid crystal electro-optic devices are generally constructed as sandwich cells with a thin layer of the liquid placed between two conducting glass plates. Because of the large electro-optic effect, the liquid layer may be very thin (3–50 μm, typically). The cells are used in two different configurations, one where the light is transmitted through the cell, the other where the incident light is reflected back at the viewer. In the latter case, the back electrode usually consists of a specularly reflecting aluminium film on a glass substrate. The front electrode is formed by a transparent conducting layer, such as SnO_2 or In_2O_3. For the transmissive cell, both electrodes must be transparent.

For practical use, the cells must be hermetically sealed, so that the liquid crystal is confined and cannot interact with the atmosphere. The major problem is water vapour which can cause decomposition and chemical reactions between the various components of the mixture. Some kinds of liquid crystal material are more susceptible to this problem than others. Chemical reactions may also take place during operation of the devices because of the electrolytic nature of the cell. This is particularly true when direct current is applied to the cell which causes transport of ions and electrochemical reactions at the electrodes. Consequently, the lifetime of the device under d.c. conditions is generally much shorter than under a.c. conditions, and most applications require the latter.

To produce any given electro-optic effect, it is necessary that the liquid crystals have a certain crystalline form or texture. The form is determined by the alignment of the molecules at the surfaces which may be perpendicular to the surface, parallel to it, or at some intermediate angle. In the parallel situation, the direction of the molecules in the plane may be required to be uniformly in a specified direction, or randomly oriented. The alignment is determined by appropriate treatments of the surface[31] in combination with chemical aligning agents.[32] In addition, the material must have a certain conductivity for the particular effect desired. The conductivity is controlled by the addition of ionic dopants.

One of the important variables in the operation of the device is its speed of response. The electro-optic effects involve a reorientation of the molecules and the response times are therefore related to the viscosity of the material, decreasing, for example, with increasing temperature. Response will be faster at higher driving fields. The equilibrium condition with no external force applied is determined by the alignment at the surface, and the speed of return to that condition increases rapidly as the cell is made thinner.

27.6.2 Hydrodynamic effects

27.6.2.1 Domains and dynamic scattering

Conductivity in liquids is generally by ionic motion. At low values of conduction, the ions move through the stationary liquid but, at higher values, instabilities are produced and the liquid is set into motion.[30] It is in this region that conduction-induced electro-optic effects occur. The effects are much more important in liquid crystals than in isotropic liquids because the optical anisotropy makes the liquid motion directly visible.

The best-understood instabilities are the Williams domains.[33] They consist of rotation vortices of liquid and are observed mainly in nematic materials of negative dielectric anisotropy. A cross-section of a cell, at right angles to the vortex axes, is shown in *Figure 27.9*. The spacing between the axes is a fixed multiple of the electrode separation. The rotating fluid causes the director to tilt, as indicated. For light polarised in the plane of the figure, the optical pathlength varies depending on the location in the cell. This causes the axes and the regions separating any two vortices to become visible as an array of parallel lines. The uniform line spacing (proportional to the cell thickness) causes the cell to act as a diffraction grating.

The threshold for onset of the domain instability is determined by the interrelationship of a number of physical effects.[34] The conductivity anisotropy produces charge separation in regions where the ions are not moving parallel to the director due to the rotating liquid (*Figure 27.8*). The resulting transverse electric

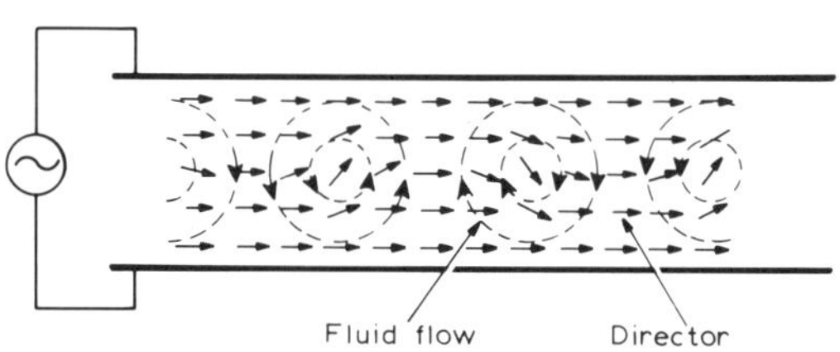

Figure 27.8 Cross section of a nematic cell exhibiting Williams' domains

fields act on the charge to produce torques in the liquid. These torques reinforce the rotation. This explains why ionic conductivity is required for domain formation.

Driving voltages can be d.c. or low frequency a.c. but, as the frequency is increased, the ionic current decreases relative to the displacement current and the instability threshold rises. Finally, near the dielectric relaxation frequency, no domains can be formed.[35] There exist other instabilities, but they are not of a hydrodynamic nature.[35]

As the driving voltage is increased beyond the threshold for domain formation, the vortices become smaller and less regular, leading finally to a turbulent state. Because the direction of the molecular axes changes rapidly from place to place, the liquid becomes highly scattering and opaque. The scattering pattern is varying continuously when observed under the microscope. This state is called *dynamic scattering*.[36]

The intensity of scattered light as a function of the applied voltage is shown in *Figure 27.9*. It can be seen that there is a

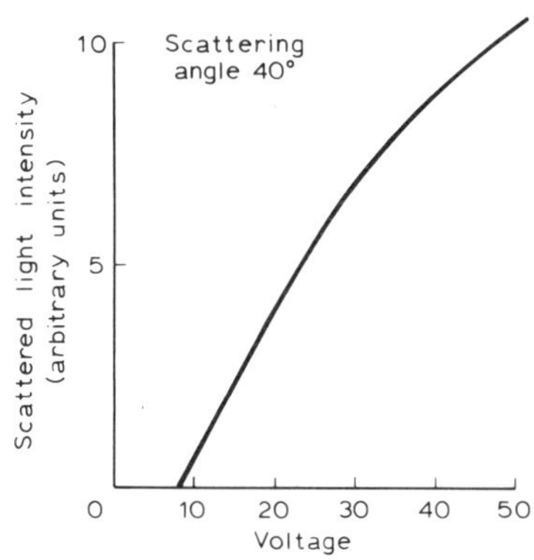

Figure 27.9 Voltage dependence of the scattered light in a dynamic scattering cell

threshold voltage, below which there is little light scattering. This voltage is slightly above the domain threshold. The amount of scattering below the threshold (off-state) depends on how well the material is aligned. Good alignment, either perpendicular or parallel to the electrodes reduces the scattered light to a negligible quantity. The scattered light intensity in the on-state depends strongly on angle, as is shown in *Figure 27.10*. A determination of the contrast ratio of a display cell therefore depends on the conditions under which it is measured. The steady state scattering properties (*Figures 27.9* and *27.10*) do not depend on the thickness of the liquid crystal film over the range of 5–10 μm. The response time to application and removal of the voltages also has a strong angular dependence. It increases proportional to the square of the film thickness if the other parameters are held constant.

Figure 27.10 shows that the scattering of light takes place predominantly in the forward direction. This produces a very good display when transmitted light is used, particularly when the geometry is arranged so that the unscattered light is not viewed by the observer. The situation is not as favourable in the reflective case because the liquid crystal must be backed by a highly reflecting mirror to produce good scattering efficiency. It becomes more difficult to prevent reflected, but unscattered, ambient light from reaching the viewer. For good results, the display must be recessed behind a set of baffles which restrict the viewing angle.

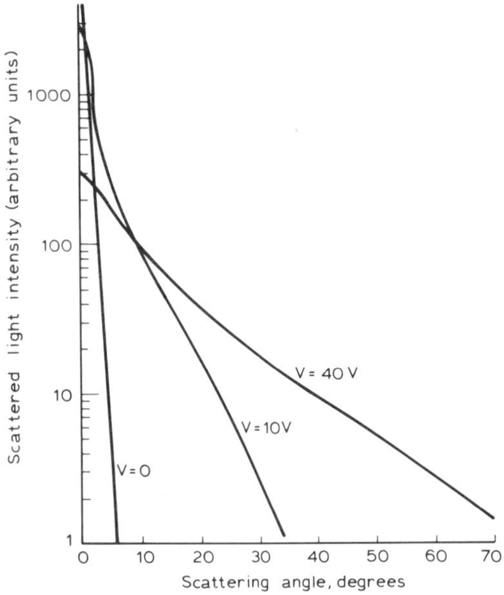

Figure 27.10 Angular dependence of the scattered light in a dynamic scattering cell

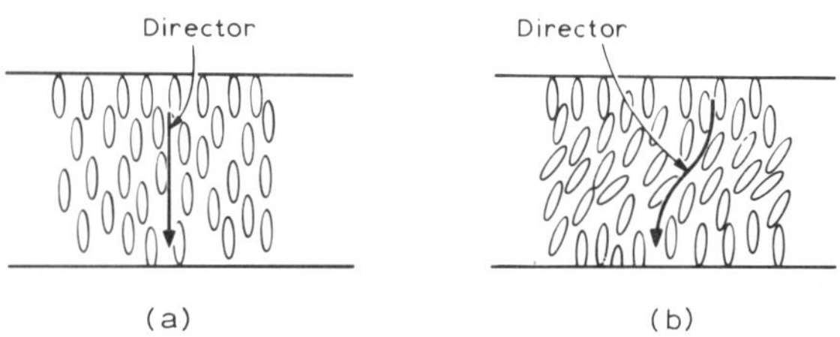

Figure 27.11 Demonstration of the field-induced birefringence effect: (a) alignment in the absence of the field; (b) when a field above the threshold field is applied between the electrodes

27.6.2.2 Storage effect

A related hydrodynamic effect[37] is observed in cholesterics of negative dielectric anisotropy. In loosely-wound materials with a long pitch, a certain current flow will also produce the dynamic scattering instability. The cholesteric structure is completely disrupted due to the turbulence. When the field is removed, the movement stops and the cholesteric forms a 'polycrystalline' state called focal-conic texture. The individual crystallites or domains are oriented randomly with respect to each other. This state is highly scattering. To make a display, a clear state is needed and this is obtained by applying a high-frequency electric field. The frequency is well above the dielectric relaxation frequency so that there is no ionic current flowing, which would re-establish the instability. Instead, the dielectric field causes the molecules to align parallel to the electrodes which, in turn, produces a uniform alignment of the helical axis at right angles to the electrodes. This cholesteric state persists after the removal of the field (storage). Thus, the sample may be switched back and forth between the scattering and clear state by applying two different kinds of driving pulses. Optically, the two states are quite similar to the on and off states of dynamic scattering but the driving voltages are higher and the transition times longer, particularly if a long-time storage is desired.

27.6.3 Field effects

27.6.3.1 Field-induced birefringence

There are a number of electric field-induced electro-optic effects in liquid crystals. The simplest one, field-induced birefringence or distortion of an aligned phase,[38] is considered first, and in some detail, as the basic principles apply to all field effects. The material used is a nematic liquid with negative anisotropy of dielectric constant and negligible conductivity. The properties of the cell surfaces are controlled so that the alignment of the molecules is uniform and at right angles to the surface (*Figure 27.11(a)*). When an electric field is applied in the perpendicular direction, it produces a torque on the individual molecules trying to turn them in a direction parallel to the electrodes because the lowest energy state of the unconstrained system is that in which the larger dielectric constant is oriented parallel to the field. However, the surface alignment is permanently fixed in the perpendicular direction and, through the elastic forces of long range order, tries to keep the bulk of the material in that same orientation. This balance between electric and elastic energy is such that below a certain threshold field E_c, the alignment remains uniformly perpendicular while, above that field, it gradually distorts towards the parallel direction as shown in *Figure 27.11(b)*. In the figure, the rotation of the molecules from the perpendicular is shown in one particular direction and the long range forces tend to align the entire sample in the same direction. There is, however, no reason to prefer one direction in the plane of the cell over any other and the direction may vary gradually throughout the sample.

Both states of the nematic fluid below and above the critical field are optically clear. Light propagating through the cell is subject to different indices of refraction and, therefore, phaseshifts in the two cases. These phase differences are converted to amplitude differences by positioning the cell between crossed polarisers. Then, light transmitted perpendicularly through the sample will be extinguished as long as the situation of *Figure 27.11(a)* applied (equivalent to light propagating along the axis of a uniaxial crystal). Above E_c, more and more light will be transmitted *Figure 27.12*,[39] until a saturation is reached when all but the surface molecules are aligned to the electrodes.

The induced birefringence effect can also be created in the opposite sense by using a material of positive dielectric anisotropy and making the surface alignment parallel to the electrodes. Then, the light transmission will be maximum at low fields and reduce to zero at high fields. The largest contrast is obtained if the polariser is located at 45° to the surface aligning direction.

The brightness and colour of this device depend on the difference in phase between the light polarised along the crystal axis and that polarised at right angles to it. This means that the colour and intensity of the transmitted light observed by the viewer will not be uniform if the cell thickness is not uniform. A further disadvantage is that the transmitted light is strongly dependent on the angle of viewing.

27.6.3.2 Twisted nematic

A field effect arrangement which overcomes this disadvantage is the so-called twisted nematic cell.[40] It is similar to the second case of induced birefringence with the positive anisotropy material. The alignment is parallel to the surfaces, but the directions of the alignment at the two surfaces are positioned at right angles to each other. The elastic forces cause the director in the interior of the crystal to remain in the plane of the sample but to twist gradually by 90° from one electrode to the other. This produces a structure which is identical to that of a cholesteric material with

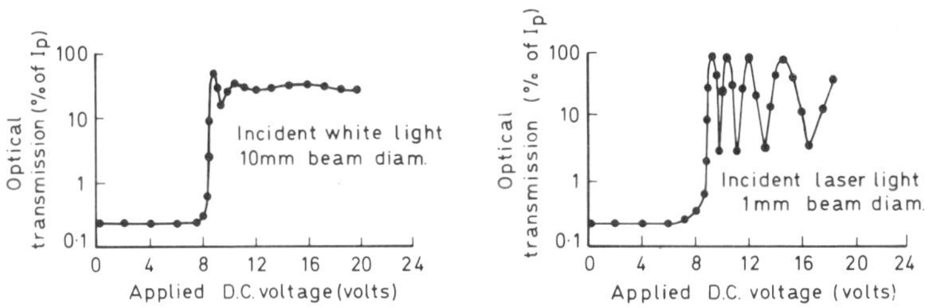

Figure 27.12 Voltage dependence of the transmitted light for a field-induced birefringence cell at normal incidence for two different kinds of light beams (courtesy American Institute of Physics)

rather long pitch (four times the thickness of the sample). If plane-polarised visible light, which has a wavelength much shorter than this pitch, propagates through the cell, its plane of polarisation is rotated at the same angular rate at which the director is rotated.[29] The total angle of rotation is 90° so that, with analyser and polariser parallel to each other, no light will be transmitted. When a field above E_c is applied to the sample, the orientation will be changed to a more or less perpendicular one and the light polarisation direction no longer rotates. The light is transmitted by the analyser. If the analyser is set at right angles to the polariser, light will be transmitted by the undisturbed cell and absorbed when the field is applied.

27.6.3.3 Guest–host effect

Another kind of field effect in nematic liquids[41] modifies the absorption of light, rather than its phase. A pleochroic dye (in which the absorption of light varies strongly depending on whether the light is polarised parallel to the molecular axis or perpendicular to it) is added to the liquid crystal. Because such dye molecules tend to be rod-like, they orient themselves parallel to the similar-shaped liquid crystal molecules. The absorption is modulated by changing the orientation of the director and, therefore, the dye molecules. For maximum absorption by the dye, the initial alignment of a liquid crystal should be in the plane of the sample, and the incident light should be polarised parallel to the director. If the liquid crystal has a positive dielectric anisotropy, then an electric field rotates the director to a perpendicular direction where the absorption is a minimum. Note that no analyser is required. To obtain a rapid response for this effect, it is necessary that the dye molecules resemble the liquid crystal molecules as closely as possible. Also, the absorption of the dye should be as high as possible to reduce the quantity of dye required.

27.6.3.4 Cholesteric field effects

A number of field effects are possible with cholesteric materials. Some produce reorientation of the existing structure as was the case with the nematics. In addition, the field may cause two kinds of changes in the cholesteric structure. These are a change in pitch of the helix and a complete disruption of the helix with a transition to the nematic phase.

A change of pitch causes optical effects if the pitch is of the order of the wavelength of light. This is because of the wavelength sensitive optical properties.[29] In particular, there is a region of total reflection for one sense of circularly polarised light propagating along the helical axis if the wavelength is close to the pitch of the helix. In such a case, the application of a field causes rapid and vivid changes of colour. However, the pitch also depends strongly on other variables, such as pressure and temperature, so that this is not a very practical electro-optic effect.

The field-induced phase change[42] is more important. In this case, use is made of the fact that, without any preferential surface alignment, the cholesteric material forms the focal-conic texture which is highly scattering and appears opaque. If the individual molecules have a positive dielectric anisotropy, then an applied field will tend to orient them parallel to the field. For a loosely wound cholesteric fluid, i.e. one with a large value of pitch, a moderate field will align the molecule into the induced nematic phase. This texture is optically clear and forms a large contrast to the zero-field opaque case. Upon removal of the field, the material returns to the scattering cholesteric phase. Cholesteric materials of large pitch are produced by mixing a short pitch cholesteric with a nematic material which, in this case, must have positive dielectric anisotropy.[43]

27.6.4 Applications

Liquid crystal displays are presently used for digital watch and clock dials, for digital panel meters and other indicators, and for calculator displays (*Figure 27.6*). They are particularly attractive for small portable devices because of their low power consumption. They can be expected to find their way into numerous other indicator applications. The information density of the displays is limited by the number of electrical connections that can be made in a practical device. Only a relatively small number of segments can be matrixed for most of the liquid crystal effects.

Whereas the first commercial displays all used dynamic scattering, most of the displays sold today use the twisted nematic field effect. They have less critical illumination requirements and a wider angle of viewability. This is particularly attractive for the watch application, where the fact that the field effects require

Other liquid crystal effects can be expected to become used in practical applications in the future.

Since electric addressing poses difficult problems in case of high information density, the only high resolution displays made so far use optical addressing schemes. In one case,[44] one electrode is replaced by a photoconducting layer with bias electrode. In the dark, the voltage drop is across the photoconductor while, under light, it appears across the liquid crystal cell and turns it on. If an optically opaque layer is placed between the liquid crystal and the photoconductor, then the same wavelength light can be used on both sides and it becomes a light amplifier.

Another optical addressing scheme[45] makes use of thermally-induced phase transitions. The effect is similar to the field-induced phase change. In the initial state, the cell is clear by having the helical axis aligned perpendicular to the plate. A focused high-intensity light beam (usually an infrared laser beam) heats the material locally above the phase transition to the liquid phase. Upon cooling, the cholesteric forms the scattering focal-conic texture which stores the written information. The entire display can be erased by applying a field across the cell, if the cholesteric has a negative molecular dielectric anisotropy. A similar device has been constructed with smectic liquid crystal material[46] where selective thermal erasure has also been found to

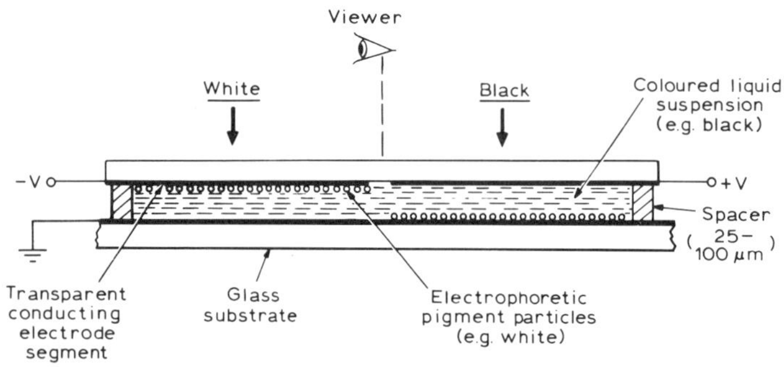

Figure 27.13 Operation of an electrophoretic display

be practical. Writing speeds of 10^4 spots/s have been found to be feasible with these displays.

27.7 Electrophoretic displays

27.7.1 Operation

Electrophoresis—the migration of charged colloidal particles in an electric field—has been used extensively for the separation and analysis of components in biological substances.[47] It is also widely used for electrostatic coating in liquids or the development process in electrophotography.

An electrophoretic display device has been developed[48] which utilises the migration of charged pigment particles suspended in a coloured liquid to produce a reflective display. An electrophoretic suspension (25–100 μm thick), composed of pigment particles and the suspending liquid, is sandwiched between a pair of electrodes, at least one of which is transparent. This basic structure is shown in *Figure 27.13*. Under the influence of a d.c. field, the pigment particles are moved towards, and deposited on, one of the electrodes; this electrode is coloured by the pigment particles while unelectroded areas or electrodes of opposite polarity retain the colour of the suspending liquid. A high-contrast reflective display results.

The pigment particles must have the same density as the suspending liquid to avoid precipitation or separation. Organic pigment particles, e.g. hansa yellow, can be provided with the same density as that of the liquid, which consists of a mixture of solvents with different densities. Inorganic pigment particles, e.g. TiO_2, are denser than normal liquids, but their average density can be reduced by encapsulation with an inert plastic. e.g. polyethylene.

27.7.2 Electrical characteristics and device performance

Electrophoretic displays, like liquid crystal displays, are passive and therefore attractive for viewing in high-brightness ambients. There are essentially no viewing angle restrictions, a distinct advantage over liquid crystal displays. Electrophoretic displays have memory, that is an element stays 'on' even if the power is removed, until the electric field excitation is reversed. Power dissipation (typically 0.4 mW cm^{-2}) is considerably higher than for liquid crystals, but low enough for moderate-sized numeric displays; the storage effect can be used to lower the average power consumption whenever the display is updated slowly.

Contrast ratios in excess of 50:1 are possible. A variety of colours is achievable by changing the pigment or suspending liquid. The addressing voltages (50–100 volts) however are a distinct disadvantage since high-voltage addressing circuitry is costly. Also, the high d.c. voltage can produce undesirable electrochemical effects which shorten the life of the display; operational life of 3000 hours has been reported[48] and can be expected to improve.

27.8 Electrochromic displays

Electrochromism is the coloration or change in colour which occurs in certain materials upon the application of an electric field. The basis for this coloration is the formation of colour centres or an electrically induced oxidation or reduction whereby a new substance is formed which absorbs visible light. The ability to colour a material reversibly by means of a locally applied electric field can be used to produce a simple reflective display, suitable for viewing in high-brightness ambients.

In general, electrochromic displays require relatively high currents (high power) or long times to achieve useful display contrasts; thus they are best suited for small alphanumeric display applications, especially those requiring infrequent updating, e.g. digital clocks, calculators.

In inorganic materials, e.g. WO_3, colour centres can be formed[49] when a d.c. electric field of $\sim 10^4$ V cm^{-1} ($\sim$3 volts across a 1 μm thick film) is applied at room temperature. Useful contrasts require $\sim 10^{18}$ colour centres/cm^3, corresponding to the passage of $\sim$100 mA cm^{-2} for a second or more. This is comparable to the charge required to deposit an opaque layer of metal by electroplating: a 50 nm thick layer of copper requires $\sim$70 mC cm^{-2} (70 mA for 1 s).

Erasibility in electrochromic displays means that irreversible electrochemical reactions must be avoided. In WO_3, electrons are trapped at colour centres reducing W^{6+} to W^{5+} while oxygen ions are given up at the anode; these ions must again be able to enter the material when the polarity is reversed. Although many electrode materials have been explored which can accept and give up oxygen ions readily, the life of these displays remains problematic.

An organic electrochromic display[50] has also been developed which produces a strongly coloured organic dye at the cathode. Since the coloration due to one dye molecule is greater than that of one metal atom or one colour centre, the electrical charge required to achieve good contrast (up to 20:1) in the organic system is considerably smaller (2 mC cm^{-2}). Life problems are claimed to be absent provided oxygen is excluded from the cell; 10^5 write–erase cycles have been achieved.[50] As with the inorganic systems, addressing voltages (1–2 volts) are compatible with integrated circuits making them of potential use for small alphanumeric applications.

References

1 NIINA, T., KURODA, S., YONEI, H., TAKESADA, H., 'A high brightness GaP green LED flat panel device for character and TV display', *Conf. Record of 1978 Biennial Display Research Conf.*, p. 18 October (1978)
See also NIINA, T., KURODA, S., YAMAGUCHI, T., YONEI, H., TOMIDA, Y. and YAGI, K., 'A multicolor GaP LED flat panel display device', *1981 SID Int. Symp. Digest of Technical Papers*, Vol. 12, p. 140, New York, April (1981)

2 BERGH, A. A. and DEAN, P. J., 'Light-emitting diodes', *Proc. IEEE*, **60**, 156 (1972)

3 NUESE, C. J., KRESSEL, H. and LADANY, I., 'Light-emitting diodes and semiconductor materials for displays', *J. Vac. Sci. Tech.*, Sept/Oct (1973)

4 DESTRIAU, G. J., *J. Chem. Phys.*, **33**, 620 (1936)

5 KAZAN, B. and KNOLL, M., *Electronic Image Storage*, Academic Press, New York, pp. 418–37 (1968)

6 LARACH, S. and SHRADER, R. E., 'Electroluminescence of polycrystallites', *RCA Rev.*, **20**, 532 (1959)

7 ARAI, H., YOSHIZAWA, T., AWAZU, K., KURAHASHI, K. and IBUKI, S., 'EL panel display', *IEEE Conf. Record of 1970 IEEE Conf. on Display Devices*, p. 52, New York, 2–3 December (1970). Also ARAI, H., 'EL panel TV', *JAEU*, **3**, 39 (1971)

8 YOSHIYAMA, M., 'Lighting the way to flat-screen TV', *Electronics*, **42**, No. 6, 114 (1969). Also YOSHIYAMA, M., OSHIMA, N. and YAMAMOTO, R., 'A television display device utilizing DC–EL panel', *National Technical Report*, **17**, No. 6, 670 (1971) (in Japanese)

9 INOGUCHI, T., TAKEDA, M., KAKIHARA, Y., NAKATA, Y. and YOSHIDA, M., 'Stable high-brightness thin-film electroluminescent panels', *1974 SID Int. Symp. Digest of Technical Papers*, Vol. 5, p. 84, San Diego, May (1974)

10 MITO, S., SUZUKI, C., KANATANI, Y. and ISE, M., 'TV imaging system using electroluminescent panels', *1974 SID Int. Symp. Digest of Technical Papers*, Vol. 5, p. 86, San Diego, May (1974)

11 SUZUKI, C., KANATANI, Y., ISE, M., MIZUKAMI, E., INAZAKI, K. and MITO, S., 'Optical writing on thin-film EL panel with inherent memory', *1976 SID Int. Symp. Digest of Technical Papers*, Vol. 7, p. 52, Beverley Hills, CA, May (1976)

12 AWANE, K., FUJII, K., YAMANO, T., TAMAKI, H., BIWA, T., HATTORI, H. and FUJIMOTO, T., 'High-voltage DSA–MOS transistor for EL display', *1978 IEEE Int. Solid-State Circuit Conf. Digest of Technical Papers*, Vol. 16, p. 224, February (1978)

13 von HIPPEL, A. R., *Dielectrics and Waves*, p. 234, Wiley, New York (1954)

14 FORMAN, J., 'Phosphor color in gas discharge panel displays', *Proc. SID*, **13**, 14 (1972). See also KANEKO, R., KAMEGAYA, T., YOKOZAWA, M., MATSUZAKI, H. and SUZUKI, S., 'Color TV display using 10″ planar positive-column discharge panel', *1978 SID Int. Symp. Digest of Technical Papers*, Vol. 9, p. 46 (1978)

15 KRUPKA, D. C., CHEN, Y. S. and FUKUI, H., 'On the use of phosphors excited by low-energy electrons in a gas discharge flat-panel display', *IEEE Proc., Special Issue on New Materials and Devices for Displays*, 1025, July (1973)

16 CHODIL, G. J., DEJULE, M. and GLASER, D., 'Cathodo-luminescent display with hollow cathodes', *US Patent No.* 3,992,633, issued 16 November (1976)

17 DE BOER, Th. J., 'An experimental 4000-picture-element gas discharge TV display panel', *Proc. 9th National Symp. on Information Display*, p. 193, Los Angeles, May (1968)

18 AMANO, Y., 'A flat-panel color TV display system', *1974 Conf. on Display Devices and Systems, Conf. Record*, 99; *IEEE Trans. Electron. Devices*, **ED-22**, 1, January (1975)

19 HOLZ, G. E., 'The primed gas discharge cell—a cost and capability improvement for gas discharge matrix displays', *1970 SID IDEA Symp. Digest of Papers*, p. 30, May (1970)

20 CHODIL, G. J., DE JULE, M. C. and MARKIN, J., 'Good quality TV pictures using a gas-discharge panel', *IEEE Conf. Record of 1972 Conf. on Display Devices*, p. 77, October (1972)

21 CHIN, Y. S. and FUKUI, H., 'A field-interlaced real-time gas-discharge flat-panel display with gray-scale', *IEEE Conf. Record of 1972 Conf. on Display Devices*, p. 70, October (1972)

22 BITZER, D. L. and SLOTTOW, H. G., 'The plasma display panel—a digitally addressable display with inherent memory', *Proc. Fall Joint Computer Conf.*, San Francisco, November (1966). Also BITZER, D. L. and SLOTTOW, H. G., 'Principles and applications of the plasma display panel', *Proc. 1968 Microelectronic Symp., IEEE*, St. Louis (1968)

23 KURAHASHI, K., TOTTORI, H., ISOGAI, F. and TSURUTA, N., 'Plasma display with gray scale', *1973 SID Int. Symp. Digest of Technical Papers*, Vol. 4, p. 72, New York, 15–17 May (1973)

24 CATANESE, C. and ENDRIZ, J., 'The physical mechanisms of feedback multiplier electron sources' and ENDRIZ, J., KENEMAN, S., CATANESE, C. and JOHNSTON, L., 'Flat TV display using feedback multipliers', *1978 SID Int. Symp. Digest of Technical Papers*, Vol. 9, pp. 122 and 124 respectively, San Francisco, April (1978)

25 CREDELLE, T., ANDERSON, C., MARLOWE, F., GANGE, R., FIELDS, J., FISHER, J., VAN RAALTE, J. and BLOOM, S., 'Cathodoluminescent flat panel TV using electron beam guides', *1980 SID Int. Symp. Digest of Technical Papers*, Vol. 11, p. 26, San Diego, April (1980)

26 HARENG, M., ASSOULINE, G. and LEIBA, E., *Proc. IEEE*, **60**, 913 (1972)

27 GREY, G. W., *Molecular Structure and the Properties of Liquid Crystals*, Academic Press, New York (1962); a general reference of all early work

28 SUSSMAN, A., *IEEE Trans. Parts, Hybrids, and Packaging*, PHP-8, 28 (1972)

29 DEVRIES, H., *Acta Cryst.*, **4**, 219 (1951)

30 FELICI, N., *Rev. Gen. Elec.*, **78**, 717 (1969)

31 CHATELAIN, P., *Bull. Soc. Fr. Miner. Cryst.*, **66**, 105 (1943); JANNING, J. L., *Appl. Phys. Lett.*, **21**, 173 (1972)

32 DREYER, J. F., in *Liquid Crystals and Ordered Fluids, 2nd edn*, JOHNSON, J. F. and PORTER, R. S., eds., Plenum Press, New York, 311 (1970)

33 WILLIAMS, R., *J. Chem. Phys.*, **39**, 384 (1963)

34 HELFRICH, W., *J. Chem. Phys.*, **51**, 4092 (1969)

35 Orsay Liquid Crystal Group, *Mol. Cryst. Liq. Cryst.*, **12**, 251 (1971)

36 HEILMEIER, G. H., ZANONI, L. A. and BARTON, L. A., *Appl. Phys. Lett.*, **13**, 46 (1968)

37 HEILMEIER, G. H. and GOLDMACHER, J. E., *Proc. IEEE*, **57**, 34 (1969)

38 FREEDERICKSZ, V. and ZWETKOFF, Y., *Acta Physicochemica URSS*, **3**, 9 (1935); SCHIEKEL, M. F. and FAHRENSCHON, K., *Appl. Phys. Lett.*, **19**, 390 (1971)

39 SOREF, R. A. and RAFUSE, M. J., *J. Appl. Phys.*, **43**, 2029 (1972)

40 SCHADT, M. and HELFRICH, W., *Appl. Phys. Lett.*, **18**, 127 (1971)

41 HEILMEIER, G. H. and ZANONI, L. A., *Appl. Phys. Lett.*, **13**, 91 (1968)

42 WYSOCKI, J. J., ADAMS, J. and HAAS, W., *Phys. Rev. Lett.*, **20**, 1024 (1968)

43 HEILMEIER, G. H. and GOLDMACHER, J. E., *J. Chem. Phys.*, **51**, 1258 (1969)

44 MARGERUM, J. D., NIMOY, J. and WONG, S.-Y., *Appl. Phys. Lett.*, **17**, 51 (1970)

45 MELCHIOR, H., KAHN, F. J., MAYDAN, D. and FRASER, D. B., *Appl. Phys. Lett.*, **21**, 392 (1972)

46 KAHN, F. J., *Appl. Phys. Lett.*, **22**, 111 (1973)

47 *Electrophoresis—Theory, Methods and Applications*, MILAN BIER, ed., Academic Press (1959)

48 OTA, I., OHNISHI, J. and YOSHIYAMA, M., 'Electrophoretic display device', *IEEE Conf. Record of 1972 Conf. on Display Devices*, p. 46, New York, October 11–12 (1972)

49 DEB, S. K., 'A novel electrophotographic system', *Appl. Opt. Suppl.*, **3**, 192 (1969)

50 SCHOOT, C. J., PONJÉE, T. J., VAN DAM, H. T., VAN DOORN, R. A. and BOLWYN, P. T., *Appl. Phys. Lett.*, **23**, 64 (1973)

28 Integrated Circuit Fabrication and Packaging

F F Mazda DFH, MPhil, CEng, MIEE, DMS, MBIM
Rank Xerox Ltd

Contents

28.1 Manufacturing processes

28.1.1 Silicon wafer preparation

The starting material for integrated circuits is a slice of single crystal silicon: the larger the slice the greater the number of identical circuits which can be made at one time. However, large slices require greater process control to ensure uniformity over the whole area. The most usual method of obtaining these slices is to pull from a silicon melt in a Czochralski puller. Impurities of p or n type can be added to the melt to give the final silicon ingots the required resistivity (typically up to about 50 Ω cm). The ingots are then cut into slices, about a millimetre thick, using a diamond impregnated saw. The saw cuts can be made along the ⟨1 1 1⟩ or ⟨1 0 0⟩ planes of the crystal. These cuts usually damage the crystal lattice near the silicon surface, resulting in poor resistivity and minority carrier lifetime. The damaged area, which is about 20 μm deep, is removed by etching in a mixture of hydrofluoric and nitric acids, and the surface is then polished with a fine diamond power to give a strain free, highly flat region.

28.1.2 Oxide growth

Silicon oxide (also called silica or SiO_2) is grown and removed from the surface of the silicon slice many times during the manufacture of an integrated circuit. It is such a fundamental substance that one of the reasons for using silicon in integrated circuits is its ability to grow a stable oxide. The oxide layer is used for diffusion masking, for sealing and passivating the silicon surface, and for insulating the metal interconnections from the silicon. Although the oxide layer may be deposited onto the slice, as for the epitaxy layer described in Section 28.1.4, it is more usual to grow it using dry or wet oxygen or steam. *Figure 28.1* shows a typical arrangement of the apparatus used for oxidation (and diffusion). The silicon slices are stacked upright in a quartz boat

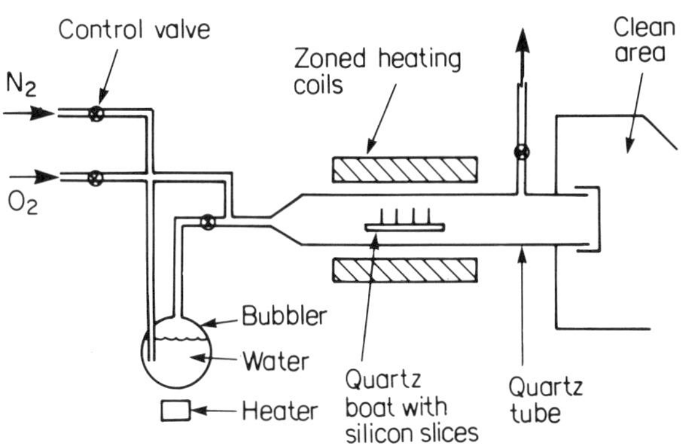

Figure 28.1 Oxidation–diffusion open tube arrangement

and inserted into a quartz tube. The tube is heated to between 1000°C and 1200°C by zoned heaters so that the boat is located in an area having a uniform temperature along a length of the tube. Nitrogen (inert atmosphere), dry oxygen, wet oxygen (oxygen bubbled into water at 95°C) or steam can be passed over the slices to grow the oxide layer. A thickness of about 1 μm takes approximately four hours to grow and consumes between 0.4 μm and 0.5 μm of the silicon. The colour of the silicon surface changes with the thickness of the oxide layer due to the shift in the wavelength of the reflected light. This effect is used as an indication of the layer thickness.

28.1.3 Photolithography

The prime use of photolithography in integrated circuit manufacture is to selectively remove the oxide from areas of the silicon slice. To do this the surface of the oxide is first covered with a thin uniform layer of liquid photoresist. This is best obtained by holding the silicon slice in a vacuum chuck and placing a fixed amount of photoresist onto its centre. The slice is then rotated at very high speeds for about one minute. The resist spreads over the slice, the excess flying off due to centrifugal forces. The amount left on the slice is clearly a function of the oxide properties and the viscosity of the resist. The slice is then heated for a few minutes in an oven to harden the resist.

The mask is next placed in contact with the oxide layer. This is usually a glass plate with the black emulsion areas on one face. It is essential that the mask is accurately aligned on the silicon slice using a microscope, and is close to it. Close physical contact is necessary to prevent the light spreading sideways on the photoresist when the mask is exposed. Unfortunately this requires considerable force with the result that particles on the slice can damage the mask. Also parts of the photoresist come away when the mask is removed. Consequently the mask is damaged and is only usable for an average of ten exposures. Special materials, such as chrome masks, give about fifty exposures, but are more expensive. The modern trend is towards projection masking where an image of the mask is projected onto the silicon slice. The mask does not now make physical contact with the silicon and consequently has a much greater life.

Once the mask is in contact with the photoresist layer it is exposed to ultraviolet light for about twenty seconds. If negative resist material is used the exposed areas become hardened by the light. The slice is now placed in a rotating chuck and developer is dropped onto it. This dissolves the unexposed resist areas. The spinning action ensures the speedy removal of dissolved parts so that all areas have equal exposure to the developer. This part of the process usually takes about thirty seconds. The slice is now placed in a bath of hydrofluoric acid which dissolves the exposed oxide areas but not the silicon. The etching time must be closely controlled; if too long the etchant spreads sideways under the remaining photoresist. For applications which require most of the oxide to be removed it is more convenient to use a positive resist material, such that the areas exposed to ultraviolet light are dissolved in the developer.

28.1.4 Epitaxy

Epitaxy means growing a single crystal silicon structure on the original slice such that the new structure is essentially a molecular extension of the original silicon. Epitaxy layers can be closely controlled regarding size and resistivity (i.e. ±10%). This compares favourably with ±30% resistivity control obtained when pulling from the silicon melt. Most of the integrated circuit structure is formed in the epitaxy layer, the rest of the slice acting purely as a ground plane.

Epitaxy apparatus is very similar to the oxide growth arrangement shown in *Figure 28.1*. However, r.f. heating coils are normally used and the silicon slices are placed in a graphite boat, which may be coated with quartz to prevent the graphite contaminating the silicon. The bubbler usually contains silicon tetrachloride ($SiCl_4$) to which may be added a controlled amount of an impurity such as PCl_3. Hydrogen gas is bubbled through this mixture before entering the quartz epitaxy tube.

Initially the slices are heated to about 1200°C and pure hydrogen and hydrochloric acid vapour are passed over them to etch away any oxide or impurities which may exist on the surface of the silicon. HCl vapour is then turned off and H_2 bubbled through the $SiCl_4$, the vapour passing over the silicon slices. When this reaches the hot silicon it dissociates and silicon atoms are deposited on the slice where they rapidly establish themselves as part of the original crystal structure. It is essential to saturate the tube with $SiCl_4$ vapour to ensure a uniform layer thickness over the whole slice. The epitaxy is about 10 to 15 μm in depth and has a resistivity which varies in the region of about 10 Ω cm. The epitaxy layer may be doped by p or n type impurities by introducing these, in the required concentration, into the vapour stream.

28.1.5 Diffusion

In epitaxy a large area is doped by a closely controlled amount of impurity. Diffusion on the other hand enables selective areas to be doped. These areas represent those which are not covered by an oxide layer so that the photolithographic stage is normally followed by diffusion.

The diffusion furnace resembles the arrangement shown in *Figure 28.1*. The bubbler contains the impurity, and nitrogen is passed through it on the way to the boat containing the silicon slices. The furnace temperature is kept close to the melting point of silicon, i.e. 1200°C, and at this value the silicon atoms are highly mobile. Impurity atoms readily move through the silicon lattice by substitution, going from a region of high concentration to that of lower density.

Diffusion can be carried out by one of two techniques, as shown in *Figure 28.2*. In the error function or one step process the concentration of impurities is kept fixed throughout the diffusion period, giving the curves shown. In Gaussian or two step diffusion a fixed amount of impurity is present so that as this moves deeper into the silicon bulk the surface concentration decreases. This gives a flatter dopant distribution of higher resistivity. Generally diffusion takes place in a slightly oxidising atmosphere. This results in the formation of a glossy layer of impurity on the silicon surface as well as a slight penetration into the silicon bulk. The silicon slice can be removed to a second furnace and heated in an inert atmosphere when the dopants diffuse out of the glassy layer to give an error function distribution. The glassy layer now not only forms a diffusion source but also protects the silicon surface from evaporation, and acts as a getter for impurities from the silicon bulk. For Gaussian diffusion the glassy layer is etched off using hydrofluoric acid prior to the slice being heated in an inert atmosphere. The critical dopants just below the surface now diffuse into the silicon bulk.

Apart from the open tube arrangement shown in *Figure 28.1* it is possible to diffuse slices by putting them in a sealed quartz container with doped silicon powder and then heating the combination in a quartz furnace. The advantage of this method is that many slices can be diffused simultaneously giving a larger throughput. However, the quartz container must be broken to remove the slices after diffusion so the process can prove expensive.

The most commonly used p and n type impurities are boron and phosphorus respectively. Both reach maximum solubility at about 1200°C and have a high diffusion constant. Arsenic, on the other hand, is a n type impurity which diffuses very slowly. It is used for making the buried layer in transistors since this must not diffuse appreciably during subsequent high temperature processes. Gold is often introduced as an impurity into integrated circuits. Gold is a lifetime killer and enables fast switching circuits to be built. To dope a slice with gold its rear surface is first coated with a thin film of gold using evaporation techniques and this is then heated in a diffusion furnace until the whole slice is saturated with gold. The gold atom is much smaller than a silicon atom. This means that it does not move through the silicon lattice by substitution as other impurities, but moves very rapidly in between the silicon atoms, i.e. intrinsically.

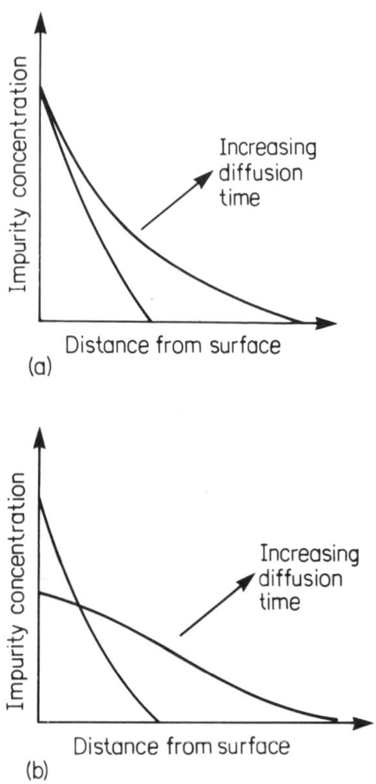

Figure 28.2 Impurity concentration during diffusion: (a) error function; (b) Gaussian

28.1.6 Vacuum deposition

During the manufacture of integrated circuits it is necessary to deposit a thin layer of metal on the silicon, for instance to form the aluminium interconnections, or to coat the back surface with a gold layer, prior to diffusion, for high speed circuits. This is carried out using vacuum deposition in an arrangement of the type shown in *Figure 28.3*. The silicon slices are placed face down around the bell jar, with the metal source in the centre. The vacuum is lowered to below 5×10^{-6} Torr before deposition is

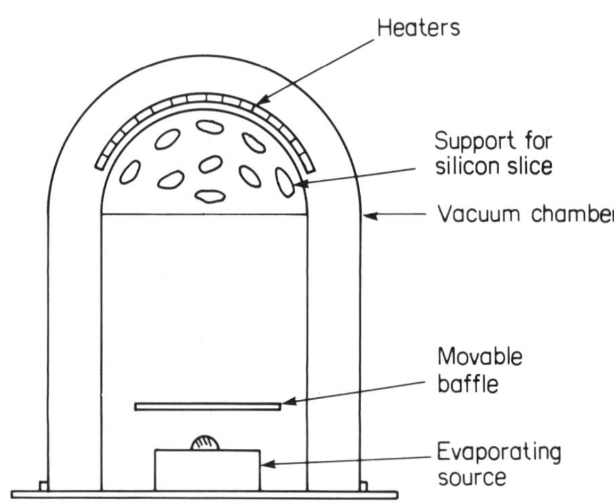

Figure 28.3 Metal evaporation equipment

commenced. The silicon is heated to between 100°C and 300°C which causes the deposited metal to form a chemical reaction with the silicon oxide and adhere to the slice. Upward evaporation is also used to prevent impurities, which may be generated by the heat source, from falling onto the slices. The metal film is usually about one micrometre thick. This thickness can be monitored by including a quartz crystal oscillator in the vacuum, whose frequency changes with the amount of metal deposited on its surface.

The metal source must have a large area. Shadowing effects can arise from point sources, which would result in weaknesses in the metal film. There are two ways of heating the metal source. It can be wrapped as a wire around a resistance heated tungsten filament. This is a cheap method but there is risk of contamination due to the evaporation of the heater metal. A cleaner solution is to use electron beam evaporation. The metal source is placed in a boat which acts as an anode to an electron gun. The high energy electrons striking the source cause it to evaporate and deposit on the surrounding silicon slices.

After evaporation it is usual to heat the slices to about 500°C in an inert (nitrogen) atmosphere. This causes the metal to alloy well with the silicon surface so that the interface between the two has a low resistance. This is referred to as a low ohmic contact joint.

28.1.7 Ion implantation

There are several basic requirements which must be met in any implantation system. These are:

(i) The impurity concentration must be uniform over a given slice and the process must be accurately reproducible over repeated slices.

(ii) The system must have a high throughput so that the expensive implantation equipment can be amortised over a large number of devices. This means that the capacity of each batch must be high and the cycle time, which is the vacuum cycle time since all such systems operate under a vacuum, must be short. A further requirement is that the doping current can be made large enough to reduce the time per slice.

(iii) The purity of the dopant must be accurately controlled. Most ion sources produce a range of dopants in addition to the one required. The impurities must be completely removed from the ion stream before it reaches the silicon slice.

(iv) The energy imparted to the ions (by the accelerating voltage) must be high enough to enable them to penetrate the maximum distance which is likely to be required.

These requirements are discussed further with reference to the basic implantation equipment shown in *Figure 28.4*. The main parts of this arrangement are as follows:

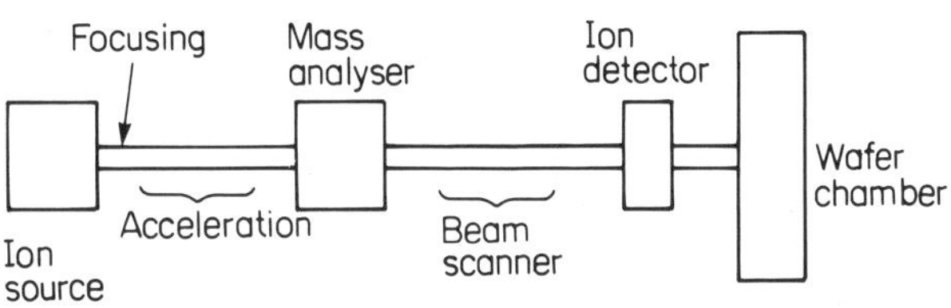

Figure 28.4 Principal parts of an ion implantation equipment

(a) An ion source which produces an abundance of the boron or phosphorus dopant required. The source output should remain constant over many days to allow reproducible devices to be made without the necessity for constant adjustment. The source current should also be variable up to about half a milliampere to give short implant times. There are several techniques used for producing ions, such as by a radio frequency discharge or by a heated filament discharge.

(b) A focusing system which is usually electrostatic since magnetic lens techniques, such as used for electron optics, give more bulky and expensive equipment due to the fact that the ions are now much heavier.

(c) An ion accelerator. This gives the ions the necessary penetration energy by applying a high voltage across the ions. It can vary from anything between 10 kilovolts to 500 kilovolts.

(d) The mass analyser is used in conjunction with the focusing system to separate out the impurities from the ion beam. First the ions are separated into different streams according to their mass and then all but the required dopant stream is blocked off before it reaches the silicon wafer chamber.

(e) The beam scanner is used to move the ion beam over the silicon surface so as to give a uniform dopant concentration. In addition it is also usual to move the slices physically past the beam for more uniformity.

(f) The doping concentration is monitored by measuring the current in the ion detector. This is quite easily done since each ion carries one positive charge unit. The dose imparted to the silicon, measured in ions/cm^2, is equal to the product of the current and the exposure time, divided by the wafer area.

(g) Finally the wafer chamber holds the silicon samples. It must be quickly accessible and large enough to hold a useful batch at each operation.

Ion implantation differs from diffusion in that the dopants are given enough energy to allow them to force their way through the silicon lattice until they reach the desired depth below the surface. The process therefore differs from diffusion where the impurity ions move gradually through the lattice and the concentration decreases away from the surface (unless epitaxy layers are grown over diffused regions, of course). Ion implantation is therefore a very precise process where the depth and concentration of impurities can be closely controlled. However, since the ions plough their way through the silicon lattice they distort the material so that implantation is usually followed by an annealing stage in which the atoms are allowed to drift into their places within the lattice.

28.2 Bipolar circuits

28.2.1 Circuit fabrication

Having looked at the different production techniques we can now see how they fit together to make an integrated circuit. The heart of any integrated circuit is its transistors. The transistor structure is very easily made in monolithic integrated circuits and it occupies relatively little space. Consequently, unlike discrete circuits, transistors are used liberally in integrated circuit designs, often to replace passive devices like resistors.

The key to integrated circuit fabrication is to build several components onto a common silicon slice without interfering with each other. There are several techniques which may be used for this in bipolar circuits. *Figure 28.5(a)* illustrates the commonest technique, known as diffusion isolation, or more correctly diode isolation. Two transistors are shown here, each completely surrounded by p type silicon. The collector of each transistor forms a diode with this p region, as shown in *Figure 28.5(b)*. If the substrate is always connected to the most negative voltage in the system then, irrespective of the voltage on each transistor, the devices are separated from each other by two reverse biased diodes.

Apart from transistors it is possible to form other electronic components in integrated circuit structures. A pn junction of base–collector or base–emitter forms a diode, a linear distance within the p base region or n emitter region of a transistor forms a resistor, and a reverse biased diode may be used as a capacitive element. All these elements have limitations, as will be evident

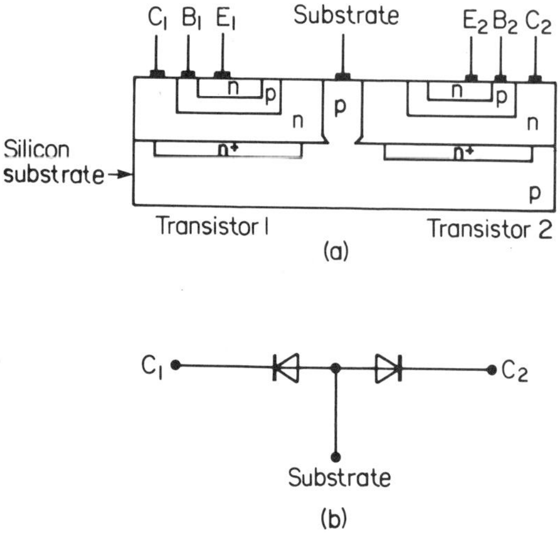

Figure 28.5 Diode isolation of two transistors: (a) device structure; (b) isolation configuration

later. For the present however let us consider how the circuit shown in *Figure 28.6(a)* can be built into an integrated circuit. This is illustrated in *Figure 28.6(b)*. Islands of n type material are first formed in a p type substrate. By taking the substrate to the most negative voltage available in the system, each island is therefore separated from the adjoining ones by the diode isolation mechanism discussed earlier. Into each of these islands it is now possible to build transistors, resistors, capacitors or diodes, as required. In *Figure 28.6(b)* the resistor is formed in what would normally be the base of a transistor and the diode utilises the base–collector of a transistor structure. The capacitor shown is a reverse diode structure although unipolar type capacitors, as described in the next section, can also be used. The various individual devices on the chip are then interconnected by metal patterns to form the required configuration shown in *Figure 28.6(a)*. The silicon oxide primarily acts as a convenient electrical insulation between the metal connectors and the rest of the silicon material.

The actual steps used in the fabrication of *Figure 28.6(b)* are shown in *Figure 28.7*. The starting material is a slice of polished

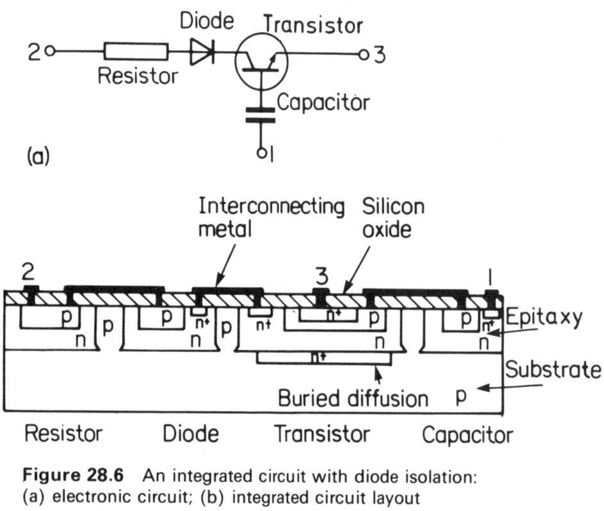

Figure 28.6 An integrated circuit with diode isolation: (a) electronic circuit; (b) integrated circuit layout

silicon as in *Figure 28.7(a)*. This slice will be capable of accommodating many hundred identical circuits or dice. Only one such die is considered here. First the transistor buried layer has to be diffused. This is done by growing an oxide on the silicon surface, and by photolithographic techniques etching a window at the required location. The mask used (first mask) will clearly be transparent except for the area covering the buried layer. The slice can now be introduced into a diffusion furnace and the n^+ layer formed (*Figure 28.7(b)*). Arsenic is used for this since, as explained earlier, it diffuses slowly and so will not move much during subsequent diffusion stages.

An n epitaxy is now grown over the whole area (*Figure 28.7(c)*). The advantage of the epitaxy is evident since it allows diffusions to be formed at different planes. Then an oxide layer is grown and photolithography is used to open up windows (second mask) through which isolation regions can be diffused (*Figure 28.7(d)*). Once again an oxide is grown over the surface and this is masked (third mask) and etched for the p diffusion of the transistor base and other areas (*Figure 28.7(e)*). The doping concentration is selected to suit the transistor requirements, and capacitors, diodes and resistors must accept this limitation. A fourth masking operation is required for the transistor emitters (*Figure 28.7(f)*) and this is followed by a fifth mask which opens up the contact areas to the silicon surface. Metal (aluminium) is now deposited over the whole surface using vacuum techniques (*Figure 28.7(g)*) and finally a sixth mask is used to etch the metal to the required interconnection pattern as shown in *Figure 28.7(h)* which corresponds to *Figure 28.6(b)*.

Two important factors should be noticed from the above description. Firstly there is a lot of to and fro movement between photolithographic, oxidation and diffusion plants. Careful handling is needed at all stages and each operation should be carried out in a clean atmosphere since dirt particles can damage the slice and could act as unwanted impurity dopants. Many different diffusion furnaces are also required since in each furnace the quartz walls become contaminated with the dopant and cannot be re-used for another purpose unless it is thoroughly saturated with the new impurity for several hours. This means that for the process just described different furnaces are needed for oxidation and for deposition and drive in (two stage, Gaussian diffusion) of three different types of impurities, i.e. arsenic (buried layer), boron (p type) and phosphorus (n type). This means a bank of eight furnaces just for one process. It is therefore very common in integrated circuit production areas to see banks of diffusion furnaces.

In addition if gold doping is used then this is usually introduced after the fourth emitter mask so that the emitter diffusion stage also drives in the gold. To prevent contamination of gold onto other devices the gold vacuum deposition plant is kept separated from that used for aluminium deposition.

The second important aspect to note is that one impurity can reverse the doping of a previous stage provided it is strong enough. Therefore the n^+ emitter diffusion is carried out into the p base diffusion. This is illustrated in *Figure 28.8* where the junctions between base, emitter and collector are defined. Note however that the transition is gradual, the impurity differences building up as one moves away from the junction. It should also be noted, as illustrated in *Figure 28.7*, that diffusions tend to spread downwards as well as sideways under the oxide masks. This must be allowed for in the mask design.

It is usual, but not always the case, to cover the completed slice with a thin layer of glass or other passive device (e.g. silicon nitride) to protect it from damage and contamination during the subsequent test and packaging stages which involve manual handling in relatively unpurified atmospheres. This step is called passivation. The glass is then etched off (seventh mask) from the bonding pad areas and in between the individual dice ready for cutting (scribing) and separation.

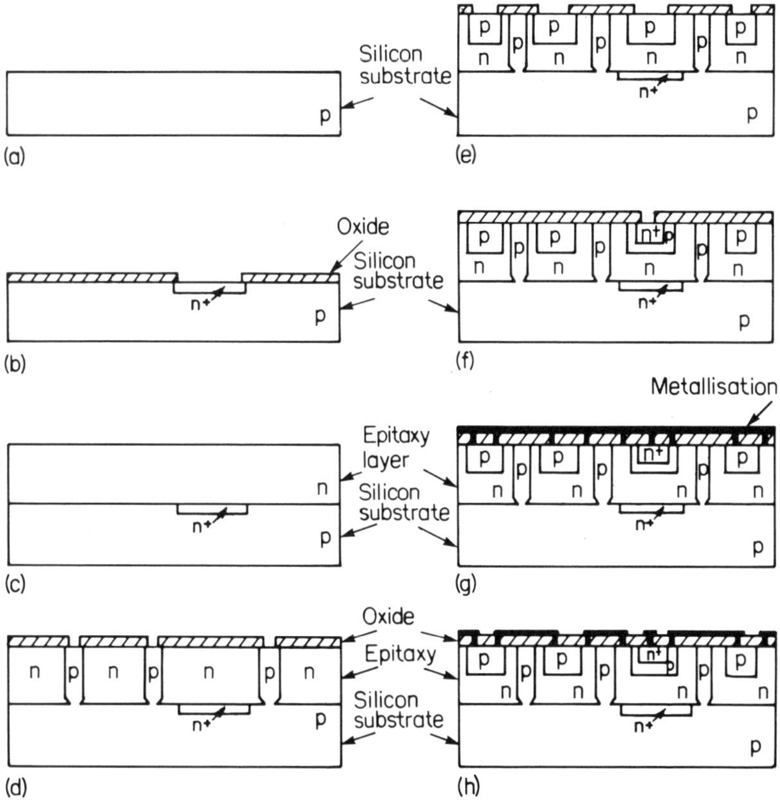

Figure 28.7 Stages in the manufacture of an integrated circuit: (a) silicon slice; (b) buried layer diffusion; (c) epitaxy; (d) isolation diffusion; (e) base diffusion; (f) emitter diffusion; (g) metal deposition; and (h) interconnection etch

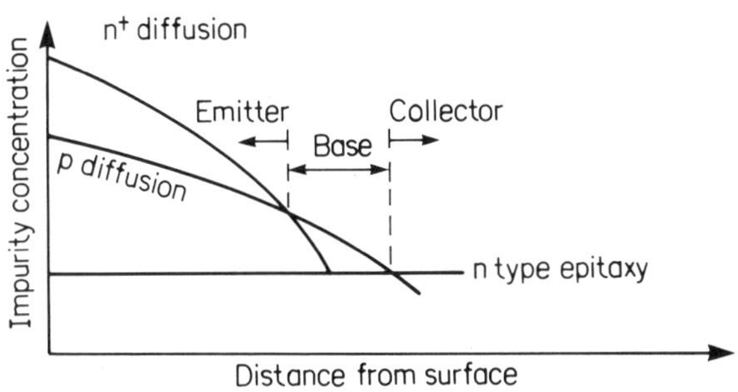

Figure 28.8 Formation of an npn transistor

28.2.2 Bipolar circuit components

Most of the different types of components which are used in integrated circuits have already been briefly introduced in the previous section. We will examine here their construction and characteristics in greater detail. There are four types of components, transistors, diodes, resistors and capacitors. As mentioned before the transistor is the single most important device in integrated circuits, all other component characteristics being dependent on it.

28.2.2.1 Transistors

In a monolithic system the substrate primarily acts as a mechanical holder to the devices and is of very high resistivity. The epitaxy forms the collector and is also of high resistivity. It is usual, as shown in *Figure 28.7*, to diffuse an n^+ region, at the same time that the emitter is diffused, at the point where the metal is to connect onto the silicon surface, to make a better joint and to prevent Schottky diode action. The high series resistance present in this collector gives poor amplifier and switching performance. However, one requires a high collector resistivity at the base junction in order to give a good collector–base breakdown voltage. Both these requirements are met by a buried n^+ diffusion as described earlier. It has also been explained that diode isolation is the most common technique in use for monolithic integrated circuits. This however presents two problems. Firstly the diode in between two devices is reverse biased and, as seen in Section 28.2.1, this acts as a capacitor. There is therefore a relatively large valued capacitor connecting the transistor collectors to their common substrate and these capacitors are voltage dependent. For low power applications this can give rise to significant leakage currents. A second problem can arise with monolithic transistors when they are biased such that the lower three layers, including the substrate, form a parasitic pnp transistor. A large current can now flow through this device from the base of the original npn transistor to the substrate. Both these effects are overcome when other isolation techniques are used, as described in the next section.

Although npn transistors are by far the most common in monolithic circuits, pnp devices can be formed if required. There are three main types, as shown in *Figure 28.9*. The substrate pnp is the simplest arrangement and only requires an emitter diffusion, although an n^+ base diffusion is also usual. It has two serious faults. Firstly all transistors on the same substrate also share the same collector, so that circuit flexibility is greatly reduced. Secondly in spite of careful process control techniques

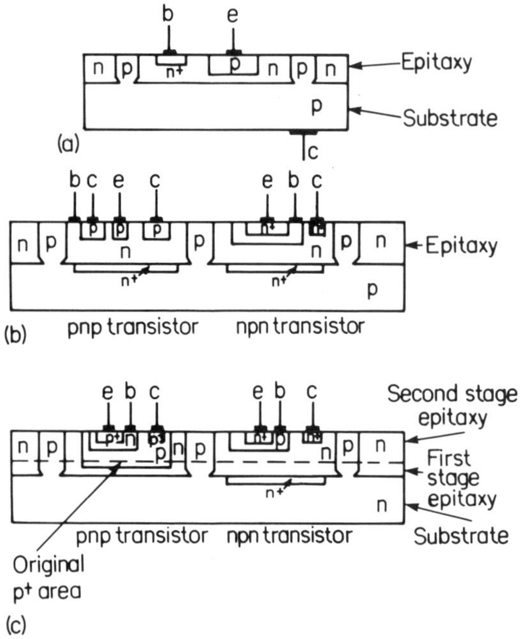

Figure 28.9 pnp transistor configurations: (a) substrate; (b) lateral; and (c) two-step epitaxy

the epitaxy layer cannot be grown to a uniform thickness over the whole slice. This means that the base width of the transistors, and consequently their gains, vary over the slice and are not predictable.

In the lateral pnp the collector is a concentric p type diffused ring around a central emitter diffusion. The base is the epitaxy between the emitter and collector diffusions. Unfortunately this again makes the base width difficult to control since it is the difference in sideways diffusion between the emitter and collector, both of which are very dependent on processing conditions. Generally to allow for process variations, the width of the base is kept relatively wide so that the gain is low. These transistors also have a low frequency response and a parasitic pnp action down to the substrate. This effect is however reduced by the presence of the n^+ buried layer. A further limitation is that the collector–base junction occurs at the surface of the silicon, instead of in the bulk as in other arrangements. This means that it is influenced by surface effects giving a low breakdown voltage.

Truely complementary pnp and npn transistors can be obtained by the two step epitaxy process. In this the epitaxy growth is interrupted half way through and a p^+ diffusion formed where the pnp transistors are due to occur. This is a mobile diffusant. The epitaxy growth is now resumed. When next a diffusion process occurs the p^+ buried diffusion moves up into the epitaxy to form the collector of the pnp transistor. Base and emitter diffusion occur as before. Quite clearly this type of pnp transistor requires more process steps. However, since the base thickness is formed by two lateral diffusions it can be accurately made to give pnp and npn devices of comparable performance.

Since base thickness in diffused monolithic transistors can be closely controlled it is possible to have very thin base high gain devices. Gains of the order of 10 000 are attainable, and are used to provide high input impedance amplifiers. Thin base regions unfortunately also give lower breakdown voltages. This can be overcome by using composite transistors, which are in effect two devices in which one provides the gain whereas the other withstands the high voltage. Using monolithic techniques both transistors are fabricated simultaneously giving minimal extra cost.

Figure 28.10(a) shows the plan view of two small signal transistors separated by a p diffusion. *Figure 28.10(b)* is the schematic of a high current transistor and illustrates the interdigitated emitter construction, which is necessary to obtain a large perimeter to area ratio and so improve the frequency performance.

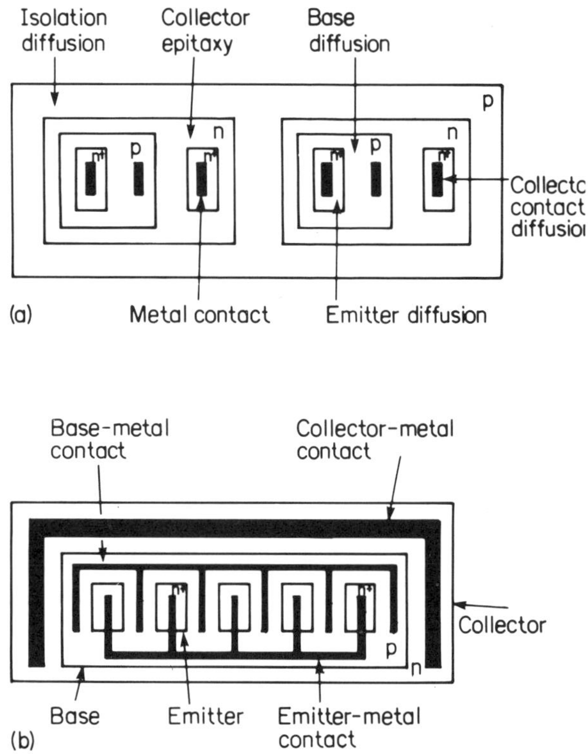

Figure 28.10 Transistor geometries: (a) two signal transistors; (b) high current transistor

28.2.2.2 Diodes

The thickness and sheet resistivity of the p and n diffusions used to make a diode are determined by those required for a transistor, since these are made at the same time as the transistor diffusions. It is also usual not to make just a p and n diffusion but to utilise two of the transistor's areas to form the diode. Clearly there are now many possibilities, and these are illustrated in *Figure 28.11*. The differences between the parameters are primarily due to the variation in the minority carriers used in the arrangements. It is also seen that there are two capacitances to be considered, that of the diode itself and of the leakage to the substrate. The fastest diode is the emitter–base arrangement since both these are heavily doped with impurities. It is also possible to obtain zener diode action, for instance by connecting the emitter and collector together and biasing it positive to the base. The zener voltage is now approximately seven volts but it will vary appreciably from slice to slice due to changes in the process parameters.

A second type of diode used in monolithic circuits is the Schottky device. Essentially it consists of a metal area such as the aluminium used in the interconnection, in contact with a lightly doped n region. The excess of free electrons in the metal cause the

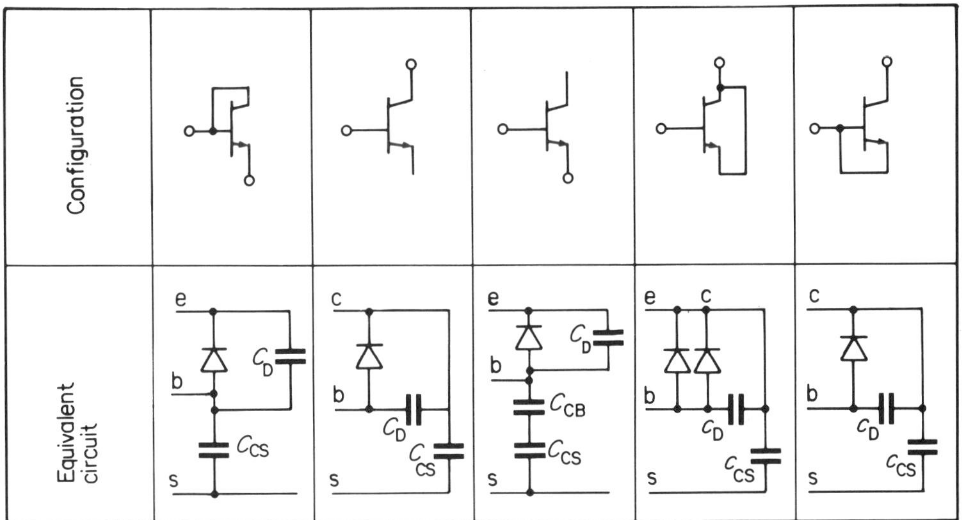

Voltage drop at 5 mA	0.7	0.7	0.7	0.7	0.7
Breakdown voltage (V)	7.0	50.0	7.0	7.0	50.0
Diode capacitance (pF)	0.5	0.8	0.5	1.0	0.8
Parasitic capacitance (pF)	3.0	3.0	1.0	3.0	3.0
Storage time (ns)	10.0	80.0	50.0	100.0	50.0

Figure 28.11 Characteristics of monolithic diodes obtained from transistors

formation of a Schottky (voltage) barrier at the junction. If the metal is now biased positive to the n region it conducts current. If it is negative it blocks, as in a traditional diode. The Schottky diode is very fast, with one nanosecond typical storage time, and has a low voltage drop when conducting, typically of about 0.3 volts. It is important to note however that the n region must be lightly doped. If it has a large impurity concentration then tunnelling takes place and the electrons in the two regions cause the formation of a good ohmic contact between the metal and the silicon. This is why an n^+ diffusion is usually made under the collector metal region.

28.2.2.3 Resistors

The resistance of a strip of diffused silicon is determined by its resistivity (sheet resistance) and dimensions. Clearly for high values of resistance the resistivity should be as large as possible. However, this implies a low impurity doping and the intrinsic carriers will now have a significant effect in reducing the temperature coefficient of the resistor. Dimensions are also limited. For instance the thinness of a line is determined by the accuracy tolerance of the masks, and the longer the resistor the greater the silicon area it occupies (long resistors are normally in the form of a zigzag pattern). Generally resistivities of 300 Ω per square are maximum for a workable temperature coefficient. The base diffusion of a transistor has a value of about 200 Ω per square and is the one most often used. The emitter has too low a resistivity and the collector too high. *Figure 28.12(a)* shows a diffused resistor in the base region (i.e. it is formed at the same time that the base of other transistors on the chip are diffused). The device is really a complex circuit, as shown in *Figure 28.12(b)*, consisting of a parasitic pnp transistor, leakage capacitor and a distributed resistor.

The absolute accuracy of a diffused resistor is limited to about 10% due primarily to the difficulty of controlling the diffusion process between slices. On a chip however the relative accuracy is about 3% being determined by tolerances in diffusion, masking and etching. *Figure 28.12(c)* shows the construction of a pinch resistor in which the emitter diffusion is used to reduce the effective width of the resistor.

28.2.2.4 Capacitors

Monolithic capacitors are not frequently used in integrated circuits since they are limited in range and performance. There are however two types available, as shown in *Figure 28.13*. The

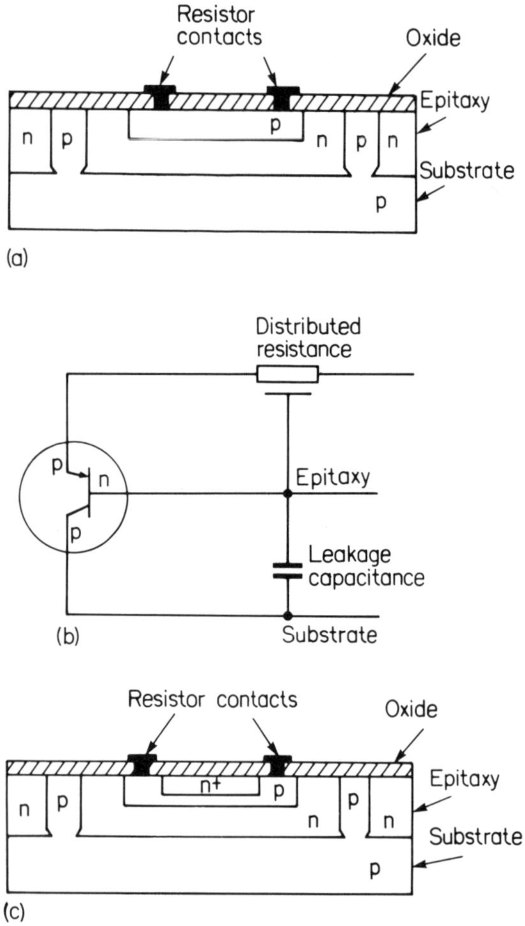

Figure 28.12 Monolithic resistor: (a) base diffused resistor; (b) equivalent circuit; and (c) pinch resistor

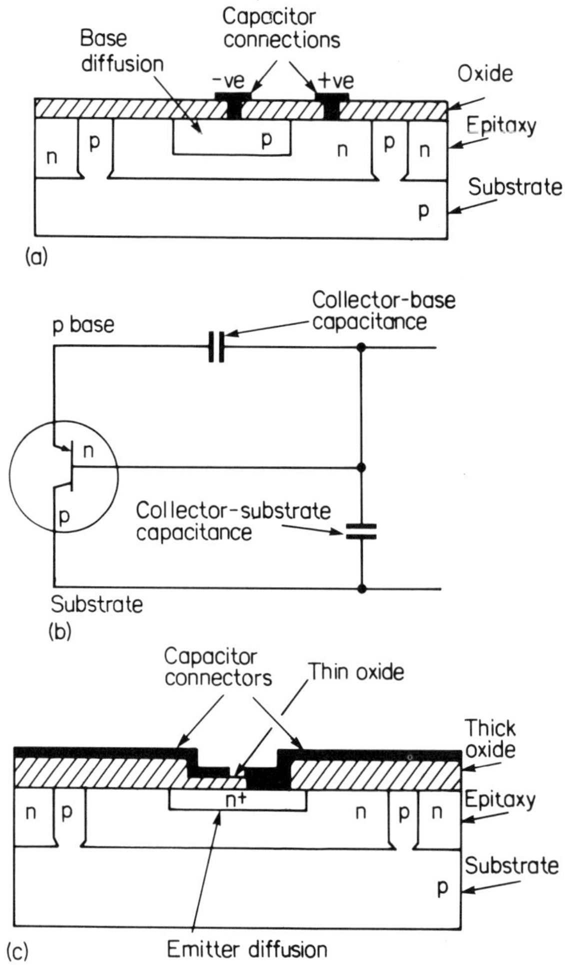

Figure 28.13 Monolithic capacitor: (a) diffused capacitor; (b) equivalent circuit; and (c) metal–oxide–silicon capacitor

diffused capacitor is a reverse biased pn junction, as explained in Section 28.2.1, formed by the collector–base or emitter–base diffusion of the transistor. The capacitance is proportional to the area of the junction and inversely proportional to the depletion thickness. This in turn varies with the resistivity of the layers so that for high impurity concentrations the capacitance is relatively large. However, the breakdown voltage of the capacitor is directly proportional to the resistivity so that one cannot simultaneously have a large capacitance value and breakdown voltage unless the junction size is made very big.

Diffused capacitors have very poor voltage coefficients since the depletion thickness is dependent on the bias voltage. Typically one can obtain a capacitance of about 1.2 nF mm^{-2} using a base resistivity of 200 Ω per square and zero bias. This value is halved at a bias of five volts. A second characteristic is low Q (equal to ratio of reactance to resistance) due to the higher resistivities involved. Generally all monolithic capacitors are inferior on this than discrete devices. *Figure 28.13(b)* shows an equivalent circuit of a diffused capacitor. In addition to the required collector–base capacitance there exists a parasitic collector–substrate capacitance and a pnp transistor. This requires the collector n region to be biased positive to cut it off.

A metal–oxide–silicon capacitor uses a thin layer of silicon oxide as the dielectric. One plate is the connecting metal and the other a heavily doped layer of silicon which is formed during the emitter diffusion. This capacitor has a lower leakage current, a much higher Q due to lower resistivitities, and is non-directional since either plate can be biased positively. The capacitance value can be varied between about 0.3 to 0.8 nF mm^{-2} and it is independent of the applied voltage provided this is below the breakdown of about 50 volts. Parasitic capacitors still exist but are lower than in the case of a diffused capacitor.

28.2.3 Bipolar circuit isolation

The standard diode isolation bipolar process suffers from many faults. The most serious of these are the leakage currents, the parasitic capacitance between collector and substrate giving low speeds, and the large size of the devices due to the fact that the epitaxy layer is relatively thick and the diffusion isolations spread considerably sideways as they move down. Many attempts have been made to devise new isolation techniques to overcome these disadvantages. They can all be divided into two groups: diffusion isolation and oxide isolation. In all cases these processes are backed by individual companies who claim various merits for their system over competitors. Therefore the company names are given in the discussion in the following sections.

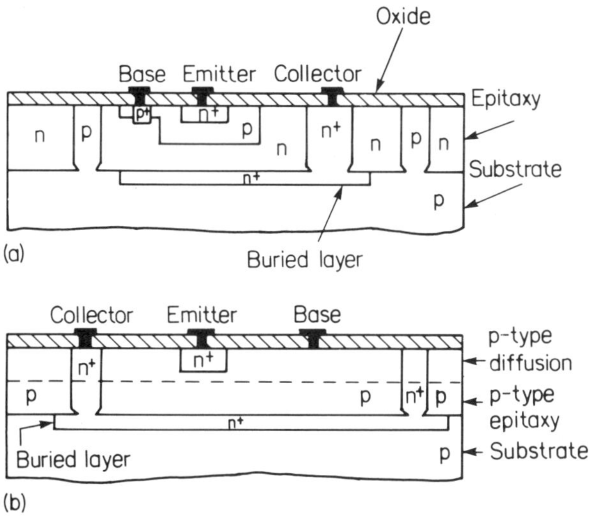

Figure 28.14 Diffusion isolated transistors: (a) Process III; and (b) Collector Diffusion Isolation (CDI)

28.2.3.1 Diffusion isolation

Two different processes are illustrated in *Figure 28.14*: Process III (Plessey Semiconductors Ltd) and Collector Diffusion Isolation (CDI) (Ferranti Semiconductors Ltd). Process III is currently the fastest system available for analogue and digital devices. It uses a very thin epitaxy layer and shallow diffusions which give narrow base regions and high frequency devices. The parasitic capacitance is also low. Diode isolation techniques are used. Process III uses an extra diffusion stage over the standard process since the collector diffusion is taken right down to the buried layer and must be made as a separate operation. This reduces the series resistance between collector junction and the buried layer, giving high frequency transistors and lower voltage drops. An extra p^+ diffusion is also used to reduce the base resistance. The thin epitaxy means that sideways diffusion is small and the devices can be closely packed.

In CDI the epitaxy is p doped and for the higher performance this may be followed by a non-mask, p^+ diffusion over the whole surface. This produces double diffused transistors different from the epitaxy base devices obtained with the usual diode isolation process. However, CDI still uses diode isolation between devices. It differs from other techniques in that the collector diffusion goes right down to the buried layer and completely surrounds the transistor. Isolation therefore occurs during the collector diffusion stage. The process therefore does not need a separate mask for the isolation diffusions. This, and the fact that the epitaxy is thin, gives very high packing densities on a chip. The thin epitaxy also gives this process, along with others which use thin layers, another advantage. Since the buried layer is close to the base region the injection efficiency of the collector is improved. This means that the transistors have high inverse gain, i.e. good gain when the roles of the collector and emitter are reversed. Not only does this result in low saturation resistance but also allows high switching speeds to be obtained without the need for gold doping. Both fast digital and linear circuits, which must not be gold doped, can therefore be fabricated on the same silicon chip.

28.2.3.2 Oxide isolation

Four oxide isolation processes are shown in *Figure 28.15*. These are Isoplanar (Fairchild Ltd), V-ATE (Raytheon Ltd), VIP (Motorola Ltd) and Polyplanar (Harris Ltd). All four processes depend on etching the silicon surface. This utilises the property that silicon nitride (Si_3N_4) when deposited on a silicon area prevents it from oxidising. Furthermore it is easy to remove the nitride by etching with phosphoric acid without affecting the silicon or oxide.

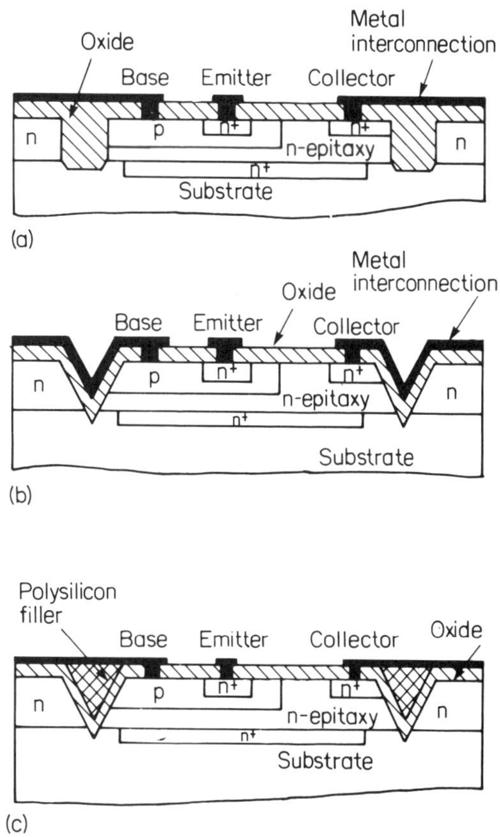

Figure 28.15 Oxide isolation techniques: (a) Isoplanar; (b) V-ATE; and (c) VIP and Polyplanar

In the Isoplanar process a buried n^+ diffusion and a thin epitaxy layer are grown as usual. This is followed by coating the silicon surface with nitride and then oxide which, using photolithographic techniques, selectively removes the nitride from the location where grooves are to be etched. The epitaxy is then etched and a thermally grown oxide used to fill up the grooves so that they are level with the surface. Base and emitter diffusions then complete the transistor. As seen in *Figure 28.15(a)* the packing density is much higher than diode isolation since the devices can be placed closer together. Furthermore since an oxide area prevents significant impurity diffusion there is a considerable amount of self-aligning effect in these transistors. The absence of diode isolation also reduces the junction capacitance giving higher operating speeds.

In the V-ATE process (vertical anisotropic etch) thin oxide and air are used as isolation between devices. It is important to use ⟨1 0 0⟩ silicon, which can be etched along the ⟨1 0 0⟩ plane some 30 to 40 times faster giving sloping edges of about 54° angle to the surface. Using nitride masking techniques V-grooves are cut into the epitaxy. The depth of the groove is determined by the initial mask width and can therefore be easily controlled. An oxide–nitride sandwich is then grown over the surface and a metallisation layer evaporated over this to form the interconnection patterns. Generally this process is similar to Isoplanar in performance but since the metallisation has to negotiate sharp bends in the V it is more likely to crack and lead to device failures.

The VIP (V isolation with polysilicon backfill) process is essentially very similar up to the point where the oxide–nitride sandwich is grown into the groove. The next stage consists in filling the grooves with polycrystaline silicon so as to slightly overfill them. The surface is then polished flat and finished with the usual diffusions and metallisations. Now however the metal does not need to pass over any sharp bends and the overall device is much more reliable.

The Polyplanar process looks very similar to VIP but differs in manufacturing technique. Principally only oxide is used in the grooves. It is claimed that this allows thicker epitaxy layers to be used giving transistors with higher voltage ratings.

28.3 Unipolar integrated circuits

28.3.1 Unipolar circuit fabrication

The manufacture of unipolar integrated circuits uses the same processes as bipolar systems. However, there is now no epitaxy layer and the most critical stage in the fabrication is the growth of the thin gate oxide.

For p channel transistors the starting material is n doped silicon which is then covered by a relatively thick oxide to a depth of about 1.5 μm. Holes are etched to define the source and drain and two p wells are next diffused in to a depth of between 2 and 4 μm. The diffusion takes place in an oxidising atmosphere so that a thin oxide layer forms over the whole surface. The oxide in the gate region is now removed and a thin, precisely controlled very pure oxide layer is regrown. This is the most important stage in the process since the gate oxide must be thin to give a low operating threshold voltage but it should also be pure, to maintain device stability, and be free from pinholes which would short the metal electrode to the silicon surface. The ohmic contact areas are then etched and metal is evaporated over the slice to a depth of 1 to 2 μm. The final masking stage defines the interconnection patterns as in bipolar integrated circuit systems.

Although bipolar and unipolar transistors derive their names from the number of charge types involved in their action, there are two other fundamental differences between them. Firstly the bipolar transistor operation occurs at its base region which is buried some distance under the silicon surface, i.e. in its bulk. Hence it is often called a 'bulk' transistor. In a unipolar transistor, on the other hand, conduction occurs due to channel formation on the silicon surface under the thin oxide. Hence it is a surface charge device. For this reason a unipolar device is very susceptible to contamination, and this initially presented many difficulties during manufacture. Most of these have now been overcome. The third difference between a bipolar and unipolar transistor is evident from an examination of their cross-sectional diagrams. Whereas the base metal connects directly to the silicon of a bipolar transistor, the gate of the unipolar device is insulated by a layer of silicon oxide. This means that the unipolar transistor has a high gate input impedance.

When unipolar devices are combined together in a silicon die a fourth difference between them and bipolar circuits becomes evident. *Figure 28.16(a)* shows two unipolar transistors on a die. Comparing this with *Figure 28.5* it is evident that the unipolar system is much less complex. Part of this is accounted for by the simpler transistor structure. Another reason however is that unipolar devices are self-isolating. There already exist surrounding areas of opposite polarity around each device so that the system resembles two reverse biased diodes shown in *Figure 28.16(b)*. By taking the substrate to the most positive potential in the system (most negative potential for n channel devices) isolation between transistors is obtained. This means that a unipolar system is usually capable of greater functional density than a bipolar system.

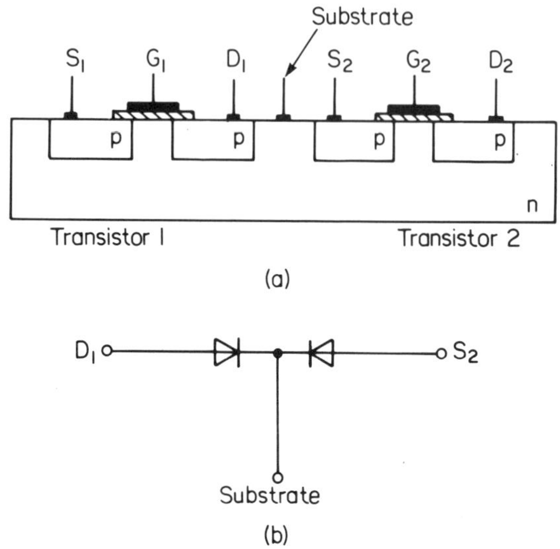

Figure 28.16 Self-isolation in unipolars: (a) two transistors on die; and (b) diode isolation of (a)

Figure 28.17 shows how the unipolar equivalent of *Figure 28.6* can be fabricated. The capacitor used in this instance consists of a dielectric (silicon oxide) separated by two conducting plates. The top plate is metal. The bottom plate is heavily doped silicon, which is a good conductor, and is eventually brought out to the metal interconnectors. The connection between the capacitor and transistor gate occurs round the side and is consequently not shown. Note also that thin silicon oxide is used as part of the capacitor and transistor gate whereas thick oxide forms an insulator between the metal interconnectors and the silicon surface, as in the bipolar circuit. It will be seen in the next section that the actual thickness of these two oxide layers is very important.

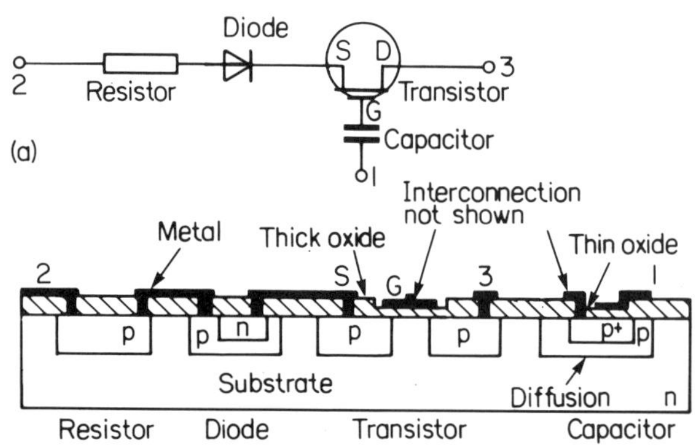

Figure 28.17 Unipolar integrated circuit: (a) functional diagram; and (b) chip construction

28.3.2 Developments in unipolars

Whereas the main battle in bipolar technology has been towards greater circuit density, unipolar systems have developed in two directions. In the early stages the prime objective was to reduce the operating voltages so that the unipolar circuit could be interfaced to traditional bipolar systems which run from 5 V supply rails. This objective having been achieved the efforts of the researchers turned towards making unipolar circuits faster. *Figure 28.18* illustrates only a few of the many systems which are currently in production.

Perhaps the easiest way of reducing the threshold voltage of a unipolar transistor, and hence reduce the operating voltage, is to use ⟨1 0 0⟩ silicon instead of the more usual ⟨1 1 1⟩ silicon. This gives a lower surface state charge and lowers the threshold from its original value of 3 to 5 V down to 1.5 to 3 V. However, the carrier mobility in ⟨1 0 0⟩ silicon is also lower so that the gain factor is seen to decrease giving lower operating speeds.

In an integrated circuit there are often metal interconnection tracks crossing two p regions, for p channel devices, where a transistor should not exist. Transistor action is prevented by ensuring that the oxide, called field oxide, is now thick enough so that the threshold voltage needed to turn the transistor on is greater than the system voltage. For a unipolar device the gate oxide is therefore made thin while the remaining areas are kept thick. Unfortunately this thinness is limited by processing difficulties since there must be no pin holes, whereas if the oxide is too thick say above 1.8 μm, then the aluminium conductors will have high steps which will cause them to crack. In ⟨1 1 1⟩ silicon devices the gate threshold is about 3 to 5 V and the field threshold 30 to 35 V. Going to ⟨1 0 0⟩ silicon reduces the gate threshold to 1.5 to 3 V but also lowers the field voltage to 15 to 20 V so that overall there is little benefit in this system and it is not frequently used.

A very popular unipolar structure is the metal–nitride–oxide–silicon (MNOS) device. The dielectric constant for silicon nitride is 7.5 compared to 3.9 for silicon so that when it is used as the gate material the threshold voltage is reduced to the 1.5 to 3 V range. Not only is the field threshold now not affected from its original 30 to 35 V value but the gain factor is doubled. Therefore the size of output transistors is reduced for the same current capability. A thin silicon oxide layer is generally used in these devices to prevent surface action between the silicon and the nitride. Pin holes are no longer a problem because of the relatively thick nitride layer so the oxide can be made very thin.

Another method of reducing threshold voltage is to use a polycrystalline silicon gate, as shown in *Figure 28.18(b)*. This reduces both the surface state charge (Q_{ss}) and the difference in work function (Q_{ms}) between the gate conductor, heavily doped

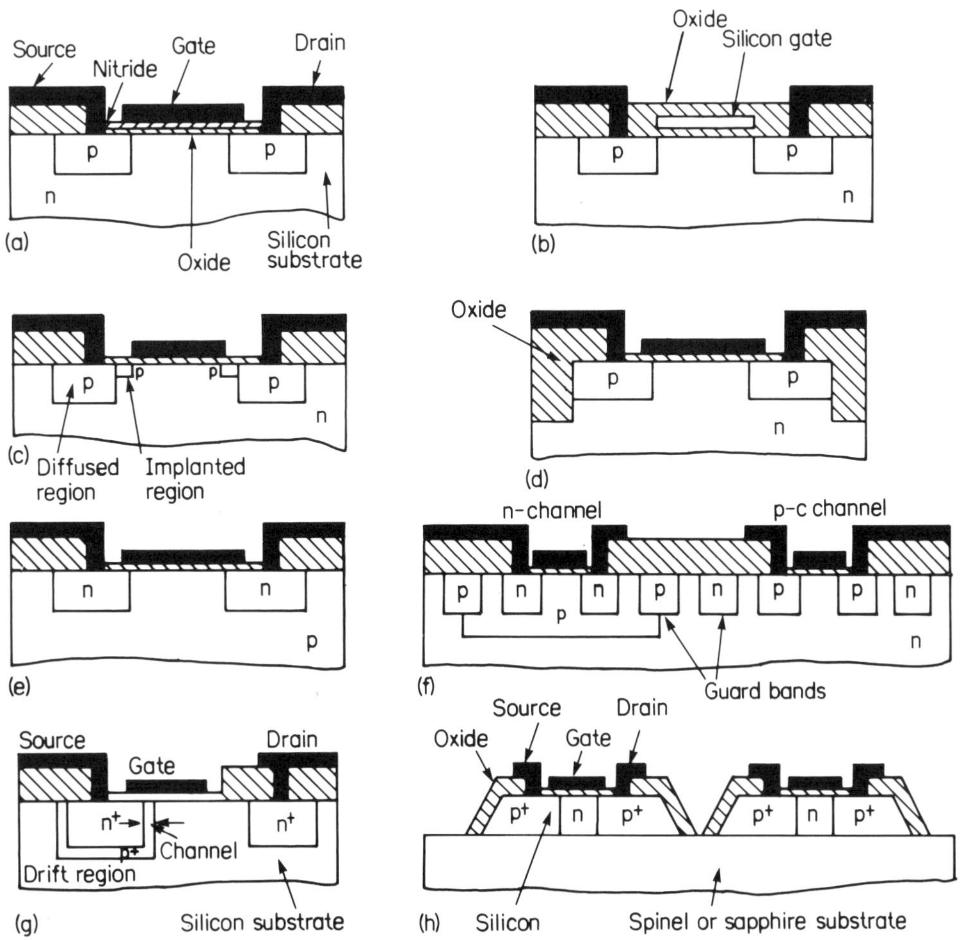

Figure 28.18 Unipolar processes: (a) MNOS; (b) silicon gate; (c) ion implant; (d) isoplanar; (e) NMOS; (f) CMOS; (g) DMOS; and (h) SOI

silicon, and the substrate. The threshold is again in the region 1.5 to 3 V. The doping concentration of the gate can be varied to give transistors with different thresholds if required in the circuit.

A very important advantage of the silicon gate structure is that it is a self-aligning process. To explain this let us look briefly at the production method used for the devices. Following the growth of the thick oxide a partial etch is used to form the thin oxide. A layer of silicon is then deposited over the entire surface, and oxide and silicon are then removed from the regions which are to form the source and drain. A non-mask diffusion is now carried out. Since the gate silicon and thick oxide will prevent impurity diffusion, the source and drain areas are very accurately defined. There is therefore very little overlap between the gate and these areas, giving low parasitic capacitance and fast operating speeds. For non-self-aligned processes it is necessary to build in tolerances for the fact that masks which define the diffusions and the gate metal cannot be exactly aligned. Therefore there can be considerable overlap between these areas. In the self-aligned processes the only overlap is that caused by the sideways diffusion under the gate, which always occurs.

There are other advantages to the silicon gate process. Reduced tolerances mean that devices can be packed closer together giving increased circuit density. The yield is high since the delicate gate oxide is covered up almost immediately by the silicon. The silicon gate can stand further high temperature processes, which the aluminium gate cannot, so that unipolar and bipolar devices can be built on the same chip if required. The silicon gate also gives a third interconnection layer, the others being the metal, which is still used to interface to the outside world, and diffused conductors on the silicon surface, so that circuit density is further increased. The major disadvantage of silicon gate is that it is a more complex process and can therefore be higher in cost.

Apart from diffusion it is also possible to introduce impurities into the silicon by ion implantation. In this the atoms, say boron for p type, are accelerated to a high energy so that they can penetrate the thin oxide and enter the silicon. The depth can be closely controlled by the accelerating voltage, and the concentration is adjustable by monitoring the boron ions. This is a low temperature process and since metal stops the impurity ions the system can be used to form devices with very little gate to source or drain overlap. As shown in *Figure 28.18(c)* the bulk of the source and drain are first diffused and the gate metal formed just short of these regions. High energy boron atoms are now implanted into the silicon to connect up to the original diffused areas without any overlap on the gate. This reduces the sizes of the devices and the parasitic capacitance even more than for silicon gate transistors since implanted ions do not diffuse sideways.

Ion implantation can also be used to vary the concentration of impurities in the channel region of selected transistors, metal being absent. This means that some devices can be made to operate in a depletion mode. This gives several advantages such as increased speed and smaller devices. It is also possible to implant resistor regions accurately so that a wider range of devices with a 2% absolute tolerance is attainable. The ion implantation process is also easily controllable and very suited to automation.

As in bipolar circuits it is possible to use oxide isolation in unipolar devices. Silicon nitride is used for selective oxidation giving the structure shown in *Figure 28.18(d)* for the Isoplanar process. Although the field oxide is relatively thick the metal steps are of much less height. This prevents cracks and allows the masks to be placed closer to the silicon surface resulting in better resolution and smaller devices—hence reduced capacitance and higher speeds. It is of course also possible to make n channel and complementary transistors (described later in this section) by this process. Two other processes resemble Isoplanar. The self-aligned thick oxide (SATO) uses nitride to define the gate and provide self-alignment for the source and drain diffusions and to grow the thick oxide. ⟨1 0 0⟩ silicon is used so that the thresholds are low, but a thick field oxide prevents parasitic transistor action. The second process, known commercially as Planox, has a thin nitride layer above the gate oxide, for lower thresholds, and uses ⟨1 1 1⟩ silicon.

N channel unipolar transistors (*Figure 28.18(e)*) have the advantage over p channel in that the electrons have twice the mobility of holes. This gives a higher gain factor and smaller transistors, with less capacitance and higher speeds. The thresholds are also lower giving much smaller operating voltages. However, initially the n channel process was difficult to make. This was due to the fact that in silicon most of the mobile contaminants is positively charged. The positive gate potential used in n channel transistors draws these to the silicon–oxide interface where they act so as to reduce threshold voltages and cause devices to behave unpredictably. In a p channel transistor the negative gate bias moves the contaminants to the aluminium–oxide interface where they have less influence on threshold voltages. However, most of the n channel processing problems have now been overcome and the device has replaced PMOS as the most frequently used unipolar transistor.

P channel and n channel transistors can be fabricated on the same slice and connected together to give complementary MOS (CMOS) devices. These have very low standby power dissipation, high operating speeds and many other circuit advantages. The disadvantage is larger size and greater processing complexity. *Figure 28.18(f)* illustrates the device. Guard band diffusions surround groups of like devices preventing leakage between them. Like NMOS the CMOS process is also rapidly establishing itself as an industry standard for many applications.

A unipolar process which is capable of very high operating speeds is DMOS (double diffused MOS). This is shown in *Figure 28.18(g)*. A very narrow gate channel is established by double diffusion of a p^+ and n^+ region into a low impurity drift region of the substrate. This drift region gives a very fast switching speed and together with the narrow gate enables operating frequencies well into several hundred megahertz. Furthermore it is capable of making high voltage devices if required. However, the process is relatively expensive since an extra diffusion must be used and all diffusions have to be precisely controlled.

Figure 28.18(h) shows the silicon on insulated (SOI) structure. In this a sapphire or spinel substrate, both of which have crystal structures very similar to silicon, is heated in an atmosphere of silicon bearing gasses. The silicon is deposited on the insulating surface in a thin layer which is an extension of the original substrate. It is now possible to etch the silicon layer so as to form isolated islands into which transistors can be diffused and connected together. Clearly the leakage and parasitic capacitance of such a system is very low so that its operating speed is comparable to that of bipolar circuits. Like other dielectrically isolated systems SOI is also very resistant to radiation. A more complex system, but one which is currently in production, uses etchants to remove silicon from the back of a slice and then fills this with glass. Clearly both bipolar and unipolar devices can be built in SOI. However, it is still not possible to grow the silicon as an exact continuation of the insulator surface. Therefore the process is best suited to lateral devices which operate on the surface of the silicon, such as in unipolar transistors.

28.4 Integrated circuit packaging

28.4.1 Die bonding

Die bonding refers to the process in which the semiconductor chip is bonded down to the thick film substrate or onto the base of its final package. This bond may be made to a metal track, when the substrate of the chip is part of the system, or it may be made to an insulating area. Therefore the bond requirement could be that of good electrical conduction or of electrical isolation. Other requirements are that it should provide good mechanical strength, good heat conduction properties and should not create any unwanted pn junctions.

There are generally three die bonding methods in use. These are eutectic alloying, soft soldering and methods using plastic adhesives.

28.4.1.1 Eutetic alloying

An eutectic alloy has a much lower melting point than that of any of its individual constituents. For die bonding a gold–silicon alloy is most commonly used. The area to which the die is to be attached is plated with gold and the silicon chip placed on it. The substrate is heated to between 400°C and 450°C and at this temperature gold and silicon form an eutectic alloy. The substrate is now removed from the heat and allowed to cool to give a strong alloy bond between the chip and the substrate. In a practical process the chip is normally held in a chuck and scrubbed ultrasonically on the bonding area in order to break down any oxides which may have formed on the silicon surface and so expose the silicon area (*Figure 28.19*). This usually takes place in an atmosphere of nitrogen since this discourages further oxide formation.

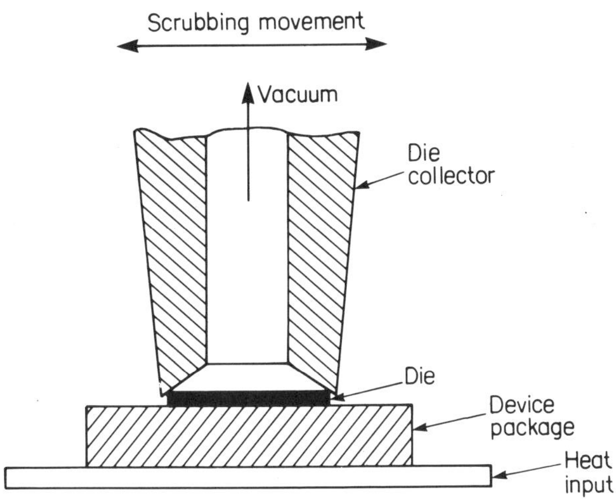

Figure 28.19 Eutectic die bonding

The gold silicon eutectic alloy melts at 370°C (melting point of gold is 1063°C and of silicon 1404°C). This is high enough to enable the bond to withstand subsequent processing steps, in particular the soldering of the completed device into a printed circuit board. The joint formed has excellent adhesion properties and heat conductivity. It is also capable of accommodating a slight amount of deformation. Its disadvantages are: (i) that it requires a metallisation area on the substrate; (ii) that the high temperatures involved can affect any thick film resistors which have been fired on the same substrate; and (iii) that due to the manual handling involved in the scrubbing process there is loss of yield caused by chips cracking or scratching. The relatively high gold content of this alloy also makes it expensive.

In spite of its limitations the gold–silicon eutectic alloy is very commonly used for die bonding. Either plain dice or devices with gold plated backing may be used. The gold plating is expensive but results in a more tenacious bond. Apart from gold other gold alloys such as gold–silicon and gold–germanium are used in this process. Gold–gallium and gold–indium can be employed with p type materials where it is required either to convert the die to a p type or to prevent n type formation. Similarly gold–antimony alloys can be used with n type dice. The melting points of all these alloys are in the region 340°C to 450°C. If a low temperature operation is required then gold–tin alloys (250°C) can be used. This also gives a very high conductivity between the chip and substrate.

28.4.1.2 Soft soldering

Lead–tin soft solders, of the type used in the assembly of printed circuit boards, can be employed to solder the silicon chip to the substrate. It is now necessary to ensure that both the pad area and the dice are gold plated for solderability. The resulting bond has good all round properties and requires a relatively low temperature, in the range 200°C to 300°C, for formation. The low temperature has the advantage of not affecting fired resistors on a thick film substrate. It does however mean that care must be taken during subsequent process steps not to exceed this value. The bond is also weak under thermal cycling conditions. It has the advantage that chips can be removed, with relative ease, from the substrate for replacement. Generally soldering is not used apart for large power chips.

28.4.1.3 Adhesives

Adhesives are used to stick the chip to the substrate when a low process temperature, below 200°C, is necessary. Silicon or epoxy adhesives may be used and the chip may be plain or gold backed. A measured amount of the adhesive is dispensed onto the bond area and the chip pressed on to it. Both conducting and non-conducting adhesives are available depending on whether or not an electrical contact is required to the bond area. The assembly is baked in an oven to cure the adhesive. This method of chip bonding has several advantages such as a strong bond formation, a low process temperature, absence of mechanical handling of the type required for eutectic systems and a thermal coefficient of expansion which can be readily controlled by modifying the composition of the adhesive. Its disadvantages are again the low temperature which subsequent steps must be limited to, the relatively high contact resistance of the bond, and the long curing times required. In addition there is always the risk of the adhesive contaminating the silicon dice.

28.4.2 Wire bonding

Having fixed the silicon dice to the substrate, wire bonding techniques must be used to electrically connect the metallisation pads on the dice to the hybrid conductor tracks or to the package leads. Thin gold or aluminium wires are most frequently used for this and the connections are made by thermocompression or ultrasonic bonding methods.

28.4.2.1 Thermocompression bonding

Thermocompression bonding depends on the fact that good molecular joints can be made between metals which are heated to well below their melting temperatures so long as the materials are of small dimension and sufficient pressure is applied to the junction area. Three methods are used to provide this force.

28.4.2.2 Ball or nail head bonding

The basic steps involved in this process are illustrated in *Figure 28.20*. The fine gold wire, about 25 μm in diameter, is led through a heated tungsten carbide capillary tube and its tip melted by a hydrogen flame. This causes meltback to occur which results in a

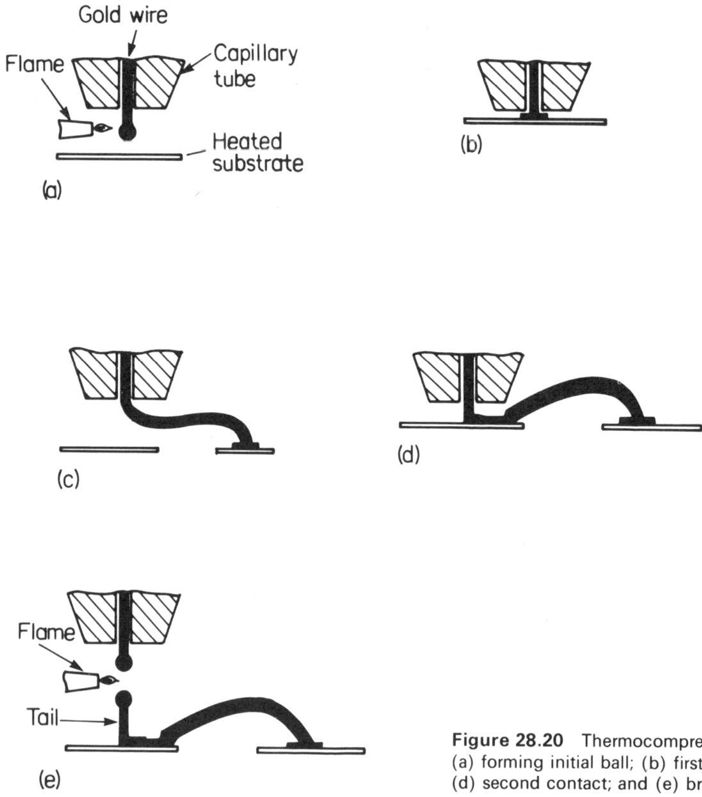

Figure 28.20 Thermocompression ball bonding of wire: (a) forming initial ball; (b) first contact; (c) move to second pad; (d) second contact; and (e) break wire and reform ball

ball of about two to three times the original wire diameter. While the ball is still hot the wire is pulled so that the ball is held tight against the end of the capillary tip and this is then brought down hard onto the contact pad. This pad is initially maintained at a temperature of about 350°C so that the effect of heat and pressure results in a strong molecular bond with the wire. The capillary tube is now moved to the second pad, and the wire is fed through during transit. Once above the pad the tube is again brought down hard on the gold wire to deform it and form the second joint. The tube is now lifted and the hydrogen flame used to cut the wire and form a ball ready for the second wire bond. This operation leaves a pigtail of gold wire on the substrate which must be removed either manually with a tweezer or by automatic machinery.

Ball bonding results in a strong bond and enables fairly high production throughputs to be obtained. The bonds are however relatively large in size and the requirement of having a heated substrate can lead to deterioration of previously fired thick film devices. The operation must also be carried out in a clean environment since the capillary tube is very fine and the wires must move through it freely. Ball bonding cannot generally be used with aluminium wire since it does not form a ball when melted.

28.4.2.3 Stitch bonding

In this system, shown in *Figure 28.21*, the bond wire is again fed through a thin capillary tube. In the starting position this wire is hooked under the edge of the tube and contact is made by pressure on the first pad. As before the substrate is heated to about 350°C. The capillary tube then moves to the second pad and contact is made by pressure from the edge of this tube. Finally the wire is broken off by thermal shock or with a mechanical cutter. In either case there is no pigtail formation and the remaining wire is hooked under the tube and ready for the next bond. Stitch bonding is particularly suitable for making successive connections in stitch fashion without breaking the wire. It is capable of moving in any direction as was also the case with ball bonding. Although both gold and aluminium wires may be used it is difficult by this method to break through the surface aluminium oxides and make reliable joints. Hence it is primarily used for gold wire bonding. Generally this method is not much used now, having given way to ultrasonic bonding techniques.

28.4.2.4 Wedge bonding

Although similar in principle to stitch bonding this system uses a capillary tube to feed out the wire, and a separate heated wedge shaped tungsten carbide tool to provide the striking force which deforms the wire and makes the join to the pad. The system is illustrated in *Figure 28.22*. After the join the wire is pulled off at the bond and no pigtail is left. The substrate must once again be heated to about 350°C and gold wire is primarily used. Wedge bonding is slower than any of the other wiring methods since two independent tools need to be positioned accurately. It is however capable of making connection to much finer wires, down to a few micrometres, and to contact pads less than 20 μm square.

28.4.2.5 Ultrasonic bonding

Although ball bonding is commonly used for interconnections in hybrid circuits, ultrasonic bonding is more popular for packaging monolithic devices. The system has the merit of requiring no direct heat to either the capillary tube through which the wire is

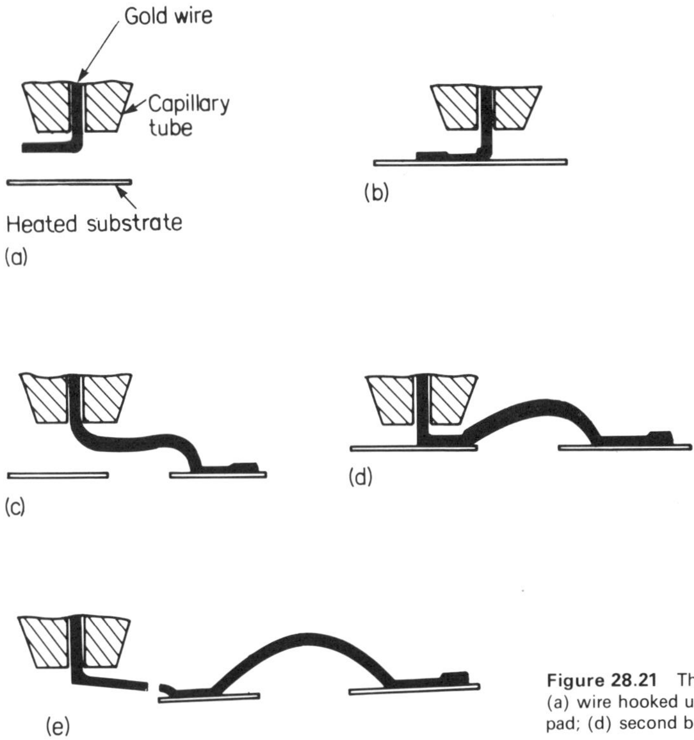

Figure 28.21 Thermocompression stitch bonding of wire: (a) wire hooked under tube; (b) first bond; (c) move to second pad; (d) second bond; and (e) break wire by mechanical shock

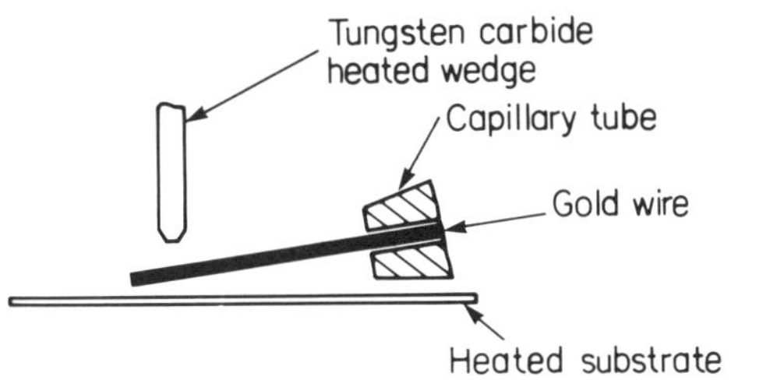

Figure 28.22 Thermocompression wedge bonding of wire

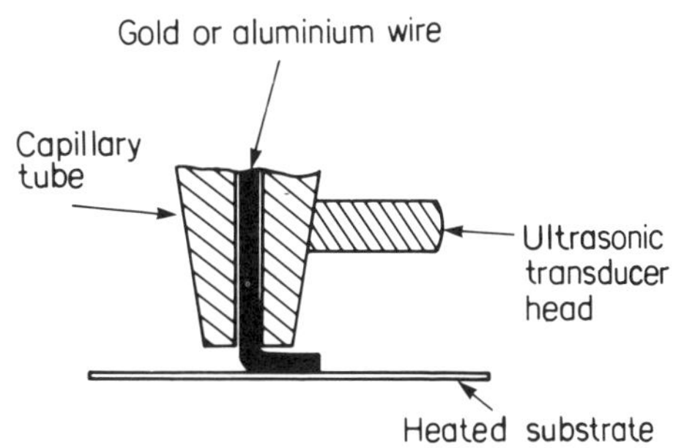

Figure 28.23 Ultrasonic bonding of wire

fed, or to the substrate. It is primarily used with aluminium wires and is shown in *Figure 28.23*. The wire is pressed down onto the contact pad and the capillary tube is vibrated at an ultrasonic frequency in the region of 20 to 60 kHz. This high frequency scrubbing action microscopically grinds down any unevenness between the two metal surfaces, removes surface oxides and results in a strong molecular joint. The absence of heat further deters the formation of aluminium oxides. It is important that the downward pressure of the tube and the chip clamping force are carefully controlled so as to optimise the dynamic stress pattern formed at the wire–pad interface by the sideways ultrasonic movement. The bond energy must exceed a given minimum but too large a value results in damage to the silicon dice.

Ultrasonic bonding tools are restricted in their movement so that constant realignment of the substrate may be necessary. Therefore they are less popular for use in hybrid circuits than ball bonding systems. However, they are generally capable of some 50% greater production rates than ball bonding and can form bonds which are two or three times more reliable.

28.4.3 Film carrier bonding

The film carrier technique uses a continuous film of polyimide which has copper lead frames mounted on its surface. The whole system resembles a film spool. A window in the film allows the inner leads of the frame to protrude. It is to these fingers that the silicon die is bonded. After bonding the lead frame and die are wound onto a second spool for storage until required. The lead frames are then removed from the film and connected into the required circuit or package by thermocompression, ultrasonic or solder bonding.

Figure 28.24 shows the steps used in bonding the lead frame to the semiconductor die. The inner leads are lowered onto the die in (a), the die having bumps as for flip chip applications and being secured to the die carrier by wax. An inert gas now flushes away the ambient air and the bonding tool descends onto the leads. A current pulse heats the tool in (b), forming the bond and melting the wax to release the die as in (c). The film carrier moves to the right as in (d) and the die carrier to the left so that a new lead

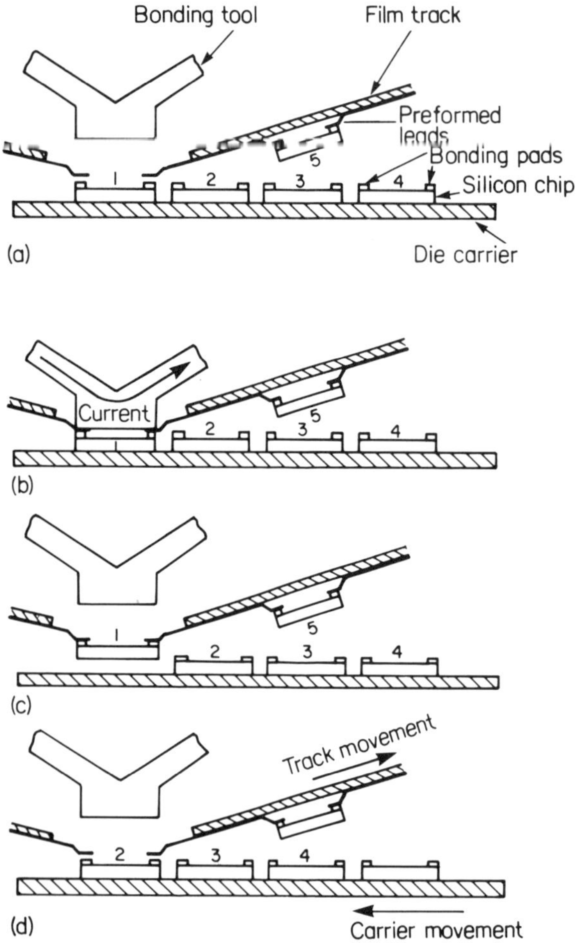

Figure 28.24 Film carrier bonding: (a) leads positioned for bond; (b) bonding operation; (c) bonded chip lifted off carrier; and (d) next chip and leads in position

frame and die are in position for another bond operation. Since all leads are bonded in one step this method is some ten times faster than wire bonding systems.

28.4.4 Plastic packages

The basic principle involved in a plastic package is to assemble the circuit on a substrate or metal lead frame and then to mould the entire structure, apart from the leads, in plastic to form the body of the device. The most commonly used plastic materials are epoxy, phenolic and silicone resins of which epoxy is the most popular. Several requirements are placed on the plastic material. It should adhere well to the lead frame so that it prevents moisture from creeping in along the package legs. The material must also have a thermal coefficient of expansion which is matched to the rest of the circuit in order to prevent stresses being set up in it which could damage the leads or the thin bonding wires. It is of course also important that the material does not contain or release any impurities which might contaminate the enclosed silicon.

Epoxy has relatively good characteristics for plastic packages. It is low in cost, chemically stable and capable of keeping out the effects of considerably hostile environments. In addition the material is capable of high mechanical strength, good adhesion properties and is an excellent electrical insulator. Silicone compounds have a much higher resistance to heat than epoxy materials. However, they are structurally weaker and are attacked by some chemicals such as salts. Phenolics are structurally very strong and relatively economical to use. But they release water vapour during curing and can contain harmful chemical impurities which attack the enclosed circuit.

Plastic packages are low cost, especially when compared to hermetically sealed devices. They are also ideally suited to volume production techniques. However, their resistance to moisture and environmental contaminants is not as good and they can be damaged under thermal cycling conditions. Plastic packages also have lower heat dissipation compared to other types, especially metal packages. Heat dissipation properties can be considerably improved by wrapping the circuit in a metal surround before plastic encapsulation and in some cases bringing out part of this metal for bolting onto an external heatsink. Plastic packages are also usually characterised for operation over the industrial temperature range of 0°C to 70°C while hermetic devices can cover the full military range from −55°C to 125°C. However, for normal industrial use plastic devices are very suitable and with chips which have been passivated with a glass or silicon nitride layer these devices can be made to operate in fairly hostile environments.

Three methods are in use for making plastic packages.

28.4.4.1 Liquid casting

In this method a mould of metal or plastic material is used into which is placed the circuit to be packaged. A premixed solution of resin and catalyst is then poured into the mould so as to completely cover the circuit but to leave a part of its lead exposed. The mould is then baked in an oven to cure the resin. After setting the moulds may be left on the devices as permanent packages or removed and reused. It is possible in this method of moulding to introduce a degree of automation in the mixing and dispension of the encapsulating liquid. However, the method is generally relatively slow and requires long curing times. It is best suited to small batch production of a few thousand devices per week or where a long bake period is essential to fully de-gas the resin and produce higher reliability.

28.4.4.2 Transfer moulding

This is a suitable for making a wide variety of packages such as dual-in-lines and flat packs. It places no restrictions on the positions of the package leads and is capable of production rates of many thousand devices per hour. The assemblies which are to be packaged are put into the bottom half of a multicavity mould, each cavity corresponding to the position of a device. The mould is closed in a transfer moulding machine and the encapsulant material injected in fine pellet form from a nozzle. The pressure at which this injection takes place is closely controlled to prevent damage to wire bonds. Under pressure and temperature the pellets melt and flow in channels within the mould so as to fill the device cavities. The resin is now cured, while still in the mould, by the applied heat and pressure. This operation takes between 1 and 3 minutes after which the mould is opened and the individual circuits separated from the lead frame assembly. If required further curing can now take place in an oven. Transfer moulding has very low operating costs although the initial price of the equipment can be high.

28.4.4.3 Fluid bed encapsulation (or conformal coating)

In this method of encapsulation only a thin layer of resin is coated over the surface and acts as the package. The coating provides a degree of protection against mechanical damage and it conforms to the outline of the encapsulated device. The system is very popularly used for low cost hybrid circuits. The substrate is heated and dipped into a bed containing epoxy powder which is fluidised by having air blown through a porous bottom in the container. On contact with the heated substrate the powder melts and adheres to it. Several dippings are used to build up a sufficiently thick layer of epoxy over the whole surface, but this is usually done using automatic equipment with many substrates undergoing simultaneous coating. After dipping the resin is cured in an oven. This method is suitable for use with substrates having leads on one side only.

Conformal coating can also be done by another encapsulation method in which the completed substrates are dipped into a liquid bed of a thixotropic resin. Many more dippings are required before a covering of sufficient thickness is built up over the substrate. The curing period is also much longer than that required with fluid bed systems and reproducibility is poorer.

28.4.5 Hermetic packages

A semiconductor chip is sensitive to the presence of contaminants such as sodium ions, hydrogen, oxygen and water vapour. These reduce the collector–base breakdown voltage of transistors, increase the leakage current and parasitic capacitance, reduce current gain, and attack the chip metallisation and the wire bonds. Hermetic sealing aims to minimise these effects by first removing the contaminants from the package, usually by heating it to about 250°C, and then sealing it in the presence of an inert gas. During sealing the package temperature is kept to as low a value as practical.

A hermetic package usually consists of a base, a body, leads and a cover. Obtaining a seal between metal surfaces is relatively easy and is done by several methods such as welding, brazing and glass frit sealing. Glass to glass seals also present little problem since both surfaces can be fused at high temperatures or cemented by means of an interconnecting lower melting point glass. The problem arises when glass to metal seals need to be made, such as when metal leads pass through a glass package. The difficulties generally arise due to the unequal heat conduction and thermal coefficients of expansion of the two materials, which make the seal unreliable under temperature cycling conditions. Two methods are generally used to produce glass to metal seals. In the first method a thin oxide layer is grown on the metal surface prior to coating with glass. The surface of this oxide is dissolved into the glass and results in a smooth transition from metal to oxide to glass. This results in a good seal since the oxide in effect acts as a intermediate buffer. The second sealing technique uses a solder glass seal. This is a special composition low melting point glass which is coated onto the two surfaces to be sealed. On heating the interconnecting glass melts and wets the two surfaces, forming a good seal on cooling.

Three types of materials are used for hermetic packages: metals, ceramics and glasses. Of the many metals the most popular is a nickel–iron–cobalt alloy called kovar. It has a relatively low electrical and thermal conductivities and is compatible with standard sealing glass. Ceramics make excellent hermetic packages and are very commonly used. They have high thermal conduction and a thermal coefficient of expansion which is compatible with that of glass. Many different ceramics can be used, the most popular being alumina having a purity between 75% and 99.5%. For high power dissipation beryllia is most common although other ceramics in production are steatite, forsterite, titanate and zircon. The thermal coefficients of all these materials are close to that of metal. Glass packages are cheap since they can give good inexpensive seals. They have poor conductivity and are generally not suitable for anything apart from very low power circuits. Hybrid packages using ceramic bases and glass walls have been used for higher power dissipations.

28.4.6 Package configurations

So many different sizes and shapes of packages have been used in the manufacture of integrated circuits that it would be folly to attempt a description of them all in this short section. Instead only a few of the more popular packages are considered here.

The TO type of package was a natural development from the transistor case. Various sizes are in use such as TO-3, TO-5, TO-8, TO-99 and TO-100. The number of pins in these packages generally varies between 8 and 14. *Figure 28.25* shows the construction of one type of TO package. The base is a gold plated header made from kovar. The leads pass into the header via glass to metal seals. The die is attached by means of a gold–silicon eutectic solder. The can or cover is usually also made from kovar and is welded onto the header flange.

Flat pack encapsulations also come in many varieties. *Figure 28.26* shows typical structures. The die is bonded to the base by eutectic solder. The gold plated leads are embedded in a glass frame and the lid is sealed with a low temperature glass frit. A metal flat pack uses kovar for most parts except for the walls

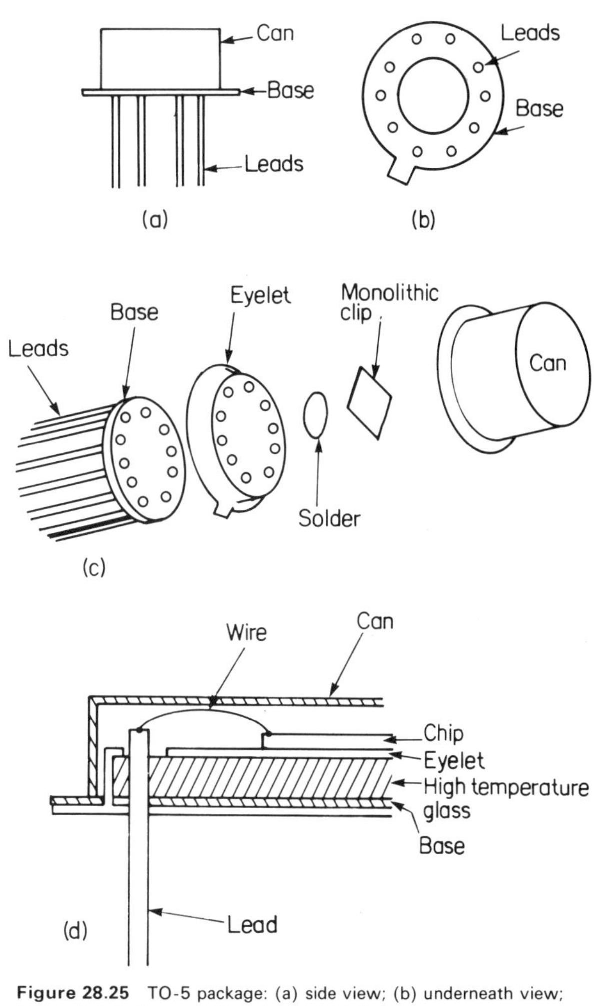

Figure 28.25 TO-5 package: (a) side view; (b) underneath view; (c) exploded view; and (d) cross-sectional view

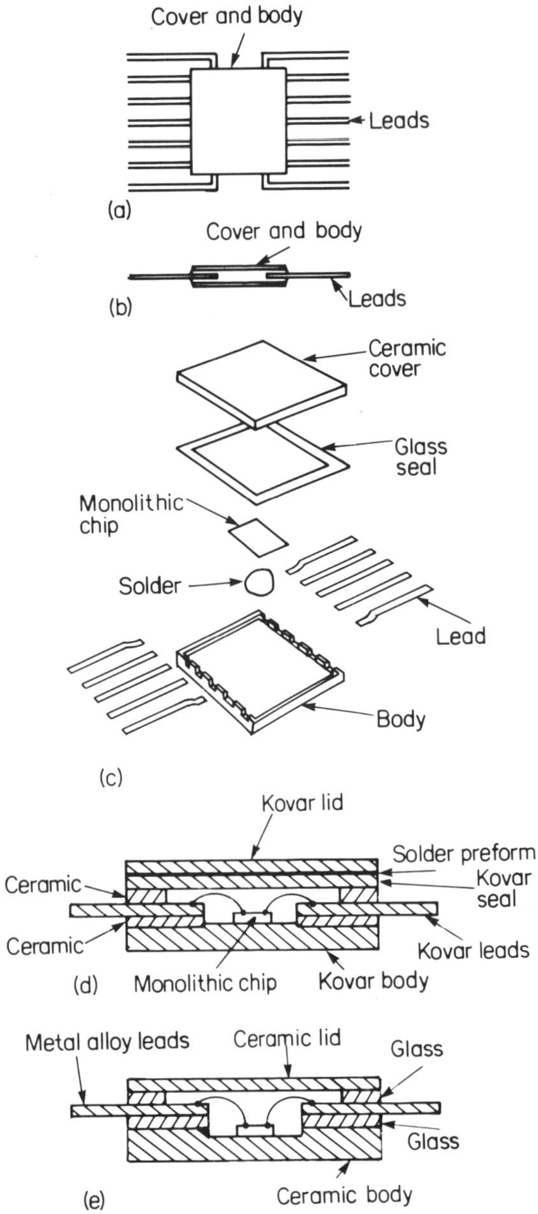

Figure 28.26 Flat package: (a) top view of 14-lead package; (b) side view; (c) exploded view of 10-lead ceramic package; (d) cross section of metal package; and (e) cross section of ceramic package

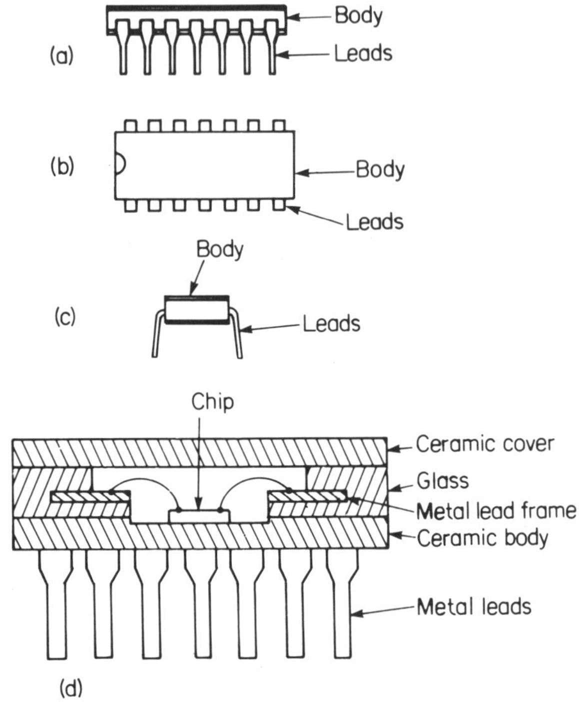

Figure 28.27 Dual-in-line package: (a) side view; (b) top view; (c) end view; and (d) cross section of ceramic package

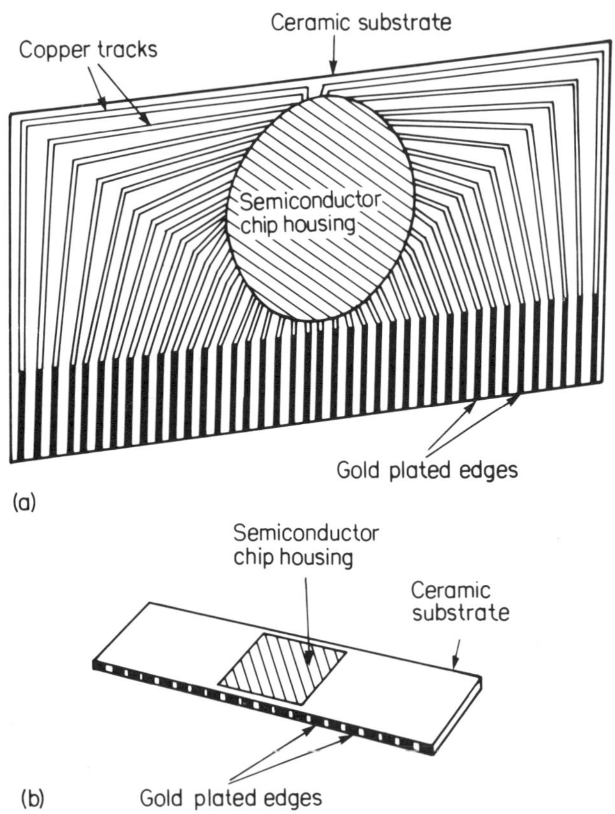

Figure 28.28 LSI packages: (a) edge mount; and (b) leadless

between the ring and base which are glass. The leads pass through this. A glass package is usually made from borosilicate and consists of a one-piece base and ring assembly in which the kovar leads are sealed. This package has good thermal and mechanical properties. The lid can be made of glass, ceramic or metal and is attached by a low melting point sealing glass. Ceramic flat packs are similar to metal devices. The base, ring and lid are now usually made from alumina or, for higher power dissipation, from beryllia.

The dual-in-line package has become very popular especially for monolithic integrated circuits, since it is convenient to handle and can be readily adapted for automatic insertion into printed circuit boards. *Figure 28.27* shows a typical device, which can be made from metal or ceramic. The chip is placed in a cavity and bonded by glass frit. The leads also pass through glass frit seals in the package. The ceramic lid is initially metallised and then brazed or solder sealed to the body, or a glass frit can again be used to connect the two together. Plastic dual-in-line packages do not have a cavity for the chip. It is essential in these instances that the moulded package completely surrounds the die to protect it from the hostile environment, and all seals between the body and the leads are reliable. The overall outlines of these packages are essentially as in *Figure 28.27*.

Packages used to house LSI devices can present a problem. These chips are relatively large and generally have many outputs, typically about 40. The most popular configuration is at present still a dual-in-line, but due to the necessity of maintaining 0.25 cm spacing between pin centres the package tends to be relatively bulky. It is difficult to align the pins and to keep the device in position during insertion into a board or socket. The pins also tend to be fragile and break, and unsoldering a board to remove a device for replacement is often a very trickly operation. To overcome these limitations several different types of packages are in use. The edge mount package, shown in *Figure 28.28*, is similar to a miniature printed card. The chip is housed in a cavity which is sealed. Connections are made to the chip and the thick film conductor tracks which are printed onto the ceramic substrate. The tracks have gold plated fingers which plug into an edge connector. Although such a package can be quickly inserted and removed from its holder the fingers tend to wear with each operation. It is also unsuitable for many external connections since the long ceramic substrate has a tendency to bow. Furthermore the assembly is relatively expensive. An alternative approach is the leadless package. The chip is once again sealed in a central cavity and connected to printed tracks. The tracks are brought out to bumps (not pins) on either the side or bottom of the package. When in use the package is placed in a receptacle which has conducting pads located near the bumps on the package. Once the cover of the receptacle is closed it squeezes the package bumps and pads into close contact, connecting the chip into circuit. This package is claimed to have several advantages such as low cost, freedom from pin damage and ease of replacement. It is however still relatively expensive and not widely used in industry.

Further reading

AHMED, A., 'Microlithography', *New Electronics*, November 25 (1980)
AHMED, A., 'Chemical vapour deposition', *New Electronics*, December 9 (1980)
AHMED, A., 'VLSI—scaling down', *New Electronics*, March 24 (1981)
ANCEAU, F. and REIS, R. A., 'Complex integrated circuit design strategy', *IEEE J. of Solid-State Circuits*, **SC-17**, No. 3 June, 459–464 (1982)
BAILEY, B., 'Ceramic chip carriers—a new standard in packaging', *Electron. Eng.*, August (1980)
BURGGRAAF, P. S., 'GaAs bulk-crystal growth technology', *Semiconductor International*, June (1982)
BURGGRAAF, P. S., 'E-beam lithography: a standard tool?', *Semiconductor International*, September (1982)
BURGGRAAF, P. S., 'Magnetron sputtering systems', *Semiconductor International*, October (1982)
DOOLEY, A., 'Choosing a reliable IC socket', *New Electronics*, January 27 (1981)
EIDSON, J. C., HAASE, W. C. and SCUDDER, R. K., 'A precision high-speed electron beam lithography system', *Hewlett-Packard J.*, May (1981)
EL REFAIE, M., 'Chip-package substrate cushions dense, high-speed circuitries', *Electronics*, July 14 (1982)
EL REFAIE, M., 'Interconnect substrate for advanced electronic systems', *Electron. Eng.*, September (1982)
FARRELL, J. S., 'Trends towards alternative packaging', *Electron. Eng.*, Mid-September (1981)
GARRETTSON, G. A. and NEUKERMANS, A. P., 'X-ray lithography', *Hewlett-Packard J.*, August (1982)
GROSS, D. R., 'Packaging complex ICs', *New Electronics*, June 15 (1982)
GROSS, D. R., 'Trends in semiconductor packaging', *Electronics & Power*, October (1982)
GROSSMAN, M., 'E-beams, new processes write a powerful legacy', *Electronic Design*, June 7 (1980)
HAYASAKA, A. and TAMAKI, Y., 'U-groove isolation technology', *JEE*, August (1982)
HAYDAMACK, W. J. and GRIFFIN, D. J., 'VLSI design strategies and tools', *Hewlett-Packard J.*, June (1981)
HESLOP, C. J., 'Reactive plasma processing in IC manufacture', *Electronic Production*, February (1980)
JONAS, A. W. and GARNER, L. E., 'Leadless chip carriers for LSI packaging', *Electronic Product Design*, April (1981)
JONES, G. A. C. and AHMED, H., 'Electron-beam lithography—a new approach to registration', *New Electronics*, August 12 (1980)
LAND, I. G., 'Mounting leadless chip carriers', *New Electronics*, January 26 (1982)
LOESCH, W., 'Custom ICS from standard cells: a design approach', *Computer Design*, May (1982)
LYMAN, J., 'Scaling the barriers to VLSIs fine lines', *Electronics*, June 19 (1980)
MARCOUX, P. J., 'Dry etching: an overview', *Hewlett-Packard J.*, August (1982)
MARSTON, P., 'Future semiconductor packaging trends', *Electron. Eng.*, September (1982)
MING LIAW, H., 'Trends in semiconductor material technologies for VLSI and VHSIC applications', *Solid State Technology*, July (1982)
MULLINS, C., 'Single wafer plasma etching', *Solid State Technology*, 88–92, August (1982)
NIXEN, D., 'Effects of materials and processes on package reliability', *Semiconductor Production*, September (1982)
RAPPAPORT, A., 'Automated design and simulation aids speed semiconductor IC development', *EDN*, August 4 (1982)
ROLFE, D., 'E-beam lithography for sub-micron geometries', *Electronic Product Design*, September (1981)
STAFFORD, J. W., 'Reliability implications of destructive gold wire bond pull and ball bond shear testing', *Semiconductor International*, May (1982)
STENGL, G., KAITNA, R., LOSCHNER, H., RIEDER, R., WOLF, P. and SACHER, R., 'Ion projection microlithography', *Solid State Technology*, 104–109, August (1982)
TEXAS INSTRUMENTS INC., 'Technology and design challenges of MOS VLSI', *IEEE J. of Solid-State Circuits*, **SC-17**, No. 3, 442–448, June (1982)
TSANTES, J., 'Leadless chip carriers revolutionize IC packaging', *EDN*, May 27 (1981)
TWADDELL, W., 'GaAs technology continues to advance, but proponents fear credibility window', *EDN*, February (1982)
VILENSKI, D. and MALTIN, L., 'Full automation of IC assembly will push productivity to new highs', *Electronics*, August 11 (1982)
WEARDEN, T., 'Applications of 111—V semiconductor materials', *Electronic Product Design*, March (1981)
WEISS, A., 'Hermetic packages and sealing techniques', *Semiconductor International*, June (1982)
WEISS, A., 'Power semiconductor packaging', *Semiconductor International*, August (1982)
WEISS, A., 'Plasma etching of aluminium: review of process and equipment technology', *Semiconductor International*, October (1982)
WEITZEL, C. E. and FRARY, J. M., 'A comparison of GaAs and Si processing technology', *Semiconductor International*, June (1982)
WINCHELL, B. and WINKLER, E. R., 'Packaging to contain the VLSI explosion', *Computer Design*, September (1982)
ZARLINGO, S. P. and SCOTT, J. R., 'Leadframe materials for packaging semiconductors', *Semiconductor International*, September (1982)

29 Hybrid Integrated Circuits

F F Mazda DFH, MPhil, CEng, MIEE, DMS, MBIM
Rank Xerox Ltd
(Sections 29.1–29.3, 29.5)

R A Mosedale
Applications Manager,
ITT Components Group
(Section 29.4)

Contents

29.1 Introduction

Integrated circuits can be classified into the major subdivisions of monolithic and hybrid. Hybrid circuits consist of two types: thick film and thin film. In physical appearance the two are very similar, but they differ in their manufacturing process, which determines their operating characteristics.

29.2 The manufacture of thin film circuits

There are five major steps in the manufacture of a thin film circuit. These are:

(i) Deposition of the film layer consisting of resistive, conductive or dielectric material.
(ii) Patterning of the layer to form the required components and interconnections.
(iii) Adjustment of the resistor and capacitor components formed in the film to give the required accuracy.
(iv) Adding on the chip components (resistors, capacitors or semiconductors) and connecting them into the circuit.
(v) Packaging the completed film circuit assembly.

29.2.1 Deposition of the film

There are three major techniques currently in use to deposit thin films. These are evaporation, sputtering and ion-plating.

29.2.1.1 Evaporation

This is perhaps the most popular method and uses equipment very similar to that employed in evaporating the metal onto monolithic circuits, as described in Chapter 28. The substrate is placed around a source which is heated above its vaporisation temperature. A vacuum of between 10^{-5} and 10^{-7} Torr is maintained and under these conditions the mean free path of the evaporated molecules is much greater than that of the distance between the source and substrate. The molecules therefore radiate in straight lines onto the substrate. On reaching it they condense and form a thin layer. The substrate is heated by an auxiliary heater so that the adhestion between it and the film is improved. The overall temperature of the substrate is determined by these auxiliary heaters, plus the radiated heat from the source, plus the energy imparted by the arriving molecules.

There are three forms of evaporation, namely resistance heated, electron beam heated and flash heated source. In the resistance heated system the source is either in the form of a wire which is wrapped round a tungsten (or tantalum or molybdenum) heater through which current is passed, or else it is a powder which is placed in a resistance heated evaporation boat. In either case this method introduces impurities from the heater into the bell jar and it is usually limited to an evaporation temperature below about 1500°C. Both these limitations are overcome by the electron beam heated system in which a high energy beam of electrons is focused onto the source material held in a water cooled container. The beam is produced from an electron source which acts as the cathode of the system and the water cooled container is its anode. High vaporisation temperatures are attainable while keeping heater contamination to a minimum.

Neither of the two heated systems is suitable for evaporating alloys. Since the constituents of the alloy will have different vaporisation pressures they will evaporate in different amounts so that the film composition will be different from that of the source. This is overcome by using flash heating. In this a small quantity of the source material is dropped in powder form into a container which is kept at a very high temperature. The powder immediately vaporises completely and deposits onto the substrate. Another method of depositing alloys is to use a non-thermal process such as sputtering.

29.2.1.2 Sputtering

A glow discharge is formed in an atmosphere of argon at between 0.01 and 1 Torr by a high voltage between the source, which forms the cathode, and the anode, which incorporates the substrate. Argon ions are formed by this discharge and are accelerated to the cathode. On striking it they cause the release of the source molecules which acquire a negative charge. This causes them to be accelerated rapidly to the anode where they impinge onto the substrate and adhere to it. The process is known as sputtering.

Sputtering is slower than evaporation and requires more elaborate equipment. An evaporated film can be deposited at a rate of about 15 μm min^{-1} whereas the rate is limited to 1 μm min^{-1} for sputtering. Sputtering however has several advantages:

(i) It is a cold process since the source does not have to be heated to anywhere near its vaporisation temperature. Hence it can be used to deposit materials with a high vaporisation temperature such as tantalum, and also for alloys, which would give incorrect film constituents if vaporised.

(ii) The molecules reaching the substrate have considerable energy so that the film density is high. The molecules on striking the substrate also ensure that all residual gas and other impurities are removed prior to and during the film formation. The adhesion is therefore very good. Both these advantages are also obtained with the ion-plated deposition method.

(iii) Since negative ions are attracted by a positive substrate the three dimensional coating capability is also much greater than for vaporisation, and equal to that of ion-plating.

29.2.1.3 Ion-plating

This may be considered to be a mixture of evaporation and sputtering. A glow discharge is formed in a low pressure gas as in the sputtering method. The substrate is now the cathode. The film molecules are introduced into the discharge by evaporating the source using resistance, electron beam or flash heated techniques. These molecules are rapidly accelerated to the substrate and impact on it to form a strong bond. Since evaporation is used film formation is rapid but there is the risk of impurities being introduced. However, all the other advantages of sputtering are now applicable.

29.2.2 Patterning of films

The previous sections described how a thin film layer can be formed on the substrate. In this section the ways in which these layers can be modified to give the required pattern, whether it be resistor, conductor or dielectric, are described. Basically there are four systems in use:

(i) *Photoresist masking* This is probably the most common process. Initially the substrate is completely covered with a film layer. Photoresist is then spun over the film and it is exposed through a glass mask, in a similar process to that used for monolithic circuits. If positive photoresist is used then the exposed areas harden and are not removed in the subsequent wash. The surface is now etched to remove the film from the unprotected regions and finally the remaining photoresist is washed away. Clearly if several layers are to be formed (e.g.

conductor and resistor) then these are both first evaporated on and then followed by photoresist masking and etchants which attack each layer separately.

(ii) *Metal masking* This is an additive process since the starting material is the bare substrate. A metal mask with the desired film pattern cut out is put into contact with the substrate and a layer of film is evaporated on. The mask is then removed to leave the film on the substrate corresponding to the cut out regions. The metal mask must be placed in close contact with the substrate during evaporation to prevent diffused edges from occurring.

Metal masking does not involve the use of photoresists and is therefore a cheaper process. It is also useful when etchants cannot be used, for instance when SiO_2 is employed for capacitors since etchants which dissolve this also attack the glass substrate. It is also much easier to reclaim the unused evaporated material, which may be a precious metal such as gold, from the mask than when they are part of the etchant chemicals. However, metal masking is not suitable for producing fine lines or when certain shapes are needed, for example a circle with a solid centre. They also become critical when several layers are to be formed requiring careful alignment of the successive masks.

(iii) *Inverse photoresist masking* In this system the bare substrate is first coated with a layer of photoresist. This is then masked and etched to form a pattern in which photoresist is left on the areas which are to have no film. As such it forms an inverse mask. The film layer is next formed over the whole substrate. It adheres to the bare substrate regions. All the remaining photoresist is then dissolved carrying away with it the film from the unwanted areas.

Inverse masking methods are capable of very good line definitions due to the close proximity of mask and substrate. They are also used when the evaporated film is difficult to etch without attacking the substrate or the photoresist. The photoresist mask is however damaged by high temperatures and so must only be used with low temperature deposition systems.

(iv) *Inverse metal masking* This is very similar to inverse photoresist masking but uses a metal mask. The bare substrate is initially covered with a metal layer which is then etched to give the inverse metal mask. The required film layer is then deposited over this and finally the original metal mask is dissolved out with a special etchant taking with it the unwanted film.

29.3 Thin film components

This section is primarily concerned with components which are formed on a thin film substrate using one of the techniques described earlier. Devices in this class include resistors, capacitors, inductors and the interconnecting conductors between them.

29.3.1 The substrate

The starting material for any thin film circuit is obviously the substrate on which it is formed. This can be made from a variety of materials such as glass or pure glazed alumina. Thin films can also be formed on silicon dioxide which is very useful as it gives a method of combining monolithic and film circuits on a single chip.

Since thin films have relatively tight tolerances they place many demands on the requirements of the substrate. It should be clean and free from cracks and scratches. The surface of the substrate should be smooth and there should be no warping or dimensional changes with time. Such changes cause poor registration of masks and bad line definition. In addition the substrate should be a good electrical insulator since many devices are placed close to each other on its surface. For power circuits it is also important that the substrate is a good thermal conductor and in these applications it is sometimes mounted on an external heatsink.

29.3.2 Conductors

The conductor tracks on film circuits should provide a low resistance path and make low loss contacts to both the resistor and capacitor films. Generally the minimum line widths and line spacings are limited to about 25 to 50 μm due to mask tolerances. The maximum thickness is determined by the available substrate area and the relative high cost of the metal used.

By far the most popular conductor metal is gold. It has low resistivity and makes an excellent low resistance joint to most resistor and capacitor plate materials. This joint takes the form of intermetallic bonds. However gold has very poor adhesion to the glass substrate due to the fact that no stable bonding oxide is formed at the glass–metal interface. It is therefore usual to cushion the gold from the glass by a thin (about 0.015 μm) nichrome layer, which does form a stable oxide with glass and therefore adheres strongly to it. The gold layer is usually about 0.5 μm in thickness. It will be seen in the next section that nichrome is the most popular thin film resistor material and the gold–nichrome combination is very frequently utilised. One of its further advantages is that the metals can be selectively etched to form the conductor and resistor patterns.

Aluminium is also used as a thin film conductor. Its obvious advantage is low material cost. However apart from the disadvantage of higher resistivity aluminium has the undesirable property of reacting with gold at relatively low temperatures. Gold wires are often used to connect add-on semiconductor devices to the conductor tracks, and if these tracks are made from aluminium then a chemical reaction occurs between them above a critical temperature. This results in the formation of a purple compound of $AuAl_2$ at the gold–aluminium interface, which eventually leads to device failure. This effect is called the 'purple plague'.

29.3.3 Resistors

Thin film resistors generally have more involved manufacturing processes than monolithic resistors but present several advantages. For instance the frequency response of a thin film resistor is several times better than a monolithic device. This is partly due to the lower parasitic capacitances associated with the resistors and partly due to their higher available sheet resistivity, which reduces the overall size. Thin film resistors can also be made to much tighter tolerances than monolithics since they may be adjusted to the required value after fabrication.

The most popular thin film resistor material is an alloy of about 80% nickel and 20% chromium, popularly known as nichrome. It has good adhesion properties and a low temperature coefficient, which is determined by several processing factors such as the rate of film deposition and the substrate composition. Typical resistor sizes are in the range 20 Ω to 50 kΩ. Nickel is unsuitable on its own for use as a film circuit material since it does not adhere adequately to the substrate. However, chromium has good adherence and is sometimes used for resistors. Its principle disadvantage is the tendency to form surface oxides which make it difficult to etch.

When high resistance values are required, cermets, which are compounds of a dielectric and a metal, are often used. They are usually formed by a two step process. The first step lays down the film and the second step anneals it to the substrate. Since cermets are usually available in powder form they are ideally suited to vacuum deposition using flash techniques. It is generally difficult to control the thickness of a cermet film during deposition but it can be subsequently trimmed to give resistor values up to about 1 MΩ with a tolerance better than ±1%.

Tantalum resistors are characterised by excellent stability, a high value of dielectric field strength and high annealing temperature. Tantalum is capable of producing uniform and reproducible film layers which can also be oxidised to form a thin passivating layer. The material has a high melting point and must therefore be deposited by sputtering. Tantalum resistors can be produced with tolerances in the range ±0.1% with stability in this range over 25 years.

Table 29.1 summarises the characteristics of a few thin film resistors and compares these with monolithic devices.

Table 29.1 Characteristic of typical thin film resistors and comparison with monolithic devices

Material	*Sheet resistance*[a] *(Ω/square)*	*Temperature coefficient*[b] *(ppm °C^{-1})*
Chromium	100–300	±50
Nichrome	100–400	±50
Cermets	1000	±100
Tantalum	10–500	±100
Diffused monolithic	2.5 (emitter) 100–300 (base)	500–2000

[a] refers to a square shaped material.
[b] parts per million per degree centigrade.

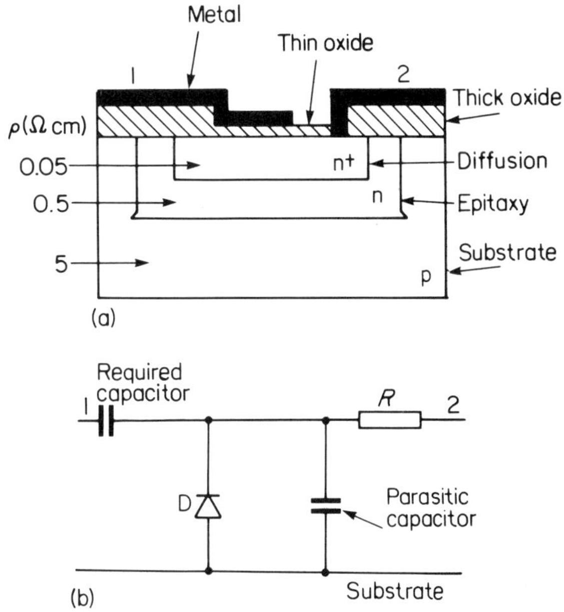

Figure 29.1 Metal–oxide–silicon capacitor: (a) cross section; (b) equivalent circuit

29.3.4 Capacitors

A thin film capacitor is basically a sandwich of a dielectric between two metal plates. The metal–oxide–silicon capacitor is shown in *Figure 29.1*. It can be used in monolithic or film circuits. It has the advantage over a diffused device of being insensitive to the polarity and the applied voltage and of having a relatively high Q due to its lower junction capacitances. *Figure 29.1(b)* shows its equivalent circuit. The parasitic capacitance is due primarily to the epitaxy-substrate diode, and the series resistance occurs in the diffused layer which forms the lower plate of the capacitor.

The thinner the oxide dielectric the higher the capitance value. However the minimum thickness is usually limited to about 0.5 μm to give a uniform layer and prevent pinholes. The typical capacitance value is in the range 0.4 to 0.7 nF mm^{-2} for an oxide thickness of 0.8 to 0.1 μm.

Tantalum oxide (Ta_2O_5) capacitors are formed by first sputtering on a tantalum layer where the capacitor is to be located. This acts as the bottom plate. The tantalum is then oxidised to give a thin tantalum oxide dielectric layer and the capacitor is completed by sputtering on the top tantalum plate. These capacitors have very high dielectric constants and a good breakdown voltage. They are sometimes troubled by the movement of substrate faults through the metal and into the dielectric, which degrades their characteristic. This is usually corrected by an electrochemical back-etch process.

In an alumina (Al_2O_3) capacitor an aluminium layer is formed on the substrate to give the lower plate of the device. A thin layer of nickel is then evaporated onto the aluminium to prevent it migrating into the dielectric. Alumina is then deposited to form the dielectric and this is followed by a second nickel layer and then the top aluminium plate.

The properties of these thin film capacitors are summarised in *Table 29.2* and compared with a diffused monolithic device. Thin film capacitors are capable of larger values and lower parasitics than monolithic devices. They are also not dependent on the polarity and magnitude of the applied voltage provided that this

Table 29.2 Thin film capacitors and comparison with diffused monolithic devices

Dielectric material	*Dielectric constant*	*Capacitance (nF mm^{-2})*	*Temperature coefficient (ppm °C^{-1})*	*Maximum operating voltage (V)*	*Q (at 10 MHz)*
Silicon dioxide	5.8	0.4–7	200	50	10–100
Tantalum oxide	20–25	2.0–3.5	150–300	20	50
Alumina (Al_2O_3)	9	0.6–2.0	400	20–50	10–100
Diffused monolithic	—	0.3	200	5–20	1–10

is below the breakdown value. However, thin film capacitors are not frequently used in practice, due to their requirement for more process steps, unless one needs a large number of low valued devices on a substrate, as for example in the decoupling of high frequency systems. In all except such applications it is more convenient to add chip capacitors.

29.3.5 Inductors

Inductors can be made quite simply in thin film technology by forming a circular or square conductor spiral onto the substrate. However, the inductance available is very low and the Q of the device is poor. Generally therefore thin film inductors are not frequently used except for some microwave hybrid applications in which devices are formed on high quality dielectric substrates. The coils in these cases operate primarily as lumped components and have inductances of the order of a few microhenries and Q values of about 50. It is much more usual in film circuits to put tiny ferrite inductors onto the substrate as add-ons, rather than to form them as a film.

29.4 Manufacture of thick film circuits

29.4.1 Thick film screen preparation

To define the pattern geometry of printed thick film inks, three basic methods exist:

(a) Stencil screens, where a woven screen mesh, typically a plain weave, either stainless steel or polyester, has been coated with a photosensitive emulsion film and subsequently exposed to ultraviolet light to polymerise, and then developed to form the screen pattern.

(b) Metal masks, in which the pattern geometry has been directly etched in a sheet of solid metal, typically molybdenum or stainless steel, 0.05 to 0.25 mm thickness.

(c) Writing-in, where the pattern geometry is directly written onto the substrate base by the application of a computer controlled X–Y table, which moves the substrate to directly beneath a stylus. The stylus is fed from an ink reservoir. The coordinates for the X–Y table movement are generated by the direct digitising of the scaled layout.

A typical flow chart for the preparation of both stencil screens and metal masks is outlined in *Figure 29.2.*

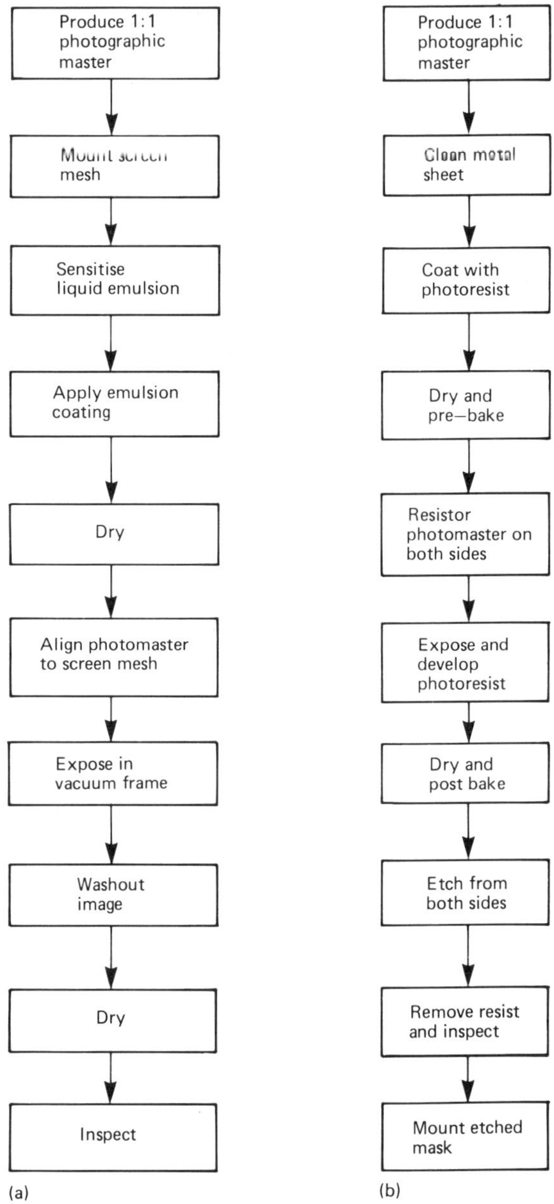

Figure 29.2 Flowchart for thick film screens: (a) stencilled screen; (b) metal mask

In both the manufacture of stencil screens and metal masks, the production of 1:1 photographic masters from scaled master artwork is required. The master artwork may be prepared by a number of techniques such as, applying opaque tapes to white paper or film, inking on a matt film, cut and peel off two-layer mylar backed film or by the direct use of computer aided design systems. The selection of master artwork scale, typically 10:1, is controlled by the design aids such as plotters and the reduction camera in terms of lens focal length and resolution.

The process of screen printing by production equipment requires that the manufactured stencil screen or metal mask be mounted onto a supporting frame, as indicated within *Figure 29.2.* This supporting frame may be fabricated from a number of materials, including wood, plastic, phenolics and metal castings, typically aluminium. The most commonly used frame material of those listed is cast aluminium due to the advantages of lightweight, strength and stability.

Attachment of the stencil screen or metal mask to the frame, is achieved by the use of one of a number of techniques including epoxy adhesive, staples, crimped tubing or metal strips. The selected dimensions of the mounting frame is totally dependant on the available printing equipment, however, sizes ranging from 7.62 cm × 7.62 cm up to 30.5 cm × 30.5 cm can be purchased for thick film printers, the most common dimensions in use being 12.7 cm × 12.7 cm and 20.3 cm × 25.4 cm.

29.4.2 Thick film deposition

Two basic methods of stencil screen or metal mask printing exist, these are: off contact printing, in which only the portion of the screen or mask directly under the squeegee is in contact with the substrate during the print cycle; and contact printing, in which the entire screen or mask remains in direct contact with the substrate during the complete pass of the squeegee and is then subsequently lifted or peeled away from the substrate. These techniques are shown in *Figure 29.3.*

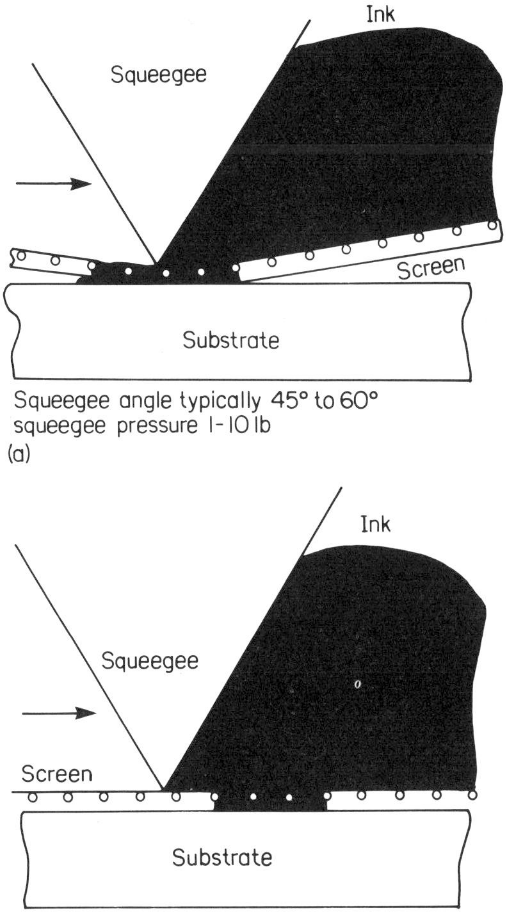

Figure 29.3 Squeegee, substrate, screen relationship during: (a) off-contact print cycle; (b) contact print cycle

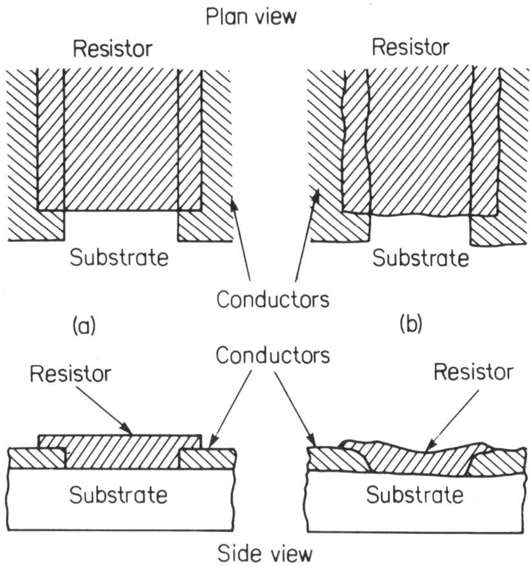

Figure 29.4 (a) Representation of an ideal thick film conductor and resistor print; (b) typical as-fired thick film conductor and resistor print

The two major requirements for thick film printing process are conformity and predictability of print as shown diagrammatically in *Figure 29.4*.

The ideal print sequence would result in the complete filling of the defined pattern on the stencil screen or metal mask with ink during the squeegee pass, concluding with all the dispensed ink being transferred directly onto the substrate base in a uniform thickness and in full compliance with the defined pattern. The inherent process variables are:

(a) Ink rheology, in which the viscosity of the ink determines to a large extent the quality of the printed pattern.
(b) Screen printer settings, where print speed, squeegee pressure and angle, screen tension and screen to substrate snap-off significantly affect the print quality.
(c) Pattern geometry as defined on the stencil screen or metal mask, which affect print thickness and definition.
(d) Screen mesh size, where ink thickness and print definition become affected.
(e) Substrate physical characteristics of flatness and parallelism, which affect print thickness and definition.

These variables prohibit the ideal print characteristics being achieved. The introduction of specialised equipment such as a lunometer to calibrate the screen mesh count and viscometer to check the viscosity of the thick film inks assist in reducing the affects of process variables, but never completely remove them.

29.4.3 Thick film inks

The thick film ink has four main constituents: active material particles, binding agents, organic binders and volatile solvents. The active material consists of metal or metal oxide particles typically 2 to 5 μm in diameter, which determine the electrical characteristics of the fired film. In general, inks with metal particles are conductive in nature and inks with metal oxide particles are resistive. The selection of an ink system with a specific active particle content is based on the availability and applicability of the system to the design parameters and required processing conditions such as soldering, where the conductor leach resistance is essential, and wire bonding, where gold or aluminium wires are directly bonded to a conductor on the base substrate. Materials such as gold, gold–platinum alloy, palladium–silver, platinum–silver alloy and copper can be added to form the conductor film. The formulation of a resistor ink system can include, palladium silver–palladium oxide, and ruthenium dioxides.

The binding ingredients consists of a glass frit, typically a glass with a low melting point, such as lead borosilicate glass. The purpose of the glass frit is twofold. Firstly, by the sintering or melting of the glass frit during the furnace firing, provide an amorphous mass within which active particle to active particle contact can occur. Secondly, to promote adhesion of the fired thick film ink to the base substrate by both the chemical and physical reaction of the glass frit with the substrate. Ink systems can be purchased where the glass frit has been replaced by a copper oxide, these systems being termed 'fritless' inks. They rely on the copper oxide providing a chemical bond between the active particles and the base substrate.

The purposes of the organic binder is to hold the active particles and glass frit in suspension such that the essential viscosity characteristics for screen printing are achieved, whilst maintaining a high rest viscosity to prevent the flow of the thick film ink prior to the squeegee pass and to subsequently retain the defined pattern on the substrate after the squeegee pass. The term 'thixotropic' is applied to those materials which exhibit such fluid characteristics.

The purpose of a volatile solvent is to reduce the 'yield point' of the thixotropic organic binder, the 'yield point' being the minimum force applied to the thick film ink by the squeegee at which the ink commences to flow through the stencil screen or metal mask. As either the 'yield point' is lowered or the applied force increased, the flow rate of the ink through the stencil screen or metal mask is increased. The volatility of the solvent system is one of the most important factors for the ease of screening. Excessively volatile solvents tend to dry out during use. This causes fluctuations in the obtained resistance values where resistor ink systems are being screened since the amount of ink deposited will vary as the solvent leaves the screened ink. Common resistor ink systems contain solvents such as terpineol, carbitol and cellosolve and their relatives and derivatives.

29.4.4 Thick film firing

When a film pattern has been screen printed onto the base substrate, the screened ink is dried by heating the substrate to about 125°C. This dries off the volatile solvent from the thick film ink. The screened substrate is then fired. The purpose of firing the ink is to burn off the organic binder and to allow the necessary physical and chemical changes to provide adhesion and the desired electrical properties to take place in the ink system. A temperature of about 350°C is required for the organic binder burn off and a temperature between 550°C to 1000°C is required to produce the changes in the ink system. The binding ingredient, if glass is used, must melt, wet the active particles and react with the base substrate. If a fritless ink system is used, the copper oxide must chemically react with the substrate to provide adhesion. The firing temperature profile affects the properties of the screened ink and must therefore be carefully controlled. In order to accomplish this, a means must be provided to gradually heat the screened substrates to the required temperature, maintain it at the temperature long enough for the desired physical and chemical reactions to take place and then to return the substrate, gradually, back to room temperature. Within this sequence, time and temperature must be precisely controlled, typically ±1°C regarding temperature. Two techniques exist for firing screen printed thick film, the moving belt furnace, which is generally the commonest production technique, and the programmable kiln, which allows the temperature time profile to be preset. *Figure 29.5* diagrammatically illustrates a moving belt furnace. It has a number of specific components. These include a muffle, generally made of mullite or fused quartz, which is designed to keep the inside of the furnace clean and to act as a heat radiator to even out the temperature directly beneath the heating coils; a continuous metal belt, typically fabricated from materials such as nichrome, which passes through the furnace at a controlled speed carrying screen substrates; baffles which are designed to provide control regarding furnace draft and atmosphere.

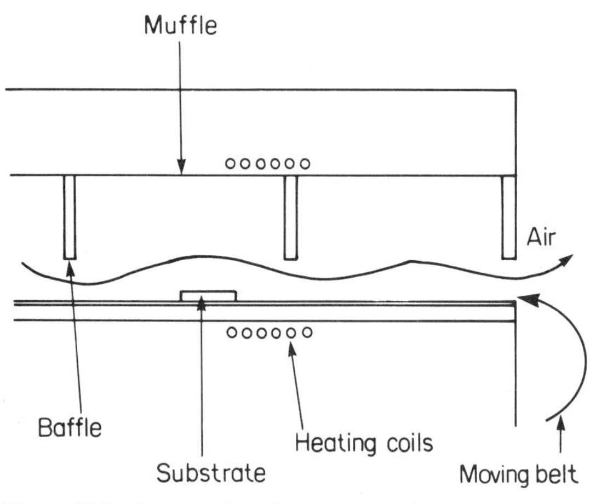

Figure 29.5 Cross section of a moving belt furnace

To achieve the essential temperature–time profile, the furnace is divided into sections, termed 'zones', each zone being surrounded by a heating coil. The minimum number of zones required for most firing profiles is four, a preheat zone, two hot zones and a cooling zone. Large furnaces may have as many as 13 independently controlled zones, in order to define the specific firing profile more accurately.

29.4.5 Thick film resistor trimming and substrate scribing

The use of thick film technology permits the designing of resistors with a wide range of ohmic values. The basic resistance, R of a screen printed resistor is given by:

$$R=\frac{\rho L}{WT} \qquad (29.1)$$

where ρ is the bulk resistivity of the ink and L, W and T are the length, width and thickness respectively of the screen printed resistor. Thus by choosing a resistor ink of suitable resistivity and by varying the relative values of L, W and T, resistors of the required values can be obtained. Practical limits are imposed by the minimum area which can be accurately screened using stencil screen or metal mask and by the maximum area of the substrates which can be allotted to a specific resistor. In addition, extremely low value resistors, typically 1 to 10 Ω, can be sensitive to screen printed conductor tracks, where they are provided an interconnection pattern to the resistor. Whilst the basic resistance, R, can be determined theoretically, the inherent process variations apparent in the screening of thick film resistors result in the actual value being a variable. Typically, with the ink as purchased exhibiting a ±10% value variation and the screening process introducing an additional ±15% value variation, resistor values as fired can be within the band of ±25% of the nominal design value. This band can be reduced by additional process controls, such that resistors with a ±15% tolerance on the design value may be achieved without any additional processing subsequent to the substrates firing. In general, tolerance requiring greater accuracy than ±15% would be achieved by the trimming of the resistor. There are two basic techniques of trimming: abrasive trimming and laser trimming.

Abrasive trimming is a technique in which material is removed by using compressed air to force finely ground alumina through a small nozzle onto the fired resistor ink. The removal of resistive material increases the aspect ratio of the resistor and increases its value. *Figure 29.6* shows a sketch of an abrasively trimmed resistor. The value of the resistor is monitored during the trimming cycle by a resistance bridge. When the values of resistance is within the design technique, a signal from the bridge is sent to the valve controlling the compressed air, shutting it off and thereby stopping the trimming. Manufacturing tolerances of ±1.0% can be readily achieved by the use of air abrasive trimming. With the introduction of a two part trimming technique which involves an initial coarse trim followed by a 24 h at 150°C stabilisation bake, prior to the second trim, manufacturing tolerances of ±0.1% can be achieved.

Laser trimming is a technique in which the heat from a laser beam is utilised to evaporate resistive material, which creates a small channel, or 'kerf' in the resistor, typically 20 μm wide. This re-routes the current through the resistor thereby effectively increasing the aspect ratio of the resistor and the resistance. The laser trim may consist of a single cut into the screened resistor or

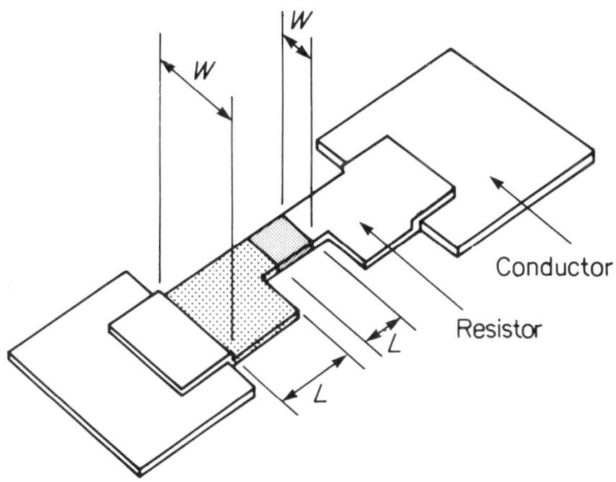

Figure 29.6 Air abrasion resistor trimming

several cuts which could include L-shaped cuts where the rate of change in resistance introduced by each laser pulse is required to be minimal. The evaporation process melts the glass in the resistor, causing it to reflow and seal the kerf, reducing the long term drift characteristics associated with abrasive trimming. In general, trimming lasers are YAG lasers (yttrium–aluminium–garnet), in preference to the gas lasers, such as the CO_2 lasers, which tend to produce too much power which in turn can produce excessive variation in trimmed resistor values. Manufacturing tolerances associated with laser trimming are comparable with those achieved with air abrasive trimming, however, speed of adjustment, cleanliness of working environment and resistor stability are considerably improved.

The requirement for substrate scribing occurs when either a multiple or individual circuits have been screen printed onto a single substrate base or when an individual circuit as printed, is less than the base substrate in area. Typically, production hybrid houses utilise 50.8 mm × 50.8 mm, 96% aluminium substrates with their screening equipment though standard substrates ranging from 25.4 mm × 25.4 mm to 102 mm × 102 mm can be purchased. The conventional technique used for scribing the individual circuits from the base substrate is the manually operated diamond scribe which scores the substrate thereby permitting the breakout of the individual substrates. This technique is however slow and can be inaccurate. The use of a laser for substrate scribing, in particular a CO_2 base system, permits a rapid scribe speed and improved accuracy to be achieved, typical figures being 17.8 cm per second and ±0.01 mm respectively.

29.5 Chip components

In the previous sections of this chapter the characteristics of thick and thin film devices have been described. This section looks at the add-on components which are available for both thick and thin film circuits. These include resistors, capacitors, wound components such as inductors and miniature transformers, and semiconductors.

29.5.1 Resistors

Generally resistors are formed as films, rather than chips, in hybrid circuits. Instances when chip components are more economical are when a few devices are required in a particular resistance range which could involve a separate screen and fire operation. Generally a given resistor paste can be used to cover a resistance range of 10:1, by adjustment of its physical dimensions, before a higher or lower resistivity ink becomes necessary. Chip resistors are also used when special characteristics have to be obtained, such as high valued resistors in thin film circuits, low TCR resistors, large wattage resistors and high frequency resistors.

There are three basic types of chips which can be bought as standard products from several suppliers. These are thin film, thick film and cermet resistors.

Thin film chip resistors are formed by thin film deposition processes on an alumina (or sometimes glass) substrate. Materials used are generally chromium or nickel–chromium. Surface resistivities up to about 500 Ω per square are available giving resistors in the region of 1 MΩ with standard tolerances of 5 to 10% or lower if specified. These resistors have a TCR better than 75 ppm $°C^{-1}$ and a temperature drift below 0.05% at 125°C for 1000 hours.

Thick film resistors are formed on alumina substrates using screen and firing techniques with noble metal pastes. Unlike thin film resistors the thick film devices are more stable at high values of resistance since the larger quantity of binder acts as an effective sealant. A very high resistance value however is difficult to reproduce (unless trimmed) and the resistors are noisy. Resistivities up to 1 MΩ per square are commonly available giving a resistance range of 10 Ω to 15 MΩ with 5% to 10% standard tolerance without trimming. The TCR is between 30 and 300 ppm $°C^{-1}$ with a temperature drift of 0.5% at 150°C for 1000 hours.

Solid cermet resistors are made from a mixture of metal, metal oxide, powdered glass and organic binder. The paste is formed into tiny blocks and fired at high temperatures to give a solid homogeneous mass. They are therefore equivalent to thick film resistors without substrates. Their electrical characteristics are very similar to those of the thick film devices described above.

Chip resistors come in a variety of different sizes and shapes, the smallest size being a block having dimensions approximately 1 mm × 1 mm × 0.2 mm. The chips can be bought with gold backing for eutectic bonding to the substrate and for interconnections into the circuit via bond wires. Alternatively they are available with metallised edges to enable them to be flow soldered. In either case the devices (or their bond wires) can often be positioned so as to provide crossovers between tracks, if these are required.

29.5.2 Capacitors

Screened capacitors are relatively difficult to produce since they require three different operations to form the two conductor plates and the intermediate dielectric. Furthermore the values of capacitance available are severely limited. For these reasons chip capacitors are very commonly used in film circuits and are consequently available in a variety of shapes.

The most popular chip capacitor dielectric material is a ceramic. The value of the dielectric constant in this type of capacitor can be varied in the range 10 to 10 000 by changing the ingredients in the ceramic while it is still in the slurry stage. Ceramic also has good physical strength and thermal tolerance to enable it to withstand the high soldering temperatures which may be attained during thick film assembly. Ceramic capacitors are basically of two types, high stability NPO (negative positive zero) and high *K*. NPO capacitors have dielectric constants in the region of 100, a linear temperature–capacitance curve, and a temperature coefficient of about ±30 ppm $°C^{-1}$ over the range −55 to +125°C. Capacitance values are fairly low since the physical size of the device is limited. However, by using multilayer devices the overall plate area is increased while the dielectric thickness is kept small, giving higher capacitance values. In such a system alternate thin layers of unfired ceramic and a conducting noble metal paste are formed and the ends of the conductors connected to external solder pads. The whole stack is then pressed and fired to yield a ceramic block which looks like a homogeneous mass. The conductor thickness is now about 2 μm and the dielectric thickness about 20 μm. Although the total height of these devices is relatively large compared to printed capacitors they occupy about the same substrate area. Capacitor values between 1 pF and about 3 μF with voltage ratings of about 50 V are common.

High K ceramic capacitors are based on a dielectric of barium titanate. The dielectric constant can be as much as 8000 giving capacitance densities in the region of 10 μF cm^{-3}. These capacitors have a lower stability and a poorer temperature coefficient compared to NPO types.

For very large capacitance values tantalum capacitors are used. The dielectric consists of a porous tantalum slug anode which is anodised and coated with a deposit of manganese dioxide to form the cathode. Capacitor values are available from about 0.1 μF to several hundred microfarads within a voltage range of 5 to 50 V. Temperature coefficients are about 50 to 2000 ppm $^{\circ}$C^{-1} for a range -50 to $+125^{\circ}$C.

Chip capacitors can be bought with tinned metal ends for reflow soldering or with wire or ribbon leads. Alternatively they are available with metal areas for thermocompression or ultrasonic bonding. It is also possible to buy capacitors with electrodes on either side such that one end is soldered down to the conducting track whilst the second side is wire bonded.

29.5.3 Wound components

Very low value inductors, in the few nanohenry region, can be formed as metal spirals in thin film or thick film. To obtain larger value inductors or transformers it is necessary to add them in chip form. Miniature toroids are available for use as inductors or pulse and broad band transformers. These are capable of relatively high inductance values, in the region of 1 mH from a volume size 6 mm × 6 mm × 1 mm. The Q of these components is about 15 which is suitable for most applications. The main limitation of these devices is that the maximum continuous power which they can handle is of the order of a few milliwatts.

For applications requiring tunable inductors the rod type of construction is used. With adjustable metal rods of high permeability inductances in the region of 20 mH are available for rod diameters of about 1 mm. These devices have a high Q and good thermal stability. They are used in applications such as r.f. and i.f. amplifiers, discriminators, modulators and tuners.

29.5.4 Semiconductors

Perhaps the largest category of add-on components is semiconductor chips. These devices can range from small signal diodes and rectifiers to complex integrated circuits. For use in hybrid circuits all devices should be passivated to prevent contamination. This means essentially that they have to be fabricated as planar devices.

Although semiconductors can be bought in the form of miniature encapsulated components specially made for hybrid circuits this is a costlier approach than using unencapsulated chips. Its merit lies in the fact that since the device is packaged it can be tested more thoroughly prior to assembly so increasing the overall yield of the system. Packaged components can also be connected into the circuit by relatively simple reflow soldering techniques.

For minimum costs unencapsulated chips are usually connected directly into the circuit. Many techniques may be used to wire these devices. Alternatively semiconductor chips are available in a form which simplify circuit connections. These include flip chip and beam lead devices.

In the flip chip bonding technique either the chip or the substrate to which it is connected are provided with contact bumps at the mounting regions. The chip is now inverted and bonded face down onto the substrate pads as shown in *Figure 29.7*. The chip is handled by a vacuum tube and the machine

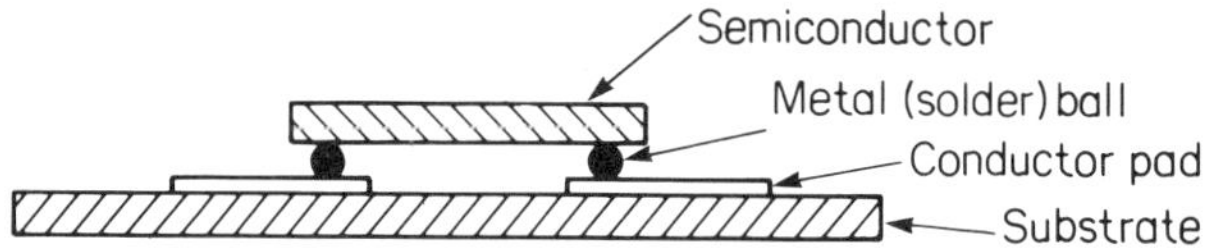

Figure 29.7 Flip chip mounted on substrate

presents to the operator a magnified split optical image of the underside of the chip, superimposed onto that of the substrate, so that alignment is simplified. The great advantage of the flip chip technique is that all bonds are made simultaneously and therefore assembly time is reduced. Furthermore electrical and mechanical connections are made in the same operation and a separate die bonding step is not required.

Flip chip bonds can be made by ultrasonic, thermocompression or solder techniques. In ultrasonic bonding the chip is pressed against the substrate and vibrated at about 60 kHz by a tungsten–carbide tool. The pressure required depends on the chip size since the larger the number of bonds the greater the force. However, the chip may be damaged if this pressure is too large so that for big chips thermocompression bonding is preferred. Essentially thermocompression bonding of flip chips is similar to that used with bond wires. Pressure is applied to the chip and substrate and both are heated to about 350°C. The system needs careful matching between the temperature coefficients of expansion of chip and substrate to prevent fracture of the bond. It is generally best applied to gold-on-gold systems.

Both thermocompression and ultrasonic bonding methods apply pressure to deform the interconnecting bumps and make reliable contacts. This deformation should be due to plastic flow, and not elastic deformation, or the reliability of the contact will suffer after the pressure has been removed, and stresses will be set up in the chip which could damage it. An alternative bonding technique uses reflow soldering in which the cutectic solder is applied to the bumps or the contact pad. The chip is placed on the pad and heated to cause solder reflow. Enough flow must occur to fill all voids, but it must not be excessive or short circuits can occur between adjacent pads. Soldering occurs at low temperatures so that unequal coefficients of expansion of the chip and substrate do not present a problem. Soldered devices can be removed relatively easily for inspection or replacement. Chips connected by thermocompression or ultrasonic bonding may also be removed by rotating the devices so as to shear off the contacts. If the bumps are on the chip, however, this will result in the removal of part of the bonding pad so that if chip relpacement is desired the pad must be large enough to accommodate it. Alternatively if the bumps are on the substrate then the bond between the bump and the substrate must be large enough to prevent it being removed with the chip.

In principle beam leads are very similar to flip chips except that in place of the contact bumps on the surface there is now a relatively thick wire lead which overhangs the surface of the dice. This lead is an extension of the normal aluminium metallisation of the pad, and since its thickness is of the order of 12 μm compared to 1 μm for the metallisation, it is called a beam. The beams are rigidly attached to the chip and act as electrical conductors as well as providing mechanical support. The chip is

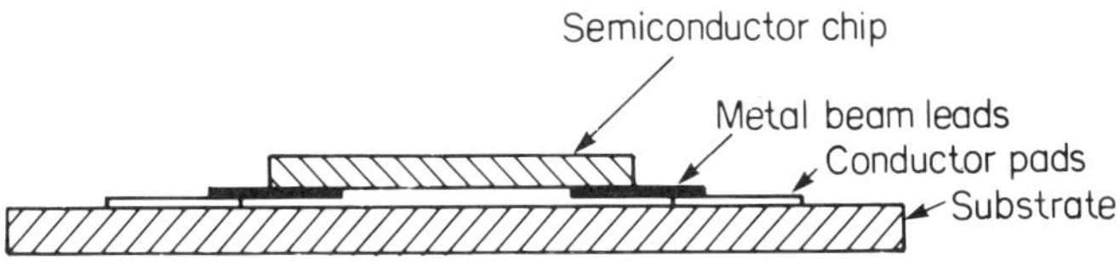

Figure 29.8 Beam lead mounted on substrate

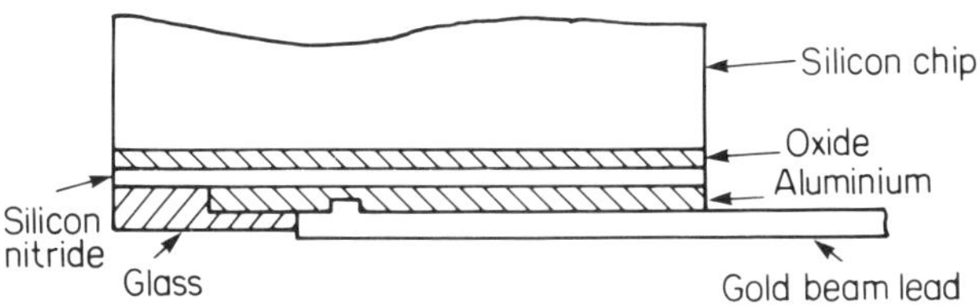

Figure 29.9 Cross section of a beam lead

usually 'flipped' to connect it into circuit, as shown in *Figure 29.8*.

Figure 29.9 shows the cross section through part of a beam lead chip. The processing of the devices, if it is bipolar, follows the usual steps right up to the stage where the emitter diffusions are made. The oxide layer is now etched to open up the metal contact areas. Silicon nitride is then deposited over the whole surface to provide a protective layer and the contact areas again etched through it. A thin layer of platinum is then sputtered on. This reacts with the exposed silicon to form platinum silicide which makes a low resistance contact. The unreacted platinum is then etched off. Alternative layers of first titanium and then platinum are next sputtered onto the surface. The titanium forms a strongly adherent surface to the nitride and the platinum acts as a barrier layer. The platinum is then etched to form the conducting paths and is plated with about 2 μm of gold. Selected areas of this gold are next etched to give beams about 70 μm wide and 200 μm long, which are further plated to a thickness of about 12 μm. The titanium layer is then etched off using gold as a mask and the silicon under the beams is also etched to give a wafer with each of its dice having cantilever gold contacts protruding over its edges. The wafer can now be probe tested and separated into individual dice using chemical etchants. To facilitate this process it is usual to use ⟨1 0 0⟩ oriented silicon for beam leads, instead of the more conventional ⟨1 1 1⟩ so that anisotropic etching can be employed. Beam leads can be attached to the substate in many ways, the most common being thermocompression bonding.

Further reading

BAILEY, B., 'Ceramic chip carriers—a new standard in packaging', *Electron. Eng.*, November (1980)

DENDA, S., 'Trends in thick-film technology for electronic components', *JEE*, June (1982)

EMBREY, D. M., 'Thick film components', *New Electronics*, May 5 (1981)

HEID, K. K. W. and STEDDOM, C. M., 'Thick film, thin film: which to use and when', *Int. J. Hybrid Microelectronics*, **4**, No. 2, 376–8, October (1981)

HUGHES, J., 'Chips for hybrids', *New Electronics*, May 5 (1981)

KEIZER, A., 'Tape automated bonding systems', *New Electronics*, May 27 (1980)

LYMAN, J., 'Tape automated bonding meets VLSI challenge', *Electronics*, December 18 (1980)

NEIDORFF, R., 'Laser-trimmed thin-film resistors hold voltage references steady', *Electronics*, June 16 (1982)

SERGENT, J. E., 'Understand the basics of thick-film technology', *EDN*, October 14 (1981)

SERGENT, J. E., 'Thin-film hybrids provide an alternative', *EDN*, November 25 (1981)

THACKRAY, I., 'Using chip carriers', *Semiconductor Production*, February (1982)

TRAVIS, W., 'Thin film hybrid circuits', *New Electronics*, October 14 (1980)

WEARDEN, T., 'The technology of thick-film hybrid circuits', *Electronic Product Design*, October (1980)

30 Digital Integrated Circuits

F F Mazda, DFH, MPhil, CEng, MIEE, DMS, MBIM
Rank Xerox Ltd

Contents

30.1 Introduction

Digital circuits may be differentiated in two ways: by the function they perform and by the technology in which they are made. Therefore a CMOS shift register will perform the same function as a TTL shift register, but it will have different parameters such as speed of operation and power consumption.

In this chapter the basic parameters used to measure the performance of digital circuits are first introduced. This is followed by a description of the different digital integrated circuit technologies, and their functional systems.

30.2 Logic parameters

Although there are many parameters of interest, such as environmental operating range, availability and cost, five key parameters are generally used in comparing digital circuit families.

30.2.1 Speed

This indicates how fast a digital circuit can operate. It is usually specified in terms of gate propagation delay, or as a maximum operating speed such as the maximum clock rate of a shift register or a flip-flop.

Gate propagation delay is defined as the time between equal events on the input and output waveforms of a gate. For an AND gate this is the delay between a point on the last changing input waveform and an equal point on the output waveform. For an OR gate the first changing input waveform is used as the reference.

30.2.2 Power dissipation

This gives a measure of the power which the digital circuit draws from the supply. It is measured as the product of the supply voltage and the mean supply current for given operating conditions such as speed and output loading. Low power dissipation is an obvious advantage for portable equipment. However, since the amount of power which can be dissipated from an integrated circuit package is limited, the lower the dissipation, the greater the amount of circuit which can be built into a silicon die.

30.2.3 Speed–power product

Compromise is generally required in the speed and power dissipation since a circuit which is designed to have high speed also has a high dissipation. *Figure 30.1* shows typical speed–power curves where the speed is measured in terms of gate delay. Curve A has an overall better speed–power product than B.

30.2.4 Current source and sink

This measures the amount of current which the digital circuit can interchange with an external load. Generally digital circuits interconnect with others of the same family and this parameter is defined as a 'fan-out', which is the number of similar gates which can be driven simultaneously from one output. The 'fan-in' of a circuit is the number of its parallel inputs.

30.2.5 Noise susceptibility and generation

Digital circuits can misoperate if there is noise in the system. This may be slowly changing noise, such as drift in the power supplies, or high energy spikes of noise. Some types of logic families can tolerate more noise than others, and generally the faster the inherent speed of the logic, the more it is likely to be affected by transient noise.

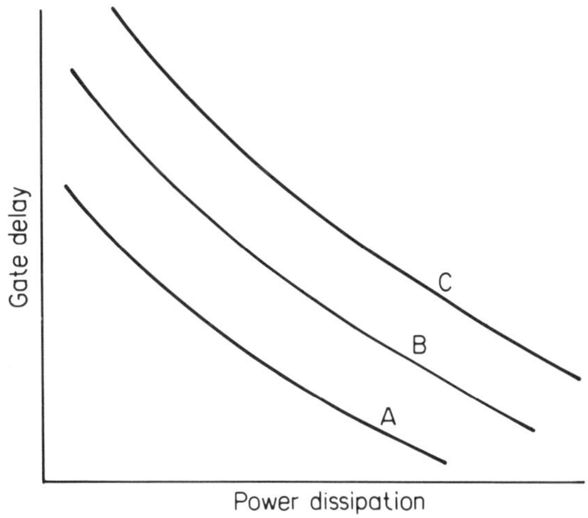

Figure 30.1 Speed–power curves

Digital circuits also generate noise when they switch and this can affect adjacent circuits. Noise generation is a direct function of the current being switched in the circuit and the speed with which it is changed.

30.3 Bipolar circuits

30.3.1 Saturating circuits

One of the earliest digital circuit families was that based on the resistor–transistor logic (RTL) gate shown in *Figure 30.2(a)*. If any of the transistors TR_1 to TR_3 is turned on by a signal on its base, then the output voltage at D falls to a low value. The disadvantage of this circuit is that it has low noise immunity, since the noise voltage has only to overcome one base–emitter junction voltage, and it is slow since input capacitors have to be charged and discharged via the input resistors.

By putting capacitors across input resistors, R_2 to R_4, the operating speed is increased, and this is known as resistor–capacitor–transistor logic (RTCL). The disadvantage of RTL and RTCL is that they occupy a large area of silicon because of the passive components.

Although diode–transistor logic (DTL) is still used, it is becoming obsolete. Input resistors are replaced by diodes.

Some DTL circuits include a second transistor, as in *Figure 30.2(b)*, to increase the circuit gain. Diodes D_1 and D_2 increase the noise immunity. For operation in very noisy environments high noise immunity logic (HNIL) is used. This also known as high threshold logic (HTL). This is similar to DTL except that a zener diode D_1 is introduced to increase the voltage levels. This circuit is also now largely obsolete.

The most popular digital family is transistor–transistor logic (TTL) and it is available in many configurations, a few of these being shown in *Figure 30.3*. If any of the inputs A, B or C is low in *Figure 30.3(a)*, then TR_3 is off. Transistors TR_1 and TR_2 form a totem-pole output stage in which the output at D swings between V_{cc} and 0 volts. Since only one of the output transistors is on at a time, resistor R_1 can be made low valued without risk of overdissipation. This enables load capacitances to be charged and discharged rapidly, so that operating speed is increased. However, due to charge storage effects, a transistor switches off more slowly than it switches on, so that for a short time both

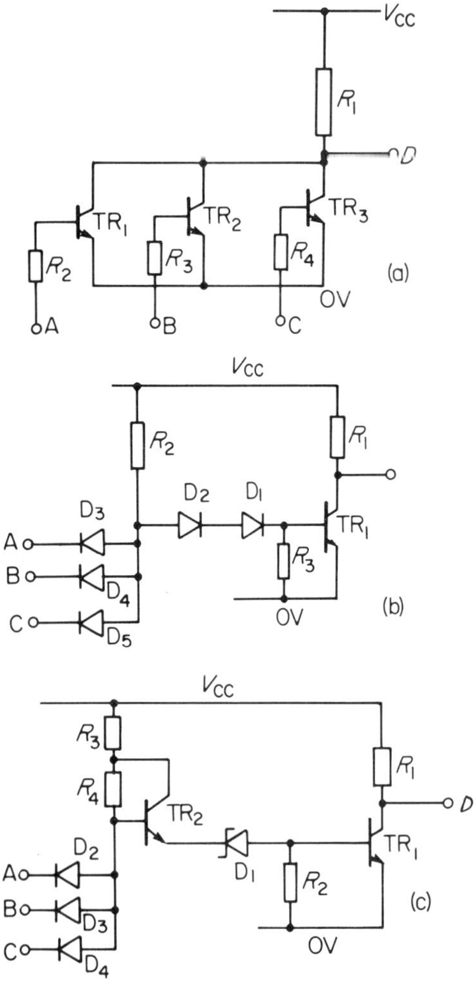

Figure 30.2 Earlier logic families: (a) RTL; (b) DTL; (c) HNIL

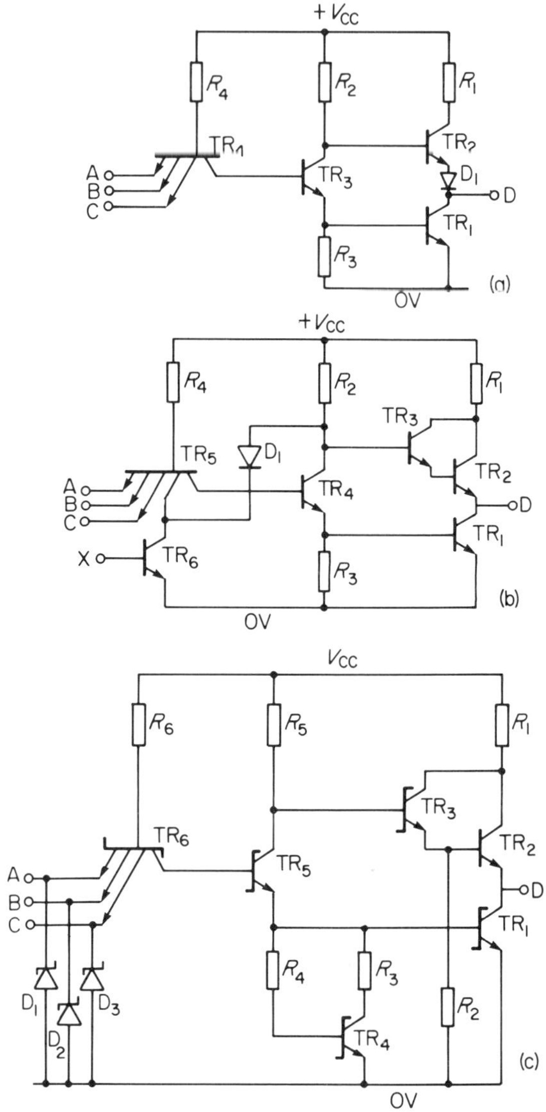

Figure 30.3 Transistor–transistor logic: (a) basic gate; (b) gate with Darlington output and tri-state control; (c) Schottky gate

transistors are on, generating current noise. This can be reduced by making R_2 and R_1 of high ohmic value, and this also reduces the power drawn from the supply. These are called low power TTL gates (LPTTL).

Figure 30.3(b) shows a Darlington output gate with tri-state control. When input X is taken to a positive voltage, both TR_2 and TR_1 are off, so that the output is at a high impedance. This feature enables the output of several TTL circuits to be connected in parallel.

30.3.2 Non-saturating circuits

The circuits illustrated in the last section are called saturating since the transistors operate either in the off mode or fully saturated. If the transistor is prevented from saturating, then hole storage effects are reduced and the operating speed can be increased.

One method of reducing the hole storage within a device is to dope it with gold. This is expensive and can result in yield loss. A better technique is to use Schottky transistors in the TTL gate. Because of the clamping effect of the collector–base diode, a Schottky transistor can be prevented from saturating even when fully on, and it can therefore be designed to have a higher gain than a conventional transistor. However, the Schottky junction introduces capacitances so that careful selection of circuit components is required.

Schottky transistor–transistor logic (TTL-S) and LPTTL-S, which compensates for some of the speed lost in LPTTL circuits, are currently popular logic families.

The fastest logic family commercially available is emitter-coupled logic (ECL), and it is also non-saturating. *Figure 30.4* shows the construction of a three-input ECL gate. A stable reference voltage is generated at point X by transistor TR_3 and its associated circuitry. Transistor TR_4 and the three input transistors form a differential amplifier. Depending on the input voltage relative to that at X, the current is switched through the circuit and either TR_1 or TR_2 is on.

In *Figure 30.4* none of the transistors saturate so the circuit has a high operating speed. The output transistors also have a low impedance when turned on so that they can provide high current drives and rapidly charge external circuit capacitances. The gate

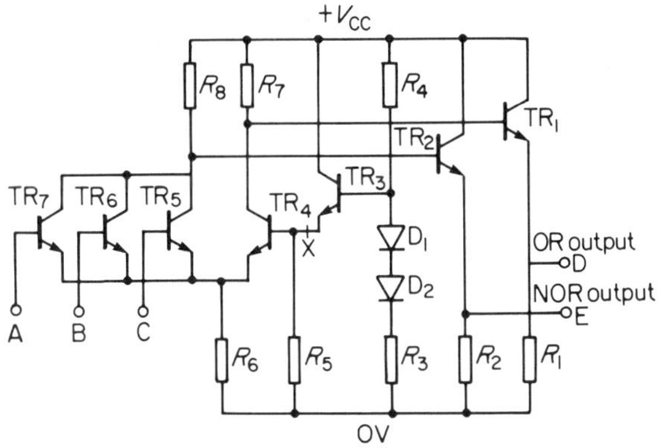

Figure 30.4 An ECL gate

is also capable of producing both OR or NOR functional outputs, and since it draws almost constant current from the supply it generates negligible switching current spikes. However, the logic voltage swings are low, usually below one volt, so that it is difficult to directly interface ECL with other types of logic.

30.3.3 Bipolar LSI

Unipolar logic is still more popular than bipolar in large scale integration (LSI). However, several circuits have now been designed which show some advantage over unipolar in certain LSI applications. Only one such system, integrated injection logic (IIL) is described here.

In IIL all emitters go to a common point and so do the bases of transistors connected to the same resistor. Only the collectors of the transistors are separate. The driving base resistor is replaced by a current source which injects current into the output transistor, as shown in *Figure 30.5*. By using lateral transistors,

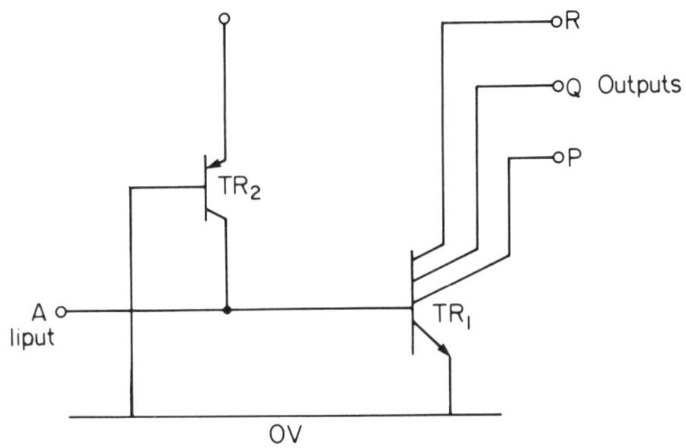

Figure 30.5 Integrated injection logic (IIL)

the pnp and npn devices can be merged together, and since no insulating regions or resistors are used, the circuit density on a die is increased, which makes this logic suitable for LSI applications.

The noise immunity and speed of an IIL gate is improved as the injection current is increased. However, power dissipation also increases. The speed–power product is low, and this means that if very high speed is not required, the power dissipation can be made low, enabling large LSI devices to be built.

Figure 30.6 shows a typical curve for the propagation delay–

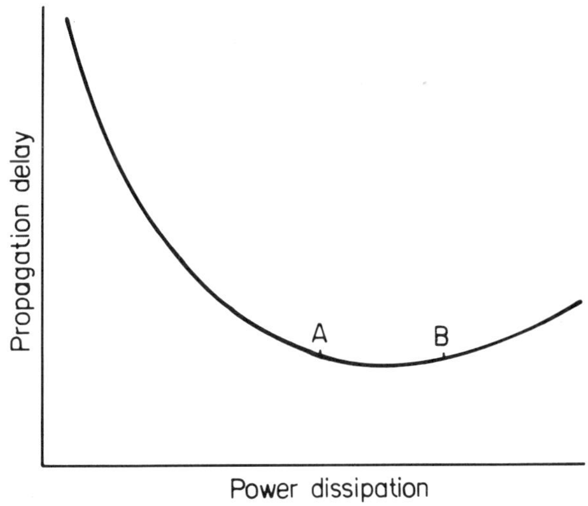

Figure 30.6 Propagation delay–power dissipation curve for IIL

power dissipation product of an IIL gate, as the injection current is increased. Up to point A this product is substantially independent of injection current, since the power dissipation increase is roughly proportional to the fall in propagation delay. Between points A and B the delay is constant since it is primarily due to active charge in the transistors. Beyond point B the base series resistance prevents rapid removal of accumulated charge so that propagation delay increases.

30.4 Unipolar logic circuits

30.4.1 CMOS logic

Complementary MOS (CMOS) is the only unipolar logic family which has found widespread use for SSI and MSI circuits. *Figure 30.7* shows the arrangement for an inverter, NAND gate and NOR gate. The circuit arrangement is similar to TTL with totem-pole outputs. Like TTL, the CMOS gate passes through a switching stage when all output transistors are on simultaneously. However, a series resistor is not now required to limit the current drawn from the power supply, since unipolar transistors have a higher inherent impedance then bipolar transistors.

CMOS logic can operate over a wide range of supply voltages. The logic 0 and logic 1 input levels are also usually guaranteed as 30% and 70% of V_{DD} so that the noise immunity band is high, as shown in *Figure 30.8*: the higher the supply voltage, the greater the absolute noise immunity.

The power dissipation of CMOS logic is low, but it increases with operating frequency since load and parasitic capacitances need to be charged and discharged.

30.4.2 Unipolar LSI logic

Unipolar circuits are generally capable of high circuit densities on a chip. They are therefore better suited to LSI rather than MSI systems, where the chip size is often limited by the bonding pad areas.

Negative logic notations are normally used in discussing logic performance. *Figure 30.9* for instance shows a NAND gate in which V_{DD} is negative with respect to V_{SS} which is at ground. TR$_1$ is a load transistor. When inputs A and B are at logic 1 (negative), the output goes to logic 0 (positive).

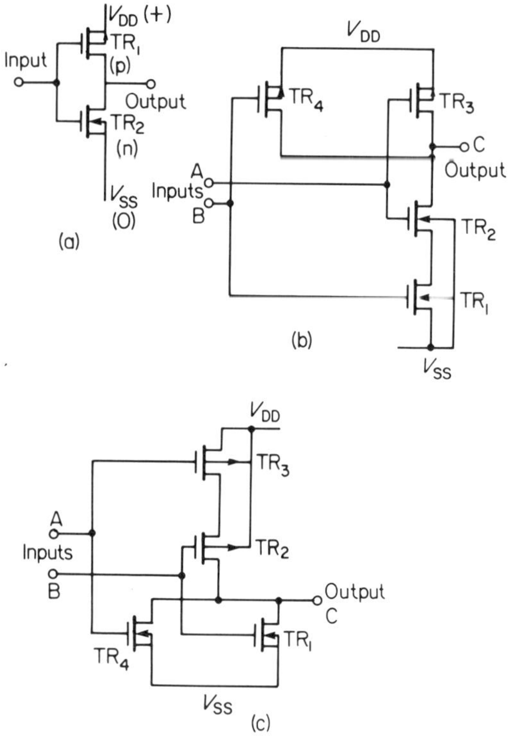

Figure 30.7 CMOS logic circuit: (a) inverter; (b) NAND gate; (c) NOR gate

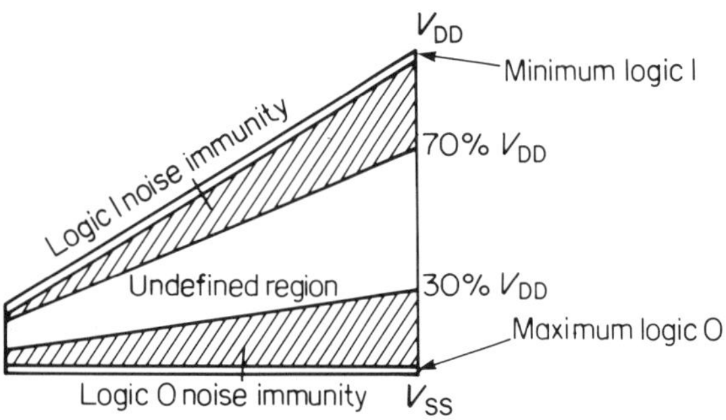

Figure 30.8 Guaranteed noise immunity bands for CMOS

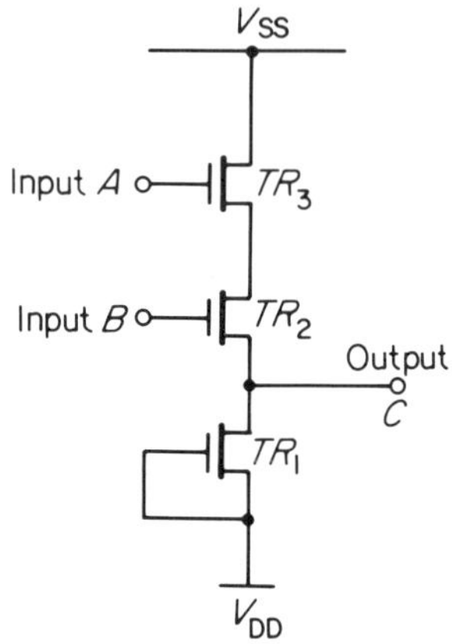

Figure 30.9 Unipolar NAND gate

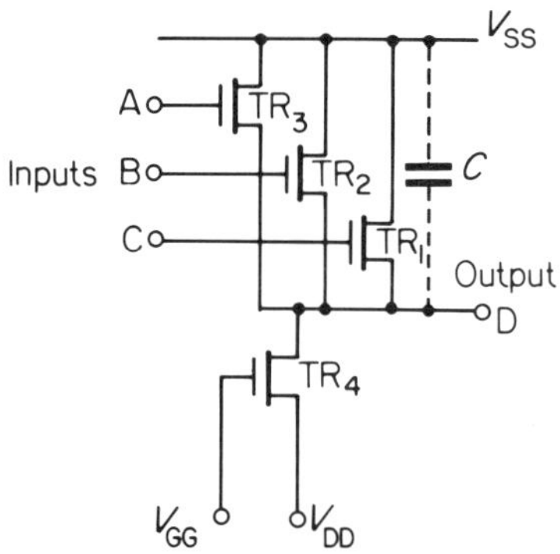

Figure 30.10 Unipolar NOR gate

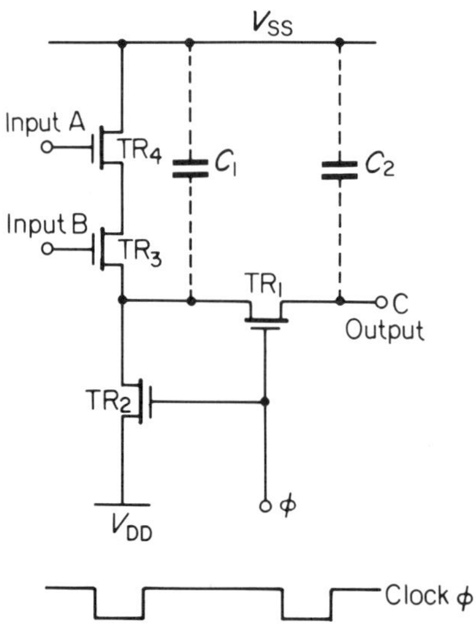

Figure 30.11 Dynamic two-phase NAND gate

A NOR gate is shown in *Figure 30.10*. A separate supply is used to bias the gate of the load transistor where $|V_{GG}| = |V_{DD} + V_T|$ to allow full logic swing on the output. The output is at logic 1 so long as none of the inputs goes to logic 1.

Both the previous figures illustrated static circuits. Dynamic circuits are commonly used in unipolar LSI systems where the logic charge is held on parasitic capacitors. This reduces dissipation since transistors are on for a short time only, and it increases the circuit density on the die.

Figure 30.11 shows a simple dynamic NAND gate. Transistors TR$_1$ and TR$_2$ come on when the clock goes negative. Depending on the state of the inputs at A and B at this time the parasitic capacitors C_1 and C_2 will be charged or discharged. TR$_1$ goes off during the positive clock period and this turns TR$_2$ off. But the charge on C_2 is maintained, provided the off period is not long.

Unipolar transistors can be enhancement mode or depletion mode, although enhancement mode devices are used in all the circuits described so far. *Figure 30.12* shows inverters with three types of loads, and *Figure 30.13* gives their characteristics. Depletion loads give larger currents and therefore shorter

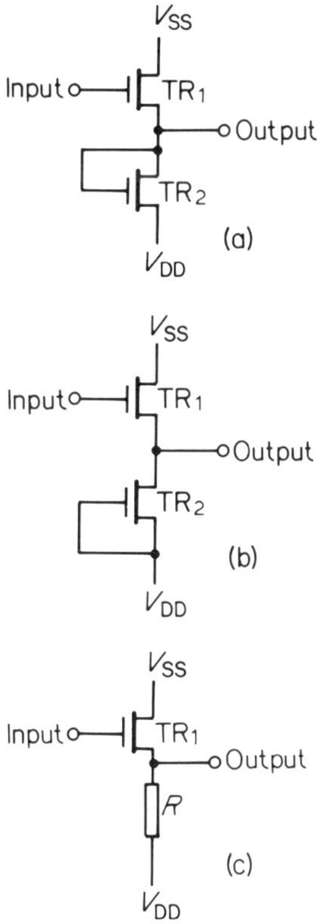

Figure 30.12 Unipolar inverters with different loads: (a) depletion; (b) enhancement; (c) resistive

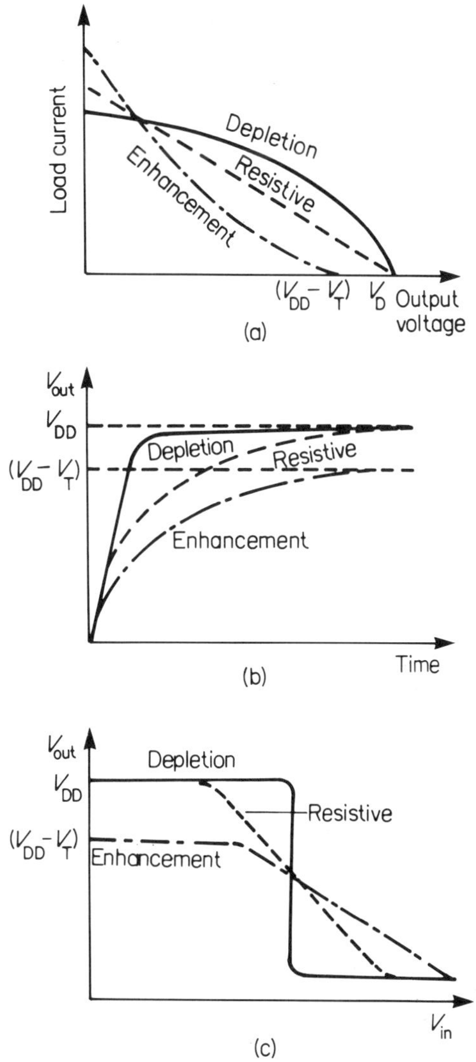

Figure 30.13 Inverter characteristics with different loads: (a) load current; (b) output risetime; (c) transfer curve

risetimes so they can generally work at higher speeds. They also have the squarest transfer characteristic, which gives the best d.c. noise immunity.

30.5 Comparisons

Table 30.1 compares the salient properties of logic families. A comparative system is used, since the absolute values can vary between vendors and in the trade-off adapted during fabrication. Currently Schottky TTL and CMOS are the most commonly used SSI and MSI families, with ECL being used for high speed systems. NMOS is used for most large memory systems.

Table 30.1 Comparison of logic families (1 = best, 6 = worst)

Logic family	*Speed*	*Power dissipation*	*Fan-out*	*Noise immunity*	*Noise generation*
DTL	4	4	3	4	2
TTL	3	4	3	3	3
TTL-S	2	5	3	3	3
LPTTL-S	3	3	3	3	3
ECL	1	7	2	3	1
NMOS	5	2	2	2	2
CMOS	6	1	1	1	2

30.6 Gates

The circuits which are described in this, and subsequent sections can be fabricated using any of the logic technologies described earlier. The electrical characteristics such as speed and power consumption will be determined by the logic family but the functional performance will be the same in all cases.

Figure 30.14 shows commonly used gates and *Table 30.2* gives the functional performance in logic 1 and 0 states. Gates are usually available as hex inverter, quad two input, treble three input, dual four input and single eight input. AND–OR–INVERT gates are also available and these are used to connect the outputs of two gates together in a wired-OR connection.

In a CMOS transmission gate a p- and an n-channel transistor are connected together. A gate signal turns on both transistors and so provides an a.c. path through them, whereas when the transistors are off the gate blocks all signals.

Gates are also made in Schmitt trigger versions. These operate as the discrete component circuits and exhibit an hysteresis between the on and off switching positions.

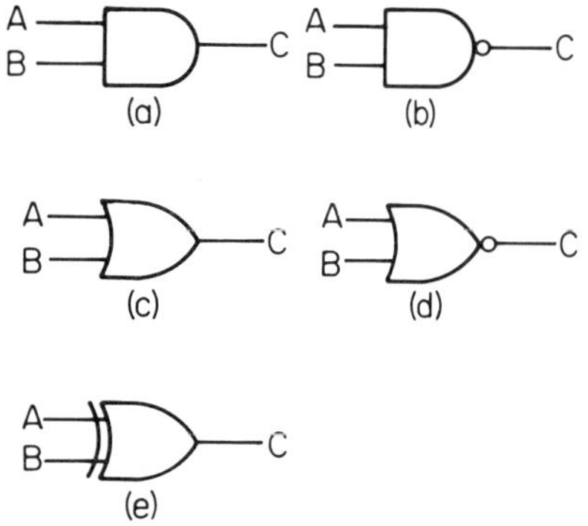

Figure 30.14 Symbols for commonly used gates: (a) AND; (b) NAND; (c) OR; (d) NOR; (e) exclusive-OR

Table 30.2 Truth table of logic gates

Inputs		*Output (C)*				
A	*B*	*AND*	*NAND*	*OR*	*NOR*	*Exclusive-OR*
0	0	0	1	0	1	0
0	1	0	1	1	0	1
1	0	0	1	1	0	1
1	1	1	0	1	0	0

30.7 Flip-flops

Flip-flops are bistable circuits which are mainly used to store a bit of information. The simpler types of flip-flops are also called latches. *Figure 30.15* shows the symbol for some of the more commonly used flip-flops. There are many variations such as master-slave *J–K*, edge-triggered and gated flip-flops.

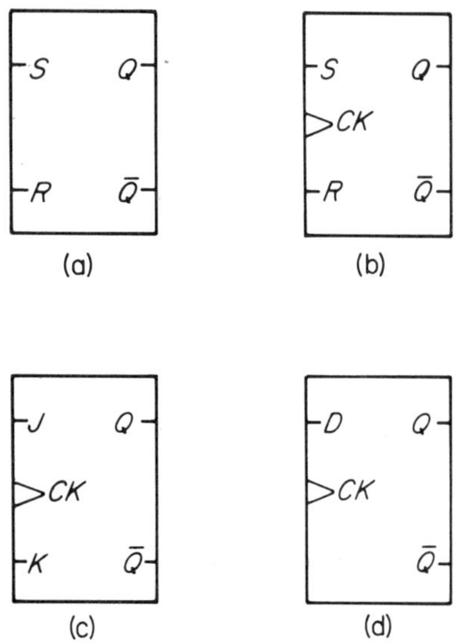

Figure 30.15 Symbols for commonly used flip-flops: (a) set-reset; (b) Master-salve set-reset; (c) *J–K*; (d) *D*-type

Table 30.3 Truth table of a set-reset flip-flop

S	*R*	Q_{n+1}	$\bar{Q}_{n+1}$
0	0	Q_n	$\bar{Q}_n$
0	1	0	1
1	0	1	0
1	1	?	?

Table 30.3 illustrates the operation of the set-reset flip-flop. Q_n represents the state of the Q output before the S and R inputs assume the logic state show, and Q_{n+1} is the state after this event. It is seen that with S and R both at logic zero, the outputs are unchanged. Otherwise the Q output will be set to the value of the S input and it will maintain this state even after the S and R inputs both return to zero. For S and R inputs both at logic 1, the output state will be forced to a logic 1, but when the S and R inputs return to zero the value of the Q and $\bar{Q}$ outputs is indeterminate.

In the master-slave set-reset flip-flop, whose operation is shown in *Table 30.4*, a clock pulse is required. During the rising edge of the clock information is transferred from the S and R inputs to the master part of the flip-flop. The outputs are unchanged at this stage. During the falling edge of the clock the inputs are disabled so that they can change their state without affecting the information stored in the master section. However, during this phase the information is transferred from the master to the slave of the flip-flop, so that the outputs change to the original S and R values. The flip-flop is disabled when no clock pulse is present.

Table 30.4 Truth table of a master-slave set-reset flip-flop (× = don't care state)

S	*R*	*CK*	Q_{n+1}	$\bar{Q}_{n+1}$
×	×	0	Q_n	$\bar{Q}_n$
0	0	⎍	Q_n	$\bar{Q}_n$
0	1	⎍	0	1
1	0	⎍	1	0
1	1	⎍	?	?

J–K flip-flops, illustrated in *Table 30.5*, are triggered on an edge of the clock waveform. Feedback is used internally within the logic such that the indeterminate state, when both inputs are equal to logic 1, is avoided. Now when the inputs are both at 1, the output will continually change state on each clock pulse. This is also known as toggling.

Table 30.5 Truth table of a *J–K* flip-flop (× = don't care state)

J	*K*	*CK*	Q_{n+1}	$\bar{Q}_{n+1}$
×	×	0	Q_n	$\bar{Q}_n$
0	0	↑	Q_n	$\bar{Q}_n$
0	1	↑	0	1
1	0	↑	1	0
1	1	↑	$\bar{Q}_n$	Q_n

D-type flip-flops have a single input. An internal inverter circuit provides two signals to the J and K inputs so that it operates as a *J–K* flip-flop, having only two input modes, as in *Table 30.6*.

Table 30.6 Truth table of a *D*-type flip–flop

D	*CK*	*Q*	$\bar{Q}$
0	↑	0	1
1	↑	1	0

30.8 Counters

Flip–flops can be connected together to form counters in a single package. There are primarily two types, asynchronous and synchronous. The synchronous counters may be subdivided into those with ripple enable and those with parallel or look-ahead carry.

In an asynchronous counter the clock for the next stage is obtained from the output of the preceeding stage so that the command signal ripples through the chain of flip–flops. This causes a delay so asynchronous counters are relatively slow, especially for large counts. However, since each stage divides the output frequency of the previous stage by two, the counter is useful for frequency division.

In a synchronous counter, the input line simultaneously clocks all the flip–flops so there is no ripple action of the clock signal from one stage to the next. This gives a faster counter, although it is more complex since internal gating circuitry has to be used to enable only the required flip–flops to change state with a clock pulse. The enable signal may be rippled through between stages or parallel (or look-ahead) techniques may be used, which gives a faster count.

Counters are available commercially, having a binary or BCD count, and which are capable of counting up or down.

30.9 Shift registers

When stored data are moved sequentially along a chain of flip–flops, the system is called a shift register. *Figure 30.16* shows a four-bit register, although commercial devices are available in sizes from four bits to many thousands of bits.

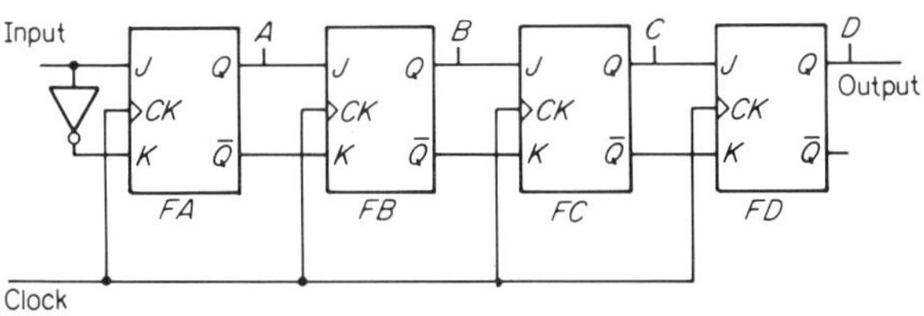

Figure 30.16 A four-bit serial-in, serial-out shift register

The shift register shown in *Figure 30.16* is serial-in, serial-out but it is possible to have systems which are parallel data input and output. The only limitation is the number of pins available on the package to accommodate the inputs and outputs. Shift registers can also be designed for left or right shift.

In the register of *Figure 30.16* input data ripple through at the clock pulse rate from the first to the last stage. Sometimes it is advantageous to be able to clock the inputs and outputs at different rates. This is achieved in a first-in, first-out (FIFO) register which is illustrated in *Figure 30.17*. Each data bit has an associated status bit and input data are automatically moved along until it reaches the last unused bit in the chain. Therefore the first data to come in will be the first to be clocked out. Data

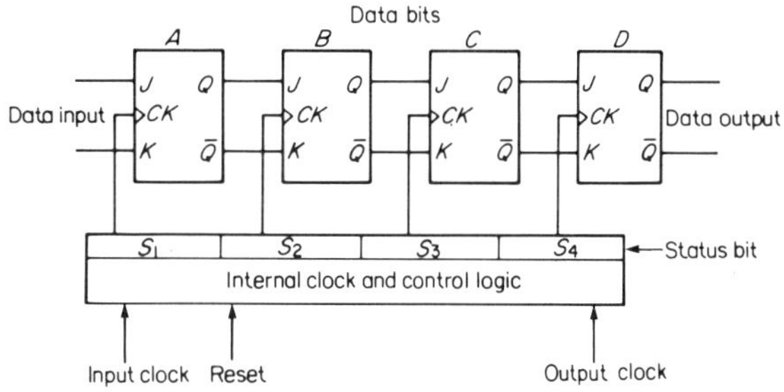

Figure 30.17 First-in, first-out shift register

can also be clocked into and out of the register at different speeds, using the two independent clocks.

30.10 Data handling

Several code converter integrated circuits are available commercially. *Figure 30.18* shows a BCD to decimal converter and *Table 30.7* gives its truth table. These converters can also be used as priority encoders. For example *Figure 30.18* may be considered to be a ten-input priority encoder. Several of the lines

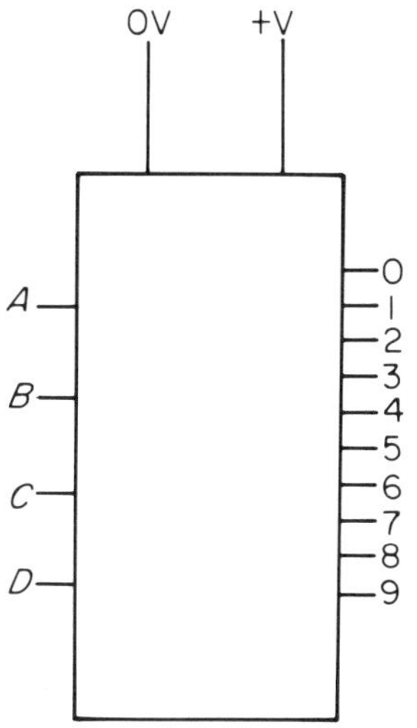

Figure 30.18 Symbol for a BCD to decimal converter

0 to 9 may be energised simultaneously but the highest number will generate a BCD output code on lines A to D.

Figure 30.19 shows an eight-channel multiplexer and *Table 30.8* gives its truth table. The channel select lines connect one of the eight data input lines to the output line using BCD code.

30.11 Timing

A variety of commercial devices are available to give monostable and astable multivibrators. Most of these incorporate control gates so that they are more versatile when used in digital systems. *Figure 30.20* illustrates a gated monostable, and its truth table, which shows the trigger mode, is given in *Table 30.9*. The monostable will trigger when the voltage at *T* goes to a logic 1.

Table 30.7 Truth table of a BCD to decimal converter

BCD code				*Outputs*									
A	*B*	*C*	*D*	*0*	*1*	*2*	*3*	*4*	*5*	*6*	*7*	*8*	*9*
0	0	0	0	1	0	0	0	0	0	0	0	0	0
0	0	0	1	0	1	0	0	0	0	0	0	0	0
0	0	1	0	0	0	1	0	0	0	0	0	0	0
0	0	1	1	0	0	0	1	0	0	0	0	0	0
0	1	0	0	0	0	0	0	1	0	0	0	0	0
0	1	0	1	0	0	0	0	0	1	0	0	0	0
0	1	1	0	0	0	0	0	0	0	1	0	0	0
0	1	1	1	0	0	0	0	0	0	0	1	0	0
1	0	0	0	0	0	0	0	0	0	0	0	1	0
1	0	0	1	0	0	0	0	0	0	0	0	0	1
1	0	1	0	0	0	0	0	0	0	0	0	0	0
1	0	1	1	0	0	0	0	0	0	0	0	0	0
1	1	0	0	0	0	0	0	0	0	0	0	0	0
1	1	0	1	0	0	0	0	0	0	0	0	0	0
1	1	1	0	0	0	0	0	0	0	0	0	0	0
1	1	1	1	0	0	0	0	0	0	0	0	0	0

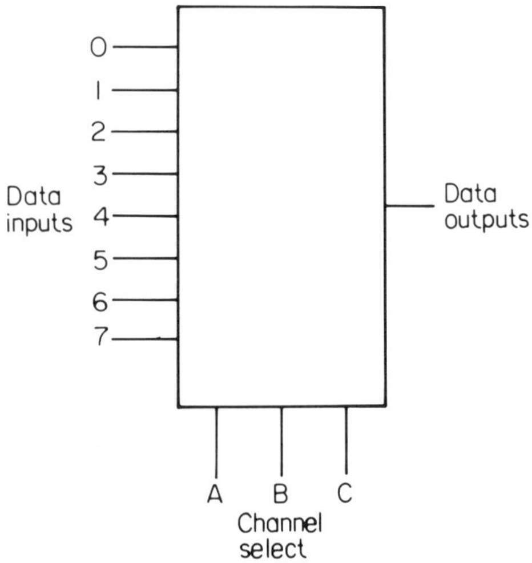

Figure 30.19 Symbol for an eight-channel multiplexer

External resistors and capacitors are used to vary the duration of the monostable pulse. By feeding the $\bar{Q}$ line back to the *D* input, this circuit can also be operated as an astable multivibrator.

Figure 30.21 shows a versatile timer circuit, which is available commercially as the 555 family. External timing components are again used, and the circuit can be interconnected to provide a variety of monostable and astable functions.

30.12 Drivers and receivers

Digital integrated circuits have limited current and voltage drive capability. To interface to power loads, driver and receiver circuits are required, which are also available in an integrated circuit package. The simplest circuit in this category is an array of transistors. Usually the emitters or collectors of the transistors are connected together inside the package to limit the package pin requirements.

Table 30.8 Truth table of an eight-channel multiplexer (× = don't care state)

Channel select			*Data inputs*								*Data outputs*
A	*B*	*C*	*0*	*1*	*2*	*3*	*4*	*5*	*6*	*7*	
0	0	0	0	×	×	×	×	×	×	×	0
0	0	0	1	×	×	×	×	×	×	×	1
0	0	1	×	0	×	×	×	×	×	×	0
0	0	1	×	1	×	×	×	×	×	×	1
0	1	0	×	×	0	×	×	×	×	×	0
0	1	0	×	×	1	×	×	×	×	×	1
0	1	1	×	×	×	0	×	×	×	×	0
0	1	1	×	×	×	1	×	×	×	×	1
1	0	0	×	×	×	×	0	×	×	×	0
1	0	0	×	×	×	×	1	×	×	×	1
1	0	1	×	×	×	×	×	0	×	×	0
1	0	1	×	×	×	×	×	1	×	×	1
1	1	0	×	×	×	×	×	×	0	×	0
1	1	0	×	×	×	×	×	×	1	×	1
1	1	1	×	×	×	×	×	×	×	0	0
1	1	1	×	×	×	×	×	×	×	1	1

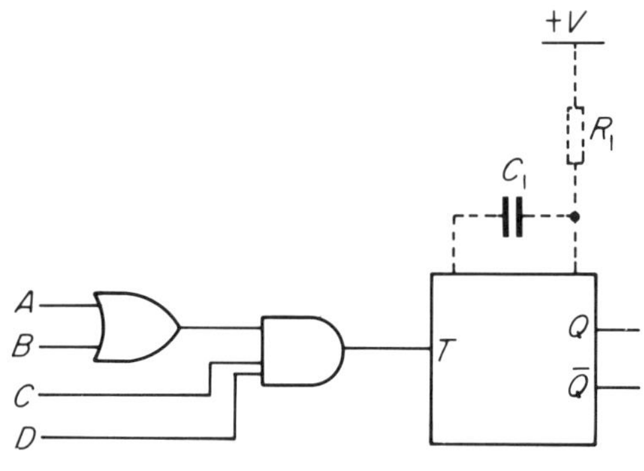

Figure 30.20 Logic diagram of a gated monostable

Table 30.9 Truth table of a monostable (↓ = 0 to 1 change, ↑ = 1 to 0 change, × = don't care)

A	*B*	*C*	*D*
↓	1	1	1
1	↓	1	1
0	×	↑	1
×	0	↑	1
0	×	1	↑
×	0	1	↑

Figure 30.22 shows a circuit used for driving inductive loads. The output transistors have high voltage ratings and freewheeling diodes are also provided inside the package. Four logic inputs are used for each output stage.

For digital transmission systems line drivers and receivers are available as in *Figure 30.23*. The digital input on the line driver controls an output differential amplifier stage, which can operate

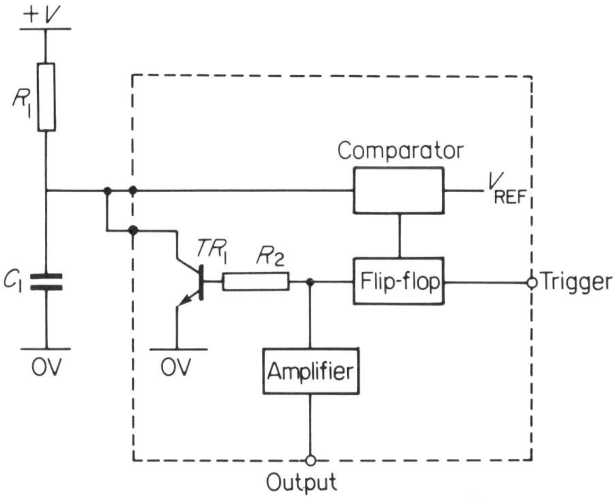

Figure 30.21 A 555 family timer

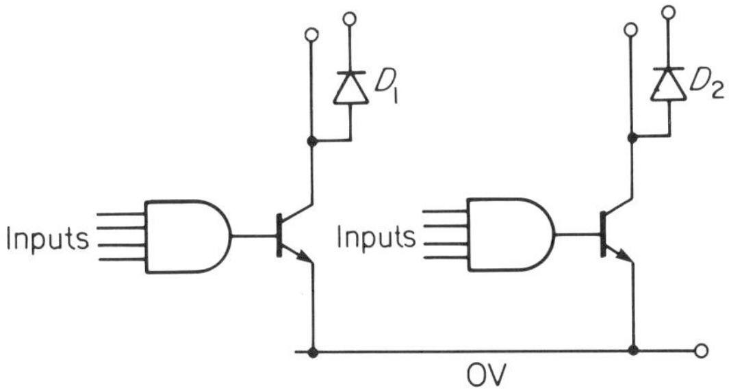

Figure 30.22 Dual driver

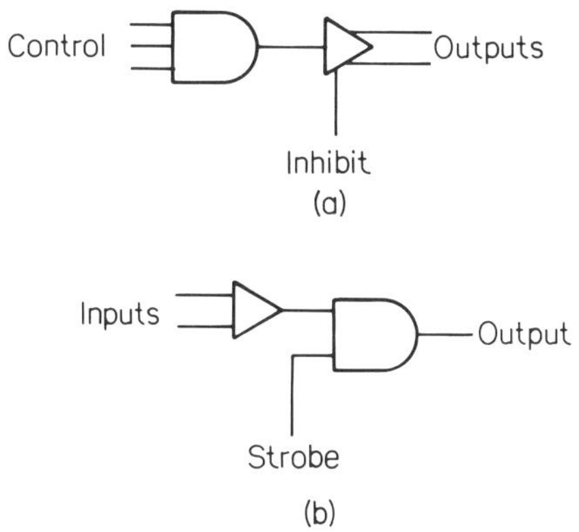

Figure 30.23 Line drivers and receivers: (a) symbol for a line driver; (b) symbol for a line receiver

into low impedance lines. The line receiver can sense low level signals via a differential input stage and provide a logic output.

30.13 Adders

Adders are the basic integrated circuit units used for arithmetic operations such as addition, subtraction, multiplication and division. A half adder adds two bits together and generates a sum

Figure 30.24 Symbol for a single bit of a full adder

Table 30.10 Truth table of a full adder

Carry in C_i	*Input bits being added*		*Sum* S	*Carry out* C_o
	A	*B*		
0	0	0	0	0
0	0	1	1	0
0	1	0	1	0
0	1	1	0	1
1	0	0	1	0
1	0	1	0	1
1	1	0	0	1
1	1	1	1	1

and carry bit. A full adder has the facility to bring in a carry bit from a previous addition. *Figure 30.24* shows one bit of the full adder and *Table 30.10* gives its truth table.

The basic single-bit adder can be connected in several ways to add multi-bit numbers together. In a serial adder the two numbers are stored in shift registers and clocked to the A and B inputs one bit at a time. The carry out is delayed by a clock pulse and fed back to the adder as a carry in.

Serial adders are slow since the numbers are added one bit at a time. *Figure 30.25* shows a parallel adder which is faster. The

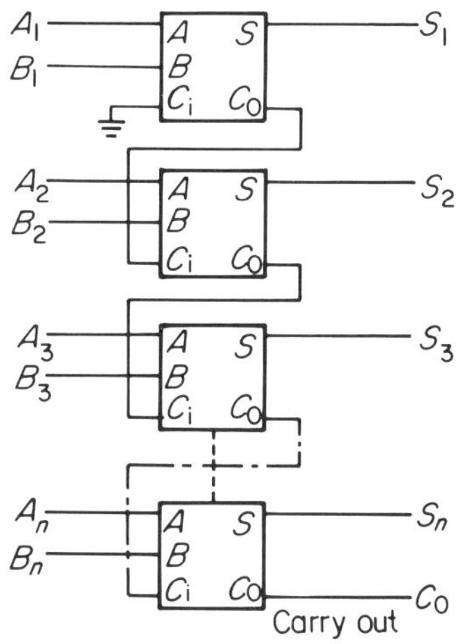

Figure 30.25 A parallel adder with ripple carry

carry output ripples through from one bit to the next so that the most significant bit cannot show its true value until the carry has rippled right through the system. To overcome this delay, look-ahead carry generators may be used. These take in the two numbers in parallel, along with the first carry-in bit, and generate the carry input for all the remaining carry bits.

Adders can be used as subtractors by taking the two's

complement of the number being subtracted and then adding. Two's complementing can be obtained by inverting each bit and then adding one to the least significant bit. This can be done within the integrated circuit so that commercial devices are available which can add or subtract, depending on the signal on the control pin.

Adders are used for multiplication by a process of shifting and addition, and division is obtained by subtraction and shifting.

30.14 Magnitude comparators

A magnitude comparator gives an output signal on one of three lines, which indicates which of the two input numbers is larger, or if they are equal. *Figure 30.26* shows one bit of a magnitude comparator and *Table 30.11* its truth table. Multi-bit numbers can be compared by storing them in shift registers and clocking them to the input of the single-bit magnitude comparator, one bit at a time, starting from the most significant bit. Alternatively, parallel comparators may be used where each bit of the two numbers is fed in parallel to a separate comparator bit and the outputs are gated together.

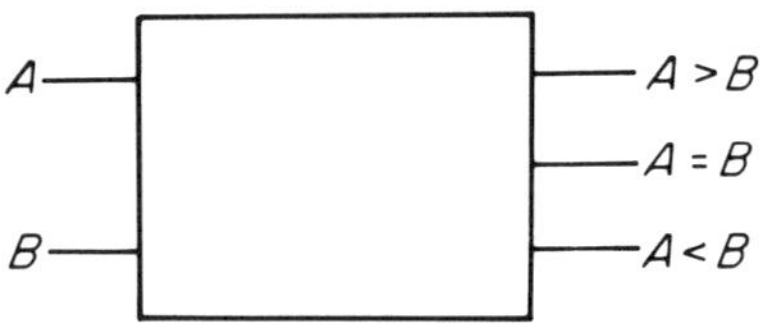

Figure 30.26 Symbol for a one-bit magnitude comparator

Table 30.11 Truth table of a one-bit magnitude comparator

A	B	$A>B$	$A<B$	$A=B$
0	0	0	0	1
0	1	0	1	0
1	0	1	0	0
1	1	0	0	1

30.15 Rate multiplier

In this device, shown in *Figure 30.27*, the input at R consists of a series of pulses and the number of pulses which appears at the output is controlled by the value of the control signal at C, where $O=RC/2^n$. Commercial devices operate in binary or BCD, giving binary or decimal rate multipliers.

Rate multipliers can be connected to give a variety of arithmetic functions. *Figure 30.28(a)* shows an adder for adding the numbers X and Y and *Figure 30.28(b)* shows a multiplier. The clock input in all cases is produced by splitting a single clock into several phases. For the adder $Z=X+Y$ and for the multiplier $Z=XY$.

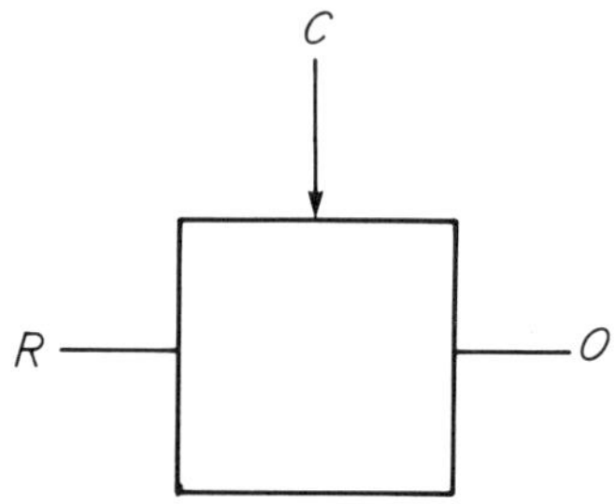

Figure 30.27 Symbol for a binary rate multiplier

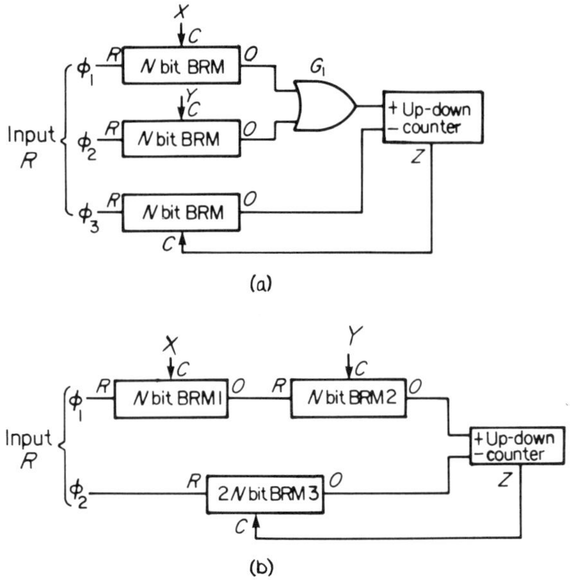

Figure 30.28 Using the binary rate multiplier: (a) addition; (b) multiplication

Acknowledgements

The contents of this chapter have been extracted from *Integrated Circuits* by F. F. Mazda, published by Cambridge Uhiversity Press.

Further reading

MIDDLETON, R. G., *Understanding Digital Logic*, Howard W. Sams, 392 pp (1982)

PATE, R. and BERG, W., 'Observe simple design rules when customizing gate arrays', *EDN* (November 1982)

PROUDFOOT, J. T., 'Programmable logic arrays', *Electronics & Power* (Nov/Dec 1980)

TWADDELL, W., 'Uncommitted IC logic', *EDN* (April 1980)

WALKER, R., 'CMOS logic arrays: a design direction', *Computer Design* (May 1982)

31 Linear Integrated Circuits

A M Pope, BSc
Product Marketing Manager,
formerly with Fairchild Cameras & Instrument (UK) Ltd,
now with The Radio Resistor Co. Ltd
(Sections 31.1–31.10)

M Trowbridge, MA
Product Marketing Engineer
Fairchild Camera & Instrument (UK) Ltd
(Section 31.11)

Contents

31.1 Introduction

The semiconductor industry today classifies almost every monolithic integrated circuit as a 'linear IC' that does not fit neatly into the categories of digital logic and memory/microprocessor circuits. Some 15 years ago, the situation was much simpler as the manufacturers were grappling with the problems of integrating basic operational amplifiers to handle analogue circuit applications in as flexible a way as possible. Then the technology in use was, of course, bipolar. Today this has expanded to embrace all the major process technologies, wherever their particular merits are advantageous, including combinations of different technologies. Hybrid integrated circuits have filled the gaps where the required technologies are impossible to combine on a single chip. The available functions have increased enormously and may currently be categorised as operational amplifiers, interface circuits, voltage regulators, data acquisition circuits and special functions. However, the operational amplifier is still the workhorse of the analogue signal processing world and forms an excellent starting point for any discussion of linear integrated circuits, or LICs, as is a convenient abbreviation.

31.2 Operational amplifiers

The operational amplifier, or op amp, was originally developed in response to the needs of the analogue computer designer. The object of the device is to provide a gain block whose performance is totally predictable from unit to unit and perfectly defined by the characteristics of an external feedback network. This has been achieved by LIC op amps to varying degrees of accuracy, largely governed by unit cost and complexity. Nevertheless, the accuracy of even low cost units has been refined to a point where it is possible to use an op amp in almost any d.c. to 1 MHz amplifier/signal processor application. Indeed, it is probably a deliberate case of 'wheel re-invention' not to use such an approach.

31.2.1 Ideal operational amplifier

The 'ideal' operational amplifier of *Figure 31.1* is a differential input, single ended output device that is operated from bipolar power supplies. As such, it can easily be used as a virtual earth

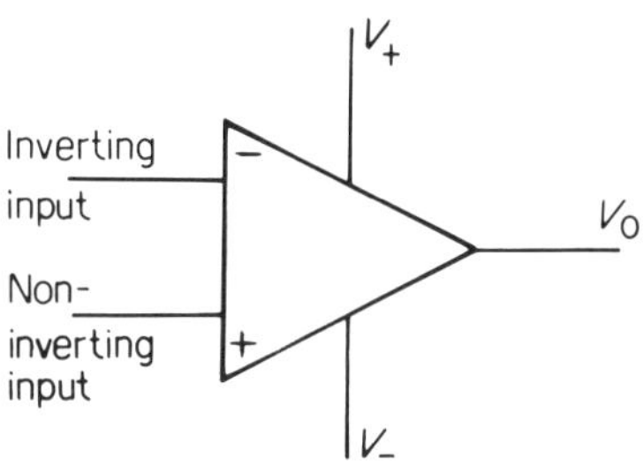

Figure 31.1 Ideal op amp

amplifier. Differential output is possible, although not common, and single supply operation will be discussed later. The ideal op amp possesses infinite gain, infinite bandwidth, zero bias current to generate a functional response at its inputs, zero offset voltage (essentially a perfect match between the input stages) and infinite input impedance. Because of these characteristics, an infinitesimally small input voltage is required at one input with respect to the other to exercise the amplifier output over its full range.

Hence, if one input is held at earth, the other cannot deviate from it under normal operating conditions and becomes a 'virtual earth' of the feedback theory definition. The shortfalls against the ideal of practical op amps are now considered together with their application consequences.

31.2.2 Input bias current

Input bias current affects all applications of op amps. In order for the op amp to function, it is necessary to supply a small current, from picoamperes for op amps with FET inputs to microamperes for junction transistor type inputs. To see the reason for this, *Figure 31.2* shows a typical op amp input stage, using a differential 'long-tailed pair' with a constant current sink in the emitters. The following gain stage is coupled out from the

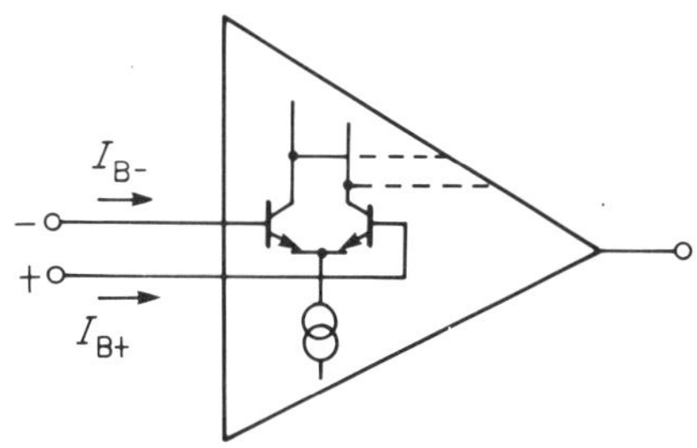

Figure 31.2 Op and amp input bias current path

collector loads. The input bias current is primarily a function of the large signal current gain of the transistors and is defined as the absolute average of I_{B+} and I_{B-}, i.e.

$$I_B = \frac{|I_{B+}| + |I_{B-}|}{2} \tag{31.1}$$

One point that is immediately obvious when considering *Figure 31.2*, but is often forgotten when designing op amp circuitry on a 'black box' approach, is that a d.c. path is always required to the op amp inputs. This is mandatory even if only a.c. amplifiers are under consideration or if op amps of the FET input variety with ultra low bias current, essentially voltage operated input devices, are being used.

31.2.3 Input offset current

Op amp fabrication uses monolithic integrated circuit construction, which can produce very well matched devices for input stages etc, by using identical geometry for a pair of devices diffused at the same time on the same chip. Nevertheless, there is always some mismatch, which gives rise to the input offset current. This is defined as the absolute difference in input bias currents, i.e.

$$I_{os} = |I_{B+} - I_{B-}| \tag{31.2}$$

31.2.4 Effect of input bias and offset current

The above indicates that the effect of input bias current will be to produce an unwanted input voltage, which can be much reduced by arranging for the bias current to each input to be delivered from an identical source resistance. *Figure 31.3* shows a simple inverting amplifier with a gain defined by R_2/R_1, in which R_3 is added to achieve to this effect. In this example the effect of amplifier input impedance is ignored, and signal input impedance is assumed to be zero. In order to balance the bias current source resistance, R_3 is made equal to R_2 in parallel with R_1, i.e.

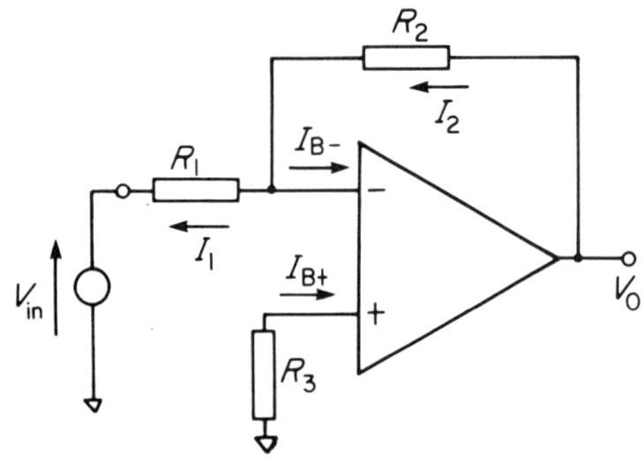

Figure 31.3 Simple inverting amplifier

$$R_3 = \frac{R_1 R_2}{R_1 + R_2} \tag{31.3}$$

Obviously, the values of the feedback network resistors must be chosen so that, with the typical bias current of the op amp in use, they do not generate voltages that are large in comparison with the supplies and operating output voltage levels. The other sources of error, after balancing source resistances, are the input offset current, and the more insidious effects of the drift of input offset and bias current with temperature, time and supply voltage, which cannot easily be corrected. Op amp designers have taken great trouble to minimise the effect of these external factors on the bias and offset currents, but, of course, more sophisticated performance is only obtained from progressively more expensive devices. As might be expected, a high precision application will require an op amp with a high price tag.

The output offset voltage due to input offset current for *Figure 31.3* under zero input signal conditions can be simply derived. The voltages around the input loop must sum to zero hence $I_{B+}R_3$ is the voltage across R_1 and the contribution to I_{B-} flowing through R_1 is $I_{B+}R_3/R_1$. The current through R_2 is then $I_{B-} - (I_{B+}R_3)/R_1$, so:

$$V_o = R_2\left[I_{B-} - \frac{I_{B+}R_3}{R_1}\right] - I_{B+}R_3 \tag{31.4}$$

where V_o is the output offset voltage. Combining equations (31.2), (31.3) and (31.4),

$$V_o = I_{os}R_2 \tag{31.5}$$

Similarly for the condition where input source resistance balancing is ignored:

$$V_o = I_B R_2 \tag{31.6}$$

By a similar process, the effect of input bias and offset current can be calculated for other op amp circuit configurations.

31.2.5 Input offset voltage and nulling

As mentioned before, a mismatch always exists between the input stages of an op amp, and the input offset voltage (V_{os}) is the magnitude of the voltage that, when applied between the inputs, gives zero output voltage. In bipolar input op amps the major contributor to V_{os} is the V_{BE} mismatch of the differential input stage. General purpose op amps usually have a V_{os} in the region of 1 to 10 mV. This also applies to the modern JFET or MOSFET input stage op amps, which achieve an excellent input stage matching with the use of ion implantation. As with the input current parameters, input offset voltage is sensitive to temperature, time and to a lesser extent input and supply voltages. Offset voltage drift with temperature is often specified as μV per mV of initial offset voltage per °C. As a general rule, the lower the offset voltage of an op amp, the lower the temperature coefficient of V_{os} will be.

Enhanced performance low V_{os} op amps are often produced today by correcting or 'trimming' the inherent unbalance of the input stage on chip before or after packaging the device. Techniques used are mainly laser trimming of thin film resistor networks in the input stage and a proprietary process known as 'zener zapping'.

Other op amps are available with extra pins connected for a function known as 'offset null'. Here a potentiometer is used externally by the user with its slider connected to V_+ or V_- to adjust out the unbalance of the device input stage (see *Figure 31.4*). A note of caution should be sounded here when using an op amp with an offset null feature in precision circuitry; the temperature coefficient of V_{os} can be changed quite significantly in some op amp types by the nulling process.

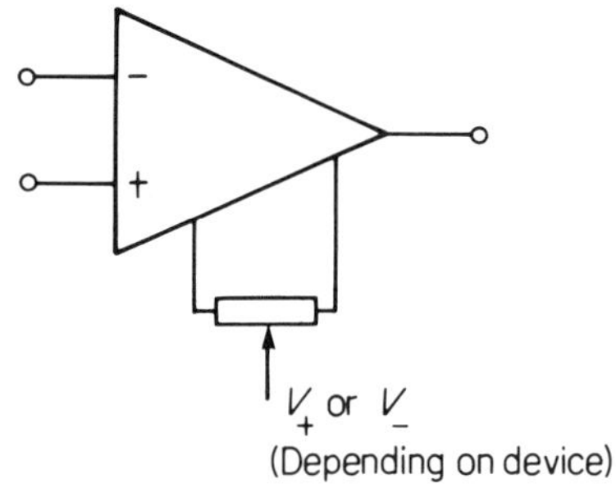

Figure 31.4 Offset nulling

31.2.6 Effect of input offset voltage

As with the input current parameters discussed previously, the input offset voltage will also produce an unwanted output voltage component. Using *Figure 31.3*, and ignoring the effect of input bias currents, then V_{os} will appear across the inputs. With no input signal, V_{os} must appear across R_1, giving I_1 as V_{os}/R_1. Since I_1 must equal I_2 in this ideal case, the output offset voltage, V_o, is given by:

$$V_o = I_2(R_1 + R_2) \tag{31.7}$$

This reduces to:

$$V_o = V_{os}\left(1 + \frac{R_2}{R_1}\right) \tag{31.8}$$

In this example of a simple inverting amplifier, it is clear that the output offset voltage derived from the input offset voltage is $V_{os}(1 + A_{cl})$ where A_{cl} is the closed loop gain. Other configurations will respond to a similar analysis.

31.2.7 Combined effect of input offsets

If the effects of the input offset current and voltage are combined to give a 'real world' equation for the output error voltage, some actual device performance figures in a simple amplifier can be evaluated.

Combining equations (31.6) and (31.8)

$$V_o = V_{os} \quad 1 + \frac{R_2}{R_1} + I_{os}R_2 \tag{31.9}$$

If the values of $R_1 = 10\ \text{k}\Omega$, $R_2 = 100\ \text{k}\Omega$ and $R_3 = 9\ \text{k}\Omega$ are taken, the result is an amplifier with inverting gain of 10 (cf. *Figure 31.3*).

Using the well known μA741 op amp,

$V_{os(max)} = 6$ mV $I_{os(max)} = 200$ nA
Output offset = 86 mV max
Using a precision type of amp (Op-07)

$V_{os(max)} = 250$ μV $I_{os(max)} = 8$ nA
Output offset = 3.55 mV max

This example shows the reduction that is possible in static errors in op amp circuits by improved grade op amps.

31.2.8 Open loop voltage gain

This is one parameter where the practical op amp approaches the infinite gain ideal very closely and typical gains of 250 000 and higher are quite common at zero frequency. However, it is also common for the gain to start to fall off rapidly at low frequencies, e.g. 10 Hz, with many internally compensated op amps (see *Figure 31.5*). The μA741 op amp, for instance, has its gain reduced

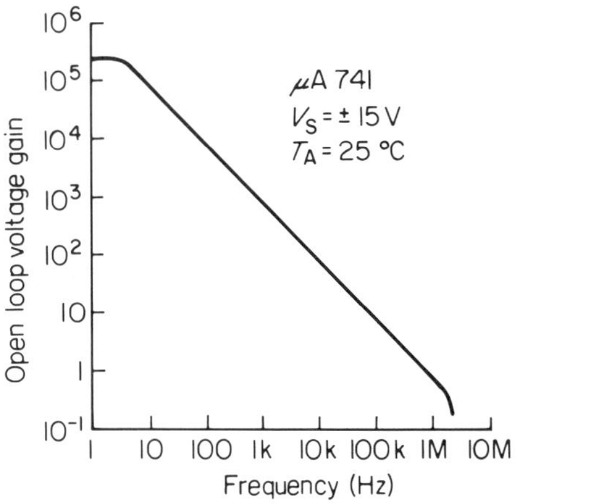

Figure 31.5 Typical op amp open loop frequency response

to unity at around 1 MHz. To assess the significance of this factor, it is necessary to develop a relationship linking the closed loop gain (A_{cl}) to the initial open loop gain (A_{ol}), in terms of accuracy. Obviously the closed loop gain is limited by the open loop gain and feedback theory indicates that the greater the difference between open and closed loop gain, the greater the gain accuracy. For the inverting amplifier of *Figure 31.3* the closed loop gain error is given by:

$$A_{cle} = \frac{100}{1 + A_{ol}R_1/(R_1 + R_2)} \%$$

Using the example of the previous section, the μA741 in an amplifier with a gain of 10 produces a remarkable 0.005% gain accuracy at d.c., with its typical A_{ol} of 2×10^5. This assumes that R_1, R_2 are exact values of course. It is a different story at 10 kHz, where, using the A_{ol} value of *Figure 31.5* gain accuracy will be 10%. To select an op amp for a given d.c. closed loop gain of Y and a gain accuracy of X per cent at a given maximum signal frequency F_{max}, a criterion is found by rearranging the above expression:

$$A_{ol} \geqslant \frac{100(1+Y)}{X} - Y + 1 \qquad (31.11)$$

Note that when the open loop voltage gain falls to the value of the feedback ratio $(R_2 + R_1)/R_1$, the closed loop gain falls by 6 dB. For the gain of ten μA741 example of *Figure 31.3*, this occurs at around 100 kHz. There are also other factors involved in specifying op amp frequency response, the most important of these being slew rate limitations. Due attention must also be paid to the overall circuit stability of the op amp and its associated feedback network impedances. Note that open loop voltage gain is strongly dependent on supply voltage in most op amps.

31.2.9 Slew rate limiting

Slew rate limiting affects all amplifiers where capacitance on internal nodes, or as part of the external load, has to be charged and discharged as voltage levels change. Indeed, the effect is clearly seen in digital logic circuitry as well, as limitations in rise and fall time. It is significant in op amps as it determines the difference between the small signal and power bandwidth. The slew rate limit in most op amps is derived from the capability of the internal circuitry to charge and discharge the compensation capacitance network, which may either be internal or external.

A general purpose op amp such as the μA741 has a typical slew rate of 0.5 V/μs. It is also capable of at least a ± 10 V swing into a 2 kΩ load resistance on ± 15 V supplies. The slew rate controls the large signal pulse response, forcing and op amp output to change from -10 V to $+10$ V in a minimum of 40 μs. Thus the large signal square wave response cannot be greater than 12.5 kHz (50% duty cycle) and, at this frequency, the output is distorted to a triangular wave. The sine wave response is affected by slew rate such that the full peak output voltage V_{pk} is available until the frequency F_{max} at which the sine wave maximum dV/dT exceeds the slew rate. Simple differentiation of a sine function, and taking the maximum rate of change of resultant cosine, leads to the following criterion:

$$\text{slew rate} = 2F_{max}V_{pk} \quad \text{at the limit} \qquad (31.12)$$

As an example, the μA741 at nominal 0.5 V/μs slew rate can produce $10V_{pk}$ (which is 20 V peak to peak or 7.07 V r.m.s.) up to a maximum of 8 kHz. The LM318 high speed op amp with 50 V/μs slew rate could extend the full power bandwidth to 800 kHz. A typical general purpose op amp output voltage capability–frequency graph is shown in *Figure 31.6*. Again note that slew rate is dependent on supply voltage to a large extent.

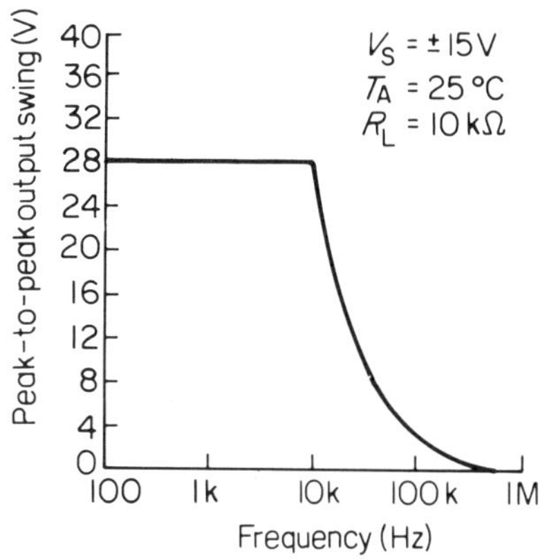

Figure 31.6 Op amp typical output swing versus frequency

31.2.10 Settling time

Frequently in op amp applications, such as digital to analogue converters, the device output is required to acquire a new level within a certain maximum time from a step input change.

The slew rate is obviously a factor in this time, but transient effects will inevitably produce some measure of overshoot and possibly ringing before the final value is achieved. This time, measured to a point where the output voltage is within a specified percentage of the final value, is termed the settling time, usually measured in nanoseconds. Careful evaluation of op amp specifications is required for critical settling time applications, since high slew rate op amps may have sufficient ringing to make their settling times worst than medium slew rate devices.

31.2.11 Output capabilities

As would be expected due to internal design limitations, op amps are unable to achieve an output voltage swing equal to the supply voltages for any significant load resistance. Some devices do achieve a very high output voltage range versus supply, notably those with CMOS FET output stages (BIMOS). All modern op amps have short circuit protection to ground and either supply built in, with the current limit threshold typically being around 25 mA. It is thus good practice to design for maximum output currents in the 5–10 mA region.

Higher current op amps are available but the development of these devices has been difficult due to the heat generated in the output stage affecting input stage drift. Monolithic devices currently approach 1 A and usually have additional circuitry built in to effect thermal overload and output device safe operating area protection, in addition to normal short circuit protection.

31.2.12 Power supply parameters

The supply consumption will usually be specified at ±15 V and possibly other supply voltages additionally. Most performance parameters usually deteriorate with reducing supply voltage. Devices are available with especially low power consumption but their performance is usually a trade-off for reduced slew rate and output capability.

Power supply rejection ratio (PSRR) is a measure of the susceptibility of the op amp to variations of the supply voltage. The definition is expressed as the ratio of the change in input offset voltage to the change in supply voltage (μV/V or dB). No self-respecting op amp ever has a PSRR of less than 70 dB. This should not be taken to indicate that supply bypassing/decoupling is unnecessary. It is good practice to decouple the supplies to general purpose op amps at least every five devices. High speed op amps require careful individual supply decoupling on a by-device basis for maximum stability, usually using tantalum and ceramic capacitors.

31.2.13 Common mode range and rejection

The input voltage range of an op amp is usually equal to its supply voltages without any damage occurring. However, it is required that the device should handle small differential changes at its inputs, superimposed on any voltage level in a linear fashion.

Due to design limitations, input device saturation etc, this operational voltage range is less than the supplies and is termed the common mode input voltage range or swing. As a differential amplifier, the op amp should reject changes in its common mode input voltage completely, but in practice they have an effect on input offset and this is specified as the common mode rejection ratio (CMRR) (μV/V or dB). Note that most general purpose op amps have a CMRR of at least 70 dB (approximately 300 μV offset change per volt of common mode change) at d.c. but this is drastically reduced as the input frequency is raised.

31.2.14 Input impedance

The differential input resistance of an op amp is usually specified together with its input capacitance. Typical values are 2 MΩ and 1.4 pF for the μA741. The input resistance of this type of device exhibits a slow decrease from about 10 kHz.

In practice, the input impedance is usually high enough to be ignored. In inverting amplifiers, the input impedance will be set by the feedback network. In non-inverting amplifiers, the input impedance is 'bootstrapped' by the op amp gain, leading to extremely high input impedance for configurations as *Figure 31.7* which shows the 100% feedback case or voltage follower. This

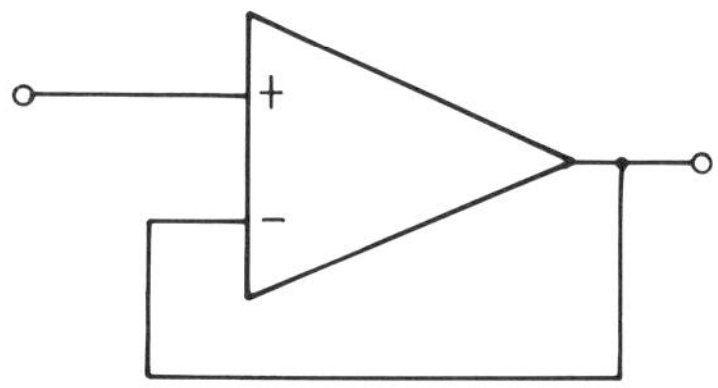

Figure 31.7 Op amp voltage follower

bootstrap effect declines as A_{ol} drops off with rising frequency, but it is safe to assume that the circuit input impedance is never worse than that of the op amp itself.

31.2.15 Circuit stability and compensation

To simplify design-in and use, many op amps are referred to as 'internally' or 'fully' compensated. This indicates that they will remain stable for closed loop gains down to unity (100% feedback). This almost certainly means that performance has been sacrificed for the sake of convenience from the point of view of users who require higher closed loop gains. To resolve this problem, many op amps are available that require all compensation components to be added externally, according to manufacturers' data and gain requirements. A compromise to this split has been the so-called 'undercompensated' op amps, which are inherently stable at closed loop gains of 5 to 10 and above.

The aim of the standard compensation is to shape the open loop response to cross unity gain before the amplifier phase shift exceeds 180°. Thus unconditional stability for all feedback connections is achieved. A method exists for increasing the bandwidth and slew rate of some uncompensated amplifiers known as feedforward compensation. This relies on the fact that the major contributors of phase shift are around the input stage of the op amp and bypassing these to provide a separate high frequency amplifying path will increase the combined frequency response (see *Figure 31.8*). Full details of the technique are available in manufacturers' literature.

31.2.16 Internal circuit design

The circuit design of actual op amps is too detailed and complex a subject to enter into here. Suffice it to say that many circuit techniques not normally available to the discrete component designer are freely used in monolithic designs. In particular, the superior component matching (differential amplifiers) and Wilson current mirrors for biasing, active loads and interstage signal coupling. An approximate internal circuit diagram of a typical μA741 device is shown in *Figure 31.9*.

In brief, the device has a differential input stage, the npn transistors Q_1 and Q_2. The emitter current sinks are formed by

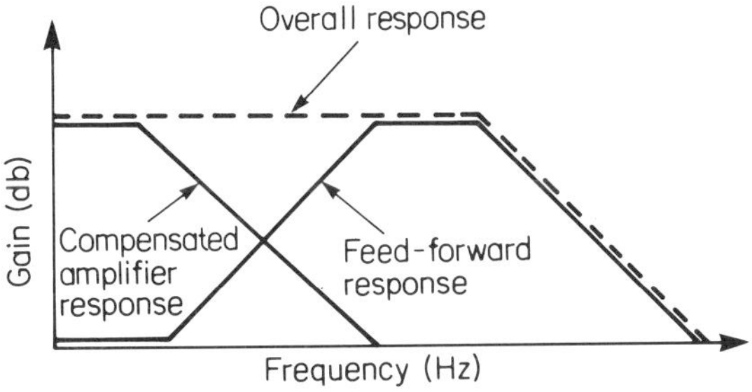

Figure 31.8 Feed-forward compensation

Q_5 to Q_7. Q_3, Q_4 are lateral pnp transistors and the signal is taken from Q_4 to the main gain stage Q_{16}, Q_{17}, which has an active collector load, Q_{13}. C_1 is the compensation capacitor. The class AB output stage is formed by Q_{14} and Q_{20}, biased by Q_{18}, Q_{19}. Short circuit limiting is provided by Q_{15}, Q_{21} and Q_{23}, Q_{24} prevent latch-up under overload conditions. The rest of the transistors are used to set bias levels. The arrangement may look strange to a discrete electronics designer, but the approach is determined by the difficulty of obtaining a wide range and number of resistor and capacitor values on a monolithic IC, compared with the relative abundance of high performance npn transistors. Elements of this circuit design may be seen in all linear ICs.

31.2.17 Operational amplifier configurations

Op amps are available ranging from general purpose to ultra-precision. The low to medium specification devices are often available in duals and quads. There is a growing trend to standardise on FET input op amps for general purpose applications, particularly those involving a.c. amplifiers, due to their much enhanced slew rate and lower noise.

Specific devices are available for high speed and high power output requirements. Additionally, some op amps are available as programmable devices. This means their operating characteristics (usually slew rate, bandwidth and output capability) can be traded off against power supply consumption by an external setting resistor, for instance, to tailor the device to a particular application. Often these amplifiers can be made to operate in the micropower mode, i.e. at low supply voltages and current.

31.2.18 Single supply operational amplifiers

It is entirely possible to operate any op amp on a single supply. However, the amplifier is then incompatible with bipolar d.c. signal conditioning. Single supply operation is entirely suitable for a.c. amplifiers. An example is shown in *Figure 31.10*. The circuit sets up the non-inverting input at $V_+/2$ to give the maximum output swing. Older design op amps such as the μA741 may not perform very well in such a circuit because of limited common mode input voltage range.

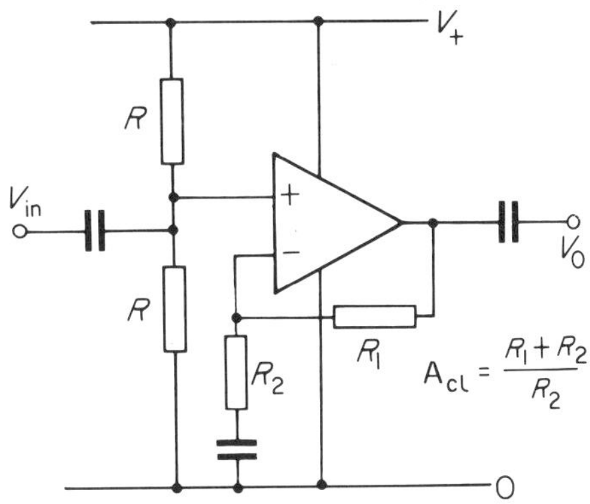

Figure 31.10 Single supply non-inverting a.c. amplifier

Most later generation op amps have been designed with single supply operation in mind and frequently use pnp differential input arrangements. They have an extended input voltage range, often including ground in single supply mode, and are referred to as 'single supply' op amps.

31.2.19 Chopper and auto zero operational amplifiers

Many attempts have been made to circumvent the offset and drift problems in op amps for precision amplifier applications. One classic technique is the chopper amplifier, in which the input d.c. signal is converted to a proportional a.c. signal by a controlled switch. It is then applied to a high gain accuracy a.c. amplifier, removing the inherent drift and offset problems. After amplification, d.c. restoration takes place using a synchronous

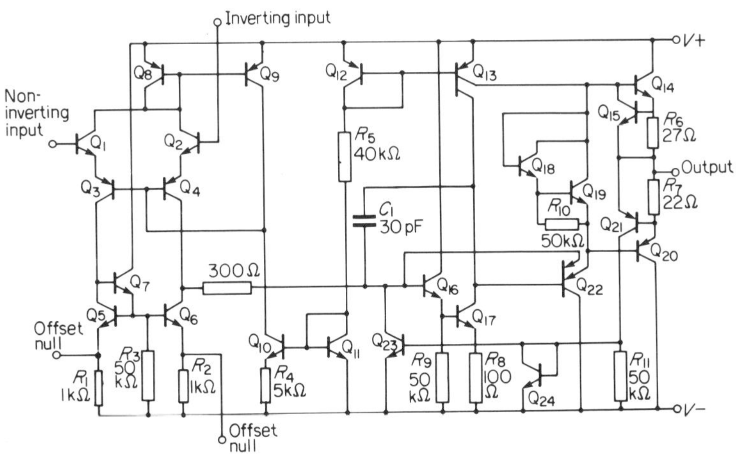

Figure 31.9 μA741 internal circuit

switching action at the chopper frequency. This is available on a single monolithic device.

Other approaches have used the fact that the output is not required continually, i.e. as in analogue to digital converters, and have used the idle period as a self-correction cycle. This again involves the use of a switch, but in this case it is usually used to ground the input of the op amp. The subsequent output is then stored as a replica of actual device error at that moment in time and subtracted from the resultant output in the measurement part of the cycle.

31.2.20 Operational amplifier applications

Op amps are suitable for amplifiers of all types, both inverting and non-inverting, d.c. and a.c. With capacitors included in the feedback network, integrators and differentiators may be formed. Active filters form an area where the availability of relatively low cost op amps with precisely defined characteristics has stimulated the development of new circuit design techniques. Filter responses are often produced by op amps configured as gyrators to simulate large inductors in a more practical fashion. Current-to-voltage converters are a common application of op amps in such areas as photodiode amplifiers, etc.

Non-linear circuits may be formed with diodes or transistors in the feedback loop. Using diodes, precision rectifiers may be constructed, for a.c. to d.c. converters, overcoming the normal errors involved with forward voltage drops. Using a transistor in common base configuration within the feedback loop is the basic technique used for generating logarithmic amplifiers. These are extensively used for special analogue functions, such as division, multiplication, squaring, square rooting, comparing and linearisation. Linearisation may also be approached by the 'piecewise' technique, whereby an analogue function is 'fitted' by a series of different slopes (gain) taking effect progressively from adjustable breakpoints.

Signal generation is also an area where op amps are useful for producing square, triangle and sine wave functions. An example of the potential simplicity of op amp designs is shown in the triangle wave generator of *Figure 31.11*. The ramp rate is set by the integrator *RC* component values. The other op amp is used as a threshold comparator with hysteresis that switches the polarity of the integrator current input as the output reaches the positive and negative limit points.

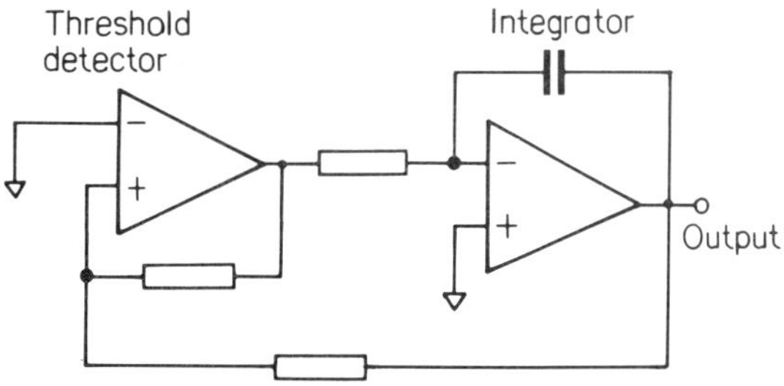

Figure 31.11 Triangle wave generator

31.2.21 Instrumentation amplifiers

Many applications in precision measurement require precise, high gain differential amplification with very high common mode rejection for transducer signal conditioning, etc. To increase the common mode rejection and gain accuracy available from a single op amp in the differential configuration, a three op amp circuit is used, which is capable of much improved performance. *Figure 31.12* shows the standard instrumentation amplifier format, which may be assembled from individual op amps or may be available as a complete (often hybrid) integrated circuit. The inputs are assigned one op amp each which may have gain or act as voltage followers. The outputs of these amplifiers are combined into a single output via the differential op amp stage. Resistor matching and absolute accuracy is highly important to maintain firstly, the CMRR, and secondly the gain precision of this arrangement. Suitable resistor networks are available in thin film (hybrid) form.

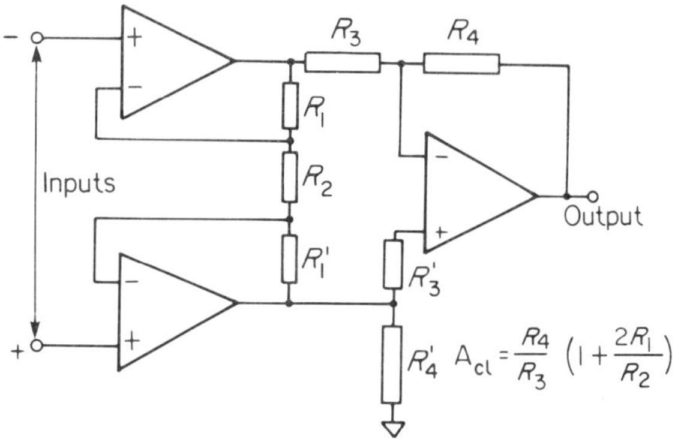

Figure 31.12 Instrumentation amplifier

31.3 Comparators

The comparator function can be performed quite easily by an op amp, but not with particularly well-optimised parameters. Specific devices are available, essentially modified op amps, to handle the comparator function in a large range of applications. The major changes to the op amp are to enable the output to be compatible with the logic levels of standard logic families (e.g. TTL, ECL, CMOS) and trade off a linear operating characteristic against speed. A wide common mode input range is also useful. Thus, all the usual op amp parameters are applicable to comparators, although usually in their role as an interface between analogue signals and logic circuits, there is no need for compensation.

A typical comparator application is shown in *Figure 31.13*. The LM311 has a separate ground terminal to reference the output swing to ground whilst maintaining bipolar supply operation for the inputs. The output TTL compatibility is achieved by a pull-up resistor to the TTL supply rail V_{cc}, as this is an open collector type comparator. Higher speed comparators will of necessity employ 'totem-pole' output structures to maintain fast output

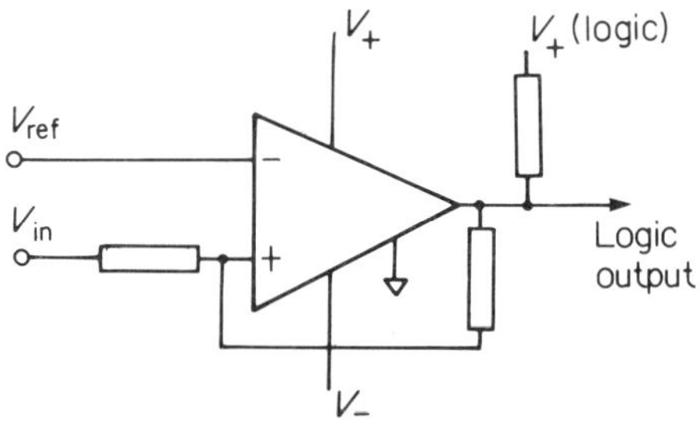

Figure 31.13 Voltage comparator using LM311

transition times. The circuit of *Figure 31.13* produces a digital signal dependent on whether the input voltage is above or below a reference threshold (V_{ref}). To clean up the switching action and avoid oscillation at the threshold region, it is quite often necessary to apply hysteresis. This is usually relatively easy to achieve with a small amount of positive feedback from the output to input as in *Figure 31.13*.

General purpose comparators such as the LM311 exhibit response times (time from a step input change to output crossing the logic threshold) in the order of 200 ns. Their output currents are limited, compared to op amps, being usually sufficient to drive several logic inputs. Often a strobe function is provided to disable the output from any input related response under logic signal control.

31.3.1 Comparator applications

Voltage comparators are useful in Schmitt triggers and pulse height discriminators. Analogue to digital converters of various types all require the comparator function and frequently the devices can be used independently for simple analogue threshold detection. Line receivers, *RC* oscillators, zero crossing detectors and level shifting circuits are all candidates for comparator use. Comparators are available in precision and high speed versions and as quads, duals and singles of the general purpose varieties. These latter are often optimised for single supply operation.

31.4 Analogue switches

Most FETs have suitable characteristics to operate as analogue switches. As voltage controlled, majority carrier devices, they appear as quite linear low value resistors in their 'on' state and as high resistance, low leakage path in the 'off' state.

Useful analogue switches may be produced with JFETs (junction field effect transistors) or MOSFETs (metal oxide semiconductor FETs). As a general rule, JFET switches can supply the lowest 'on' resistances and MOSFET switches the lowest 'off' leakage. Either technology can be integrated in single or multi-channel functions, complete with drivers in monolithic form.

The switched element is the channel of a FET or channels of a multiple FET array, hence the gate drive signal must be referred to the source terminal to control the switching of the drain-source resistance. It can be seen that the supply voltages of the gate driver circuit in effect must contain the allowable analogue signal input range. The FET switched element has its potential defined in both states by the analogue input voltage and additionally in the off stage by the output potential. The gate drive circuit takes an input of a ground referred, logic compatible (usually TTL) nature and converts it into a suitable gate control signal to be applied to the floating switch element. The configuration will be such that the maximum analogue input range is achieved whilst minimising any static or dynamic interaction of the switch element and its control signal. *Figure 31.14* shows a typical analogue switch in block diagram form. This is a dual SPST function. Because of their close relationship to mechanical switches in use, mechanical switch nomenclature is often used, e.g. SPST is single pole, single throw, and DPDT is double pole, double throw, both poles being activated by the same control signal. Break-before-make switching is frequently included to avoid shorting analogue signal inputs by a pair of switch elements.

The equivalent of single pole multi-way switches in monolithic form are referred to as analogue multiplexers.

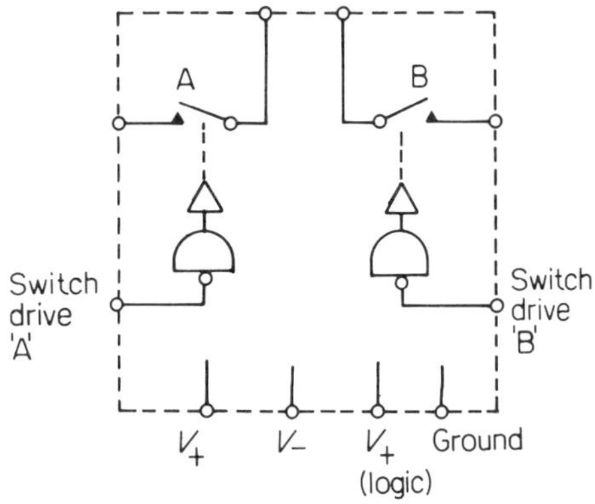

Figure 31.14 Dual SPST analogue switch (DG181)

31.4.1 Analogue switch selection

Selection is based on voltage and current handling requirements, maximum on resistance tolerable, minimum off resistance and operating speed. Precautions have to be taken in limiting input overvoltages and some CMOS switches were prone to latch up, a destructive SCR effect that occurred if the input signal remained present after the supplies had been removed. Later designs have since eliminated this hazard.

31.5 Sample and hold circuits

A sample and hold takes a 'snapshot' of an analogue signal at a point in time and holds its voltage level for a period by storing it on a capacitor. It can be made up from two op amps connected as voltage followers and an analogue switch.

This configuration block diagram is as shown in *Figure 31.15*.

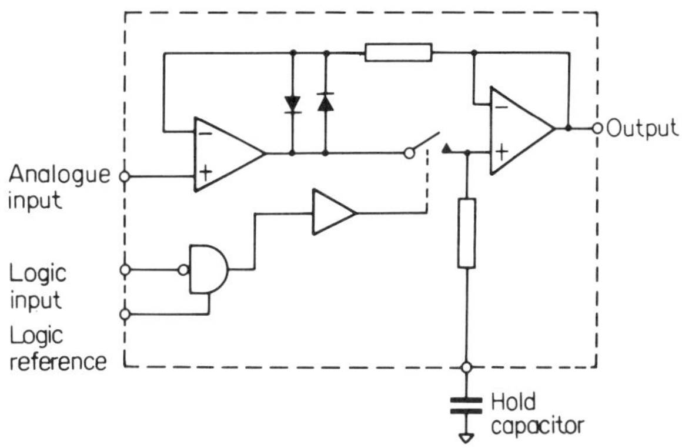

Figure 31.15 Sample and hold circuit (LF398)

In operation, the input follower acts as an impedance buffer and charges the external hold capacitor when the switch is closed, so that it continually tracks the input voltage (the 'sample' mode). With the switch turned off, the output voltage is held at the level of the input voltage at the instant of switch off (the 'hold' mode). Eventually, the charge on the hold capacitor will be drained away by the input bias current of the follower. Hence the hold voltage 'droops' at a rate controlled by the hold capacitor size and the leakage current, expressed as the droop rate in mV s^{-1} μF^{-1}.

When commanded to sample, the input follower must rapidly achieve a voltage at the hold capacitor equivalent to that present at the input. This action is limited by the slew rate into the hold capacitor and settling time of the input follower and it is referred to as the acquisition time. Dynamic sampling also has other sources of error. The hold capacitor voltage tends to lag behind a moving input voltage, due primarily to the on-chip charge current limiting resistor. Also, there is a logic delay (related to 'aperture' time) from the onset of the hold command and the switch actually opening, this being almost constant and independent of hold capacitor value. These two effects tend to be of opposite sign, but rarely will they be completely self-cancelling.

31.5.1 Sample and hold capacitor

For precise applications, the hold capacitor needs most careful selection. To avoid significant errors, capacitors with a dielectric exhibiting very low hysteresis are required. These dielectric absorption effects, are seen as changes in the hold voltage with time after sampling and are not related to leakage. They are much reduced by using capacitors constructed by polystyrene, polypropylene and PTFE.

The main applications for sample and hold circuits are in analogue to digital converters and the effects of dielectric absorption can often be much reduced by effecting the digitisation rapidly after sampling, i.e. in a period shorter than the dielectric hysteresis relaxation time constant.

31.6 Digital to analogue converters (DAC)

Digital to analogue converters are an essential interface from the digital world into the analogue signal processing area. They are also the key to many analogue to digital converter (ADC) techniques that rely on cycling a DAC in some fashion through its operating range until parity is achieved between the DAC output and the analogue input signal. All DACs conform to the block diagram of *Figure 31.16*. The output voltage is a product of the digital input word and an analogue reference voltage. The output can only change in discrete steps and the number of steps is immediately defined by the digital inputs. Should this be eight, e.g. an eight-bit DAC, the number of steps will be 256 and the full scale output will be 256 times some voltage increment related to the reference.

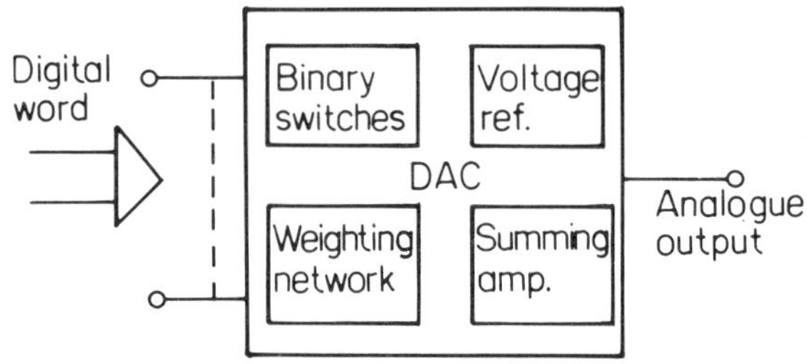

Figure 31.16 Digital to analog converter block diagram

Although various techniques can be used to produce the DAC function, the most widely used is variable scaling of the reference by a weighting network switched under digital control. The building blocks of this kind of DAC are: (a) reference voltage; (b) weighting network; (c) binary switches; and (d) an output summing amplifier. All may be built in, but usually a minimum functional DAC will combine network and switches.

31.6.1 R–2R ladders

The weighting network could be binary weighted as in *Figure 31.17(a)*, with voltage switching and summation achieved with a normal inverting op amp. However, even in this simple four-bit example, there is a wide range of resistor values to implement and speed is likely to suffer due to the charging and discharging of the network input capacitances during conversion. *Figure 31.17(b)* shows the R–$2R$ ladder network employing current switching to develop an output current proportional to the digital word and a reference current. The current switching technique eliminates the transients involving the nodal parasitic capacitances. Significantly, only two resistance values are required and the accuracy of the converter is set by the ratio precision rather than of the absolute value. Linear IC fabrication is capable of accommodating just such resistance value constraints quite

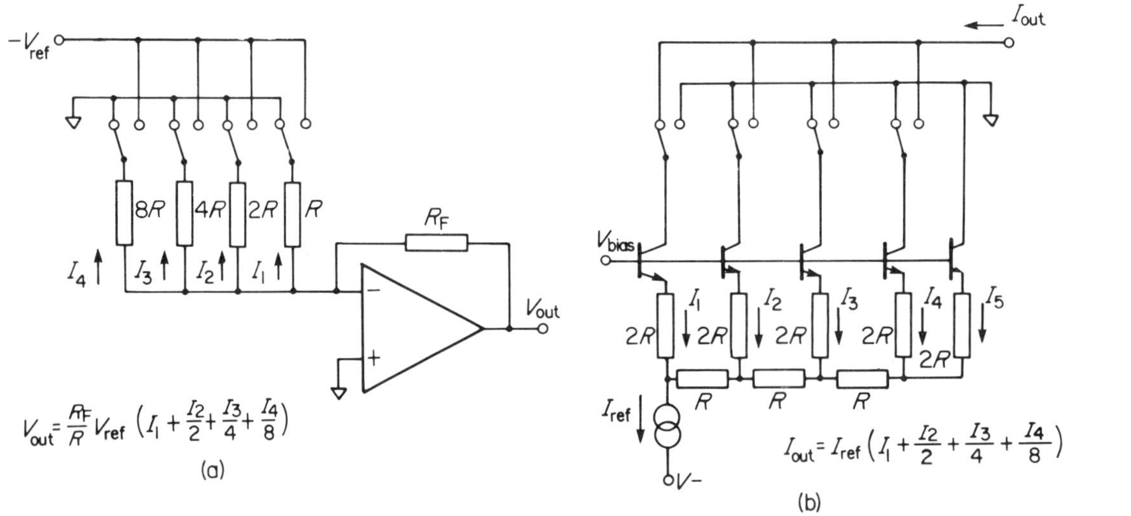

Figure 31.17 Digital to analogue converter circuits. (a) Binary weighted ladder—voltage switching. (b) R-2R ladder—current switching

conveniently. Monolithic DACs are available in 8 and 12 bit versions and recently in 16 bit. 16 bit and above DACs are also available in hybrid form.

Notice that the output of the DAC shown in *Figure 31.17(b)* is in a current form. It is converted to a voltage output quite simply by an external op amp current-to-voltage converter. *Figure 31.18* shows the standard internal layout of a typical 8 bit DAC. The op

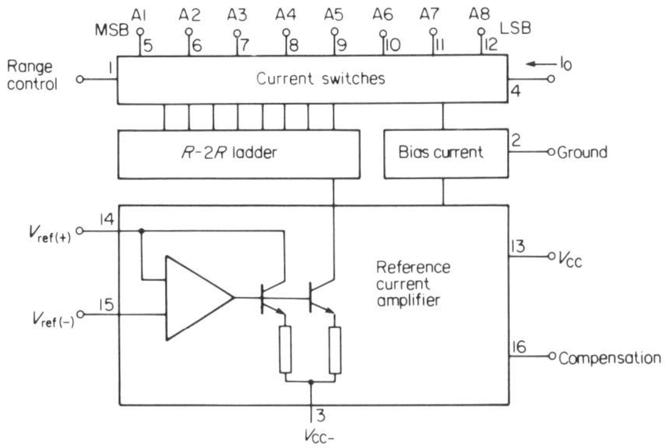

Figure 31.18 8-bit digital to analogue converter IC (MC1408)

amp and current mirror circuit are used to produce a current sink related to the supply from a voltage reference of either polarity referred to ground. True or complementary current outputs are available and both may be used in differential mode, in single-ended mode by grounding one output.

The output of such a DAC is directly proportional to the reference current and where it is able to operate over a significant dynamic reference current range, it is referred to as a 'multiplying' DAC.

31.6.2 Resolution, linearity and monotonicity

Resolution has already been touched upon as being defined by the digital word length and reference voltage increment. However, it is important to note that it is quite possible to have, say, a 12 bit DAC in terms of resolution which is incapable of 12 bit accuracy. This is because of non-linearities in the transfer function of the DAC. It may well be accurate at full scale but deviate from an ideal straight line response due to variations in step size at other points on the characteristic. Linearity is specified as a worst case percentage of full scale output over the operating range. To be accurate to n bits, the DAC must have a linearity better than $\frac{1}{2}$LSB (LSB = least significant bit, equivalent to step size) expressed as a percentage of full scale (full scale = $2^n \times$ step size).

Differential linearity is the error in step size from ideal between adjacent steps and its worst case level determines whether the converter will be monotonic. A monotonic DAC is one in which, at any point in the characteristic from zero to full scale, an increase in the digital code results in an increase in the absolute value of the output voltage. Non-monotonic converters may actually 'reverse' in some portion of the characteristic, leading to the same output voltage for two different digital inputs. This obviously is a most undesirable characteristic in many applications.

31.6.3 Settling time

The speed of a DAC is defined in terms of its settling time, very similar to an op amp. The step output voltage change used is the full scale swing from zero and the rated accuracy band is usually $\pm\frac{1}{2}$LSB. Settling times of 100–200 ns are common with 8 bit R–$2R$ ladder monolithic DACs. The compensation of the reference op amp will have a bearing on settling time and must be handled with care to maintain a balance of stability and speed.

31.6.4 Other DAC techniques

There are many other methods that have been proposed and used for the DAC function. There is insufficient space here to make a full coverage. However, mention should be made of the pulse width technique that is frequently used in consumer ICs. The digital inputs are assigned time weighting so that in, say, a 6 bit DAC, the LSB corresponds to 1 time unit and the MSB to 32 time units. The time units are derived by variable division from a master clock and usually clocked out at a rate which is some sub-multiple of the master clock.

The duty cycle will then vary between 0 and 63/64 in its most simple form. This pulse rate is integrated by an averaging filter (usually *RC*) to produce a smooth d.c. output. The full scale is obviously dependent on the pulse height and this must be related back to a voltage reference, albeit only a regulated supply. This type of DAC is often used for generating control voltages for voltage controlled amplifiers used for volume, brightness, colour, etc. (TV application) and it is possible to combine two 6 bit pulse width ratio outputs together for 12 bit resolution (not accuracy) to drive voltage controlled oscillators, i.e. varactor tuners. This is one application (tuning) where non-monotonicity can be acceptable.

31.7 Analogue to digital converters (ADC)

ADCs fall into three major categories: (a) direct converters; (b) DAC feedback; and (c) integrating.

31.7.1 Direct or 'flash' converters

Flash converters may have limited resolution but are essential for very high speed applications. They also have the advantage of continuous availability of the digitised value of the input signal, with no 'conversion time' waiting period. They consist of a reference voltage which is subdivided by a resistor network and applied to the inputs of a set of comparators. The other inputs of the comparator are common to the input voltage. A digital encoder would then produce, in the case of a 3 bit converter for example, a weighted 3 bit output from the 8 comparator inputs. The method is highly suitable to video digitisation and is usually integrated with a high speed digital logic technology such as ECL or advanced Schottky TTL. Flash converters in monolithic form of up to 9 bit resolution have been built.

31.7.2 Feedback ADCs

Feedback ADCs use a DAC within a self-checking loop, i.e. a DAC is cycled through its operating range until parity is achieved with the input. The digital address of the DAC at that time is the required ADC output. Feedback ADCs are accurate with a reasonably fast conversion time. Their resolution limitations are essentially those of the required DAC cost, performance and availability.

Feedback ADCs require a special purpose logic function known as a successive approximation register (SAR). The SAR uses an iterative process to arrive at parity with the ADC input voltage in the shortest possible time. It changes the DAC addresses in such a way that the amplitude of the input is checked

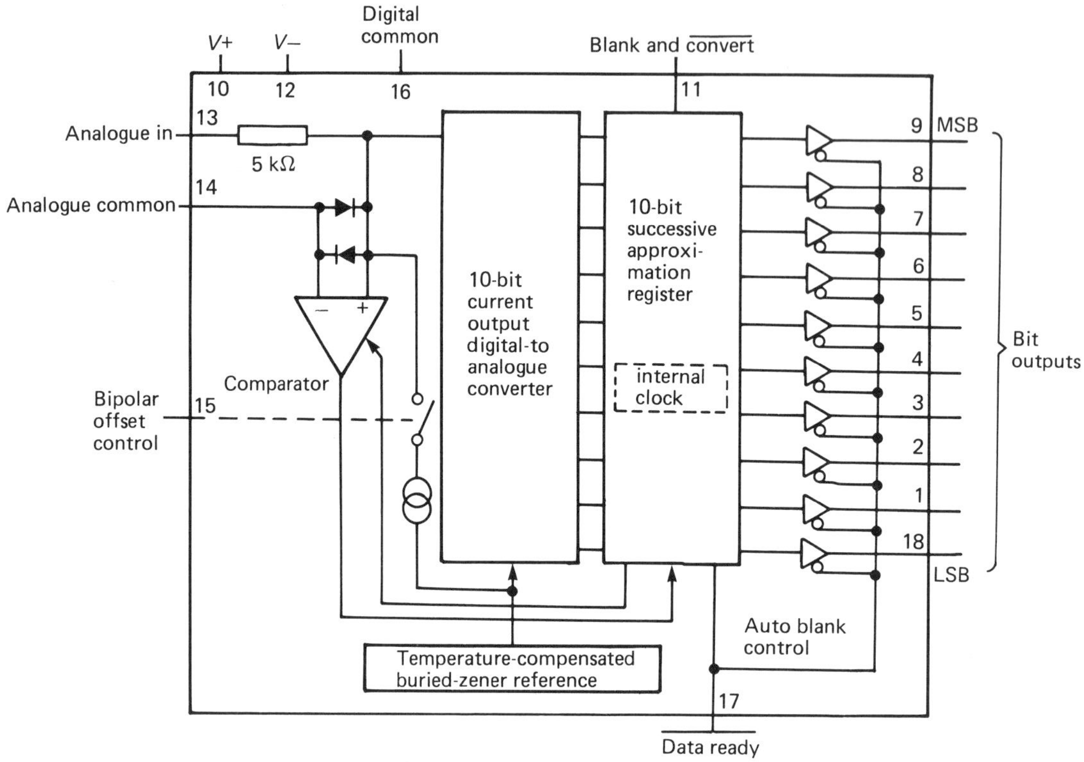

Figure 31.19 10-bit analog to digital converter (AD571)

to determine whether it is greater or smaller than the input on a bit sequential basis, commencing with the MSB. This implies continuous feedback from a comparator on the analogue input and DAC output connected to the SAR.

Figure 31.19 shows a block diagram of a 10 bit ADC available on a single chip. The 10 bit DAC is controlled by 10 bit SAR with internal clock and the DAC has an onboard reference. The digital output is fed through tri-state buffers to ease the problems of moving data onto a bus structured system. These are high impedance (blank) until the device is commanded to perform a conversion.

After the conversion time, typically 25 μs for this type of device, a data ready flag is enabled and correct data presented at the ADC output. An input offset control is provided so that the device can operate with bipolar inputs. Because of the nature of the conversion process, satisfactory operation with rapidly changing analogue inputs may not be achieved unless the ADC is preceded by a sample and hold circuit.

31.7.3 Integrating ADCs

Integrating ADCs use a variety of techniques, but all usually display high resolution and linearity. They also have a much greater ability to reject noise on the analogue input than other types of ADC. The penalty that is paid for this is a much longer conversion time which is also variable with actual input voltage. These factors make integrating ADCs very suitable for digital voltmeter, panel meter and other digital measuring applications. LICs to perform this function often have direct display (LED, LCD, etc.) drive capability built in and, because of their BCD coding and subsequent display segment/digit formatting, will be unsuitable for data acquisition systems.

31.7.4 'Dual slope' integrating ADC

The dual slope ADC will be explained in some detail, as the understanding of this technique leads to familiarity with most other integrating converter principles. The dual slope technique has advantages in that a large number of the possible error sources are inherently self-cancelling. The basic principle consists of ramping down a capacitor from a known voltage level with a discharge current directly proportional to the input voltage for a fixed period of time. *Figure 31.20* shows the effect on the integrating capacitor. At the end of the fixed period the capacitor

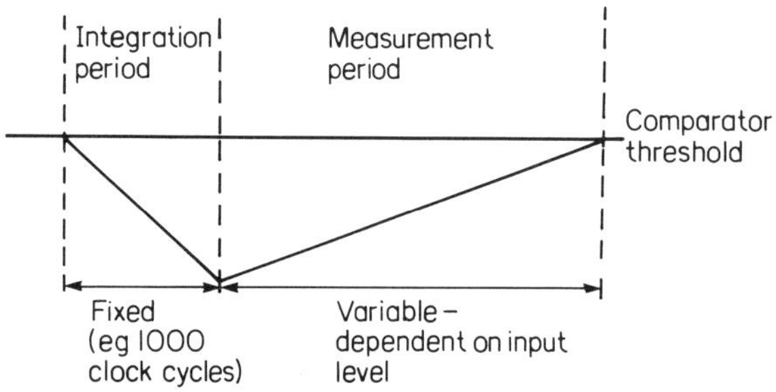

Figure 31.20 Capacitor voltage ramp—integrating ADC

is recharged at constant current defined by a voltage reference. During this second period, the clock used to generate the fixed period timing is counted until a comparator indicates the original voltage level has been reached. This then stops the count, which is

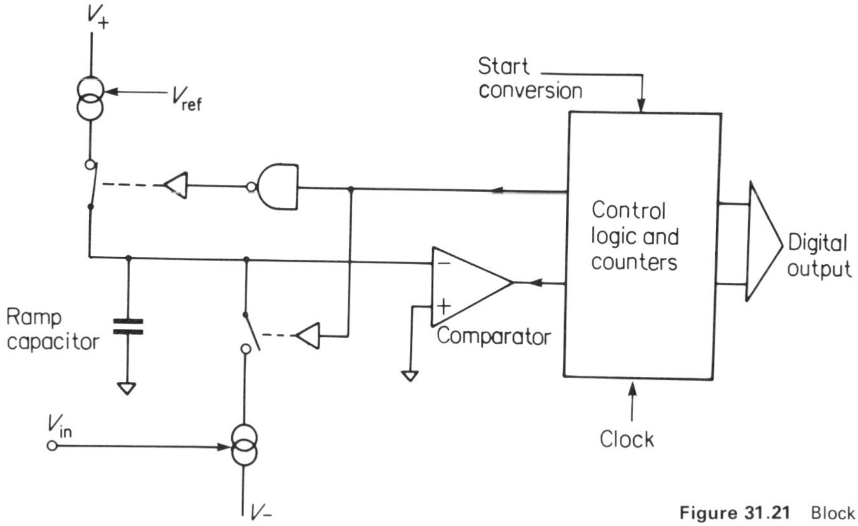

Figure 31.21 Block diagram—dual slope integrating ADC

then latched and presented to the output in binary or BCD form. With suitable scaling, the BCD output can be directly in volts.

Figure 31.21 shows a block diagram of the implementation of the dual slope ADC. The switch positions will be reversed during the integration period. In a typical application, say a $3\frac{1}{2}$ digit DVM, full scale count is 1999, the control logic using BCD counters. After the previous conversion, the counters are reset to 0999. The integration period is defined from this start point to when the counter reaches its maximum count of 1999, this point being arranged to reverse the switches and the next counter state is 0000. The counter continues during the measurement period until the comparator changes state. Exceeding 1999 during measurement will latch an overload indication. Note that this example ($3\frac{1}{2}$ digit BCD) is equivalent to 11 bit binary resolution. Integrating ADCs are capable of 18 bit accuracy and resolution with careful design and application.

Dual slope ADCs can perform this accurately due mainly to the tolerance of long term drift of various components. The clock frequency is not critical as long as its short term stability is good, i.e. relative to the conversion period, since the integration period changes are corrected by counting the same clock during the measurement period.

The absolute value of the ramp capacitor is not important as long as it possesses short term capacitance stability and those dielectric parameters as outlined for sample and hold circuits. The actual comparator threshold and its error parameters, e.g. offsets, are not a source of error as long as they do not change over the dynamic ramp range and in the short term. Primary error sources revolve around the reference current source (or sink) and the input amplifier, acting as a voltage-to-current converter. The op amp errors can often be corrected by auto zero techniques applied during the measurement period. This will involve switching the op amp to measure (and store for correction during integration) its own offset, during the period in which its output is redundant. The effects of noise on the input voltage are averaged out during the integration, and in mains powered equipment the conversion rate can be synchronised to the mains frequency to avoid jitter induced at mains frequency.

31.7.5 Voltage-to-frequency converters (VFC)

The voltage-to-frequency converter is a form of direct ADC that produces a serial rather than parallel digital output. Converters of this type have advantages in their fast response, good linearity and continuous output that may be used for transmission of analogue values in serial digital form over cables, fibre optic and radio channels. Direct counting of the output with normal digital frequency counter techniques will produce a read-out in BCD or binary format. Additionally, most monolithic VFCs can have their functional elements configured by external connections to perform as frequency-to-voltage converters (FVCs). 12 bit resolution is possible with monolithic VFCs, but greater resolution and accuracy is obtained with hybrid devices.

Figure 31.22 shows the basic principle of a VFC operating on positive input voltages. If the input voltage is greater than the

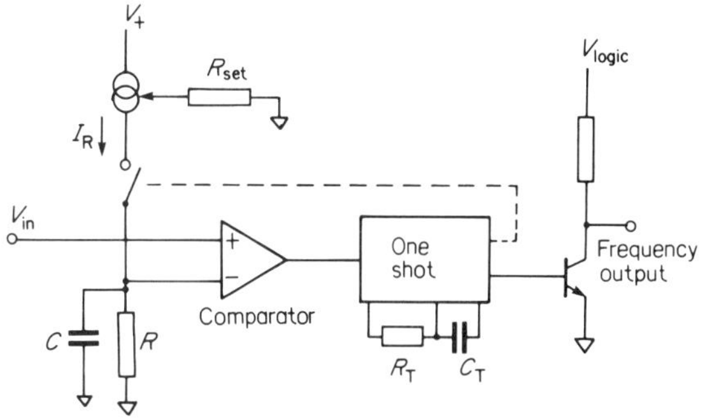

Figure 31.22 Voltage to frequency converter (outline)

voltage stored on the capacitor, the comparator will trigger the monostable multibrator (one-shot). The output of the one-shot switches the logic compatible output stage and also switches on the current source. The current source is on for a period determined by the one-shot *RC* time constant, and is programmed by an external resistor in conjunction with an internal voltage reference. The charge increment delivered to the comparator input capacitor will usually be sufficient to raise its voltage to a level in excess of the input voltage. At the end of the one-shot period, the current source is switched off and the voltage on the input capacitor decays at a rate determined by the parallel resistor *R*. When it falls below the input, the comparator will retrigger the one-shot again and the cycle repeats. At balance, the average current flowing into and out of the capacitor must be equal. The reference current, I_R, and the one-shot period, t, are constants, so that the average capacitor current is $I_R tF$, where F is the VFC frequency. Average capacitor current is also, to a good approximation, given by V_{in}/R. So it can be seen that frequency output is directly proportioned to input voltage, and a relatively simple VFC can be linear over a wide range.

31.8 Voltage regulators

LICs have an important part to play in power supplies for all forms of electronic circuitry, digital or analogue. They have, for a considerable period, dominated linear regulator applications and, with the increasing tendency to space and weight reduction and energy conservation, are emerging as self-contained switching regulators or control circuits for switchers. The widest usage at present in linear regulators is for series pass types with some shunt devices for special applications.

31.8.1 Series regulators

The block diagram of a series regulator is shown in *Figure 31.23*. This equates functionally to the μA723 type regulator IC with the exception of thermal and SOA protection. This was an early

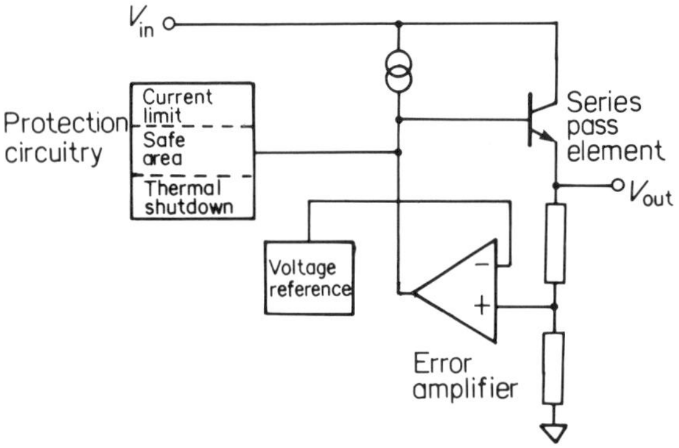

Figure 31.23 Voltage regulator block diagram

approach to a flexible 'universal' voltage regulator that has remained an industry standard to this day. The essential parts are a voltage reference, an error amplifier (op amp), a series pass element, and facilities for overload protection. The μA723 contains a temperature compensated zener diode reference and the series pass element is an npn Darlington, capable of about 150 mA collector current. An additional npn transistor is available to act as a current limiter in conjunction with an external sense resistor. The device connections allow it to be configured as a series or shunt linear regulator or various types of switching regulator using different external components to achieve the desired output voltage and boost the current handling capability. This amount of flexibility is not always the optimum solution and other fixed and adjustable concepts will be discussed.

31.8.2 Load and line regulation

The ideal voltage regulator produces its desired output voltage without any change when the input voltage and load current fluctuate. The line regulation is defined as the change in output voltage, in percentage or absolute terms, for a specified change in input voltage. Similarly, a load regulation is defined as the output voltage change for a specified change in load current. Usual applications of voltage regulator ICs will be to produce regulated d.c. lines from normal a.c. mains, i.e. regulating the raw d.c. voltage produced by a step-down transformer, rectifier and reservoir capacitor combination.

The dynamic performance of the device line regulation is important here to attenuate the mains frequency ripple present at its input. This parameter is usually specified as ripple rejection and the frequency at which it is guaranteed is usually 100–120 Hz (full wave a.c. rectification). Line and load regulation will be better than 1% and ripple rejection around 60–70 dB on a general purpose regulator IC.

31.8.3 Drop out voltage

Drop out voltage is significant in power supply design as it identifies the minimum input–output voltage differential under which a regulator can still function. This is usually set by the saturation voltage of the series pass element and its circuit configuration. The user must ensure that the input voltage is in excess of the required output plus the worst case drop out voltage. This must allow, in the case of a mains power supply, for the troughs of the waveform of unregulated d.c. input caused by ripple effects.

31.8.4 'Three-terminal fixed' regulators

As the electronics industry has standardised on particular supply voltages, e.g. +5 V for TTL, ±15 V for analogue signal processing (op amp) circuitry, the demand has grown for fixed output regulators with minimum external components requirements and simplicity of use.

This has resulted in the families of so-called fixed voltage, three-terminal regulators. The feedback resistor network sampling the output and the current limit resistor have been integrated into these devices, along with a more substantial series pass transistor to boost current availability. This has dictated the choice of various power packages, e.g. TO3, TO220, TO202, to dissipate the extra heat. It has also made mandatory the inclusion of additional protection circuitry as in *Figure 31.23* to prevent 'blowout' of the regulator in case of overdissipation (internal thermal sensing and shutdown) and simultaneous excess voltage and current on the pass element (safe operating area protection).

Fixed voltage three-terminal regulators are available, usually with 5% initial voltage tolerance, in positive and negative versions, at various voltages from 2.6 V to 24 V. Widely available current levels are 0.1 A, 0.25 A, 0.5 A and 1 A to 1.5 A. The current rating usually decides the package options that are used. Monolithic fixed voltage regulators are available up to around 5 A at certain popular voltages (+5 V) and hybrid devices can achieve higher currents (10 A–20 A).

Figure 31.24 shows the circuit arrangement of a three-terminal regulator. The quiescent (stand-by) current, I_Q, of the regulator

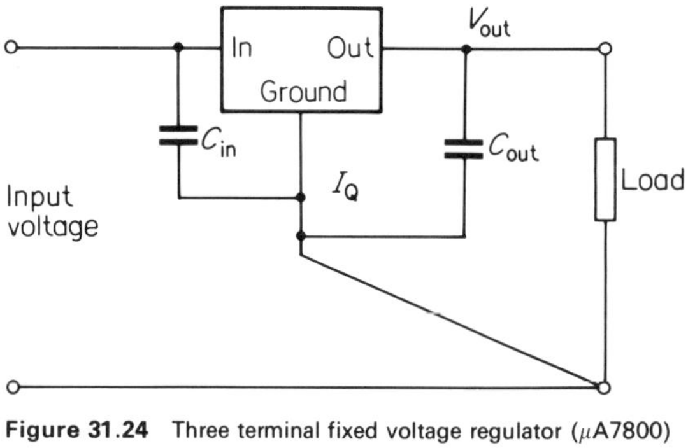

Figure 31.24 Three terminal fixed voltage regulator (μA7800)

flows down the ground leg and is of the order of a few milliamperes to avoid excessive power loss during no-load periods. To ensure stability, a capacitor, C_{in}, is usually required, although some regulators may tolerate its absence if the power supply reservoir capacitor is physically close enough to the device.

Negative regulators are usually much more susceptible to oscillation without complete input bypassing and the use of the recommended C_{in} (value and type) is necessary. C_{out} improves the transient response of the regulator to step load changes but is not usually essential.

A disadvantage of the three-terminal configuration if that the device can only regulate to its terminal and hence uncompensated voltage drops can occur if the load is remote. Because I_Q is low, the ground terminal can be returned to a load sense point easily. However, heavier gauge wire must be used to reduce this effect on the positive terminal, as in *Figure 31.24*.

Since the I_Q of this type of regulator is quite constant with load changes, alterations of the output voltage are possible with resistance added in the ground lead. However, three-terminal regulators specifically designed as adjustable parts are available with better performance.

31.8.5 Adjustable regulators

The proliferation of fixed voltage regulator ranges have satisfied many applications, but there is always a demand for different voltages, greater accuracy, etc. LICs such as the μA723, LM104 and LM105 can handle a variety of voltage and current combinations with a high level of precision. Specialised three-terminal adjustable regulators have been developed to offer a simple solution to both non-standard and standard voltages.

These have a 'floating' configuration as in *Figure 31.25*, and always maintain a nominal 1.2 V reference across the V_{out} and ADJ terminals. The output voltage is defined by the voltage across R_2 plus the reference voltage. Since the constant current

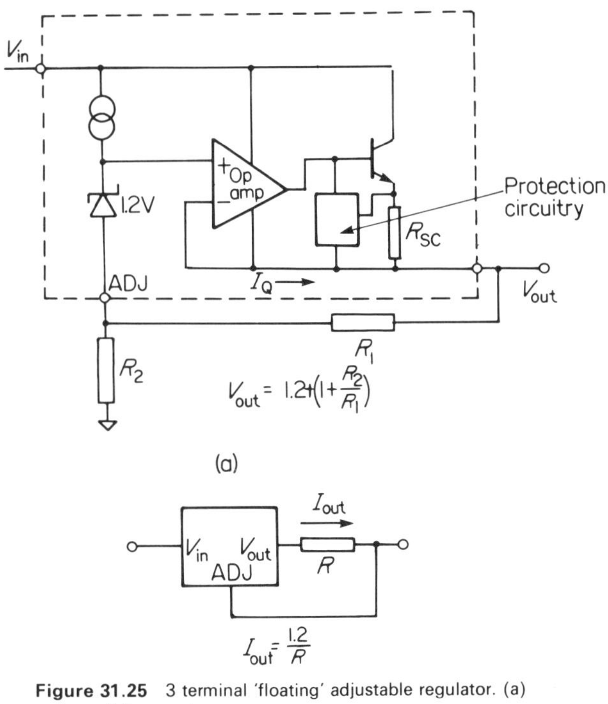

Figure 31.25 3 terminal 'floating' adjustable regulator. (a) voltage; (b) current

source biasing of the reference is low, typically around 50 μA, the current through R_2 is very close to $1.25/R_1$. Errors resulting from the reference bias current can be swamped by selecting as much value for the current through R_1/R_2. This current has to be equal to or greater than the regulator quiescent current, which flows out of the V_{out} terminal, in any case. The reference used in this, as in other three-terminal regulators, is of the 'band-gap' variety. This type of three-terminal regulator also has the same comprehensive overload protection circuitry. The three-terminal adjustable regulator is also capable of implementing very simply the constant current source function as in *Figure 31.25(b)*.

31.8.6 Foldback limiting

Merely limiting the current output of a regulator during overloads is rarely sufficient, since in the case of short circuit, input–output differential increases greatly and hence so does power dissipation. To avoid exceeding the thermal design limitations, regulators using devices, such as the μA723, can be arranged to progressively reduce their limit current as the output voltage is forced below its regulated level. Complete regulator subsystems, i.e. three-terminals, achieve foldback by thermal sensing on the chip and progressive shutdown (output current restriction), commencing at around 150°C.

31.8.7 Overvoltage protection

One regulator may well be supplying many ICs in a system and it is often necessary to ensure that in the event of a regulator failure, the output voltage cannot go significantly higher than the nominal level. TTL, in particular, has a relatively small range between normal operating ($5\ V \pm 5\%$) and absolute maximum (7.0 V) supply levels.

Overvoltage protection uses an additional voltage reference and comparator to sample the regulator output and, should it exceed a preset (higher than nominal) limit, to rapidly fire a thyristor which short circuits or 'crowbars' the output. The regulator current limiting may also have failed, of course, and a fuse in series with the regulator input will be the terminal interruption of the fault condition. LIC subsystems are available containing the elements of overvoltage protectors, although they do not usually contain the thyristor device.

31.8.8 Voltage references

All regulator ICs contain some form of voltage reference. Originally references used in voltage regulators were zener diodes, either selected for low temperature coefficient by having a breakdown voltage close to the transition between true zener operation and avalanche operation (5.1 V–5.6 V), or an avalanche diode compensated with the addition of a forward biased junction (around 7 V). This latter type has been extensively used in ICs, typified by the μA723. Its disadvantages are noise and shortcomings in long term stability. For low voltage regulators, it dictates a minimum V_{in} of 8 V to 9 V to activate the reference, even if the load related dropout parameter is lower.

The 'band-gap' reference produces a highly predictable, stable voltage of nominally 1.2 V with low noise and temperature coefficient. 1.2 V is the band-gap of silicon, corresponding to the V_{BE} of a silicon transistor at 0 K. The operation of the reference depends on the ability to fabricate transistors adjacent to each other within an IC whose V_{BE} difference can be controlled by area (geometry) ratio during diffusion.

Because of the number of individual components making up the reference, the band-gap can be more susceptible to thermal gradients on the chip. High precision references therefore have tended to continue to use the zener technique with improvements being made by employing ion implantation. The subsurface breakdown thus produced has much better long term stability, and temperature stability effects can be minimised by a constant temperature oven formed by an on-chip heater, thermal sensing and feedback control. This kind of reference is frequently employed for precision DACs and ADCs.

31.8.9 Shunt regulators

Shunt regulators are commonly seen using zener diodes to regulate supply lines. Series resistance is essential with a shunt regulator, and, if the load current falls to zero, all the power previously supplied to the load is dissipated in the regulator. This makes the shunt regulator cumbersome and inefficient for high power loads, particularly if they are variable. LIC shunt regulators are more precise, adjustable versions of the standard zener diode, with the ability to be programmed over a wide voltage range with two resistors. *Figure 31.26* shows a typical device and its basic circuit application.

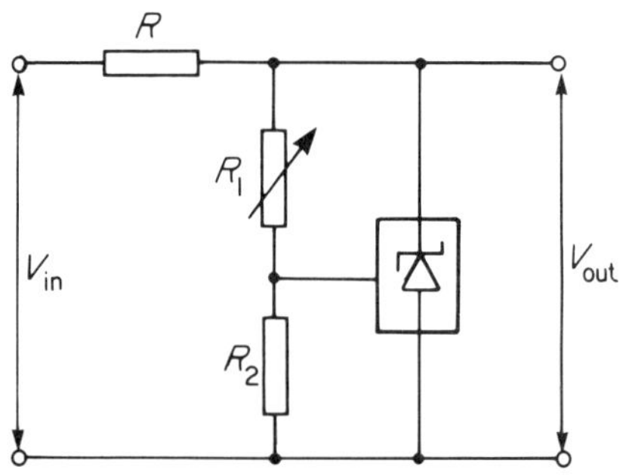

$$V_{out} = 2.5\left(1 + \frac{R_1}{R_2}\right)$$

Figure 31.26 Adjustable shunt regulator (μA431)

31.8.10 Switching regulators

Switching regulators differ from the series and shunt regulators discussed previously in two major areas. Firstly, they are capable of much greater efficiency since the minimum possible power is dissipated in the switching elements. Secondly, they are not limited to regulating a voltage to some lower level than its original form (step-down) but are capable of stepping up and inverting the polarity of the d.c. voltages.

LICs for switching regulators basically fall into two categories, those intended as controllers for switching regulators utilising transformer coupling from output to input, and those intended as switching inverters/converters with relatively low d.c. input voltages (around 50 V and below) and no input/output isolation. The former are often used for direct off-line (mains) switch mode power supply (SMPS) controllers.

31.8.11 SMPS control ICs

The typical off-line SMPS reduces the bulk, weight and heat dissipation of a linear supply by full wave rectifying and filtering and a.c. line input and then switching it at a frequency, say 25 kHz, which is much higher than that of the mains supply. This means a dramatic reduction in size from the normal mains frequency transformer and reservoir capacitors. The primary of the SMPS transformer is switched, either single-ended or in push–pull, by one or more high speed bipolar or MOSFET power transistors. The output voltage is rectified at the secondary with fast recovery, e.g. Schottky rectifiers. The SMPS control circuit contains the necessary oscillator and, in addition to the normal regulator components, a comparator, pulse steering logic and drivers to interface to the power switch devices. The switching drives are pulse width modulated to regulate the output. To preserve isolation the control circuits will be coupled to the output via some form of isolator, e.g. optocouplers.

31.8.12 Single inductor switching regulators

Many useful voltage conversion/inversion and regulator functions can be achieved with LIC switching regulators of the single-ended, single-inductor variety. Such circuits allow the provision of multiple supplies (of either polarity) from a single voltage input, such as a battery, whilst ensuring low power consumption through switching efficiency. A typical LIC able to perform such d.c./d.c. translations is shown in *Figure 31.27*. The op amp is a 'bonus' that can be used to provide additional series regulated supplies as it has high current output capability. The remaining circuitry is used to compare a fraction of the output voltage with a reference (band-gap) and commit the switch to a cycle of the oscillator waveform, depending on the result.

Figure 31.28 shows the three basic switching configurations with a single inductor, and all of these, with the exception of inverting, which requires an external switch transistor, can be handled by the device of *Figure 31.27*. The oscillator free runs and the comparator enables the output switch on a cycle-by-cycle basis. During an active cycle, the risk of inductor current rising too high is contained by an active current limit circuit within the oscillator. At this point, the switch-on time is aborted, and the inductor allowed to discharge into the output load and storage capacitor. Thus the frequency of operation of this type of switching regulator is variable with load and input voltage change.

31.9 IC timers

Timers may be considered as digital circuits, since their basic functions are as astable and monostable multivibrators. However, the function is extremely useful in linear systems. Most timer circuits are derivatives of the basic NE555 type as in *Figure 31.29*, with two comparators controlling a set/reset flip–flop. In continuous (astable) operation, an external capacitor is charged and discharged between two thresholds, $\frac{2}{3}V_{cc}$ and $\frac{1}{3}V_{cc}$, via an external resistor network. In the monostable mode, the circuit times out when the external capacitor charges from zero to $\frac{2}{3}V_{cc}$. Quad and dual versions of the circuit are available and also versions in CMOS technology. LICs are also available containing a 555 type timer coupled to an *n*-bit programmable counter. These can produce long delays very accurately and repeatably.

31.10 Consumer LICs

Consumer LICs are usually devices for which the prime volume usage is within the consumer radio and TV industries. Certain functions may be usable in the professional communications, telecommunications and, of course, broadcasting areas. However, some functions are quite specialised and are only really useful for their designed application. A good example of this is the chrominance processing circuitry for a television receiver.

The range of available devices is very wide and cost pressures on consumer equipment have led to some of the largest scale linear ICs on the market. *Figure 31.30* gives basic outlines of a VHF/FM radio receiver and a UHF colour TV. All of the

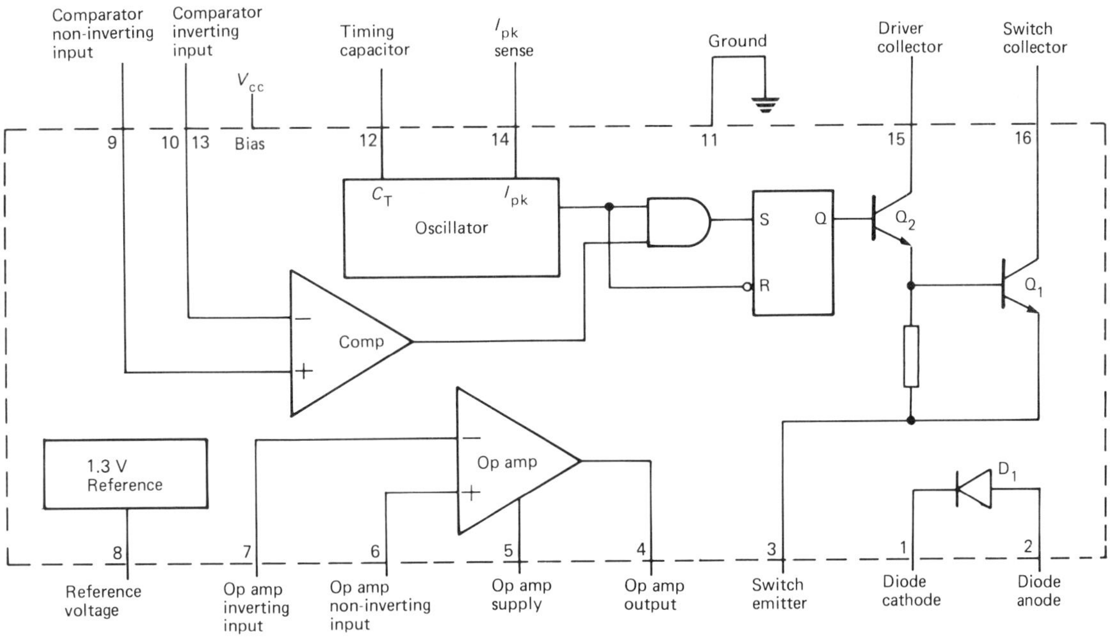

Figure 31.27 Single inductor switching regulator sub-system (μA78S40)

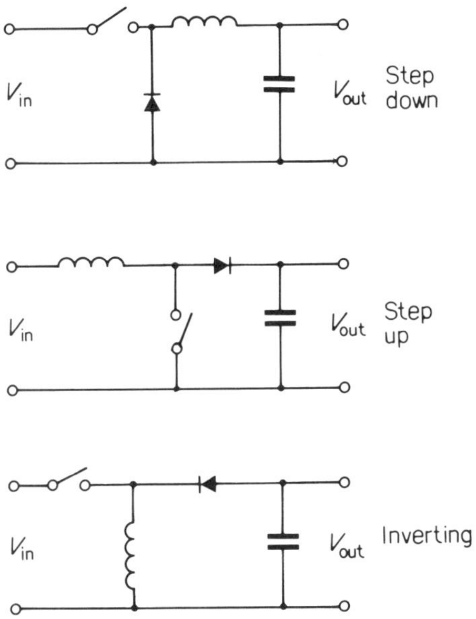

Figure 31.28 Single inductor switching regulator configurations

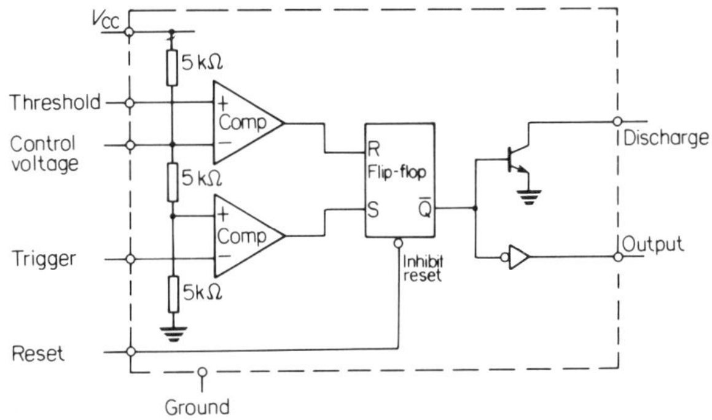

Figure 31.29 IC timer block diagram (NE555)

functions of the radio can be integrated and the present state of the art for minimum chip count is four, partitioning being as shown. The tuner IC would use external tuned circuits for selectivity (with varactor tuning) and ceramic (block) filtering prior to the 10.7 MHz limiting IF amplifier and detector IC. Multiplex stereo decoding is achieved in another IC which consists of a phase locked loop. This locks an oscillator (often 76 kHz) to the incoming 19 kHz pilot tone to regenerate the correct phase 38 kHz signal for matrix decoding. The dual audio power amplifier then drives the loudspeakers. Monolithic audio power amplifiers are currently capable of upwards of 20 W output. The complete receiver may of course require voltage regulators, tuning control circuits, etc.

The TV block diagram is an even more graphic indication of available LIC complexity. The UHF tuner is so far resistant to monolithic integration, but most other functions are heavily integrated. The video IF amplifier and detector IC handles the incoming video signal at 38–39 MHz and provides the amplitude modulation detection. Intercarrier sound at 6 MHz is handled in the sound channel IC, which additionally provides FM detection and audio power amplification. The composite video is applied to three ICs, two of which separate the synchronising signals and process the line and field scan oscillators. Field scan drive is direct, but line scan and associated EHT (25 kV) generation for the CRT are usually handled by an external high voltage power semiconductor. The chrominance information is decoded in another IC and the resultant red, green and blue (RGB) drives are applied to discrete video amplifiers, since ICs cannot currently achieve the breakdown voltage necessary to drive the CRT grids.

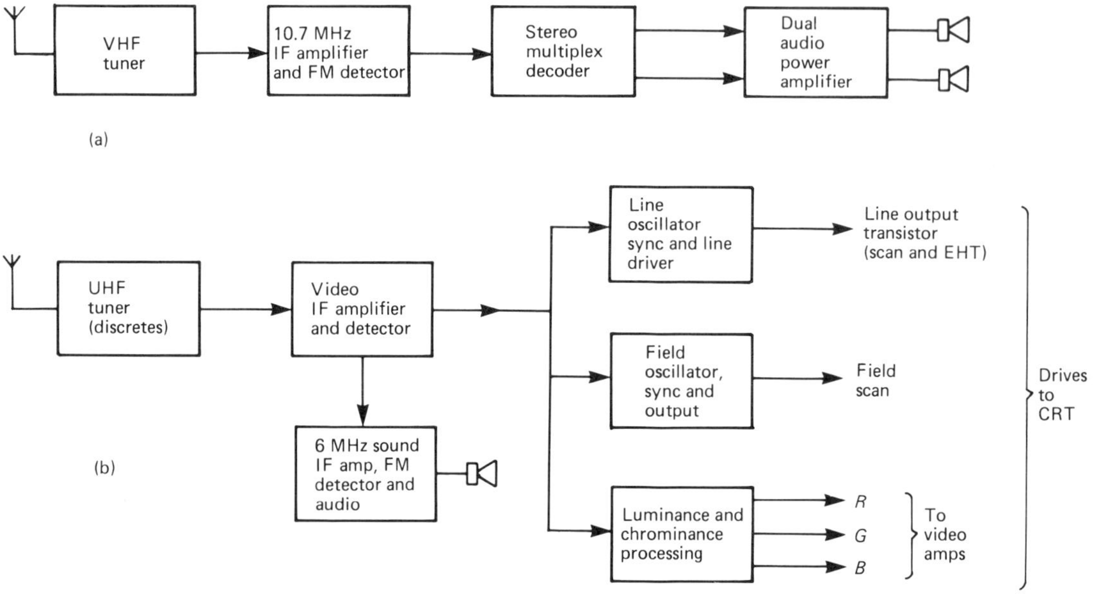

Figure 31.30 Consumer LIC applications: (a) VHF/FM stereo radio (four ICs); (b) UHF colour TV (basic five ICs)

Single chip chroma processors are probably the densest linear ICs available. Colour TVs also contain many other ICs for power supply control (SMPS controllers), tuning, remote control, etc.

31.11 Charge transfer devices

A charge transfer device (CTD) is a semiconductor structure in which discrete charge packets are moved. CTDs find application in shift registers, high speed filtering, imaging systems and dynamic memories. The two main methods of constructing CTDs result in bucket brigade devices and charge coupled devices.

31.11.1 Bucket brigade devices (BBD)

BBDs are formed by connecting a series of capacitors with switches, normally FET or bipolar transistors. A single storage element consists of two capacitor-switch units. Charge held on the capacitors is transferred along the device by turning odd and even switches on alternately. Modern commercial BBDs can transfer at 1 MHz.

Non-linearities encountered at small signal levels are avoided by making the zero signal correspond to a fixed offset. Charge deficit represents the signal amplitude.

FETs as switches have a slower transfer rate than bipolar transistors. The latter lose charge through the base current of each transistor. Compensating by amplifiers reduces the dynamic range of the device.

BBDs incorporate a simple concept, but integration and performance are better in a CCD structure.

31.11.2 Charge coupled devices (CCD)

31.11.2.1 Construction

In a CCD a voltage applied to an electrode placed on, but insulated from, a p-type silicon substrate creates a potential maximum to which electrons are attracted. It is useful to consider the potential profile as a well, and charge as water which is held in it. Clocking a series of electrodes on the surface creates a moving potential profile which 'pours' the charge along the device. This avoids the spurious leakage paths in the transistor switching of BBDs.

Losses incurred by charge being trapped while it is transferred along an interface are overcome by moving the potential maximum away from that interface by selective diffusion. This buried channel CCD (BCCD) structure, shown in *Figure 31.31* has a faster, less noisy transfer than its surface channel equivalent. The typical efficiency of a BCCD transfer is 0.999995 at up to 20 MHz.

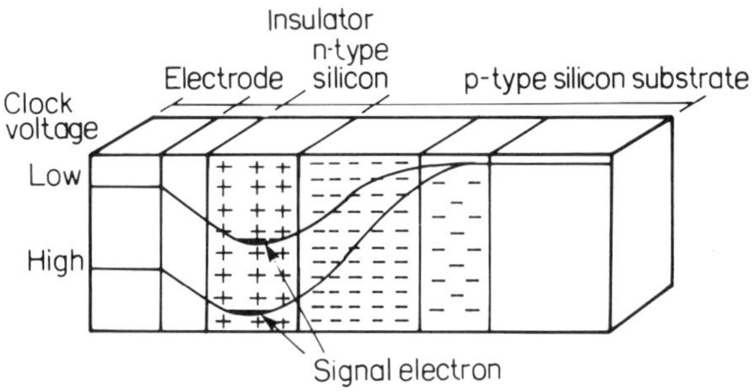

Figure 31.31 Buried channel CCD structure

31.11.2.2 Charge transfer

Early CCDs used three phase clocking to transfer charge while maintaining potential barriers to separate the charge packets. Doping creates an asymmetrical potential profile along the device, allowing two phase clocking to transfer charge, while still keeping the charge packets separate. By applying a d.c. level to one set of electrodes charge may be transferred with only one external clock ($1\frac{1}{2}$ phase clocking). This is the basis of most modern CCD structures.

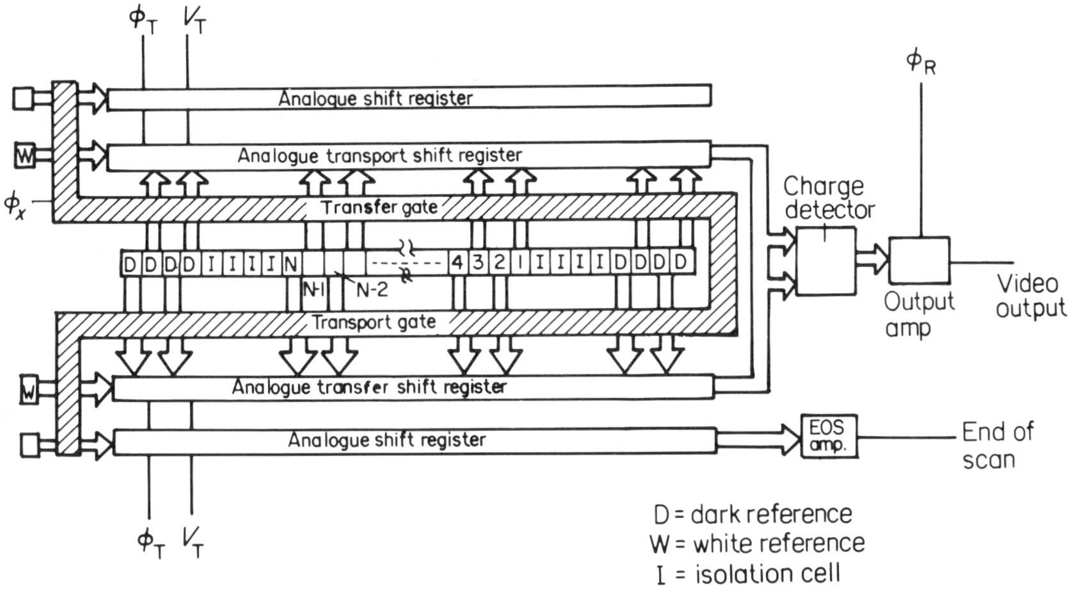

Figure 31.32 Block diagram of a typical CCD line-scan sensor

31.11.2.3 Charge input

Charge may be injected to a CCD shift register electrically or optically. Silicon is sensitive to light in the wavelength range 400 nm to 1200 nm, which includes the visible region of the spectrum. Light falling on the depletion region of a CCD creates an electron–hole pair. The electron moves to the potential well. A charge packet is created in proportion to the intensity of the light falling on the device.

31.11.2.4 Charge output

Charge sensing methods vary according to whether the charge can be read destructively, as at the end of an image sensor, or must be read without destruction, such as part-way along a tapped delay line, or in a CCD memory. In the former, charge enhances the leakage current of a reverse biased diode. In the latter, charge packets produce mirror image charges in a floating gate electrode. In both, the charges are then amplified for output.

31.11.3 CCD applications

31.11.3.1 Imaging

Line scan CCD sensors In the Fairchild CCD142 image sensor (*Figure 31.32*), 2048 photosites, 13 μm square on 13 μm centres, accumulate charge which is linearly dependent on the intensity of the light entering the device, and its integration time. Eight shielded elements, separated from the photosites by isolation cells, represent the random generation of charge in the device—the dark signal.

After a suitable exposure time (typically 100 μs to 100 ms) a pulse applied to the transfer clock (φ_x) causes charge packets to be transferred to the transport registers. By transferring odd photosites to one side, and even photosites to the other, the number of packet transfers is halved.

The d.c. voltage on V_T and the waveform on the transport clock, φ_t, provide the $1\frac{1}{2}$ phase clocking which moves charge in the CCD shift registers. Charge packets are delivered to the output circuit alternately to re-establish their original sequence. Here, the charge alters the potential of a precharged diode. The change is amplified, sampled, held and output. The reset clock (φ_R) recharges the diode.

A white reference charge is injected into the ends of the shift registers to be output with the video and used in an external automatic gain control circuit.

Two further shift registers provide an end-of-scan pulse, and reduce incursion of stray electrons into the inner shift registers.

The dynamic range of this device is typically 2500 to 1. Its saturation exposure is 0.4 μJ cm^{-2} and its responsivity 5.0 V μJ^{-1} cm^{-2}. The efficiency of each shift register transfer is 0.999995.

The CCD142 has an output data rate of 2 MHz. Other sensors offer resolution of up to 3456 elements and data rates up to 20 MHz.

The precise manufacture of the photosites means that objects may be resolved accurately and repeatedly. These attributes are exploited in facsimile, telecine and industrial inspection applications. By moving objects past the sensor a two-dimensional picture is built up.

Area CCD sensors Area image sensors are fabricated with their photosites arranged in a matrix. The sensor turns a picture into precise electrical signals. Suitable clocking of the sensor results in a full interlaced video output signal. Currently available production sensors have 185 440 photosites in a 488 × 380 matrix, although development devices up to 800 × 800 elements have been produced.

The low noise, high sensitivity, low power consumption, high accuracy and inherent reliability make CCD cameras ideal for replacement of tube cameras, for applications such as broadcasting, robot vision and image analysis.

31.11.3.2 Signal processing

With the charge input, shifting and output elements, a CCD dealy line may be constructed. In its simplest form, an input waveform appears at the output after a delay dependent on the number of shift register elements and the clock rate. The horizontal line delay needed in PAL TV receivers can be created in this way.

In a tapped delay line, portions of a signal can be sensed after variable delays and used for adding to or subtracting from the input signal. The matched filters which may be constructed in this way find application in communications and radar to detect weak signals in high background noise. Here CCDs offer longer delays than surface acoustic wave (SAW) devices.

31.11.3.3 Memory

A CCD shift register may be viewed as a digital, rather than an analogue, storage device by arbitrarily defining the sizes of binary '0' and '1' charge packets. Advantages over other technologies are that the CCD structure is very compacr and so cost and power consumption will be low. However, because its serial access is slower than random access memories it is regarded as a replacement for magnetic and bubble memories.

Further reading

BOARDMAN, C. M. and DESCURE, P., 'CCDs—Injection, detection, operation and use', *Electronic Engineering* (September 1982)

DANCE, M., 'Relent developments in op amps', *Electronics Industry* (June 1981)

FAULKENBERRY, L. M., *An Introduction to Operational Amplifiers with Linear IC Applications*, John Wiley & Sons (1982)

JUNG, W. G., 'Stable FET—input op amps achieve precision performance', *EDN* (November 1982)

KRAUS, K., 'Designing with programmable operational amplifiers', *Electronics Industry* (October 1982)

MUTO, A. and NEIL, M., 'ADC dynamic performance testing', *Electronic Product Design* (June 1982)

OHR, S., 'Converters show off linear LSI processing', *Electronic Design* (June 1980)

SHOREYS, F., 'New approaches to high-speed high-resolution analogue-to-digital conversion', *Electronics & Power* (February 1982)

TSANTES, J., 'Data converters', *EDN* (August 1982)

ZUCH, E. L., 'Understanding and applying sample and hold circuits', Part 1, *New Electronics* (October 1978); Part 2, *New Electronics* (October 1978)

32 Semiconductor Memories

S Young BSEE, MBA
Marketing Manager,
SEEQ Corporation

F Jones BSEE
Memory Component Applications Manager,
MOSTEK Corporation

Contents

32.1 Dynamic RAM

Dynamic RAM is the lowest cost, highest density random access memory available. Since the 4k generation DRAM (dynamic RAM) has held a 4 to 1 density advantage over static RAM, its primary competitor. Dynamic RAM also offers low power and a package configuration which easily permits using many devices together to expand memory sizes. Today's computers use DRAM for main memory storage with memory sizes ranging from 16 kbytes to many megabytes. With these large size memories very low failure rates are required. The metal oxide semiconductor (MOS) DRAM has proven itself to be more reliable than previous memory technologies (magnetic core) and capable of meeting the failure rates required to build huge memories. Dynamic RAM is the highest volume and revenue product in the industry.

32.1.1 Cell construction

The 1k and early 4k memory devices used the three-transistor (3-T) cell shown in *Figure 32.1*. The storage of a 'one' or a 'zero'

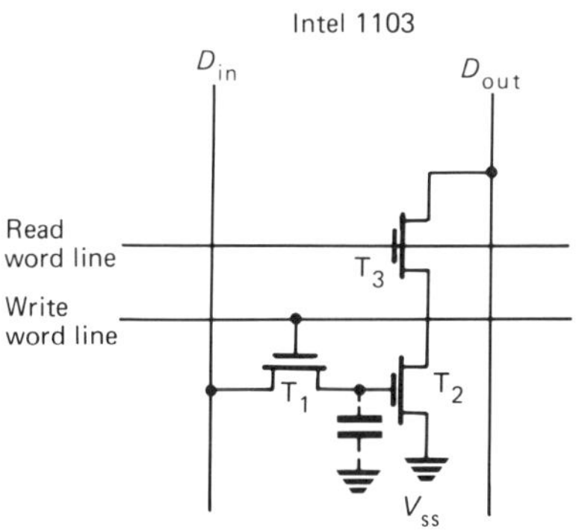

Figure 32.1 Three-transistor cell used in 1 k bit RAM device

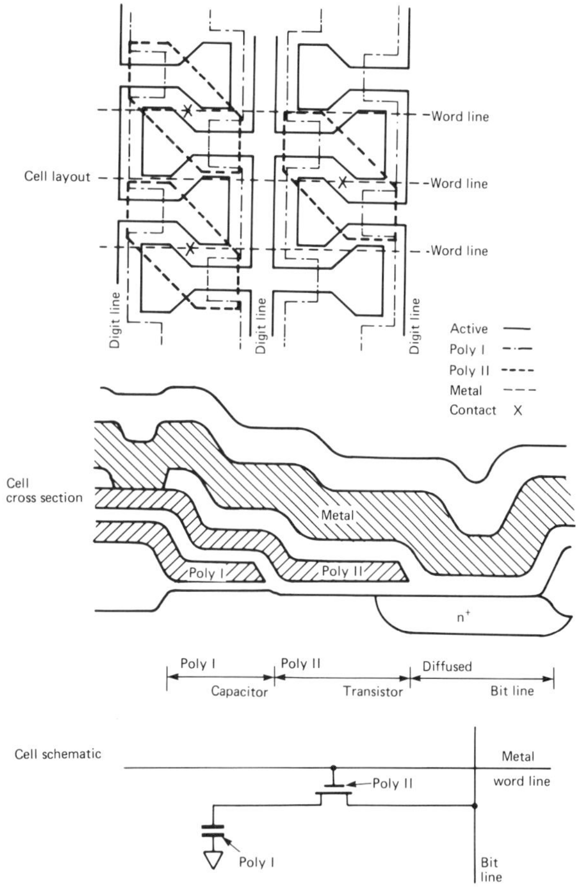

Figure 32.2 Storage cell configuration used in most 16 k RAM devices

occurred on the parasitic capacitor formed between the gate and source of T_2. Each cell had amplification thus permitting storage on the very small capacitor. Due to junction and other leakage paths the charge on the capacitor had to be replenished at fixed intervals; hence the name dynamic RAM. Called refresh, a maximum duration was specified as 2 ms.

The next evolution, the 1-T cell, was the breakthrough required to make MOS RAM a major product. The 1-T cell is shown in *Figure 32.2*. This is the 16k RAM version of the cell. The 4k is similar except that only one level of polysilicon (poly 1) is used. The two-level poly process improves the cell density by about a factor of two at the expense of process complexity. The transistor acts as a switch between the capacitor and the bit line. The bit line carries data into and out of the cell. The transistor is enabled by the word line which is a function of the row address. The row address inputs are decoded such that one out of N word lines is enabled. N is the number of rows which is a function of density and architecture. The 16k RAM has 128 rows and 128 columns. The storage cell transistor is situated such that its source is connected to the capacitor, its drain is connected to the bit line and the gate is connected to the word line.

32.1.2 Cell sensing

Two types of sense amplifiers have been used historically with 1-T cells. Since one approach draws significantly less power, it has become the standard of the 16k RAM generation. Sensing of the signal from a 1-T cell is explained in the following paragraphs while contrasting the two approaches.

When the row decoder enables a row, charge is transferred from each storage capacitor in that row to its respective digit/sense line, destructively reading data. Each column has its own sense amplifier, the function of which is to detect and amplify the signal caused by this charge. To maximise the signal into the sense amplifier, the bit line capacitance must be minimised. This is accomplished by cutting the bit line in half. The sense amplifier is then placed in the centre of the bit line, where it senses a differential voltage between the two halves of the line.

To each side of the sense amplifier is added one additional column of storage cells. These additional cells, commonly called 'dummy cells' have a capacitance equal to approximarely one half of a data storage cell. The dummy cell is used to establish a reference voltage on the side of the sense amplifier not containing the selected cell. Since the dummy cell is half the capacitance of a storage cell, a voltage reference level halfway between the voltage difference between a '1' and '0' is established. This reference voltage is commonly called the cell's 'trip point' and may be varied by changing the physical size of the 'dummy cell'. In actual practice this 'trip point' varies due to alignment tolerances in the manufacturing process.

Between cycles, the two halves of each bit line are equilibrated to precisely the same voltage and the 'dummy cell' is restored to 0 volts (*Figure 32.3*). After the row decoder has been activated, a

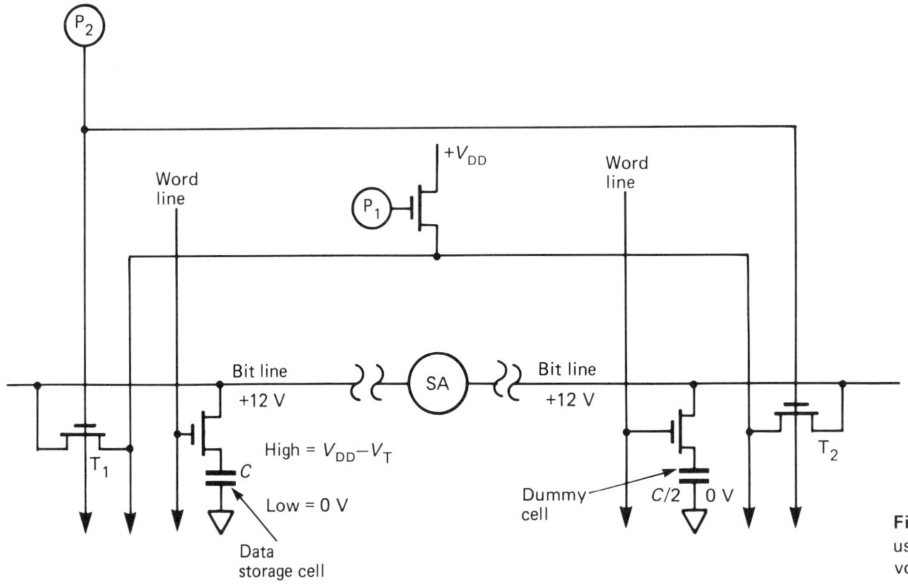

Figure 32.3 Sensing scheme uses dummy cell to establish voltage reference

single cell in each column turns on, including the 'dummy cell'. If we assume the bit lines were initially set to 12 volts then the following occurs. The 0 volts stored in the 'dummy cell' pulls the side of the sense amplifier not containing the selected storage cell to approximately 11.75 volts. (This is determined by the ratio of the bit line capacitance to that of the 'dummy cell's' capacitance.) If the storage cell contained 0 volts, then charge sharing between the bit line and the storage cell would bring the bit line voltage to approximately 11.5 volts (remember the storage cell is twice as large as the dummy cell). A voltage difference between 11.5 and 11.75 volts can now be sensed. Had 12 volts been stored, then the bit line would remain at 12 volts and a voltage difference between 12 and 11.75 volts would be sensed. In practice the storage cell stores 12 V minus the threshold voltage of the access transistor thereby somewhat moving the bit line away from 12 volts.

The sense amplifier is now turned on and via the regenerative action of the cross-coupled amplifier the lower voltage side of the sense amplifier is pulled to ground. The higher voltage side remains at its previous level. At the end of a cycle the word lines are deselected and the preceding transistors, T_1 and T_2, are turned on. This equilibrates the bit lines to the V_{DD} supply potential. The node P_1 is placed at a voltage at least one threshold above V_{DD} to permit the bit lines to reset to the V_{DD} potential.

Two types of sense amplifiers have been used in commercially available products—variations of the static design of *Figure 32.4(a)* and of the dynamic amplifier of *Figure 32.4(b)*. Both are about equal in their ability to detect and amplify small signals.

The load resistors (R_1 and R_2) employed in the static amplifiers consume a substantial amount of power, typically half or more of total chip power. Since dynamic amplifiers do not employ these resistors, their power consumption is considerably reduced. However, there are formidable design and layout problems associated with dynamic amplifiers, and many designers in the early days chose to incorporate power-consuming static sense amplifiers into their designs.

We must look at the write cycle—or more accurately the read–modify–write cycle—to understand the differing circuit requirements for static and dynamic sense amplifiers. As an example, suppose cell 64 (*Figure 32.4(a)*) had originally stored a low voltage and was read.

Upon detecting a lower voltage on node B than that on node A, the sense amplifier will drive node B to ground and node A to power supply V_{DD}. Transistor T_3 then turns ON and the data from the cell becomes available to the output buffer at one end of the data bus.

Now assume we want to write the opposite data back into the cell. This requires forcing a high voltage onto node B and the storage capacitor, C_{64}. To do this, the data input buffer drives the input/output data bus to ground. T_3 then forces node A to ground, overpowering R_1. When node A goes to ground, T_2 turns OFF allowing R_2 to pull node B to V_{dd} as required to write the high level into the storage cell.

Without R_2, node B would remain at ground, and it would be impossible to write a high voltage into the cell. With the resistors present, data can be written into a cell in either matrix half with a single input/output data bus.

A trade-off exists in the choice of values for R_1 and R_2. Since either R_1 or R_2 dissipates power in all sense amplifiers, low resistance values result in very high power consumption. Then again, digit line capacitance of node B is quite large, and a high-value resistor would lead to excessively long write time. Unfortunately, there is no good compromise, and circuits employing static sense amplifiers suffer from high power consumption and long write times.

On paper, the dynamic amplifier solves the problem very well. Having read a low voltage from cell 64 (*Figure 32.4(b)*) assume that we again want to write a high voltage back into the cell. As before, the input buffer drives the true data bus to ground, with T_3 causing node A to follow. However, the input buffer also forces the complement data bus to V_{DD} with T_4 causing node B to follow.

The complement data bus thus performs the job previously done by the resistor. With cell 64 still selected, the high voltage on node B is transferred into the cell and the write operation is complete.

Note that T_3 and T_4 function only as switches and can have very low resistances to speed up write time. This involved no speed/power trade-off. Therefore, memory designs using dynamic amplifiers consume far less power and write much faster than do designs using static sense amplifiers.

The previous was the technique used in the 4k/16k generations. This generation permitted use of a 12 V (V_{DD}) power supply and a

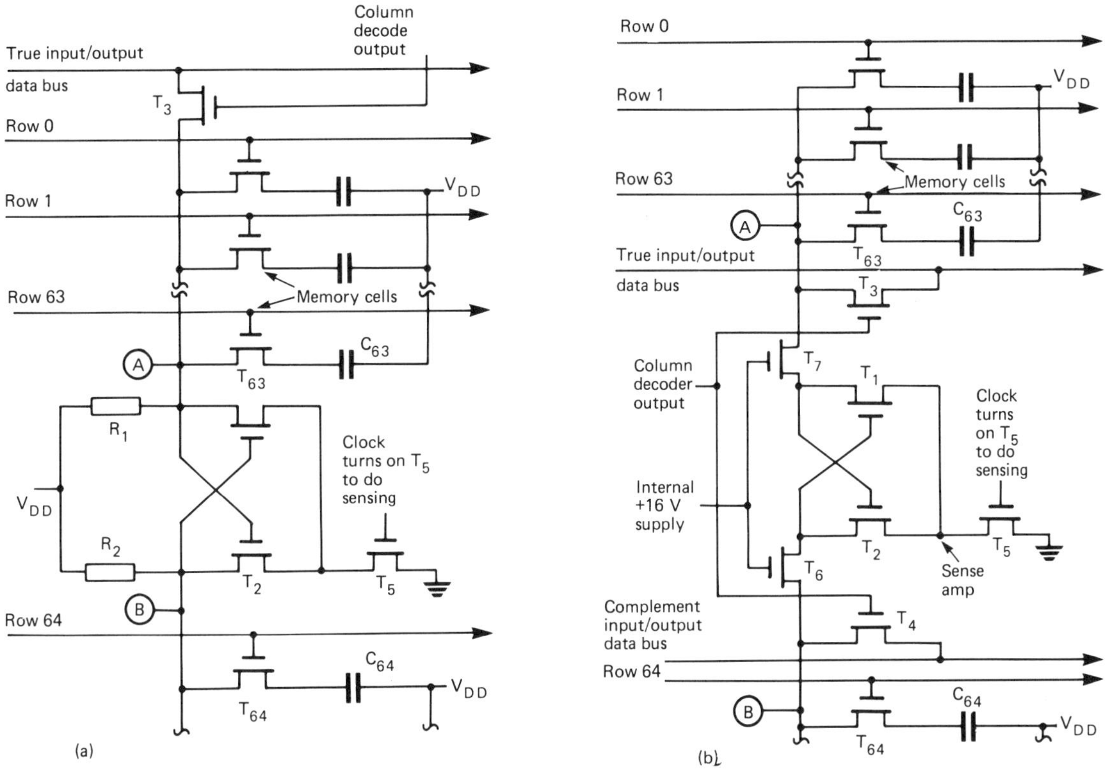

Figure 32.4 Sense amplifiers: (a) static; (b) dynamic

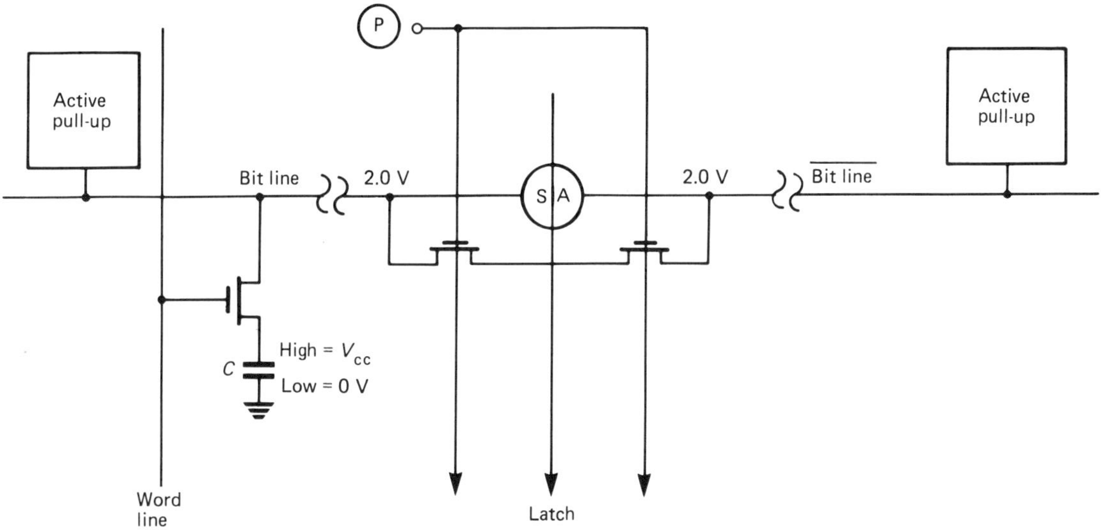

Figure 32.5 64 k DRAM sensing approach

-5 V (V_{BB}) bias supply. The 64k generation offers neither of these to the designer. The designer must generate a bias supply on chip and also has only a 5 V supply to work with. One obvious ramification is that the signal is decreased to 5/12 of the 16k if circuit techniques and geometries are held constant. To permit manufacture of a 64k RAM both dimensions and circuit techniques were changed.

Figure 32.5 illustrates one approach used at the 64k density level. The following describes the operation of this approach.

This sensing scheme equilibrates to about half the supply voltage, $V_{CC}/2$, and uses active pull-ups in order to write a full power supply level back into the cell. This method eliminates the neccessity of a dummy cell; no longer is a dummy cell required in order to establish a reference voltage to guarantee a differential voltage across the sense amplifier during sensing.

At the start of the cycle, when RAS goes low, both sides of the sense amplifier have been equilibrated to about $V_{CC}/2$. A high is stored as a full V_{CC} level (not a $V_{CC}-V_T$) and a low is stored as ground or 0 volts. After the row decoder has determined which row to select, the appropriate word line is enabled turning on the selected row of storage cells. If a zero had been stored in the storage cell, the bit line previously equilibrated to $V_{CC}/2$, will be pulled down several hundred milivolts. If a high had been stored, the bit line would be pulled up several hundred milivolts. (The exact change on the bit lines is dependent on the capacitance of the bit line and the capacitance of the storage cell. The higher the ratio of storage cell capacitance to the bit line capacitance the greater the change on the bit line due to a stored high or a stored low level.) Note that in either case a differential voltage is established, without using dummy cells, and this voltage is sufficient to be detected by the sense amplifier.

When the differential voltage on the bit lines has settled following the selection of the storage cells, the sense amplifier is activated via the latch signal. The regenerative action of the sense amplifier pulls the lower side down to 0 volts. The higher side remains unchanged.

At the end of the cycle, when RAS goes inactive, three events occur. First, the word line is boot-strapped up to a level greater than V_{CC} plus a V_T (threshold voltage). This assures that a full V_{CC} is stored in the storage cell for a high level. Second, the active pull-ups are enabled resulting in the higher side of the sense amplifier being reinforced to a full V_{CC} level while leaving the lower side at ground. Third, the word lines are deselected, after the storage cell has been written into, and the equilibration transistors are turned on. This allows charge sharing to occur between the bit lines on both sides of the sense amplifier to about half the supply voltage. This equilibrated level remains until the next RAS active cycle and the above process is repeated. The sequence of events described above is summarised in the timing diagram shown in *Figure 32.6*.

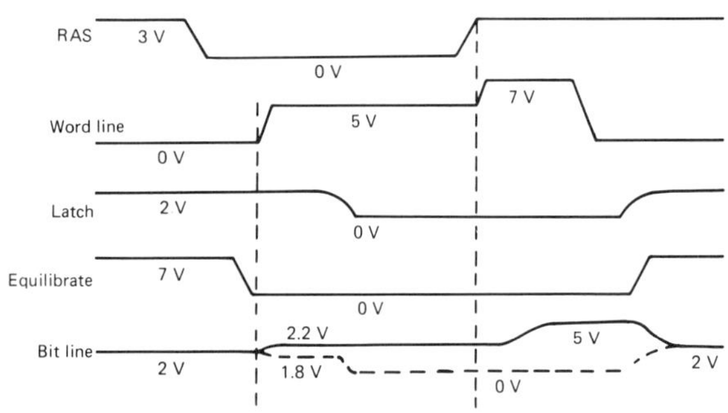

Figure 32.6 Timing diagram for the system of *Figure 32.5*

32.1.3 Cell process

The principal process used in current generation DRAM is the double-level poly-silicon gate process. The process uses n-channel devices which rely on electron mobility. The 1k RAM (1103) used a p-channel process which relied on hole mobility. The p-channel process although simpler to manufacture (it is more forgiving to contamination) is by nature slower than n-channel (electrons are much faster than holes) and cannot operate at 5 V levels. Therefore once manufacturing technology advanced enough to permit n-channel to yield it quickly displaced p-channel techniques for memory.

32.1.4 System design considerations using dynamic RAMs

Dynamic RAMs provide the advantages of high density, low power, and low cost. These advantages do not come for free; the dynamic RAM is considered to be more difficult to use than static RAMs. This is because dynamic RAMs require periodic refreshing in order to retain the stored data. Furthermore, although not generic to dynamic RAMs, most dynamic RAMs multiplex the address bits which requires more complex edge-activated multi-clock timing relationships. However, once the system techniques to handle refreshing, address multiplexing, and clock timing sequences have been mastered, it becomes obvious that the special care required for dynamic RAMs is only a minor inconvenience.

A synopsis of the primary topics of concern when using dynamic RAMs follows.

32.1.5 Refreshing

Dynamic RAMs use a 'sample-and-hold' principle to store binary data; the presence of a charge in the storage capacitor can represent a stored logical '1' and the absence of charge can represent a stored logical '0'. A characteristic of a stored level in a 'sample-and-hold' type circuit is that it can retain data for only a short period of time and that the retention time or storage time interval is limited by the size of the storage capacitor and leakage paths present. A further limitation on data retention time is the junction temperature. The storage time of any dynamic RAM may be expressed by the empirical equation:

$$t_s = A\exp(-BT) \tag{32.1}$$

where T is the junction temperature, B is a variable relating the magnitude of the generation–recombination current to the junction temperature, and A is a scaling constant reflecting such variables as junction area, bulk defect density, and sense amplifier design.

Storage time typically doubles each time the junction decreases by about 10°C. This is illustrated in *Figure 32.7*. Typically, dynamic RAM vendors guarantee the retention of stored data for about 2 ms at a maximum ambient temperature of 70°C.

The architecture of the RAM determines the number of refresh cycles required to refresh all storage cells. The most common refresh requirement is 128 refresh cycles each 2 ms interval. This is accomplished by cycling through 128 ROW locations (addresses clocked into the RAM during RAS assertion) each and every 2 ms interval. A less popular refresh scheme which appeared at the 64k density level is 256 refresh cycles each 4 ms. This scheme requires that each cell retain the stored charge twice as long as with the previous method.

As indicated, dynamic RAMs require periodic refresh cycles. The refresh cycles may be performed either in a distributed fashion or in a burst mode. With distributed refreshing, single refresh cycles are scheduled evenly throughout the refresh interval. With burst refreshing, all the required refresh cycles are performed in a back-to-back fashion. Since dynamic RAMs use not only dynamic circuitry in the storage array circuitry but also in the peripheral clock circuits, these clocks must be cycled periodically or the device will not operate properly; a long interval between active RAM cycles results in degraded levels within the clock circuitry. For this reason, distributed refreshing is the preferred refreshing mode.

There are a number of methods that can be used to schedule the required periodic refresh cycles. Some microprocessors provide a signal or status bits that can easily be decoded to indicated when the refresh cycle can be safely performed.

Many systems require arbitration circuitry to select between normal memory cycles and pending refresh cycles. The

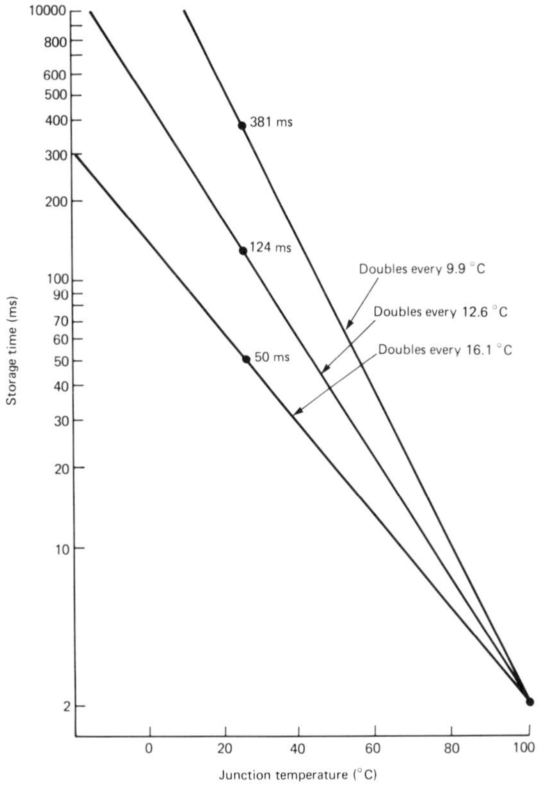

Figure 32.7 Effect of temperature on cell storage time

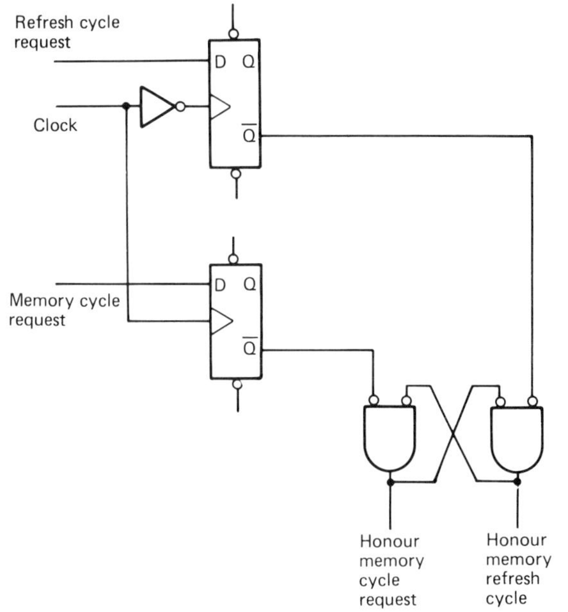

Figure 32.8 Cross-coupled flip-flop arbitration between a refresh request and a memory access request

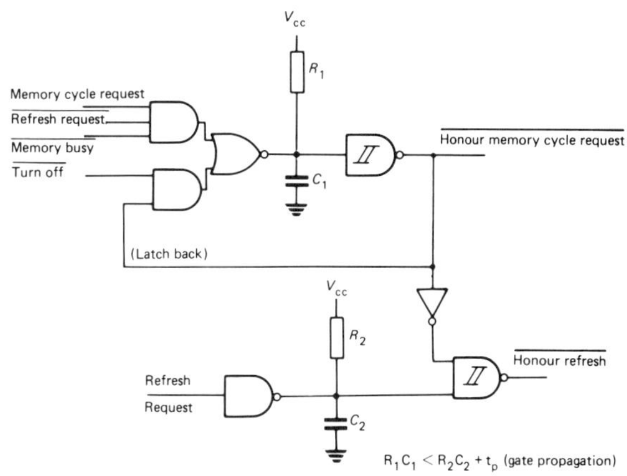

Figure 32.9 Arbitration between a synchronous memory cycle request and refresh request

arbitration task is not trivial. This circuitry must decide between either a normal access or a refresh and start a memory cycle. Furthermore, the arbitration must be done such that glitching on the RAM clocks (RAS, CAS, or W) does not occur and that a memory cycle is not prematurely aborted. *Figures 32.8* and *32.9* show two different schemes that have been used successfully to arbitrate between memory access activity and refresh cycles. *Figure 32.8* indicates a technique when a memory 'start' timing signal is available with specific timing relative to a system clock which is also available to the arbitrator. *Figure 32.9* illustrates a technique which has been successful when the memory request signal is asynchronous and a synchronising clock is not available.

The purpose of the arbitration circuit is to schedule processor (or user cycle) requests and required refresh requests. Once a refresh has been given priority by the arbitrator. the refresh circuitry must provide the refresh address (during row address time) to the RAM and generally a 'RAS-only' type cycle is executed. A read cycle (RAS and CAS active, W inactive) will also accomplish the refresh operation. However, when two or more devices have their data outputs connected together, 'RAS-only' type cycles should be executed in order to avoid data output contention. Note that if normal memory access cycles can guarantee that all combinations of the ROW addresses within the refresh field are cycled within the specified refresh interval, additional refreshing is not required.

32.1.6 Address multiplexing

The use of address multiplexing reduces the number of address lines and associated address drivers required for memory interfacing by a factor of 2. This scheme, however, requires that the RAM address space be partitioned into an X–Y matrix with half the addresses selecting the X or ROW address field and the other half of the addresses selecting the Y or column address field. The ROW addresses must be valid during the RAS clock and must remain in order to satisfy the hold requirements of the on-chip address latches. After the ROW address requirements have been met, the addresses are allowed to switch to the second address field, column addresses, a similar process is repeated using CAS (column address strobe) clock to latch the second (column) address field. Note, for RAS-only refreshing the column address field phase is not required. (See *Figure 32.10* for address multiplexing timing relationships for read cycles and *Figure 32.11* for write cycles.)

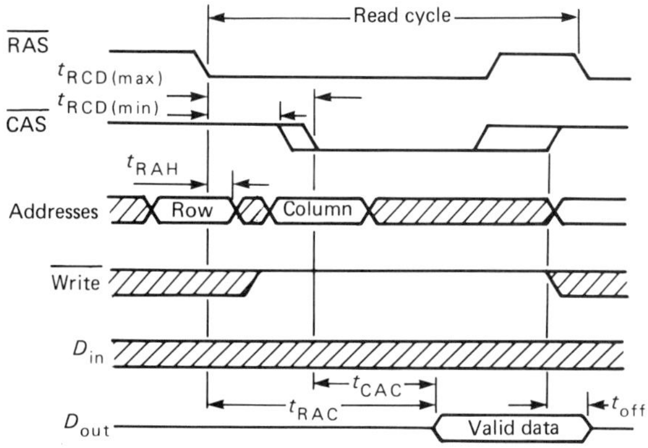

Figure 32.10 Read cycle timing for 16 pin multiplexed dynamic RAM

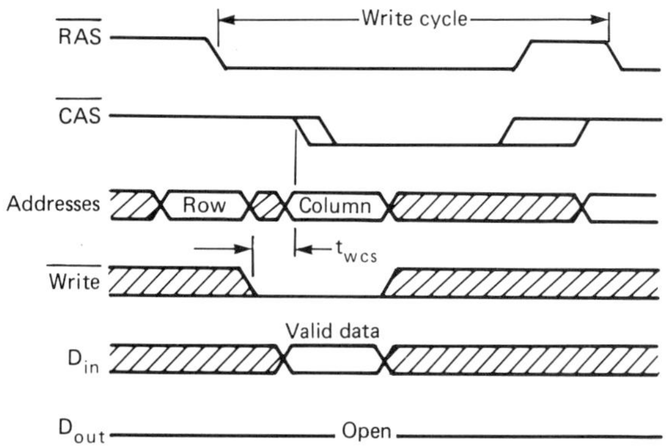

Note: If, write command occurs before CAS, then D_{out} remains high impedance.

Figure 32.11 Write cycle timing for 16 pin multiplexed dynamic RAM

32.1.7 Dynamic RAM system timing generation

When using dynamic RAMs, the system performance is very much dependent on the timing relationships of the interface signals. RAM vendors have attempted to simplify the system timing problem. This is done by first reducing the time that the ROW address must be held valid following the RAS clock going active. This allows the column addresses to be presented to the RAM as soon as practical. After the column addresses are valid, the CAS clock can be asserted. CAS can be activated at any time during the t_{RCD} minimum-to-maximum interval without affecting the worst case data access time relative to the RAS active edge. If, however, the CAS signal cannot be asserted prior to the t_{RCD} maximum point, storage errors or reading errors will not result. The only penalty that results from pulling CAS active beyond the t_{RCD} maximum limit is that the access time is pushed out such that the access time is now entirely dependent on the access time from CAS or t_{CAC} (see *Figure 32.12*).

Figures 31.13, *32.14* and *32.15* illustrate three popular schemes that are commonly used to generate the required interface timing signals. A combination of these techniques can be used. The method shown in *Figure 32.13* is generally used where low cost is the primary consideration. This method uses the inherent propagation delays of the devices themselves to generate the require timing relationships. *Figure 32.14* shows how delay lines

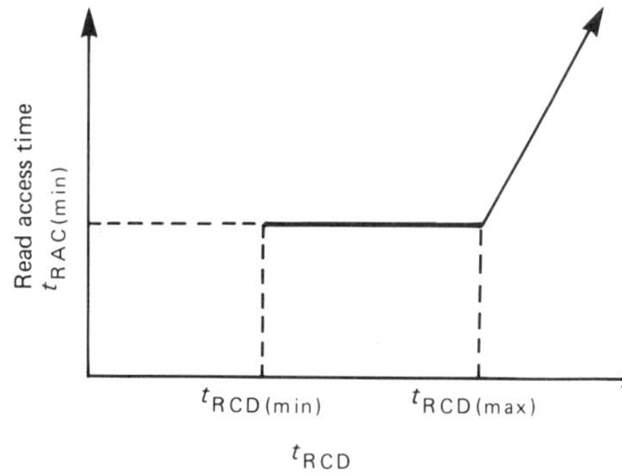

Figure 32.12 Gated CAS timing relationship

can be used to generate the required relationships. 'Digital' delay modules have recently found widespread acceptance in this application. *Figure 32.15* shows a synchronous technique that can be used when high frequency timing edges are available. Regardless of the method used, worst case timing calculations should be made in order to ensure operation within the limits specified by the RAM vendor.

32.1.8 Board layout considerations

Several guidelines should be followed when laying out printed circuit boards for memory arrays using dynamic RAMs. The four areas of primary concern are power distribution, power decoupling, placement of array driver chips, and treatment of signal and clocks through the memory array.

Careful attention to power distribution and power decoupling is important since dynamic RAM current requirements are transient in nature. Dynamic RAMs require very little quiescent current during periods of inactivity. However, large current spikes occur during any active memory or refresh cycle. Due to the transient current requirements of these devices, low inductance power distribution techniques should be used within the memory array. This can be done by dedicating 'voltage' planes when using multilayer printed circuit boards or by using gridding techniques with double-sided boards. High frequency decoupling capacitors should be judiciously placed close to the memory devices and connected to the low impedance power bus. Generally, 0.1 μF decoupling capacitors are used and will adequately suppress power bus transients in the array. Additionally, low frequency-capacitors should be placed around the perimeter of the memory array to provide further smoothing of the supply voltages by minimizing the effects of inductive and resistive voltage drops.

32.1.9 Soft errors

With dynamic RAMs, particular care must be taken to provide a friendly operating environment in order to avoid the occurrence of intermittent errors. When dynamic RAMs are operated in an 'unfriendly' environment, the occurrence of intermittent errors may exist. These errors are referred to as 'soft' errors. Board related soft errors can be prevented by making certain that all operating conditions are within the RAM manufacturer's specification limits. Some general design guidelines that should be observed when using MOS dynamic RAMs include:

(1) Always design well within the vendor's specification limits. Adequate voltage (supply and signal), timing, and temperature margins are essential for trouble free operation.

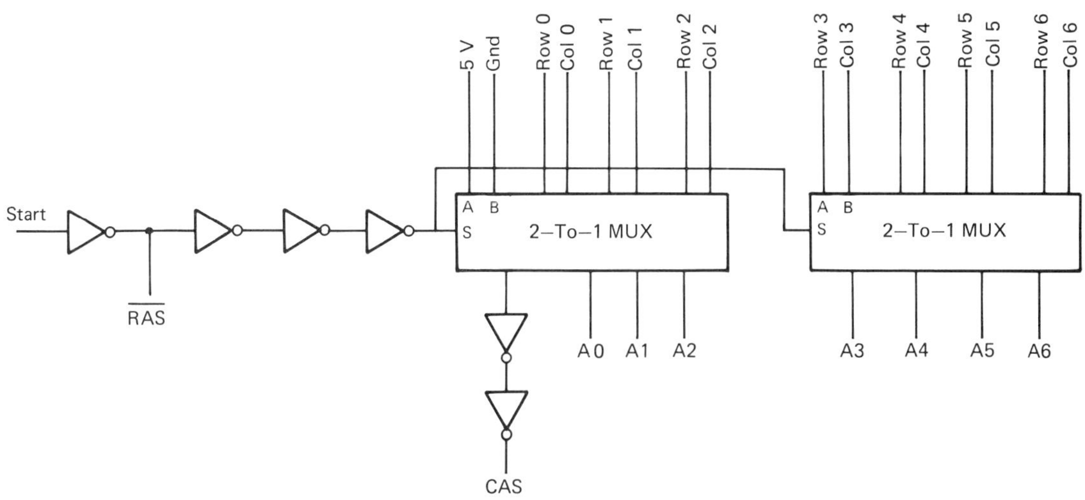

Figure 32.13 Device delays are used to generate timing

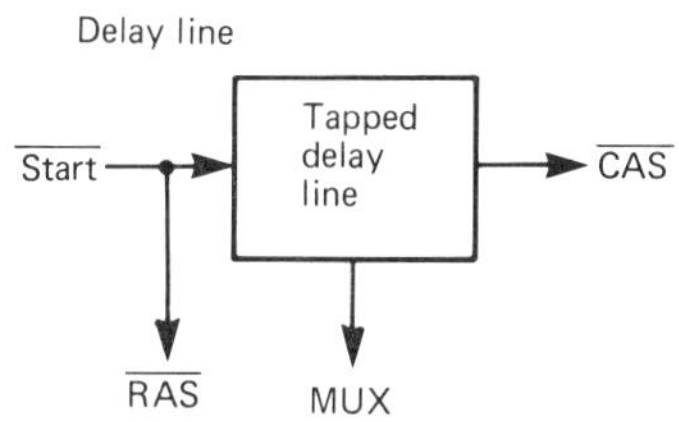

Figure 32.14 Tapped delay line is used to generate timing signals

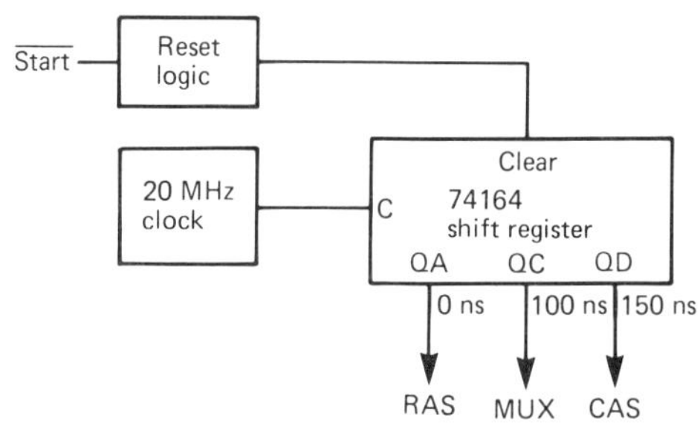

Figure 32.15 Shift register is used to generate timing signals

(2) Layout array as densely as reasonably possible while maintaining adequate power busses and array power decoupling.

(3) Use low impedance power distribution through the array. Gridding is essential with double-sided PC boards.

(4) Place high frequency decoupling capacitors within the array matrix. At least one capacitor per critical supply for every two devices (preferably each device) is suggested. A 0.1 μF capacitor has been found to be suitable for this purpose.

(5) Where possible, place bulk decoupling capacitors around the perimeter of the memory matrix.

(6) Use transmission line terminations on all signals to the RAMS. Conventional termination techniques are effective when used with constant impedance transmission lines that have predictable impedances. Many of the single supply (+5 V) RAMs will tolerate as much as −2.0 V negative excursions below ground on input signals. This does not indicate that line terminations are not required. When excessive negative overshoot occurs, even within the specification limit, the positive excursions resulting from the damped ringing will reduce noise margins and will generally result in a violation of the $V_{IL(max)}$ specification limit. The user should be cautioned that high speed dynamic RAMs can respond to nanosecond spikes and glitches which may result in unpredictable behaviour.

(7) Avoid long parallel runs with signals that cannot tolerate crosstalk. If long runs are necessary, provide an effective ground plane by routing power or ground runners on the adjacent layer. This reduces the crosstalk by *E*-field cancellation and results in a nearly constant impedance transmission line (inclusion of a power grid yields an effective microstrip characteristic) for signals transversing the array. Also, crosstalk can be reduced by placing decoupled power ground runners between the signal traces.

(8) When possible, avoid stubbing of array signal lines. This represents an impedance discontinuity with resulting energy reflections which can reduce noise margins or possibly violate the vendor's specification limits.

(9) Locate the driver circuits as close to the array as possible with short direct routing of interface signals between the drivers and the RAMs.

(10) Place both high frequency caps and bulk decoupling caps near the array driver circuits.

(11) Provide sufficient ground bussing for drive circuits and between the drivers and the array to minimise ground bounce and ground offset conditions.

(12) Avoid using driver gates in a common package when crosstalk due to ground bounce or other internal coupling mechanisms cannot be tolerated.

(13) If any device is being used in an unconventional way, verify with the device vendor that the device will support the desired operation.

Although careful board/system design can prevent soft errors, other soft error mechanisms exist. These soft errors are due to traces of radioactive contaminants which emit alpha particles. These alpha emiters are contained in the packaging material used for semiconductor devices. The most common elements with alpha emiting isotopes are uranium, thorium, radium, polonium, bismuth, and radon.

When the alpha particle passes into the RAM it creates a positive charge in the oxide. The alpha continues to travel into the silicon to a depth of approximately 25 μm resulting in electron–hole pairs being generated. These electrons are attracted to the positively charged oxide. Electrons generated in depletion regions drift toward the surface and electrons generated in the substrate diffuse toward the surface. This process can result in a modification of the charge on the storage cell capacitor or can affect levels during the sensing process. In either case, a soft error can occur if the energy levels associated with the alpha particle passing through the silicon are high enough. Seldom does a single alpha particle cause more than a single location to err and the mechanism is non-catastrophic. The location affected by the alpha 'hit' will properly store new (rewritten) data and this location is no more susceptible to another soft error than any other location on the device.

Fortunately, alpha-related soft errors occur very infrequently. Device manufacturers have attempted to make these devices as insensitive to alpha radiation as possible. Methods such as using radioactively 'clean' packaging materials, protective die coatings, design improvements (for more signal margins), and new processing techniques have reduced alpha-type soft errors to a very low level.

D_2	D_1	D_0
0	0	0
0	0	1
0	1	0
0	1	1
1	0	0
1	0	1
1	1	0
1	1	1

(a)
Complementing any single bit results in another valid data word (Hamming distance = 1)

D_2	D_1	D_0	$P_{(even)}$	D_2	D_1	D_0	$P_{(odd)}$
0	0	0	0	0	0	0	1
0	0	1	1	0	0	1	0
0	1	0	1	0	1	0	0
0	1	1	0	0	1	1	1
1	0	0	1	1	0	0	0
1	0	1	0	1	0	1	1
1	1	0	0	1	1	0	1
1	1	1	1	1	1	1	0

(b)
Complementing any single bit does not result in another valid data word (Hamming distance = 2)

Figure 32.16 Error detection

32.1.10 Error detection and error correction

In a large memory array there is a statistical probability that a soft error(s) or device failure(s) will occur resulting in erroneous data being accessed from the memory system. The larger the system, the greater the probability of a system error.

In memory systems using dynamic RAMs organised N by 1 configuration, the most common type of error/failure is single bit oriented. Therefore, error detection and error correction schemes with limited detection/correction capabilities are ideally suited for these applications.

Single bit detection does little more than give some level of confidence that an error has not occurred. Single bit detection is accomplished by increasing the word width by one bit. In order to understand how this provides error detection capability refer to *Figure 32.16*. In the example shown a small word length of three bits has been used. Note that three bits provides eight binary combinations as shown in *Figure 32.16(a)*. With the three bit word, complementing any single bit in the word results in another valid binary word combination.

By adding a parity bit and selecting the value of this bit such that the four bit word has an even number of 'ones' (*Figure 32.16(b)*) or an odd number of 'ones' (*Figure 32.16(c)*), complementing any single bit in the word no longer results in a valid binary combination. Notice that it is necessary to complement at least two bit locations across the word in order to achieve a new valid data word with the proper number of 'ones' or 'zeros' in the word. In order to extract the error information from the stored data word. This concept was introduced by Richard Hamming in 1950. The scheme of defining the number of positions between any two valid word combinations is called the 'Hamming distance'. The Hamming distance determines the detection/correction capability.

Correction/detection capability can be extended by additional parity bits in such a way that the Hemming distance is increased. *Figure 32.17* shows how a 16-bit word with five parity bits (commonly referred to as 'check bits') can provide single bit detection and single bit correction. In this case, a single bit error in any location across the word, including the check bits, will result in a unique combination of the parity errors when parity is checked during an access.

As indicated above, the ability to tolerate system errors is

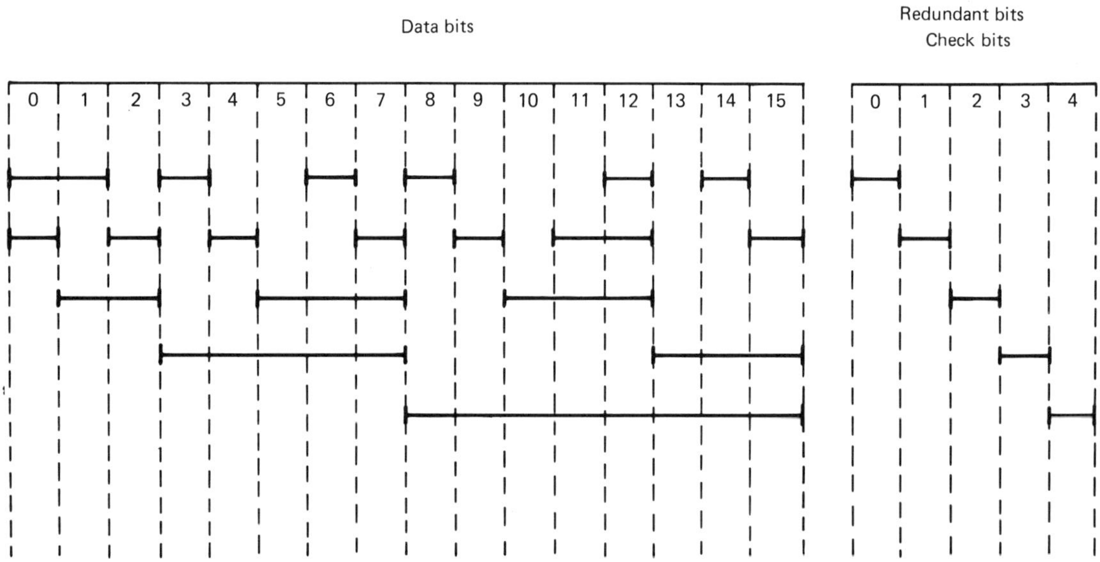

Figure 32.17 An algorithm for determining check bits (parity from subsets of data word) for providing single error bit detection and single bit correction

bought at the expense of using additional memory devices (redundant bits). However, the inclusion of error correction at the system level has been greatly simplified by the availability of error correction chips. These chips provide the logic functions necessary to generate the check bits during a memory write cycle and perform the error detection and correction process during a memory read cycle.

32.2 Static RAM

Static random access memory is a memory device that utilises a storage vehicle that requires only that power remain on to retain data. Typically it requires no synchronous timing edges and is easy to use. Due to its characteristics it can be an extremely fast memory device. Speed and/or ease of use are therefore the primary motivations for designing with this class of device. This contrasts with density, low cost, and lower power which are the key motives for using a dynamic RAM.

32.2.1 Organisation

Static RAMs in the 1k (k = 1024) and 4k generations are organised as a square array (number of rows of cells equals the number of columns of cells). For the 1k device there are 32 rows and 32 columns of memory cells. For a 4096 bit RAM there are 64 rows and 64 columns. The RAMs of the seventies and beyond include on-chip decoders to minimise external timing requirements and number of input/output (I/O) pins.

To select a location uniquely in a 4096 bit RAM 12 address inputs are required. The lower-order six addresses decode one of 64 to select the row while the upper-order 64 addresses decode to select one column. The intersection of the selected row and the selected column locates the desired memory cell.

Static RAMs of the late seventies and early eighties departed from the square array organisation in favour of a rectangle. This occurred for two primary reasons: performance and packaging considerations.

Static RAM has developed several application niches. These applications often require different organisations and/or performance ranges. Static RAM is currently available in one bit I/O, four bit I/O, and eight bit I/O configurations. The one and four bit configurations are available in a 0.3 inch wide package with densities of 1k, 4k, and 16k bits. The eight bit configuration is available in a 0.6 inch wide package in densities of 8k and 16k bits. The eight bit I/O device is pin compatible with ROM (read only memory), and EPROM (electrically programmable ROM) devices. All of the static RAM configurations are offered in two speed ranges. One hundred nanosecond, typically is used as the dividing line for part numbering.

32.2.2 Construction

The static random access memory (RAM) uses a six device storage cell. The cell consists of two cross-coupled transistors, two I/O transistors and two load devices. There are three techniques that have been successfully employed for implementing the load devices. These are: enhancement transistors, depletion transistors, and high impedance polysilicon load resistors.

Enhancement transistors are normally off and require a positive gate voltage (relative to the source) to turn the device on. There is a voltage drop across the device equal to the transistors' threshold. Depletion transistors are normally on and require a negative gate voltage (relative to the source) to turn them off. There is no threshold drop across this device. Polysilicon load resistors are passive loads which are always on. Formed in the poly level of the circuit they are normally undoped and have very high impedances (5000 MΩ typically). They have the advantage of very low current (nanoamperes) and occupy a relatively small area. All present large density, state of the art, static RAMs now use this technique for the cell loads. Further enhancements have been made by manufacturing a two-level poly-process thereby further decreasing cell area.

The operation of the three-cell approaches is shown in *Figure 32.18*.

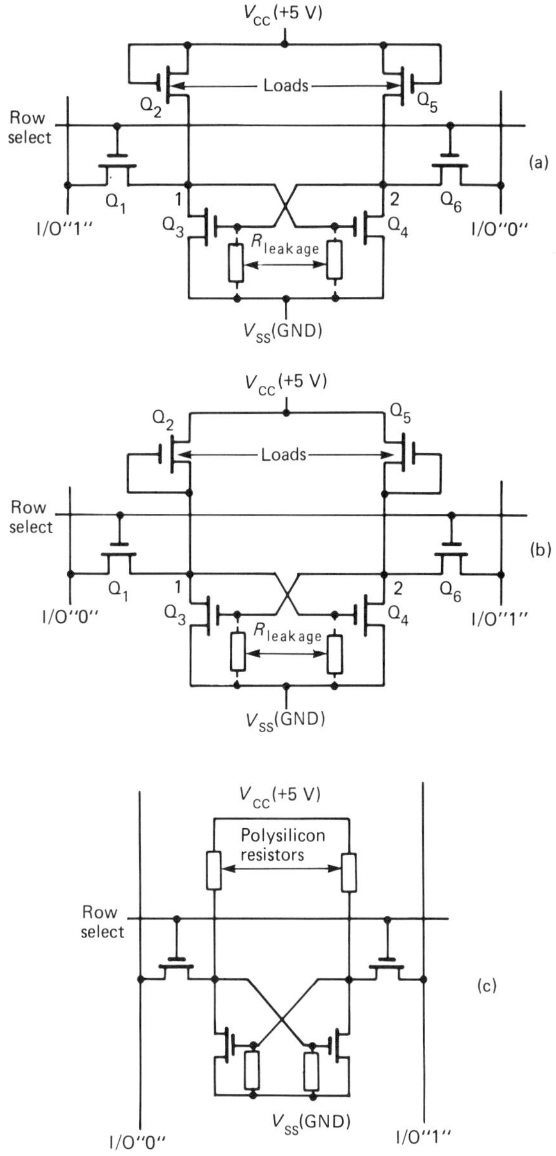

Figure 32.18 Static RAM storage cells

32.2.2.1 Enhancement cell

Consider the storage cell shown in *Figure 32.18(a)*. Data are stored as a charge on the gate of either Q_3 or Q_4 (which determines the logic state of the cell). The voltage on the charged node is approximately $V_{CC}-V_{TH}$ (where V_{TH} is the effective threshold of the load devices) and turns Q_3 or Q_4 on. By definition a logic '0' is stored in the cell if Q_3 is on and a logic '1' is stored if Q_4 is on.

If it is assumed that Q_3 is on (logic '0' stored) then current will flow from the load on Q_3 (device Q_2) through Q_3 to ground (V_{SS}). This current will cause the voltage at node 1 to assume a value near V_{SS} (the voltage is proportional to the effective on resistance of Q_2 and Q_3). The resultant low voltage on node 1 turns device Q_4 off. Device 5 maintains the charge on the gate of Q_3 by replacing charge leaked off through the high impedance parasitic leakage resistor $R_{leakage}$ (this leakage is typically in the picoampere range). The storage cell will remain in this logic state until an external forcing function is applied (write cycle).

32.2.2.2 Depletion cell

Operation of the storage cell is as follows: assume the gate of Q_3 is high turning Q_3 on causing current to flow in Q_3 and Q_2. Since devices Q_2 and Q_3 are ratioed (that is, Q_2 has a higher impedance than Q_3) the voltage at node 1 will drop close to V_{SS}. Note that the gate of Q_2 is tied to node 1; therefore as node 1 decreases in voltage, the voltage drive on Q_2 is reduced, making the effective impedance of Q_2 higher. This allows the voltage at node 1 to move even closer to V_{SS}.

Since node 1 is low and is tired to the gate of Q_4, device Q_4 is off. The charge on Q_3 is maintained by the load device Q_5. Note that only leakage currents flow through device Q_5 which has a minimal effect on the voltage at node 2. Since increased positive voltage at node 2 increases the voltage drive on Q_5, device Q_5 turns on hard. The voltage at node 2 is therefore equal to V_{CC} (note that there is no threshold drop across device Q_5 since it is a depletion mode device).

32.2.2.3 Resistor cell

Operation of the storage cell is as follows: assume the gate of Q_3 is high turning Q_3 on causing current to flow in Q_3 and Q_2. Since device Q_2 has a much higher impedance than Q_3 (orders of magnitude higher) the voltage at node 1 will drop close to V_{SS}. Since node 1 is low and is tied to the gate of Q_4, device Q_4 is off. The charge on Q_3 is maintained by the load resistor Q_5. Note that only leakage currents flow through resistor Q_5 which has a minimal effect on the voltage at node 2. Since the current through Q_5 is very low the voltage at node 2 asymptotically approaches V_{CC}.

32.2.3 Accessing the storage cell

The storage cell is interrogated for a read or write operation by activating the proper row select line which turns devices Q_1 and Q_6 on (*Figure 32.18*). For a read operation, a sense amplifier connected to both the I/O '0' and I/O '1' outputs of each column detects the state of the selected storage cell in that column. If Q_3 is on (logic '0') then current will flow in the I/O '0' line. If Q_4 is on (logic '1'), current will flow in the I/O '1' line. A write buffer places a high level (approximately V_{CC}) on the I/O '0' line to write a logic '0', and a high level on the I/O '1' to write a logic '1'. For both write conditions, the opposite line is held low (V_{SS}).

Figure 32.19 is typical of circuitry used in static RAM. Data is gated to/from the appropriate columns by column select. Note that chip enable(s) gate the output data to a three state buffer and then to the output pin. Therefore, if a chip is not selected, the output pin goes to a high impedance state (allowing the output pins to be OR tied).

32.2.4 Characteristics

Static RAM is typified by ease of use. This comes at the expense of power dissipation, density, and cost. To service a broad spectrum of applications many variations of the peripheral circuitry

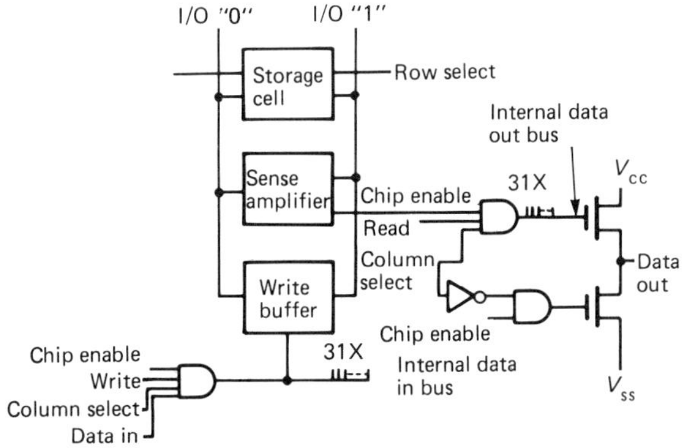

Figure 32.19 Typical circuitry used in static RAM

surrounding the static cell have been implemented. This section will analyse the key device types available.

The fundamental device is often referred to as a ripple through static RAM. Its design objective is ease of use. This device offers one power dissipation mode 'on'. Therefore its selected power (active power) is equal to its deselected power (standby power). One need simply present an address to the device and select it (via chip select) to access data. Note that the write signal must be inactive. This sequence is shown in *Figure 32.20*. In this RAM type the chip select access is typically 50% of the address access.

The write cycle can be executed in a simple manner as shown in *Figure 32.21*.

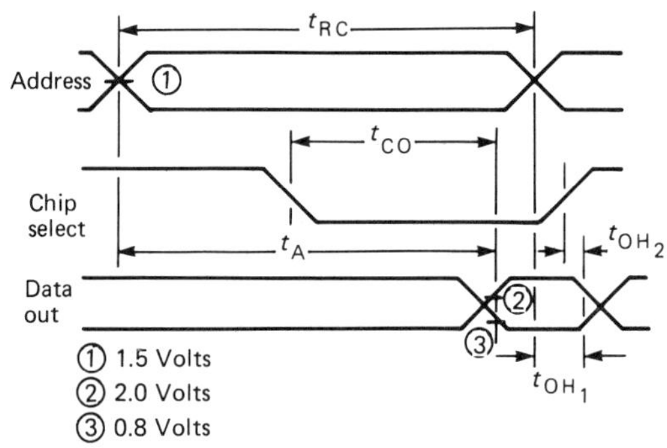

Figure 32.20 Ripple through read cycle

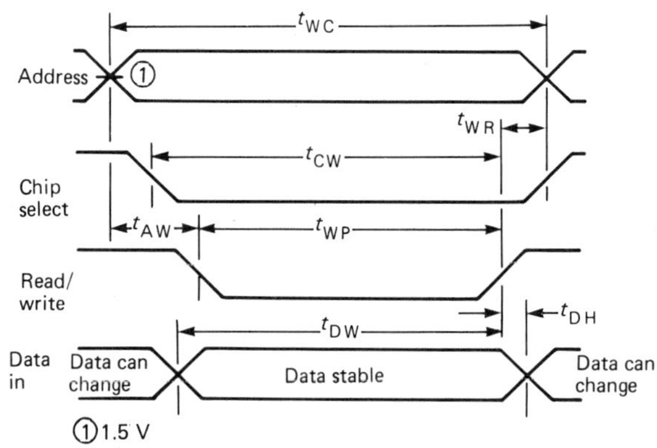

Figure 32.21 Write cycle ripple through interface

The address must be valid before and after read/write active to prevent writing into the wrong cell. Chip select gates the write pulse into the chip and must meet a minimum pulse width which is the same as the read/write minimum pulse width.

Read modify write cycles can be performed by this as well as other static RAMs. Read modify write merely means combining a read and write cycle into one operation. This is accomplished by first performing a read then activating the read/write line. The status of the data out pin during the write operation varies among vendors. The product's data sheet should therefore be read carefully.

The device achieves its ease of use by using very simplistic internal circuitry. To permit a non-synchronous interface (no timing sequence required) all circuitry within the part is active at all times. The static 'NOR' row decoder shown in *Figure 32.22* is

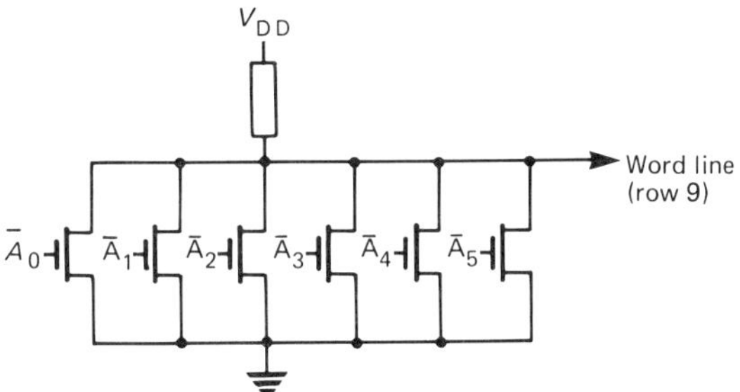

Figure 32.22 Static NOR row decoder

typical of the circuitry used. For a 4k RAM with 64 rows, this circuit is repeated 64 times with the address lines varing to identify each row uniquely. The decoder operates as follows.

In the static row decoder (*Figure 32.22*) a row is selected when all address inputs to the decoder are low making the output high. The need to keep the power consumption of this decoder at a minimum is in direct conflict with the desire to make it as fast as possible. The only way to make the decoder fast is to make the pull-up resistance small so that the capacitance of the word line can be charged quickly. However, only one row decoder's output is high; while the output of each of the other 63 decoders is low, causing the resistor current to be shunted to ground. This means that the pull-up resistance must be large in order to reduce power consumption.

Figure 32.23 is a typical input buffer. In this circuit device A and C act as loads and are always on. If the address is a logic '0'

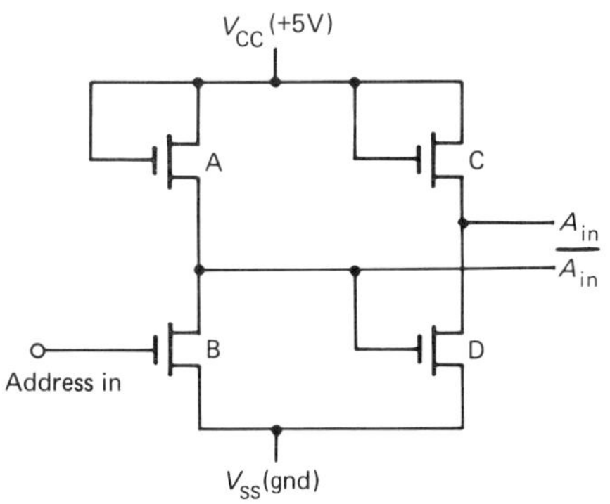

Figure 32.23 Address input buffer

then B is off and D is on; vice versa if the address is a logic '1' then B is on and D is off. There is always a current path to ground. This circuit is repeated for each address (12 times for 4k).

32.2.5 Power gated static RAM

On the previous static circuit the chip select function was used to select the device for write cycles and enable the output for read cycles. A type of static RAM, introduced at the 4k density level uses the chip select to reduce standby power. A ratio of active to standby of 5 to 1 is achieved. The timing for this device differs from the previous RAM in that the chip select access time is now equal to address access. An anomaly in timing occurs when the chip select is inactive for a short interval of time (approximately 30 ns or less). For this situation the access time is increased by about 20 ns. This anomaly was caused by a design problem and should be eliminated as a specification requirement in future product generations.

The effect of the power down feature is clearly shown in *Figure 32.24*.

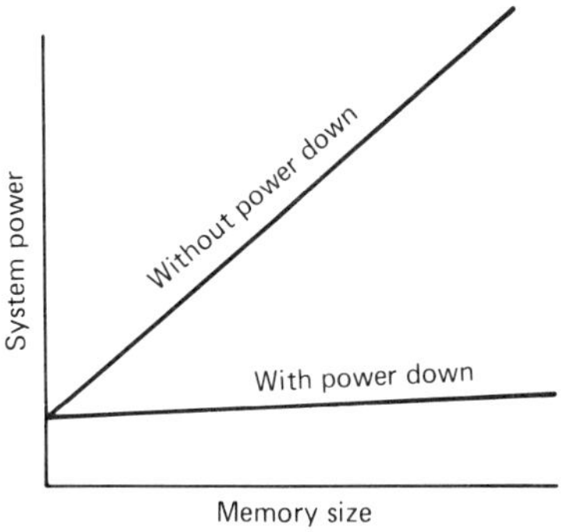

Figure 32.24 Effect of power down at the system level

This technique is also a compromise between ease of use and power. If we build a 64k × 9 system the average power is:

without power down
$P_{npd} = 144$ devices × 5 V × 180 mA/device = 129.6 W where P_{npd} is power with no power down.

with power down

$$\begin{aligned}P_{pd} &= 9 \text{ devices (active)} \times 5 \text{ V} \times 180 \text{ mA/device}\\ &\quad + 135 \text{ devices (standby)} \times 5 \text{ V} \times 30 \text{ mA/device}\\ &= 8.100 + 20.250 \text{ W}\\ &= 28.350 \text{ W}\end{aligned}$$

where P_{pd} is power with power down.

The power reduction is achieved with the minor complication of requiring chip select to be valid at the start of a cycle. If only 4k of memory is used *CS* can be grounded, raising power levels but simplifing the timing requirements. The internal circuitry is slightly more complicated in order to implement the power gate feature.

A radically different static RAM was designed to maintain the static cell (no periodic activation needed) and dramatically reduce the devices power consumption. The power dissipation of the static RAM can be reduced by using a 'clocked' interface. The clocked interface required synchronisation of timing and trades ease of use for power dissipation. Clocked static rams are commonly known as 'edge-activated' RAMs. The device is designed to work easily with microprocessors thereby minimising the difficulty of use. The need to synchronise timing edges to the RAM will cause a speed penalty in some applications. Therefore this RAM was designed for the microprocesser market and is not applicable to the sub-100 ns application. In microprocessor systems the system timing of the micro usually masks the speed penalty due to synchronisation.

The edge-activated device is configured as a 4k × 1. It is best used in applications where the system requires a memory more than 4k deep. Since it is a by 1 organisation, parity can be implemented easily. Parity is the addition of 1 extra bit per word to verify the validity of the word.

The edge-activated concept can best be explained by reference to *Figure 32.25*. The edge-activated RAM requires all address inputs to be valid prior to initiation of a negative going edge-on the chip enable input. The chip enable signal must remain valid for a specified duration, equivalent to the minimum access time of the component. A recovery time between cycles, 50% of specified access is required for proper operation. Data out becomes available, t_{acc} after CE and remains valid until chip enable is deactivated.

The requirements of the edge-activated interface is one which is readily available from most microprocessor chips or can be readily obtained from processor/memory controller timing.

The benefits of this RAM is its very low power dissipation. Active power is typically 20% of a ripple through static's. Standby power is 1/5 to 1/30 that achieved with other static techniques. For example, a 64k × 9 (144 devices) static RAM memory implemented with edged-activated circuits dissipated 4.7 W while a 'ripple through' static dissipation 129 W. In a similar application, the power P_{ea} required for an edge-activated static RAM is given by:

$$\begin{aligned} P_{ea} &= \text{devices (active)} \times \text{power (active)} \\ &\quad + \text{devices (standby)} \times \text{power (standby)} \\ &= 8 \text{ devices} \times 5 \text{ V} \times 30 \text{ mA} \\ &\quad + 135 \text{ devices} \times 5 \text{ V} \times 5 \text{ mA} \\ &= 4.7 \text{ W} \end{aligned}$$

$$\begin{aligned} P_{ripple} &= \text{devices} \times \text{power} \\ &\quad 144 \text{ devices} \times 5 \text{ V} \times 180 \text{ mA} \\ &= 129 \text{ W} \end{aligned}$$

where P_{ripple} is power ripple through static.

The low power dissipation of the edge-activated RAM is achieved by using many circuit techniques commonly associated with dynamic RAM. It is designed as is a dynamic RAM to eliminated d.c. current paths. The circuit of *Figure 32.26* is typical of the techniques used. Its operation is as follows. The decode is accomplished by presetting (precharging) node A, on the decoders, to a high state during the chip inactive time. The precharge clock is turned off at chip enable time and the addresses are strobed into the decoders causing node A on all but the one selected decoder to go low. The only power consumed by this circuit is the transient power drawn during capacitor charge time at node A.

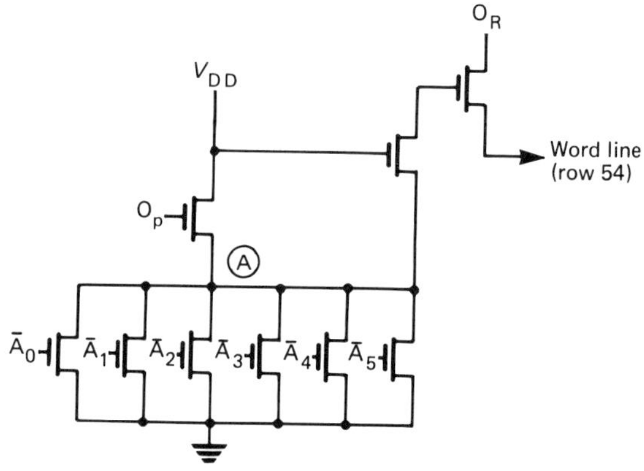

Figure 32.26 Dynamic NOR row decoder

The last technique to be described combines the best features of the static RAMs previously discussed. Called address activated, it uses many of the clocked circuitry techniques of edge activated but obtains the 'edge' from any one of the addresses or clocks (CS, WE). Since all inputs must be active between cycles, looking for a transition, the device has a power specification that is less than ripple through static but greater than edge activated. Power gate, techniques can be applied to further reduce power if the application the part is designed for warrants it. This device uses clocked circuitry primarily for speed enhancement, not power.

If we analyse the static NOR row decode its performance is a function of the load device's ability to charge the word line. Also performance is a function of the transistors (between the word line and ground) ability to discharge the word line. As we discussed earlier in a 4k RAM 63 out of 64 decoders are shunting current to ground to prevent selection. This means that the pull-

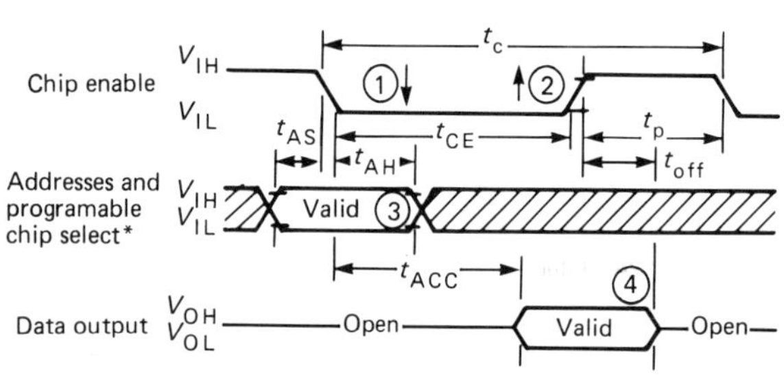

1. A simple high to low transistion at the chip enable $(\overline{CE})$ input activates this entire family of memory devices.
2. Returning to CE input to a high level is all that is required to achieve a 75% reduction in device operating power.
3. Address information is strobed and latched into a set of on-chip registers.
4. You have full control of the data output; this is determined by the CE pulse width.

Figure 32.25 Low voltage, edge-activated RAM/ROM family

up (load) resistance must be large in order to reduce power consumption which impacts speed. Since the dynamic decoder has no d.c. path the trade-off between power and speed is minimised. Other areas of the circuit also benefit in speed if a clock can be used. These all add up to make address activated a useful high performance lower power circuit technique.

The timing specifications for the address activated RAM resemble that of the ripple through static RAM. In by 1 devices power gating is used while with the by 8 devices the applications do not warrant it.

Static RAM applications are segmented into two major areas based on performance. Slower devices, greater than 100 ns access time, are generally used with microprocessors. These memories are usually small (several thousands of bytes) with a general perference for wide word organisation (by 8 bit). When by one organised memories are used, the applications tend to be 'deeper' (larger memory), and desire lower power or parity. The by 4 bit product is an old design which is currently being replaced by 1k × 8 and 2k × 8 static RAM devices. Statics are used in these applications due to their ease of use and compatability with other memory types (RAM, EPROM). The by 4 and by 8 devices use a common pin for data in and data out. In this configuration, one must be sure that the data out is turned off prior to enabling the data in circuitry to avoid a bus conflict (two active devices fighting each other drawing unnecessary current). Bus contention can also occur if two devices connected to the same bus are simultaneously on fighting each other for control. The by 8 devices include an additional function called output enable (OE) to avoid this problem. The output enable function controls the output buffer only and can therefore switch it on and off very quickly. The by 4 devices do not have a spare pin to incorporate this function. Therefore, the chip select (CS) function must be used to control the output buffer. Data out turns off a specified delay after CS turns off. One problem with this approach is that it is possible for bus contention to occur if a very fast device is accessed while a slow device is being turned off. During a read/modify/write cycle the leading edge of the read write line (WE) is normally used to turn off the data out buffer to free the bus for writing.

The second segment is high performance applications. Speeds required here are typically 55–70 ns. This market is best serviced by static RAM due to their simple timing and faster circuit speeds. The requirement of synchronising two edges as required in clocked part typically costs the system designer 10–20 ns. Therefore, fast devices are always designed with a ripple through interface. The by 1 devices have been available longer, and tend to be faster than available by 8 devices. As a result the by 1 device dominates the fast static RAM market today. Applications such as cache and writeable control store tend to prefer wide word memories. The current device mix should shift over time as more suppliers manufacture faster devices.

32.2.6 Process

The pinouts and functions tend to be the same for fast and slow static devices. The speed is achieved by added process complexity and the willingness to use more power, for a given process technology.

The key to speed enhancement is increasing the circuit transistor's gain, minimising unwanted capacitance and minimising circuit interconnect impedance. Advanced process technologies have been developed to achieve these goals. In general, they all reduce geometries (scaling) while pushing the state of the art in manufacturing equipment and process complexity. These process technologies have been given identification names such as HMDS and scaled poly 5. *Figure 32.27* shows several of the key parameters involved.

The figure shows the cross section of a scaled device and lists

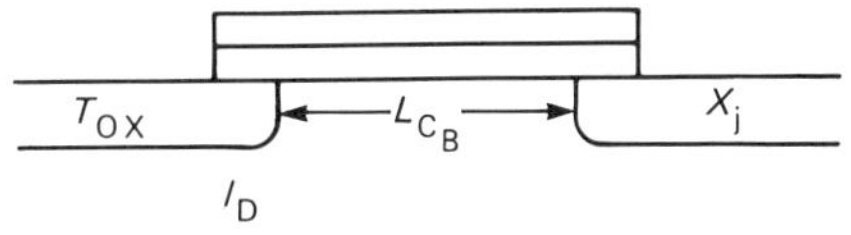

Performance factors
- High device gain
- Low diffusion capacitance
- Low Miller capacitance
- Low body effect

Gain $\propto 1(T_{ox} l)$
$C_p \propto X_j$
$C_m \propto l_D$
$\Delta V_r \propto \sqrt{C_B} T_{ox}$

Limits
- Punch through voltage
- Threshold voltage

$V_{PT} \propto C_B l^2$
$V_r \propto \sqrt{C_B} T_{ox}$

Result
- Decrease l, T_{ox}, X_j, l_D
- Increase C_B

l = Channel length
T_{ox} = Oxide thickness
X_j = Diffusion depth
l_D = Gate overlap
C_B = Concentration

Figure 32.27 HMOS scaling

the parameters of scaling, one of which is device gain. The slew rate of an amplifier or device is proportional to the gain. Because faster switching speeds occur with high gain, the gain is maximised for high speed. Device gain is inversely proportional to the oxide thickness (T_{ox}) and device length (l), consequently, scaling these dimensions increases the gain.

Another factor which influences performance is unwanted capacitance which appears in two forms: diffusion and Miller capacitance. Diffusion capacitance is directly proportional to the overlap length of the gate and the source (l_D). Capacitance on the input shunts the high frequency portion of the input signal so that the device can only respond to low frequencies. Secondly, capacitance from the drain to the gate forms a feedback path creating an integrator or low pass filter which degrades the high frequency performance. This effect is minimised by reducing l_D.

One of the limits on scaling is punch through voltage, which occurs when the field strength is too high, causing current to flow when the device is 'turned off'. Punch through voltage is a function of channel length (l) and doping concentration (C_B) thus channel shortening can be compensated by increasing the doping concentration. This has the additional advantage of balancing the threshold voltage which was decreased by scaling the oxide thickness for gain.

32.2.7 Design techniques

32.2.7.1 By 1 statics

The by 1 static memories are offered in 18 pin or 20 pin packages for 4k and 16k densities respectively.

The 4k/16k by 1 static RAM can be easily integrated into large memory configurations in highly compact board layouts. The devices can use standard TTL logic to achieve the desired signal characteristics. A typical design for a 16k × 9 (using 4k devices) is shown in *Figure 32.28*. Chip select is decoded in order that it occurs on only one memory word simultaneously. Address lines and write enable (read/write line) go to all chips simultaneously.

Power supplies were placed in the corners of the devices to permit use of two-sided printed circuit boards. The RAM chips draw significant transient currents as well as d.c. current. To supply the transient current at the very high frequencies required it is necessary to place high frequency (ceramic) capacitors in close proximity to the memory chip. The rule of thumb is to use one 0.1 μF ceramic capacitor every other RAM.

32.2.7.2 By 8 static

The byte wide (8 bit) statics offer pin compatability with RAM and EPROM circuits. This gives the designer the ability to utilise

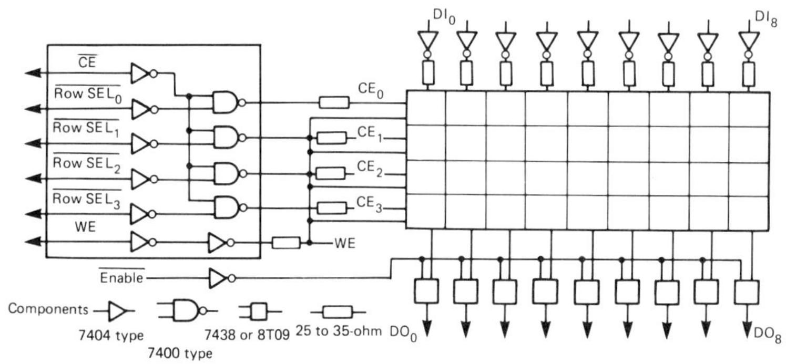

Figure 32.28 Design of 16 k × 9 memory

on board design for his RAM/ROM/EPROM needs. The pinout also permits flexibility in defining the type of memory to be implemented in a given application. The compatability is most useful in the lower performance application geared at microprocessor memory. The pinouts chosen are expandable well beyond the 2k × 8 level. To permit this expansion a 28 pin package pinout was defined to be compatible with the 24 pin used by current generation statics, EPROMs, and ROMs.

32.3 Non-volatile memories

With today's faster, more powerful microcomputer chips emerging in abundance, and larger, more memory-intensive programs being written, semiconductor memory requirements for larger storage capacities, faster access times, and lower subsequent costs have become dominant system design factors. Basic semiconductor memory-chip technology involves variations of random access memory (RAM) and read only memory (ROM). RAM allows binary data to be written in, and to be read out. New and different programs and data can be loaded and stored in RAM as needed by the processor. Because information is stored electrically in RAM its contents are lost whenever power goes down or off. When fixed or unchanging programs and data are needed by the processor, they are loaded into some form of ROM. In ROM, information is physically (permanently) embedded; therefore, its contents are preserved whenever power is off or interrupted momentarily.

Depending on the type and quantity of microprocessor systems to be produced, a decision has to be made as to whether ROM, PROM, or EPROM will be used for permanent program storage. If only a few systems are to be manufactured, it may be more cost-effective to use either PROM or EPROM. EPROM-based storage also allows the main program to be changed at any time, even in the field by the end-user. The PROM based system requires replacement; however, it is field programmable. If the main requirement is a minimum parts configuration and many microprocessor systems must be produced the decision should be to use ROM-based storage.

32.3.1 ROM operation

In mask-programmed ROM, the memory bit pattern is produced during fabrication of the chip by the manufacturer using a masking operation. The memory matrix is defined by row (X) and column (Y) bit-selection lines that locate individual memory cell positions.

For example, in *Figure 32.29* refer to column C_2 and row 127 as the storage cell location of interest. When the proper binary inputs on the address lines are decoded, the cell at R_{127}, C_2 will be selected. If the drain contact of this cell is connected to bit line L_2, then L_2 will be pulled below threshold, turning off device C_2; note that devices C_0, C_1, and C_3 through C_{15} will also be off since they are not addressed. Therefore, device A pulls the OUT line to V_{CC} for a logic 1 output when cell R_{127}, C_2 is selected.

Alternatively, consider when cell R_{127}, C_2 is masked it does not have a drain contact to bit line L_2. Then when this cell is addressed, device C_2 is not connected to V_{CC} and will be turned on.

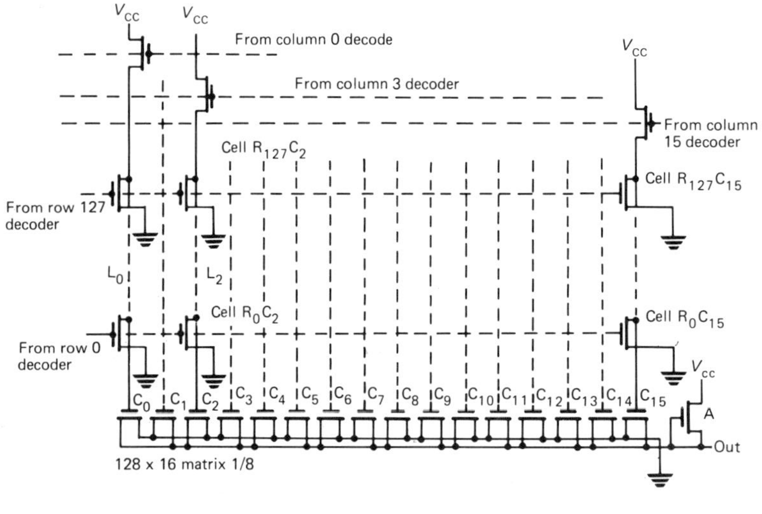

Figure 32.29 Portion of ROM matrix and output circuitry

Thus, the OUT line will be pulled to ground through device C_2 and will appear as a logic 0 output. To program a 1 or a 0 into a ROM storage cell, the drain contact will or will not be connected, respectively, to the particular bit line. Note that this type of programming is permanent. An alternative method of performing the same operation would be to eliminate the gate of the storage cell.

32.3.2 ROM process

For many designs, fast manufacturing turnaround time on ROM patterns is essential for fast entry into system production. This is especially true for the consumer 'games' market. Several vendors now advertise turnaround times that vary from two to six weeks for prototype quantities (typically small quantities) after data verification. Data verification is the time when the user confirms that data have been transferred correctly into ROM in accordance with the input specifications.

Contact programming is one method that allows ROM programming to be accomplished in a shorter period of time than with gate mask programming. In mask programming, most ROMs are programmed with the required data bit pattern by vendors at the first (gate) mask level, which occurs very early in the manufacturing process. In contact programming, actual programming is not done until the fourth (contact) mask step, much later in the manufacturing process. That technique allows wafers to be processed through a significant portion of the manufacturing process, up to 'contact mask' and then stored until required for a user pattern. Some vendors go one step further and program at fifth (metal) mask per process. This results in a significantly shorter lead time over the old gate-mask programmable time of 8 to 10 weeks; the net effect is time and cost savings for the end user.

32.3.3 ROM applications

Typical ROM applications include code converters, look-up tables, character generators, and non-volatile storage memories. In addition, ROMs are now playing an increasing role in microprocessor-based systems where a minimum parts configuration is the main design objective. The average amount of ROM in present microprocessor systems is in the 10k to 32k byte range. In this application, the ROM is often used to store the control program that directs CPU operation. It may also store data that will eventually be output to some peripheral circuitry through the CPU and the peripheral input/output (P I/O) device.

In a microprocessor system development cycle, several types of memory (RAM, ROM and EPROM or PROM) are normally used to aid in the system design. After system definition, the designer will begin developing the software control program. At this point, RAM is usually used to store the program, because it allows for fast and easy editing of the data. As portions of the program are debugged, the designer may choose to transfer them to PROM or EPROM while continuing to edit in RAM. Thus, he avoids having to reload fixed portions of the program into RAM each time power is applied to the development system.

32.3.4 Electrically erasable programmable ROM

The ideal memory is one that can perform read/write cycles at speeds meeting the needs of microprocessors (200 ns typical) and store data in the absence of power. The EEPROM (electrically erasable PROM) is a device which meets two of the three requirements. It is non-volatile (retains data in the absence of power) and can be read at speeds compatible with today's microprocessors. However, its write cycle requires typically 20 ms to execute. The EEPROM has two major advantages over the EPROM. The EEPROM can be programmed in circuit and can selectively change a byte of memory instead of all bytes. The process technology to implement the EEPROM is quite complex and the industry is currently on the verge of mastering production.

32.3.5 EEPROM theory

EEPROMs use a floating gate structure, much like the ultraviolet erasable PROM (EPROM), to achieve non-volatile operation. To achieve the ability to electrically erase the PROM a principle known as Fowler–Nordheim tunnelling was implemented. Fowler–Nordheim tunnelling predicts that under a field strength of 10 MV cm^{-1} a certain number of electrons can pass a short distance, from a negative electrode, through the forbidden gap of an insulator entering the conduction band and then flow freely toward a positive electrode. In practice the negative electrode is a polysilicon gate, the insulator is a silicon dioxide and the positive electrode is the silicon substrate.

Fowler–Nordheim tunnelling is bilateral, in nature, and can be used for charging the floating gate as well as discharging it. To permit the phenomenon to work at reasonable voltages (e.g. 20 V), the oxide insulator needs to be less than 200 angstroms thick. However, the tunnelling area can be made very small to aide the manufacturability aspects (20 nm oxides are typically one half the previously used thickness).

Intel Corporation produces an EEPROM based on this principle. Intel named the cell structure used, FLOTOX. A cross section of the FLOTOX device is shown in *Figure 32.30*.

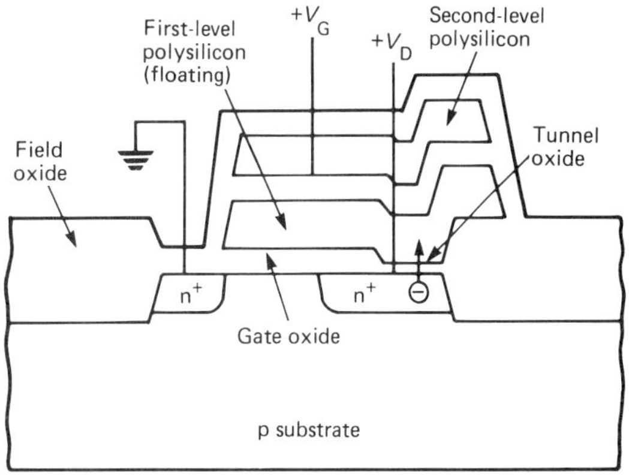

Figure 32.30 EEPROM cell using Fowler-Nordheim tunnelling mechanism

The FLOXTOX structure resembles the FAMOS structure used by Intel for EPROM devices. The primary difference is in the additional tunnel-oxide region over the drain. (*Figure 32.31* for EPROM cell configuration.) To charge the floating gate, of the FLOTOX structure of *Figure 32.30*, a voltage V_G is applied to the top gate and with the drain voltage V_D at 0 V, the floating gate is capacitively coupled to a positive potential. Electrons will then flow to the floating gate. If a positive potential is applied to the drain and the gate is grounded the process is reversed and the floating gate is discharged.

32.3.6 EEPROM circuit operation

The electrically erasable cell is the building block for EEPROM memories. One manufacturers approach is to assembly the Fowler–Nordheim concept cell into a memory array using two transistors per cell as shown in *Figure 32.32*. The floating gate cell

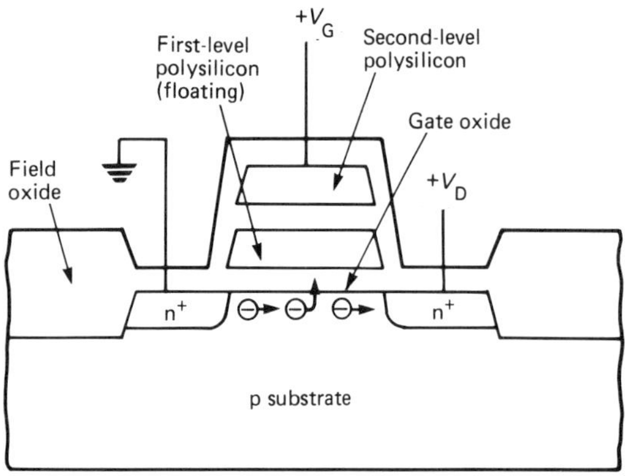

Figure 32.31 EPROM ultraviolet eraseable cell. Similar to the EEPROM cell

(bottom device) is the actual storage device. The upper device is used as a select transistor to prevent devices on non-selected rows from discharging when a column is raised high. Before information is entered the array must be cleared. This returns all cells to a charged state as shown schematically in *Figure 32.32(a)*. To clear the memory all the select lines and program lines are raised to 20 V while all the columns are grounded. This forces electrons through the tunnel oxide to charge the floating gates on all of the selected rows. The Intel Corporation device offers the user the option of chip-clear or byte-clear. When byte-clear is initiated, only the select and program lines of an addressed byte are raised to 20 V.

To write a byte of data, the select line for the addressed byte is raised to 20 V while the program line is grounded (see *Figure 32.32(b)*). Simultaneously, the columns of the selected byte are raised or lowered according to the incoming data pattern. The byte on the left in *Figure 32.32(b)*, for example, has its column at a high voltage, causing the cell to discharge, whereas the bit on the right has its column at ground so its cell will experience no change. Reading is accomplished by applying a positive bias to the select and program lines of the addressed cell. A cell with a charged gate will remain off in this condition but a discharged cell will be turned on.

The EEPROMs being introduced today are configured externally to be compatible with ROM and EPROM standards which already exist. Devices typically use the same 24 pin pinout as the generic 2716 (2k × 8 EPROM) device. A single 5 V supply is all that is needed for read operations. For the write and clear operations an additional supply (V_{PP}) of 20 V is necessary. The device reads in the same manner as the EPROM it will eventually replace.

The timing for writing a byte is shown in *Figure 32.33*. The chip is powered up by bringing CE low. With address and data applied, the write operation is initiated with a single 10 ms, 20 V

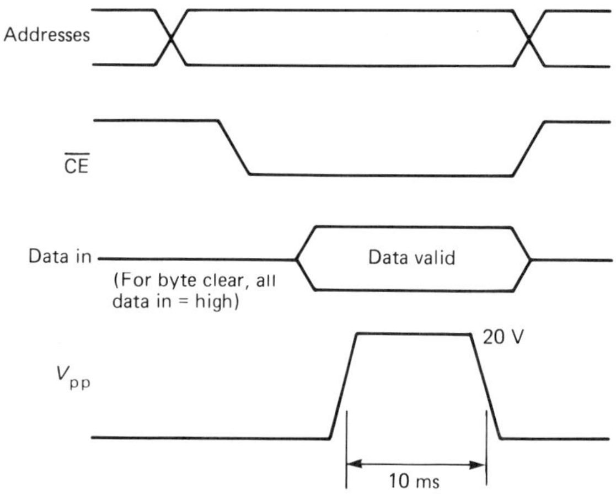

Figure 32.33 Byte write operation for an EEPROM. Two cycles are required

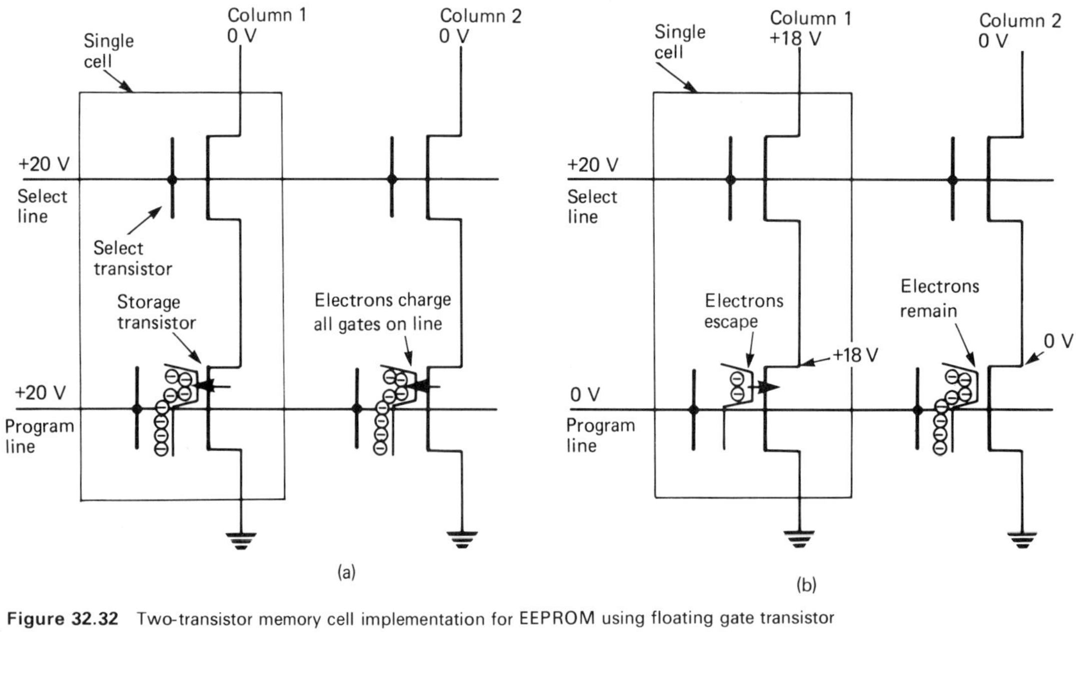

Figure 32.32 Two-transistor memory cell implementation for EEPROM using floating gate transistor

pulse applied to the V_{PP} pin. During the write operation, OE is not needed and is held high.

A byte clear is really no more than a write operation. A byte is cleared merely by being written with all 1's (high). Thus altering a byte requires nothing more than two writes to the address byte, first with the data set to all 1's and then with the desired data. This alteration of a single byte takes only 20 ms. In other non-volatile memories, changing a single byte requires that the entire contents be read out into an auxiliary memory. Then the entire memory is rewritten. This process not only requires auxilary memory; for a 2 kbyte device it takes about one thousand times as long (20 ms versus 20 s).

The only difference between byte clear and chip clear is that OE is raised to 20 V during chip clear. The entire 2 kbytes are cleared with a single 10 ms pulse. Addresses and data are not all involved in a chip-clear operation.

32.3.7 EEPROM applications

The electrically erasable PROM (EEPROM) has the non-volatile storage characteristics of core, magnetic tape, floppy and Winchester disks but is a rugged low power solid state device and occupies much less space. Solid state non-volatile devices, such as ROM and EPROM have a significant disadvantage in that they cannot be deprogrammed (ROM) or re-programmed in place (EPROM). The non-volatile bipolar PROM blows fuses inside the device to program. Once set the program cannot be changed, greatly limiting their flexibility. The EEPROM therefore, has the advantages of program flexibility, small size, and semiconductor memory ruggedness (low voltages and no mechanical parts).

The advantages of the EEPROM create many applications that were not feasible before. The low power supports field programming in portable devices for communication encoding, data formatting and conversion, and program storage. The EEPROM in circuit change capability permits computer systems whose programs can be altered remotely, possibly by telephone. It can be changed in circuit to quickly provide branch points or alternate programs in interactive systems.

The EEPROMs non-volatility permits a system to be immune to power interruptions. Simple fault tolerant multiprocessor systems also become feasible. Programs assigned to a processor that fails can be reassigned to the other processors with a minimum interruption of the system. Since a program can be backed up into EEPROM in a short period of time key data can be transferred from volatile memory during power interruption and saved. The user will no longer need to either scrap parts or make service calls should a program bug be discovered in fixed memory. With EEPROM this could even be corrected remotely. The EEPROM's flexibility will create further applications as they become available in volume and people become familiar with their capabilities.

32.3.8 Erasable programmable ROM

The EPROM like the EEPROM satisfies two of our three requirements. It is non-volatile and can be read at speeds comparable with today's microprocessors. However, its write cycle is significantly slower like the EEPROM. The EPROM, has the additional disadvantage of having to be removed from the circuit to be programmed as contrasted to the EEPROM's ability to be programmed in circuit.

EPROM is electrically programmable, then erasable by ultraviolet (UV) light, and programmable again. Erasability is based on the floating gate structure of an n- or p-channel MOSFET. This gate, situated within the silicon dioxide layer, effectively controls the flow of current between the source and drain of the storage device. During programming, a high positive voltage (negative if p-channel) is applied to the source and gate of a selected MOSFET, causing the injection of electrons into the floating silicon gate. After voltage removal, the silicon gate retains its negative charge because it is electrically isolated (within the silicon dioxide layer) with no ground or discharge path. This gate then creates either the presence or absence of a conductive layer in the channel between the source and the drain directly under the gate region. In the case of an n-channel circuit, programming with a high positive voltage depletes the channel region of the cell; thus a higher turn-on voltage is required than on an unprogrammed device. The presence or absence of this conductive layer determines whether the binary 1-bit or the 0-bit is stored. The stored bit is erased by illuminating the chip's surface with UV light. The UV light sets up a photocurrent in the silicon dioxide layer which causes the charge on the floating gate to discharge into the substrate. A transparent window over the chip allows the user to perform erasing, after the chip has been packaged and programmed, in the field.

32.3.9 Programmable ROM

The PROM has a memory matrix in which each storage cell contains a transistor or diode with fusible link in series with one of the electrodes. After the programmer specifies which storage cell positions should have a 1-bit or 0-bit, the PROM is placed in a programming toll which addresses the locations designated for a 1-bit. A high current is passed through the associated transistor or diode to destroy (open) the fusible link. A closed fusible link may represent a 0-bit, while an open link may represent a 1-bit (depending on the number of data inversions done in the circuit). A disadvantage of the fusible-link PROM is that its programming is permanent; that is, once the links are opened, the bit pattern produced cannot be changed.

32.3.10 Shadow RAM

A recently introduced RAM concept is called the shadow RAM. This approach to memory yields a non-volatile RAM (data are retained even in the absence of power), by combining a static RAM cell and an electrically erasable cell into a single cell. The electrically erasable programable read only memory (EEPROM) shadows the static RAM on a bit by bit basis. Hence the name shadow RAM. This permits the device to have read/write cycle times comparable to a static RAM, yet offer non-volatility.

The non-volatility has a limit in that the number of write cycles to the EE cell is typically one thousand to one million maximum. The device can however be read indefinitely. Currently two product types are offered. One permits selectively recalling bits stored in the non-volatile array while the other product recalls all bits simultaneously.

32.3.11 Shadow RAM operation

The static RAM portion of the shadow RAM operates the same as the depletion mode static cell discussed earlier in the static RAM section. The EEPROM portion is formed by adding a transistor with a floating gate and a capacitor as shown in *Figure 32.34*. While operating as a static RAM the capacitor and transistor have no effect. The device will automatically move the contents of the EEPROM into the static cell upon power up. This operates as follows: during switching of the internal +5 V supply, the word-select lines isolating nodes N_1 and N_2 from the bit lines are turned off. If the floating gate transistor was programmed on, N_2 will have a larger capacitance than N_1. This causes N_2 to rise more slowly than N_1, and the cell will latch with N_2 low and N_1 high. Had the floating gate transistor been programmed off, then the capacitance load on N_1 will be larger than that on N_2. As the supply switches from ground to +5 V, node N_2 will rise more quickly than N_1 and N_2 will latch high with N_1 low.

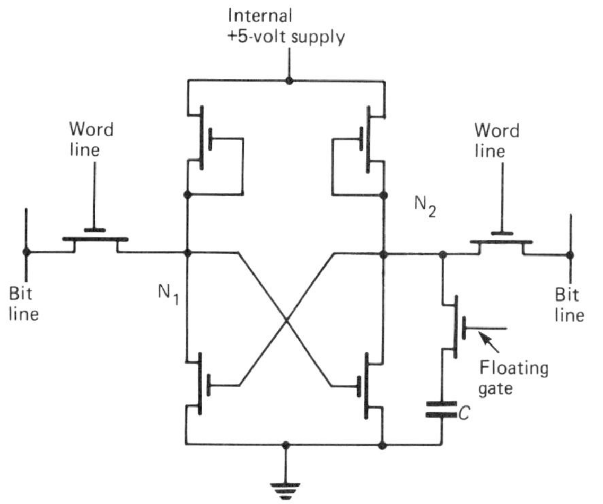

Figure 32.34 Shadow RAM storage cell comprised of a classical depletion load static cell with a transistor and capacitor added

During operation data are moved between the RAM and EEPROM positions under the control of two TTL signals, recall and store. In the case of storing into the EEPROM all bits must be transferred. For recall two parts are available to give the designer the option of either recalling all bits or selectively recalling one bit at a time. The bits are recalled in less than 1 ms. The operation in non-power-up conditions is described in the next section.

32.3.12 Shadow RAM construction

The shadow RAM currently produced by Xicor Coporation uses a triple-poly-process. A schematic of the cell is shown in *Figure 32.35*. The polysilicon electrodes are separated by oxide layers

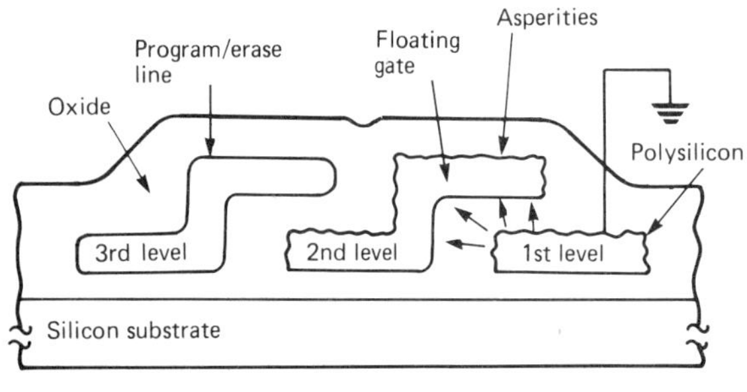

Figure 32.35 Schematic of shadow RAM cell

approximately 100 nm thick. Asperities (projections) that exist uniformly across the surface of the polysilicon layer, help to create a reproducible flow of electrons. The asperities greatly reduce the voltage potential needed. The part is programmed by bringing the program/erase line to a positive voltage thereby coupling the floating gate to a positive potential. The programing electrode (on poly 1 level) is held at ground and, with a large enough electric field, emits electrons that are captured by the floating gate making it negative. For erasure, the floating gate is held close to ground potential while the program/erase line is brought high. To avoid the need for an extra polysilicon electrode, a switching transistor holds the gate to ground. Having established an electric field between the program/erase electrode and the floating gate, the asperities on the gate emit electrons to the electrode so that the gate becomes positively charged.

32.3.13 Shadow RAM theory

It was observed for many years that a fairly low voltage would cause electrons to flow between polysilicon layers (electron tunnels) in unexpectedly large numbers. This enhanced electron tunnelling was found to be due to projections, called asperities, that exist uniformly across the surface of a polysilicon layer. The shadow RAM uses a triple-poly-process aimed at enhancing and exploiting this tunnelling effect.

The shadow RAM uses a floating gate transistor whose ability to be left either charged or uncharged, in the absence of power, as the basis for operation of the non-volatile RAM. A floating gate is an island of conducting material surrounded by oxide and coupled capacitively to the silicon substrate to form a transistor. The presence or absence of charge on the gate determines whether the transistor is on or off. Charge is induced into the gate region by overcoming the oxide barrier by applying a relatively low voltage between the floating gate itself and the other two polysilicon layers. Since the charge is trapped in the oxide, it acts as a barrier to the flow of current, and the charge once placed on the gate should remain indefinitely as long as the charging voltage is removed. When the memory is being programmed, the floating gate is charged with electrons, turning the transistor off; during erasure, electrons are removed from the gate turning the transistor on. Hence a programmed floating gate is at a negative potential and an erased gate is positive.

32.4 Charge coupled devices

A CCD (charge coupled device) is, in essence, a shift register formed by a string of closely spaced MOS capacitors. (See Chapter 31). A CCD can store and transfer analogue signals—either electrons or holes—that may be introduced electrically or optically. Charge coupled devices are being used in photosensor arrays, large-storage memories, and such signal processing components as variable delay lines, transversal filters and signal correlators. Today the memory applications are the least successful effort due to the competition from dynamic RAM.

32.4.1 Theory of operation

To retain data, CCDs rely on dynamic charge storage on a semiconductor surface. *Figure 32.36* shows a metal–oxide–semiconductor structure that could be part of a CCD shift register or the storage capacitor and address transistor of a one-transistor RAM. The storage region is the silicon surface beneath

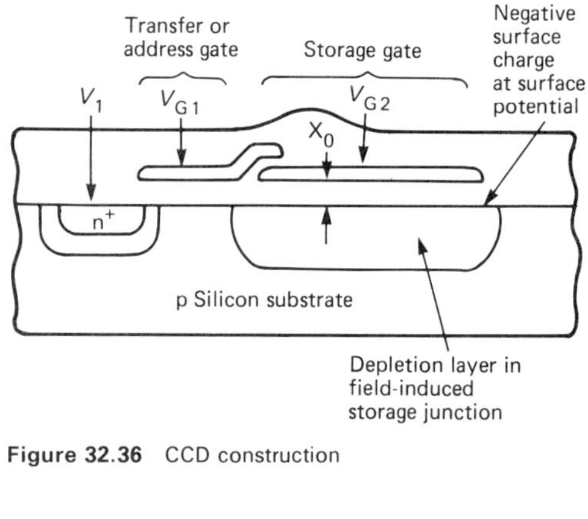

Figure 32.36 CCD construction

the storage gate where a field-induced junction contains a negative charge of mobile electrons.

The surface of the semiconductor behaves like a capacitive voltage divider—the upper capacitor is the fixed gate capacitance, while the lower one is the space-charge region of the field-induced junction; the connection between the capacitors is the field-induced junction itself.

The transfer or address gate forms a temporary conductive path between the two junction regions. If the voltage applied to the storage gate (V_{G2}) exceeds the threshold voltage, then the surface-potential of the field-induced junction will equal the diffuse-junction potential when the transfer or address gate (V_{G1}) is turned on.

Figure 32.37 shows the dependence of surface charge in the field-induced storage region on surface potential (with respect to

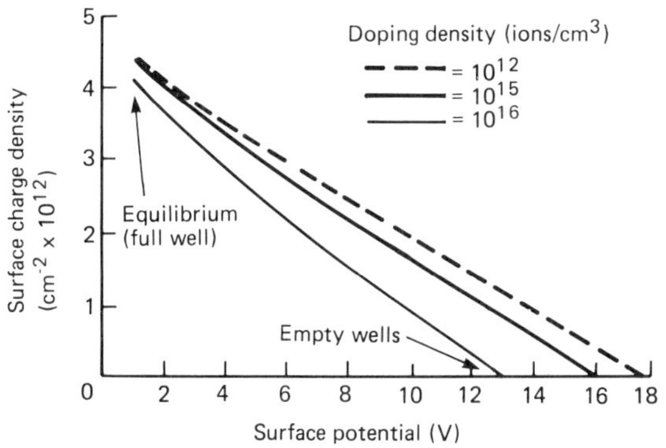

Figure 32.37 Relationship between surface charge and potential in a CCD

the substrate) for three different values of substrate doping density. (The plot has been made for values typical of three-supply 16k dynamic RAMs.) The 16k RAM uses a similar storage principle. The graph shows that the surface charge is inversely proportional to the surface potential.

The minimum surface potential corresponds to inversion and to maximum charge in the inversion layer. If the transfer gate is turned off, minimum surface potential is in a condition of equilibrium. With the transfer gate turned off, any other surface potential is not at equilibrium, and leakage currents arising from generation of current in the space charge region, plus diffusion current from the adjacent neutral region (plus any other source of electron–hole pairs, including light and other ionising radiation) cause a decay in surface potential toward the equilibrium value. As surface potential decays, the surface charge density increases toward its maximum value; thus the potential well tends to fill with charge as a result of leakage current.

High surface potentials are empty wells and correspond to logic level ones. The time required for a logic level 1 to decay to logic level 0 (for an empty well to fill due to leakage) is a measure of the dynamic storage time. In actual circuits, the dominant leakage current at high temperatures is from diffusion current in the substrate. Minimum frequency in CCDs is determined by how much decay in surface potential will be allowed before the sense amplifier detects logic errors.

Both the maximum charge density and maximum surface potential increase with more positive gate voltages and more negative threshold voltages since the size and capacity of the potential well is proportional to the difference of the gate voltage and the threshold voltage. In CCDs, charge (or surface potential) stored in one cell is transferred to an adjacent cell by increasing its gate voltage—which in turn increases its surface potential—making it more attractive to charge. Charge then flows from the present stored well into the well with the higher driving force.

32.4.2 Organisation

There are several techniques that have been utilised to move the CCD charge packet, as in *Figure 32.38*. One technique uses two

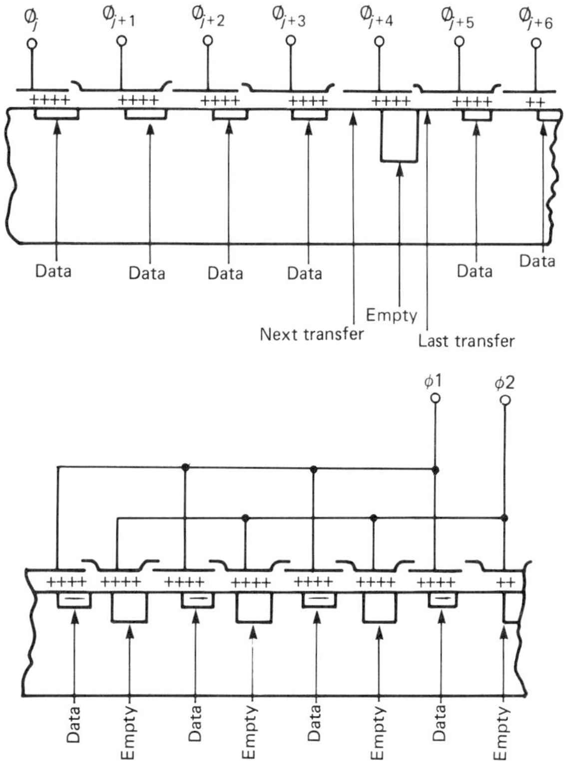

Figure 32.38 Types of CCD organisation: (a) electrode/bit CCD; (b) standard two-electrode/bit CCD

electrodes and two storage wells (locations) per bit. In this technique all data bits are transferred to the next location simultaneously. This requires that two locations be available for each bit to prevent overwriting.

Another technique uses one electrode per bit and has one additional storage location for the entire register. This requires a data rippling effect to function. Each electrode has a unique clock that is time sequenced to all other clocks. This permits moving data one bit at a time permitting the next location to be vacated before data is moved in. This technique almost halves the number of storage locations needed.

The primary disadvantage of this technique is that a clock decoder is required to provide the ripple through clock signals necessary. This decoder adds complexity and area to the device.

32.4.3 Device construction techniques

In CCD devices not all the available charge survives a transfer. The fraction of charge transferred from one well to the next is referred to as the charge transfer efficiency, n. The fraction left behind per transfer is the transfer loss, or transfer inefficiency,

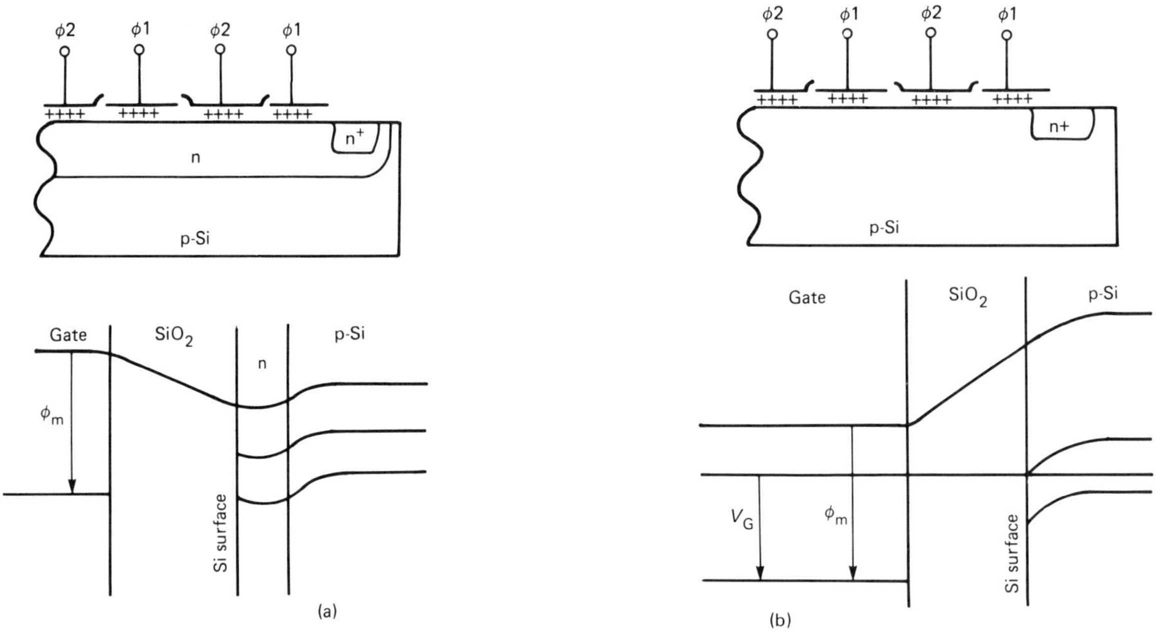

Figure 32.39 CCD construction and band diagrams: (a) buried channel; (b) surface channel

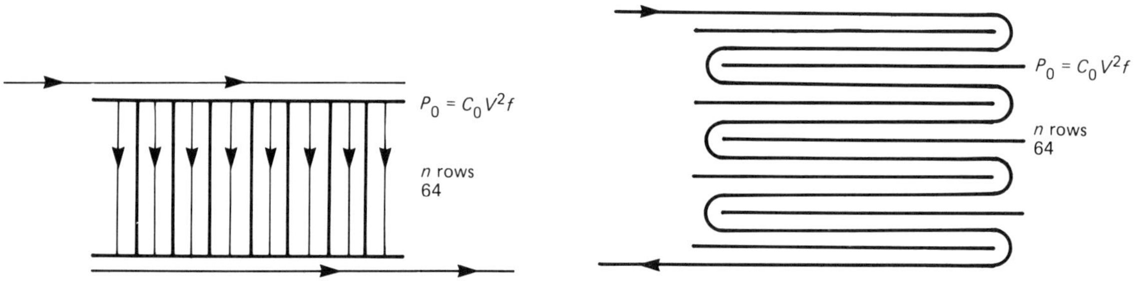

Capacity : $m \times n$ 4096	Number of transfers : $2 \times (m+n-2)$ 252
Clocks : 4	Power : $2 \times (m+n-2)\ CV^2f$ 252 P_0

(a)

Capacity : $m \times n$ 4096	Number of transfers : $2 \times m \times n$ 8192
Clocks : 2	Power : $2m \times n \times CV^2f$ 8192 P_0

(b)

Figure 32.40 CCD memory architecture: (a) series–parallel–series; (b) serpentine

denoted by e, so that $n+e=1$. Because n determines how many transfers can be made before the signal is seriously distorted and delayed, it is a very important figure of merit for a CCD. To give some perspective, assume a loss of 10% of charge is the maximum to be tolerated, then a 330 stage shift register requires a transfer efficiency of 99.99%. The maximum achievable value for n depends on two factors: how fast free charge can transfer between adjacent gates and how much of the charge gets trapped at each gate location by stationary states.

The charge coupled device can be constructed as a surface-channel device (SCCD) or as a buried-channel (BCCD), as in *Figure 32.39*. The potential wells of an SCCD are formed at the Si–SiO_2 interface. In contrast, the BCCD forms wells below the silicon surface to avoid charge trapping by surface states. Therefore a BCCD device can achieve an efficiency greater than 0.9999.

Enhancements in SCCD devices have permitted their efficiencies to reach the $n=0.9999$ range while similar enhancements let BCCD devices reach 0.99999.

The choice between surface and bulk channel depends on the application. The surface channel offers somewhat simpler processing, more conventional on-chip MOS circuits, and higher charge handling capability per unit area. Bulk channel devices can be designed for very high frequency operation. The potential for very low noise operation is another advantage. Noise impacts the operating capabilities of the CCD.

However, the maximum charge signal in a BCCD is up to three times smaller than that of SCCD. Also the BCCD has a higher drain current (a type of leakage) than the SCCD device.

32.4.4 Memory architecture

CCD registers can be layed out in two structures as shown in *Figure 32.40.* The first is called serpentine and requires all bits to propagate through all possible locations before leaving. This increases the device latency (access to any bit) dramatically. It also significantly increases power. The serpentine's major advantage is its simplistic layout and minimal clocks needed. This saves area and reduces cost. An alternative is the SPS (serial-parallel–series) architecture. The SPS architecture permits use of smaller loops thus decreasing the average access time (latency). The access of the individual loops is achieved by using a decoder.

32.4.5 Applications

The CCD was perceived to have several applications. They can be used for various types of signal processing applications such as electronically variable delay lines, and a variety of transversal filters. CCDs can also be constructed to operate as very effective self-scanned photosensor arrays (charge coupled imagers) (see Chapter 31). Because of their small size CCDs were perceived as a low cost alternative for memory. That is low cost compared to RAM but higher cost than serial memory such as disk. They did offer a significant speed advantage over disk (microsecond versus milliseconds) and it was thought that they could fill a cost performance gap. This gap is shown in *Figure 32.41.* To use the CCD in this manner the host computer would need to be designed for a 'gap filler' memory architecture and CCD would have to cost 25% of DRAM. The technology never supported this cost structure, thereby all but eliminating this niche. Another application was for disk replacement with solid state memory.

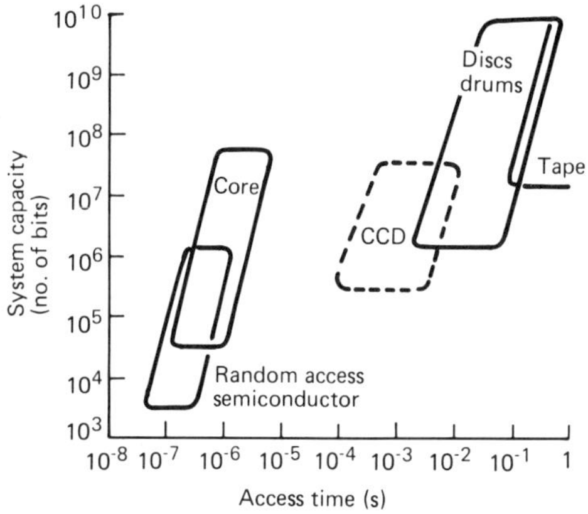

Figure 32.41 Memory hierarchies

High speed fixed head per track disk was the primary target, CCDs were perceived to be more durable and reliable than the mechanical disk system. CCDs could come close to competing on cost and the durability and performance enhancements made this a viable niche.

Further reading

JOHNSON, W. S., KUHN, G. L., RENNIGER, A. L. and PERLEGOS, G., '16k EE-PROM Relies on Tunnelling for Byte Erasable Program Storage', *Electronics*, February 28 (1980)

ROCHERS, G. D., 'EEPROM Elipses Other Reprogrammable Memories', *Electronics Design*, 22 November (1980)

BROWN, J. R. Jr, 'Timing Peculiarities of Multiplexed RAMs', *Computer Design*, July (1977)

KLEIN, R., OWEN, W. H., SIMKO, R. T. and TCHON, W. E., '5-Volt-Only Nonvolatile RAM Owes It All To Polysilicon', *Electronics*, 11 October (1979)

YOUNG, S., 'Evolution of MOS Technology', Presented at *Wescon* (1978)

YOUNG, S., 'Memories Have Not a Density Ceiling but New Process Will Push Through', *Electronics Design*, 25 October (1978)

PROEBSTING, R. and SCHROEDER, P., 'A 16k X 1 Bit Dynamic RAM', *ISCC*, February (1977)

PROEBSTING, R., 'Dynamic MOS RAM's. An Economic Solution for Many System Designs', *Electronics Design News*, 20 June (1977)

PROEBSTING, R. and GREEN, R., 'A TTL Compatible 4096 N-channel RAM', *ISSCC*, February (1973)

FOSS, R. C. and HARLAND, R., 'Standards for Dynamic MOS RAMs', *Electronics Design*, 16 August (1977)

GREENE, R., 'Pinout Standard Amplifies Variable Density Memory Design', *Electronics Design*, 6 December (1980)

BURSKY, D., 'UV EPROMs and EEPROMs Crash Speed and Density Limits', *Electronics Design*, 22 November (1980)

DROIR, J., OWEN, W. H. and SIMKO, R. T., 'Computer Systems Acquire Both RAM and EEPROM From One Chip With Two Memories', *Electronics Design*, 15 February (1980)

YOUNG, S., 'Uncompromising 4-K Static RAM Runs Fast on Little Power', *Electronics*, 12 May (1977)

OLIPHANT, J., 'Designing Non-Volatile Memory Systems with Intel's 5101 RAM', *Intel Applications Note* AP-12

GOSNEY, M., 'Reappraising CCD Memories: Can They Stand Up to RAMs?', *Electronics*, 7 June (1979)

GOSNEY, M., 'CCDs, Production Device or LAB Experiment', *Mostek Corporation Technology Brief*

OWEN, R., 'A Testing Philosophy for 16K Dynamic Memories', *Mostek Corporation, Applications Note*

'Mostek's BYTEWYDE(tm) Memory Products', *Mostek Corporation, Applications Note*

'Designing Memory Boards For RAM/ROM/EPROM Interchange', *Mostek Corporation, Applications Note*

'Resolving Microprocessor Memory Bus Contention', *Mostek Corporation, Applications Note*

'N-Channel MOS—Its Impact on Technology', *Mostek Corporation, ApplicationsNote*

'Which Way For 4K ... 16, 18, or 22 Pin?', *Intel Corporation, Application Brief* AP-11

JONES, F. and LAUTZENHEISER, D., 'Printed Circuit Board Layouts for Compatible Dynamic RAM Family', *Computer Design*, December (1980)

BURSKY, D. 'Memories Pace Systems Growth', *Electronic Design*, 27 September (1980)

HNATEK, E. R., 'Semiconductor Memory Update: EEPROMs', *Computer Design*, December (1981)

HNATEK, E. R., 'Semiconductor Memory Update: DRAMs', *Computer Design*, January (1982)

THREEWITT, B., 'A VLSI Approach to Cache Memory', *Computer Design*, January (1982)

EATON, S. S. and WOOTON, D., 'Circuit Advances Propel 64 K RAM across the 100 ns Barrier', *Electronics*, 24 March (1982)

DONNELLY, W., 'Memories—New Generations Push Technology Forward', *Electronics Industry*, October (1982)

WILCOCK, J. D., 'Semiconductor Memories', *New Electronics*, 17 August (1982)

WHITTIER, R. J., 'Semiconductor Memories', *Mini-Micro Systems*, December (1982)

33 Microprocessors

A Shewan, BSc
Systems Sales Engineer, Motorola Ltd
(Sections 33.1–33.5)

Robin K Saxby, BEng
Microprocessor Manager, Motorola Ltd
(Sections 33.6, 33.7)

Contents

33.1 Basic structure

The basic principles of operation of a microprocessor are identical to those that control the operation of any larger computer, and consist of four parts, as shown in *Figure 33.1*: (1) the memory; (2) the arithmetic-logic unit (ALU); (3) the input–output system; (4) the control unit.

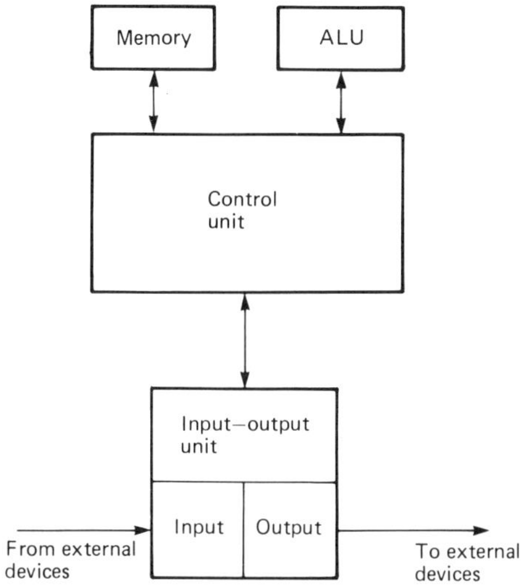

Figure 33.1 Basic structure of a microprocessor

The *memory* is an indispensable part of a microprocessor. It contains the *data* used in a program as well as the *instructions* for executing the program. A *program* is a group of instructions telling the microprocessor what to do with the data. The *arithmetic-logic unit* (ALU) is part of the microprocessor that performs those arithmetic or logic operations required by the routine or program, and generates the status bits (condition codes) that are the heart of the decision-making capabilities of any computer. The *input–output system* controls communication between the microprocessor and its external devices. The *control unit* consists of a group of flip-flops and registers that regulate the operation of the microprocessor itself. The function of the control unit is to cause the proper sequence of events to occur during the execution of each computer instruction.

33.1.1 Memory concepts

The binary bits of a block of memory can be organised in various ways. It is usual for most microprocessors to be arranged so that the bits are grouped into *words* of 4, 8 or 16 bits each. In most present-day microprocessors, an 8-bit word length (or byte) is used. Modern microprocessors are *word oriented* in that they transfer one byte (or word) at a time by means of the data bus. This bus, the internal registers, and the ALU (arithmetic logic unit) are *parallel* devices that handle all the bits of a word *at the same time*.

Microprocessor instructions require 1, 2, or 3 bytes (or words) and typical routines include 5 to 50 instructions. Since IC memories are physically small and relatively inexpensive, they can include thousands of words and thereby provide very comprehensive computer programs.

Each word in memory has *two parameters*; its *address*, which locates it within memory, and the *data*, which is stored at that location. The process of accessing the contents of the memory locations requires two registers, one associated with address and one with data. The memory address register (MAR) holds the *address of the word currently being accessed*, and the memory data register (MDR) holds the *data being written into or read out of the addressed memory location*. These registers can be considered part of the memory or of the control unit.

In most microprocessors the memory consists of two parts with different memory addresses, a *ROM area* used to hold the *program*, *constant data*, and *tables*, and a *RAM area*, used to hold *variable data*. Generally, the data on which the program operates must be rewritten every time the system is started again (restarted), and the system must always be restarted (going through a start-up procedure) after any power failure occurs. This is not a severe drawback because data is usually invalid after a power failure. Fortunately, the program, if it is contained in ROM, can be restarted immediately because a power failure does not affect a ROM.

33.1.2 The arithmetic logic unit

The ALU performs all the arithmetic and logical operations required by the microprocessor. It accepts two operands as inputs (each operand contains as many bits as the basic word length of the computer) and performs the required arithmetic or logical operation upon them. ALUs are readily available as ICs but microprocessors contain their ALUs within their chip.

Most ALUs perform the following arithmetic or logical operations: (1) addition; (2) subtraction; (3) logical OR; (4) logical AND; (5) EXCLUSIVE OR; (6) complementation; (7) shifting.

Microprocessors are capable of performing more sophisticated arithmetic operations such as multiplication, division, extracting square roots, and taking trigonometric functions; however, in most devices these operations are performed as *software subroutines*. A multiplication command, for example, is translated by the appropriate subroutine into a series of add and shift operations that can be performed by the ALU.

33.1.3 The control section

The function of the *control section* is to regulate the operation of the microprocessor. It decodes the instructions and causes the proper events to occur in the correct order.

The control section of a microprocessor consists of a group of registers and flip-flops and the timing circuitry necessary to make them operate properly. In a rudimentary device the following registers might be part of the control section:

(1) The memory address register (MAR)

(2) The program counter (PC) This is a register that contains as many bits as the MAR. It holds the memory address of the next instruction word to be executed. It is usually *incremented* during the execution of an instruction so that it contains the address of the next instruction to be executed.

(3) The instruction register This register holds the instruction while it is in the process of being executed.

(4) The instruction decoder This decodes the instruction presently being executed. Its inputs come from the instruction register.

(5) The accumulator The accumulator contains the basic operand used in each instruction. In ALU operations where two operands are required, one of the operands is stored in the accumulator as a result of previous instructions. The other operand is generally read from memory. The two operands form the inputs to the ALU and the result is normally sent back to the accumulator.

The control unit usually contains several flip-flops. The flags or condition codes are flip-flops and most microprocessors also have FETCH and EXECUTE flip-flops.

These determine the state of the microprocessor. The instructions are contained in the microprocessor's memory. The microprocessor starts by fetching the instruction. This is the FETCH portion of the computer's cycle. It then executes, or performs the instruction. At this time, the computer is in EXECUTE mode. When it has finished executing the instruction, it returns to FETCH mode and reads the next instruction from memory. Thus the computer alternates between FETCH and EXECUTE modes and the FETCH and EXECUTE flip-flops determine its current mode of operation.

33.2 The Motorola 6800 microprocessor

In this chapter the Motorola 6800 microprocessor is used to describe features of operation, which are also common to many other devices currently available from other manufacturers.

Figure 33.2 shows the block diagram of a 6800 microprocessor. An 8-bit internal bus interconnects the various registers with the instruction decode and control logic. The nine control lines that communicate with the external devices can be seen on the left, with the address bus at the top and the data bus at the bottom of the figure.

33.2.1 ALU and registers

The Arithmetic Logic Unit included in the 6800 is an 8-bit, parallel processing, 2s complement device. It includes the condition code (or processor status) register. The system has a 16-bit address bus, but as seen in *Figure 33.2*, the internal bus is 8-bits wide, and the index and stack pointer registers (and the address buffers) are implemented with a high (H) and low (L) byte. The two 8-bit accumulators, A and B, speed up program execution by allowing two operands to remain in the microprocessor. Instructions that can be performed using both accumulators (e.g. ABA and SBA) are very fast because they do not require additional cycles to fetch the second operand.

33.2.2 Vectored interrupts

An interrupt is a signal to the microprocessor that causes it to stop execution of the normal program and to branch (or jump) to another location that is the beginning address of an interrupt service routine. These routines are written to provide whatever action is necessary to respond to the interrupt. Four types of interrupts are provided in the 6800, and each has its unique software service routine and vector. Three of them also are implemented by pins on the 6800. The four types are:

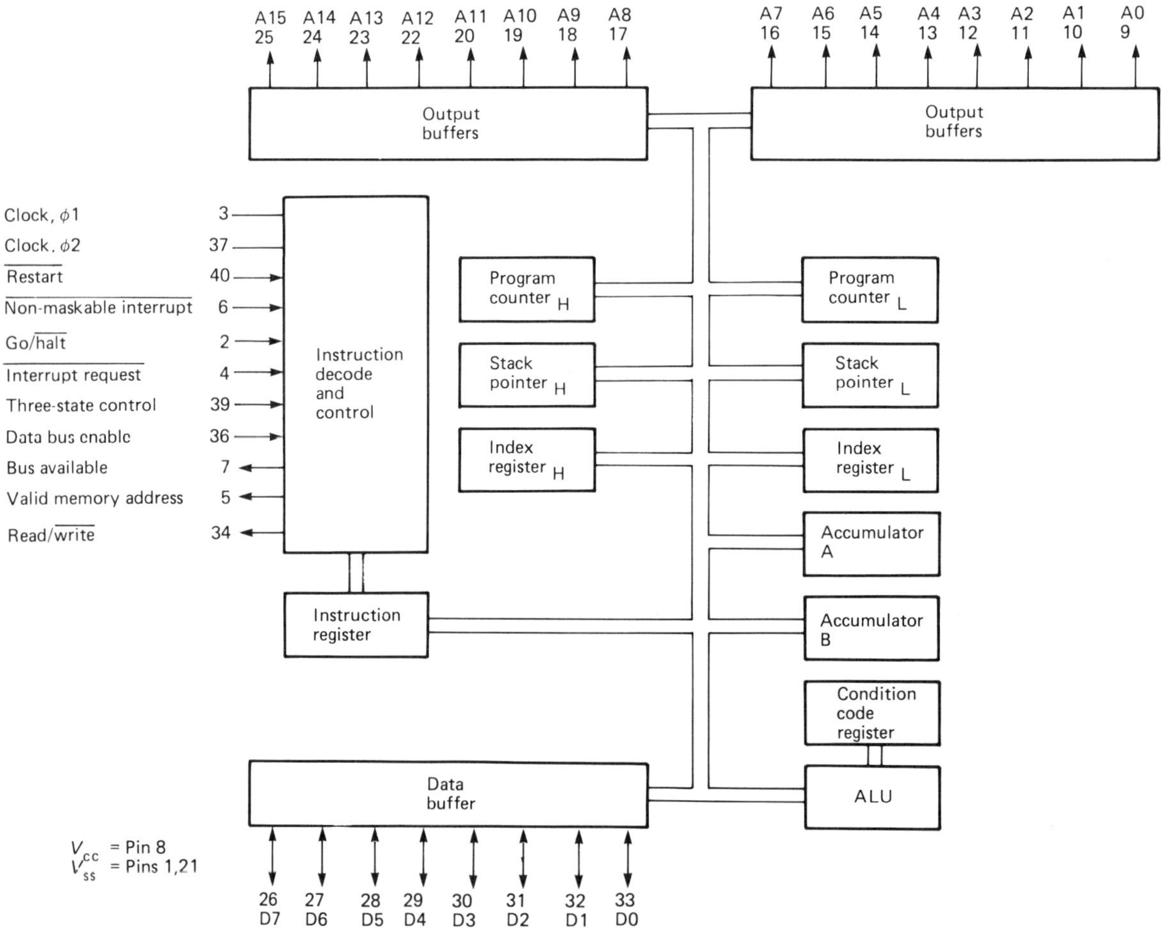

Figure 33.2 Block diagram of 6800 microprocessor

RESTART (RST)
NONMASKABLE INTERRUPT (NMI)
SOFTWARE INTERRUPT (SWI)
INTERRUPT REQUEST (IRQ)

In order for the program to be branched to the appropriate routine, the top eight locations of the ROM or PROM that is highest in memory, are reserved for these interrupt vectors. The contents of these locations contain the 16-bit addresses where the service routines begin. When activated, the program is 'vectored' or pointed to the appropriate address by the microprocessor logic. This is called a 'vectored interrupt.'

33.2.3 Control lines and their functions

The hardware aspects of system design involve interconnecting the components of the system so that the data is properly transferred between them and to any external hardware that is being utilised. In addition to the obvious requirement for power and ground to each component, the address bus, and the bidirectional data bus, numerous control signal lines are required. These control lines are in effect a third bus called the control bus. The lines are identified in *Figure 33.2* which also shows their pin numbers.

To design a system that performs properly, it is necessary to understand each signal's characteristics and function. These are described in the following paragraphs.

33.2.3.1 Read/Write (R/W)

This output line is used to signal all external devices that the microprocessor is in a READ state (R/$\overline{\text{W}}$ = HIGH), or a WRITE state (R/$\overline{\text{W}}$ = LOW). The normal standby state of this line, is HIGH. This line is three-state. When three state control (TSC) goes HIGH, the R/$\overline{\text{W}}$ line enters the high impedance mode.

33.2.3.2 Valid memory address (VMA)

The VMA output line (when in the HIGH state) tells all devices external to the microprocessor there there is a valid address on the address bus. During the execution of certain instructions, the address bus may assume a random address because of internal calculations. VMA goes LOW to avoid enabling any device under those conditions. Note also that VMA is held LOW during HALT, TSC, or during the execution of a WAIT (WAI) instruction. VMA is not a 3-state line and therefore direct memory access (DMA) cannot be performed unless VMA is externally opened (or gated).

33.2.3.3 Data bus enable (DBE)

The DBE signal enables the data bus drivers of the microprocessor when in the HIGH state. This input is normally connected to the phase 2 clock but is sometimes delayed to assure proper operation with some memory devices. When HIGH, it permits data to be placed on the bus during a WRITE cycle. During a microprocessor READ cycle, the data bus drivers within the microprocessor are disabled internally. If an external signal holds DBE LOW, the microprocessor data bus drivers are forced into their high impedance state. This allows other devices to control the I/O bus (as in DMA).

33.2.3.4 HALT and RUN modes

When the HALT input to the 6800 is HIGH, the microprocessor is in the RUN mode and is continually executing instructions. When the HALT line goes LOW, the microprocessor halts after completing its present instruction. At that time the microprocessor is in the HALT mode. Bus Available (BA) goes HIGH, VMA becomes a 0, and all 3-state lines enter their high impedance state. Note that the microprocessor does not enter the HALT mode as soon as HALT goes LOW, but does so only when the microprocessor has finished execution of its current instruction. It is possible to stop the microprocessor while it is in the process of executing an instruction by stopping or 'stretching the clock.

Bus available (BA)

The bus available (BA) signal is a normally LOW signal generated by the microprocessor. In the HIGH state it indicates that the microprocessor has stopped and that the address bus is available. This occurs if the microprocessor is in the HALT mode or in a WAIT state as the result of the WAI instruction.

33.2.3.6 Three-state control (TSC)

TSC is an externally generated signal that effectively causes the microprocessor to disconnect itself from the address and control buses. This allows an external device to assume control of the system. When TSC is HIGH, it causes the address lines and the READ/WRITE line to go to the high impedance state. The VMA and BA signals are forced LOW. The data bus is not affected by TSC and has its own enable (DBE).

The microprocessor is a dynamic IC that must be periodically refreshed by its own clocks. Because TSC stops the clocking of the internal logic of the microprocessor, it should not be HIGH longer than 3 clock pulses, or the dynamic registers in the microprocessor may lose some of their internal data. (Up to 19 cycles at 2 MHz for the new 68B00.)

33.2.4 Clock operation

The 6800 utilises a *two-phase clock* to control its operation. The waveforms and timing of the clock are critical for the proper operation of the microprocessor and the other components of the family.

The clock synchronises the internal operations of the microprocessor, as well as all external devices on the bus. The program counter, for example, is advanced on the falling edge of phase 1 and data are latched into the microprocessor on the falling edge of phase 2. All operations necessary for the execution of each instruction are synchronised with the clock.

In a typical instruction table the number of bytes and clock cycles are listed for each instruction. The number of bytes for each instruction determines the size of the memory, and the number of cycles determines the time required to execute the program. LDA A $1234 (which is the extended addressing mode), for example, requires three bytes, one to specify the operation (OP) code and two to specify the address, but requires four cycles to execute. Often an instruction requires the processor to perform internal operations in addition to the fetch cycles. Consequently, for any instruction the number of cycles is generally larger than the number of bytes.

33.3 Hardware configuration of a microprocessor system

Several different types of devices are connected to a microprocessor to form a total control system. A few such devices are described here and illustrated by reference to the Motorola 6800 family.

The basic family of microcomputer components consists of five parts: (1) the microprocessor; (2) masked programmable read only memory (ROM) (1024 bytes of 8 bits each); (3) static random access memory (RAM) (128 bytes of 8 bits each); (4) peripheral interface adapter (PIA) (for parallel data input/output); (5) asynchronous communications interface adapter (ACIA) (for serial data input/output).

As shown in *Figure 33.3*, a complete microcomputer can be built using the components listed above, plus a clock, which is needed to control the timing of the system. Several clock devices

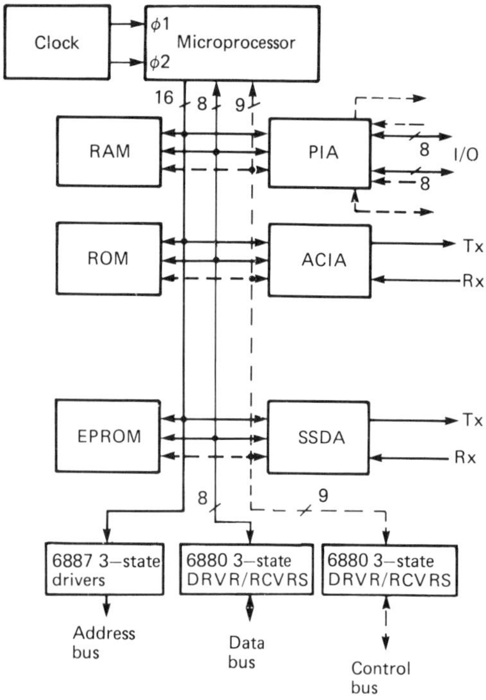

Figure 33.3 The 6800 family components

are available. These are *crystal controlled oscillators* that provide the necessary two-phase nonoverlapping timing pulses and are equipped with output circuits suitable for driving MOS circuitry.

The suitability of the family of components as elements of a microcomputer system depends on several factors. First, is the requirement that *all elements of a computer be present* and, second, that they be partitioned into the various packages so that they are *modular* and a variety of configurations can be easily assembled. Finally, there is a need for a simple way to interconnect them.

In the system of *Figure 33.3*, the program instructions for the system would typically be stored in ROM or EROM, and all variable data would be written into or read from RAM. The input/output (I/O) of data for the system would be done via the PIA, ACIA or SSDA.

To understand the ways these units work together, it is necessary to know the hardware features of each part. *Figure 33.3* also shows that the system components are interconnected via a 16-wire address bus, an 8-wire data bus, and a 9-wire control bus. For a component to be a member of a microcomputer family and to insure compatibility, it must meet specific system *standards*. The standards must also make it convenient for external devices to interface (or communicate) with the microprocessor. These standards, which apply to all components, are as follows:

(1) 8-bit bidirectional data bus;
(2) 16-bit address bus;
(3) 3-state bus switching techniques;
(4) TTL/DTL level compatible signals;
(5) 5 volt n-channel MOS silicon gate technology;
(6) 24 and 40 pin packages;
(7) clock rate 100 kHz to 2 MHz (MC68B00);
(8) temperature range of 0 to 70°C.

Since the basic word length of the 6800 is 8-bits (one byte) it communicates with other components via an 8-bit data bus. The data bus is *bidirectional*, and data are transferred into or out of the 6800 over the same bus. A read/write line (one of the control lines) is provided to allow the microprocessor to control the direction of data transfer.

An 8-bit data bus can also accommodate ASCII (American standard code for information interchange) characters and packed BCD (two BCD numbers in one byte).

A 16-bit address bus was chosen for these reasons:

(1) for programming ease, the addresses should be multiples of 8 bits;
(2) an 8-bit address bus would only provide 256 addresses but a 16-bit bus provides 65 536 distinct addresses, which is adequate for most applications.

33.3.1 Peripheral interface adapters

The peripheral interface adapter (PIA), shown in *Figure 33.4* provides a simple means of interfacing peripheral equipment on a parallel or byte-wide basis to the microcomputer system. This device is compatible with the bus interface on the microprocessor side, and provides up to 16 I/O lines and 4 control lines on the peripheral side, for connection to external units. The PIA outputs are TTL and CMOS compatible.

33.3.2 The asynchronous communications interface adapter

The ACIA provides the circuitry to connect serial asynchronous data communications devices (such as a teletypewriter terminal TTY) to bus organised systems such as the 6800 microcomputer system.

The bus interface includes select, enable, read/write, and interrupt signal pins in addition to the 8-bit bidirectional data bus lines. The parallel data of the 6800 system is serially transmitted and received (simultaneously) with proper ASCII formatting and error checking. The control register of the ACIA is programmed via the data bus during system initialisation. It determines word length, parity, stop bits, and interrupt control of the transmit or receive functions.

The PIA and ACIA are the most used I/O components in the 6800 family. Other I/O devices are available however, and include:

(1) synchronous serial data adapter (SSDA);
(2) advanced data link controller (ADLC);
(3) general purpose interface adapter (IEEE 488-1975 bus);
(4) CRT controller (CRTC);
(5) floppy disk controller (FDC);
(6) direct memory access controller (DMAC).

The reader should consult the manufacturers literature for information on these peripheral controller products.

33.3.3 The synchronous serial data adapter

Asynchronous communications via the ACIA are used primarily with slow speed terminals where information is generated on a keyboard (manually) and where the communication is not necessarily continuous. Synchronous communications are usually encountered where high speed continuous transmission is required. This information is frequently read to MODEMs, disk or tape systems at 1200 BPS or faster. Syncrhronous systems

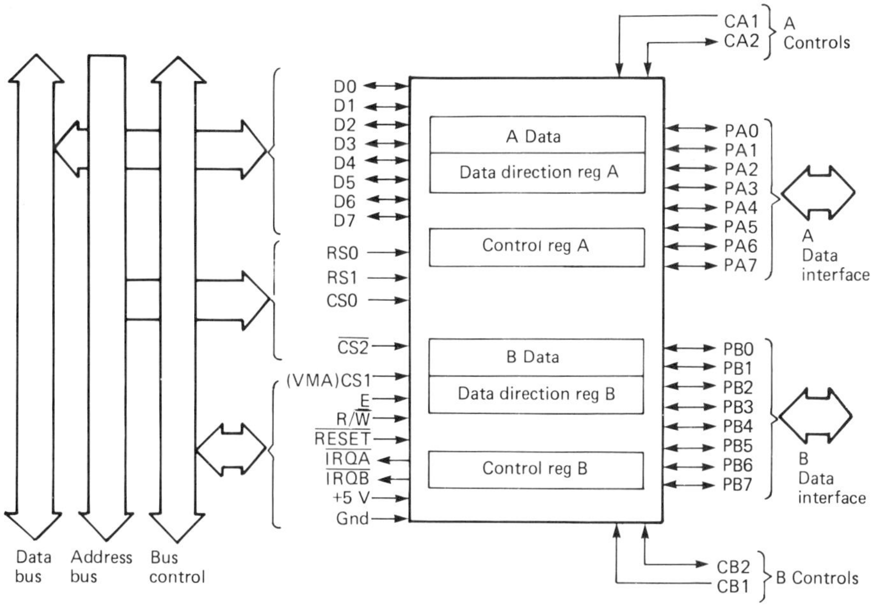

Figure 33.4 PIA bus interface and registers

transmit a steady stream of bits even when no characters are available (a *sync* character is substituted). Because there are no start and stop bits to separate the characters, care must be taken to synchronise the receiving device with the transmitted signal so that the receiver end of the circuit can determine which bit is bit 1 of the character.

Synchronous systems usually use a preamble (all 1s, for example) to establish synchronisation between the receiver and transmitter and will then maintain sync by transmitting a sync pattern until interrupted. Because start and stop bits are not needed, the efficiency of transmission is 20% better for 8-bit words (8 instead of 10 bits per character).

The SSDA provides a bidirectional serial interface for synchronous data information interchange with bus organised systems such as the 6800 microprocessor. It is a complex device containing seven registers. Although primarily designed for synchronous data communications using a 'Bi-sync' format, several of its features and, in particular, the first-ins, first-out (FIFO) buffers, make it useful in other applications where data are to be transferred between devices that are not being clocked at precisely the same speed, such as tape cassettes, tape cartridges, or floppy disk systems.

33.3.4 Memory space allocation

The I/O devices such as the PIA, ACIA, and SSDA all have internal registers that contain the I/O data or control the operation of the device. Each of these registers must be allocated a unique address on the address bus and is communicated with just as if it were memory. This technique is called *memory-mapped* I/O and, in addition to allowing the use of all memory referencing instructions for the I/O functions, it eliminates the need for special I/O instructions.

The process of addressing a particular memory location, includes not only selecting a cell in a chip, but also selecting that chip from among all those on the same bus. The low order address lines are generally used to address registers within the chips, and the high order lines are available to single out the desired chip. These high order lines could be connected to a decoder circuit with one output line used to enable each chip but, since most of the 6800 family devices have several chip select pins available, these are frequently connected directly to the high order address lines (A15, A14, A13,...) and separate decoders are not needed, particularly in simple systems.

Table 33.1 shows the number of addresses required and the number of chip select (CS) pins available for typical 6800 family components.

Table 33.1 Addresses required and chip select pins needed

Component	*Addresses required*	*Positive CS pins*	*Negative CS pins*
RAM	128	2	4
ROM	1024	*	*
PIA	4	2	1
ACIA	2	2	1
SSDA	2	0	1

* Four programmable enables are defined as positive or negative when the mask is made.

A component is selected only if all its CS lines are satisfied. When unselected, a component places its outputs in a high impedance state and is effectively disconnected from the data bus.

When setting up a microcomputer system, each component must be allocated as much memory space as it needs and must be given a unique address so that no address selects more than one component. In addition, one component (usually a ROM), must contain the vector interrupt addresses.

33.3.5 Addressing techniques

Most systems do not use all 64K of available memory space. Thefefore, not all of the address lines need to be used, and *redundant addresses* will occur (i.e. components will respond to two or more addresses). It is important, however, to choose the lines used so that no two components can be selected by the same address. Since the chip selects serve to turn ON the bus drivers (for a microprocessor READ), only *one* component should be on at any time or something may be damaged.

For proper system operation, most memory and peripheral devices should only transfer data when phase 2 and VMA are

HIGH. Consequently one CS line on each component is usually connected to each of these signals or a derivative of these signals. The PIA, for example, must have its E pin connected to phase 2 or it will not respond to an interrupt and can lock up the system following the execution of a WAI instruction.

When allocating memory, it is wise to place the 'scratch pad' RAMs at the bottom of the memory map, since it is then possible to use the direct address mode instructions throughout the program when referencing these RAM locations. Because two-byte instead of three-byte instructions are used, a savings of up to 25% in total memory requirements is possible.

33.4 Software

This section introduces the software features of microprocessor, again using the 6800 as an example, the mnemonics or assembly language concept, and then the accumulator and memory referencing instructions.

The first byte of each instruction is called the *operation code* (op code) because its bit combination determines the operation to be performed. The second and third bytes of the instruction, if used, contain address or data information. They are the *operand* part of the instruction. The op code also tells the microprocessor logic how many additional bytes to fetch for each instruction. There are 197 unique instructions in the 6800 instruction set. Since an 8-bit word has 256 bit combinations, a high percentage of the possible op codes are used.

When an instruction is fetched from the location in memory pointed to by the program counter (PC) register, and the first byte is moved into the instruction register, it causes the logic to execute the instruction. If the bit combination is 01001111 (or 4F in hex), for example, it CLEARs the A accumulator (RESETs it to all 0s), or if the combination is 01001100 (4C), the processor increments the A accumulator (adds 1 to it). These are examples of *one-byte* instructions that involve only one accumulator. An instruction byte of 10110110 (B6) is the code for the *extended addressing mode* of LOAD A. It also commands the logic to fetch the *next two bytes* and use them as the *address* of the data to be loaded into the A accumulator.

33.4.1 Assembly language

When the op codes for each instruction are expressed in their binary or hex format, as in the previous paragraph, they are known as *machine language instructions*. The bits of these op codes must reside in memory and be moved to the instruction decoder for analysis and action by the microprocessor. Programs and data can be entered into a microprocessor in machine language using data switches or keyboards.

Writing programs in machine language is very tedious, and trying to follow even a simple machine language program strains the ability of most people. As a result, various techniques have evolved in an effort to simplify the program documentation. The first step in this simplification is the use of *hexadecimal notation* to express the codes. This reduces the digits from 8 to 2 (10100011 = A3, for example) or, in the case of an address, from 16 binary digits (bits) of 4 hex digits. Even this simplification, however, is insufficient and the concept of using *mnemonic language* to describe each instruction has therefore been developed. A *mnemonic* is defined as a device to help the memory. For example, the mnemonics LDA A (load A) and STA A (store A) are used for the LOAD and STORE instructions instead of the hex OP codes of 86 and B7 because the mnemonics are descriptive and easier to remember.

Not only are mnemonics easy to remember, but the precise meaning of each one makes it possible to use a computer program to translate them to the machine language equivalent that can be used by the computer. A program for this purpose is called an *assembler* and the mnemonics that it recognises constitute an *assembly language program.*

When a program is written in mnemonic or assembly language, it is called a *source program.* After being assembled (or translated) by the computer, the machine language codes that are produced are known as the *object program.* An assembler also produces a *program listing*, which is kept as a record of the design.

The 6800 assembler uses several symbols to identify the various types of numbers that occur in a program. These symbols are:

(1) A blank or no symbol indicates the number is a *decimal* number;
(2) A $ immediately preceding a number indicates it is a *hex* number ($24, for example, is 24 in hex or the equivalent of 36 in decimal);
(3) A # sign indicates an *immediate* operand;
(4) A @ sign indicates an *octal* value;
(5) A % sign indicates a *binary* number (01011001, for example).

33.4.2 Accumulator and memory instructions

The accumulator and memory instructions can be broken down roughly into the following categories:

(1) Transfers between an accumulator and memory (LOADs and STOREs);
(2) Arithmetic operations—ADDITION, SUBTRACTION, DECIMAL ADJUST ACCUMULATOR;
(3) Logical operations—AND, OR, XOR;
(4) Shifts and rotates;
(5) Test operations—bit test and compares;
(6) Other operations—CLEAR, INCREMENT, DECREMENT, and COMPLEMENT.

In order to reduce the number of instructions required in a typical program and, consequently, the number of memory locations needed to hold a program, microprocessors feature several ways to address them. They are called *addressing modes* and many of them only use one or two bytes.

Instructions can be addressed in one or more of the following modes: *immediate*, *direct*, *indexed*, *implied*, or *extended.*

These modes are explained in the following paragraphs and illustrated in *Figure 33.5*, where each instruction is assumed to start at location 10. The figure shows the op code at location 10 and the following bytes. It explains the function of each byte and gives a sample instruction for each mode.

33.4.2.1 Immediate addressing instructions

All the *immediate instructions* require two bytes. The first byte is the op code and the *second byte* contains the *operand* or *information to be used.* If, for example, the accumulator contains the number 2C and the following instruction occurs in a program:

ADD A #$23

The # indicates that the hex value, $23 is to be added to the A accumulator. The immediate mode of the instruction is indicated by the # sign. After the instruction is executed, A contains 4F (23 + 2C).

Immediate instructions are used if the variable or operand is known to the programmer when he is coding and need not reside in memory. For example, if the programmer knows he wants to add 5 to a variable, it is more efficient to add it immediately than to store 5 in memory and do a direct or extended ADD. When one of the operands must reside in memory, however, direct or extended instructions are required.

Location		Example of code	Instruction and effect
10	OP code	8B	ADD A $33
11	Immediate value	33	Adds $(33)_{16}$ to A.

(a) Immediate addressing

Location		Example of code	Instruction and effect
10	OP code	9B	ADD A $33
11	Direct address	33	Adds the contents of location 0033 to A.

(b) Direct addressing

Location		Example of code	Instruction and effect
10	OP code	BB	ADD A $0133
11	High address byte	01	Adds the contents of location 0133 to A.
12	Low address byte	33	

(c) Extended addressing

Location		Example of code	Instruction and effect
10	OP code	AB	ADD A $06, X
11	Offset	06	Adds the contents of the location given by the sum of the index register +6 to A.

(d) Indexed addressing

Location		Example of code	Instruction and effect
10	OP code	1B	ABA Adds the contents of A to B. The results go into A.

(e) Inherent

Figure 33.5 Examples of the various addressing modes of the 6800

33.4.2.2 Direct instructions

Like immediate instructions, direct instructions require two bytes. The second byte contains the address of the operand used in the instruction. Since the op code identifies this as a two-byte instruction, only 8 address bits are available. The microprocessor contains 16 address lines, but for this instruction, the 8 MSBs of the address are effecrively set to 0. The memory locations that can be addressed by a direct instruction are therefore restricted to 0000 to 00FF. It is often wise to place variable data in these memory locations because this data is usually referenced frequently throughout the program. The programmer can then make maximum use of direct instructions and reduce memory requirements by up to 25%.

33.4.2.3 Extended instructions

Extended instructions are 3-byte instructions. The OP code is followed by two bytes that specify the address of the operand used by the instruction. The second byte contains the 8 high order bits of the address. Because 16 address bits are available, any one of the 65 536 memory locations can be selected. Thus, extended instructions have the advantage of being able to select any memory locations, but direct instructions only require two bytes in the program instead of three. Direct instructions also require one less cycle for execution so they are somewhat faster than the corresponding extended instructions.

33.4.2.4 Indexed instructions

As its name implies, an indexed instruction makes use of the *index register* (X). For any indexed instruction, the address referred to is the sum of the number (called the OFFSET) in the second byte of the instruction, plus the contents of X. The 6800 provides instructions to LOAD, STORE, INCREMENT, and DECREMENT X as well as the stack pointer (SP).

Index registers are useful when it is necessary to relocate a program. For example, if a program that originally occupied locations 0000 to 00CF must be moved or relocated to locations 0400 to 04CF, all addresses used in the original program must be changed. In particular, direct instructions cannot be used because the program no longer occupies lower memory. If X is loaded with the base address of the program (400 in this example), then all direct instructions can be changed to indexed instructions and the program will function as before.

The index register is often used by programs that are required to perform *code conversions*. Such programs might convert one code to another (ASCII to EBCDIC, for example), or might be used for trigonometric conversions where the sine or cosine of a given angle may be required.

33.4.2.5 Implied addressing

Implied instructions are used when all the information required for the instruction is already within the CPU and no external operands from memory or from the program (in the case of immediate instructions) are needed. Since no memory references are needed, implied instructions only require one byte for their OP code. Examples of implied instructions are CLEAR, INCREMENT and DECREMENT the accumulators, and SHIFT, ROTATE, ADD, or SUBTRACT accumulators.

33.4.3 Logic instructions

AND, OR, and EXCLUSIVE OR instructions allow the programmer to perform Boolean algebra manipulations on a variable, and to SET or CLEAR specific bits in a byte. They can also be used to test specific bits in a byte, but other logic instructions such as BIT, TEST, or COMPARE may be more useful for these tests.

33.4.3.1 Setting and clearing specific bits

AND and OR instructions can be used to SET or CLEAR a specific bit or bits in an accumulator or memory location. This is very useful in those systems where each bit has a specific meaning, rather than being part of a number. In the control and status registers of the PIA or ACIA, for example, each bit has a distinct meaning.

33.4.3.2 Testing bits

In addition to being able to SET or CLEAR specific bits in a register, it is also possible to test specific bits to determine if they are 1 or 0. In the peripheral interface adapter (PIA), for example, a 1 in the MSB of the control register indicates some external event has occurred. The microprocessor can test this bit and react appropriately. Typically the results of the test sets the Z or N bit. The program then executes a conditional branch and takes one of two different paths depending on the results of the test.

Accumulator bits can be tested by the AND and OR instructions, but this modifies the contents of the accumulator. If the accumulator is to remain unchanged, the BIT TEST instruction is used. This ANDs memory (or an immediate operand) with the accumulator without changing either.

33.4.3.3 Compare instructions

A COMPARE instruction essentially subtracts a memory or immediate operand from an accumulator, leaving the contents of both memory and accumulator unchanged. The actual results of the subtraction are discarded; the function of the COMPARE is to SET the condition code bits.

There are two types of COMPARE instructions; those that involve accumulators, and those that use the index register. Effectively, the two numbers that are being compared are subtracted, but neither value is changed. The subtraction serves to SET the condition codes. In the case of the CPX instruction, only the Z bit is significant, but for the COMPARE accumulator (CMP A or B) instruction, the carry, negative, zero and overflow bits are affected and allow us to determine which of the operands is greater.

The COMPARE INDEX REGISTER instructions compare the contents of the Index Register with a 16-bit operand. They are often used to terminate loops.

33.4.3.4 The TEST instruction

The TEST (TST) instruction subtracts 0 from an operand and therefore does not alter the operand. Its effect, like that of compares or bit tests, is to set the N and Z bits. It differs in that it always CLEARs the overflow and carry bits. It is used to set the condition codes in accordance with the contents of an accumulator or memory location.

33.4.4 Branch and Jump instructions

BRANCH and JUMP instructions allow programs requiring *decisions*, *branches*, and *subroutines* to be written.

33.4.4.1 Jump instructions

One of the simplest instructions is the JUMP instruction. It loads the PC with a new value and thereby transfers or jumps the program to a new location.

The JUMP instruction can be specified in one of two modes, indexed or extended. The JUMP INDEXED (JMP 0,X) is a two-byte instruction.

The second byte or *offset* is added to the index register and the sum is loaded into the PC.

Extended jumps are three-byte instructions, where the last two bytes are a 16-bit address. Since a 16-bit address is available, they allow the program to jump to any location in memory. They are very easily understood because they require no calculations.

33.4.4.2 Unconditional branch instructions

Branch instructions are two-byte relative address instructions. The second byte contains a displacement. Normally, when a branch instruction is executed, the contents of the PC are incremented twice to point to the address of the next instruction. When the branch is taken, the PC is altered by the displacement, and the next op code is found at the address that equals the address of the branch instruction plus two, plus the displacement. The displacement is treated as an 8-bit signed number that is added to the PC. Displacements with MSBs of 1 are negative numbers, which cause the program to branch backward. Since the maximum positive number that can be represented by an 8-bit signed byte is 127_{10} and the most negative number is -128, the program can branch to any location between PC $+129$ and PC -126, where PC is the address of the first byte of the BRANCH instruction.

The unconditional branch, BRA, causes the program to branch whenever it is encountered. It is equivalent to a JUMP instruction, but since it is only a two-byte instruction, it is used when the location being jumped to is within the range of $+129$ or -126 bytes relative to the current PC address.

33.4.4.3 Conditional branch instructions

A conditional branch instruction branches only when a particular condition code or combination of condition codes are SET or CLEAR. Therefore the results of instructions preceding the branch determine whether or not the branch is taken. Conditional branches allow the user to write programs that make decisions and give the microprocessor its ability to compute.

33.4.5 Subroutines

A *subroutine* is a small program that is generally used more than once by the main program. Multiplications, 16-bit adds, and square roots are typical subroutines.

Figure 33.6 illustrates the use of the same subroutine by two different parts of the main program. The subroutine located at

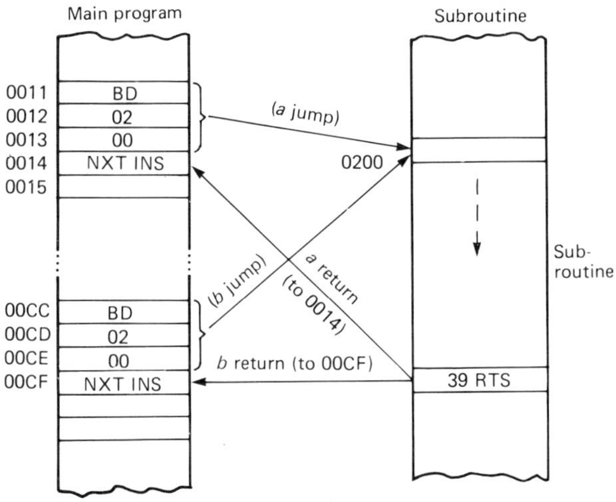

Figure 33.6 Use of a subroutine

200 can be entered from either location 0011 or 00CC by placing a jump to subroutine (AD, BD, or 8D, depending on the mode) in these addresses. The PC actions, as a result of the subroutine jump at 0011, are identified by the *a* in *Figure 33.6* and *b* identifies jumps from 00CC.

After the subroutine is complete, the program resumes from the instruction following the location where it called the subroutine. Because the subroutine must return to one of several locations, depending on which one caused it to be entered, the original contents of the PC must be preserved so the subroutine knows where to return.

33.4.5.1 Jumps to subroutines

The JUMP TO SUBROUTINE (JSR) instruction remembers the address of the next main instruction by writing it to the stack before it takes the jump. The JSR can be executed in the indexed or extended mode. There is a branch to subroutine (BSR) that can be used if the starting address of the subroutine is within +129 to −126 locations of the program counter. The advantage of the BSR is that it requires one less byte in the main program.

33.4.5.2 Return from subroutine

The JSR instructions preserve the contents of the PC on the stack, but a return from subroutine (RTS) instruction is required to properly return. The RTS is the last instruction executed in a subroutine. It places the contents of the stack in the PC and causes the SP to be incremented twice. Because these bytes contain the address of the next main instruction in the program (put there by the JSR or BSR that initiated the subroutine), the program resumes at the place where it left off before it entered the subroutine.

33.4.5.3 Nested subroutines

In some sophisticated programs the main program may call a subroutine, which then calls on a second subroutine. The second subroutine is called a nested subroutine because it is used by and returns to the first subroutine.

33.4.5.4 Use of registers during subroutines

During the execution of a subroutine, the subroutine will use the accumulators; it may use X and it changes the contents of the CCR. When the main program is re-entered, however, the contents of these registers must often be as they were before the jump to the subroutine.

The most commonly used method of preserving register contents during a subroutine is to write the subroutine so that it PUSHes those registers it must preserve onto the stack at the beginning of the subroutine and then PULLs them at the end of the subroutine, thus restoring their contents before returning to the main program.

33.4.6 Other instructions

33.4.6.1 CLEAR, INCREMENT, and DECREMENT instructions

These allow the user to alter the contents of an accumulator or memory location as specified. Those instructions referring to memory locations can be executed in extended or indexed modes only. They are simple to write, but require 6 or 7 cycles for execution because the contents of a memory location must be brought to the CPU, modified, and rewritten to memory.

33.4.6.2 SHIFT and ROTATE instructions

The drawings in *Figure 33.7* show diagrammatically how these work. Note that they all use the carry bit, either for input, outpit, or both. Rotates are 9-bit rotates that combine the 8-bit accumulator and the carry bit.

33.4.6.3 COMPLEMENT and NEGATE instructions

The COMPLEMENT instructions invert all bits of a memory location. They are useful as logic instructions and in programs requiring complementation, such as the BCD subtraction program.

The NEGATE instructions take the 2s complement of a number and therefore negates it. The negate instruction works by subtracting the operand from 00. Since the absolute value of the operand is always greater than the minuend, except when the operand itself is 00, the negate instruction SETs the carry flag for all cases, except when the operand is 00.

33.4.7 Assembler directives

When writing a program, options are available to enable the programmer to reserve memory bytes for data, specify the starting address of the program, and select the format of the assembler output. These options are called assembler directives. Assembler directives are written into the source program and interpreted during the assembly process.

33.4.8 The two-pass assembler

To convert a source program to an object program, this assembler reads the source program twice. It is known as a two-pass assembler. When it reads through the program the first time (called the first pass or pass 1), the number of bytes required for each instruction is determined and the PC is incremented accordingly. As each label or symbol is encountered, its address is stored away in the symbol table but nothing is printed (except errors if they occur). During the second reading (pass 2), the values of these labels are inserted in the object code and the offsets are calculated by the assembler for each branch instruction. It then prints the assembly listing. It also produces the object code in a form that allows it to be entered into the microcomputer.

33.4.9 Error indication

If the source program is not written in accordance with the rules specified for each type of statement, error lines will be printed. Note that each error has a number. The user's manual provides an explanation of each type of error as an aid to debugging. The most common causes of errors are improper spacing or the use of illegal characters.

Most of the errors are printed during pass 1 and allow the operator to abort the assembly to make corrections. If there are only a few errors and they are not serious, the operator may elect to let the assembler complete the listing anyway, so an object machine code file can be obtained and tested. Usually, other revisions are also required. In addition to the printing of an error number, which identifies the type of error, the original line is reprinted (unformatted) so the programmer can find any mistake.

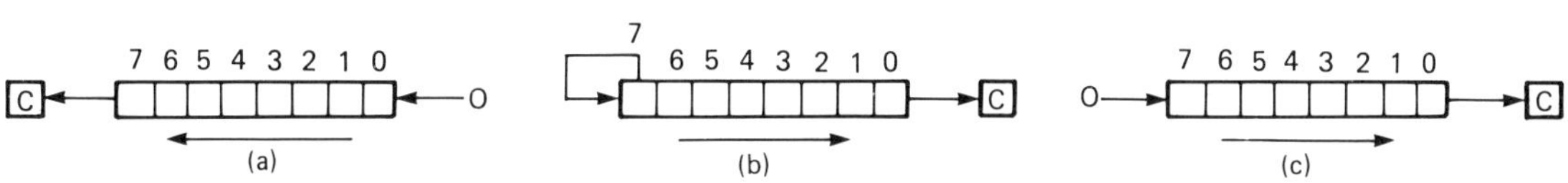

Figure 33.7 The 6800 shift instructions: (a) arithmetic shift left (ASL); (b) arithmetic shift right (ASR); (c) logical shift right

Many times the error is not obvious, and the rules may have to be reviewed before the reason is found.

A summary schematic of what an assembler does is shown in *Figure 33.8*.

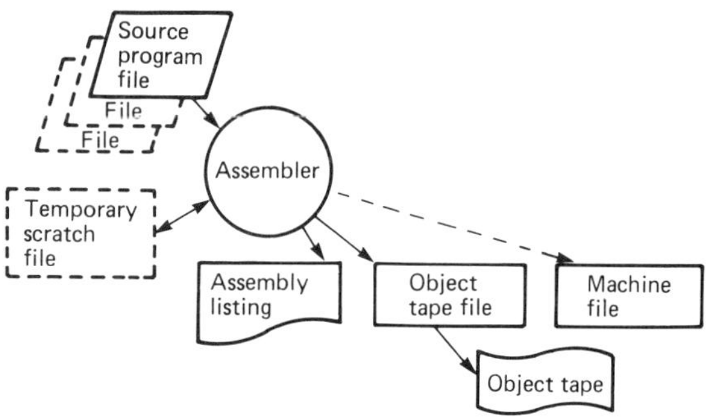

Figure 33.8 Assembler operation summary

33.5 Interrupts

One of the most important features of a microprocessor is its ability to control and act on feedback from peripheral devices such as line printers or machinery controllers. It must be able to sense the operation of the system under its control and respond quickly with corrective commands when necessary.

When conditions that require fast response arise, the system is wired so as to send a signal called an interrupt to the microprocessor. An interrupt causes the microprocessor to stop execution of its main program and jump to a special program, an interupt service routine, that responds to the needs of the external device. The main program resumes when the interrupt service routine is finished.

Important aspects of the 6800 interrupt structure are the stack concept, the use of vectored interrupts, and the interrupt priority scheme provided by the microprocessor logic.

The stack is an area in memory pointed to by the stack pointer (SP) register. The stack has three basic uses:

(1) to save return addresses for subroutine calls;
(2) to move or save data;
(3) to save register contents during an interrupt.

Use of the stack during interrupts is discussed in this chapter.

The 6800 uses four different types of interrupts: reset (RST), non-maskable (NMI), software (SWI), and hardware interrupt request (IRQ). Unique interrupt servicing routines must be written by the system designer for each type of interrupt used, and they can be located anywhere in memory. Access to the routines is provided by the microprocessor that outputs a pair of addresses for the appropriate interrupt. The two locations addressed must contain the address of the required interrupt service routine.

It should be noted that the actual ROM (or PROM) accessed may appear to be at some lower address as long as it also responds to the addresses of the vectors. When one of the four types of interrupts occurs, the microprocessor logic fetches the contents of the appropriate two bytes and loads them into the program counter. This causes the program to jump to the proper interrupt routine. The fetched addresses are commonly called vectors or vector addresses since they point to the software routine used to service the interrupt.

Three of the interrupts (RST, NMI, and IRQ) are activated by signals on the pins of the microprocessor, and the fourth (SWI) is initiated by an instruction. Each of these interrupts have similar, but different sequences of operation and each will be described.

33.5.1 Reset (RST)

A reset is used to start the program. A LOW pulse on the RESET pin of the microprocessor causes the logic in the microprocessor to be reset and also causes the starting location of the program to be fetched from the reset vector locations.

The RESET line is also connected to any hardware devices that have a hardware reset and need to be initialised, such as the PIA. Grounding of the RESET line clears all registers in the PIA, and the restart service routine reprograms the control and direction registers before allowing the main program to start. The ACIA has no RESET pin and depends entirely on software initialisation. All software flags, or constants in RAM, must also be preset. If the system, includes power-failure sensors and associated service routines, additional steps will be needed in the RESET service routine to provide the automatic restart function.

33.5.2 The IRQ interrupt

The IRQ interrupt is typically used when peripheral devices must communicate with the microprocessor. It is activated by a LOW signal on the IRQ pin (pin 4) of the microprocessor. Both the ACIA and PIA have IRQ pins that can be connected to the microprocessor when desired. Even though this line is pulled LOW the IRQ interrupt does not occur if the I bit of the Condition Code (CC) register is SET. This is known as masking the interrupt. The I bit is SET in one of three ways:

(1) by the hardware logic of the microprocessor as a part of the restart procedure;
(2) whenever the microprocessor is interrupted;
(3) by an SEI (set interrupt mask) instruction.

Once SET, the I bit can be cleared only by a CLI (clear interrupt mask) instruction. Therefore, if a program is to allow interrupts, it must have a CLI instruction near its beginning.

33.5.2.1 Interrupt action

When an interrupt is initiated, the instruction in progress is completed before the microprocessor begins its interrupt sequence. The first step in this sequence is to save the program status by storing the PC, X, A, B, and CC registers on the stack. These seven bytes are written into memory starting at the location in the stack pointer (SP) register, which is decremented on each write. When completed, the SP is pointing to the next empty memory location. The microprocessor next sets the interrupt mask bit (I), which allows the service program ro tun without being interrupted. After setting the interrupt mask, the microprocessor fetches the address of the interrupt service routine from the IRQ vector location and inserts it into the PC. The microprocessor then fetches the first instruction of the service routine from the location now designated by the PC.

33.5.2.2 Nested interrupts

Normally an interrupt service routine proceeds until it is complete without being interrupted itself, because the I flag is SET. If it is desirable to recognise another IRQ interrupt (of higher priority, for example), before the servicing of the first one is completed, the interrupt mask can be cleared by a CLI instruction at the beginning of the current service routine. This allows 'an interrupt of an interrupt,' or nested interrupts. It is handled in the 6800 by storing another sequence of registers on the stack. Because of the automatic decrementing of the stack pointer by each interrupt, and subsequent incrementing by the RTI instruction when an interrupt is completed, they are serviced in the proper order. Interrupts can be nested to any depth, limited only by the amount of memory available for the stack.

Return from interrupt (RTI)

The interrupt service routine must end with an RTI (return from interrupt) instruction. It reloads all the microprocessor registers with the values they had before the interrupt and, in the process, moves the stack pointer to where it was before the interrupt. The RTI essentially consists of seven steps that write the contents of the stack into the microprocessor registers and an additional one that allows the program to resume at the address restored to the PC (which is the same place it was before the interrupt occurred). Note that the RTI restores the CC register as it was previously, and interrupts will or will not be allowed as determined by the I bit.

33.6 Development tools

Because the microprocessor or microcomputer is a programmable device requiring suitably generated machine code to cause it to function in a desired manner, development tools are available to generate this code. The range of tools available today spans the low cost evaluation boards, medium priced single user systems usually having high level language capabilities as well as assemblers and in-circuit emulators, up to multi-user stations with hard disk stores, Pascal compilers and cross software and hardware development stations to support a range of microprocessors.

Generally speaking, the greater the cost of the development system, the more powerful it is in terms of monitoring, editing and debugging facilities; ease of operator interface; flexibility in program, storage and retrieval; and variety of high level languages available.

Using the more sophisticated tools will usually cut down development time, whilst low cost evaluation boards are often used in education and for first time users to become acquainted with a new device.

Today development tools are supplied by the semiconductor manufacturers and some of the instrument manufacturers. The tools from the instrument houses support a range of products from various semiconductor vendors, whilst the semiconductor suppliers are generally able to offer more comprehensive and up to date support for their own devices and future product enhancements. Many software houses offer support for various devices on a range of manufacturers' hardware.

A few example of Motorola tools are described here to show what can be acquired.

33.6.1 The Motorola MEK6802D5 microcomputer evaluation board

This is a basic, low cost system to evaluate the capability of the MC6800 microcomputer family of components. It is useful as a tool to learn programming techniques, and to develop custom applications for microprocessors.

The system consists of an MC6802 microprocessor, D5 BUG firmware in a 2K byte ROM, 1K byte of RAM, 7 segment multiplexed displays, a 25-key scanned keyboard and a 300 baud Kansas City cassette interface, uncommitted user peripheral interface adapter, wire wrap area and on board power supply.

33.6.2 The Motorola Exorset 100 system

This is based on the very powerful 8-bit microprocessor MC6809. It has two floppy disks providing 320K bytes of storage and 56K bytes of dynamic RAM. There is a full ASCII keyboard with 16 user definable function keys, a 12 inch CRT screen with 22 lines of 80 characters or 16 lines of 40 characters with a 320×256 dot-matrix for graphic display which can be overlaid by the alpha numeric display. Position-independent self-test programs on diskette can be transferred to EPROM on the system main board. They include RAM diagnostics, keyboard test, serial port test and disk read/write tests. Disk operation and file management are handled by the XDOS operating system which is fully compatible with the Exorciser and MDOS system. The programming language is BASICM (and extended function interface compiler, designed for microprocessor applications). A trace function is provided as a program test aid. Pascal can also be provided. A software communications program for file transfer between Exorciser and Exorset is also available. The Exorset is compatible with a large range of micromodules which can be used to adapt the system to specialised tasks. Powerful statements and data types (bit access, byte integer, 31 character stringing, and real data types with digit accuracy options, real time capabilities, matrix operations) are provided as are easy control of special peripheral circuits and generations of a ROMable re-entrant and position independent code that can execute ultimately on the application with a runtime ROM package. Compatible in-circuit emulators for a variety of Motorola microprocessors can connect to the Exorset from a variety of suppliers.

33.6.3 Motorola Exormacs for 16-bit and 32-bit microprocessor

The Exormacs is a development system for designing and developing advanced 16-bit microprocessor based systems using Motorola families of microprocessors, microcomputers, and peripheral parts. It is also ideally suited for developing applications using the Versamodule family of 16-bit borard level application products and accessories.

Exormacs is a third generation development system providing support for Motorola's M68000 microprocessor family. The chassis houses the development system card complement, a switching power supply, cooling fans and a front control panel. The internal card cage accommodates 15 cards. A switching power supply handles power needs for all system configurations. The front panel is unique in that it provides the user with a manual control input capability as well as visual indication of current system status and existing error conditions.

In the first card slot resides the debug module containing macsbug firmware, bus arbitration logic, a parallel printer port and two RS-232C terminal ports.

The second card slot of the chassis is the MPU module. This contains the MC68000 MPU chip, its clock system, a four-segment memory management unit (MMU), primary and secondary map switching logic, and firmware that provides module diagnostics. The MMU allows the system to allocate memory under control of the Versados operating system, and provides multitasking operation. This realtime multitasking operation system helps speed programme development by allowing concurrent tasks. For example, when an assembly is in progress using the printer, the CRT console is free for program editing.

Two 128K byte dynamic memory modules provide Exormacs with 128K 16-bit words of RAM which include byte parity. Parity is read during memory access, providing the MC68000 MPU with soft error status such that a memory re-try may be initiated. The base address may be set by the user through switch inputs. The chassis can support up to eight RAM modules providing the user with a megabyte of directly addressable resident memory.

The universal disk controller supplied with the M68KMACSH2 supports up to two 32 megabytes hard disk subsystems and up to two megabytes of floppy disk mass storage. An MC68120 microcomputer intelligent peripheral controller is used to handle data requests from the M68000 system and provides self contained module diagnostics for the disk system. The use of this multiprocessing technique offers increased system performance which results in more efficient utilisation of multiple users time. The industry standard SMD interface is used and provides for a 1.25 megabytes/second transfer rate. It requires two Exormacs.

The MC68000 software deve.opment package provides the user with a cost-effective and efficient operating system, structured macroassembler, linkage editor, CRT text editor, symbolic debugger and PASCAL compiler.

33.6.4 Typical development cycle

The development system enables the engineer to construct a working model of a prototype system. This is accomplished by the ability of the system to emulate user system hardware and to debug the interface of external devices and user software. Rather than immediately designing and building a prototype system, the engineer sets up the development system to functionally represent the system using plug-in modules. The process is depicted in *Figures 33.9* and *33.10*.

(a) Define the system and determine the software and hardware functions to be performed.

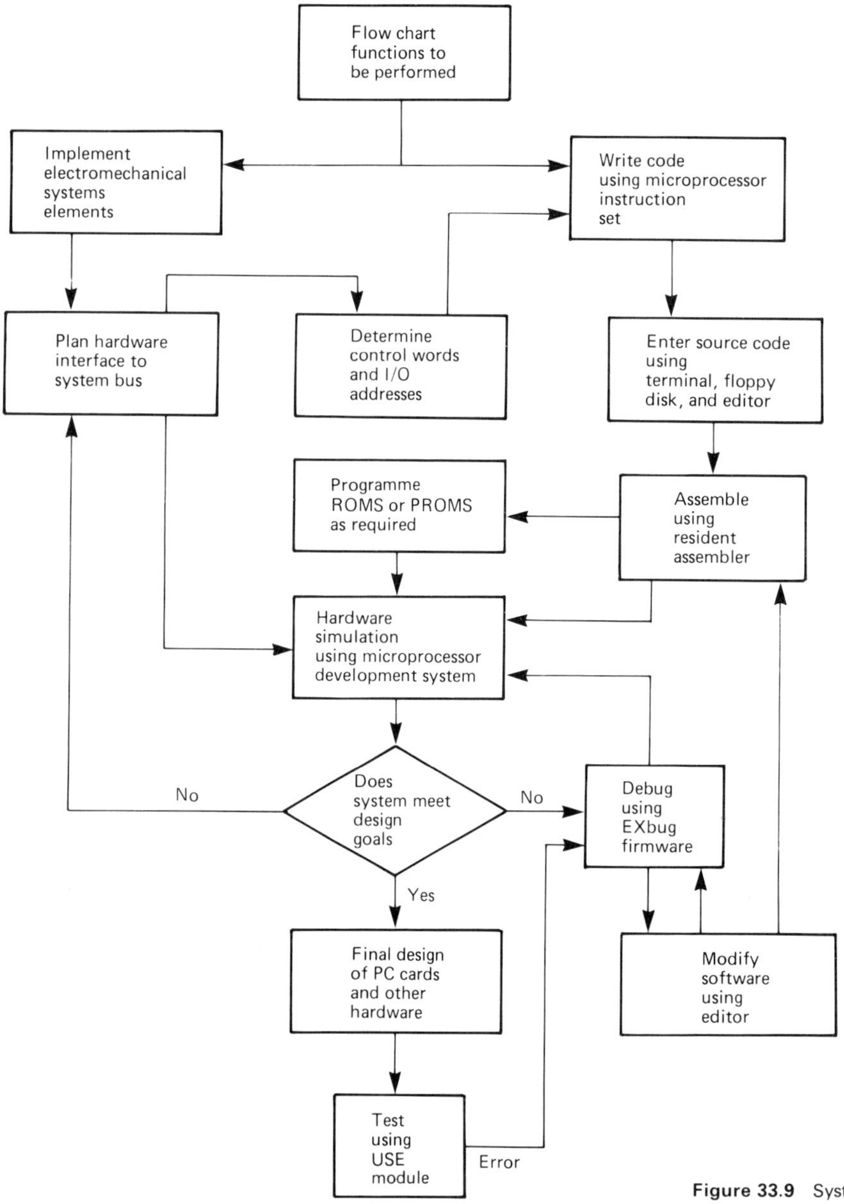

Figure 33.9 System designing and verifying procedure

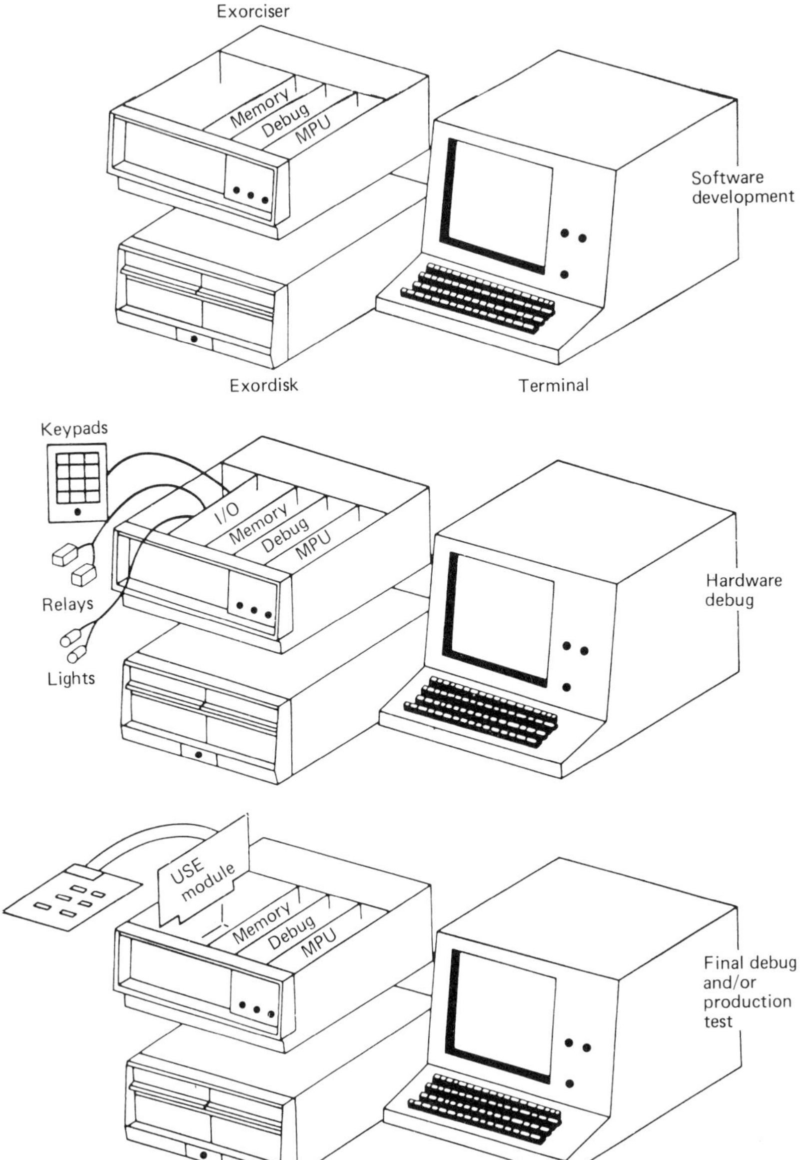

Figure 33.10 Exorciser, the development tool

(b) Set up the Exorciser to emulate the system hardware and if required build any special hardware interface.
(c) Prepare software programs on the Exorciser after testing the hardware required to accomplish the intended function.
(d) Load the software into the Exorciser and using the 'Exbug' firmware, debug both the hardware and the software.
(e) Design and build a pre-production model of the system.
(f) Combine the pre-production hardware with the user system software and using the Exbug firmware and user system emulator* (USE) module, debug and make any hardware and software adjustments.
(g) Extensively test and evaluate the system in an actual working environment. At this point the program may be stored in PROM or EPROM using the optional PROM programmer module.
(h) Build the production hardware and the ROM's. The USE* can be used to test and debug the production systems.
(i) Combine the production hardware with the system software and make final adjustments.
(j) Release the system to production.
(k) Analyse problems in production hardware using the Exorciser and USE.

* Firmware a 'debug' and 'monitor' program resident in ROM within the development system. USE. The user system evaluator is an optional module with an interconnecting cable (and buffer) which plugs into the processor socket of the user system such that all the debug capabilities of the Exorciser are usable in the user hardware.

33.7 Choosing a microprocessor

Of primary importance in the choice of a microprocessor is the cost of the device to be used and the cost of developing prototypes and production volumes. The cost of a device is inversely related to the volume of product used and the age of the device since its first introduction. Most manufacturers today supply families of compatible devices for applications of varying volumes. For prototyping and production volumes up to say 1000 per annum, one would use either microprocessor units with program stored in external EPROM or microprocessors with their own EPROM on chip. For volumes greater than 1 to 2 thousand per annum, one would freeze the software into the silicon by use of on-chip ROM. For specialist high volume applications, say greater than 500 thousand per annum, special devices are designed providing extra features such as phase-locked loop, analogue-to-digital conversion, display driving, power fail back-up, random access memory on chip.

Other primary considerations are:

(a) Availability of compatible interface circuits, such as cathode ray tube controllers, floppy disk drivers, dedicated input/output devices and special maths chips like floating point arithmetic.
(b) The way the device is packaged, such as plastic dual in-line for high volume low cost, chip carriers for applications with limited physical space, ceramic for high reliability and high humidity and temperature.
(c) The power consumption of the device in both the running and standby mode. This is a function of the device technology used in manufacture. CMOS devices tupically consume one hundredth the power of equivalent NMOS or HMOS devices.

Normally of secondary importance are:

(a) The architecture of the microprocessor, such as 4-bit, 8-bit, 16-bit, 32-bit.
(b) Availability of support software such as high level language compilers, real-time operating systems and application packages.
(c) Existence of good support tools for program writing in assembler, and for end system emulation and debugging both hardware and software. A lack of such tools could of course cause a major problem in affecting the development time scale and hence the cost of a project.
(d) Good manufacturer support in the form of available training courses, data sheets, programming manuals, and worked examples. Some manufacturers of microcomputers supply worked software examples and standard routines to short-cut the design process. Often consultancy support programs and user group libraries exist to assist the system designer.
(e) An awareness of the long term plans of device family enchancement to ensure that designs can be sensibly and cost effectively upgraded.
(f) Existence of second sources to support the prime manufacturer.

Microprocessors are often supplied as board level products. Examples are single board computers, memory modules, universal disk controllers, intelligent peripheral controllers, Winchester disk controllers, standard input/output boards, computer grade interface modules and semiconductor memory to replace the traditional hard disks. The system designer may find it more cost effective to interface these standard boards rather than to design his own product at the component level. Equally important as the boards themselves, is the availability of back-up software—operating systems and utilities to support them.

Further reading

BAUER, J. R., 'Alterable microprocessors: tailoring chip design to meet applications', *Digital Design* (August 1982)

CLEMENTS, A., 'An introduction to bit-slice microprocessors', *Electronics & Power* (March 1981)

CUSHMAN, R. H., 'CMOS microprocessor and microcomputer ICs', *EDN* (September 1982)

DONNELLY, W., 'Single chip microcomputers and customised LSI', *Electronics Industry* (March 1982)

GRAPPEL, R. D., 'Design powerful systems with the newest 16-bit microprocessors', *EDN* (September 1982)

GRAPPEL, R. D., 'Instruction-set power makes a microprocessor's job easier', *EDN*, November 10 (1982)

JOHNSON, R. C., 'Operating systems hold a full house of features for 16-bit microprocessors', *Electronics*, March 24 (1982)

34 Electron Valves and Tubes

M J Rose, BSc(Eng)
Mullard Ltd
(Sections 34.1, 34.6, 34.7.1, 34.10, 34.11, 34.12)

A P O Collis, BA, CEng, MIEE
Manager, Broadcast Tubes Division,
English Electric Valve Co. Ltd
(Section 34.2)

G T Clayworth, BSc
Manager, High Power Klystron Dept.,
English Electric Valve Co. Ltd
(Section 34.3)

F J Weaver, BSc, MIEE
English Electric Valve Co. Ltd
(Section 34.4)

P M Chalmers, BEng, AMIEE
English Electric Valve Co. Ltd
(Section 34.5)

E W Herold, BSc, MSc, DSc, FIEE
(Sections 34.7-2 34.7.5)

W E Turk, BSc, CEng, FIEE, FRTS, MInstM
Manager, Phototube Marketing,
English Electric Valve Co. Ltd
(Section 34.8)

L W Turner, CEng, FIEE, FRTS
Consultant Engineer
(Section 34.9)

Contents

34.1 Small valves

34.1.1 Triode

The triode contains three electrodes: a grid is placed between the anode and cathode. The circuit symbol for a triode is shown in *Figure 34.1*.

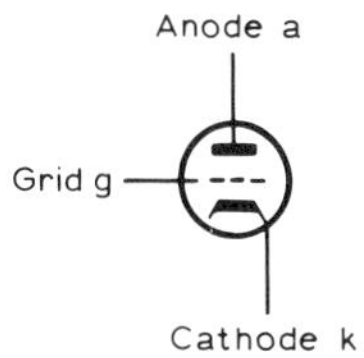

Figure 34.1 Circuit symbol for triode

The grid is in the form of a fine wire spiral surrounding the cathode. Because the grid is closer to the cathode than is the anode, small changes in grid voltage V_g have a significant effect on the electron flow and hence on the anode current. If the grid is highly negative with respect to the cathode, all the electrons from the cathode are repelled and no anode current flows. If the grid is highly positive, all the electrons are attracted to the grid and again no anode current flows. If no voltage is applied to the grid, the triode acts as a diode. This is shown by the $V_g = 0$ curve in the anode characteristic of *Figure 34.2*.

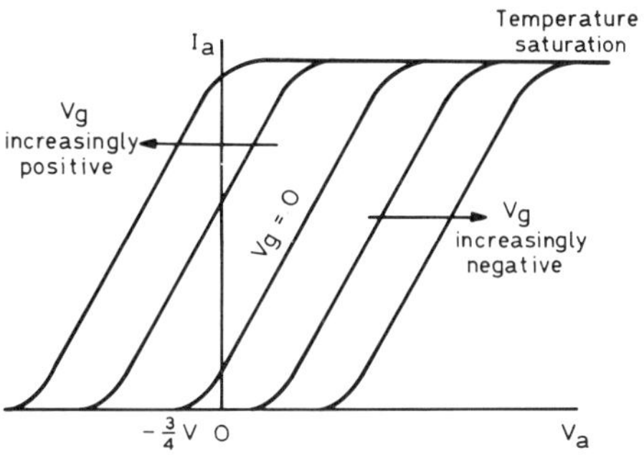

Figure 34.2 Anode characteristic of triode

If the grid is slightly positive, the electrons from the cathode are accelerated towards the anode and the triode starts to conduct at an anode voltage more negative than $-\frac{3}{4}$ V. This is shown by the curve to the left of the $V_g = 0$ curve in *Figure 34.2*. If the grid is slightly negative, the anode voltage has to overcome the effect of the grid voltage and a larger anode voltage is required before conduction starts. This is shown by the curves to the right of the $V_g = 0$ curve in *Figure 34.2*. Because the triode is never operated in circuits with a negative anode voltage, the practical anode characteristic is shown in *Figure 34.3*.

The triode has three principal characteristics. Besides the anode characteristic just described, there are the grid or mutual characteristic, the change of anode current with grid voltage with anode voltage as parameter, and the constant-current characteristic, the change of anode voltage with grid voltage with anode current as parameter. The mutual characteristic is shown in *Figure 34.4* and the constant-current characteristic in *Figure 34.5*.

The slopes of the three characteristics are used to give the small-signal parameters defining the performance of the triode.

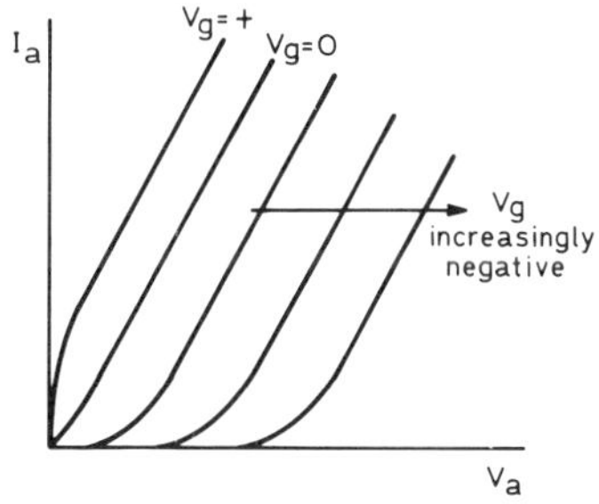

Figure 34.3 Practical anode characteristic of triode

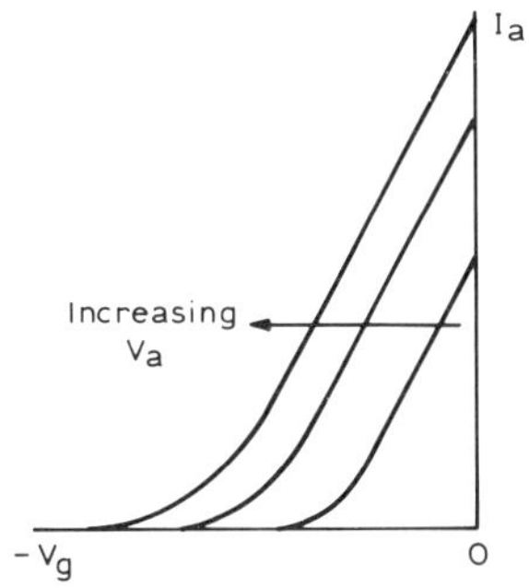

Figure 34.4 Mutual characteristic of triode

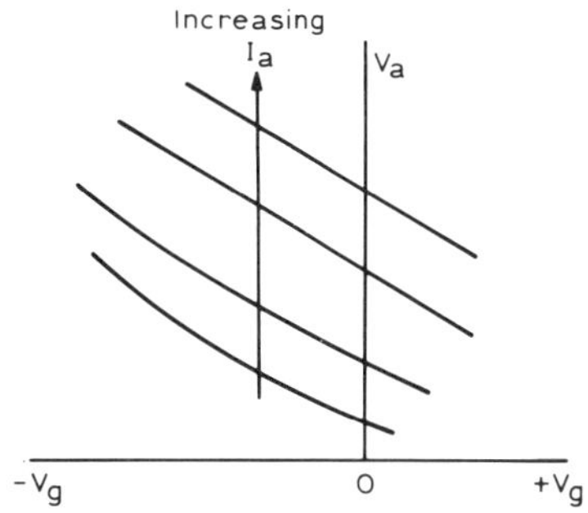

Figure 34.5 Constant-current characteristic of triode

Because the slopes vary over the length of the characteristics, the value of the parameter is qualified by the operating point at which it applies. The slope of the mutual characteristic defines the mutual conductance g_m of the triode, also called the slope. The slope of the constant-current characteristic defines the amplification factor μ, while the reciprocal of the slope of the anode characteristic defines the anode slope resistance r_a.

34.1.2 Tetrode

The performance of a triode is limited by the interelectrode capacitances. The grid-to-anode capacitance can be reduced by placing a second grid, the screen grid, between the grid and the anode. The valve now becomes a tetrode; the circuit symbol is shown in *Figure 34.6*.

The control grid is operated at a low negative voltage, typically up to -10 V, while the screen grid is held at a moderate positive voltage, typically 80 V, to accelerate the electrons from the

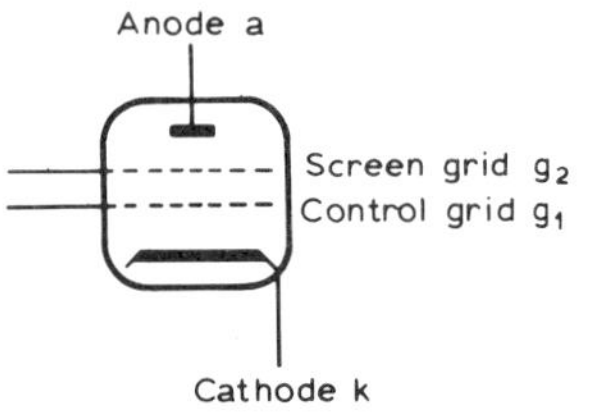

Figure 34.6 Circuit symbol for tetrode

cathode towards the anode. If the anode voltage is zero, all the electrons flow to the screen grid and a high screen-grid current flows. If the anode voltage is increased but is still lower than the screen-grid voltage, anode current flows and the screen-grid current falls. If the anode voltage is increased further so that it is higher than the screen-grid voltage, the anode current increases further while the screen-grid current is further reduced. This idealised operation of the tetrode is shown by the characteristic of anode current I_a and screen-grid current I_{g2} plotted against anode voltage in *Figure 34.7(a)*.

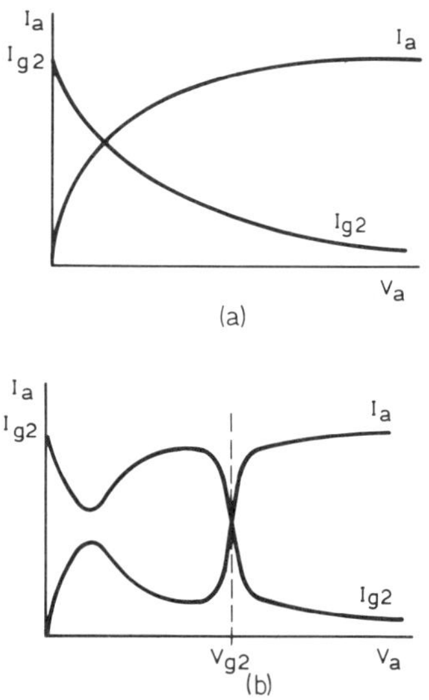

Figure 34.7 Anode current and screen-grid characteristics of tetrode: (a) ideal characteristics neglecting secondary emission; (b) practical characteristics

The operation is ideal because it neglects the effects of secondary emission. The effect of secondary emission is to produce kinks in the characteristic as shown in *Figure 34.7(b)*.

The anode characteristic of the tetrode can be varied by the control-grid voltage V_{g1} to give a family of curves. The mutual characteristic of the tetrode, the variation of anode current with control-grid voltage with screen-grid voltage constant, is similar to that of the triode.

34.1.3 Pentode

The secondary emission can also be suppressed by inserting a third grid into the valve, the suppressor grid. This grid is wound to a larger pitch than the others, and is placed between the screen grid and anode. The suppressor grid is held at or near cathode potential, and is sometimes connected inside the valve envelope

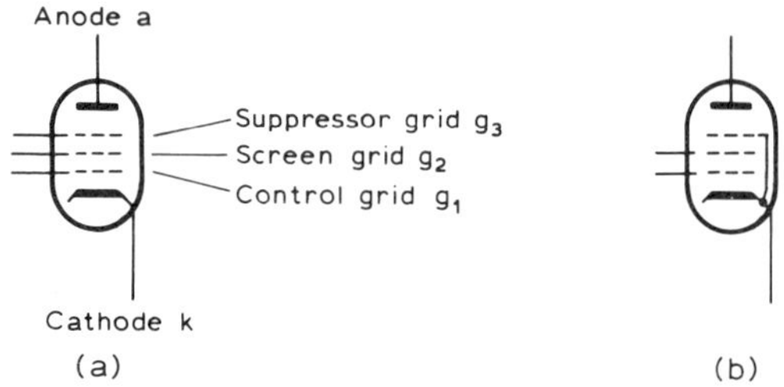

Figure 34.8 Circuit symbol for pentode: (a) separate suppressor grid; (b) suppressor grid connected to cathode

to the cathode. The valve now becomes a pentode; the circuit symbol is shown in *Figure 34.8*.

The control grid is operated at a low negative voltage and the screen grid at a moderate positive voltage as in the tetrode. If the anode voltage is lower than the screen-grid voltage, the screen grid acts as an anode and a high screen-grid current flows. As the anode voltage is increased, the electrons from the cathode flow to the anode because the large pitch of the suppressor grid has little effect on primary electrons. Any secondary electrons from the anode, however, are repelled by the suppressor grid and return to the anode. There is therefore a sharp rise in anode current and a corresponding fall in screen-grid current, as shown in *Figure 34.9*. The effect of control-grid voltage on the anode characteristic is shown in *Figure 34.10*, while the effect of screen-grid voltage on the mutual characteristic is shown in *Figure 34.11*.

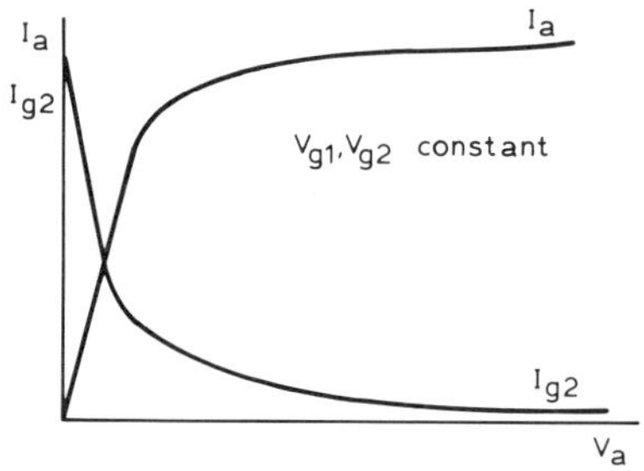

Figure 34.9 Anode current and screen-grid current characteristics of pentode

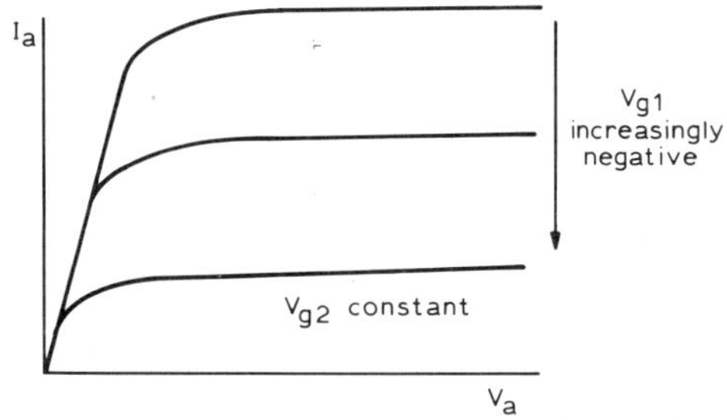

Figure 34.10 Anode characteristic of pentode with control-grid voltage as parameter

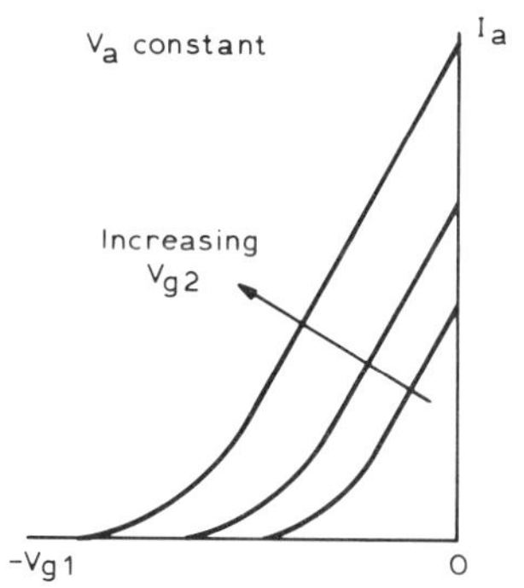

Figure 34.11 Mutual characteristic of pentode with screen-grid voltage as parameter

34.2 High power transmitting and industrial tubes

The earliest vacuum tubes had glass envelopes with an internal anode cooled only by the radiation of heat through the glass envelope. This limited the maximum anode dissipation to a few kilowatts before the glass became too hot. The development of a method of making large diameter glass to metal seals overcame this limit by making it possible for the anode to be part of the vacuum envelope. Its surface could then be easily and efficiently cooled.[1] This opened the way for the development of the modern high power triode and tetrodes and although glass has been replaced by ceramics the principal remains the same. Tubes are manufactured with r.f. power outputs ranging from several kilowatts to over 2 MW continuous rating.

The applications of these tubes can be divided into two chief categories of roughly equal importance. One category includes communications, radio and television broadcasting where the tubes are used almost exclusively as power amplifiers. Modern transmitters use tetrodes because, with their high gain the minimum number of stages are needed, and solid state amplifiers can easily supply the drive power. The second category covers industrial applications where it is usual to use triode oscillators which are simple and reliable, such as in induction heating and dielectric heating of non-conductors.

34.2.1 Construction of tubes

A vacuum tube consists of: (a) the cathode which is heated to a high temperature and is the electron source; (b) one or more grids which closely surround the cathode and control the flow of electrons to the anode; and (c) the anode on which the electrons are collected. Each of these electrodes has its own special requirements and particular problems (*Figure 34.12*).

34.2.1.1 Cathode

The properties required of the cathode are that it should be mechanically stable at high temperatures, that it should be a good emitter of electrons and that this emission should be stable and last a long time.[2,3]

In the lastest designs of high power tube the cathode is cylindrical in shape and made from a mesh of fine wires of tungsten to which 1–2% of thorium oxide has been added. The mesh is directly heated to the required operating temperature of 2000 K by passing current through each wire. This design has been developed to overcome the problem of distortion caused by the stresses which occur each time the filament is 'heat-cycled' from room temperature to its working temperature.

Figure 34.12 (a) A water cooled anode capable of dissipating 120 kW; (b) complete electrode assembly for the BW1185J2 (Photo by courtesy of English Electric Valve Company Limited)

To obtain a stable emission the filament is heated in a hydrocarbon atmosphere during manufacture in order to change the outer surface of the wire to tungsten carbide. The thickness of this layer is limited to the equivalent of 20–30% of the cross-sectional area. If the degree of carburisation is greater than this the wires become brittle and there is a risk of them breaking. At the working temperature thorium produced by the reduction of the thorium oxide diffuses to the surface of the carbide to form an emissive layer. During running, some of the throium is lost by evaporation and some is also lost as the result of bombardment by positive ions which are formed by the collision of high energy electrons with the molecules of the residual gases left in the tube after evacuation. Therefore the residual gas pressure must be kept as low as possible and the filament must be run sufficiently hot for the rate of thorium diffusion to be fast enough to keep the emissive surface replenished. If this is not done the emission will fail. At the same time, however, carbon from the carbide is also being lost partly by evaporation, and partly by diffusion into the core of the filament wire. When the carbide has completely diffused and no longer exists as a distinct layer, the emission fails.

This process cannot be reversed. The rate of carbon diffusion like that of thorium is a rapid function of temperature, and while the filament temperature must be high enough to maintain the emission it must not be too high or the rapid diffusion of carbon will drastically shorten the life. The importance of this is illustrated by the fact that a 10% rise in filament voltage will increase the operating temperature by about 30 K, doubling the rate of carbon diffusion and halving the life.

34.2.1.2 Grids

The characteristics of the tube are determined by the geometric dimensions of the grids and its distance from cathode and anode.

The grid and anode currents may be calculated using appropriate formula[4] although computer programs are now often used where either greater accuracy or information on the electron trajectories is required.[5]

The grids must have good mechanical stability at high temperatures but, unlike the cathode, they must have low thermionic emission and low secondary emission.

The grid is heated not only by radiation from the very hot cathode, but when it is driven positive it will attract electrons to itself producing a grid current, which dissipates power (*Figure 34.13(a)*). The combined effect results in the grid being heated to between 1000 and 1500°C and it may itself emit thermionic electrons which are attracted to the neighbouring electrodes (*Figure 34.13(b)*). The impact of the incident or primary electrons in the grid may knock electrons from the surface (*Figure 34.13(c)*).

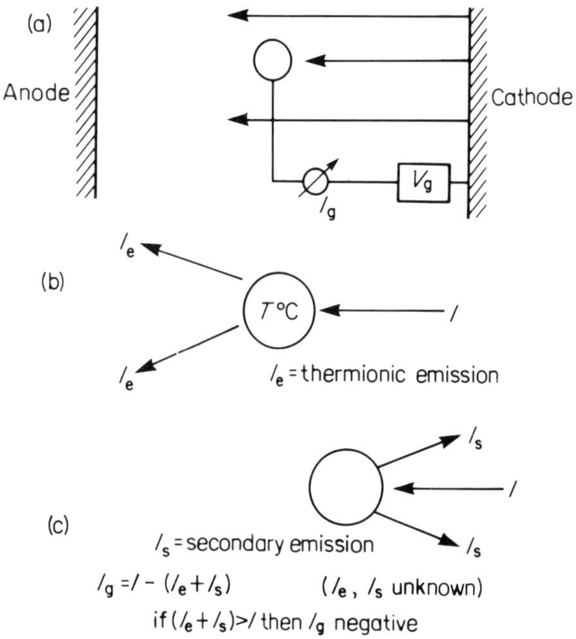

Figure 34.13 Diagram showing components of the grid current

These secondary electrons will be collected on nearby electrodes. The grid current measured externally to the tube therefore has three components: the incident current I, the thermionic current I_e and the secondary electron current I_s, the last two being in the opposite direction from the first.

In general therefore the measured current is less than the incident current and in extreme cases may be negative. The variation from tube to tube, the change during life, plus the possibility of negative currents and hence negative resistance, make circuit design difficult and it is preferable to reduce the effect until it is negligible.

For reasons of mechanical stability and hot strength, metals such as molybdenum and tungsten must be used for making the grid, but these are good thermionic emitters particularly if covered with thorium evaporated from the cathode. A reduction in grid emission can be obtained by coating the grid with a high work function metal such as platinum and this is often done. However platinum has a high secondary emission coefficient. In addition if the grid becomes too hot metallurgical changes occur due to diffusion between the platinum and the core metal, which renders the suppression of thermionic emission ineffective and causes distortion.

One method of solving this problem is to provide a barrier layer between the platinum and core which is done in the so-called K grid.[5,7] This is made by sintering a thin coating of zirconium carbide onto the grid wire at 1900°C and then platinum plating. The black coating gives higher thermal emissivity thus reducing the operating temperature while the rough surface lessens secondary emission. The final result is a grid which can run with a dissipation of 25 W cm^{-2} of surface compared with 7 W cm^{-2} for an ordinary platinum coated grid.

In a more recent development[8] the grid is made from pure carbon, thus making use of that element's desirable properties of high hot strength, low thermionic and secondary emission. A thin shell of carbon, the shape of the final grid, is first made by vapour deposition on to a mandrel at 2000°C. The grid apertures are then machined into the shell either by abrasive machining through a mask or by cutting with a numerically controlled laser machine. High quality grids can be made in this way which can be operated in excess of 50 W cm^{-2}.

34.2.1.3 Anodes

The anode must have good mechanical strength, good electrical and thermal conductivity and is therefore invariably made of copper (*Figure 34.12(a)*). It can be cooled externally in several ways. When determining the power to be dissipated due allowance must be made for the increase due to mismatching or mistuning of the output and other causes.

34.2.2 Air cooling

This is a very convenient and popular method of cooling but because of the size and weight of the radiator and problems of noise from the air ducting and blowers it is difficult to make a forced air cooled valve with an anode dissipation greater than 50 kW. The temperature distribution over the cooling fins and anode surface can be calculated approximately for a given dissipation and airflow by using a method described by Mouromsteff.[9] The determination of the air flow required is done using the airflow curves given in the manufacturer's tube data such as those shown (*Figure 34.14*). From these the flow and

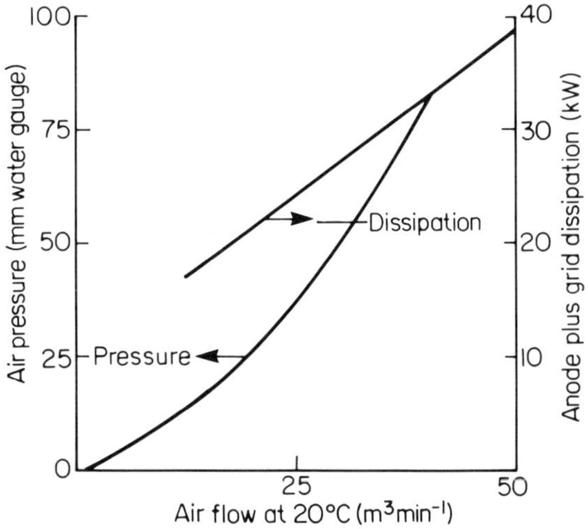

Figure 34.14 Typical air flow characteristics for a forced air cooled anode

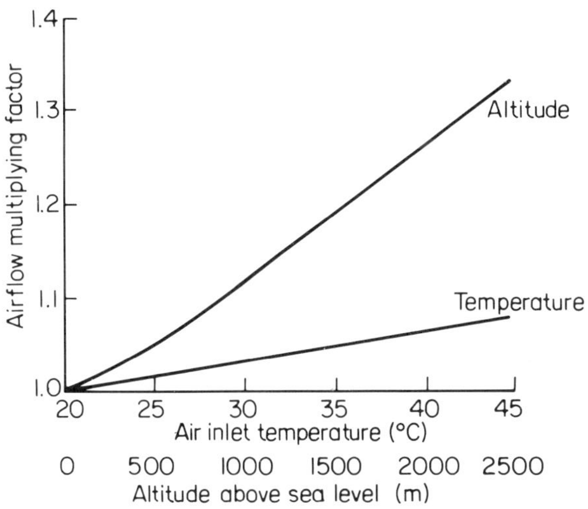

Figure 34.15 Factors to correct air flow for altitude and ambient temperature

pressure head across the radiator are obtained but it may be necessary to correct these for altitude or ambient temperature if these are different from the data (*Figure 34.15*). To the head thus obtained must be added the pressure head in all the other parts of the cooling system including bends, changes in duct size, filters etc. This gives the total pressure head and flow required.[10] Consideration must also be given to the direction of flow. It is recommended that the air should flow from the tube cathode, over the envelope and then through the anode radiator because it means that the cool air will first flow over other components cooling them and improving their reliability.

The cooling air may be sucked through the cabinet, the inside of which will be at a slightly lower pressure than the outside, when air and dust may leak in. The fan will also be blowing air heated to over 100°C during its passage through the anode radiator. Alternatively air may be blown into the cabinet. The inside pressure will then be greater than outside, which will prevent dust getting in, but the opening of a door or the removal of a panel will severely reduce the cooling of the tube. Once a decision has been made a blower can be selected from data supplied by the manufacturer.

On installation of the tube it is advisable to measure the airflow, and this may be done by commercial instruments or running the tube with a known anode dissipation. The temperature rise of the cooling air is measured and then the airflow can be calculated from the relation:

$$\text{airflow} = \frac{0.173 \times \text{inlet temperature} \times \text{total dissipation}}{\text{temperature rise}} \tag{34.1}$$

where airflow is measured in $m^3\ min^{-1}$, temperature and temperature rise in K, and dissipation in kW. Once satisfied that the airflow is sufficient the flow switches should be set to switch off all supplies if the flow is reduced. It is good practise to include a temperature switch to detect if the outlet temperature is too high, indicating excessive anode dissipation.

34.2.3 Water cooling

Similar care to ensure adequate flow is equally important with water cooling. However, there are two important additional constraints. Firstly the tube manufacturer will specify a maximum outlet temperature for the water. If this is exceeded it is possible for local boiling to occur, forming a layer of steam over the surface of the anode and severely reducing the heat transfer. A hot spot will form which may cause the anode to distort or in a severe case the anode may melt.

Secondly if the quality of the water, is not adequate, corrosion and scaling of the anode can occur. The dissolved solids should be less than 30 ppm to prevent scale formation which reduces heat transfer from the anode to the water, restricts the water flow, and may block small water channels used in some tube types. To minimise corrosion the electrical conductivity must also be low, at worst 300 μmho cm^{-1}. It is usual to run the anode at high potential above earth, and the cooling water has therefore to be fed through insulating pipes whose length must be about 1 metre per kV. Leakage currents flowing through the water will erode the jacket or connecting pipes by electrolysis. Special sacrificial anodes may be fitted in the water pipes and arranged so that they are attacked preferentially, but it is important that these are checked regularly and replaced if necessary. Another solution is to run the anode at earth potential despite the increased circuit complexity. If no suitable water supply is available a closed circuit system with heat exchanger must be installed.

Since the water flow may be measured easily it provides a simple method of determining the anode dissipation and efficiency of the tube. The temperature rise and flow rate are measured and the dissipation is given by:

$$\text{dissipation} = \frac{\text{flow} \times \text{temperature rise}}{15} \tag{34.2}$$

where flow is in litre min^{-1} and the other quantities as in equation (34.1).

34.2.4 Vapour cooling[12]

The danger of allowing the water to boil at the surface of the anode has been described above, but it is possible to design an anode where it is quite safe for this to happen. The steam produced is taken via an insulating pipe to a steam condenser, thus utilising the high value of the latent heat of vapourisation to transport the heat (*Figure 34.16*). The first requirement is that

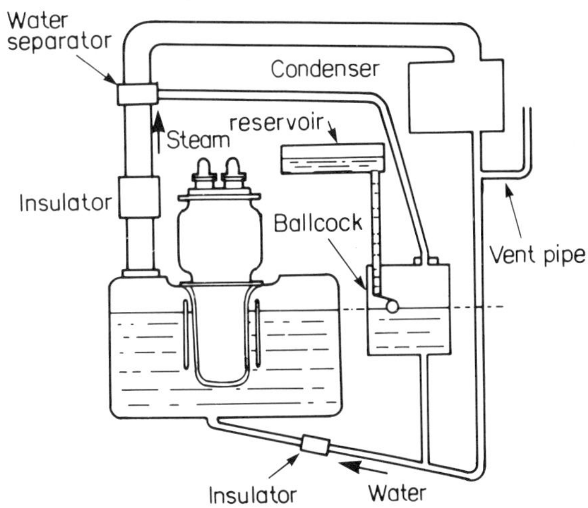

Figure 34.16 Vapour cooling system with external condensor

adequate water must always be maintained in the boiler and that the level must never drop below the specified minimum level even at full dissipation. To achieve this the pressure in the steam pipe must be cancelled by means of a balance pipe returning to the top of the sealed level tank. There is also a pressure fall in the water return pipe due to flow of water back to the boiler and this must be kept as little as possible. Although the flow is only 27 cm^3 kW^{-1} min^{-1} the return pipe should be of generous size. A temperature switch included in the vent pipe will detect any steam blow off due to condenser failure or excessive dissipation. The system must be filled with distilled or deionised water and it is important that it does not become contaminated with oil or grease as this will result in a severe reduction in permissible dissipation by forming a hydrophobic film over the anode surface (*Figure 34.17*). A simpler system particularly suited for industrial

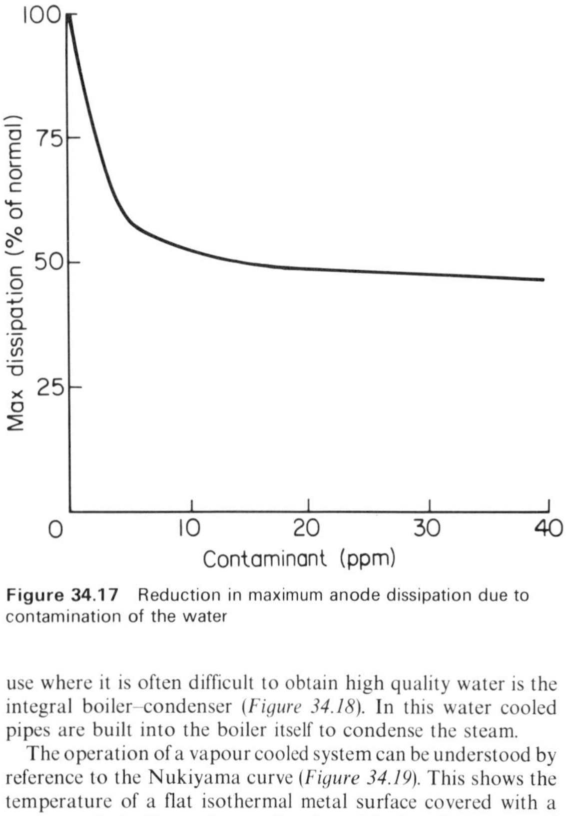

Figure 34.17 Reduction in maximum anode dissipation due to contamination of the water

use where it is often difficult to obtain high quality water is the integral boiler–condenser (*Figure 34.18*). In this water cooled pipes are built into the boiler itself to condense the steam.

The operation of a vapour cooled system can be understood by reference to the Nukiyama curve (*Figure 34.19*). This shows the temperature of a flat isothermal metal surface covered with a coolant at its boiling point as a function of the heat flux expressed in W cm^{-2}. As the flux is increased the temperature of the surface

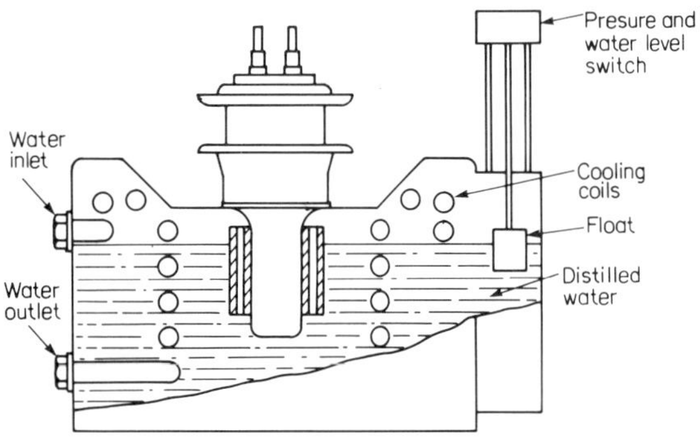

Figure 34.18 Boiler–condenser unit

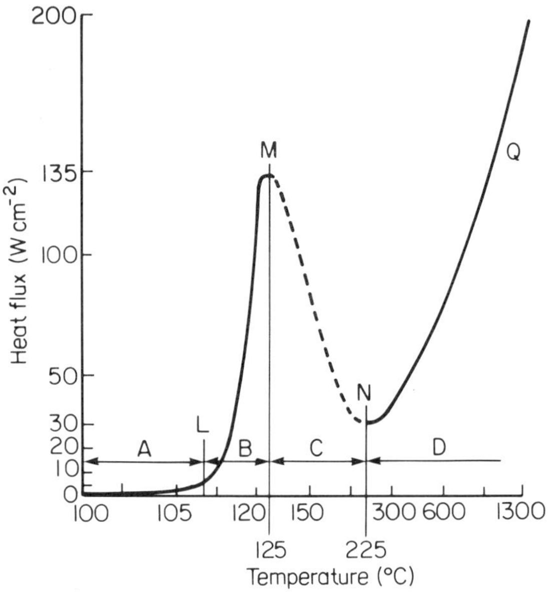

Figure 34.19 Nukiyama curve: variation of surface temperature with heat flux

will follow the curve until point M is reached. Any attempt to increase the heat flux further will force the operating point to jump to Q where the temperature will destroy the anode. Anodes designed for vapour cooling have thick ribs and grooves machined into their surface, and are operated so that, although part of the surface may be in the region MN, the temperature is stabilised by heat conduction to other parts operating in regions ML and A.[11,12] If part of the rib is operating in region A the anode is in thermal equilibrium and an increase in dissipation increases the temperature in that region only.

The Hypervapotron[R] system (trademark of Thomson-Brandt)[13,14] is a recent improvement whereby water cooled tubes may be run with dissipations up to 2 kW cm^{-2}, the water inlet temperature being as high as feasible on the equipment using the tube, while outlet temperature can reach up to 90°C at atmospheric pressure or 100°C to 105°C at 1 or 2 bars pressure. The anodes for this system have grooves at right angles to the water flow formed in the surface. Steam formed in the grooves is ejected at high velocity into the water stream where it is rapidly condensed, fresh water rushing back into the groove. This pulsating action prevents the formation of a steam film and allows very high dissipations to be obtained. The variation of the mean heat transfer coefficient with heat flux plays a crucial role in the operation. It has been found that the temperature at the root of the groove tends to be a constant value of 225–300°C, the heat transfer coefficient increasing to maintain this temperature[13] (*Figure 34.20*). All these systems offer the advantages of a smaller heat exchanger the size of which is governed by the average temperature difference between the primary and secondary circuits and since the steam from a vapour cooled system is at 100°C the heat exchanger can be correspondingly smaller (*Table 34.1*).

34.2.5 Ratings

It is normal practice to express the ratings of high power tubes as absolute values. This means that the equipment designer has to make sure that no limit is ever exceeded, whatever the variation in

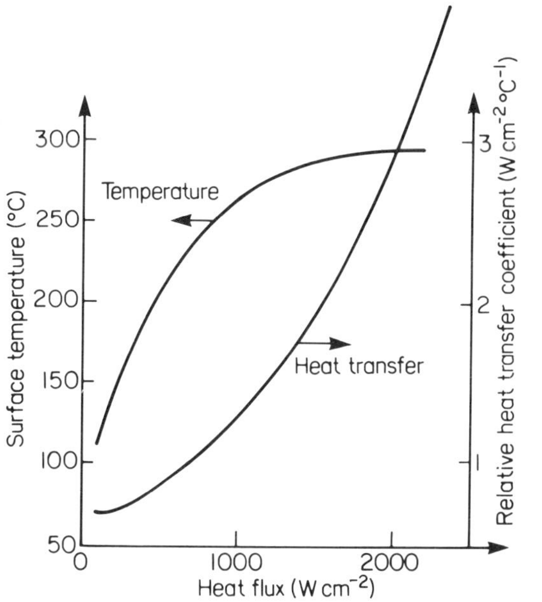

Figure 34.20 Surface temperature and relative heat transfer coefficient for the Hypervapotron system (Trade Mark of Thomson Brandt)

power supply voltage, transients, mistuning, maladjustment or other causes.[15,16]

34.2.6 Calculation of performance

The characteristics of high power tubes are most conveniently shown by constant current curves, where the x and y axes are the instantaneous anode and grid voltages respectively, and the plot is the locus of the values of these voltages which give a constant anode or grid current (*Figure 34.21*). For sine waves the instantaneous grid and anode voltages are:

$$v_g = V_g \sin \omega t + V_{bias} \qquad (34.3)$$

$$v_a = V_a \sin(\omega t + \pi + \phi) + V_{dc} \qquad (34.4)$$

where

V_g = peak r.f. grid voltage

V_{bias} = d.c. grid bias voltage

V_a = peak r.f. anode voltage

V_{dc} = d.c. anode voltage

$\omega = 2\pi \times$ frequency

ϕ = phase angle difference

Coincident values of the voltages v_a and v_g may also be plotted on the constant-current characteristics, and the locus of these points is called the load line, which is in the general case an ellipse. The

Table 34.1 Typical operating conditions for heat exchangers

Method of cooling	*Temperature (°C)*					*Ratio*
	Ambient	*Inlet*	*Outlet*	*Mean*	*Differential*	
Vapour and Hypervapotron ᴿ	45	60	100	80	35	1
Water	45	50	70	60	15	2.3

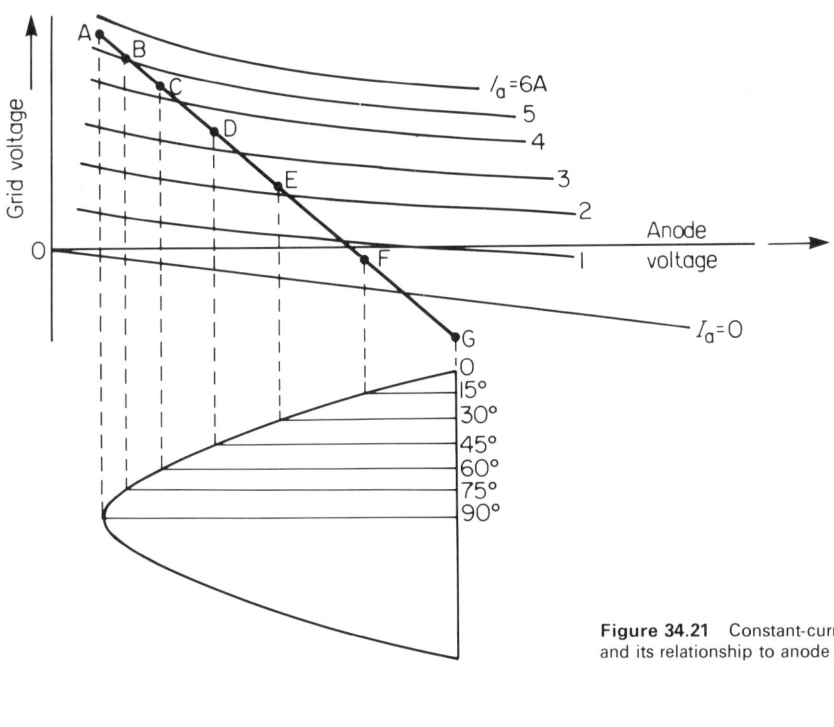

Figure 34.21 Constant-current characteristics showing load line and its relationship to anode and grid voltages

corresponding electrode currents may be read from the characteristics and plotted against time to give the pulses of current taken by the anode and grid. A Fourier analysis may be carried out on these to give the d.c., fundamental r.f. and harmonic components from which the power output and load impedance may be calculated. The analysis can be greatly simplified if the times at which the current values are found are suitably chosen. This is particularly so if the load is a pure resistance, i.e. ϕ is zero in equation (34.4), when the load line becomes a straight line.

Once the load line has been fixed the values of grid and anode currents at, say, 15° intervals may be read from the curves with the aid of a transparent overlay (*Figure 34.22*).[18,19] Approximate design factors for various operating classes are given in *Table 34.2*.

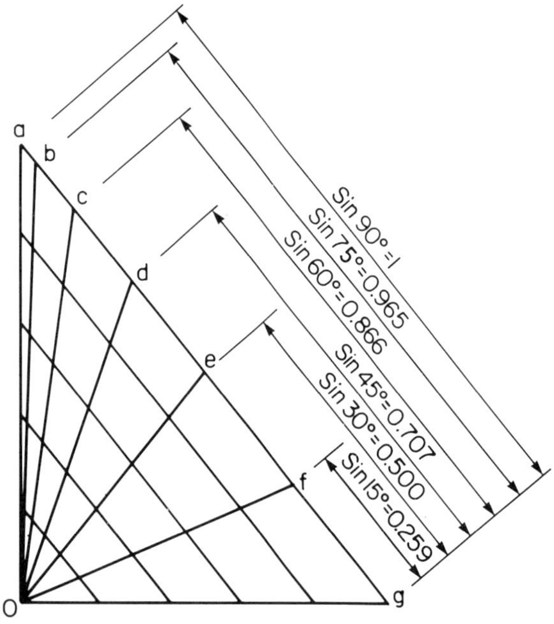

Figure 34.22 Construction of overlay to assist in calculation of tube performance

Table 34.2 Approximate design factors

	Efficiency (%)	*Conduction angle (deg)*	*Ratio of peak to mean anode current*
Class C	75	120–140	4 to 4.5
Class B	65	180	π
Class AB	60	180	3

34.2.7 High frequency effects

The time taken for the electrons to travel from cathode to anode is short enough not to have an important effect on the operation of the type of tubes discussed at frequencies below 100 MHz. Special tubes are made which are satisfactory to 1000 MHz or more. However, there is a very important high frequency effect due to the inductance of the cathode structure inside the tube (*Figure 34.23*).[4]

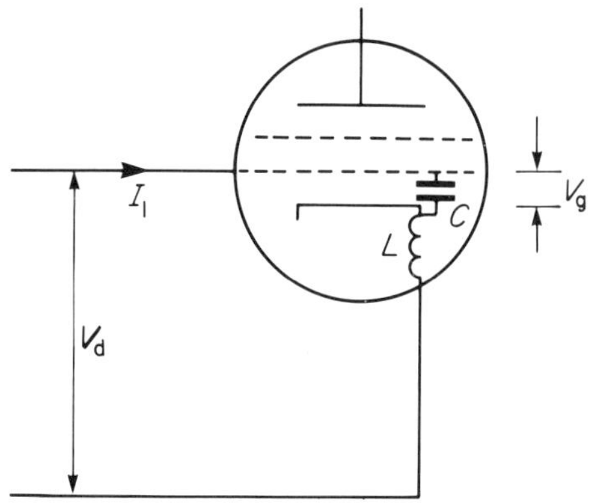

Figure 34.23 Equivalent circuit at high frequencies

The grid cathode voltage V_g will be less than the applied voltage V_i by the voltage across the cathode inductance L:

$$V_i = V_g - j\omega L I_k \tag{34.5}$$

At a fixed anode voltage the tube current may be approximated by

$$I_k = G_m V_g \tag{34.6}$$

Where G_m is the slope expressed in amperes per volt.

The input current I_g to the tube will be given by

$$I_g = V_g \times j\omega C \tag{34.7}$$

Substituting these in equation (34.5) and rearranging gives the input admittance Y of the tube as

$$Y = j\omega C \frac{1 - j\omega L G_m}{1 + \omega^2 L^2 G_m^2}$$

$$= \frac{j\omega C}{1 + \omega^2 L^2 G_m^2} + \frac{\omega^2 L C G_m}{1 + \omega^2 L^2 G_m^2} \tag{34.8}$$

The second term on the right-hand side of equation (34.8) represents a conductance and will absorb power. Although this power is not dissipated, being transferred to the output circuit, it must be supplied by the driver.

Typical values for a large tube are:

$L = 5 \times 10^{-9}$ H

$C = 185$ pF

$G_m = 0.3$ A V^{-1}

At 30 MHz for example the input conductance is 0.01 mho, i.e. the input resistance is 100 ohm. If the peak r.f. drive voltage is 1200 V then the additional drive power required will approximately be 7 kW, and the drive source must be designed accordingly.

34.3 Klystrons

34.3.1 Types of klystron available

Klystrons are a type of electron tube using the principle of velocity modulation to avoid the transit time problems which produce difficulties for triodes and tetrodes as operating frequencies increase to above 1 GHz.

Klystron oscillators, sometimes called reflex kylstrons, produce microwave r.f. energy at power levels up to a few watts at the lower microwave frequencies and up to several hundred milliwatts at the highest microwave frequencies. In some applications klystron oscillators have been replaced by solid-state devices but their inbuilt robust tolerance of supply transients and relatively low cost has recently led to their ousting solid-state devices in some applications.

Klystron amplifiers begin to be useful from about 400 MHz and are available up to around 20 GHz. The majority of applications are below 2 GHz where CW klystron amplifiers can produce power outputs in the range 1–60 kW at gains of 30–40 dB. At frequencies above 2 GHz CW power outputs of up to 25 kW are available. At the lowest frequencies (a few hundred MHz) CW powers of up to 1 MW are obtainable.

Pulsed klystron amplifiers are usually operated above 2 GHz where peak output powers in the range 1–20 MW are available at gains of 30–50 dB. Pulsed amplifiers are less well known at lower frequencies than 1 GHz but tubes are available giving peak output powers in the range 0.5–20 MW.

The klystron was first described by the Varian Brothers[20] in 1939 and by Hahn and Metcalf.[21] The klystron oscillator first appeared in 1940 but successful klystron amplifiers were not available until the early 1950s.

The size and weight range of klystrons is astonishing. A small oscillator klystron might be 10 cm high and weigh 150 g whilst a large CW amplifier klystron could be more than 2 m long and weigh 100 kg. Some special purpose tubes are much larger and heavier even than this.

34.3.2 Principle of operation

Klystrons use an r.f. voltage to velocity modulate an initially uniform velocity electron beam. An electron beam travels down a metal tunnel which has gaps along its wall across which r.f. voltages are applied. Individual electrons have their velocity shifted by interaction with the r.f. electric fields across the tunnel gaps according to the instantaneous value of the r.f. electric field at the instant when the electron arrives at the gap. Some electrons are slowed, some accelerated, some unaffected according to their arrival time at the gap. In the subsequent tunnel section slowed electrons will be caught by faster electrons and accelerated electrons will catch up slower electrons. Thus the uniform electron beam is converted by velocity modulation into a beam of non-uniform density which may then be used to perform the desired function of either amplification or oscillation.

The tunnel is usually called the *drift tube* and the gaps *interaction gaps*. The r.f. voltages across the interaction gaps are produced by resonant metal boxes called *cavities* suitably coupled to the gaps.

34.3.2.1 Klystron amplifiers

A klystron amplifier having two cavities is shown in *Figure 34.24*. An electron beam (1) is formed by the electron gun (2). The electrons in this beam are accelerated to a constant velocity U_0 determined by the relation:

$$U_0 = \left(\frac{2e}{m} V_0\right)^{1/2} \tag{34.9}$$

where

U_0 = electronic velocity
e = electronic charge (1.6×10^{-19} C)
m = electronic mass (9.1×10^{-31} kg)
V_0 = beam voltage

The electrons enter the drift tube (3) and continue to travel at velocity V_0 until they enter the input interaction gap (4) which is surrounded by the input resonant cavity (5). When this cavity is adjusted to resonate at the same frequency as that of an r.f. signal coupled into the cavity by means of a small loop antenna an r.f. electric field is developed across the interaction gap parallel to the beam axis.

In each successive half-cycle the electrons crossing the gap are either accelerated or decelerated depending on the phase of the gap voltage. If the instantaneous gap voltage is $V \sin \omega t$, where $\omega = 2\pi f$, then the velocity of the emerging electrons will be given by

$$U = U_0\left(1 + m\frac{V_1}{V_0}\sin \omega t\right)^{1/2}$$

$$\simeq U_0\left(1 + M\frac{V_1}{2V_0}\sin \omega t\right) \tag{34.10}$$

where M is the gap coupling coefficient, dependent on the gap geometry and beam diameter.

The velocity modulated beam enters a field free region (6) called the drift tube and here the fast electrons overtake those slowed down during the preceding half cycle. In this way electron bunches are formed and the beam becomes density modulated.

The first theoretical work on klystron behaviour was published by Webster,[22] who showed that the value of r.f. current on the beam as a function of drift tube length is given by the relation:

$$i_{rf} = 2I_0 J_1(K) \tag{34.11}$$

where K, the bunching parameter, is given by

$$K = \frac{\omega z V_1}{2U_0 V_0} \tag{34.12}$$

where

I_0 = beam current
J_1 = a Bessel function of the first kind
z = distance measured from the input gap
V_1 = input gap voltage

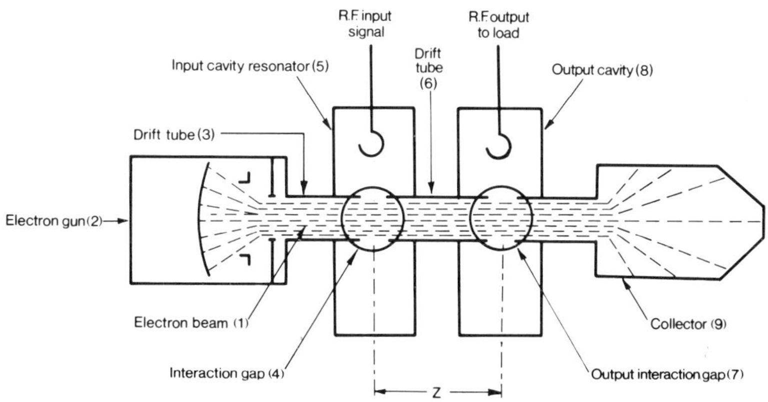

Figure 34.24 Two-cavity klystron

The maximum value of $J_1(K)$ is 0.58 and occurs when $K=1.84$. It follows that for any given value of V_1 the value of i_{rf} can be made to be a maximum by making z the appropriate length. In theory, therefore, it is possible to achieve very high gain by making z very long, but in practice space charge de-bunching[23] limits the useful value of z.

The output gap (7) is positioned at a distance z which corresponds to the maximum value of i_{rf}. On crossing this gap the bunched beam induces currents in the output cavity (8) which contains a loop antenna which delivers power to an external load via a transmission line.

The voltage developed across the output gap is in anti-phase with the electron bunches. Consequently, more electrons cross the gap during the retarding voltage half cycle than do during the positive cycle. There is, therefore, a nett transfer of energy during each cycle. The electrons are slowed down as they cross the gap and the maximum useful peak voltage is that which just stops the slowest electrons and is approximately the beam voltage V.

The spent beam that emerges from the output gap travels into the collector (9) where the remaining energy is dissipated as heat.

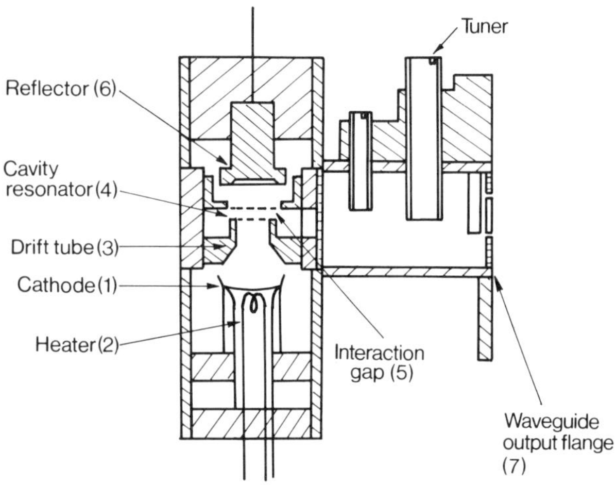

Figure 34.25 Structure of typical reflex klystron

34.3.2.2 Klystron oscillator

A klystron oscillator is shown in *Figure 34.25*. A cathode (1) is heated by a heater (2) and an electron beam is projected via a drift tube (3) into a cavity resonator (4) which surrounds a gridded interaction gap (5). The grids serve to couple more strongly the r.f. electric fields generated by the resonant cavity to the electrons within the beam. Having traversed the interaction gap the beam enters a region of decelerating electric field produced by the reflector electrode (6).

The beam is velocity modulated on its first transit of the interaction gap and density modulation, called bunching, subsequently takes place. The electron bunches are reflected at the reflector electrode and returned to the interaction gap. For maximum r.f. output via the waveguide output (7) the electron bunches must arrive back at the interaction gap so as to experience a maximum r.f. retarding voltage. This requires a drift transit time of $n+\frac{3}{4}$ r.f. cycles, where n is an integer.

Modulation of the electron beam is initiated within the tube by a noise process and oscillation builds up to give full output provided the $n+\frac{3}{4}$ relationship is satisfied. The reflector electrode voltage is adjusted to produce the required relationship and oscillation is sustained. The reflex klystron will, therefore, oscillate in a number of modes determined by the reflector voltage.

Since the bunched beam presents a reactive component to the resonant cavity and the sign of this reactance is dependent upon the phase of the returning bunches, the frequency of oscillation can be adjusted over a limited range by altering the reflector voltage.

The use of a single resonant cavity simplifies mechanical tuning arrangements and most reflex klystron oscillators can be tuned over relatively large frequency ranges with a single control. Electronic tuning is achieved by varying the reflector voltage and as the reflector draws no current very little drive power is required for this type of tuning.

34.3.3 Multicavity klystron amplifiers

Most commercially available klystron amplifiers have at least four resonant cavities and tubes having 5–7 cavities are not uncommon. The two-cavity tube shown in *Figure 34.24* will amplify a single frequency but has a narrow bandwidth. In most real applications for klystron amplifiers significant instantaneous bandwidth is required and this can be achieved by the use of a stagger tuning technique where multiple cavities are tuned to frequencies lower and higher than the centre frequency to build up the required bandpass characteristic.

A typical four-cavity klystron amplifier is shown diagrammatically in *Figure 34.26*. *Figure 34.27* shows an actual four-cavity tube used in a UHF television transmitter. The

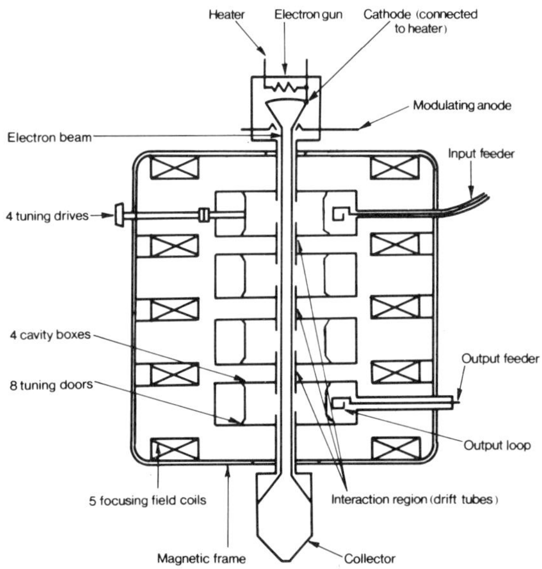

Figure 34.26 Four-cavity klystron amplifier

penultimate and output cavities of the tube can be considered as a power amplifier stage and the earlier cavitities (two in this case) as a driver stage for the two-cavity power amplifier stage. The power amplifier stage determines the linearity of the whole tube whilst the driver stage largely determines the gain and bandwidth of the whole tube.

A hot cathode in the electron gun generates an electron beam which is held cylindrical as it traverses the drift tubes by means of the axial magnetic field of a series of coils retained in a supporting structure which also forms a flux return path of low reluctance. The electron gun and collector are external to the magnetic field,

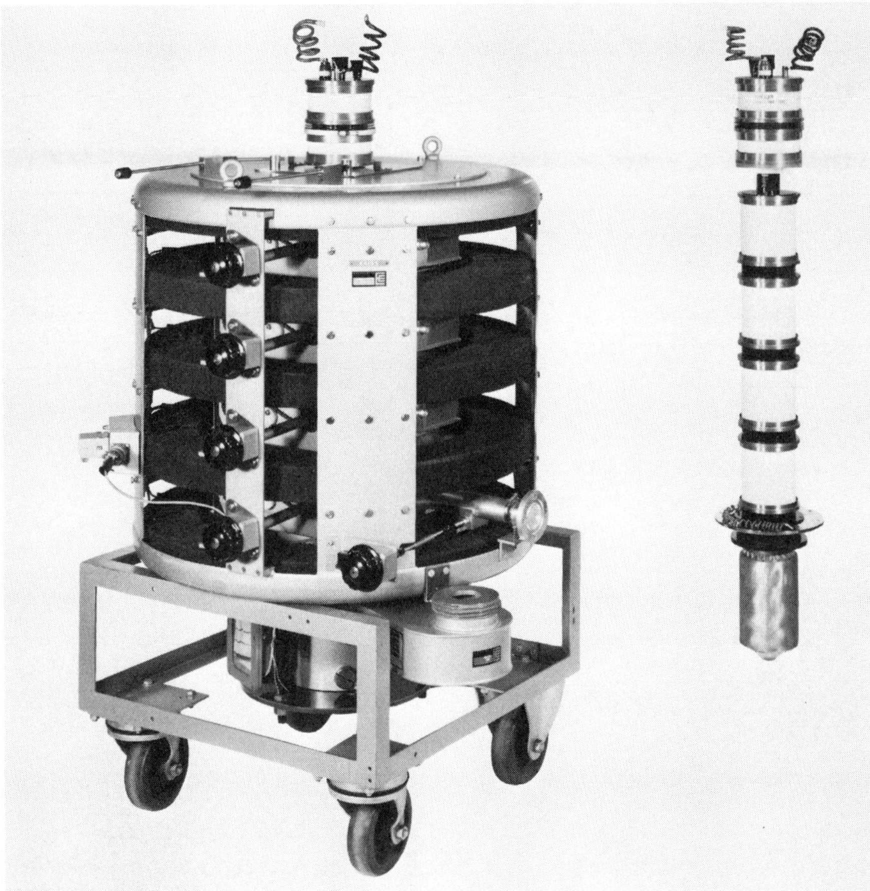

Figure 34.27 10 kW UHF television klystron amplifier and circuit assembly (left). The actual klystron tube used (right). (Photo by courtesy of The English Electric Valve Company Limited)

though some flux is allowed to link the cathode to ensure correct injection of the electron beam into the drift tube.

Metal cavity boxes surround each of the interaction gaps. Each cavity is mechanically tuned by moveable doors and the input and output cavities have loop antennas.

The beam is velocity modulated at the input cavity by a small r.f. voltage. The signal level and hence the electron velocities are insufficient to overcome the space charge repulsion forces between the electrons. Under these conditions the space charge wave theory of Hahn[24] and Ramo[25]is applicable and the r.f. current maximum occurs at a distance of a quarter plasma wavelength z given by

$$z=\frac{U_0}{2\omega_q} \tag{34.13}$$

where

z = distance from the gap
U_0 = initial electron velocity
ω_q = reduced plasma angular frequency

A second, lightly loaded, cavity is positioned at this distance and because of its higher impedance a significantly higher voltage is developed across the second gap. This voltage now remodulates the beam and produces a higher r.f. current maximum at a third gap situated a further 0.25 plasma wavelength away. The same process is repeated at this gap resulting in a higher voltage still. Depending on the number of cavities this process continues until the penultimate gap is reached. By now the signal level at the gap will be large and the space charge forces will be overcome so that the basic two-cavity klystron mechanism will apply. Webber[26] has shown that under large signal conditions the optimum drift tube length is 0.1 of a reduced plasma wavelength and this figure is normally used in klystron design.

In the driver stages the gap voltages are small relative to the beam voltage and the amplification is linear. However, in the output stage the 'drive' level ($V_{\text{pen.gap}}$) is sufficiently high for large signal conditions to apply and the r.f. current at the output gap is given by

$$i_{\text{rf}}=2I_0J_1(K_p) \tag{34.14}$$

where K_p, the bunching parameter at the penultimate gap, is given by

$$K_p=\frac{\omega z V_{\text{pen.gap}}}{2U_0V_0} \tag{34.15}$$

The output voltage as a function of drive has, therefore, the form of a Bessel function (*Figure 34.28*). At signal levels well below saturation the behaviour is approximately linear but on raising the power level the gain falls smoothly to a saturation value 6.0 dB below the small signal gain.

As the output power increases so kinetic energy is extracted from the beam. This results in a reduction in the velocity of the

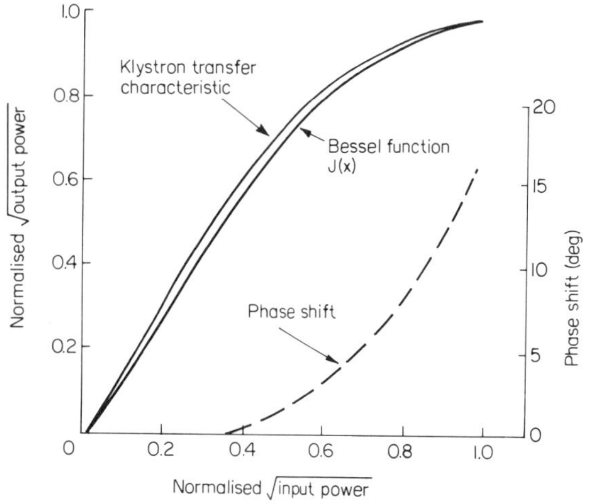

Figure 34.28 Typical klystron transfer characteristic

electrons which introduces a delay and hence a phase shift. This level-dependent phase shift is known as a.m./p.m. conversion (*Figure 34.28*).

Amplifier klystrons require a long electron beam of constant diameter. It is, therefore, necessary to provide some means of constaining the beam which would otherwise spread as a result of the electron forces. By far the most common solution is to apply a magnetic field along the axis of the beam.

For any beam there is a value of magnetic field B_B, known as the Brillouin field, at which the centripetal forces generated by the electrons rotating in the magnetic field balance exactly the space charge repulsion forces. This value is given by the relation

$$B_B = \frac{8.32 \times 10^{-4}(I_0)^{1/2}}{r_0 V_0^{1/2}} \qquad (34.16)$$

where r_0 is the beam radius.

Unfortunately Brillouin flow requires that the current density in the beam remains constant—a condition which is not satisfied in practical klystrons. It is more usual to employ field values twice to three times Brillouin. Under these conditions the beam diameter perturbations caused by changes in current density are much reduced.

Klystrons decrease in size as the operating frequency increases (because the cavities get smaller and the drift lengths shorter). Large low frequency tubes are usually solenoid-focused but higher frequency tubes often use permanent magnets.

34.3.4 Construction techniques

The construction of a typical external cavity klystron is shown in *Figure 34.29*. The major parts are: the electron gun; the body section, made up of the interaction gaps and drift tubes; and the collector.

34.3.4.1 The electron gun

The cathode and its surrounding electrodes are shaped and positioned to produce an electrostatic field which will focus the electrons through the hole in the anode. A focus electrode usually surrounds the cathode.

The shaped anode, with its central hole, may be made integral with the body or insulated from it in which case it is known as a modulating anode. Some guns employ beam current control grids or electrodes in front of the cathode.

The electron gun geometry controls the beam current for any given voltage. Under normal operating conditions the beam current is given by the relation:

$$I_{beam} = K V_A^{3/2} \qquad (34.17)$$

where

I_{beam} = the beam current (in amperes)
V_A = the anode voltage (in volts)
K = a constant known as the (micro)perveance

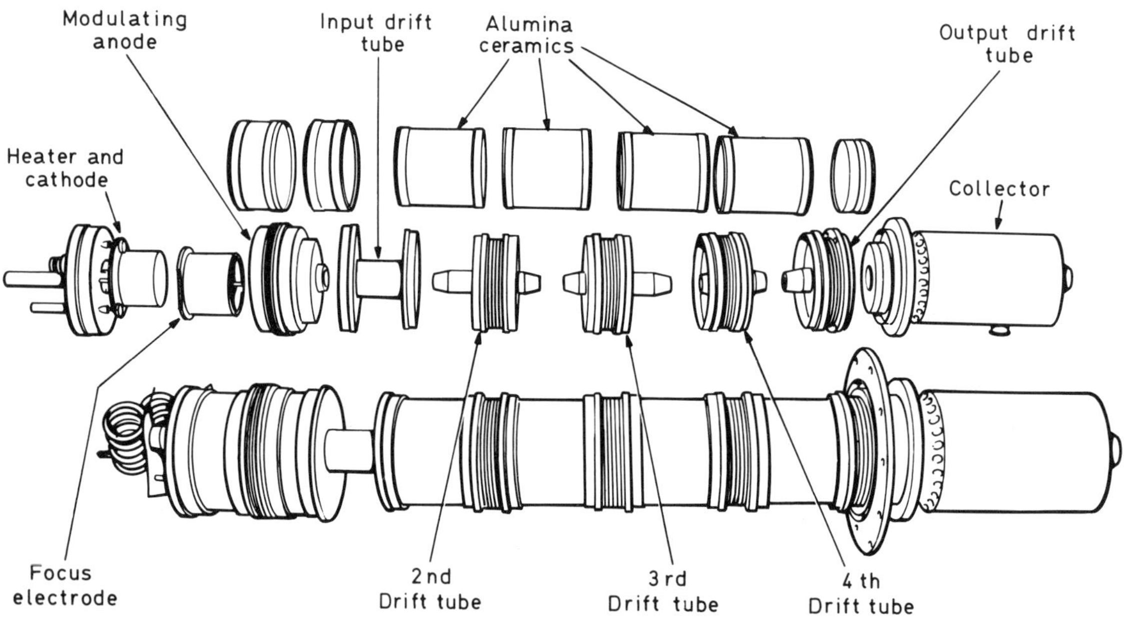

Figure 34.29 Construction of a typical external cavity amplifier klystron (By courtesy of The English Electric Valve Company Limited)

34.3.4.2 The body

The beam formed by the gun and focused by the magnetic field passes into the drift tubes and the interaction gaps. At the higher microwave frequencies the body usually consists of a series of cylindrical cavities brazed together or milled from solid material, with dividing walls supporting drift tubes. The tuner mechanism, usually an adjustable capacitance plate, is fitted through a vacuum bellows. The whole body is cooled by either liquid or air blowing.

At lower frequencies, particularly between 400 and 1000 MHz external cavity klystrons are widely used. In this construction part of each cavity is outside the vacuum envelope, which then consists of a ceramic cylinder joined to robust flanges supporting the drift tubes. The external part of the cavity is made in two halves, joining around the ceramic on a plane across which no r.f. current flows. Two opposing doors or cavity walls can be moved transverse to the klystron axis to provide tuning, while a coupling loop can be rotated to adjust the coupling to external circuitry.

34.3.4.3 The collector

The power density of the beam is usually much too high to be allowed to impinge directly on a metal surface. The density is, therefore, reduced by allowing the beam to spread out on leaving the focusing field beyond the output gap.

34.3.4.4 Cooling

Three methods of heat removal are in common use for high power klystrons: forced air, liquid and vapour (evaporative) cooling. The first of these is relatively simple but is generally limited to mean beam powers less than 30 kW by the need to keep the collector size reasonable.

High power tubes may be water or vapour cooled. Water cooled collectors require large flows and high pressure so the much more efficient vapour cooling technique is usual on the high power CW tubes.

34.4 Magnetrons

The magnetron is a microwave oscillator with a high efficiency requiring only modest operating voltages for the generation of high-power levels. In particular it has found an enduring place in the transmitters of pulsed radar systems and it is with this type of tube that we will be mainly concerned, other types of magnetron such as the low-power voltage tunable tubes and CW tubes for r.f. heating described more briefly are fully listed in the bibliography.

The magnetron concept, a cylindrical diode immersed in a uniform magnetic field parallel to its axis, was first reported by A. W. Hull in 1921[45] and was later developed as a high-frequency oscillator.

The development of pulsed radar led to a pressing need for a high-power microwave source in order to improve angular and range resolution and permit the use of physically smaller antennae. The work of Prof. J. R. Randal and Dr H. A. H. Boot[46] at Birmingham University in 1939 satisfied that need with the generation of powers at 3 GHz several orders of magnitude greater than those previously achieved.

Further war-time development in Britain and the USA provided a profusion of types at frequencies between 1 GHz and 30 GHz operating with output powers ranging from a few kilowatts to several megawatts.[46–49] Methods of mechanical tuning were devised and deployed and a considerable effort was applied, though with less success, to electronic tuning. The post-war years have seen vast improvements in operating stability and life expectancy of magnetrons, while the introduction of the long anode magnetron and the coaxial magnetron, together with the concept of rapid tuning or frequency agility, have further extended the range of application.

34.4.1 Principles of operation

A simplified representation of the main components of the multicavity magnetron in cross section is given in *Figure 34.30.*

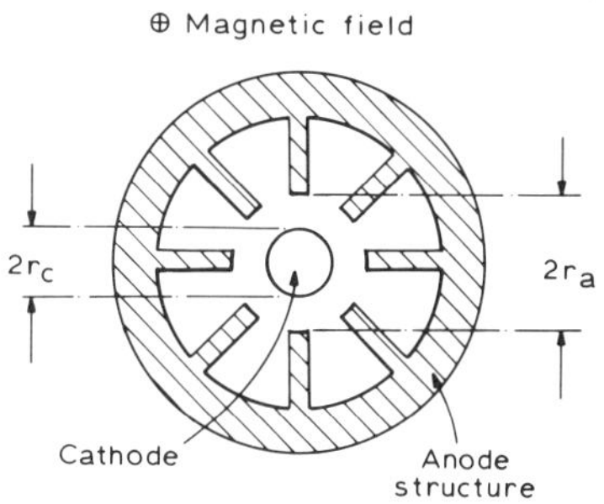

Figure 34.30 Main components of the multicavity magnetron

The anode, in addition to being the positive electrode and collector of electrons, also forms a structure resonant near the desired r.f. output frequency. This resonant structure is made up of a number of inter-coupled resonant cavities which provide, in the interaction space, a rotating electromagnetic wave of phase velocity below that in free space so as to give interaction with a rotating cloud of electrons emitted from the cathode.

The magnetic field, which is perpendicular both to the r.f. electric field and to the d.c. electric field (the magnetron is a 'crossed field' device), causes the electrons to rotate round the cathode in the interaction space. The strength of the magnetic field is such that, in the absence of an r.f. field, electrons would be unable to reach the anode.

With an r.f. field present, electrons in a favourable position in the rotating electron cloud will give up energy to the r.f. field and move towards the anode, thereby transforming potential energy arising from the d.c. electric field into r.f. energy. Finally they will be collected on the inner surfaces of the anode structure. Electrons in an unfavourable position in the electron cloud will take up a small amount of energy from the r.f. field and will be quickly returned to the cathode surface where secondary electrons will be emitted, some of which will be useful in the generation of further r.f. power. These returning electrons expend their energy at the cathode surface, an effect known as back bombardment, thus heating the cathode and hence limiting the average power handling capacity. Electrons between the favourable and unfavourable positions described above will be accelerated or retarded into one of these groups. This process, whereby electrons are selected and phased for optimum generation of r.f. power is known as 'phase focusing' and leads to the high electronic efficiency of the magnetron.

R.F. output power is coupled from the resonant anode structure, through a vacuum window and into an external waveguide or coaxial line.

The anode voltage at which oscillation can build up is known as the threshold voltage and is given by the following expression:

$$V_{th} = 1.01\left(\frac{r_a}{\lambda}\right)^2 \frac{T\lambda(1 - r_c^2/r_a^2)}{535N} - \frac{4\times10^7}{N^2} \qquad (34.18)$$

where

V_{th} = the threshold voltage (V)
r_a = the anode bore radius (m)

r_c = the cathode radius (m)
T = the magnetic density (T)
N = the number of resonators
λ = the free space wavelength (m)

The maximum electronic efficiency is given by

$$\eta_{max} = 1 - \frac{1}{NX - 1} \tag{34.19}$$

where

$$X = T\lambda\left(1 - \frac{r_c^2}{r_a^2}\right) \times 4.67 \times 10^{-5} \tag{34.20}$$

34.4.2 Magnetron construction

34.4.2.1 R.F. structure

The r.f. structure which acts as the magnetron anode is normally made of copper and forms part of the vacuum envelope, thus providing good heat dissipation. However, temperature rise along the vane structure can be a limiting factor on power rating and in high power CW tubes hollow water-cooled vanes are used. In the case of magnetrons at higher frequencies the power density at the vane tips facing the cathode may be great enough to lead to quite large temperature rises during the operating pulse time.

The r.f. output system is coupled to one resonator of the r.f. structure in conventional types through an appropriate impedance matching system, either waveguide or coaxial line.

In long anode and coaxial magnetrons,[50,51] the r.f. output is symmetrically coupled to the resonator system, in the latter case through a stabilising resonator which can materially improve the frequency pulling and pushing phenomena.

Figure 34.31 Vapour-cooled L-band tunable magnetron with anode and cathode assemblies (photo: English Electric Valve Company Limited)

34.4.2.2 Cathode

The cathode which operates in a mixed thermionic and secondary emitting fashion is normally equipped with non-emitting end shields or *hats* to prevent axial spread of the electron stream.

The axial magnetic field required for operation is usually provided by permanent magnets frequently with soft iron pole pieces built through the vacuum envelope in order to reduce the reluctance of the magnetic circuit. In the case of long anode magnetrons the magnetic field is normally provided by an electromagnet due to the length of uniform field required.

34.4.2.3 Typical magnetron structures

A representative selection of magnetrons with their r.f. structures and cathodes are shown in *Figures 34.31* to *34.34*. *Figure 34.31* shows a high-power tunable L-band magnetron with conventional strapped r.f. structure giving a 60 MHz tuning range and a power rating of 2.3 MW peak, 3.3 kW mean, with an operating voltage of 39 kV. The tube is vapour cooled, providing excellent frequency stability and permitting the use of a very compact heat exchange system. An external permanent magnet is used.

A *Q*-band magnetron is shown in *Figure 34.32*. The r.f. structure is of the rising sun variety which permits relatively high power ratings at high frequencies, in this case 50 kW peak and 20 W mean: the operating voltage is 14 kV. The magnetic field is provided by an integral permanent magnet.

The *S*-band long anode magnetron shown in *Figure 34.33* is capable of generating a peak output of 2.5 MW with 3.75 kW mean. The operating voltage is 39 kV, the magnetic field being provided by an external solenoid, the water-cooling system of which also cools the magnetron. The length of the anode

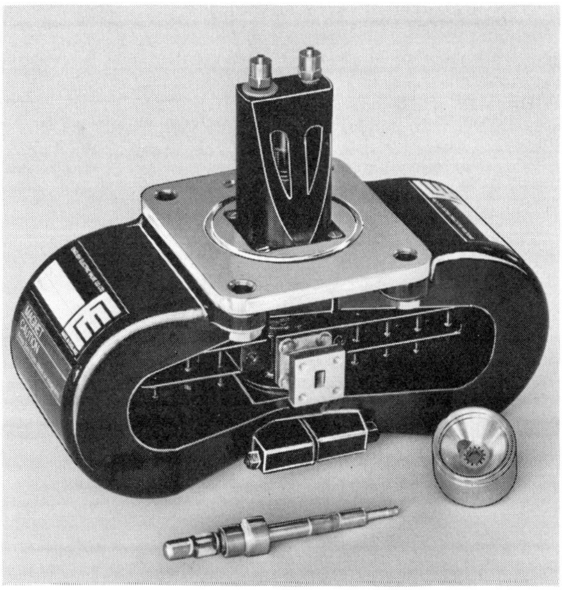

Figure 34.32 Frequency-agile Q-band magnetron anode and cathode assemblies (photo: English Electric Valve Company Limited)

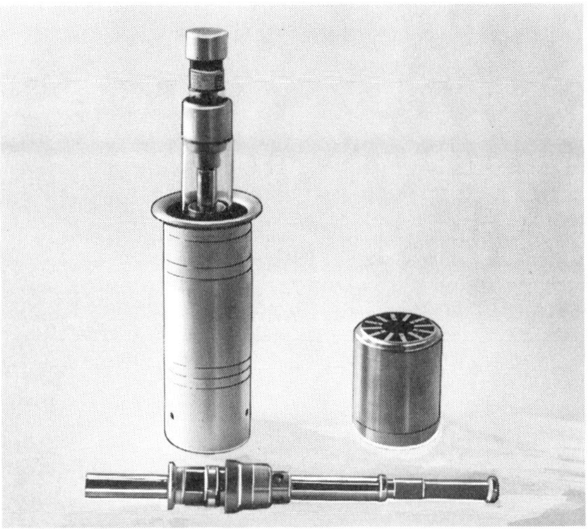

Figure 34.33 Long anode S-band magnetron with anode and cathode assemblies (photo: English Electric Valve Company Limited)

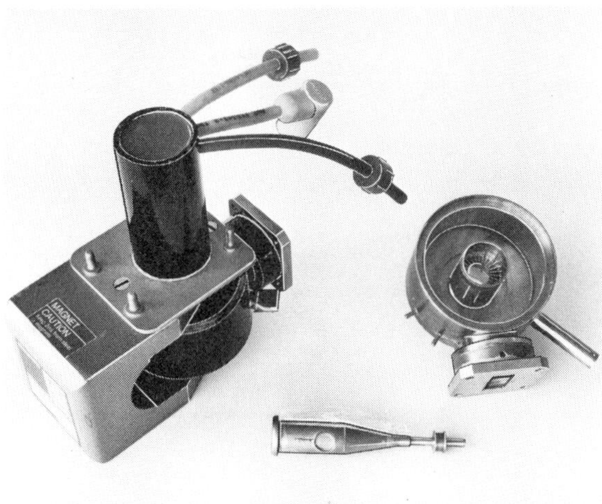

Figure 34.34 Co-axial X-band magnetron with anode and cathode assemblies (photo: English Electric Valve Company Lkmited)

compared with the operating wavelength and the large cathode area will be noted.

Figure 34.34 shows an *X*-band coaxial magnetron with a power rating of 75 kW peak and 75 kW mean. The operating voltage is 13 kV, the magnetic field being provided by an integral permanent magnet. This design provides large dimensions for both anode and cathode for a given operating frquency. The stabilising resonator and its coupling slots surround the typical magentron resonator system.

34.4.3 Methods of tuning magnetrons

In many applications it is necessary that the operating frequency of the magnetron be adjustable or variable. In some cases very rapid tuning is required, this is known as frequency agility. Successful tunable magnetrons have been realised using both mechanical and electronic methods.

34.4.3.1 Mechanical methods

Tuning may be accomplished by variation of either the inductance or capacitance of the anode resonator system directly or by tuning a separate resonant cavity coupled to it. In its simplest form the magnetron tuner consists of a conducting ring placed near the end of the magnetron resonator system. When the ring is moved towards the anode the r.f. magnetic field coupling adjacent davities is modified, the inductance of the resonator is decreased and thus the resonant frequency is increased. The tuner ring may be moved by a simple mechanism passing through a vacuum bellows. The frequency range of the inductive tuner may be increased by attaching conducting pins to the ring; these pins protrude into the resonators of the magnetron anode.

For rapid tuning (frequency agility) the ring may be driven by a transducer capable of rapid movement; both electromagnetic and piezoelectric transducers have been used, the latter being particularly successful in the millimetre wavelength applications. In another manifestation, the spin tuned magnetron, an inductive tuning element takes the form of a toothed ring or tube which rotates in the magnetic r.f. field providing rapid tuning.

Magnetrons with tuners which interact with the electric component of r.f. field, capacitative tuners, have been made. In this case the tuning element may be either a conductor or a dielectric but it must move in regions of high electric field and hence such tuners are prone to arcing when used directly in the anode resonator system.

Coupled cavity tuning has found considerable application in coaxial magnetrons in which the operating frequency may be varied by tuning the stabilising cavity. This has the advantage of separating the tuner from the anode resonator system; it may even be outside the vacuum envelope. In one type of coaxial magnetron frequency agility has been achieved by rapidly rotating dielectric paddles within the external cavity.

34.4.3.2 Electronic methods

Practical electronic tuning has been realised in two ways:

(a) By variation of the anode voltage of specially designed tubes known as voltage tuned magnetrons (VTM).
(b) By utilising a multipactor electron discharge to change the frequency of resonators coupled to the magnetron anode system.

Of these approaches only the latter has been applied to the high power pulsed tubes considered here and which have applications in frequency agile radar systems.

While it is possible to achieve frequency agility using the electromechanical systems described above, these devices suffer from two major limitations. Firstly the variation of frequency with time is essentially cyclic, and although this may be concealed to some extent by varying the radar pulse repetition frequency with time, a modern electron counter measures (ECM) system should be capable of detecting and predicting the inherently cyclic tuning pattern. Secondly the tuning rate is low, leading to a restriction of the pulse to pulse frequency change. Multipactor tuned magnetrons are free from both these limitations permitting improvements in the performance of radar systems to be obtained by reduction of glint, improved subclutter visibility, and better resolution.

Multipactor tuning is achieved by control of a multipactor electron discharge in a number of secondary resonant cavities coupled to the magnetron anode system. The discharge, which is produced by r.f. voltages between cold secondary emitting surfaces, may be inhibited by the application of a switching voltage of 3–4 kV with a negligible current drain if it is applied during the inter pulse period. Thus by switching, say, three tuning elements between 'on' and 'off' states eight discrete operating ffequencies can be obtained and the frequency of each operating pulse can be selected at random with a total tuning range of about 1%, 100 MHz at X-band.

The system does not significantly affect the other characteristics of the magnetron and there are only small increases in size and weight. Switching power consumption is very low and variation of power output over the range of frequencies is small, about 1 dB.

34.4.4 Operating characteristics

The major operating characteristics of the magnetron follow from two facts:

(1) There is a sharply non-linear relation between applied voltage and current through the tube.
(2) It is a self-oscillator.

34.4.4.1 Input characteristics

The input voltage/current relationship results from the fact that no current flows through the device for applied voltages below the threshold value; as soon as the threshold value is exceeded oscillation starts to build up accompanied by a rapid rise in current, the operating voltage rises only slowly above the threshold point. Thus the magnetron input impedance is very high below the threshold point and very low above that point; this fact mube be borne in mind when designing magnetron drive circuits. In particular the performance of drive circuits on resistive load will in general differ markedly from performance with a magnetron load.

The performance chart (*Figure 34.35*), shows this non-linear relation between applied voltage and current for several values of magnetic field, together with contours of constant efficiency. The

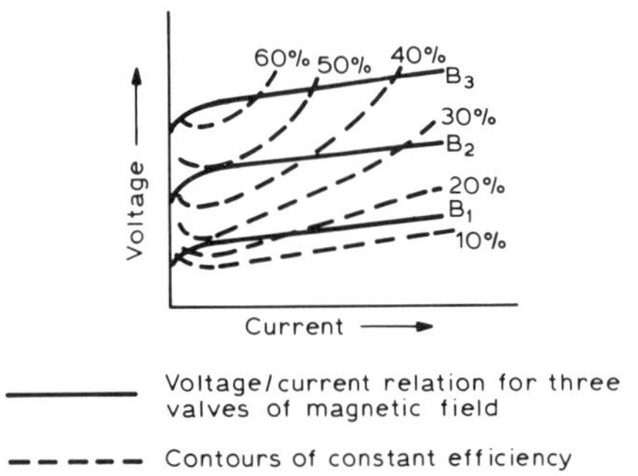

Figure 34.35 Performance chart for magnetron with separate magnet

choice of operating point in this chart may be limited by the onset of various malfunctions in certain regions. It is therefore essential that any deviation from the recommended operating point should be queried with the manufacturer.

In pulsed applications the oscillation must build up from noise levels at the start of each pulse. Thus there is a finite rate of build up of current and r.f. and hence the rate of build up of applied voltage must not exceed the recommended value if malfunction is to be avoided.

The front edge of the r.f. output pulse is subject to an inevitable time jitter which is of the order of 5 r.f. cycles in conventional magnetrons but may be an order of magntiude greater in the case of coaxial magnetrons.

34.4.4.2 Output characteristics

The magnetron is a self-oscillator and hence the output power and frequency of operation are affected by the r.f. load impedance, the variation of frequency is referred to as frequency pulling. Output power and frequency are also functions of the input current (the operating voltage remaining nearly constant), the variation of frequency with input current is referred to as frequency pushing.

The effects of both frequency pushing and pulling may be reduced by the use of coaxial magnetrons with their high degree of frequency stabilisation, in particular the low pushing figures obtained by these tubes may in certain circumstances be useful in ensuring minimum r.f. output bandwidth when operating with long duration pulses. The effects of frequency pulling may be reduced by the use of ferrite isolators or circulators. The use of these devices is particularly important when the susceptance of the r.f. load is a rapid function of frequency as in the case of linear accelerators or where the feeder is very long.

Thermal drift of operating frequency occurs as the magnetron warms up or if the ambient temperature changes. In this context the high-power vapour-cooled magnetrons are particularly useful as the temperature of the resonator system is stabilised at the boiling point of the cooling fluid.

34.4.5 Magnetron performance: practical limits

Figures 34.36 and *34.37* indicate the present limits of peak and mean power output in commercially available magnetrons as a function of r.f. frequency.

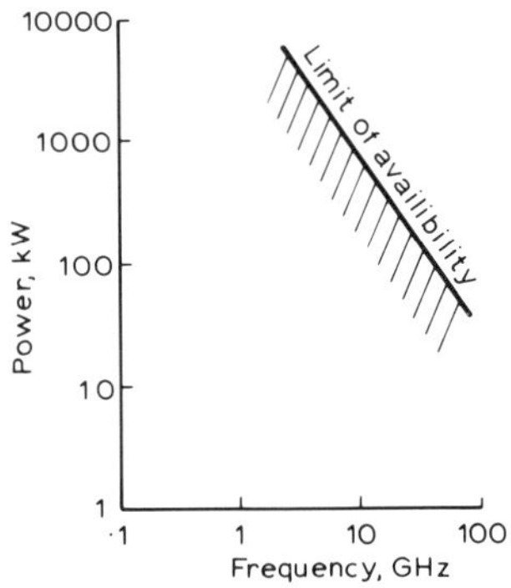

Figure 34.36 Peak output power

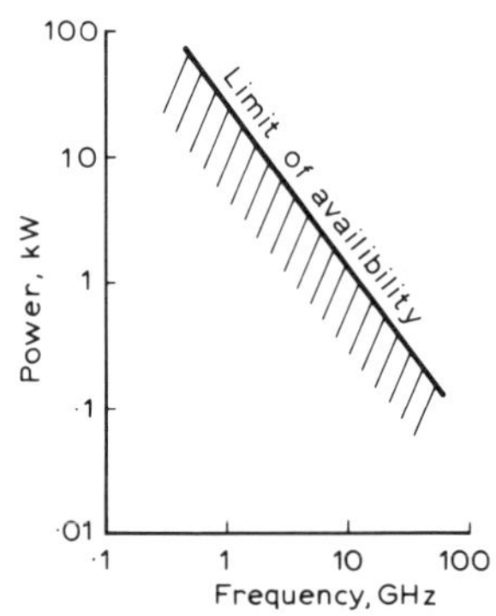

Figure 34.37 Mean output power

Available tunability exceeds 10% of the r.f. frequency for slow tuning and 5% for rapid tuning or frequency agility.

Both conventional and coaxial magnetrons have been used successfully in moving target indication radars. In this application there are demands on magnetron pulse-to-pulse frequency jitter and pulse-to-pulse starting jitter. Rugged magnetrons have been specially developed for airborne and missile applications where the environment can be one of high levels of shock and vibration.

34.5 Travelling-wave tubes

The travelling-wave tube (t.w.t.) is a microwave amplifier capable of amplifying over very wide frequency bands. The amplification process takes place by continuous interaction between an electron beam and an electromagnetic wave propagatin along a slow-wave structure. The principle was invented by Kompfner[74] in 1943 who used a simple wire helix as a slow-wave structure. Similar tubes were then developed and first used as microwave amplifiers in microwave relay link systems. During the last 25 years travelling-wave tubes have been continuously developed using other slow-wave structures such as coupled cavities to provide CW output powers of tens of kilowatts and pulse powers of several megawatts with power gains of up to 60 dB. Travelling-wave tubes are usefully employed from u.h.f. to centrimetric wavelengths. However, the original helix slow-wave structure is still one of the most useful due to its great bandwidth. Tubes employing helices have been made with useful amplification properties over a bandwidth greater than two octaves.

A travelling-wave tube consists of three main parts as illustrated in *Figure 34.38*. An electron gun[75] which is capable of producing a beam of electrons. A slow-wave structure which is in close proximity to the electron beam and which can propagate microwave signals at approximately the same speed as the electrons usually at a fraction of the speed of light. The slow-wave structure has an attenuator region approximately half way along its length to absorb r.f. energy which may propagate in the reverse direction. Without an attenuator the tube could be unstable and self oscillate. The final part is a collector which is capable of trapping the spent electron beam and dissipating this remaining energy as heat.

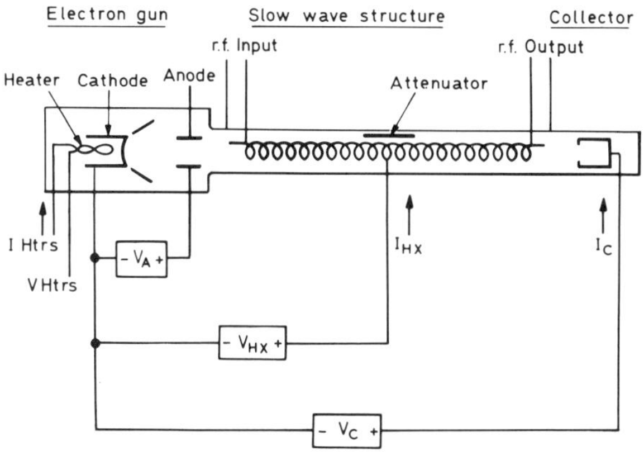

Figure 34.38 Schematic outline of a travelling-wave tube

An axial magnetic field is required to confine the beam in the structure region and prevent it diverging under the repulsive forces between the electrons. This focusing field can be provided either by a solenoid as in all early tubes or by using permanent magnets. Most travelling-wave tubes with power outputs up to a few hundred watts use a periodic permanent magnet focusing system[76] known as p.p.m. which is lighter, more compact and has less leakage magnetic field than a uniform system.

The velocity of the electron beam in the travelling-wave tube can be adjusted by varying the voltage applied between the cathode and slow-wave structure. The magnitude of the electron beam current is generally controlled independently by varying the voltage on an additional grid or anode in the electron gun.

34.5.1 Amplifier mechanism

At the input and output terminals of the travelling-wave tube there are transitions which allow the r.f. signal to be coupled to the slow-wave structure. The axial component of the r.f. electric field at the start of the slow-wave structure accelerates or decelerates electrons as they enter. This causes a velocity modulation on the beam which gradually gives rise to a density jodulation or electron bunching as the beam progresses along the slow-wave structure. As the beam and r.f. wave are in near synchronism the electron bunches induce voltages on the slow-wave structure which re-enforce those already present and thereby promote a rapid increase in the rlf. signal. To maintain this interaction the electron beam is initially travelling slightly faster than the r.f. signal eave. However, as the r.f. wave increases the average velocity of the electron bunches is reduced as more energy is extracted from the beam. Finally the electron bunches lose synchronism with the r.f. wave and no further energy is extracted from the beam. At this point the tube is said to be saturated and gives maximum power output. Any further increase of input signal level causes the output to fall. A typical travelling-wave tube input/output or transfer characteristic is shown in *Figure 34.39*.

34.5.2 Gain

The gain depends directly on the length of the slow-wave structure. As the helix or structure voltage is varied above or below the synchronous value the gain decreases rapidly. This

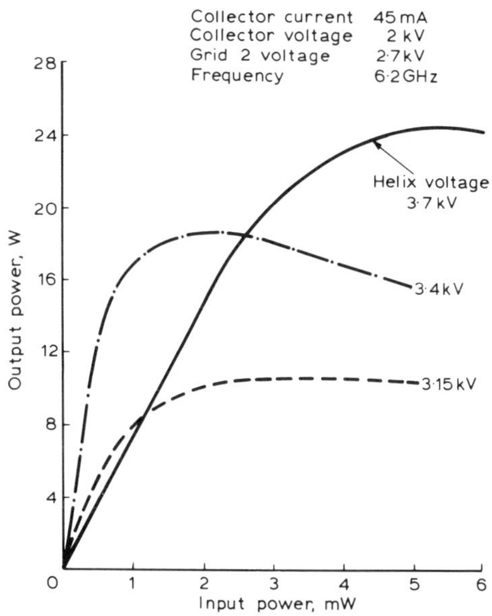

Figure 34.39 Transfer characteristics of a typical N1072 t.w.t. (courtesy English Electric Valve Company Limited)

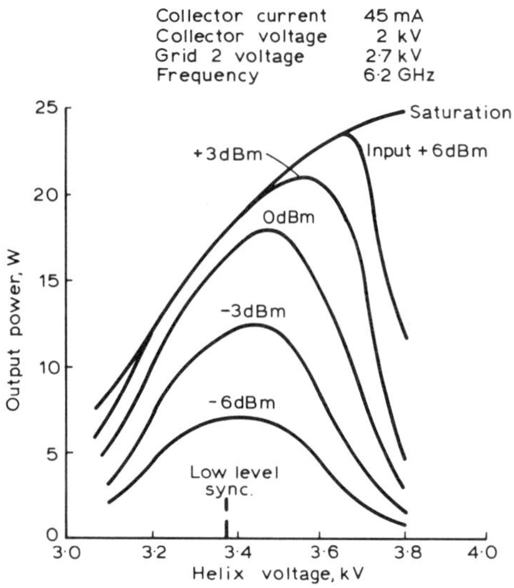

Figure 34.40 Outline power against helix voltage for a typical N1072 t.w.t. (courtesy English Electric Valve Company Limited)

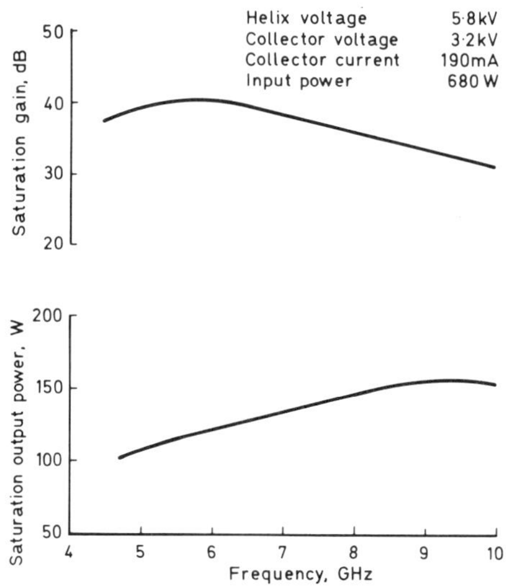

Figure 34.41 Typical performance characteristics of the N1077 t.w.t. (courtesy English Electric Valve Company Limited)

characteristic is illustrated in *Figure 34.40*. The gain also increases at a rate proportional to the cube root of the beam current.

The small signal gain G in decibels is given by

$$G = BCN - A - L \tag{34.21}$$

where A, B and L are constants for a given design and they have typical values of 10, 40 and 6 respectively. C is Pierce's[77] gain parameter and N is the number of wavelengths along the slow-wave structure:

$$C = \left(\frac{KI_0}{4V_0}\right)^{1/3} \tag{34.22}$$

where K is called the coupling impedance and depends on the geometry of the slow-wave structure, I_0 is the beam current and V_0 is the beam voltage.

For a helix slow-wave structure the phase velocity of the r.f. wave is largely independent of frequency therefore the wave and beam velocity remain in synchronism and large bandwidths are possible. Typical gain and power/frequency characteristics of a broad band t.w.t. are shown in *Figure 34.41*.

34.5.3 Efficiency

The highest power which can be obtained from a given travelling-wave tube is given by

$$P_{\mathrm{rf}} = kI_0V_0C \tag{34.23}$$

where k depends on the slow-wave structure and electron beam diameters and typically has a value between 2 and 3. Generally the beam efficiencies (100 kC) of the t.w.t.s. lie in the range 5% to 10% for low-power tubes and are greater than 35% for high-power pulsed tubes.

One method of improving the beam efficiency which has proved successful is the introduction of a taper[78] in the slow wave structure towards the output. The r.f. circuit wave is then slowed down at a similar rate to the beam and more energy can be extracted before synchronism is lost. Alternatively the beam can be accelerated in the output region by means of a voltage jump to re-synchronise the beam and wave velocities.

The overall efficiency of a travelling wave can be improved by a factor of 2 or 3 over the beam efficiency by making the cathode-collector potential smaller than the beam voltage and therefore reducing the wasted energy dissipated in the collector. Some travelling-wave tubes used in space applications have overall efficiencies greater than 50%. Such tubes make use of two or three collectors capturing the spent electron beam at progressively lower voltages. The overall efficiency of a travelling-wave tube is

$$\frac{P_{\mathrm{rf}}}{I_cV_c + V_{\mathrm{HX}}I_{\mathrm{HX}} + V_{\mathrm{HRTS}}I_{\mathrm{HRTS}}} \tag{34.24}$$

34.5.4 Noise

The main sources of noise in a travelling-wave tube arise from the shot noise or current density variations as electrons leave the cathode and secondly the random velocity fluctuations of the emitted electrons. In low noise tubes[79] the electron gun has several electrodes designed so that the potential profile between the input to the helix and the cathode can be optimised so that minimal noise in the beam is coupled to the helix. Low noise tubes have noise figures between 6 dB and 10 dB with power outputs of a few milliwatts.

By using a high magnetic focusing field over the cathode and introducing a special beam-forming electrode to produce a divergent electric field at the cathode ultra low noise tubes[80,81] are produced with noise figures between 2.5 dB and 5.0 dB.

34.5.5 Linearity

Since travelling-wave tube amplifiers are generally operated in the region of maximum output non-linearity of amplitude and phase distortion can occur. In some cases the saturation characteristic is put to good use and the travelling-wave tube may be employed as a limiter. However, in most other amplifier uses where linearity is important the tube is operated 2 dB to 3 dB below the saturated output level.

The conversion of amplitude modulation to phase modulation (a.m./p.m. conversion) occurs in a t.w.t. due to the reduction in the beam velocity as the input signal is increased. At saturation the a.m./p.m. conversion may rise to 5 °/dB. T.W.T. microwave link amplifiers are generally designed to have an a.m./p.m. conversion of about 1 °/dB at their operating power level.

When several signals are simultaneously amplified by a travelling-wave tube a mixing or intermodulation[82] occurs. This results in intermodulation products (i.p.) at the output with power levels dependent on the relative levels of the original signals. As in the case of a.m./p.m. conversion intermodulation distortion is significantly reduced by operating the tube in the small signal or linear region. With two equal carrier signals operating at saturation the third order i.p. is typically 10 dB below the carriers.

34.5.6 The construction and types of travelling-wave tube

34.5.6.1 The electron gun

The electron gun shown in *Figure 34.42* produces a controlled diameter beam of electrons from the cathode. The cathode has a surface with a low work function so that electrons are emitted when the cathode is heated. This surface is commonly obtained by depositing a layer of barium and strontium oxides on to a nickel base, stable emission is obtained at temperatures in the region of 750°C. Several other types of cathode construction are in common use but one which is now finding wide application is the barium aluminate cathode. This consists of barium aluminate impregnated into a porous tungsten pellet. The temperature of operation is higher than for the oxide cathodes but higher electron emission densities are possible up to two amps/cm^2, these cathodes are also less susceptible to ion damage. The operational life of the tube is often governed by cathode life and with care in designs and production this can be in excess of 20 000 hours. The beam current may be switched on and off by modulating the voltage between the slow-wave structure and the cathode. It is also possible to introduce control grids close to the cathode that switch the beam with much smaller voltages.

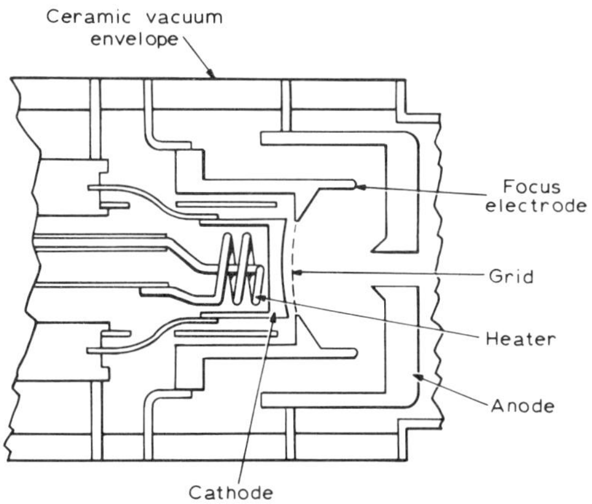

Figure 34.42 The electron gun

34.5.6.2 The slow-wave structure

The helix The helix was used in the first t.w.t.s developed and still is commonly used today. The great virtue of the circuit is its capability of producing bandwidths in excess of an octave. This capability has not been equalled by recently developed structures. The helix is however not easily cooled and this generally limits the maximum power output to below 2 kW.

The helix is usually made of tungsten and is wound with great accuracy to give a constant propagation velocity. It is supported by three or sometimes four insulating rods as shown in *Figure 34.43*. In higher-power tubes these are made of beryllia ceramic which having good thermal conductivity keep the structure cool. For power levels below 20 watts the cheaper alumina ceramics or quartz are used.

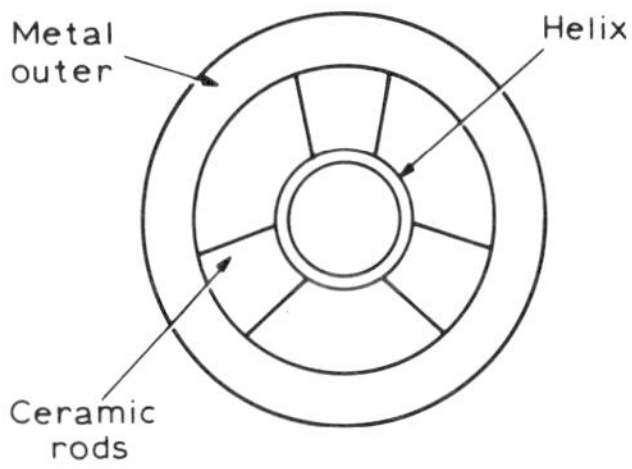

Figure 34.43 The helix structure

In low-power non-rugged tubes the outer vacuum envelope is made of glass but with higher power levels a metal cylinder is shrunk or brazed on to the tube rod assembly to ensure a good radial thermal path.

The ring and bar. In order to overcome the power limitations of the simple helix the contra wound helix was investigated and now finds use in the formalised version called the ring and bar structure, *Figure 34.44*. The internal beam diameter is larger than a comparative helix and therefore allows beams of lower current density to be used for a given output. The periodic nature of the structure however leads to a reduction in bandwidth. The basic construction is the same as for the helix tubes but power levels of 10 kW have been obtained with 30% bandwidth.

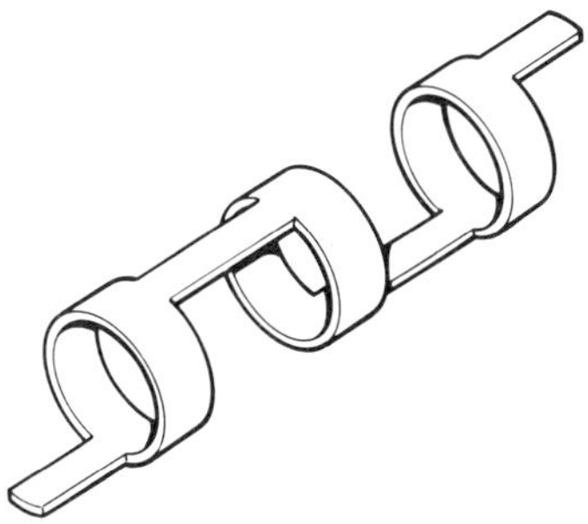

Figure 34.44 The ring and bar structure

The Hughes structure. A series of cavities may be coupled together by slots or loops to form a band pass structure. These structures are normally made of copper with geometrics such that good thermal paths allow high-power operation. One of the more successful arrangements is the Hughes structure shown in *Figure 34.46*. This structure gives bandwidths of 20% when designed for use in the region of 20 kW CW output power. At higher powers the usable bandwidth is less.

The clover-leaf circuit. The clover-leaf circuit has four or six wedges projecting into each pill box cavity and radial slots form

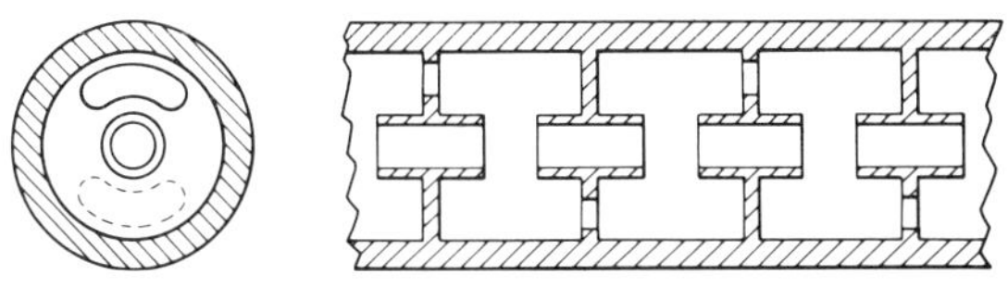

Figure 34.45 The Hughes structure

coupling elements into the adjacent cavities, producing a structure propagating a forward fundamental wave. The structure has found application at output power levels up to 1 megawatt peak with bandwidth of 10%. A typical tube[83] is shown in *Figure 34.46*.

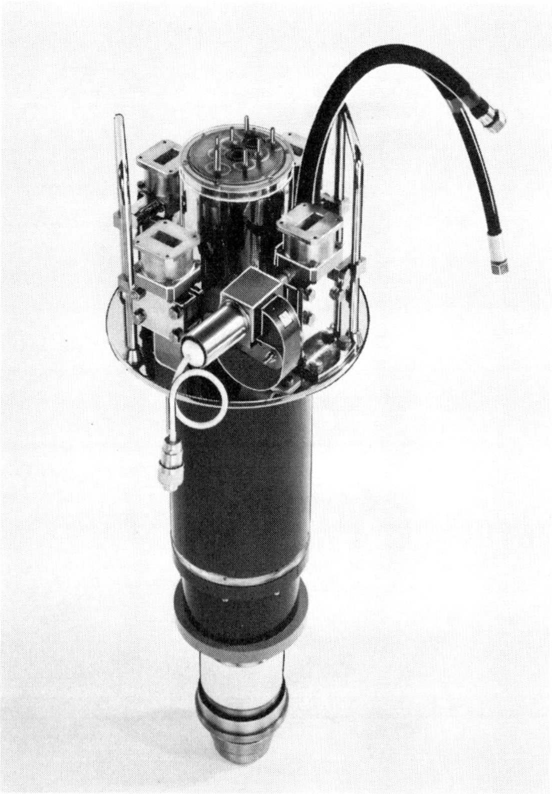

Figure 34.46 A high-power cloverleaf travelling-wave tube type N1061 (photo: English Electric Valve Company Limited)

34.5.6.2 The collector

The collector is an electrode designed to collect the spent beam of electrons after they have left the slow-wave structure. The energy of the electrons is converted into heat on impact with the collector surface and care is taken in the design of collectors for high-power tubes to ensure that internal surfaces of the collector are not overheated. The collector may be cooled by thermal conduction to an external heat sink, by air blown over the collector or by liquid cooling. In tubes where the overall efficiency needs to be high the collector is operated at a reduced voltage relative to the cathode. The electrons are slowed down between the slow-wave structure and the collector and thus have less energy to convert to heat on impact. Operation in this manner is called depressed collector operation.

34.5.6.3 Magnetic focusing

A magnetic field is required to contain the beam as it travels through the slow-wave structure. This field is provided by two main methods, the solenoid and by periodic permanent magnets. The solenoid is used today on low-noise receiver tubes and on some high-power coupled cavity tubes. It is however bulky and heavy and also requires a separate power source. The periodic permanent magnet focusing systems (p.p.m.) do not suffer from these drawbacks and find wide application wherever weight, size and efficiency are important.

A diagram of a periodic permanent magnet structure is shown in *Figure 34.47*. The polarity is reversed with each cylindrical

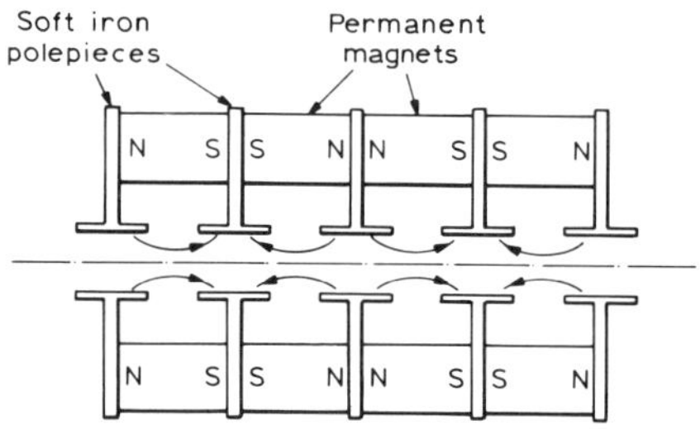

Figure 34.47 Periodic permanent magnet structure

magnet so producing a stack with an alternating magnetic field profile on the beam axis. The focusing system may be an integral part of the tube construction as illustrated in *Figure 34.48* or it may form a separate mount into which tubes are plugged.

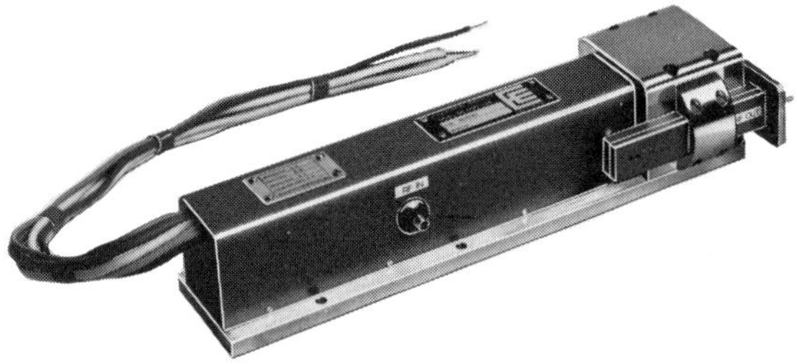

Figure 34.48 A broad-band helix travelling wave tube type N1081 (photo: English Electric Valve Company Limited)

34.5.7 Application of travelling-wave tubes

Travelling-wave tubes are widely used as amplifiers in all types of pulse and CW radars. Where larger bandwidths are required, with greater frequency agility, travelling-wave tubes are used as the final output stage of the radar transmitter as well as in driver stages. Low-noise travelling-wave tubes are used as the first stage in radar receivers. For use in airborne radars where high gain small size and rugged construction is important the travelling-wave tube finds application. These features also make the travelling-wave tube attractive for space use.

Other applications in the military field include uses in guidance and blind landing systems and control in missile weapon systems. The broad bandwidth of the helix t.w.t. makes it particularly attractive in electronic counter measures (ECM) and jamming applications.

In the civil communications field the t.w.t. is used as a transponder for television broadcasting both on the ground and in satellites. One of the widest uses is in microwave link communication systems where there may be several hundred travelling-wave tubes in one system.

34.6 Cathode-ray tube

The electrons from a heated cathode can be focused into a narrow beam. If this beam strikes a luminescent screen, a visible indication of the position of the beam is obtained. An electric field applied to the beam will deflect it, and so the screen will show a visible representation of the applied electrical signal. This is the principle of the cathode-ray tube, used in such instruments as the oscilloscope.

34.6.1 Construction

A schematic cross section of the cathode-ray tube is shown in *Figure 34.49*. The tube can be considered as having three sections:

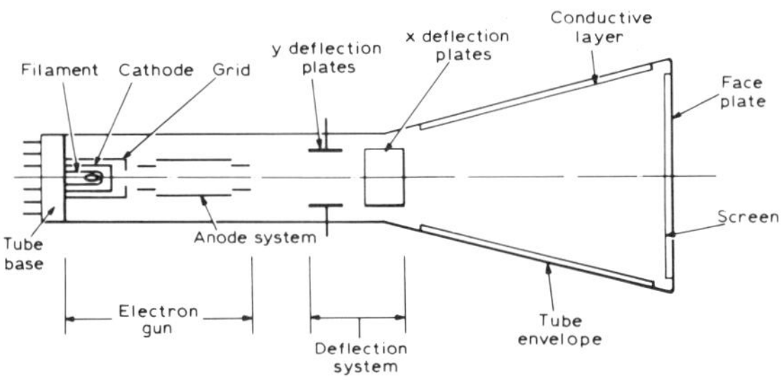

Figure 34.49 Schematic cross section of cathode-ray tube

the electron gun, deflection system, and the screen. The electron gun and deflection system are housed in the narrow neck of the tube; the screen terminates the cone part of the tube which is coated on the inside with a conductive layer.

34.6.1.1 Electron gun

The cathode of a cathode-ray tube is a cylinder closed at one end. The outside of the closed end is coated with a mixture of barium and strontium oxides which emits electrons when heated. The cathode is heated by a tungsten filament inside the cylinder, the filament being coated with alumina to insulate it electrically from the cathode. The cathode can then be held at earth potential.

The cathode is surrounded by a second cylinder, again closed at one end but containing an aperture through which the electrons from the cathode can pass towards the anodes. This second cylinder forms the grid of the cathode-ray tube. If the grid is made highly negative with respect to the cathode, no electrons can pass through and no display is obtained on the screen. If the grid is only slightly negative, electrons can pass through. As the brightness of the display depends on the number of electrons stiking the screen, and the number of electrons flowing through the tube is inversely proportional to the grid voltage, so this voltage can be used to control the brightness of the display. The range of grid voltage is typically 0 to −50 V. The cutting off of the electron beam by a highly negative grid voltage is called blanking.

A more detailed cross section of a typical electron gun is shown in *Figure 34.50*. The first anode is in the form of a disc with an aperture in the centre, or a short closed cylinder with apertures at both ends. This anode is held at a constant positive voltage with respect to the cathode, typically 1 to 2 kV. The electric field produced by the first anode brings the electron beam to a focus at point P between the grid and first anode. From this point the beam diverges again, but it is brought to a second focus at a point on the screen by the action of the second and third anodes. The spot on the screen is an image of the electron beam at point P, and so the spot diameter on the screen depends on the beam at P rather than on the cathode diameter or on any aperture in the electron gun. The spot diameter on the screen, also known as the line width, is typically between 0.2 and 0.4 mm.

The second anode is the focus electrode, and is an open cylinder to which a variable voltage about a mean value of typically +250 V is applied. The third anode is held at a constant positive voltage similar to that of the first anode and accelerates the electrons towards the screen. The aperture in this anode limits the beam diameter.

The first focus point of the beam, point P, is affected by changes in grid voltage (adjusting the brightness of the display) which may cause defocusing of the spot on the screen. This defocusing can be corrected by adjusting the voltage on the focus anode. Because of the screening effect of the first anode, adjusting the focus voltage does not affect the first focus of the beam.

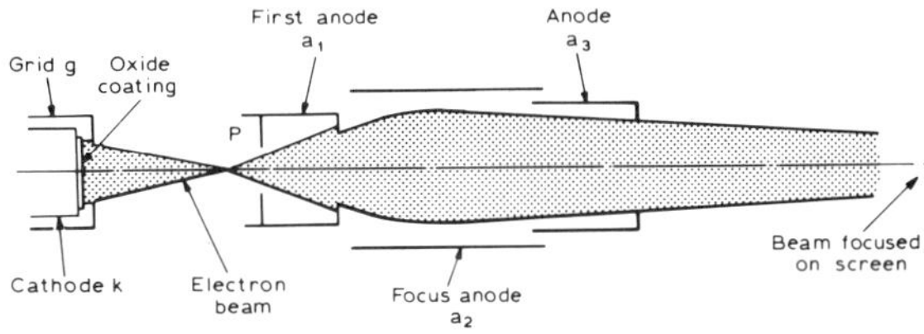

Figure 34.50 Cross section of electron gun

34.6.1.2 Deflection system

The electron beam in a cathode-ray tube can be deflected by a magnetic or electric field. For cathode-ray tubes used in television (picture tubes), magnetic deflection is used but for instrument tubes such as those in oscilloscopes electrostatic deflection is used. The beam passes between two parallel plates and a voltage is applied across the plates. The direction in which the beam is deflected and the distance through which it is deflected depend on the polarity and magnitude of the voltage across the plates. To prevent the deflected beam striking the plates, in a practical tube the plates are not parallel but either diverge towards the screen or are shaped to follow the path of the deflected beam. As a result, the deflecting force is not always at right angles to the beam, causing a variation in the force acting over the cross section of the beam. The spot on the screen may therefore appear elliptical rather than circular at the edges of the screen, an effect called deflection defocusing.

The deflection of the beam in a particular cathode-ray tube is related to the voltage applied to the deflection plates by a quantity called the deflection factor. Typical values of deflection factor lie between 4 and 45 V cm^{-1}.

Two sets of deflection plates are used in a cathode-ray tube, the sets being at right angles to each other. The sets are known as the x- and y-deflection plates, as shown in *Figure 34.51*, referring to the direction in which the spot is deflected on the screen. To prevent interaction between the two deflecting fields, a screen known as the interplate shield is placed between the x and y plates. This shield is usually held at the mean deflection-plate voltage, typically between 0.5 and 2 kV, although sometimes porivision is made to vary this voltage to correct distortion of the display. A shield may also be placed around the deflection plates to screen them from external fields.

34.6.1.3 Luminescent screen

The function of the screen of a cathode-ray tube is to provide a visible indication of the position of the deflected electron beam at any instant. This is achieved by forming the screen from

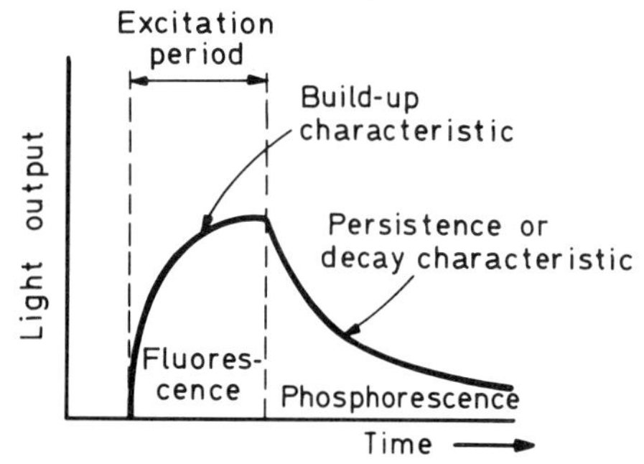

Figure 34.51 Light output characteristic of phosphor

phosphors, materials that emit visible light when struck by electrons. This effect is called luminescence. Practical phosphors are manufactured from 'host' materials to which are added 'activators'. Typical host materials include the oxides, sul'hides, silicates, selenides, and halides of aluminium, cadmium, manganese, silicon, and zinc; typical activators are copper, magnesium, and silver. The activator prolongs the time for which the light is emitted, but other materials called 'killers' (such as nickel) can be added to reduce the emission time. By choosing the host material and the activator or killer, phosphors of different characteristics can be obtained. The screen of a cathode-ray tube is formed by depositing a thin layer of the chosen phosphor on the inside of the faceplate of the tube.

When a point on the screen is struck by the electron beam, the light output from the phosphor rises to a constant level over the build-up time. The level of light output is determined by the energy of the incident beam, proportional to the square of the velocity of the electrons which, in turn, is proportional to the accelerating voltage of the tube. The light output remains constant while the phosphor is bombarded, but decays when the beam is removed. The emission of light when the phosphor is bombarded is called fluorescence and that during the decay is called phosphorescence. A typical light output characteristic is shown in *Figure 34.51*. The build-up and decay times depend on the phosphor used. The decay time is the more important characteristic, and the decay of light from the screen is called the persistence.

Screens of difference colours and characteristics can be obtained by choosing the appropriate phosphor. *Table 34.3* shows a classification of phosphors on the fluroescent and phosphorescent colours and the persistence.

When the electron beam strikes the screen, secondary electrons may be released. Some of these secondary electrons may leave the screen to be collected by the conductive layer inside the cone section of the tube (shown in *Figure 34.49*) which is held at the same voltage as the third anode. The ratio of secondary electrons leaving the screen to primary electrons arriving, the secondary

Table 34.3 Designation of phosphors

Pro-electron designation	*Fluorescent colour*	*Phosphorescent colour*	*Persistence**	*Equivalent JEDEC designation*
BA	Purplish blue	—	Very short	—
BC	Purplish blue	—	Killed	—
BD	Blue	—	Very short	—
BE	Blue	Blue	Medium short	P11
BF	Blue	—	Medium short	—
GB	Purplish blue	Yellowish-green	Long	P32
GE	Green	Green	Short	P24
GH	Green	Green	Medium short	P31
GJ	Yellowish-green	Yellowish-green	Medium	P1
GK	Yellowish-green	Yellowish-green	Medium	—
GL	Yellowish-green	Yellowish-green	Medium short	P2
GM	PurOlish-blue	Yellowish-green	Long	P7
GN	Blue	Green (Infra red excited)	Medium short (fluorescence)	—
GP	Bluish-green	Green	Medium short	P2
GR	Green	Green	Long	P39
GU	White	White	Very short	—
KA	Yellow-green	Yellow-green	Medium	P20
LA	Orange	Orange	Medium	—
LB	Orange	Orange	Long	—
LC	Orange	Orange	Very long	—
LD	Orange	Orange	Very long	P33
W	White	—	—	P4
X	Tricolour screen	—	—	P22
YA	Yellowish-orange	Yellowish-orange	Medium	—

* *Persistence of phosphors*

less than 1 μs	very short
1 μs to 10 μs	short
10 μs to 1 ms	medium short
1 ms to 100 ms	medium
100 ms to 1 s	long
greater than 1 s	very long

emission ratio, depends on the beam energy. At certain energies the secondary electrons may remain on the screen, and because the screen is a poor conductor, a negative voltage builds up and effectively decreases the accelerating voltage on the beam. This effect is called sticking, and results in a reduced light output from the screen.

Sticking can be prevented by depositing a thin layer of aluminium over the phosphor layer and connecting it to the conductive layer on the inside of the tube. Any electrons remaining on the screen are then conducted away. The aluminium backing layer also reflects forward the light from the phosphors that would otherwise shine down the tube and thus be lost.

34.6.2 Post-deflection acceleration

To ensure the brightest possible display on the screen, the accelerating voltage on the beam (the voltage on the third anode) should be as high as possible. Unfortunately the higher this voltage, the higher the deflection voltage must be to maintain the deflection factor (each part of the beam remains for a shorter time in the deflection region). There is therefore a limit to the accelerating voltage, and this can affect the brightness of the display for high-speed operation with the type of tube just described, the mono-accelerator cathode-ray tube. The beam can, however, be accelerated after deflection to overcome this limitation. A fourth anode is placed between the deflection plates and the screen to form the post-deflection acceleration cathode-ray tube.

This fourth anode is provided by the conductive layer on the inside of the tube envelope. The layer is now made in the form of a helix, as shown in *Figure 34.52*. The end of the helix nearest the

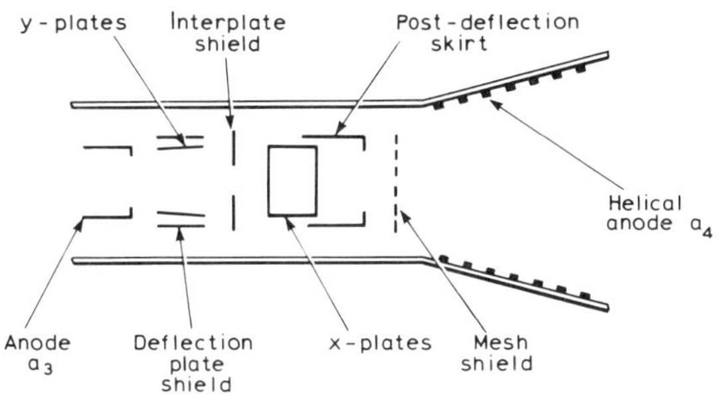

Figure 34.52 Electrode structure of cathode-ray tube with post-deflection acceleration

deflection plates is held at the third-anode voltage, while the end nearest the screen is held at a higher voltage. The ratio of these two voltages is the post-deflection acceleration (p.d.a.) ratio, and for a helical (spiral' p.d.a. tube this ratio is typically 6.

Because the electric field produced by the fourth anode can penetrate into the deflection region, a shield is placed between the deflection plates and the end of the helix. This shield is known as the mesh shield and is shown in *Figure 34.52*.

The electric field produced by the p.d.a. helix without a mesh shield is shown in *Figure 34.53(a)*. The modified field when a mesh shield is incorporated in the tube is shown in *Figure 34.53(b)*. The field now forms a low-magnification diverging lens, and a p.d.a. ratio as high as 10 can now be obtained. The voltage applied to the screen end of the helix can be as high as 10 to 20 kV.

The electron gun of a typical cathode-ray tube with p.d.a. is shown in *Figure 34.54*.

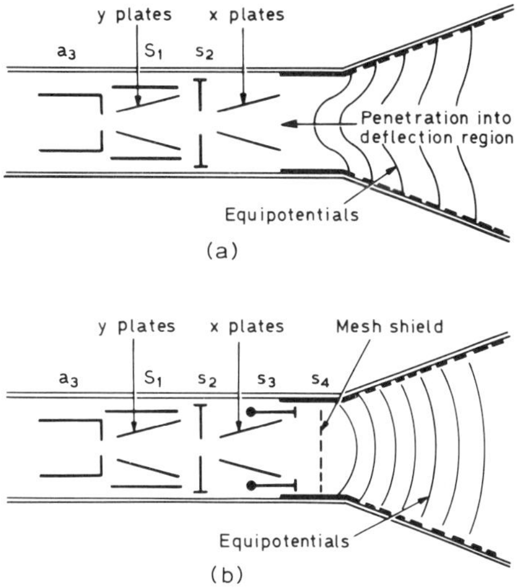

Figure 34.53 Electric field produced by p.d.a. voltage: (a) no mesh shield; (b) with mesh shield

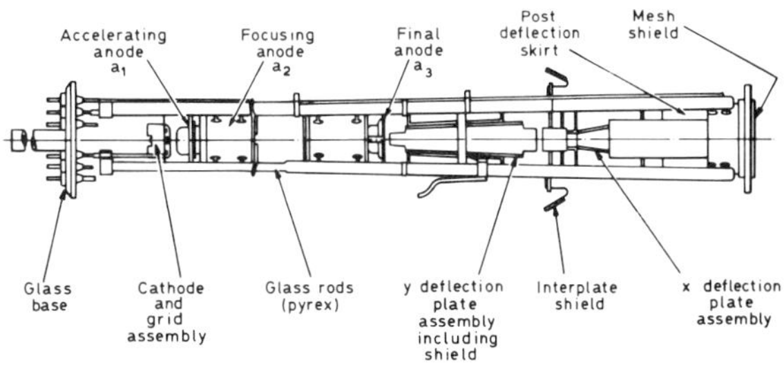

Figure 34.54 Electron gun for cathode-ray tube with post-deflection acceleration

34.6.3 Multidisplay cathode-ray tubes

It is often useful in an oscilloscope to display two traces simultaneously, particularly if the display is of related waveforms in a circuit. Two electron beams are required in the cathode-ray tube to achieve this, each with its own *y*-deflection system. The two beams may have a common *x*-deflection system, representing time for example, or can have separate *x* deflections. Two methods can be used in the tube to obtain the two beams. In one, the double-gun tube, two electron guns are used to provide the two beams. In the other, the split-beam tube, one electron gun is used and the beam is split into two so that each part can be deflected separately.

The splitting of the beam is done by placing a horizontal plate, the splitter plate, between the third anode and the *y* plates. The splitter plate is held at the same voltage as the third anode, and it is continued along the tube to form a screen between the two *y*-deflection plates. A deflection voltage applied to the upper plate will therefore operate only on the upper electron beam, while a voltage applied to the lower plate operates only on the lower beam. The two beams can then be deflected simultaneously in the horizontal direction by the *x*-deflection plates.

Because the beam is split into two, the brightness of the displays on the screen is half that of a single display. This loss of brightness can be a disadvantage when the tube is operating at high frequencies. An alternative method of splitting the beam to overcome this disadvantage is to split it in the electron gun. The third anode has two apertures instead of one, and so two beams emerge.

Post-deflection acceleration can be applied to both beams so the performance of a split-beam tube is comparable to that of a single-beam tube. A problem that may arise in the use of a split-beam tube is that when the two displays have widely different duty cycles, there will be a considerable difference in the brightness of the displays that may affect photography of the screen. (The brightness control will affect both beams). This disadvantage can be overcome by using a double-gun tube.

Because two guns are used in the tube, the brightness and focus of each beam can be adjusted independently. The beams can have a common x-deflection system or separate ones to give two independent displays in one tube. Because there are two electron guns, the double-gun is more bulky than the split-beam tube.

34.6.4 Higher frequency operation

There are two limits to the operating frequency of the cathode-ray tube previously described. As the frequency of the signal applied to the deflection plates increases, so the reactance of the plates presented to the deflecting signal increases. Each part of the electron beam takes a finite time to pass through the deflection region, the beam transit time, and if the deflecting signal varies over this time, the deflecting force will vary and the deflection of the beam will be less than expected. If the period of the deflecting signal is the same as the beam transit time, the beam will experience equal and opposite deflecting forces and the net deflection is zero.

The limit set by the reactance of the deflection plates depends on the construction of the tube. If the connections to the deflection plates are made by pins in the tube base, the inductance of leads the limits the operating frequency to about 10 MHz. If the connections are made through the sidewalls of the tube, and are therefore shorter, the frequency limit is between 50 and 80 MHz. Above this frequency, the transit-time limit starts to be significant.

The operating frequency can be increased further by effectively reducing the beam transit time. This is done by using a series of short deflection plates connected as a transmission line. The deflecting signal is propagated along this line at a velocity equal to that of the electron beam. Each plate deflects the beam, the individual deflections adding to give the final deflection. The frequency limit with this type of tube is about 500 MHz.

Higher operating frequencies can be obtained by using a helical deflection system. The axial propagation velocity of the deflecting signal along the helix is the same as that of the electron beam. Each part of the beam, therefore, remains in a constant field as it travels through the helix but with a steadily increasing deflecting force. Cathode-ray tubes with helical deflextion systems can operate up to frequencies of, typically, 800 MHz.

34.7 Television picture tubes

A television picture tube can be regarded as a c.r.t. adapted for the special requirements of displaying a picture. Because a large rectangular picture is to be displayed, a rectangular rather than a circular screen is required to avoid unused screen area. The large display area requires a large deflection angle if the tube is not to be excessively long. This large deflection angle, together with the high beam current when white areas are beind displayed on the screen, means that magnetic deflection has to be used instead of electrostatic deflection. It is through considerations such as these that the present-day black-and-white television picture tube has evolved.

34.7.1 Construction of black-and-white picture tube

The schematic cross-section of a present-day television picture tube is shown in *Figure 34.55*. Electrostatic focusing is used, combined with magnetic deflection. Magnetic focusing was used on some post-war picture tubes, but is not used on present-day tubes.

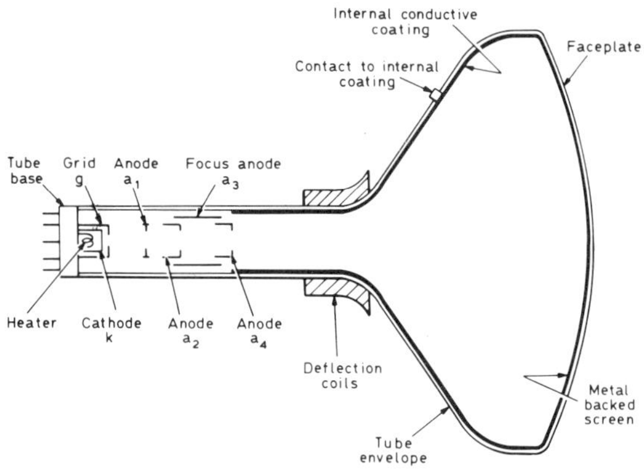

Figure 34.55 Schematic cross-section of black-and-white television picture tube (neck exagerated for clarity)

Electrons are emitted from the heated oxide-coated cathode k, and are brought to a focus between the grid g and the first anode a_1 by the effect of the potentials applied to these two electrodes. It is normal practice in present-day television receivers to drive the picture tube by modulating the cathode rather than the grid. The d.c. potential of the cathode is determined by the conditions of the video output stage, and this potential is modulated by the video signal. The grid is held at a potential which can be caried to adjust the brightness of the picture on the screen. The d.c. voltage applied to the grid controls the electron flow (beam current) as in any thermionic tube, and so controls the electron bombardment of the screen. When the brightness of the picture on the screen. The d.c. voltage applied to the grid controls the electron flow (beam current) as in any thermionic rube, and so controls the electron bombardment of the screen. When the brightness of the picture has been adjusted, the potential on the grid remains consrant. If the video signal were applied to the grid, the grid potential would vary and so alter the point at which the beam was focused. Cathode drive avoids this change of focus, while the varying cathode-to-grid voltage in response to the video signal modulates the electron beam.

The screen of the black-and-white picture tube is formed by a fine uniform layer of phosphor deposited on the inside of the faceplate of the tube. The electron beam striking the phosphor layer produces light by luminescence. The phosphor used is type W which produces a white fluorescence. A thin layer of aluminium is deposited over the phosphor layer to form a 'metal-backed' screen. This aluminium layer has two functions. It reflects forward light from the phosphors that would otherwise shine back down the tube and therefore be lost. The light output from the screen is therefore increased. The second function of the aluminium layer is to prevent *sticking* of the screen. The electron bombardment of the phosphor layer can produce secondary electrons. Because the phosphor is a poor conductor of electricity, these secondary electrons could remain on the screen. A negative charge would accumulate, decreasing the accelerating voltage on the electron beam, and so reducing the light output from the screen. The aluminium layer is held at a high positive potential through its connection with the internal coating and anode a_4, and so secondary electrons are conducted away and no negative charge can accumulate.

An external conductive coating is applied to part of the cone of the tube. This coating is connected to the chassis of the receiver, and the capacitance between this layer and the internal layer can be used to smooth the e.h.t. voltage applied to anodes a_2 and a_4 through the connector in the cone.

34.7.2 Colour television picture tube

In a black-and-white cathode-ray picture tube, a single, pencil-like beam of electrons is accelerated, deflected by a magnetic field into a rectangular scanning pattern and then strikes a phosphor material deposited on the inside of the front glass surface. The high-velocity electrons cause the phosphor to emit white light. There are many inorganic materials that have this light-emitting characteristic, known as cathodo-luminescence. The so-called 'white' phosphor is actually a balanced mixture of colour-emitting phosphors. A colour picture tube uses three colour phosphors, red, green and blue emitting which, if mixed in the right proportion, would also produce white. However, in a colour tube they are not mixed but are deposited on the inside of the front glass surface in many small triplet or triad groups of dots or lines. Instead of a single pencil-like electron beam, the shadow-mask tube uses three such beams, closely spaced. They are accelerated and deflected just as in the black-and-white tube. The three beams converge on each other to meet at the phosphor screen. However, by use of a perforated metal mask close to the phosphor screen, with one perforation for each triplet phosphor group, each beam is cut off or 'shadowed' from two of the phosphor colours and can strike only one of them. Thus, there is one beam that produces only red, one only green and one only blue.

With any colour-television system, a receiver can be designed to demodulate or decode the received signal into its three primary component signals, red, green and blue. If each colour component signal is connected to the picture tube to control the corresponding electron beam, separate red, green and blue pictures are produced that appear to the viewer to be superimposed because his eye cannot discern the very small, closely-spaced phosphor elements. If all three beams are simultaneously excited, white can be produced, or any colour or combination of colours. Most of the entire gamut of colours distinguishable to the eye are available over a wide range of brightness. Such a colour-picture tube can be used with any system of television. (However, there are minor modifications of aperture spacing in the shadow mask that minimise moiré effect with a particular number of scanning lines.)

Figure 34.56 shows schematically a small section of the two most common forms of shadow mask. One form uses small round holes and round phosphor dots, *Figure 34.56(a)*. The three electron beams originate from electron guns in a triangular, or 'delta' arrangement. In the other form, *Figure 34.56(b)*, the mask openings are narrow vertical slits, and the phosphor triplets are narrow vertical lines. In the latter case, the electron guns are on a horizontal line. In both forms, mask and screen are at the same potential and electrons travel in this region in straight lines, as shown. The angle between the beams is exaggerated in *Figure 34.56*, in reality it is only about 1°. The figure does not show the deflecting system or the overall shape of the tube, both of which closely resemble those of a black-and-white tube. However, the figure does show how the shadow effect is used to prevent each beam from striking more than its own colour phosphor. To be noted is that each mask aperture must be exactly in register with a trio of phosphor elements; this is a major problem that makes a colour picture tube so much more difficult to fabricate than its black-and-white counterpart.

34.7.3 Characteristics and limitations

A colour picture tube is characterised by colour fidelity, brightness (luminance), contrast, picture size, resolution and sharpness. Each characteristic usually involves compromise with the ideal because of various limitations, both theoretical and practical.

Colour fidelity is best understood in terms of the CIE colour diagram[93] shown in *Figure 34.57* and derived for an average observer. All visible colours are found by x and y coordinates that lie within the horseshoe-shaped figure. One can think of the x coordinates as one of three colour stimuli, the y coordinates as another and z, not plotted, as the third, the sum being unity. Thus, 'white' is in the neighbourhood of $x=\frac{1}{3}$, $y=\frac{1}{3}$, as shown by the central open circle. Extending a radial line out from the white point, the rotational angle determines the hue, and the distance out from white is a measure of the saturation or purity. Colours and spectral wavelengths are marked on the figure. A three-colour employs primaries that determine a triangle, and only colours within that triangle are reproduced.

In the colour-television systems used in most of the world, the

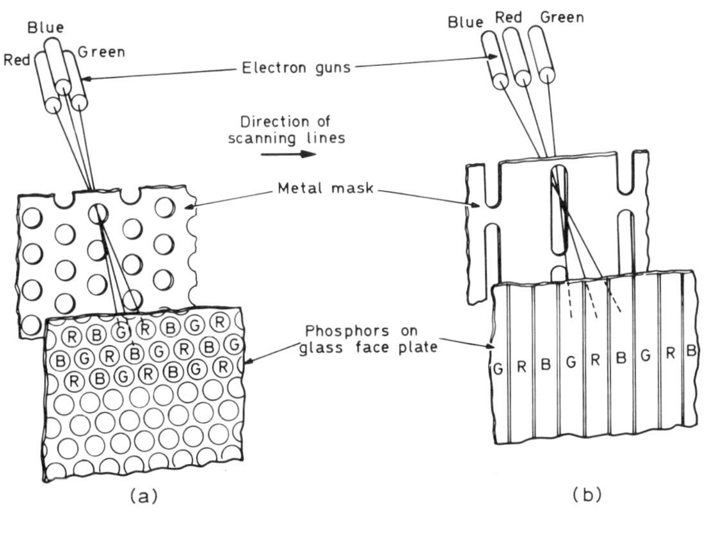

Figure 34.56 Schematic of two shadow-mask systems: (a) with round holes, and guns in a delta configuration; (b) with slit openings and in-line guns. In actual tubes, mask and face-plate are slightly curved

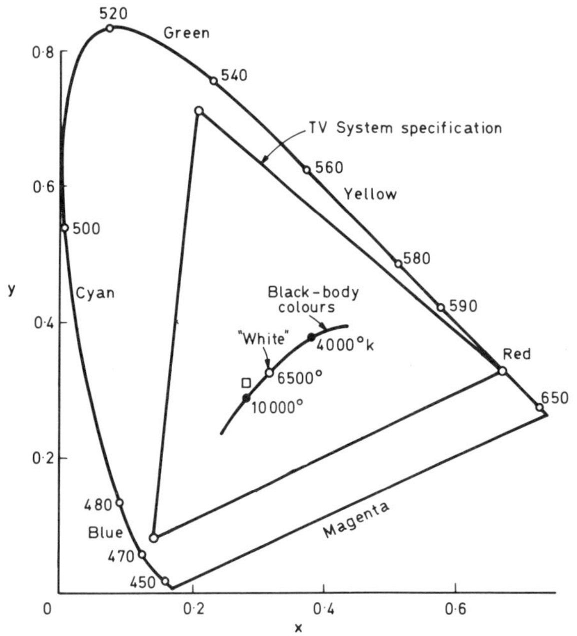

Figure 34.57 The CIE colour diagram. The specified transmitted colour system primaries and white point are shown by open circles. In the USA the receiver 'white' point is often set at the point shown by the square

colour cameras are arranged to produce the three chosen primaries that are approximated by the open circles. A 'white' point is set up, approximately as shown by the central open circle, and actually at $x=0.31$, $y=0.33$. The transmitted colour fidelity is better than that achieved by printing or photography and can be described as excellent. Ideally, the receiver picture tube should have phosphors that match these circles and, in fact, such phosphors are possible but they are not the most efficient. Fortunately, the compromise is hardly noticed by most viewers. Nevertheless, as phosphor research continues, one may expect closer approaches to the ideal, i.e. to the open circles.

A receiver using a colour shadow-mask picture tube, for example with the phosphors indicated by the crosses, can be adjusted to match the transmitted 'white' (open circle), by balancing the three individual beam currents. The open circle represents a *warmer* 'white' than is common in so-called black-and-white tubes, and in the USA many receivers are adjusted to a bluer 'white' shown by the square at $x=0.28$, $y=0.31$. Even this departure from ideal has been accepted although, once again, passage of time will undoubtedly bring about a closer approach to the transmitted standard. In summary, man's colour perception is so adaptive that departures from ideal colour fidelity are well tolerated,[95] and the compromises that are possible are manifold.

The maximum brightness or luminance of a colour picture is primarily determined by the phosphor efficiencies, the electron-beam power in watts per unit area, and the transmission of the shadow-mask and of the faceplate glass of the tube. Bright pictures are highly desirable but there is an increasingly noticeable 'flicker' above a certain brightness, particularly in television systems using 50 field/second picture sequence. Brightness is measured in candelas/m^2, also known as *nits* (for those accustomed to older units 1 candela/m^2 = 0.2918 foot lamberts). The limits to brightness, contrast picture size and sharpness or resolution, are interrelated.

The maximum contrast in a picture can be considered as the ratio of the highlight brightness to the brightness of a 'black' portion. The contrast ratio is a maximum when the picture is viewed in a completely dark room, but generally some ambient illumination is preferred. When such an ambient is present, the maximum contrast becomes the ratio of the highlight brightness to the reflected diffused ambient. Phosphors in use have a white body colour, i.e. they are almost fully diffusing. For this reason, means must be employed to minimise the light diffused back to the viewer. In a colour tube, the diffused light will also add white to a pure colour, thereby reducing colour saturation as well as contrast. Two methods have been developed for reducing the diffused light and are used either singly, or both together.

If light-absorbing material of a netural tint is incorporated in the faceplate glass to reduce its transmission to a fraction, T, then the highlight brightness is also reduced by T. The ambient light diffused back, however, goes through the 'grey' glass once, is diffused by the phosphor and again goes through the 'grey' glass on its return to the viewer. Thus, the diffused light is reduced by the factor T^2. The contrast is then improved by the amount, $1/T$, at the expense of a brightness loss. Values of T in the range of 0.5 have been employed when this is the only method used.

A second method incorporates an apertured black coating (called a matrix) that is deposited on the rear of the faceplate glass before the phosphor is deposited.[96] The apertures in the black coating exactly correspond to the central portion of the phosphor dots. With this second method, the phosphor area that is visible from the front remains white but it now covers only a fraction, F, of the screen area, thereby cutting the diffused ambient by the same fraction. In practical designs, it is not possible to excite the entire screen area with the electron beam without danger of electron spot overlap on to undesired colours. In a conservative design (i.e. tolerance of the order of ± 0.1 aperture diameters in spot position) the useful phosphor portion is of the order of 0.5 of the entire area, i.e. F is also about 0.5, the remainder being black. This design improves contrast by just as much as grey glass with $T=0.5$, but without the loss of highlight brightness by absorption. In practice, some matrix-type shadow-mask tubes have used $F=0.53$ and $T=0.85$ in combination. Compared with grey glass only, of $T=0.5$, such matrix tubes have a brightness improvement of $0.85/0.50=1.7$ and a contrast improvement of $0.5/(0.85\times 0.53)=1.1$. More commonly, matrix types are designed for greater brightness by increasing the useful phosphor portion to $F=0.7$, at the expense of tighter tolerance and lower contrast. In a dark room, contrast of practical shadow-mask tubes approaches 50:1, which is excellent, and in a reasonably well-lit room (250 lux illumination) the contrast is still of the order of 10:1 or more, which is acceptable. A black matrix increases tube manufacturing cost and is used where maximum brightness is desired; because of flicker, 50 field/s systems are less likely to need matrix screens.

There are limits to brightness and, consequently, to contrast that are imposed by the electron-beam power per unit area and the necessary sharpness of the picture. To keep X-ray radiation below safe limits, even with heavy elements added to the faceplate glass, the upper voltage limit is of the order of 25–35 kV. A focused electron beam produces a spot on the screen that gets larger as the current is incrreased, but the effect is less in shorter tubes with large deflection angles. Thus, a 100° deflection-angle tube will be brighter than a 90° one, at the same beam spot size, because the beam current can be larger. In a given tube, if an attempt is made to increase the beam current excessively for highlights, the spot gets so much larger that picture sharpness and resolution are lost and the picture appears fuzzy. The beam spot is described as *blooming*. In general, it is desirable to have a small-enough spot to show the scanning lines, although a more important characteristic for a sharp picture is the edge appearance of large-area objects.[97,98]

If all else is equal, large picture sizes are much more pleasing than small ones. The picture shape is fixed by the transmission standards, i.e. rectangular with 4:3 ratio of horizontal width to vertical height. The faceplate is ordinarily made in approximately this shape to conserve space. A practical limit to the size of the picture tube is imposed by cost, and by the awkwardness of length.

The shortest tubes are those with the largest deflection angle; the angle was once 70°, but by 1973 only 90°, 110° and 114° were being used. The larger of these angles are close to the limit imposed by deflection difficulties as well as by practical fabrication techniques.

Both black-and-white and colour television systems compress the signal amplitudes at the transmitter by a fractional power law; usually the 'gamma' or fractional exponent is 0.46. This compression considerably improves the signal-to-noise ratio. In a colour receiver, it is particularly important that the transmitted exponent be corrected by its inverse, i.e. the light output for each colour should be approximately proportional to the 2.2-power of the signal voltage. Fortunately, it is a characteristic of cathode-ray guns that they do have about this power law and no special receiver circuits are necessary.

34.7.4 The round-hole shadow-mask tube

The shadow mask with round holes, as in *Figure 34.58(a)*, can be thought of as the traditional type and was the only type in use prior to 1968. The principles have already been outlined, but many details need amplification.[87,99] *Figure 34.58* shows a cutaway view of such a shadow-mask tube with a 90° deflection angle and a 630-mm picture diagonal. The mask is spherically

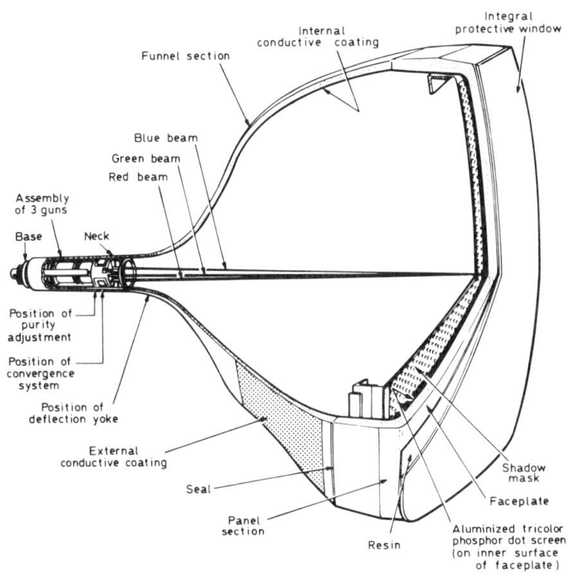

Figure 34.58 Cutaway view of a 90° deflection angle, delta-gun, shadow-mask tube

curved and rests behind a similarly curved glass faceplate at a spacing of about 14 mm. There are three separate electron guns in a delta arrangement, tilted slightly to form beams at an angle of about 1° from each other; when undeflected, the beams point at the centre of the faceplate. Corresponding to each mask hole is a triad of red, green and blue light-emitting phosphor dots superposed on a black matrix.

The figure suggests how such a tube is put together. The funnel section represents one part and the faceplate or panel section, with an integrally mounted shadow mask, representing a second part. The two are sealed together with a vacuum-tight low-melting-point glass frit and the electron gun is then sealed into the neck. The figure also shows an integral protective window on the faceplate to prevent flying glass in the event of an implosion. This window is bonded to the faceplate with a resin. Other types of protection are also in use, including a tight metal band around the rim of the faceplate.[100]

Inside the faceplate, the tri-colour arrays of phosphor dots are covered with a thin, electron-permeable aluminium layer to reflect the phosphor light in a forward direction, as with black-and-white tubes.

In operation, the deflection yoke is placed over the neck, up against the funnel section. In the tube shown, an external magnetic shield is placed around the funnel section, but some types have an internal shield. In either case, an automatic demagnestising field coil (degaussing coil) is placed around the whole, and is actuated for a short time with alternating current each time the receiver is turned on.[101]

Operating voltages are connected to the gun, and anode high voltage to the mask, phosphor screen and internal conductive coating. In most types, focusing is done electrostatically using a bi-potential design that requires an adjustable voltage for focusing of the order of 15% to 20% of the anode voltage; the beam focus electrodes for all three guns are internally connected so that only a single focus voltage is required. In some smaller-screen types, with somewhat less critical performance requirements, a so-called 'enizel-lens' focus is used; this is an electrostatic system with one element at approximately cathode potential and requires little or no adjustment.

In any three-gun colour tube it is important that the three beams coincide (converge) at the phosphor screen over the entire deflection raster, or else the colour picture would not be properly superposed. In the central region of the screen convergence is achievable by aiming the guns at one point, making minor correcting adjustments with four external magnets, combined with internal pole pieces built into the gun. Such central-region coincidence is called static convergence. When the three beams of a delta-gun-type tube are simultaneously deflected, for example to the corners of the picture, convergence errors occur. These are corrected by a number of special current wave shapes that are applied to a dynamic magnetic-field convergence system that uses internal pole pieces. These current wave shapes are derived from the deflecting system so as to be in correct time synchronism with scanning. In a typical receiver with a delta-gun shadow-mask tube, each tube must be separately adjusted for static and dynamic convergence; this requires four static and 12 dynamic adjustments. One other adjustment is required, known as 'purity'; this is also done by an external magnet system, and serves to shift the centre-of-deflection of the three beams to the position that leads to maximum colour uniformity over the entire screen. The convergence adjustments bring the three beams together over the entire picture area, and the purity assures that each beam strikes only its correct colour phosphor dot over the entire picture area. In spite of the apparent complexity, the use of appropriate dot and cross-hatch test patterns permits these adjustments to be made by an experienced person in a matter of minutes.

In most delta-gun shadow-mask tubes there are separate cathode (*K*), control-grid (*G*1) and screen grid (*G*2) electrodes for each of the three guns. Such designs can be connected to the receiver circuits in many ways. Some receivers use a luminance signal (equivalent to a black-and-white signal) on the cathodes, and three colour-difference signals on the control grids. In such circuits, the picture tube itself is part of the colour matrixing system. Other receivers fully separate the red (R), green (G) and blue (B) video signals, and these are applied to either the three cathodes or the three control grids.

In the design of the round-hole shadow-mask tube, consideration must be given to the moiré pattern that results from the scanning lines interacting with the periodic shadow-mask openings.[87] There is a minimum moiré effect for any given number of scanning lines; a 625-line t.v. system will sometimes use a different shadow-mask periodicity than a 525-line system in order to stay at the minimum.

34.7.5 The Trinitron

The Trinitron[90,102] was the first major variation from the round-hole shadow-mask tube. In the Trinitron, the mask consists of a grill of vertical strips from top to bottom. Such a mask is not self-supporting and also cannot be spherically curved. For this reason, it is mounted under tension on a relatively massive frame and the curvature is cylindrical. The mask is used with a similarly curved faceplate and the phosphors are deposited in triads of R, G and B vertical lines (similar to *Figure 34.56(b)*). To prevent vibration, very small horizontal wires are placed on the convex surface of the mask, but these are so fine as not to be noticeable.

The three electron beams lie on a horizontal line, as with in-line guns sometimes used with the round-hole shadow mask, or as with the slotted mask of *Figure 34.56(b)*. The Trinitron tube is considered relatively costly to manufacture. However, an important advantage of the Trinitron grill mask is that it has no structure in the vertical direction.[98] Thus, the sharpness of brightness changes in the vertical direction is comparable to that of a black-and-white tube, and is limited only by the scanning lines and the beam spot size and shape. This feature, together with the larger diameter lens that is employed in the gun, may explain the oft-expressed viewer reaction that Trinitron pictures are exceptionally sharp.

The Trinitron gun is unusual because it is not an assembly of three separate in-line guns but, instead, uses three electron beams in a single-gun system. The advantage of this lies in the larger lens diameter (with lower spherical aberration) and smaller electron-beam spot than is possible when three individual guns must be placed in the same size neck. *Figure 34.59* shows the gun configuration of a 114°-deflection Trinitron.[102] Only one

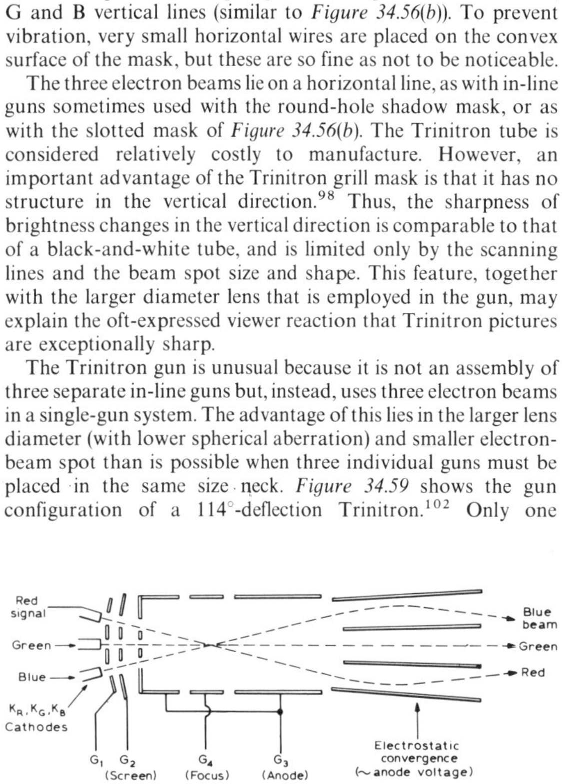

Figure 34.59 The single-gun, three-beam configuration of the 114° deflection Triniton. The three beams are in line and are used with a vertical line screen. By crossing the beams at the focus electrode, a single large-diameter lens can be used and this reduces the spherical aberration

electrode is used for the first grid and another one for the second grid, in distinction to the three each used in the traditional designs. To enable the gun to produce three beams, three separate cathodes are used, one for each colour signal. In the gun, the three beams are tilted to cross each other at the centre of the focus electrode; this focus system then acts in common on all three. Beyond this convergence point, the beams separate again but are electrostatically converged. When undeflected, the beams strike a common point at the centre of the screen, as indicated in the figure.

By proper yoke design, it is possible to simplify dynamic convergence in in-line tubes compared with delta-gun types. A delta-gun tube uses 12 dynamic convergence adjustments. An in-line gun reduces this to four adjustments and the 114° Trinitron gun and yoke uses two preset corrections to reduce the adjustable number to two. The predicion in-line system, to be described in the next section, eliminates *all* dynamic convergence adjustment.

34.8 Television camera tubes

34.8.1 General principles

The camera tube has been the subject of continuous research since its inception and many varied designs have seen commercial operation. The television camera provides a lens system, usually magnetic electron beam focus and deflection units. Electrical power supplies, and signal processing circuits complement the camera tube.

The tube has three essential components. These are:

(a) A photosensitive surface located in the image plane of the camera lens and upon which the incoming light produces an electronic charge.
(b) A charge storage target.
(c) An electron gun which produces an electron beam to scan the target and evaluate its charge pattern to produce the video signal.

(a) and (b) are often combined in a single component.

The tube, usually of cylindrical form and highly evacuated, can rely upon electromagnetic, electrostatic or a combination of both for its operating principles.

In operation, the photosensitive surface receives the optically focused image from the camera lens and in the case of photoemitters the resulting electrons are directed on to the storage target. Alternatively, for photoconductors, electrons are generated to produce internal conduction. If the sensor material is a polar crystal material then a change in dipole polarisation intensity occurs.

In either case, the changed electrical state of the target is restored to the former 'steady state' by the scanning electron beam and it is the recharging current as a function of position, and therefore time, which constitutes the video signal. The precise mechanism of signal generation is peculiar to each tube type. Furthermore, tube constructional detail varies among the various manufacturers.

34.8.2 The image orthicon

The sensitivity of the orthicon was inadequate for many purposes because of basic limitations in the storage target. Its high capacitance produced only a small operational voltage swing causing beam discharge lag which resulted in unacceptable handling of moving objects. On the other hand higher potentials due to picture highlights tended to build up into values above the control of the scanning beam.

Both these deficiencies were overcome and combined with an image section similar to that of the image iconoscope to produce the image orthicon—announced in 1946[136] by RCA. With this tube the majority of television services throughout the world were established and it is still the preferred camera tube for monochrome television.

Seen in *Figure 34.60* the tube incorporates three major sections:

(a) an image intensifier;
(b) a scanned storage target; and
(c) an electron multiplier.

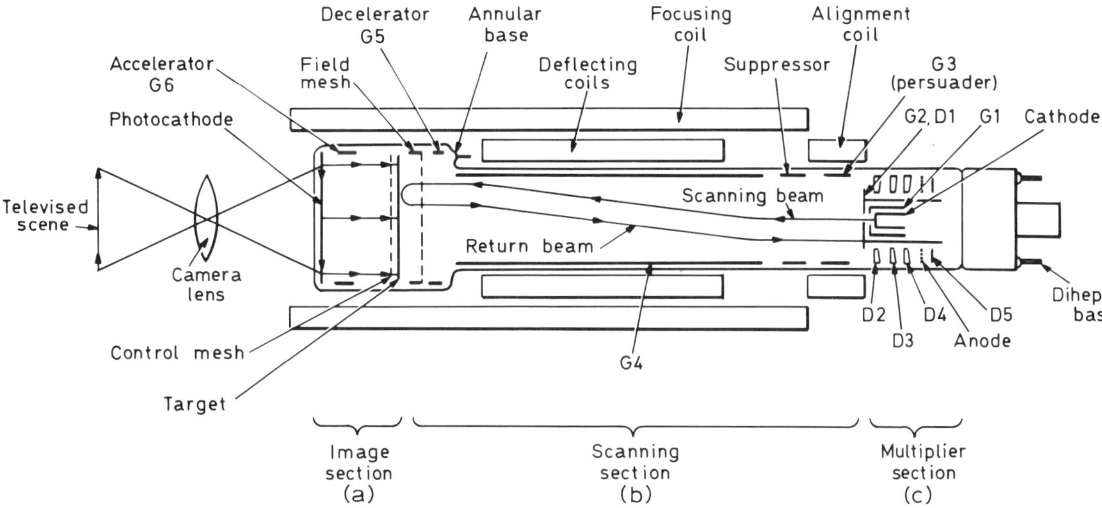

Figure 34.60 Schematic of 3″ image orthicon tube. The $4\frac{1}{2}$″ version uses the same diameter optical image but incorporates a magnifying electron lens in the image section in order to completely fill the larger target

The optical image projected on to the continuous pyotocathode generates an electron image which is focused by the surrounding solenoid and the electrostatic field in the image section upon the micro-thin glass target. In modern tubes this glass has very slight ohmic conductivity[137] compared with the ionic process of early examples. At the target surface secondary electron emission takes place—the secondary electrons being collected by the slightly positive and highly transparent mesh. The loss of secondaries produces a nominal electron gain and a positive charge pattern on the glass of a density varying in proportion to the original pattern of scene brightness. Charging continues in the picture highlight areas until the glass surface slightly exceeds the mesh in potential. The charge pattern is transferred to the opposite target surface for evaluation by the scanning beam. As the target is scanned some beam electrons drift back randomly ro the electron gun. Other beam electrons then neutralise the target charges and the then zero potential elements resistively reduce the input face of the target to near zero potential so that a new value of charge may be assumed. The remaining beam electrons are reflected by the zero potential elements roughly along their outgoing path to the limiting aperture of the electron gun. Due to the electron-optical conditions at the target, the returning electrons impinge on the dynode surface rather than disappear into the aperture. At the dynode, secondary electrons are produced and are guided into the electron multiplier by the negative potential of the persuader. In the 5-stage electron multiplier with about 300 volts per stage a total gain of about 1000 is achieved to give a final output signal of several microamps—well above the noise level of the head amplifier. The signal amplpitude is measured from maximum output corresponding to full beam returned from picture black, to lo lower values created by the various shades of grey, with a minimum generated by peak white in the original scane. This negative polarity signal is one grave disadvantage of the image orthicon since the maximum return beam from picture black has highest noise content and this is most objectionable in those dark areas of the received picture.

Signal-to-noise ratio is optimised by making the stored target charge per picture element as high as is practicable consistent with maximum read out efficiency to give minimum lag. This is achieved by using targets with high storage capacitance—i.e. with mesh and target close together and/or large area. In this respect the series of tubes of $4\frac{1}{2}$″ diameter[138] is superior to the original 3″ series. The latter is retained in high sensitivity applications where the lower capacitance target can become fully charged at lower light levels. As mentioned above, charge accumulates on the target until the potential of its input face has risen to a value just above that applied to the mesh such that the net secondary emission is zero. Clearly, this saturation value is reached at lower light levels for wider spaced, smaller diameter systems which have lower capacity.

Graphically the Signal (S)/Illumination (I) relationship is shown in *Figure 34.61*, but variations occur according to tube type and mode of operation. With logarithmic coordinates this

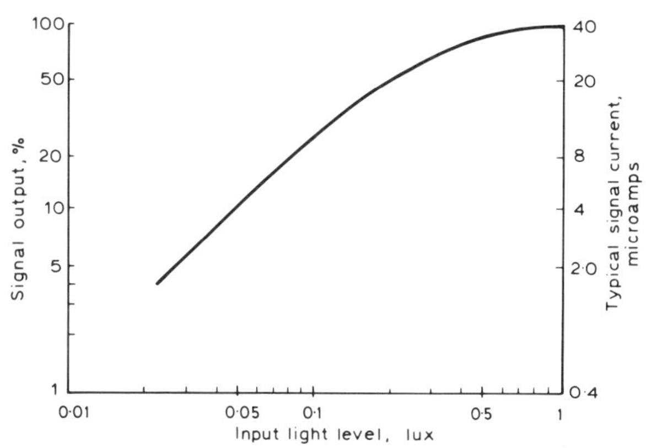

Figure 34.61 Typical transfer characteristic of an image orthicon tube. It may be noted that, according to the light level and contrast ranges for the input values, a gamma of 1 or 0.5 or a combination may be obtained. This is of value to camera operation

curve can be represented by the formula $S = \gamma I + \text{constant}$, γ or gamma, the index of proportionality is unity at lower levels of illumination and reduces at higher levels—an operational attribute of the image orthicon.[139–142]

34.8.3 The image isocon

The scanning of the target in an image orthicon gives rise to scattered electrons, in quantity proportional to the scattering charge. By using a selective electrode system, diagrammatically shown in *Figure 34.62*, these can be guided into the electron multiplier and the reflected electrons discarded to form a video signal of correct polarity and with none of the disadvantages of the image orthicon signal. By the end of 1973 such tubes began to assume great importance in low light level television systems due to the possibility of using a lower capacity storage target than is practically feasible in the image orthicon, to produce higher sensitivity without the attendant noise problems.

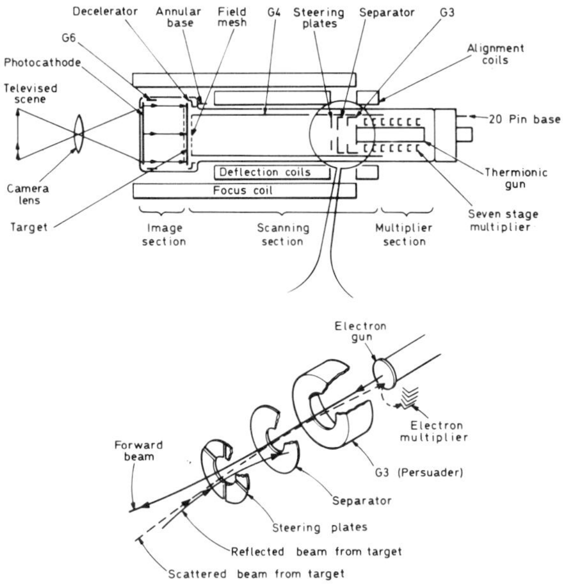

Figure 34.62 Schematic diagram of the image isocon

34.8.4 The vidicon

This name, originally coined by RCA[143] is now often used as a generic term for all orthogonally scanned camera tubes using a photoconductive target (*Figure 34.63*).

In operation the laminar target suffers a proportionate reduction in resistance according to the local brightness of the incoming light image. The applied steady potential to the transparent conducting backing layer then gives rise to a potential pattern on the vacuum side of the target. When scanned to cathode potential, capacitive coupling acorss the target layer produces a video output current of varying amplitude proportional to the potential of the element from which it was derived and hence proportional to the elemental brightness of the original scene.

Materials for use as photoconductive targets must satisfy fairly well defined requirements.[144] There must be a maximum conversion of light into available charge carriers and to achieve this, absorption of the incoming radiation must be high and the effective band-gap of the marerial should correspond in energy to the longest wavelength in its spectrum. For the visible range a band-gap of less than 2 eV is required. Freedom from trapping centres is also necessary in order that the constituent molecules can return rapidly to the unexcited state.

For satisfactory operation the completed target must have a satisfactory time constant, so imposing upon the basic material a lower-limit of resistivity of 10^{12} Ω cm and certain restrictions upon thickness when used as a continuous layer. In this regard adequate thickness is necessary to ensure maximum light absorption but on the other hand it must be limited to minimise charge diffusion, optical scattering and working capacitance.

Various materials can be used as vidicon targets. They include selenium, antimony trisulphide, cadmium selenide, lead oxide and silicon. The latter two materials have low resistivities and must be processed to incorporate blocking contacts in the target layer for satisfactory operation. Details are given below.

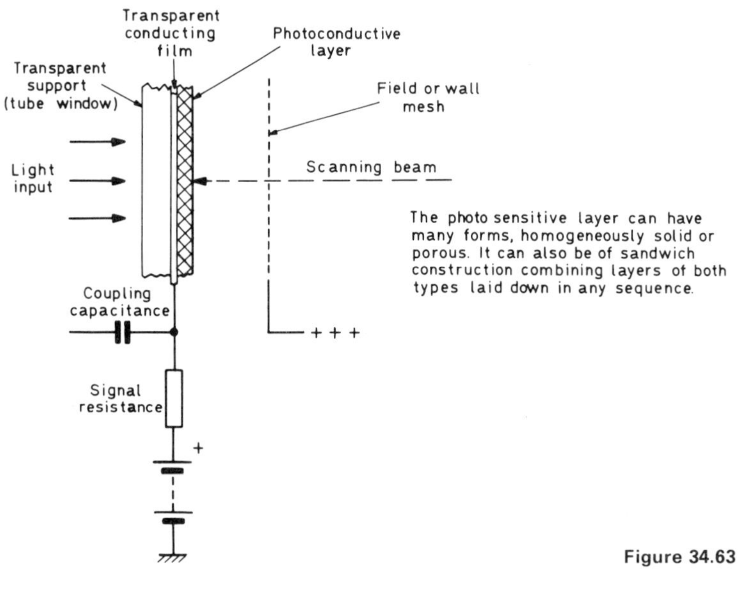

Figure 34.63 Schematic diagram of vidicon target arrangement

34.8.4.1 Selenium

Selenium was used in first examples[143] of the vidicon-type tube. It was unstable and had a tendency to revert to an electrically conducting form after some hours of use. Resurrected briefly in the form of a Se/Te alloy in the USA in the mid-1960s it was eventually abandoned because no lasting remedy could be discovered for its instability.

34.8.4.2 Antimony trisulphide

Basically of the formula Sb_2S_3, antimony trisulphide has seen extensive use in vidicons.[145] It is the least costly material to process but has limited sensitivity and suffers from movement lag.

As a target layer, antimony sulphide can be vacuum evaporated to be thin and dense or, in an atmosphere of inert gas such as argon, to a porous film of greater thickness. The operating capacity of the latter is desirably lower as also is its sensitivity. In practise, combinations of hard and porous coatings about 1 μm to 3 μm thick are used. Controlled exposure to air is permissible to allow convenient assembly methods. Antimony trisulphide vidicons have a standing output current or dark current which is temperature dependent when not illuminated, a severe disadvantage in colour cameras. Operational gamma varies from 0.7 at low illumination to 0.4 at higher levels.

Resolving power of early antimony sulphide vidicons was inadequate and was shown to result from the presence of heavy positive ions in the end region of the beam focus electrode near to its terminating field mesh.

A considerable improvement is obtained when the latter is electrically separated from the beam focus electrode and the two operated at slightly different potentials, see *Figure 34.64*.[146,147]

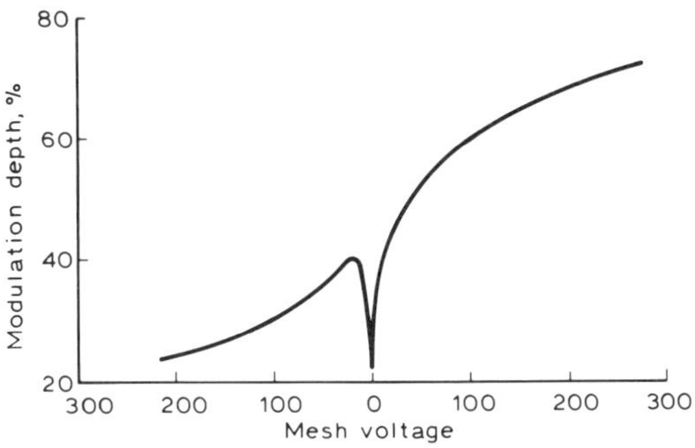

Figure 34.64 Graph showing relation between resolution and mesh potential of a separate mesh vidicon type tube

The separate mesh tube allows a correction to be made for the non-linear magnetic field at the target end of the focusing solenoid. Furthermore, due to the higher field gradient between the mesh and target, higher beam currents can be used for better control of picture highlights.

In spite of several operational drawbacks such as smearing of moving objects, high temperature dependent dark current, etc. the antimony sulphide vidicon is widely used in a variety of closed-circuit television applications.

34.8.4.3 Silicon

Pure silicon exhibits very high photo-conductivity. However, due to its low ohmic resistance, a mosaic of discrete diodes must be formed[148,149,150] for it to serve as a camera tube target.

One technique is to first heavily oxidise the eventual mosaic surface of an optically worked lamina of single crystal silicon. Then, by photo-lithographic techniques the continuous silicon dioxide layer is converted into a fine grid a few microns in pitch—so exposing the base silicon material as a conglomerate of holes. Next a p-type material such as boron glass is fused over the whole surface so that each silicon hole now becomes a pn junction, surface insulated from its neighbours. To minimise the self-biasing action of the insulating grid when scanned, a metal, or reasonably conducting material is applied to each diode surface to reduce the effective bar width of the grid but still maintaining mutual insulation between the diode surfaces. *Figure 34.65* indicates this arrangement.

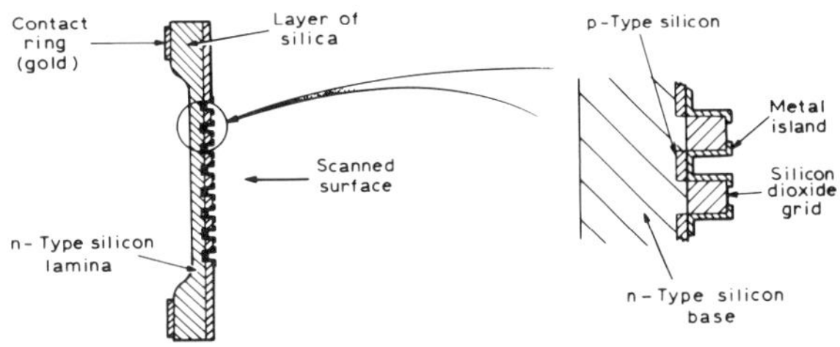

Figure 34.65 Diagrammatic cross-section of a silicon diode array target

The final manufacturing operation is to chemically etch the back of the intended picture raster area to its operational thickness—about 25 μm. Sometimes this surface is given dichroic treatment to enhance a particular spectral sensitivity. The completed target is mechanically mounted in conventional vidicon position with minimum clearance between it and the inside of the tube faceplate. The inevitable small gap is of no optical consequence but does preclude the use of fibre-optic techniques.

In operation, the continuous silicon backing plate is biased some 10–20 volts positive with respect to the scanning beam cathode. Scanning the mosaic stabilises it at cathode potential so that each diode assumes a reverse bias condition—the p-type islands just below cathode potential. Behind each one, and probably forming a continuous layer, is a region depleted of current carriers to insulate it from its neighbours.

When illuminated, current carriers, electrons and holes, are photoelectrically produced in the silicon layer. The electrons are attracted by the positive signal plate potential and the holes migrate to charge the p-type islands for evaluation in normal manner by the scanning beam.

The silicon vidicon has very high sensitivity and extends into the near infra-red. It is of use in this spectral band for simple surveillance work. Its large scale use was intended to be for the television telephone where its ability to survive electrical and light overloads and its anticipated long life is of extreme advantage compared with other vidicon-type tubes. The tube is used to a limited extent in the red channel of some vidicon colour cameras.

34.8.4.4 Lead oxide

In recent years lead oxide of the form PbO, has become the most important photoconductor used in tubes of the vidicon format.[151–155] Lead oxide is a true semiconductor and can exist in p, n or intrinsic forms. In the lead oxide vidicon a porous target, structured as shown diagramatically in *Figure 34.66* exists, having been produced by evaporation under carefully controlled conditions.

In operation, the positively biased signal plate and the negative

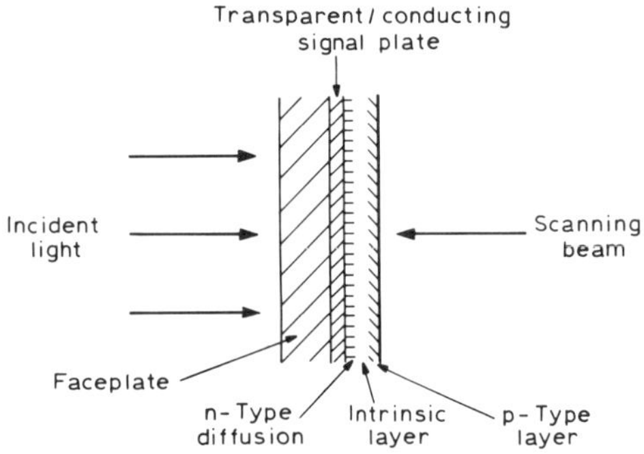

Figure 34.66 Schematic section of a lead-oxide target

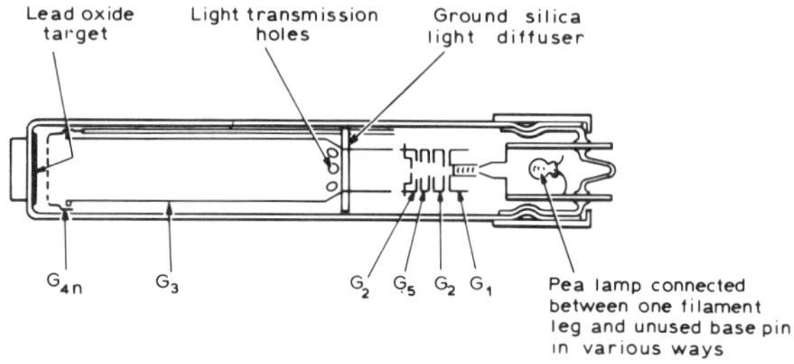

Figure 34.67 Schematic diagram of English Electric HOP/Light Bias Leddicon tube. Gun electrode G2 is, in effect, divided and an additional one, GS, interposed

scanned surface stabilised slightly below cathode potential effectively remove all current carriers from the body of the layer to give the tube a zero output current when un-illuminated, a most valuable asset for colour signal processing. When illuminated, strict proportionality prevails over the operational range of scene brightness between incident illumination and signal current.

Tubes incorporating a principle known as light bias, became available[156] in the early 1970s. 'Light bias'—a small amount of target illumination—can be generated either in the optical system, in the tube socket or in an internal appendix of the tube itself. Its function is to raise the potential of the scanned surface at very low light levels so that the efficiency of the scanning beam is improved at the resulting low signal currents. Without light bias this surface stabilises slightly below cathode potential at a value at which only the higher energy beam electrons can reach the target so that discharge effieicncy is seriously impaired.

Tube light bias effectively enables differential lag between individual camera tubes to be minimised.

Other relatively undesirable properties of early lead oxide vidicons have led to improved variants each of which has particular application in the television field.

Tetragonal lead oxide (litharge) has a band gap of 2 eV which gives it a wavelength threshold sensitivity at approximately 640 nm. This prevents it from 'seeing' long wavelength red colours. The deficiency has been overcome, where required, by a partial doping of the lead oxide, with sulphur for example, to produce a material with a band gap significantly below 2 eV. The sensitivity increase in red is roughly doubled. In ganeral, electron trapping in the modified material tends to be higher so that image retention can be slightly greater than in the standard tube. However, the higher sensitivity and better colourimetry are advantages to be offset against this.

The light transfer characteristic of the lead oxide target is essentially a linear function ideally suited to the mathematical signal processing required in colour television systems. However, unlike simple vidicon and image orthicon tubes, there is no high-light saturation—a situation leading to unwanted 'ballooning' of picture high-lights. Some allveiation can be achieved by careful camera and lighting application but up to five lens stops improvements results from special tube designs designated ACT (anti-comet-tail)[155] or HOP (high-light overload protection).[156] These types use a modified electron gun which incorporates an additional electrode (see *Figure 34.67*).[156] A voltage pulse applied to this electrode during the line scan retrace time causes a high density beam to reach the target and remove the high surface charges resulting from high input illuminations. This technique also removes the 'comet-tail' produced when a high-light in a televised scene has movement relative to the camera.

Alternative and somewhat simpler systems use either the signal current or that part of the scanning beam remaining after target recharge which returns to and is collected by the second grid of the electron gun to modulate the reading beam in accordance with the highlight density. Ih either case the control is achieved during active line time via the usual beam control grid—often in conjunction with a modified gun in which this grid operates at a positive potential instead of the negative value conventionally used.

Lead oxide vidicon tubes under various trade names and of several diameters are vailable. The 30 mm tube is manufactured in both integral and separated field mesh construction—the latter being most efficient. 25 mm and 18 mm lead oxide tubes are also commercially available and a developmental 13 mm format was announced in 1981. These tubes are all of separated mesh construction—the smaller sizes being applied almost exclusively to the electronic news gathering (ENG) requirements of television broadcasters. A domestic and industrial market of large magnitude is forecast for cameras using tubes of the smaller sizes, both for colour and monochrome reproduction.

34.8.4.5 Other target materials

Lead oxide, properly processed and incorporated into a vidicon-type camera tube meets practically all the demands of the television broadcaster and is, at the time of writing, the best all-round compromise. However, better sensitivity and resolution are always welcome and research, mainly in Japan, has resulted in the evolution of new target layers especiallygood in one or two operational parameters. Among these are Chalnicon[157] using cadmium selenide with antimony trisulphide, Newvicon with zinc selenide and zinc cadmium telluride, Saticon[158] using a layered structure of selenium, arsenic and tellurium.

34.8.5 Low light level camera tubes

These are of two basic forms—either a combination of a conventional type as described above with an image intensifier[142] or a tube of image orthicon concept but incorporating an intensifier target,[159] *Figure 34.68*. Potassium chloride has been successfully used[160] as an intensifier target. Operated at approximately 15 kV, a cascade of secondary electrons is produced in the thickness of the target from the incoming photoelectrons such that a charge gain of approximately 50 times is achieved. Such tubes carry the generic name of SEC vidicons—the prefix denoting secondary electron conduction.

Arsenic trisulphide,[161] the silicon diode array[150] and zinc sulphide[162] are also used in this manner but in these cases the target conductivity is produced by high energy electron bombardment. Target gains of 100–500 times are typical of tubes of the electron bombarded silicon target—variously known as SIT (silicon intensifier target) tubes or of the EBSICON type (electron bombarded silicon) type.

In each of the above tubes the video signal is taken from the

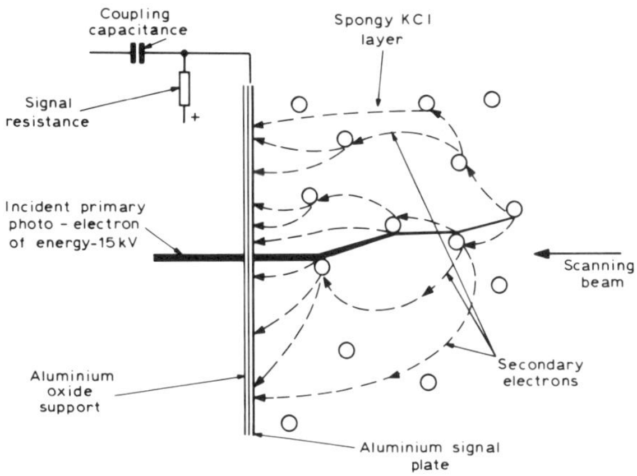

Figure 34.68 Diagrammatic representation of the secondary electron cascade in a target of spongy potassium chloride

conducting support for the target mosaic—for the SEC vidicon a mesh supported aluminium film and, in the case of the EBSICON, the continuous silicon backing layer.

For tube combinations, coupling can be by lens or direct between the output phosphor and fibre optic window of the image intensifier to the fibre optic input window of the camera tube. The photocathode of the latter is specially processed to match the spectral output of the phosphor. The increase ih sensitivity of the system is, naturally, the gain of the intensifer. Some reduction in resolution is introduced by the fibre optics.

34.8.6 Multicolour tubes

Colour television requires the generation of three primary signals in the red, blue and green spectral regions and these must be originated in the television camera. For this purpose, apart from farme sequential systems, according to the coding principle adopted one, two, three or four video channels are employed requiring a corresponding number of camera tubes each conforming to a particular monochrome specification.

In this section tubes which, in themselves, generate composite 'coloured' signals are described. These require quite different signals processing to that used with combinations of monochrome tubes.

The simple vidicon was adapted to produce red, blue and green signals in 1960.[163] It used a target system which was in effect a superimposition of three. A sequence of colour filter stripes was applied to the inside face of the entrance window. On each stripe was deposited a transparent signal plate and the photoconductor applied as a continuous layer over the whole. A common contact was applied to each set of signal plate stripes so that when exposed and scanned in conventional manner signals from each of the three colours were produced to be analysed and reconstituted by appropriate circuitry on a shadow-mask kinescope. The tube did not reach commercial production due, probably, to the extreme precision necessary for the construction of the target and also to its very low sensitivity.

Modern tubes use similar principles but rely less on extreme mechanical accuracy and only brief constructional details are available.

Several arrangements[164,165,166] of internal stripe dichroic filter systems with a common signal plate have been announced but with little detailed information. One tube uses in addition an indexing signal plate in combination with the striped filters. An alternative possibility is to apply the filters to the outside face of a fibre-optic entrance window to the tube.

In all cases, when scanned these colour indexed targets give rise to composite video signals from which the required colour components are extracted by phase or frequency discriminating circuitry.

Two tube cameras[167] with an optical splitter use on e tube to supply the luminance signal and the second, possessing a simple frequency coded filter system, to provide the red and blue signals. Again electronic circuitry enables the desired chrominance signal components to be derived.

In spite of their low relative cost, application of these colour tubes has been in fields other than broadcasting due to their marginal performance. Howeverk in the education world, microscopy and limited single shot interview work their use is increasing significantly.

34.9 X-ray tubes

It is now known that the X-ray is an electromagnetic radiation lying in the wavelength region of 10^{-10} metres (10^{12} MHz) between ultraviolet and γ radiation.

Proof of the electromagnetic nature of X-rays came from early experiments which showed that X-rays travel in straight lines, that their intensity at a distance obeys the inverse square law, that they are not deflected by magnetic fields and that they can be transmitted through a vacuum. More recent investigations proved that X-rays can be polarised and refracted and can produce interference and diffraction.

34.9.1 Generation of X-rays

The simple X-ray tube comprises an evacuated envelope containing a thermionic cathode and an anode or target. If a very high potential difference is applied between anode and cathode, with the anode positive, electrons emitted from the cathode by thermionic emission are accelerated towards the anode and strike it with considerable energy, producing X-rays, the intensity of the X-rays being a function of the current in the tube. These X-rays emanate from the surface of the anode in all directions.

Study of these X-rays reveals that a number of different frequencies are emitted and with the aid of a spectrometer a graphy of intensity versus frequency can be built up. A typical graph is shown in *Figure 34.69.*

As will be seen, the typical spectrum consists of a continuous or 'white' range of frequencies upon which are superimposed a number of sharply defined spectral lines. These two types of spectral are believed to originate in different ways and will be dealt with separately.

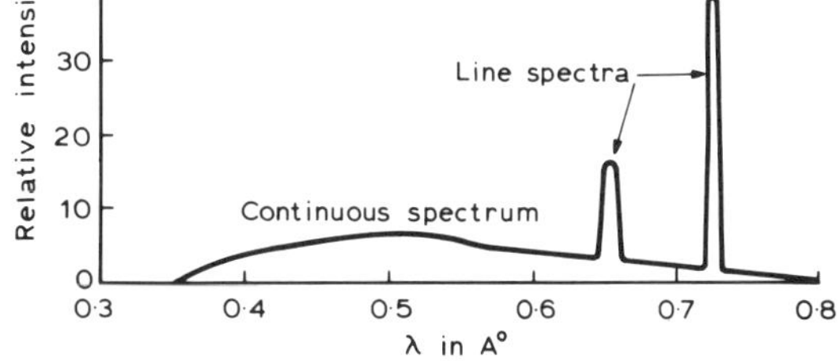

Figure 34.69 Spectrum of a typical X-ray tube

34.9.1.1 The continuous spectrum

When a fast moving electron strikes an atom under the surface of the target, its energy of motion is converted into radiant energies. The electron may lose all its energy in one collision in which case a radiation of frequency given by the quantum relation $E_e = hf$ is produced, where E_e is the original energy of the electron, h is Planck's constant and f is the frequency. Thus if the original energy E_e is sufficiently high, i.e. the electron has sufficiently high velocity, then a high frequency radiation is given off as an X-ray.

Thus the highest frequency of radiation given out as a result of the electron losing all its energy in a single collision, being dependent upon the electron velocity, is a function of the accelerating force, i.e. the anode-to-cathode potential.

Thus for any given applied voltage V, a radiation of frequency f_{max} is produced where

$$V = \frac{hf_{max}}{e} \qquad (34.25)$$

where e is the charge on the electron. Alternatively

$$V = \frac{12.4}{\lambda_{min}} \qquad (34.26)$$

where λ is the wavelength in Å and V is in kilovolts. For example, if $V = 10$ kV, $\lambda = 1.24$ Å.

However, it is possible for an electron to lose its energy not in a single collision but as a result of a number of glancing blows with a successive number of target atoms and, in this case, each collision produces a radiation, the frequency depending upon the energy lost.

This does not necessarily mean that X-rays are produced at each successive collision—some of the collisions may produce electromagnetic waves of much lower frequency, i.e. light or heat.

Thus, if an electron of energy E_e, say, strikes three target atoms before coming to rest and loses energy E_1 at the first collision E_2 at the second and E_3 at the third, then electromagnetic waves of three different frequencies f_1, f_2 and f_3 will be produced.

The continuous spectrum may therefore be considered to be caused by two different effects—firstly the sharp cut off at the maximum frequency which results from electrons in a single collision and secondly the broad band of frequencies (the 'white' spectrum) resulting from electrons losing their energy in a number of separate collisions.

34.9.1.2 Line spectra

The frequencies of the sharply defined line spectra, however, have been found experimentally to be a function of the element from which the target is manufactured and these X-rays are believed to emanate from deep within the target atoms, mostly from the electrons in the three inner shells.

In the normal atom each of these shells contains its normal complement of electrons. It is possible, however, for a cathode-emitted electron to penetrate deep into the atom and to eject an electron which is contained in one of the inner shells, thus leaving a vacancy.

If this should occur, an electron from the next shell (higher energy) falls into the vacancy, giving up energy as a quantum of X-radiation. This leaves another vacancy which is filled by an electron from the next higher energy shell producing another quantum of radiation at a different frequency.

The energy given up by an electron falling from one shell to the next is given by the energy difference between the two shells. Thus the frequency of the radiation is calculated from:

$$f = \frac{E_2 - E_1}{h} \qquad (34.27)$$

where E_2 and E_1 are the energy levels of the two shells.

In the case of the K, L and M shells the energy difference is sufficient to produce X-ray frequencies and, since this energy difference varies from element to element, the frequency of the line spectra is a characteristic of the element. The production of line spectra requires a minimum potential difference which varies according to the atomic number of the target element. The higher the atomic number, the greater is the voltage required to give the electron sufficient energy to enter the inner shells of the element and to eject an orbital electron.

For example, the minimum potential differences to eject K shell electrons in copper and molybdenum are as follows:

copper (atomic number 29) $V_{min} = 8.9$ kV
molybdenum (atomic number 42) $V_{min} = 29$ kV

Of all the electrons leaving the cathode of an X-ray tube, less than 1% actually produce X-radiation. The energy of the remaining electrons is dissipated as heat and massive cooling systems must be used in practice.

34.9.2 Properties of X-rays

X-rays, like light rays, are electromagnetic in character and many of the properties of light are also properties of X-radiation subject to the fact that X-rays have a much higher frequency and are therefore more energetic.

Many substances which are transparent to light are also transparent to X-rays but, owing to their higher energy, X-rays can also pass through heavier or denser materials. Different elements absorb X-rays to a different degree and in fact the amount of absorption is proportional to the atomic weight and the density.

Röntgen proved that X-rays would pass more easily through flesh than bone and thus provided the medical profession with a means of investigating the human body without resorting to surgery.

X-rays also have the property of ionising gases. The X-ray energy produces a few high speed electrons in the gas and these, in turn, liberate a large number of low energy electrons by a collision process. From the point of impact on the material under bombardment, reflected or scattered X-rays are found to leave the material in such a manner as to obey the optical laws of reflection.

Other X-rays, however, are also evident. These—the characteristic X-rays—are particular in nature to the properties of the material under bombardment. They are, however, only produced if the energy of the incident X-ray beam is greater than the minimum energy required to displace K, L or M electrons in the material under bombardment.

Finally, β particles or high speed electrons are also evident but can only be detected in the area of impact as a result of their limited range in a medium. As stated, these arise mainly from an inverse X-ray effect. Some of the particles, however, may result from a photo-emissive effect on the surface of the material.

X-rays themselves are affected by the medium, the attenuation as already stated being proportional to the atomic weight or density. Even in air there is a measurable attenuation.

Intense X-rays also affect living tissue by inhibiting the growth of body cells, destroying tissue and causing inflammation as a result of the production of intense heat. Although the biological action of X-rays is fundamentally injurious, it is this very property which is used in the annihilation of diseased body cells. This treatment is called radiotherapy.

34.9.2.1 Secondary radiation

A most interesting phenomenon associated with X-ray bombardment is the complex character of a secondary X-radiation which occurs. This secondary radiation comprises scattered X-rays, characteristic X-rays and β particles. The scattered X-rays may be assumed to be analogous to reflected or refracted light and the β particles, having the same velocity as the electron beam in the X-ray tube, probably result from a type of inverse X-ray effect.

The characteristic X-rays, however, arise from the ejection of electrons deep in the inner shells of the atoms of the scattering material and are produced by a process similar to the production of line spectra. The secondary radiations are therefore a characteristic of the scattering material and provide a convenient method of analysis.

34.9.3 Diffraction and X-ray crystallography

Two properties of X-rays which are common to all electromagnetic rays are diffraction and interference and it was in 1912 that the scientist von Laue suggested that X-rays might be diffracted if the regular arrangements of atoms in a crystal were used as a diffraction grating. In 1913 W. H. Bragg used crystal diffraction to measure the wavelengths of X-rays, calculating the crystal spacing from Avogadro's number. X-ray diffraction by crystals has led to an entirely new field of research—X-ray crystallography—three principal methods being in present day use.

34.9.3.1 Laue method

In tne Laue method a continuous spectrum of X-rays is allowed to pass through a single crystal by way of a collimator which narrows the beam. X-ray diffraction occurs within the crystal and a number of reinforcement spots appear on a photographic screen placed beyond the crystal.

34.9.3.2 Bragg method

In the Bragg method a narrow X-ray beam is reflected by a crystal. Interference occurs and the positions of the resultant spectral lines are investigated using an ionisation chamber or a photographic plate.

The principle of Bragg's appratus, known as an X-ray spectrometer, is illustrated in *Figure 34.70(a)*.

The incident beam of X-rays from the source A is limited by the collimator B. The reflecting crystal C is mounted in wax on the table which rotates, its position being determined by the vernier V. An arm D, rotating about the same axis as the table, carries the ionisation chamber E. In practice the ionisation chamber and the crystal are rotated slowly and the magnitude of ionisation in the chamber is noted. A graph of ionisation versus angle of rotation can then be plotted.

Figure 34.70(b) shows a typical spectrum obtained in this manner. The maxima correspond to line spectra in the original beam.

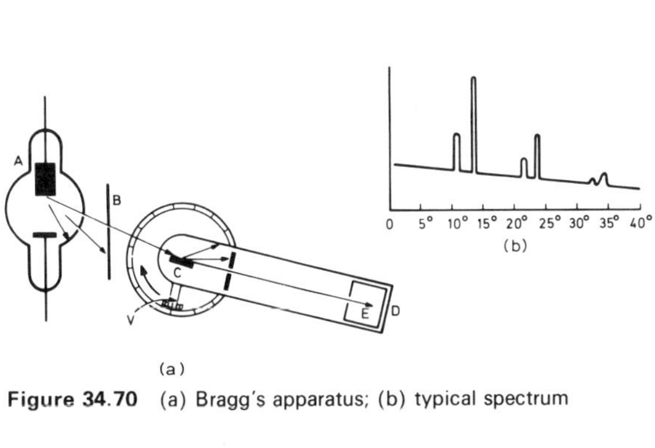

Figure 34.70 (a) Bragg's apparatus; (b) typical spectrum

34.9.3.3 Powder method

The powder method has the advantage of requiring only a very small quantity of the material under test. Crystals are ground to a very fine powder through which an X-ray beam is allowed to pass. The method is essentially that of Laue except that monochromic X-radiation is used. The random orientation of the crystal planes in the powder produces a circular pattern which is characteristic for a particular element of compound as each circle corresponds to reflection from a given set of crystal planes.

34.9.4 X-ray tubes

The most difficult electrode to design is the anode or target. At the focal point, i.e. the small area of the anode bombarded by cathode-emitted electrons, less than 1% of the electron energy actually produces X-rays. The remaining energy is given up in the form of intense heat. A favoured technique employed is the use of an anode which comprises a massive copper block into which is embedded a small tungsten target. The high thermal conductivity of the copper together with the very high melting point of the tungsten make this combination very suitable.

All X-ray pictures are shadows and so cannot be sharp unless the source is of small dimensions. In the older tubes the cathode rays were focused on a small circular area but many modern tubes have what is called a line focus, i.e. the focal area is a narrow line. This prevents excessive heating and enables the X-ray output of the tube to be increased.

Figure 34.71 illustrates the principle of line focus. If A is the target, F is the focal area and CF is the axis of the tube, it is seen

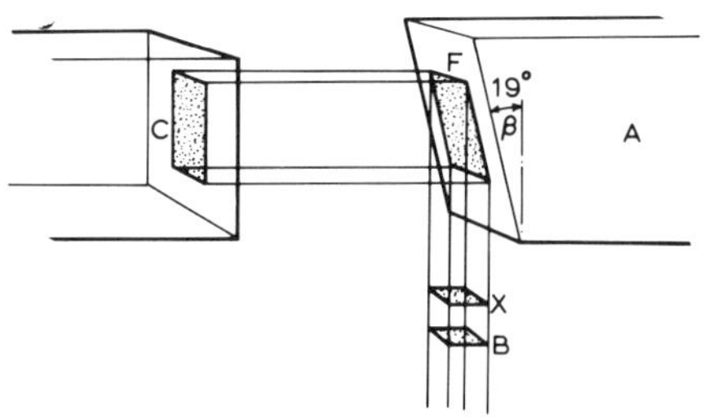

Figure 34.71 Circuit symbol for photoemissive tube

that the electron beam travelling along the axis CF is rectangular in cross-section. The face of the target A is cut at an angle of 19° to the direction from which the X-rays are viewed, BF. At B the X-rays appear to have originated from a small square-shaped area of dimensions as shown at *X*. (In this diagram the cross-sectional areas of the electron and X-ray beams have been considerably enlarged for the purpose of clarity.)

X-tay tubes can effectively be divided into two types—those used for treatment (therapy) and those used for diagnosis.

Therapeutic tubes give rise to demands for an efficient cooling system owing to the necessity of operation over long periods of time.

Figure 34.72 shows the schematic diagram of a typical X-ray tube used for diagnosis. This comprises a substantial heater surrounded by a cup-shaped assembly K which is negatively charged with respect to the anode so as to form an electron beam that focuses on the target T. The anode comprises a massive block of copper *A*, into which a piece of tungsten T is embedded, this acting as the actual target for the electrons. All the electrodes are situated in an evacuated tube G which is itself encased with a lead shield having a small aperture from which the X-rays radiate.

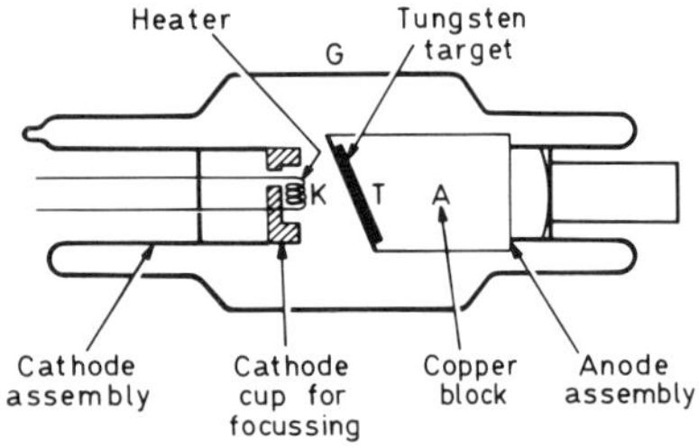

Figure 34.72 Typical X-ray tube

34.9.5 X-ray image intensifier

Before the X-ray image intensifier was introduced, the only method of viewing X-rays directly was by means of a fluorescent screen and, unfortunately, unless the X-ray intensity is maintained at a dangerously high level, the intensity of illumination provided by the fluorescent screen is extremely low indeed.

The image intensifer provides an illumination bright enough not only for direct viewing but also for cinephotography under conditions of very low X-ray intensities.

As illustrated in *Figure 34.73*, the image intensifier comprises an evacuated envelope E containing a complex cathode structure S, K and a simple fluorescent screen A.

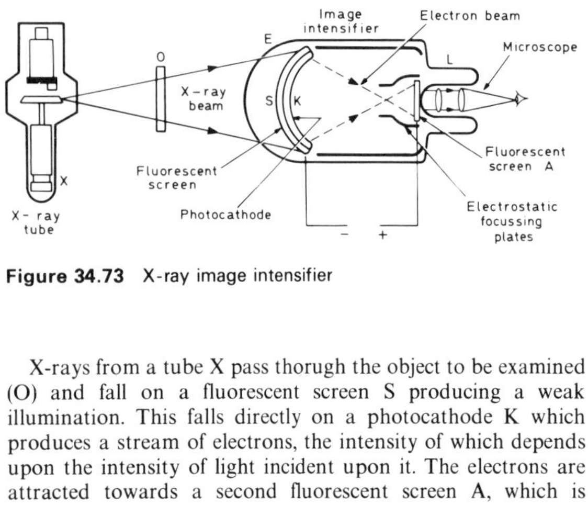

Figure 34.73 X-ray image intensifier

X-rays from a tube X pass thorugh the object to be examined (O) and fall on a fluorescent screen S producing a weak illumination. This falls directly on a photocathode K which produces a stream of electrons, the intensity of which depends upon the intensity of light incident upon it. The electrons are attracted towards a second fluorescent screen A, which is maintained at a positive potential with respect to the photocathode, and upon impact produce a light image of small size but of high intensity. This image can be viewed by a simple lens system L, a 'still' camera or a cine camera.

34.10 Photoemissive tubes

Photoemissive tubes are pf two types: vacuum tubes and gas-filled tubes. Both types have the same electrode structure sealed in a glass envelope; the difference is that the vacuum type has a high vacuum inside the envelope while the gas-filled type has a low-pressure filling of an inert gas.

The cathode of a photoemissive tube, called the photocathode, is made from a material which emits electrons when illuminated. The photocathode area is large so that the largest amount of radiation is collected and the largest number of electrons emitted. The anode is a thin rod so as not to obstruct the radiation falling on the photocathode. The circuit symbol for a photoemissive tube is shown in *Figure 34.74*.

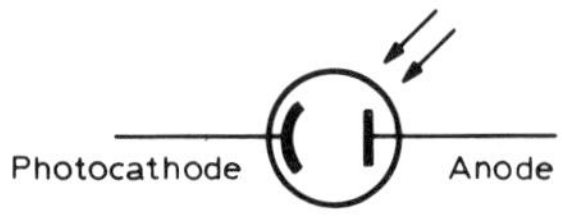

Figure 34.74 Simplified cross section of photomultiplier

In a vacuum photoemissive tube, if the anode-to-cathode voltage exceeds a value called the saturation voltage V_s, all the electrons emitted when the photocathode is illuminated are attracted to the anode. For a fixed level of radiation, the photocurrent remains constant with increasing anode voltage. As the radiation increases, so a greater number of electrons is emitted from the photocathode and a larger photocurrent flows. Thus a family of anode voltage/cureent curves with luminous flux as parameter is obtained, as shown in *Figure 34.75*.

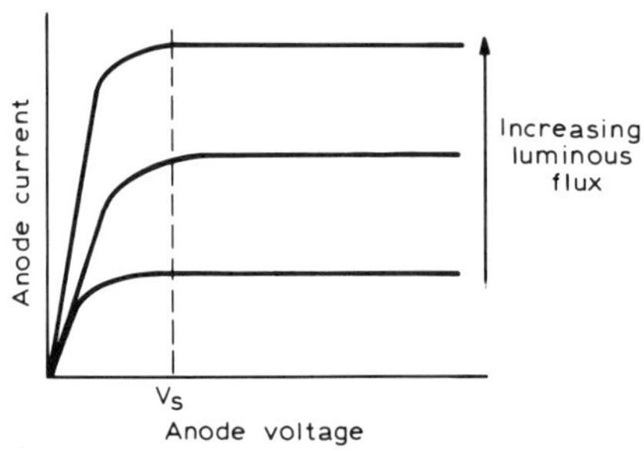

Figure 34.75 Linear cascade dynode system

In a gas-filled photoemissive tube, the photocurrent is initially the same as in a vacuum photoemissive tube up to the saturation voltage V_s. The photocurrent remains reasonably constant with increasing anode voltage until the ionisation voltate v_i is reached, at which voltage the electrons emitted from the photocathode have sufficient energy to ionise the gas filling. There will therefore be an increase in anode current as 'gas amplification' occurs, the gas amplification increasing as the anode voltage is increased. The anode voltage/current characteristic with luminous flux as parameter for a gas-filled photoemissive tube is shown in *Figure 34.76*.

When the photocathode is not illuminated, a small current still flows, the 'dark current'. The magnitude of this current varies

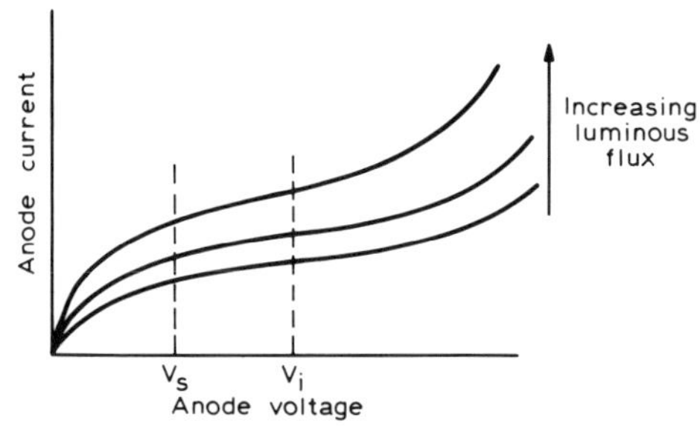

Figure 34.76 Venetian blind dynode system

with the type of tube, being as low as 0.15 nA typically for a vacuum photoemissive tube and up to 0.1 μA for a gas-filled tube. The spectral response of the tube is determined by the material of the photocathode although this can be modified by using different filters in the window of the tube.

The published data for a photoemissive tube gives the spectral response and sensitivity, together with the operating and limiting voltages and currents. The anode voltage/current characteristic may be given with loadlines so that suitable external circuits can be designed.

Photoemissive tubes are used in applications where a current proportional to the incident radiation is required. Many of the applications are similar to those of photodiodes and phototransistors but with the advantages of larger photocurrents being produced and a wider range of spectral responses through the choice of photocathode material.

34.11 Photomultiplier

A photomultiplier is a special form of photoemissive tube in which multiplication of the electrons emitted from the photocathode occurs to produce a larger anode current. The electron multiplication is produced by secondary emission in an electrode structure between the photocathode and the anode.

34.11.1 Construction of photomultiplier

A simplified cross section of a photomultipl8er is shown in *Figure 34.77*. The photocathode is deposited on the inside of the window

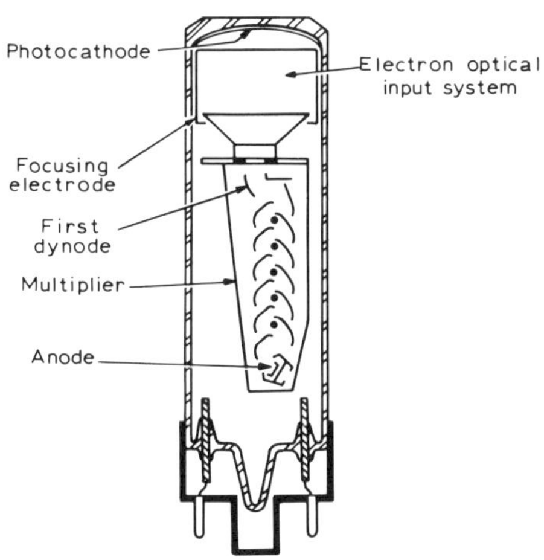

Figure 34.77 Voltage divider networks for dynodes of photomultiplier: (a) for maximum gain; (b) for best linearity and time response

of the tube. The electrons emitted by the photocathode when illuminated are directed by the electron–optical input system onto the first dynode. Secondary electrons are emitted by this dynode, and these electrons strike the second dynode. Electron multiplication occurs through successive impacts of the secondary eleectrons through the dynode system so that the anode current of the photomultiplier is considerably larger than the initial current emitted by the photocathode.

The spectral response of the photomultiplier is determined by the photocathode material. The response can be modified by the choice of material for the window of the photomultiplier. It is therefore possible to adapt a photomultiplier to respond to a particular spectrum by the choice of photocathode and window materials.

The transit time of electrons emitted from the photocathode to their striking the first dynode should be independent of the point on the photocathode at which they were emitted. This is particularly important for photomultipliers used in very fast applications. Such differences in transit time are minimised by making the window and photocathode part of a sphere. When a photomultipler is not intended for very fast operation, the window is flat.

The electron–optical system ensures that as many as possible of the electrons emitted from the photocathode reach the first dynode. The focusing electrode is an aluminium ring or a metalised coating on the inside of the tube, and held at the same voltage as the photocathode. An accelerating electrode held at the same voltage as the first dynode accelerates the electrons towards that dynode so that they have sufficient energy to release secondary electrons on impact.

Two forms of dynode structure are used in a photomultiplier: the linear cascade and the venetian blind structures. The linear cascade structure is shown in *Figure 34.78*. Each dynode acts as a

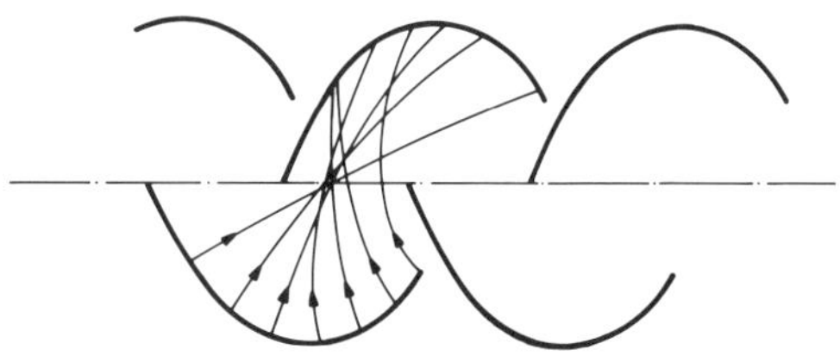

Figure 34.78 Simplified cross-section of image intensifier tube

reflective element, directing the emitted secondary electrons to the next dynode. The voltage between successive dynodes must be large enough to ensure a reasonable collection efficiency and to prevent the formation of a space charge. If a space charge were allowed to form, it would affect the linearity of the anode current with respect to the initial photocathode current. A progressively higher voltage is applied to each dynode through the multiplier twoares the anode to obtain the required voltage graidient. The venetian blind dynode structure is shown in *Figure 34.79*. This is

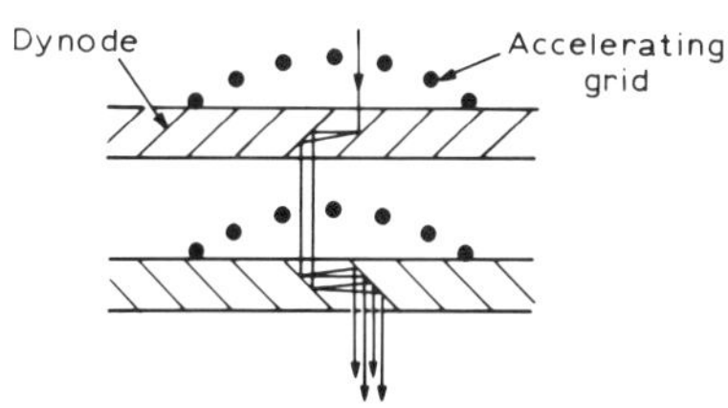

Figure 34.79 Schematic cross-section of three-stage image intensifier with fibreoptic coupling

a regular array of dynodes at 45° preceded by an accelerating grid. All the electrons from the photocathode strike the first dynode, to release secondary electrons which strike the first dynode again ro release more secondary electrons which are attracted to the second dynode. Because of the 'double impact', the venetian blind structure is a relatively slow one and can only be used when the transit time of electrons through the dynode structure is not important.

The material from which the dynodes are made must be one which has a good secondary emission characteristic. Typical materials are magnesium and caesium with a silver oxide layer or copper and berylium, the later combination being more widely used as it has a higher stability.

The anode of a photomultiplier is in the form of a grid placed in front of the final dynode. The secondary electrons from the preceding dynode pass through the anode grid to strike the final dynode. The secondary electrons released from the final dynode then are attracted to the anode.

34.11.2 Supplies for photomultipliers

The dynodes of a photomultiplier require a progressively higher voltage to ensure the transmission of secondary electrons through the multiplier section of the tube. This voltage supply is provided by a resistive voltage divider network. A stabilised high-voltage power supply is required, the voltage depending on the number of dynodes in the tube. Typical values lie between 1.2 and 7 kV. The minimum voltage between successive dynodes is determined by the need to prevent a space charge forming so that a linear relationship exists between the initial photocathode current and the anode current. The maximum value is determined by flashover. Good linearity between currents also requires the electrode currents to be maintained in the saturation region by the dynode voltages. To prevent excessive variations in the dynode voltages, the current through the voltage divider network should be high compared with the electrode currents themselves. A minimum value of at least 100 times the maximum average anode current is required.

For pulse operation, for example in scintillation counting, a high anode current occurs for the duration of the pulse. The voltage on the dynodes can be kept constant for the duration of the pulse by connecting capacitors acrosss the resistors at the high-voltage end of the divider network. These bypass capacitors reduce the values of the currents through the divider network required for constant dynode voltage. However, the capacitors increase the time-constant of the divider network. At high pulse repetition rates where the intervals between pulses are short, the capacitors may not have time to discharge and considerable voltage variations may occur. The errors produced by these variations increase with the repetition frequency. For applications where a short time-constant is required at a high repetition frequency, high currents in the divider network must be accepted.

Typical voltage divider networks for the photomultiplier supplies are shown in *Figure 34.80*. The divider in *Figure 34.80(a)* is designed for maximum gain with a given supply voltage; that in *Figure 34.80(b)* is designed for good linearity between the initial photocathode current and the final anode current, and for the best time response. The progressive increase in voltage across the final stages of the photomultiplier in the network of *Figure 34.80(b)* prevents premature saturation, and so improves the linearity further. In both networks, the voltage on the accelerating electrode (acc) is made adjustable to allow the photomultiplier to be set up for optimum collection efficiency or

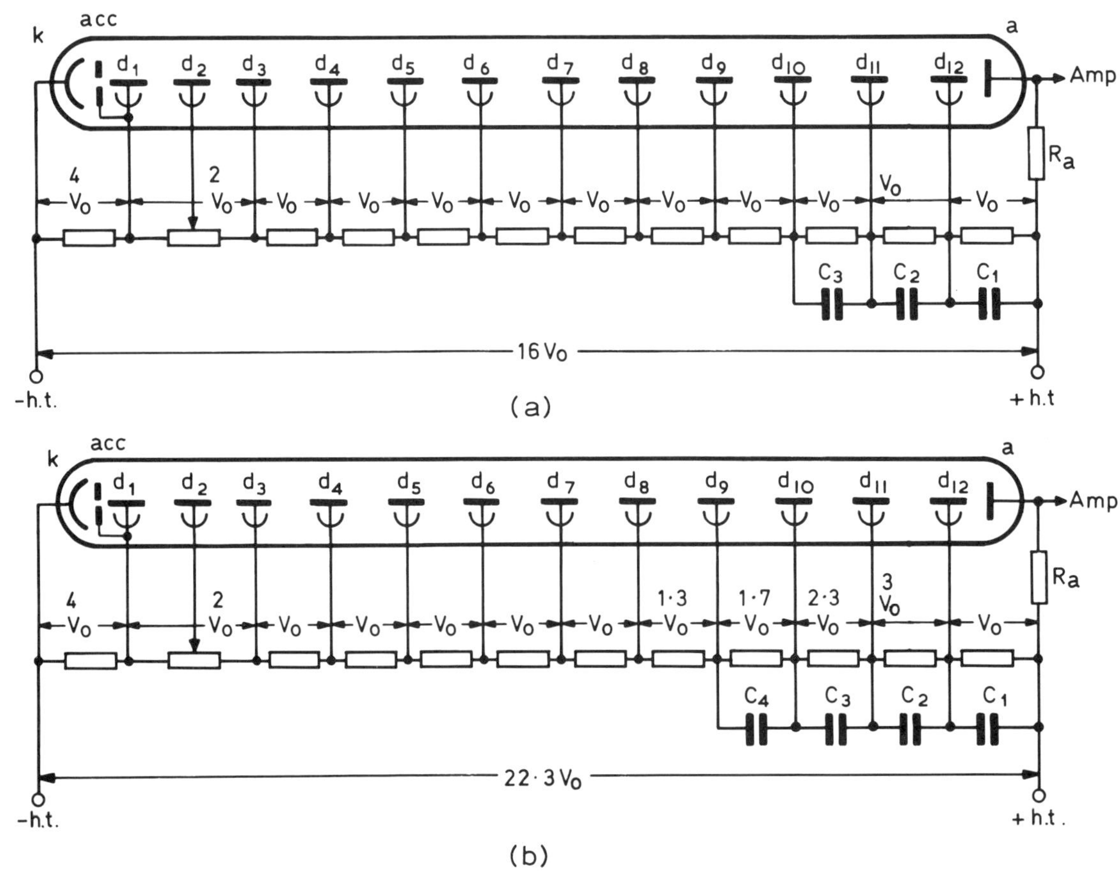

Figure 34.80 Voltage multiplier for use with three-stage image intensifier

optimum time response. Other divider networks can be designed to optimise different parameters to give compromise performances other than those of the two networks shown.

From the overall supply voltage, the maximum and minimum values of interelectrode voltage given in the published data, the maximum anode current, and the relative values of the interelectrode voltages, suitable resistor values for the divider network can be calculated.

34.11.3 Data for photomultipliers

The published data for photomultipliers has some items in common with that for photoemissive tubes already mentioned. The material of the photocathode and of the window are given, together with the spectral response. Luminous and spectral sensitivities under specified conditions are also given. The circuits of the two types of voltage divider network, sometimes with recommended component values, are given.

For the divider network of *Figure 34.80(a)* (maximum gain), the characteristics given include the anode dark current, and either the supply voltage required for a particular gain or the gain at a specified supply voltage. For the other type of divider network shown in *Figure 34.80(b)* (best linearity and time response), information on linearity and transit time is given. The gain of a photomultiplier, the current amplification, is of the order of 10^6 or 10^7.

From the user's point of view, the important dimension of a photomultiplier is the useful diameter of the photocathode. This diameter is typically between 10 and 110 mm.

34.11.4 Applications of photomultipliers

Typical applications of pyotomultipliers include their use in scintillation counters, flying-spot scanners, low-lefel light detectors, photometry, and spectrometry. In scintillation counters, the incident radiation produces small flashes of light in the scintillation material which are amplified by the photomultiplier whose output is used in a counting or detection circuit. In a flying-spot scanner the photomultiplier provides a convenient method of producing an electrical signal proportional to the light transmitted from the spot on the cathode-ray tube through the slide or film being scanned, while the low-level detecti9n, prhotometry, and spectrometry applications all make use of the high luninous sensitivity and spectral response of the photomultiplier.

34.12 Image intensifier and image converter

The image intensifier and image converteer operate on the same principle. An image of the screen to be viewed is focused on the photocathode of the tube. The photocathode emits electrons in fresponse to the incident photons corresponding to the illumination of the scene. The photogenerated electrons are accelerated to strike a luminescent screen. Because each incident photon on the photocathode gives rise to many photons on the screen, a gain in intensity is obtained.

The visible image on the screen is determined by the spectral response of the photocathode. If the response lies within the visible region of the spectrum, and the sensitivity of the photocathode is greater than that of the human eye, an intensified image of a dimly-lit scene is obtained. The device is therefore called an image intensifier. If the spectral response of the photocathode lies outside the visible region, say in the infrared part of the spectrum, the image on the photocathode will produce a visible image on the screen. The device now becomes an image converter.

In both type of device, any area of the scene that cannot produce sufficient illumination of the photocathode to release electrons cannot produce an image on the screen, however greater the gain of the device.

34.12.1 Proximity image intensifier

The original form of image intensifier, developed in the 1930s and still in use today, is the proximity-focused image intensifier. It consists of a flat photocathode placed close to and parallel with a luminescent screen. The screen is held at a high positive voltage with respect to the photocathode, typically 5 to 6 kV and the spacing between the photocathode and screen is approximately 0.6 mm. The electrons released from the photocathode by the incident radiation therefore travel in streiahg tlines to the screen. This system produces a distortion-free image on the screen, and this is the advantage of this type of image intensifier over the more modern types.

34.12.2 Single-stage image intensifier

A simplified cross section of a single-stage image intensifier tube is shown in *Figure 34.81*. A spherical photocathode is used, deposited on the inside face of the window. The emitted electrons are focused by the cylindrical focusing electrode, which is held at the same voltage as the photocathode, at the apex of the cone-shaped electrode. The electrons are accelerated by the cone onto the luminescent screen. Ideally the screen should be spherical with the same radius of curvature as the photocathode, but in practice a flat screen is used. Slight distortion at the edges of the image will therefore occur. The cone-shaped electrode is held at the same voltage as the screen, and the voltage between the photocathode and screen is typically 12 to 15 kV. Typical dimensions for an image intensifier of this type are a length of 50 mm and useful photocathode and screen diameter of 20 to 25 mm.

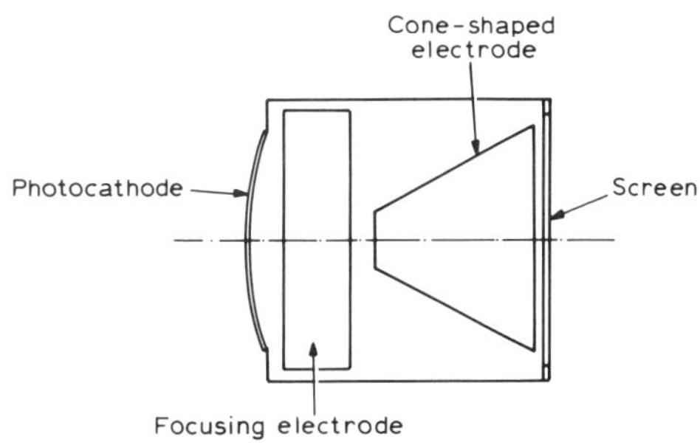

Figure 34.81 Simplified cross section of channel image intensifier

The phosphor used for the luminescent screen is chosen to suit the response of the eyes for direct-viewing systems, or to suit films in systems where the final image is to be photographed. The back of the phosphor on the screen (away from the viewing side) is coated with a layer of aluminium, as in a television picture tube, to increase the brightness of the display. The image with this type of device is inverted with respect to the photocathode. If an upright image is required, however, the input optical system can produce an inverted image on the photocathode.

The gain of an image intensifier is the ratio of the output luminance from the screen to the input illuminance of the photocathode. The gain will depend on the luminous sensitivity of the photocathode and the efficiency of the screen. A typical value for this type of image intensifier is 75.

The published data for this type of image intensifier have elements in common with photoemissive tubes. The material, sensitivity, and spectral response of the photocathode are given, combined with details of the luminescent screen. Magnification, resolution, and distortion under specified operating conditions are given. The gain of the device is given, together with a quantity 'background equivalent illumination'. With the device operating but with no input illumination on the photocathode, there will be abackground luminance on the screen. This is caused by thermionic or field emission from the photocathode, electron or ion scintillation, or long-term phosphorescence of the screen from previous operation. The background luminance is equivalent to the dark current of photoemissive tubes and photomultipliers, and can be regarded as the noise of the system, determining the minimum input signal that can be amp.ified by the device.

34.12.3 Multistage image intensifier

Higher gains than can be achieved by the single-stage image intensifier can be obtained by, in effect, connecting several image intensifiers in series. The screen of the first intensifier is used as the input to the photocathode of the second intensifier, and so on. The schematic cross section of such a multistage image intensifier is shown in *Figure 34.82*.

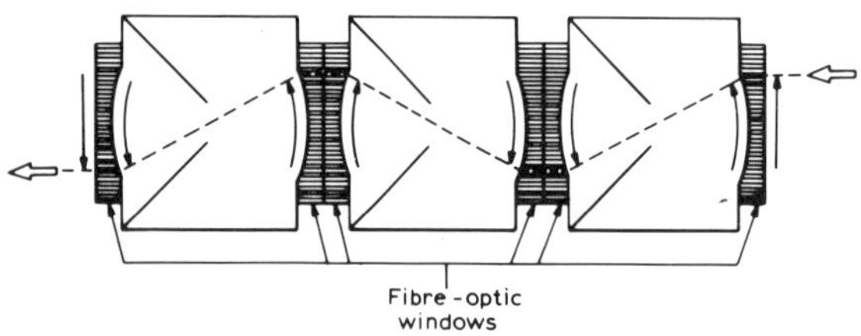

Figure 34.82 Voltage supply for channel image intensifier

A three-stage image intensifier is shown, with the input photocathode on the right. The screen of one stage is coupled to the photocathode of the next stage by plano-concave fibre-optic lenses. The radius of curvature of the concave side of the lens on which the photocathode or screen is deposited is chosen to optimise the focusing of the electrons between the photocathode and screen. The plane surface of the lens enables simple coupling between adjacent lenses to be achieved.

The voltages required for the operation of the three-stage image intensifier can be obtained from a voltage multiplier such as the one shown in *Figure 34.83*. This operates from a supply voltage of 2.8 kV at 1.5 kHz. The connections to the second and third stages are made through high-value resistors. These resistors are included to limit the brightness of the final display in the event of a sudden and unexpected increase in the illumination of the scene. The components of the voltage multiplier can be encapsulated with the image intensifier to form a portable device requiring only the 2.8 kV supply.

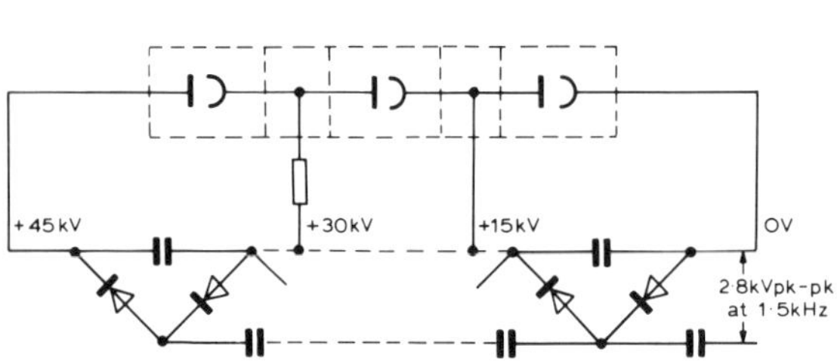

Figure 34.83

The gain of a typcial three-stage image intensifier is 50 000. Typical dimensions of a practical device are a length of 200 mm and a useful photocathode diameter and useful screen diameter of 25 mm.

34.12.4 Channel image intensifier

A gain comparable to that of the three-stage intensifier in a device of similar size to that of the single-stage intensifier can be obtained from the channel image intensifier. A simplified cross section of the channel image intensifier is shown in *Figure 34.84*.

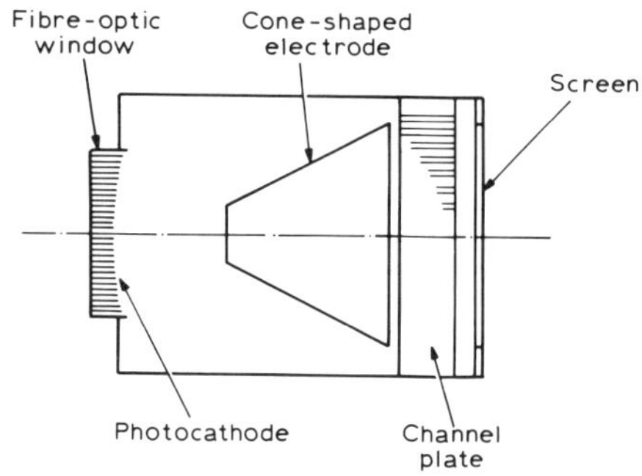

Figure 34.84

The high gain results from the use of the channel plate. This may be regarded as a large number of channel electron multipliers packed closely together. A channel electron multiplier is a thin tube, the inside wall of which is coated with a high-resistivity material with good secondary emission characteristics. With a high voltage across the length of the tube, the resistive material on the inside of the tube acts as a continuous dynode. An electron entering the tube and striking the side wall will release secondary electrons. Successive impacts of these secondary electrons result in a large number of electrons emerging from the end of the tube. If the initial electron entering one tube of the channel plate is emitted from a photocathode, and the emerging electrons from the opposite end of the tube are made to strike a luminescent screen, then a high-gain image intensifier can be obtained. The large number of tubes in the channel plate enable all the electrons from the photocathode to be multiplied.

In the cross section of *Figure 34.84*, the photocathode is deposited on the concave surface of the fibre-optic input window. The electrons emitted from the photocathode are accelerated to the channel plate by the cone-shaped electrode. The channel plate itself is approximately 1 mm thick, and a voltage of 1 kV is maintained across its thickness. The electrons emerging from the channel plate are accelerated to the screen by the high voltage between the plate and the screen, approximately 5 kV. Because of the close spacing of the channel plate and screen, about 0.5 mm, the electrons travel in a striaght lines to the scfreen, and so the resolution of the device is high.

The gain of a channel image intensifier is typically 30 000. The gain can be varied by varying the gain of the channel plqte, by varying the voltage across it. The voltages in a channel image intensifier are shown in *Figure 34.85*. The voltage between the photocathode and the channel plate is V_1, between the channel plate and the screen V_3, and that across the channel plate itself V_2.

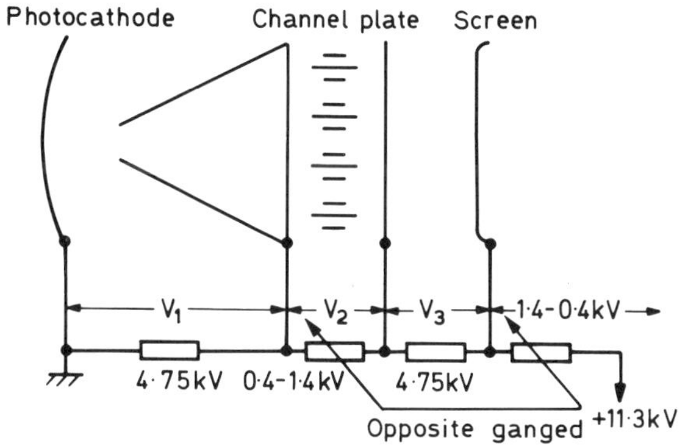

Figure 34.85

Ideally, varying voltage V_2 should not affect V_1 and V_3 so that the electrostatic focusing of the electrons from the photocathode and the proximity focusing of the electrons stiking the screen are not affected. A resistive voltage divider network to achieve this, with typical voltage values and ranges marked, is shown in *Figure 34.85*.

The published data for the channel image intensifier are similar to that of the other types of image intensifier. One characteristic that is particular to the channel device is that a curve of gain plotted against channel plate voltage may be included in the data.

34.12.5 Image converter

As mentioned previously, an image converter is a device used to produce a visible image of an 'invisible' scene. The various types of image intensifier described can be used as image converters, the photocathode material being one that responds to radiation outside the visible spectrum. A typical use of an image converter is the viewing of a scene lit by 'invisible light' such as infrared radiation so that the inhabitants of that scene would be unaware of the illumination and of being viewed.

Applications of image converters include military and security uses, as well as astronomical and scientific ones.

Acknowledgement

The material for Section 34.9 on X-ray tubes was kindly made available by Mullard Limited.

References

1 HOUSEKEEPER, W. G., 'The art of sealing base metals through glass', *J. Am. Inst. EE*, **42**, 870 (1923)
2 SCHNEIDER, P., 'Thermion emission of thoriated tungsten', *J. Chem. Phys.*, **28**, 675 (1958)
3 AMBRUS, J., KEREKES, L. and WALDHAUSER, I., 'The relation between the structure and properties of transmitting tube cathodes', *Tungsram Technische Mitteilungen*, 32 (1977) (German)
4 SPRANGENBURG, K. R., *Vacuum Tubes*, McGraw-Hill (1948)
5 DUNN, D. A., HAMZA, V. and JOLLY, J. A., 'Computer design of beam pentodes', *IEEE Trans.*, **ED12**, 6 June (1965)
6 Australian Patent 220414, 26 February (1959)
7 PAPENHUIJZEN, P. J., 'A transmitting triode for frequencies up to 900 MHz', *Philips Tech. Rev.*, **19**, 4 (1957/1958)
8 CFTH British Patent 1,206,049
9 MOUROMTSEFF, I. E., 'Temperature distribution in vacuum tube coolers with forced air cooling', *J. Appl. Phys.*, **12**, 491 (1941)
10 PANNET, W. E., *Radio Installations*, Chapman & Hall, London (1951)
11 BEUTHERET, C., 'The vapotron technique', *Rev. Tech. Thomson-CSF*, 24 (1956)
12 CFTH British Patent 940,984
13 BEUTHERET, C., 'A breakthrough in anode cooling systems: the hypervapotron', *IEEE Conf.—Electron Devices and Techniques* (Sept. 1970)
14 CFTH British Patent 1.194,249
15 SPITZER, E. E., 'Principles of electrical ratings of vacuum tubes', *Proc. IRE*, **39**, 60 (1951)
16 British Standards Code of Practice CP1005 (1962)
17 TYLER, V. J., 'A new high efficiency high power amplifier', *Marconi Review*, **21**, 96, 3rd quarter (1958)
18 SARBACHER, R. I., 'Graphical determination of high power amplifier performance', *Electronics*, **15**, 52, Dec. (1942)
19 THOMAS, H. P., 'Determination of grid driving power in radio frequency power amplifiers', *Proc. IRE*, **21**, 1134 (1938)
20 VARIAN, R. H. and VARIAN, S. F., 'A high frequency oscillator and amplifier', *J. Appl. Phys.*, **10**, 321, May (1939)
21 HAHN, W. C. and METCALF, G. F., 'Velocity modulated tubes', *Proc. IRE*, **27**, 106, February (1939)
22 WEBSTER, D. L., 'Cathode ray bunching', *J. Appl. Phys.*, **10**, 501, July (1939)
23 WEBSTER, D. L., 'Theory of klystron oscillators', *J. Appl. Phys.*, **10**, 864, December (1939)
24 HAHN, W. C., 'Small signal theory of velocity modulated electron beams', *GEC Rev.*, **42**, 258 (1939)
25 RAMO, S., 'Space charge and field waves in an electron beam', *Phys. Rev.*, **56**, 276 (1939)
26 WEBBER, S., 'Ballistic analysis of a two-cavity finite beam klystron', *Trans. IRE*, **ED-5**, 98 (1958)
27 BECK, A. H. W., *Space-Charge Waves*, Pergamon Press (1958)
28 BECK, A. H. W., *Thermionic Valves*, Cambridge University Press (1953)
29 CHALK, G. O. and O'LOUGHLIN, C. N., *Klystron Amplifiers for Television*, English Electric Valve Co. (1965)
30 EDGCOMBE, C. J. and O'LOUGHLIN, C. N., 'The television performance of the klystron amplifier', *Radio and Electronic Engineer*, **41**, 405 (1971)
31 GITTINS, J. F., *Power Travelling Wave Tubes*, The English University Press Ltd, Chapters 4 and 5 (1965)
32 HAMILTON, KUPER and KNIPP, *Klystrons and Microwave Triodes*, McGraw-Hill (1948)
33 HARMAN, W. W., *Fundamentals of Electronic Motion*, McGraw-Hill (1953)
34 PIERCE, J. R., *Theory and Design of Electron Beams*, Van Nostrand (1954)
35 PIERCE, J. R. and SHEPHERD, W. G., 'Reflex oscillators', *Bell Syst. Tech. J.*, **26**, 460 (1947)
36 SIMS, C. D. and STEPHENSON, I. M., *Microwave Tubes and Semiconductor Devices*, Blackie and Sons Ltd (1963)
37 SPANGENBERG, K. R., *Vacuum Tubes*, McGraw-Hill (1948)
38 STAPRANS, A., MCCUNE, E. W. and REUTZ, J. A., 'High power linear beam tubes', *Proc. IEEE*, **61**, 299 (1973)
39 STRINGALL, R. L. and LEBACQZ, J. V., 'High power klystron development at the Stanford linear accelerator centre', *Proc. 8th MOGA Conf.*, Kluver-Deventer (Amsterdam), 14–13 (1970)
40 WARNECKE, R. R., CHODOROW, M., GUENARD, P. R. and GINZTON, E. L., 'Velocity modulated tubes', *Advances in Electronics and Electronic Physics*, **3**, Academic Press, 43 (1958)
41 JAPANESE PATENT, UK 1328546, 'An apparatus for modulating and amplifying high frequency carrier waves' (1970)
42 HEPPINSTALL, R. and CLAYWORTH, G. T., 'The importance of water purity in the successful operation of vapour cooled television klystrons', *The Radio and Electric Engineer*, **45**, No. 8, Aug. (1975)
43 KONRAD, G. T., 'Performance of a high efficiency high power UHF klystron', *SLAC. Publ. 1896*, March (1977)
44 BOHLEN, H., 'Properties of a TV transmitter employing grid modulated klystrons', *Proc. 12th Int. TV Symp., Montreux* (1981)
45 HULL, A. W., *Phys. Rev.*, 18–31 (1921)
46 RANDAL and BOOT, *J. Inst. Elect. Engrs.*, **93**, pt. IIIA, No. 5, 928 (1946)
47 WILLSHAW, RUSHFORTH, STAINSBY, LATHAM, BALLS and KING, *J. Inst. Elect. Engrs.*, **93**, pt. IIIA, No. 5, 985 (1946)
48 FISK, HAGSTRUM and HARTMAN, 'The magnetron as a

generator of centimetre waves', *Bell Syst. Tech. J.*, 25–167 (1946)
49 COLLINS, G. B. (ed), *Microwave Magnetrons*, Radiation Laboratory Series, McGraw-Hill (1948)
50 BOOT, FOSTER and SELF, *Proc. IEE*, 105B, Suppl. No. 10 (1958)
51 OKRESS, E. C. (ed), *Crossed Field Devices*, 2 Vols, Academic Press (1953)
52 COOPER, B. F. and PLATTS, D. C., 'Frequency agile magnetrons using piezo electric tuning elements', *Proc. European Microwave Conference*, Brussels (1973)
53 BARKER, D., 'An experimental investigation of the energy distribution of returning electrons in the magnetron', *Proc. 8th MOGA Conf.*, Amsterdam (1970)
54 BARTON, D. K., 'Simple procedures for radar detection calculations', *IEEE Trans. on AES*, Sept. (1969)
55 BECK, A. H. W., *Thermionic Valves, Their Theory and Design*, Cambridge University Press (1953)
56 BENSON, F. A., *Millimetre and Sub-Millimetre Waves*, Iliffe Books Ltd (1969)
57 BIRKENEIMIER, W. P. and WALLACE, N. D., 'Radar tracking accuracy improvement by means of pulse to pulse frequency modulation', *IEE Trans. on Comm. and Electronics*, Jan. (1968)
58 BRODIE, I. and JENKINS, R. O., 'Secondary electron emission from barium dispenser cathodes', *Br. J. Appl. Phys.*, **8**, May (1957)
59 JEPSON, R. L. and MULLER, M. W., 'Enhanced emission from magnetron cathodes', *J. Appl. Phys.*, **22**, 9, Sept. (1951)
60 LATHAM, R., KING, A. H. and RUSHFORTH, L., *The Magnetron*, Chapman & Hall (1952)
61 KIND, G., 'Reduction of tracking errors with frequency agility', *IEE Trans. on AES*, May (1968)
62 MALONEY, C. E. and WEAVER, F. J., 'The effect of gas atmospheres on the secondary emission of magnetron cathodes', *Proc. 7th MOGA Conf.*, Hamburg (1968)
63 OKRESS, E. C., *Microwave Power Engineering*, 2 Vols, Academic Press (1968)
64 PICKERING, A. H. and COOPER, B. F., 'Some observations of the secondary emission of cathodes used in high power magnetrons', *Proc. 5th MOGA Conf.*, Paris (1964)
65 PICKERING, A. H. and LEWIS, P., 'Microwave sources for industrial heating', *Proc. Bradford Conference on Microwave Heating*, Oct. (1970)
66 RAY, H. K., 'Improving radar range and angle detection with frequency agility', *Microwave Journal*, May (1966)
67 SIMMS, G. D. and STEPHENSON, I. M., *Microwave Tubes and Semiconductor Devices*, Blackie & Son (1963)
68 SLATER, J. C., *Microwave Electronics*, D. Van Nostrand (1950)
69 TWISLETON, J. R. G., 'Twenty-kilowatt 890 MHz continuous-wave magnetron', *Proc. IEE*, **3**, 1, Jan. (1964)
70 PICKERING, A. H., LEWIS, P. F. and BRADY, M., 'Multipactor tuning in pulse magnetrons', *Commun. Int. (GB)*, March (1977)
71 VYSE, B. and LEVINSON, H., 'The stability of magnetrons under short pulse conditions', *IEEE Trans.*, **MTT-29**, No. 7, July (1981)
72 RUDEN, T. E., 'Design and performance of one megawatt 3.1–3.5 GHz coaxial magnetron', *Proc. 9th European Microwave Conf.* (1979)
73 PICKERING, A. H., 'Electronic tuning of magnetrons', *Microwave J.*, July (1979)
74 KOMPFNER, R., *The Invention of the Travelling-Wave Tube*, San Francisco Press (1964)
75 PIERCE, J. R., *Theory and Design of Electron Beams*, Van Nostrand (1954)
76 STERRETT, J. E. and HEFFNER, H., 'The design of periodic magnetic focussing structures', *Trans. IRE*, **ED-5**, 1, 35 (1958)
77 PIERCE, J. R., *Travelling-Wave Tubes*, Van Nostrand (1950)
78 SAUSENG, O., 'Efficiency enhancement of travelling-wave tubes by velocity re-synchronisation', *7th International Conference on Microwave and Optical Generation and Amplification, Hamburg*, 16 (1968)
79 PETER, R. W., 'Low noise travelling-wave amplifier', *RCA Review*, XIII, 3, 344 (1952)
80 CURRIE, M. R. and FORSTER, D. C., 'New mechanism of noise reduction in electron beams', *J. Appl. Phys.*, **30**, 1, 94 (1959)
81 CHALK, G. O. and JAMES, B. F., 'A wide dynamic range ultra low noise TWT for S-Band', *5th International Conference on Microwave and Optical Generation and Amplification, Paris*, 14 (1964)
82 KUNZ, W. E., LAZZARINI, R. F. and FOSTER, J. H., 'TWT amplifier characteristics for communications', *Microwave Journal*, **10**, 3, 41 (1967)
83 CHALK, G. O. and CHALMERS, P. M., 'A 500 kW travelling-wave tube for X-Band', *6th International Conference on Microwave and Optical Generation and Amplification, Cambridge*, 54 (1964)
84 GOLDMARK, P. C., *et al.*, 'Color television', *Proc. IRE*, **30**, 162–182 (1942) and **31**, 465–478 (1943)
85 HEROLD, E. W., 'Methods suitable for television color kinescopes', *Proc. IRE*, **39**, 11, 1177–1185 (1951). See also HEROLD, E. W., *Proc. Soc. Inform. Displays*, **15**, 141–159 (1974)
86 LAW, H. B., 'A three-gun shadow-mask color kinescope', *Proc. IRE*, **39**, 11, 1186–1194 (1951)
87 MORRELL, A. M., *et al.*, *Color Television Picture Tubes*, Academic Press, New York (1974)
88 THIERFELDER, C. W., 'Large screen narrow-neck 110° color television system', *IEEE Trans.*, **BTR-17**, 141–147 (1971)
89 NARUSE, Y., 'An improved shadow-mask design for in-line, three-beam color picture tubes', *IEEE Trans*, **ED-18**, 697–702 (1971)
90 YOSHIDA, S. and OHKOSHI, A., 'The trinitron—a new color tube', *IEEE Trans.*, **BTR-14**, 19–27 (1968)
91 RAMBERG, E. G., 'Focusing-grill color kinescopes', *IRE Convention Record*, Vol. 4, Part 3, 128–134 (1956)
92 DE HAAN, E. F. and WEIMER, K. R. U., 'The beam-indexing colour display tube', *R. Television Soc. Journal*, **11**, 278–282 (1967)
93 WRIGHT, W. D., *The Measurement of Colour*, 4th edn, Adam Hilger, Bristol (1969)
94 LARACH, S. and HARDY, A. E., 'Cathode-ray-tube phosphors: principles and applications', *Proc. IEEE*, Vol. **61**, 915–926 (1973)
95 BARTLESON, C. J., 'Color perception and color television', *J. SMPTE*, **77**, 1–12 (1968)
96 FIORE, J. P. and KAPLAN, S. H., 'The second-generation color tube providing more than twice the brightness and improved contrast', *IEEE Trans.*, **BTR-15**, 267–275 (1969)
97 HIGGINS, G. C. and PERRIN, F. H., 'The evaluation of optical images', *Photog. Science and Eng.*, **2**, 66–76 (1958)
98 MACHIDA, H. and FUSE, Y., 'Gain in definition of color CRT image displays by the aperture grill', *IEEE Conf. Record on Display Devices*, 101–108, Oct. 11–12 (1972)
99 MORRELL, A. M., 'Development of the RCA family of 90-degree rectangular color picture tubes', *IEEE Trans.*, **BTR-11**, 90–95 (1965)
100 SPEAR, B. W. and POWELL, D. E., 'KIMCODE, a method for controlling devacuations of TV tubes', *IEEE Trans.*, **BTR-9**, **1**, 25–31 (1963)
101 BLAHA, R. F., 'Degaussing circuits for color TV receivers', *IEEE Trans.*, **BTR-18**, 7–10 (1972)
102 YOSHIDA, S., 'A wide-deflection angle (114°) trinitron color picture tube', *IEEE Trans.*, **BTR-19**, 231–238 (1973)
103 BARBIN, R. L. and HUGHES, R. H., 'New color picture tube system for portable TV receivers', *IEEE Trans.*, **BTR-18**, 193–200 (1972)
104 BECK, A. H. W., *Thermionic Tubes*, Cambridge University Press, London (1953)
105 FONDA, G. R. and SEITZ, F., *Preparation and Characteristics of Solid Luminescent Materials*, Wiley/Chapman and Hall, London (1948)
106 GARLICK, G. F. J., *Luminescent Materials*, O.U.P., Oxford (1949)
107 GOLDING, J. F. (editor), *Measuring Oscilloscopes*, Iliffe, London (1971)
108 HARMAN, W. W., *Fundamentals of Electronic Motion*, McGraw-Hill, New York (1953)
109 KNOLL, M. and KAZAN, B., *Storage Tubes and their Basic Principles*, Wiley, New York (1952)
110 LEVERENZ, H. W., *An Introduction to the Luminescence of Solids*, Wiley/Chapman and Hall, London (1950)
111 PARR, G. and DAVIE, O. H., *The Cathode Ray Tube and Its Applications*, Chapman and Hall, London (1959)
112 SHAW, D. F., *An Introduction to Electronics*, 2nd edn, Longman, London (1970)

113 SOLLERT, T., STARR, M. A. and VALLEY, G. E., *Cathode Ray Tube Displays*, M.I.T. Series No. 22, McGraw-Hill, New York (1948)
114 SOWAN, F. A. and REID, I. A. (editors), *Screen Phosphors and Industrial Cathode Ray Tubes*, Mullard Limited, London (1964)
115 SPANGENBERG, K. R., *Vacuum Tubes*, McGraw-Hill, New York (1948)
116 THOMSON, J. and CALLICK, E. B., *Electron Physics and Technology*, Physical Science Texts, English Universities Press, London (1959)
117 ZWORYKIN, V. K. and MORTON, G. A., *Television*, 2nd edn, Wiley/Chapman and Hall, London (1954)
118 NIPKOW, P., 'Electrische Teleskop', *Ger. Pat.* 30105 (1884)
119 CAREY, G., *Design & Work*, **8**, 569 (1880)
120 GARRAT, G. R. M. and MUMFORD, A. H., 'The history of television', *Proc. IEE*, Part IIIA, 25–42 (1952)
121 CAMPBELL SWINTON, A. A., *Nature*, **78**, 151 (1908) also 'Presidental Address', *J. Rontgen Soc.*, **8**, 1–15 (1912)
122 ZWORYKIN, V. K., 'Television system', *U.S. Pat.* 1691324 (1928) also *U.S. Pat.* 2141059 (1938) (application date December 1923)
123 ZWORYKIN, V. K., 'Television with cathode-ray tubes', *J. IEE*, **73**, 438 (1933)
124 SCHADE, O. H., 'Electro-optical characteristics of television systems', *RCA Review*, **9**, 34 (1948)
125 MCGEE, J. D. and LUBSZYNSKI, H. G., 'EMI cathode-ray television transmission tubes', *J. IEE*, **84**, 468–482 (1939)
126 (*a*) LUBSZYNSKI, H. G. and RODDA, S., 'Improvements in or relating to television', *Brit. Pat.* 442666 (1934);
(*b*) CAIRNS, J. E. L., 'A small high-velocity scanning television pick-up tube', *Proc. IEE*, Part IIIA, 89–94 (1952)
127 IAMS, H., MORTON, G. A. and ZWORYKIN, V. K., 'The image iconoscope', *Proc. IRE*, 541–547 (1939)
128 (*a*) STRUBIG, H., 'Die Superikonoscop-Kamerarohre der Fernseh G.m.b.H.', Kurzmitteilungen, Fernseh G.m.b.H., Darmstadt, *Sonderheft*, 1 (1955)
(*b*) Editorial, 'The mode of operation of the Rieseliko', *Fernseh—Communications*, **13/14**, 229–240 (1957)
129 Manufactured by Cathodeon Ltd, Cambridge
130 COPE, J. E., GERMANY, L. W. and THEILE, R., 'Improvements in design and operation of image iconoscope type camera tubes', *J. Br. IRE*, **12**, 139–149 (1952)
131 SCHAGEN, P., BOERMAN, J. R., MAARTENS, J. H. J. and VAN RUSSEL, T. W., 'The Scenioscope, a new television camer tube', *Philips Tech. Rev.*, **17**, 189–198 (1956)
132 BLUMLEIN, A. D. and MCGEE, J. D., 'Mosaic stabilisation to cathode potential by scanning', *Brit. Pat.* 446661 (1934)
133 LUBSZYNSKI, H. G., 'Improvements in or relating to television and like systems', *Brit. Pat.* 468965 (1936) also *Brit. Pat.* 522458 (1958)
134 ROSE, A. and IAMS, H., 'Television pick-up tubes using low velocity scanning', *Proc. Inst. Radio Eng.*, **27**, 547–555 (1939)
135 MCGEE, J. D., 'A review of some television pick-up tubes', *Proc. IEE*, Pt. III, **97**, 377–392 (1950)
136 ROSE, A., WEIMER, P. K. and LAW, H. B., 'The image orthicon—a sensitive television pick-up tube', *Proc. Inst. Radio Eng.*, **34**, 424–432 (1946)
137 BANKS, P. B., 'Improvements in or relating to television camera cathode-ray tubes', *Brit. Pat.* 1048390 (1964)
138 HENDRY, E. D. and TURK, W. E., 'An improved image orthicon', *J. Soc. Motion Picture and Television Engineers*, **69**, 88–91 (1960)
139 WEIMER, P. K., 'The image isocon—an experimental television pick-up tube based on the scattering of low velocity electrons', *RCA Review*, **10**, 366–386 (1949)
140 KLEM, A., and KINGMA, R. V., 'Low light level systems developed by N.V. Optische Industrie de Oude Delft', *Proc. Electro-Optics Int.*, **71**, 304–312 (1971)
141 KLEM, A., 'Delcalix with isocon', *Odelca Mirror*, **9**, 1–4.
142 NIXON, R. D. and TURK, W. E., 'The image isocon for low light level operation', *J. Soc. Motion Picture Televis. Eng.*, **81**, 454–458 (1972)
143 WEIMER, P. K., FORGUE, S. V. and GOODRICH, R. R., 'The vidicon photoconductive camera tube', *Electronics*, **23**, 70–74 (1950)
144 TURK, W. E., 'Photoconductive TV camera tubes—a survey', *Journal of Science and Technology* (General Electric Company Ltd), *37*, No 4, 163–170 (1970)
145 FORGUE, S. V., GOODRICH, R. R. and COPE, A. D., 'Properties of some photoconductors, principally antimony trisulphide', *RCA Review*, **12**, No 3, 335–349 (1951)
146 LUBSZYNSKI, H. G., 'Improvements in and relating to television and like systems', *Brit. Pat.* 468965 (1936)
JEPSON, H. B., 'Improvements in or relating to photoconductive devices', *Brit. Pat.* 1030173 (1961)
147 DAWE, A. C., 'Special types of vidicon camera tubes', *Industrial Electronics*, November (1963)
148 CROWELL, M. H. and LABUDA, E. F., 'The silicon diode array camera tube', *Bell System Tech. J.*, **48**, 1481–1528 (1969)
149 WOOLGAR, A. J. and BENNETT, E. F., 'Silicon diode array tube and targets', *J.R. Television Society*, **13**, 53–58 (1970)
150 SANTILLI, V. J. and CONGER, III, G. B., 'TV camera tubes with large silicon diode array targets operating in the electron bombarded mode', *Advances in Electronics and Electron Physics*, **33A**, 219–228 (1972)
151 DE HAAN, E. F., 'The Plumbicon, a new television camera tube', *Philips Tech. Rev.*, **24**, 57–58 (1962–63)
152 DE HAAN, E. F., VAN DER DRIFT, A. and SCHAMPERS, P. P. M., 'The Plumbicon, a new television camera tube', *Philips Tech. Rev.*, **25**, 133–151 (1964)
153 VAN DE POLDER, L. J., 'Target stabilisation effects in television pick-up tubes', *Philips Res. Rep.*, **22**, 178–207 (1967)
154 DOLLEKAMP, J., SCHUT, TH. G. and WEIJLAND, W. P., 'Advances in Plumbicon camera tube design', *Electron Appl.*, **30**, 18–32 (1971)
155 VAN DOORN, A. G., 'The Plumbicon compared with other television camera tubes', *Philips Tech. Rev.*, **27**, 1–4 (1966)
156 BAILEY, P. C., 'New lead oxide tubes', *Sound and Vision Broadcasting*, **11**, 19–21 (1970)
156 DOLLEKAMP, J., 'One-inch diameter Plumbicon camera tube type 19XQ', *Mullard Technical Communications*, **109**, 196–200 (1971)
157 YOSHIDA, O., 'Chalnicon—a new camera tube for colour TV use', *Japan Electronic Engineering*, 40–44 (1972)
158 HAOHIRO GOTO, YUKINAO ISOZAKI and KEIICHI SHIDARA, 'New photoconductive camera tube, Saticon', *NHK Laboratories Note No. 170*, Sept. (1973)
159 GOETZE, G. W., 'Transmission secondary emission from low density deposits of insulators', *Advances in Electronics and Electron Physics*, **XVI**, 145–153 (1962)
160 GOETZE, G. W. and BOERIO, A. H., 'SEC camera-tube performance characteristics and applications', *Advances in Electronics and Electron Physics*, **28A**, 159–171 (1969)
161 SCHNEEBERGER, R. J., SKORINKO, G., DOUGHTY, D. D. and FEIBELMAN, W. A., 'Electron bombardment induced conductivity including its application to ultra-violet imaging in the Schumann region', *Advances in Electronics and Electron Physics*, **XVI**, 235–245 (1962)
162 LODGE, J. A., 'A review of television pick-up tubes in the United Kingdom', *Proceedings of Electro-Optic Conference, Brighton, England*, 253–264 (1971)
163 WEIMER, P. K., 'A development tri-color vidicon having a multi-element target', *IRE Trans.*, **ED-7**, 147–153 (1960)
164 BRIEL, L., 'A single-vidicon television camera system', *J.S.M.P.T.E.*, **79**, 326–330 (1970)
165 KUBOTA, Y. and KAKIZAKI, T., 'An ENG colour camera using a single pick-up tube', Technical publication by *Sony Corporation, Tech. Publ., Tokyo, Japan*
166 ATTEW, J. E., 'A single-tube colour camera', *J. Royal Televis. Soc.*, **14**, No. 5, 123–125 (1972)
167 (*a*) JESTY, L. C., 'Recent developments in colour television', *J. Royal Televis. Soc.*, **7**, 488–508 (1955)
(*b*) For example Sony colour camera DXC-5000CE

35 Transducers

J E Harry BSc(Eng), PhD
Senior Lecturer, University of Technology
Loughborough

Contents

35.1 Introduction

A transducer is a device which is used to convert a measured quantity (pressure, flow, distance, etc) into some other parameter. Although transducers include devices that convert one form of mechanical input to a different form, for example the manometer for pressure measurement, we are here concerned almost exclusively with electrical transducers which have become almost universally used because of the ease of transmitting and processing the output signal. The ready availability of low cost electronic systems and increasing degree of automation has resulted in ever increasing applications of transducers in all areas of industry, research and development for direct measurement and process control.

Transducers use a large number of different electrical phenomenon, some of which are listed in *Table 35.1*. The different

Table 35.1 Physical effects used in instrument transducers

(1) Active transducers
Electromagnetic
Piezoelectric
Magnetostrictive (as a generator)
Thermoelectric
Photoelectric (photoemission)
Photovoltaic (photojunction)
Electrokinetic (streaming potential)
Pyroelectric
(2) Passive transducers
Resistance
Inductance
Capacitance
Mechanical-resistance (strain)
Magnetoresistance
Thermoresistance
Photoresistance
Piezoresistance
Magnetostrictive (as a variable inductance)
Hall effect
Radioactive ionisation
Radioactive screening
Ionisation (humidity in solids)

transducers can be classified as, active in which energy conversion occurs, or passive in which energy is controlled. The overall operation is similar and a complete measurement system normally comprises the transucer or sensor, a signal conditioner or convertor, the output of which may be used to supply an indicator or for control purposes, and a supply in the case of passive transducers. Of the many effects that are used in transducers the principal effects used are variation of resistance, inductance, capacitance, piezoelectric effect and thermal effects which are described here.

35.2 Transducers for distance measurement

35.2.1 Resistance transducers

Resistance transducers using linear or rotating potentiometers for measuring linear or angular movement are both simple and versatile. Simple measurement circuits can be used enabling low cost systems to be devised. The potentiometer track is manufactured with a linear resistance and may be either wire-wound or continuous using a conducting plastic. The d.c. resistance R of a conductor is given by

$$R = \rho l/A \tag{35.1}$$

where ρ is the resistivity, l the length and A the cross-sectional area.

The potentiometer is usually supplied with an a.c. input, overcoming problems of d.c. bias of the output and the need to amplify a d.c. signal. The a.c. resistance of the conductor is greater than the d.c. resistance due to the skin effect. At high frequencies it is necessary to take into account the increase in a.c. resistance.

The inductance of wound tracks will be higher than for moulded tracks and where high-frequency excitation is used or high frequency response is required it may need to be taken into account. Both the inductance and capacitance can normally be ignored when excitation is d.c. or at a low frequency, but capacitive and inductive effects limit the response at higher frequencies.

The resolution of a wire-wound track is limited by the winding density (turns/mm) and the dimensions of the wiper contact. If the wiper is in contact with only one turn at a time then the resolution is R/n, where R is total resistance and n the number of turns. However, for a continuous output the wiper must overlap at least two windings reducing the resolution to $Rn/(n-1)$.

Moulded tracks are normally used enabling very high resolutions up to 0.0035% with a linearity of $\pm 0.2\%$ over distances in excess of 1 m to be obtained. A typical standard resistance of $40\ \Omega\ \text{mm}^{-1}$ at an operating voltage of 10 V is used. Where required a second track can be incorporated to give two independent electrical outputs.

The conducting plastic track has the advantage of stepless operation and therefore virtually infinite resolution. Imperfect contact between the potentiometer wiper and the track results in contact resistance and hence noise, however, this will only affect dynamic measurements. Wear of the track occurs due to movement of the wiper, however, the operating life is normally in excess of 10^6 cycles.

The accuracy is expressed as a percentage at full scale deflection and therefore the absolute accuracy varies inversely with the deflection. The output voltage signal limits the sensitivity of the system (V mm^{-1}) and depends on the voltage supplied to the potentiometer. This in turn is limited by the maximum permissible power dissipation of the potentiometer winding, $W = V^2/R$ if the resistance is high the input impedance of the conditioning circuit following the transducer must be high in order to avoid errors due to loading of the track. The error can be calculated from the circuit shown in *Figure 35.1*. The voltage ratio of the unloaded transducer is given by

$$V_o/V_i = xR_0 \tag{35.2}$$

when the transducer is loaded this becomes

$$\frac{V_o}{V_i} = \frac{x}{1 + mx(1-x)} \tag{35.3}$$

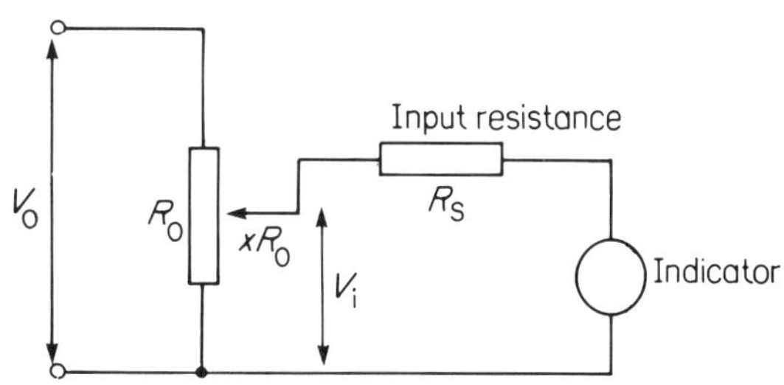

Figure 35.1 Resistance transducer loaded by measuring circuit

where $m = R_o/R_s$ and the error

$$e = \left(1 - \frac{1}{1 + mx(1-x)}\right)100\% \tag{35.4}$$

The error will be zero at both zero and full scale deflection of the transducer and will increase to a maximum when the transducer resistance is 50% of the maximum value. In general the load resistance must be of the order of 10 to 20 times greater than the transducer to obtain an accuracy of 1–2% of full scale. The error results in a non-linear response and to achieve a linear output accurate to within 1% the impedance of the conditioning circuit should be greater than $25R_s$. The principal limitation of potentiometers is the relatively high inertia of the moving parts which limit the rate of response typically to about 0.5 m s^{-1}.

35.2.2 Strain gauges

Strain is defined as extension/original length and is dimensionless. Strain gauges utilise the change in resistance of an electrical conductor that occurs when a stress is applied corresponding to a change in resistivity.

The sensitivity of a strain gauge along its axes of measurement can be defined in terms of a gauge factor K:

$$K = \frac{\text{electrical strain}}{\text{mechanical strain}} = \frac{\Delta R/R}{\Delta l/l} \tag{35.5}$$

For alloy strain gauges $2 < K < 5$ while for semiconductor gauges $K > 100$. Since the strain gauge will have a conducting path on an axis transverse to the axial path the cross sensitivity is typically less than 1% for a given axial strain.

The allowable percentage elongation of a gauge depends on the metallurgical condition of the gauge material, the insulating support material, and the adhesive used. It varies from 0.5% for high temperature applications in which ceramic cements are used up to 5% for polyamide and paperbacked foil gauges.

Strain gauge transducers are extensively used for the measurement of small displacements and, by deduction, of force, pressure and weight. The change in resistance is directly proportional to the strain and is measured with the strain gauge acting as one arm of a Wheatstone bridge. Strain gauges are available in a number of different forms ranging from wirewound gauges on thin insulating supports to thin film gauges obtained by sputtering and semi-conductor gauges.

The measurement of static strain imposes the greatest demands on strain gauge performance and copper nickel alloys are generally used because of their low and controllable temperature coefficients. Nickel iron alloys are used for dynamic measurements, where the larger temperature coefficient is not so important and the higher gauge factor (sensitivity) is an advantage when a high output is required. Nickel chrome and platinum alloys are used where higher operating temperatures are required. Temperature compensation can be achieved by incorporating a dummy strain gauge in the Wheatstone bridge. Temperature limits for dynamic strain measurements are generally much higher than for static conditions.

An important property of a strain gauge material is a very high degree of stability, particularly important for measurement of static strain or in load cells, necessitating the use of high grade resistance alloy materials, principally nickel chrome alloys and platinum in the form of wire or a thin film, both of which have a low temperature coefficient of resistance. Unbonded strain gauges use fine resistance wire wound in one plane on a former, stress being transmitted to the gauge by an elastic deflection system. Unbonded strain gauges are by nature bulky and cumbersome and are therefore limited in application to areas where they are part of an integral unit such as load cells and accelerometers.

The resistance element of a bonded gauge is attached along its length to the substrate. Various substrate materials are used depending on the application including paper and plastics. The simplest form of bonded strain gauge utilises flat wire formed in a grid (*Figure 35.2(a)*) in the one plane, but this method of construction is relatively bulky. A helically wound grid enables a longer effective strain gauge length to be obtained and consequently higher sensitivity by winding the wire on both sides of the substrate. Very high sensitivities and complex geometrical arrangements are obtainable using photoetched techniques (*Figure 35.2(b)*) to produce miniature strain gauges and composite strain gauges and rosettes to measure strain on more than one axis (*Figures 35.2(c)* and *(d)*). Where high sensitivities are required semiconductor strain gauges formed by deposition can be used.

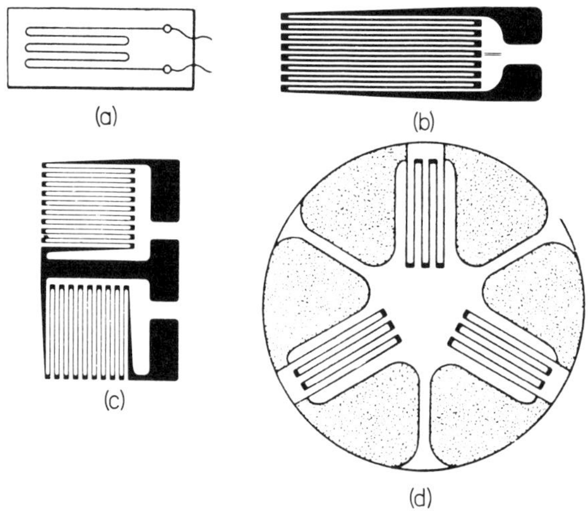

Figure 35.2 Examples of different strain gauges: (a) wire grid; (b) photoetched grid; (c) strain gauge for twin axes measurement; and (d) strain gauge rosette

The entire Wheatstone bridge may be incorporated in a single gauge thereby eliminating errors due to temperature variations, etc., and enabling compact transducers suitable for load cells to be constructed.

The attachment of the strain gauge to the assembly for which the strain is to be measured is the critical part of the use of the strain gauge and a wide variety of cements and adhesives have been developed for this purpose. Creep over long periods and shear under high stress must be avoided and the insulation resistance must be high, typically of the order of 10^4 MΩ.

The movement measured in strain measurements is normally very small and as a result the change in resistance of the gauge is correspondingly small and sensitive measuring circuits are required. The most commonly used circuit is the Wheatstone bridge shown in *Figure 35.3*. The bridge is balanced when

$$\frac{R_1}{R_2} = \frac{R_3}{R_4} \tag{35.6}$$

indicated when no current flows in the null detector. The bridge has the highest sensitivity when the resistors in the bridge arms are equal. Typical values of strain gauge resistance are 100–500 Ω. The bridge is normally balanced by varying a resistance in series with the strain gauge. Since the change in resistance is normally small the effect of temperature variation on the resistance may be

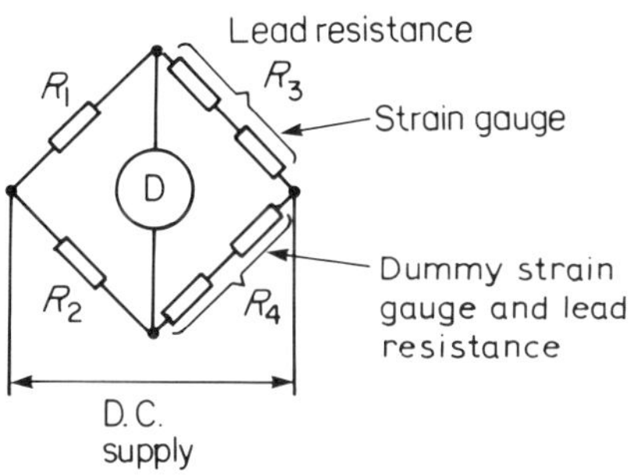

Figure 35.3 Wheatstone bridge with dummy gauge for temperature and lead compensation

significant and it is normal practice to incorporate a dummy strain gauge in the opposite arm of the bridge but which is not subjected to any strain. Any changes in temperature cause a corresponding change in resistance of this gauge therefore minimising the error. The effect of variation in resistivity with temperature will not normally be important for dynamic measurements for which one strain gauge will be adequate. Temperature compensation coupled with increased sensitivity can be obtained where two strain gauges are used, one in tension and one in compression. For example in the case of a cantilever, or the measurement of torque, the two strain gauges are connected in opposite arms of the bridge, while two active gauges in adjacent arms can be used for measurement of bending strain. The various bridge circuits that are used are referred to as quarter bridge (one strain gauge) half bridge (two strain gauges) and full bridge in which all four arms are strain gauges.

35.2.3 Capacitive transducers

The capacitance between two parallel plates is given by

$$C=\varepsilon_0\varepsilon_r\frac{A}{d} \qquad (35.7)$$

where ε_0 is the permittivity of free space, ε_r is the relative permittivity, A is the effective area of the plates, and d is the separation between the plates. Differentiating gives

$$\frac{dC}{dd}=-\varepsilon_0\varepsilon_r\frac{A}{d} \qquad (35.8)$$

The response of a capacitive transducer is shown in *Figure 35.4*.

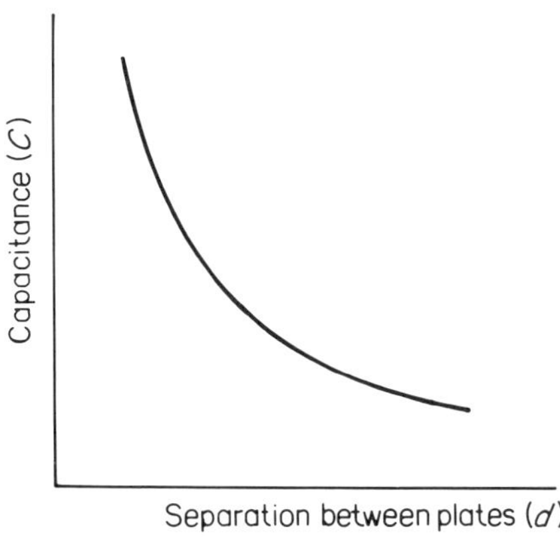

Figure 35.4 Variation of capacitance with separation for parallel plate capacitor

From this it can be seen that the resolution is high when the separation between the plates is small. A number of different capacitor configurations are used for distance measurement some of which are illustrated in *Figure 35.5*. Flat plate

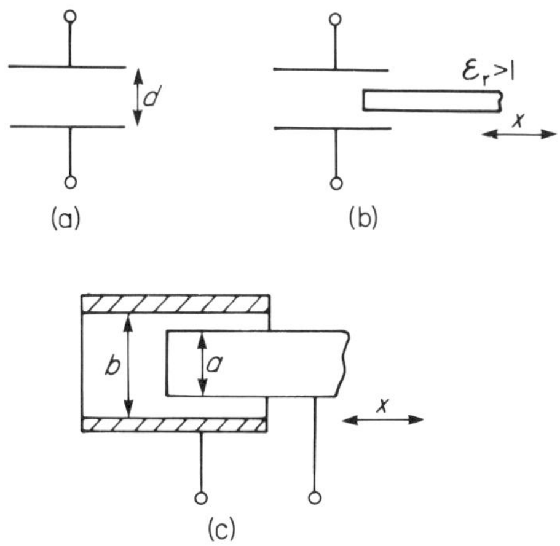

Figure 35.5 Examples of capacitive transducers: (a) parallel-plate, variable separation; (b) parallel-plate, variable permittivity; and (c) co-axial tube

transducers are limited to very small but accurate measurement of distance with almost infinite resolution of distances less than 1 mm.

The capacitance of a concentric tube capacitor is given by

$$C=\frac{2\pi\varepsilon_0\varepsilon_r}{\ln(b/a)} \qquad (35.9)$$

It varies linearly with distance, and is capable of measurements up to 0.4 m with a resolution of 10^{-4} mm.

Capacitive transducers may also be used for proximity detection in which the capacitance in the ancillary circuit is varied by the proximity of a non-metallic material or by stray capacitance to unearthed metallic objects, and may be used in applications such as counting, etc. The signal conditioning circuit usually uses either amplitude modulation of a high frequency signal, an a.c. bridge circuit, or measures the change in frequency of a tuned circuit.

35.2.4 Inductive transducers

The inductance of a solenoid is given by

$$L=\mu_0\mu_r\frac{AN^2}{l} \qquad (35.10)$$

where μ_0 is the permeability of free space, μ_r the relative permeability, A the area of cross section of the solenoid, N the number of turns and l the length of the solenoid. To achieve adequate sensitivity a core of ferrous material is normally used, typical arrangements being shown in *Figure 35.6*. The inductance may be made to change by varying the reluctance of the magnetic circuit, i.e.

$$L=\frac{\mu_0}{A}\left(\frac{l_i}{\mu_r}+l_a\right) \qquad (35.11)$$

where l_i is the length of the path in the magnetic material and l_a the length of the air gap. In practice this is achieved by varying the

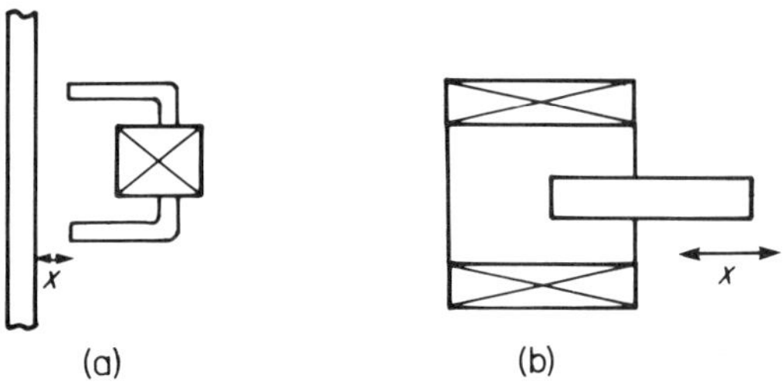

Figure 35.6 Examples of variable reluctance inductive transducers

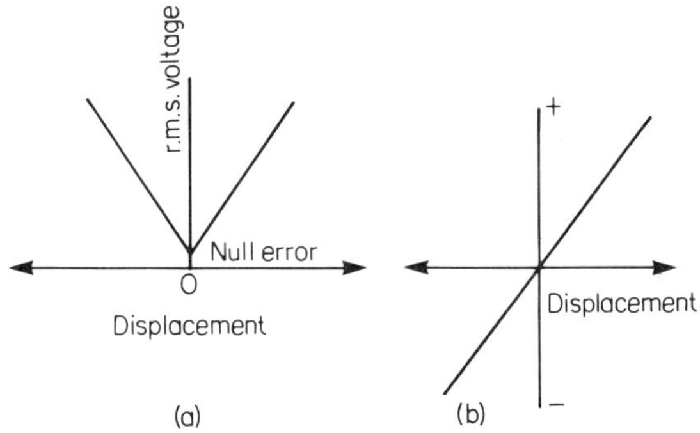

Figure 35.8 Response of LVDT: (a) a.c. r.m.s. output; (b) phase sensitive rectified output

distance between ferromagnetic core and the coil so as to vary the air gap, i.e.

$$\frac{dL}{dl} = -\mu_0 A N^2\left(\frac{\mu_r}{l_i^2} + \frac{1}{l_a^2}\right) \tag{35.12}$$

Inductive transducers can be used for distance measurement with a high degree of resolution. In one form known as the linear variable-differential transformer (LVDT), illustrated in *Figure 35.7*, a core of ferrous material is moved between two coils

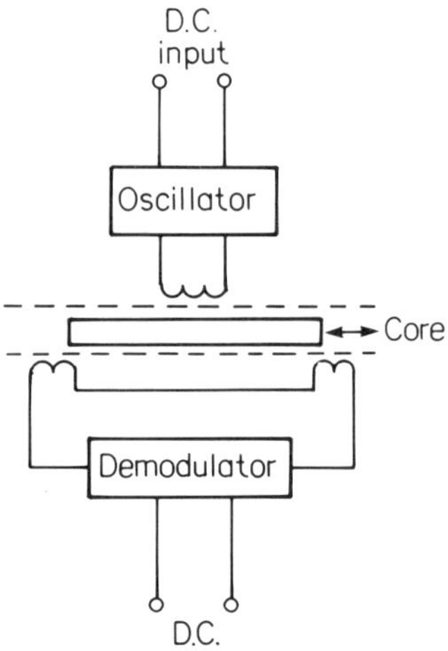

Figure 35.7 Linear variable differential transformer (LVDT) with integral oscillator and de-modulator

connected in anti-phase and supplied with a high frequency excitation voltage. The a.c. input to the primary is normally derived by an ancilliary circuit incorporated within the transducer and the secondary output demodulated so that a d.c. output voltage, directly proportional to a displacement of the armature, is obtained. The windings are connected in anti-phase so that the output voltage is at a null when the core is equally positioned in both coils (*Figure 35.8*), although small imperfections in winding, etc, will result in the null voltage being slightly offset from zero. The principal advantage is the absence of contact and virtually infinite resolution. Typical ranges of measurement are from 0.125 mm to 75 mm with a sensitivity of 0.25 mV mm^{-1}. A similar construction can be used to measure angular displacement up to 300°.

Inductive sensors may also be used for proximity measurement to given an output signal when a metallic object approaches the front surface of the sensor. The output of the sensor varies non-linearly as the distance from the sensors decreases and sensors of this kind are normally limited to simple systems such as proximity measurement and counting. The inductance is made part of a resonant circuit in which the natural frequency of the circuit is changed by the proximity of the metal object. Similar circuits to those used for capacitive transducers are used.

35.2.5 Optical encoders

Encoders consist of a glass disc in the case of rotary encoders or a strip in the case of linear encoders with accurately generated lines at regular intervals. A narrow beam of light passes through the encoder and relative movement between the encoder and light source generates pulses which are detected by a photoelectric detector and amplified to give a square-wave output. Two signals can be obtained for each output pulse corresponding to the increase and decrease of output, giving twice the resolution. The resolution can be further improved by using a second track (*Figure 35.9*) phase shifted by 90° which automatically gives a count pulse rate of four times the number of lines. The direction of

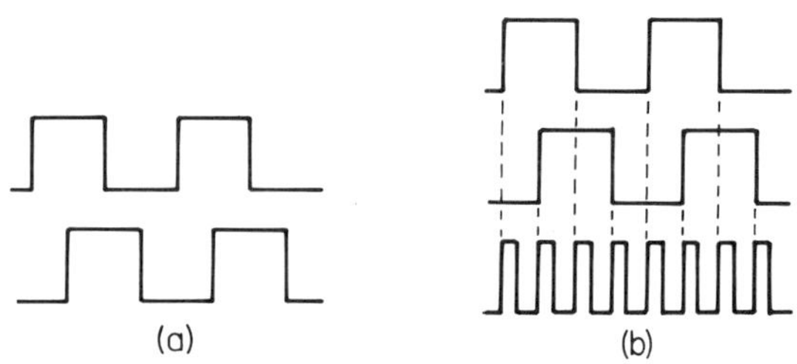

Figure 35.9 Output signals derived from optical encoder: (a) single pulse; (b) pulse output corresponding to leading and trailing pulse edges

movement can be derived by using a direction sensing logic circuit which differentiates each edge of the signals and checks their relationship with the other track. The overall accuracy is better than ± half bit (one bit being equivalent to number of lines times four).

The same principle can be used for absolute measurement of distance. However an absolute encoder has a discrete pattern or code for each line on the disc such as a binary or BCD code which requires decoding. To avoid ambiguity in reading an absolute pattern, cyclic gray codes are normally used in absolute encoders. A grey code is where only one bit changes from line to line. To illustrate the reasons for a cyclic grey code consider the transition from seven to eight in a four bit binary code. All four bits change 'simultaneously'. As it is impossible to obtain a simultaneous change a momentary erroneous reading could be obtained. To overcome this problem a grey code is generated.

A higher degree of precision can be obtained by the use of Moiré fringes which are formed when two diffraction gratings are superimposed. Instead of two complete plates only one plate and a small part of a plate are required if the plates are tilted. This produces alternating light and dark bands across the grating, illustrated in *Figure 35.10*. These move as the gratings

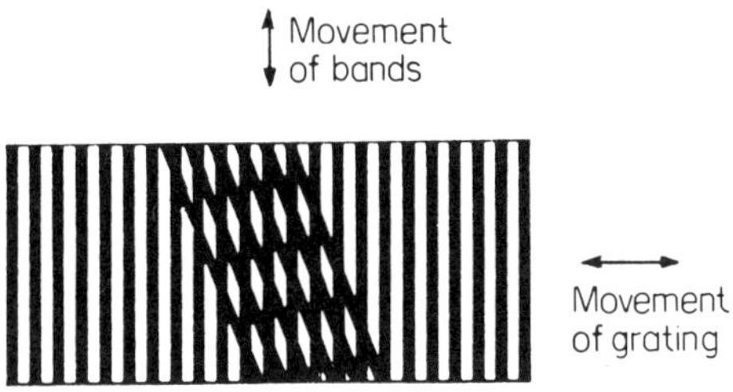

Figure 35.10 Formation of Moiré fringes by crossed gratings

move with respect to each other and can be detected by photocells. The gratings can be made to very high accuracies, a common degree of resolution being 254 lines/mm. Optical encoders enable non-contact measurement of position with very high resolution to be obtained. Resolution up to 1/40 000 of a revolution is possible. Absolute encoders use a second track with associated light source and detector.

Applications of encoders include measurement of distance and velocity in such applications as machine tools, weighing machines, automatic cut to length processes and remote control of position.

The ultimate (at least at present) in precision is obtainable by optical methods. The interferometer which relies on the constructive and destructive interference of light from a coherent source have been used as a laboratory measurement tool for many years. The availability of lasers with depth of coherence of more than a metre has enabled interferometer methods to be taken from the laboratory and these are now in use in such applications as precision machining and calibration. Measurements to accuracies of a fraction of a wavelength are possible enabling measurements of distance over several metres to within 0.1 μm to be made.

35.3 Transducers for velocity measurement

Inductance transducers may be used for measurement of linear velocity which is related to the induced e.m.f., which for a single conductor moving in a magnetic field is given by

$$V = Blu \qquad (35.13)$$

where B is the magnetic flux density, l the length of the conductor and u the velocity of the conductor perpendicular to the magnetic field.

One type of linear velocity transducer uses a magnet coupled to the moving object which moves along the axis of a solenoid coil. This generates an e.m.f. proportional to the velocity but is limited to relatively short path lengths. An advantage of moving the magnet rather than the coil is the elimination of the need for flexible leads. Measurement of rotational velocity is possible using small generators or tachometers (sometimes referred to as tacho generators). The generated e.m.f. is directly proportional to the rotational speed.

The a.c. tachogenerator, essentially an a.c. generator using a rotating permanent magnet to produce an output voltage which is dependent on the rotational frequency, has the advantage that there are no connections required through carbon brushes or slip rings and it is therefore simple and robust. A simple rectifying circuit is required to smooth the output. One of the disadvantages of the a.c. tachogenerator is that the output impedance changes with the frequency which will result in an error unless a high impedance measuring circuit is used.

The d.c. tachogenerator is similar in construction to the a.c. generator except that the magnet is stationary and the coil rotates. Carbon brushes provide contacts to the rotating commutator and a rectified d.c. output voltage proportional to speed is obtained. The impedance is not dependent on the rotational velocity, and a smoothing circuit is still required to reduce the ripple output.

The drag cup tachometer uses relative motion between the aluminium cup and the rotating magnet to induce eddy currents in the cup which provide an electromagnetic torque proportional to the relative speed. The tachometer can be made direct reading and can also act with a considerable degree of inherent damping which in many cases is an advantage.

Two phase induction generators may be used to provide an output voltage directly proportional to the rotational frequency, at a constant frequency, thereby eliminating some of the problems associated with the permanent magnet a.c. generator. The two windings are wound at right angles to each other one of which is excited at a constant frequency. Eddy currents are induced in the rotor, which may be either of the squirrel cage or drag cup type. This links the second field coil resulting in an induced voltage which is independent of frequency but varies in amplitude with the speed of the rotor.

35.4 Measurement of force and pressure

Strain gauges are frequently used for measurement of force either directly in newtons (N) or for weighing (kilograms, kg). The strain gauges are incorporated in elastic members such as shackles and links (tension), and load cells (compression) connected to half or full bridges. A very high degree of accuracy and stability is required for weighing equipment with accuracies in excess of 0.5%.

of charge with pressure can be used to measure very rapid changes in pressure, torque and acceleration over very large ranges. The construction of a piezoelectric transducer is illustrated in *Figure 35.11*. The transducer element may be quartz

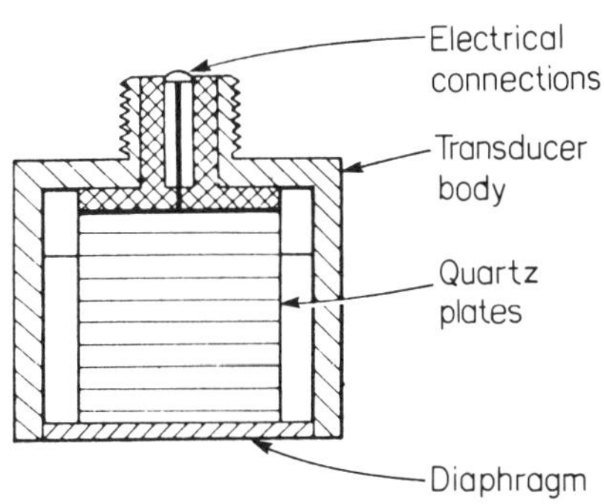

Figure 35.11 Piezoelectric transducer

or a ferroelectric ceramic such as barium titanate. The transducer acts as a capacitor of varying charge whose output is expressed either in terms of charge as C/g or voltage (V/g) sensitivity. The transducer has a very high output impedance and therefore it is necessary to use conditioning circuits with very high input impedances, typically of the order 10^{14} Ω to limit discharge of the capacitor. Since the capacitance is small (typically 20 pF) an amplifier with a high input impedance is necessary. Resolution of 1 part in 10^6 is possible. Differential pressure transducers using an inductive transducer to measure the deflection of a diaphragm are suitable for use up to about 1 bar.

Measurement of very low pressure is more difficult. The McLeod gauge traps a sample of the gas in the system which is compressed by a column of mercury of known height. The pressure is related to the volume occupied by the compressed gas. Measurement down to about 10^{-2} Pa are possible but the operation is difficult to carry out automatically. The thermal gauge is suitable for operation over the range 10^{-2} to 10^2 Pa. A resistance heater is inserted in the vacuum; the power dissipation by convection is a function of the pressure and gas. Alternatively the resistance may also be used to indicate the pressure. Thermocouple gauges can be used over the range 10^{-2} to 10^2 Pa.

Ionisation gauges are capable of measuring very low pressure from 0.1 Pa to 10^{-8} Pa. As the gas pressure is reduced through the Paschen minimum the voltage gradient increases and can be used to measure the pressure.

35.5 Accelerometers

Various different forms of accelerometers are based on the acceleration of a known mass, sometimes referred to as the seismic or proof mass used in conjunction with a displacement transducer for measurement of acceleration. The principle of operation of a seismic accelerometer is defined by Newton's second law:

$$ma = kx \tag{35.14}$$

in which a is the accelerating force, m the seismic mass, x the deflection of the restoring force, e.g. spring, and k the force due to the deflection. The operating principle is illustrated in *Figure 35.12*. A system based on this alone would result in oscillation at

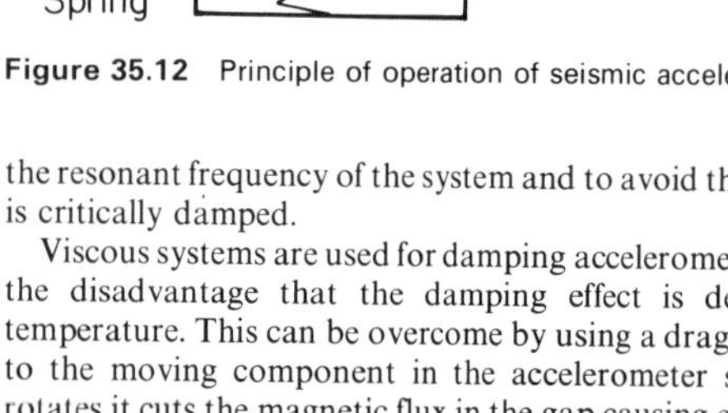

Figure 35.12 Principle of operation of seismic accelerometer

the resonant frequency of the system and to avoid this the system is critically damped.

Viscous systems are used for damping accelerometers but have the disadvantage that the damping effect is dependent on temperature. This can be overcome by using a drag cup coupled to the moving component in the accelerometer so that as it rotates it cuts the magnetic flux in the gap causing eddy currents to flow in the cup. This results in a force which acts in a direction so as to slow the motion of the cup proportional to the current and therefore the velocity of the cup.

The piezoelectric accelerometer utilises the shear displacement of a piezoelectric crystal (*Figure 35.11*) which generates a capacitive charge. No restoring spring or damping is required and the damping factor which is of the order of 0.01 can be treated as zero over the useful frequency range of the accelerometer which extends to several kHz. Full scale outputs are obtained for accuracies of from 10^{-10} g to more than 10^4 g linear over a range of 10^4:1.

35.6 Vibration

Vibration is the oscillatory motion of a body which is characterised by the frequency of oscillation and the amplitude. The oscillation may be sinusoidal or non-sinusoidal with higher harmonics present. Vibration is measured in units of acceleration (m s^{-2}), velocity (m s^{-1}) and displacement (m). Levels of amplitude of vibration vary over a very wide range and an international decibel scale has been agreed for velocity (reference 10^{-3} m s^{-1}) and acceleration (reference 10^{-5} m s^{-2}). The vibration level measured in terms of velocity or acceleration is expressed as

$$\text{vibration level} = 20 \log_{10}(A_1/A_0) \text{ dB} \tag{35.15}$$

The variation of the amplitude of the measured parameter with frequency is referred to as the vibration spectrum which in many cases varies uniformly with frequency (response), i.e. no resonance occurs. As the acceleration increases the displacement tends to decrease whilst the velocity remains constant.

The amplitude of vibration can be measured using a seismic mass transducer similar in construction to an accelerometer. If the transducer is vibrated at frequencies greater than its resonant frequency a voltage is induced in the coil by the relative motion between the coil and the magnet, which is proportional to the velocity of the coil with respect to the stationary magnet. The induced voltage is often adequate without further amplification. The piezoelectric accelerometer can be used to measure vibration acceleration.

Vibration displacements can be measured using capacitive or inductive transducers in the same way as distance measurement but are usually limited to the measurement of relative displacement only. Other techniques are occasionally used including stroboscopic methods in which the vibration can be viewed by freezing the motion with a rapidly pulsed lamp and the reed vibrometer in which the effective length of the reed is varied until resonance occurs enabling the frequency to be determined. This method is applicable over the range of 5 Hz to 10 kHz.

35.7 Transducers for fluid flow measurement

35.7.1 Variable area flow meter

The variable area flow meter is illustrated in *Figure 35.13*. The tube is mounted vertically and the fluid flowing through the tube raises the float until the force due to the momentum of the fluid is balanced by the downward force of the float. The variable area flow meter is a simple and effective way of measuring small flows of gases or liquids with a high degree of accuracy and wide range of flow rates. Although normally used as direct indicating flow meters, by using a non-magnetic flow tube combined with a magnetic float the output can be coupled to give an output signal. Variable area flow meters are also used in conjunction with orifice plates at high flow rates.

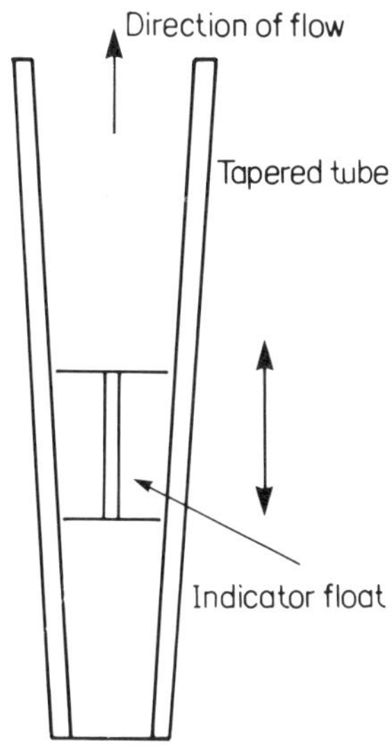

Figure 35.13 Principle of operation of variable area flow meter

35.7.2 Differential pressure flow meters

The rate of flow of a fluid can be deduced from the pressure drop across an orifice. The two principal types of orifice used are the orifice plate, used where the fluid has a high viscosity and/or flow rate, and the Venturi tube for which the total pressure drop is a minimum (*Figure 35.14*). The pressure drop across an orifice may be measured with a manometer or with a variable area flow meter or by a differential pressure transducer generally using capacitive transducer elements. Flow meters of this kind are capable of operation over flow ranges of up to 3:1 and the square root characteristic produced by orifice meters results in a high sensitivity and accuracy at flow rates above 50% of full scale.

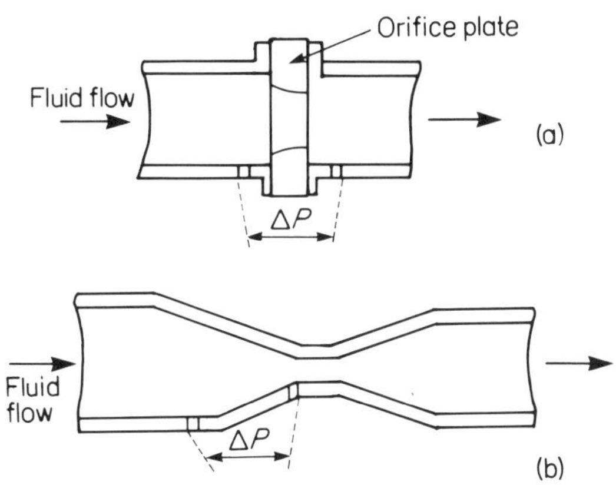

Figure 35.14 Differential pressure flow meters: (a) orifice plate; (b) Venturi tube

35.7.3 The turbine flow meter

The rurbine flow meter uses a turbine wheel in the flow channel which is rotated by the flow. The angular velocity is ideally proportional to the velocity of the liquid and hence the flow rate. The turbine blade itself may be used to generate a voltage by using a paramagnetic material and incorporating a pickup coil in the side walls of the flow meter. Errors may result due to swirling or turbulence of the flow but this can be reduced by incorporating axial vanes at the entrance to the meter. The effects due to viscosity are small for values of Reynolds number greater than about two to three thousand which set a lower limit to the linear range of the meter. The output of the meter is nominally linear and accurate over quite a wide range typically ±1% over a range of 20:1.

35.7.4 Magnetic flow meters

The magnetic flow meter relies on the voltage induced in a conductor passing through a magnetic field. The principle of operation is illustrated in *Figure 35.15*. Its application is limited to liquids with an electrical conductivity above 5×10^{-6} S. No obstruction occurs in the flow line and provided the electrical conductivity is adequate the magnetic flow meter has important applications in metering fluid materials in which pressure drops must be minimised or a substantial particulate material exists such as in slurries. The effect of non-conducting deposits on electrodes can be reduced by using capacitive coupling to the electrodes. The magnetic field is obtained by a.c. excitation of the field coils so as to eliminate the build-up of a polarising potential between the electrodes.

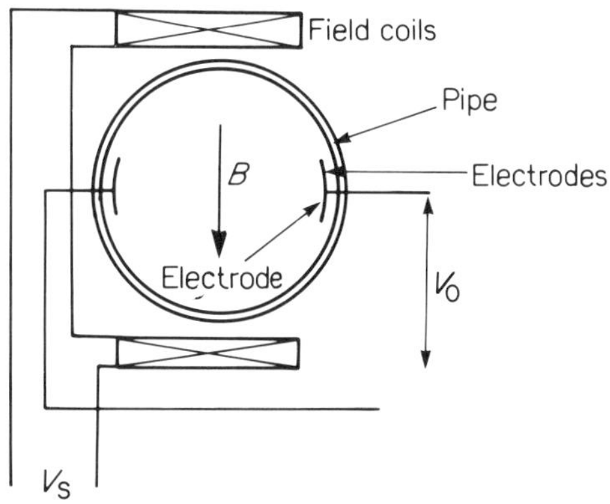

Figure 35.15 Principle of operation of magnetic flow meter

Leakage currents from other sources may also flow and although the error can be reduced by rejecting the quadrature components, the in phase component may cause an error. Provided this is constant it can be eliminated during setting up of the flow meter; however when it changes the zero setting will need to be changed. This can be overcome by using a pulsed d.c. magnetic field which elininates problems due to polarisation and errors due to voltage drops from other sources.

35.7.5 Vortex flow meter

The principle of operation of the vortex flow meter is to generate pressure variations using the curvature of the blades at the inlet to produce a velocity component tangential to the initial axial flow. This results in an unsteady pressure distribution and secondary rotation of the vortex core and local velocity fluctuations. The fluctuations can be detected by modulating an ultrasonic signal or using a piezoelectric detector to respond to the pressure change generated by the vortex rotation, or a thermistor to sense the temperature fluctuation caused by the vortex. A pulsed output signal is obtained in this way which is directly proportional to the volumetric flow. The vortex meter may also be used for measuring total flow rates since the digital output is directly proportional to the volumetric flow. The piezoelectric sensor may be totally encased making it extremely rugged and corrosion resistant, however the thermistor sensor offers greater sensitivity but is more susceptable to thermally insulating contaminants.

35.7.6 Thermal transducers for flow measurement

Alternative methods of measuring fluid velocity include thermistors and hot wire anemometers. Both have an electric current passed through them and rely on the rate of cooling due to the liquid flowing past them which is a function of its velocity. Both methods are capable of measuring instantaneous flow patterns.

35.8 Transducers for temperature measurements

35.8.1 Resistance thermometers

The resistance of most electrical conductors varies with temperature according to the relation

$$R = R_0(1 + \alpha T + \beta T^2 + \cdots) \quad (35.16)$$

where

R_0 = resistance at temperature T_0
R = resistance at T
α, β = constants
T = rise in temperature above T_0.

Over a small temperature range, depending on the material we may write

$$R = R_0(1 + \alpha T) \quad (35.17)$$

where α is the temperature coefficient of resistance.

Important properties of materials for resistance thermometers include a high temperature coefficient of resistance, stable properties so that the resistance characteristic does not drift with repeated heating and cooling or mechanical strain and a high resistivity to permit the construction of small sensors.

The variation of resistivity with temperature of some of the materials used for resistance thermometers is shown in *Figure 35.16*. Tungsten has a suitable temperature coefficient of resistance but is brittle and difficult to form. Copper has a low resistivity and is generally confined to applications where the sensor size is not restricted. Both platinum and nickel are used extensively because they are relatively easy to obtain in a pure state but platinum has an advantage over nickel in that its temperature coefficient of resistance is linear over a larger temperature range.

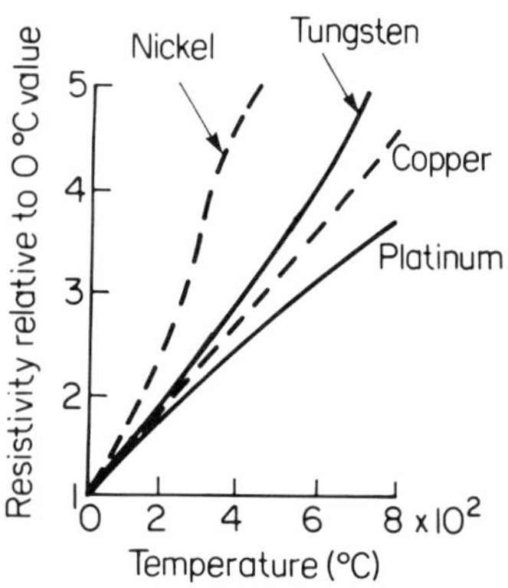

Figure 35.16 Variation of resistivity with temperature of materials used for resistance thermometers

In addition to the temperature range, temperature coefficient and interchangeability characteristics, other important parameters include accuracy, stability, repeatability, rate of response, self-heating effect, insulation resistance and resistance to vibration. Most resistance thermometers adhere to the performance characteristics set out in BS1904 and DIN3760.

The specifications are given based on a typical 100 Ω platinum resistance thermometer.

Accuracy The accuracy of calibration is defined as the ability of a thermometer to conform to its predetermined resistance–temperature relationship and is expressed in terms of % of actual temperature reading. The majority of industrial thermometers fall with the 0.1 to 0.5% range.

Stability Stability is defined as the ability of a thermometer to maintain and reproduce its specified resistance–temperature characteristics for long periods of time within its specified temperature range. The drift in the ice-point resistance after 10 000 hours of operation at 600°C must be less than 0.05% (approximately 0.15°C).

Repeatability Repeatability is defined as the conformity of consecutive temperature measurements for a thermometer at selected temperatures within its specified range of operation. Consecutive temperature measurements should agree within 0.02% (approximately 0.05°C).

Time response The time response is the time required for a thermometer to react to a step change in temperature and reach the resistance corresponding to 63.2% of the total temperature change. The time response can vary from more than two seconds for an industrial encapsulated thermometer to less than 0.5 seconds for the wafer-type sensing element and as low as 0.2 second for a platinum-film sensing element.

Self-heating The heat generated by Joule heating of the resistance element can be a source of error, which is specified as the rise in indicated temperature due to the power dissipated through the sensor over the full range of operating current. The maximum self-heating error over a current range of 0 to 10 mA must be less than 0.1°C based on a minimum dissipation factor of 100 mW °C^{-1}.

Platinum is the material most generally used in the construction of precision laboratory standard thermometers for calibration work and is used for the international thermometer scale from the liquid oxygen point (−182.96°C) to the antimony point (630.74°C).

The resistance–temperature relationship for platinum resistance elements is determined from the Callendar equation

$$T = \frac{100(R_T - R_0)}{R_{100} - R_0} + \delta\left(\frac{T}{100} - 1\right)\frac{T}{100} \quad (35.18)$$

where T is temperature and R_T is resistance at temperature T, R_0 is resistance at 0°C, R_{100} is resistance at 100°C, and δ is the Callendar constant (approximately 1.5).

Subsequent errors from the Callendar equation result in temperature differences of less than ±0.1°C at temperature below 500°C, however these differences approach ±1.0°C at

850°C and should be taken into account with industrial thermometers specified for use at high temperatures.

The temperature coefficient of pure nickel is almost twice that of platinum thus offering the advantage of better sensitivity. Very high purity of nickel is difficult to achieve and different batches of nickel have different temperature coefficients of resistance of the order 0.0062 $\Omega\,^\circ C^{-1}$ to 0.00617 $\Omega\,^\circ C^{-1}$. This is compensated for with a coil constructed from a negligible temperature coefficient alloy, such as constantan, connected in series or parallel with the sensing element and the resultant coefficient and hence the sensitivity is effectively lowered to approximately that of platinum. Construction of nickel resistance thermometers, unlike those of platinum, has been based on a common design, however the recent widespread use of fully encapsulated platinum thermometers with its superior performance characteristics has replaced nickel in most industrial applications.

Copper of the highest purity is readily available commercially, having a temperature coefficient of resistance slightly higher than platinum. The usable temperature range is confined to −200°C to 150°C as copper oxidises at higher temperatures. Construction methods for copper resistance thermometers are very similar to those used in the manufacture of nickel thermometers but as high purity copper is easily obtained no compensating resistor is required. The relatively low resistivity of copper necessitates the use of fine wire to avoid excessively large sensors with subsequent slow response times.

The temperature coefficient of resistance is highly linear, over the range −50°C to 150°C and this is useful when applied to applications where the measurement of temperature difference is required. In many instances copper themometers with their high degree of reproducibility and interchangeability, linear temperature coefficient of resistance, and relatively simple mode of construction are prefereable to the more expensive platinum thermometers for applications near ambient temperature.

35.8.2 Thermistors

The thermistor (temperature sensitive resistor) is a semiconductor whose resistance is a known function of temperature. The variation of resistance with temperature is highly non-linear, with a high temperature coefficient of resistance of the order of 3% to 5% per °C compared with 0.4% per °C for platinum. They are normally used up to about 350°C but are available for operation up to 600°C. The temperature coefficient of resistance is normally negative. The resistance, at any temperature T is given approximately by

$$R_T = R_0 \exp \beta\left(\frac{1}{T} - \frac{1}{T_0}\right) \tag{35.19}$$

where

R_T = thermistor resistance at temperature T (K)
R_0 = thermistor resistance at temperature T_0 (K)
β = a constant determined by calibration

At high temperatures this reduces to

$$R_T = R_0 \exp(\beta/T) \tag{35.20}$$

The resistance is normally high eliminating errors due to the lead resistance. Thermistors are available in various forms. One form extensively used is a bead of the semiconductor coated with glass for protection. A small size and fast response are obtained. Values of resistance lie between 100 Ω to over 10^7 Ω and bead sizes vary from 0.2 mm to 2.5 mm diameter. The bead may be sealed into a

Table 35.2 Commonly used thermocouples†

Thermocouple	*Maximum continuous operating temperature (°C)*	*Typical output ($\mu V\,^\circ C^{-1}$)*	*Comments*
Platinum–rhodium			
$Pt–Pt_{87}Rh_{13}$ (Type R)*	1500	12(1600)	Stable; good corrosion resistance
$Pt_{94}Rh_6–Pt_{70}Rh_{30}$	1600	11.6(1600)	
$Pt_{80}Rh_{20}–Pt_{60}Rh_{40}$	1700	4.5(1600)	
Palladium			
$Pt_{90}Ir_{10}–Pd_{40}Au_{60}$	1000	60(800)	Good resistance to corrosion; higher outputs than Pt/Rh alloys
$Pt_{12.5}Pd–Au_{54}Pd_{46}$	1200	35(400)	
Iridium			
$Ir–Ir_{40}Rh_{60}$	2100	6.6(2000)	Fragile at high temperatures
Tungsten–rhenium			
$W–W_{74}Re_{26}$	2700		Operation in inert atmosphere vacuum only
$W_{95}Re_5W_{14}Re_{26}$	2700		
Chromel–alumel (type K)			
$Ni_{90}Cr_{10}–Ni_{94}$, Al, Si, Mn	1300	25(150–1300)	Deteriorates rapidly in H_2S or CO_2; most commonly used up to 1100°C
Iron–constantan (type J)			
$Fe–Cu_{57}Ni_{35}$	800	63(800)	Can be used in oxidising and reducing atmospheres; high output
Copper constantan	350	60(350)	Copper oxidises above 350°C;
Chromel constantan	700	81(500)	Very high sensitivity

* Supersedes $Pt–Pt_{90}Rh_{10}$ (type S).
† E.m.f. tables for the following thermocouple combinations are given in BS4937: 1973
Part 1, Platinum/10% rhodium–platinum (type S)
Part 2, Platinum/13% rhodium–platinum (type R)
Part 3, Iron/copper–nickel
Part 4, Nickel–chromium/nickel–aluminium.

Table 35.3 Comparison of temperature transducers

Property	*Platinum resistance thermometer*	*Thermistor*	*Thermocouple*
Repeatability	0.03°C to 0.05°C	0.1°C to 1°C	1°C to 10°C
Stability	<0.1% drift in 5 years	0.1°C to 2.5°C drift per year	0.5°C to 1°C drift per year
Sensitivity	0.2 to 10 Ω °C^{-1}	100 to 1000 Ω °C^{-1}	10–50 μV °C^{-1}
Temperature range	−120°C to 850°C	−100°C to 350°C	−200°C to +1600°C
Signal output	1–6 V	1–3 V	0–60 mV
Minimum size	7.5 mm dia. × 6 mm long	0.44 mm dia.	0.4 mm dia.
Linearity	good	poor	good
Special features	greatest accuracy over wide range: highly stable	greatest sensitivity lead effects minimised by high impedance	largest operating range

solid glass rod which can be easily mounted and can be used for measuring liquid temperatures. Discs are made by pressing the thermistor material into round discs up to 25 mm diameter which are sintered. Discs enable higher power dissipation to be achieved and are used for surface temperature measurement where space is not a problem. Washers may be manufactured in the same way to enable mounting on a bolt.

35.8.3 Thermocouples

Various thermocouple materials and methods of construction are used depending on the temperature, environment and sensitivity. Examples of some typical thermocouple materials and conditions of usage are given in *Table 35.2.* Where a continuous output from the temperature sensor is required for indication or control the thermocouple is the most widely used method of temperature measurement and the chromel–alumel thermocouple is almost solely used today for measuring temperatures up to about 1300°C. Above this temperature, depending on the application, platinum/platinum–rhodium alloy and tungsten/tungsten–rhenium thermocouples are employed. Only a relatively small output of the order of 10–20 μV °C^{-1} is obtained and it is necessary to amplify the output. Although over short ranges of temperature the output may be approximately proportional to the temperature over wide ranges it is non-linear and linearising circuits are necessary for different thermocouple materials.

The output voltage is measured with respect to the cold junction e.m.f. which must be held constant. This is normally carried out by a built-in temperature compensated reference voltage. The advent of stable high sensitivity solid state amplifiers with built in linearisation has enabled temperatures to be measured to within 0.1% and has extended the application of thermocouples into many areas where resistance thermometers were formerly used.

35.8.4 Comparison of temperature transducers

A comparison of the advantages and disadvantages of resistance thermometers, thermistors and thermocouples is shown in *Table 35.3.*

The temperature range of thermocouples is the largest and the small mass of the thermocouple junctions means a rapid response time and a sensitive reading at a point.

Reproducibility, stability and accuracy are the main advantages of resistance thermometers. The wire type thermometers have a large surface area useful in area sensing applications. They suffer the disadvantages of limited miniaturisation and relatively high cost but these factors are counterbalanced by linear output which permits the use of less expensive instrumentation. The film type thermometers are not as yet as accurate, but they offer the possibility of substantial size reductions.

Thermistors combine the small size of thermocouples and are relatively cheap. They suffer from the problems of non-interchangeability of element, non-linearity of output and poor stability. The most appropriate applications are in small temperature ranges due to their high sensitivity.

Further reading

BENEDICT, R. P., *Fundamentals of Temperature, Pressure and Flow Measurements*, Wiley, New York, p. 58 (1977)

DROMGOOLE, W. V., 'Thermistor linearisation permits wide range of temperature measurements', *Elec. Engng OCT*, pp 68–70 (1972)

DROMS, C. R., 'Thermistors for temperature measurements', *Temperature: Its Measurement and Control in Science and Industry*, **3**, (2) 339–46 (1962)

GRANT, D. A. and HICKENS, W. F., 'Industrial temperature measurement with nickel resistance thermometers', *Temperature: Its Measurement and Control in Science and Industry*, **3**, (2) 305–15 (1962)

HALL, J. A., *The Measurement of Temperature*, Chapman and Hall, London, pp. 13–18 (1966)

HASLAM, J. A., SUMMERS, E. R. and WILLIAMS, D., *Engineering Instrumentation and Control*, Edward Arnold, London (1981)

JONES, E. B., *Instrument Technology*, Vol. 1, 3rd edn, Newnes-Butterworth, Sevenoaks (1974)

NEUBERT, H. K. P., *Instrument Transducers*, Clarendon Press, Oxford (1963)

OLIVER, F. J., *Practical Instrumentation Transducers*, Pitman, London (1972)

SHEPHERD, I. E., 'Temperature measurement with low cost thermistors', *Elec. Engng*, Sept. p. 21 (1974)

WERNER, F. D., 'Some recent developments in applied platinum resistance thermometry', *Temperature: Its Measurement and Control in Science and Industry*, **3**, (2) 297–305 (1962)

Part 4

Electronic Design

36

Filters

P J Cottam, BSc
Group Leader,
STC Components

Contents

36.1 Types of filter

The classification of electric wave filters can be done in several ways. They may be grouped in terms of the frequency spectrum against the realisation (*Figure 36.1*), or they may be grouped according to the elements that make up the filters.

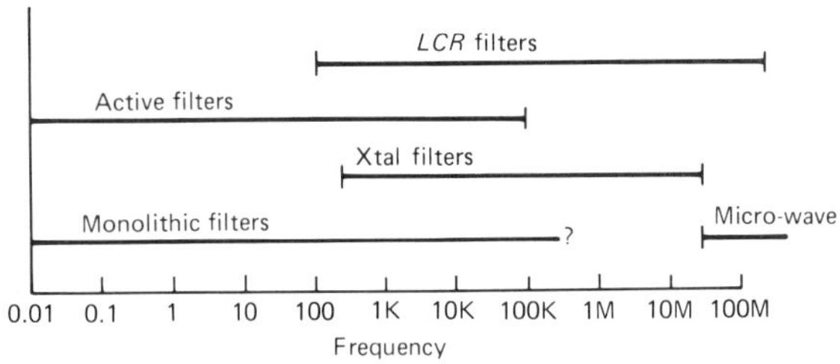

Figure 36.1 Filter frequency guide

There are five basic types of filters used for frequency discrimination in electronic circuits.

(1) The low pass filter (*Figure 36.2(a)*). This type of filter passes all signals in the frequency band from zero frequency up to the required cut-off frequency.

(2) The high pass filter (*Figure 36.2(b)*). This type of filter rejects all signals in the frequency band from zero frequency up to the required cut-off frequency.

(3) The band pass filter (*Figure 36.2(c)*). This type of filter passes all signals in the frequency band defined by a lower and an upper frequency. This is the most common form of filter.

(4) The band reject filter (*Figure 36.2(d)*). This type of filter rejects all signals in the fequency band defined by a lower and an upper frequency. This type of filter is often used to take out unwanted side tone.

(5) The all pass filter (*Figure 36.2(e)*). This type of filter is not normally amplitude sensitive, but provides a controlled phase response. It may be used for phase correction in digital transmission systems, phase splitting of signals, or expansion of signals in the time domain.

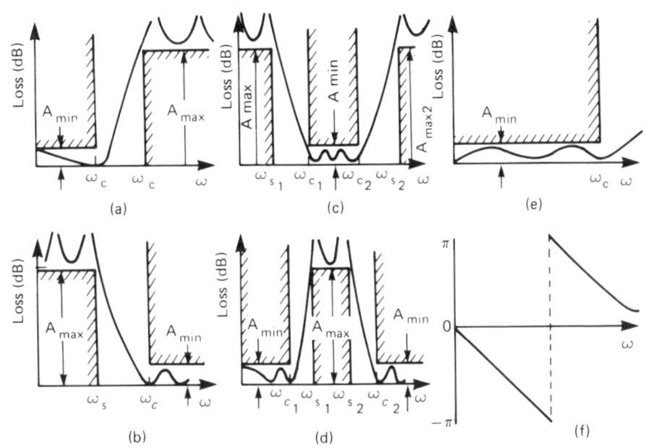

Figure 36.2 Different types of filter: (a) low pass; (b) high pass; (c) band pass; (d) band reject; (e) all pass; (f) all pass

36.2 Filter design using image parameters

The theory of image parameter design was proposed by O. Zobel in the 1920s, but has in more recent times become less popular as more modern synthesis techniques have been developed.

The basic filter sections used in image parameter design are given in *Figures 36.3, 36.4, 36.5* and *36.6*. These figures give only the simpler elements normally used.

The basic filter circuits are shown in *Figure 36.7*. The basic equations for the image impedances Z_T and Z_π for the $\frac{1}{2}T$ sections are derived by assuming that the network is terminated with impedances that change with frequency in accordance with the following image impedance equations.

Z_T = mid-series image impedance

= impedance looking into the input 1–2 with Z_π across output 3–4.

Z_π = mid-shunt image impedance

= impedance looking into the output 3–4 with Z_T across input 1–2.

Therefore

$$Z_T=\frac{Z_1}{2}+\frac{2Z_2Z_\pi}{2Z_2+Z_\pi} \tag{36.1}$$

and

$$Z_\pi=\frac{[(Z_1/2)+Z_T]2Z_2}{2Z_2+(Z_1/2)+Z_T} \tag{36.2}$$

This gives

$$Z_TZ_\pi=Z_1Z_2 \tag{36.3}$$

Solving for Z_T and Z_π gives

$$Z_T=(Z_1Z_2)^{1/2}[1+(Z_1/4Z_2)]^{1/2}\,\Omega \tag{36.4}$$

and

$$Z_\pi=(Z_1Z_2)^{1/2}/[1+(Z_1/4Z_2)]^{1/2}\,\Omega \tag{36.5}$$

The general equation for the transfer constant of a network is given as

$$\theta=a+\mathrm{j}b \tag{36.6}$$

where a is the image attenuation constant (nepers) and b is the image phase constant (radians).

In the pass band, for a full section, the following equations apply:

$$\cosh\theta=1+(Z_1/2Z_2) \tag{36.7}$$

$a=0$ nepers for frequencies giving $-1\leqslant Z_1/4Z_2\leqslant 0$, $b=\cos^{-1}[1+(Z_1/2Z_2)]$ rad.

In the stop band, for a full section, the following equations apply:

$$a=\cosh^{-1}|1+(Z_1/2Z_2)| \quad \text{for } Z_1/4Z_2>0 \tag{36.8}$$

$b=0$ rad

$$a=\cosh^{-1}|1+(Z_1/2Z_2)| \quad \text{for } Z_1/4Z_2<-1 \tag{36.9}$$

$b=\pm\pi$ rad

36.2.1 Realisation of a low pass filter using image parameter design

It is required to design a typical telecommunications low pass filter with the following requirements:

(1) cut-off frequency 3.4 Hz;
(2) peak attenuation 4.5 kHz and 6.5 kHz;
(3) output load resistance 600 Ω.

We will construct this filter using a constant-*k* mid-section and an *m*-derived section with *m*-derived terminating half-sections. The basic sections are shown in *Figure 36.8*.

(1) Constant-*k* mid-section (*Figure 36.8(a)*)

$$L_k=\frac{R}{\omega_c}=600/(2\pi\times 3400)=28.1\times 10^{-3}\,\mathrm{H} \tag{36.10}$$

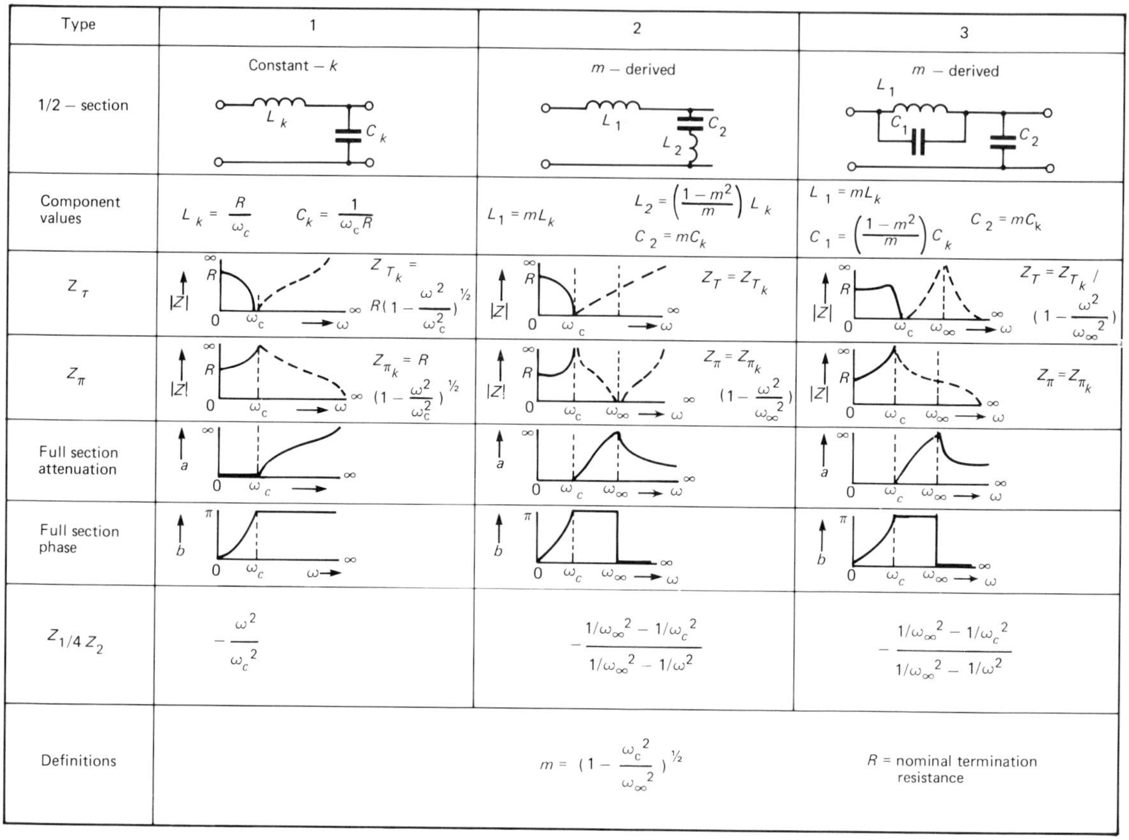

Figure 36.3 Low pass structures

$$C_k=\frac{1}{\omega_c R}=(2\pi\times 3400\times 600)^{-1}=0.078\times 10^{-6}\ \text{F} \quad (36.11)$$

$$a=2\cosh^{-1}(\omega/\omega_c)=2\cosh^{-1}(f/3400) \quad (36.12)$$

$$b=2\sin^{-1}(\omega/\omega_c)=2\sin^{-1}(f/3400) \quad (36.13)$$

(2) *m*-derived mid-section (*Figure 36.8(b)*). We shall use this section to derive the peak attenuation at 6.5 kHz:

$$m=\left(1-\omega_c^2/\omega_\infty^2\right)^{1/2}=\left(1-\frac{3400^2}{6500^2}\right)^{1/2}=0.852 \quad (36.14)$$

$$L_1=mL_k=0.852\times(28.1\times 10^{-3})=23.9\times 10^{-3}\ \text{H} \quad (36.15)$$

$$L_2=[(1-m^2)/m]\times L_k=9.0441\times 10^{-3}\ \text{H} \quad (36.16)$$

$$C_2=mC_k=0.852\times(0.078\times 10^{-6})=0.066\times 10^{-6}\ \text{F} \quad (36.17)$$

$$a=\cosh^{-1}\left(1-\frac{2m^2}{(\omega_c/\omega)^2-(1-m^2)}\right)$$

$$=\cosh^{-1}\left(1-\frac{1.45}{(3400^2/f^2)-0.27}\right) \quad (36.18)$$

$$b=\cos^{-1}\left(1-\frac{2m^2}{(\omega_c/\omega)^2-(1-m^2)}\right)$$

$$=\cos^{-1}\left(1-\frac{1.45}{(3400^2/f^2)-0.27}\right) \quad (36.19)$$

3. *m*-derived half-sections (*Figure 36.8(c)*). We shall use these sections to derive the peak attenuation at 4.5 kHz:

$$m=[1-(\omega_c^2/\omega_\infty^2)]^{1/2}=0.655 \quad (36.20)$$

$$L_1=mL_k=0.655\times(28.1\times 10^{-3})=18.4\times 10^{-3}\ \text{H} \quad (36.21)$$

$$L_2=[(1-m^2)/m]\times L_k$$

$$L_k=[(1-0.655^2)/0.655]\times(28.1\times 10^{-3})=24.5\times 10^{-3}\ \text{H} \quad (36.22)$$

$$C_2=mC_k=0.655\times(0.078\times 10^{-6})=0.051\times 10^{-6}\ \text{F} \quad (36.23)$$

The final filter is shown together with its response curve in *Figure 36.9*.

36.3 Filter design using synthesis

In modern filter design the image parameter theory is giving way to the design technique which is based on the use of complex polynomials to define the transfer function. All filter synthesis using this approach is done with reference to a normalised low pass filter, i.e. $\omega_c=1$ rad s^{-1}. Many publications now exist with comprehensive 'look up' tables for normalised low pass filters. The most commonly used approach to filter synthesis the Butterworth, Chebyshev, elliptic and Bessel polynomial approximations.

36.3.1 Butterworth approximation

The well-known Butterworth approximation is given by:

$$K(s)=k(s/\omega_c)^n \quad (36.24)$$

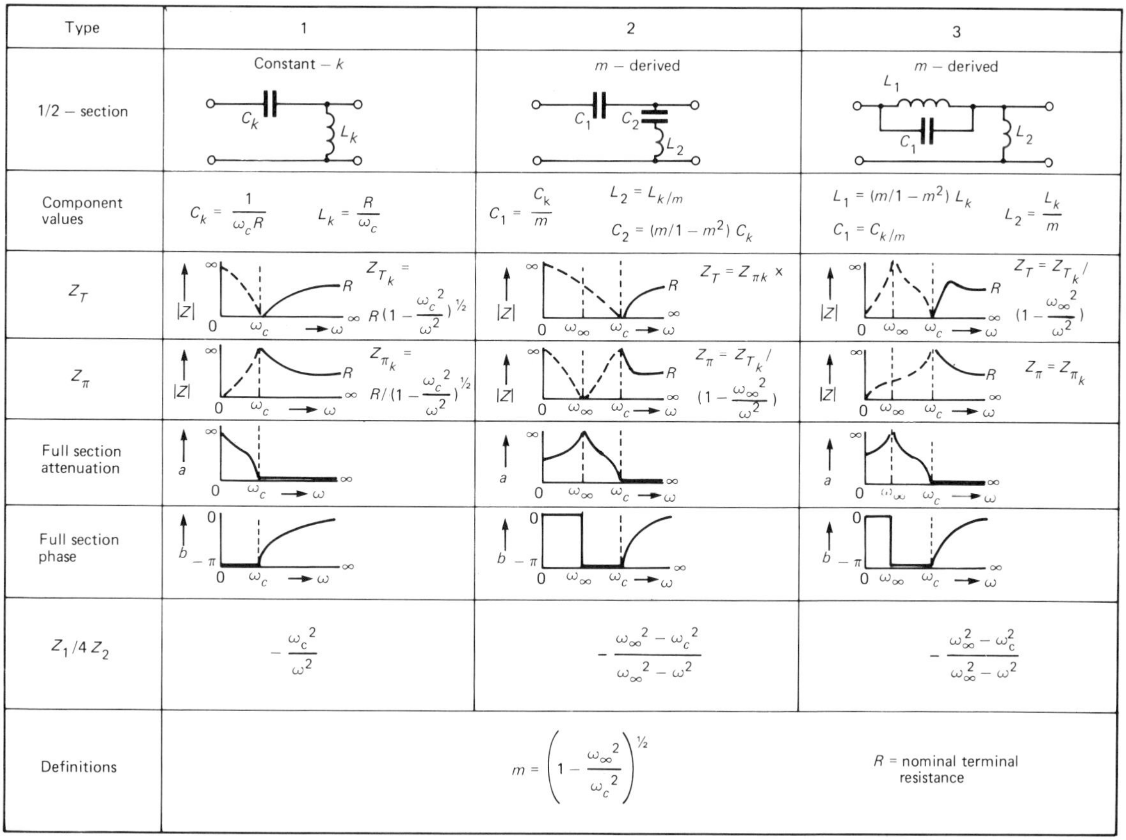

Figure 36.4 High pass structures

where k is a constant, n is the order of the polynomial, and ω_c is the passband cut-off frequency. This gives the loss function for a low pass filter as:

$$H(s)=V_i(s)/V_o(s)=\sqrt{1+k^2(\omega/\omega_c)^{2n}} \tag{36.25}$$

The loss in dB is given by

$$A=10\lg[1+k^2(\omega/\omega_c)^{2n}] \tag{36.26}$$

This response has the flattest possible characteristic at the centre of the pass band. At the pass band cut-off frequency the loss is given by (again in dB)

$$A=10\lg(1+k^2) \tag{36.27}$$

In the filter tables this is given as A_{max}. As the frequency ω becomes much greater than ω_c then the loss is given by (dB)

$$A=20\lg[k(\omega/\omega_c)]^n \tag{36.28}$$

This indicates that the loss increases at $6n$ dB/octave.

A graph of the first eight Butterworth responses is given in *Figure 36.10*. The first eight Butterworth polynomials are given in *Table 36.1*. These are derived as follows:

$$|H(s)|^2=1+(-s^2)^n \tag{36.29}$$

where $\omega_c=1$ rad, $k=1$ (hence to denormalise a function we multiply s by $(k^{1/n}/\omega_c)$).

The roots of $|H(s)|^2$ are given by:

$$s_N=\exp\left(\frac{j\pi}{2}\,\frac{2N+n-1}{n}\right) \tag{36.30}$$

where $N=1, 2, \ldots, 2n$. This gives the roots located on a unit circle and equally spaced at π/n rad.

The roots of a fourth-order Butterowrth approximation are given in *Figure 36.11*.

36.3.2 Realisation of a Butterworth low pass filter

The characteristics of the filter are as follows: $A_{max}=1.0$ dB, $A_{min}=12$ dB, $f_c=3.4$ kHz, $f_s=4.5$ kHz. From equation (36.27)

$$k^2=10^{A_{max}/10}-1 \qquad \text{i.e. } k=0.51 \tag{36.31}$$

The attenuation per octave is given by:

$$(12-1)/(4.5-3.4)\times 3.4 \text{ dB/octave}=34 \text{ dB/octave}$$

The order of the filter is 34/6 which equals 5.7, therefore a sixth-order filter is required.

The normalised function is

$$(s^2+0.5176s+1)(s^2+1.4144s+1)(s^2+1.9305s+1)=H(s) \tag{36.32}$$

To denormalise the function we multiply s by $(k^{1/n}/\omega_c)$ and therefore

$$\frac{k^{1/n}}{\omega_c}=\frac{0.51^{1/6}}{2\pi\times 3400}=0.0000418$$

and

$$H(s)=\frac{[(s^2+12382s+1)(s^2+33835s+1)(s^2+46181s+1)]}{5.723\times 10^8} \tag{36.33}$$

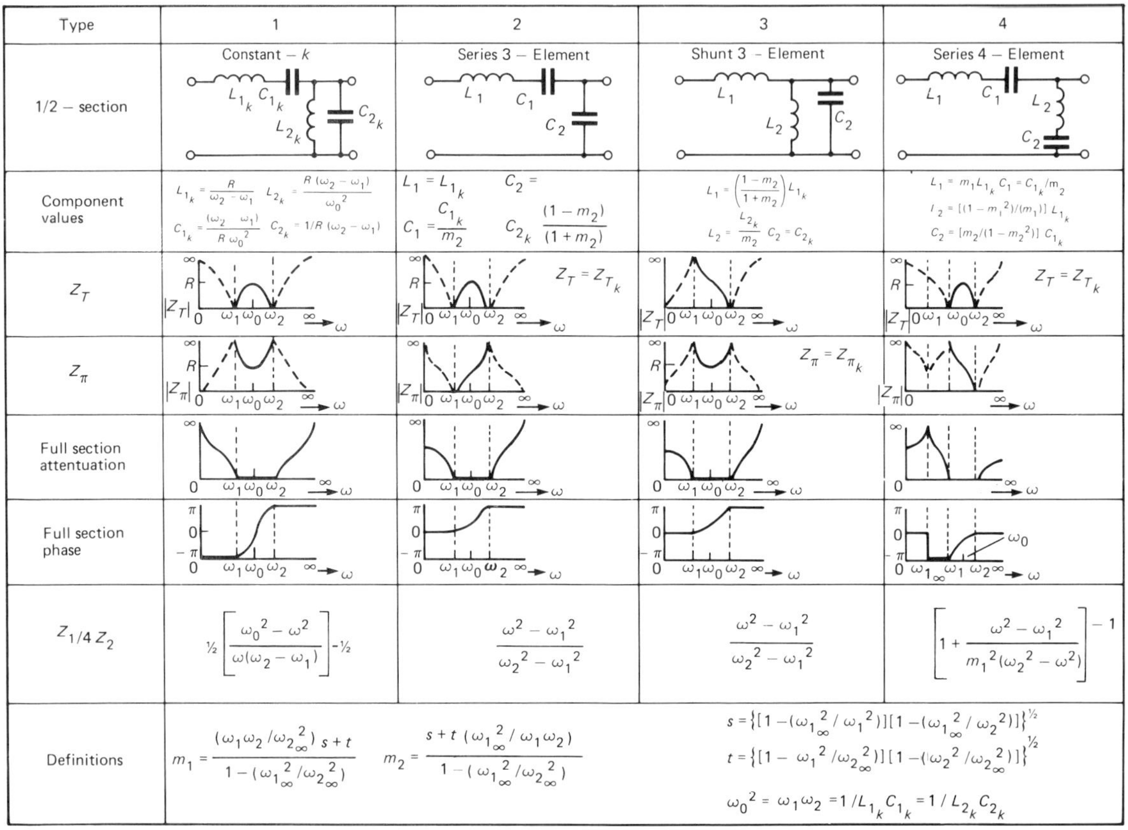

Figure 36.5 Band pass structures

36.3.3 The Chebyshev approximation

The Butterworth approximation provided a response that was maximally flat at $\omega = 0$ but got progressively poorer as ω approached ω_c. Also the stop band attenuation is not as good as can be achieved by other types of polynomial approximations. The Chebyshev approximation is a polynomial which can achieve good stop band attenuation. It sacrifices flatness in the pass band in favour of equi-ripple, and gives a sharper roll-off in the stop band.

The Chebyshev function is definee as:

$$C_n(\omega/\omega_c) = \begin{cases} \cos(n\cos^{-1}\omega/\omega_c) & |\omega/\omega_c| \leqslant 1 \quad (36.34) \\ \cosh(n\cosh^{-1}\omega/\omega_c) & |\omega/\omega_c| > 1 \quad (36.35) \end{cases}$$

In the pass band we aim to use the approximation which gives equi-ripple, and the Chebyshev function can be rewritten as a polynomial:

$$C_{n+1}(\omega/\omega_c) + C_{n-1}(\omega/\omega_c)$$

$$= \cos[(n+1)(\cos^{-1}\omega/\omega_c)] + \cos[(n-1)(\cos^{-1}\omega/\omega_c)]$$

$$= 2\cos(\cos^{-1}\omega/\omega_c)\cos(n\cos^{-1}\omega/\omega_c) = 2\omega/\omega_c C_n(\omega/\omega_c)$$

therefore

$$C_{n+1}(\omega/\omega_c) = 2\omega/\omega_c C_n(\omega/\omega_c) - C_{n-1}(\omega/\omega_c) \quad (36.36)$$

For $n = 0$

$$C_0(\omega/\omega_c) = 1 \quad (36.37)$$

and for $n = 1$

$$C_1(\omega/\omega_c) = \omega/\omega_c \quad (36.38)$$

Then

$$C_2(\omega/\omega_c) = 2(\omega/\omega_c)C_1(\omega/\omega_c) - C_0(\omega/\omega_c) = 2(\omega/\omega_c)^2 - 1 \quad (36.39)$$

$$C_3(\omega/\omega_c) = 2(\omega/\omega_c)C_2(\omega/\omega_c) - C_1(\omega/\omega_c) = 4(\omega/\omega_c)^3 - 3(\omega/\omega_c) \quad (36.40)$$

$$C_4(\omega/\omega_c) = 8(\omega/\omega_c)^4 - 8(\omega/\omega_c)^2 + 1 \quad (36.41)$$

Plots of these polynomial relationships on the (ω/ω_c) axis are given in *Figure 36.12*, for $-1 < \omega/\omega_c < +1$, and shows that in the pass band the response is equi-ripple. It is also apparent that at $\omega/\omega_c = 1$, $C_n(\omega/\omega_c) = 1$.

The Chebyshev loss function is given by

$$H(s) = V_i(s)/V_0(s) = \sqrt{1 + k^2C_n^2(\omega/\omega_c)} \quad (36.42)$$

The loss (dB) is given by

$$A = 10\lg[1 + k^2C_n^2(\omega/\omega_c)] \quad (36.43)$$

At the pass band cut-off frequency that loss (in dB) is given by

$$A_{\max} = 10\lg(1 + k^2) = \text{pass band ripple} \quad (36.44)$$

As the frequency ω becomes much greater than ω_c then the loss becomes

$$A = 20\lg[k\,C_n(\omega/\omega_c)] = 20\lg[k\,2^{n-1}(\omega/\omega_c)^n] \quad (36.45)$$

The Butterworth approximation gave $A = 20\lg K(\omega/\omega_c)^n$, therefore the Chebyshev polynomial gives an additional attenuation of $20\lg(2)^{n-1} \sim 6(n-1)$ dB for the same value of n. *Table 36.2* gives the first six Chebyshev polynomials and *Figure 36.13* gives graphs of some of the Chebyshev responses. The multiplying factor in *Table 36.2* is the value necessary to provide a minimum loss in the pass band of 0 dB.

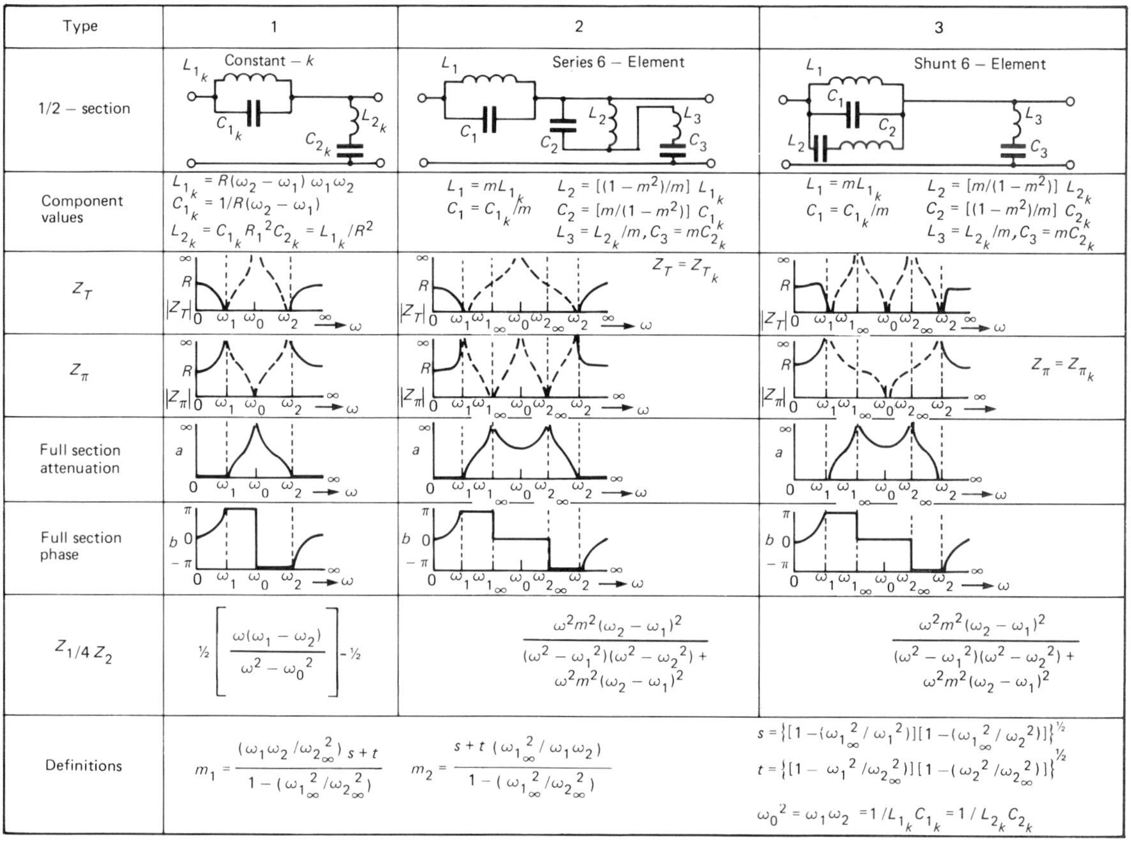

Figure 36.6 Band stop structures

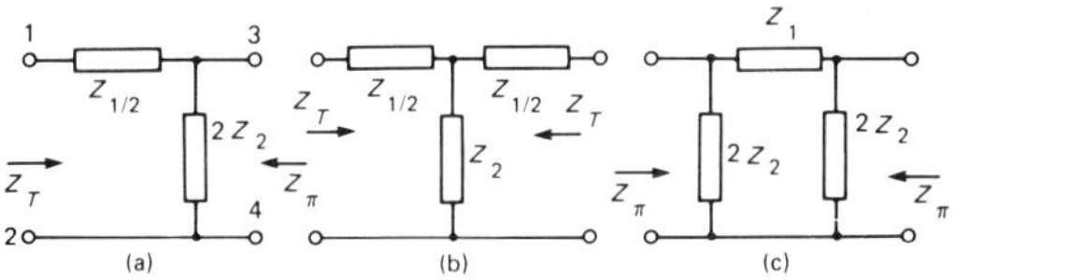

Figure 36.7 Image parameter sections: (a)$\frac{1}{2}$-section; (b) full-T; (c) full-π

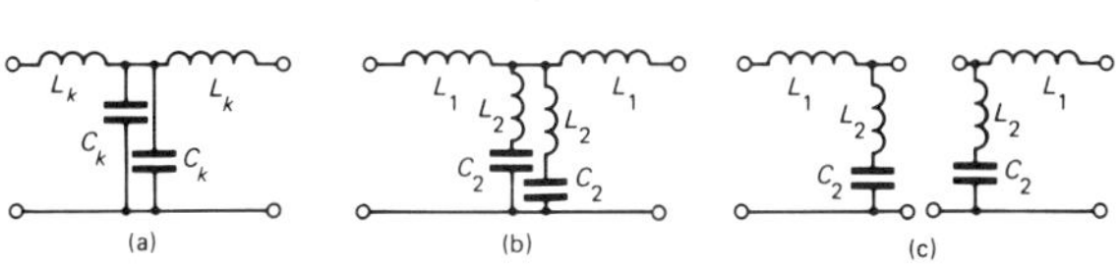

Figure 36.8 Sections for worked example in Section 36.3.1: (a) constant-k mid-section; (b) m-derived mid-section; (c) m-derived end section

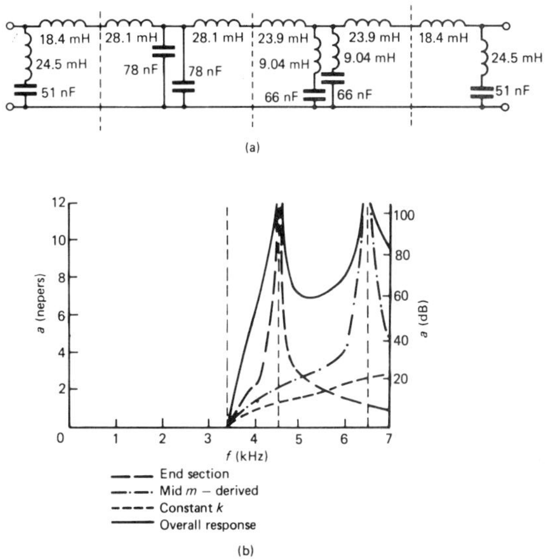

Figure 36.9 Circuit and response of worked example in Section 36.2.1: (a) circuit; (b) response

From the equation

$$|H(s)|^2 = 1 + k^2 C_n^2(\omega/\omega_c) \tag{36.46}$$

It can be shown that the roots of $H(s)$ are $s_N = \sigma_N \pm j\omega_N$, where $N = 0, 1, 2, \ldots, 2n-1$. In this expression

$$\sigma_N = \pm \sin\frac{\pi}{2}\frac{(1+2N)}{n}\sinh\left(\frac{1}{n}\sinh^{-1}\frac{1}{k}\right) \tag{36.47}$$

and

$$\omega_N = \cos\frac{\pi}{2}\frac{(1+2N)}{n}\cosh\left(\frac{1}{n}\sinh^{-1}\frac{1}{k}\right) \tag{36.48}$$

and it can be shown that the roots lie on an ellipse whose equation is

$$\left\{\sigma_N\left[\sinh\left(\frac{1}{n}\sinh^{-1}\frac{1}{k}\right)\right]^{-1}\right\}^2+\left\{\omega_N\left[\cosh\left(\frac{1}{n}\sinh^{-1}\frac{1}{k}\right)\right]^{-1}\right\}^2$$

as shown in *Figure 36.14*.

The values of the Chebyshev polynomials can now be found in a similar manner to that used in the Butterworth approximation.

36.3.4 Realisation of a Chebyshev low pass filter

We shall use the same characteristics as in Section 36.3.2.

From *Figure 36.13(b)* it can be seen that for an attenuation of 34 dB per octave we need a value of n just greater than 4, and so we must choose $n=5$. This is an order less than the Butterworth filter. The normalised $H(s)$ is as below

normalised $H(s)$

$$=(s^2+0.17892s+0.98831)(s^2+0.46841s+0.42930)\times\frac{(s+0.28949)}{0.12283} \tag{36.49}$$

The denormalising multiplier is $(s)\omega_c^{-1}$, the k multiplier having been taken care of in *Table 36.2.* It follows that

$H(s)=$

$$\frac{(s^2+608.3s+11.42\times10^6)(s^2+1592.6s+4.96\times10^6)(s+984.3)}{5.581\times10^{16}} \tag{36.30}$$

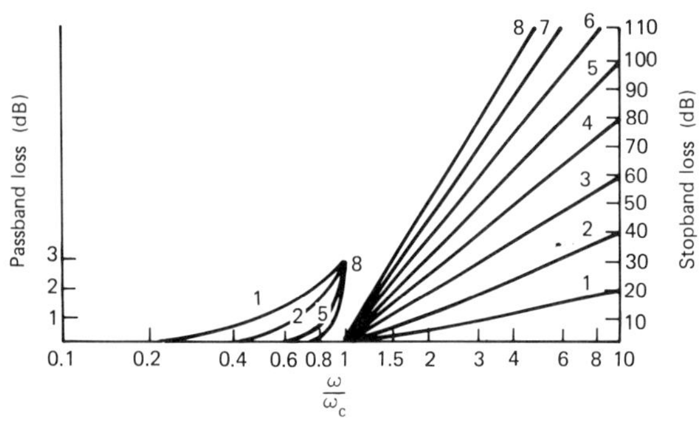

Figure 36.10 Butterworth response

36.3.5 Elliptic approximation

Both the Butterworth and the Chebyshev approximations have stop band losses that increase by 6n dB/octave for an nth order response. Therefore to get a very sharp roll-off after ω_c needs a filter with a very high order. One way to overcome the use of such complex filters is to provide finite poles of attenuation in the stop band. Butterworth and Chebyshev approximations were both monotonic in the stop band with a pole at infinite frequency. This new type of characteristic is an elliptic function. The most common type of approximation not only gives equi-ripple in the pass band, but also gives equi-ripple in the stop band, and this is known as the Cauer approximation.

To define an elliptic response the following parameters are needed:

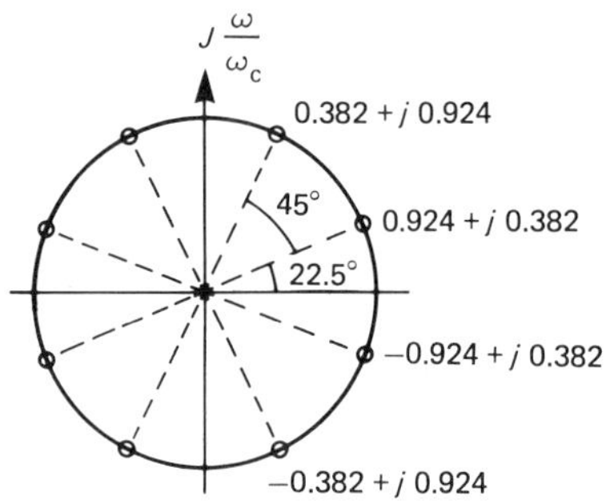

Figure 36.11 Roots of a fourth-order Butterworth

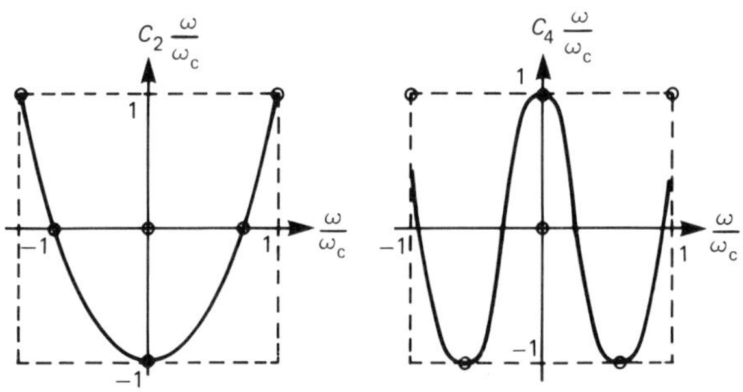

Figure 36.12 Plots of a Chebyshev second- and fourt-order

Table 36.1 First eight Butterworth polynomials

n	$H(s)$	$H(s)$ (*factorised*)
1	$s+1$	$s+1$
2	$s^2+1.414s+1$	$s^2+1.414s+1$
3	$s^2+2.000s^2+2.000s+1$	$(s^2+s+1)(s+1)$
4	$s^4+2.1631s^3+3.414s^2+2.6131s+1$	$(s^2+0.7654s+1)(s^2+1.8478s+1)$
5	$s^5+3.236s^4+5.236s^3+5.236s^2+3.236s+1$	$(s^2+0.618s+1)(s^2+1.618s+1)(s+1)$
6	$s^6+3.864s^5+7.464s^4+9.142s^3$	$(s^2+0.518s+1)(s^2+1.414s+1)$
	$7.464s^2+3.864s+1$	$(s^2+1.931s+1)$
7	$s^7+4.494s^6+10.098s^5+14.592s^4+14.592s^3$	$(s^2+0.445s+1)(s^2+1.247s+1)$
	$+10.098s^2+4.494s+1$	$(s^2+1.802s+1)(s+1)$
8	$s^8+5.153s^7+13.137s^6+21.846s^5+25.688s^4$	$(s^2+0.390s+1)(s^2+1.111s+1)$
	$+21.846s^3+13.137s^2+5.153s+1$	$(s^2+1.664s+1)(s^2+1.961s+1)$

Table 36.2 First six Chebyshev polynomials

n	A_{max}	$H(s)$	*Denominator* × *constant*
1	0.5 (dB)	$s+2.8628$	2.8628
2		$s^2+1.4256s+1.5162$	1.4314
3		$(s^2+0.6246s+1.1425)(s+0.6265)$	0.7157
4		$(s^2+0.3507s+1.0635)(s^2+0.8467s+0.3564)$	0.3579
5		$(s^2+0.2239s+1.0358)(s^2+0.5863s+0.4768)(s+0.3623)$	0.1789
6		$(s^2+0.1553s+1.0230)(s^2+0.4243s+0.5901)(s^2+0.5796s+0.1570)$	0.0948
1	1.0 (dB)	$s+1.9652$	1.9652
2		$s^2+1.0977s+1.1025$	0.9826
3		$(s^2+0.4942s+0.9942)(s+0.4942)$	0.4913
4		$(s^2+0.2791s+0.9865)(s^2+0.6737s+0.2794)$	0.2457
5		$(s^2+0.1789s+0.9883)(s^2+0.4684s+0.4293)(s+0.2895)$	0.1228
6		$(s^2+0.1244s+0.9907)(s^2+0.3398s+0.5577)(s^2+0.4641s+0.1247)$	0.0689

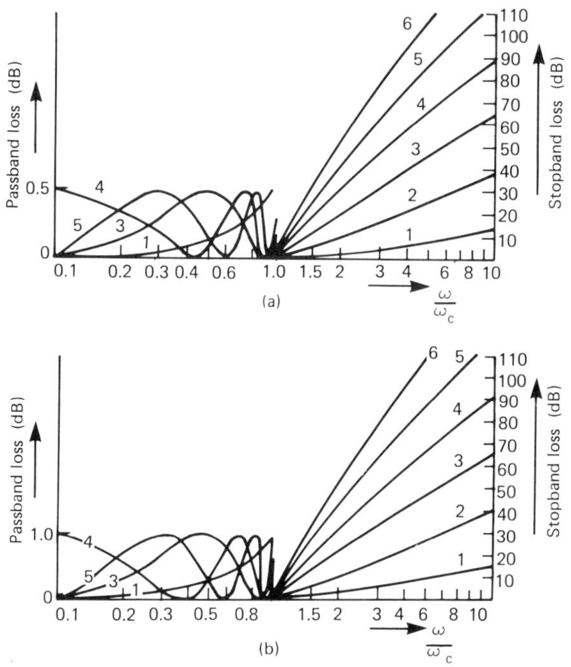

Figure 36.13 Chebyshev responses to sixth-order

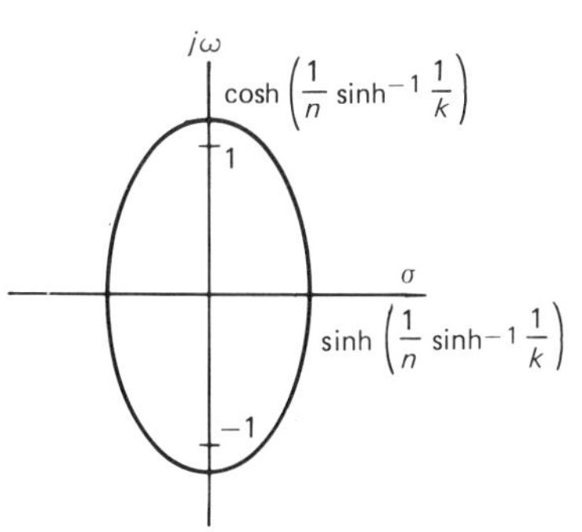

Figure 36.14 Ellipse for roots of Chebyshev polynomial

(1) Frequency transition ratio ω_s/ω_c. This is sometimes given as θ, where $\theta=\sin^{-1}\omega_c/\omega_s$;

(2) Pass band ripple A_{max};

(3) Stop band minimum attenuation A_{min},

(4) The order of the filter n.

Instead of A_{max} the table will sometimes refer to the reflection coefficient $\rho(\%)$, where

$$A_{max}=-10\lg(1-\rho^2)\,\text{dB}$$

The mathematical proof behind the calculations needed to define the elliptic functions is to be found in R. W. Daniels text on approximation methods.[4]

The general loss function is given by:

$$H(s)=\prod_{n=2,4,6,\ldots}\frac{(s^2+b_{1_n}+b_{0_n})}{k(s^2+a_{0_n})}\quad \text{even function} \qquad (36.51)$$

$$\prod_{n=1,3,5,7,\ldots}\frac{(s^2+b_{1_n}+b_{0_n})(s+c_0)}{k(s^2+a_{0_n})}\quad \text{odd function} \qquad (36.52)$$

36.3.6 Bessel approximation

The approximations discussed so far have been concerned with amplitude response, and have taken no account of the associated phase shift. An approximation that gives maximally flat group delay in the frequency domain is the Bessel approximation (sometimes referred to as the Thomson filter).

The ideal delay characteristic is given by the equation

$$V_o(t)=V_i(t-T_0) \qquad (36.53)$$

where T_0 is the delay constant.

In the s-plane this equation becomes

$$V_o(s)=V_i(s)\exp(-sT_0)$$

$$H(s)=V_i/V_o(s)=\exp(sT_0) \qquad (36.54)$$

If we normalise this function for $T_0=1$ then $H(s)=e^s$. The Bessel function for this is

$$H(s)_n=B_n(s)/B_n(0) \qquad (36.55)$$

where $B_n(0)=[1+\overline{2\times 0}][1+\overline{2\times 1}][1+\overline{2\times 2}]\ldots[1+\overline{2\times(n-1)}]$

$B_n(s)$ is the nth order Bessel polynomial which is defined by

$$B_0(s)=1,\quad B_1(s)=s+1 \qquad (36.56)$$

and

$$B_n(s)=2(n-1)B_{n-1}(s)+s^2B_{n-2}(s) \qquad (36.57)$$

Table 36.3 The first six Bessel polynomials

n	$H(s)$	*Denominator × constant*
1	$s+1$	1
2	s^2+3s+3	3
3	$(s^2+3.678s+6.459)(s+2.322)$	15
4	$(s^2+5.792s+9.140)(s^2+4.208s+11.488)$	105
5	$(s^2+6.704s+14.273)(s^2+4.649s+18.156)(s+3.647)$	945
6	$(s^2+8.497s+18.801)(s^2+7.471s+20.853)(s^2+45.032s+26.514)$	10395

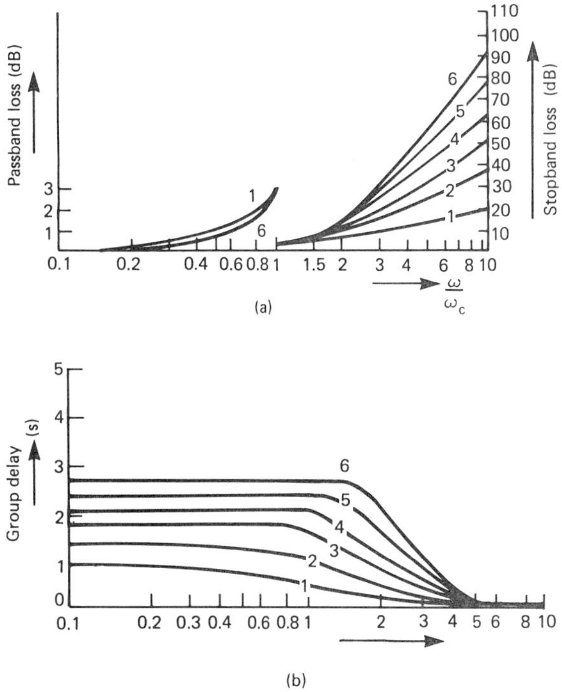

Figure 36.15 Bessel response to sixth-order

The first six Bessel approximations are given in *Table 36.3* and the first six Bessel characteristics are shown in the graphs in *Figure 36.15*.

36.3.7 Delay equaliser

The Bessel approximation in Section 36.3.6 gives a maximally flat delay response in the pass band but gives a poor amplitude rejection response in the stop band. The alternative approach is to use a filter with the correct attenuation response and then use a phase correcting section to level the delay response in the pass band. As the amplitude response of the filter is correct then the amplitude response of the phase correcting section should be flat over both the pass band and stop band of the preceding filter.

This type of phase correction circuit is called a delay equaliser. The transfer characteristic is given by

$$T(s)=\frac{1}{H(s)}=\frac{V_o(s)}{V_i(s)}=\frac{s^2-as+b}{s^2+as+b} \tag{36.58}$$

The gain in dB of this section for all values of ω is

$$\begin{aligned}20\lg\left|\frac{V_o(s)}{V_i(s)}\right|_{s=j\omega} &= 10\lg[(b-\omega^2)^2+(-a\omega)^2] \\ &\quad -10\lg[(b-\omega^2)+(+a\omega)^2] \\ &= +0\,\text{dB}\end{aligned} \tag{36.59}$$

The cut-off frequency is given by

$$\omega_c=\sqrt{b} \tag{36.60}$$

The quality factor for $T(s)$ is given by

$$Q=\sqrt{b}/a \tag{36.61}$$

The delay can then be expressed:

$$\text{delay}=\frac{2s(b+\omega^2)}{(b-\omega^2)^2+a^2\omega^2}\ \text{s} \tag{36.62}$$

The plots of the delay characteristics for different values of Q are given in *Figure 36.16*.

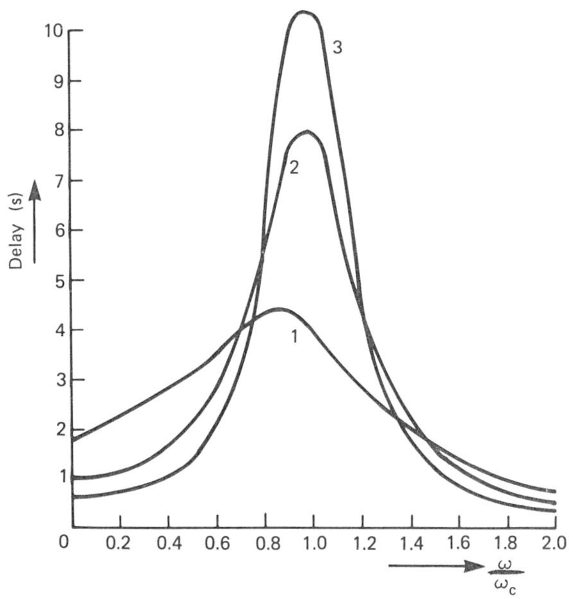

Figure 36.16 Delay equaliser plot for Q=1, 2, 3

36.3.8 Filter transformations

All the characteristics discussed so far in Section 36.3 have all been with reference to the low pass characteristic. All the design tables for filter characteristics are given with reference to the low pass characteristic. We therefore need to know the frequency transforms to convert the low pass filter to a high pass, band pass, and band stop filter.

36.3.8.1 High pass

Figure 36.17 shows a low pass characteristic with a lower axis

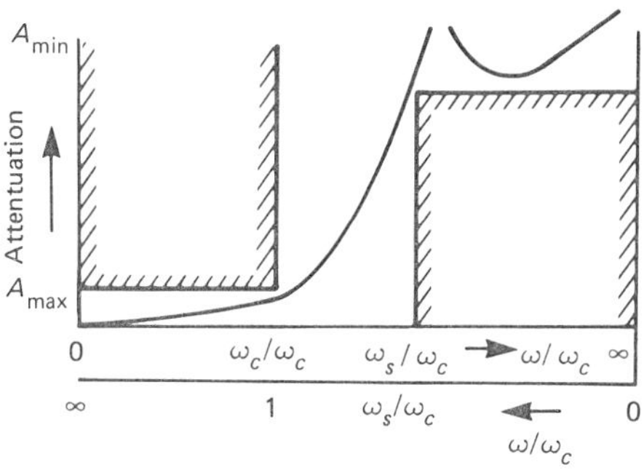

Figure 36.17 Filter transformation—high pass

which is the inverse of the frequency axis. If the lower axis is now used as the frequency axis the following factors are apparent.

At $\omega=0$	loss = maximum
At $\omega=\infty$	loss = 0 dB
At $\omega=1$	loss = A_{max}
At $\omega=\omega_s/\omega_c$	loss = A_{min}

This is the characteristic of a high pass filter. Therefore if we take a low pass loss characteristic and replace s by $1/s$ we get the complementary high pass filter.

If the characteristic of a high pass filter is

$A_{min}=30$ dB, $A_{max}=1$ dB, $\omega_c=1000$, $\omega_s=800$

then the characteristic of a low pass complement is

$A_{min}=30$ dB, $A_{max}=1$ dB, $\omega_s=1000$, $\omega_c=800$

The normalised low pass polynomial can now be found. This can be denormalised to a high pass polynomial by replacing s by ω_c/s.

36.3.8.2 Band pass

Figure 36.18 shows a low pass characteristic with a lower axis shifted in frequency to realise a band pass characteristic. The

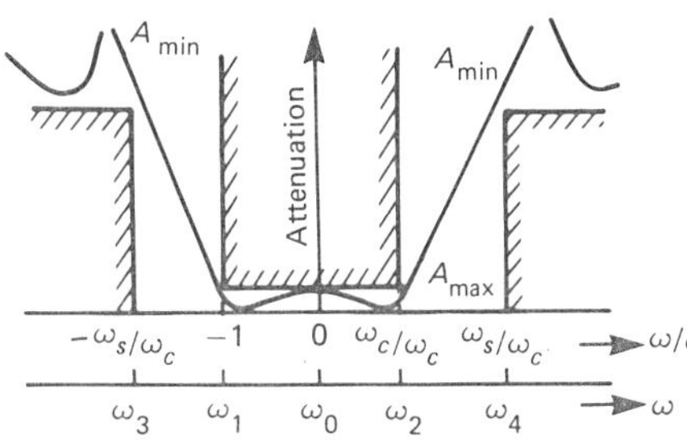

Figure 36.18 Filter transformation—band pass

characteristic for the band pass is symmetrical. If the desired band pass characteristic is not symmetrical then the A_{min} requirement must be made equal on either side to the maximum attenuation, and the frequencies of the stop band corners must be made symmetrical by moving the frequency which is furthest away from the centre frequency nearer to the centre frequency to regain symmetry. The relationship between the four frequencies of a symmetrical band pass filter is given by

$$\omega_1\omega_2=\omega_3\omega_4=\omega_0^2 \tag{36.63}$$

The frequency transformation to convert the low pass to a symmetrical band pass is accomplished by feplacing s by

$(s^2+\omega_0^2)/s(\omega_2-\omega_1)$

To characterise the filters the following low pass information is required

$$A_{max},\ A_{min},\ \omega_c=1,\qquad \frac{\omega_s}{\omega_c}=\frac{\omega_4-\omega_3}{\omega_2-\omega_1}$$

From this can be realised the normalised low pass polynomial. This is denormalised and converted to a band pass by replacing s as above.

36.3.8.3 Band stop

The band stop characteristic can be derived from the low pass characteristic by inserting the frequency scale as we did for the high pass transform. The band stop transform is therefore the inverse of the band pass, s being replaced by

$$\frac{(\omega_2-\omega_1)s}{s^2+\omega_0^2}$$

To characterise the filter the following low pass information is required

$$A_{max},\ A_{min},\ \omega_c=1,\ \frac{\omega_s}{\omega_c}=\frac{\omega_2-\omega_1}{\omega_4-\omega_3}$$

From this can be realised the normalised low pass polynomial. This is denormalised and converted to a band pass by replacing s as above.

36.4 Passive realisation of filters derived by synthesis

Figure 36.19 shows the impedance functions which can be created using R, L and C combinations. It is apparent that by combining

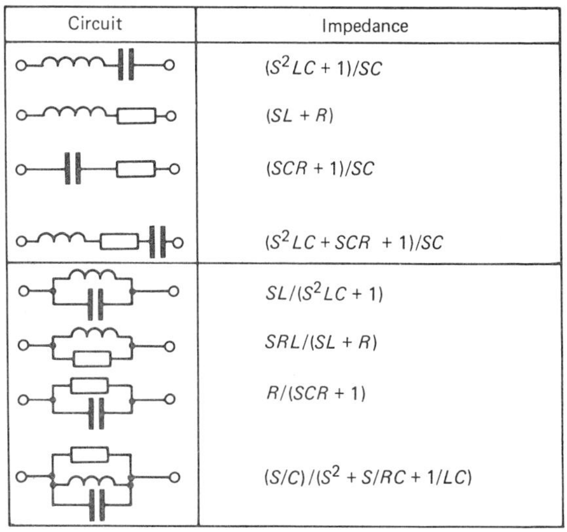

Figure 36.19 Passive *RLC* circuits

these elements any of the transfer characteristics in Section 36.3 can be realised.

The usual method of passive realisation is carried out by consulting filter design tables where not only will the polynomial constants be given, but also the normalised values for R, L and C.

36.4.1 Active realisation of filters derived by synthesis

Because of the high value of inductors necessary for the frequency range below 30 kHz, active realisation has received a great deal of attention. Initially attention was focused on the single amplifier realisatiin for each section of a filter, but with the arrival of dual, quad and hex operational amplifiers, two and three amplifier realisations were produced.

36.4.2 Single and multiple amplifier filters

The advantages of the modern operational amplifier are that it has a very low output impedance and a high input impedance. This makes impedance matching very simple, and so the cascade approach is easily implemented.

36.4.2.1 Sallen and Key filters[5]

The basic low pass circuit shown in *Figure 36.20* has the transfer function

$$T(s)=(\alpha/R_1R_2C_1C_2)\left[s^2+s\left(\frac{1}{R_1C_1}+\frac{1}{R_2C_1}+\frac{1-\alpha}{R_2C_2}\right)+\frac{1}{R_1R_2C_1C_2}\right]^{-1} \tag{36.64}$$

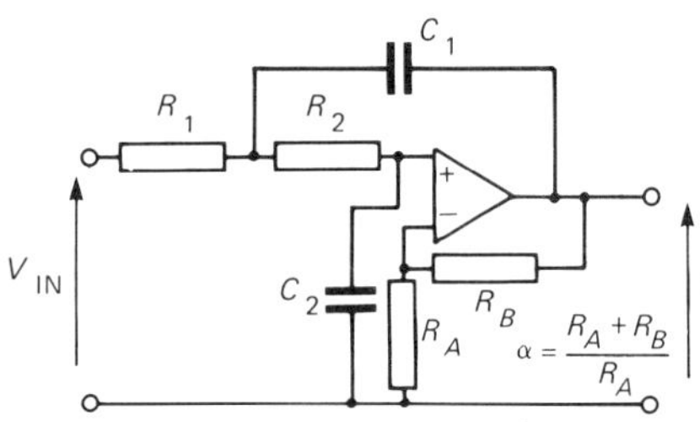

Figure 36.20 Sallen and Key low lass filters

This is equivalent to the general low pass function

$T(s)=k\,.\,b/(s^2+as+b)\equiv\omega_c^2/(s^2+(\omega_c/Q_c)s+\omega_c^2)$

where

$b=1/R_1R_2C_1C_2$

$$a=\frac{1}{R_1C_1}+\frac{1}{R_2C_1}+\frac{1-\alpha}{R_2C_2}$$

By comparison we get

$\omega_c=(1/R_1R_2C_1C_2)^{1/2} \qquad k=\alpha$

$$Q_c=\frac{\omega_c}{a}=(1/R_1R_2C_1C_2)^{1/2}\left(\frac{1}{R_1C_1}+\frac{1}{R_2C_1}+\frac{1-\alpha}{R_2C_2}\right)^{-1}$$

There are three possible solutions to this. They are as follows.

(1) $R_1=R_2 \qquad \alpha=1$

Then

$C_1=2Q_c/\omega_c \qquad C_2=(2\omega_cQ_c)^{-1}$

This has the disadvantage that $C_1/C_2=4Q_c^2$, and this may produce an awkward value for one of the capacitors.

(2) $C_1=C_2=C \qquad R_1=R_2=R$

Then

$\omega_c=1/RC \qquad \text{and} \qquad \alpha=3-(Q_c)^{-1}$

This approach however produces a circuit which is more susceptible to component variations than (1). For optimum sensitivity we adopt the Saraga[6] approach in solution (3).

(3) $C_1/C_2=\sqrt{3}\,Q_c \qquad R_2/R_1=Q_c/\sqrt{3}$

Then

$\alpha=4/3$

This approach produces the optimum sensitivity to component changes with regard to the stability of Q_c.

The basic high pass function shown in *Figure 36.21* has the transfer function

$$T(s)=\alpha s^2\left[s^2+s\left(\frac{1}{R_1C_1}+\frac{1}{R_2C_1}+\frac{1-\alpha}{R_2C_2}\right)+\frac{1}{R_1R_2C_1C_2}\right]^{-1} \tag{36.66}$$

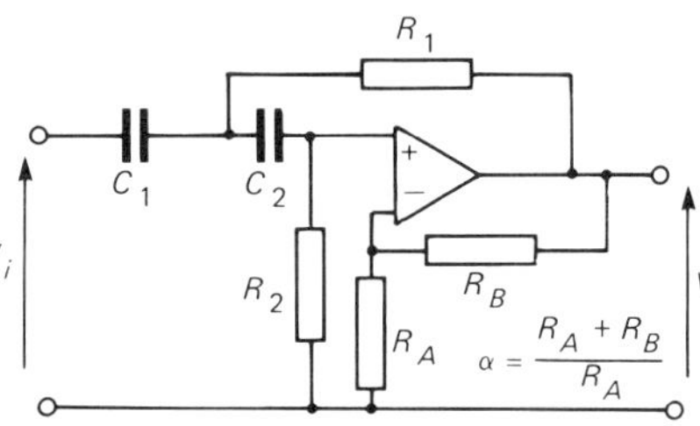

Figure 36.21 Sallen and Key high pass filters

The method of solving for the component values is exactly the same as for the low pass function.

By cascading the Sallen and Key high pass and low pass sections it is relatively easy to synthesise any Butterworth or Chebyshev filter characteristic which has been derived as in previous sections.

The basic band pass function shown in *Figure 36.22* has the transfer function

$$T(s)=\frac{\alpha s}{R_1C_1}\left[s^2+s\left(\frac{1}{R_1C_1}+\frac{1}{R_3C_2}+\frac{1}{R_3C_1}+\frac{1-\alpha}{R_2C_1}\right)+\frac{R_1+R_2}{R_1R_2R_3C_1C_2}\right]^{-1} \tag{36.67}$$

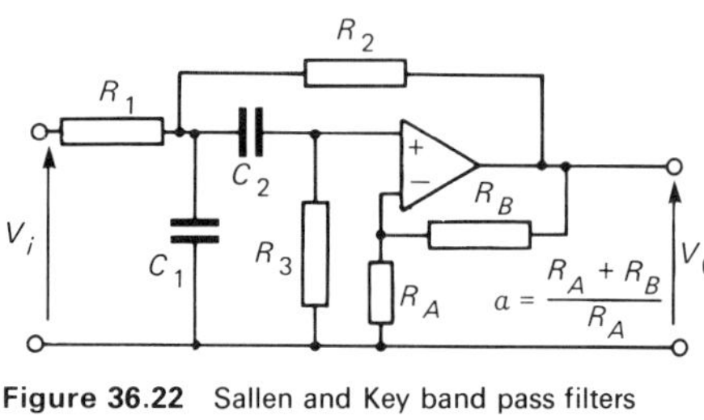

Figure 36.22 Sallen and Key band pass filters

This is equivalent to the general band pass function

$$T(s)=Ks[s^2+(\omega_c/Q_c)s+\omega_c^2]^{-1} \tag{36.68}$$

The usual solution to this is to let

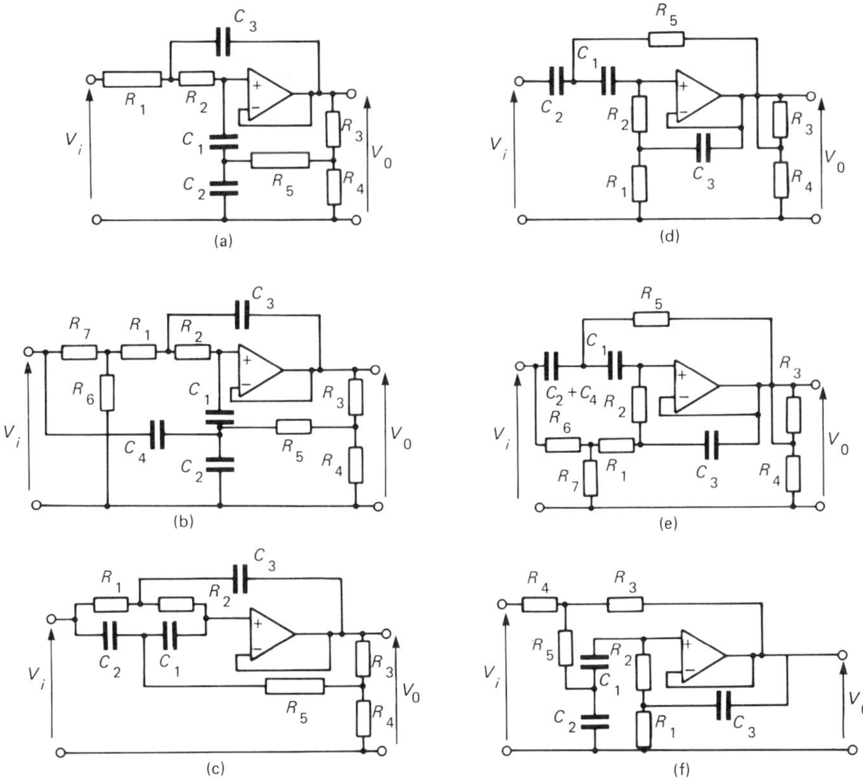

Figure 36.23 Lim filters: (a) Lim low pass filter; (b) Lim low pass biquad; (c) Lim band stop; (d) Lim high pass; (e) Lim high pass biquad; (f) Lim band stop

(a) $T(s) = \dfrac{\omega_c^{\,2}}{S^2 + S\dfrac{\omega_c}{Q} + \omega_c^{\,2}}$

(b) $T(s) = \dfrac{S^2 + \omega_\infty^{\,2}}{S^2 + S\dfrac{\omega_c}{Q} + \omega_c^{\,2}}$

(c) $T(s) = \dfrac{S^2 + \omega_c^{\,2}}{S^2 + S\dfrac{\omega_c}{Q} + \omega_c^{\,2}}$

(d) $T(s) = \dfrac{S^2}{S^2 + S\dfrac{\omega_c}{Q} + \omega_c^{\,2}}$

(e) $T(s) = \dfrac{S^2 + \omega_\infty^{\,2}}{S^2 + S\dfrac{\omega_r}{Q} + \omega_r^{\,2}}$

(f) $T(s) = \dfrac{S\omega_c/Q}{S^2 + S\dfrac{\omega_c}{Q} + \omega_c^{\,2}}$

$R_1 = R_2 = R_3 = R \qquad C_1 = C_2 = C$

Then

$\omega_c = \sqrt{2}/RC \qquad \text{and} \qquad \alpha = 4 - \sqrt{2}/Q$

This leaves

$$K = \frac{\alpha}{R_1 C_1}$$

36.4.2.2 Lim filters

The circuit configurations shown in *Figure 36.23* were first proposed by J. T. Lim.[7] The circuit is essentially a balanced twin-*T* network round a unity gain amplifier. The advantage of the twin-*T* approach is that it gives a much lower sensitivity to variations in the amplifier open loop gain than does the Sallen and Key filter.

The design equations which apply to the low pass, high pass, band pass and band stop are:

$\omega_c = 1/CR$

where

$C = C_1 = C_2 = C_3$

and

$R = R_2/2 = R_1$

$$\frac{R}{3} = R_5 + \frac{R_3 R_4}{R_3 + R_4}$$

$$Q = \frac{1}{3}\left(\frac{R_4}{R_3} + 1\right)$$

The two biquadratic realisations are useful, as it provides us with a means of realising Caver eliptic responses.

The design equations which apply to these filters are

$\omega_c = 1/CR$

$C = C_1 = C_2 + C_4 = C_3$

$$R = R_1 + \frac{R_6 R_7}{R_6 + R_7} = \frac{R_2}{2}$$

$$\frac{R}{3}=R_5+\frac{R_3R_4}{R_3+R_4} \qquad Q=\frac{1}{3}\left(\frac{R_4}{R_3}+1\right)$$

and

$$\frac{\omega_\infty}{\omega_c}=\left[\left(1+\frac{C_2}{C_4}\right)\left(\frac{R_6}{R_6+R_7}\right)\right]^{1/2} \qquad \text{for } \omega_\infty>\omega_c \tag{36.69}$$

and

$$\frac{\omega_c}{\omega_\infty}=\left(\frac{R_6+R_7}{R_7}\right)^{1/2} \qquad \text{for } \omega_\infty<\omega_c \tag{36.70}$$

where ω_∞ is the frequency of maximum attenuation.

The adjustment technique for these filters is worth noting at this point. The adjustment is quite elaborate as it requires three different arrangements of the circuit components.[8] This procedure is shown in *Figure 36.24*. The first two stanges are

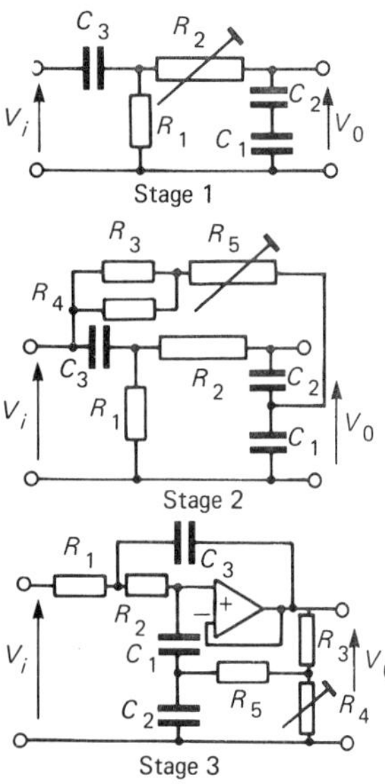

Figure 36.24 Lim low/high pass filter tuning

adjusted for a 0° phase shift at ω_c. The third stage of adjustment is to correct for any error in the Q factor of the filter.

36.4.2.3 *Wien-bridge notch filter*[9]

The typical second-order notch filter response is given by

$$T(s)=(s^2+\omega_c^2)(s^2+s(\omega_c/Q_c)+\omega_c^2)^{-1} \tag{36.71}$$

The circuit shown in *Figure 36.25* gives the response

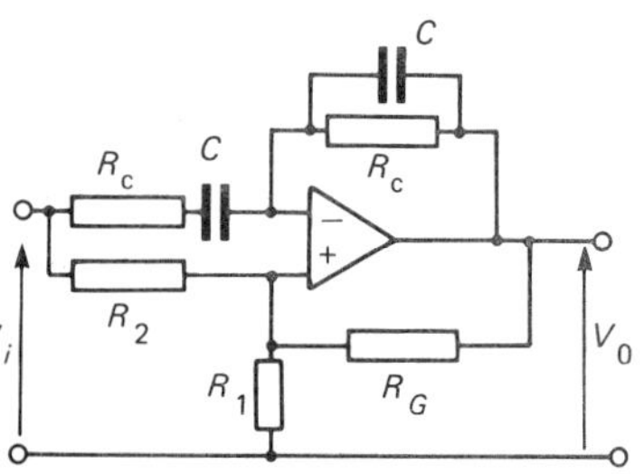

Figure 36.25 Notch filter

$$T(s)=K\frac{s^2+(3-R/R_2)\omega_c s+\omega_c^2}{s^2+(3-R_G/(R_G-R))\omega_c s+\omega_c^2} \tag{36.72}$$

where

$$R=\left(\frac{1}{R_1}+\frac{1}{R_2}+\frac{1}{R_G}\right)^{-1}$$

$$K=\frac{RR_G}{(R_G-R)R_2}$$

and

$$\omega_c=\frac{1}{R_c C}$$

To obtain equivalence we require that $R/R_2=3$ and

$$Q_c=\left(3-\frac{R_f}{R_f-R}\right)^{-1}$$

The advantage of this type of filter is that the resonant frequency controlling components are independant of the null-determining components.

36.4.2.4 *The three amplifier biquadratic filter*[10]

The circuit shown in *Figure 36.26* will produce low pass and band pass functions.

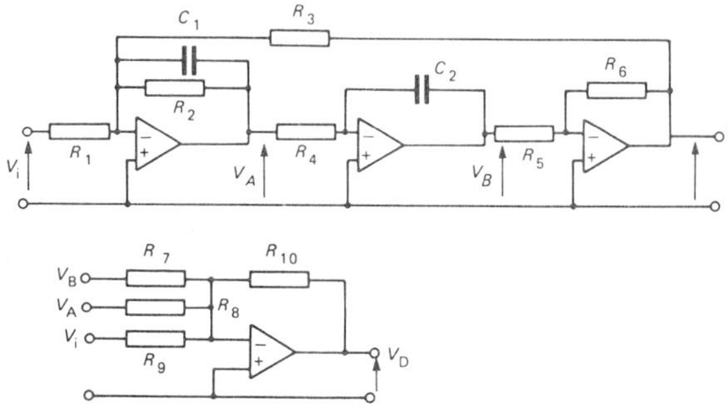

Figure 36.26 The three amplifier biquadriatic and summing amplifier

The transfer function between V_i and V_c is given by

$$T(s)V_c/V_i=-(R_1R_4C_1C_2)^{-1}\left(s^2+\frac{1}{R_2C_1}s+\frac{1}{R_4R_3C_1C_2}\right)^{-1}$$

$$=T(s)_{V_B/V_i} \qquad \text{if } R_5=R_6 \tag{36.72}$$

which is a second-order low pass function.

The transfer function between V_i and V_A is given by

$$T(s)_{V_A/V_i}=-s(R_1C)^{-1}\left(s^2+\frac{1}{R_2C_1}s+\frac{1}{R_4R_3C_1C_2}\right)^{-1} \tag{36.73}$$

which is a second-order band pass function.

If we now add another amplifier and sum the points V_i, V_A, V_B, we then generate a new transfer function

$$T(s)_{V_D/V_i}=-T(s)_{V_B/V_i}\frac{R_{10}}{R_7}-T(s)_{V_A/V_i}\frac{R_{10}}{R_8}-T(s)_{V_i}\frac{R_{10}}{R_9}$$

$$=\frac{R_{10}}{R_7}\frac{1}{R_1R_4C_1C_2}+\frac{R_{10}}{R_8}\frac{s}{R_1C_1}$$

$$-\frac{R_{10}}{R_9}\left(s^2+\frac{1}{R_2C_2}s+\frac{1}{R_4R_3C_1C_2}\right)$$

$$\times\left(s^2+\frac{1}{R_2C_1}+\frac{1}{R_4R_3C_1C_2}\right)^{-1}$$

$$=-K\frac{s^2+cs+d}{s^2+as+b} \tag{36.74}$$

the general biquadratic function.

The normal conditions used to design with this filter are to make $C_1=C_2$, $R_3=R_4$ and $R_8=R_{10}$.

A further use of this filter is to realise the delay equaliser function

$$T(s)=-K\frac{s^2-as+b}{s^2+as+b} \tag{36.75}$$

This is achieved by making $R_8=\infty$.

36.4.2.5 Inductance simulation and fdnr simulation

The gyrator is a two-part circuit that inverts an impedance (*Figure 36.27*). K is the gyration impedance multiplier.

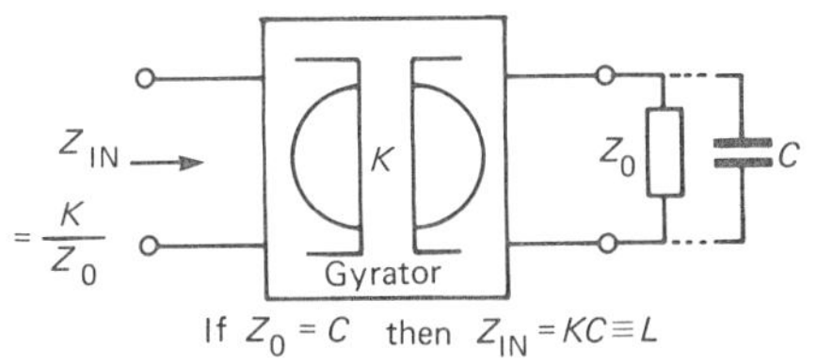

Figure 36.27 Gyrator

Two circuits were proposed for the realisation of the gyrator by Riordan,[11] and Antoniou[12] and Bruton.[13] The circuits are shown in *Figure 36.28*. The input impedance formula for both circuits is the same. It is apparent that if either Z_2 or Z_4 were a capacitor and all other impedances were resistors then the input impedance would be inductive. With this technique values of Q in excess of 1000 may be obtained, even when using amplifiers with gains as low as 40 dB.

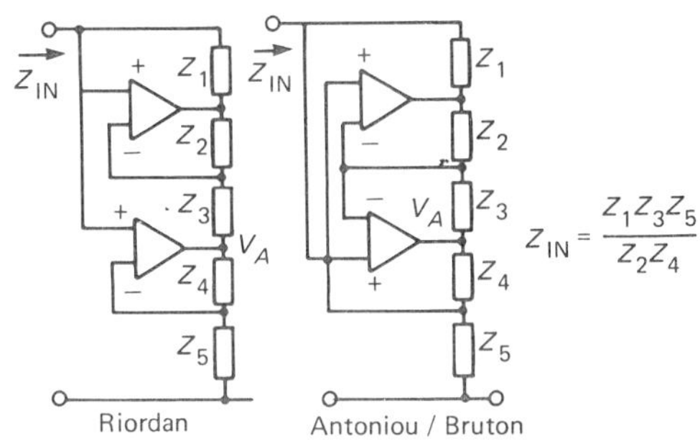

Figure 36.28 Gyrator circuits

If Z_1 or Z_3 are replaced by a capacitor, and all other impedances are resistors then we generate an input impedance given by the expression

$$Z_{\rm IN}=\frac{R_3}{R_2R_4}\frac{1}{s^2C_1C_3}=\frac{K}{s^2}\ (\text{symbol}=\text{⫨}) \tag{36.76}$$

This is a frequency dependant negative resistor (FDNR), which varies with the square of the frequency.

When using this type of circuit it should be remembered that the output at V_A is given by

$$V_A=V_i[(Z_1+Z_2)/Z_1] \tag{36.77}$$

This can cause the amplifier to limit without the input voltage approaching the supply lines.

36.4.2.6 Applications of the FDNR and inductance simulations

Consider the circuits shown in *Figure 36.29*.

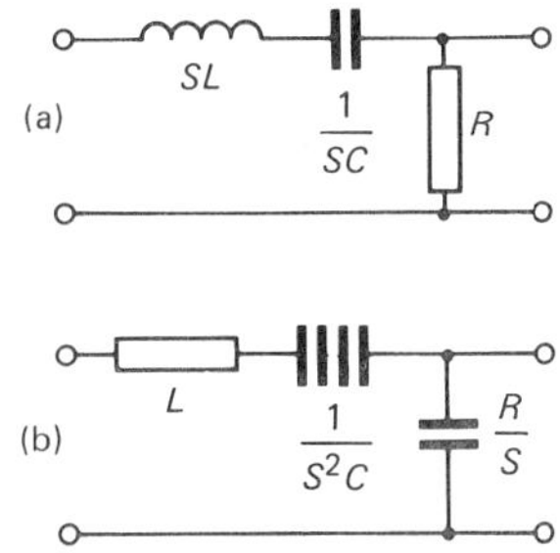

Figure 36.29 Illustration of impedence scaling

$$T(s)=\frac{s(R/L)}{s^2+s(R/L)+1/LC} \tag{36.78}$$

for circuit (a) and

$$T(s)=\frac{s(R/L)}{s^2+s(R/L)+1/LC)} \tag{36.79}$$

for circuit (b).

This technique is known as impedance scaling and holds for all orders of circuits. Therefore, for any LCR circuit there are two possible options for the elimination of the inductor.

Consider a typical fifth-order Butterworth /Chebyshev LCR network as shown in *Figure 36.30*. Circuit (a) can be converted to circuit (b).

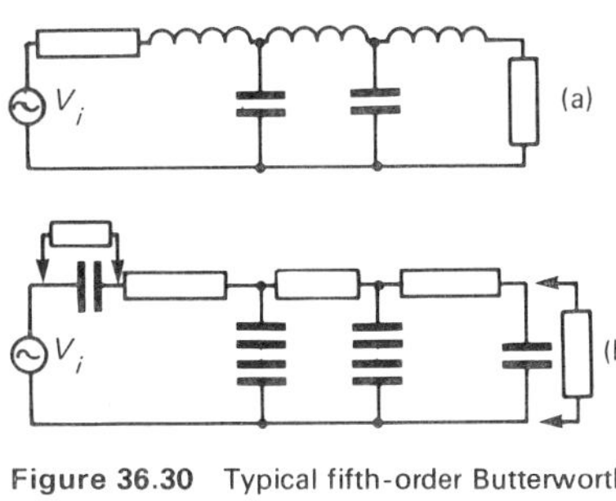

Figure 36.30 Typical fifth-order Butterworth/Chebyshev *LCR* network

From the equivalent circuit it will be seen that the circuits will no longer work down to zero frequency, and there is no path to ground for the amplifier bias currents. A high value resistance is connected across the source and load impedances as shown to overcome these limitations. Because of the capacitive input and output impedances this type of circuit is normally buffered at its input and output.

Now consider a typical fifth elliptic LCR network as in *Figure 36.31*. This may be realised by direct inductance simulation.

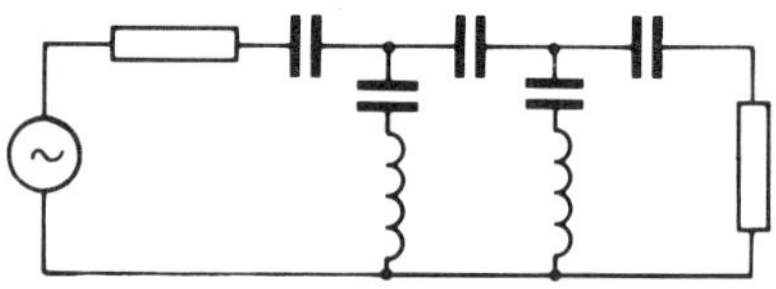

Figure 36.31 Typical fifth elliptic *LCR* network

Therefore, provided the *LCR* normalised circuit can be derived from published tables, the gyrator approach provides a quick and easy means of realising an active filter which will closely follow the theoretical filter characteristic. It should be remembered, though, that a filter is only as good as the componants used, and due consideration should be given to the type of capacitors used. The preferred choice of capacitor is the NPO type of ceramic capacitor.

This type of realisation has the disadvantage that it requires two amplifiers to produce a second-order response, but it does produce an active filter with very low sensitivity to componant variations.

The problem with the circuits shown in *Figures 36.29–36.32* is that they are grounded. For floating inductance simulation it can be shown that the circuit of *Figure 36.32* realises a floating inductor between terminals 1 and 2.

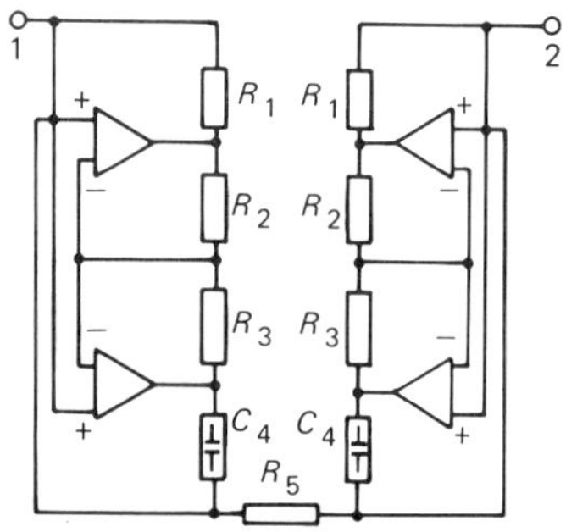

Figure 36.32 Floating inductor simulation

This type of inductance simulation has opened up the way for the production of very low frequency filters as very large values of inductance may be realised with very high values of *Q*.

36.5 Crystal filters

As the need for more precise control of frequency became necessary, considerable effort was put into developing a frequency conscious component with a very tight frequency selectivity component. The result of this work was the quartz crystal. The crystal structure is shown in *Figure 36.33* and defines the *X* and *Y* axis. The *Z* axis is the optical axis and exhibits no piezoelectric characteristics.

The crystal is cut according to two groups, the *X* group and *Y* group.

For the *X* group the thickness dimension is parallel to the *X* axis, and for the *Y* group the thickness dimension is parallel to the *Y* axis. The plates are cut at different angles to *Z* axis, the angle being classified by the name of the cut *AT, BT, MT,* etc. The frequency range associated with some of these cuts is given in *Table 36.4*.

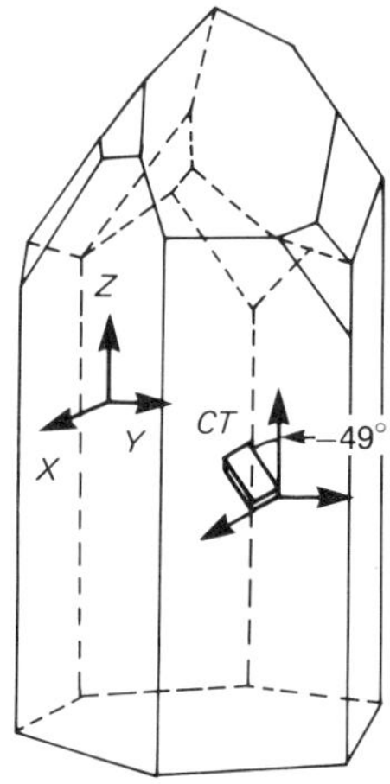

Figure 36.33 Quartz crystals

Table 36.4

Cut	*Group*	*Frequency (kHz)*
X		40–20 000
5°*X*	*X*	0.9–500
MT		50–100
NT		4–50
Y		1000–20 000
AT		500–100 000
BT		1000–75 000
CT		300–1100
DT	*Y*	60–500
ET		600–1800
FT		150–1500
GT		100–550

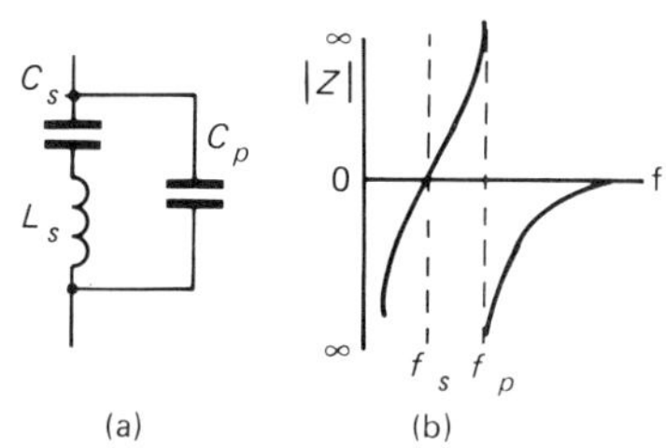

Figure 36.34 Crystal equivalent circuits: (a) circuit; (b) impedence characteristic

The equivalent circuit is given in *Figure 36.34* together with the impedance characteristic. The stability of this circuit is dependant on the *Q* factor which is the ratio of the reactance to the resistance. The *Q* of a crystal can lie between 10000 and over a million, against which the *Q* of most coils lie between 10 and 500. From the impedance plot it is seen that the crystal has two resonant frequencies, a series and a parallel resonance.

The main use of crystals is in band pass filters, especially where very narrow bands are required. Two types of band pass filters are given in *Figure 36.35*.

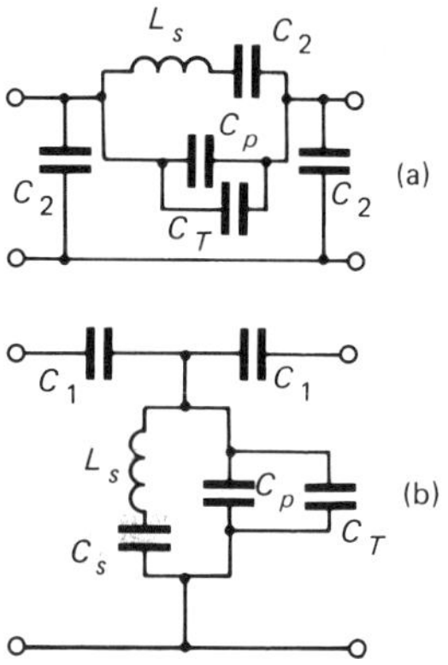

Figure 36.35 Crystal filters: (a) -section; (b) *T*-section

For a π network

$$C_1 \approx \frac{1-m^2}{4\pi m f_a R_0} \qquad C_2 \approx \frac{m}{2\pi f_a R_0} \tag{36.80}$$

$$C_s \approx \frac{\Delta f}{2\pi m f_a R_0} \qquad L_s \approx \frac{m R_0}{2\pi \, \Delta f}$$

where f_a is the mid-band frequency $=\frac{1}{2}(f_1+f_2)$, Δf the bandwidth (f_2-f_1), $C_1=C_T+C_P$ and $m=[(f_2^2-f_\infty^2)/(f_1^2-f_\infty^2)]^{1/2}$.

For a *T* network

$$C_1 \approx \frac{m}{2\pi f_a R_0} \qquad C_2 \approx \frac{m}{\pi(m^2-1) f_a R_0}$$

$$C_s \approx \frac{2m^3 \, \Delta f}{\pi(m^2-1)^2 f_a R_0} \qquad L_s \approx \frac{(m^2-1)^2 R_0}{(8\pi m^3 \, \Delta f)}$$

where $C_2=C_P+C_T$.

The *T* and π sections can be cascaded into ladder networks to form wider bandwidth pass band filters and tables[2] exist to give different orders of filter with defined ripple in the pass band.

36.6 Monolithic approach to filters

The filters described so far have been composed of discrete components. A lot of research has been applied in order to realise these filters in monolithic form. There have been two approaches which have predominated: the switched-capacitance filter and the charge coupled device filter.

In 1977 several papers[14] were published which suggested replacing the resistor in an integrator circuit by a capacitor and two switches (*Figure 36.36*). The output voltage is given by

$$V_o = -V_i \frac{C_1}{C_2} \frac{1}{j\omega T}$$

where *T* is the clock time and ω the input signal frequency. This is equivalent to the conventional integrator using a capacitor and a resistor. The time constant is controlled by the switching

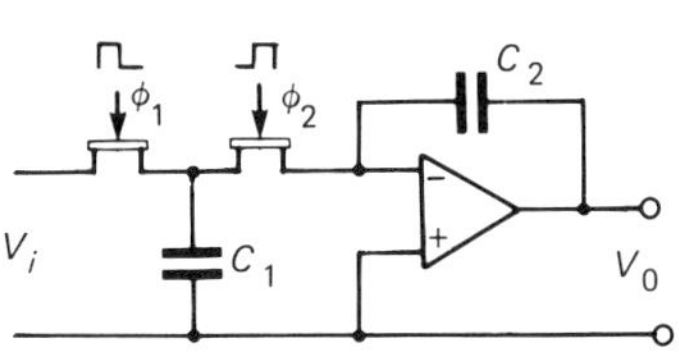

Figure 36.36 Switched-capacitor integrator

frequency and the capacitive ratio, not by the capacitive values. Capacitance ratios are more easily achieved in monolithic form than accurate values of capacitance. Positive and negative integrators are shown in *Figure 36.37*. The extra switches are

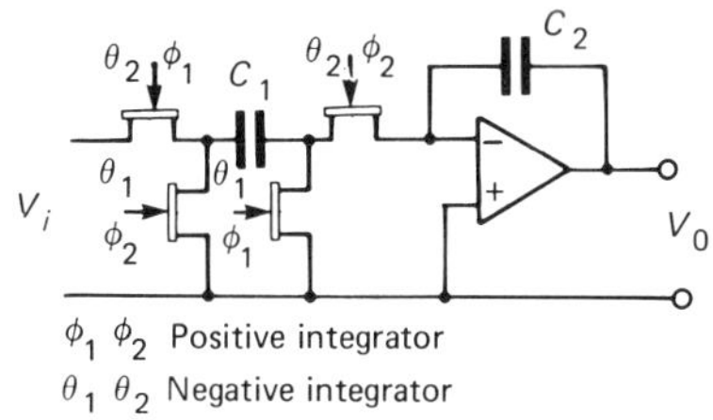

Figure 36.37 Stray capacitor modification

there to minimise the effects of stray capacitance in the switches. Also required for state variable filters are integrators with a damping coefficient (*Figure 36.38*). With these first-order functions we can generate a tri-state filter (*Figure 36.39*). From

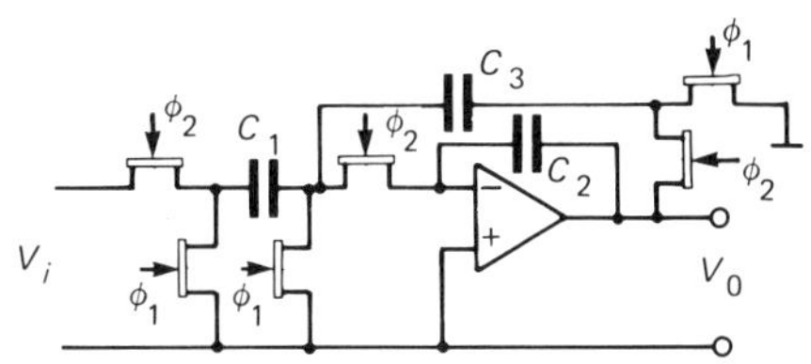

Figure 36.38 Integrator with damping coefficient

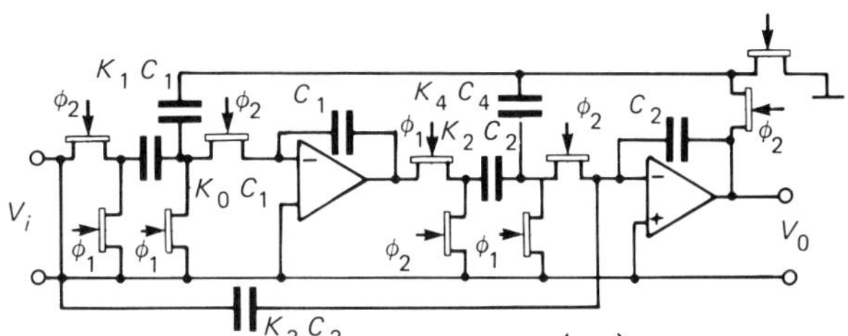

$$H(\omega) = \frac{V_0(\omega)}{V_i(\omega)} \left[\frac{1 - \dfrac{4K_3}{K_0K_2} \sin^2\left(\dfrac{\omega T}{2}\right)}{1 - 2\dfrac{(2+K_4)}{K_1K_2} \sin^2\left(\dfrac{\omega T}{2}\right) + j\dfrac{K_4}{K_1K_2} \sin(\omega T)} \right]$$

Figure 36.39 Second-order notch filter

this can be generated an integrated circuit realisation of a second-order filter. The next extension of this form of realisation of filters will be the UFA (uncommitted filter array). This will be a monolithic chip with the final top mask performing the interconnections between the standard filter elements. The oscillator to provide the correct timing for these switches will be incorporated on the chip, with possibly only one external timing capacitor or crystal being needed to set the oscillator frequency.

References

1 I.T.T., *Reference Data for Radio Engineers*, 5th edn, Howard W. Sams and Co. Inc., New York (1968)

2 Zverev, A. I., *Handbook of Filter Synthesis*, Wiley, New York (1967)

3 Christian, E. and Eismann, E., *Filter Design Tables and Graphs*, Wiley, New York (1966)
4 Daniels, R. W., *Approximation Methods for Electronic Filter Design*, McGraw-Hill, New York (1974)
5 Daryanani, G., *Principles of Active Network Synthesis and Design*, Wiley, New York (1976)
6 Saraga, W., 'Sensitivity of 2nd order Sallen–Key type active *R.C.* filters', *Electron. Lett.*, **3**, 442–4 (1967)
7 Lim, J. T., 'Improvements in or relating to active filter networks', *British Patent Application 9657* (dated 16 April, 1971)
8 Jeffers, R. and Haigh, D. G., 'Active *RC* lowpass filters for FDM and PCM systems', *Proc. IEE*, **120**, 945–53 (1973)
9 Darilek, G. and Tranbarger, O., *Electron. Design*, **3**, 80–1 (1978)
10 Thomas, L. C., 'The biquad: part 1—some practical design considerations', *IEEE Trans. Cct Theory*, **CT-18**, 350–7 (1971)
11 Riordon, R. H. S., 'Simulated inductors using differential amplifiers', *Electron. Lett.*, **3**, 50–1 (1967)
12 Antoniou, 'A realization of gyrator using operational amplifiers and their use in *RC*-active network synthesis', *Proc. IEE*, **116**, 1838–50 (1969)
13 Bruton, L. T., 'Network transfer functions using the concept of frequency dependent negative resistance', *IEEE Trans. Cct Theory*, **CT-16**, No. 3, 406–8 (1969)
14 Martin, K., 'Improved circuits for the realisation of switched-capacitance filters', *IEEE Trans. Cct Syst.*, **CAS-27**, 237–44 (1980)
15 Bowron, P. and Stephenson, F. W., *Active Filters for Communications and Instrumentation*, McGraw-Hill, Maidenhead (1979)
16 Chen, C., *Active Filter Design*, Hayden, Rechelle Park (1982)
17 Christansen and Fink, *Electronic Engineer's Handbooks*, 2nd edition, McGraw-Hill (1982)
18 Lane, G., 'Introduction to digital filters', *Electronic Product Design*, Part 1 (October 1982); Part 2 (November 1982); Part 3 (December 1982)

37 Attenuators

V J Green BSc
Senior Engineer,
Standard Telephones and Cables PLC

Contents

37.1 Introduction

An attenuator is a network designed to introduce a known loss when inserted between resistive impedances Z_1 and Z_2 to which the input and output impedances of the attenuator are matched. Z_1 and Z_2 can be interchanged as the source and load with the loss, expressed as a power ratio, remaining the same.

Three forms of resistance network that can conveniently be used are the T section, the π section, and the bridged T section. The equivalent balanced sections are also shown. Also included in this chapter are networks designed to match a source impedance Z_1 to a load impedance Z_2 $(Z_1 > Z_2)$ with the minimum possible loss.

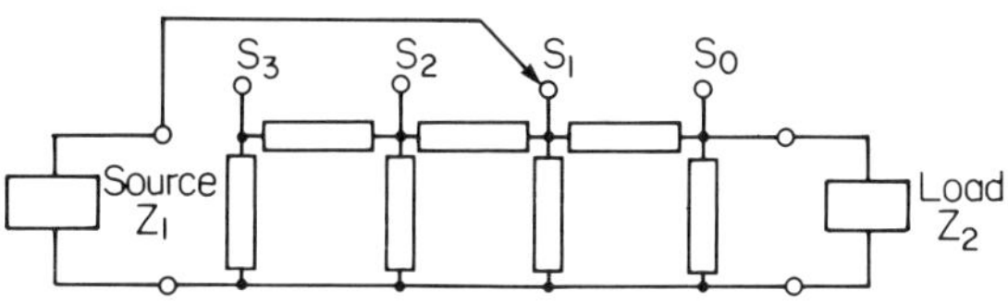

Figure 37.1 A ladder attenuator

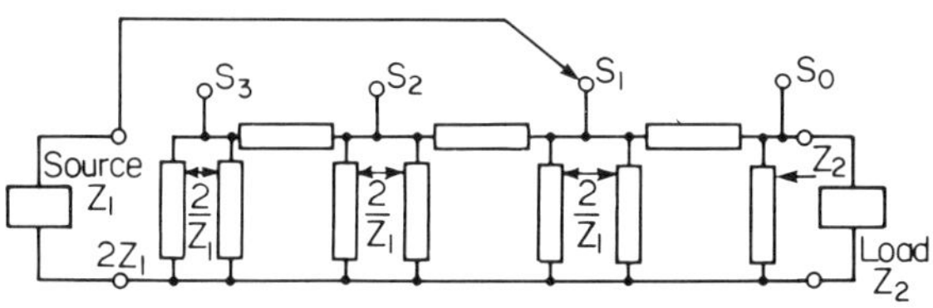

Figure 37.2 A ladder attenuator resolved into π sections

37.2 Symbols used

Z_1 and Z_2 are the image impedances of the network, generally: Z_1 is the input impedance and Z_2 the load impedance. N is the ratio of power delivered to the attenuator by the source to the power delivered to the load. K is the ratio of the input current into the attenuator to the output current into the load.

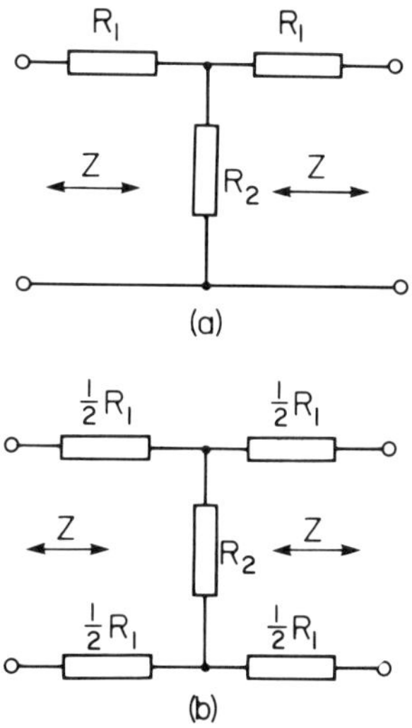

Figure 37.3 Symmetrical T and H attenuators: (a) T attenuator; (b) H attenuator

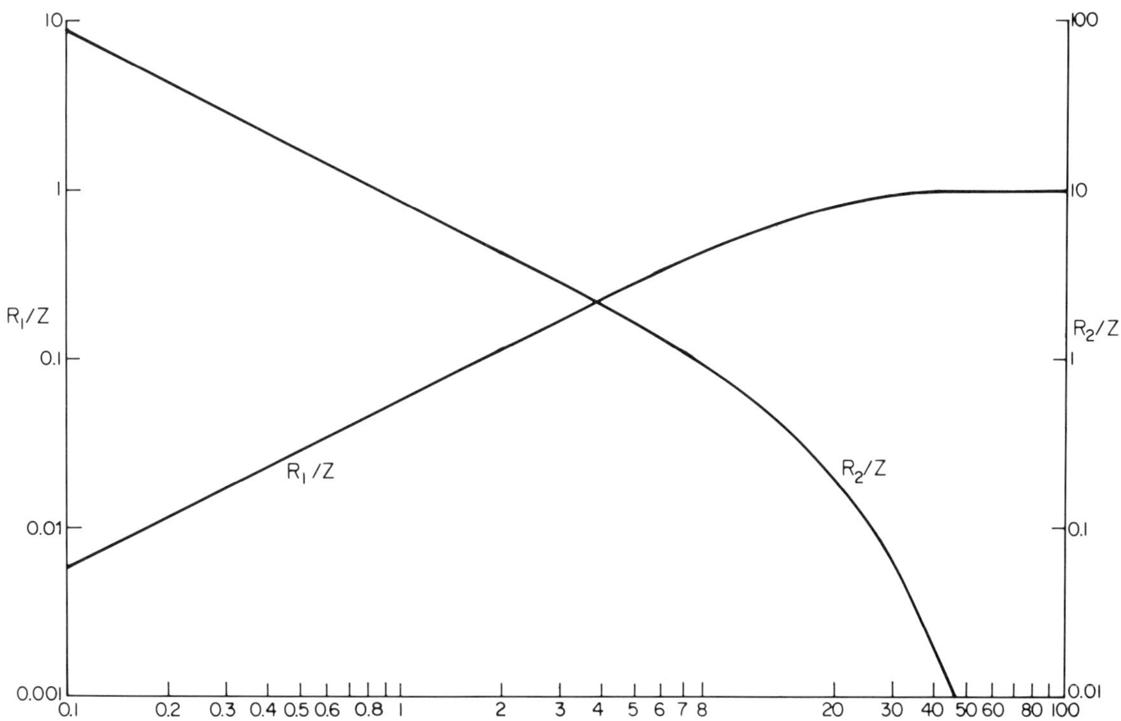

Figure 37.4 T-section design chart

Attenuation in dB $=10\lg N$ (37.1)

Attenuation in nepers $=\theta=\frac{1}{2}\ln N$ (37.2)

37.3 Ladder attenuator

A ladder attenuator (*Figure 37.1*) allows the input to be switched between shunt arms (S_3, S_2, S_1, S_0). A ladder attenuator can be designed by resolving it into a cascade of π sections (*Figure 37.2*) splitting the shunt arms into two resistors. The last section matches Z_2 to $2Z_1$ with all other sections being symmetrical, matching impedance $2Z_1$. There is a terminating resistor of $2Z_1$ on the first section. Design each section for the loss required between the switch points at the ends of each section.

For input connected to S_0:

Loss in dB $=10\lg(2Z_1+Z_2)^2/4Z_1Z_2$ (37.3)

Input impedance $=Z_2/2$ (37.4)

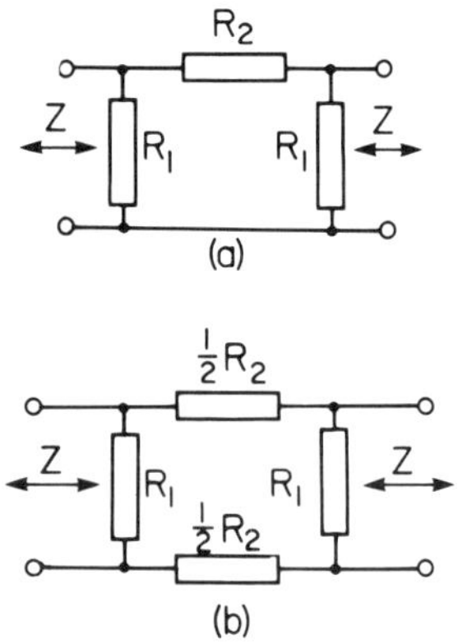

Figure 37.5 Symmetrical π and 0 attenuators: (a) π attenuator; (b) 0 attenuator

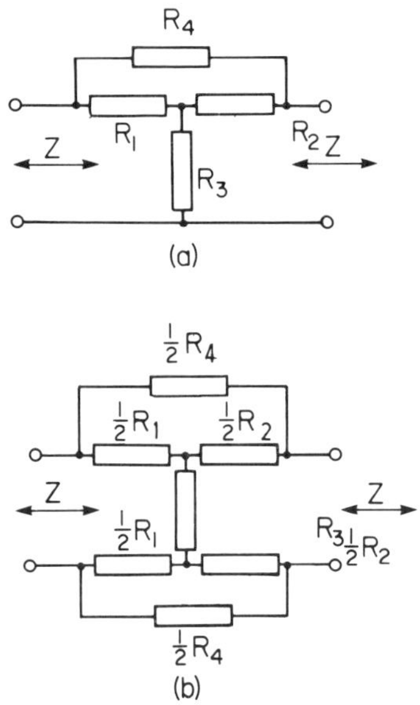

Figure 37.7 Bridged *T* and *H* attenuators: (a) *T* attenuator; (b) *H* attenuator

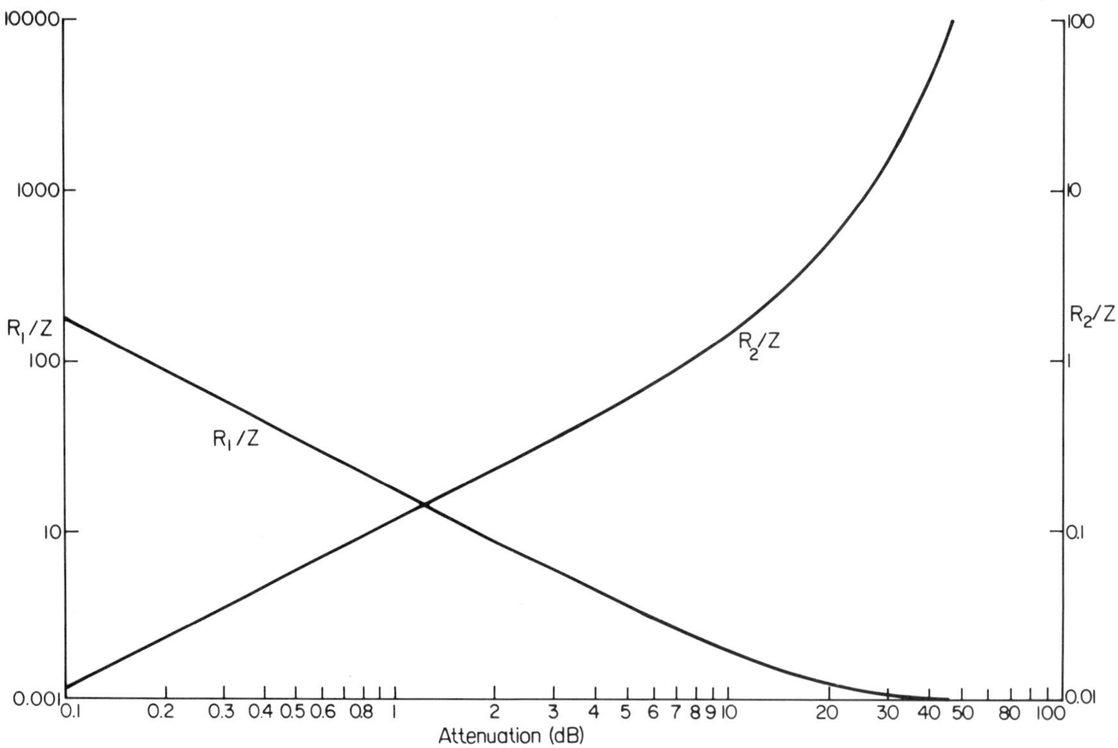

Figure 37.6 π-section design chart

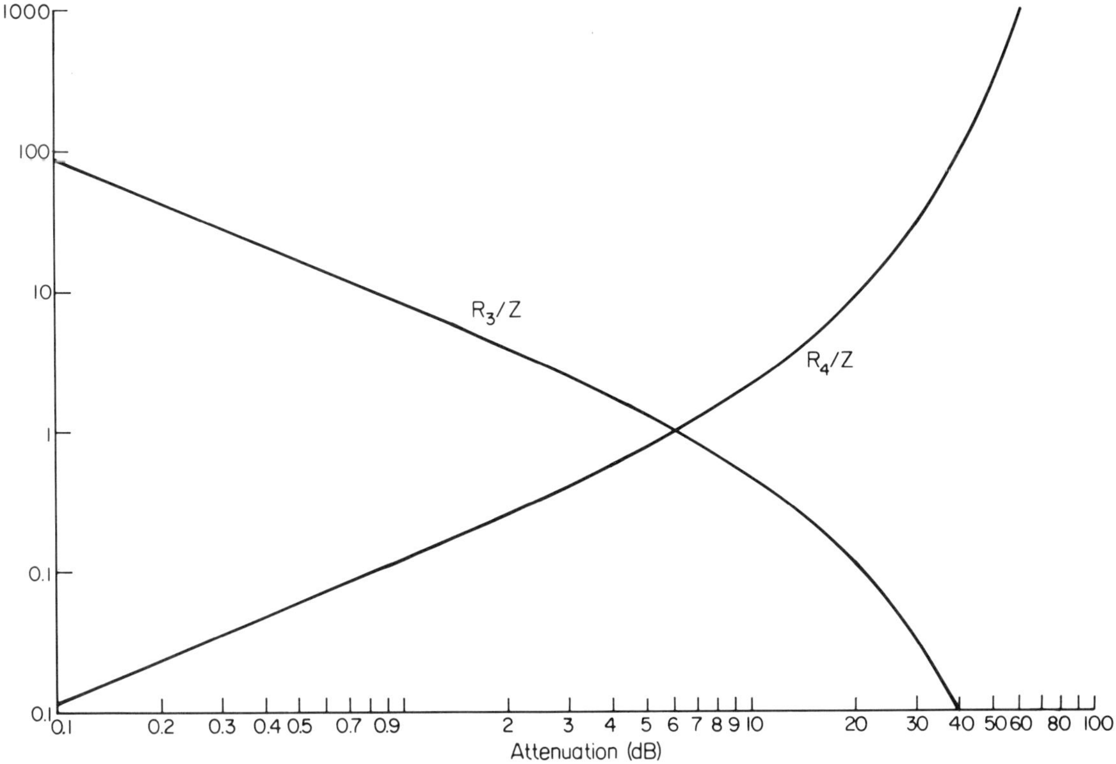

Figure 37.8 Bridged *T* design chart

Output impedance $=Z_1Z_2/(Z_1+Z_2)$ (37.5)

For input connected to S_1, S_2, S_3:
Loss in dB $=3+$ sum of losses of π sections between input and output.

Input impedance $=Z_1$

37.3.1 Effect of incorrect load impedance on the operation of an attenuator

Let

$Z_2+\Delta Z_2=$ actual load impedance terminating the attenuator (ΔZ_2 need not be purely resistive).

$Z_1+\Delta Z_1=$ resulting input impedance

$K+\Delta K=$ resulting current ratio.

The relationships between these quantities are:

$$\frac{\Delta Z_1}{Z_1}=\frac{2\,\Delta Z_2/Z_2}{2N+(N-1)(\Delta Z_2/Z_2)} \tag{37.6}$$

and

$$\frac{\Delta K}{K}=\left(\frac{N-1}{2N}\right)\frac{\Delta Z_2}{Z_2} \tag{37.7}$$

37.4 Symmetrical *T* and *H* attenuators

37.4.1 Configuration

Configurations are shown in *Figure 37.3*.

37.4.2 Design equations

For the *T* and *H* attenuators

$$R_1=Z\tanh(\theta/2) \tag{37.8}$$

$$R_2=Z/\sinh\theta \tag{37.9}$$

$$R_1=Z\left(\frac{\sqrt{N}-1}{\sqrt{N}+1}\right) \tag{37.10}$$

$$R_2=\frac{2Z\sqrt{N}}{N-1} \tag{37.11}$$

37.4.3 Design chart

This is shown in *Figure 37.4*. Values of R_1/Z can be read from the appropriate curve off the left-hand vertical axis. Values of R_2/Z can be read from the appropriate curve off the right-hand vertical axis. Attenuation in dB is on the horizontal axis.

37.5 Symmetrical π and 0 attenuators

37.5.1 Configuration

These are shown in *Figure 37.5*.

37.5.2 Design equations

For the π and 0 attenuators:

$$R_1=Z/[\tanh(\theta/2)] \tag{37.12}$$

$$R_2 = Z \sinh \theta \quad (37.13)$$

$$R_1 = Z\left(\frac{\sqrt{N}+1}{\sqrt{N}-1}\right) \quad (37.14)$$

$$R_2 = Z\left(\frac{N-1}{2\sqrt{N}}\right) \quad (37.15)$$

37.5.3 Design chart

This is shown in *Figure 37.6*. Values of R_1/Z can be read from the appropriate curve off the left-hand vertical axis. Values of R_2/Z can be read from the appropriate curve off the right-hand vertical axis. Attenuation in dB is on the horizontal axis.

37.6 Bridged *T* or *H* attenuators

37.6.1 Configuration

These are shown in *Figure 37.7*.

37.6.2 Design equations

$$R_1 = R_2 = Z \quad (37.16)$$

$$R_3 = Z/(\sqrt{N}-1) \quad (37.17)$$

$$R_4 = Z(\sqrt{N}-1) \quad (37.18)$$

37.6.3 Design chart

This is given in *Figure 37.8*. Values of R_3/Z and R_4/Z can be read from the appropriate curve off either vertical axis. Attenuation in dB is on the horizontal axis.

37.7 Minimum-loss pads

37.7.1 Configuration

This is shown in *Figure 37.9*.

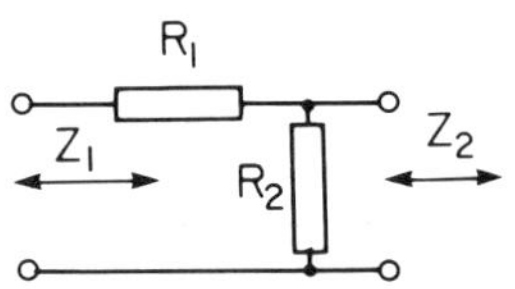

Figure 37.9 Minimum-loss pad

37.7.2 Design equations

$$\cosh \theta = (Z_1/Z_2) \quad (37.19)$$

$$R_1 = Z_1\{[1-(Z_2/Z_1)]^{1/2}\} \quad (37.20)$$

$$R_2 = Z_2/\{[1-(Z_2/Z_1)]^{1/2}\} \quad (37.21)$$

37.7.3 Design charts

Design chart (1) is shown in *Figure 37.10*. The minimum loss in dB can be read from the vertical axis. The ratio Z_1/Z_2 is the horizontal axis.

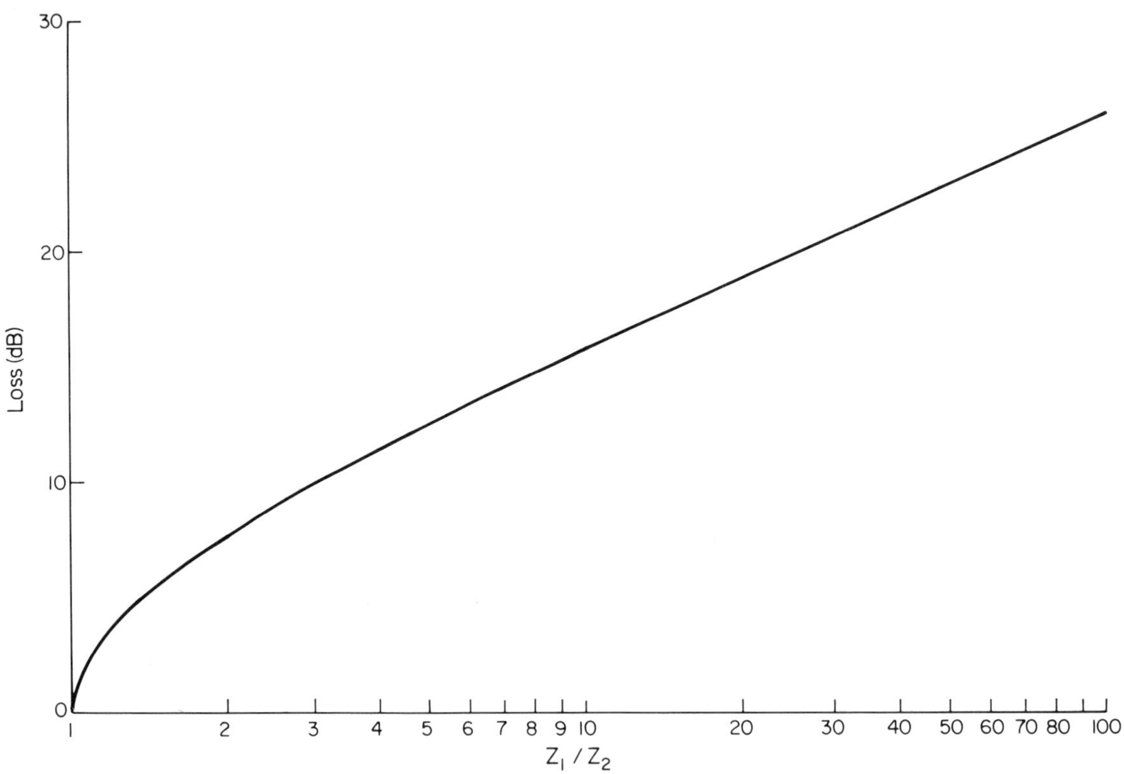

Figure 37.10 Minimum-loss pad design chart (1)

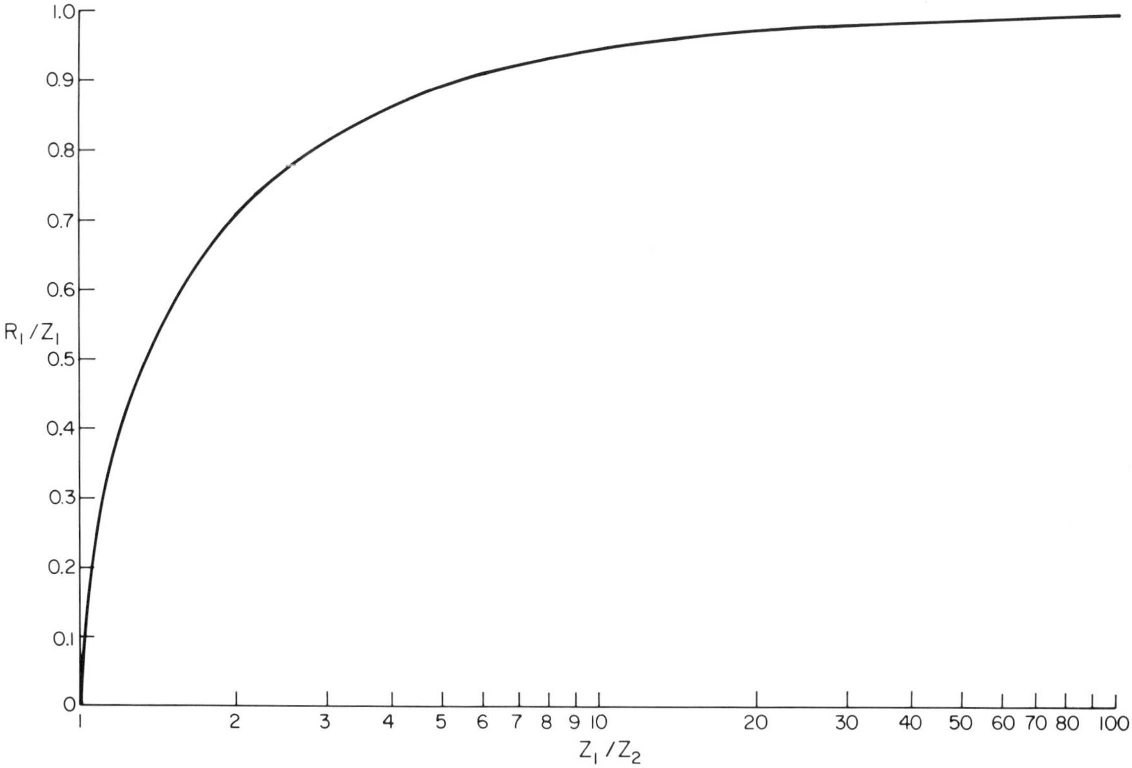

Figure 37.11 Minimum-loss pad design chart (2)

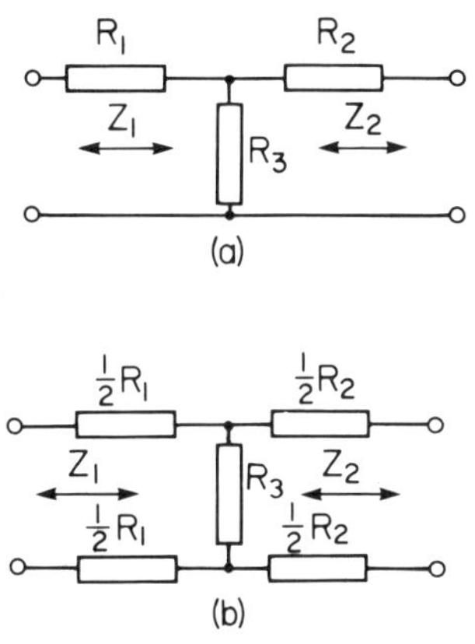

Figure 37.12 *T* and *H* pads: (a) *T* pad; (b) *H* pad

Design chart (2) is shown in *Figure 37.11*. The value of R_1/Z_1 can be read from the vertical axis. The ratio Z_1/Z_2 is on the horizontal axis.

Let ratio $Z_1/Z_2 = M$ (37.22)

$R_2 = R_1/(M-1)$ (37.23)

37.8 Miscellaneous *T* and *H* pads

37.8.1 Configuration

These are shown in *Figure 37.12*.

37.8.2 Design equations

For the *T* and *H* pads:

$R_3 = (Z_1Z_2)^{1/2}/\sinh\theta$ (37.24)

$R_1 = (Z_1/\tanh\theta) - R_3$ (37.25)

$R_2 = (Z_2/\tanh\theta) - R_3$ (37.26)

$$R_3 = \frac{2(NZ_1Z_2)^{1/2}}{N-1} \quad (37.27)$$

$R_1 = Z_1[(N+1)/(N-1)] - R_3$ (37.28)

$R_2 = Z_2[(N+1)/(N-1)] - R_3$ (37.29)

38 Amplifiers

S W Amos BSc(Hons), CEng, MIEE
Freelance Technical Editor and Author

Contents

38.1 Introduction

An amplifier is essentially an assembly of active and passive devices designed to increase the amplitude of an electrical signal.

The amplifier may be required to deliver appreciable power output, for example to operate a loudspeaker system. This is an example of a large-signal amplifier and the output signal has significant voltage and current components. In many amplifiers, however, e.g. the r.f. amplifier in a radio or television receiver, the output power is small because one component is small compared with the other. In such small-signal amplifiers the signal is best regarded as a voltage or current waveform. Active devices are by their very nature best suited for operation by a specific type of signal. For example valves and field-effect transistors are controlled by an input voltage and bipolar transistors by an input current. But, irrespective of the type of active device employed, an amplifier can be designed to act as a voltage or current amplifier by suitable choice of input resistance and output resistance. It is, in fact, the ratio of signal source resistance to input resistance and of output resistance to load resistance which determines whether the signal at input or output is best regarded as a voltage or a current. If the ratio is small compared with unity, the signal is best treated as a voltage, and if the ratio is large compared with unity it is more conveniently regarded as a current (*Figure 38.1*).

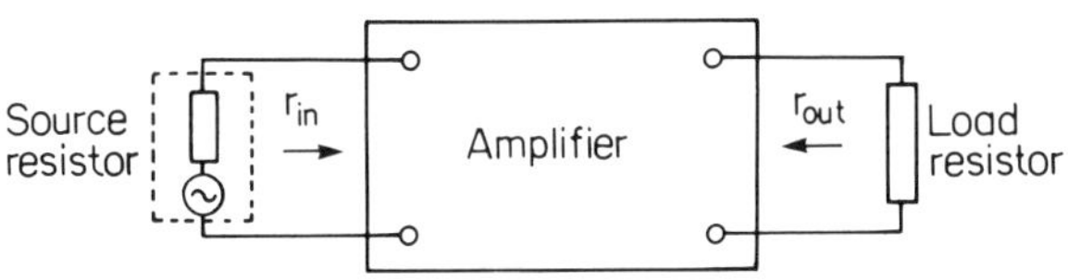

Figure 38.1 External and internal terminating resistances of an amplifier

38.2 Analogue amplifiers

Certain types of signal, notably audio and video signals, are characterised by the fact that at any instant they may have any value within certain limits. They are examples, in fact, of analogue signals, and amplifiers for such signals must accurately reproduce the way in which the input signal varies with time. In other words the output waveform of an analogue amplifier must be a substantially accurate copy of that of the input signal. This is only possible if the input–output characteristic of the amplifier is linear. The linearity of the input–output characteristics of active devices themselves is seldom good enough for many purposes, but the required degree of linearity can be achieved by the use of negative feedback and some of the circuits used for this purpose are illustrated later.

The frequency response of an analogue amplifier depends on its purpose. Certain d.c. amplifiers need a response extending from zero to a few kHz. A high-fidelity audio amplifier is likely to have a response level between 30 Hz and 15 kHz which is approximately the frequency range of the average human ear. The upper frequency limit of an amplifier for video or pulse signals is determined by the steepness of the vertical edges in the signals to be reproduced. The steepness of the leading (or tailing) edge of a pulse is measured by the time taken for its amplitude to change from 10% to 90% of its final steady value as shown in *Figure 38.2*. This is known as the rise time (or fall time) and is related to the upper frequency limit of the amplifier by the approximate expression

$$f_{max} = (2 \times \text{rise (fall) time})^{-1} \qquad (38.1)$$

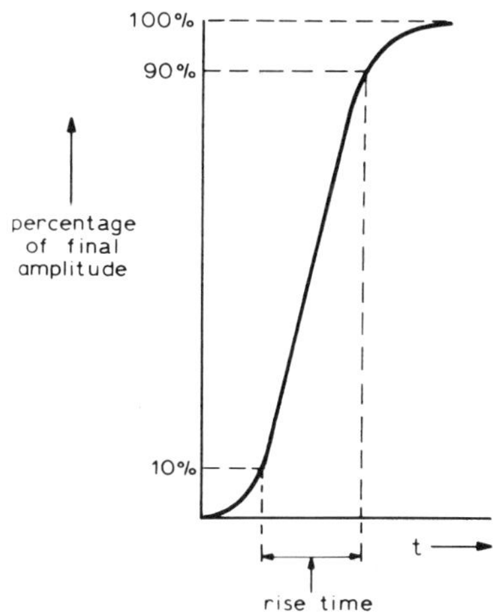

Figure 38.2 The rise time of an edge

Thus if an amplifier is required to reproduce pulses with a rise (or fall) time of 0.1 μs, a response up to at least 5 MHz is needed.

The low-frequency limit of the amplifier is determined by the levelness of the horizontal sections of the signals to be reproduced. The degree of levelness is measured by the amount of sag in, for example, a pulse top as indicated in *Figure 38.3*. The

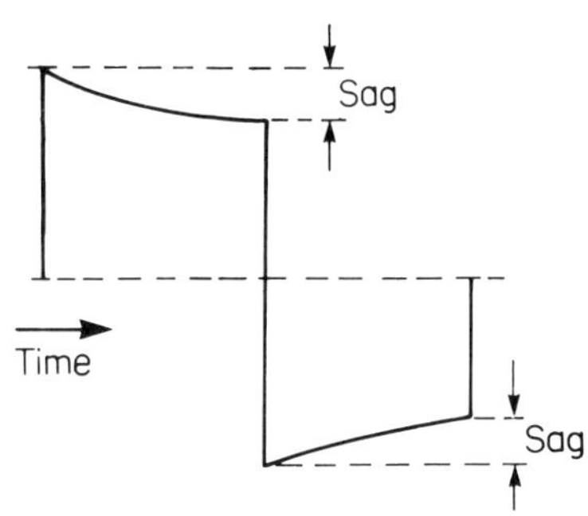

Figure 38.3 Sag in the reproduction of a square wave

low-frequency extreme f_{min} is given approximately by the relationship

$$f_{min} = \frac{\text{percentage sag}}{100\pi} \times \text{square-wave frequency} \qquad (38.2)$$

Thus to keep the sag in the reproduction of a 50 Hz square wave to less than 2% (representing high-grade performance) requires a low-frequency response extending down to 0.3 Hz!

38.3 Limiting amplifiers

There is another way of amplifying pulses in which the input signal is deliberately made large enough to overload the amplifier so that the output signal alternates between the two states corresponding to cut-off or saturation of the output stage. The output signal is thus in pulse form and the pulse amplitude is independent of the amplitude (or the waveform) of the input pulses provided that this is sufficient to cause overloading. For an amplifier used in this manner the shape of the input–output characteristic (which is so important in analogue amplifiers) is of little concern because the changes of state of the amplifier normally occur very rapidly in order to give short rise and fall times in the output pulses. In this way an amplifier can generate large-amplitude pulses from input pulses of smaller amplitude; it is a pulse amplifier which operates on principles quite different from those of analogue amplifiers.

38.4 Class A, B, and C operation

Whether an amplifier is linear depends on the shape of the characteristics relating output current with input signal of the active devices, the amplitude of the input signal and the way in which the signal swing is accommodated on the characteristics.

For example if the characteristic has a linear or near-linear section, and if the input signal has a small amplitude compared with the extent of the linear section, then the signal can be accommodated on any part of the linear section and the output current is linearly related to the input signal as shown in *Figure 38.4*, which illustrates a sinusoidal input signal. This form of operation is known as class A and is extensively used in small-signal amplifiers where linearity is required. It is a most inefficient form of operation because the current taken by the device from the supply is independent of the signal amplitude and remains constant even when the input signal is removed. This constancy of mean current is utilised in certain automatic biasing circuits as illustrated later in this chapter. By making use of the full extent of the linear section of the characteristic of suitable devices it is possible to obtain considerable output power. Early radio receivers and audio amplifiers used triodes and pentodes operating in class A but the efficiency, measured by the ratio of output power to the power taken from the h.t. supply, was limited to a theoretical maximum of 25% and in practice is considerably smaller.

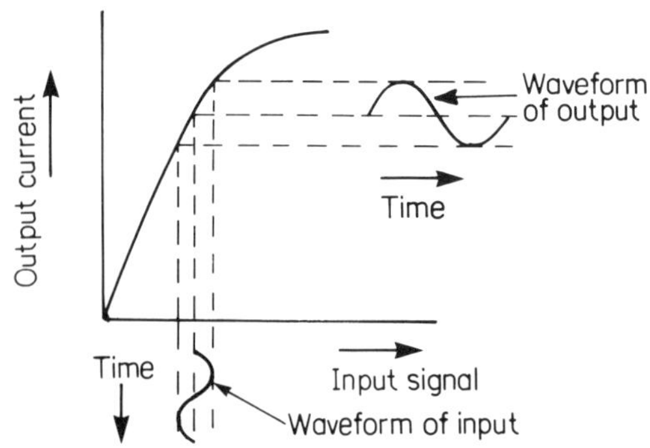

Figure 38.4 Class A operation

If the input signal is accommodated at the bottom end of the linear section of the characteristic near the point of output current cut-off, as shown in *Figure 38.5*, only one half of each cycle of input signal is reproduced. This difficulty can be overcome and linear amplification achieved by using two matched devices in push–pull, each biased to cut-off as shown in the figure. A great advantage of this form of operation, known as class B, is that the output current is very low in the absence of an input signal and is proportional to the amplitude of the input signal. Theoretically the efficiency of a class B stage is 78% and in practice figures approaching this value can be obtained. This is the form of output stage, using matched bipolar transistors, employed in most modern amplifiers and radio receivers.

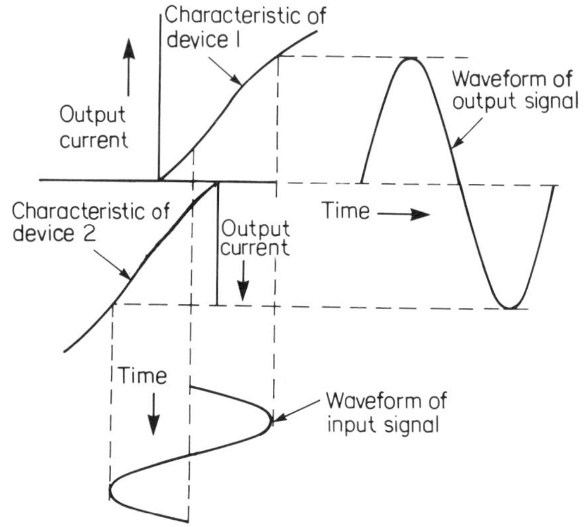

Figure 38.5 Class B operation using a push–pull pair

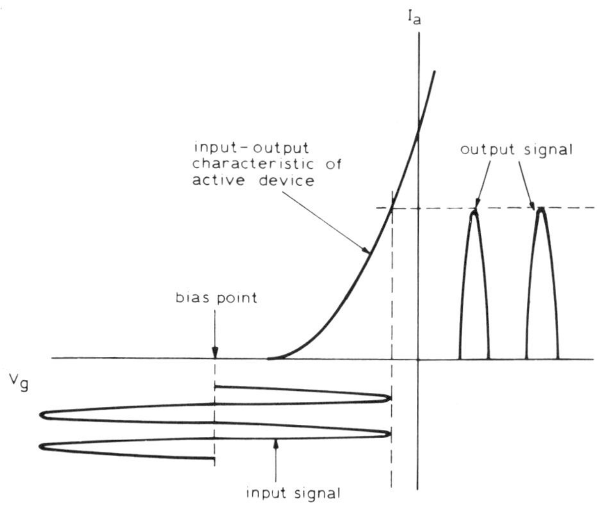

Figure 38.6 Class C operation

In a further type of operation, known as class C, only the peaks of the input signal are reproduced, as shown in *Figure 38.6*. The output current thus consists of a series of pulses at the frequency of the input signal and the waveform of the output current is not a replica of that of the input signal. This type of operation cannot, therefore, be used for linear amplification and is used where a constant-amplitude sinusoidal signal is to be amplified. Because of the non-linearity of the input–output characteristic the output contains a wealth of components including a strong fundamental component at the frequency of the input signal and a number of components at multiples of this frequency. These harmonics can be eliminated by using an *LC* circuit, resonant at the fundamental frequency, as the output load. The output signal is thus an amplified and undistorted version of the sinusoidal input signal. Very high efficiencies approaching 90% can be obtained from a class C amplifier and they are used in the high-power stages of radio transmitters where a constant-amplitude sinusoidal signal has to be amplified.

38.5 Frequency range

As already mentioned some amplifiers are required to have a uniform response over a wide frequency range beginning at a very low frequency. For example a high-grade audio amplifier with a response from 30 Hz to 15 kHz covers a range of nine octaves and for a video amplifier the range can approach 20 octaves. To achieve such wide bandwidths requires the use of active devices with suitably high upper frequency limits and the inter-device coupling circuits must be aperiodic. Resistive loads are generally used and sometimes small inductors are employed to offset the effects of the inevitable shunt capacitance which tends to reduce amplifier gain as the upper frequency limit is approached. Examples of such circuits are given later.

Amplifiers may alternatively be required to have a level response over a more restricted frequency range. For example the r.f. stage of an f.m. radio receiver requires a response extending from 87.5 to 100 MHz. For the i.f. amplifier of a medium wave or long wave a.m. radio receiver the passband may be as small as 10 kHz centred on 465 kHz. To achieve such restricted frequency ranges resonant circuits are used to couple the active devices, the resonance frequencies and the Q values of the circuits being chosen to give the required passband and possibly a particular rate of fall off of response outside the passband.

38.6 Fundamental amplifying circuits of active devices

The three types of active device, valve, bipolar transistor and field-effect transistor in their simplest forms have three terminals. The input signal is applied between two of them, of which one must be the base (gate or grid). The output signal is generated between two terminals, of which one must be the collector (drain or anode). It follows that one terminal must be common to input and output circuits and the basic amplifier configurations are classified by this common terminal (*Figure 38.7*).

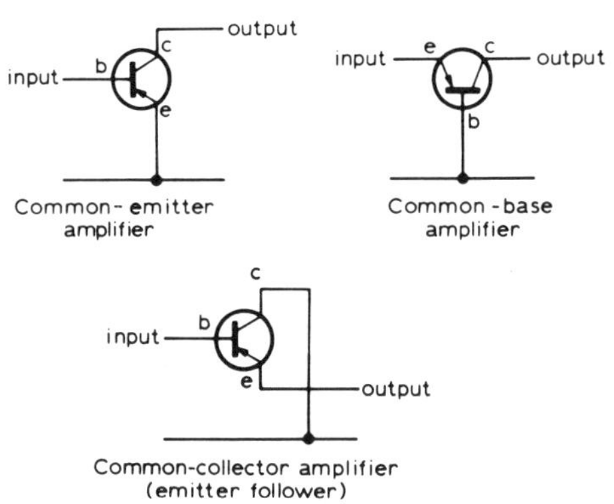

Figure 38.7 Basic forms of bipolar-transistor amplifying circuits

For example if the common terminal of a bipolar transistor is the base, the circuit is known as the common-base amplifier. It is characterised by very low input resistance, very high output resistance and a current gain of unity. The output signal is in phase with the input signal. Voltage gain can be high. Because the base acts to some extent as an earthed screen between emitter and collector there is little capacitance linking input and output circuits and the amplifier can be used with stability at v.h.f. and u.h.f.

The common-collector circuit has high input resistance, low output resistance and a voltage gain of approximately unity. Current gain is considerable. The output signal is in phase with the input signal and, because the gain is unity, is almost equal to it. The output signal thus follows the input signal and the circuit is more popularly known as an emitter follower. The circuit is used where a high input resistance or a low output resistance is required or as a buffer between stages where it is important to minimise unwanted coupling.

The third fundamental amplifying circuit is the common-emitter. The input resistance is low, the output resistance high (though the ratio of the two is much less than in the common-base circuit) and the voltage gain and current gain can both be considerable. This probably accounts for the fact that this is by far the most used of the three basic circuits. The output signal is inverted with respect to the input signal.

Similar circuit arrangements are possible, of course, with valves and field-effect transistors and the properties listed above for the bipolar transistor apply also to the valve and the field-effect transistor except for the references to current gain which do not apply to these voltage-controlled devices.

38.7 General amplifying circuits

Figure 38.8 gives the circuit diagram of a common-emitter class A amplifying stage consisting of a single bipolar transistor. One of

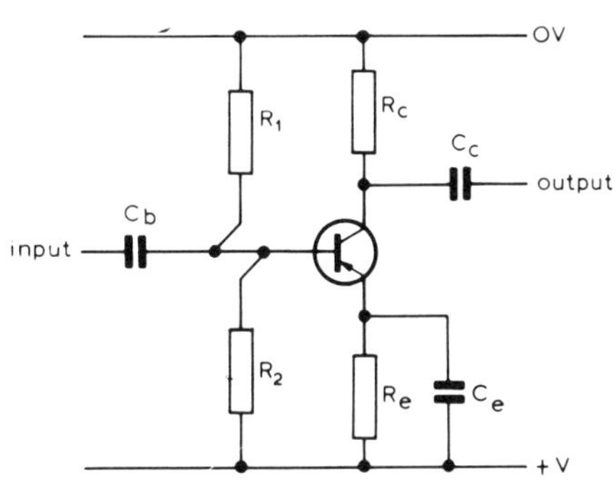

Figure 38.8 Common-emitter stage of amplification with d.c. stabilisation by potential divider and emitter resistor

the most important features of the circuit is the method employed to bias the transistor and to ensure constancy of mean emitter current.

It is important to keep the mean emitter current constant because a number of the transistor properties depend on it. For example the current gain h_{fe} varies with emitter current and often has a maximum at a particular value of current. The input resistance r_{in} also depends on emitter current, being given approximately by the expression

$$r_{in} = h_{fe}\frac{25\ (\text{mV})}{I_e\ (\text{mA})} \tag{38.3}$$

Thus if the current amplification factor h_{fe} is 100 and if I_e is 2 mA, r_{in} is 1250 Ω.

The maintenance of constancy of mean emitter current is often termed d.c. stabilisation of the operating point and is necessary because, for a given base bias, emitter current can vary, from specimen to specimen of the same transistor type, over a range as great as 3:1, often termed the manufacturing spread. Moreover with germanium transistors there is a leakage component of the emitter current which is not controlled by the base current and which, at high ambient temperatures, can become comparable with the useful (controlled) component of emitter current, so degrading the performance of the amplifier by limiting the signal output. At normal operating temperatures leakage current is not troublesome in silicon transistors but d.c. stabilisation is still desirable to achieve constancy of mean emitter current in spite of the manufacturing spreads in transistor parameters.

In the circuit diagram of *Figure 38.8* stabilisation is achieved by the resistors R_1, R_2 and R_e. The potential divider R_1R_2 applies a particular value of steady voltage V_b to the base. Suppose V_b is 3 V. Between the base and the emitter terminals of the transistor, when it is operating normally as an amplifier, there is a voltage difference of approximately 0.7 V for a silicon transistor so that the voltage across R_e is 2.3 V. If R_e is made 1 kΩ, the emitter current is stabilised at 2.3 mA. R_e introduces negative feedback which, if undesired, can be eliminated by decoupling R_e by a capacitor C_e with a low reactance at the signal frequencies employed.

Stabilisation is achieved in the following manner. If for any reason, such as a rise in temperature, the emitter current tends to rise, the increased voltage drop across R_e causes a decrease in base-emitter voltage which limits (but does not completely neutralise) the rise in current. Similarly if the emitter current tends to fall, the reduced voltage drop across R_e causes an increase in base-emitter voltage which offsets but does not completely neutralise the fall in current. Better compensation can be obtained by feeding the potential divider from the collector terminal instead of the positive supply terminal. Now when emitter current tends to rise the collector voltage falls, causing the base voltage also to fall and this, combined with the rise in emitter voltage, gives improved d.c. stabilisation. It may be necessary to take precautions to avoid signal-frequency negative feedback from collector to base via the upper resistor of the potential divider. This improved method of d.c. stabilisation is used in some of the two-stage voltage and current amplifiers described later. Enhancement-type field-effect transistors also require a bias voltage which lies between the source and the drain voltages, and these can also be d.c. stabilised by use of a potential divider to feed the gate and by including a resistor in the source circuit.

For valves and depletion-type field-effect transistors where the gate voltage lies outside the range of the source-drain voltage the biasing circuit shown in *Figure 38.9* can be used. Again class A

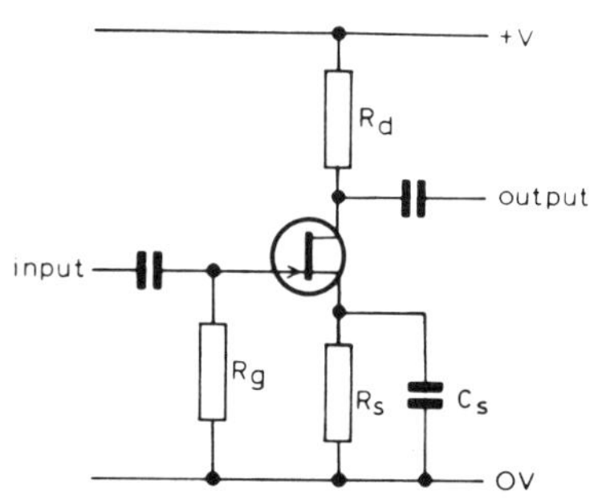

Figure 38.9 A depletion-type junction-gate field-effect transistor with bias and d.c. stabilisation provided by a source resistor

operation is assumed. The voltage generated across the source resistance R_s by the source current is applied between gate and source, the gate resistance R_g being included to permit the application of an external input signal. C_s is included to minimise negative feedback. The required value of R_s is given by

$$R_s = \frac{\text{gate bias voltage}}{\text{mean source current}} \qquad (38.4)$$

Thus if the mean current is 5 mA and the gate bias 2 V, R_s should be 400 Ω.

38.7.1 Use of negative feedback

The two biasing circuits just described enable a consistent value of mean current to be achieved but in spite of this there are inevitably differences in gain between nominally identical circuits because of the variations in characteristics from specimen to specimen of the same transistor type. The application of negative feedback is the usual way of minimising these differences and one method of applying negative feedback is by using an un-decoupled emitter or source resistor. The effect of negative feedback is to reduce gain and improve linearity and it also makes the gain of the stage more dependent on the values of the feedback components and less dependent on the characteristics of the transistor itself. This form of feedback circuit, shown in basic form in *Figure 38.10(a)*, also increases the input resistance

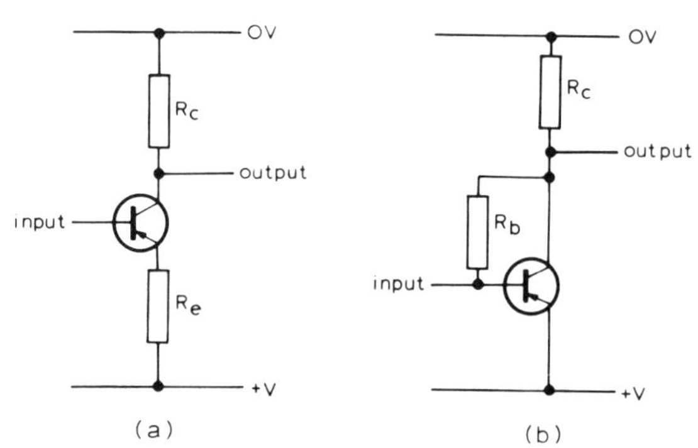

Figure 38.10 Two basic methods of applying negative feedback to a common-emitter amplifier

and the output resistance of the amplifier. The performance of an amplifier with a high input resistance is best expressed in terms of input voltage as mentioned at the beginning of this chapter. The high output resistance implies that the output signal is best regarded as an output current. This type of circuit can thus be treated as a voltage-to-current converter and the ratio of the output current to the input voltage is given approximately by the expression

$$\frac{i_{out}}{v_{in}} = \frac{1}{R_e} \qquad (38.5)$$

An alternative method of applying negative feedback to a single transistor is illustrated in *Figure 38.10(b)*. The resistor R_b returns to the input circuit a feedback current proportional to the output voltage. The effect of this type of feedback circuit is, as before, to reduce gain and improve linearity and to make the gain of the stage more dependent on the value of R_b than on the properties of the transistor. However this form of feedback reduces both the input resistance and the output resistance of the stage. The behaviour of an amplifier with a low input resistance is best explained by assuming the input signal to be a current. The low output resistance implies that the output signal is best regarded as a voltage. This circuit is thus a current-to-voltage converter and the ratio of the output voltage to the input current is given approximately by the expression

$$\frac{v_{out}}{i_{in}} = R_b \tag{38.6}$$

By combining a voltage-to-current converter with a current-to-voltage converter it is possible to produce a two-stage voltage or current amplifier.

38.8 Two-stage voltage amplifier

If the voltage-to-current stage is placed first as indicated in *Figure 38.11* then the two-stage amplifier has a high input resistance and a low output resistance as required in a voltage amplifier. The high output resistance of the first stage feeds directly into the low input resistance of the second stage and these are the conditions required to transfer the current output of the first stage into the following stage with negligible loss. Thus equating i_{out} with i_{in} in the above two equations gives the approximate relationship

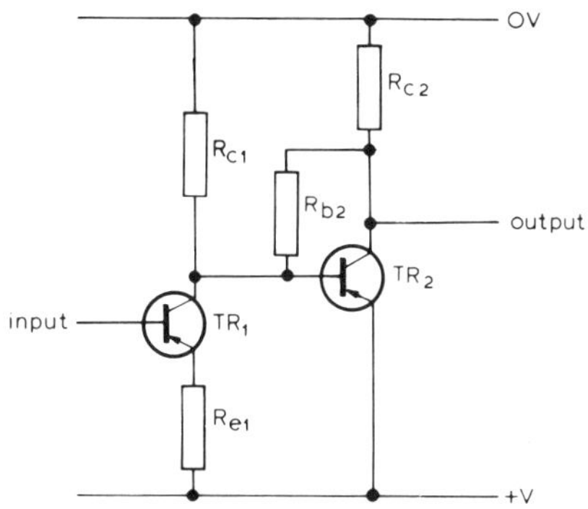

Figure 38.11 Basic form of two-stage voltage amplifier

$$\frac{v_{out}}{v_{in}} = \frac{R_{b2}}{R_{e1}} \tag{38.7}$$

i.e. the voltage gain is given by the ratio of the two feedback resistors and is independent of the characteristics of the two transistors.

In practical versions of the voltage-amplifying circuit R_{b2} is usually returned to the emitter circuit of TR_1 as shown in *Figure 38.12.* This has little effect on the performance because the current injected into TR_1 emitter by R_{b2} emerges with negligible loss from the collector and so enters TR_2 base, R_{c1} being large compared with TR_2 input resistance. R_{b2} and R_{e1} now form a voltage divider across TR_2 with the junction connected to the base. This helps in the d.c. stabilisation of the circuit, all that is necessary to complete it being the inclusion of an emitter resistor R_{e2} as shown in *Figure 38.12.* C_{e2} is needed to minimise signal-frequency negative feedback.

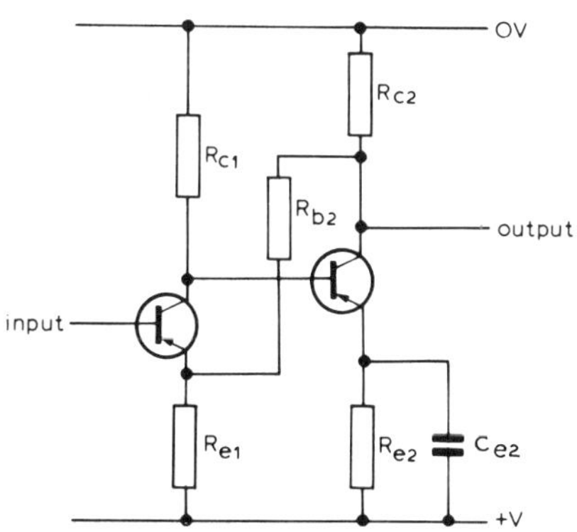

Figure 38.12 Practical circuit for two-stage voltage amplifier incorporating means of ensuring d.c. stability

38.9 Two-stage current amplifier

If the current-to-voltage converter is placed first, as shown in *Figure 38.13*, then the two-stage amplifier has a low input resistance and a high output resistance as required in a current amplifier. The low output resistance of the first stage feeds directly into the high input resistance of the second stage. These are the conditions required to transfer the voltage output of the first stage into the input of the second stage with negligible loss. Thus equating v_{in} and v_{out} in the above two equations gives the approximate relationship

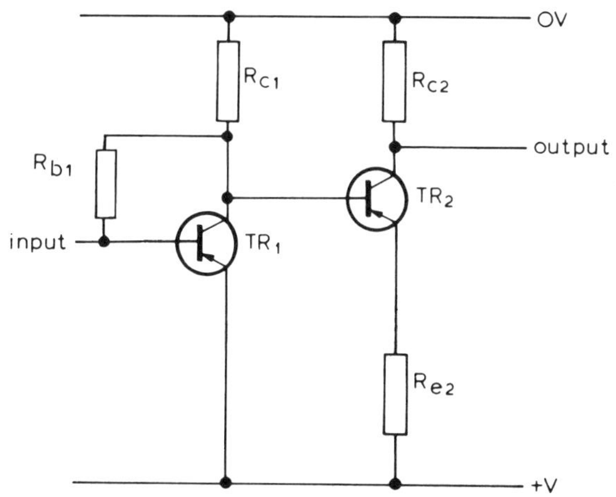

Figure 38.13 Basic form of two-stage current amplifier

$$\frac{i_{out}}{i_{in}} = \frac{R_{b1}}{R_{e2}} \tag{38.8}$$

i.e. the current gain is given by the ratio of the two feedback resistors and is independent of the characteristics of the two transistors.

In practical versions of the current amplifying circuit R_{b1} is usually fed from TR_2 emitter as shown in *Figure 38.14.* This has little effect on the performance of the circuit because TR_1 collector is connected to TR_2 base and TR_2 emitter follows the signal voltage at the base. To achieve d.c. stability a common technique is to add a resistor $R_{e2'}$ in series with R_{e2} to form a voltage divider to feed TR_1 base via R_{b1}. R_{e1} is then included in TR_1 emitter circuit being decoupled by C_{e1} to eliminate signal frequency negative feedback. $R_{e2'}$ should be decoupled as shown for the same reason.

It is possible to effect economies in components and so produce simpler circuits by using complementary transistors in these two-stage amplifiers. As an example *Figure 38.15* gives the circuit diagram of a complementary voltage amplifier. As before the gain is given by

$$\frac{v_{out}}{v_{in}} = \frac{R_3}{R_2} \tag{38.9}$$

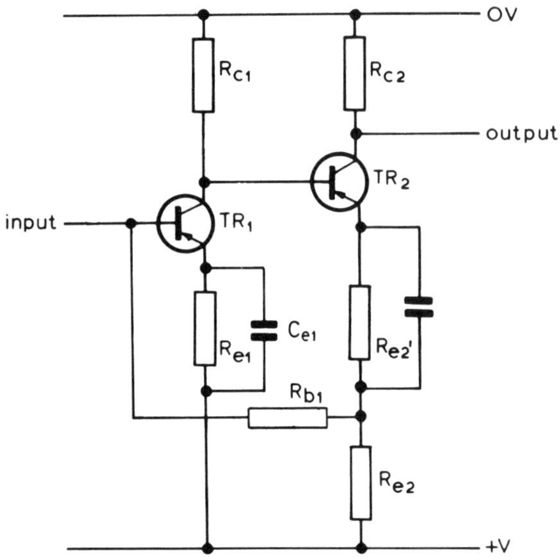

Figure 38.14 Practical circuit for two-stage current amplifier incorporating means of ensuring d.c. stability

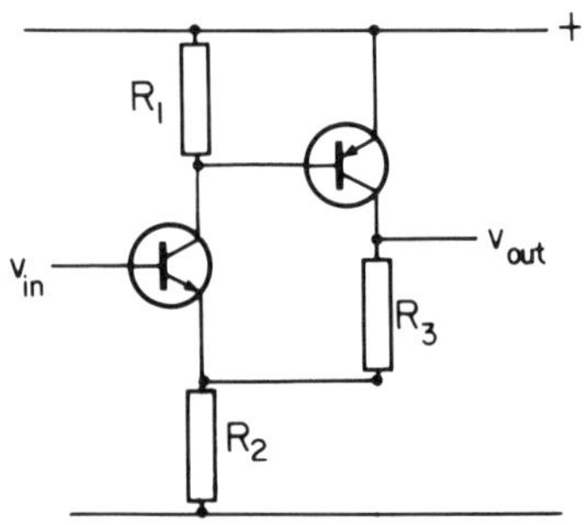

Figure 38.15 Basic form of complementary two-stage voltage amplifier

38.10 D.C. amplifiers

All the amplifier circuits so far discussed have included a direct inter-transistor coupling. Such amplifiers have a frequency response extending to zero and are thus capable of amplifying steady currents and voltages.

It would not, however, be practical to use a succession of such stages to achieve high zero-frequency gain. In such an arrangement the inevitable slight drift in collector current in the early stages could, after amplification by all the following stages, give rise to a serious spurious output. To minimise this effect it is customary to construct d.c. amplifiers of long-tailed pair stages of the type shown in *Figure 38.16*. Each pair consists of two identical transistors with identical collector loads and a common external emitter resistor. Any drift in the collector current of one transistor, caused for example, by a change in ambient temperature, is accompanied by an equal change in collector current of the other transistor which shares the same environment. The output of the stage is developed between the collector terminals and is thus zero for all current changes which affect both transistors equally. In a d.c. amplifier the signal to be amplified is applied to one base of the long-tailed pair and, if required, a negative feedback voltage can be applied to the other base to improve linearity.

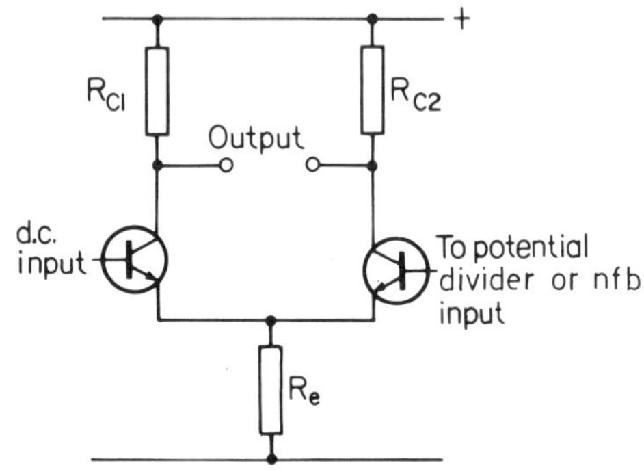

Figure 38.16 Basic form of long-tailed pair

The common emitter connection also introduces a measure of negative feedback, the emitter of each transistor being effectively terminated by the internal emitter impedance of the other transistor, causing 6 dB loss in gain. The common external impedance can be high and in long-tailed pairs forming part of an integrated circuit is often the collector-emitter path of a third bipolar transistor, the current through which (and hence in the long-tailed pair) is stabilised by the potential divider and emitter resistor technique.

38.11 Bipolar transistor output stages

A common-emitter stage of the type illustrated in *Figure 38.8* can, by a suitable choice of transistor, be used to deliver several watts of power, e.g. to operate a loudspeaker. If the transistor operates in class A there is considerable dissipation in the transistor itself. The resulting temperature rise must be limited otherwise the transistor can be damaged or even destroyed. The usual precaution is to mount the transistor in intimate thermal contact with a heat sink, usually a metal structure with cooling fins. Suitable output transistors may have a mean collector current of 1 A and the optimum load is low, commonly only a few ohms, so that the speech coil can be used as a direct-coupled load and there is no need for a matching transformer.

In general, however, transistors required to deliver appreciable undistorted power are operated in push–pull. This reduces distortion by the cancellation of even harmonics and permits the use of class B operation which economises on power supply. A symmetrical push–pull circuit is illustrated in *Figure 38.17*. The

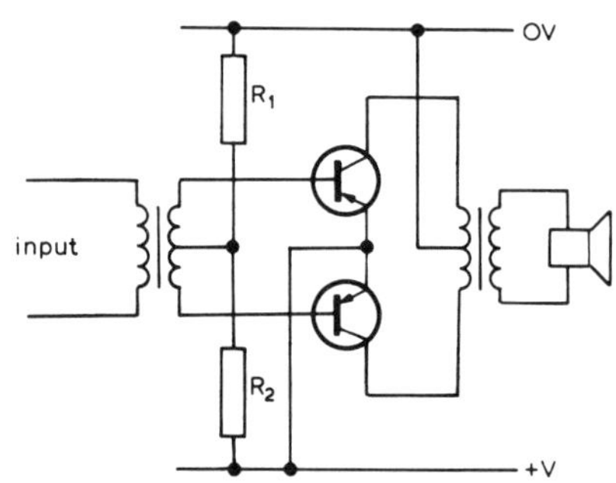

Figure 38.17 Symmetrical push–pull bipolar-transistor output amplifier

potential divider R_1R_2 provides base bias and is adjusted to give sufficient collector current in the absence of an input signal to minimise crossover distortion. To keep general distortion at a satisfactorily low level the characteristics of the two transistors must be closely matched and negative feedback is usually employed.

The need for a phase-splitting transformer is one of the disadvantages of the symmetrical circuit but such a transformer is not needed if a pair of complementary transistors is used in the output stage and the essential features of the complementary circuit are shown in *Figure 38.18*. TR_1 and TR_2 are npn and pnp transistors of closely matched characteristics. They are connected in series across the supply, their common-emitter terminals being returned to the midpoint of the supply via the output load shown as a loudspeaker. The transistors operate in class B and have a common base connection forming the input terminal. Positive-going signals drive TR_1 into conduction but cut TR_2 off so that the voltage of the common-emitter terminal approaches supply positive value. Negative-going input signals drive TR_2 into conduction but cut TR_1 off so that the voltage of the common-emitter terminal approaches supply negative value. An alternating input signal thus causes the common-emitter potential to swing between the limits of the supply voltage and the circuit operates in push–pull without need of any form of phase-splitting device.

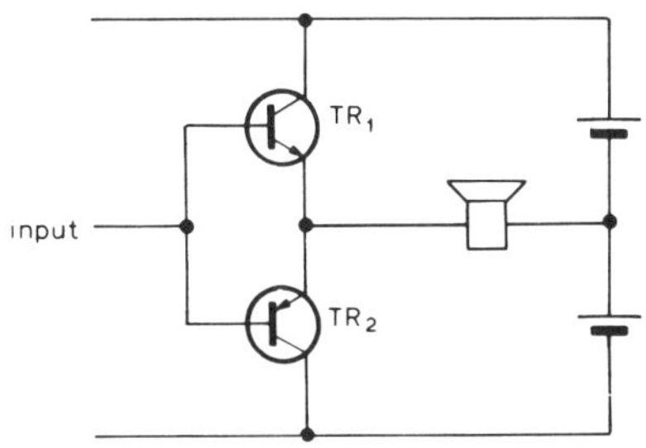

Figure 38.18 Essential features of a complementary push–pull single-ended output stage

A difficulty of the basic circuit of *Figure 38.18* is that the load is in the emitter circuit of both transistors which therefore operate as emitter followers with unity voltage gain. A large input voltage is needed to deliver maximum output into the load. This disadvantage is overcome by the modifications shown in the circuit diagram of *Figure 38.19* which also includes components for biasing the transistors. The load is now fed from the common-emitter terminal via a capacitor. This enables the other terminal of the load to be connected to the positive supply line so that the junction of load and capacitor can be used as the supply source for the driver transistor TR_1. This arrangement ensures that the signal voltage generated across the collector load resistor for TR_1 is applied between base and emitter of both output transistors. Expressed differently, the top end of the collector load of TR_1 is decoupled to the common-emitter terminal of the output pair. Thus TR_2 and TR_3 operate as common-emitter amplifiers which have much greater voltage gain than the emitter-followers of *Figure 38.18*.

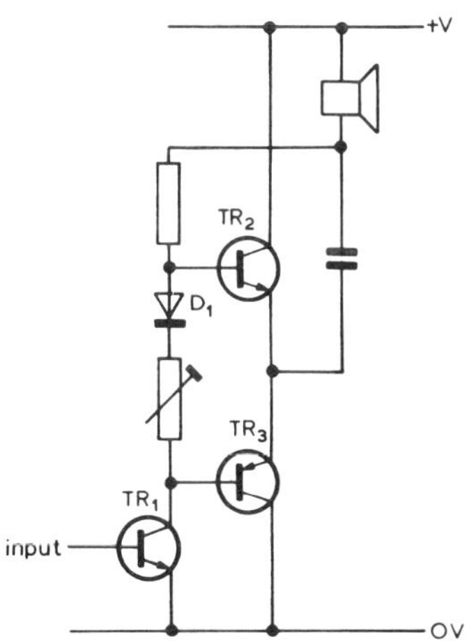

Figure 38.19 Modification of previous circuit to ensure that the output transistors operate in common-emitter mode

The forward-biased diode D_1 is included to compensate for the changes in base-emitter voltage of the output transistors which occur with alterations in temperature. The diode thus ensures greater constancy in the standing current of TR_2 and TR_3. The preset resistor in the base circuit enables this current to be set to the desired value. *Figures 38.18* and *38.19* are examples of the asymmetrical or single-ended push–pull circuit.

It may be difficult to find a pair of complementary transistors with closely matched characteristics to deliver considerable output power but it is possible to use matched transistors of the same type in the output stage, these being driven by a complementary push–pull pair as shown in *Figure 38.20*. The output transistors are also connected in series across the supply and each is driven from one of the complementary pairs. As TR_4 and TR_5 are of the same type their input signals must be in antiphase to ensure push–pull operation and this is achieved by arranging for TR_2 to act as an emitter follower (which does not invert the signal) and for TR_3 to act as a common-emitter amplifier (which does invert the signal). The current gain of an emitter follower and a common-emitter amplifier are approximately equal and this ensures that the output pair receive equal-amplitude input signals. The circuit diagram of *Figure 38.20* differs from that of *Figure 38.19* in that separate capacitors are used for loudspeaker coupling and driver collector circuit decoupling.

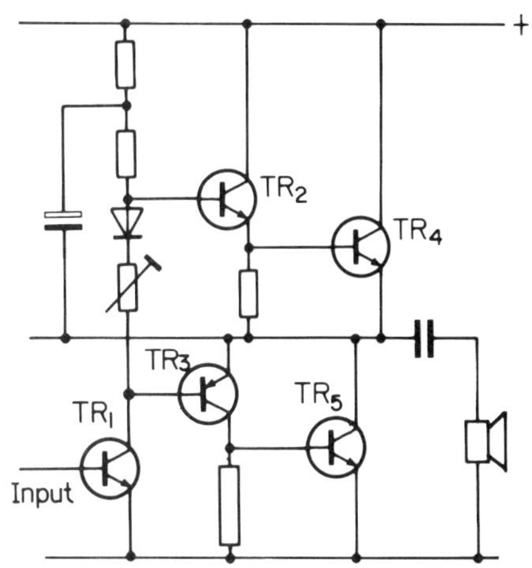

Figure 38.20 Elaboration of the previous circuit to permit the use of similar transistors in the output stage

38.12 Wideband amplifiers

The two-transistor circuits illustrated in *Figures 38.12* and *38.14* are capable of a wide frequency response. In fact the upper frequency limit is primarily determined by the transistion frequency f_T of the transistors. There is no difficulty in finding bipolar transistors with values of f_T of several hundred MHz and f_T measures the gain-bandwidth product of the device. Thus a transistor for which $f_T = 200$ MHz can give a gain of 10 up to 20 MHz and of 3 up to 67 MHz. To achieve these values of gain, however, it is necessary to offset the effects of the inevitable shunt capacitance which degrades the high-frequency response. There are, in general, two ways of doing this; by neutralising the capacitance by use of inductance and by the use of negative feedback.

Figure 38.21 shows the collector circuit of a wideband amplifier using shunt inductance to extend the high-frequency response.

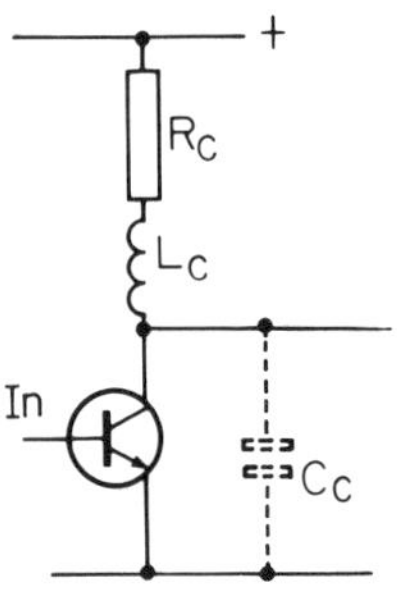

Figure 38.21 Essential features of the shunt-inductance method of extending high-frequency response

The value of the inductance L_c must be chosen with care. If it is too small it will not be effective in extending the frequency response sufficiently, and if it is too large it will cause overshoots in reproducing signals with sharp transitions. A suitable value is given by

$$L_c = 0.5 R_c^2 C_c \qquad (38.10)$$

where R_c is the collector load resistance and C_c is the total shunt capacitance in parallel with R_c (including any contribution from the input circuit of the following stage). The value of R_c to use can be calculated from the expression

$$R_c = \frac{1}{2\pi f_{max} C_c} \qquad (38.11)$$

where f_{max} is the highest frequency to be reproduced. Thus if the upper limit is 5 MHz and if C_c is 30 pF we have that R_c is approximately 1 kΩ. From this the required inductance value is given by $L_c = 0.5 \times 10^6 \times 30 \times 10^{-12} = 15$ μH.

Figure 38.22 is an example of an amplifier circuit in which

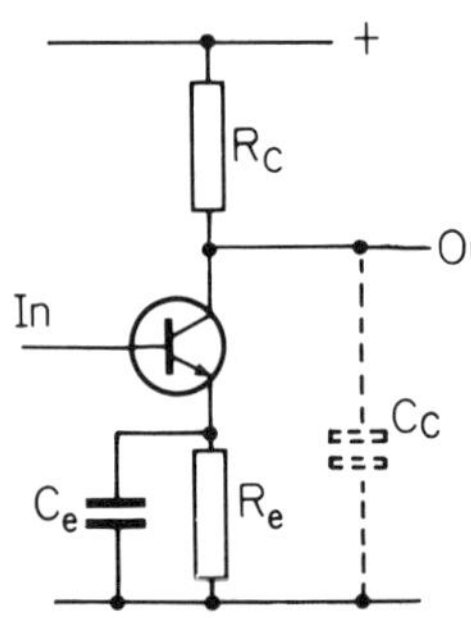

Figure 38.22 Method of extending high-frequency response by frequency-discriminating negative feedback in the emitter circuit

extension of the high-frequency response is achieved by negative feedback. The circuit is basically a common-emitter stage of the type shown in *Figure 38.8* but the bypass capacitor C_e is made too small to provide effective decoupling over the whole of the frequency band to be covered. It is effective at the high-frequency end so giving maximum gain from the transistor at these frequencies but is ineffective at lower frequencies where negative feedback due to R_e reduces the gain. Thus C_e gives a high-frequency boost which can be used to offset the effect of the shunt capacitance in the collector circuit. In fact the best results are achieved when the collector and emitter time constants are equal, i.e.,

$$R_e C_e = R_c C_c \qquad (38.12)$$

and this gives the following simple expression for calculating the required value of C_e

$$C_e = \frac{R_c C_c}{R_e} \qquad (38.13)$$

As a numerical example if $R_c = 1$ kΩ and if $C_c = 30$ pF (as in the above example) and if $R_e = 200\,\Omega$

$$C_e = \frac{10^3 \times 30 \times 10^{-12}}{200} \text{F} = 150 \text{ pF}$$

Figure 38.23 gives the circuit diagram of a transistor pulse amplifier also using negative feedback. TR_1 is an emitter follower

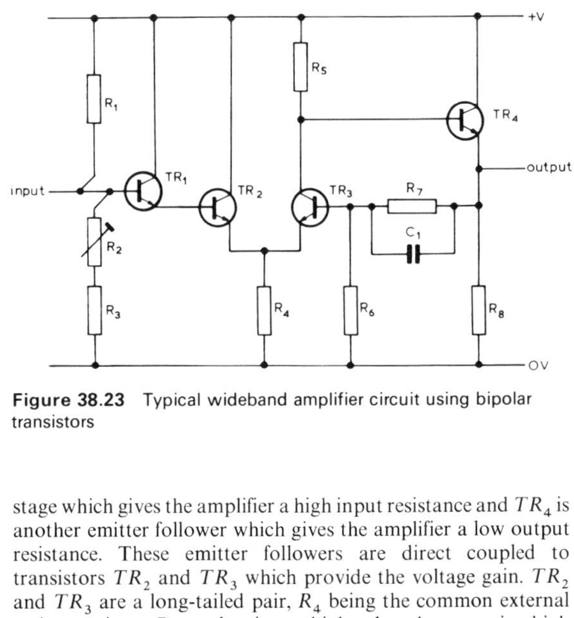

Figure 38.23 Typical wideband amplifier circuit using bipolar transistors

stage which gives the amplifier a high input resistance and TR_4 is another emitter follower which gives the amplifier a low output resistance. These emitter followers are direct coupled to transistors TR_2 and TR_3 which provide the voltage gain. TR_2 and TR_3 are a long-tailed pair, R_4 being the common external emitter resistor. R_4 can be given a high value, thus ensuring high d.c. stability, without at the same time losing considerable gain as a result of signal-frequency negative feedback in this resistor. R_4 is shunted by the internal emitter resistance of TR_3 and this reduces the signal frequency loss to only about 6 dB.

Another useful feature of the long-tailed pair is that the base of the second transistor provides a convenient point at which to inject a negative feedback signal. In *Figure 38.23* the feedback is made frequency-dependent by inclusion of the network $R_6 R_7 C_1$ which is designed to make the frequency response of the amplifier fall off above the required passband. The voltage gain of $TR_2 TR_3$ (and hence of the amplifier) is given by $g_m R_5$ where g_m is the mutual conductance of each transistor and R_5 is the value of TR_3 collector load resistance. Typical values for g_m and R_5 are 40 mA

V^{-1} and 2.5 kΩ giving a voltage gain of 100. This is reduced to the required value by the feedback network R_6R_7. For example if a gain of 20 is required then R_7 is made equal to $19R_6$ and if the passband limit is required to be 10 MHz, then C_1 is chosen to have a reactance equal to R_7 at this frequency.

The resistor R_2 is used to set the no-input-signal standing output voltage to the required value.

38.13 Wideband amplifier using field-effect transistors

Because of their low noise, field-effect transistors are most likely to be used in the input stage of a wideband amplifier such as a camera head amplifier. The high input resistance makes this type of transistor particularly suitable for such an application because camera tubes have a high, predominantly capacitive, output impedance which needs to be loaded by a high resistance.

A single field-effect transistor as a common-source amplifier is not ideal because feedback via the internal drain-gate capacitance causes a fall off in response and in signal-to-noise ratio at high frequencies. The usual solution to this problem is to feed the output of a common-source amplifier into a second field-effect transistor connected as a common-gate amplifier. Direct coupling is commonly used and the two transistors can be connected across the supply as shown in *Figure 38.24*, in a circuit arrangement known as a cascode amplifier. The voltage gain of

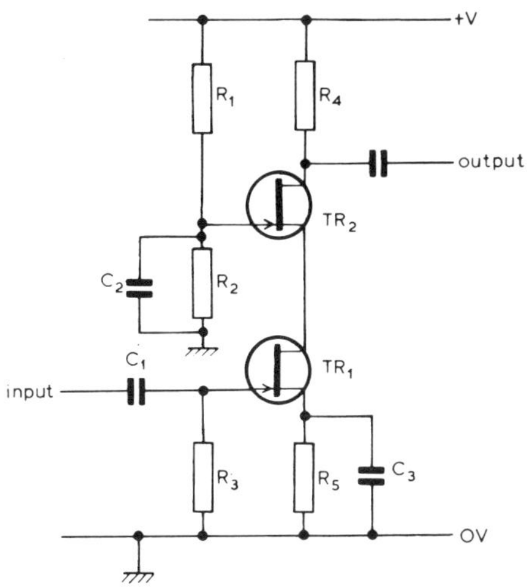

Figure 38.24 A cascode input stage for a wideband amplifier using two junction-gate field-effect transistors

the cascode from the gate of TR_1 to the drain of TR_2 is that of TR_1 feeding the load resistor R_4 but feedback via the drain-gate capacitance is eliminated because the voltage gain of TR_1 is limited to unity, its effective load resistance being the internal source resistance of TR_2. Feedback cannot occur in TR_2 via the drain-gate capacitance because the gate is decoupled at signal frequencies by C_2. The cascode is a well-known arrangement which has been used with thermionic valves and with bipolar transistors. It is also used in r.f. amplifiers at v.h.f. and u.h.f. and an example is given later in this chapter.

38.14 Tuned amplifiers

As mentioned earlier tuned circuits are used as inter-device coupling circuits where the passband of the amplifier is small compared with the centre frequency. The signal to be amplified may be a constant-amplitude sine wave. This applies, for example, to the carrier-wave stages in an a.m. transmitter and to many of the stages in an f.m. transmitter. A linear input-output characteristic is not necessary in such amplifiers and class C operation can be used, giving very high efficiencies.

The obvious form for a tuned amplifier for this application is that of an active device, biased to ensure class C operation, with input and output LC circuits resonant at the frequency of operation. In the high-power stages of an a.m. transmitter the active device is likely to be a triode valve and its anode-grid capacitance even at frequencies as low as hundreds of kHz is likely to be sufficient to cause instability. The circuit operates, in fact, as a tuned-grid tuned-anode oscillator. It is necessary, therefore, to offset the positive feedback occurring via the anode-to-grid capacitance by an equal amount of negative feedback to secure stability of operation. The basic form of a typical circuit is illustrated in *Figure 38.25*. The inductor of the anode-tuned

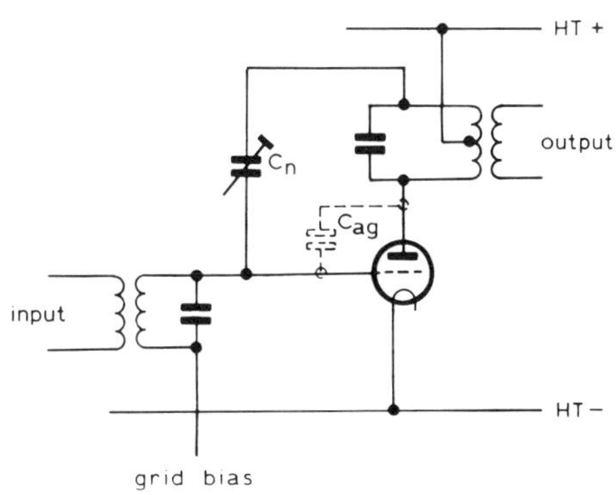

Figure 38.25 Basic form of neutralised-triode r.f. amplifying circuit used in transmitters

circuit is centre-tapped and the h.t. feed is introduced via this point which is, therefore, effectively at earth potential. The r.f. signal at the top end of the inductor is thus in antiphase with that at the anode. Thus the signal fed back to the grid via the neutralising capacitor C_n is in antiphase with that fed back via the anode-grid capacitance (shown in broken lines). In practice C_n is made variable and is adjusted empirically to secure stable amplification.

38.15 Frequency multipliers

The output tuned circuit of an amplifier can be kept in oscillation by regular bursts of current at $\frac{1}{2}$, $\frac{1}{3}$ or in general $1/n$ of its resonance frequency. In other words the output circuit can be tuned to a harmonic of the input frequency. This is the basis of one form of frequency multiplier. The danger of instability in such an arrangement is not so serious as that when input and outputs circuits are tuned to the same frequency.

38.16 Tuned amplifiers for a.m. signals

In amplifying amplitude-modulated signals, changes in carrier amplitude which represent the modulating signal must be faithfully reproduced. The amplifiers used must therefore have a linear input–output characteristic. Class A stages are used in the r.f. and i.f. amplifiers of receivers. In high-power a.m. transmitters, where economy of operation is important, class B stages are used, often in push–pull to minimise radiation of even harmonics of the carrier frequency.

38.16.1 Use of single tuned circuits

One form of tuned amplifier which can be used to amplify a.m. signals consists of a succession of LC circuits all tuned to the same centre frequency and separated by active devices. The i.f. amplifiers of sound and television receivers are often examples of such amplifiers. The centre frequency of each tuned circuit is determined by the product LC: the rate of fall-off of the response on each side of the centre frequency is determined by the effective Q value of the inductor. The relationship between response and mistuning Δf is illustrated in *Figure 38.26*. This shows that the loss is 3 dB when $2Q\,\Delta f/f_c = 1$. This gives

$$2\,\Delta f = \frac{f_c}{Q} \qquad (38.14)$$

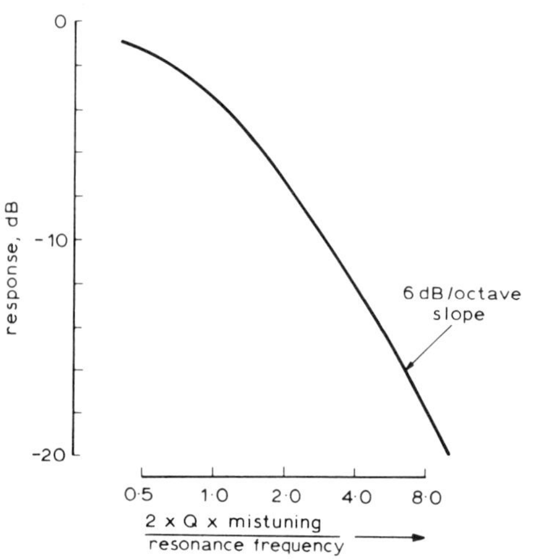

Figure 38.26 Universal curve illustrating the frequency response of a single tuned circuit

$2\,\Delta f$ is the frequency difference between the two 3-dB down points and is hence the passband. Thus we have

$$Q = \frac{\text{centre frequency}}{\text{passband}} \qquad (38.15)$$

This is a useful relationship which enables Q values to be assessed for each application. For example the standard centre frequency for the i.f. amplifier of a 625-line television receiver is 36.5 MHz and the bandwidth must be 6 MHz to include vision and sound signals. The required Q value in an LC circuit for such an amplifier is thus given by

$$Q = \frac{\text{centre frequency}}{\text{passband}} = \frac{36.5}{6} \approx 6$$

It would be difficult to construct an inductor to have precisely this value of Q and the usual solution to this difficulty is to use coils of higher Q value (say 100) and to damp them by parallel resistance to give the required Q value. The input resistance of transistors is usually sufficiently low at this value of centre frequency that it can conveniently be used to give the required damping. The precise damping effect can be controlled by adjusting the L to C ratio of the tuned circuit, keeping the LC product constant to maintain the value of the centre frequency.

As an example a typical value for the input resistance of a bipolar transistor is 2 kΩ. If the transistor base and emitter terminals are effectively in parallel with the LC circuit then the input resistance dictates the dynamic resistance R_d of the circuit. R_d is equal to $L\omega Q$ and thus we can calculate the required value of inductance from the relationship

$$L = \frac{R_d}{\omega Q} \qquad (38.16)$$

Substituting for R_d, ω and Q we have that the inductance for the vision i.f. stage is given by

$$L = \frac{2 \times 10^3}{6.284 \times 36.5 \times 10^6 \times 6}\ \text{H} = 1.4\ \mu\text{H}$$

C can be calculated from the relationship

$$\omega = \frac{1}{\sqrt{(LC)}} \qquad (38.17)$$

Substituting for ω and L we have

$C \approx 14$ pF

In practice it is sometimes more convenient to use values of L and C which give a higher dynamic resistance and to reduce this to the required value by suitable choice of tapping point on the inductor for the base connection of the transistor.

Figure 38.27 gives the circuit diagram of a typical stage of i.f. amplification using a pnp bipolar transistor. The single resonant

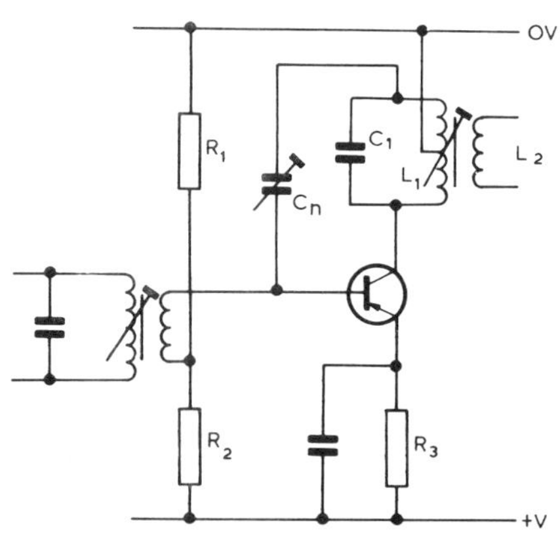

Figure 38.27 Stage of i.f. amplification using a neutralised pnp transistor

circuit L_1–C_1 is connected in the collector circuit of the transistor and the output to the next stage is taken from a secondary winding L_2 tightly coupled to the primary winding. D.C. stability is ensured by the potential divider R_1–R_2 and the emitter resistor R_3.

In a receiver the next stage is likely to be a further common-emitter amplifying stage or a diode detector, both with an input resistance of the order of 2 kΩ. For maximum gain from the stage under examination the turns ratio of the r.f. transformer L_1–L_2 should be chosen to match the output resistance r_o of the transistor to the input resistance r_i of the following stage. For unity coupling between the windings the turns ratio n:1 is given by the usual formula

$$n=\sqrt{\frac{r_o}{r_i}} \tag{38.18}$$

With such a ratio the transistor is presented with a collector load resistance equal to its own output resistance. To calculate the gain available we can regard the transistor as a current generator I $(=g_m V_{in})$ feeding into its own output resistance r_o in parallel with the external load resistance (also equal to r_o). Thus the current entering the external load is equal to $\frac{1}{2}g_m V_{in}$ and the signal at the collector is equal to $\frac{1}{2}g_m V_{in} r_o$. Because of the step-down ratio of the r.f. transformer the signal at the secondary winding is given by

$$V_{out}=\frac{\frac{1}{2}g_m V_{in} r_o}{n} \tag{38.19}$$

But $n=\sqrt{(r_o/r_i)}$

Therefore

$$V_{out}=\frac{\frac{1}{2}g_m V_{in} r_o}{\sqrt{(r_o/r_i)}} \tag{38.20}$$

giving

$$\frac{V_{out}}{V_{in}}=\tfrac{1}{2}g_m\sqrt{(r_o r_i)} \tag{38.21}$$

This can give surprisingly large gains. Typical values for a silicon planar transistor for g_m, r_i and r_o are 40 mA/V. 2 kΩ and 200 kΩ. Substitution in Equation (38.21) shows the gain as 400. The internal collector-base capacitance of the transistor (typically 1 pF) would certainly be sufficient to cause r.f. instability and oscillation unless the stage is neutralised. One way of achieving neutralisation is illustrated in *Figure 38.27*. A disadvantage of neutralisation, particularly for mass-produced i.f. amplifiers, is that the neutralising capacitors ideally require individual adjustment for optimum results. Moreover the alignment of a succession of neutralised stages is not easy.

The need for neutralisation is however not the only disadvantage of this circuit. A possibly more serious disadvantage is that the signal-handling capacity of a stage with so high a collector load resistance is very limited. If the supply voltage is 12 V then a collector current swing of less than 0.1 mA will cause the collector voltage to swing the full extent of the supply. Thus the output power available from the stage is very low, certainly insufficient to drive a diode detector adequately and probably insufficient for a following common-emitter stage.

To obtain good power output from the amplifier it is better to present it with a load resistance which makes good use of the current and voltage swings available. If the supply is 12 V and the mean collector current 1 mA it is best to regard the optimum load as the quotient of these: 12 kΩ or perhaps 10 kΩ to allow for an emitter voltage of 2 V. A repeat of the above calculation for the amended value of collector load resistance will show that the gain is reduced to 90, still high enough to justify neutralisation.

The following are some of the techniques which may be employed to avoid the need for neutralisation:

(*a*) Use a common-base stage instead of common emitter. This helps because the internal feedback capacitance now responsible for any possible instability is that between collector and emitter and this is much smaller than the collector-base capacitance. This technique is extensively used in the r.f. stages of v.h.f. and u.h.f. receivers.

(*b*) Use a cascode stage, i.e. a common-emitter stage feeding directly into a common-base stage. A circuit diagram of a cascode stage using two f.e.t.s. is given in *Figure 38.24*. This technique gives the same gain as the common-emitter stage but with stability.

(*c*) Limit the gain of the common-emitter stage to a value at which the collector-to-base capacitance cannot cause instability. It may well be that this expedient increases the number of stages necessary to give the required overall i.f. gain but there will be no neutralising capacitors to be adjusted and the alignment of the i.f. tuned circuits will be straightforward.

If a transistor has tuned circuits connected to base and collector terminals (both tuned to the same frequency) instability will result if $\omega c_{bc} g_m R_b R_c$ exceeds 2 where R_b and R_c are the resistances effectively in parallel with the input and output circuits.

As an example suppose, as in the previous example, $c_{bc}=1$ pF, $g_m=40$ mA/V, $R_b=2$ kΩ and $R_c=200$ kΩ. At the standard a.m. sound intermediate frequency of 465 kHz we have

$$\omega c_{bc} g_m R_b R_c = 6.284\times 465\times 10^3\times 1\times 10^{-12}\times 40\times 10^{-3}\times 2 \times 10^5\times 2\times 10^3 = 46.7$$

which is greater than 2 and hence instability, even at the low frequency of 465 kHz is inevitable unless neutralising is used.

To avoid instability, without neutralisation, the design technique used is to equate expression 17 to say 0.5 giving a protection factor of 4:1 against instability. Suppose we wish the stage to operate at 10.7 MHz the standard f.m. sound intermediate frequency. We have

$$R_b R_c=\frac{0.5}{\omega c_{bc} g_m}=\frac{0.5}{6.284\times 10.7\times 10^6\times 1\times 10^{-12}\times 40\times 10^{-3}}$$
$$=1.86\times 10^5$$

The product of R_b and R_c should now be chosen so that their product does not exceed this value. R_c should preferably be equal to the optimum load for the transistor, i.e. (collector voltage swing)/(collector current swing). Suppose, in a mains-driven receiver where current economy is not a serious consideration, the transistor has a mean collector current of 4 mA and that the collector voltage swing can be 12 V. R_c is then 3 kΩ and the maximum value of R_b is given by

$$R_b=\frac{1.86\times 10^5}{R_c}=\frac{1.86\times 10^5}{3\times 10^3}=62\ \Omega$$

The input circuit of the transistor should be designed so that the resistance between base and emitter at 10.7 MHz does not exceed 62 Ω. The base can be connected to a tapping point on the inductor so that this value of resistance is not exceeded.

38.16.2 Use of coupled tuned circuits

A better approximation to the ideal square-topped frequency response required for the amplification of modulated r.f. signals can be obtained by using *LC* circuits in coupled pairs. The response of two identical coupled circuits depends on the coupling between the coils. If the mutual inductance is M, the coupling coefficient is given by M/L and is usually represented by k. If k is less than $1/Q$ (where Q is the ratio of reactance to resistance for each coil) the response is low and has a single peak at the resonance frequency of the tuned circuits. Maximum output occurs when $k=1/Q$, known as optimum coupling and the response has a flattened peak at the resonance frequency. If k is greater than $1/Q$ the response is still a maximum but now has two peaks at frequencies given by

$$f_1 = \frac{f_c}{\sqrt{(1-k)}} \quad (38.22)$$

$$f_2 = \frac{f_c}{\sqrt{(1+k)}} \quad (38.23)$$

where f_c is the resonance frequency of both LC circuits. The peak separation measures the passband of the coupled coils and is given by

$$f_1 - f_2 = f_c k \quad (38.24)$$

Thus the coefficient of coupling required to give a required passband at a particular centre frequency is given by

$$k = \frac{\text{passband}}{\text{centre frequency}} \quad (38.25)$$

As an example consider the i.f. amplifier of an a.m. broadcast receiver. The carrier-frequency spacing on the medium waveband is 9 kHz and this figure is often taken as the bandwidth required. The centre frequency is standardised at 465 kHz and thus the coefficient of coupling required in an i.f. transformer is, from Equation (38.25), given by

$$k = \frac{9}{465} \approx 0.02$$

Thus if the inductors have a Q value of 50 or less, this value of k corresponds to critical coupling or less and a single-peaked response is obtained. If Q exceeds 50 the response is double-humped and the two peaks become more marked as the Q value is increased.

38.16.3 Automatic gain control

The gain of the early stages of the i.f. amplifiers of sound and television receivers is usually automatically controlled (a) to combat the effects of signal fading and (b) to ensure that all received signals are reproduced at approximately the same amplitude irrespective of their level at the input to the receiver. A.G.C. is achieved by deriving the d.c. bias for these early stages from the detector output or a post-detector point, modulation frequencies being filtered out, commonly by an RC circuit as illustrated elsewhere in this book. There are two ways in which the gain of a bipolar transistor amplifying stage can be controlled by a d.c. bias.

In one method the d.c. bias is polarised so as to reduce collector current; this is, of course, the method which was used with valve amplifiers and is known as reverse control. Positive-going bias is required to control pnp stages and negative bias for npn stages. A circuit using reverse control is illustrated in *Figure 38.28*. This method of applying a.g.c. is not very satisfactory with transistors because increase in bias, as occurs on strong received signals, reduces the signal-handling capacity of the stage when, in fact, an increase in signal-handling capacity is desirable.

Better results can be obtained from the second method of automatic gain control known as forward control. In this method the transistor is forward biased by the a.g.c. signal so that collector current is increased. An essential feature of the circuit, shown in *Figure 38.29*, is the resistor R_c in the collector circuit. As the collector current increases so does the voltage across R_c and the voltage across the transistor falls. With bipolar transistors which are specially designed for this application, the characteristics become more crowded implying reduced gain as the voltage across the transistor decreases.

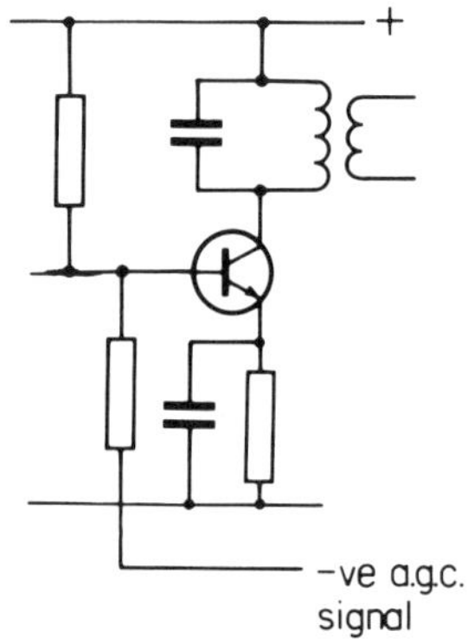

Figure 38.28 A bipolar transistor i.f. stage designed for reverse automatic gain control

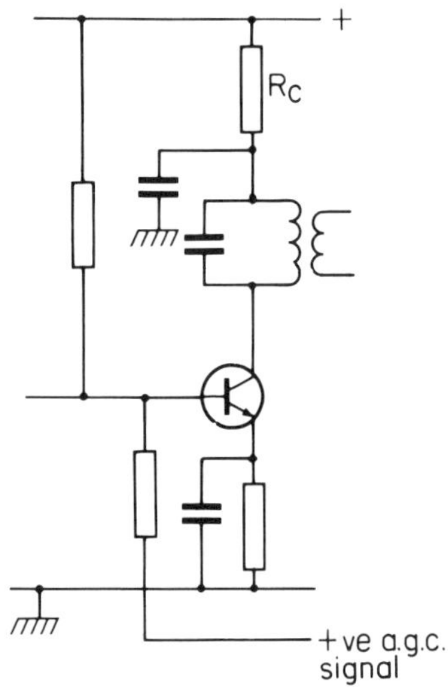

Figure 38.29 A bipolar transistor i.f. stage designed for forward automatic gain control

38.16.4 V.H.F. and u.h.f. tuned amplifiers

At v.h.f. and u.h.f. it is common practice to use common-base and common-gate stages as amplifiers. The low input resistance is no disadvantage and the stability of the amplifier is better than that of the common-emitter and common-source amplifier because the base or gate acts to some extent as a screen between input and output terminals. Often the common-base or common-gate stage forms part of a cascode circuit.

In the circuit of *Figure 38.24* the cascode is formed of two discrete field-effect transistors, the lower acting as common-source and the upper as common-gate amplifier. It is significant that there is no external connection to the link between the lower drain and the upper source terminals. It is thus possible to replace the two transistors by one dual-gate type as shown in *Figure 38.30* and this form of amplifier, using a dual-insulated-gate field-effect transistor is often used as the r.f. stage of a v.h.f. or u.h.f. receiver. The upper gate is connected to a potential divider across the supply and is decoupled by a low-reactance capacitor and this makes the circuit very similar to that of an r.f. tetrode or pentode used as an r.f. amplifier; in fact the dual-gate transistor is sometimes described as a tetrode.

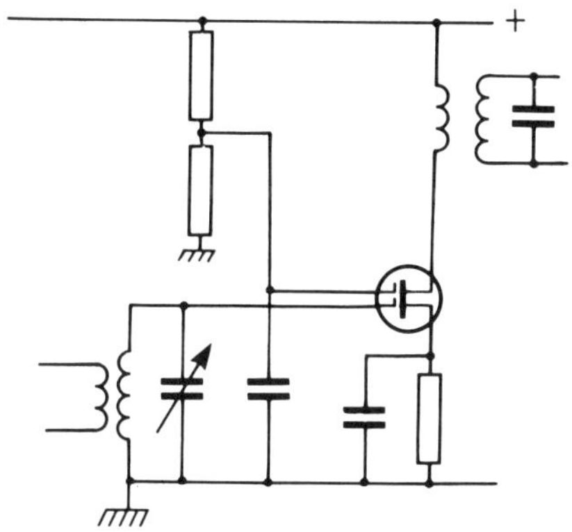

Figure 38.30 A dual-insulated-gate field-effect transistor used as an r.f. amplifier

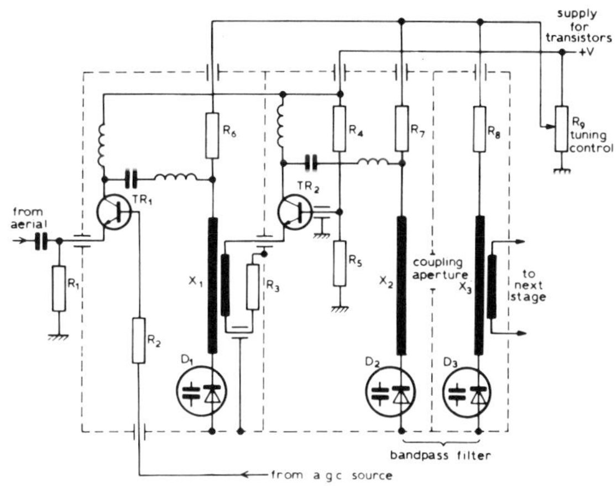

Figure 38.31 A v.h.f. trough-line tuner using varicap diodes

A modern tendency in the r.f. stages of f.m. and television receivers is to use semiconductor diodes for tuning. A reverse-biased pn junction behaves as a capacitance and the value of the capacitance can be controlled by adjustment of the current through the diode. This idea is attractive to receiver designers because it enables all the r.f. circuits (including the oscillator) to be tuned remotely by adjustment of a single variable resistance carrying only d.c. It also simplifies the provision of preset tuning for only one preset potential divider is required for each station to be selected.

A simplified diagram of the r.f. circuits of a u.h.f. television tuner is given in *Figure 38.31*. The tuning inductors used in this circuit are lecher lines each consisting of a central conductor mounted within a conducting rectangular box, an arrangement known as a trough line. One way of obtaining the coupling between inductors, which is necessary to form a bandpass filter, is to provide an aperture in the common trough wall. The capacitance between the lines then provides the required coupling.

A transmission line less than $\lambda/4$ in length and shortcircuited at one end behaves as an inductance at the other end and can be tuned by a variable capacitor (or a varicap diode) connected across the open end. Similarly transmission lines between $\lambda/4$ and $\lambda/2$ in length and open-circuited at one end behave as an inductance and can be tuned by a variable capacitance connected across the other end. Both systems are in use in v.h.f. and u.h.f. tuners but the longer lines are preferred for use with varicap diodes because the current which controls the diode capacitance can be easily introduced via the centre conductor of a trough line as shown in *Figure 38.31*.

The aerial input is applied to the emitter of the common-base stage TR_1 which is controlled from the a.g.c. line. The trough-line inductor X_1 forms the collector load for TR_1 this being tuned by the varicap diode D_1 which is controlled from the potential divider R_9 which is in fact the tuning control. X_1 is coupled to a secondary inductor in close proximity to it and this feeds the amplified signal from TR_1 to the emitter of the second common-base stage TR_2. This is stabilised by the potential divider R_4–R_5 and the emitter resistor R_3. The collector load for TR_2 is the trough line X_2, tuned by varicap diode D_2. X_2 is capacitively coupled to a similar trough line X_3 via the aperture in the common trough-line wall to form a bandpass filter. X_3 is tuned by varicap diode D_3 and all three diodes are controlled from the tuning control R_9. X_3 is coupled to a further inductor close to it and this conveys the output of the two r.f. stages to the next stage, the frequency changer, the oscillator of which is similarly tuned by a varicap diode controlled from R_9.

39 Oscillators and Frequency Changers

S W Amos BSc (Hons), CEng, MIEE
Freelance Technical Editor and Author

Contents

39.1 Oscillators

Sinusoidal signals have wide application in radio and electronics. They are used to test equipment, e.g. a.f. amplifiers, to provide the carrier source in transmitters, in the frequency changers of superhet receivers, to provide bias and wiping in tape recorders.

Oscillators, i.e. generators of sinusoidal waveforms contain three basic sections:

(*a*) a frequency-determining section which is normally an *LC* network but can also be an *RC* network;
(*b*) an amplifying section which is usually a transistor;
(*c*) a limiting section which restricts the amplitude of the generated sine wave to a value at which it can be handled without overloading by the amplifying section. This is essential if a particularly pure waveform is required.

39.1.1 Hartley oscillator

In order to generate oscillations the amplifying section must contain a signal path between the output and the input circuits by virtue of which the amplifier is capable of supplying its own input: in order words the amplifier must have positive feedback. In many oscillators the frequency-determining section provides the positive feedback and if the amplifier is a single transistor which inverts the input signal the correct phase relationship between input and output connections can be obtained by use of a tapping point on the inductor as shown in *Figure 39.1*.

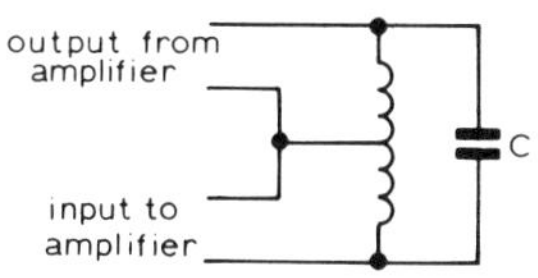

Figure 39.1 The three essential connections to the inductor of an *LC* oscillator

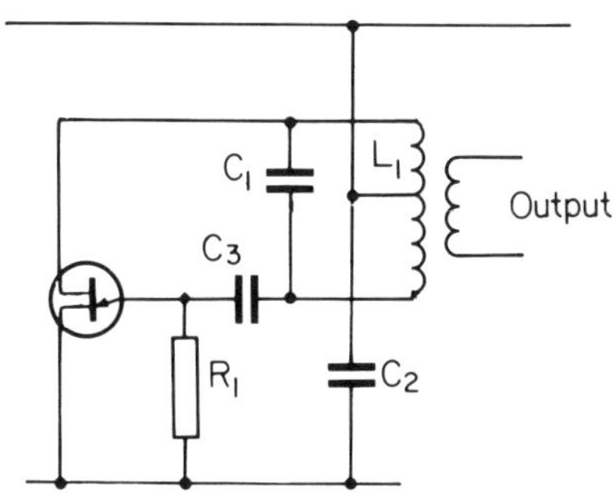

Figure 39.2 A Hartley oscillator circuit using a junction-gate field effect transistor

By making these three connections to the source, gate and drain of a junction-gate field-effect transistor, the oscillator circuit of *Figure 39.2* is obtained. The input to the transistor is applied between gate and source, and the output is generated between drain and source. The source is thus the common connection and goes to the centre tap on the inductor (via the low-reactance capacitor C_2). The drain is connected directly to one end of L_1 and the positive supply for this electrode is introduced via the tapping point. The gate is connected to the other end of L_1 via the low-reactance capacitor C_3 which is necessary to isolate the gate from the positive supply. The gate is connected to the source by R_1.

C_3 and R_1 provide automatic gate bias. At the moment of switching on there is zero gate bias and the transistor takes considerable drain current. Oscillation thus starts immediately and, as a result, the gate is driven alternately positive and negative with respect to the source. However when the gate is driven positive the input junction becomes forward-biased and current flows in the gate circuit. This current charges C_3, the polarity of the charge being such as to bias the gate negatively with respect to source. In the final state of equilibrium considerable amplitude of oscillation can result, the gate being driven positive for only a brief period in each cycle. During the remainder of each cycle, when the gate is negative with respect to the source and there is no gate current, the charge on C_3 leaks away through R_1 but, provided R_1 is large enough, very little of it is lost before the next cycle of oscillation restores the charge again. The burst of gate current momentarily applies a low resistance across the tuned circuit and it is this which limits the amplitude of oscillation generated. Normally the amplitude is so large that the transistor is cut off for a large part of each cycle. In other words the amplifier operates in class C. Thus it can be said that the resonant circuit is kept in oscillation by bursts of drain current at the resonance frequency.

For successful results the time constant R_1C_3 should be long compared with the periodic time of the oscillation and the frequency of oscillation is given by the expression

$$f = \frac{1}{2\pi\sqrt{(L_1C_1)}} \qquad (39.1)$$

This type of oscillator in which the inductor is centre-tapped is known as the Hartley. The sinusoidal oscillation is set up, of course, in L_1C_1 and the output must be taken from this circuit: a convenient method of doing this is by coupling an inductor to L_1 as shown in *Figure 39.2*.

39.1.2 Colpitts oscillator

It is alternatively possible to tap the capacitive branch of an *LC* circuit in order to provide the three connections essential for positive feedback: this is shown in *Figure 39.3* and provides the

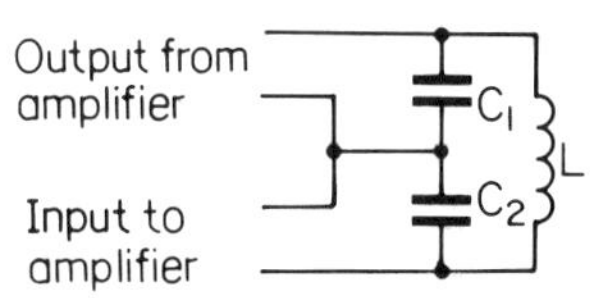

Figure 39.3 The three essential connections to the capacitive branch of an *LC* oscillator

basic feature of the Colpitts oscillator which thus enables oscillation to be set up in an inductor without a tapping point. If we attach these three connections to the emitter, base and collector of a pnp transistor we obtain the practical circuit shown in *Figure 39.4*.

The connection which is common to the input and output of the transistor is the emitter and this is connected directly to the centre point of the two capacitors. One end of the inductor is connected directly to the collector and the other to the base via the capacitor C_3 which isolates the base from the supply. The collector cannot be connected directly to the negative supply because this would result in an effective short-circuit of C_1 and so the resistor R_1 is introduced. This damps the circuit L–C_1–C_2

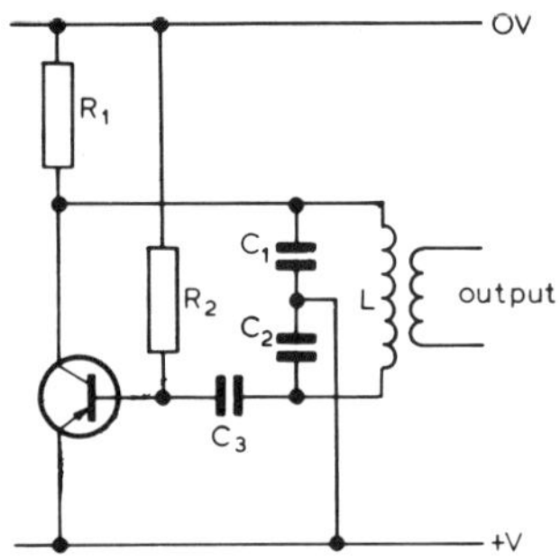

Figure 39.4 A Colpitts oscillator circuit using a bipolar transistor

and thus cannot be too small. The base is biased by R_2 from the negative supply line. R_2–C_3 provide automatic bias, the transistor being driven into conduction during negative half cycles applied to the base and remaining cut off during positive half cycles, again an example of class C operation. Oscillation occurs at the frequency

$$f = \frac{1}{2\pi\sqrt{(LC)}} \qquad (39.2)$$

where

$$C = \frac{C_1 C_2}{C_1 + C_2} \qquad (39.3)$$

The output is again taken from L–C_1–C_2 by a coupling coil.

39.1.3 RC phase-shift oscillator

Generators governed by RC circuits can be designed to produce a wide variety of different waveforms and the production of sawtooth and rectangular waves is described elsewhere in the book. However an RC network can also form the frequency-determining element of a sinusoidal oscillator: two examples of such oscillators are described; the first is the RC phase-shift oscillator.

In the Hartley and Colpitts oscillators the LC circuit introduced a phase inversion between its input and output terminals. This, combined with the phase inversion of the amplifier gives the positive feedback essential to sustain oscillation. For a symmetrical wave such as a sinusoid the effect of phase inversion is the same as that of altering the phase by 180°. If therefore we can find a network which shifts the phase by 180° at a particular frequency and if the gain of the amplifier can compensate for the attenuation introduced by the network at this frequency, then we have the basis for an oscillator. A single RC network can at best give 90° phase shift and its attenuation is then infinite. However a network of three RC sections, each introducing 60° phase shift is a practical possibility: if the resistance values are equal and if the capacitance values are also equal the attenuation introduced by the 3-section network is 29 (i.e. this is the value of i_{in}/i_{out}) which can be made good by a single transistor. The circuit diagram of an oscillator operating on such principles is given in *Figure 39.5*. The frequency-determining network is R_1–C_1–R_2–C_2–R_3–C_3 in which $R_1 = R_2 = R_3 = R$ and $C_1 = C_2 = C_3 = C$. D.C. stabilisation is ensured by the potential divider R_4–R_5 and the emitter resistor R_6. The frequency of oscillation is given by $f = 1/2\pi\sqrt{6}\,RC$.

No means is shown in *Figure 39.5* of limiting the amplitude of oscillation and this is essential to preserve the purity of the waveform generated. It is possible, by adjustment of the value of R_6, to set the emitter current at a value which only just gives

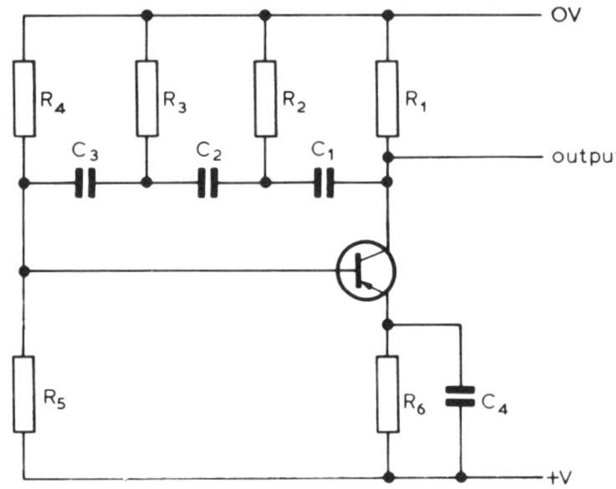

Figure 39.5 *RC* phase-shift oscillator

sufficient current gain to make up for the attenuation in the RC network but this is not a good method and it would be better to use an automatic system in which the transistor is biased back when the output amplitude exceeds a predetermined value.

39.1.4 Wien bridge oscillator

It would be difficult to make a variable-frequency oscillator based on the circuit of *Figure 39.5*. A circuit much better suited to this purpose is the Wien bridge oscillator which is used as the basis of a number of a.f. test oscillators.

The basic Wien bridge network is shown in *Figure 39.6*. It contains two equal resistors and two equal capacitors. At the frequency for which

$$f = \frac{1}{2\pi RC} \qquad (39.4)$$

the network has zero phase shift between input and output, and the voltage attenuation is 3, i.e. $v_{in}/v_{out} = 3$. To use such a network as the frequency-determining element in an oscillator the amplifier must also introduce zero phase shift and have a voltage gain of 3. The amplifier must, moreover, have a very high input resistance (as shown later) for it is essential to minimise any shunting effect on the parallel RC branch. Two stages at least are required in the amplifier to give the required zero phase shift and it is usual to include a third, an emitter follower, to provide a low-resistance output for the frequency-determining network and for the oscillator itself.

By using two sections of a two-gang variable capacitor for the two capacitors in *Figure 39.6* it is possible to vary the frequency of oscillation over a range of say 10:1. The two resistors can be

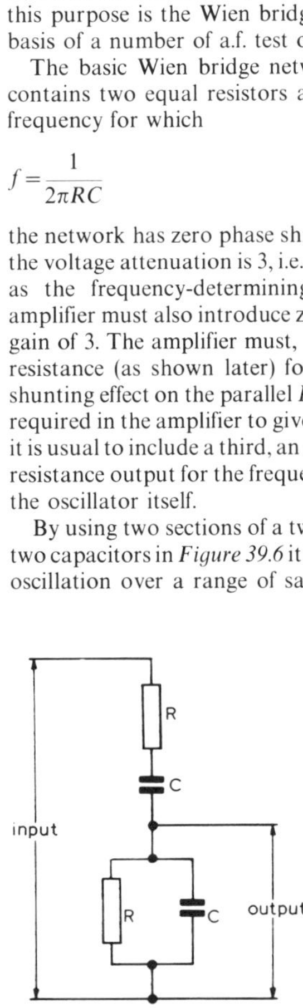

Figure 39.6 Basic Wien bridge network

varied in decade steps to produce a number of frequency ranges. Three ranges could thus cover the frequency band from 30 Hz to 30 kHz which is suitable for an a.f. test oscillator. If the variable capacitors are of 500-pF maximum capacitance then the value of R required for the lowest-frequency range can be obtained from Equation (39.4) by rearranging it thus

$$R=\frac{1}{2\pi fC} \qquad (39.5)$$

Substituting $f=30$ and $C=500\times10^{-12}$ we have

$$R=\frac{1}{6.284\times30\times500\times10^{-12}}\ \Omega\approx10\ \text{M}\Omega$$

To avoid shunting this appreciably (which would increase the frequency and prevent accurate alignment of the two sections of the Wien bridge network) the input resistance of the amplifier must be very high indeed and an f.e.t. is the obvious choice for the first stage. The second and third stages can be bipolar transistors as indicated in the circuit diagram of *Figure 39.7*.

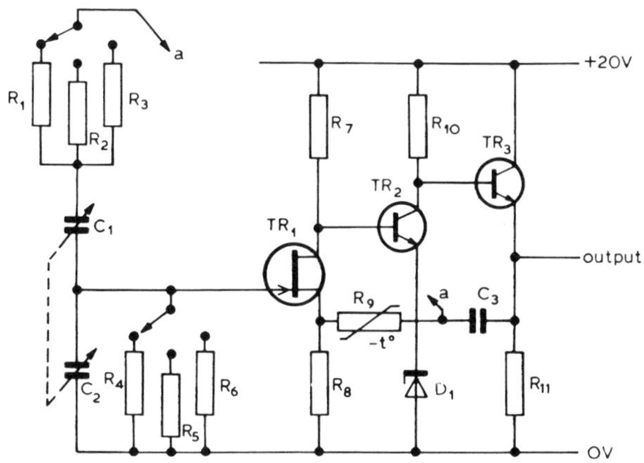

Figure 39.7 A Wien bridge oscillator with three frequency ranges

Negative feedback is applied between the emitter of TR_3 (in effect the collector of TR_2 since TR_3 is an emitter follower) and the source of TR_1 via the non-linear resistor R_9 which has a negative temperature coefficient. This has the effect of maintaining the output of the oscillator substantially constant in spite of frequency or range changes. Immediately after switching on the resistance of R_9 is high because there is no signal in it and hence no heat. The gain of the amplifier is determined by the feedback components and is given by R_9/R_8 and, at the instant of switch on, this is high, much higher than the value of 3 necessary to ensure oscillation. Oscillation therefore begins and builds up rapidly. As soon as the oscillation reaches R_9 via C_3 the resistance of R_9 falls due to the heating effect of the signal in it. The fall continues and the resistance of R_9 settles around a value approximately equal to twice R_8. The type of resistor chosen for R_9 should be such that its resistance will equal twice R_8 with the required value of oscillation amplitude across it.

The amplifier is direct-coupled throughout and an interesting design point is the inclusion of the zener diode D_1 in the emitter circuit of TR_2. The drain voltage of TR_1 is probably around 10 V, i.e. half the supply voltage which means that TR_2 emitter voltage must be approximately the same. A resistor could be included in TR_2 emitter circuit to permit such an emitter voltage to be realised but this would give negative feedback and reduce the gain of TR_2 to a very low value. This feedback could be minimised by decoupling the emitter resistor by a low-reactance capacitor but it is desirable at some point in a multistage d.c. amplifier to stabilise the voltage. By including a zener diode of a suitable voltage rating TR_2 emitter voltage can be stabilised so eliminating feedback and obtaining maximum gain from TR_2. It is true that the overall gain of the amplifier is reduced to 3 by the negative feedback due to R_8 and R_9 but the constancy of output amplitude improves as the gain of the individual transistors is increased.

39.2 Frequency changers

In many examples of electronic equipment it is necessary to combine two signals of frequency f_1 and f_2 to produce new signals with frequencies of (f_1+f_2) and (f_1-f_2). Perhaps the most obvious example occurs in superhet receivers where the received r.f. signal is combined with the output of the local oscillator in the frequency-changer stage to produce the difference-frequency signal which is amplified in the i.f. amplifier.

39.2.1 Additive mixers

If the two signals f_1 and f_2 are connected in series or in parallel and applied to a linear amplifier, both are amplified without distortion and the output of the amplifier contains only amplified versions of f_1 and f_2: there are no components at the sum and difference frequencies. Such an arrangement is therefore no use as a generator of sum and difference frequencies: for this it is essential that the amplifier should have a non-linear characteristic and that this should be used in the mixing process. In stages of this type therefore the normal design technique is to bias the amplifier to a non-linear part of the input–output characteristic and then to ensure that one of the signals, normally that from the local oscillator, has sufficient amplitude to sweep over this non-linear region. The amplitude of the second signal, the r.f. input from the aerial or r.f. stage, is then non-critical. A frequency changer can therefore consist of an oscillator stage and a non-linear mixing stage. These stages may be separate transistors or they can be combined: it is possible to use an oscillator as a frequency changer by arranging to inject an r.f. signal into it and to abstract an i.f. signal from it.

39.2.2 Self-oscillating mixers

As mentioned earlier many oscillators operate in class C. The characteristic around the point of collector-current cut off is therefore used in the oscillation process and the non-linearity essential for successful mixing is present. Provided care is taken in the design of the circuit so that r.f. can be put in and i.f. can be taken out without significant detriment to the oscillating process a self-oscillating mixer can be a good frequency changer and have a conversion efficiency little short of that of a separate mixer and oscillator.

Figure 39.8 gives the circuit diagram of a self-oscillating mixer used in an f.m. receiver. The npn transistor is d.c. stabilised by the potential divider R_1–R_2 and the emitter resistor R_3. It operates as a Colpitts oscillator (*c.f. Figure 39.4*) the two fundamental capacitors being C_1 and C_2 while C_3, in parallel with them, provides oscillator tuning. C_3 is ganged with other variable capacitors tuning r.f. circuits and it is desirable therefore for its moving vanes to be earthed. This necessarily earths the base of the transistor and clearly the emitter cannot also be earthed. The inductor L_2 is therefore introduced: it has high reactance at the oscillator frequency and thus permits the emitter potential to fluctuate at r.f. and also provides a convenient point at which the r.f. signal can be introduced via C_4. C_5 and C_6 are d.c. blocking

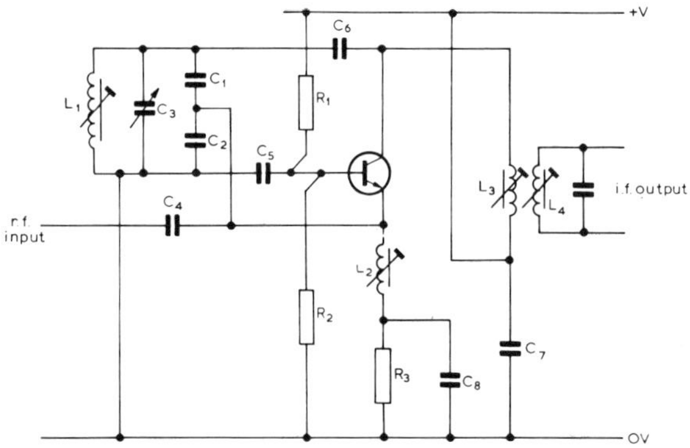

Figure 39.8 Self-oscillating mixer circuit suitable for use in an f.m. receiver

capacitors permitting L_1 and C_3 to be earthed while allowing working d.c. voltages to be applied to the base and the collector. L_3 and L_4 constitute an i.f. transformer coupling the frequency changer to the first i.f. stage. C_7 is a decoupling capacitor and C_6 acts as the tuning capacitor for L_3, L_1 having negligible reactance at the intermediate frequency. To avoid negative feed-back at i.f. due to L_2 in the emitter circuit L_2 is made variable and is adjusted to resonate with C_8 at the intermediate frequency.

39.2.3 Multiplicative mixers

An alternative approach in designing a mixer stage is to adopt the multiplicative principle. If a device can be found which multiplies two inputs together instead of adding them, then the difference frequency output is obtained directly and without any need for non-linearity in the device. This can be seen directly from the trigonometrical identity

$$2\sin\omega_1 t\,.\,\sin\omega_2 t=\cos(\omega_1-\omega_2)t+\cos(\omega_1+\omega_2)t \qquad (39.6)$$

Multiplication of two signals requires an active device with two input terminals, such that the signal applied to one terminal controls the gain of the signal applied to the other. One such device is a dual-insulated-gate field-effect transistor. The alternating output current is given approximately by $g_m V_{g2}$ where g_m is the mutual conductance and V_{g2} is the alternating voltage applied to gate 2. The mutual conductance is controlled by the voltage on gate 1. Thus if signals are applied to the two gates, the output current contains a strong component at the product of the two signals.

40 Modulation Theory and Systems

K R Sturley BSc, PhD, FIEE, FIEEE
Telecommunications Consultant

Contents

40.1 Introduction

The transmission of information, whether speech, music, vision or data, over long distances requires the use of a carrier channel at least equal to that of the frequency spectrum of the information components. The carrier frequency must have one of its characteristics varied (modulated) by the information and the receiver must contain an information extractor (detector) designed to react to the carrier characteristic that is varied and to produce an output, which is as close a copy of the original information as possible.

If the carrier frequency is continuous the information may be employed to modulate either the amplitude (a.m.) or the time characteristic, frequency (f.m.) or phase (p.m.). The information components generally have two properties, amplitude or intensity, and frequency. *Table 40.1* shows how each is conveyed.

Table 40.1

Information property	*Amplitude modulation*	*Time modulation*	
		Frequency	*Phase*
Amplitude	change of carrier amplitude	change of carrier frequency	change of carrier phase
Frequency	rate of change of carrier amplitude	rate of change of carrier frequency	rate of change of carrier phase

The carrier may consist of a pulsed carrier, whose amplitudes (p.a.m.), positions (p.m.m.) or duration (p.d.m.) are varied by the information as shown in *Figure 40.1*. The pulsed form of carrier operation allows a number of information channels to be transmitted on the same carrier frequency by regularly allocating a pulse to a given channel. For example 8 channels can be accommodated using pulses 1, 10, 19, etc. for channel 1, pulses 2, 11, 20, etc. for channel 2. Pulses 9, 18, 27, etc. are required for synchronising a distribution gate at the receiver. The gate separates the information channels and directs them to their required destinations. Such a system is known as time division multiples (t.d.m.).

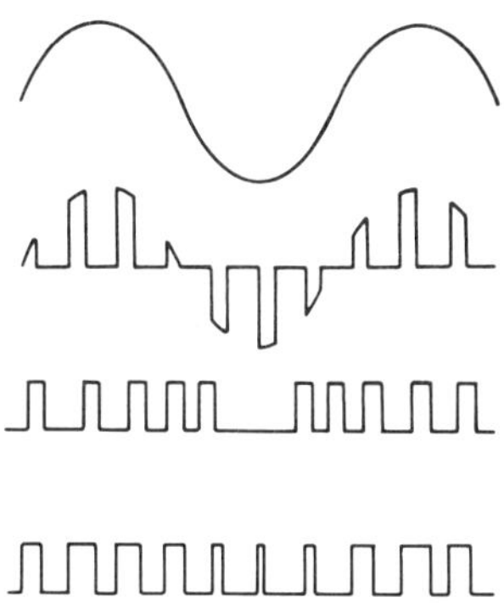

Figure 40.1 Examples of the three types of pulse modulation

Frequency division multiplex (f.d.m.) is used with continuous carrier in telephony, each speech channel being shifted in frequency before being added to the others occupying different frequency bands. The combined signals modulate the carrier; at the receiver the channels are filtered from each other and their frequency spectra returned to normal.

All the above systems are known as analogue modulation because the information controls the carrier directly. Considerable message-to-noise ratio advantage is gained by converting the amplitude characteristic of the information into a digital form before modulation. This is known as pulse code modulation (p.c.m.). The digital code has to be converted back to its analogue form when it is desired to interpret the information. Time division multiplex is used with p.c.m. when a number of different information channels have to be accommodated.

40.2 Continuous carrier modulation

40.2.1 Amplitude modulation

The carrier amplitude may be represented by $E_c \cos \omega t$ and the modulation by $E_m \cos pt$, so that the a.m. carrier expression is

$$E_c(1 + M\cos pt)\cos \omega t \tag{40.1}$$

where M, the modulation ratio, $\leqslant 1 \propto E_m$.

This is analysed into three components:

$$E_c \cos \omega t + \tfrac{1}{2}E_c M \cos(\omega \pm p)t \tag{40.2}$$

i.e. the unmodulated carrier and a pair of sidebands. Information normally consists of several frequencies, and each frequency has its own M proportional to its amplitude (the sum of all M values must not exceed unity) and produces a sideband pair.

Double sideband (d.s.b.) transmission is the simplest form of a.m. requiring only a very simple unidirectional detector at the receiver and it is therefore used for broadcasting. It requires a transmission channel bandwidth twice the highest frequency of the information components.

To obtain d.s.b. amplitude modulation the information voltage is used to control the gain of a carrier amplifier, and an example with a transistor amplifier is shown in *Figure 40.2*. The information could have been inserted in series with the carrier in the base, and the variation of gain is due to the non-linear characteristic of the transistor. A similar circuit can be used for an electronic tube amplifier. In a triode tube modulation can be achieved by inserting the modulating signal in the grid, cathode or anode circuit. The advantage of grid or cathode modulation is that a relatively low voltage or power is required from the modulator, but modulation envelope distortion occurs at high values of M. Anode modulation requires relatively large power from the modulator but distortion is low up to modulation percentages of 90% ($M = 0.9$). Thus broadcasting transmitters using triodes in the modulated amplifier employ anode

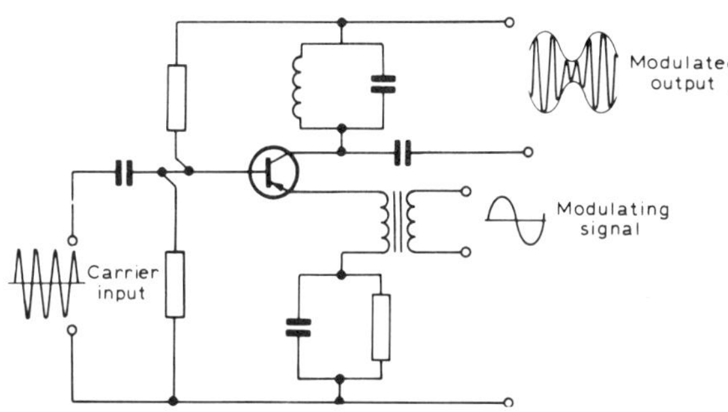

Figure 40.2 A modulated amplifier circuit

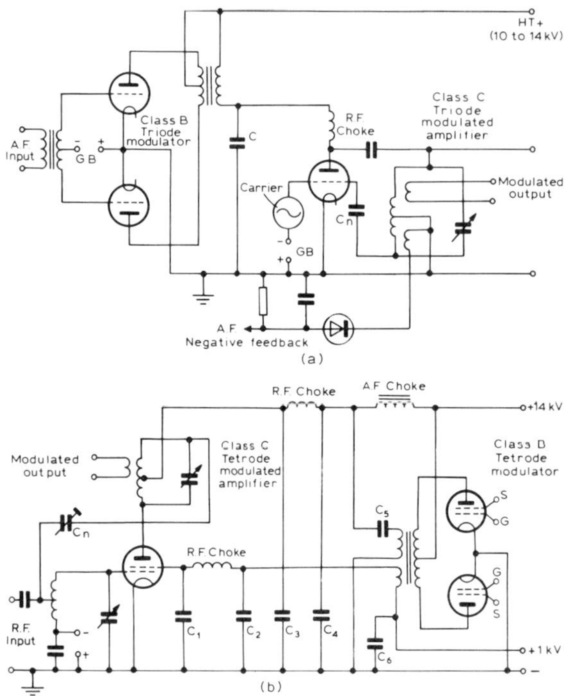

Figure 40.3 (a) A high-power modulator stage using triodes. (b) A high-power modulator stage using tetrodes

modulation. A typical high-power modulator stage is shown in *Figure 40.3(a)*; speech or music is amplified in a high power class B push–pull stage, which must be capable of supplying half the carrier power. Its transformer coupled a.f. varies the anode voltage of the carrier amplifier tube from almost zero to twice the d.c. supply voltage (about 14 kV). The transformer secondary may be a.f. choke-coupled to the r.f. stage so as to divert the d.c. current of the latter from the a.f. transformer. The carrier amplifier tube in the modulated amplifier stage operates at high efficiency in class C, and envelope distortion is low up to $M=0.90$; r.f. currents are bypassed from the a.f. side by the capacitor C. Envelope distortion is often reduced still further by rectifying the modulated output and applying the resulting a.f. wave as negative feedback to an earlier a.f. stage.

A single class A a.f. amplifier may be used in series with the r.f. carrier tube anode but it suffers from low d.c. efficiency and complications due to the high d.c. voltage.

Since the modulated amplifier r.f. stage in *Figure 40.3(a)* uses a triode, there will be r.f. feedback from the anode into the grid, and self oscillation is likely to occur. The stage must be neutralised by feeding into the grid circuit an inverse voltage from the end of the anode tuning coil opposite from the anode via the neutralising capacitance C_n. If the tuning coil is centre tapped $C_n \triangleq C_{ag}$, the interelectrode capacitance from anode to grid. If the r.f. output stage is in push–pull, neutralisation is secured by neutralising capacitors from the anode of one of the push–pull pair to the grid of the other.

The problem of neutralisation is much reduced if the grid of the modulated amplifier is earthed and the r.f. drive inserted in the cathode. The power output of the r.f. driver into the modulated amplifier stage adds to the total power output, and if this is unmodulated it reduces the output modulation percentage below that of the modulated amplifier. Thus the driver stage itself must be modulated.

Tetrode and pentode tubes can be used as modulated amplifiers with the modulating signal applied to the screen-grid or for the pentode to the suppressor-grid. Suppressor-grid modulation suffers from the same disadvantage as grid modulation, viz., modulation envelope distortion with high modulation.

In modern practice tetrodes using combined screen and anode modulation are tending to replace triodes in modulated amplifier stages, because they have higher r.f. gain, often higher a.c./d.c. efficiency and, due to the r.f. earthing of the screen, reduced neutralisation problems. Their higher r.f. gain reduces the number of r.f. stages required between carrier master oscillator and modulated amplifier. Screen modulation alone is unsatisfactory because of envelope distortion but combined screen and anode modulation gives results comparable to that of anode modulation of a triode. The modulation transformer requires two secondary taps having a.f. voltage ratios approximately equal to the ratio of d.c. screen and anode voltages on the tetrode. A typical circuit is given in *Figure 40.3(b)*, where C_1–C_4 are r.f. capacitors and C_5–C_6 are a.f. capacitors.

When the modulated amplifier supplies its output direct to the aerial, the system is called high power modulation; if modulation is carried out before the final r.f. power stage, and envelope distortion is to be avoided, the modulated signal must thereafter be amplified in class B amplifier stages, which have an appreciably lower a.c./d.c. efficiency than the high power class C modulated amplifier.

There have been three successful attempts, known as Chireix, Ampliphase and Doherty, to use low-power modulation and approach the efficiency of the high-power modulator.

The Chireix method employs two r.f. class C amplifying chains each carrying oppositely phase-modulated carriers of almost constant amplitude. The phase modulation is obtained by adding sidebands in phase quadrature to the unmodulated carrier as shown by the vector diagram of *Figure 40.4(a)*. A balanced

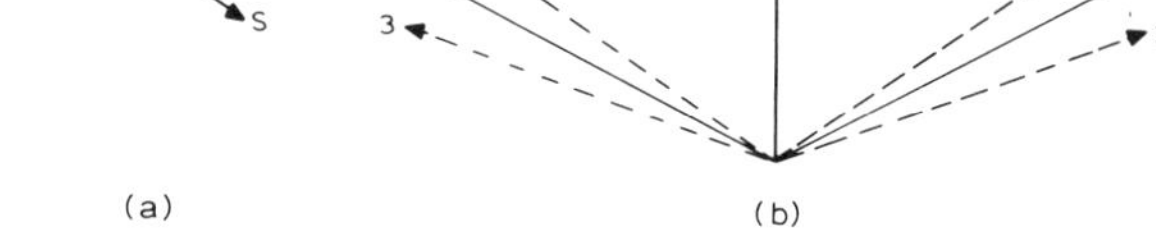

Figure 40.4 The Chiriex (high efficiency) method using low power phase modulation

modulator permits the sidebands to be developed separately from the carrier. The two oppositely phase-modulated carriers after amplification are added at the output to generate an a.m. wave as shown in *Figure 40.4(b)*. The unmodulated carriers must be given a phase shift $<180°$ in order that the required unmodulated carrier amplitude (E_c) appears at the output; this is achieved by a phase shifting network in one of the amplifying chains.

The Ampliphase system operates on the Chireix principle of combining at the output of two amplifying chains phase-modulated signals in such a way as to produce a.m. One system uses $\lambda/4$ transformers to couple the power from the anodes of the final power output stages into the common load, which is the aerial system. The two power valves operate in class C to develop constant r.f. voltages, and these are converted by the $\lambda/4$ transformer into constant currents in the load. Phase variations of the currents cause an a.m. output in the load as shown in *Figure 40.4(b)*.

The Doherty high efficiency output stage uses two valves, whose grid circuits are driven by modulated inputs in phase quadrature and whose anodes are coupled by a $\lambda/4$ network of characteristic impedance R. Both supply power into a load of $R/2$ as shown in *Figure 40.5*. The $\lambda/4$ network introduces a phase

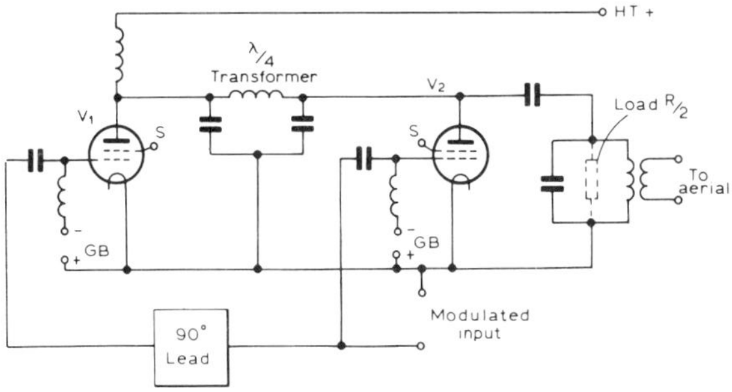

Figure 40.5 The Doherty high efficiency amplifier

delay of 90° so that the modulated input to V_1 must be given a compensating 90° phase lead. When the carrier is equal to or less than its unmodulated value V_1 only is operative, V_2 being cut off. During positive modulation cycles, anode voltage limitation of V_1 occurs and this is transferred by the $\lambda/4$ network as a constant current into the load; V_2 begins to supply power and this increases the effective terminating impedance $R/2$. The $\lambda/4$ network, whose characteristic impedance is R, transfers this as a decreasing load seen by V_1 at the input to the network, and this changes from $2R$ when V_2 is cut off, to R for 100% positive peak of modulation when V_2 is fully operative and supplying half the total power. Due to the change from $2R$ to R and the constant output voltage the power delivered from V_1 at 100% positive peak modulation is double the unmodulated carrier power and V_2 provides an equal power into the load, effectively increased to R by the action of V_1.

Television transmitters generally employ grid modulation since it requires much less power than anode modulation—an important consideration with wide band signals—and non-linear distortion of video signals is less serious than with speech or music. The synchronising pulse part of the video signal can be operated over the most non-linear part of the transmitter characteristic and can be predistorted to compensate for the non-linearity.

40.2.2 Vestigial sideband transmission

Channel bandwidth should be as small as possible to ensure economic use of the transmission frequency spectrum, and vestigial sideband transmission can represent an appreciable saving of spectrum with television signals having information components up to about 5.5 MHz. Only a part of one of the sidebands, that within about 1 MHz of the carrier is transmitted; if a simple detector circuit is employed distortion of the modulation content of all information frequencies exceeding about 1 MHz results. Fortunately their amplitudes are small and the degree of distortion is quite acceptable on vision signals. The receiver has to be detuned to half carrier amplitude so as to reduce the low-frequency sideband energy to the original level in relation to the high-frequency sidebands. Vestigial sideband operation is unsuitable for speech or music because of the distortion produced by the simple unidirectional detector used in broadcasting receivers.

40.2.3 Single sideband carrier system

One sideband of a d.s.b. amplitude modulated audio transmission can be removed if a more complicated product detector, having a square law characteristic, is employed in the receiver. The unidirectional peak detector can function with relatively low distortion if the modulation sideband power is low. The main advantage is the halving of the transmission bandwidth though if power from the suppressed sideband is transferred to the remaining one, signal-to-noise ratio is increased by 3 dB because halving the bandwidth halves the noise power.

40.2.4 Single sideband with pilot carrier

In a double sideband transmission the ratio of sideband to carrier power is $M^2/2$, which is 50% at 100% modulation. The same is true of s.s.b. when the power from the suppressed sideband is transferred to the other. The carrier power therefore accounts for a very large proportion of the total transmitted power and a worthwhile saving in running costs in a point-to-point communication system is realised by reducing the carrier to a low value. The carrier cannot be entirely eliminated because detection cannot be achieved in its absence. The carrier is restored at the receiver by having an oscillator locked in frequency and phase by the transmitted residual pilot carrier. If almost all the carrier power is transferred to the sideband the transmitted information content can be increased 3 times relative to the 100% modulation sideband power so that s.s.b. with pilot carrier gives a signal-to-noise ratio approaching 8 dB better than d.s.b. with full carrier (5 dB increase in signal and 3 dB decrease in noise due to halving bandwidth).

A balanced modulator permits the sidebands to be generated and the carrier to be eliminated; it may be a push–pull type r.f. modulator stage with carrier frequency in the same phase to the input base circuits and the information applied in opposite phase as in *Figure 40.6*. If the characteristics of each half stage are

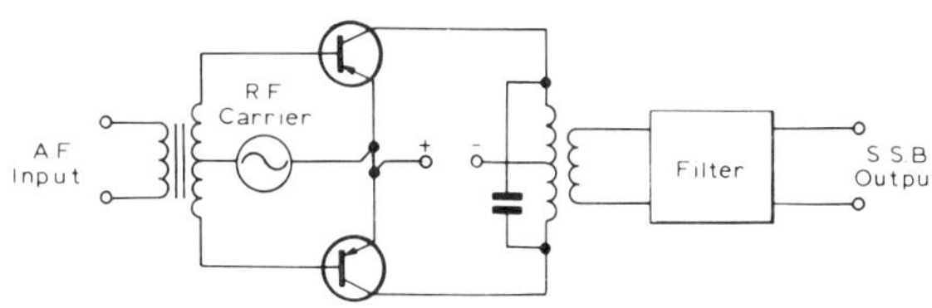

Figure 40.6 A push–pull balanced modulator for carrier elimination

identical the carrier component is cancelled at the output but the two sidebands are unaffected. The wanted sideband is selected by passing through a filter rejecting the unwanted.

The second sideband channel may be used to carry separate information on the same pilot carrier by means of another balanced modulator and filter; the two independent sidebands and the correct proportion of carrier are combined before power amplification to the information amplitude.

40.2.5 Phase modulation

In phase modulation the phase angle of the carrier is advanced and retarded in proportion to the information amplitude. The mathematical representation is

$$E_c \cos(\omega_c t + \varphi \cos pt) \tag{40.3}$$

where $\varphi \cos pt \propto E_m \cos pt$.

Like a.m. it can be resolved into carrier and sideband components but there is more than one pair of sidebands per information frequency component, and the general expression is

$$E_c\left[J_0(\varphi)\cos\omega_c t + \sum_{n=1}^{\infty} J_n(\varphi)\cos\left[\omega_c + n\left(\frac{\pi}{2} \pm p\right)\right]t\right] \tag{40.4}$$

where J_o, etc. are Bessel Functions.

Theoretically there are an infinite number of sidebands spaced $\pm f_m$, $\pm 2f_m$, etc. from the carrier, but in practice the amplitudes of the higher order sidebands fall off rapidly if φ is not large. Thus when $\varphi = 45°$ (0.787 rad), only the first two pairs ($\pm f_m$, $\pm 2f_m$) have significant amplitudes.

Phase modulation is produced by variation of the inductive or capacitive element of a buffer amplifier stage following a master oscillator. The variable element may be a varactor diode biased by the information voltage. Phase change must not exceed about $\pm 25°$ if linear modulation is to be achieved, but it can be increased by applying it as an input to a multiplier stage; thus a tripler stage multiplies the phase change by 3. A frequency changer can restore the carrier to its original frequency leaving the multiplied phase untouched.

40.2.6 Frequency modulation

Frequency modulation requires the carrier frequency to be varied in accordance with the information so that carrier instantaneous frequency is

$$f = f_c + f_d \cos pt$$

where $f_d \propto E_m$ and is the frequency derivation of the carrier. Since

$$\varphi = 2\pi \int f.dt$$

$$= \omega_c t + \frac{f_d}{f_m}\sin pt$$

and the expression for a f.m. carrier is

$$E_c \cos\left(\omega_c t + \frac{f_d}{f_m}\sin pt\right) \tag{40.5}$$

The carrier and sideband representation is

$$E_c\left(J_o\frac{f_d}{f_m}\cos\omega_c t + \sum_{n=1}^{\infty} J_n\left(\frac{f_d}{f_m}\right)[\cos(\omega_c + (np)t + (-1)^n\cos(\omega_c - np)t]\right) \tag{40.6}$$

The ratio f_d/f_m is an angle inversely proportional to f_m, so that for a given information amplitude f_d/f_m is much larger when the information frequency component is low. It is similar to phase modulation with low-frequency boost. Thus if $f_d = 10$ kHz, $f_d/f_m = 0.785$ rad., when $f_m = 12.75$ kHz and only two sidebands are significant, whereas when $f_m = 500$ Hz, $f_d/f_m = 20$ rad. and about 23 pairs of sidebands have significant amplitudes. This gives frequency modulation an advantage over noise because the latter phase-modulates a carrier whose amplitude is maintained constant by a limiter in the receiver. Hence the low frequency components of thermal noise have much less effect than the high frequency components at the output of a f.m. receiver. The normally flat frequency spectrum of thermal noise is 'triangulated' with minimum noise at low frequencies.

The carrier and sideband component amplitudes vary from maximum to minimum through zero as f_d/f_m varies, but the root mean square value of the amplitudes is always constant and equal to the unmodulated carrier amplitude.

A feature of f.m. is that an increase in f_d increases the information modulation content of the signal in direct proportion, so that an increase of f_d from 15 kHz (the maximum broadcast audio frequency) to 75 kHz increases information amplitude content by 5 and is equivalent to 14.25 dB increase in signal strength. Thus unlike a.m. information content is determined by the product of bandwidth and power, and there are advantages in increasing transmission bandwidth beyond $2f_m$. Channel spacing of v.h.f. transmission does not permit $f_d > 75$ kHz and the bandwidth required is found to be 180 kHz or $2\,[f_d(\max) + f_m(\max)]$.

Frequency modulation is obtained by varying the inductive or capacitve element of the tuned circuit of a master oscillator. The average carrier frequency must be maintained constant otherwise adjacent transmission channels will suffer interference. One way of reducing this problem is to use balanced reactance modulators, which have adding reactance changes but whose resistance components change in opposite directions so maintaining a constant total resistance component. A further improvement is gained by comparing average frequency against a crystal oscillator and converting any error into a d.c. voltage controlling the reactance modulators, or into a driving force to operate a motor controlling a tuning capacitance across the f.m. oscillator. A block schematic of a motorised correction circuit is shown in *Figure 40.7*. The f.m. oscillator frequency is changed to about 7.5 MHz with the aid of a crystal oscillator; this frequency is divided down to 15 kHz and applied to a pair of balanced modulators together with the output from a 15 kHz crystal oscillator phase-shifted by 90° to produce a 2 phase field from the balanced modulators. This field drives a synchronous motor mechanically coupled to a trimmer capacitor across the f.m. oscillator. The field is only present when there is an error in the divider frequency. A crystal oscillator[1] can be frequency-modulated and a highly stable average frequency maintained.

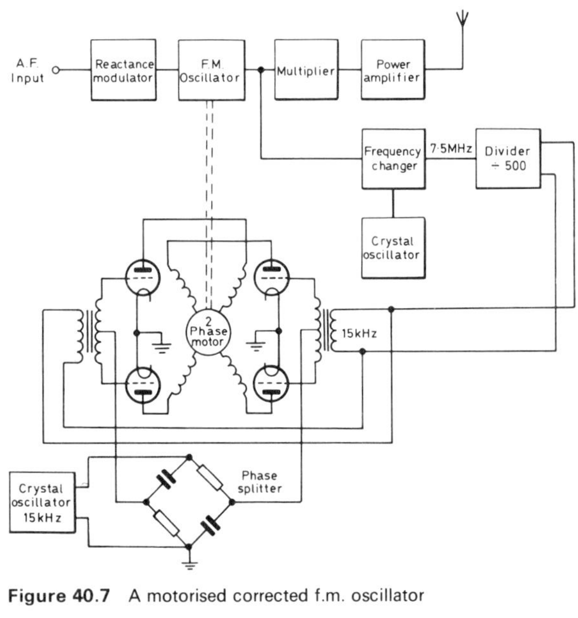

Figure 40.7 A motorised corrected f.m. oscillator

40.3 Pulsed carrier modulation

Pulsed carrier modulation is a form of double modulation in which the information is used to modulate a pulse, which is in turn used to modulate the final carrier. It requires a pulse repetition frequency (p.r.f.) at least about 2.3 times the maximum information frequency component. A series of unmodulated pulses can be resolved into a d.c. component and a theoretically infinite number of integer multiples of the p.r.f. When the pulses are amplitude-modulated the frequency spectrum can be resolved into a d.c. component and the full range of base band information frequencies, and sideband pairs of the base band components associated with each integer multiple of the p.r.f. as indicated in *Figure 40.8*. The sideband ranges must not overlap if the signal is to be detected without distortion at the receiving point. This cannot occur with a.m. if the p.r.f. exceeds $2f_m$(max).

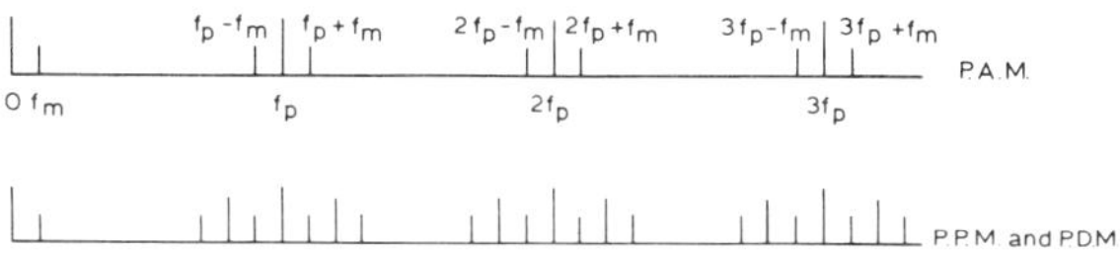

Figure 40.8 Frequency spectrum of p.a.m., p.p.m. and p.d.m. pulses

When the pulse is position or duration modulated, harmonic sidebands of the pulse repetition integer frequencies ($nf_p \pm mf_m$) are produced but there are no harmonic frequencies of the base band, *Figure 40.8*. Detection at the receiver requires that no overlap shall occur, and if this is probable either the p.r.f. must be increased or the degree of modulation reduced. Pulsed carrier modulation is valuable for multiplex operation with t.d.m. but for simplex operation it has no advantages over continuous carrier modulation.

40.3.1 Pulse amplitude modulation

Pulse amplitude modulation is a sampling procedure at the p.r.f. The pulse amplitude may change during the 'on' period in accordance with the information amplitude, or it may remain constant resulting in a stepped envelope version of the original signal. The stepped form is an important stage in the conversion to a digital signal in pulse code modulation. The varying pulse amplitude is used to modulate the carrier amplitude in the normal way. Detection of p.a.m. at the receiving point is achieved by a peak diode followed by a low pass filter which selects the base band from the pulse spectrum. This type of modulation cannot give any better signal-to-noise ratio than normal amplitude modulation, and it can give a much worse ratio if the receiver is not 'muted' during the pulse off periods.

40.3.2 Pulse position modulation

P.P.M. can be produced by using the information amplitude to frequency or phase modulate the p.r.f. Frequency modulation can be realised by a variable reactance, controlled by the information amplitude, across the p.r.f. oscillator. A limiter in the receiver removes the amplitude noise and allows the signal-to-noise improvement to be achieved. Detection by peak diode followed by a low pass filter allows the base band frequencies to be recovered. An alternate method is to use a frequency discriminator tuned to the p.r.f.

40.3.3 Pulse duration modulation

Modulation of pulse duration is obtained by varying the level at which a sawtooth voltage derived from the p.r.f. oscillator is sliced. Detection may be by peak diode followed by a low pass filter, or the pulse duration may be converted to an amplitude variation by using the pulses to control the duration of a ramp voltage.

40.3.4 Telegraphic communication

Telegraph signals themselves are in pulse form and when used to modulate a carrier generate pulsed carrier modulation. Originally the carrier was switched on for mark and off for space, but this is unsatisfactory because the receiver reverts to full gain on space, and noise may cause interference and spurious signals. Frequency shift keying (f.s.k.) is the method now adopted and the carrier is shifted by about 100–200 Hz from mark to space.

The Morse code is satisfactory for manual operation and visual observation but is less useful when teleprinter or automated punched paper tape or magnetic tape operation is employed. A 5-unit code, which is a binary on–off system, allows 2^5 or 32 characters to be defined but gives no error indication. The 7 unit code, made up of 3 mark and 4 space elements permits 35 characters ($7C_3 = 7.6.5/1.2.3$) to be defined, and at the same time provides a degree of error detection. The marks are counted and unless 3 are registered in each character the message must be in error and a request for a repeat can be sent from the receiver to the transmitter. The process can be automated and is then called automatic error correction (a.r.q.) A.R.Q. equipment generally records groups of 3 characters, checks for the correct number of marks in each character and sends back to the transmitter a repeat request or go-ahead signal. After checking and finding no error, the recorded signals are passed to the teleprinter. The transmitter cuts off at every 3 characters and remains inoperative until the repeat or go-ahead signal is received.

Only a single error is detected and a double error which drops a mark and inserts interference or noise simulating a mark into a space passes uncorrected. If more than 35 characters have to be accommodated a higher unit code is required, thus an 8 unit code of 3 marks and 4 spaces could cover 56 characters or with 4 marks and 4 spaces 70 characters are possible.

The term baud is used in telegraph communication, 1 baud signifying 1 on–off period/second; a 50 baud signal has 50 pulses/second, the duration of each on–off period and of the pulse being 0.02 second and 0.01 second respectively, and the required transmission bandwidth being 100 Hz. A single telephonic speech channel can accommodate 24 such telegraph signals.

40.4 Pulse code modulation

With pulse code modulation the information amplitude is divided into a number of levels, each of which is designated by a number; it is regularly sampled and given the number of the specified level nearest the given sample. Normally a binary system of numbering is employed because this is the simplest to achieve electronically, and the digital output consists of a number of pulses per unit time corresponding to the binary number given to the sampled amplitude level. In a binary system the number of levels is 2^n where n is the maximum number of pulses per unit time; thus 8 levels require 3 pulses with zero regarded as a level, whereas 1024 levels require 10 pulses. Since the coded signal represents a number of discrete levels, the reproduced information at the receiving point after decoding will be a stepped copy of the original information, *Figure 40.9*. The steps introduce a spurious signal, and when there are many steps of small amplitude the interference appears as a hiss like random noise

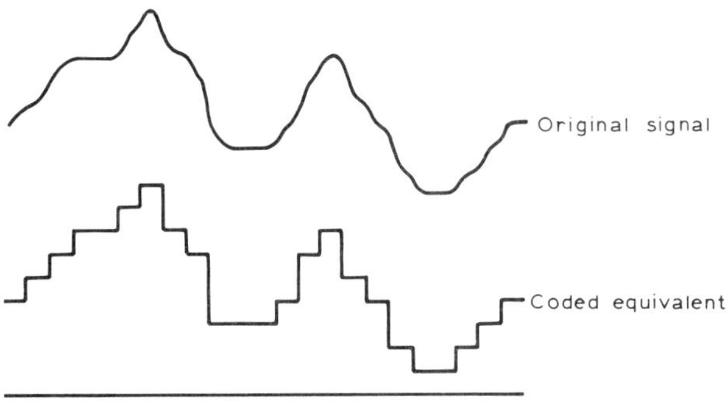

Figure 40.9 Pulse coded equivalent of the original signal

superimposed on the original information amplitude, and it is worst when the information amplitude lies midway between two levels. The hiss is termed quantising noise since it is caused by the discrete level quanta; its effect can clearly be reduced by increasing the number of levels, but this requires more digital pulses in a given time interval and therefore a greater transmission bandwidth.

When the number of level steps is small, i.e. when the information amplitude is very small, the step interference changes from a hiss to a seriously distorted signal and this condition is known as granular distortion. In telephonic speech, amplitude variations can be smoothed out by compression, and satisfactory communication can be achieved by a 7 digit binary code (128 levels). The same degree of compression is not permissible for high quality sound broadcasting, and subjective tests have shown that a 13 digit[2] coding (8192 levels) is required. The 13 digit coding provides a peak signal to r.m.s. quantising noise ratio of 83 dB and a 7 digit 47 dB. It is possible to reduce the audibility of granular distortion by adding a 'dither' voltage to the information before coding. The dither voltage has two components, one a square wave of peak-to-peak amplitude equal to half a level step and at half the sampling frequency, and the other white (thermal) noise at a much lower level (about 4 dB below quantising noise). At the receiver the half sampling frequency square wave is removed by the filter selecting the information frequencies. A slightly improved signal-to-noise ratio is obtained by substituting a pseudo random signal for the white noise and at the receiver cancelling this by subtracting an identical sequence obtained from another pseudo random generator synchronised by the transmitted pseudo random signal.

Conversion of the analogue signal into its digital equivalent involves three processes, sampling, quantising and coding. *Figure 40.10* illustrates this with reference to an 8 level 3 digit coding. A

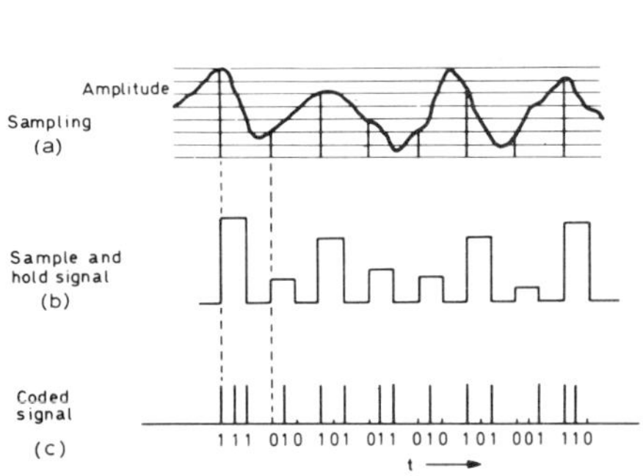

Figure 40.10 The process of pulse coded modulation

d.c. component must be added to the analogue signal before sampling to make it unidirectional; the combined signal is sampled, *Figure 40.10(a)* and converted to a p.a.m. signal but the amplitude of each pulse is held constant at the analogue plus d.c. value at the start of the pulse, until the digit number corresponding to the amplitude has been generated, *Figure 40.10(b)*. Quantising involves comparing each pulse amplitude sample with a number of equally spaced reference amplitude levels and either determining to which level its amplitude is nearest or, what is often easier, selecting the level one lower than that which it just fails to reach. Each level is assigned a binary number and the output of the coder provides the binary pulse equivalent of the level determined from the unidirectional p.a.m. analogue signal, *Figure 40.10(c)*.

A simplified block diagram of the method of sample and hold, producing constant amplitude pulses, is shown in *Figure 40.11*.

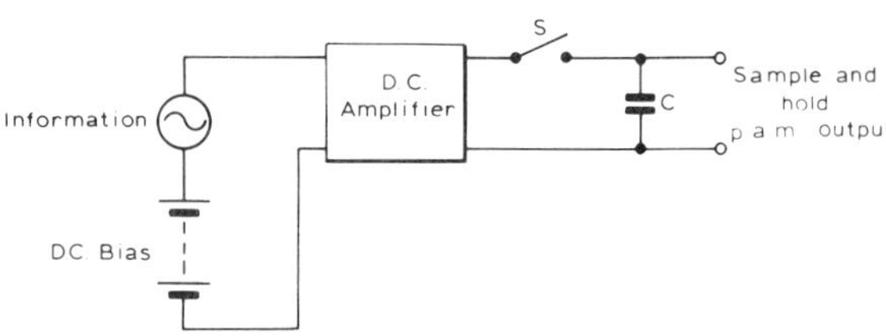

Figure 40.11 Schematic diagram of sample and hold circuit

The d.c. and analogue combined signal is amplified and sampled by closing of the switch *S*, which charges the capacitor *C* to the sample voltage. The switch, generally a diode bridge, is operated from the sampling oscillator, whose p.r.f. is about $2.3f_m$(max). The closure time for the switch is very short compared with the period of the analogue signal and after opening, the capacitor voltage remains substantially constant at the sample amplitude until the level assessment has been made. Shortly afterwards the capacitor is fully discharged and is then ready to accept the next amplitude sample.

There are a number of systems for converting the p.a.m. signal into a digital equivalent; one uses a counter coder and another employs successive approximations. A simplified block diagram of the counter coder is illustrated in *Figure 40.12*. The held

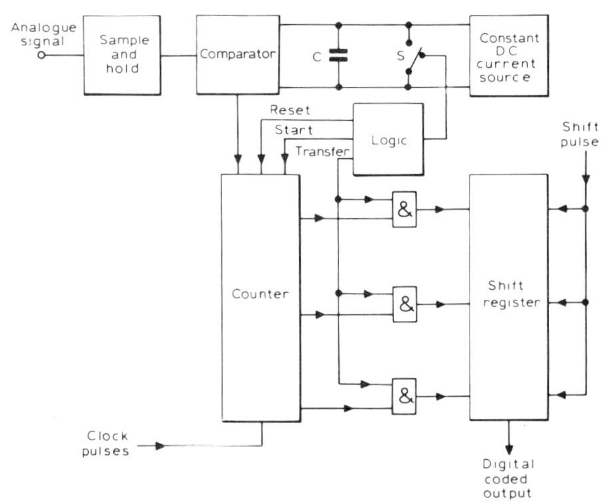

Figure 40.12 Block diagram of counter encoder

sample of information amplitude is compared against a ramp voltage of constant slope provided by charging a capacitor at a constant current. When an assessment is to be made a short circuit S across the ramp generator capacitor C is removed and simultaneously clock pulses are fed to a counter. When the voltage across capacitor C equals that of the sample, the comparator sends a pulse which stops the counter. Since the voltage across C increases linearly with time until equality is reached with the sample, the counter registers the binary equivalent of the sample amplitude. The state of the counter is transferred to the shift register via AND gates. The counter is cleared and the ramp capacitor short circuited in readiness for coding the next sample. Shift pulses are fed into the register to release the digital information in time sequence; the digital representation may be transmitted direct over coaxial cable or may be used to modulate a u.h.f. or s.h.f. carrier for transmission over a microwave link.

At the receiving point the reverse process may be carried out to convert the digital signal back to its analogue original. A block diagram of the counter decoder is shown in *Figure 40.13*. The

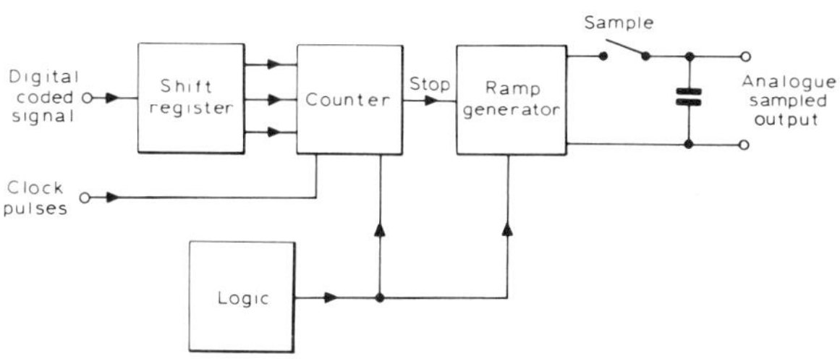

Figure 40.13 Block diagram of counter decoder

coded digital signal is stored in a shift register from which it is transferred to the counter. A range generator—a capacitor charged from a constant current—is started simultaneously with the counter which counts back to zero. When all counter outputs register zero, the charging current to the capacitor is cut off, and the voltage across the latter is proportional to the quantised value of the original sample. All these operations are controlled by a sampling oscillator synchronised with that at the sending end, and after each digital conversion the capacitor of the ramp generator is short circuited ready for recharging on the next set of digital pulses. A stepped p.a.m. copy of the original information signal is generated across the capacitor.

The successive-approximation coder compares the sampled analogue signal with a set of successively presented reference voltages corresponding to one digit position in the code. If the maximum reference voltage is less than the sample the comparator gives a logic 1 output; the sample voltage minus the reference voltage is passed to another comparator connected to the next digit voltage reference and if the reference is more than the former the comparator gives a logic 0 output. If the maximum reference voltage is greater than the sample, the comparator registers a logic 0 output and the sample is passed on unchanged to the next reference voltage. This process is repeated until all the digit positions have been defined.

Time division multiplex can readily be applied to allow several p.c.m. signals to share the same transmission channel. With the aid of electronic switches the binary coded samples of each information signal are connected in turn to the transmission system. All information channels are sampled once per cycle of the multiplex operation and extra binary digits called parity bits are added for synchronisation and error detection to complete the whole cycle or frame. The frame synchronising signal maintains the multiplexing and demultiplexing operations in synchronism. The demultiplexer is prevented from responding to a frame synchronising pattern occurring accidentally during the information sequence.

Errors in p.c.m. signals may be detected by transmitting an extra parity bit—with the information. The parity bit is inserted at the sending end to bring the total binary digits to an even or an odd number in each transmitted information sample. Failure at the receiver to register an odd or even number (whichever is chosen) of bits in each sample indicates an error.

References

1 MORTLEY, W. S., 'Frequency-modulated quartz oscillators for broadcasting equipment', *Proc. IEE*, Part **B**, **104**, 239 (1957)
2 SHORTER, D. E. L. and CHEW, J. R., 'Application of P.C.M. to sound signal distribution in a broadcasting network', *Proc. IEE*, **119**, 1442 (1972)

41 Amplitude Modulated and Frequency Modulated Detectors

S W Amos, BSc(Hons), CEng, MIEE
Freelance Technical Editor and Author

Contents

41.1 Introduction

According to BS 4727 the function of a detector is to abstract information from a radio wave. The information may be the modulation waveform and the detector is then alternatively termed a demodulator. All detectors are not, however, demodulators. For example the a.g.c. detector in an a.m. receiver is required to give an output related to the unmodulated amplitude of the signal, and a filter is incorporated to remove modulation-frequency components from the detector output.

The output of an f.m. detector can similarly be a copy of the modulation waveform and such a circuit may also be termed a discriminator. Alternatively an f.m. detector may be required to give an output for automatic frequency control of an oscillator and modulation-frequency components in the output are suppressed by use of a filter. A further complication in the nomenclature of f.m. detectors is that they are sometimes called phase detectors or phase discriminators.

The relationship between frequency modulation and phase modulation is simple. In frequency modulation, for a constant-amplitude modulating signal, the phase shift of the carrier is swept between limits which are inversely proportional to the modulating frequency whereas in phase modulation the limits are fixed. Similarly in phase modulation, for a constant-amplitude modulating signal, the frequency of the carrier is swept between limits directly proportional to the modulating frequency whilst in frequency modulation the limits are fixed. In practice this means that one form of modulation can be converted to the other by including a 6 dB per octave filter in the modulating-signal path and, by use of such a filter, the same circuit can be used for the detection of f.m. or p.m. signals. For simplicity all the circuits mentioned in this chapter are referred to as f.m. detectors or discriminators.

41.2 A.M. detectors

If the mode of operation of the various a.m. detectors is considered in detail it is found that each conforms to one of four basic modes. There are minor variations in the details of operation but all a.m. detectors conform to one of the following four types.

1. Those in which the detector output is made up of samples of the peak value of the modulated r.f. input.
2. Those in which the detector clamps the peaks of the modulated r.f. input at a constant potential so that the mean value of the signal varies at modulation frequency.
3. Those in which the output stems from the interaction between the side frequencies and the carrier of the modulated r.f. input, the interaction being caused by the non-linearity of the transfer characteristic.
4. Those in which the output results from the effective multiplication of the modulated r.f. input and a second input at the carrier frequency.

We shall now examine this classification in detail.

41.2.1 Sampling detectors

41.2.1.1 Series-diode circuit

The simplest example of a sampling detector is the series-diode circuit shown in *Figure 41.1*. It is similar to a half-wave rectifier circuit and the capacitor C_1 can be called a reservoir capacitor. Operation of the circuit relies on the rapid charging of C_1 through the low value forward resistance and the subsequent discharge through the high value diode load resistor R_1.

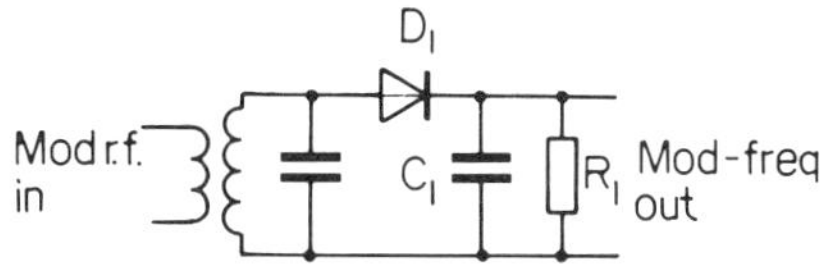

Figure 41.1 The simple series-diode detector circuit (example of a sampling detector)

Diode D_1 conducts during positive half-cycles of r.f. input and charges C_1 to the peak value of the input signal. During negative half-cycles the diode is cut off and C_1 begins to discharge through R_1. The ratio of the time constant R_1C_1 to the period of the carrier is, however, so chosen that very little of the charge on C_1 is lost before D_1 begins to conduct on the next positive half-cycle of input and C_1 is again charged to the peak value. Thus C_1 maintains a positive voltage which keeps D_1 cut off except for the instants when the input signal passes through its positive peaks. In practice the period of conduction is only a small fraction of the positive half cycle.

Thus the load circuit R_1C_1 is connected to the modulated r.f. source by the low forward resistance of the diode for only a brief fraction of each input cycle and during this time the capacitor voltage is 'topped up' to the peak value of the r.f. input. For the remainder of each cycle the diode is cut off, isolating the load circuit from the r.f. input so that the voltage across R_1C_1 begins a small exponential fall. Thus the diode acts as a switch which is turned on and off by the carrier component of the input signal. This is an example of a sampling process in which the modulated r.f. input signal is sampled once per cycle when it is passing through its positive peak. As the peak value changes as a result of modulation, the voltage across R_1C_1 changes to give a simulation of the modulating signal waveform made up of a number of 'topping up' increases separated by exponential falls. These constitute an r.f. ripple of small amplitude superposed on the modulating-frequency waveform and which is easily removed by an r.f. filter to make the output waveform a good approximation to the modulating signal.

This type of detector is widely used in a.m. receivers and gives a good performance provided that the input signal is large enough to switch the diode effectively, i.e. so that it has a low forward resistance and a high reverse resistance. For an r.f. input of small amplitude the forward resistance is higher and the reverse resistance lower than could be wished and thus detection of small-amplitude signals is inefficient. Better results could be obtained for small inputs if the diode could be switched by a large-amplitude signal synchronised with the carrier component. This is possible using a synchronous detector of the type normally used for the demodulation of suppressed-carrier a.m. signals: such detectors are described later. The switching signal can be obtained from a local oscillator as in synchrodyne receivers or from the received signal by removing the modulation as in the homodyne receiver. I.cs are available with limiter stages suitable for use in a homodyne receiver.

41.2.1.2 Synchronous detectors

Circuits of the type so far considered are used to detect modulated signals in which the carrier is present. They take samples of the positive peaks of the modulated r.f. input and are not affected by variations in the timing or phase of the peaks. To detect carrier-suppressed a.m. signals the detector must be sensitive to the phase as well as the amplitude of the peaks of the input signal; the reason for this will be made clear in the discussion of *Figure 41.2*. Thus the detector must have a reference signal of constant frequency against which it can compare the phase of the modulated r.f. input. To this end the detector is

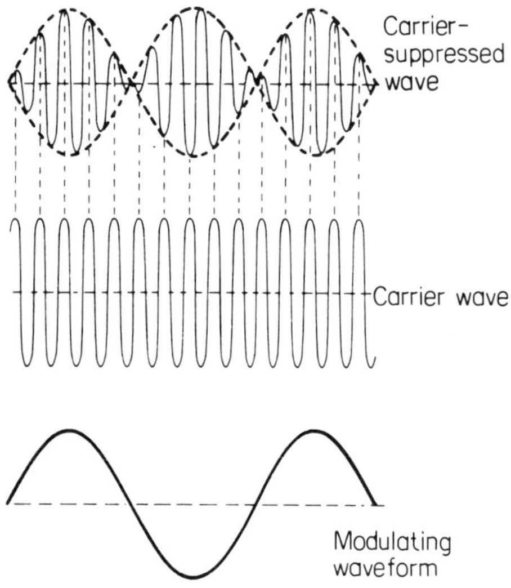

Figure 41.2 Action of a synchronous sampling detector in detecting a carrier-suppressed signal (the dashed lines indicate the sampling periods)

provided with a second input in the form of a constant-amplitude sinusoidal signal synchronised with the (suppressed) carrier frequency of the modulated r.f. signal to be detected.

41.2.1.3 Synchronous sampling detector

One possible circuit for a synchronous sampling detector is given in *Figure 41.3*. The single series diode of the prototype a.m.

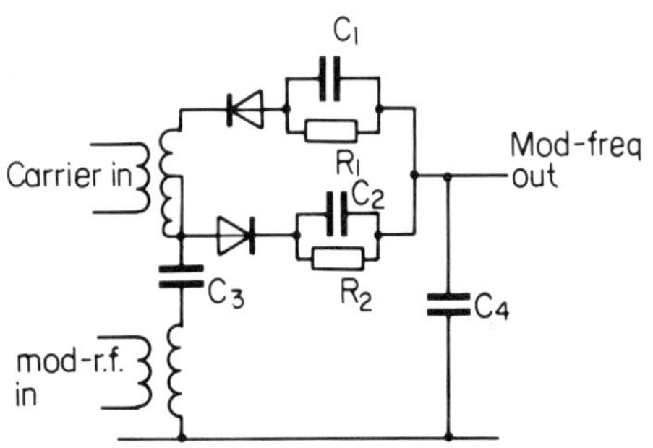

Figure 41.3 Synchronous sampling detector using a diode bridge

detector is replaced by two diodes and a centre-tapped transformer. Both diodes conduct together to produce the low-impedance path which connects the source of modulated r.f. to the capacitor C_4. When the diodes are non-conductive the path is open-circuited and C_4 retains its charge. The diodes must be driven into conduction and non-conduction by the carrier input and not by the modulated r.f. input and thus the carrier input must be large compared with the other input signal. The balanced form of the carrier circuit is adopted to minimise any carrier component which may reach C_4. The time-constant circuits R_1C_1 and R_2C_2 are included as diode loads to ensure that the diodes conduct for only a small fraction of each cycle, i.e. when sampling is required.

The way in which such a detector demodulates a double-sideband suppressed-carrier signal is illustrated in *Figure 41.2*, in which the vertical dashed lines indicate the sampling periods. A non-synchronous a.m. detector, being insensitive to phase, would sample all the positive peaks and would thus produce a grossly distorted output. The synchronous detector operates stricly at carrier-frequency intervals and samples the positive peaks during one half-cycle of the modulating signal and negative peaks during the other half-cycle, thus correctly reconstituting the waveform of the modulating signal. The output has positive and negative swings and, for a symmetrical modulating signal such as a sine wave, has a mean value of zero, i.e. there is no d.c. component as in the output of the prototype non-synchronous series-diode detector.

This type of circuit can be used to demodulate the quadrature-modulated colour-difference signals in a colour television receiver. Here the modulated signal has two carrier components in quadrature, each amplitude modulated by a different signal. The circuit of *Figure 41.3* can demodulate one of these signals without interference from the other because, during the time it is sampling the peaks of one signal, the other is passing through zero and so has no effect on the detector output. A second detector with its carrier input in quadrature with that of the first is required to demodulate the second colour-difference signal.

For some applications the components R_1C_1 and R_2C_2 can be omitted. The diodes then connect C_4 to the source of modulated r.f. for the whole of one carrier half-cycle.

41.2.2 Clamping detectors

41.2.2.1 Shunt-diode circuit

In the circuit of *Figure 41.1* the output of the detector is taken from the reservoir capacitor, but it could alternatively be taken from the diode, the circuit being re-arranged as shown in *Figure 41.4* to enable one leg of the output to be earthed. In this version

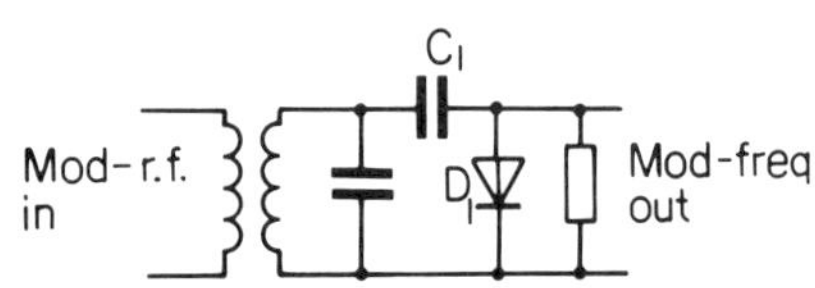

Figure 41.4 The simple shunt-diode detector circuit (example of a clamping detector)

of the circuit, known as the shunt-diode detector, the reservoir capacitor is series-connected, which makes the circuit convenient when d.c. isolation is required between the output terminals and the source of modulated r.f. input.

There is, however, no r.f. isolation between r.f. input and the output as in the series-diode circuit. The reservoir capacitor provides a low-reactance path at r.f. and transfers the modulated-r.f. input signal with little attenuation to the detector output terminals. The output is, in fact, made up of the modulation-frequency signal generated across the reservoir capacitor in series with the modulated-r.f. signal transferred from the input. Thus the output of the shunt-diode detector has a much greater r.f. ripple content than that of the series-diode circuit. The waveform of the output from the shunt-diode circuit can be deduced in the following way.

The diode conducts during a small fraction of each r.f. input cycle to charge the reservoir capacitor and for this brief period it acts as a short circuit across the output terminals. Thus for the duration of the charging period the output of the detector is zero: this occurs at each of the positive peaks of the input signal. The detector output therefore consists of a version of the modulated-r.f. input waveform in which each r.f. cycle is so displaced vertically that all positive peaks touch the zero volts line as shown in *Figure 41.5*. The mean value of such a signal varies with modulation and, if the r.f. ripple is suppressed, consists of the modulation waveform superposed on a negative zero-frequency component proportional to the amplitude of the unmodulated r.f. input.

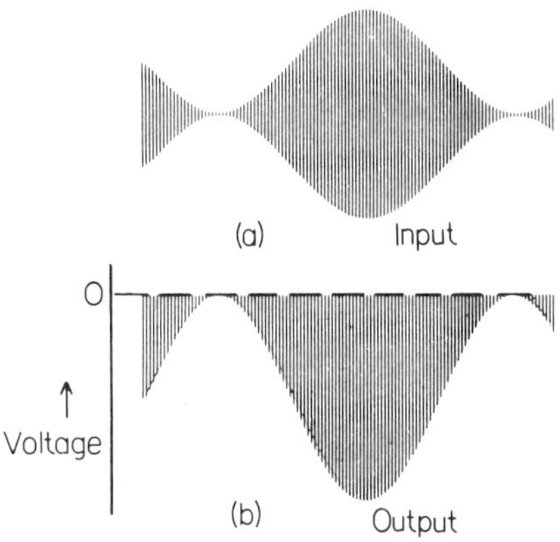

Figure 41.5 Waveforms for the shunt-diode detector (a) Input (b) Output

The action of this form of detector is an example of clamping in which positive peaks of the input signal are clamped at zero volts. The circuit is often used in television to clamp the sync tips of a video waveform at a particular voltage: in this application the circuit is known as a d.c. restorer.

41.2.2.2 Synchronous clamping detector

Figure 41.6 gives the circuit diagram of a synchronous clamping detector. It has much in common with the synchronous sampling

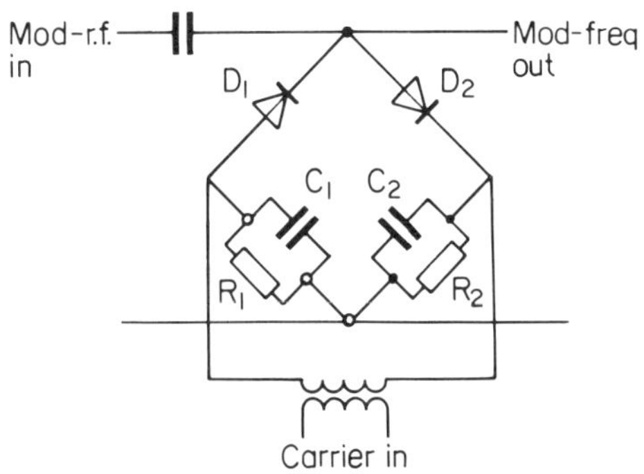

Figure 41.6 A synchronous clamping detector using a diode bridge

detector of *Figure 41.3* except, of course, that the diodes are arranged to produce a shunt short circuit once per carrier cycle. The diodes and their load circuits form a balanced circuit chosen to minimise carrier content in the detector output and the time constant of the load circuits R_1C_1 and R_2C_2 is made long compared with the carrier period so that the diodes conduct for only a small fraction of each cycle. At each conduction period that part of the modulated-r.f. input waveform which coincides in time with it is clamped at zero volts.

The way in which the detector demodulates a suppressed-carrier signal is illustrated in *Figure 41.7*, in which the vertical dashed lines indicate the conduction periods. For a correctly

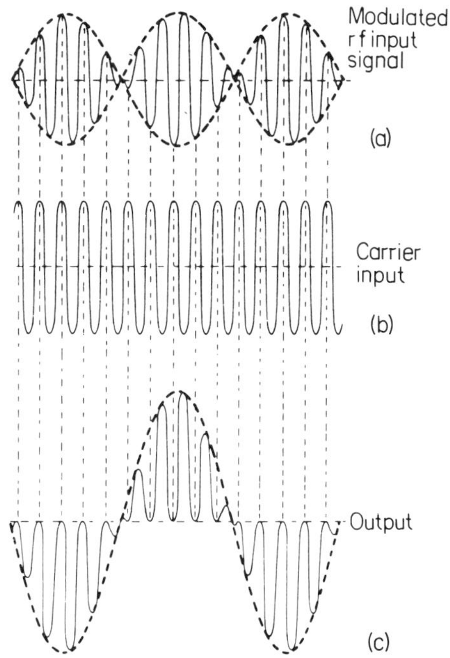

Figure 41.7 The action of a synchronous clamping detector in detecting a carrier-suppressed signal

synchronised carrier these coincide with positive peaks of the modulated-r.f. signal during one half-cycle of the modulating signal and with negative peaks during the other half-cycle. Thus the output signal has positive and negative swings as shown in *Figure 41.7(c)*. As for the prototype non-synchronous shunt-diode detector there is a very large r.f. ripple content in the output but, for a symmetrical modulating signal such as a sine wave, there is no d.c. component.

The diodes can be replaced by a shunt-connected bipolar transistor which is switched on and off by the carrier signal applied to the base. The circuit diagram (*Figure 41.8*) includes an *RC* combination in the base circuit which determines the duration of each clamping period. If the transistor is a symmetrical type transmission of the carrier signal to the detector output can be minimised.

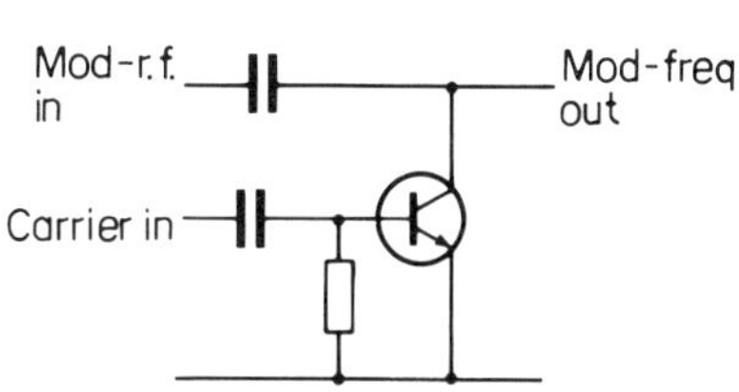

Figure 41.8 A simple synchronous clamping detector using a symmetrical bipolar transistor

41.2.3 Additive (non-linear) detectors

In all the detectors so far considered, a reservoir capacitor has played an essential part: it is charged during part of each cycle of carrier component and discharges during the remaining part of the cycle. Thus the shape of the input–output characteristic of the charging device has only a second-order effect.

There is, however, a type of detector in which the shape of the input–output characteristic is all-important because it is in use for most if not the whole of each cycle of input signal. Any device with a non-linear relationship between input and output behaves as this type of detector.

Detection is achieved because of the unequal response to positive and negative half-cycles of the input signal and this is a consequence of the non-linearity of the characteristic as shown in *Figure 41.9*. Clearly the mean value of the output signal varies

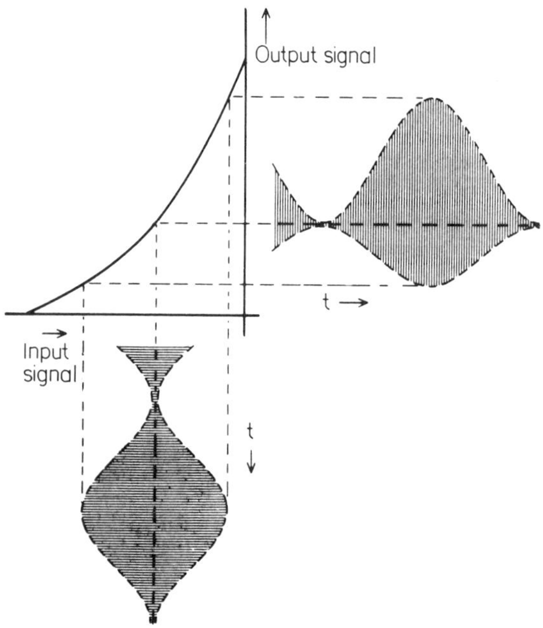

Figure 41.9 Typical non-linear input–output characteristic showing detection of an amplitude-modulated signal

with the modulation and the magnitude of the modulation-frequency output depends on the severity of the non-linearity of the characteristic. The mean output also varies with the magnitude of the input signal.

There is an alternative method of explaining the operation of this type of detector. When two sinusoidal signals with different frequencies are applied to a device with a linear characteristic, the output has only two components and these are at the frequencies of the two input signals. If, however, two such signals are applied to a device with a non-linear characteristic, the output contains components not only at the frequencies of the two input signals but also at multiples of these two frequencies (harmonics) and at the sums and differences of the various harmonics. The last mentioned are known as combination frequencies and are given by $(mf_1 \pm nf_2)$ where f_1 and f_2 are the frequencies of the two input signals, m and n being integers. Perhaps the most interesting of the combination frequencies is $(f_1 - f_2)$ the difference frequency. Non-linear devices are often used as r.f. mixers in superheterodyne receivers, the inputs from oscillator and the r.f. circuit being connected in parallel or series and applied to the single input terminal: it is the difference term which is selected from the output of the mixer for amplification in the i.f. amplifier. In a non-linear detector the input, assumed amplitude-modulated by a single sinusoidal signal, has three components—the carrier, the upper side frequency and the lower side frequency. The difference term resulting from interaction between the upper side frequency and the carrier, and between the carrier and the lower side frequency, both yield the required modulation-frequency output. But interaction between the upper and lower side frequencies yields an unwanted second harmonic of the modulating signal and interaction between the harmonics of the side frequencies and the carrier yields a complex of other unwanted terms. Thus the non-linearity of the characteristic on which the action of the detector depends inevitably causes considerable harmonic and intermodulation distortion.

41.2.4 Multiplicative (product) detectors

As shown in the previous section one method of achieving a.m. detection is by use of a non-linear device which generates an output at the difference between the frequencies of two components of the input signal. An alternative method is to use a device with two input terminals and which in effect multiplies the two inputs to form the output. This process yields an output at the sum and difference frequencies directly as shown by the identity:

$$\sin \omega_1 t \sin \omega_2 t = \tfrac{1}{2}[\cos(\omega_1 - \omega_2)t - \cos(\omega_1 + \omega_2)t] \quad (41.1)$$

The difference term is thus obtained without need of non-linearity.

There are a number of r.f. mixers and synchronous detectors which use this principle in which, as the identity implies, current is assumed to flow in the device throughout each cycle of both input signals. In all these examples both input terminals control the current through the device and one of them can be regarded as controlling the mutual conductance g_m of the device. The output current is given by $g_m v_{in}$ approximately (where v_{in} is the signal applied to the second input terminal) and is thus proportional to the product of the two inputs.

One method of producing a circuit in which two inputs control the same current is by connecting two transistors in series across the supply as indicated in *Figure 41.10*. A number of circuits of

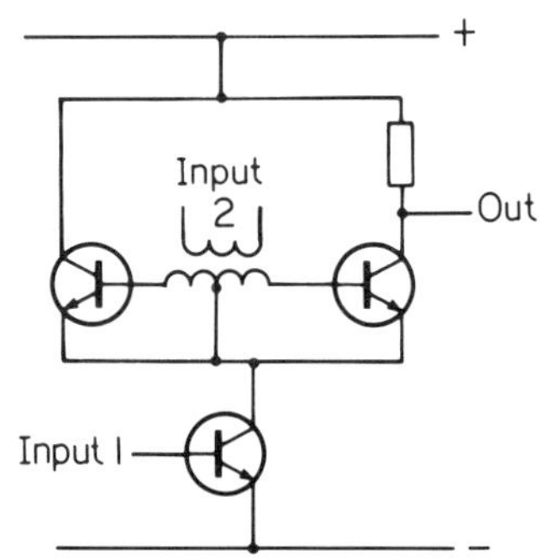

Figure 41.10 Simplified circuit diagram of a product detector using bipolar transistors

this type are in common use, particularly in integrated circuits, and frequently the upper transistor is replaced by a parallel push–pull pair, the input being applied to their bases in push–pull, the output being taken from only one of the transistors. The advantage of using push–pull is that the currents of the paralleled transistors are in antiphase so that alternating currents at the frequency of the push–pull input are confined to the push–pull stage and do not stray into the supply circuits or to the lower transistor which controls the current to the push–pull pair.

A third type of multiplicative device is the dual-gate, field-effect transistor. Both gates control the channel current and thus if two signals are applied to the two gates, sum and difference signals are available in the drain current.

41.3 F.M. detectors

An examination of the various types of f.m. detector suggests that they all belong to one of the following four categories.

(a) Those consisting essentially of an f.m.-to-a.m. converter followed by an a.m. detector.
(b) Those using phase comparators, i.e. circuits in which the output is dependent on the degree of overlap of two sets of carrier frequency pulses.
(c) Those using a counter circuit as a discriminator.
(d) Those using the locked-oscillator principle.

This classification will now be examined in detail.

41.3.1 F.M. detectors incorporating an f.m.-to-a.m. converter

Perhaps the most obvious way of detecting an f.m. signal is to convert the frequency variations into corresponding amplitude variations of the carrier which is then applied to an a.m. detector. A number of types of f.m. detector operate on this principle which is illustrated in *Figure 41.11*.

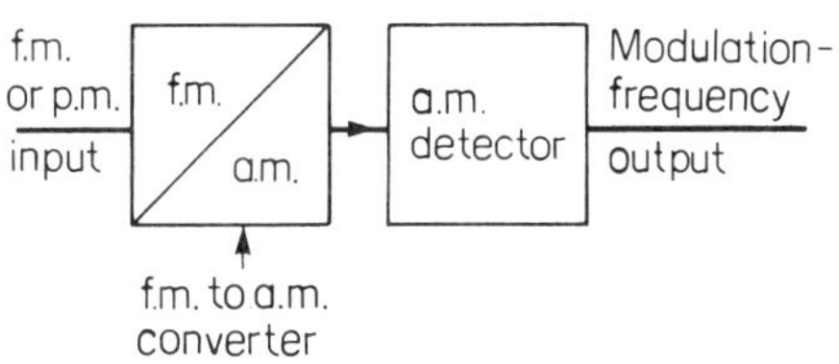

Figure 41.11 Block diagram illustrating the form of a number of types of f.m. detector

41.3.1.1 Slope detector

A simple way of achieving f.m.-to-a.m. conversion is to make use of the slope of the skirts of the amplitude/frequency characteristic for a tuned circuit. If the resonance frequency of the tuned circuit is so chosen that the centre frequency of the signal falls on a suitable part of the characteristic, as shown in *Figure 41.12*, then the output is a signal which is amplitude-modulated and frequency-modulated by the same modulating signal. If this output is applied to a simple a.m. detector, the frequency modulation will be ignored but the amplitude modulation will give an output at the modulation frequency. The curvature of the skirts of the resonance curve causes harmonic distortion which can be minimised by choice of Q value and resonance frequency for the tuned circuit but the distortion is still serious.

41.3.1.2 Round–Travis detector

In this form of detector the distortion caused by curvature of the tuned-circuit characteristic is reduced by use of the push–pull

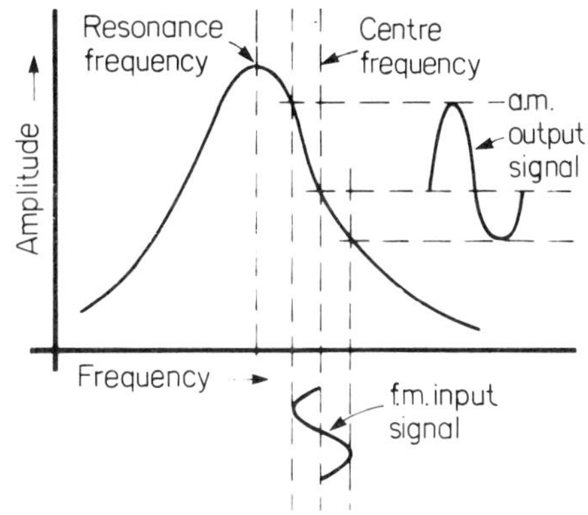

Figure 41.12 Simple f.m. slope detector

principle. Two similar tuned circuits are used, one (L_1C_1) resonant at a frequency f_1 above the centre frequency and the other (L_2C_2) resonant at f_2 an equal amount below the centre frequency. The signals developed across L_1C_1 and L_2C_2 are detected by separate a.m. detectors, their outputs being connected in series opposition. One possible circuit diagram for a Round–Travis detector is shown in *Figure 41.13* in which simple sampling-type detectors are shown.

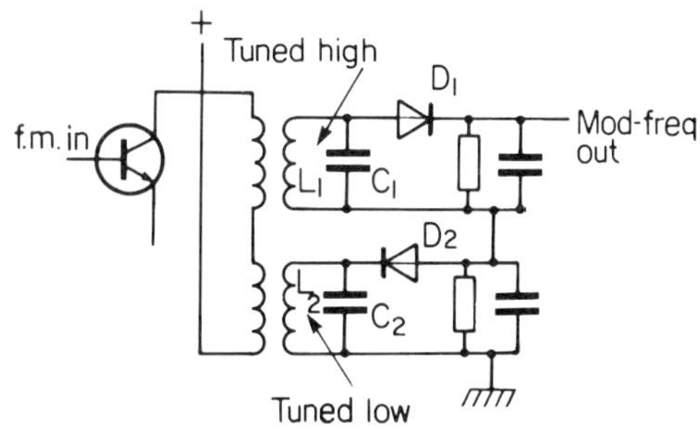

Figure 41.13 Round–Travis f.m. detector

The operation of the detector is illustrated in *Figure 41.14*. At the centre frequency equal outputs are received from the two diodes so that the net output is zero. At frequencies above the centre frequency D_1 gives a larger output than D_2 and the combined output is positive: at frequencies below the centre frequency D_2 gives a larger output than D_1 and the combined output is negative. Thus the net output indicates by its polarity whether the instantaneous frequency of the input is above or below the centre value and by its magnitude the extent of the deviation.

Figure 41.14 shows that the complementary curvature of the characteristics for L_1C_1 and L_2C_2 yields a straighter overall amplitude/frequency relationship than is possible from a single tuned circuit. The overall relationship shown in *Figure 41.14* has the S-shaped form characteristic of that of many f.m. detectors.

The Round–Travis detector was at one time used in f.m. receivers but has long since been abandoned in favour of some of the alternative types described later. It has two main disadvantages:

(1) L_1C_1 and L_2C_2 must be so adjusted that their resonance frequencies f_1 and f_2 are symmetrically disposed about the centre

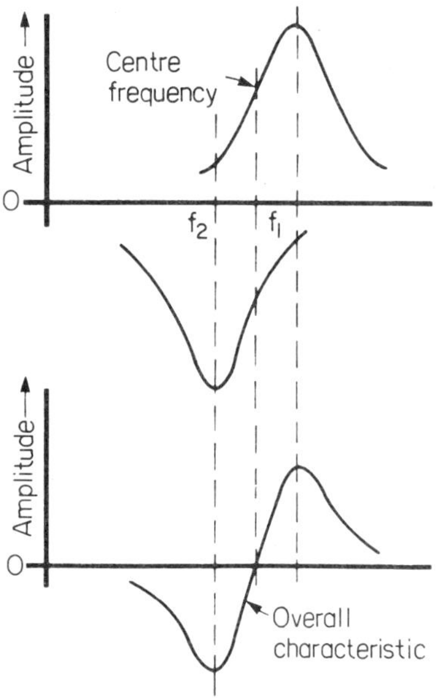

Figure 41.14 Derivation of overall characteristic of Round–Travis f.m. detector

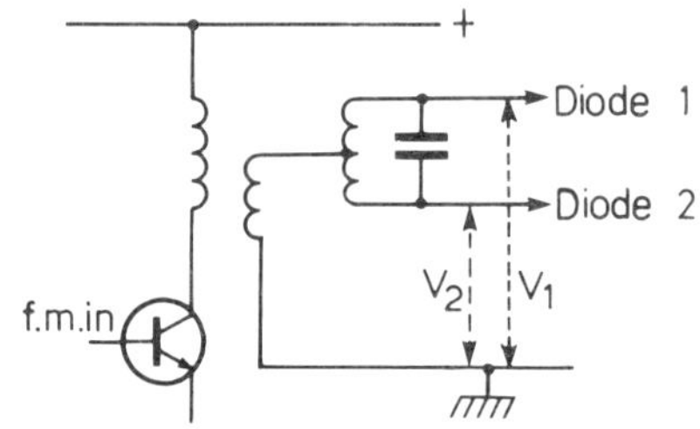

Figure 41.15 Method of deriving the two diode inputs in the Seeley–Foster and ratio detectors

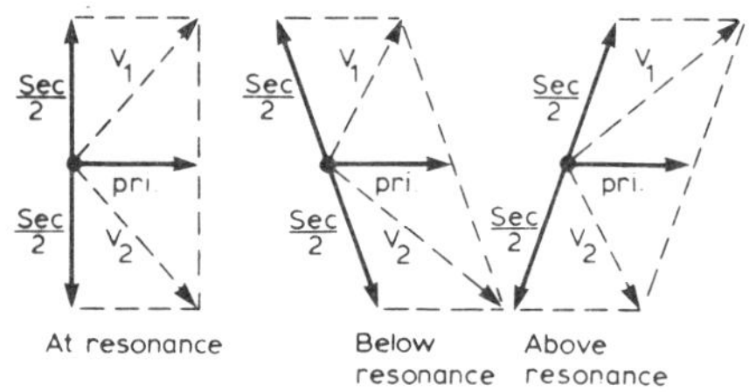

Figure 41.16 Vector diagram for the circuit of *Figure 41.15* showing how the voltages applied to the diodes vary with frequency

frequency. Thus alignment of the detector circuit is more complicated than for a number of the alternative types which require alignment only at the centre frequency.

(2) It responds to any amplitude modulation of the input signal. To obtain maximum signal-to-noise ratio, an f.m. receiver should respond only to frequency modulation of the input signal and should ignore any amplitude modulation which may be present. Some f.m. detectors can be designed to have inherent a.m. rejection and these are naturally preferred.

41.3.1.3 Seeley-Foster discriminator

This f.m. detector uses an arrangement of diodes similar to that of the Round–Travis circuit but the method of providing the diode input signals is different. The method makes use of the phase relationship between the voltage across the tuned secondary winding of a transformer and that across the primary winding. Whether the primary winding is tuned or not, these two voltages are in quadrature when the applied signal is at the resonance frequency of the secondary winding. At frequencies above resonance the secondary voltage lags the quadrature condition to an extent dependent on the frequency deviation and at frequencies below resonance the secondary voltage leads to the quadrature condition to an extent depending on the deviation.

If therefore the secondary winding is centre-tapped and if a sample of the primary voltage is injected into the centre tap, as shown in *Figure 41.15*, the voltages V_1 and V_2 at the two ends of the secondary winding vary with frequency in the same way as those from the two tuned circuits in the Round–Travis circuit. This is shown in the vector diagram of *Figure 41.16* which illustrates the relative magnitudes of V_1 and V_2 at resonance, above and below resonance. These diagrams apply when the primary voltage is equal to half the secondary voltage.

Thus a Seeley-Foster circuit could be made up from the circuit shown in *Figure 41.15* feeding into two simple diode circuits as shown in *Figure 41.17*. An alternative circuit which simplifies the design of the transformer is to use a capacitive link between primary winding and secondary centre tap as shown in *Figure 41.18*. By this means the whole of the primary voltage is injected

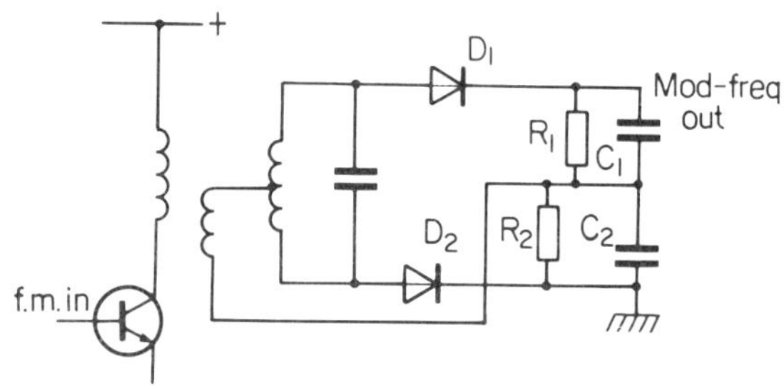

Figure 41.17 One circuit for a Seeley–Foster discriminator

into the secondary circuit. However, the introduction of the capacitor C_p interrupts the diode circuit. Normally when a diode detector is fed via a series capacitor the diode and its load resistor are both shunt-connected to ensure that the capacitor can be charged once per cycle when the diode conducts and can discharge through the load resistor when the diode is cut off by the input signal. In the circuit of *Figure 41.18(a)* the series capacitor can certainly charge when the diodes are driven into conduction by the input signal but, for the periods when the diodes are cut off by the input signal, a discharge path must be provided between the right-hand plate of C_p and the junction between R_1C_1 and R_2C_2. Moreover this path must not introduce significant damping of the primary circuit. There are two techniques which are commonly adopted to achieve this end.

(1) As shown in *Figure 41.18(a)* an inductor can be introduced between the secondary centre tap and R_1R_2 junction. This should have an inductance such that its reactance is large compared with that of C_1 and C_2 at the operating frequency.

(2) If the link between R_1R_2 and C_1C_2 is cut a direct connection can be made between the coupling capacitor and R_1R_2 junction as shown in *Figure 41.18(b)*. Damping of the primary circuit can be minimised by using sufficiently large values for R_1 and R_2. As

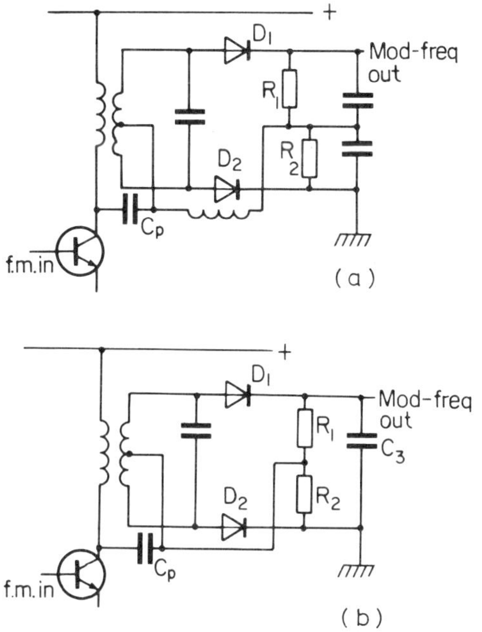

Figure 41.18 Two forms of Seeley–Foster circuit using a capacitive link between primary and secondary windings

shown C_1 and C_2 can be replaced by a single equivalent capacitor, C_3.

The Seeley-Foster discriminator was extensively employed in early f.m. receivers. Alignment is straightforward, needing only a signal source at the centre frequency and linearity can be made acceptable. Its chief disadvantage, shared with the Round–Travis circuit, is that it responds to any amplitude modulation of the input signal. Thus to obtain the high signal-to-noise ratio of which an f.m. receiver is capable it is necessary to precede the Seeley-Foster circuit by one or more amplitude-limiting stages to minimise any a.m. content in the received signal.

41.3.1.4 Ratio detector

By a simple modification the Seeley-Foster discriminator can be made capable of a useful degree of a.m. suppression. The detector circuit so produced is known as the ratio detector and it is not surprising that it rapidly displaced the Seeley-Foster discriminator. The way in which the ratio detector operates can be approached in the following way.

If one of the diodes in the circuit of *Figure 41.17* or *Figure 41.18(a)* is reversed, the net output is the sum of the voltages across the individual diode loads (not the difference as in the Seeley-Foster circuit). Thus for an input to the circuit at the centre frequency there is a voltage at the combined output approximately equal to the sum of the peak input voltages to the diodes: this compares with zero output from the Seeley-Foster circuit.

If the frequency of the input is displaced from the centre value the output across one diode load increases whilst that across the other decreases as shown for V_1 and V_2 in *Figure 41.16* and the combined voltage output tends to be independent of frequency and thus of frequency modulation. This combined output is proportional to input signal amplitude and can be used to operate a tuning indicator.

Even though the voltage across (C_1+C_2) is constant (for a given input amplitude) the voltages across the individual reservoir capacitors C_1 and C_2 vary with the frequency of the input signal and either capacitor can be used as the source of modulation-frequency output from the detector. In a balanced ratio detector circuit the junction of C_1 and C_2 is earthed and the detector output is taken from the non-earth terminal of C_1 (as shown in *Figure 41.19(a)*) or C_2. In an unbalanced ratio detector one end of the combined diode load is earthed as shown in *Figure 14.19(b)* and the detector output is taken from C_1C_2 junction. In

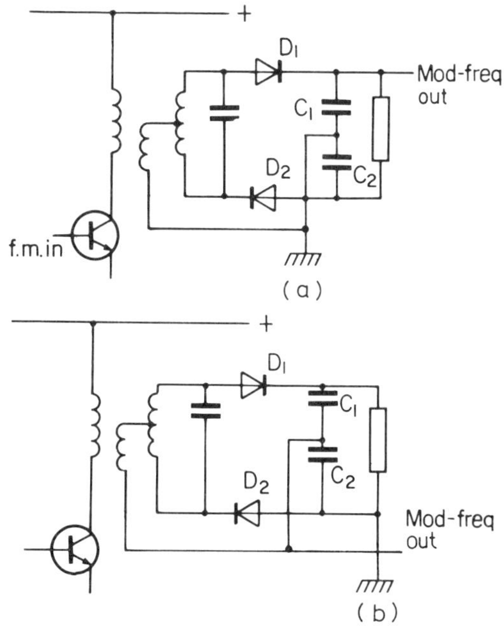

Figure 41.19 Simplified circuits of ratio detector with no provision for a.m. rejection (a) balanced form (b) unbalanced form

both types of circuit the constant voltage across the series-connected reservoir capacitors C_1 and C_2 is divided in a ratio determined by the peak inputs to D_1 and D_2: this is the origin of the name of the circuit.

To make the circuit capable of a useful degree of a.m. rejection the diode load resistor(s) are given low value(s) so that the tuned circuit feeding the detector is heavily dampled. A large value capacitor is then connected across the load resistors to give a time constant approaching one second. *Figure 41.20* illustrates these modifications applied to an unbalanced circuit. The voltage across the long-time-constant network is in practice

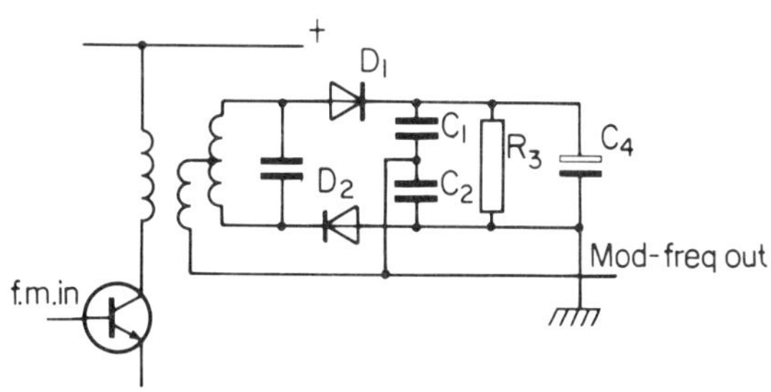

Figure 41.20 The circuit of *Figure 41.19(b)* modified to give a measure of a.m. rejection

approximately equal to the peak value of the input signal to the diodes and adjusts itself to any permanent change in the value of the peak input. As already mentioned this voltage can be used to operate a tuning indicator.

Suppose as a result of amplitude modulation of the input signal there is a momentary increase in the peak amplitude of the signal input to the ratio detector. The voltage across the long time constant diode load circuit cannot instantaneously adjust itself to equal the peak value of the spike and as a result the diodes are driven heavily into conduction and their forward resistance increases the already heavy damping on the tuned circuit thus momentarily reducing the voltage gain of the previous stage, minimising the effect of the spike.

Similarly if there is a momentary reduction in the peak value of the input signal to the detector, the long time constant network again cannot register the change and the diodes are cut off so removing the damping imposed by the diode load on the tuned circuit. Thus the gain of the previous stage is momentarily increased, offsetting the effect of the change in input signal. In fact the removal of the diode load dampling can result in overcompensation and a common technique is to include low-value resistors in series with the diodes as shown in *Figure 41.21*,

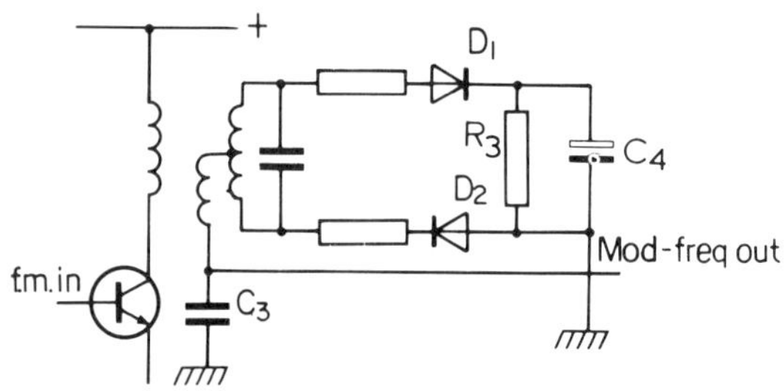

Figure 41.21 An unbalanced ratio detector with a single reservoir capacitor (C_3)

the resistance being adjusted empirically to give optimum a.m. rejection. Thus the inclusion of the long time constant circuit enables very short term changes in input signal amplitude to be minimised: in fact the ratio detector operates as a dynamic limiter.

Figure 41.21 gives the circuit diagram of an unbalanced ratio detector which differs from that described earlier in that it contains only a single reservoir capacitor C_3 in place of the two shown in earlier circuits. The way in which the modulation-frequency output is developed across C_3 can be explained as follows.

If we replace the secondary and tertiary windings of the transformer by equivalent generators V_1 and V_2, the essential features of *Figure 41.21* take the form shown in *Figure 41.22*. Both diodes conduct together once per carrier cycle and, because of the long time constant R_3C_4, the period of conduction is very brief and occurs as the combined diode input signal $(V_1 + V_2)$ reaches its peak value. As a result of this conduction C_4 is charged to the peak value of $(V_1 + V_2)$. During this brief conduction period D_1 and D_2 can be regarded as short circuits and D_2 effectively connects C_3 across the generator V_2. C_3 thus charges to the peak value of V_2. For an input signal at the centre frequency V_1 is equal to V_2 and thus C_3 is charged to a voltage equal to one half that across C_4. For the remainder of each carrier cycle when D_1 and D_2 are non-conductive the charge on C_3 remains except for a small leak through any resistor in parallel with it.

One cycle later, during the next period of conduction of D_1 and D_2, the voltage across C_3 is adjusted by charge or discharge to

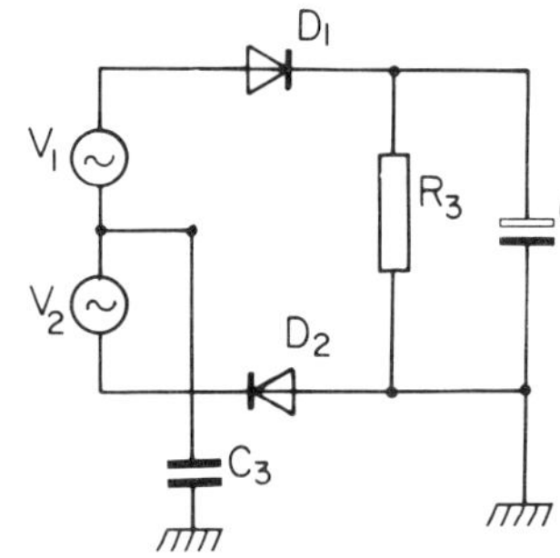

Figure 41.22 Equivalent circuit of *Figure 41.21*

agree with any change in the peak value of V_2. Thus a copy of the changing value of V_2 is built up across C_3 and this is, of course, a representation of the changing phase relationship between primary and secondary voltages which, in turn, represents the frequency-modulated waveform of the input signal.

41.3.2 Phase-comparator detectors

This type of detector also makes use of the varying phase relationship between two input signals, nominally in quadrature, such as the voltages across the primary and tuned secondary windings of a transformer but it does so in a manner quite different from that of the detectors described earlier. In Seeley-Foster and ratio detectors the two input signals are added to produce resultant voltages (the amplitude of which varies with the phase difference) which are applied to amplitude detectors, the combined output giving the required modulation-frequency signal.

In phase-comparator detectors the two input signals are limited so as to form rectangular pulses. Limiting may be carried out in separate stages preceding the phase comparator or in the phase comparator itself. The degree of overlap of these pulses varies with the phase difference between the two inputs and determines the output current of the comparator which is therefore a copy of the modulation waveform. The output of the comparator thus depends on the relative timing of the two sets of pulses and is independent of the amplitude of the input signals provided this is sufficient to give satisfactory limiting. To summarise, in the ratio and Seeley-Foster detectors the amplitude of the primary and secondary voltages is the significant quantity whereas in the phase comparator it is the timing of these voltages which matters.

The general form of a phase-comparator detector is illustrated in the block diagram of *Figure 41.23*.

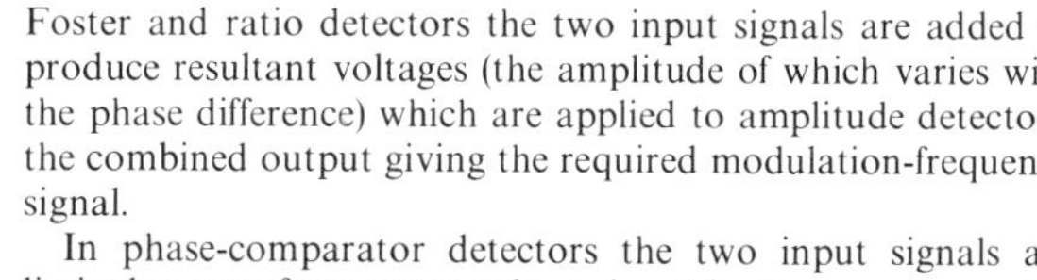

Figure 41.23 General form of phase-comparator detector

41.3.2.1 Transistor phase-comparator detectors

In its simplest form a transistor phase comparator could take the form shown in *Figure 41.24*. One of the disadvantages of such a simple circuit is that the output would contain a large component at the input frequency in addition to the wanted modulation-

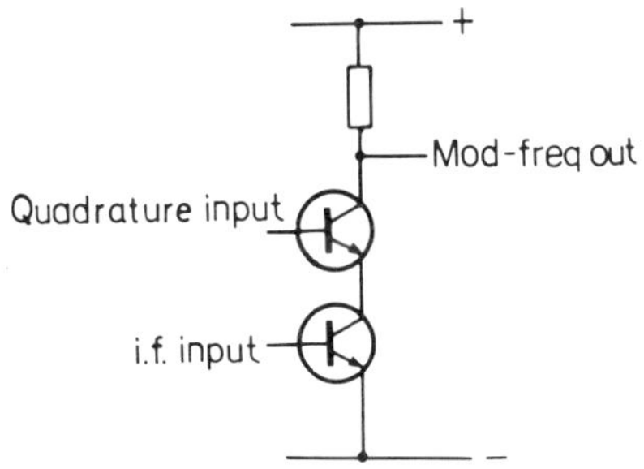

Figure 41.24 Basic form of bipolar-transistor coincidence detector

frequency component and in practical forms of phase-comparator detector precautions are taken to minimise this unwanted component.

In integrated circuits, for example, extensive use is made of the push–pull principle and a simplified version of a typical circuit is given in *Figure 41.25*. The output of the i.f. amplifier (also included in the i.c.) is applied in the form of push–pull pulses to the bases of TR_5 and TR_6 so that when one of these transistors is driven into conduction the other is cut off. The quadrature signal is derived from the i.f. output by use of an external *LC* circuit and associated reactance (one possible arrangement is shown in dashed lines) and is applied also in pulse form to two push–pull pairs TR_1TR_2 and TR_3TR_4 in a balanced circuit which ensures that none of the quadrature component appears between the output terminals. Suppose TR_1 base is driven positive by the quadrature signal at an instant when TR_5 is conductive. The effect is to promote conduction in TR_1 and thus to cut off TR_2, producing a net output between the output terminals. Half a cycle later, when TR_6 is conductive, TR_3 and TR_4 behave similarly and again there is a net output. The duration of these outputs depends, of course, on the extent of the overlap between the i.f. and quadrature inputs and varies with the phase difference between the two inputs. The output can be used as a.f. in an f.m. receiver or for a.f.c. purposes.

TR_7 is included to stabilise the mean current through the detector and is one of the many auxiliary components included in i.cs to ensure that the performance is substantially unaffected by variations in ambient temperature or in supply voltage.

A number of i.cs designed for use in f.m. receivers incorporate detectors with a circuit similar to that of *Figure 41.25* and they are often described as balanced, symmetrical, quadrature coincidence or product detectors.

41.3.3 Counter discriminator

This uses a principle quite different from those employed in the detectors so far described. If an f.m. signal is rectified the result is a succession of half-sinewave pulses the frequency of which varies according to the modulation. At periods where the pulses are crowded the mean value per unit time is greater than at instants when they are less crowded. This variation in mean value represents the modulation waveform and if the rectified signal is passed through a low pass filter to suppress all but a.f. the output consists of the wanted modulation-frequency component superposed on a direct component. The change in frequency in the signal radiated from an f.m. broadcast transmitter is, however, very small compared with the centre frequency, typically ± 75 kHz maximum at a carrier frequency of, say, 90 MHz—a variation of less than $\pm 0.1\%$ representing a very small change in mean value of the rectified signal and thus a very small a.f. output from the low pass filter. The relative change in frequency is greater in the i.f. circuits, ± 75 kHz in 10.7 MHz being approximately $\pm 0.7\%$. It is usual, however, in receivers using pulse-counter discriminators to employ an i.f. of 455 kHz or even lower. At 455 kHz the maximum change in frequency is nearly $\pm 17\%$ which gives a worthwhile modulation-frequency component in the rectified signal.

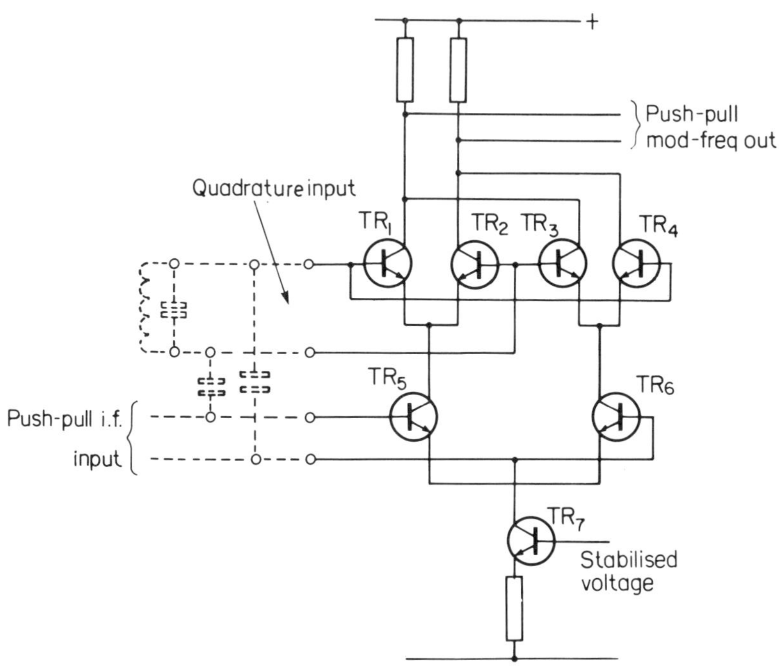

Figure 41.25 Simplified form of coincidence detector used in integrated circuits

The signal presented to the low pass filter must be free of amplitude variations because these would give a spurious output. Moreover all the input pulses must be of identical shape because variations in shape could also give unwanted components in the output. The problem is, therefore, to generate from the i.f. signal a series of pulses all of identical form and amplitude, the number per unit time varying according to the modulation.

Early pulse-counter discriminators were fed with square waves from the final limiting stage in the i.f. amplifier. The square wave was differentiated in an *RC* circuit which, as shown in *Figure 41.26*, incorporated diodes to eliminate negative-going blips. The resulting train of positive-going blips was passed through a low pass filter with a cut-off frequency of, say 30 kHz. A simple *RC* filter is shown in *Figure 41.26* which is based on an article published by M. G. Scroggie.

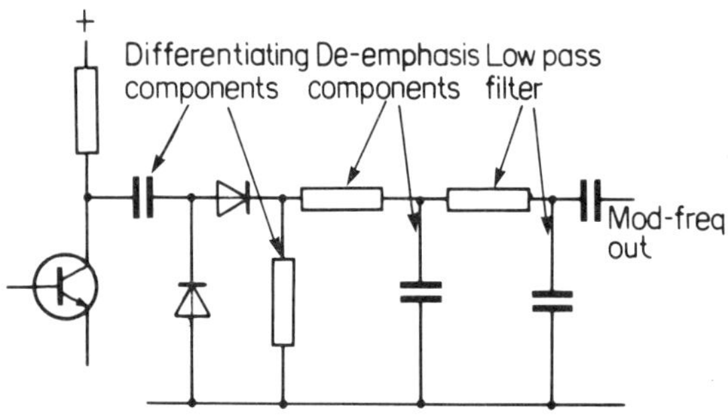

Figure 41.26 Basic form of pulse-counting discriminator

In more recent pulse-counter discriminators the positive-going blips are used to trigger a multivibrator giving, for example, 1 μs pulses which are passed through a squarer stage (to eliminate any overshoots) before being applied to the low pass filter.

Pulse-counter discriminators are used in applications where linearity is important, e.g. in f.m. rebroadcast receivers and in f.m. deviation meters.

41.3.4 Locked-oscillator discriminators

As the title suggests this last type of f.m. discriminator is based on an oscillator which is synchronised by the f.m. signal so that its frequency follows any changes in that of the input signal. Such a system can be expected to have two useful properties. Firstly the amplitude of the oscillator output can be many times that of the input signal, implying a useful degree of voltage gain. Secondly the oscillator output amplitude is independent of that of the input signal provided this is sufficient to give effective synchronising; in other words the system should give effective amplitude limiting. Thus the oscillator can be used as a source of amplified and amplitude-limited f.m. signals which can be followed by any of the types of discriminator described above. Used in this way, of course, the oscillator is not itself a discriminator but a source of input signal for a discriminator. Circuits of this type were described as early as 1944.

The synchronised oscillator can, however, act as a discriminator. If it operates in class C, taking one burst of current from the supply per cycle of oscillation, the frequency of the bursts follows that of the input signal and so contains a modulation-frequency component which can be used as detector output. For the reasons given under the previous section, however, a low value of intermediate frequency (and hence oscillator frequency) is necessary to give a worthwhile performance from such a circuit.

41.3.4.1 Phase-locked-loop circuits

In this more recent application of the principle the frequency of the oscillator is controlled not by direct application of the f.m. signal but by a control voltage dependent on the difference between the phase of the oscillator and that of the f.m. signal. The circuit, illustrated in principle in *Figure 41.27*, is so designed that the effect of the oscillator control voltage is to minimise the phase difference between the two signals applied to the phase comparator. Thus the phase of the oscillator is locked to that of the input signal and follows any variations in it. As in the circuits described earlier the output of the phase comparator contains the required modulation-frequency component but the output is usually passed through a low pass filter to suppress any radio-frequency components.

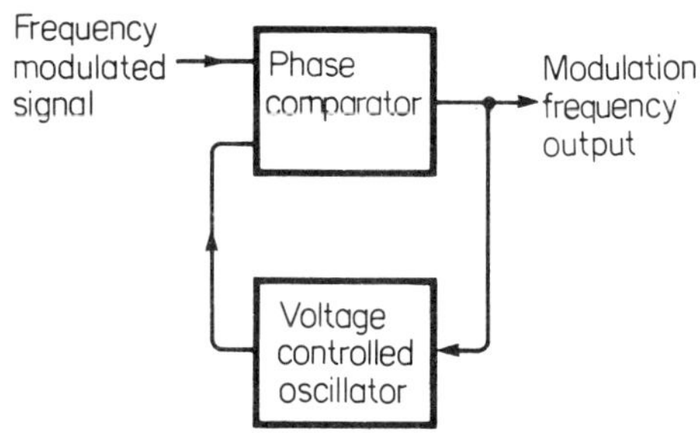

Figure 41.27 Block diagram of phase-locked-loop f.m. detector

This type of detector has something in common with those described under 'phase comparator detectors'. Comparison of *Figure 41.23* and *Figure 41.27* shows that in phase-comparator circuits the quadrature input is derived from the other input by use of a phase-shifting network whereas in the phase-locked-loop the second input is derived from the output of the comparator so introducing negative feedback.

The oscillator must be such that its frequency can be readily controlled by a voltage applied to it. It can be a Hartley or Colpitts type in which part of the capacitance of the frequency determining network is provided by a varactor, the d.c. input to which therefore controls the operating frequency. Alternatively, and this arrangement avoids any need for an *LC* circuit, the oscillator can be an astable multivibrator, the control voltage being applied to the resistors of the *RC* circuits which determine the free-running frequency of the oscillator. The phase comparator may have a circuit similar to that shown in *Figure 41.25*.

Further reading

HERBERT, J. W., 'A homodyne receiver', *Wireless World* (September 1973)

SCROGGIE, M. G., 'Low-distortion f.m. discriminator', *Wireless World* (April 1956)

AMOS, S. W., 'A.M. detectors', *Wireless World* (April 1980)

AMOS, S. W., 'F.M. detectors', *Wireless World* (January and February 1981)

42 Waveform Generators and Shapers

S W Amos BSc(Hons), CEng, MIEE
Freelance Technical Editor and Author

Contents

42.1 Introduction

A number of particular waveforms are extensively used in electronic equipment. One is the sinusoidal form and circuits for generating sine waves were described in Chapter 39. A second waveform in widespread use is the rectangular pulse and a third is the sawtooth or ramp waveform. From pulses and sawtooth forms other waveforms can be derived by use of circuits known as shapers and this chapter describes the basic methods of generating pulses and sawtooth waveforms and of waveform shaping.

42.2 Pulse generators

The rectangular pulse, shown in idealised form in *Figure 42.1*, is the fundamental form of signal in computers and digital equipment generally and is also extensively used in television, radar and other equipment; it features in the radiated television waveform.

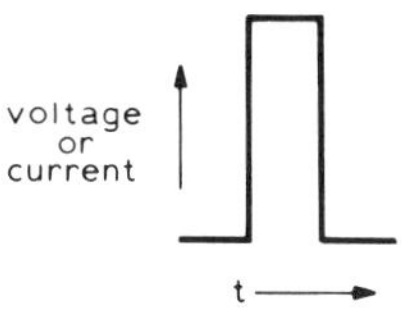

Figure 42.1 Ideal pulse waveform

One way of generating rectangular pulses is by shaping other waveforms such as sine waves. For example if a sinusoidal signal is applied to the circuit of *Figure 42.2(a)* the diode will conduct and hence attenuate all parts of the signal which exceed the diode cathode bias in value, giving an output of the form shown. If the negative-going excursions of this output are eliminated (for example by the use of a second diode) we are left with an approximation to a rectangular pulse: this becomes a nearer approach to the ideal form the greater the ratio of the sine-wave amplitude to the diode bias.

Better results can be obtained by using transistors in place of diodes as limiting devices and the circuit diagram of a two-transistor circuit which limits on positive and negative peaks is given in *Figure 42.2(b)*. The two transistors are emitter-coupled by a common resistor, an arrangement known as a long-tailed pair. TR_1 is an emitter follower driving the common-base stage TR_2. Positive-going signals applied to TR_1 base appear at substantially the same amplitude at TR_2 emitter and, if large enough, cut TR_2 off, causing its collector potential to rise to supply positive value. Negative-going signals, if large enough, cut TR_1 off so that TR_2 takes a steady collector current (determined by the potential divider R_5–R_6 and the emitter resistor R_3) and the collector potential is at a steady value below that of the positive supply rail. Thus a large-amplitude signal at the input to this circuit gives an output at TR_2 collector which alternates between two steady values, i.e. it is a pulse output at the frequency of the input signal.

A simple common-emitter amplifier can, of course, be used as a limiter. By giving it a very large input signal (which, in linear amplifier operation would be described as a gross overload) the transistor is driven into cut-off on one half-cycle of the input signal and into saturation on the other half cycle. Thus the collector potential is either at the positive or the negative supply value. The larger the input is made, the shorter is the time taken for the collector potential to switch from one extreme voltage to the other and the better is the shape of the output pulses. The input signal can be sinusoidal or it can be a pulse waveform which has overshoots, ripples or other undesirable features on it. These unwanted features can, with proper design, be removed in the limiting circuit which can thus be regarded as a 'cleaner stage'.

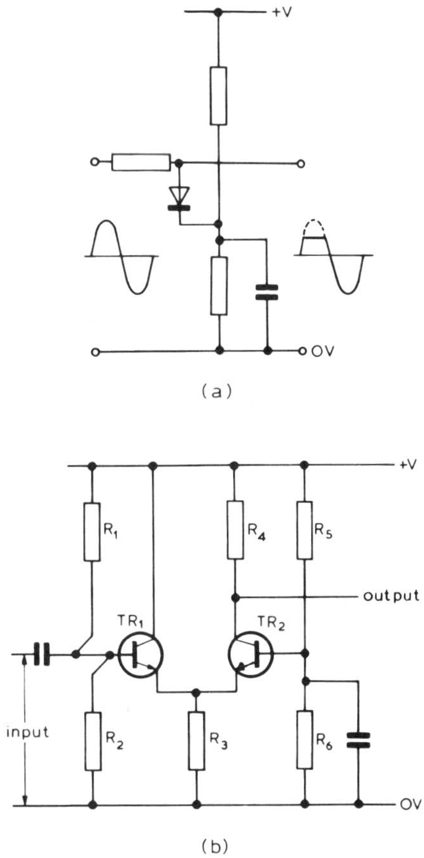

Figure 42.2 Limiting circuits: (a) diode; (b) two-transistor

There are other circuits which generate rectangular pulses and do not require any signal input waveform for their operation. These are therefore true rectangular pulse generators and the first example of such a generator is the blocking oscillator. The basic form of the circuit is given in *Figure 42.3*. TR_1 is a common emitter stage with base bias provided by the resistor *R*. There is phase inversion between the base and collector signals and *T* is a transformer which also gives phase inversion between the signals delivered to the base and collector circuits. Thus a state of positive feedback exists and this results in oscillation at the resonance frequency of the collector winding, usually the larger of the two windings. One of the aims in the design of blocking oscillators is to provide considerable positive feedback so that the oscillation amplitude builds up very rapidly. As it does so TR_1 takes a burst of base current which charges up the capacitor *C*, the polarity being such as to drive the base negative and to cut the transistor off. This is, of course, the basis of the automatic system of biasing. Ideally TR_1 should be cut off within the first half cycle of oscillation. *C* now discharges through *R* and TR_1 remains cut off until the voltage across *C* has fallen sufficiently for TR_1 to take base current again. This promotes another half-cycle of oscillation as a result of which TR_1 is again cut off by the negative voltage generated at the junction of *R* and *C* by the burst of base current.

Thus bursts of base current occur regularly and, of course, there are associated bursts of collector current. Negative-going

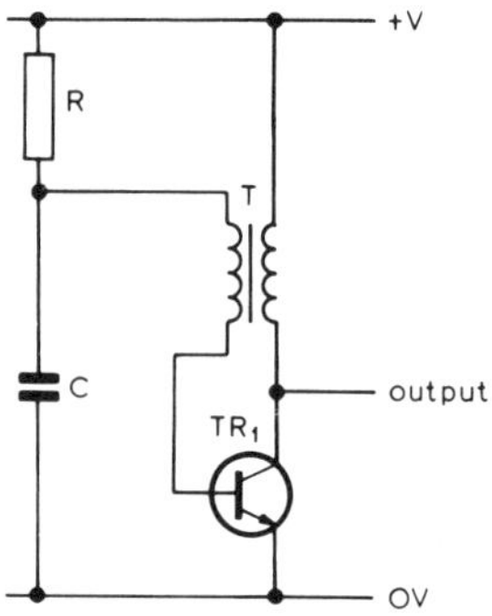

Figure 42.3 Basic circuit for a blocking oscillator

voltage pulses are hence generated at the collector terminal and these can be taken as the output of the circuit. The interval between the pulses is governed by the time constant RC so that either R or C can be made variable to provide a pulse frequency control. The duration of the pulses is a function of the transformer design being primarily dependent on the inductance and self-capacitance of the primary winding.

The natural frequency of a blocking oscillator of the type illustrated in *Figure 42.3* is given by the approximate expression

$$f=\frac{n+1}{RC} \tag{42.1}$$

where $n:1$ is the step-down ratio of the transformer. It is possible to control the natural frequency of a blocking oscillator by adjustment of the voltage to which R is returned and this method is illustrated in the circuit diagram of *Figure 42.4*. This method of

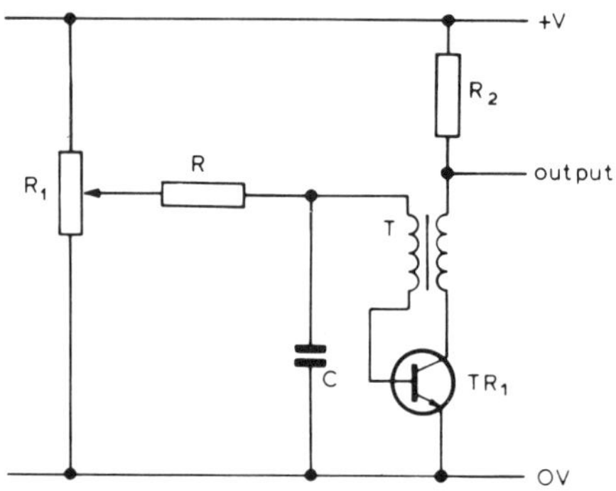

Figure 42.4 Blocking oscillator circuit with frequency control

frequency control has the advantage that the potential divider R_1 carries only d.c. and can thus be situated at some distance from the blocking oscillator itself: this could be useful in a television receiver for example in which R_1 is the line-hold or field-hold control and must be readily accessible to the user. The output from the circuit of *Figure 42.4* is taken from the resistor R_2: this avoids the inclusion of any inductive effects which may be present in the output of the circuit shown in *Figure 42.3*.

42.2.1 Synchronisation of blocking oscillators

Although the blocking oscillator does not need an input signal the circuit can readily be synchronised at the frequency of any regularly occurring signal applied to it. The synchronising signal is arranged to terminate the relaxation period earlier than would occur naturally. Thus to synchronise the circuits of *Figures 42.3* or *42.4* a synchronising signal could be in the form of positive-going pulses applied to the base circuit or negative-going pulses applied to the collector circuit. A blocking oscillator intended for synchronised operation should be designed to have a natural frequency slightly lower than that of the synchronising pulses.

If the natural frequency is made slightly below one half the frequency of the synchronising pulses then the blocking oscillator will be triggered into oscillation by every second synchronising signal, i.e. it will run at precisely one half of the synchronising frequency. This idea can be extended and it is possible to arrange for a blocking oscillator to run at 1/3, 1/4, 1/5, etc., of the synchronising frequency. In other words the blocking oscillator can be used as a frequency divider. Ratios much above 1/5 are a little difficult to achieve with reliability because such ratios demand close control over synchronising-pulse amplitude and natural frequency.

42.2.2 Unijunction transistor

The unijunction transistor (sometimes known as the double-base diode) makes possible the simple pulse-generating circuit shown in *Figure 42.5*. This semiconducting device has a filament of n-type silicon with ohmic connections at each end and a p-type

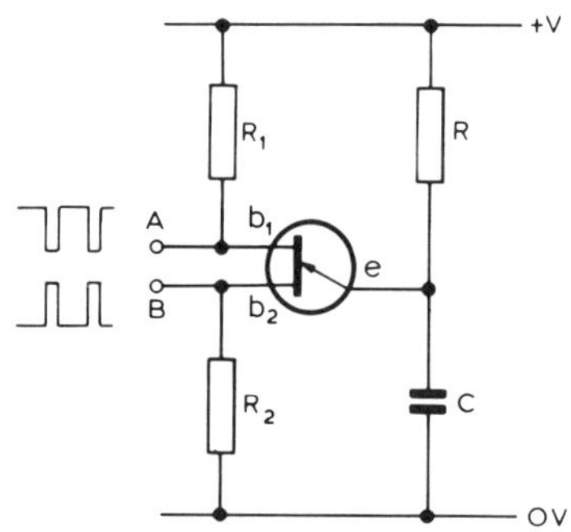

Figure 42.5 Unijunction transistor pulse-generating circuit

junction near the centre. The device has something in common with the field-effect transistor in that the filament has a high resistance when the pn junction is reverse biased but a low resistance when the junction is forward biased.

Initially C is discharged and the pn junction is reverse biased. The filament resistance is high and little current therefore flows through R_1 and R_2. As C charges through R the voltage across C rises. When it reaches the potential of that part of the filament which is in contact with the p-type emitter, the pn junction becomes conductive and the filament resistance abruptly drops. The enhanced conductivity is chiefly due to the injection of holes from the emitter which move towards the negative end of the filament thus increasing the forward conduction of the pn junction. A regenerative process is thus set up which culminates in a very rapid increase of current in the filament which discharges C. The discharge C causes the pn junction to become reverse biased again with the result that the circuit just as suddenly reverts to the state originally assumed. C now begins to charge again and the cycle recommences. Each sudden burst of filament current causes a negative pulse at terminal A and a positive pulse at B. Thus the circuit produces pulses of opposite polarity at a rate determined by the time constant RC.

42.3 Multivibrators

If the transformer in the circuit of *Figure 42.3* is replaced by any other device which introduces a phase inversion between the signal received from the collector and that fed to the base, then the circuit still has positive feedback and is capable of oscillation. Such a device is a transistor connected in the common-emitter mode and the two-transistor circuit so obtained is known as a multivibrator. The behaviour of the circuit depends on the nature of the coupling circuits, i.e. whether these are direct or capacitive. There are, in fact, three basic versions of the multivibrator circuit:

(*a*) with two direct couplings;
(*b*) with one direct and one capacitive coupling;
(*c*) with two capacitive couplings.

42.3.1 Bistable circuit

Figure 42.6 shows the circuit diagram of a multivibrator with two direct inter-transistor couplings. The base circuit of each transistor is fed from the collector circuit of the other via a potential

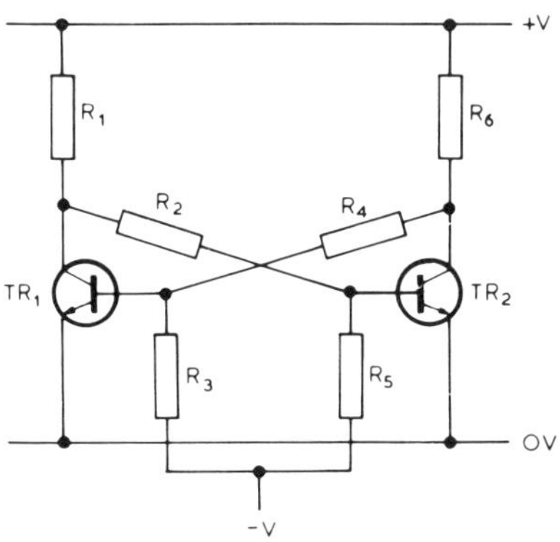

Figure 42.6 Basic bistable multivibrator circuit

divider and the resistor values are so chosen that when one transistor is conducting, its low collector potential ensures that the other is cut off. Moreover the high collector potential of the cut-off transistor ensures that the other transistor is maintained in the conductive state. For example TR_1 may be cut off and TR_2 conducting. When the circuit is placed in this state, it will remain in it indefinitely unless compelled to change by an external triggering signal. Such a signal may consist of a positive-going pulse applied to TR_1 base to make it conduct or a negative-going pulse applied to TR_2 base to cut it off. Either form of signal will cause a change of state in both transistors. The circuit will now remain in this new state (TR_1 on and TR_2 off) indefinitely unless compelled to change it by an external signal. Persistent states such as those described are known as stable and the direct-coupled multivibrator thus has two stable states: such a circuit is known as bistable. The changes of state, once initiated by the external signal, are very fast, being accelerated by the positive feedback inherent in the circuit. The transitions from one state to the other can be made even faster by connecting capacitors in parallel with R_2 and R_4 as shown in *Figure 42.7*(*a*): C_3 and C_4 are known as speed-up capacitors.

One of the principal features of the bistable is its ability to maintain a particular state, i.e.; it has a memory. For this reason bistables are extensively used in the stores and registers of computers and similar equipment.

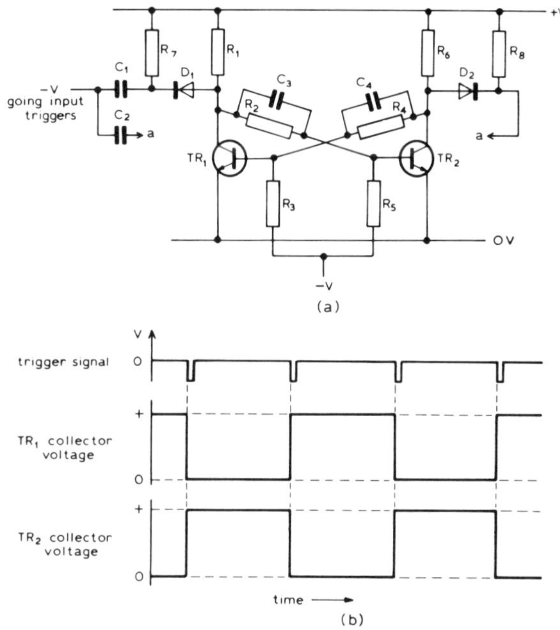

Figure 42.7 Bistable circuit with speed-up capacitors and diode input gate: (a) circuit; (b) associated waveforms

The bistable can also be used as a pulse generator for if it is fed with a regular stream of triggering signals it will change state with each received signal and will thus generate square waves at each collector. If the triggering signals are negative-going blips and if the first cuts TR_1 off then the next must be fed to TR_2 base to cut this off and so on, i.e. the triggers must be directed alternately to the two bases. This routing of triggers can be achieved by a diode gate circuit such as that illustrated in *Figure 42.7*(*a*). If TR_1 is cut off the diode D_1 will conduct a negative trigger to TR_1 collector and thus to TR_2 base via the speed-up capacitor C_3. TR_2 is conductive and its low collector potential biases D_2 off so that the trigger cannot reach TR_1 base. Thus the trigger can only affect TR_2 which is cut off by it, causing TR_1 to be turned on. D_2 can now direct the next trigger to TR_1 base. The signals generated at the collectors are thus square waves at half the frequency of the applied triggers as shown in *Figure 42.7*(*b*). The signals at the bases are also square waves in antiphase to the collector signals and are of smaller amplitude, but the speed-up capacitors can give some overshoots on the square waves which are illustrated in idealised form.

It is significant that the transistors in this circuit are either on (with a substantial collector current) or off (with zero collector current). Except for a very short period at each transition the collector currents never have any values other than these two. The transistors are used, in fact, as switches and the shape of the input–output characteristic, which is of considerable importance in analogue applications, is generally of little consequence in pulse circuits.

42.3.2 Monostable multivibrator

Suppose a multivibrator has one direct coupling and one capacitive coupling as shown in *Figure 42.8*(*a*). The introduction of the coupling capacitor C_1 makes a fundamental difference to the behaviour of the circuit because there is now no means of preserving a negative voltage on the base of TR_2 to keep it non-conductive. Because the base resistor R_3 is returned to the

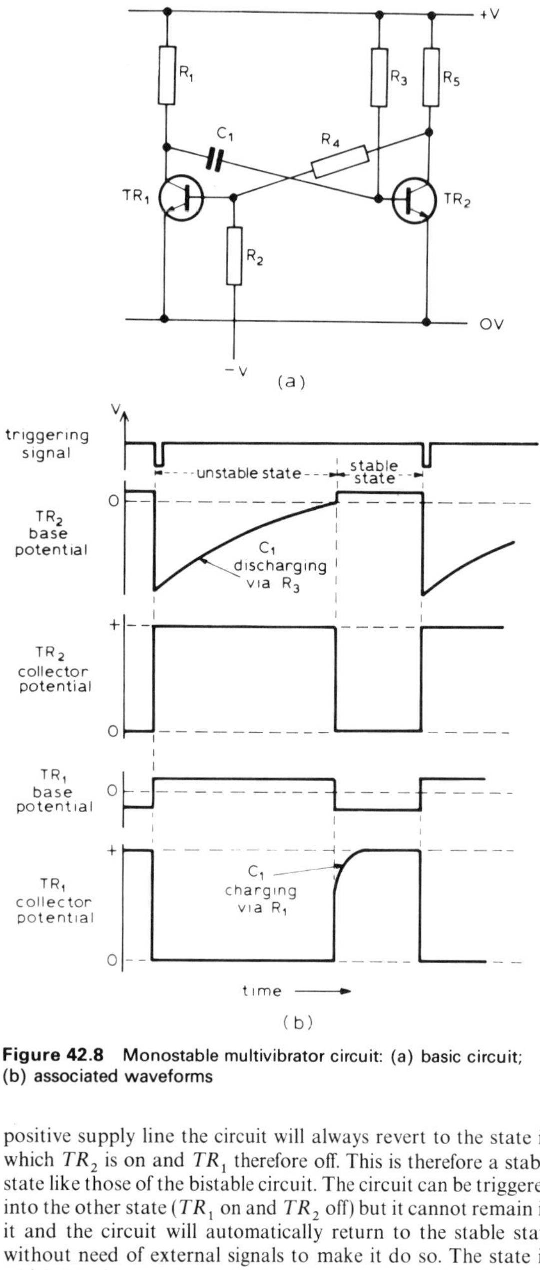

Figure 42.8 Monostable multivibrator circuit: (a) basic circuit; (b) associated waveforms

positive supply line the circuit will always revert to the state in which TR_2 is on and TR_1 therefore off. This is therefore a stable state like those of the bistable circuit. The circuit can be triggered into the other state (TR_1 on and TR_2 off) but it cannot remain in it and the circuit will automatically return to the stable state without need of external signals to make it do so. The state in which TR_1 is on and TR_2 is off is therefore an unstable state and circuits such as this which have one stable state and one unstable state are known as monostables.

In the stable state TR_2 is on and TR_1 is off. TR_2 has base current and the base potential is near supply negative value. The collector potential of TR_1 is near that of the supply positive line and C_1 is hence charged to the supply voltage. A negative-going trigger applied to TR_2 base cuts this transistor off and turns TR_1 on so that TR_1 collector potential falls abruptly to negative supply value, carrying the base potential of TR_2 to a considerable negative voltage. C_1 now begins to discharge through R_3 and as it does so TR_2 base potential rises towards zero. But for the connection to TR_2 base the potential at the junction of R_3 and C_1 would rise to the supply positive value. However as soon as this junction becomes slightly positive with respect to supply negative TR_2 begins to conduct and this initiates another transition, accelerated by positive feedback, which ends with TR_2 on and TR_1 off. The rise in TR_1 collector potential cannot be instantaneous because it is controlled by C_1 which charges via R_1. The waveforms for this circuit are therefore as shown in *Figure 42.8(b)*.

During the unstable state the collector potential of TR_1 is low and that of TR_2 is high. Thus the circuit develops negative-going pulses at TR_1 collector and positive-going pulses at TR_2 collector during this period the duration of which is determined by the time constant R_3C_1. The main application of the monostable is the generation of pulses of a predetermined duration on receipt of a triggering signal. To obtain an estimate of the duration of the pulses we can assume that TR_2 conducts when C_1 has discharged to half the voltage it had across it immediately after triggering. The voltage V_t across a capacitor discharging into a resistance falls exponentially according to the relationship $V_t = V_0 e^{t/RC}$ where V_o is the initial voltage. Since $V_t = V_o/2$ we have

$$t = \log_e 2R_3C_1 = 0.6931R_3C_1 \qquad (42.2)$$

Thus if we require 100-μs pulses $R_3C_1 = 144.3$ μs. R_3 supplies TR_2 with base current in the stable state and if TR_2 is to take, say, 5 mA collector current at this time then the base current should preferably not be less than 100 μA. For a supply voltage of 10 V therefore R_3 should not be less than 100 kΩ. This gives C_1 as

$$C_1 = \frac{144.3}{10^5}\ \mu\text{F} = 1.44\ \text{nF}$$

42.3.3 Astable multivibrator

The circuit diagram of a multivibrator with two capacitive couplings is given in *Figure 42.9(a)*. There is no means of maintaining a permanent negative voltage on the base of either transistor and this circuit has therefore no stable state. Both states are unstable and the circuit oscillates between them continuously and automatically without need for triggering signals. It is, in fact, a free-running circuit but a multivibrator with two unstable states is generally described as astable.

The waveforms can be deduced from those of the monostable circuit and are given in *Figure 42.9(b)*. The duration of one unstable state is given approximately by $0.6931R_3C_1$ and of the other $0.6931R_2C_2$ so that the natural frequency of the astable circuit is given by

$$f = \frac{1}{0.6931(R_3C_1 + R_2C_2)} \qquad (42.3)$$

Although the circuit does not require external signals for its operation it can readily be synchronised at the frequency of a regular external signal. Normally synchronisation is achieved by terminating the unstable periods earlier than would occur naturally and therefore positive-going synchronising signals are required at the bases of the npn transistors and the natural frequency of the astable circuit should be slightly lower than that of the synchronising signal.

42.3.4 Emitter-coupled multivibrator circuits

All the multivibrator circuits so far described have included two collector-to-base couplings. It is possible to replace one of these with an emitter-to-emitter coupling: in a direct coupling the emitters are simply bonded and returned to the supply via a common resistor as shown in the monostable circuit of *Figure 42.10*. Alternatively a capacitive emitter coupling can be obtained by using individual emitter resistors bridged via a capacitor. An advantage of emitter coupling is that one collector terminal is

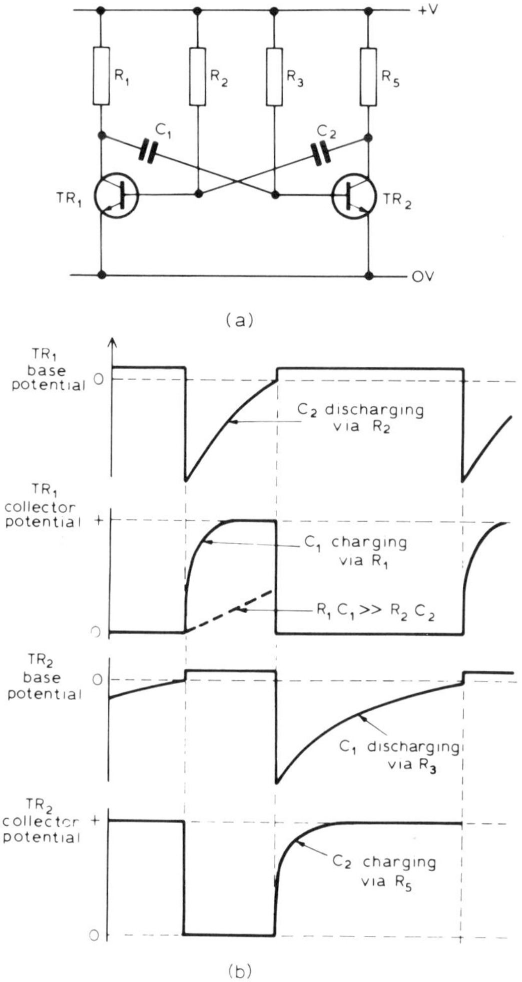

Figure 42.9 Astable multivibrator: (a) basic circuit; (b) associated waveforms

free, i.e. not involved in the provision of positive feedback and can thus be used as a convenient output point as suggested in *Figure 42.10*.

42.3.5 Complementary multivibrator circuits

By using a combination of pnp and npn transistors the particularly-simple bistable circuit of *Figure 42.11* is possible. When the npn transistor TR_1 is cut off there is no voltage drop across R_1 and therefore the pnp transistor TR_2 is also cut off. The absence of collector current in R_2 ensures that TR_1 remains cut off. Alternatively if TR_1 is on, there is a large voltage drop across R_1 which biases TR_2 on and the consequent large voltage drop across R_2 keeps TR_1 on. A feature of this circuit therefore is that in one state both transistors are off and in the other state both transistors are on.

42.4 Sawtooth generators

Waves of the shape shown in *Figure 42.12*, known as sawtooth or ramp waveforms, are used for electron beam deflection in

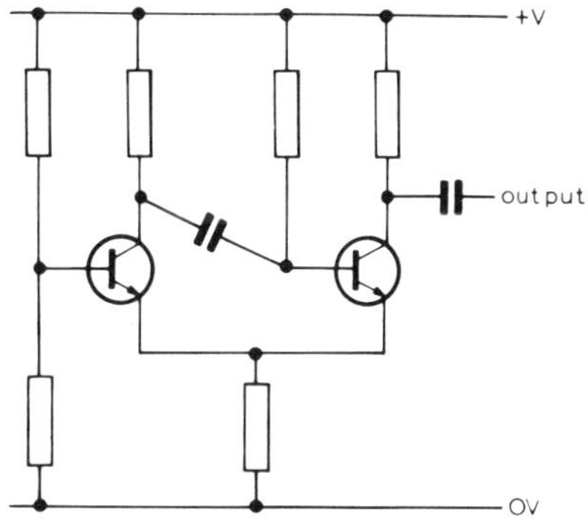

Figure 42.10 Circuit for a monostable emitter-coupled multivibrator

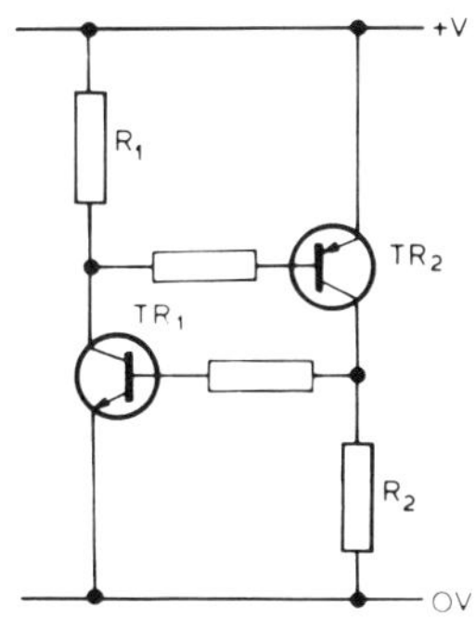

Figure 42.11 Complementary bistable multivibrator circuit

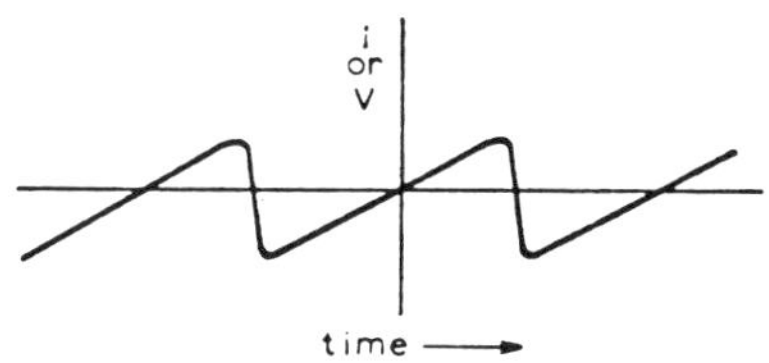

Figure 42.12 Sawtooth waveform

oscilloscopes, in television transmitting and receiving equipment, in digital to analogue conversion equipment and in measuring equipment. Normally the slow rise is required to be linearly related to time for this is the working stroke of the waveform but the shape of the rapid fall (the return stroke) need not be linear and the shape of this transition is not usually significant.

42.4.1 Production of sawtooth voltages

An approximation to a sawtooth voltage waveform can be obtained from a simple circuit such as that shown in *Figure 42.13*. The required output is generated across the capacitor C as it charges from the supply via R. The flyback voltage is obtained by discharging C by the transistor which is turned on for the duration of the flyback period. Such a simple circuit has a number of serious limitations: the most important is that the rise of voltage across C is exponential not linear. The departure from linearity is not serious provided that the rise in voltage is restricted to a small fraction of the supply voltage. Normally the performance of the circuit is unsuitable and methods of improv-

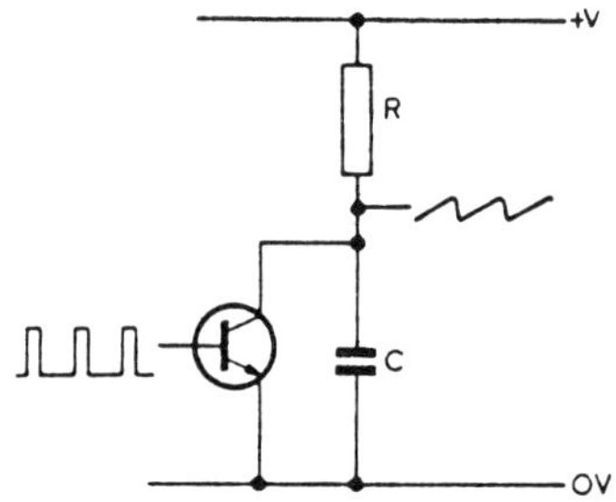

Figure 42.13 Simple discharger circuit for production of sawtooth voltages

ing the linearity of the working stroke are necessary. The reason for the lack of linearity is that the rise in voltage across C causes an equal fall in voltage across R and hence a fall in the charging current. Ideally, to achieve linearity, the charging current must be kept constant as charging proceeds. This is shown in the fundamental relationship:

$$V = \frac{Q}{C} \tag{42.4}$$

But $Q = It$

Therefore

$$V = \frac{I}{C} \cdot t \tag{42.5}$$

I must be constant to make V directly proportional to t.

Two methods of achieving a constant charging current are in common use. The first is the bootstrap circuit illustrated in *Figure 42.14*. An emitter follower is connected across the capacitor C

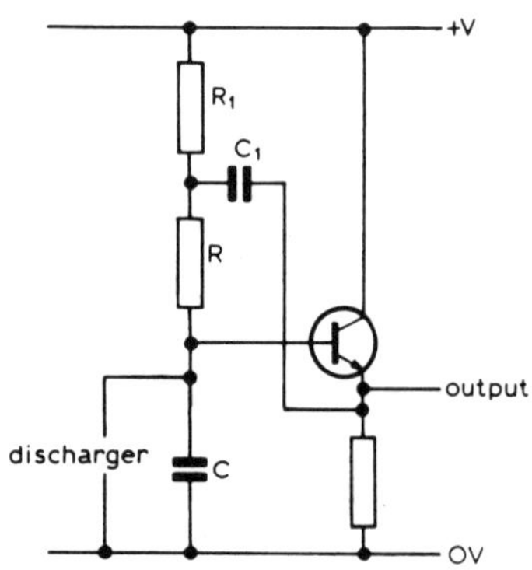

Figure 42.14 Bootstrap circuit

and delivers, at its emitter terminal, a copy of the voltage across C. The transistor therefore provides a low-resistance output terminal for the sawtooth generator. The emitter follower output is transferred via the long-time-constant circuit R_1–C_1 to the supply point for the charging resistor R. Thus as the voltage across C rises so does the voltage at the junction of R_1 and R. The voltage across R remains substantially constant throughout the charging process and thus the charging current is maintained constant.

The second method of achieving linearity is to use a source of constant current in place of the charging resistor R. A d.c.-stabilised common-emitter transistor circuit is one possible source of constant current and a circuit using this is given in *Figure 42.15*. The transistor circuit is stabilised by the potential divider R_1–R_2 connected to the base and by the emitter resistor R_3. Suppose as a numerical example it is required to generate a voltage across C which rises linearly to 10 V in 100 μs. The transistor circuit could readily be designed to give a charging current of 5 mA. We can calculate the required value for C from the simple relationship of Equation (42.5), $C = (I/V)t$.

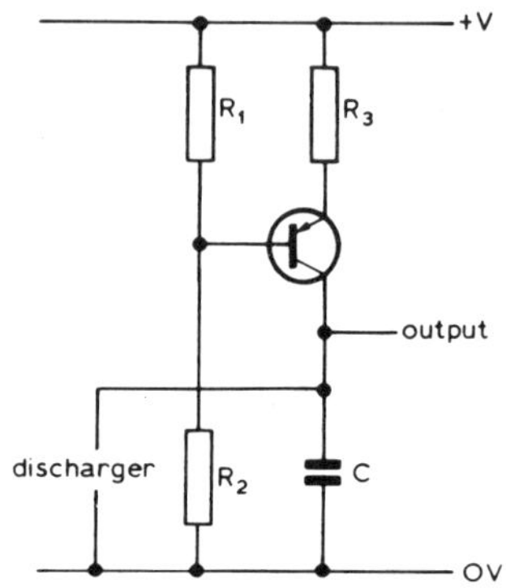

Figure 42.15 Method of achieving linearity using a transistor as a constant-current source

Substituting for I, V and t

$$C = \frac{5 \times 10^{-3} \times 100 \times 10^{-6}}{10} \text{ F} = 0.05\ \mu\text{F}$$

All the circuits so far described for the production of sawtooth voltages and currents have required a pulse waveform to operate them. Mathematically the circuits have generated the time integral of the input waveform and can thus be described as integrators. Because of the need for input signals such circuits are known as *driven circuits*. By the addition of other components sawtooth generators can be made to provide their own control pulses and can thus generate sawtooths without need for external triggering signals. Such generators are known as *free-running* and a multivibrator can be used as a free-running sawtooth generator, as described in the next section.

42.4.2 Multivibrator giving sawtooth output

Fundamentally a multivibrator is a generator of approximately rectangular current pulses and if such pulses are fed to a capacitor a sawtooth waveform is developed across the capacitor. There is no need to introduce an additional capacitor into the circuit for this purpose (indeed in collector-coupled circuits this might affect the positive feedback on which multivibrator action depends) because one of the coupling capacitors can be used.

It has already been pointed out that when a transistor is cut off in a multivibrator circuit, its collector potential cannot rise instantaneously to supply voltage value because the capacitor coupling the collector to the base of the other transistor must charge during this period via the collector load resistor. Thus in *Figure 42.9(a)* when TR_1 is cut off C_1 charges via R_1. If the time constant R_1C_1 is made long compared with the period of non-conduction of TR_1 (determined by the time constant R_2C_2) C_1 is not fully charged during this period and the voltage generated at TR_1 collector is exponential in form. The initial part of an exponential rise is almost linear as shown in dotted lines in *Figure 42.9(b)*. Thus the condition to be satisfied to obtain such an output is: $R_1C_1 \gg R_2C_2$. The choice of values for R_1 and C_1 is limited because the time constant R_3C_1 determines the period of non-conduction of TR_2 and hence of conduction of TR_1.

42.4.3 Production of sawtooth currents

For television camera and picture tubes a sawtooth current is required in the deflection coils to deflect the beam horizontally at line rate and vertically at field rate. At the line frequency the deflector coils are predominantly inductive and the problem thus is to generate a sawtooth current in an inductive circuit. There is a very easy solution for if an inductor is connected across a constant voltage supply the current in the inductor rises linearly with time. The voltage across an inductor is related to the rate of change of current in it according to the expression $E = -L(\mathrm{d}i/\mathrm{d}t)$. Thus if L and E are constant then $\mathrm{d}i/\mathrm{d}t$ must be constant. A very simple sawtooth current generator can thus have the circuit diagram shown in *Figure 42.16*. The transistor is held in the conductive state for the required duration of the working stroke by the positive-going pulse waveform applied to the base. This gives the transistor a low-resistance path between collector and emitter so that, in effect, L is connected directly across the supply and current therefore grows in it linearly. At the end of the working stroke the transistor is turned off by the negative-going signal at the base and the current in L falls rapidly to zero. This is the very simple basis of the line deflection circuits used in television receivers. In practice the circuits are considerably more complex as illustrated in *Figure 42.17(a)*.

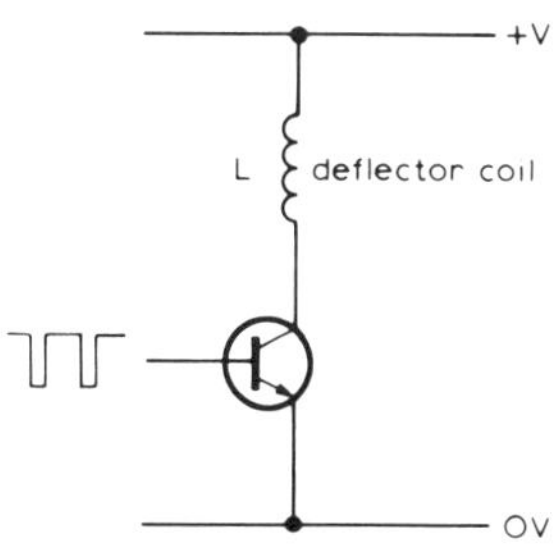

Figure 42.16 Basic circuit for the generation of sawtooth current waveforms

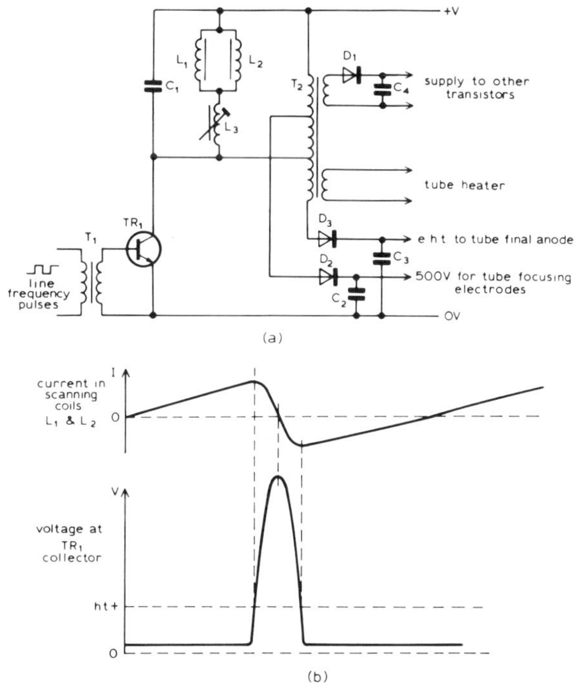

Figure 42.17 More practical line output circuit for a television receiver: (a) circuit; (b) associated waveforms

The line scan coils L_1 and L_2 are fed from a line output transformer to ensure that no d.c. flows in the coils (this would produce an undesirable static deflection of the beam) and via the variable inductor L_3 which enables the magnitude of the line-scan current (and hence picture width) to be adjusted. The line-scan circuit is tuned by capacitor C_1: this plays a vital part in the operation of the circuit which will now be described and is illustrated in the curves of *Figure 42.17(b)*.

TR_1 is driven into conduction by pulses from the line oscillator and driver stages and is held conductive while a linearly growing current flows into the coils from the supply. At the end of the working stroke TR_1 is cut off but the current in an inductive circuit cannot cease instantaneously and continues to flow, being taken from C_1 which was previously charged to the supply voltage. C_1 is rapidly discharged by this current which continues to flow and charges C_1 in the reverse direction. As energy flows into C_1 the current in the line-scan coils falls, reaching zero when the voltage across C_1 is a maximum. These exchanges in energy between the scan coils and C_1 are the beginning of free oscillation in the resonant circuit L_1–L_2–L_3–C_1 and in the next stage current begins to flow in the scan coils again but in the reverse direction and this reaches a maximum at the moment when C_1 is again discharged. Scan current continues to flow and C_1 begins to acquire charge again of the polarity assumed initially. When the voltage across C_1 is approximately equal to the supply voltage it forward biases the collector-base junction of TR_1 and this, together with the secondary winding of transformer T_1, provides a low-resistance path from the lower end of L_3 to the negative terminal of the supply. Thus the scanning-coil circuit is again connected directly across the supply. Now however the current in the coils is a maximum and in the reverse direction to that assumed initially. The current therefore falls and the constant voltage across the coils ensures that the rate of fall of current is also constant, i.e. the scan current falls linearly to zero. This current flows into the supply via TR_1 collector-base junction and the secondary winding of T_1. When the current has reached zero the circuit is in the state assumed initially and TR_1 is now turned on again by the external signal to provide another period of linear growth of scan current. With proper design the linear fall and subsequent linear growth of current in the scanning coils combine to produce an uninterrupted linear change of current which constitutes the working stroke. This 'resonant return' circuit is extensively used in scanning systems and is very efficient because almost as much energy is returned by the coils to the supply during the first half of the working stroke as is taken by the coils from the supply during the second half of the working stroke. The transistor TR_1 is turned on only during the second half of the working stroke. The resonant circuit L_1–L_2–L_3–C_1 performs half a cycle of oscillation during the flyback period and this requirement enables suitable values of L_1–L_2–L_3 and C_1 to be calculated.

The rapid change of scan current during flyback causes a high voltage peak to be generated in the line output stage. A sample of this voltage is obtained from a winding on the line output transformer and is rectified by D_3 to provide the e.h.t. voltage for the final anode of the picture tube. The rectifier polarity is such that it conducts during the flyback period and it is commonly a multiplier type to give the high voltage, typically 11 kV, which is required.

A tapping on the primary winding of the line output transformer similarly provides a supply of approximately 500 V for the focusing electrodes of the picture tube. The rectifier is D_2. A secondary winding on the same transformer provides a supply, not rectified, for the tube heater. This is a convenient method of obtaining a low-voltage supply for the heater and avoids the provision of a mains transformer for the purpose.

Another winding on the line output transformer provides a d.c. supply of say 25 V for most of the transistors in the receiver. The rectifier D_1 is arranged to conduct during forward strokes of the line output stage and C_4 provides smoothing. Clearly this 25 V supply cannot be used for the line output transistor itself and the associated driver and oscillator stages and these are usually powered from mains rectifying equipment.

42.5 Waveform shaping circuits

42.5.1 Differentiating and integrating circuits

Bistable and monostable circuits need triggering signals for their action and astable circuits are usually controlled by synchronising signals. It was assumed in the descriptions of these circuits that the external signals had the form of pulses. Certainly pulses can be used for this purpose but it is clear from the descriptions that it is the leading or trailing edge of the pulse which is effective in the triggering or synchronising process: the horizontal part of the pulse is unimportant in this application.

For this reason it is common practice to feed pulses to multivibrators and other circuits via an *RC* circuit of the form shown in *Figure 42.18(a)*. Such an arrangement of series capacitance and shunt resistance is also commonly encountered in the intertransistor coupling circuits of linear amplifiers where the time constant must be so chosen that the signal is transmitted through the *RC* circuit with negligible change in waveform. For a pulse signal a very long time constant would be necessary to transmit the horizontal sections with negligible sag. However the horizontal sections are of no interest if the pulses are used for triggering or synchronising and it is therefore permissible to use a short time constant in the network used to transmit them. The effect of a time constant which is short compared with the pulse repetition period is illustrated in *Figure 42.18(b)*. The vertical leading and trailing edges are reproduced without distortion but the horizontal section is badly distorted. If the time constant is reduced sufficiently the output becomes simply a succession of alternate positive-going and negative-going spikes. Such a waveform is, of course, quite suitable for triggering or synchronising purposes. Mathematically such a waveform is the first derivative of the input waveform and for this reason the network of *Figure 42.18(a)* is often called a differentiating circuit. Thus the time constants R_7C_1 and R_8C_2 in *Figure 42.7(a)* could both be small compared with the period of the input signal.

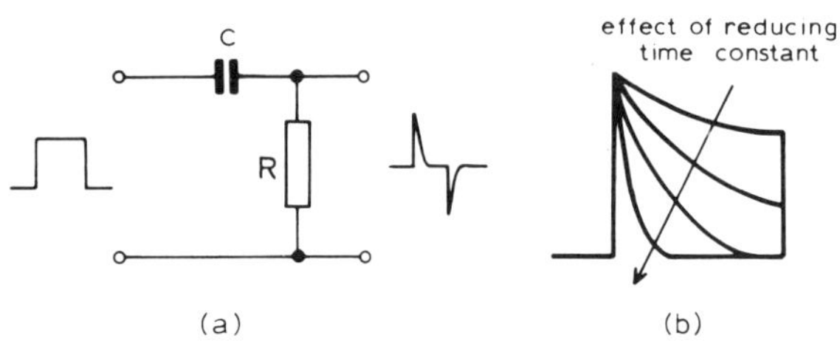

Figure 42.18 A simple differentiating circuit: (a) circuit; (b) the effect on the output waveform of reducing the time constant

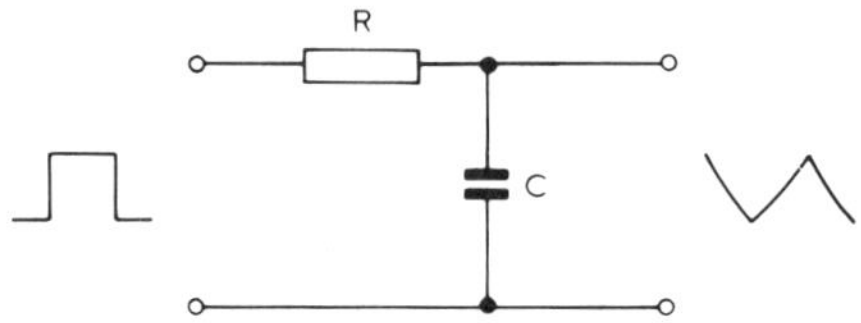

Figure 42.19 A simple integrating network. *RC* must be large compared with the period of the input signal

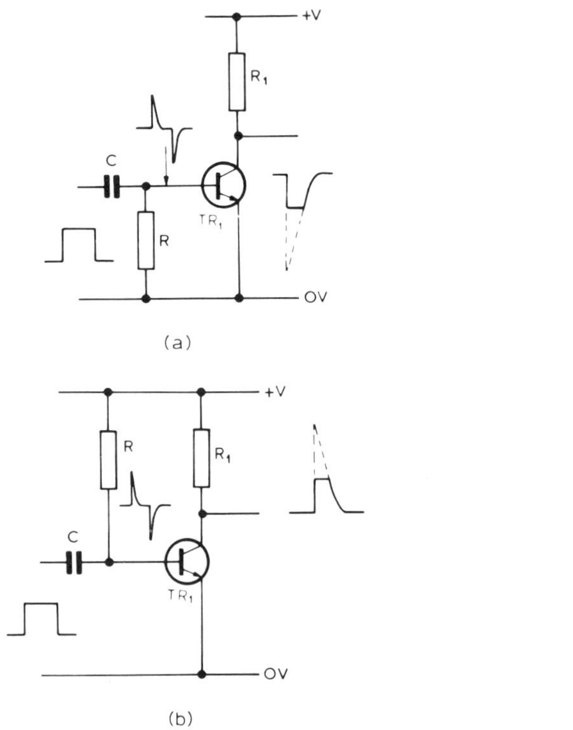

Figure 42.20 Pulse-shortening circuit giving output pulses with a leading edge coincident with: (a) the leading edge; and (b) the trailing edge of the input pulses

The circuit of *Figure 42.19*, on the other hand, gives an output waveform similar to the time integral of the input waveform, provided the time constant *RC* is long compared with the period of the input signal.

42.5.2 Pulse-shortening circuit

The circuit illustrated in *Figure 42.20(a)* makes use of a differentiating network to produce short pulses. Input pulses are differentiated by *RC* to produce positive-going and negative-going spikes as shown. TR_1 is normally biased off by resistor *R* and its collector potential is therefore at supply positive value. The positive-going spike of the differentiated input signal triggers TR_1 into conduction and produces a negative-going pulse at the collector. The leading edge of this pulse is coincident with the positive-going leading edge of the pulse input to the circuit. Provided the input pulse amplitude is large enough TR_1 will reach saturation when triggered and the width of the collector pulse waveform will depend on the time constant *RC*, increasing with increase in the time constant.

If the resistor *R* is returned to the positive supply line as shown in *Figure 42.20(b)*, the transistor is normally biased to saturation and the collector potential is at supply negative value. The transistor now ignores the positive-going spikes of the differentiated input waveform but is cut off by the negative-going spikes so that a positive-going pulse is generated at the collector and its leading edge is coincident with the trailing edge of the input waveform. Again the width of the pulse so generated can be increased by increases in the time constant *RC*.

43 Rectifier Circuits

J P Duggan
Logic Product Officer
Mullard Ltd

Contents

43.1 Single-phase rectifier circuits with resistive load

The commonly used single-phase rectifier circuits, and the output voltage waveforms for these circuits when used with a resistive load, are shown in *Figures 43.1* and *43.2* respectively.

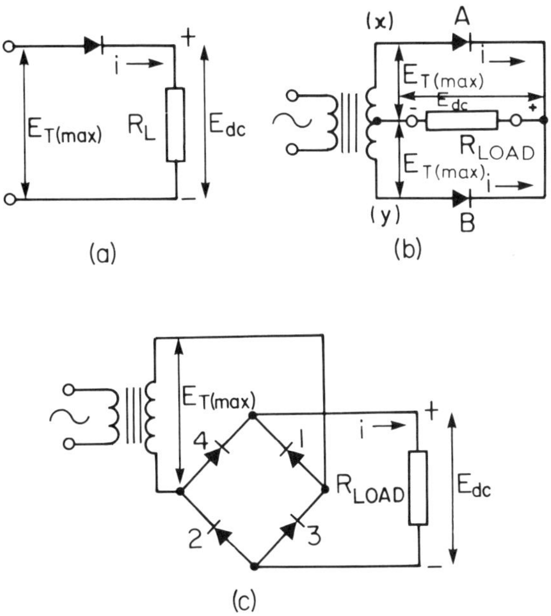

Figure 43.1 Single-phase rectifier circuits: (a) half-wave; (b) full-wave centre-tap; (c) full-wave bridge

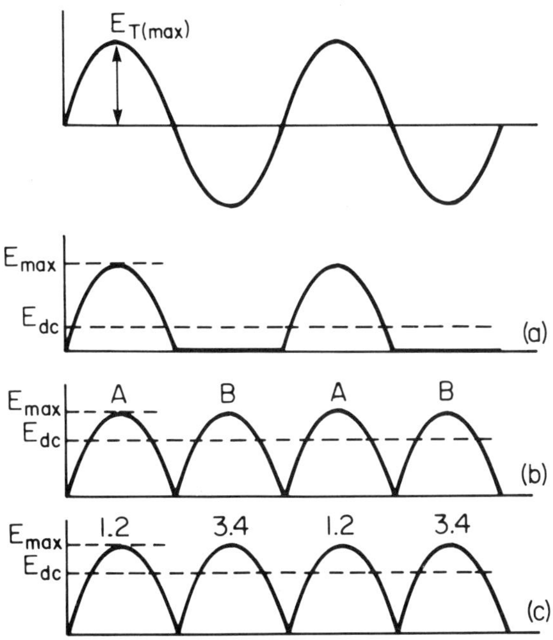

Figure 43.2 Waveforms for single-phase circuits. The sine wave input is shown at the top, and the output waveforms below: (a) half-wave; (b) full-wave centre-tap; (c) full-wave bridge

The secondary input voltage applied to the circuit is sinusoidal and has a crest value $E_{T(\max)}$. For all three circuits of *Figure 43.1*, the crest output voltage $E_{\max}=E_{T(\max)}$.

The half-wave rectifier circuit conducts during the positive half cycle and blocks during the negative half cycle of the applied alternating voltage.

In the full-wave centre-tap circuit, the rectifiers are mounted so that rectifier *A* conducts when point (*x*) goes positive, and rectifier *B* conducts when point (*y*) goes positive.

In the full-wave bridge circuit, rectifiers 1 and 2 conduct during the positive half cycle, and rectifiers 4 and 3 during the negative half cycle.

The current through the load in each of the three circuits is unidirectional.

If it is assumed that the rectifiers and transformer used are ideal, the performance of any of these circuits can be calculated. The values obtained for each of the above circuits are given in *Table 43.1*.

43.1.1 Percentage ripple

If it is assumed that the amplitudes of the higher harmonics are small compared to that of the harmonic at fundamental frequency f_r, then

$$V_R\% = \frac{\text{Fundamental r.m.s. ripple voltage}}{E_{dc}} \times 100 \qquad (43.1)$$

From Equation (43.1), the r.m.s. harmonic component at fundamental frequency (which is twice the supply frequency for this circuit) is

$$\frac{4}{3\pi}\frac{E_{\max}}{\sqrt{2}}$$

therefore

$$V_R\% = \frac{4}{3\pi}\frac{E_{\max}}{\sqrt{2}}\left(\frac{2E_{\max}}{\pi}\right)^{-1} \times 100 = 47.2$$

43.2 Single-phase circuits with capacitor input filter

Single-phase half-wave, full-wave, and voltage-doubler circuits are discussed in this section.

43.2.1 Half-wave circuit

The half-wave circuit, *Figure 43.3*, is the simplest rectification circuit giving continuous load current. In the absence of the capacitor *C* the rectifier will deliver power to the load R_{LOAD} during the positive half cycle, and will block during the negative half cycle. This leads to discontinuous voltage and current in the load.

With the capacitor *C* in circuit, the capacitor charges to the crest value of the applied voltage on the first positive half cycle. When the applied voltage falls below the crest value, the capacitor voltage is higher than the applied voltage, and thus the rectifier is reverse biased. The capacitor now discharges into the load until such time as the applied voltage exceeds the capacitor voltage again. The rectifier is then forward biased and charges the capacitor to the crest applied voltage again. The rectifier then ceases to conduct, as previously explained, and the cycle is repeated.

The idealised current waveforms for this circuit, after a steady state has been established, are shown in *Figure 43.4*. The current through the rectifier does not rise instantaneously in practice, because of the time-constant formed by the capacitor *C* and the

Table 43.1 Idealised rectifier circuit performances

Type of rectifier circuit

Secondary input voltage per phase

Output voltage across a–b

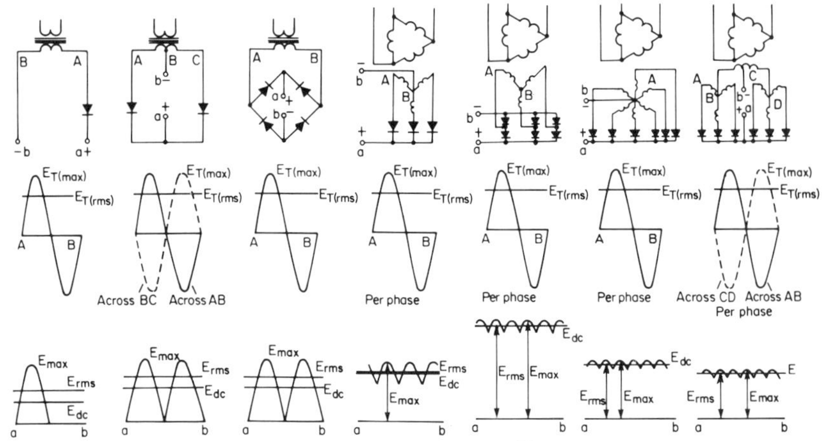

	Single-phase			Three-phase			
	Half-wave	*Centre-tap full-wave*	*Full-wave bridge*	*Half-wave*	*Full-wave bridge*	*Centre-tap*	*Double-star*
Output voltage across a–b	$E_{max}=E_{T(max)}$ $E_{rms}=0.707E_{T(rms)}$	$E_{max}=E_{T(max)}$ $E_{rms}=E_{T(rms)}$	$E_{max}=E_{T(max)}$ $E_{rms}=E_{T(rms)}$	$E_{max}=E_{T(max)}$ $E_{rms}=1.2E_{T(rms)}$	$E_{max}=\sqrt{3}E_{T(max)}$ $E_{rms}=2.34E_{T(rms)}$	$E_{max}=E_{T(max)}$ $E_{rms}=1.35E_{T(rms)}$	$E_{max}=0.866E_{T(max)}$ $E_{rms}=1.17E_{T(rms)}$
Number of output voltage pulses per cycle (N)	1	2	2	3	6	6	6
Output voltage							
E_{dc} in terms of r.m.s. input voltage per phase $E_{T(rms)}$	$0.45E_{T(rms)}$	$0.90E_{T(rms)}$	$0.90E_{T(rms)}$	$1.17E_{T(rms)}$	$2.34E_{T(rms)}$	$1.35E_{T(rms)}$	$1.17E_{T(rms)}$
E_{dc} in terms of r.m.s. output voltage E_{rms}	$0.636E_{rms}$	$0.90E_{rms}$	$0.90E_{rms}$	$0.98E_{rms}$	E_{rms}	E_{rms}	E_{rms}
E_{dc} in terms of peak output voltage E_{max}	$0.318E_{max}$	$0.636E_{max}$	$0.636E_{max}$	$0.826E_{max}$	$0.955E_{max}$	$0.955E_{max}$	$0.955E_{max}$
R.M.S. output voltage E_{rms} in terms of E_{dc}	$1.57E_{dc}$	$1.11E_{dc}$	$1.11E_{dc}$	$1.02E_{dc}$	$1.00E_{dc}$	$1.00E_{dc}$	$1.00E_{dc}$
Peak output voltage E_{max} in terms of E_{dc}	$3.14E_{dc}$	$1.57E_{dc}$	$1.57E_{dc}$	$1.21E_{dc}$	$1.05E_{dc}$	$1.05E_{dc}$	$1.05E_{dc}$
Output current							
Average current per rectifier leg I_0	I_{dc}	$0.5I_{dc}$	$0.5I_{dc}$	$0.33I_{dc}$	$0.33I_{dc}$	$0.167I_{dc}$	$0.167I_{dc}$
I_{rms} per rectifier leg R	$1.57I_{dc}$	$0.785I_{dc}$	$0.785I_{dc}$	$0.588I_{dc}$	$0.577I_{dc}$	$0.408I_{dc}$	$0.293I_{dc}$
L		$0.707I_{dc}$	$0.707I_{dc}$	$0.577I_{dc}$	$0.577I_{dc}$	$0.408I_{dc}$	$0.289I_{dc}$
I_{pk} per rectifier leg R	$3.14I_{dc}$	$1.57I_{dc}$	$1.57I_{dc}$	$1.21I_{dc}$	$1.05I_{dc}$	$1.05I_{dc}$	$0.525I_{dc}$
L		I_{dc}	I_{dc}	I_{dc}	I_{dc}	I_{dc}	$0.5I_{dc}$
Transformer rating							
Secondary r.m.s. voltage per transformer leg $E_{T(rms)}$	$2.22E_{dc}$	$1.11E_{dc}$ (to centre-tap)	$1.11E_{dc}$ (total)	$0.855E_{dc}$ (to neutral)	$0.428E_{dc}$ (to neutral)	$0.74E_{dc}$ (to neutral)	$0.855E_{dc}$ (to neutral)
Secondary r.m.s. current per transformer leg $I_{T(rms)}$ R	$1.57I_{dc}$	$0.785I_{dc}$	$1.11I_{dc}$	$0.588I_{dc}$	$0.816I_{dc}$	$0.408I_{dc}$	$0.293I_{dc}$
L		$0.707I_{dc}$	I_{dc}	$0.577I_{dc}$	$0.816I_{dc}$	$0.408I_{dc}$	$0.289I_{dc}$
Secondary volt-amp VA_s R	$3.48E_{dc}I_{dc}$	$1.74E_{dc}I_{dc}$	$1.23E_{dc}I_{dc}$	$1.50E_{dc}I_{dc}$	$1.05E_{dc}I_{dc}$	$1.81E_{dc}I_{dc}$	$1.50E_{dc}I_{dc}$
L		$1.57E_{dc}I_{dc}$	$1.11E_{dc}I_{dc}$	$1.48E_{dc}I_{dc}$	$1.05E_{dc}I_{dc}$	$1.81E_{dc}I_{dc}$	$1.48E_{dc}I_{dc}$
Secondary utility factor U_s R	0.287	0.574	0.813	0.666	0.95	0.552	0.666
L		0.636	0.90	0.675	0.95	0.552	0.675
Primary voltage per transformer leg (Transformer ratio 1:1)	$2.22E_{dc}$	$1.11E_{dc}$	$1.11E_{dc}$	$0.855E_{dc}$	$0.428E_{dc}$	$0.74E_{dc}$	$0.855E_{dc}$
Primary current per transformer leg (Transformer ratio 1:1) R	$1.57I_{dc}$	$1.11I_{dc}$	$1.11I_{dc}$	$0.588I_{dc}$	$0.816I_{dc}$	$0.577I_{dc}$	$0.408I_{dc}$
L		I_{dc}	I_{dc}	$0.471I_{dc}$	$0.816I_{dc}$	$0.577I_{dc}$	$0.408I_{dc}$
Primary volt-amp VA_p R	$3.48E_{dc}I_{dc}$	$1.23E_{dc}I_{dc}$	$1.23E_{dc}I_{dc}$	$1.50E_{dc}I_{dc}$	$1.05E_{dc}I_{dc}$	$1.28E_{dc}I_{dc}$	$1.05E_{dc}I_{dc}$
L		$1.11E_{dc}I_{dc}$	$1.11E_{dc}I_{dc}$	$1.21E_{dc}I_{dc}$	$1.05E_{dc}I_{dc}$	$1.28E_{dc}I_{dc}$	$1.05E_{dc}I_{dc}$
Primary utility factor U_p R	0.287	0.813	0.813	0.666	0.95	0.78	0.95
L		0.90	0.90	0.827	0.95	0.78	0.95
Fundamental ripple frequency f_r	f	$2f$	$2f$	$3f$	$6f$	$6f$	$6f$
% Ripple = $\frac{\text{r.m.s. fundamental ripple voltage}}{E_{dc}}\times 100$	111	47.2	47.2	17.7	4.0	4.0	4.0
Crest working voltage							
In terms of E_{dc}	$3.14E_{dc}$	$3.14E_{dc}$	$1.57E_{dc}$	$2.09E_{dc}$	$1.05E_{dc}$	$2.09E_{dc}$	$2.42E_{dc}$
In terms of $E_{T(rms)}$	$1.41E_{T(rms)}$	$2.82E_{T(rms)}$	$1.41E_{T(rms)}$	$2.45E_{T(rms)}$	$2.45E_{T(rms)}$	$2.83E_{T(rms)}$	$2.83E_{T(rms)}$

R = Resistive load L = Inductive load f = Supply frequency Hz

In the calculation of the above circuit performances, the rectifier forward voltage drop and the transformer impedance have been ignored. The primary volt-amp rating of the transformer does not take primary magnetising current into account.

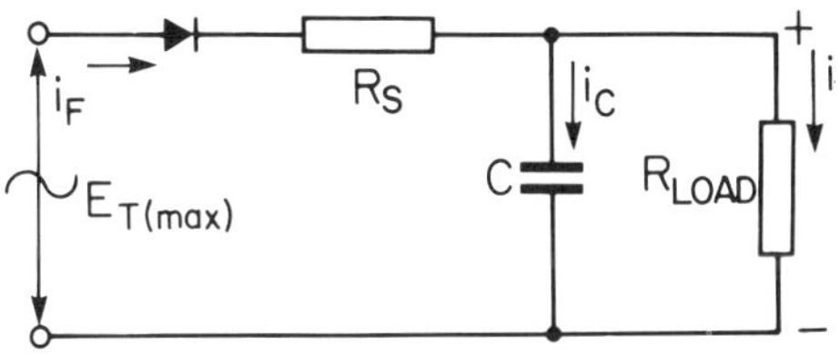

Figure 43.3 Single-phase half-wave circuit

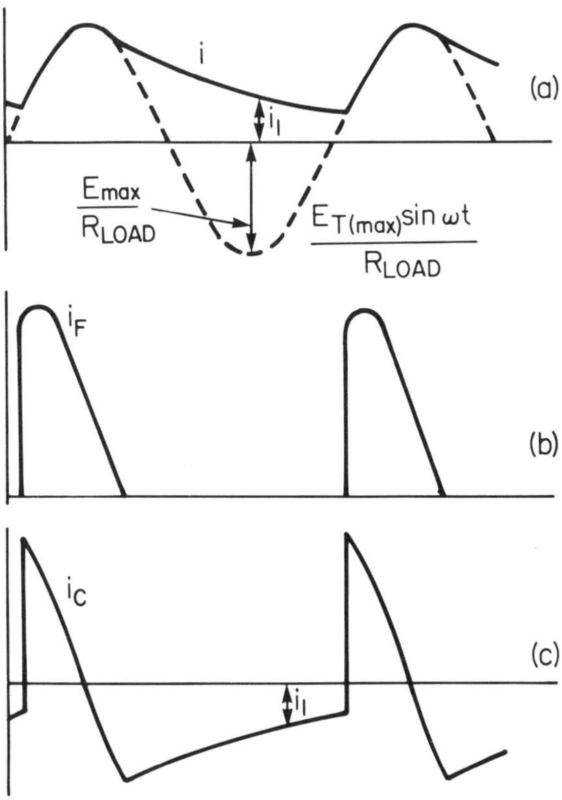

Figure 43.4 Waveforms for single-phase half-wave circuit, after establishment of steady state: (a) load current; (b) rectifier current; (c) capacitor current

source resistance of the supply plus the rectifier and any series resistance.

The capacitor acts as a reservoir, storing up energy during the period that the rectifier conducts. The rectifier current is thus the sum of the capacitor and load currents. The capacitor loses part of its charge during the period that the rectifier is non-conducting, because it discharges into the load during this period. The load current is then equal to the capacitor current i_1. Because of this action, the voltage across the capacitor does not remain constant. The ripple voltage is at the same frequency as that of the applied voltage.

The series resistor R_s is included in the circuit to limit the peak current through the rectifier on initial switch-on.

43.2.2 Performance of half-wave circuit

The charging current from the rectifier to the capacitor flows in pulses which are large in amplitude. The ripple frequency is the same as that of the applied voltage, and an expensive capacitor filter must be used to reduce the ripple to a reasonable value.

If a transformer is used to supply the power to the circuit, then the secondary of the transformer carries unidirectional current each time the rectifier conducts. The transformer has to be rated at the maximum r.m.s. current that flows through the rectifier.

The unidirectional current through the secondary winding of the transformer can lead to core saturation, which in turn leads to increases in magnetising current and hysteresis loss, and introduction of harmonics in the secondary voltage.

The regulation and conversion efficiency of the circuit is low. If a transformer is used, the utility factor of the transformer is also low. Because of the above disadvantages, this circuit is normally only used direct from the mains, and where efficiency is of secondary importance to cost.

43.2.3 Full-wave circuits

There are two types of full-wave rectifier circuit: the full-wave bridge circuit (*Figure 43.5*) and the full-wave centre-tap circuit (*Figure 43.6*). The performance of each is the same, except that,

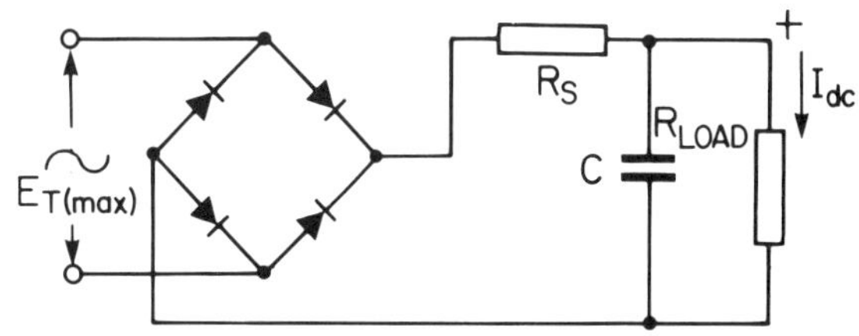

Figure 43.5 Single-phase full-wave bridge circuit

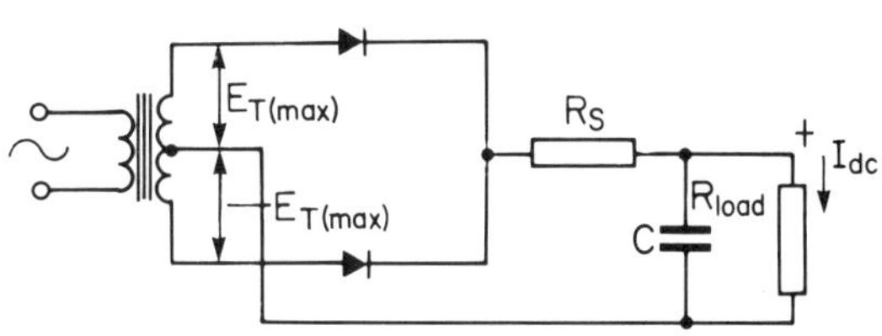

Figure 43.6 Single-phase full-wave centre-tap circuit (also known as two-phase half-wave)

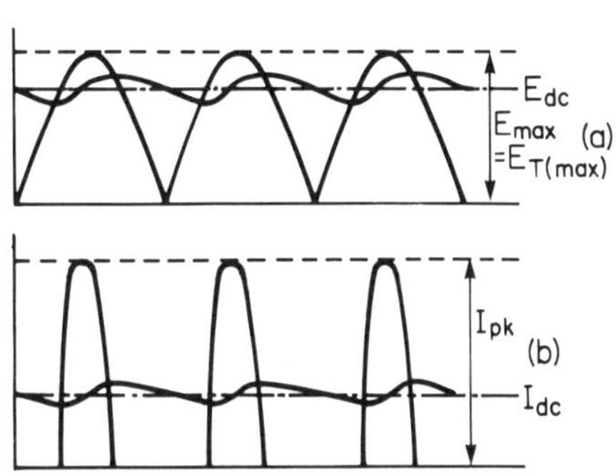

Figure 43.7 Waveforms for single-phase full-wave circuits: (a) voltage; (b) current

with rectifiers of specified crest working voltage, the d.c. voltage available in a bridge circuit is twice that of the centre-tapped transformer circuit. The voltage and current waveforms for both are shown in *Figure 43.7*.

In the full-wave bridge circuit the applied alternating voltage is rectified by the bridge, and the output from the bridge is smoothed by the capacitor filter in a similar manner to that described for the half-wave circuit. More efficient smoothing is obtained in this case, because the capacitor maintains the load current for a shorter period, and therefore the capacitor voltage will change by a smaller amount. This means that the d.c. voltage available at the output is greater than that for the half-wave circuit, and the ripple voltage is smaller. The ripple frequency is twice that of the applied voltage.

The centre-tap circuit operates in a similar manner. The rectifiers conduct alternately, and therefore the current flows through each half of the transformer secondary alternately. The rectifiers must withstand a crest working voltage which is equal to the peak value of the applied voltage across both halves of the transformer secondary.

43.2.4 Comparison of single-phase full-wave circuits

The two full-wave circuits are compared in *Table 43.2*.

Table 43.2 Comparison of single-phase full-wave circuits

	Bridge (*Figure 43.5*)	*Centre-tap* (*Figure 43.6*)
No. of rectifiers	4	2
Ripple frequency f_r	$f_r = 2f$	$f_r = 2f$
Ripple amplitude	Small compared to half-wave circuit	Small compared to half-wave circuit
Smoothing	Relatively easy	Relatively easy
Crest working voltage	$E_{T(\max)}$	$2E_{T(\max)}$
Conversion efficiency	Relatively high, but slightly lower than for centre-tap circuit because of voltage drop across additional rectifier	High
Transformer	Low transformer secondary volt-amp rating	High transformer secondary volt-amp rating

43.2.5 Applications of single-phase full-wave circuits

The principal drawback of the centre-tap circuit is the cost of the transformer. The circuit can never be used without a transformer, whereas in certain circumstances the full-wave bridge circuit may be directly operated from the mains, if the rectifiers used are rated to withstand the crest working voltage. On the other hand, it is easy to obtain a three-wire d.c. supply from a single transformer with the centre-tap circuit.

The bridge circuit is the more widely used of the full-wave circuits. It is generally used wherever the desired output voltage is approximately equal to the r.m.s. applied voltage. The centre-tap full-wave circuit is used for low-power and low-voltage applications, where low ripple is desired.

43.3 Voltage-doubler circuits

There are two types of voltage-doubler circuit: the symmetrical and common-terminal circuits.

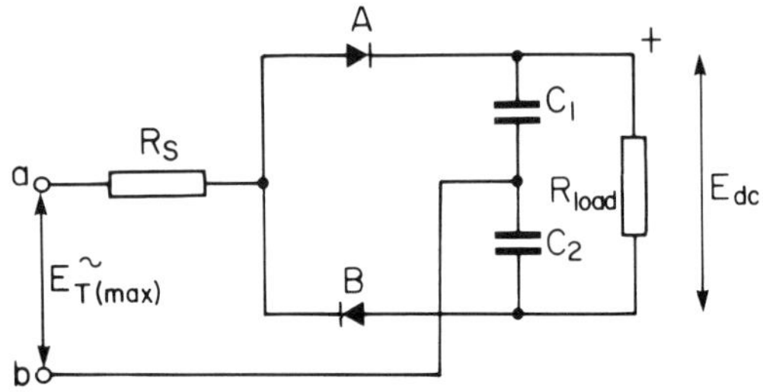

Figure 43.8 Symmetrical voltage-doubler circuit

43.3.1 Symmetrical voltage-doubler

The symmetrical voltage-doubler (*Figure 43.8*) is essentially a combination of two half-wave rectifier circuits, with smoothing filters connected in series, but supplied from the same power source. The output voltage waveform is shown in *Figure 43.9*.

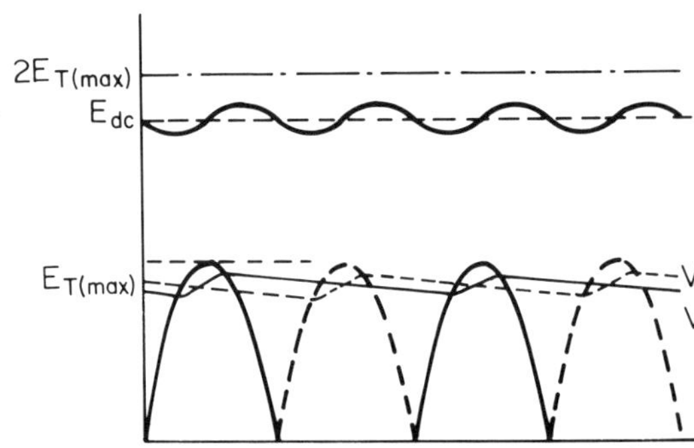

Figure 43.9 Output voltage waveform for symmetrical voltage-doubler

When (a) is positive, current flows through R_s and rectifier A to charge C_1, whose other terminal is connected to (b). When (b) is positive, C_2 is charged, the return to (a) being via rectifier B and R_s. Each capacitor charges to the peak applied voltage. The capacitors continually discharge through the load and also act as smoothing elements. The output voltage therefore tends towards twice the peak applied voltage, but cannot achieve it unless the load R_{LOAD} is disconnected.

The rectifiers must be able to withstand twice the peak applied voltage in the reverse direction. The capacitors must be rated at the peak applied voltage. The ripple frequency of the circuit is twice that of the applied voltage.

43.3.2 Common-terminal voltage-doubler

The common-terminal voltage-doubler is shown in *Figure 43.10*, and the output voltage waveform in *Figure 43.11*.

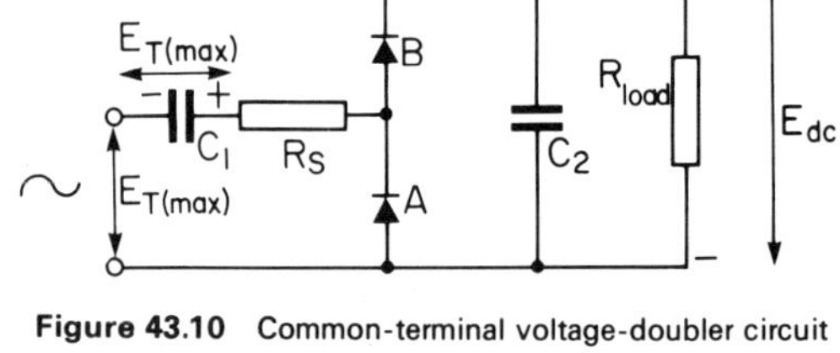

Figure 43.10 Common-terminal voltage-doubler circuit

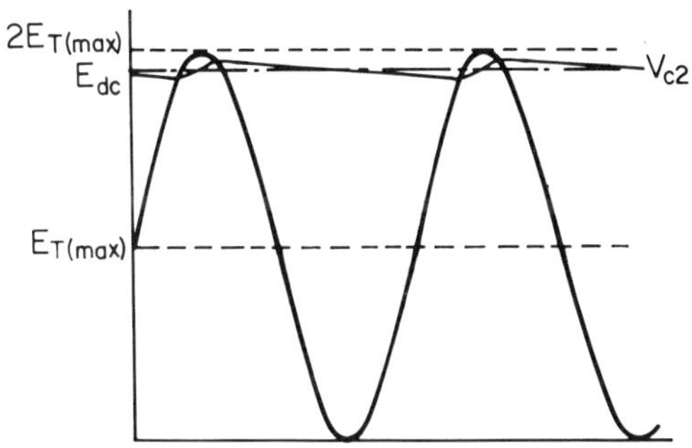

Figure 43.11 Output voltage waveform for common-terminal voltage-doubler

During the first negative half cycle of the applied voltage, C_1 charges to the peak voltage $E_{T(\max)}$ through rectifier A. During the next positive half cycle, the voltage across C_1 is in series with the applied voltage and aids it to charge C_2 to $2E_{T(\max)}$ through rectifier B. Capacitor C_1 loses part of its charge during this process, but charges up to $E_{T(\max)}$ again during the next negative half cycle. The cycle is then repeated.

The voltage across C_2 does not remain constant at approximately $2E_{T(\max)}$ because it discharges into the load R_{LOAD} when rectifier B is not supplying the load current.

The ripple frequency is the same as that of the applied voltage. Capacitor C_2 must be rated at twice the peak applied voltage, and the rectifiers must withstand twice the peak applied voltage.

Applying similar reasoning, it is relatively simple to construct a voltage tripler, a voltage quadrupler, or a circuit with an output voltage which is any other multiple of the peak applied voltage. The essential points to bear in mind are (a) the maximum crest working voltage that the rectifier must withstand, and (b) the rating of the capacitors.

43.3.3 Comparison of voltage-doubler circuits

The two voltage-doubler circuits are compared in *Table 43.3.*

Table 43.3 Comparison of single-phase voltage-doubler circuits

	Symmetrical	*Common-terminal*
Crest working voltage	$2E_{T(\max)}$	$2E_{T(\max)}$
Ripple frequency f_r	$f_r = 2f$	$f_r = f$
Capacitor rating	Rating of C_1 and C_2 must be equal to the peak applied voltage	Rating of C_1 must be equal to the peak applied voltage, and that of C_2 twice the peak applied voltage. C_1 must be rated to carry the r.m.s. load current
Regulation	Poor, but better than for common-terminal doubler	Poor

43.4 Design of single-phase circuits using capacitor input filters

The graphical solution of a capacitor filter rectifier circuit, as put forward by Schade,[6] is presented in *Figures 43.12* to *43.17*. The peak resistance $\hat{R}_s$ introduced by Schade to include the peak tube resistance is replaced by the source resistance R_s, which includes the transformer winding resistance, the rectifier resistance, and the series resistance added to limit the initial peak rectifier current.

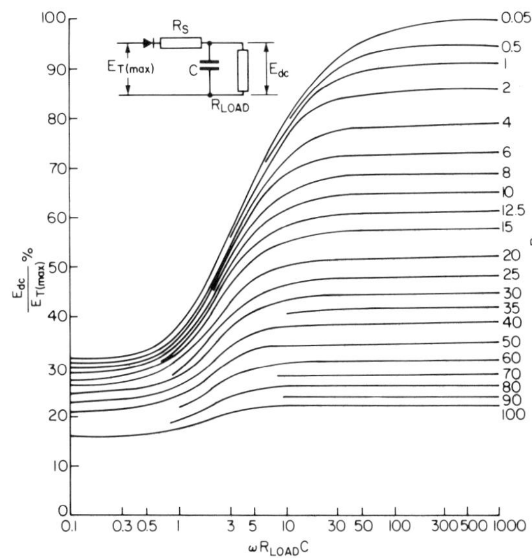

Figure 43.12 $E_{dc}/E_{T(\max)}\%$ as a function of $\omega R_{\text{LOAD}}C$ for half-wave circuits. C in farads, R_{LOAD} in Ω, $\omega = 2\pi f$

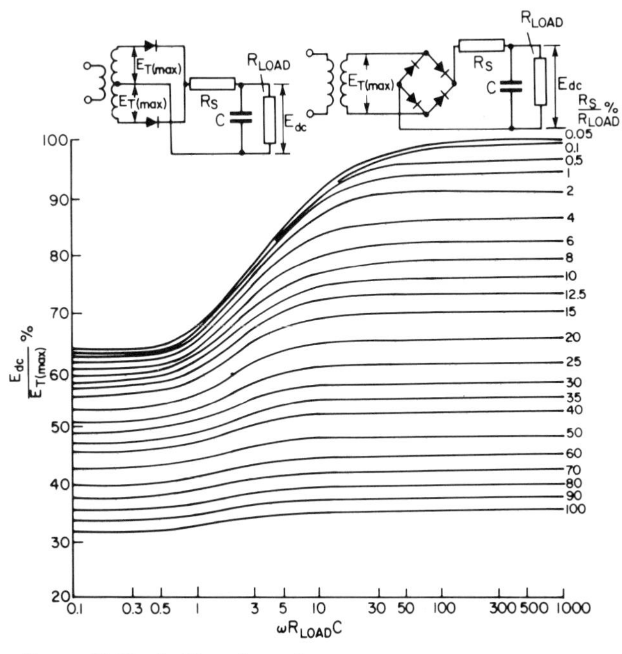

Figure 43.13 $E_{dc}/E_{T(\max)}\%$ as a function of $\omega R_{\text{LOAD}}C$ for full-wave circuits. C in farads, R_{LOAD} in Ω, $\omega = 2\pi f$

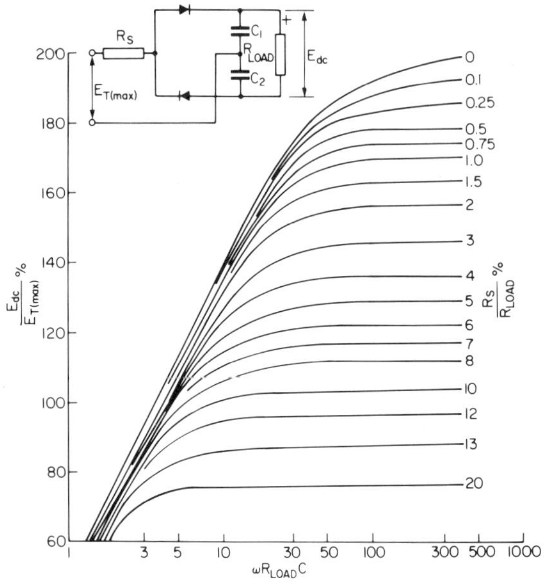

Figure 43.14 $E_{dc}/E_{T(max)}$% as a function of $\omega R_{LOAD}C$ for voltage-doubler circuits. C in farads, R_{LOAD} in Ω, $\omega = 2\pi f$

Figures 43.12, *43.13* and *43.14* give the conversion ratio $E_{dc}/E_{T(max)}$ as a function of $\omega R_{LOAD}C$ for half-wave, full-wave, and voltage-doubler circuits respectively. The conversion ratio depends on the value of (R_s/R_{LOAD}%). For reliable operation the value of $\omega R_{LOAD}C$ should be selected to allow operation on the flat portion of the curves.

Figure 43.15 gives information on the minimum value of $\omega R_{LOAD}C$ that must be used to reduce the percentage ripple to a desirable figure. *Figures 43.16* and *43.17* give, respectively, the ratio of r.m.s. rectifier current to average current per rectifier and

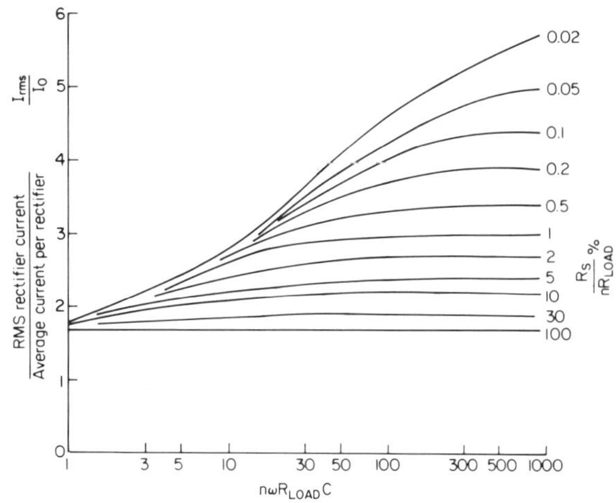

Figure 43.16 The ratio r.m.s. rectifier current/average current per rectifier plotted against $n\omega R_{LOAD}C$. C in farads, R_{LOAD} in Ω, $n=1$ for half-wave, $n=2$ for full-wave and $n=0.5$ for voltage doubler

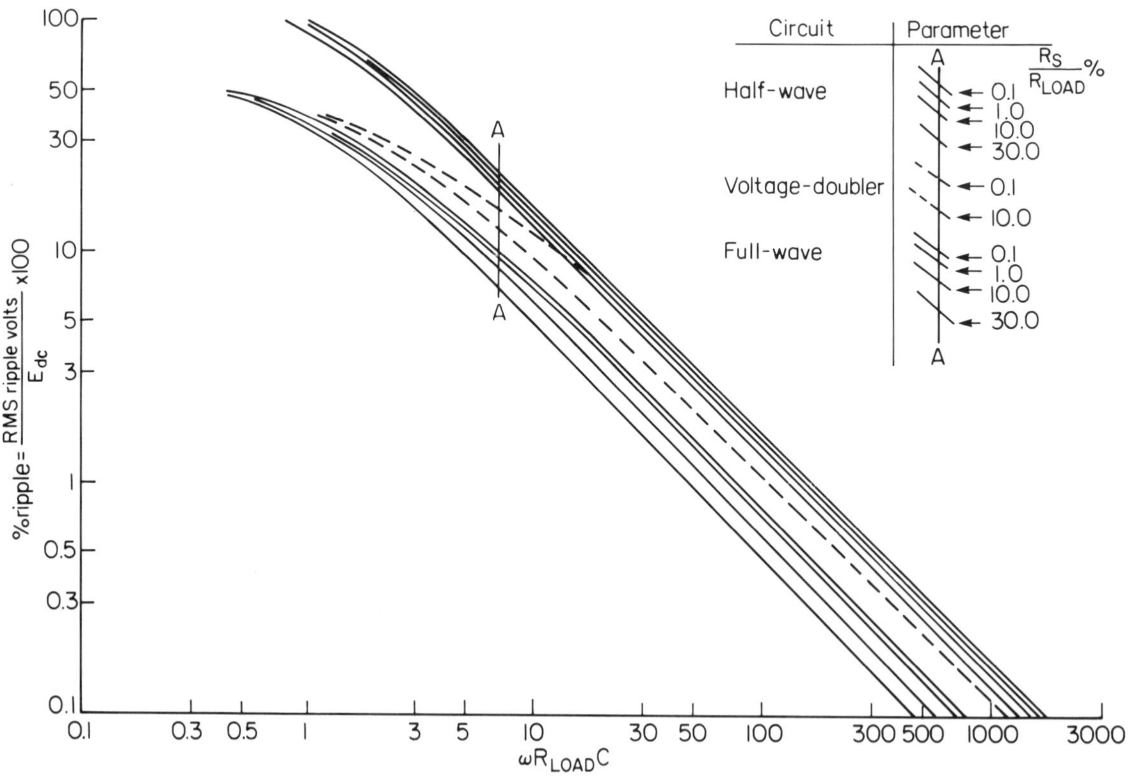

Figure 43.15 Percentage ripple as a function of $\omega R_{LOAD}C$ for capacitor input filter. C in farads, R_{LOAD} in Ω, $\omega = 2\pi f$ and f is the line frequency

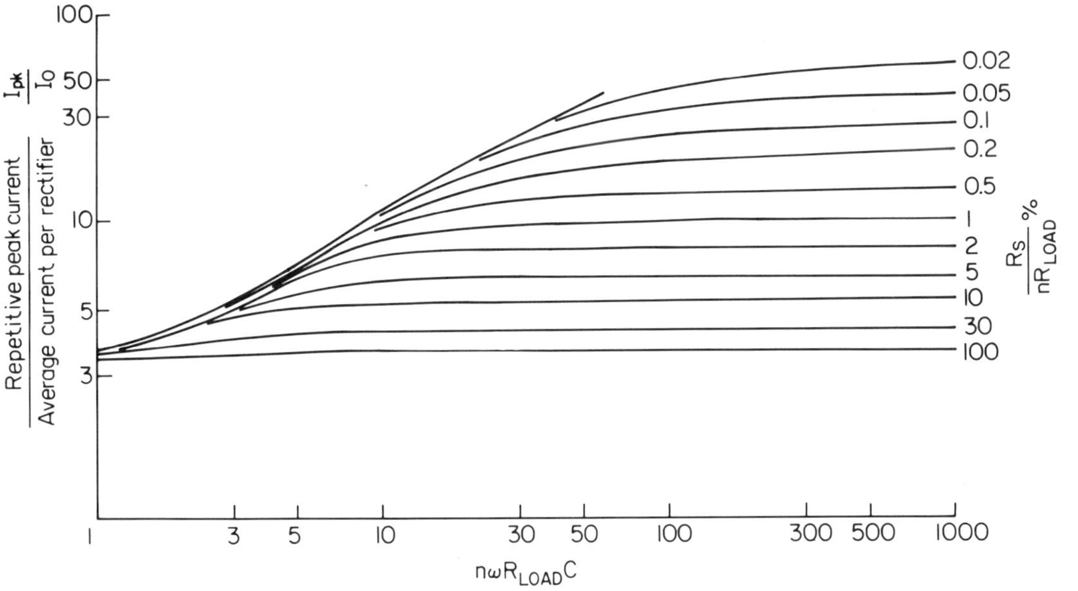

Figure 43.17 The ratio repetitive peak current/average current per rectifier plotted against $n\omega R_{LOAD}C$. C in farads and R_{LOAD} in Ω. ω=2 f, f is the line frequency. n=1 for half-wave, n=2 for full-wave and n=0.5 for voltage doubler

the ratio of peak repetitive rectifier current to average current per rectifier, plotted as functions of $n\omega R_{LOAD}C$. These ratios are dependent on the value of $R_s/nR_{LOAD}\%$.

The transformer leakage reactance has not been taken into account in the design procedure. However, it tends to reduce the peak rectifier current, and therefore assists in limiting the peak current.

43.4.1 Design procedure

The following design procedure is recommended in the design of single-phase silicon rectifier circuits with capacitor input filter.

(1) Determine the value of R_{LOAD}.
(2) Assume a value of R_s (usually between 1 and 10% of R_{LOAD}).
(3) Calculate $R_s/R_{LOAD}\%$.
(4) From the percentage ripple graph against $\omega R_{LOAD}C$ (*Figure 43.15*), determine the value of $\omega R_{LOAD}C$ required to reduce the ripple to a desired value for $R_s/R_{LOAD}\%$ determined in (3). Calculate the value of C required.
(5) From the $E_{dc}/E_T(\max)\%$ against $\omega R_{LOAD}C$ curves for the appropriate circuit (*Figures 43.12, 43.13* or *43.14*) determine the conversion ratio for the value of $\omega R_{LOAD}C$ determined in (4) and $R_s/R_{LOAD}\%$ determined in (3).
(6) Determine the $E_{T(\max)}$ and $E_{T(\mathrm{rms})}$ that must be applied to the circuit, using information derived in (5).
(7) Determine the crest working voltage that the rectifiers must withstand.
(8) Determine the r.m.s. current per rectifier from *Figure 43.16*.
(9) Decide on the rectifiers to be used.
(10) Check the peak repetitive current per rectifier from *Figure 43.17*.
(11) Check the initial switch-on current I_{on} given by $E_{T(\max)}/R_s$. If the value obtained exceeds that specified for the rectifier, then R_s must be increased and the design procedure repeated.
(12) Design the transformer and adjust the value of R_s accordingly, taking into account the transformer resistance and the forward resistance of the rectifier at the average current.
(13) Check the r.m.s. ripple current through the capacitor.
(14) Design the RC damping circuit as recommended in the published data of the rectifier.
(15) Determine the size of heatsink to be used to allow operation at the desired ambient temperature (from published data).

43.5 Design procedure for single-phase rectifier circuits with choke input filter

The analysis of the capaacitor input filter rectifier circuits has shown that for any high-current conversion, the circuit requires a large value of smoothing capacitor, which has to carry a large ripple current, and large initial and repetitive peak currents flow through the rectifiers. These limitations may be overcome by the use of choke input filters.

The single-phase half-wave circuit (*Figure 43.3*) cannot be used with a choke input filter, as it would require an infinite value of inductance to cause current to flow throughout the cycle.

For the full-wave bridge circuit of *Figure 43.5* and the full-wave centre-tap circuit of *Figure 43.6*, R_s is replaced by a series choke L.

The full-wave bridge circuit with choke input filter is shown in *Figure 43.18*, and the voltage and current waveforms in *Figure*

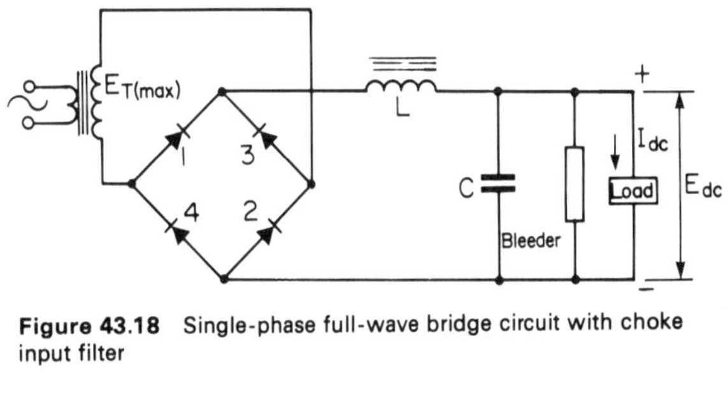

Figure 43.18 Single-phase full-wave bridge circuit with choke input filter

43.19. The action of the choke is to reduce both the peak and the r.m.s. value of current and to reduce the ripple voltage. The choke input filter circuit, however, requires a higher applied voltage than the capacitor input filter circuit, to produce the same output voltage.

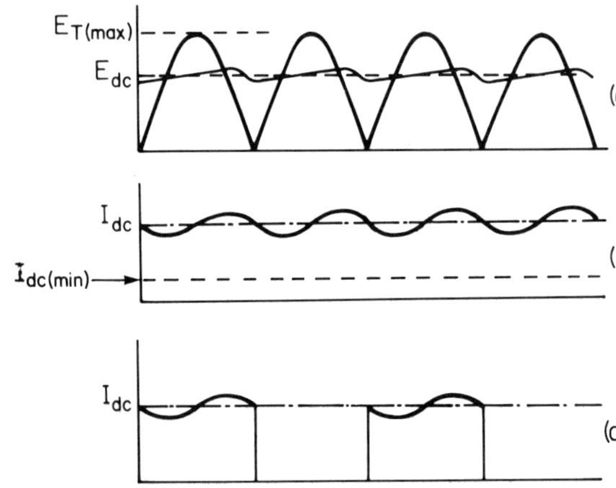

Figure 43.19 Waveforms for full-wave bridge circuit with choke input filter (a) output voltage (b) current through choke (c) current through rectifiers (1 and 2) or (3 and 4)

43.5.1 Smoothing circuit

The choke input filter must, ideally, pass only one frequency, which is zero, and attenuate all others. The filter must allow direct current to flow to the load without much power loss, and at the same time present a high impedance to the fundamental and other ripple frequencies. The capacitor shunts the load so as to by-pass the harmonic currents.

The attenuation factor K of the filter with series choke L and shunt capacitor C, is defined as the ratio of the total input impedance of the filter to the impedance of the parallel combination of the shunt capacitor C and load R_{LOAD}. For the choke input filter to function efficiently, the choke reactance at fundamental ripple frequency f_r should be much greater than its d.c. resistance, and the capacitor reactance much lower than the minimum load resistance. If it is assumed that, the inductance of the choke being L,

$$2\pi f_r L \gg \text{choke resistance } R_L$$

and

$$\frac{1}{2\pi f_r C} \ll R_{LOAD(min)}$$

then

$$K = \frac{2\pi f_r L - (2\pi f_r C)^{-1}}{(2\pi f_r C)^{-1}}$$

therefore

$$K = 4\pi^2 f_r^2 LC - 1 \tag{43.2}$$

The value of the inductance L used in the circuit must be such as to allow the rectifiers to conduct over one cycle of the fundamental ripple frequency. If the rectifier conducts for a period less than this, then the choke input filter will behave more and more like a capacitor input filter. This will give rise to a higher repetitive peak current through the rectifiers, and will also result in poor regulation.

The use of sufficient inductance allows the rectifier to conduct over the complete cycle; whereas the capacitor input filter allows the rectifier to conduct over only a fraction of a cycle. It follows that, for a given current, the rectifier will switch off before the cycle is completed, for a certain value of inductance. This value is termed the critical inductance L_{crit}.

43.5.2 Output voltage

Consider the single-phase full-wave bridge circuit shown in *Figure 43.18* and the waveforms shown in *Figure 43.19*. The rectified voltage is applied to the choke input filter. This voltage may be expressed as a series containing a d.c. component and harmonic components. The crest value of the output voltage is E_{max}, and it is equal to $E_{T(max)}$ in this circuit.

The rectified voltage can be approximated to a d.c. term plus a harmonic at the fundamental ripple frequency, assuming that the amplitudes of the higher harmonics are negligible. Therefore,

$$e \simeq \frac{2}{\pi} E_{max} - \frac{4}{3\pi} E_{max} \cos 2\omega t \tag{43.3}$$

43.5.3 Critical inductance

From *Figure 43.19* it can be seen that for the rectifier to conduct throughout the fundamental ripple cycle, the negative-going peak ripple current delivered by the rectifier must not exceed the minimum d.c. current, which occurs with a load of $R_{LOAD(max)}$. Thus

$$I_{dc(min)} = \frac{E_{dc}}{R_{LOAD(max)}} = \frac{2E_{max}}{\pi} \frac{1}{R_{LOAD(max)}} \tag{43.4}$$

If $2\pi f_r L \gg R_L$, and

$$\frac{1}{2\pi f_r C} \ll R_{LOAD(min)}$$

$$\text{peak a.c. current} = \frac{4}{3\pi} E_{max} \frac{1}{2\pi f_r L} \tag{43.5}$$

The critical inductance is reached when the peak a.c. current equals the direct current. That is,

$$\frac{4}{3\pi} E_{max} \frac{1}{2\pi f_r L_{crit}} = \frac{2E_{max}}{\pi} \frac{1}{R_{LOAD(max)}}$$

therefore

$$L_{crit} = \frac{R_{LOAD(max)}}{3\pi f_r} \tag{43.6}$$

For 50 Hz supply frequency and full-wave rectification, $f_r = 100$ Hz, so that

$$L_{crit} = \frac{R_{LOAD(max)}}{943} \tag{43.7}$$

Because of the approximations made, it is necessary to use a somewhat higher value of inductance than L_{crit}. In practice, it is found that for reliable and satisfactory operation the optimum value of inductance that should be used is twice the value of L_{crit}.

It is obvious from the nature of the circuit that it is not possible to maintain the critical value of the inductance over all values of load current. This would require an infinite inductance at zero load current. Two methods are available to ensure that current flows throughout the cycle, and that good regulation is maintained over a wide range of load currents. These are the use of a bleeder resistance or a swinging choke.

43.5.4 Bleeder resistance

A bleeder resistance of a suitable value is connected across the shunt capacitor to maintain the minimum current that will satisfy the critical inductance condition, even when no load is connected. The use of a bleeder will prevent the output voltage from rising to the peak applied voltage in the absence of the load.

43.5.5 Swinging choke

The swinging choke method is based on the fact that the inductance of an iron-cored inductor partly depends on the amount of direct current flowing through it. The swinging choke is designed so that it has a high inductance value at low currents, and this decreases as the d.c. current is increased. The use of such a choke is therefore very satisfactory for maintaining good regulation over a range of load current, and it is also more efficient than the bleeder resistance method.

Since the inductance is continually varying with the load current, the ripple voltage is no longer independent of the load current. When using a swinging choke, it is necessary to ensure that the inductance does not fall to a very low value at the maximum load current, as this will lead to high repetitive peak currents. In practice, the inductance at full load L_F should be such that

$$L_F = 2R_{\text{LOAD(min)}}/943 \qquad (43.8)$$

43.5.6 Ripple current and voltage

If $2\pi f_r L \gg R_L$,

$$\frac{1}{2\pi f_r C} \ll R_{\text{LOAD(min)}}$$

and

$$2\pi f_r L \gg \frac{1}{2\pi f_r C}$$

then

$$\text{r.m.s. ripple current } I_{\text{c(rms)}} = \left(\frac{4}{3}\frac{E_{\max}}{\pi}\frac{1}{\sqrt{2}}\right)\frac{1}{2\pi f_r L} \qquad (43.9)$$

Since

$$E_{\text{dc}} = \frac{2}{\pi} E_{\max}$$

$$I_{\text{c(rms)}} = \frac{\sqrt{2}}{3} E_{\text{dc}} \frac{1}{2\pi f_r L} \qquad (43.10)$$

% ripple = % ripple before filtering $\times 1/K$

From Table 43.1, % ripple before filtering = 47.2%.

From Equation (43.2), if $4\pi^2 f_r^2 LC \gg 1$ then

$$K \simeq 4\pi^2 f_r^2 LC$$

$$\%\ \text{ripple} = \frac{47.2}{4\pi^2 f_r^2 LC} = \frac{1.193}{f_r^2 LC} \qquad (43.11)$$

For 50 Hz supply frequency and full-wave rectification, $f_r = 100$ Hz, therefore

$$\%\ \text{ripple} = 119.3/LC \qquad (43.12)$$

where L is in henries and C is in μF.

43.5.7 Minimum value of shunt capacitance

In evaluating the percentage ripple and the attenuation factor of the filter, it has been assumed that the reactance of the capacitor at the fundamental ripple frequency is very much lower than the minimum load resistance. In practice, it is found that satisfactory performance is obtained when the reactance of the capacitor is made less than one-fifth the minimum load resistance. That is,

$$\frac{1}{2\pi f_r C} \leqslant R_{\text{LOAD(min)}}/5$$

Therefore

$$C \geqslant \frac{5\times 10^6}{2\pi}\frac{1}{f_r R_{\text{LOAD(min)}}}\ \mu\text{F}$$

$$\geqslant \frac{796\,000}{f_r R_{\text{LOAD(min)}}}\ \mu\text{F} \qquad (43.13)$$

Because of the nature of the circuit, the capacitor will resonate with the inductor at a certain frequency. At this frequency the output impedance will be greater than the capacitor reactance. Therefore, when a non-linear loading is applied, precautions must be taken to ensure that the output impedance of the filter is small at the load current frequency.

43.5.8 Additional filter sections

When it is required to reduce the ripple voltage across the load to a very low value, a single-stage choke input filter may require large values of inductance and capacitance, which may lead to an uneconomic filter design. In this case, the same results may be achieved by using a multi-stage filter with small value inductors and capacitors. It can be shown that the optimum smoothing is achieved when all stages are identical.

Figure 43.20 shows the attenuation factor K plotted against $f_r^2 LC$ for one-, two-, and three-stage filters. A suitable arrangement can therefore be selected by studying the filter characteristics. For a K factor between 23 and 160, the two-stage filter is the most economic. For K above 160, a three-stage filter is more suitable.

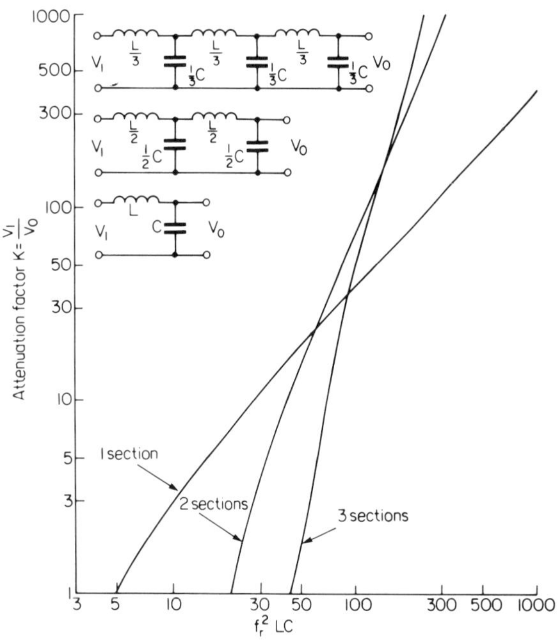

Figure 43.20 Characteristics of choke input filters. L in henries and C in microfarads

43.6 Three-phase rectifier circuits

There are many advantages in using a polyphase rectifier system when high-power conversion is required. The object is to superimpose more voltages of the same peak value but in different time relation to each other. An increase in the number of phases leads to the following improvements.

(i) Higher output voltage E_{dc} for the same voltage input.
(ii) Higher fundamental ripple frequency and lower amplitude ripple voltage.
(iii) Higher overall efficiency.

43.6.1 Comparison of three-phase circuit performances

Table 43.1 includes the performance of the commonly used three-phase rectifier circuits. In evaluating the results in this table, it has been assumed that the transformer and rectifiers are ideal. The table, however, gives a good indication of the relative merits of the circuits, and may be used to select the best circuit for any particular application. It may also be used for comparing the kilowatts per rectifier available from various circuits. This is best illustrated by an example.

Consider the single-phase and three-phase full-wave bridge circuits, with rectifiers rated at a crest working voltage of 400 V and with a current rating of 20 A. The attainable performances are compared in *Table 43.4*.

From the above calculation, it follows that better use of rectifiers is made in the three-phase bridge circuit.

Table 43.4 Comparison of three-phase circuits

	Single-phase bridge	*Three-phase bridge*
Number of rectifiers in circuit from *Table 43.1*	4	6
Output voltage E_{dc}	$\frac{400}{1.57}=255$ V	$\frac{400}{1.05}=380$ V
Output current I_{dc}	$2\times 20=40$ A	$3\times 20=60$ A
Power available $E_{dc}I_{dc}$	$255\times 40=10.2$ kW	$360\times 60=22.8$ kW
Kilowatts per rectifier	$\frac{10.2}{4}=2.55$ kW	$\frac{22.8}{6}=3.8$ kW

References

1. CROWTHER, G. O. and SPEARMAN, B. R., 'Mains overvoltages: protection and monitoring circuits', *Mullard Technical Communications*, Vol. 5, No. 47, March 1961, pp. 301 to 304, and Vol. 6, No. 51, Sept. 1961, pp. 12 to 21.
2. TULEY, J. H., 'Design of cooling fins for silicon power rectifiers', *Mullar Technical Communications*, Vol. 5, No. 44, June 1960, pp. 118 to 130.
3. WAIDELICH, D. L., 'Diode rectifying circuits with capacitance filters', *Trans. AIEE*, Vol. 60, 1941, pp. 1161 to 1167.
4. WAIDELICH, D. L., 'Voltage multiplier circuits', *Electronics*, Vol. XIV, No. 5, May 1941, pp. 28 to 29.
5. ROBERTS, N. H., 'The diode as half-wave, full-wave and voltage-doubling rectifier, with special reference to the voltage output and current input', *Wireless Engineer*, Vol. XIII, No. 154, July 1936, pp. 351 to 362, and Vol. XIII, No. 155, August 1936, pp. 423 to 430.
6. SCHADE, O. H., 'Analysis of rectifier operation', *Proc. IRE*, Vol. 31, No. 7, July 1943, pp. 341 to 361.
7. SAY, M. G., 'The performance and design of alternating current machines', Pitman, London, 1948 (reprinted 1963).
8. TOBISCH, G. J., 'Parallel operation of silicon diode rectifiers'. To be published in *Mullard Technical Communications*.
9. CORBYN, D. B. and POTTER, N. L., 'The characteristics and protection of semiconductor rectifiers', *Proc. IEE*, Vol. 107, Part A, No. 33, June 1960, pp. 255 to 272 (Originally Paper 3135U, November 1959).
10. GUTZWILLER, F. W., 'The current-limiting fuse as fault protection for semiconductor rectifiers'. *Trans. AIEE* (Part 1, Communication and Electronics), No. 35, November 1958, pp. 751 to 755.
11. GUTZWILLER, F. W., 'Rectifier voltage transients: causes, detection and reduction', *Electrical Manufacturing*, Vol. 64, No. 12, Dec. 1959, pp. 167 to 173.

44 Power Supply Circuits

M Burchall, CEng, FIERE
Previously Engineering Manager, Gould Advance Ltd
Now Engineering Manager, CELAB Ltd

Contents

44.1 Introduction

A power supply can essentially be considered as a matching device converting energy from a source to that needed by a load usually including some regulation. The source may be a.c. such as 50/60 Hz or 400 Hz mains, in which case rectification is required, or it may be d.c. from a battery, photovoltaic cell etc. in which case only d.c.–d.c. conversion is required. In most cases where the source is a.c. isolation is required for safety and signal isolation purposes. This isolation is carried out in a transformer either working at the mains frequency or some other internally generated frequency which is usually much higher. In many cases where the source is d.c., isolation is not required and the power supply only has to provide voltage conversion and/or inversion. If isolation is required when operating from a d.c. supply then a transformer is provided with a frequency which is generated internally in the power supply.

In *Figure 44.1* a generalised chain of most of the possible stages from a.c. or d.c. input to d.c. output is shown. After the a.c. input has been isolated and rectified to become d.c. the circuitry becomes common with the d.c. input and all discussion of the following circuit will start from the assumption of an unregulated d.c. input. Some of the circuits such as the linear regulator or the switching regulator can be utilised in different parts of the chain.

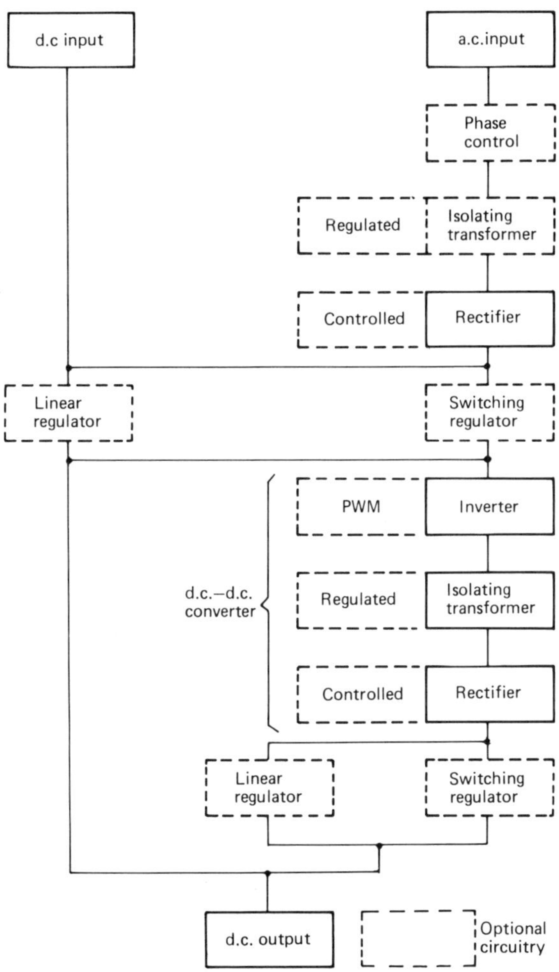

Figure 44.1 Power supply systems

The linear regulator and switching regulator are defined in this context as non-isolated four-terminal networks and the d.c.–d.c. converter as an isolated four-terminal network. There are some exceptions to this general rule in that a d.c.–d.c. converter may be fitted with an autotransformer or have a common path through a d.c. feedback loop but the general principles are still valid.

44.2 Performance

The various factors which affect the performance of a power supply are defined in BS5654 (IEC478).[1] The major items of normal concern are:

(1) line regulation;
(2) load regulation;
(3) temperature regulation;
(4) ripple and noise.

In a regulated power supply the line and load regulation are controlled by the d.c. loop gain of the feedback loop which can be of any desired level within the bounds of loop stability and bandwidth. Additionally feedforward techniques can be used. Temperature regulation is a function of the quality of the reference voltage, the resistive network sampling the output voltage and the input stages of the feedback amplifier. Ripple and noise are functions, firstly of the type of circuit, e.g. linear or switching and, secondly, the loop gain of the control amplifier at the relevant frequencies and the magnitude and quality of the circuit components.

To a first approximation power supplies can be categorised in four performance brackets (*Table 44.1*).

Table 44.1 Performance of power supplies

Type of power supply	*Total deviation of O/P voltage*
Unregulated	±10–20%
Semi-regulated	±5–10%
Regulated	±1–5%
Precision	±1%

44.2.1 Unregulated power supplies

Unregulated power supplies include all the basic a.c. to d.c. rectification circuits, with or without a transformer, comprised of half-wave, full-wave, multi-phase and voltage multiplication circuits. These are described in Chapter 43 and will not be discussed further except to say that because of their open-loop nature the line regulation approximates to the line voltage variation (usually specified in the range ±6% to ±15%) and the load regulation is highly dependant on the load variation and type of rectification used.

44.2.2 Semi-regulated power supplies

Semi-regulated power supplies fall into two main classes, the phase-regulated thyristor rectified type which is also described in Chapter 45 and the ferro-resonant transformer[2,3] (C.V.T.) regulated power supply. This is an extension of the unregulated a.c. to d.c. rectifier circuits where the line transformer is replaced by a magnetic regulating transformer which removes most of the input line variation. For fixed load applications this is a very adequate circuit possessing high reliability due to its inherent simplicity. It is only recently that it is starting to be replaced by the direct-off-line type of switching power supply on the grounds of cost and weight.

44.2.3 Regulated power supplies

Regulated power supplies comprise all types with feedback and/or feedforward to improve line and load regulation. They are either of the linear or switching type or combinations of both and will be discussed in detail in subsequent sections.

44.2.4 Precision regulated power supplies

Precision regulated power supplies are a special group ranging from high quality general purpose bench power supplies to the most accurate precision voltage sources which are used for calibration purposes. They are normally linear types with very high feedback loop gains and will not be discussed further.

44.3 Linear or switching

In recent years there has been a significant change in usage away from the linear power supply to the switching power supply[4] as a result of the growth to maturity of the direct-off-line type of switching power supply which is now cost-competitive with linear power supplies down to a few watts. The reasons for this can be summarised as improvements in weight, size and efficiency[5] with some deterioration in the overall performance notably in the area of ripple and noise. For most applications except the most sensitive of linear amplifiers this deterioration is acceptable.

Table 44.2 lists the comparison between linear and switching power supplies for a 5V 20A application. Although the improvement in efficiency from 50% to 75% does not, in itself, look large it leads to a reduction in losses from 100 W to 33 W. It is this reduction, in addition to the reduction in size of magnetic components and capacitors, which enables a switching power supply to be so much smaller than the equivalent linear power supply.

Table 44.2 Comparison between linear and switching power supplies at 5V 20A

	Linear power supply	*Switching power supply*
Efficiency	50%	75%
Power loss (100 W load)	100 W	33 W
Total system power	200 W	133 W
kW/m^3	30	90–180
W/kg	5	15–30

44.4 Protection

It is usual in most power supplies to provide some form of protection against external and internal fault conditions. The most important of these are over-current protection and over-voltage protection.

44.4.1 Over-current protection

If the load current increases beyond its designed level due to some external fault, then, unless the power supply is self-limiting on current (e.g. the shunt regulator, the ferro-resonant regulator and some forms of switching regulators) damage will be caused to the power supply. In the unregulated or phase-regulated supply this protection can be a fuselink, but for most electronic power supplies the protection has to be fast acting and usually consists of extra electronic circuitry which detects the over-current and over-rides the voltage control. The three commonest forms of this are (a) current limiting, (b) constant current and (c) re-entrant or fold-back limiting. These are shown in *Figure 44.2*. The commonest type of protection is (a). Type (b) is only usually found in precision power supplies with constant voltage constant current (CVCC) characteristics where the constant current performance is comparable with the constant voltage performance. Type (c) is often used on linear series regulators because it reduces the maximum dissipation seen by the series transistors under shortcircuit conditions. It is sometimes used with switching power supplies to reduce the magnitude of current flowing into the external circuit fault. It suffers from problems with non-linear loads (such as tungsten filament lamps) in that it can lock-out at a stable low voltage condition. It can also lock-out when two power supplies are connected in series.

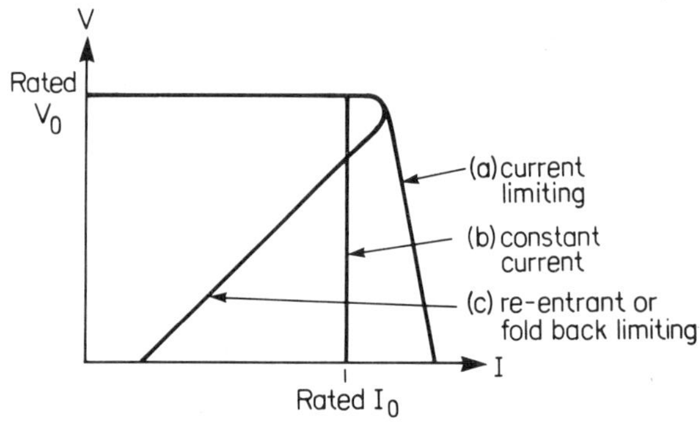

Figure 44.2 Over-current protection

44.4.2 Over-voltage protection

Over-voltages can be caused either internally or externally. It is possible for an internal fault (such as a short-circuit series transistor) to cause a loss of control of the output voltage. Alternatively, it is possible for some external voltage, perhaps from a higher voltage within a multi-voltage system, to be connected to the power supply by an external fault. In either case the output voltage rises and will eventually cause damage to the connected load. To prevent this an over-voltage detection circuit continually monitors the output voltage and if it rises beyond predetermined limits some form of short-circuit is applied to the output terminals. This is usually a thyristor and is commonly called over-voltage crowbar protection. For security of protection the over-voltage circuit is preferably of a two-terminal configuration, i.e. independant of any external power to the control circuits which might be missing at the particular time when protection is needed. *Figure 44.3* shows a typical example of a two-terminal crowbar circuit where the reference voltage is the V_{be} of TR_1 thermally compensated for by a thermistor R_3.

44.5 Linear regulators

Most linear regulators are of the series type, i.e. a transistor connected in series with an unregulated d.c. source which absorbs and dissipates the surplus power. The other type of linear

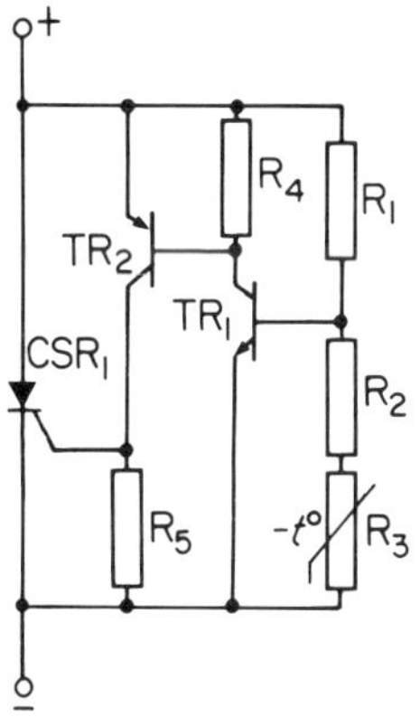

Figure 44.3 Over-voltage crowbar circuit

regulator is the shunt type, but applications for these are mainly confined to low powers because of their inefficiency.

44.5.1 Zener regulator

The zener diode regulator[6] is an open-loop version of the shunt regulator, the simplest form being shown in *Figure 44.4*. To a first

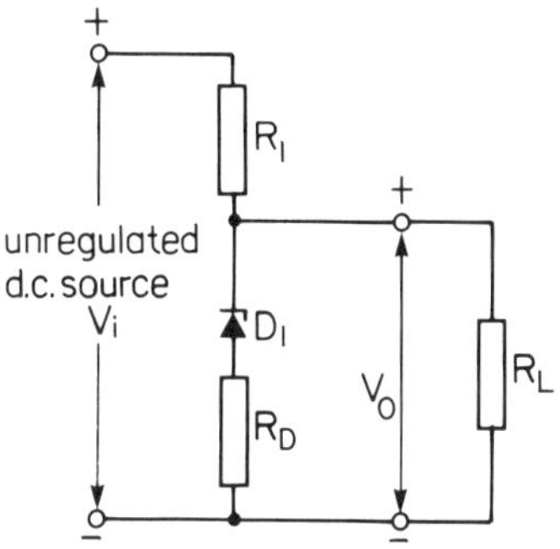

Figure 44.4 Zener diode regulator

approximation the stability ratio S is defined as

$$S=\frac{\%\text{ input voltage change}}{\%\text{ output voltage change}} \tag{44.1}$$

is

$$S=\frac{R_1V_0}{R_DV_i} \tag{44.2}$$

R_D is the incremental resistance of the zener diode at the chosen operating point and is the a.c. or d.c. value depending on the conditions being analysed.

For simple zener diodes where R_D is in the range 5 to 20 Ω values of S of 20 to 50 are typical. Extra stability can be achieved by cascading circuits when the stability ratio increases to the product of the individual stage ratios. With the introduction of i.c. band-gap references, which have an R_D value of the order of 0.2 Ω, values of S of 500 to 1000 are achievable which are sufficient for most applications.

44.5.2 Shunt regulator

The basic circuit of a shunt regulator is shown in *Figure 44.5*. If the output voltages rises for any reason (such as a reduction in load current) the sampling resistors R_2 and R_3 apply an increasing potential to the base of TR_3. This reduces the potential

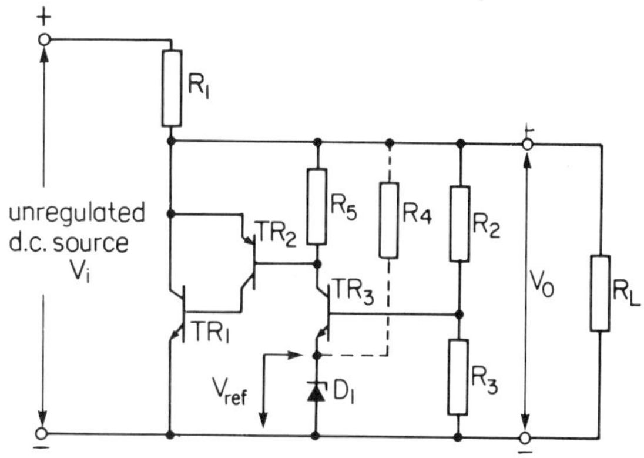

Figure 44.5 Shunt regulator

at the base of TR_2, which turns on transistors TR_2 and TR_1 causing them to draw more current through resistor R_1, counteracting the original tendency for V_0 to rise. The output voltage is given by the equation

$$V_0=\frac{R_2+R_3}{R_3}(V_{\text{REF}}+V_{be}) \tag{44.3}$$

Resistor R_4 is usually necessary to provide D_1 with an optimum working current and lower its inherent incremental resistance. D_1 is usually chosen to be around 6.2 V to have a positive temperature coefficient to balance the negative V_{be} coefficient of TR_3. The shunt regulator has one important characteristic not shared by the series regulator in that it can accept current as a sink as well as supply current as a source.

A useful application of this is as an over-voltage protection clamp where the shunt regulator is used to absorb the energy from an over-voltage fault or interference spike in a power system before deciding that the over-voltage is of long duration and applying a crowbar short-circuit in the form of a thyristor. The shunt regulator is also automatically protected against overload currents and needs no extra over-current protection providing the supply resistor R_1 can withstand the extra power of the overloaded condition.

44.5.3 Series regulator

The simplest form of series regulator is shown in *Figure 44.6*. This has a series pass transistor TR_1, a reference source D_1 and a control amplifier TR_2. All other circuits are elaborations of this. As in the shunt regulator, if V_0 rises for any reason the potential at the base of TR_2 rises causing it to draw more current through R_1. This lowers the potential at the base of TR_1 causing it to pass less current which counteracts the original tendency for V_0 to rise.

The major improvements in this circuit are to increase the current capability of TR_1 by connecting 2 or 3 transistors in a Darlington connection, and providing a separate stabilised source for R_1 to provide more gain and independence from variations in the source V_i. This circuit is shown in *Figure 44.7*. This provides a better line and load regulation than the basic circuit. Middlebrook[7] gives a detailed analysis of the perfor-

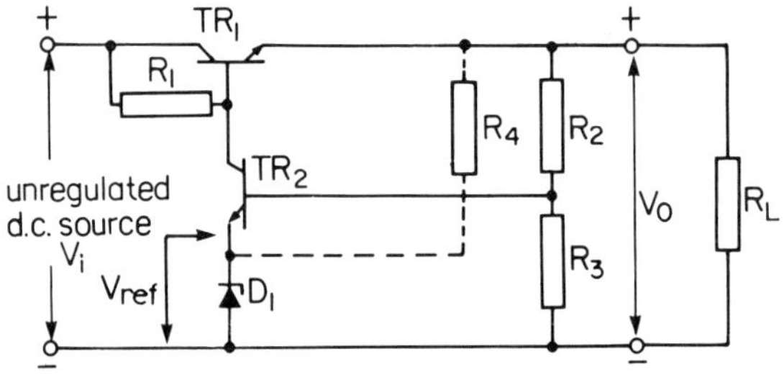

Figure 44.6 Series regulator

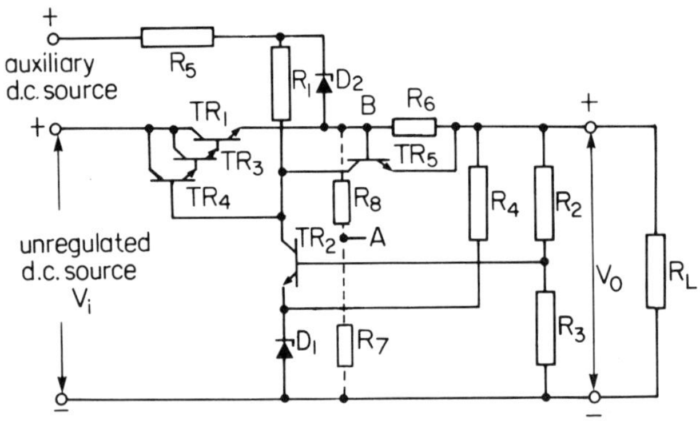

Figure 44.7 Series regulator with current-limiting protection

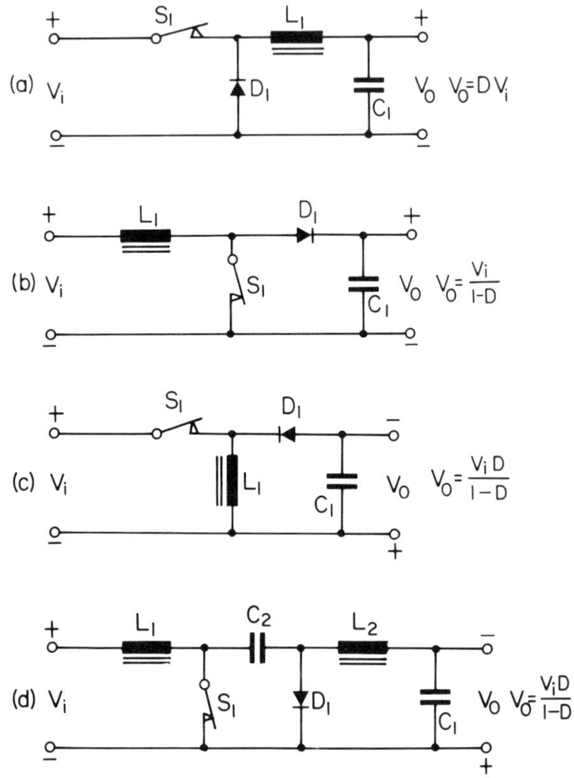

Figure 44.8 Switching regulator: (a) buck; (b) boost; (c) inverting; (d) Cuk

mance of this circuit. This circuit also provides current limiting protection. A voltage is developed across R_6 by the load current and when it exceeds the V_{be} of TR_5 it turns TR_5 on and draws current from R_1 diverting it from the series transistors. This reduces the output voltage to whatever level is necessary to satisfy the load resistance.

The circuit shown in *Figure 44.7* can be modified for foldback current protection by moving the base of TR_5 from point B to A. Now the voltage developed by the load current has to overcome both the V_{be} of TR_5 and the fraction of the output voltage developed across R_8 by the divider network R_7 and R_8. As the output voltage reduces in the overloaded condition the voltage across R_8 and the amount of current required in R_6 to turn on TR_5 also reduce. This continues all the way down to the short-circuit condition where the load current is determined by R_6 and the V_{be} of TR_5. By the correct selection of R_6, R_7 and R_8 the load current can be made to fold back to any desired short-circuit level.

Series regulators are available in IC and hybrid versions ranging from devices which can provide tens of mA of current through fixed-output TO3 regulators, to variable three-pin and multi-pin regulators and high power hybrid regulators capable of supplying several amps of current.

44.6 Switching regulators

The two basic switching regulators are the buck (step-down) regulator[8] and the boost (step-up) regulator shown in *Figure 44.8*. They have the generalised property of d.c. voltage conversion where the output voltage can be controlled or varied by modulation of the duty ratio *D*. This is known as pulse-width modulation (PWM). These can be combined as the buck-boost regulator which, by circuit simplification, becomes the inverting regulator; or the boost-buck regulator which becomes the Cuk converter.[9] An extensive analysis of various converter topologies and references is given by Severns.[10]

Both the inverting and Ćuk converters have the extra property of continuous variation of the output voltage above and below the input voltage. The Ćuk converter also has the benefit of non-pulsating input and output currents and, because the energy storage transfer is by capacitor instead of inductor, can be made smaller and lighter. Capacitors can store many more J/m^3 and J/kg than inductors.

The switching action in *Figure 44.8* can be carried out by the usual switching devices such as a thyristor, GTO[11] (gate turn-off thyristor), bipolar transistor or FET[12] (field-effect transistor).

44.6.1 Buck regulator

A block diagram of a bipolar transistor buck regulator is shown in *Figure 44.9*. This is used to illustrate the principle of pulse-width modulation (PWM) which is common to all switching stabilisers and controlled d.c. to d.c. converters. It contains the necessary features of error comparison in amplifier A_1, a ramp generator A_3 (this can be a single or double-sided ramp as convenient) and a modulator in amplifier A_2. Amplifiers A_2 and A_3 need a wide bandwidth to obtain fast switching edges for driving TR_1 with fast 'on' and 'off' transition times particularly at oscillation frequencies above 20 kHz.

If the output voltage increases for any reason the output of amplifier A_1 also increases. This moves the input to amplifier A_2 further up the ramp waveform as shown between *Figures 44.10(a)* and *44.10(b)* and causes the output of A_2 to be high for shorter

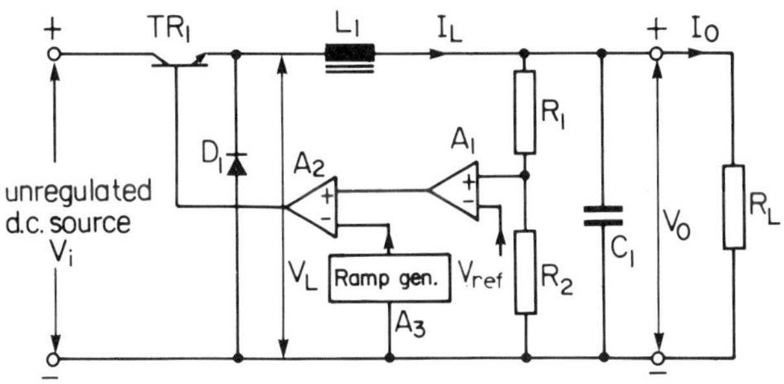

Figure 44.9 Block diagram of a buck regulator

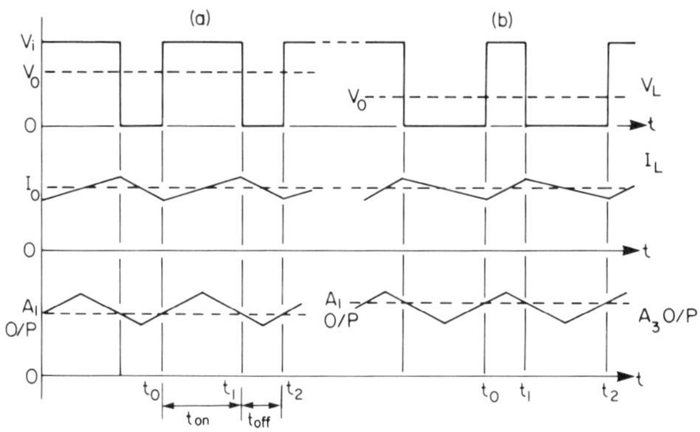

Figure 44.10 Waveforms for a buck regulator: (a) high duty ratio *D*; (b) low duty ratio *D*

periods of time, thus lowering the duty ratio *D* of the switch TR_1. This reduces the output voltage to counteract the original tendency to rise.

When transistor TR_1 turns 'on' at time t_0 the current in the inductor rises at the rate

$$\frac{dI_L}{dt}=\frac{V_i-V_0}{L} \tag{44.4}$$

When TR_1 turns 'off' at time t_1 the current in the inductor decays at the rate

$$\frac{dI_L}{dt}=\frac{V_0}{L} \tag{44.5}$$

therefore

$$t_{on}(V_i-V_0)=t_{off}V_0 \tag{44.6}$$

where

$$D=\frac{t_{on}}{t_{on}+t_{off}} \tag{44.7}$$

therefore

$$V_0=DV_i \tag{44.8}$$

This is for a theoretically dissipationless system and in practice losses in TR_1, D_1 and L_1 modify this to some extent.

All the above functions of A_1, A_2 and A_3 can be obtained in purpose built ICs.

44.6.2 Boost regulator

The boost regulator will not be examined in detail since the principles are similar to that of the buck regulator, i.e. energy is stored in the input inductor while the switch is closed and released to the output when the switch is open.

44.7 D.C. to d.c. converters[13,14,15,16]

There are five basic types of square wave d.c. to d.c. converter available and these are shown in *Figures 44.11–44.15*. The

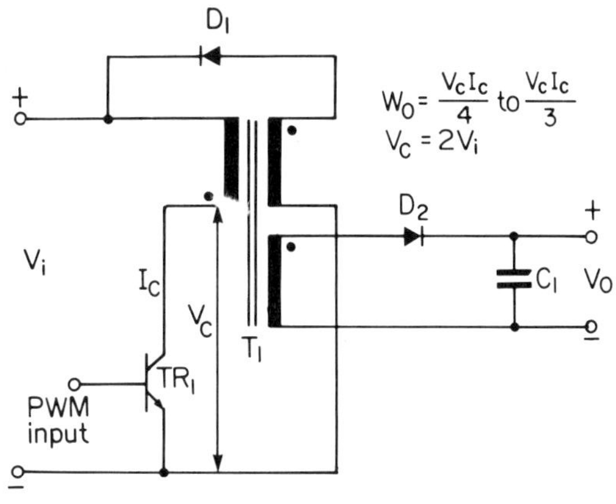

Figure 44.11 Flyback converter

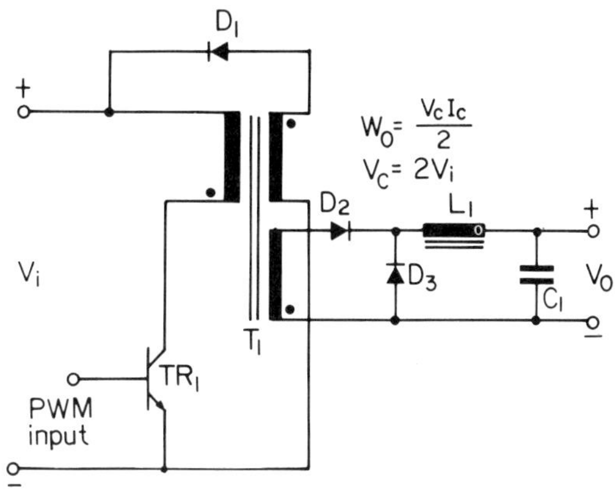

Figure 44.12 Forward converter

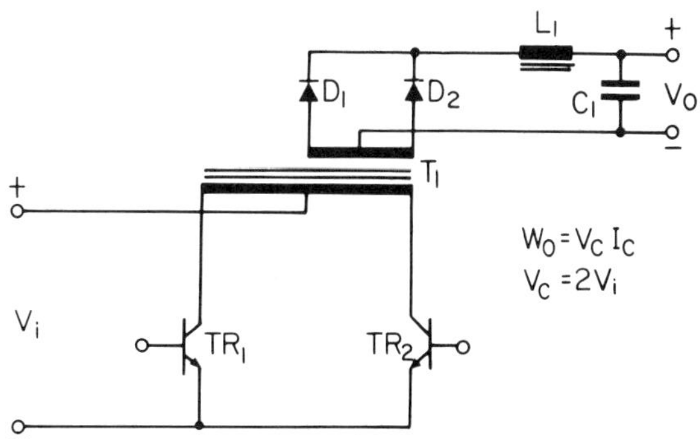

Figure 44.13 Push–pull converter

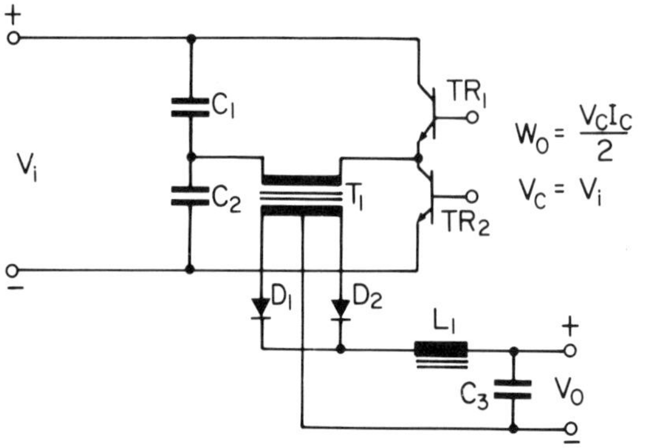

Figure 44.14 Single-ended push–pull or half-bridge converter

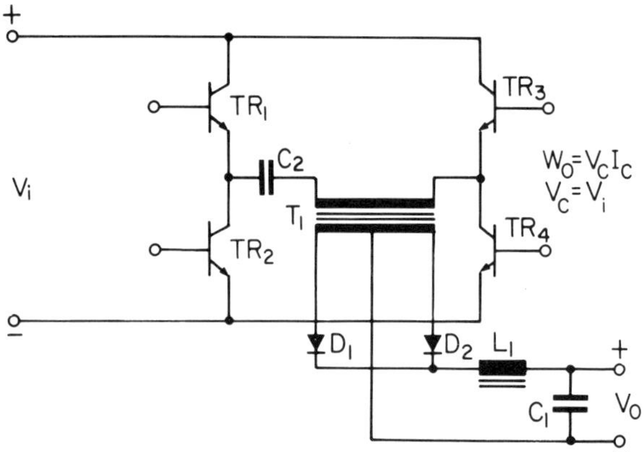

Figure 44.15 Full bridge converter

simplest forms of these are self-oscillating with either voltage or current feedback or both. These give uncontrolled d.c. to d.c. conversion. Externally driven versions can be provided with PWM drive waveforms which give a variable d.c. transformation ratio, in the same manner as the switching regulators, but with input–output isolation.

The two single-transistor converters have poor transformer utilisation since only one half of the *B–H* loop is used because of the d.c. polarisation. It has been proposed to overcome this by placing a small permanent magnet in series with the magnetic circuit but this has not found any commercial usage so far.

44.7.1 Flyback converter

This is one of the two single-transistor converters and is shown in *Figure 44.11*. The winding polarities are shown by the conventional dot notation. This circuit can be regarded as the isolated version of a boost regulator in that energy is stored (in this case in the transformer primary inductance) during the transistor 'on' time and released to the output during the 'off' time. The extra winding connected through D_1 is only necessary under off-load conditions to return the surplus energy back to the supply and limit the voltage on TR_1 collector to twice V_i.

This circuit is not normally used at powers above 100 W because of the poor utilisation of the peak current capacity of the switching transistor, due to the triangular nature of the collector current waveform, and also because the output capacitor C_1 has to carry ripple currents which are of the same order of magnitude as the output current. These ripple currents can also produce a large output-voltage ripple, not only because of the ramp voltage produced in the capacitance by the charge and discharge currents, but also because of the inherent series resistance and series inductance of the output capacitor. In most practical low voltage applications the capacitor is of an electrolytic type.

The flyback converter is often used for multiple output applications[17] due to the simplicity of coupling extra diodes and capacitors to extra windings. It has superior cross-regulation under these conditions to the forward converter due to the absence of integrating inductors with their attendant resistance.

44.7.2 Forward converter

This is the second of the single-transistor converters and is shown in *Figure 44.12*. In this case power is transferred to the output winding during the 'on' time of the transistor. The winding connected through D_1 back to the input voltage is now an essential part of the operation and serves to carry the transformer magnetising current while the transformer core re-sets. Since the transistor is now carrying essentially square waves of current the output power can be twice as high as in the flyback converter using a transistor of the same voltage and current ratings. There is no practical limit to the output power of this circuit and multi-transistor and multi-phase versions have been developed up to several kW.

44.7.3 Push–pull converter

This circuit shown in *Figure 44.13* is one of the most popular of the two-transistor converters particularly for low voltage sources where V_i is below 50 V. This allows transistor collector voltage ratings to be around 100 V giving a good selection of fast switching types. In its self-oscillating form (discussed in Section 44.7.6) it is economic and simple. It is not often used where V_i is greater than 300 V, partly because this requires 600 V rating transistors, but also because there is no self-balancing action against asymmetric switching times which cause the transformer flux density to move cycle by cycle into saturation. This saturation is normally followed by the transistor destruction.

44.7.4 Single-ended push–pull or half bridge converter

This circuit shown in *Figure 44.14* has two important benefits. Firstly, because the transistors are only switching between the positive and negative of the input source V_i the collector voltage rating can be of the same magnitude as V_i. This is important for direct-off-line power supplies where V_i is the rectified and smoothed mains supply and can reach over 370 V off-load. Secondly, because the output transformer T_1 is connected to the centre point of C_1 and C_2 any direct current flowing in T_1 as a result of asymmetric switching times causes this centre point to move up or down from $\frac{1}{2}V_i$ in such a direction that a compensating action takes place.

44.7.5 Full bridge converter

This circuit shown in *Figure 44.15* has one of the benefits of the half bridge converter in that the transistor only has to have a collector voltage rating equal to V_i but it has no asymmetric switching time compensation. This can be provided by an optional series capacitor C_2 which develops a compensating d.c. bias voltage in a similar fashion to C_1 and C_2 in the half-bridge circuit. Unfortunately this requires a very specialised capacitor capable of handling high a.c. powers at high frequencies, and these are quite rare.

44.7.6 Self-oscillating converters

All of the above d.c.–d.c. converters can be produced in a self-oscillating form. For example the flyback converter of *Figure 44.11* when provided with a feedback winding becomes the 'ringing choke' or 'blocking oscillator' of *Figure 44.16*. The 'on' period is ended either when the transformer T_1 saturates or when the base current supplied by R_1 is insufficient to maintain the transistor TR_1 in saturation.

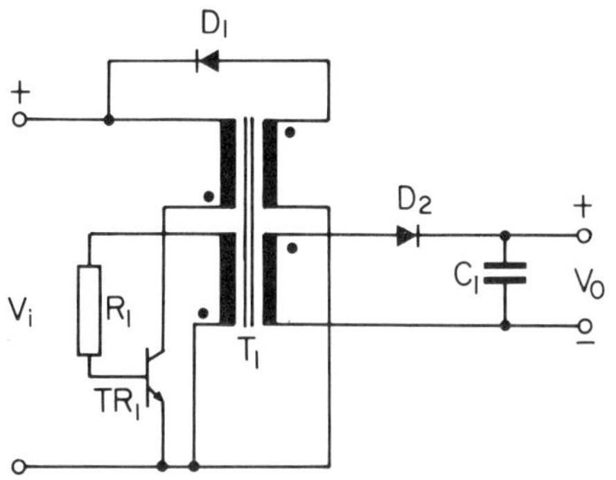

Figure 44.16 Blocking oscillator or ringing choke converter

Figure 44.17 shows a self-oscillating push–pull converter with both voltage[18] and current feedback. The circuit will work with either type of feedback but has limitations. For example the circuit with voltage feedback tends to be inefficient at partial loads since the drive current has to be designed to be large enough to supply the maximum load condition whereas the circuit with current feedback cannot work at no load since there is no collector current in transformer T_2 primary and, therefore, no available base current. Note that since this is a square wave converter with no PWM the output inductor is not required to integrate the rectified waveform and only a smoothing capacitor C_1 is required.

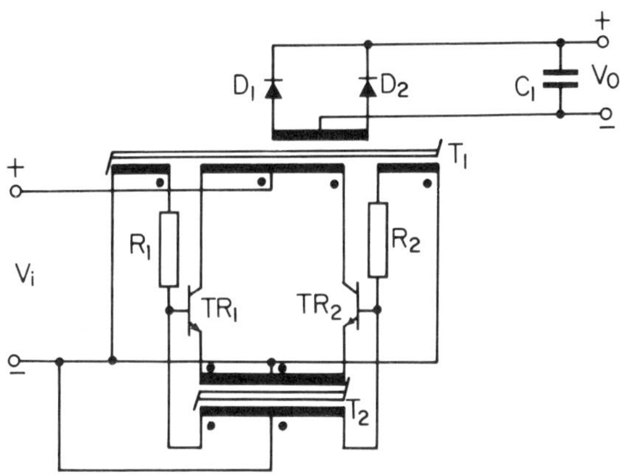

Figure 44.17 Self-oscillating push–pull converter

The oscillation frequency is determined by the saturation of the output transformer T_1 (in the voltage feedback case). This is only suitable at low powers since the core losses of a transformer swinging between positive and negative peaks of flux density at a high frequency become unacceptable. At higher powers the frequency is determined by a small auxiliary drive transformer (such as T_2 in the current feedback case) or a small saturating voltage drive transformer.[19] It is also possible for the frequency to be determined by *RC*, *RL* or *LC* circuits in the feedback path or by a saturating base inductor.

44.7.7 Sinusoidal converters

It is also possible for the d.c. to d.c. conversion process to be carried out in parallel[20,21] or series[22] *LC* resonant circuits. These are similar to those used in d.c. to a.c. inversion except that the oscillation frequency can be optimised at any convenient level instead of at some fixed level, e.g. 400 Hz. There are some benefits to sinusoidal operation mainly in respect of the dV/dt on the switching devices which can be a major source of electromagnetic interference. A typical parallel *LC* converter is shown in *Figure 44.18*.

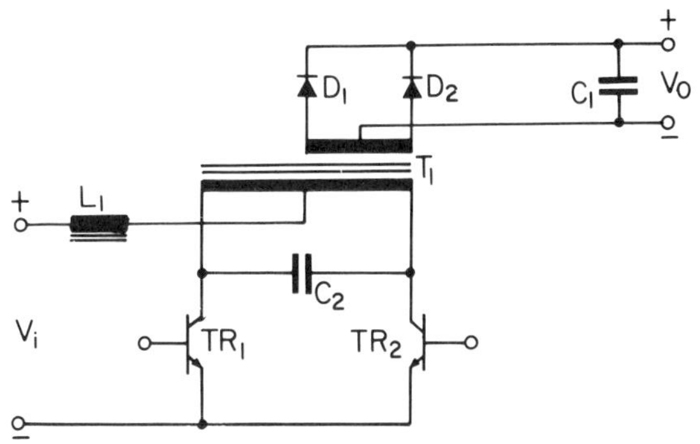

Figure 44.18 Sinusoidal converter

Sinusoidal converters are particularly useful at very high frequencies, above 100 kHz, where there are practical difficulties in making transformers with low leakage inductance and stray capacity. This is even more so for high voltage power supplies[23] where the winding capacity of the multi-turn output winding makes square wave operation impractical.

44.8 Direct off-line regulators

Until recently the major areas of application of d.c. to d.c. converters have been in low power or low voltage (up to 50 V) circuits. In the last few years a major change has taken place in that direct-off-line switching power supplies have started to replace linear power supplies for the reasons outlined in Section 44.3.

In essence a direct-off-line power supply consists of an unregulated supply derived from the a.c. input (normally 115 V, 220 V or 240 V) providing a 'raw' d.c. voltage of about 300 V. This is connected to one of the standard d.c. to d.c. converters detailed in Section 44.7, the particular converter being selected according to the voltage, current and power requirements. Regulation against line and load changes can be provided by one of the systems shown in *Figure 44.1* the commonest being by direct PWM of the d.c. to d.c. converter.[24]

Switching frequencies are normally in the region 20 to 50 kHz although the introduction of the power FET and combinations of FET and bipolar transistors[25] are pushing practical frequencies above 100 kHz.[26]

References

1 'Stabilised power supplies. D.C. output', *BS5654* (1979), *IEC478* (1974)
2 'Ferroresonant transformer bibliography', *Magnetic Metals Inc.*
3 HART, H. P. and KAKALEC, R. J., 'The derivation and application of design equations for ferroresonant voltage regulators and regulated rectifiers', *IEEE Trans. on Magnetics*, MAG-7, No. 1, 205–11, March (1971)
4 BURCHALL, M., 'A guide to the specification and use of switching power supplies', *Gould Advance Ltd* (1977)
5 BURCHALL, M., 'Why switching power supplies are rivalling linears', *Electronics*, 141–43, September 14 (1978)
6 CHANDLER, J. A., 'The characteristics and applications of Zener (voltage reference) diodes', *Electron. Eng.*, 78–86, February (1960)
7 MIDDLEBROOK, R. D., 'Design of transistor regulated power supplies', *Proc. IRE*, **45**, 11, 1502–9 (1957)
8 MIDDLEBROOK, R. D., 'Switching regulator design guide', *Unitrode Publication* No. U-68
9 CUK, S. and MIDDLEBROOK, R. D., 'A new optimum topology switching DC–DC converter', *IEEE Power Electron. Spec. Conf.*, 160–79, June (1977)
10 SEVERNS, R., 'Switchmode converter topologies—make them work for you', *Intersil. Appl. Bull.*, A035 (1980)
11 BURGUM, F., NIJHOF, E. B. G. and WOODWORTH, A., 'Gate turn-off switch', *El3ctron. cpmp. appl.*, **2**, 4, 194–202, August (1980)
12 CLEMENTE, S., PELLY, B. and RUTTONSHA, R., 'A universal 100 kHz power supply using a single HEXFET', *Int. Rect. Appl. Note*, AN-939 (1981)
13 JANSSON, L. E., 'A survey of converter circuits for switched-mode power supplies', *Mullard Technical Note 24*, TP1442/1 (1975)
14 JANSSON, L. E., 'The design and operation of transistor d.c. converters', *Mullard Technical Communications*, **2**, 17, February (1956)
15 PALMER, M., 'The ABC's of DC to AC inverters', *Application Note AN-222*, Motorola Semiconductor Products, Inc.
16 TOWERS, T. D., 'Practical design problems in transistor DC/DC converters and DC/AC inverters', *Proc. IEE*. Pt. B, Suppl. 18, 1373–83, May (1959)
17 BASELL, M. C., '50 W multiple-output switched-mode power supply', *Mullard Technical Note 33*, TP1520 (1975)
18 ROYER, G. H., 'A switching transistor AC to DC converter', *Trans. AIEE*, July (1955)
19 JENSEN, J. L., 'An improved square wave oscillator circuit', *Trans. IRE*, Vol. CT4 No. 3, September (1957)
20 WAGNER, C. F., 'Parallel inverter with resistance load', *Trans. AIEE*, **54**, 1227–35 (1935)
21 RIDGERS, C., 'Voltage stabilised sinusoidal inverters using transistors', *Radio Eledtron. Eng.*, 109–27, August (1967)
22 EBBINGE, W., 'Designing very high efficiency converters with a new high frequency resonant GTO technique', *Powercon 8 Proc.*, A-1, 1–8, April (1981)
23 SCHADE, O. H., 'Radio-frequency-operated high-voltage supplies for cathode-ray tubes', *Proc. IRE*, 158–63, April (1943)
24 WOOD, P. N., 'Design of a 5 volt, 1000 watt power supply', *TRW Application Note*, No. 122A
25 TAYLOR, B. and FARROW, V., 'A new switching configuration improves the performance of off-line switching converters', *Powercon 8 Proc.*, G-2, 1–7, April (1981)
26 SEVERNS, R., 'The design of switchmode converters above 100 kHz', *Intersil. Appl. Bull.*, A034 (1980)

Further reading

PRESSMAN, A. I., *Switching and Linear Power Supply, Power Converter Design*, Heyden Book Co, NJ (1977)
MIDDLEBROOK, R. D. and CUK, S., *Advances in Switched-Mode Power Conversion*, Teslaco, California (1981)
MCLYMAN, C. W. T., *Transformers and Inductor Design Handbook*, Marcel Dekker Inc., NY (1978)
BEDFORD, B. D. and HOFT, R. G., *Principles of Inverter Circuits*, Wiley, NY (1964)
HNATEK, E. R., *Design of Solid-State Power Supplies*, Van Nostrand Reinhold, NY (1971)
FINK, D. G. (ed), *Electronic Engineer's Handbook*, McGraw-Hill, NY (1975). Also *Solid-State Power Circuits*, RCA Technical Series, SP-52 (1971)
RODDAM, T., *Transistor Inverters and Converters*, D. Van Nostrand Co. Inc., NJ (1963)
SNELLING, E. C., *Soft Ferrites, Properties and Applications*, Iliffe Books Ltd, London (1969)
NIJOFF, E. B. G. and EVERS, H. W., 'Introduction to the series resonant power supply', *Electronic Product Design* (September 1982)

45 Naturally Commutated Power Circuits

Tony Clark, DipEE, CEng, MIEE
Electrical Engineering Consultant

Contents

45.1 The principles of natural commutation in rectifier circuits

Natural commutation occurs when the circuit provides the means by which the current is reduced to zero and transferred from one thyristor to another allowing the conducting thyristor to turn off.

The simplest and most widely used method makes use of the alternating voltage waveform to effect the current transfer.

The current in a thyristor can be reduced to zero without interrupting the current in the load by switching on another thyristor connected to a higher-voltage supply source. In *Figure 45.1*, thyristor SCR_2 is triggered whenever voltage V_2 is larger than voltage V_1, the current in thyristor SCR_1 will transfer to thyristor SCR_2 and thyristor SCR_1 will turn off. If the voltages V_1 and V_2 are sinusoidal then everytime voltage V_1 is larger than voltage V_2, thyristor SCR_1 will switch on and resume taking the current.

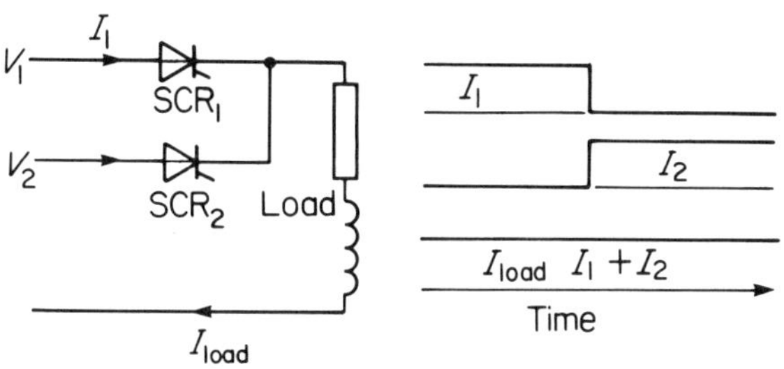

Figure 45.1 Basic circuit for natural commutation

For a simple half-wave circuit as shown in *Figure 45.2*, thyristor SCR_1 is triggered every positive half cycle of voltage V_1, and the current flows through the load as I_1. When voltage V_2 becomes greater in value, thyristor SCR_2 is triggered and the load current is transferred to it as current I_2. The load voltage is unidirectional and the level is controlled by altering the point at which the thyristors are triggered.

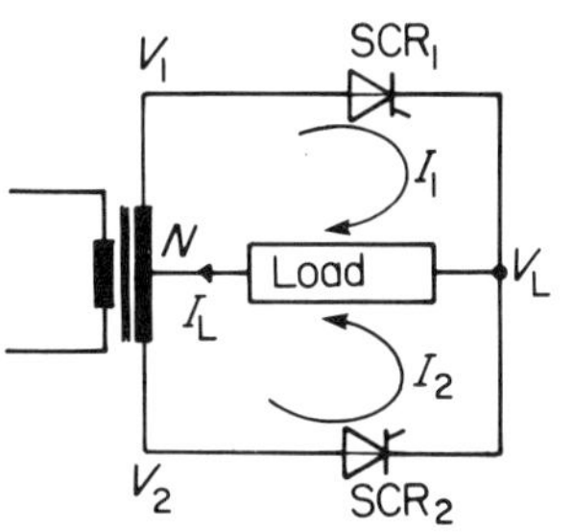

Figure 45.2 Simple half-wave circuit

45.1.1 Waveforms

For a resistive load, the current will cease to flow when the voltage reaches a zero (*Figure 45.3*), but with inductance in the load circuit it will continue for a period after the voltage zero. The period is dependent upon the magnitude of the impedance, and the load voltage will no longer follow the input waveform. As the current becomes continuous the output voltage will remain constant. For a capacitive load, the current will only flow for the period when the d.c. output voltage exceeds the capacitor voltage.

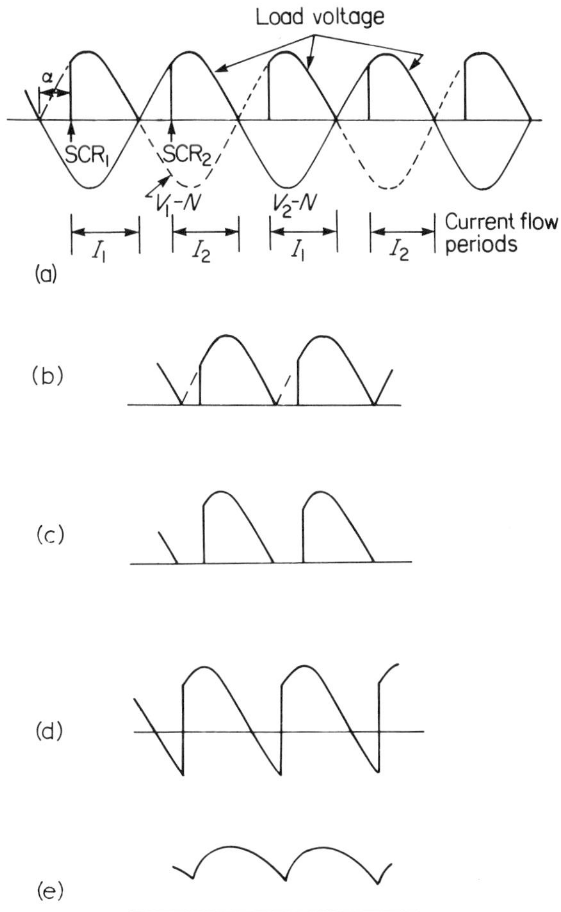

Figure 45.3 Effect of load inductance: (a) current flow periods; (b) output voltage, resistive load; (c) load current, resistive load; (d) output voltage, inductive load; (e) load current, inductive load

45.1.2 Overlap angle

Due to the stray circuit impedance in each arm, two thyristors can be conducting during the commutation period and this period is called the overlap angle (u). It is dependent on the magnitude of the load current and the value of the transformer leakage impedance. *Figure 45.4* shows the distortion of the a.c. voltage waveform due to overlap.

45.1.3 Regulation

This is defined as the reduction of the output voltage as the load increases. It comprises of thyristor and diode voltage drops, and transformer leakage impedance.

45.1.4 A.C. harmonics

The a.c. current is non-sinusoidal and contains harmonics related to the supply frequency and the circuit arrangement. The

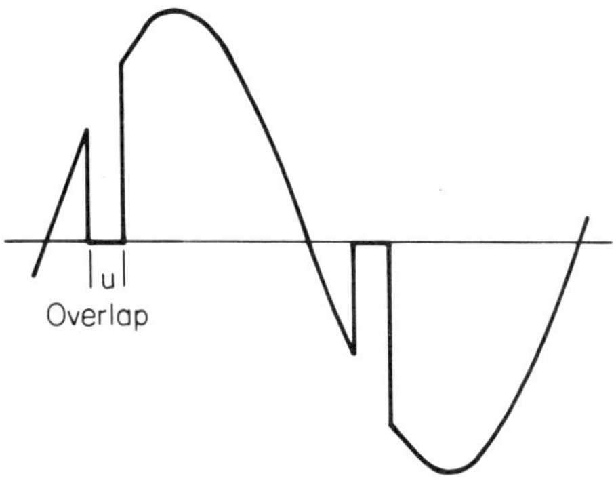

Figure 45.4 Effect of overlap on a.c. voltage waveform

magnitude of these harmonics can be found using Fourier analysis of the waveform.

For smooth d.c. current, and minimum overlap angle, the harmonic content will be constant for all control angles.

$$\text{RMS harmonic} = \frac{100}{m}\ \%\ \text{of fundamental} \qquad (45.1)$$

where m = order of harmonic.

In practical circuits the harmonic content changes with overlap angle and the d.c. ripple content, and changes in circuit impedances will also increase differences.

45.1.5 Power factor of input current

The power factor changes with the delay angle α and the effects of the overlap angle.

For a fully controlled bridge circuit the power factor is given by

$$\cos\varphi = \cos\left(\alpha + \frac{u}{2}\right) \qquad (45.2)$$

The power factor can be measured using the two wattmetre method and is normally defined as the power factor of the fundamental current.

45.2 D.C. output harmonics

The d.c. waveform of the output for all circuits is easily analysed and the harmonics present in the waveform for each circuit are dependent upon the number of commutations per cycle.

Figures 45.5 to *45.8* illustrate the d.c. harmonic content for continuous d.c. load current.

45.3 Inversion

If the load is sufficiently reactive or has a back e.m.f. such as a d.c. rotating machine, the delay angle can be advanced to 90 degrees where the mean d.c. voltage becomes zero. Providing the current remains continuous the delay angle can be increased beyond 90° and the load voltage will become negative.

With a negative load voltage and the same direction of current flow, the load power is fed back into the supply. The conditions for a single phase half wave circuit are shown in *Figure 45.9* and *45.10*. The reverse power flow will continue until the load energy is dissipated.

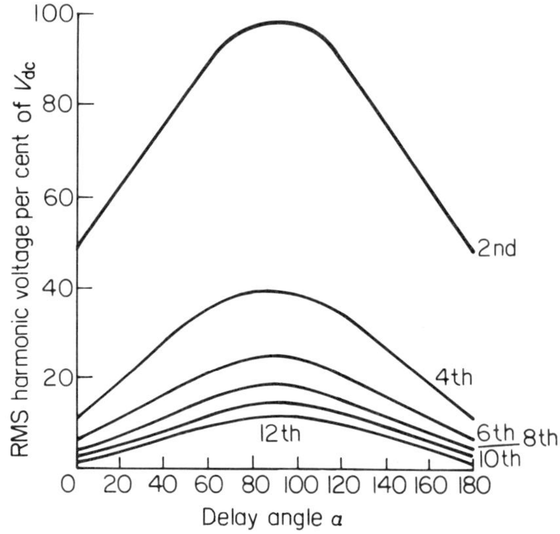

Figure 45.5 Single-phase half-wave and fully controlled bridge circuit

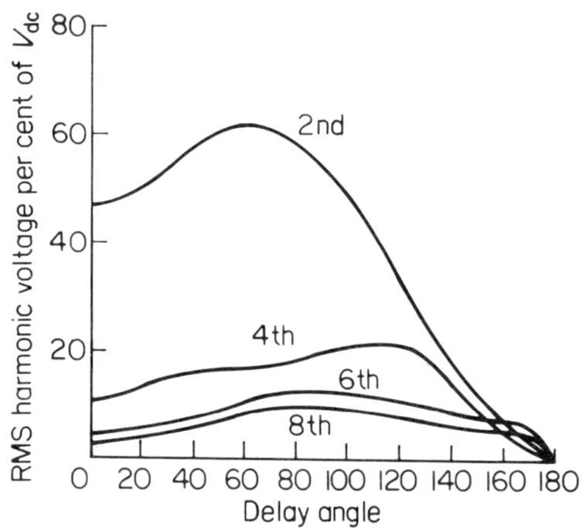

Figure 45.6 Single-phase half controlled bridge

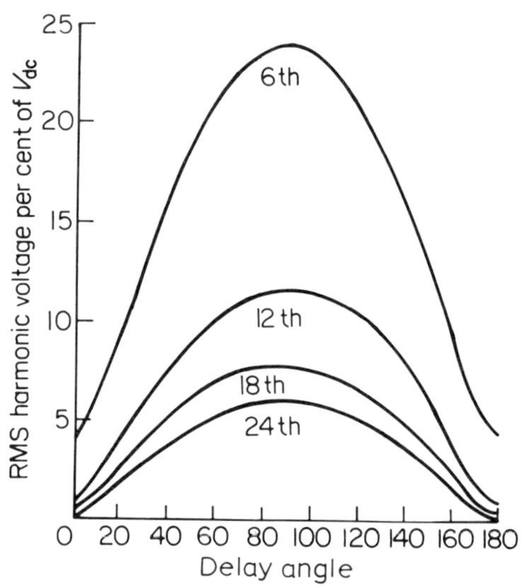

Figure 45.7 Three-phase fully controlled bridge and six-phase half-wave

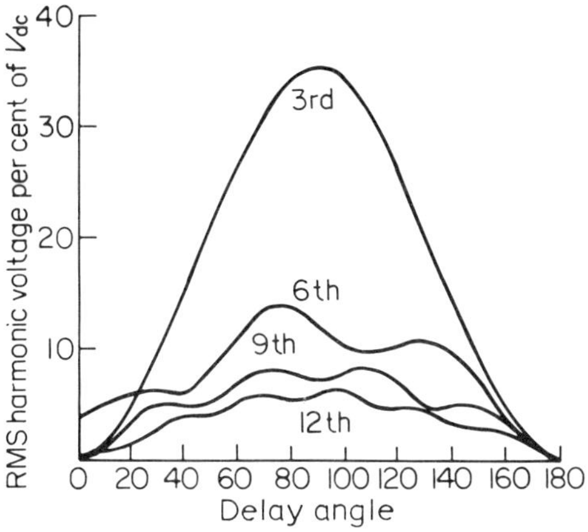

Figure 45.8 Three-phase half controlled bridge

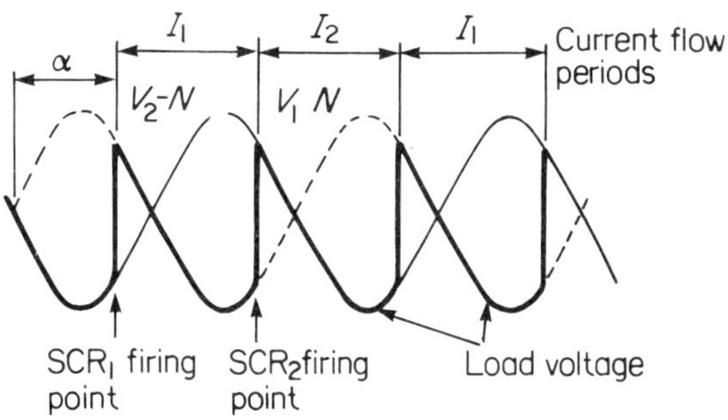

Figure 45.9 Power reversal during inversion

45.4 Circuit parameters for various a.c. configurations

These are shown in *Table 45.1*.

45.5 Typical applications of naturally commutated circuits

A number of circuits are available for controlling a.c. voltages with thyristors. Reverse blocking thyristors may be connected in anti-parallel or single-phase bridge may be used to rectify the a.c. line current, so that one thyristor can control both halves of the a.c. waveform. This is accomplished by replacing the d.c. load on the single-phase bridge with a short circuit and locating the load in the a.c. line. In some polyphase circuits, the same control of the voltage can be obtained when one thyristor is replaced with a rectifier diode giving a more economical arrangement.

45.5.1 Phase control of the a.c. voltage

When the load is inductive, current flows as a sinusoid which lags the supply voltage by the angle θ, which is a measure of the power factor. If each thyristor is triggered at this angle, the load current will be unaffected. If the triggering angle is made to lag behind θ, the load current will flow in a series of non-sinusoidal pulses of less than 180 electrical degrees duration.

As the angle of phase retard is increased, the pulse becomes increasingly shorter until at 180 degrees retard, they cease to exist and the voltage across the load is zero. Thus the voltage across the load will be reduced by phase retard in much the same manner as with a resistive load, except that voltage control will take place over a narrower range of triggering angles from θ to 180 degrees. At all triggering angles, the power factor of the load does not depart significantly from the value observed with no

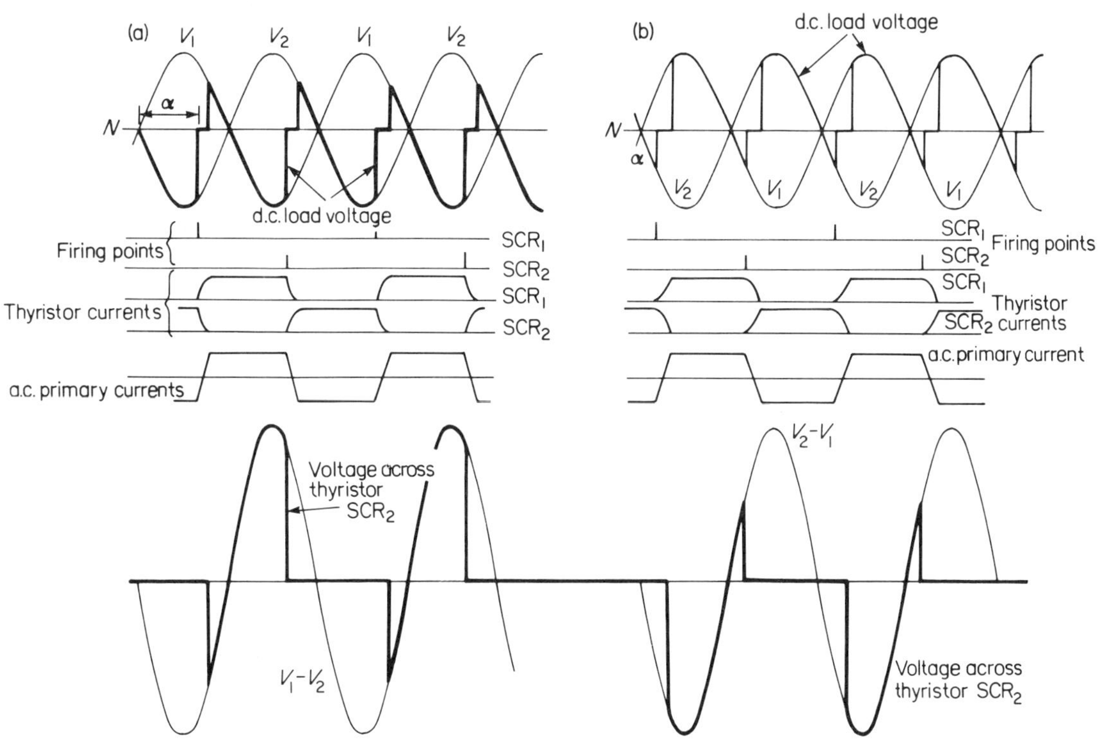

Figure 45.10 Waveforms for single-phase half-wave circuit during inversion. (a) delay angle 120; (b) delay angle 30

Table 45.1

Circuit name	Circuit connection	Load voltage waveform	Peak forward voltage on thyristor	Peak reverse voltage: Thyristor	Peak reverse voltage: Diode	Max. load voltage ($\alpha=0$) E_{dc} = av. d.c. E_{ac} = r.m.s. d.c. value	Load voltage (V_S) trigger delay angle (α)	Trigger delay angle range	Maximum thyristor current: I_{AV}	Maximum thyristor current: Conduction angle	Maximum diode current: I_{AV}	Maximum diode current: Conduction angle	Regenerative capability	Ripple frequency of load voltage	Notes
1. Half-wave resistive load	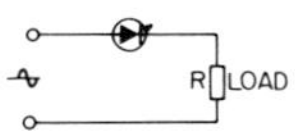	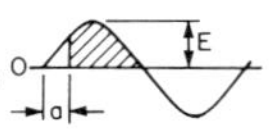	E	E	—	$E_{dc}=\frac{E}{\pi}$ $E_{ac}=\frac{E}{2}$	$E_{dc}=\frac{E}{2\pi}(1+\cos\alpha)$ $E_{ac}=$ $\frac{E}{2\sqrt{\pi}}\left(\pi-\alpha+\frac{\sin 2\alpha}{2}\right)^{1/2}$	180°	$\frac{E}{\pi R}$	180°	—	—	—	f	
2. Half-wave inductive load with free wheel diode	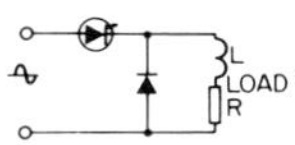	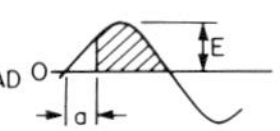	E	E	E	$E_{dc}=\frac{E}{\pi}$	$E_{dc}=\frac{E}{2\pi}(1+\cos\alpha)$	180°	$\frac{E}{2\pi R}$	180°	$0.54\frac{E}{\pi R}$	210°	No	f	
3. Centre-tapped with resistive or inductive load	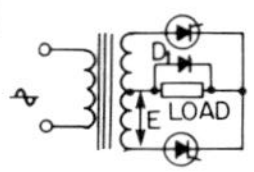	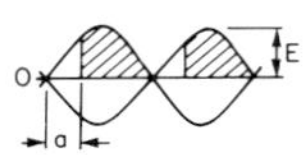	E	$2E$	E	$E_{dc}=\frac{2E}{\pi}$	$E_{dc}=\frac{E}{\pi}(1+\cos\alpha)$	180°	$\frac{E}{\pi R}$	180°	$0.26\frac{E}{\pi R}$	148°	No	$2f$	
4. Centre-tapped with resistive or inductive load with thyristor in d.c. circuit	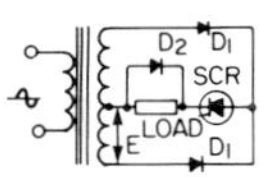	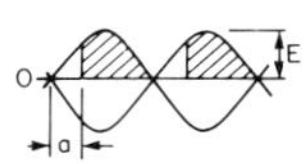	E	0	$2E$ on D_1 E on D_2	$E_{dc}=\frac{2E}{\pi}$	$E_{dc}=\frac{E}{\pi}(1+\cos\alpha)$	180°	$\frac{2E}{\pi R}$	360°	$D_1=\frac{E}{\pi R}$ $D_2=0.26\frac{E}{\pi R}$	180° -- 148°	No	$2f$	Recovery of thyristor in d.c. load limits frequency
5. Centre-tapped with inductive load	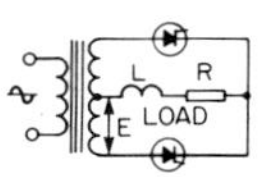	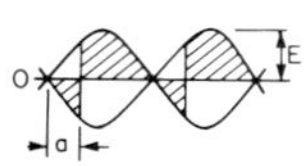	$2E$	$2E$	—	$E_{dc}=\frac{2E}{\pi}$	$E_{dc}=\frac{2E}{\pi}\cos\alpha$ for continuous current	180°	$\frac{E}{\pi R}$	180°	—	—	Yes	$2f$	
6. Hybrid single-phase bridge with free wheel diode	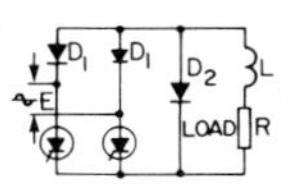	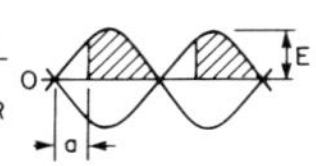	E	E	E	$E_{dc}=\frac{2E}{\pi}$	$E_{dc}=\frac{E}{\pi}(1+\cos\alpha)$	180°	$\frac{E}{\pi R}$	180°	$D_1=\frac{E}{\pi R}$ $D_2=0.26\frac{E}{\pi R}$	180° 148°	No	$2f$	D_2 necessary to ensure turn off on inductive loads

7. Hybrid single-phase bridge thyristor arm common	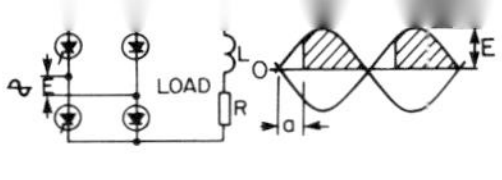	E	E	E	$E_{dc}=\frac{2E}{\pi}$	$E_{dc}=\frac{E}{\pi}(1+\cos\alpha)$	180°	$\frac{E}{\pi R}$	180°	$\frac{E}{\pi R}$	180°	No	$2f$	Rectifier diodes provide free wheel path. Conducts $(\pi+\alpha)°$
8. Single-phase bridge	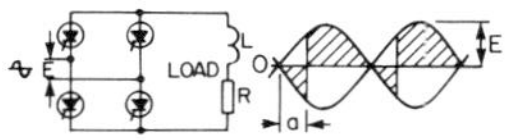	E	E	—	$E_{dc}=\frac{2E}{\pi}$	$E_{dc}=\frac{2E}{\pi}\cos\alpha$ for continuous current	180°	$\frac{E}{\pi R}$	180°	—	—	Yes	$2f$	Recovery of thyristor in d.c. load limits frequency
9. Single-phase with thyristor in d.c. circuit	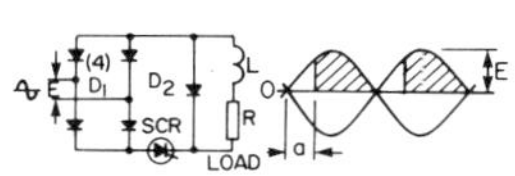	E	0	E on D_1 and D_2	$E_{dc}=\frac{2E}{\pi}$	$E_{dc}=\frac{E}{\pi}(1+\cos\alpha)$	180°	$\frac{2E}{\pi R}$	360°	$D_1=\frac{E}{\pi R}$ $D_2=0.16\frac{2E}{\pi R}$	180° 148°	No	$2f$	
10. Three-phase half wave (inductive load free wheel diode)	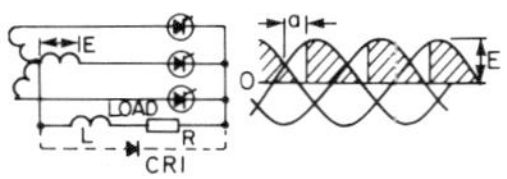	E	$\sqrt{3}E$	E	$E_{dc}=3\sqrt{3}\frac{E}{2\pi}$	$E_{dc}=\frac{3\sqrt{3}}{2\pi}\cos\alpha$ $(0<\alpha<30°)$ $E_{dc}=\frac{3E}{2\pi}(1+\cos(\alpha+30°))$ $(30°<\alpha<150°)$	150°	$\frac{\sqrt{3E}}{2\pi R}$	120°	$0.16\frac{3\sqrt{3E}}{2\pi R}$	134°	No	$3f$	
11. Three-phase half-wave	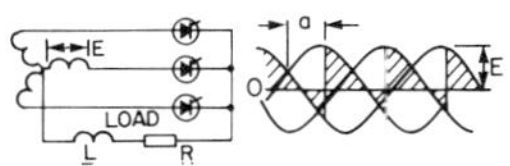	$\sqrt{3}E$	$\sqrt{3}E$	—	$E_{dc}=3\sqrt{3}\frac{E}{2\pi}$	$E_{dc}=\frac{3\sqrt{3E}}{\pi}\cos\alpha$ for continuous current	150°	$\frac{\sqrt{3E}}{2\pi R}$	120°			Yes	$3f$	
12. Three-phase hybrid bridge	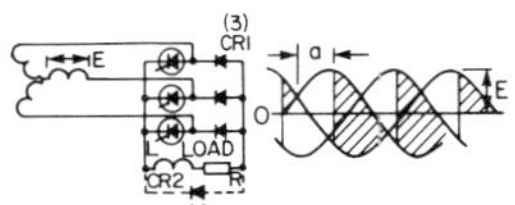	$\sqrt{3E}$	$\sqrt{3E}$	$\sqrt{3E}$	$E_{dc}=\frac{3\sqrt{3E}}{\pi}$	$E_{dc}=\frac{3\sqrt{3E}}{2\pi}(1+\cos\alpha)$	180°	$\frac{\sqrt{3E}}{\pi R}$	120°	$D_1=\frac{\sqrt{3E}}{\pi R}$ $D_2=0.14\frac{3\sqrt{3E}}{\pi R}$	120° 132°	No	$3f$	D_2 necessary to ensure turn off on inductive loads
13. Three-phase full controlled bridge	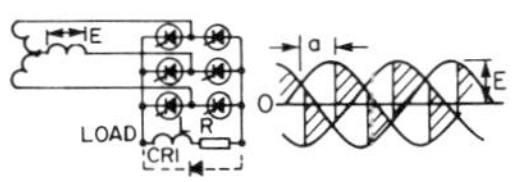	$\sqrt{3E}$	$\sqrt{3E}$	$\sqrt{3E}$	$E_{dc}=\frac{3\sqrt{3E}}{\pi}$	$E_{dc}=\frac{3\sqrt{3E}}{\pi}\cos\alpha$ $(0<\alpha<60°)$ $E_{dc}=\frac{3\sqrt{3E}}{\pi}\left(1+\frac{\cos\alpha}{2}-\frac{\sqrt{3}}{2}\sin\alpha\right)$ $(60°<\alpha<120°)$	120°	$\frac{\sqrt{3E}}{\pi R}$	120°	$0.056\frac{3\sqrt{3E}}{\pi R}$	212°	No	$6f$	Thyristors require two gate signals 60° apart each cycle

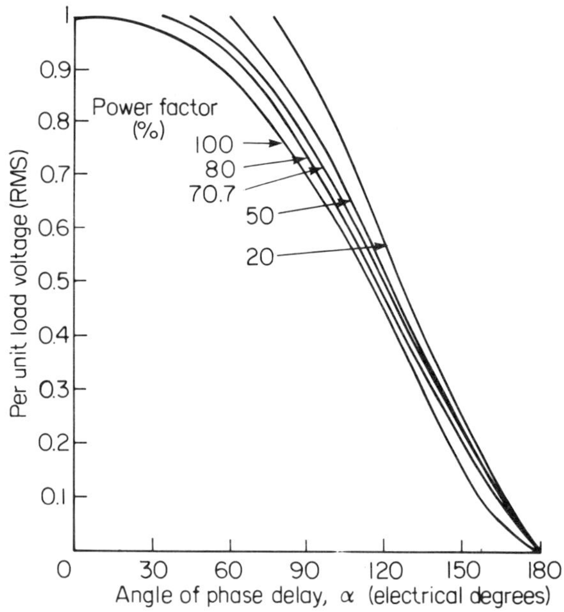

Figure 45.11 Load current variation with angle of phase retard

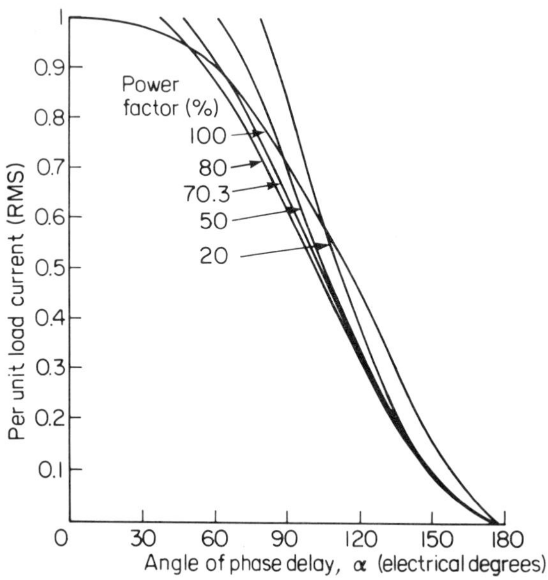

Figure 45.12 Load voltage variation with angle of phase retard

phase control. Typical transfer characteristics of a.c. phase control circuits are shown in *Figures 45.11*, *45.12* and *45.13*.

45.6 Gate signal requirements

When thyristors are used to control resistive loads, almost any of the varied forms of triggering circuits may be used with satisfactory results. Synchronization of the triggering pulses may be accomplished from either the line voltage or the voltage across the thyristor.

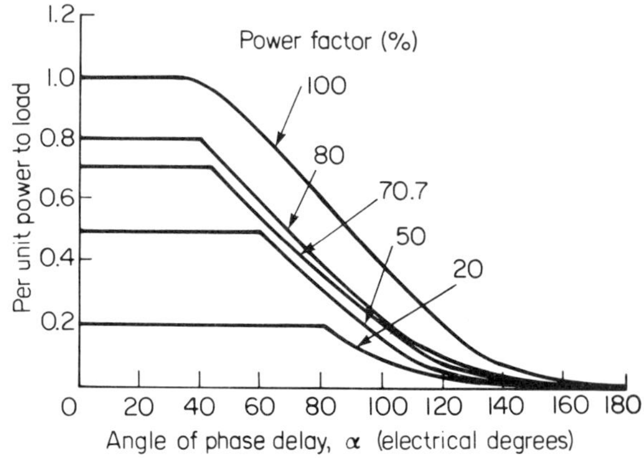

Figure 45.13 Angle of phase retard variation with power to load

However, when the load is inductive, several precautions must be observed in order to achieve optimum performance.

(1) If a triggering circuit is used which produces a narrow spike of gate signal, the charge injected into the thyristor may not be sufficient to maintain the thyristor in the conducting state until the load current has built up to a magnitude larger than the latching current. This may result in mistriggering and erratic control, or no load current whatsoever. One solution is to shunt the inductive load with a small resistive load drawing a current somewhat larger than the maximum value of thyristor holding current. An alternative (and usually more satisfactory) solution is to provide a triggering circuit which produces a square-wave gate signal which lasts from the time at which triggering is initiated until the time when the thyristor conducts a significant amount of current.

(2) For inductive loads, it is mandatory to obtain line synchronization for the triggering circuit from the line voltage, not from the voltage across the thyristor. If the synchronizing signal is obtained from across the thyristor, a large unbalance between the positive and negative half-cycles of current is probable. The result may be overloading of one thyristor, saturation of a transformer core (if a transformer is being controlled), and unstable control. (See *Figure 45.14*.)

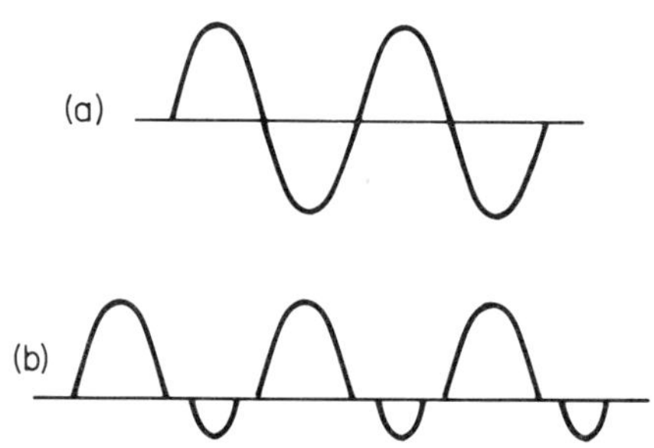

Figure 45.14 Triggering circuit synchronisation: (a) Symmetrical line current when triggering circuit synchronisation is taken from line. (b) Asymmetrical (undesirable) line current which results when triggering circuit synthronisation is taken across *SCR* assembly

For highly inductive loads, triggering the thyristor full on can result in no output if the triggering pulse is so short that it disappears and the thyristor regains its off-state blocking ability before current can flow in the load circuit. To avoid this, it is necessary to make the triggering pulses long enough so that the thyristor will always be able to conduct whenever circuit conditions are right for conduction, once it has been triggered. *Figure 45.15* defines the necessary triggering pulse duration as a function of the load power factor, when the pulse is initiated with zero phase retard.

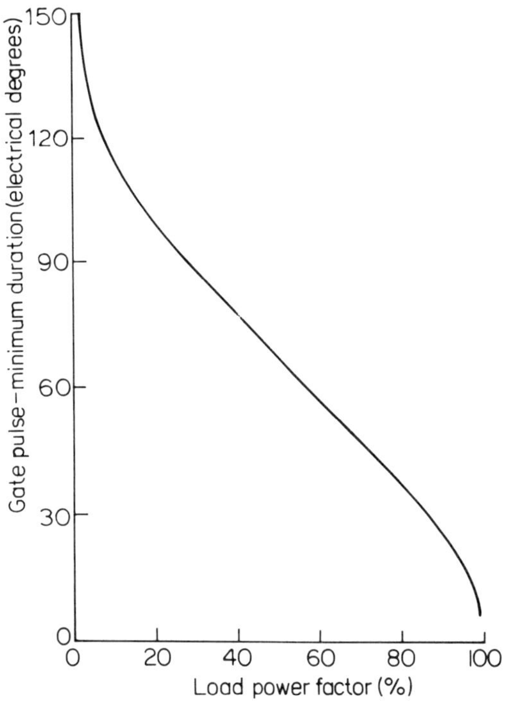

Figure 45.15 Pulse duration variation with load power factor

45.7 Other ways of controlling the a.c. voltage

There are numerous types of equipment or systems which can be controlled readily and advantageously by the use of thyristors as switches (as opposed to using them in the phase control mode for continuous variability of voltage). Of course, many of the switching mode circuits may be readily modified so that they incorporate phase control to provide a fine voltage adjustment to supplement the switching mode of control.

Some of the advantages of the switching mode of operation of thyristors for controlling voltage are as follows.

(1) Switching of load voltage (either from zero to full voltage or from partial to full voltage) is accomplished without mechanical contacts, thus eliminating common maintenance problems which result from burning, pitting, and welding of contacts. 'Contact bounce' is also eliminated, thereby reducing radio frequency interference (RFI) caused by the repetitive shock excitation of reactive circuit elements.

(2) 'Zero voltage switching' may be used to essentially eliminate radio frequency interference (RFI) often encountered when voltage is controlled by phase control.

(3) The use of the switching mode of voltage control eliminates the reduction in power factor which inherently occurs when voltage is reduced by phase control.

Many of these switching control circuits will find use in controlling heating elements for ovens, furnaces, hot plates, crucibles, and space heaters. They may also be used for speed controls of squirrel cage and wound rotor induction motors where the motor and/or load inertia is high. Still other types of load, for example resistance welders, stud welders, flashers and flashing beacons, and magnetic hammers or pulsers, demand this type of control as an inherent element in their mode of operation.

Most heating elements have a high thermal inertia and are easily adaptable to control by the switching mode, either by being switched from zero to full voltage or from a low voltage tap to full voltage.

When circuits with a tapped transformer winding are compared to those with a tapped load, it will be seen that the final control result is very similar. However, where the transformer is tapped, the load voltage is initially low and is switched to a higher value. In the tapped load circuit, the initial voltage is high across one section of the load and zero across the other section. After switching, the voltage decreases across one section while increasing across the other. This tapped load 'shunt controller' performs equally well for resistive loads (such as heating elements) and for speed control of wound rotor induction motors, by varying the resistance in the rotor circuit.

Phase control is also very effective (although radio frequency interference filtering may be necessary) using the shunt mode of control. *Figure 45.16* illustrates the control characteristic for both a resistive load and an inductive (70 degree lagging) load. When R_B is small compared to R_A, the range of load current variation is small, but precision of adjustment is very good. When R_B is large compared to R_A, the swing in load current can be very large, but with a reduction in adjustment precision. In either case, the voltage is smoothly adjustable over the design range, and is readily adaptable to automatic control.

In the cases of flashers and beacons, it often becomes possible to substantially lengthen filament life of the lamps by switching from full voltage to a lower transformer tap (reducing the voltage below the incandescence level), thus minimizing the range of filament temperature excursion. A similar improvement in the life of heating elements may also be expected.

To obtain good temperature regulation, or speed regulation, with the switching mode of control, the control system may employ pulse-burst modulation. This mode of control is usually based on a fixed time control period, for example, 20 cycles at power frequency. The number of cycles during this period when the thyristor switch is in conduction is made variable and is adjusted by temperature or speed feedback controls. The switch always conducts for essentially an integral number of cycles, and is either off, or conducts at a reduced voltage level, for the balance of each period (*Figures 45.17* and *45.18*).

Pulse-burst modulation can be applied to the control of most heating systems and motor drives due to the high thermal or mechanical inertia of these systems. Systems with very high inertia can tolerate relatively long control periods, whereas systems with lower inertia, or requiring a fine control resolution, will demand a short control period.

As mentioned earlier, the use of pulse-burst modulation with thyristor switches will minimise radio frequency interference, but this, by itself, will not completely eliminate it. The fast turn-on of a thyristor when the supply voltage is at any value other than zero, still causes one pulse of shock excitation each time the thyristor is turned on to carry a burst of pulses. However, with thyristor switches, it is possible and often desirable to incorporate zero voltage switching. When controlled in this mode, the thyristors are always turned on at zero voltage (they turn off at zero current), thus eliminating the interference caused by fast switching of heavy currents.

All of the circuits which switch from one tap to another (either tapped transformer or tapped load) may incorporate phase

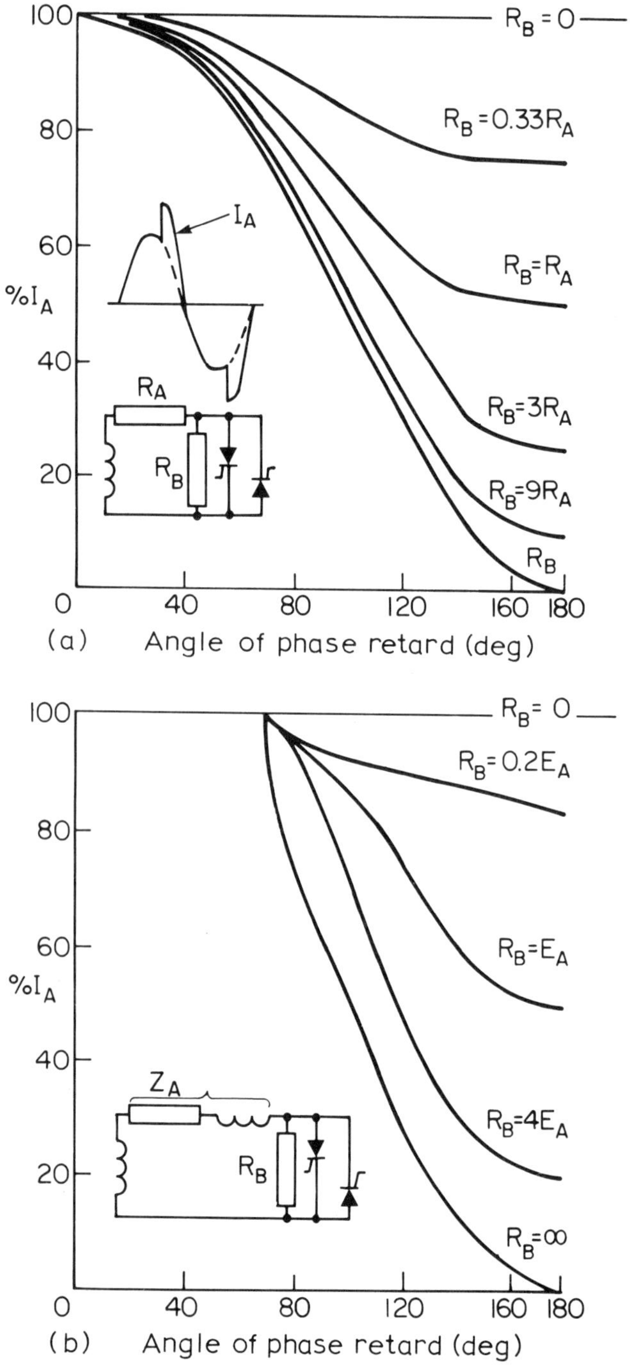

Figure 45.16 A.C. control using shunt thristor: (a) resistive load; (b) inductive load

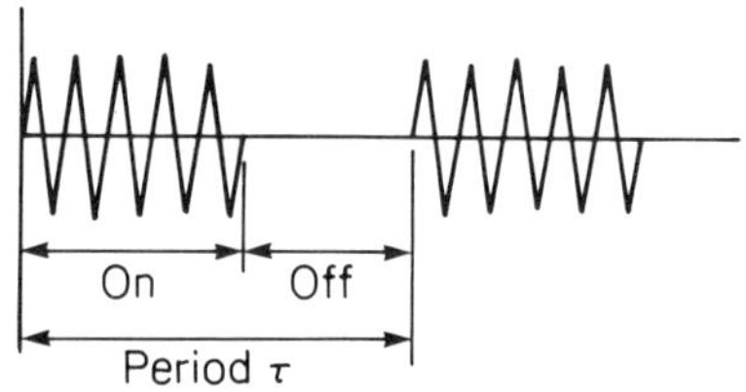

Figure 45.17 Pulse burst modulation voltage control—full on to off.

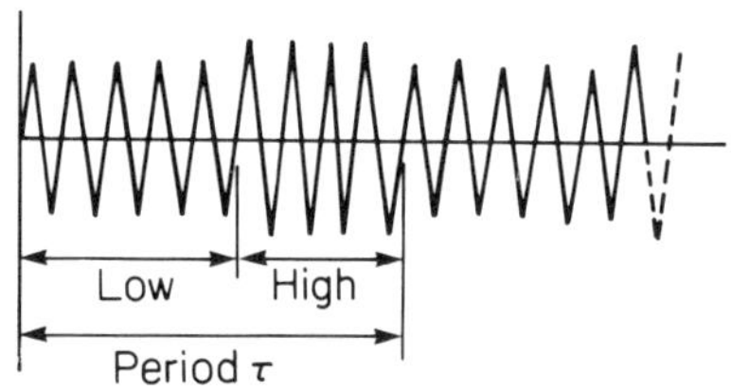

Figure 45.18 Pulse burst modulation voltage control—transformer tap switching. Control of a.c. load (circuits 4A, 6A and 15A)

control to supplement the switching type of control to obtain voltage or current regulation of power supplies. Methods of applying phase control have been discussed earlier in this chapter. It is possible to achieve such regulation by phase control alone and eliminate the taps. However, where only a limited range of adjustability is required, the combination of phase control with switching mode operation greatly reduces the peak-to-RMS current ratio in a.c. power supplies. For instance, output voltage can be smoothly varied between the voltages obtained from any two transformer taps by first gating a thyristor connected to the lower voltage tap, and then later in the cycle gating a thyristor connected to the higher voltage tap. This type of voltage control reduces the peak-to-average current ratio. It also minimises the change in power factor as voltage is varied.

45.8 The cycloconverter

A cycloconverter is a means of changing the frequency of alternating power using thyristors which are commutated by the a.c. supply. It can be used as an alternative to a rectifier followed by an inverter.

Cycloconverters utilise any fully controlled, naturally commutated rectifier circuits, using a number of separate circuits in series or parallel. Thyristor ratings will follow the normal practice for mains connected naturally commutated circuits. Anti-parallel operation using two converters in each phase is preferable allowing freedom of choice on the connection of the separate converters employed in the circuit. (Detailed circuit and mathematical analysis given in References 1, 2 and 3.

45.8.1 Basic concept of a single-phase circuit

Figure 45.19 shows the circuit discussed in this section. The arms of a single-phase full-wave rectifier circuit are replaced by two pairs of thyristors connected in anti-parallel which can be triggered to produce opposite polarities. When thyristor SCR_1 and SCR_3 are triggered the d.c. output is sensed in one direction, and reversed when thyristors SCR_2 and SCR_4 are triggered instead. By alternatively triggering each pair of thyristors at a frequency lower than the supply frequency a square wave of current is produced and flows in the load. A filter would be required to reduce or eliminate the ripple produced.

A crude sine wave can be synthesised by triggering the individual thyristors at different delay angles.

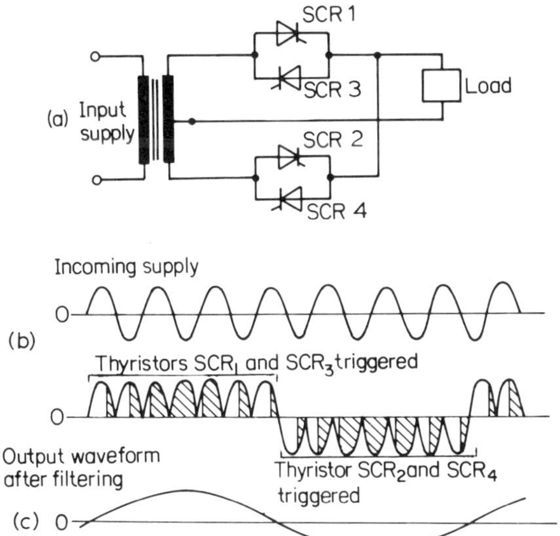

Figure 45.19 Simple single-phase cycloconverter: (a) circuit diagram; (b) output waveforms; (c) output waveform after filtering

45.8.2 Practical single-phase circuit

Figure 45.20 shows the circuit discussed in this section. To achieve an improved output waveform shape, two three-phase fully controlled bridges are connected in anti-parallel. This arrangement of the bridges is capable of feeding power to the load

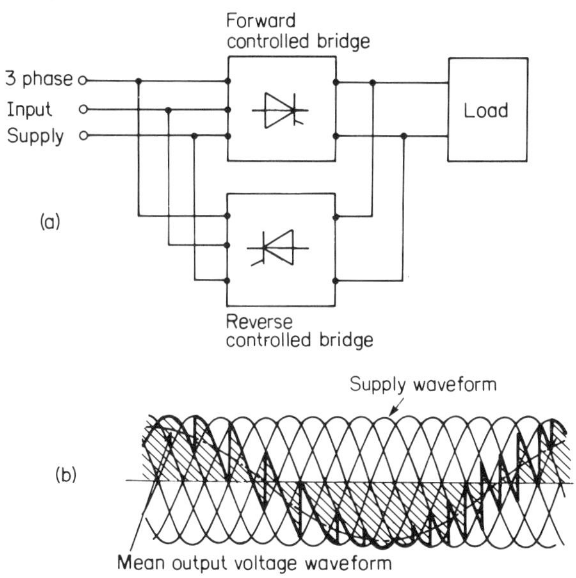

Figure 45.20 Practical single-phase cycloconverter: (a) circuit diagram; (b) basic output waveform

or returning power to the supply from the load. Both bridges produce full voltage control in either direction and providing the current is correctly sensed, any combination of output voltage and current is possible.

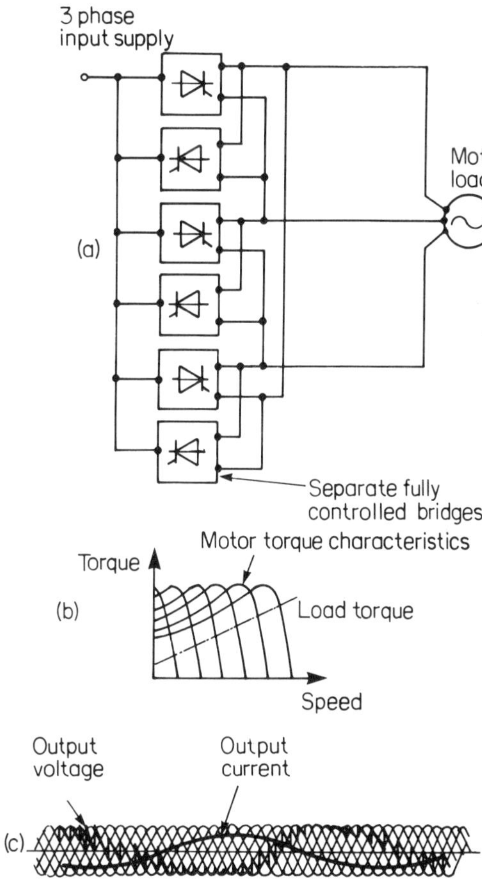

Figure 45.21 Typical three-phase cycloconverter driving a motor load: (a) circuit diagram; (b) load characteristics; (c) output waveforms

45.8.3 Practical three-phase circuit

Figure 45.21 shows the circuit discussed in this section. Three single-phase cycloconverters are usually connected in star or delta to produce a three-phase output, the choice of connection depending on the load characteristics. An example of a delta-connected motor load is illustrated, the circuit producing a wide range of frequency and voltage control to enable the highest torque to be maintained over the speed range. The circuit provides a.c. power in both directions as required by the load.

45.8.4 Characteristics of cycloconverters

(1) A large number of thyristors are required for practical circuits. Three-phase circuits usually need a minimum of eighteen devices.

(2) They can only be operated at sub-supply frequency.

(3) Output voltage can be characterised up to maximum cycloconverter voltage. In excess of this level the waveform shape is trapezoidal without voltage control. On motor loads, the input power factor can be improved at the expense of additional harmonic content.

(4) The input power factor is low for most circuits and changes with variation of output frequency. For motor load, output voltage is proportional to frequency to maintain good speed control parameters. There is a low power factor at low speed.

(5) Although cycloconverters have more complicated electronic systems, the output is easily characterised by changing signals to give: (a) changes in the direction of power flow from feeding the load to regenerating back onto the supply lines; (b) changes of phase rotation on multi-phase types; (c) output waveform can easily be characterised to suit load.

References

1 McMURRAY, W., *Theory and Design of Cycloconverters*, MIT Press, Cambridge, Mass. (1972)
2 PELLY, B. R., *Thyristor Phase Controlled Converters and Cycloconverters*, John Wiley (1971)
3 GYUGYI, L. and PELLY, B. R., *Static Power Frequency Changers*, John Wiley (1976)

Further reading

BLODFORD, B. D. and HOFT, R. C., *Principles of Inverter Circuits*, Chapter 3, John Wiley (1964)
BOWLER, P., 'The application of a cycloconverter to the control of induction motors', *IEE Conf. Publication* No. 17, 137–45, November (1965)
MAZDA, F. F., *Thyristor Control*, Newnes-Butterworths (1973)
READ, J. C., 'The calculation of rectifier and inverter performance characteristics', *IEE Proc.*, **92**, 29 (1945)
SALZMAN, T., 'Cycloconverters and automatic control of ring motors driving tube mills', *Siemens Rev.*, **45**, 3–8 (1978)
I.G.E., *SCR Manual*
I.R., *SCR Applications Handbook*
I.E.E., 'Semiconductor converters', *IEE Publication* No. 146, revised edition (1975)

46 Forced Commutated Power Circuits

F F Mazda, DFH, MPhil, CEng, MIEE, DMS, MIBM
Rank Xerox Ltd

Contents

46.1 Introduction

The design aspects of d.c. to d.c. converters (choppers) and d.c. to a.c. converters (inverters) are considered in this chapter. The effect of forced commutation circuits, as required by thyristors, is initially ignored so that the results are equally applicable to other forms of switching, such as using power transistors.

46.2 D.C. to d.c. converter without commutation

46.2.1 Voltage control methods

Figure 46.1 shows a simplified circuit arrangement of a chopper, and the waveforms for mark-space and frequency control. With

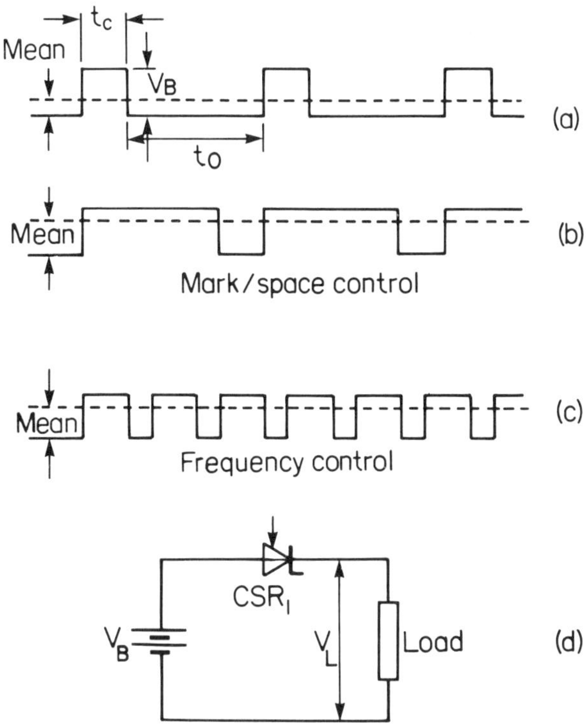

Figure 46.1 Operating principle of a d.c. to d.c. converter (chopper). (a) Output waveform; (b) output waveform with mark-space control; (c) output waveform with frequency control; (d) simplifier circuit arrangement

thyristor CSR_1 conducting the load voltage is equal to V_B (loss across the thyristor being assumed small). When CSR_1 is non-conducting load voltage is zero. Varying the ratio of thyristor on to off times, either in a variable mark-space or controlled frequency mode, the mean load voltage is changed, being given by:

$$V_{L(mean)} = V_B \frac{t_c}{t_c + t_0} \tag{46.1}$$

46.2.2 Load and device currents

Figure 46.2 shows an equivalent circuit of a chopper operating into a load of voltage V_F. L, R and C form filter components and

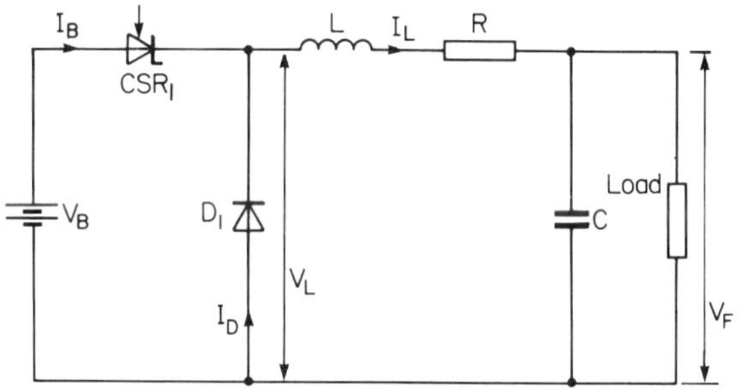

Figure 46.2 Equivalent circuit of a chopper

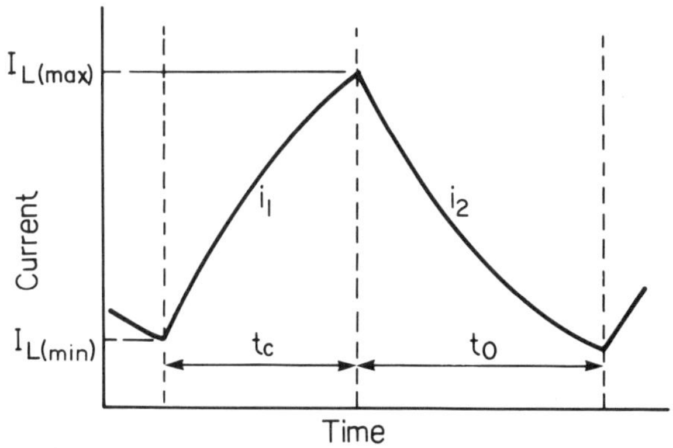

Figure 46.3 Steady-state load-current waveform of a chopper

D_1 is a free-wheeling diode, which carries the inductive load current during the off period of thyristor CSR_1. *Figure 46.3* gives this steady state current waveform. Time t_c corresponds to the period for which CSR_1 is on and t_0 for that when it is off. The following assumptions are made.

(1) The thyristors and diodes have zero voltage drop when conducting.

(2) The devices have infinite resistance when non-conducting.

(3) Turn-on time of the thyristors is short compared to the switching period. Therefore switching losses can be neglected.

(4) The d.c. source impedance is negligible so that energy can flow in either direction through it without affecting terminal voltage.

(5) V_F is constant during a cycle of operation.

(6) Load current is continuous.

With these assumptions the values of $I_{L(max)}$ and $I_{L(min)}$ can be written from *Figure 46.3* as:

$$\frac{I_{L(min)}}{V_B/R} + \frac{V_F}{V_B} = \exp\left(-\frac{R}{L}t_0\right)\left[1 - \exp\left(-\frac{R}{L}t_c\right)\right] \left[1 - \exp\left(-\frac{R}{L}(t_c + t_0)\right)\right]^{-1} \tag{46.2}$$

$$\frac{I_{L(max)}}{V_B/R} + \frac{V_F}{V_B} = \left[1 - \exp\left(-\frac{R}{L}t_c\right)\right]\left[1 - \exp\left(-\frac{R}{L}(t_c + t_0)\right)\right]^{-1} \tag{46.3}$$

The mean current I_B through thyristor CSR_1 and I_D through diode D_1 can be found as the mean of i_1 over time t_c and i_2 over t_0, respectively.

$$\frac{I_B}{V_B/R}=\frac{t_c}{t_c+t_0}-\frac{L/R}{t_c+t_0}\left[1-\exp\left(-\frac{R}{L}t_c\right)\right]\left[1-\frac{I_{L(min)}}{V_B/R}\right]$$
$$-\frac{V_F}{V_B}\left[\frac{t_c}{t_c+t_0}-\frac{L/R}{t_c+t_0}\left(1-\exp\left(-\frac{R}{L}t_c\right)\right)\right] \qquad (46.4)$$

$$\frac{I_D}{V_B/R}=\frac{L/R}{t_c+t_0}\left[1-\exp\left(-\frac{R}{L}t_0\right)\right]\left[\frac{I_{L(max)}}{V_B/R}+\frac{V_F}{V_B}\right]$$
$$-\frac{V_F t_0}{V_B(t_c+t_0)} \qquad (46.5)$$

Mean load current I_L is given by:

$$I_L=\frac{I_V+I_D}{V_B/R}$$

or

$$\frac{V_L-V_F}{R}$$

Therefore

$$\frac{I_L}{V_B/R}+\frac{V_F}{V_B}=\frac{t_c}{t_c+t_0} \qquad (46.6)$$

Equations (46.3) to (46.6) are shown plotted in *Figures 46.4* to *46.7*. The ripple current, derived from Equations (46.2) and (46.3)

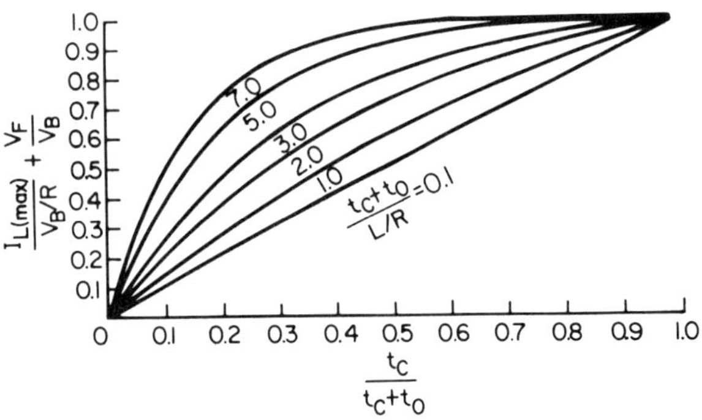

Figure 46.4 Variation of $I_{L(max)}/(V_B/R)+V_F/V_B$ with $t_c/(t_c+t_0)$

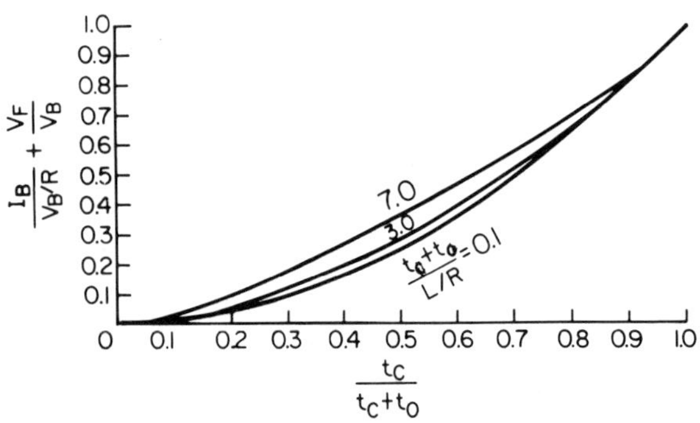

Figure 46.5 Variation of $I_B/(V_B/R)+V_F/V_B$ with $t_c/(t_c+t_0)$

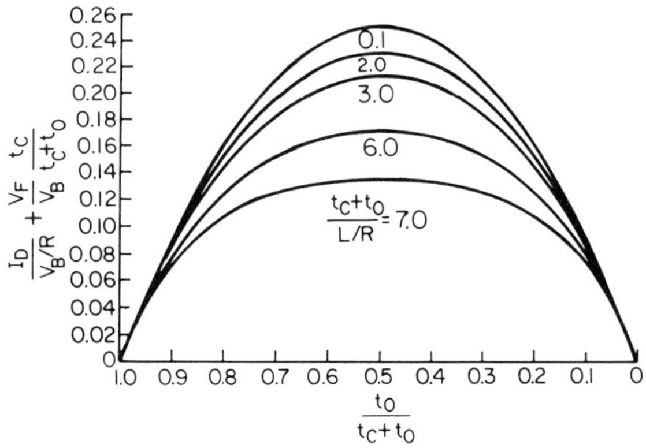

Figure 46.6 Variation of $I_D/(V_B/R)+(V_F/V_B)[t_c/(t_c+t_0)]$ with $t_0/(t_c+t_0)$

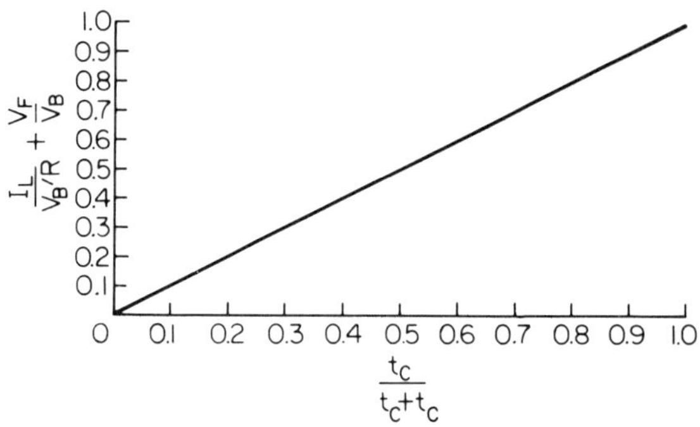

Figure 46.7 Variation of $I_L/(V_B/R)+V_F/V_B$ with $t_c/(t_c+t_0)$

is plotted in *Figure 46.8*. These curves allow device ratings, at any operating frequency, determined by the ratio of $(t_c+t_0)/(L/R)$ to be obtained. The peak load current, $I_{L(max)}$, is that value which has to be commutated in thyristor CSR_1. This varies with operating frequency and pulse width, raching a maximum value of $(V_B-V_F)/R$ when the thyristor is on all the time. Peak current should be as low as possible for any pulse width. Therefore it is clearly advantageous to operate at higher frequencies. *Figure 46.8* shows that the current ripple is then also minimal and from *Figure 46.5* the thyristor mean current rating is reduced. Although a penalty is paid in terms of increased diode current, a diode is cheaper than an equivalent rated thyristor so that this is

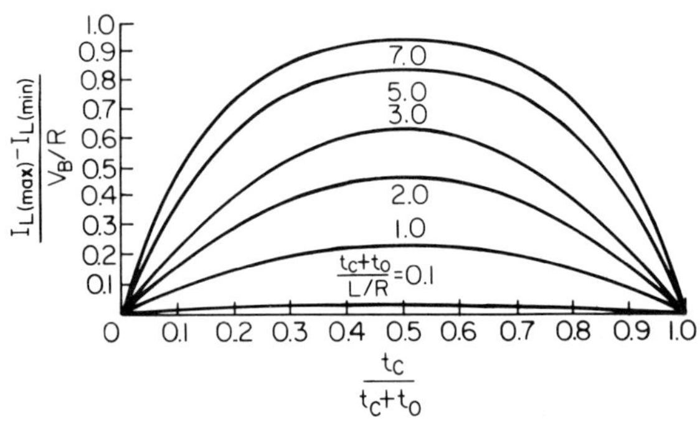

Figure 46.8 Variation of $(I_{L(max)}-I_{l\,(min)})/(V_B/R)$ with $t_c/(t_c+t_0)$

acceptable. It is interesting to note that ripple current is independent of the value of V_F and reaches a peak at half control width.

46.3 D.C. to a.c. converter without commutation

46.3.1 Operating principle

All inverter circuits fall into two broad groups, push–pull and bridge. A push–pull inverter is shown in *Figure 46.9*, the commutation components for the thyristors being omitted.

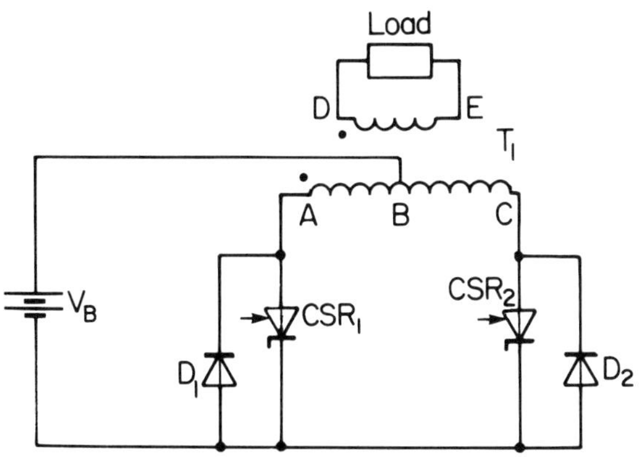

Figure 46.9 Fundamental push–pull inverter

When CSR_1 is turned on, end A of the output transformer T_1 goes negative so that end D of the load is made negative. When CSR_1 is turned off and CSR_2 fired the load polarity reverses. Diodes D_1 and D_2 carry the inductive load current when the thyristors are first turned on. Clearly T_1 is not always essential. It may be possible to make the load itself centre-tapped. For instance ABC could be the stator winding of a single-phase induction motor. In either case it is important to note that when CSR_1 conducts a voltage V_B is impressed across AB. If the two halves of the load are closely coupled, this raises the anode of CSR_2 to a potential $2V_B$ above its cathode. Similarly CSR_1 and D1 must have a voltage rating of at least $2V_B$.

A bridge inverter without commutation components is shown in *Figure 46.10*. It is evident that such an inverter does not need the load to be centre-tapped or connected via an intermediate centre-tapped transformer. Firing CSR_1, CSR_2 and CSR_3, CSR_4 in pairs gives an alternating load polarity. D_1 to D_4 carry the

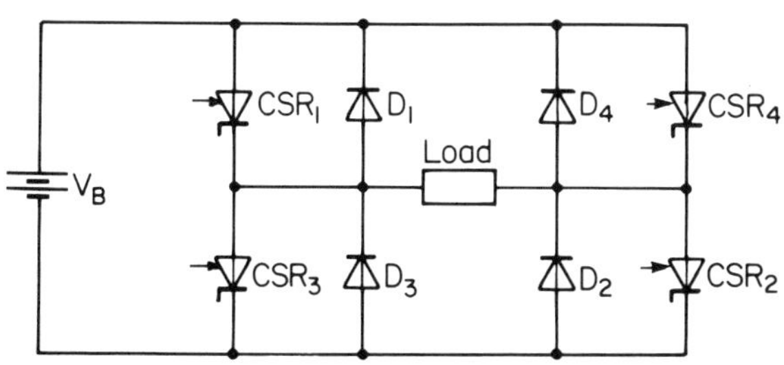

Figure 46.10 Basic bridge inverter

inductive load current. Although there are twice as many components in a bridge inverter than in a push–pull inverter, the devices now need only be rated to a voltage V_B.

Generally the choice between bridge and push–pull inverters is not difficult. If the load is not centre-tapped, or does not require a transformer, bridge inverters should be used. If the load is transformer coupled, bridge inverters are used at high supply voltages, say above 200 V, and push–pull inverters at lower voltages, which would not necessitate using unduly high voltage rated and expensive devices.

46.3.2 Voltage control methods

There are two instances when voltage control within an inverter is required:

(*a*) when the output is to be kept at a fixed value in spite of regulation within the inverter, or fluctuations of the supply voltage;

(*b*) when the output is to be varied in a given manner, e.g. proportional to frequency for variable-frequency drives.

There are several ways in which this voltage control can be achieved. In all cases the a.c. output will be made up of a fundamental component and a band of harmonic frequencies. The various control methods all contrive to reduce the harmonic voltages while avoiding excessive circuit complexity.

Voltage control in inverter circuits is important. All too often the alternatives are not considered during the design stage, and traditional methods such as mark-space (or quasi-square wave) are employed, even though another technique could be much more advantageous.

Perhaps the most popular method of controlling the a.c. voltage is to vary its mark-space ratio, as in *Figure 46.11*. Fourier

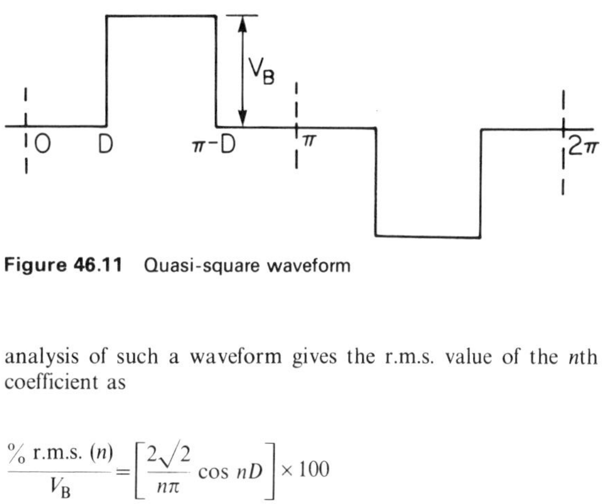

Figure 46.11 Quasi-square waveform

analysis of such a waveform gives the r.m.s. value of the nth coefficient as

$$\frac{\%\ \text{r.m.s.}\ (n)}{V_B} = \left[\frac{2\sqrt{2}}{n\pi}\cos nD\right] \times 100$$

The total r.m.s. voltage of the waveform, including all harmonics, is obtained as

$$\frac{\%\ \text{r.m.s.}\ (T)}{V_B} = \left[1 - \frac{2D}{\pi}\right]^{1/2} \times 100 \qquad (46.8)$$

Equations (46.7) and (46.8) are shown evaluated in *Table 46.1* up to the 15th harmonic. From this table it is evident that the harmonic content of the output increases rapidly as the mark-space ratio of the waveform is reduced. This is illustrated more clearly by *Figure 46.12*. At low voltages various harmonics are almost equal in value to the fundamental, the total harmonic content being about ten times larger. This represents the greatest

Table 46.1 Harmonic composition of a quasi-square wave

	R.M.S. voltage as per cent of d.c. supply								
2D/T	*1*	*3*	*5*	*7*	*9*	*11*	*13*	*15*	*Total*
0.00	90.0	30.0	18.0	12.9	10.0	8.18	6.92	6.00	100
0.02	89.8	29.5	17.1	11.6	8.44	6.30	4.74	3.53	98.0
0.04	89.3	27.9	14.6	8.20	4.26	1.53	0.43	1.85	95.9
0.06	88.4	25.3	10.6	3.20	1.25	3.94	5.33	5.71	93.8
0.08	87.2	21.9	5.56	2.41	6.37	7.61	6.87	4.85	91.7
0.10	85.6	17.6	0.00	7.56	9.51	7.78	4.07	0.00	89.4
0.12	83.7	12.8	5.56	11.3	9.69	4.38	1.30	4.85	87.2
0.14	81.4	7.46	10.6	12.8	6.85	1.03	5.85	5.71	84.9
0.16	78.9	1.88	14.6	12.0	1.87	5.96	6.71	1.85	82.5
0.18	76.0	3.76	17.1	8.80	3.68	8.17	3.34	3.53	80.0
0.20	72.8	9.27	18.0	3.97	8.09	6.62	2.14	6.00	77.5
0.22	69.3	14.5	17.1	1.61	9.98	2.03	6.26	3.53	74.8
0.24	65.6	19.1	14.6	6.89	8.76	3.48	6.44	1.85	72.1
0.26	61.6	23.1	10.6	10.9	4.82	7.40	2.55	5.71	69.3
0.28	57.4	26.3	5.56	12.8	0.63	7.92	2.95	4.85	66.3
0.30	52.9	28.5	0.00	12.2	5.88	4.81	6.58	0.00	63.2
0.32	48.2	29.8	5.56	9.37	9.30	0.51	6.07	4.85	60.0
0.34	43.4	29.9	10.6	4.73	9.82	5.60	1.72	5.71	56.6
0.36	38.3	29.1	14.6	0.81	7.29	8.12	3.71	1.85	52.9
0.38	33.1	27.1	17.1	6.19	2.49	6.91	6.80	3.53	49.0
0.40	27.8	24.3	18.0	10.4	3.09	2.53	5.60	6.00	44.7
0.42	22.4	20.5	17.1	12.6	7.71	3.01	0.87	3.53	40.0
0.44	16.9	16.1	14.6	12.5	9.92	7.17	4.41	1.85	34.5
0.46	11.3	11.0	10.6	9.91	9.05	8.04	6.91	5.71	28.3
0.48	5.65	5.62	5.56	5.47	5.36	5.22	5.05	4.85	20.0
0.50	0.00	0.00	0.00	0.00	0.00	0.00	0.00	0.00	0.00

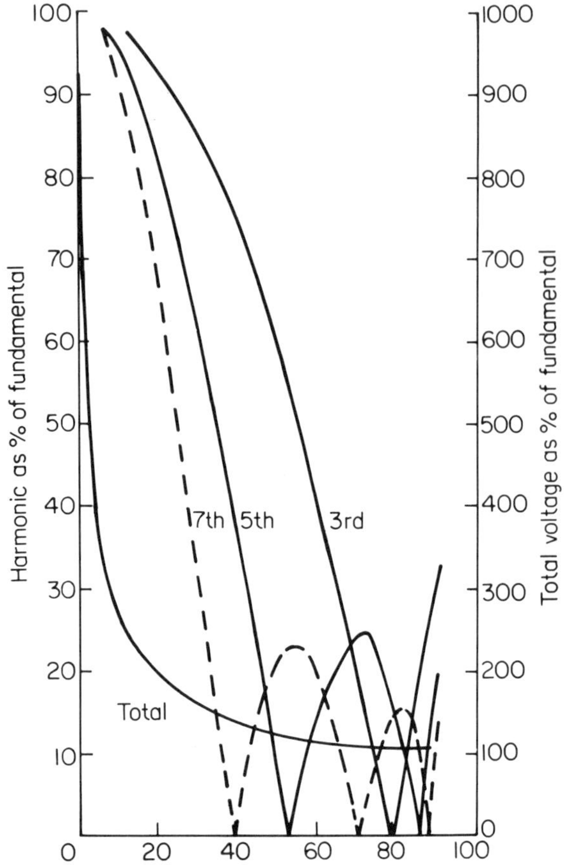

Figure 46.12 Harmonics in a quasi-square waveform

disadvantage of the quasi-square voltage control system, and normally limits the output voltage range to between 30 and 90% of the d.c. supply, i.e. a frequency change of 3:1. *Figure 46.11* may also be referred to as a single-pulse, unidirectional wave. It is unidirectional since in any half cycle the output is either positive or negative, never both. There is also a single pulse in each half cycle.

A second extension to quasi-square control is to have several equally spaced unidirectional pulses in a half cycle. This is conveniently obtained by feeding a high-frequency carrier triangular wave and a low-frequency reference square wave into a comparator as in *Figure 46.13(a)*. The output voltage swings between V_B and zero volts according to the relative magnitudes of the two waveforms, as in *Figure 46.13(b)*. The width of the output pulses is determined by the ratio of A_S/A_T and there are as many pulses per cycle as the ratio f_T/f_S. The Fourier coefficient is given by

$$\frac{\%\ \text{r.m.s.}\ (n)}{V_B} = \sum_{M=0,2,4,\ldots} \frac{2\sqrt{2}}{n\pi}[\cos nx(M) - \cos nx(M+1)] \tag{46.9}$$

and

$$\frac{\%\ \text{r.m.s.}\ (T)}{V_B} = 100\left[\frac{f_T}{2\pi f_S}\{x(0) - x(1)\}\right]^{1/2} \tag{46.10}$$

The third and seventh harmonics derived from Equations (46.9) and (46.10) are plotted in *Figures 46.14* and *46.15* respectively for $f_T/f_S = 2$, 4, 6, 10 and 20. This shows the improvement in attenuation of the lower harmonics, compared to quasi-square

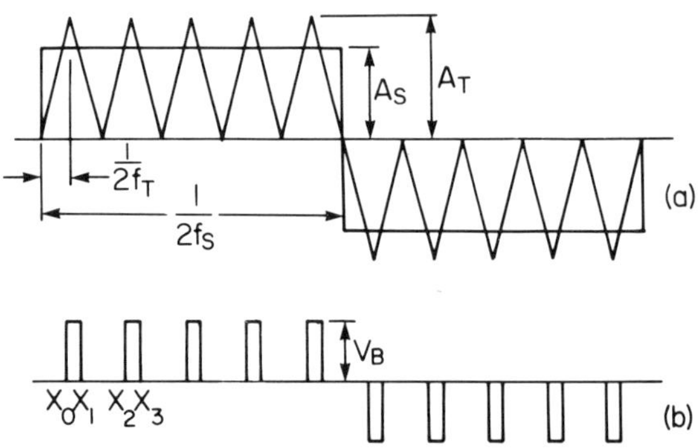

Figure 46.13 Pulse width modulation with a square wave using unidirectional switching: (a) control waveforms; (b) output waveform

wave control methods, although for higher harmonics it is necessary to go to a larger pulse number to obtain appreciable harmonic reduction.

In *Figure 46.13* a square wave was used as the reference source. A greater reduction in harmonic content would be obtained if a sine wave was used as the reference as in *Figure 46.16*. The output pulses are not of constant width, but provided the values of X_1, X_2, X_3, etc., are known the magnitude of the harmonic coefficient

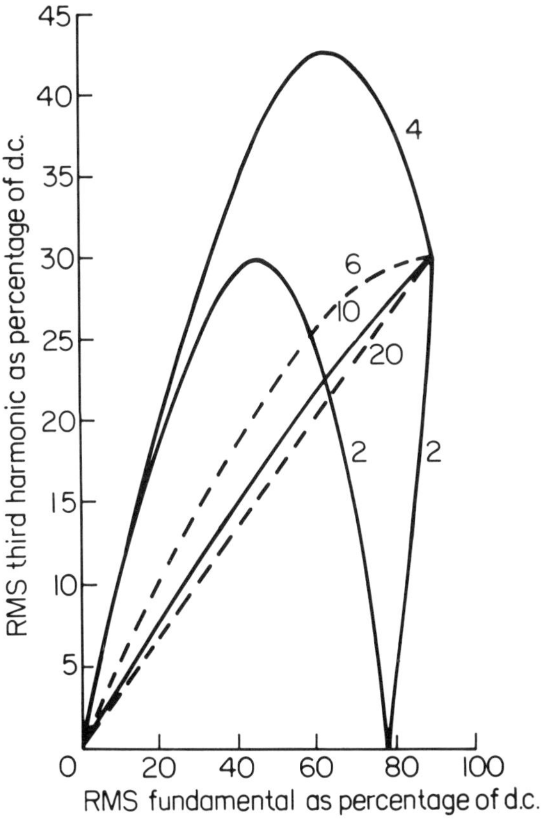

Figure 46.14 Third-harmonic content of a square-modulated unidirectional wave

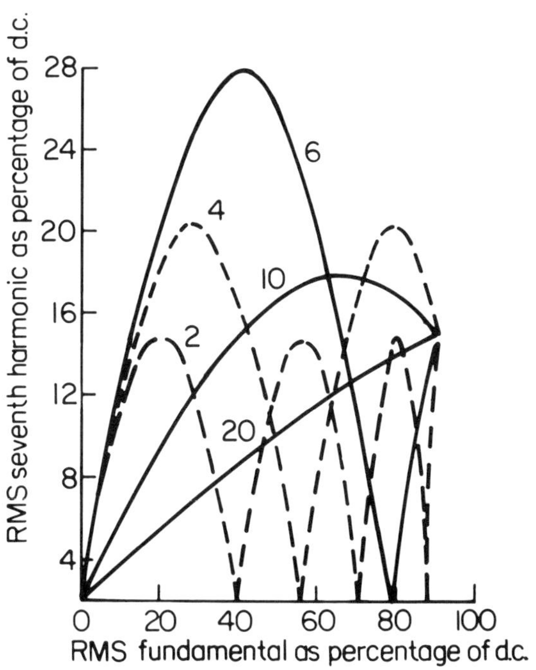

Figure 46.15 Seventh-harmonic content of a square-modulated unidirectional wave

Figure 46.16 High frequency pulse-width modulation with unidirectional switching

can again be found from Equation (46.9). However, Equation (46.10) is now modified to take account of the unequal pulse widths and is given by

$$\frac{\%\ \text{r.m.s.}\ (T)}{V_B} = \left[\frac{2}{\pi}\sum\{x(M+1)-x(M)\}\right]^{1/2} \tag{46.11}$$

Equations (46.9) and (46.11) are shown evaluated in *Tables 46.2* to *46.5*. Voltage control is again effected by changing the ratio of A_S/A_T. The maximum value of this is limited to 0.98 rather than unity to prevent adjacent pulses from merging into each other. The following factors can be noticed.

Table 46.2 Harmonic content of a sine modulated unidirectional wave with $f_T/f_S=4$

A_S/A_T	*R.M.S. voltage as per cent of d.c. supply*								
	1	*3*	*5*	*7*	*9*	*11*	*13*	*15*	*Total*
0	0	0	0	0	0	0	0	0	0
0.1	9.93	9.77	9.46	8.99	8.40	7.69	6.88	6.00	26.5
0.2	16.6	15.9	14.4	12.4	9.97	7.32	4.64	2.14	34.4
0.3	23.2	21.1	17.4	12.5	7.18	2.26	1.65	4.13	40.7
0.4	29.6	25.4	17.9	9.19	1.25	4.24	6.48	5.68	46.2
0.5	33.9	27.5	16.9	5.54	3.19	7.27	6.60	2.87	49.5
0.6	40.0	29.5	13.4	1.02	8.41	7.66	2.01	3.52	54.2
0.7	44.0	30.0	10.0	5.36	9.91	5.05	2.41	5.90	57.0
0.8	47.9	29.8	6.01	9.03	9.44	0.95	5.85	5.13	59.7
0.9	51.6	29.0	1.63	11.6	7.08	3.44	6.92	1.61	62.3
0.98	53.4	28.3	0.62	12.4	5.33	5.33	6.38	0.62	63.6

(1) Unlike the previous system this method of voltage control results in severe attenuation of frequencies below a certain value. On examining the tables it will be seen that the harmonic numbers with the largest amplitude occur at $f_T/f_S \pm 1$. Therefore, for example, with $f_T/f_S = 10$ the harmonics are largest at the ninth and eleventh. This is logical since the tenth harmonic is the 'carrier' wave itself, and no attempt is made to eliminate it. Quite clearly the higher the ratio of f_T/f_S the more effective this system becomes. The same statement was made when considering modulation with a square wave, but there the effect of higher carrier frequencies was to keep the proportion of the harmonics constant (and equal to the value for a square wave) as the fundamental was varied, and not to eliminate it.

Table 46.3 Harmonic content of a sine modulated unidirectional wave with $f_T/f_S=6$

A_S/S_T	*R.M.S. voltage as per cent of d.c. supply*								
	1	*3*	*5*	*7*	*9*	*11*	*13*	*15*	*Total*
0	0	0	0	0	0	0	0	0	0
0.1	6.98	0.29	7.20	6.57	0.84	7.20	5.98	1.32	24.0
0.2	14.7	0.60	14.7	13.1	1.60	13.1	10.2	2.09	34.8
0.3	21.2	0.25	19.9	18.0	0.59	14.4	11.4	0.55	41.7
0.4	27.8	0.45	23.9	21.5	0.88	12.7	9.08	0.39	47.7
0.5	34.3	1.46	26.7	23.4	2.17	8.47	3.59	0.38	52.9
0.6	40.8	2.74	28.1	23.1	2.72	3.07	3.48	2.69	57.7
0.7	48.8	5.52	27.2	20.4	2.22	2.73	10.9	6.50	63.0
0.8	55.4	7.28	25.7	15.6	0.32	5.67	13.7	6.41	67.1
0.9	60.8	10.4	22.5	10.8	3.59	5.06	12.6	4.51	70.2
0.98	66.3	13.6	18.7	5.12	7.86	2.74	8.82	0.21	73.1

Table 46.4 Harmonic content of a sine modulated unidirectional wave with $f_T/f_S=10$

A_S/A_T	*R.M.S. voltage as per cent of d.c. supply*								
	1	*3*	*5*	*7*	*9*	*11*	*13*	*15*	*Total*
0	0	0	0	0	0	0	0	0	0
0.1	7.26	0.28	0.24	0.88	7.53	6.70	0.43	0.68	25.0
0.2	13.6	0.54	0.12	0.11	13.7	12.3	1.48	0.29	34.6
0.3	20.8	0.95	0.18	0.12	19.9	17.4	2.72	0.27	42.8
0.4	27.5	0.49	0.57	0.02	24.1	21.2	3.63	0.92	49.1
0.5	34.7	0.98	0.69	0.80	27.7	23.0	5.70	0.66	55.2
0.6	42.1	0.60	0.35	3.67	27.8	24.7	6.21	1.71	60.7
0.7	48.9	0.17	0.93	4.90	27.0	23.2	8.51	2.16	65.2
0.8	55.9	1.31	0.91	8.17	24.9	20.8	9.71	2.89	69.9
0.9	62.5	1.43	1.57	11.0	20.7	17.5	10.5	4.11	73.9
0.98	68.5	0.93	1.57	13.4	16.6	12.7	11.9	4.70	77.3

Table 46.5 Harmonic content of a sine modulated unidirectional wave with $f_T/f_S=20$

A_S/A_T	*R.M.S. voltage as per cent of d.c. supply*								
	1	*3*	*5*	*7*	*9*	*11*	*13*	*15*	*Total*
0	0	0	0	0	0	0	0	0	0
0.1	7.58	0.68	1.04	0.22	0.33	0.26	0.38	1.26	25.9
0.2	14.9	0.60	0.09	0.50	0.35	0.30	0.62	0.48	36.2
0.3	21.5	0.15	0.15	0.77	0.53	0.52	0.88	0.24	43.7
0.4	28.2	0.03	0.56	0.89	0.00	0.01	1.02	0.55	50.2
0.5	35.3	0.05	0.00	0.41	0.53	0.15	0.47	0.50	56.2
0.6	42.4	0.16	0.05	0.21	0.22	0.13	0.82	0.53	61.6
0.7	49.2	0.10	0.22	0.39	0.29	0.15	0.85	0.97	66.3
0.8	56.8	0.53	0.59	0.26	0.06	0.36	0.28	0.09	71.2
0.9	64.0	0.55	0.02	0.15	0.49	0.21	0.13	0.76	75.5
0.98	69.0	0.26	0.42	0.19	0.10	0.16	0.32	1.25	78.6

(2) This system has two disadvantages. Firstly the high inverter frequency required to give effective lower-harmonic reduction leads to smaller efficiencies due to inverter losses. Secondly the maximum output is well below 90% of d.c. supply, as obtained with a square wave. This would limit the maximum operating frequency when running with some types of loads which need to be fully fluxed, as described before.

(3) There is a characteristic increase in total harmonic content, with higher operating frequencies. This is not normally serious since higher harmonics can be more easily filtered out than lower-order harmonics.

In the above discussions all the waveforms looked at have been unidirectional, that is the instantaneous voltage in any half cycle has been either positive or negative, as in *Figure 46.16*.

Knowing the value of intersection points X_0, X_1, X_2, etc. the r.m.s. of the nth harmonic can readily be obtained by the arithmetic sum of the individual pulses over the 0 to π interval. Therefore

$$\frac{\%\ \text{r.m.s.}\ (n)}{V_B}=\frac{\sqrt{2}}{n\pi}100\Big\{1-\cos nx(0)+\sum[\cos nx(2M+1)-\cos nx(2M+2)]-\sum[\cos nx(2M)-\cos nx(2M+1)]\Big\} \tag{46.12}$$

The r.m.s. value of the total harmonic is constant, irrespective of the depth of modulation and the operating frequency, and is equal to that of a square wave since there are no zero periods in the output.

For bidirectional switching, even harmonics are absent from the output, apart from the inverter operating frequency and its triplen harmonics. For unidirectional switching all even harmonics are missing.

The lower harmonic content with bidirectional switching, when the inverter operating frequency is higher than the output frequency, is less than for a quasi-square wave. It is also less than that obtained with unidirectional switching and is due to the absence of any 'zero' periods in the waveform. For maximum modulation the output reduces to a square wave and the harmonic content is identical to comparable quasi-square and unidirectional systems.

With bidirectional switching the odd-harmonic content generally increases as the operating frequency is raised. The reverse was true for unidirectional switching. However, the presence of the large-valued even harmonics still makes it desirable to operate this system at an inverter frequency of the order of twenty times the desired output frequency.

The square reference wave may of course be replaced by a sine wave. The output is very similar to *Figure 46.17* except that the

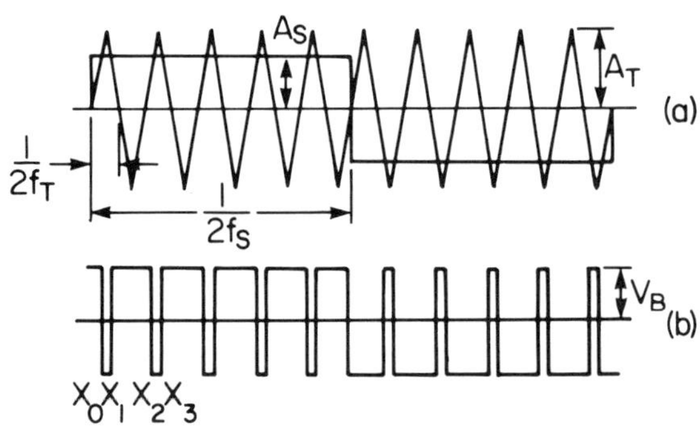

Figure 46.17 Pulse-width modulation with a square wave using bidirectional switching: (a) control waveforms; (b) output waveform

modulation depth varies linearly along the cycle. This waveform will also contain odd and even sine terms, and since $X_0=0$ the r.m.s. voltage of the nth harmonic can be derived from Equation (46.13).

$$\frac{\%\ \text{r.m.s.}\ (n)}{V_B}=\frac{\sqrt{2}}{n\pi}100\Big\{\sum[\cos nx(2M+1)-\cos nx(2M+2)]$$

$$-\sum[\cos nx(2M)-\cos nx(2M+1)]\Big\} \qquad (46.13)$$

The harmonic content of the waveform is very similar to unidirectional switching as in *Tables 46.2* to *46.5*. The harmonic with the largest amplitude is that which occurs close to the chopping frequency f_T, both odd and even harmonics being considered. As an example when operating at $f_T/f_S=6$ the sixth harmonic is very large. For zero modulation depth the fundamental is zero and the sixth harmonic has a value 90% of the d.c. supply, since the output is a square wave at this frequency. As the modulation depth increases the fundamental also increases in value and the sixth harmonic reduces (whereas adjacent even harmonics, i.e. the fourth and the eighth increases in value). When operating with $f_T/f_S=20$ the harmonic content up to the 15th is very similar to unidirectional methods except that there are now odd and even terms. However, the total harmonic content, which for bidirectional switching is 100% of the d.c. supply irrespective of the modulation depth, is much higher. As in previous methods it is clear that the inverter frequency should be several orders larger than the output frequency for effective harmonic reduction. This is therefore a disadvantage of the system as it can lead to lower efficiencies.

Before leaving this section it would be useful to look at two other techniques for voltage control which have been used specifically as methods for harmonic reduction. The first of these is called staggered phase carrier cancellation. It essentially consists in combining several high-frequency modulated waves in which the carriers are out of phase, whereas the modulating (low-frequency) wave as in phase. This results in a strengthening of the low-frequency and a weakening of the high-frequency carrier.

This system can be extended. For instance combining four waveforms with their low frequencies in phase but carriers phase shifted by 90°, 180° and 270° would result in the carrier and its first, second and third harmonics being eliminated from the output.

The disadvantage of staggered phase-carrier cancellation is the extra hardware needed. Its prime use is in fixed-frequency sine wave inverters where the extra cost can be justified on account of the reduction in size of the output filter.

The second technique worth mentioning is that of waveform synthesis. This is frequently used, especially in larger installations, where several inverters are run in parallel but phase shifted, their outputs being summed by a transformer to produce a stepped waveform with reduced harmonic content. The same effect can be obtained by using a tapped supply. Alternatively, instead of tapping the supply, the primary or secondary of the transformer can be tapped. *Figure 46.18* shows an inverter arrangement with a tapped secondary, the primary being the normal form of a push–pull inverter. The firing sequence of the thyristors is also shown. Provided the tappings on the secondary are such as to give the waveform indicated, it can be calculated that the output contains no harmonics below the eleventh.

46.3.3 Load and device currents

Figure 46.19 shows the load voltage and current waveforms for the inverter of *Figure 46.10*. This also indicates the device conducting periods. Using the same assumptions as made for the chopper, the characteristics of the inverter can be determined as before. If $T=t_c+t_0$ is the chopping period, then:

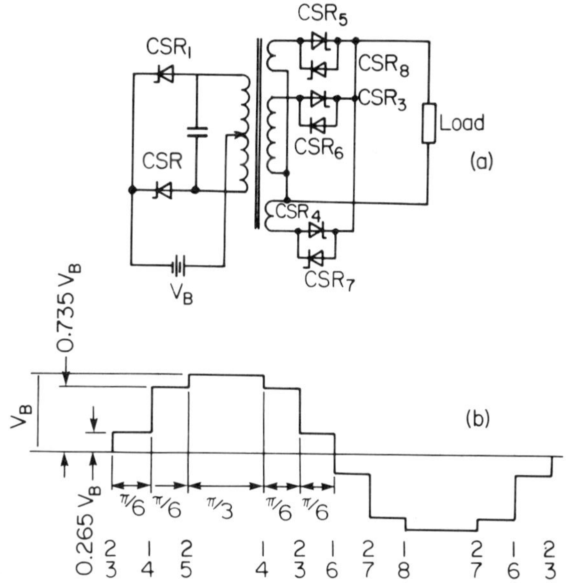

Figure 46.18 Waveform synthesis with a tapped secondary transformer: (a) circuit diagram; (b) load waveform

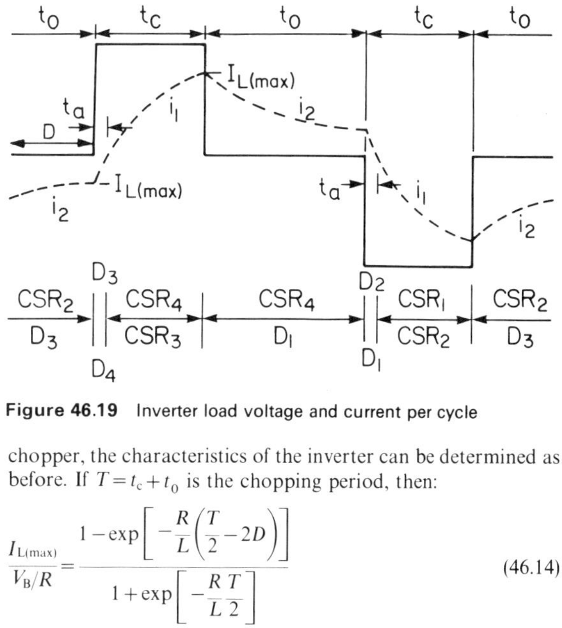

Figure 46.19 Inverter load voltage and current per cycle

$$\frac{I_{L(max)}}{V_B/R}=\frac{1-\exp\left[-\frac{R}{L}\left(\frac{T}{2}-2D\right)\right]}{1+\exp\left[-\frac{R}{L}\frac{T}{2}\right]} \qquad (46.14)$$

The r.m.s. current rating of the devices is obtained by considering the current and voltages indicated in *Figure 46.19*. It is important to note that one arm of the bridge carries a larger current than the other, and the 'worst case' should be considered. Therefore the rating of the thyristors can be taken as i_1 over t_c-t_a or as the sum of i_1 over t_c-t_a and i_2 over t_2. The latter gives the 'worst case'.

The values of thyristor (I_T), diode (I_D) and load (I_{LR}) r.m.s. currents are given by

$$\frac{I_T}{V_B/R}=\left[\left(\frac{I_{L(max)}}{V_B/R}\right)^2\frac{L/R}{2T}\left\{1-\exp\left[-\frac{T}{L/R}\cdot\frac{4D}{T}\right]\right\}+\left(\frac{I_{TO}}{V_B/R}\right)^2\right]^{1/2} \qquad (46.15)$$

where

$$\frac{I_{\mathrm{TO}}}{V_{\mathrm{B}}/R}=\left[\frac{L/R}{T}\left\{\frac{T/2-2D}{L/R}-\log P+2P\exp\left(-\frac{T/2-2D}{L/R}\right)\right.\right.$$
$$\left.\left.-\frac{P^2}{2}\exp\left(-\frac{T-4D}{L/R}\right)-\frac{3}{2}\right\}\right]^{1/2}$$

and

$$P=1+\frac{I_{\mathrm{L(max)}}}{V_{\mathrm{B}}/R}\exp\left(-\frac{Rt_0}{L}\right)$$

$$I_{\mathrm{D}}/(V_{\mathrm{B}}/R)=\left[\left(\frac{I_{\mathrm{T}}}{V_{\mathrm{B}}/R}\right)^2-\left(\frac{I_{\mathrm{TO}}}{V_{\mathrm{B}}/R}\right)^2\right.$$
$$\left.+\frac{L/R}{T}\left\{\log P-2P+\frac{P^2}{2}+\frac{3}{2}\right\}\right]^{1/2} \qquad (46.16)$$

$$\frac{I_{\mathrm{LR}}}{V_{\mathrm{B}}/R}=\left[2\left\{\left(\frac{I_{\mathrm{D}}}{V_{\mathrm{B}}/R}\right)^2+\left(\frac{I_{\mathrm{TO}}}{V_{\mathrm{B}}/R}\right)^2\right\}\right]^{1/2} \qquad (46.17)$$

Although Equations (46.16) and (46.17) could be plotted, it would be more informative to show the current variations, with fundamental output voltage, as a ratio of the load current, using the curves of *Figure 46.12* for a quasi-square wave. This is shown in *Figures 46.20, 46.21* and *46.22* and allows peak current and the r.m.s. currents of the thyristor or diode to be determined at any load voltage and r.m.s. current, for a given operating frequency.

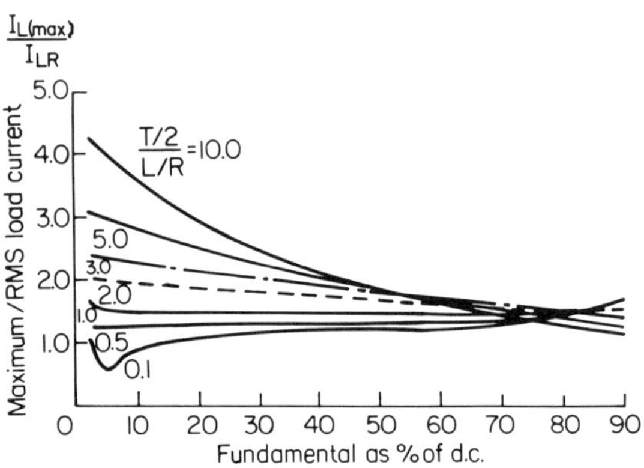

Figure 46.20 Variation of (peak load/r.m.s. load) current with fundamental load voltage

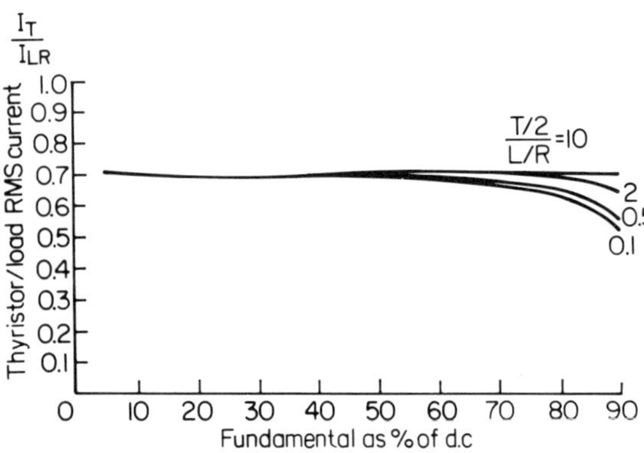

Figure 46.21 Variation of (r.m.s. thyristor/r.m.s. load) current with fundamental load voltage

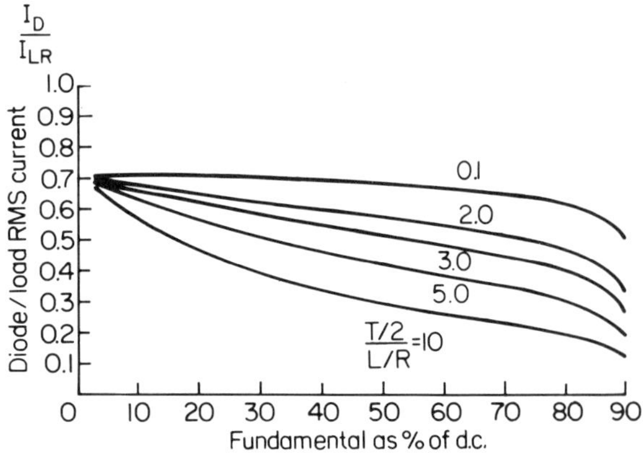

Figure 46.22 Variation of (r.m.s. diode/r.m.s. load) current with fundamental load voltage

Figure 46.20 shows the large increase in commutation current requirements at low frequencies and voltages, illustrating the unsuitability of this control method for low mark-space operation. The thyristor r.m.s. current is almost constant at $1/\sqrt{2}$ times the load current indicating that the devices conduct for approximately half the load current period. The diode ratings also tend to this value at low voltages.

46.4 Commutation in choppers and inverters

46.4.1 Methods of commutation

Although there are many different commutation circuits for choppers and inverters they can be divided into four groupings. These are illustrated with reference to a chopper in *Figure 46.23*.

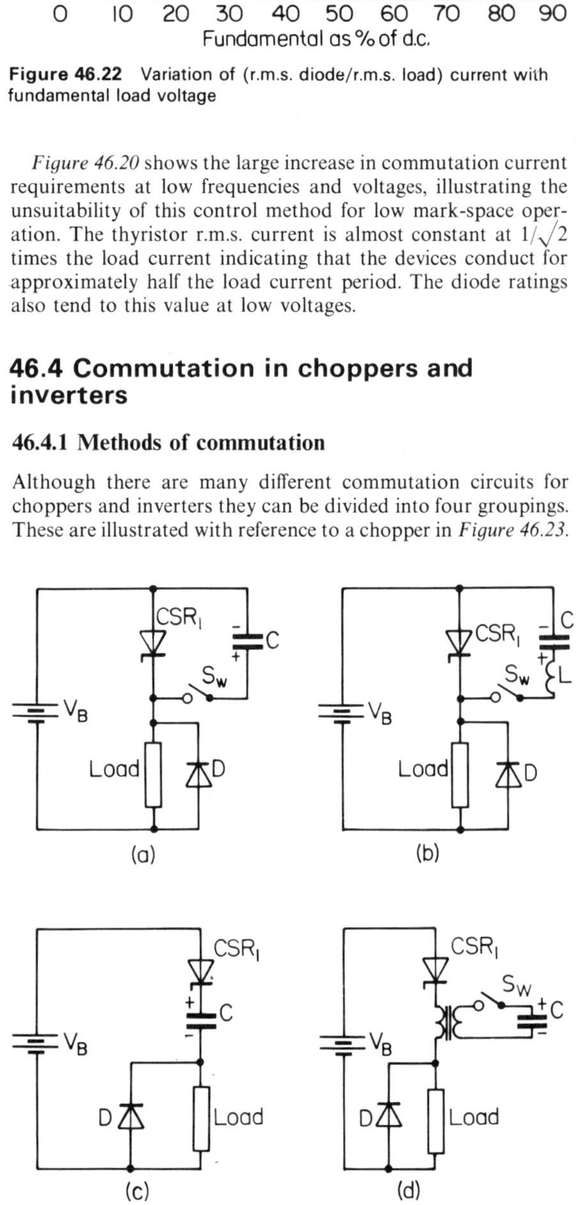

Figure 46.23 Commutation methods: (a) parallel capacitor; (b) parallel ocapacitor-inductor; (c) series capacitor; (d) coupled pulse

(1) *Parallel-capacitor commutation*, the charged capacitor being placed directly across the thyristor to be turned off as in *Figure 46.23(a)*. Switch Sw forms part of the commutation circuit.

(2) *Parallel capacitor-inductor commutation*, an inductor being connected in series with the commutation capacitor as in *Figure 46.23(b)*.

(3) *Series-tocapacitor commutation*, the commutation capacitor being connected in series with the thyristor to be turned off as in *Figure 46.23(c)*.

(4) *Coupled-pulse commutation*, the turn-off pulse being coupled to the thyristor through a transformer or auto-transformer as in *Figure 46.23(d)*.

Figure 46.23 indicates only the basic principles of the commutation methods. The auxiliary circuits used to prime the commutation capacitor, ready to turn off the main thyristor CSR_1, are not shown. Typical chopper and inverter circuits in the above four groups are given below.

46.4.1.1 Parallel-capacitor commutation

For the circuit shown in *Figure 46.24* thyristor CSR_2 is fired initially, charging C to voltage V_B with plate b positive. CSR_1 is

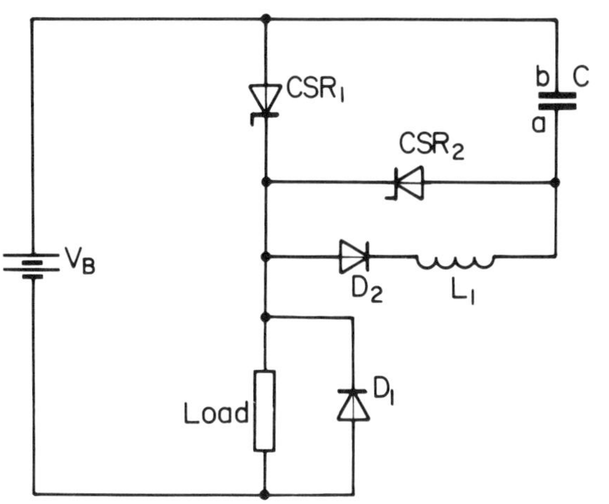

Figure 46.24 Parallel capacitor commutation in a chopper circuit

now turned on. C discharges via CSR_1, D_2, L_1 and resonates, recharging to V_f with plate a positive. For negligible resonant loss V_f is equal to V_B.

To turn CSR_1 off CSR_2 is fired. The voltage across CSR_1 is reversed, C discharges through the load (current assumed constant) and recharges with plate b positive. CSR_2 then turns off.

In the circuit shown in *Figure 46.24* assume C to be charged to a voltage V_f. The load current just prior to CSR_2 fixing is $I_{L(max)}$ and is assumed constant during the short discharge period of C. If the reverse recovery current of CSR_1 is neglected then:

$$I_{L(max)} = \frac{CV_f}{t_F} \tag{46.18}$$

where t_F is the time during which CSR_1 is reverse-biased. For commutation to be successful, this time must be greater than the device turn off time at the rate of re-applied forward voltage given by:

$$\frac{dV}{dt} = \frac{I_{L(max)}}{C} \tag{46.19}$$

Equations (46.18) and (46.19) are applicable to all parallel-capacitor commutated circuits and enable the correct capacitor value to be chosen for a given load and device characteristic. However, the magnitude of V_f will be determined by the auxiliary commutation circuit adopted.

Figure 46.25 shows two examples of commutation in an inverter. In *Figure 46.25(a)* firing CSR_1 charges capacitor C with

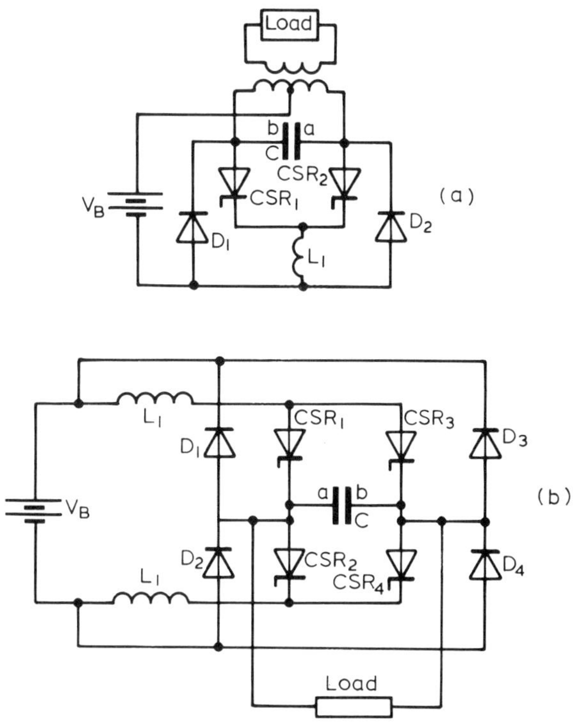

Figure 46.25 Parallel capacitor commutation in inverter circuits: (a) push–pull; (b) bridge

plate a positive to a voltage $2V_B$. When CSR_2 is fired capacitor C is connected across CSR_1 turning it off. The capacitor now discharges to zero voltage, its stored energy then being dissipated in the L_1-D_1-C-CSR_2 conduction path. After this C charges to $2V_B$ with plate b positive, ready to turn CSR_2 off when CSR_1 is fired.

Figure 46.25(b) shows a single-phase bridge inverter circuit. With CSR_1 and CSR_4 conducting capacitor C is charged to V_B with plate a positive. When CSR_2 and CSR_3 are fired to commence the next step of the output, capacitor C is connected across CSR_1 and CSR_4 and turns them off. The circuit is therefore an example of parallel-capacitor commutation. Inductor L_1 prevents the supply from being instantaneously short-circuited during commutation.

46.4.1.2 Parallel capacitor-inductor commutation

Figure 46.26(a) shows a chopper circuit in which C is initially charged through L_1, D_3 and the load with plate 'a' positive. The load cycle is commenced when CSR_1 is turned on. C cannot discharge since D_3 is reverse biased and CSR_2 is non-conducting. To turn CSR_1 off CSR_2 is fired. C resonates through CSR_2 and L_1 and discharges through L_1 and D_3 turning CSR_1 off. D_2 limits the maximum discharge period, even with the load open circuit.

The values of C and L_1 required to commutate a thyristor of

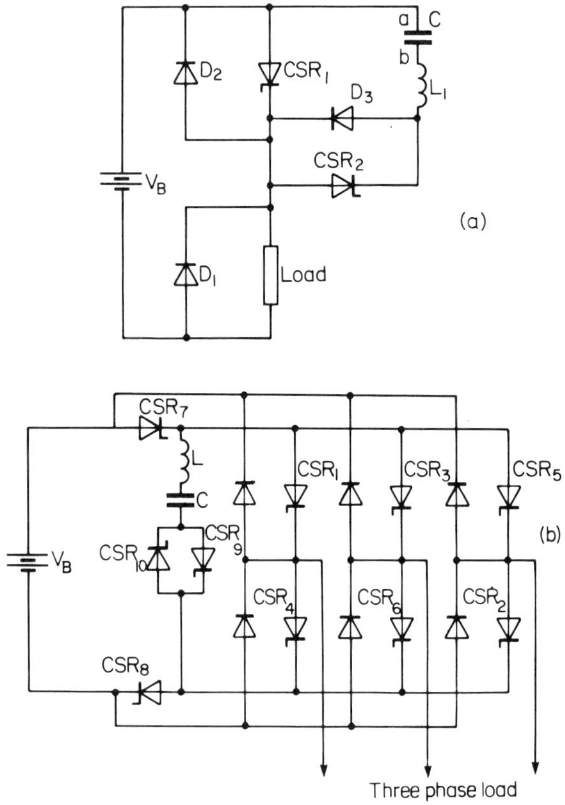

Figure 46.26 Parallel capacitor-inductor commutation: (a) chopper; (b) inverter

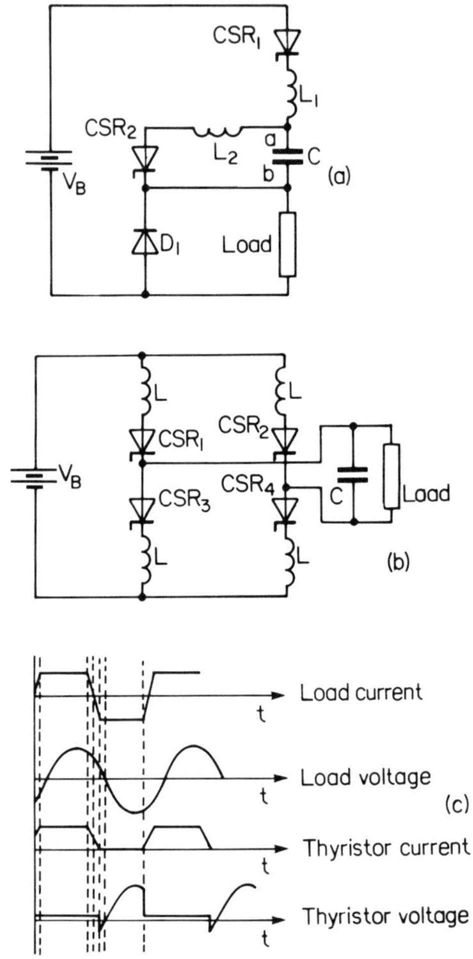

Figure 46.27 Series capacitor commutation: (a) chopper; (b) inverter circuit; (c) inverter waveforms

turn-off time t_F in parallel capacitor-inductor circuits is given by:

$$C = \frac{2\sqrt{2t_F I_{L(max)}}}{\pi V_B} \tag{46.20}$$

$$L_1 = \frac{\sqrt{2t_F V_B}}{\pi I_{L(max)}} \tag{46.21}$$

From this it can be seen that parallel capacitor-inductor commutation is unsuitable for use in systems controlling large currents from low supply voltages. For example, to turn off 250 A from 40 V using thyristors with $t_F = 50$ μs would give $C = 280$ μF and $L_1 = 2.5$ μH. In all probability the inductance of the connecting leads would exceed this value.

Figure 44.26(b) shows an inverter circuit which has been developed for aerospace applications. CSR_9 is fired to prime C ready for commutation. When CSR_{10} is now turned on capacitor C discharges via its inductor L, turning CSR_7 and CSR_8 off and so eventually commutating the inverter thyristors. This circuit also has high operating efficiencies.

46.4.1.3 Series-capacitor commutation

Referring to the basic series-capacitor commutator circuit of *Figure 46.23(c)* it is seen that CSR_1 will turn off once C has charged to V_B volts. Before CSR_1 can be refired the capacitor must be reset. There are several ways in which this may be done. In *Figure 46.27(a)* CSR_1 is fired to commence the load cycle. This causes C to charge through L_1 and turn CSR_1 off. CSR_2 is fired which resets C through L_2 after which CSR_2 turns off and CSR_1 can be refired.

Series-capacitor commutation circuits have a very limited operating range, due to the capacitor reset time, and are rarely used as choppers. They find much greater application in sine wave inverter circuits. *Figure 46.27(b)* shows a typical circuit with its waveforms (*Figure 46.27(c)*). The inverter is operated at a frequency determined by the oscillatory frequency of the load and C, so that these inverters find most frequent application for producing fixed frequency sine wave output.

46.4.1.4 Coupled-pulse commutation

A chopper circuit using an auto-transformer to couple the turn-off pulse to the thyristors is shown in *Figure 46.28(a)*. CSR_1 is the main thyristor and CSR_2 an auxiliary device used to commutate CSR_1. L_1 L_2 is a tapped auto-transformer. The circuit operation is as follows. CSR_1 is fired to commence the load cycle. This also causes C_1 to charge through L_1 to supply voltage V_B. The clamping action of D_2 prevents C_1 charging to a higher voltage. To turn CSR_1 off CSR_2 is fired. Capacitor C_1 discharges through L_2 coupling a voltage pulse via L_1 to CSR_1 so turning it off. The voltage on C_1 falls to zero after which D_1 conducts, carrying the current due to energy stored in L_2 and the load. CSR_2 turns off when the current in L_2 has fallen to below the device holding value. C_2 is not normally required. By including it CSR_1 can be

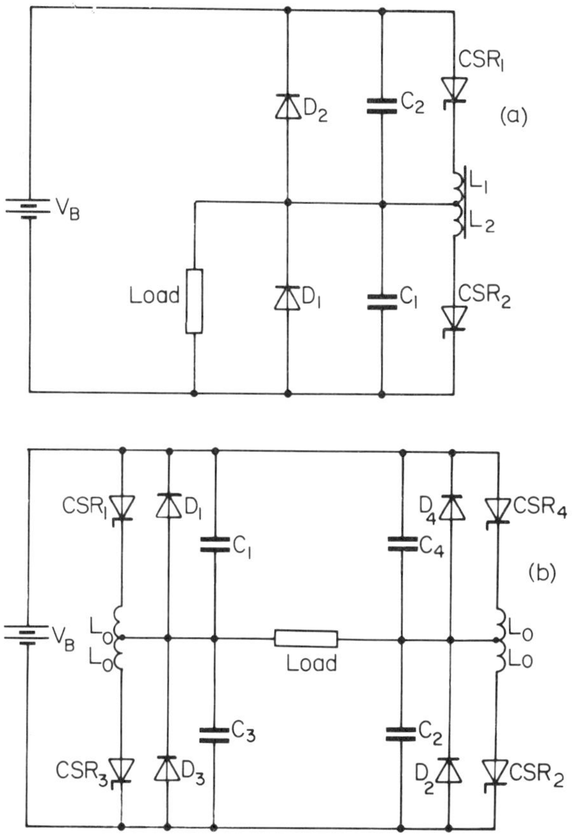

Figure 46.28 Coupled-pulse commutation: (a) chopper; (b) inverter

fired before CSR_2 has turned off since then C_2 discharges through L_1 coupling a pulse to CSR_2 via L_2 and so turning it off. This allows a wider voltage control range.

Assuming a symmetrical system with $C_1 = C_2 = C$ and $L_1 = L_2 = L$ the values of L and C can be found as

$$L = \frac{2.76 \times V_B \times t_F}{I_{L(max)}} \tag{46.22}$$

$$C = \frac{2.15 \times t_F \times I_{L(max)}}{V_B} \tag{46.23}$$

This again shows the unsuitability of this commutation method for low-voltage high-current operation. The circuit of *Figure 46.28(a)* can be extended to give the inverter circuit of *Figure 46.28(b)*, which is perhaps the most reliable and frequently used inverter circuit.

With any device conducting the commutation capacitor in the corresponding half of the bridge is charged to V_B. When the thyristor in that leg is fired the capacitor discharges through half of the auto-transformer, coupling a turn-off pulse to the conducting thyristor via the other half of the transformer. The circuit is versatile in that firing one device automatically commutates the other half of the bridge. It can be used to produce outputs with zeros in the waveform with no additional circuitry. It does, however, suffer from low efficiency, due to commutation losses at higher frequencies, and is therefore not suited for pulse-width modulated voltage control methods. It cannot be used in circuits which have d.c. supply variations to change the output, since the commutation voltage is now affected.

46.4.2 Effect on load and device currents

In this section a typical chopper and inverter are considered along with their commutation components. A method of specifying the device ratings and the value of the commutation inductor and capacitor will also be studied. Although these will differe according to the circuit adopted, the calculation procedure remains unchanged.

In *Figure 46.24*, the mean current rating of diode D_1 is clearly still given by Equation (46.5), its peak rating being $I_{L(max)}$, the current just before CSR_1 turns off (assuming the commutation interval to be short relative to the chopper operating period). The diode voltage rating should exceed $2V_B$ which is obtained when $L_F = 1$. Mean current rating of thyristor CSR_1 is I_B, as given by Equation (46.4), plus the resonant current due to C and L_1. This gives a total current

$$I_{CSR1(mean)} = I_B + \frac{2CV_B}{t_c + t_0} \tag{46.24}$$

The peak current is either $I_{L(max)}$ or the resonant value of $V_B\sqrt{(C/L_1)}$ whichever is greater. The voltage rating must exceed V_B. Assuming C to discharge at constant load current, CSR_1 must have a turn-off time shorter than $CV_B/I_{L(max)}$.

Thyristor CSR_2 carries a current of $I_{L(max)}$ during the charge period of C only. Its mean current rating is $2CV_B/(t_c + t_0)$ although its size is normally determined by its peak current capacity of $I_{L(max)}$. Similarly diode D_2 has a mean rating of $2CV_B/(t_c + t_0)$ but a peak resonant current of $V_B\sqrt{(C/L_1)}$ which can be high. The value of C is fixed by the commutation requirements and L_1 is chosen such that the resonant time $\pi\sqrt{(L_1C)}$ is small compared to the operating frequency (between 5 and 10%).

Figure 46.29 shows the instant of commutation for CSR_3 to an enlarged scale, for the inverter of *Figure 46.28(b)*. Assuming CSR_3

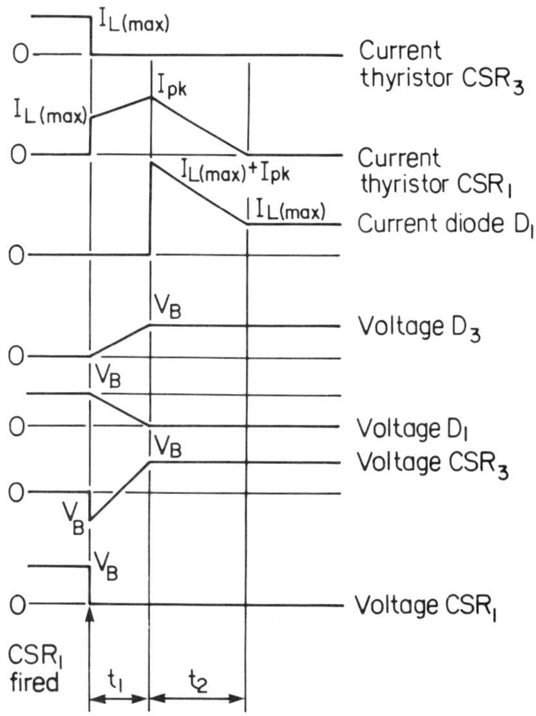

Figure 46.29 Circuit waveforms for *Figure 46.28(b)* during the commutation of thyristor CSR_3

and CSR_4 to be conducting, the load current being at its peak value of $I_{L(max)}$. C_3 and C_4 are at zero voltage (device volt-drops and d.c. resistance of centre-tapped chokes L_0L_0 being neglected). CSR_1 is now fired. If the leakage inductance of the chokes be ignored, current $I_{L(max)}$ transfers instantaneously from CSR_3 to CSR_1, capacitor C_1 discharging to support both this and the load current.

The current through CSR_1 increases, reaching a peak of I_{pk} after time t_1 when C_1 has completely discharged. Energy stored in L_0 now free-wheels through diode D_1 and is dissipated, falling to zero after a further time t_2. Load current is carried by CSR_4 and D_1.

When CSR_2 is fired at a later interval to commutate CSR_4 the load current has decayed to $I_{L(min)}$. Therefore commutation requirements and increased device ratings are not as severe as before. In the 'worst' case, when there are no zero dwell periods in the output voltage waveform, the ratings of all devices are equal. This will be considered here.

If t_F denotes the turn-off time seen by the thyristor being commutated, and W_L is the watts loss per commutation, caused by the dissipation of the energy $\frac{1}{2}L_0(I_{pk})^2$ stored in L_0 then *Figure 46.30* shows the variation of turn-off, peak current and watts-loss

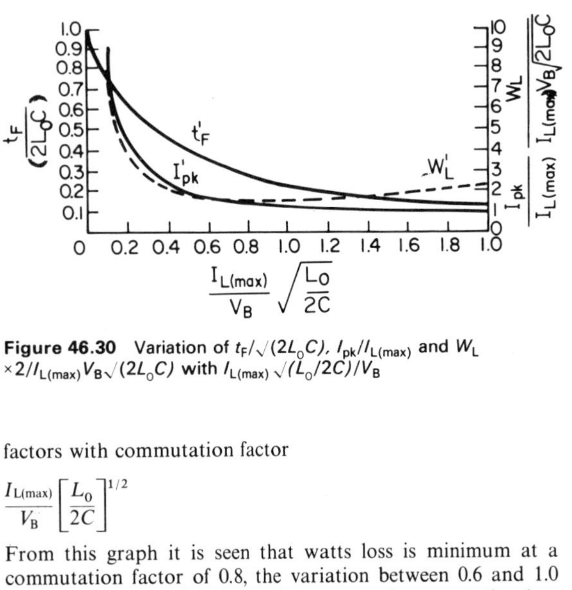

Figure 46.30 Variation of $t_F/\sqrt{(2L_0C)}$, $I_{pk}/I_{L(max)}$ and $W_L \times 2/I_{L(max)}V_B\sqrt{(2L_0C)}$ with $I_{L(max)}\sqrt{(L_0/2C)}/V_B$

factors with commutation factor

$$\frac{I_{L(max)}}{V_B}\left[\frac{L_0}{2C}\right]^{1/2}$$

From this graph it is seen that watts loss is minimum at a commutation factor of 0.8, the variation between 0.6 and 1.0 being slight. Below 0.6 the peak current and commutation loss increases steeply, although available turn-off time also increases. Working on the minimum loss point the value of L_0 and C can be found as:

$$L_0 = \frac{2.76 V_B t_F}{I_{L(max)}} \qquad (46.25)$$

$$C = \frac{2.15 t_F I_{L(max)}}{V_B} \qquad (46.26)$$

The contribution to device ratings by the commutation interval is directly dependent on time t_2 in *Figure 46.29*. This is the period required for the current in CSR_1 to decay from I_{pk} to zero due to losses across CSR_1 and D_1 (assumed a constant and equal to ΔV), and due to the 'effective loss' resistance of the CSR_1, L_0, D_1 loop (equal to R_e). To reduce this time to a minimum means introducing an external resistance in series with D_1. This gives greater losses during normal operation, with a subsequent reduction in efficiency. There is no one acceptable solution for all cases. Some inverters may be using devices which are overrated, so that higher r.m.s. currents can be accepted. In others efficiency may not be important so losses across a larger R_e are tolerated. In any case it is always important to ensure that t_2 is less than half a cycle of inverter operation, to prevent commutation failures.

If I_{D_1} and I_{T_1} represent the commutation current, in r.m.s. values, through the diode and thyristor respectively, and T is the inverter periodic time, *Figures 46.31* and *46.32* show plots of

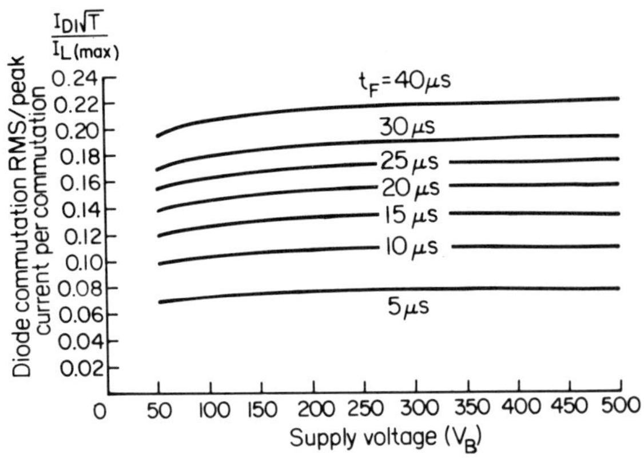

Figure 46.31 Variation of $I_{D_1}\sqrt{T}/I_{L(max)}$ with supply voltage

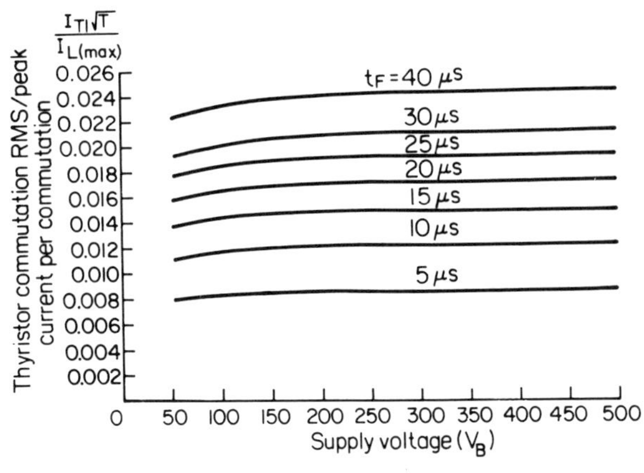

Figure 46.32 Variation of $I_{T_1}\sqrt{T}/I_{L(max)}$ with supply voltage

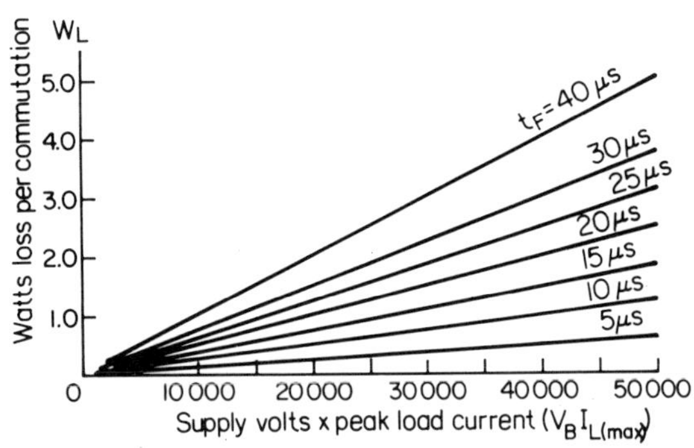

Figure 46.33 Power loss per commutation

$I_{D1}\sqrt{T}/I_{L(max)}$ and $I_{T1}\sqrt{T}/I_{L(max)}$ against supply voltage. These are calculated for values of L_0 and C given by Equations (46.25) and (46.26) and for $\Delta V = 2.5$ V and $R_e = 0.2V_B/I_{L(max)}$. These graphs show the advantage of using fast turn off time devices. The total thyristor and diode ratings are given by the geometric sum of values read off from *Figures 46.21*, *46.22*, *46.23* and *46.24*. Therefore when operating at an inverter frequency of 10 Hz, $I_{T1}/I_{L(max)}$ is nearly constant at $0.0248/\sqrt{0.1} = 0.079$ whereas *Figure 46.21* gives the thyristor current without commutation as $0.35I_{L(max)}$. The total current is also approximately this value. For 1 kHz operation, however, $I_{T1} = 0.79I_{L(max)}$ and *Figure 46.22* still gives the same value as before so that total thyristor r.m.s. current with commutation is $\sqrt{(0.79^2 + 0.35^2)}I_{L(max)} = 0.745I_{L(max)}$. Therefore this inverter is not suitable for higher-frequency operation. The same conclusions can be derived from *Figure 46.33* which gives the watts loss per commutation. Now for 40 μs devices and $V_B I_{L(max)} = 50\,000$, total energy loss per second at 10 Hz is 200 W whereas at 1 kHz it is 20 kW.

Further reading

MURRAY, R. (Ed.), *Silicon Controlled Rectifier Designer's Handbook*, Westinghouse Electric Corporation (1963)

GENTRY, F. E. *et al.*, *Semiconductor Controlled Rectifiers*, Prentice-Hall Inc. (1964)

BEDFORD, B. D. and HOFT, R. G., *Principles of Inverter Circuits*, John Wiley & Sons (1964)

GRIFFIN, A. and RAMSHAW, R. S., *The Thyristor and its Applications*, Chapman and Hall Ltd (1965)

SEYMOUR, J. (Ed.), *Semiconductor Devices in Power Engineering*, Pitman Books Ltd (1968)

DAVIS, R. M., 'Power diode and thyristor circuits', *IEE Monograph No. 7* (1971)

PELLY, B. R., *Thyristor Phase-Controlled Converters and Cycloconverters*, Wiley-Interscience (1971)

MAZDA, F. F., *Thyristor Control*, Butterworths (1973)

47 Instrumentation and Measurement

G L Bibby, BSc, CEng, MIEE
University of Leeds
(Sections 47.7–47.10)

D K Bulgin, PhD
Cambridge Instruments Ltd
(Sections 47.13–47.17)

C L S Gilford
MSc, PhD, FInstP, MIEE, FIOA
Independent Acoustic Consultant
(Sections 47.11–47.12)

F F Mazda
DFH, MPhil, CEng, MIEE, DMS, MBIM
Rank Xerox Ltd
(Sections 47.3–47.5 and 47.6.7)

J A Smith, CEng, MIM
Cambridge Instruments Ltd
(Sections 47.13–47.17)

R C Whitehead, CEng, MIEE
Polytechnic of North London
(Sections 47.1–47.2, 47.6.1–47.6.6 and 47.6.8)

Contents

47.1 Analogue voltmeters

These instruments may be designed for the measurement of d.c. and/or a.c. In addition, resistance measuring facilities may be added. In *Table 47.1*, comparisons are made between the performances of typical modestly priced non-electronic and electronic meters.

Table 47.1 Performance comparison between non-electronic and electronic meters

	Non-electronic	*Electronic*	*Units*
Maximum sensitivities for full-scale deflection	100	0.1	mV
Maximum operating frequencies	10 kHz	Highly variable from MHz to GHz	
Input resistances for d.c. operation	50 000	10 MΩ	Ω/V

The values of input resistances of a.c. operation are not amenable to simple tabulation, but the values for electronic meters normally exceed by very large margins the values for non-electronic meters.

Since these instruments take less power from the circuit under test than is required to operate the moving-coil meter used for the final display, they must incorporate amplifiers. Power sources for the amplifiers may be mains units and/or batteries. The batteries may be of the primary or secondary type, and if they are the latter then charging facilities are usually provided. Provision is made for checking battery voltages using the same moving-coil meters that are used for the display of signal amplitude.

47.1.1 D.C. measurement

Consider first operation at d.c., for which a basic schematic and simplified circuit are shown in *Figure 47.1*. R_1 is a switch-controlled resistor chain used for range changing. R_2 is first

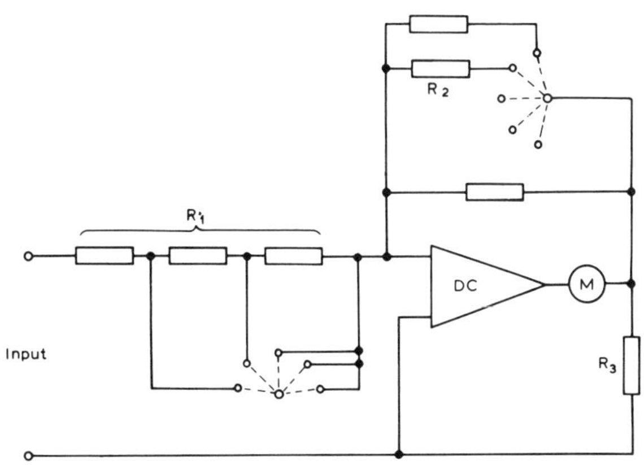

Figure 47.1 An improved voltmeter employing an operational amplifier

adjusted to standardise the effective supply voltage. The voltage gain of the d.c. amplifier, as modified by negative feedback, is given by R_2/R_1, and the current through the meter M is further controlled by R_3. Typically the value of R_2 is fixed for all except the high voltage ranges at a value of 100 MΩ, while the value of R_1 is switched in accordance with the range in use, so that the overall sensitivity of the complete circuit is 1 μA/V. For ranges above 100 V however this would involve the use of an inconveniently high value of R_1, so for these higher voltage ranges the value of R_1 may be kept fixed at 100 MΩ and reduced values of R_2 be switched into circuit, giving thereby a relatively reduced value of input resistance in terms of Ω/V.

For high values of sensitivity, amplifier stability must be correspondingly high and this is commonly achieved by the use of chopper-stabilised amplifiers. In these, the d.c. signal is modulated to produce a low-frequency signal which is amplified and then rectified. The amplifier should have a narrow bandwidth. For operation at the lowest levels it is advisable to use synchronous detection to make the indication phase sensitive. The signal will produce deflection in one direction only, but noise will produce recrified components in both directions, causing the meter to dither about its correct indication. If a long time-constant circuit be added to the rectifier output, the noise component can be reduced, but at the expense of sluggish operation. A basic circuit is shown in *Figure 47.2*.

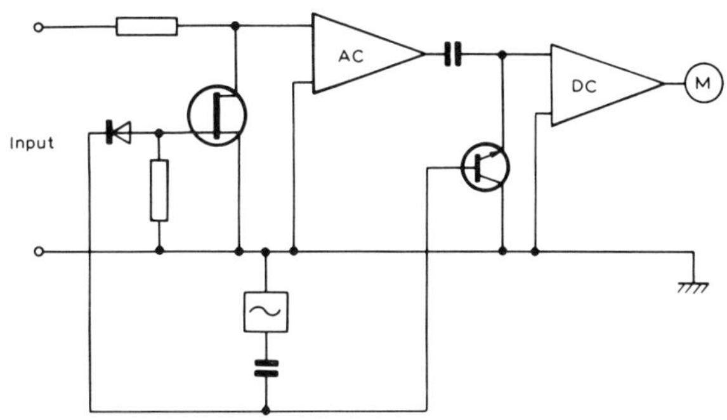

Figure 47.2 Chopped operation with synchronous detection. The a.c. amplifier has an input signal of low level, the d.c. amplifier an input signal of high level

When measuring d.c. it is immaterial whether the positive or negative or both terminals are at high potential, but this is normally an important point which must be considered when making a.c. measurements, particularly at high frequencies and particularly when the voltmeter is being operated from a mains supply. Normally one input terminal is to be operated at high potential and has only a low value of capacitance (2–20 pF) to all other bodies. The other terminal is at low potential and has a significant value of capacitance to earth and to other bodies, particularly if the instrument is mains operated. When the instrument case is metallic, this should, for safety reasons, be connected to the earth of the mains supply when the latter is being employed. The low-potential terminal is connected to the metal case in some instruments, but not in others, although significant capacitance to the case normally exists. For use where *both* input terminals are at high potential, a special instrument with a balanced input circuit should be used.

47.1.2 A.C. measurement

In the measurement of alternating voltages, attention must first be given to the quantity which is to be measured. Although peak-to-peak values may be required for special purposes (especially when the meter is used in conjunction with an oscilloscope), normally these voltmeters are used for the measurement of the root-mean-square values of the fundamental components of waveforms which are nominally sinusoidal, but which may contain small proportions of harmonics, noise, hum and/or other components. The influence of these additional components upon the accuracy of measurement depends upon the measurement circuit employed and sometimes upon the amplitude of the signal. The meter deflection may be proportional to: the arithmetic mean value of the waveform; the peak value; the peak-to-peak value; the root-mean-square (r.m.s.) value. Due notice must be taken of the fact that the relationships between the r.m.s. value of the waveform and the quantities to which the meter readings are proportional are respectively $\pi/2$, $1/\sqrt{2}$, $1/2\sqrt{2}$ and 1. (Manufacturers take account of these factors when calibrating the meter scales.)

Examples of the results of adding an even-order (second) harmonic, and odd-order (third) harmonics to sinusoids are illustrated in *Figure 47.3*. At (*a*) a second harmonic has been added. During any half-period of the fundamental, the arithmetic mean value of the second harmonic is zero. The total arithmetic value of the waveform is thus unchanged when an even-order harmonic of small amplitude is added to the fundamental. This is illustrated by the two shaded areas in (*a*), which as shown are equal for the fundamental and the sum. Variation of the phase of the harmonic relative to the fundamental will not change this.

The additions of third harmonics to the fundamentals, but with different phase relationships, are shown in (*b*) and (*c*). In (*b*) the sum waveform encloses a smaller area than the fundamental, while in (*c*) the reverse is true. Thus the circuit which responds to the arithmetic mean value of the waveform performs satisfactorily in the presence of even-order harmonics, but less so in the presence of odd-order harmonics, the sign of the error being unknown unless the phase relationship is known.

Consider now the measurement of peak values. Examination of *Figure 47.3*(*a*) shows that for even order harmonics the sign of the error is dependent upon which peak is measured. Examination of *Figure 47.3*(*b*) and (*c*) shows that while the sign of the error is independent of which peak is measured, the magnitude of the error is dependent upon the phase relationship.

Examination of the peak-to-peak values shows that these are not affected by the addition of small proportions of even-order harmonics, but that positive or negative errors may occur when odd-order harmonics are added.

When root-mean-square operation is employed, the fundamental and harmonic and noise amplitudes are added in quadrature. For instance a 10% harmonic will produce an error of $100\,(\sqrt{1^2+0.1^2}-1)\%=0.5\%$. The error is always positive.

Two typical basic forms of a.c. voltmeters are shown in *Figure 47.4*. In (*a*) the signal is first rectified, then, when necessary, attenuated to keep the amplitude within the operating range of the d.c. amplifier–meter combination. In (*b*), the signal is, when necessary, attenuated, amplified by the a.c. amplifiers, rectified, and a value displayed on the meter. These alternative schemes are now considered in greater detail.

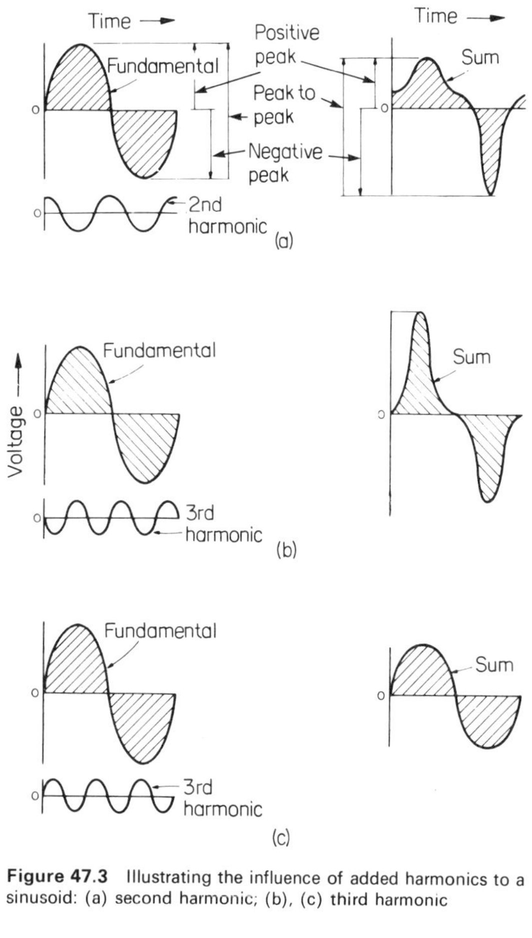

Figure 47.3 Illustrating the influence of added harmonics to a sinusoid: (a) second harmonic; (b), (c) third harmonic

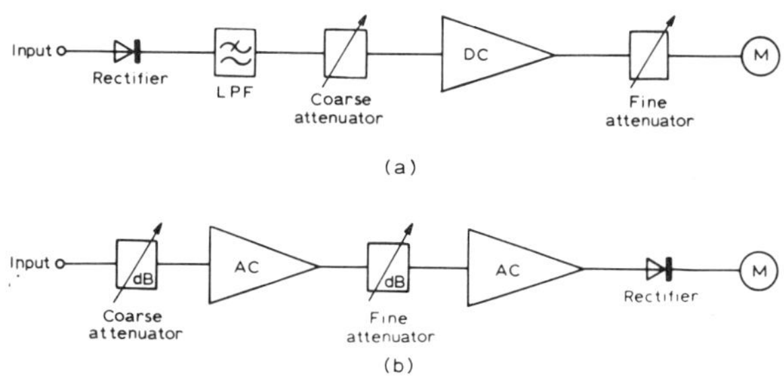

Figure 47.4 Schematic diagrams of a.c. voltmeters: (a) rectifier–amplifier type; (b) amplifier–rectifier type

A basic rectifier circuit is shown in *Figure 47.5*(*a*). Initially the following assumptions are made.

(1) Source resistance $R_s \ll$ the load resistance R.
(2) Diode resistance r when highly conductive $\ll$ the load resistance R.
(3) Resistance R_f in low pass filter $\gg$ the load resistance R.
(4) Amplitude of the source is high, e.g. >10 V so that the diode D acts as an on-off switch.
(5) Time constants CR and $C_fR_f \gg$ the time period of the source t.

If the r.m.s. value of the input voltage is V_i then the capacitor voltage V_c reaches a maximum value of $\sqrt{2}V_i$, falling due to the discharge through R during the non-conductive period of the diode to $\sqrt{2}V_i-\delta V_c$ where δV_c is the change in capacitor voltage as is shown in (*b*). When $CR \gg t$ then the conductive period of the diode is $\delta t \ll t$, thus the *average* value of V_c is approximately $\sqrt{2}V_i-(\delta V_c/2)$. The circuit is nominally peak indicating so that the fractional error

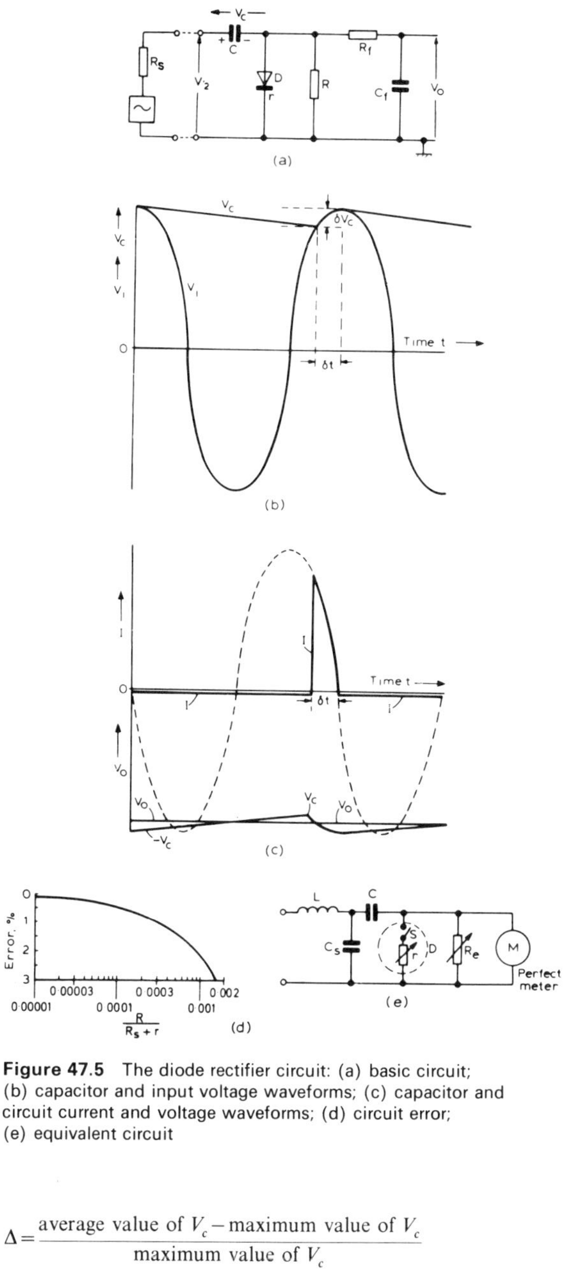

Figure 47.5 The diode rectifier circuit: (a) basic circuit; (b) capacitor and input voltage waveforms; (c) capacitor and circuit current and voltage waveforms; (d) circuit error; (e) equivalent circuit

$$\Delta = \frac{\text{average value of } V_c - \text{maximum value of } V_c}{\text{maximum value of } V_c}$$

$$\simeq \frac{-\delta V_c}{2V_{c\max}}$$

$$= \frac{-\delta Q}{2V_{c\max}C}$$

where Q is the change in capacitor charge.

If the capacitor discharge period approximates to the period t and the capacitor current is I as in (c) then

$$\Delta \simeq \frac{-It}{2V_{c\max}C}$$

$$= -\frac{t}{2CR}$$

or for design purposes

$$CR = \frac{1}{-\Delta 2f} \qquad (47.1)$$

The required time constant is very much longer than that which is necessary for a similar value of Δ in an amplifier coupling circuit.

When the diode is a thermionic valve, there is hardly any practical limit to the value of capacitance that may be used and hence the minimum frequency of operation. When a semiconductor diode is used, a limit occurs due to the magnitude of the charge that must be passed by the diode when a connection is made to a point where a high value steady potential exists, e.g. at the collector of a transistor or anode of a valve. The output voltage V_o is shown in *Figure 47.5(c)* to be the conjugate of the average value of the capacitor voltage V_c. This is due entirely to the polarity definition which has been adopted for V_c. The capacitor current I is also shown in *Figure 47.5(c)*. During the charge it takes on a shape which is a small portion of a sine wave, and the areas enclosed below and above the axis must be equal. The infinitely sharp wave-front of the current waveform during the charging period implies that the source and diode resistances have zero value. With finite values, the rise-time will of course be finite also, and the peak value of V_c will be less than that of the source e.m.f.

Because the capacitor charging current flows for only a very small fraction of the period t, its peak magnitude must be extremely high in comparison with its average value, and this results in significant errors unless the sum of the source resistance plus diode resistance is extremely small in comparison with the value of the load resistance R. The percentage error, as given by Scroggie is as shown in *Figure 47.5(d)*. In addition, a combination of high value source resistance with intermittancy of current flow causes distortion of the waveform under investigation. Because of this short duty cycle of the diode, the value of input resistance cannot be specified by any simple figure. Arbitrarily, it could be defined as say 50 times the value of source resistance in connection with which there was a -2% error, but the value would change with the tolerance. However, when the circuit is operated from a resonant circuit of high magnification factor Q, then the input resistance approximates to $R/2$. This is because R is the only component in which there is significant power dissipation at low and medium frequencies, and current is taken from the source only when the instantaneous amplitude is close to $\sqrt{2}V_i$, so that the dissipation in R is $(\sqrt{2}V_i)^2/R$ so that the apparent value of R is $R/2$.

When measuring signals of high amplitude, the output voltage V_o is proportional to the input voltage V_i, so that for this condition the moving-coil meter used for display may carry, say, two scales only and multiplication factors may be used as for a simple milti-meter. For lower amplitudes the signal waveform is accommodated on the curved portion of the characteristic of the diode so that the response changes from peak, through arithmetic mean, to square law, and the period of diode conduction increases. This has the advantage that errors due to harmonics are reduced and the connection of the meter introduces less distortion of the signal under test. One disadvantage, however, is due to the fact that the diode is not operating as an on-off switch, but is conducting moderately for a large proportion of the operating cycle, the dissipation within the diode increases so that the effective value of the input resistance falls. Another disadvantage is that due to the gradual change from linear to square law operation additional meter scales are required such that simple multiplying factors can no longer be applied. The proportional reading accuracy of square-law scales is less than that on linear scales. Some improvement may however be achieved by switching into circuit non-linear devices to improve the scale law, but such devices tend to be temperature sensitive.

The problem of amplifier stability becomes all important at low levels of input signal.

An approximate equivalent circuit for a diode rectifier circuit is given in *Figure 47.5(e)*. Here *S* and *r* represent the diode. At high amplitude, and when driven from low impedance or high *Q* circuits, the duty of cycle of *S* is low as also is the value of *r*. If the amplitude is low and/or the source resistance is high, then the reverse is true. At low frequencies the actual and effective values of R_e are high, and the CR_e product limits the performance at low frequencies. At high frequencies, e.g. above 10 MHz, dielectric losses cause the value of R_e to be reduced drastically. Stray capacitance C_s and stray inductance *L* commonly cause a rise to occur in the response of the circuit at very high frequencies, followed by a progressive fall of response.

Extension of the linear law down to about 100 mV can be achieved by the use of the sharper changes of conductivity of the zener diode using a circuit employing two such diodes of similar voltage rating, one D_1 for rectification, and another D_2 for biasing as in *Figure 47.6*. The output will of course be superimposed upon a steady potential. Due to the absence of hole storage, the circuit is operable up to hundreds of megahertz.

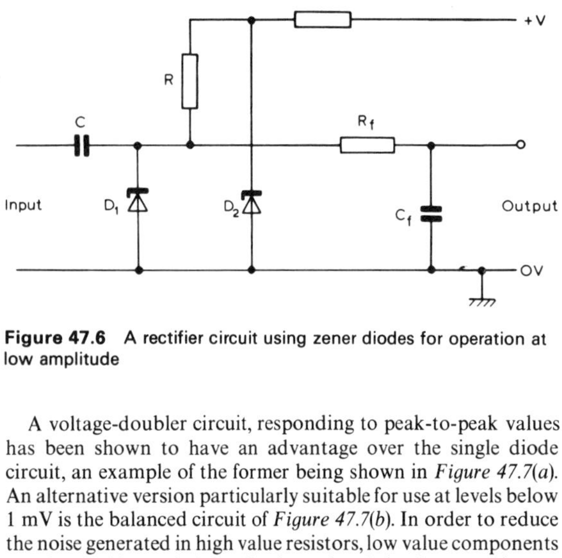

Figure 47.6 A rectifier circuit using zener diodes for operation at low amplitude

A voltage-doubler circuit, responding to peak-to-peak values has been shown to have an advantage over the single diode circuit, an example of the former being shown in *Figure 47.7(a)*. An alternative version particularly suitable for use at levels below 1 mV is the balanced circuit of *Figure 47.7(b)*. In order to reduce the noise generated in high value resistors, low value components may be used instead, their effective values being increased by use of positive feedback bootstrapping circuits.

Reference to *Figure 47.4(a)* shows that d.c. amplifiers are needed. For moderate degrees of sensitivity it is necessary either to employ frequent electrical zero setting or else a chopper-stabilised circuit in order to compensate for drift.

In order to minimise capacitance loading of the circuit under test, the diode and other components closely associated with it, are commonly housed in a probe extension from the main unit. In one type a thermostatically controlled heater prevents the operating temperature of the probe from falling below 33°C, because rectification efficiency falls at low temperatures.

The maximum sensitivities which may be achieved using the circuit of *Figure 47.4(a)* are less than those for circuit (*b*). The former type of circuit is thus only popular for use at very high frequencies for which it would be difficult to design an a.c. amplifier for use in circuit (*b*).

For operation below a few tens of megahertz, amplification first and rectification afterwards is to be preferred, as exemplified in the circuit of *Figure 47.4(b)*. The main advantages which this circuit has over the rectifier–amplifier combination are as follows.

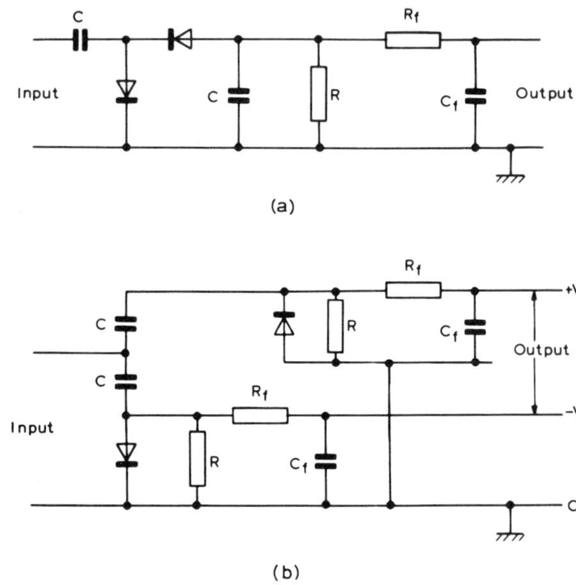

Figure 47.7 Voltage doubler circuits: (a) unbalanced; (b) balanced

(1) An output connection may be taken off just ahead of the rectifier circuit so that the instrument may be used as a wide-band amplifier of adjustable and known value of gain.
(2) The value of the input resistance can be expressed as a simple quantity which does not change with change of amplitude of the signal.
(3) The rectifier circuit always operates at high amplitude, thereby producing high and consistent rectification efficiency.
(4) There is no need for zero setting or the use of chopper-stabilised amplifiers.
(5) Deflection is proportional to the arithmetic mean value of the signal so that the scale law is linear. This minimises the number of scales which appears on the meter face and thus permits scale multiplication factors to be used. Without overcrowding the meter face, a decibel scale may be added.

The detailed design of any instrument in this category centres largely upon the attenuator arrangement. In this connection, steps of 10 decibels are almost standard, giving a meter face as shown in *Figure 47.8(a)*. The zero decibel value is normally taken as 775 mV r.m.s., corresponding to 1 mW in a 600 Ω circuit. The scale for a corresponding attenuator switch, which might control either both attenuators or only the fine attenuator, is shown in *Figure 47.8(b)*. Where a single attenuator switch is used, the coarse attenuator may provide steps of 20 dB, the setting of this

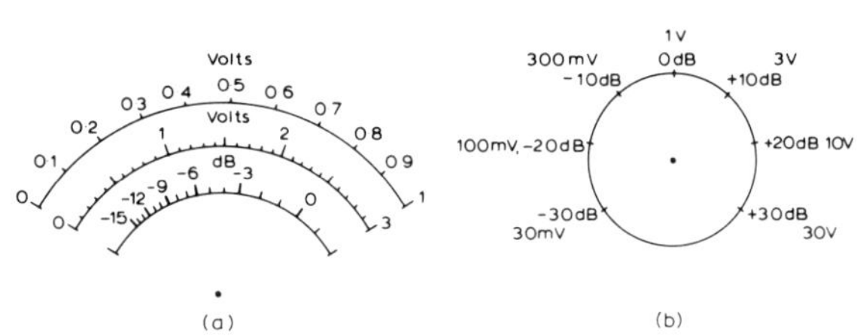

Figure 47.8 Typical voltmeter scales (a) and corresponding range switch engravings (b)

attenuator changing on *alternate* positions of the switch, while steps of 10 dB are alternately brought in and out of circuit in the fine attenuator. This makes for efficient use of the amplifiers. An alternative arrangement more suitable for use where a very wide range of amplitudes is encountered, or where a probe must be used, is to have separate controls for the attenuators. In this case the coarse attenuator usually has two positions, one marked VOLTS/O dB and the other marked mV/−60 dB. The fine attenuator will then cover the 60 dB range in steps of 10 dB. Where a probe is provided the coarse attenuator is usually located within the probe. A basic circuit for such an attenuator is given in *Figure 47.9*.

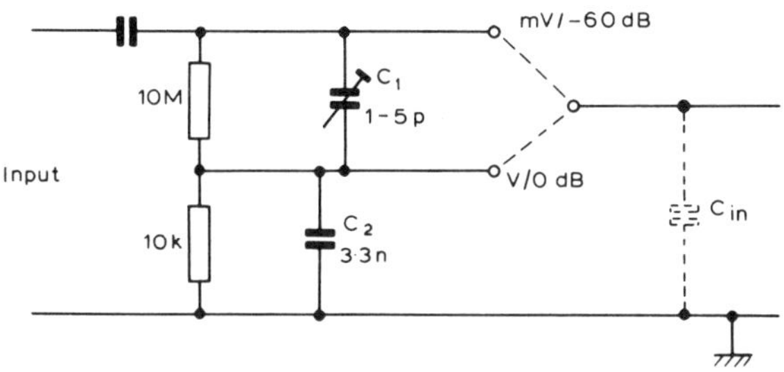

Figure 47.9 A 60 dB attenuator circuit

If the two amplifiers be designed with stages having very high values of input impedance (using say f.e.t. source-followers or compound-emitter followers), then the attenuators may take the form of frequency-compensated potential dividers, rather than constant-impedance networks.

Where the required gain × bandwidth product is small, the first amplifier may be a simple f.e.t. source-follower, feeding a potential-divider-type fine attenuator. Otherwise an f.e.t. input stage and an emitter-follower output are appropriate, preferably with overall feedback. Since the forward resistances of the rectifier diodes change both with changes of current and changes of temperature, they should be fed from a high impedance source. This can be achieved in two ways. First the metering circuit is incorporated into a current negative-feedback loop which thus raises the effective value of the output impedance of the amplifier. Secondly, the output stage of the amplifier may be supplied with constant current load.

47.2 Digital voltmeters

The digital voltmeter provides a digital display of d.c. and/or a.c. inputs, together with coded signals of the visible quantity, enabling the instrument to be coupled to recording or control systems. Depending on the measurement principle adopted, the signals are sampled at intervals over the range 2–500 ms. The basic principles are: (i) linear ramp; (ii) successive approximation/potentiometric; (iii) voltage to frequency, integration; (iv) dual-slope; (v) some combination of the foregoing. Techniques (iii) and (iv) are described in Chapter 31 and (i) and (ii) are described below.

47.2.1 Linear ramp

This is a voltage/time conversion in which a linear time base is used to determine the time taken for the internally generated voltage v_s to change by the value of the unknown voltage V. The block diagram, *Figure 47.10(b)*, shows the use of comparison networks to compare V with the rising (or falling) v_s; these

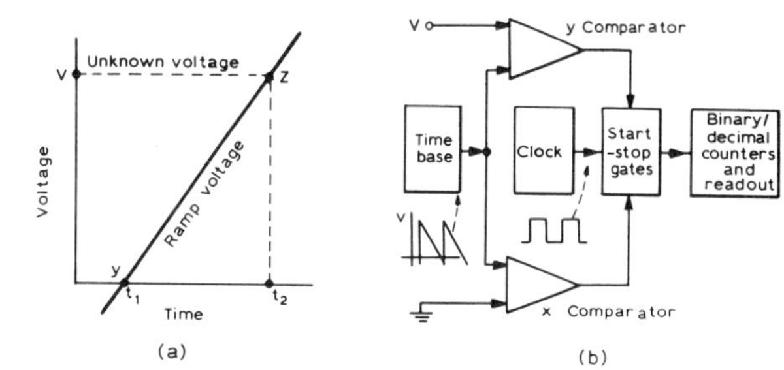

Figure 47.10 Linear-ramp digital voltmeter: (a) ramp voltage; (b) block diagram

networks open and close the path between the continuously running oscillator, which provides the counting pulses at a fixed clock rate, and the counter. Counting is performed by one of the binary coded sequences, the translation networks give the visual decimal output. In addition a binary coded decimal output may be provided for monitoring or control purposes.

Limitations are imposed by small non-linearities in the ramp, the instability of the ramp and oscillator, imprecision of the coincidence networks at instants y and z, and the inherent lack of noise rejection. The overall uncertainty is about ±0.05%, and the measurement cycle would be repeated every 200 ms for a typical 4-digit display.

Linear 'staircase' ramp instruments are available in which V is measured by counting the number of equal voltage 'steps' required to reach it. The staircase is generated by a solid-state diode pump network, and linearities and accuracies achievable are similar to those with the linear ramp.

47.2.2 Successive approximation

As it is based on the potentiometer principle, this class produces very high accuracy. The arrows in the block diagram of *Figure 47.11* show the signal-flow path for one version; the resistors are selected in sequence so that, with a constant-current supply, the test voltage is created within the voltmeter.

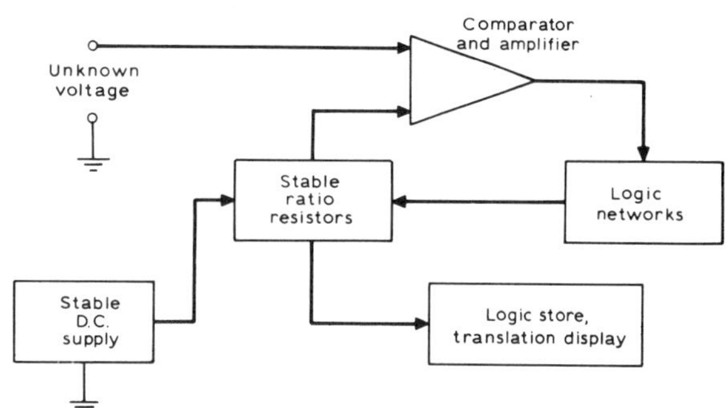

Figure 47.11 Successive-approximation digital voltmeter

Each decade of the unknown voltage is assessed in terms of a sequence of accurate stable voltages, graded in descending magnitudes in accordance with a binary (or similar) counting scale. After each voltage approximation of the final result has been made and stored, the residual voltage is then automatically re-assessed against smaller standard voltages, and so on to the smallest voltage discrimination required in the result. Probably four logic decisions are needed to select the major decade value of the unknown voltage, and this process will be repeated for each lower decade in decimal sequence until, after a few milliseconds, the required voltage is stored in a coded form. This voltage is then translated for decimal display.

It is necessary to sense the initial polarity of the unknown signal, and to select the range and decimal marker for the read-out; the time for the logic networks to settle must be longer for the earlier (higher voltage) choices than for the later ones because they must be of the highest possible accuracy; off-set voltages may be added to the earlier logic choices, to be withdrawn later in the sequence; and so forth.

The total measurement and display takes about 5 ms. When noise is present in the input, the necessary insertion of filters may extend the time to about 1 s. As noise is more troublesome for the smaller residuals in the process, it is sometimes convenient to use some different techniques for the latter part. One such is the voltage-frequency principle which has a high noise rejection ratio. The reduced accuracy of the technique can be tolerated as it applies only to the least significant figures.

47.2.3 Digital multimeters

Any digital voltmeter can be scaled to read d.c. or a.c. voltage, current, impedance or any other physical property provided that an appropriate transducer is inserted. The trend with instruments of modest accuracy (0.1%) is to provide a basic digital voltmeter with separate plug-in converter units as required for each parameter. Instruments scaled for alternating voltage and current normally incorporate one of the a.c./d.c. converter units mentioned in a previous section, and the quality of the result is limited by the characteristics inherent in such converters. The digital part of the measurement is more accurate and precise than the analogue counterpart, but is more expensive.

For systems application, command signals can be inserted into, and binary-coded or analogue measurements received from, the instrument through multi-way socket connections, enabling the instrument to form an active element in a control situation.

Resistance, capacitance and inductance measurements depend to some extent on the adaptability of the basic voltage measuring process. The dual-slope technique can be easily adapted for two-, three- or four-terminal ratio measurements of resistance by using the positive and negative ramps in sequence; with other techniques separate impedance units are necessary.

47.2.4 Input and dynamic impedance

The high precision and small uncertainty of digital voltmeters makes it essential that they have a high input impedance if these qualities are to be exploited. Low test voltages are often associated with source impedances of several hundred kilohms: for example, to measure a voltage with source resistance 50 kΩ to an uncertainty of ±0.005% demands an instrument of input resistance 1 GΩ, and for a practical instrument this must be 10 GΩ if the loading error is limited to *one-tenth* of the *total* uncertainty.

The dynamic impedance will vary considerably during the measuring period, and it will always be lower than the quoted null, passive, input impedance. These changes in dynamic impedance are coincident with voltage 'spikes' which appear at the terminals due to normal logic functions; this noise can adversely affect components connected to the terminals, e.g. Western standard cells.

Input resistances of the order of 1–10 GΩ represent the conventional range of good quality insulators. To these must be added the stray parallel reactance paths through unwanted capacitive coupling to various conducting and earth planes, frames, chassis, common rails, etc.

47.2.5 Noise limitation

The information signal exists as the potential difference between the two input leads. Each can have unique voltage and impedance conditions superimposed on it with respect to the basic *reference* or *ground* potential of the system, as well as another and different set of values with respect to a local *earth* reference plane.

An elementary electronic instrumentation system will have at least one ground potential and several earth connections—possibly through the (earthed) neutral of the main supply, the signal source, a read-out recorder or a cathode ray oscilloscope. Most true earth connections are at different electrical potentials with respect to each other due to circulation of currents (d.c. to u.h.f.) from other apparatus, through a finite earth resistance path. When multiple earth connections are made to various parts of a high-gain amplifier system, it is possible that a significant frequency spectrum of these signals will be introduced as electrical noise. It is this interference which has to be rejected by the input networks of the instrumentation, quite apart from the concomitant removal of any electrostatic/electromagnetic noise introduced by direct coupling into the signal paths. The total contamination voltage can be many times larger (say 100) than the useful information voltage level.

Electrostatic interference in input cables can be greatly reduced by 'screened' cables (which may be 80% effective as screens), and electromagnetic effects minimised by transposition of the input wires and reduction in the 'aerial loop' area of the various conductors. Any residual effects, together with the introduction of 'ground and earth-loop' currents into the system are collectively referred to as *series* and/or *common-mode* signals.

47.2.6 Series and common-mode signals

Series-mode (normal) interference signals V_{sm} occur in series with the required information signal. Common-mode interference signals V_{cm} are present in both input leads with respect to the reference potential plane: the required information signal is the difference voltage between these leads. The results are expressed as *rejection ratio* (in dB) with respect to the input error signal V_e that the interference signals produce, i.e.

$$K_{sm} = 20 \log (V_{sm}/V_e) \qquad (47.2)$$

and

$$K_{cm} = 20 \log (V_{cm}/V_e) \qquad (47.3)$$

where K is the rejection ratio.

Consider the elementary case in *Figure 47.12*, where the input lead resistances are unequal, as would occur with a transducer

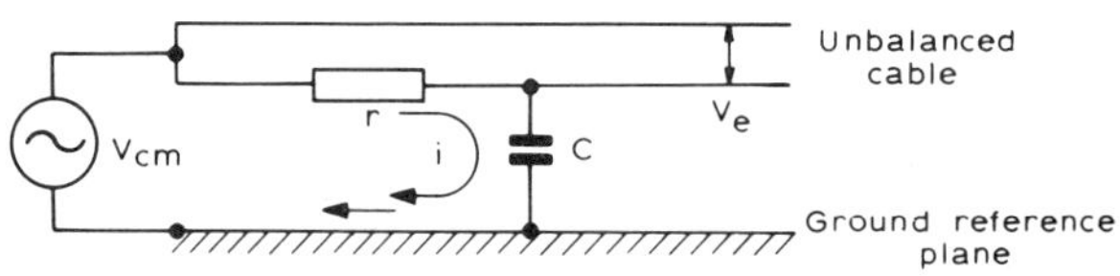

Figure 47.12 Common-mode effect in unbalanced network

input. Let r be the difference resistance, C the cable capacitance, with common-mode signal and error voltages V_{cm} and V_e respectively. Then the common-mode numerical ratio is

$$V_{cm}/V_e = V_{cm}/ri = 1/2\pi fCr \qquad (47.4)$$

assuming the cable insulation to be ideal, and $X_C \gg r$. Clearly for a common-mode statement to be complete it must have a stated frequency range and include the resistive unbalance of the source. (It is often assumed in c.m.r. statements that $r = 1$ kΩ.)

The c.m.r. for a digital voltmeter could be typically 140 dB (corresponding to a ratio of $10^7/1$) at 50 Hz with a 1 kΩ line unbalance, and leading consequently to $C=0.3$ pF. As the normal input cable capacitance is of the order of 100 pF/m, the situation is not feasible. The solution is to inhibit the return path of the current i by the introduction of a guard network. Typical guard and shield parameters are shown in *Figure 47.13* for a six-

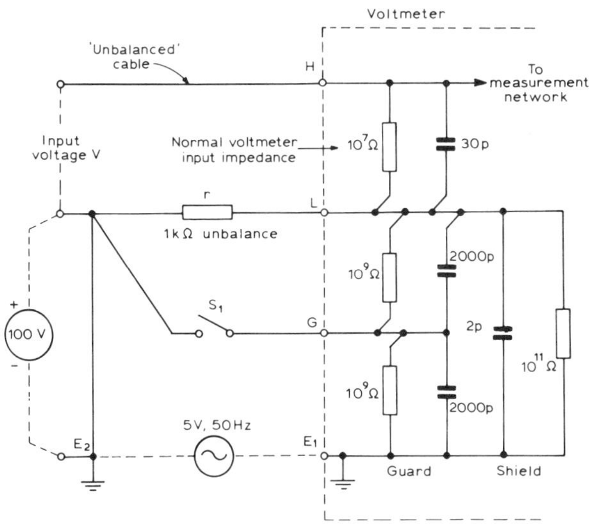

Figure 47.13 Typical guard and shield network for digital voltmeter

figure digital display on a voltmeter with $\pm 0.005\%$ uncertainty. Consider the magnitude of the common-mode error signal due to a 5 V 50 Hz common-mode voltage between the shield earth E_1 and the signal earth E_2:

Switch S_1 open. The a.c. common-mode voltage drives current through the guard network and causes a change of 1.5 mV to appear across r as a series-mode signal; for $V=1$ V this represents an error V_e of 0.15% for an instrument whose quality is $\pm 0.005\%$.

Switch S_1 closed. The common-mode current is now limited by the shield impedance, and the resultant series-mode signal is 3.1 μV, an acceptably low value that will be further reduced by the noise-rejection property of the measuring circuits.

47.2.7 Floating-voltage measurement

If the voltage difference V to be measured has a potential difference to E_2 of 100 V, as shown, then with S_1 open the change in potential difference across r will be 50 μV, as a series-mode error of 0.005% for a 1 V measurement. With S_1 closed, the change will be 1 μV, which is negligible.

The interconnection of electronic apparatus must be carefully made to avoid systematic measurement errors (and shortcircuits) arising from incorrect screen, ground or earth potentials.

In general it is preferable, wherever possible, to use a signal common reference mode, which should be at the zero signal reference potential to avoid leakage current through r. Indiscriminate interconnection of the shields and screens of adjacent components can increase noise currents by shortcircuiting the high-impedance stray path between the screens.

47.2.8 Instrument selection

The most precise 7-digit voltmeter, when used for a 10 V measurement, has a discrimination of ± 1 part in 10^6 (i.e. ± 10 μV), but has an uncertainty ('accuracy') of about ± 10 parts in 10^6. The distinction is important with digital read-out devices least a higher quality be accorded to the number indicated than is in fact justified. The quality of any reading must be based upon the time-stability of the total instrument since it was last calibrated against external standards, and the cumulative evidence of previous calibrations of the like kind.

Selection of a digital voltmeter from the list of types given in *Table 47.2* is based on the following considerations.

(1) No more digits than necessary, as the cost per digit is high.
(2) High input impedance, and the effect on likely sources of the dynamic impedance.
(3) Electrical noise-rejection, assessed and compared with (*a*) the common-mode rejection ratio based on the input and guard terminal networks and (*b*) the actual inherent noise-rejection property of the measuring principle employed.
(4) Requirements for binary-coded decimal facilities.
(5) Versatility and suitability for use with a.c. or impedance converter units.
(6) Use with transducers (in which case (3) is the most important single factor).

47.2.9 Calibration

It will be seen from *Table 47.2* that digital voltmeters should be recalibrated at intervals between three and twelve months. Built-in self-checking facilities are normally provided to confirm satisfactory operational behaviour, but the 'accuracy' check cannot be better than that of the included standard cell or zener diode reference voltage and will apply only to one range. If the user has available some low-noise stable voltage supplies (preferably batteries) and some high-resistance helical voltage-dividers, it is easy to check logic sequences, resolution and the approximate ratio between ranges. The accuracy of the 1 V range can be tested with an external Weston standard cell provided that the cell voltage is known to within ± 3 μV by recent NPL calibration.

47.3 Signal generators

In this section signal generators represent those instruments which generate a waveform whose frequency and amplitude can be varied. This includes sine wave oscillators, square, triangular and sawtooth waveform generators, and pulse generators.

The basic circuits used for generating waveforms are described in Chapter 42. A simplified block diagram of a commercial generator is shown in *Figure 47.14*. The master oscillator usually produces sine waves, for a sine wave oscillator. For low frequencies it is not possible to use a tuned LC oscillator and integrating circuits are used to generate a triangular waveform, and this is synthesized to alter the shape of the triangular wave as its amplitude changes, and so produce a sine wave.

Generally, different waveshapes are generated in different parts of the system, and these can be brought out in a function generator. Commercial instruments are capable of operating over a range of frequencies from 0.001 Hz to 1 GHz.

The output frequency can be varied in steps by a range switch, and continuously within any step. The oscillator can free-run, or it can be synchronised to an external signal, or it can be f.m. or a.m. modulated by another signal. The amplitude of the output can be varied independently of the frequency.

Sometimes several outputs are available having different impedances for matching. A pulse generator, like a sine wave

Table 47.2 Typical characteristics of digital voltmeters and multimeters

1: Operating principle—
DS, dual slope;
DDS, inductive divider + dual slope;
I, integration;
ISA, mixed I and SA:
SA, potentiometer/successive approx. R, ramp.

2: Number of display digits.
3: Operating time.
4: Instrument ranges.
5: Uncertainty, smallest digits to be added in ± parts per million of maximum reading.
6: Maximum discrimination, in ± p.p.m. of maximum reading.

7: Parallel input resistance.
8: Parallel input capacitance.
9: Common (or series) mode rejection.
10: Common (or series) mode rejection, 50 Hz.
11: Recalibration period.

1	2	3 ms	4	5 p.p.m.	6 p.p.m.	7 MΩ	8 pF	9 dB	10 dB	11 months
Single-purpose Voltmeters										
R	3	500	100 mV–1 kV	5000	1000	10	—	90–10	40	3
I	4	2–200	100 mV–1 kV	300	10	10^5, 10	—	150	150	12
I	6	2–200	20 mV–1 kV	40	10	10^5, 10	—	—	160	6
ISA	7	1000	1 V–1 kV	60	1	10^4, 10	40	—	160	3
DD	7	—	1 V–1 kV	10	1	10^5, 10	—	120	150	3
Small Multimeter or Panel-meter										
			0.2 V–1 kV							
			0–20 kHz	1000–4000	100	10	110	—	90	12
DS	4	200	200 μA–2A							
			0.2–2000 kΩ							
Modular Meters										
SA	5	2–250	0.1 mV–1 kV (d.v.)	200	10	10^4, 10	—	—	100	3
	5	2–250	1 V–1 kV	1000–5000	10	1	100	—	60	3
			50 Hz–100 kHz							
SA	5	2–250	1 kΩ–10 MΩ	500	10	—	—	—	—	3

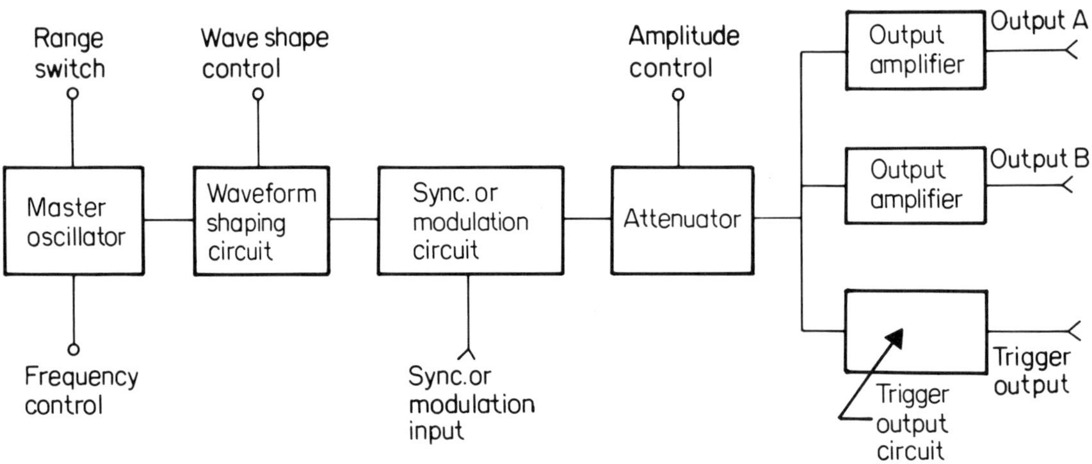

Figure 47.14 Simplified block diagram of a signal generator

generator, must also be properly terminated at its operating frequency, with its characteristic impedance, to prevent reflections. If a pulse generator has a rise time of t_r ns then this is equivalent to a sine wave frequency of f_s MHz where

$$f_s = \frac{350}{t_r} \tag{47.5}$$

In selecting a signal generator the quality of the output waveform is an important consideration. For example a sine wave should have minimum distortion; a pulse generator should have clean rise and fall times with no overshoot, ringing or sag; a square wave generator should have equal mark to space periods and often the slope of the edges of the waveform should be variable.

Sine wave generators are often used for testing receivers. Pulse generators are used to test radar and communication systems, for device testing such as the recovery time of transistors, and for circuit analysis such as the transient analysis of an amplifier.

47.3.1 Frequency synthesiser

This is a sine wave generator which uses the phase-locked loop

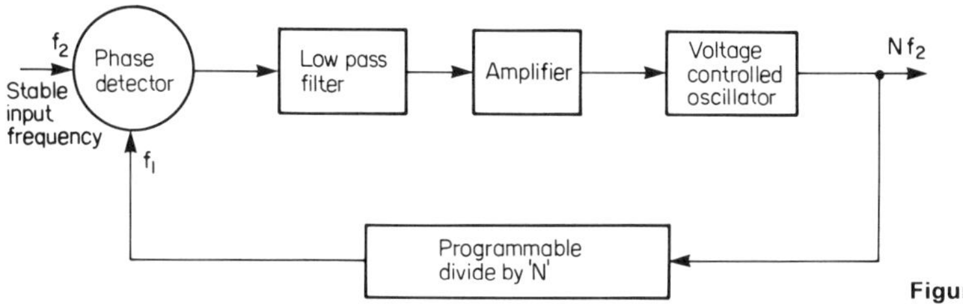

Figure 47.15 Block diagram of a frequency synthesiser

principle to generate a range of output frequencies. *Figure 47.15* shows the operating principle of the instrument.

The input frequency is generated by a stable source such as a voltage controled crystal oscillator. This is compared in a phase detector whose output consists of $f_2 \pm f_1$. The low pass filter selects the difference frequency $f_2 - f_1$. This is amplified and fed to the voltage-controlled oscillator. This drives f_1 towards f_2 so that they lock together, and only a phase difference exists between f_1 and f_2 which gives sufficient signal to f_1 to keep it locked to f_2.

To vary the output frequency the value of 'N' is changed in the programmable divider. Usually the input frequency f_2 is obtained from a high frequency oscillator. Since it is expensive to design a programmable divider to run at high frequencies, a fixed count divider, called a pre-scalor, is introduced in series with the programmable divider.

47.3.2 Sweep frequency generator

A sweep frequency generator is usually a signal generator which produces a sine wave output in the r.f. range, whose frequency can be continuously and smoothly varied, usually at a low audio frequency rate. *Figure 47.16* shows a schematic diagram of a sweep frequency generator.

The frequency can be swept electronically or manually. Electronically tuned oscillators include backward wave oscillator tubes. Example of a manual method is the use of a motor driven variable capacitor in a tuned *LC* circuit of an oscillator. The frequency sweeper gives a synchronously varying voltage which can also be used for the horizontal deflection in a cathode ray tube display or an XY recorder. This enables a plot to be obtained giving the response of a device which is fed by the sweep generator.

Manual controls can be used for adjusting the frequency of the master oscillator. The range switch operates in several bands, and the frequency sweeper covers each band usually at 10 to 30 sweeps per second. The level control circuit monitors the r.f. level at some point in the system and holds the power to the load constant as the frequency and load impedance change. This ensures a fixed readout calibration with frequency and prevents source mismatch.

47.3.3 Random-noise generator

A random-noise generator gives a signal whose instantaneous amplitude varies at random, and contains no periodic frequency component. It has many uses such as testing radio and radar systems for signal detection in the presence of noise, and intermodulation and crosstalk tests in communication systems.

Random noise is usually generated within the instrument by a semiconductor noise diode which gives an output frequency in the band 80 kHz to 220 kHz. This signal is amplified and modulated down to the audio-frequency band. It is then passed through a filter which gives an output noise signal in one of three spectrums, white noise, pink noise and USASI noise, as in *Figure 47.17*.

Pink noise is so called by analogy to red light since it emphasises lower frequencies. It has a voltage spectrum which varies inversely as the square root of frequency, and is used in bandwidth analysis. USASI noise is used for testing audio systems since its spectrum approximately equals the energy distribution of speech and music frequencies.

47.4 Waveform analysers

These instruments are primarily used to determine the composition of a waveform. Two types of instruments are described below, harmonic analysers and the spectrum analyser. Harmonic analysers are further divided into harmonic distortion

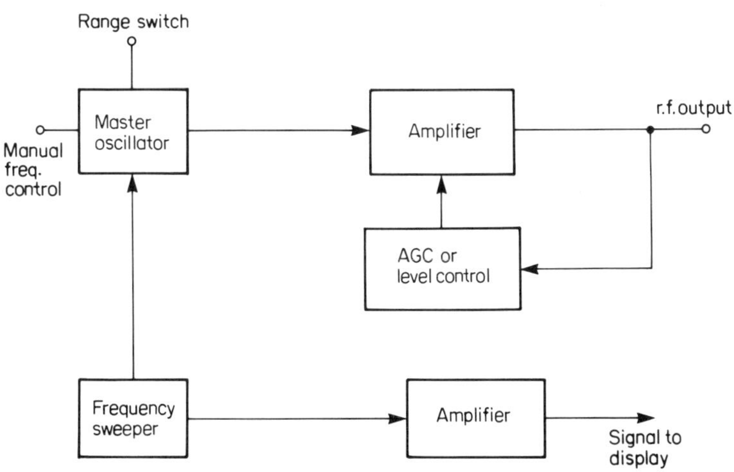

Figure 47.16 Sweep frequency generator block (schematic)

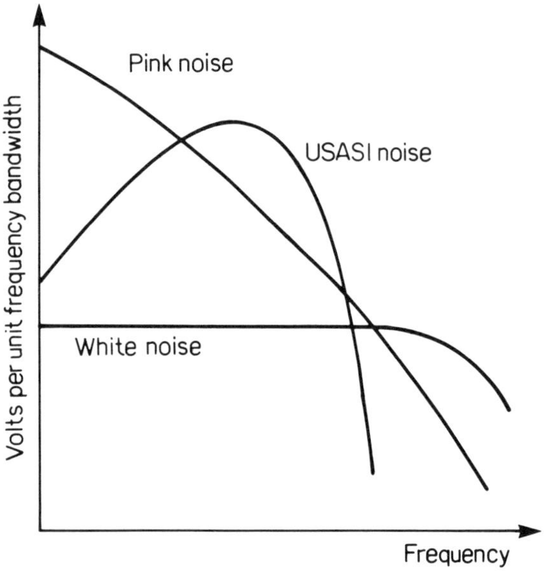

Figure 47.17 Noise spectra

analyser and intermodulation distortion analyser. Transient intermodulation distortion is also described in this section.

47.4.1 Harmonic distortion analyser

This is shown in *Figure 47.18*. The amplifier under test is fed from a signal source having very low distortion. The harmonic output of this source can of course be checked separately by the meter. The output from the amplifier is fed into a notch filter. This is tunable and is designed to cut off the fundamental frequency and to let other signal frequencies through. The voltmeter measures a.c. r.m.s. voltage. Both the total voltage E_T of the harmonics and fundamental, and the harmonic r.m.s. voltage E_H can be read by means of the voltmeter switch. The total harmonic distortion is given by

$$\text{IHD}=\frac{E_H}{E_T}\times 100 \qquad (47.6)$$

In practice the meter is calibrated to a percentage scale. It is set to E_T and adjusted for full-scale deflection. When the switch is now moved to E_H the total harmonic distortion can be read off directly on the meter.

There are several types of distortion. Amplitude distortion occurs when the output amplitude is not proportional to its input. When the amplifier is overdriven it will clip and distortion will increase. Cross over distortion occurs in push–pull output amplifiers, and this increases as the test signal level decreases. The distortion meter will also read noise in the system. This noise increases as the signal level decreases so it can be confused with certain types of distortion. Care must therefore be taken to minimise noise.

47.4.2 Intermodulation distortion analyser

A non-linear system will generate frequencies in its output which are equal to the sums and differences of the frequencies present in the input signal. When a high frequency f_1 signal and a low frequency signal f_2 are mixed in a linear network the output will contain the two original frequencies only. If the signals are mixed in a non-linear system, such as an amplifier which has distortion, modulation will occur, and the output will contain at least f_1, f_2, f_1+f_2 and f_1-f_2, and usually also several harmonics and their sum and difference frequencies.

Amplitude modulation is caused by the system distortion as in *Figure 47.19*. The intermodulation distortion is given by

$$\text{IMD}=\frac{C-M}{M}\times 100 \qquad (47.7)$$

Figure 47.20 shows how an intermodulation distortion analyser may be used to measure amplifier distortion. The test frequency is usually a two-tone wave in which one frequency is at least 50 times larger than the other. For example 50 Hz and 5 kHz signals may be used in which the 50 Hz signal has an amplitude several times larger than the 5 kHz signal. When the two signals are processed by an amplifier any amplitude non-linearity will cause the 50 Hz signal to amplitude modulate the 5 kHz signal. The intermodulation distortion amplifier has filters and demodulators for separating and measuring the amplitude modulated signal.

47.4.3 Transient intermodulation distortion

This occurs when there are sudden changes in the input signal and the amplifier cannot respond quickly. The amplifier under test is usually checked by a test signal which is a square wave with a superimposed high frequency sine wave, as in *Figure 47.21(a)*. If transient intermodulation distortion is present then the output

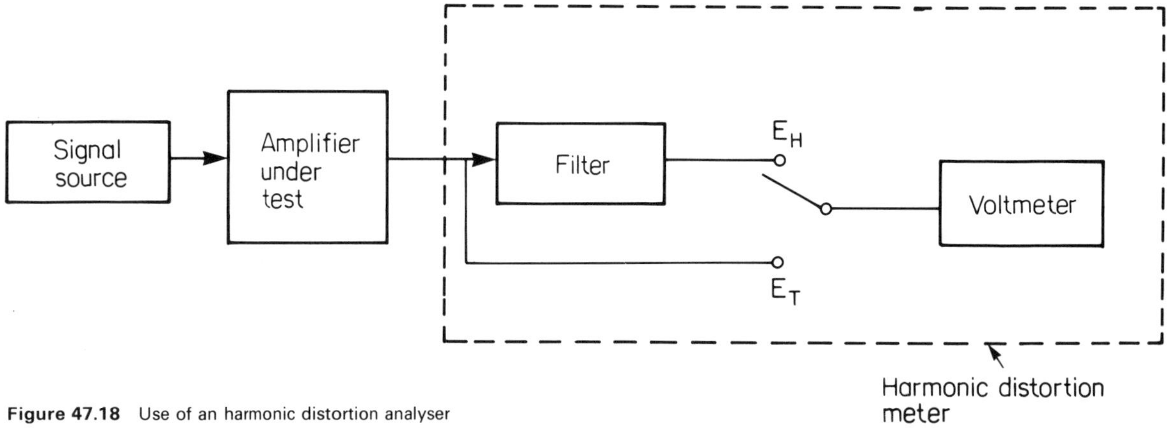

Figure 47.18 Use of an harmonic distortion analyser

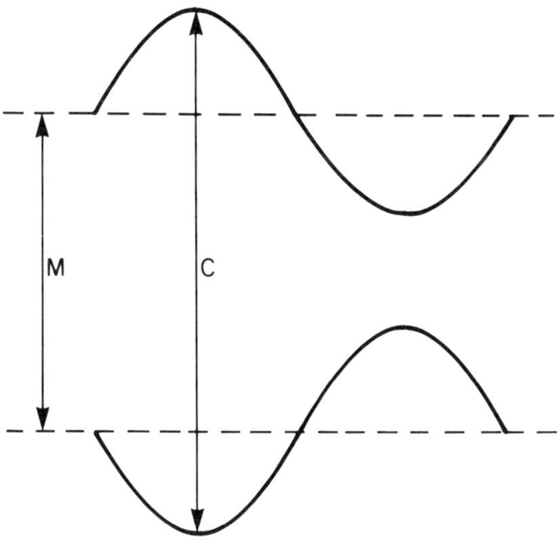

Figure 47.19 Amplifier modulation in a non-linear system

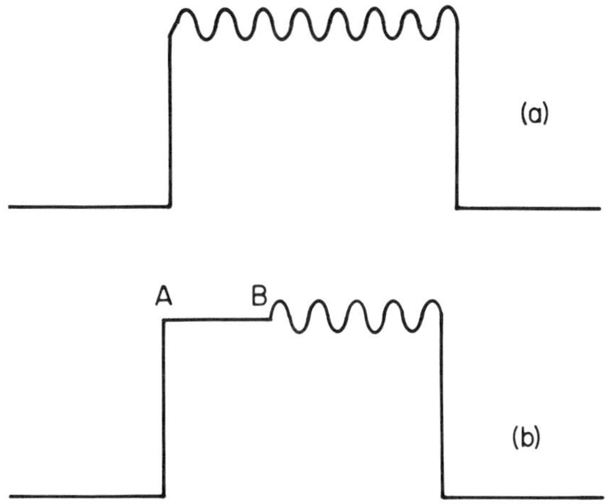

Figure 47.21 Illustration of transient intermodulation distortion: (a) input signal; (b) output signal

from the amplifier will have parts of the high frequency component of the input missing, as at AB in *Figure 47.21(b)*. This can be checked on an oscilloscope.

47.4.4 Spectrum analyser

Figure 47.22 shows a simplified diagram of a spectrum analyser. The output from the amplifier under test is broken down into its individual harmonics and the frequency and amplitude of each is displayed on an internal CRT display, as in *Figure 47.23*.

The input signal to a spectrum analyser is used to modulate an internal high frequency oscillator, producing sideband frequencies corresponding to the input frequency. The oscillator is swept by a sweep generator, which also provides the signal for the X plates of the internal cathode ray tube display. The sideband frequencies generated at the mixer are also swept and pass sequentially through the filter. They are then demodulated and applied to the Y plates of the CRT display.

47.5 Counters

The electronic circuits which go into making a counter, and the different types of counters, were described in Chapter 30. These circuits will be referred to as the counter module. *Figure 47.24* shows how this module is used in a commercial instrument. The instrument can operate in several modes, such as totalising, frequency measurement, period measurement, ratio measurement and averaging. These modes are selected by a mode select switch and the path which the signals follow in the instrument are varied by internal control circuitry.

The input signals being measured are usually not clean square waves, and may sometimes not have sufficient amplitude to operate other circuits within the instrument. They are therefore amplified (or attenuated if the signals are too large) and sharpened up using waveform shapers and a Schmitt trigger. *Figure 47.25* shows the operation of the circuits. The Schmitt trigger has hysteresis and the input waveform must cross both the upper and lower trigger lines in order to register on the instrument.

In some counters the trigger levels can be varied in order to control the trigger points on the input waveform, and to change the width of the hysteresis band.

The counter has a very stable internal crystal oscillator which generates an internal measurement time base. The crystal is usually temperature controlled. Programmable dividers can be used to divide the frequency of the internal time base, or the input signals, before they are used for measurement within the instrument.

47.5.1 Counter measurement modes

The counter usually operates in one of six modes.

47.5.1.1 Totaliser

In this mode of operation the counter simply adds the number of pulses received and displays the total. Counting starts and stops

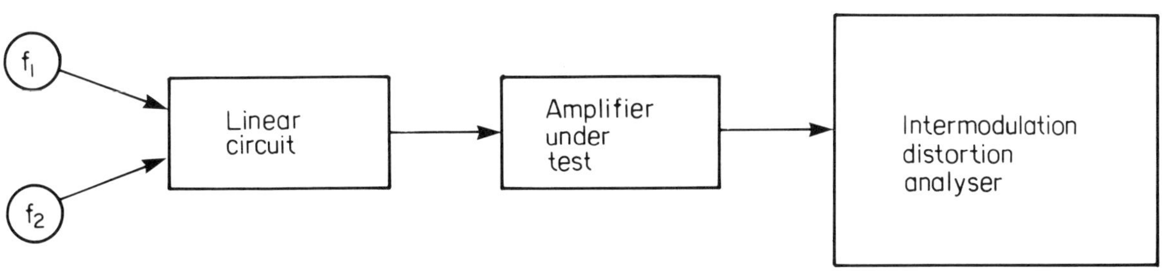

Figure 47.20 Measurement of intermodulation distortion

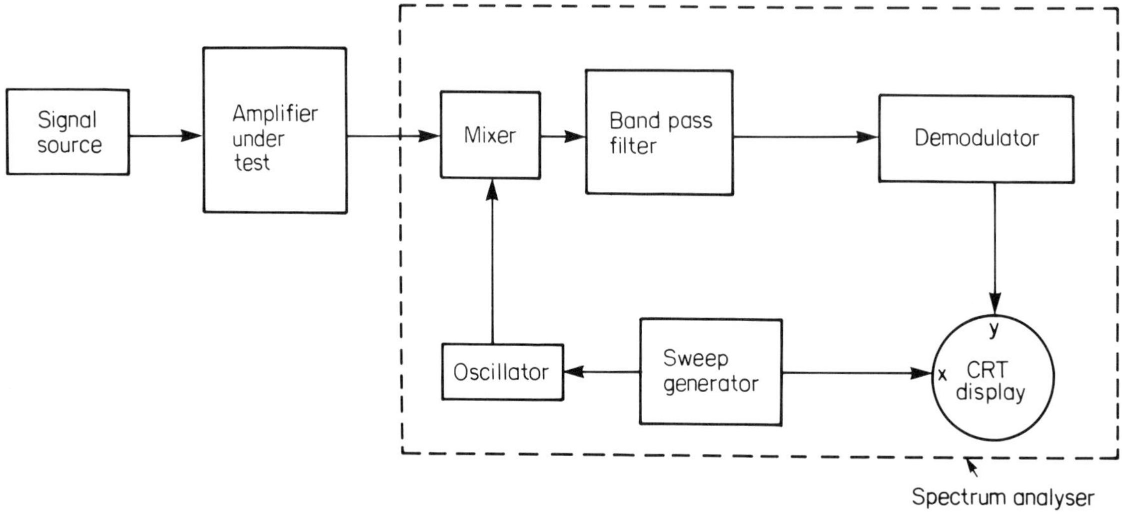

Figure 47.22 Block diagram of a spectrum analyser

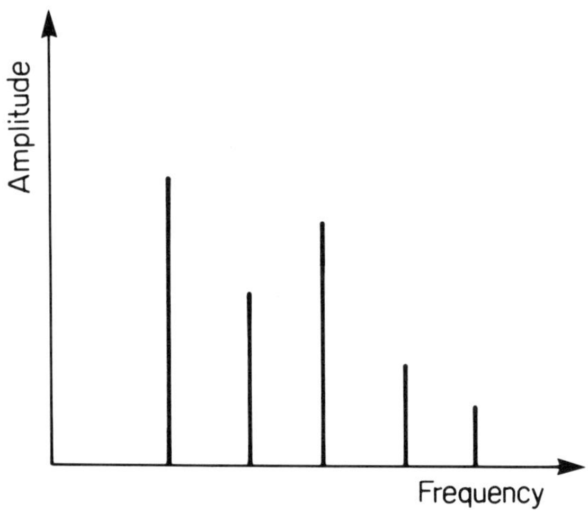

Figure 47.23 Screen display of a spectrum analyser

between two periods which are usually determined by external signals, such as switches on the instrument.

47.5.1.2 Frequency measurement

This measures the frequency of the input signal, which is connected to the signal 1 terminal of the instrument. The control gate to the counter module is kept open for a known period of time (a fixed number of internal time base cycles) and the number of input signal pulses is counted. The frequency of the signal is given by

$$f_s = \frac{\text{counter reading}}{\text{time}} \tag{47.8}$$

47.5.1.3 Period and time measurements

A period is the time between two identical points on the input waveform. Now the roles of the signal input and internal time base are reversed. The gate to the counter module is controlled by the input signal and it is kept open for one cycle of the signal. During this time the number of pulses (cycles) from the internal oscillator are counted. The period is given by

$$T_s = \frac{\text{counter reading}}{\text{frequency of time base}} \tag{47.9}$$

It is also possible to use this technique to measure the time between any two points on the input signal, such as its pulse width, by operating the on-off control to the counter module from these points.

47.5.1.4 Ratio measurement

This is used to find the ratio between two input signals. The slower of the two signals is connected to the signal 2 terminal and it replaces the internal time base. The instrument now measures the number of input cycles of signal 1 for a single cycle of signal 2, and this gives the ratio between the two signals.

47.5.1.5 Averaging measurements

This measurement mode is used with the frequency, period or ratio measurements. By using the programmable dividers the counter gate is kept open for several cycles of the control waveform and the average over one cycle is found.

47.5.1.6 High frequency measurement

For measurements of very high frequency, in the gigaHertz range, it is usual to divide down the input frequency by using prescalers before counting. This increases the measurement time, and it is also not possible to count a number which is smaller than the divider ratio. An alternative technique is to use a heterodyne counter where the input frequency is mixed with another internal frequency and the difference frequency is selected and measured.

47.5.2 Counter errors

Errors arise in the use of the counter due to three main causes.

(1) Time base error. This is caused by a change in the frequency of the internal oscillator. It affects all measurements of frequency and period. The error is constant irrespective of the frequency

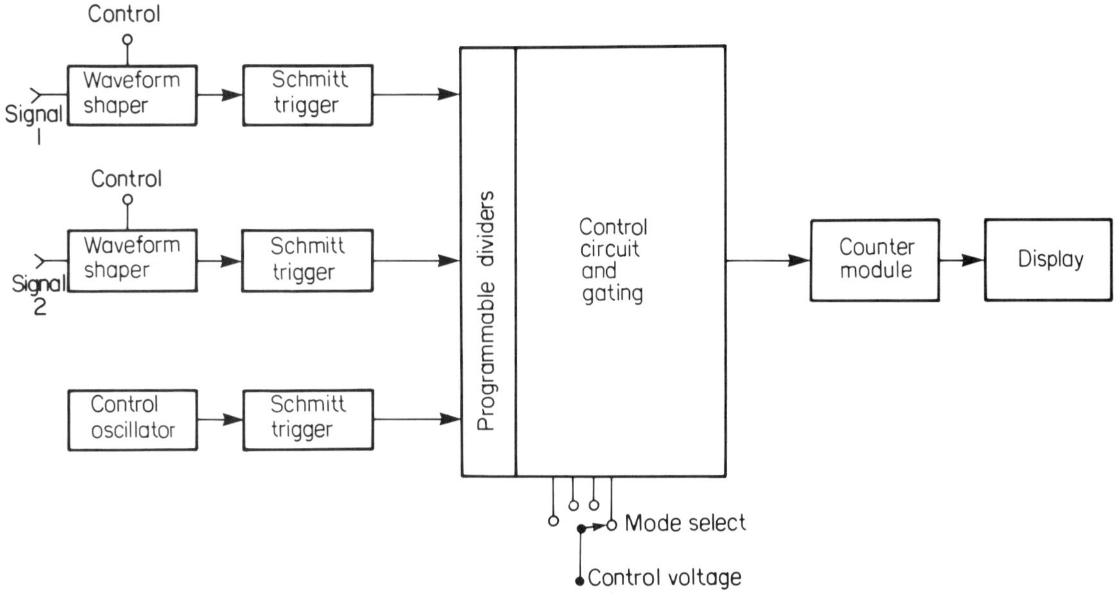

Figure 47.24 Block diagram of a counter

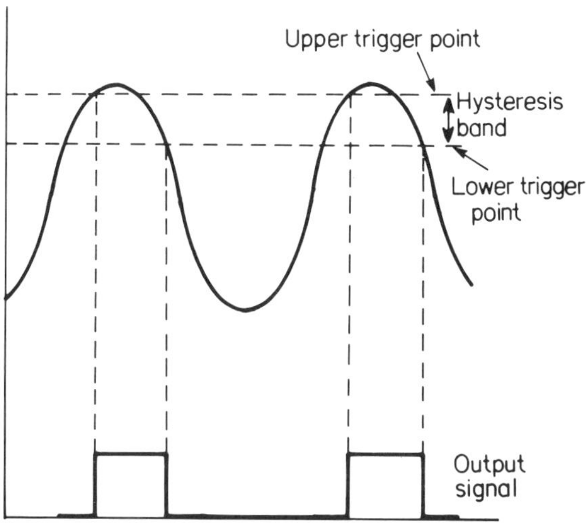

Figure 47.25 Operation of a waveform shaper and Schmitt trigger

being measured, for example a 0.001% error will occur whether 1 kHz or 1 MHz is being measured. There are several causes for the time base error.

(*a*) Initial error due to faulty setting up during calibration. Counters are most frequently calibrated by comparing them with a standard frequency broadcast.

(*b*) Short term stability error. These result in momentary frequency variations caused by shock, vibration, voltage transients, etc. The effect of this error can be minimised by taking measurements over a long time and using the averaging measurement mode described in Section 47.5.1.

(*c*) Long term stability error. This is due to the drift in the crystal oscillator frequency over a long period of time, and is referred to as its ageing rate. As a crystal is temperature cycled, and kept oscillating, its frequency gradually increases with time. The frequency will eventually settle down to a low rate of change, but if the crystal is allowed to cool and start again it will show a jump in frequency of several Hertz from its previous value. In many instruments the crystal is kept at a stable temperature inside a temperature controlled oven. Back up rechargeable batteries are used to keep the oven going if the main power is turned off.

(2) Gating error. This occurs when measuring frequency or period and is illustrated in *Figure 47.26*, where, depending on the phase of the input signal relative to the gating waveform, the counter reading can vary by ± 1. The error is inversely proportional to the frequency being measured, and at low frequencies it can be very large. It is therefore better to take period measurements at low frequencies. Generally if f_i is the frequency of the internal time base then for signal frequencies lower than $(f_i)^{1/2}$ period measurement should be used, and for input frequencies above this value frequency measurements should be used. The gating error is also reduced by using the averaging measurement mode.

shown in *Figures 47.27*, *47.28* and *47.29*. The input signal must

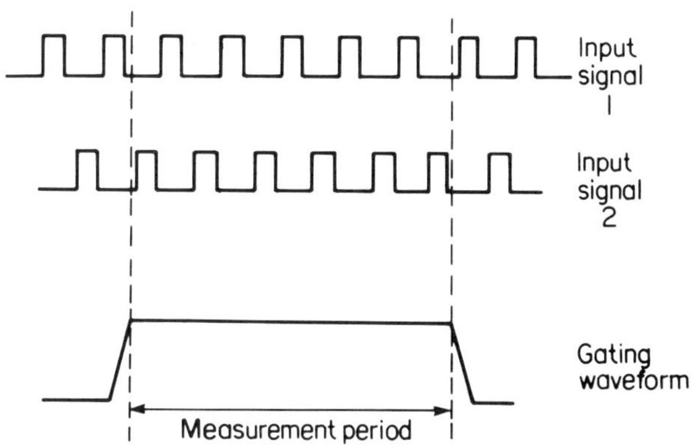

Figure 47.26 Illustration of gating error

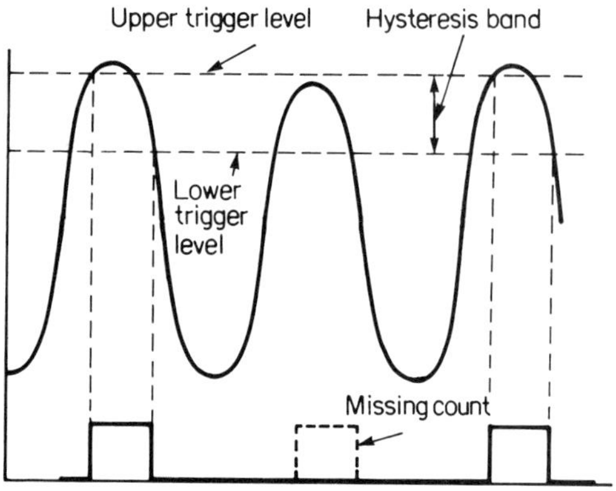

Figure 47.27 Trigger error giving low count

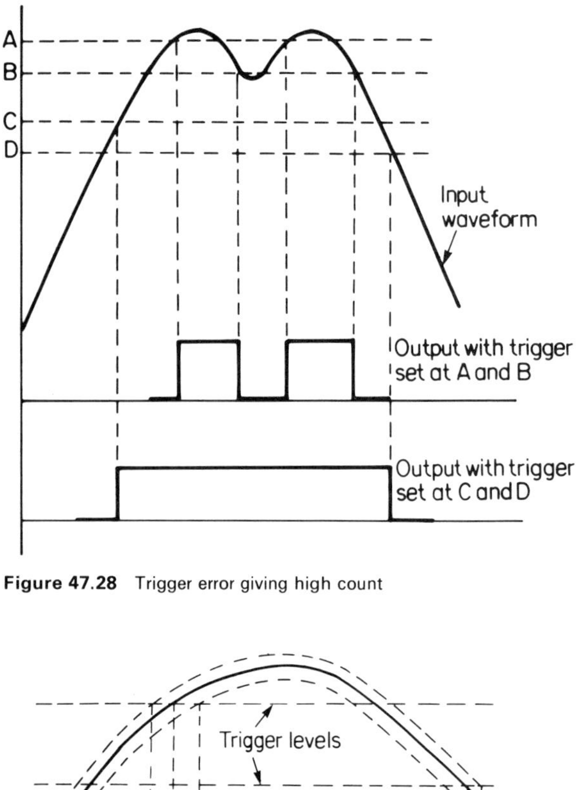

Figure 47.28 Trigger error giving high count

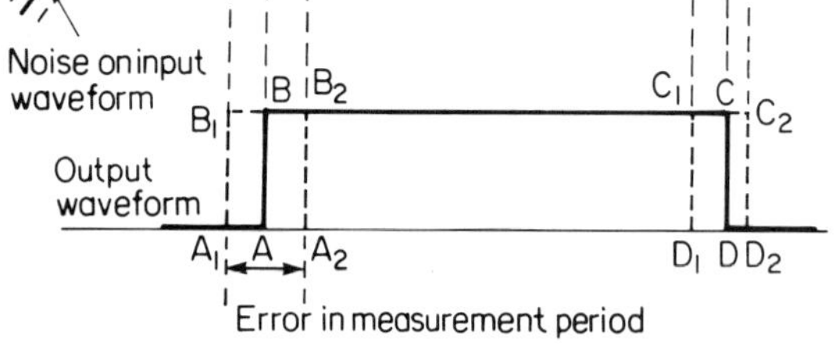

Figure 47.29 Trigger error giving uncertain period of measurement

cross both the trigger level lines of the Schmitt trigger in order to register on the counter. Variation in amplitude, as in *Figure 47.27*, can cause counts to be missed. This can be overcome by amplifying the input signal or by narrowing the hysteresis band and lowering the trigger levels.

Figure 47.28 illustrates the effect of harmonics on the signal waveform which can result in a counter reading which is too high. This can also occur due to rings or transients on the waveform. It can be avoided by moving the trigger levels so that they do not cross any of the distorted parts of the wave.

Noise on shallow input signal waveforms can cause jitter in the measurement period, and false readings, as in *Figure 47.29*. Its effect can be reduced by making the signal go quickly through the hysteresis band, by sharpening up the waveform edges, and by narrowing the hysteresis band.

47.6 Cathode ray oscilloscopes

A cathode ray oscilloscope is an electronic equipment designed to display two-dimensional information on the fluorescent screen of a cathode ray tube in a non-pictorial form. If brightness variations are employed, then to a limited extent, it can also present three-dimensional information. The horizontal axis usually represents time and the vertical axis voltage, but the latter may represent one of many quantities such as loudness or magnetic field strength.

The cost of this type of equipment varies considerably and it follows that the range of functions, facilities, performances and techniques employed is also very wide. The high degree of flexibility of the cathode ray tube and the many versions of equipment currently available make it difficult to decide where the boundary line lies between oscilloscopes and various other specialised instruments employing cathode ray tubes as visual display devices. In this section the emphasis is on the medium-priced conventional instrument. Some information is however provided on the more expensive and specialised instruments. The cathode ray tube itself is described in detail in Chapter 34.

47.6.1 Basic oscilloscope

A block schematic diagram of a typical instrument is given in *Figure 47.30*. The voltage waveform to be examined is applied to the input. A capacitor, which may be shortcircuited by means of a switch, is located in the high-potential input connection. If the waveform is to be examined in its entirety, then the switch should be set to the d.c. position. If the signal contains d.c. and a.c. components, but only the latter is to be examined, then the switch should be set to the a.c. position.

A variable attenuator *ATT* now follows, the function of which is to enable the required magnitude of the vertical component of the display on the oscilloscope screen to be achieved, with a wide range of possible input amplitudes.

The attenuator is followed by the vertical deflection amplifier *Y*, the function of which is to deliver a signal of tens or hundreds of volts peak-to-peak to the *Y* or vertical-deflection plates of the cathode ray tube. Normally a positive rate of change of voltage at the input causes the fluorescent spot on the face of the tube to rise. The *Y* shift control adjusts the position of the reproduced pattern to the required location in the vertical plane.

The horizontal-deflection amplifier *X* usually delivers to the *X* plates a waveform having a constant rate of change of voltage during its active period, causing the spot to move from left to right. The *X* shift control adjusts the position of the reproduced pattern to the required location in the horizontal plane.

The *X* amplifier normally derives its input from the sweep circuit. This is triggered to commence its active period of operation during which time the spot on the cathode ray screen

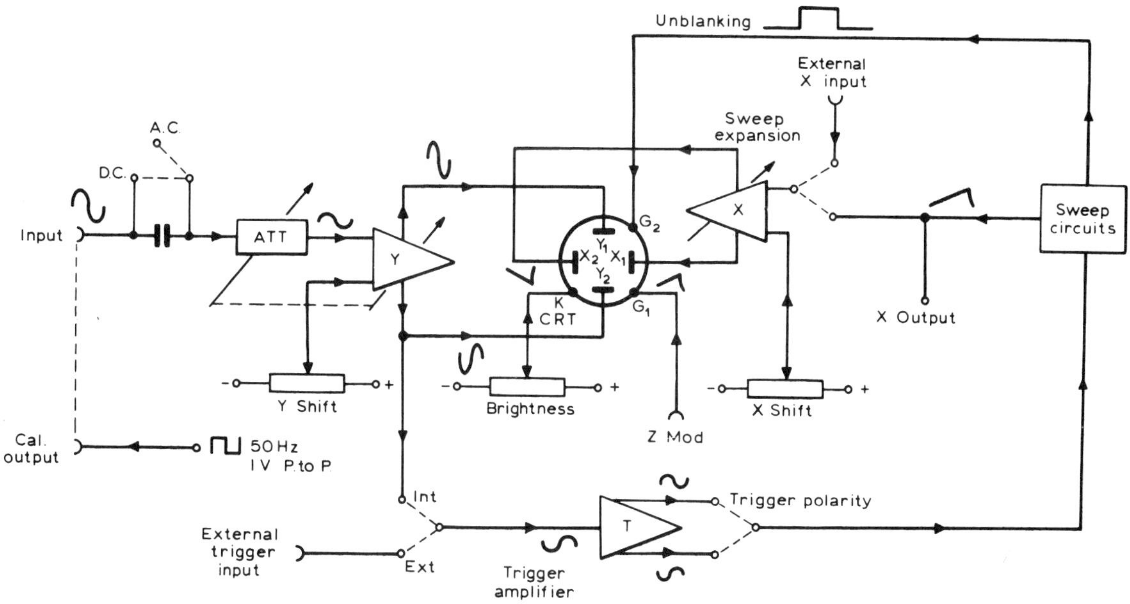

Figure 47.30 Schematic diagram of a typical oscilloscope

travels from left to right. The spot velocity is governed by the setting of the velocity control(s) on the instrument.

An auxiliary output from the sweep circuit is usually available to operate other equipment such as varactors in frequency-sweeping generators, thus allowing response/frequency characteristics of circuits to be displayed.

It will be seen that triggering may be effected indirectly from the normal input signal via the *Y* amplifier or from some external source. When experimenting upon or adjusting a piece of equipment under examination it is usually more convenient to trigger the oscilloscope directly from the signal generator in use. Some signal generators have an auxiliary output circuit especially for this purpose, from which a signal is available at a fixed amplitude. This minimises the need for adjustments of the sweep circuits.

Figure 47.30 reveals three further inputs to the cathode ray tube, at the cathode *K*, and the grids G_1 and G_2. The potential difference between *K* and G_1 may be varied continuously by means of the brightness control and by means of an external a.c. signal via the *Z* modulation socket. The beam is deflected to prevent it reaching the screen until G_2 receives an unblanking signal which coincides with the active left-to-right movement of the beam.

47.6.2 Oscilloscope amplifiers

The main features of oscilloscope amplifiers which need to be considered are:

(1) Gain.
(2) Maximum amplitude of output signal.
(3) Bandwidth.
(4) Input impedance.
(5) Balanced operation.

In the case of the *Y* deflection amplifier, the magnitude of the input impedance is kept as high as possible in order to minimise loading of the circuit under test.

The required magnitudes of the output signals from the deflection amplifiers vary from tens of volts to hundreds of volts peak-to-peak. The difference between the *X* and *Y* deflection potentials required for a given cathode ray tube is due to the following causes. The *X* deflection plates are nearer to the screen than the *Y* deflection plates and therefore must bend the beam through a wider angle. In many oscilloscopes provision is made for a longer deflection in the horizontal direction than in the vertical direction, typically in the ratio 4 : 3. Due to this, the two *X* deflection plates must be further apart than the two *Y* deflection plates which further reduces their deflection sensitivity proportionally.

A moderately priced oscilloscope might have its sensitivity quoted as 50 mV/cm. (More strictly, this is its *inverse* sensitivity.) Instruments having high sensitivity and using tubes of low deflection sensitivities may use amplifiers having gains up to 5000, at which value stabilisation becomes a major problem. In this connection one or more of the following techniques may be employed:

(1) Use of stabilised supply potentials.
(2) Use of a chopper/stabilised amplifier in a feedback network.
(3) Use of balanced stages throughout.

Although the potential required for full screen horizontal deflection is always larger than that which is required for vertical deflection, the amplitude of the available amplifier input potential is normally greater.

In the case of oscilloscopes which are specifically designed for the production of Lissajous figures (Section 47.6.8) the *X* amplifier gain should be higher than the *Y* amplifier gain by a factor which is inversely proportional to the relative deflection sensitivities of the cathode ray tube. The number of inverting stages and the manner of the connection to the *X* plates should be such that, in this particular type of oscilloscope, the application of a positive potential to the *X* input causes the spot to move from left to right.

An economically designed amplifier will have its maximum amplitude of output signal and its bandwidth determined almost entirely in the final stage, with earlier stages much more conservatively rated, i.e. these earlier stages will have greater overload capacity and greater bandwidth.

The relative bandwidths of the *X* and *Y* deflection amplifiers need consideration. If the instrument is to be used mainly for Lissajous figures and for the examination of unmodulated waveforms then there is little logic in making the bandwidths different. Most manufacturers, however, take the view that when the *Y* deflection circuit is operating near to its maximum frequency, then the instrument is being used either as a peak-to-peak voltmeter, or is being used for viewing only *modulated* waveforms, so that some economy may be effected by providing an *X* amplifier with a lower bandwidth.

The stray capacitances in the circuit however, particularly those of the active devices, effectively shunt the load resistors and thus limit the operation of the amplifiers at the high frequency end of the spectrum. The effective bandwidth of the amplifiers may be extended by the use of one or more of the following techniques:

(1) Stages may have circuit forms which alternate between voltage-amplifier and current-amplifier, e.g. between common-emitter and common-collector types as is shown in *Figure 47.31*.

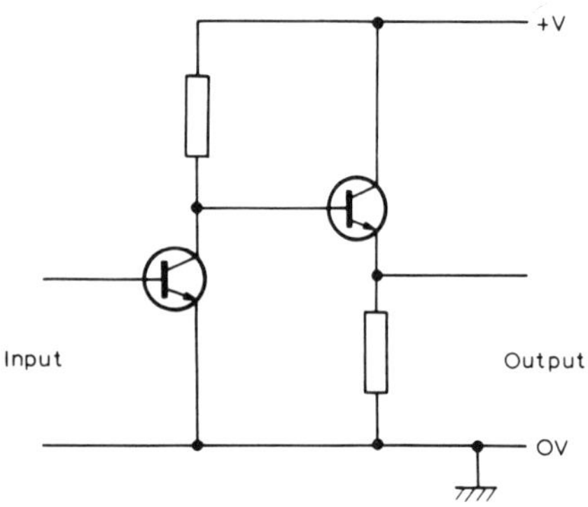

Figure 47.31 A voltage-amplifier current-amplifier combination

(2) A correction inductor may be added to the circuit, its effects offsetting partially the shunting effects of the stray capacitancies. An inductor may be connected in series with the load, this being known as shunt correction. Alternatively it may be connected between the output electrode of one active device and the input electrode of the succeeding active device; this is known as series correction.
(3) Emitter (or source), correction circuits may be employed.
(4) Distributed amplifiers may be employed. (These are normally used at frequencies exceeding 20 MHz.)

Some subsidiary (but nevertheless very important) aspects of oscilloscope amplifiers that need consideration are:

(1) Gain control.
(2) Nature of input circuit, balanced or unbalanced. (If the former arrangement is used then the common-mode rejection ratio is important.)
(3) Protection against damage by input signal of excessive amplitude.
(4) Signal delay to permit examination of the leading edge of a pulse.

Y deflection amplifiers are normally provided with continuously variable gain controls which cover the ranges that lie between the discrete steps that are provided by their associated switch controls. The maximum range is usually 1:2.5 and the control may be either a preset or an operational one.

A common requirement is examination of the leading edge of a pulse which is triggering the oscilloscope. To this end it is necessary that a delay circuit be introduced into the *Y* deflection circuit after the point at which a connection is made to the trigger circuit. The characteristic impedance of a typical balanced delay line is of the order of 200 Ω. For an oscilloscope having a bandwidth of the order of 50 MHz this value of impedance may be suitable for interposition between the *Y* deflection amplifier and the *Y* plates, allowing a signal of high amplitude to be fed to the trigger circuit. However, in an instrument of lower bandwidth, most load impedances have higher values for reasons of current economy and hence the delay line may be located at an earlier stage. The *Y* deflection amplifier is thus split into two sections, the delay line being operated at a lower level of amplitude. This necessitates an increase of gain in the trigger circuit. Schematics for these two conditions are given in *Figure 47.32*.

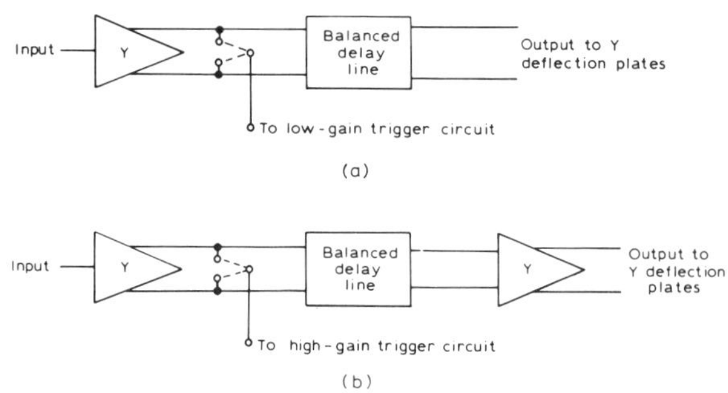

Figure 47.32 Locations for delay lines in: (a) a wide-band oscilloscope and (b) narrow-band oscilloscope

47.6.3 Attenuators

The main requirements of an attenuator for inclusion in a deflection circuit of an oscilloscope are:

(1) The attenuation must be accurate at all frequencies within the specified band.
(2) The magnitude of the input impedance should be high.
(3) The impedance of the attenuator–amplifier combination should not alter with any change of overall sensitivity. (This is to facilitate the design of probe units and also to maintain a constant loading upon the circuit under test.)

The attenuations, expressed as voltage ratios, are normally in the sequence 1, 2, 5, 10, 20, 50, 100, 200, 500, etc., as far as is necessary. The nine values quoted are adequate for a simple oscilloscope, but further values are necessary for sensitive instruments. In the latter case, variation of overall sensitivity is sometimes achieved by a combination of attenuation adjustment and gain adjustment of the following amplifier, the latter being used at maximum gain only when all attenuation has been switched out. There may be a reduction of bandwidth at maximum sensitivity.

The overall attenuation required is normally provided by the cascade connection of a number of attenuating sections in separately screened compartments. The nine values of attenuation are usually provided by a selection of 1, 2 or 5 in one screened compartment, cascaded with a selection of 1, 10 or 100 in another compartment. An example of a circuit giving unity or 10:1 attenuation is illustrated in *Figure 47.33*(*a*).

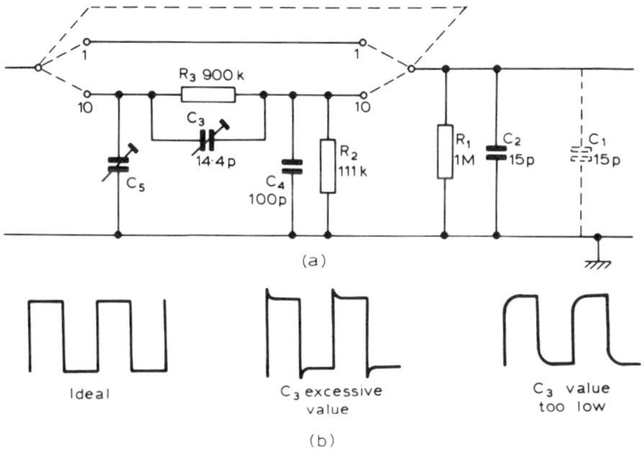

Figure 47.33 Attenuators (a) basic attenuator circuit and (b) adjustment of high frequency response

The input impedance of the amplifier is represented by the dotted C_1. Since its value is likely to be somewhat unstable it may be shunted by a capacitor C_2. The input impedance of the oscilloscope with all attenuators switched out of circuit is thus 1 MΩ shunted by 30 pF, and all these values must be maintained when the attenuators are switched into circuit. The required attenuation at low frequencies is given by

$$\frac{(R_1 \| R_2)+R_3}{R_1 \| R_2}=10 \quad (\| \text{ signifies parallel connection}) \qquad (48.10)$$

While $(R_1 \| R_2)+R_3=1$ MΩ as required to maintain constant input resistance. To ensure that this attenuation is maintained over the specified frequency band it is necessary that the time-constants of the series and parallel combinations should be identical. Since $(C_1+C_2)/9$ would produce the inconveniently low value of 3.3 pF for the capacitor C_3, a capacitor $C_4=100$ pF is added, so now the equality of time-constants can be realised by $C_3R_3=(C_1+C_2+C_4)(R_1 \| R_2)$. This produces $C_3=14.4$ pF. The value of input capacitance which has been considered so far is 14.4 pF in series with 130 pF, i.e. 13.5 pF. To maintain a constant value of input capacitance it is therefore necessary to add $C_5=30-13.5=16.5$ pF. Other attenuating circuits are designed on a similar basis.

In the course of production, and possibly at long-term intervals thereafter, the values of C_3 and C_5 (and similar capacitors in other attenuator circuits), should be checked using a square-wave input to the oscilloscope. C_3 will be adjusted with only this attenuator circuit operative. An excessively high value of C_3 will produce overshoot spikes on the displayed waveform, while too low a value will produce prolonged risetimes, examples being shown in *Figure 47.33(b)*.

Attenuators are expensive items, so that in the case of oscilloscopes having input circuits that are balanced, and hence requiring the more expensive balanced attenuators, it is usual for the manufacturers to put greater reliance on the use of amplifiers of variable gain and thus to economise on the number of circuits in the balanced attenuators.

An example of a complete attenuator circuit is shown in *Figure 47.34*. The components C_{16} and R_{10} are to safeguard the amplifier input circuit against damage from an input signal of excessively high amplitude. The dual input sockets facilitate comparison of two input signals.

47.6.4 Probes

Probes are designed for use with oscilloscopes and other instruments for one or more of the following purposes:

(1) To isolate the circuit under test from the effects (principally shunt capacitance) caused by normal connecting leads or cable to the oscilloscope and the input circuit of the oscilloscope itself.
(2) To increase or decrease the gain of the measuring system.
(3) To provide detection of an amplitude-modulated waveform.

Unless it is known to the contrary, it is unsafe to take a probe provided by one manufacturer and to use it with another manufacturer's instrument. However, some probes are adjustable in this respect.

The length and type of cable which is provided with the probe is critical and should not be changed. Whereas the capacitance of

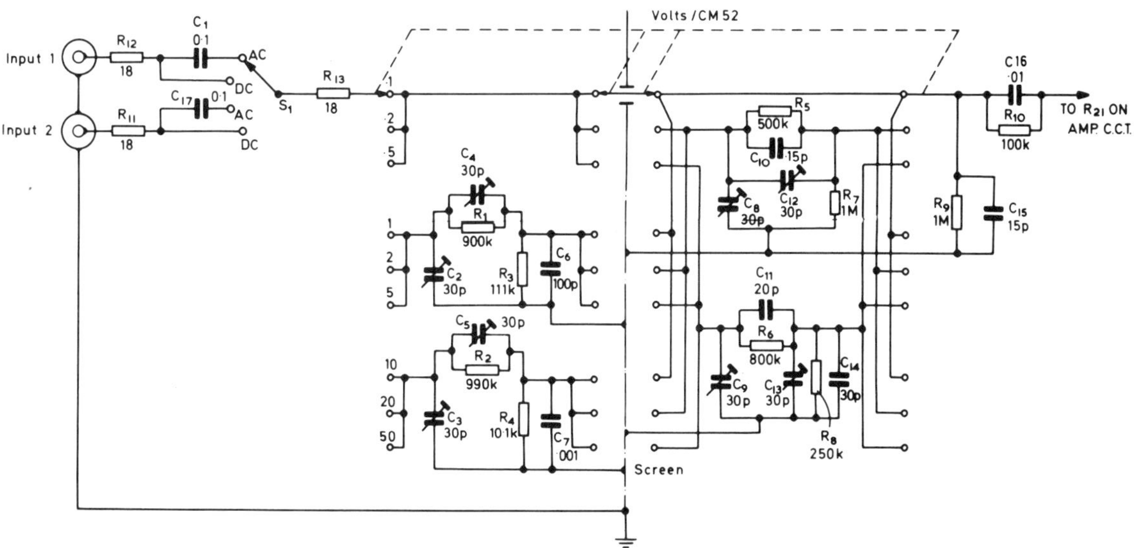

Figure 47.34 Attenuator circuit type H (courtesy Telequipment)

normal co-axial cable is typically 100 pF per metre, special low-capacitance cables employing very thin inner conductors are normally used. These conductors are usually kinked so that the dielectric is mostly air. The inductance of the cable, combined with its capacitance tends to cause ringing on signals having abrupt transitions, as components are reflected backwards and forwards along the cable. Since such a simple circuit cannot be terminated in its characteristic impedance while maintaining the other properties required, a compromise is reached by using resistance wire for the inner conductor, the resistance value being several hundreds of ohms. Careful design of the cable is necessary if microphony is to be avoided.

Some probes contain only passive components, while others contain active components which must be supplied with power along the interconnecting cable from the main instrument. Probes containing only passive components, and some containing active components also, normally attenuate the signal. In the simpler systems the attenuation is a simple round number such as 2 or 10. Since this may lead to operational errors, more complex system are available which operate when the probe plug is inserted into the socket on the oscilloscope. This action may:

(1) Raise the gain of the vertical deflection amplifier; or
(2) change the circuit of an optical display system using light-emitting diodes or fibre-optic devices, thus indicating a change of sensitivity.

The simplest probe, containing only two components, is shown in *Figure 47.35*. Here the time-constant of the probe

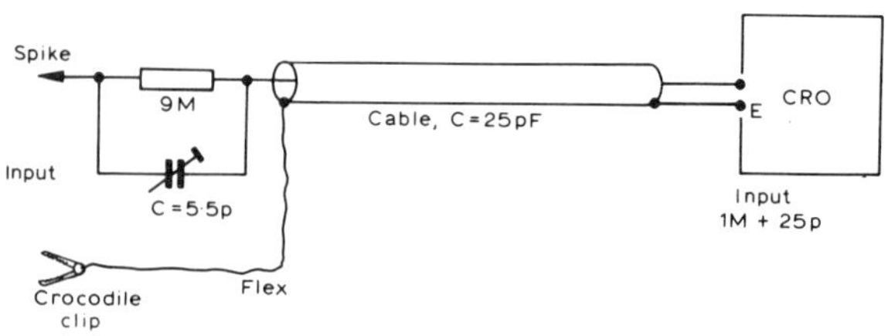

Figure 47.35 Simple probe giving an attenuation of 10

$5.5\times10^{-12}\times9\times10^{6}=50\times10^{-6}$ is made to equal the time-constant of the cable-plus-oscilloscope input circuit, i.e. $1\times10^{6}\times(25+25)\times10^{-12}=50\times10^{-6}$.

The capacitor in the probe is normally adjusted using a square-wave generator as previously explained in the section on attenuators.

Where much loss of deflection sensitivity cannot be tolerated, where the input impedance of the measuring circuit must be very high, or where an abnormally long cable must be used between probe and oscilloscope, it is usual to use active devices in the probe. The circuit forms are either cathode-follower thermionic valves or circuits which are based upon the field-effect transistor.

A cathode-follower circuit and its equivalent circuit are given in *Figure 47.36*. It will be seen that the attenuation of such a probe is given by

$$\left(R+\frac{1}{g_m}\right)\Big/R \tag{47.11}$$

It is often convenient to make $1/g_m=R$ to produce an attenuation of 2. The length of the grid/base of an idealised valve is V_{aa}/μ, so that the equivalent for a cathode-follower circuit is

$$\frac{V_{aa}\left(R+\dfrac{1}{g_m}\right)}{\mu R} \tag{47.12}$$

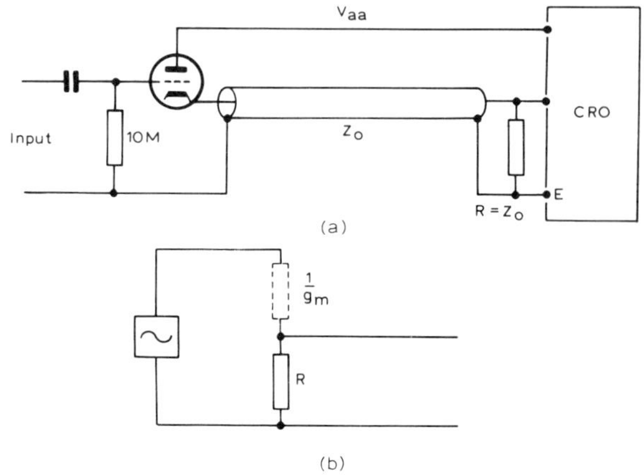

Figure 47.36 A cathode-follower probe: (a) basic circuit; (b) equivalent cathode circuit

In practice the length of the grid-base is, for low-distortion operation, only about half this quantity.

It should be noted that while the value of input impedance is raised to about 10 MΩ shunted by 5 pF, operation at frequencies below about 10 Hz is sacrificed.

An alternative scheme using semiconductors is illustrated in *Figure 47.37*. Due to the lower mutual conductance of the field-

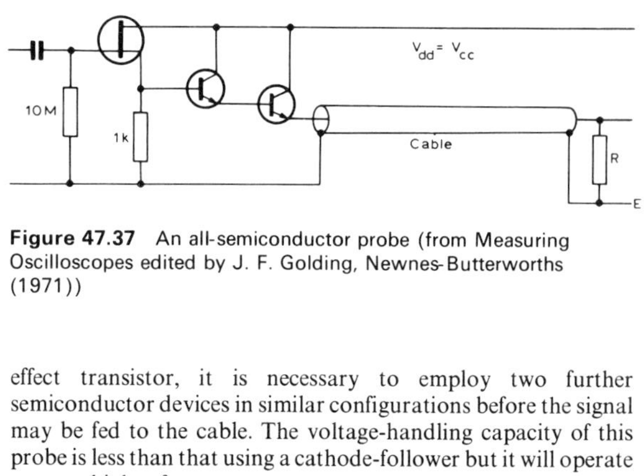

Figure 47.37 An all-semiconductor probe (from Measuring Oscilloscopes edited by J. F. Golding, Newnes-Butterworths (1971))

effect transistor, it is necessary to employ two further semiconductor devices in similar configurations before the signal may be fed to the cable. The voltage-handling capacity of this probe is less than that using a cathode-follower but it will operate up to a higher frequency.

Current measuring probes are also available. These are basically current transformers. The current carrying wire acts as a primary, a split ferrite core has jaws which can be opened to clip over the wire and then closed. A winding on the core acts as the secondary winding. Electrostatic and electromagnetic shielding are necessary. Such probes operate satisfactorily up to about 50 MHz, but performance is less satisfactory below 1 kHz, although frequency-compensated amplifiers may be supplied with the probe.

47.6.5 Multiple display facilities

Most multiple display facilities are provided by oscilloscopes specially designed for the purpose; alternatively they are provided by means of additional units used in conjunction with conventional oscilloscopes. In either case the screen size should be large and the aspect ratio preferably 1 : 1 rather than the 4 : 3 ratio commonly encountered.

Each channel must be provided with its own sensitivity control(s), its own Y shift control, and its own input switch. Some systems have separate horizontal sweep and shift, brightness and focus controls for the two channels, while in some cases they are common.

Multiple displays usually provide dual display facilities, sometimes with the additional facility of displaying either the sum or the difference of two quantities along a single trace. Quadruple display instruments are also available. These are particularly useful for teaching and demonstration purposes, but they are not usually precision instruments. They normally employ very large cathode ray tubes of types designed for television or radar display purposes, and as such they are magnetically deflected, so that the bandwidth of the Y deflection circuits is commonly limited to about 8 kHz. Instruments employing long after-glow cathode ray tubes are available having sweep circuits which operate with repetition frequencies as low as 2 cycles/minute. Some 4-channel oscilloscopes can be switched to provide 2-channel or 1-channel operation when required.

The multiplexing facility may be provided either by the use of a special type of cathode ray tube or alternatively by time-division-multiplex electronic circuitry. Several sub-divisions of these two techniques are available.

Where a single-beam cathode ray tube is used for multiple display purposes, alternative time-division-multiplex sampling systems are available. Some oscilloscopes use one technique, some the other. Some instruments provide facilities for the use of both techniques, the choice being in certain cases left to the user and in other cases being governed by the sweep adjustment control.

In one case the switching action in the signal circuits is synchronised with the horizontal sweep, so that each sweep of the spot is devoted entirely to one signal, the various signals being written on to the screen in sequence. This is known as synchronous-mode operation, or where only two channels are provided, as alternate-mode operation. In the other case the switching action is unsynchronised, occurring usually at about 100 kHz. A typical schematic diagram is shown in *Figure 47.38*.

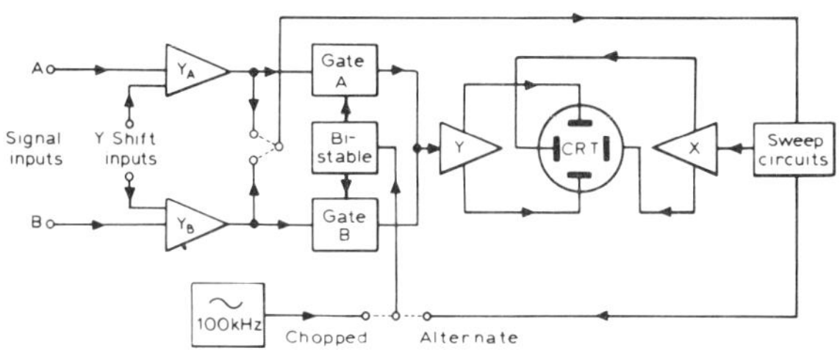

Figure 47.38 A dual-trace oscilloscope

In this circuit the gates open alternately under the control of the bistable circuit, so that the signals are displayed in turn. The bistable circuit may be triggered either from the internal sweep circuit or from the 100 kHz internal generator.

Consider as an example two waveforms which are to be examined, a sinusoid being applied to input A and a triangular waveform to input B. To the former, a positive Y shift potential is added, producing V_A, and to the latter a negative Y shift potential, producing V_B, both being illustrated in *Figure 47.39*. The horizontal sweep waveform is shown as V_X. If alternate mode operation is employed then the input to the common Y amplifier will appear as V_{Y_1}. If however chopped operation is employed, then using the chopping waveform V_{CH} the input to the common Y amplifier will be as shown at V_{Y_2}. (The vertical transitions of this waveform are so rapid that there is a negligible brightening of the screen in between the two waveform patterns.) The gaps which appear in both waveforms in any one horizontal sweep are filled by successive sweeps, so that irrespective of the switching employed (alternate or chopped), the final result appears as shown at the bottom of *Figure 47.39*.

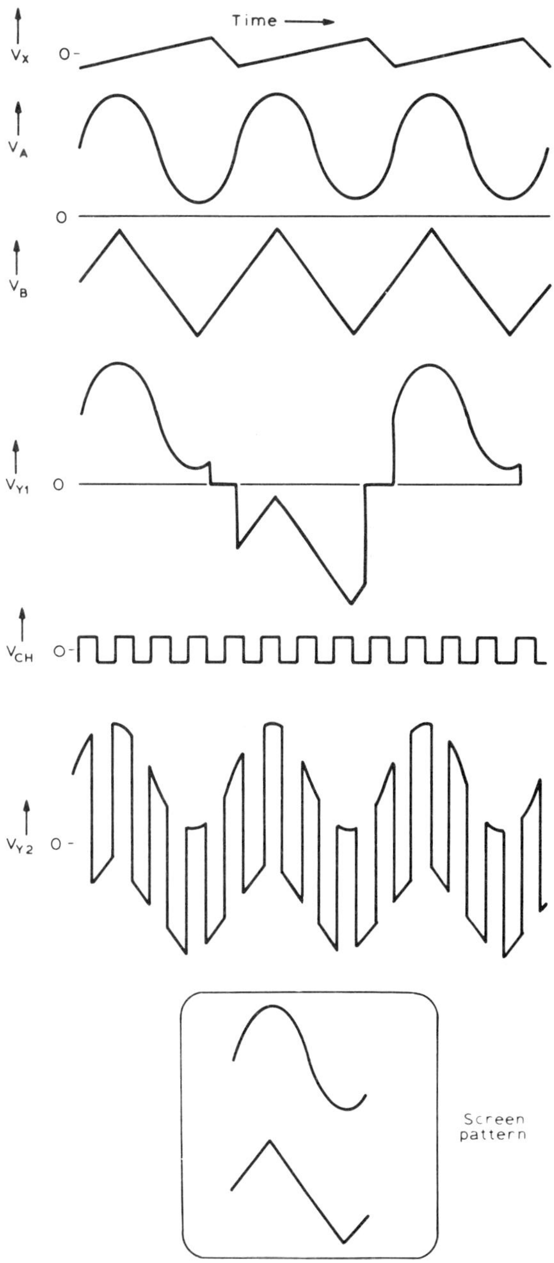

Figure 47.39 Typical waveforms and screen pattern for a single beam dual-channel oscilloscope using either alternate-mode or chopped-mode operation

47.6.6 Sampling

For the examination of waveforms having repetition frequencies between about 100 MHz and 1 GHz, or having rise-times of the

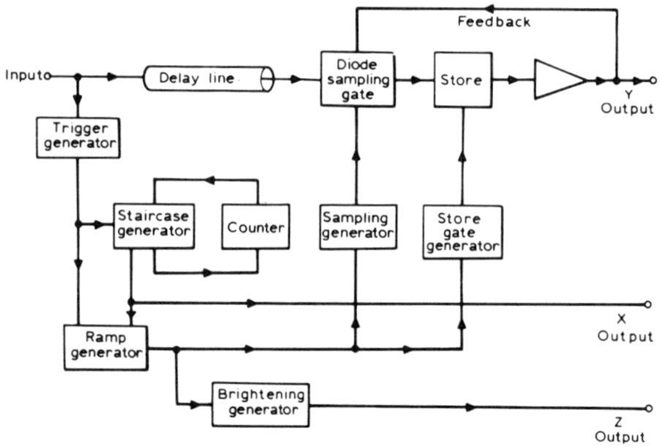

Figure 47.40 A sampling unit

order of nanoseconds it is usual to employ a sampling technique, using either a complete sampling oscilloscope or else a sampling adaptor in conjunction with a conventional oscilloscope. A simplified schematic diagram appears in *Figure 47.40* with waveforms in *Figure 47.41*.

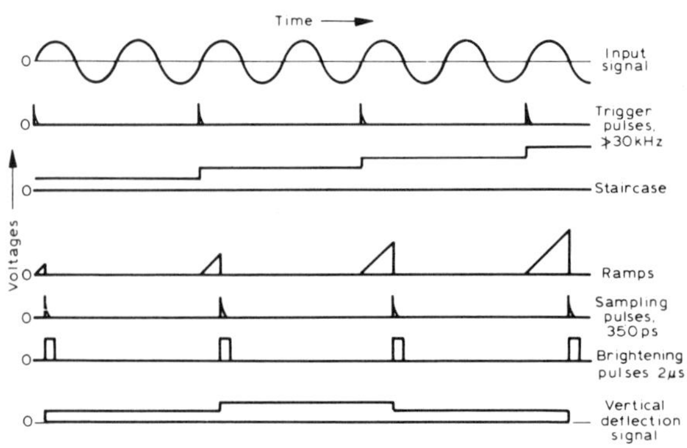

Figure 47.41 Waveforms in a sampling oscilloscope

The input signal is applied to a trigger generator which delivers a series of pulses, normally at a frequency which is of the order of 30 kHz but at an integral submultiple of the input frequency. (This presupposes that the signal frequency exceeds 30 kHz. If however the signal frequency is less than this, then the sampling frequency is made to equal the signal frequency.)

Each trigger pulse initiates one step in the operation of a staircase generator, the output of which has its steps counted to determine the number of dots per cycle of the waveform which is to appear on the oscilloscope screen in any one horizontal sweep. After the required count, say between 200 and 1000, the counter resets the staircase generator, which then commences the generation of a new staircase.

Each trigger pulse also causes the ramp generator to commence the generation of a very fast ramp of fixed velocity. The duration of each ramp is, however, caused to be proportional to the instantaneous magnitude of the staircase, so that during the staircase cycle the ramps are of progressively increased durations. The *termination* of each ramp now causes the sampling generator to produce a sampling pulse, typically of 350 ps, these sampling pulses occurring progressively later and later in individual periods of the test waveform.

The staircase waveform also causes the spot of the oscilloscope to move horizontally across the screen in a series of rapid movements, and at the same time the spot is brightened by a pulse (typically of 2 μs duration) which coincides in time with the sampling pulse.

The signal now passes via a delay line (producing typically 50 ns delay) to the sampling gate. The function of this delay line is to facilitate the examination of the leading edge of a pulse as explained previously.

Samples of the pulse are taken and stored in a capacitor store, access to which is controlled by the store-gate generator. From the store, an output is taken via an amplifier and feedback path of unity gain, to reverse-bias the diodes in the sampling gate so that during each sampling period the component which is passed into the store is proportional only to the *change* of signal amplitude which has taken place since the last sample was taken. The output from the amplifier is then used to provide vertical deflection of the oscilloscope beam.

47.6.7 Storage oscilloscope

A non-storage oscilloscope needs a periodic signal of a relatively high frequency in order to display a steady trace. The persistance of the screen can vary from milliseconds to a few seconds, depending on the type of phosphor. Storage oscilloscopes are used to display transient events, and also for signals with very low frequencies in which, on a conventional oscilloscope, the first bit of the trace would begin to fade before the end was finished.

Two storage techniques are used in oscilloscopes, analogue and digital. The first uses a special storage cathode ray tube and the second digitises the analogue signal and stores it within the oscilloscope circuitry.

47.6.7.1 Storage tube

Figure 47.42 shows a schematic of one type of storage tube. The flood guns have no associated deflection plates and they flood the entire area of the target. Their function is to maintain the state of

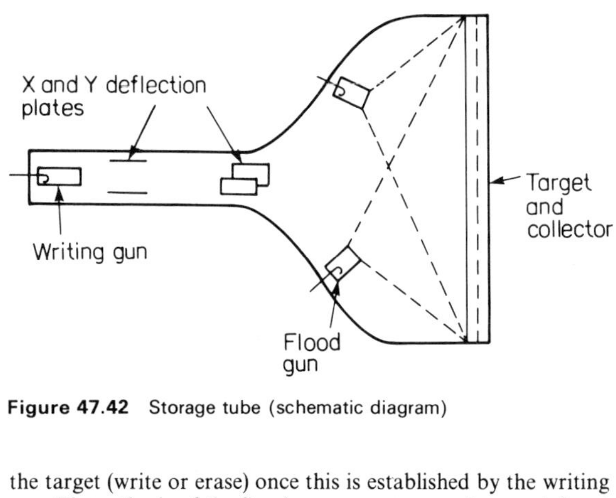

Figure 47.42 Storage tube (schematic diagram)

the target (write or erase) once this is established by the writing gun. The cathode of the flood guns are at ground potential and they are kept continuously on. The cathode of the writing gun is at a high negative voltage and it can be deflected onto any point on the screen which is to be written into.

The screen consists of scattered phosphor particles in which any area can be written into without affecting an adjacent area. This is the target and it is placed on a conductive coated glass face plate which acts as the collector.

When primary electrons from the writing or flood guns strike the target they result in secondary emission. The ratio of the secondary emission current to primary emission current is called the emission ratio. *Figure 47.43* shows a plot of this ratio as the target voltage is chaged. The collector is at a positive potential of about V_2. Point A represents the erased position and point C the written position. Assume that the writing gun is off and the target

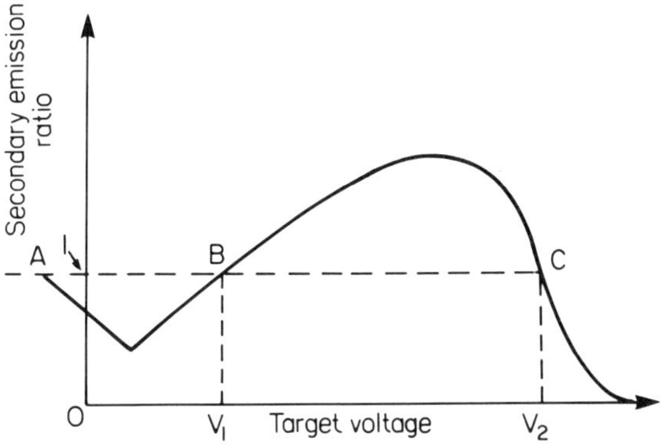

Figure 47.43 Secondary emission due to flood gun

is at point A. The flood guns have insufficient energy to cause the phosphor to move away from A. When the writing gun is switched on it causes a large number of electrons to strike the target and this increases the secondary emissions. From A to B the flood guns oppose the writing gun but beyond B they aid the writing gun in moving the phosphor to point C. The writing gun can now be switched off, the flood guns having sufficient energy to maintain the phosphor in the written position at C.

To erase information stored on a storage tube the collector may be pulsed negative. This causes the secondary emission from the target to be repelled back into the target. The target voltage reduces until it reaches the erased point A after which the collector pulse can be slowly reduced to zero.

47.6.7.2 Digital storage

Digital storage oscilloscopes use electronic circuitry to store the signal, and a conventional cathode ray tube which is periodically refreshed by the stored information. *Figure 47.44* shows the schematic of a digital storage oscilloscope. The input signal is amplified and the resulting waveform is sampled and converted to a digital signal, by the analogue to digital converter, before being stored. A digital to analogue converter changes this stored information back to an analogue signal before it is amplified and fed to the *Y* plates of the cathode ray tube. When the control switch is set to position A the storage circuitry is by passed and the instrument behaves like a conventional oscilloscope.

The analogue to digital converter is usually the limiting component of a digital storage oscilloscope and its speed determines the frequency response of the instrument. The memory capacity is another important parameter and it determines the resolution of the instrument, and the number of dots used to create a waveform. For example if an 8000 bit memory is available and 8 bits are used to represent a word then the instrument has a resolution of 1 in 2^8, or 1 in 256, and it can use up to 1000 dots to create a waveform. If both channels use the same memory for storage then the number of dots available is halved.

The accuracy of a oscilloscope is usually worse than its resolution, since it includes errors due to resolution of the analogue to digital converter and due to non-linearity of the amplifiers. Modern storage oscilloscopes have resolutions of the order of 0.025% and accuracies of 0.25%. Some oscilloscopes have a dot joining feature which uses simple lines to join between the dots on the waveforms.

The digital storage oscilloscope stores information electronically. This information can therefore be manipulated, such as reproduced on an expanded or reduced time base. It can also be output at different speeds, for example to suit a pen recorder. In this instance the output speed should match the inertia of the pen so that lines between dots are smooth.

47.6.8 Lissajous figures

A less familiar use of the oscilloscope is for the formation of Lissajous figures by a process illustrated in *Figure 47.45*. In this case the *Y* plates receive their signals in the normal manner but the *X* plates receive theirs via the external *X* socket. The internal horizontal sweeping generator and the unblanking operation are inoperative. In the example given, it is shown that the *Y* plates are receiving a sinusoid while the *X* plates are receiving a cosinusoid of the same frequency. The two graphs in *Figure 47.45* employ

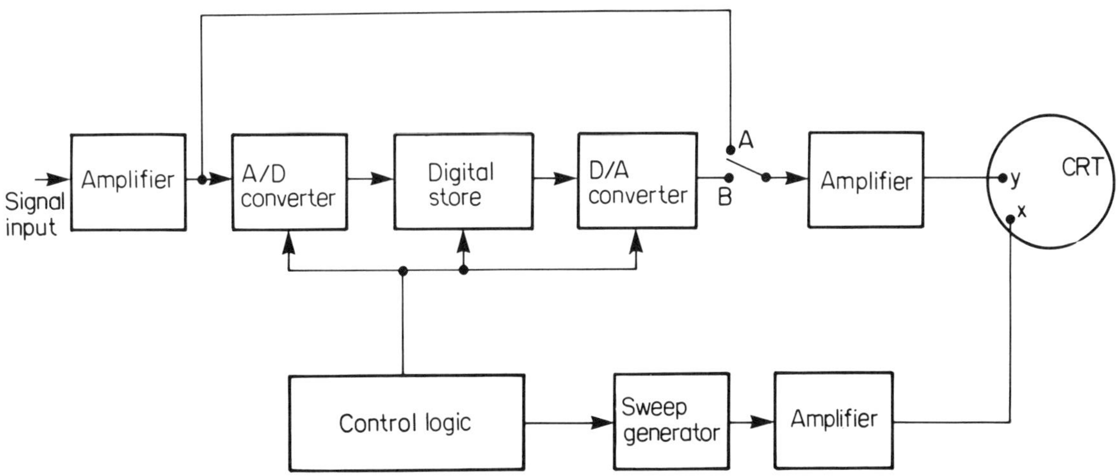

Figure 47.44 Block diagram of a digital storage oscilloscope

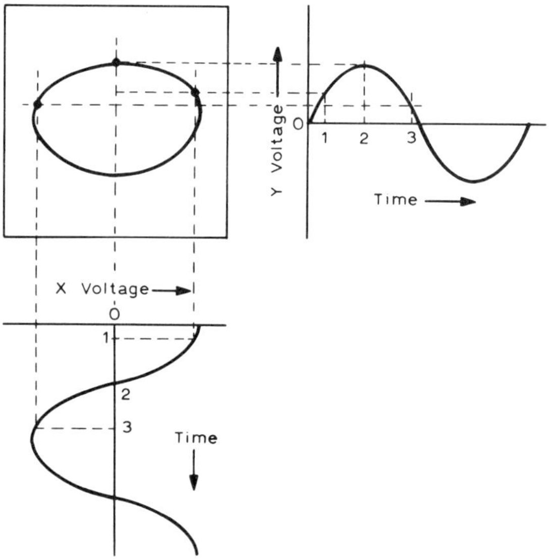

Figure 47.45 Formation of a simple Lissajous figure

similar time scales. A succession of times 1, 2, 3, etc., are similarly selected on both graphs and construction lines drawn to indicate the resulting positions of the spot. The result is a pattern which consists of an ellipse having its axes horizontal and vertical, the spot travelling in an anti-clockwise direction. Careful consideration shows that two *similar* waveforms would produce a straight line with a positive slope. If one waveform be inverted, the straight line has a negative slope. By employing this principle of construction, the pattern which results may be predicted for any two waveforms of any two repetition frequencies, but in practice the technique is normally restricted to simple waveforms having some simple frequency relationship.

The phase relationship between two sinusoidally shaped waveforms of the same frequency may be determined as follows: First one of the waveforms is applied simultaneously to both X and Y inputs. At least one of these circuits should have a gain control which is continuously variable. The gain controls are now adjusted to produce on the screen a straight line with a slope of either $+45°$ or $-45°$. (The 45° slope indicates that the deflection sensitivities are identical. The straightness of the line indicates that the deflection circuits have identical phase shift at this frequency.) The preliminary adjustment having been made, the two signals are now fed *separately* to the horizontal and vertical deflection input sockets. The result is likely to be an ellipse, the major and minor axes Z_{maj} and Z_{min} being shown in *Figure 47.46*. The phase difference is now given by:

$$\phi = 2\tan^{-1}\frac{Z_{min}}{Z_{maj}} \tag{47.13}$$

Lissajous figures are also used for frequency comparison purposes, e.g. for checking the operation of frequency multipliers and dividers and for calibrating multifrequency oscillators from single-frequency sources. (It should be noted that the technique can be used only when the relative frequencies are stable to within about 1 Hz.) Examples where $f_X = 2f_Y$, and $f_X = f_Y/3$, are shown in *Figures 47.47(a)* and *(b)* respectively. For any frequency relationship however, an infinite number of patterns is available depending upon *phase* relationships. Where the frequency relationship is high, e.g. greater than six, it is more convenient to employ a technique which includes Z modulation, the necessary circuit arrangements and resulting screen pattern are shown in

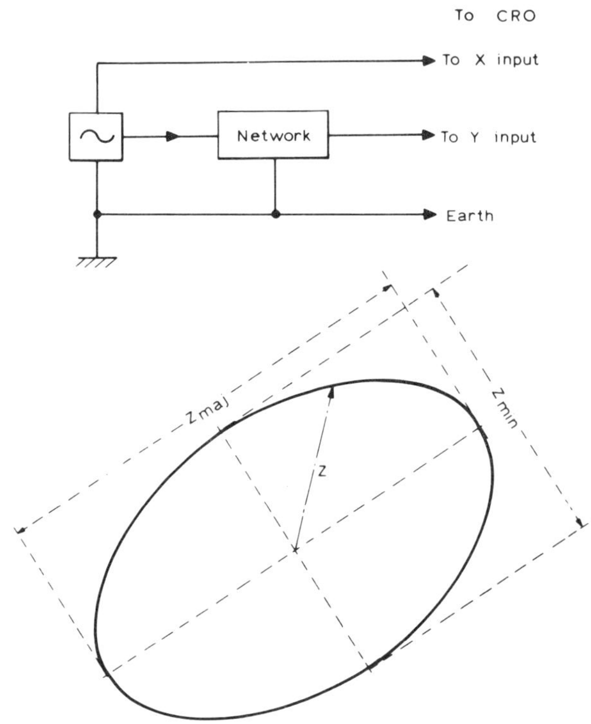

Figure 47.46 Measurement of phase shift

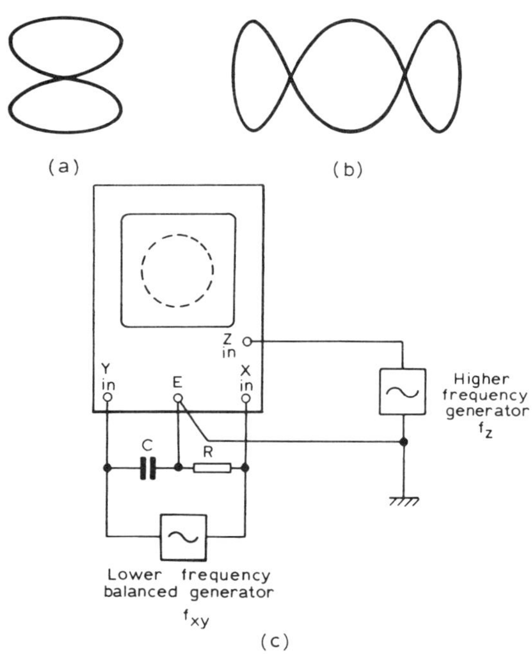

Figure 47.47 Frequency comparison using Lissajous figure: (a) $f_x = 2f_y$; (b) $f_x = f_y/3$; (c) $f_x = 15f_{xy}$

Figure 47.47(c). Here the lower frequency signal is applied from a *balanced* source to a phase-splitting circuit consisting of C and R in series. Assuming that the horizontal and vertical deflection systems have similar sensitivities, then the magnitude of the reactance of C should have approximately the same value as the resistance of R. This will produce an ellipse, the axes of which are horizontal and vertical. Adjustment of X or Y sensitivity or C or R will enable a circle to be produced. The higher frequency signal, having an amplitude of several volts, will cause the pattern to be broken up as shown, the number of breaks equallying f_Z/f_{XY}.

Lissajous figures may be employed for the measurement of time delay, using a simple arrangement such as is shown in *Figure 47.48*. It will be realised that whenever the time delay t_d of the

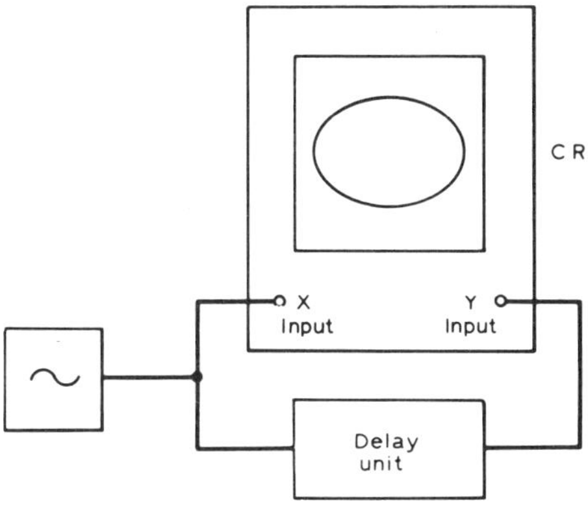

Figure 47.48 Measurement of delay by means of a Lissajous figure

network is equal to an odd number of quarter periods of the source, then the pattern on the screen will be an ellipse, the axes of which are horizontal and vertical. The most convenient procedure to adopt is to set the generator frequency f to a low value where the delay t_d is only a very small fraction of the generator period $1/f$. The screen pattern should then be a sloping straight line. The value of f is now raised until, for the *first* time, the axes are horizontal and vertical, then $t_d = 1/4f$.

A further use of Lissajous figures is for checking the amplitude modulation of a carrier by means of a trapezium figure. Here the modulated high-frequency signal is applied to the Y deflection circuit (because this normally has higher gain and bandwidth) and the modulating signal is applied to the X deflection circuit. An example is shown in *Figure 47.49(a)* where the percentage modulation is given by

$$100\,\frac{a-b}{a+b} \qquad (47.14)$$

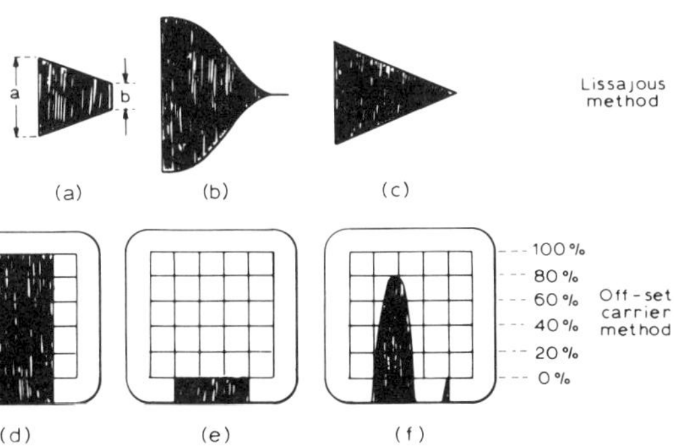

Figure 47.49 Modulation measurements

An example of over modulation, including peak-amplitude limiting and carrier-cutting is shown in *Figure 47.49(b)*. Perfect 100% modulation is illustrated in *Figure 47.49(c)*.

A more accurate method of measuring modulation is illustrated in *Figure 47.49(d)*, (*e*) and (*f*). First, the sweep circuit of the oscilloscope is set to operate normally. Then, in the absence of any signal, the Y shift control of the oscilloscope is set to produce a horizontal line to coincide with the bottom line of the oscilloscope graticule. The signal is now applied to the Y deflection circuit in the normal manner, and while it is *un*modulated, the Y deflection sensitivity (or signal amplitude), is adjusted to produce a pattern which rises by 5 or 10 divisions of the graticule as shown at (*d*). This action calibrates the vertical scale. The Y shift control is now operated to lower the pattern so that its upper extremity coincides with the bottom line of the graticule as shown at (*e*). When modulation is applied, the upper extremity of the pattern indicates the percentage of the modulation, as shown in the example (*f*). It is convenient, but not essential, to synchronise the operation of the sweep circuit with the modulating signal.

47.7 Temperature measurement

A transducer is a device that can sense changes of one physical kind and transpose them systematically into a different physical kind which are compatible with a signal processing system. This section is concerned with the exploitation of thermal changes in practical sensors that can produce proportional alterations in the output conditions of a transducer; the output changes may be variations in voltage, resistance or position. Reference is made to some of the associated metrological techniques and instrumentation used in control systems and data acquisition.

The application of heat energy to any substance will modify the extant kinetic and potential energies associated with the molecular, atomic and electronic structure of the material; these changes will be different for each substance. The useful alterations in the output phenomena of these tranducers which can be measured conveniently, may be classified, very approximately, as follows:

(1) Thermo-electric effects; due to the separation of electric charges between the conducting or semi-conducting surfaces of different materials which are in contact. E.M.F. is the output parameter which can change.
(2) 'Thermo-resistive effects', due to alterations (from a non-quantum point of view) in the mean free path of the conduction electrons in a material. Resistance is the output parameter which can change.
(3) 'Thermo-mechanical effects', due to significant changes in the external dimensions of solids caused by alterations in both the amplitude of atomic vibrations and in the elastic forces holding the molecules together. The resultant change in dimension can be sensed externally.

The macroscopic thermal behaviour of substances is sensed by changes in the mean temperature which occurs due to a combination of the inherent thermal inertia of the material and the changing thermal environment in which the sensor resides. The three thermal responses above are included in many of the sensors and instruments listed in BS 1041 *Code of Temperature Measurement*. In this Code the sensors and instruments are classified according to temperature range as follows:

(1) Specialised thermometers for measurements near absolute zero. One such, based on the magnetic susceptance of certain paramagnetic salts, is suitable for temperatures below 1.5 K; another, employing an acoustic resonant-cavity technique, can be used up to 50 K. The associated

instruments are a mutual inductance bridge for the former, and electronic measurement of length for the latter.
(2) Ge, Si and GaAs *p–n* junction diode and carbon resistor thermometers are used with d.c. potentiometers and Wheatstone bridges for temperatures up to 100 K.
(3) Vapour pressure thermometers are used for standard readings below 5 K and for general measurements up to 370 K.
(4) Thermistor (resistance) and quartz (resonance) methods, with Wheatstone bridge and electronic counter respectively, cover the range 5–550 K.
(5) Electrical resistance sensors (usually of Pt) are used with d.c. bridges throughout the range 14–1337 K.
(6) Thermocouple e.m.f.s are measured by d.c. potentiometer. The method, widely employed in industry, has ranges between 100 and 3000 K with various combinations of metals.
(7) The expansion of mercury in glass or steel capillary tubes is applied to measurements from 230 to 750 K. The range is 80–300 K if the mercury is replaced by toluene.
(8) Radiation and optical thermometers employ thermopiles, photodiodes or photo-multipliers for measurements up to 5000 K, using d.c. potentiometric or optical balance methods. These provide the only practical methods for very high temperatures.

Mercury-in-glass and optical thermometers are indicating instruments only, but the remainder can be made to furnish graphical records. The electrical resistance and thermoelectric thermometers are especially suitable for multi point recording as for heated solid surfaces, points in a mass, or inaccessible places in electrical machines.

The scientific scale of temperature is based upon the degree Kelvin(K), and the unit temperature difference on this scale is, by definition, the same as that for the degree Celsius (°C, previously called degree centigrade). The fundamental fixed point for temperature is the triple point of water, to which is assigned the thermodynamic temperature of 273.16 K exactly. On this scale the temperature of the ice point of water is 273.15 K (or 0°C). The platinum resistance and thermocouple thermometers are the primary standard instruments up to 1336 K (gold point), above which the standard optical pyrometer is used to about 4000 K.

The International Practical Temperature Scale (IPTS-68) is a very close approximation to the fundamental thermodynamic temperature scale. The Practical Scale is used since fixed points on this scale can be realised and disseminated much more readily, and with higher precision, than can those of the thermodynamic scale.

47.8 Thermoelectric devices

In 1822 it was discovered by J Seebeck that if a loop be made by joining together the ends of two conductors (composed of different materials) and a temperature difference is maintained between the two junctions then (due to the thermal e.m.f.) a current will flow around the conducting loop. This 'Seebeck effect' is the dominant phenomenon in the thermocouple although there will be minor contributions due to the Peltier and Thomson effects as described below.

In 1834 J C Peltier demonstrated that, when a source of current is connected by two conductors of the same material to a third conductor of a different material, then the temperature at one end of the third conductor will rise and that of the other end will fall. Useful thermo-electric cooling may be achieved with Peltier cooling units, although large d.c. currents are required due to the inefficiency of this process.

In 1854, W Thomson (Lord Kelvin) showed that, if there is a variation in temperature along the length of a conductor, then an electrical potential difference is produced. The nett Thomson effect is often negligibly small except when very large temperature gradients are encountered.

47.8.1 Thermocouples

A thermocouple consists of two wires of dissimilar material (*Figure 47.50*) which are electrically insulated from each other

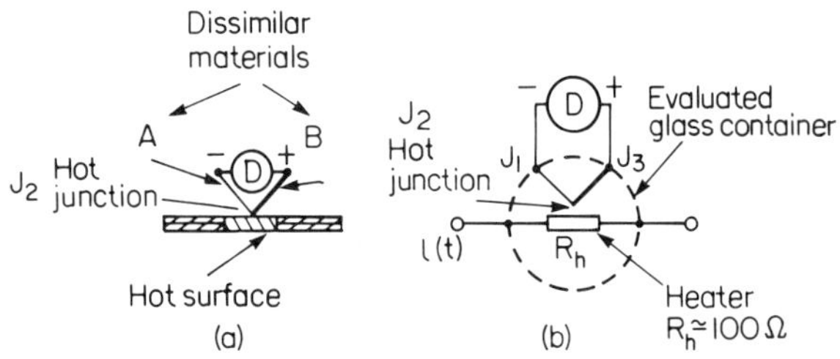

Figure 47.50 Thermocouple arrangements

and connected at one end, J_2, where the changing thermo-e.m.f.'s will occur due to alterations in the temperature at J_2. Provided J_1 and J_3 are at the same temperature, the insertion of a detector D will not affect the resultant e.m.f. The detector could be, for example, a high input-impedance d.c. digital voltmeter with data readout facility (e.g. BCD conversion, IEEE 488 bus etc) or a balanced d.c. potentiometric measurement.

Figure 47.50(a) is a symbolic representation of the wide range of direct-contact thermocouples used in industrial applications, many of which will be electrically grounded or earthed. These thermocouple probes are available in a wide variety of shapes and sizes from 6 mm diameter fast response low mass heads with excellent repeatability, and low mass beryllium copper clip probes for transistor thermal sink tests, to stainless steel needle probes for semi-solids and stainless steel, sheathed, mineral insulated probes for general applications in hostile environments at more elevated temperatures. Exposed junction thermocouples with PTFE or fibreglass shields have a very quick response.

A thermal radiation pyrometer can incorporate a sensor consisting of a sensitive thermopile formed from a group of similar thermocouples connected in series; the thermocouples are grouped together on a small annulus which constitutes the focus area of an optical system.

Figure 47.50(b) represents an instrumentation version of a thermocouple which would be used for a.c./d.c. comparisons of small currents (e.g. 5 mA maximum with a 100 Ω heater) since the device behaves similarly for d.c. or a.c. currents up to 100 MHz for special designs. The mean of the positive and negative d.c. readings must be used. A recent international investigation between the National Physical Laboratory and other national laboratories has shown that the a.c./d.c. comparison uncertainty is about ±5 ppm up to 50 kHz; such tests constitute the present fundamental metrological link for a.c./d.c. transfer measurements. This type of thermocouple has been included in direct reading r.f. moving coil milliammeters and in the secondary windings of ferrite-core current transformers for higher current r.f. measurements. Inevitably the scales follow a square law and will be cramped at one end. The thermal inertia of the heater is very small and it can easily be destroyed by even a disturbance of short duration (e.g. a switching surge), hence a quick-acting protection network (e.g. diode) must be incorporated although care should be taken to ensure that the characteristics of the protection circuits do not influence the required response.

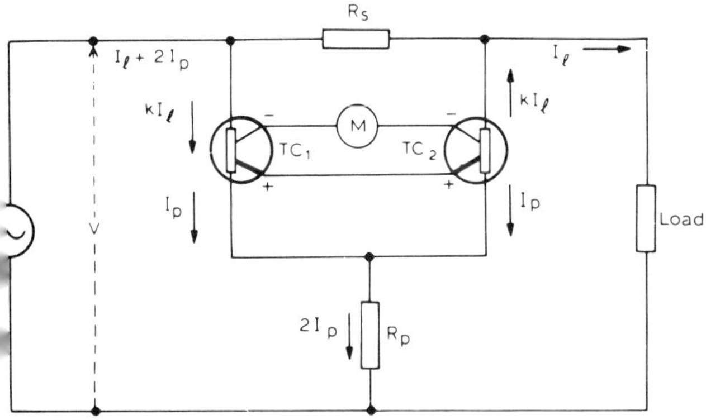

Figure 47.51 A wattmeter circuit using two thermocouples

A network employing two thermocouples has been employed to measure power as shown in *Figure 47.51*. The resistor R_s has a relatively low value of resistance and it corresponds to the shunt in a conventional ammeter. The resistor R_p has a relatively high value of resistance and it corresponds to the series resistor in a conventional voltmeter. Other symbols have the following significance:

V = supply voltage
I_l = load current
kI_l = a fixed proportion of the load current
I_p = half of the current in R_p
$kI_l + I_p$ = current in the heater of thermocouple 1
$kI_l - I_p$ = current in the heater of thermocouple 2
E_1 = e.m.f. produced by thermocouple 1
E_2 = e.m.f. produced by thermocouple 2
θ = angle of deflection of moving coil meter M

The thermocouples being connected in opposition

$$\theta \propto E_1 - E_2 \qquad (47.15)$$

therefore

$$\theta \propto (I_l + I_p)^2 - (kI_l - I_p)^2 \qquad (47.16)$$

and since $I_p \propto$ supply voltage V

$$\theta \propto I_l V \qquad (47.17)$$

i.e. to the power dissipated in the load.

The e.m.f./temperature characteristics of thermocouples are known, empirically, for all the useful combinations of materials each being quoted, normally, with respect to a 'cold' junction (0°C) formed from the same materials: general information about the calibration and characteristics of these materials is available in British Standard BS 4937-1973 (1981).

The e.m.f./temperature relationship is non-linear hence, when the 'cold' junction is at a temperature θ, above 0°C, then the tabulated junction e.m.f. e_1, corresponding to θ_1 must be added to the measured e.m.f. e_2 corresponding to the unknown much higher temperature θ_2; then read off the tabulated value of θ_2 which corresponds to $e = e_1 + e_2$. For less accurate measurements the input terminals of the instrument, being at ambient temperature, could constitute the 'cold' terminals without incurring unacceptable errors provided compensating leads are connected between the thermocouple and the detector. For difference measurements, at elevated temperatures over a short range of temperature differences, the error could be unimportant in some cases even without compensation, although it would be normal to include some 'cold junction compensation' such as a thermistor circuit. For direct reading instruments the (reference) cold junction can be maintained at 0°C by immersion within chipped melting ice in a thermos flask, or by using a commercial ice-point apparatus working upon the Peltier effect.

A typical value of source resistance for a thermocouple is 10 Ω. To achieve maximum sensitivity the resistance of the thermocouple load should equal the source resistance. Since the power available from a source S equals $E_s^2/4R_s$, the power available from a combination of couples is proportional to the number of couples.

The two connections between a thermocouple and its load (typically a sensitive moving-coil meter) are usually made with the aid of special conductors having thermo-voltaic characteristics similar to the materials used in the couples. This is to ensure that the effective hot and cold junctions are well separated thermally. It is important to connect each individual wire to its appropriate conductor in the couple, i.e. the two wires are *not* interchangeable. A standard colour code for identifying the types of cable for use with a given type of couple and for identifying the individual wires is given in British Standard BS 1843/1952 (1981).

The e.m.f./temperature coefficients, if considered to be linear, would be constant over the working range for any pair of materials; the values lie, generally, between 10–100 μV/K with the lower coefficients applying to the materials used for the higher temperatures. Some typical values are given in *Table 47.3*.

Table 47.3 Characteristics of thermocouples

Materials	*Approximate e.m.f. in μV/°C of temperature difference*	*Operating temperature range °C*
Chromel/Eureka	41	0 to 1000
Iron/Eureka	59	−200 to +1382
Chromel/Alumel	40	−200 to +1200
Platinum/Platinum–Rhodium	6.5	0 to +1450
Copper/Constantan	42	−200 to +300
Carbon/Silicon–Carbide	292 in the range 1210°C to 1450°C	0 to +2000

Sensitivity up to 1000 μV/°C can be achieved when one element is a thermistor.

An approximate equation for e.m.f. (e μV) for a hot junction θ(°C) using a cold junction of 0°C as reference is $\lg(e) = A \lg(\theta) + B$ with, for example, constants $A = 1.14$, $B = 1.36$ for copper/constantan. A closer approximation to the true e/θ curves for each material can be achieved in terms of higher-order polynomial equations covering various temperature bands. Such equations can be stored in digital form for each material, as part of an automatic data curve-fitting program for the direct interpretation of temperature from the measured e.m.f.

A recent microprocessor-controlled digital readout thermometer, sensed by thermocouples, is a mains-driven instrument with a 1000-hour battery for off mains use. Storage in memory of the characterisation parameters includes those of six thermocouple specifications (to ±0.1°C) defined in BS 4937, HBS 125, NFC 42-321 and DIN 43710 (−200°C to +1767°C). Measurements are effected, without change of range, to an overall uncertainty (including characterisation) of less than ±0.2%. The cycle of measurements includes short circuit, cold junction and internal 'calibration', and the display is refreshed every 0.25 s. Scaled displays are μV, °C, °F or K (°A) and facilities include programmable multiplying factors, alarms, dwell times for up to 99 minutes per channel with pause between scans for up to four days, 10 input channels and parallel BCD or high-resolution analogue outputs.

Many portable and bench digital thermometers are available and usable with a range of thermocouple probes giving 0.1°C resolution plus auto-ranging and analogue output. Cold junction compensation is based upon a precision thermistor and the overall uncertainty is about ±0.3% reading.

47.8.2 Pyrometers

Radiation thermometers respond to the total radiation (heat and light) of a hot body, while optical thermometers make use only of the visible radiation. Both forms of pyrometer are specially suited to the measurement of very high temperatures because they do not involve contact with the source of heat.

(1) Radiation thermometers. Both the Fery variable focus and the Foster fixed focus types comprise a tube containing at the closed end a concave mirror which focuses the radiation on to a sensitive thermocouple. In use the tube is 'sighted' on to the hot body. The sighting distance is not critical provided that the image formed is large enough to cover the thermocouple surface. Calibration is by direct sighting on to a body of known surface temperature.

(2) Optical thermometers. The commonest form is the disappearing filament type, which consists of a telescope containing a lamp, the filament brightness of which can be so adjusted by circuit resistance that, when viewed against the background of the hot body, the filament vanishes. The lamp current passes through an ammeter scaled in temperature. The telescope eyepiece contains a monochromatic glass filter to utilise the phenomenon that the light of any one wavelength emitted by the hot body depends on its temperature. Although calibration is based on the assumption that the hot body is a uniform radiator, departure from this condition involves less error than in the radiation pyrometer.

47.9 Thermo-resistive devices

A resistive sensor based upon a substantially linear resistance/temperature coefficient (α) should have a small thermal capacity to provide a rapid response as well as to avoid a temperature gradient across the sensor. Hence materials of high resistivity and high α values ensure a small volume device with adequate sensitivity. Networks which included thermistors can be made to fulfil these requirements although these materials are inherently non-linear.

47.9.1 Pure metal sensors

Pure conductors such as Pt, Ni and W have relatively low resistance/temperature coefficients but these are stable and fairly lihear. Pt is generally adopted for precision measurements and in particular for the definitive experiment within the range 90–903.5 K, of the International Practical Temperature Scale. Platinum and other metals can be deposited on ceramics to produce metal film resistors suitable (if individually calibrated) for temperature measurement.

Let the resistance values of a pure conductor be R at θ°C and R_o at 0°C; then the relationship can be written as $R = R_o(1 + \alpha\theta + \beta\theta^2)$. For most materials (nickel excepted) the second-order coefficient β is very small and $R = R_o(1 + \alpha\theta)$ is reasonably accurate. Examples of the properties of some commonly used materials are given in *Table 47.4*.

47.9.2 Platinum resistance thermometer (see BS 1904)

To a reasonable approximation the resistance R of a platinum wire at temperature θ°C in terms of its resistance R_o at 0°C and R_{100} at 100°C is

$$R = R_o(1 + \alpha\theta) \quad (47.18)$$

where

$$\alpha = (R_{100} - R_o)/100R_o \quad (47.19)$$

The temperature θ_p for a given value R is

$$\theta_p = 100(R - R_o)/(R_{100} - R_o) \quad (47.20)$$

and is known as the platinum temperature. Conversion to true temperature is found from the difference formula

$$\theta - \theta_p = \delta[(\theta/100)^2 - (\theta/100)] \quad (47.21)$$

The value of δ depends upon the purity of the metal. It is obtained

Table 47.4 Thermal properties of some commonly used materials

Material	*Symbol*	*Coefficient of thermal expansion* $\times 10^{-6}$	*Resistivity in ohm–metres* $\times 10$	*Resistance/temperature coefficient* $\times 10^{-3}$	*Melting point* °C
Aluminium	Al	28.7	2.82	3.9	660
Advance, Constantan, Eureka (55 Cu, 45 Ni)	—	14.8	49.1	0.2	1210
Brass (66 Cu, 34 Zn)	—	20.2	6.72	2.0	920
Copper, annealed	Cu	16.1	1.72	3.93	1083
Curalumin	—	—	5.76	2.0	500 to 637
Iron, pure	Fe	12.1	9.65	5,2 to 6.2	1535
Manganin (84 Cu, 12 Mn, 4 Ni)	—	—	44.8	0.02	910
Mercury	Hg	—	96.0	0.89	−38.87
Molybdenum	Mo	6.0	5.7	4.5	2630
Nichrome (65 Ni, 12 Cr, 23 Fe)	—	—	112.0	0.17	1350
Nickel	Ni	15.5	9.7	4.7	1455
Phosphor-bronze (95.5 Cu, 0.5 P, 4 Sn)	—	16.8	9.4	3.0	1050
Platinum	Pt	9.0	10.6	3.0	1774
Tungsten	W	4.6	5.6	4.5	3370

by measurement of the resistance of the sensor at 0, 100 and 444.67°C, the last value being the boiling point of sulphur. The value of δ is typically 1.5. A practical sensor, as shown in *Figure 47.52* usually consists of a coil of pure platinum wire wound on a mica or steatite frame, the coil being protected by a tube of steel

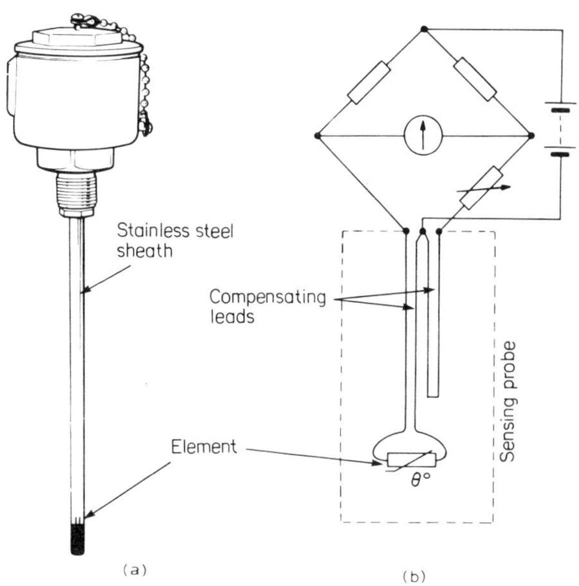

Figure 47.52 (a) Platinum resistance thermometer (courtesy Sangamo-Weston Ltd); (b) a suitable circuit

or refractory material. The resistance measurement is carried out by connecting the coil in a Wheatstone bridge network which usually has equal ratio arms, a pair of compensating leads being connected in the fourth arm. These leads run in parallel to the actual leads from the sensor coil and compensate for their resistance changes. If the initial resistance and coefficient α of the sensor are large, the compensating leads may not be required. The range of resistance values available for different sensors is 10 Ω to 25 kΩ and the overall response time to temperature changes is 5–10 s. The current in the sensor should be limited to 10 mA to avoid self-heating effects. The best accuracy obtainable (from NPL calibration) would be to ±1 m°C with a possible linear interpolation between given test points to provide a precision of ±0.1 m°C.

47.9.3 Semiconductor sensors

47.9.3.1 P–N junctions

Certain semiconductor p–n diodes are suitable for temperature measurement over a wide range; Ge, for example is useful from 1.5–100 K with excellent sensitivity. One differential thermometer has two matched p–n diodes which, when linearly amplified, give a discrimination better than 0.0001°C. Such instruments have several indirect applications, such as sensing thermal gradients in 'constant temperature' enclosures or vats, monitoring load changes, and displaying the input–output liquid temperature differences in fluid pumps.

47.9.3.2 Thermistors (thermally sensitive resistors)

These semiconductor sensors are made from sintered compounds of the metallic oxides of Cu, Mn, Ni and Co and formed into beads, rings or discs; they are usually enclosed in epoxy resin or glass. A high resistivity is achieved with a very large resistance/temperature coefficient, thus the beads can be as small as 0.1 mm in diameter giving a very exact location of the temperature and coupled with a fast response (of a few milliseconds) to temperature changes; but responses are much slower for discs and coated specimens. The devices can be designed for resistance values, at 20°C, from 1 Ω to several MΩ together with one outstanding property, which is a negative temperature coefficient of about −4% per °C. If the thermistor carries appreciable current, of even a few milliamperes, then one possible consequence which must be avoided is the cumulative effect of self-heating.

Since polarisation effects are absent the thermistor can be used in a.c. or d.c. circuits. Wheatstone bridge networks are widely used with the thermistor in one arm and the out-of-balance detector current as an indication of temperature. Although the characteristics are strongly non-linear, it is possible to obtain thermistors in matched pairs which can be applied to differential temperature measurement. Another application, not directly a temperature measurement, is to the indication of air flow in pipes, or as anemometers. One thermistor is embedded in a metal block acting as a thermal reservoir, the other senses the air speed as a cooling effect. If the two thermistors occupy adjacent arms of a bridge then the out-of-balance signal is a measure of the fluid rate of flow.

47.9.3.3 Bolometer bridge

A thermo-sensitive resistor TH may be used for the measurement of power (particularly at very high frequencies), using the bridge circuit of *Figure 47.53*. The operating range of TH must include

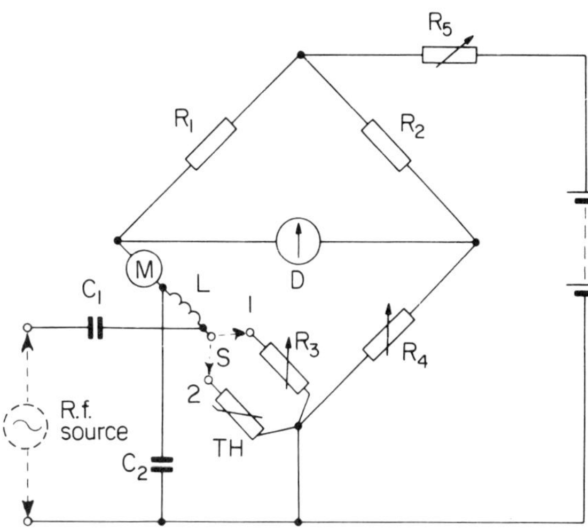

Figure 47.53 A bolometer bridge

the intended load resistance R for the a.c. source. The combination of C_1–C_2–L enables a.c. and d.c. to be mixed in TH without the sources feeding power to each other. With no a.c. input initially, the switch S is first set to position 1, R_3 is set to the value of the intended load resistance and R_4 is adjusted to produce a balance as indicated by the detector D. S is then set to position 2 and R_5 is adjusted to produce a balance. TH now has the value R of the intended load resistance. The value I_1 of the current in the meter M is noted. The a.c. input is now applied and this changes the resistance value of TH thereby unbalancing the bridge. The value of R_5 is then increased to reduce the power fed in from the d.c. source and thus rebalance the bridge. The new value I_2 of the current in M is noted. The *reduction* of power from the d.c. source will equal the power which has been supplied by the a.c. source. Let P_t be the total power which is required to drive TH to the value R, and P_{ac} be the a.c. power which is to be measured, than,

$$P_t = I_1^2 R \qquad (47.22)$$
$$= I_2^2 R + P_{ac}$$

By subtraction

$$O = I_1^2 R - I_2^2 R - P_{ac} \qquad (47.23)$$

Therefore

$$P_{ac} = R(I_1 + I_2)(I_1 - I_2) \qquad (47.24)$$

47.10 Thermo-mechanical devices

The thermo-mechanical effect constitutes a change of dimension of a material (itself not necessarily an electrical conductor), caused by a change of temperature. Normally an increase of temperature causes an increase of physical dimensions. The coefficients of expansion of some commonly used materials are given in *Table 47.4*.

47.10.1 Contact thermometers

One simple example is the mercury contact thermometer, *Figure 47.54*. This is used as a sensing element in thermostatically

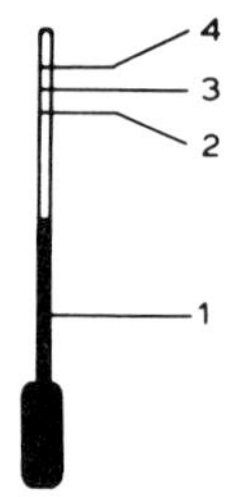

Figure 47.54 A contact thermometer

controlled ovens and refrigerated chambers. A large bulb, full of mercury, is eonnected to a thin stem, thus producing a large change in the length of the mercury column for a small change of temperature. Four thin platinum wires, numbered 1–4 in the diagram, are sealed into the stem. In normal operation the upper limit of the mercury column oscillates slowly between contact 3 and a point which is just below this contact. When the column fails to reach contact 3, the breaking of an electric circuit between contacts 1 and 3 causes either a heater to be switched on or a refrigerator to be switched off, as may be appropriate to the application under consideration. In the event of the temperature passing outside the permissible limit, then completion of a circuit between contacts 1 and 4, or the breaking of a circuit between contacts 1 and 2, causes an alarm to be operated. Slight changes of temperature, e.g. ± 0.1°C are to be expected with such an on-off form of temperature control, best results being achieved when the power of the heater or refrigerator is such as to cause the on and off periods to be approximately equal. Since the mercury column is allowed to carry currents of small magnitude only, the thermometer circuit should operate to control the input signals to high impedance devices such as field-effect transistors. See British Standard BS 1704 (1951/1973) for general information about liquid-in-glass thermometers.

47.10.2 Thermostat (liquid sensor)

One means of thermostat operation makes use of the expansion with temperature of a volatile liquid enclosed in a container which is exposed to the heated medium being controlled. The expansion extends a metal bellows or diaphragm against the action of a spring and the resulting movements with rising and falling temperature operate the contact mechanism. In the case of room thermostats of this type the expansion bellows are self-contained within the instrument. For immersion thermostats a phial or bulb containing the volatile liquid forms the sensitive member and may be connected to the bellows in the instrument by means of metallic tubing.

The construction of a thermostat of this type is shown in *Figure 47.55(b)*, which also shows the type of phial used. In this case the

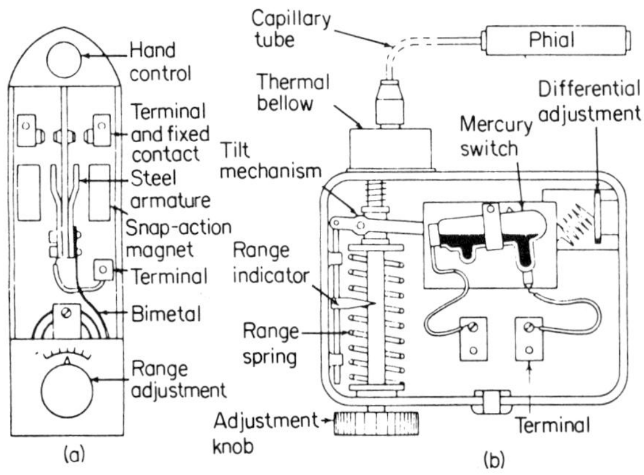

Figure 47.55 Thermostats

switch is of the mercury type and is tilted on and off by the operating level actuated by the movements of the diaphragm. The temperature setting is changed by turning the adjusting knob after removing its locking screw. Differential adjustment is provided by a milled knob inside the case.

47.10.3 Thermostat (bimetal sensor)

There are several forms of this device, corresponding to the intended application. In the case of room thermostats a typical construction makes use of a coiled bimetal strip which winds and unwinds with change of temperature. A contact carried on the free outer end of the strip makes and breaks contact with a fixed member, and a means of adjustment is provided for determining the operating temperature. A typical range is between the limits of 5 and 25°C for heating installations and -8 to $+15$°C for cooling, in each case with a differential (between opening and closing temperatures) of 1–2°C. A 'snap' switching action is provided by magnets which hold the moving contact until the accumulated operating force overcomes the magnetic attraction. If the load current is too large to be carried by the bimetal, the thermostat operates a separate relay or contactor.

The arrangement of a typical room thermostat with high and low temperature contacts is shown in *Figure 47.55(b)*. The hand control knob for closing either contact is seen at the top of the instrument.

47.10.4 Bimetallic strips

Temperature sensitive bimetals, or thermometals, are produced by firmly bonding together two or more metals or alloys having different expansion coefficients, the composite product being rolled into strip. A change of temperature produces a change of curvature, utilisable in several forms such as the deflection of a straight strip, the opening or closing of U-shaped pieces, the rotation of spirals and helices, or the dishing of washers and discs. Thermometals may be used to give temperature indication, thermostatic control, process control or temperature compensation. The metal may itself be heated, e.g. by the passage of a current as in overload circuit-breaker tripping.

Materials used in bimetal devices include brass, various steels, nickel, manganese alloy, etc.

For a given bimetal in strip form there is a range of temperature θ over which the radius of curvature R is given by $(1/R)=2K\theta/t$, where $2K$ is the change in curvature per unit change of temperature per unit of thickness t. A straight strip of length L, fixed at one end and free to move at the other, will have a free-end deflection of approximately $d=K\theta L^2/t$. If the strip is prevented from deflecting it will develop a force $p=4KE\theta t^2/L$ per unit width, where E is its elastic modulus.

In many applications bimetals must do work in moving or supporting a load. The work capacity depends on E, K^2, θ^2 and the volume of the bimetal. Overstressing will produce a permanent set in the material.

47.11 Noise and sound measurements

47.11.1 Fundamental quantities

The words *sound* and *noise* will be treated here as synonymous from the viewpoint of measurement. Two quantities are directly measurable, the sound pressure and the particle velocity, other quantities such as sound power being derived from one or both of these.

It is assumed, unless otherwise stated, that all sound measurements are carried out in air, so that it is necessary for only one of the basic quantities, normally the pressure, to be measured, since pressure and particle velocity in a parallel wave motion are related by the equation

$$p=\rho cv \tag{47.25}$$

where ρ is the density of the air, c is the velocity of sound in air and v is the particle velocity.

Sound pressures are, for practical purposes, converted into sound pressure levels, defined in decibels (dB) as

$$L=20\ \lg_{10}(p/p_{\text{ref}}) \tag{47.26}$$

where L is the sound pressure level and p_{ref} is a reference pressure of 2×10^{-5} N/m^2.

Alternatively, intensity level (which is the decibel equivalent of intensity, the rate of flow of sound energy across unit cross-section at a point in a sound field) is defined as

$$L_1=10\ \lg_{10}(I/I_{\text{ref}}) \tag{47.27}$$

where

$$I_{\text{ref}} \text{ is } 10^{-12}\ \text{W/m}^2 \tag{47.28}$$

and

$$I=p^2/\rho c \tag{47.29}$$

These two reference levels are nearly equal for air at normal temperature and pressure, differing by only about 0.14 dB. For most purposes they are regarded as equivalent.

Subjective measures of sound sensation are derived from objective measurement of pressure level or spectrum pressure level (pressure per unit bandwidth) throughout the spectrum.

47.11.1.1 Measurement of particle velocity

Particle velocity may be measured absolutely by means of a Rayleigh disc,[1] which is a thin disc suspended from a torsion fibre in the sound field. The torque on the disc when held at an angle to the direction of propagation is related to the particle velocity. This method of measurement is unsuitable for use outside a specialised laboratory, and it is therefore better to measure pressure.

47.11.1.2 Measurement of sound pressure

For the measurement of sound pressure a calibrated microphone is needed, together with an amplifier of known gain with rectifier and meter. The microphone may be calibrated by the method of reciprocity,[2] by Rayleigh disc or by comparison with a standard which has been calibrated by one of these methods. The comparison must be carried out in a free-field room (i.e. one in which all surfaces are covered with sound-absorbing material so that only sound directly from the loudspeaker or other source of test sound reaches the microphones).

For calibration of the amplifier, a steady alternating voltage is applied through a calibrated attenuator to its input and the attenuation is varied until the output voltage of the amplifier is equal to the input of the attenuator. *Figure 47.56* shows a circuit for noise measurement embodying these components. The calibrated microphone is connected to the input of the amplifier and the output is rectified and applied to a d.c. meter. The rectification characteristic should give an accurate indication of the root-mean-square (r.m.s.) value of the signal, which represents the power irrespective of the waveform. British Standard 3489:1962 makes use of an r.m.s. meter obligatory for all noise measurements. To achieve a true r.m.s. reading some form of thermal meter is often used, but simple square-law rectification gives sufficient accuracy even with complex waveforms. The amplitude of the waveform can be measured with the use of a peak-rectifying meter, and the ratio of this to the amplitude of a sinusoid having the same r.m.s. value is known as the crest factor.

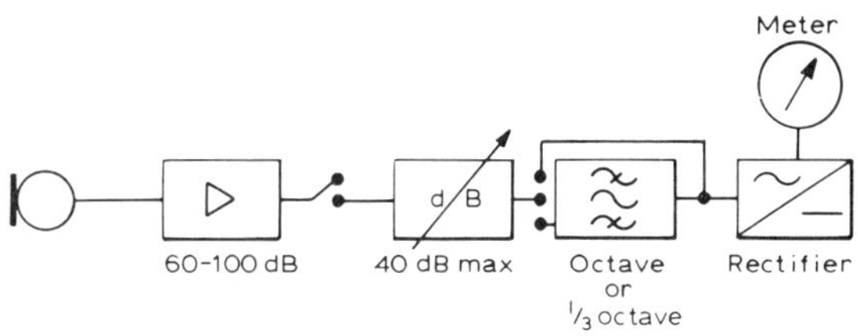

Figure 47.56 Noise and sound measurement

As the rectifier is necessarily limited in dynamic range, it is usual to insert switched attenuators into the amplifier, by which the rectifier input and consequently the meter indication, can be brought within a range of 10–20 dB. For the rapid measurement of steady or s.owly varying sounds, a calibrated attenuator with 1 dB steps may be used to bring the meter to a reference reading.

47.11.1.3 Measurement of sound spectra

The spectrum of a complex sound may be measured by inserting band pass filters into the amplifier chain of a sound-pressure measuring circuit, so as to obtain a series of band pressure levels.

The range of audible frequencies is divided into octave or one-third octave bands and filters for the purpose are specified in British Standard 2475:1964. The one-third octave bands, except at very low frequencies, correspond closely to the critical bandwidths of hearing. Thus, measurements in one-third octave bandwidths are used for the evaluation of loudness by the method of Zwicker[3] and also for the great majority of development work on the silencing of machines.

Octave bandwidths are used in the Stevens' method[4] of loudness evaluation and for many applications in connection with industrial and community noise. Noise containing discrete frequency components requires narrow-band heterodyne filters, typically 10% of the centre frequency in width.

47.11.2 Practical equipment for the measurement of sound pressure level

47.11.2.1 Sound level meters

A sound level meter consists of a portable battery-operated noise measuring circuit with an r.m.s. meter calibrated in decibels; most sound level meters are provided with a variety of facilities such as band pass filters and weighting networks. A typical block diagram is shown in *Figure 47.57*. The requirements of such an

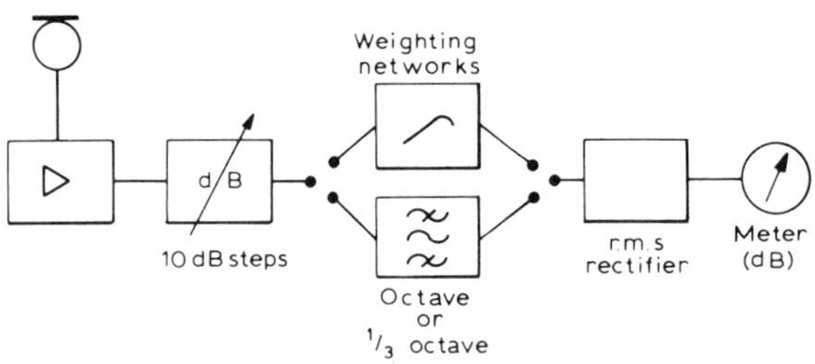

Figure 47.57 Block diagram of a typical sound-level meter

instrument are specified in British Standards 3489:1962 (industrial grade meters) and 4197:1967 (precision grade), the difference between the types being mainly in the tolerances.

The microphone, which is usually of a capacitor or electret type, has an omnidirectional polar diagram in which the off-axis sensitivity deviation is held within specified limits. The main amplifier is followed by an attenuator with switched 10 dB steps by means of which the indicating meter may be brought within the scale for a wide range of microphone inputs. Until recently, the meter scale was usually 10 dB wide with perhaps a small extension at the lower end. Recently it has become possible to provide meters with very much wider scales to facilitate reading of rapidly changing levels. In use, the meter reading is added to the number shown on the attenuator control, which represents the sound pressure level when the meter is at zero.

Weighting networks and other filters are inserted after the attenuators, either built-in and selected by switches or external and connected by input and output terminals.

An overall calibration of sound level meters is carried out by exposing the microphone to a sound generator with a known output. In one type of generator, small steel balls are allowed to fall freely on to a metal diaphragm inclined to the horizontal from where they bounce away into a reservoir. The microphone is placed at a stated distance from the diaphragm and the instrument is adjusted to bring the meter to a marked point on the scale. Another type consists of a metal tube designed to fit snugly on the microphone and containing a small diaphragm driven by an oscillator.

The gain of the amplifier chain is separately checked by noting the meter reading produced by an inbuilt stable oscillator.

A meter reading of sound pressure level by means of an instrument such as the above, in which equal weight is given to all frequencies within the range of measurement, is not well correlated with the subjective loudness or other subjective effects of the sound. This is because the human ear is relatively insensitive to sounds of very high and, more particularly, low sounds.[5] Weighting networks have therefore been devised for insertion into noise measuring chains by which the low and very high frequency components of complex sounds are progressively attenuated. *Figure 47.58* shows four commonly used networks as specified in British Standards 4197:1967 and 5647:1972. The A

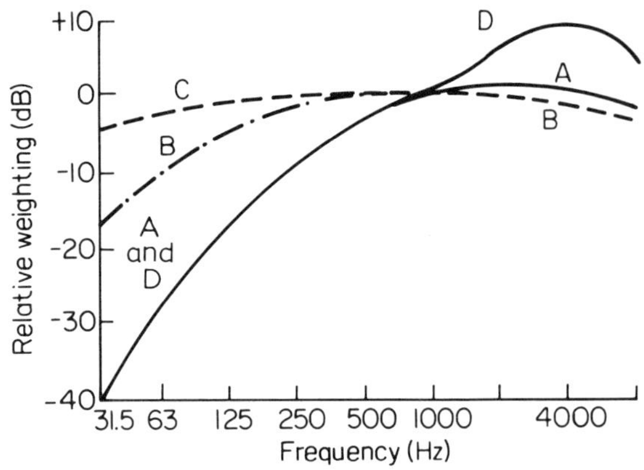

Figure 47.58 Weighing networks (A, B, C, D) (from British Standards Nos 4197:1967 and 5647:1972)

network is the most commonly applied and is always provided in sound level meters, being used in connection with road traffic noise, hearing damage risk and assessment of neighbourhood noise.

The proliferation of integrated circuits in the last five years has enabled the versatility of sound level meters to be increased in several important respects. Facilities now available include the following.

(a) Very fast time constants in conjunction with peak-hold enabling the maximum value of a fluctuating signal to be accurately determined.

(b) Continuous integration or averaging of the acoustic energy input for any desired time.

(c) Real-time display of octave or one-third octave band pressure levels.

(d) A.C. and d.c. outputs corresponding to the signal reaching the indicating meter, for driving tape recorders or chart level recorders.

(e) Inputs for alternative microphones or for acceleraometers for vibration measurement.

47.11.2.2 Chart level recorders

Similar in principle to the measurement of noise levels by measurement of the attenuation necessary to reduce the output of the amplifier to match a reference level, as described in section 47.11.1.2 is the chart level recorder, by which permanent records of level can be made over a range of 75 dB or more. In this instrument, the first of which was marketed by Neumann, a high speed servo system is actuated by the difference between the amplified signal and an arbitrary value. The servo drives the wiper of a potentiometer in the amplifier chain and comes to rest at the balance point. The position of this point is continuously recorded by a pen attached to the wiper arm. The potentiometer is usually designed to give a plot in dB, but any other law may be produced.

For most purposes the chart is driven at constant speed to give a record of level against time, but by gearing the chart drive to the dial of an oscillator, response/frequency curves can be obtained.

47.11.2.3 Real time displays of sound spectra

Equipment is now obtainable from several manufacturers, consisting of a noise-measurement chain and a set of narrow band pass filters which are scanned at short intervals and their outputs displayed on an oscilloscope. The filters are usually one-third octave width or narrower, up to about 400 channels being included in one instance. These instruments are designed as complete systems with a variety of facilities such as the freezing of individual scans as a steady spectrum, the brightening of any selected filter position on the display with a simultaneous readout of its value, or the digital storage of the spectrum of a transient event for subsequent replay. In one instrument, using one-third octave filters, an automatic loudness summation is presented by the method of Zwicker.[3] Time-compression and digital processing may be used to speed up analysis and increase the possible number of channels.

47.11.3 Recording of sound for measurement and analysis

47.11.3.1 Analogue recording

Recording of sound for subsequent analysis may often greatly speed up work on site and also allow short-lived events which cannot be repeated exactly, such as the flyover of an aircraft, to be submitted to frequency analysis by replaying through band pass filters.

The output of the amplifier stages of a sound level meter are recorded on a magnetic tape recorder, taking the signal from a point after any switched attenuators, filters or weighting networks so that the recording will be able to handle the whole range of levels for which the meter is designed to be used. The range may include, for instance, a calibrating signal of over 124 dB followed by a sound of 30 dB or less.

To make use of the maximum available dynamic range, the recorder should have a good linear characteristic of 40 dB or more and its maximum input should be matched to the highest reading on the meter scale.

47.11.3.2 Digital and computer systems for noise measurement

Sound signals, being continuous functions of time, lend themselves readily to digital recording and analysis, either in real time or in digital recordings. The output voltage of a microphone amplifier is converted by means of an analogue-to-digital converter into digital characters which are stored on magnetic tape, with associated information such as time or filter channel. The magnetic tape, usually in the form of a domestic type cassette, is subsequently analysed by computer—a process for which the microcomputers now on the market are in widespread use. Other applications of computers and microprocessors to acoustic measurements are dealt with in the section below on general acoustic measurements.

47.11.3.3 Continuous integration of noise power

During the last decade, increasing importance has been attached to the integrated intensity of noise or its average power over a period. The equivalent level L_{eq} of a noise for an averaging time T is defined as

$$L_{eq} = 10 \lg(T^{-1}) \int_0^T (p^2/p_0^2)\,dt \qquad (47.30)$$

and the noise dose is the integral of the energy alone, usually expressed as the percentage of the integral of 90 dB (relative to 10^{-12} W/m^2) for 8 hours.

The introduction of microprocessors has enabled sound level meters to produce a direct reading of either noise dose or L_{eq} for such varied purposes as assessing the hearing damage risk in a discotheque or measuring the transmission loss through the facade of a building, using varying traffic noise as the test sound. Reference will be made to other applications below.

47.11.4 Use of equipment for noise measurement

47.11.4.1 Road traffic noise

A-weighted levels are used for all traffic noise measurements whether of the noise emission of industrial vehicles or of the noise climate caused by traffic as a whole, because it is found to correlate more closely with subjective reaction than any other weighting.[6]

The noise emitted by a vehicle is subject to a legal limit, depending on the type of vehicle. British Standard 3425:1967 specifies details of the test site and conditions and of the special sound level meter with a wide scale and no switched attenuators, the settings of which could be questioned in the event of legal proceedings.

Compensation for sound insulation may be claimed by householders living in a climate of excessive traffic noise. The criterion for compensation is that the facade of the house should be subject to a level over 68 dB for more than 10% of the 18 hour period from 6 a.m. to 10 p.m. Apparatus for deriving this so-called L_{10} level has been available for many years. It consists of a number of mechanical counters corresponding to 5 dB intervals on the potentiometer of a chart recorder. The counter system is pulsed at regular intervals and each counter records the number of samples falling within the corresponding 5 dB range of level. After an adequate period of time a histogram or ogive is plotted from the total counts, from which L_{10} or any other similar index can be calculated to within about 1 dB.

Recent versions of this equipment use microprocessor techniques to give direct read-out of L_{10} etc and of L_{eq}.

47.11.4.2 Machinery noise

The noise output from a machine may be measure in free-field conditions by means of a sound level meter placed in succession at a number of points distributed over a hemispherical solid angle at a constant distance from the surface of the machine. British Standard 4196:1967 describes the method. A choice of four distances from 0.3 m up to 10 m is given. Alternatively, diffuse field measurements of noise power may be made in a highly reverberant enclosure. For measurements on a machine *in situ*, where the conditions are intermediate between free-field and diffuse field, the free-field method of measurement is used and correction is made for the effect of the reverberant sound. For the case of a substantially diffuse field, the sound power level, H, is given by

$$H = (L) - 10 \lg T + 10 \lg V - 14 \text{ dB} \qquad (47.31)$$

where (L) is the mean sound pressure level, T is the reverberation time of the enclosure and V is its volume. (SI units).

47.11.4.3 Hearing damage risk

Modern practice in hearing conservation for those exposed to industrial noise at work is summarised in the Department of Employment's *Code of Practice for Reducing the Exposure of Employed Persons to Noise.*[7] It is recommended that exposure to steady sound exceeding 90 dB(A) should not be continued for more than 8 hours in any day. For varying or intermittent exposure, the A-weighted energy level integrated for the whole working period in a day should not exceed the same integral of 90 dB(A) over 8 hours.

It is usual to make measurements in terms of the equivalent energy level for the working period (see Section 47.11.3.3). The value of L_{eq} for an exposure period may be computed from a record of sound level against time, but instruments giving a total integrated energy level for a period of exposure, or even a value of L_{eq} are how in general use. These are known as noise dose meters.

47.11.4.4 Industrial noise in residential areas

The application of measurement to the acceptability of noise from industrial premises in a residential area is covered by a rather complicated set of rules given in BS4142:1967. A-weighting is specified, as with road traffic, and the interpretation of level readings must take into account the type of area, e.g. rural, or residential with some industry, the time of day, a notional background level for the area, the actual background level if it can be measured in the absence of the specific interference, the intermittency of the interfering sound and its character.

47.11.4.5 Aircraft noise

Assessment of community disturbance by aircraft noise requires a knowledge of the number of flyovers as well as the mean level of the peak sound level during individual flyovers. The two factors are combined in an index known as the noise and number index (NNI).

$$\text{NNI} = \text{Average level of peak noise} + 15 \lg N - 80 \qquad (47.32)$$

where N is the number of flyovers per day. The peak level is expressed in *PN* dB, which is a weighted sum of the spectral components having a high correlation with disturbance.[8]

In place of *PN* dB it is now usual to substitute a sound level meter reading with the D-weighting (see *Figure 47.58*). Compared with A-weighting, this gives prominence to frequencies above 1 kHz.

47.12 Acoustic measurements

47.12.1 Definitions

Sound is transmitted from one point in a building to another by various paths through the air or the structure. The point of origin is described as the *source* and the other is usually in a room known as the *receiving room.* Airborne sound transmission is that in which the sound is generated in the air at the first point and received in the air at the second. The shortest path of transmission, usually through one or more intervening partitions is called the *direct path*; sound that travels by indirect paths, often including substantial distances through solid structures, is said to travel by *flanking transmission.* Sound travelling through the solid material of a building, particularly from a vibrating source in contact with it, is known as *structure borne sound. Figure 47.54* shows examples of these forms of transmission.

The sound insulating characteristic of a partition is its *transmission coefficient.* This is the ratio of the sound power radiated into the receiving room to that falling on the source room side. It is usually denoted by τ. The difference in sound power levels is known as the *sound reduction index* (SRI), or *transmission loss* (TL), and is the figure usually quoted. The two quantities are related by the equation

$$\text{SRI} = 10 \lg 1/\tau \qquad (47.33)$$

The reduction of sound pressure level between two rooms is known as the *sound level reduction* or *sound level difference.*

47.12.2 Measurement of airborne sound transmission

To measure the sound level difference between two rooms, it is simply necessary to place a sound source in one of them and to measure the difference between the equivalent sound pressure levels in the two rooms. To derive the SRI of the partition, a correction must be applied to the sound level reduction to take into account the absorption of sound in the receiving room. If the sound transmission between the two rooms is entirely by the direct path through the intervening wall, the sound level reduction will depend not only on the SRI of the partition but also on its area and the increase of sound pressure level due to reverberation in the receiving room. The relationship between the SRI and the sound level difference is[9]

$$\text{SRI} = \text{Sound level difference} + 10 \lg \left(\tfrac{1}{4} + A/S\alpha\right) \qquad (47.34)$$

where A is the area of the partition, S the total area of the interior surfaces of the receiving room and α is their average absorption coefficient (see section 47.12.4).

47.12.2.1 Standard conditions

Standards for measurement of level reduction and derivation of SRI are given in British Standard 2750:1956. In that document, the $\frac{1}{4}$ within the correction term is ignored, so that it becomes $10 \lg(A/S\alpha)$. This simplification causes little error except when the receiving side is in the open air or a room with very great absorption. A normalising correction is recommended when assessing walls between dwellings or rooms within a dwelling where there may be multiple flanking paths. The correction term becomes $10 \lg(10/S\alpha)$ or $10 \lg(T/0.5)$ where T is the reverberation time of the room (see section 47.12.3).

To make a full assessment of the SRI of a partition, narrow bands of noise at one-third octave intervals are radiated by one or more loudspeakers into the source room. At each frequency, the sound-pressure level is measured at five microphone positions in each room by one of the methods described in section 47.11, and the mean energy level computed for each room. The microphones should be in positions more than half a wavelength from any wall. The reverberation time also is measured (see section 47.12.3), and the total absorption calculated for substitution in equation (47.34).

The mean energy level is computed for each room by averaging the power ratios derived from the pressure levels in the five microphone positions and reducing the average to decibels. The measured levels are expressed as decibels relative to any convenient reference, e.g. the multiple of 10 to the next below the lowest, and the power ratios read from a table such as *Table 47.5* (one place of decimals is usually sufficient).

If the spread of the five SPL does not exceed 5 dB, they may be averaged directly, giving results less than 1 dB in error.

Table 47.6 gives the central frequencies recommended for bands of noise used for test of sound insulation and other acoustic quantities.

The mean SRI of the partition is defined as its average value in one-third octave bands from 100 Hz to 3150 Hz.

Table 47.5 Power ratios for sound pressure levels (SPL)

SPL	*Power ratio*	*SPL*	*Power ratio*
0	1.00	7	5.01
1	1.26	8	6.31
2	1.58	9	7.94
3	2.00	10	10.00
4	2.51	11	12.59
5	3.16	12	15.85
6	4.00		etc.

Table 47.6 Standard testing frequencies for acoustic measurements

Octave bands (Hz)	*One-third octave bands (Hz)*		
63	50	63	80
125	100	125	160
250	200	250	315
500	400	500	630
1000	800	1000	1250
2000	1600	2000	2500
4000	3150	4000	5000
8000	6300	8000	10000

47.12.2.2 Practical details

Any method may be used for the measurement of the band pressure levels in the two rooms. Readings for all frequency bands and microphone positions may be made in one room before passing to the other. A chart recorder may be used for giving a permanent record of the levels, the source and receiver microphones being switched in alternately. If the dial of the oscillator providing the test signal is driven by gearing from the recorder and the filters are switched by the same mechanism, each filter position gives a pair of levels corresponding to the two rooms.

As five microphone positions are normally necessary in each room to give satisfactory accuracy, one must either go to the expense of a number of matched microphones of high quality or interrupt the measurements to change microphone positions in both rooms. An alternative method is to mount each microphone on a driven rotating arm so that it traces out a circular path along which the pressure level can be sampled.

An important point is that, whichever method is used, checks must be made frequently in measuring the receiving room level to confirm that the wanted signal is sufficiently above the noise level, whether due to electrical or ambient noise, to give reliable measurements. For this purpose the loudspeakers are momentarily switched off and the measured level should fall by at least 6 dB.

For extensive routine measurements at the BBC, digital equipment developed by Moffat[10] has been used, the test signals being played from magnetic tapes which also carry trigger signals for actuating digital processing equipment. The outputs of microphones in the two rooms are then recorded on a second tape which is then processed to yield the sound level reduction.

For laboratory measurement of sound reduction index of partitions and walls the sample must be fixed in an opening in a heavy wall separating two rooms, constituting a so-called transmission suite, in which every precaution has been taken to avoid transmission of sound by any path other than that through the sample partition. Each room must be of adequate size to allow the establishment of a diffuse field and the edge constraints of the sample should be similar to those in its normal situation. It should be rectangular in shape with an area of 10 m^2.

47.12.2.3 Separation of airborne from structure-borne transmission

The measurement of airborne sound transmission by the British Standards method described above gives no information about the path or paths of transmission. These may include structural paths and parallel transmission paths through the partition between the rooms. Part of the energy may be transmitted directly through holes, cracks, thin doors, etc., which will be greatly inferior to the rest of the partition with respect to transmission loss. *Figure 47.59* shows some of the paths by which sound may travel from a source room to an adjacent one. There

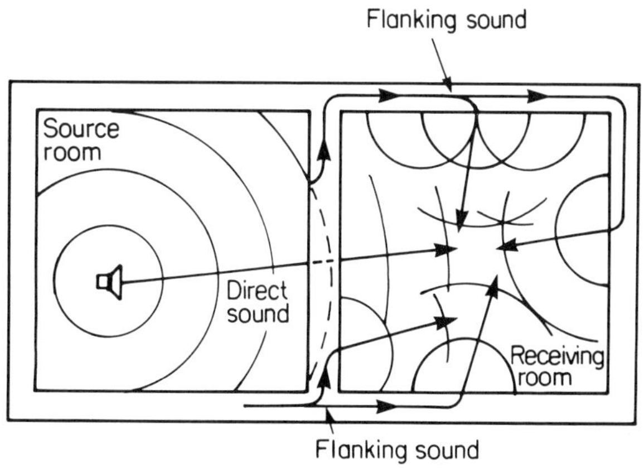

Figure 47.59 Sound transmission between adjacent rooms

are several ways in which the indirect paths may be separated from the principal path for diagnostic purposes or to determine the true sound reduction index of a partition in the presence of other paths even if the energy transmitted through them is the greater.

A non-permeable partition cannot transmit sound except by the bodily movement of its mass and the inherent SRI of a partition can therefore be determined by measuring the velocity amplitude of the surface on the receiving room side with the pressure on the source room surface. The velocity amplitude may be derived from the acceleration amplitude of the surface measured by means of accelerometers attached to it.[11] The velocity amplitude is related to the acceleration amplitude by the equation $v = q/f$, where q is the acceleration and f is the frequency of the sound. Small accelerometers are available in which the sensitive element consists of a steel disc between two discs of piezoelectric material. The acceleration of the whole causes a force difference between the steel and the two piezoelectric discs and the resulting voltage is a measure of the acceleration. The equivalent near-field sound pressure level corresponding to an accelerometer voltage level V relative to 1 volt is given by

$$L = V + 207 - 20 \lg kf \qquad (47.35)$$

where f is the frequency and k is the sensitivity in mV/g acceleration.

This equation breaks down below the frequency at which the speed of bending waves in the surface is equal to that of sound in air.

The accelerometer is in other ways also a powerful tool for the tracing of sound and vibration through the structure of a building, and for predicting the near-field sound pressures to be expected in a building subject to vibration.

A second very powerful method is to derive the cross-correlation function between the outputs of microphones in the source and receiver rooms using a comparatively broad band (about an octave) of noise as the signal. A variable time delay is inserted into the source-room microphone circuit and as this is slowly increased the correlation function reaches a maximum value when the delay is equal to the time taken by sound to travel from the source room microphone to the receiving room microphone by any one of the paths. There may be a succession of such maxima, their amplitudes being proportional to the sound pressure levels at the receiving room attributable to the corresponding paths. The transmission losses can therefore be calculated. Details of this method have been given by Goff[12] and Burd.[13]

The radiation of sound in or out of a surface may be reduced by covering the surface with additional flexible layers such as plasterboard supported over a layer of sound absorbing material. BS 2750 points out that this method may be used to reduce the contributions from individual paths of the flanking transmission and so effectively eliminate them from the measurements. This, like the correlation method, is mainly of use in laboratory conditions.

47.12.2.4 Measurement of impact sound transmission

The transmission of sound from impacts within a building cannot be measured in terms of a level reduction. Instead, standard impacts are delivered to a point on the structure and the band pressure level is measured at the standard test frequencies at a number of positions in another room. Most commonly, the blows are delivered to a floor immediately above the receiving room to test the effects of footsteps in producing sound in the room below. The impacts are produced by an impact machine or footsteps machine specified in British Standard 2750:1956, consisting of a number of hammer heads each of 500 g weight and a mechanism for allowing the hammers to fall freely in sequence on to the floor at a rate of between five and ten per second. The band pressure levels in the receiving room are plotted against frequency as a spectrum which is compared with a standard form of spectrum. A disadvantage of the tapping machine described above as a representation of footsteps is that the hammers are very much lighter than legs and consequently may give misleading comparisons between structures of widely differing mechanical form, e.g. between a carpet and a floating floor, as means of reducing sound transmission. Cremer and Gilg,[14] have found that an electromagnetic shaker gives more reliable results.

47.12.3 Measurement of reverberation time

When a steady sound is radiated into an enclosure and then suddenly cut off, the sound pressure decays according to Franklin's equation[15]

$$P=P_0 e^{-kt} \tag{47.36}$$

where P_0 is the initial pressure amplitude and k is a constant.

The reverberation time of the enclosure is defined as the time required for the sound pressure to fall to 1/1000 of its initial value, i.e. for the level to fall by 60 dB. Thus equation (47.36) yields

$$L=L_0-60\,t/T \qquad \text{(dB)} \tag{47.37}$$

where L, L_0 are the instantaneous and initial levels and T is the reverberation time with a value of $6.9/k$. This is a linear function of time and the term decay curve is usually taken to refer to this form, i.e. the curve of sound pressure level against time.

A straight decay curve will be obtained only in small rooms at the frequency of a strong isolated-room mode; in all other cases it will be modified by beats, rapid fluctuations or changes in general slope as the decay proceeds. Measurement of reverberation time consists of recording or displaying a decay curve and measuring the slope of the best-fit straight line.

47.12.3.1 Measurement from decay curves

The block diagram of *Figure 47.60* shows the commonest method of measuring reverberation time by using a chart recorder. A test

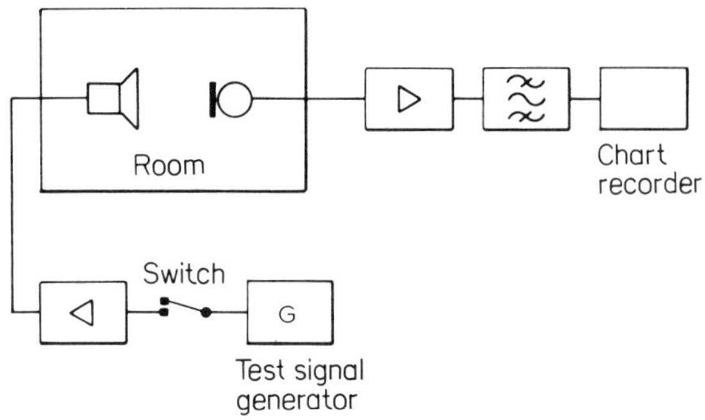

Figure 47.60 Block diagram of equipment for measurement of reverberation time

signal from an oscillator or noise generator is radiated by a loudspeaker into the enclosure under test and received, together with the resulting reverberant sound, by a microphone. The output voltage of the microphone is amplified and fed to a logarithmic chart recorder through band pass filters which remove noise and harmonics. The chart is started and the sound is cut off so that a decay curve is recorded. This process is repeated at a series of test frequencies, the standard frequencies being those listed in *Table 47.6*. The series is repeated with the microphone in at least four other positions, and in greater numbers in the case of large halls.

The slopes of the decay curves are measured or directly converted to reverberation times by means of a calibrated portractor, and average for all microphone positions at each frequency.

Figure 47.61 shows the way in which an oscillograph may be used in place of the chart recorder, with advantage for the speed and conveience of measurements.[16]

The on/off switch for the sound source is replaced by an electronic switch which releases short bursts of tone or bands of noise at regular intervals. The end of each burst triggers the time base of the oscilloscope. The microphone signal, after amplification, is converted to its logarithm which is displayed on the oscilloscope as its Y-deflection. The reverberation time is read from the screen with the aid of a rotatable scale. A screen with a phosphor having a persistence of about a second greatly facilitates readings.

47.12.3.2 Signals for reverberation time measurement

Owing to the presence of strong standing-wave systems in a room, the shapes and mean slopes of decay curves change rapidly with small variations of frequency. It is therefore essential to use a signal of finite bandwidth so that a number of room modes are simultaneously excited and a decay pattern representative of the frequency region is produced. Random noise may be used, bands of one octave or one-third of an octave being selected in the

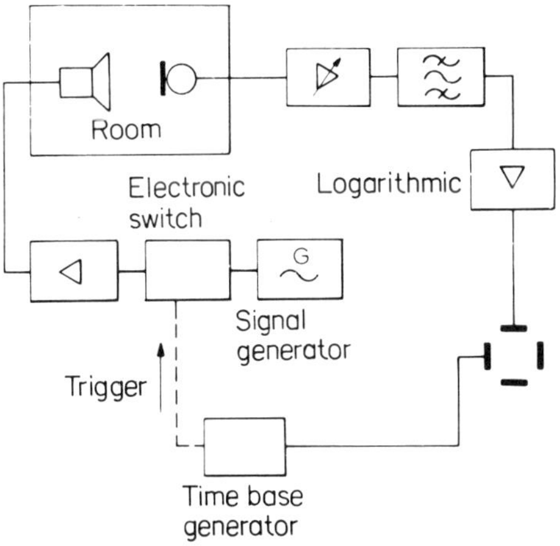

Figure 47.61 Block diagram of reverberation measuring equipment using oscilloscope

microphone circuit. A disadvantage is that full use is not made of the power handling capacity of the loudspeaker since only a small fraction of the total radiated power is selected for measurement. Frequency modulated tone may be used instead in which the oscillator frequency is varied at a rate of about 5 Hz to 10 Hz and the depth of modulation is $\pm 5\%$.

Pistol shots or chords from an orchestra may be used to obtain tape recordings of wide-band decay curves for subsequent analysis through band pass filters to obtain a graph of reverberation time against frequency. This method is frequently employed to measure the reverberation time of a concert hall in the presence of an audience; pistol shots should not be used if other methods are available as they are found by the author to give unreliable results in comparison with steady test signals.

It has been noted that the decay curves of an enclosure are seldom straight unmodulated lines and therefore some human judgement is required in assessing the best-fit slope. Moreover, background noise and the decreasing gradient, which often occur at the end of a decay, may influence judgement severely. By general agreement, it is now usual to quote the slope between levels 5 and 35 dB below the steady-state level reached before cut-off of the sound.

With either warble tone or bands of noise, successive repetitions of a decay vary noticeably, and a great advantage of the oscilloscope method is that the graticule can be set to agree with the mean slope of a large number of decays which follow one another on to the screen.

47.12.3.3 Schroeder's method of processing decay curves

It has been mentioned that a series of decay curves obtained under identical conditions when using finite-band noise such as bands of noise or warble tone, will all differ from each other. This is because, although the successive test pulses possess a common spectrum, they represent different functions of time and hence excite the room differently. It is possible to produce identical time-functions having a predetermined effective bandwidth by the use of short trains of waves started and stopped at fixed points in their cycles. With such decays, however, it is found that the variability of interpretation is greater than with more usual types of test signals with which a number of successive traces can be superimposed and the average slope estimated.

Schroeder[17] showed that by using a short tone-burst and integrating the energy in the reverberant signal from present time to infinity, a decay time was derived, representing the ensemble average of all possible decays resulting from a test signal with that particular spectrum. Curves thus obtained are free from the adventitious fluctuations which characterise ordinary decay curves.

In mathematical terms, one plots the function $\int_t^\infty p^2\,\mathrm{d}t$ where p is the instantaneous sound pressure and t the time. To obtain this integral, the decay curve can be recorded on magnetic tape which is then reversed and replayed into a squaring and integrating circuit. The output is converted to its logarithm and recorded by a chart recorder. The reverberation time is measured in the usual manner from the slope and divided by two since it represents integrated energy level instead of pressure level.

As an alternative to reversing a tape recording to obtain the integral, the first of two identical decays can be squared and integrated over its whole course and the second continuously integrated and subtracted from the whole integral before recording on the chart.

Thus,

$$\int_t^\infty p^2\,\mathrm{d}t = \int_0^\infty p^2\,\mathrm{d}t - \int_0^t p^2\,\mathrm{d}t \qquad (47.38)$$

Equipment for carrying out this method of measurement has been marketed, but it cannot be regarded as a suitable method for field use.

47.12.3.4 Computer derivation of reverberation time

The intrusion of human judgement into the assignment of decay curve slopes led Moffat and Spring[18] to develop completely objective methods, based on recorded test tapes, for routine use in the BBC studios.

The test tape carried bursts of signals consisting of one-third octave bands of noise and trigger signals. The tape was played into the studio through a loudspeaker and re-recorded, with the added reverberation of the studio. The signals on the tape were subsequently digitised and analysed to extract

(a) the reverberation time of each decay;
(b) the mean reverberation time and variance for each frequency;
(c) the level difference between start of the analysed part of the decay and noise level.

47.12.3.5 Steady-state determination of reverberation time

The methods of reverberation time determination described above yield accurate results but need dynamic equipment which may not be readily available to a non-specialised laboratory requiring to make occasional measurements, as for example in connection with transmission loss.

Two steady-state methods are possible. In one, due to Schroeder,[19] a loudspeaker is fed with a signal of slowly rising frequency f, and the phase difference φ between sound at any point in the room and at the loudspeaker is continuously plotted. It may be shown that

$$T = 2.2\,\mathrm{d}\varphi/\mathrm{d}f \qquad (47.39)$$

Another method is to measure the average sound level in the room produced by a sound generator of known power, or to compare, in effect, the SPL in the near and far fields. The necessary equipment consists of a loudspeaker with low directivity and a signal generator, sound level meter and band pass filters.

47.12.4 Measurements of acoustic absorption coefficients

The absorption coefficient of a material at a stated frequency is the proportion of sound energy incident on it which is absorbed or lost by transmission. This may be quoted for normal or random incidence. The absorption or absorbing cross section of a finite area of a material is the product of its area and its mean effective absorption coefficient.

47.12.4.1 Measurement by reverberation room

The principle of the reverberation method of measuring absorption coefficient, which yields the random-incidence coefficient, is to measure the reverberation time of a room at a series of frequencies as listed in *Table 47.6*, and then to repeat the measurements after fixing a suitable area of the sample material on to one or more surfaces of the room. The total absorption in the room is calculated from the reverberation time by the formula of Eyring[20] with and without the specimen in place. If $\bar{\alpha}$ is the mean absorption coefficient of all the room surfaces and S their area

$$-\ln(1-\bar{\alpha})=0.162V/TS \qquad (47.40)$$

where V is the volume of the room, and absorption $=S\bar{\alpha}$.

According to British Standard 3638:1963, the volume of the room should be between 180 and 250 m^3 and it should be surfaced with hard sound-reflecting surface finishes. Its reverberation time when empty should be at least 5 s up to 500 Hz, the lower limit diminishing by 1 s per octave above this frequency. The walls and ceiling should be heavy and rigid to avoid absorption at low frequencies by structural vibrations.

The sound should be radiated from a loudspeaker either near one corner of the room or one-third the way along a diagonal of the room, as these positions alone ensure a satisfactorily uniform excitation of all room modes.

The state of diffusion in the sound field should be enhanced by the use of sheets of sound-reflecting material hung from the ceiling. The sizes and orientations of the sheets should be distributed to obtain directional uniformity.

A single sample of 10 m^2 area is recommended for general tests of commercial materials, since this arrangement was found by Kosten[21] to result in the greatest measure of agreement between different testing laboratories. Divided samples, distributed on three or four surfaces of the room are to be recommended for tests in small areas to promote good diffusion. Subdivision increases the absorption coefficient of most materials especially at frequencies around 500 Hz. It should also be noted that subdivided distributed samples improve the diffusion of a reverberation room to such an extent that hanging sheets are unnecessary.[22] At least five measurements, from different positions of the microphone should be averaged, and it is an advantage to have two loudspeakers which are used alternately for each microphone position to increase the number of replications at each frequency. Bands of noise or warble tone are suitable as test signals.

It is usual not to make any correction for the loss of absorption of the test room surface when covered by the sample. This effect is kept small by the use of hard materials for the room surfaces, and attempts to make corrections may actually result in larger errors.

47.12.4.2 Measurements by standing-wave tube

Measurements of absorption coefficient by the reverberation method described above requires a large specially treated room, large test samples and facilities for the accurate measurement of reverberation time. A few establishments with a major interest in the development or use of sound absorbers are able to maintain these facilities, but most measurements of coefficients are necessarily carried out by two or three specialist consultant firms. The cost of this service makes it more suitable for checking production prototypes than as an aid for experimental product development. Useful information can, however, be obtained for

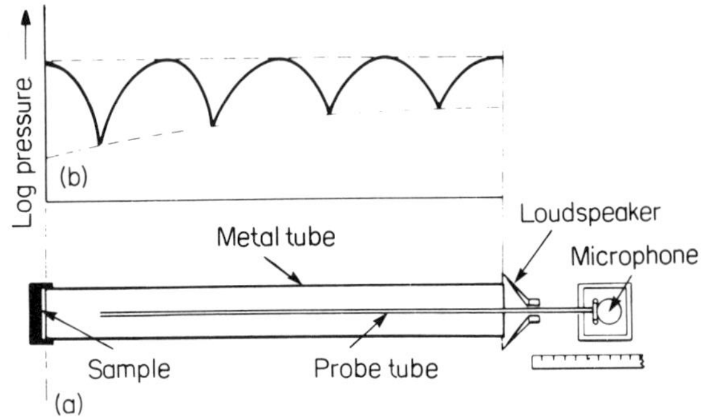

Figure 47.62 Standing wave tube apparatus for measuring sound absorption coefficient at normal incidence: (a) diagram of equipment; (b) variation of pressure along tube

the latter purpose by measurements of the normal-incidence absorption coefficient in a standing-wave tube.[23]

Figure 47.62(a) shows the principle of this apparatus. The sample, a small disc of 30 to 100 mm diameter, is cut from the material and held in a cap fitting tightly over one end of a tube about 1 m long. The other end of the tube is closed by a loudspeaker diaphragm and the magnet of the loudspeaker is bored centrally to permit the insertion of a probe tube attached to a microphone. The probe tube is long enough to reach the surface of the sample and the microphone is mounted on a sliding or wheeled carriage so that the sound pressure may be measured at any point along the tube.

With the sample in place and pure tone radiated by the loudspeaker, the sound pressure along the tube shows a series of maxima and minima at intervals of a half wavelength as shown in *Figure 47.62(b)*. The ratio of maximum to minimum pressures diminishes as distance from the sample increases owing to losses in the tube; if its value, extrapolated to the face of the sample, is n, the normal-incidence absorption coefficient is

$$\alpha_N=4/(n+n^{-1}+2) \qquad (47.41)$$

The theory of this apparatus is given by Beranek.[24]

If the exact distance of the first minimum from the surface of the sample is measured, the real and imaginary parts of the acoustic impedance at the face of the sample can be calculated, and the random-incidence absorption coefficient derived from these parameters for many types of material.[25,26] The calculation depends on the assumption that the impedance at the surface is independent of the angle of incidence. Although this is approximately true for a large proportion of absorbers, there is an element of doubt in the accuracy of the random-incidence coefficients arrived at in this way, and it is essential that final tests on a material to be used in acoustic treatment should be made by the reverberation method.

47.13 The electron microscope

The limitations of traditional light microscopy have been well known for many years. The resolution of such microscopes is determined by diffraction effects due to the wavelength of the illumination, and is of the order of 100–200 nm. This corresponds to a useful magnification limit of about $\times 5000$.

In 1924, de Broglie[27] showed wave particle duality and hence that the wavelength of an electron is a function of its energy, $E=h\nu$ where h is Planck's constant. Energy can be imparted to a charged particle by means of an electric accelerating field. Thus, at a sufficiently high voltage, say 50 kV, electrons of extremely short wavelength ($\lambda=0.0055$ nm) can be produced.

Since electrons can also be focused by electrostatic and electromagnetic fields, the potential for use of electrons as a source for microscopy was quickly realised. In addition, electromagnetic lenses are extremely versatile since the focal length of the lens may be changed by altering the current through it.

Two types of electron microscope were originally developed, the transmission electron microscope (TEM) and the scanning electron microscope (SEM). However in recent years, combination of these techniques to form the scanning transmission electron microscope (STEM) has made rapid advancement.

The transmission electron microscope is applied to ultra-thin or sectioned material, through which the electron beam is projected to form an image. The whole field of view is illuminated simultaneously and the enlarged image observed on a fluorescent screen incorporated in the microscope.

The scanning electron microscope images solid surfaces. The electron beam is focused to a small spot and is then scanned over the surface. The resultant electron signal is collected and displayed as a brightness modulated image on a cathode ray tube.

47.14 The transmission electron microscope

Transmission electron microscopes were first built in the early 1930s and were soon developed into commercial instruments for research. Modern TEM have resolutions of the order 0.2 nm, making them powerful instruments in the study of crystalline materials and investigation of the structure of fine particles.

47.14.1 Instrument construction

Figure 47.63 shows a schematic diagram of the main components of a TEM. The electron gun contains a directly heated cathode and a Wehnelt cylinder, acting as a bias shield, mounted on an insulator. The accelerating potential between gun and anode can be varied from about 50 kV to 200 kV depending on the type of instrument, thereby varying the penetrating power of the electron beam. A condenser lens system makes it possible to reduce the cross section of the beam emitted from the gun and is used to illuminate the area of interest on the sample. This illumination can be varied by adjustment of the condenser lens current for differing working conditions. The current in the objective lens controls the focus of the image on the flourescent viewing screen, and the projector lens is used to vary the magnification typically $\times 1000$ to $\times 500\,000$.

In order to allow the accelerated electrons to reach the fluorescent screen, the electron optics and viewing screen or photographic plate must be maintained in a vacuum. Improved vacuum systems giving working pressures of 10^{-6} to 10^{-7} Torr (mm Hg) are obtained using oil diffusion pumps in conjunction with cold traps. This reduces any effect of contamination of the specimen due to residual oil vapours.

The high resolving power of modern TEM requires that the power supplies for the lenses and high voltage applied to the gun be extremely stable.

Brightness variation in the TEM image is caused by scattering of the electron beam due to different densities within the specimen. The electrons pass through the specimen which must be thin enough to allow the beam to be transmitted, approximately 20–40 nm. When the focus is correct, electrons screate a projected image of the sample on the fluorescent screen. Images can be recorded on photographic plate or film by swinging away the fluorescent screen and allowing the image to fall onto the film which is housed within the vacuum environment of the microscope.

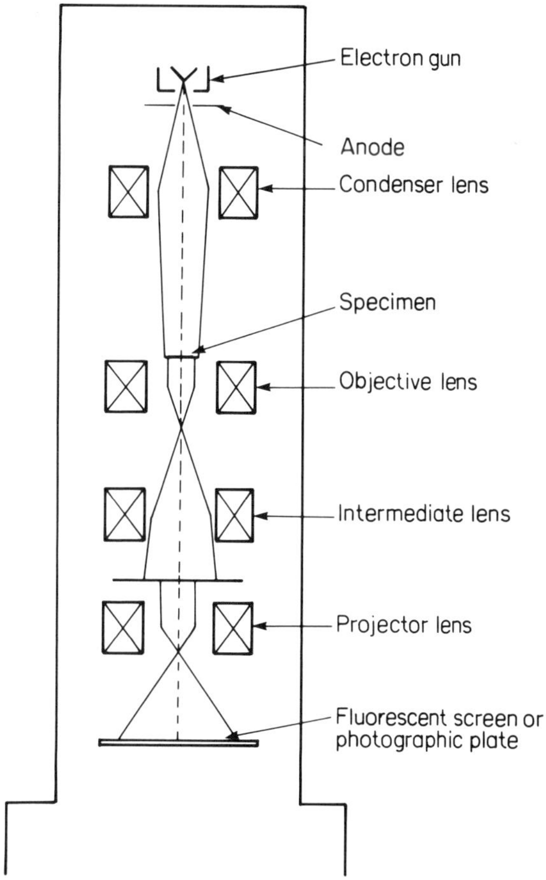

Figure 47.63 Schematic diagram of a transmission electron microscope

47.14.2 Specimen handling in the TEM

Specimen preparation is an important consideration for transmission electron microscopy. Because contrast is formed by electron scattering within the specimen, fine detail will only be observed in very thin specimens. Thick specimens show overlapping of detail from different height levels within the specimen and reduced resolution due to chromatic effects. Scattered radiation is either prevented by the objective aperture from reaching the image plane or merely contributes to the background intensity.

47.14.2.1 Replication

The preparation of replicas of specimens which are opaque to electrons and which cannot be thinned enables their surface structure to be studied in the TEM. Replicas are most widely used for the examination of bulk specimens such as polished and etched metals.

A replica consists of a thin film of material which is electron transparent, the material corresponding exactly to the topography of the specimen surface. One technique for the production of replicas is to vacuum evaporate carbon over the surface to be examined. The contrast produced in the TEM by replicas is often very low and can be improved by shadowing using an electron-dense material evaporated at an angle to the replica surface (*Figure 47.64*). The shadowed replica can then be

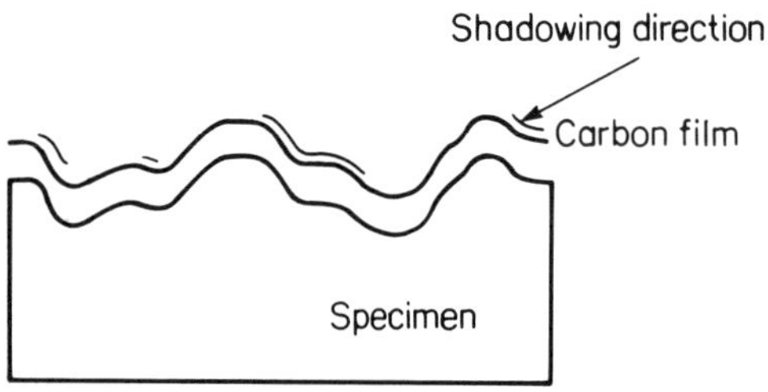

Figure 47.64 Schematic diagram of replication process

stripped from the specimen and mounted for examination in the TEM.

47.14.2.2 Preparation of materials

Thick samples whose structure and composition is constant throughout the thickness can be prepared for TEM by thinning. Controlled removal of material from the surface to a suitable thickness can be obtained by electro-polishing or ion-beam thinning.

47.14.2.3 Embedding and sectioning

This technique is suitable for soft materials, particularly biological tissue. The specimen is embedded in a resin that will remain stable under the electron beam, and thin sections are shaved from the surface using a microtome. Specimens about 40 nm thick can be cut by repeatedly moving the sample past a sharp cutting edge whilst making a small advance towards the knife.

47.14.2.4 Mounting of specimens

Specimens are usually mounted on 3.0 or 2.5 mm diameter metal grids. The grid materials may be of copper or nickel and varying mesh sizes are available. A thin support film may be used to hold the specimen in place on the grid. The grid, complete with specimen, may then be mounted in a special holder and inserted into the vacuum system of the TEM.

47.15 The scanning electron microscope

47.15.1 Construction and performance characteristics

The SEM was postulated in the 1930s by Knoll[28] and Von Ardenne,[29] and serious design study started in 1948 under Oatley in Cambridge, resulting in commercial production in 1965.

A schematic diagram of the SEM is shown in *Figure 47.65*. An electron source is used to provide electrons which are accelerated to a high potential (0.5–50 kV). The beam of electrons is focused by two or three electromagnetic lenses to form a small spot on the sample surface. Double deflection coils housed inside the final lens cause the spot to be scanned over the specimen surface in a raster.

The beam of electrons can interact with the specimen surface in a variety of ways, generating radiations from the sample which are characteristic of its composition and topography; a proportion of this radiation is emitted from the surface and can be used to characterise it. Furthermore, as the beam is scanned, the signal level of each characteristic radiation may vary with surface composition and topography. By applying these signals as brightness modulation to a cathode ray tube scanned in synchronisation with the electron beam, images of the surface characteristics can be derived.

The electron source is in general a heated tungsten hairpin,

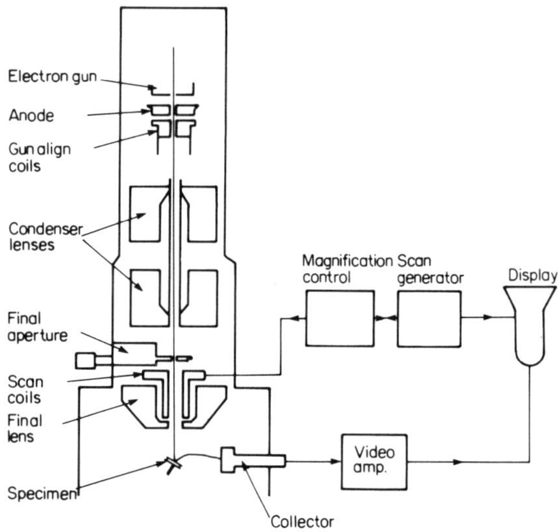

Figure 47.65 Schematic diagram of scanning electron microscope

operating over accelerating voltages of 1–50 kV, but alternative sources such as lanthanum hexaboride and field emission sources are also in use, and these give higher brightness. However the alternative sources generally have greater vacuum requirements than that for a tungsten filament, and so ancillary pumping must be provided. The required vacuum level in an SEM is usually provided by an oil diffusion pump backed by a rotary pump, but contamination-free environments are now being required. Liquid nitrogen cooled traps or baffles fitted about the diffusion pump help reduce the contamination. However, oil-free pumps such as turbo-molecular pumps are considered essential to minimise contamination.

The principal imaging mode of the SEM uses secondary electrons. These are generated by electron–atom interactions and have energies sufficient to travel only some 2–20 nm within the substrate material. Thus only those generated close to the surface can be re-emitted and even these are vulnerable to absorption by surface topography.

The spatial resolution currently attainable in this mode is of the order of 3–7 nm. The detector is commonly a positively biased scintillator accelerating electrons into the active area, and thence transmitting a signal via a light guide and photomultiplier to amplifiers and signal processors and finally to a cathode ray tube. Particular properties of the secondary electron image are the large depth of focus available and the flexible magnification range. Magnification derives simply from the ratio of a length scanned on the sample to that scanned on the cathode ray tube, and hence can be varied over a range of typically $\times 5$–$\times 300\,000$ simply by changing the currents through the scanning coils whilst maintaining the scanned area on the cathode ray tube.

Magnification is independent of lens focus and can be zoomed rapidly through its range centring on a fixed point on the specimen surface. Although using a two-dimensional display, the images produced are characterised by their three-dimensional appearance and relative ease of interpretation by non-specialist staff. Images may be viewed at TV rates on the visual display, and interfacing to video recorders is also possible. A wide variety of image processing techniques are available (e.g. differentiation, expanded contrast) which can be used to highlight important features of images to aid interpretation.

Image recording uses a second cathode ray tube and conventional camera system. The record tube is of high resolution (typically 2500 lines) to give optically high quality pictures. Other imaging signals available in the SEM include the following:

(1) Backscattered electrons. Electrons from the primary beam may be scattered through large angles by atomic nuclei and re-emitted with only small losses in energy. Since the backscattering process is dependent on the mean atomic number of the material, the contrast obtained gives compositional detail of the sample rather than topographic detail, provided that the detector is placed symmetrically about the electron beam and the sample surface is normal to the electron beam. If the detector is orientated such that there is a preferred direction for detection, then topographic detail may be enhanced.

(2) Absorbed electrons. Under most operating conditions in the SEM, the rate of input of electrons from the beam is greater than the total rate of emission of secondary and backscattered electrons. The resultant charge is usually dissipated by earthing the specimen. However if the sample is connected to a current amplifier rather than earth, it is possible both to measure the absorbed current and produce a specimen current image. The absorbed current image produced is complementary to the backscattered image. Certain materials exhibit beam induced conductivity effects also, and these are particularly useful in semiconductor research.

(3) X-rays. The action of the electron beam to create vacancies in material electronic structure leads to electron transitions accompanied by emission of the appropriate quanta of electromagnetic radiation. In general, this radiation is in the x-ray region of the spectrum, and may be monitored using an appropriate detector. The x-rays so produced are characterised for each element, and this forms the basis of the technique of x-ray microprobe analysis. Detection may be either by the conventional crystal spectrometer (wavelength dispersive systems) or by lithium drifted silicon solid state devices (energy dispersive system). These give information about the elemental composition of the specimens and can yield qualitative and quantitative analyses.

(4) Cathodoluminescence. Some materials exhibit fluorescence in the visible region of the spectrum under electron bombardment as illustrated by the phosphors of cathode ray tubes. The emitted radiation, known as cathodoluminescence may be used for imaging in the SEM using an appropriate detector. The radiation is in many cases a function of impurity levels within materials and is used both in semiconductor materials research and in many minieralogical investigations.

47.15.2 Specimen handling in the SEM

The necessity of a vacuum environment in the SEM precludes the examination of gases and liquids, and special precautions must be taken with solid materials containing high proportions of liquid or gas. In addition, bombardment by negatively charged electrons ultimately builds up a surface charge and surfaces which are not normally electrically conducting must be rendered so.

Large specimen chambers and specimen stages with wide ranging movements allow samples of up to 175 mm diameter to be examined in a variety of orientations.

Metals may be mounted directly or after cleaning, either chemically or ultrasonically, to remove surface debris. Minerals, ceramics etc are coated with a thin (10–20 nm) conductive layer usually of gold, gold–palladium or carbon. This is done either by evaporation or sputtering. Carbon layers are most frequently used where microanalysis is required and no light elements are to be detected. Organic materials are treated similarly, but are generally examined at lower beam voltages to prevent bombardment damage. Biological material contains a large amount of fluid and is generally subjected to some drying procedure such as freeze drying or critical point drying. In some cases quench freezing is used and material is then examined in the frozen state on a cold stage.

Because of the large specimen chamber available, the SEM is particularly well suited to *in situ* sample processing. Behaviour of materials at high and low temperatures, during tensile testing etc are studied using videotape to record the dynamic sample changes as they occur.

The SEM is applied in both fundamental and applied research to many problems involving the physical structures of solid surfaces and particularly those requiring greater depth of focus or higher magnification than is obtainable with the light microscope. Other uses include fractography, failure mechanics, the study of powders and compacted materials, tribology and corrosion science. In biology, geology and pathology the SEM is exploited both as a research tool and as a means of presentation of micrographical information to the non-specialist and student. The SEM has found many uses in forensic sciences. It has also gained a wide acceptance throughout the electronics industry, both in research and quality control. Its range of applications in this field are detailed in section 47.17.

47.16 Scanning transmission electron microscopy (STEM)

STEM has made rapid advances over recent years. It is possible to generate variations on STEM by adding scanning attachments to transmission microscopes or transmission detectors to scanning microscopes. However such is the growth of the technique that specially designed STEMs are now being built.[30]

Thin sectioned material is used as in the TEM but scanning image collection and processing of the transmitted electrons are retained. Single-atom imaging is obtainable using STEM optimised electron optics. A variety of imaging modes is available for STEM and finds application in materials science particularly in combination with microanalytical techniques.

47.17 Electron microscope applications in electronics

The scanning electron microscope is particularly suited to examination of electronic components and semiconductors. The lack of sample preparation required makes the SEM invaluable in providing rapid examination of failed components and product monitoring which are essential processes in the electronics industry.

The interaction of an SEM beam with a semiconductor sample can generate many phenomena worthy of study, including some that cannot be observed by any other technique. Electron beams are easy to control, to deflect in scan patterns, switch on and off, and to adjust in energy with which they bombard the specimen. Secondly, the alternative operating modes of the SEM quickly answer many of the questions which arise during semiconductor research and production. The use of electron beams in the microelectronics field has grown rapidly in the past ten to fifteen years. Large depth of focus is one of the most important features of the SEM for examination of integrated circuits, allowing detailed study of step coverage at very high tilt angles. As circuits become more complex, with circuit elements reaching the limits of photolithography the superior resolving power and depth of focus of the SEM has become a necessary part of the evaluation of new circuits, construction techniques and quality control.

A considerable reduction in the size of circuit elements can be made by using an electron beam technique instead of conventional photolithography. The electron beam microfabricator (EBMF) uses an electron beam to write patterns on semiconductor wafers, and the system is computer controlled.

The EBMF allows the minimum linear dimension of a circuit element to be reduced from 2 μm (using conventional photolithography) to 100 nm therefore increasing the available packing density.

47.17.1 Semiconductor investigations in the SEM

Very little sample preparation is needed for examination of semiconductor material in the SEM. Part-processed or completed wafers can be observed in most modern SEM without breaking and with no preparation. Part-processed wafers can thus be examined at intervals during manufacture to aid quality assessment. Finished devices need to be opened such that the circuit can be seen, and mounted in the SEM with at least one termination grounded.

For the purposes of SEM imaging, a semiconductor device is conductive, even silicon oxide and passivation layers applied to circuits exhibit induced conductivity at the beam energies normally used. Several types of SEM imaging can be used on semiconductor samples to give information which is unobtainable by other techniques.

47.17.1.1 Secondary electron imaging

Imaging using secondary electrons is used to show surface details of topography and also contrast due to differences in atomic number. If a semiconductor wafer is tilted with respect to the incident electron beam, the large depth of focus of the SEM is ideal for monitoring features such as oxide steps formed by multiple diffusions and the integrity and profile of metal tracks (*Figure 47.66*).

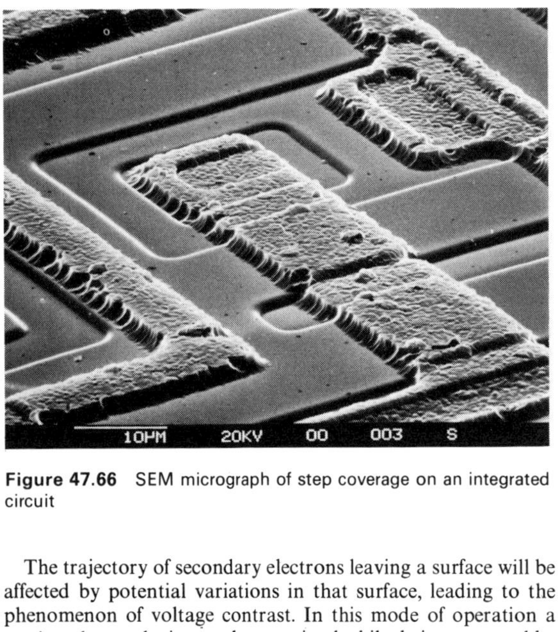

Figure 47.66 SEM micrograph of step coverage on an integrated circuit

The trajectory of secondary electrons leaving a surface will be affected by potential variations in that surface, leading to the phenomenon of voltage contrast. In this mode of operation a semiconductor device can be examined whilst being operated by an external circuit. The potential of any part of the circuit will affect the generation of secondary electrons and therefore the brightness of the resultant image. Thus voltage contrast can be used to observe electrical failure beneath the surface of a device or reverse biased junctions because of discontinuity in contrast. The contrast change is proportional to the value of the applied voltage and therefore it is possible to measure potential at various points in the circuit.

The study of integrated circuits using voltage contrast is often done dynamically, with the circuit under simulated working conditions, e.g. an MOS shift register can be run at high frequency and a beam switching system used to blank the electron beam at some sub-harmonic of the fundamental logic frequency applied to the device. This technique effectively freezes the device in a particular logic state. Dynamic voltage contrast studies can be recorded on videotape, the stroboscopic effect of the beam blanking system effectively slowing down the operating frequency of the device.

Voltage contrast effects are most easily observed if there is no passivation layer over the surface of the device, but it is still possible to observe contrast on passivated devices provided the energy of the primary electron beam is carefully selected with reference to the passivation thickness (*Figure 47.67*(*a*) and (*b*)).

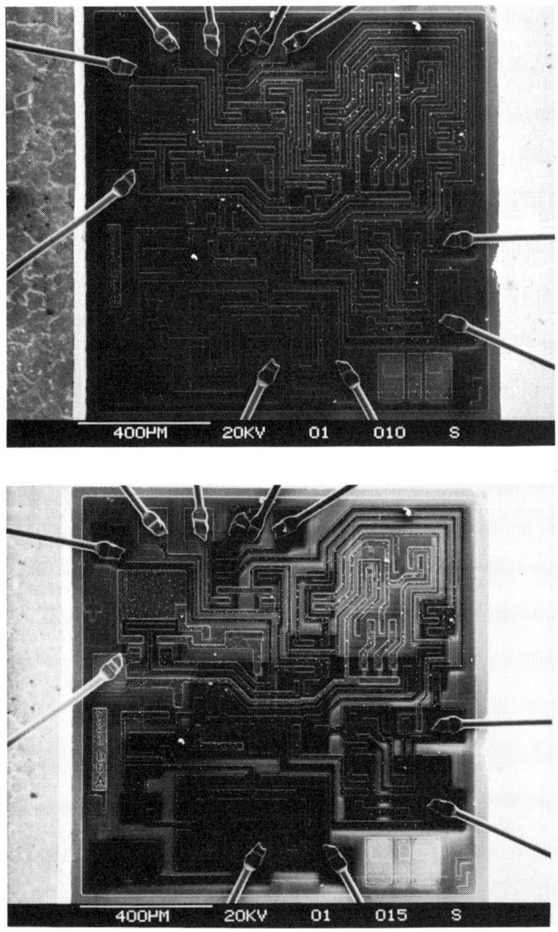

Figure 47.67 Secondary electron image of standard voltage regulator, +9 volts applied

47.17.1.2 Absorbed electron and conductive imaging

A specimen current amplifier can be used to produce an image by using the absorbed current to modulate the CRT display. The absorbed current image gives contrast which is generally complementary to the backscattered electron image. Whilst the technique can be applied to many types of samples, it is often semiconductor materials which yield most information.

If some pins of a completed semiconductor device are earthed and one or more pins connected to a specimen current amplifier, the specimen current flowing in the device as a result of conductivity induced by the beam can be used to modulate the brightness of the CRT display. The conduction mode depends on the conductivity increase of the target region due to the generation of extra carriers. The primary electrons entering the semiconductor create electron–hole pairs, available for conduction processes until removed by recombination.

The conduction mode (called EBIC, electron beam induced conductivity) can be used to study depletion region boundaries both with and without an externally applied bias (*Figure 47.68*).

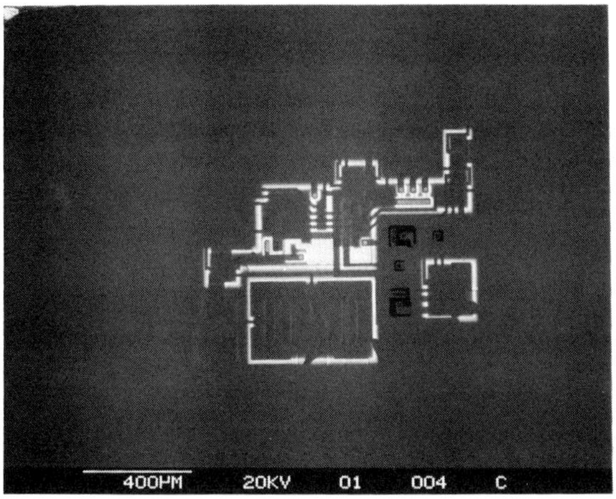

Figure 47.68 EBIC image of voltage regulator, −0.25 volts applied

EBIC imaging is an important technique for monitoring depletion layer spreading, junction breakdowns and oxide pinholes which do not produce compositional or topographic contrast and therefore cannot be observed in any other SEM imaging mode. Since the electron beam will easily penetrate semiconductor materials, careful choice of primary beam voltage allows information in the EBIC mode to be obtained from junctions buried beneath a surface layer. A 10 keV electron beam penetrates about 1 μm of silicon or silicon dioxide, but the penetration depth increases to about 5 μm for a 30 keV electron beam.

The recombination of carriers can also lead to the emission of light photons from the sample. The emission of light under the influence of a primary electron beam is known as cathodoluminescence. Phosphors, light emitting diodes and semiconductor laser materials all exhibit cathodoluminescence which can be detected and used to modulate the CRT display brightness or analysed for wavelength using a spectrometer. Luminescent efficiency of many materials is highly temperature dependant; special purpose specimen stages available for the SEM allow controlled temperature variation, above and below ambient during examination.

47.17.1.3 X-ray analysis

Analysis of x-rays produced by interaction of the primary electron beam and the specimen is a powerful technique for characterising the composition of all types of electronic components. Concentrations of elements as low as 0.01% from depths as small as 0.5–1 μm in the specimen surface can be analysed non-destructively on complete components, devices or wafers. X-ray analyses can be displayed in the form of mappings which show the distribution of a chosen element over the surface of a specimen.

The technique is particularly useful in troubleshooting electronic components which have failed due to contamination occurring in production or operating lifetime.

47.17.1.4 Electron channelling

Modification of the operation of the lenses of an SEM allows the beam to be rocked about a point on the surface. The contrast arising from backscattered electrons varies due to the arrangement of atomic planes in the crystal lattice and electron beam channelling patterns are formed. These patterns relate to crystallographic structure and can be used to identify orientation of single-crystal material such as silicon, germanium and gallium arsenide. Selected area channelling patterns can be obtained from grains having a minimum size of about 2 μm.

References

1 KING, L. V., *Proc. R. Soc. A.*, **153**, 17 (1935)
2 COOK, R. K., *J. Res. N.B.S.*, **25**, 489 (1940)
3 ZWICKER, E., *Acustica*, **10**, 304 (1960)
4 STEVENS, S. S., *J. Acoust. Soc. Am.*, **28**, 807 (1956)
5 ROBINSON, D. W. and DADSON, R. S., *Brit. J. Appl. Phys.*, **7**, 166 (1956)
6 MILLS, C. H. G. and ROBINSIN, D. W., *Engineer*, **211**, 1070 (1961)
7 DEPARTMENT OF EMPLOYMENT, *Code of Practice for Reducing the Exposure of Employed Persons to Noise*, H.M.S.O. (1972)
8 KRYTER, K. D., *Noise Control*, **6**, 12 (1960)
9 GILFORD, C. L. S., *Acoustics for Radio and Television Studios*, p. 52, Peter Peregrinus, London (1972)
10 MOFFAT, M. E. B., *BBC Research Department Report*, **PH 8** (1967)
11 WARD, F. L., *Proc. 4th Internat. Congress on Acoustics, Copenhagen*, Paper L 11
12 GOFF, K. W., *J. Acoust. Soc. Am.*, **27**, 233 (1955)
13 BURD, A. N., *J. Sound Vib.*, **7**, 13 (1968)
14 CREMER, L. and GILG, J., *Acustica*, **23**, 54 (1970)
15 FRANKLIN, W. S., *Phys. Rev.*, **16**, 372 (1903)
16 SOMERVILLE, T. and GILFORD, C. L. S., *BBC, Q.*, **8**, 41 (1952)
17 SCHROEDER, M. R., *J. Acoust. Soc. Am.*, **37**, 409 (1965)
18 MOFFAT, M. E. B. and SPRING, N. F., *B.B.C. Eng.*, **80** (1969)
19 SCHROEDER, M. R., *Proc. 3rd Internat. Congress on Acoustics, Stuttgart*, **2**, 897 (1959)
20 EYRING, C. F., *J. Acoust. Soc. Am.*, **1**, 217 (1930)
21 KOSTEN, C. W., *Proc. 3rd Internat. Congress on Acoustics, Stuttgart*, **2**, 815 (1959)
22 GILFORD, C. L. S., *Acoustics for Radio and Television Studios*, p. 178, Peter Peregrinus, London (1972)
23 SCOTT, R. A., *Proc. Phys. Soc. London*, **58**, 253 (1946)
24 BERANEK, L. L., *J. Acoust. Soc. Am.*, **12**, 3 (1940)
25 ATAL, B. S., *Acustica*, **9**, 27 (1959)
26 DUBOUT, P. and DAVERN, E., *Acustica*, **19**, 15 (1959)
27 DE BROGLIE, L., *Phil. Mag.*, **47**, 466 (1924)
28 KNOLL, M., 'Aufladepotential und Sekundar-emission elektronbestrahlter Oberflachen', *Z. Techn. Phys.*, **2**, 467 (1935)
29 VON ARDENNE, M., 'Das Elektronen raster mikroskop', *Z. Techn. Phys.*, **19**, 407–416 (1938)
30 WIGGINS, J. W., ZUBIN, J. A. and BEER, M., *Rev. Sci. Instrum.*, **50**, 403 (1979)

Further reading

WELLS, O. C., *Scanning Electron Microscopy*, McGraw-Hill (1974)
GOLDSTEIN, J. I., YAKOWITZ, H., *Practical Scanning Electron Microscopy*, Plenum Press (1975)
GLAUERT, A. M. (Editor), *Practical Methods in Electron Microscopy*, North Holland (1974 onwards)
REIMER, L., *Scanning*, **1**, 3 (1977)
JOHARI, O., *Scanning Electron Microscopy* (Proceedings of Annual Conferences)

MORGAN, D., 'Digital storage oscilloscopes', *New Electronics* (April 1979)
PARISH, D., 'New developments in digital storage oscilloscopes', *Electronic Engineering* (November 1980)
FARNDALE, L., 'Developments in digital storage oscilloscopes', *New Electronics* (April 1981)
BLATCH, H., 'The increasing capabilities of modern DMMS', *New Electronics* (April 1981)
NELSON, R., 'High frequency instruments', *EDN* (February 1981)
TEJA, E. R., 'Voice-input equipment', *EDN* (May 1981)
PARISH, R., 'Programming digital storage oscilloscopes', *New Electronics* (November 1981)
THOMPSON, T., 'Temperature testing electronic components', *Electronic Production* (January 1982)
CHRONES, C., 'Capable analog testers handle a variety of devices', *EDN* (January 1982)
CHESNUTIS, G. F., 'Intelligent instruments', *EDN* (March 1982)
ROLFE, D., 'The measurement of RF levels', *Electronic Product Design* (March 1982)
CONNEL, R. J., 'IEEE—488 bus test systems', *New Electronics* (April 1982)
TEGEN, A. and WRIGHT, J., 'Oscilloscopes: the digital alternative', *New Electronics* (April 1982)
MILLS, J. and STAMATIADES, A., 'Aids for the digital designer', *New Electronics* (April 1982)
MUTO, A. and NEIL, M., 'ADC dynamic performance testing', *Electronic Product Design* (June 1982)
BUONO, J. A., 'Scanning electron microscopes: matching the instrument to the application', *Test & Measurement World* (June 1982)
KRANS, M., 'Intelligent oscilloscopes', *New Electronics* (June 1982)
TILDEN, M. and RAMIREZ, R., 'Understanding IEEE—488 basics simplifies system integration', *EDN* (June 1982)
CHESNUTIS, G. F., 'Logic analyzers', *EDN* (June 1982)
KIKUCHI, Y., 'Digital LCR measurement instruments', *Electronics Industry* (July 1982)
ROARK, D. and FRENCH, H., 'Bringing true 3D imagery to color raster graphics', *Mini-Micro Systems* (July 1982)
SAMPL, S., 'HP-IL data-acquisition, control system opens new measurement territory', *Electronics* (October 1982)
DONNELLY, W., 'Logic test instruments—choice extends from probes to complex analysers', *Electronics Industry* (October 1982)
CHESNUTIS, G. F., 'Modular instruments', *EDN* (November 1982)

48 Control Systems

J E Harry, BSc (Eng), PhD
Senior Lecturer
Department of Electronic and Electrical Engineering
University of Technology
Loughborough

Contents

48.1 Open-loop systems

Control systems are used in all areas of industry. The process may encompass almost any conceivable operation ranging from operation of a machine tool to filling milk bottles. The components comprising a simple control system can be categorised as controller, final actuator or servo, the process and the sensor. The components vary depending on the process and some examples of different components used in different processes are listed in *Table 48.1*. The process parameters are varied by the final actuator. The final actuator may be controlled in turn by a secondary actuator, e.g. the speed of a hydraulic drive may be increased with a motorised valve. The control setting may be varied with the controller which may be mechanical, electromechanical, hydraulic or electronic. The setting (set point) of the controller is fixed by the operative in charge of the process.

Such a system is known as an open-loop system and is illustrated schematically in *Figure 48.1(a)*. The sensor indicates the state of the process. Open-loop systems are used in all areas of industry where fluctuations in the process variables occur within acceptable limits or occur slowly and can be corrected by the operator adjusting the set point. If now a feedback signal is obtained from the sensor this can be used to maintain the process within set limits. The feedback path is shown in *Figure 48.1(b)*. This now becomes a closed-loop control system and is known as a servomechanism. The action of the controller is now to compare a reference signal corresponding to the set point with the feedback signal, the difference being the error signal E. This is normally amplified and used to restore the process to the set point.

48.2 Closed-loop systems

A simple example of a closed-loop control system is the self balancing potentiometer illustrated in *Figure 48.2*. Depending on the polarity of the out-of-balance voltage the servomotor varies the tapping position on the slide wire. The error signal tends to zero as the balance position is reached so that in practice a small error exists corresponding to a region either side of the balance position over which no control action occurs known as the dead space. This is reduced by the amplifier. If the gain A of the amplifier is high and the feedback is large so that the amplification is small the sensitivy of the system is increased by a factor $1/A$.

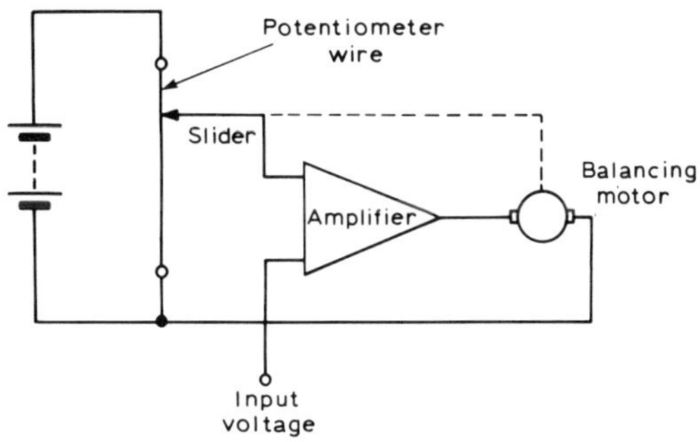

Figure 48.2 Servo-assisted self-balancing potentiometer

48.3 Control equations

Each element of a control system can be typified by its response in terms of the change in output over the input which will be a time-dependent function and can be expressed in terms of first or higher order differential equations.

By taking the Laplace transform of the response the transfer function of a system can be readily determined by combining the transfer functions of the individual elements of the system in

Table 48.1 Examples of elements in control systems

Process	*Control variable*	*Controller*	*Servo*	*Sensor*
Milling	position	on-off	stepper motor	encoder
Oil flow to burner	flow	proportional	d.c. shunt motor	turbine flowmeter
Rolling mill	velocity	proportional and derivative	Ward-Leonard drive	tachometer
Batch furnace for heat treatment	temperature	proportional, derivative and integral control	triac	thermocouple

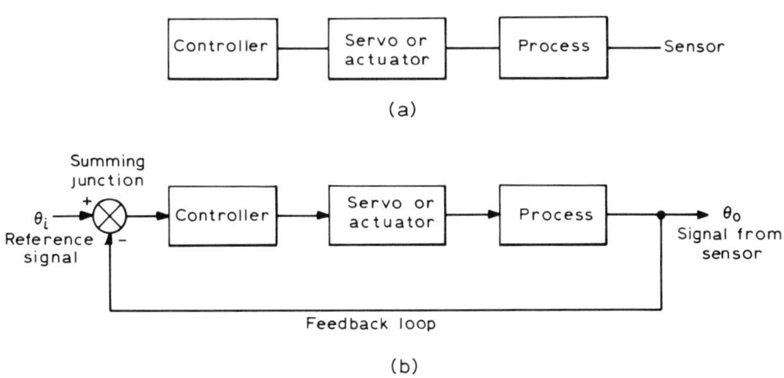

Figure 48.1 Basic components of a control system: (a) open loop; (b) closed loop

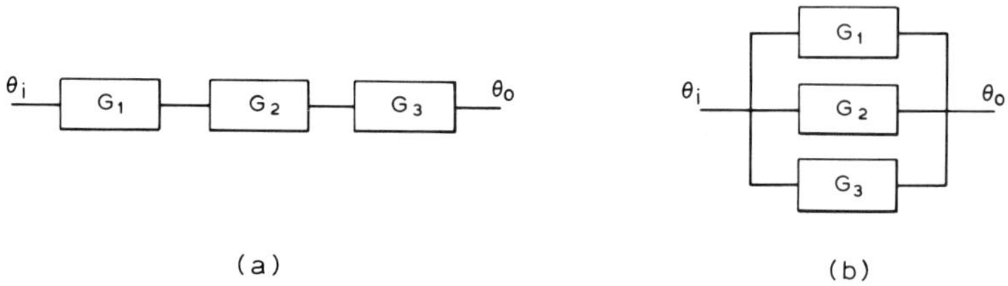

Figure 48.3 Series and parallel connection of system components: (a) series; (b) parallel

Figure 48.3. The transfer function of elements connected in series is equal to their product

$$G = G_1 \,.\, G_2 \,.\, G_3 \qquad (48.1)$$

and in parallel

$$G = G_1 + G_2 + G_3 \qquad (48.2)$$

The servo system can now be represented by the block diagram in *Figure 48.4*.

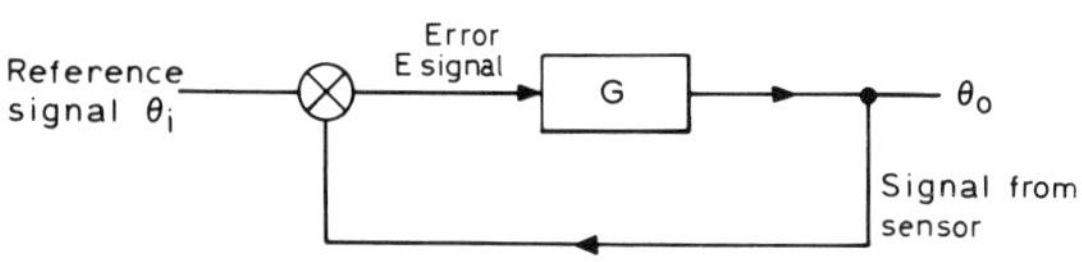

Figure 48.4 Closed-loop control system (servo or batch operation)

The system can be described by the system transfer function, the control ratio and the error function. The system transfer function is given by

$$G = \frac{\theta_o}{E} \qquad (48.3)$$

The control ratio is given by

$$\frac{\theta_o}{\theta_i} = \frac{G}{1+G} \qquad (48.4)$$

and the error ratio

$$\frac{E}{\theta_i} = \frac{1}{1+G} \qquad (48.5)$$

The response of any single control element can be resolved in terms of the damping coefficient ξ the time constant T and ω critical frequency.

For a first-order system the response is given by

$$T\frac{d\theta_o}{dt} + \theta_o = \theta_i \qquad (48.6)$$

and

$$a_2 \frac{d^2\theta_o}{dt^2} + a_1\, d\theta_o + a_o\theta_o = a_o\theta_i \qquad (48.7)$$

where

$$T = \frac{2a_2}{a_1} \qquad \xi = \frac{a_1}{2\sqrt{a_o a_2}}$$

and

$$\omega_n = a_o/a_2$$

from which it can be seen that

$$T = \xi\omega_n \qquad (48.8)$$

By combining the Laplace transforms of each element it is possible to determine the response of the complete system. The Laplace transforms corresponding to various common control system elements are tabulated in *Table 48.2*. The response of the system can be solved analytically for first- and second-order systems. Since the order of the system is equal to the sum of the order of each component in the system, systems with third and higher order responses are often encountered which need to be solved by iterative techniques.

Table 48.2 Examples of control systems and their transfer functions

System or process	*Transfer function*
First-order	$\frac{1}{Ts+1}$
Second-order	$\frac{1}{T^2\xi^2 s^2 + 2T\xi^2 s + 1}$
Integral controller	$\frac{1}{Ts}$
Proportional controller	K_c
Rate controller	Ts
Linear final control element with a a first-order lag	$\frac{K_v}{Ts+1}$

Another example of a simple closed-loop control system is the variable load or regulator system shown in *Figure 48.5*. This corresponds to a process with varying but controlled parameters,

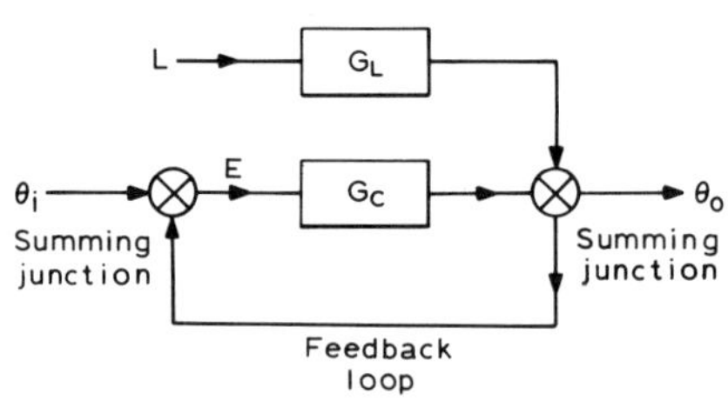

Figure 48.5 Regulator or variable load operation

e.g. throughput. In this case assuming the control level θ_i is fixed the transfer function becomes

$$\frac{E}{L}=\frac{G_2}{1+G} \quad (48.9)$$

$$\frac{\theta_o}{L}=\frac{G_2}{1+G} \quad (48.10)$$

48.4 Control system characteristics

In any closed-loop control system a high degree of precision in terms of the error ratio is required. One factor that limits this is the system stability which decreases as the amplifier gain is increased. If a system becomes unstable it goes out of control which may have disastrous effects.

For any control system there is a small region either side of the set point over which no control action occurs. This may be due to mechanical backlash, thermal inertia, etc. If the dead space is made very small instability may occur. This is known as hunting and occurs where the control element is in effect oscillating about the set point, and is an important consideration in the design of precision control systems.

The stability of a system can be determined if the damping coefficient time constant and critical frequency are known. By varying the relative magnitudes of these parameters the stability can be changed. The cause of instability can be considered in terms of the open-loop frequency response of a system. If delays which exist in the system result in phase lags which are in anti-phase with the input, the output signal is fed back to the input so as to increase the error rather than decreasing it in the same way as positive feedback causes instability in an amplifier.

The magnitude of the delays in a control system is often an important factor governing the stability of the system as well as the degree of precision of control that may be obtained. Delays occur with each element of the system, e.g., sensor controller, actuator and process. Delays also occur in transmission between each element although this will often be small. Depending on the system involved the relative magnitudes of the delays may be important or not. In complex systems the response of individual elements may not be known or the overall response of the entire system may be too complex to analyse. In these cases the response may be measured by determining the response of the system or a model of it to standard input signals. These are usually the response to a continuous sinusoidal input, a ramp function and a square wave input.

The response to a sinusoidal input as a function of frequency enables the amplitude frequency response to be obtained which when plotted on a log scale of frequency is referred to as a Bode diagram. The plot of the phase shift as a function of frequency in polar coordinates is referred to as the Nyquist diagram from which the system stability can be obtained.

48.5 Control modes

The application of control systems can be illustrated by an electric oven operating in an on/off mode as in *Figure 48.6*. The power is supplied until the set point is reached. Delays exist due to the thermal capacity of refractories, the load and the temperature sensor. When the power is switched off as the set point is reached, heat continues to flow resulting in overshoot and when the temperature falls through the set point there is a delay before the power is switched on. The delay is known as lag. The relative positions of the sensor and load with respect to the heating element are critical. If the sensor is remote from the heating element, large relatively slow swings of temperature with possible overheating of the load will occur, while if it is close to the heating element rapid oscillations will occur and the load may not reach the required temperature.

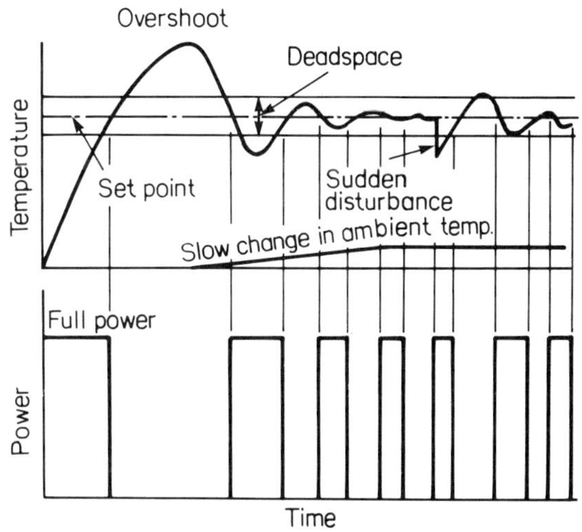

Figure 48.6 Response to on-off control

Oscillation about the set point is inevitable with on/off control since switching occurs only at the set point, when full power is applied. A small difference between the set point and the temperature at which the power is connected often exists and this may be increased to prevent rapid wear of switch contacts.

Proportional action is the basis of continuous control. The power input at any temperature θ within the proportional band is given by

$$W=\frac{\theta-\theta_0}{\theta_2-\theta_1}W_0 \quad (48.11)$$

and the response is

$$p=\frac{e}{b} \quad (48.12)$$

where θ_2 is the temperature of the upper limit of the proportional band, θ_1 at the lower limit, W_0 is the maximum power input at θ_2, e is the error signal and b is a constant for the system.

The power can be varied by controlling the mark/space ratio by relays or contactors, or by using thyristors to give rapid sequence or phase-shift control. The overshoot is reduced by decreasing the power input as the set point is approached so that it is equal to the heat losses when the set point is reached. This is known as proportional control (*Figure 48.7*). The proportional band is confined to only part of the range so that initially full power is available at low temperatures. The position and width of the proportional band can be easily varied.

The position of the set point in the proportional band is preset so that when the set point is reached the power input is exactly equal to the heat losses. This will occur only for the precise conditions existing when the system is set up governed by the position of the set point in the proportional band. Any changes such as in ambient temperature, heat losses and any other factors affecting the power requirements at the set point will result in a change in the power required and hence in the actual control temperature (*Figure 48.7*). The difference is known as offset or droop and is typically up to half the width of the proportional band. The offset can be reduced by increasing the controller gain, i.e. by decreasing the width of the proportional band. However,

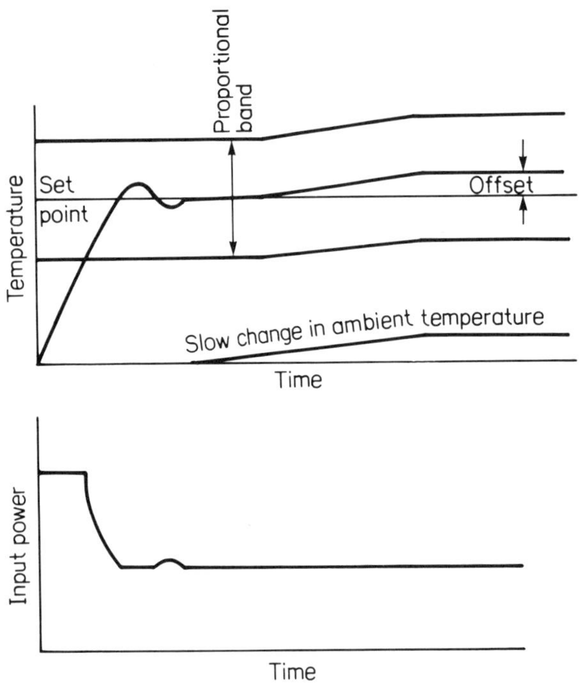

Figure 48.7 Proportional control showing effect of offset

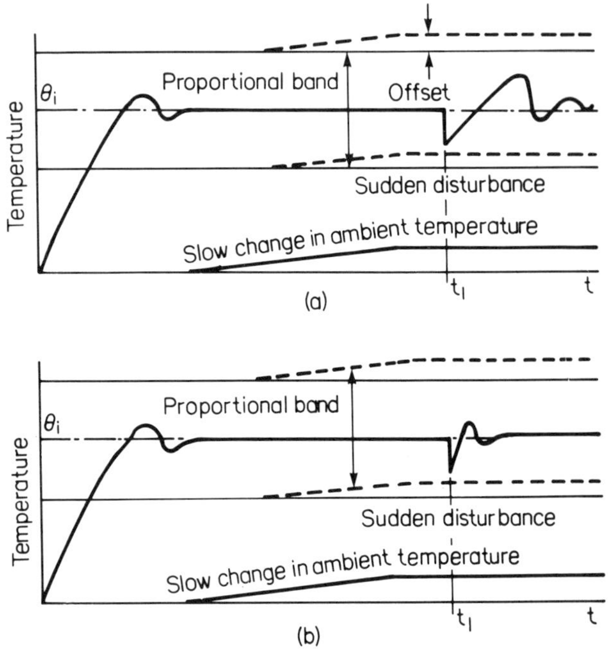

Figure 48.8 Effects of derivative and integral control: (a) proportional and integral control response; (b) proportional, integral and derivative (3-term) control response

this results in increased initial overshoot, reduced stability and in the limit uncontrolled oscillations equivalent to on/off operation with no dead space. Proportional control alone is used quite extensively for temperature control systems in which a high degree of precision is not required. Fluctuations in the position of the proportional band occur due to changes in the ambient temperature θ_0 and operating conditions in the furnace such as heat losses due to deterioration of the refractories. As a result, the proportional band is offset and the power available at the set point changes since it is critical in governing the control characteristic of the process. Slowly changing fluctuations in variables outside the control system can be reduced by integral control. The function of integral control is to integrate the deviation of the control temperature from the temperature at the set point over a long period and to correct the input power *W*. The effect of integral control compensating for changes in ambient temperature by restoring the proportional band to its correct level is illustrated in *Figure 48.8(a)* and the response is given by $p=f\int e\,dt$. The integral time constant can be adjusted to suit the process parameters so that it compensates only for long-term variations and does not respond to small changes.

In any temperature control system, since control only occurs within the proportional band, the integral action may move the proportional band so far up-scale that the set point may be reached before the proportional band. To offset the effect of integral control during start-up, the proportional band may be automatically shifted to a lower temperature.

A sudden fluctuation about the set point can only be slowly corrected by proportional control. This can be compensated for by differential control which responds to fluctuations over only a short period and results in a restoring signal much greater than that due to proportional control alone. The effect of derivative control in response to a sudden change is shown in *Figure 48.8(b)* and is given by

$$p=r\frac{de}{dt} \tag{48.13}$$

The derivative time constant can be adjusted so as to increase the power input in the proportional band over a period sufficiently long to compensate for a rapid fluctuation.

48.6 Analog control techniques

48.6.1 The electronic controller

The electronic analog controller is based on proportional control to which derivative or integral action may be added. Other functions may also be incorporated to modify the response.

A typical electronic 3-term temperature controller with adjustable set point of the kind used for temperature control during plastics extrusion is shown in *Figure 48.9*. As well as proportional, derivative and integral (PDI) control additional functions may be included such as a ramp function generator to gradually increase or lower the set point during the process as required.

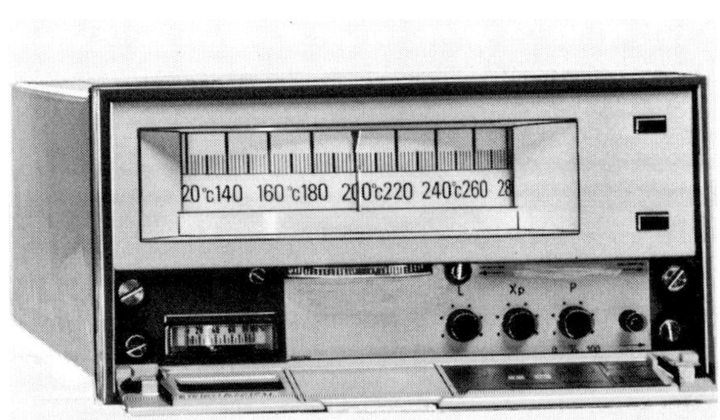

Figure 48.9 A 3-term controller showing pre-set controls for proportional, integral and derivative controls and set point deviation meter (courtesy Eurotherm Ltd)

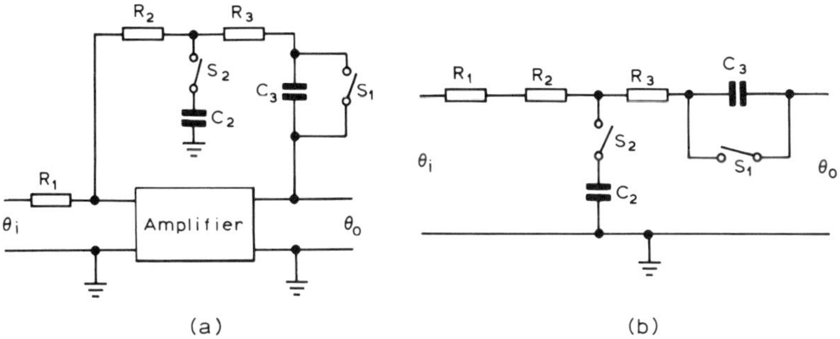

Figure 48.10 (a) Schematic diagram of 3-term controller and (b) simplified equivalent circuit

A schematic diagram of a 3-term controller is shown in *Figure 48.10* together with an equivalent circuit. The proportional signal is obtained from the resistors R_1, R_2 and R_3. Derivative control can be added by including a capacitor C_2 so that the transfer function becomes

$$\frac{V_o}{V_c}=\frac{Ts+1}{Ts+\frac{R_1}{R_2}+1} \tag{48.14}$$

Integral action is obtained by connecting a capacitor across R_2 so that

$$\frac{V_o}{V_i}=\frac{1}{Ts+\frac{R_1}{R_2}+1} \tag{48.15}$$

which if $\frac{R_1}{R_2}<Ts$ tends to

$$\frac{V_o}{V_i}=\frac{1}{Ts} \tag{48.16}$$

By using an amplifier the load on the reference and sensor input is reduced and the sensitivity increased. If the gain of the amplifier is large the input to the amplifier A can be considered as a virtual earth. The transfer function of the controller is

$$\frac{\theta_o}{\theta_i}=-\left|\frac{R_2+R_3}{R_o}+\frac{R_2C_2}{R_1C_3}+\frac{1}{R_1C_{3p}}+\frac{R_3}{R_1}R_2C_{2p}\right| \tag{48.17}$$

If $R_2+R_3=R_1$, $R_3=R_2$, and $C_2 >> C_3$ then

$$\frac{\theta_o}{\theta_i}\simeq-\left(1+\frac{1}{T_i}+T_d\right) \tag{48.18}$$

where T_i and T_d are the integral and derivative time constants

$$T_D\simeq\frac{R_2C_2}{2} \qquad T_i\simeq R_3C_3$$

i.e. the integral and derivative action can be adjusted by varying C_3 and C_2 independently.

48.6.2 Analog computers in control systems

The analog computer is based on the use of operational amplifiers to carry out arithmetical operations. These include summation, integration, differentiation and scaling. Multiplication and division and non-linear functions can also be obtained.

Electronic analog computers may be used to simulate complex systems. Changes in the parameters may be made by varying resistance and capacitance. These may be at preset values or varied according to a preset programme. Analog computers are usually designed for one specific application although some degree of versatility may be achieved by varying the interconnection of the computer modules.

The analog computer is capable of dealing with continuous quantities rather than by discrete steps and operates in real time. Complex functions can be simulated without the necessity for the large stores and high number of iterative operations required by digital computers. A further advantage is the ability to relate directly the performance of the analog computer with the process itself and the simplicity of programming.

Notwithstanding these advantages digital computers have in many areas replaced analog computers and the falling price of digital computers is likely to increase this trend. The greater accuracy, increased flexibility and the capability of dealing with large numbers of inter-related variables tend to outweigh many of the advantages of analog computers except for relatively simple applications. Nevertheless, analog computers still have and will retain an important part in electronic control systems.

48.7 Servomotors

Servomotors are used in an enormous variety of applications ranging from miniature motors used in data recorders to motors with ratings in excess of 1 MW for winding machines used in mines. Both d.c. and a.c. motors are used.

The d.c. commutator motor is the most commonly encountered servomotor drive. Capable of continuous variation in speed from zero to full speed, the torque and power output can be continuously controlled over wide ranges. Although the a.c. squirrel cage induction motor is most extensively used in industry where constant speed drives are required it is not suitable for applications where wide variations in speed are needed. For speed control of a.c. motors a wound rotor with slip ring outputs is necessary or special field coil designs such as those used in the Schrage motor. For low-power low cost applications the 2-phase synchronous motor is used.

48.7.1 Control of speed and torque

Field connections of the d.c. commutator motor are illustrated schematically in *Figure 48.11*. The field may be separately excited, shunt excited, series excited or, a combination of series and shunt excitation (compound excitation). Where very small motors are used the field coil is replaced by a permanent magnet.

The principal of operation of each is similar although the interaction of field and armature currents varies enabling different operating characteristics to be obtained. The simplest machine to consider is the separately excited motor which is extensively used in control systems. The flux from the field coil is proportional to the field current, i.e. $\varphi \propto I_f$. The induced voltage (back e.m.f.) in the armature is proportional to

$$E\propto\frac{N\,\mathrm{d}\varphi}{\mathrm{d}t} \tag{48.19}$$

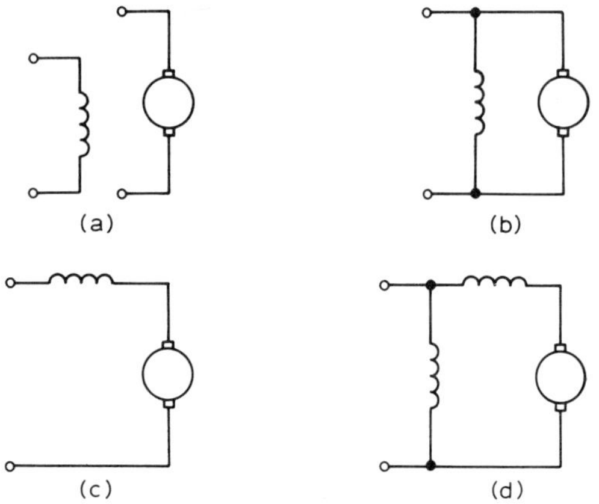

Figure 48.11 Field connections of d.c. motors: (a) separately excited; (b) shunt excited; (c) series excited; and (d) compound excited

which in turn is proportional to $V-I_aR_a$. Hence

$$N \propto \frac{V-I_aR_a}{\varphi} \qquad (48.20)$$

and the torque $T \propto \varphi I_a$.

The rotor should be capable of being rapidly accelerated where the acceleration is given by

$$\frac{d^2\theta}{dt^2}=\frac{Tg}{J}\ \text{rad/s}^2 \qquad (48.21)$$

where θ is the angular displacement and J is the moment of inertia.

The rotor should be sufficiently damped so as to reduce overrun where the damping coefficient is given by

$$D=\frac{dT}{dN}gs \qquad (48.22)$$

And the time constant of the rotor should be sufficiently short where

$$t_n=\frac{J}{D}s \qquad (48.23)$$

The time required to reverse is approximately $1.7t_n$.

Speed control of motors can be derived directly from the armature current at constant field current. Speed control to an accuracy of 5–15% can be achieved in this way. If armature compensation is used accuracies of $2\frac{1}{2}$ to 5% are possible.

Change of the direction of rotation is obtained by reversing the direction of either the armature current or field current. Split field windings are sometimes used to enable continuous variation of speed through zero. The motor is stationary when the currents in the field are such that the fluxes oppose each other. The speed and direction is a function of the magnitude and direction of the out of balance current.

48.7.2 Position control

The synchro or resolver normally consists of a polyphase stator with a single or polyphase rotor. A single polyphase stator with single phase rotor is shown in *Figure 48.12.* The stator is excited and the voltages in the rotor are a function of its angular position.

If the receiver (which is identical) is also excited by the same source the rotor will be driven by the transmitter until the induced e.m.f. is equal but in opposition to that from the transmitter. A resolution corresponding to about 0.3° is possible, however as the receiver torque decreases as the required position is approached the error signal decreases and effects due to friction and load torque limit the accuracy obtainable. Amplification of the error signal from the rotor enables this to be minimised.

Synchronous motor drives may be used where a very high degree of precision is not necessary and only a small torque is required. The rotational speed is a function of the frequency and power and high frequency supplies may be used. Resolution to about 1/100 of a revolution is possible. A multiple tooth permanent magnet rotor is driven by a 2-phase field winding at power or higher frequencies. The 2-phase supply can be derived from a single-phase supply as shown in *Figure 48.13* or with a Scott-connected transformer. A 100-tooth rotor operated at 50 Hz has a slew speed of 60 r.p.m.

A stepper motor translates a digital signal into angular movement. Variable reluctance motors use a soft iron rotor in which teeth and slots are cut. A stepper motor is illustrated schematically in *Figure 48.14.* The wound stator has corresponding teeth and slots. When the stator is energised the motor aligns so as to minimise the reluctance of the air gap. Rotation is controlled and synchronised by input pulses to the stator winding. Variable reluctance motors are capable of very high stepping rates at small stepping angles making them ideally suited for control of position to a high degree of precision. An additional advantage of variable reluctance motor drives is the small compact dimensions that can be achieved. Higher speeds with lower rotor inertia than permanent magnet motors are obtainable, however, the efficiency and power output are lower. Typical step angles vary between 0.9° to 90°.

Another form of construction uses a permanent magnet rotor and wound stator. The rotor rotates until it reaches an equilibrium position opposite to an opposing pole. If the stator windings are excited in sequence the rotor rotates. An alternative construction uses a toothed soft iron rotor and a stator comprising a permanent magnet and control winding. The control winding is used to alter the distribution of the field resulting in stepwise rotation. A feature of this design is the high

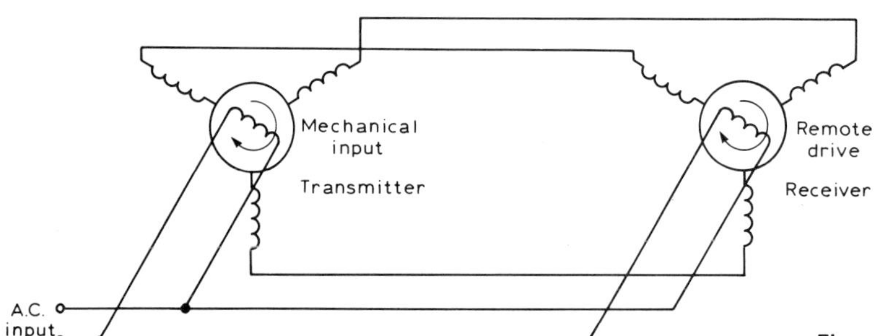

Figure 48.12 Synchronous transmitter and receiver

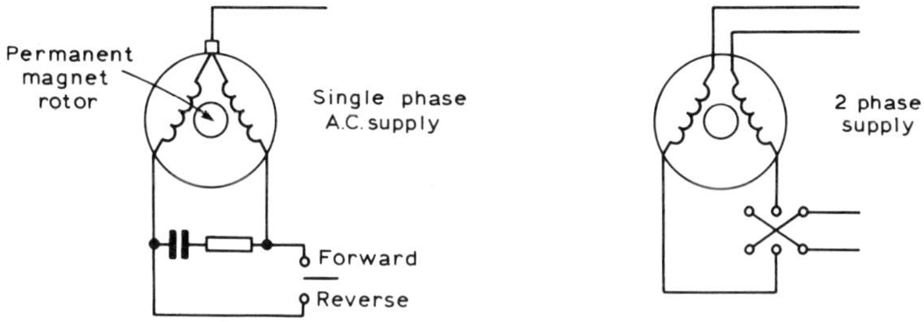

Figure 48.13 Synchronous motor drives

Figure 48.14 Variable reluctance stepper motor

residual torque obtained when the control winding is not energised.

The drive signals for stepper motors are usually obtained from pulses derived from logic circuits operating in sequence. In its simplest form where the motor is operated in on-off mode within its synchronous range the drive circuit is in the form of a transistor switch for each motor coil. This is controlled by a logic circuit to switch the coils in the correct sequence. The complexity of the logic circuit will depend on the number of stator windings. The voltage and current required to drive the motor may vary from a few volts and 50 mA to more than 20 A at 100 V. Stepping speeds from zero to more than 10 000 revs/second are possible.

Further reading

ELGEROD, O. L., *Control Systems Theory*, McGraw-Hill, New York (1967)

GUTZWILLER, F. W., *Silicon Controlled Rectifier Manual*, General Electric, USA (1967)

HEALEY, M., *Principles of Automatic Control*, 2nd edn., English Universities Press, London (1970)

LUKE, H. D., *Automation for Productivity*, Wiley, New York (1972)

PELLEY, B. R., *Thyristor Phase Controlled Converters and Cycloconverters*, Wiley-Interscience (1971)

PIKE, C. H., *Automatic Control of Industrial Drives*, Newnes-Butterworth, London (1971)

PRIME, H. A., *Modern Concepts in Control Theory*, McGraw-Hill, London (1969)

RAVEN, F. H., *Automatic Control Engineering*, 2nd edn., McGraw-Hill, New York (1968)

SPEISS, R., 'Hybrid computer aids process control development', *Mini-Micro Systems* (August 1982)

TABAK, D. and KUO, B. S., *Optimal Control by Mathematical Programming*, Prentice-Hall, New Jersey (1971)

YOUNG, A. J., *An Introduction to Process Control System Design*, Longmans, London (1955)

'Electrical Variable Speed Drives', *I.E.E. Conference Publication*, Number 93 (1972)

'Power Thyristors and their Application', *I.E.E. Conference Publication*, Number 53 (1969)

Proceedings of Conference, Industrial Static Power Conversion, *Institution of Electronic and Electrical Engineers*, New York (1965)

49 Antennas and Arrays

A D Monk, MA, CEng, MIEE
Section Chief,
Antenna Division,
Marconi Research Centre,
GEC Research Laboratories

Contents

49.1 Fundamentals

49.1.1 Antenna function and performance

The function of an antenna is to provide a transition between a transmission line or a voltage or current feed point, and a plane wave (or more exactly a spectrum of plane waves) propagating in free space. The most important description of the performance of an antenna is its radiation pattern or polar diagram, which quantifies the directional characteristics of the interface with free space provided by the antenna. For the majority of antenna types the most important parameters which may be obtained from the radiation pattern are the peak gain, the beamwidth of the main lobe between half power (3 dB) points, and the levels of the sidelobes relative to the main lobe peak. Other parameters which may be of interest are the operational bandwidth, VSWR or impedance match, front-to-back ratio, and field polarisation properties.

49.1.1.1 Antenna gain

The power gain (generally abbreviated to gain) of an antenna in a specified direction is defined as 4π times the ratio of the power per unit solid angle radiated in that direction, to the net power accepted by the antenna from its generator. If the direction is not specified then it is generally assumed that the 'gain' refers to the maximum or peak value. The mathematical form of this definition is

$$G(\theta, \varphi) = \frac{4\pi\Phi(\theta, \varphi)}{P_0} \tag{49.1}$$

An alternative but exactly equivalent definition is that the gain is the ratio of the power per unit solid angle radiated by the antenna, to the power per unit solid angle which would be radiated by a lossless isotropic radiator with the same power accepted at its terminals. From this definition the reason for the choice of dBi or 'dB above isotropic' as the unit of antenna gain is apparent.

49.1.1.2 Antenna directivity

The antenna gain G may be expressed as the product

$$G = \eta D \tag{49.2}$$

where η is the antenna efficiency and accounts for all the losses in the antenna (ohmic loss, tranmission line loss, impedance mismatch, etc.), and D is the antenna directivity, which is a function only of the antenna radiation pattern shape. Except in the hypothetical case of a lossless antenna, the gain will always be less than the directivity.

As with antenna gain, antenna directivity may be defined in two exactly equivalent ways. Either as the ratio of the radiated power density in the given direction, to the average radiated power density

$$D(\theta, \varphi) = \frac{\Phi(\theta, \varphi)}{\Phi_{av}} \tag{49.3}$$

or as the ratio

$$D(\theta, \varphi) = \frac{P_t'(\theta, \varphi)}{P_t} \tag{49.4}$$

where P_t is the total power radiated by the antenna, and $P_t'(\theta, \varphi)$ is the power which would have to be radiated by an isotropic antenna to provide the same power density in the given direction.

From the first definition, the antenna directivity in the given reference direction may be written in terms of the radiated power density in the reference direction Φ_r, and the radiated power density at other angles $\Phi(\theta, \varphi)$

$$D_r = \frac{\Phi_r}{(P_t/4\pi)} = \frac{4\pi\Phi_r}{\int_0^{\pi}\int_0^{2\pi}\Phi(\theta, \varphi)\,d\varphi \sin\theta\,d\theta} \tag{49.5}$$

Thus the directivity may be computed by integrating a measured or theoretical radiation pattern $\Phi(\theta, \varphi)$ over the surface of a sphere.

49.1.2 Field strength and power density relationships in a plane wave

For a plane wave propagating in free space, if the electric and magnetic field strengths E and H are expressed in V m^{-1} and A m^{-1} respectively, they are related by

$$E = 120\pi H \tag{49.6}$$

where $120\pi\,\Omega \simeq 377\,\Omega$ is known as the impedance of free space.

The power density p expressed in W m^{-2} is given in terms of the field strengths as

$$p = E^2/120\pi = 120\pi H^2 \tag{49.7}$$

49.1.3 Radiated power density and field strengths

If a power P_0 is accepted by an antenna whose gain in a given direction is G, then the radiated power density and field strengths in that direction in the antenna far-field at a distance r are

$$p = \frac{P_0 G}{4\pi r^2} \tag{49.8}$$

$$E = \frac{\sqrt{30 P_0 G}}{r} \tag{49.9}$$

$$H = \sqrt{\frac{P_0 G}{30}}\,(4\pi r)^{-1} \tag{49.10}$$

(all units are m.k.s. as before).

49.1.4 Received power

The power received by an antenna at its terminals when illuminated by a plane wave of power density p is

$$P_r = p A_e \tag{49.11}$$

where A_e is known as the 'effective area' of the antenna, and is related to the antenna gain thus

$$A_e = \frac{G\lambda^2}{4\pi} \tag{49.12}$$

The received power in terms of the power density and field strengths of the incident plane wave is therefore

$$P_r = \frac{pG\lambda^2}{4\pi} = \left(\frac{E\lambda}{4\pi}\right)^2 \frac{G}{30} = 30H^2\lambda^2 G \tag{49.13}$$

49.1.5 Aperture efficiency

The gain of a lossless antenna with a uniformly illuminated planar aperture of area A is

$$G_0 = \frac{4\pi A}{\lambda^2} \tag{49.14}$$

The aperture efficiency of an aperture antenna is the ratio of the actual gain to the gain if the same aperture were uniformly illuminated, assuming that the losses are the same in both cases. The aperture efficiency is given in terms of the aperture field distribution f as

$$\eta_a = \frac{1}{A}\frac{|\int_A f\,dA|^2}{\int_A |f|^2\,dA} \tag{49.15}$$

49.2 Radiation from elemental sources

For the analysis of many antenna types, it is convenient to describe the radiation as occurring from elemental sources disposed across the radiating area or along the radiating length of the antenna. The two basic elemental sources are the electric and magnetic dipoles. The electrical dipole may be considered as an electric current I flowing along a length δL, where δL is much smaller than λ, and has a moment equal to $I\,\delta L$. Similarly the magnetic dipole may be considered as a closed loop of electric current I with an area δA, where the maximum loop dimension is much smaller than λ, and has a moment equal to $I\,\delta A$.

One example of an antenna which can be analysed as a distribution of elemental sources is the linear wire antenna, in which the continuous current distribution along its length is represented as an infinite number of electric dipole sources, each of vanishingly small length and locally constant current. The net radiated field is then equal to the sum or integral of the fields radiated by the elemental sources. Another example would be an aperture antenna such as an electromagnetic horn, with the aperture E and H fields replaced by a combination of magnetic and electric dipoles respectively, by invoking the equivalence principle.

49.2.1 The electric dipole

49.2.1.1 General solution

For a short length δL of electric current I, i.e. an electric dipole, orientated along the z axis of a standard spherical coordinate system as shown in *Figure 49.1(a)*, it may be shown that the components of the radiated electric and magnetic fields in free space are

$$E_r = 60k^2 I\,\delta L\left[\frac{1}{(kr)^2} - \frac{j}{(kr)^3}\right]\cos\theta\, e^{-jkr} \tag{49.16}$$

$$E_\theta = j30k^2 I\,\delta L\left[\frac{1}{kr} - \frac{j}{(kr)^2} - \frac{1}{(kr)^3}\right]\sin\theta\, e^{-jkr} \tag{49.17}$$

$$H_\phi = j\frac{k^2}{4\pi} I\,\delta L\left[\frac{1}{kr} - \frac{j}{(kr)^2}\right]\sin\theta\, e^{-jkr} \tag{49.18}$$

$$E_\phi = H_r = H_\theta = 0 \tag{49.19}$$

where r is the distance from the dipole to the observation point, and k is the free-space wave number $2\pi/\lambda$. If I is in amperes and δL, r and λ are in metres, then E will be given in volts/metre and H in amperes/metre.

49.2.1.2 Near-field solution

Very close to the dipole, nominally $r < 0.01\lambda$, only the E_r and E_θ fields are significant, and these become

$$E_r \simeq -j60I\,\delta L\cos\theta(e^{-jkr}/r^3) \tag{49.20}$$

$$E_\theta \simeq -j30I\,\delta L\sin\theta(e^{-jkr}/r^3) \tag{49.21}$$

Under this condition these two components are in time phase, with amplitudes simply related

$$E_r/E_\theta \simeq 2\cot\theta \tag{49.22}$$

49.2.1.3 Far-field solution

Very far from the dipole, nominally $r > 5\lambda$, the radial E_r field component vanishes, and E_θ and H_ϕ become

$$E_\theta \simeq j(60\pi/\lambda)I\,\delta L\sin\theta(e^{-jkr}/r) \tag{49.23}$$

$$H_\phi \simeq (j/2\lambda)I\,\delta L\sin\theta(e^{-jkr}/r) = E_\theta/120\pi \tag{49.24}$$

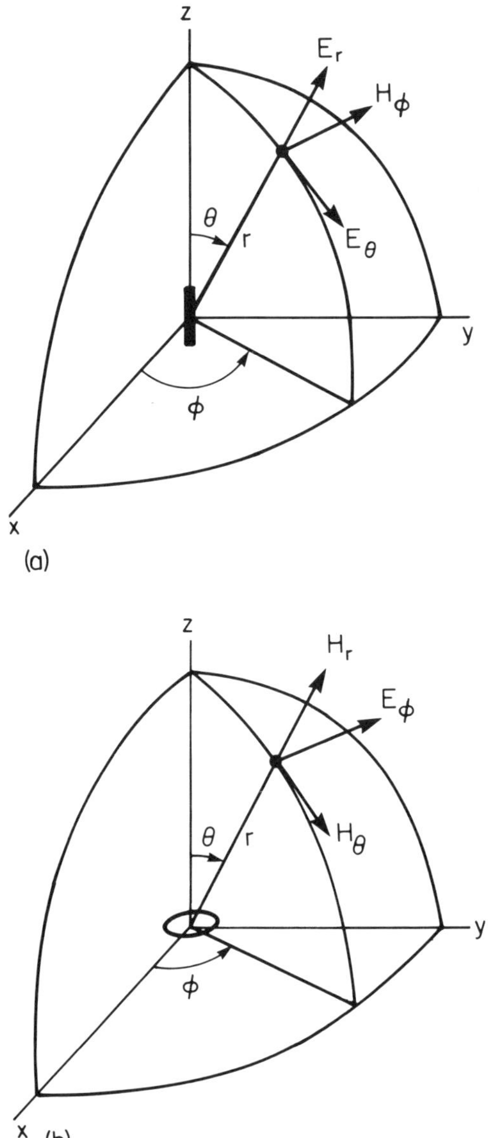

Figure 49.1 Radiation from elemental electric and magnetic dipoles. (a) Electric dipole; (b) magnetic dipole

The far-field amplitude is thus proportional to $\sin\theta$, giving the familiar 'figure of eight' voltage pattern in the planes containing the current element, when plotted in polar form. It is also apparent that the far-field amplitude is proportional to $1/r$, so that the power density is proportional to $1/r^2$, giving the inverse square law relationship with distance.

49.2.2 The magnetic dipole

49.2.2.1 General solution

For a small loop area δA of electric current I, i.e. a magnetic dipole, in the xy plane of a standard spherical coordinate system as shown in *Figure 49.1(b)*, it may be shown that the components of the radiated electric and magnetic fields in free space are

$$E_\phi = 30k^3 I\,\delta A\left(\frac{1}{kr} - \frac{j}{(kr)^2}\right)\sin\theta\, e^{-jkr} \quad (49.25)$$

$$H_r = j\frac{k^3}{2\pi} I\,\delta A\left(\frac{1}{(kr)^2} - \frac{j}{(kr)^3}\right)\cos\theta\, e^{-jkr} \quad (49.26)$$

$$H_\theta = \frac{k^3}{4\pi} I\,\delta A\left(\frac{1}{kr} - \frac{j}{(kr)^2} - \frac{1}{(kr)^3}\right)\sin\theta\, e^{-jkr} \quad (49.27)$$

$$E_r = E_\theta = H_\phi = 0 \quad (49.28)$$

where the units and symbols are the same as for the electric dipole.

49.2.2.2 Near-field solution

Very close to the dipole, nominally $r < 0.01\lambda$, only the H_r and H_θ fields are significant, and these become

$$H_r \simeq \frac{1}{2\pi} I\,\delta A \cos\theta(e^{-jkr}/r^3) \quad (49.29)$$

$$H_\theta \simeq -\frac{1}{4\pi} I\,\delta A \sin\theta(e^{-jkr}/r^3) \quad (49.30)$$

so that H_r and H_θ are related in a way similar to E_r and E_θ in the near-field of an electric dipole

$$H_r/H_\theta \simeq -2\cot\theta \quad (49.31)$$

49.2.2.3 Far-field solution

Very far from the dipole, nominally $r > 5\lambda$, the radial H_r field vanishes, and E_ϕ and H_θ become

$$E_\phi \simeq 30k^2 I\,\delta A \sin\theta(e^{-jkr}/r) \quad (49.32)$$

$$H_\theta \simeq \frac{k^2}{4\pi} I\,\delta A \sin\theta(e^{-jkr}/r) = \frac{E_\phi}{120\pi} \quad (49.33)$$

The $\sin\theta$ factor corresponds to the same 'figure of eight' polar diagram as for the electric dipole, in this case in the planes perpendicular to the plane of the loop.

49.2.3 Elemental aperture field source

49.2.3.1 The equivalence principle

The equivalence principle[1,2] is a powerful technique for analysing the radiation from apertures within which the tangential **E** and **H** field distribution is known. Stated briefly, if the field distribution is known across a closed surface surrounding the antenna, then it is possible to define a distribution of electric and magnetic currents, **J** and **M** respectively, across the surface such that the fields everywhere inside that surface are zero

$$\mathbf{J} = \hat{\mathbf{n}} \times \mathbf{H} \quad (49.34)$$

$$\mathbf{M} = \hat{\mathbf{n}} \times \mathbf{E} \quad (49.35)$$

where $\hat{\mathbf{n}}$ is the outwards unit vector normal to the surface. The radiated fields due to the antenna are then exactly equal, except for a change of sign, to the fields radiated by the electric and magnetic currents.

This principle may be applied to a large class of aperture antennas. For example for an electromagnetic horn, if it is assumed that the field distribution at the aperture has the same form as that of the waveguide modes exciting the horn at its throat, then the radiated fields may be obtained from that aperture field distribution. For a reflector antenna it may be assumed that the fields at the reflector surface due to either illumination by the feed (transmit mode analysis) or by an incoming plane wave (receive mode analysis), are the same as would exist without the reflector present, except that reflection of the wave at the surface cancels the tangential **E** field and doubles the tangential **H** field. The radiation from an element on the reflector surface is then identical to the radiation from the elemental electric current source

$$\mathbf{J} = 2\hat{\mathbf{n}} \times \mathbf{H} \quad (49.36)$$

For an element $dx\,dy$ of a radiating aperture in which the tangential **E** and **H** fields E_x^a and H_y^a are known (see *Figure 49.2*),

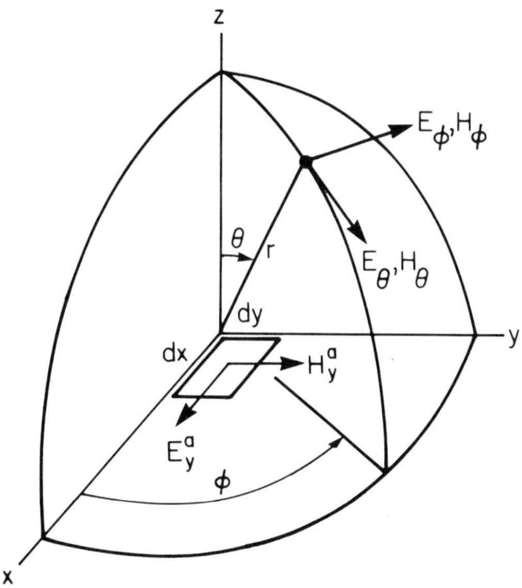

Figure 49.2 Radiation from elemental aperture

the far-field components of the radiated **E** field are

$$E_\theta = -j(E_x^a + 120\pi H_y^a \cos\theta)\frac{\cos\varphi}{2\lambda}\,dx\,dy(e^{-jkr}/r) \quad (49.37)$$

$$E_\phi = j(E_x^a \cos\theta + 120\pi H_y^a)\frac{\sin\varphi}{2\lambda}\,dx\,dy(e^{-jkr}/r) \quad (49.38)$$

For the special case in which the E and H aperture field amplitudes are related by the impedance of free-space

$$E_x^a = 120\pi H_y^a \quad (49.39)$$

so that the equivalent current source is a balanced Huygens source, the far-fields become

$$E_\theta = -jE_x^a(1+\cos\theta)\frac{\cos\varphi}{2\lambda}\,dx\,dy(e^{-jkr}/r) \quad (49.40)$$

$$E_\phi = jE_x^a(1+\cos\theta)\frac{\sin\varphi}{2\lambda}\,dx\,dy(e^{-jkr}/r) \quad (49.41)$$

49.3 Antenna polarisation

49.3.1 The polarisation ellipse

The polarisation attributed to an antenna is the polarisation or orientation in space of the field vectors it radiates into space. It has long been a universal convention that antenna or wave polarisation refers to the orientation of the electric or **E** field rather than the magnetic or **H** field. Thus when for example an antenna is referred to as being vertically polarised, then the radiated **E** field vector is in the vertical plane and the **H** field vector in the horizontal plane.

For an arbitrarily polarised wave, in a fixed plane normal to the direction of propagation the locus of the tip of the **E** field vector with respect to time will be an ellipse—the polarisation ellipse. This is shown in *Figure 49.3*. The wave polarisation may

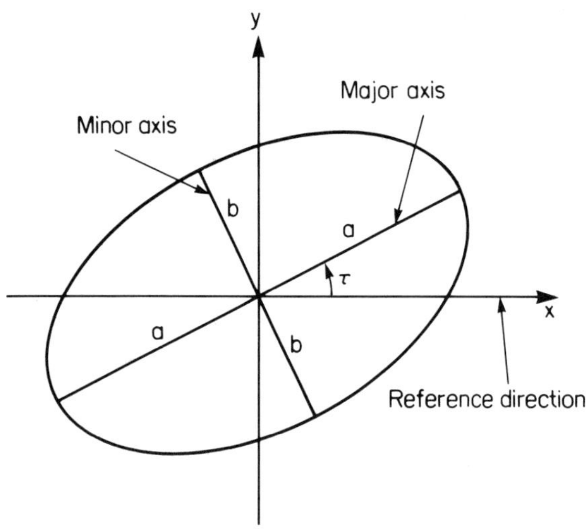

Figure 49.3 The polarisation ellipse

be defined completely by three parameters, all of which are directly related to the geometry of the polarisation ellipse. The first of these is the voltage axial ratio A, which is equal to the ratio of the lengths of the major and minor axes of the ellipse, i.e.

$$A = a/b \tag{49.42}$$

The second is the angle τ between the ellipse major axis and the reference direction, in this case the x axis. The third is the sense of rotation of the tip of **E** field vector around the ellipse. A clockwise rotation as viewed in the direction of propagation is termed right-hand elliptical polarisation, and rotation anticlockwise is termed left-hand elliptical pollarisation. It is conventional to restrict the axial ratio to the range of values $1 < A < \infty$. A wave polarisation with axial ratio A, which is less than unity, and tilt angle τ is identical to the polarisation with axial ratio $1/A$, which is greater than unity, and tilt angle $\tau \pm \pi/2$. The axial ratio is also often expressed in dB where

$$\mathrm{dB} = 20 \lg A \tag{49.43}$$

so that an axial ratio of unity is 0 dB.

The two special cases are linear and circular polarisation. For linear polarisation the axial ratio is infinity, with the tilt angle τ defining the orientation (the sense of rotation is redundant). For circular polarisation the axial ratio is unity or 0 dB, and the sense of circular polarisation is either right-hand (RHCP) or left-hand (LHCP) (the angle τ is redundant). A circularly polarised wave or field may be considered to comprise two equal and orthogonal linearly polarised components in phase quadrature. The sense of phase quadrature determines the hand of circular polarisation.

49.3.2 Polarisation coupling loss

The power coupling between two antennas or between a receiving antenna and an incident plane wave, will depend on the relative polarisation of the two antennas or antenna and incident wave. Maximum coupling with zero polarisation coupling loss will occur when the two polarisations are identical. Zero coupling will occur when the two polarisations are orthogonal.

Quantitatively the polarisation power coupling loss P $(0 \leqslant P \leqslant 1)$ is given by

$$P = \tfrac{1}{2}[1 + \sin 2\theta_1 \sin 2\theta_2 + \cos 2\theta_1 \cos 2\theta_2 \cos 2(\tau_1 - \tau_2) \tag{49.44}$$

where the two axial ratios are $\tan \theta_1$ and $\tan \theta_2$, and the polarisation ellipse tilt angles (referred to the same reference direction) τ_1 and τ_2. The sign of θ_1 and θ_2 defines the rotation sense or hand.

The maximum and minimum power coupling factors, which occur for $\tau_1 - \tau_2 = 0$ and $\pi/2$ respectively, are

$$P_{\max} = \cos^2(\theta_1 - \theta_2) \tag{49.45}$$

$$P_{\min} = \sin^2(\theta_1 + \theta_2) \tag{49.46}$$

When one polarisation is linear so that $\theta_1 = \pi/2$, the coupling becomes

$$P_{\text{lin}} = \tfrac{1}{2}[1 - \cos 2\theta_2 \cos 2(\tau_1 - \tau_2)] \tag{49.47}$$

which has maximum and minimum values of $\sin^2 \theta_2$ and $\cos^2 \theta_2$ respectively.

When one polarisation is circular so that $\theta_1 = \pi/4$, the coupling is independent of the orientation of the second polarisation ellipse, and is given by

$$P_{\text{cp}} = \tfrac{1}{2}(1 + \sin 2\theta_2) \tag{49.48}$$

49.3.3 Orthogonal polarisations and cross polarisation

For every polarisation state with axial ratio A and tilt angle τ there is a unique orthogonal polarisation state which has the same axial ratio, tilt angle $\tau \pm \pi/2$ and the opposite sense of rotation. For linear polarisation the orthogonal polarisation is also linear but with the orientation rotated through 90°, and for circular polarisation the orthogonal polarisation is circular of the opposite hand. There is zero coupling between two orthogonally polarised antennas or between a plane wave and an orthogonally polarised receiving antenna.

The concept of polarisation orthogonality is sometimes applied to provide two separate channels from a single antenna, by using two space-orthogonal linear polarisations, or LHCP and RHCP. The most notable example of such an application is frequency reuse in satellite communications systems. A more widely used application of the concept is in a quantitive description of the polarisation purity of an antenna. In any direction in space it is possible to define a 'wanted' and 'unwanted' or 'co' and 'cross' orthogonal polarisation pair, so that the polarisation purity in that direction may be expressed as the ratio of the cross-polar to co-polar components of the vector radiation pattern in the same direction.

For a nominally circularly polarised antenna the co-polar and cross-polar polarisations are simply LHCP and RHCP in all directions. For a nominally linearly polarised antenna, although in the principal planes the definition of co-polar and cross-polar polarisations is obvious, there is some ambiguity in other directions. Ludwig[3] has examined this problem in some detail and has identified three alternative definitions for linear co- and cross-polarisation—Ludwig's first, second and third definitions. The three definitions for an antenna nominally linearly polarised in the y direction are shown graphically in *Figure 49.4* and are described mathematically by

$$\hat{\mathbf{a}}_{\text{co}}^{(1)} = \sin\theta \sin\varphi \hat{\mathbf{r}} + \cos\theta \sin\varphi \hat{\boldsymbol{\theta}} + \cos\varphi \hat{\boldsymbol{\varphi}} \tag{49.49}$$

$$\hat{\mathbf{a}}_{\text{cross}}^{(1)} = \sin\theta \cos\varphi \hat{\mathbf{r}} + \cos\theta \cos\varphi \hat{\boldsymbol{\theta}} - \sin\varphi \hat{\boldsymbol{\varphi}} \tag{49.50}$$

$$\hat{\mathbf{a}}_{\text{co}}^{(2)} = (\sin\varphi \cos\theta \hat{\boldsymbol{\theta}} + \cos\varphi \hat{\boldsymbol{\varphi}})(1 - \sin^2\theta \sin^2\varphi)^{-1/2} \tag{49.51}$$

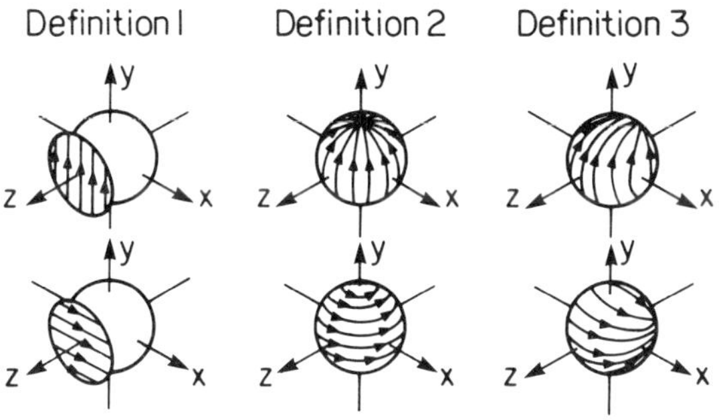

Figure 49.4 Ludwig's three definitions of cross polarisation. Top, direction of the reference polarisation. Bottom, direction of the cross polarisation

$$\hat{\mathbf{a}}_{\text{cross}}^{(2)} = (\cos \hat{\varphi\theta} - \cos\theta \sin\varphi\hat{\varphi})(1 - \sin^2\theta \sin^2\varphi)^{-1/2} \quad (49.52)$$

$$\hat{\mathbf{a}}_{\text{co}}^{(3)} = \sin\varphi\hat{\boldsymbol{\theta}} + \cos\varphi\hat{\boldsymbol{\varphi}} \quad (49.53)$$

$$\hat{\mathbf{a}}_{\text{cross}}^{(3)} = \cos\varphi\hat{\boldsymbol{\theta}} - \sin\varphi\hat{\boldsymbol{\varphi}} \quad (49.54)$$

where $\hat{\mathbf{a}}_{\text{co}}$ and $\hat{\mathbf{a}}_{\text{cross}}$ are the unit vectors in the co- and cross-polar directions expressed in terms of the unit vectors $\hat{\mathbf{r}}$, $\hat{\boldsymbol{\theta}}$ and $\hat{\boldsymbol{\varphi}}$ in a standard spherical coordinate system. The co- and cross-polarised components of the antenna field $\mathbf{E}$ are

$$E_{\text{co}} = \mathbf{E} \cdot \hat{\mathbf{a}}_{\text{co}} \quad (49.55)$$

$$E_{\text{cross}} = \mathbf{E} \cdot \hat{\mathbf{a}}_{\text{cross}} \quad (49.56)$$

Of the three definitions the third is the most widely used, and has the advantage that a reflector antenna feed having zero cross polarisation according to Ludwig's third definition, will produce parallel field lines in the reflector aperture, and will hence contribute no cross polarisation to the secondary pattern. It also gives zero cross polarisation for a balanced Huygens source. The second definition in contrast gives zero cross polarisation for an electric dipole source, and has the disadvantage that the co-polar field line pattern does not become orthogonal to itself when rotated about the z axis by 90°.

49.3.4 Circular polarisation from orthogonal linear polarisations

Circular polarisation may be obtained by combining two equal amplitude linear polarisations in phase quadrature. Any amplitude imbalance between the two linear components, or departure from the quadrature phase relationship will yield elliptical polarisation with a non-unity axial ratio. This is equivalent to saying that the errors will introduce a circularly polarised component of the opposite hand to that desired, i.e. circular cross polarisation.

If the two linear polarisation amplitudes are E_1 and E_2 and the relative phase difference is δ, then the ratio of RHCP and LHCP amplitude is

$$\frac{E_R}{E_L} = \left(\frac{E_1^2 + E_2^2 + 2E_1E_2 \sin\delta}{E_1^2 + E_2^2 - 2E_1E_2 \sin\delta}\right)^{1/2} \quad (49.57)$$

so that pure RHCP is obtained when $E_1 = E_2$ and $\delta = \pi/2$. The voltage axial ratio is given as

$$A = \frac{(E_R/E_L) + 1}{(E_R/E_L) - 1} \quad (49.58)$$

For the special case where the two linear components are in exact phase quadrature but are imbalanced in amplitude, the net axial ratio is simply equal to the ratio of the linear component amplitudes E_1/E_2.

For the other special case where the two linear components are equal in amplitude but are misphased by an angle Δ from the ideal quadrature relationship so that $\delta = \pi/2 \pm \Delta$, the next axial ratio is given by

$$A = \frac{(1+\cos\Delta)^{1/2} + (1-\cos\Delta)^{1/2}}{(1+\cos\Delta)^{1/2} - (1-\cos\Delta)^{1/2}} \underset{\Delta\to 0}{\simeq} \frac{1+\frac{1}{2}\Delta}{1-\frac{1}{2}\Delta} \quad (49.59)$$

where Δ is in radians.

A set of curves is plotted in *Figure 49.5*, showing how the axial ratio and circular polarisation purity vary with combinations of amplitude and phase imbalance between the linear components.

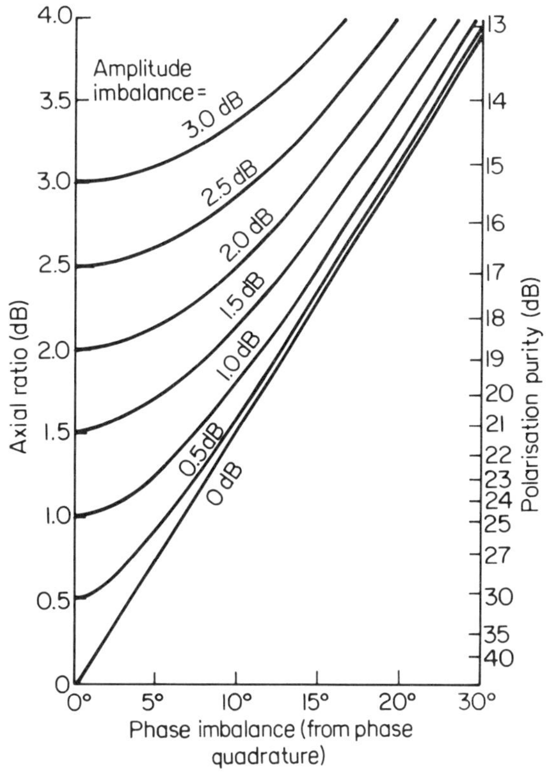

Figure 49.5 Circular polarisation purity versus orthogonal linear component amplitude and phase imbalance

49.4 Linear and planar arrays

Linear and planar arrays comprise elements positioned along a line or across a plane respectively. The excitation function or distribution for a planar array is generally separable in the x and y directions in rectangular coordinates, so that the principal plane patterns are also separable. Because of this, most planar arrays may be analysed in terms of the linear distributions in the two planes.

For an array with a fixed distribution, the main beam position will be fixed in space. If however a linear phase taper is introduced across the array, the main beam will squint, and if the distribution phase can be continuously or nearly continuously varied, then the beam can be continuously scanned in space without physical movement of the array. Such an antenna is known as a phased array, and is widely used in both radar and other applications.

49.4.1 Continuous line source distributions

The spacing between elements in an array is usually of the order of half a wavelength, so that the array pattern is to a fair approximation given by the product of the element pattern and the pattern of a continuous array of isotropic sources with the same aperture extent and distribution as the real array. It is therefore convenient when comparing the properties of different array distributions, to consider and compare the properties of the equivalent line source distributions.

49.4.1.1 The Fourier transform

For a continuous linear aperture distribution $f(x)$ $(-a/2 \leqslant x \leqslant a/2)$ the far-field radiation pattern is given by

$$g(a)=\int_{-a/2}^{a/2} f(x)\,\mathrm{e}^{-\mathrm{j}2\pi ux}\,\mathrm{d}x \tag{49.60}$$

where

$$u=\frac{\sin\theta}{\lambda} \tag{49.61}$$

which may be recognised as the Fourier transform.[4]

For a distribution with even symmetry, i.e. $f(-x)=f(x)$, the transform reduces to

$$g(u)=2\int_{0}^{a/2} f(x)\cos(2\pi ux)\,\mathrm{d}x \tag{49.62}$$

which is known as the cosine Fourier transform. For a distribution with odd symmetry, i.e. $f(-x)=-f(x)$, the transform reduces to

$$g(u)=2j\int_{0}^{a/2} f(x)\sin(2\pi ux)\,\mathrm{d}x \tag{49.63}$$

which is known as the sine Fourier transform. The aperture efficiency of an even symmetrical distribution $f(x)$ is given by

$$\eta_a=\frac{2}{a}\left|\int_{0}^{a/2} f(x)\,\mathrm{d}x\right|^2\left[\int_{0}^{a/2}|f(x)|^2\,\mathrm{d}x\right]^{-1} \tag{49.64}$$

and the relative power in the aperture by

$$P=\frac{2}{a}\int_{0}^{a/2}|f(x)|^2\,\mathrm{d}x \tag{49.65}$$

The boresight field $g(0)$ according to equations (49.60) and (49.62) is given by

$$g(0)=\sqrt{\eta_a P}\;a \tag{49.66}$$

so that the pattern normalised to $g'(0)=1$ is simply

$$g'(u)=\frac{g(u)}{g(0)}=\frac{g(u)}{\sqrt{\eta_a P}\;a} \tag{49.67}$$

49.4.1.2 Patterns of common equiphase distributions

In *Table 49.1* the radiation pattern equations and the values of the most important associated parameters, for eight common even symmetrical equiphase distributions are tabulated. The beamwidth constant K is the ratio of the 3 dB beamwidth to the aperture width in wavelengths, for large apertures

$$\theta_3 \simeq K/(a/\lambda) \quad \text{(degrees)} \tag{49.68}$$

The exact relationship, valid for lage and small apertures is

$$\theta_3=2\sin^{-1}\left(\frac{\pi K}{360a/\lambda}\right) \tag{49.69}$$

49.4.1.3 Cosine-squared on a pedestal distribution

A widely used linear distribution is the 'cosine-squared on a pedestal' distribution

$$f(x)=e+(1-e)\cos^2 \pi x/a \tag{49.70}$$

where $-a/2 \leqslant x \leqslant a/2$, and e is the relative pedestal height or edge taper. The absolute pattern $g(u)$ for this distribution is given by

$$\frac{g(u)}{a}=\frac{\sin\pi au}{\pi au}\left[e+\frac{(1-e)}{2}\,\frac{\pi^2}{\pi^2-(\pi au)^2}\right] \tag{49.71}$$

with a boresight level $g(0)/a=(1+e)/2$. By varying the edge taper parameter e, a varying trade-off between side lobe levels, aperture efficiency and beamwidth may be obtained. The aperture efficiency is given by

$$\eta_a=\left[1+\tfrac{1}{2}\left(\frac{1-e}{1+e}\right)^2\right]^{-1} \tag{49.72}$$

and is plotted against the edge taper e together with the beamwidth constant K and the first four side lobe levels in *Figure 49.6*.

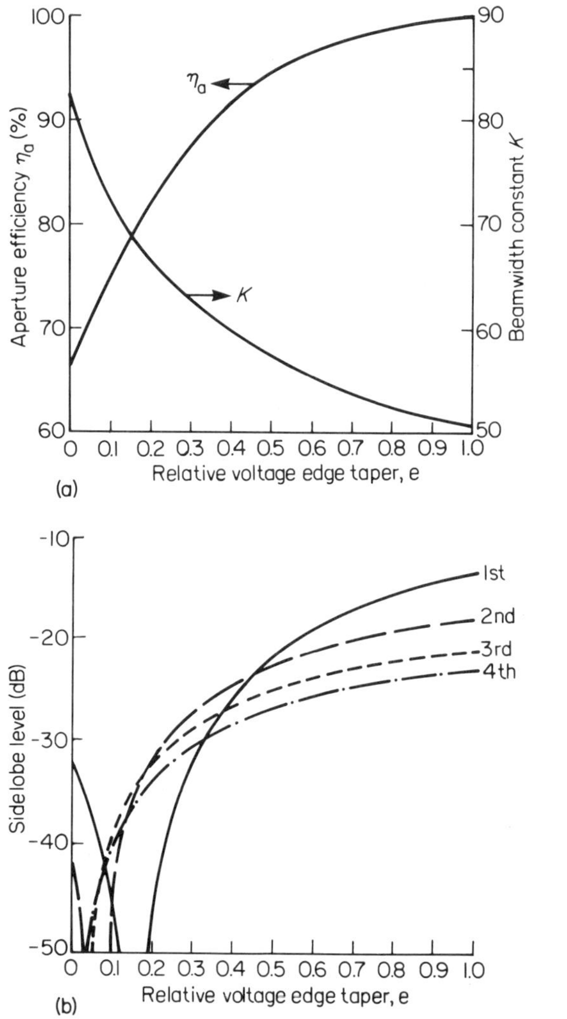

Figure 49.6 Properties of the cosine-squared on a pedestal distribution. (a) Aperture efficiency and beamwidth constant; (b) first four sidelobe levels

Table 49.1 Patterns of common equiphase distributions

Distribution type	$f(x)$	($z=\pi au$) $g(u)/a$	*Aperture efficiency* η_a	*Relative power (P) in the aperture*	*3 dB beamwidth constant K*	*1st null position (au)*	*1st side-lobe level (dB)*	*2nd side-lobe level (dB)*	*1st side-lobe position (au)*	*2nd side-lobe position (au)*
Uniform	1	$\frac{\sin z}{z}$	1 (100%)	1	50.8	1.00	−13.3	−17.8	1.43	2.46
Cosine	$\cos\left(\pi\frac{x}{a}\right)$	$\left(\frac{\pi}{2}\right)\frac{\cos z}{(\pi/2)^2-z^2}$	$8/\pi^2$ (81.1%)	$\frac{1}{2}$	68.1	1.50	−23.0	−30.7	1.89	2.93
Cosine^2	$\cos^2\left(\pi\frac{x}{a}\right)$	$\frac{1}{2}\frac{\sin z}{z}\frac{\pi^2}{\pi^2-z^2}$	2/3 (66.7%)	$\frac{3}{8}$	82.5	2.00	−31.5	−41.5	2.36	3.41
Parabolic	$1-\left(\frac{2x}{a}\right)^2$	$\frac{2}{z^2}\left(\frac{\sin z}{z}-\cos z\right)$	5/6 (83.3%)	$\frac{8}{15}$	66.2	1.43	−21.3	−29.0	1.83	2.90
Parabolic^2	$\left[1-\left(\frac{2x}{a}\right)^2\right]^2$	$\frac{8}{z^2}\left[\frac{3}{z^2}\left(\frac{\sin z}{z}-\cos z\right)-\frac{\sin z}{z}\right]$	7/10 (70.0%)	$\frac{128}{315}$	78.8	1.83	−27.7	−37.8	2.23	3.32
Double cosine	$\left\|\sin\left(2\pi\frac{x}{a}\right)\right\|$	$\pi\frac{1+\cos z}{\pi^2-z^2}$	$8/\pi^2$ (81.1%)	$\frac{1}{2}$	53.4	1.00	−7.5	−23.3	1.65	3.89
$1\frac{1}{2}$ cycle cosine	$\cos\left(3\pi\frac{x}{a}\right)$	$-\left(\frac{3\pi}{2}\right)\frac{\cos z}{(3\pi/2)^2-z^2}$	$8/9\pi^2$ (9.01%)	$\frac{1}{2}$	29.7	0.50	+7.6	−9.2	1.40	2.91
Gable	$1-\left\|\frac{2x}{a}\right\|$	$\frac{1-\cos z}{z^2}=\frac{1}{2}\left[\frac{\sin(z/2)}{(z/2)}\right]$	3/4 (75.0%)	$\frac{1}{3}$	73.1	2.00	−26.5	−35.7	2.86	4.92

Table 49.2 Hamming distribution sidelobes

	1st	*2nd*	*3rd*	*4th*	*5th*	*6th*	*7th*
Level (dB)	−44.0	−56.0	−43.6	−42.7	−43.2	−44.1	−45.0
Position (au)	2.22	2.79	3.53	4.50	5.49	6.49	7.49

The value of the maximum sidelobe level is minimised for $e=0.08$, and this special case of the cosine-squared on a pedestal distribution is known as the Hamming distribution. The levels and angular positions of the first seven sidelobes are tabulated in *Table 49.2*.

49.4.1.1 The Taylor distribution

It may be shown that the distribution producing the narrowest beamwidth for a specified maximum sidelobe level is that which produces a pattern in the form of a modified Chebyshev polynomial, with all sidelobes of equal amplitude.[5] Such a distribution has become known as the Dolph–Chebyshev distribution. In practical cases this distribution has two disadvantages. Firstly that the distribution tends to require high amplitudes at the edge of the aperture, and secondly that for large aperture sizes the directivity will be limited, asymptotically approaching 3 dB above the specified main beam to sidelobe level.

A modified set of distributions was proposed by Taylor,[6] producing equal amplitude sidelobes at the specified level out to the $\bar{n}$th sidelobe, with subsequent side lobes tapering off as $1/u$. The distributions of this form are more realizable in practice, and do not suffer from the gain limitation of the Dolph–Chebyshev distributions.

For a linear aperture of width a, the distribution is given by

$$f(p, A, \bar{n}) = 1 + 2 \sum_{n=1}^{\bar{n}-1} F(n, A, \bar{n}) \cos np \tag{49.73}$$

where

$$F(n, A, \bar{n}) = [(n-1)!]^2 \prod_{m=1}^{\bar{n}-1} 1-(n/z_m)^2 \times [(\bar{n}-1+n)!\,(\bar{n}-1-n)!]^{-1} \tag{49.74}$$

$$p = 2\pi x/a \qquad (-a/2 \leqslant x \leqslant a/2) \tag{49.75}$$

$$A = \pi^{-1} \cosh^{-1} \eta \tag{49.76}$$

where η is the sidelobe voltage ratio, z_n is the nth pattern zero

$$z_n = \pm\sigma[A^2 + (n-\tfrac{1}{2})^2]^{1/2} \qquad 1 \leqslant n \leqslant \bar{n} \tag{49.77}$$

$$= \pm n \qquad \bar{n} \leqslant n < \infty \tag{49.78}$$

and σ is the pattern stretchout factor (slightly >1)

$$\sigma = n[A^2 + (\bar{n}-\tfrac{1}{2})^2]^{-1/2} \tag{49.79}$$

The resulting pattern is given by

$$g(z, A, \bar{n}) = \frac{\sin \pi z}{\pi z} \prod_{n=1}^{\bar{n}-1} \frac{1-(z/z_n)^2}{1-(z/n)^2} \tag{49.80}$$

where $z = (a/\lambda) \sin \theta$. The 3 dB beamwidth is given by

$$\theta_3 = 2 \sin^{-1}\left(\frac{\lambda\sigma\beta_0}{2a}\right) \tag{49.81}$$

where

$$\beta_0 = 2 \sin^{-1}\left\{\frac{1}{\pi}\left[(\cosh^{-1}\eta)^2 - \left(\cosh^{-1}\frac{\eta}{\sqrt{2}}\right)^2\right]^{1/2}\right\} \tag{49.82}$$

Comprehensive tables of the aperture distributions for various sidelobe levels and values of $\bar{n}$ are available.[7]

49.4.2 Discrete arrays

For a discrete equispaced array of N elements excited by a voltage or current distribution A_i $(i=1, N)$, the far-field pattern is

$$E(\theta) = \sum_{i=1}^{N} A_i\, e^{j\pi(s/\lambda)\sin\theta(2i-N-1)} \tag{49.83}$$

where s is the element spacing.

For element spacings in the range $\lambda/2 \leqslant s \leqslant \lambda$ the directivity is given by

$$D = 2\eta_a(N-1)s/\lambda \tag{49.84}$$

where the aperture or illumination efficiency η_a is given by

$$\eta_a = \left(\sum_{i=1}^{N} A_i\right)^2 \left(N \sum_{i=1}^{N} A_i^2\right)^{-1} \tag{49.85}$$

49.4.2.1 The uniform distribution

For a distribution uniform in both amplitude and phase, the far-field pattern is given by

$$E(\theta) = \frac{\sin[N\pi(s/\lambda)\sin\theta]}{N \sin[\pi(s/\lambda)\sin\theta]} \tag{49.86}$$

$$\underset{s/\lambda\to 0}{=} \frac{\sin[N\pi(s/\lambda)\sin\theta]}{N\pi(s/\lambda)\sin\theta} \tag{49.87}$$

where s is the element spacing. For small element spacings therefore the pattern approaches the Fourier transform of a continuous uniform distribution with a −13.3 dB first side lobe level, and beamwidth constant $K=50.8$.

The directivities or gains for uniform broadside arrays with 2–10 elements are plotted in *Figure 49.7* for isotropic elements, and for short dipoles aligned normal to the array and collinear with the array. In all cases the gain is approximately linear with element spacing for $0.1\lambda < s < 0.9\lambda$, and for isotropic elements the gain is equal to the element number N when $s=\lambda/2$. For both the isotropic and normally orientated dipole elements the gain drops rapidly at $s \simeq \lambda$ due to the emergence of the grating lobe into real space. This effect is largely absent for colinear dipole elements, since the $\cos\theta$ dipole pattern suppresses the grating lobe as it appears in the end-fire mode. The gain of arrays with isotropic or collinear dipole elements are roughly equal for large element numbers, whereas the gain is significantly increased with normally orientated dipole elements.

49.4.3 Beam scanning

An equiphase linear array will produce a broadside main beam normal to the array. The beam may however be scanned away from the broadside direction by an angle θ_0 by applying a linear phase taper of α between consecutive elements to the distribution, where α is given by

$$\alpha = 2\pi(s/\lambda) \sin\theta_0 \tag{49.88}$$

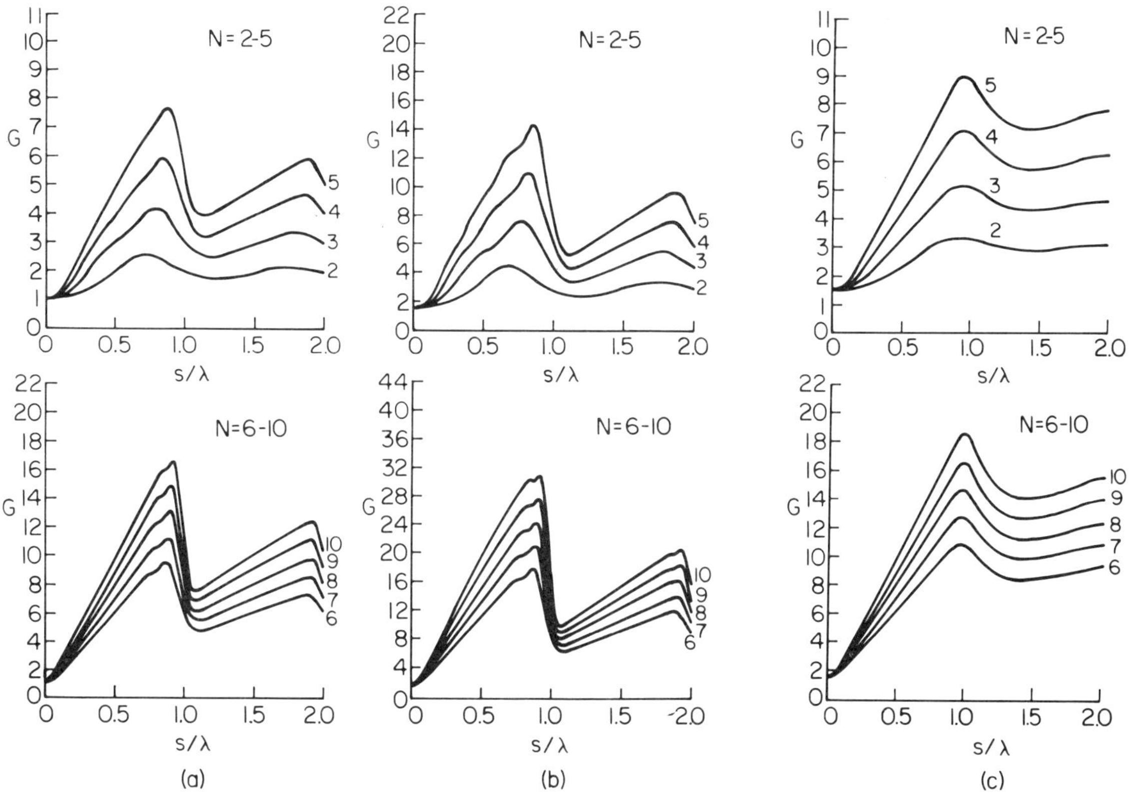

Figure 49.7 Uniform broadside linear array directivity. (a) isotropic elements; (b) normally oriented Hertzian dipole elements; (c) collinear Hertzian dipole elements

where s is the element spacing. The array pattern variable u, normally equal to $\sin\theta/\lambda$, will then be replaced by $\sin(\theta-\theta_0)/\lambda$ with the pattern shape $g(u)$ as a function of this variable remaining the same.

One consequence of scanning the beam away from the broadside direction is that the grating lobe will appear in real space for a smaller electrical spacing.

The spacing just placing the grating lobe peak into real space is given by

$$s/\lambda=(1+|\sin\theta_0|)^{-1} \tag{49.89}$$

For the broadside case ($\theta_0=0$) this condition is $s/\lambda=1$, and for the end-fire case ($\theta_0=\pm\pi/2$) the condition is $s/\lambda=1/2$.

The other main consequence of beam scanning is that the main beam will broaden as the scan angle is incrreased. For a uniform distribution and array length a, the 3 dB beamwidth is given by

$$\theta_3=\cos^{-1}\left(\sin\theta_0-\frac{0.443}{a/\lambda}\right)-\cos^{-1}\left(\sin\theta_0+\frac{0.443}{a/\lambda}\right) \tag{49.90}$$

$$\underset{a/\lambda\to\infty}{\simeq} 0.886\,\frac{\sec\theta_0}{a/\lambda} \quad \text{(at or near broadside)} \tag{49.91}$$

up to the scan limit where the two beams either side of the array axis start to merge.

At end-fire ($\theta_0=\pm\pi/2$) the beamwidth is given by

$$\theta_3=2\cos^{-1}\left(1-\frac{0.443}{a/\lambda}\right) \tag{49.92}$$

$$\underset{a/\lambda\to\infty}{\simeq} 2\left(\frac{0.886}{a/\lambda}\right)^{1/2} \tag{49.93}$$

These beamwidth curves are plotted against array length in *Figure 49.8* . To obtain the beamwidths for other distributions, the uniform distribution beamwidth should be multiplied by $K/50.8$, where K is the beamwidth factor in degrees for the distribution in question.

49.5 Circular arrays and loop antennas

To generate useful radiation patterns, it is possible to distribute radiating elements around the circumference of a circle—the circular array. The loop antenna is a special case of the circular array, since it may be considered to comprise an infinite number of infinitesimal elementary dipole sources aligned tangentially to the circumference of the circle. Similarly the annular slot (see section 49.7.3).

It is possible to vary the radiation pattern of a circular array by varying the array excitation function, the array radius and the element orientation. Because of this there are effectively two extra degrees of freedom when compared with the linear and planar array geometries, for which only the array excitation may be varied to attempt to form the desired radiation pattern. An additional advantage of the circular array is its ability to scan a directional radiation pattern about its axis through a full 360° with little or no pattern change.

49.5.1 Element orientation

For a circular array of linearly polarised elements, there are three principal orientations—axial, tangential and radial, as depicted in *Figure 49.9*. For arrays of discrete elements, the axial

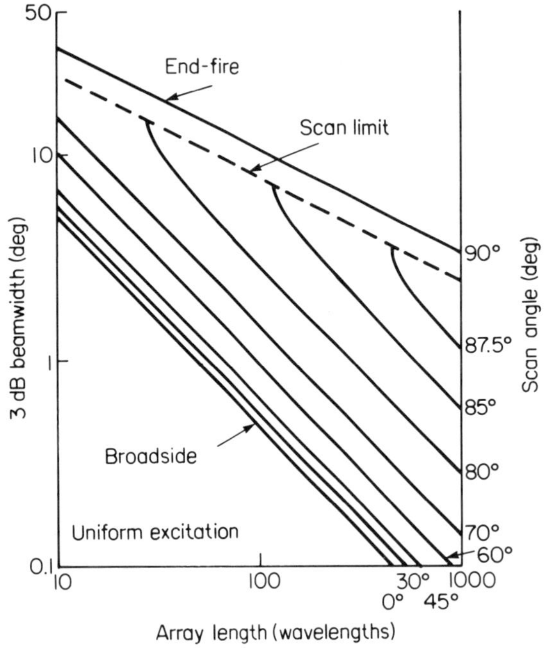

Figure 49.8 Scanned uniform linear array beamwidth

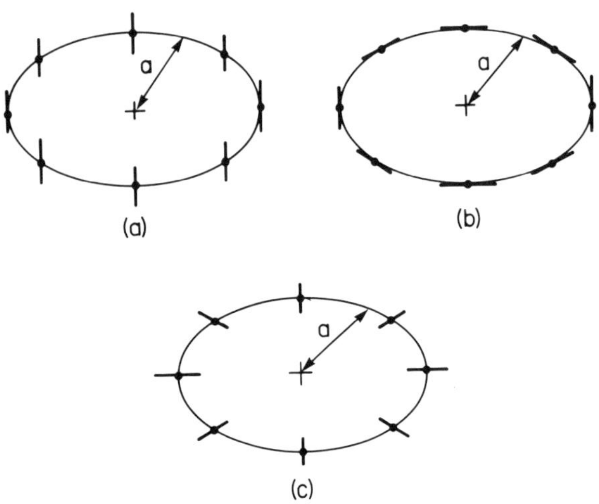

Figure 49.9 Circular array element orientations. (a) Axial; (b) tangential; (c) radial

orientation is the most widely used, and the polarisation of the radiated field is everywhere linear and $\hat{\boldsymbol{\theta}}$ directed. The tangential and radial orientations produce elliptical polarisation, and are used, for example, when a circularly polarised radiation maximum is required in the direction of the array axis, which may be achieved with a particular excitation, as will be described later. The analyses for these two orientations are also of use when applied to continuous radiating structures, the circular loop and annular slot already referred to being two examples for the tangential orientation. A centre-fed circular disc would be an example for the radial orientation.

49.5.2 Phase modes for circular arrays

The concept of phase modes in circular arrays not only simplifies the analysis of arrays when excited with particular distributions, but also may be applied directly to the synthesis of radiation patterns. Indeed in many cases the antenna feed network which is actually used directly reflects this.

Any array voltage or current distribution $F(\varphi)$ where φ is the azimuthal angle around the array, may be analysed as a Fourier series of 'phase modes'

$$F(\varphi)=\sum_{-\infty}^{\infty} a_m \, e^{jm\phi} \quad (49.94)$$

where a_m is the complex coefficient for the mth mode. From this equation it may be seen that the phase of the mth mode excitation varies linearly with φ, passing through $2\pi m$ radians for each complete revolution around the array. It may be shown that the far-field pattern radiated by the mth mode excitation exhibits exactly the same property.

49.5.2.1 Continuous array radiation patterns

Adopting the phase mode concept just described, the far-field radiation patterns for continuous arrays of radius a of elementary or Hertizian dipole sources are

$$F_\theta=-\sin\theta J_m(a)\, e^{jm(\phi-\pi/2)} \quad (49.95)$$
$$F_\phi=0 \quad (49.96)$$
for axial orientation

$$F_\theta=\frac{m}{z} J_m(z)\cos\theta\, e^{jm(\phi-\pi/2)}$$
$$F_\phi=jJ'_m(z)\, e^{jm(\phi-\pi/2)} \quad (49.98)$$
for tangential orientation (49.97)

$$F_\theta=jJ'_m(z)\cos\theta\, e^{jm(\phi-\pi/2)} \quad (49.99)$$
$$F_\phi=-\frac{m}{z} J_m(z)\, e^{jm(\phi-\pi/2)} \quad (49.100)$$
for radial orientation

$$z=2\pi\frac{a}{\lambda}\sin\theta \quad (49.101)$$

These equations apply where the array is aligned in the xy plane in the same spherical coordinate system as for the elemental loop shown in *Figure 49.1(b)* such that at $\varphi=0$ the phase of the excitation voltage or current is zero for each mode. F corresponds to the radiated **E** or **H** field, depending on whether the elemental dipole sources are electric or magnetic respectively. In all cases the $[\exp(-jkr)]/r$ factor is suppressed.

The only cases for which the radiated field is non-zero in the $\theta=0$ direction are the tangential and radial orientations with $m=\pm1$ excitations. In these cases the field in that direction is perfectly circularly polarised, with the hand of polarisation (LHCP or RHCP) changing between the $m=+1$ and $m=-1$ excitations.

The higher the order of the phase mode, the more the radiated energy will be concentrated towards the $\theta=\pi/2$ (array) plane. This is demonstrated in *Figure 49.10*, which shows the gain at $\theta=\pi/2$ (and $\theta=0$ for $m=\pm1$) for tangentially and radially orientated arrays.

49.5.2.2 Discrete array radiation patterns

The continuous array patterns will be closely approximated by those of a discrete array of a finite number of elements, provided two conditions are met.

(a) The circumferential element spacing is a half-wavelength or less, i.e. $2\pi a/s<\lambda/2$, where a is the array radius and s the number of elements.

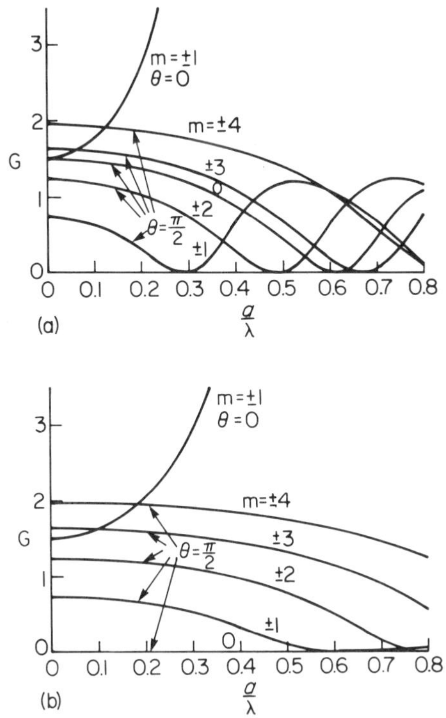

Figure 49.10 Circular array directivity. (a) Tangential orientation; (b) radial orientation

(b) The number of elements is much greater than the mode order, i.e. $s >> m$. As a general rule three times the mode order represents a minimum.

If either or both these conditions are not met, the resulting patterns in the azimuthal plane (θ constant) will exhibit amplitude and phase ripple with a dominant period equal to $2\pi/s$, i.e. with one cycle per element. For a discrete array with axially orientated elements the radiation pattern will be modified to

$$F_\theta = -\frac{\sin\theta}{2}\sum_{q=0}^{\infty}(2-\delta_{oq}) \times [J_{m+qs}(z)\,e^{j(m+qs)(\phi-\pi/2)} + J_{-m+qs}(z)\,e^{-j(-m+qs)(\phi+\pi/2)}] \quad (49.102)$$

This may be broken down into the continuous array solution plus an infinite series of perturbation terms consisting of harmonic phase modes. Similar forms apply to discrete arrays with tangentially or radially orientated elements.[8]

49.5.2.3 Array radiation patterns with directional elements

All of the preceding analysis has assumed that the radiating elements are short elemental electric or magnetic dipoles with $\sin\theta$ radiation patterns. When the directional patterns of the elements do not conform to this ideal, the array patterns will be modified from the previously given forms.

One case of interest is that in which the elements are linear dipoles of finite length, so that the element pattern is still symmetrical about its axis, but has a θ dependence other than $\sin\theta$, with a narrower beamwidth. For axially orientated array elements the effect of the element pattern on the net array pattern is straightforward, simply replacing the $\sin\theta$ in equation (49.95) by the particular element pattern. The basic radiation properties of the array and their dependence on the array radius are unaffected. For the tangential and radial orientations the modifications are more complex,[8] since for these cases the element pattern is not separable from the array pattern. For dipole lengths of less than $\lambda/2$ however the changes to the elemental dipole array patterns are small.

The other case of interest is the axially orientated circular array in which the element pattern is no longer symmetrical about the vertical array axis. Examples of this would be arrays with vertical dipoles backed by reflecting ground planes, or with small horns pointing radially outwards. For moderate element directivity in the φ plane, for example with an element pattern of the form $1+\cos\varphi$, the effects of this directivity are generally beneficial. For a continuous dipole array equations (49.95) and (49.96) show that in a given θ direction, the relative radiated field will vary with frequency according to the Bessel function $J_m(z)$. If z is such that $J_m(z)=0$ then it will be impossible to radiate any energy in that direction at that frequency. Elements which are directional in the φ plane tend to 'smooth over' the Bessel function zeros, so that the relative field in any given direction will vary by only a few dB over a greater than one octave frequency range. This is of great importance for broad-band circular arrays.

49.5.3 Oblique orientation—the Lindenblad antenna

All the preceding arrays have used elements axially, tangentially or radially orientated with respect to the array circle. In the most general case the elements may be arbitrarily orientated with three degrees of freedom—the direction cosines $\cos\alpha$, $\cos\beta$ and $\cos\gamma$ to the $\hat{\mathbf{r}}$, $\hat{\boldsymbol{\theta}}$ and $\hat{\boldsymbol{\varphi}}$ unit vectors respectively in the array or xy plane.

There are several special cases of element orientation apart from the three principle orientations just discussed. One of the more useful of these is with $\alpha=\pi/2$ so that the elements are normal to the radius vector and inclined at an angle γ to the array plane. When excited with the $m=0$ uniform amplitude and phase distribution, and γ is chosen to satisfy

$$\gamma = \pm\tan^{-1}[J_1(ka)/J_0(ka)] \quad (49.103)$$

where k is the free-space wavenumber $2\pi/\lambda$ and a is the array radius, the resulting pattern is azimuthally omni-directional and circularly polarised—the Lindenblad antenna.[9,10] The relative radiation pattern is of the form

$$F_\theta = \pm[J_1(ka)/J_0(ka)]J_0(z)\sin\theta\cos\gamma \quad (49.104)$$

$$F_\phi = -jJ_1(z)\cos\gamma \quad (49.105)$$

where $z=ka\sin\theta$. For small radii ($ka<<1$) these reduce to

$$F_\theta \simeq \tfrac{1}{2}ka\sin\theta \quad (49.106)$$

$$F_\phi \simeq -\tfrac{1}{2}jka\sin\theta \quad (49.107)$$

so that the field is circularly polarised everywhere in space with the radiation pattern of a short dipole.

49.5.4 Beam cophasal excitation

To produce a pencil beam in the $\theta=\pi/2$ azimuth plane with an axially orientated array of linear elements, the obvious excitation is that which produces equal phase signals from all the elements in the wanted direction—the beam cophasal excitation. For convenience, particularly when scanning the beam electronically, the excitation amplitudes are often uniform. Under these conditions for a continuous array the relative azimuth and elevation plane radiation patterns are

$$F_{az} = J_0(4\pi a\lambda^{-1}\sin\tfrac{1}{2}\varphi) \quad (49.108)$$

$$F_{el} = J_0[2\pi a\lambda^{-1}(1-\sin\theta)] \quad (49.109)$$

where a is the array radius, φ is the azimuthal angle from the beam peak, and θ is the elevation angle from the zenith (beam

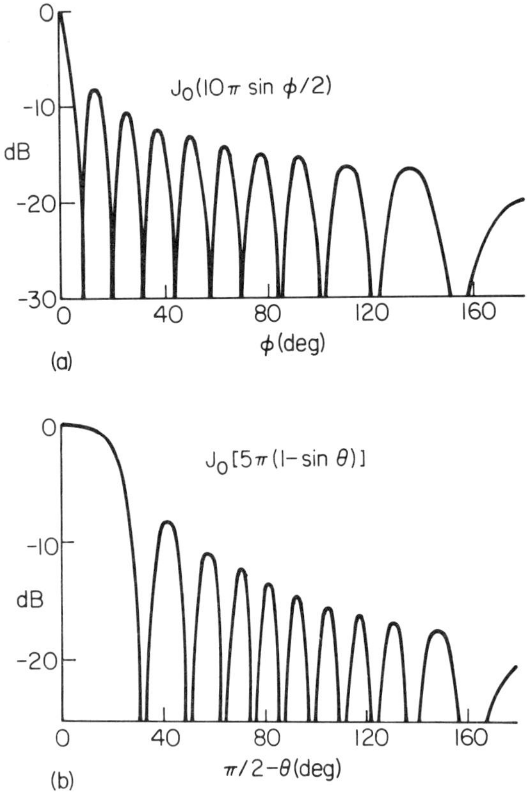

Figure 49.11 Radiation patterns for 5 diameter beam cophasal circular array. (a) Azimuth; (b) elevation

peak at $\theta=\pi/2$). These patterns are plotted for a 5λ diameter array in *Figure 49.11*. The azimuth pattern beamwidth is narrower and the side lobes higher than for a uniform linear array whose length is equal to the circular array diameter, because the geometry of the circular array tends to concentrate the power towards the edges of the aperture. The elevation pattern has the same side lobe levels as the azimuth pattern, but the main beam is broader because the beam in the elevation plane is effectively formed from the end-fire array mode.

To achieve lower side lobes it is necessary to excite only the elements along the half circular arc directly visible from the beam maximum direction, and to taper the amplitude from centre to edge. This makes electronic beam scanning more difficult to implement, since the phase *and* amplitude of the excitation to each element must be changed as the beam is scanned.

In all cases the circumferential element spacing must be less than $\lambda/2$ to closely approximate the continuous array patterns and avoid the emergence of grating lobes. For elements which are directional in the azimuth plane, the spacing must be made somewhat smaller than what would be allowable for omnidirectional elements.

49.5.5 Loop antennas

An example of a continuous tangentially orientated circular array is the current-fed circular wire loop. In most cases it may be assumed that the current in the loop is uniform in amplitude and phase, so that the excitation function corresponds to the $m=0$ phase mode.

49.5.5.1 Radiation patterns

For a circular loop of radius a aligned in the xy plane of a standard spherical coordinate system, the radiated far-fields are

$$E_\phi=120\pi^2 Ia\lambda^{-1}J_1(2\pi a\lambda^{-1}\sin\theta) \tag{49.110}$$

$$H_\theta=\pi Ia\lambda^{-1}J_1(2\pi a\lambda^{-1}\sin\theta) \tag{49.111}$$

where I is the loop current (the $[\exp(-jkr)]/r$ factor is suppressed). For a small loop of arbitrary shape and area A ($<\lambda^2/100$) the far-field pattern is

$$E_\phi\simeq 120\pi^2 IA\lambda^{-2}\sin\theta \tag{49.112}$$

$$H_\theta\simeq \pi IA\lambda^{-2}\sin\theta \tag{49.113}$$

i.e. the pattern for a short magnetic dipole.

Far-field polar diagrams for circular loops of 0.1, 1 and 5 wavelengths diameter are shown in *Figure 49.12(a)*. *Figure 49.12(b)* shows the polar diagram for a square loop of side length 4.44λ. Compared with the 5λ diameter (equal area) circular loop pattern, the lobes are of uniform amplitude, rather than decreasing from $\theta=0$ to $\theta=\pi/2$.

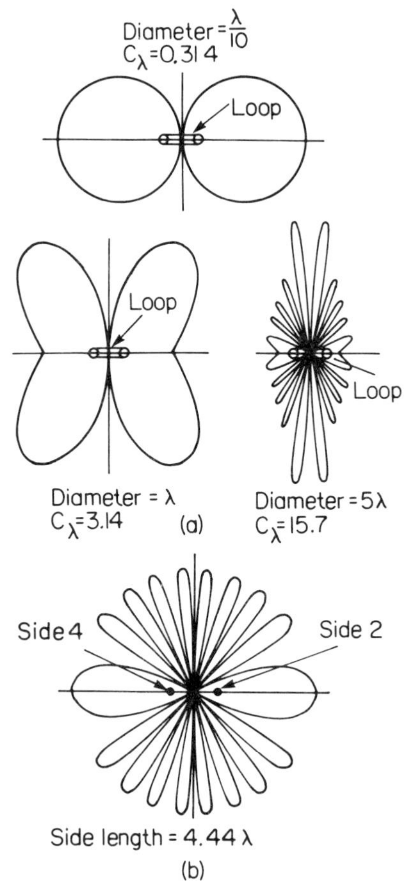

Figure 49.12 Loop radiation patterns. (a) Circular loops; (b) square loop

49.5.5.2 Directivity

The directivity of a circular loop of radius a is given by

$$G=\frac{4\pi a\lambda^{-1}J_1^2(2\pi a\lambda^{-1}\sin\theta)}{\int_0^{4\pi a/\lambda}J_2(y)\,dy} \tag{49.114}$$

For a small loop ($A<\lambda^2/100$) of arbitrary shape, the directivity in the $\theta=\pi/2$ plane is

$$G\simeq 3/2 \tag{49.115}$$

and for a large circular loop ($2\pi a/\lambda>5$) the peak directivity approaches an asymptotic solution

$$G\simeq 4.25a\lambda^{-1} \tag{49.116}$$

The peak directivity is plotted against loop circumference in *Figure 49.13(a)* showing also the small and large loop approximations.

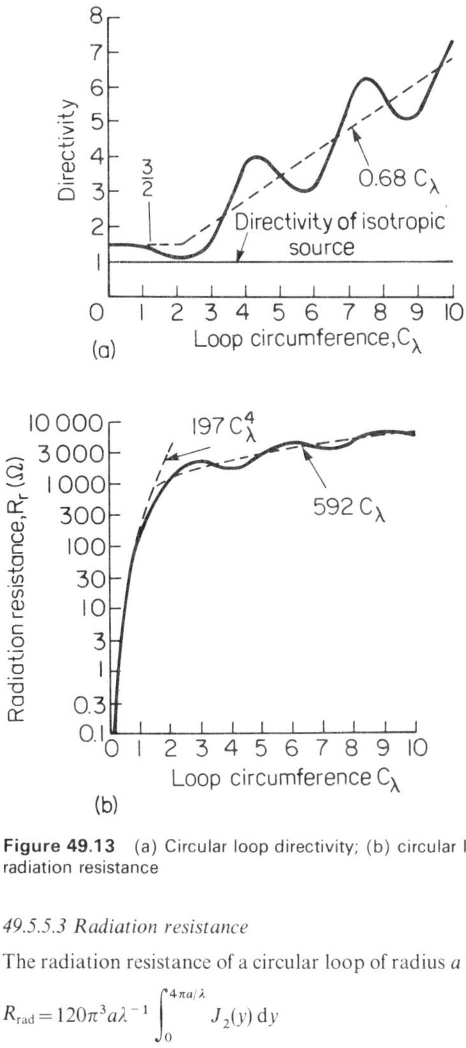

Figure 49.13 (a) Circular loop directivity; (b) circular loop radiation resistance

49.5.5.3 Radiation resistance

The radiation resistance of a circular loop of radius a is given by

$$R_{rad}=120\pi^3 a\lambda^{-1}\int_0^{4\pi a/\lambda}J_2(y)\,dy \tag{49.117}$$

For a small loop of arbitrary shape and area A ($A<\lambda^2/100$) the radiation resistance is approximately

$$R_{rad}\simeq 320\pi^4A^2\lambda^{-4} \tag{49.118}$$

and for a large circular loop ($2\pi a/\lambda>5$) the asymptotic solution is

$$R_{rad}\simeq 3720a\lambda^{-1} \tag{49.119}$$

The radiation resistance of a circular loop is plotted against its circumference in *Figure 49.13(b)*, showing also the small and large loop approximations.

49.5.5.1 Multi-turn loops

If the loop contains more than a single turn, the relative radiation pattern and directivity are unchanged. For a given loop current the radiated fields are N times what they would be for a single turn (N = number of turns), and conversely the induced voltage is N times what it would be for a single turn when illuminated by a plane wave of a given power density. The radiation resistance is raised by a factor N^2.

The applications of multi-turn loops are primarily to raise the radiation resistance of a small loop for matching purposes, and for producing a voltage multiplication effect for maximum sensitivity when feeding a high impedance receiver input.

49.6 Wire antennas

49.6.1 Linear wire antennas (linear dipoles)

A linear wire antenna comprises a straight length of wire whose diameter is much less than its length, fed at some point along its length. A balanced dipole antenna is obtained by feeding the antenna at its centre, which will provide a good non-reactive impedance match to a 50 Ω or 75 Ω transmission line, for a dipole length slightly shorter than a half wavelength. A wire may also be end-fed against earth or a ground plane, which is often a particularly convenient configuration.

49.6.1.1 Radiation patterns

For antenna lengths of up to a few wavelengths, it is a reasonable assumption that the current distribution along the wire is a sinusoidal standing wave with a period equal to one free space wavelength, and with zero current at the free end(s) of the wire. Based on this assumption it is possible to derive the radiated far-field pattern for a centre-fed linear dipole of length L

$$E_\theta(\theta)=j60I\left(\frac{\cos(\pi L\cos\theta/\lambda)-\cos(\pi L/\lambda)}{\sin\theta}\right)\frac{e^{-jkr}}{r} \tag{49.120}$$

where θ is measured from the antenna axis, and I is the current at the point of current maximum. For the special cases of $L=\lambda/2$, λ, $3\lambda/2$, the pattern becomes

$$\underset{L=\lambda/2}{E_\theta(\theta)} = j60I\left(\frac{\cos(\frac{1}{2}\pi\cos\theta)}{\sin\theta}\right)\frac{e^{-jkr}}{r} \tag{49.121}$$

$$\underset{L=\lambda}{=} j60I\left(\frac{\cos(\pi\cos\theta)+1}{\sin\theta}\right)\frac{e^{-jkr}}{r} \tag{49.122}$$

$$\underset{L=3\lambda/2}{=} j60I\left(\frac{\cos(\frac{3}{2}\pi\cos\theta)}{\sin\theta}\right)\frac{e^{-jkr}}{r} \tag{49.123}$$

These patterns are plotted in polar form in *Figure 49.14*. The main lobe in the direction $\theta=\pi/2$ narrows with increasing dipole length, and the secondary anti-phase lobes only appear for lengths greater than one wavelength.

The widely used half-wave dipole has a 3 dB beamwidth of 78° and a gain of 1.64 (2.15 dBi). These values may be compared with the 90° 3 dB beamwidth and 1.5 (1.76 dBi) gain for a vanishingly small Hertzian dipole.

49.6.1.2 Impedance

The radiation resistance of a thin linear dipole of length L, at the point of the current maximum is given by

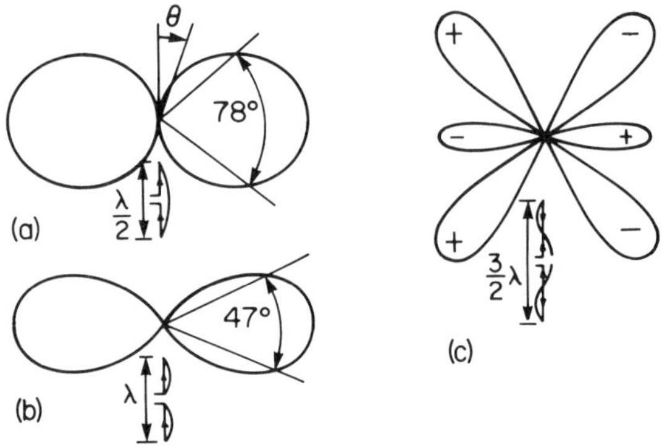

Figure 49.14 Linear dipole patterns. (a) $\frac{1}{2}$ wavelength; (b) full wavelength; (c) $\frac{3}{2}$ wavelength. The antennas are centre-fed, and the current distribution is assumed to be sinusoidal

$$R_{rad} = 60\{\gamma + \ln(kL) - C_i(kL) + \tfrac{1}{2}\sin(kL)[S_i(2kL) - 2S_i(kL)] + \tfrac{1}{2}\cos(kL)[\gamma + \ln(kL/2) + C_i(2kL) - 2C_i(kL)]\} \quad (49.124)$$

where S_i and C_i are the standard sine and cosine integrals and γ is Euler's constant (0.5772...). If the dipole is fed symmetrically at the centre the radiation resistance becomes

$$R'_{rad} = R_{rad}\,\mathrm{cosec}^2(kL/2) \quad (49.125)$$

which predicts an infinite resistance at the dipole centre when its length is equal to an odd number of quarter-wavelengths, and a value of 73.1 Ω for an exact half-wave. For a short ($L << \lambda$) centre-fed dipole the radiation resistance is given by

$$R_{rad} \simeq 20\pi^2(L/\lambda)^2 \quad (49.126)$$

The radiation resistance is plotted against dipole length for both feed points in *Figure 49.15*.

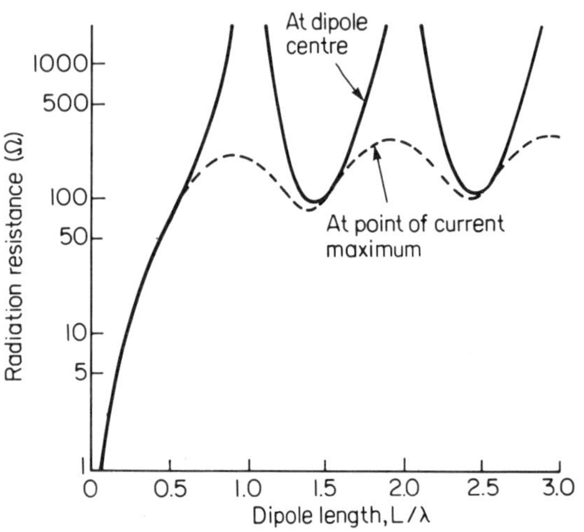

Figure 49.15 Radiation resistance of a thin linear dipole

The reactive component of the antenna impedance is dependent on the radius to length ratio a/L for the wire, and at the current maximum feed point is given by

$$X_a = 30\{2S_i(kL) + \cos(kL)[2S_i(kL) - S_i(2kL)] - \sin(kL)[2C_i(kL) - C_i(2kL) - C_i(2ka^2/L)]\} \quad (49.127)$$

A resonant design is achieved when this reactance is zero, and this occurs for lengths slightly shorter than a half-wavelength. At resonance the radiation resistance will be in the approximate range 50 Ω–75 Ω, the exact value depending on the wire radius, and such a dipole will therefore present a reasonably good match to a 50 Ω or 75 Ω transmission line.

49.6.2 Linear monopoles

A vertical linear monopole of length L and fed against a ground plane, will exhibit identical radiation patterns in the half space above the ground plane to those of a balanced dipole of length $2L$ radiating into free space. Because, for a given drive current, the total radiated power is reduced by a factor of two, the radiation resistance compared to the equivalent dipole will be exactly halved. A quarter-wave monopole for example will have a radiation resistance of 36.6 Ω.

49.6.3 The rhombic antenna

The linear dipole and end-fed antennas just described are members of the class of 'standing wave' antennas. The current along the length of such antennas is formed from two equal waves travelling in opposite directions, such that it is uniform in phase and sinusoidal in amplitude. If one end of a linear wire antenna is correctly terminated and the other end is excited, then there will be only a single travelling wave along the length of the wire, propagating at, or close to the velocity of light in free space. The linear phase progression of the current associated with such a wave will radiate a unidirectional pattern tilted in the direction of the wave propagation.

One of the more widely used forms of wire travelling wave antenna is the horizontal rhombic.[11,12,13] Its main application is in long distance h.f. communication, and whether excited in free space or a wavelength or so above ground, it has the useful property that the position of the beam peak in the elevation plane may be adjusted by choosing the geometry and dimensions correctly.

The basic rhombic is shown in *Figure 49.16*, orientated in the horizontal or xy plane such that the azimuth pattern beam peak

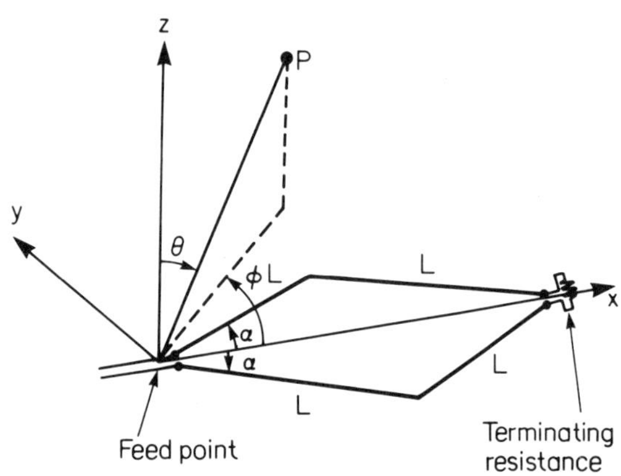

Figure 49.16 The rhombic antenna

is in the direction away from the feed point. along the x axis. Note the terminating resistor connected at the acute corner opposite to the feed point. For a drive current I the far field radiation pattern for an isolated rhombic is

$$E(\theta, \varphi) = 4(15/\pi)^{1/2} I \sin\alpha \frac{\sin[\frac{1}{2}kL(1-\cos\psi_1)(1-\cos\psi_2)]}{[(1-\cos\psi_1)(1-\cos\psi_2)]^{1/2}} \quad (49.128)$$

where

$$\cos \psi_1 = \sin \theta \cos(\varphi + \alpha) \quad (49.129)$$

$$\cos \psi_2 = \sin \theta \cos(\varphi - \alpha) \quad (49.130)$$

The pattern is expressed in a form such that the radiated power per unit solid angle is equal to $E(\theta, \varphi)^2$. When the rhombic is located at a height h above perfectly conducting ground, the pattern is modified to

$$E'(\theta, \varphi) = 2E(\theta, \varphi) \sin (kh \cos \theta) \quad (49.131)$$

To maximise the radiated field strength at an angle θ in the elevation plane ($\theta = 90° -$ elevation angle above the horizon) it may be shown that the optimum height above ground, arm length and included semi-angle are given by

$$h_{opt} = \lambda/4 \cos \theta \quad (49.132)$$

$$L_{opt} = \lambda/2 \cos^2 \theta \quad (49.133)$$

$$\alpha_{opt} = \tfrac{1}{2}\pi - \theta \quad (49.134)$$

For a typical design the arm length would be in the range 2–7 λ, the height above ground between 1 λ and 2 λ, and the included semk-angle α in the range 15–30°. The input impedance is generally relatively high, say 700–800 Ω, the side lobes are also rather high at −10 dB or worse, and the 3 dB beamwidth is typically 10–15°.

49.7 Slot antennas

Slots cut in conducting surfaces often provide convenient antennas or antenna elements, since they are 'flush-mounting', and require no protrusion above the ground plane surface. A possible disadvantage is that their radiation patterns are strongly influenced by the size or shape of the ground planes into which they are cut, and the pattern with a ground plane even 10λ across will be significantly different to that of the same slot on an infinite ground plane. A slot may be fed in a number of different ways, either from coaxial line or from waveguide. For example as shown in *Figure 49.13* a small rectangular slot can be fed from a coaxial connected across the narrow dimension of the slot at some point, from the open end of a rectangular waveguide whose axis is normal to the ground plane, or by cutting the slot in either the broad or the narrow face of the rectangular waveguide itself. The method of excitation will have little effect on the radiation pattern of the slot, but will affect primarily the impedance seen by the feeder and the bandwidth. The fourth feed method shown in *Figure 49.17*, and using a coaxial T-fed cavity to excite the slot, offers a relatively wide bandwidth, with a VSWR less than 2:1 over an octave bandwidth.

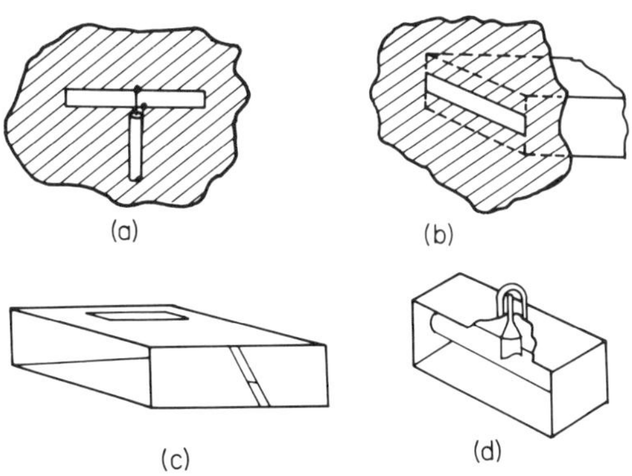

Figure 49.17 Some alternative rectangular slot excitation methods

The simple slot in a ground plane will radiate equally on both sides with a symmetric radiation pattern. In most applications a uni-directional pattern is required, so that the radiation on one side of the ground plane must be suppressed, usually by enclosing the slot on that side with a cavity. Apart from increasing the forward radiated field uniformly by 3 dB, the addition of the cavity will usually have little effect on the radiation pattern. Again the effect will be primarily on the impedance and bandwidth. Suppressing the radiation on one side of the ground plane will double the slot impedance.

49.7.1 Complementary antennas

A useful theory of equivalence between slot and wire antennas based on Babinet's principle has been developed.[15] The basis of this theorem is that any slot antenna has an equivalent wire antenna formed by transposing the areas of conductor and free space. The radiation patterns of these two antennas will be identical except that the **E** and **H** fields will be transposed. The rectangular slot is thus complementary to a linear wire dipole antenna of the same length and width, the former vertically polarized and the latter horizontally polarised, both with the same figure-of-eight pattern in the vertical plane. The annular slot is complementary to a wire loop of the same diameter and width.

This equivalence theory allows the impedance z_{slot} of a slot to be simply related to the impedance z_{comp} of its complementary antenna

$$z_{slot} z_{comp} = \eta_0^2/4 = 3600\pi^2 \quad (49.135)$$

where η_0 is the impedance of free space ($=120\pi$).

49.7.2 The rectangular slot

49.7.2.1 Radiation patterns

The radiation pattern of a slot radiating into free space, and orientated in the xy plane of a standard spherical coordinate system with the **E** field across the narrow dimension parallel to the x axis, may be computed in two ways. The slot may be converted to its complementary form, and the radiation pattern computed for that using one of the standard forms. Alternatively the field equivalence principle may be invoked and the x directed electric fields **E** in the aperture replaced by equivalent y directed magnetic currents **M** from the relationship

$$\mathbf{M} = -\hat{\mathbf{n}} \times \mathbf{E} \quad (49.136)$$

where $\hat{\mathbf{n}}$ is the z directed unit vector normal to the aperture. The fields radiated by the equivalent magnetic currents are those radiated by elementary magnetic dipole sources, as described in section 49.2.2. The radiated electric fields from a vanishingly small element $dx\,dy$ of a rectangular slot are thus

$$E_\theta = -jE_x \frac{\cos \varphi}{2\lambda} \, dx\, dy \frac{e^{-jkr}}{r} \quad (49.137)$$

$$E_\phi = jE_x \frac{\cos \theta \sin \varphi}{2\lambda} \, dx\, dy \frac{e^{-jkr}}{r} \quad (49.138)$$

where E_x is the x directed field in the slot aperture, and k is the free space wavenumber $2\pi/\lambda$. This pattern is uniform in the xz plane, and has a $\cos \theta$ variation in the yz plane. The net field radiated by the slot is found as an integral across the aperture involving the preceding equations for the elemental radiation field. This analysis is only valid if the aperture field in the slot is all or virtually all electric, with little or no magnetic or H component. For this reason the slot width w must be very much less than a wavelength.

The rectangular slot is often made half a free space wavelength long, to provide a resonant design with a purely resistive impedance. For the half-wave resonant slot the radiation pattern in the plane of the slot is the same as the radiation pattern of a half-wave dipole, i.e.

$$E_\phi = \frac{\cos(\frac{1}{2}\pi \sin \theta)}{\cos \theta} \tag{49.139}$$

As mentioned earlier, the size of the ground plane into which the slot is cut will have a considerable effect on the overall radiation pattern, and the pattern for an infinite ground plane will only be approached for dimensions in excess of 10λ. Most effect will be observed in the plane perpendicular to the slot, and qualitatively the pattern will exhibit a degree of ripple and an amplitude taper in the direction $\theta = \pi/2$ along the ground plane. This is demonstrated in *Figure 49.18*, which shows patterns for a half-wave slot on circular ground planes of various diameters.

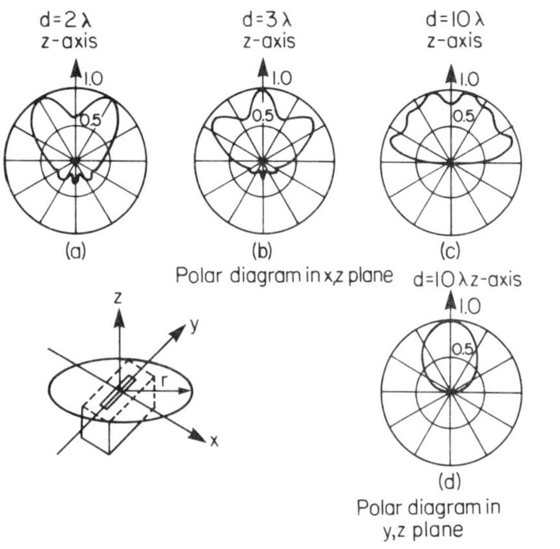

Figure 49.18 Radiation patterns of a narrow half-wavelength slot on a finite circular ground plane

49.7.2.2 Impedance

By using the principle of complementary antennas already described, if a slot is transformed to its complementary antenna and its impedance found, then the slot impedance is given directly by equation (49.135). This method is applied to half-wave, resonant half-wave, and full-wave rectangular slots in *Figure 49.19*. The complementary cylindrical dipole has the same length as the slot and a diameter equal to half the slot width, since it may be shown that a cylindrical conductor of diameter d is equivalent to a flat strip conductor of width $w = 2d$.[16] It is worth noting that a centre-fed full-wave slot will be almost exactly matched to a 50 Ω coaxial transmission line. Although the centre impedance of a resonant half-wave slot is non-reactive, to match it to a 50 Ω feed line it is necessary to move the feed point to a distance $\lambda/20$ from one end.

Measured slot impedances (*Figure 49.20*) generally confirm the predictions of the preceding analysis, with some slight discrepancies. The most notable discrepancy is that the resonant length appears from the measurements to be independent of the slot width, whereas the complementary antenna analysis suggests that it will be reduced as the slot width is increased.

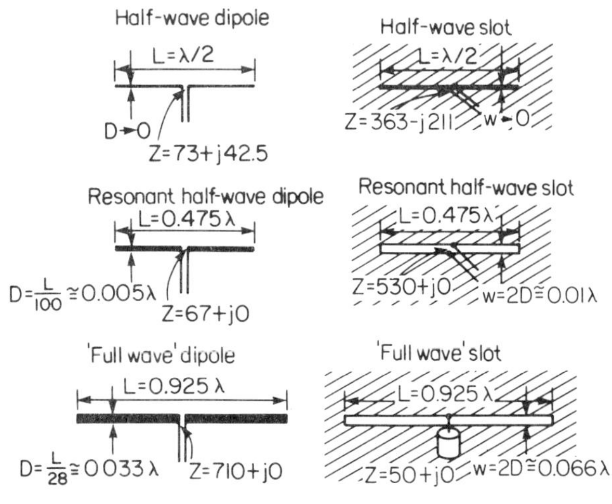

Figure 49.19 Rectangular slot and complementary wire dipole impedances

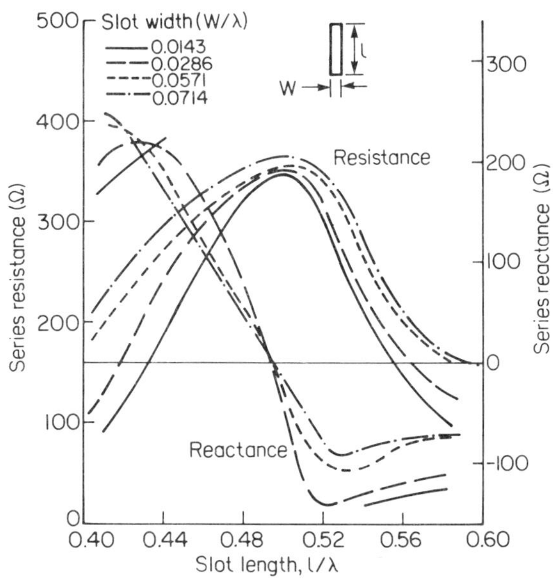

Figure 49.20 Measured impedances for a narrow rectangular slot

The radiation conductance of short rectangular slot of length l may be found, since it may be shown that for the complementary short dipole the radiation resistance is

$$R_{rad} = 20\pi^2(l/\lambda)^2 \tag{49.140}$$

and hence from equation (49.135)

$$G_{rad} = (l/\lambda)^2/180 \tag{49.141}$$

49.7.3 The annular slot

49.7.3.1 Radiation patterns

An annular slot in a ground plane may be excited as a radiator by appling a voltage radially across the slot. The radiation pattern is identical to that of the complementary wire loop, with **E** and **H** fields transposed. With the slot in the xy plane of a standard spherical coordinate system, the radiation pattern is symmetrical about the z axis, and is zero along that axis in the direction $\theta=0$.

The radiation pattern for a narrow circular slot of radius a is

$$E_\theta=j2\pi aVJ_1(ka\sin\theta)\,e^{-jkr}/r \qquad (49.142)$$

where V is the voltage applied across the slot, k is the free space wave number $2\pi/\lambda$, and J_1 is the first-order Bessel function of the first kind. The radiation pattern is of exactly the same form as that of the circular loop (section 49.5.5), for which patterns are plotted for various radii in *Figure 49.12(a)*. The field along the ground plane in the direction $\theta=\pi/2$, is a function of the electrical radius of the slot, and, for example, for $ka=3.83$ (the first zero of J_1) there is zero radiated field along the ground plane.

For small radii ($a<<\lambda$) equation may be approximated by

$$E_\theta\simeq j2\pi a^2\lambda^{-1}V\sin\theta\,e^{-jkr}/r \qquad (49.143)$$

which is the radiation pattern of a short electric dipole aligned along the z axis.

The annular slot radiation pattern is affected qualitatively by finite ground plane size, in a similar way to the rectangular slot. The pattern of a small annular slot on a finite ground plane is identical to that of a short stub monopole on the same ground plane (apart from the field transposition).

49.7.3.2 Impedance

The impedance of an annular slot may be computed in the same way as for the rectangular slot. In this case the complementary antenna is a thin wire loop of the same radius.

For a slot fed from a TEM-mode coaxial line with inner diameter and outer diameters equal to the inner and outer diameters, respectively, of the slot, the feed impedance normalized to the coaxial characteristic impedance has been computed.[17] This is plotted in *Figure 49.21* as normalised conductance and susceptance against slot radius, for a number of slot widths. For small radii the conductance is proportional to $(ka)^4$, and the susceptance directly proportional to ka. Although the slot susceptance falls for ka, greater than unity, it does not actually become zero for any radius, so that there is no true resonance (c.f. the rectangular slot case).

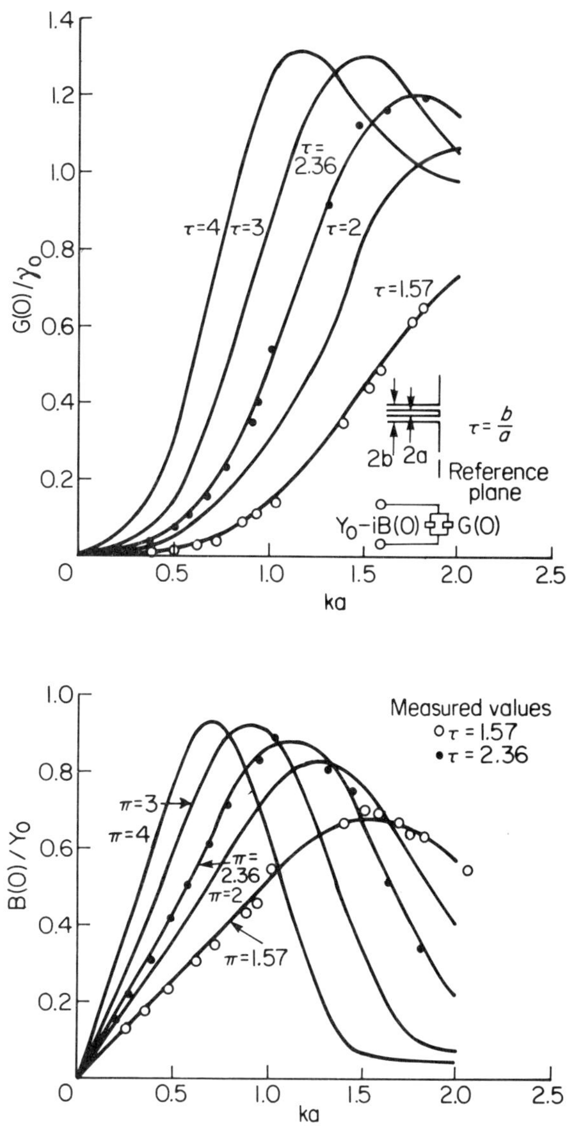

Figure 49.21 Normalised impedance of an annular slot

49.8 Electromagnetic horns

Electromagnetic horns find application as antennas in the frequency range covering VHF to millimetric frequencies, where a moderate gain in the approximate range 10–35 dBi is required over bandwidths of up to an octave (multi-octave bandwidths are also available using special techniques). They can be used as antennas in their own right, for example as earth-coverage satellite antennas, as feeds for reflector antennas, and as accurate and stable antenna gain standards.

Some of the more commonly used horn types are shown in *Figure 49.22.* The pyramidal and conical horns are used to produce beamwidths of the same order in the two principal planes, the sectoral horns produce fan beams which are much narrower in one plane than in the other, and the biconical horn produces a pattern which is omni-directional in the azimuth plane. In nearly all cases the horn is fed from rectangular waveguide (pyramidal and sectoral types) or from circular waveguide (conical and biconical types).

49.8.1 Radiation from open-ended waveguide

The radiation pattern of a horn may be computed from the tangential aperture fields (see section 49.2.3). For a horn with no major discontinuities in the flare it is a reasonable assumption that the fields at the aperture are of the same form as that of the exciting waveguide mode(s) at the throat, with a quadratic phase taper across the aperture due to the wavefront curvature introduced by the flare (see *Figure 49.23*). The edge-to-centre or peak value ψ_0 of this phase error is given quite accurately by the expression

$$\psi_0=2\pi\delta_0/\lambda\simeq\tfrac{1}{2}\pi w\alpha/\lambda \qquad (19.144)$$

where w is the aperture width in the plane being considered, α is the semi-flare angle in the same plane, and ψ_0 and α are in the same units (degrees or radians).

The effect of the flare-induced quadratic aperture phase error is slight for small to moderate phase errors, certainly over the top 15 dB of the main beam, and it is possible to correct out the phase error with a lens in the aperture. It is therefore useful to consider as a basis the radiation patterns of open-ended waveguides with zero aperture phase error.

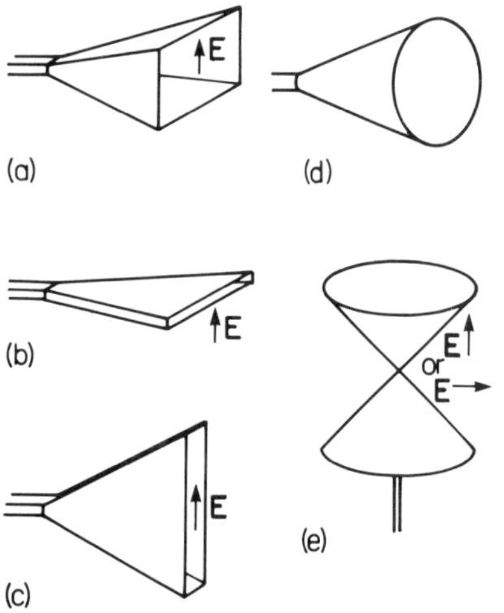

Figure 49.22 Some commonly used electromagnetic horn types. (a) Pyramidal; (b) sectoral *H* plane; (c) sectoral *E* plane; (d) conical; (e) biconical

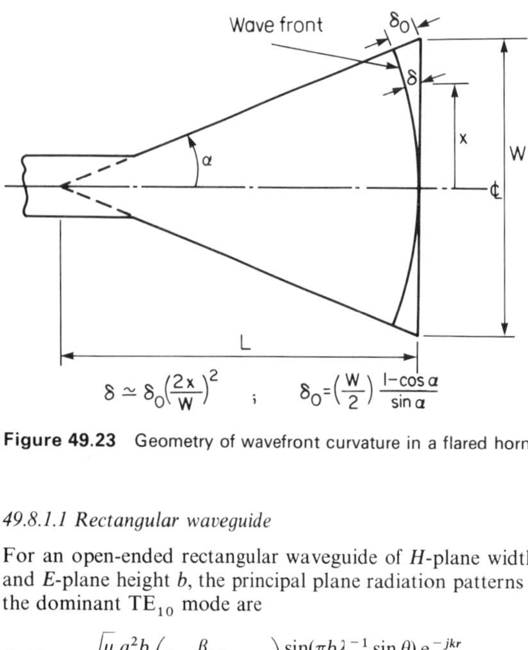

Figure 49.23 Geometry of wavefront curvature in a flared horn

49.8.1.1 Rectangular waveguide

For an open-ended rectangular waveguide of *H*-plane width a and *E*-plane height b, the principal plane radiation patterns for the dominant TE_{10} mode are

$$E_E(\theta)=2\sqrt{\frac{\mu}{\varepsilon}\frac{a^2b}{\pi\lambda^2}}\left(1+\frac{\beta_{10}}{k}\cos\theta\right)\frac{\sin(\pi b\lambda^{-1}\sin\theta)}{(\pi b\lambda^{-1}\sin\theta)}\frac{e^{-jkr}}{r} \tag{49.145}$$

$$E_H(\theta)=-\sqrt{\frac{\mu}{\varepsilon}\frac{\pi a^2b}{2\lambda^2}}\left(\frac{\beta_{10}}{k}+\cos\theta\right)\times\frac{\cos(\pi a\lambda^{-1}\sin\theta)}{(\pi a\lambda^{-1}\sin\theta)-(\frac{1}{2}\pi)^2}\frac{e^{-jkr}}{r} \tag{49.146}$$

where k is the free space wavenumber $2\pi/\lambda$, β_{10} is the guide wavenumber for the TE_{10} mode

$$\beta_{10}=k\left[1-\left(\frac{\lambda}{2a}\right)^2\right]^{1/2} \tag{49.147}$$

and r is the distance to the observation point.

The beamwidths between first nulls are given by

$$\theta_n^E=2\sin^{-1}\left(\frac{\lambda}{b}\right) \tag{49.148}$$

$$\theta_n^H=2\sin^{-1}\left(\frac{3\lambda}{2a}\right) \tag{49.149}$$

and for large apertures (narrow beams) the 3 dB beamwidths in degrees are

$$\theta_3^E\simeq 50.8\,\frac{\lambda}{b} \tag{49.150}$$

$$\theta_3^H\simeq 68.1\,\frac{\lambda}{a} \tag{49.151}$$

For a given aperture width the pattern is narrower in the *E* plane than in the *H* plane, because the aperture field amplitudes in those planes are uniform and cosine-tapered respectively. The sidelobes are higher in the *E* plane for the same reason.

Equations (49.145) and (49.146) are based on several assumptions. Most notable of these are that the radiation occurs from the fields across the aperture alone and there are no fields outside the aperture or currents flowing on the outside waveguide walls, and that the aperture is matched and there are no higher-order modes generated there. Even though these assumptions are not strictly valid, particularly for small aperture sizes, the results are usefully accurate in practice.

The gain of an open-ended rectangular waveguide or a pyramidal or sectoral horn with small aperture phase error is given by

$$G\simeq\frac{32}{\pi}\frac{ab}{\lambda^2} \tag{49.152}$$

48.8.1.2 Circular waveguide

For an open-ended circular waveguide of radius a, coaxial with the z axis of a standard spherical coordinate system, and supporting the dominant TE_{11} mode polarised in the y direction, the θ and φ components of the far-field pattern are given by

$$E_\theta(\theta,\varphi)=\frac{\omega\mu}{2}\left(1+\frac{\beta_{11}}{k}\right)J_1(\chi'_{11})\frac{J_1(u)}{\sin\theta}\sin\varphi\frac{e^{-jkr}}{r} \tag{49.153}$$

$$E_\phi(\theta,\varphi)=-\frac{ka\omega\mu}{2}\left(\frac{\beta_{11}}{k}+\cos\theta\right)J_1(\chi'_{11})\times\frac{J'_1(u)}{1-(u/\chi'_{11})^2}\cos\varphi\frac{e^{-jkr}}{r} \tag{49.154}$$

where all the variables are as defined for the rectangular waveguide, together with $u=ka\sin\theta$, χ'_{11} is the first root of $J'_1(x)$ $(=1.84118)$, and β_{11} is the guide wavenumber

$$\beta_{11}=\left[k^2-\left(\frac{\chi'_{11}}{a}\right)^2\right]^{1/2} \tag{49.155}$$

The E_θ pattern represents the *E*-plane pattern with $\varphi=\pi/2$, and the E_ϕ pattern represents the *H*-plane pattern with $\varphi=0$.

The *E*- and *H*-plane beamwidths between the first nulls are

$$\theta_n^E=2\sin^{-1}\left(\frac{3.832}{2\pi a}\right) \tag{49.156}$$

$$\theta_n^H=2\sin^{-1}\left(\frac{5.331}{2\pi a}\right) \tag{49.157}$$

and for large apertures (narrow beams) the 3 dB beamwidths in degrees are

$$\theta_3^E\simeq 29.4\lambda/a \tag{49.158}$$

$$\theta_3^H \simeq 37.2\lambda/a \quad (49.159)$$

As with the rectangular waveguide the pattern is narrower in the E plane than the H plane, and the side lobes are also higher in the E plane.

The gain of an open ended circular waveguide, or a coinical horn with small aperture phase error is given by

$$G \simeq 33.0(a/\lambda)^2 \quad (49.160)$$

49.8.2 Finite flare angles

The radiation patterns given for open-ended rectangular and circular waveguide will also be representative for pyramidal and conical flared horns, provided the flare is narrow enough to introduce little or no wavefront curvature at the aperture. As a raule of thumb, the edge-to-centre peak aperture phase error ψ_0 given by equation should be less than 20° for this to be satisfied. For quadratic phase errors greater than this, the main beam will broaden, the sidelobes will be increased, the nulls will fill in, and the gain will be reduced. Normalised E- and H-plane pyramidal horn patterns are plotted for phase errors in the range 0 to 200° in *Figure 49.24*. The pattern changes are similar for the conical horn.

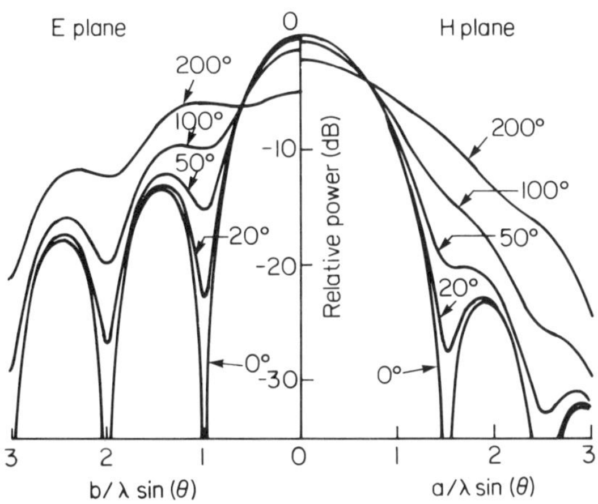

Figure 49.24 Principal plane patterns for a flared pyramidal or sectoral horn with varying aperture phase error

49.8.2.1 Optimum horns

If the axial length L of a flared horn is kept constant, and the flare angle increased, the gain will at first increase due to the increasing aperture size, and then beyond some point start to decrease again as the gain loss due to the increasing quadratic aperture phase error becomes dominant. A horn with the geometry giving the maximum gain for the given flare length is known as an optimum horn.

It is found that for a pyramidal horn the gain maximum occurs when the edge-to-centre path length difference δ_0 is approximately 0.25λ in the E plane and 0.4λ in the H plane (peak phase errors of 90° and 144° respectively). The gain of an optimum pyramidal horn is

$$G \simeq 6.43ab/\lambda^2 \quad (49.161)$$

where a and b are the H- and E-plane aperture dimensions.

For an optimum conical horn the edge-to-centre path length difference δ_0 is 0.375λ (peak phase error 135°), and its gain is

$$G \simeq 20.6(a/\lambda)^2 \quad (49.162)$$

where a is the aperture radius.

The flare semi-angle α is given in terms of the axial length L and edge-to-centre path difference δ_0 as

$$\alpha = \cos^{-1}[L/(L+\delta_0)] \quad (49.163)$$

This equation allows the design of an optimum horn in terms of its length L, using the values of δ_0 just given. Experimentally determined design data together with 3 dB beamwidths for a pyramidal horn are plotted in *Figure 49.25*.

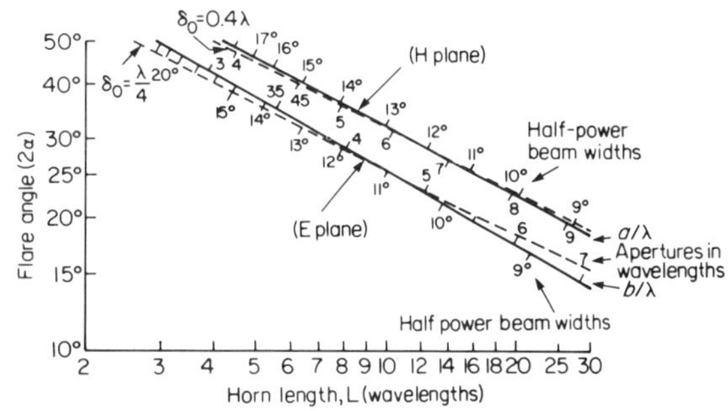

Figure 49.25 Optimum pyramidal horn design data

49.9 Reflector antennas

A reflector antenna comprises a feed and one or more reflecting surfaces, configured so that the rays from the feed are redidrected in space to achieve a particular radiation pattern. In the most commonly used type—the focusing pencil beam reflector, the feed rays are collimated, producing a narrow pencil beam with a gain considerably in excess of the gain of the feed itself. In other types the reflector profile is chosen to produce a broader secondary radiation pattern of a given shape. In most cases a single reflecting surface is used, however in the Cassegrain and Gregorian types two reflectors are used—one large main reflector, and one much smaller subreflector. Antennas with 'beam waveguide' feed systems use as many as six reflectors, although four of these are effectively used only to replace the main section of the waveguide run to the feed horn.

Despite the advances made in array technology, the reflector antenna remains the most attractive solution as a large aperture antenna in many cases. Its relative simplicity and reliability provide great advantages, and yet very high levels of performance are available, particularly with the benefits of modern design techniques.

49.9.1 Focusing pencil beam reflector systems

49.9.1.1 The parabolic reflector

The geometry of the parabolic reflector is shown in *Figure 49.26(a)*. The reflector profile is a parabola defined by

$$y^2 = 4fx \quad (49.164)$$

where f is the focal length, and collimates the diverging rays from a feed placed at the focus into a parallel beam. The reflector is completely defined by its diameter D and focal length, although the focal length is generally expressed indirectly as the ratio f/D, which will in most cases lie within the range 0.25–2.0. The semi-angle α subtended by the reflector at the focus is given in terms of the f/D ratio as

$$\alpha = 2\tan^{-1}[(4f/D)^{-1}] \quad (49.165)$$

which is plotted in *Figure 49.27*. The geometrical relationships for a ray emerging from the focus at an angle ε to the axis are

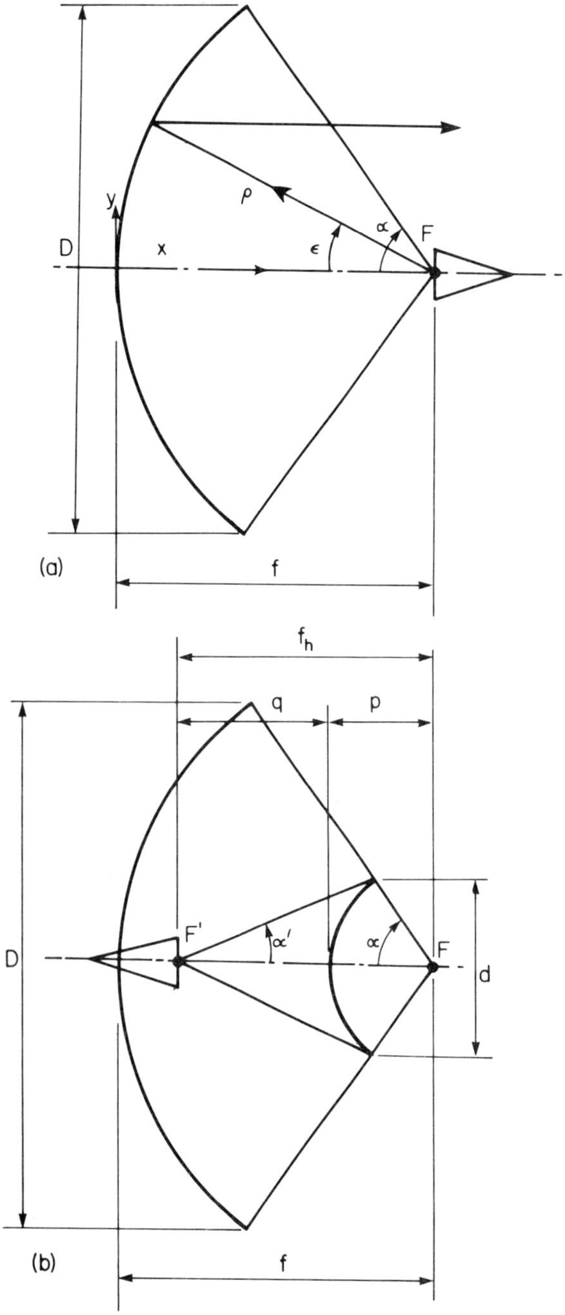

Figure 49.26 (a) Parabolic reflector geometry; (b) Cassegrain (dual reflector) geometry

$$\rho = f\sec^2(\tfrac{1}{2}\varepsilon) \tag{49.166}$$

$$y = 2f\tan(\tfrac{1}{2}\varepsilon) \tag{49.167}$$

The most common parabolic reflector (the paraboloid) is a surface of revolution formed by rotating the parabola about the axis, and has a point focus. The two dimensional equivalent with a line focus is parabolic in one plane but unshaped in the other plane, and is used with a line source feed to illuminate the reflector with a cylindrical wave.

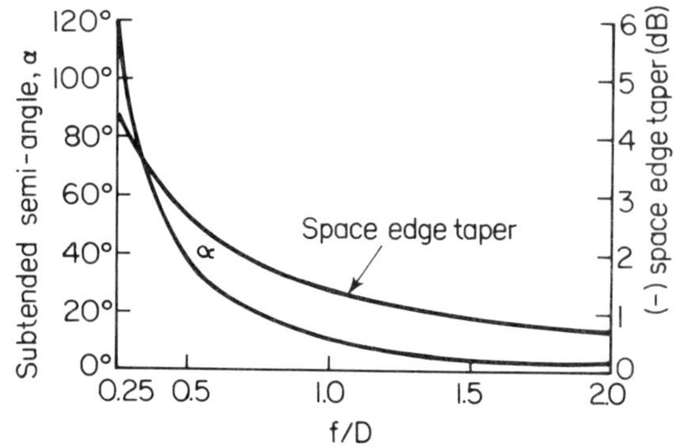

Figure 49.27 Subtended semi-angle and space edge taper for a parabolic reflector

The secondary radiation pattern of a reflector antenna may be computed in two principal ways. The more rigorous of these invokes the physical optics principle described briefly in section 49.2.3 to compute the electric currents **J** induced in the reflector surface by the incident magnetic field **H**

$$\mathbf{J} = 2\hat{\mathbf{n}} \times \mathbf{H} \tag{49.168}$$

($\hat{\mathbf{n}}$ = unit normal vector), and then to compute the fields radiated into space in the form of an integral over the reflector surface. The second method first computes the fields projected onto the aperture plane of the antenna due to illumination by the feed, and then uses the Kirchoff principle (equations (49.37) and (49.38)) to compute the secondary radiation pattern as the Fourier transform of the aperture fields.

For a feed radiating a perfect spherical wave originating at the focus, the fields in the aperture plane of a paraboloid will be uniform in phase. The amplitude distribution of the aperture plane fields will be due both to the amplitude taper of the feed radiation pattern, and an additional 'space taper' introduced by the paraboloid geometry, due to the fact that the diverging rays from the feed must travel further before being collimated, near the aperture edge than near the centre. The space taper (in dB) is given in terms of the ray angle ε as

$$e_{\text{space}} = 20\lg[\cos^2(\tfrac{1}{2}\varepsilon)] \tag{49.169}$$

The space taper at the reflector edge ($\varepsilon = \alpha$) is plotted against f/D in *Figure 49.27*.

The beamwidth of the feed radiation pattern will affect the edge taper of the aperture fields, and hence both the secondary pattern shape and boresight gain. The boresight directivity for a circular aperture reflector of diameter D is given by

$$G = (\pi D/\lambda)^2 \eta_a \eta_s \tag{49.170}$$

where η_a and η_s are the aperture and spillover efficiencies respectively

$$\eta_a = \frac{4}{\pi D^2}\left(\left|\int_0^{2\pi}\int_0^{D/2} f(y,\varphi)y\,\mathrm{d}y\,\mathrm{d}\varphi\right|^2\right) \times \left(\int_0^{2\pi}\int_0^{D/2} |f(y,\varphi)|^2 y\,\mathrm{d}y\,\mathrm{d}\varphi\right)^{-1} \tag{49.17.1}$$

$$\underset{\text{azimuthal symmetry}}{=} \frac{8}{D^2}\left(\left|\int_0^{D/2} f(y)y\,\mathrm{d}y\right|^2\right)\left(\int_0^{D/2} |f(y)|^2 y\,\mathrm{d}y\right)^{-1} \tag{49.172}$$

$$\eta_s = \left(\int_0^{2\pi}\int_0^{\alpha} |g(\theta,\varphi)|^2 \sin\theta\,\mathrm{d}\theta\,\mathrm{d}\varphi\right) \times \left(\int_0^{2\pi}\int_0^{\pi} |g(\theta,\varphi)|^2 \sin\theta\,\mathrm{d}\theta\,\mathrm{d}\varphi\right)^{-1} \tag{49.173}$$

$$\underset{\substack{\text{azimuthal}\\ \text{symmetry}}}{=} \left(\int_0^{\alpha} |g(\theta)|^2 \sin\theta \, d\theta \right) \times \left(\int_0^{\pi} |g(\theta)|^2 \sin\theta \, d\theta \right)^{-1} \quad (49.174)$$

where $f(y, \varphi)$ and $g(\theta, \varphi)$ are the fields in the reflector aperture, and feed pattern fields respectively, with other variables as defined earlier.

For a given reflector f/D the aperture efficiency will increase with increasing feed pattern beamwidth, whereas the spillover efficiency will decrease with increasing feed pattern beamwdith. The antenna gain will therefore maximise for a particular value of feed pattern beamwidth, and this will usually occur when the net edge taper is between -8 dB and -12 dB. For most feed patterns the main beam will be approximately Gaussian down to the -12 dB points, so that if a feed pattern edge taper of $-E$ dB is required, the 3 dB beamwidth is approximately given by

$$\theta_3 \simeq 2\alpha(3/E)^{1/2} \quad (49.175)$$

49.9.1.2 Cassegrain (dual reflector) geometry

There are two dual reflector geometries using a parabolic main reflector—the Cassegrain and Gregorian geometries, of which the former is the more widely used. The subreflector of a Cassegrain antenna system (shown in *Figure 49.26(b)*) is hyperbolic in profile and has the effect of creating a virtual focus at F', between the two reflectors.

The subreflector is completely defined by its diameter d, focal length $f_{h'}$ and magnification factor M. The relationships between the various geometrical parameters for the Cassegrain system are

$$M = q/p \quad (49.176)$$

$$e = (M+1)/(M-1) \qquad (e = \text{eccentricity}) \quad (49.177)$$

$$f_h = p + q \quad (49.178)$$

$$\alpha = 2\tan^{-1}[(4f/D)^{-1}] \quad (49.179)$$

$$\alpha' = 2\tan^{-1}[(4Mf/D)^{-1}] \quad (49.180)$$

If the main reflector diameter and focal length, together with the subreflector focal length and magnification factor are already defined, and if, as is usually the case, the subreflector edge is required to subtend the same angle at the prime focus F as the main reflector edge, the subreflector diameter d is then no longer independent

$$d = 2f_h/(\cot\alpha + \cot\alpha')^{-1} \quad (49.181)$$

It may be shown that for minimum aperture blockage, when the blockage shadows of the feed horn and subreflector are equal, the subreflector diameter is given approximately by

$$d \simeq \sqrt{2\lambda f} \quad (49.182)$$

49.9.2 Reflector shaping

The basic 'optical' Cassegrain geometry using hyperbolic and parabolic reflector profiles, has two disadvantages when applied to microwave antennas. Firstly the illumination of the main reflector will be determined by the shape of the feed pattern (generally amplitude tapered), so that it is not possible to fully maximise the antenna efficiency, except by using special feed types. Secondly internal diffraction effects due to the finite electrical size of the reflectors will come into play, so that the simple geometric or ray tracing analysis will break down, generally reducing the gain below that expected.

To overcome the first of these disadvantages it is possible to synthesise 'shaped' reflector profiles, slightly perturbed from the basic parabolic/hyperbolic pair, which according to geometric optics will produce a uniform aperture field distribution and hence increase the antenna gain.[18,19] This synthesis is generally known as the Williams method. A rigorous analysis of the aperture fields for a geometrically shaped Cassegrain antenna of finite size will show that the field distribution achieved is not exactly uniform, and exhibits amplitude and phase ripple. More sophisticated profile synthesis techniques are available which will further improve the aperture field uniformity and increase the gaink by taking account of the internal diffraction effects.[20] Such Cassegrain designs are sometimes termed 'diffraction optimised'.

A modification of both the geometrically shaped (Williams) and diffraction optimised designs is available, in which only the subreflector is shaped and the main reflector is a best-fit paraboloid. Except for large main reflector diameters with geometrically shaped subreflectors, the modified design will provide almost as much gain as a fully shaped system.

To show how the efficiencies of the various shaped designs compare, these are plotted against main reflector diameter in *Figure 49.28*.

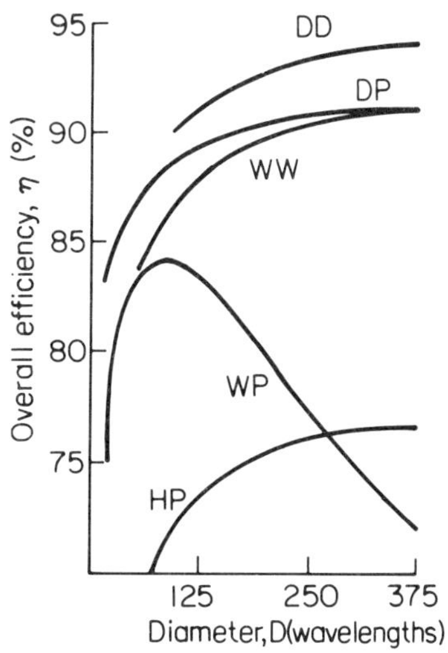

Figure 49.28 Efficiencies for dual reflector antennas with different profiles. HP = hyperboloid/paraboloid; WP = Williams subreflector/paraboloid; WW = Williams subreflector/Williams main reflector; DP = Diffraction optimised subreflector/paraboloid; DD = Diffraction optimised subreflector/diffraction optimised main reflector

49.9.3 Aperture blockage

The aperture of a rotationally symmetric reflector antenna will be blocked by the feed for a front-fed (single reflector) design, and by the subreflector and feed for a Cassegrain (dual reflector) design. The effect will be to create a central shadow in the aperture which will reduce the antenna gain and increase the sidelobe levels.

If the aperture field distribution is assumed to be circularly symmetric, and parabolic tapered in the radial direction with an edge taper (voltage) e, then it may be shown that the absolute secondary pattern relative to an isotropic source is

$$e(\theta) = 2\pi \frac{D}{\lambda} \sqrt{\frac{3}{1+e+e^2}} \left\{ 2(1-e)\frac{J_2(u)}{u^2} + e\frac{J_1(u)}{u} - \left(\frac{d}{D}\right)^2 \left[2(1-f)\frac{J_2(v)}{v^2} + f\frac{J_1(v)}{v} \right] \right\} \quad (49.183)$$

where d and D are the blockage shadow and aperture diameters respectively, and

$$u=\pi D\lambda^{-1}\sin\theta \tag{49.184}$$

$$v=(d/D)u \tag{49.185}$$

$$f=1-(1-e)(d/D)^2 \tag{49.186}$$

For a uniform distribution ($e=1$) equation (49.183) reduces to

$$e(\theta)=(2\pi D/\lambda)\left[\frac{J_1(u)}{u}-\left(\frac{d}{D}\right)^2\frac{J_1(v)}{v}\right] \tag{49.187}$$

The boresight gain loss due to the aperture blockage is given by

$$\frac{G}{G_{\text{unblocked}}}=\left[1-\left(\frac{d}{D}\right)^2\frac{(1+f)}{(1+e)}\right] \tag{49.188}$$

which for a uniform distribution becomes

$$\frac{G}{G_{\text{unblocked}}}=\left[1-\left(\frac{d}{D}\right)^2\right] \tag{49.189}$$

The blockage gain loss is plotted against normalised blockage diameter and edge taper in *Figure 49.29(a)*.

For small blockage diameters the main effect of the aperture blockage on the radiation pattern will be to increase the levels of the odd sidelobes (1st, 3rd, ...) and to decrease the levels of the first few even sidelobes (2nd, 4th, ...). The first sidelobe level for a parabolic on a pedestal distribution is plotted against normalised blockage diameter and edge taper in *Figure 49.29(b)*. For $d/D>0.3$ it is apparent that the edge taper has little influence on the sidelobe level.

49.9.4 Reflector profile errors

If the reflector surfaces of a focusing pencil beam antenna deviate from the required profiles, phase errors will be introduced into the aperture field distribution. The most significant effects of these will be a reduction of the boresight gain and an increase in the sidelobe levels.

For a random distribution of profile errors across the reflector surface, the error distribution may be described by two parameters[21]

ε (the normalised r.m.s. profile error)
c (the error distribution correlation interval)

ε is simply half the r.m.s. path length error, and is related to the r.m.s. profile error measured normal to the surface of, or parallel to the axis of a parabolic reflector (Δ_n and Δ_z respectively) by

$$\frac{\varepsilon}{\Delta_n}=4\frac{f}{D}\left\{\ln\left[1+\left(\frac{D}{4f}\right)^2\right]\right\}^{1/2}\underset{f/D\to\infty}{=1} \tag{49.190}$$

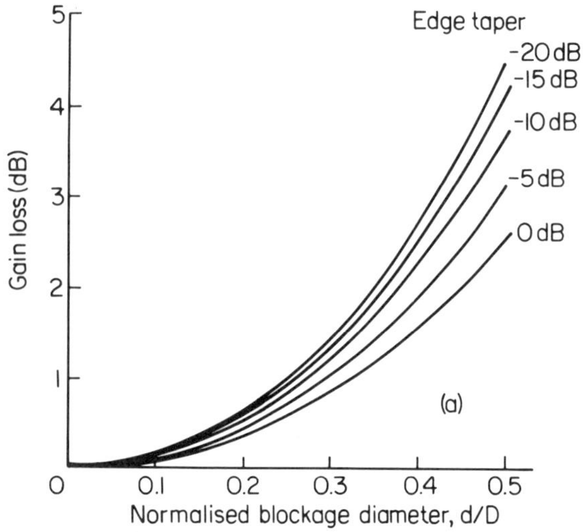

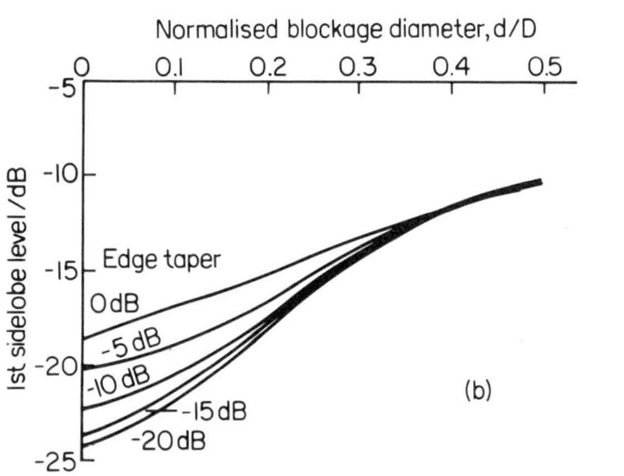

Figure 49.29 Effects of circular central aperture blockage. (a) Gain loss; (b) first side lobe level

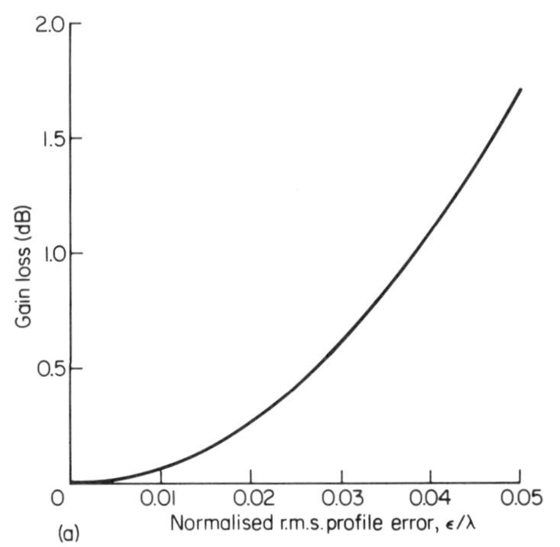

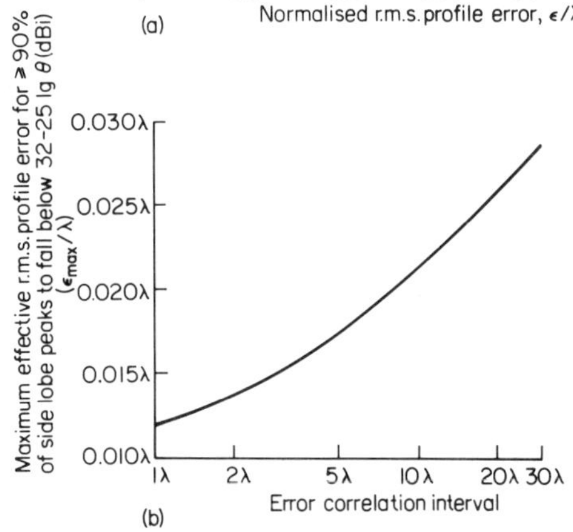

Figure 49.30 Effects of reflector profile errors. (a) Gain loss; (b) profile accuracy to meet wide-angle side lobe specifications

$$\frac{\varepsilon}{\Delta_z}=\left[1+\left(\frac{4f}{D}\right)^{-2}\right]^{-1/2} \underset{f/D\to\infty}{=} 1 \qquad (49.191)$$

$$\frac{\varepsilon}{\Delta_n}=4\frac{f}{D}\left\{\ln\left[1+\left(\frac{D}{4f}\right)^{2}\right]\right\}^{1/2} \underset{f/D\to\infty}{=} 1 \qquad (49.190)$$

where f and D are the reflector focal length and diameter. The correlation interval c is the radius over which the errors remain correlated.

For small errors and small correlation intervals compared to the wavelength and antenna diameter respectively, the boresight gain loss (in dB) is given approximately by

$$\Delta G \simeq 686(\varepsilon/\lambda)^2 \qquad (49.192)$$

This is plotted in *Figure 49.30(a)*.

The expected mean side lobe power pattern due to the profiles errors is given by

$$G(\theta)=(2\pi c/\lambda)^2\,e^{-\overline{\delta^2}}\sum_{n=1}^{\infty}\frac{(\overline{\delta^2})^n}{n\cdot n!}\,e^{-(\pi c\sin\theta/\lambda)^2/n} \qquad (49.193)$$

where

$$\overline{\delta^2}=(4\pi\varepsilon/\lambda)^2$$

As an example of the profile error tolerances necessary to meet realistic side lobe specification, *Figure 49.30(b)* plots the maximum permissible r.m.s. error against correlation interval, to meet a widely used satellite earth station antenna side lobe specification.

References

1 SCHELKUNOFF, S. A., *Electromagnetic Waves*, Van Nostrand (1943)
2 SCHELKUNOFF, S. A. and FRIIS, H. T., *Antennas: Theory and Practice*, Wiley, New York (1952)
3 LUDWIG, A. C., 'The Definition of Cross Polarization', *I.E.E.E. Trans.*, **AP-21**, 116–119 (1973)
4 RAMSEY, J. F., 'Fourier Transforms in Aerial Theory', *Marconi Rev.*, **83–89** (1946–48)
5 DOLPH, C. L., 'A Current Distribution for Broadside Arrays which Optimizes the Relationship between Beam Wdith and Sidelobe Level', *Proc. I.R.E.*, **34**, 335 (1946)
6 TAYLOR, T. T., 'Design of Line-Source Antennas for Narrow Beamwidth and Low Sidelobes', *I.R.E. Trans.*, **AP-3**, 16–28 (1955)
7 SPELLMIRE, R. J., *Tables of Taylor Aperture Distributions*, TM581, Hughes Aircraft Co., Culver City, California (1958)
8 KNUDSEN, H. L., 'Radiation from Ring Quasi-Arrays', *I.R.E. Trans.*, **AP-4**, 452–472 (1956)
9 LINDENBLAD, N. E., 'Antennas and Transmission Lines at the Empire State Television Station, Part 2', *Communications*, **21**, 10–14 and 24–26 (1941)
10 BROWN, G. H. and WOODWARD, O. M., 'Circularly-Polarized Omnidirectional Antenna', *RCA Rev.*, **8**, 259–269 (1947)
11 BRUCE, E., BECK, A. C. and LOWRY, L. R., 'Horizontal Rhombic Antennas', *Proc. I.R.E.*, **23**, 24–46 (1935)
12 FOSTER, D., 'Radiation from Rhombic Antennas', *Proc. I.R.E.*, **25**, 1327–1353 (1937)
13 HARPER, A. E., *Rhombic Antenna Design*, Van Nostrand, New York (1941)
14 Radio Research Laboratory Staff, *Very High Frequency Techniques*, McGraw-Hill, New York (1947)
15 BOOKER, H. G., 'Slot Aerials and Their Relation to Complementary Wire Aerials', *J. I.E.E.*, **93**, Part III A, No. 4 (1946)
16 HALLEN, E., 'Theoretical Investigations into the Transmitting and Receiving Qualities of Antennae', *Nova Acta Regide Soc. Sci. Upsaliensis*, **Ser IV**, **11**, No. 4, 1–44 (1938)
17 LEVINE, H. and PAPAS, C. H., 'Theory of the Circular Diffraction Antenna', *J. App. Phys.*, **22**, No. 1, 29–43 (1951)
18 GALINDO, V., 'Design of Dual-Reflector Antennas with Arbitrary Phase and Amplitude Distributions', *I.E.E.E. Trans.*, **AP-12**, 403–408 (1964)
19 WILLIAMS, W. F., 'High Efficiency Antenna Reflector', *Microwave J.*, **8**, 79–82 (July 1965)
20 WOOD, P. J., 'Reflector Profiles for the Pencil-Beam Cassegrain Antenna', *Marconi Rev.*, **XXXV**, No. 185, 121–138 (1972)
21 RUZE, J., 'Antenna Tolerance Theory–A Review', *Proc. I.E.E.E.*, **54**, 633–640 (1966)

50 Noise Management in Electronic Hardware

Joseph M Camarata
Manager, Electronic Design,
Electronics Division,
Xerox Corporation

Contents

50.1 Introduction

50.1.1 Noise susceptibility

Susceptibility of the electronic subsystem to noise is a common concern in the design of electronic hardware. In this chapter some of the important considerations and techniques relative to the control of noise susceptibility are examined.

The rapidly advancing state-of-art in microelectronics has created many new opportunities for the use of electronics. The proliferation of electronics into new product areas has allowed dramatic product improvements; but it has also brought problems. Often these applications contain relatively noisy environments for the electronics. Consequently, one important area of concern in any product which utilises electronics is the management of electrical noise. The term 'management' is used because noise generation is a practical consequence from the operation of electronic/electrical subsystems. Of course, techniques are used to limit the amount of noise generated consistent with other product objectives. But, for the noise which inevitably remains, proper design (management) is necessary to ensure an acceptable level of performance.

Noise is loosely defined to be any form of electromagnetic energy (radiated or conducted) except for intentional signals along intended signal paths. Noise susceptibility refers to the sensitivity or response of electronic circuits and subsystems to noise. The susceptibility threshold for a particular electronic subsystem under specified operating conditions is normally defined as the point at which functional problems occur.

This chapter focuses on the problem of noise control within and between electronic subsystems and subsystem interactions with the remainder of the machine electrical environment. However, most of the information provided is equally applicable to noise whose origin is external to the machine. Although this information is relevent to any type of equipment, it is oriented towards electronic systems in commercial equipment using digital techniques.

50.1.2 Noise management philosophy

Noise management is an essential part of the design process and must be considered early when design concepts are being formulated. When establishing a noise management strategy, the designer recognises that there are three elements to the problem: noise source, coupling mechanism and receiving circuit/subsystem.

The strategy should address each of these elements to determine the most cost effective way to prevent harmful interactions. The task for the designer is to recognise and understand potential problems, identify design alternatives and to select the most cost effective combination of techniques. Most of the techniques utilised will be nothing more than good design practice which have little or no cost impact to the product. Occasionally, difficult problems will require special hardware or software solutions which increase product cost. In this chapter, we will only examine the problem relative to hardware design. The total noise management strategy for digital equipment, however, should also include software design. In particular, software filtering and timers, data refresh and recovery techniques, error detection and correction, etc, should be included as appropriate. The number of special hardware solutions required and their cost can be minimised by properly addressing noise management at the beginning and throughout the product design cycle. Solutions that must be patched into an existing design will often carry cost/schedule penalties that could have been avoided.

50.2 Background information

In this section, a brief summary of useful formulas and concepts is presented. For the reader wishing to explore the subject in more depth, there are many good reference books on this material, some of which are listed in the references.

50.2.1 Waveform analysis

Normally, designers closely monitor and control the time domain characteristics of signal waveforms. However, most circuit and system modeling and analysis relative to the control of interference is done in the frequency domain (i.e. field generation, coupling phenomena, and receptor sensitivity). Understanding the frequency content of potentially disturbing signals greatly aids in the task of interference control.

50.2.1.1 Fourier series

Trigonometric form Any normally encountered periodic waveform of period T may be represented as the sum of a series of sinusoidal components. The components will consist of sinusoids at the fundamental and multiples of the fundamental frequency (harmonics). This is called a Fourier series and is given by:

$$f(t)=\tfrac{1}{2}A_0+\sum_{n=1}^{\infty}(A_n\cos\omega_n t+B_n\sin\omega_n t) \qquad (50.1)$$

Where the coefficients A_n and B_n are given by:

$$A_n=2/T\int_{-T/2}^{T/2}f(t)\cos\omega_n t\,dt \qquad (50.2)$$

$$B_n=2/T\int_{-T/2}^{T/2}f(t)\sin\omega_n t\,dt \qquad (50.3)$$

In equations (50.1–50.3), $\omega_n=n\omega_1$ where $\omega_1=2\pi/T$ and the limits of integration may be any values that are one period apart. Substituting $n=0$ into equation (50.2) yields A_0 where $A_0/2$ is the average value of the waveform.

The frequency spectrum of a periodic signal consists of discrete line spectra occurring at fundamental and harmonic frequencies. The sine and cosine terms in equation (50.1) can be combined into the alternate form of equation (50.4)

$$f(t)=\frac{C_0}{2}+\sum_{n=1}^{\infty}C_n\cos(\omega_n t+\varphi_n) \qquad (50.4)$$

where

$$C_0=A_0 \qquad (50.5)$$

$$C_n=\sqrt{A_n^2+B_n^2} \qquad (50.6)$$

$$\varphi_n=\tan^{-1}(-B_n/A_n) \qquad (50.7)$$

C_n is the amplitude of the frequency line spectra and φ_n the phase.

Exponential form Rather than compute A_n and B_n, it will often be more convenient to use the exponential form of the Fourier series. This form is given by equations (50.8 and 50.9)

$$f(t)=\sum_{n=-\infty}^{\infty}D_n^{jn\omega t} \qquad (50.8)$$

$$D_n=1/T\int_{-T/2}^{T/2}f(t)\,e^{-jn\omega t}\,dt \qquad (50.9)$$

The appearance of negative frequencies in the summation interval of equation (50.8) is a mathematical requirement associated with the complex notation. It should be noted that in general, D_n is a complex quantity. Also, D_n and D_{-n} are always complex conjugates of one another and $D_n = C_n/2$. The following general symmetry relationships are useful;

(a) if a time function possesses half wave symmetry, i.e.: $f(t) = f(t + T/2)$, $D_n = 0$ for even n.
(b) all D_n are real for even functions.
(c) all D_n are imaginary for odd functions.

(section 50.2.1.2)

To demonstrate the usefulness of the exponential form of Fourier series, it will be applied to the general rectangular pulse train of *Figure 50.1*.

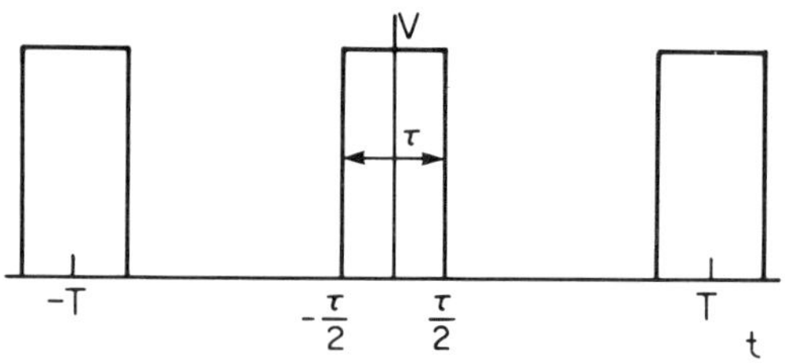

Figure 50.1 Rectangular pulse waveform

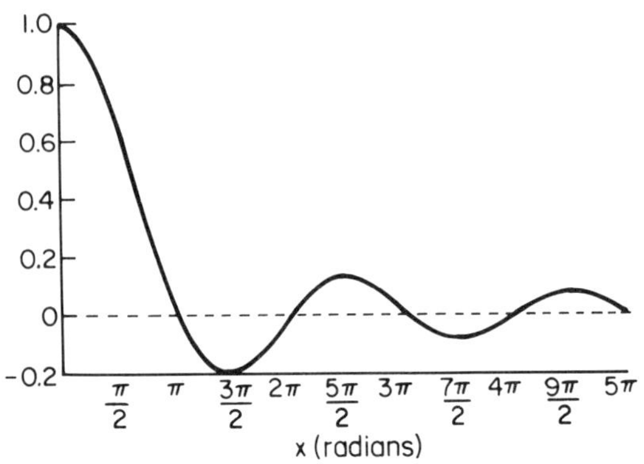

Figure 50.2 sin x/x curve. From *Electronic Designers Handbook* by R W Landee *et al.* Copyright 1957, McGraw-Hill Book Company. Used with the permission of McGraw-Hill Book Company

From equation (50.9)

$$D_n = 1/T\int_{-T/2}^{T/2} f(t)e^{-jn\omega_1 t}\,dt = \frac{V\tau}{T}\left(\frac{\sin\frac{1}{2}n\omega_1\tau}{\frac{1}{3}n\omega_1\tau}\right)$$

$$= \frac{V\tau}{T}\left(\frac{\sin n\pi\tau/T}{n\pi\tau/T}\right) = \frac{V\tau}{T}\left(\frac{\sin n\pi f_1\tau}{n\pi f_1\tau}\right) \qquad (50.10)$$

The envelope of the amplitude spectra is of the form sin x/x which frequently occurs in signal analysis. The following characteristics of sin x/x can be immediately deduced;

(a) for x very small ($x\to 0$), sin $x \to x$ and sin $x/x \to 1$.
(b) sin $x/x = 0$ for $x = n\pi$ where $n = 1, 2, 3 \ldots$. For other values of x, sin x/x oscillates between positive and negative values.
(c) For x very large ($x\to\infty$), sin $x/x \to 0$. The curve of sin x/x is plotted in *Figure 50.2*.

The frequency spectrum corresponding to the pulse train of *Figure 50.1* is plotted in *Figure 50.3* for the case of $T/\tau = 4$.

The spectrum consists of discrete line spectra at the fundamental frequency and harmonics. The amplitude of the line spectra is $2|D_n|$. D_n is given by equation (50.10), and is plotted in *Figure 50.3* as a function of frequency (nf_0). The spectrum envelope is zero at $nf_0 = 1/\tau$, $2/\tau$, e/τ, etc. The number of line spectra between zero crossings of the envelope is equal to $(T/\tau - 1)$.

Equations for the Fourier coefficients for some common waveforms are provided in *Figure 50.4*.

Spectrum envelope approximation Often, one is not interested in obtaining the exact value of the frequency spectrum components but only needs a straight line approximation to the envelope which passes through the maximum values of line spectrum components. Such straight line approximations are shown in *Figure 50.5* on log–log scale for rectangular and trapazoidal pulses. From this graph it can be seen that the low frequency spectrum for these equal area pulses is the same. The first break frequency occurs at $f = 1/\pi\tau$ and the maximum amplitude spectra is falling off at -20 dB/decade. At $f = 1/\pi t_r$ for the trapezoidal pulse the envelope takes another break and it falls off at -40 dB/decade.

50.2.1.2 Even/odd functions

A given function may be even, odd or the combination of even and odd functions. An even function is defined by

$$f(t) = f(-t) \qquad (50.11)$$

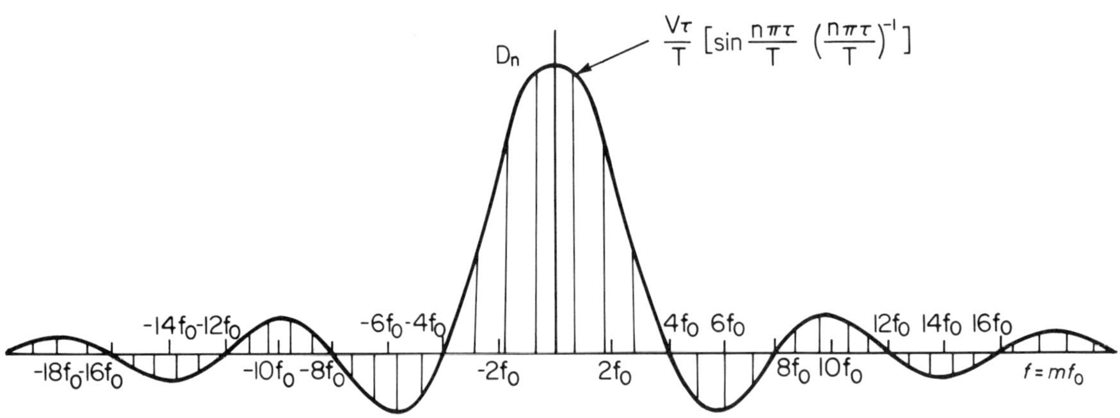

Figure 50.3 Spectral lines for rectangular pulse train with $T/\tau = 4$. Reproduced from reference 4 with permission

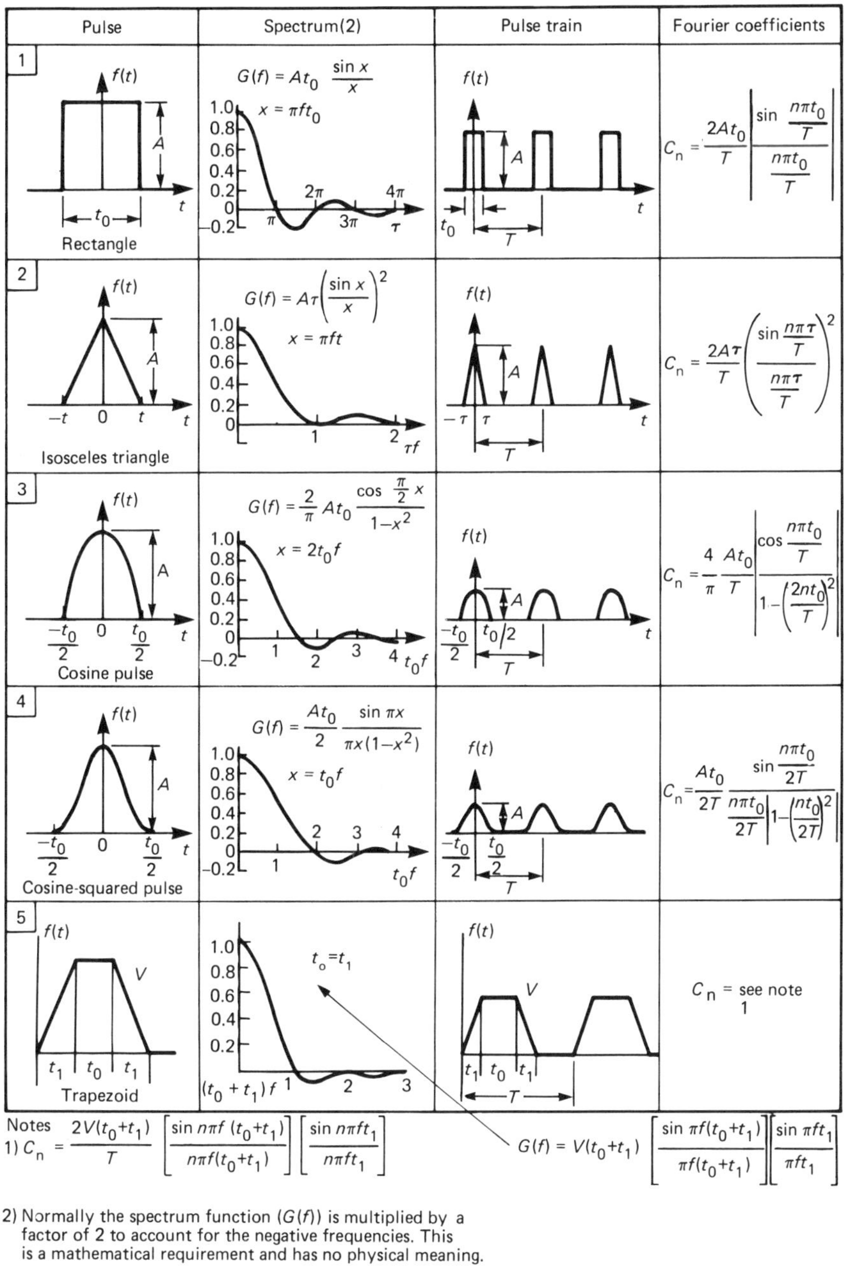

Figure 50.4 Spectrum and Fourier coefficients of some common waveforms. Reproduced from reference 3 with permission

The definition of an odd function is given by

$$f(t) = -f(-t) \tag{50.12}$$

Often, a function can be made either even or odd, depending on the choice of the axis. Note that even/odd functions should not be confused with even or odd harmonics which are discussed in the next section. The even or odd harmonic content cannot be affected by the axis choice.

50.2.1.3 Even/odd harmonics

A waveform with a period T will contain only odd harmonics if

$$f(t) = -f(t + T/2) \tag{50.13}$$

Similarly, it will only contain even harmonics if

$$f(t) = f(t + T/2) \tag{50.14}$$

The above conditions are sufficient but not necessary.

50.2.1.4 Fourier integral

As the period of the pulse train increases without limit, the spacing between line spectra ($2\pi/T$) decreases to zero. The discrete line spectrum becomes a continuous frequency spectrum. In equation (50.9), the ratio D_n/ω is defined as g_n and as $T \to \infty$ this becomes a continuous function $g(\omega)$ given by equation (50.15).

$$g(\omega) = 1/2\pi \int_{-\infty}^{\infty} f(t)\,e^{-j\omega t}\,dt \tag{50.15}$$

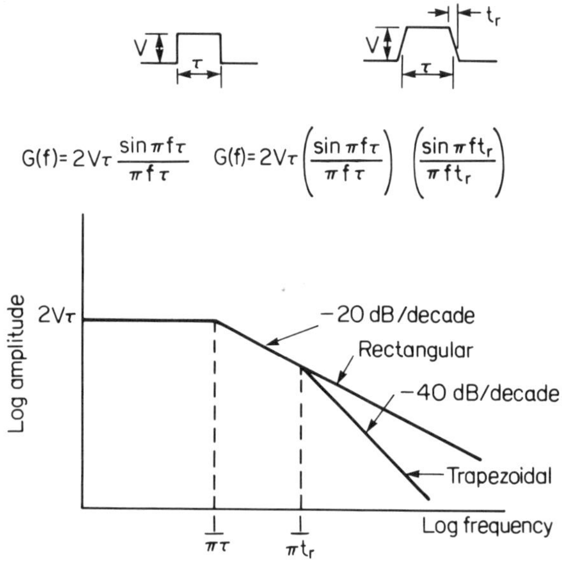

Figure 50.5 Asymptote approximations to spectrum envelope for rectangular and trapezoidal pulses

Also, the summation indicated in equation (50.8) becomes an integral in the limit and is given by equation (50.16)

$$f(t)=\int_{-\infty}^{\infty} g(\omega)\, e^{j\omega t}\, d\omega \tag{50.16}$$

These two equations define the Fourier integral and allow the frequency spectrum for a single pulse to be obtained analogous to the way that the Fourier series did for a periodic waveform.

Applying equation (50.15) to a single rectangular pulse (*Figure 50.6*) yields

$$g(\omega)=V\tau\left(\frac{\sin \pi f\tau}{\pi f\tau}\right) \tag{50.17}$$

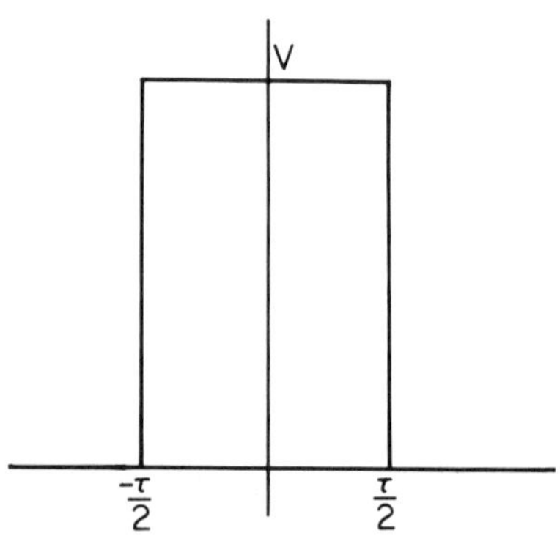

Figure 50.6 Rectangular pulse

Note that the frequency spectrum envelope shape depends only on the pulse's time domain waveform. It is independent of whether the pulse is periodic or non-repetitive. The periodic pulse train has a discrete line spectrum while the non-recurring pulse has a continuous spectrum. It should also be noted that the units for $g(\omega)$, (equation 50.17), are volts/frequency or volt-seconds while for a periodic waveform, (equation 50.10) they are volts. Spectrums for some common pulse waveshapes are provided in *Figure 50.4*.

50.2.2 Field theory

Most electronic hardware contain many elements which are capable of antenna-like behaviour, either as a transmitter or receiver (e.g. cables, printed wiring board interconnect, electronic components, ground or power distribution interconnect, etc). Under certain conditions, these elements may emit or receive sufficient energy to cause operational problems. This energy transfer takes place via electric, magnetic, or electromagnetic fields which couple the source and receiver circuits. The fields are generated as a result of time varying voltages and currents that exist within or external to the electronic system. An understanding of the nature of these fields can be obtained from an examination of the field equations corresponding to an elementary electric dipole (current filament) and an elementary magnetic dipole (current loop). The following equations utilise the spherical coordinate system and assume that: (1) the size of the differential element is small compared to wavelength and the distance to the observation point, (2) current is uniformly distributed over the element length, and (3) the diameter of the elemental conductor is small compared to its length.

50.2.2.1 Elementary electric dipole source

Assume an elemental current filament as shown in *Figure 50.7*. Using spherical coordinates, the field will have Er, E_θ and H_ϕ

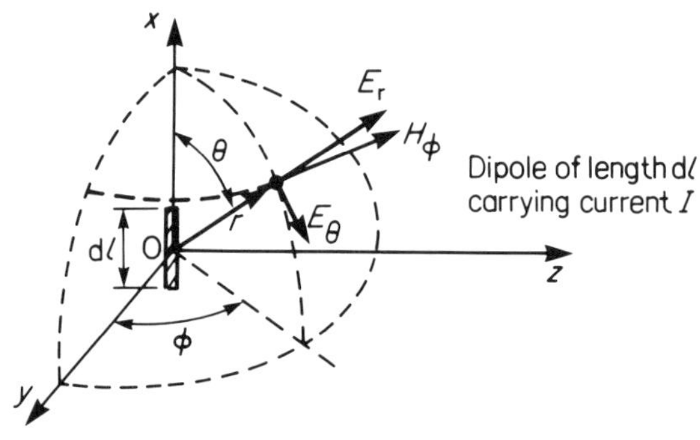

Figure 50.7 Field from an electric dipole. Reproduced from reference 4 with permission

vectors. The magnitudes of these vectors are given by equations (50.18, 50.19, 50.20)

$$E_r=\frac{I\,dl\,\beta^3 e^{-j\beta r}}{\omega\quad 2\pi\varepsilon_0}\left[\frac{1}{(\beta r)^2}-\frac{j}{(\beta r)^3}\right]\cos\theta \quad \text{(V/m)} \tag{50.18}$$

$$E_\theta=\frac{I\,dl\,\beta^3 e^{-j\beta r}}{\omega\quad 4\pi\varepsilon_0}\left[\frac{j}{(\beta r)}+\frac{1}{(\beta r)^2}-\frac{j}{(\beta r)^3}\right]\sin\theta \quad \text{(V/m)} \tag{50.19}$$

$$H_\phi=\frac{I\,dl\,\beta^2 e^{-j\beta r}}{\omega\quad 4\pi}\left[\frac{j}{(\beta r)}+\frac{1}{(\beta r)^2}\right]\sin\theta \quad \text{(A/m)} \tag{50.20}$$

where

dl = differential length of current element (m)
I = current in element (A)
ω = angular frequency of I (rad/sec)
v = velocity of propagation (in free space $v=c=3\times10^8$ m/s)
β = phase constant ω/V

Also,

$$\beta=\frac{2\pi f}{C}=\frac{2\pi}{\lambda}, \text{ where } \lambda=\text{wavelength (m)}$$

ε_0 = permittivity of free space = $1/36\pi\times10^9$ f/m
r = distance from elemental source to observation point (m)

An examination of equations (50.18–50.20) yields the following information.

(1) For $\beta r<<1$, the higher-order terms dominate with the **E** field components varying as $1/r^3$, $1/r^2$ and the **H** component as $1/r^2$. The field very close to the source is dominated by the $1/r^3$ terms and is called the electrostatic field. This field term corresponds to that of an electric dipole and represents a constant energy field. The $1/r^2$ terms are called the induction field. This total field region where these higher-order terms predominate is called the near field.
(2) For $\beta r>>1$, E_r becomes insignificant compared to the transverse term (E_θ) and E_θ and H_ϕ vary as $1/r$. This is called the radiation or far field. In this field energy is being propagated from the current dipole.
(3) The transition from the far to the near field is defined as $\beta r=1$ or $r=1/\beta=\lambda/2\pi$, approximately at a distance corresponding to one sixth of a wavelength.

50.2.2.2 Elementary current loop

Similar to the previous section, the fields corresponding to the elementary current loop of *Figure 50.8* are given.

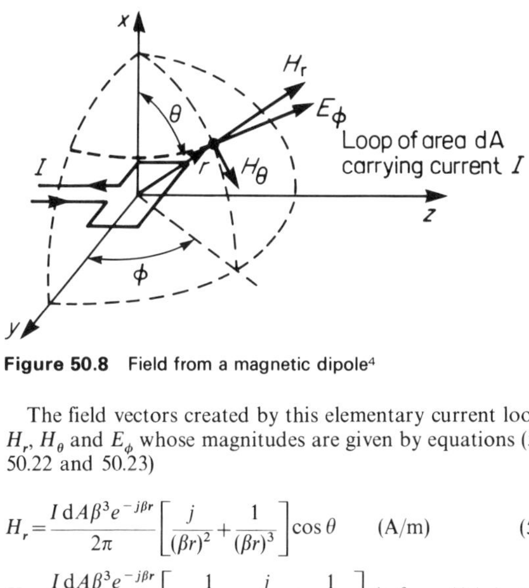

Figure 50.8 Field from a magnetic dipole[4]

The field vectors created by this elementary current loop are H_r, H_θ and E_ϕ whose magnitudes are given by equations (50.21, 50.22 and 50.23)

$$H_r=\frac{I\,dA\beta^3e^{-j\beta r}}{2\pi}\left[\frac{j}{(\beta r)^2}+\frac{1}{(\beta r)^3}\right]\cos\theta \quad \text{(A/m)} \tag{50.21}$$

$$H_\theta=\frac{I\,dA\beta^3e^{-j\beta r}}{4\pi}\left[-\frac{1}{(\beta r)}+\frac{j}{(\beta r)^2}+\frac{1}{(\beta r)^3}\right]\sin\theta \quad \text{(A/m)} \tag{50.22}$$

$$E_\phi=\frac{I\,dA\beta^4e^{-j\beta r}}{\omega 4\pi e_0}\left[\frac{1}{(\beta r)}-\frac{j}{(\beta r)^2}\right]\sin\theta \quad \text{(V/m)} \tag{50.23}$$

where dA = differential area of the current loop (m^2). All other parameters are the same as defined for the electric dipole.

An analysis of these equations will produce results similar to those obtained for the electric dipole.

50.2.2.3 Wave impedance

At any point in a medium containing an electromagnetic field, the field has an impedance η defined as

$$\eta=\mathbf{E}/\mathbf{H} \tag{50.24}$$

In the far field, this ratio equals the characteristic impedance of the medium which for air is given by equation (50.25)

$$\eta=\sqrt{\mu/\varepsilon}\simeq120\pi\simeq377\ \Omega \tag{50.25}$$

In the near field, the impedance is either higher or lower than 377 Ω depending on the nature of the source. Sources with high voltage and low current generate high impedance fields and low impedance fields result from high current, low voltage sources. For a high impedance field (i.e. electric dipole) the near field impedance is

$$|Z|=\frac{\eta\lambda}{2\pi r}=\frac{1}{2\pi f\varepsilon r} \quad (\Omega) \tag{50.26}$$

For a low impedance field (i.e. current loop), the near field impedance is given by

$$|Z|=\frac{\eta 2\pi r}{\lambda}=2\pi f\mu r \quad (\Omega) \tag{50.27}$$

Figure 50.9 plots wave impedance as a function of distance from the source (normalised to $\lambda/2\pi$). It can be seen that $r=\lambda/2\pi$ represents the transistion from the near to far field.

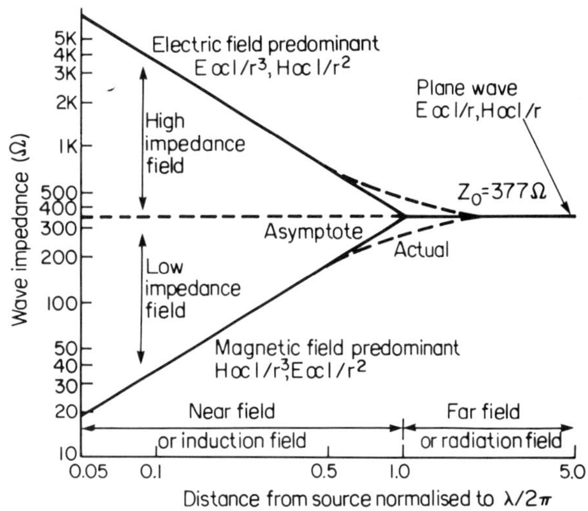

Figure 50.9 Wave impedance against distance from source[1]

50.2.3 Skin effect

A conductor carrying a d.c. current will have the current uniformly distributed across the conductor cross section. An a.c. current, however, is not uniformly distributed but will tend to concentrate on the conductor surface. As frequency increases, the effect becomes more pronounced until the current flow is confined to a very thin layer on the conductor surface. This phenomenon is called the skin effect and significantly increases the effective resistance of conductors carrying high frequency signals. It is also an important factor relative to shielding effectiveness.

The current density magnitude decreases exponentially with distance from the surface and is reduced to $1/e$ the surface value at a distance δ (in metres) given by

$$\delta=\sqrt{\frac{1}{\pi f\mu\sigma}} \tag{50.28}$$

where

δ = skin depth
f = frequency
μ = conductor permeability
σ = conductivity (mhos/metre)

For a copper conductor, $\mu=\mu_0=4\pi\times10^{-7}$, and $\sigma=5.75\times10^7$ mhos/metre at 20°C and equation (50.28) becomes

$$\delta=\frac{0.0664}{\sqrt{f}} \qquad (50.29)$$

Figure 50.10 shows the current distribution in round solid copper wire of 1 mm diameter (radius a=0.5 mm). *Figure 50.11* shows the depth of current penetration against frequency for some common conductor materials.

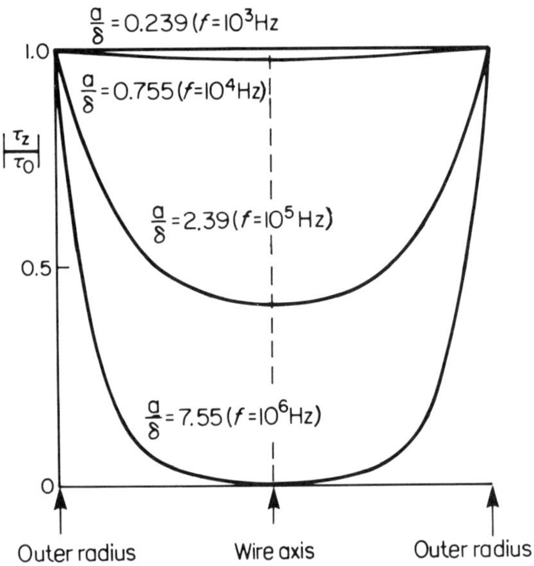

Figure 50.10 Current distribution in cylindrical 1 mm diameter cu wire against frequency[9]

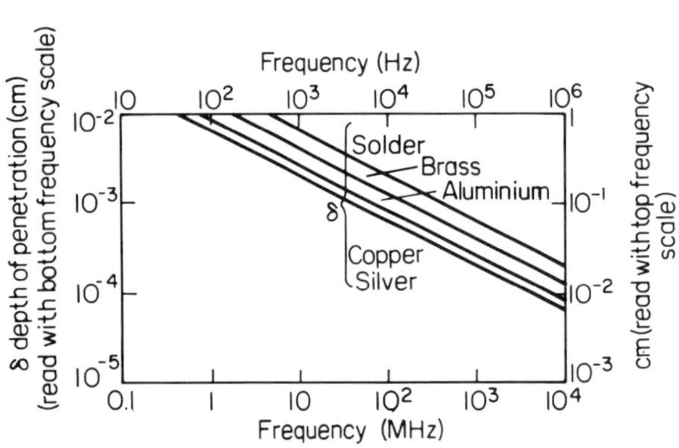

Figure 50.11 Skin effect against frequency for plane conductors[9]

The skin depth δ, is an important parameter in determining the a.c. resistance of a conductor. Consider a round conductor for which δ is small relative to the conductor radius, a. The equivalent conductor geometry for current conduction may be shown to be a cylindrical shell of thickness δ which carries a uniformly distributed current. The a.c. resistance is proportional to $1/2\pi a\delta$.

The ratio of a.c. to d.c. resistance is

$$\frac{R_{\text{a.c.}}}{R_{\text{d.c.}}}=\frac{a}{2\delta}=\frac{a\sqrt{\pi f\mu\sigma}}{2} \qquad (50.30)$$

which for copper becomes

$$\frac{R_{\text{a.c.}}}{R_{\text{d.c.}}}=7.53\,a\sqrt{f} \qquad (50.31)$$

50.2.4 Shielding

Shielding is the use of a conducting and/or permeable barrier between a potentially disturbing field source and the circuitry to be protected. Shields may be used to protect interconnect (e.g. cables) or packaged circuit functions. The effectiveness of shielding depends on the characteristics of: (a) the field (type of field, strength, polarisation, angle of incidence, frequency), (2) the shielding material (conductivity, permeability), and (3) the physical geometry of the shield (thickness, openings in the shield). Although the following discussion is strictly true for shields in the form of a large flat sheet, the principles are true in general.

Mathematically, shielding effectivness (S) for electric and magnetic fields is defined by equations (50.32) and (50.33) respectively.

$$S=20\lg E_1/E_2 \qquad (50.32)$$

$$S=20\lg H_1/H_2 \qquad (50.33)$$

in dB, where E_1, H_1 are the incident fields and E_2, H_2 are the attenuated fields.

Shielding attenuates incident fields via two loss mechanisms—absorption (A) and reflection (R). With A and R expressed in dB, the total shielding loss is

$$S=A+R+B \qquad (50.34)$$

The B term is a correction factor required if there are multiple reflections within the shield. Normally A is sufficiently high (greater than 10 dB) so that B can be neglected. This is especially true for electric fields which have a substantial reflection loss at the incident surface. However, thin shields in the presence of low frequency magnetic fields may require the use of this term. The effect of B reduces the overall shielding effectiveness as indicated in the following expression[1]

$$B=20\lg(1-e^{-2t/\delta}) \qquad (50.35)$$

50.2.4.1 Absorption loss

When time varying electric and magnetic fields impinge on a conducting medium, current flow is induced. In accordance with section 50.2.3 (skin effect), the magnitude of current at distance x from the surface (I_x) is given by

$$I_x=I_s e^{-x/\delta} \qquad (50.36)$$

I_s = current at surface
δ = skin depth (see equation (50.28)).

Electromagnetic field attenuation in the medium follows the same relationship which yields

$$A=x/\delta\ (\text{nepers})=8.69x/\delta\ (\text{dB}) \qquad (50.37)$$

From equation (50.37) it can be seen that absorption loss is directly proportional to the thickness of the material and equals 8.69 dB for each skin depth of distance into the shielding material.

50.2.4.2 Reflection loss

The characteristic impedance of any medium is given by equation (50.38)

$$Z_0=\left(\frac{j\omega\mu}{\sigma+j\omega\varepsilon}\right)^{1/2} \quad (50.38)$$

In an insulator $\sigma << j\omega\varepsilon$ and

$$Z_0=(\mu/\varepsilon)^{1/2} \quad (50.39)$$

In a conductor $\sigma >> j\omega\varepsilon$ and

$$Z_0=\left(\frac{\pi f\mu}{\sigma}\right)^{1/2}(1+j) \quad (50.40)$$

In section 50.2.2, a field was shown to have an impedance which depends on the electrical distance from the source. The far field impedance is 377 Ω. The near field impedance could be either higher or lower depending on whether the source generator is predominantely electric or magnetic field respectively. When such a field impinges on a conducting barrier, the impedance mismatch will cause some of the incident field to be reflected The portion that penetrates is attenuated by absorption loss. The penetrating field which reaches the exit surface will undergo another reflection loss. The total reflection loss is given (in dB) by

$$R=20\lg\frac{|Z_W|}{4|Z_S|} \quad (50.41)$$

where Z_W is the impedance of wave just outside the shield and Z_S is the shield impedance.

Equation (50.41) is valid for both electric and magnetic fields. For electric fields, the biggest reflection loss occurs at the incident air/conductor boundary so that very little electric field penetration occurs. Conversely, magnetic fields more readily penetrate the shield material at the incident boundary and suffer large reflection at the exit boundary.

Reflection loss (in dB) in the near field for electric and magnetic fields is given by equations (50.42) and (50.43) respectively.

$$R_e=320.2-10\lg\left[f^3r^2\left(\frac{\mu_r}{\varepsilon_r}\right)\right] \quad (50.42)$$

$$R_m=14.6+10\lg\left[fr^2\left(\frac{\sigma_r}{\mu_r}\right)\right] \quad (50.43)$$

where

f = frequency (Hz)
r = distance between source and shield (m)
μ_r = relative permeability of shield material (relative to air)
ε_r = relative permitivity of shield material (relative to air)
σ_r = relative conductivity of shield material (relative to copper)

From equations (50.42) and (50.43) the following facts are evident.

(1) *Electric field*
(a) Reflection loss decreases 30 dB/decade of frequency.
(b) Reflection loss decreases 20 dB/decade of distance; loss is greater for shield close to the source.

(2) *Magnetic field*
(a) Reflection loss increases 10 dB/decade of frequency.
(b) Reflection loss increases 20 dB/decade of distance; loss is greater for shield farther from source.

Equation (50.42) shows that it is possible to achieve relatively high reflection losses for electric fields. Low frequency magnetic fields, however, have low reflection losses and must depend on absorption loss as the primary loss mechanism.

50.2.5 Interconnect formulae

The distribution of electrical signals and power between physically separated points is a characteristic of all electronic equipment. The interconnect system is the means by which this distribution is accomplished. In general, the interconnect system must handle many different types of signals. Signal frequency content can range from d.c. frequency to hundreds of MHz and power from μW to hundreds of watts. Often these incompatible signals will be in close proximity to each other creating a high potential for interference. The physical/electrical length of the interconnect, effect on signal quality and coupling to other circuits (either as a disturbing source or victim) are factors which the designer must consider.

Depending on signal characteristics, a signal line and its return path may be viewed as a short line (lumped element line) or a long line (distributed parameter transmission line). The important criterion is the electrical length of the line relative to the signal frequency content. If the line is carrying digital signals, a long line is defined as a line whose one way delay time (t_d) is larger than one half the transition time of the signal (i.e. $t_d > t_r/2$ or $t_f/2$).

Regarding interconnect electrical characteristics, two categories can be defined: self and mutual. Self parameters are those associated with a given signal path itself while mutual parameters concern the coupling between different signal paths. For either type, there can be a lumped or distributed case depending on signal frequencies and line lengths.

The basic electrical parameters of interest are the inductance (L) and capacitance (C) per unit length. For distributed lines, the characteristic impedance (Z_0) and propagation delay (t_d) for a lossless line are given by equations (50.44) and (50.45)

$$Z_0=(L/C)^{1/2} \quad (\text{in } \Omega) \quad (50.44)$$

$$t_d=v^{-1}=(LC)^{1/2} \quad (\text{in s/m}) \quad (50.45)$$

where

v = velocity of propagation (m/s)
L = inductance (H/m)
C = capacitance (F/m)

Figure 50.12 provides formulae for several useful transmission line configurations.

In *Figure 50.13*, Z_0, L and C are plotted as a function of D/d to illustrate the sensitivity of these parameters to physical spacing.

50.3 Interference mechanisms

When undesirable energy from a source function finds its way into a victim circuit or subsystem, the potential for interference exists. In this section, the primary mechanisms by which this energy transfer can occur are examined. The two predominant types of interference crosstalk are common impedance and induction coupling.

50.3.1 Common impedance coupling

Shared conductive paths between circuits and/or subsystems create the potential for common impedance crosstalk. The two most common places for this to occur are in the power and ground distribution systems. A simple example illustrates how this can occur. Assume a linear amplifier circuit with three stages of amplification implemented as shown in *Figure 50.14*. Further assume that the power handling capacility increases from stage 1 to 3 (i.e. $I_3 > I_2 > I_1$).

From *Figure 50.14* it can be seen that crosstalk between stages is occurring through both the ground and power distribution systems. The input signal to each stage is given by

$$e_{in(n)}=e_{0(n-1)}+e_{gn}=e_{0(n-1)}(1+\varepsilon) \quad (50.46)$$

Transmission line configuration		Characteristic impedance (Z_0)	Capacitance (f/m)	Inductance (H/m)	Conditions
Parallel wires (air)	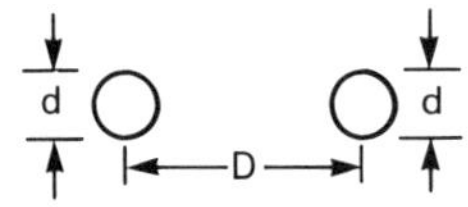	$\frac{(\mu/\epsilon)^{1/2}}{\pi}\cosh^{-1}(D/d)$	$\frac{\pi\epsilon}{\cosh^{-1}(D/d)}$	$\mu/\pi \cosh^{-1}(D/d)$	
Wire over ground	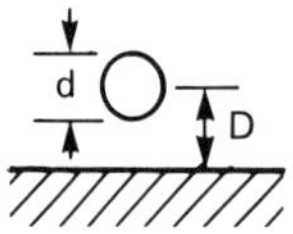	$\frac{(\mu/\epsilon)^{1/2}}{2\pi}\cosh^{-1}(2D/d)$	$\frac{2\pi\epsilon}{\cosh^{-1}(2D/d)}$	$\mu/2\pi \cosh^{-1}(2D/d)$	
Microstrip	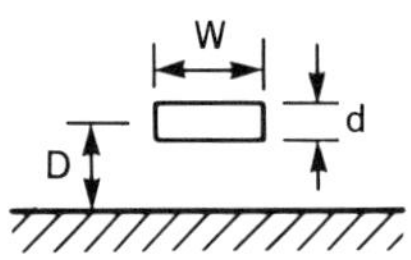	$(\mu/\epsilon)^{1/2}D/W$	$\epsilon W/D$	$\mu D/W$	$2D < W$, $d << 2D$
		$\frac{(\mu/\epsilon)^{1/2}}{2\pi}\ln(2\pi D/W+d)$	$\frac{2\pi\epsilon}{\ln[(2\pi D/W+d)]}$	$\mu/2\pi \ln(2\pi D/W+d)$	$2D >> W$
Stripline	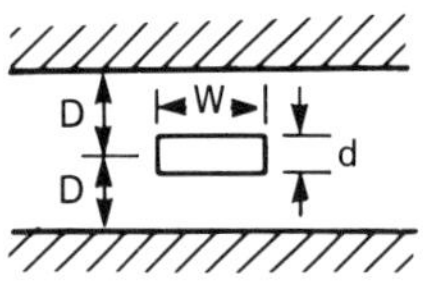	$(\mu/\epsilon)^{1/2}D/2W$	$2\epsilon W/D$	$\mu D/2W$	$D << W$, $d << D$
Parallel strip	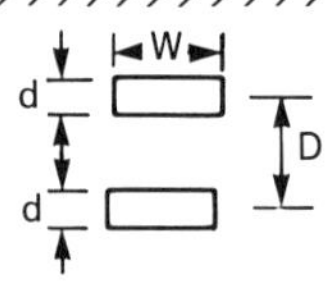	$(\mu/\epsilon)^{1/2}D/W$	$\epsilon W/2\pi$	$\mu D/W$	$D << W$, $d << D$
		$\frac{(\mu/\epsilon)^{1/2}}{\pi}\ln(\pi D/W+d)$	$\frac{\pi\epsilon}{\ln[(\pi D/W+d)]}$	$\mu/\pi \ln(\pi D/W+d)$	$D >> W$
Coaxial		$\frac{(\mu/\epsilon)^{1/2}}{2\pi}\ln(D/d)$	$\frac{2\pi\epsilon}{\ln(D/d)}$	$\mu/2\pi \ln(D/d)$	$W >> d$

Figure 50.12 Transmission line formulae[2]

Where $\varepsilon = e_{gn}/e_{0(n-1)}$ is the error term, and is indistinguishable from signal. The ground offset error voltage, e_{gn}, contains a component generated from the nth stage itself and a component from each of the stages which follow. The letter terms are particularly serious due to their higher operating currents. In the case of digital circuitry, some reflief exists since one is working against a threshold instead of linear gain stages. Here the individual logic stages are roughly identical and there is no difference if the current is flowing towards either stage 1 or stage 3. The power distribution system is subject to the same type of problem, however, and the effect of power distribution noise on circuit performance would need to be characterised for a particular circuit. From *Figure 50.14*, design philosophies for controlling conduction crosstalk coupling are self evident and can be summarised as follows:

(1) Keep the impedances of the ground and power distribution systems as low as practicable.

(2) Design ground and power distribution systems such that undesirable currents do not flow through current paths which couple noise into critical circuits. This is essentially a task of understanding and managing current flow paths. In particular, preventing currents from high noise sources from flowing through the ground structure of lower power more sensitive circuits. For example, in *Figure 50.14*, connecting the power supply to the high power stage would have currents flowing away from the input stage.

(3) Utilise adequate bypass filtering and decoupling on the power distribution system.

50.3.2 Induction field coupling

In section 50.2.2 the field equations for an electric dipole and elemental current loop were examined. The field characteristics as a function of distance from the source allowed two regions to be defined. In the near (induction) field region the field impedance is determined by the source characteristics and field intensity varies as $1/r^n$ ($n=2, 3$). In the far (radiation) field region, field impedance is constant at 377 Ω and field intensity varies as $1/r$. The boundary between these two regions is commonly defined to be approximately 1/6 wavelength ($\lambda/2\pi$) from the source.

At 30 MHz (corresponding to a pulse rise time of about 10 ns), the wavelength in free space is 10 m corresponding to a near/far field transition boundary of approximately 1.8 m. This is still quite a large distance relative to the major crosstalk problems likely to be encountered within most equipment. Consequently, internal machine field coupling interactions are usually from the induction field where the electric and magnetic fields are considered separately. The most likely vehicle for coupling interactions within the equipment environment is the interconnect (signal, power, and ground).

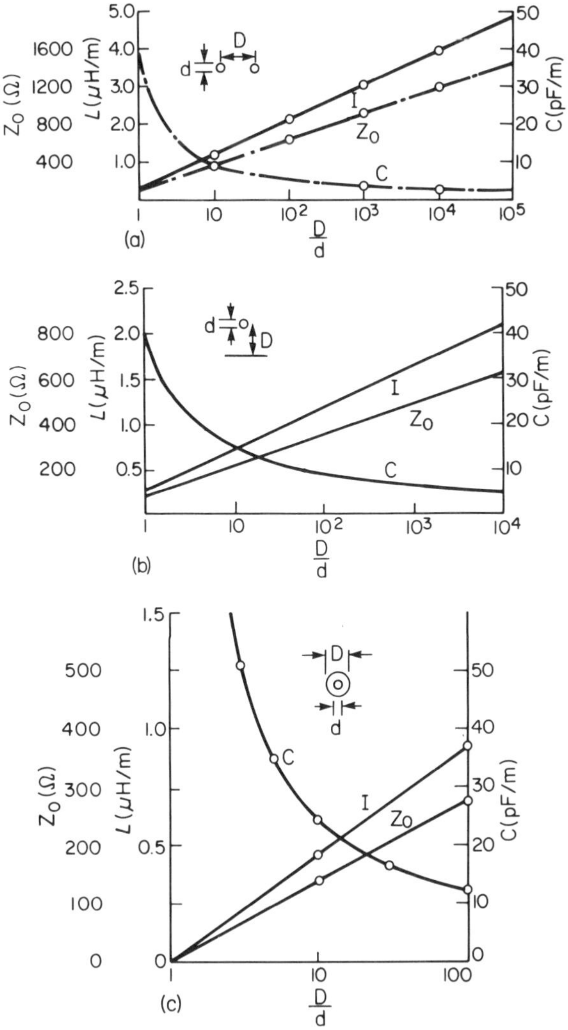

Figure 50.13 Characteristic impedance (Z_0), inductance (L), and capacitance (C) against D/d (all with air dielectric). (a) Parallel wires; (b) wire over ground; (c) coaxial cable[8]

50.3.2.1 Crosstalk equations—electrically short lines

Assume a pair of electrically short parallel wires (two-way delay time less than rise or fall time) over a ground plane which are coupled by mutual capacitance (C_m) and mutual inductance (L_m). Because these are electrically short lines, C_m and L_m are the total lumped mutual capacitance and inductance respectively. One of the wires is actively driven by the generator V_S and the other wire is the receiver or victim circuit (*Figure 50.15*). It is common in wire-to-wire crosstalk situations for the following relationship to exist

$1/j\omega C_m >> R_{R1},\ R_{R2} >> j\omega L_m$

The capacitive and inductive coupling may then be considered separately and then combined by superposition to obtain the total effect.

Under these conditions, the equivalent circuit of *Figure 50.16* can be developed. It consists of a series voltage source of

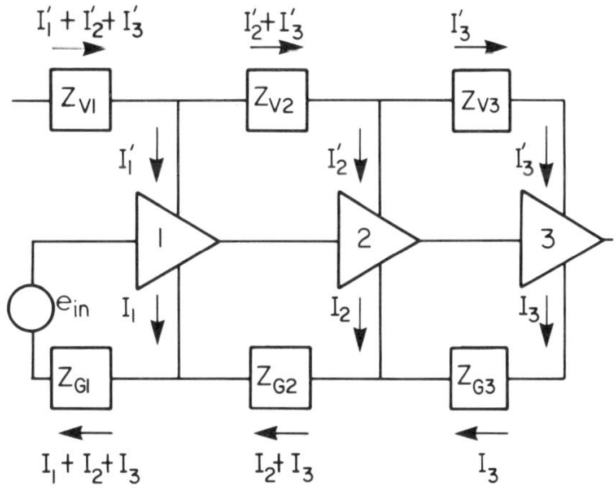

Figure 50.14 Example of common impedance coupling in power and ground distribution

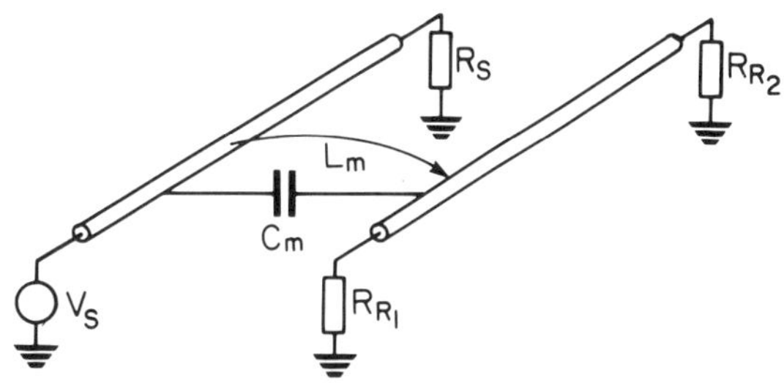

Figure 50.15 Parallel coupled signal lines

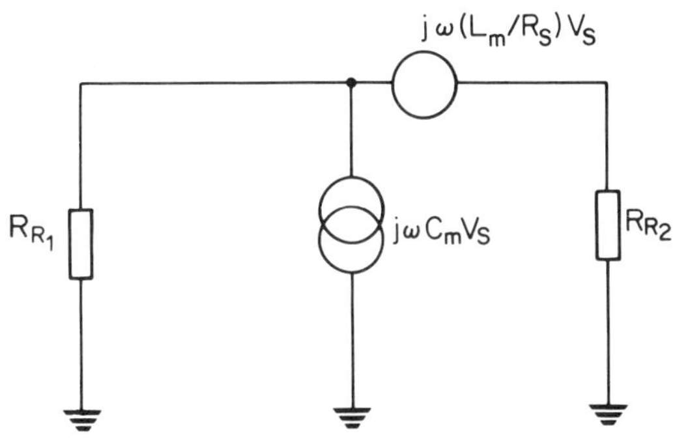

Figure 50.16 Equivalent circuit of victim line

magnitude $j\omega L_m V_S/R_S$ from the mutual inductance and a parallel current source of magnitude $j\omega C_m V_S$ from the mutual capacitance. It can be seen that the capacitive and inductive components of crosstalk are in phase at the end closest to the generator and out of phase at the far end. The near and far end crosstalk equations for the receiver line are

$$V_{R1} = C_m \frac{dV_S}{dt}\left(\frac{R_{R1}R_{R2}}{R_{R1}+R_{R2}}\right) + \frac{L_m}{R_S}\frac{dV_S}{dt}\left(\frac{R_{R1}}{R_{R1}+R_{R2}}\right) \quad (50.47)$$

$$V_{R2} = C_m \frac{dV_S}{dt}\left(\frac{R_{R1}R_{R2}}{R_{R1}+R_{R2}}\right) - \frac{L_m}{R_S}\frac{dV_S}{dt}\left(\frac{R_{R2}}{R_{R1}+R_{R2}}\right) \quad (50.48)$$

If $R_{R1}=R_{R2}=R_S=R$, equations (50.47) and (50.48) reduce to

$$V_{R1}=\tfrac{1}{2}\left(C_mR+\frac{L_m}{R}\right)\frac{dV_S}{dt} \quad (50.49)$$

$$V_{R2}=\tfrac{1}{2}\left(C_mR-\frac{L_m}{R}\right)\frac{dV_S}{dt} \quad (50.50)$$

From equations (50.49) and (50.50), the following observations can be made.

(1) Crosstalk is directly proportional to the mutual coupling terms (L_m, C_m) which for electrically short lines is also proportional to the length of the coupled lines.

(2) High impedance circuits $R>(L_m/C_m)^{1/2}$ favour capacitive crosstalk.

(3) Low impedance circuits, $R<(L_m/C_m)^{1/2}$, favour inductive crosstalk.

(4) V_{R2} is zero for $R=(L_m/C_m)^{1/2}$.

Both capacitive and inductive crosstalk are directly proportional to the rate of change or frequency of the noise generator.

50.3.2.2 Reducing induction field coupling

The observations relative to equations (50.49) and (50.50) in section 50.3.2.1 suggest the following techniques for reducing induction field coupling in interconnecting lines:

(1) Reduce mutual coupling terms, L_m and C_m. This reduction can be achieved by increasing the physical separation of coupled lines, segregating high noise lines from lines associated with sensitive circuits (in cable runs) or, if inductive coupling predominates, minimising loop areas of source and victim circuits especially common area circuits (i.e. use twisted pair, ground plane, etc.). Shield conductors and for microstrip lines, interpose grounded line(s) between source and victim lines.

(2) Adjust circuit impedances in accordance with the type of coupling predominating. For capacitive coupling, reduce circuit and line impedances (using close proximity ground planes, etc) or for inductive coupling increase circuit impedances to reduce currents.

(3) Reduce signal speed, i.e. increase transition times of pulses, use lower frequencies to the extent possible.

50.3.2.3 Coupled transmission line

In high speed digital equipment the fast transition times of pluse signals often require a distributed transmission line model for the coupled lines. Distributed analysis techniques should be used whenever pulse transition times are less than twice the line's propagation delay. The disturbed line will have two crosstalk components. Forward crosstalk propagates in the same direction as the disturbing signal while backward crosstalk propagates in the opposite direction. Assuming lossless lines terminated in Z_0, the crosstalk equations for near and far ends of the disturbed line are

$$V_{R1}=K_B[V_S(t)-V_S(t-2T_d)] \quad (50.51)$$

$$V_{R2}=K_f l(d/dt)V_S(t-T_d) \quad (50.52)$$

$$K_B=[C_mZ_0+(L_m/Z_0)]/4t_d \quad (50.53)$$

$$K_f=\tfrac{1}{2}[C_mZ_0-(L_m/Z_0)] \quad (50.54)$$

where

L_m = mutual inductance per unit length
C_m = mutual capacitance per unit length
Z_0 = characteristic impedance (Ω)
l = length of coupled region
t_d = delay/unit length

The crosstalk coefficients (K_B, K_f) depend on the geometry of the signal lines and the material properties. *Figures 50.17* and *50.18* provide values of K_B and K_f for different microstrip line geometries.

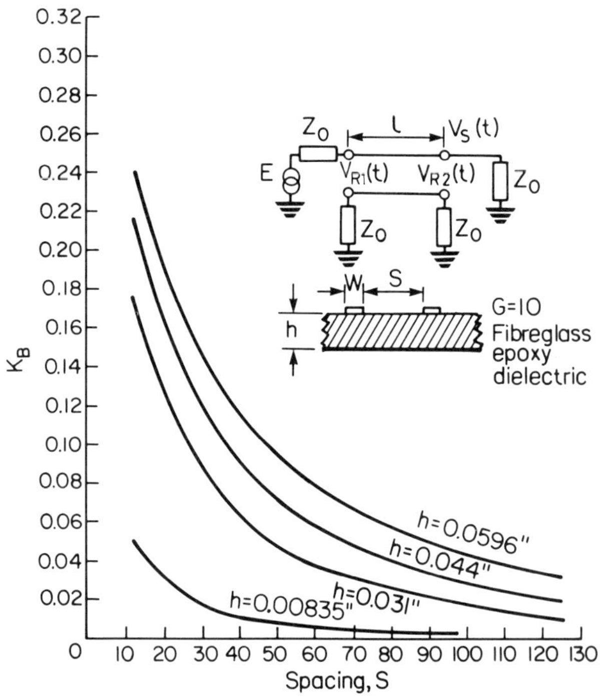

Figure 50.17 Back crosstalk constant against spacing. The following points should be noted. (1) Lines are terminated in Z_0, the characteristic impedance of each line. (2) Curves are valid for $0.0075'' < W < 0.025''$. (3) Copper weight is 2 ounces: 0.0025" thick. (4) Time delay of line (T_d) is 1.8 nS/ft (~5.4 nS/m). **Reproduced from reference 7 with permission**

From equations (50.51) and (50.52), the following information can be deduced regarding transmission line crosstalk.

(1) The amplitudes of both the backward and forward crosstalk components are directly proportional to the respective crosstalk coefficient, K_B and K_f.

(2) The amplitude of the forward crosstalk component is directly proportional to the length of the coupled lines (l), and the rate-of-change of the disturbing signal. The amplitude of backward crosstalk does not depend on the length, l.

(3) The backward crosstalk component is the summation of two parts, each an attenuated replica of the disturbing signal. The first part appears at the same instant as the disturbing signal and is of the same polarity. The second part is inverted and appears after a delay equal to twice the one way delay of the coupled lines.

(4) The forward crosstalk component waveshape is the derivative of the disturbing signal and it appears one line delay time after the disturbing signal.

(5) The forward crosstalk component is 0 when $Z_0=(L_m/C_m)^{1/2}$. This occurs when the signal lines are embedded in a homogeneous dielectric.

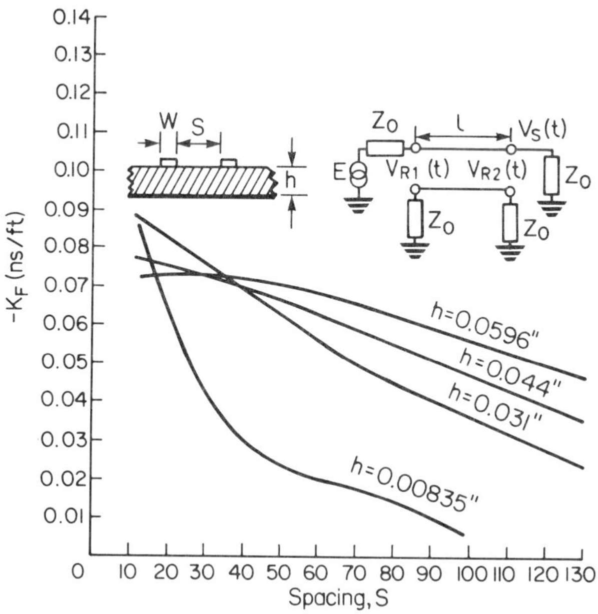

Figure 50.18 Forward crosstalk constant against spacing. Lines terminated in characteristic impedances. Notes (2) and (4) for *Figure 50.17* also apply. Reproduced from reference 7 with permission

50.4 Design techniques for interference control

50.4.1 Grounding

A large percentage of noise problems experienced in electronic equipment result from inadequacies in the ground system design. A well considered ground strategy should be established early in the hardware conceptualisation stage of any product.

50.4.1.1 Grounding—objectives/definitions

At the outset, it is necessary to understand that the term 'ground' is not completely descriptive. Different concerns often result in several distinct ground system identifications. Essentially, there are three objectives relative to ground system design. They are:

(1) Safety.
(2) Providing an equipotential reference for electronic circuits/subsystems.
(3) Providing current paths for the control of internal machine crosstalk and emissions into the environment (regulatory limits).

The purpose of safety ground is to prevent shock hazards to operating personnel. Other common names for safety ground are primary ground, green wire ground, hardware ground, a.c. ground, and earth ground.

Signal ground is needed as a common reference for electronic circuitry and as a return medium for currents. Operational considerations (e.g. preventing common impedance crosstalk between incompatible circuits) will often dictate several distinct and separated signal grounds. The grounds within a system are normally electrically connected at some point(s). Occasionally, special performance requirements dictate the use of electrically isolated grounds. This approach tends to be expensive, however, and it is difficult to maintain the isolation at high frequencies.

Other names for signal ground are secondary ground, circuit ground, and d.c. ground.

50.4.1.2 Types of grounds

Safety ground Equipment operating from power mains must be designed such that a fault condition does not cause conducting surfaces (accessible to an equipment user) to become electrically 'hot'. The technique predominantly in use for electrically powered equipment is to provide a safety ground connection to conducting surfaces accessible to the operator. Under normal operating conditions, the current carried by the safety ground conductor is very small and limited by the safety agencies of various countries. Under a fault condition (i.e. short to equipment cabinet), the safety ground conducts the fault current and triggers the circuit breaker or ground-fault-interrupter. In general, safety ground should not be used to perform signal ground functions.

Signal ground Normally, electronic circuits and subsystems require at least one common voltage reference called signal ground. Circuit requirements usually dictate that signal ground closely approximates equipotential characteristics. Because the ground may carry significant current in normal operation, this establishes a limitation on ground impedance. The degree to which the signal ground system must approximate ideal zero impedance conditions is dependent on many factors. For example, the type of circuitry utilised, ground current characteristics, system physical characteristics, and circuit/system susceptibility to ground noise are important considerations.

These factors should be identified early in the design cycle and the knowledge used to establish ground performance specifications. The following sections provide a general framework and guidelines for ground system design.

50.4.1.3 Performance considerations

Performance of the ground system can be defined in terms of the following criteria:

(1) Ground system potential differences (both d.c. and transient).
(2) Conduction crosstalk magnitude.
(3) Susceptibility to external noise sources.
(4) Noise fields emanating from ground system.

Fortunately, these requirements do not normally present conflicting requirements on the ground system design.

Ground system potential differences There are two basic mechanisms for inducing potential differences in the ground system: conduction induced potential differences, i.e. ground current flowing through ground impedance and induction field coupling. Conduction induced offsets can be minimised by keeping the impedance of the ground system and ground currents as low as possible.

The equivalent circuit of a ground system is a series *R*, *L* network. At high frequencies, the a.c. resistance must be used. Both the resistance and inductance of the ground system can be reduced by using ground planes, large diameter wire and minimising physical size. The inductance of the ground system can be kept low by making signal currents return in close proximity to the path from which they came.

Of course, there are practical limitations as to how low the impedance of the ground system can be made. Consequently, current path control is an important technique which must also be utilised.

The primary mechanism relative to induction field induced offsets in the ground system is **H** field coupling into ground loops. The magnitude of the induced voltage is directly proportional to the magnitude of the **H** field, the area of the ground loop and the rate of change of the **H** field.

Common impedance crosstalk In the previous section, conduction induced potential differences were discussed. These ground potential differences may be particularly troublesome when several circuits share a common ground path. Problems can occur when currents from one circuit induce ground potential differences in another. Apart from reducing ground impedance, the best way to control this problem is to force currents to remain within the ground structure of their associated circuits. By constantly attempting to visualise probable current return paths, the designer can usually manipulate physical layout and impedances to make ground currents follow the desired path. In particular, two approaches which can be used to control ground currents to selected paths are (a) redundant parallel ground returns, and (b) segregated grounds. An example of redundant parallel returns is shown in *Figure 50.19*. The intent here is to make the redundant return path impedance (primarily inductance) much lower than the ground return path. Accomplishing this will cause a large portion of the return current to use this path. In difficult cases, a balun transformer can be used to achieve good confinement to the desired path at signal frequencies.

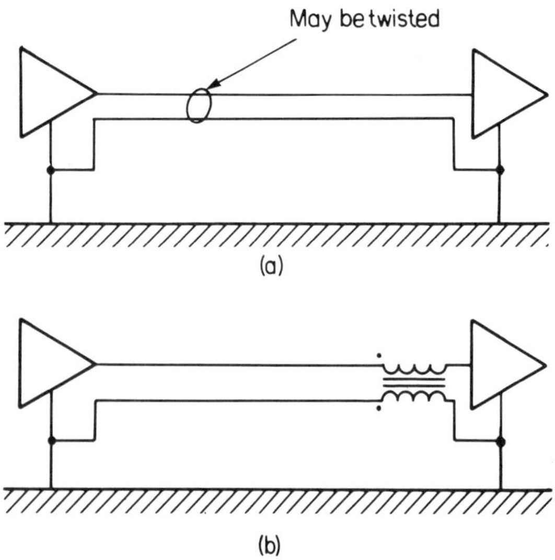

Figure 50.19 Single ended circuit configuration. (a) Circuit with redundant current return path; (b) circuit with redundant current return path and Balun transformer

Segregated ground trunks (*Figure 50.20*) are particularly useful when the equipment contains different types of hardware which tend to be incompatible (e.g. electromechanical devices like motors, solenoids, clutches, etc, low level analogue signals and digital logic circuitry). However, it is often necessary for functions on one ground to communicate with functions on a different ground. Isolated techniques could be used but often aren't because of cost and/or performance issues. If direct coupled circuits are used, signal loop requirements will often require the communicating circuits to use a local ground return. These ground ties introduce ground loops which work against the original intent for ground segregation. They also raise the possibility of additional problems occurring (i.e. circulating currents and induction field coupling).

Ground loops Whenever a circuit/subsystem has more than one path to the common ground point, a ground loop exists.

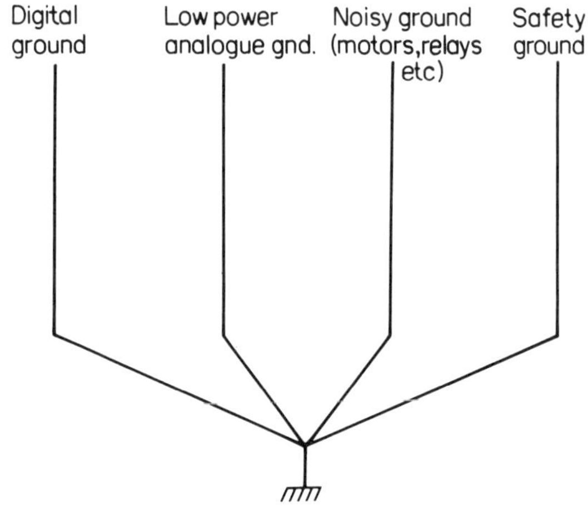

Figure 50.20 Example of ground segregation

Normally, the designer tries to prevent ground loops but practical considerations often result in their presence (particularly in a large piece of equipment). Ground loops are susceptible to **H** field induced voltages and circulating currents. In addition to the effect of the voltage offsets introduced by these circulating currents, they can be a source of noise.

50.4.1.4 System ground philosophies

There are basically four design philosophies for ground systems;

(1) single-point ground;
(2) multipoint ground;
(3) hybrid ground;
(4) isolated ground.

In general, it is not practicable (or desirable) to utilise a given philosophy at all hardware levels. Unavoidable physical effects (i.e. parasitic coupling) often enter the picture and tend to modify a given grounding philosophy.

The following factors should be considered in the selection of a ground system approach:

(1) Physical size and complexity of equipment.
(2) Frequency content of signals.
(3) Incompatibility between functional elements in the equipment (e.g. high power drivers and sensitive amplifier circuits).
(4) Sensitivity of functional elements to ground noise (common mode noise susceptibility).
(5) Equipment noise environment.
(6) Communication requirements between circuits/subsystems.

Single-point ground In a single-point ground system (*Figure 50.21*), circuits/subsystems have only one path to a common system ground point. This approach has the advantage of well controlled ground current paths and no ground loops but it also has disadvantages. It is difficult to maintain a low impedance ground (particularly at high frequency) when the individual ground trunks must all terminate on a common ground point. There is an opportunity for common impedance crosstalk to occur in the trunks with multiple circuit branches. It tends to be physically cumbersome with regard to wiring, especially if an attempt is made to minimise branches. It tends to break down at high frequencies due to parasitic coupling and when it is necessary for circuits/subsystems on different ground trunks to communicate.

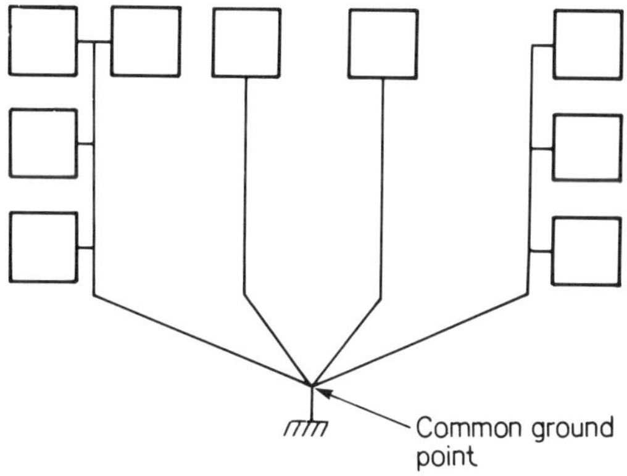

Figure 50.21 Example of single-point ground

Because of the degradation of the single point ground system at high frequencies, it is normally considered to be best at the lower frequencies (i.e. $f < \sim 5\text{–}10$ MHz).

Multipoint ground *Figure 50.22* illustrates a multiground system. This approach is usually able to maintain a lower ground

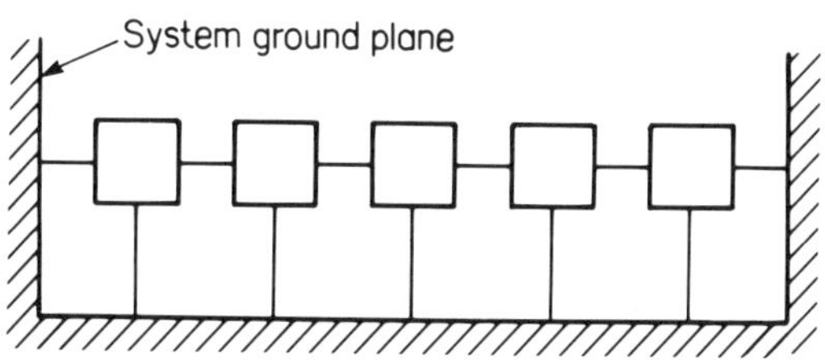

Figure 50.22 Example of multipoint ground

system impedance and thus is used at higher frequencies ($f > 5\text{–}10$ MHz). Care must be taken in multipoint ground systems to insure that the resulting ground loops and common impedance paths do not create interference conditions. These concerns often limit the size of a system using multiground techniques.

Hybrid ground As previously mentioned, a pure single-point or multipoint ground system is normally not practical or desirable. The factors identified in section 50.4.1.4 need to be considered when contemplating the ground system design. For example, it will often be convenient to use multipoint ground techniques at the lower packaging levels like the printed wiring board assembly (PWBA). Here the ground loops are small and the multipoint ground structure provides a low impedance local ground for high frequency digital circuitry. Depending on the type of equipment, it might be appropriate to extend the multipoint ground structure to a ground of PWBA in a card chassis and then to utilise single-point ground techniques between card chassis.

Isolated ground On occasion, difficult performance requirements require isolated ground systems for subsystems which are communicating.

Figure 50.23 shows three ways that this is commonly accomplished. *Figure 50.23(a)* uses an opto coupler and has response down to d.c. *Figure 50.23(b)* uses transformer coupling which cannot pass d.c. and requires that the signal characteristics (i.e. pulse width, duty cycle) be compatible with the transformer circuit capability. *Figure 50.23(c)* shows a balun connected

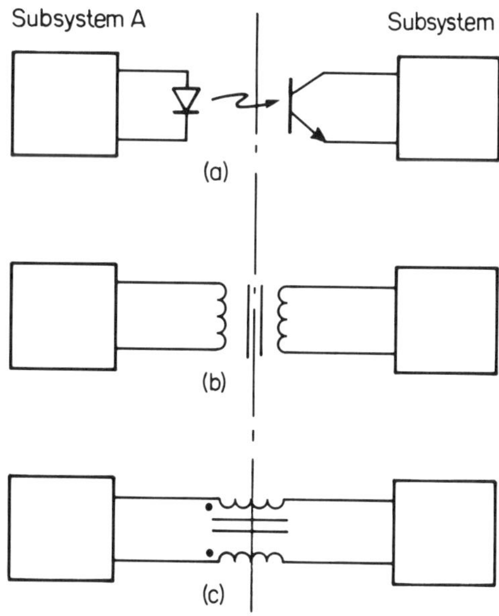

Figure 50.23 Isolation techniques. (a) Opto couler; (b) transformer; (c) Balun transformer

transformer which passes d.c. and thus does not provide low frequency isolation. It does, however, provide reasonably good high frequency isolation. An electromagnetic relay is another alternative for achieving isolation if the slower speed can be tolerated.

50.4.2 Power distribution/decoupling

Most electronic hardware requires well regulated d.c. voltages at the point of usage. Normally, these voltages are created from the a.c. mains via a power conversion function. The power conversion may be completely performed within one centralised function or alternatively, it may be distributed. *Figure 50.24* illustrates these two options. In this discussion only the effect of distribution on power regulation is examined. The technique of

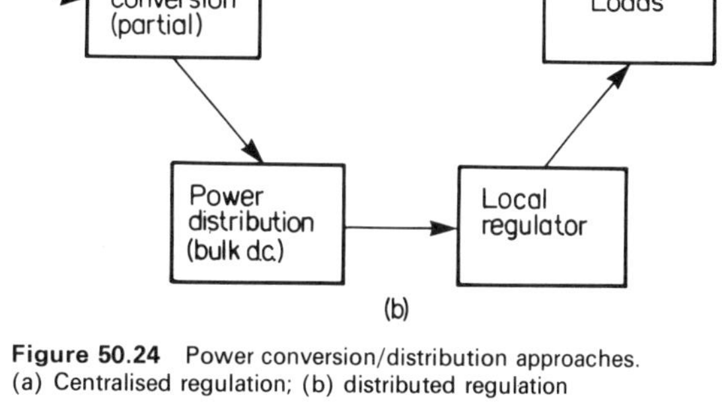

Figure 50.24 Power conversion/distribution approaches. (a) Centralised regulation; (b) distributed regulation

Figure 50.24(a) is the one most commonly used and has the more difficult distribution problem. The objective of the distribution system is to distribute power from point-of-generation to point-of-use without the addition of unacceptable noise or losses. The maximum allowable voltage variations (d.c. and transient) should be specified and budgeted to the various distribution system loss/noise mechanisms.

50.4.2.1 Performance considerations

The power distribution system can modify the distributed voltage in any of three ways:

(1) Static d.c. drop due to distribution system resistance and d.c. load current.
(2) Transient voltage changes due to load current and distribution system impedance.
(3) Induction field coupling into the distribution lines.

Static losses in the distribution system are the easiest to predict and control once the maximum load current (I_{LM}) is defined. Generally, it is most satisfactory to use a statistical approach to calculate I_{LM} as opposed to a worst case calculation. In a system of any size, the latter approach tends to be overly conservative. Once I_{LM} is determined, the maximum resistance (R_{DM}) allowed for the distribution system lines is given by

$$R_{DM}=\frac{\Delta V_{\text{d.c.}}}{I_{LM}}-R_X \quad (50.55)$$

where $\Delta V_{\text{d.c.}}$ is the allowable d.c. loss and R_X the resistance of other elements in the power distribution path beside wire (i.e. fuses, connectors, etc). Note that R_{DM} is the total resistance of the distribution lines and includes the resistance of both the source and return lines.

Transient load current changes act on the distribution system impedance to create voltage excursions. In digital equipment, current demand changes/spikes can be associated with logic state changes, capacitive charging, line driving and overlap conduction spikes from logic circuits (i.e. TTL). Distributed filtering with high frequency bypass capacitors supply charge for high speed local current transients. The capacitor value (in farads) is calculated as follows

$$C=\Delta I\,\Delta t/\Delta V \quad (50.56)$$

where ΔI is the current transient magnitude, ΔV the allowable change in voltage at the filtered node and Δt the time interval corresponding to the ΔI current change.

This capacitor effectively lowers the impedance at specific supply nodes. A low loop impedance is essential for good results and requires both the loop area and capacitor's inductance to be kept small. These capacitors should be selected for high frequency performance. Typically they have values from 0.01 to 0.1 μf. Larger value bypass capacitors (10–50 μf) are usually used to handle the lower frequency variations for a large group of circuits (e.g. PWBA). The general philosophy is to contain high frequency current transient and current demand variations as close to the source as possible (*Figure 50.25*).

Low impedance laminated bus-bars are often used for low impedance distribution systems. These systems consist of rectangular cross section parallel copper strips separated by a thin dielectric layer (e.g. see parallel strip Z_0 formula in *Figure 50.12*). Characteristic impedance values down to a few ohms are possible with this configuration. Physically distributing lumped capacitors along a distribution bus is another way to lower bus impedance within the limits of capacitor and distribution inductance.

50.4.3 Circuit techniques

The net effect of noise on electronic systems is to create extraneous voltages and/or currents. When these voltages and currents interfere with signals, an interference problem exists.

A noise management strategy has two basic thrusts. They are to keep noise out of the electronic system and to utilise circuit techniques which allow satisfactory operation in the noise environment that remains.

Viewing a signal circuit to consist of a transmitter, receiver and associated source and return lines, two types of noise can be defined. The first is called common mode noise and causes a potential difference to exist between the common terminals of the transmitter and receiver circuits. A signal circuit containing a source and return line connected between these two points would have the noise currents flowing in the same direction on these lines. The other noise type is called differential mode and causes currents to flow in opposite directions in the sense source and return lines.

50.4.3.1 Common mode noise

Figure 50.26 shows a circuit operating in the presence of a common mode voltage. This single ended open wire configuration has very poor performance in the presence of common mode noise. The problem is that the noise voltage appears in the signal loop. It can either detract from a valid signal or be falsely detected as a valid signal. This circuit configuration is very poor relative to common mode to differential mode conversion (the entire V_N appears as a differential mode signal).

Figure 50.27 shows a balanced system using a differential amplifier and twisted pair interconnect. The characteristics of the

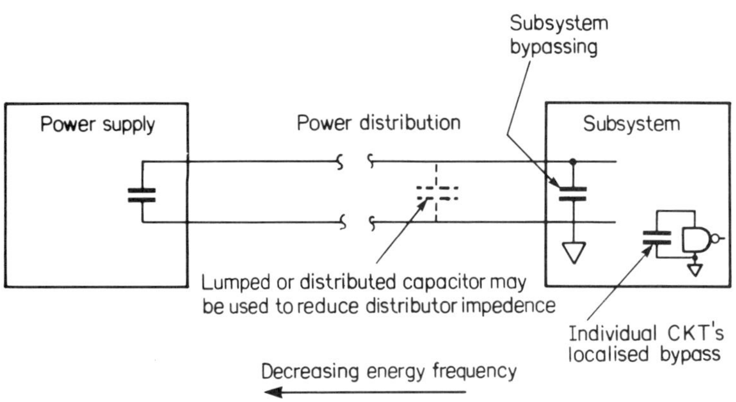

Figure 50.25 Decoupling the power distribution system

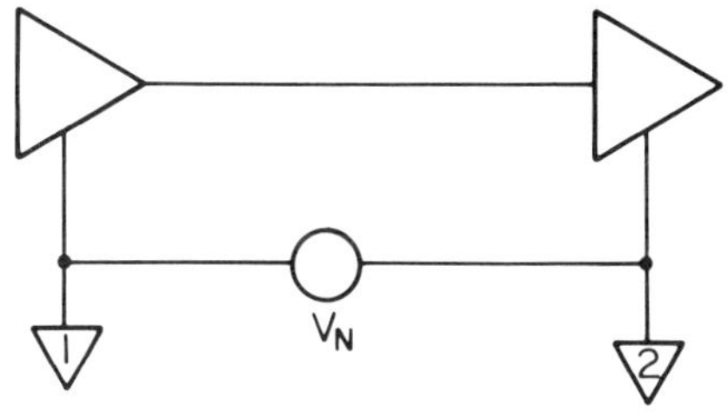

Figure 50.26 Common mode noise on a single-ended circuit

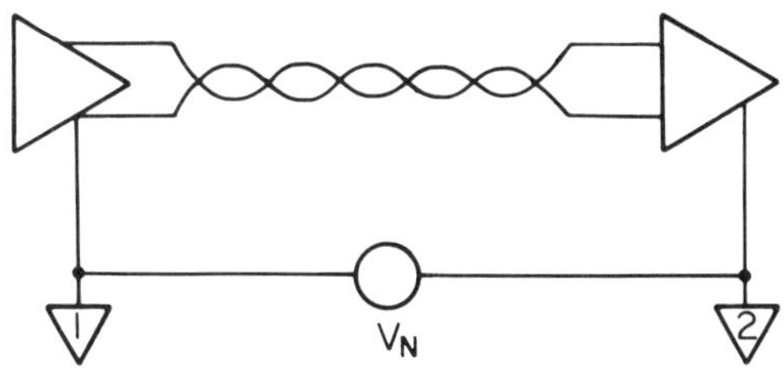

Figure 50.27 Common mode noise on a balanced differential circuit

differential amplifier are to sense the differential signal and reject common mode noise. Because of the balanced structure there is very little common mode to differential mode conversion. Thus, this circuit configuration can tolerate common mode noise quite well.

In *Figure 50.28*, a high impedance current source configuration is used for the transmitting circuit. The current drive signal (and

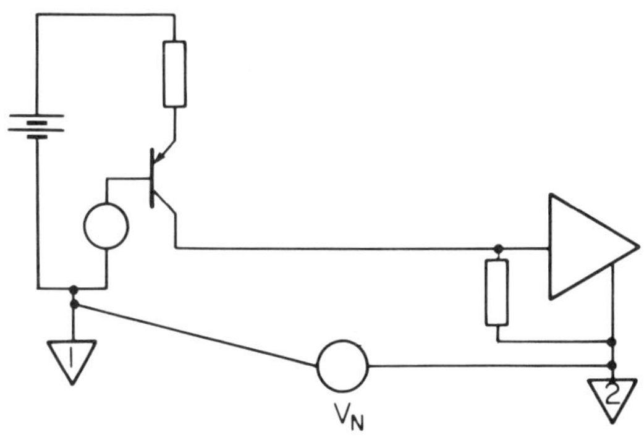

Figure 50.28 Cirduit technique to provide common mode rejection

thus the sense voltage) is independent of ground potential difference.

Figure 50.29 shows a single ended configuration similar to *Figure 50.26* but with coax grounded at both ends used. V_N causes a current (I_S) to flow through the shield which creates a voltage across the shield. Neglecting the resistive component of sheild voltage, the transfer function for the error voltage, V_E, generated from the V_N noise source is

$$V_E = V_N\left[\frac{1}{1+(j\omega L_S/R_S)}\right] \tag{50.57}$$

where R_S is the shield resistance and L_S the shield inductance.

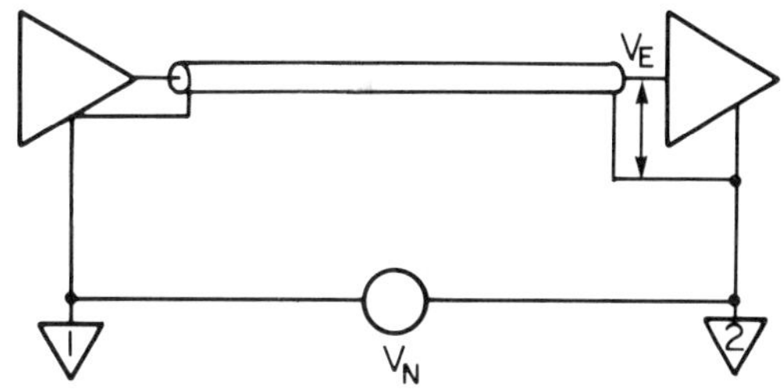

Figure 50.29 Common mode noise on a single ended system using coaxial cable

The ratio R_S/L_S is called the shield cut-off frequency, f_c^1 where the error signal is reduced to 0.707 of its low frequency value. From that point on the error voltage decreases at the rate of 6 dB/octave (*Figure 50.30*). Since the values of f_c for most coaxials are quite low (e.g. 0.5–2 kHz), the coaxial cable itself can effectively reject high frequency common mode noises. The shield resistance drop, however, is not cancelled and care must be taken to keep it small.

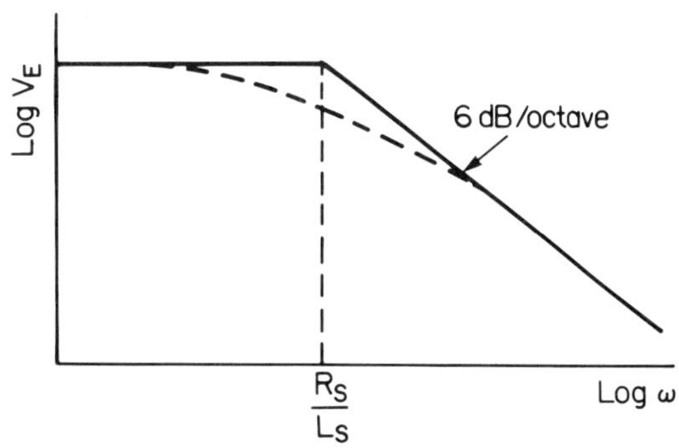

Figure 50.30 Error voltage against frequency for circuit of *Figure 50.29*

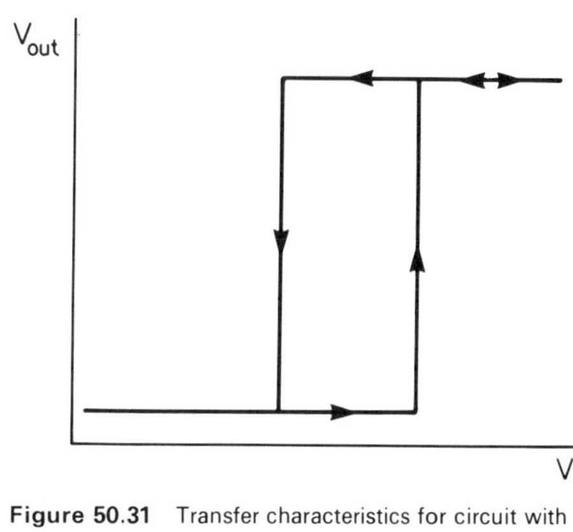

Figure 50.31 Transfer characteristics for circuit with hysteresis

Isolation is another technique often used to reject common mode potential differences (see *Figure 50.23*). However, parasitic coupling across the isolation elements will tend to degrade the isolation at high frequencies and must be evaluated.

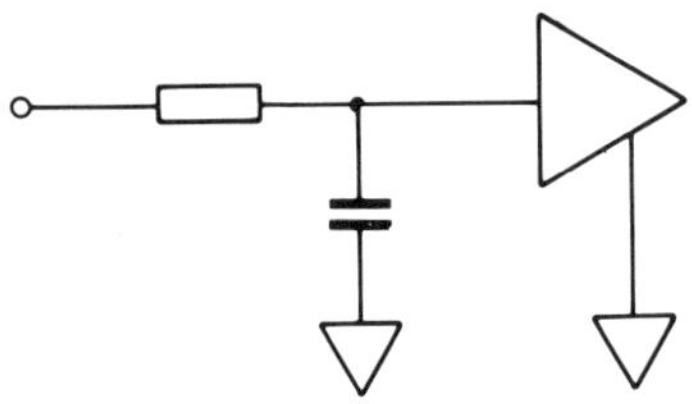

Figure 50.32 Low pass filter on circuit input

50.4.3.2 Differential mode noise

The normal signal mode is differential, thus, this type of noise is particularly troublesome. Amplitude discrimination is the first line of defence against differential noise. The approach is to utilise sense thresholds for circuits and to keep noise amplitudes below these levels. Valid signals, of course, will exceed these thresholds with adequate margin. Input hystersis (*Figure 50.31*) can further enhance the circuits noise rejection capability.

Time discrimination (strobe) can also be an effective tool to reject noise (particularly when the noise is synchronised to some machine event). In this approach, the sense circuits are only activated when they are being used.

Frequency discrimination (filtering) is another effective approach to reject differential noise. Because of the derivative nature of most noise mechanisms, the noise will often be of relatively high frequency compared to signals. This fact can be used to advantage through the use of low pass filters at signal inputs likely to be subjected to noise such as cable receiver circuits (*Figure 50.32*).

References

1 OTT, H. W., *Noise Reduction Techniques in Electronic Systems*, John Wiley & Sons, pp. 35, 36, 37, 139 (1976)

2 BELL TELEPHONE LABORATORIES INC., *Physical Design of Electronic Systems; Volume 1, Design Technology*, Prentice-Hall Inc., pp. 362, 363 (1970)

3 ENGELSON, M. and TELEWSKI, F., *Spectrum Analyzer Theory and Applications*, Avtech House Inc., p. 66 (1974)

4 EVERETT, W. W. Jr., *Topics in Intersystem Electromagnetic Compatibility*, Holt, Rinehart and Winston Inc. (1972)

5 GRAY, H. S., *Digital Computer Engineering*, Prentice-Hall Inc., pp. 273, 274, 275, 276 (1963)

6 INTERNATIONAL TELEPHONE AND TELEGRAPH CORP., *Reference Data for Radio Engineers*, 4th Edition, International Telephone and Telegraph Corp. (1956)

7 KAUPP, H. R., *Pulse Crosstalk between Microstrip Transmission Lines*, Symposium Record, Seventh International Electronic Circuit Packaging Symposium, pp. 9, 10 (1966)

8 LANDEE, R. W., DONOVAN, C. D., ALBRECHT, A. P., *Electronic Designer's Handbook*, McGraw-Hill Book Co. Inc., pp. 20.19, 20.20, 20.23 (1957)

9 RAMO, S., WHINNEY, J. R., and DANDUZER, T., *Fields and Waves in Communication Electronics*, John Wiley & Sons Inc., pp. 252, 293 (1965)

10 RYDER, J. D., *Networks Lines and Fields*, Prentice-Hall Inc., Third printing (1958)

11 WHITE, D. R. J., *EMC Handbook*, Vol. 3, Don White Consultants Inc. (1973)

12 WHITE, D. R. J., *EMI Control in the Design of Printed Circuit Boards and Backplanes*, Don White Consultants Inc. (1981)

13 MORRISON, R., *Grounding and Shielding Techniques in Instrumentation*, John Wiley & Sons Inc. (1977)

14 ZAIS, A., 'RF shielding becomes more necessary', *Microwave Systems News* (December 1982)

51

Noise and Communication

K R Sturley, PhD, BSc, FIEE, FIEEE
Telecommunications Consultant

Contents

51.1 Interference and noise in communication systems

Information transmission accuracy can be seriously impaired by interference from other transmission systems and by noise. Interference from other transmission channels can usually be reduced to negligible proportions by proper channel allocation, by operating transmitters in adjacent or overlapping channels geographically far apart, and by the use of directive transmitting and receiving aerials. Noise may be impulsive or random. Impulsive noise may be man-made from electrical machinery or natural from electrical storms; the former is controllable and can be reduced to a low level by special precautions taken at the noise source, but the latter has to be accepted when it occurs. Random (or white) noise arises from the random movement of electrons due to temperature and other effects in current-carrying components in, or associated with, the receiving system.

51.2 Man-made noise

Man-made electrical noise is caused by switching surges, electrical motor and thermostat operation, insulator flash-overs on power lines, etc. It is generally transmitted by the mains power lines and its effect can be reduced by:

(i) Suitable r.f. filtering at the noise source;
(ii) Siting the receiver aerial well away from mains lines and in a position giving maximum signal pick-up;
(iii) Connecting the aerial to the receiver by a shielded lead.

The noise causes a crackle in phones or loudspeaker, or white or black spots on a monochrome television picture screen, and its spectral components decrease with frequency so that its effect is greatest at the lowest received frequencies.

Car ignition is another source of impulsive noise but it gives maximum interference in the v.h.f. and u.h.f. bands; a high degree of suppression is achieved by resistances in distributor and spark plug leads.

51.3 Natural sources of noise

Impulsive noise can also be caused by lightning discharges, and like man-made noise its effect decreases with increase of received frequency. Over the v.h.f. band such noise is only evident when the storm is within a mile or two of the receiving aerial.

Cosmic noise from outer space is quite different in character and generally occurs over relatively narrow bands of the frequency spectrum from about 20 MHz upwards. It is a valuable asset to the radio astronomer and does not at present pose a serious problem for the communications engineer.

51.4 Random noise

This type of noise is caused by random movement of electrons in passive elements such as resistors, conductors and inductors, and in active elements such as electronic valves and transistors.

51.4.1 Thermal noise

Random noise in passive elements is referred to as thermal noise since it is entirely associated with temperature, being directly proportional to absolute temperature. Unlike impulsive noise its energy is distributed evenly through the r.f. spectrum and it must be taken into account when planning any communication system. Thermal noise ultimately limits the maximum amplification that can usefully be employed, and so determines the minimum acceptable value of received signal. It produces a steady hiss in a loudspeaker and a shimmering background to a television picture.

Nyquist has shown that thermal noise in a conductor is equivalent to a r.m.s. voltage Vn in series with the conductor resistance R, where

$$Vn=(4kTR\,\Delta f)^{1/2} \tag{51.1}$$

k = Boltzmann's constant, $1.372\times10^{-23}J/K$
T = absolute temperature of conductor
Δf = pass band (Hz) of the circuits after R

If the frequency response were rectangular the pass band would be the difference between the frequencies defining the sides of the rectangle. In practice the sides are sloping and bandwidth is

$$\Delta f=\frac{1}{E_o^2}\int_0^\infty [E(f)]^2\,df \tag{51.2}$$

where E_o = midband or maximum value of the voltage ordinate and $E(f)$ = the voltage expression for the frequency response.

A sufficient degree of accuracy is normally achieved by taking the standard definition of bandwidth, i.e. the frequency difference between points where the response has fallen by 3 dB.

Figure 51.1 allows the r.m.s. noise voltage for a given resistance and bandwidth to be determined. Thus for

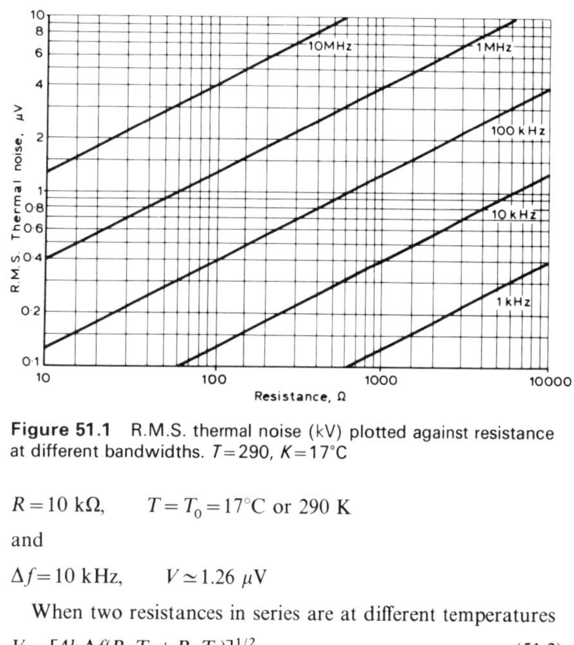

Figure 51.1 R.M.S. thermal noise (kV) plotted against resistance at different bandwidths. $T=290$, $K=17°C$

$R=10\ k\Omega$, $\quad T=T_0=17°C$ or 290 K

and

$\Delta f=10$ kHz, $\quad V\simeq1.26\ \mu V$

When two resistances in series are at different temperatures

$$V_n=[4k\,\Delta f(R_1T_1+R_2T_2)]^{1/2} \tag{51.3}$$

Two resistances in parallel at the same temperature *Figure 51.2(a)* are equivalent to a noise voltage

$$V_n=[4kT\Delta f\,.\,R_1R_2/(R_1+R_2)]^{1/2} \tag{51.4}$$

in series with two resistances in parallel, *Figure 51.2(b)*.

The equivalent current generator concept is shown in *Figure 51.2(c)* where

$$I_n=[4kT\,\Delta f(G_1+G_2)]^{1/2} \tag{51.5}$$

If R is the series resistance of a coil in a tuned circuit of Q factor, Q_o, the noise voltage from the tuned circuit becomes

$$V_{no}=V_nQ_o=Q_o(4kTR\,\Delta f)^{1/2} \tag{51.6}$$

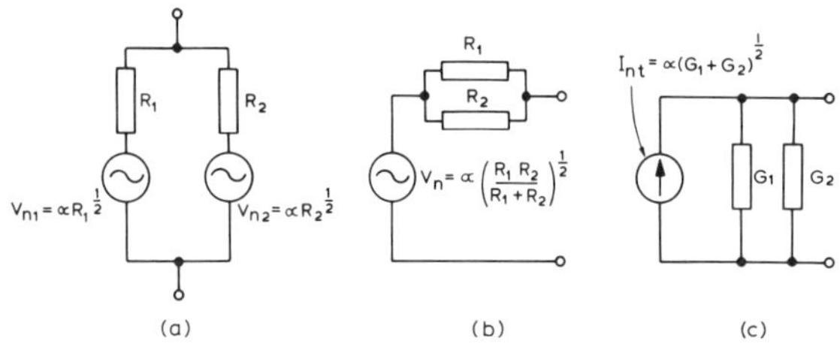

Figure 51.2 (a) Noise voltages of two resistances in parallel; (b) an equivalent circuit and (c) a current noise generator equivalent $\alpha = (4kT'f)^{1/2}$

The signal injected into the circuit is also multiplied by Q_o so that signal-to-noise ratio is unaffected.

51.5 Electronic valve noise

51.5.1 Shot noise

Noise in valves, termed shot noise, is caused by random variations in the flow of electrons from cathode to anode. It may be regarded as the same phenomenon as thermal (conductor) noise with the valve slope resistance, acting in place of the conductor resistance at a temperature between 0.5 to 0.7 of the cathode temperature.

Shot noise r.m.s. current from a diode is given by

$$I_n = (4k\alpha T_k g_d \Delta f)^{1/2} \qquad (51.7)$$

where

T_k = absolute temperature of the cathode
α = temperature correction factor assumed to be about 0.66

$g_d = \dfrac{dI_d}{dV_a}$, slope conductance of the diode.

Experiment[1] has shown noise in a triode is obtained by replacing g_d in equation (51.7) by g_m/β where β has a value between 0.5 and 1 with a typical value of 0.85, thus

$$I_{na} = (4k\alpha T_k g_m \Delta f/\beta)^{1/2} \qquad (51.8)$$

Since $I_a = g_m V_g$, the noise current can be converted to a noise voltage at the grid of the valve of

$$V_{ng} = I_{na}/g_m = (4k\alpha T_k \Delta f/\beta g_m)^{1/2}$$
$$= [4kT_o \Delta f \,.\, \alpha T_k/\beta g_m T_o]^{1/2} \qquad (51.9)$$

where T_o is the normal ambient (room) temperature.

The part $\alpha T_k/\beta g_m T_o$ of expression (51.9) above is equivalent to a resistance, which approximates to

$$R_{ng} = 2.5/g_m \qquad (51.10)$$

and this is the equivalent noise resistance in the grid of the triode at room temperature. The factor 2.5 in R_{ng} may have a range from 2 to 3 in particular cases. The equivalent noise circuit for a triode having a grid leak R_g and fed from a generator of internal resistance R_1 is as in *Figure 51.3*.

51.5.2 Partition noise

A multielectrode valve such as a tetrode produces greater noise than a triode due to the division of electron current between screen and anode; for this reason the additional noise is known as partition noise. The equivalent noise resistance in the grid circuit becomes

$$R_{ng}(tet) = (I_a/I_k)(20I_s/g_m^2 + 2.5g_m) \qquad (51.11)$$

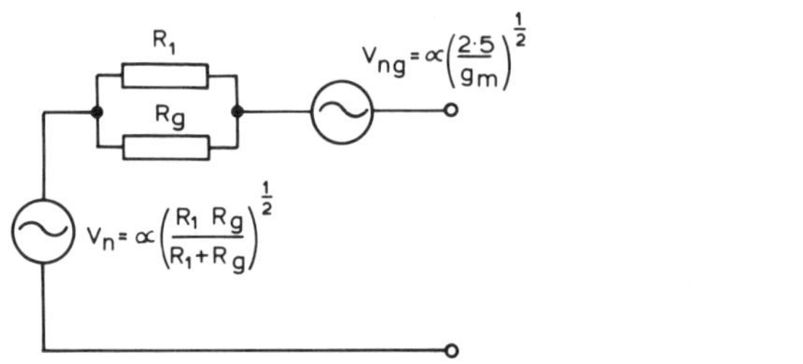

Figure 51.3 Noise voltage input circuit for a valve $\alpha = (4kT'f)^{1/2}$

Where I_a, I_k and I_s are the d.c. anode, cathode and screen currents respectively. I_s should be small and g_m large for low noise in tetrode or multielectrode valves. The factor $20I_s/g_m^2$ is normally between 3 to 6 times $2.5/g_m$ so that a tetrode valve is much noisier than a triode.

At frequencies greater than about 30 MHz, the transit time of the electron from cathode to anode becomes significant and this reduces gain and increases noise. Signal-to-noise ratio therefore deteriorates. Partition noise in multielectrode valves also increases and the neutralised triode, or triodes in cascode give much better signal-to-noise ratios at high frequencies.

At much higher frequencies (above 1 GHz) the velocity modulated electron tube, such as the klystron and travelling wave tube, replace the normal electron valve. In the klystron, shot noise is present but there is also chromatic noise due to random variations in the velocities of the individual electrons.

51.5.3 Flicker noise

At very low frequencies valve noise is greater than would be expected from thermal considerations. Schottky suggested that this is due to random variations in the state of the cathode surface and termed it *flicker*. Flicker noise tends to be inversely proportional to frequency below about 1 kHz so that the equivalent noise resistance at 10 Hz might be 100 times greater than the shot noise at 1 kHz. Ageing of the valve tends to increase flicker noise and this appears to be due to formation of a high resistance barium silicate layer between nickel cathode and oxide coating.

51.6 Transistor noise

Transistor noise exhibits characteristics very similar to those of valves, with noise increasing at both ends of the frequency scale. Resistance noise is also present due to the extrinsic resistance of the material and the major contributor is the base extrinsic resistance r_b'. Its value is given by expression (51.1), T being the absolute temperature of the transistor under working conditions.

Shot and partition noise arise from random fluctuations in the movement of minority and majority carriers, and there are four sources, viz.

(i) Majority carriers injected from emitter to base and thence to collector.
(ii) Majority carriers from emitter which recombine in the base.
(iii) Minority carriers injected from base into emitter.
(iv) Minority carriers injected from base into the collector.

Sources (i) and (ii) are the most important, sources (iii) and (iv) being significant only at low bias currents. Under the latter condition which gives least noise, silicon transistors are superior to germanium because of their much lower values of I_{co}.

A simplified equivalent circuit for the noise currents and voltages in a transistor is that of *Figure 51.4* where

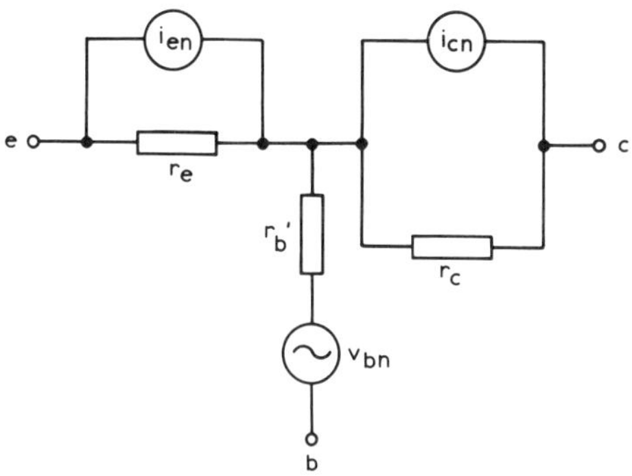

Figure 51.4 Noise circuit equivalent for a transistor

$i_{en} = (2eI_e\,\Delta f)^{1/2}$ the shot noise current in the emitter
$i_{cn} = [2e(I_{co} + I_c(1-\alpha_o))\,\Delta f]^{1/2}$, the shot and partition noise current in the collector
$v_{bn} = (4kT\,.\,r_b'\,\Delta f)^{1/2}$, the thermal noise due to the base extrinsic resistance
e = electronic charge = 1.602×10^{-19} coulomb

Since transistors are power amplifying devices the equivalent noise resistance concept is less useful and noise quality is defined in terms of noise figure.

Flicker noise, which is important at low frequencies (less than about 1 kHz), is believed to be due to carrier generation and recombination at the base–emitter surface. Above 1 kHz noise remains constant until a frequency of about $f_\alpha(1-\alpha_o)^{1/2}$ is reached, where f_α is the frequency at which the collector–emitter current gain has fallen to $0.7\alpha_o$. Above this frequency, which is about $0.15f_\alpha$, partition noise increases rapidly.

51.7 Noise figure

Noise figure (F) is defined as the ratio of the input signal-to-noise available power ratio to the output signal-to-noise available power ratio, where available power is the maximum power which can be developed from a power source of voltage V and internal resistance R_s. This occurs for matched conditions and is $V^2/4R_s$.

$$F = (P_{si}/P_{ni})/(P_{so}/P_{no}) = P_{no}/G_a P_{ni} \tag{51.12}$$

where G_a = available power gain of the amplifier.

Since noise available output power is the sum of $G_a P_{ni}$ and that contributed by the amplifier P_{na}

$$F = 1 + \frac{P_{na}}{G_a P_{ni}} \tag{51.13}$$

The available thermal input power is $V^2/4R_s$ or $kT\,\Delta f$, which is independent of R_s, hence

$$F = 1 + P_{na}/G_a kT\,\Delta f \tag{51.14}$$

and

$$F(\text{dB}) = 10 \log_{10}(1 + P_{na}/G_a kT\,\Delta f) \tag{51.15}$$

The noise figure for an amplifier whose only source of noise is its input resistance R_1 is

$$F = 1 + R_s/R_1 \tag{51.16}$$

because the available output noise is reduced by $R_1/(R_s+R_1)$ but the available signal gain is reduced by $[R_1/(R_s+R_1)]^2$. For matched conditions $F = 2$ or 3 dB and maximum signal-to-noise ratio occurs when $R_1 = \infty$. Signal-to-noise ratio is unchanged if R_1 is noiseless because available noise power is then reduced by the same amount as the available gain.

If the above amplifier has a valve, whose equivalent input noise resistance is R_{ng},

$$F = 1 + \frac{R_s}{R_1} + \frac{R_{ng}}{R_s}\left[1 + \frac{R_s}{R_1}\right]^2 \tag{51.17}$$

Noise figure for a transistor over the range of frequencies for which it is constant is

$$F = 1 + \frac{r_b' + 0.5r_e}{R_s} + \frac{(r_b' + r_e + R_s)^2(1-\alpha_o)}{2r_e R_s \alpha_o} \tag{51.18}$$

At frequencies greater than $f_\alpha(1-\alpha_o)^{1/2}$, the last term is multiplied by $[1 + (f/f_\alpha)^2/(1-\alpha_o)]$. The frequency f_T at which collector–base current gain is unity is generally given by the transistor manufacturer and it may be noted that $f_T \simeq f_\alpha$, the frequency at which collector–emitter current gain is $0.7\alpha_o$.

Expression (51.18) shows that transistor noise figure is dependent on R_s but it is also affected by I_c through r_e and α_o. As a general rule the lower the value of I_c the lower is noise figure and the greater is the optimum value of R_s. This is shown in *Figure 51.5* which is typical of a r.f. silicon transistor. Flicker noise causes

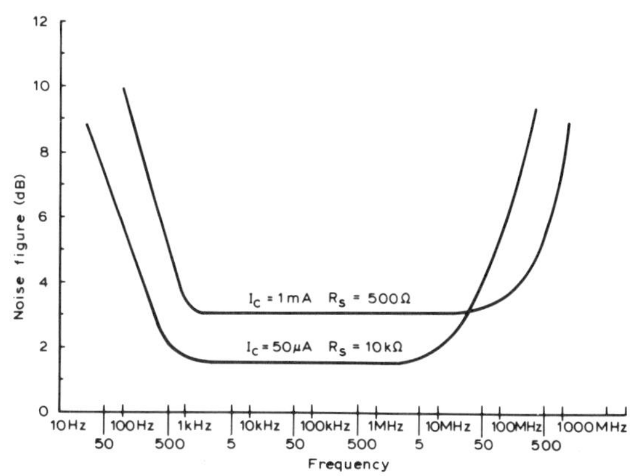

Figure 51.5 Typical noise figure—frequency curves for a r.f. transistor

the increase below 1 kHz, and decrease of gain and increase of partition noise causes increased noise factor at the high frequency end. The high frequency at which F begins to increase is about $0.15f_\alpha$; at low values of collector current f_α falls, being approximately proportional to I_c^{-1}. The type of configuration, common emitter, base or collector has little effect on noise figure.

Transistors do not provide satisfactory noise figures above about 1.5 GHz, but the travelling-wave tube and tunnel diode can achieve noise figures of 3 to 6 dB over the range 1 to 10 GHz.

Sometimes noise temperature is quoted in preference to noise figure and the relationship is

$$F = (1 + T/T_o) \tag{51.19}$$

T is the temperature to which the noise source resistance would have to be raised to produce the same available noise output power as the amplifier. Thus if $T = T_o = 290$ K, $F = 2$ or 3 dB.

The overall noise figure of cascaded amplifiers can easily be calculated and is

$$F_t = F_1 + \frac{F_2 - 1}{G_1} + \frac{F_3 - 1}{G_1 G_2} + \ldots \frac{F_n - 1}{G_1 \ldots G_{n-1}} \tag{51.20}$$

where $F_1, F_2, \ldots, F_n$ and $G_1, G_2, \ldots, G_n$ are respectively the noise figures and available gains of the separate stages from input to output. From equation (51.20) it can be seen that the first stage of an amplifier system largely determines the overall signal-to-noise ratio, and that when a choice has to be made between two first-stage amplifiers having the same noise figure, the amplifier having the highest gain should be selected because increase of G_1 reduces the noise effect of subsequent stages.

51.8 Measurement of noise

Noise measurement requires a calibrated noise generator to provide a controllable noise input to an amplifier or receiver, and a r.m.s. meter to measure the noise output of the amplifier or receiver. The noise generator generally consists of a temperature-limited (tungsten filament) diode, terminated by a resistance R as shown in *Figure 51.6*. The diode has sufficient anode voltage to ensure that it operates under saturation conditions and anode saturation current is varied by control of the diode filament current. A milliammeter reads the anode current I_d and the shot noise current component of this is given by $(2I_d e\,\Delta f)^{1/2}$ where e is electronic charge, 1.602×10^{-19} coulomb. The shot noise has the same flat spectrum as the thermal noise in R, and the meter is calibrated in dB with reference to noise power in R and so provides a direct reading of noise factor. R is generally selected to be 75 Ω, the normal input impedance of a receiver.

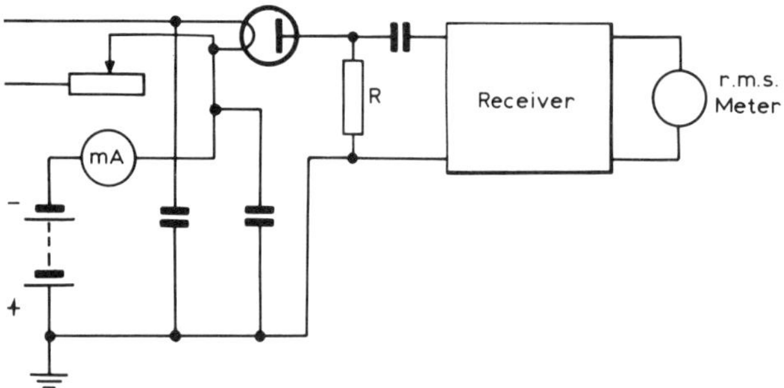

Figure 51.6 Noise figure measurements

When measuring, the diode filament current is first switched off, and the reading of the r.m.s. meter in the receiver output noted. The diode filament is switched on and adjusted to increase the r.m.s. output reading 1.414 times (double noise power). The dB reading on the diode anode current meter is the noise figure, since

Noise output power diode off $= GP_{nR} + P_{na}$
Noise output power diode on $= G(P_{nR} + P_{nd}) + P_{na} = 2GP_{nd}$
Noise figure $= 10\log_{10}[(GP_{nR} + P_{na})/GP_{nR}] = 10\log_{10} P_{nd}/P_{nR}$

The diode is satisfactory up to about 600 MHz but above this value transit time of electrons begins to cause error. For measurements above 1 GHz a gas discharge tube has to be used as a noise source.

51.9 Methods of improving signal-to-noise ratio

There are five methods of improving signal-to-noise ratio, viz.,

(i) Increase the transmitted power of the signal.
(ii) Redistribute the transmitted power.
(iii) Modify the information content before transmission and return it to normal at the receiving point.
(iv) Reduce the effectiveness of the noise interference with signal.
(v) Reduce the noise power.

51.9.1 Increase of transmitted power

An overall increase in transmitted power is costly and could lead to greater interference for users of adjacent channels.

51.9.2 Redistribution of transmitted power

With amplitude modulation it is possible to redistribute the power among the transmitted components so as to increase the effective signal power. As described in Chapter 40 suppression of the carrier in a double sideband amplitude modulation signal and a commensurate increase in sideband power increases the effective signal power, and therefore signal-to-noise ratio by 4.75 dB (3 times) for the same average power or by 12 dB for the same peak envelope power. Single sideband operation by removal of one sideband reduces signal-to-noise ratio by 3 dB because signal power is reduced to $\frac{1}{4}$ (6 dB) and the non-correlated random noise power is only halved (3 dB). If all the power is transferred to one sideband, single sideband operation increases signal-to-noise ratio by 3 dB.

51.9.3 Modification of information content before transmission and restoration at receiver

51.9.3.1 The compander

A serious problem with speech transmission is that signal-to-noise ratio varies with the amplitude of the speech, and during gaps between syllables and variations in level when speaking, the noise may become obtrusive. This can be overcome by using compression of the level variations before transmission, and expansion after detection at the receiver, a process known as companding. The compressor contains a variable loss circuit which reduces amplification as speech amplitude increases and the expander performs the reverse operation.

A typical block schematic for a compander circuit is shown in *Figure 51.7*. The input speech signal is passed to an amplifier

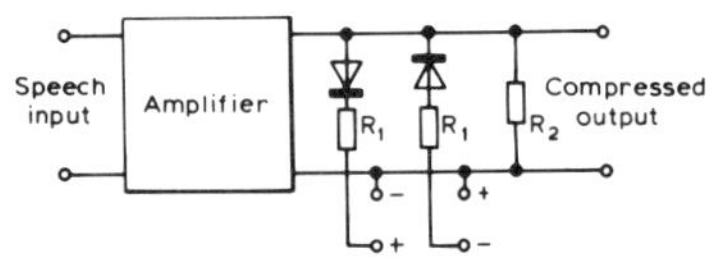

Figure 51.7 A compressor circuit

across whose output are shunted two reverse biased diodes; one becoming conductive and reducing the amplification for positive going signals and the other doing the same for negative-going signals. The input–output characteristic is S shaped as shown in *Figure 51.8*; the diodes should be selected for near identical

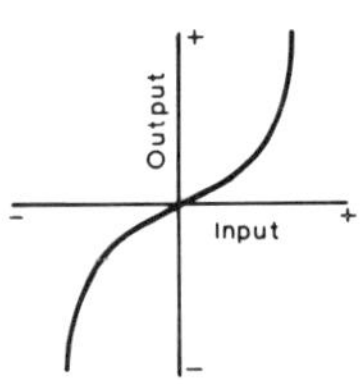

Figure 51.8 Compressor input–output characteristic

shunting characteristics. Series resistances R_1 are included to control the turn-over, and shunt resistance R_2 determines the maximum slope near zero.

A similar circuit is used in the expander after detection but as shown, *Figure 51.9*, the diodes form a series arm of a potential divider and the expanded output appears across R_3. The expander characteristic, *Figure 51.10*, has low amplification in the gaps between speech, and amplification increases with increase in speech amplitude.

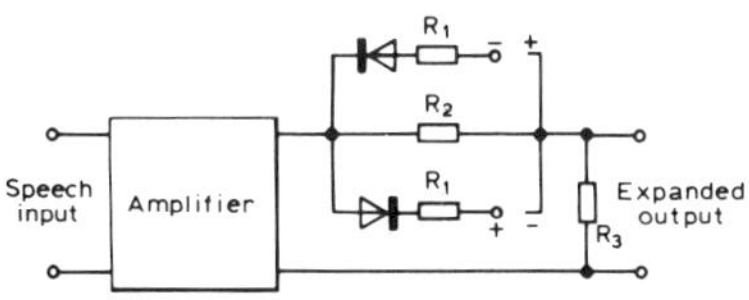

Figure 51.9 An expander circuit

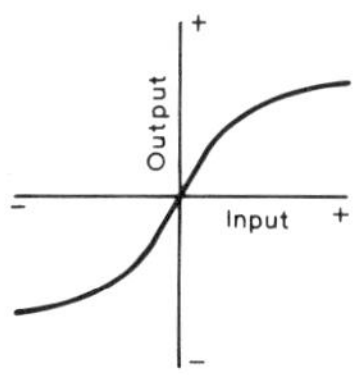

Figure 51.10 Expander input–output characteristics

The diodes have a logarithmic compression characteristic, and with large compression the dB input against dB output tends to a line of low slope, e.g., an input variation of 20 dB being compressed to an output variation of 5 dB. If greater compression is required two compressors are used in tandem.

The collector–emitter resistance of a transistor may be used in place of the diode resistance as the variable gain device. The collector–emitter resistance is varied by base-emitter bias current, which is derived by rectification of the speech signal from a separate auxiliary amplifier. A time delay is inserted in the main controlled channel so that high-amplitude speech transients can be anticipated.

51.9.3.2 Lincompex

The compander system described above proves quite satisfactory provided the propagation loss is constant as it is with a line or coaxial cable. It is quite unsuitable for a shortwave point-to-point communication system via the ionosphere. A method known as Lincompex[2] (linked compression expansion) has been successfully developed by the British Post Office. *Figure 51.11* is a block diagram of the transmit–receive paths. The simple form of diode compressor and expander cannot be used and must be replaced by the transistor type, controlled by a current derived from rectification of the speech signal. The current controls the compression directly at the transmitting end and this information must be sent to the receiver by a channel unaffected by any propagational variations. This is done by confining it in a narrow channel (approximately 180 Hz wide) and using it to frequency-modulate a sub-carrier at 2.9 kHz. A limiter at the receiver removes all amplitude variations introduced by the r.f. propagation path, and a frequency discriminator extracts the original control information.

The transmit chain has two paths for the speech signals, one (A) carries the compressed speech signal, which is limited to the range 250 to 2700 Hz by the low-pass output filter. A time delay of 4 ms is included before the two compressors in tandem, each of which has a 2 to 1 compression ratio, and the delay allows the compressors to anticipate high amplitude transients. The 2:1 compression ratio introduces a loss of $x/2$ dB for every x dB change in input, and the two in tandem introduce a loss of $2(x/2)=x$ dB for every x dB change of input. The result is an almost constant speech output level for a 60 dB variation of speech input. Another time delay (10 ms) is inserted between the compressors and output filter in order to compensate for the control signal delay due to its narrow bandwidth path.

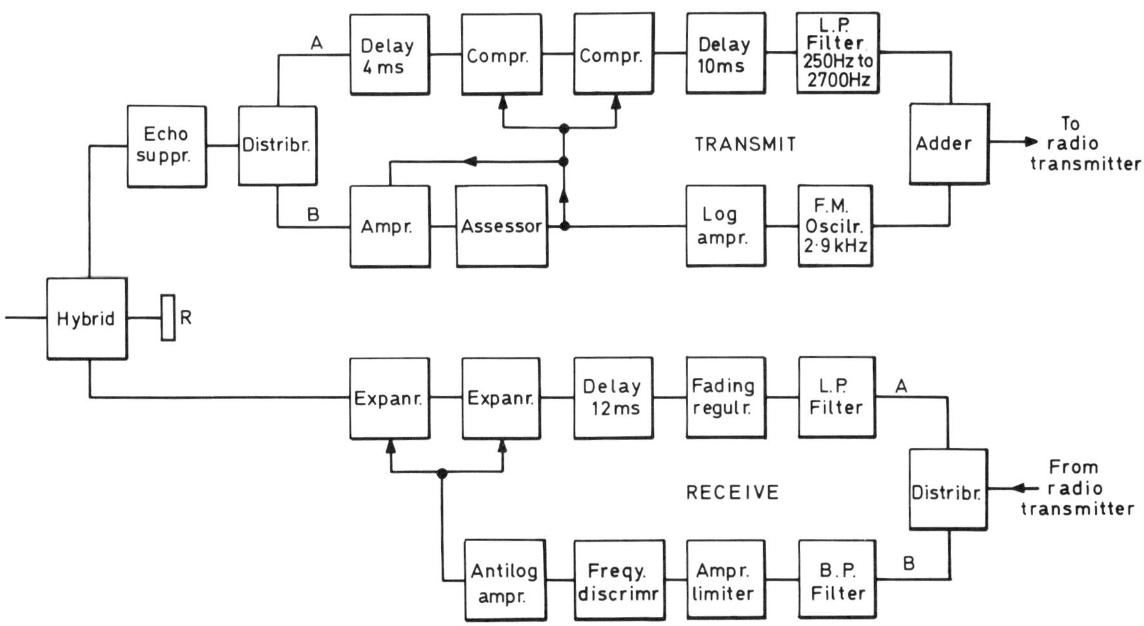

Figure 51.11 Block schematic of Lincompex Compander system for radio transmissions

The other transmit path (B) contains an amplitude-assessor circuit having a rectified d.c. output current proportional to the speech level. This d.c. current controls the compressors, and after passing through a logarithmic amplifier is used to frequency modulate the sub-carrier to produce the control signal having a frequency deviation of 2 Hz/dB speech level change. The time constant of the d.c. control voltage is 19 ms permitting compressor loss to be varied at almost syllabic rate, and the bandwidth of the frequency-modulated sub-carrier to be kept within ±90 Hz. The control signal is added to the compressed speech and the combined signal modulates the transmitter.

The receive chain also has two paths; path (A) filters the compressed speech from the control signal and passes the speech to the expanders via a fading regulator, which removes any speech fading not eliminated by the receiver a.g.c., and a time delay, which compensates for the increased delay due to the narrow-band control path (B). The latter has a band pass filter to remove the compressed speech from the control signal and an amplitude limiter to remove propagational amplitude variations. The control signal passes to a frequency discriminator and thence to an antilog amplifier, the output from which controls the gain of the expansion circuits. The time constant of the expansion control is between 18 ms and 20 ms.

51.9.3.3 Pre-emphasis and de-emphasis

Audio energy in speech and music broadcasting tends to be greatest at the low frequencies. A more level distribution of energy is achieved if the higher audio frequencies are given greater amplification than the lower before transmission. The receiver circuits must be given a reverse amplification-frequency response to restore the original energy distribution, and this can lead to an improved signal-to-noise ratio since the received noise content is reduced at the same time as the high audio frequencies are reduced. The degree of improvement is not amenable to measurement and a subjective assessment has to be made. The increased high-frequency amplification before transmission is known as pre-emphasis followed by de-emphasis in the receiver audio circuits. F.M. broadcasting (maximum frequency deviation ±75 kHz) shows a greater subjective improvement than a.m., and it is estimated to be 4.5 dB when the pre- and de-emphasis circuits have time-constants of 75 μs. A simple *RC* potential divider can be used for de-emphasis in the receiver audio circuits, and 75 μs time constant gives losses of 3 and 14 dB at 2.1 and 10 kHz respectively compared with 0 dB at low frequencies.

51.9.4 Reduction of noise effectiveness

Noise, like information, has amplitude and time characteristics, and it is noise amplitude that causes the interference with a.m. signals. If the information is made to control the time characteristics of the carrier so that carrier amplitude is transmitted at a constant value, an amplitude limiter in the receiver can remove all amplitude variations due to noise without impairing the information. The noise has some effect on the receiver carrier time variations, which are phase-modulated by noise, but the phase change is very much less than the amplitude change so that signal-to-noise ratio is increased.

51.9.4.1 Frequency modulation

If the information amplitude is used to modulate the carrier frequency, and an amplitude limiter is employed at the receiver, the detected message-to-noise ratio is greatly improved. F.M. produces many pairs of sidebands per modulating frequency especially at low frequencies, and this 'bass boost' is corrected at the receiver detector to cause a 'bass cut' of the low frequency noise components. This triangulation of noise leads to 4.75 dB signal-to-noise betterment. Phase modulation does not give this improvement because the pairs of sidebands are independent of modulating frequency. The standard deviation of 75 kHz raises signal-to-noise ratio by another 14 dB, and pre-emphasis and de-emphasis by 4.5 dB, bringing the total improvement to 23.25 dB over a.m.

The increased signal-to-noise performance of the f.m. receiver is dependent on having sufficient input signal to operate the amplitude limiter satisfactorily. Below a given input signal-to-noise ratio output information-to-noise ratio is worse than for a.m. The threshold value increases with increase of frequency deviation because the increased receiver bandwidth brings in more noise as indicated in *Figure 51.12.*

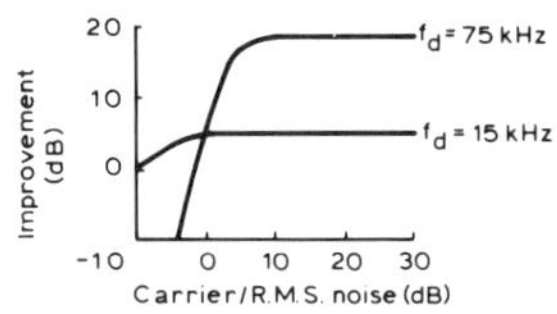

Figure 51.12 Threshold noise effect with f.m. compared with a.m.

51.9.4.2 Pulse modulation

Pulse modulated systems using change of pulse position (p.p.m.) and change of duration (p.d.m.) can also increase signal-to-noise ratio but pulse amplitude modulation (p.a.m.) is no better than normal a.m. because an amplitude limiter cannot be used.

51.9.4.3 Impulse noise and bandwidth

When an impulse noise occurs at the input of a narrow bandwidth receiver the result is a damped oscillation at the mid-frequency of the pass band as shown in *Figure 51.13(a)*. When a wide bandwidth is employed the result is a large initial amplitude with a very rapid decay, *Figure 51.13(b)*. An amplitude limiter is much more effective in suppressing the large amplitude near-single pulse than the long train of lower amplitude oscillations. Increasing reception bandwidth can therefore appreciably reduce interference due to impulsive noise provided that an amplitude limiter can be used.

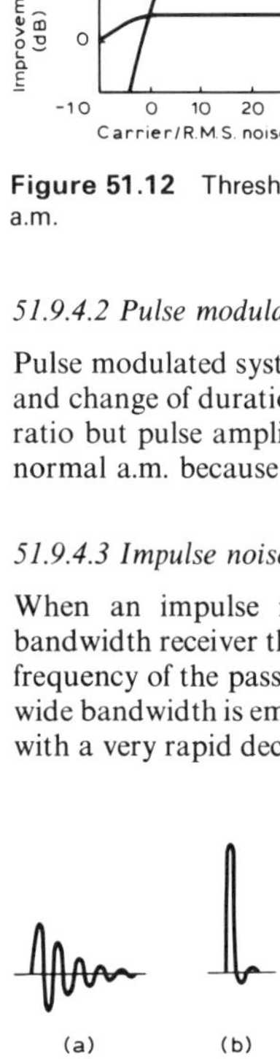

Figure 51.13 Output wave shape due to an impulse in a (a) narrow band amplifier; (b) a wideband amplifier

51.9.4.4 Pulse code modulation

A very considerable improvement in information-to-noise ratio can be achieved by employing pulse code modulation[3] (p.c.m.). P.C.M. converts the information amplitude into a digital form by sampling and employing constant amplitude pulses, whose presence or absence in a given time order represents the amplitude level as a binary number. Over long cable or microwave links it is possible to amplify the digital pulses when signal-to-noise ratio is very low, to regenerate and pass on a freshly constituted signal almost free of noise to the next link. With analogue or direct non-coded modulation such as a.m. and f.m., noise tends to be cumulative from link to link. The high signal-to-noise ratio of p.c.m. is obtained at the expense of much increased bandwidth, and Shannon has shown that with an ideal system of coding giving zero detection error there is a relationship between information capacity C (binary digits or bits/sec), bandwidth W (Hz) and average signal-to-noise thermal noise power ratio (S/N) as follows:

$$C = W \log_2 (1 + S/N) \qquad (51.21)$$

Two channels having the same C will transmit information equally well though W, and S/N may be different. Thus for a channel capacity of 10^6 bits/s, $W = 0.167$ MHz and $S/N = 63 \equiv 18$ dB, or $W = 0.334$ MHz and $S/N = 7 \equiv 8.5$ dB. Doubling of bandwidth very nearly permits the S/N dB value to be halved, and this is normally a much better exchange rate than for f.m. analogue modulation, for which doubling bandwidth improves S/N power ratio 4 times or by 6 dB.

In any practical system the probability of error is finite, and a probability of 10^{-6} (1 error in 10^6 bits) causes negligible impairment of information. Assuming that the detector registers a pulse when the incoming amplitude exceeds one half the normal pulse amplitude, an error will occur when the noise amplitude exceeds this value. The probability of an error occurring due to this is

$$Pe = \frac{1}{(2\pi)^{\frac{1}{2}}} \frac{2Vn}{V_p} \exp(-V_p^2/8Vn^2) \qquad (51.22)$$

where V_p = peak voltage of the pulse
and V_n = r.m.s. voltage of the noise.

The curve is plotted in *Figure 51.14*.

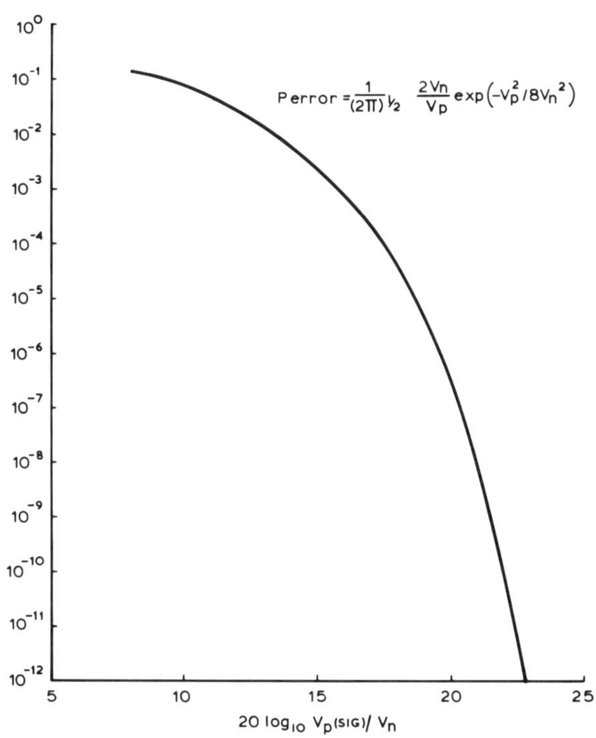

Figure 51.14 Probability of error at different V_p(sig)/r.m.s. noise ratios

An error probability of 10^{-6} requires a V_p/V_n of approximately 20 dB, or since $V_{av} = V_p/2$ a V_{av}/V_n of 17 dB. In a binary system 2 pulses can be transmitted per cycle of bandwidth so that by Shannon's ideal system

$$\frac{C}{W} = 2 = \log_2 (1 + S/N) \quad \text{and} \quad S/N = 3 \simeq 5 \text{ dB} \qquad (51.23)$$

Hence the practical system requires 12 dB greater S/N ratio than the ideal, but the output message-to-noise ratio is infinite, i.e. noise introduced in the transmission path and the receiver is completely removed. There will, however, be a form of noise present with the output message due to the necessary sampling process at the transmit end. Conversion of amplitude level to a digital number must be carried out at a constant level, and the reconstructed decoded signal at the receiver is not a smooth wave but a series of steps. These quantum level steps superimpose on the original signal a disturbance having a uniform frequency spectrum similar to thermal noise. It is this quantising noise which determines the output message-to-noise ratio and it is made small by decreasing the quantum level steps. The maximum error is half the quantum step, l, and the r.m.s. error introduced is $l/2(3)^{1/2}$. The number of levels present in the p.a.m. wave after sampling are 2^n where n is the number of binary digits. The message peak-to-peak amplitude is $2^n l$, so that the

$$\text{Message (pk-to-pk)/r.m.s. noise} = \frac{2^n l}{l/2(3)^{1/2}} = 2(3)^{1/2} 2^n \qquad (51.24)$$

$$\begin{aligned} M/N \text{ (dB)} &= 20 \log_{10} 2(3)^{1/2} 2^n \\ &= 20n \log_{10} 2 + 20 \log_{10} 2(3)^{1/2} \\ &= (6n + 10.8) \text{ dB} \end{aligned} \qquad (51.25)$$

Increase of digits (n) means an increased message-to-noise ratio but also increased bandwidth and therefore increased transmission path and receiver noise; care must be exercised to ensure that quantising noise remains the limiting factor.

Expression (51.25) represents the message-to-noise ratio for maximum information amplitude, and smaller amplitudes will give an inferior noise result. A companding system should therefore be provided before sampling of the information takes place.

51.9.5 Reduction of noise

Since thermal noise power is proportional to bandwidth, the latter should be restricted to that necessary for the objective in view. Thus the bandwidth of an a.m. receiver should not be greater than twice the maximum modulating frequency for d.s.b. signals, or half this value for s.s.b. operation.

In a f.m. system information power content is proportional to (bandwidth)^2 so that increased bandwidth improves signal-to-noise ratio even though r.m.s. noise is increased. When, however, carrier and noise voltages approach in value, signal-to-noise ratio is worse with the wider band f.m. transmission (threshold effect).

Noise is reduced by appropriate coupling between signal source and receiver input and by adjusting the operating conditions of the first stage transistor for minimum noise figure.

Noise is also reduced by refrigerating the input stage of a receiver with liquid helium and this method is used for satellite communication in earth station receivers using masers. The maser amplifies by virtue of a negative resistance characteristic and its noise contribution is equivalent to the thermal noise generated in a resistance of equal value. The noise temperature of the maser itself may be as low as 2 K to 10 K and that of the other parts of the input equipment 15 K to 30 K.

Parametric amplification, by which gain is achieved by periodic variation of a tuning parameter (usually capacitance), can provide the relatively low noise figures 1.5 to 6 dB over the range 5 to 25 GHz. Energy at the 'pump' frequency (f_p) operating the variable reactance, usually a varactor diode, is transferred to the signal frequency (f_s) in the parametric amplifier or to an idler frequency ($f_p \pm f_s$) in the parametric converter. It is the resistance component of the varactor diode that mainly determines the noise figure of the system. Refrigeration is also of value with parametric amplification.

References

1 NORTH, D. O., 'Fluctuations in Space-Charge-Limited Currents at Moderately High Frequencies', *RCA. Rev.*, **4**, 441 (1940), **5**, 106, 244 (1940)

2 WATT-CARTER, D. E. and WHEELER, L. K., 'The Lincompex System for the Protection of HF Radio Telephone Circuits', *P.O. Elect. Engrs. J.*, **59**, 163 (1966)

3 BELL SYSTEM LABORATORIES, *Transmission Systems for Communications* (1964)

52 Broadcasting System Characteristics

K R Sturley, PhD, BSc, FIEE, FIEEE
Telecommunications Consultant

Contents

52.1 Introduction

Broadcasting is concerned with the generation, control, transmission, propagation and reception of sound and television signals. Sound broadcasting with its relatively small bandwidth presents fewer technical problems than television. Adequate reproduction of speech and music can be achived within a frequency range from about 100 Hz to 8 kHz, but high fidelity monophonic and stereophonic programmes require a bandwidth from about 30 Hz to 15 kHz. Television broadcasting generates a signal to control the light output of the receiver picture tube and this covers a frequency range of the order of d.c. to 5.5 MHz for 625 line interlaced scanning. This means that the carrier frequency must be much greater than 5.5 MHz.

52.2 Carrier frequency bands

The carrier frequency bands allocated internationally to sound and television broadcasting are shown in *Table 52.1*.

Radio propagation may be accomplished by means of a ground wave, sky wave or space wave.

Table 52.1 Carrier frequency band allocation

Frequency	*Wavelength*	*Frequency range*	*Purpose*
Low	long wave	150–285 kHz	Sound
Medium	medium wave	525–1605 kHz	Sound
High	short wave	Bands approx. 259 kHz wide located near 4, 6, 7, 9, 11, 15, 17, 21, 26 MHz	Sound
Very high	Band I	47–68 MHz	Television
	Band II	87.5–100 MHz	F.M. Sound
	Band III	174–216 MHz	Television
Ultra-high	Bands IV and V	470–960 MHz	Television

52.3 Low frequency propagation

The low frequency range is propagated by the ground wave which follows the contour of the earth; signal strength tends to be inversely proportional to distance, the losses in the ground itself being small, and reception is possible at considerable distances though marred by noise interference. Any sky waves are absorbed in the ionospheric layers.

52.4 Medium frequency propagation

The medium frequencies are propagated by ground wave during the day, any sky wave being absorbed by the ionospheric *D*-layer. The ground wave generates currents in the ground, and the energy loss is greatest over poor conductivity ground such as granite rocks, and is least over sea water, where propagation approaches the inverse square law. The greater the frequency the greater is the loss at a given distance. Curves of ground wave attenuation with distance for sea water and ground of good, moderate and poor conductivity, for a 1 kW transmitter and a zero-gain aerial are published by the CCIR.[1] These permit the field strength (referred to 1 μV/m as zero level) of low and medium frequencies to be calculated for distances from 1 to 1000 km and are reproduced in *Figures 52.1* to *52.4*. When using these curves the following points should be especially noted.

(1) They refer to a smooth homogeneous earth.
(2) No account is taken of tropospheric effects at these frequencies.
(3) The transmitter and receiver are both assumed to be on the ground. Height-gain effects can be of considerable importance in connection with navigational aids for high-flying aircraft, but these have not been included.
(4) The curves refer to the following conditions:
   (*a*) They are calculated for the vertical component of electric field from the rigorous analysis of van der Pol and Bremmer.
   (*b*) The transmitter is an ideal Hertzian vertical electric dipole to which a vertical antenna shorter than one quarter wavelength is nearly equivalent.
   (*c*) The dipole moment is chosen so that the dipole would radiate 1 kW if the Earth were a perfectly conducting infinite plane, under which conditions the radiation field at a distance of 1 km would be 3×10^5 μV/m.
   (*d*) The curves are drawn for distances measured around the curved surface of the Earth.
   (*e*) The inverse-distance curve *A* shown in the figures, to which the curves are asymptotic at short distances, passes through the field value of 3×10^5 μV/m at a distance of 1 km.
(5) The curves should, in general, be used to determine field strength only when it is known that ionospheric reflections at the frequency under consideration will be negligible in amplitude—for example, propagation in daylight between 150 kHz and 2 MHz and for distances of less than about 2000 km. However, under conditions where the sky wave is comparable with, or even greater than, the ground wave, the curves are still applicable when the effect of the ground wave can be separated from that of the sky wave, by the use of pulse transmissions, as in some forms of direction-finding systems and navigational aids.

As an example of the use of the curves, consider a 100 kW transmitter feeding into a 0.4 λ medium wave aerial at 1 MHz. The field strength at a point 50 km away over ground of medium conductivity from a 1 kW transmitter feeding a very short aerial is from *Figure 52.3*, 53 dB (ref. 1 μV/m). The 0.4 λ aerial has a gain of about 1 dB and the transmitter is +20 dB with reference to 1 kW, so that the field strength is 74 dB or 5.5 mV/m. When the signal path conductivity varies, field strength can be calculated by changing from one curve to another. Suppose the receiving point for the above were moved to 100 km with the additional 50 km over ground of poor conductivity. *Figure 52.4* shows that the loss from 50 km to 100 km at 1 MHz is $(40-26)=14$ dB, so the field strength at 100 km would appear to be $74-14=60$ dB (1 mV/m). It has been found in practice that the actual field strength is always less than this and that a more correct value is obtained by determining the field strength with the transmitter and receiver positions interchanged, and then taking the average of the direct and reversed dB values. Thus for the transmitter at the receiving site *Figure 52.4* shows a field strength at 50 km of 40 dB for 1 kW, i.e. a field strength for the 100 kW transmitter and 1 dB gain aerial of 61 dB. The attenuation from 50 to 100 km from *Figure 52.3* is 53 dB to 38 dB, or 15 dB. Hence by this reciprocal method field strength would be 46 dB. Average field strength is $(60+46)/2=53$ dB, and this is taken as the correct estimated field strength.

Sky wave propagation occurs from the ionospheric *E* layer after nightfall due to the disappearance of the absorbing *D* layer. This can be a serious disadvantage because it leads to greatly increased service area often well beyond national boundaries, causing interference in the service areas of distant transmitting stations using the same or adjacent frequencies. Sky wave propagation is subject to considerable fading due to changes in

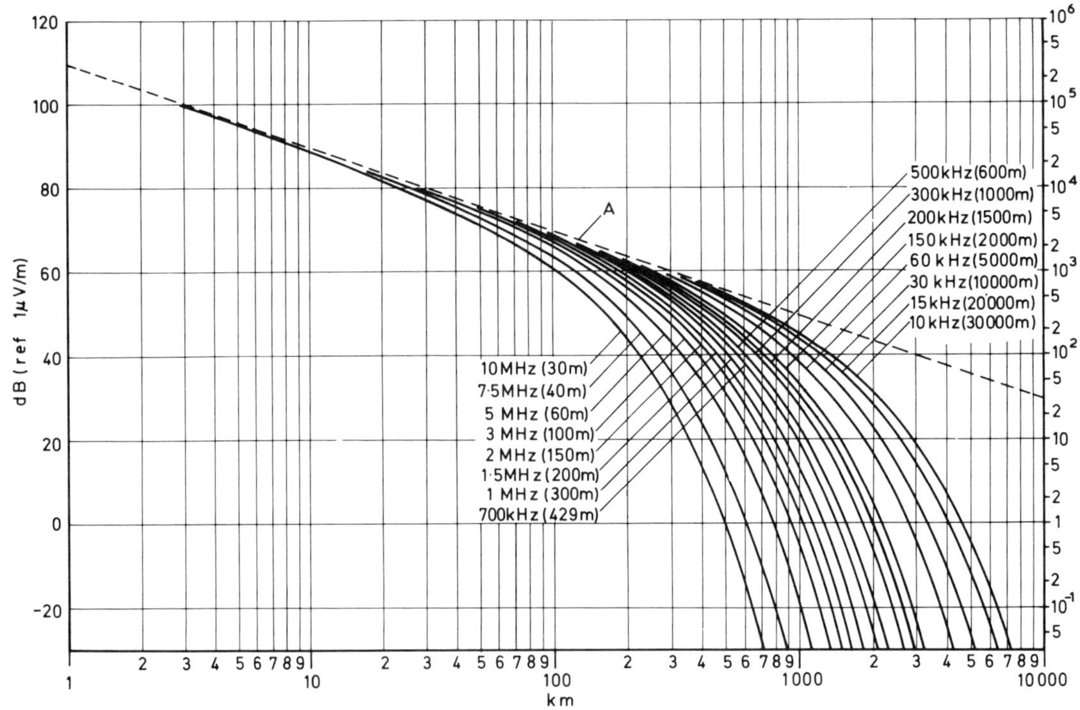

Figure 52.1 Ground-wave propagation curves over sea, σ=4 mho/m, ε=80 (Courtesy C.C.I.R.)

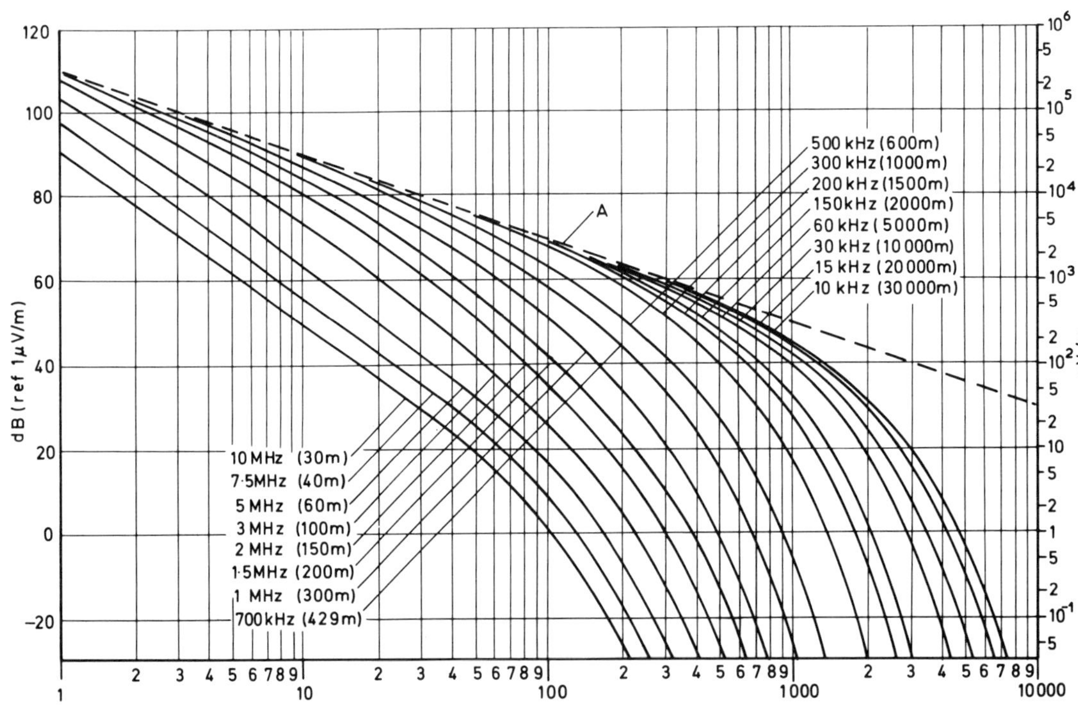

Figure 52.2 Ground-wave propagation curves over land of good conductivity, $\sigma=10^{-2}$ mho/m, ε=4 (Courtesy C.C.I.R.)

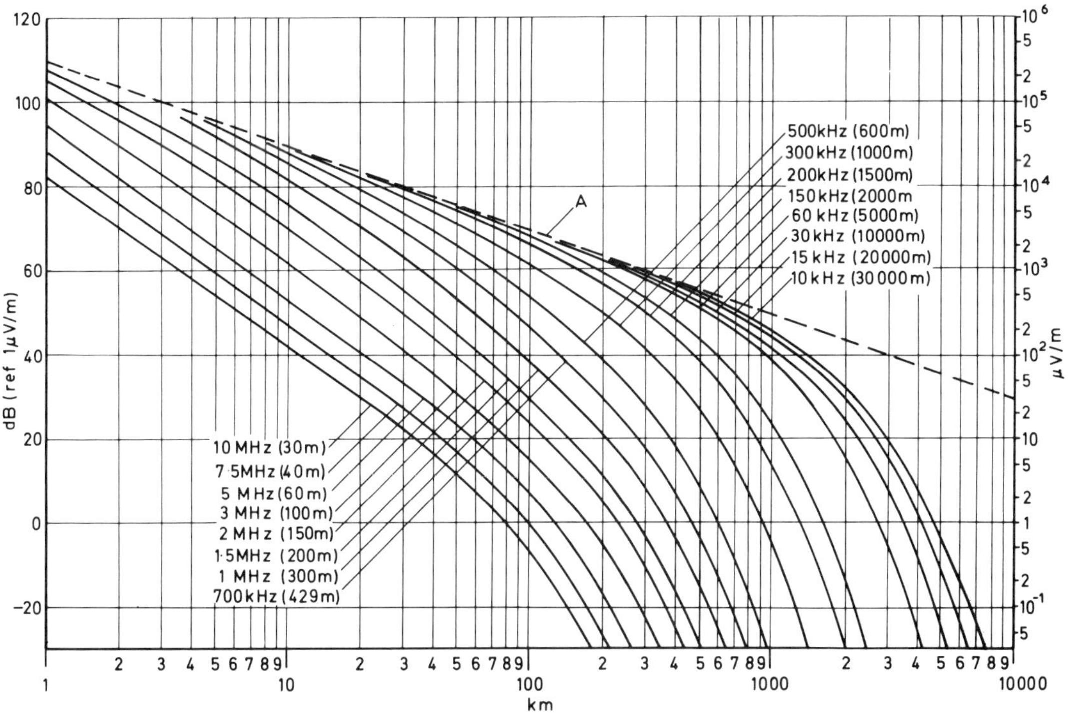

Figure 52.3 Ground-wave propagation curves over land of moderate conductivity, $\sigma=3\times10^{-3}$ mho/m, $\varepsilon=4$ (Courtesy C.C.I.R.)

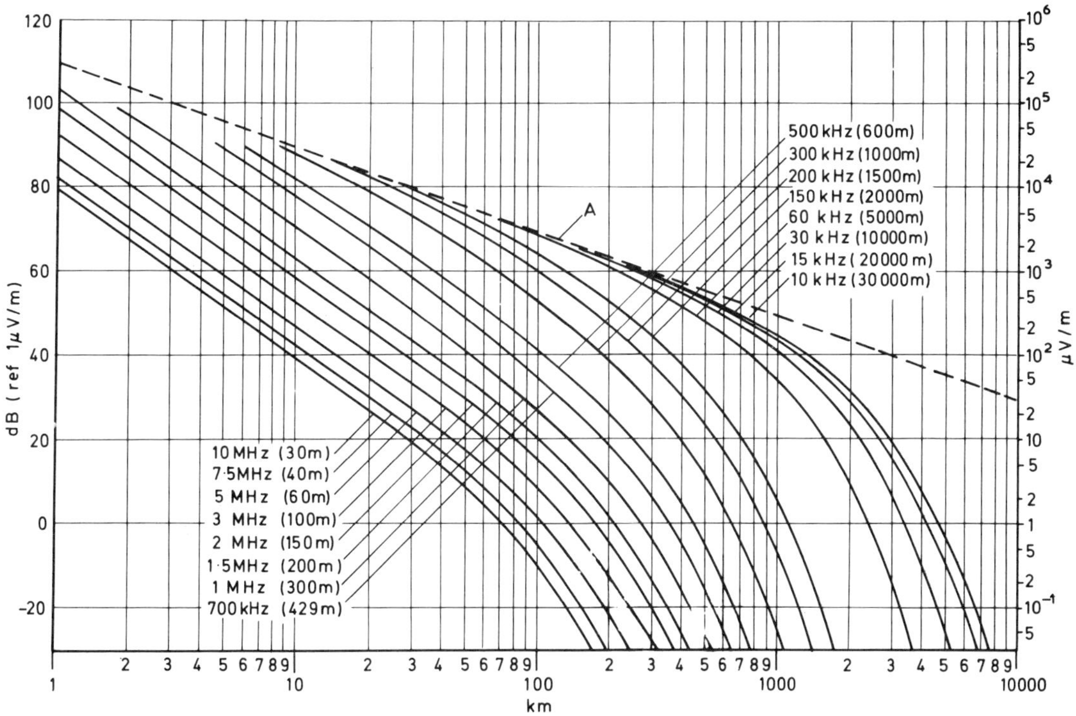

Figure 52.4 Ground-wave propagation curves over land of poor conductivity, $\sigma=10^{-3}$ mho/m, $\varepsilon=4$ (Courtesy C.C.I.R.)

the *E* layer, and at a distance of 300 km to 400 km it may produce a field strength comparable to that of the ground wave. Mutual interference between the two is then greatest and severe fading can occur. The height of a vertical medium-wave aerial largely determines the amount of sky wave reflected by the *E* layer. The greater the height (up to about 0.6 λ) the less the sky-wave reflection.

52.5 High frequency propagation

As the ground wave from a high frequency transmitter is rapidly attenuated by ground absorption which increases with increase of frequency, h.f. propagation is by ionospheric reflection from the *E* and *F* layers. The range of frequencies reflected by each layer is determined by the angle of projection from the ground—(the shallower this is the greater the reflection frequency)—the time of day, season and sun-spot activity. Above a given frequency, which is very variable, there is no reflection because the wave penetrates beyond the layer. Highest reflection frequency from *E* and *F* layers occurs about midday at the centre of the propagation path in the summer season and at sun-spot maximum. Greatest useable reflection frequency is about 30 MHz from the *F* layer during maximum sun-spot years falling to about 20 MHz during sun-spot minimum years.

Field strength median values are amenable to calculation.[2]

52.6 Very high frequency propagation

Very high frequency propagation[1] is achieved by the space wave; near to the ground energy is quickly lost but the effect disappears with increase of height above ground. Thus at heights of 2 λ or greater, the space wave is little affected by ground losses. Some bending of the v.h.f. wave occurs in the troposphere, but satisfactory propagation is normally limited to line-of-sight. An intervening hill produces a radio shadow on the side furthest from the transmitter, with much reduced field strength in the shadow. The v.h.f. signal has the advantage of much lower impulsive noise interference and it is normally unaffected by the vagaries of the ionosphere. However, fixed or moving objects comparable or greater in size than the v.h.f. wavelength can act as reflectors to cause interfering wave patterns. A fixed object may increase or decrease the field strength at a given receiving point, depending on the phase relationship between the direct and reflected wave. A moving reflecting object, such as an airliner, causes the phase relationship of the reflected wave to vary, and considerable fading and distortion of the received signal occurs.

With a.m. sound signals the change in field strength due to interference from the reflected wave of a fixed object such as a tall building or water tower would not be serious, but when f.m. is used the phase relationship of the reflected varying carrier frequency is constantly changing and this causes serious distortion (known as multi-path) of the sound content. Multipath a.m. television signals produce multiple images (ghosting) on the receiver picture tube due to time delay differences between the received signals. The smaller size of v.h.f. receiving aerials (due to the shorter wavelength) permits the design of highly directional aerials, which can be angled to reduce the undesirable reflections from fixed objects.

The energy from a v.h.f. transmitting aerial is projected at other angles as well as horizontal; some is projected upwards to penetrate the ionosphere and be lost, but some is projected downwards to the ground. Part of the energy is absorbed in the ground at the point of incidence but a good deal may be reflected to cause interference at the receiving aerial in the same manner as reflections from a fixed object. The extent of the interference depends on the amplitude and phase of the reflection coefficient and this is determined by the conductivity and permittivity of the ground at the point of reflection, the angle of incidence to the surface and the polarisation of the wave. Typical examples of the variation of the reflection coefficient of relative magnitude (ρ, the ratio of the reflected to incident wave amplitude) and phase (ϕ) with respect to angle to the surface (ψ) are given in *Figure 52.5*.

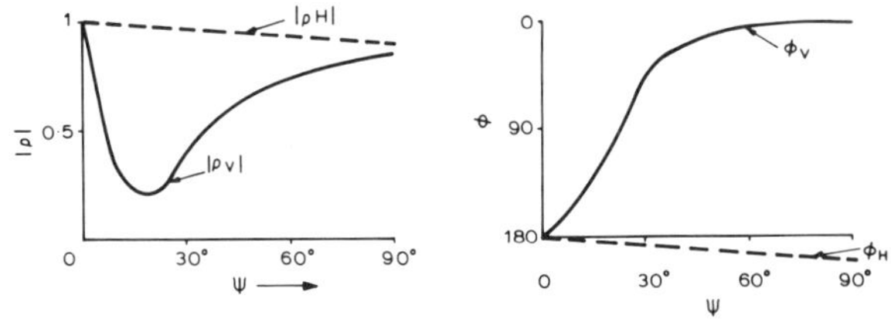

Figure 52.5 Reflection coefficient relative magnitude ρ and phase ϕ at different angles of incidence ψ to the surface (V is vertical polarisation and H horizontal polarisation)

The reflection coefficient magnitude is maximum at low values of ψ, i.e. at glancing incidence, and for the horizontally-polarised wave decreases slowly as ψ is increased. For the vertically polarised wave the magnitude decreases relatively rapidly to a minimum and then rises again towards the horizontally polarised value. With a perfectly conducting ground the magnitude would fall to zero at an angle ψ known as the Brewster angle. At the low angles of surface incidence and for both polarisations the reflected wave suffers a phase reversal, which changes only to a small extent at higher angles of incidence for horizontal polarisation, but decreases rapidly for vertical polarisation.

High values of reflection magnitude are only obtained when the reflecting surface is smooth, for when this is not so the reflected energy is scattered in all directions. A surface may be considered as rough if the variations in height of the surface multiplied by sin ψ are greater than $\lambda/8$, so that a land or sea surface appears rougher as the transmission frequency is increased.

The total field (E_t) at the receiving aerial is the vector sum of direct (E_d) and reflected (E_r) waves with the phase of the latter suitably modified by the extra path length it has had to travel. The extra path length is seen from *Figure 52.6* to be

$$[d^2+(h_t+h_r)^2]^{1/2}-[d^2+(h_t-h_r)^2]^{1/2}$$

$$=d\left\{\left[1+\left(\frac{h_t+h_r}{d}\right)^2\right]^{1/2}-\left[1+\left(\frac{h_t-h_r}{d}\right)^2\right]^{1/2}\right\}$$

$$\simeq\frac{2h_th_r}{d} \quad \text{since } d \gg (h_t+h_r)$$

This corresponds to a phase angle of

$$\frac{2h_th_r}{d}\cdot\frac{2\pi}{\lambda}=\frac{4\pi h_th_r}{\lambda d} \tag{52.1}$$

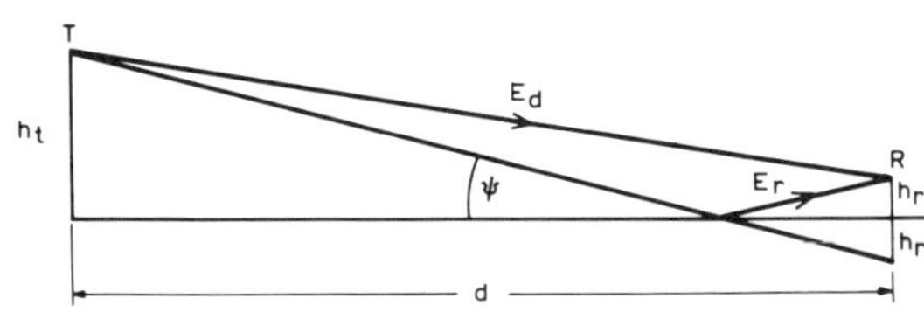

Figure 52.6 Combination of direct and reflected ground ray with v.h.f. propagation (E_d is direct field and E_r reflected field)

Hence the phase angle of the reflected signal at the receiver aerial is ($\phi = 4\pi h_t h_r/\lambda d$) and its amplitude is $k\rho E_d$ where k is the correction factor for the directional characteristic of the transmitter aerial. If $k=\rho=1$ and $\phi=-180°$ the vector sum of E_d and E_r becomes

$$E_t = 2E_d \sin(2\pi h_t h_r/\lambda d) \tag{52.2}$$

so that the field at the receiver aerial varies from zero at the ground ($h_r=0$) through a series of maxima when

$$2\pi h_t h_r/\lambda d = (2n-1).\pi/2 \tag{52.3}$$

or

$$h_r = (2n-1)\lambda d/4h_t$$

and minima when

$$h_r = 2n\lambda d/4h_t \tag{52.4}$$

Under the above conditions a transmitting aerial at a height of 300 m would give maximum field strength at a distance of 30 km when the receiving aerial is at a height of 75 m. For low height receiving aerials $2\pi h_t h_r/\lambda d$ is small and $\sin(2\pi h_t h_r/\lambda d) \rightarrow 2\pi h_t h_r/\lambda d$; since the direct wave $\propto 1/d$, the received field strength will tend to be $\propto 1/d^2$.

In practice due to tropospheric refraction it is found that propagation is always greater than line-of-sight by an amount which would be obtained with a ground profile radius of about 1.33 earth radius, and it is usual to use this value in calculations. Temperature inversions with lowest temperatures near the ground can occur from time to time to produce a ducting effect permitting propagation over considerable distances. The effect is more noticeable over sea than over land.

Diffraction of the transmitted wave can occur due to a hill between the transmitting and receiving aerials, and at grazing incidence a loss of about 6 dB in signal strength occurs. The loss increases rapidly when the hill blocks the line of sight. In order to reduce the loss to a low value, the direct path between the two aerials should clear the obstacle by

$$H = \left[\frac{\lambda d_1(d-d_1)}{d}\right]^{1/2} \tag{52.5}$$

where d_1 = distance from the receiving aerial to the obstacle.

Thus the summit of a hill halfway between transmitting and receiving aerials spaced apart 30 km should be at least

$$H = \left[\frac{3 \times 30 \times 10^3}{2}\right]^{1/2} = 210 \text{ m}$$

below the direct path if $f = 100$ MHz.

52.7 Ultra-high frequency propagation

The propagation of ultra-high frequencies follows the same principles as with very high frequencies. The reduction in wavelength means that smaller objects can produce reflections and the absorption from, for example, trees in leaf is greater; on the other hand aerials are smaller and their directivity and gain can be increased.

References

1 C.C.I.R., *Documents of the XIIth Plenary Assembly*, Vol. II, 'Propagation', 217 to 223 (1970)

2 PIGGOTT, W. R., 'The calculation of Median Sky-Wave Field Strength in Tropical Regions', *UK Department of Scientific and Industrial Research*, No. 27 (1959)

53

Digital design

M D Edwards, BSc, MSc, PhD
Department of Computation,
The University of Manchester
Institute of Science and Technology

Contents

53.1 Number systems

53.1.1 General representation of numbers

In everyday life we normally represent numbers in the decimal radix, or base, for example, the current year, 1983. Different number systems are employed, however, for other uses; base 12 for counting inches, base 60 for seconds and minutes, base 7 for days of the week and base 3 for feet. In general any number system can represent the integer number N, as

$$N = a_{n-1}r^{n-1} + a_{n-2}r^{n-2} + \ldots + a_1 r^1 + a_0 r^0 \qquad (53.1)$$

where

r = radix or base
r^i = digit weighting value
a_i = value of the digit in the ith position, where $0 \leqslant a_i \leqslant r-1$. Thus a radix k number system requires k different symbols to represent the digits 0 to $k-1$.
n = number of digits in the representation of the number.

The digits are written in order so that the position of the digit implies the weighting to which it corresponds.

53.1.2 Decimal numbers

We can represent the decimal number, 1983, in the above format as:

$$1983 = (1.10^3) + (9.10^2) + (8.10^1) + (3.10^0)$$

where $r = 10$ and the digit values are either 0, 1, 2, 3, 4, 5, 6, 7, 8, 9.

53.1.3 Binary numbers

The binary, base 2, number system is used in digital systems and consists of two digits, 0 and 1. Digital systems employ the binary system because it is a straightforward task to decide if an electrical device (or logical element) is either ON (1) or OFF (0). Physical devices exist in one of either two states, e.g. a light bulb is either on or off, a switch is either open or closed. Thus it is logical to use the binary number system in digital systems. Naturally, devices could be constructed to handle numbers with larger bases, but, they are ruled out due to their greater complexity.

A number, N, in the binary radix may be represented as:

$$N = b_{n-1}2^{n-1} + b_{n-2}2^{n-2} + \ldots + b_1 2^1 + b_0 2^0 \qquad (53.2)$$

where $r = 2$ and the digit values $(b_0 \ldots b_{n-1})$ are either 0 or 1 and are known as bits.

Thus the binary representation of the decimal number 1983 is 11110111111 in BINARY NOTATION.

This is equivalent to:

$$\begin{aligned} 1983 &= (1.2^{10}) + (1.2^9) + (1.2^8) + (1.2^7) + (0.2^6) + (1.2^5) \\ &\quad + (1.2^4) + (1.2^3) + (1.2^2) + (1.2^1) + (1.2^0) \\ &= 1024 + 512 + 256 + 128 + 32 + 16 + 8 + 4 + 2 + 1 \end{aligned}$$

53.1.4 Conversion from binary to decimal numbers

There are two methods for converting binary numbers to decimal numbers. The first method consists of summing up the powers of 2 corresponding to the 'one' bits in the number. For example:

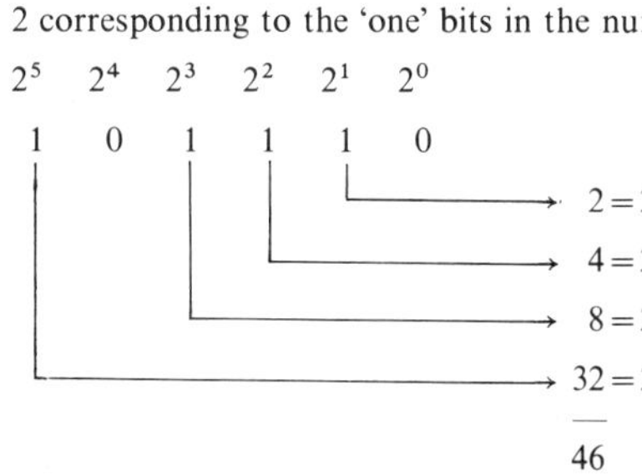

The second method consists of writing the binary number vertically, one bit per line, with the left-most bit on the bottom line. The bottom line is called line 1, the one above it line 2, and so on. The decimal number will be constructed in a column next to the binary number. The procedure to construct the decimal number is to begin by writing a 1 on line 1, the entry on line x consists of two times the entry on line $x-1$ plus the bit on line x (either 0 or 1), the entry on the top line is the answer. This method is known as 'successive doubling' and an example is given below.

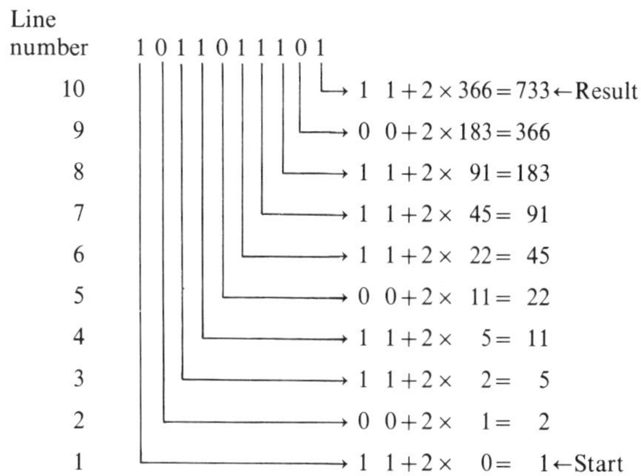

53.1.5 Conversion from decimal to binary numbers

There are also two methods for converting decimal numbers to binary numbers. The first method involves subtracting powers of 2 from the decimal number. The largest power of two smaller than the number is subtracted from the number. The process is then repeated on the remainder. When the number has been decomposed into powers of two, the binary number can be assembled with 'ones' in the bit position corresponding to the powers of 2 used and 'zeros' elsewhere. *Table 53.1* gives powers of 2.

Table 53.1 Powers of two

Power of 2	*Decimal*	*Power of 2*	*Decimal*
2^0	1	2^8	256
2^1	2	2^9	512
2^2	4	2^{10}	1024
2^3	8	2^{11}	2048
2^4	16	2^{12}	4096
2^5	32	2^{13}	8192
2^6	64	2^{14}	16384
2^7	128	2^{15}	32768

The second method consists of successively dividing the decimal number by 2. The quotient is written directly underneath the original number, and the remainder, 0 or 1, is written next to the quotient. The quotient is then halved and the process repeated until the number 0 is reached. The binary number can now be obtained directly from the remainder column, starting at the bottom. As an example the decimal number 483 is converted to binary below.

Quotient	Remainder
483	
241	1
120	1
60	0
30	0
15	0
7	1
3	1
1	1
0	1

1 1 1 1 0 0 0 1 1

53.1.6 Octal numbers

The binary number system is too cumbersome for human usage especially when large binary numbers containing lots of bits are encountered. It is easy to make errors when reading and writing large binary numbers and hence it is usual to encode the binary digits into a more human readable form. The octal, or base 8, number system helps to alleviate this problem. A number, N, in the octal radix may be represented as:

$$N = l_{n-1}8^{n-1} + l_{n-2}8^{n-2} + \ldots + l_1 8^1 + l_0 8^0 \qquad (53.3)$$

where $r = 8$ and the digit values $(l_0 \ldots l_{n-1})$ are either 0, 1, 2, 3, 4, 5, 6, 7 (see *Table 53.2*).

The octal number system reduces the number of digits when handling binary numbers. To convert binary numbers to octal numbers, the binary digits are divided into 3-bit groups which can be represented by a single octal digit. For example, the binary number,

1 1 0 0 1 1 0 1 0 1 0 1 0

is separated into 3-bit groups, starting at the right-hand end of the number and supplying leading zeros if necessary:

0 0 1   1 0 0   1 1 0   1 0 1   0 1 0

The 3-bit groups may be replaced by their octal equivalents:

1 4 6 5 2

and the binary number is converted to its octal equivalent, 14652. Conversely, an octal number can be expanded into a binary number by replacing the octal digits by their 3-bit binary equivalents:

612 = 110 001 010

Octal numbers may be converted to their decimal equivalents, and vice versa, by similar processes used to interchange binary and decimal numbers.

In order to avoid ambiguity between decimal, octal and binary numbers, a subscript of 2, 8 or 10 is used to indicate the radix being used; for example 27(10) = 33(8) = 11011(2).

53.1.7 Hexadecimal numbers

Another convenient number system is the hexadecimal, or base 16, system. The hexadecimal number system, like octal, is used to represent binary numbers in a more readily understandable form. A number $N(16)$, in the hexadecimal radix, may be represented as:

$$N(16) = h_{n-1}16^{n-1} + h_{n-2}16^{n-2} + \ldots + h_1 16^1 + h_0 16^0 \qquad (53.4)$$

where $r = 16$ and the digit values $(h_0 \ldots h_{n-1})$ are either

0, 1, 2, 3, 4, 5, 6, 7, 8, 9, *A*, *B*, *C*, *D*, *E*, *F*

A table of hexadecimal numbers together with their equivalents is given in *Table 53.2*.

Table 53.2 Decimal, hexadecimal, octal and binary number equivalents

Decimal	*Hex*	*Octal*	*Binary*	*Decimal*	*Hex*	*Octal*	*Binary*
0	0	0	0	8	8	10	1000
1	1	1	1	9	9	11	1001
2	2	2	10	10	*A*	12	1010
3	3	3	11	11	*B*	13	1011
4	4	4	100	12	*C*	14	1100
5	5	5	101	13	*D*	15	1101
6	6	6	110	14	*E*	16	1110
7	7	7	111	15	*F*	17	1111

To convert binary numbers to hexadecimal numbers, the binary digits are divided into 4-bit groups. The 4-bit groups can be represented by a single hexadecimal digit. For example, the binary number,

1 0 1 1 0 1 0 0 1 1 1 0 1

is separated into 4-bit groups, starting at the right-hand end of the number and supplying leading zeros if necessary:

0 0 0 1   0 1 1 0   1 0 0 1   1 1 0 1

The 4-bit groups are replaced by their hexadecimal equivalents:

1 6 9 *D* (16)

Conversely, a hexadecimal number can be expanded into a binary number by replacing the hexadecimal digits by their binary equivalents:

A 4 *F* (16) = 1 0 1 0   0 1 0 0   1 1 1 1

Hexadecimal numbers may be converted to decimal numbers, and vice versa, in the usual manner.

53.1.8 Binary addition and subtraction

Binary arithmetic operations, addition and subtraction, are very similar to those in decimal arithmetic. In fact, the rules are much simpler. Consider the addition of two bits, *A* and *B*:

A	B	$A+B$
0	0	0
0	1	1
1	0	1
1	1	0 → carry 1

A carry (the next highest power of two) is only propagated if both bits are 'one'.

Binary subtraction follows the same basic rules as binary addition. Consider the subtraction of two bits, *A* and *B*:

A	B	$A-B$
0	0	0
0	1	1 → borrow 1
1	0	1
1	1	0

Subtraction is carried out in the normal manner except that a borrow (the next highest power of two) is only propagated if the minuend is 'zero' and the subtrahend is 'one'.

53.1.9 Binary multiplication and division

The remaining two basic arithmetic operations are multiplication and division. There are numerous methods for multiplying and dividing binary numbers, however, only the simplest methods are considered here.

Binary multiplication is similar to decimal multiplication and, again, the rules are simpler. Binary multiplication is performed in the normal manner by multiplying and then shifting one place to the left and finally adding together the partial products. Note that the multiplier digit can only be 0 or 1 and thus the partial product is either zero or equal to the multiplicand. Consider the example below:

```
Multiplicant    1 1 0 1  (13)
Multiplier      0 1 0 1   (5)

      1 1 0 1
      0 1 0 1
      -------
      1 1 0 1
    0 0 0 0
  1 1 0 1
0 0 0 0
-------------
1 0 0 0 0 0 1                 (65)
-------------
```

The process of binary division is very similar to that of decimal arithmetic. Binary division is performed by dividing and then shifting the divisor one place to the right and subtracting from the dividend. Note that the division process is simplified to either divide once or not at all. Consider the example below:

```
Dividend:        1 0 1 1 1 1  (47)
Divisor:             1 0 0 0   (8)
                     1 0 1 ←— Quotient (5)
               -------------
        1 0 0 0) 1 0 1 1 1 1
                 1 0 0 0
                 -------
                   1 1 1 1
                   1 0 0 0
                   -------
                     1 1 1 ←— Remainder (2)
                   -------
```

53.1.10 Positive and negative binary numbers

The normal way to represent numbers is to use the 'sign and magnitude' method, for example, +83 and −72. In the binary number system, the most significant bit is used to represent the sign and the remaining bits are used to denote the magnitude. If the most significant bit is '0', the number is considered to be positive and if the bit is '1' then the number is negative.

We will assume that all numbers are 4-bits long for simplicity. Having one bit for the sign reduces the magnitude of the largest numbers that can be represented. If only positive numbers were used then we could represent numbers in the range 0 to 15 ($2^4=16$), but we now have to have approximately half for negative numbers, so the range of numbers is reduced to roughly −8 to +8.

There are three common methods used to represent positive and negative binary numbers; sign and magnitude, one's complement and two's complement.

53.1.10.1 Sign and magnitude

In the sign and magnitude method, the negative number is simply the positive number with the sign but reversed. In general, the range of numbers that may be represented in n-bits is

$$-(2^{n-1}-1) \leqslant N \leqslant (2^{n-1}-1)$$

Therefore the negative equivalent of an n-bit positive sign and magnitude number, a, is $2^{n-1}+a$.

The sign and magnitude method makes arithmetic complex because there are two values for zero and the binary sum of a positive number and its sign and magnitude negative is non-zero (see *Table 53.3*). For example:

```
  6−6≡6+(−6)

  0 1 1 0
 +1 1 1 0
 ---------
1 0 1 0 0 = −4 (or +4 if the carry is ignored)
```

The sign and magnitude method is used, for example, in analog to digital and digital to analog conversion, but the numbers have to be converted to another form before any arithmetic operations can be performed.

53.1.10.2 One's complement

In the one's complement representation the negative equivalent of a number may be given by:

$$-N=(2^n-1)-\mathrm{N} \qquad (53.5)$$

where

$$N=(b_{n-1}2^{n-1}+b_{n-2}2^{n-2}+\ldots+b_1 2^1+b_0 2^0)$$

Thus, the negative equivalent of +4 in one's complement may be found by subtracting 4 from 15 where ($15=2^n-1$, $n=4$ bits):

```
 1 1 1 1    (15)
−0 1 0 0    (4)
 -------
 1 0 1 1    (−4)
 -------
```

The negative representation of a number is the logical complement of the number; that is, replacement of all the zeros in the binary representation by ones and all the ones by zeros. In general, the range of one's complement numbers that may be represented in n-bits is

$$-(2^{n-1}-1) \leqslant N \leqslant (2^{n-1}-1)$$

It is possible to perform meaningful arithmetical operations using one's complement numbers. For example, consider the following addition and subtraction sums:

```
(a)     3+2          0 0 1 1
                   + 0 0 1 0
                     -------
        =5           0 1 0 1
                     -------

(b)     7−5≡7+(−5)     0 1 1 1
                     + 1 0 1 0
                     ---------
                     (1) 0 0 0 1
             carry    └──────→1
                     ---------
             =2          0 0 1 0
                     ---------
```

In example (b), a carry is generated and in order to produce the correct answer the carry must be added back into the sum. Care must be taken when performing arithmetic operations on one's complement numbers to ensure that numerical overflow does not occur. Overflow is the condition which occurs when the result of an arithmetic operation cannot be correctly expressed. For example, when adding or subtracting 4-bit numbers the result must lie within the range -7 to $+7$ otherwise the result will be interpreted wrongly. For example:

```
7+6      0 1 1 1
         0 1 1 0
         -------
 = -2    1 1 0 1      (incorrect)
         -------
```

The fact that there are two representations of 'zero' in one's complement arithmetic, and addition and subtraction are relatively complex, has made the one's complement representation of negative numbers almost obsolete.

53.1.10.3 Two's complement

In the two's complement representation, the negative equivalent of a number may be given by:

$$-N = 2^n - N \qquad (53.6)$$

where

$N = (b_{n-1}2^{n-1} + b_{n-2}2^{n-2} + \ldots + b_1 2^1 + b_0 2^0)$

Thus, the negative equivalent of $+5$ in two's complement may be found by subtracting 5 from 16; where ($16 = 2^n$, $n = 4$ bits):

```
           1 0 0 0 0     (16)
             0 1 0 1      (5)
          ----------
ignore →  (0) 1 0 1 1    (−5)
          ----------
```

The negative representation of a number is the logical complement of the number plus 1. For example:

$3 = 0\ 0\ 1\ 1$

```
1's complement  =  1 1 0 0
          +1    +        1
                   -------
                   1 1 0 1  =  −3
                   -------
```

In general, the range of two's complement numbers that may be represented in n-bits is

$-(2^{n-1}) \leqslant N \leqslant (2^{n-1} - 1)$

It is possible to perform meaningful arithmetical operations using two's complement numbers. For example:

```
(a)   4+3              0 1 0 0
                     + 0 0 1 1
                       -------
       =7              0 1 1 1
                       -------

(b)   5−2≡5+(−2)           0 1 0 1
                         + 1 1 1 0
                         ---------
                   ┌───→(1) 0 0 1 1  =  3
                   │     ---------
                   └ignore the carry

(c)   −3−4≡−3+(−4)          1 1 0 1
                          + 1 1 0 0
                          ---------
                       ↗(1) 1 0 0 1  =  −7
                          ---------
            ignore the carry
```

As with one's complement arithmetic care must be taken to avoid overflow conditions occurring.

The use of two's complement arithmetic is widespread due to the simplicity of the basic arithmetic operations and the fact that there is only one representation of 'zero'.

The three methods for representing positive and negative binary integers are summarised in *Table 53.3.*

53.1.11 Shifting and binary fractions

In addition to the binary arithmetic operations already mentioned there is an additional operation known as the 'shift' operation. The first type of shift operation is known as a logical shift and may be used to multiply or divide a positive number by two. The logical shift left operation, shown below, is equivalent to multiplying by two:

Logical shift left	before	S_3	S_2	S_1	S_0	
	after	S_2	S_1	S_0	0	
Example:	before	0101	(5)			
	after	1010	(10)			

The logical shift right operation, shown below, is equivalent to dividing by two:

Logical shift right	before	S_3	S_2	S_1	S_0	
	after	0	S_3	S_2	S^1	
Example:	before	0100	(4)			
	after	0010	(2)			

Table 53.3 Representation of positive and negative binary integers

Representation method	*Representation of +5 and −5*	*−a*	*Zero*	*Max positive*	*Max negative*
Sign and magnitude	+5 0:101 −5 1:101	$2^{n-1}+a$	0:000 1:000	$2^{n-1}-1$	$2^{n-1}-1$
One's complement	+5 0:101 −5 1:010	$2^{n-1}-a$	0:000 1:111	$2^{n-1}-1$	$2^{n-1}-1$
Two's complement	+5 0:101 −5 1:011	2^n-a	0:000	$2^{n-1}-1$	2^{n-1}

The above shift operations will not produce the correct result if two's complement numbers are shifted left or right. Thus, an extra type of shift operation, known as the arithmetic shift, is employed. The arithmetic shift left and right operations also multiply and divide by two respectively, and they maintain the sign of the number. Both arithmetic shifts are illustrated below:

Arithmetic shift left		before	S_3	S_2	S_1	S_0
		after	S_3	S_1	S_0	0
			preserve the sign bit			
Examples:	(a)	before	0010	(2)		
		after	0100	(4)		
	(b)	before	1101	(-3)		
		after	1010	(-6)		
Arithmetic shift right		before	S_3	S_2	S_1	S_0
		after	S_3	S_3	S_2	S_1
Examples:	(a)	before	0100	(4)		
		after	0010	(2)		
	(b)	before	1000	(-8)		
		after	1100	(-4)		

53.1.12 Binary fractions

So far we have only considered integer binary numbers, however, it is also possible to represent binary fractions. A binary number, integer plus fraction may be represented by:

$$N=(b_{n-1}2^{n-1}+\ldots+b_1 2^1+b_0 2^0)+(b_{-1}2^{-1}+b_{-2}2^{-2}+\ldots+b_{-m}2^{-m})$$

where the term in the first bracket is the integer part, the + sign is the binary point and the term in the second bracket is the fraction part.

In the decimal number system the fractional part of the number is represented by tenths, hundredths, thousandths, etc whereas in the binary number system the fractional part is represented by halves, quarters, eighths, etc. For example, the following are valid binary fractions represented in two's complement form:

1.0	=0001.0000	-1.0	=1111.0000
1.5	=0001.1000	-1.5	=1110.1000
3.75	=0011.1100	-3.75	=1100.0100
6.4375	=0110.0111	-6.4375	=1001.1001

53.1.13 Floating point numbers

In many calculations a large dynamic number range is required. It is possible to use fixed point arithmetic to represent large numbers but this incurs the penalty of needing a large number of bits; for example, 20 bits are required to express the decimal number 1 million. Therefore, a system of representing numbers is required where the range of possible numbers is independent of the number of significant digits. The scientific, or floating point, notation may be used, where a floating point number may be expressed in the form:

$$N=f\times b^e$$

where b is the base (or radix), f is the fraction (or mantissa) and e is the exponent.

For example, the decimal number 5280 may be represented in floating point format as 0.528×10^4. The range of the number is determined by the number of digits in the exponent and the precision by the number of digits in the fraction. In floating point format numbers will nearly always have to be rounded to the nearest expressible number. However, the loss of precision is compensated for by a greater dynamic range.

For binary floating point numbers, the number of available bits is divided into a fraction part and an exponent part. The exponent is expressed as a fixed point binary integer and the fractional part as a normal binary fraction. Consider the value 3.125 held in floating point format, base 2, in 16 bits; with 10 bits for the fraction and 6 bits for the exponent:

3.125 = 011.001 (integer + fraction)

Equivalent to:

	Exponent	*Fraction*
0.390625×2^3	000011	0110010000
0.78125×2^2	000010	1100100000
0.1953125×2^4	000100	0011001000

Note that there are several possible ways to represent the number. In order to maximise the number of significant digits in the fraction it is usual to represent all non-zero floating point numbers in *Normalised* form. This is achieved by adjustment of the fraction and exponent until the fraction has a non-zero most significant bit; that is, the fraction lies in the range $0.5\leqslant f<1$.

It is usual practice to be able to represent both positive and negative values. However, it is necessary to have both positive and negative exponents, so that negative and normalised numbers of less than 0.5 may be represented. Thus, a 16-bit floating point number may be postulated as:

m.s.b.		l.s.b.
X	X X X X X	X X X X X X X X X X
sign bit 0 = +ve 1 = −ve	exponent $-16\leqslant e\leqslant 15$	fraction $0.5\leqslant f<1$

Thus, the range of numbers which may be represented is:

$$\pm(0.5\times 2^{-16}\text{ to }0.999\times 2^{15})$$
$$=\pm(7.6\times 10^{-6}\text{ to }3.27\times 10^4)$$

Example:

	Sign	*Exponent*	*Fraction*
$5280=0.645\times 2^{13}$	0	01101	1010010100
0	0	00000	0000000000
$-1.5\times 10^{-5}=0.5\times 2^{-15}$	1	10001	1000000000
$47.0=0.734375\times 2^6$	0	00110	1011110000

Any floating point number with a zero fraction represents zero (see above). However, the standard representation of zero (true zero) has a zero fraction and the smallest possible exponent; which means that the exponent should be 10000.

53.2 Basic mathematics of digital systems

In order to design digital systems it is necessary to understand the basic mathematics of the subject, known as Boolean algebra. The relevant mathematical theory is presented so that the engineer may readily appreciate the fundamentals of Boolean algebra and apply it to the design of digital systems.

53.2.1 Boolean algebra

Logic is a tool used in the study of deductive reasoning, also known as propositional logic. In 1854, the mathematician Geroge Boole made a significant advance in this field when he postulated that symbols may be used to represent the structure of logical thought. The work of Boole may be readily applied to the design of digital systems, where a digital circuit may be represented by a set of symbols and the circuit function expressed as a set of relationships between the symbols. The basic symbols used in digital systems are '1' for *true* and '0' for *false*. There are additional symbols used to represent the relationship between the basic symbols. The logical function is the relationship between a set of assertions, which are represented by symbols, or letter groups.

Boolean algebra is based on five postulates which may be applied to switching circuits. The five postulates are given below:

Let a symbol X be associated with the state of an element (or circuit), then

(1) $X=0$ if $X \neq 1$ and
(1a) $X=1$ if $X \neq 0$
(This is the definition of a binary variable)

(2) 0 AND 0=0
(2a) 1 OR 1=1

(3) 1 AND 1=1
(3a) 0 OR 0=0

(4) 1 AND 0=0 AND 1=0
(4a) 0 OR 1=1 OR 0=1

(5) NOT 0=1
(5a) NOT 1=0

The basic postulates use three logical relationships which are unique to boolean algebra, AND, OR, NOT. In addition, a number of theorems may be derived, which are applicable to the algebra of switching circuits. The basic theorems are given below, where the switching variables are represented by the symbols X, Y, Z and may take the value 0 or 1. The AND operator is represented by the symbol '.', the OR operator by '+' and the NOT operator by $\bar{A}$, (*not A*):

(6) $X+0=X$
(6a) $X.1=X$

(7) $1+X=1$
(7a) $0.X=0$

(8) $X+X=X$

(8a) $X.X=X$

(9) $(\bar{X})=\bar{X}$
(9a) $\overline{(\bar{X})}=X$

(10) $X+\bar{X}=1$

(10a) $X.\bar{X}=0$

(11) $X+Y=Y+X$
(11a) $X.Y=Y.X$

(12) $X+X.Y=X$
(12a) $X.(X+Y)=X$

(13) $Y.(X+\bar{Y})=X.Y$
(13a) $X.\bar{Y}+Y=X+Y$

(14) $X+Y+Z=(X+Y)+Z=X+(Y+Z)$

(14a) $XYZ=(XY).Z=X.(YZ)$

(15) $XY+XZ=X.(Y+Z)$
(15a) $(X+Y).(X+Z)=X+YZ$

(16) $(X+Y).(Y+Z).(Z+\bar{X})=(X+Y).(Z+\bar{X})$

(16a) $XY+YZ+Z\bar{X}=XY+Z\bar{X}$

(17) $(X+Y).(\bar{X}+Z)=XZ+\bar{X}Y$

(18) $\overline{X+Y}=\bar{X}.\bar{Y}$
(18a) $\overline{X.Y}=\bar{X}+\bar{Y}$

A principle of duality exists between the basic theorems; if the values 0, 1 and the operations (.), (+) are interchanged then the alternative representation is obtained (cf 6, 6a; 7, 7a, etc). The general form of the duality principle is given in theorems 18 and 18a, and is known as De Morgan's theorem.

The above theorems may be represented in terms of simple electrical switching circuits and the physical representation of some of the theorems is given in *Figure 53.1*.

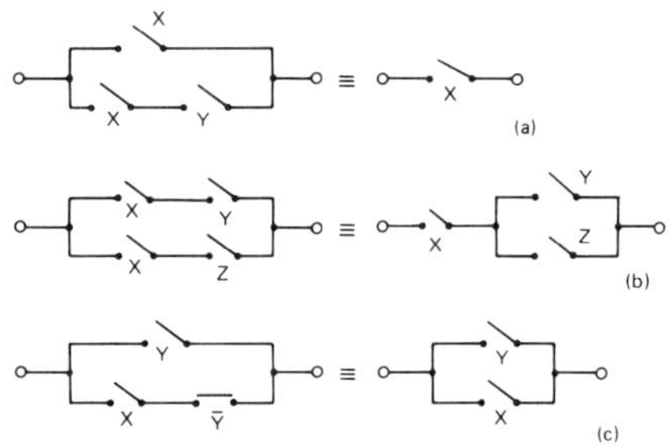

Figure 53.1 Electrical representation of logic theorems (a) $X+X.Y=X$, (b) $X.Y+X.Z=X.(Y+Z)$, (c) $X\bar{Y}+Y=X+Y$

53.2.2 Theorem proving

The theorems of switching algebra may be proved by the use of two techniques; algebraic manipulation and perfect induction.

To prove a theorem (or the equivalence of two logical functions) it is necessary to invoke a subset of the other theorems. Consider the proof of the following relationship: theorem 15a:

$(X+Y).(X+Z)=X+YZ$

Consider the l.h.s. of the equation:

$=(X+Y).(X+Z)$
$=X.X+XZ+YX+YZ$ (by theorem 15)
$=X+XZ+YX+YZ$ (by theorem 8a)
$=(X+XZ)+YX+YZ$
$=X+YX+YZ$ (by theorem 12)
$=(X+YX)+YZ$
$=X+YZ$ (by theorem 12)

To prove a theorem (or the equivalence of two logical functions), by the method of perfect induction, the following steps must be performed:

(1) Write out all the combinations of the variables in tabular form.
(2) Deduce the result of each expression for all combinations of the variables.
(3) If the result of each expression for all combinations of the variables is the same, then the two expressions are equivalent.

The table of all combinations of the variables and the result of each expression is also known as a *truth table*.

The truth table for theorem 15a is given below in *Table 53.4*.

Table 53.4 Truth table for theorem 15a

X	Y	Z	$(X+Y).(X+Z)$	$X+YZ$
0	0	0	0	0
0	0	1	0	0
0	1	0	0	0
0	1	1	1	1
1	0	0	1	1
1	0	1	1	1
1	1	0	1	1
1	1	1	1	1

53.2.3 Functions of binary variables

A number of functions may be derived from n binary variables. For n binary variables there are 2^n possible combinations of the variables and 2^{2n} different functions. Thus, for two binary variables (X, Y) there are four possible combinations and sixteen unique functions, as shown in *Table 53.5*.

Table 53.5 Combinations of two binary variables

X	Y	f_0	f_1	f_2	f_3	f_4	f_5	f_6	f_7	f_8	f_9	f_{10}	f_{11}	f_{12}	f_{13}	f_{14}	f_{15}
0	0	0	0	0	0	0	0	0	0	1	1	1	1	1	1	1	1
0	1	0	0	0	0	1	1	1	1	0	0	0	0	1	1	1	1
1	0	0	0	1	1	0	0	1	1	0	0	1	1	0	0	1	1
1	1	0	1	0	1	0	1	0	1	0	1	0	1	0	1	0	1

There are four important functions: f_1 (AND), f_7 (OR), f_8 (NOR), f_{14} (NAND), together with the complement functions f_{12}, f_{10} for X and Y respectively. In general a function may be expressed in terms of the AND/OR operations or its dual in terms of the NAND/NOR/OR operations by application of De Morgan's theorem.

$F = A \,.\, B + C$ (AND/OR)

The following are equivalent to F

$$F = \overline{\overline{A \,.\, B} \,.\, \bar{C}} \qquad \text{(NAND)}$$
$$= \overline{(\bar{A} + \bar{B})} + C \qquad \text{(NOR/OR)}$$
$$= \overline{(\bar{A} + \bar{B}) \,.\, \bar{C}} \qquad \text{(OR/NAND)}$$

Thus, any switching system may be expressed in terms of a combination of the AND/OR/NAND/NOR functions (see also Chapter 30).

53.2.4 Minterms and maxterms

Logic design problems are normally presented in truth table form. The mathematical description of the function may be extracted from the truth table and then manipulated into a form suitable for implementation by an interconnected set of logic gates. Consider *Table 53.6*.

Table 53.6 Example of a truth table

X	Y	Z	F
0	0	0	1
0	0	1	0
0	1	0	0
0	1	1	1
1	0	0	0
1	0	1	1
1	1	0	1
1	1	1	0

The function F is true (equal to 1) for certain combinations of the variables X, Y, Z. The value of F is false (equal to 0) for the remaining combinations. The function may be described in mathematical terms by taking the logical OR of all the combinations of the variables which produce a 'true' result, thus:

$$F = \bar{X}\bar{Y}\bar{Z} + \bar{X}YZ + X\bar{Y}Z + XY\bar{Z} \qquad (53.7)$$

This form of the equation is known as the *sum of products*. The dual of this equation may be obtained by taking the logical OR of all combinations of the variables which produce a 'false' result thus:

$$\bar{F} = \bar{X}\bar{Y}Z + \bar{X}Y\bar{Z} + X\bar{Y}\bar{Z} + XYZ \qquad (53.8)$$

By applying de Morgan's theorem to equation (53.8), we have:

$$F = (\overline{\bar{X}\bar{Y}Z + \bar{X}Y\bar{Z} + X\bar{Y}\bar{Z} + XYZ})$$
$$= (X + Y + \bar{Z}) \,.\, (X + \bar{Y} + Z) \,.\, (\bar{X} + Y + Z) \,.\, (\bar{X} + \bar{Y} + \bar{Z}) \qquad (53.9)$$

Thus, equations (53.7 and (53.9) are identical and the form of equation (53.9) is known as the *product of sums*.

The sum of products is also known as the *minterm* form of an equation and the product of sums the *maxterm* form. The possible minterms and maxterms for three binary variables, X, Y, Z are given in *Table 53.7*.

Table 53.7 Minterms and maxterms of three variables

X	Y	Z	*Minterms*	*Maxterms*
0	0	0	$m_0 = \bar{X}\bar{Y}\bar{Z}$	$M_0 = \bar{X} + \bar{Y} + \bar{Z}$
0	0	1	$m_1 = \bar{X}\bar{Y}Z$	$M_1 = \bar{X} + \bar{Y} + Z$
0	1	0	$m_2 = \bar{X}Y\bar{Z}$	$M_2 = \bar{X} + Y + \bar{Z}$
0	1	1	$m_3 = \bar{X}YZ$	$M_3 = \bar{X} + Y + Z$
1	0	0	$m_4 = X\bar{Y}\bar{Z}$	$M_4 = X + \bar{Y} + \bar{Z}$
1	0	1	$m_5 = X\bar{Y}Z$	$M_5 = X + \bar{Y} + Z$
1	1	0	$m_6 = XY\bar{Z}$	$M_6 = X + Y + \bar{Z}$
1	1	1	$m_7 = XYZ$	$M_7 = X + Y + Z$

Thus equations (53.7) and (53.9) may be expressed in their minterm and maxterm forms as:

$$F = (m_0, m_3, m_5, m_6) \qquad \text{(minterms)} \qquad (53.10)$$
$$F = (M_0, M_3, M_5, M_6) \qquad \text{(maxterms)} \qquad (53.11)$$

The following relationships exist between minterms and maxterms:

(1) For n binary variables there are 2^n maxterms and 2^n minterms.
(2) The logical OR of all minterms equals 1.
(3) The logical AND of all maxterms equals 0.
(4) The complement of any minterm is a maxterm and vice versa.

53.2.5 Canonical form

If all the binary variables, or their complements, appear once and only once in each term of a logical function then the function is said to be in its *canonical* form. The canonical form of a function is useful in the comparison and simplification of boolean functions.

To express a function in its canonical form logically multiply (AND) each term by the absent variables in the term, expressed in the form $(A + \bar{A})$.

For example:

$$F_1 = XY + YZ + \bar{X}Z$$
$$= XY(Z + \bar{Z}) + YZ(X + \bar{X}) + \bar{X}Z(Y + \bar{Y})$$
$$= XYZ + XY\bar{Z} + XYZ + \bar{X}YZ + \bar{X}YZ + \bar{X}\bar{Y}Z$$
$$= XYZ + XY\bar{Z} + \bar{X}YZ + \bar{X}\bar{Y}Z$$

$$F_2 = XY + \bar{X}Z$$
$$= XY(Z + \bar{Z}) + \bar{X}Z(Y + \bar{Y})$$
$$= XYZ + XY\bar{Z} + \bar{X}YZ + \bar{X}\bar{Y}Z$$

Thus, when expanded into their canonical forms, F_1 and F_2 are equivalent (see theorem 16a).

53.3 Minimisation of logic functions

A boolean function may be represented in either a sum of products or a product of sums form. In the design of digital systems the function must be implemented using the least number of logical units (AND, NAND, OR, NOR gates).

It is possible to simplify most boolean functions by algebraic manipulation. However, for complex functions which involve many variables (greater than five), the simplification process becomes too difficult and tedious. Techniques exist for overcoming this deficiency, which are based on graphical and algorithmic (or tabular) minimisation methods. This section describes both graphical and algorithmic methods for the minimisation of boolean functions.

53.3.1 Graphical minimisation of boolean functions

There are numerous ways which may be used to represent a boolean function graphically. One of the most commonly used approaches is to use a *Venn diagram* which represents variables as areas inside a rectangle, as in *Figure 53.2*.

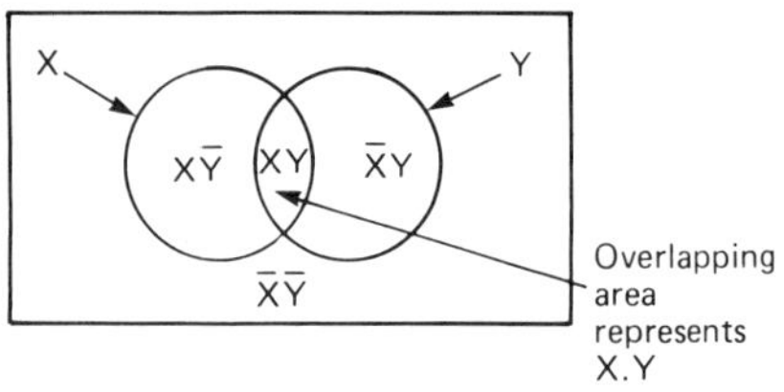

Figure 53.2 A Venn diagram

The rectangle is known as the 'universe of discourse' and the variables are represented by circles inside it. For two variables, the Venn diagram can represent all the possible combinations of the variables and this can be extended to any number of variables, although the diagram becomes too complex to understand for more than four variables.

An alternative representation of boolean functions may be obtained by the use of 'geometric figures', where variables are represented by nodes, arcs and faces of the geometric figure. Single-, two- and three-variable geometric figures may be drawn, as shown in *Figure 53.3*.

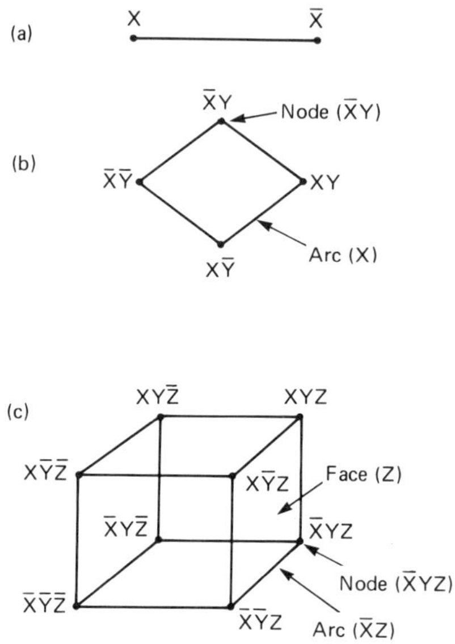

Figure 53.3 Geometrical interpretation of boolean functions. (a) Single variable, (b) two variables, (c) three variables

A boolean function may be simplified by marking each term at an appropriate node and then joining the nodes by the appropriate arcs. The simplified function is then the logical OR of all the arcs. Consider the following example:

$$F=XY\bar{Z}+XYZ+X\bar{Y}Z+\bar{X}\bar{Y}Z \tag{53.13}$$

The terms may be marked on the cube, as shown in *Figure 53.4*, and the four nodes joined by the three arcs.

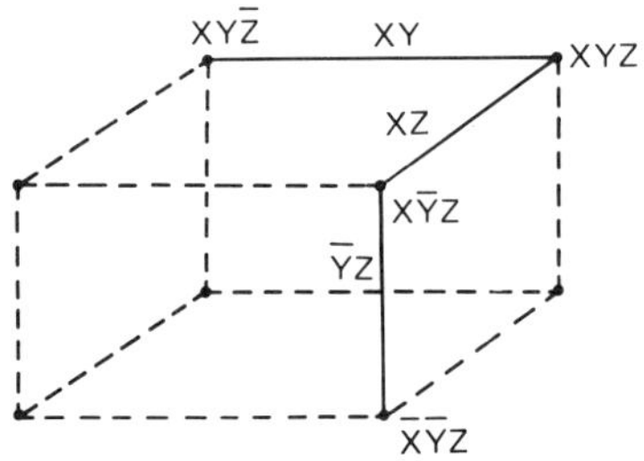

Figure 53.4 Example of a three-variable function

The four nodes are joined by the three arcs, XY, XZ, $\bar{Y}Z$ and the function thus simplifies to

$$F=XY+XZ+\bar{Y}Z \tag{53.14}$$

However, the function may be further simplified. The XY arc includes the $XY\bar{Z}$ and XYZ nodes and the $\bar{Y}Z$ arc includes the $X\bar{Y}Z$ and $\bar{X}\bar{Y}Z$ nodes; thus, the XZ arc is redundant as it also contains the XYZ and $X\bar{Y}Z$ nodes.

The use of geometric figures to simplify (minimise) boolean functions is not widely used as the optimum minimisation is not readily apparent and it is difficult to visualise a geometric figure to represent four or more variables.

53.3.2 Minimisation using Karnaugh maps

The Karnaugh map method for representing boolean functions graphically is based on the Venn diagram and geometric figure techniques. An extension to the Venn diagram was proposed by Veitch, where all the possible combinations of n variables are represented as areas inside a rectangle. The areas are arranged in such a manner that a move from one overlapping area to another, in a horizontal or vertical direction, changes only one variable. Thus, all adjacent areas can be related according to the switching theorem $XY+X\bar{Y}=X$. The Veitch diagram for four variables is shown in *Figure 53.5*.

The Karnaugh map is a matrix representation of a Veitch diagram, where each combination of the binary variables is assigned to a square in the matrix. The squares (also known as cells) are assigned binary codes such that each square in a horizontal or vertical direction differs from its neighbour in the value of a single variable only. Thus, terms in a boolean function may be combined using the relationship $XY+X\bar{Y}=X$. Two, three and four variable Karnaugh maps are shown in *Figure 53.6*.

It must be emphasised that a similar relationship exists between the cells in the top and bottom rows and the leftmost and rightmost columns in the Karnaugh maps.

It is possible to plot Karnaugh maps for n variables, however, the number of cells become large above four variables and it is

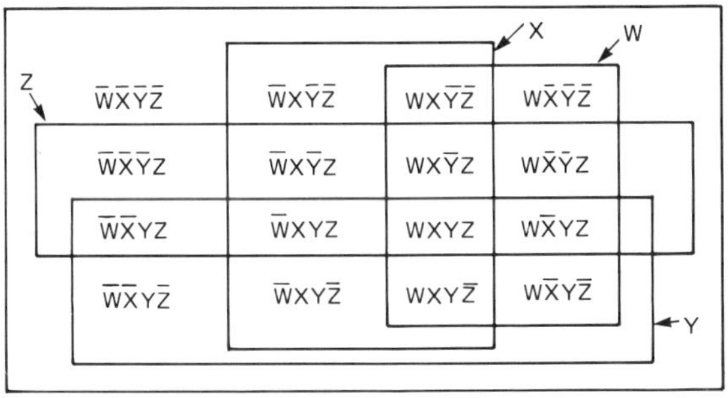

Figure 53.5 Veitch diagram for four variables

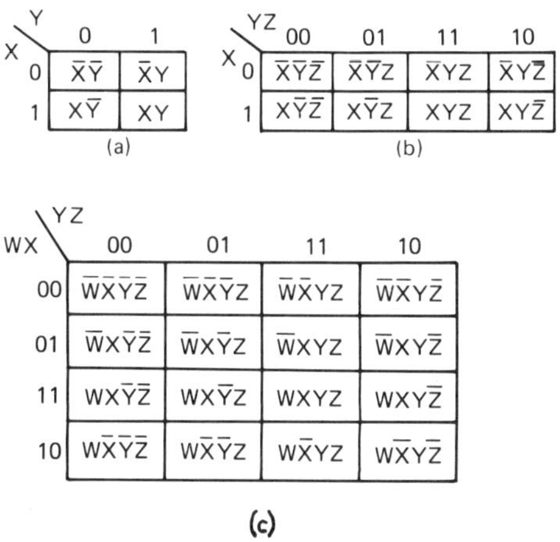

Figure 53.6 Karnaugh maps (a) two variable, (b) three variable, (c) four variable

difficult to see the relationships between cells. Note that for n variables, there are 2^n cells in the Karnaugh map.

Whereas single adjacent cells differ by a single variable, it is possible to group together larger number of cells (known as sub-cubes) which differ by more than one variable but still correspond to the general theorem $XY+X\bar{Y}=X$. A sub-cube may be defined as a set of cells of the Karnaugh map within which one or more of the variables have constant values.

A sub-cube of 2 cells differs by a single variable and thus the variable is redundant. A sub-cube of 4 cells has two redundant variables and similarly a sub-cube of 8 cells has three redundant variables.

To plot a boolean function on a Karnaugh map, the function must normally be in its canonical sum of products form. If, originally, it is not then it must be expanded into the canonical form by the method described previously. A '1' is then entered into each map cell corresponding to each term in the function. The function may then be minimised (if possible) by grouping together cells to form sub-cubes and thereby eliminating redundant variables. For example:

$$\begin{aligned}F &= \bar{Y}Z+\bar{W}\bar{X}\bar{Y}\bar{Z}+\bar{W}\bar{X}Y\bar{Z}+W\bar{X}\bar{Y}\bar{Z}+W\bar{X}Y\bar{Z}\\ &= WX\bar{Y}Z+W\bar{X}\bar{Y}Z+\bar{W}X\bar{Y}Z+\bar{W}\bar{X}\bar{Y}Z+\bar{W}\bar{X}\bar{Y}\bar{Z}\\ &\quad +\bar{W}\bar{X}Y\bar{Z}+W\bar{X}\bar{Y}\bar{Z}+W\bar{X}Y\bar{Z}\end{aligned} \quad (53.15)$$

The Karnaugh map is shown in *Figure 53.7*.

The sub-cubes enclosing the terms of the function are shown by solid lines and thus by inspection the function may be minimised to yield:

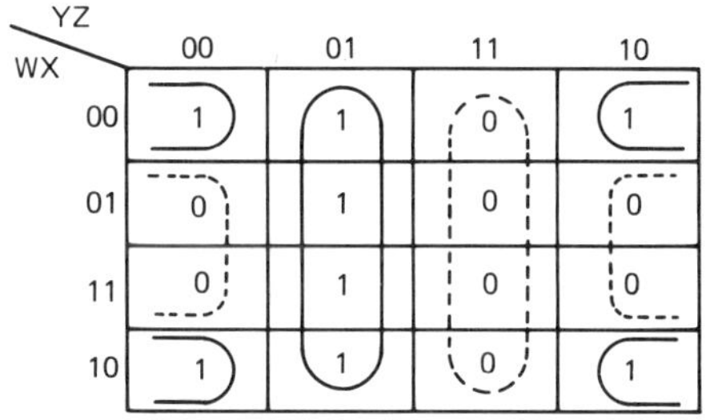

Figure 53.7 Karnaugh map of equation (53.15)

$$F=\bar{Y}Z+\bar{X}\bar{Z} \quad (53.16)$$

The looped terms are known as *prime implicants*. The cells containing 0's (that is, the terms not in the function) may also be combined as shown by the dotted lines. When reading the result the variables must be inverted and combined in the product of sums form to give

$$F=(\bar{Y}+\bar{Z})(\bar{X}+Z) \quad (53.17)$$

The dotted looped terms are known as *prime implicates*. It should be noted that it is conceptually simpler to combine the 1's rather than the 0's, even though the two results are identical.

When combining cells on a Karnaugh map the following rules should be obeyed:

(1) Every cell that contains a 1 must be included in the minimisation at least once.

(2) The largest possible group of cells must be formed. Sub-cells containing a power of 2 single cells may only be used; that is, groups of 2, 4, 8,... cells. This ensures that the maximum number of redundant variables is eliminated.

A significant advantage of the Karnaugh map minimisation technique is that it enables 'don't care' conditions to be used in the minimisation process. In some digital systems it is not necessary (or desirable) to specify the output function for all possible combinations of the input variables. This fact may be used to aid minimisation of the function as it is immaterial what value the output function takes for the input variable combination thus it may be assigned the value 0 or 1 in the Karnaugh map. Don't care conditions are assigned the value X (either 0 or 1) in the Karnaugh map. For example, consider the following function, F given by *Table 53.8*.

Table 53.8 Function F

W	X	Y	Z	F
0	0	0	0	0
0	0	0	1	1
0	0	1	0	1
0	0	1	1	0
0	1	0	0	0
0	1	0	1	1
0	1	1	0	1
0	1	1	1	0
1	0	0	0	X
1	0	0	1	X
1	0	1	0	X
1	0	1	1	X
1	1	0	0	X
1	1	0	1	X
1	1	1	0	X
1	1	1	1	X

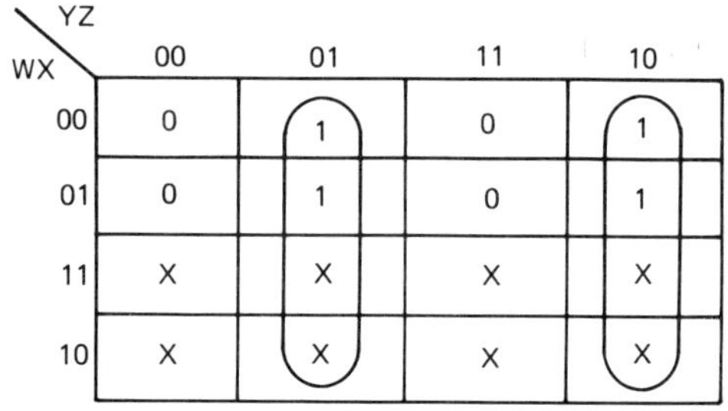

Figure 53.8 Karnaugh map of *F* defined by *Table 53.8*

The Karnaugh map is shown in *Figure 53.8*.
The function in its minimum form is:

$$F = Y\bar{Z} + \bar{Y}Z \qquad (53.18)$$

It is immediately obvious that the use of don't cares has enabled maximum sized sub-cubes to be derived thus eliminating the maximum number of redundant variables.

53.3.3 Tabular minimisation of boolean functions

The Karnaugh map minimisation technique becomes unwieldy if more than four variables are used. For problems with a greater number of variables a tabular method may be used. The technique most widely used is due to Quine and McCluckey and is based on the technique used in the Karnaugh map method; that is, examining each term and its reduced derivatives, exhaustively and systematically, applying the theorem $X\bar{Y} + XY = X$ at each stage. The result is a minimisation of the function. Although the method is normally performed by hand, it is amenable for programming into a digital computer.

The Quine-McCluskey tabular minimisation method is exemplified below. The following function is to be minimised, expressed in minterm form:

$$F = (2, 9, 11, 13, 15, 16, 18, 20, 21, 22, 23)$$

The first step is to tabulate the terms into groups according to the number of 1's contained in each term as in *Table 53.9(a)*.
The first group contains a single binary digit (2,16) the second group two binary digits (9, 18, 20) and so on. Each term in a group is then compared with each term in the group below it. If terms differ by one variable only, then they may be combined; for example, 00010 is compared with 10010 and found to differ by one variable and the term −0010 (the dash representing the redundant variable) is used to start a new list (*Table 53.9(b)*). Both terms are then 'ticked off' from *Table 53.9(a)* and the comparisons are continued until no more are possible.

The process continues by comparing each term in *Table 53.9(b)*. This time the redundant variables must also correspond.

The process terminates when no terms may be combined. The uncombined terms (unticked list entries) are the prime implicants of the function. Thus, the function *F* has been minimised to:

$$F = \bar{B}\bar{C}D\bar{E} + A\bar{B}\bar{E} + \bar{A}BE + A\bar{B}C \qquad (53.19)$$

During the minimisation process, the repeated terms in a table may be ignored in succeeding tables.

53.4 Sequential systems

A description of combinatorial logic was given in previous sections. A combinatorial logic circuit may be defined as a circuit whose outputs are a function of the present inputs only, as in *Figure 53.9*.

Table 53.9 Example of Quine–McCluskey tabular minimisation method. (a) step I, (b) step 2, (c) step 3

(a)	*A*	*B*	*C*	*D*	*E*	
2	0	0	0	1	0	√
16	1	0	0	0	0	√
9	0	1	0	0	1	√
18	1	0	0	1	0	√
20	1	0	1	0	0	√
11	0	1	0	1	1	√
13	0	1	1	0	1	√
21	1	0	1	0	1	√
22	1	0	1	1	0	√
15	0	1	1	1	1	√
23	1	0	1	1	1	√

(b)	*A*	*B*	*C*	*D*	*E*		
(2,18)	–	0	0	1	0		(1
(16,18)	1	0	0	–	0	√	
(16,20)	1	0	–	0	0	√	—
(9,11)	0	1	0	–	1	√	
(9,13)	0	1	–	0	1	√	
(18,22)	1	0	–	1	0	√	
(20,21)	1	0	1	0	–	√	
(20,22)	1	0	1	–	0	√	
(11,15)	0	1	–	1	1	√	
(13,15)	0	1	1	–	1	√	
(21,23)	1	0	1	–	1	√	
(22,23)	1	0	1	1	–	√	

(c)	*A*	*B*	*C*	*D*	*E*	
16,18/20,22	1	0	–	–	0	(2
16,20/18,22	1	0	–	–	0	
9,11/13,15	0	1	–	–	1	(3
9,13/11,15	0	1	–	–	1	
20,21/22,23	1	0	1	–	–	(4
20,22/21,23	1	0	1	–	–	

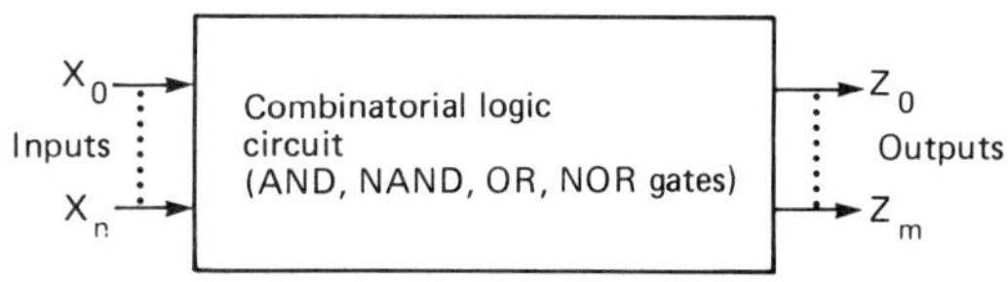

Figure 53.9 Combinatorial logic [$Z = f(X)$]

In order to describe practical digital systems it is often necessary to design circuits whose outputs are a function of the present and past inputs. For example, a combination lock must be aware of the sequence of coded digits which are input to it before the lock will open, and in order to dial a telephone number the digits must be input to the telephone system in the correct sequence. This type of digital system is known as a sequential system.

A sequential system is basically a combinatorial circuit with some of the outputs fed-back to the inputs through memory storage elements, as in *Figure 53.10*.

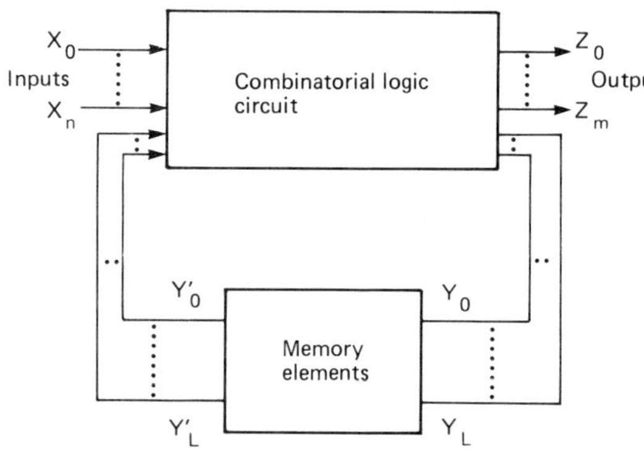

Figure 53.10 Sequential logic [$Z = f(X, Y')$, $Y = f(X, Y')$]

The memory elements are used to remember a particular function of the previous inputs to the combinatorial logic circuit. The values of these variables $Y'_0, \ldots, Y'_L$ are also known as the 'system state' of the sequential system.

There are two types of sequential system—synchronous and asynchronous. In synchronous systems, the inputs, outputs and system state are sampled at regular time intervals. The sampling is controlled by a *clock* which determines the 'frequency' of the system. In general, the inputs, outputs and system state only change on a particular phase of the clock.

In asynchronous systems, the logic circuits proceed at their own 'speed' regardless of any basic regular timing. As a result of asynchronous systems is relatively complex and the operation of such systems is difficult to analyse.

53.4.1 Finite state machines

In order to design sequential systems it is necessary to adopt a rigorous design methodology. It is normal to represent a sequential system by means of a state transition table and a state transition graph. A state transition graph is a graphical representation of a sequential system, where the system state is represented by circles, and the lines connecting the circles representing state transitions. For example, consider the state transition graph for a 2-bit counter as given in *Figure 53.11*.

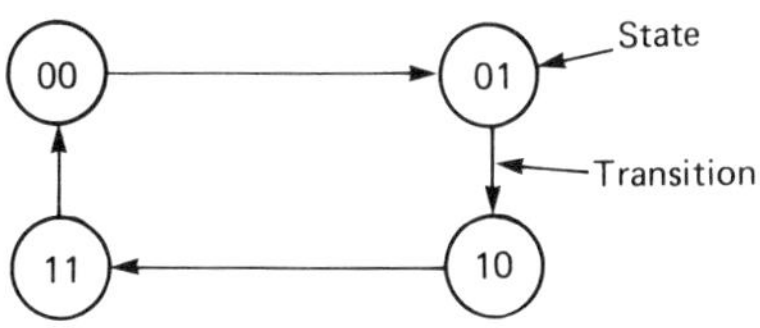

Figure 53.11 State transition graph for a two-bit counter

The stage transition graph indicates that the counter counts in the sequence 00, 01, 10, 11, 00...; thus the graph illustrates the operation of the sequential system. The state transition table may be obtained directly from the state transition graph as shown in *Table 53.10*, where *A* and *B* are the system state variables. The *Table 53.10* indicates the next state reached when the sequential system is in a particular state. The sequential system may be realised directly from the state transition table. From *Table 53.10*:

$$A' = \bar{A}B + A\bar{B}$$
$$B' = \bar{B}$$

Table 53.10 State transition table for *Figure 53.11*

Present state		*Next state*	
A	*B*	*A'*	*B'*
0	0	0	1
0	1	1	0
1	0	1	1
1	1	0	0

Thus, the sequential system may be implemented as shown in *Figure 53.12* (see also Chapter 30 for flip-flop description).

The above sequential system is known as a 'finite-state machine' as the system can only exist in one of 2^n states, where *n* is equal to the number of bits in the system state (in this case $n = 2$).

It is possible to formalise a general finite-state machine which may be used to implement any sequential system. Consider the state transition graph shown in *Figure 53.13*.

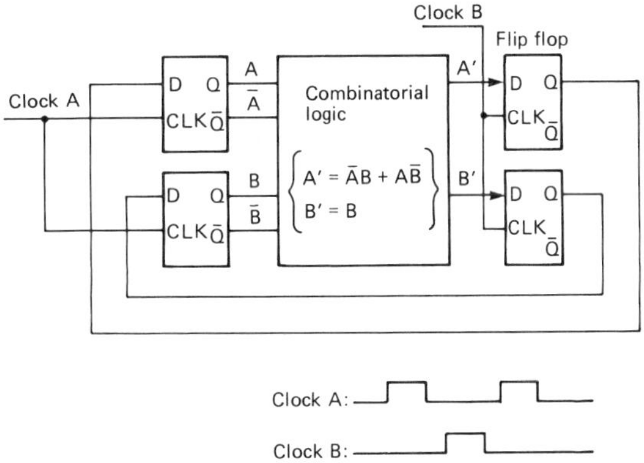

Figure 53.12 Circuit implementation of *Table 53.10*

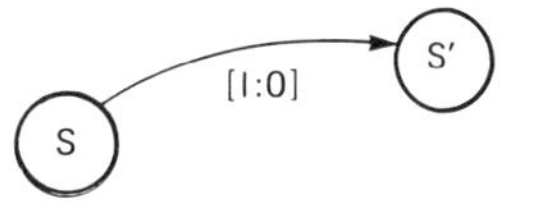

Figure 53.13 A state transition graph

The circles contain a coded representation of the system state. The transition line, or directed arc, shows the transition between states. The arc leaves one state and terminates on the next state. Alongside the directed arc is displayed the value of any inputs [I] which existed at the beginning of the state transition sequence and also the value of the outputs [O] produced by the state transition sequence. The inputs cause certain 'conditional' stage transitions to occur, which are controlled by the state of the input variables. Thus, the general finite-state machine must be able to generate outputs and state transition sequences which are controlled by inputs and the present system state. The general finite-state machine may be implemented as shown in *Figure 53.14*

In *Figure 53.14* the following symbols are used:

R = collection of *n* bistables
$mf = f(I, S)$ = machine function
$sf = f(I, S)$ = (state function)
S = system state
I = input variables
O = output variables

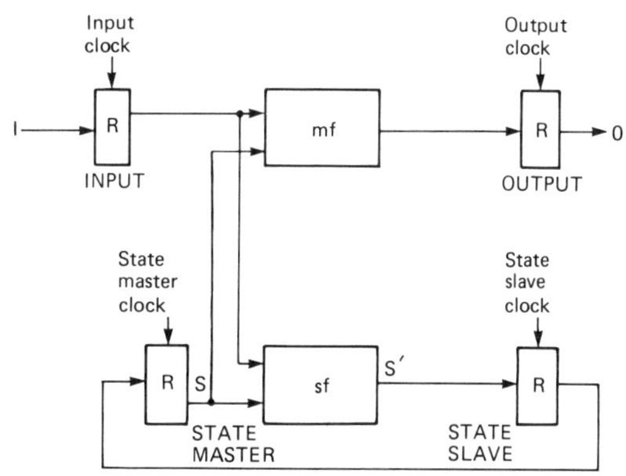

Figure 53.14 General finite-state machine

Note that four non-overlapping clocks are required for the correct sequencing of the finite-state machine. A practical example will now be considered:

It is required to design a finite-state machine which recognises the 3-bit pattern 110, to produce an output Z, in a continuous serial bit stream. A typical serial bit stream input is 1100101011001....

The first step is to derive the state transition graph for the finite-state machine and then to generate the state transition table from the graph. The state transition graph is shown in *Figure 53.15*.

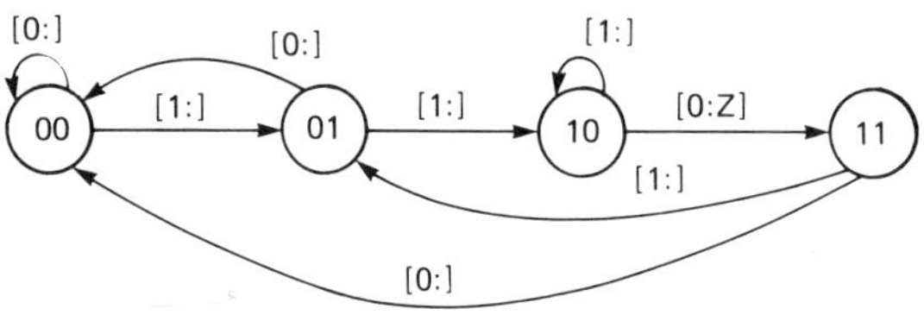

Figure 53.15 State transition graph for the text example

The state transition table is given in *Table 53.11*.

Table 53.11 State transition table for *Figure 53.15*

Input X	*Present state* A	B	*Next state* A'	B'	*Output* Z
0	0	0	0	0	0
0	0	1	0	0	0
0	1	0	1	1	1
0	1	1	0	0	0
1	0	0	0	1	0
1	0	1	1	0	0
1	1	0	1	0	0
1	1	1	0	1	0

The equations for the state function and machine function may be derived from the state transition table:

$$\text{State function} = sf = f(I, S)$$
$$= A' = A\bar{B} + X\bar{A}B$$
$$B' = \bar{X}A\bar{B} + X\bar{A}\bar{B} + XAB$$

$$\text{Machine function} = mf = f(I, S)$$
$$= Z = \bar{X}A\bar{B}$$

Thus, the finite-state machine may be implemented as in *Figure 53.16*.

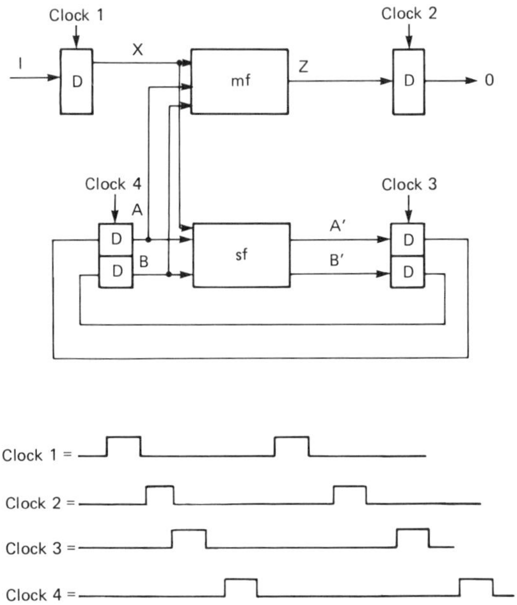

Figure 53.16 Implementation of finite-state machine of the text example

53.5 Threshold logic

Threshold logic is the term given to a special type of boolean function, where the function is determined by a set of weighted inputs to a threshold logic element.

A boolean function $F(X)$ of n binary variables is defined to be a threshold function if the following conditions are met:

$$F(X) = 1 \quad \text{if} \quad f(X) = \sum_{i=1}^{n} W_i x_i \geq T$$

$$F(X) = 0 \quad \text{if} \quad f(X) = \sum_{i=1}^{n} W_i x_i < T$$

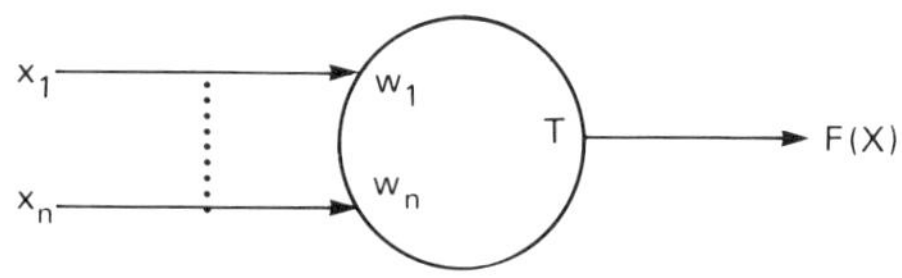

Figure 53.17 Threshold logic element

where

$X = (x_1, x_2, \ldots, x_n)$
x_i = binary variable
w_i = weight of x_i
T = threshold value
$f(X)$ = algebraic function of X

Thus, the output of a threshold logic element is 'true' if the weighted sum of its inputs is equal to or exceeds a predetermined threshold value, T.

A threshold logic element is a physical device whose inputs consist of n binary variables and whose output is a threshold function of the input variables. A threshold logic element (or threshold gate) may be represented by *Figure 53.18*.

Consider the following threshold function:

$$F = A\bar{C} + A\bar{B} \qquad (53.20)$$

Table 53.12 indicates the combinations of the input variables which produce a 'true' output and those which produce a 'false' output.

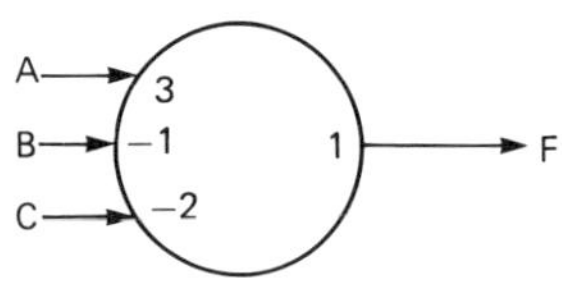

Figure 53.18 Threshold function implementation of equation (53.20)

Table 53.12 True and false output for equation (53.20)

	'True' output				'False' output		
	A	*B*	*C*		*A*	*B*	*C*
(1)	1	0	0	(5)	0	0	1
(2)	1	0	1	(6)	0	1	0
(3)	1	1	0	(7)	0	1	1
(4)	0	0	0	(8)	1	1	1

Thus three input combinations generate a 'true' output and five do not. The problem is to assign weightings to each of the input variables and a value for the threshold such that the terms (1) to (3) produce a ptrue' output only. It is normal practise to assign integer weightings to each of the binary input variables and it is possible to have both positive and negative weighting values.

A possible set of weightings, together with the corresponding threshold value, is shown below:

$W_A = 3 \qquad W_B = -1 \qquad W_C = -2 \qquad T = 1$

This is normally expressed in terms of a threshold function, $R[F]$, where:

$R[F] = (3, -1, -2{:}1)$

Thus, input term (3) produces a 'true' output as the sum of its weighted values exceeds the threshold value $(3+(-1)+0=2>T(1))$ and input term (8) does not produce a 'true' output as the sum of its weighted values is not equal to nor does it exceed the threshold value $(3+(-1)+(-2)=0<T(1))$. The threshold function may be implemented as in *Figure 53.18.*

For n binary variables there are 2^{2n} possible switching functions. This, however, is not the case for threshold functions. In general, there are fewer threshold functions than switching functions; for example, for three binary variables there are 256 possible switching functions but only 104 possible threshold functions. Thus, before a boolean function may be implemented as a single threshold function, it must be determined if the boolean function is expressible as a threshold function. If this is the case then the relative weightings of the input variables and the threshold value must be determined for the threshold logic element. This is a difficult process and involves complex mathematics, which is outside the scope of this discussion.

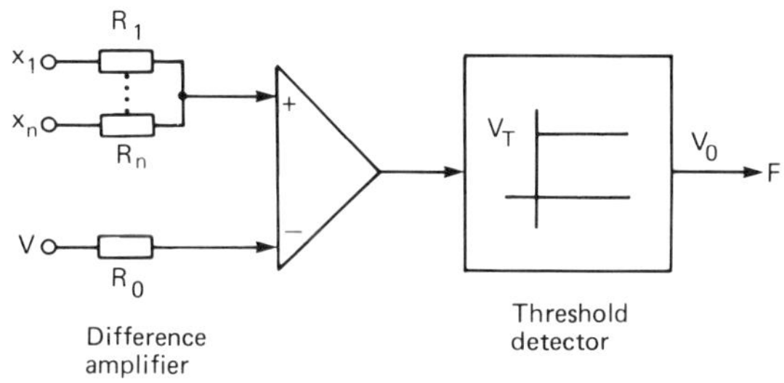

Figure 53.19 Circuit implementation of threshold logic elements

53.5.1 Implementation of threshold logic elements

Threshold logic elements may be physically realised by an operational amplifier and threshold detector circuit, as shown in *Figure 53.19.*

The input variables are either 0 or V volts. The output of the difference amplifier is either a positive or negative voltage depending upon the weighted sum of the inputs (where $w_i \propto 1/R_i$). If negative weightings are required then the input variable should enter the negative input terminal of the difference amplifier. The threshold detector produces an output (V_0 volts) if its input voltage exceeds or is equal to a threshold voltage, V_T; otherwise it produces an output of zero volts. It must be noted that there will be practical limits to the number of inputs to the difference amplifier and the relative sizes of the input variable weightings.

Threshold logic elements may be used as general logic elements as the basic boolean operations may be readily realised:

Boolean operation $[F]$	*Threshold function* $R[F]$
AND of n variables	$[1, 1, \ldots, 1{:}n]$
OR of n variables	$[1, 1, \ldots, 1{:}1]$
NAND of n variables	$[-1, -1, \ldots, -1{:}(1-n)]$
NOR of n variables	$[-1, -1, \ldots, -1{:}0]$
Inverter	$[-1{:}0]$

Thus, complex boolean functions may be implemented by a single type of threshold logic element by merely adjusting the weights and threshold value.

The use of threshold logic elements is not widespread at present, however, their usage may become more popular if the threshold logic elements may be implemented without recourse to the use of basic analog circuitry.

Part 5

Applications

54 Television and Sound Broadcasting

J L Eaton, BSc, MIEE, CEng
Consultant (Broadcasting)
(Sections 54.13–54.17)

S M Edwardson, CEng, FIEE
BBC Research Department
(Sections 54.18–54.20)

R P Gabriel, BSc, CEng, FIEE, MIERE, FIEE
Formerly Chief Engineer,
Rediffusion Ltd
(Section 54.6)

P Hawker
Independent Broadcasting Authority
(Sections 54.1–54.5)

C R Spicer, AMIEE
Senior Design Engineer,
BBC Designs Department
(Sections 54.7–54.12)

Contents

54.1 Fundamentals of television transmission

For television broadcasting, as for sound radio, the first step is the transmission of a carrier frequency from a transmitter on which the picture information can be radiated and picked up by the receivers. The carrier frequency is selected in accordance with the appropriate channels used in the country concerned. The wanted transmission, provided that the receiver is located within its service area, may be selected by tuning the receiver, although in practice this may be done by means of pre-tuned push or touch buttons.

The carrier frequency links transmitter to receiver and is modulated with the picture information. For terrestrial television broadcasting the picture or *video* signal is transmitted using *amplitude modulation*. However for space satellite broadcasting, where transmitter power is at a premium, it may be preferable to use FM for transmitting the video signal.

54.1.1 Scanning

Television systems are based on translating a scene into a series of small points, termed *picture elements*, by methodically tracing out or *scanning* the picture, and then transmitting the intensity or light value of each picture element in sequence. In practice the first step in deriving the video signal for a particular scene is to form an optical image on the photocathode of a camera tube. This image is scanned by the electron beam which moves across the image in much the same way as a printed page is read: i.e. from left to right across the top of the image, with a rapid return to the left-hand side of the image, to 'read-off' a second line just below the first, until the entire 'page' has been read. When the bottom of the image is reached the electron beam returns to the initial top left-hand position to read off the next 'page' as shown in *Figure 54.1*.

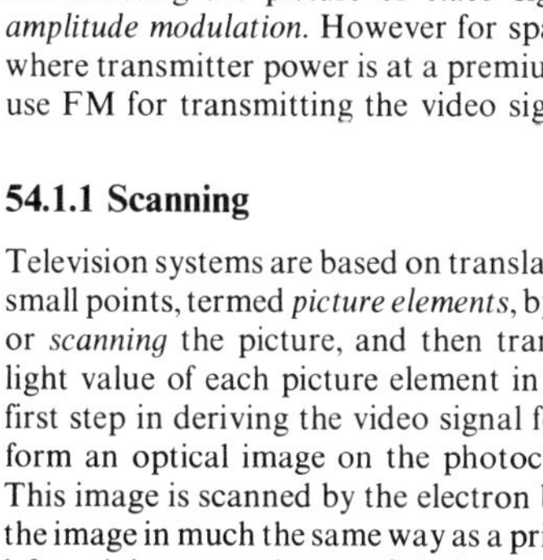

Figure 54.1 Simplified representation of sequential scanning

At the receiver a reproduction of the image is built up on the screen of the picture tube by an electron beam moving in precise synchronism with the beam in the camera tube. The screen glows with an intensity dependent upon the beam current and the glow persists for a short time: this combined with the persistence of vision of the eye gives the viewer the impression that the picture tube face is presenting a complete picture; moreover if the subsequent pictures follow sufficiently rapidly the illusion of movement, as in a cinematograph film, appears natural.

The movement of the camera-tube beam across the optical image, or picture elements, from left to right is termed a line scan, and the rapid return to the left-hand side of the picture the line flyback. In a simple form, supposing a complete picture was transmitted as a series of 100 lines to each picture then it would be a 100-line television system. In practice however certain periods of time are required for synchronising purposes, reducing the number of *active* lines to a lower figure than the theoretical.

A series of still images presented to the eye at the rate of about 10/s provides an illusion of continuous motion but accompanied by pronounced flicker. If the rate is increased to 25/s, flicker is reduced but still noticeable, particularly when the images are bright. A repetition rate of 50/s provides motion virtually without flicker, although for very bright pictures some 60 images are desirable. The rate at which images follow one another is called the *field* or *frame* rate.

54.1.1.1 Interlaced scanning

To transmit a complete high-definition picture at a sequential field rate of 50 or 60/s would require an excessively large bandwidth. An alternative technique is called *interlacing*: instead of transmitting each horizontal line in sequence, alternate lines are scanned first, with the missing lines traced out subsequently as shown in *Figure 54.2* thus providing two half-scans for each complete picture. Since the eye hardly perceives individual lines,

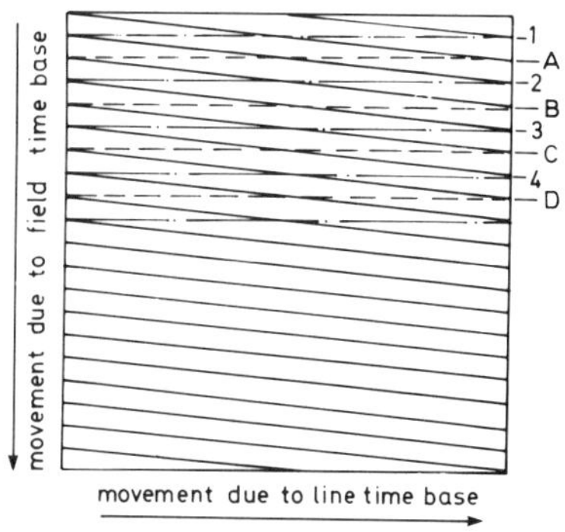

Figure 54.2 Simplified representation of double interlaced scanning used for all current television broadcasting

the effect is to provide 50 fields or images per second with 25 complete pictures per second, even though the total detail, and hence the bandwidth, is that of a 25 pictures-per-second system. However it is possible with modern techniques to provide roughly equivalent pictures with sequential scanning, though interlacing is universally incorporated in broadcast standards. With interlacing the final line of the first field and the first line of the second field are half lines, the line scan starting or ending at the centre of the top and bottom of the picture.

54.1.2 Video and sync signals

Since the scanning process must be carried out synchronously at both the studio and the receiver it is necessary to provide *synchronising signals* to ensure exact correspondence between the two scanning processes. This is achieved by transmitting timing pulses which are used by the receiver as a reference to which to lock its horizontal and field times bases (it would be feasible in modern practice to provide only one set of timing pulses from which the other set could be derived using digital counting techniques). The way in which the picture signal is combined with the line sync pulses is illustrated in *Figure 54.3* and the waveform of the field sync signals is given in *Figure 54.4*.

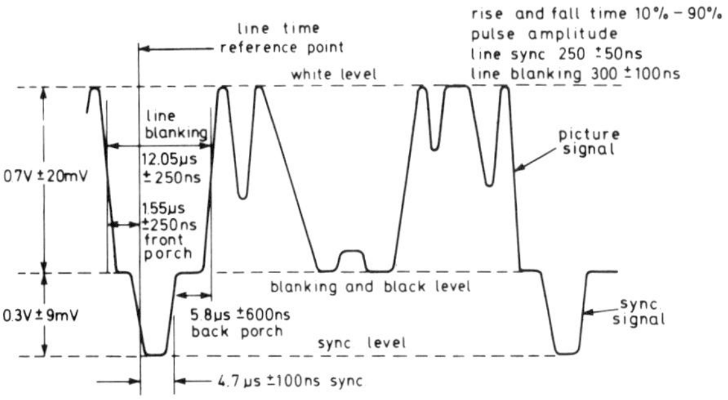

Figure 54.3 The waveform of a typical line showing synchronising signals

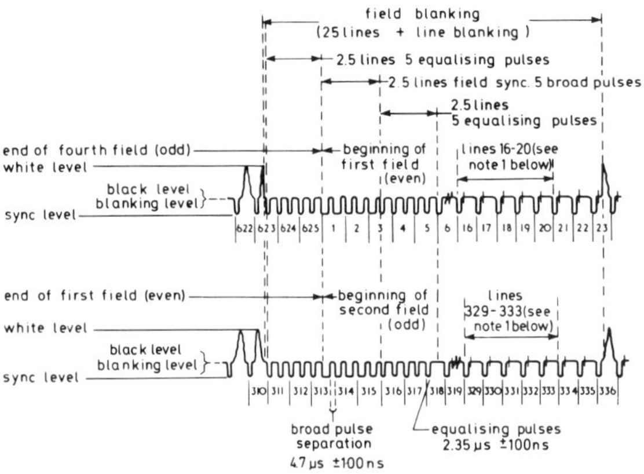

Figure 54.4 Field synchronising and blanking waveforms for a typical signal. Lines 7 to 14 and 320 to 327 have been omitted. Lines 16 to 20 may contain identification, control, or test signals or teletext data. The first and second fields are identical with the third and fourth in all respects except burst blanking

The signals required to transmit the proportional brightness of each picture element of a high-definition system are of much higher frequency than for speech or music. As each line is traced out, the output of the camera varies in voltage, changing with changes in light value. If the line consists entirely of white or black or the same shade of grey the output is steady DC or 0 Hz. If the line represents a fine network of vertical lines of alternate black and white picture elements the output is at a frequency of several MHz.

It is common practice to term composite picture and synchronising information as *video* signals, but during radio transmission when these frequencies are modulated on a carrier wave, it is termed a *vision* signal.* Video signals are normally distributed as 1 V peak-to-peak signals, across an impedance of 75 Ω.

A high-definition television system thus requires:

(1) Transmission of picture information at frequencies extending to several MHz.
(2) Transmission of synchronising pulses to keep the received time bases in step with those used in the studio; the transmission of these pulses equally requires wide bandwidth.
(3) Transmission of an accompanying sound channel.

* The AM may be positive or negative as indicated in *Figure 54.4*.

The precise way in which these requirements are met and the associated frequencies and tolerances together form a television *standard*. There are a number of different standards in current use, including 525-line 30 pictures (60 fields) and several variations of 625-line 25 pictures (50 fields) are the most significant, although 405-line and 819-line standards are still in use in the UK and France, but are being phased out.

54.1.2.1 Bandwidth

The maximum upper frequency of a video signal is governed by the picture content and the scanning standard being used. For the UK System I this is normally taken as 5.5 MHz.

Were conventional double-sideband AM to be used, this would imply a maximum vision bandwidth of $2 \times 5.5 = 11$ MHz. In addition, further bandwidth would be required for the sound signal. In view of the restricted frequency spectrum at VHF and UHF, it has for many years been the practice to reduce the bandwidth of one of the two sidebands to produce *vestigial sideband* or *asymmetric sideband* vision transmission. In System I the full upper sidebands to 5.5 MHz are transmitted but the bandwidth of the lower sideband is restricted to 1.25 MHz (in 625-line System G as used in many countries the lower sideband is restricted even more to 0.75 MHz).

This means that for System I the total vision bandwidth is 1.25 + 5.5 or 6.75 MHz and a gap of 0.5 MHz is left before the sound carrier, which is thus 6 MHz above the vision carrier, as shown in *Figure 54.5*.

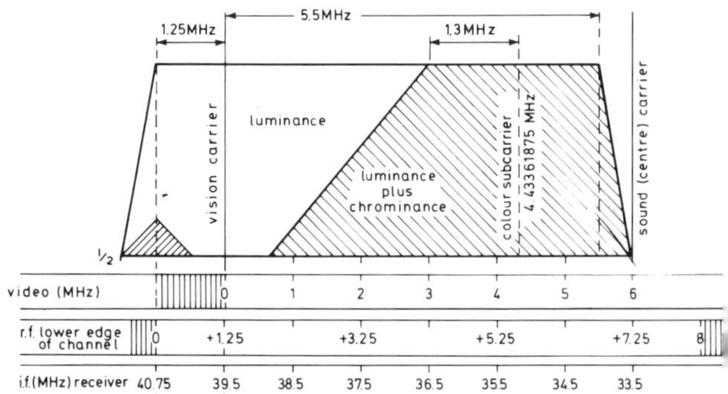

Figure 54.5 The frequency bands occupied by colour picture components and sound signal from an ideal 625-line system-I transmitter as related to video, vision, and the i.f. of a receiver (BREMA standard i.f.)

For both Systems I and G in the European region the agreed international UHF channels occupy 8 MHz; in System I the vision carrier is always +1.25 MHz from the lower end of the channel; the FM sound carrier +7.25 MHz. *Figure 54.5* shows the vision response curve of System I as related to the video signal, the RF channel, and the standard BREMA receiver IF channel. The FM sound is transmitted with a maximum AF signal of 15 kHz and with a peak carrier deviation (corresponding to a 400 Hz tone at a level of +8 dBm at the modulator input) of ±50 kHz. The pre-emphasis time constant is 50 μs.

The use of an asymmetric sideband vision signal results in a degree of quadrature distortion when the signal is envelope demodulated: this form of distortion (which can usually be detected only with some difficulty, even on a test card) can be eliminated by the use of synchronous demodulation.

54.2 Colour television

54.2.1 Development

Demonstrations of low-definition colour television were made by Baird as early as 1928 and later during World War II he was able to demonstrate a system of potentially greater value. The first public service in colour was inaugurated in the United States in 1951, although this system occupied a large bandwidth and could not be satisfactorily received either in colour or black-and-white except on a receiver designed for the system. It was soon appreciated that to achieve wide acceptability a public colour system needed to be *compatible*, that is the transmission should be receivable as a black-and-white picture on a monochrome receiver, and have *reverse compatibility*, that is allow the reception of black-and-white pictures on the colour receiver. These requirements were carefully studied and a specification for a colour system was drawn up by the National Television System Committee (NTSC) in the United States, and transmissions to this specification began on January 1, 1954. A further important feature of this system was that the colour information is encoded into the video signal in such a way that the transmissions occupy the same bandwidth as the black-and-white transmission. During the period from 1949 onwards, RCA successfully developed the shadow-mask colour display tube which, although difficult to manufacture, was an important and far-reaching advance on earlier colour display systems.

The early spread of colour in the United States was relatively slow and some engineers believed this was due to the variable quality of the colour, primarily due to instrumental and transmission shortcomings. It was not until about 1964 to 1966 that demand for colour receivers became widespread.

In Europe the slow growth of American colour television was attributed to some degree to the susceptibility of NTSC signals to relatively small changes of *differential gain* and *differential phase* anywhere within the transmission chain; further, the development in 1956 of quadruplex videotape black-and-white recording appeared to pose particular problems if it was also to meet the more stringent colour requirements. Differential phase refers to changes of phase of the colour subcarrier as a result of changing brightness levels; similarly differential gain is any change of gain with brightness.

As a result alternative encoding systems were developed and studied, though most of the other techniques of NTSC were retained. Two systems proved of particular importance: SECAM (sequential and memory) which introduced the concept of a delay line; and PAL (phase alternation line) which combined the delay line of SECAM with the suppressed subcarrier techniques of NTSC.

All three systems were found to have advantages and disadvantages, and no common agreement could be reached. The outcome was that NTSC is used in North America, Japan, and some other countries; PAL is used in the UK, many European countries, South Africa, Australia, New Zealand, and some others; SECAM is used in France, USSR and East Europe, and some African and Middle East countries. With the improved techniques now available each system is capable of giving roughly similar results, though PAL and SECAM remain less susceptible to transmission errors; SECAM presents rather more problems in the studio, although relatively easy to record on tape. It is possible to transcode between systems and the development of digital standards conversion also allows the signals to be readily converted between the main world standards. The development of fully-digital and time-multiplexed analogue systems based on component video rather than composite video is of increasing practical importance for studio and satellite transmissions.

54.2.2 Colour television principles

The sensation of colour derives from the different reactions of the eye/brain system to visible electromagnetic radiation at different frequencies: a colour is what we feel when we look at light of a predominant wavelength: see *Figure 54.6*. The eye is more sensitive to radiation in the middle range of the visible spectrum (green) than to radiation towards the extremities (violet and red, orange and blue).

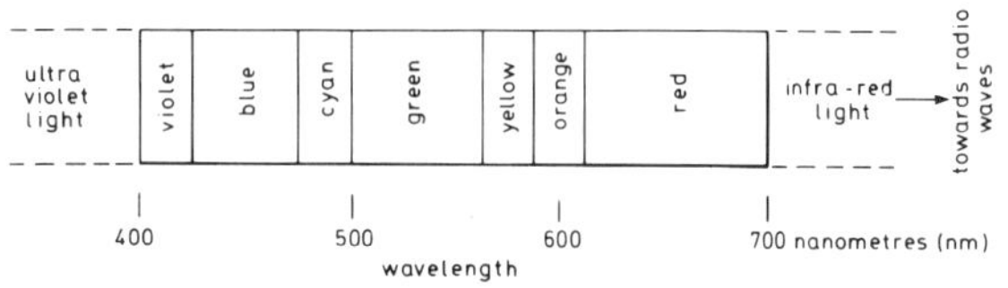

Figure 54.6 The electromagnetic spectrum of visible light. Short wavelengths such as these may be quoted in Angstrom units (10^{-10} m)

White light is the reflection of a uniform radiation throughout the visible spectrum; if the brightness is reduced such radiation appears grey; if little or no light is reflected the object appears black. Colour may be experienced by looking directly at a source of radiation or at reflected light.

The eye is less sensitive to the fine detail of colour than to fine detail in black and white: this fact allows colour information to be transmitted more crudely (less bandwidth) than black-and-white.

The experience of almost any colour can be obtained by adding (*additive mixing*) specific proportions of the three primary colours red, green, and blue. In painting and printing, colour mixing is a *subtractive process* and the primary colours are then red, yellow, and blue.

Any colour can thus be defined by three characteristics: its brightness (*luminance*); its *hue* (dominant electromagnetic wavelength or frequency); and its intensity (*saturation*) which corresponds to its colourfulness.

The standard adopted to define whiteness is known as the *colour temperature* (which determines the amount of blue or green in peak white). The standard adopted in the UK is *Illuminant D* corresponding to 6500 K.

Luminance, the characteristic used in monochrome television, conveys details of the varying levels of brightness of the picture elements. In any compatible colour television system luminance is transmitted, allowing a monochrome picture to be received.

In addition, for colour display, it is necessary to radiate additional information about the *hue* and *saturation* characteristics: such information is termed the *chrominance* (often abbreviated to *chroma*) signal. Saturation is a measure of the intensity or colourfulness of a colour: 0 per cent saturation is entirely grey with a complete absence of colour; 100 per cent saturation has no dilution of the colour with white. An example is that pink and red light may both have the same predominant wavelength (i.e. the same hue) but the red is the more fully saturated colour.

Chrominance information defines the hue and saturation of the picture independently of the luminance, so that theoretically any distortion of the chrominance information does not affect the detail of the picture: a system designed to fulfil this is termed a *constant-luminance* system; in practice some departure from this ideal is to be found in most systems.

To define completely the luminance and chrominance information, a colour television camera analyses the light from the scene in terms of its red (*R*), green (*G*) and blue (*B*) components by means of optical filters and then gamma corrects to take into account differences between camera and display tube characteristics; gamma-corrected signals are termed *R′*, *G′*, *B′* signals. For closed-circuit sequential colour systems, these three signals may be kept separate, but for a compatible broadcast system the signals are processed to provide the basic luminance signal (Y); in some cameras a fourth pick-up tube is used for operational reasons to obtain a luminance signal although it should be appreciated that the basic R, G, B signals contain all the information required to define luminance, hue and saturation. The green and red signals contribute more to luminance than blue and in practice a matrixing network is designed to the following specification.

$$Y' = 0.3R' + 0.6G' + 0.1B'$$

The four signals, Y', R', G', B' are related mathematically and it is therefore unnecessary to transmit all four. Since for compatibility we require a Y' luminance signal, this is transmitted. Two other signals are obtained by taking the red and blue signals and subtracting from them the luminance signal: that is $(R' - Y')$ and $(B' - Y')$. $(G' - Y')$ can be derived within the receiver by matrixing.

The transmitted signals are thus Y' (i.e. $0.3R' + 0.6G' + 0.1B'$); $(R' - Y')$ and $(B' - Y')$. These allow us to recover Y', R', G', B' in the receiver.

The current colour systems depend on this process but differ in the way that the three signals are transmitted within a single vision channel. Both NTSC and PAL transmit all three signals simultaneously; SECAM transmits luminance continuously but radiates on any given line only one of the two chrominance signals, changing over between each line. This indicates that in SECAM the vertical colour information is only one-half that of the other two systems, although this is of only minor practical significance since the eye is unable to detect fine colour detail. Delay lines used in SECAM decoders can be more tolerant than those required for PAL.

Although the chrominance signals generated in the colour camera occupy the full range of video frequencies, it is only the luminance signal that requires to be transmitted to this degree of resolution. The ability to resolve fine detail depends on *visual acuity*, and our ability to resolve colour in small details of a picture is inferior to that for corresponding black-and-white or grey pictures. Since the human eye does not resolve colour in small areas there is no need to reproduce this, even for high-quality television pictures. This influences many aspects of colour television, not least the ability to limit the bandwidth of chrominance information relative to that required for the luminance signal; in practice chrominance information in a 625-line system can begin to roll off at about 1.3 MHz. The restriction of chrominance bandwidth makes possible *inband* transmission of chrominance information, although this technique gives rise to some loss of compatibility (dots and crawl being seen on a monochrome picture) and also cross-colour effects on colour reproduction (flaring of patterned jackets is a common example), due to luminance signals appearing in the colour channels.

Basically the information in a monochrome signal is distributed in a series of packets separated by the line frequency, with only little spectrum energy in the gaps between as shown in *Figure 54.7*. By choosing a colour subcarrier frequency (see below) that is accurately placed between multiples of the line frequency, and noting that chrominance modulation energy is similarly in the form of packets it is possible to interleave the basic energy spectra of the luminance and chrominance signals: for 625-line transmission the colour subcarrier frequency is maintained very precisely at 4 436 618.75 Hz $\pm$1 Hz, with a maximum rate of change of subcarrier frequency not exceeding 0.1 Hz/s.

The relationship between subcarrier and line frequency is

$$f_{sc} = (284 - \tfrac{1}{4})f_h + \tfrac{1}{2}f_v \text{ Hz} \tag{54.1}$$

where f_h is the line frequency and f_v is the field frequency.

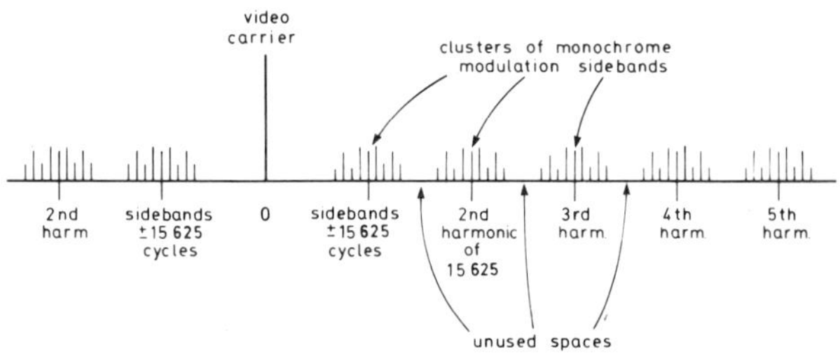

Figure 54.7 In a monochrome television transmission the sidebands are in groups around harmonics of 15 625 Hz leaving spaces in which colour information may be inserted

54.2.3 Chrominance transmission

In the PAL and NTSC colour systems, both of the colour-difference signals are transmitted simultaneously. Both channels of information are transmitted using a single subcarrier frequency with the subcarrier itself suppressed by means of balanced modulators except for a brief synchronising burst (the *colour burst*) which is transmitted following each line synchronising pulse, that is to say it is radiated only during the period known as the back porch of the line pulses as shown in *Figure 54.8*. This brief colour burst is used to correct an accurate

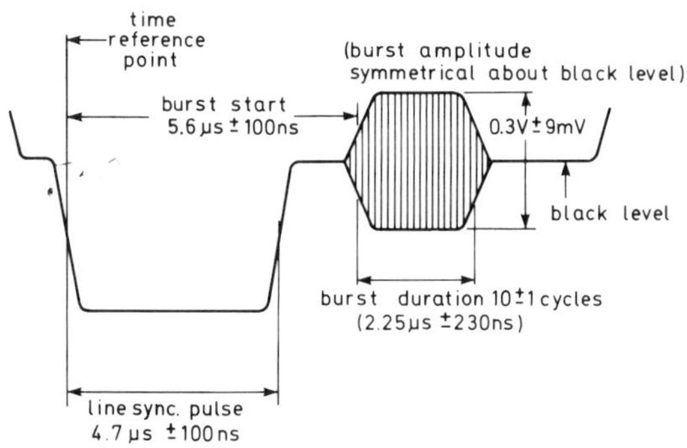

Figure 54.8 The colour burst on a standard level signal (700 mV white level)

reference oscillator in the colour receiver; by this means not only can the subcarrier be suppressed during the transmission of picture information, but it allows the receiver to take advantage of the technique of *quadrature modulation* and to recover separately two channels of colour information using the process known as *synchronous demodulation*: see *Figure 54.9*. One of the two chrominance signals is shifted through 90° of phase so that the peaks of signal amplitude on one channel occur precisely at the zero cross-over points of the other (a similar technique is used in stereo disc recordings).

Since this means that the subcarrier is modulated in both amplitude and phase, the two chrominance signals may be represented by a single phasor of which the length represents the saturation and angular direction (0–360°) the hue as shown in *Figure 54.10*.

The amplitude of the phasor, since it represents saturation, reduces to zero for grey/black tones and is low when representing low saturated colour. This indicates that with PAL and NTSC colour systems little chrominance energy is radiated except for

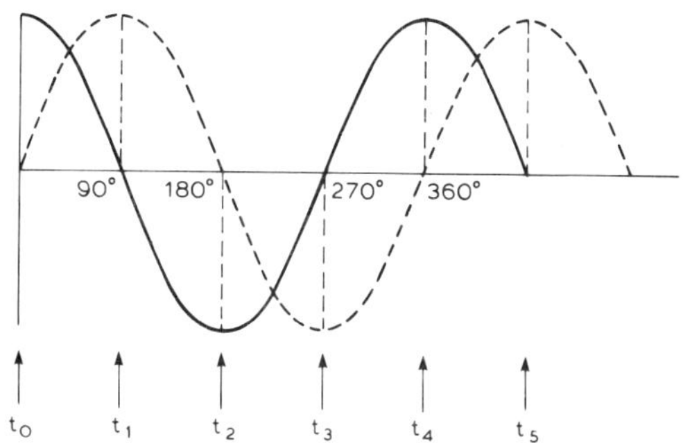

Figure 54.9 The principle of synchronous demodulation of quadrature-modulated signals. Two carrier waves are shown of the same frequency but in quadrature (i.e. with 90° phase difference between them). If each is amplitude modulated by different information and if two demodulators, one synchronised with each carrier, sample the waves only at the times when one is at a peak and the other at null, then the two sets of information can be separately retrieved at the receiver. The solid-line waveform is sampled at times t_0, t_1, t_4, etc. and the dashed-line curve at times t_1, t_3, t_5, etc.

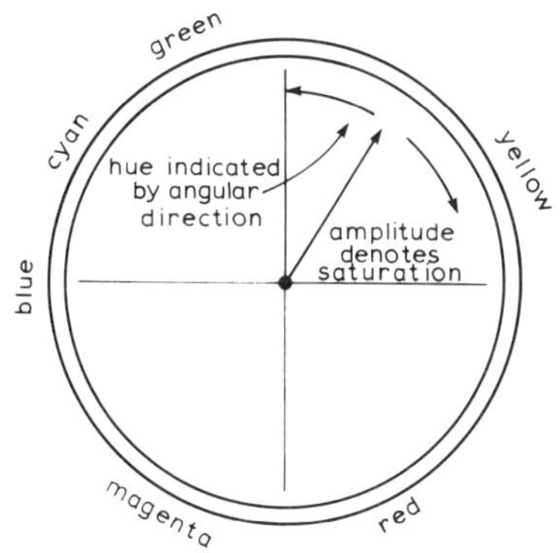

Figure 54.10 In the NTSC and PAL systems any hue can be represented by a phasor having a specific phase angle and an amplitude representing the degree of saturation

heavily saturated colours, and this feature improves the compatibility of the black-and-white picture.

In the NTSC system the phase of the colour phasors remains constant throughout transmission, the hue information being determined by the degree of synchronism between the studio source and the crystal-controlled reference oscillator for the synchronous demodulators in the receiver. Any disturbance of this synchronism anywhere along the transmission path results in a change of hue, unless corrected or compensated for in the receiver. It was this susceptibility of NTSC to differential phase that led to the development of SECAM and PAL, although many of the colour errors noted in early receivers were due to camera drift and instrumentation faults that have today been largely overcome.

54.2.4 PAL transmission

The PAL system differs from NTSC in that the phase of one of the chrominance signals is reversed during alternate line periods. There are in effect two different colour circles in use, and phase errors induced in one of these colour circles has an equal but opposite effect on the other.

In practice the chrominance signal corresponds to the sideband components of two AM subcarriers in phase quadrature, identified by the letters *U* and *V*, the phase of the *V* axis component being electronically reversed after every line. Colours are thus represented by the amplitude and phase appropriate for the particular line. The burst of subcarrier following the line synchronising pulse is used to establish the reference subcarrier phase in the receiver and *also* to identify and synchronise the switching of the *V* axis (*swinging burst*). This is done by making the phase of the colour burst $\pm 135°$ on odd lines

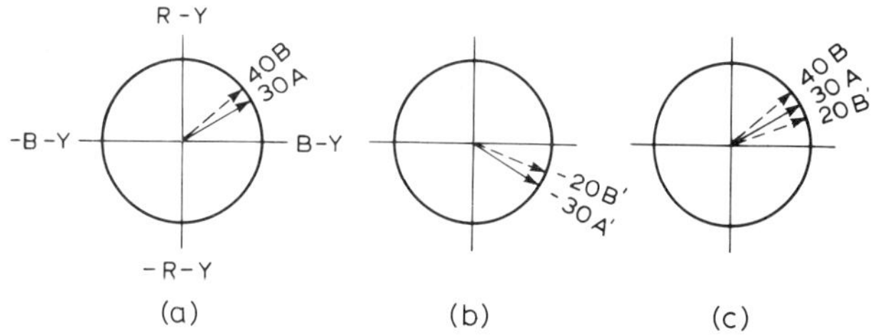

Figure 54.11 Principle of PAL automatic line correction by averaging. (a) Hue *A* transmitted with a positive (*R–Y*) component; received as *B* because of a phase shift; (b) same hue transmitted with negative (*R–Y*) component *A'*: received as *B'* due to a phase error equivalent to that at (a); (c) after reversing the polarity of (b) the two received signals *B* and *B'* represent *B* and *B''* which when averaged give the correct hue *A*

of the first and second fields and even lines of the third and fourth fields. On even lines of the first and second fields and odd lines of the third and fourth fields, the phase of the colour burst is $+225°$. The mean burst phase is held within $180 \pm 2°$ of the reference axis. This allows the PAL receiver to identify whether the *V* axis $(R' - Y')$ signal is being transmitted in positive or negative form. Receivers incorporating a delay line can provide tolerable colour on signals subjected to as much as 40° differential phase, although every effort is made during transmission to keep radiated errors far below this figure; multipath propagation, however, can induce errors after the signals have left the transmitting antenna.

Between a programme originating source, such as a camera, and the transmitter output, many items of equipment in the signal path are capable of introducing distortion into a colour signal and for this reason broadcasting authorities may specify codes of practice covering studio and outside-broadcast equipment, transmission networks, transmitters, and distinguishing between direct path and worst path conditions, anticipating that, for example, many signals pass more than once through a video-tape recorder in the process of production and editing. Such codes normally specify the permissible tolerances of output signal level, non-linearity distortion (including total phase errors and differential gain), linear distortion, noise and accuracy of synchronising signals.

54.2.5 Insertion test signals

During the field blanking interval, certain of the unused lines may carry additional information, for example, teletext (Oracle/Ceefax) data transmissions, identification, control and test signals. Lines used for such information on the 625-line standard are 16 to 20 inclusive on first and third (even) fields and 329 to 333 inclusive on second and fourth (odd) fields.

Operational teletext transmissions in the UK used lines 15 (328), 16 (329), 17 (330) and 18 (331) and may be further extended later.

UK national insertion test signals are normally transmitted on lines 19 (332) and 20 (333).

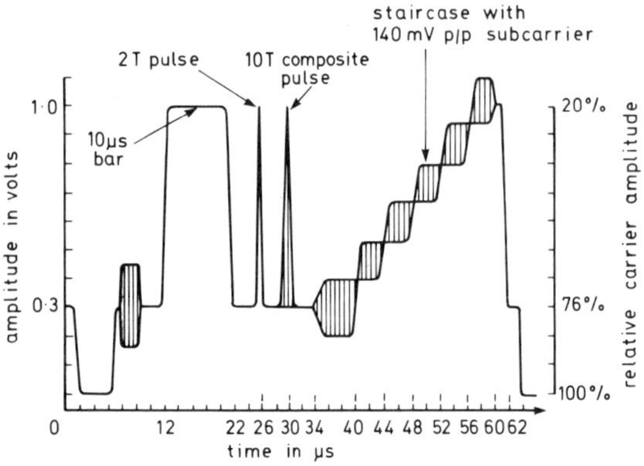

Figure 54.12 Insertion test signal 1 (lines 19 and 332)

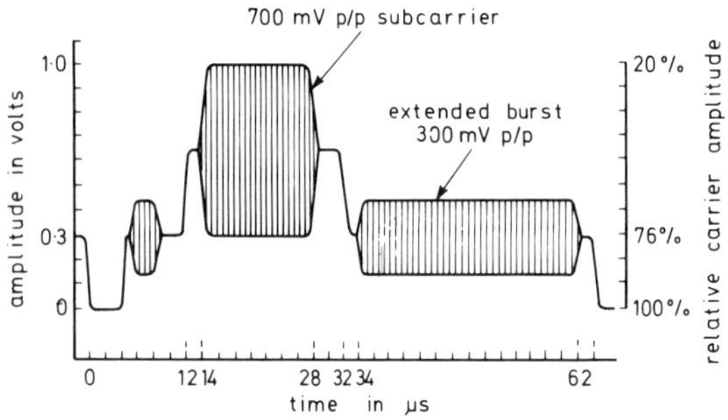

Figure 54.13 Insertion test signal 2 (lines 20 and 333)

The UK test signals 1 and 2 are shown in *Figures 54.12* and *54.13*.

Measurements of the test signals allow various distortions to be measured where required throughout the transmission chain. The advantage of the insertion test signal technique is that it allows optimum performance to be achieved without in any way interrupting normal picture transmission. Both manual and automatic monitoring of insertion test signal is possible.

The 2T pulse and 10 μs bar on test signal 1 enable the *K* rating to be found, while measurement of line–time non-linearity may be made by passing the staircase waveform through a filter of restricted bandwidth. The 10T composite pulse allows chrominance-to-luminance gain and delay inequalities to be assessed.

Chrominance-to-luminance crosstalk can be measured from the bar waveform on test signal 2. The extended burst on test signal 2 is of constant phase and amplitude and is intended for demodulating the subcarrier on the previous line in order to measure the differential phase.

The following are illustrations of the relevance of measurements made using interval test signals:

Measurement	*Relevance*
Width of 2T pulse	Resolution
Tilt on 10 μs bar	Smear
Pulse/bar ratio	Amplitude/frequency characteristics
2T echoes and rings	Phase response; echoes
10T pulse	Luminance/chrominance gain and delay inequalities
Chrominance staircase	Differential gain and phase, also luminance linearity
Mini-bar, line 20	Luminance–chroma crosstalk

Different organisations tend to use the interval test signal facilities in slightly different ways.

An important feature of the test signals is the 2T sine-squared pulse having a half-duration equal to the reciprocal of the nominal bandwidth and the bar which is used both as an amplitude reference for the pulse and also as a means of indicating sag or droop (caused by a falling off in response to lower middle frequencies).

An accepted way of making waveform measurements involves the use of special graticules in conjunction with oscilloscopes. For example measurements with the 2T pulse are made by suitably positioning the pulse within the graticule. Distortions represented by changes in amplitude, width, or resulting from overshoots and ringing, can be expressed by assessing the displayed pulse as one of a series of *K* ratings, with the least favourable rating considered to represent overall performance. A typical *K*-rating graticule is shown in *Figure 54.14*.

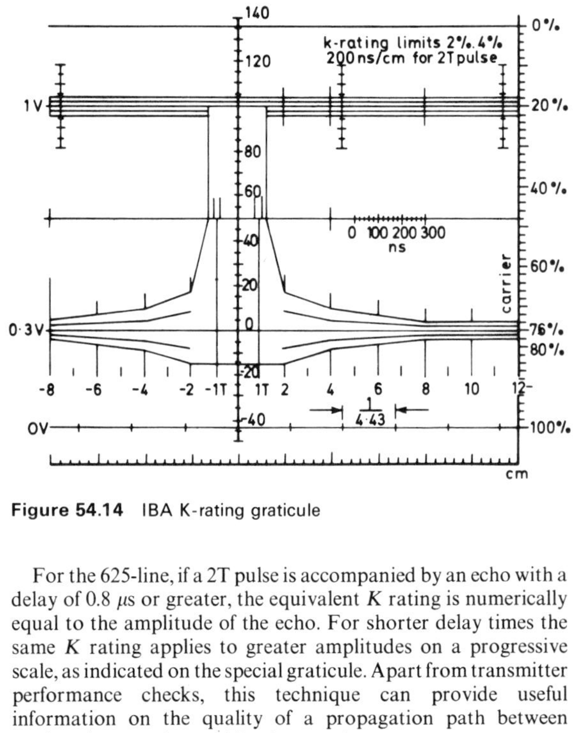

Figure 54.14 IBA K-rating graticule

For the 625-line, if a 2T pulse is accompanied by an echo with a delay of 0.8 μs or greater, the equivalent *K* rating is numerically equal to the amplitude of the echo. For shorter delay times the same *K* rating applies to greater amplitudes on a progressive scale, as indicated on the special graticule. Apart from transmitter performance checks, this technique can provide useful information on the quality of a propagation path between stations in assessing multipath reception.

The staircase waveform can be used to determine linearity of the transmission chain.

To facilitate the use of interval test signals for colour transmission a broader pulse, consisting of subcarrier signal, modulated to provide a unilateral sine-squared envelope, was added to permit measurement of luminance-to-chrominance gain and delay inequalities.

The interval test signal system is widely used in Europe for transmitter propagation and link checking and monitoring; it is also used for studio checking although seldom on a continuous basis.

54.2.6 Colour impairments

Although modern colour transmissions provide a high standard of colour fidelity, subjective impairments may occasionally occur. The following notes are based on experience of the

Australian Broadcasting Commission during test transmissions prior to the introduction of 625-line (System G) PAL colour television.

(1) *Saturation variations* in received pictures are caused by an incorrect amplitude relationship between the subcarrier burst and the chrominance component of the video signal. Receivers incorporate automatic colour control circuits (ACC) which vary the gain of the chrominance circuits in accordance with the amplitude of the burst on the assumption that any variation in the chrominance level (such as that caused by frequency response) equally affects the burst level. Thus if the ratio of burst to picture is correctly set at the signal source, the saturation of the original signal is maintained at the receiver output. If, however, some variation in frequency response occurs and the burst amplitude only is corrected prior to transmission, the output of the receiver will be incorrect. This problem may not be detected prior to transmission because picture monitors do not usually include ACC circuits. It is essential that the amplitude of burst should always be set accurately at signal sources, that it should never be varied independently of the picture chrominance, and that all sources employ the same standard burst to picture ratio.

(2) *Colour balance* errors may be caused by lack of matching between colour cameras, by the inherent differences in colorimetry between electronic and film cameras, by inherent colour casts in colour film, by poor grading of sections of colour film, or by variations in subcarrier phasing. It is essential for colour matching that all colour monitors be adjusted to the same colour temperature, that the colour scheme of viewing areas be standardised and that a white point reference of the correct colour temperature be available.

(3) *Streak and smear* are the descriptive names given to types of transient distortion and are explanations in themselves of the subjective effect. They are caused by transient distortions of comparatively low frequency and may not always be obvious with pulse and bar when viewed on a picture monitor. A window signal is a suitable test, and as a compromise can be combined with pulse and bar signals by windowing the bar. This fault can be troublesome with long-distance inter-city relay circuits where the effect of a number of small distortions becomes cumulative. It is also a common fault of monitor and receiver circuitry.

(4) *Transient errors* anywhere within the range of video frequencies can cause subjective impairment of pictures. The standard monochrome practice of pulse and bar testing is equally applicable to colour practice, and to explore the response of the chrominance channel a composite pulse is usually added. Other transient test signals are also used for particular applications.

Transient errors may also arise from injudicious use of operational facilities. For example, use of the crispener circuit on a particular colour camera to sharpen the edge detail may produce very pleasing pictures on the transmission monitor of an outside-broadcast van, but can result in an objectionable edge ringing after passing through a mediocre link circuit and transmitter.

(5) *Resolution* may be affected by many faults, from incorrect focusing of optical components to unsatisfactory frequency response of electronic equipment. A small amount of group delay, causing small displacement of the chrominance with respect to the luminance information may also be perceived as a loss of resolution, as may transient errors.

(6) *Crosstalk* in this context refers to the effect of unwanted coupling between two circuits. Usually this is frequency dependent, and increases with higher frequencies. Thus crosstalk becomes a more serious problem with the introduction of colour because the presence of the chrominance channel centred on 4.43 MHz has greatly increased both the HF components of the video signal and their importance. If the crosstalk is synchronous with the wanted signal it will be visible as a faint picture in the background. In the more general case it will cause variations in the colour of at least parts of the picture, usually as a slow pulsing of the colour, as the unwanted signal beats with the chrominance components of the wanted signal.

(7) *Video tape headbanding and velocity errors.* Quad video tape machines have four individual rotating heads which each scan 16 lines of the video signal in turn. Any difference between individual heads or head amplifiers can result in visible bands across the picture. Any differences in performance between the first line scanned by one head and later lines scanned by the same head will also cause banding. Where the errors affect the amplitude of the chrominance signal (e.g. differences in high-frequency equalisation) there is a change in colour saturation. Where the errors affect the phase of the chrominance signal (e.g. velocity errors) there is a change in colour hue. Sophisticated electronic compensators reduce these errors. These problems are overcome in modern helical scan recorders and by the use of digital timebase correctors.

(8) *Chrominance luminance delay inequalities* cause the chrominance information to be displaced from the luminance information. This fault arises from unsatisfactory phase response at chrominance frequencies. It is often present where older equipment, designed for monochrome is used.

(9) *Chrominance noise.* In general HF noise is less objectionable than LF noise. However, in colour receivers or monitors the chrominance circuits take the HF noise components from around 4.43 MHz and demodulate them down to low frequencies resulting in a subjective increase in noise when operating in colour. To overcome this either the noise performance of the equipment must be improved and/or notch filters fitted to luminance channels just prior to encoding. Difficulties also occur where the level of lighting is inadequate, e.g. in some outside-broadcast situations. This results in an increase in noise in each camera channel and a consequent increase in chrominance noise.

(10) *PAL ident problems.* The standards for the PAL system lay down that the phase of the subcarrier burst shall be 135° for the first line after the burst blanking sequence. For a continuous picture there is no problem if this is incorrect (225°) because the oscillator in the receiver locks up satisfactorily. Normally there is no problem even if a cut is made between two signals having different burst phasing because the receiver oscillator changes phase and locks to the second phasing very quickly. However, if two such signals are mixed, where the chrominance information of both is compared with the burst of one only, one signal appears in incorrect colours. This situation can occur unless all coders at picture sources operate with the correct ident or where any mechanism can change the PAL ident (e.g. some abnormal video tape conditions).

To check that the PAL ident is correct it is necessary to examine the phase of the burst on the first line after burst blanking and this can be done with a PAL vectorscope with the facility for displaying a single identifiable line (similar to an insertion-test-signal facility). The most convenient line to check is line 6 of field 3.

(11) *Hum and control tones for mains.* The mains supply can be a source of interference to television pictures if any component from it becomes combined with the video signal.

One source of such interference arises from the inter-winding capacitance between the primary and secondary of power transformers. A small current flows via this capacitance to the earth side of the secondary, and then back to the power supply earth via the earth connections of the equipment. It is inevitable that part of this path is common to the signal earth connections, and so a small interfacing signal is injected into the signal paths.

In theory, any hum or other interfering signals below line frequency can be removed by clamp amplifiers, and this method is often used. However, it is desirable to avoid the use of clamp amplifiers where possible as they can add other distortions to the signal. Interference arising from the inter-winding capacitance of power transformers can be overcome by using electrostatic shields and returning all electrostatic shields to mains earth independently of signal earth paths.

(12) *Cross-colour* is an effect whereby luminance information in the region near 4.43 MHz is decoded and appears as LF chrominance information. For example a check pattern on clothing, at a critical distance from a camera, produces coloured effects. One way of minimising this problem is to use a notch filter in the luminance channel just prior to encoding.

54.3 International television standards and systems

Although a measure of international agreement exists, different countries have adopted different television standards, and these are recognised by the CCIR. The two main standards (1) 625-line 50 fields and (2) 525-line 60 fields are now used in most countries but within these two standards a number of different characteristics are to be found. For example the UK 625-line standard (system I) differs in the frequency spacing between vision and sound carriers (6 MHz) from the 625-line systems B and G which are widely used elsewhere and which have 5.5 MHz vision–sound spacing: this means that receivers are not interchangeable without modification or adjustment. Similarly, three main colour encoding systems are used (PAL, NTSC and SECAM) including 525-line PAL in some South American countries although most 525-line systems use NTSC.

The PAL system is widely used on 625-line systems in Western Europe, Australia, South Africa, New Zealand, Hong Kong, etc.

The following countries have adopted SECAM: France, East Germany, Hungary, Lebanon, Poland, USSR, Bulgaria, Czechoslovakia, Egypt, Tunisia, Cuba, Haiti, Ivory Coast, Luxembourg, Monaco, Zaire, Iran, Iraq, Morocco and Saudi Arabia.

54.3.1 Channels

Because of the different channel-widths the channel numbering adopted by different countries differs, particularly on bands I and III (VHF).

The UK channels for bands I and III include for historical reasons a double-sideband allocation for channel I. This channel will be lost to broadcasting by 1986 when frequencies 41 to 47 MHz are deleted from this band. The future use in the UK of bands I and III after the phasing out of 405-line transmissions by 1986 will not include television broadcasting.

Table 54.1 Television channels in bands I and III (being phased out)

Channel No	*Sound (MHz)*	*Vision (MHz)*	*Channel No*	*Sound (MHz)*	*Vision (MHz)*
1	41.50	45.00	8	186.25	189.75
2	48.25	50.75	9	191.25	194.75
3	53.25	56.75	10	196.25	199.75
4	58.25	61.75	11	201.25	204.75
5	63.25	66.75	12	206.25	209.75
6	176.25	179.75	13	211.25	214.75
7	181.25	184.75			

UK channels (21 to 68) for bands IV and V conform with a common Western European designation, each channel being 8 MHz wide. Channel 21 starts at 470 MHz, 22 at 478 MHz, etc. Channels 35 to 38 are not used and form the gap between band IV and V. In System I the vision carrier is always +1.25 MHz from the lower limit to the channel and the FM sound channel +7.25 MHz from the lower edge. To find the lower frequency edge of any channel between 21 and 68, multiply the channel number by 8 and then add 302. For example channel $63 = (63 \times 8) + 302 = 806$. Therefore for channel 63 the nominal vision carrier is $806 + 1.25 = 807.25$ MHz and sound carrier $806 + 7.25 = 813.25$ MHz.

In the USA there is no VHF channel 1, and each channel is 6 MHz wide. Channel 2 begins at 54 MHz with vision carrier at 55.25 MHz and sound carrier at 59.75 MHz. Channels 3 to 6 follow consecutively. Channel 7 has lower limit at 174 MHz, and channels 8 to 13 follow consecutively. The American UHF band begins with channel 14 at a lower limit of 470 MHz and each channel is again 6 MHz wide, following consecutively to channel 83 between 884 to 890 MHz.

In Japan there is no television allocation below 90 MHz and similar exceptions and non-standard frequency allocations can be found in a number of countries.

54.3.2 System M (525-line 60 field) plus NTSC characteristics

Tips of sync pulses 100 per cent modulation
Blanking or pedestal level (front and back porches)
75 per cent ±2.5 per cent of peak value

Reference black level	70 per cent of peak level
Reference white level	12.5 per cent of peak level
Intercarrier frequency	4.5 MHz
Colour subcarrier frequency	3.579 545 MHz
$(B-Y)$ signal (I)	1.5 MHz
$(R-Y)$ signal (Q)	0.5 MHz
Line frequency	15 734.26 Hz
Field frequency	59.94 Hz

54.4 Digital television

For more than a decade the application of digital techniques to video processing and transmission has been under investigation, and progressively used in specialised operational applications since 1971. The target is an all-digital studio environment with transmission in digital form to the transmitter, but this still seems a few years away. In the mid-1970s most applications used digital coding of the composite colour waveform in 'stand-alone' units with input and output in analogue form. More recently attention has turned to component-coding which offers many attractions but requires higher bit rates.

Several factors contribute to this paradox: the more rugged nature of the digital signal which requires a far lower signal-to-noise ratio, and its greater immunity to most forms of noise and interference. The ability to make use of picture redundancy and complex coding strategies can provide very large degrees of bit-rate reduction without excessively increasing the harmful effects of bit errors. For composite coded 625-line pictures, transmission bit rates of the order of 34 Mbit s^{-1} (17 MHz bandwidth) have been shown experimentally by the BBC, and even less bandwidth may be required for 525-line systems. Digital transmission also provides better energy dispersal.

In Europe, an improved 60 Mbit s^{-1} experimental system, using small-dish terminals for both the up and down link, has been demonstrated by the IBA, who have also shown that the

Table 54.2 Main characteristics of television standards

CCIR designation	*A*	*M*	*(N)*	*B*	*C*	*G*	*(H)*	*I*	*D, K (K1)*	*L*	*F*	*E*
Frequency bands	I, III	I, III	IV, V	I, III	I, III	IV, V		IV, V	I, III, IV, V	IV, V	I, III	I, III
Lines per picture	405	525	(625)	625	625	625		625	625	625	819	819
Fields per second (Hz)	50	60	(50)	50	50	50		50	50	50	50	50
Line frequency (Hz)	10 125	15 750		15 625	15 625	15 625		15 625	15 625	15 625	20 475	20 475
Video bandwidth (MHz)	3	4.2		5	5	5		5.5	6	6	5	10
Channel bandwidth (MHz)	5	6		7	7	8		8	8 (8.5)	8	7	14
Sound vision carrier spacing (MHz)	−3.5	+4.5		+5.5	+5.5	+5.5		+6	+6.5	+6.5	+5.5	+11.15
Width of vestigial sideband (MHz)	0.75	0.75		0.75	0.75	0.75	(1.25)	1.25	0.75 (1.25)	1.25	0.75	2
Vision modulation polarity	Pos	Neg		Neg	Pos	Neg		Neg	Neg	Pos	Pos	Pos
Sound modulation	AM	FM		FM	AM	FM		FM	FM	AM	AM	AM
Maximum deviation (kHz)	—	±25		±50	—	±50		±50	±50	—	—	—
Pre-emphasis time-constant (μs)	—	75		50	50	50		50	50	—	—	—
Relative transmitter power (vision sound)	4:1	10–5:1		5–10:1	4:1	5:1		5:1	2–5:1	8:1	4:1	—
Where used (selected countries)	Ireland UK (Obsolescent) Monochrome only	Canada Iran Japan Korea Mexico Panama Saudi Arabia USA Venezuela (*N*) Argentine Uruguay	Australia Austria Cyprus Denmark Egypt Finland W. Germany E. Germany Greece Iceland India Indonesia Iraq Israel Libya Malaysia	Malta Netherlands New Zealand Nigeria Norway Pakistan Portugal Spain Sweden Switzerland Syria Tunisia Turkey Uganda Yugoslavia Zambia Zimbabwe	Algeria Belgium Luxembourg	Austria Finland W. Germany E. Germany Israel Italy Malaysia	Portugal Spain Sweden Switzerland Uganda Zambia Zimbabwe (*H*) Belgium Cyprus Yugoslavia	Ireland Nigeria South Africa (VHF/UHF) UK	Czechoslovakia (*D*, *K*) Hungary (*D*, *K*) Poland (*D*, *K*) Roumania USSR (*K*1) Central African countries	France Luxembourg Monaco		France Monaco (Obsolescent) monochrome only

highest quality 14/7/7 component-coded signals can be reduced from around 230 Mbit s^{-1} to fit the 140 Mbit s^{-1} hierarchical structure of the telecommunications authorities, virtually without visible impairment.

The advantages of digital transmission include freedom from the ill effects of differential phase and differential gain and the ability to regenerate an exact replica of the input data stream at any point in the chain, thus avoiding cumulative signal-to-noise degradation. It was Shannon's communication theory that first underlined mathematically the outstanding efficiency of digitally encoded transmission systems.

However it is important to realise that digits do not eliminate all problems. From the earliest days of manual and machine cable telegraphy it has been recognised that the transmission of high-speed pulses within a channel of restricted bandwidth can present severe practical problems, including inter-symbol interference and susceptibility of the error rate to all forms of 'echoes' and multipath propagation.

Digital systems are inevitably subject to quantising noise, which depends upon the number and arrangement of the levels at which the original analogue signal is digitised, and also to aliasing foldover distortion. Aliasing represents spectral components arising from the process of sampling not in the original signal; when these fall within the spectrum of the sampled signal they result in foldover distortion. Aliasing can be minimised by effective filtering, although the ease and cost with which such filtering can be accomplished is very much a factor of the sampling rate.

A digital system is where all waveforms are selected from a restricted number, as opposed to the infinite number of shapes and amplitudes of an analogue signal. A digital signal has far cruder and more rugged time relationships and unlike an analogue signal is far less susceptible to differential phase and differential gain distortions. A binary digital system has only two states and so is a 'go/no-go' or 'on-off' system. Analogue systems degrade gradually through all operations; digital systems can show imperceptible degradation until they reach a certain level of errors and then degrade very rapidly.

Digital operations can be of calculable, repeatable, consistent standard whereas analogue systems require accurate adjustment and alignment to achieve optimum results. 'Line-up time' is essentially an analogue situation, whereas an 'on-off' switch may be the controlling factor of a digital system.

Digital systems score where the user needs to store masses of information or perform accurately calculable operations on the information: standards conversion, graphics storage, synchronisation, compression, expansion, noise reduction, image enhancement can all be accomplished more readily in digital form; indeed digital techniques may allow users to do things which cannot be done satisfactorily or at all in analogue form.

Nevertheless such a radical change as an all-digital approach to studio and distribution operations has to be justified on many grounds: economics, performance, reliability, stability and ease of operation amongst them.

The basis of any digital video technique involves quantising the picture into a number of discrete brightness levels, sampling the signal at a repetition rate normally greater than the Nyquist figure of twice the highest frequency component in the signal. Each level is allotted a unique code in the form of the presence or absence of pulses, usually in the form of linear pulse code modulation (PCM). In practice some 256 brightness levels may be used, corresponding to an 8-bit data word. Signal processing is then achieved by arithmetical operations on these data words.

Most of the early work on the 625-line standard, involved a sampling rate of 13.3 MHz on the composite video signal, resulting in a 106.4 Mbit s^{-1} bit stream in serial form, although normally during processing each of the eight bits is in parallel form. For 525-line NTSC the basic bit rates for component coding fall between 75 to 114 Mbit s^{-1} depending upon which multiple of the colour sub-carrier frequency is used (in the USA as early as 1974–75 it began to be recognised that $4f_{sc}$ is to be preferred to $3f_{sc}$).

For component coding of a 625-line picture in *YUV* form the total baseband becomes significantly higher, a 12/4/4 system (i.e. a system in which the luminance signal is sampled at 12 MHz, the colour-difference signals at 4 MHz) results in a bit-stream of about 160 Mbit s^{-1}; while a 14/7/7 system represents a bit stream of about 230 Mbit s^{-1}.

Digital video processing enables picture information to be stored for as long as required in a digital memory and then read out either in real time or at higher or lower speeds. This means that digital signals can readily be delayed, stretched or compressed in time, facilitating the correction of timing errors, synchronisation and conversion of signals between different television standards.

The digital tape recorder opens the way to multiple generations of tape without significant degradation: the 20th certainly, even perhaps the 50th generation of tape remains of full broadcast standard. With component coding the way would be opened, in some cases, to complex post-production editing and assembly: for example the ability to insert chroma keyed backgrounds in post-production would enormously simplify the problems and expense of 're-takes'.

54.4.1 Digital transmission

A curious paradox appears to be developing in the field of digital transmission, particularly over distribution satellites. As noted above, the very high bit rates associated with digital video implies a significantly greater bandwidth; a 100 Mbit s^{-1} bit rate requires a 50 MHz bandwidth. Yet one is beginning to see in North America the exploitation of digital transmission techniques to allow *more* video channels to be put through a bandwidth-limited transponder. This possibility arises from the fact that an analogue video signal is normally transmitted in the form of wide-deviation FM with a bandwidth in the order of 27 MHz. By reducing the digital signal using the very effective bit-rate reduction techniques, it is possible to put two digital channels through a satellite transponder intended for one FM circuit.

An analogue signal can be converted to digital form, processed and then reconverted to analogue form several times in tandem with little visible degradation. However if this tandem chain is extended too far it will tend to lead to a marked increase in the quantisation noise caused by the use of a restricted number of amplitude levels. Other impairments in digital video may be due to aliasing, clock jitter, error rates and the impossibility of transmitting perfect pulses. Digital video signals may also call for added complexity for monitoring and measuring the impairments, and some still uncertain cost factors in equipment and maintenance.

These points are stressed to indicate why not all engineers expect to see, even when digital standards are finally agreed, any immediate rush to convert all existing studios to an all-digital approach, or to see any early 'retirement' of the current composite-coded stand-alone digital machines.

The reasons are partly technological, partly economic. Even in countries where capital expenditure can be largely offset against taxable profits, the high cost of studio equipment demands that equipment is not replaced prematurely unless it can be shown beyond question that the new units will provide a real competitive advantage over those organisations without them.

During the past few years much the early colour equipment in British studios has been replaced. This includes the important change from 2-inch quadruplex VTR machines increasingly to the more flexible and more economic 1-inch helical scan machines. These machines, even when they incorporate digital time base correction, still record the signal in analogue form. Together with the adoption of time-coding systems they are bringing about a situation where the editing of tape is becoming as flexible and as convenient and precise as the traditional editing of film. The provision of fast and slow motion viewing, stop-motion, etc., means that the whole operation is coming more and more to resemble post-production film editing and assembly, using either multiple or, in future, single electronic cameras in both studio and field production.

In the early 1970s, the VTR was undoubtedly still the limiting factor in so many ways, a situation that is less evident with the 1-inch analogue machine.

In 1981 an international standard for studio operations was finally agreed, although final details of some parameters (including sound) were left in abeyance. The agreed sampling rates are 13.5/6.75/6.75 MHz, a sampling rate that can be applied both to 625-line and to 525-line systems. Digital equipment suitable for operation at this sampling rate became available in 1982, but the final format of an all-digital video tape recorder has yet to be determined.

54.5 Teletext transmission

During the 1970s, a data transmission system, riding 'piggy-back' on conventional television transmission was developed in the UK and introduced under the service designations Ceefax (BBC) and Oracle (IBA–ITV). An agreed technical specification was introduced in 1974. Teletext transmission systems have since been introduced in a number of countries, and basically similar systems but with rather different technical specifications have also been developed in some countries. Teletext uses the broadcast television signal to carry extra information. These extra signals do not interfere with the transmission and reception of normal programmes. A teletext receiver, a television receiver with additional circuits, is capable of reconstructing written information and displaying it on the screen. The system allows the transmission of very many bulletins of information, and the viewer can choose any one by selecting a three-figure number on a set of controls, usually push buttons or thumbwheel switches. After a short interval the information appears and remains for as long as it is needed.

A major use of teletext is as an information service, but it can also supplement normal television programmes with sub-titles or linked pages. The entire signal is accommodated within the existing 625-line television allocation, and so costs nothing in terms of radio frequency spectrum space. It is, effectively, the first broadcasting system to transmit information in digital form.

Teletext pages look rather like pages of typescript, except that they can also include large-sized letters and simple drawings. The standard-sized words can use upper or lower case, and be in any one of six colours—red, green, blue, yellow, cyan and magenta—or white. The shape of the characters is usually based on a 7 × 5 dot matrix, with a refinement known as character rounding. As many as 24 rows of the standard-sized characters can be fitted on a page, and each row can have up to 40 characters. Each page can carry about 150–200 words.

The specification also allows the option of characters of twice the height of standard characters. Larger-sized characters and also drawings are made by assembling small illuminated rectangles, each one-sixth the size of the space occupied by a standard character, and these too can be in any of the six colours or in white. Part or all of each page can be made to flash on-and-off (usually once per second), to emphasise any particular item.

The page background is usually black, although the specification allows the editor to define different background colours for part or all of the screen. Also, at the teletext editor's discretion, text can be enclosed in a black window and cut into the normal picture. Further, certain receiver designs allow the whole page (sometimes in white only) to be superimposed upon the picture.

It is potentially possible for the system to carry up to 800 single pages; but because of the way the pages are transmitted, this could mean an appreciable waiting time between page selections. So for the moment not all 800 pages are used at any one time.

Additional circuits are needed in a television receiver to decode teletext. First, data extraction and recognition circuits examine the incoming 'conveyer belt' of teletext data lines, and extract those signals which make up the page which has been selected. The data from these lines is then stored, usually in a semiconductor memory, so that the page can be displayed at the same rate as a normal picture. The binary number codes are then translated to their corresponding characters or graphics patterns. Finally a video raster scan representation of the page is switched onto the screen.

The reception of teletext, like normal television, is susceptible to various kinds of interference. IBA trials, however, indicate that the great majority of those who receive television pictures well, are also able to receive teletext.

Figure 54.15 shows one teletext data organisation.

54.5.1 Enhanced graphics

Work is continuing in the UK aimed at defining and specifying broadcast teletext systems in such a way that public broadcast services can be introduced or progressively up-dated at the lowest-possible receiver-decoder costs. Each hierarchical step embraces the properties of lower levels and permits subsequent extensions to higher levels and a fine-line drawing set. Fine-line graphics have the general appearance of alphageometric graphics but demand less 'storage' in the decoder, though with the penalty of providing rather less scope than true alphageometric systems.

The following 'levels' for broadcast teletext systems have been proposed:

(1) *Level 1.* This is the current operational teletext specification incorporating the joint BBC–BREMA–IBA specification and subsequent minor amendments. Fully operational teletext magazines of news and information have been available in the UK on all three programme channels for a number of years. By 1983 there were some 750 000 teletext receivers and adaptors in use, with the number doubling every year. The specification includes optional extensions for linked pages, basic page check word, programme or network label and data for equipment control, including time and data in UTC with local offset.

(2) *Level 2.* This level offers Polyglot 'C' for multi-language text and incorporates a wider range of display attributes which may be non-spacing.

(3) *Level 3.* This level introduces dynamically redefined character sets, permitting the use of non-Roman characters such as Arabic or Hindustani, of alphageometric instructions (AGI) in a closed-circuit mode. Tests are shortly to start using similar AGI pages in the operational Oracle magazine.

The efficiency of the system is underlined by its ability to contain within two normal length teletext pages sufficient AGI to provide the following three separate displays: (*a*) multi-coloured pie chart, (*b*) a chess board with a game in progress, and (*c*) an engineering announcement incorporating a large IBA logo. A further one-and-a-half pages of AGI provide enough data for a fully coloured detailed weather map, with symbols representing sunshine and rain clouds, and carrying typical weather forecast captions and titles.

Level 4. This covers the form of alphageometric coding which the IBA is currently developing. Unlike lower levels it requires the

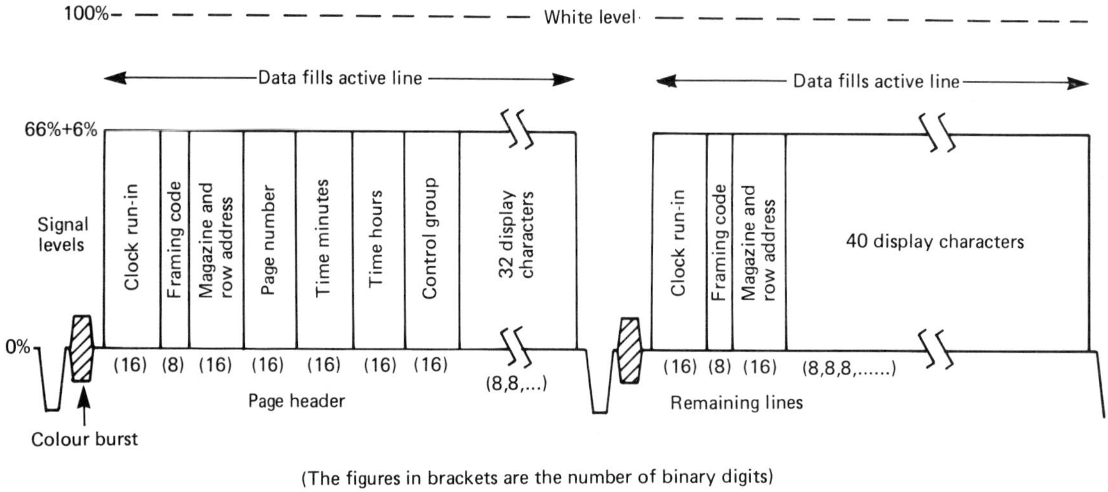

Figure 54.15 Teletext data organisation

use of some computing power within the decoder, and some additional storage capacity.

Level 5. This level covers the transmission of high-definition still pictures. The limitation to the quality of these still pictures, which may be superior to that of the associated television system, becomes that of the receiver display device. Experimental transmissions at this level have included the Oracle-adapted broadcast transmissions of 'Picture-Prestel' during IBC80 (September 1980) and the high-quality still-picture system demonstrated by the BBC.

The level 4 coding as demonstrated by the IBA differs from previously proposed alphageometric systems in two significant ways:

Firstly, 'incremental coding' is used which is particularly suited for drawing maps and other irregular outlines and, secondly, a 'delay' command has been introduced. This latter feature is regarded as of particular importance where information is to be transmitted in the form of a series of connected explanatory pages. For example, the IBA system may be explained by a series of 10 pages, the contents of which appear to be drawn as the viewer watches and the complete sequence of pages involves the transmission of only one or two pages of teletext (one or two thousand bytes).

The experimental level 4 system could be introduced into the existing UK system without affecting existing decoders when it becomes economically feasible to build consumer-type decoders having sufficient storage. The system is capable of receiving the currently broadcast level 1 teletext pages, in addition to displaying a series of specially coded pages.

54.6 Cable television

54.6.1 Historical

The use of cables to convey entertainment to the home goes back to the earliest days of the telephone: Gilbert installed it in 1882 and advised Sullivan to do the same to listen to the stage performance of their masterpieces.[1]

In Budapest in 1892 a network carried news services to homes and in the same period the Electrophone Company in London enabled subscribers to listen to theatres, balls and churches.[1]

These rather specialised services did not, however, achieve a wide acceptance and it was not until the advent of the popular demand for broadcasting in the 1920s that wired distribution set off on the path which now seems certain to lead ultimately to the fully integrated communication network carrying any kind of signals – voice, data or vision – to and from each member of the community.

In April 1924 the first commercial radio relay service was started by A. L. Bauling in Koog aan de Zaan in Holland[1]. He was followed independently by Mr. A. W. Maton[2] of Hythe, a village in Hampshire, who also started experimenting in 1924 and opened his commercial service in January 1925.

It was upon this foundation that cable television systems grew in Europe, although their recent rapid growth in North America had a different origin.

54.6.2 Present systems

Wired distribution systems vary considerably in their size and complexity. Perhaps the simplest possible concept is that of a single receiving aerial serving a pair of semi-detached houses or a small block of three or four flats. In a very strong signal area such a system may not need any active equipment, the signal picked up being shared between the users by means of passive dividing networks. In weaker signal areas, or where a somewhat higher number of outlets is required, it may be necessary to incorporate an amplifier before the signal from the aerial is divided up between outlets. In still more complex systems, in weak signal areas or to serve large blocks of flats, it may be necessary to change the frequency of the signal derived from the aerial before it is distributed to individual users. This is to ensure that at a television receiver interference does not occur between the signal carried over the wired system and the signal available directly from the broadcasting station.

The most sophisticated wired systems may comprise an elaborate directional aerial system, sited at a high vantage point to gather in weak signals from a distant transmitter; a device to change the frequencies of the incoming signals to other channels more suitable for distribution by wire; together with an extensive network of underground and/or overhead cables carrying the signals to many thousands of households in a given locality.

Some of the transmission systems currently in use for this type of system are described in the following sections. In most cases the input television signals are picked up at suitably sited stations receiving normal off-air broadcast signals; however, there are some cases in which signals are received by direct cable connection from the broadcasting authority. There would be no technical problem in feeding signals from other sources, such as local studios or video tape recorders, directly into the wired distribution system.

There are two main types of distribution system representing a difference of approach which arose for historical reasons. In the early 1930s a substantial business developed in the wired relay of sound radio programmes at audio frequency and the total number of subscribers reached about 1 million in 1950. This business was confined to Europe, mainly Holland and the UK and developed in large provincial cities, e.g., Hull, Leeds, Newcastle, Nottingham, Bristol. The cost of the cable connection, programme switch and loudspeaker was competitive with the radio receivers of the day and this, with the convenience and reliability of the system, provided the economic base for its growth. With the advent of television in the early 1950s it became necessary for the relay operators, if they were to stay in business, to develop television distribution methods which would, as with sound radio, be competitive with direct reception off-air in the generally good reception conditions of the large provincial cities. The result was the HF multipair system, *Figure 54.16*, which can serve simplified receivers directly, or conventional off-air receivers through an 'inverter'. Since the programmes are carried on physically separate channels these systems may be described as space division multiplex.

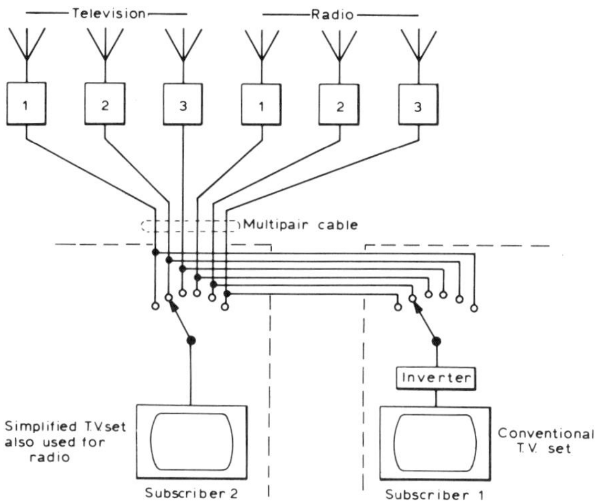

Figure 54.16 H.F. multi-pair cable television systems

The other type of system was developed, principally in the USA for different reasons. There is no obligation on the United States broadcasting industry to serve the smaller towns where the available advertising revenue may be insufficient to support a plurality of broadcasting stations. This led to the development of systems which were purely concerned with the improvement of reception of weak signals from distant stations. In these circumstances the system does not have to compete as an alternative means of reception of local transmissions. The most suitable system for this purpose uses a single co-axial cable to which standard television receivers can be directly connected. The different programmes are carried on different frequency channels in a manner exactly analogous to over the air broadcasting *Figure 54.17*, and may be referred to generally as frequency division multiplex.

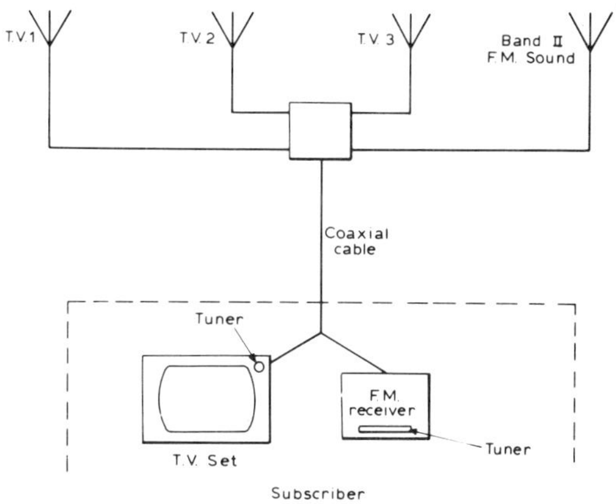

Figure 54.17 V.H.F. coaxial cable television system

54.6.2.1 HF systems

An HF system uses a single channel in the HF band for all programmes and each programme is carried on a separate pair of wires in a multi-pair cable. The vision signals lie in the band between 4 and 10 MHz where cable attenuation is relatively low but the frequency is high enough to avoid difficulties in the demodulation process in the receivers. The result is that a very large number of subscribers can be fed from one amplifier installation; the number ranges from several hundred to as many as 2500. The sound accompanying the vision signal is carried at audio frequency and operates the loudspeaker directly. The sound radio programmes are also carried at audio frequency on separate pairs. Modern installations of this type employ a 12-pair cable of special construction which has capacity for six television programmes with a further six sound radio programmes. In an alternative system a 4-pair cable is used and the sound accompanying the vision signal is carried on an amplitude modulated carrier at approximately 2 MHz, enabling the same pairs to carry a further four sound programmes at audio frequency.

The distribution of sound direct to loudspeakers has represented the most economic solution for many years but with an increase in the number of programmes to be carried and the reduction in the cost of integrated circuits for FM demodulation and audio amplification, the position is now changing and HF systems in future are likely to abandon it. This will leave the audio and supersonic bands up to about 2 MHz on each pair free for two-way data or voice services.

Many occasions arise when it is desirable to serve a conventional television receiver from an HF system and this is done by means of a frequency changing unit which takes the vision signal from the cable and converts it into one of the standard broadcast channels in the VHF or UHF bands; the sound signal is fed directly to the loudspeaker.

The principal advantage of the HF system is its ability to serve receivers which are simpler and cheaper and more reliable than conventional television receivers. Further advantages are simplicity in operation, since no tuning is required, and the fact that sound radio programmes are reproduced by the loudspeaker in the television receiver. The major drawback is that the number of programmes which can be distributed is limited by the number of pairs in the cable and the number can only be increased by providing additional cable or by carrying more than one programme on each pair which, though possible, leads to a more complex receiver, thus largely sacrificing the main advantage of the system.

54.6.2.2 VHF systems

Modern VHF systems employ wide band amplifiers covering the range 40–300 MHz. Owing to the relatively high cable attenuation at these frequencies amplifiers must be inserted at fairly frequent intervals. One amplifier may serve up to 150 households and a total system gain up to 800 dB can be achieved with satisfactory performance in respect of noise and cross-modulation, enabling large towns to be served from a single reception site. The system may also carry FM sound programmes in band II exactly as broadcast. Although the cable and amplifiers are capable of carrying a large number of channels it is usual to locate the television channels in broadcasting bands I and III for direct operation of standard VHF television receivers. The design of these receivers is such as to place a practical limit on the number of channels that may be used in these bands. The main factors which set a limit to the number of programmes are:

(1) The receivers are not manufactured with adequate adjacent channel selectivity, hence adjacent channels cannot be used on the same system.

(2) Receivers are not well enough screened to avoid direct pick-up of strong signals from a television transmitter; hence the channels used for broadcasting in an area cannot be used in a cable system in the same area.

(3) Local oscillator interference from a receiver connected to the cable to a nearby receiver working from an aerial precludes the use of certain channels.

As a result of these limitations it is rare for systems to be able to distribute more than six television programmes with high technical quality in areas with strong ambient signals.

The programme required is selected by means of the selector switch or tuner fitted on the television receiver. Since the distribution system is required to serve standard domestic television receivers, some of which are only capable of receiving 405-line signals, it is usually necessary to distribute the 405-line programmes as well as those of 625-lines. A problem which has arisen in this country since UHF became the primary service is that new receivers capable of operating on VHF are almost unprocurable and the connection of a modern receiver to a 625-line VHF system now requires the use of a frequency changer or 'up-converter' to convert the VHF signal received to a UHF channel to be fed to the receiver.

This problem is temporary and peculiar to the UK but it illustrates the difficulties which are met in devising a cable distribution system to serve receivers which were designed for a different purpose, namely, reception from an aerial.

54.6.2.3 Mixed systems

The need to serve 405-line VHF and 625-line UHF receivers on the same system has led to the development of VHF/UHF systems in which 405-line programmes are fed to the subscriber at VHF 625-line programmes at UHF, and FM sound programmes on band II. Individual 200-subscriber areas are served by a primary distribution system which may be operated at HF (in space division) or VHF (in frequency division), the appropriate signals being converted to UHF at distribution points.

54.6.2.4 UHF systems

Modern UHF systems employ wideband amplifiers covering frequencies up to 860 MHz. Because of the high cable attenuation, such systems are at present limited to areas containing about 200 subscribers. Single standard UHF receivers are used and the programme required is selected by means of the selector switch or tuner fitted on the receiver. As with the VHF system FM sound programmes are usually provided in band II exactly as broadcast.

UHF frequencies are unsuited by their nature to large cable systems and are unlikely to find more than limited use in such conditions as now exist in the UK.

54.6.2.5 Switched systems

Higher numbers of programmes can be handled by HF systems with a different arrangement of the distribution network in which pairs in the cable are allocated to individual subscribers to connect them to a switching centre or programme exchange.[3,4] Selection of the desired programme is then carried out by a switch in the programme exchange by remote control over a separate pair from the subscriber's premises, in a manner somewhat similar to that used in an automatic telephone exchange. The higher frequencies to be transmitted place a strict economic limit on the length of the subscriber's connection to the exchange so that a larger number of exchanges, each covering a smaller area, is required than in the case of the telephone system. A switched system can be extended to two-way working, the return path to the exchange being provided over the same subscriber's pair of wires at a higher frequency than the outgoing signal. The system is illustrated in *Figure 54.18*.

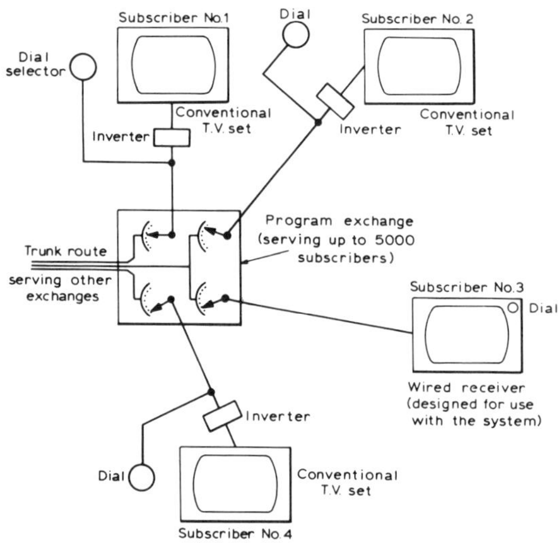

Figure 54.18 Switched systems for cable television

54.6.2.6 Pay TV

Pay-TV in which the recipient pays directly for the programmes which he takes, is technically possible. The simplest form is payment on a per-channel basis; a number of channels each carrying programmes on a particular subject or group of subjects might be offered and subscribers would elect to pay a monthly subscription for the channels in which they are interested in the same way as people subscribe to magazines. This is very simply arranged on HF systems since the circuit carrying the chosen programme can be connected or not to each subscriber in the junction box or programme exchange. In coaxial systems the simplest scheme is to use channels to which ordinary receivers will not tune and to supply simple frequency converters to subscribers requiring the pay channels. Security against illicit use of converters is a problem and an improvement can be obtained at extra cost by scrambling the signal as transmitted and descrambling for each Pay-TV subscriber.

Payment on a per channel basis is satisfactory in many circumstances but it is necessary to achieve payment on a per programme basis if the full potential of Pay-TV is to be realised. This is because the owners of the most suitable programmes such as first-run films and important sporting events will not make them available to television unless they can be sure of receiving their share of the money paid by the public to see their particular programme, as they would from a cinema or sports arena.

Because HF systems lend themselves readily to the backward transmission of control signals on one or more pairs from the subscriber to the distribution centre, payment per programme is fairly simple and was demonstrated on a substantial scale in South London and Sheffield in 1966. These pilot schemes showed promise of commercial success but were discontinued because of political difficulties which prevented their expansion to a much wider basis.

On coaxial systems it is necessary to prevent unauthorised reception by the use of 'out of band' channels and scrambling and to release the programme to the subscriber by sending a control signal with an address code for the subscriber in question. The request from the subscriber may be made either by telephone or as the result of interrogation and response signals from a central computer using the two-way facility provided on some modern networks.[5] Systems of these types were beginning commercial operation in the USA in 1973.

54.6.3 The future

The future of wired distribution systems must be set against a wide range of possibilities. At one extreme lies a universal communications network, wideband, two-way and switched, capable of conveying any kind of information, visual, voice or data from anywhere to anywhere. Besides television and sound for entertainment purposes the range of services which might be provided by such a universal network, include:

- Telephone and viewphone facilities.
- Electricity, gas and water meter reading.
- Facsimile reproduction of newspapers, documents, etc.
- Electronic mail delivery.
- Business concern links to branch offices.
- Access to computers.
- Information retrieval (library and other reference material).
- Computer-to-computer communications.
- Special communications to particular neighbourhoods or ethnic groups.
- Surveillance of public areas for protection against crime.
- Traffic control.
- Fire detection.
- Educational and training television and sound programmes.

At the other extreme one can envisage some small advance on the present situation with a minor extension in the number of programmes distributed on wired broadcasting systems operating side by side with entirely separate switched networks for telephone and data as at present. There is no theoretical limit to the number of television programmes that could be delivered to a subscriber's premises by cable but the developments in the UK during the next ten years will depend upon the customer demand for additional television and other services, on the cost of providing them, and on political and legal decisions regarding licensing, copyright and other complicated issues. In the USA, the Federal Communications Commission has laid down certain rules for the licensing of new wired distribution systems which call for at least 20-television-channel distribution capability, together with some capacity for return communication.

The technical possibilities for future wired communication systems, incorporating the distribution of sound and television programmes, are very wide. Initial studies of the various forms of these networks, now being conducted in this country and abroad, are based on the use of cable as the only transmission medium but, in the future, the transmission medium might include millimetric radio links and optical fibre guides. One must also consider as a future possibility the carriage of full-bandwidth television to the home in digital form. The ultimate pattern of network connections will probably be the result of an evolutionary process whose timing will be influenced by demand considerations and by the extent and type of past investment.

For HF systems which employ a separate pair of wires for each programme, the practical limit to the number of programmes is about 12 because of the large number of pairs that must be handled in junction boxes and selector switches.

For greater numbers of programmes a switched HF system may be used. Because this is a four-wire system (one programme pair and one control pair) it has the advantage that control of the use of certain channels (e.g., pay-television channels) is easily arranged. On the other hand only one programme at a time is available on the subscriber's premises unless a second exchange line is provided or arrangements are made to transmit a second programme over the same line by frequency division multiplex. Switched HF systems employing four wires can be extended to provide for all the services listed at the beginning of this chapter.

For systems operating at VHF or higher frequencies a single coaxial cable should be capable of delivering up to 24 programmes. For example, the channels might be spaced at 10.5 MHz as in the Post Office educational television system used by the Inner London Education Authority. With this spacing the 24 channels might be accommodated in the band 40–400 MHz but, as the spacing is different from that of the broadcast channels, the system would be unable to serve directly ordinary receivers in use for off-air reception and either special receivers or tunable converters would be required by every subscriber. Also with 24 channels the linearity of the long chains of repeaters required in these systems needs to be of the highest order if the distortion products, particularly those of the third order, are not to degrade the picture quality to an unacceptable degree.

In order to provide many of the services mentioned at the beginning of this section systems must be capable of passing messages in both directions. In coaxial systems messages in the return direction can use the frequency space below the lowest television channel transmitted, i.e., below about 50 MHz. Initially there may be sufficient capacity on the normal simple tree topology at present used with the systems but looking farther into the future, schemes are in consideration for services other than full bandwidth television in which digital transmission would be used in conjunction with coaxial cable laid in a ring formation beginning and ending at communications exchanges. Transmission over the cable would be in one direction only and subscribers would be connected to points on the ring. The subscriber's equipment would pick out from the total digit stream that information appropriate to his needs and would re-inject into the stream appropriate return communication in digital form. A limited number of entertainment and other distributed television programmes in FDM analogue form could occupy the bandwidth in the ring above the highest frequency (possibly 80 MHz) required for the both-way digital services. With a ring topology it would be possible to provide a two-way television service to a limited number of subscribers by using the remainder of the ring for the return direction of transmission.[6]

54.7 The equipment required for sound broadcasting

The basic equipment required for sound broadcasting would seem relatively simple, namely, a microphone, AF amplifier with volume control, programme distribution to transmitter and a feeder from transmitter to aerial. In practice the technique is much more complicated and involves a smooth change from one programme source to another, the insertion of special effects, e.g., crowd noises, motor car starting etc, the reduction of the dynamic range of programme volume without destroying artistic values, the recording and reproduction of programme, the monitoring of output, and the provision of a communication system to co-ordinate and supervise the distribution to and from studios and to transmitters. A block schematic diagram illustrating a possible grouping of programme requirements is shown in *Figure 54.19*. The programme connection from each source to the

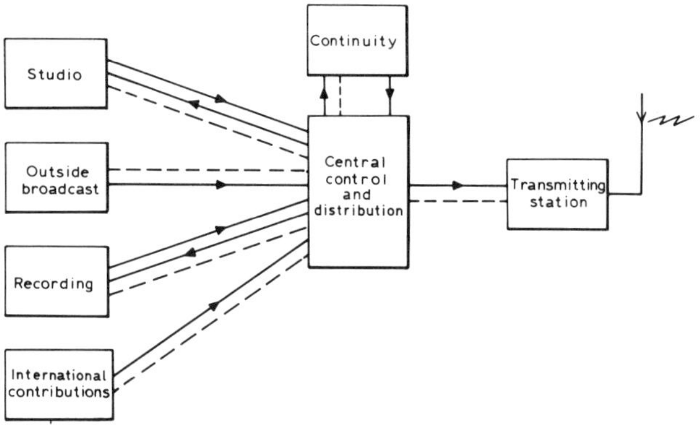

Figure 54.19 Programme collection and distribution (full lines – programme links, dotted lines – two-way communication control links)

control room, from which the composite programme is distributed to the transmitters, is paralleled by a link enabling communication between each source and the distribution centre to be established and to allow a cue feed from the preceding programme to be given when necessary.

54.7.1 The studio

The studio will need to be treated acoustically to give the right 'atmosphere' for performers and listeners, and this will depend on whether it is required for speech, drama or music. Reverberation time, which is time taken for sound to die away to 10^{-6} of its original energy intensity after it has been cut off, is important and should be reasonably constant over the AF range. For speech it should be of the order of 0.2 s and for large concert halls a value of about 1.5 s would be normal. Drama may call for considerable variations from a very low value, less than 0.1 s simulating an open air scene, to a high value, about 3 s simulating a large room with many reflections from relatively bare walls. Artificial reverberation is often used to cover the latter situation.

The range of audio frequencies encompasses about nine octaves and separate studio treatment has to be applied to control the low, middle and high frequencies. Acoustic design is too specialised for inclusion here.

54.7.2 The studio apparatus

The requirements of one particular broadcasting organisation can differ appreciably from another; one may need a relatively simple operation of recorded disc or tape programmes, interspersed with news and advertisements, whereas another may have to cater for live programmes using many performers in large orchestral, chorus and soloist ensembles or in drama productions. Nevertheless there may be a minimum of three microphone channels (rising to perhaps twelve), a disc and a tape reproducing channel together with facilities for adding artificial reverberation or echo, for aural monitoring and for measuring programme volume. An independent channel may be included to permit an emergency announcement to be superimposed on the programme, or a narrator for a drama programme to be inserted.

Outside sources are fed at zero level from the central control room to the studio cubicle, where they can be heard on headphones by using prefade keys. They are attenuated by about 70 dB to bring to about the same level as a microphone output before being plugged into the source amplifier associated with one of the microphone channels.

Talk-back may be provided from a microphone in the studio control cubicle to a loudspeaker in the studio so that the producer can issue instructions during rehearsal. During transmission the studio loudspeaker is automatically cut off, but talk-back can be provided to artists via headphones. The studio loudspeaker can be used for acoustic effects, which reproduce, for example, crowd noises so that studio background atmosphere is preserved during the insert. A change of background by fading out the studio to insert the effect may destroy an illusion.

Special distortion effects such as the simulation of telephonic speech quality may sometimes be required, and this is achieved by inserting variable frequency-response networks in a channel.

When a discussion is to be broadcast between two or more speakers in geographically distant locations, each must be able to hear the others; unless special precautions are taken 'howl-back' is a possibility. This is prevented by feeding back the output of one studio to the other from a point prior to the insertion of the other contributor. The 'clean' feed is supplied on headphones to the latter.

54.7.3 Continuity control

When a programme is normally broadcast nationwide, special steps may be required to maintain it at all normal broadcasting times, and at some point in the main chain before distribution to the transmitters a continuity control, *Figure 54.19*, is inserted. The apparatus is a modified version of the studio equipment with a microphone which can be switched into the circuit by an announcer. The latter can step in and take control with a special announcement or explanation for an over-run of programme or a technical fault. He will also have facilities for inserting a recorded item as a fill-in for an under-running programme.

54.7.4 The main control centre

The chief purpose of the main control centre is to accept programmes from many sources and route them to their destinations, either the studios or the transmitters. A degree of aural and visual monitoring is carried out but no gain control is normally performed except when a channel is being set up with test signals. Communication circuits will be available to all sources originating programme, and there will be facilities for carrying out engineering tests such as those for noise, distortion, frequency response, etc. If the majority of programmes are recorded the distribution network can be made to operate automatically using a memory storage system.

54.8 Stereophonic broadcasting

Monophonic broadcasting suffers from the disadvantage that no spatial sense can be given because the sound source is effectively a single ear. The aim of stereo is to make the sounds appear to come from the position they originally occupied. To do this completely would require many microphone channels with as many loudspeakers. In practice it is not normal to use more than two channels to convey the stereo sound. The illusion of space has to be created by a difference in sound amplitude from two loudspeakers, which should have identical characteristics and operate in phase. Out-of-phase sources give a diffuse sound image appearing to come from behind the head. An amplitude unbalance of about 20 dB shifts the sound image from the centre line between the loudspeakers to the extreme edge.

The two stereo signals are designated A (left-hand side) and B (right-hand side), and since monophonic receivers must be able to use the stereophonic signals it is necessary to form an addition of A and B to give a compatible monophonic version.

Accordingly the A and B signals are converted at the transmitter into $(A+B)$ and $(A-B)$ signals by the transformer shown in *Figure 54.20*. A special multiplex system (the GE Zenith system used in many countries including the UK and known as the 'pilot tone' system) is employed to permit both signals to be transmitted on the same carrier and to be separated at the receiver, *Figure 54.21*.

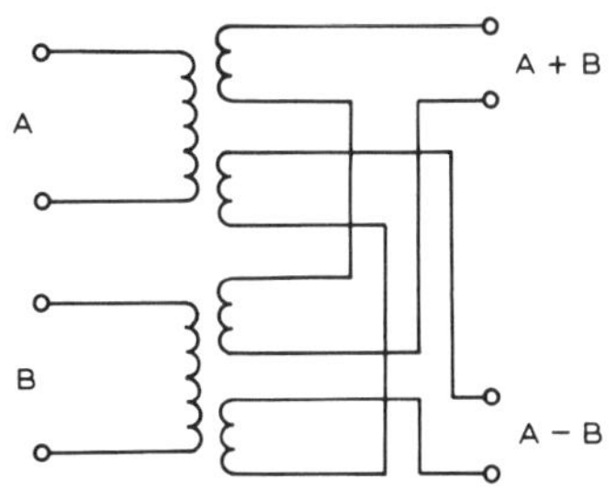

Figure 54.20 Combination of A and B signals to provide $(A+B)$ and $(A-B)$ outputs

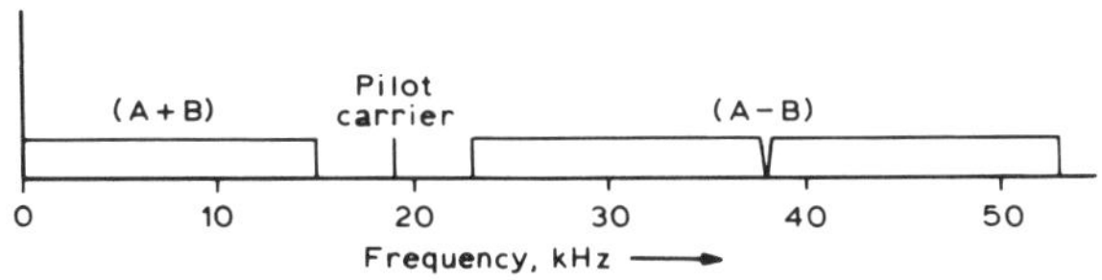

Figure 54.21 Pilot sub-carrier coded stereo signal

The $(A+B)$ signal is left unchanged covering the AF range 30 Hz to 15 kHz but the $(A-B)$ signal together with a sub-carrier at 38 kHz is applied to a balanced modulator whose output contains only the amplitude-modulated sidebands covering the range 23 to 53 kHz. A low-amplitude pilot tone at half the sub-carrier frequency, i.e. 19 kHz, is added to the $(A+B)$ and the $(A-B)$ sidebands and this composite signal is used to frequency modulate the transmitter. The $(A+B)$ signal is separated at the receiver from the pilot sub-carrier (19 kHz) and the $(A-B)$ sidebands. The 19 kHz sub-carrier is filtered from the other signals and is amplified and multiplied by two before being applied with the $(A-B)$ sidebands to a detector which extracts the $(A-B)$ signal. Addition and subtraction of the recovered $(A+B)$ and $(A-B)$ signals recovers the original left (A) and right (B) signals. The monophonic receiver rejects the pilot tone and $(A-B)$ sidebands and only reproduces the $(A+B)$ signal.

54.9 Distribution to the transmitter

The sound component of television and radio programmes is distributed between the studio centres and the transmitters by means of a network of cable and microwave links. Pulse-code modulation (PCM) techniques[7,8,9] are tending to replace older analogue methods of distribution. This provides a higher standard of performance and in the case of stereo radio perfect matching between left and right channels.

Pulse-code modulation ensures that the system is immune to all but the most severe noise and distortion on the bearer circuit. The presence or absence of the pulses may be detected up to a threshold point above which the effect on the audio channel rapidly becomes catastrophic. It can be shown[7] that the threshold for an ideal detector occurs with white noise at a peak signal/RMS noise ratio of approximately 20 dB. Taking this into account a PCM distribution system may be designed so that the audio performance is determined by the coding and decoding equipment, and this performance maintained throughout the complete distribution system.

54.10 Television sound distribution, sound-in-syncs

Television sound can be distributed by the sound-in-syncs (SIS) system. This combines the sound in PCM form with the video signal in such a way as to enable the complete programme to be routed via the same video link. At the transmitter site the signal is split back into separate audio and video signals and transmitted in the usual way to the viewer. The composite SIS signal offers advantages financially due to economies in audio links, and also operationally since the audio cannot become separated from the video between terminal sites.

The SIS system is essentially a time-division multiplex system in which the audio signal is coded in pulse-code modulated form and the pulses inserted during the line-synchronising period of the video waveform (*Figure 54.22*).

The video may be any 625-line colour signal using a field waveform comprising of five equalising pulses, five broad pulses

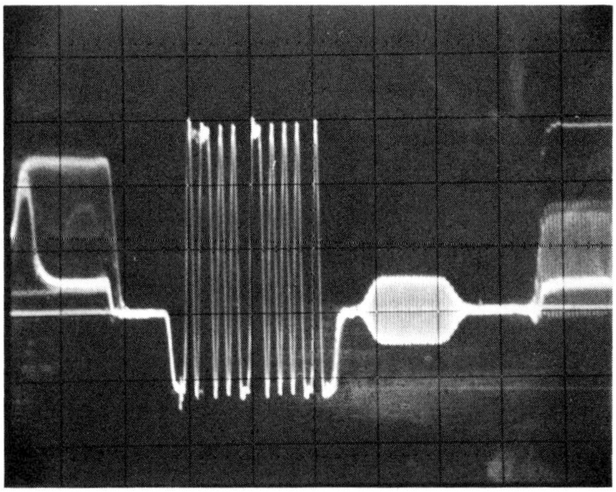

Figure 54.22 Sound-in-syncs waveform showing sound pulses inserted into the line synchronising period

and then a further five equalising pulses. The equipment with modification, may be used with 525-line System M video signals.

The audio channel obtained has the following performance:

Frequency response	50 Hz–13.5 kHz (±0.7 dB reference 1 kHz)
	25 Hz–14 kHz (−3 dB reference 1 kHz)
Signal-to-noise ratio	53 dB CCIR Rec 468-2 quasi-peak meter and weighting network, bandwidth 20 Hz–20 kHz
Non-linear distortion	0.25% (1 kHz at full modulation).

The quality of the audio channel is independent of the noise and distortion on the bearer circuit up to a threshold point above which the audio channel rapidly deteriorates. The threshold point for SIS is to a large extent determined by the ability to separate the sync pulses from the SIS waveform. This is because the leading edge of the line-sync pulse is used as the timing reference for the insertion of the sound pulses.

Typical distortions permissible on the bearer circuits are:

White noise	23 dB (peak signal/RMS unweighted noise)
Pulse K rating	8%
High-frequency loss	−8 dB at 4.3 MHz.

54.10.1 SIS equipment description

The simplified block diagram of the complete equipment is shown in *Figure 54.23.* The audio input is first compressed in dynamic range by means of a circuit that is basically a fast acting audio limiter. The signal is then sent to an analogue-to-digital converter (ADC) where it is sampled at a frequency equal to twice the video line frequency (31.25 kHz) and each sample converted into a 10-bit binary code. This 10-bit code is then temporarily stored while the next sample is being converted. The two codes are combined to form a 20-bit word, a marker bit added making 21 bits total, and the complete message inserted into the following line sync pulse at a peak-to-peak amplitude of 0.7 V for a nominal 1 V video signal, *Figure 54.24.* The individual pulses are shaped into sine-squared form with a half-amplitude duration of 182 ns, and a spacing of 182 ns between the middle of the pulses.

These pulses have negligible energy above a frequency of 5.5 MHz and so are quite suitable for transmission on a video link. During the field-blanking period alternate equalising pulses are widened so as to accommodate the pulse groups, *Figure 54.25.* At the decoder the sync pulses are blanked out and a regenerated sync waveform, which is standard in all respects, gated in.

In order to minimise variations in the mean level of the pulse group, the 10-bit words are interleaved and one word

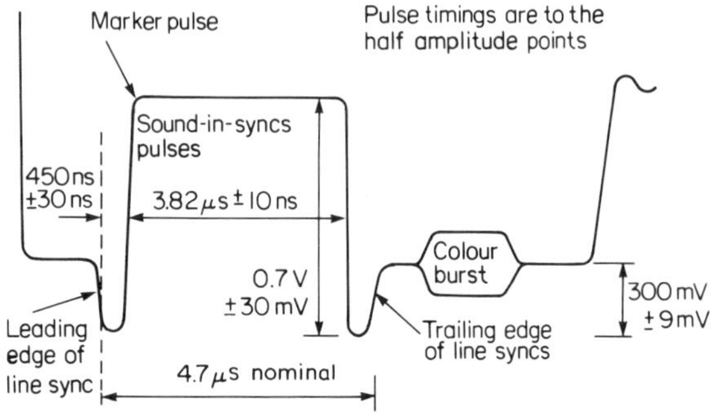

Figure 54.24 Idealised waveform showing combined pulse group, in the line-sync period

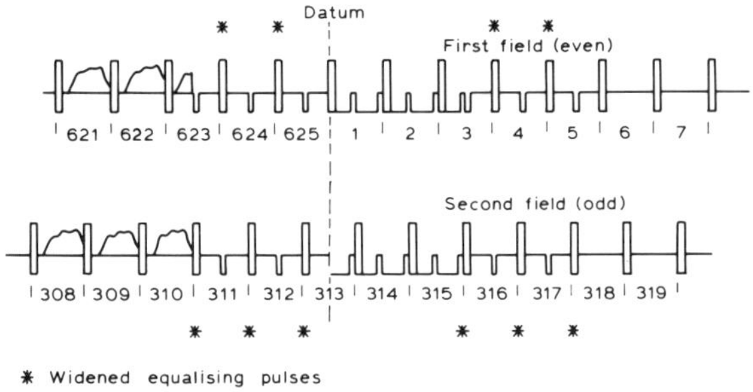

Figure 54.25 Modified field-blanking waveform

complemented (logic 1's exchanged for a logic 0 and conversely), *Figure 54.26.* In addition, the least significant bits are arranged to come first so as to keep the most rapidly changing pulses as far away as possible from the video back porch. The reason for this is, if the video bearer circuit has low-frequency amplitude distortion, variations in mean level during the sync pulse can be impressed on to the back porch, and if a clamp is used, these variations may then be transferred to the picture period and produce sound-on-vision effects.

At the receiving terminal the video signal is separated from the combined SIS signal and restored to a standard form. The 21-bit message is sent to a digital-to-analogue converter (DAC), where it is split into the two 10-bit codes, each code is then separately decoded. The analogue output of the converter is then fed to the expander where the dynamic range of the audio signal is restored.

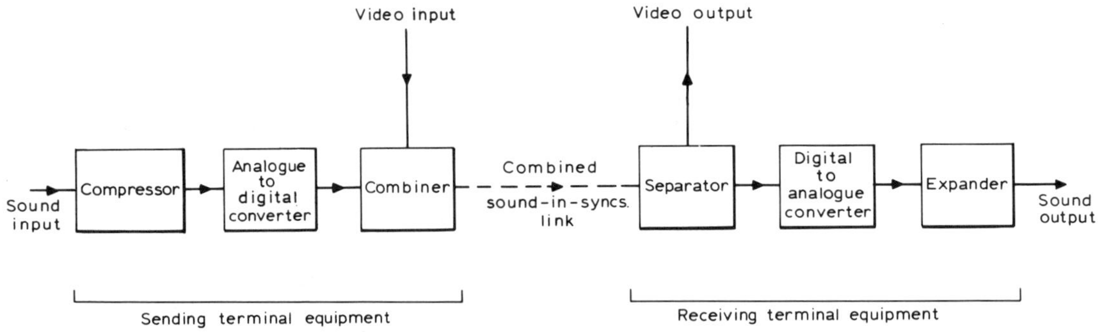

Figure 54.23 Simplified block diagram of the sound-in-syncs system

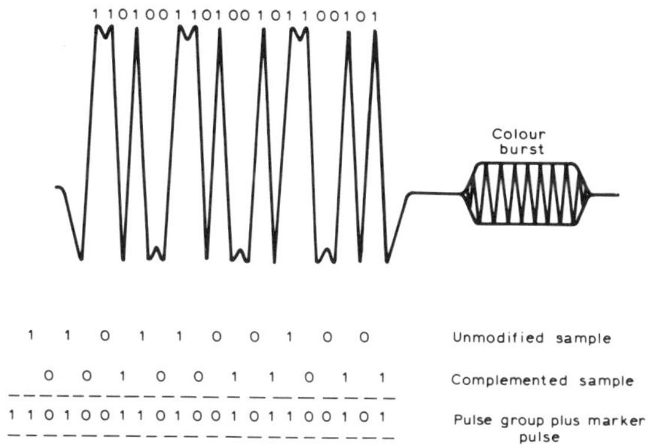

Figure 54.26 Details of pulse group waveform

54.11 The 13-channel PCM radio distribution system

The 13-channel PCM system was developed in order to meet the demand for high-quality sound circuits capable of carrying stereo programme material between the studio centres and the main transmitters. All channels provide the same high standard of performance, and any two channels may be used as a stereo pair over long distances without loss of quality. The performance achieved on a single codec is as follows:

Frequency response	50 Hz–14.5 kHz ($\pm$0.2 dB with respect to 1 kHz, typically -1.5 dB at 30 Hz and 15 kHz).
Signal-to-noise ratio	57 dB CCIR Rec 468-2 quasi-peak meter and weighting network, bandwidth 20 Hz–20 kHz.
Non-linear distortion	0.1% (1 kHz at full modulation).

54.11.1 Bit rate

The bit rate was chosen to be 6336 kbit s^{-1}. This enables the bit stream to be conveyed on a bearer circuit designed to carry 625-line television signals. The individual pulses are shaped into sine-squared form with a half-amplitude duration of 158 ns, and the pulses are spaced 158 ns apart, *Figure 54.27*. These pulses therefore have negligible energy about 6.336 MHz; the bandwidth may be reduced in exchange for a worsening of the immunity to noise and distortion. With a bandwidth up to 6.336 MHz the PCM system can withstand a peak signal/RMS white noise ratio of 20 dB. If the bandwidth is restricted to about 4.5 MHz then the noise immunity becomes approximately 22 dB.

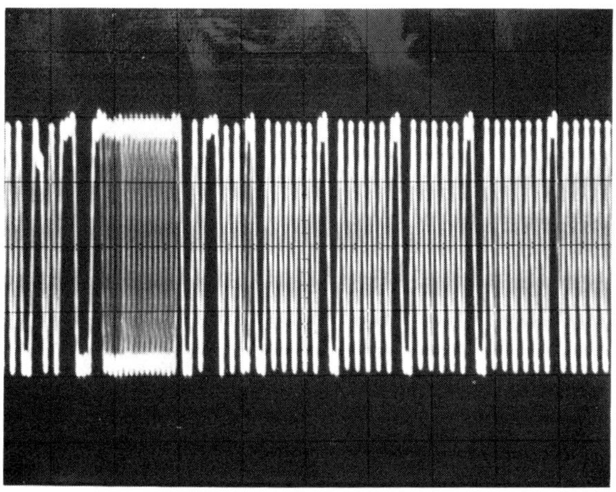

Figure 54.27 Part of PCM bit stream

54.11.2 Coding equipment

The simplified block diagram of the coding equipment is shown in *Figure 54.28*.

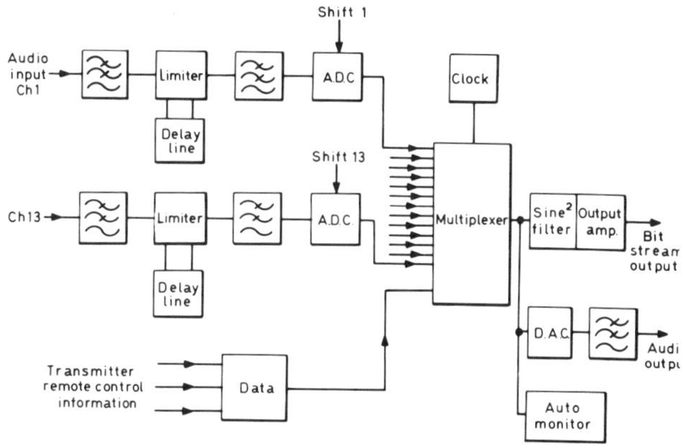

Figure 54.28 Simplified block diagram of PCM coder

Each audio input is fed to a delay-line type limiter via a 15 kHz low-pass filter. The limiter is adjusted so as to operate at a level 2 dB above the nominal peak programme level. This prevents the occasional high programme peak over-modulating the PCM system. If two channels are used as a stereo pair, then the limiters are interconnected so that if either limiter operates both change gain by the same amount, thus preventing shift of the stereo image. A 50 μs pre-emphasis network is associated with the limiter. This enables the channel to be used to feed a FM transmitter without an additional limiter at the transmitter site. The PCM limiter will then prevent over-deviation of the transmitter.

The audio signal passes via a second 15 kHz low-pass filter to the ADC. The low-pass filters are designed to pass 15 kHz with little attenuation while frequencies at 16 kHz and above are greatly attenuated.

The ADC uses a sampling frequency of 32 kHz. This enables an upper frequency limit of 15 kHz to be achieved without aliasing components causing distortion. Each sample is converted into a 13-bit binary code using a converter with a linear quantising law. In order to protect the system against serious bearer circuit disturbances, a single parity bit is added to each 13-bit word. This provides a check on the five most significant bits. If an error is detected then the word is discarded and the previous word substituted in its place. If the error rate becomes high enough to impair seriously the quality of the signal the audio output is muted.

The output of each ADC is fed to the multiplexer. The multiplexer carries out the time-division process, combining the thirteen channels into a stream of pulses at a bit rate of 6.336 Mbit s^{-1}. After pulse shaping into sine-square form the signal is passed to the output, the bit stream is shown in *Figure 54.27*. A bit-rate of 6.336 Mbit s^{-1} along with a sampling rate of 32 kHz gives a total of 198 bits per frame. The 13 audio channels each use 14 bits including the parity bit, the remaining 16 bits are used as a framing pattern (11 bits) and a data control channel (5 bits).

The data channel provides the facility for remote switching of equipment at decoder sites. The data signal comprises of two parts, an address and a switching message. It is possible to switch either equipment at all the decoder sites simultaneously or to send a message to an individual site. The basic capacity of the system is 128 messages but this may be extended if the time to send a message is increased. Typical uses of the data messages are for mono/stereo switching and transmitter remote control.

Automatic monitoring is used to give warning of any faults occurring in the coder. The monitor is sequential and dwells on each audio channel in turn for 256 ms. During this time 256 checks are made by sampling the audio input to the channel and coding by means of a 5-bit ADC. The output of this ADC is then compared with the 5 most significant bits of the appropriate word appearing in the bit stream at the output of the multiplexer. If more than 50% of the samples are in error then the monitor will register a fault on that particular channel.

For aural monitoring a DAC and a low-pass filter are included in the equipment. These can be manually switched so as to provide a means of quality monitoring any audio channel in the bit stream.

54.11.3 Decoding equipment

The simplified block diagram is shown in *Figure 54.29*.

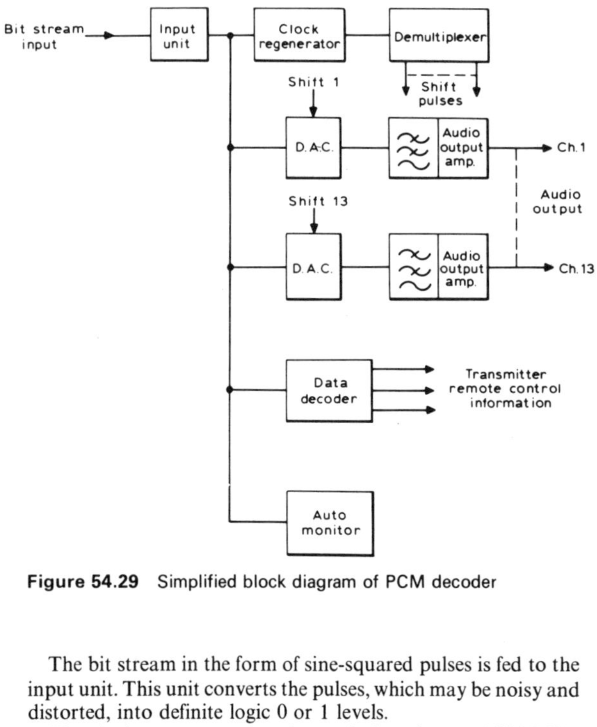

Figure 54.29 Simplified block diagram of PCM decoder

The bit stream in the form of sine-squared pulses is fed to the input unit. This unit converts the pulses, which may be noisy and distorted, into definite logic 0 or 1 levels.

The clock regenerator produces clock pulses at 6.336 MHz locked to the incoming bit stream. The synchronisation is achieved by comparing the timing of the framing pattern occurring in the bit stream with a locally generated framing pattern, any timing error produces a change in a voltage used to control the frequency of the local timing reference.

The regenerated clock pulses are fed to a demultiplexer where groups of bit-rate shift pulses are produced. The shift pulses are used to enter the individual 14-bit pulse groups, present in the continuous bit stream, into the correct DAC.

Initially the parity bit is checked and if correct the conversion is allowed to proceed in a DAC using the dual ramp-counter technique.

The analogue output from the DAC is then fed to a low-pass filter with a cut-off frequency of 15 kHz. This filter removes the unwanted high-frequency components associated with a sampled waveform. The audio signal then passes through an amplifier to the channel audio output of the decoder.

The automatic monitoring is similar to the sequential system used in the coder. A 5-bit ADC is used, with the audio samples taken from the output of the decoder. The output of the ADC is then compared with the five most significant bits occurring in the appropriate word in the input bit stream. If necessary the monitor may be used to check other equipment in the programme chain placed after the decoder.

54.12 Digitally companded PCM systems

Multi-channel digital circuits are being increasingly used for telephony with the result that analogue circuits will in the future become scarce.

This has resulted in developments designed to use the new digital circuits for high-quality sound programme distribution. Several bit-rates have been recommended by the CCITT,[10] the primary and secondary rates proposed in Europe being 2.048 Mbit s^{-1} and 8.448 Mbit s^{-1}. To make the best use of digital circuits companding techniques are used to conceal the audibility of background noise. Companding has the effect of varying the noise depending on the signal level, the noise being least at low signal levels and highest at maximum signal level. A system known as[11] NICAM 3, using digital companding has been developed by the BBC. This system codes the samples to 14-bit accuracy and then digitally compresses to 10 bits per sample for transmission. By this means it is possible to obtain six high-quality sound channels on a 2.048 Mbit s^{-1} bit stream.

54.13 Low frequency and medium frequency transmitters for sound broadcasting

Figure 54.30 shows a typical schematic for high-power low-frequency (LF) and medium-frequency (MF) transmitters.

54.13.1 The drive unit

The drive unit produces a stable output at the carrier frequency. It often employs a crystal oscillator circuit housed in a temperature-controlled enclosure to achieve maximum stability. Crystal drives can be stable to within one part in 10^9 per month. If greater stability is required a rubidium vapour standard can be employed. This unit makes use of an atomic resonance of rubidium 87 at about 9 GHz. A vapour cell is irradiated with a signal near to the resonance frequency derived from a 10 MHz crystal oscillator. The light from a rubidium spectral lamp also passes through the cell onto a photodetector. When the frequency of the signal equals the resonance frequency the light is absorbed. The output of the photodetector is arranged to control the precise frequency of the crystal oscillator. The unit provides an output at 10 MHz from the crystal oscillator to a stability of about 3 parts in 10^{11} per month. The output is applied to a frequency synthesiser to produce the required carrier frequency.

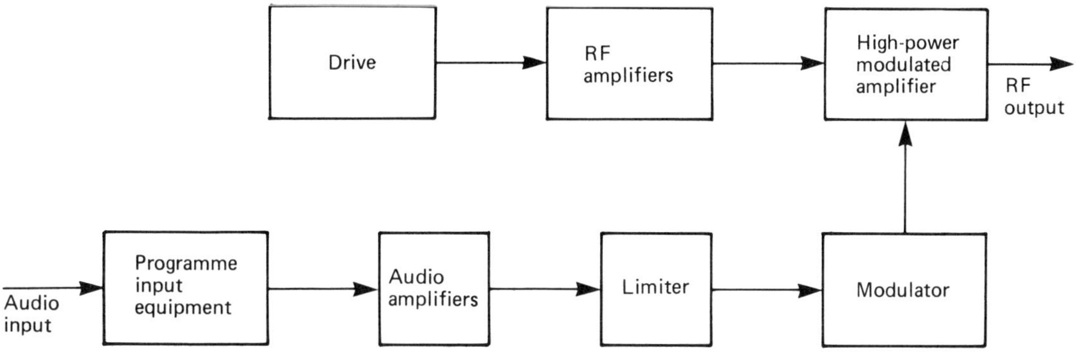

Figure 54.30 Schematic of high-power low- and medium- frequency sound transmitter

54.13.2 The modulated amplifier

In high-power transmitters this is a valve stage employing triodes or tetrodes. It may have a single valve or a pair operating in push–pull. The load for the valve or valves comprises a tuned circuit inductively coupled to the output through a matching circuit. In modern transmitters the output stage is often operated in Class D. A square voltage waveform at the carrier frequency is applied to the grid which drives the valve into saturation. The tuned anode circuit responds to the fundamental component of the train of anode current pulses thus giving a sinusoidal carrier output. If the duration of the pulses is varied in accordance with the audio signal the output will be an amplitude modulated carrier. The fundamental component of a square wave increases according to a sinusoidal law as the mark/space ratio increases from zero, having a maximum at unity mark/space ratio. When fully modulated, therefore, the amplifier can theoretically impose a compression of about 4 dB. As an alternative to control grid modulation, screen grid modulation is possible in tetrodes. With valve circuits output powers of up to 500 kW are typical and greater powers can be obtained by paralleling the outputs of several amplifiers.

The Doherty amplifier is commonly found in high-power LF/MF transmitters. The basic circuit of the Doherty amplifier is shown in *Figure 54.31*. V1, the carrier valve, delivers power alone

Figure 54.31 The Doherty amplifier

at carrier amplitudes up to the mean level (unmodulated level) through a quarter-wave network. In this range V2, the peaking valve, is biased back and inoperative. Above the mean carrier level, V2 begins to contribute; its grid drive is delayed so that its output is in phase with that from V1. A property of a quarter-wave network (or quarter-wave line) is that it transforms a constant voltage source into a constant current source. At mean carrier amplitude, V1 is made to saturate when it behaves like a constant voltage source. The power delivered by V1 is then dictated by the apparent impedance of the load circuit which, in turn, increases with increasing output from V2. From mean carrier level to peak level the outputs from V1 and V2 increase. The circuit constants are arranged so that, at peak amplitude (100% modulation) both valves are contributing equal power. Clearly this circuit can be operated in the Class D mode and in push–pull.

Its principal advantages compared with conventional arrangements are that smaller valves can be used for a given peak power and the efficiency is more constant over the modulation cycle; usually about 60%.

For output powers of up to 1 kW, solid state transmitters are available. The output transistors are operated in a switching mode and since they are either 'on' or 'off' the dissipation is low. In a typical arrangement the power amplifier comprises four transistors in a bridge configuration switched at the carrier frequency across an audio modulated supply rail. These transmitters offer good efficiency (about 50% overall), reliability and easy adjustment of the output power level.

54.14 Short-wave transmitters

Broadcasting on short waves generally calls for regular changes of frequency to match ionospheric conditions and, either a number of crystals must be available to be switched as desired, or a frequency synthesiser must be used. The advantage of the latter is that any frequency can be selected in decade steps from about 4 to 30 MHz, although its stability may be lower than that of the crystal oscillator. The transmitter schematic is similar to *Figure 54.30*. The low-power radio frequency (RF) amplifiers are broadband to accommodate the range of frequencies required and the high-power amplifier can be automatically tuned to a series of spot frequencies; the tuning being under the control of a memory system and a pre-selector switch. An output power of 250 kV is typical.

54.15 VHF frequency-modulated sound transmitters

Frequency-modulated transmissions in Band II (88–108 MHz) offer high-quality services with stereo capability. The schematic of a VHF sound transmitter is shown in *Figure 54.32*. In the first stages of the stereo transmitter the incoming left-hand (A) and right-hand (B) audio signals are passed through pre-emphasis networks and then applied to a stereo encoder which produces a

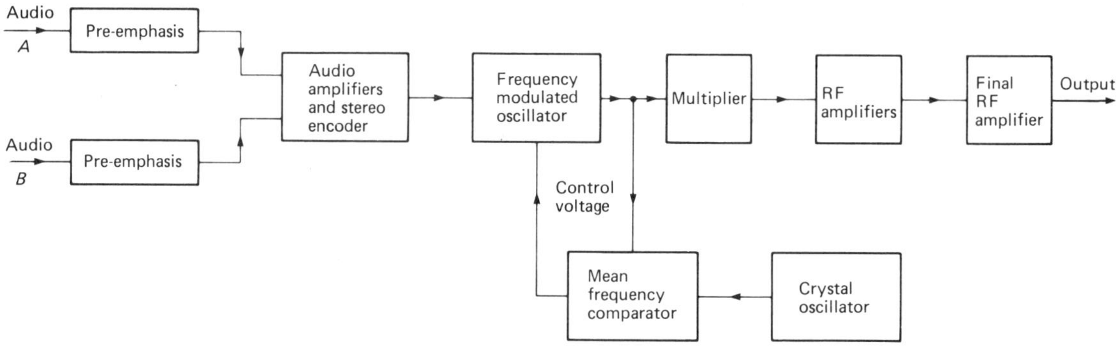

Figure 54.32 VHF frequency modulated sound transmitter

composite baseband signal prior to modulation. In the 'pilot tone' system the first 15 MHz of the spectrum of the composite signal is occupied by the sum signal $(A+B)/2$. The difference signal $(A-B)/2$ is amplitude suppressed carrier modulated on a sub-carrier frequency of 38 kHz and added to the sum signal. A pilot tone at 19 kHz is also added to the composite signal. The multiplier stages following the oscillator increase the frequency deviation by the same ratio as the carrier multiplication. A maximum frequency deviation of ± 75 kHz is usual. The following RF amplifier stages are wide-band. Output powers of up to 20 kW are typical and for lower powers all solid-state equipments are available.

54.16 UHF television transmitters

The schematic for one type of UHF television transmitter is shown in *Figure 54.33*. As shown the video signal amplitude modulates the carrier at the final frequency. The diode circuit presents a variable reflection coefficient at the middle port of the circulator thus imposing an amplitude variation on the carrier in accordance with the video signal. An equivalent type of modulator uses a directional coupler with variable reflection coefficient circuits on two ports. The modulation may alternatively be carried out at an intermediate (IF) frequency, the modulator being followed by a frequency changer. The IF modulation method is tending to replace the absorption modulator because pre-correction and vestigial side-band (VSB) filtering can be carried out in the IF stages and this gives a better overall performance. *Figure 54.34* illustrates the vision carrier envelope (one line) when modulated with a colour bar signal (PAL system).

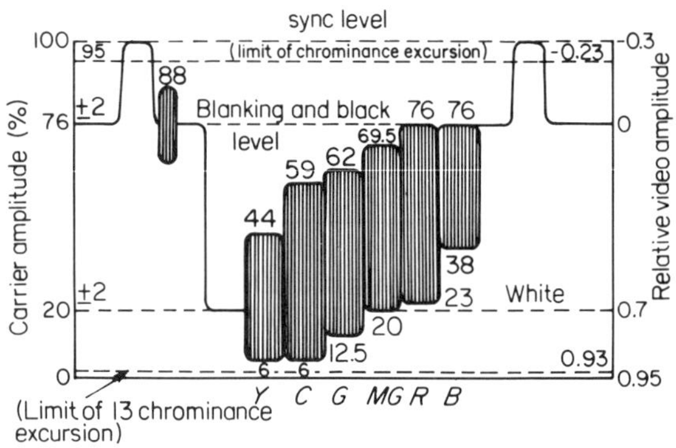

Figure 54.34 Vision carrier envelope (one line) when modulated with a colour bar signal

54.16.1 Television sound transmitter (UHF)

This is a separate transmitter, often employing a klystron output stage. The type of modulation depends on the television system in use. For system I (in the United Kingdom) frequency modulation is employed on a carrier frequency that is 6 MHz above the vision carrier frequency. The sound carrier power can be between 1/10 and 1/5 of the peak vision carrier power. The outputs of the vision and sound transmitters are combined in a diplexer. Many stations have pairs of sound and vision transmitters operating in parallel which can be switched to single-transmitter operation in the event of a breakdown.

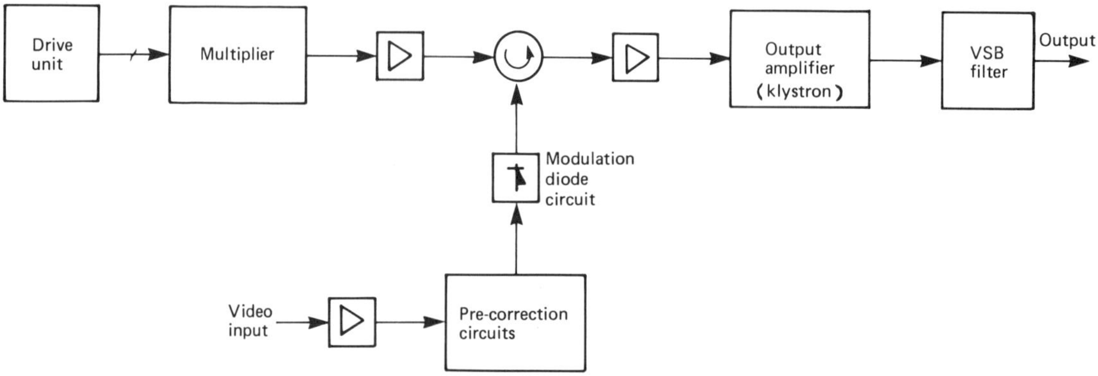

Figure 54.33 UHF television transmitter

54.17 Television relay stations: transposers

Television services in the UHF bands can require a number of low-to-medium-power transmitters to fill 'holes' in the service area of a main transmitter. These relay stations are often of the non-demodulating transposer type in which a complete television channel, comprising sound and vision signals, is received off air, frequency translated to another channel, and re-radiated. The range of output powers is from a few watts to 1 kW and output amplifiers may be solid-state, travelling-wave tubes or klystrons depending on the power required.

54.18 General aspects of radio and television receivers

Broadcast radio and television receivers have remained substantially unchanged in their signal flow and general principles of operation; in fact, the introduction of stereo radio and the addition of colour and teletext to television represent the only two basic changes that have been made. With few exceptions, the use of the superheterodyne receiver has been retained, although many of the circuit functions are now combined in a single integrated circuit. The broadcast radio signals themselves have, of necessity, also remained unchanged over the years (except for stereo) and the broadcast chain through which the signal passes is also much the same.

Most broadcast radio and television receivers work on the superheterodyne principle, with a local oscillator in step with the incoming signal to provide a constant heterodyne frequency, the so-called intermediate frequency (IF). It is in the IF band that the main amplification of the received signal takes place, prior to demodulation and final audio or video amplification. On the LF, MF and HF bands and television bands, local oscillators work above the frequency of the incoming signal but, in band II (VHF/FM), receivers vary and oscillators operate both above and below the RF input signal frequency. The IF frequency for band II FM sound receivers has been standardised world-wide for many years at 10.7 MHz. However, the value used for LF, MF and HF receivers is now found to vary from the older UK standard of 470 kHz to 455 kHz, most receivers marketed in Britain now being manufactured in the Far East. The values of 39.5 MHz (vision IF) and 33.5 MHz (sound IF) are standard for British television receivers.

While the bulk of gain and selectivity is obtained in the fixed-frequency IF amplifier, some first-stage gain and selectivity is needed to achieve a good signal-to-noise ratio and to provide some selectivity to reduce the interference from out-of-band strong signals that can cause, for example, image-frequency breakthrough and unwanted signals due to intermodulation between harmonics of the oscillator and the incoming signals. Some examples of radio receiver IF responses are shown in *Figures 54.35* and *54.36*, although it should be realised that the performances of practical radio receivers vary widely; moreover,

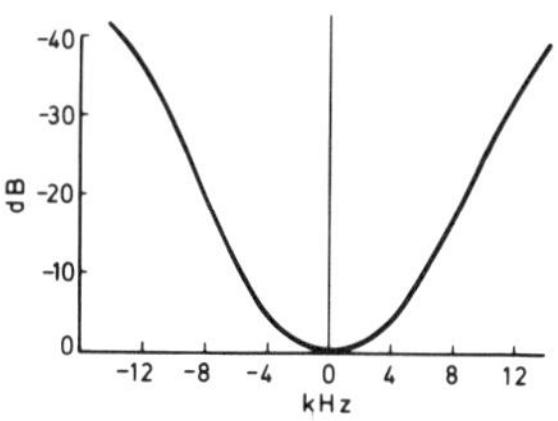

Figure 54.35 Typical MF receiver, 470 kHz IF amplifier response using 5 tuned circuits

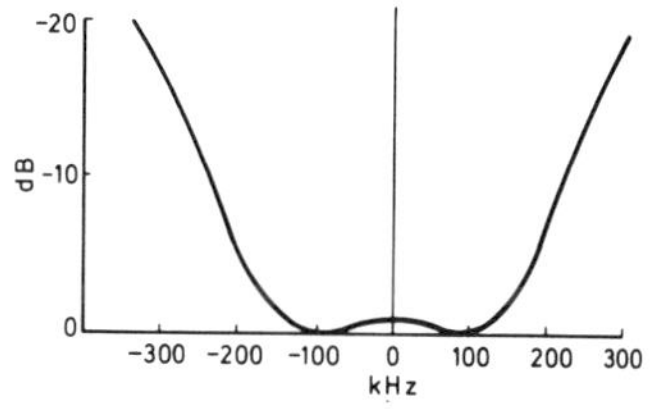

Figure 54.36 Typical VHF, FM receiver, 10.7 MHz IF amplifier response using 5 tuned circuits

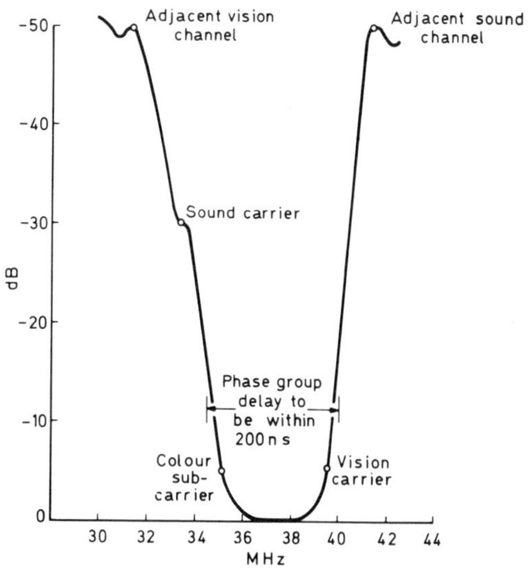

Figure 54.37 Typical television receiver IF amplifier response

improved (and cheaper) IF filters are now being introduced using ceramics in place of conventional tuned circuits.

Figure 54.37 shows the kind of IF response obtained in modern television receivers using surface-acoustic-wave (SAW) IF filters[12,13,14], rather than the earlier tuned-circuit IF filters. The response of television receivers is complicated by the need to attenuate the adjacent channel vision and sound carriers and to place the wanted sound carrier on the 'intercarrier-shelf' about 23 dB below the point of maximum gain. This is in order to obtain the best sound performance with the least amount of interference from intermodulation with low-frequency video signal components (causing 'buzz') and from intermodulation with the chrominance sub-carrier which can cause unwanted patterning on the picture. The phase/frequency or group-delay response is also of crucial importance; poor phase response causes 'plastic' picture effects, overshoots and smearing of outlines as well as increasing the likelihood of errors in the reception of teletext signals. The phase shift through the receiver should vary linearly with frequency over the pass-band or, in other words, the relative timings of all the spectral components of the picture signal should remain undisturbed. However, in the vicinity of the vision carrier, and at the other band edge, the phase response changes rapidly with errors in alignment and automatic tuning facilities of good stability are of great importance. Moreover, the amplitude and phase responses should remain essentially unchanged with gain variations brought about by the application of automatic gain control (AGC); AGC is an integral feature of most receivers and is usually achieved by utilising the rectified demodulator output potential to provide control of the bias to the preceding IF amplifier srages. The introduction of the television SAW IF filter,

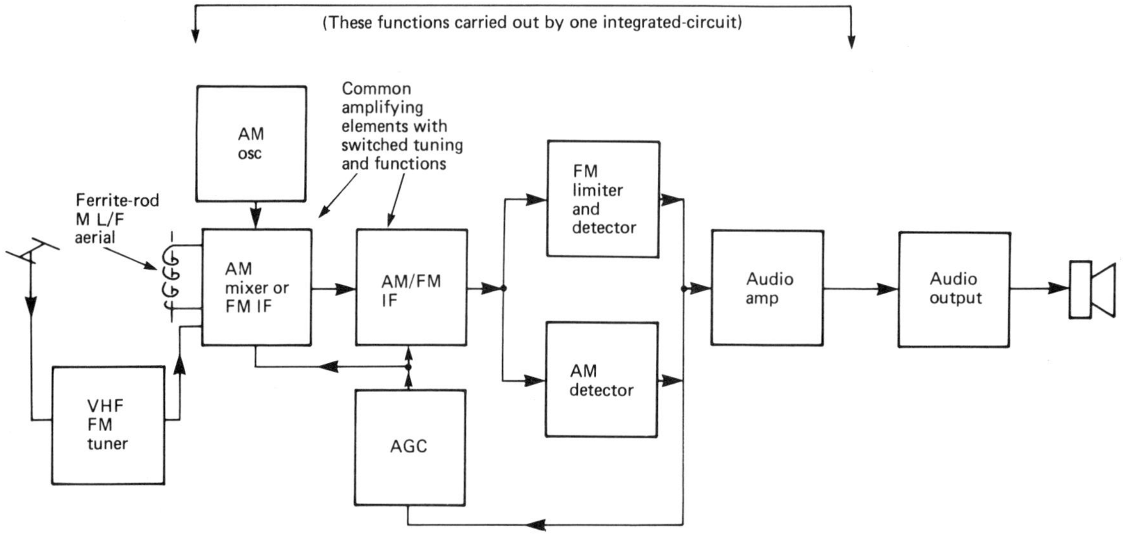

Figure 54.38 AM/FM receiver

whose group-delay response is controllable independently of the amplitude response, has done much to alleviate many of the difficulties mentioned above.

54.19 Radio receivers

Figure 54.38 shows a functional diagram of an AM/FM receiver.

54.19.1 Radio aerials

Long and medium-wave domestic radio receivers used long-wire aerials for many years, usually in the form of an inverted 'L'. This aerial, short compared with a wavelength, presents a capacitive source impedance to the receiver and can form part of the input circuit turning. During the 1950s, the development of ferrites led to the emergence of the ferrite-rod aerial which has been almost universally adopted for use in medium- and long-wave portables. Essentially a loop or frame aerial, the ferrite-rod concentrates the magnetic (or *H*) component of the incident wave within a coil wound on a rod of high-permeability ferrite; the same coil normally provides the inductance for the first-stage tuning in the receiver. The 'figure-of-eight' directional properties of the ferrite-rod aerial can often be put to good use in reducing co-channel or other interference.

When the VHF FM service was first introduced in Britain, it was assumed that reception would be via a fixed aerial at roof-top level and it was for this purpose that horizontal polarisation was chosen both here and in most other countries. One important exception is Radio Telefis Eireana, of Eire, which makes exclusive use of vertical polarisation. In more recent years, other forms of reception such as the portable and car receiver using vertical 'whip' aerials have become important and the choice of polarisation has been the subject of review. Experiments have been carried out in a number of European countries and, in the mid-1970s, there began the addition of vertical components of equal power to some existing horizontally-polarised transmissions. New stations were constructed using this so-called 'mixed polarisation' from the start.

Fixed VHF receivers are often operated with a short-wire indoor aerial but, if signal strength is lacking or if there is multipath interference, a roof-mounted aerial with some directivity is to be preferred. Reception of VHF FM signals on portable receivers is still achieved by the extendable rod or 'whip' aerial. This is awkward and inconvenient, however, and would be better superseded, perhaps by the ferrite-rod, although attempts to develop ferrite aerials for VHF have not been very successful so far, mainly because of sensitivity problems[15].

54.19.2 Stereophonic FM radio receivers

The pilot tone system of stereophonic (stereo) sound broadcasting is now well established in many parts of the world[16].

The first requirement of a broadcast stereo sound system is that the signal should give satisfactory reception on a monophonic (mono) radio receiver. Thus, the left hand (L) and right hand (R) stereo signals are combined to form the sum or the mono (M) signal and the difference or S signal is modulated on to a 38 kHz subcarrier. The 38 kHz subcarrier is modulated using double-sideband, suppressed-carrier, amplitude modulation and, at the receiver, a synchronous detector is used for its demodulation. A 19 kHz pilot tone is transmitted to provide a reference or synchronising signal for the synchronous detector.

A useful way[17] of looking at the pilot tone stereo system is shown in *Figure 54.39(a)*. Here it will be seen that the signal fed along the transmission path is formed from both the L and R signals on a time-sharing or time-division multiplex basis, the fast changeover switches spending equal periods of time in each position. The mean value of the waveform shown dotted in *Figure 54.39(b)* corresponds to the $\frac{1}{2}(L+R)$ (or M) signal whilst the high-frequency component of the waveform has an amplitude of $\frac{1}{2}(L-R)$ and corresponds to the modulated subcarrier signal. If this whole waveform were passed through a suitable (53 kHz) low-pass filter, harmonics of the subcarrier frequency would be removed and the resultant sinusoidal waveform would be substantially identical to that of the combined M and S signals in pilot tone stereo. To complete the picture, it is necessary to add the 19 kHz pilot tone and to reduce the level of the subcarrier by a factor of $\pi/4$ to restore its sinusoidal amplitude to that of the square wave.

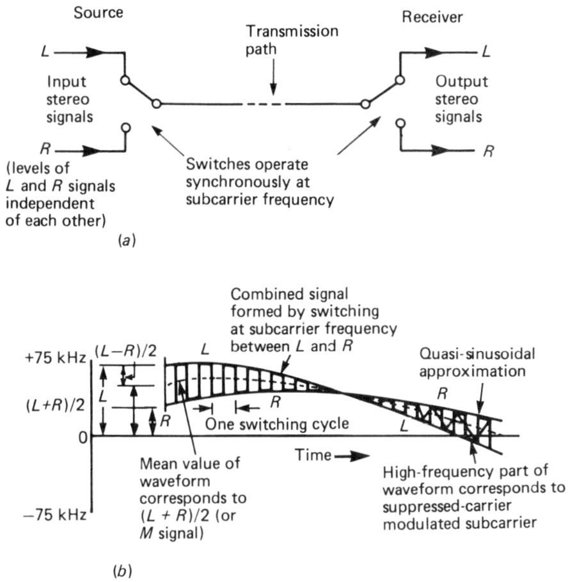

Figure 54.39 Time-division multiplex of *L* and *R* signals. (a) Circuit; (b) waveforms

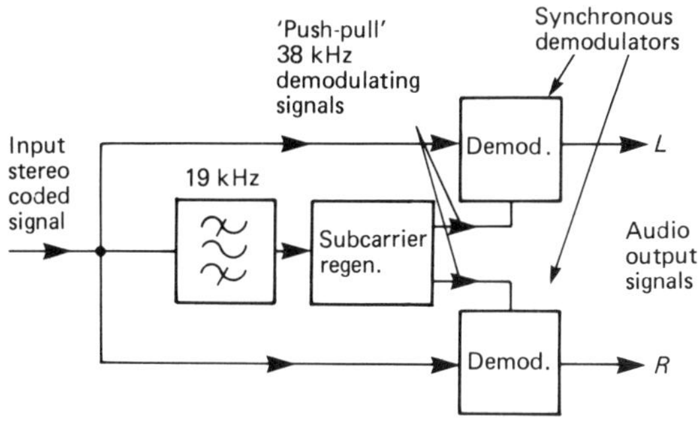

Figure 54.40 Switching stereo decoder

54.19.2.1 Stereo decoders

The basic requirements of a stereo decoder may be listed:

(*a*) To regenerate the 38 kHz subcarrier which is suppressed in transmission; this is usually accomplished using the 19 kHz pilot tone.
(*b*) To demodulate the subcarrier and thereby recover the stereo difference signal.
(*c*) To recombine the stereo sum and difference signals so as to recover the original *L* and *R* radio signals.

The switching decoder is the most commonly used and is popular because it is amenable to embodiment in integrated-circuit form and requires a minimum of other circuits and adjustments. It also fulfils requirements (*b*) and (*c*), above, as one operation. The basic principle of the switching decoder can be described using the time-division multiplex approach and with further reference to *Figure 54.39*, where a synchronous switch is used at the receiver to separate the time-multiplexed *L* and *R* signals. Two push–pull sampling waveforms select the positive and negative maximum values of the subcarrier. *Figure 54.40* shows the essential details. The action of the 38 kHz demodulating signals can, in a practical circuit, be that of a square wave by which the synchronous demodulators are switched 'on' or 'off', or that of a sine wave which multiplies the stereo signal in a linear multiplier. The former case is the easier to achieve in practice but does not give perfect decoding, whilst the sine-wave linear-multiplication process is difficult to achieve in practice but gives, theoretically, perfect decoding[17]. Square-wave on/off switching results in the generation of different multiplying coefficients or gains for the $L+R$ and $L-R$ signals which, in turn, cause the output *L* channel to contain demodulated *R* signal components, and vice versa. This can easily be compensated by employing a common *L* and *R* signal amplifier with a degree of common-mode rejection and, in practical receivers, this is often built-in to the demodulating circuits.

Figure 54.41 shows the functional block diagram of a commonly used integrated-circuit switching decoder. The upper chain of functions is designed to lock the internal 76 kHz oscillator to the incoming 19 kHz pilot tone signal and thereby to provide a 38 kHz reference signal for demodulation of the stereo signal. In input multiplier 1, the returned 19 kHz signal and the incoming 19 kHz pilot tone are multiplied together and the low-frequency output from the multiplier is used to control the frequency of the 76 kHz oscillator via the loop filter and hence to

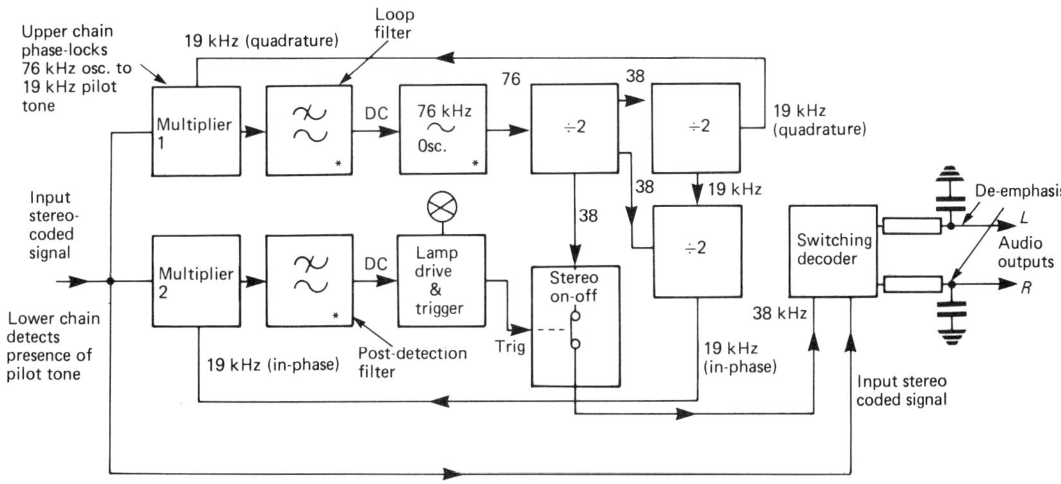

Figure 54.41 Integrated-circuit switching stereo decoder. * denote IC functions that involve external components

pull in and phase lock it to the pilot tone. When the oscillator is locked, the 19 kHz returned signal is in phase quadrature with the incoming stereo pilot tone signal. Other incoming signal components are also subjected to multiplication by the returned 19 kHz signal, but the resulting products are removed by the low-frequency loop filter following multiplier 1. The third frequency divider receives 38 kHz and 19 kHz signals from the first two dividers and is so connected as to deliver a 19 kHz output signal that is in phase with the incoming pilot tone. This is fed to multiplier 2 which therefore delivers a DC output signal, the value of which is proportional to the level of the incoming pilot tone. After filtering, this DC signal is applied to the lamp drive and trigger circuit so that, when a sufficiently strong pilot tone is present, both the lamp and the stereo-on switch are activated.

The integrated-circuit switching decoder just described contains no filtering elements other than the loop filter, the post-detection filter and the two de-emphasis circuits. These are all low frequency, low pass in characteristic and replace the bandpass and higher-frequency low-pass filters that might have been used, by operating on the signals after they have been reduced to baseband frequencies. However, noise and other interfering signals above 53 kHz accompanying the stereo signal can result in harmful intermodulation, particularly with harmonics of the subcarrier signal, to produce audio frequency output components via the difference channel; the noise triangulation effect of FM tends to enhance the level and importance of these unwanted high-frequency terms. It is preferable, therefore, to use a low-pass input filter to remove all components with frequencies above 53 kHz.

Unwanted cross talk between the L and R signal components can also occur and is usually due to non-uniformities in the response of the IF filter. The tolerance for left to right cross talk is given by the CCIR[18] as 30 dB separation between the L and R signals (between 300 Hz and 4000 Hz) and this degree of cross talk can easily occur due to deficiencies in the amplitude-response of practical receiver IF amplifier circuits. For example an amplitude 'roll off' resulting in a response of about -1 dB at 30 kHz could, typically, result in a separation between L and R signals of about 30 dB, and it is the practice of some designers to include some amplitude/frequency compensation to increase this figure. Improved IF filters[19], however, can reduce these and other deficiencies.

The basic requirements of a stereo decoder may also be carried out separately, as in a frequency-multiplex decoder. The main virtue of this method of decoding is that de-emphasis and bandwidth limiting are applied before demodulation of the difference signal and the unwanted noise and interfering signals are reduced in level before intermodulation effects can occur. It is unfortunate that this high-performance circuit has not found favour with designers on account of the filters required and its unsuitability for large-scale integration.

54.19.3 Future radio receivers

In attempting to forecast the future trends in radio receiver design, it seems reasonable to look at the more important shortcomings of present receivers and to consider ways and means that might be available to overcome them. Certainly the shapes, sizes and general appearances of future radio receivers will vary widely and there is also little doubt that it will become possible to make receivers that are very small indeed. These are important aspects but, in one sense, they will be only cosmetic in their effect unless the receiver of the future provides additional facilities as well.

It is widely recognised that many radio listeners experience difficulties in tuning their receivers to a desired station or programme and are thus unable to make the best use of the broadcast programme available; indeed it is thought that some listeners leave undisturbed the tuning of their receivers, from fear of being unable to recover the signal once it has been lost. Although there are problems with tuning in the UK, they are even more severe in some other European countries. In using the word 'tuning', here, we are not concerned with the receiver adjustment necessary to obtain best reception of a station already identified, but rather to consider the actual station identification process itself. Indeed, we can take the meaning further and use it to include the selection of a particular network, such as Radio 3, or even a category of programme such as news, sport, etc.

The problem of station and programme identification seems to be most acute at VHF, mainly because of the less predictable nature of propagation in this band. In many areas, there is often a large choice of transmitters from which a given programme can be received and there may be uncertainty as to which signal should be chosen for best reception or to obtain the desired local programme variations. Car radio receiver designers were the first to get to grips with this problem by introducing sophisticated storage and search facilities in the receiver. In one car radio, for example, it is possible to store the values of seventy different frequencies, covering VHF, HF, MF and LF. On VHF, sixty frequencies are stored under six programme headings, thus allowing a choice of ten different carrier frequencies for each radio programme. When one of the six programmes is selected by the user, the receiver automatically selects the strongest signal that can be found on any of its ten pre-set frequencies. If the user wishes, he can instruct the receiver to provide the second strongest signal instead; this can be valuable in those cases where the strongest signal may not be the best, due to multipath interference for example. As the car moves out of the service area of one transmitter into that of another, the receiver will automatically switch to the new stronger transmitter. Automatic or search tuning is also possible, as well as conventional manual tuning. Other new designs of car radio provide facilities that are different in detail from these, but are similar in general principle.

54.19.3.1 Radio data signals

Some broadcasting organisations in Europe are tackling the same problem by adding inaudible data signals for identification and other purposes to their otherwise normal broadcast radio signals. These additional signals are intended primarily as tuning aids, whereby it will be possible to identify a signal once it has been received[20,21]. Although final agreement on an international standard for radio data has not yet been reached in Europe, it seems very likely that a low-level subcarrier with a frequency in the region of 57 kHz will be used. Some (compatibility) problems have been encountered with interference to the normal programme from the data signal but these difficulties have been almost entirely concerned with older stereo receivers and reports of interference have been few over a three-year period. It is expected that any residual problem will disappear as these older receivers are replaced.

It seems likely that radio data will do much to facilitate tuning simply by providing identification information. For example, a manually tuned receiver can carry an alphanumeric display so that, as the band is explored, the display indicates the identity of each station found. In a motor car a voice synthesiser (as now used in some cars) could announce the identity of the station. In another example, a receiver is provided with a key-pad or other device through which the user instructs the receiver to find a particular station or programme. In a search operation (which can be automatic or manual) the receiver then examines the successive signals, stopping correctly tuned as soon as the wanted signal is identified.

Taking the argument further, fairly low-cost storage can be provided in the receiver so that, once acquired, the location (i.e. the carrier frequency) of a particular station or programme can be 'remembered' for use in the future. This has led to the idea of a 'learning' receiver. In one form, such a receiver would have a quiescent mode, with the receiver apparently switched off but, in fact, scanning the band, locating and storing the frequency locations of all stations that could be detected. In another form there could be two receiver circuits in the one unit. The second receiver would carry out the band-scanning and storing operation, providing a continuous up-to-date supply of information to enable instant retuning of the main receiver, on demand.

Many other options are possible, both with regards to the data to be transmitted and the use to which it may be put in the receiver.

54.20 Television receivers

The practical advances in television since the inception of the British 405-line interlaced system in the 1930s have been extensive. However, as far as the picture transmission process itself is concerned, these advances have not been brought about by any fundamental change, apart from the addition of colour.

Throughout the world, there are several different scanning standards in use for broadcast television and all have in common the use of interlaced scanning. The two most commonly used scanning standards are the 625/50 standard (i.e. 625 lines per television picture, each picture being formed from two interlaced television fields occurring at a repetition rate of 50 per second) which is predominantly European, and the 525/60 standard which is predominantly American. However, it should be realised that the use of a particular scanning standard does not necessarily mean that television receivers for one country would operate successfully in another country using the same scanning standard. For example, the form and polarity of vision signal modulation, the channel bandwidths and spacings, the modulation used for the sound signal and the frequency spacing between the vision and sound carriers vary throughout the world, quite apart from the colour system and many other aspects.

Full details may be found in the appropriate CCIR texts[22].

The development of practical colour television broadcasting began in the United States of America with the introduction of the NTSC (National Television Systems Committee) colour system[23]. Two other systems, SECAM[24] (Séquentiels Couleurs A Memoire) and PAL (Phase Alternation by Lines[25]) were developed in Europe during the 1960s.

All broadcast colour television systems have a number of characteristics in common.

(*a*) Pictures can be received on black-and-white receivers as well as in colour on colour receivers. This property of 'compatibility' was an important aspect during the development of colour systems. 'Reverse compatibility' is also preserved, whereby a black-and-white transmission can be received in black and white on a colour receiver.

(*b*) Additional 'colouring' information (the chrominance signal) is transmitted simultaneously with the normal brightness (or luminance) signal.

(*c*) The colour television signal occupies the same bandwidth as the black-and-white equivalent. This is achieved by using an in-band chrominance subcarrier to carry the additional colouring information.

(*d*) The chrominance channel is of low resolution and does not, in principle, carry information describing fine detail in the picture; this results from exploitation of the properties of the eye which does not demand the portrayal of fine colour detail, only brightness detail.

(*e*) The transmitted chrominance and luminance signals are derived from red, green and blue primary signals. At the receiver, a picture tube is used with red, green and blue phosphors which accurately match the optical colour analysis characteristics of the original signal source.

In an outline description, such as this, it is not possible to give much detail and many simplifications have been made. For example, the fact that all television picture tubes have a non-linear transfer characteristic, and the consequential fact that 'gamma correction' is applied to television signals at source, have both been ignored although they have a considerable practical effect.

The PAL colour television system is often described as a variant of the American NTSC system, which was developed first during the early 1950s. The red, green and blue primary colour signals (R, G and B) are combined in weighted proportions to form a full bandwidth luminance or brightness signal (Y) corresponding to the normal black-and-white television signal. In addition, two 'colour-difference' colouring signals, $B-Y$ and $R-Y$, are formed: these signals become the 'U' and 'V' signals, respectively, when they have been attenuated to form the chrominance modulating signals.

The simplified equations relating the various signals are as follows:

$$\begin{aligned} Y &= 0.3R+0.59G+0.11B \\ R-Y &= 0.7R-0.59G-0.11B \\ B-Y &= -0.3R-0.59G+0.89B. \end{aligned}$$

The chrominance subcarrier modulating signals U and V are derived from the colour-difference signals thus:

$$\begin{aligned} U &= 0.49(B-Y) \\ V &= 0.88(R-Y). \end{aligned}$$

Since the Y signal is formed from a mixture of the original R, G and B signals, it is possible to recover these original signals at the receiver from the Y, U and V signals by the appropriate linear matrix operation. However, it will be remembered that colour television operates by adding colour signals of limited resolution (the colour-difference signals) to a brightness or luminance signal of a higher resolution corresponding to the full video bandwidth of the system. Moreover, in a picture containing little or no colour, the colour-difference signals become vanishingly small and only the higher resolution luminance signal is generated; when there is colour in the picture, the appropriate colour-difference signals appear. At the receiver, the colour picture tube display is balanced to produce a black-and-white (i.e. colourless) picture when the colour-difference signal input is zero. When the picture contains colour, the action of the (narrow-band) colour-difference input signal is to upset this balance, appropriate to the colouring signal information but, because of the narrow bandwidth, the effect is 'broad-brush' and fine colour changes are not portrayed. Of course, areas containing fine detail can still be portrayed with an overall colour.

In the NTSC system, the two colour-difference signals are each modulated onto one of two quadrature components of an in-band subcarrier, using double-sideband, suppressed-carrier amplitude modulation. The frequency of this colour subcarrier is carefully chosen to cause minimum interference to the black-and-white picture and, in fact, is chosen to be an odd multiple of half the line frequency. A colour-synchronising burst is also transmitted during the line-blanking period and, at the receiver, quadrature synchronous detectors are used to recover the individual colour-difference signals for display.

A disadvantage of the NTSC system was found to be its sensitivity to errors in the phase of the chrominance signal. It will be appreciated that, since the colour subcarrier signal is formed from the vector sum of two quadrature components each carrying different colouring information, any shift in the phase of the colour subcarrier relative to the synchronising burst will alter the proportions of the detected colour-difference signals and give rise to a change in the hue of the picture. An error of a few degrees can be noticeable and a particular form of phase error, known as differential phase distortion, can cause the hue of the scene to be dependent also upon the level of the video signal; this can produce disturbing effects such as the colour of a face altering as the person moves in and out of shade. The PAL system was designed to overcome this particular defect of NTSC. Protection against phase errors is achieved in PAL by reversing the polarity of the transmitted '*V*' component of the colour subcarrier on successive television lines (known as *V*-axis switching) and recovering the colour-difference signals at the receiver by combining the signals from successive lines. The obvious penalty for this change in operation is a reduction in picture sharpness along horizontal colour transitions because each output line signal is now formed from two adjacent scanning lines instead of one. However, this is not serious because, as stated earlier, high resolution is not required to describe colouring information. Phase distortion now results in a slight reduction in the level of the colour signal (loss of saturation) and this can sometimes combine with differential-gain distortion to produce undesirable effects. For example, the deepness of colour of a face can vary unrealistically with light and shade (under the chin for instance).

54.20.1 Outline of receiver operation

The diagram shown in *Figure 54.42* does not represent a particular television receiver but rather a collection of building blocks from which a receiver might be formed. Indeed it is unlikely that there is a receiver anywhere that conforms to *Figure 54.42.* A colour receiver is shown because of its greater importance. Black-and-white receivers may, in one sense, be regarded as obsolescent. Here, we may consider them as receivers without either colour circuits or a colour display tube and, in many areas, with less stringent performance requirements.

54.20.1.1 Channel selection

Referring to *Figure 54.42*, tuning or channel selection can be achieved by means of conventional pre-set varicap controls or by one of several 'synthesiser' methods. In the synthesiser method shown, the frequency of the receiver's local oscillator is steered to match the (reference) frequency provided by an accurate and stable crystal oscillator through a chain of frequency dividers, one of which is variable or programmable. The programmable divider is controlled by the channel selector which also feeds an appropriate varicap tuning control voltage to the tuner RF stage. When a particular channel is selected, the programmable divider is set to the appropriate value of division ratio so that, allowing for the intermediate-frequency offset, the local oscillator frequency becomes locked to the correct frequency for the channel selected. To provide a push-button facility, it is a relatively simple matter to store the values of the division ratio applicable to each pre-selected channel. Alternatively, a key-pad may be provided so that an instruction can be entered into the receiver to cause it to tune to a particular channel; however, this facility is more applicable to radio than television receivers since the latter are usually static and the number of available channels is small and their identities are known locally. Although the use of 'synthesiser tuning' is undoubtedly advantageous, it has so far found favour only in those countries where the number of available television channels is large. The number might amount to eight or more (for example in across-border listening on the mainland of Europe) and, under these circumstances, the costs of pre-set varicap controls and synthesiser operation become comparable.

54.20.1.2 IF stage

The IF stages of modern television receivers comprise an amplifier and a bandpass filter, the latter in the form of a surface-acoustic-wave (SAW) filter[12,13,14]. As stated earlier, this is a small device that gives a response accurately tailored to the requirements of television reception. It has the advantages that its performance is repeatable in production and is stable to close tolerances over a wide range of temperatures. Moreover, its bandpass amplitude/frequency characteristic can be independent of the phase characteristic, thus avoiding the need for equalisation circuits whilst still realising a phase characteristic much closer to the ideal than with LC circuits; this is of special advantage for teletext reception where the preservation of pulse shape is important. There are no trimming adjustments to make during manufacture and the small unit is hermetically sealed in a transistor type of metal can giving immunity to interference from external fields. The only important disadvantage of the SAW filter is its insertion loss which requires an amplifier with a gain of about 20 dB to compensate for the loss.

54.20.1.3 Intercarrier sound reception

Most modern television receivers employ intercarrier sound reception. In an intercarrier sound receiver, both the vision and sound IF signals are amplified in a common IF amplifier and are passed to the detector. This is an amplitude demodulator to which the frequency-modulated sound IF signal appears simply as another sideband of the vision carrier (i.e. a subcarrier) resulting in an output of an FM 'intercarrier' signal to feed the FM sound demodulator at a frequency equal to the difference between the vision and sound carrier frequencies (6 MHz in UK system I).

The advantages of intercarrier sound reception are fairly well known and may be summarised:

The intercarrier frequency is wholly independent of the local oscillator frequency, the allowable drift of which is thereby increased by several times being restricted only by changes in the effective IF response and the relative levels of the vision and sound carriers.

Because of the foregoing, the tuning of the receiver by the user is simplified.

The use of common amplifiers and components brings about a useful economy.

Unwanted frequency modulation of the local oscillator is not transferred to the FM intercarrier signal.

The disadvantages of intercarrier reception are less well known. However, most intercarrier sound receivers give rise to some degree of 'buzz on sound' interference from the vision signal and some of the more important causes are described below.

(i) Overmodulation of the vision carrier, either (rarely) at the transmitter, or due to propagation effects.

(ii) Phase modulation of the received vision carrier (and hence of the 6 MHz intercarrier signal) due to the receiver vestigial sideband amplitude response.

(iii) Phase modulation of the FM sound signal due to (varicap) variations in receiver circuit impedances caused by the vision signal.

(iv) The effects of receiver non-linearity on video signal components at sub-harmonics of 6 MHz, causing interference to the 6 MHz intercarrier signal.

(v) Mixing of chrominance components with video signal components, causing interference to the 6 MHz intercarrier signal.

(vi) Unwanted audio frequency interfering signals generated by various less well defined mechanisms within television receivers.

The severity of these effects is found to vary with receivers and receiving locations but, in general, the problems encountered in

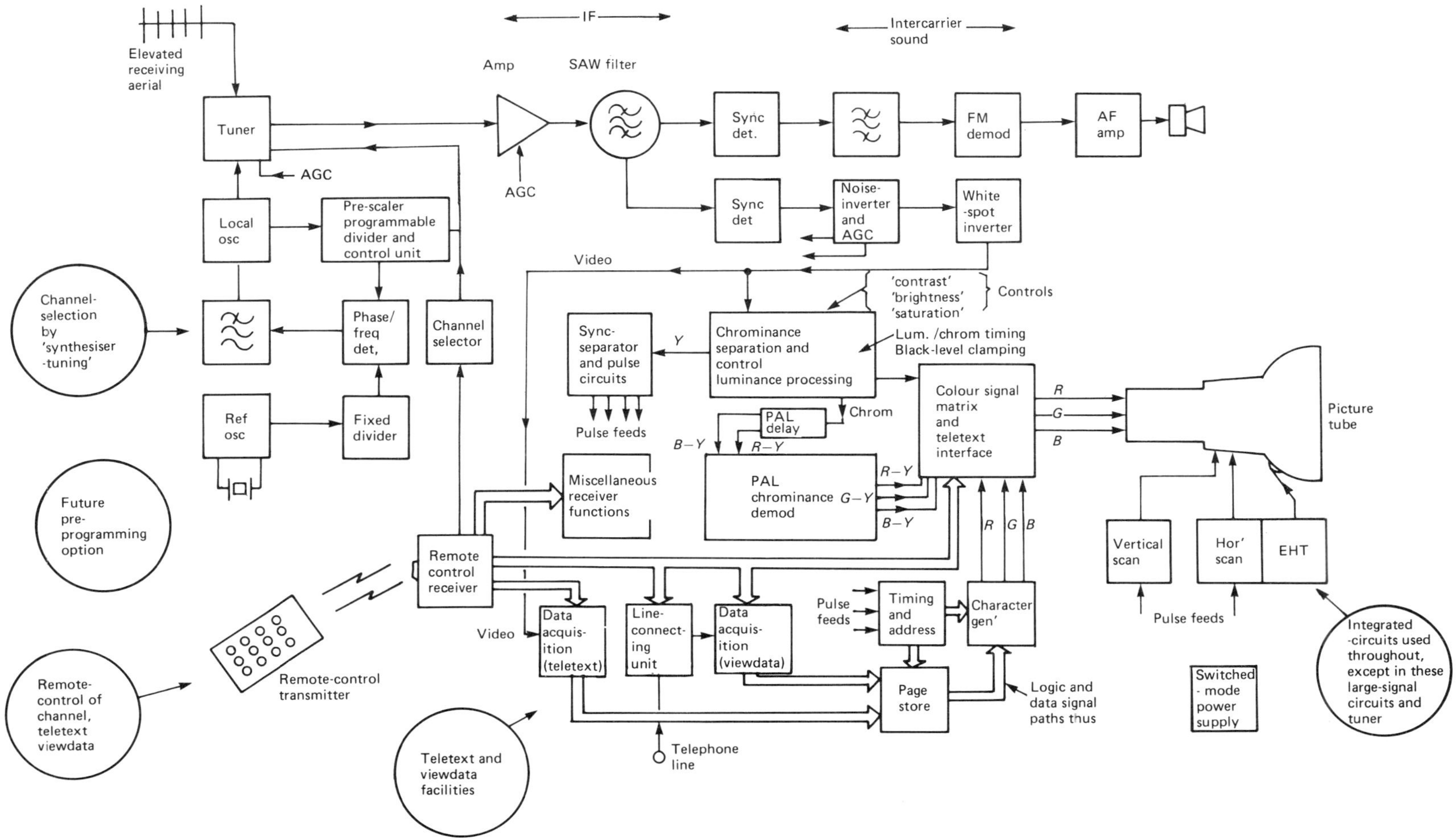

Figure 54.42 Domestic television receiver

modern installations are not serious. Further discussion is outside the scope of this section.

54.20.1.4 Synchronous detectors

It will be seen that the IF signal is demodulated by means of synchronous detection. This is a well known technique in which the input signal is multiplied by a demodulating reference frequency signal; here, the reference signal is obtained from the carrier itself by a process of amplifying, filtering and limiting. The chief advantages of a synchronous detector, compared with the earlier envelope detector, are its excellent linearity and the fact that it eliminates the quadrature distortion that is characteristic of vestigial sideband transmission and reception. The good linearity results in reduced 'buzz on sound' interference from the vision signal and the removal of quadrature distortion improves both the reception of teletext signals and the quality of the displayed television picture. However, care in design is necessary to protect against phase errors in the reference signal and consequent mal-operation of the circuit, in particular the emergence of intermodulation products which may cause interference to picture or sound.

54.20.1.5 White spot and noise inverters and AGC

A disadvantage of the synchronous detector is found under conditions of interference from impulsive noise. With negative modulation (i.e. white level near to minimum carrier level), an envelope detector cannot produce an interference 'spike' greater than the modest excess level of video signal corresponding to zero carrier level. A synchronous detector, however, because of its inherent ability to handle suppressed-carrier signals, can produce 'super-white' interference spikes which can appear on the picture as bright de-focused spots. This has necessitated the use of limiters and other circuitry, such as the white-spot inverter, a fast-acting circuit which detects a white spot and holds its level at a less noticeable grey value. The noise inverter is required for protection against interference or noise spikes that would otherwise excurse below black level and might easily exceed the level corresponding to the tips of the synchronising pulses. The action of the noise inverter is twofold. First, it is a fast-acting circuit that detects and holds noise pulses to a value close to the black level and thus prevents interference with picture synchronisation. Secondly, the detector supplying automatic gain control (AGC) signals is also protected against the effects of noise pulses that might otherwise cause erroneous changes in gain, by holding the level of the AGC signals to a reference or standby value during the noise pulse.

The AGC system itself operates by.

(*a*) Comparing the (maximum) level of the vision signal during the line-synchronising pulse periods with a reference level.

(*b*) Applying gain control signals to the IF amplifier and tuner. At low input signal levels, gain control is applied to only the IF amplifier. At higher levels of input signal, the 'tuner takeover' point is reached, after which the IF gain is stabilised and gain control is applied to the tuner RF stages. This two-stage AGC arrangement is intended to maximise the signal-to-noise ratio and to avoid overload at high levels of input signal.

54.20.1.6 Chrominance and luminance separation and control

The video signal is next applied to the chrominance separation unit. This unit, which also includes the processing of the luminance signal, provides the means for control of 'contrast' (simultaneous, ganged adjustment of the levels of the luminance and chrominance signals), 'brightness' (the potential corresponding to black level of the luminance signal) and 'saturation' (the separate adjustment of the level of the chrominance signal, sometimes described as the 'colour' control). The ganged adjustment of the luminance and chrominance signal levels is provided so that contrast may be adjusted without affecting the saturation of the colour picture. Separation of the chrominance subcarrier signal is achieved by means of a bandpass filter after which the signal is subjected to automatic level control. This is an AGC process (known as ACC for automatic colour control) in which the amplitude of the colour burst is sampled as a measure of the chrominance signal level and compared against a reference. This is an important precaution because of the fluctuations or errors in chrominance signal level that can occur due to multipath interference or echoes. These can cause a significant change in the levels of the (low-frequency) signals close to the carrier, relative to the level of the chrominance sidebands (at 4.43 MHz from the carrier). For example, consider an echo or delayed signal with a level of 25% relative to the main signal and a delay of about 0.1 μs. Depending upon the the phase of the delayed carrier, the colour subcarrier level and hence the saturation of the displayed picture could vary over a range of about 8 dB around the correct value. Moreover, if the chrominance signal level were not automatically controlled, disturbing fluctuations in saturation could occur when switching between channels, because of the varying influences of multipath propagation at different carrier frequencies.

Before being subjected to level adjustment by the 'contrast' control, the luminance signal is first passed through the sound trap to remove interference from the 6 MHz sound signal and is then adjusted in timing by means of a short delay line. This re-timing process is necessary to compensate for the additional delay suffered by the colour signals during the bandpass filtering and demodulation processes and to ensure that, when the two signals are combined for display, the colouring signal will be in register on the picture with the brightness or luminance signal. The luminance signal processing also includes black-level clamping. As well as being a convenient way for a designer to change the DC reference level of a video signal, black-level clamping is essential in a good-quality television picture display to ensure that the (low) level of brightness corresponding to 'black' is not disturbed by fluctuations in the mean level of the picture. As an example, if a picture contains a small area of nominal black, the brightness level in this area should not change if the picture is faded down. Indeed if the picture is faded down fully to black, then the whole screen should assume the same low value of brightness as was previously held by the small black area.

54.20.1.7 PAL delay line

The level-stabilised chrominance signal is next passed to the chrominance demodulator through a PAL one-line delay circuit. The function of this circuit, shown in *Figure 54.43*, is to derive chrominance signals for demodulation from two successive television lines and thus provide the phase error compensation that is inherent in the PAL system. For simplicity in analysis, similar signals are assumed to occur on successive lines and circuits with unity gain are also assumed. It will be seen that, because of PAL ('*V*' axis) switching, the polarity of the '*V*' signal is reversed on successive lines and hence the outputs of the adder and subtractor will consist primarily of *B*-*Y* and *R*-*Y* signals. The main object of this process is to provide the first step at the receiver in deriving one pair of colour-difference signals from each pair of television lines. If the chrominance signals on successive lines are different, this will result in a *B*-*Y* quadrature component of subcarrier appearing on the *R*-*Y* output and vice versa. This is not very important, however, because the synchronous detectors that follow will discriminate against these unwanted quadrature terms.

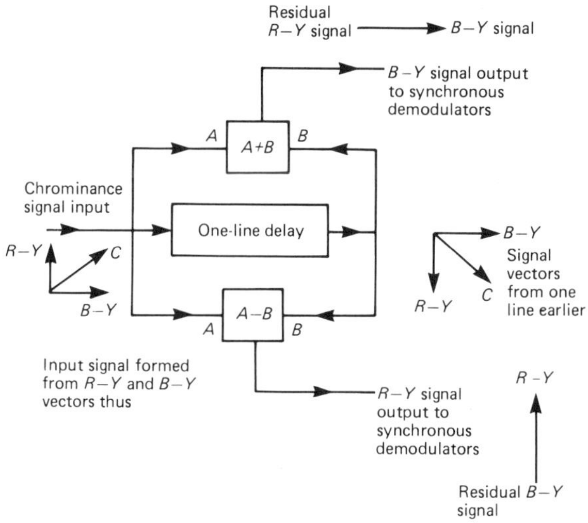

Figure 54.43 PAL one-line delay circuit

54.20.1.8 Chrominance demodulator

Referring again to *Figure 54.42*, in the chrominance demodulator quadrature reference signals are obtained from an oscillator locked to the colour-burst signal. In some receivers, the locked oscillator operates at twice the colour subcarrier frequency (~8.87 MHz), so that its output frequency may be divided by two to produce demodulating reference signals that are accurately in quadrature without the need for extra components and phase adjustment. In the case of the *R-Y* signal, this suffers a line-by-line polarity inversion (due to the action of PAL V-axis switching) which is corrected in the chrominance demodulator, usually by 0°/180° switching of the associated demodulating reference signal. A 'colour-killer' circuit is also provided that detects the presence of the colour burst before allowing the colour circuits to become active. This is desirable for best reception of black-and-white television signals on colour receivers, because it removes the undesirable effects of cross colour and noise that would otherwise appear as spurious colouring of the picture.

Having recovered the *R-Y* and *B-Y* baseband signals, the corresponding *B-Y* signal is derived from them by a linear matrix operation.

54.20.1.9 Colour signal matrix and teletext interface unit

At one time, television receivers only rarely contained a unit of this kind. The three colour-difference signals were fed individually to the three cathodes of the colour picture tube and the luminance or *Y* signal was fed to the three control grids; thus the net drive to the picture tube was formed from the difference between the two. In modern television receivers it has been necessary to provide for the display of teletext alphanumeric and graphics characters and this has proved most satisfactory and convenient with the direct use of *R*, *G* and *B* signals. Thus the unit accepts inputs of *RGB* signals, or of *Y* and colour difference signals, and delivers *RGB* drive signals to the picture tube.

54.20.1.10 Teletext and viewdata

The teletext[26] and viewdata[27] units consist almost entirely of logic and data signal circuits and it will be seen that, apart from

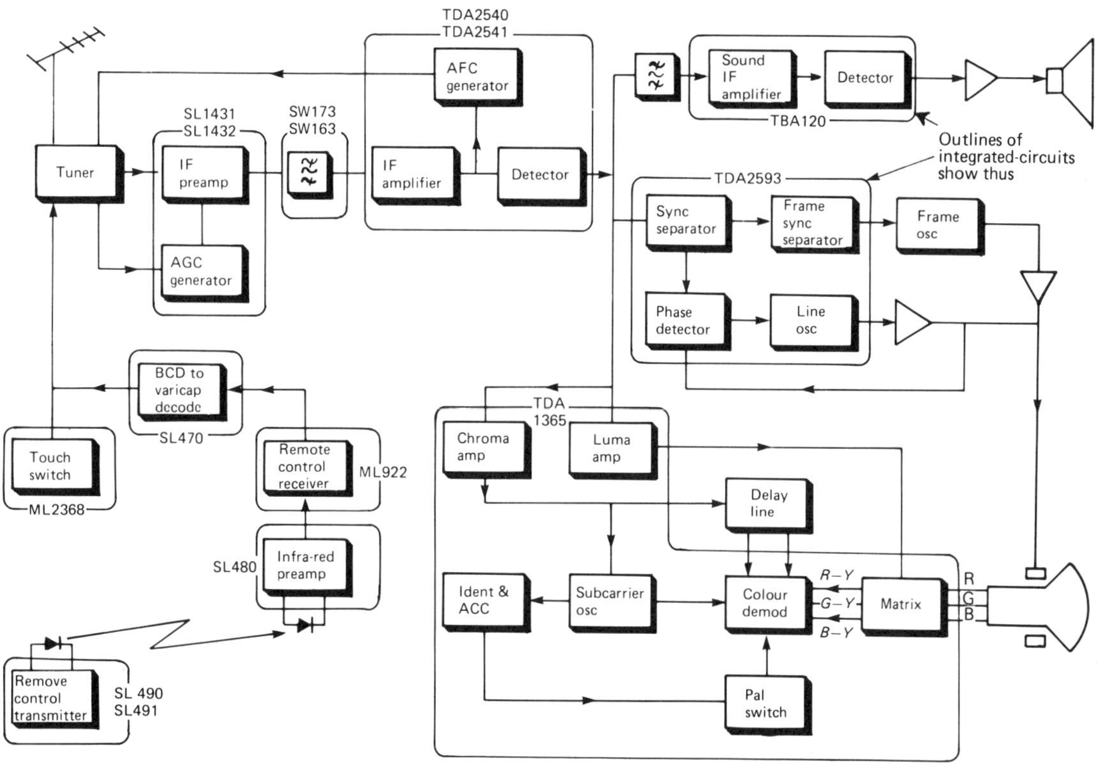

Figure 54.44 Alternative receiver schematic showing use of integrated circuits

separate data acquisition units, the teletext and viewdata signals share the same circuits in the receiver; this is possible because the same coding and formatting standard is used for both. The received video signal is applied to the teletext data acquisition circuit where the data signal occurring during the field blanking period is detected. The selected teletext page is recognised on arrival by its coded page number and is acquired and stored row-by-row in the page store; the stored data is then used to control a character generator for display. In a similar way, when using the viewdata facility, the data acquisition circuit acquires and stores the received data. Different data acquisition circuits are needed because of the different forms of data transmission. Teletext uses a very high bit rate for a short burst of data that is transmitted during the television field-blanking period, whilst viewdata uses a continuous, low bit rate transmission method.

54.20.1.11 Remote control

A remote-control unit is shown whereby operation of the receiver, including the use of data facilities, is made possible from an armchair. Ultrasonic acoustic signals (at a frequency of a few tens of kHz) are used in this case, although other media such as infra-red signals are also used. The kind of facilities that are possible depend primarily upon those provided by the television receiver itself. In the case described here, adjustment of all normal control settings (by raise/lower 'instructions'), teletext/viewdata/television operation, teletext page selection, viewdata number entry etc., television channel selection (12 channels), numerical display of selected channel, display of clock time, and several other lesser facilities could be available.

54.20.2 Influence of LSI

The schematic of an alternative, and somewhat simpler, receiver is shown in *Figure 54.44*. Its purpose is to illustrate the way in which the use of integrated circuits had become almost total by 1981. The horizontal-scanning output stage represents the one remaining area that has not been invaded by the integrated circuit. During the 1970s, more and more small-signal functions were performed by integrated circuits and, by 1980, even some large-signal applications such as field scanning were being met by them in some receivers.

The receiver shown in *Figure 54.44* is typical of the early 1980s. The introduction of integrated circuits has, broadly, improved both the reliability and performance of receivers, and has helped to keep down costs because of the simpler procedures at the assembly stages.

References

1 EXWOOD, M., 'Cable Television', *Paper to Royal Television Society Convention*, 7 September 1973
2 COASE, R. H., 'Wire Broadcasting in Great Britain', *Economica*, Vol. XV, No. 59, August 1948
3 GABRIEL, R. P., 'Cable Television and the Wired City', *I.E.E. Electronics and Power*, April 1972
4 GARGINI, E. J., 'The Total Communication Concept of the Future', *Royal Television Society Journal*, March/April 1973
5 *Canadian Radio-Television Commission*, 'The Integration of Cable Television in the Canadian Broadcasting System', 26.2.1971
6 HARE, A. G., 'Telecommunication 20 Years On—A Look Into Local Distribution', *Post Office Research Dept.*, May 1971
7 OLIVER, B. M., PIERCE, J. R. and SHANNON, C. E., 'The Philosophy of P.C.M.', *Proc. I.R.E.*, **36** (1948)
8 SHORTER, D. E. L. and CHEW, J. R., 'Applications of Pulse-Code Modulation to Sound-Signal Distribution in a Broadcasting Network', *Proc. I.E.E.*, **119**, No 10 (1972)
9 HOWARTH, D. and SHORTER, D. E. L., 'Pulse-Code Modulation for High-Quality Sound Signal Distributions: Appraisal of System Requirements'; *BBC Research Department Report* EL10, 12 (1967)
10 'CCITT, Sixth Plenary Assembly', *Recommendations* C731 and C741, 111–2, p. 425 and pp. 444–46, OCITT, Geneva (1976)
11 CAINE, C. R., ENGLISH, A. R., O'CLAREY, J. W. H., 'NICAM 3: Near-Instantaneously Companded Digital Transmission System for High-Quality Sound Programmes', *The Radio and Electronic Engineer*, **50**, No 10, pp. 519–530, October 1980
12 ZYPKEMA, DEVRIES and BANACH, 'Engineering Aspects of the Application of Surface-Wave Filters in Television I.F.'s', *IEEE Trans. on Consumer Electronics*, **CE-21**, No 2, May 1975
13 MAINES and PAIGE, 'Surface-Acoustic-Wave Components, Devices and Applications', *Proc. I.E.E.*, **210**, No 10R, October 1973
14 PLESSEY, 'Plessey Surface-Acoustic-Wave Filters and Pre-Amplifiers for TV IF Systems', *Consumer News*, **1**, No 2, May 1978
15 THODAY, R. D. C., 'Band II Ferrite Aerial Unit', *Wireless World*, September 1977
16 CCIR, 'Recommendations and Reports of the CCIR, 1978, XIVth Plenary Assembly, Kyoto, Vol. X, Rec. 450, 86 (1978)
17 SPENCER, J. G. and PHILLIPS, G. J., 'Stereophonic Broadcasting and Reception', *The Radio and Electronic Engineer*, **27**, No 6, June 1964
18 CCIR, 'Recommendations and Reports of the CCIR, 1978, XIVth Plenary Assembly, Kyoto', Vol. X, Report 293–4, 155 (1978)
19 GROSJEAN, J. P., 'Phase Locked Loops and Surface Wave Filters an Ideal Combination, *I.E.E.E. Trans. on Consumer Electronics*, **CE-26**, August 1980
20 WHYTHE, D. J. and ELY, S. R., 'Data and Identification Signalling for Future Radio Receivers', International Broadcasting Convention, 1978, *I.E.E. Conference Publication* No 166 (1978)
21 ELY, S. R., 'The Impact of Radio-data on Broadcast Receivers', *I.E.R.E. Conf. on Radio Receivers and Associated Systems*, July 1981
22 CCIR, 'Recommendations and Reports of the CCIR, 1978, XIVth Plenary Assembly, Kyoto', Vol. XI (1978)
23 BROWN, G. H. and LUCK, D. G. C., 'Principles and Development of Colour Television Systems', *RCA Review*, No 14 (1953)
24 DE FRANCE, H., 'Le Système de Télévision en Couleurs Séquentiels-Simultanis', *L'Onde Electrique* (1958)
25 BRUCH, W., 'PAL—A Variant of the NTSC Colour Television System', *Selected Papers* I and II, Telefunken (1966)
26 *Broadcast Teletext Specification*, BBC, IBA and BREMA joint publication, September 1976
27 FEDIDA, S., 'Viewdata: An Interactive Information Service for the General Public', *Post Office Research Department Report* No 553, October 1976

Further reading

JURGEN, R. K., 'Two-way Applications for Cable Television Systems in the 1970s', *I.E.E.E. Spectrum*, November 1971
'Communication Technology for Urban Improvement', June (1971), *Proc. I.E.E.E. Special Issue on Cable Television*, July 1970
MASON, W. F., 'Urban Cable Systems', *The Mitre Corporation*, National Academy of Engineering, Washington D.C. (May 1972)
REPORT OF THE SLOAN COMMISSION, *On the Cable—The Television of Abundance*, McGraw-Hill
TELEVISION ADVISORY COMMITTEE 1972, papers of the Technical Sub-Committee, Chapter 4, H.M.S.O.
WARD, JOHN E., 'Present and Probable CATV/Broadband Communication Technology', *Report of the Sloan Commission*
KLEIN, A. B., *Colour Cinematography*, Chapman and Hall
WRIGHT, W. D., *Measurement of Colour*, Hilger
HARDY, A. C., *Handbook of Colorimetry*, M.I.T. Mass. U.S.A.
LUCKIESH and MOSS, *Science of Seeing*, Macmillan
WRATTEN, *Colour Filters*, Kodak
BENSON, J. E., 'A survey of the methods and colorimetric principles of colour TV', *J. Brit. I.R.E.*, Jan. (1953), [Reprinted from *Proc. I.R.E.* (Australia), July, Aug. (1951)]
CARNT, P. S. and TOWNSEND, G. B., *Colour Television*, Iliffe Books (1961)
HENDERSON, H., *Colorimetry*, BBC Engineering Training Supplement No 14

LIVINGSTON, D. C., 'Colorimetric analysis of the NTSC colour television system', *Proc. I.R.E.*, **42**, January (1954)
LIVINGSTON, D. C., 'Reproduction of luminance detail by NTSC colour television systems', *Proc. I.R.E.*, **42**, January (1954)
NEUHAUSER, R. G., ROTOW, A. A. and VEITH, F. S., 'Image orthicons for colour cameras', *Proc. I.R.E.*, **42**, January (1954)
Proc. I.R.E., numerous papers, **42**, January (1954)
SIMS, H. V., 'Black level in television', *Wireless World*, **68**, No 1 (1962)
SIMS, H. V., *Principles of PAL Colour Television and Related Systems*, Newnes-Butterworths (1969)
CCIR, 'Recommendations and Reports of the CCIR, 1978, XIVth Plenary Assembly, Kyoto', Vol. X, Rec. 415 (1978)
MORRELL, LAW, RAMBERG and HEROLD, 'Colour Television Picture Tubes', Academic Press, New York and London (1974)
MULLARD LIMITED, '30AX Self-Aligning 110° in-line Colour TV Display', *Mullard Technical Note* No 119, Mullard Limited, London (1979)
HINDIN, H. J., 'Videotext looks brighter as developments mount', *Electronics*, August 25 (1982)
TSANTES, J., 'AM-stereo technology gains momentum, but no industry standard is in sight', *EDN*, September (1982)

55 Communication Satellites

A K Jefferis, BSc, CEng, FIEE
Chief Executive,
Satellite Systems Division,
British Telecom International

S C Pascall, BSc, PhD, CEng, MIEE
Head of UNISAT Launch and Propulsion Group,
Satellite Systems Division,
British Telecom International

Contents

55.1 Introduction

The geostationary satellite concept was first proposed by Arthur C Clarke who recognised the potential of rocket launches and the advantage of the geostationary orbit.[1] The major uses of communication satellites at present are (*a*) point-to-point international communications for telephony, data and telegraph traffic and TV program distribution, (*b*) regional and domestic telecommunications and TV program distribution and (*c*) communication with mobile terminals mainly for maritime telecommunication services. Another use of growing importance is direct television broadcasting and systems for commercial international fixed services.[2]

There are two systems at present, the INTELSAT system and the INTERSPUTNIK system. The INTELSAT system is owned by the 106-member INTELSAT Consortium and uses geostationary satellites. The INTERSPUTNIK organisation[3] members are the USSR and its allied countries. The system is based on the use of satellites in highly inclined elliptical orbits as well as geostationary satellites.

55.2 Frequency bands

Of the bands allocated to communication satellites the following, or parts of them, are the ones used or likely to be used for commercial systems:

Up paths (GHz)	*Down paths (GHz)*
	3.4–4.2
5.85–7.075	4.5–4.8
8.025–8.4 (use depends on region)	
12.7–12.75 (R2)	10.7–11.7
	11.7–12.3 (R2)
12.5–12.75 (R1)	12.5–12.75 (R1, R3)
12.75–13.25	
14.0–14.5	
	17.7–18.8
	19.7–20.2
27.5–30	
22.55–23.55 (Inter-satellite links)	
32–32.3 (Inter-satellite links)	

(Note: R1, R2, R3 refer to the three ITU Regions)

In addition the bands 10.7 to 11.7 GHz (in Europe and Africa), 14.5 to 14.8 GHz (except Europe) and 17.3 to 18.1 GHz may be used for up links but only as feeder links to broadcasting satellites. Most of these bands are shared with terrestrial services. When this occurs, there are limits on transmitted powers and the power flux density a satellite may set up at the Earth's surface.[4] In the 3.4–7.75 GHz band the limit is -152 dBW m^{-2} (4 kHz)$^{-1}$ below 5° arrival angle, rising to -142 dBW m^{-2} (4 kHz)$^{-1}$ at 25° arrival angle and above. In the 12.2–12.75 GHz band, it is 4 dB greater than in the 3.4–7.75 GHz band and in the shared part of the 17.7–19.7 GHz 33 dB greater still, but specified for a 1 MHz rather than 4 kHz bandwidth.

55.3 Orbital considerations

By far the most useful orbit for communication satellites is the geostationary satellite orbit, a circular equatorial orbit at approximate height 36 000 km for which the period is 23 hours 56 mins, the length of the sidereal day. A satellite in this orbit remains approximately stationary relative to points on the Earth's surface. There are two main perturbations to the orbits of such satellites:

(1) A drift in orbit inclination out of the equatorial plane due to effects of the Sun and Moon. This is at the approximately linear rate of 0.86° a year for small inclinations. If not corrected it causes the satellite to move around a progressively increasing daily 'figure of eight' path as viewed from the Earth.

(2) Acceleration of the satellite in longitude towards one of the two stable points at 79°E and 101°W longitude. This is caused by non-uniformity of the Earth's gravitational field.

55.4 Launching of satellites

The launcher generally places the satellite first into a low-altitude circular inclined orbit and then into an elliptical transfer orbit with apogee at the altitude of the geostationary orbit. At an approximate apogee the apogee boost motor, generally solid fuelled and forming part of the satellite itself, is fired to circularise the orbit and remove most of the remaining orbit inclination. The satellite's own control system is used to obtain the final desired orbit and maintain it.

Satellites can be launched using either expandable rockets or the reusable Space Shuttle. Some US launchers suitable for communication satellites, together with their approximate geostationary orbit payload capabilities, are the Thor Deltas 2914 (360 kg), 3914 (465 kg) and 3910 PAM (550 kg), the Atlas Centaur (1092 kg) and the Titan 34D (1909 kg). Ariane I (1078 kg) is being developed by the European Space Agency. Satellites will be launched from the US Space Shuttle with its various upper stages. STS//SSUS-D (682 kg), STS/SSUS-A (1260 kg) and STS/IUS (2860 kg). The Shuttle was used in November 1982 to launch SBS-3 and ANIK-C3 satellites.

55.5 Satellite stabilisation and control

Stabilisation of a satellite's altitude is necessary since, for high communications efficiency, directional aerials must be used and pointed at the Earth. Geostationary communication satellites are either spin stabilised or three-axis body stabilised. INTELSAT IV and INTELSAT IVA are spin-stabilised whilst INTELSAT V is three-axis body stabilised.

In a spinning satellite, the body of the spacecraft spins at 30–100 rpm about an axis perpendicular to the orbit plane, but the aerial system is generally de-spun, that is, located on a platform spinning in the opposite direction so that the net effect is a stationary antenna beam relative to the Earth. A reference for the control system is usually obtained primarily by infra-red Earth sensors supplemented by Sun sensors. Antenna pointing accuracy of $\pm 0.2°$ or better[5] is obtained through the antenna de-spin control electronics and by occasional adjustments to the direction of the satellite's spin axis.

Body-stabilised designs generally employ an internal momentum wheel with axis perpendicular to the orbit plane. Control about the pitch axis is through the wheel's drive motor electronics, while control about the yaw and roll axes (necessary because of perturbations to the wheel axis direction) may be by gimballing the wheel or by use of hydrazine monopropellant thrusters to correct the axis direction. In any case, thrusters must be used for occasional dumping of momentum.

The orbit of the satellite must also be controlled and this is achieved by ground command of hydrazine thrusters. The mass of hydrazine required for a 7-year life, expressed as a percentage of total satellite mass, is approximately 15–20%[6] for N–S station keeping (inclination control) and less than 1% for E–W station keeping assuming a conventional hydrazine system is employed. In a spinning satellite, axial thrusters are used for inclination corrections and radial thrusters, operated in a pulsed mode, for longitude corrections. Methods such as the use of bi-propellant, power augmented electrothermal hydrazine decomposition or electrical propulsion,[7,8] exist to improve on the specific impulse of conventional thrusters. In electrical propulsion, a propellant (an alkali metal such as caesium) is ionised by passing it through a heated grid. The ions are then accelerated by an electric field and ejected through a nozzle.

55.6 Satellite power supplies

Silicon solar cells are the accepted source of primary power, except during eclipse when power is maintained by nickel–cadmium cell batteries. Full shadow eclipse occurs on 44 nights in the spring and 44 nights in the autumn. The longest eclipses occur at the equinoxes and last for 65 minutes (full shadow). The nickel–hydrogen battery is intended to replace the nickel–cadmium batteries now used on operational communications satellites. The nickel–hydrogen battery offers significant advantages in terms of reliability, life expectancy and improved energy density.

Spinning satellites have body-mounted solar cells producing at the end of their operating life 9–11 W per kg mass of solar cell array. Body-stabilised satellites using extendible arrays rotated so as to always face the Sun can give 18–22 W per kg mass. A disadvantage of any extendible array for synchronous orbit missions is the inability to provide power during transfer orbit when the array is still stored.

55.7 Telemetry, tracking and command (TT&C)

Telemetry is used to monitor and evaluate the satellite performance and provide the ground control with data for the operation and failure diagnosis of the satellite. The command function is required for the initiation of manoeuvres by the satellite as part of routine operations, such as to switch paths, in the communications payload or to respond to an emergency by initiating corrective action. Tracking is necessary for the execution of manoeuvres or corrections. TT&C signals generally occupy a narrow channel within the communication band but are outside the communication channels. For example, in the INTELSAT V spacecraft 6175 MHz is used for the command functions and 3947.5 MHz and 3952.5 MHz for telemetry.

55.8 Satellite aerials

A circular radiation pattern of 17.5° beamwidth is just adequate to cover the area of the Earth visible from a geostationary satellite. This corresponds to a beam edge gain, relative to isotropic, of approximately 16 dB and is generally provided by means of a horn antenna. Spot beam aerials, which may be steerable, are used for increasing the gain and hence the effective isotropically radiated power (e.i.r.p.). On more advanced satellites spot beam aerials are used to permit reuse of the frequency band by relying on the high degree of isolation possible if the beams have adequate angular separation. Spot beam aerials are generally front-fed paraboloids. The beam edge gain of such aerials is approximately $41-20\log_{10}$ (beam width in degrees) dB, regardless of frequency.

The aerial types used in INTELSAT satellites to date have been as follows. The INTELSAT I (Early Bird) and INTELSAT II satellites used linear arrays of dipoles (not de-spun) producing toroidal radiation patterns. The INTELSAT III satellite has a de-spun global coverage horn while INTELSAT IV has global horns and two 4.5° spot beam paraboloids all mounted on a de-spun platform which also carries the transponders.[10]

A particular aerial implementation combining large area coverage with frequency re-use between east and west has been used on the INTELSAT IV-A satellite. The coverage areas in this case are obtained by composite spot beams formed with multiple feeds illuminating large offset paraboloids, selected feeds being fed in quadrature with others.

INTELSAT V uses six communication aerials, two conventional global coverage horns (6/4 GHz), two hemispherical hemi/zone offset-fed reflectors (6/4 GHz) and two offset fed spot beam reflectors (14/11 GHz), *Figure 55.1*.[11] Each hemi/zone aerial has 88 feed horns, each feed horn producing a small spot beam that can be set to any amplitude and phase. A combination of these spot beams forms the required footprint shape. The hemi beams are shaped to cover whole continents, such as the Americas, whilst the zone beams, operating in the opposite sense of circular polarisation, cover smaller areas, such as Europe, *Figure 55.2*.

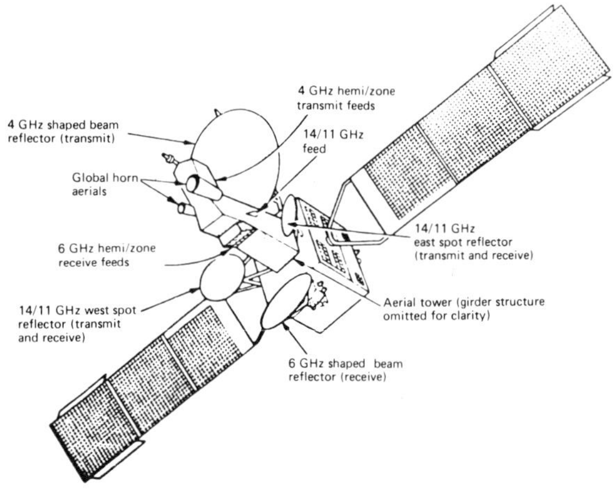

Figure 55.1 INTELSAT V satellite

The two 14/11 GHz spot beam aerials are provided to service high traffic areas such as Western Europe and North-East USA. Each aerial receives and transmits into the same spot area. Beam isolation is achieved by beam shaping and linear orthogonal polarisation separation.

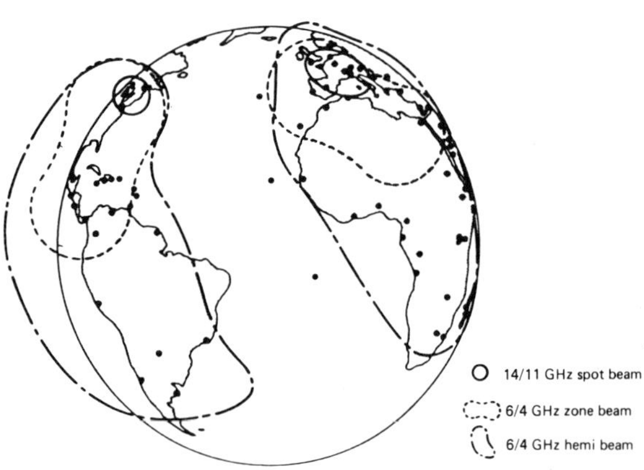

Figure 55.2 Approximate aerial beam coverage of INTELSAT V spacecraft for the Atlantic Ocean region

55.9 Satellite transponders

The essential purpose of the transponder is to translate the received signals to a new frequency band and amplify them for transmission back to the Earth. In early satellites this process was carried out broadband but modern satellites generally separate the signals into a number of channels at least for part of the amplification. Although the INTELSAT IV satellites are nearing the end of their useful life the transponder arrangement used is typical of many current satellites that do not employ frequency re-use and it is therefore the INTELSAT IV repeater[12] that will be described in order to illustrate typical techniques. *Figure 55.3* shows the basic elements. Signals in the band 5.932 to 6.418 GHz from the receive aerial are first amplified in a 6 GHz tunnel diode amplifier (TDA) and then frequency translated by 2225 MHz for further broadband amplification in the 4 GHz TDA and low-level travelling-wave tube (TWT). This part of the transponder sub-system has fourfold redundancy to achieve high reliability. The signals are then split by a circular filter dividing network into 12 channels of 36 MHz bandwidth and 40 MHz spacing between centre frequencies. The channels have gain control attenuators with eight 3.5 dB steps, except channels 9 to 12 where there are only 4 steps. These are followed by the high-level (6 watt) TWT (and standby) and then the switches selecting the transmit aerials. Channels 1, 3, 5 and 7 can be connected to global beam aerial or spot beam aerial No 1, channels 2, 4, 6 and 8 can be connected to the other global beam aerial or to spot beam aerial No 2 and channels 9, 10, 11 and 12 are permanently connected to the global beam aerials. No two adjacent channels are ever connected to the same aerial which simplifies the output multiplexer filter design.

The principal characteristics of the transponder and aerial sub-system are as follows:

Receive system gain-to-noise temperature ratio $(G/T) - 17.6$ dB K^{-1}
Flux density for TWT saturation -73.7 to -55.7 dBW m^{-2}
Transmit e.i.r.p. global beam 22 dBW per channel
spot beam (4.5°) 33.7 dBW per channel
Receive aerials are left-hand circularly polarised
Transmit aerials are right-hand circularly polarised.

Other specifications relate to gain stability, gain slope, group delay response, amplitude linearity and amplitude to phase modulation transfer (am to pm).

The above relates to INTELSAT IV but alternative arrangements are possible. In the European Communication Satellite (ECS), for example,[13] besides operating in the 14/11 GHz rather than 6/4 GHz frequency bands, the receiver employs a parametric amplifier instead of a TDA and a double conversion transponder is used, the bulk of the amplification taking place in a broadband intermediate frequency amplifier operating at UHF. The channels are connected to one of three 3.7° spot beams or the elliptical (5.2° × 8.9°) European coverage beam referred to as Eurobeam. There are 12 channels of 72 MHz nominal bandwidth each, with the later models incorporating two 72 MHz channels at 12 GHz to be used for specialised services. Finally, the transponder sub-system is completely duplicated since the satellite employs frequency re-use through polarisation discrimination.

In many modern satellites, the transponder arrangements are further complicated by frequency re-use between spot beams as well as, or instead of, re-use on separate polarisations. Furthermore, the use of more than one set of frequency bands in the same satellite is practised, e.g. 6/4 GHz and 14/11 GHz interconnected with each other. The INTELSAT V spacecraft operates in both 6/4 GHz and 14/11 GHz and utilises these bands in seven distinct coverage areas, five at 6/4 GHz and two at 14/11 GHz, *Figure 55.2*. Re-use of the frequency bands between coverage areas is achieved by means of spatial and/or polarisation isolation.[14]

In all these cases the main complication to the transponders arises through the need to achieve an adequate degree of connectivity between receive and transmit beams, together with flexibility to meet changing requirements.

Satellite transponders are being revolutionised by the use in the satellite of time division switching between beams. A

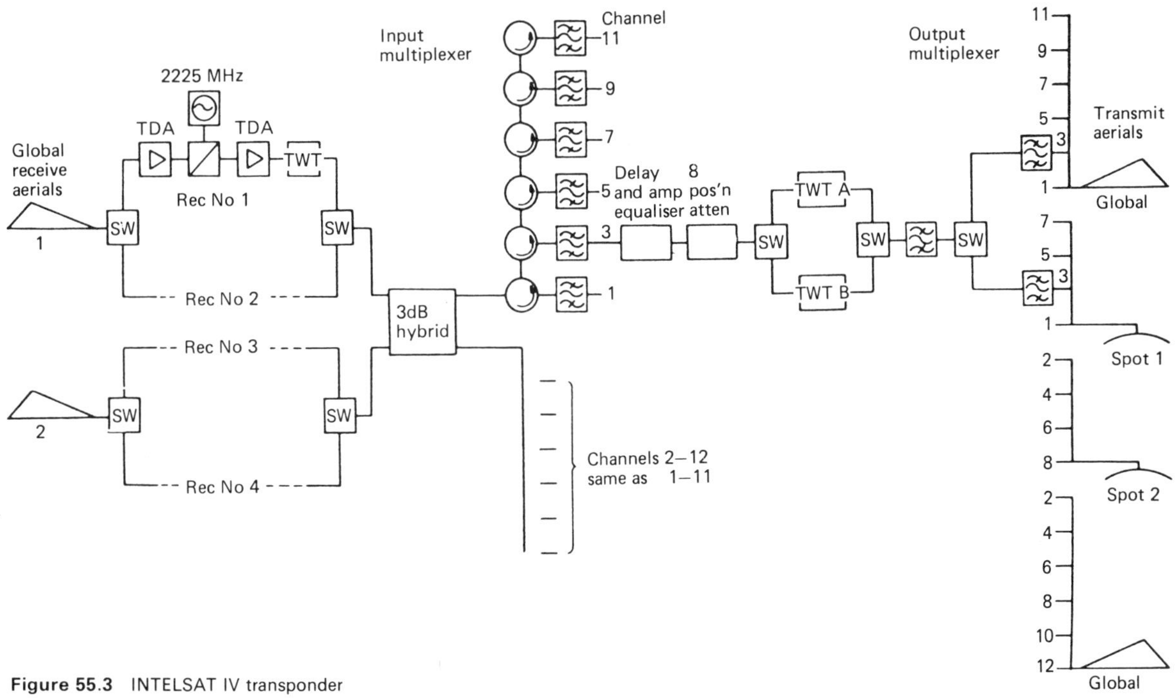

Figure 55.3 INTELSAT IV transponder

switching capability in the satellite permits interconnections between different up- and down-link beams to be dynamically changed, thus ensuring the desired connectivity and flexibility for re-allocation of capacity. The Advanced Westar[15] satellite due to be launched in the near future is equipped with a 4 × 4 satellite switch in the 14/11 GHz band operating exclusively in the satellite switched/time division multiple access (SS/TDMA) mode. The satellite switch permits interconnection of all its four channels on a sequential basis. The four satellite channels consist of four independent receiver and transmitter combinations. The switch matrix connects the output of each receiver to the input of each transmitter sequentially.

55.10 Multiple access methods

In a communication satellite system many Earth stations require access to the same transponder sub-system and usually also to the same RF channel in a channellised satellite. Currently frequency division multiple access (FDMA) is used in which the RF carriers from the various stations are allocated frequencies according to an agreed frequency plan. When several carriers are passed through the same TWT the non-linearity and am to pm conversion cause intermodulation which adds considerably to the noise level.[16,17] It becomes necessary to reduce the TWT drive level well below the saturation point, so reducing the output power. At the optimum operating level the intermodulation noise is usually half the down-path thermal noise. Typical output power back-off for INTELSAT V is 4.5 dB for global beam channels, 6 dB for hemi/zone channels, 6 dB for 80 MHz spot beam channels.

Alternatively, for digital signals, time division multiple access (TDMA) can be used,[18,19] each station transmitting its digital traffic in a short burst during the portion of the overall time frame allocated to that station. The frame format of the INTELSAT TDMA system is shown in *Figure 55.4*. A station maintains its correct burst position by applying information provided by the reference station. The reference station derives this information from a comparison of the burst position relative to the reference burst. Initial acquisition is achieved in open loop by stations transmitting a shortened burst at an instant determined from the time at which the reference burst is received and the ranging information supplied from the reference station.

Most multiple access systems have fixed assignment of telephone channels, with enough channels allocated to meet the peak demand on each route. For low-capacity routes it may be more economic to use 'demand assignment' systems in which channels from a pool are allotted to pairs of Earth stations as the demand arises. The first system of this type developed for commercial use is known as 'SPADE'.[20] In the INTELSAT system, SPADE uses a 36 MHz transponder bandwidth containing a pool of 800 channels. Individual 4 kHz voice channels are PCM encoded using 8 kHz sampling rate and 7 bits of quantisation. Each digital voice signal is transmitted on a separate carrier using four-phase PSK (phase shift keying) modulation. Systems of this type are referred to as single-channel per carrier (SCPC) systems. The main difference between SPADE and other SCPC systems is that SPADE incorporates a demand assignment system. Pre-assigned SCPC systems using FM modulation with and without companding (SCPC/FM, SCPC/CFM) have been developed to operate at carrier to noise values lower than those required by the standard SPADE channel units.

55.11 Transmission equation

The carrier-to-thermal noise power ratio, C/N, in bandwidth B Hz, for either the up path or down path is given by:

$$C/N = \text{e.i.r.p.} - L - LA + G/T - 10\log_{10} k - 10\log_{10} B \,\text{dB} \tag{55.1}$$

where

e.i.r.p. = transmitter power × aerial gain, dB W
L = free-space path loss, $20\log_{10}(4\pi d/\lambda)$, dB

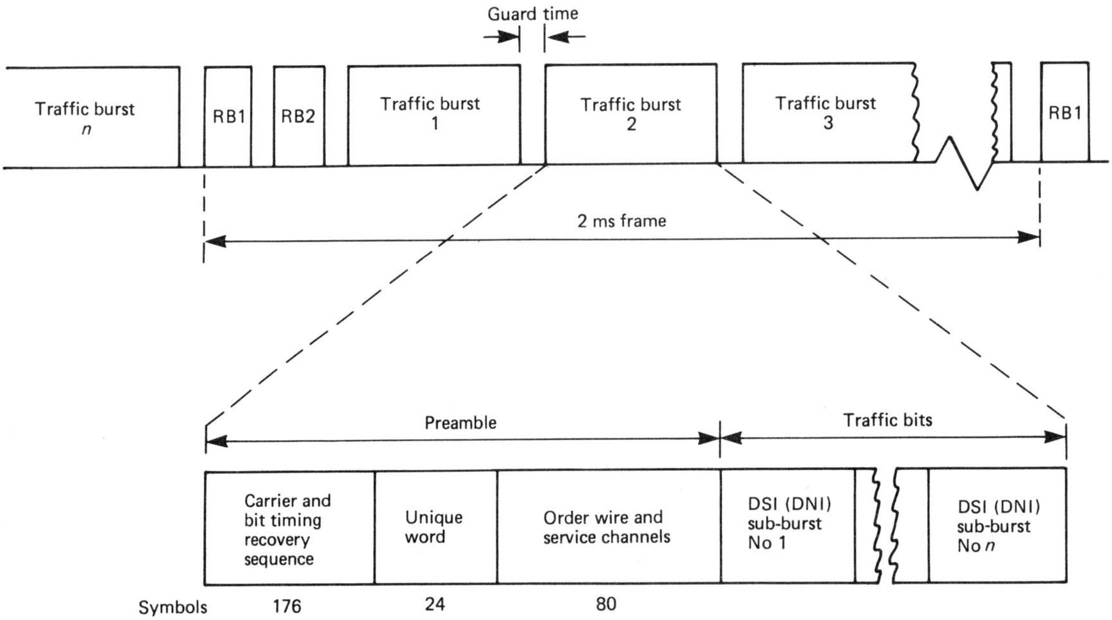

Figure 55.4 Frame and burst format of the INTELSAT TDMA system. RB1 and RB2 are the reference bursts from reference stations 1 and 2 respectively. The drawing is not to scale

LA	=	atmospheric loss (0.3 dB at 5° elevation in clear weather at 4 GHz)
G/T	=	receive system aerial gain-to-noise temperature ratio, dB K^{-1}
k	=	Boltzmann's constant, 1.37×10^{-23} W Hz^{-1} K^{-1}.

The C/N ratio available in a satellite link is used to determine the channel capacity and performance of the link. Typical contributions to the overall C/N ratio for INTELSAT V are listed in *Table 55.1*.

Table 55.1 Typical INTELSAT V beam edge C/N budgets for a 6/4 GHz hemibeam and a 14/11 GHz spot beam for multicarrier FM transmission

	Hemi-beam C/N (dB)	*Spot beam* C/N (dB)
Up-path thermal	27.4	26.3
Up-path interference (frequency re-use/spatial and polarisation isolation)	22.5	33.0
Satellite intermodulation	20.6	24.0
Down-path thermal	18.3	19.8
Down-path interference (frequency re-use/spatial and polarisation isolation)	22.5	33.0
Total C/N	14.4	17.5

55.12 Performance objectives

The CCIR recommends performance objectives for international telephone and television connections via satellite. For a telephone channel the most important objective for the system designer is that the psophometrically weighted noise power must not exceed 10000 pW for more than 20% of any month.[21] For television the corresponding objective is 53 dB signal-to-weighted-noise ratio for the 2500 km hypothetical reference circuit for 99% of any month measured in the band 0.01 to 5 MHz.[22]

For PCM telephony systems bit error rates of 10^{-6} (10 minute mean value) 10^{-4} (1 minute mean value) and 10^{-3} (1 second mean value) for more than 20% of any month, 0.3% of any month and 0.01% of any year, respectively, are recommended by CCIR.[23]

55.13 Modulation systems

Frequency modulation is the most widely used at present. In the case of telephony most traffic is carried in large FM carriers but single-channel per carrier systems are also used in some cases. For the multiple channel per carrier case the channels are first assembled in a frequency division multiplex (FDM) baseband.

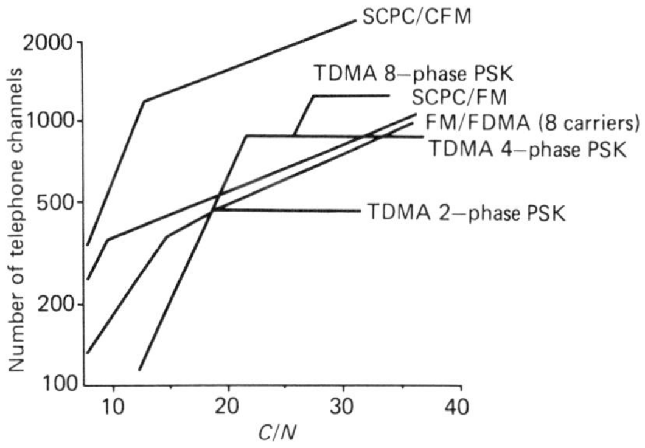

Figure 55.5 Telephone channel capacity in 36 MHz channel

Modulation indices are chosen according to the available ratio of carrier power to noise density. *Figure 55.5* shows the number of channels (one telephone circuit requires two channels) which can be obtained in a 36 MHz bandwidth as a function of total carrier power to noise ratio. The knee in the curve is the point below which the full 36 MHz cannot be occupied because the C/N ratio would be below threshold, even assuming the use of threshold extension demodulators. Methods of calculating performance, etc are given in references 24 and 25.

A typical INTELSAT allocation of the 10000 pW of noise allowed in the telephone channel would be as follows:

Up path thermal	350	Down path thermal	2900
Up-path intra-system interference co-channel	585	Down-path intra-system interference co-channel	585
cross polarisation	690	cross polarisation	690
Satellite intermodulation	1700	Terrestrial interference	1000
		Other satellite system interference	2000
		Margin	500

In the case of television, frequency modulation is again the universally adopted method at present. Reference 25 gives a method for calculating performance.

Table 55.2 gives the modulation parameters used in INTELSAT IVA and V for typical carrier sizes

Television is normally transmitted in 17.5 MHz bandwidth allowing two 49 dB S/N ratio video signals to be passed through a single global beam transponder. 30 MHz bandwidth, giving 54 dB S/N ratio is also used in certain cases.

The modulation method for digital signals is generally phase shift keying (PSK). The theoretical minimum C/N ratio for 4 phase PSK, the most commonly preferred system, is 11.4 dB in a bandwidth equal to the symbol rate (half the bit rate) for a bit

Table 55.2 Example of modulation parameters used with INTELSAT IVA and V

Carrier-type No of channels	*Bandwidth allocated (MHz)*	*Bandwidth occupied (MHz)*	*Top baseband limits (kHz)*	*Deviation (rms) for 0 dBmO test tone kHz*	*Overall C/N ratio (dB)*
24	2.5	2.0	108	164	12.7
432	20.0	18.0	1796	616	16.1
972	36.0	36.0	4028	802	17.8

error rate of 10^{-4}. Practical bandwidth requirements are greater than the symbol rate by typically 20% and the practical C/N ratio is greater than the theoretical minimum value by typically 2 to 3 dB. *Figure 55.5* shows also the capacity obtainable from 36 MHz using PSK modulation for PCM, telephony.

The parameters of the TDMA system[18] adopted by INTELSAT are:

Channel encoding at 64 kbit s^{-1} (8 bits per sample) with or without digital speech interpolation (DSI).
4 phase PSK modulation at 120 Mbit s^{-1}.
Approximately 1600 channels per 72 MHz (3200 with DSI).
2 ms frame period.

The capacity of a TDMA transponder can be approximately doubled by employing a digital speech interpolation (DSI) system.[26,27,28] DSI enables an increase in capacity to be achieved by utilising the quiet periods of a speech channel. Any active input channel is able to switch to any available channel in a pool of satellite channels. Hence, the number of terrestrial channels that can be connected is approximately double the number of satellite channels available.

The FDM/FM and PCM/TDMA systems described above are used mainly for large traffic streams. For thin route applications single-channel per carrier (SCPC) operation is frequently adopted. The main parameters of typical SCPC PSK and companded FM systems are as follows:

SCPC CFM/PSK 4ϕ parameters for SPADE and pre-assigned (INTELSAT standard A station)[29]

Audio channel input bandwidth	300–3400 Hz
Encoding	7 bit PCM $A = 87.6$ companding law 8 kHz sampling rate
Modulation	4 phase coherent PSK
Carrier control	voice activated for voice channels
Channel spacing	45 kHz
IF noise bandwidth	38 kHz
C/N at threshold	13.5 dB
Threshold bit error rate	10^{-4}

SCPC/CFM[30] (typical thin route domestic system)

Audio channel processing	2:1 companding
Modulation	FM
Carrier control	voice activated
Channel spacing	22.5 kHz
Noise bandwidth	18.75 kHz
C/N	16.8 dB for 10 000 pW (subjective equivalent) for a median talker.

55.14 Summary of international, regional and domestic satellites

Table 55.3 Main features of communication satellites

Name	*Number successfully launched by end 1981*	*Date of first launch*	*Mass (orbit) (kg)*	*No*	*Transponder characteristics Bandwidths (MHz)*	*Saturation output e.i.r.p. (dB W)*	*Total telephone channel capacity*
INTELSAT							
IV	7	Jan 1971	730	12	36	22, 33.7	8 000 (typ)
IVA	5	May 1975	750	20	36	22, 26, 29	12 000 (typ)
V	2	Dec 1980	1012	27	78, 72, 36 38, 240	23.5, 26.5, 26 29, 41.1, 44.4	24 000 (typ)
RCA	1	Dec 1980	461	24	34	32, 36	28 800
SBS	2	Nov 1980	555	10	43	40–43.7	27 800
ECS	—	Sept 1982	461	9	72	34.8, 40.8	16 000

55.15 Earth stations for international telephony and television[31,32,33,34]

55.15.1 Aerial system

Several classes of Earth station are currently in use, namely the fixed, mobile and transportable stations. INTELSAT recommends only the standard Earth stations for fixed satellite communication services. Three standard Earth stations (standard A, B and C) are currently authorised by INTELSAT. A fourth standard for use at shore stations in the maritime mobile service is under study. Domestic systems using leased INTELSAT transponders generally use Earth stations comparable with the standard B. *Table 55.4* lists the standard INTELSAT earth station parameters.

Modern large Earth stations adopt the Cassegrain construction in which the use of a sub-reflector permits the feed to be located at the centre of the main reflector where it is readily accessible and from where the waveguide runs to the transmitters and receivers are relatively short. The recent introduction of beam waveguide systems allows the feed to be located off the moving parts of the aerial close to the low-noise and high-power amplifiers. This eliminates long waveguide runs and awkward flexible or rotary joints.

Aerial steering, generally in azimuth and elevation axes, is necessary because the satellites are not perfectly stationary when considered in terms of the very narrow beamwidth (approximately 0.2° at 6 GHz) of a 30 m aerial. Small beam-pointing adjustments can also be obtained by moving the feed. The satellite beacon is generally used to derive the control signal for the auto-track steering system, a common method being by the detection in the feed system of higher order modes which undergo rapid rate of change as the received signal direction varies near the axis.

55.15.2 Low-noise receiving system[35]

The wide-bandwidth and low-noise temperature of the parametric amplifier makes it the universal choice as the first-stage amplifier. The full satellite bandwidth of 500 MHz can be readily covered. Cooling to 20 K or below using closed-cycle cryogenic systems with gaseous helium was used in the past to produce amplifier noise temperatures of the order of 15 K. Now, thermoelectrically(Peltier)-cooled low-noise amplifiers (LNA), producing amplifier temperatures of 40 to 55 K (at 6/4 GHz), are replacing the cryogenically-cooled low-noise amplifiers. The total receiving system noise temperature at 4 GHz, however, is typically 70 K and 90 K for cryogenically- and thermo-electrically-cooled LNA, respectively at 5° elevation angle. The receiving system noise temperature of 70 K for cryo-genically-cooled systems is made up of about 25 K due to feed and waveguide losses and 15 K due to the LNA.

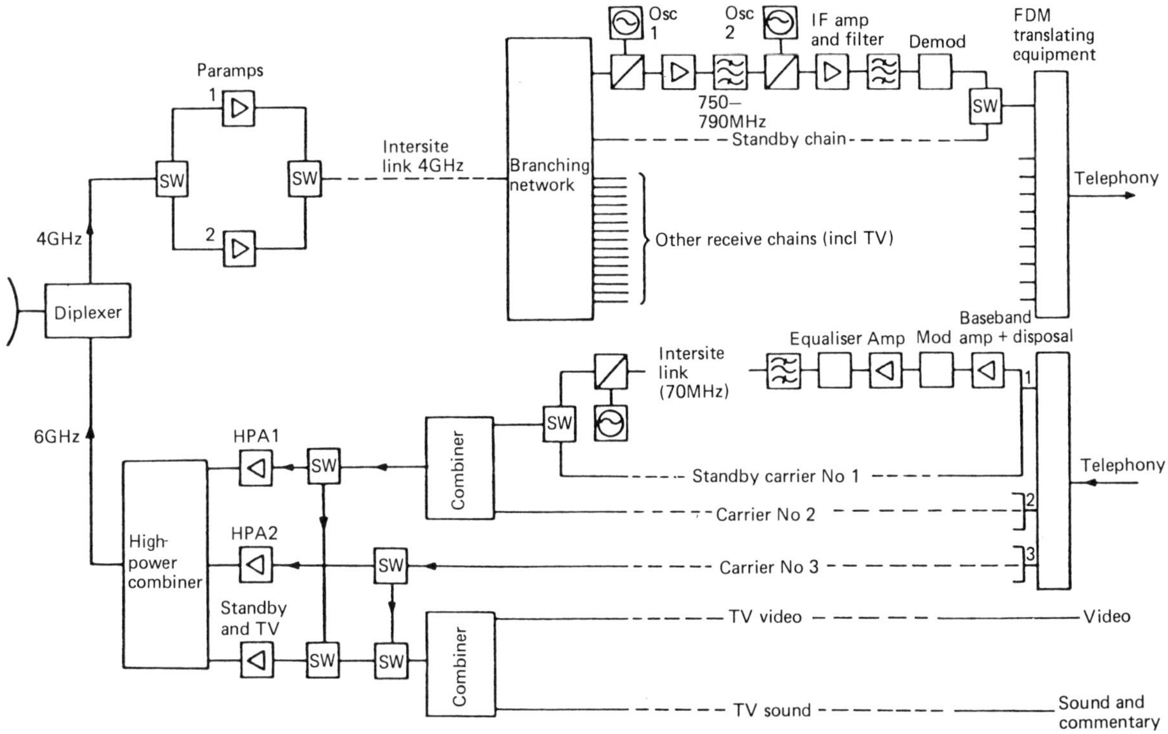

Figure 55.6 Typical Earth station equipment using TWT

Table 55.4 INTELSAT Earth station parameters

Standard	*Uplink (GHz)*	*Downlink (GHz)*	*G/T(min) (dB K^{-1})*	*Antenna diameter (m)*	*e.i.r.p. range (dBW)*
A	5.925–6.425	3.7–4.2	40.7	30	73–99
B	5.925–6.425	3.7–4.2	31.7	11	63–85
C	14.0–14.5	10.95–11.2 11.45–11.7	39.0 (nom)	13	63–92
Shore station	6.417–6.425	4.1925–4.2005	32	11	69.0 max

55.15.3 High-power amplifiers

Transmitter powers are relatively modest but the transmitter arrangements are frequently complicated by the need to transmit more than one carrier. In a small or medium size station transmitting only one or two carriers, a klystron of 50 MHz bandwidth for each carrier may be the best arrangement. For a large station, klystrons with low-loss combiners (loss from 1.5 to 4.5 dB) or TWT covering the whole 500 MHz band are used. A TWT may be used with multiple carriers but a large station can still need several TWT, connected through a high-power combining network to the transmit waveguide.

The transmitter power requirements can be judged by the following examples for INTELSAT V 6/4 GHz band operation, assuming 60 dB aerial gain.

Global beam carriers 132 ch. 91 W, 972 ch. 813 W, TV 501 W
Hemi/zone beam carriers 132 ch. 129 W, 972 ch. 575 W
TDMA one carrier per 80 MHz transponder 3200 ch. with DSI 316 W.

If two or more carriers are passed through the same TWT or klystron, a back-off of 3 to 8 dB is necessary to reduce intermodulation products to the allowed levels. At 6/4 GHz band, TWT and klystrons with power outputs of 10 kW and 3 kW, respectively, are commonly used although TWT with power outputs of 20 kW are commercially available.

55.15.4 Intermediate frequency equipment, combining and branching equipment

On the transmit side the link between the main equipment building and the aerial site is usually at the IF of 70 MHz. Up conversion to 6 GHz is followed by either low-power combining of carriers to be amplified in the same TWT or by intermediate TWT amplifications followed by high-power klystron amplification and combining. On the receive side the frequently large number of received carriers and the need for flexibility make the use of an SHF link from the aerial to the equipment building more convenient. A flexible elliptical waveguide can be used for this. A large station may need to extract perhaps 30 separate carriers, one for each station from which traffic is to be received. A modern approach to this problem uses stripline branching techniques. The signal path is first broken down to the required number of receive chains, which are all broadband at this stage. A first downconverter, fed with a local oscillator frequency selected to pick out the 40/80 MHz band in which the carrier falls, translates to a 770 MHz IF from which a second downconverter translates to 70 MHz for feeding to the demodulator. For TDMA the first downconverter selects a 80 MHz bandwidth which is translated in the second downconverter to a 140 MHz IF. After equalisation the IF signal is passed to the PSK demodulator. Redundant receive chains and automatic changeover are provided for reliability. Group delay equalisation is carried out at 70 MHz (140 MHz for TDMA), the satellite group delay being normally dealt with by the transmit side equalisation. Carefully designed IF filters are necessary in transmit and receive paths to eliminate adjacent channel interference.

55.15.5 Modulators, demodulators and baseband equipment

The design of frequency modulators follows conventional techniques. The main requirement is for a high degree of linearity.

Demodulators also follow conventional techniques except when the carrier-to-noise ratio is close to the FM threshold (approximately 10 dB). Then threshold extension demodulators[36] are used, the FM feedback demodulator being the commonest of this type.

The telephone channels are assembled into the baseband signals required for the particular transmitted carriers using conventional FDM translating equipment. All these basebands commence at 12 kHz but the band below 12 kHz is used for engineering service circuits and for the energy dispersal signal, a triangular-wave signal which keeps the RF spectrum well spread during the light traffic loading conditions to reduce interference problems. On the receive side those channels destined for the particular station are extracted from the received basebands using FDM translating equipment. They are then reassembled for onward transmission.

Where TDMA is employed, the 120 Mbit s^{-1} four-phase PSK signal is demodulated into two 60 Mbit s^{-1} data streams that are passed to the TDMA terminal proper. The TDMA terminal extracts the appropriate data bursts containing the relevant telephone channels. These are reformatted as standard 30-channel PCM circuits for onward transmission. The technology exists for satellite communications using small (3 to 5 m) Earth terminals.[37] A number of organisations mainly in the USA are offering specialised business services and TV distribution using small Earth terminals.[38,39] Small terminals located at the user premises can provide voice, live colour TV for two way video conferences, reduced bandwidth freeze frame colour TV for slide presentation and various digital computer and facsimile transmissions.

55.16 Regional and domestic communication satellite systems

Regional and domestic satellites can concentrate all their available radiated power into relatively small areas, thus achieving higher e.i.r.p. compared to international service satellite covering large areas of the globe. Typically, the e.i.r.p. of a global beam is 23 dB W whilst a beam using the same TWT but illuminating mainland USA or Western Europe achieves about 33 dB W. This tenfold increase in incident power can be used to reduce ground segment costs by the use of much smaller aerials than the ones used in international communications, typically 3–9 m.

Some general principles concerning regional and domestic systems can be stated. Such systems are likely to be justified where:

(i) The terrain is very difficult for terrestrial communications.
(ii) There are no existing facilities and distances are large.
(iii) Terrestrial facilities exist but large volumes of traffic have to be carried over large distances and/or where a wideband service is needed.

Table 55.5 lists basic characteristics of selected regional/domestic service satellites.

ORBITA (USSR).[40,41,42] The first purely domestic satellite system was the ORBITA network of the USSR based on the use of MOLNIYA satellites placed in a highly elliptical 12 hour orbit (538 and 39 300 km at 65.5° inclination) providing mainly TV distribution. The first MOLNIYA (Lightning) satellite was launched in April 1965. A network of ground terminals was set up near big cities with parabolic aerials of 12 m which allowed reception of TV and reception/transmission of voice and telegraph signals. Initially, the frequency band around 800 MHz was used but in 1971 after the launch of MOLNIYA-2 the 6/4 GHz was used. A portable ground station was also deployed which has an aerial of 7 m and works to both MOLNIYA and RADUGA (Rainbow) satellites. The ORBITA system has about 50 Earth stations in the USSR at present.

ANIK (Canada).[43,44] Telesat of Canada with the launch in 1972 of the first of three ANIK-A satellites established the world's second domestic system. The second generation ANIK-B satellites[45] have been added to the original ANIK-A and the first of the third generation ANIK-C[46] was launched in 1981 (*Table 55.5*).

WESTAR (USA).[47,48] The first USA domestic satellite, WESTAR I (*Table 55.5*) was launched by the Western Union Telegraph Co in 1972 followed by WESTAR II in the same year. The WESTAR satellite system provides voice, telegraph and data communications in the USA. The earth segments consist mainly of five major Earth terminals utilising 15 m aerials located at major cities. Western Union also leases out capacity to other common carriers. The American Satellite Corporation (AMSAT) leases transponders on WESTAR and provides services to its users via about forty 5 to 11 m Earth terminals owned and operated by AMSAT. A WESTAR III was launched in 1979 after having been held in ground storage for 5 years and a WESTAR IV is planned to be launched in 1982.

RCA SATCOM (USA).[49] RCA satellites SATCOM 1 and SATCOM 2 were launched in 1975 and 1976, respectively (*Table 55.5*). Each satellite has twenty-four 36 MHz transponders with 5 W TWT, operates in the 6/4 GHz band and utilises polarisation diversity for frequency re-use. The SATCOM system services the USA and provides voice, data and TV services to business, media and government. A new data, voice, facsimile, slow-scan TV and teleprint service called '56 plus' is also being offered.

COMSTAR (USA).[50] Comsat General Inc established the COMSTAR satellite system with the launch of the COMSTAR D1 satellite in 1976 (*Table 55.5*). The system was designed to meet the specific requirements of the American Telephone and Telegraph Co (AT&T) for long-distance voice communications. The space segment is leased entirely to AT&T and GT&E who operate 7 INTELSAT standard A equivalent Earth stations. The

Table 55.5 Basic characteristics of selected domestic service satellites

	ANIK-A	*ANIK-B*	*ANIK-C*	*WU WESTAR*	*RCA SATCOM*	*SBS*	*TELECOM 1*	*COMSTAR*	*INDONESIA PALAPA*
Service	Fixed voice	Fixed voice	Fixed voice, TV, data	Fixed voice, TV, data	Fixed voice, TV	Fixed voice, TV, data	Fixed voice, TV, data	Fixed voice	Fixed voice, TV
Frequencies (GHz)	6/4	6/4, 14/11	14/12	6/4	6/4	14/12	6/4, 8/7, 14/12	6/4	6/4
Mass in geostationary orbit (kg)	272	440	567	297	461	555	640	810	200
Coverage	Canada	Canada	Canada	USA	USA	Continental USA	France, Overseas Dept and Territories	USA	Indonesia
No of transponders (bandwidth (MHz))	12(36)	12/6 (36/72)	16(54)	12(36)	24(34)	10(43)	6(36), 2(120), 4(40)	24(36)	12(36)
No of beams	1	1 4/1	4/1	2	2	1	4	4	1
G/T (dB K^{-1})	−7	−6, −1	+1	−6	−5, −10	2 to −2	−8.6 to −4.6, +6.3	−8.8	−7
E.i.r.p. (dB W)	33	36, 47.5	48	33	32, 26	43.7 to 40	27.3–31.5, 47.0	33	32
Modulation	FDM/FM QPSK, SCPC	FDM/FM QPSK, SCPC	FDM/FM QPSK, SCPC	FM QPSK	FDM/FM QPSK, SCPC	QPSK	FDM/FM BPSK, SCPC	FDM/FM	FM/SCPC
Multiple access	FDMA TDMA	FDMA TDMA	FDMA TDMA Re-use	FDMA TDMA	FDMA	TDMA	FDMA TDMA	FDMA Re-use	FDMA
Primary power (W)	300	840	900	300	770	914	920	600	300

satellites employ polarisation isolation frequency re-use in the 6/4 GHz band to provide twenty four 36 MHz transponders capable of handling 28 800 telephone channels. Three COMSTAR D satellites have been launched to date covering 51 states in the USA.

SATELLITE BUSINESS SYSTEMS (USA).[51,39] The most ambitious domestic satellite communications system in the USA so far is that of the Satellite Business System (SBS), a partnership of COMSAT General Corporation, Aetna Life and Casualty and IBM. The SBS provides an all-digital private-line switched network for integrated voice, data and video services.[38] The overall system consists of the satellites, Earth stations, facilities for TT&C and a system management function which brings all these together under computer control. The system operates four-phase PSK/TDMA in the 14 to 11 GHz band (*Table 55.5*).

The space segment comprises three satellites, two operational in orbit (one also acting as a spare) and one ground spare. The first operational satellite was successfully launched in November 1980 and service began in early 1981. Each satellite contains one active and three spare wideband receivers and ten communication channels. The offset parabolic satellite antennae with maximum gain of 35 dB (at 14/12 GHz) provide shaped beams that cover the contiguous 48 states.

The SBS Earth stations[52,53] which are located in the customers' premises consist of (*a*) RF terminals employing 5.5 m or 7.6 m symmetrically fed parabolic reflectors with Cassegrain feeds; (*b*) burst modems that perform modulation/demodulation functions for the high-speed transmission bursts; (*c*) satellite communications controllers for a series of functions including multiple access control and port adaptor systems for connecting the customers' communications and business equipment to the SBS system.

TELECOM 1 (France).[54,55] The French Government is planning to set up a domestic satellite communications system, the TELECOM 1, to become operational in 1983. Three satellites will be employed, two in orbit (one operational and one spare) and one ground spare. The operational satellite will be positioned 10°W above the Gulf of Guinea.

TELECOM 1 will operate in the 14/12 GHz, 8/7 GHz and 6/4 GHz frequency bands (*Table 55.5*). At 14/12 GHz, coverage will be provided over the territory of metropolitan France and neighbouring countries. The six 14/12 GHz transponders will provide two different services, namely, an intra-company service for the business communications and a video service from a point of origin towards community viewing receive points. The 8/7 GHz band will be used exclusively for French Government communications. A near-global beam provides the required coverage. The 6/4 GHz will be used for telephony and television communications between metropolitan France and the Overseas Departments and Territories.

The intra-company communications service will include data, telephony and video conferencing and will operate in the TDMA mode. Data rates will range from 2.4 kbit s^{-1} to 2 Mbit s^{-1}. Easy to install, unmanned Earth stations with a 3 to 4 m aerial and 150 W rated transmitter power will be installed at or near the business premises subscribing to the service. The satellite and Earth stations will employ permanent pooling of transmission capacities, flexibly and dynamically allocated on demand to any system user, according to his varying transmission speed requirements. Five transponders of 36 MHz bandwidth each will provide a total of 125 Mbit s^{-1} (about 370 channels of 64 kbit s^{-1} per transponder).

The video service will provide live TV transmissions to cinemas, schools etc using small receiving aerials.

Communications between metropolitan France and the Overseas Departments and Territories will include about 1800 voice channels and one TV channel. INTELSAT standard A and B equivalent Earth stations will be used in metropolitan France and overseas territories, respectively.

INTELSAT In addition to its global system INTELSAT is currently leasing out about 17 INTELSAT satellite transponders, mostly on spare satellites on a low-tariff pre-emptable basis, to individual signatories for their domestic communications. Countries such as Algeria, Brazil, Chile, Colombia, France, India, Norway, Nigeria, Oman, Peru, Sudan, Saudi Arabia, Spain and Zaire lease whole or partial transponders from INTELSAT for domestic communications. Most domestic systems use 11 to 14 m Earth stations and operate their voice channels in the voice activated SCPC mode. Some countries lease INTELSAT transponders for domestic communications before launching their own domestic satellites. This was done by Indonesia and is planned by Australia, China, India and Mexico among others.

PALAPA (Indonesia).[56,57,58] In 1976, Indonesia initiated a domestic communications service on its PALAPA A1 Satellite (*Table 55.5*). Two in-orbit satellites, one operational and one spare, nearly identical to ANIK-A and WESTAR, form the space segment. Fifty 10 m Earth terminals strategically located on the 13 677-island archipelago (only 1000 islands inhabited) relay satellite traffic to telephone exchanges in the cities and communities served, and interconnect with microwave and cable network for distribution of voice TV and data traffic. The Indonesian Government also leases transponders to neighbouring countries for domestic use.

The second generation PALAPA B satellites scheduled for launch in 1983 will replace the PALAPA A. PALAPA B will be twice as big, with twice the capacity and four times the electrical power of PALAPA A.

55.16.1 Future regional and domestic systems

RCA new satellites replacing the SATCOM 1 and 2 will feature 8.5 W solid state transponders with improved aerial design and will provide 50% increase in voice and data capacity, reduce ground segment costs and have an expected in-orbit life of 10 years.

Western Union's next satellite generation, the Advanced WESTAR,[15,59] will be the first dual-purpose communications satellite handling communications between NASA, ground stations and NASA spacecraft as well as Western Union's own commercial traffic. The Advanced WESTAR will provide twelve 36 MHz transponders at 6/4 GHz band and four 250 Mbit s^{-1} SS–TDMA transponders in the 14/12 GHz band.

EUTELSAT, the European Consortium, will provide in 1983 European regional satellite services using two in-orbit ECS satellites, one operational and one spare. The system will operate in the 14/11 GHz band using TDMA for voice data and FM for TV distribution. In addition, EUTELSAT will provide specialised business services via small Earth stations using the ECS satellites (flight F2 onwards) and on leased TELECOM1 capacity.

The European Space Agency is planning an experimental regional European satellite, temporarily abbreviated to LSAT,[60,61,62] which will carry mainly satellite multi-service and direct broadcasting payloads. LSAT will be about twice the size of the ECS and is due for launching in 1985.

55.17 Satellite systems for aeronautical and maritime mobile communications[63,64,65]

An important advantage of satellite communications over all other modes of long-distance wideband communications is the ability to operate with mobile terminals. A number of satellite systems have been proposed to serve mobile platforms, one of which is the now fully operational 'MARISAT' system.[66] MARISAT was developed to serve mobile maritime users in the three ocean regions, Atlantic, Indian and Pacific, respectively. The system consists of three operational geostationary satellites (launched in 1976), three shore stations and 542 ship terminals.

Each satellite contains one 500 kHz and two 25 kHz channels in the 250/300 MHz bands for use by the US Navy and two 4 MHz transponders for commercial use. One transponder translates shore-to-ship signals from 6 GHz to 1.5 GHz while the other translates ship-to-shore signals from 1.6 GHz to 4 GHz.

The MARISAT communications system provides voice, telegraph and signalling channels. The voice channels are transmitted on a SCPC/CFM/FDMA mode. The telegraph channels are two-phase coherent PSK modulated and use TDM and TDMA for shore-to-ship and ship-to-shore directions, respectively. Each satellite supports one voice and 44 telex channels (low-power mode).

In February 1982, the International Maritime Satellite Organisation (INMARSAT) system replaced MARISAT in providing the space segment for maritime communications. Two European Space Agency MARECS satellites, three INTELSAT V satellites with a maritime services subsystem and an existing MARISAT satellite will provide the required three-ocean coverage for INMARSAT.

For economy and continuity of service, INMARSAT has retained the modulation and multiple access methods used by MARISAT. The basic INMARSAT system characteristics are as follows:

(*a*) Satellite communication system

	Shore to ship	Ship to shore
Receive G/T	−21 to −17 dB K^{-1}	−17.5 to −21.1 dB K^{-1}
Total e.i.r.p.	27 to 34.2 dB W	14.5 to 20 dB W
Transponder bandwidth	4–7.5 MHz	

(*b*) Ship terminal characteristics

G/T	−4 dB K^{-1}
Gain	23 dBi
Aerial diameter	1.2 m
E.i.r.p.	37 dB W

(*c*) Shore Earth station characteristics

G/T	32.2 dB K^{-1}
Gain	43 to 46 dBi (at 60% efficiency)
Aerial diameter	12.8 m
E.i.r.p.	60 dB W (voice), 57 dB W (TDM)

(*d*) Frequencies used

	Ship to satellite	Satellite to shore
Uplink	1636.5 to 1644 MHz	6417.5 to 6425 MHz
Downlink	1535 to 1542.5 MHz	4192.5 to 4200 MHz

A number of aeronautical satellite communications systems have been proposed in the last decade but due to international, institutional and economic problems, progress has been slow. Communication with aircraft would probably require a small number of voice channels and perhaps some channels for automatic data transmission as a complement to a radio determination system which would allow closer spacing between aircraft on heavily loaded oceanic routes. Voice channels would probably use narrow-band FM or delta modulation with PSK. The largest single factor influencing system performance is the aircraft antenna which necessarily has very wide beamwidth and hence low gain since it cannot be steerable, except perhaps for rudimentary steering with a phased array or switching between elements. This constraint means that only a handful of channels can be obtained from a satellite which would give thousands of channels to standard INTELSAT Earth stations.

The frequency bands of prime interest for maritime and aeronautical communication satellites[4] are:

	Space to Earth	Earth to space
Maritime mobile satellites	1530 to 1544 MHz	1626.5 to 1645.5 MHz
Aeronautical mobile satellites	1545 to 1559 MHz	1646.5 to 1660.5 MHz

For the links between the satellites and the land-based stations frequencies in the 6/4 GHz bands are used although use of the above mentioned UHF bands is permitted.

55.18 Television broadcasting satellite systems[67,68]

A wide diversity in the technical devices and the performance characteristics of TV broadcasting satellite systems exist. Besides the coverage areas and the frequencies used, the main differences between systems result from the service definition. For example, the Europeans are planning a purely direct to home reception requiring high flux density; others specify community-type reception which requires less powerful satellite transmission.

The WARC 77[69] drew up a plan in which every country in regions 1 (Europe/USSR/Africa) and 3 (Asia/Australia) was allocated an orbital position with 5 TV channels of 27 MHz necessary bandwidth and specified e.i.r.p. and footprint characteristics. A regional conference to conduct a similar assignment plan for region 2 (Americas) is scheduled for 1983.

Typical parameters for television broadcasting direct to home and community receivers are given in *Table 55.6*.

Table 55.6 Examples of parameters for FM television broadcast satellite at 12 GHz. (Satellite aerial efficiency of 55% is assumed.)

Service type	*Receive aerial diameter (m)*	*Receive aerial G/T (dB K^{-1})*	*System noise factor (dB)*	*Overall C/N required (dB)*	*Bandwidth per channel (MHz)*	*Received p.f.d. (dB W m^{-2})*	*Satellite e.i.r.p. (dB W)*	*Satellite RF power (W) for 3.4°*	*1.15°*
Home	0.9	6	5.9	14	27	−103	62	631	71
Community	1.8	14	4.2	14	27	−111	54	100	11

At present there are three national television broadcasting satellite systems,namely, CTS/HERMES (Canada/USA),[70] EKRAN (USSR) and BSE (Japan).[71] The main parameters of these systems are listed in *Table 55.7*.

Table 55.7 Television broadcasting satellite system parameters

	CTS/HERMES	*EKRAN*	*BSE*
Country	Canada/USA	USSR	Japan
Launch date	1976	1976–1980	1978
Launcher	DELTA-2914	PROTON-D	DELTA-2914
Lifetime (years)	3	Unknown	3
Frequency (GHz)	12	0.7	12
TWTA RF power (W)	200	200	100
Coverage	2.5°	9×10^6 km^2	Japan
E.i.r.p. (dB W)	59.5	47.5	55
Aerial diameter (m)	0.6 to 3	YAGI Array	16
G/T (dB K^{-1})	5.8	−6	15.7
C/N_0 (dB Hz)	86.4	87.6	92.5
C/N (dB)	13	13.8	18.5

Future TV broadcasting satellite systems are the INSAT[72] (Indian community TV 1981), the Franco–German program (direct to home TV 1983–1984), the Australian multipurpose communication system (community TV 1984).

The bands allocated to satellite broadcasting[4] are 2.5 to 2.69 GHz (community reception only) and 11.7 to 12.5 GHz, 12.1 to 12.7 GHz (subject to detailed planning at RARC) and 11.7 to 12.2 GHz for regions 1, 2 and 3, respectively.

55.19 Future developments

The trend in the 1980s will be for higher capacity satellites with lower cost per channel making more efficient use of the frequency spectrum and offering better inter-satellite connectivity.

Frequency re-use with multi spot beams[73] and the use of 30 to 20 GHz[74,75,76,77] frequency bands will offer the higher capacity needed.

As the number of satellite beams increases, on-board switching[78] and processing will become increasingly important. Regenerative transponders will demodulate the received digital carrier, reconstruct the signal and remodulate it for transmission to Earth. On-board processing eliminates the up-link noise component of the retransmitted signal making it particularly useful for small terminal links where the up-link noise can be a large portion of the overall C/N.

High transmission efficiency will be achieved by the use of efficient digital coding and processing. Inter-satellite connectivity can be achieved by the use of inter-satellite links (ISL). ISL can interconnect Earth stations operating to different satellites without additional aerials or double hops. This facility will significantly influence the capacity, connectivity, coverage, orbit-utilisation and overall cost of a satellite system.

The Space Shuttle with its 18.3 m by 4.6 m payload bay and 30 000 kg launch capability into low Earth orbit will greatly influence the design of communications satellites. The Shuttle could place very large stable platforms in geostationary orbit, that can support complex large antenna and payload configurations serving different missions.[79] Alternatives to the platform concept include (*a*) very large satellites launched by the Shuttle; (*b*) numerous small special-purpose satellites; (*c*) several smaller satellites connected via ISL; (*d*) clusters of satellites[80] interconnected and collocated within the beam of an Earth station.

Information on the following further topics can be found in the references given.

Propagation factors in satellite communication.[81]
Interference with terrestrial or other satellite systems.[82,83]
Efficiency of use of the geostationary orbit.[84]
Effects of propagation delay and echo.[85,86]

References

1 CLARKE, A. C., 'Extra terrestrial relays', *Wireless World*, 305 (1945)
2 EDELSON, B. I. and DAVIS, R. C., 'Satellite communications in the 1980s and after', *Phil. Trans. R. Soc. London*, **289**, 159 (1978)
3 KRUPIN, Y. U., 'The INTERSPUTNIK International Space Communications Organisation', *Radio Telev. (Czechoslovakia)*, **1**, 23 (1978)
4 WARC, 'Final acts of the World Administrative Radio Conference', ITU Geneva (1979)
5 YOSHIDA, N., SHIOMI, T. and OKAMOTO, T., 'On-orbit antenna pointing performance of Japanese communications satellites CS', *AIAA Paper 82-0441* (1982)
6 COLLETTE, R. C. and HERDAN, B. L., 'Design problems of spacecraft for communication missions', *Proc. of I.E.E.E.*, **65** (No 3), 342–356 (1977)
7 HAYN, D., BRAITINGER, M. and SCHMUCKER, R. H., 'Performance prediction of power augmented electrothermal hydrazine thrusters', *Technische Universitaet Lehrstuhl für Raumfahrttechnik, Munich, W Germany* (1978)
8 FREE, B. A., 'North–south station keeping with electric propulsion using on-board battery power', *1980 JANNAF Propulsion Meeting*, **5**, 217 (1980)
9 RICARDI, L. J., 'Communication satellite antennas', *Proc. of I.E.E.E.*, **65** (No 3), 356–359 (1977)
10 HALL, G. C. and MOSS, P. R., 'A review of the development of the INTELSAT system', *Post Office Electrical Engineers Journal*, **71**, 155–163 (1978)
11 EATON, R. J. and KIRKBY, R. J., 'The evolution of the INTELSAT V system and satellite, part 2, spacecraft design', *Post Office Electrical Engineers Journal*, **70** (No 2), 76–80 (1977)
12 JILG, E. T., 'The INTELSAT IV Spacecraft', *COMSAT Technical Review*, **2** (No 2), 271 (1972)
13 BARTHOLOME, P., 'The European communications satellite programme', *ESA Bulletin (France)*, **14**, 40–47 (May 1978)
14 FUENZALIDA, J. C., RIVALAN, P. and WEISS, H. J., 'Summary of the INTELSAT V communications performance specifications', *COMSAT Technical Review*, **7** (No 1), 311–326 (1977)
15 RAMASASTRY, J., CALLANAN, W., MARKHAM, R., KAUL, P. and GOLDING, L., 'Advanced WESTAR SS/TDMA system', *IVth Int. Conf. on Digital Satellite Communications, Montreal, Canada 23–25 Oct 1978* p. 36–43 (1978)
16 WESTCOTT, R. J., 'Investigation of multiple FM/FDM carriers through a satellite TWT operating near to saturation', *Proc. of I.E.E.E.*, **114** (No 6), 726 (1972)
17 CHITRE, N. K. M. and FUENZALIDA, J. C., 'Baseband distortion caused by intermodulation in multicarrier FM systems', *COMSAT Technical Review*, **2** (No 1), 147 (1972)
18 INTELSAT 'INTELSAT TDMA/DSI system specification (TDMA/DSI traffic terminals), *INTELSAT Document BG 42/65 Rev.*, **1** (June 1981)
19 EUTELSAT, 'EUTELSAT TDMA/DSI system specification', *EUTELSAT Document ECS/C-11-17 Rev.*, **1** (September 1981)
20 EDELSON, B. I. and WERTH, A. W., 'SPADE system progress and application', *EUTELSAT Document ECS/C-11-17*, **2** (No 1) 221 (1972)
21 CCIR, 'Allowable noise power in the hypothetical reference circuit for frequency-division multiplex telephony in the fixed satellite service', *CCIR Rec. 568, ITU Geneva*, **12**, 39 (1978)

22 CCIR, 'Single value of the signal-to-noise ratio for all television systems', *CCIR Rec. 568, ITU Geneva*, **12**, 39 (1978)

23 CCIR, 'Allowable bit error rates at the output of the hypothetical reference circuit for systems in the fixed satellite service, using pulse-code modulations for telephony', *CCIR Rec. 522, ITU Geneva*, **4**, 61 (1978)

24 HILLS, M. T. and EVANS, B. G., '*Telecommunications System Design, Vol. 1 Transmission Systems*', 176–198, George Allen and Unwin (1973)

25 BARGELLINI, P. L., 'The INTELSAT IV communication system', *COMSAT Technical Review*, **2** (No 2), 437 (1972)

26 TERRELL, P. M., 'Application of digital speech interpolation', *Communs. Int.* (*GB*), **6** (No 2), 22, 24, 26 & 30 (February 1979)

27 CAMPANELLA, S. J., 'Digital speech interpolation techniques', 1978 National Telecommunications Conf., Birmingham AL USA, *Conf. Record of the IEEE*, 14.1/1.5 (December 1978)

28 SEITZER, D., GERHAUSER, H. and LANGENBUCHER, G., 'A comparative study of high quality digital speech interpolation methods', *Int. Conf. on Communications, Toronto, Canada*, 50.2/1.5 (June 1978)

29 INTELSAT 'Standard A performance characteristics of Earth stations in the INTELSAT IV, IVA and V systems having a *G/T* of 40.7 dB/K (6/4 GHz frequency bands), *INTELSAT DOCUMENT BG28/72*, 75 (August 1977)

30 FERGUSON, M. E., 'FM—The new single-channel-per-carrier technique', *Technical Memorandum, California Microwave Inc*, Sunnyvale CA, USA

31 THE INSTITUTE OF ELECTRICAL ENGINEERS LONDON (UK), 'Antennas and propagation, Part 1 Antennas', *IEE Conf. Publ.*, IEE, Savoy Place, London WC2, Publication No 195 (April 1981)

32 INTELSAT, 2nd Earth Station Technology Seminar, Athens, Greece (October 1977)

33 LOVE, A. W. (Ed), '*Reflector Antennas*', I.E.E.E. Press, Institute of Electrical and Electronic Engineers Inc, New York (1978)

34 CLARRICOATS, P. J. B., 'Some recent advances in "microwave reflector antennas"', *Proc. of I.E.E.*, **126** (No 1), 9–25 (January 1979)

35 KAJIKAWA, M., HAGA, I., FUKUDA, S. and AKINAGA, W., 'Development status of low-noise amplifiers', *AIAA 8th Communications Satellite System Conf., Orlando FA, USA*, 309–316 (April 1980)

36 VAN DASLER, G., VAN LAMBALGEN, H. and VAN DAAL, P. M., 'A threshold extension demodulator', *Phillips Telecom Rev.* (*Netherlands*), **31** (No 3), 131–146 (October 1973)

37 KAISER, J., VEENSTRA, L., ACKERMANN, E. and SEIDEL, F., 'Small Earth terminals at 12/14 GHz', *COMSAT Technical Review*, **9** (No 2), 549–601 (Autumn 1979)

38 SCHNIPPER, H., 'The SBS system and services', *I.E.E.E. Commun. Mag.* (*USA*), **18** (part 5), 12–15 (September 1980)

39 BARNLA, J. D., 'The SBS digital communications satellite system', *1978 Wescon Technical Paper*, Los Angeles CA, USA, 31.2/1-9 (September 1978)

40 MERCADER, L., 'Direct broadcasting of TV signals from satellites. A description of the Russian systems', *Mundo Electron* (*Spain*), No 88, 71–77 (September 1979)

41 FISCHER, C., 'Satellite communications systems ORBITA-2', *Radio Fernbehen Elektron.* (*Germany*), **23** (No 17), 548–550 (September 1974)

42 KANTOR, L. YA., POLUKHIN, U. A. and TALYZIN, N. V., 'New relay stations of the ORBITA-2 satellite communications system', *Elektrosvyaz* (*USSR*), **27** (No 5), 1–8 (May 1978)

43 ALMOND, J., 'Commercial communication satellite systems in Canada', *I.E.E.E. Commun. Mag.* (*USA*), **19** (No 1), 10–20 (January 1981)

44 ROSCOE, O. S., 'Direct broadcast satellites — the Canadian experience', *Satell. Commun.* (*USA*), **4** (No 8), 22–23, 29–32 (August 1980)

45 GOTHE, G., 'The ANIK-B slim TDMA pilot project', NTC '80, *IEEE 1980 National Telecommunications Conference*, Houston, Texas, USA, 71.4/1-9 (1980)

46 CHAN, K. K., HUANG, C. C., CUCHANSKI, M., RAAB, A. R., CRAIL, T. and TAORMINA, F., 'ANIK-C antenna system', *1980 Int. Symp. Digest*, Antennas and Propagation, Quebec, Canada, 2–6 June 1980, 89–92 (1980)

47 VERMA, S. N., 'US domestic communication system using WESTAR satellites', *World Telecommunication Forum Technical Symp., Geneva*, 6–8 October 1975, 2.4.3/1-6 (1975)

48 SCHNEIDER, P., 'WESTAR today and tomorrow (satellite communications)', *Signal* (*USA*), **34** (No 3), 43–45 (December 1979)

49 KEIGLER, J. E., 'RCA SATCOM: An example of weight of optimised satellite design for maximum communications capacity', *Acta Astronautica*, **5**, 219–242 (1978)

50 BRISKMAN, R. D., 'The COMSTAR program', *COMSAT Tech. Rev.*, **7** (No 1), 1–34 (1977)

51 WHITTAKER, P. N., 'Satellite business system (SBS) — A concept for the 80s', *Policy Implications of Data Network Development in the OECD*, 35–39 (1980)

52 WEISCHADLE, G. M. and KOURY, A., 'SBS terminals demand advanced design', *Microwave Syst. News* (USA), **9** (No 4), 70 (April 1979)

53 WESTWOOD, D. H., 'Customer premises RF terminals for the SBS system', *ICC 79 Int. Conf. on Communications*, 6.3/1–5 (1979)

54 FLEURY, L., GUENIN, J. P. and RAMAT, P., 'The TELECOM 1 system', *Echo Rech.* (*France*), No 101, 11–20 (July 1980)

55 GRENIER, J., POPOT, M., LAMBARD, D. and PAYET, G., 'TELECOM 1, A national satellite for domestic and business services', ICC '79, *1979 Int. Conf. on Communications, Boston, MA* (*USA*), 10–14 June 1979, 49.5/1 (1979)

56 HOGWOOD, P., 'PALAPA — Indonesia to the fore', *J. Br. Interplanet Soc.* (*GB*), **30** (No 4), 127–130 (1977)

57 SOEWANDI, K. and SOEDARMADI, P., 'Telecommunications in Indonesia', *I.E.E.E. Trans. Commun.* (*USA*), **COM-24** (No 7), 687–690 (1976)

58 TENGKER, J. S., 'Indonesian domestic satellite system', EASCON '76, *Proc. of Eastern Electronics Conf., Washington D.C.* (*USA*), 11-A to 11-U (1976)

59 BLYTH, R. and HALDEMAN, D., 'TDRSS multiple access telecommunications service', *AIAA 8th Communications Satellite Systems Conf., Orlando FL* (*USA*), 20–24 April 1980, 317–327 (1980)

60 WATSON, J. R., 'L-SAT: A multipurpose satellite for the 1980s and 1990s', *IEE Colloquium on Satellite Broadcasting, London, England, 20 November 1980*, 4/1 (1980)

61 BARTHOLOME, P. and DINWIDDY, S., 'European Satellite Systems for Business Communications', ICC '80, *1980 Int. Conf. on Communications, Seattle WA* (*USA*), 8–12 June 1980, 51.5/1–7 (1980)

62 LORIORE, M., 'Communication satellite payloads: A review of past, present and future ESA developments', Technology Growth for the 80s, *IEEE MTT-S International Microwave Symposium Digest, Washington DC* (*USA*), 28–30 May 1980, 189–191 (1980)

63 IEE LONDON, 'Maritime and aeronautical satellite communication and navigation', *Int. Conf. I.E.E. London* (March 1978)

64 CCIR, 'Conclusions of the Interim Meeting of Study Group 8 (Mobile Services), *ITU Geneva* (1980)

65 CCIR, 'Recommendations and Reports of the CCIR (Mobile Service), Vol 8, *ITU Geneva* (1978)

66 LIPKE, D. W. *et al.*, 'MARISAT-A maritime satellite communications system', *COMSAT Technical Review*, **7** (No 2), 351–391 (1977)

67 HMSO London, 'Direct broadcasting by satellite', Report of Home Office Study London, *Her Majesty's Stationery Office* (1980)

68 CCIR Report 215-4, 'Systems for the broadcasting-satellite services (sound and television), *CCIR*, **11**, 163–185, ITU Geneva (1978)

69 ITU, Final Acts, 'World Broadcasting – satellite administrative radio conf.', *ITU Geneva* (1977)

70 SIOCOS, C. A., 'Broadcasting satellite reception experiment in Canada using high power satellite HERMES', *IBC 78 I.E.E. Cont. Publ.* (*UK*), **166**, 197–201 (1978)

71 ISHIDA, *et al.*, 'Present situation of Japanese satellite broadcasting for experimental purposes', *I.E.E.E. Trans.*, **BC-25** (No 4), 105 (1979)

72 KALE, P., and GRAUL, D. W., 'India's multi-purpose satellite', *Telecommunications*, 60–63 (1980)

73 REUDINK, D. O., 'Spot beams promise satellite communications breakthrough', *I.E.E.E. Spectrum*, 36–42 (1978)

74 FUKETA, H., KURAMOTO, M., INOVE, T. and KATO, E., 'Design and performance of 30/20 GHz band Earth stations for domestic satellite communication system', *AIAA Communications*

Satellite Systems Conference, 361–369 (1980)
75 KRIEGL, D. O. W. and BRAUN, H. M., '20/30 GHz satellite business system in Germany', *AIAA Communications Satellite Systems Conf.*, 634–639 (1980)
76 BERRETTA, G., 'Outlook for satellite communications at 20/30 GHz', *ESA Journal*, **3** (1979)
77 TIRRO, S., 'Utilisation of the 20/30 GHz spectrum part for high capacity national communications', *Int. Conf., Space Telecommunications and Radio Broadcasting*, Toulouse, France (1979)
78 FORDYCE, S. W. and JAFFE, L., 'Future communications concepts, switchboard in the sky', *Satellite Communications, Feb/March 1978* (1978)
79 MORGAN, W., 'Large geostationary communications platform', *XXX Congress IAF, Munich FR6, September 1979* (1979)
80 VISHER, P., 'Satellite clusters', *Satellite Communications (USA)*, 22–27 September (1979)
81 IEE London, '2nd Int. Conf. on antennas and propagation, Part 2, Propagation', I.E.E. London (April 1981)
82 JOHNS, P. B. and ROWBOTTOM, T. R., *Communication Systems Analysis*, Butterworths, London (1972)
83 CCIR, 'Methods for determining interference in terrestrial radio-relay systems and systems in the fixed satellite service', *CCIR Report 388-3*, Vol. IX, 340–357, ITU Geneva (1978)
84 CCIR, 'Technical factors influencing the efficiency of use of the geostationary satellite orbit by radiocommunication satellites sharing the same frequency bands', *CCIR Report 453-2*, Vol. IV, 181–206, ITU Geneva (1978)
85 HUTTER, J., 'Customer response to telephone circuits routed via a synchronous-orbit satellite', *Post Office Elec. Engr. J.*, **60**, part 3, 181 (1967)
86 CAMPANELLA, S. J., SUYDERHOUD, H. G. and ONUFRY, M., 'Analysis of an adoptive impulse response echo cancellor', *Comsat Technical Review*, Vol. 2, No 1, 1 (1972)

Bibliography

PELTON, J. N. and SNOW, M. S., *Economic and Policy Problems in Satellite Communications* (*Economic and Political Issues of the First Decade of INTELSAT*, Praeger Publications (1977)
SCHMIDT, W. G. and LAVEAN, G. E., *Communication Satellite Development Technology*, American Institute of Aeronautics (1976)
SNOW, M. S., *International Commercial Satellite Communications* (*Economic and Political Issues of the First Decade of INTELSAT*). Praeger Publications (1976)
UNGER, J. H. W., *Literature Survey of Communication Satellite Systems and Technology*, I.E.E.E. Press (1976)
VAN TREES, H. L., *Satellite Communications*, I.E.E.E. Press (1979)
MIYA, K. (Ed), *Satellite Communications Engineering*, Lattice Co, Tokyo (1975)
LOVE, A. W. (Ed), *Reflector Antennas*, I.E.E.E. Press, The Institute of Electrical and Electronic Engineers Inc, New York (1978)
CHAYES, A. *et al.*, *Satellite Broadcasting*, Oxford University Press (1973)

56 Point-to-point Communication

P J Howard, BSc, CEng, MIEE
Transmission Products Division,
Standard Telephones and Cables PLC
(Sections 56.5–56.11)

T Oswald, BSc, MIEE
Submarines Systems Division,
Standard Telephones and Cables PLC
(Sections 56.19–56.20)

Sydney F Smith, BSc(Eng), CEng, MIEE
Switching Main Exchange Products Division,
Standard Telephones and Cables PLC
(Sections 56.1–56.4)

D B Waters, BSc(Eng)
Design Manager,
Transmission Products Division,
Standard Telephones and Cables PLC
(Sections 56.12–56.18)

Contents

56.1 Telephone instrument (subset)

The speech transmission elements of a telephone instrument, *Figure 56.1*, consist of a receiver and transmitter (microphone) usually assembled in a common mounting to form a handset and connected to the line through an anti-sidetone induction coil (hybrid transformer). The complete instrument also incorporates signalling elements, the exact form of which depends upon the type of exchange with which it is designed to interwork. Generally they comprise a dial or push buttons and associated gravity switch for signalling to the exchange and a ringer (bell) for receiving calling (ringing) signals from the exchange.

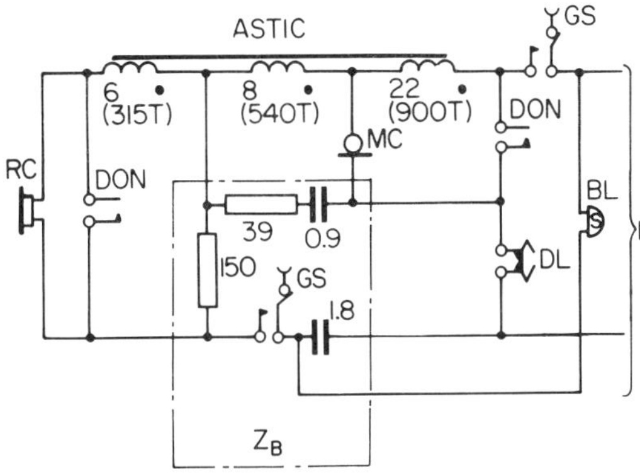

Figure 56.1 Typical telephone instrument circuit

56.1.1 Transmitter

Most instruments employ a carbon microphone. Sound waves cause the diaphragm to vibrate, varying the pressure exerted on the carbon granules. This produces corresponding variations in the electrical resistance between the granules and hence in the current flowing through them when connected to a d.c. supply.

Power may be derived from a battery or other d.c. supply locally at the instrument but in public systems it is now the universal practice for a single central battery to provide a supply to all lines on the exchange. The a.c. component of the line current produced in this way is known as the speech current and the d.c. component derived from the battery is the microphone feed current (transmitter feed) or polarising current.

The feed to each line is decoupled by individual inductors, often in the form of relay coils, and speech is coupled from one line to another through the exchange switching apparatus by means of capacitors as shown in *Figure 56.2* or through a transformer. The combination of d.c. feed, battery decoupling and speech coupling components is known as a transmission bridge or feeding bridge. The relays providing the battery decoupling also serve to detect line signals to control the holding and releasing of the connection.

The microphone resistance is non-linear and is subject to wide variations due to temperature, movement of granules and, of course, the effects of sound. Typically it is between 40 and 400 Ω under normal working conditions.

Some telephone instruments employ other types of microphone (e.g. moving coil) which do not themselves require a d.c. feed. Such microphones, however, lack the inherent amplification of the carbon microphone. They are usually supported by a transistor amplifier driven by the line current and designed to operate with similar feeding bridge conditions to carbon microphone instruments.

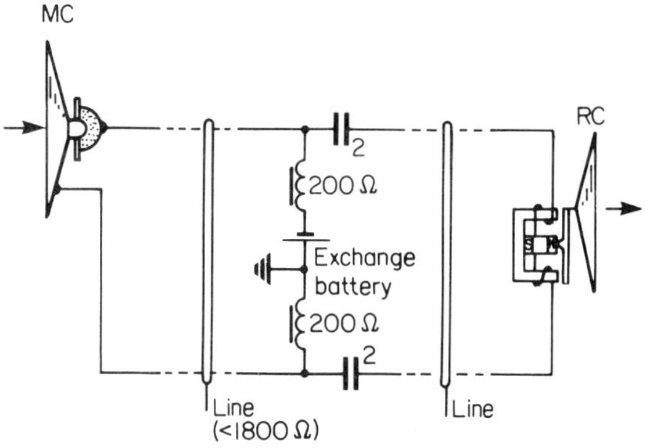

Figure 56.2 Principle of transmission bridge

56.1.2 Receiver

Most types operate on the moving-iron principle employed by Alexander Graham Bell for both transmitting and receiving in his original telephone in 1876. In some cases the diaphragm is operated on directly by the magnets but a more efficient arrangement is shown in *Figure 56.2*, in which the design of armature and diaphragm can each be optimised for its own purpose without conflict of mechanical-acoustic and magnetic requirements. A permanent magnet is included in the simple receiver to prevent frequency doubling due to the diaphragm being pulled on both the positive and the negative half cycle. In the rocking armature arrangement shown in *Figure 56.1*, no movement of the armature would occur at all without the permanent magnet.

Unlike the carbon microphone, the moving iron receiver requires no direct current for its operation and is usually protected against possible de-polarisation by a series capacitor.

56.1.3 Anti-sidetone induction coil (ASTIC)

This is a hybrid transformer which performs the dual function of matching the receiver and transmitter to the line and controlling sidetone.

Sidetone is the reproduction at the receiver of sound picked up by the transmitter of the same instrument. It occurs because the transmitter and receiver are both coupled to the same two-wire line. The most comfortable conditions for the user are found to be when he hears his own voice in the receiver at about the same loudness as he would hear it through the air in normal conversation. Too much sidetone causes the talker to lower his voice, reduces his subsequent listening ability and increases the interfering effect of local room noise at the receiving end. The complete absence of sidetone, however, makes the telephone seem to be 'dead'.

The use of a hybrid transformer divides the transmitter output between the line and a corresponding balance impedance. If this impedance exactly balanced that of the line there would be no resultant power transferred to the receiver. In practice there is some transfer to produce an acceptable level of sidetone.

56.1.4 Gravity switch (switch hook)

To operate and release relays at the exchange and thus indicate calling and clearing conditions, a contact is provided to interrupt the line current. This contact is open when the handset is on the rest and closes (makes) when the handset is lifted. Because it is operated by the weight of the handset it is known as the gravity switch. It is also known as a switch hook contact and the terms off-hook and on-hook are often used to describe signalling conditions corresponding to the contact closed or open respectively.

56.1.5 Dial

This is a spring operated mechanical device with a centrifugal governor to control the speed of return. Pulses are generated on the return motion only, the number of pulses up to ten depending upon how far round the finger plate is pulled.

The pulses consist of interruptions (breaks) in the line current produced by cam-operated pulsing contacts. A set of off-normal contacts operate when the finger plate is moved and are used to disable the speech elements of the telephone during dialling. These dial pulses are usually specified in terms of speed and ratio and the standard values of these parameters are designed to match the performance of electromechanical exchange switching equipment, *Figure 56.3*.

56.1.6 Push-button keypad

For use with exchanges designed for dial pulse signalling there are telephones in which a numerical keypad is used to input numbers to an l.s.i. circuit which then simulates the action of a dial. The output device for signalling to line is usually a mercury-wetted reed relay.

For electronic exchanges, and some types of electromechanical exchange, a faster method of signalling known as multifrequency (m.f.) can be used. In this, a twelve-button keypad is provided with oscillators having frequencies as shown in *Figure 56.4*. Pressing any button causes a pair of frequencies to be generated, one from each band.

56.1.7 Ringer (bell)

In the idle state, the handset is on its rest, the gravity switch is open and no direct current flows in the line. The ringer is therefore designed to operate on alternating current. This is connected from a common supply at typically 75 V r.m.s. and 25 Hz connected through a 500 Ω resistance at the exchange.

56.2 Telephone networks

Some use is made of telephones interconnected to form self-contained private networks but the majority of the world's telephones are connected to the public switched telephone network (PSTN).

56.2.1 Types of customer (subscriber) installation

Telephones are installed at customers' (subscribers') own premises or made available to the general public at call offices (paystations) with coin collecting boxes (CCB). Each of these installations is normally connected to a local exchange by a pair of wires but in isolated locations radio links may be used. In some cases more than one customers' instrument may be connected to the same line which is then described as a party line. In some

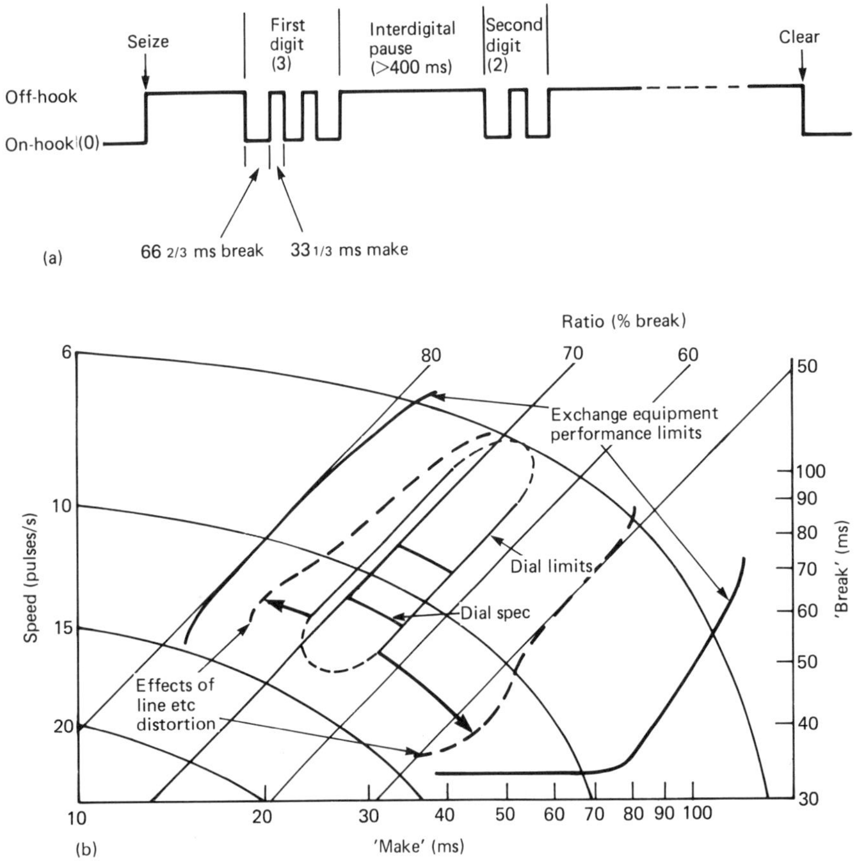

Figure 56.3 Telephone subscriber's line signalling: (a) waveform; (b) performance curves

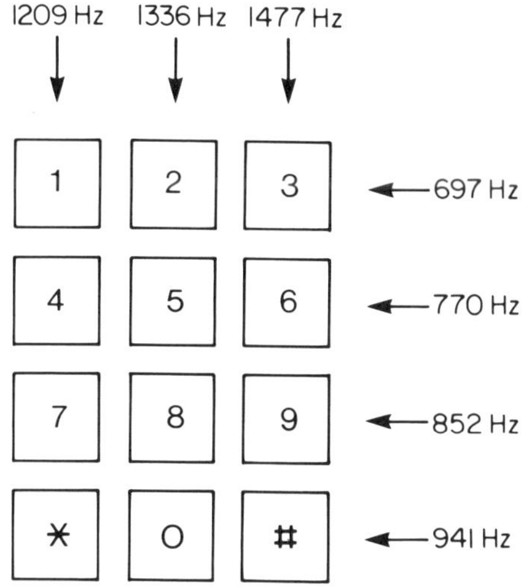

Figure 56.4 Push-button (multi-frequency) dialling

countries a party line, especially in remote rural areas, may serve as many as twenty customers with a system of coded ringing signals. In the UK the only party lines are those serving two customers on a system known as shared service. The case of one customer per line is known as exclusive service.

At the customer's premises there can be more than one telephone. Residential and small business users often have two or more instruments with one of several single methods of interconnection known as extension plans. Larger businesses generally have private branch exchanges (PBX) on their premises providing connection between their extensions either under control of a switchboard operator (private manual branch exchange—PMBX) or by dialling (private automatic branch exchange—PABX).

These private exchanges are connected to their local exchange by a group of exchange lines and the public exchange selects a free line within the group (PBX hunting) when a caller dials the first line of the group.

56.2.2 Hierarchy of exchanges

Each local public exchange provides the means for setting up calls between its own customers' lines. It is normally also connected to other local exchanges in the same area (e.g. the same town) by junctions. For calls outside this area, junctions are provided to trunk exchanges which are in turn interconnected by trunk lines enabling calls to be established to customers in other areas. Trunk lines are also provided to one or more international exchanges which enable calls to be established to customers in the national networks of other countries.

A single exchange can sometimes combine two or more of these functions and the larger national networks have more than one level of trunk exchange in their hierarchy (e.g. three levels in the UK, four in the USA). A trunk call in the UK therefore passes through up to six trunk exchanges in addition to the local exchanges at each end.

In some large areas, some local exchanges may be interconnected only by having calls switched through another local exchange, a technique known as tandem switching which can be independent of the trunk switching hierarchy used for calls outside the area.

56.2.3 Signalling

To connect calls through a network of exchanges it is necessary to send data referring to the call between the exchanges concerned.

In the simplest case this consists of loop/disconnect signalling similar to that used on customers' lines, *Figure 56.3*. Each speech circuit consists of a pair of copper wires which have direct current flowing when in use on a call and this current is interrupted to form break pulses corresponding to those produced by the telephone dial. When the called party answers, his exchange sends back an 'answer' signal to the calling exchange by reversing the direction of current flow.

Longer circuits are unsuitable for direct current signalling, e.g. due to the use of amplifiers, multiplexing equipment or radio links. Voice frequency (VF) signalling is then used in which pulses of alternating current are used at frequencies and levels compatible with circuits designed for handling speech currents (typically 600, 750, 2280 and 2400 Hz). This use of frequencies within the speech band (300–3400 Hz) is termed in-band signalling.

An alternative is the use of out-of-band (outband) signalling in which a signalling frequency outside the speech band, e.g. 3825 Hz is separately modulated on to the carrier in a frequency division transmission system.

In PCM transmission systems, the signals are digitally encoded and certain timeslot(s) are reserved for this purpose separate from the speech timeslots.

These are all channel-associated signalling systems in which the signalling is physically and permanently associated with the individual speech channel even though no signalling information is transmitted during the greater part of the call. The use of stored program control has given rise to a more efficient signalling means known as common-channel signalling. In this, one signalling channel serves a large number of speech channels and consists of a direct data link between the control processors (computers) in the exchanges concerned. Each new item of information relating to a call is sent as a digital message containing an address label identifying the circuit to which it is to be applied.

56.2.4 Manual exchanges

The early exchanges were all manually operated and many are still in service around the world. The exchange equipment consists of a switchboard at which an operator (telephonist) sits. The simplest type is the magneto board on which the caller gains the attention of the operator by using a hand generator to send an a.c. signal which is detected by an electromechanical indicator at the switchboard. An improved design is the central battery (CB) board which uses gravity switch contacts at the telephone to operate a relay at the exchange to light a lamp at the switchboard. In both cases the caller then tells the operator verbally what call he wants to make.

56.2.5 Automatic and auto-manual exchanges

The term automatic is used to describe exchanges where the switching is carried out by machine under the remote control of the caller who could be either a customer or an operator. Even on these systems there are some calls which customers either cannot or will not set up for themselves. To provide for these, numbers are allocated (e.g. 100) which can be dialled for assistance. Such calls are connected through the automatic exchange to a switchboard where the operator can establish the required connection also through the automatic equipment. These switchboards are called automanual exchanges.

56.3 Automatic exchange switching systems

The first public automatic exchange opened in 1892 at La Porte, Indiana (USA), and used rotary switches invented in 1891 by Almon B Strowger, an undertaker in Kansas City, after whom the system is named. Since then several systems have been developed. The most successful of these use mechanical devices called crossbar switches or arrays of reed relays. The most recent of these systems use stored program control (s.p.c.) in which the operations to set up each call are determined by software in the exchange processors (computers).

The new generation (digital) systems employ totally solid-state switching as well as stored program control.

56.3.1 Strowger (step by step)

Ratchet-driven switches remotely controlled directly by pulses from the calling telephone are used to simulate the operator's actions on a plug and cord switchboard. The wipers (moving contacts) are connected by short flexible cords and perform a similar function to the switchboard plugs. The place of the jack field is taken by a bank of fixed contacts and the wipers are moved one step from each contact to the next for each pulse, as in *Figure 56.5*, by means of a ratchet and pawl mechanism.

At first, Strowger's exchange provided each customer with his own 100 outlet two-motion selector. The expense of providing one large switch or selector for each customer is now avoided by using a stage of smaller cheaper uniselectors to connect the callers to the selectors as and when they require them. Expansion of the exchange beyond one hundred lines is by adding further stages of selectors as shown in *Figure 56.5*.

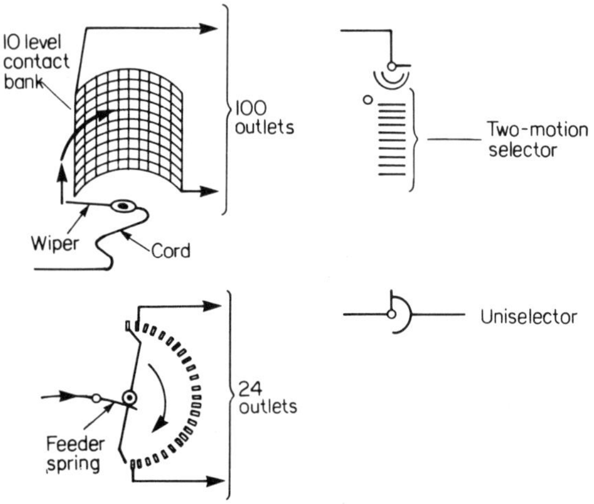

Figure 56.5 Strowger (step-by-step) selectors

The setting up of a call progresses through the exchange stage by stage. Each selector connects the call to a free selector in the next stage in time for the train of pulses to set that selector. The wipers step vertically to the required level under dial pulse control at 10 pulses per second then rotate automatically at 33 steps per second to select a free outlet by examining the potential on each contact. This is step-by-step operation and when used with Strowger selectors, it needs only simple control circuits which can be provided economically at each selector.

A large city often has hundreds of exchanges with overlapping service areas where it would be unacceptable for callers to have to consult a different list of dialling codes according to which exchange the telephone they happened to be using is connected. To allow the same list of three-digit codes to be used anywhere in the area, director equipment is provided, *Figure 56.7*, which translates the code dialled by the caller into the actual routing digits required.

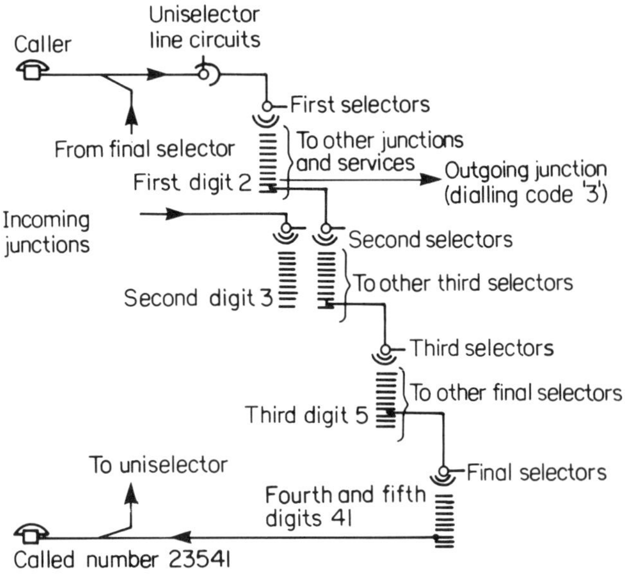

Figure 56.6 Strowger (non-director) exchange trunking

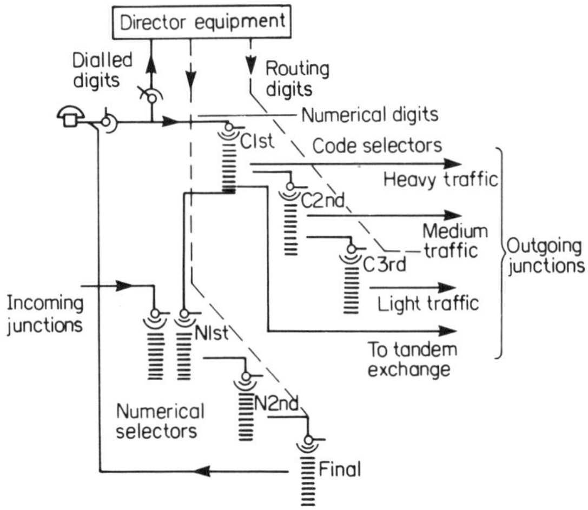

Figure 56.7 Strowger director exchange trunking

56.3.2 Crossbar

The crossbar switch is mechanically simpler than the rotary switches of the Strowger system and can be designed for greater reliability and smaller size. They cannot be directly controlled by dial pulses, however, so more complex control circuits are required which are therefore shared over more than one switch, a method of operation known as common control. Translation facilities like those provided by the director system, but not limited by considerations of mechanical switch sizes, are readily obtained because there is no inherent relationship between number allocation and particular switch outlets.

The switch mechanism consists of a rectangular array of contact sets or crosspoints which can be selected by marking their vertical and horizontal co-ordinates.

Figure 56.8 shows a much simplified diagram of a single crosspoint and a general view of a complete switch. The flexible

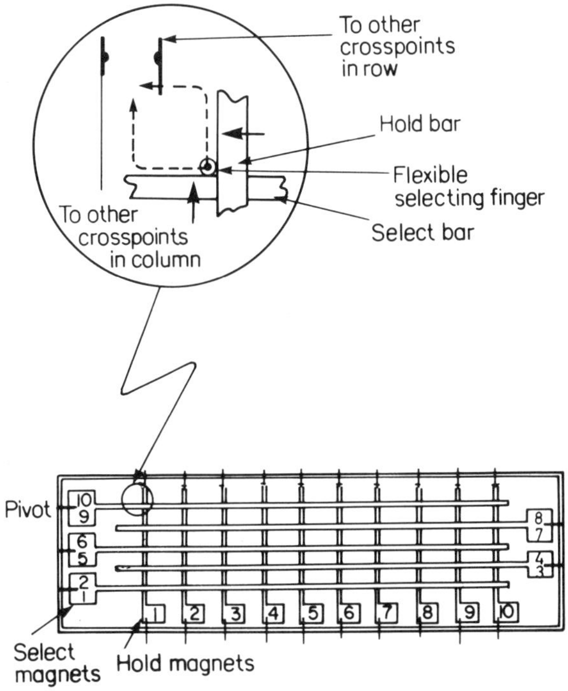

Figure 56.8 Crossbar switch

selecting finger should be thought of as a stiff piece of wire sticking up out of the paper. The horizontal (select) and vertical (hold) bars move in the direction shown by the arrows when their associated coils are energised. It can be seen that the crosspoint contact will be operated only when the horizontal bar 'selects' before the vertical bar operates. Once operated, the contact remains held by the vertical bar which traps the finger and holds it even when the horizontal bar has been restored. The horizontal bar can be used again to select another crosspoint in the row without disturbing those already in use.

In the most usual form of construction, the movement of the bars is produced by a slight rotation about the pivot in a manner similar to the armature of a telephone relay. The spring finger is attached to the select bar and swings across when the bar rotates.

56.3.3 Crosspoint system trunking

To provide an exchange capable of serving say 10 000 customers it would not be economic either to construct a single large cross bar switch or simply to gang a large number of smaller switches to form a square matrix of 10 000 × 10 000 crosspoints. A more practical arrangement is to connect the switches in two or more stages. To illustrate this *Figure 56.9* shows first a single square matrix of 81 crosspoints providing for nine paths between nine inlets and nine outlets and then how to meet the same requirement with only 54 crosspoints. The advantage of this

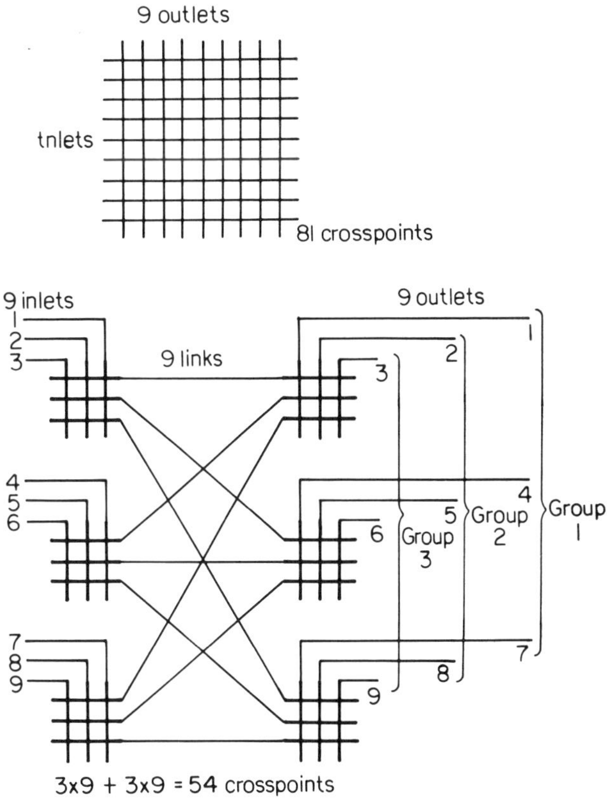

Figure 56.9 Trunking principle of distribution stage for crossbar and electronic exchanges

approach is even greater with larger numbers of circuits and practical crossbar switches having typically 10 × 20 crosspoints or more, rather than only 3 × 3 as shown in the example.

This simple example provides only for the distribution of calls from the inlets to an equal number of outlets through the same number of links. In practical exchanges it is only necessary to provide as many paths through the switching network as the number of calls which are expected to be in progress at one time. The concentration of lines on to a smaller number of paths or trunks can be achieved using the principle illustrated by another simple example in *Figure 56.10*.

A complete local exchange then consists of a combination of a distribution (group selection) stage and one or more line concentration units as shown in *Figure 56.11*. To establish a call, the line circuit LC detects the loop calling or 'Seize' condition (*Figure 56.3*) and signals the identity of the calling line to the exchange control. In analogue systems, the usual method of operating is that the control sets the crosspoints to connect the line to a register which detects the dial (or key) pulses and signals them to the control. In digital electronic systems, the dial pulses are usually detected at the line circuit. In either case the control, which is duplicated or sectioned in some way for security, analyses the digits received and sets the appropriate crosspoints to establish the call.

56.3.4 Analogue electronic exchanges

Typical systems of this sort are TXE2 and TXE4 used in the UK for small (up to 2000 lines) and large exchanges respectively.

This type of system employs a switching matrix having the

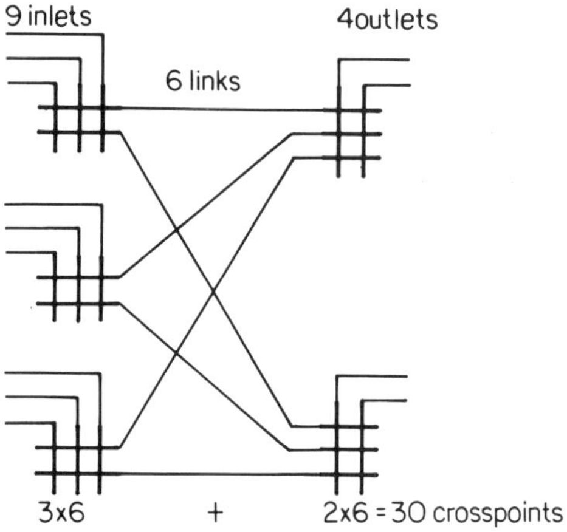

Figure 56.10 Trunking principle of concentration stage for crossbar and electronic exchanges

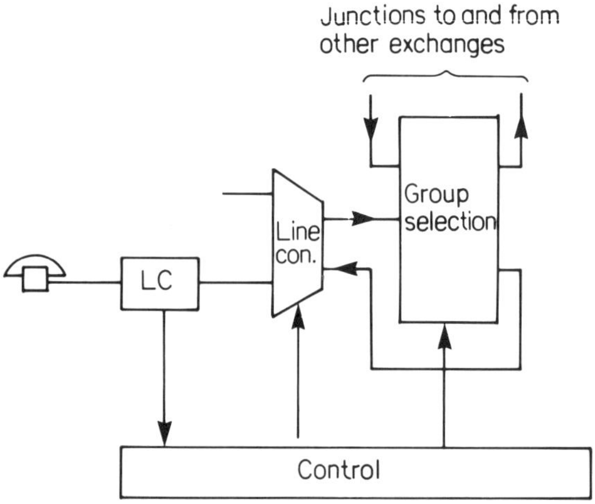

Figure 56.11 Generalised architecture of a complete crossbar or electronic exchange

same structure as crossbar systems (*Figures 56.9*, *56.10* and *56.11*). Public exchanges usually employ reed relays for the crosspoints. These have the advantage of being sealed against corrosion and dirt and they operate in only about 2 ms making them more amenable to electronic control than Strowger or crossbar switches. They also have a separate operating coil for each crosspoint enabling switch sizes to be optimised for system requirements.

The higher speed of operation and the use of electronic control permits greater use of self-checking features to improve reliability and quality of service. The greater use of electronics enables these systems to benefit from advancing technology in that field in terms of size and versatility in the provision of customer facilities.

There are systems using solid state crosspoints such as thyristors but these are confined mainly to PABX applications where crosstalk, transmission and overvoltage requirements are often less onerous than on public systems.

56.3.5 Space division and time division switching

Exchanges such as Strowger, crossbar and reed relay systems employ space division switching in which a separate physical path is provided for each call and is held continuously for the exclusive use of that call for its whole duration.

Digital and other pulse modulated systems employ time division switching in which each call is provided with a path only during its allocated time slot in a continuous cycle. Each switching element or crosspoint is shared by several simultaneous calls each occupying a single time slot in which a sample of the speech is transmitted. This sample may be a single pulse as in pulse amplitude modulated systems or a group of pulses as in pulse code modulated systems.

The range of exchangers known as System X recently introduced in the UK are in this category. Speech is pulse code modulated and transmitted through the network in 8-bit bytes at 64 k bit s^{-1}.

56.3.6 Digital trunk exchange

Digital transmission systems are multiplexed in groups usually of 24 or 30 speech channels on each link. The trunk exchange has to be able to connect an incoming channel in one PCM link to an outgoing channel in another PCM link. Received speech samples must therefore be switched in space to the appropriate link and translated in time to the required channel time slot. *Figure 56.12* shows in simplified form a typical arrangement known as T–S–T (time-space-time). A practical system would have some 1024 time slots per speech store instead of the four shown and a space switch of perhaps 96 × 96 instead of the 5 × 5 in the diagram.

The first time switch connects to a cross-office time slot through the space switch which itself operates in a time division switching mode. The second time switch connects to the time slot required on the outgoing PCM link. The connection illustrated in *Figure 56.12* is from incoming channel 2 to outgoing channel 4 via crosspoint row 3 at cross-office time slot '*t*'.

Each time switch operates by storing received speech samples in the order in which they are received and reading them out of store in a different order according to which time slot they are to be connected.

The arrangement shown carries speech in one direction only. All receive channels are connected to one side of the switch, regardless of whether the circuits are incoming or outgoing in the traffic sense, and all 'transmit' channels are connected to the other side. Each call requires transmission in each direction and these two paths occupy the same crosspoint in the space switch but at different time slots usually 180° out of phase. In the simple example shown time slot $t+2$ would be used for the corresponding opposite direction of transmission.

56.3.7 Digital local exchange

The T–S–T switching network of *Figure 56.12* provides only for distribution, like the crosspoint network of *Figure 56.9*. A complete local exchange needs also a concentration stage. Telephone customers' lines are generally on individual pairs employing analogue transmission. First generation electronic exchanges have analogue concentration stages, *Figure 56.10*, so that the relatively expensive digital encoding and multiplexing is provided after concentration. Cost trends in electronic components are now beginning to favour the provision of the conversion on a per line basis so that digital concentration stages can be provided in the form of an additional time switching stage in which the available time slots to the main T–S–T switch is less than the total number of input time slots and store locations in the concentrator speech stores they serve. This might consist typically up to 16 groups of 256 lines, each written cyclically into a 256 word store but with only 256 time slots available for allocation to the acyclic read out from all these stores.

The interface to the analogue customer's line includes certain functions which require voltage and power levels incompatible

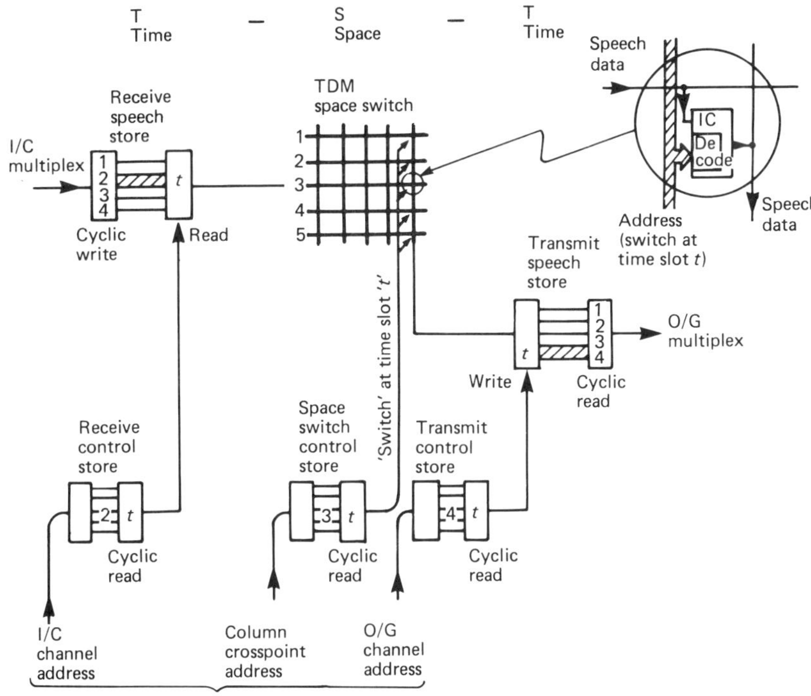

Figure 56.12 Principle of time–space–time switching

with the digital concentrator switch and which therefore have to be provided at the individual line circuit (line card). The line interface functions are generally referred to as the Borscht functions:

Battery feed to line.
Overvoltage protection.
Ringing current injection and ring trip detection.
Supervision (on-hook/off-hook detection).
Codec for analogue to digital conversion.
Hybrid for two-wire to four-wire conversion.
Test access to line and associated line circuit.

56.3.8 Integrated networks

The combination of digital exchanges and digital (PCM) junctions and trunk lines constitutes an integrated digital network (IDN) and has advantages of economy and improved transmission quality by avoiding the need for frequent analogue to digital conversions.

The availability of end-to-end digital circuits enables data at rates up to 64 kbit s^{-1} to be transmitted over the same network as speech without the use of modems. The joint use of the network in this way to provide voice and non-voice services constitutes an integrated services digital network (ISDN).

56.4 Teletraffic

56.4.1 Grade of service

It would be uneconomic to provide switching equipment in quantities sufficient for all customers to be simultaneously engaged on calls. In practice, systems are engineered to provide an acceptable service under normal peak (i.e. busy hour) conditions. For this purpose, a measure called the grade of service (g.o.s.) is defined as the proportion of call attempts made in the busy hour which fail to mature due to the equipment concerned being already engaged on other calls. A typical value would be 0.005 (one lost call in 200) for a single switching stage. To avoid the possibility of serious deterioration of service under sudden abnormal traffic it is also usual to specify grades of service to be met under selected overload conditions.

56.4.2 Traffic units

Traffic is measured in terms of a unit of traffic intensity called the Erlang (formerly known as the traffic unit or TU) which may be defined as the number of call hours per hour (usually, but not necessarily the busy hour), or call seconds per second etc. This is a dimensionless unit which expresses the rate of flow of calls and, for a group of circuits, it is numerically equal to the average number of simultaneous calls. It also equates on a single circuit to the proportion of time for which that circuit is engaged, and consequently to the probability of finding that circuit engaged (i.e. the grade of service on that circuit). The traffic on one circuit can, of course, never be greater than one Erlang. Typically a single selector or junction circuit would carry about 0.6 Erlangs and customers lines vary from about 0.05 or less for residential lines to 0.5 Erlang or more for some business lines.

It follows from the definition that calls of average duration t seconds occurring in a period of T seconds constitute a traffic of A Erlangs given by:

$$A = Ct/T \qquad (56.1)$$

An alternative unit of traffic sometimes used is the c.c.s. (call cent second) defined as hundreds of call seconds per hour (36 c.c.s. = 1 Erlang).

56.4.3 Erlang's full availability formula

Full availability means that all inlets to a switching stage have access to all the outlets of that stage. Trunking arrangements in which some of the outlets are accessible from only some of the inlets are said to have limited availability.

For a full availability group of N circuits offered a traffic of A Erlangs, the grade of service B is given by

$$B = \frac{A/N!}{1 + A + A^2/2! \ldots A^N/N!} \qquad (56.2)$$

(This formula assumes pure chance traffic and that all calls originating when all trunks are busy are lost and have zero duration.)

56.4.4 Busy hour call attempts

An important parameter in the design of stored program control systems in particular and common control systems in general is the number of call attempts to be processed, usually expressed in busy hour call attempts (BHCA). Typically a 10 000 line local exchange might require a processing capacity of up to 80 000 BHCA.

56.4.5 Blocking

If two free trunks cannot be connected together because all suitable links are already engaged the call is said to be blocked. In *Figure 56.9*, if one call has already been set up from inlet 1 to outlet 1, an attempted call from inlet 2 to outlet 2 will be blocked The effects of the blocking are reduced here by allocation of the outlets to the three routes.

Blocking can be reduced by connecting certain outlets permanently to the inlet side of the network to permit a blocked call to seek a new path. Alternatively a third stage of switching may be provided.

It is possible to design a network having an odd number of switching stages in which blocking never occurs. Such non-blocking arrays need more crosspoints than acceptable blocking networks and are uneconomic for commercial telephone switching applications.

56.5 Analogue transmission by cable

56.5.1 International standardisation

All transmission facilities in public switched telephone networks are planned to achieve the recommended standards of performance set by the CCITT (the International Telephone and Telegraph Consultative Committee of the ITU). In principle, compliance with these recommendations relates only to international connections, but in order to ensure that any call set up from within a country to a point in another country is satisfactory, national connections are operated to the same standards.

The basic transmission qualities recommended by the CCITT include bandwidth, interface levels, noise, return loss and level stability. For specific equipments the CCITT recommend many other parameters to facilitate the setting up of international connections. It is important to recognise that the CCITT is constantly being requested to study aspects of the transmission network and seeks to reflect technological improvements in its recommendation. The current status can be found in the appropriate volumes of the preceding Pleanary Assembly. The current issue is the Yellow Book published by the ITU in Geneva, 1981, of which volume III relates to line transmission.

56.5.2 Terminology

The telecommunication terms used in this section can be explained as follows:

56.5.2.1 Decibel (dB)

This is the unit of power ratios, e.g.

$$P_1/P_2 = 10 \log_{10}(P_1/P_2) \text{ dB} \qquad (56.3)$$

If the impedance under discussion is the same it follows that

$$P_1/P_2 = V_1^2/V_2^2 = 20 \log_{10}(V_1/V_2) \text{ dB} \qquad (56.4)$$

dB is, therefore, the unit for expressing amplifier gains, cable losses etc.

56.5.2.2 Absolute power

The decibel becomes an absolute unit by reference to a power of 1 mW.

Thus 1 mW corresponds to 0 dBm
1 W corresponds to +30 dBm
1 μW corresponds to −60 dBm
1 pW corresponds to −90 dBm

56.5.2.3 Relative level

When analysing a chain of tandem connected equipments, the level at any one input will depend on prior conditions. However, the equipment in question will have been designed for a particular working level and tolerance range. The concept of relative level, dBr, is a means of expressing the designed operating conditions. Conventionally, an 0 dBm signal at the two-wire audio connection, would produce a signal of −10 dBm at a −10 dBr point. Likewise a −15 dBm signal would produce a −25 dBm signal at the same −10 dBr point.

56.5.2.4 Level ratios

It is convenient to express noise levels, pilot levels etc, relative to the nominal traffic level (i.e. the dBr level) at a point in the transmission path; e.g. a pi.ot level might be at −10 dBm0, which means 10 dB below traffic signal. Thus a −10 dBm0 pilot measured at a −33 dBr point would read −43 dBm. Similarly, interfering signals and noise can be expressed as dBm0 with respect to traffic levels. Thus an interfering signal of −50 dBm0 would measure −60 dBm at a −10 dBr point.

56.5.2.5 Psophometric weighting

The human ear does not respond equally to all audio frequencies and speech intelligence is primarily conveyed in the midrange frequencies of 500 to 1600 Hz. Thus interfering signals which fall outside this band can be higher than the mid-range frequencies. This property is quantified in the internationally agreed psophometric weighting network. Such a network is always included in voice frequency noise measuring equipment. The effect of the network on randomly distributed noise (white noise) is to reduce its power by 2.6 dB. Also, the noise in a 4 kHz band will be reduced by 1.1 dB when measured in the standard 3.1 kHz bandwidth (0.300 to 3.4 kHz).

Noise results, inclusive of psophometric weighting and bandwidth correction, are expressed as dBm0p. e.g. Noise in a 4 kHz bandwidth of −60 dBm0 is equivalent to −63.6 dBm0p in a 3.1 kHz telephony channel.

56.5.2.6 Hypothetical reference circuit

To assist in the allocation of noise between types of equipment a 'hypothetical reference circuit' has been conceived which is representative of a continental call. The circuit is 2500 km long and is made of nine homogeneous line sections, each 280 km long, with various combinations of carrier multiplex equipment at the intermediate points. The noise in any 3.1 kHz channel operating end to end over this circuit must not exceed 10 000 pW0p (−50 dBm0p). 2500 pW0p are allowed in the multiplex equipment and

7500 pW0p for the line. Thus all line transmission plant is designed to achieve a noise performance of not more than 3 pW0p per km. The multiplex equipment noise is allocated between the various translating equipments. For example, the first translation stage, the primary group equipment, is allocated 200 pW0p.

56.5.2.7 Four-wire working and phantom circuits

Local exchange connections to subscribers are two-wire line carrying both the 'go' and 'return' traffic.

For amplified and carrier circuits, four-wire working is normal, using a separate 'pair' for each direction of transmission.

A third independent 'phantom' circuit can be derived from two pairs as shown in *Figure 56.13*.

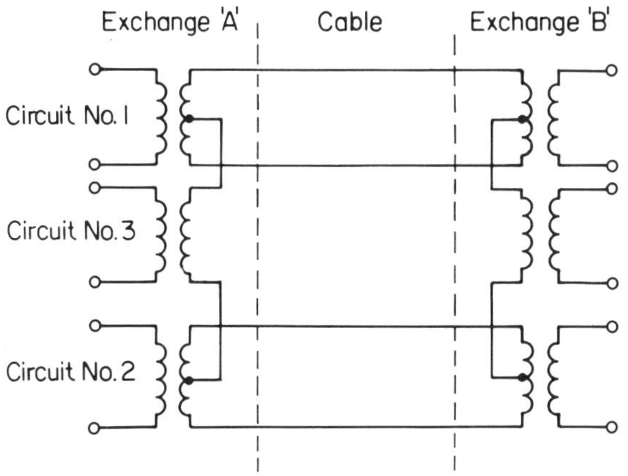

Figure 56.13 Derivation of phantom circuit (No. 3) using centre-tapped line coils

56.5.3 Voice frequency transmission

In addition to subscriber connections, voice frequency circuits are commonly used between nearby exchanges. Loading, the application of coils as a lumped inductance, is normally applied to reduce circuit attenuation and equalise attenuation.

56.5.3.1 Effect of loading

Attenuation is a function of the four primary cable parameters, *R*, *C*, *L* and *G* (leakance). In modern cables *L* and *G* are small and the attenuation approximates to

$$\tfrac{1}{2}R(C/L)^{1/2} \qquad (56.5)$$

By increasing *L*, the attenuation is reduced. The effect of loading is to reduce and flatten the low-frequency response, but attenuation rises rapidly beyond a cut-off frequency, given by

$$f_c = \pi^{-1}(LC)^{-1/2} \qquad (56.6)$$

The effective transmission bandwidth is taken as $0.75f_c$, so for a 3.400 kHz circuit f_c needs to be 4.500 kHz. Loading also increases the characteristic impedance to the value

$$Z_0 = (L/C)^{1/2} \qquad (56.7)$$

(where *L* and *C* relate to the loading interval).

56.5.3.2 Standard loading

Standard loading coil inductances are 22, 44, 66 and 88 mH, and a coil spacing of 1830 m is commonly used. *Figure 56.14* shows the

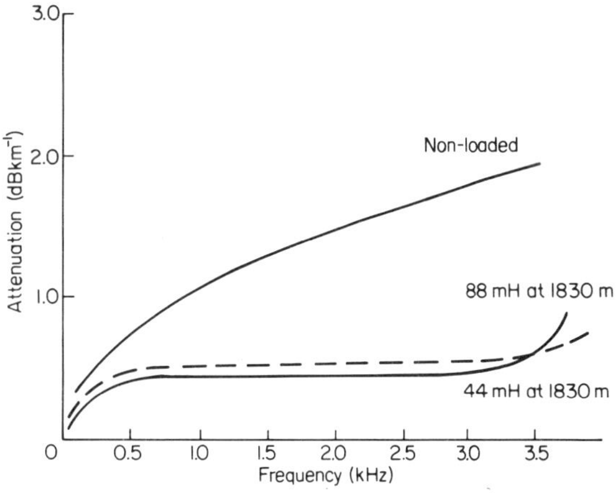

Figure 56.14 Effect of loading: 0.63 mm conductors of mutual capacitance 45 nF km^{-1}

loss of 0.63 mm paired conductors of mutual capacitance 45 nF km^{-1}, unloaded and with 44 mH and 88 mH loading.

56.6 Frequency division multiplex equipment

For the more efficient utilisation of buried copper cables, carrier techniques have been used extensively for very many years in both national and international telephone networks. The particular technique universally adopted is amplitude modulation. However, only one side band is transmitted and the carrier is suppressed. By this means a maximum number of channels can be contained within a frequency band and the absence of high-level carrier reduces the power loading on the equipment in the transmission path.

At the receiving end the incoming side band is modulated with a locally generated carrier. Consequently, both the transmitting and receiving carriers have to be generated to a great accuracy, such that the reconstituted voice frequency signal does not differ by more than one Hz from the original signal. The internationally defined primary, or basic, group comprises twelve 4 kHz channels assembled into a (60–108) kHz frequency band. Each channel provides an audio path extending from 300 to 3400 Hz.

In some designs a 3825 Hz outband signalling channel is also provided, but this clearly imposes a severer requirement on the channel separating filters.

56.6.1 Primary group equipment designs

The 12-channel group equipment is by definition the most commonly employed equipment in carrier networks and, hence, design cost effectiveness is particularly important. Consequently, several design approaches have evolved exploiting the economic advantages of technological developments. However, all these designs comply with the international recommendations and can, therefore, be interconnected.

56.6.2 Requirements

The CCITT recommendations (Yellow Book, Volume III, Rec. G232) are very comprehensive. Key features are as follows:

(*a*) The average attenuation/frequency characteristic measured through send and receive equipment connected back-to-back shall lie within the mask shown in *Figure 56.15*.

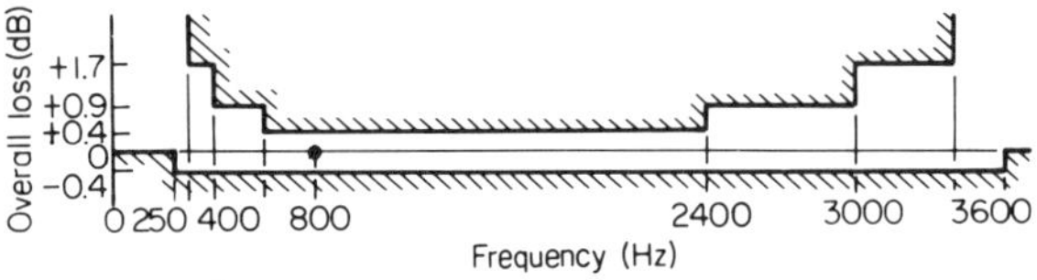

Figure 56.15 Limits for the average variation of overall loss of 12 pairs of equipment of one 12-channel terminal equipment

(*b*) Measuring at the HF out, no carrier signal in the (60–108) kHz band shall be greater than −26 dBm0. The aggregate shall be less than −20 dBm0, and no carrier signal greater than −50 dBm0 shall be present outside the (60–80) kHz band.

(*c*) Intelligible crosstalk between channels shall be less than 65 dB. Unintelligible crosstalk from an adjacent channel shall be suppressed by at least 60 dB.

(*d*) The permitted noise contribution is discussed in Section 56.6.2.

56.6.3 Crystal filter design approach

This design approach exploits the characteristics of quartz crystal filters. This equipment requires the provision of twelve carrier frequencies spaced at 4 kHz intervals from 64 to 108 kHz. The design of each channel unit is identical, apart from the pass band of its crystal filters.

The modulator is switched hard by the carrier signal, effectively by a square-law function.

If the

input signal $= E_1 \sin \omega t$ (56.8)

and the

carrier signal $= E_2 \sin pt$ (56.9)

then the output signal is given by

$E_1 \sin \omega t[\frac{1}{2} + 2\pi^{-1}(\sin pt + \frac{1}{3}\sin 3pt + \ldots)]$
= modulating signal + upper and lower side bands of the carrier frequency + upper and lower side bands of 3 times the carrier frequency ... (56.10)

An active modulator (namely, a transistor) requires lower carrier powers and provides good linearity over a wide range of signal power levels.

Audio filtering stages prior to the modulator restrict the traffic band to the range 0.3 to 3.4 kHz. The output crystal filters select the lower side band. In the receive direction, the incoming crystal filter selects the wanted channel and the post modulator filters suppress the unwanted side bands and carrier.

Typical crystal filter characteristics are shown in *Figure 56.15*. Clearly the most damaging crosstalk component would be impurity in the carrier supply. Typically, spurious signals which are multiples of 4 kHz, need to be 80 dB below carrier level.

56.6.4 Alternative design approaches

One technique exploits the stability, small size and low cost of modern coils and capacitors. In order to reduce the number of designs, use can be made of a pregroup, and, to facilitate filter design this 'pregroup' may be first modulated into a frequency spectrum most appropriate to the technology employed.

Figure 56.17 shows one method by which the standard basic group can be derived using these techniques. It will be noted that different carrier supplies are required, and as these are non-standard they are normally generated within the channel equipment from incoming standard carriers.

Several other methods are currently available, including the use of mechanical filters, crystal filters operating at 2 MHz etc. All equipments meet the internationally agreed requirements and can, therefore, be interworked.

56.7 Higher-order multiplex equipment

The same concepts used in the generation of the primary group of 12 channels in the 60–108 kHz band have been applied to build up a hierarchy of frequency allocations. Thus five 60–108 kHz primary groups are assembled into a 312–552 kHz basic supergroup of 60 channels.

The basic supergroup can in turn be modulated into the sixteen supergroup assembly extending from 60 to 4028 kHz. This is a

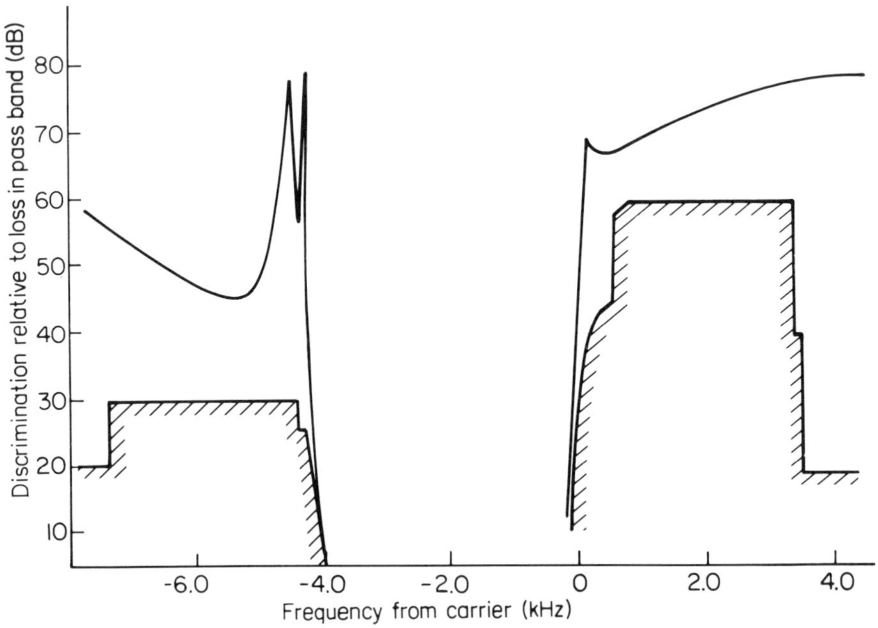

Figure 56.16 Typical channel filter characteristic

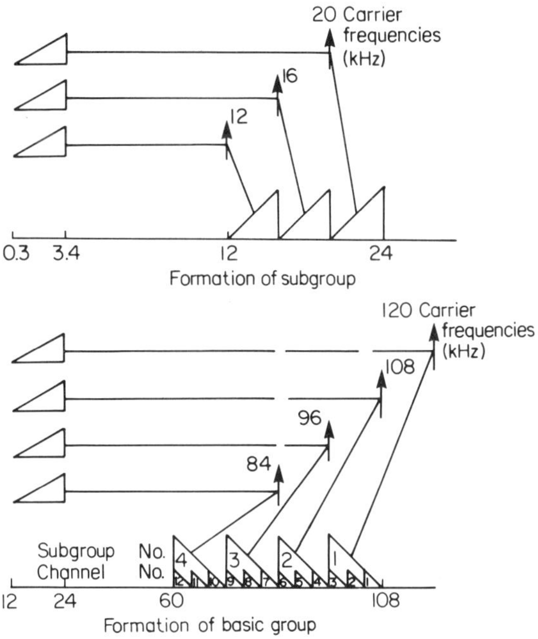

Figure 56.17 Example of basic group formation via a subgroup

commonly used hierarchical level for application to 4 MHz line systems.

The supergroup can also be assembled into a basic master group from 812 to 2204 kHz of 300 channels. Three master groups are assembled into the basic super master group from 8516 to 12 388 kHz of 900 channels. In the United States very similar allocations are used, except that the term master group applies to a 600 channel assembly.

In many countries, including the United Kingdom, it was not advantageous to adopt the master group and super master group levels. Instead a 15 supergroup assembly, extending from 312 to 4028 kHz is modulated up into the next stage.

The purpose of these hierarchical structures is to provide a modular approach to the assembly of increasing numbers of telephony channels. At most points in the hierarchy, line transmission systems are available to provide interconnections between stations.

56.8 Line transmission systems

The motivation for the development of line transmission system is to achieve lower cost per telephone circuit. The primary constraints in attaining this objective are the need to provide a transmission bandwidth conforming to one of the hierarchical steps previously discussed, and the need to use one of the cable designs which is internationally accepted.

A system which requires special translating equipment or a special cable will inevitably attract additional costs cancelling any 'per circuit' advantage. An exception is in the field of submarine cable systems where the unique environmental and commercial factors justify special designs almost on a per system basis. For land lines, cables are installed with a life expectancy of forty years and the transmission engineers task is to exploit this medium with ever increasing efficiency.

56.8.1 Cable designs

Copper pair cables have been used for low-capacity (12, 60, 120 channel) connections, but these techniques are obsolete. Modern FDM transmission systems are designed for use on coaxial pair cables. The coaxial cores standardised by the CCITT are large core, which has a 9.5 mm inside diameter of the outer conductor and a 2.6 mm diameter inner conductor, and small core for which the figures are 4.4 and 1.2 mm respectively. Both these cables are effectively air spaced as the inner is typically supported on relatively widely spaced polyethylene discs.

These standardised coaxial cores can be laid up with other conductors to form a great variety of cable to suit individual administration's needs. Typical designs would comprise some 6, 8 or 12 coaxial cores together with twisted pairs for supervisory use laid up in the interstitial spaces, with perhaps an outside layer of twisted pairs to provide local audio circuits. In some territories such a cable would be armoured and directly buried, in others, and in most urban areas, the armouring is omitted and the cable is installed in cable ducts.

56.8.2 Cable characteristics

The characteristics of these cables for analogue transmission is well defined, and recent work has extended this knowledge further in recognition of the need to apply high-bit-rate digital transmission systems to existing networks.

For FDM applications, the key characteristics are as follows:

(*a*) Large core cable:
Loss $=2.355\sqrt{F}+0.006F$ dB km^{-1} at 283 K (10°C)
Impedance $=75\pm3\ \Omega$ at 1 MHz
Temp coefficient of attenuation $=0.2\%$ per K (°C)
Far end crosstalk = greater than 130 dB

(*b*) Small core cable:
Loss $=5.32\sqrt{F}$ dB km^{-1} at 283 K (10°C)
Impedance $=75\pm1\ \Omega$ at 1 MHz
Temperature coefficient of attenuation $=0.2\%$ per K (°C)
Far end crosstalk = greater than 130 dB

It will be seen that the loss increases with the square root of the frequency. The linear element is very small. A useful approximation is that small core cable has $2\frac{1}{4}$ times more loss. Hence an equipment designed for small-core cable will operate on large-core cable at a spacing $2\frac{1}{4}$ times longer.

Both cables have a characteristic impedance of 75 Ω. The crosstalk performance between cores is excellent and not a practical limitation on system design. Such cables also have good immunity against external interference.

56.9 System design considerations

The principal design considerations are illustrated in the following paragraphs by reference to a 2700 channel system. The bandwidth of this system extends from nominally 300 kHz to 12.5 MHz. The design objective is to provide a stable transmission path with a flat frequency characteristic and negligible noise. This is accomplished by providing amplifying points (repeaters) at regular intervals along the cable.

56.9.1 Traffic path

56.9.1.1 Gain shaping

The recommended repeater or amplifier spacing for a 12 MHz system on 4.4 mm cable is 2 km, corresponding to a loss of about 40 dB at 12.5 MHz and 6 dB at 300 kHz. The usual design technique is to compensate for some of this slope with a shaped negative feedback network, i.e. minimum NFB is provided at the highest frequency, but not less than 15 to 20 dB in order to preserve the virtues that NFB brings.

It is not practical to put the whole cable shape in the feedback path and additional fixed networks, with zero loss at the top frequency, are used. By splitting these between input and output, good return losses are obtained at low frequencies and buffering to lightning surges from the cable is provided.

A long tandem connection of amplifiers requires equalisation at the terminal stations to compensate for any systematic error accumulating along the line.

56.9.1.2 Noise contribution

The noise contribution of each amplifier must never exceed 6 pW0p in any channel and a design target of 4 pW0p is set to ensure this. This will comprise say two thirds basic or thermal noise, and one third intermodulation noise.

$$\begin{aligned}\text{Thermal noise contribution} &= 2.66\ \text{pW0p}\\ &= -90+4.25=-85.75\ \text{dBm0p}\\ \text{Amplifier noise} &= -139+\text{NF}+\text{gain dBm0p}\\ &= -99+\text{NF dBm0p}\end{aligned}$$

Hence the amplifier noise figure must be better than 13 dB at a point of zero relative level. The amplifier cannot be operated at such a level because intermodulation would be too high.

For an output level of −10 dBr, the noise figure would need to be −3 dB. Thus knowing the intermodulation performance and noise figure it is possible to select line levels to give optimum performance. The calculation is complicated because some intermodulation products add on a square law basis. Hence the calculation is most conveniently done with computer and the form of a typical print out is indicated in *Table 56.1*.

The optimum line levels required at the line amplifier outputs can be set by introducing an appropriate network in the transmit terminal (the pre-emphasis network) and taking this slope out at the receiving station (de-emphasis network). The computer program can also be used to test the sensitivity of the design to level errors.

56.9.1.3 Overload performance

In addition to fulfilling the requirements for noise performance the amplifier must also tolerate the equivalent r.m.s. sine-wave power of the peak of the composite multiplex signal. This is given by:

$$P_{eq}=-5+10\log_{10}N+10\log_{10}(1+15/N^{1/2})\ \text{dBm0} \qquad (56.11)$$

where N in this case is 2700 channels and

$$P_{eq}=+30.4\ \text{dBm0}.$$

On a pre-emphasised system, this would apply at the frequency of mean power, typically about 8 MHz where the line level is −17 dBr.

Thus the overload requirement becomes +13 dBm. However, several decibels margin must be added to allow for misalignments and a better specification would be +20 dBm.

It is important to recognise that both the noise and overload requirements have to be met with any secondary lightning protection components connected. These typically comprise high-speed diodes connected across the signal transistors and can contribute to the intermodulation products.

56.9.2 Level regulation

The coaxial cable attenuation is subject to a temperature coefficient of 0.21% per K (°C). Cable is buried at about 80 cm and statistics gathered over many years show a mean annual temperature variation at this depth of ±10 K (°C). Thus a 2 km repeater section of small core cable will change by

$$40\times 0.0021\times 10=\pm 0.8\ \text{dB at 12.5 MHz.}$$

and by

$$6\times 0.0021\times 10=\pm 0.12\ \text{dB at 300 kHz.}$$

After a tandem connection of several repeaters this could accrue to an unacceptable level.

It is necessary to compensate for this effect. One solution is to include a temperature sensor in the repeater which controls an appropriately shaped network. This assumes that the local temperature is representative of the whole cable length and that the sensing element has very stable characteristics. Another technique is to send a pilot signal the whole length of the system, and send back a command signal which adjusts the gain of the intermediate amplifiers. The most adaptable solution, however, is to use the line pilot signal to directly control the intermediate amplifiers.

A block schematic of a regulated repeater is shown in *Figure 56.18* and a typical route in *Figure 56.19*. It will be noted that only

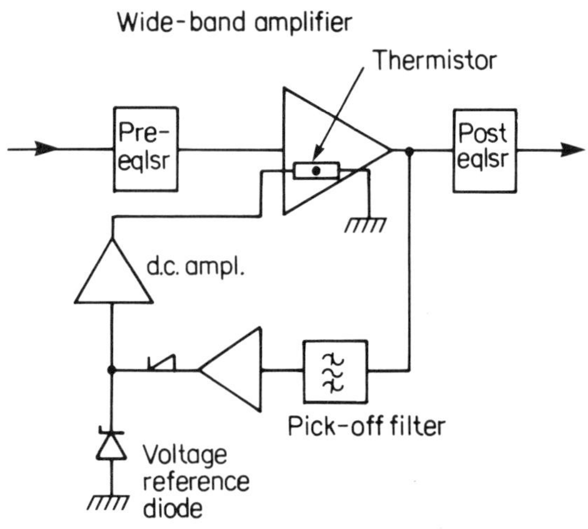

Figure 56.18 Regulated repeater

Table 56.1 Calculated noise performance 500 km route (9.5/2.6 mm cable)

Frequency (MHz)	*Relative level (dBr)*	*Thermal*	*A+B*	*A−B (pW0p km⁻¹)*	*A+B−C*	*Total*
0.4	−22	0.092	0	0.003	0.001	0.096
2.3	−22	0.177	0	0.003	0.013	0.193
5.0	−20	0.280	0	0.002	0.029	0.311
7.6	−18	0.396	0	0.001	0.088	0.485
10.4	−15	0.487	0.001	0.001	0.223	0.712
12.3	−12	0.501	0.003	0	0.154	0.658

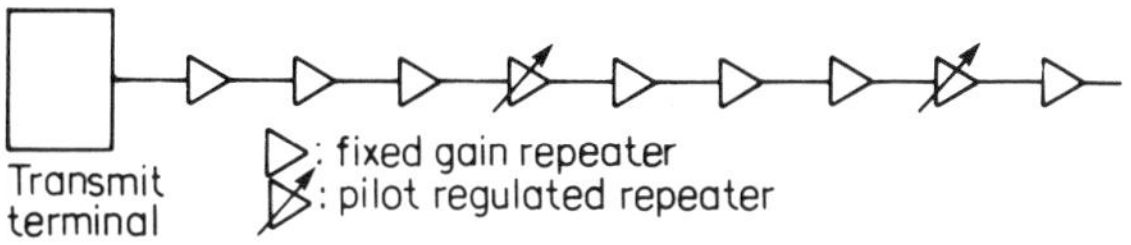

Figure 56.19 Allocation of regulated repeaters

every fourth amplifier needs be regulated, the rest operate on fixed gain.

A line pilot frequency of 12 435 kHz is allocated for this purpose and transmitted 10 dB below the virtual traffic level. It is picked off by a crystal filter in the regulator, amplified and rectified. The d.c. level is compared with a reference signal, the difference signal is amplified and controls the current into a thermistor. The resistance of this thermistor, which changes with current, determines the loss of the Bode network inserted in the amplifier interstage network, or in the negative feedback path. When the resistance is nominal, this network has a nominal loss. If the resistance falls, the loss is reduced, if it rises the loss is increased. The network configuration is such that the loss frequency characteristic is identical to the cable shape.

56.9.2.1 Control ratio

If the pilot level at the amplifier output has changed by 1 dB and the regulator reduces this to 0.1 dB, the control ratio is ten. The control ratio error does not accumulate, because, for example, the next regulator in these circumstances would see a change of 1.1 dB and reduced it to 0.11 dB. The next 1.11 dB reduces to 0.111 dB and so on.

56.9.2.2 Control range

The control range is the range over which the regulator can correct and is determined by the limits of resistance through which the thermistor can be driven and by the network design. Typical practical values are ± 6 dB at 12.5 MHz.

56.9.2.3 Simulation error

In all gain regulating methods the objective is to compensate for the actual cable loss change with temperature. The difference between the characteristic of the compensating network and the cable change is the simulation error. This can be made very small, but nevertheless on a long route it will accumulate in a systematic fashion and require secondary equalisation. Another source of level instability is a change of gain of the repeater with ambient temperature, which can be almost unmeasurable on an amplifier, but significantly accumulative on a long route.

56.9.2.4 Dynamic performance

The dynamic performance of a tandem connection of many pilot gain controlled amplifiers needs analysing as this is a limiting factor on the number of amplifiers, and, therefore, route length, controlled from one station.

The dynamic behaviour can be evaluated in two ways. Firstly, by stepping the send pilot level in say 0.5 dB increments and observing the resultant level change at the far end of the system. The requirement is that the overshoot shall always be less than the input change.

A second method is to amplitude modulate the pilot signal with a low-frequency (e.g. 10 Hz) signal and measure the modulation amplitude at the receive end. This is gain enhancement and a typical requirement is that on the longest regulated section

$$20 \log_{10}(B/A) \leqslant 6 \text{ dB}$$

where A is the amplitude modulation depth applied, and B the modulation depth measured at the receiving end.

56.9.3 Secondary regulation

A second pilot, 308 kHz, is available to control networks compensating for these very small errors. The use of such a network is generally restricted to every hundredth repeater say, but the difficulty in practice is determining the real requirement and then controlling it, when the maximum may indeed not be at 308 kHz. For telephony applications, additional automatic regulating pilots are commonly transmitted within each group or super group. These control flat gain networks which is a reasonable assessment of the gain drift which is likely to occur across a limited bandwidth. Another method is to periodically readjust the residual equalisers provided at the ends of a route which notionally compensates for the systematic fixed equalisation error. Equipment is available to do this automatically by means of inter-supergroup pilots, which as implied, are transmitted for maintenance surveillance in the frequency gaps between adjacent supergroups. The solution adopted depends on the lengths of route involved and the maintenance policies of the administration. For route lengths of about 300 to 650 km of large core cable, modern coaxial line equipment has sufficient stability to require no maintenance attention in this respect.

56.9.4 Crosstalk

The most stringent crosstalk requirement stems from the need for the transmission of broadcast programme material over coaxial systems. In this case three or four telephony channels are replaced with a 12 kHz programme modulation equipment.

Assuming unidirectional programme circuits are operating in opposite directions over a 300 km coaxial link, an 86 dB crosstalk margin can be allocated between the two directions of transmission. It is known that crosstalk can add in phase, that is, over narrow bands voltage addition can occur.

Thus, in the 300 km route, with 2 km spaced repeaters there are 150 crosstalk sources voltage adding, i.e. an addition of $20 \log_{10} 150$ or 44 dB. The crosstalk requirement per repeater is then (86 + 44) or 130 dB. Referring to *Figure 56.20* this means that the signal loss from the output of one amplifier to the input of the other needs to be 170 dB. Achievement of such a performance consistently in a production environment requires a design in which this need has been recognised from the outset. By careful attention to screening and by minimising common path earth currents, crosstalk values of this order can be realised.

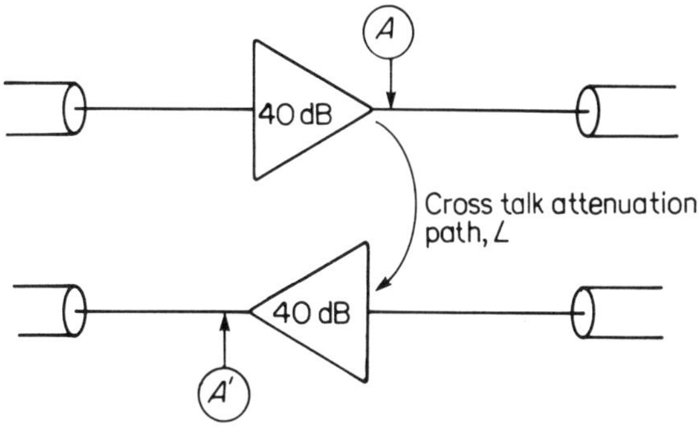

Figure 56.20 Crosstalk. If level $A' = A - 130$ dB then $L = 130 + 40 = 170$ dB

56.9.5 Power feeding

The intermediate line repeaters are installed in underground housings at 2 km intervals. The power for this equipment is fed over the inner conductor of the coaxial cable. The most common technique is to transmit a constant current d.c. as this requires least equipment at the buried repeater point.

Figure 56.21 shows the configuration at a repeater and *Figure 56.22* a complete power feeding section. The spacing between power feeding stations can be up to 300 km or more depending on the system and cable type. The power separating filter requirements are calculated on the need to control the effect of the output signal passing back through the power path and adding to the input signal. Such an effect could accumulate along the line and produce an unequalisable ripple.

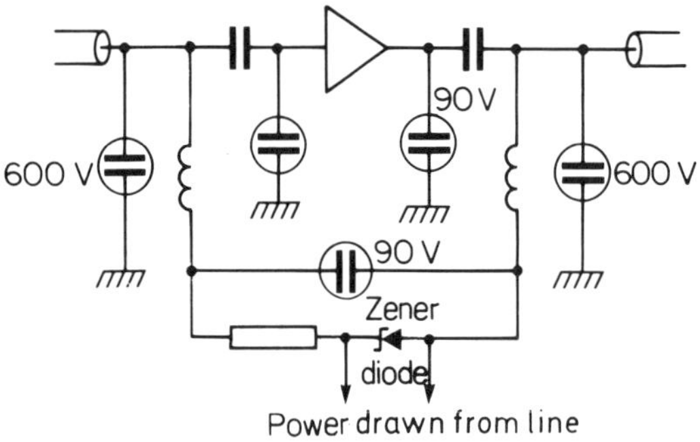

Figure 56.21 Repeater power feed circuit and surge protection components. Voltages indicate nominal breakdown voltages

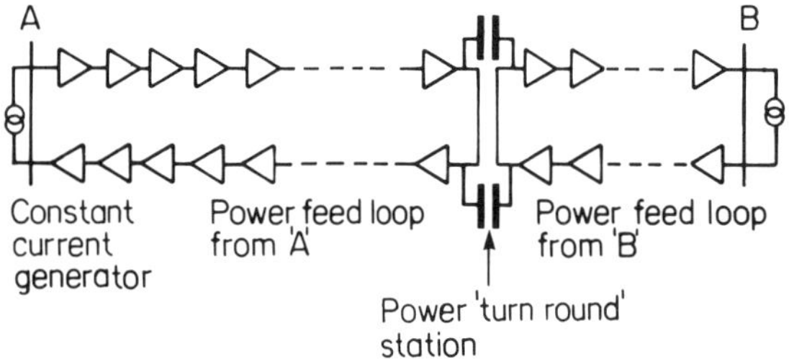

Figure 56.22 'Power-feeding' sections

Suppose the ripple signal is to be restricted to 0.1 dB after 280 km

Maximum level of interfering signal	= −38.6 dBm0
Voltage addition of 140 repeaters	= 43 dB
Therefore	
Maximum contribution per repeater	= −82 dBm0
Gain at 12.5 MHz, 0.3 MHz	= 40 dB, 6 dB
Then	
Suppression required in power path	= 122 dB, 86 dB
or	
Suppression in filter at 12.5 MHz, 0.3 MHz	= 61 dB, 43 dB.

At the power turn round point, i.e. where the d.c. power feeding loop is completed, the requirement is the same as the repeater crosstalk requirement. Any extra attenuation can be provided in the turn round connection. The repeater circuits are connected in series and parallel configurations to make best use of the incoming current. In the UK, and some other countries, considerable operating benefit is obtained by restricting the current to 50 mA which is recognised as being inherently safe to personnel. However, it is not uncommon for higher values to be used, but, beyond about 300 mA, the voltage drop in the cable becomes an increasingly dominant factor. A modern repeater consumes less than 2 W. The line voltage drop is of the order of 15 V at 50 mA for each amplifier. With modest sending end voltages, ±350 V say, very long spans between power feeding points can be achieved.

56.9.6 Surge protection

The zener diode shown in *Figure 56.21* is not used as for power stabilisation. Power stability can easily be set by the tolerance on the power feeding equipment constant current source. The purpose of the zener diode is to limit the maximum voltage which could appear across the repeater power input terminals in the event of an induced current surge into the cable.

Surges can result from either lightning storms, railway electric traction or electricity power line failures. The problem is to protect the transmission equipment from damage over a very wide range of operational conditions. International agreement has been reached on the level of immunity required (CCITT Rec. K17) although special circumstances may demand higher values. The normal conditions are survival from 5 kV discharges of 10/700 pulse shape, applied transversely to the input and output terminals and longitudinally through the repeater. Additionally, the repeater must withstand 50 Hz currents (10 A for 1 s) applied longitudinally. Several techniques are used to obtain surge immunity of this order. The primary protection comprises gas discharge tubes connected as shown in *Figure 56.21*. These need to be low capacitance designs with very fast (<1 μs) striking characteristics. It is also necessary to ensure that the power feeding current will not maintain the arc once it has been struck. Lower voltage gas tubes are used within the repeater and the pre-and-post attenuation equalisers also buffer the amplifier. Consequently, the components in the equalisers need to tolerate surges. Finally, semiconductor diodes etc are distributed within the repeater to absorb residual surges, but these need to be used with care to avoid additional intermodulation of signal products.

56.9.7 Supervisory systems

Analogue line transmission systems are extremely reliable and failure is a rare event. Nevertheless, it is necessary to provide maintenance staff with a means of unambiguously identifying a faulty repeater when the situation arises. The repair team can then be sent to the correct geographical location and replace the equipment in a minimum of time.

In the past, use has been made of the conductors laid up in the interstitial spaces between the coaxial tubes. However, these techniques are not effective on systems with significant numbers of repeaters, or on cables with a small number of interstice pairs. The traffic path can be utilised as the bearer for supervisory information. One such technique is to provide a fixed frequency oscillator at each repeater point, using a different frequency at each location. At the receiving terminal the level of signal received from each oscillator can be used to deduce amplifier conditions along the route. The disadvantages are that a large number of frequencies have to be used at close spacing and the oscillator is, therefore, expensive. In addition, by virtue of its allocated frequency, the repeater becomes tailored to a location.

An alternative method is to use the same oscillator frequency at all locations and interrogate each oscillator in turn from the receiving terminal. This can be done by stepping an interrogation signal from repeater to repeater as described by Howard.[1]

56.9.8 Cable breaks

The system described above will locate faults in the HF path, but if the cable is cut, power is lost to all repeaters in that power fed section and the oscillator monitoring system is also inoperative. In practice cable breaks are invariably caused by external events which are self evident in their own right, e.g. road works in the vicinity of telephone cables. Nevertheless, it can be postulated that the break might be due to hidden causes and a location scheme should be provided. A commonly used solution is to provide diodes at each repeater connected across the two cable inner conductors. If the power feeding supply is reversed, these diodes become conducting, the zener diodes in the through path are also forward biassed and hence only drop one volt. Consequently, a resistance measurement indicates which repeater section is broken. The application of this technique can be extended by grading the resistors connected in series with the transverse diodes.

Location of the actual break point within a repeater section is made by time domain reflectometer measurement on the coaxial core.

56.9.9 Engineering order wire circuit

It is common practice to provide an engineering telephone circuit from end to end of the line system with access at the repeater points. This is particularly useful during system installation and is an important maintenance aid.

Such systems are normally worked as conventional four-wire circuits over loaded interstice pairs. VF amplifiers are provided at the intermediate power feeding stations.

The underground repeater housings are often pressurised and provided with a pressure sensitive contactor which closes when the pressure falls, indicating the danger of water ingress. These contacts can be paralleled up on the phantom of the speaker circuit. Then any closure can be located by measuring the resistance of the phantom loop.

56.10 Non-telephony services

The transmission equipment described is also designed to cater for non-voice services, e.g. voice frequency telegraph and data modems of various rates. More significant users of bandwidth are programme circuit networks provided for the broadcasting authorities. Television can also be transmitted over FDM line systems. Because the television signals extend down to almost d.c. SSB suppressed carrier modulation is not practical. Instead vestigial sideband modulation is used. For a 12 MHz system the carrier frequency is approximately 7 MHz and the television information extends from about 6.3 to 12.3 MHz. The only change which has to be made to the line system is that it has to be equalised for group delay as well as for attenuation distortion.

56.11 Current status on analogue transmission

The principal characteristics of some analogue line systems currently available are given in *Table 56.2*. Of these the commercially most important are probably the 960 channel and 2700 channel systems. The 10 800 channel (60 MHz) system is in operation in several European countries and a very similar system (Bell L5) is used extensively in the USA. Application of the 40 MHz system is limited and it is not recommended by the CCITT. Technological improvements in recent years have made it possible to introduce an 18 MHz system which provides a fifty per cent increase in channel capacity with the same repeater spacing as older 12 MHz systems. It is primarily intended for application to large core cable.

Table 56.2 Some currently available systems

Bandwidth (MHz)	*Channel capacity*	*Cable (mm)*	*Repeater spacing (km)*	*Notes*
4	960	9.5	9.1	
12	2700	9.5	4.55	Also TV transmission
18	3600	9.5	4.55	Also TV transmission
60	10800	9.5	1.5	Also TV transmission
4	960	4.4	4	
12	2700	4.4	2	
18	3600	4.4	2	
40	7200	4.4	1	Not standardised

Systems of this type in conjunction with the multiplex hierarchy discussed in Section 56.7 provide the great majority of national and international telecommunication connections. Proposals for higher capacity analogue transmission systems, e.g. 200 MHz offering some 30 000 channels on a single coaxial cable, have been made but it is very unlikely that such developments will now take place. As more administrations make the decision to change from analogue space-switched networks, more design centres are directing their energies to the design of digital transmission systems. This trend will probably be quickened by the availability of optical fibre transmission technology, which certainly, at the present time, is more suited to digital than analogue transmission.

Thus the systems listed in *Table 56.2* may well be the last developments in analogue trunk transmission systems. However, such systems will continue to be brought into service for many years to come and analogue cable plant will form the basis of most telecommunication administrations' transmission networks for the foreseeable future.

56.12 Digital transmission

The past few years has seen a major shift by telecommunication authorities away from analogue techniques for transmission and switching to digital ones. There are a number of economic reasons for this of which the most important are:

(*a*) The ability to increase the capacity of existing paired transmission cables.

(*b*) The availability of cheap digital processing.

(*c*) The ability to carry different types of information, such as speech, data or facsimile, over the same system without mutual interference.

(*d*) Transmission quality independent of distance and unaffected by digital manipulation such as multiplexing or switching.

(*e*) The ability to exploit non-linear media such as fibre optical systems.

Most telecommunication authorities are aiming towards an integrated digital network with digital transmission and switching in which it will be possible to have a digital path from subscriber to subscriber.

The elements of digital transmission are:

(*a*) Primary multiplex—converting analogue voice signals to digital form and multiplexing several channels together.

(*b*) Higher-order multiplexes—combining lower-speed digit streams to form a higher-speed one for higher-capacitity transmission.

(*c*) Digital transmission systems of various capacities and using various media.

(*d*) Coders and multiplexes for non-speech traffic.

56.13 Basic digital techniques

56.13.1 Sampling

If a band-limited analogue signal is sampled at a rate equal to at least twice its highest frequency then the original signal may be recovered from the sampled signal without impairment by filtering. The upper and lower sidebands about each harmonic of the sampling rate will not overlap if the sampling frequency (f_s) is greater than twice the maximum signal frequency (f_{max}). Under these conditions the original can be recovered by ideal filtering, with usual low-pass filtering to recover the baseband information. To allow for practical filtering f_s has to be in excess of $2f_{max}$. If the signal to be sampled contains frequencies in excess of half the sampling rate then the sidebands overlap and interfere with correct demodulation, and this is known as aliasing. To prevent this signals are band-limited by low-pass 'anti-aliasing' filters prior to sampling.

Sampling means that continuous (analogue) information can be transmitted in a discontinuous form which enables signals to be multiplexed by time division. A sequence of samples is known as a pulse-amplitude-modulated (PAM) waveform.

For telephone quality speech with a signal band limited to a frequency of 3400 Hz, a sampling rate of 8 kHz (125 μs interval) is standard.

56.13.2 Quantising

Instead of continuously variable amplitudes the PAM samples can be assigned the nearest values on a vertical scale with a finite number of steps; this is known as quantising, see *Figure 56.23*.

The difference between the actual and quantised value is the quantisation error and in use gives rise to a multiplicative, noise-like, error signal known as quantisation distortion (QD). The finer the resolution of the quantised scale the less the QD.

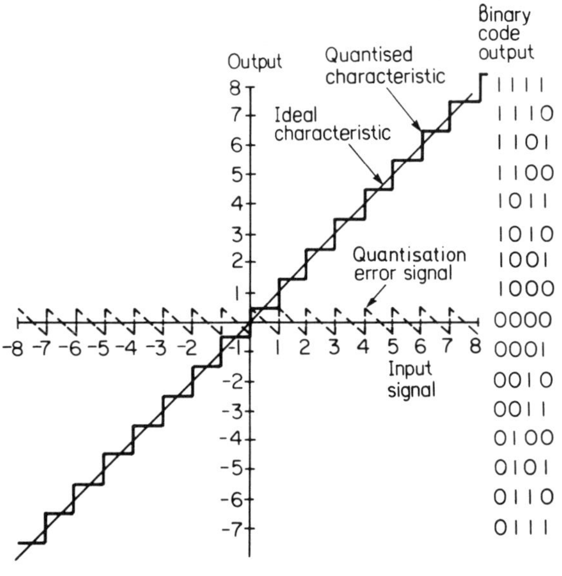

Figure 56.23 Illustration of quantising

56.13.3 Coding

For transmission the quantised samples can be represented by binary numbers, the number of bits per sample corresponding to the resolution of the quantiser, e.g. N-bit coding would correspond to 2^N quantising levels. In *Figure 56.23*, 16 levels are represented by 4 bits. The resulting bit stream is the pulse-code-modulated signal. Telephone quality speech is typically coded as 8 bits per sample.

56.13.4 Companding

From the description of quantising it can be seen that small amplitude samples are quantised with relatively less accuracy than large ones. Therefore, a low-level signal will have a worse signal to quantising noise ratio than a large one. Ideally the signal to noise ratio would be independent of amplitude. This can be approximated to by the use of a non-linear relationship between the input and output of the coder. The CCITT recommended law for 30 CH systems is:

$$y=\frac{1+\log_{10}(Ax)}{1+\log_{10}A} \qquad A^{-1}\leqslant x\leqslant 1 \qquad (56.12)$$

$$y=\frac{Ax}{1+\log_{10}A} \qquad 0\leqslant x\leqslant A^{-1} \qquad (56.13)$$

where y is the ideal output, x the input and $A=87.6$. In practice a segmented approximation to this law is used in which the slopes of adjacent segments are in ratios of 2. CCITT recommend 13 segments as shown in *Figure 56.24* for the positive quadrant. The first two segments are co-linear. The full range of output codes is 8 bits in 256 levels. 16 of these are allocated to each of the outer segments and 32 to the centre one.

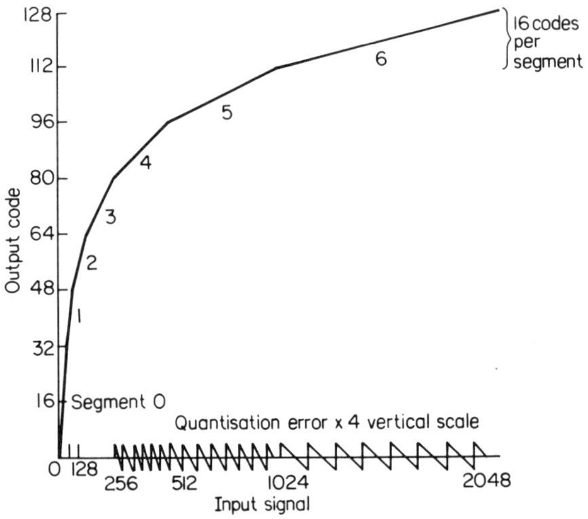

Figure 56.24 Segmented *A* law companding curve

The non-linear output is said to be compressed. A decoder has a complementary characteristic (expanded) so that the analogue to analogue characteristic is linear. The process is known as companding from COMpression and exPANsion.

56.14 Time division multiplexing

If the groups of bits representing successive PAM samples are much shorter in duration than the internal between samples then it is possible to use the remainder of the time to transmit other signals. This enables a multiplicity of signals to be carried in one bit stream so obtaining time division multiplexing.

An arrangement standardised by CCITT is for 32 groups of 8 bits to be transmitted every 125 μs corresponding to the 8 kHz sampling rate of each channel. This gives an aggregate bit rate of $32 \times 8 \times 1$ kHz = 2048 kbit s^{-1} for the multiplexed signal.

In modern practice the anti-aliasing filter PAM sampling and analogue to digital conversion are performed on a per channel basis in a single integrated circuit. The output from these integrated circuits is 64 kbit s^{-1} in bursts of 8 bits at an instantaneous rate of 2048 kbit s^{-1}. These signals are digitally combined to give the aggregate 32 time slot 2048 kbit s^{-1} signal.

56.14.1 Framing

In order for the multiplexed information in a bit stream to be recovered at the receiver it must be contained within a recognisable frame structure.

For example, the CCITT European frame structure is shown in *Figure 56.25*. It consists of 32 by 8 bit time slots labelled 0 to 31 and repeats at an 8 kHz rate or 125 μs period. Thirty of the time slots are used for speech, these are 1 to 15 and 17 to 31. Time slot 0 is reserved for alignment. That is, it carries fixed bit patterns which can be recognised at the receiver so that the time slot and bit counters in the receiver can be aligned with those of the transmitter (apart from transmission delay) thus enabling the information bits in the frame to be allocated to their correct receive channels.

To enable alignment to be maintained in the presence of digital errors between transmit and receive multiplexes a loss of alignment is not recognised until three successive frame alignment words are corrupted. When alighment has been lost all digit positions are checked until an alignment pattern is found. Confirmation that this is the correct time slot zero (TS0) is achieved by checking for a 1 in digit two of the following TS0 and that the alignment pattern re-occurs in the next TS0.

Time slot 16 (TS16) is reserved for signalling information, that is, information passed between telephone exchanges concerned with the setting up and breaking down of calls on the speech channels. The detailed contents of TS16 depends on the exchange types. For typical electro–mech types it would include seize and release signals and dialling pulses. More modern types of exchange would use multi-frequency signalling over speech circuits for numerical information. Stored programme controlled exchanges can use TS16 as a data link between processors (common channel signalling).

56.15 Higher-order multiplexing

The frame structure described in the previous section is for a primary multiplex. Where higher numbers of channels are required the outputs of primary multiplexes can be combined to form higher-order multiplexes. In the CCITT European multiplex hierarchy (see *Figure 56.26*) four primary bit streams are combined to form a secondary multiplex with a bit rate of 8448 kbit s^{-1}. Four secondary outputs can be combined to give a third order of 34.368 Mbit s^{-1} and four of these combined to give a fourth order at 139.264 Mbit s^{-1} corresponding to 1920 speech channels.

The basic technique for each stage of higher-order multiplexing is identical. *Figure 56.27* is a block diagram of a second-order system while *Figure 56.28* shows the frame structure. The four 2048 kbit s^{-1} inputs are assumed pleisiochronous, that is, with the same nominal rate, but not locked together (the primary multiplex clock tolerance is 50 ppm).

In each tributary the 2048 kbit s^{-1} data are written into a store. Data are read out from the store at a maximum rate of 206 digits/frame, equal to 2052.2 kbit s^{-1} (see *Figure 56.27*) and a minimum rate of 205 digits/frame, equal to 2042.3 kbit s^{-1}, the 206th 'justifiable' bit being filled arbitrarily. By comparing the phases of the write and read clocks to the store the choice of using or not using the justifiable digit for data is made so that the mean data rate out of the store exactly equals the input rate. This process is known as justification. Other regular fixed bits are added in the store so that the total digit rate out of the store is 2112 kbit s^{-1}. These bits are used for the justification control bits and alignment words.

The outputs from all four tributaries are then bit interleaved to give the final 8448 kbit s^{-1} output. The function of the justification control bits is to indicate to the receive demultiplex whether or not the justifiable bit contains valid information.

At the receiver the 8448 kbit s^{-1} bit stream is aligned and distributed to the receive tributaries where the the justification control bits are used to remove redundant bits. The resulting bit stream has a clock rate of 2112 kbit s^{-1} with clock periods

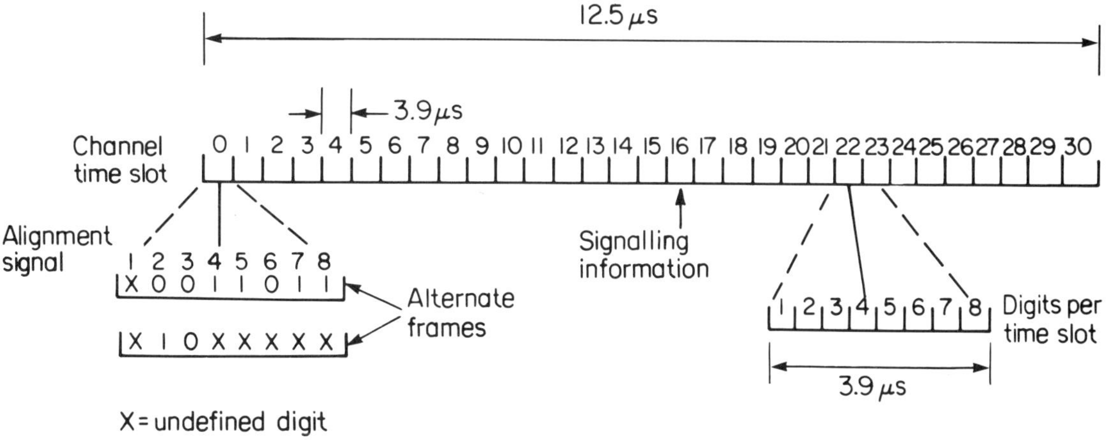

Figure 56.25 CCITT European frame structure

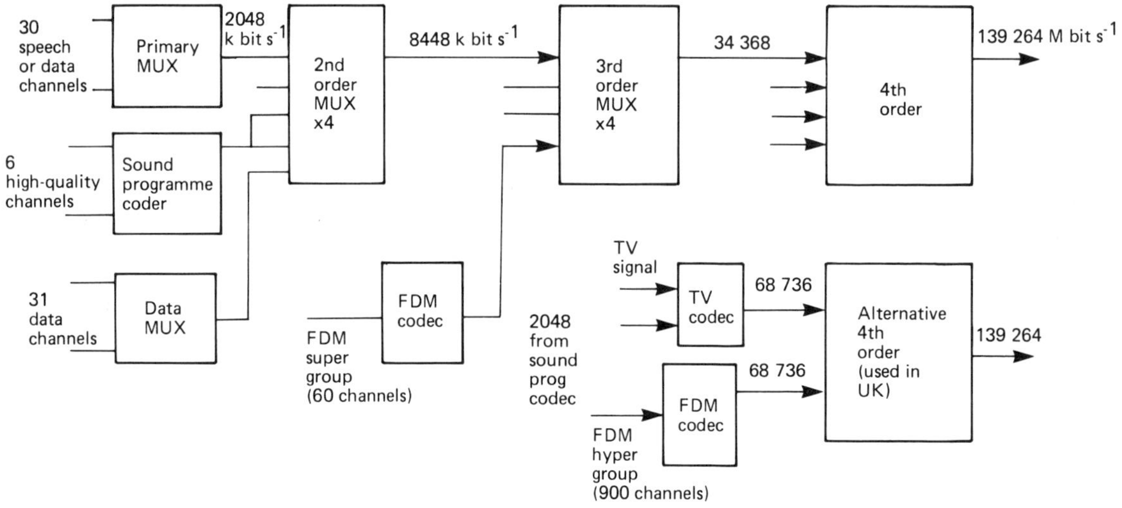

Figure 56.26 Multiplexing hierarchy showing various types of traffic

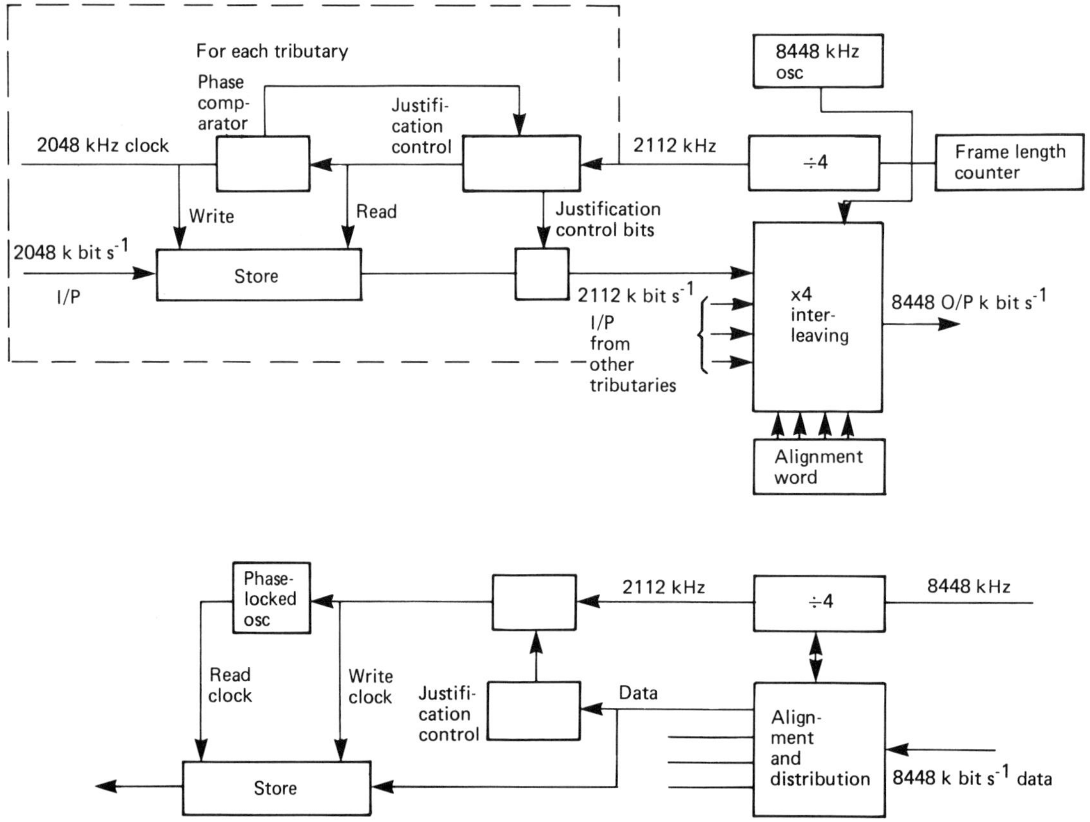

Figure 56.27 Multiplex block diagram

omitted so that the mean rate is exactly the same as that of the original 2048 kbit s^{-1} input. A phase-locked oscillator and buffer store are used to smooth out this bit stream to recover the original clock for output. The frame structure of *Figure 56.28* provides more detail of the operation. One justifiable bit is provided per tributary per frame of bits. Three justification control bits, J_c, are used per frame. This enables a majority decision to be made at the receiver to protect the process against transmission errors.

Note that no use is made of any information contained in the

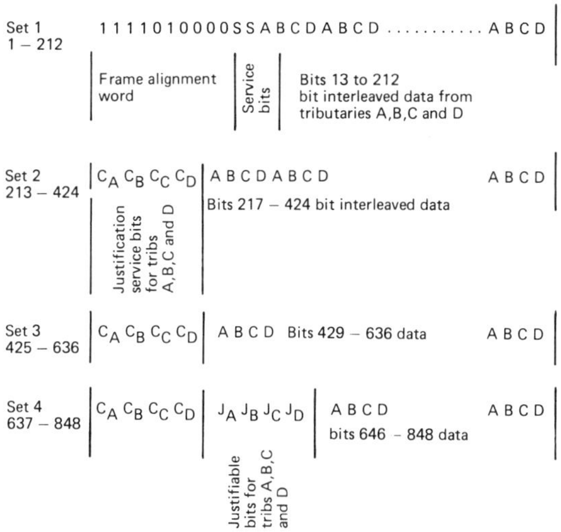

Figure 56.28 Frame structure

2048 kbit s^{-1} input data. Thus any 2048 kbit s^{-1} bit stream can be multiplexed.

56.16 Digital transmission

A major reason for converting information to digital form is to take advantage of regenerative digital transmission. In addition to employing linear amplification to overcome attenuation a regenerative repeater recognises what the original digital signal would have been and transmits a signal identical to that original. This avoids the build up of signal imperfections along the length of the route.

Figure 56.29 shows a regenerative system for use with paired cable at say 2048 kbit s^{-1}. Power for the intermediate regenerators is provided by a constant direct current (typically 50 mA) which is fed longitudinally over each of the two pairs used for go and return transmission. Transformers at the input and output of each repeater separate and combine the d.c. and signal. Since two pairs are used instead of the 30 pairs which would be required for audio transmission the capacity of the original cable is greatly increased.

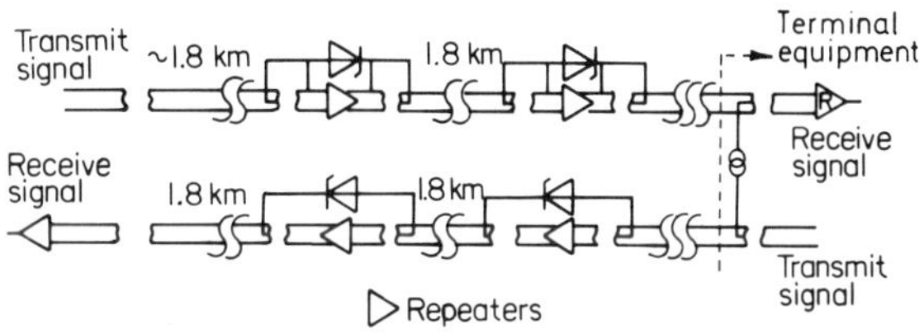

Figure 56.29 Regenerative system for use with paired cable at say 2048 kbit s^{-1}

Figure 56.30 shows a block diagram of a paired cable repeater for a three-level signal. The input signal consists of half width pulses transmitted from the previous repeater, attenuated and distorted by the cable pair plus interference from systems on other pairs in the same cable (crosstalk coupled interference). The automatic line-building out (ALBO) section corrects for the effect of various line lengths between repeaters. A passive input equaliser reduces the distortion and the amplifier produces a signal at 'D' the 'decision point' which is the signal to be regenerated. The pulses are approximately full width and the equalised bandwidth up to the digit rate. This signal is split four ways. One path is to the timing extraction where the signal is sliced and rectified to give a spectral component at the digit rate. This is extracted by a tuned circuit and limited to reduced-pattern-induced amplitude variations. The resulting square wave provides a clock to govern the time at which the positive and negative regenerators decide on the presence or absence of pulses.

Two other paths for the amplified signal are to the positive and negative regenerators. The signals are sliced against thresholds to determine whether they exceed the minimum acceptable pulse amplitudes. If these thresholds are exceeded during the sampling

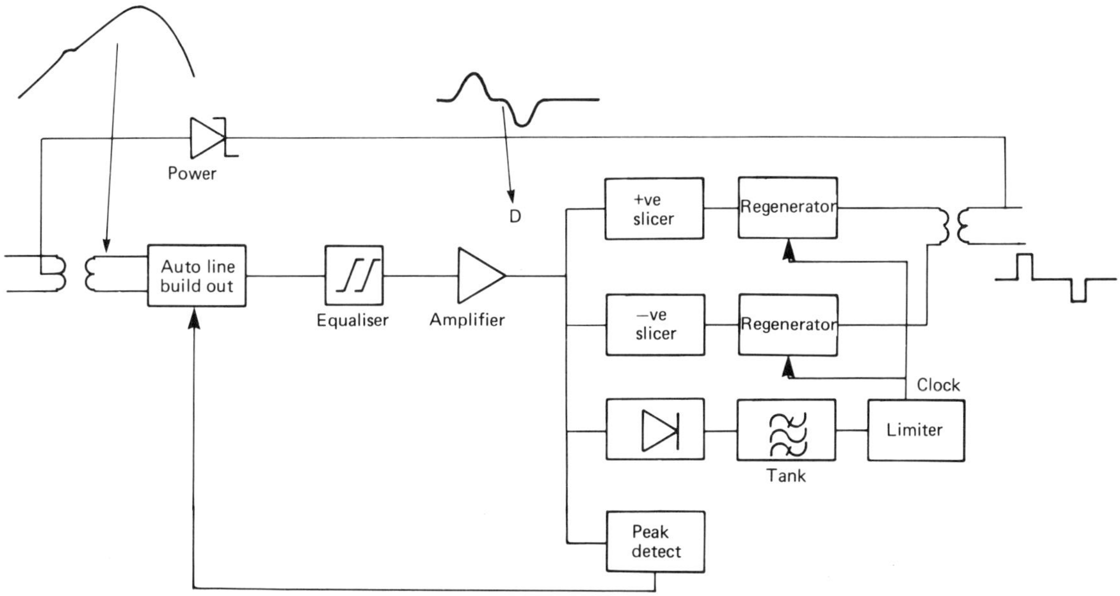

Figure 56.30 Block diagram of a paired cable repeater for a three-level signal

instant determined by the timing waveform then a new half width output pulse is generated, the pulse width is governed by the timing waveform.

Outputs from the positive and negative regenerators are combined in a transformer and sent to line. The block diagram for a coaxial cable repeater would be similar except for the addition of filters at the input and output to separate d.c. power from signal power. Optical repeaters generally use two-level transmission and so have only one regenerator. The input transformer would be replaced by an optical power detector (avalance photodiode) and the output current would drive an optical transmitter, LED or laser. Other forms of timing extraction use phase-locked loops or crystal filters.

Because of the a.c. coupling in the signal path and the need to have frequent signal transitions to maintain the timing waveform non-redundant binary is not suitable for transmission and it is necessary to use a 'line code' with added redundancy to have the necessary properties. Redundancy can be added by using more levels (typically three-level ternary codes) or by increasing the bit rate. A simple code originally used for paired cable systems is AMI (alternate mark inversion), a three-level code in which binary ones are transmitted alternately as +1 or −1. HDB3 (high-density bi-polar) is a modification of AMI which avoids lack of timing during long strings of binary zeros. For coaxial systems more efficient codes can be used such as the 4B3T types where blocks of 4 binary digits are translated to blocks of 3 ternary digits. The reduced symbol rate reduces attenuation and noise bandwidth for a given repeater spacing. Optical systems are restricted to two levels by the non-linearity of the transducers, binary codes such as 7B8B, where 7 information bits are transmitted in 8 bit words, provide sufficient redundancy.

56.17 Coding of other types of information

56.17.1 High-quality sound

A linear coder for broadcast quality sound would require 32 kHz sampling and 14-bit resolution, this would allow 4 channels in a 2048 kbit s^{-1} system. Bit rate reduction by *A* law companding can give rise to undesirable distortion. In a system developed by the BBC, NICAM III, near instantaneous companding is used. Initial analogue to digital conversion is 32 kHz sampling and 14 bits, giving an audio bandwidth of 15 kHz. In every period of 1 ms (32 samples) the amplitude of the largest sample is used to determine which of four available linear coding scales will be used. The maximum amplitudes of the four scales are in 6 dB steps and the one chosen is the lowest which can accommodate the largest sample. Each scale has 10-bit resolution, thus the original 14 bits per sample is reduced to 10 giving 320 kbit s^{-1} channel. A data channel multiplexed in the 2048 stream indicates to the receiver which scale is required to decode each block of 32 samples. With the reduced bit rate due to compression six sound channels can be included within the 2048 kbit s^{-1} rate together with data channels and alignment words.

56.17.2 TV coder

A TV transmission system for a composite PAL signal must handle more than 5.5 Mbit s^{-1} bandwidth and have 8-bit linear resolution. It is convenient to lock the sampling rate to three times the colour subcarrier (subcarrier frequency $f_{sc}=4.43$ MHz), that is 13.3 MHz. This gives a bit rate of 106 Mbit s^{-1}. Bit rate reduction can be achieved by the use of differential PCM (DPCM). In this technique an estimate is made of the next sample based on previous samples, then instead of sending the next sample with 8-bit resolution the difference between the predicted and actual values is sent with 5-bit resolution. At the receiver the same prediction is made as at the transmitter, and the incoming DPCM sample used to correct this value to reconstruct the original sample. Because small prediction errors are more common than large ones, and because the eye is less critical of prediction errors occurring at large changes of luminousity, it is possible to use a tapered quantising law for the DPCM samples. Thus small differences between predicted and actual signals are coded with finer resolution than large ones. For a TV signal sampled at three times f_{sc} sample 3 before the current one gives a good prediction. The resulting signal has a bit rate of 66.5 Mbit s^{-1}, this can be multiplexed with a 2048 kbit s^{-1} channel for high-quality sound, etc, and an alignment signal to give a total

Table 56.3

System *bit rate and medium*	*Line code* *symbol rate*	*Repeater* *spacing*	*Notes*
2048 kbit s^{-1} paired cable	HD83 2048 kbaud	Depends on cable ~1.8 km	Widely used to increase traffic capacity of existing cables
2048 kbit s^{-1} paired cable	MS43 (4B3T) 1536 kbaud	Depends on cable ~1.8 km	Allows substantially more systems in existing cables than HDB3 systems
8448 kbit s^{-1} paired cable	MS43 6336 kbaud		Exploits cables originally installed for FDM 24 channel carrier
34	MS43 27	4 km on 1.2/4.4 mm coax	
139 Mbit s^{-1}	6B4T 692.8	2 km on 1.2/4.4 mm coax	
8448 Mbit s^{-1} optical	2B3B	~12 km	Highly redundant code allows cheap terminals. Used for junction systems without intermediate repeaters
34 Mbit s^{-1} optical, 139 Mbit s^{-1} optical	7B8B	~10 km depending on fibre	

rate of 68.736 Mbit s^{-1}. This occupies half the capacity of a 140 Mbit s^{-1} transmission system.

56.18 Submarine telecommunications systems

Submarine telecommunications systems are designed primarily. to carry telephone traffic. They use analogue transmission methods assembling single side-band channels by frequency division multiplex. Circuit capacities range from 120 to 5500 and larger capacities could be developed should the need arise. Systems at present installed are up to about 3000 n.m. between terminals and the total route length is very large: British supplied complete systems alone total over 60 000 n.m. at the end of 1980.

Submarine systems are designed to be very reliable. The methods of reliability assessment are similar in principle to those ordinarily in use for high-quality equipment, differing largely in the extensiveness, care and cost devoted to such matters as elimination of known failure mechanisms, accelerated life tests, burn-in, degree of inspection and production control, yield rates, etc.

Apart from a few early designs, submarine telephone systems use a single cable for both directions of transmission. This is because, for a certain bandwidth, the lowest attenuation is given by a single coaxial tube. This fact, which is in contrast with land-line practice, leads to greater economy in submerged plant and dictates the configuration of the system. The relevant parts of the system are the cable, the submerged apparatus, which includes both repeaters and equalisers, the power feeding equipment and the terminal translation apparatus.

56.19 Submarine telephone cables

The cable is now invariably a single coaxial tube of a size and structure depending on the bandwidth of the system and the circumstances of the route.

56.19.1 Deep-sea cable

This cable is used in deep water, greater than about 500 fathoms,* where there is no risk of damage due to fishing activity or ship's anchors.

A bottom survey is normally undertaken so that unsuitable conditions such as rocky under-water valley transits or fast bottom currents can be avoided. This cable needs no mechanical protection and a design has evolved (see *Figure 56.31*) which comprises:

(*a*) A central strength member of high tensile steel wires whose function is to bear the static load of some miles of suspended cable, including that of any suspended repeaters, and the dynamic forces due to inertia and water drag. Up to ten tons load or more can be experienced in recovering repeatered cable from deep water in a rough sea.

This member must have a low twist/load characteristic and torsionally balanced designs are often used.

(*b*) A copper tape which is formed around the strength member with a welded longitudinal seam. It is electrically thick at the lowest transmission frequency and is swaged to an accurate diameter. This forms the inner conductor.

(*c*) A dielectric core of polyethylene. This is the least expensive practical dielectric and has excellent properties for the purpose, being almost incompressible and having a low permittivity and

* The standard of length in most submarine systems is still the telegraph nautical mile (n.m.) of 6087 feet (1.855 km) so that 1000 fathoms, of 6 feet each, is about $1\frac{1}{2}\%$ less than 1 n.m.

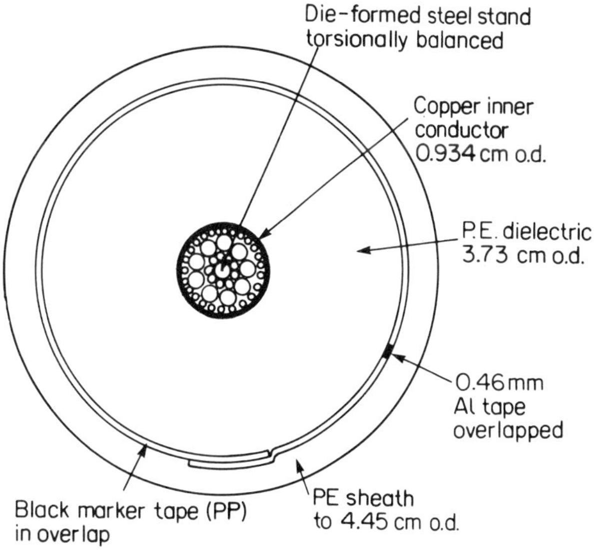

Figure 56.31 3.73 cm/0.934 cm (1.47 in/0.368 in) lightweight cable

a very low loss angle. The core is shaved to an accurate diameter, to give predictability of attenuation and to give a good base for the outer conductor.

(*d*) An outer conductor of a single metal tape having a longitudinal overlap. This can be of copper but aluminium is often used, in which case the overlap is insulated by a polypropylene or polythene tape, which guards against intermittent contact, gives a good seal under pressure and forms a marker to indicate twist.

(*e*) An overall polythene sheath. This supplies compression to the outer conductor, some abrasion resistance and makes it possible to eliminate water from the core/outer-conductor structure when, as is usual, water barriers are fitted at the repeaters.

At a repeater, the load is transferred from the strength member to the repeater housing, sometimes by a gimbal but, more often, by applying armour wires over a few fathoms at the end. This forms a flexible stocking anchored at one end to the housing and drawing down on the cable, so that the load is transferred from the strength member to the armour wires, through the layers of the cable structure, over a considerable length.

56.19.2 Shallow-water cable

In shallow seas, such as the North Sea, and on continental shelves, the cable is subject to hazards—such as abrasion in a tide way, fouling by ship's anchors and, more importantly, by the devices which spread the net of heavy trawlers. The ideal cable here is heavy, strong and impenetrable, with good abrasion resistance properties, which are conferred by one or two layers of mild steel armour wires. In severe conditions the outer armour can be applied with a quite short lay, to give improved resistance to crushing and penetration. In difficult situations, it is possible to bury the cable in a trench on the sea floor, made by a plough at the same time as the cable lay. Armoured cable now usually uses the lightweight, deep-water cable as a basis, although mild steel is usually substituted for the high tensile steel centre member (see *Figure 56.32*).

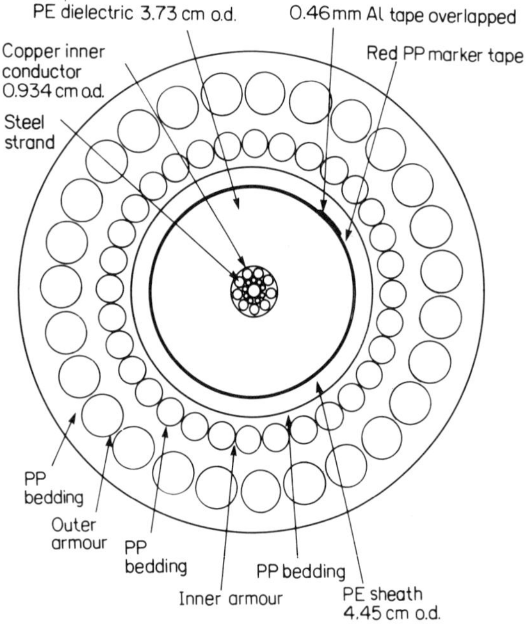

Figure 56.32 Type AB armoured cable

56.19.3 Screened cable

Some extension overland of submarine systems, sometimes including repeaters, is common. A submarine cable outer conductor is unsuitable as a screen and, in the past, protection against radio interference was given by an expensive screen of multiple iron tapes added to a small diameter cable. With the advent of higher frequency systems, however, a thick copper outer conductor with a generous, uninsulated, overlap is suitable and behaves like a theoretical homogeneous tube.

56.20 Performance of cable

The attenuation of the cable, α, is given by

$$\alpha=\tfrac{1}{2}RvC+(\pi f/v)\tan\delta \qquad (56.14)$$

in units of nepers per unit length at all frequencies, if the loss angle δ is small. The first term is the conductor loss, due to the resistance R, where C is the cable capacitance (constant) and v is the propagation velocity. At high frequencies, v approaches the velocity of a plane wave in the dielectric so that R is the only cause of significant variability. At high frequencies R is proportional to the square root of the frequency f, because of the diminishing penetration of the magnetic field into the metal. The second term is independent of dimensions and is directly proportional to frequency if δ is invariant. It is commonly about 10% of the first term at the highest frequency of modern (say 45 MHz) systems. The conductor loss is least, for a given overall diameter, where the outer-conductor to inner-conductor diameter ratio is 3.6/1,* which corresponds to a characteristic impedance $(1/vC)$ of 51* Ω for a polythene dielectric. This optimum is flat and impedances of 43 to 60 Ω have been used giving a wide range of properties for the central member. Although the use of an aluminium outer conductor increases the attenuation by nearly 6% over that for an all-copper cable, and increases the number of repeaters by the same amount, it can lead to a significant overall cost advantage. Such cables show great predictability of initial performance and usually have negligible ageing rates.

* For conductors of the same metal. For copper inner, aluminium outer conductors, we have 3.8/1 and 53 Ω.

The cables have a pressure coefficient of attenuation of about $\frac{1}{2}$% per n.m. of depth and a temperature coefficient of attenuation of from 0.17 to 0.1% or less, reducing as the diameter and frequency increase, where the effect of the loss angle of the dielectric affects the attenuation to a greater extent.

The optimum cable size for minimum system cost occurs where cable and repeater costs are roughly equal, but this optimum is quite flat and the 3.73 cm (1.47 in) diameter design is suitable for both 1840 circuit (13.7 MHz) and 5520 circuit (44.3 MHz) systems. The attenuation-frequency response of this cable is, at 283 K (10°C) zero depth:

MHz	1	5	15	45
dB/n.m.	1.69	3.84	6.84	12.54

56.21 Submerged repeaters

A repeater must perform several functions:

(*a*) Separate the directions of transmission.
(*b*) Provide amplification in each direction.
(*c*) Separate the power feed from the signal currents.
(*d*) Protect itself from power surges on the cable.
(*e*) Provide a supervisory facility for fault location.
(*f*) Protect itself against the ingress of sea water at high pressure.

Figure 56.33 shows the schematic diagram of a repeater in which functions (*a*) and (*b*) are combined; most early and some modern repeaters use this method. *Figure 56.34* shows the most common modern arrangement in which a different amplifier is used for each direction of transmission.

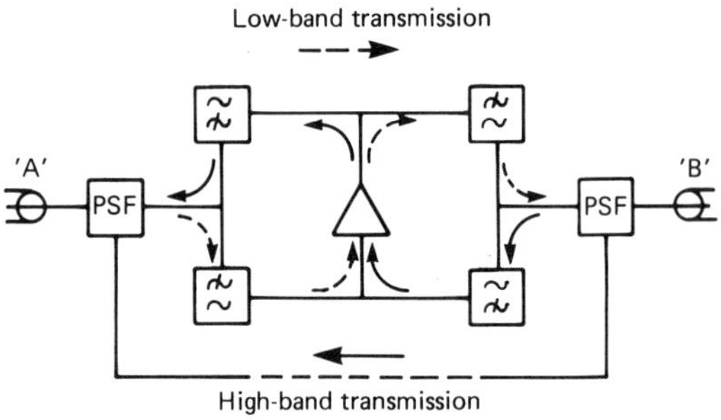

Figure 56.33 Single amplifier repeater

56.21.1 Directional filters

To enable gain to be inserted in each direction, it is necessary to use two frequency bands, one above the other with a space between them. These bands are separated at the repeater, the signal levels in the bands are amplified, and the bands are then combined for onward transmission. The band spacing (1.2/1 to 1.4/1, according to design and frequency) is required to enable the filters to develop enough stop-band loss in the opposite band to suppress not only oscillation but small gain ripples. The loop loss should exceed 50 to 60 dB, in which is included the effect of the amplifier gain, to suppress ripples to below ±0.03 to ±0.01 dB per repeater.

In *Figure 56.33* a single amplifier provides gain for both bands.

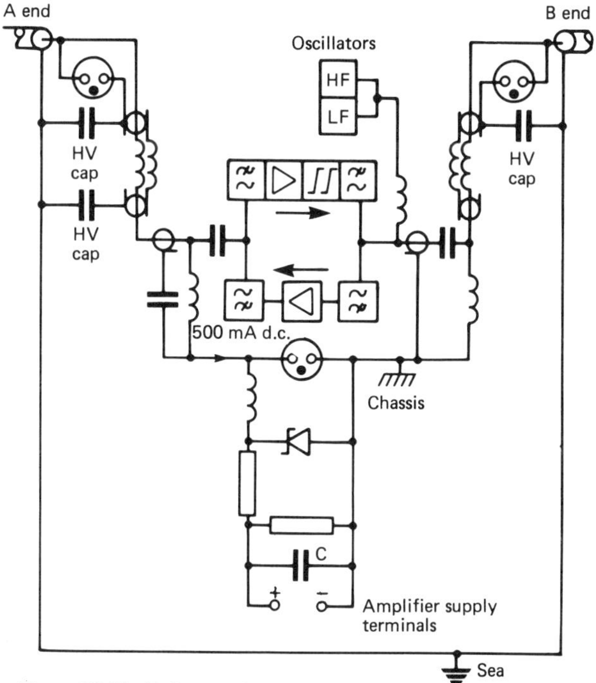

Figure 56.34 Twin amplifier repeater

Thus intermodulation produced by non-linearity in the amplifiers can inject noise into the other band. This can lead to a form of oscillation, 'noise singing', on long-term systems, in which the iterative effect of a large number of noise transfers from band to band is large enough to be self-sustaining. This effect is much aggravated by a gain step in the system, or some other types of misalignment, caused, for example, by a succession of repairs, which inserts gain into the transfer mechanism. This led to the configuration of *Figure 56.34*, to which all modern British repeaters conform, in which a separate amplifier is used for both directions of transmission; this virtually eliminates the problem.

The filters must have a very good return loss. 30 dB is necessary to reduce ripples, due to reflections, to a level comparable with those caused by finite-loop loss.

56.21.2 Amplification

Figure 56.34 as well as eliminating 'noise singing', improves the performance obtainable from the amplifiers, especially the high-band amplifier which sets the system performance. Because its bandwidth is now less than 2/1, second-order intermodulation products are eliminated, feed-back loop design, especially if the loop includes the transformers, is made easier and there is a small load power advantage. These advantages are enough to justify the additional amplifier even on economic grounds.

Negative feed-back is used, from 20 to 30 dB over the band, to stabilise the gain and input/output impedances and, more importantly, to reduce third-order intermodulation. This sets the performance (and hence the repeater spacing) on long systems, rather than the overload point. The build up of intermodulation noise is less on submarine links than on land lines. Because of the phase dispersion introduced by the directional filters, the power level of the products, in the worst part of the spectrum, is about $N^{1.7}$ (where N is the number of repeaters) and the index is less at other frequencies (by contrast, in land lines, where the phase dispersion is small, the level approaches N^2). This effect is sometimes enhanced by the addition of phase dispersion ('all-pass') networks to the high band of the repeater.

The amplifier components, especially those in the feed-back network which set the gain, have very small parameter drifts. In particular, the reliability specification for the transistors includes a parameter drift requirement which gives a tolerable drift for as many as 1000 sets of triples in tandem.

56.21.3 Power separation

The power to a repeater is fed via the inner conductor and is at direct current. Because of the enormously high bulk resistivity of polyethylene, the attenuation to d.c. is zero, so that the feed current level does not alter along the route. A combination high-pass, low-pass filter is used at the input and output of a repeater to separate the signal from the feed current. A capacitor, at each end, must withstand the full system voltage and, because the repeater chassis is at high voltage with respect to the sea, this capacitor must be in series with the earthed outer conductor of the sea cable. The impedance of this capacitor, especially at the higher frequencies where its inevitably large size causes it to be rather inductive, gives rise to a cross talk path across the repeater. To increase the loss of this path to acceptable proportions requires the use of coaxial chokes and, sometimes, an additional high-voltage capacitor to augment the natural capacitance between the repeater chassis and the sea.

All the repeaters are fed in series with the same current (about 500 mA is common). Both repeater amplifiers are in parallel and are shunted by a resistor whose function is to provide an accurate value to the repeater resistance. This is independent of the characteristics of the semiconductors when the repeater current is reduced to a value where the semiconductors are below their operating threshold voltage. This is necessary for fault location by the d.c. or quasi-d.c. tests which have been inherited from telegraph cable practice. For surge suppression, the repeater d.c. supply is shunted by a large 'Zener' diode. This diode is often arranged to be in its avalanche condition, shunting about 50 mA or so, and thus has the advantage of reducing the sensitivity of repeater gain to feed current changes.

About 15 to 20 V is used for the voltage drop across a repeater, and the cable voltage drop does not add much to this, so that, for a 1000 repeater system, voltages of $\pm 10\,000$ V, are possible. A repeater rating is up to $12\frac{1}{2}$ kV to allow a margin for contingencies, such as magnetic storms which add a voltage in series with the cable. An important requirement for the power feed arrangements, and all the parts of the system at high voltage, is immunity from impulse noise, such as is produced by ionisation, which can seriously affect data circuits. The oil-filled paper dielectric high-voltage capacitors, and then the repeater as a whole, are rigidly inspected for this property, for which the specification can be very severe, e.g. $1/N$ counts in 15 minutes which exceed a level of 12 dB below the long-term signal power nominally loading a speech channel band. This ionisation specification is more severe than that required by the reliability of high-voltage components.

56.21.4 Surge suppression

The most vulnerable place for an earth fault on the cable is inshore in shallow water, i.e. close to a terminal station where the system voltage is highest. Thus the peak current which can flow into the fault path, which may include a repeater, is about 200 A. Also, there may be a step change of about 10 kV in voltage. To protect against surges of this magnitude, several measures are taken.

(*a*) Referring to *Figure 56.34*, the gas tubes protect the input/output components and bypass the long-term cable discharge current around the repeater.

(*b*) The amplifier voltage supply is protected by the avalanche diode, the peak current through which is diminished by the several inductors in series with it. The capacitor C, which is fitted primarily to extend the band of low-frequency fault location, further reduces surge impulses.

(*c*) The gas tubes allow a great deal of high-frequency energy through to the amplifiers via the signal paths. The amplifier inputs are protected by back-to-back connected diodes. The amplifier outputs are similarly protected, but here the diodes are biassed back to avoid lowering the overload point to signal voltages.

(*d*) Each transistor emitter is protected by a shunt diode, connected in the opposite sense to that of the emitter/base junction.

The nature of the surge problem, because of the non-linearities involved, is not amenable to calculation, and much empirical testing is required, using artificial lines and careful analysis of components and devices after test, before a design can be approved.

56.21.5 Supervisory facilities

In the past, supervisory facilities have been comprehensive and have used the two-band nature of submarine systems to allow a supervisory signal in one direction to be converted into a signal in the other, i.e. to be returned to the originating station. Both frequency domain and time domain signals have been used. The latter was most common and the set-up was similar to a radar method. A short pulse was emitted by one station then modulated by a carrier, also transmitted from the station, into a pulse transmitted in the opposite direction. The time delay and magnitude of each return pulse was peculiar to a particular repeater.

However, it has come to be recognised that simpler facilities are satisfactory. A modern repeater has merely an oscillator generating one or two tones, of precise frequency just outside the signal traffic bands, which are unique to that repeater.

The level of this tone establishes the relative level at that repeater, and changes in level between repeaters give the changes of level along the route. The tones are at fairly high level (while still contributing a negligible addition to system loading) so that if an additional tone, carried by all repeaters, is sent from a terminal station at high level, an intermodulation product is produced, e.g. third order, in the same band, which first appears at a unique repeater. If this repeater has poor non-linearity, this product stands above that produced by the normal repeaters, by the process (mentioned in *56.21.2*) by which phase dispersion reduces the build-up of products from normal repeaters along the route. A transistor which produces an abnormally high shot noise is likely to affect the gain of a repeater, so that it could be found by tone level measurements. But, as an insurance, a method to locate basic noise faults has been developed. This relies on the principle that the attenuation/frequency response, observed at a terminal from a noise source, depends on the number of repeaters which have been traversed. Measurements are made of received noise power level against system attenuation in a part of the spectrum where the attenuation/frequency response is both rapid and precise, e.g. at the band edges of the directional filters. Analysis of this response, using computer techniques, can indicate the approximate position of the noise source. Field trials have shown that a location accuracy of 1 to 2% of the observed length of system can be achieved, which means that a faulty repeater could be recovered at the second trial.

The supervisory measures are almost always sufficient, but they rely on a supply of power to the repeaters. This is usually possible, even with a cable fault, because the fault can be so balanced that the voltage drop to earth at the sea is close to zero, the voltage ratio incidentally giving an additional indication of the fault position. If necessary, the fault can carry the whole feed current from one end of the system. Occasionally, an open circuit fault occurs, e.g. if the cable is damaged by stretching so that the central member retracts within the polythene core, the resultant gap being eliminated by hydrostatic pressure. Here quasi-d.c. capacitance tests can give quite precise locations. A whole variety of d.c. tests, from telegraph experience, can be used, in other cases, if the repeater has proper bypass circuitry. These tests can be supplemented by impedance/frequency tests, which measure resonances in the faulty length, or by pulse-echo tests (which are their equivalent in the time domain) using the low-frequency path through the power feed circuit of the repeaters.

56.21.6 Repeater pressure housing and assembly

Many housing arrangements have been, and still are, used; the following describes the methods used today in Britain.

The internal apparatus is contained in circular or octagonal cans carried by four longitudinal methyl-methacrylate bars, mounted on two circular brass end plates. These end plates, holding the apparatus, are sealed into a brass shell, at one end by soldering and at the other end by a lead seal. Flexibility is built into the bar mounting to allow some longitudinal expansion. Clamped polythene glands allow the passage of the signal cables, here reduced to about 8 mm in diameter, through the end plates. Silica-gel dessicators are enclosed in this capsule. These stabilise the humidity to around 10% irrespective of temperature or the original humidity of the gas space, since the gas is merely a transfer device to interchange the water content of the dessicators and the methacrylate bars. Dry, oxygen-free, nitrogen is used as the encapsulant.

The pressure housing is a thick cylindrical shell, of high-tensile nickel chrome molybdenum steel, hardened and tempered. This casing is closed by forged steel bulkheads of similar material. The gap between the bulkheads and the casing is sealed by rubber 'O' rings located in grooves in the bulkheads. A polythene gland is used to withstand the sea pressure at signal cable entry. Machined on the bulkhead is a protrusion carrying a central threaded bore and external castellations. Through this is passed the 8 mm cable with a large projection on both faces. The whole is covered by polythene, after the exposed metal surfaces have been sensitised by a separate application of hot polythene, using a high-pressure injection moulding process. This also forces the polythene of the cable into the threaded bore, effectively preventing extrusion of the cable under pressure. Sea pressure intensified the seal of the moulding to the castellations. These cable entries are made before offering up the bulkheads to the casing and, at this time, a joint is made to the internal unit cables. Flexible braid, covering the cables and pushed back to enable the inner conductor joints to be made, are drawn over the joints, coupled together and taped down. The sea side of the bulkhead is protected from corrosion by polyisobutylene enclosed in a neoprene bell which forms a diaphragm transferring pressure to this viscous material from the sea.

The large sea cable is fitted with 8 mm cable terminations which include a seal to exclude water under hydrostatic pressure from the spaces between the outer conductor and the core and sheath of the sea cable. This 8 mm cable is covered with braids which are fitted with couplers and are retractable to enable the joint between the repeater and sea cable (8 mm cables) to be made. All joints are tested in a water bath at 40 kV d.c. They are made by brazing and polythene injection mouldings which are subject to inspection by X-ray photography. The braids are protected and consolidated by hot polyisobutylene and self-amalgamating and adhesive tapes. The terminations include a cone through which the cable armour wires are taken, laid over the outside surface of the cone and bound to the armour. The small core cables are formed into a coil which is inserted into short joint chambers which are screwed to the end of the housing proper. These chambers are fitted with an anchor plate, to which the cones are clamped by split collets, and which is covered by a screwed end cap.

The housing is corrosion protected by 0.4 mm of zinc spray and three layers of vinyl paint. If the sea cable is of aluminium, zinc anodes of about 9 kg are fitted, outboard of the end caps and electrically connected to the housing, to provide additional cathodic protection.

56.22 Submerged equalisers

It is not practicable to match the gain of an individual repeater to the loss of its associated cable section with an accuracy sufficient to give the performance required for long systems.

It is possible to measure both repeaters and cable in the factory with great accuracy; then the accumulation of systematic errors from a number of repeater/cable sections can be equalised out in a factory-built network installed in a repeater or submerged equaliser, once per group of repeaters.

The non-systematic changes of loss, even if slightly unknown, occurring in cable during the lay, are of smooth shapes and can be corrected by switchable networks during the lay, the switches being permanently set by fusible links. Such methods have been successfully used, but have fallen out of favour because of the unfavourable production sequences involved (which cannot always be followed) and a lack of flexibility for unexpected behaviour.

Another method uses factory-built networks but which are not installed until the lay is under way. A large number of pre-built simple networks are available to choose from and, in addition, simple changes to loss and resonant frequency can be made. This method is also used successfully but suffers rather in flexibility, accuracy of equalisation of awkward shapes, and results in a rather high equaliser loss, requiring additional repeaters.

The method for dealing with non-systematic loss favoured in Britain is to design and construct networks, in the light of measurements made during the lay, using a comprehensive kit of components, fine adjustments of resonant frequency being made by variable inductance cores. Almost any circuit configuration can be used, giving great ability to deal with unpredicted contingencies, which simplify the production specifications for repeaters. A mixture of empirical methods and programmed computer methods are used to design networks and extract their performance requirements from transmission measurements. A clean room on the laying ships, similar to that used in the repeater factory, is provided for the construction of networks, which are generally plugged-in to pre-wired boards (to test their response) before they are finally soldered-in by inspected joints.

A housing, sealed by 'O' rings, following the techniques used for repeater assembly, is finally closed after tests have given assurance of the performance.

Equalisers of this type are used in the highest frequency systems (5520 channels, 44.3 MHz) of which the longest one has 271 repeaters with 13 equalisers, i.e. one per 20 repeaters. This has about 9500 dB total loss which was equalised to within $\pm 1\frac{1}{2}$ dB of the planned response at any frequency. Some lower frequency systems have been equalised to similar accuracies with about 20 000 dB of cable loss.

The above equalisation arrangements allow laying speeds of about 5 knots.

56.23 Terminal power feed

Although on short routes, where the total feed voltage is less than the repeater rating and systems can be fed from one end, it is desirable to feed from both ends, with opposite polarities, to minimise the voltage stress on the repeater capacitors. Two constant current power units in series are inherently unstable but one of them can be given two constant current characteristics, one slightly above nominal current and one slight below, with a finite slope between them, i.e. with a finite output impedance. The centre point of this crossover is arranged to be at approximately half of the system voltage, whereupon the system stabilises at nominal current and at approximately half voltage per unit. If one power unit fails, a voltage limit operates to prevent the other power unit feeding the entire voltage, if this is not permissible.

Current alarms are given at $\pm 1\%$ and $\pm 10\%$, and the unit is usually switched out at $+10\%$. Voltage alarms are at $\pm 3\%$ with a switch off settable at 10% to 40%. These facilities are duplicated.

500 mA is the usual power feed for high-capacity (British) systems. The long-term accuracy is a small fraction of the minimum alarm current, although the need for this is less with systems having repeaters protected by diodes in the avalanche mode.

A 'hot transfer' facility can be provided, in which a spare power unit, normally operating on a dummy load and adjusted to give the local system voltage, can be switched in place of the normal power unit, if it becomes faulty or requires maintenance, with insignificant change to the line current. On short systems, this is unnecessary as either end can maintain line current if the other end fails.

Primary power supply is provided by the station battery, converted to a.c. (in the audio frequency band or above) and controlled by transductors before rectification to give the d.c. line voltage.

56.24 Termination translation apparatus

Frequency allocations are based on internationally agreed spectra at the base band so that extension over a land network is possible where the bandwidth is sufficient. This base band is transmitted in one direction of transmission; and in the other direction a band formed from the base band by modulation by a single carrier is transmitted.

However, special multiplexing equipment enables one to obtain a higher capacity by interconnection at one of the standard levels (see *Figure 56.35* which illustrates a 45 MHz system). Only the higher multiplexing levels require non-standard frequency translations so that standard designs may be used for supergroup and mastergroup translations. Again, the normal gaps are retained in the spectrum for working at super master group and master group level. However, only 8 kHz is required for super group access and special filters are available for through-group working with 3 kHz spaced channels.

Up to $7\frac{1}{2}\%$ of the spectrum is used for ancillary services, like engineering speaker circuits, noise monitor bands, pilots and supervisory signals.

On the widest band systems, group delay distortion (despite the effects of the directional filters) can be low enough, in some parts of the spectrum, to permit colour television transmission. On 45 MHz systems, a 625 line PAL, $5\frac{1}{2}$ MHz service is available, in either or both directions, for up to 2500 km to AA/CMTT standards. This is in addition to half the normal telephony usage. On short systems (up to about 100 n.m.) a single TV channel can be sent above the high-band spectrum while permitting the full telephony traffic to be carried. The primary translation from video to carrier transmission uses equipment developed for land coaxial systems. It is in the band 6.3 to 12.3 MHz of which the lower 0.5 MHz is a vestigial side band. Further translations place this band into the required parts of the line spectra. TV transmission requires very comprehensive group delay and/or waveform distortion correction but in some other respects, e.g.

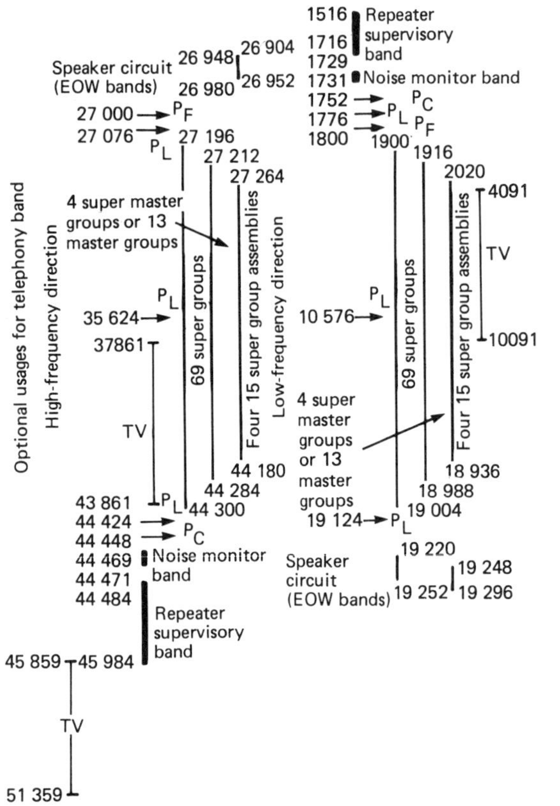

Figure 56.35 NG-1 system line frequency spectrum. All frequencies are in kHz. P_L, line pilot; P_C, cable pilot; P_F, frequency comparison pilot

differential gain and phase, problems can be less than for microwave systems.

56.25 Recent developments

Two refinements have recently become available to reduce the effects of misalignment of levels on a system. These eliminate the reduction øf repeater gain otherwise necessary to maintain signal-to-noise ratio requirements.

One is the use of a network, containing thermistors, in the feedback path of the repeater amplifiers, which changes the gain according to the ambient temperature in such a way as to largely cancel the temperature coefficient of loss of its cable section. Reductions of at least 3/1 on the apparent temperature coefficient are normal.

The other refinement is to control the gain of the several parts of a system from shore. In British practice this takes the form of an electronic switching device which may be linked to the supervisory frequency (for identification) at every 20th repeater. Automatic gain and equalisation control is never used in the submerged plant, because of fail-safe requirements, although it is increasingly used at terminal stations to avoid manual adjustments.

The use of speech traffic concentrators, such as 'TASI'* and 'CELTIC'†, which can increase the traffic capacity of an existing system of sufficient loading tolerance, by at least 2/1, will be increasingly common. These continually switch traffic into the otherwise quiet periods of channels.

* TASI: time-assigned speech interpolation.

† CELTIC: concentrateur exploitant les temps d'inactivite des circuits (telephone channel speech concentrator).

56.26 Future development

56.26.1 Coaxial cable systems

Although coaxial cable FDM systems are a mature technology, some future development is possible.

Improved production methods and components, such as the use of hybrid thick film technology, can offer some cost reduction with no relaxation of standards and, on high-traffic routes, wider bandwidths can reduce the cost per circuit mile. Laboratory work has shown that 150 MHz bandwidths are quite practicable, probably using a combination of feed-back and feed-forward principles, if the demand for high-traffic systems should warrant it.

Repeater gains can be increased (reducing the numbers required in proportion) by measures to reduce intermodulation. These include the use of increased phase dispersion in repeaters and partial cancellation of the intermodulation products at the terminals. A very effective method, as yet untried, is to use a submerged frequency frogger to transpose the high and low bands in the middle of a route. Since the low-band noise can be made negligible with respect to the high-band noise, an improvement of 7 dB in repeater gain (giving about 20% fewer repeaters in high-capacity systems) and $3\frac{1}{2}$ dB in overload point is theoretically possible.

However the advent of optical fibre systems will probably render coaxial cable systems increasingly obsolescent.

56.26.2 Optical fibre systems

It is likely that the future of high capacity submerged systems lies in the exploitation of optical fibres, see Chapter 57. This is because of the extremely low attenuation which is applicable to appropriate wavelengths of near monochromatic weakly guided light waves in high-purity silica glass.

While there may be many vicissitudes on the way, the probable nature of optical systems in the middle future is fairly clear, and a typical configuration will comprise:

(*a*) *Monomode fibre*

Multimode fibre is unattractive because of the limited frequency response due to phase interference between modulation signals arriving via different path lengths. On a well designed single-mode fibre, the phase dispersion penalty can be made very low so that only the attenuation limits repeater spacing. With doped silica glass the lowest attenuation occurs at about 1.55 μm wavelength, where it may be only $\frac{1}{3}$ dB km^{-1} including the effects of cabling and splices.

(*b*) *Digital transmission*

It is hardly practicable to design an optical system with the signal-to-noise ratio and linearity required for wide-band FDM transmission, so that digital transmission is required. Binary PCM will be used, with a code having some redundancy, such as 7B/8B, at 140 or 280 Mbit s^{-1} (1920 or 3840 voice channels per fibre).

(*c*) *Laser light source*

Lasers are necessary for maximum light output and it is important to couple the light source efficiently to the small diameter fibre. The probable material is GaInAsP with a mean power output at 1.55 μm of upwards of −8 dBm with lens launching.

(*d*) *Detector*

Avalanche photodiodes are not attractive at long wavelengths so that the detector is likely to be a PIN quaternary device using GaInAsP. This must feed into an extremely high-impedance, low-capacitance preamplifier for which a specially developed FET will be used. The nett capacitance at the interface dominates the frequency response which is equalised by a rising gain characteristic at 6 dB/octave, over the minimum band necessary to control pulse spreading. A minimum received mean power of about −45 dBm, or less, should be possible.

(*e*) *Repeater section length*

With the above arrangements, and allowing about 8 dB margin for contingencies, the section length is about 90 km. A six fibre cable at 140 Mbits s^{-1} has then a capacity of 3840 circuits. This compares with a 45 MHz coaxial system of 3900 circuits, but which has a repeater spacing of only $5\frac{1}{2}$ km on 3.73 cm (1.47 in) cable. The long section is the attraction of optical systems. Assuming that the cable cost is much the same, and neglecting all but submerged apparatus costs, a regenerator could cost several times that of a coaxial repeater to break even. Although there could be about 50 transistors in a regenerator, compared with seven in a coaxial repeater, and other components are also rather more numerous, the circuitry is much more amenable to integration techniques, using large numbers of active devices on a single chip combined with hybrid thick film technology for the passive components. With these techniques it is expected that suitable cost targets can be met with no sacrifice to the reliability considerations traditional to submerged systems.

Reference

1 HOWARD, P. J. *et al.*, '12 MHz Line Equipment', *Electrical Communication*, **48**, No. 1/2, 27–37 (1973)

Further reading

Telephone systems and equipment

ATKINSON, J., *Telephony*, Pitman (1950)
BEAR, D., *Principles of Telecommunication Traffic Engineering*, Peter Peregrinus (1976)
BERKELEY, G. S., *Traffic and Trunking Principles in Automatic Telephony*, Benn (1949)
FLOOD, J. E. (Ed), *Telecommunications Networks*, Peter Peregrinus (1976)
SMITH, S. F., *Telephony and Telegraphy*, 3rd Edn, Oxford University Press (1978)

General

BELL TELEPHONE LABORATORIES, *Transmission Systems for Communications* (1965)
Electrical Communication, **48**, No 1/2, Special issue on Transmission (1973)
Siemens Review, **38**, Special issue *Communications Engineering* (1971)
SCHNURR, L., 'Mobile communications', *Telecommunications*, **16**, No 11, October (1982)
MAYO, J. S., 'Communications at a distance', *Mini-Micro Systems*, December (1982)

Line systems

Bell Systems Technical Journal, **32**, No 4, Special issue *L3 Coaxial System* (1953)
Bell Systems Technical Journal, **48**, No 4, Special issue *L4 System* (1969)
Bell Systems Technical Journal, **53**, No 10, Special issue *L5 Coaxial Carrier System* (1974)
BAX, W. G., '1 pW per km on a 3600 Channel Coaxial Line', *Philips Telecommunication Review*, **35**, No 4 (1977)
HERMES, W. *et al.*, 'Level Regulation of Coaxial Line Equipment', *Philips Telecommunication Review*, **30**, No 1 (1971)
KALLGREN, OVE, 'New Generation of Line Systems', *Ericsson Review*, No 2, 48–53 (1974)
MILLER, J. R. *et al.*, 'Recent Development in FDM Transistor Line Systems', *Post Office Engineers Journal*, **63**, Part 4, 234–241 (1971)
NORMAN, P. *et al.*, 'Coaxial System for 2700 Circuits', *Electrical Communication*, **42**, No 4 (1967)

Multiplex

HALLENBACH, F. J. *et al.*, 'New L multiplex', *Bell Systems Technical Journal*, **42**, No 2 (1963)

Standardisation

RECOMMENDATIONS OF THE CCITT, 'Line Transmission', *Yellow Book*, Vol. 3, ITU, Geneva (1981)
RECOMMENDATIONS OF THE CCITT, 'Protection', *Yellow Book*, Vol. 9, ITU, Geneva (1981)

57 Fibre-optic Communication*

H J M Otten
N.V. Philips Gloeilampenfabrieken,
Electronic Components and Materials Division

Contents

* Based on an article which appeared in *Electronic Components and Applications*, **3**, No. 2 (1981).

57.1 Introduction

The idea of using light as a medium of communication is by no means new: indeed, the technology has now celebrated its centenary.[1] For most of its history, the development of optical communications centred on the use of beams of light, but success was limited. Suitable light sources were a problem until the advent of the laser; practical ranges were limited by atmospheric conditions and beam alignment accuracy.[2] Thus, although the idea was sporadically revived for short-range applications, it was not until the rapidly developing technologies of semiconductor optoelectronics and fibre optics were married that the full potential of optical communications could be realised.

As with most new technologies, interest in fibre optics has centred on the most technically demanding applications: long-distance wideband links for telephony, data and TV signals. However, as is often the case, technological spin-off from the main endeavour has provided a basis for the use of fibre-optic communication techniques in a range of applications that are likely, in total, to represent the dominant market.

The reason for the widening interest in fibre-optic communications is not hard to find since an optical fibre is almost the ideal transmission line. It is made from abundant, low-cost materials, and its bandwidth is larger than that of any other line. It has a low weight per unit length, and it is an electrical insulator. It is immune to electromagnetic interference and, thus it is also immune to crosstalk from adjacent fibres. It is virtually immunte to tapping. Several different types of optical fibre, in various forms, are now available commercially, most having attenuations well below 10 dB km^{-1}. To accompany them, connector systems and splicing aids have been developed.

Unlike electrical transmission lines, optical fibres require light sources and photodetectors to interface them with electronic equipment. These are, however, little if any more complicated than the drivers and multiplexers used with wideband cables. A variety of semiconductor generators and detectors have been developed that are very suitable for short- and medium-range applications. Most are in packages designed for easy connection to optical fibres.

It is now possible to assemble from commercially available components optical communications systems that are ideal for use in hostile or hazardous environments, or where large potential differences exist, or where earth loops must be avoided. Applications range from factory instrumentation, through multi-terminal data systems to vehicle control. It is with such practical, local fibre-optic systems that this chapter is concerned.

Optical fibres rely for their operation on the phenomenon of total internal reflection. If a ray of light propagating within a medium of refractive index n_1 approaches the boundary with a second medium of refractive index n_2 at an angle less than a critical angle,

$$\theta = \cos^{-1}(n_2/n_1) \qquad \text{(Snell's law)} \tag{57.1}$$

it will not pass through the boundary but will be reflected back into medium n_1, provided that $n_1 > n_2$, as shown in *Figure 57.1*.

Thus, light propagating within a rod of transparent material will be reflected from the walls of the rod if it approaches them at an angle less than the critical angle for the material and the surrounding medium (usually air). Unfortunately, such a simple arrangement is unsuitable for most fibre-optic applications since if two rods touch, as would happen within a cable, light passes from one to the other, *Figure 57.2*. This would lead to excessive losses, crosstalk and lack of security.

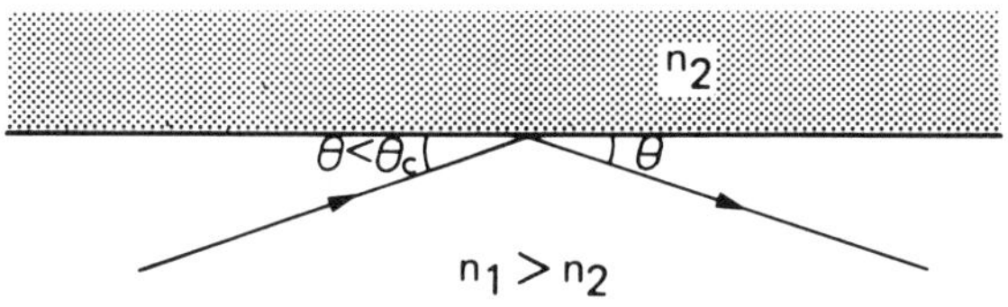

Figure 57.1 Reflection at the boundary of two media

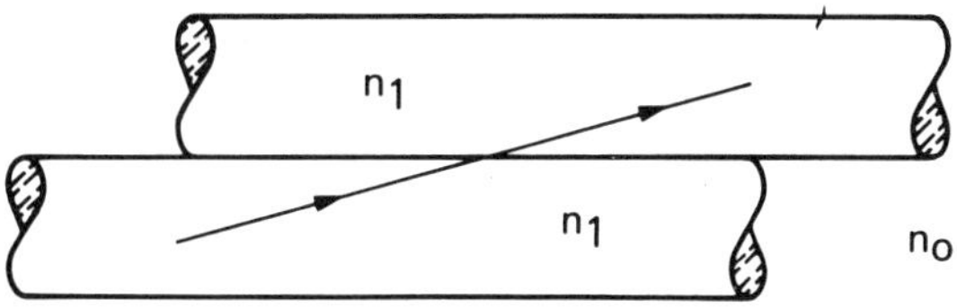

Figure 57.2 Light passing through two single fibres which touch

57.2 Practical optical fibres

In practice, optical fibres for communications purposes are made with a core of higher refractive index than that of the material near the walls, so that they are self contained. The angle at which light can enter the fibre and be trapped is reduced, but losses due to light escaping and crosstalk are eliminated, and the fibre is almost proof against tapping.

Two basic types of optical fibres are commercially available. The simplest is the step-index fibre which consists of a central homogeneous core surrounded by a cladding layer of lower refractive-index material. Light will be trapped within the core if it approaches the boundary with the cladding at an angle less than

$$\theta_c = \cos^{-1}(n_{cl}/n_{co}) \tag{57.2}$$

where n_{cl} is the refractive index of the cladding and n_{co} the refractive index of the core, *Figure 57.3*. Two fibres of this construction can touch without light escaping from one to the other.

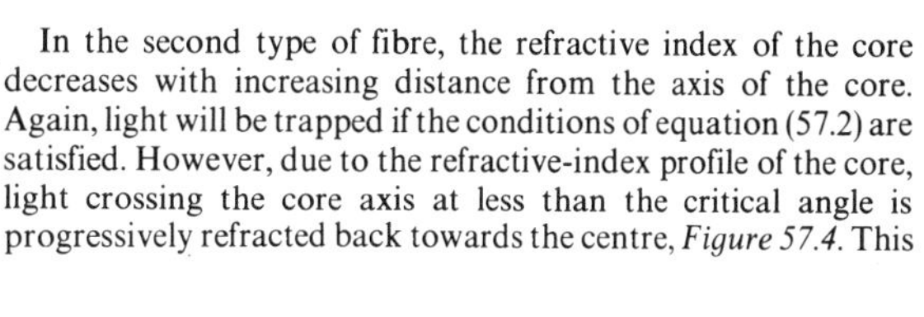

Figure 57.3 Step-index optical fibre

In the second type of fibre, the refractive index of the core decreases with increasing distance from the axis of the core. Again, light will be trapped if the conditions of equation (57.2) are satisfied. However, due to the refractive-index profile of the core, light crossing the core axis at less than the critical angle is progressively refracted back towards the centre, *Figure 57.4*. This

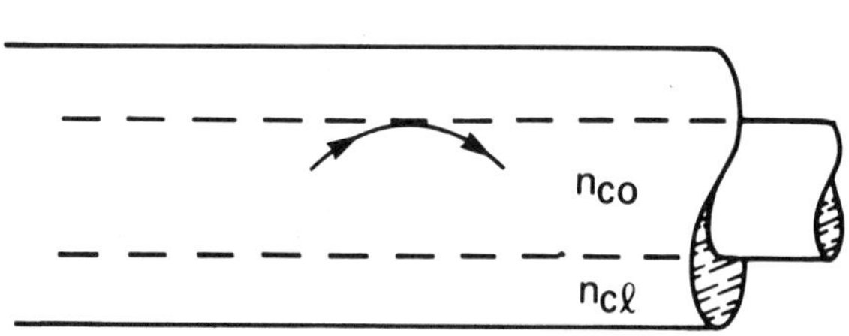

Figure 57.4 Graded-index optical fibre

property of graded-index fibres is called self-focusing. The difference between the refractive index at the axis and at the boundary of the core is about 1%.

Depending on the radius a of the fibre core and refractive indices of core and cladding, light waves propagate through a fibre in one or more modes, much as microwaves in a waveguide.[3] The number of modes N that can be sustained by a fibre at a given free-space wavelength λ_0 is a function of the normalised fibre frequency V:

$$N \simeq \tfrac{1}{2}V^2 \tag{57.3}$$

$$V = \frac{2\pi a}{\lambda_0}(n_{co}^2 - n_{cl}^2)^{1/2} = \frac{2\pi a n_{co}}{\lambda_0}(2\Delta)^{1/2} \tag{57.4}$$

where $\Delta = (n_{co} - n_{cl})/n_{co}$ is the relative refractive index difference.

Figure 57.5 shows the modes present at different V numbers,[4] and the V number as a function of core radius for $\lambda_0 = 900$ nm and n_{co} greater than n_{cl} by 1%. As the radius a and normalised frequency V of a fibre decrease, so does the number of modes of propagation possible within it. Note that hybrid mode HE_{11} has no cut-off frequency. Thus, for $V < 2.405$ there exists only one propagatin mode; the corresponding fibre is called a monomode fibre.

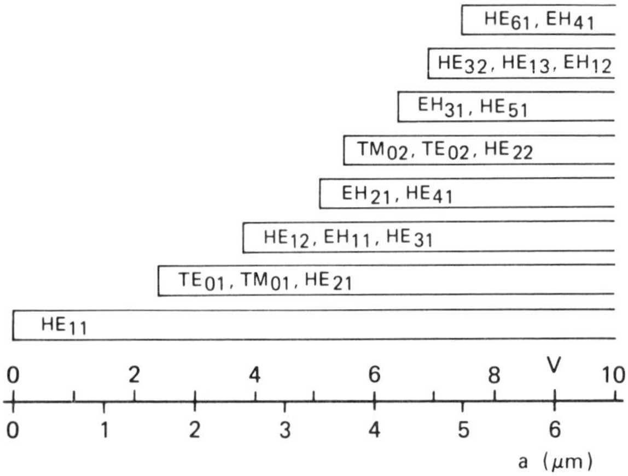

Figure 57.5 Modes present at different values of V and a. λ=900 nm, Δ=0.01, n_{co}=1.50

Figure 57.6 shows the relation between the refractive index difference and core radius a of this kind of waveguide. For a 1% index difference $a < 1.8$ μm. To avoid such very small core size and thus extreme connector accuracy the index difference is usually reduced to about 0.1%, bringing the maximum core radius to about 5 μm.

On the other hand, when $V > 2.405$, many modes propagate and the fibre is termed multimode. *Figure 57.7* shows cross sections and index profiles of usual fibre types.

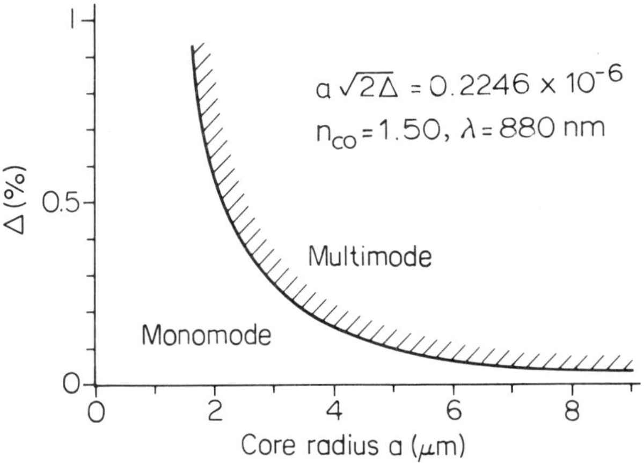

Figure 57.6 The boundary between monomode and multimode fibre operation as a function of refractive index difference

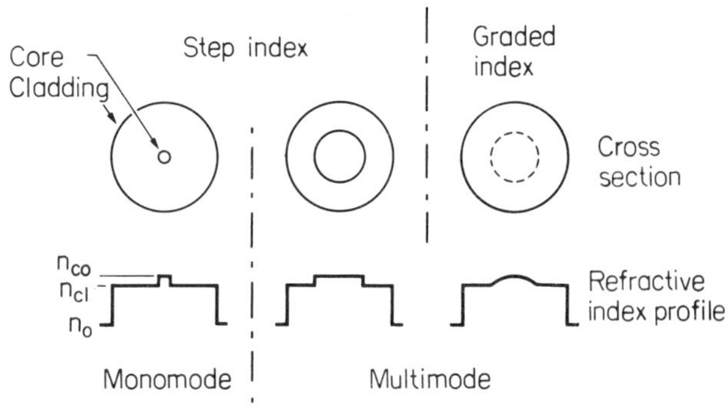

Figure 57.7 Refractive index profiles of the three basic types of optical fibre

57.3 Optical-fibre characteristics

From the point of view of the system designer, the most important characteristics of an optical fibre are material dispersion, modal dispersion, numerical aperture and attenuation. Material and modal dispersion together set the maximum fibre length or maximum repeater spacing as a function of signal frequency or data bit rate. Numerical aperture and attenuation set the maximum fibre length in terms of the available light power.

57.3.1 Modal dispersion

In step-index fibres, different modes correspond to different angles of reflection from the core/cladding boundary, and, thus, different speeds of propagation through the fibre because path length depends on reflection angle. Since refractive index is the ratio of the velocity of light in a medium to its velocity in free space, c, it follows that phase velocities due to the different propagation modes lie between two limits:

$$\frac{c}{n_{co}} \leqslant v_{ph} \leqslant \frac{c}{n_{cl}} \tag{57.5}$$

where v_{oh} is the phase velocity. The spread of phase velocities results in pulse spreading which, in turn, limits the upper frequency of modulation for a given fibre length.

In practice, due to deviations from the ideal, the degree of pulse spreading in step-index fibres is always less than that predicted by equation (57.5). Irregularities in the core/cladding boundary lead to mode conversion; fast propagation modes, corresponding to low angles of reflection, are converted to slower, higher-angle, modes; slow, high-angle, modes are converted to faster, lower-angle modes. The net result is a narrowing of the phase velocity spread (*Figure 57.8*).

Perfect step transitions in refractive index cannot be achieved in practice either, so that a continuous transition from n_{co} to n_{cl} takes place about the core/cladding boundary. Within this

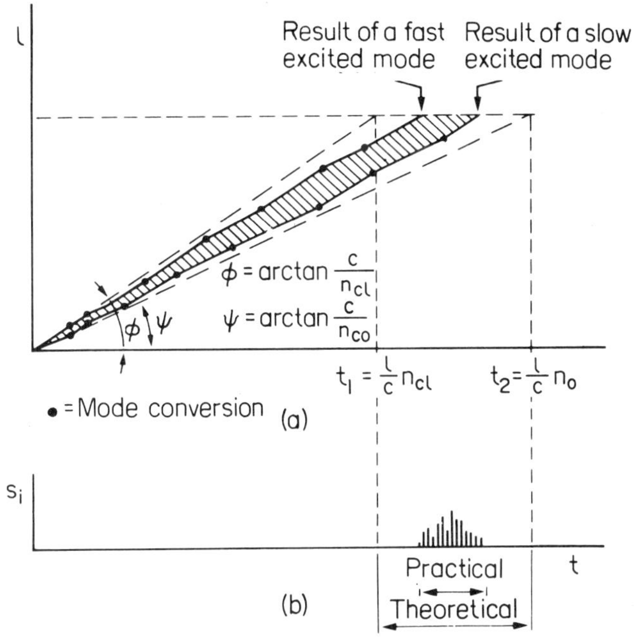

Figure 57.8 (a) Phase velocity and (b) pulse dispersion in step-index fibres. $\phi = \tan^{-1}(c/n_{Cl})$, $\psi = \tan^{-1}(c/n_{CO})$

transition zone, some self-focusing similar to that in a graded-index fibre takes place.

Modal dispersion in graded-index fibres is less than that in step-index fibres due to path-length differences being reduced by the self-focusing action. Calculation shows that the phase-velocity spread is least when the core refractive-index profile approximates a parabola.[5]

For step-index fibres, the maximum pulse spread per unit length due to modal dispersion is given by

$$\tau_m = n_{co}\Delta/c \tag{57.6}$$

In the case of graded index fibres, the maximum modal dispersion amounts to

$$\tau_m \simeq n_{co}\Delta^2/c \tag{57.7}$$

Modal dispersion can result in pulse spreads up to 50 ns km^{-1} for step-index fibres, but only about 0.5 ns km^{-1} for graded index fibres, both for multimode waveguides.

57.3.2 Material dispersion

Material dispersion is the variation of group velocity with wavelength in a transparent medium whose refractive index depends on wavelength. Pulse dispersion τ_c per unit length as a result of this phenomenon equals:

$$\tau_c = -\frac{\lambda}{c}\frac{d^2n}{d\lambda^2}\Delta\lambda \tag{57.8}$$

where $\Delta\lambda$ indicates the spectral width of the applied light source. In silica fibres, the behaviour of the refractive index n is such that the material dispersion shows a real zero at a wavelength determined by the chemical composition.

Reference 6 describes a silica fibre with zero material dispersion at 1.27 μm. In reference 7 a fibre with a different composition is presented showing a zero at 1.55 μm.

Figure 57.9 depicts the material dispersion as a function of wavelength for these reported fibres. Both curves pass through points of zero material dispersion, offering the prospect of greatly improved optical fibre performance at the longer wavelengths at which the zeros occur. The shaded bands indicate the wavelength regions of two semiconductor light source materials.

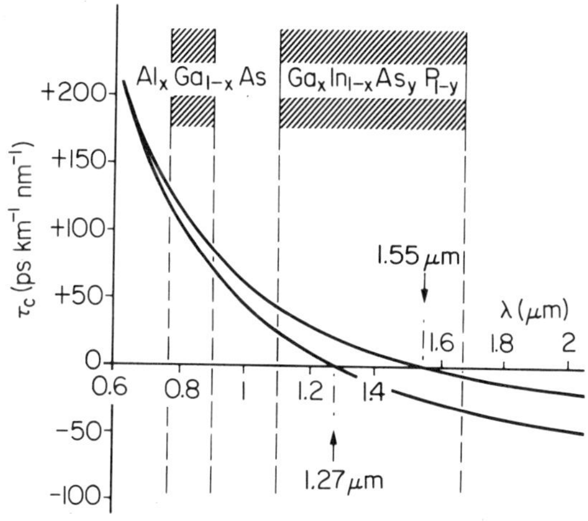

Figure 57.9 Material dispersion τ_c as a function of wavelength for two types of optical fibre

57.3.3 Waveguide dispersion

Waveguide dispersion arises principally because of the wavelength dependence of the modal V number. This waveguide dispersion is negligibly small for all modes not close to cut-off. The dispersion contribution from this source is generally not significant and can be disregarded. It is evident that monomode fibres, operated at the zero material dispersion wavelength, behave as an extremely wideband transmission line. At other wavelengths, spectral width of the light source sets the bandwidth of a single-mode fibre. A survey of typical modal and material dispersion data of silica fibres is given in *Table 57.1*.

Table 57.1 Typical dispersion values for silica fibres (Δ=0.01, λ=880 nm)

	Monomode	*Multimode*	
		Graded index	*Step index*
Modal dispersion	0	0.5 ns km^{-1}	50 ns km^{-1}
Material dispersion	60 ps km^{-1} nm^{-1}	115 ps km^{-1} nm^{-1}	60 ps km^{-1} nm^{-1}

57.3.4 Numerical aperture

Since there is a maximum angle at which light can strike the core/cladding boundary and be reflected back, there must also be a maximum angle, called the acceptance angle, at which light can enter the core of an optical fibre and be trapped. This is illustrated by *Figure 57.10*, where it can be seen that, with θ_c as defined in equation (57.1), and from Snell's law,

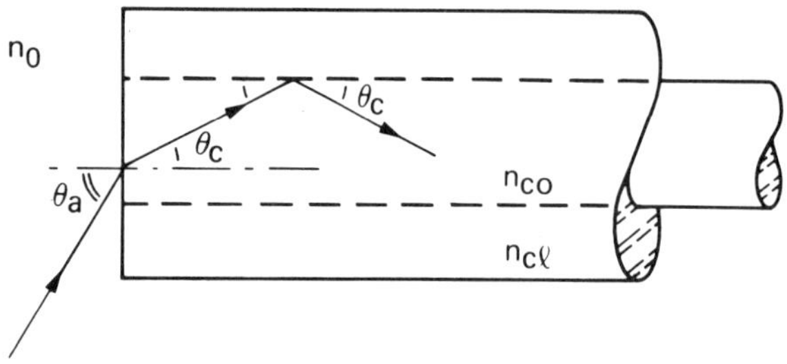

Figure 57.10 Illustration of acceptance angle, $\theta_c = \cos^{-1}(n_{cl}/n_{co})$ and $\theta_a = \sin^{-1}(n_{co} - n_{cl})^{1/2}$

$$\frac{\sin \theta_a}{\sin \theta_c} = \frac{n_{co}}{n_0} \tag{57.9}$$

For a fibre-to-air interface, $n_0 = 1$, so

$$\sin \theta_a = n_{co} \sin \theta_c = (n_{co}^2 - n_{cl}^2)^{1/2} \simeq n_{co}(2\Delta)^{1/2} \tag{57.10}$$

Angle θ_a is termed the acceptance angle of the fibre; the quantity $\sin \theta_a$ is the numerical aperture NA.

The numerical aperture determines the power than can be coupled from a light source into an optical fibre. Where the diameter of the emitting area of the source D_e is less than or equal to the diameter of the fibre core D_c, and source and fibre are in contact, the theoretical optical power P_o coupled into a step-index fibre is, assuming a Lambertian source,

$$P_o = \tfrac{1}{4}\pi^2 D_e^2 R_0 (NA)^2 \tag{57.11}$$

When D_e is greater than D_c, so that the emitting area overlaps the fibre core,

$$P_o = \tfrac{1}{4}\pi^2 D_c^2 R_0 (NA)^2 \tag{57.12}$$

where R_0 is the on-axis radiance of the source in both cases.

Equations (57.11) and (57.12) apply only when the light rays lie in a plane containing the axis of the fibre (meridional rays). However, for an extended source, some rays not in a plane with the fibre axis will also be accepted by the fibre (helical rays). This about doubles the power actually coupled[8] for a fibre of $NA = 0.2$.

In practice, source and fibre will not be in perfect contact and there will be a power loss in consequence. In most cases, this loss will be at least 2 dB.

57.3.5 Attenuation

Attenuation, or rather its reduction, is the key to the current success of fibre-optic techniques. Materials research has reduced the attenuation of optical fibres from well over 1000 dB km^{-1} to less than 10 dB km^{-1}. Specimens have been produced with attenuations approaching 0.2 dB km^{-1}. There are two main causes of loss in optical fibres, absorption and scattering.

Absorption is still the principal cause of loss. It is caused by metallic ion impurities in the core material. Many of these ions have electron transition energies corresponding to light wavelengths in the region 0.5 to 2 μm.

Light passing through a fibre will be scattered by inclusions and dislocations in the material that are small compared to the wavelength of the light. Scattering can also be caused by local temperature gradients. Rayleigh scattering due to the molecules of the material themselves is independent of light intensity but varies with $1/\lambda^4$.

Very high grade glasses have Rayleigh scattering losses of about 0.9 dB km^{-1} at a wavelength of 1 μm. This must be regarded as close to the practical limit at that wavelength. *Figure 57.11* shows the attenuation of a low-loss monomode fibre as a function of wavelength.[9] The peaks are due to absorption by metallic or hydroxyl ions.

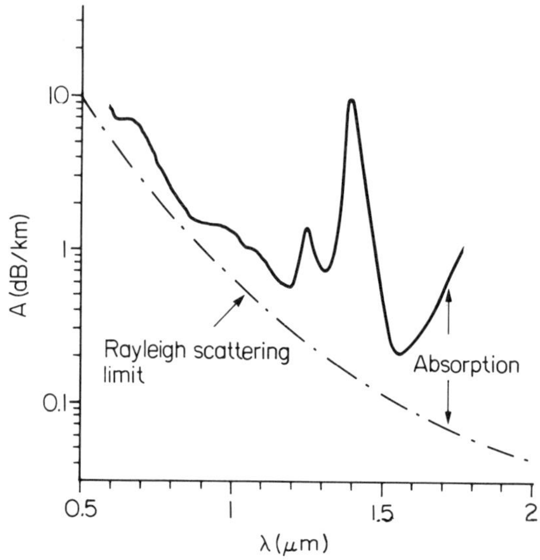

Figure 57.11 The attenuation A of a low-loss monomode fibre as a function of wavelength

57.3.6 Optical-fibre specifications

We are now able to list those optical-fibre properties that must be known if basic quantitative design is to be undertaken. They are:

- dimensions
- mechanical properties
- optical attenuation
- dispersion
- numerical aperture
- minimum modulation bandwidth
- wavelength at which optical characteristics are measured.

By way of example, *Table 57.2* lists the principal characteristics of graded-index optical fibre, type 21183 manufactured by Philips' Glass Division by a chemical vapour-deposition process.[10]

Table 57.2 Characteristics of Philips' graded-index optical fibre type 21183

Bandwidth	⩾800 MHz km
Attenuation (850 nm)	⩽3 dB km^{-1}
Pulse dispersion (FWHM)	⩽0.6 μs km^{-1}
Numerical aperture	0.21 ± 0.01
Core diameter	50 ± 2 μm
Cladding diameter	125 ± 2 μm

57.4 Light sources

Fibre-optic communications systems require sources of light of adequate brightness capable of being modulated at the desired frequency. The choice, in practice, lies between the semiconductor laser and the light-emitting diode. These devices are attractive in that their drive requirements are compatible with semiconductor practice, they are compact and robust, and, of great importance, they have the long-life and high-reliability potential that characterises well made semiconductor devices. In the case of the LED this potential is well on the way to being realised and useful lives of the order of 100 000 h can now be achieved. Chapter 26 describes the characteristics of light sources.

57.5 Photodetectors

For the purposes of fibre-optic system design, the principal characteristics of a photodetector are:

- quantum efficiency, the proportion of incident photons converted into current
- response time, which sets bandwidth
- internal multiplication factor, the ratio between primary photocurrent and output current
- noise
- practical considerations, such as operating voltage and encapsulation.

For the reasons given in connection with light sources, semiconductor devices are also the preferred choice for detectors in fibre-optic systems.

Table 57.3 lists response time and internal current multiplication factor M for the main types of semiconductor photodetectors. It is apparent that the best available performance for wideband links is currently that offered by silicon p-i-n and avalanche photodiodes.

Table 57.3 Internal multiplication factor M and response time for various types of semiconductor photodetector

Detector type	*M*	*Response time* (*s*)
Photoconductor	10^5	10^{-3}
p-n diode	1	10^{-6}
p-i-n diode	1	10^{-9}
Phototransistor	10^2	10^{-5}
Avalanche photodiode	10^3	10^{-9}
Field-effect transistor	10^2	10^{-7}

Figure 57.12 shows the responsivity and quantum efficiency of various types of semiconductor photodetectors. Silicon is evidently the most suitable material for use with GaAs–AlGaAs light sources. For longer-wavelength sources, InP–InGaAsP seems the most promising material.

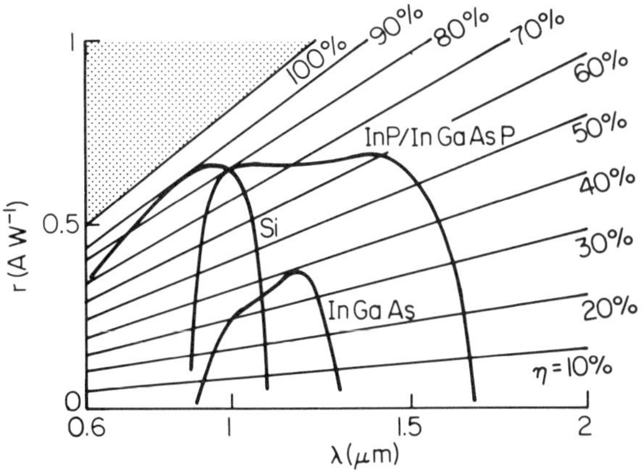

57.12 Responsivity r and quantum efficiency of some semiconductor photodetectors in various semiconductor materials

57.5.1 p-i-n photodiodes

Silicon p-i-n diodes, *Figure 57.13*, are sufficiently sensitive for most short-range applications. They have the additional advantages of being both comparatively cheap and rugged devices. Moreover, they operate with low bias (5 to 10 V).

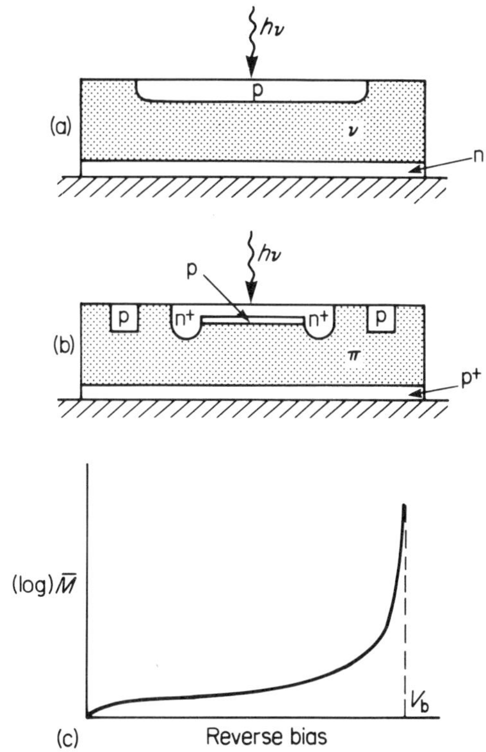

Figure 57.13 Cross sections of (*a*) p-i-n photodiode and (*b*) avalanche diode photodetectors with (*c*) typical avalanche multiplication factor M as a function of reverse bias characteristic

For fibre-optic applications, photodetector structures are generally optimised for the LED or laser wavelength to be used. In this way, quantum efficiencies

$$\eta = I_p h\nu / eP_p \qquad (57.13)$$

where I_p is the photocurrent and P_p the optical power, approaching 90% can be achieved in the region 0.75 to 0.9 μm wavelength, with modulation bandwidths of several hundred megahertz.

57.5.2 Avalanche photodiodes

Avalanche photodiodes (APD) combine detection of light with internal amplification by avalanche multiplication. They are operated with a reverse bias close to the breakdown voltage (about 200 V) so that photon-liberated electrons passing through the high field gradient about the junction gain sufficient energy to create new electron–hole pairs by impact ionisation. The amplification of an avalanche diode is strongly dependent on the bias.[11]

Since the process by which gain is achieved is stochastic, excess noise is generated by random gain fluctuations. The excess-noise factor F_e of an avalanche diode is a function of average gain $\bar{M}$,

Figure 57.13. For this reason, there is a value of $\bar{M}$ and, hence, of operating bias, for which detector signal-to-noise ratio is optimum for a given bandwidth, modulation method and optical power combination.

Due to their very high sensitivity, avalanche photodiodes are the preferred detectors in long-distance fibre-optic communication systems using semiconductor light sources.

57.6 Device packages

Optimum coupling between an optical fibre and a laser, LED, or photodetector requires that the end of the fibre be positioned accuractly close to the active region of the semiconductor die. This is a job for which the device manufacturer is best equipped, especially since positioning the fibre involves exposing the die to possible contamination.

Light sources and detectors are now available in packages that already incorporate a length of optical fibre one end of which is in close proximity to the active region of the die. Two of these are shown in *Figure 57.14*. That of *Figure 57.14*(*a*) has a pigtail of fibre about 350 mm long designed to be terminated at an equipment panel or internal bulkhead, where connection to the main fibre system is made.

In the encapsulation of *Figure 57.14*(*b*) an optical-fibre bar extends to the end of the metal ferrule; this package forms part of a standard BNC, TNC or RIM-SMA optical-fibre connector. Connection may be made to devices in either encapsulation with a loss of about 1.5 dB.

57.7 Coupling optical fibres

In order to assemble a practical fibre-optical communications system, it must be possible to make connections to and between optical fibres. For maintenance purposes, some of the connections will have to be demountable.

In essence, the optical coupling of two fibres is achieved by ensuring that their ends are clean and of adequate optical quality, and then holding them concentrically close together. *Figure 57.15* shows a number of the fibre-to-fibre joint defects that can contribute to connection loss.

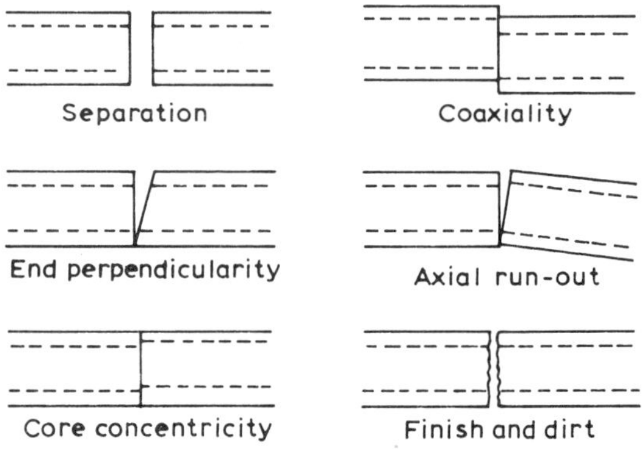

Figure 57.15 Factors affecting losses in fibre-to fibre joints

57.7.1 Fibre-to-fibre connectors

Demountable connectors are available for both single- and multiple-fibre cables. Their essential features are two aligning and holding terminations for the fibres, and a connecting bush. Most commercially available fibre connectors are based on the mechanical arrangements of coaxial types, such as those of *Figure 57.16*. Connectors for thick step-index fibres made from plastic materials are also available (see also Chapter 18).

A low-loss connection between graded-index or monomode fibres requires an extremely precise mechanical arrangement. Simple ferrule constructions in standard SMA or DIN housings are generally insufficient.

Figure 57.16 shows two types of fibre-to-fibre connectors suitable for step-index fibres, one for a fibre cable and the other for a single- or double-coated fibre for use inside equipment. Both connectors achieve alignment accuracy by means of a watch jewel at the ends of the male parts.

Welding (fusing) together fibres is an attractive, low-cost solution in cases where demountability is not required, as in buried trunk cables. Losses as low as 0.1 dB can be achieved by this method.[12]

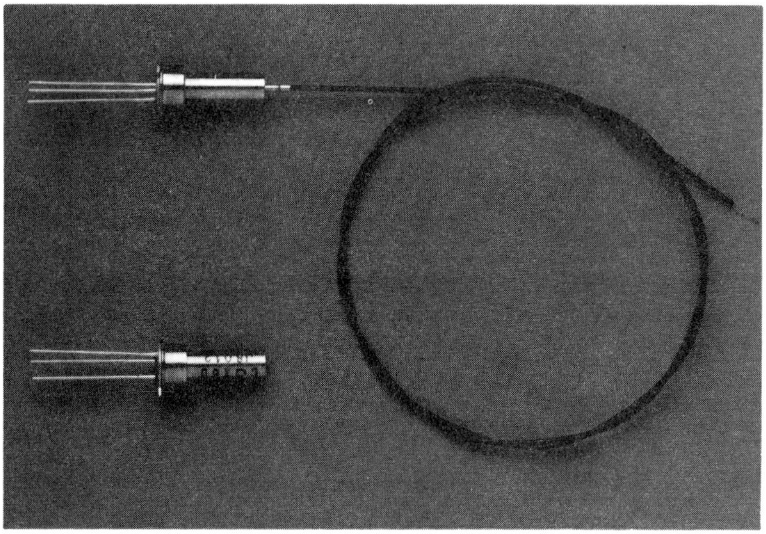

Figure 57.14 Encapsulations used for active fibre-optic devices. (*a*) Package with fibre 'pigtail'; (*b*) package with short thick fibre which is designed to form part of an active connector

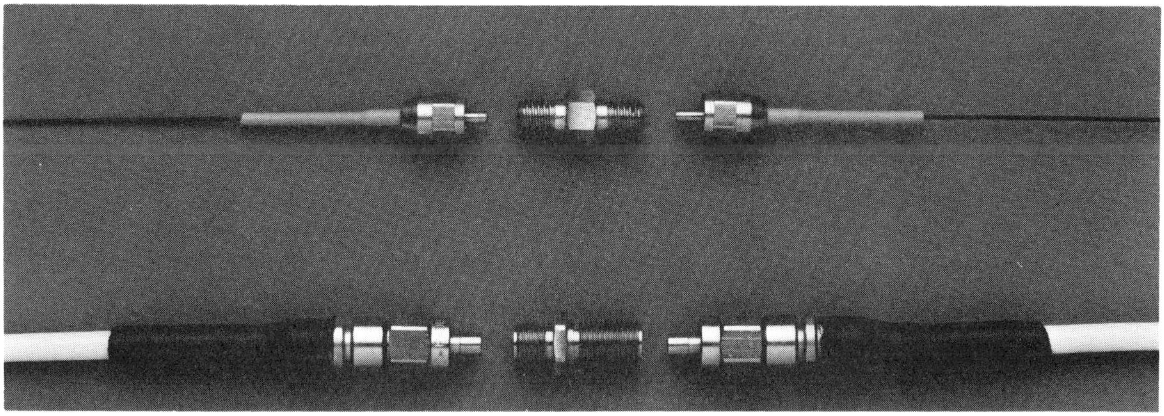

Figure 57.16 Demountable connectors for optical fibres. (*a*) Connector for coated fibres; (*b*) connector for fibre cable with strain-relieving sleeve

57.7.2 Coupling fibres to active elements

Efficient coupling of light sources and detectors to system fibres has been greatly simplified by the introduction of device packages with integral coupling fibres as shown in *Figure 57.14*. All active optical elements can, in principle, be mounted in one of these two encapsulations.

The fibre pigtail package allows the designer to position a device freely on a printed-circuit board. This is a major advantage where parasitic inductances and stray capacitance can affect system performance. Such a situation would be found on racked p.c. boards in telephone exchanges.

The fibre bar package, *Figure 57.14*(*b*), for active connectors, *Figure 57.17*, is mainly used in short-haul data links where the active elements are LED and p-i-n diodes.

Efficiency of coupling between LED or lasers and optical fibres can be improved by the use of small lenses.

57.8 System design considerations

The essential properties of a fibre-optic communication system are bandwidth and attenuation.

57.8.1 Bandwidth

Assuming Gaussian pulses, the bandwidth can be estimated simply by summing the various system time constants:

$$\tau_{tot}^2 = \tau_s^2 + \tau_f^2 + \tau_d^2 \qquad (57.14)$$

Here, τ_{tot} is the overall system time constant, τ_s is the time constant of the light source and driver, τ_d is the time constant of the system fibre and τ_f is the time constant of the detector and receiver input circuit.

The light source time constant is the rise or fall time, whichever is the longer. The optical-fibre time constant is the sum of the time constants due to material and modal dispersion:

$$\tau_f^2 = l^2(\tau_c^2 + \tau_m^2) \qquad (57.15)$$

where l is the fibre length, τ_c is the phase delay per unit fibre length due to material dispersion and τ_m is the phase delay per unit length due to modal dispersion.

Once the time constant has been calculated, the overall modulation bandwidth of the system is

$$B = 1/(2\tau_{tot}) = 1/2T \qquad (57.16)$$

where T is the minimum pulse length available for digital transmission.

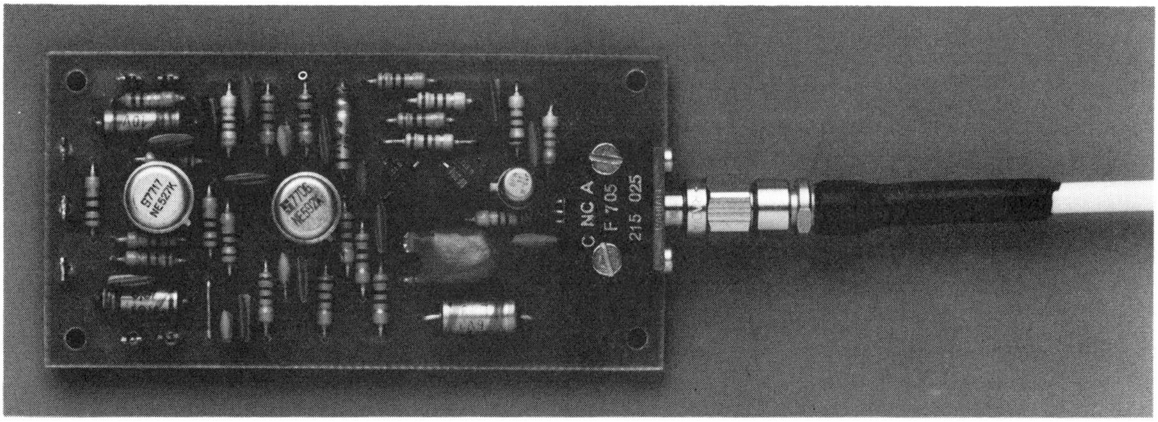

Figure 57.17 LED-equipped active fibre connector using package of *Figure 57.14*(*b*)

57.8.2 Signal attenuation

The maximum permissible signal attenuation in a fibre-optic system is

$$A_{max} = P_{s\,max}/P_{d\,min} \quad (57.17)$$

where P_s is the optical power available from the light source and P_d is the power coupled to the photodetector. A more powerful light source is equivalent to a more sensitive detector.

About 1 to 3 mW of optical power can be coupled into a graded or step-index fibre from a semiconductor laser, depending on coupling efficiency, fibre numerical aperture and laser performance.

The power coupled into a fibre from a LED can be calculated from equation (57.11) for a step-index fibre. Due to the Lambertian character of the source, the influence of the numerical aperture is much greater than with a laser. Coupled power varies, in practice, from 10 to 200 μW.

The main limitations on the sensitivity of a p-i-n diode equipped receiver are set by shot and thermal noise. Optimum performance is obtained using a high-resistance transimpedance amplifier, a silicon FET input stage, and a differentiator to compensate for the low-pass characteristic of the input stage.[13] At modulation frequencies above a few tens of megahertz, better results will be obtained with a good bipolar or GaAs FET input stage.

Figure 57.18 gives the sensitivity of a silicon FET equipped p-i-n diode receiver as a function of a diode load resistor R_L for a bit error rate of 10^{-10} at a transmission speed of 20 Mbits s^{-1}. An input capacitance of 2 pF and a FET noise equivalent series resistance of 300 Ω are assumed. R_{eq} is the equivalent series noise resistance of the FET.

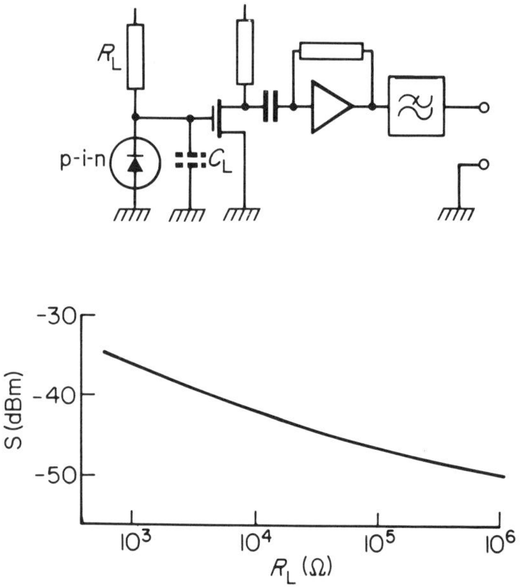

Figure 57.18 p-i-n diode receiver. *(a)* Circuit diagram; *(b)* variation of sensitivity S with load resistor R_L

Due to the internal multiplication factor M, an avalanche photodiode is much more sensitive than a p-i-n diode. The sensitivity of a receiver equipped with an avalanche diode detector is plotted in *Figure 57.19* as a function of avalanche multiplication factor M. A load resistor of 100 kΩ and an excess noise factor of $2+0.02M$ are assumed; other details are as for *Figure 57.18*.

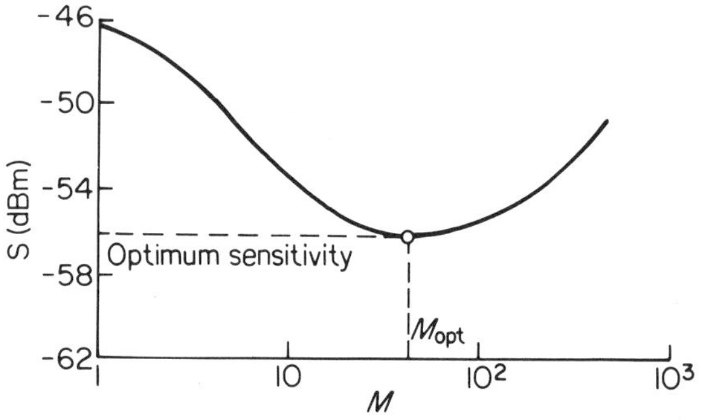

Figure 57.19 Sensitivity of receiver using an avalanche photodiode

In order to achieve a bit error rate of 10^{-10} in digital transmission, a signal-to-noise ratio of 22 dB is required. For analogue TV transmission, acceptable picture quality requires a signal-to-noise ratio of at least 45 dB and, consequently, more optical power.

57.8.3 Range

The maximum range of a fibre-optic system is set by either bandwidth or attenuation. The ranges achievable with various source, fibre and detector combinations are plotted as a function of bit rate in *Figure 57.20.*

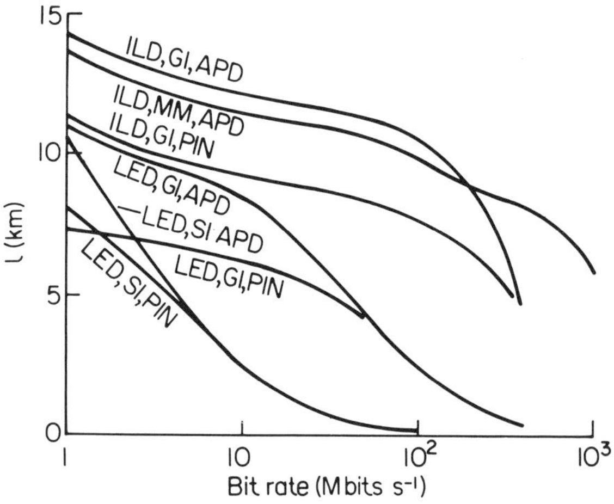

Figure 57.20 Bit rate as a function of range/repeater spacing l for fibre-optic systems using various source–fibre–detector combinations; $A=5$ dB km^{-1}, $\lambda=880$ nm. SI, step-index fibre ($\tau_m=50$ ns km^{-1}); GI, graded-index fibre ($\tau_m=0.5$ ns km^{-1}); MM, monomode fibre; ILD, laser diode; LED, light-emitting diode; PIN, p-i-n photodiode; APD, avalanche photodiode

57.9 Applications

Optical communications systems compete with millimetre waveguide, coaxial and twisted-pair transmission lines. Optical fibres are certainly attractive in prospect compared with moderate and high-quality coaxial cables. Whether they will replace twisted-pair lines depends on the price of copper.

The extra cost of the optoelectronic interface devices often outweighs the cost advantage of the glass-fibre cable, particularly in short-haul systems. However, an optical system should not necessarily be regarded as a one-to-one replacement for a copper line system. The use of multiplexing techniques to take advantage of the large information-carrying capacity of glass fibre cables can greatly reduce the cost per channel.

In hazardous or noisy environments, glass-fibre cables often do not require the coding and decoding equipment, interference suppressors, heavy shielding or isolation transformers necessary with copper lines. Fibre-optic systems are inherently interference insensitive. Due to their low weight and small size, a given cable duct can accommodate more optical cables than metal ones.

System feasibility can best be judged economically. Compare the value in use of the existing system with the actual cost of an optical replacement, taking all costs, direct and indirect, into account.

Telephone trunk lines and television distribution require levels of performance best achieved with lasers, graded index fibres and avalanche diode detectors. Particularly on the basis of cost per channel, fibre-optic cables will be cheaper than coaxial cables. In addition, due to the very low attenuation of optical fibres, few or no repeaters will be required.

A 140 Mbits s^{-1} optical link has been demonstrated that has a range of 102 km between repeaters, compared to the 1.5 km of an equivalent coaxial system.[14]

The immunity to interference, low weight, compactness and difficulty of tapping make fibre-optics especially attractive for both military and general high-security applications.

Fibre-optic systems are ideal for linking computers to peripherals. Here, speed is the most important property.

Systems using LED, step-index fibres and p-i-n diode detectors are ideal for monitoring and control purposes in industrial plants, where distances are short and information is transferred at relatively low speeds.

The small size and immunity to crosstalk of optical fibres make them attractive in prospect for internal wiring in such applications as automobiles, data-processing equipment and telephone exchanges.

Using optical fibres, a wideband network to deal with the increasing flow of information to and from the home becomes a practical possibility. One or two fibres could be used to carry TV and radio programmes, telephony, viewdata, telebanking, public utilities metering and similar services. Several projects with this in view are already in progress or under consideration: Hi-Ovis in Japan, BIGFON in the Federal Republic of Germany, DIVAC in The Netherlands, and Biarritz in France. If the idea is widely taken up, its impact on the market for fibre-optic components will be dramatic.

References

1 BELL, A. G., 'On the Production and Reproduction of Sound by Light', *Proc. Am. Assn Adv. Sci.*, **29**, 115–136 (1881)

2 CADE, C. M., 'Eighty Years of Photophones', *Brit. Comm. Electron.*, **9**, No 2, 112–115 (1962)

3 SNITZER, E., 'Cylindrical Dielectric Waveguide Modes', *J. Opt. Soc. Am.*, **51**, No 5, 491–505 (1961)

4 MILLER, S. E., *et al.*, 'Research Towards Optical Fibre Transmission Systems', *Proc. I.E.E.E.*, **61**, No 2, 1703–1751 (1973)

5 KAWAKAMI, S. and NISHIZAWA, J. I., 'An Optical Waveguide with Optimum Distribution of the Refractive Index with Reference to Waveguide Distortion', *I.E.E.E. Trans. MTT*, **16**, No. 10, 814–818 (1968)

6 PAYNE, D. N., *et al.*, 'Zero Material Dispersion in Optical Fibres', *Electron. Lett.*, **11**, No 8, 176–178 (1975)

7 CHANG, C. T., 'Minimum Dispersion at 1.55 μm for Single Mode Step Index Fibres', *Electron. Lett.*, **15**, No 23, 765–767 (1979)

8 ALLAN, W. B., *Fibre Optics, Theory and Practice*, Plenum Press, London and New York, 29–33 (1973)

9 MIYA, T., *et al.*, 'Ultimate Low Loss Single Mode Fibre at 1.55 μm', *Electron. Lett.*, **15**, No 4, 106–108 (1979)

10 VAN ASS, H. M. J. M., *et al.*, 'The Manufacture of Glass Fibres for Optical Communication', *Philips Tech. Rev.*, **36**, No 7, 182–189 (1976)

11 WEBB, P. P., *et al.*, 'Properties of Avalanche Photodiodes', *RCA Rev.*, **35**, No 6, 234–278 (1974)

12 FRANKEN, A. J. J., *et al.*, 'Experimental Semi-Automatic Machine for Hot Splicing Glass Fibres for Optical Communication', *Philips Tech. Rev.*, **38**, No 6, 158–159 (1978/1979)

13 PERSONICK, S. D., 'Receiver Design for Digital Fibre Optic Communications Systems', *Bell Systems Tech. J.*, **52**, No 6, 843–886 (1973)

14 CAMERON, K. K., *et al.*, '102 km optical fibre transmission experiment at 1.52 μm using an external cavity controlled laser transmitter module', *Electron Lett.*, **18**, No 15, 650–651 (1982)

Further reading

LOMBAERDE, R., 'Fiber-optic multiplexer clusters signals from 16 RS—232-C channels', *Electronics* (March 1982)

CHALLANS, J., 'Connector for optical trunk transmission system', *Electronic Engineering* (September 1982)

ARCHER, J. D., 'Fibre optic communications', *Electronic Product Design* (November 1982)

SUEMATSU, Y. and IGA, K., *Introduction to Optical Fibre Communications*, Wiley Interscience (1982)

58 Videotape Recording

S Lowe
British Broadcasting Corporation

Contents

58.1 Introduction

The frequency spectrum occupied by a video signal extends from a few cycles per second to around 5.5 MHz. It covers an 18 octave range, well in excess of the 10 octave limit of magnetic recording. Video recorders, therefore, modulate the signal onto a carrier and thereby reduce the bandwidth, at the expense of increasing the recorded frequencies, to about 4 octaves. Over this range it is possible to equalise the signal to give a flat frequency response.

Both frequency and amplitude modulation are used. Broadcast machines use f.m. and record the composite video directly, industrial and domestic machines use f.m. for the chrominance and therefore have to separate and recombine these signals appropriately.

Frequency modulation is chosen because of its relative simplicity compared with pulse-coded systems and its tolerance of amplitude variations compared with amplitude modulation. In the direct system there is the added advantage that no bias signal is required, while in the combined system the f.m. signal can be used to bias the a.m. component. *Figure 58.1* shows the spectrum of typical systems. It will be noted that in the combined

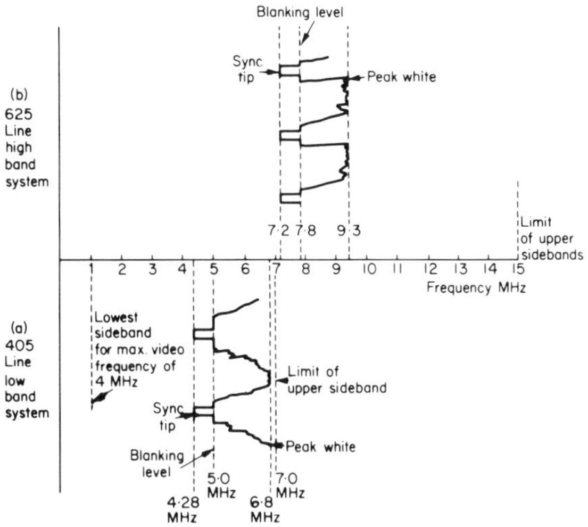

Figure 58.1 Frequencies used in (*a*) low-band and (*b*) high-band videotape recording systems

system, the colour component has been down converted to a lower frequency. It is recorded as an a.m. signal below the luminance f.m. This has given rise to the name colour under recording commonly applied to industrial and domestic machines.

58.2 Tape transport

The video head has to record and replay signals with sidebands up to 15 MHz for broadcast machines, 8 MHz for industrial and domestic machines.

Current heads have gaps in the region of 6 μm (0.235 mil) and tape is capable of recording wavelengths of 2 μm (0.067 mil). The head to tape speed is therefore determined by the head and is found to be around 0.5 m s^{-1} (900 ips) for broadcast, 0.8 m s^{-1} (300 ips) for industrial and 0.5 m s^{-1} (200 ips) for domestic machines. Longitudinal tape transport at these speeds is impractical since one hour of recording would require 82 000 m (90 000 yards) of tape travelling at 82 km h^{-1} (51 mph) for a broadcast machine.

To overcome this problem, video tape recorders employ heads mounted on drums which rotate so that the head travels transversely across the tape which can then move at normal tape recorder speeds, e.g. 38.1 cm s^{-1} (15 ips).

58.3 Quadruplex machines

Early VTR machines were of the quadruplex type, i.e. with four heads, using a tape 50.8 mm wide running at 19.05 cm s^{-1} and 38.1 cm s^{-1}, the latter being the normal standard. The tape is transported from reel to reel by means of a capstan and pinch roller in the manner of a professional tape sound recorder. For video recording the four heads are mounted in quadrature on the periphery of a head drum (sometimes called a headwheel) approximately 50 mm in diameter which rorates at right angles to the length of the tape, *Figure 58.2*. The tape is made to curve around the head drum by a guide of similar diameter into which the tape is sucked by a vacuum, *Figure 58.3*. The guide is slotted

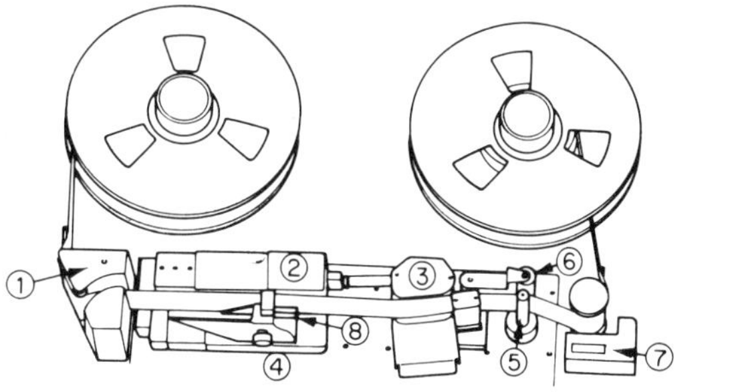

Figure 58.2 Videotape transport system: 1 master erase head, 2 head-drum, incorporating motor, head-drum, tachometer and pick up transformer or sliprings, 3 sound and cue recording heads, 4 vacuum tape guide, 5 capstan, 6 pressure roller, 7 tape timer, 8 control track head

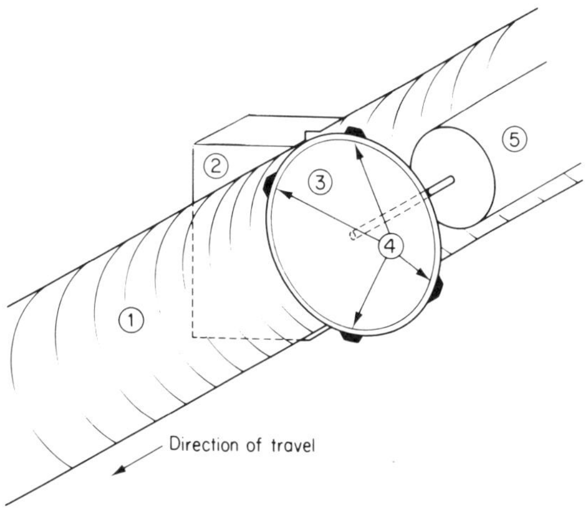

Figure 58.3 Head drum arrangement for transverse-scan recorder: 1 tape, formed into a curve around head-drum by vacuum guide, 2 vacuum guide, 3 head-drum (or head wheel), 4 recording heads projecting from head-drum, 5 head-drum motor

to enable the tip of the head drum to penetrate the normal line of the tape, stretching it slightly at the point of contact. This stretching action ensures consistent head-to-tape congtact and makes it possible to compensate for head wear, *Figure 58.4*. The

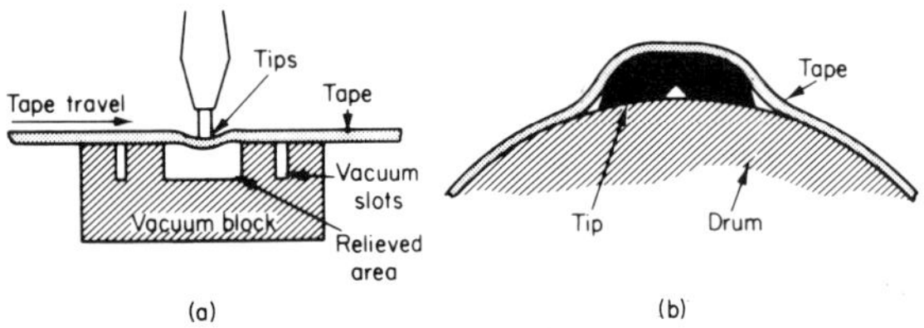

Figure 58.4 Arrangement of tape and head-drum, (*a*) cross section of tape guide vacuum block, showing penetration of recording head tips below normal level of tape, (*b*) enlarged side view of head drum showing tip projection

head drum rotates at five times the television field frequency, i.e. 250 rps for the 625-line standard, or four times field frequency, i.e. 240 rps for the 525-line system.

For the 625-line television standard the heads sweep across the tape in turn covering an arc of 114° at a head-to-tape speed of 4013.2 cm s^{-1} laying down 4 × 250 tracks per second. The tracks are 0.25 mm wide, and due to the movement of the tape at 38.1 cm s^{-1} are spaced apart 0.14 mm, i.e. the centre-to-centre spacing is 0.4 mm, *Figure 58.5*.

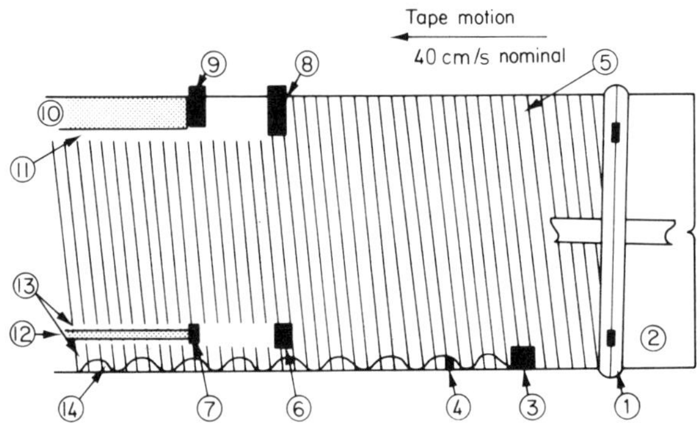

Figure 58.5 Quadruplex videotape recorder-arrangement of tracks, 1 head-drum, 2 tape, 3 control track head, 4 edit pulse, 5 video tracks 0.25 mm wide with 15.9 mm guard band, 6 cue track erase head, 7 cue track record/reproduce head, 8 audio trace erase head, 9 audio track record/replay head, 10 audio track, 1.78 mm wide, 11 guard band, 0.5 mm wide, 12 cue track 0.5 mm wide, 13 cue track bands 0.25 mm wide, 14 control track 1.27 mm wide

In the recording mode all four heads are fed with the signal continuously so that recording takes place over almost the whole width of the tape, the information recorded at the bottom of one track being duplicated at the top of the next. On reproduction each head must be switched in sequence otherwise the noise produced by the non-productive heads would be excessive. Although only 90° of rotation is required of each head in turn representing a head-to-tape sweep of 4.01 cm in fact a sweep of 4.62 cm is used so as to allow an overlap at the beginhing and end of each track in which the switching can take place. This is done in the line blanking periods of the video wave form to prevent flashing due to transients. With the 625-line standard each 90° of the head drum corresponds to 15.625 lines.

58.4 Helical machines

The present generation of machines is of the helical type, i.e. with one or two heads on a drum rotating at frame or field rate around which the tape is wrapped in the form of a helix, *Figure 58.6*. Broadcast machines use a 25.4 mm (1 in) wide tape running at a longitudinal speed of 23.98 cm s^{-1} (9.44 ips) for the C format and 24.30 cm s^{-1} (9.57 ips) for the B format, industrial and domestic machines use 19 mm ($\frac{3}{4}$ in) and 12.8 mm ($\frac{1}{2}$ in) tapes running at speeds of 9.5 cm s^{-1} ($3\frac{3}{4}$ ips) and 25 mm s^{-1} (1 ips).

58.4.1 C format

The C format machines record one field for each rotation of the head drum on a track at an angle of approximately $2\frac{1}{2}$°. When the video head reaches the end of the track it loses contact with the tape while it moves to the other edge and the start of the next track. This period during which no signal is recorded is called the vertical drop-out or format drop-out and lasts 12 lines on the 625 standard, 10 lines on the 525 standard.

If it is required that all lines should be recorded, then an extra sync head is added, spaced 30° further round the drum, which records the missing lines and some of those also recorded by the main head to allow for switching between them.

The rotational speed of the head drum, or scanner is fixed by the field rate of the video signal to 50 rpm for 625 use or 60 rpm for 525 use. The diameter of the scanner, 13.4 cm (5.3 in) is then chosen to give the required head-to-tape speed of 21.146 m s^{-1} (832.56 ips) for 625 and 25.375 m s^{-1} (999.06 ips) for 525. The video head is 0.16 mm (0.006 in) wide and record tracks 411 mm (16.2 in) long at a pitch of 0.2144 mm (0.0084 in) for 625 and 0.1823 mm (0.0072 in) for 525.

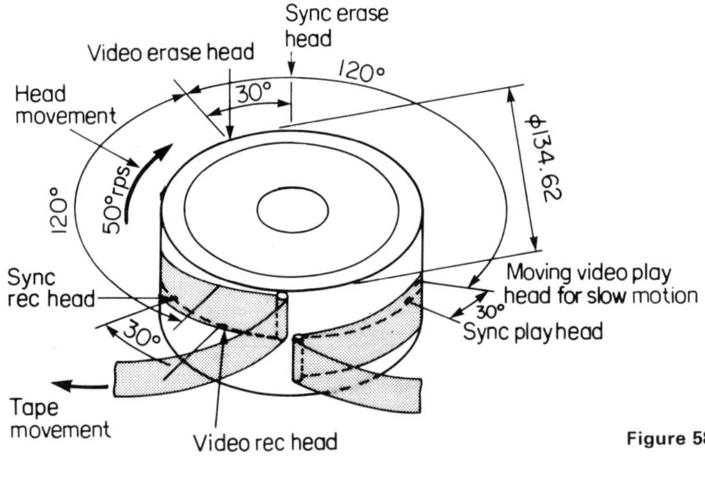

Figure 58.6 C format helical tape path

In record, the scanner is synchronised with the video signal so that the start of each track coincides with the beginning of the field. The combined effect of tape movement and the track angle ensure that the line synchronising pulses are aligned on adjacent tracks taking into account the half line offset between fields. This improves the quality of still frames.

The signal being recorded may be monitored either by viewing the f.m. signal feed to the heads via the replay demodulator and signal system, or by placing a second video head on the scanner and viewing the signal replayed from the tape. This simultaneous or confidence replay is one of the major advantages of this type of machine since it removes the need for back-up recording and spot checks.

The second advantage of the C format is the facility for still picture and slow motion. If the longitudinal movement of the tape is stopped, the video head will scan along the same path on each rotation. The angle of the scan will be altered by the lack of tape movement and the head will cross from one recorded track to another. There will be a momentary loss of picture, seen as a horizontal noise bar about one tenth of the picture height, but because of the sync line-up referred to above, the disturbance will be minimal.

These pictures are adequate for viewing during a non-broadcast operation such as editing, but to achieve undisturbed pictures the head must be made to follow the recorded track. This is done by mounting a replay-only head on a pizeoelectric crystal and driving it to follow the peak replayed r.f. This automatic scan tracking, AST, or dynamic tracking, DT, enables full broadcast quality pictures to be replayed at speeds from 3 reverse to $\frac{1}{5}$ play × normal forward.

Before the pictures are suitable for transmission it is necessary to stabilise the signal timing and to replace the pulses lost in the format drop-out. This is done in the signal processor known as the time base corrector, TBC, discussed later.

There are four audio tracks available, all recorded along the edges of the tape and 0.8 mm (0.031 in) wide. Tracks 1 and 2 may be used as a stereo pair, track 3 is usually reserved for the 80-bit frame identifying code, time code, used for sequence logging, machine synchronisation and during editing. Track 4 is an optional feature and occupies the part of the tape set aside for the sync head of full information machines (*Figure 58.7*).

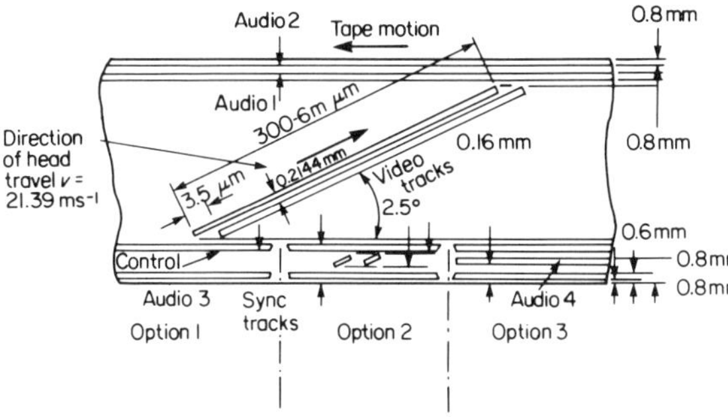

Figure 58.7 Positions of recorded tracks in the three options of C format

58.4.2 B format

The B format occupies a place between the quadruplex and C formats in that it uses a scanner with two video record replay heads and 2.54 cm (1 in) tape upon which pictures are recorded in segments.

The rotational speed of the scanner, 150 rps, is the same for both the 625 line and the 525 line systems. Each 625 line picture is recorded in six segments, each 525 line picture in 5 segments.

The tape is wrapped in a helix around the 5.08 cm (2 in) scanner so that it is in contact for 190° of the circumference to allow overlap in the recording. The track angle is 14.289° and their length is 80 mm which is traversed by the head at a speed of 23.73 m s^{-1} (933.76 ips). The tracks are 160 μm wide and the longitudinal tape speed of 24.3 cm s^{-1} causes them to be laid at a pitch of 0.2 mm. This leaves a guard band between tracks of 40 μm (*Figure 58.8*).

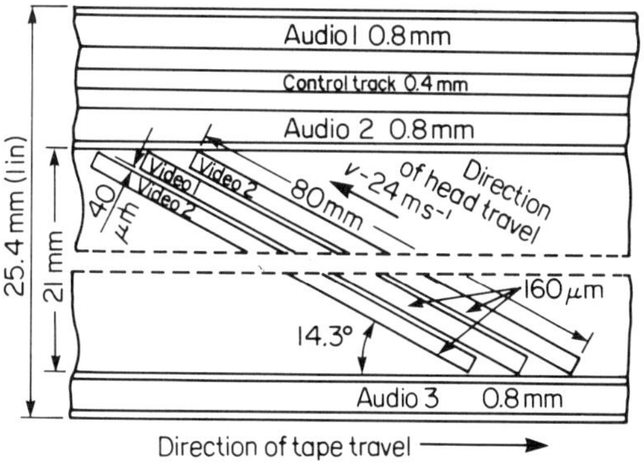

Figure 58.8 B format tupe layout

There is no provision for off-tape replay during recording although there is space on the scanner should an extra pair of heads be required at a later date. The segmentation of the picture prevents slow motion or still frame except with the addition of a field store. On the other hand, the overlap of the recorded video means that no special provision is required to record the field interval.

The small-diameter low-mass scanner is virtually immune to speed changes caused by gyroscope forces when the macine is moved rapidly. The size and the 190° wrap make it relatively easy to produce a cassette machine, either for portable use or for random access machines used to transmit advertisements.

As with all colour VTR for broadcast use the output has to be stabilised by a time base corrector. In the case of the B format a range of half a line (33 μs) is adequate because of the good mechanical stability of the scanner. The larger diameter scanner of the C format can cause gyro-induced errors requiring a range of ten to twenty lines although the slow motion facility also requires an extended range.

58.5 Modulation

The video signal is frequency modulated in order to reduce the bandwidth. The carrier frequency is limited by head and tape development and for early machines was barely greater than the video signal. As better heads and tape became available, it was possibly firstly to accommodate colour and secondly to improve picture quality. More recently, the trend has been to reduce tape consumption.

58.5.1 Low band

The original (low-band) modulation system used for video tape recording of 405 line black-and-white pictures had black level based on 5.0 MHz with peak white extending to 6.8 MHz and sync tips at 4.28 MHz, i.e. a maximum deviation of 2.52 MHz. With such a low-modulation index and comparatively little energy at high frequency the side bands are virtually limited to a single pair. For a bandwidth of 4 MHz the lowest sideband is 1 MHz (5–4) and the upper sideband can be limited to just above the upper deviation frequency, i.e. 7 MHz so that the whole video signal can be contained within the range 1 to 7 MHz which is less than three octaves, *Figure 58.1(a)*.

58.5.2 High band

The low-band system is not suitable for 625 line colour television recording because of the extended bandwidth (a flat response up to 5.5 MHz is required) and the large amount of h.f. energy in the colour subcarrier signal at 4.43 MHz. If a low carrier frequency is used for the recording system, spurious noise pattern effects can be produced by the colour subcarrier because the lower sidebands, which contain substantial energy extending to below zero frequency (relative to the carrier), can become 'folded back' in the modulation process to beat with the carrier, *Figure 58.9*.

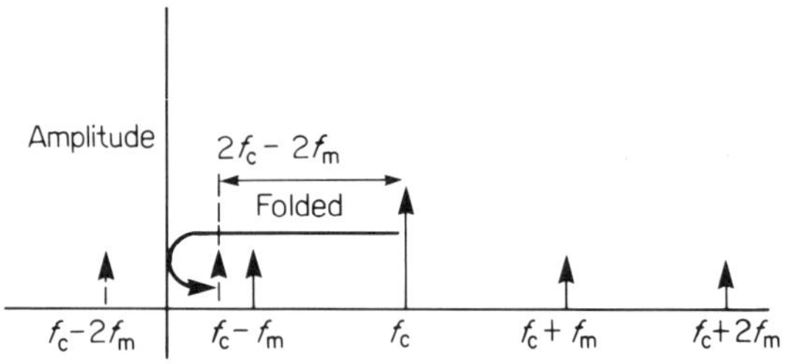

Figure 58.9 The folded sideband. f_c is the carrier frequency and f_m the modulating frequency

For this reason the 'high-band' system is used for 625 line colour picture recording. The system is based on blanking level clamped at 7.8 MHz with sync tips at 7.16 MHz and peak white 9.3 MHz for quadruplex machines. The helical formats use slightly different frequencies, *Figure 58.10*.

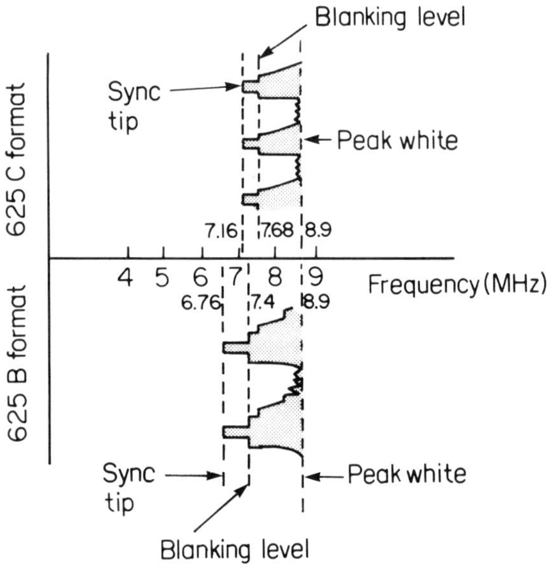

Figure 58.10 Frequencies used in (*a*) EBU C format and (*b*) EBU B format videotape recording systems

In practice, it is necessary for the video replay head to repdocuce frequencies up to above 15 MHz in order to recover the necessary upper sidebands, *Figure 58.10(b)*.

58.5.3 Super high band

The high-band system suffers from a picture degradation known as Moiré. It is visible as a moving patterning and results from the sidebands which have been folded back. By increasing the carrier frequencies to 9.9 MHz for blanking, 9 MHz for sync tips and 12 MHz for peak white, 9.9 , 9.58 and 10.7 MHz for 525 line systems, Moiré can be practically eliminated.

The super high-band standard also proposed the use of an extra subcarrier, pilot tone, which could be used during replay to give continuous correction of amplitude and phase errors in the chrominance component of the signal. The pilot tone frequency is 6.65 MHz for 625 and 5.37 MHz for 525.

58.6 Colour under

For some applications it is desirable to reduce tape consumption at the expense of the best picture quality. This may be achieved for a black-and-white system simply by restricting the bandwidth, thereby losing fine detail. For colour signals, any reduction in bandwidth would eliminate the colour subcarrier with catastrophic consequencies.

The solution is to change the frequency of the colour subcarrier so that it occupies a space in the spectrum below the luminance, hence colour under. Because the eye cannot perceive colour in fine detail, it is possible to restrict the bandwidth of the chrominance signal to about $\pm\frac{1}{2}$ MHz. The chrominance is then recorded as an amplitude modulated signal using the luminance f.m. as bias (*Figure 58.11*).

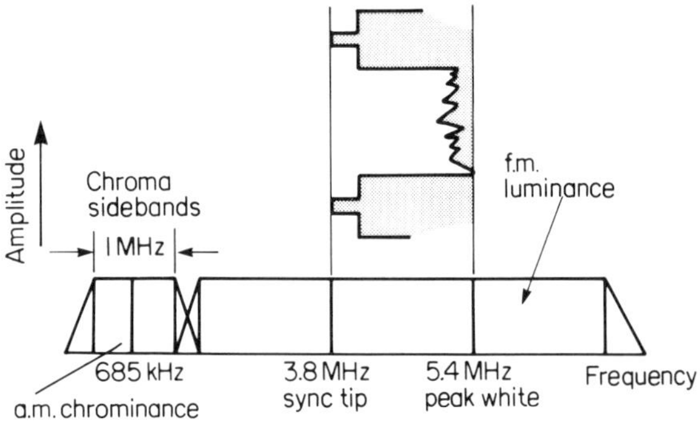

Figure 58.11 Colour under spectrum

58.7 Sound tracks

The sound tracks on all formats are recorded longitudinally using conventional audio recorder techniques.

On the quadruplex system the magnetic particles of the tape are oriented transversely to favour the video recording and consequently there is a 6 dB drop in the audio sensitivity. This is compensated by the relatively high tape speed and width of the tracks. The main audio track, placed at the top edge of the tape is 1.7 mm (70×10^{-3} in) wide. Near the bottom of the tape there is a second audio track 0.6 mm (20×10^{-3} in) wide which is poorer quality and is intended mainly for recording cue tones to control the edit process. It is also used to record the sound effects of sporting events which may be used later to add to commentaries in various languages.

On the helical systems, the video tracks are nearly parallel to the tape edge and so the magnetic particle orientation favours both video and audio. The C format has three audio tracks with the option of a fourth one. All the tracks are 0.8 mm (31×10^{-3} in) wide, and may be used jointly or as a stereo pair. Track 3 is reserved for the 80-bit frame identifying code, time code, used to synchronise machines and locate frames during editing.

The fourth track is an option on 625 machines only. The area of tape it occupies is used on 525 machines to record the vertical interval lost in the format dropout. On 625 machines the four audio configuration is chosen because of the advantage it offers as a dubbing and intermediate mixing channel during editing.

The B format has three audio tracks, two are located near to the top of the tape separated by the control track described below. They may be used in parallel, for stereo or for separate programme material. Track 3 may be used for dubbing but is mainly intended to carry time code and it is located near the bottom of the tape. All the tracks are 0.8 mm (31×10^{-3} in) wide.

58.8 Control track

In order to replay a recording, the video head has to be made to follow the video tracks. This is achieved by recording pulses derived from the head drum rotation on to a longitudinal track similar to the audio tracks. Included in the control track waveform are pulses to identify the points where edits may be made without disturbing the colour subcarrier to line sync relationship.

On quadruplex machines, the control track waveform is a 250 Hz sinc wave which is recorded without bias. On the B and C formats, there are field rate pulses recorded with bias. Additionally, edit pulses are added every four or eight fields to identify the PAL frame sequence.

58.9 Record

In the record mode the video is fed to the rotating heads via slip rings or rotary transformers. The rotation of the drum is controlled by the incoming video signal so that the field synchronising pulses are recorded at a specific point on the video track. In the case of the quadruplex machine, one had is chosen for this task and the pulses are recorded in the middle of the tape. On helical machines where one head records a complete field, the field pulses occur at the beginning of each track. The drum is further controlled in conjunction with tape movement to ensure that the line sync pulses are aligned on adjacent tracks. This can be an advantage during the replay mode especially for still frames.

It is usual to apply the same signal to all heads simultaneously during record. Provision then has to be made to erase the space for the audio tracks, although on helical machines it is possible to arrange the tape path so that the video heads do not contact the area used by the audio tracks.

When editing, however, it is necessary to arrange for the video head record current to be turned off as it passes over the audio and control tracks. This is done by sensing the position of the drum using the tachometer pulses which form part of the drum servo loop.

58.10 Reply

In replay only the head which is in contact with the tape is fed to the replay electronics. This is to obtain the best signal-to-noise ratio, which would otherwise be degraded by the output of the non-relevant heads.

The signals are again routed via slip rings or rotary transformers to the pre-amplifiers and replay electronics. The rotation of the drum and the movement of the tape are controlled so that the signal is replayed at the correct rate and can be synchronised to other sources if this is required. If the machine is replaying in isolation it may be locked to the mains supply frequency or an internal crystal.

On non-segmented helical machines the switch from one head to another occurs once every field and it is placed in the field-blanking period. This allows time for any timing discontinuities to be corrected before the active picture starts. On quadruplex and B format machines arrangements have to be made to switch during the much shorter line-blanking period. It is also necessary to correct any timing or colour phase errors which would show as horizontal bands across the screen in areas of highly saturated colour.

The replayed r.f. signal is amplified, limited to remove any variation in amplitude and demodulated. Before the pictures may be viewed the phase and timing errors introduced by the mechanical instability of the tape transport have to be removed. This may be done in one of two ways depending on how the signal is to be viewed.

For broadcast use the signal, sync pulses, luminance and chrominance must be retimed. This involves passing the signal through a variable-length delay line, either in its analogue form or after conversion to an 8-bit digital form. The duration of the delay is varied in accordance with the error between the off-tape signal timing and a suitable reference to obtain a fully stabilised signal.

For non-broadcast use only the colour component of the signal need be processed. The timing errors present in the sync pulses and luminance component are easily accepted by black-and-white picture monitors or receivers. In order to decode the colour signal correctly there has to be a stable relationship between the reference colour burst at the start of each line and the colour subcarrier along the line. This may be achieved on the colour under recording system used for domestic and industrial machines by modulating the reference oscillitor used in the heterodyne process. By varying its frequency in sympathy with the offtape jitter a stable colour signal may be obtained, but at the expense of losing the fixed subcarrier to line sync relationship. This means that the output may not be mixed or added to another signal, say, from a camera or caption generator.

58.11 Time base correction

All broadcast machines and any that are used as feeds to a mixer during editing require time base correction to overcome the effects of the mechanical instability of the tape and head movements (*Figure 58.12*).

The head drum servo on a quadruplex machine is capable of synchronising the video output to within $\pm 100\ \mu s$. This is further improved to $\pm 4°$ of a cycle of the subcarrier by varying the length of an analogue delay line in oppoistion to the error.

It is necessary to produce a large range with a high degree of resolution, which in its simplest form would require many short delay line sections. To reduce the number of sections required they are arranged to have lengths increasing in a binary series. The resolution is then equivalent to the duration of the shortest section while the range may be almost doubled by adding just the largest delay element. This is especially important on quadruplex machines where there may be large error changes from one line to the next as a result of head switching.

The offtape signal for a helical machine may have errors of several microseconds in normal play and several lines when being used in a portable mode. To provide time base correction with a suitably large range, the signal is first converted to a digital form,

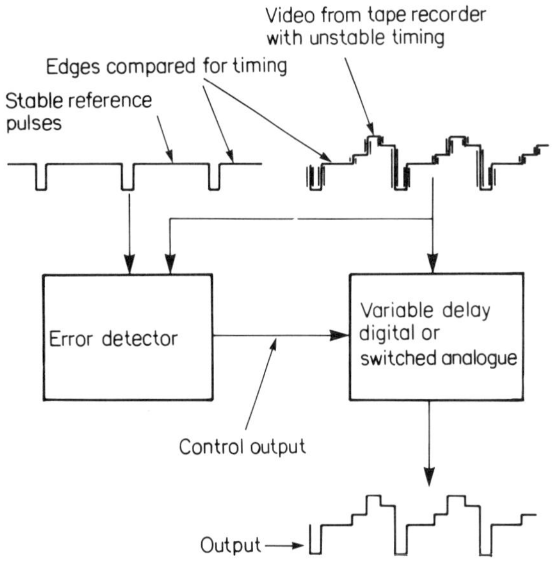

Figure 58.12 Action of a time base corrector

using eight bits for each sample. It is then possible to load the off-tape signal into a store, either a shift register or random access memory, at a rate determined by the off-tape sync rate. The signal is then read out at a steady rate, thereby accomplishing the desired correction.

To correct signals which arrive early with respect to the reference it is merely necessary to add some delay to the signal path. On the other hand, to correct late signals, delay must be removed and therefore a reserve has to be built in. It is usual to have half the total delay in circuit when the error is zero. On a helical machine with a ten line correction range this would mean a built-in delay of five lines. The head drum of the replay machine would then be synchronised to a reference five lines early so that the output of the time base corrector would match the studio. The early reference signal may be provided as an advanced syncs output from the time base corrector or internally from the replay VTR.

Time base correctors for use with colour under signals have to restore the subcarrier to sync relationship that was destroyed by the colour stabilising circuitry. This is done either by decoding the signal or by replacing the chrominance jitter before converting the signal to a digital form.

The luminance and chrominance components may be fed separately to the time base corrector so that the chrominance correction may be carried out at the same time as the frequency conversion. This saves signal processing and improves the picture quality.

58.12 Drop out compensation

Mechanical handling and manufacturing defects results in minute holes in the coating. The loss of signal which results from such holes is called drop out and is visible on the screen as black lines. They usually last a fraction of a line but as the head to tape speed reduces with more modern machines, they may extend up to a line's length. Their existence may be detected by sensing the drop in r.f. level and the resultant pulse is used to activate the compensation circuitry. Perfect repair is not possible but a high degree of concealment may be achieved by replacing the missing part of the signal with part of a close previous line.

58.13 Velocity errors

Timing errors, such as may be caused by head drum speed jitter or head/tape misalignment can accumulate over the length of each line depending upon the rate of change of timing (i.e. velocity) error. Velocity errors due to head drum jitter tend to be small, as the rate of change is small, and random in position so that they tend to cancel out.

58.13.1 Quadruplex machines

Errors due to head/tape guide geometry are related to head drum rotation which is locked to the field frequency so that they appear in the same position repeatedly and are therefore visible. If the tape and head drum are not properly concentric, due, for example, to incorrect adjustment of the tape guide height on a quadruplex machine, the head/tape velocity, and hence the frequency, will vary along the length of each track consisting of some 16 lines (*Figure 58.13*). Although correction is applied at the start of each line there can still be a considerable rate of change of phase resulting in a change of hue across each line of picture. The hue error will vary according to the position of each line in relation to the sweep of the head with the result that the picture is broken up into bands of approximarely 16 lines with a marked change of hue between them. With the delay line PAL colour system the hue errors are translated into saturation errors which, due to the effect of the banding, are equally objectionable.

58.13.2 Helical machines

The velocity compensators used for helical machines are built into the associated time base correctors. They act to adjust the timing of the write clocks in response to the error voltage obtained by comparing off-tape line length with a reference.

Velocity errors on helical machines are virtually linear and constant across the whole field. This makes them rather easier to compensate than the rapidly changing errors seen on quadruplex systems. Because the velocity errors occur at the field rate, banding is not introduced.

The error detection circuitry is already incorporated for use when the tape is running at non-play speeds for slow motion or still frame.

By adjusting for velocity errors at the same time as instability errors, a set of delays and its circuitry is saved. Velocity errors become particularly significant when dubbing from one machine to another is required, e.g. for editing purposes, because the velocity errors build up and because the sync pulses and colour bursts are replaced, they are not easily corrected.

When tapes recorded on one machine are played on another it is unlikely that the recording and reproducing heads will match exactly. If there is any difference in the length of the individual head tips this is equivalent to incorrect head tip insertion with consequent velocity error and banding. To make possible complete interchangeability of tapes as well as dubbing between machine it is necessary to employ velocity compensation.

58.14 Capstan and head drum servos

As with an audio recorder the longitudinal movement of the tape is caused by a capstan and pinch roller (*Figure 58.14*). The speed is controlled by the capstan servo lock which has two modes of operation. In record, there is no means of measuring the amount of tape which passes, and so the capstan rotates at a fixed speed. The tape speed is then determined by the capstan diameter.

The head drum speed is determined by the number of heads and the format, e.g. 250 Hz for a quadruplex, 50 Hz for C format.

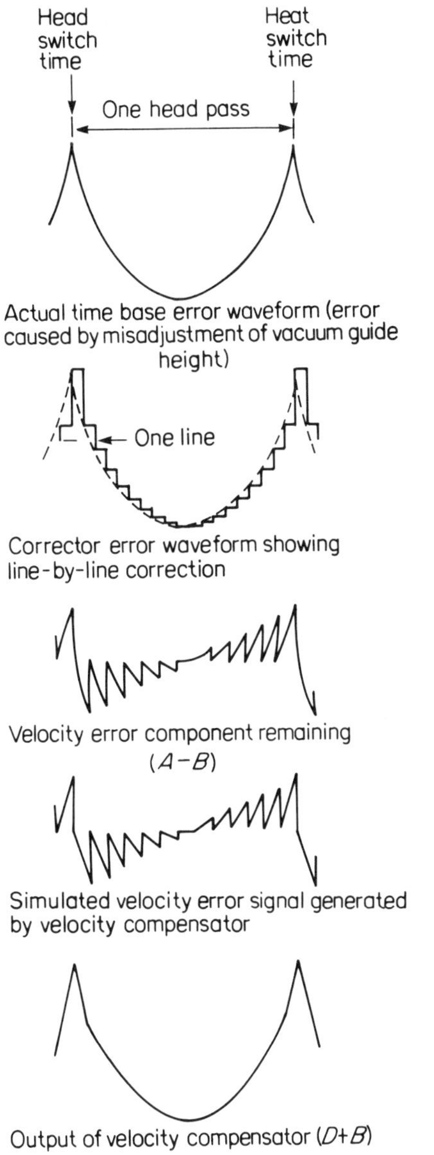

Figure 58.13 Quadruplex velocity error correction

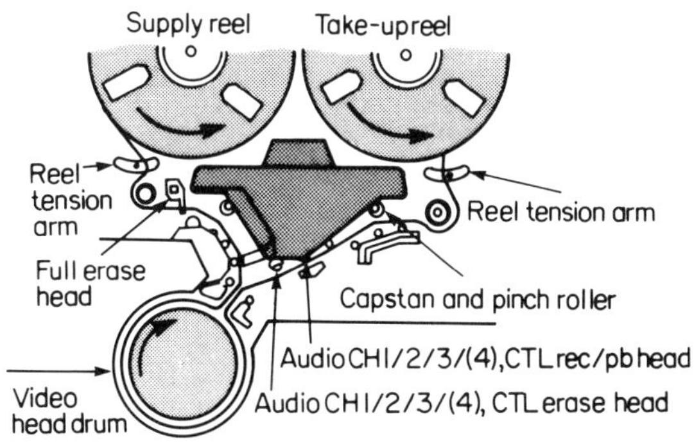

Figure 58.14 Typical helical tape path

In record, the speed of the drum is measured using an integral tachometer whose output is compared with the input video. There is also a phase comparator which ensures that the start of each track is laid down at the appropriate point on the tape.

To enable the head and capstan servos to repeat their relative movements during replay a set of pulses, the control track, is laid down along the tape parallel to its edge during the recording.

In replay, the head drum is again referred to a video input so that the output of the VTR can be synchronised to a studio or broadcast chain. The capstan speed is then controlled, comparing the control track output with a suitable reference derived from the head drum servo. In this way, the speed and phase of the capstan may be adjusted to ensure that the video heads follow the recorded tracks.

The tracking control found on most machines alters the phase relationship to enable non-standard recordings to be accurately replayed. The errors may creep in through misadjustment, mechanical tolerances or tape stretch.

It is possible to detect the position of the tracks by measuring the amplitude of the replayed r.f. The error voltage may then be used to adjust automatically the tracking for output. There is a consequential increase in the time taken for the machine to lock up when replay is initiated.

58.15 Tape editing

Tape editing with quadruplex video tape machines can be achieved either by physically cutting the tape or by electronic means.

58.15.1 Physical editing

Two-inch (5.08 cm) video tape can be physically cut and joined using adhesive tape on the back as in a sound tape recorder, but the tape must be cut and joined very accurately during the vertical blanking interval between picture frames. With the tape travelling at approximately 19.05 cm s^{-1}. These occur every 0.762 cm at 25 pictures per second (European standard).

To enable the correct point at which to cut the tape to be located, edit pulses are recorded on the control track. These occur at a frequency of 25 Hz for monochrome (European standard) of 12.5 Hz for PAL colour pictures. In the latter case it is necessary, when making to join, to preserve the colour burst phase relationship which, in the PAL system is a four-field sequence. A cut can therefore only be made at intervals of two complete interlaced pictures. Thus editing cannot be quite so fine with colour as with monochrome pictures.

The cutting points are located by marking the tape at the precise moment of the programme cue and, when the machine is stopped, painting over the area of the control track with a liquid containing powered iron in suspension. The edit pulse can then be seen with the aid of a microscope and by means of a special splicer the tape is cut and joined to the next sequence which has also been cut at an edit pulse.

The operation is complicated by the fact that the sound recording head is approximately 23 cm in advance of the vision recording so that if a cut is made on a vision cue approximately 0.6 s of the sound from the cut material will remain on the wanted tape. Similarly, if the cut is made at the start of a vision cue 0.6 s of the new sound will be lost. Whenever possible, therefore, for cut editing, the cutting point should be chosen to coincide with a gap in sound of at least 0.6 s. If this is not possible it will be necessary to 'lift off' that part of the sound by recording it on to 6.25 mm tape and re-recording it over the join.

58.15.2 Electronic editing

Editing of video tape by physical cutting of tape has been superceded by electronic dubbing methods. In its simplest form one machine replays the uncut tape and another machine records it. Because of the time needed for the VTR servos to lock up, each machine is parked ten seconds before the chosen edit point. They are then started simultaneously, both in play. At the chosen point, the edit machine is switched to record either manually or by a pre-recorded tone pulse on the audio cue track. It is possible to rehearse each edit and adjust its timing if necessary. There is no longer any problem of sound offset, but it is common practice to use a separate audio recorder to blend the pre- and post-edit sound. The audio is transferred to the $\frac{1}{4}$ in (0.635 cm) audio tape recorder, ATR, during the dubbing of the scene which follows the previous edit. It is sometimes necessary to introduce a frame or two of delay during the remix to overcome phasing problems which distort the sound, applause is a typical case.

The use of only two machines limits editing to simple cuts. To enable post-production mixing and effects it is necessary to provide a replay VTR for each source to the vision mixer. Three and four machine editing requires much more sophisticated synchronising of the play and record machines and where possible mixer and audio recorders.

A modern edit suite will achieve the necessary synchronising action with the aid of an eighty-bit frame-identifying-code time code. The time code signal is recorded on one of the audio tracks using a code which contains only two frequencies. A continuous stream of logic ones would be recorded as a 2 kHz square wave, a stream of logic zeros as a 1 kHz signal. This bi-phase mark signal is self-clocking and can be read at speeds from less than one-fifth play slow motion to greater than 50 times play during wind.

There are many commercial units which use this time code to log edit points and then to control the VTR via microprocessors during rehearsal, adjustment and printing of the edit.

The ability to log the times of chosen scenes has led to a development in video tape post-production known as 'off-line' editing. The time code numbers are displayed in the associated pictures which are then recorded on a low-cost VTR, such as U-Matic, VHS or Beta. The producer may then choose scenes and edit points without the need for expensive broadcast machine time. A rough cut may be made on the cheap format so that several different verions may be tried economically before the final edit on the broadcast machines. The time codes of the final version of the off-line may then be logged on paper or magnetic tape and used to control the machines making the final print.

There are two main modes of edit, assemble and insert (*Figure 58.15*). In assemble editing each new scene is added to the end of the last recording and an unrecorded tape is used. Insert editing involves putting new scenes into an existing recording. The main difference as far as the VTR is concerned is that in an assemble edit a new control track is recorded while for an insert edit the capstan servo remains in the control track replay mode. This is to ensure that the new recorded tracks exactly cover the ones which are to be replaced.

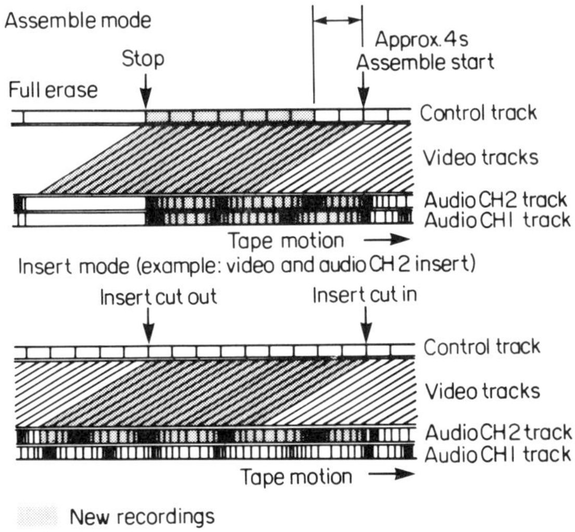

Figure 58.15 Track patterns for edit modes, U-Matic example shown

58.16 Domestic systems

There are two systems that are within the economic reach of domestic users, tape and disc.

58.16.1 Tape

There are several video tape recorder formats using $\frac{1}{2}$ in (1.27 cm) tape which have developed from the colour under system developed for educational and industrial use.

The bandwidth is severely limited and cannot cover the 4.43 MHz colour subcarrier and its sidebands. The signal is therefore split into separate luminance and chrominance components by suitable low-pass and notch filters. The luminance component is then recorded in the usual way, by frequency modulation of a suitable carrier. The lower sidebands are limited to a minimum of 1.5 MHz to allow for the inclusion of the modified colour signal. The chrominance is limited to a 1 MHz bandwidth and amplitude modulated on to a new carrier in the region of 700 kHz which is placed in the spectrum below the luminance.

To further increase the packing density of the recorded signal the usual guard band between tracks is eliminated. To prevent crosstalk during replay, adjacent tracks are recorded by heads whose azimuths are angled from the perpendicular by 6 to 15°. This causes adequate rejection of the high frequency f.m. but not the a.m. chrominance.

On alternate fields, the phase of the chrominance is advanced by 90° per line. Suitable addition or subtraction of adjacent lines during replay allows the separation of the wanted and crosstalk signals.

Several track and lacing patterns have been developed, led by the Philips VCR, but VHS, Beta and VCC are the only ones to achieve any degree of acceptance currently. They are all two-head formats recording one field per head pass. A crude form of still picture may be obtained by stopping the tape movement as for C format. The lack of a guard band means that there is no loss of signal due to the change in track angle, but a double-width head is fitted on one side of the drum which ensures continuity of the signal for the whole track.

Longitudinal control and sound tracks are added in the usual way but with tape speeds as low as one inch per second (2.54 cm s^{-1}) the quality is somewhat limited.

58.16.2 The Philips VLP system

The system uses discs of similar size and material to an ordinary LP record but instead of the usual groove there is a spiral track of minute indentations. These are all of equal depth and width but their length and the distance between them along the track varies such that a type of pulse modulation system containing all the necessary information to reproduce a colour television picture with synchronous sound is provided. The disc revolves at 25 resolutions per second so that each revolution contains one complete image. Playing time is 60 min. The surface of the disc is coated with a thin reflective metallic layer after pressing. A very small spot of light from a helium–neon laser source is focused on to the track and is reflected back, modulated in accordance with the pattern of indentations, into a photodiode. The resulting signal after amplification and suitable processing can be fed directly to the inmput of a television receiver.

The light beam is centred on to the track by means of an opto-electronic control system. Thus there is no need for a mechanical groove and the track pitch can be made extremely small (*Figure 58.16*). The opto-electronic mechanism does not have to remain synchronised to the track pitch and it is possible to speed up or slow down or reverse the motion or to hold a still frame. As there is no mechanical contact between the pick-up and the disc this can be accomplished, as can normal replay, without any wear of the record or the pick-up.

(a)

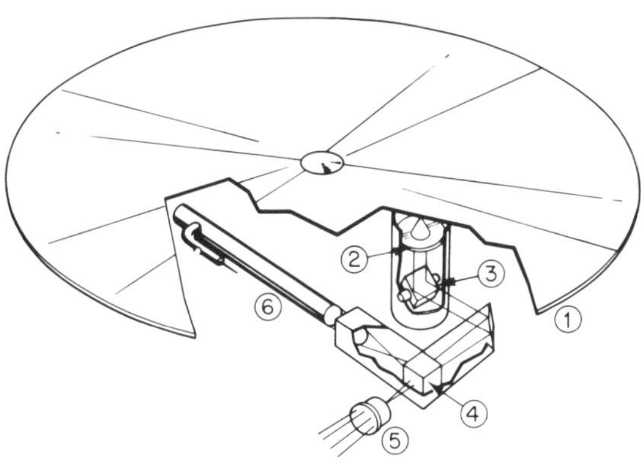

(b)

Figure 58.16 Philips VLP disc system (*a*) surface of the long-playing record through a scanning electron microscope, (*b*) schematic representation: 1 video long-playing record, 2 spring suspended lens with automatic focusing of the light beam, 3 hinged mirror for following the track, 4 beam-splitting prism, 5 photodiode (detector), 6 light source (Courtesy Philips Electrical Limited)

58.17 Slow and stop motion

Before the advent of the broadcast helical recorders, the segmented nature of video recording meant that a field store was needed for still frame or slow motion. The first method of providing the store was the Ampex HS 100 which was developed from computer disc memory systems. There were two discs mounted of a common axle and rotating at field rate. One video field was recorded on each of the four surfaces with each field completing a single circular track. It was then possible to replay the fields repeatedly to obtain a still frame. By moving the recording heads it was possible to record up to 36 s on the disc.

The slow and still picture functions have now been taken over by the C format machines or by digital frame stores in conjunction with B format machines.

58.18 Composite recorders

The need for lightweight recorders for news gathering has led to the introduction of the half-inch tape machines based on the Betamax and the VHS formats which were developed for domestic use. The tape transports of the broadcast machines are refined and strengthened versions of the domestic machines while the tape cassettes are identical. This is important because it ensures a world wide availability of tape.

In order to obtain the necessary increase in bandwidth the tape consumption has had to be increased by a factor of eight. This has been done by increasing the longitudinal speed and by increasing the width of tape used by each video head track. As can be seen from *Figure 58.17* there is a separate track for the luminance and

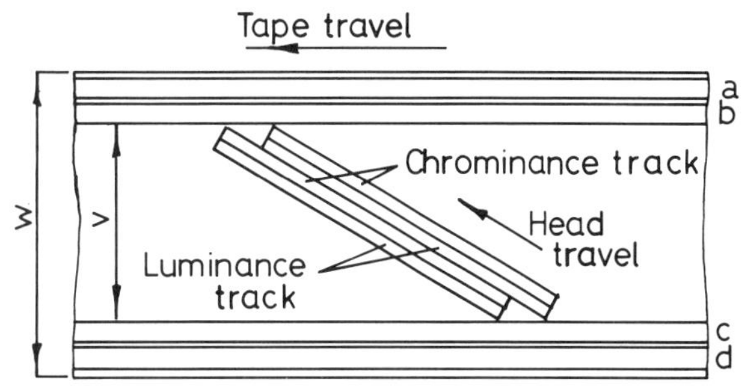

Figure 58.17

chrominance signals. There are three main types of recorder using this technique, two on half-inch and one on quarter-inch tape.

The method by which the different designs encode their separate signals varies, but in all cases the luminance is directly recorded using an FM carrier. The chrominance is either modulated on to two separate FM carriers and applied to the chrominance track, M format, or the chrominance from one line is compressed to half duration and recorded as B–Y and R–Y in sequence, Betacam. The third method, Lineplex from Bosch, multiplexes the Y and C signals after modifying the Y to occupy 1.5 times its real time and compressing the B–Y and R–Y by a factor of 2.

As can be seen from *Figure 58.17*, there are several longitudinal tracks for audio and for control signals. The different formats are summarised below.

B	*Betacam*	*M format*	*Lineplex*
Tape width (W)	$\frac{1}{2}$ in	$\frac{1}{2}$ in	$\frac{1}{4}$ in
Recording width (V)	9.38 mm	10.07 mm	4.6 mm
Video track angle	4.68°	5.97°	2.7°
Control track position	c	d	a & b
Audio track 1	b	b	d
Audio track 2	a	a	c
Timecode track	d	none	a & b

All these formats are intended for use in recorders which are integrated with cameras and give an all up weight less than 7 kgs. The Beta and M formats are not interchangeable and will co-exist as standards.

58.19 Digital recording

At the time of writing there is no agreed standard for a digital recorder although it is recognised that this is the last and essential link in the video chain. However, it appears that the original intention to digitise the composite video signal is no longer valid. Instead it is probable that the three signals Y, R–Y and B–Y will be separately digitised and recorded. This component format fits well with the systems already in use for special effects devices and offers the best packing density.

In general digital signals require more bandwidth than analogue signals and so tape economy may be sacrificed. The main discussions are concerned with the number of bits required. Although eight are adequate for time base correctors it is not enough for a signal which might be expected to have its gamma changed. The quantising levels would then become visible. It is in this area that most of the degradation is seen since the process of changing from analogue-to-digital and back introduces errors which fix the maximum signal-to-noise that may be achieved.

Current predictions are that digital machines will not be available in large quantities before 1986.

Further reading

The Fourth International Conference on Video and Data Recording, *IERE Conference Proceedings*, No 54 (April 1982)

International Broadcast Convention, *IEE Proceedings*, No 195 (September 1982)

Videodisc Special Issue, *RCA Review*, **43**, No 1 (March 1982)

59

Radar Systems

H W Cole
Principal Radar Systems Engineer,
Marconi Radar Systems Ltd

Contents

59.1 Introduction

Radar techniques quickly spread into a number of branches different from their original application in early pre-World War II days. The word 'RADAR' is taken generally to be an acronym of USA origin of the phrase *RA*dio *D*etection *A*nd *R*anging. The word 'detection' appears to have been replaced, at the time of its being first coined, by the term 'direction-finding' in the cause of security. The primary radar technique, from which others have grown, is based upon two phenomena:

(*a*) That radio energy impinging upon a discontinuity in the atmosphere is reflected by the discontinuity.
(*b*) That the velocity of propagation of radio waves is constant.

Exploitation of these phenomena led to transmission of regular pulses of radio energy, range being obtained from measurement of the round-trip time of transmitted pulses. Further pulse transmission and reception engineering led to systems which although radar-based, or inspired, are not true radar, e.g. Oboe, Gee, Omega, etc—since these are pure distance measuring systems they will not be treated here. A close relative of primary radar, which does not use the reflection of transmitted energy, is, however, so important that it merits separate treatment. This is the secondary surveillance radar (SSR) system wherein pulse transmissions from the ground are received in aircraft, detected and decoded in the aircraft's transponder. The aircraft's transponder then transmits coded pulses back to the ground after a short fixed delay. Thus the operational benefit of primary radar is maintained and many bonuses accrue which are explained later.

59.1.1 History

Radar's history is currently incompletely written. The most prolific historian to date is the generally acknowledged 'father of radar' Sir Robert Watson-Watt.[1,2,3,4] He rightly observes the radar principle as being 'often discovered and always rejected'. Early workers, even Heinrich Hertz himself, made observation of the reflection of radio waves from metallic surfaces. In 1922 Marconi introduced publicly the notion of this effect being put to use in the detection and location of the direction of ships. If there is any one point in time at which we can say radar was invented it comes from Sir Robert's memorandum of 27 February 1935. This laid down the basic principle we know as primary radar and also pointed to the need and possibility of SSR by realising how necessary it would be to distinguish 'friend from foe'. Emphasis was placed also on the need for reliable ground–air communications. This memorandum was stimulated by a previous UK Air Ministry request to investigate the possibility of destroying the attacking power of aircraft by radiation. It was concluded that the aim could not be achieved at that time—but in any case the aircraft had first to be located. This would be entirely practicable and Sir Robert elaborated upon this in the second memorandum.

In the few months following February 1935, remarkable progress was made, culminating in detection and ranging of aircraft to half mile accuracy out to nearly 60 miles. Height measurement was also achieved out to 15 miles using range and elevation angle. All this early work was conducted on wavelengths between 50 and 25 metres. No measurement of bearing was attempted at this time, it being thought adequate to rely upon independent range measurements taken from a chain of stations of known location. However, around January 1936, the application of receiver DF techniques, using goniometer principles provided a crude facility which was immediately successful. Similar work was carried out on the European continent, and in the USA. Various reports ascribe the lower level of progress, in Germany for instance, to the German High Command view that the bomber was unstoppable and bombing, a very quick way to victory. In the USA, the defence needs were held to be much less urgent than in the UK. This inhibited progress in the early days of radar history.

By 1939 in the UK a radar chain (CH) was established, operating eastwards and ranging from the Orkney Isles to Portsmouth. By September 1941 the chain encircled the whole of England, Scotland and Wales. By this time, the needs and possibilities of radar led to the production of equipments for airborne use, rotatable aerials and the plan position indicator (PPI). In 1940, stimulated by the need for small equipment Randall and Boot successfully operated their resonant cavity magnetron and thus revolutionised the radar technique. Their device,[5] producing at that time 50 kW peak power at 10 cm wavelength, is now the most widely used generator of microwave power. The use of very short wavelengths made possible the development of small aerials with high discrimination, and equipments with a wide range of power weight and size. The study of radar applications in defence and aircraft navigation continued in peacetime, the latter blossoming into the field of civil aviation electronics, air traffic control, satellite communications. By the end of World War II, radar engineering had produced many different equipments with differing attributes. Almost immediately the concept of systems engineering took concrete shape and became virtually a separate discipline. The possibility of mixing computer technology into signal and radar data processing has led to the present point of the next revolution in radar.

59.2 Primary radar

59.2.1 Fundamentals

The radar system can be represented in universal form as the functional diagram of *Figure 59.1*.

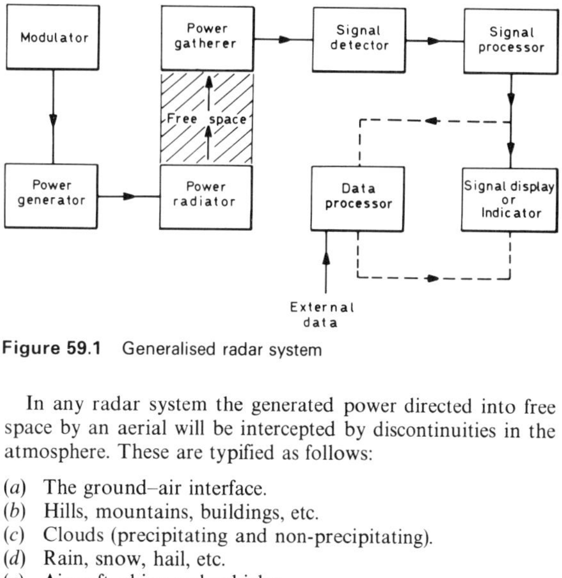

Figure 59.1 Generalised radar system

In any radar system the generated power directed into free space by an aerial will be intercepted by discontinuities in the atmosphere. These are typified as follows:

(*a*) The ground–air interface.
(*b*) Hills, mountains, buildings, etc.
(*c*) Clouds (precipitating and non-precipitating).
(*d*) Rain, snow, hail, etc.
(*e*) Aircraft, ships and vehicles.
(*f*) Birds, insects, dust clouds, atmospheric discontinuities (i.e. sharp changes of refractive index).

Although some power will be absorbed on impact, most is reflected and some will travel back to the power gatherer, the receiving aerial. The received power is amplified in a receiver, the signal competing with random noise gathered by the aerial and added to that generated at the receiver input (galactic noise, interference signals from electric devices and receiver noise). After frequency selective amplification, the signal plus noise is passed to a signal processor. Here various characteristics of the numberous types of received signal are exploited to separate the wanted from unwanted, e.g. moving targets from stationary targets, large from small, long from short, etc.

The filtered signals are then passed to a display equipped to register the signals in such a manner as to allow their position to be measured by means of range, bearing and sometimes height, scales. The signal processor is being increasingly used to serve a data processor which also accepts external data on targets, situations and various other criteria. Its output can furnish modified display presentations to augment the radar display, and categorise and clarify the total radar-sensed situation.

59.2.1.1 *Monostatic, bistatic and adaptive radars*

Referring to *Figure 59.1* the modulator and power generator constitute the transmitter in all of the above categories. Monostatic systems differ from the bistatic in that a single aerial is used for power radiation and power gathering, *Figure 59.2*. A duplexer is used to separate the transmit and receive functions, both in the time and power amplitude domains. In the bistatic system, *Figure 59.3*, two separate aerials are used, one for the transmit and the other for the receive function. These aerials may sometimes be many miles apart.

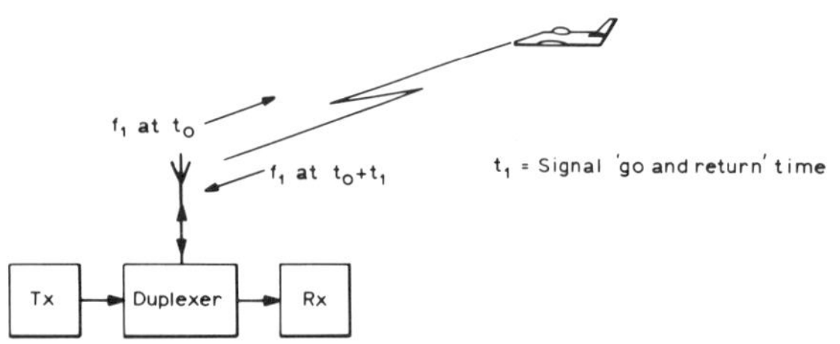

Figure 59.2 Monostatic radar

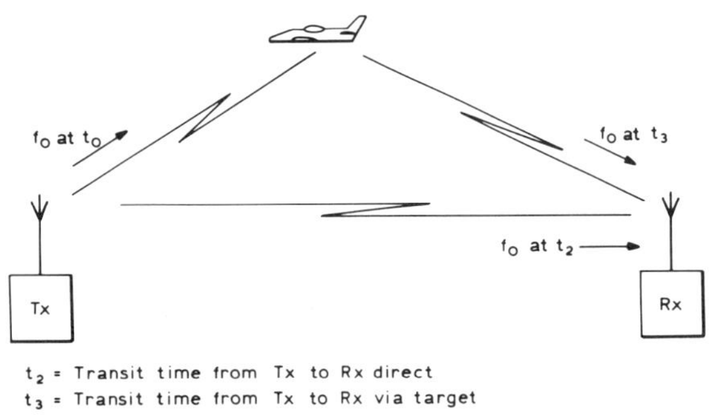

Figure 59.3 Bistatic radar

The range of a reflecting object (aircraft, ship, mountain, etc) is given by measuring the pulse 'round trip' time from transmission to reception using the leading edges of both transmitted and received pulses as reference.

Round trip time for 1 nautical mile $=12.36\ \mu s$
Round trip time for 1 km $=6.65\ \mu s$.

Normally the operating parameters of a radar system are fixed or there can be a few selectable changes during operation, e.g. type of modulation, data gathering rate, power output, receiver sensitivity and selectivity

In the so-called adaptive radar system, parameters of operation (within limits) are varied as a function of the radar's performance on a number of targets. For example, feedback is generated by the processor or data processor to increase information on specific targets by, for instance, directing the aerial to targets in a specific order and governing the dwell time on each to improve data quality; varying the output pulse rate or spectrum. These systems are not yet in general use.

The most commonly used radar system is the monostatic and this section therefore concentrates upon this. Further information on bistatic technique and adaptive radar will be found in references 6–10 and 38.

59.2.1.2 *The radar equation*

There are many forms of the radar equation to be found in the literature. It is therefore considered helpful to develop it from fundamental ideas in order that these various forms, each with their own subtleties and idiosyncrasies, can be better understood. Taking the simple case of free-space performance, the maximum range of radar is a function of:

P_t	peak transmitted power
G	gain of the aerial in the direction of the target
σ	effective reflecting area of the target
λ	wavelength of radiation
S_{min}	minimum received signal power required to be detectable above system noise
A_r	effective area of receiving aerial.

Consider a target at range R. The power density at R will be equal to

$$\frac{P_t G}{4\pi R^2} \tag{59.1}$$

The target will intercept this power over its equivalent area of σ and reflect it back over the same distance, R, to the aerial which now becomes the receiver in the monostatic system. Thus the power gathered by the aerial equals

$$\frac{P_t G \sigma A_r}{4\pi R^2 4\pi R^2} \tag{59.2}$$

Now A is related to the aerial gain in the following way:

$$A=\frac{G\lambda^2}{4\pi} \tag{59.3}$$

Maximum gain is obtained when the full aperture of the aerial is available to gather the returned energy and since the same aerial is used for transmission and reception, equation (59.2) can be rewritten:

$$P_r=\frac{P_t G^2 \lambda^2 \sigma}{(4\pi)^3 R^4} \tag{59.4}$$

Postulating the maximum range as that achievable when P_r reduces to S_{min}, then

$$R_{max}^4=\frac{P_t G^2 \sigma \lambda^2}{(4\pi)^3 S_{min}} \tag{59.5}$$

The dynamic performance of a radar system cannot, however, be directly calculated by this formula since various statistical factors have yet to be accounted for. One already appears in the notion of S_{min}, since noise is a random phenomenon and is further complicated by the behaviour of σ, the effective target's echoing area. An idea of this can be gained from study of *Figure 59.4* which gives information on the typical values found in practice.

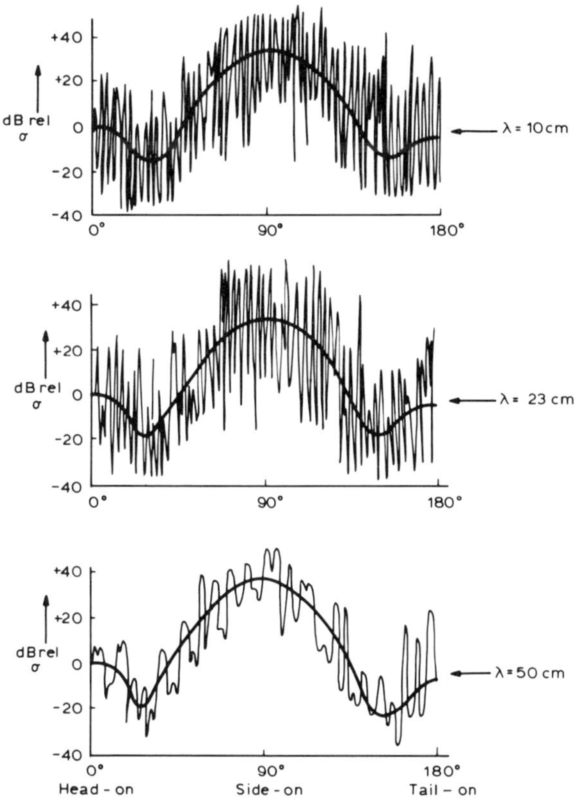

Figure 59.4 Showing typical variation in effective echoing area (σ) of aircraft targets

Table 59.1 Target effective echoing areas

Target	*Attitude*	*σ (typical mean)* (m^2)
Small fighter jet	Head-on	2
	Tail on	3
	Side on	500
Large transport jet	Head-on	10
	Tail on	15
	Side on	1000

Table 59.1 gives typical values of target effective echoing area. At microwave frequencies there is no extrapolable relationship between wavelength and effective echoing area. The value of σ can change violently over very small azimuth increments at microwave frequencies. As the target attitude to the radar can change even during interpulse periods, there is a range of probabilities that the echoing area will produce a returned signal of a given strength. The distribution of these progabilities is taken variously as Gaussian, Raleigh or exponential. Radar engineers are generally concerned with probability of target detection of the order 80% and so for all practical purposes the difference in these distributions is negligible. These probabalistic factors, together with others associated with display, operator, atmospheric loss and system loss factors are all combined to modify the final calculation. Two commonly used formulations are those due to L. V. Blake[9] and W. M. Hall.[11]

59.2.1.3 Propagation factor

In all radar systems, theory begins with free-space propagation conditions. These almost always do not pertain since radiations, from and to the aerial via the target and its environment, are seldom via a single path. The most common effect is that of the ground above which the radar aerial is mounted. Consider the situation in *Figure 59.5(a)* where the phase centre of the aerial is at height h above the ground. Energy reaches the distant target by

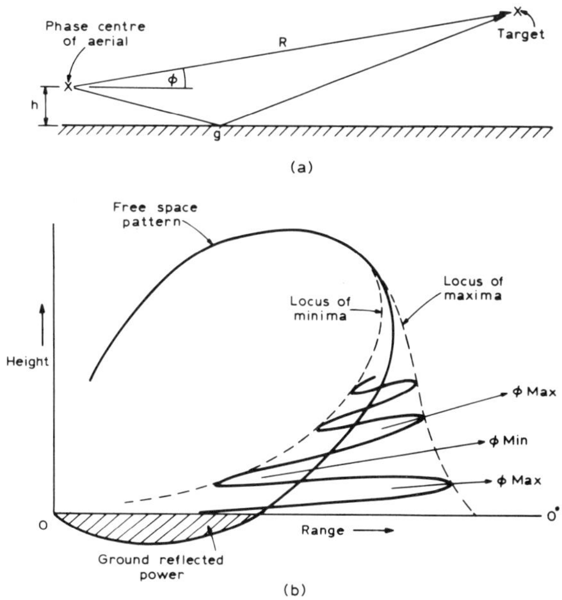

Figure 59.5 (*a*) Ground reflection mechanism (*b*) Showing lobes and gaps in vertical polar pattern

the direct ray R and that reflected at g. Over the elevation range of which ϕ is one example, the direct and reflected rays will be alternatively in and out of phase with each other. The free-space pattern of *Figure 59.5(b)* will thus be modulated, a series of lobes being formed. The position of these is governed by h and the wavelength in the following way:

$$\phi_{\max}=\frac{\lambda n_1}{4h} \qquad \phi_{\min}=\frac{\lambda n_2}{4h} \tag{59.6}$$

where λ is the wavelength, h the height, n_1 are odd integers and n_2 are even integers. Max and min indicate peaks and troughs of lobes respectively.

This model and theory holds for very flat ground or water but becomes a complicated calculation as ground roughness and departures from flat plane increase. Vincent and Lynn[12] have made the problem tractable.

Obviously these lobes are useful providers of extra range but the gaps are less welcome. The modulations can be reduced by several means:

(*a*) Putting less power into the ground.
(*b*) Increasing the aerial tilt in elevation.
(*c*) Intercepting the reflected power and dissipating or scattering it in incoherent fashion.

An interesting alternative is to be found in 50 cm radars such as are used by the Civil Aviation Authority for almost all of the UK airways surveillance system. Here the wavelength and mean aerial height combine to give a long low lobe embracing most of

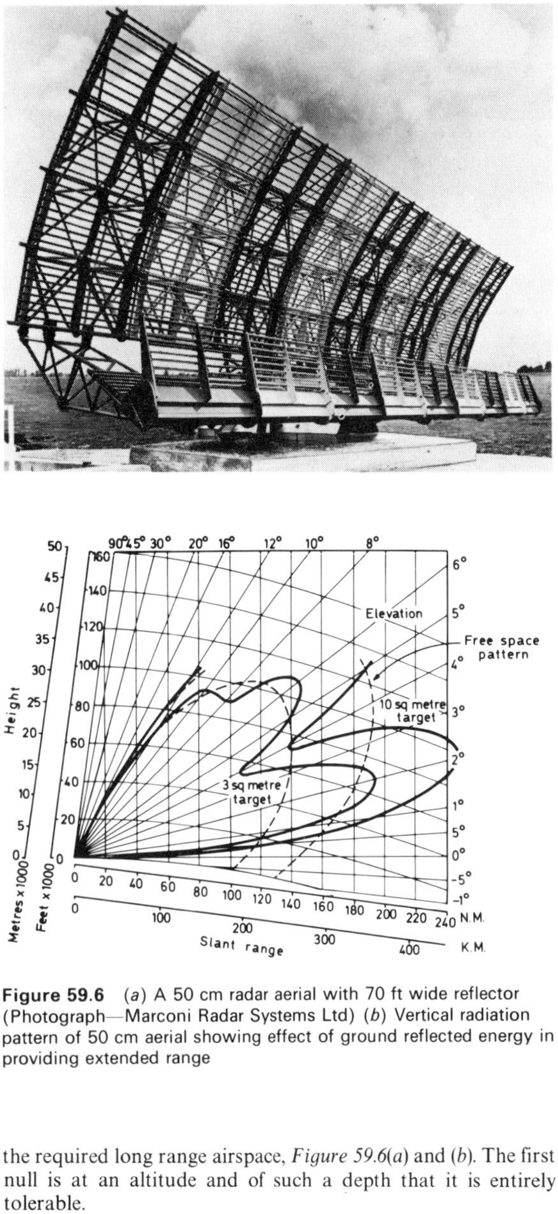

Figure 59.6 (*a*) A 50 cm radar aerial with 70 ft wide reflector (Photograph—Marconi Radar Systems Ltd) (*b*) Vertical radiation pattern of 50 cm aerial showing effect of ground reflected energy in providing extended range

the required long range airspace, *Figure 59.6*(*a*) and (*b*). The first null is at an altitude and of such a depth that it is entirely tolerable.

Dependence is placed upon a very flat site, the tolerable roughness being great because of the longer (in the microwave sense) wavelength.

A useful comparison of ground reflection effects across the microwave band is to be found in reference 13. The pattern propagation factor which thus modulates the free-space pattern is also embodied in range calculation.

The magnitude of the lobes and gaps is a function of the effective reflection coefficient of the reflecting surface. Reference 14 treats the subject in great depth and reference 15 gives tables from which specific cases can be calculated.

59.2.1.4 Anomalous propagation

Radar coverage in the vertical plane is usually expressed by a slant range/height diagram drawn for equi-probability of detection of a given effective echoing area and probability of false alarm. A 'standard atmosphere' is generally taken wherein the refractive index of the atmosphere varies with height as the density of the air changes. In such a standard atmosphere containing water vapour, rays are bent downwards as range increases. To counteract this and allow rays in the elevation domain to be drawn as straight lines the range/height diagrams assume an Earth curvature $\frac{4}{3}$ times the normal radius. Naturally local variations in atmospheric condition distort the calculated radar cover diagram.

Sometimes these variations are severe enough to cause a discontinuity in the rate of change of refractive index with height. The consequent ray bending can be large enough to equal the Earth's curvature. Under these conditions ('super-refraction') the radar has no horizon and excessive ground ranges can be reached. Apart from difficulties, under these conditions, in obtaining target height data at low elevation angles from height finding radars, normal surveillance radars experience ground clutter and target signals due to transmissions previous to the current one. These are called 'second/third/*n*th-time around' signals. The effect is illustrated in *Figure 59.7*.

In conditions causing anomalous propagation all microwave frequencies are affected. However the onset and degree of effect is later and less severe as wavelength increases. The region of concern can be considered for all practical purposes to be at angles of elevation up to 1°.

59.2.2 Types of radar

Figure 59.1 represents many types of radar; some of the more important of these, starting with the most simple, are now considered.

59.2.2.1 C.W. radar

Here there is no modulation provided in the transmitter. The movement of the target provides this in its generation of a Doppler frequency. The system is illustrated in *Figure 59.8*. The presence of the target is indicated by the presence of the Doppler frequency of value

$$f_D = \frac{89.4V_r}{\lambda} \tag{59.7}$$

where V_r is the radial speed, λ the wavelength and f_D at 10 cm is approximately 9 Hz per mph (5.6 Hz per km h).

Range is not measurable with one c.w. element; but certain modern equipments based on this principle do not need to indicate range, e.g. police speed traps using radar techniques. The technique is widely used for velocity measurements and provided no great range is required, very modest power output can be used, e.g. 10 to 15 W at 10 cm with a small dish aerial of some 90 cm (3 ft) diameter can effectively operate up to 16 to 24 km (10–15 miles) on aircraft.

A.M./C.W. radar

This variant of the c.w. technique employs two slightly differing c.w. transmissions. With two frequencies f_1 and f_2 a single target at give range will return $f_1 \pm f_D$ and $f_2 \pm f_D$. The Doppler frequency difference can be made small if $f_1 \sim f_2$ is small. However the range domain from the radar is characterised by a phase difference between f_1 and f_2 which is linearly proportional to distance and unambiguous up to phase differences of 2π. By detecting the phase difference of the two almost identical Doppler components of the returned signals the range can be measured, being:

$$R = \frac{C\phi}{4\pi(f_1 - f_2)} \tag{59.8}$$

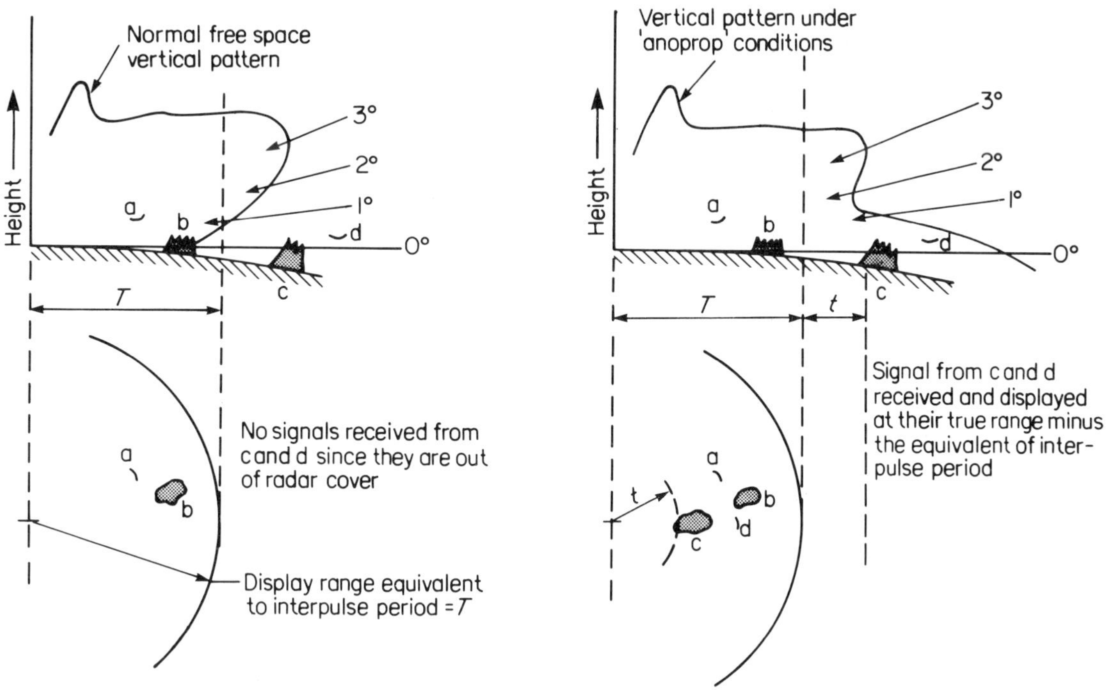

Figure 59.7 Illustrating the effect of anomalous propagation, a and d=aircraft, b and c=ground clutter

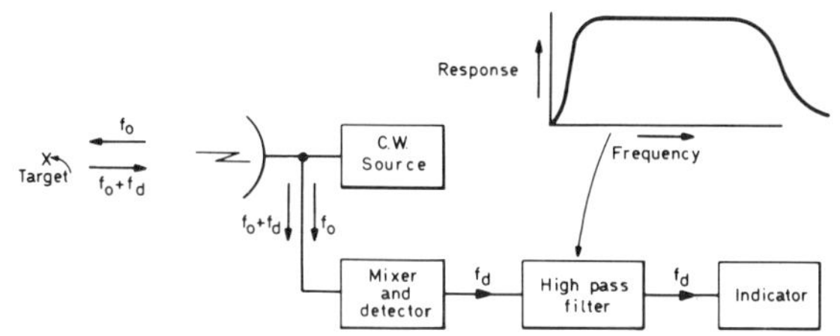

Figure 59.8 Simple c.w. radar

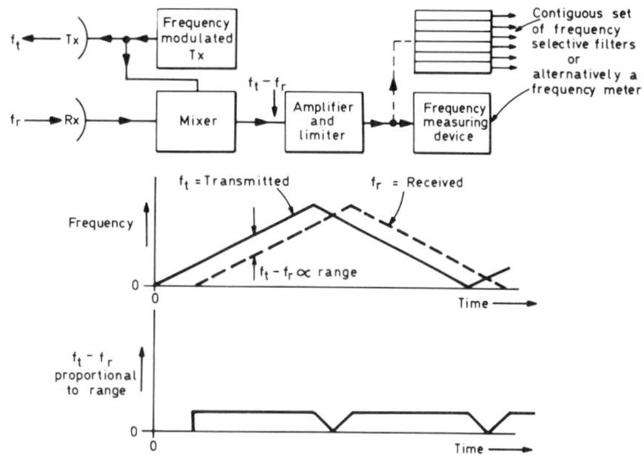

Figure 59.9 An f.m./c.w. radar system

where R is the range to target, C the velocity of propagation, $f_1 - f_2$ the frequency difference of the two transmitted c.w. and ϕ the phase difference between the two Doppler frequencies.

The complication of dealing with a number of targets simultaneously present in the system is resolved by erecting a number of Doppler filters, each with its own phase measuring element.

F.M./C.W. radar

A simple form of this is illustrated in *Figure 59.9*. Here the range continuum has its analogue in frequency deviation from a starting point. Comparison of the frequency of the returned signal and the transmission frequency gives a difference which is directly proportional to range. Range accuracy is determined largely by the bandwidth of the frequency measuring cells into which the range scale is broken and the stability of frequency deviations. The technique has been used for a radio altimeter, the Earth being the radar's 'target', and the range being the height of the aircraft above the reflecting surface.[16]

In all the systems described above there was a practical limitation of performance due to the limited c.w. power that could be generated at very high frequencies so that small aerials of high directivity could be used. This limitation of power was eventually overcome by using pulse modulation techniques. In this way very high peak power could be produced within the device's mean power capability. When pulse modulation is used the transit time of the pulses to and from targets is the direct measure of range. A pulse modulated system is illustrated in *Figure 59.10(a)* and *(b)*.

59.2.2.4 Pulse-modulated Doppler system

In the normal Doppler system, target detection is based upon sensing the Doppler frequency generated by target movement. Straight pulse modulation will detect and indicate both fixed and moving targets. Bandwidth and other considerations allow the latter to be used as a surveillance radar scanning regularly in

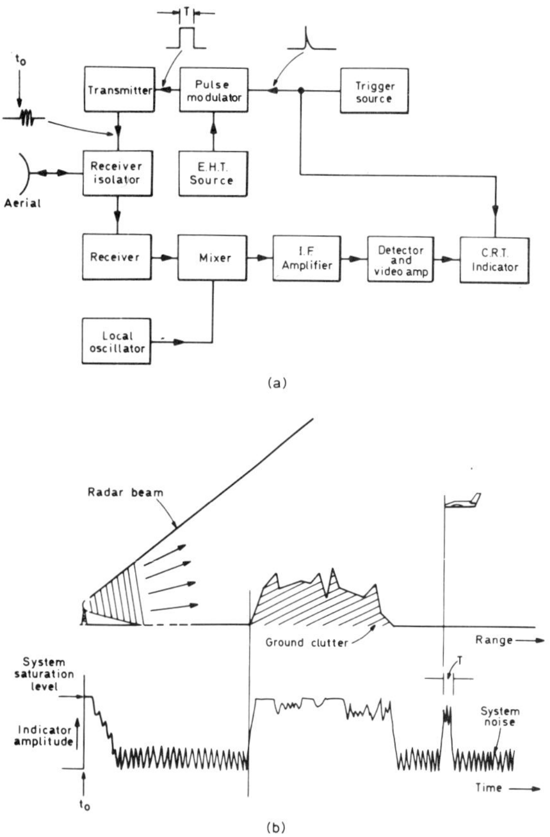

Figure 59.10 (*a*) Pulse modulated radar system, (*b*) signals in the pulse modulated radar

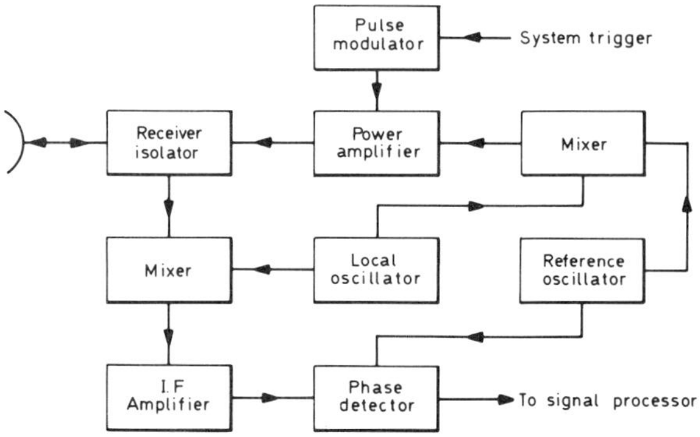

Figure 59.11 Self-coherent pulse radar

azimuth since target dwell time can be short. In order to create a surveillance system that can reject stationary targets at will and preserve moving targets, the pulse modulated system can be made to include the necessary frequency and phase coherence of the Doppler system. This produces the most common form of moving target indicator (MTI) system in use today.

59.2.2.5 MTI systems

Basically these are of two main types:

(*a*) Self coherent.
(*b*) Coherent by phase-locking (commonly called the 'coho-stalo' technique).

Both types operate as follows.

Targets at a fixed range (mountains, etc) will produce signals which exhibit the same r.f. phase from pulse to pulse when referred to the phase of the transmitted r.f. pulse. Moving targets, by the same token, will produce different r.f. phase relationships from pulse to pulse, because of their physical displacement in inter-pulse periods. By use of a phase sensitive detector these phase relationships can be used in processors to separate fixed from moving targets. The importance of maintaining phase coherence can be seen, for, in order to measure the phase of the received signals relative to that transmitted, a reference has to be laid down.

In the self-coherent system the transmitter is a power amplifier whose r.f. source is crystal controlled by a reference oscillator. The same reference is taken for the receiver's local oscillator. Thus coherence is assured and maintained at i.f. level at which point the phase information is extracted. The system is illustrated in *Figure 59.11*.

In the 'coho-stalo' system, the transmitter is a self-oscillating magnetron whose phase at the onset of oscillating is varying from pulse to pulse. Coherence is achieved as follows.

The receiver has an extremely stable local oscillator (stalo). At each transmission, a sample of the r.f. output power is taken at low level. This, after mixing with the stalo, forms an i.f. locking pulse which is used to prime an oscillator which becomes the coherent reference (the coho). This oscillator is switched off during the time the lock pulse is injected. Upon switching on, the stored energy of the lock pulse starts the oscillation in a controlled manner. Thus the coho preserves the phase reference for a pulse period. The coho is then switched off just prior to the arrival of the next lock pulse and the process is repeated. The system is illustrated in *Figure 59.12*. Signal processing is described in Section 59.2.9.

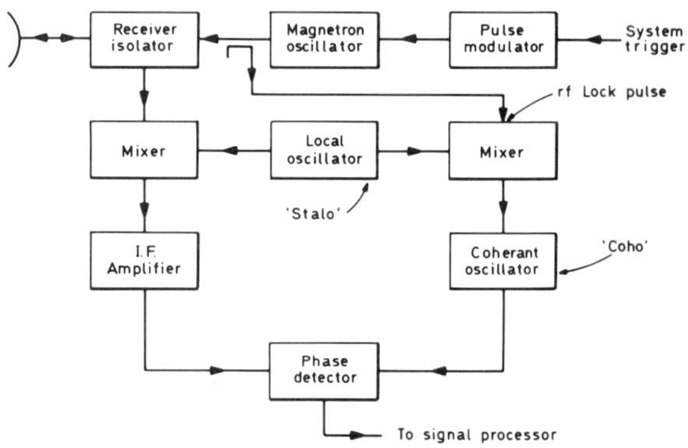

Figure 59.12 Coho-stalo system

59.2.2.6 Chirp

Radar performance is dependent upon the mean power of the system. If this is increased by lengthening the output pulse duration, immediately the range discrimination is reduced (12.36 μs is equivalent to 1 nautical mile in primary radar terms). The *chirp* or pulse compression technique is a means of obtaining high mean power by use of longer pulse output while retaining high range discrimation. It operates as follows.

The transmitter output pulse is frequency modulated, usually linearly, throughout its duration. Received signals which bear this modulation are passed through a pulse compression filter which has a delay characteristic which is frequency dependent.

This characteristic is made to be of opposite sense to that used in transmission so that the frequencies occurring later in the transmitted pulse are delayed least. Those occurring early in the output pulse are delayed most. By this means the received energy is compressed in time. If the original pulse duration is τ the output pulse duration from the compression filter is $1/B$ where $B = f_{max} - f_{min}$ used in the frequency modulation. The peak power using this technique is effectively increased by a factor of $B\tau$ which is called the pulse compression ratio or dispersion factor.[7,8]

An unfortunate by-product of this technique is the generation of 'time sidelobes' (see references 9, 10, 17) and an effect where small signals in close range proximity to large signals are masked until range separation of pulse widths has been established.

59.2.3 Operational roles

Radar's ability to 'see without eyes' has placed the technique at the service of a wide variety of users. A resumé of operational roles with pertinent data is given in *Table 59.2*. It will be seen that a wide range of engineering is required to furnish the requisite hardware.

59.2.4 Transmitters

Across the band of microwaves from 8 mm to 50 cm transmitters of peak power from watts up to 10 MW are found. For the higher powers required, magnetrons are almost exclusively used for wavelengths up to 23 cm. Klystrons and other forms of power amplifiers have been employed at wavelengths of 10 and 23 cm. For 50 cm wavelengths high-power klystrons are universally employed. The advantage of klystrons and power amplifiers is their ability to be driven with crystal-controlled sources. This produces automatic frequency and phase coherence in the system which is called for in all MTI systems. Magnetrons have to be excited by a very high voltage pulse.

This is produced from a source the energy of which can be transferred as a pulse with a controlled rate of rise and fall, at a known time, for a known duration. This is usually effected by a pulse-forming network, the stored energy of which is released to the magnetron by a trigger device via a pulse transformer (usually a thyratron, although hard valve modulators are still used). The released energy, in pulses of some 30–60 kV of up to 10 μs duration, is stored by a charging circuit which can be resonant, for greater efficiency, at the repetition frequency of the transmitted pulses.Care has to be exercised in pulse shaping to avoid magnetron moding and unwanted frequency modulations. Purity of the frequency spectrum is difficult to achieve unless great care in modulator design is exercised.[6] Klystrons are virtually power amplifiers and still need high-voltage pulses to provide the amplification during the required pulse output.

Other methods of providing high peak power in the microwave range are possible but very rarely found in practice, examples are as follows.

High-power klystrons. These are available for operation at B (250–500 MHz), L (0.5–2.0 GHz), S (1.5–5.0 GHz) and C (4.0–6.0 GHz) band with output power up to 20 MW at L band. Power gain typically 40 dB.

High-power travelling-wave tube (TWT). These can be operated at L, S and C band giving output of up to 2 MW peak with gains typically 45 dB. They have very wide bandwidth.

Twystron. So called because it uses a klystron-type cathode gun assembly and the TWT technique of slow wave restriction to provide gain of up to 40 dB at S band. It has the very desirable property of some 200 MHz instantaneous bandwidth, allowing frequency agility to be used without transmitter complication.

Crossed-field amplifiers (CFA). These are found for use across the X to C (4.0–11.0 GHz) band and provide typically 10 dB gain. They are a form of amplitron and use cold cathodes.[7,8]

Solid-state devices. By use of pulse compression techniques the requisite mean power can be provided by power amplifiers consisting of batteries of transistors whose output is combined in proper phase relationship. This has great advantages in high reliability. No modulator is required, pulses being formed by low-level c.w. input switching.

59.2.5 Receivers

Across the microwave band, receivers almost always use the superheterodyne technique. This is to simplify signal handling in processors, etc. The design aim is to preserve dynamic range to prevent amplitude and phase distortion outside the designer's control. Parametric amplifier techniques are now common which produce noise figures of some 3 dB at 50 cm and 6 dB at 3 cm wavelength.

These commonly employ variable capacitance diodes which 'noiselessly' extract power from a separate microwave c.w. oscillator source called the *pump* and convert it to the signal frequency, thus achieving amplification of signal without introduction of extra noise. The gain of a parametric amplifier of this kind is proportional to the ratio of pump to signal frequency. When integrated into a high-power radar system care has to be used to protect the sensitive diodes from damage due to the leakage of power during transmission.

Very nearly theoretical limits of sensitivity have been reached in some receiver designs (i.e. noise figures approaching 0 dB indicating the receiver operating at the limit of KTB*). This is achieved by super-cooling the receiver input elements incidentally producing a novel engineering problem embracing physics, electronics, mechanics, chemistry and refrigeration disciplines. These types of parametric amplifiers are to be found in radio telescopes the powers of which are determined to a large degree by receiver performance.[6] In new radar systems the parametric amplifier is being displaced by a simple transistor 'front end' of very low noise figure.

The radar system sensitivity is given by the sum of all losses between the antenna and receiver plus the receiver noise figure itself; typical figures (in dB) are as follows:

Receiver noise figure	2.5
Rotating joint	0.5
Diplexer (diversity system)	0.25
Duplexer	0.5
Receiver protecting device	0.25
Wave guide and coaxial cable losses	0.25
Total effective noise figure	4.25

In radar performance calculations these losses are entered into the radar equation when deriving S_{min}.

I.F. amplifiers

It is common in modern transistorised equipment to find 100 MHz bandwidth easily achieved and bandwidths subsequently restricted to the optimum by the design of a filter at the input or output stages. Dynamic characteristic forming is usually done at i.f. and can produce linear, logarithmic, limiting or compression characteristics fairly readily over 80 dB of dynamic range above noise.

* K = Boltzmann's constant, T = temperature in degrees absolute, B = bandwidth.

Table 59.2 Operational roles

Operational role	*Typical wave-length*	*Typical peak power and pulse length*	*Deployment*	*Characteristics*
Infantry manpack	8 mm	f.m.–c.w.	Used for detecting vehicles, walking men	Has to combat clutter from ground and wind-blown vegetation etc. Hostile environment. Battlefield conditions
Mobile detection and surveillance	3 cm	20 kW at 0.1 μs to 0.5 μs	Used in security vehicles moving in fog. Also military tactical purposes	Needs high resolving power for target discrimination. Hostile environment. Mobile over rough ground
Airfield surface movement indicator	8 mm	12 kW at 20 ~ 50 ns	Used on airfields to detect and guide moving vehicles in fog and at night	Resolving power has to produce almost photographic picture. Can detect walking men
Marine radar (civil and military)	3 cm 10 cm 23 cm	50 kW at 0.2 μs (see Defence below)	Used for navigation and detection of hazards. Longer wavelength for aircraft detection, i.e. 'floating surveillance' radars	3 cm needs good resolution and very short minimum range performance. All have to contend with sea clutter and ships' movement
Precision approach radar (PAR)	3 cm	20 kW at 0.2 μs	Used to guide aircraft down glidepath and runway centreline to touchdown	Very high positional accuracy required together with very short minimum range capability and high reliability
Airfield control radar (ACR)	3 cm 10 cm	75 kW at $\frac{1}{2}$ μs 0.4 MW at 1 μs	Usually 15 rpm surveillance of airfield area. Guides aircraft into PAR cover or on to instrument landing system	Needs anti-clutter capability. 10 cm has MTI. Good range accuracy. High reliability required
Airborne radar	3 cm	40 kW at 1 μs	Used for weather detection and storm avoidance	Forward-looking sector scanning. Storm intensity measurement system usually incorporated
Met. radar	3 cm 6 cm 10 cm	75 kW to $\frac{1}{2}$ MW $\frac{1}{2}$ ~ 1 μs	Surveillance and height scanning gives range/bearing/height data on weather. Also balloon-following function	Good discrimination and accuracy. Rain intensity measuring capability
Air traffic control—terminal area surveillance (TMA)	10 cm 23 cm 50 cm	$\frac{1}{2}$ MW to 1 MW 1 μs to 3 μs	Surveillance of control terminal areas. Detection and guidance of aircraft to runways and navigational aids. 60 n miles range	Needs good discrimination and accuracy. MTI system necessary. Display system incorporates electronic map as reference
Air traffic control—long range	10 cm 23 cm 50 cm	$\frac{1}{2}$ MW to 2 MW 2 μs to 4 μs	Surveillance of air routes to 200 n miles range. Monitoring traffic in relation to flight plans	Has to combat all forms of clutter. MTI essential. Circular polarisation necessary except at 50 cm
Defence—tactical	10 cm 23 cm	$\frac{1}{2}$ MW to 1 MW 1 μs to 3 μs	Mobile or transportable. Used for air support and recovery to base	As for TMA radar plus ability to combat jamming signals
Defence—search	10 cm 23 cm	2 MW to 10 MW 2 μs to 10 μs	Used for detection of attacking aircraft monitoring of defending craft including direction for interception. Usually a height finder operated together with search	As for ATC long range plus ability to combat all forms of jamming. Forward stations report data to defence centre
Defence—3D search	23 cm	5 MW at 60 μS compressed to 0.6 μs	Used for defence search but multiple elevation beams allow automatic calculation of target height	As for defence—search, polarisation is horizontal

59.2.6 Transmitter–receiver devices

Until very recent times high-power radars used gas discharge tubes as a means of producing isolation between transmitter and receiver. A simple form is illustrated in *Figure 59.13*. When the transmitter fires, the gap in the ATR cell breaks down and allows power to progress to aerial and receiver. Power across the TR cell will cause the cell's tube to strike and impose a short circuit across the receiver input terminals. This short circuit is so placed that matching to the aerial is not disturbed. Upon cessation of the transmission the ATR device will revert to quiescence and thus disconnect the transmitter from the system. The TR cell's short circuit is also removed and thus returned signals can reach the receiver. Disadvantages of this technique are:

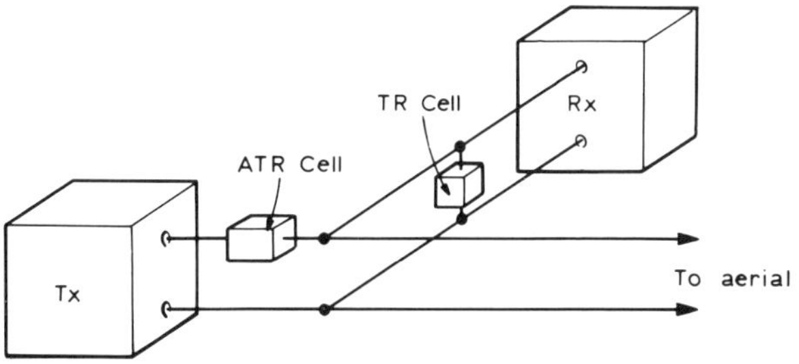

Figure 59.13 Simplified diagram showing receiver protection devices

(*a*) Leakage past the TR device into the receiver.
(*b*) Long recovery time after firing which provides attenuation to received signals. This limits minimum range performance.

These devices are being replaced in modern equipment by high-power-handling diodes using the same technique of open and short circuit lines. They have better recovery times and less insertion loss but produce problems associated with the need to switch bias voltages in synchronism with transmission in order to prevent destruction of the diodes.

59.2.7 Aerials

The gain of an aerial is related to its area. However the maximum gain is not the only, nor even the prime, consideration. The radar designer attempts to achieve gain in desired directions and thus beam shape in both horizontal and vertical planes becomes extremely important. The beam may be formed in a variety of ways as follows:

(*a*) Use of groups of radiators (dipoles, unipoles, radiant rods, helices, etc).
(*b*) Reflector illuminated by various elements, e.g. dipole, horn, poly-rod, linear feed, etc.
(*c*) Slotted waveguides.
(*d*) Microwave lens system.

By far the most common is group (*b*). There are here many choices of illumination, distribution and reflector shape to permit practically any radiation pattern to be produced.[18]

The beamwidth to the half-power points (θ) in either azimuth or elevation planes is approximately given by

$$\theta = 70\lambda/D$$

where λ is the wavelength and D the physical aperture of antenna in the plane considered, expressed in the same units as λ. The beam is formed at the Rayleigh distance r from the antenna, given by

$$r = fD^2/\lambda$$

where f is a factor not less than unity and up to two. Taking $f=2$ ensures that the transition between near- and far-field effects has been passed.

Further information on antenna will be found in Chapter 49.

59.2.8 Radar displays

Almost exclusively these use cathode ray tubes to represent various related dimensions on its screen, e.g. range and azimuth, range and height, range and relative amplitude, etc. Two types of modulation are used: amplitude and intensity. Typical examples are given in *Figure 49.14*. Of these the most important is the PPI (plan position indicator).

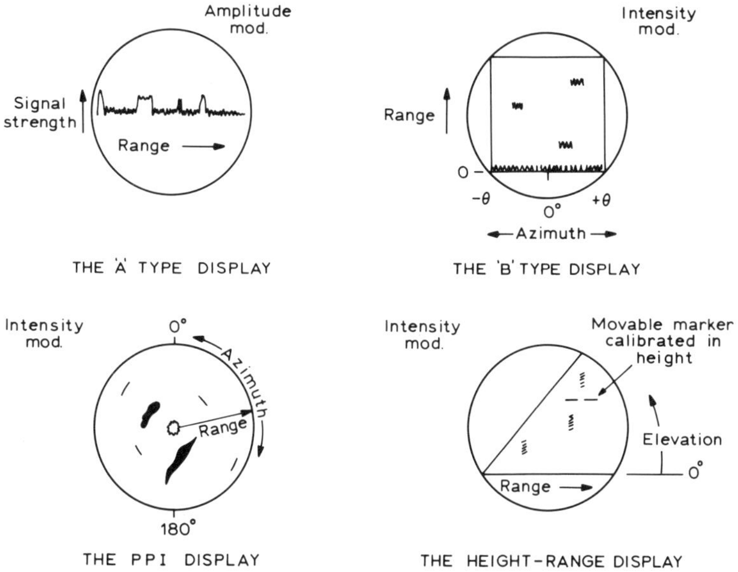

Figure 59.14 Various radar displays

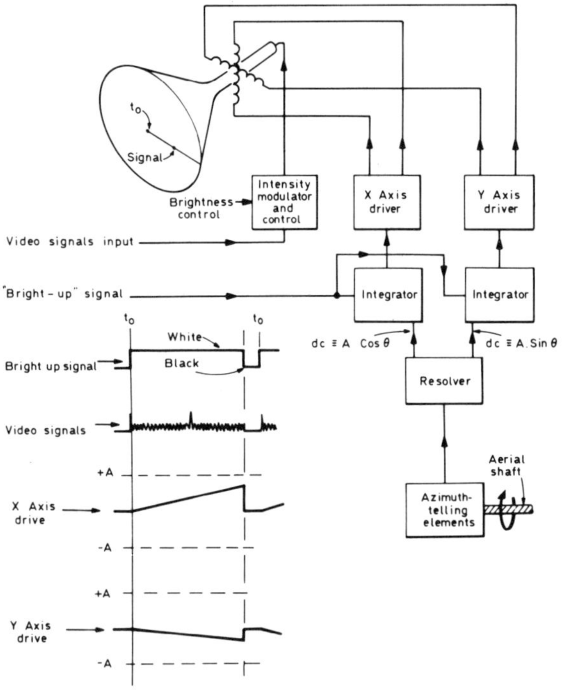

Figure 59.15 PPI display system

59.2.8.1 The PPI

Although *Figure 59.15* shows magnetic deflection, electrostatic deflection can also be used. The deflection coils, nowadays of sophisticated design with correction for many types of error and aberration due to tube geometry and manufacture, are mounted around the neck of the tube and produce orthogonal fields across the path of the electron beam. Variation of these fields deflect the beam across the face of the tube which it strikes and excites the phosphor on its surface producing an intense glow at the point of contact. This glow is visible through the glass of the screen.

Assume the tube centre is to represent the position of the radar and its edge, the radar's maximum range. The display system generates a *bright-up* waveform synchronised to the radar transmission time t_0, and with its amplitude and level set to give a threshold producing very small excitation of the phosphor in the absence of modulating signals. The waveform is also used to govern the period over which the time-base integrators operate. There are identical chains for the X and Y axes and both perform integration of a d.c. voltage expressive of the sine and cosine of the bearing to be indicated. These d.c. voltages can be generated in a number of ways, a common method being a 'sin/cos' potentiometer. The resolver is made to rotate in synchronism with azimuth tell-back elements mechanically geared to the aerial turning system. This may be done, for example, with torque-transmitting rotary transformers ('selsyns') or a servo-drive synchro system, again using rotary transformer technique. As the aerial turns, the values of the resolver's d.c. outputs will change, going through one cycle of maxima and minima per aerial revolution. The action of the integrators will therefore produce X and Y saw-tooth waveforms whose amplitude and polarity change in sympathy with the aerial's rotation thus automatically reproducing the fields necessary to cause the electron beam to move across the tube at the azimuth of the aerial. By making the X and Y drive waveforms highly linear and governing the overall gain of the deflection system, the tube face can be made to represent areas to calculable scales in a very linear fashion. Since the time base is synchronised with radar transmission time, received signals appear $12.36 \times \gamma$ μs later, where γ is their real range in nautical miles. Signals of detectable amplitude are amplified and used as sources of intensity modulation of the c.r.t. beam. Thus as the radar's narrow beam is rotated in azimuth the signals returned are 'written' on to the tube face. By varying the chemical properties of the phosphor the length of after-glow time can be varied. By this means the display system can add to the radar's fulfilment of its operational role, e.g. if track history is required, long after-glow can provide this; if small changes in signal pattern are required in a fast-scan system, a short after-glow is used.

The example in *Figure 59.15* is capable of many variations, e.g. the resolution into X and Y saw-tooth deflection waveforms can be done by a rotary transformer element with two fixed orthogonal field coils and the rotor fed with a single fixed amplitude integrated saw-tooth. The rotor would be turned in sympathy with the aerial by a servo-driven azimuth transmitting system. Some displays still exist where the deflection coils are rotated around the c.r.t. tube neck; still others are used with a group of coils fixed around the tube which produce a rotating field in sympathy with the aerial rotation. The displays described above are all real-time systems. In non-real time systems, increasingly to be found, data in the form of symbols or alpha-numeric characters are written upon the screen in machine time. Here it is common to find deflection systems consisting of fast and slow elements working in conjunction. In this case no saw-tooth integration is needed. The 'slow' deflection coils are made to take upfield values which give the beam a desired position at a desired machine time. The 'fast' deflection coils have analogue waveforms of small amplitude to cause the beam to 'write' desired characters. Typical speeds are 20 μs to move from one tube edge to another and 5 μs per character for data writing. It is unusual to find more than 25 characters (5×5) associated with individual targets.

Modern systems utilise digital techniques for both azimuth telling and time-base generation. They may be organised in real or machine time. Two main systems of azimuth telling are to be found, both producing high resolution of the circle into a 13- or 14-bit structure. Digital shaft encoders of magnetic or optical type can produce a multi-bit expression of the aerial's position as a parallel multi-bit word; alternatively, for reporting the position of a continuously rotating shaft (as is common in most search radars), the encoder reports only changes to the least significant digit of the multi-bit word. Integration into the full word is done by a digital resolver at the receiving end of the data link. This produces great economy in data transmission.

59.2.9 The radar environment and signal processing

Wanted signals are always in competition with unwanted signals. The total environment can be appreciated by reference to *Figure 59.16*. This is typical of the general case and illustrates need to separate the wanted aircraft signals from those simultaneously arriving from, in this radar's role, unwanted or clutter sources. The radar system also suffers total blindness due to shadowing by solid objects and attenuation due to gases, dust, fog and similar particular media. e.g. rain and clouds. The radar aerial attempts to concentrate its radiation in a well defined beam. However there are always side lobes present which produce radiation simultaneously with the main beam in unwanted directions. These also gather unwanted signals.

Signal processing can be used to discriminate in favour of wanted signals. The basic radar system is also designed to provide some bias in favour of the wanted signals, e.g. by using as narrow a radar beam as possible, by the use of the longest

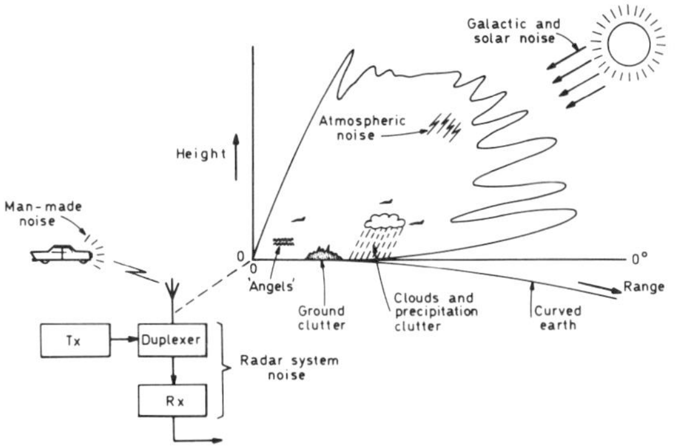

Figure 59.16 The radar environment

wavelength possible to minimise signals from cloud and rain, by the use of the shortest pulse possible at the highest repetition rate, etc.

The radar's resolution capability is largely determined by three parameters

(*a*) The azimuth beamwidth.
(*b*) The elevation beamwidth.
(*c*) The transmitter pulse length.

This is illustrated in *Figure 59.17*.

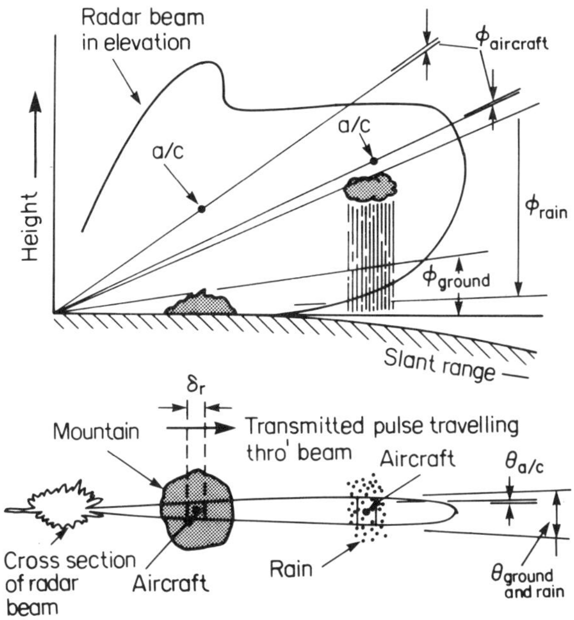

Figure 59.17 The significance of resolution cell. The ground and rain occupy a larger volume of the resolution cell than do aircraft

The desirability of reducing the size of the resolution cell to improve wanted 'target-to-clutter' ratio can be appreciated by considering the following practical case:

Azimuth beamwidth	1.5°
Target range	60 n.m.
Pulse duration	3 μs
Elevation beamwidth	10°

(approx. 0 to 70 000 ft at 60 n. miles)

The cross-sectional area of the cell is about $\frac{1}{2}$ (n. miles)^2 and rises to 11 (n. miles)^2. If the wanted target is an aircraft it is very much smaller than the resolution cell which can gather unwanted targets (ground clutter, rain and clouds, angels, etc) throughout its volume.

59.2.9.1 Discrimination against ground clutter

A number of techniques are available ranging from the very simple to the highly sophisticated.

Map blanking. An electronic video map of the clutter pattern from a fixed site is used to produce suppression signals which inhibit the radar display, thus producing a 'clean' picture containing only moving targets. This method is not effective with moving targets in regions of clutter and for this reason is seldom used.

Pulse length discrimination (*PLD*). Here use is made of the difference in the range continuum in the duration between aircraft signals and the general mass of ground clutter. All clutter greater in duration than twice the transmitted pulse length is totally rejected, those up to this limit are displayed thus preserving aircraft targets and rejecting almost all ground clutter. This is a relatively simple video signal process which has the advantage of providing supra-clutter visibility. Thus if the wanted signal is of sufficient strength to show above the clutter, it is displayed in clutter regions. A logarithmic receiver is usually employed ahead of the process to increase the dynamic range before signal amplitude limiting takes place.

Moving target indicators (*MTI*). In this case the processing is closely linked with modulator and transmitter design and is therefore more of a radar system than pure signal processing. The purpose of an MTI system is to discriminate between fixed and moving targets, and system coherence is required (see Section 59.2.2.5).

The discrimination is made in the following manner. Consider the radar aerial stationary and illuminating a fixed target, e.g. the face of a mountain. Successive signals received from regular pulse transmissions will be at fixed range R. In the continuum this range can be expressed as $R = n\lambda + d\lambda$, where λ is the wavelength of transmission, n the number of whole wavelengths to the target and d a fraction.

By establishing a phase reference at each transmission and preserving it for a whole reception period it is possible to give coherent meaning to the above expression using a phase-sensitive detector, the output of which for any target is characterised by three factors:

(i) Its occurrence in time relative to transmission (due to the value of $n\lambda$) to the nearest whole wavelength.
(ii) Its peak-to-peak amplitude due to the usual radar parameters, e.g. target echoing area, etc.
(iii) Its amplitude and polarity within the peak-to-peak range due to the value of $d\lambda$.

In the case cited the output would be as illustrated in *Figure 59.18*(*a*). A moving target would produce different values of $d\lambda$ for relatively slow speeds and different values of $n\lambda$ as speed increased. Thus the phase detected output might behave as illustrated. By the storage and comparison technique, successive received signals can be subtracted from each other. Both analogue and digital techniques are employed.

In the example given it will be seen that the signals from fixed targets can be reduced to zero and moving targets will produce non-cancelling outputs. The rate at which the moving targets change their amplitude from pulse to pulse is analogous, and equal in value, to the Doppler frequency referred to in equation (59.7). A simple pulse Doppler MTI is illustrated in *Figure 59.18*(*b*). It has a number of limitations which tend to offset its

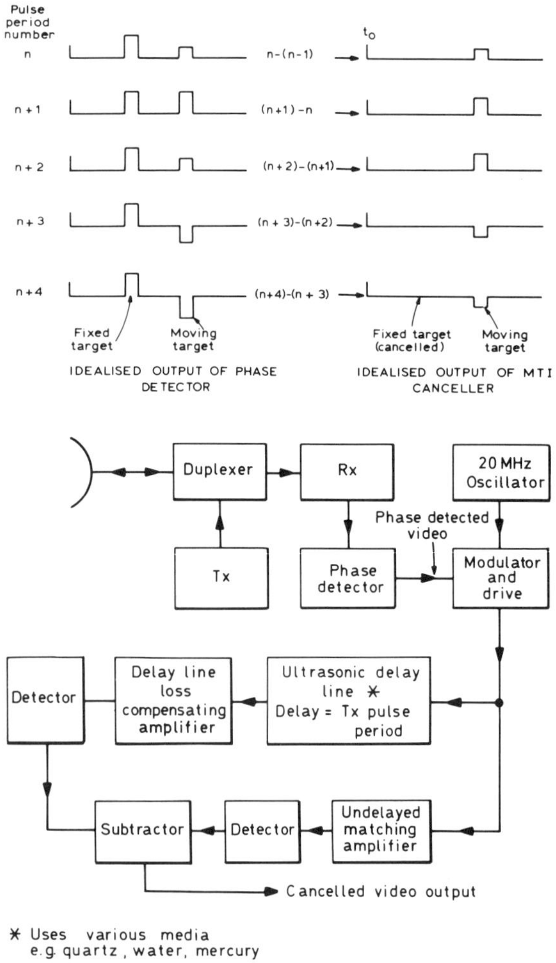

Figure 59.18 (*a*) The MTI system's phase detector-canceller principle (*b*) A simple MTI radar

major advantage of providing sub-clutter visibility (SCV, the ability to see moving targets superimposed upon fixed targets where the moving target is smaller than the fixed). Values of 20 dB for SCV are common in many surveillance radars using this type of MTI. Some of the limitations referred to are as follows:

(*a*) Blind speeds—where a moving target behaves as though it was stationary, i.e. $n\lambda$ changes by an integer in a pulse period. In this condition the Doppler frequency is a multiple of the pulse repetition frequency of the transmitter.
(*b*) Blind phases—where a moving target produces the same output from the phase detector whose characteristic is symmetrical in the phase axis.
(*c*) Phase difference masked by noise or swamped by phase distortion in the receiver.

The system disadvantages can be overcome in varying degrees, e.g. blind speeds by staggered p.r.f. systems; blind phases by dual phase detectors, whose references phases I and Q are 90° displaced.

Other disadvantages are systematic and less tractable. For example 'fixed' targets are seldom found. The 'mountain' referred to above commonly has trees growing on it and wind causes fluctuations which spoil complete cancellation. Also the aerial is rotating in practice and its horizontal polar pattern modulates the target's amplitude again spoiling cancellation.[7,8]

Digital techniques. These are now almost always used. The principles described above remain the same. Differences from the analogue technique begin at the output of the phase detector. By quantifying the amplitude and polarity of the phase-detected output at a regular clock rate, typically 1 to 4 MHz, the resultant output is able to be put into a range-ordered store. It remains in store until the next transmission and, range cell by range cell, the current video is compared with that in store. Differences are taken and either used or stored again for second differences to be taken before release as cancelled output. It is a real time process.

Staggered p.r.f. The simple two-pulse canceller MTI radar illustrated in *Figure 59.18* can be improved by carrying out successive cancellations. The system response for cancellers with regular p.r.f. is illustrated in *Figure 59.19*. The 'blind speeds' are

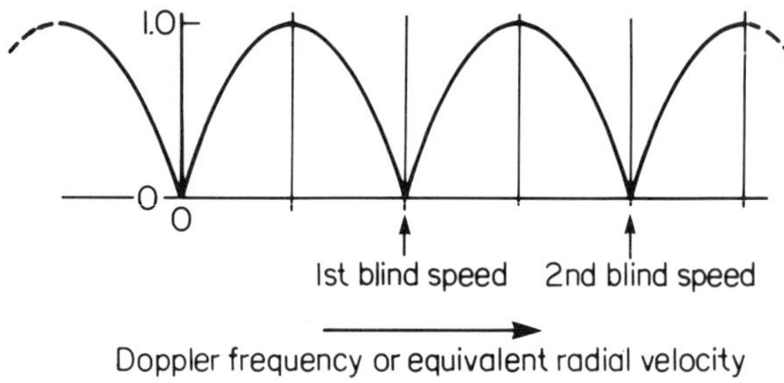

Figure 59.19 MTI response of two-pulse comparator

equivalent to a target movement of integral numbers of half wavelengths in an interpulse period. In a typical L-band radar a target radial speed of 150 knots gives a Doppler frequency of about 600 Hz. From equation (59.7) and *Figure 59.20* it is seen the relationship is linear. As seen from *Figure 59.19* a regular p.r.f. results in repeated 'blind speeds'. By varing the interpulse period at successive transmissions the overall system response is a compound or average of a number of responses shown in *Figure 59.21*, each slightly displaced because of differences of blind speeds. The first true blind speed will occur when all the zeros of those in the group coincide. Thus if two p.r.f. were used in the

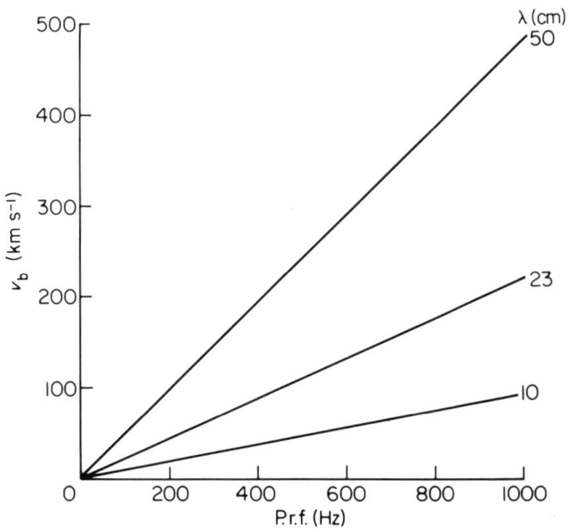

Figure 59.20 MTI blind speeds at three wavelengths. v_b (km s^{-1}) = λ (m) × p.r.f. (Hz) × 0.968

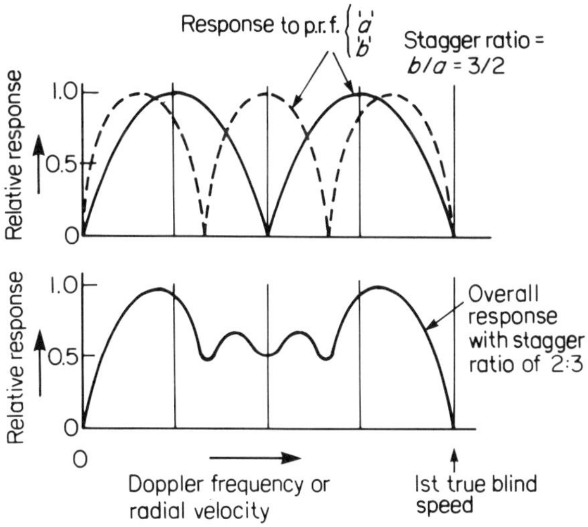

Figure 59.21 MTI response with staggered p.r.f.

ratio of 4:5 then v_{btrue} is given where $4v_{b4} = 5v_{b5}$ where v_{b4} and v_{b5} are the respective blind speeds of the two p.r.f. Note that by making the 'stagger ratio' small the value of v_{btrue} increases but vestigial blind speed losses increase.

The response in *Figure 59.19* is a series of half sine waves, resulting from a two-pulse comparison MTI. If a three-pulse comparator is used the response becomes a series of halfsine waves squared. A general expression for the response shape of an 'n' pulse comparison system is:

$$A = \sin^{n-1} \omega t \qquad (59.9)$$

where n is the number of pulses compared and ωt the equivalent Doppler frequency.

When staggered p.r.f. are used the general form of the modulus of the overall response is:

$$y = A \sin^{n-1} \omega_1 t + A \sin^{n-1} \omega_2 t + \cdots + A \sin^{n-1} \omega_m t \qquad (59.10)$$

where n is the number of pulses compared and $\omega_1 t, \omega_2 t, \ldots, \omega_m t$ are the individual responses of each p.r.f. used in the stagger pattern.

Doppler filtering. As described above, MTI can be regarded as a time sampling and comparison system. The coherence of receptions with transmissions allows the wavelength to be used as a vernier of the range scale.

The target movement is detected by comparing the phase of returned signals from successive transmissions as well as their range (which changes very little between the samples at the p.r.f. rate).

It happens that the method produces signals whose amplitude changes from the phase detector occur at the Doppler frequency created by the target's movement.

It is possible, using digital techniques, to construct a series of contiguous passband filters in the Doppler domain and by these to discriminate (to the degree governed by the filter bandwidths) between fixed and moving targets. Phase detector outputs form the filter input.[8,37] Such an arrangement is illustrated in *Figure 59.22.* Points of note are:

(*a*) Each filter has to be followed by some form of TTI. The zero Doppler filter (f_0) output will contain all ground clutter and thus provides targets only when they exhibit super-clutter visibility.

(*b*) Because the passband characteristics of filters overlap and each has side lobes, heavy ground clutter can make clutter also appear in filter f_1 output. The same will be true of weather clutter at radial velocities higher than zero.

(*c*) Because the phase detector cannot distinguish range changes between samples (successive transmissions) of greater than a wavelength, the zero Doppler filter repeats itself at f_D, the sample rate (i.e. the radar p.r.f.). This is directly analogous to 'blind speed'. Thus any target flying at a non-zero blind radial velocity will be masked by the 'alias' of the ground clutter, if at the same range, and not be seen unless it has super-clutter visibility. For this reason it is necessary to operate with staggered p.r.f. as in the pulse comparison system.

(*d*) The Doppler filter technique has the distinct advantage of giving target visibility in ground and weather clutter where radial velocities of clutter and target are separated by a filter passband width.

(*e*) Filters formed by digital techniques require samples at a regular rate equal in number to the number of filters. Thus staggered p.r.f. can only be executed in blocks (i.e. if four filters then four samples at one p.r.f. have to be executed before changing to another p.r.f.). This can have the effect of causing half the target data to disappear if the target happens to be at the blind speed of one of the p.r.f.

Data on Doppler filtering techniques can be found in reference 19.

Integration techniques. As far as MTI is concerned ground clutter containing wind blown vegetation is characterised by exhibiting movement but no displacement over successive aerial revolutions. This is used in the technique of temporal threshold integration (TTI) to remove clutter signals in the following way.

The surveillance area is covered by a large mass of contiguous storage cells (typically 65 000) each of the same azimuth and range size. At successive aerial revolutions the contents of each cell is peak detected and a proportion of the value is stored. Thus over a number of aerial revolutions the true peak value of clutter amplitude would be achieved for non-fluctuating cell contents. The stored value for each cell thus establishes a threshold which must be exceeded if detection is to produce an output. To account for clutter fluctuations a margin is added to the stored values. Thus if an aircraft were to be detected over clutter it requires to exhibit 'super-clutter visibility', i.e. to be of such an amplitude that its addition to the clutter results in exceeding the established clutter threshold pulus the margin. The integration time is such that aircraft dwell time in any cell is too short effectively to modify the cell's established threshold.

A variant of this technique performs its integration over batches of cells in the range domain only. The threshold for each cell is now set by the average contents of cells either side of it. It is used in environments which could contain barrage or noise jamming signals which can be rapidly switched on and off. The TTI integration time is many aerial revolutions therefore any such jamming, having set high thresholds in many cells, when switched off could result in needless desensitisation for long times. In range integration the reaction time is very much less than one aerial revolution.

Adaptive signal processing (ASP). A compound of the principles of digital MTI, Doppler filtering and temporal threshold integration is found in the adaptive signal processor. The principle of operation can be seen from the block diagram in *Figure 59.23.* Phase detected signals are used as input to a four-pulse digital canceller. Staggered p.r.f. gives a Doppler filter passband as shown in *Figure 59.24*(*b*). This rejects signals with zero and near-zero Doppler components. Moving weather clutter will still appear at its output and must be rejected. Dual phase detectors whose references are 90° phase displaced express the Doppler frequency of the input signal as a phase angle relative to the reference. The measured phase angle is used in a further phase detector to govern the phase shift of the reference

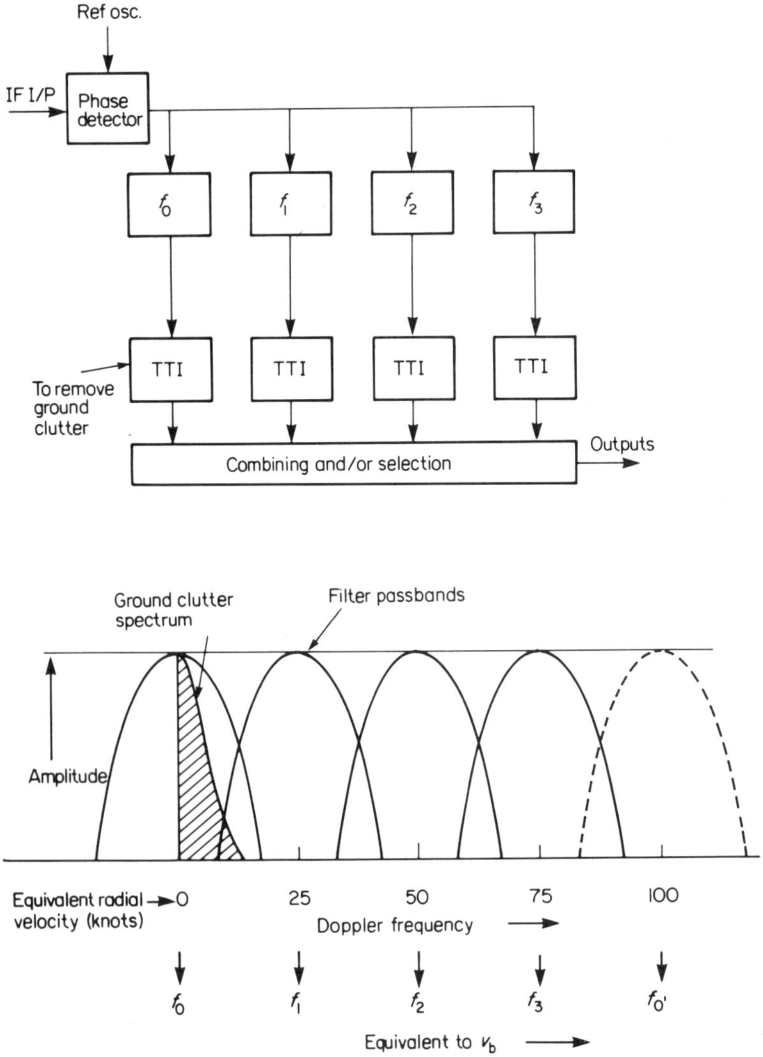

Figure 59.22 A typical four-filter arrangement

oscillation. Subsequent four-pulse cancellation with staggered p.r.f. gives an output characteristic with a rejection 'notch' in the Doppler domain centred upon the measured Doppler frequency. Integration constants are such that aircraft targets of non-zero Doppler components do not cause the phase shift to take place. Thus signals outside the passband of the rejection notch are allowed as output. Simultaneously with the above two detection processes, non-coherent detection is performed.

Temporal threshold integration is applied to each of the three detection processes' outputs and allows any available wanted signal to be gated out in real time. The whole processor thus automatically adapts itself to deal with fixed and moving clutter.

59.2.9.2 Discrimination against interference

P.R.F. discrimination. Use is made of the fact that radar transmission is at regular rates, i.e. a known repetition frequency. Using a storage technique similar to that of the MTI system, successive signals are compared. Those due to the station's own transmission will correlate in range, those due to interference will do so only if harmonically or randomly related to the station's p.r.f. Instead of a subtraction process, coincidence gating is used and so wanted signals are released and unwanted interference rejected. In systems not using a fixed p.r.f. use is made of the known time of transmission. Range correlation of wanted targets is used as detailed above, de-correlation again provides rejection of unwanted interference.

59.2.9.3 Discrimination against cloud and precipitation

To a certain extent the PLD technique can be used, but only when clutter is very severe. It is more general to use the circular polarisation (CP) technique. Here the radiation from the radar is changed from linear to circular by either waveguide elements or quarter-wave plates.[20] Rain drops and cloud droplets return nearly equal components in vertical or horizontal planes. When put into the aerial system these components of the signals from rain are of opposite phase and if of equal amplitude, will cancel. Signals from irregular targets have inequality of vertical and horizontal components and thus do not cancel. However, there will be some loss of wanted target strength (typically 2–3 dB). Rejection of rain signals can be as high as 20 dB thus

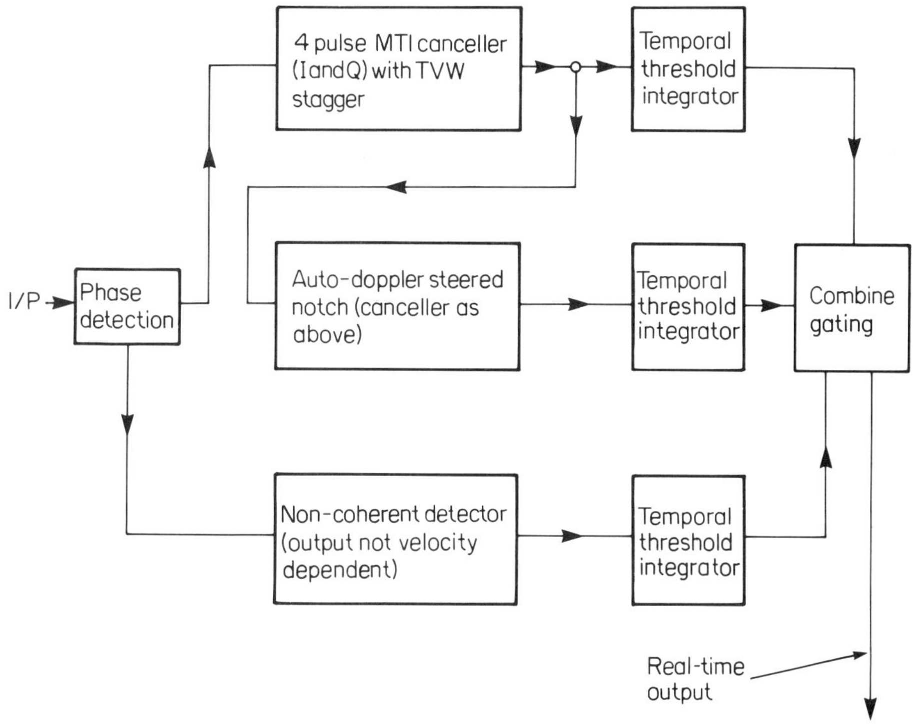

Figure 59.23 The adaptive signal processor

discrimination in favour of wanted targets is of the order 17–18 dB.

The choice of wavelength is critical in achieving discrimination against rain, the longer wavelengths increasingly provide protection due to an inverse fourth power law relating echoing area and wavelength.[6,21]

59.2.9.4 Discrimination against angels

Again, wavelength is an important factor, angels being more apparent at the shortest wavelengths. Angels are now generally recognised as being due to birds, in isolation and in great flocks.[22] Use is made of the target signal amplitude difference between angels and aircraft, angels being smaller in general. Range-dependent signal attenuation or receiver desensitisation is used to reduce angel signals to noise level; the coincident aircraft signals are also reduced at the same time but being stronger are retained above noise.

59.2.10 Plot extraction

Radar signals from given targets exist, in real time, for a few microseconds in periods of milliseconds. From these data the operator reads the signal's centre of gravity as displayed upon his p.p.i. The data on a given target are repeated each time the radar beam illuminates the aircraft—typically, every 4 to 15 s. Radar data commonly have to be sent over long distances, e.g. from the aerial's high vantage point to a control centre sometimes many miles away. To transfer these data in real time, inordinately large bandwidths must be wastefully employed.

In order to avoid this, and simultaneously to provide positional data in a form more suitable to computer handling, plot extraction equipment is fast emerging and being operationally used. The plot extractor accepts radar signals in real time from the processor, together with the digital expression of the aerial azimuth. The extractor performs logical checks such as range correlation, azimuth contiguity, etc, after signals cross a pre-set threshold and become digital. The extractor derives the range and azimuth centre of all plots which meet the logic criteria and stores these in azimuth and range order ready for release as range and azimuth words in digital form; typically 12 bits for range of 200 miles and 13 bits for 360° of azimuth.

The assembled data are read from store at rates suitable for transmission over telephone bandwidths. By this means 100 to 200 plots can be reported in a few seconds. Computer programmes have been constructed which convert plot data into track data, deriving speed and heading as by-products.[23,24] Track data may be fed into p.p.i. type displays at a high refresh rate (approximately 40 times per second) in machine time producing a picture which may be viewed in high ambient lighting.

59.3 Secondary surveillance radar (SSR)

59.3.1 Introduction

The need for automatic means of distinguishing friend from foe was foreseen by Sir Robert Watson-Watt in his original radar memorandum. The SSR system, which gives automatic aircraft identity, started its life as the wartime IFF system (identification friend or foe); the term IFF is used universally for the military application of the technique. The civil version is known variously as SSR and, more commonly in the USA, as the radar beacon system (RBS). Both IFF and SSR use the same system parameters these days and no distinction is made in the following text unless necessary.

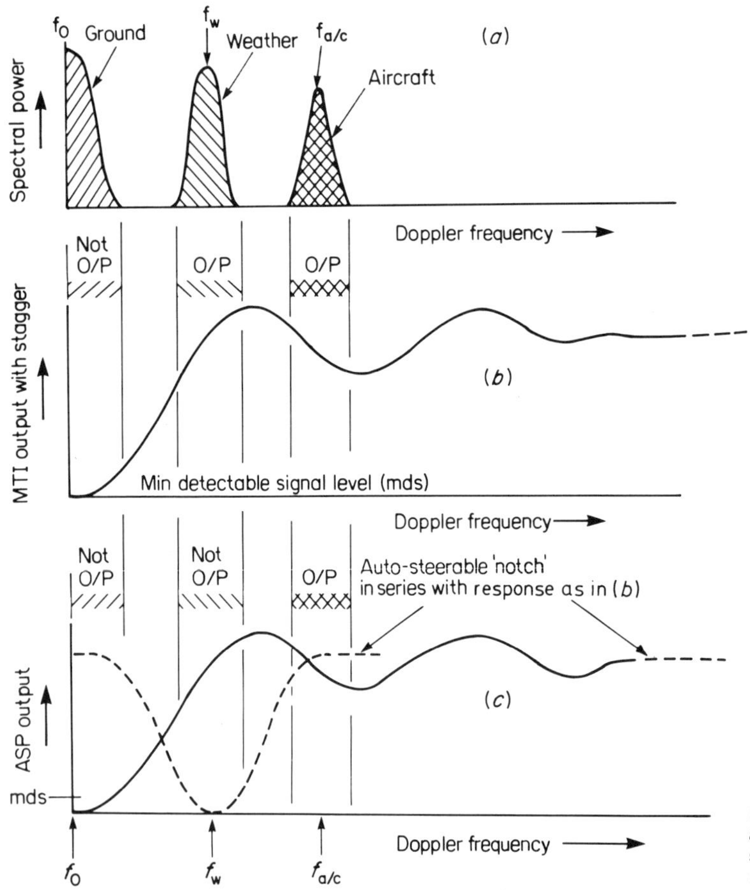

Figure 59.24 Signals and responses in adaptive signal processor (ASP). (a) Signal spectra; (b) traditional MII response; (c) ASP response

59.3.2 Basic principles

Regular transmissions of pairs of pulses are made from the ground station at a frequency of 1030 MHz via a rotating aerial with a beam shape narrow in azimuth and wider in elevation. A pulse pair constitutes an interrogation and modes of interrogation are characterised by coding of the separation of the pair, internationally designated P_1 and P_3. Aircraft carry transponders which detect and decode interrogation. When certain criteria are met the transponders transmit a pulse position coded train of pulses at 1090 MHz via an omni-directional aerial. The ground station receives these replies and decodes them to extract the data contained. This process of interrogation and reply is illustrated in *Figure 59.25*.

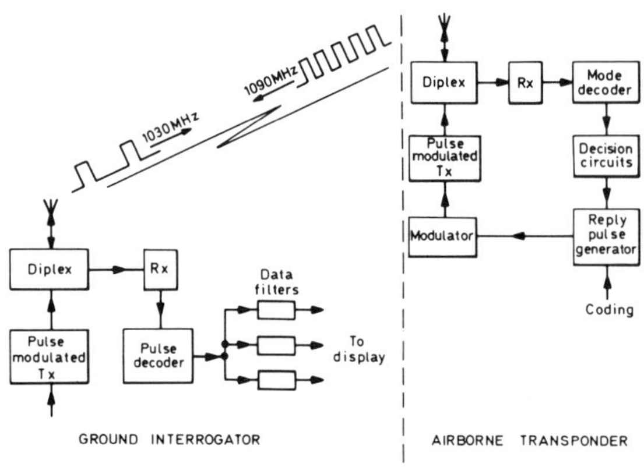

Figure 59.25 The SSR principle

59.3.2.1 *Modes of interrogation*

These are shown in *Figure 59.26*. Through the agency of the International Civil Aviation Organisation (ICAO) there is internationally agreed spacing and connotation for the civil modes. Various military agencies treat the military modes in the same fashion:

Military modes:
- Mode 1—Secure
- Mode 2—Secure
- Mode 3—Joint military and civil identity

Civil modes:
- Mode A—Joint military and civil identity (same as Mode 3)
- Mode B—Civil identity
- Mode C—Altitude reporting
- Mode D—System expansion (unassigned).

The dwell time of the ground stations beam on target is generally of the order of 30 ms, dependent upon beamwidth, aerial rotation rate and p.r.f. There is thus time enough to make some 15 to 20 interrogations of a given target. Advantage is taken of this to execute repetition of different modes by mode

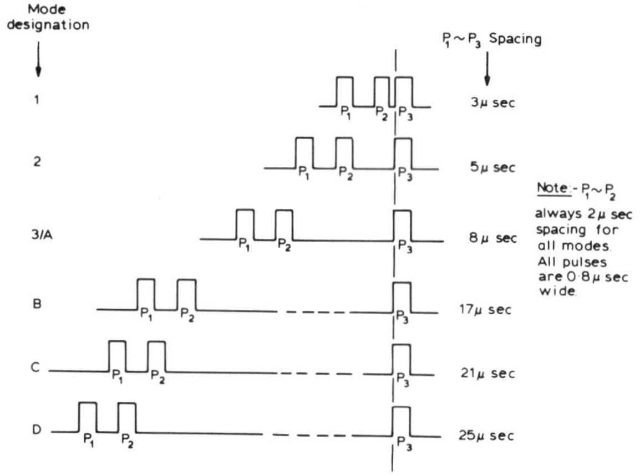

Figure 59.26 Interrogation pulse structure in IFF/SSR systems

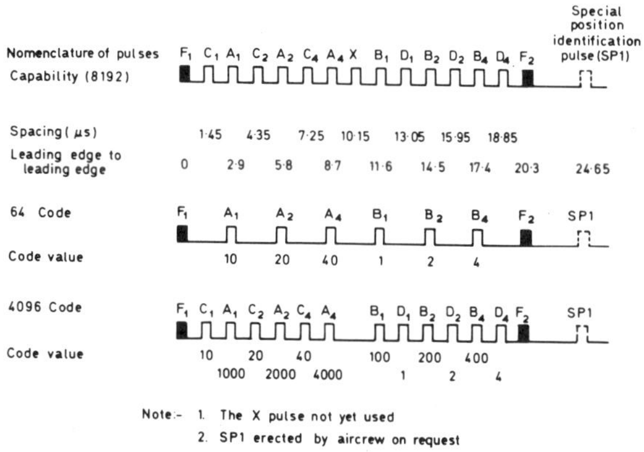

Figure 59.28 The ICAO reply format in SSR

interlacing. Repetition confers the necessary redundancy of data in the system so that data accuracy is brought to a high level. By this means, for example, the identify of an aircraft and its altitude can be accurately known during one passage of the beam across the target. Hence in one aerial revolution these data are gathered on all targets in the radar cover.

59.3.2.2 Replies

Aircraft fitted with transponders will receive interrogation during the dwell time of the interrogating beam. The transponder carries out logic checks on the interrogations received. If relative pulse amplitudes, signal strength, pulse width and spacing criteria are met, as specified in reference 25 replies are made within 3 μs of the receipt of the second pulse (P_3) of the interrogation pair. Replies are made at a frequency of 1090 MHz. This is 60 MHz higher than the interrogation frequency. In air traffic control terms this provides an important advantage with SSR vis-a-vis primary radar as can be seen from *Figure 59.27* illustrating SSR's clutter freedom.

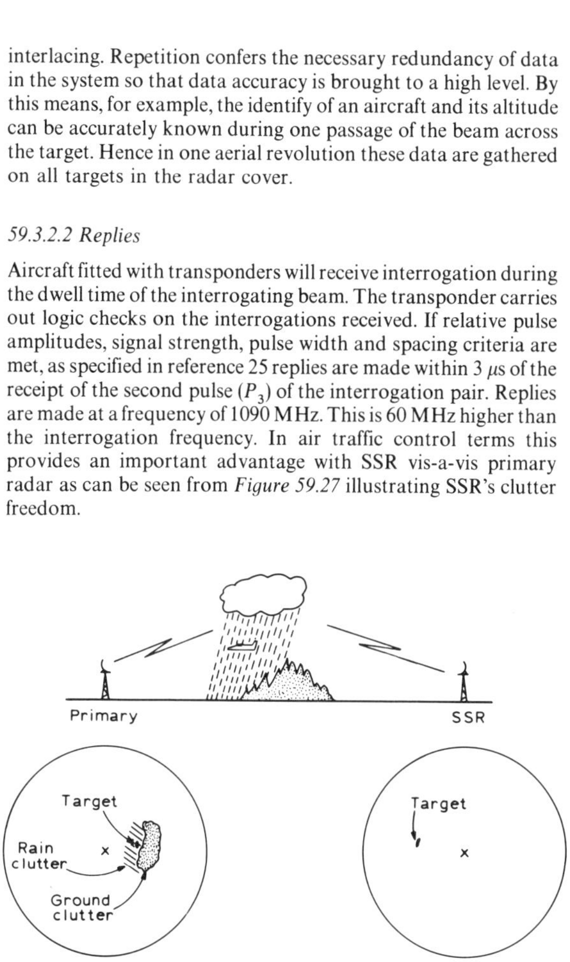

Figure 59.27 The clutter freedom of SSR

The pulse train of a reply is constructed, again according to internationally agreed standards, and allows codification of any or all of 12 information pulses contained between two always-present 'bracket' or 'framing' pulses designated F_1 and F_2. The pulse structures possible may be deduced from *Figure 59.28*. The code is octal based; the four groups of three digits are designated in order of significance as groups A, B, C and D. Of the 4096 possible codes, ICAO specifies that code 7700 is reserved for signalling emergencies, code 7600 to signal communication failure and code 7500 to signal 'Skyjack'.

The codification which signals altitude is formulated by a Gilham code pattern giving increments of 100 ft (30.5 m). The tabulation can be found in the ICAO document Annex 10[25] and covers altitudes from −1000 ft to +100 000 ft (−300 m to +30 000 m).

59.3.2.2 Side lobe suppression

The interrogation pair P_1 and P_3 is transmitted from an aerial with a radiation pattern narrow in azimuth and inevitably produces side lobes carrying sufficient signal strength to stimulate replies from aircraft transponders at short and medium range. This impairs azimuth discrimination and causes transponders to reply for longer than necessary thus generating unwanted interference (fruit) to other stations receiving replies. ICAO specifies a system of interrogation side lobe suppression (ISLS) to prevent replies being made to interrogations not carried by an interrogator's main beam. It operates as follows. The interrogation pair, P_1 and P_3, both of equal amplitude are accompanied by a control pulse P_2 of the same amplitude. The aerial is arranged to produce two different horizontal radiation patterns, one of which carries P_1 and P_3 in a narrow beam and the other P_2 at lesser amplitude in the main beam region and greater amplitude in side lobe regions. A method of achieving this is by means of the 'sum and difference' aerial technique.[26] The control pulse P_2 is always 2 μs after P_1 for any mode of interrogation. The transponder's logic circuits compare the amplitude of P_1 and P_3 with that of P_3. If P_1 and P_3 are less than P_2, no reply is made. If P_1 and P_3 are 9 dB greater than P_2 a reply must be made. Between these limits a reply may or may not be made. We have then the so-called 'grey region' of 9 dB allowed by ICAO. In practice this is currently more like 6 dB with civil transponders and individual transponders can produce quite stable system beamwidths of the order of ±10% of the mean value. The ISLS system is illustrated in *Figure 59.29*.

59.3.2.3 Fruit and defruiting

All interrogations from any ICAO station are made at 1030±0.2 MHz and all replies from all ICAO standard transponders are made at 1090±3 MHz.

Interrogations are made via beams narrow in azimuth, replies are made omni-directionally. Thus replies to one interrogator

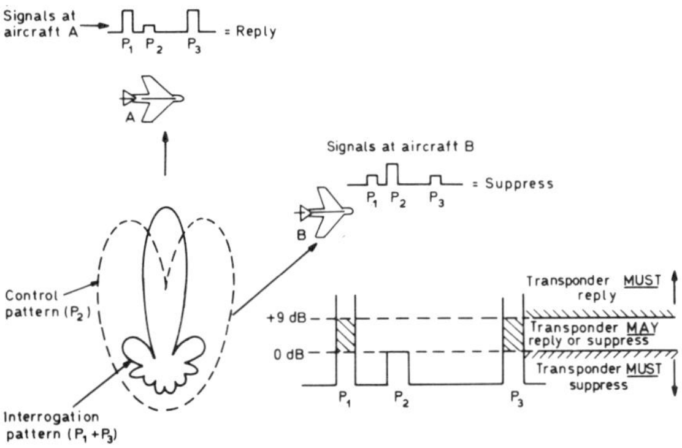

Figure 59.29 The ISLS system

station can be received by other stations via their main beam or receiver side lobes (unless receiver side lobe suppression is used). To prevent ambiguous range and bearing information being gathered by these means, each interrogator is assigned a specified p.r.f. or interrogation rate. A synchronous detector at each station can then separate replies due to its own interrogations from those due to other interrogators. Asynchronous replies are known as *fruit* and the process of filtering them out is known as *defruiting*. There are various defruiting techniques but range oriented digital comparators are mostly used.

59.3.2.4 Garbling and degarbling

If two aircraft on one bearing are close enough in slant range, their replies can overlap because of their simultaneity. This condition is known as synchronous garbling. It is common to find SSR decoders with fast dual registers operated by a commutating clock. These enable all but the most severe pulse masking to be tolerated. In view of the dubiety of data under garbled conditions it is common to find data labelled in such a way that garble conditions are indicated to the operator.

59.3.2.5 Real time decoding

Replies are standardised as shown in *Figure 59.28*. All pulses are 0.45 ± 0.1 μs in duration separated by 1.45 μs and contained within bracket pulses spaced 20.3 μs apart. Thus reply detectors of various sorts can be constructed which sense pulse width, position and spacing. Decoders operating in this manner pass all valid code data to operators who have control units upon which wanted codes may be set by mode/code switches. Every reply contains F_1/F_2 bracket pulses. Decoders are generally arranged to seek coincidence of F_1 and F_2 by delaying F_1 by 20.3 μs. When coincidence is found the decoder produces an output pulse similar to that of a primary radar signal. Thus p.p.i. display responses from SSR appear similarly to the operator, giving range and azimuth data.

59.3.2.6 Real-time active decoding

The signals indicating aircraft 'presence' and position can be used by the operator to call out the specific code data of any aircraft reply. By use of a light-pen or real-time gate associated with a steerable display marker, any individual target can be isolated and code data of its replies examined by the decoder. The code structure and the mode of interrogation are indicated to the operator on separate small alpha-numeric displays as four octal digits representing the four groups of information pulses referred to above. A similar indicator is used to show automatically reported altitude of targets in decimal form in units of 100 ft (30.5 m) flight levels.

59.3.2.7 Real-time passive decoding

Aircraft identity codes can be discovered by use of active decoding as described above. Commonly, identities will be known by either flight planning or radio telephone request. Operators are usually given sets of controls upon which they can set any mode and code combinations, again using the four-digits octal structure. When the control is activated the decoder looks for correspondence between the set values and all replies. When correspondence occurs a further signal is generated by the decoder to augment that due to detection of F_1/F_2. Thus unique identity of a number of targets can be established.

59.3.2.8 Automatic decoding

The processes described above operate in real time and require the operator to filter wanted data from that given, by use of separate controls and indicators. Automatic decoding carries out the process of active decoding on all targets continually, storing in digital form all the mode and code data, together with positional data. Using modern fast p.p.i. display techniques this data can be converted into alpha-numeric form with other symbols for direct display upon the p.p.i. screen, the position of the data being associated with the position symbol of the relevant target. By this means a fully integrated primary and secondary radar system can be produced. This concept is widely being implemented for air traffic control use.

59.3.3 Important distinctions between primary and secondary radar

The following is a brief resumé of the advantages and disadvantages of SSR relative to primary radar.

Positional data. Both systems give this, although quality of data in terms of resolution is better with primary radar. SSR reply time (20.3 μs) and wide horizontal beamwidth precludes high range and azimuth resolution. The SSR pulse width (0.45 μs), however, provides high range accuracy potential.

Height data. In primary radar systems either a separate height-finding equipment or a multi-beam surveillance system must be deployed. In SSR the height-finding facility is in-built but dependent upon aircraft transponders being fully equipped with height reporting elements (this is not yet mandatory but may become so).

Identify of targets. This is inherent in SSR giving 4096 individual identities in a service area. In primary radar, identity has to be established by request of manoeuvre or unreliably inferred from position reports.

Dependence upon aircraft size. In SSR all transponders have ICAO standard performance hence the signals from small craft are as strong as those from large aircraft. In primary radar the equivalent echoing area is dependent upon target size and shape. Although both systems suffer from target 'glint', SSR suffers least and can be improved by use of multiple aerials.

Constraints of ground clutter. Primary radar has to include clutter rejecting systems with 'in-clutter' target detection capability. This always has a limit which can inhibit target detection in regions of high-level clutter. SSR is completely free from this constraint since the ground station is not tuned to receive on its transmitting frequency; reflected energy is thus rejected.

Constraints of weather clutter. The same is true here as for ground clutter, i.e. SSR will not receive energy reflected from weather clutter. Two-way attenuation of microwaves is a factor for consideration in both primary and secondary radar but is less significant in the SSR system which, by ICAO specification, is organised with power in hand and thus more able to cope with losses.

Constraints of angel clutter. SSR has complete freedom from angel clutter. This freedom can be achieved in primary radar only by sacrifice of detection of very small targets, or complication of signal processing.

Dependence upon target cooperation. SSR is entirely dependent upon target cooperation and is totally impotent unless aircraft carry properly working transponders. Primary radar is dependent only upon the presence of a target within its detection range.

Range capability. The range performance of SSR is amenable to calculation as with primary radar. A major difference is that SSR range is the compound of two virtually independent inverse square law functions and primary radar range, when a single aerial is used, is an inverse fourth power law function. With SSR it is required to calculate

(*a*) Maximum interrogation range ($R_{i\,\max}$).
(*b*) Maximum reply range ($R_{r\,\max}$).

$$R^2_{i\,\max} = \frac{P_i G_i G_r \lambda_i^2}{(4\pi)^2 S_{a/c} L_i L_{a/c}} \qquad (59.9)$$

and

$$R^2_{r\,\max} = \frac{P_r G_r G_i \lambda_r^2}{(4\pi)^2 S_r L_i L_{a/c}} \qquad (59.10)$$

where

P_i = Peak of interrogator power output
G_i = Gain of the interrogation aerial
G_r = Gain of the aircraft transponder aerial
$S_{a/c}$ = Signal necessary at aircraft aerial to produce satisfactory interrogation
L_i = Losses between interrogator and its aerial terminals
$L_{a/c}$ = Losses between aircraft aerial and transponder receiver terminals
λ_i = Wavelength of interrogation (29 cm)
P_r = Transponder peak output power
G_r = Gain of aircraft aerial
S_r = Reply signal level necessary at responser (ground receiver) aerial for satisfactory decoder operation
λ_r = Wavelength of reply (27.5 cm)

The following points should be noted relating to the above:

(1) *Interrogator aerial gain G_i.* It is usual to find aerials of very small vertical dimension, of the order of 46 cm to 61 cm ($1\frac{1}{2}$ to 2 wavelengths approximately). This produces beams which are very wide in the elevation plane and consequently a great deal of power is directed into the ground. The resultant vertical polar diagram is therefore subject to very large modulation about the free-space pattern by the mechanism described earlier. The system uses vertically polarised radiation.

(2) *Aircraft aerial gain G_r.* The aircraft aerial is intended to be omnidirectional. Due to aerodynamic restrictions and varying shape of the airframe, the polar pattern of the aerial is modulated away from the desired shape and, more importantly, subject to shadowing and sometimes total obscuration during aircraft manoeuvring. Average reported gain performance is ± 7 dB about a mean of zero. There is evidence of wide variations over small solid angles of -40 dB and $+20$ dB. For these reasons it is usual to find figures of unity gain assumed in calculations.

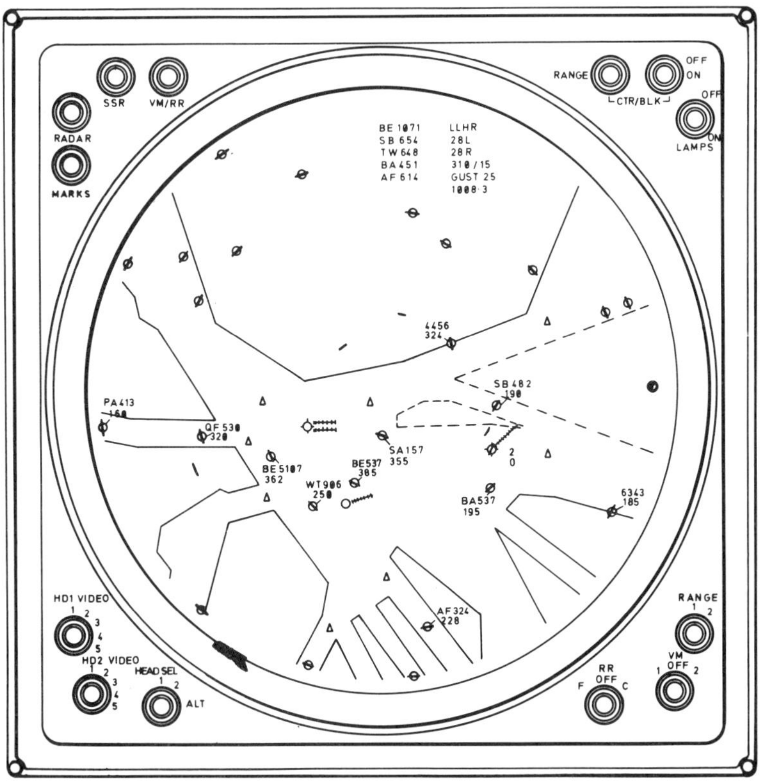

Figure 59.30 Air traffic control system radar display. Example of modern PPI display capable of presenting real time signals (arcs) together with machine time data (circle, alpha-numerics, video map and symbols). (Marconi Radar Systems Ltd)

(3) *Range performance.* If an interrogator is operated at the maximum permitted power ($52\frac{1}{2}$ dBW ERP) and a transponder of nominal performance is taken, maximum interrogation free-space ranges of 300 to 400 natutical miles can be achieved. Allowing the system to have some beamwidth (say equivalent to that at −3 dB relative to the horizontal polar pattern peak), the range is still of the order 200 to 280 nautical miles.

Taking practical values for system loss and ground receiver (responser) sensitivity, reply ranges in excess of 300 naturical miles are obtained. Thus the system has gain in hand on the reply path, ensuring high signal strength of received data.

59.3.4 Future extension of SSR

The SSR system, as with any other, suffers limitation and has its own special systematic problems, e.g. over-interrogation of transponders; over-suppression of transponders; corruption of data by fruit; and poor azimuth discrimination. These have been variously reported and analysed.[27–32]

In an attempt to lessen the effects of these shortcomings a number of variations on the beacon theme are currently under intensive investigation with a view to future implementation on the same internationally agreed basis as current SSR.

These variants (notably DABS and ADSEL; acronyms for 'direct addressed beacon system' and the 'addressed selected system') postulate the ability of a ground station to send an interrogating train of pulses to selected aircraft of known position. The interrogation would contain the selected aircraft's code of identity. If the aircraft found correspondence between the addressed code and its own identity it would issue its reply data which, it is claimed, can be more intense than at present and contain for instance speed, heading, altitude, fuel state, etc. ADSEL and DABS interrogations have been designated as Mode S.[33,34,35]

All the currently proposed variants are organised to be compatible with existing SSR specifications (which doubtless will be protected by ICAO for many years to come) and use normal SSR to obtain the necessary positional information. In automated radar systems this form of SSR is ideally suited to operational needs since, under computer control, aircraft data may be oeganised to be obtained when the system needs it and in the form required. All SSR systems are eminently suited for integration into data handling systems because of the digital nature of its signals.

59.3.5 Modern operational methods

Radar art and technology have reached a point where practically any civil operational requirement can be met. In the military sphere electronic countermeasures are developed to great sophistication. As an example of the former, *Figure 59.30* illustrates the way in which primary and secondary radar, display and data handling systems can be combined to give all the information an air traffic control system needs for the safe and expeditious handling of air traffic. Data on selected aircraft are available to operators by simple controls organised in such a way that displays are not cluttered with masses of available data but which is on call at the operator's discretion.

References

1 WOOD, D. and DEMPSTER, D., *The Narrow Margin*, Arrow Books, Revised Illustrated Edition (1969)
2 WATSON-WATT, SIR ROBERT, *Three Steps to Victory*, Odhams Press (1957)
3 WATSON-WATT, SIR ROBERT, 'Radar in War and Peace', *Nature*, **156**, 319 (1945)
4 PRICE, A., *Instruments of Darkness*, McDonald & Jane (1977)
5 RANDALL, J. T. and BOOT, H. A. H., 'Early Work on Cavity Magnetrons', *J.I.E.E.*, **93**, 997 (1946)
6 SKOLNIK, M. I., *Introduction to Radar Systems*, McGraw-Hill, International Student Edition (1962)
7 SKOLNIK, M. I., *The Radar Handbook*, McGraw-Hill (1970)
8 NATHANSON, F. E., *Radar Design Principles*, McGraw-Hill (1969)
9 BLAKE, L. V., 'Recent Advances in Basic Radar Range Calculation Technique', *I.R.E. Trans.*, **MIL.5**, 154 (1961)
10 BARTON, D. K., *Radars*, Vol. 3 Pulse Compression, Artech House 1(975)
11 HALL, W. M., 'Prediction of Pulse Radar Performance', *I.R.E.*, **44**, 224 (1956)
12 VINCENT, N. and LYNN, P., 'The Assessment of Site Effects of Radar Polar Diagrams', *The Marconi Review*, **28**, No 157 (1965)
13 HANSFORD, R. F., *Radio Aids to Civil Aviation*, Heywood & Co., 112 (1960)
14 BECKMAN, P. and SPIZZICHINO, A., *The Scattering of Electromagnetic Waves from Rough Surfaces*, Pergamon (1963)
15 *Atlas of Radio Wave Propagation Curves for Frequencies Between 30 and 10 000 MHz*, Radio Research Labs, Min. of Postal Services, Japan (1955)
16 RIDENOUR, L. N., *Radar System Engineering*, McGraw-Hill, 143 (1947)
17 BERKOWITZ, R. S., *Modern Radar—Analysis, Evaluation and System Design*, Wiley, chap 2 (1967)
18 SILVER, S., *Microwave Antenna Theory & Design*, McGraw-Hill (1949)
19 SKOLNIK, M. I., *Introduction to Radar Systems*, McGraw-Hill and Kogakusha, Issue No 2, para 4.7, 328–331 (1980)
20 RIDENOUR, L. N., *Radar Systems Engineering*, McGraw-Hill, 81–86 (1947)
21 BARTON, D. K., *Radar Systems Analysis*, Prentice Hall, 105 (1964)
22 EASTWOOD, SIR ERIC, *Radar Ornithology*, Methuen (1967)
23 HOWICK, R. E., 'A Primary Radar Automatic Track Extractor', *I.E.E. Conference Publication*, No 105, 339 (1973)
24 'A Simple Automatic Track Extraction System', *I.E.E. Conference Publication* No 155 (1977) (this volume contains numerous related papers)
25 Annex 10 to the Convention on International Civil Aviation (International Standards & Recommended Practices—Aeronautical Telecommunications) H.M.S.O., **2**, April (1968)
26 SKOLNIK, M. I., *The Radar Handbook*, McGraw-Hill, 38.10 (1970)
27 COLE, H. W., 'S.S.R.—Some Operational Implications in Technical Aspects', *World Aerospace Systems*, **1**, No 10, 468, Oct. (1965)
28 ULLYAT, C., 'Secondary Surveillance Radar in ATC', *I.E.E. Conference Publication*, No 105 (1973)
29 ARCHER, D. A. H., 'Reply Probabilities in SSR', *World Aviation Electronics*, Dec. (1961)
30 HERMANN, J. E., 'Problems of Broken Targets and What to Do About Them', *Report on the 1972 Seminan on Operational Problems of the ATC Radar Beacon System*, National Aviation Facilities Experimental Centre, Atlantic City, NJ, 31 (1973)
31 GORDON, J., 'Monopulse Technique Applied to Current SSR', *GEC Journal of Science & Technology*, **47**, No 3, February (1982)
32 COLE, H. W., 'The Suppression of Reflected Interrogation in SSR', *Marconi Review*, **43**, No 217 (2nd Qtr (1980))
33 BOWES, R. C., GRIFFITHS, H. N. and NICHOLS, T. B., 'The Design and Performance of an Experimental Selectivity Addressed (ADSEL) SSR System', *I.E.E. Conference Publication*, No 105, 32 (1973)
34 STEVENS, M. C., 'New Developments in SSR', *I.E.E. Conference Publication*, No 105, 26 (1973)
35 AMLIE, T. S., 'A Synchronised Discrete-Address Beacon System', *I.E.E.E. Transaction on Communications*, **COM-21**, No 5, 421 (1973)
36 BEAN, B. R., CAHOON, B. A., SAMSON, C. A. and THAYER, G. O., *A World Atlas of Atmospheric Radio Refractivity*, US Dept of Commerce, Environmental Science Services Administration (ESSA Monograph No 1), US Govt Printing Office (1966)
37 TAYLOR, J. W. Jr., 'Sacrifices in Radar Clutter Suppression Due to Compromises in Implementation of Digital Doppler Filters', Radar '82', *IEE Conf. Publ.*, No 216, p. 46
38 GASKELL, S. and FINCH, M., 'Fixed Beam Multi-lateration Radar System for Weapon Impact Scoring', Radar '82', *IEE Conf. Publ.*, No 216, p. 130

60 Computers and their Application

Paul F Musson, BA
Digital Equipment Co. Ltd
(Sections 60.17 and 60.18)

Ian G Robertson, BSc
Digital Equipment Co Ltd
(Sections 60.1–60.16)

Contents

60.1 Introduction

Although the advent of computers in our everyday lives may seem very recent, the principles of the modern computer were established before the existence of any electronic or electromechanical technology as we know them today, and electronic computers were beginning to take shape in laboratories in 1945.

The work of Charles Babbage, a Cambridge mathematician of the 19th century, in attempting to build an 'analytical engine' from mechanical parts, remarkably anticipated several of the common features of today's electronic computers. His proposed design, had he been able to complete it and overcome mechanical engineering limitations of the day, would have had the equivalent of punched card input, storage registers, the ability to branch according to results of intermediate calculations, and a form of output able to set numeric results in type.

Many purely mechanical forms of analogue computer have existed over the last few centuries. The most common of these is the slide rule, and other examples include mechanical integrators and even devices for solving simultaneous equations.

Much of the development leading to modern electronic computers, both analogue and digital, began during World War II with the intensified need to perform ballistics calculations. The development of radar at this time also provided the stimulus for new forms of electronic circuit which were to be adopted by the designers of computers.

A further development of momentous importance to the technology of computers, as it was for so many branches of electronics, was that of the transistor in 1949. Continued rapid strides in the field of semiconductors have brought us the integrated circuit which allows a complete digital computer to be implemented in a single chip.

60.2 Types of computers

Although there are two fundamentally different types of computer, analogue and digital, the former remains a somewhat specialised branch of computing, completely eclipsed now, both in numbers of systems in operation and in breadth of applications, by digital computers. Whilst problems are solved in the analogue computer by representing the variables by smoothly changing voltages on which various mathematical operations can be performed, in a digital computer all data is represented by binary numbers held in two-state circuits, as are the discrete steps or instructions for manipulating the data, which make up a program.

60.2.1 Analogue computers

An analogue computer consists of a collection of circuit modules capable of individually performing summation, scaling, integration or multiplication of voltages, and also function generating modules. On the most up to date systems, these modules contain integrated circuit operational amplifiers and function generators. Several hundred amplifiers are likely to be used on a large analogue computer.

To solve a given problem in which the relationship of physical quantities varying with time can be expressed as differential equations, the inputs and outputs of appropriate modules are interconnected, incorporating scaling, feedback and setting of initial conditions as required, with voltages representing the physical quantities. In this way single or simultaneous equations can be solved, in such applications as engineering and scientific calculation, modelling and simulation.

The interconnection and setting of coefficients and initial conditions required is normally done by means of a patch panel and potentiometers. An analogue computer may also have a CRT display or chart recorder for display or recording of the results of a computation. Where an analogue computer is being used for design work, the designer may choose to observe immediately the effect of changing a certain design parameter by varying the appropriate potentiometer setting thus altering the voltage representing that parameter. In a simulation application, outputs from the analogue computer may be used as input voltages to another electrical system.

Except in some specialised areas, the work formerly done by analogue computers is now most likely to be carried out on the modern high-speed, cost-effective digital computer, using numerical methods for operations such as integration. Digital computers can work to high precision completely accurately, whereas using analogue techniques there are inherent limits to precision, and accuracy suffers through drift in amplifiers.

60.2.2 Hybrid computers

The analogue part of a hybrid computer is no different in its circuitry and function to that of a stand alone analogue system. However, the task of setting up the network required to solve a particular problem is carried out by a digital computer, linked to the analogue machine by analogue/digital converters and digital input/output. Thus, the digital computer sets potentiometers and can read and print their values, and can monitor and display or log voltages at selected parts of the analogue network.

Linking the two technologies in this way brings the advantage of programmability to the mathematical capabilities of the analogue computer. Set up sequences for the analogue circuitry can be stored, and quickly and accurately reproduced at will. Tasks can be carried out in the part of the system best adapated to performing them. In particular, the ability of the digital computer to store, manipulate and present data in various ways is particularly advantageous.

The use of hybrid systems in, for example, the simulation of aircraft and weapon systems, has helped to perpetuate analogue computing as a technique applicable in certain narrow areas.

60.2.3 Digital computers

Digital computers in various forms are now used universally in almost every walk of life, both in business and in public service. In many cases unseen, computers nonetheless influence people in activities such as travel, banking, education and medicine.

As can be imagined from the variety of applications, computers exist in many different forms, spanning a range of price, from the smallest personal system to the largest supercomputer, of more than 10 000:1. Yet, there are certain features which are common to all digital computers:

(*a*) Construction from circuits which have two stable states, forming binary logic elements.

(*b*) Some form of binary storage of data.

(*c*) Capability to receive and act on data from the outside world (input), and to transmit data to the outside world (output).

(*d*) Operation by executing a set of discrete steps or instructions, the sequence of which can be created and modified at any time to carry out a particular series of tasks. This ability to be programmed, with a program stored in the system itself, is what gives great flexibility to the digital computer.

It has also given rise to a whole new profession, that of the computer programmer and the systems analyst, as well as the industry of designing, building and maintaining computers themselves.

60.3 Generations of digital computers

Beginning with circuits consisting of relays, the history of the digital computer can be seen as having fallen into four generations between the 1940s and today.

(*a*) *First generation:* built with valve circuits and delay-line storage, physically very massive, taking up complete rooms, requiring very large amounts of electricity with corresponding heat dissipation, low overall reliability requiring extensive maintenance. Input/output was rudimentary (teleprinters, punched cards) and programming very laborious.

(*b*) *Second generation:* developed during the 1950s with transistorised circuits, faster, smaller and more reliable than first generation, but still large by today's standards, magnetic core main store with magnetic drums and tapes as back-up, line printers for faster printed output. Programming language translators emerged.

(*c*) *Third generation:* heralded by the integrated circuit, allowing more compact construction and steadily improving speed, reliability and capability. The range and capabilities of input/output and mass storage devices increased remarkably. In the software area, high-level languages (e.g. FORTRAN, COBOL, BASIC) became commonplace and manufacturers offered operating system software developed, for example, to manage time-sharing for a large number of computational users, or real-time process control.

Most significantly, a trend of downward cost for given levels of performance was established, and the minicomputer, aimed at providing a few or even one single user with direct access to and control over their own computing facilities, began to gain in numbers over the large, centrally managed and operated computer system per organisation.

(*d*) *Fourth generation:* while a great many third-generation computer systems are in use, and will remain so for some time, the semiconductor technology of large-scale integration (LSI) has brought complete computers on a chip known as microprocessors allowing further refinement and enhancement of third-generation equipment.

Semiconductor memory has almost completely replaced core memory, and the continuing reduction in size and cost has brought the 'personal computer', numbered in hundreds of thousands of units supplied, truly within the reach of the individual in his own home or office.

60.4 Digital computer systems

A digital computer system is a collection of binary logic and storage elements combined in such a way as to perform a useful task. Any computer system, whether a microcomputer based on just a few integrated circuit packages incorporated within a laboratory instrument, or a large data processing system consisting of many cabinets of equipment housed in a specially-built air-conditioned computer room, invariably contains a combination of the following parts, described in detail in the following sections and shown in *Figure 60.1*.

(*a*) *Central processor unit.* The CPU is where instructions forming the stored program are examined and executed, and is therefore in control of the operation of the system. Instructions and data for immediate processing by the CPU are held in main memory, which is linked directly to the CPU.

(*b*) *Input/output.* This is the structure which provides optimum communication between the CPU and other parts of the system.

(*c*) *Peripherals.* These are the devices external to the CPU and memory which provide bulk storage, man/machine interaction and communication with other electronic systems.

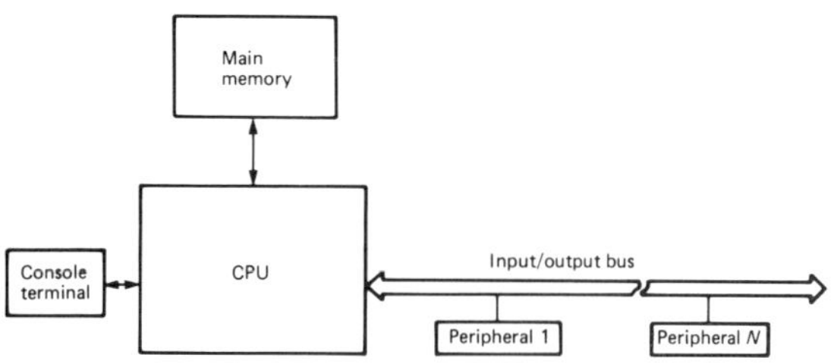

Figure 60.1 Components of a computer system

In going from microcomputer to large data processing systems there are many differences in complexity, technology, interconnection, capacity and performance. At any given level of system cost or performance, however, there are many similarities between systems from different manufacturers.

Several district categories of computer system can be identified. Going from least to most comprehensive and powerful, these are described in subsequent sections.

60.4.1 Microcomputers

Apart from microprocessor chips incorporated into other equipment and dedicated to performing a fixed application such as control or data acquisition, the smallest type of digital computer system is the desk top microcomputer.

These are the so-called personal computers, built around a microprocessor chip, normally with a small amount of memory available to the user, and some simple software stored permanently in read-only memory to allow the user to write programs in a language such as BASIC. A keyboard for entering programs and data, a CRT display or drive circuitry for a domestic television set, storage in the form of an audio tape recorder, or small magnetic discs, a character printer, capable of a print rate in the region of 100 characters per second, are the other components of a typical microcomputer.

Such a system is capable of a range of data processing and scientific applications, but is essentially only capable of one such task at a given time. Speed of operation is relatively low.

60.4.2 Minicomputers

Since its introduction as a recognisable category of system in the mid 1960s, with machines such as the Digital Equipment Corporation PDP8, the minicomputer has evolved rapidly. It has been the development which has brought computers out of the realm of specialists and large companies into common and widespread use by non-specialists.

The first such systems were built from early integrated circuit logic families, with core memory. Characteristics were low cost, ability to be used in offices, laboratories and even factories, and simplicity of operation allowing them to be used, and in many cases programmed, by the people who actually had a job to be done, rather than by specialist staff remote from the user.

These were also the first items of computer equipment to be incorporated by original equipment manufacturers (OEM) into other products and systems, a sector of the market which has contributed strongly to the rapid growth of the minicomputer industry.

Applications of minicomputers are almost unlimited, in areas such as laboratories, education, commerce, industrial control, medicine, engineering and government.

With advancing technology, systems are now built using medium-scale and large-scale integrated circuits, and memory is now almost entirely semiconductor. While earlier systems had a very small complement of peripherals, typically a teleprinter and punched paper tape input and output, there has been great development in the range and cost effectiveness of peripherals available. A minicomputer system will now typically have magnetic disc and tape storage holding tens of millions of characters of data, a printer capable of printing 300 lines of text per minute, and a number of CRT display terminals or teleprinter terminals.

60.4.3 Midicomputers

The word midicomputer, or supermini, has been coined to describe a type of system which has many similarities in implementation to the minicomputer, but by virtue of architectual advances, has superior performance. These advances include:

(*a*) Longer word length. The amount of information processed in one step, or transferred in a single operation between different parts of the system, is usually twice that of a minicomputer.

(*b*) As well as increasing the rate of information handling, the longer word length makes it possible to provide a more comprehensive instruction set. Some common operations, such as handling strings of characters or translating high-level language statements into CPU instructions, have been reduced to single instructions in newer midicomputers.

(*c*) Longer word length provides larger memory addressing. A technique called virtual memory (see section 60.6.6) gives further flexibility to addressing in some midicomputers.

(*d*) Higher data transfer speeds on internal data highways, which allow faster and/or larger numbers of peripheral devices to be handled, and larger volumes of data to be transmitted between the system and the outside world.

Despite providing substantial power, even when compared with the mainframe class of system described below, midicomputers fall into a price range below the mainframe. This is because they have almost all originated from existing minicomputer manufacturers, who have been able to build on their volume markets, including in most cases the OEM market.

60.4.4 Mainframes

The mainframe is the class of system typically associated with commercial data processing in large companies, or handling a large base of time-sharing terminal users. Today's mainframes, all products of large, established companies in the computer business (except for systems which are software compatible emulators of the most popular mainframe series) are the successors to the first and second generations as described in section 60.3. They inherit the central control and location, emphasis on batch card readers and line printers, COBOL programming, and the need for specialised operating staff.

Mainframes are capable of supporting large amounts of on-line disc and magnetic tape storage as well as large main memory capacity, and more recent versions have data communications capabilities supporting remote terminals of various kinds.

Although some of the scientific mainframes have extremely high operating rates (over 10 million instructions per second), most commercial mainframes are distinguishable more by their size, mode of operation and support than by particularly high performance.

60.5 Central processor unit

This part of the system controls the sequence of individual steps required for the execution of instructions forming a program. These instructions are held in storage, and when executed in the appropriate order, carry out a task or series of tasks intended by the programmer.

Within any particular computer system, the word length is the fixed number of binary digits of which most instructions are made up. Arithmetic operations within the CPU are also performed on binary numbers of this fixed word length, normally 8, 12, 16, 24, 32, 36 or 64 binary digits or bits. The CPU is connected via a

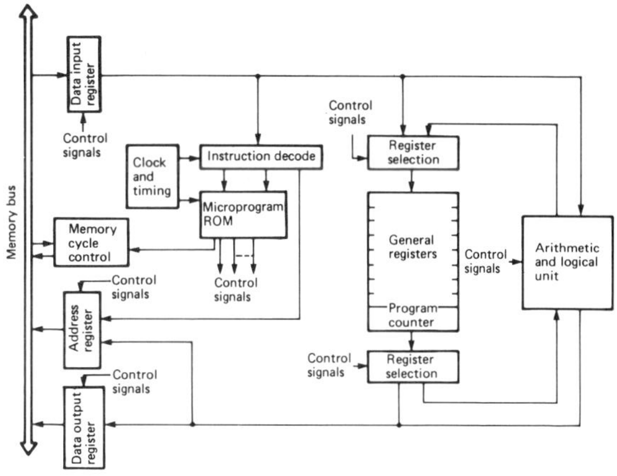

Figure 60.2 CPU block diagram

memory bus, as in *Figure 60.2*, to a section of fast core or semiconductor memory organised as a number of randomly accessible words, each of which can be written to or read from individually. Time for reading one word from or writing one word into main memory is typically 0.5 to 1.0 μs. Each word or location of memory can contain either an instruction or data. Apart from simple systems, some form of magnetic tape or disc memory peripheral is present on a system as file storage and back-up to main memory.

Control and timing circuits in the CPU cause instructions and data to be fetched from memory for processing, and write into memory the results of any instructions which require to be stored for further processing. The program counter holds the memory address of the next instruction to be fetched after each instruction has been processed. Frequently, the next instruction is held in the next location in memory and the counter need simply be incremented by one, but at other times the sequence of the program dictates that a new value be written into the program counter. Instructions which alter the sequence of a program calculate and insert a new value into the program counter for the next instruction.

In order to start the CPU when no programs are already in memory, the program counter is loaded with a predetermined address by external means, initiated by, for example, a push button or the action of switching power on to the system. The initial program which is entered is a simple, short loader program either held in ROM or hand loading into memory using the front panel switches. Its function is to load a more comprehensive general purpose loader, which automatically loads user or system programs. This process is known as bootstrapping or booting the system and the initial ROM program is known as the bootstrap loader. Alternatively, or where programs are already in memory, a start address can be set up on front panel switches through a keyboard.

60.6 Memory

In order to provide storage for program instructions and data in a form where they can be directly accessed and operated upon, all CPUs have main memory which can be implemented in a variety of technologies and methods of organisation.

60.6.1 Memory organisation

Memory is organised into individually addressable words into which binary information can be loaded for storage, and from which the stored data pattern can be read. On some systems, memory is arranged in such a way that more than one word at a time is accessed. This is done to improve effective memory access rates, on the basis that by accessing say two consecutive words, on most occasions the second word will be the one which the CPU requires next.

A memory controller is required between the memory arrays and the CPU, to decode the CPU requests for memory access and to initiate the appropriate read or write cycle. A controller can only handle up to a certain maximum amount of memory, but multiple controllers can be implemented on a single system. This can be used to speed up effective memory access, by arranging that sequentially addressed locations are physically in different blocks of memory with different controllers. With this interleaved memory organisation, in accessing sequential memory locations the operation of the controllers is overlapped, i.e. the second controller begins its cycle before the first has completed. Aggregate memory throughput is thus speeded up.

In some more complex computer systems, all or part of the memory can be shared between different CPUs in a multiprocessor configuration. Shareable memory has a special form of controller with multiple ports, allowing more than one CPU to content for access to the memory, allocated among them on a cycle by cycle basis.

It is sometimes appropriate to implement two types of memory in one system; random access or read/write memory (RAM), and read-only memory (ROM). Programs have to be segregated into two areas:

(*a*) Pure instructions, which will not change, can be entered into ROM.

(*b*) Areas with locations which require to be written into, i.e. those containing variable data or modifiable instructions, require to occupy RAM.

Read-only memory is used where absolute security from corruption of programs, such as operating system software or a program performing a fixed control task, is important. It is normally found on microprocessor based systems, and might be used, for example, to control the operation of a bank cash dispenser.

Use of ROM also provides a low cost way of manufacturing in quantity a standard system which uses proven programs which never require to be changed. Such systems can be delivered with the programs already loaded and secure, and do not need any form of program loading device.

60.6.2 Memory technology

The commonest technologies for implementing main memory in a CPU are:

(*a*) MOS RAM. This technology is very widely used, with the abundant availability, from the major semiconductor suppliers, first of 1K bit chips, then advancing through 4K and 16K chips to 64K units now available. In the latter form, very high density is achieved, with up to 1024K words of memory available on a single printed circuit board.

Dynamic MOS RAMs require refresh circuitry which automatically at intervals re-writes the data in each memory cell. Static RAMs, which does not require refreshing, can also be used. This is generally faster, but also more expensive, then dynamic RAM memory.

Semiconductor RAMs are volatile, i.e. lose contents on powering down. This is catered for in systems with back-up storage by re-loading programs from the back-up device when the system is switched on, or by having battery back up for all or part of the memory.

In specialised applications requiring memory retention without mains for long periods, or battery operation of the complete CPU, CMOS memory can be used because of its very low current drain. It has the disadvantage of being more expensive than normal MOS memory, but where it is essential to use it, circuit boards with on-board battery and trickle charger are now available.

(*b*) Read only memories, used as described in section 60.6.1 can be either erasable ROMs, or a permanently loaded ROM such as fusible-link ROM.

(*c*) Bubble memories are at present used for small amounts of bulk memory replacing, for example floppy disc storage in areas such as intelligent terminals. Apart from its increasing cost advantage over magnetic tape or disc storage, bubble memory has the advantage that its contents are non-volatile in the event of power being removed.

(*d*) Core memory remains in some applications, but although it has come down substantially in cost under competition from semiconductor memory, more recently MOS RAMs of higher capacity have been much cheaper and have largely taken over from core memory.

Core memories are made up of planes of tiny toroidal ferrite cores in a square array, threaded by wires at right angles, arranged so that each core is threaded by one each of these *X* and *Y* wires. The operation of core memory relies upon the hysteresis property of ferrite. A core, storing one bit of one word, can be magnetised to saturation in one direction by driving half the current required to produce the necessary field for switching along one *X* and one *Y* drive wire. Only the core at the intersection of *X* and *Y* wires will experience the full field necessary to switch. It will remain in that state until driven to saturation in the opposite direction by half currents of the opposite sense in the chosen *X* and *Y* lines.

In one of the several possible arrangements, each plane represents one bit position in memory, *X* and *Y* lines being selected by decoding the address for the required location. A single sense wire is threaded through every core in the plane. When it is required to read from a location, the action of writing zeros into each bit position, i.e. negative half currents on the selected *X* and *Y* lines, is performed. In a core which previously held a 1, the large flux change which takes place causes a small voltage pulse to be induced on the sense wire for that bit position. Sense amplifiers amplify and shape this pulse, so that it can be used to set the corresponding bit in a memory data register to 1. If the core was already storing a 0, there is no flux change sufficient to cause a pulse to be obtained.

The action of read out is destructive, so the data read into the register has to be automatically written back into that memory location.

60.6.3 Registers

The CPU contains a number of registers accessible by instructions, together with more which are not accessible, but are a necessary part of its implementation. Other than single-digit status information, the accessible registers are normally of the same number of bits as the word length of the CPU.

Registers are fast-access temporary storage locations within the CPU and implemented in the circuit technology of the CPU. They are used, for example, for temporary storage of intermediate results or as one of the operands in an arithmetic instruction.

A simple CPU may have only one register, often known as the accumulator, plus perhaps an auxiliary accumulator or quotient register used to hold part of the double-length result of a binary multiplication.

Larger word length, more sophisticated CPUs typically have 8 or more general purpose registers which can be selected as operands by instructions. Some systems such as the PDP11 use one of its 8 general purpose registers as the program counter, and can use any register as a stack pointer. A stack in this context is a temporary array of data held in memory on a 'last in, first out' basis. It is used in certain types of memory reference instructions, and for internal housekeeping in interrupt and subroutine handling. The stack pointer register is used to hold the address of the top element of the stack. This address, and hence the stack pointer contents, is incremented or decremented by one at a time as data are added to or removed from the stack.

60.6.4 Memory addressing

Certain instructions perform an operation in which one or more of the operands is the contents of a memory location, for example arithmetic, logical and data movement instructions. In a simple machine such as the PDP8, the other operand is invariably the contents of the accumulator register, but in a more sophisticated CPU, various addressing modes are available to give, for example, the capacity of adding together the contents of two different memory locations and depositing the result in a third.

In such a CPU as, for example, the PDP11 all such instructions are double operand, i.e. the programmer is not restricted to always using one fixed register as an operand. In this case any two of the general purpose registers can be designated either as each containing an operand, or through a variety of addressing modes, each:

(*a*) Containing the memory address of an operand.

(*b*) Containing the memory address of an operand, and the register contents then incremented following execution.

(*c*) Containing the memory address of an operand, and the register contents then decremented following execution.

(*d*) Containing a value to which is added the contents of a designated memory location, known as indexed addressing.

(*e*)–(*h*) All of the above, but where the resultant operand is itself the address of the final operand, known as indirect or deferred addressing.

This richness of addressing modes is one of the benefits of more advanced CPUs, as, for example, it provides an easy way of processing arrays of data in memory, or of calculating the address portion of an instruction when the program is executed.

Further flexibility is provided by the ability on the PDP11 for many instructions to operate on half words of 8 bits (known as a byte), and on some more comprehensive CPUs such as that of the VAX11 to operate on double and quadruple length words and also arrays of data in memory.

60.6.5 Memory management

Two further attributes may be required of memory addressing. Together they are often known as memory management.

(1) The ability, particularly for a short word length system (16 bits or less), for a program to use addresses greater than those implied by the word length. For example, with the 16 bit word length of most minicomputers, the maximum address which can be handled in the CPU is 65 536.

As applications grow bigger this is often a limitation, and extended addressing functions by considering memory as divided into a number of pages. Associated with each page at any given time is a relocation constant which is combined with relative addresses within its page to form a longer address. For example, with extension to 18 bits, memory addresses up to 262 144 can be generated in this way.

Each program is still limited at any given time to 65 536 words of address space, but these are physically divided into a number of pages which can be located anywhere within the larger memory. Each page is assigned a relocation constant, and as a particular program is run, dedicated registers in the CPU memory management unit are loaded with the constant for each page, *Figure 60.3*.

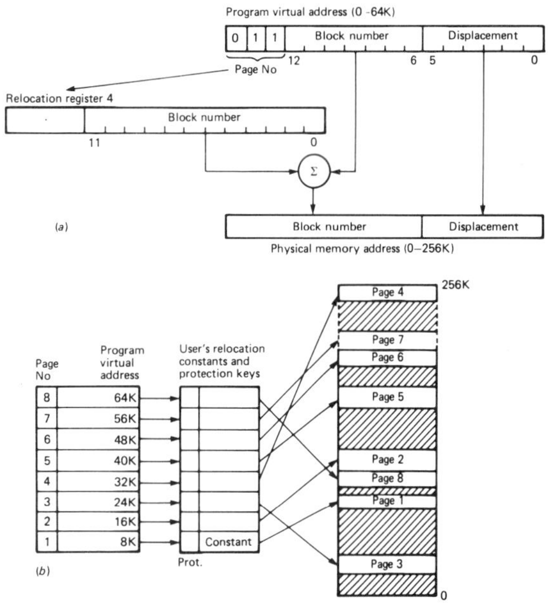

Figure 60.3 Memory management for a 16 bit CPU. (*a*) Generation of a physical address in the range 0 to 256K by combination of user's program virtual address in the range 0 to 64K with a relocation constant for the page concerned. Memory is handled in 64 byte blocks, with 8 relocation registers, giving segmentation into 8 pages located anywhere in up to 256K of physical memory. (*b*) The user's program is considered as up to 8 pages, up to 8K bytes each. Relocation constants for that program map these pages anywhere in up to 256K bytes of physical memory. Protection per page can also be specified

Thus many logically separate programs and data arrays can be resident in memory at the same time, and the process of setting the relocation registers, which is performed by a supervisory program, allows rapid switching between them in accordance with a time scheduling scheme. This is known as multiprogramming. Examples of where this is used are a time-sharing system for a number of users with terminals served by the system, or a real-time control system where programs of differing priority need to be executed rapidly in response to external events.

(2) As an adjunct to the hardware for memory paging or segmentation described above, a memory protection scheme is readily implemented. As well as a relocation constant, each page can be given a protection code to prevent it being illegally accessed. This would be desirable for example for a page holding data that is to be used as common data among a number of programs. Protection can also prevent a program from accessing a page outside its own address space.

Both of the above features are desirable for systems performing multiprogramming. In such systems, the most important area to be protected is that containing the supervisory program or operating system, which controls the running and allocation of resources for users' programs.

60.6.6 Virtual memory

Programmers, particularly those engaged in scientific and engineering work, frequently have a need for very large address space within a single program for instructions and data. This allows them to handle large arrays, and to write very large programs without the need to break them down to fit a limited memory size.

One solution is known as virtual memory, a technique of memory management by hardware and operating system software whereby programs can be written using the full addressing range implied by the word length of the CPU, without regard to the amount of main memory installed in the system. From the hardware point of view, memory is divided into fixed length pages, and the memory management hardware attempts to ensure that pages in most active use at any given time are kept in main memory. All of the current programs are stored on disc backing store, and an attempt to access a page which is not currently in main memory causes paging to occur. This simply means that the page concerned is read into main memory and an inactive page written out to disc to make room for it.

A table of address translations holds the virtual physical memory translations for all the pages of each program. The operating system generates this information when programs are loaded on to the system, and subsequently keeps it updated. Memory protection on a per page basis is normally provided, and a page can be locked into memory as required to prevent it being swapped out if it is essential for it to be immediately executed without the time overhead of paging.

When a program is scheduled to be run by the operating system, its address translation table becomes the one in current use. A set of hardware registers to hold a number of the most frequent translations in current use speeds up the translation process when pages are being repeatedly accessed.

60.7 Instruction set

The number and complexity of instructions in the instruction set or repertoire of different CPU varies considerably. The longer the word length, the greater is the variety of instructions which can be coded within it. This means generally that for a shorter word length CPU, a larger number of instructions will have to be used to achieve the same result, or that a longer word length machine with its more powerful set of instructions needs fewer of them and hence should be able to perform a given task more quickly.

Instructions are coded, according to a fixed format, allowing the instruction decoder to determine readily the type and detailed function of each instruction presented to it. The simple instruction set of the PDP8 is shown as an example in *Figure 60.4*.

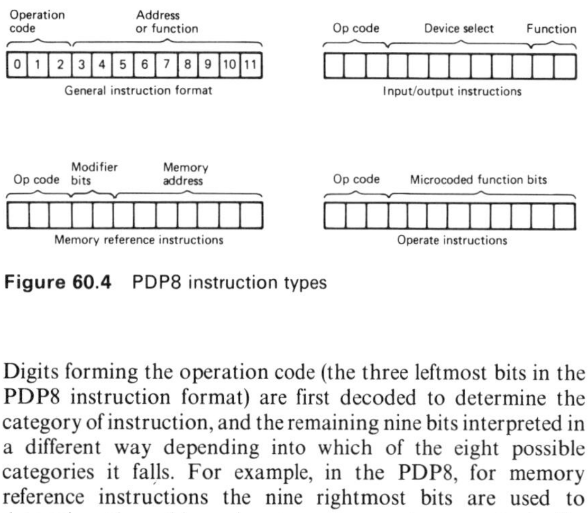

Figure 60.4 PDP8 instruction types

Digits forming the operation code (the three leftmost bits in the PDP8 instruction format) are first decoded to determine the category of instruction, and the remaining nine bits interpreted in a different way depending into which of the eight possible categories it falls. For example, in the PDP8, for memory reference instructions the nine rightmost bits are used to determine the address in memory on whose contents the instruction operates, while for an instruction to transfer data between an input/output device and the CPU, the two rightmost bits specify the operation to be carried out by the input/output device and the other seven bits a selection address to identify which device is required.

Most systems have a considerably more elaborate range of instructions than the 12-bit word length PDP8, but all employ the principle of decoding a certain group of digits in the instruction word to determine the class of instruction and hence how the remaining digits are to be interpreted.

The contents of a memory location containing data rather than an instruction are not applied to the instruction decoder. Correct initial setting of the program counter, and subsequent automatic setting by any branch instruction to follow the sequence intended by the programmer, ensures that only valid instructions are decoded for execution.

Logical and arithmetic instructions perform an operation on data (normally one or two words for any particular instruction) held in either the memory or registers in the CPU. The addressing modes available to the programmer (see section 60.6.4) define the range of possible ways of accessing the data to be operated on. This ranges from the simple single operand type of CPU, where the accumulator is always understood to contain one operand while the other is a location in memory specified by the addressing bits of the instruction to a multiple operand CPU with a wide choice of how individual operands are addressed.

In some systems such as the PDP11, instructions to input data from and output data to peripheral devices are the same as those used for manipulating data in memory. This is achieved by implementing a portion of the memory addresses at the high end as data and control registers in peripheral device controllers. More detail is given in section 60.9.

There are certain basic data transfer, logical, arithmetic and controlling functions which must be provided in the instruction sets of all CPUs. This minimum set allows the CPU to be programmed to carry out any task which can be broken down and expressed in these basic instructions. However, it may be that a program written in this way will not execute quickly enough to perform a time-critical application such as control of an industrial plant or receiving data on a high-speed communicatins line. Equally the number of steps or instructions required may not fit into the available size of memory. In order to cope more efficiently with this sort of situation, i.e. to increase the power of the CPU, all but the very simplest CPUs have considerable enhancements and variations to the basic instruction set. The more comprehensive the instruction set, the fewer are the steps required to program a given task, and the shorter and faster in execution are the resulting programs.

Basic types of instruction, with examples of the variations to these, are:

(*a*) *Data transfer*. Load accumulator from a specified memory location and write contents of accumulator into specified memory location. Most CPUs have variations such as add contents of memory location to accumulator and exchange contents of accumulator and memory location.

CPU with multiple registers also have some instructions which can move data to or from these registers, as well as the accumulator. Those with 16-bit or greater word lengths may have versions of these and other instruction types which operate on bytes as well as words.

With a double operand addressing mode (see section 60.6.4) a generalised 'Move' instruction allows the contents of any memory location or register to be transferred to any other memory location or register.

(*b*) *Logical* (i) AND function on a bit by bit basis between contents of a memory location and a bit pattern in the accumulator. Leaves ones in accumulator bit positions which are also one in the memory word. Appropriate bit patterns in accumulator allow individual bits of chosen word to be tested. (ii) Branch (or skip the next instruction) if the accumulator contents are zero/non-zero/positive/negative.

Many more logical operations and tests are available on more powerful CPUs, such as OR, exclusive OR, complement, branch if greater than or equal to zero, branch if less than or equal to zero branch if lower or the same. The branch instructions are performed on the contents of the accumulator following a subtraction or comparison of two words, or some other operation which leaves data in the accumulator. The address for branching to is specified in the address part of the instruction. With a skip, the instruction in the next location should be an unconditional branch to the code which is to be followed if the test failed, while for a positive result, the code to be followed starts in the next but one location.

Branch or skip tests on other status bits in the CPU are often provided, e.g. on arithmetic carry and overflow.

(*c*) *Input/output*. CPUs like the PDP11, with memory mapped input/output do not require separate instructions for transferring data and status information between CPU and peripheral controllers. For this function, as well as performing tests on status information and input data, the normal data transfer and logical instructions in (*a*) and (*b*) are used.

Otherwise, separate input/output instructions provide these functions. Their general format is a transfer of data between the accumulator or other registers, and addressable data, control or status registers in peripheral controllers. Some CPUs also implement special input/output instructions such as (i) Skip if 'ready' flag set. For the particular peripheral addressed, this instruction tests whether it has data awaiting input, or whether it is free to receive new output data. Using a simple program loop, this instruction will synchronise the program with the transfer rate of the peripheral. (ii) Set interrupt mask. This instruction outputs the state of each accumulator bit to an interrupt control circuit of a particular peripheral controller, so that, by putting the appropriate bit pattern in the accumulator with a single instruction, interrupts can be selectively inhibited or enabled in each peripheral device.

(*d*) *Arithmetic*. (i) Add contents of memory location to contents of accumulator, leaving result in accumulator. This instruction, together with instructions in category (*b*) for handling a carry bit from the addition, and for complementing a binary number, can be used to carry out all the four arithmetic functions by software subroutines. (ii) Shift. This is also valuable in performing other arithmetic functions, or for sequentially testing bits in the accumulator contents. With simpler instruction sets, only one bit position is shifted for each execution of the instruction. There is usually a choice of left and right shift, and arithmetic shift (preserving the sign of the word and setting the carry bit) or logical rotate.

Extended arithmetic capability, either as standard equipment or a plug in option, provides multiply and divide instructions and often multiple bit shift instructions.

(*e*) *Control*. Halt, no operation, branch, jump to sub-routine, interrupts on, interrupts off, are the typical operations provided as a minimum. A variety of other instructions will be found, peculiar to individual CPUs.

60.7.1 CPU implementation

The considerable amount of control logic required to execute all the possible CPU instructions and other functions is implemented in one of two ways:

(*a*) With random logic using the available logic elements of gates, flip-flops etc, combined in a suitable way to implement all the steps for each instruction, using as much commonality between instructions as possible. The various logical combinations are invoked by outputs from the instruction decoder.

(*b*) Using microcode—a series of internally programmed steps making up each instruction. These steps or micro-instructions are loaded into ROM using patterns determined at design time, and for each instruction decoded, the micro program ROM is entered at the appropriate point for that instruction. Under internal clock control, the micro-instructions cause appropriate control lines to be operated to effect the same steps as would be the case if the CPU were designed using method (*a*).

The great advantage of microcoded instruction sets is that they can readily be modified or completely changed by using an alternative ROM, which may simply be a single chip in a socket. In this way, a different CPU instruction set may be effected.

In conjunction with microcode, bit slice microprocessors may be used to implement a CPU. The bit slice microprocessor contains a slice or section of a complete CPU, i.e. registers, arithmetic and logic, with suitable paths between these elements. The slice may be one, two or four bits in length, and by cascading a number of these together, any desired word length can be achieved. The required instruction set is implemented by suitable programming of the bit slice microprocessors using their external inputs controlled by microcode.

The combination of microcode held in ROM and bit slice microprocessors is used in the implementation of many CPU models, each using the same bit slice device.

60.8 CPU enhancements

There are several areas in which the operating speed of the CPU can be improved with added hardware, either designed in as an original feature or available as an upgrade to be added in field. Some of the more common ones are described below.

60.8.1 Cache memory

An analysis of a typical computer program shows that there is a strong tendency to access instructions and data held in fairly small contiguous areas of memory repetitively. This is due to the fact that loops—short sections of program re-used many times in succession—are very frequently used, and data held in arrays of successive memory locations may be repetively accessed in the course of a particular calculation.

This leads to the idea of having a small buffer memory, of higher access speed than the lower cost technology employed in main memory, between CPU and memory. This is known as cache memory. Various techniques are used to match the addresses of locations in cache with those in main memory, so that for memory addresses generated by the CPU, if the contents of that memory location are in cache, the instruction or data is accessed from the fast cache instead of slower main memory. The contents of a given memory location are intially fetched into cache by being addressed by the CPU. Precautions are taken to ensure that the contents of any location in cache which is altered by a write operation is re-written back into main memory so that the contents of the location, whether in cache or main memory, are identical at all times.

A constant process of bringing memory contents into cache, thus overwriting previously used information with more currently used words takes place completely transparently to the user. The only effect to be observed is an increase in execution speed. This speeding up depends on two factors, hit rate, i.e. percentage of times when the contents of a required location are already in cache, and the relative access times of main and cache memory. The hit rate, itself determined by the size of cache memory and algorithms for its filling, is normally better than 90%. This is dependent, of course, on the repetitiveness of the particular program being executed.

60.8.2 Fixed and floating point arithmetic hardware

As far as arithmetic instructions go, simpler CPUs only contain add and subtract instructions, operating on single-word operands. Multiplication, of both fixed and floating point numbers, is then accomplished by software subroutines, i.e. standard programs which perform multiplication or division by repetitive use of the add or subtract instructions, which can be invoked by a programmer who requires to perform a multiplication or division operation.

By providing extra hardware to perform fixed point multiply and divide, which also usually implements multiple place shift operations, a very substantial improvement in the speed of multiply and divide operations is obtained. With the hardware techniques used to implement most modern CPU, however, these instructions are wired in as part of the standard set.

Floating point format (*Figure 60.5*) provides greater range and precision than single-word fixed point format. In floating point representation, numbers are stored as a fraction times 2 raised to a power which can be positive or negative. The fraction (or mantissa) and exponent are what is stored, usually in two words for single-precision floating point format, or four words for double precision.

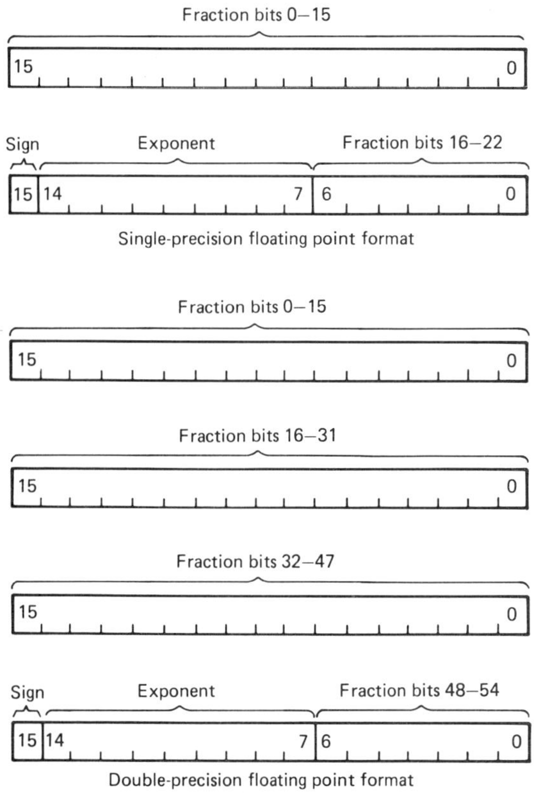

Figure 60.5 16-bit word floating point formats

Hardware to perform add, subtract, multiply and divide operations is sometimes implemented as a floating point processor, an independent unit with its own registers to which floating point instructions are passed. The floating point processor can then access the operands, perform the required arithmetic operation and signal the CPU, which has been free to continue with its own processing meantime, when the result is available.

An independent floating point processor clearly provides the fastest execution of these instructions, but even without that, implementing them within the normal instruction set of the CPU using its addressing techniques to access operands in memory, provides a significant improvement over software subroutines.

60.8.3 Decimal instruction set

For commercial data processing work such as that involved in processing large monetary values, the ability to do arithmetic operations on variables held precisely in decimal format is important, to avoid accumulated rounding errors in sums of money from using a binary floating point representation.

While this can be handled by software subroutines, decimal arithmetic instructions incorporated in the CPU instruction set or as a plug-in option are important in systems engaged in financial work.

60.8.4 Array processors

Similar to an independent floating point processor described above, an optional hardware unit which can perform complete computations on data held in the form of arrays of data in memory, independent from the CPU and at high speed, is known as an array processor.

These are used in specialised technical applications such as simulation, modelling and seismic work. An example of the type of mathematical operation which would be carried out by such a unit is matrix inversion.

60.8.5 Timers and counters

For systems which are used in control applications, or where elapsed time needs to be measured for accounting purposes as, for example, in a time-sharing systems where users are to be charged according to the amounts of CPU time they use, it is important to be able to measure intervals of time precisely and accurately. This measurement must go on while the system is executing programs, and must be 'real time', i.e. related to events and time intervals in the outside world.

Most CPUs are equipped with a simple real-time clock which derives its reference timing from 50 or 60 Hz mains. These allow a predetermined interval to be timed by setting a count value in a counter which is decremented at the mains cycle rate until it interrupts the CPU on reaching zero.

More elaborate timers are available as options, or are even standard items on some CPUs. These are driven from high-resolution crystal oscillators, and offer such features as:

(*a*) More than one timer simultaneously.
(*b*) Timing random external events.
(*c*) Program selection of different time bases.
(*d*) External clock input.

The system supervisory software normally keeps the date and time of day up to date by means of a program running in the background all the time the system is switched on and running. Any reports, logs or printouts generated by the systems can then be labelled with the date and time they were initiated. To overcome having to reset the date and time every time the system is stopped or switched off, some CPUs now have a permanent battery driven date and time clock which keeps running despite stoppages and never needs reloading once loaded initially.

Counters are also useful in control applications to count external events or to generate a set number of pulses, for example to drive a stepping motor. Counters are frequently implemented as external peripheral devices, forming part of the digital section of a process input/output interface (see section 60.9).

60.9 Input/output

In order to perform any useful role, a computer system must be able to communicate with the outside world, either with human users via keyboards, CRT screens, printed output etc, or with some external hardware or process being controlled or monitored. In the latter case, where connection to other electronic systems is involved, the communication is via electrical signals.

All modern computer systems have a unified means of supporting the variable number of such human or process input/output devices required for a particular application, and indeed for adding such equipment to enhance a system in the field. The impetus to develop this common way of connecting all external or peripheral parts of a computer system to the CPU arose as the number and variety of such peripheral units increased during the 1960s and 1970s.

As well as all input/output peripherals and external mass storage in the form of magnetic tape and disc units, some systems also communicate with main memory in this common, unified structure. In such a system, for example the PDP11, there is no difference between instructions which reference memory and those which read from and write to peripheral devices.

The benefits of a standard input/output bus to the manufacturer are:

(*a*) It provides a design standard allowing easy development of new input/output devices and other system enhancements.
(*b*) Devices of widely different data transfer rates can be accommodated without adaptation of the CPU.
(*c*) It permits development of a family concept.

Many manufacturers have maintained a standard input/output highway and CPU instruction set for as long as a decade or more. This has enabled them to provide constantly improving system performance and decreasing cost by taking advantage of developing technology, whilst protecting very substantial investments in peripheral equipment and software.

For the user of a system in such a family, benefits are:

(*a*) The ability to upgrade to a more powerful CPU whilst retaining existing peripherals.
(*b*) Retention of programs in moving up or down range within the family.
(*c*) In many cases the ability to retain the usefulness of an older system by adding more recently developed peripherals.

60.9.1 Input/output bus

The common structure for any given model of computer system is implemented in the form of an electrical bus or highway. This specifies both the number, levels and significance of electrical signals and the mechanical mounting of the electrical controller or interface which transforms the standard signals on the highway to ones suitable for the particular input/output or storage device concerned.

Some attempts at standardisation between different manufacturers have been successful, e.g. the S100 bus available on a number of microcomputer systems, which now forms the basis for an I.E.E.E. standard. Even without standardisation, any data highway or input/output bus needs to provide the following functions.

(*a*) *Addressing.* A number of address lines is provided, determining the number of devices which can be accommodated on the system. For example, 6 lines would allow 63 devices. Each interface on the bus decodes the address lines to detect input/output instructions intended for it.

(*b*) *Data.* The number of data lines on the bus is usually equal to the word length of the CPU, although it may alternatively be a sub-multiple of the word length, in which case input/output data is packed into or unpacked from complete words in the CPU. In some cases data lines are bi-directional, providing a simpler bus at the expense of more complex drivers and receivers.

(*c*) *Control.* Control signals are required to synchronise transactions between the CPU and interfaces, and to gate address and data signals to and from the bus. Although all the bits of an address or data word are transmitted at the same instant, in transmission down the bus, because of slightly different electrical characteristics of each individual line, they will arrive at slightly different times. Control signals are provided to gate these skewed signals at a time when they are guaranteed to have reached their correct state.

60.9.2 Types of input/output transaction

Three types of transaction via the input/output bus between CPU and peripheral devices are required, as described below.

(*a*) *Control and status.* This type of transfer is initiated by a program instruction to command a peripheral device to perform a certain action in readiness for transferring data or to interrogate the status of a peripheral. For example, a magnetic tape unit can be issued with a command to rewind; the read/write head in a disc unit to be positioned above a certain track on the disc; the completion of a conversion by an analogue-to-digital converter verified; or a printer out of paper condition, may be sensed.

Normally, a single word of control or status information is output or input as a result of one instruction, with each bit in the word having a particular significance. Thus multiple actions can be initiated by a single control instruction, and several conditions be monitored by a single status instruction. For the more complex peripheral devices, more than one word of control or status information may be required.

(*b*) *Programmed data transfer.* For slow and medium speed devices, for example small floppy disc units or line printers, data are input or output one word at a time with a series of program instructions required for every word transferred.

The word or data are transferred to or from one of the CPU registers, normally the accumulator. In order to effect a transfer of a series of words forming a related block of data, as is normally required in any practical situation, a number of CPU instructions per word transferred is required. This is because it is necessary to take the data from, or store it into, memory locations. As a minimum, in a simple case, at least six CPU instructions are required per word of data transferred.

In a system such as the PDP11 where instructions can reference equally easily memory locations, peripheral device registers and CPU registers, the operation is simplified since a MOVE instruction can transfer a word of data directly from a peripheral to memory, without going through a CPU register. This applies equally to control and status instructions on the PDP11, with a further advantage that the state of bits in a peripheral device status register can be tested without transferring the register contents into the CPU.

The rate of execution of the necessary instructions must match the data transfer rate of the peripheral concerned. Since it is usually desired that the CPU continue with the execution of other parts of the user's program while data transfer is going on, some form of synchronisation is necessary between CPU and peripheral to ensure that no data are lost. In the simplest type of system, the CPU simply suspends any other instructions and constantly monitors the device status word awaiting an indication that the peripheral has data ready for input to the CPU or is ready to receive an output from it. This is wasteful of CPU time where the data transfer rate is slow relative to CPU instruction speeds, and in this case the use of 'interrupt' facilities (see section 60.9.3) provides this synchronisation.

(*c*) *Direct memory access.* For devices which transfer data at a higher rate, in excess of around 20 000 words per second, a different solution is required.

At these speeds, efficiency is achieved by giving the peripheral device controller the ability to access memory autonomously without using CPU instructions. With very fast tape or disc units which can transfer data at rates in excess of 1 million bytes per second, direct memory access (DMA) is the only technique which will allow these rates to be sustained.

The peripheral controller has two registers which are loaded by control instructions before data transfer can begin. These contain (*a*) the address in memory of the start of the block of data and (*b*) the number of words which it is desired to transfer in the operation.

When the block transfer is started the peripheral controller, using certain control lines in the I/O bus, sequentially accesses the required memory locations until the specified number of words has been transferred. The memory addresses are placed on address lines of the I/O bus, together with the appropriate control and timing signals, for each word transferred. On completion of the number of words specified in the word count register, the peripheral signals to the CPU that the transfer of the block of data is completed.

Other than the instructions required initially to set the start address and word count registers and start the transfer, a DMA transfer is accomplished without any intervention from the CPU. Normal processing of instructions therefore continues. Direct memory access (more than one peripheral at a time can be engaged in such an operation) is, of course, competing with the CPU for memory cycles, and the processing of instructions is slowed down in proportion to the percentage of memory cycles required by peripherals. In the limit, it may be necessary for a very-high-speed peripheral to completely dominate memory usage in a burst mode of operation, to ensure that no data are lost during the transfer through conflicting requests for memory cycles.

60.9.3 Interrupts

The handling of input/output is made much more efficient through the use of a feature found in varying degrees of sophistication on all modern systems. This is known as 'automatic priority interrupt', and is a way of allowing peripheral devices to signal an event of significance to the CPU (e.g. a terminal keyboard having a character ready for transmission, or completion of DMA transfer) in such a way that the CPU is made to suspend temporarily its current work to respond to the condition causing the interrupt.

Interrupts are also used to force the CPU to recognise and take action on alarm or error conditions in a peripheral, e.g. printer out of paper, error detected on writing to a magnetic tape unit.

Information to allow the CPU to resume where it was interrupted, e.g. the value of the program counter, is stored when an interrupt is accepted. It is necessary also for the device causing the interrupt to be identified, and for the program to branch to a section to deal with the condition which caused the interrupt (*Figure 60.6*).

Examples of two types of interrupt structure are given below, one typical of a simpler system such as an 8-bit microprocessor or an older architecture minicomputer, the other representing a more sophisticated architecture such as the PDP11.

In the simpler system, a single interrupt line is provided in the input/output bus, on to which the interrupt signal from each peripheral is connected. Within each peripheral controller, access to the interrupt line can be enabled or disabled, either by means of a control input/output instruction to each device separately, or by a 'mask' instruction which, with a single 16-bit word output, sets the interrupt enabled/disabled state for each of up to 16 devices on the input/output bus. When a condition which is defined as able to cause an interrupt occurs within a peripheral, and interrupts are enabled in that device, a signal on the interrupt line will be sent to the CPU. At the end of the instruction currently being executed, this signal will be recognised.

In this simple form of interrupt handing, the interrupt servicing routine always begins at a fixed memory location. The interrupt forces the contents of the program counter, which is the address of next instruction which would have been executed had the interrupt not occurred, to be stored in this first location and the program to start executing at the next instruction. Further interrupts are automatically inhibited within the CPU, and the first action of the interrupt routine must be to store the contents of the accumulator and other registers so that on return to the main stream of the program these registers can be restored to their previous state.

Identification of the interrupting device is done via a series of conditional instructions on each in turn until an interrupting device is found. Having established which device is interrupting, the interrupt handling routine will then branch to a section of program specific to that device. At this point or later within the interrupt routine, an instruction to re-enable the CPU interrupt system may be issued, allowing a further interrupt to be received by the CPU before the existing interrupt handling program has completed. If this 'nesting' of interrupts is to be allowed, each interruptable section of the interrupt routine must store the return value of the program counter elsewhere in memory, so that as each section of the interrupt routine is completed, control can be returned to the point where the last interrupt occurred.

A more comprehensive interrupt system, that on the PDP11 for example, differs in the following ways from that described above.

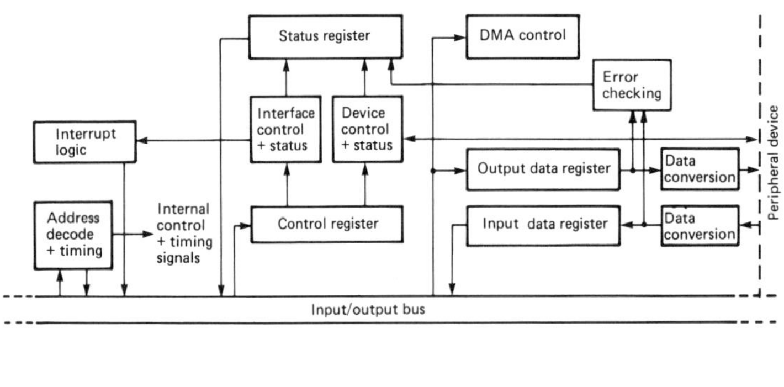

Figure 60.6 Block diagram of peripheral interface

(*a*) Four separate interrupt lines are provided, and any number of devices can be on each line or level. The CPU status can be set to different priority levels, corresponding to the four interrupt lines. Only interrupts on a level higher than the current priority are accepted by the CPU.

This provides a more adaptable way of dealing with a wide range of devices of different speeds and with different degrees of urgency.

(*b*) When an interrupt is accepted by the CPU, the interrupting device sends a vector or pointer to the CPU on the I/O bus address lines. This points to a fixed memory address for each device, which holds the start address of its interrupt routine, and in the following memory word, a new status word for the CPU, defining its priority level and hence its ability to respond to other levels of interrupt during this interrupt routine.

By avoiding the need for the CPU to test each device until it finds the interrupting one, response to interrupts is much faster.

(*c*) The current value of the program counter and processor status word are automatically placed on a push-down stack when an interrupt occurs. A further interrupt accepted within the current interrupt routine will cause the program counter and status word to be stored on the top of the stack, and the existing contents to be pushed down into the stack. On return from an interrupt routine, the program counter and status word stored when that interrupt occurred are taken from the top of the stack and used by the CPU, allowing whatever was interrupted to continue as before.

This can take place for any number of interrupts, subject only to the capacity of the stack. Thus 'nesting' to any level is handled automatically without the need for the programmer to store the program counter at any stage.

60.10 Peripherals

Peripheral devices fall into three categories:

(*a*) Those which are designed to allow humans to interact with the system by outputting information in the form of readable alphanumeric text or graphics, either on paper or on a display screen, and accepting information from humans through a keyboard or voice recognition device, or by scanning printed text.

The general function performed by devices in this class is sometimes referred to as man–machine interaction.

(*b*) Those which act as a back-up form of storage to supplement the main memory of the system. The most simple of these, now largely superseded, was probably punched paper tape or cards, and the most complex now ranges up to very large discs or other specialised forms of storage, each capable of holding hundreds of millions of characters of information. Peripherals of this type are generally known as mass storage devices.

(*c*) Interfaces between the computer system and other forms of system or electronic device. Analogue/digital converters, digital input/output and communication line interfaces are good examples.

Either (*a*) or (*c*) or both are required in every system. The existence of (*b*) depends on the need for additional storage over and above main memory. All peripheral devices in a system are connected via the input/output structure (as described in section 60.9) to the CPU, memory, and in some systems to a separate I/O processor.

The throughput rates and flexibility of the input/output structure determine the number and variety of peripheral devices which can be handled in a system before the input/output requirements begin to saturate the system and prevent any processing of instructions being done by the CPU. In deciding on the configuration of a particular system, it is important to analyse the throughput requirement dictated by peripheral devices, to ensure the system does not become I/O bound, and that data from any peripheral device is not lost due to other devices taking too many of the I/O resources.

Historically in the computer industry, independent manufacturers as well as the larger computer systems companies have developed and manufactured peripherals. The products of the independent manufacturers are either bought by system manufacturers for design into their systems or are sold by the independent manufacturers direct to users of the more popular computers, with an interface providing compatibility with the I/O bus of the system.

This has fostered the development of many of the widely-used, cost-effective peripherals available today, such as floppy discs and printers.

Certain storage devices with removable storage media, where the formats for recording data on the media have been standardised, can be used for exchanging data between systems from different suppliers. This is important where data may be gathered on one system and need to be analysed on a different system more suitable for that purpose. Punched tape and cards, and principally magnetic tape are used for this purpose. To a lesser extent, floppy discs are also used, but although they and other forms of discs may be physically and mechanically the same on different systems, there is little standardisation of recording formats.

Interchange between systems of the same model using removable media is, of course, not subject to problems of different formats, and is used extensively for the exchange of data and programs. Another method of exchange of data between systems irrespective of manufacturer is data transmission via communication lines, as described in the section on communications.

60.11 Terminals

A data terminal is essentially a device at which a computer user sits, either to receive data in alphanumeric or graphic form, or to input data through a keyboard or other form of manual input, or both. Terminals range in cost and complexity from a 30 characters per second serial printer to a full graphics terminal with a large, perhaps colour, CRT screen. Most are connected to the CPU by data communications interfaces and can therefore be situated at some distance from the system (e.g. in various offices within a building), or indeed remote from the computer site communicating over the public telephone network.

60.11.1 Teleprinter

In the early days of computing, one of the first peripheral devices to be used was the teleprinter operating at around 10 characters per second, with 5-level punched tape, as used in telegraphy and telex at the time. Subsequently, the computer industry has devised its own keyboard/printer devices. These print characters serially on continuous paper, typically at rates of either 30 characters per second or in the region of 150 to 180 characters per second. Devices in the higher-speed range are sometimes offered without a keyboard, for use simply as a medium-speed printer.

The print head consists of usually seven needles held in a vertical plane in the head assembly which is positioned with the needles perpendicular to the paper and spaced a short distance from it, with a carbon ribbon interposed between. Each needle can be individually driven by a solenoid to contact the paper through the ribbon, thus printing a dot. A complete character is formed by stepping the head through 5 positions horizontally, and at each position energising the appropriate solenoids. The head is then stepped on to the position for the next character. When the end of a line is reached, the paper is advanced one line and the print head either returns to the left margin position or, in some faster printers, prints the next line from right to left. This is possible where the printer is provided with storage for a line or more of text, and the characters can be extracted from this store in the reverse order. Throughput speed is improved where this technique is used, by saving redundant head movement.

The 7×5 dot matrix within which this type of printer forms each character allows an acceptable representation of alpha and numeric characters. Better legibility, particularly of lower case characters with descenders, can be achieved by using a larger matrix such as 9×7, i.e. a head with nine needles stepping through seven positions for each character.

Character codes are received for printing and sent from the keyboard in asynchronous serial form, on a line driven by a data communications interface in the computer, whose transmission speed determines the overall printing and keying throughput. Alternatively, connection to the computer by an interface which transmits all bits of a character in parallel is sometimes used with higher speed printer only devices. Buffer storage for up to a line or more of characters is provided in printers which use serial data communications, to make the most efficient use of communication lines, and in printers with built in intelligence, to allow look-ahead so that the print head can skip blanks and take the shortest route to the next printable character position.

Character sets can be readily changed by replacing the ROM chip which contains the dot patterns corresponding to each character code.

Other variants of serial dot matrix printer include:

(*a*) 'Silent' printers using special sensitised paper which is marked either by heat or by electrical discharge. In this type of printer, the needles in the print head are either electrically heated or electrically charged, in the appropriate sequence to form characters as the head is advanced along the print line.

(*b*) Plotting printers, in which the head movement is controlled in a special way allowing a simple form of graphic output to be produced. Lines and shaded or solid areas can be produced, as well as normal characters.

60.11.2 Letter quality printer

Where improved print quality is required in applications such as word processing, printers sometimes known as daisy wheel or petal printers are used (see section 61.8.6).

Another type of serial printer, which gives high print quality is the ink jet printer (see Chapter 61).

60.11.3 Visual display unit

A VDU is a terminal in which a CRT is used to display alphanumeric text. It is normally also equipped with a keyboard for data input, and occasionally a printer may be slaved to the VDU to produce a permanent paper copy of the information displayed on the screen at a particular time.

The format for the layout of text on the screen is most commonly 24 lines of text, each with up to 80 character positions. A few VDUs allow 132 columns to be displayed in a line, using a special compressed character set. This latter feature is useful for compatibility with computer printouts which normally have up to 132 columns of print. Displaying of new text on the screen takes place a character at a time, starting in the top left-hand corner and continuing line by line. When the screen is full, the page of text moves up one line allowing a new line to be added at the bottom.

Characters are caused to be displayed either by outputting data from the CPU through a serial communications interface, or by the operator typing on a typewriter-like keyboard.

VDUs are classified depending on which of the following modes of message composition they use:

(*a*) *Block mode.* A full screen of text is composed and, if need be, edited by the operator, and the corresponding character codes held in a buffer store in the VDU. Transmission of the text from the buffer store to the CPU is then done as a continuous block of characters, when a 'transmit' key is pressed.

(*b*) *Character mode.* Each character code is transmitted to the CPU as the corresponding key is pressed. Characters are 'echoed' back from the CPU for display on the screen. This function, plus editing of entered data, is performed by program in the CPU.

To assist with character positioning on the screen, a cursor is displayed, showing the position in which the next character will appear. On most VDUs, its position can be altered by control characters sent by program to the VDU, or by the operator using special keys, so that inefficient and time consuming use of 'space' and 'new line' controls is unnecessary.

Other functions, some or all of which are commonly provided depending on cost and sophistication, are: blinking; dual intensity; reverse video (black characters on a white or coloured background); underline; alternate character sets and protected areas. Each of these can be selected on a character by character basis. Additionally, the following are attributes of the whole screen: bidirectional scrolling; smooth scrolling (instead of jumping a line at a time); split screen scrolling; enlarged character size; compressed characters.

The most commonly used colours for displayed characters are white, green and amber, all on a dark background, or the reverse of these where reverse video is available. Some VDUs are physically constructed with the keyboard as an integral part of the unit, whilst others, like the Digital VT100 have a detached keyboard.

As with serial printers, characters are formed as a series of dots within a matrix.

VDUs with attributes such as multicolour display, symbol characters for the creation of mimic-type diagrams and simple graphics, and point addressable (or bit-map) displays for higher resolution graphics, are available for use in industrial control and monitoring, or in training or graphic design. Their characteristics are otherwise similar to those of standard VDUs, except that physically they are frequently implemented with industrial-grade colour monitors, which can, if required, be rack mounted.

The colour bit-map display is found on many personal microcomputer systems, often using a domestic colour TV set as the display screen.

60.11.4 Application terminals

Specialised terminals exist for use in areas such as:

(*a*) Shops, as point of sale terminals, i.e. sophisticated cash registers linked to an in-store computer.

(*b*) Banking for customer cash dispensing, enquiries and other transactions, or for teller use including the ability to print entries in pass books.

(*c*) Manufacturing for shop floor data collection and display.

These typically use features found in the terminals described above, and in addition may have the capability to read magnetic stripes on credit cards, punched plastic cards or identity badges, or bar codes on supermarket goods or parts in a factory.

60.11.5 Graphics terminals

CRT terminals which provide full graphics capabilities, i.e. the ability to generate a display with lines, curves, shaded areas etc, are in common use for applications known generically as computer-aided design (CAD).

Industries and professions such as aircraft and vehicle manufacture, electronics, structural and civil engineering and architecture use CAD for interactive design of their products such as car bodies, printed circuit boards and building frameworks, by creating and modifying two or three dimensional representations of these objects on the CRT screen. Keyboards and devices such as joysticks, rolling balls and light pens are provided for modification of displayed information, as in *Figure 60.7*.

Figure 60.7 Graphic terminal with keyboard and joystick. (Reproduced with kind permission of Digital Equipment Co Ltd)

Graphics terminals may communicate with the CPU either via a serial character interface, or where very high performance is required by a special high-speed interface outputting data in parallel form. In either event, the terminal normally has its own graphics processor to interpret picture information output from the CPU and held in a picture store. Thus, the program in the CPU need only supply parameters such as the start and end points of a vector, or the centre and radius of a circle, and the necessary processing is done within the terminal to display them on the screen.

60.12 High-speed printers and plotters

60.12.1 Line printer

For greater volumes of printed output than can be achieved with serial printers, line printers which can produce a whole line of characters almost simultaneously are available. Using impact techniques, speeds up to 2000 full lines (usually of 132 characters each) per minute are possible. Continuous paper in fan fold form, which may be multi-part to produce carbon copies, is fed through the printer a line at a time by a transport system consisting of tractors which engage sprocket holes at the edges of the paper to move it upwards and through the printer from front to rear. A paper tray at the rear allows the paper to fold up again on exit from the printer.

As well as advancing a line at a time, commands can be given to advance the paper to the top of the next page, or to advance a whole page or line. This is important, for example where pre-printed forms are being used.

Two types of line printer are in common use, drum printers and band printers. Both use a horizontal row of hammers, one per character position or in some cases shared between two positions. These are actuated by solenoids to strike the paper through a carbon film against an engraved representation to print the desired character. In a drum printer, a print drum the length of the desired print line rotates once per print line. In each character position, the full character set is engraved around the circumference of the drum. A band printer has a horizontal revolving band or chain of print elements, each with a character embossed on it. The full character set is represented on the band in this way. To implement different character fonts involves specifying different barrels in the case of a drum printer, whereas a change can be made readily on a band printer by an operator changing bands, or individual print elements in the band can be replaced.

The printer has a memory buffer to hold a full line of character codes. When the buffer is full, or terminated if a short print line is required, a print cycle is initiated automatically. During the print cycle, the stored characters are scanned and compared in synchronism with the rotating characters on the drum or band. The printer activates the hammer as the desired character on the drum or band approaches in each print position.

60.12.2 Other printers/plotters

Laser printers and electrostatic printer/plotters are used for high-speed outputs (see section 61.8.6). A similar end result to that of the electrostatic printer/plotter is available with a device which uses conventional impact techniques to generate patterns of overlapping dots. Normal paper, which may be multi-part, is used.

60.12.3 Pen plotter

Another form of hard copy output is provided by pen plotters. These are devices aimed primarily at high complexity graphics with a limited amount of text. Their uses range from plotting graphs of scientific data to producing complex engineering drawings in computer-aided design applications such as drawings used in integrated circuit chip design.

The plotter has one or more pens held vertically above a table on which the paper lies. These can be of different colours, and as well as being raised or lowered on to the paper individually by program commands, they can be moved in small steps, driven by stepping motors. Plotting in X and Y directions is achieved either by having a single fixed sheet of paper on the plotting table with the pen or pens movable in both X and Y directions, or, achieving control in one axis by moving the paper back and forth between supply and take-up rolls under stepping motor control, and in the other axis by pen movement. Diagonal lines are produced by combinations of movements in both axes.

With step sizes as small as 0.1 mm, high-accuracy plots can be produced, in multiple colours where more than one pen is used, and annotated with text produced in a variety of sizes and character sets. Supporting software is usually provided with a plotter. This will, for example scale drawings and text, and generate alphanumeric characters.

60.13 Direct input

Other forms of direct input, eliminating the need for typing on the keyboard of a teleprinter or VDU, are described in this Section.

60.13.1 Character recognition

Although the ultimate goal of a machine which can read any handwriting is a long way off, the direct conversion of readable documents into codes for direct computer input is a reality. These are used for capture of source data as an alternative to keyboard input, and for processing documents such as cheques. Several types of device exist, with varying capabilities and functions:

(*a*) Page and document readers, with the capability to read several special OCR fonts, plus in some cases lower quality print including hand printing, and hand marked forms as opposed to written or printed documents. Most character readers have some form of error handling, allowing questionable characters to be displayed to an operator for manual input of the correct character.

A wide range of capabilities and hence prices is found, from simple, low-speed (several pages per minute) devices handling pages only, to high-speed devices for pages and documents, the former at up to one page per second, with the latter several times faster.

(*b*) Document readers/sorters which read and optionally sort simple documents such as cheques and payment slips with characters either in magnetic ink or OCR font. These are geared to higher throughputs, up to 2000 documents per minute, of standard documents.

(*c*) Transaction devices, which may use both document reading and keyboard data entry, and where single documents at a time are handled.

60.13.2 Writing tablets

Devices using a variety of techniques exist for the conversion of hand printed characters into codes for direct input to a CPU. The overall function of these is the same—the provision of a surface on which normal forms, typically up to A4 size, can be filled in with hand printed alphanumeric characters, using either a normal writing instrument in the case of pressure sensitive techniques, or a special pen otherwise.

The benefits of this type of device include:

(*a*) Immediate capture of data at source, avoiding time consuming and error prone transcription of data.

(*b*) By detecting the movements involved in writing a character, additional information is gained compared with optical recognition, allowing characters which are easily confused by OCR to be correctly distinguished.

60.14 Disc storage

The rapidly declining cost of semiconductor memory has meant that main memory capabilities of 1 megabyte for minicomputers and up to 10 or more megabytes for midi and mainframe computers are not uncommon. However, computer systems of all categories from microcomputer to mainframe are now almost invariably equipped with magnetic disc storage ranging in capacity from one quarter megabyte to over 1000 megabytes per disc unit or drive.

The reasons for the almost universal application of the current generation of magnetic discs, having eclipsed earlier forms of backing storage such as magnetic drums, fixed head per track discs and block addressable magnetic tape systems of various kinds, are:

(*a*) Lower cost per byte of storage, particularly with large capacity disc drives.

(*b*) Random access to any data on the disc within milliseconds, due to the fact that the disc is rotating past read and write heads which can be moved quickly to span the whole recording surface.

(*c*) In most cases, the magnetic storage medium can be removed from the drive. This allows copies to be made for distribution, or for security reasons. Indefinitely large amounts of data can be stored off-line on media recorded by the system, which can subsequently be re-loaded and processed on the system, one at a time.

Discs are connected to a CPU by a controller which is normally a DMA device attached to the input/output bus or to a high-speed data channel, except in the case of the slowest of discs which may be treated as a programmed transfer device. The controller is generally capable of handling a number of drives, usually up to 4 or 8. Having multiple disc drives on a system, as well as providing more on-line storage, allows copying of information from one disc to another and affords a degree of redundancy since, depending on application, the system may continue to function usefully with one less drive in the event of a failure.

A disc controller is relatively complex, since it has to deal with high rates of data transfer, usually with error code generation and error detection, a number of different commands, and a large amount of status information.

Four types of disc drive will be described in the following sections: floppy disc, cartridge disc, removable pack disc and 'Winchester Technology' disc. The following elements are the major functional parts common to all the above types of drive, with differences in implementation between the different types.

(*a*) *Drive Motor*. This drives a spindle on which the disc itself is placed, rotating at a nominally fixed speed. The motor is powered up when a disc is placed in the drive, and powered down, normally with a safety interlock to prevent operator access to rotating parts, until it has stopped spinning, when it is required to remove the disc from the system.

(*b*) *Disc medium*. The actual recording and storage medium is the item which rotates. It is coated with a magnetic oxide material, and can vary from a flexible diskette of one quarter megabyte capacity recording on one surface only (floppy disc) to an assembly of multiple discs stacked one above the other on a single axle (disc pack) holding hundreds of megabytes of data.

(*c*) *Head mechanism*. This carries read/write heads, one for each recording surface. The number of recording surfaces ranges from only one on a single-sided floppy disc to ten or more for a multi-surface disc pack. In the latter case, the heads are mounted on a comb-like assembly, where the 'teeth' of the comb move together in a radial direction between the disc surfaces.

During operation, the recording heads fly aerodynamically extremely close to the disc surface, except in the case of floppy discs where the head is in contact with the surface. When rotation stops, the heads either retract from the surface or come to rest upon it, depending on the technology involved.

The time taken for the read/write head to be positioned above a particular area on the disc surface for the desired transfer of data is known as the access time. It is a function partly of the rotational speed of the disc, which gives rise to what is known as the average rotational latency, namely one half of the complete revolution time of the disc. Out of a number of accesses, the average length of time it is necessary to wait for the desired point to come below the head approaches this figure. The second component of access time is the head positioning time. This is dependent upon the number of tracks to be traversed in moving from the current head position to the desired one. Again, an average figure emerges from a large number of accesses. The average access time is the sum of these two components. In planning the throughput possible with a given disc system, the worst case figures may also need to be considered.

(*d*) *Electronics.* The drive must accept commands to seek, i.e. position the head assembly above a particular track, and must be able to recover signals from the read heads, and convert these to binary digits in parallel form for transmission to the disc controller. Conversely, data transmitted in this way from the controller to the disc drive must be translated into appropriate analogue signals to be applied to the head for writing the desired data on to the disc.

Various other functions concerned with control of the drive and sensing of switches on the control panel are performed. On some more advanced drives, much of the operation of the drive and electronics can be tested off-line from the system, allowing fault diagnosis to be performed without affecting the rest of the system.

Information is recorded in a number of concentric, closely spaced tracks on the disc surfaces, and in order to write and thereafter read successfully on the same or a different drive, it must be possible to position the head to a high degree of accuracy and precision above any given track. Data are recorded and read serially on one surface at a time, hence transfer of data between the disc controller and disc surface involves conversion in both directions between serial analogue and parallel digital signals. A phase-locked loop clock system is normally used to ensure reliable reading by compensating for variations in the rotational speed of the disc.

Data are formatted in blocks or sections on all disc systems, generally in fixed block lengths pre-formatted on the disc medium at time of manufacture. Alternatively 'soft sectoring' allows formatting into blocks of differing length by program. The drive electronics are required to read sector headers, which contain control information to condition the read circuitry of the drive, and sector address information, and to calculate, write and check an error correcting code—normally a cyclic redundancy check—for each block.

Finding the correct track in a seek operation, where the separation between adjacent tracks may be as little as 0.02 mm, requires servo-controlled positioning of the head to ensure accurate registration with the track. All rigid disc systems have servo-controlled head positioning, either using a separate surface pre-written with position information and with a read head only, or with servo information interspersed with data on the normal read/write tracks being sampled by the normal read/write head. Floppy disc systems, where the tolerances are not so fine, have a simpler stepping motor mechanism for head positioning.

60.14.1 Floppy disc

The floppy disc, whilst having the four elements described above, was conceived as a simple, low-cost device providing a moderate amount of random access back-up storage to microcomputers, word processors and small business and technical minicomputers.

As the name implies, the magnetic medium used is a flexible, magnetic oxide coated diskette, which is contained in a square envelope with apertures for the drive spindle to engage a hole in the centre of the disc and for the read/write head to make contact with the disc. Diskettes are of two standard diameters, approximately 203 and 133 mm, with an 89 mm microfloppy emerging as a likely third. The compactness and flexibility of the disc makes it very simple to handle and store, and possible for it to be sent by post.

One major simplification in the design of the floppy disc system is the arrangement of the read/write head. This runs in contact with the disc surface during read/write operations, and is retracted otherwise. This feature, and the choice of disc coating and the pressure loading of the head are such that, at the rotational speed of 360 rpm, the wear on the recording surface is minimal. Eventually, however, wear and therefore error rate are such that the diskette may have to be replaced, copying the information on to a new diskette.

Capacities vary from the 256 kilobytes of the earliest drives, which record on one surface of the diskette only, to a figure of over one megabyte on more recent units, most of which use both surfaces of the diskette. Access times, imposed by the rather slow head positioning mechanism using a stepping motor, are in the range of 200 to 500 ms. Transfer rates are below 100 kilobytes per second.

Another simplification is in the area of operator controls. There are generally no switches or status indicators, the simple action of moving a flap on the front of the drive to load or remove the diskette being the only operator action. The disc motor spins all the time that power is applied to the drive.

60.14.2 Cartridge disc

This type of disc system is so-called because the medium, one or two rigid discs on a single spindle, of aluminium coated with magnetic oxide and approximately 350 mm in diameter, is housed permanently in a strong plastic casing or cartridge. When the complete cartridge assembly is loaded into the drive, a slot opens to allow the read/write heads access to the recording surfaces. As well as providing mechanical mounting, the cartridge provides protection for the disc medium when it is removed from the drive.

Drives are designed either for loading from the top when a lid is raised, or from the front when a small door is opened allowing the cartridge to be slotted in. Power to the drive motor is removed during loading and unloading, and the door is locked until the motor has slowed down to a safe speed. On loading and starting up, the controller cannot access the drive until the motor has reached full speed. Operator controls are normally provided for unload, write protect and some form of unit select switch allowing drive numbers to be reassigned on a multiple drive system. Indicators typically show drive on-line, error and data transfer in progress.

Access times are normally in the region of 50 to 75 ms, aided by a fast servo-controlled head positioning mechanism actuated by a coil or linear motor, the heads being moved in and out over the recording surface by an arm which operates radially. Heads are light weight, spring loaded to fly aerodynamically in the region of 0.001 mm from the surface of the disc when it is rotating at its full speed, usually 2400 or 3600 rpm. Because of the extremely small gap, cleanliness of the oxide surface is vital, as any particle of debris or even smoke will break the thin air gap causing the head to crash into the disc surface. In this rare event, permanent damage to the heads and disc cartridge occurs. Positive air pressure is maintained in the area around the cartridge, in order to minimise the ingress of dirt particles. Care should be taken to ensure cleanliness in the handling and storage of cartridges when not mounted in the drive.

The capacity of cartridges is in the range from 2 to 30 megabytes, with data transfer rates in the region of 0.5 megabytes per second. Up to four or eight drives can be accommodated per controller, and because of the data transfer rate, direct memory access is necessary for transfer of data to or from the CPU.

60.14.3 Disc pack

The medium used in this type of drive has multiple platters (5 or more) on a single spindle, and is protected when removed from the drive by a plastic casing. When loaded on the drive, however, the casing is withdrawn. The drives are top loading, and unlike cartridge discs which can generally be rack mounted in the cabinet housing the CPU, are free standing units.

Other than this difference, most of the design features of disc pack drives follow those of cartridge units. The significant difference is the larger capabilities (up to 1000 megabytes) and generally high performance in terms of access times (35 to 50 ms) and transfer rates (in the region of 1 megabyte per second).

60.14.4 Winchester drive

So-called from a name local to the laboratory in the United States where it was developed, This is a generic name applied to a category of drive where the disc medium itself remains fixed in the drive. The principal feature of the drive, the fixed unit is known as a head–disc assembly. By being fixed and totally sealed, with the read/write heads and arm assembly within the enclosure, the following benefits are realised:

(*a*) Contaminant free environment for the medium allows better data integrity and reliability, at the same time having less stringent environmental requirements. Simpler maintenance requirements follow from this.

(*b*) Lighter weight heads, flying to tighter tolerances closer to the recording surface, allow higher recording densities. Since the disc itself is never removed, instead of retracting, the heads actually rest on special zones on the disc surface when power is removed.

(*c*) The arrangement of read/write heads is two per surface, providing lower average seek times, by requiring less head movement to span the whole recording area.

The head positioning arrangement differs mechanically from that of the previously described drives by being pivoted about an axis outside the disc circumference.

Three general types of Winchester drive exist, with approximate disc diameters of 133, 203 and 355 mm, providing capacities from 5 megabytes to over 500 megabytes. Performance, for the reason described above, can exceed that for disc cartridge or pack drives of corresponding capacity.

The smallest versions of Winchester drive are becoming popular as the storage medium for microcomputers and smaller configurations of minicomputer, offering compact size with very competitive prices and fitting above the top end of the floppy disc range. Operationally, the fact that the discs are not removable from the drive means that a separate form of storage medium which is removable must be present on a system using a Winchester drive. Back-up and making portable copies is done using this separate medium—another type of disc drive, or a magnetic type system matched to the disc speed and capacity.

60.15 Magnetic tape

Reliable devices for outputting digital data to and reading from magnetic tape have been available for a considerable time. The use of this medium, with agreed standards for the format of recorded data, has become an industry standard for the interchange of data between systems from different manufacturers.

In addition to this, low-cost magnetic tape cartridge systems exist providing useful minimal-cost back-up storage for small systems plus a convenient medium for small volume removable data and the distribution of software releases and updates.

60.15.1 Industry standard tape drive

These allow reels of 12.7 mm wide oxide-coated magnetic tape, which are normally 731 m in length on a 267 mm diameter reel, or 365 m on a 178 mm reel, to be driven past write and read head assemblies for writing, and subsequent reading, at linear densities from 200 to 6250 bits per inch. Tapes are written with variable length blocks or records with inter-record gaps in the region of 12.7 mm. Each block has lateral and longitudinal parity information inserted and checked, and a cyclic redundancy code is written and checked for each block. The latter provides a high degree of error correction capability.

The tape motion and stop/start characteristics are held within precise limits by a servo-controlled capstan around which the tape wraps more than 180 degrees for sufficient grip. Earlier drives used a capstan and pinch roller mechanism, which imposes greater strain and damage to the tape and has inferior speed control compared with a single capstan.

Correct tape tension and low inertia is maintained by motors driving the hubs of the two tape reels in response to information on the amount of tape in the path between the two reels at any time. One of the following forms of mechanical buffering for the tape between the capstan and reels is used:

(*a*) *Tension arm.* This uses a spring loaded arm with pulleys over which the tape passes, alternating with fixed pulleys such that when loaded, the tape follows a W-shaped path. The position of the arm is sensed and the information used to control the release and take up of tape by the reel motors.

(*b*) *Vacuum chamber.* This technique, used on modern, higher-performance tape drives, has between each reel and the capstan, a chamber of the same width as the tape, into which a U-shaped loop of tape of around 1 to 2 m is drawn by vacuum in the chamber. The size of the tape loops is sensed photoelectrically to control the reel motors.

To prevent the tape from being pulled clear of the reel when it has been read or written to the end or rewound to the beginning, reflective tape markers are applied near each end of the reel. These are sensed photoelectrically, and the resulting signal used to stop the tape on rewind, or to indicate that forward motion should stop on reading or writing.

Three different forms of encoding the data on the tape are encountered, dependent upon which of the standard tape speeds is being used. Up to 800 bits per inch, the technique is called 'non return to zero' (NRZ), while at 1600 bits per inch 'phase encoded' (PE) and at 6250 bits per inch 'group code recording' (GCR) are used. Some drives can be switched between 800 bits per inch NRZ and 1600 bits per inch PE.

Block format on the tape is variable under program control between certain defined limits, and as part of the standard, tape marks and labels are recorded on the tape, and the inter-block gap is precisely defined. Spacing between write and read heads allows a read-after-write check to be done dynamically to verify written data. Writing and reading can only be done sequentially. These tape units do not perform random access to blocks of data.

From one to four or eight tape drives can be handled by a single controller. For PE and GCR, a formatter is required between controller and drive, to convert between normal data representation and that required for these forms of encoding.

Tape drives can vary in physical form from a rack mountable unit occupying around 250 mm of panel height to a floor standing unit around 1.5 m in height. The smallest units can only handle smaller sized tape reels, while the floor standing units are normally at the high end of the performance range and their size may be dictated by the length of vacuum columns required.

Operator controls for on-line/off-line, manually controlled forward/reverse and rewind motion, unit select and load are normally provided. To prevent accidental erasure of a tape containing vital data by accidental write commands in a program, a write protect ring must be present on a reel when it is to be written to. Its presence or absence is detected by the drive electronics. This is a further part of the standard for interchange of data on magnetic tapes.

60.15.2 Cartridge tape

Low-cost tape units storing in the region of 0.5 megabytes of data on a tape cartridge are sometimes used for back-up storage on microcomputer systems, or for low-volume program loading on larger systems. Cartridge units using block formatted tapes which can search to locate numbered blocks can be used for operations requiring random access, such as supporting operating system software. This is only feasible, however, on a microcomputer system in view of the very slow access times.

Cartridge tape units are normally operable over a serial asynchronous line, and can therefore be connected by any communications interface to a system. Data transfer rates are normally below 5 kilobytes per second.

At even lower cost and performance levels, drives using standard cassettes of the dimensions of the normal audio cassette, are also found. With some of the personal microcomputers, standard domestic tape recorders are even used as a very simple storage device.

60.15.3 Streamer tape unit

The emergence of large capacity, non-removable disc storage in the form of Winchester technology drives has posed the problem of how to make back up copies of complete disc contents for security or distribution to another similarly equipped system. One solution is a tape drive very similar to the industry standard units described in section 60.15.1 but with the simplification of writing in a continuous stream, rather in blocks. The tape controller and tape motion controls can, therefore, be simpler than those for the industry standard drive.

A streamer unit associated with a small Winchester drive can accept the full disc contents on a single reel of tape.

60.16 Digital and analogue input/output

One of the major application areas for minicomputers and microcomputers is direct control of and collection of data from other systems by means of interfaces which provide electrical connections directly or via transducers to such systems. Both continuously varying voltages (analogue signals) and signals which have discrete on or off states (digital signals) can be sensed by suitable interfaces and converted into binary form, for analysis by programs in the CPU. For control purposes, binary values can also be converted to analogue or digital form by interfaces for output from the computer system.

In process control and monitoring, machine control and monitoring, data acquisition from laboratory instruments, radar and communications to take some common examples, computer systems equipped with a range of suitable interface equipment are used extensively for control and monitoring. They may be measuring other physical quantities such as temperature, presssure and flow converted by transducers into electrical signals.

60.16.1 Digital input/output

Relatively simple interfaces are required to convert the ones and zeros in a word output from the CPU into corresponding on or off states of output drivers. These output signals are brought out from the computer on appropriate connectors and cables. The output levels available range from TTL levels for connection to nearby equipment which can receive logic levels to over 100 V DC or AC levels for industrial environments. In the former case, signals may come straight from a printed circuit board inside the computer enclosure, while in the latter, they require to go through power drivers and be brought out to terminal strips capable of taking plant wiring. This latter type of equipment may need to be housed in separate cabinets.

In a similar manner, for input of information to the computer system, interfaces are available to convert a range of signal levels to logic levels within the interface, which are held in a register and can be input by the CPU. In some cases, input and output are performed on the same interface module.

Most mini and micro systems offer a range of logic level input/output interfaces, while the industrial type of input and output equipment is supplied by manufacturers specialising in process control. Optical isolators are sometimes included in each signal line to electrically isolate the computer from other systems. Protection of input interfaces by diode networks or fusible links is sometimes provided to prevent damage by overvoltages. In industrial control where thousands of digital points need to be scanned or controlled, interfaces with many separately addressable input and output words are used.

Although most digital input and output rates of change are fairly slow (less than 1000 words per second), high-speed interfaces at logic levels using direct memory access are available. These can in some cases transfer in burst mode at speeds up to the region of 1 million words per second. High transfer rates are required in areas such as radar data handling and display driving.

60.16.2 Analogue input

Analogue to digital converters, in many cases with programmable multiplexers for high- or low-level signals and programmable gain preamplifiers covering a wide range of signals (from microvolts to 10 V), allow conversion commands to be issued, and the digital results to be transferred to the CPU by the interface.

Industrial grade analogue input subsystems typically have a capacity of hundreds of multiplexer channels, low-level capability for sources such as thermocouples and strain gauges, and high common mode signal rejection and protection. As with digital input/output, this type of equipment is usually housed in separate cabinets with terminal strips, and comes from specialised process control equipment or data logger manufacturers.

For laboratory use, converters normally have higher throughput speed, lower multiplexer capacity and often direct cable connection of the analogue signals to a converter board housed within the CPU enclosure. Where converters with very high sampling rates (in the region of 100 000 samples per second) are used, input of data to the CPU may be by direct memory access. Resolution of analogue to digital converters used with computer systems is usually in the range 10 to 12 bits, i.e. a resolution of 1 part in 1024 to 1 part in 4096. Resolutions of anything from 8 to 15 bits are, however, available. Where a programmable or auto-ranging preamplifier is used before the analogue to digital converter, dynamic signal ranges of one million to one can be handled.

60.16.3 Analogue output

Where variable output voltages are required, for example to drive display or plotting devices or as set points to analogue controllers in industrial process control applications, one or more addressable output words is provided, each with a digital to analogue converter continuously outputting the voltage represented by the contents of its register. Resolution is normally no more than 12 bits, with a usual signal range of ± 1 V or ± 10 V. Current outputs are also available.

60.16.4 Input/output subsystems

Some manufacturers provide a complete subsystem with its own data highway separate from the computer system input/output bus, with a number of module positions into which a range of compatible analogue and digital input/output modules can be plugged. Any module type can be plugged into any position to make up the required number of analogue and digital points.

60.17 Data communications

60.17.1 Introduction

The subject of data communications is one which is currently receiving a great deal of attention from hardware manufacturers, software designers and users alike—perhaps more so than any other branch of computers. Indeed, many large organisations have individuals or whole departments whose sole function is to explore ways of connecting the systems in existence within the organisation and to design and manage the resulting networks. Furthermore, these systems will consist increasingly of types of equipment other than computers, equipment such as facsimile transmission devices, telex machines, photocopiers, typesetting systems and so on.

The requirement for the communication of data is not, of course, new. Interest has increased and will continue to increase, simply because improving technology is constantly opening up new possibilities. These technological improvements come in many forms.

Large-scale integration and consequent lower costs have made much more readily available very powerful computers which can contain the sophisticated software required to handle complex networks and overcome complex problems such as finding alternative routes for messages when a transmission line is broken. With the computer 'space' thus available, programmers can produce complex programs required to transmit data from a terminal connected to one computer to a program running in another, without the operator being aware of the fact that two computers are involved. Interface devices between the computer and the data network are becoming increasingly powerful, indeed they are frequently microcomputers in their own right, and are thus relieving the main computer of much of the previous load. Of course, the PTTs (Post, Telephone and Telegraph Authority) such as British Telecom, recognising the enormous potential of the data communications market, are developing both the range and sophistication of their offerings.

Computers have always been able to communicate with their peripheral devices such as card readers, mass storage devices and printers, but in the 1960s it was not typical for the communications to extend beyond this. Data was transcribed onto 'punching documents' by functional departments within an organisation. A punching document would contain a number of boxes each of which held one character which would subsequently be converted into a hole on a punched card by a key punch operator. There was frequently an entire department of key punch operators situated next to the computer room, and punch documents would be delivered to the punch room with the processed data being returned to the functional department some days later, in the form of a printed report.

This method of operation frequently proved time-consuming and slow, and one of the earliest common forms of data communications consisted of the automation of this process. It involved installing remote job entry (RJE) terminals in the department generating the original data (*Figure 60.8*). The RJE terminal usually consisted of an 80 column card reader which was fed with cards punched in the user department itself. As the terminal was connected by public or private telephone lines to the computer, the delay in entering the punched data into the computer system was reduced to the time required to transmit the batch over the telephone line. In addition, it was common to attach a printer to the RJE terminal, so that reports could be retransmitted back up the line, thereby effecting further savings in time. Large print runs, however, were still typically carried out on the main computer, since the line costs for the transmission of these may well have proven prohibitive.

However, in the late 1960s and 1970s the development of both hardware and software technology made it increasingly attractive to replace these terminals with more intelligent remote systems. These systems varied in their sophistication. At one end of the spectrum were interactive screen based terminals which could interrogate files held on the central computer. Greater sophistication was found in data validation systems which held sufficient data locally to check that, for example, part numbers on a customer order really existed, before sending the order to the computer for processing. Most sophisticated still were complete minicomputers carrying out a considerable amount of local data processing before updating central files to be used in large 'number crunching' applications such as production scheduling and materials planning.

Figure 60.8 IBM 2780 R.J.E. terminal, comprising card reader and printer. (Reproduced by kind permission of IBM UK Ltd)

From these systems have grown a whole range of requirements for data communications. We have communications between mainframe computers, between minis, between computers and terminals, between terminals and terminals and so on.

60.17.2 Data communications concepts

Computers communicate data in binary format, the bits being represented by changes in current or voltage on a wire. Various code systems have been developed in attempts to standardise the way characters are represented in binary format. One of the earliest of these was the 5-bit Baudot code, invented towards the end of the nineteenth century by Emile Baudot for use on telegraphic circuits. Five bits can be used to represent 32 different characters and whilst this was adequate for its purpose it cannot represent enough characters for modern data communications. Nonetheless Baudot gave his name to the commonly used unit of speed 'baud' which, although strictly speaking meaning signal events per second, is frequently used to denote 'bits per second'.

Nowadays, one of the most commonly used codes is the ASCII code (American Standard Code for Information Interchange). This consists of seven information bits plus one parity (error checking) bit. Another is EBCDIC (Extended Binary Coded Decimal Interchange Code), an 8-bit character code used primarily on IBM equipment.

Within the computer and between the computer and its peripheral devices such as mass storage devices and line printer, data are usually transferred in parallel format (*Figure 60.9*). In parallel transmission a separate wire is used to carry each bit, with an extra wire carrying a clock signal. This clock signal indicates to the receiving device that a character is present on the information wires. The advantage of parallel transmission is, of course, speed, since an entire character can be transmitted in the time it takes to send one bit. However, the cost would prove prohibitive where the transmitter and receiver are at some distance apart. Consequently, for sending data between computers and terminal devices and between computers which are not closely coupled, serial transmission is used.

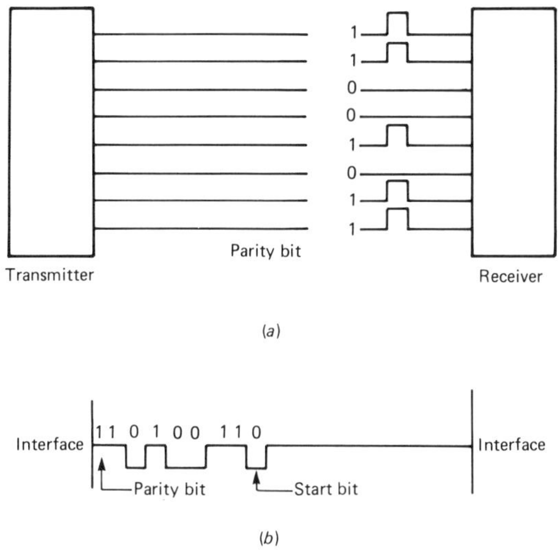

Figure 60.9 Data transmission; (*a*) parallel, (*b*) serial

Here a pair of wires is used, with data being transmitted on one wire whilst the second acts as a common signal ground. As the term implies, bits are transmitted serially and so this form of transmission is more practical for long-distance communication because of the lower cost of the wiring required. In addition, it is simpler and less expensive to amplify signals rather than use multiple signals in order to overcome the problem of line noise, which increases as the distance between the transmitter and receiver grows. Data transmission frequently makes use of telephone lines designed for voice communication, and since the public voice networks do not consist of parallel channels, serial transmission is the only practical solution.

Parallel data on multiple wires are converted to serial data by means of a device known as an interface. In its simplest form, an interface contains a register or buffer capable of storing the number of bits which comprise one character. In the case of data going from serial to parallel format, the first bit enters the first position in the register and is 'shifted' along, thereby making room for the second bit (*Figure 60.10*). The process continues until the sampling clock which is strobing the state of the line indicates that the correct number of bits has been received and that a character has been assembled. The clock then generates a signal to the computer which transfers the character in parallel format. The reverse process is carried out to convert parallel to serial data.

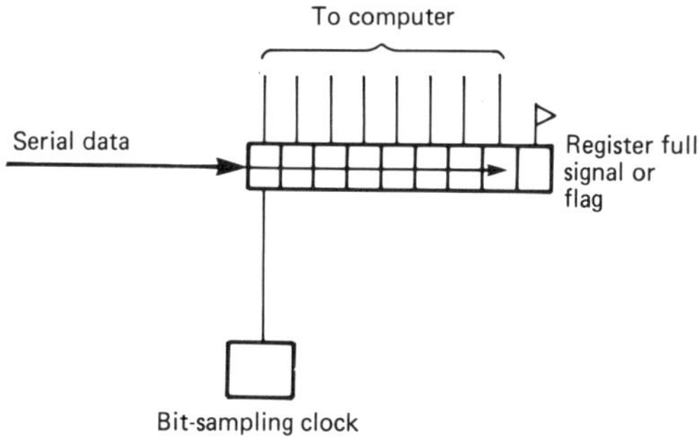

Figure 60.10 Serial-to-parallel interface

This 'single buffered' interface does have limitations, however. The computer effectively has to read the character immediately, since the first bit of a second character will be arriving to begin its occupation of the register. This makes no allowance for the fact that the computer may not be available instantly. Nor does it allow any time to check for any errors in the character received.

To overcome this problem, a second register is added creating a 'double buffered' interface (*Figure 60.11*). Once the signal is received indicating that the requisite number of bits have been assembled, the character is parallel transferred to the second, or holding register, and the process can continue. The computer now has as much time as it takes to fill the shift register in order to check and transfer (again in parallel format) the character.

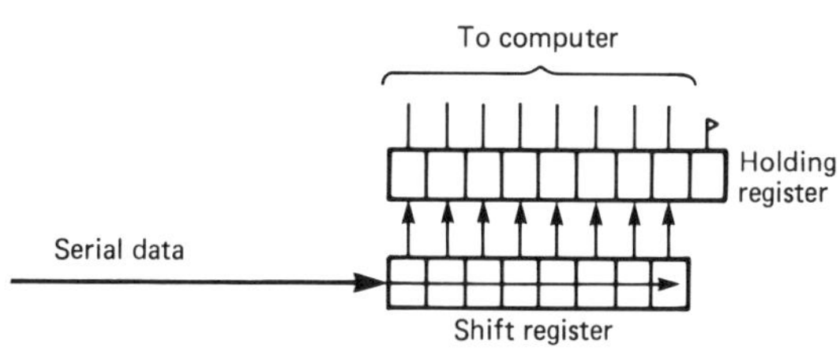

Figure 60.11 Double-buffered interface

60.17.3 Multiline interface

With the development of technology, the transmitter and receiver functions are now carried out by an inexpensive chip and so the major costs in the interface are the costs of the mechanism used to interrupt the CPU when a character has been assembled and the connection to the computer's bus used to transmit the received data to the CPU, or in some cases direct to memory. The interrupt mechanism and the bus interface are not heavily used. Indeed, they function only when a character is received or transmitted. These facilities are shared in a multiline interface, sometimes (though not strictly correctly) known as a 'multiplexor' (*Figure 60.12*). To achieve this the device has several receivers and transmitters and a first in, first out (FIFO) buffer for received characters. The receivers are scanned and when a flag is found indicating that a character has been received the character is transmitted into the FIFO buffer, along with its line number. An interrupt tells the CPU that there are characters in the buffer and they are communicated over the bus to the computer. Similarly, a scanner checks the transmitters and when it discovers a flag indicating that a transmitter buffer is empty, it interrupts the CPU. Typically, the number of lines supported by a multi-line interface increases by powers of two for convenient binary representation, 4, 8, 16, 32, 64, 128 and 256 being common.

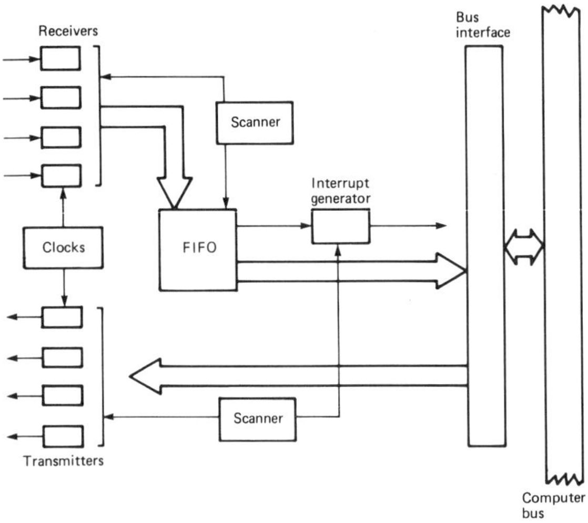

Figure 60.12 Schematic diagram of a multiline interface

The economies of scale in such an interface mean that further sophistications can be included such as program selectable formats and line speed, and modem control for some or all of the lines.

However, the term 'multiplexing', strictly speaking, actually refers to the function of sharing a single communications channel across many users. There are two commonly used methods of achieving this. One is a technique called time division multiplexing. It consists of breaking down the data from each user into separate messages which could be as small as one or two characters and meaningless when taken individually. The messages, together with identifying characters, are interleaved and transmitted along a single line. They are separated at the other end and the messages reassembled. This is achieved by use of devices known as concentrators or multiplexors.

The second technique used to achieve this objective of making maximum use of a communication line is frequency division multiplexing. The concept is similar to that of time division multiplexing. It is achieved by transmitting complete messages simultaneously but at different frequencies.

60.17.4 Modem

A significant complication of using public voice networks to transmit data is that voice transmission is analogue whereas data generated by the computer or terminal is digital in format. Thus an additional piece of equipment is required between the digital sender/receiver and the analogue circuit. This device modulates and demodulates the signal as it enters and leaves the analogue circuit, and is known by the abbreviated description of its functions, a modem (*Figure 60.13*). Modems are provided by the common carrier such as British Telecom or by private manufacturers. In the latter case, however, they must be approved by the carrier and must contain or be attached to a device which provides electrical isolation (*Figure 60.14*).

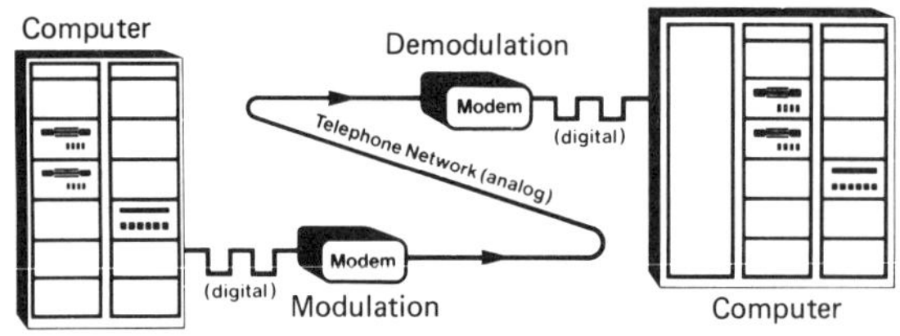

Figure 60.13 The use of modems in a communications link

60.17.5 Transmission techniques

There are two techniques commonly used to transmit data on serial lines. One varies the current and the other varies the voltage in order to indicate the presence or absence of bits on the line.

The current based technique communicates binary data by turning on and off a 20 mA current flowing through both the transmitter and receiver. Current on indicates a 'mark' or '1' bit and current off signifies a 'space' or '0' bit. This technique of turning on and off a current is less susceptible to noise than the technique of varying the voltage. However, it does have some drawbacks. Optical isolators are needed to protect logic circuits from the high voltages which may be required to drive the loop. Since there is one current source, an active interface and a passive interface are required, and finally, since a 20 mA system cannot carry the necessary control information it cannot be used with modems.

However, the EIA (Electronic Industries Association) and CCITT (Comité Consultatif Internationale de Télégraphie et Téléphone) systems do contain specifications and recommendations for the design of equipment to interface data terminal equipment (computers and terminals) to data communication equipment (modems). The specific EIA standard to which most modem equipment is designed is RS-232-C. The CCITT, being formed by the United Nations to consider all aspects of telecommunications across several national boundaries, was unable to publish firm standards, and instead produced a list of recommendations. Its equivalent of RS-232-C is known as 'V.24—List of Definitions for Interchange Circuits between Data Terminal Equipment and Data Circuit Terminating Equipment'. The EIA/CCITT systems communicate data by reversing the polarity of the voltage; a '0' is represented by a positive voltage and a '1' by a negative voltage.

The signals in the EIA/CCITT specifications are not

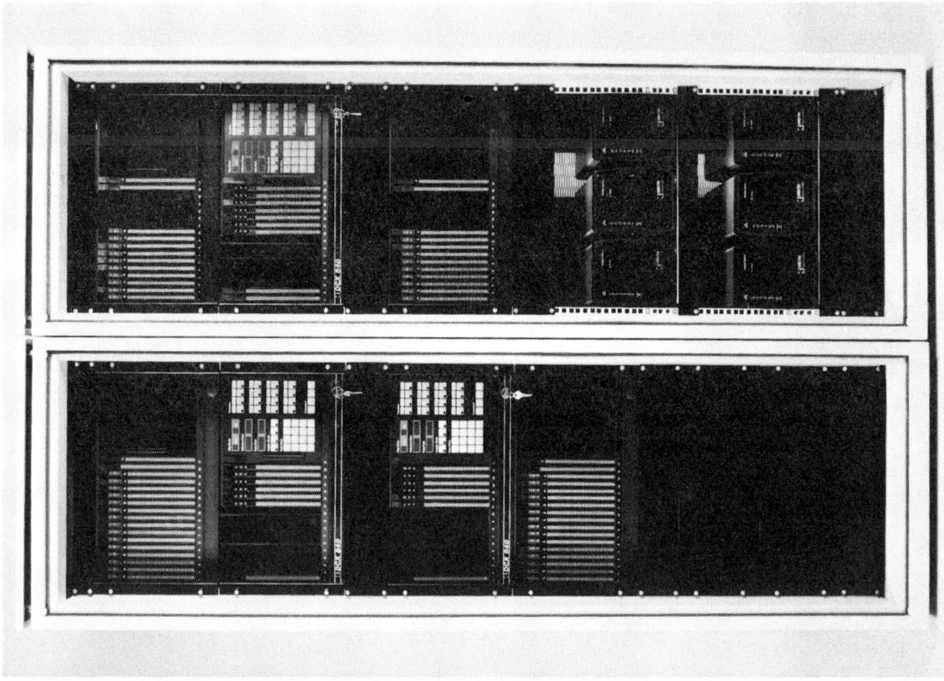

Figure 60.14 Racks containing networking multiplexors and modems. (Reproducted by kind permission of Computer and Systems Engineering Ltd)

recommended for use over distances greater than 50 feet. Consequently, the modem and interface should not be more than 50 feet apart.

60.17.6 Transmission types

Different communications applications use one of two types of transmission, asynchronous or synchronous. Slower electromechanical devices such as teleprinters typically use asynchronous (or 'start–stop') transmission in which each character is transmitted separately. In order to tell the receiver that a character is about to arrive, the bits representing the character are preceded by a start bit, usually a zero. After the last data bit and error checking bit the line will return to the 1-bit state for at least one bit time—this is known as the stop bit.

Asynchronous transmission has the advantage that it requires relatively simple and therefore low-cost devices. Since characters can be sent at random times it is also appropriate for interactive applications which do not generate large amounts of data to be transmitted. It is, however, inefficient, since at least two extra bits are required to send eight data bits, and so would not be used for high-speed communication.

In synchronous transmission, characters are assembled into blocks by the transmitter and so the stream of data bits travels along the line uninterrupted by start and stop bits. This means that the receiver must know the number of bits which make up a character so that it can re-assemble the original characters from the stream of bits. Preceding the block of data bits synchronisation characters are sent to provide a timing signal for the receiver and enable it to count in the data characters. If the blocks of data are of uniform length, then this is all that is required to send a message. However, most systems would include some header information which may be used to indicate the program or task for which the data are destined and the amount of data in the block. In addition, if the messages are of variable length, some end of message characters will be required.

Because it does not contain start and stop bits for every character, synchronous transmission is more efficient than asynchronous. However, it can be inappropriate for some character-orientated applications and the equipment required to implement it is more expensive.

60.17.7 Direction of transmission

There are three types of circuit available for the communication of data and correspondingly, there are three direction combinations, simplex, half duplex and full duplex. However, it is possible to use a channel to less than its full potential.

Simplex communication is the transmission of data in one direction only, with no capability of reversing that direction. This has limitations and is not used in the majority of data communications applications. It can be used, however, for applications which involve the broadcasting of data for information purposes in a factory for example. In this instance, there is neither a need nor a mechanism for sending data back to the host. The simplex mode of operation could not be used for communication between computers.

Half duplex, requiring a single, two-wire circuit, permits the user to transmit in both directions, but not simultaneously. Two-wire half duplex has a built in delay factor called turnaround time. This is the time taken to reverse the direction of transmission from sender to receiver and vice versa. The time is required by line propagation effects, modem timing and computer response time. It can be avoided by the use of a four-wire circuit normally used for full duplex. The reason for using four wires for half duplex rather than full duplex may be the existence of limitations in the terminating equipment.

Full duplex operation allows communication in both directions simultaneously. The data may or may not be related depending on the applications being run in the computer or computers. The decision to use four-wire full duplex facilities is usually based on the demands of the application compared to the increased cost for the circuit and the more sophisticated equipment required.

60.17.8 Error detection and correction

Noise on communications lines will inevitably introduce errors into messages being transmitted. The error rates will vary according to the kind of transmission lines being used. In-house lines are potentially the most noise free since routing and shielding are within the users control. Public switched networks, on the other hand, are likely to be the worst as a result of noisy switching systems and dialling mechanisms.

Whatever the environment, however, there will be a need for error detection and correction. Three systems are commonly used: VRC, LRC and CRC.

VRC, or vertical redundancy check, consists of adding a parity bit to each character. The system will be designed to use either even or odd parity. If the parity is even, the parity bit is set so that the total number of ones in the character plus parity is even. Obviously, for odd parity the total number will be odd. This system will detect single bit errors in a character. However, if two bits are incorrect the parity will appear correct. VRC is therefore a simple system designed to detect single bit errors within a character. It will detect approximately 9 out of 10 errors.

A more sophisticated error detection system is LRC, longitudinal redundancy check, in which an extra byte is carried at the end of a block of characters to form a parity character. Unlike VRC, the bits in this character are not sampling an entire character but individual bits from each character in the block. Thus the first bit in the parity character samples the first bit of each data character in the block. As a result, LRC is better than VRC at detecting burst errors which affect several neighbouring characters.

It is possible to combine VRC and LRC and increase the combined error detection rate of 99% (*Figure 60.15*). A bit error can be detected and corrected, because the exact location of the error will be pinpointed in one direction by LRC and the other by VRC.

Even though the combination of LRC and VRC significantly increases the error detection rate, the burst nature of line noise means that there are still possible error configurations which could go undetected. In addition, the transmission overhead is relatively high. For VRC alone, in the ASC11 code, it is 1 bit in 8, or $12\frac{1}{2}\%$. If VRC and LRC are used in conjunction it will be $12\frac{1}{2}\%$ plus one character per block.

A third method which has the advantage of a higher detection rate and, in most circumstances, a lower transmission overhead is CRC, cyclic redundancy check. In this technique the bit stream representing a block of characters is divided by a binary number. In the versions most commonly used for 8-bit character format, CRC-16 and CRC-CCITT, a 16-bit divisor is used. There are no carry overs in the division and a 16-bit remainder is generated. When this calculation has been completed, the transmitter sends these 16 bits—two characters—at the end of the block. The receiver repeats the calculation and compares the two remainders. With this system, the error detection rises to better than 99.99%. The transmission overhead is less than that required for VRC/LRC when there are more than 8 characters per block, as is usually the case.

The disadvantage with CRC is that the calculation overhead required is clearly greater than for the other two systems. The check can be performed by hardware or software but, as is usually the case, the higher performance and lower cost of hardware is making CRC more readily available and more commonly used.

Once bad data have been detected, most computer applications require that it be corrected and that this occurs automatically. Whilst it is possible to send sufficient redundant data with a message to enable the receiver to correct errors without reference to the transmitter, the calculation of the effort required to achieve this in the worst possible error conditions means that this technique is rarely used. More commonly, computer systems use error correction methods which involve retransmission. The two most popular of these are 'stop and wait retransmission' and 'continuous retransmission'.

'Stop and wait' is reasonably self explanatory. The transmitter sends a block and waits for a satisfactory or positive acknowledgement before sending the next block. If the acknowledgement is negative, the block is retransmitted. This technique is simple and effective. However as the use of satellite links increases, it suffers from the disadvantage that these links have significantly longer propagation times than land based circuits and so the long acknowledgement times are reducing the efficiency of the network. In these circumstances, 'continuous retransmission' offers greater throughput efficiency. The difference is that the transmitter does not wait for an acknowledgement before sending the next block, it sends continuously. If it receives a negative acknowledgement it searches back through the blocks transmitted and sends it again. This clearly requires a buffer to store the blocks after they have been sent. On receipt of a positive acknowledgement the transmitter deletes the blocks in the buffer up to that point.

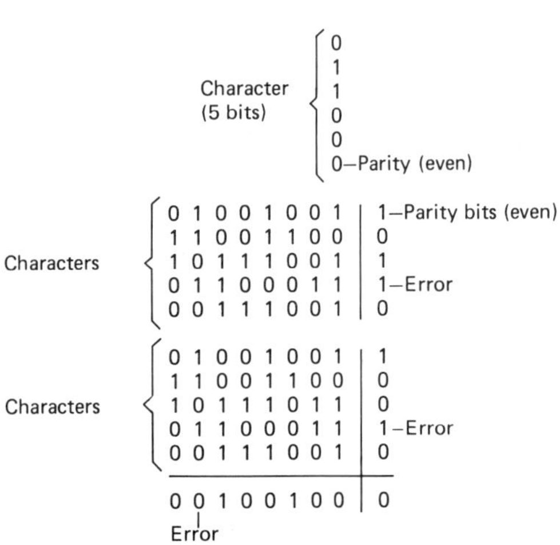

Figure 60.15 VRC, LRC and VRC/LRC combined (with acknowledgements to Digital Equipment Co Ltd)

60.17.9 Communication protocols

The communications protocol is the syntax of data communications. Without such a set of rules, a stream of bits on a line would be impossible to interpret. Consequently, many organisations, notably computer manufacturers, have created protocols of their own. Unfortunately, however, they are all different and consequently yet another layer of communications software is required to connect computer networks using different protocols. Examples of well known protocols are Bisync and SDLC from IBM, DDCMP from Digital Equipment Corporation, ADCCP from the American National Standards Institute (ANSI) and HDLC from the International Standards Organisation (ISO). The differences between them, however, are not in the functions they set out to perform, but in the way they achieve them. Broadly speaking, these functions are as follows.

Framing and formatting. These define where characters begin and end within a series of bits, which characters constitute a message and what the various parts of a message signify. Basically, a transmission block will need control data, usually contained in a 'header' field, text—the information to be transmitted—held in the 'body', and error checking characters, to be found in the 'trailer'. The actual format of the characters is defined by the information code used such as ASCII or EBCDIC.

Synchronisation. It involves preceding a message or block with a unique group of characters which the receiver recognises as a synchronisation sequence. This enables the receiver to frame subsequent characters and fields.

Sequencing. It numbers messages so that it is possible to identify lost messages, avoid duplicates and request and identify retransmitted messages.

Transparency. Ideally all of the special control sequences should be unique and, therefore, never occur in the text. However, the widely varied nature of the information to be transmitted, from computer programs to data from instruments and industrial processes, means that occasionally a bit pattern will occur in the text which could be read by the receiver as a control sequence. Each protocol has its own mechanism for preventing this, or achieving 'transparency' of the text. Bisync employs a technique known as 'character stuffing'. In Bisync the only control character which could be confusing to the receiver if it appeared in the text is DLE (data link escape). When the bit pattern equivalent to DLE appears within the data a 'second' DLE is inserted. When the 2 DLE sequences are read, the DLE proper is discarded and the original DLE-like bit pattern is treated as data. This is character stuffing. SDLC, ADCCP and HDLC use a technique known as 'bit stuffing' and DECMP uses a bit count to tell the receiver where data begins and ends.

Start-up and time-out. These are the procedures required to start transmission when no data have been flowing and recovering when transmission ceases.

Line control. It is the determination, in the case of half-duplex systems, of which terminal device is going to transmit and which to receive.

Error checking and correction. As described in the section on error checking techniques, each block of data is verified as it is received. In addition, the sequence in which the blocks are received is checked. For data accuracy all the protocols discussed in this section are capable of supporting CRC (cyclic redundancy check). The check characters are carried in the trailer or block check character (BCC) section.

60.17.10 Computer networks

In the early days of data communications, and to some extent this is still true today, information travelled along a single, well defined route between the remote computer to the 'host'. The reason for this was that the remote computer was fairly restricted in its computing and data storage capabilities and so the 'serious' computing was carried out at the data centre. Most large organisations have data centres and few have plans to do away with this strategy. However, the development of powerful mini and so-called 'midi' systems have given rise to the 'specialist' computer, that is, the machine bought for one specific application. Within large organisations, it is increasingly common for departments to want to share these specialist machines.

In the motor industry, for example, the European headquarters of a US corporation would have its own design and engineering department with a computer capable of processing, displaying and printing design calculations. However, it may still require access to the larger US machine for more complex applications requiring greater computer power. In addition, there may be a number of test units, testing engines and transmissions, each controlled by its own mini and supervised by a host machine. If there is a similar engineering department in Germany, it may be useful to collect and compare statistical data from test results. And since people require paying it may be useful to have a link with the mainframe computer in the data centre for the processing of payroll records. So it goes on. The demand for the linking of computers and the sharing of information and resources is increasing constantly.

A communications network may exist within a single site. Provided it is not necessary to cross a public thoroughfare an organisation may connect its systems together to form its own internal network. Because of the cost and disruption associated with laying cables, many companies are using internal telephone circuits for data communications.

For the factory environment many computer manufacturers offer proprietory networks for connecting terminal equipment to circuits based on 'tree' structures or loops. Connections to the circuit may be from video terminals for collection of, for example, stores data, special purpose card and badge readers used to track the movement of production batches, or transducers used in the control of industrial processes.

Current developments of such internal networks are aimed at extending the range of equipment types which can be connected to such circuits. The ultimate aim is to be able to connect all types of electronic equipment to a single network—photocopiers, word processors, telex machines, facsimile machines and so on. Perhaps the best known is the Ethernet system, being promoted as a potential standard for internal networks by Xerox, Digital and Intel (see section 61.9).

There are a number of network types:

Point to point. This is the simplest form of network and involves the connection of two devices—two computers or a computer and a terminal. If the communication line goes down for any reason then the link is broken, and so it is usual to back up leased lines with dial-up facilities (*Figure 60.16*).

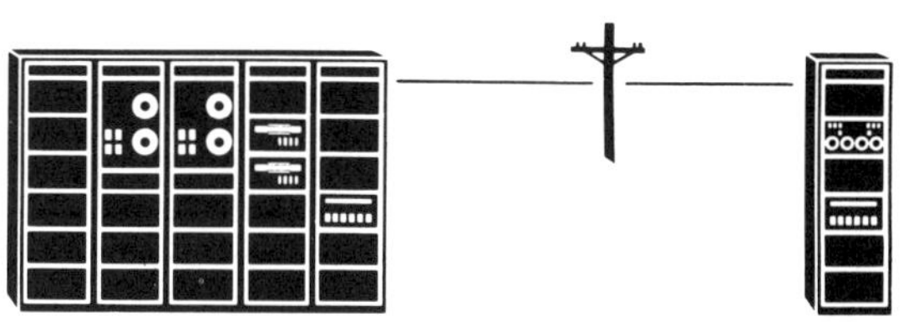

Figure 60.16 A point-to-point link (with acknowledgements to Digital Equipment Co Ltd)

Multi-point. As the name implies, Multi-point describes the connection of several tributary stations to one host. It is usual for the host to 'poll' the tributary stations in sequence, requesting messages, and for the network to be based on leased lines. In the case of one 'spur' being disconnected, the tributary station will dial into the host using a port received for the purpose (*Figure 60.17*).

Centralised. In this type of network, the host exercises control over the tributary stations all of which are connected to it. The host may also act as a message-switching device between remote sites (*Figure 60.18*).

Hierarchical. A hierarchical structure implies multiple levels of supervisory control. For example, in an industrial environment,

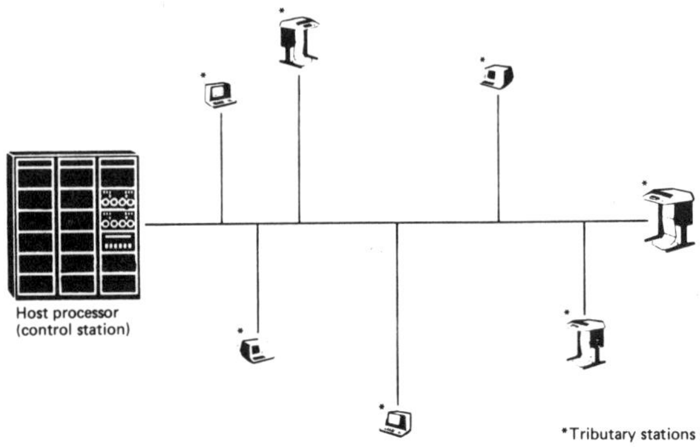

Figure 60.17 A multi-point network (with acknowledgements to Digital Equipment Co Ltd)

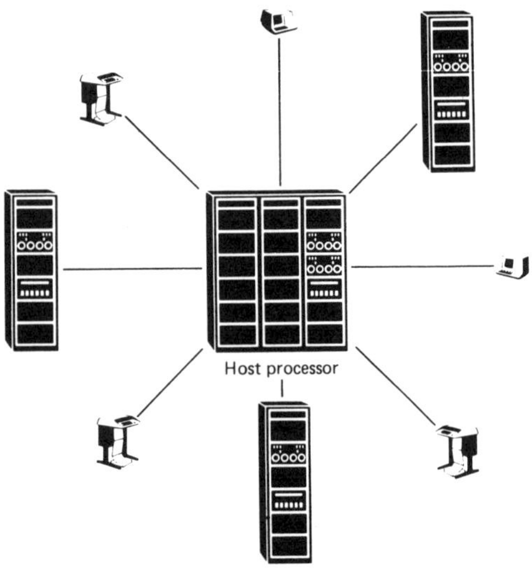

Figure 60.18 A centralised network (with acknowledgements to Digital Equipment Co Ltd)

special purpose 'micros' may be linked to the actual process equipment itself. Their function is to monitor and control temperature and pressure.

These 'micros' will then be connected to supervisory 'minis' which can store the programs and set points for the process computers and keep statistical and performance records (*Figure 60.19*).

The next link in the chain will be the 'resource management computers', keeping track of the materials used, times taken, comparing these with standards, calculating replenishment orders, adjusting forecasts and so on.

Finally, at the top of the network, the financial control system records costs and calculates the financial performance of the process.

Fully distributed. Here each station may be connected to several others in the network. The possibility then exists to share resources such as specialised peripheral devices or large memory capacity and to distribute the data base to the systems which access the data most frequently. It also provides alternative routes for messages when communication lines are broken or traffic on one link becomes excessive (*Figure 60.20*).

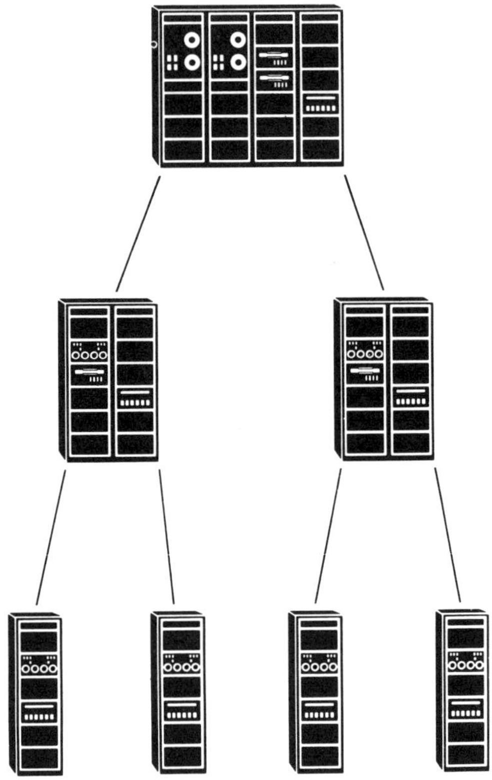

Figure 60.19 A hierarchical network (with acknowledgements to Digital Equipment Co Ltd)

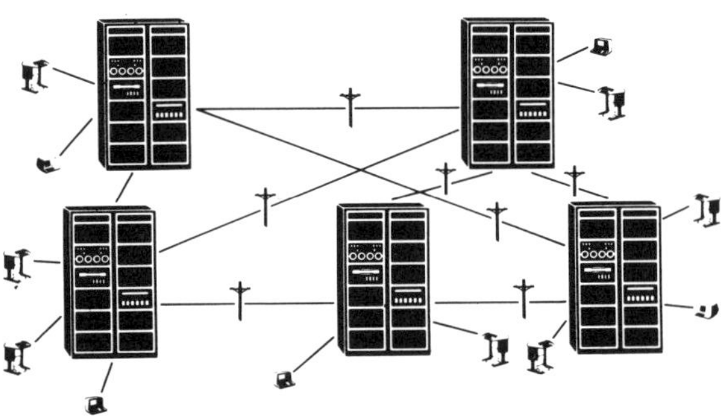

Figure Figure 60.20 A fully distributed network (with acknowledgements to Digital Equipment Co Ltd)

However, the design of such systems requires sophisticated analysis of traffic and data usage and even when set up is more difficult to control than less sophisticated networks.

60.17.11 Network concepts

Whatever the type of network, there are a number of concepts which are common.

File transfer. A network should have the ability to transfer a file, or a part file, from one node to another without the intervention of programmers each time the transfer takes place. The file may contain programs or data and since different types, and possibly generations, of computers, and different applications are involved, some reformatting may be required. This requires a set of programs to be written to cover all foreseen transfer requests and requires a knowledge of all local file access methods and formats.

One good exmple of the need for this is the application known as archiving. It involves the transmission of copies of files held on computer to another system in another location. In the event of original files being lost as a result of fire, the files can be recreated using the archived information.

Resource sharing. It may be more cost effective to set up communication links to share expensive peripheral devices than to duplicate them on every computer in the network. For example, one computer may have a large sophisticated flatbed printer/plotted for producing large engineering drawings. To use this the other computers would store the necessary information to load and run the appropriate program remotely. This would be followed by the data describing the drawing to be produced.

Remote file access/enquiry. It is not always necessary or desirable to transfer an entire file, especially if only a small amount of data is required. In these circumstances what is required is the ability to send an enquiry from a program (or task) running in one computer and remotely load, to the other system. This enquiry program will retrieve the requisite data from the file and send them back to the original task for display or processing. This comes under the broad heading of task-to-task communications.

Logical channels. The user of a computer network will know where the programs and data, which he wants to access, exist. He does not want to concern himself with the mechanics of how to gain access to them. He expects there to be a set of pre-defined rules in his system which will provide a 'logical channel' to the programs and data he wishes to reach. This logical channel will use one or more logical links to route the users request and carry back the response efficiently and without errors. It may be that there is no direct physical link between the user's computer and the machine he is trying to access. In these circumstances the logical channel will consist of a number of logical links. The physical links, in some cases, may be impossible to define in advance, since in the case of 'dial up' communication using the public switched network, the route will be defined at connection time by the PTT.

Virtual terminal. This is a very simple concept. It describes a terminal physically connected to computer A but with access (via A) to computer B.

The fact that he is communicating via A should be invisible to the user. Indeed, to reach his ultimate destination, he may unknowingly have to be routed through several nodes.

In reality, the vast number of different terminals available on the market, each with different functions and characteristics, together with the possibility of several computer systems being involved, plus multiple-line types, can make the concept difficult to implement. In practice, the network designer will define a set of basic specifications to which all connected terminals must conform if they wish to use the virtual terminal facility.

Emulator. As the name implies, this consists of one device performing in such a way that it appears as something different. For example, a network designer wrestling with the 'virtual terminal' concept may define that any terminal or computer to be connected to his network should be capable of looking like a member of the IBM 3270 family of video terminals for interactive work and the IBM 2780/3780 family for batch data transfer. In other words, they must be capable of 3270 and 2780/3780 emulation. Indeed, these two types of emulation have been amongst the most commonly used in the computer industry.

Routing. As soon as we add a third node, C, to a previously point-to-point link from A to B, we have introduced the possibility of taking an alternative route from A to B, namely via C. This has advantages. If the physical link between A and B is broken, we can still transmit the message. If the traffic on the AB link is too high we can ease the load by using the alternate route.

However, it does bring added complications. The designer has to balance such factors as lowest transmission cost versus load sharing. Each computer system has to be capable of recognising which messages are its own and which it is required merely to transmit to the next node in the logical link. In addition, when a node recognises that the physical link it was using has, for some reason, been broken, it must know what alternative route is available.

Network design is a complicated and specialist science. Computer users do not typically want to re-invent the wheel by writing all the network facilities they require from scratch. They expect their computer supplier to have such software available for rent or purchase, and, indeed, most large computer suppliers have responded with their own offerings. IBM clearly dominates the computer industry with its systems network architecture (SNA) designed to support networks hosted by the many IBM mainframe systems installed throughout the world. Amongst the minicomputer manufacturers probably the best known network offering is Digital's 'Decnet', a peer-to-peer network system.

The fact remains, of course, that there is still no standard network architecture permitting any system to talk to any other. Hence the need for emulators. However, the PTTs of the world have long recognised this need and are uniquely placed as the suppliers of the physical links to bridge the gap created by the computer manufacturers. They have developed the concept of public 'packet-switched networks' to transmit data between private computers or private networks. .

Public packet-switched networks (PPSNS). Packet switching involves breaking down the message to be transmitted in to 'packets' which are 'addressed' and introduced into the network controlled by the PTT. Consequently, the user has no influence over the route the packets take. Indeed, the complete contents of a message may arrive by several different routes. Users are charged according to line usage and the result is generally greater flexibility and economy. The exception is the case where a user wants to transmit very high volumes of data regularly between two points. In this instance, a high-speed leased line would probably remain the most viable option.

What goes on inside the network should not concern the subscriber, provided the costs, response times and accuracy meet his expectations. What does concern him is how to connect to the network. There are basically two ways of doing this.

(*a*) If he is using a relatively unintelligent terminal he needs to connect to a device which will divide his message into packets and insert the control information. Such a device is known as a PAD (packet assembler/disassembler) and is located in the local packet-switching exchange. Connection between the terminal and the PAD may be effected using dedicated or dial-up lines.

(*b*) More sophisticated terminals and computer equipment may be capable of performing the PAD function themselves in which case they will be connected to the network via a line to the exchange, but without the need to use the exchange PAD.

The CCITT has put forward recommendation X25 ('Interface between data terminal equipment for terminals operating in the packet mode in public data networks') with the aim of encouraging standardisation. X25 currently defines three levels within its recommendations.

(*a*) The *physical* level defines the electrical connection and the hand-shaking sequence between the data terminals equipment (DTE) and the data communications equipment (DCE).

(*b*) The *link* level describes the protocol to be used for error-free transmission of data between two nodes. It is based on the HDLC protocol.

(*c*) The *packet* level defines the protocol used for transmitting packets over the network. It includes such information as user identification and charging data.

Packet-switched networks have been in use for some years in the USA. In other countries they are in various stages of development, from experimental to operation. They include TRANSPAC in France, DataPac in Canada and PSS in the UK.

60.17.12 Data terminal equipment

The most basic all-round terminal is the teleprinter and although its popularity is waning in favour of the video terminal, it remains an attractive low-cost device, still widely used as a control console.

The most widely used terminal is the video display or VDU. VDUs may be clustered together in order to optimise the use of a single communications line. In this instance a controller is required to connect the screens and printers to the line (*Figure 60.21*).

Batch terminals are used when a high volume of non-interactive data is to be transmitted. Most commonly the input medium is punched cards with output on high-speed line printers. Transmission speeds can reach 22 000 bits per second. As with VDUs it is quite feasible to build intelligence into batch terminals in order to carry out some local data verificiation and local processing.

In addition to these commonly found terminals there are a host of special purpose devices including various types of optical and magnetic readers, graphics terminals, hand-held numeric (keypad) devices, badge readers, audio response terminals, point of sale terminals and more.

Finally, of course, computers can communicate directly with each other without the involvement of any terminal device.

60.18 Software

60.18.1 Introduction

Software is the collective name for programs. Computer hardware is capable of carrying out a range of functions represented by the instruction set. A program (the American spelling is usually used when referring to a computer program) simply represents the sequence in which these instructions are to be used to carry out a specific application. However, this is achieved in a number of ways. In most cases the most efficient way of using the hardware is to write in a code which directly represents the hardware instruction set. This is known as machine code and is very machine dependent. Unfortunately, it requires a high level of knowledge of the particular type of computer in use, and is time consuming. In practice, therefore, programmers write in languages in which each program instruction represent a number of machine instructions. The programs produced in this 'high-level language' clearly require to

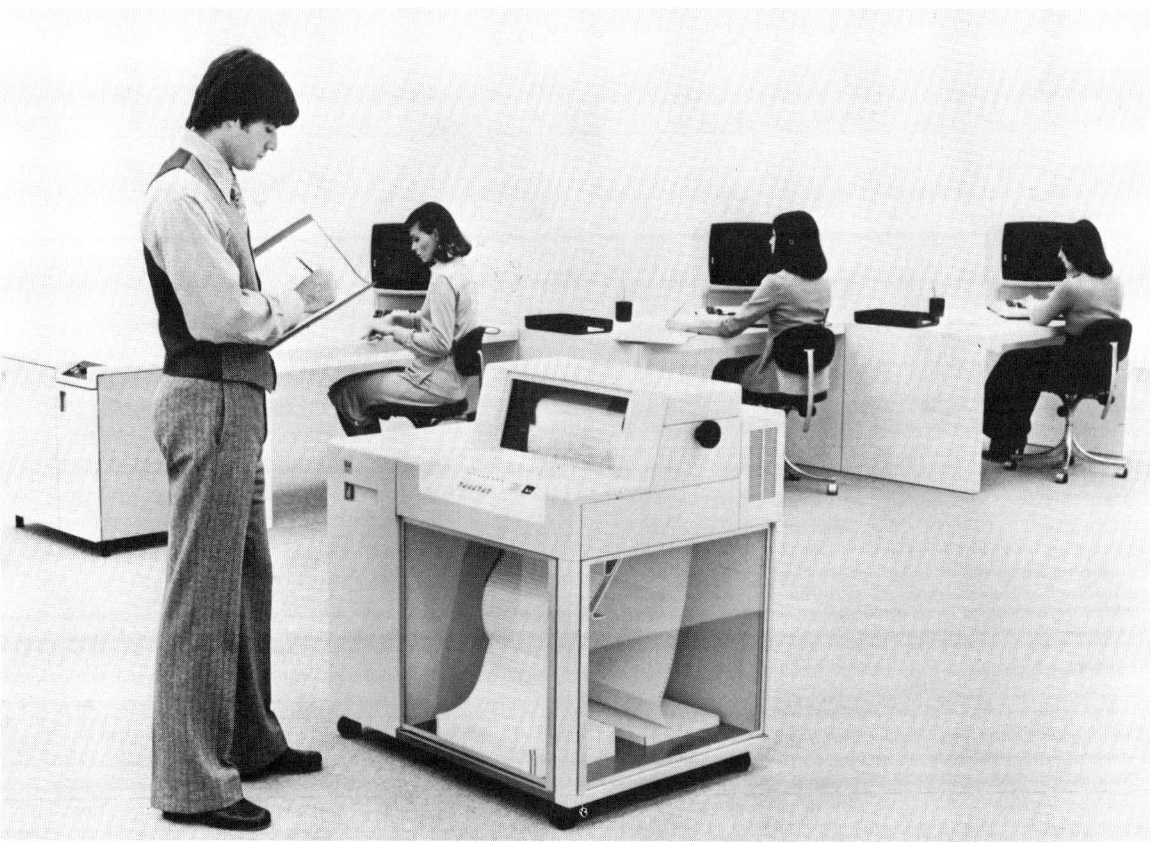

Figure 60.21 A cluster of video terminals and printers, with controller (reproduced by kind permission of IBM UK Ltd)

be translated into code which can operate upon the computers' instruction set.

It would be possible, of course, to buy computer hardware and then set out to write every program you needed. However, this would take a very long time indeed. Most users require their system to perform the same set of basic functions such as reading, printing, storing and displaying data, controlling simultaneous processes, translating program and many others. Consequently, most computers are supplied with pre-written programs to carry out these functions. They fall into three basic categories: operating systems, data management systems and language translators.

60.18.2 The operating system

The operating system sits between the application program designed to solve a particular problem and the general purpose hardware. It allocates and controls the system's resources such as the CPU memory, storage and input/output, and allocates them to the application program or programs.

Part of the operating system will be permanently resident in main memory and will communicate with the operator and the programs that are running. The functions it will carry out will typically be:

(*a*) The transfer into memory of non-resident operating system routines.

(*b*) The transfer into memory of application programs or parts of application programs. In some cases, there is insufficient memory to hold an entire program and so little used portions are held on disk and 'overlayed' into memory as they are required.

(*c*) The scheduling of processor time when several programs are resident in memory at the same time.

(*d*) The communication between tasks. For ease of programming, a large program may be broken down into sections known as tasks. In order to complete the application it may be necessary to transfer data from task to task.

(*e*) Memory protection, ensuring that co-resident programs are kept apart and are not corrupted.

(*f*) The transfer of data to and from input and output devices.

(*g*) The queuing of input/output data until the appropriate device or program is ready to accept it.

There are several ways to use the resources of a computer system and each makes different demands on an operating system. The four main distinctions are as follows:

Batch processing. This was the original processing method and is still heavily used where large amounts of data have to be processed efficiently. Data are transcribed onto some input medium such as punched cards or magnetic tape, along with some checking data such as column to tab, and then run through the system to produce, typically, a printed report. Classical batch jobs include such applications as payroll and month and statement runs.

Batch operating systems require a command language (often known as JCL—job control language) which can be embedded between the data, and which will load the next program in the sequence. Jobs are frequently queued on disc before being executed and the operating system may offer the facility of changing the sequence in which jobs are run, either as a result of operator intervention or as a result of pre-selected priorities.

The advantage of batch processing is its efficiency in processing large amounts of data. The major disadvantage is that once a user has committed his job he must wait until the cycle is completed before he receives any results at all. If they are not correct he must re-submit his job with the necessary amendments.

Interactive processing. Ths involves continuous communication between the user and the computer—usually in the form of a dialogue. The user frequently supplies data to the program in response to questions printed or displayed on his terminal whereas in batch processing all data must be supplied, in the correct sequence, before the job can be run.

A single operator using a keyboard does not use the power of a computer to any more than a fraction of its capacity. Consequently, the resources of the system are usually shared between many users in a process known as *time sharing.* This should not be apparent to the individual user who should receive a response to his request in two or three seconds. Time sharing, as the name suggests, involves the system allotting 'time slices', in rotation, to its users, together with an area of memory. Some users may have a higher priority than others and so their requests will be serviced first. However, all requests will be serviced eventually.

Requirements of interactive time-sharing operating systems are efficient system management routines to allocate, modify and control the resources allocated to individual users (CPU time and memory space) and a comprehensive command language. This language should be simple for the user to understand and should prompt the inexperienced operator whilst allowing the experienced operator to enter his commands swiftly and in abbreviated format.

Transaction processing. This is a form of interactive processing which is used when the operations to be carried out can be predefined into a series of structured transactions. The communication will usually take the form of the operators 'filling out' a form displayed on the terminal screen, a typical example being a sales order form. The entered data are then transmitted as a block to the computer which checks the data and sends back any incorrect fields for correction. The options available to the operator will always be limited and he or she may select the job to be performed from a 'menu' displayed on the screen.

Typical requirements of a transaction processing operating system are:

(*a*) Simple and efficient forms design utilities.

(*b*) The ability to handle multidrop terminals (terminals 'dropped off' a communications line). This may be achieved by a time-sharing approach or by the central processor 'polling' the terminals in search of a 'message ready' signal.

(*c*) Efficient file management routines, since many users will be accessing the same files at the same time.

(*d*) Comprehensive journaling and error recovery. Journaling is the recording of transactions as they occur, so that in the event of a system failure the data files can be updated to the point reached at the moment of failure.

Real time. It is an expression sometimes used in the computer industry to refer to interactive and transaction processing environments. Here it is used to mean the recording and control of processes. In such applications, the operating system must respond to external stimuli in the form of signals from sensing devices. The system may simply record that the event has taken place, together with the time at which it occurred, or it may call up a program which will initiate corrective action, or it may pass data to an analysis program.

Such a system can be described as 'event' or 'interrupt' driven. As the event signal is received, it will interrupt whatever processing is currently taking place, provided that it has a higher priority. Interrupt and priority handling are key requirements of a real-time operating system. Some operating systems may offer the user 16 possible interrupt levels and the situation can arise in which a number of interrupts of increasing priority occur before the sysrem can return to the program that was originally being executed. The operating system must be capable of recording the point reached by each interrupted process so that it can return to each task according to its priority level.

There are some concepts which are common to most operating systems.

(a) *Foreground/background.* The simplest form of processing is 'single user', either batch or interactive. However, a more effective use of the computer resources is to partition the memory into two areas. One, background, is used for low priority, interruptable programs. The other, foreground, is occupied by a program requiring faster response to its demands. The latter will therefore have higher priority.

(b) *Multiprogramming* is an extension of foreground/background in which many jobs compete for the system's resources rather than just two. Only one task can have control of the CPU at a time. However, when it requires an input or output operation, it relinquishes control to another task. This is possible because CPU and I/O operations can take place simultaneously. For example a disk controller, having received a request from the operating system, will control the retrieval of data thus releasing the CPU until it is ready to pass on the data it has retrieved.

(c) *Bootstrapping* (booting). The operating system is normally stored on a systems disk drive used solely for this purpose. Its heavy usage makes it unwise to mix operating system modules and data on the same device. When the computer is started up, the monitor (the memory resident position of the operating system) must be read from storage into memory. The routine which does this is known as the bootstrap.

(d) *System generation* (sysgen). When a computer is installed or modified, the general purpose operating system has to be tailored to the particular hardware configuration on which it will run. A sysgen defines such things as the devices which are attached to the CPU, the optional utility programs which are to be included and the amount of memory available and the amount to be allocated to various processes.

It is unlikely that any single operating system can handle all the various processing methods if any of them is likely to be very demanding. An efficient batch processing system would not be able to handle the multiple interrupts of a real-time operating system. There are, however, multipurpose systems which can handle batch and interactive, interactive and real time.

(e) *Data management software.* Data to be retained are usually held in auxilliary storage rather than in memory, since if it were held in memory it would be lost when the system was turned off and, moreover, even with current memory prices, the cost would prove prohibitive. To write and retrieve the data quickly and accurately requires some kind of organisation and this is achieved by data management software. This is usually provided by the hardware manufacturer, although independent software houses do sell such systems which they claim are more efficient or more powerful or both.

The most commonly used organisational arrangement for storing data is the file structure. A file is a collection of related pieces of information. An inventory file, for example would contain information on each part stored in the warehouse. For each part would be held such data as the part number, description, quantity in stock, quantity on order and so on. Each of these pieces of data is called a field. All the fields for each part form a record and, of course, all the inventory records together constitute the file.

The file is designed by the computer user, though there will usually be some guidelines as to its size and structure to aid swift processing or efficient usage of the storage medium. With file management systems the programs using the files must understand the type of file being used and the structure of the records with it. There are four types of file organisation, sequential, relative, direct and indexed.

(a) *Sequential file organisation.* Before the widespread use of magnetic storage devices, data were stored on punched cards. The program would cause a record (punched card) to be read into memory, the information was updated and a new card punched. The files thus created were sequential, the records being stored in numeric sequence. A payroll file, for example would contain records in employee number sequence.

This type of file organisation still exists on magnetic tapes and disks. However, the main drawback is that to reach any single record, all the preceding records must be read. Consequently, it has application only when the whole file is processed from beginning to end, and random enquiries to individual records are rarely made.

(b) *Relative file organisation.* Relative files permit random access to individual records. Each record is numbered according to its position relative to the first record in the file and a request to access a record must specify its relative number. Unfortunately, most user data, such as part number, order number, customer number and so on, does not lend itself to such a simplistic numbering system.

(c) *Direct (hashed) file organisation.* This is a development of the relative file organisation and is aimed at overcoming its record numbering disadvantage. The actual organisation of the file is similar. However, a hashing algorithm is introduced between the user number identifying a particular record and the actual relative record number which would be meaningless to the user. The algorithm is created once and for all when the system is designed and will contain some arithmetic to carry out the conversion (*Figure 60.22*).

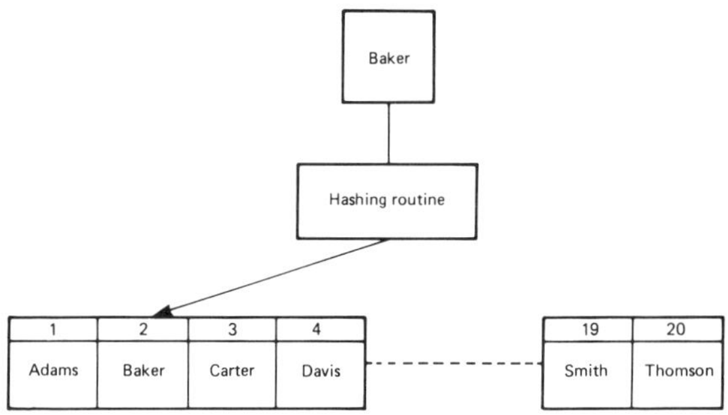

Figure 60.22 Direct (hashed) file organisation

This file organisation permits very fast access but it does suffer from the disadvantage that most algorithms will occasionally arrive at the same relative record number from different user record identification numbers, thus creating the problem of 'synonyms'. To overcome this problem the file management software must look to see if the record position indicated by the algorithm is free. If it is a new record can be stored there. If it is not a synonym has occurred and the software must look for another available record position. It is, of course, necessary to create a note that this has occurred so that the synonym can subsequently be retrieved. This is usually achieved by means of pointers left in the original position indicating the relative record number of the synonym.

The user numbering possibilities permitted with direct files may be more acceptable to the user since they are not directly tied to the relative record number. However, the need for an algorithm means that these possibilities are limited. In addition the design of the algorithm will affect the efficiency of recording and retrieval since the more synonyms that occur the slower and more cumbersome will be these operations.

(d) *Indexed file organisation.* The indexed method of file organisation is used to achieve the same objectives as direct files, namely the access to individual records by means of an identifier known to the user, without the need to read all the preceding records.

It uses a separate index which lists the unique identifying fields (known as keys) for each record together with a pointer to the location of the record. Within the file the user program makes a request to retrieve part number 97834, for example. The indexed file management software looks in the index until it finds the key 97834 and the pointer it discovers there indicates the location of the record. The disadvantage of the system is fairly apparent; it requires two accesses to retrieve one record and is therefore slower than the direct method (assuming a low incidence of synonyms in the latter). However, there are a number of advantages.

(i) It is possible to access the data sequentially as well as randomly, since most data management systems chain the records together in the same sequence as the index by maintaining pointers from each record to the next in sequence. Thus we have indexed sequential or ISAM (indexed sequential access method) files.

(ii) Depending on the sophistication of the system, multiple keys may be used, thus allowing files to be shared across different applications requiring access from different key data (*Figure 60.23*).

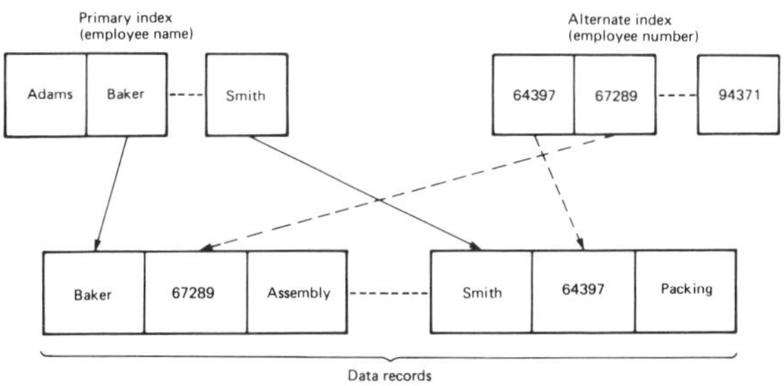

Figure 60.23 Multi-key 'ISAM' file organisation

(iii) Additional types of keys can be used. Generic keys can be used to identify a group of like records. For example, in a payroll application, employee number 74639 may identify K Jones. However, the first two digits (74) may be used for all employees in the press shop. It is therefore possible to list all employees who work in this department by asking the software to access the file by generic key.

Another possibility is that of asking the system to locate a particular record that contains the key value requested, or the next highest, if the original cannot be found. This is known as using approximate keys.

(iv) Most computer manufacturers provide multi-key ISAM systems and so the user does not need to concern himself about the mechanics of data retrieval.

60.18.3 Data base management

Files tend to be designed for specific applications. As a result, the same piece of information may be held several times within the same system. This has two disadvantages. Firstly it is wasteful of space and effort. Secondly it is very difficult to ensure that the information is held in its most recent form in every location.

It is, of course, possible to share files across applications. However, a program usually contains a definition of the formats of the data files, records and fields it is using. Changes in these formats necessitated by the use of the data within new programs will result in modifications having to be made in the original programs.

The data base concept is designed to solve these problems by separating the data from the programs which use it. The characteristics of a data base system are: a piece of data is held only once; the data are defined so that all parts of the organisation can use them; it separates data and their description from application programs; it provides definitions of the logical relationships between records in the data so that they need no longer be embedded in the application programs; it should provide protection of the data from unauthorised changes and from hardware and software. The data definitions and the logical relationships between pieces of data (the data structures) are held in the schema (*Figure 60.24*).

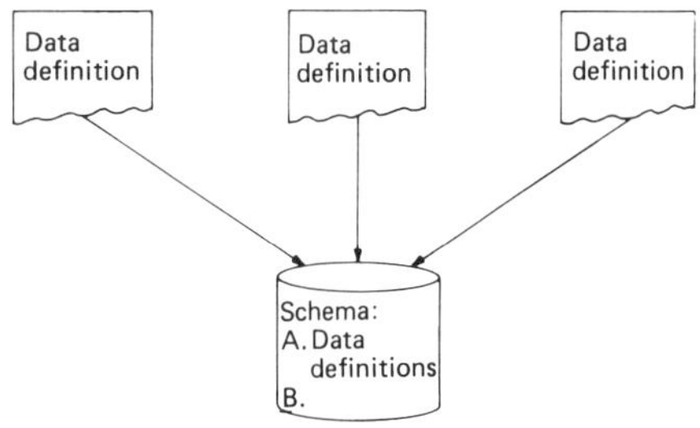

Figure 60.24 The schema

The database is divided into realms—the equivalent of files, and the realms into logical records. Each logical record contains data items which may not be physically contiguous.

Records may be grouped into sets which consist of owner and member records. For example a customer name and address record may be the owner of a number of individual sales order records.

When an application is developed a subschema is created defining the realms to be used for that application. The same realm can appear in other subschemas for other applications (*Figure 60.25*).

There are three major definitions of the logical relationships between the data (*Figure 60.26*).

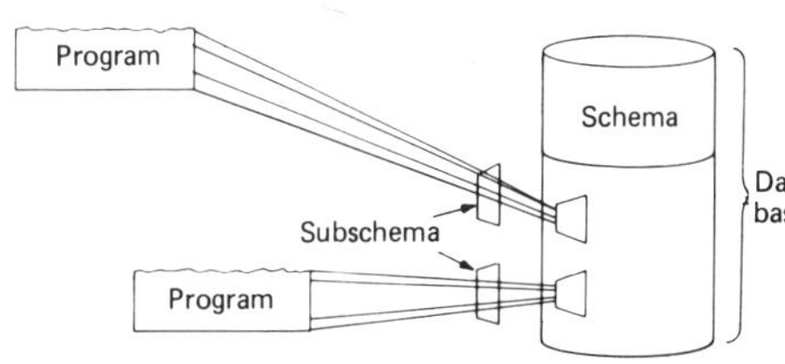

Figure 60.25 The subschema

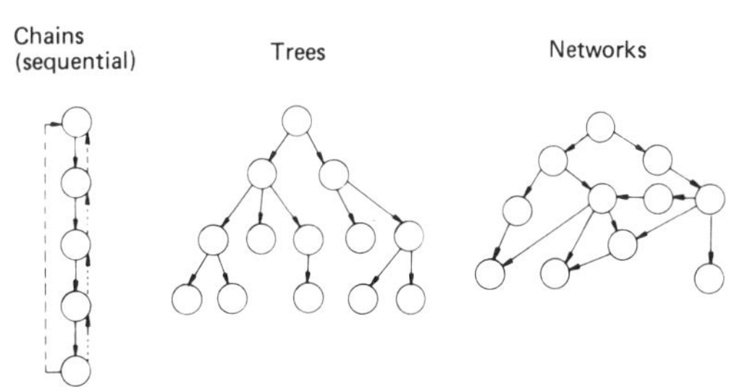

Figure 60.26 The three types of logical relationships

(*a*) *Sequential* (sometimes known as chain). Here each record is related only to the immediately preceding and following records.

(*b*) *Tree or hierarchical.* In this structure each record can be related to more than one record following it. However, records along separate 'branches' are not directly linked with each other and the relationship can be traced only by travelling along the branches.

(*c*) *Networks.* These are the most complex structures. They are effectively groups of trees where records can be related across branches. Any record can, in fact, be related to any other.

Because, within a data base management system, data are separated from the programs which use it, the data are regarded as a corporate asset. Management of this asset is in the hands of a data base administrator. He maintains the schema and works with application programmers to define the parts of the data base to which they may have access and to help them create subschemas for their particular applications.

Codasyl began to take an interest in developing data base standards in 1965 when the data base task group was formed. There now exist Codasyl standards for data base design.

60.18.4 Language translators

A programming language is a convention comprising words, letters and symbols which have a special meaning within the context of the language. However, programs have to be translated into the binary language understood by computers. The programmer writes a 'source' program which is converted by the language translator into an 'object program'. Usually during this process, checks are made on the syntax of the source program to ensure the programmer has obeyed the rules. The errors discovered will be noted, usually in two categories—terminal and warning. Terminal errors are those which will prevent the program running. Warning errors indicate that the translator would have expected something different, but that it may not actually be an error. Program errors are known as 'bugs' and the process of removing them, as 'debugging'.

Programs are stored in libraries and usually a program will be stored in both its object source and formats. The working program is the object program but when changes have to be made, these will be made to the source program which will then be translated to produce a new object program.

There are two kinds of language translators, assemblers and compilers.

Assemblers. An assembler is a language processor designed for use on a particular type of computer. In assembly language there is a one-to-one relationship between most of the language mnemonics and the computer binary instructions, although pre-defined sets of instructions can be 'called' from the assembly program.

There are four parts to an assembly language instructions:

(*a*) *Label.* This is a name defined by the programmer. When he wants to refer to the instruction he can do so by means of the label.

(*b*) *Operation code.* This will contain a 'call or an instruction mnemonic. If a call is used the assembler will insert a pre-defined code during the assembly process. If the programmer used a mnemonic this will define the operation to be carried out.

(*c*) *Operand.* This represents the address of the item to be operated on. An instruction may require one or two operands.

(*d*) *Comments.* This is an optional field used for ease of interpretation and correction by the programmer.

Assembly languages are generally efficient and are consequently used for writing operating systems and routines which require particularly rapid execution. However, they are machine dependent, slow to write and demanding in terms of programmer skill.

Compilers. These are used to translate high-level languages into binary code. These languages are relatively machine independent, though some modifications are usually required when transferring them from one type of computer to another. The instructions in the high-level language do not have a one-for-one relationship with the machine instructions.

Most compilers read the entire source program before translating. This permits a high degree of error checking and optimisation. An An incremental compiler, however, translates each statement immediately into machine format. Each statement can be executed before the next is translated. Although it does not allow code optimisation it does check syntax immediately and therefore permits the programmer to correct errors as they occur.

60.18.5 Languages

Basic (beginners' all-purpose symbolic instruction code). It is an easy to learn, conversational programming language which enables beginners to write reasonably complex programs in a short space of time. The growth in the popularity of time-sharing systems has increased its use to the point where it is used for a whole range of applications from small mathematical problems, through scientific and engineering calculations and even to commercial systems.

A Basic program consists of numbered statements which contain English words, words, symbols and numbers such as 'LET', 'IF', 'PRINT', 'INPUT', * (multiply), + and so on.

Basic was developed at Dartmouth College and while there is a standard Basic there are many variations developed by different manufacturers.

Fortran (formula translation). It originated in the 1950s and was the first commercially available high-level language. It was designed for technical applications and its strengths lie in its mathematical capabilities and its ability to express algebraic expressions. It is not particularly appropriate when the application requires a large amount of data editing and manipulation.

In 1966 an attempt was made by ANSI to standardise the Fortran language. However, manufacturers have continued to develop their own extensions.

A Fortran program consists of four types of statement:

(*a*) Control statements (such as GOTO, IF, PAUSE and STOP) control the sequence in which operations are performed.

(*b*) Input/output statements (such as READ, WRITE, PRINT, FIND) cause data to be read from or written to an I/O device.

(*c*) Arithmetic statements, such as * (multiplication), ** (exponentiation), / (division), perform compulation.

(*d*) Specification statements define the format of data input or output.

Cobol (common business-oriented language) is the most frequently used commercial language. The first Codasyl (Conference of Data Systems Languages) specifications for Cobol were drawn up by 1960, the aims of which were to create a language that was English-like and machine independent.

Cobol is a structured language with well defined formats for individual statements. A Cobal program consists of four divisions:

(*a*) Identification division which names and documents the program.

(*b*) Environment division which defines the type of computer to be used.

(*c*) Data division which names and describes data items and files used.

(*d*) Procedure division which describes the processing to be carried out.

The 'sentences' within Cobol can contain 'verbs' such as ADD, SUBTRACT, MULTIPLY and DIVIDE and are readable in their own right. For example, in an invoicing program you may find the line:

IF INVOICE TOTAL IS GREATER THAN 500 THEN GO TO DISCOUNT ROUTINE

As a result, by intelligent use of the language the programmer can produce a program which is largely self documenting. This is a significant advantage when modifications have to be made subsequently, possibly by a different programmer.

A general purpose, structured language like Cobol is not as efficient in terms of machine utilisation as assembly language or machine code. However, in a commercial environment, programmer productivity and good documentation are generally the most important factors. The calculations are usually not complex and therefore do not require great flexibility in terms of number manipulation.

PL1 (programming language one). Primarily an IBM language, PL1 was introduced to provide greater computational capabilities than Cobol but better file handling than Fortran.

APL (a programming language). Introduced through IBM, APL is generally used interactively. At one end of the scale it allows the operator to use his terminal as a calculator. At the other, it enables him to perform sophisticated operations such as array manipulation with the minimum of coding.

RPG (report program generator) was introduced by IBM for 360 and System 3 machines originally. A commercially orientated, very structured language, it processes records according to a fixed cycle of operations, although developments have been made in the language to make it more suitable for transaction processing environments.

In borad terms, the increased speed of processing and data handling currently available, and the low cost of memory have reduced the pressures on programmers to code for maximum speed and efficiency. It is frequently more economic to spend more money on hardware than to allow programmers to spend time optimising the performance of their programs.

This, coupled with the shortage of trained programmers has resulted in increased emphasis on simple languages, good program development tools and a general emphasis on programmer productivity.

Further reading

RADIN, H. and RAUTENBERG, L., 'Interactive Software Automatically Translate Tasks into Programs', *Electronics*, August 25 (1982)

FREEDMAN, D., 'Optical Disks; Promises and Problems', *Mini-Micro Systems*, October (1982)

KESTER, W. A., 'Design of Raster Scan Graphics Systems', *Digital Design*, August (1982)

MANUEL, T., 'Molding Computer Terminals to Human Needs', *Electronics*, June 30 (1982)

SUZUKI, J., 'Recent Trends in Raster-Scan Graphic Displays', *JEE*, June (1982)

LERNER, E. J., 'Automating Programming', *IEEE Spectrum*, August (1982)

BELCASTRO, R. J., 'Specification Template Speeds Software Design', *EDN*, October 27 (1982)

KENEALY, P., 'Personal Computers: a New Generation Emerges', *Mini-Micro Systems*, August (1982)

CHANG, S. S. L., *Fundamentals Handbook of Electrical Computer Engineering*, John Wiley & Sons (1982)

MARTINI, N. J., ELLEMENT, D. M. and MA, P. L., 'Low-Cost Plotter Electronics Design', *Hewlett-Packard Journal*, December (1982)

MALLACH, E. G., 'Computer Architecture', *Mini-Micro Systems*, December (1982)

REED, J. S., 'Computer Graphics', *Mini-Micro Systems*, December (1982)

61 Office Communication

J A Dawson, CChem, MRSC
Manager Special Materials Section
Rank Xerox Ltd
(Sections 61.1–61.3)

R F G Linford, BSc, CEng, MIEE
Engineering Manager, Information Processing
Rank Xerox Ltd
(Sections 61.8 and 61.9)

R C Marshall, MA, CEng, FIEE
Engineering Manager, Facsimile Products
Rank Xerox Ltd
(Sections 61.4–61.7)

Contents

61.1 Basic principles of xerography

The successful, commercial exploitation of xerography has resulted from the integration of a number of electrostatic process steps. The creation of an electrostatic image, which allows development by charged, pigmented particles forms the basis of the xerographic process. These particles can then be transferred to plain paper and subsequently fixed to provide a permanent copy. The generation of the electrostatic image in all xerographic copiers, depends upon the phenomenon of photoconductivity, whereby a material's electrical conductivity will increase by many orders of magnitude under the influence of light. By initially charging the photoconductor to a high, uniform surface potential, light reflected from the background areas of the original cause charge decay in the corresponding area of the image. Thus, a charge pattern remains on the surface of the photoconductor corresponding to the original document.

The basic steps in the xerographic process are outlined in the following paragraphs and depicted in *Figure 61.1*.

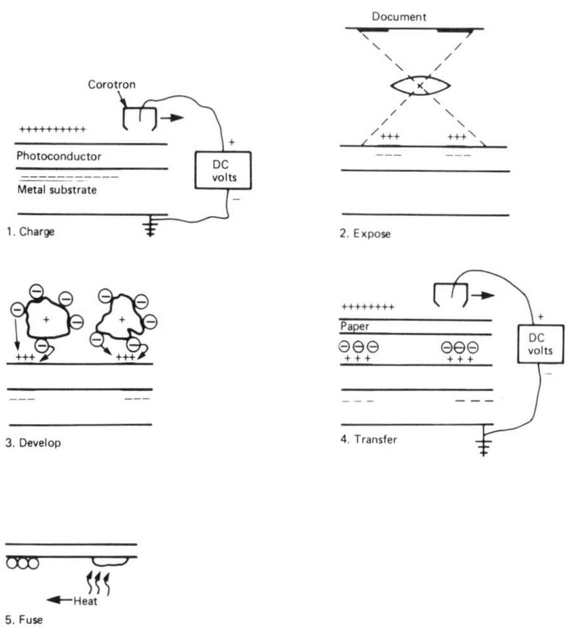

Figure 61.1 The basic steps of xerography

(i) Charging, or sensitisation of the photoconductor to a uniform surface potential of 800–1000 V is usually accomplished by a corotron. Literally a thin wire, stretched between terminals parallel to the photoconductor surface, and driven at approximately 8000 V, the corotron emits a corona of ions which deposit on the photoconductor surface.[1] The charged photoconductor must be kept in the dark to prevent discharge.

(ii) Exposure of the sensitised photoconductor to light reflected from the original to be copied generates the required electrostatic image. Voltage decay occurs via photon absorption by the photoconductor surface with the creation of an electron–hole pair. These separate under the influence of the electrostatic field, the electron neutralising a surface charge, and the hole, transported through the photoconductor, neutralises the corresponding image charge at the photoconductor–substrate interface. Selenium obeys the reciprocity law in that it responds to the product of light intensity and time, regardless of their individual values.[2] Typical light discharge and electrostatic contrast characteristics are shown in *Figure 61.2*.

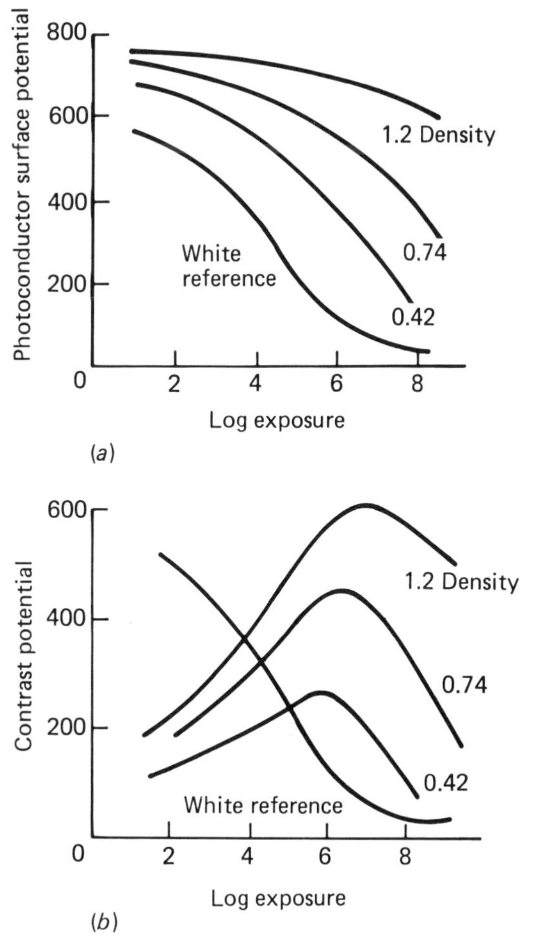

Figure 61.2 Photoconductor characteristics, (*a*) light discharge, (*b*) electrostatic contrast

(iii) Xerographic development of the latent electrostatic image renders the image pattern visible. To ensure selective development of the image, the black toner particles are charged to a polarity opposite to that of the photoconductor surface. Toner consists of finely dispersed carbon black in a thermoplastic polymer matrix. The toner particles are small to ensure that reasonable image, edge definition and resolution performance are obtained. Usually transported to the development zone via a carrier, it is the careful selection of carrier–toner pairs that ensures the correct charging, and hence development characteristics.

(iv) Transfer of the developed image from the photoconductor surface to plain paper is effected by a corotron. Similar in design and process to the charge corotron, positive charge is sprayed onto the back side of the paper which itself is in contact with the developed photoconductor surface. Sufficient fields are generated to ensure that most, but not all, of the toner will transfer to the paper.

(v) Fusing the image into the surface of the paper is accomplished by heat from a radiant fuser, or, a combination of heat and pressure from a fuser roll/back-up roll combination. It is this step that will dictate the rheological requirements of the thermoplastic resins used in toner manufacture.

61.2 Extension to a dynamic copier

The processes outlined in section 61.1 allow us to visualise the xerographic steps required to produce a copy via a plate photoconductor in the static mode. Dynamic operation of a copier places many constraints of space and geometry on the elements of process design and this in turn requires significant demands being made on the xerographic developers and photoconductors in use today. The trend towards faster, less expensive and smaller copiers will ensure that technology development will continue for some time to come. A schematic representation of a copier is given in *Figure 61.3*.

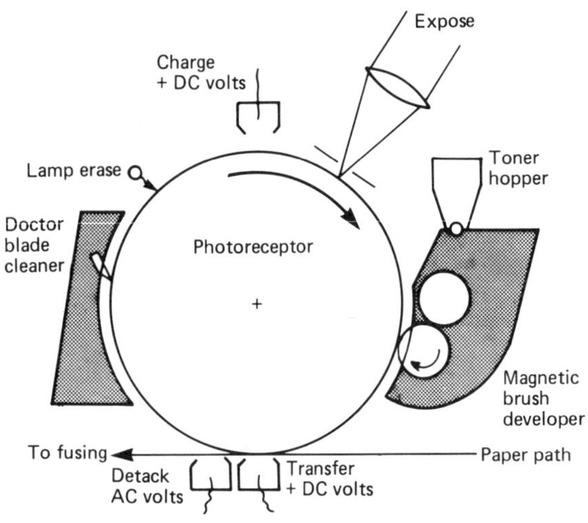

Figure 61.3 Xerographic copier schematic

Charging remains as discussed earlier with the rotation of the photoreceptor now providing the linear motion underneath the corotron. Exposure, represents more of a challenge. With stationary platen copiers, the original document is scanned and the reflected light is transmitted via lens, mirrors and exposure slit to the photoreceptor. Some small copiers today incorporate a moving platen and strip optics for cost considerations.

Development of the electrostatic image has been extensively studied following the trend towards faster, smaller copiers.[3] Initial copier designs utilised a cascade development system which poured developer over the rotating photoreceptor. Increasing process speeds demanded developers with smaller particle size (increasing surface area to mass ratio), to increase the toner carrying capacity. Cascading developer over the photoreceptor in the against mode brought some relief for further speed enhancement but the resultant development system remained bulky. The trend towards magnetic brush development systems commenced approximately ten years ago and immediately offered the advantages of compactness and enhanced developability. The latter, resulting mainly from the fibrous nature of the brush, and, the density of developer at time of photoreceptor contact, has led to almost all modern copiers using this form of development. A very recent extension of this concept is the single-component magnetic brush copiers used by some Japanese manufacturers and these are discussed in the section on materials.

Apart from the optimisation of currents and geometry, the process steps on transfer (and often detack), are direct analogues of the static environment.

Fusing the toner image onto the paper is one of the major contributors to power consumption within the copier and this has led to the abandonment of the early, inefficient, radiant or oven fusers. Centrally-heated polymer-coated, steel or aluminium rolls now effect fusing by a combination of heat and pressure. Release agent fluids are often used on these rollers to prevent toner offsetting from the paper to the polymeric surface.

In the dynamic mode, the photoreceptor has to be returned to a virgin condition prior to reaching the charge corotron and hence embarking on another cycle. If this is not the case, residual voltages will build up within the transport layer of the photoconductor and cyclic problems will rapidly manifest themselves and cause print to print variations. Following the transfer of image toner to paper, a residual image (some 5–20% of the developed toner), will remain on the photoreceptor surface. (Total transfer is avoided as it would increase print background levels.) Removal is effected by rotating brushes in some machines, soft webbed fabric in others but the popular choice for all small and mid-volume copiers today is the so-called doctor blade. A straight-edged polyurethane blade is angled against the photoreceptor and this scrapes the residual toner from the surface. The clean photoreceptor subsequently passes underneath an AC corotron or an erase lamp to remove all vestiges of voltage fluctuations prior to charging once again.

61.3 Xerographic materials

61.3.1 Photoreceptors

The xerographically active part of the photoreceptor is the photoconductor, generally amorphous selenium with one or two minor components, supported on a cylindrical drum of aluminium or a flexible metallic belt. The photoconductor layer is deposited onto the metallic substrates in vacuum coaters. An electrical barrier layer is essential between the 60 μm thick photoconductor and the metallic substrate to prevent charge injection from the latter when the photoconductor has reasonable electron mobility.

Early experimental photoreceptor[2] plates used a variety of photoconductors such as sulphur, anthracene, zinc oxide or sulphide, cadmium sulphide or selenide. The first xerographic copiers were based on extremely pure selenium photoreceptors and these performed adequately at the low cycling speeds. As copiers became faster and smaller, two major improvements were needed in terms of photoreceptor characteristics; greater photosensitivity, and more stable cycling characteristics.

Photosensitivity increases were obtained relatively easily by incorporating small percentage additions of arsenic or tellurium to the deposited film of selenium. Sensitivity increases result from an extension in the photoconductor's spectral response beyond the 550 nm limit of selenium, and enhanced response at smaller wavelengths (see *Figure 61.4*).

Cyclic stability proved more difficult to overcome but is generally improved by halogen dopants at the ppm level in the deposited photoconductor layer. However, careful balancing is required. The presence of chlorine (the most commonly used dopant), at the photoconductor/substrate interface layer, reduces the cycle up of residual potentials as desired. However, if the dopant migrates to the outward surface of the photoconductor (often referred to as the generator layer), increasing fatigue occurs. In this case, halogen-rich photoconductor can trap electrons in the upper layers and partially discharge the photoreceptor during the charging and expose stations in the copier. The lower apparent charging voltage will manifest itself as loss of density on the final print.

Organic photoconductors are in use today and present their own unique problems but remain a very poor second or third in usage compared with the selenium- or selenide-based

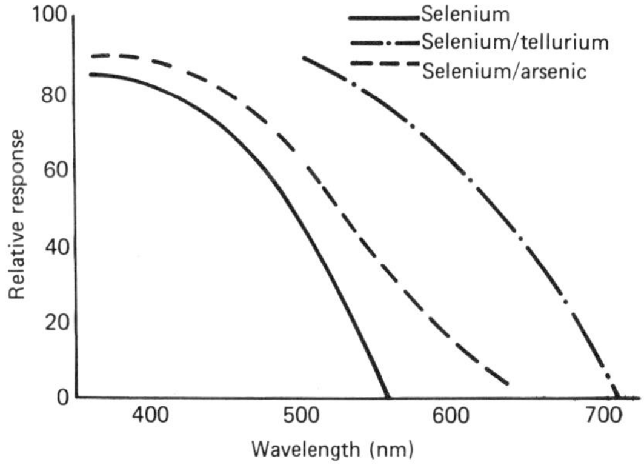

Figure 61.4 Spectral response of typical photoconductors

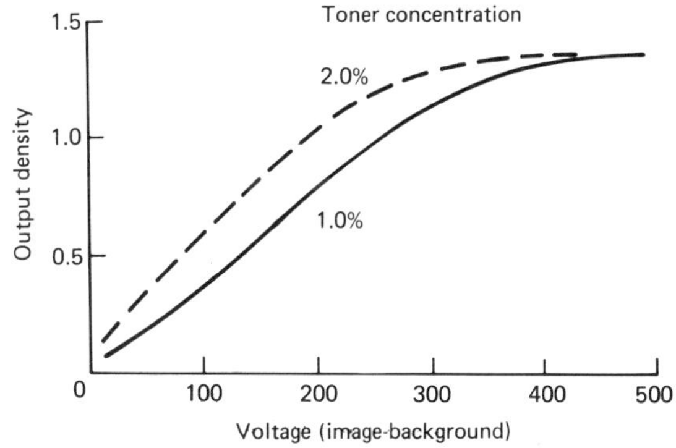

Figure 61.5 Development characteristic as a function of toner concentration

photoconductors. The ultimate advantage of cost will probably justify a trend towards organic materials and away from selenium-based photoreceptors.

61.3.2 Toners

Dry xerographic toners have not changed appreciably since the introduction of copiers with an almost universal dependence upon carbon black for pigmentation, finely dispersed in a styrene/methacrylate polymer matrix. Other polymers (polystyrene, styrene/butadiene, polyesters) have been used, the choice often following the desire for different fusing systems to be employed.

Requirements for toner are that it will charge acceptably, to allow development and transfer; it can be cleaned from the photoreceptor surface, and it will fuse readily into the paper. The advent of doctor blade cleaning systems has necessitated additives in the toner to lubricate the photoreceptor surface and prevent blade chatter. Small amounts of zinc stearate are often added to solve this problem.

Toners are melt mixed either in Z-blade mixers or extruders, mechanically crushed to a small size and air-micronised to a final average particle size distribution of 10–15 μm by weight, with virtually all particles being between 5 and 35 μm.

61.3.3 Carriers

Performing two basic functions, the carrier acts as a conveying system for the toner, and provides a charging surface to effect triboelectrification. It is this triboelectric charge that ensures selective image development on the surface of the photoconductor. Developers usually contain 1–2% by weight of toner, the development capability being a function of toner concentration (*Figure 61.5*).

In early copiers, xerographic carriers were based on smooth, round sand or glass beads up to 600 μm in diameter. The resultant developer was cascaded over the electrostatic image. Classically, xerographic development is electrostatic field, and not surface potential, dependent and these cascade systems (open development), suffered two major limitations. As there were only fringe fields present, solid areas on originals would not reproduce (apart from the edges), and contrast enhancement occurred on development. The imposition of a development electrode above the surface of the photoconductor during the development process significantly improves the solid area performance.[4] The electrode, a positively biassed plate in cascade development, or, the development roll itself in magnetic brush systems, is set to a value some 100–200 V above the background potential on the

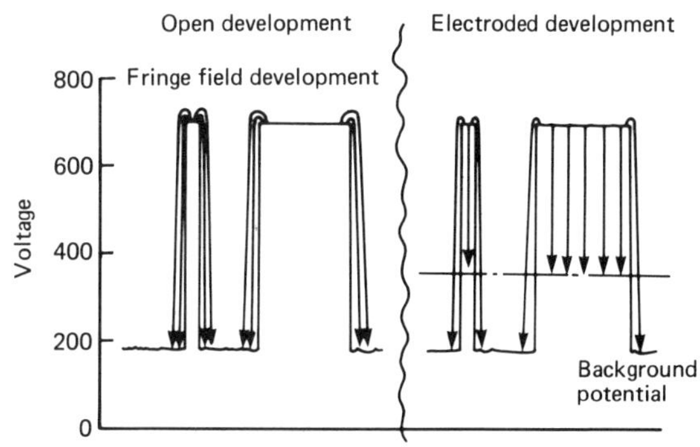

Figure 61.6 Open and electroded development

photoreceptor, see *Figure 61.6*. Significant gains in latitude are apparent with the electrode in place as background development is now suppressed by a cleaning field and image development is enhanced.

In order to benefit from the electroded development systems almost universally used today, and to enable increased development speed, carriers have become smaller and more conductive. Steel, or iron shot, in the 80–200 μm size range is now the most common core material in use although magnetite, sponge iron and ferrites have been employed. To produce carriers, the cores are lacquer coated with polymeric films to provide the triboelectric charging interface for the toners. Lacquers have been based on ethyl cellulose but methacrylates are more widely used today. Carrier lacquers and the toner formulation are chosen such that they charge to opposite polarities through physical contact. A triboelectric table can be generated such that any material on the list will acquire a negative charge when in contact with a material above it on the list.

Triboelectric series[5]
Polymethylmethacrylate
Ethylcellulose
Nylon
Cellulose acetate
Styrene-butadiene copolymer
Polystyrene
Polyvinylchloride
Polytetrafluoroethylene

Triboelectrification is a poorly understood phenomenon and in copiers that go faster and faster, has a limitation in terms of time dependence. Incoming toner to the developer housing can be transported to the development zone in less than 5 s. If inadequately charged or wrong sign toner is produced this will develop in background areas. Quantities of such poorly charged toner are minimised by design approaches in the developer housing to ensure adequate dynamic mixing. In order for any development system to function, the electrical forces attracting the toner to the electrostatic image must exceed all other forces acting on the toner. Toner deposition occurs by three-body contact (electrostatic image/toner/carrier) or by separation of toner from carrier particles on impact generating a toner cloud within the developer.

61.3.4 Single-component developers

A recently introduced modification of the magnetic brush development system utilises a single-component developer. By loading standard toner polymers with up to 50% magnetite, transport to the development zone is enabled magnetically with the magnetite also acting as the pigment. Charging of the toner particles can be by direct induction from a biassed interference bar or by allowing electrostatic induction from the electrostatic image itself. Some problems are immediately apparent with these technologies in that tight control of background and image voltages is needed to maintain development latitude. Also, the requirement for a given conductivitiy in the toner can diminish the transfer efficiency and lead to low-density prints and high toner consumption rates. Many other development processes are available for specific uses.[3]

61.3.5 Paper

The ability to use plain paper has perhaps been one of the major reasons for the success of xerographic copiers. However, the characteristics of the paper are by no means unimportant. Machines that feed sheets of paper often rely on the beam strength of the paper to effect stripping from the photoreceptor or fuser roll. Surface roughness is important in affecting the fixedness (fusing) of the toner image on the copy. Conductivity (especially in humid conditions) can markedly affect transfer.

61.4 Raster scanning

61.4.1 Introduction

The electronic office brings together two different kinds of technology. On the one hand there is the electronic computer, with its emphasis on data processing, employing a limited set of alphanumeric characters. On the other hand is the photocopier, capable of producing replicas of first-class appearance whether the original be English or Arabic, machine-generated or manuscript. A way has to be found to add this flexibility to work with any sort of image to an electronic office system intended for non-specialist use.

This problem was recognised and solved by those pioneers who devised raster scanners to encode and decode written characters and so allow their communication via a telegraph circuit—a process now known as facsimile communication. Today, raster-scanning document readers are being used for character and signature coding and recognition, whilst the raster-scanning output principle allows multi-font character and graphics printers to be designed for computers and word processors. These scanning devices and their associated electronic data processing and communication systems are the subject of this section.

61.4.2 Principles

An image can be dissected into picture elements (pixels or 'pels') for transmission, processing or storage, and reassembled from such elements for display or printing. Such conversion between a two-dimensional image and a serial data stream is usually accomplished by scanning in two orthogonal directions as shown in *Figure 61.7*. This principle is shared with television; the

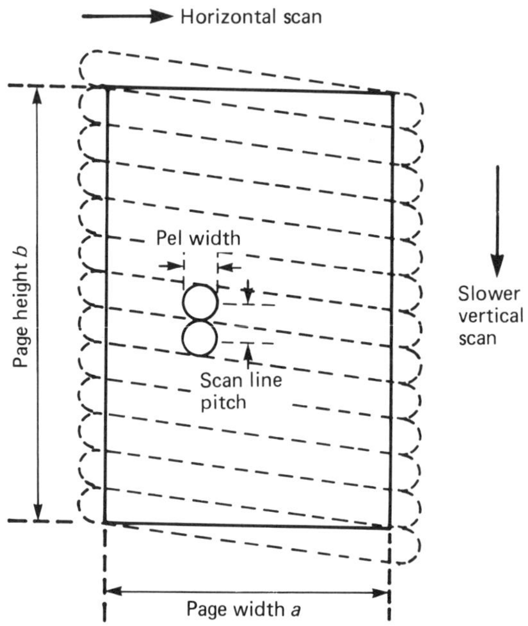

Figure 61.7 The raster scan principle. Horizontal resolution, $r_h \simeq 1$/pel width; vertical resolution, $r_v \simeq 1$/scan line pitch

essential differences lie in the use of paper or film-based input and output media and a frame scanning time increase of several orders of magnitude.

A mechanism that feeds the medium through the machine can also provide scanning in one direction. This has led to a multitude of economical designs in which the horizontal and vertical scanning mechanisms are quite different—a further contrast to TV practice. Very similar raster-scanning mechanisms can be used for image input and output and in many cases this leads to very cost-effective combined mechanisms. A wide variety of configurations are possible and will be discussed later, but first it is necessary to discuss the requirements of various applications.

61.4.3 Resolution

The spatial precision with which an image can be encoded or printed by a raster scanner is commonly measured in lines per millimetre or lines per inch. This contrasts with photographic usage, where line *pairs* per millimetre are normally quoted. The subjective appearance of an image depends on several factors. Firstly, since the scanning methods used for the horizontal and vertical directions are, in general, different the corresponding resolution figures will also be different. It is found that a worthwhile improvement may result if one dimension has up to twice the resolution of the other. This improvement is useful as it allows quality to be traded for speed by changing just the vertical resolution. *Table 61.1* lists typical resolution specifications for various applications. The performance of the human eye is such

Table 61.1 Typical resolution figures for raster scanners

Application	*Lines/mm (horizontal × vertical)*
Facsimile transmission: cost of prime importance	3.2 × 2.6
Facsimile, CCITT group 3	3.8 × 3.8
Office systems printer	12 × 12
Office printer with half-tones and colour	24 × 24
Graphic arts image correction	80 × 80

that about 8.5 lines/mm can be resolved at 400 mm reading distance.[6]

61.4.4 Kell factor

Suppose that a black dot is scanned by an aperture of the same width. There is then a chance that the dot will lie partly in the tracks of each of two successive scan lines and so be subsequently reproduced at twice its original width. Averaged over the whole document this effect reduces the resolution by a factor of about 0.7; this factor is known as the Kell factor.[6] It must be applied in the vertical direction (as defined in *Figure 61.7*) of all raster-scan systems. It also applies to the horizontal direction when lines of fixed reading elements such as the photodiode array of *Figure 61.10(d)* are used.

61.4.5 Grey scale

If shades of grey in the original document are to be faithfully reproduced, then the system optical density transfer characteristic must approximate to the form shown in *Figure 61.8* curve A. This will require the transmission and perhaps storage of

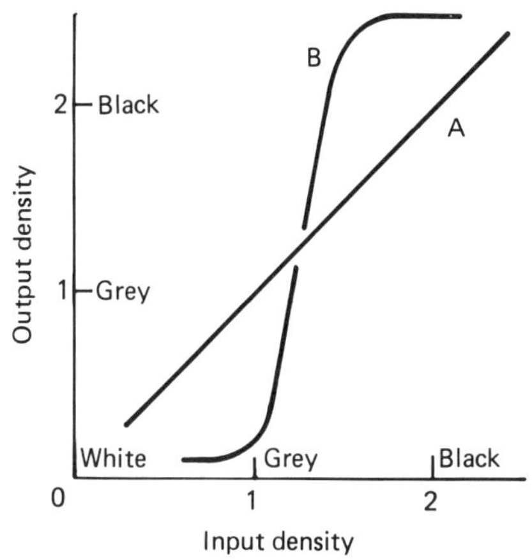

Figure 61.8 Optical density transfer characteristics. A, with grey scale; B, no grey scale

analogue or digitised analogue signals. The problem of reproducing analogue grey tones is usually avoided by substituting a pattern of variable-size black dots ('half-tones') to give variable percentage cover of the surface. However if the spatial frequency of these dot patterns is of the same order as that of the raster scan, 'More patterns' will result from the beating of these two frequencies. To avoid this higher resolution must be used. Half-tone grey scale print can be produced by raster printers.[7]

If, as is generally the case, shades of grey are not required, then imperfections in both white and black may be removed by non-linear electronic processing such as that shown by curve B in *Figure 61.8*. This corresponds to a 'high gamma' photographic system. The ease with which this processing can be adjusted to clean up an image is one of the advantages that raster-scanned electronic copiers can offer over page-parallel photocopiers.

61.4.6 Colour

An input scanner may deliberately be made blind to certain colours, such as the boxes that define the layout of a business form, by a suitable optical filter. The more general case of reading full-colour information requires the use of three appropriately filtered photodetectors.[8] Output printing in colour requires separate printing devices for each primary colour. When intermediate shades and combinations of colours are desired then the printer resolution must be such as to permit electronic half-tone generation.[7]

61.4.7 Multiple-element scanning

If, as shown in *Figure 61.7*, an image of dimensions $a \times b$ is to be converted into a serial data stream in time t by a single optical detector, and resolutions of r_h and r_v elements per unit length are required respectively in the horizontal and vertical scanning directions, then it follows that the scanning velocity (assuming that no time is wasted) equals abr_v/t. When high performance is required this velocity can become too high for mechanical scanners. Furthermore, the 'exposure time' available for each element of the image equals t/abr_hr_v which also constrains the design of fast, high-resolution scanners. Both restraints can be relaxed by configurations in which a number of reading or writing elements are in use at the same time, with electronic conversion to or from the serial bit stream. For example, in *Figure 61.10(d)* the exposure time of each element can approach t/bt_v, that is ar times greater than in the single-detector system.

61.4.8 Drum scanners

If the original document is wrapped around a drum (so that dimension a in *Figure 61.7* is circumferential) then the required scanning pattern becomes a spiral. This may easily be implemented by rotating the drum whilst moving the elemental scanner axially by a screw mechanism as shown in *Figure 61.9*. This simple principle has dominated raster-scanner design since its invention in 1850. Its optical arrangements are simple because the field of view need only cover one pel, and by using one of the direct writing processes to be discussed later, the same scanning mechanism can operate as a raster printer. The principle is

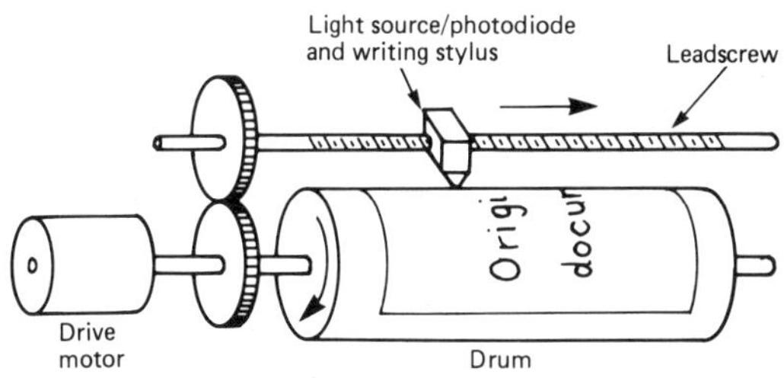

Figure 61.9 Drum scanner principle

therefore of considerable value to facsimile communication. However, the maximum scanning rate is limited by the need for precision balancing of mechanical components, and mechanisms must be provided to load, retain and subsequently unload the paper from around the drum.

The rotational speed of the drum must be tightly controlled. It is usual to employ a synchronous motor driven from an electronic power invertor whose frequency is derived from a quartz-crystal source. When varying resolution is to be provided by an adjustable 'gear ratio' between the drum and the leadscrew the latter is turned by a separate stepper motor driven from an adjustable frequency divider from the same source.

61.4.9 Flat-bed scanners

A basic alternative to the drum scanner uses the movement of the paper through the machine to provide the slower 'vertical' scanning action. The horizontal line actually being scanned is kept flat; hence the name 'flat bed'. A great variety of line scanning methods have been proposed for image input and printing, and some of these are discussed in the next two sections.

61.5 Image input scanners

61.5.1 Optics

Images are invariably sensed using visible light and a photodetector, with optical filtering added in order to approximate to the response of the human eye. Optical systems are angled so as to avoid the detection of light reflected from the surface of smooth documents. Commonly used light sources include xenon arc lamps and lasers as well as incandescent and fluorescent lamps. It is usually necessary to supply these with smoothed direct current or pulses synchronised to the scanning process in order to avoid modulation of the image. At least in high-resolution or high-speed scanners, the need to obtain sufficient illumination to give an adequate signal-to-noise ratio from the photodetector is a basic design constraint. The designer of lamp power supplies therefore aims to get the maximum possible useful output consistent with the required lamp life.

61.5.2 Line scan methods

Figure 61.10 shows a variety of methods by which one line of an input document may be scanned.

In *Figure 61.10(a)* the whole area of the document is illuminated (typically by one or two tubular fluorescent lamps) and the photodetector is arranged to 'see' a single elemental area by the lens. A rotating mirror then scans this elemental field of view across the page. Just as for the drum scanner, precise control of the scanning motor is required.

In *Figure 61.10(b)* the optical arrangement is reversed, and an elemental area of light is scanned across the page and one or more photodetectors arranged to view the whole area. Electronic scanning of a light-spot imaged from a cathode ray tube is shown in this example. This is an alternative that can out-perform rotating mirror scanning at high speeds. However, it is bulky, expensive and requires specialised high-voltage and electronic scanning supplies.

It is possible to scan both the light source *and* the photodetector aperture as shown in *Figure 61.10(c)*.

An attractive alternative is to electronically scan the outputs of an line array of CCDs or photodiodes. This has the advantages of multiple-element scanning discussed earlier. Typically, a monolithic array of 1728 elements is used to scan the width of an A4 page.

61.6 Output printers

61.6.1 Direct printers

Figure 61.11 shows examples of writing methods that produce a hard-copy by a single process step. Although these are inherently cheap they are difficult to optimise. In particular, most use special papers that look and feel wrong and produce images that may lack permanence.

The electro-percussive stylus, the raster-scanned equivalent of the typewriter, is shown in *Figure 61.11(a)*. It is slow and noisy.

The electroresistive principle shown in *Figure 61.11(b)* is widely used with drum scanners, particularly those used in facsimile. Its disadvantage is the ash and smell resulting from the erosion of the surface. It is limited to a resolution of about 4 lines/mm.

Ink jet printing[9,10] is capable of printing high-resolution images onto ordinary plain paper. The liquid ink is forced through a small aperture and is thus broken into fine droplets. *Figure 61.11(c)* shows how these droplets may be produced by ultrasonic vibration. Electrostatic fields may be used to accelerate and deflect the droplets.

The thermal process outlined in *Figure 61.11(d)* has recently come into widespread use for cheap printers for calculators, small computers and facsimile machines. Since the transfer of heat to blacken the paper is an inherently slow process, multiple-element arrays are usual. These arrays use resistors or diodes as heating elements and are energised by some sort of multiplexing arrangement to reduce the number of connections and driver circuits. Resolution is limited by constructional difficulties to about 8 lines/mm but arrays of up to 1728 elements are in common use.

61.6.2 Electrostatic printers

Electrostatic printers use 'dielectric paper' which acts as a capacitor. A pattern of electric charge corresponding to the image is laid onto the paper surface and this is 'developed' with liquid or powder ink ('toner') and then 'fixed' by heat or pressure just as in a photocopier. *Figure 61.12* shows the operating process of a roll-fed liquid-ink printer widely used as a graphic printer. The 'writing head' consists of an array of styli that are selectively driven to a voltage that ensures ionisation and consequent surface charging of the paper. The ionisation process provides a threshold (no charge is deposited at 400 V, full charge is deposited at 600 V) which allows the use of simple but specialised multiplexing techniques. This same threshold does, however, mean that analogue grey scale printing is hardly possible.

61.6.3 Optical intermediate printers

Light-sensitive materials such as photographic film and paper can be used as printout media by exposing them to an elemental spot of light that moves in the required raster pattern (*Figure 61.7*) whilst being modulated in intensity by the required video signal. Since subsequent development and fixing will be required the machine process is at least as complex as that needed for an electrostatic printer, but very high resolution and a good analogue grey scale can be obtained.

A crater lamp is a neon glow discharge tube shaped so that light from the crater in its cathode provides a source that can be focused onto an elemental area of the printout medium. *Figure 61.13(a)* shows how this may be applied to a drum scanner. The intensity is modulated by controlling the current and typically 1 to 75 mA might be required. Although this arrangement is very simple the scanning speed is limited by the low light output.

Figure 61.13(b) shows how light from a laser or xenon arc lamp may be modulated, and caused to scan along a line by a rotating

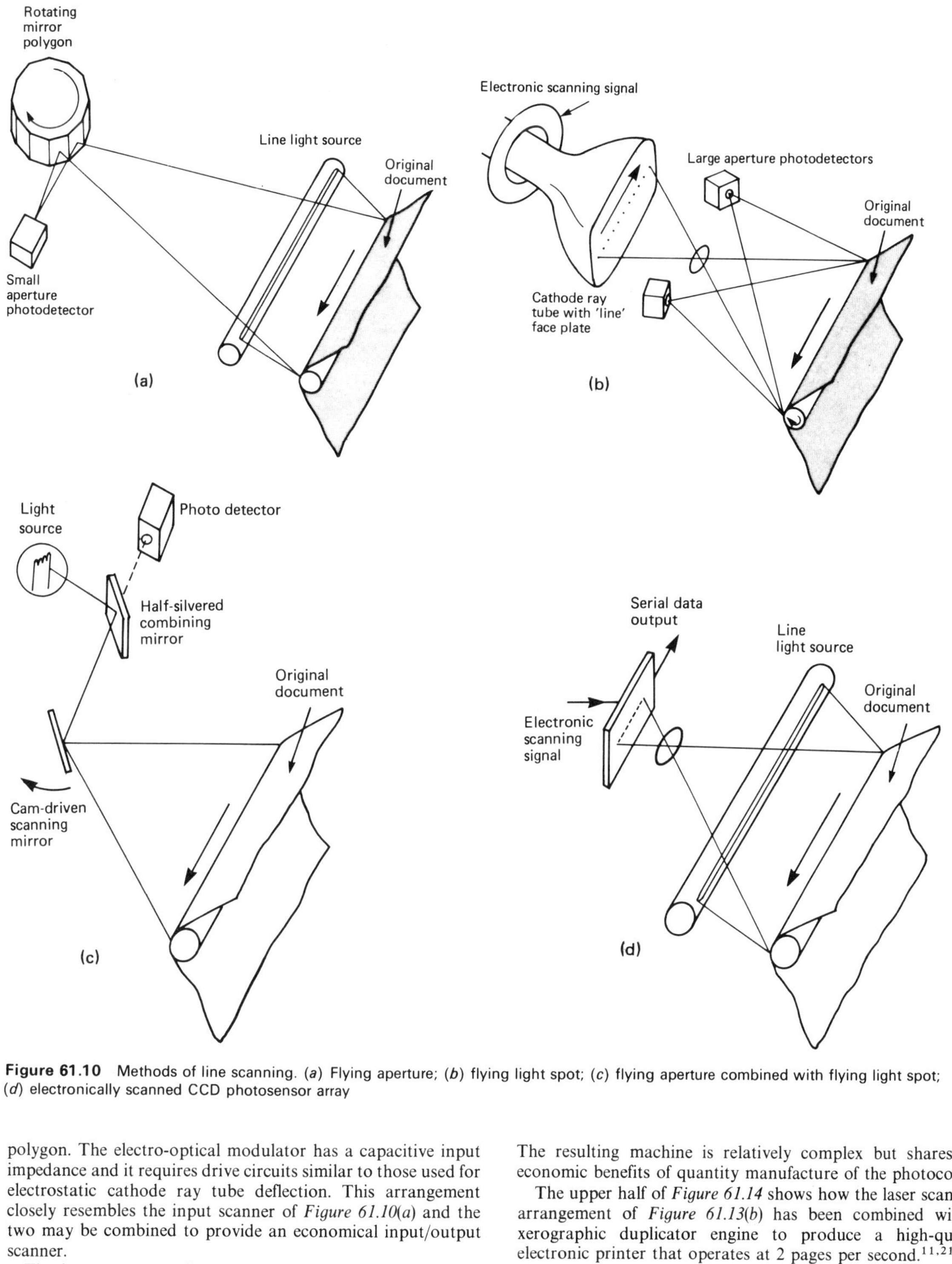

Figure 61.10 Methods of line scanning. (*a*) Flying aperture; (*b*) flying light spot; (*c*) flying aperture combined with flying light spot; (*d*) electronically scanned CCD photosensor array

polygon. The electro-optical modulator has a capacitive input impedance and it requires drive circuits similar to those used for electrostatic cathode ray tube deflection. This arrangement closely resembles the input scanner of *Figure 61.10*(*a*) and the two may be combined to provide an economical input/output scanner.

The input scanner of *Figure 61.10*(*b*) is also adaptable for output printing, by modulating the electron beam of the cathode ray tube by the video signal, and substituting light-sensitive copy paper for the original document shown.

Photocopiers may be adapted to work as raster-scan printers by using one of the above methods to expose the photoreceptor. The resulting machine is relatively complex but shares the economic benefits of quantity manufacture of the photocopier.

The upper half of *Figure 61.14* shows how the laser scanning arrangement of *Figure 61.13*(*b*) has been combined with a xerographic duplicator engine to produce a high-quality electronic printer that operates at 2 pages per second.[11,21]

61.6.4 Printer systems

The laser modulator of such a printer must be supplied with some 10 megabits of video in each half-second page exposure time. To store this whole page in bit-mapped form is inconveniently

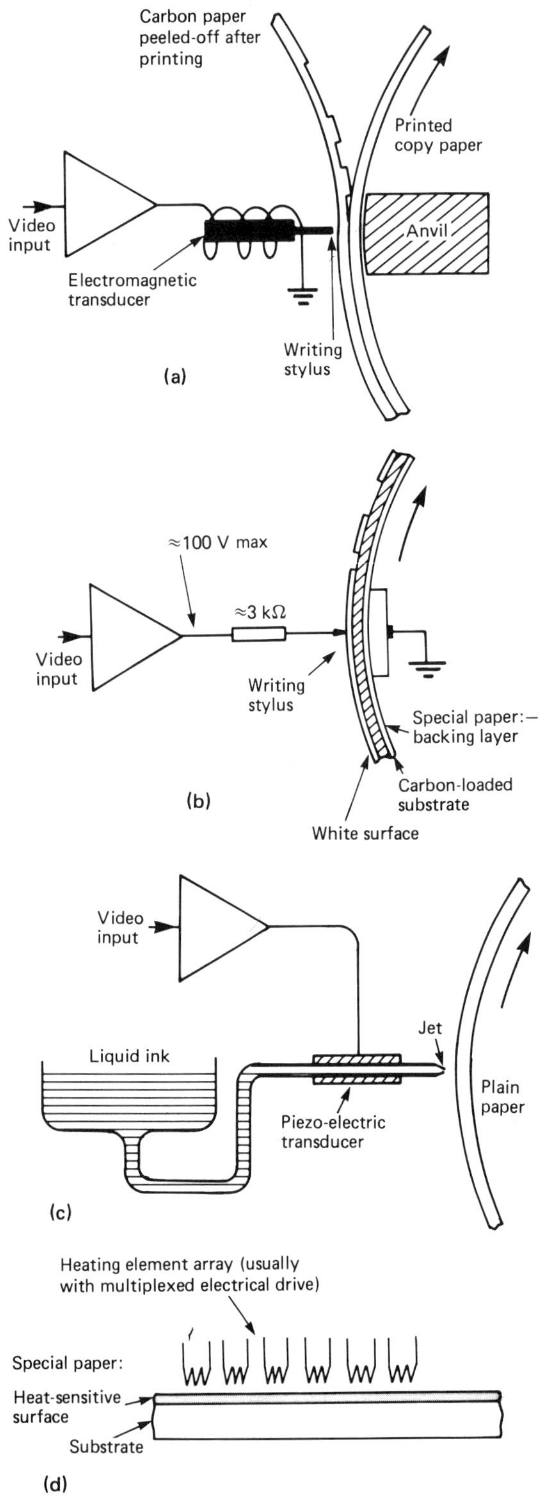

Figure 61.11 Direct writing processes. (*a*) Electro-percussive: a vibrating hammer transfers pigment from carbon paper to the copy paper. (*b*) Electro-resistive: the white surface layer is removed by spark erosion to uncover the black substrate. (*c*) Ink jet: the ink is held in the tube by capilliary action until driven by the shock wave from the transducer. (*d*) Thermal: the paper coating darkens when the threshold is exceeded

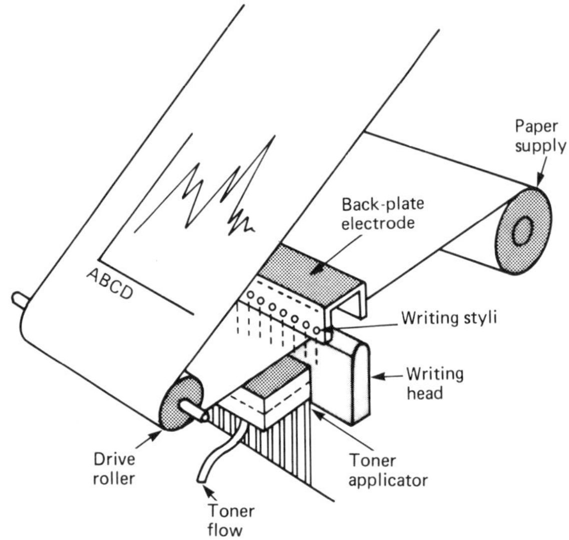

Figure 61.12 The electrostatic writing process (courtesy Versatec Inc.)

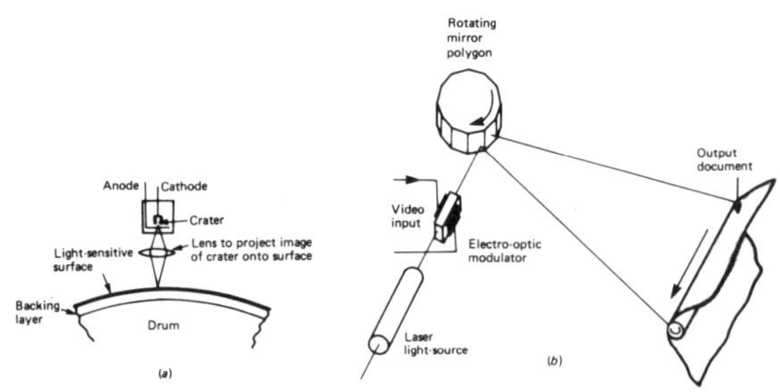

Figure 61.13 Printing arrangements that use light-sensitive copy paper. (*a*) Crater lamp arranged for use with a drum scanner. (*b*) Electro-optically modulated laser with rotating mirror scanner

expensive, so techniques have been developed for storing the information in a more compact form, and translating it into the line-by-line image signal when required.[12] Such compression requires the discarding of combinations that will not be required, or accepting a small statistical risk of being unable to store a desired combination. Alpha-numeric characters and features such as vertical and horizontal lines and logos may be stored in a 'font memory' and fetched when required to build up the raster-scanning line.

The hardware required to do this is included in the lower half of *Figure 61.14*, which outlines the electronic subsystem that must be added to the laser scanned raster printer to make a general-purpose phototypesetter/printer for use in an electronic office system. Input data are received from magnetic tapes produced by other systems or from a communication channel (such as Ethernet), and stored on the systems disk. Job description information, such as format of pages, choice of fonts, details of forms and logos, is prepared and edited by the operator. When a job has been completely specified the print machine is run by the machine controller, whilst the image generator, a specialised hard-wired electronic unit, fetches and assembles the information from the font generator as it is needed to produce the video signal that modulates the writing laser.

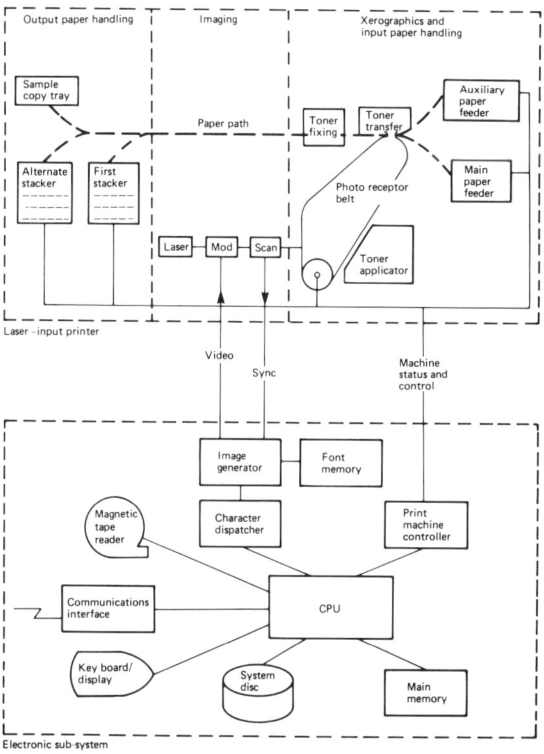

Figure 61.14 Printer system

61.7 Facsimile

61.7.1 Systems

An input scanner can be used to transmit images over a telecommunications link to a remote printer. This principle has been in use commercially for news photograph transmission[13] since the 1930s. Combined send/receive machines suitable for office use became widely available in the 1960s, and by 1980 there were about a quarter of a million such machines connected to the world-wide telephoe network. Facsimile provides very fast transmission of almost any documentary material without specialist preparation. However, between 2 million and 10 million bits are required for a raster scan of an A4 page; this has to be compared with 20 000 bits for a similar-sized page of ASCII-coded characters. This added transmission burden is somewhat alleviated by the ability to tolerate very high error rates.

The design of facsimile machines is strongly influenced by the use of the PSTN (public-switched telephone network). Firstly, the network was designed for speech, not data. Therefore the power/frequency/time characteristics of the transmitted facsimile signal must be chosen to suit the network. Secondly, transmission time is expensive, so effective data compression and channel modulation methods must be used.[14] Thirdly, since the public network is *switched*, every facsimile machine can, in principle, be connected to every other. Rigorous development and application of international standards is therefore necessary to ensure that this potential for interconnection is not wasted.[15]

Figure 61.15 shows a block diagram of a typical facsimile machine. When data are read from an input document it is first compressed and then modulated onto an audio-frequency carrier prior to being coupled to the line. The receive path is the reverse of this.

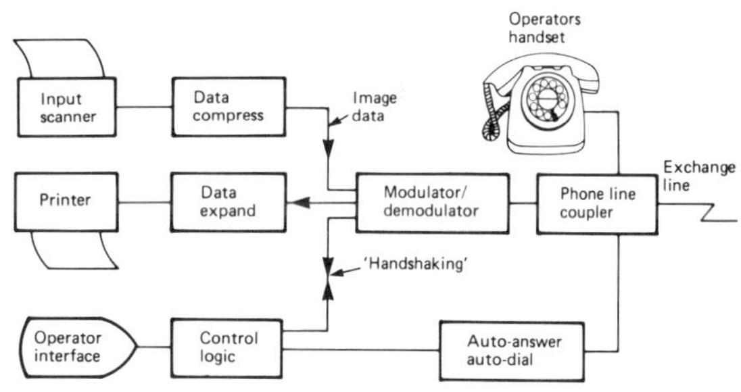

Figure 61.15 A facsimile machine

61.7.2 Handshaking

Control logic in each machine carries out a 'handshaking' process with the other machine designed to ensure that the two machines, which may be supplied by different manufacturers to different specifications, carry out the operator's instructions using the highest possible level of performance that is within their joint capability.

This handshaking is accomplished by an exchange of audio-frequency tone bursts. When the called machine first comes on line it announces its own capability. The calling machine compares this with its own capability and any special requirements of the operator, such as alternative resolution standards, and instructs the called machine to prepare itself accordingly. The sender then sends a line scan phasing signal. Once the receiver has achieved synchronisation with this signal it sends a 'confirmation for receive' signal. Upon receipt of this confirmation the sender commences image transmission. When this is complete another handshake sequence takes place to confirm safe delivery of the image and arrange for the transmission of the next page, if any.

The more sophisticated machines have automatic facilities for unattended reception and transmission to bridge time differences between users and allow the exploitation of cheaper telecommunication tariffs.

Operating standards of facsimile machines are recommended by the International Telegraph and Telephone Consultative Committee (CCITT), which has over the years defined a succession of 'groups' of increasing performance.[16] The principal characteristics of these groups are summarised in *Table 61.2*, and will be discussed below. It should be noted that many commercial facsimile machines are capable of operating in more than one group, and the necessary decisions are made as part of the 'handshaking' process.

61.7.3 Group 1 operation

Early systems for video transmission by wire used frequency modulation, chosen in the belief that it was the best method of overcomimg the distortion of a telephone circuit. Since telecommunications authorities restrict the power in certain frequency bands to avoid interference with control functions[17] the range 1300 to 2100 Hz was selected for the CCITT group 1 recommendation in 1971. This modulation system limits the maximum data rate, so that at a resolution of 3.8 lines/mm the transmission time of an A4 page is 6 minutes. Manufacturers therefore provided lower resolution modes to reduce transmission time (e.g. 4 minutes at 2.6 lines/mm) but these were not standardised. Margin stops are used to reduce the transmission time of short documents but since the recommendations were intended to apply to mechanical scanners whose speed could not be varied no other sort of picture compression was proposed.

Table 61.2 A comparison of CCITT facsimile group characteristics

CCITT 'group'	*A4 page time (minutes)*	*Horizontal resolution (pels/mm)*	*Vertical resolution (lines/mm)*	*Half-tone capability*	*Handshake method (T30)*	*Video transmission modulation*	*Picture data compression compression*
1 (T2)	6	4	3.85	Common option	Tonal	FM Black = 2100 Hz White = 1300 Hz	None
2 (T3)	3	4	3.85	Option	Tonal FSK option	LSB AM Carrier = 2100 Hz	None
3 (T4)	less than 1	8	Std: 3.85 Option: 8	Rare option	FSK (V21) 300 bit s^{-1} 1650 Hz +1850 Hz	MPSK Std: (V27ter) 2.4/4.8 Kbit s^{-1} Opt: (V29) 7.2/9.6 Kbit s^{-1}	Horizontal: run-length Vertical: optional relative-address Channel: modified Huffman

Note: numbers in parenthesis refer to CCITT recommendations.[18]

61.7.4 Group 2 operation

By 1976 an amplitude modulation system had been developed that offered adequate signal/noise performance at substantially higher transmission rates. Recommendations were prepared to standardise the use of a lower-sideband vestigial carrier system using 'alternate phase encoding', in which bandwidth is reduced by arranging that successive periods of full carrier are sent with reversed phase. The recommendation also defined improved 'handshaking' protocols that allow the use of unattended machines.

Some machines in this group employ 'white-space skipping' to reduce transmission time without compromise of quality. When a completely white area is identified a special code is sent and both sender and receiver scan quickly until further data are encountered. This technique can halve the transmission time of a simple business letter.

61.7.5 Group 3 operation

In 1980 agreement was reached on recommendations for a facsimile service that employs developments in digital modem technique, and image processing algorithms, to transmit an A4 page in under one minute.[22]

The standard modem makes use of multiphase shift keying (MPSK) to provide 8 (or alternatively 4) modulation levels, so allowing a binary data transmission rate of 4800 bit s^{-1} (or alternatively 2400 bit s^{-1}) whilst using only an 1800 Hz carrier. An optional modem for use on lines of above-average quality provides data rates of 9600 bit s^{-1} or 7200 bit s^{-1} by combined multiphase and amplitude modulation.

The handshaking procedures are extended to provide for automatic testing of line quality and subsequent selection of modem data rate from the alternatives described above, together with standardisation of methods for indicating a wide variety of special machine features.

A substantial reduction in transmission time has been achieved by the use of three steps of digital data compression that exploit the correlation between nearby pels. The statistical nature of this process requires raster scan mechanisms that can operate at a rate that varies with the information content of the immediate image area, and so the phasing process used with earlier machines to achieve line synchronisation is no longer appropriate. Run length coding is applied along each horizontal scan line by converting the data to a list of the lengths of each of the black and white sections. Optionally, further lines may be encoded by reference to the previous line 'relative address coding'. Although such coding is not continued beyond four lines, to limit the image degradation due to an error being propagated from one line to the next, it can nevertheless provide about 30% reduction in transmission time. The above two coding processes provide a number of different signals, with widely differing probabilities of occurrence, which have to be sent over the channel to the receiver. 'Channel codes' are therefore chosen so that the most common signals are sent using the shortest codes. This 'modified Huffman' algorithm[19] when combined with the run length coding provides overall compression of data in a typical page[20] by a factor of 5 to 10.

61.7.6 The future

The increasing availability of long-distance communications and the decreasing cost of digital electronics will make facsimile even more attractive as an electronic alternative to the postal service. CCITT group 4 recommendations are expected to define standards for the intermixing of raster scan and character-based data, so binding together the hitherto separate worlds of facsimile and word processing.

61.8 Word processors

61.8.1 Introduction

The concept of word processing has been recognised for some 10–15 years. It refers both to the application of electronic techniques within office equipment to prepare and manipulate textual material, and to the associated disciplines and procedures of the office environment. The word processor, as a discrete office machine, has developed from the typewriter, and shares the typewriter's basic function of creating typed script of defined, uniform appearance under the command of an operator using a keyboard. In addition word processors can store the textual material in electronically coded form, permitting subsequent manipulation, correction or editing. In all but the simplest forms of word processor the output printing process, 'print-out', is usually separate from input keying, and retrieval of stored text and the creation of single or multiple print-out may be executed at will.

Word processing is now regarded as a vital first step in the evolution of the 'electronic office' or 'office of the future'. These concepts anticipate substantial movement towards the purely electronic creation, storage and distribution of commercial or business information, thereby alleviating the growing burden of paperwork encountered in most offices today.

To achieve this radical change the office of the future will be one in which the component items of equipment—word processors, electronic files, high-speed printers, facsimile machines, scanner-copiers, optical character readers, etc—are electronically connected to one another and, if need be, to those in other offices. The facilities for the high-speed exchange of information so created will be highly automated, permitting many different types of equipment from different manufacturers to interact without fundamental compatibility constraints.

The impressive growth over the last decade in the range and capability of word processors has been stimulated by the dramatically falling cost of digital electronic control and storage technology, and in particular by the development of the integrated microprocessor. The same trend is influencing the development of all the components of the electronic office.

61.8.2 Development from the typewriter

The typwriter has provided the means for local preparation of printed text, and has been in existence in a compact and easily usable form for almost a century. Although many detailed refinements have been introduced, the typewriter's prime function of linking legended keys on a one-for-one basis to typing elements which carry the typeface remains essentially intact. The earliest and most basic word processor to emerge was the so-called editing typewriter in the mid 1960s. In its commonest form it consists of an electric typewriter in which the individual key depressions are encoded onto paper tape. This record of the initial typed draft can be replayed, causing direct actuation of the printing mechanism, but the print-out may be manually interrupted at any point for the execution of simple amendments or corrections to the draft. At the same time a corrected tape is produced, and may be replayed on demand in order to produce any number of correctly amended typescripts.

These early machines have many of the vital features of a modern word processor—a keyboard, a printing mechanism which is no longer directly linked to the keyboard, and a means for recording and retrieving the sequence of keystrokes input by the operator.

The next major innovation was the introduction of a magnetic means of recording the keyed text. Magnetic tape and card drives use inherently far denser and faster media than paper tape, and are relatively silent in operation. Furthermore, the increasingly sophisticated digital electronic control techniques employed, and particularly microprocessor-based systems, introduced various means of electronically logging sections of recorded textual data. This eliminated the need to playout the tape sequentially in order to reach a particular part of the text for editing or print-out, and greatly speeded up the creation of accurate and well presented typescript.

Word processors at the level of development so far described generally retain a superficial physical resemblance to the electric typewriter, i.e. the keyboard and printing mechanism reside in a shared desk-top unit, although the major parts of the electronic controller, together with the magnetic tape or card drive(s), are contained in a separate free-standing console, electrically connected to the 'typewriter' by a multi-way cable. There are, moreover, additional controls which allow the operator to manipulate pre-recorded text in units of characters, words, lines and even paragraphs. Indeed, the word processor has rapidly progressed from the automatic typewriter stage to become a dedicated computer system which processes words rather than data in general.

All but the simplest of modern word processing systems now include some form of electronic display on which initial preparation and subsequent editing of the text is carried out. Magnetic card and magnetic tape storage systems have been largely superseded by 'floppy discs'. These discs have one-to-two orders of magnitude more storage than tape cassettes, and average access times of less than one second rather than the tens of seconds, or even minutes of tape systems.

The term 'word processor' now includes an increasing variety of system configurations, other than the mid-range stand-alone one. The main divisions of equipment types are discussed below.

61.8.3 Types of word processors

In view of the rapid pace of innovation in all branches of office automation it is extremely difficult to categorise word processors in other than very broad terms. The following descriptions of the various types in current use should therefore be regarded as merely a general guide. There are many designs of equipment which span more than one category, and the falling cost of complex digital integrated circuits for control and data storage is enabling more and more word processing capability to be included for a given cost, or position in the market for equipment.

61.8.3.1 Electronic typewriters

These systems bear some resemblance to the electric typewriter, but are not directly derived from the latter. They are desk-top units, with all their components included within one assembly the superficial appearance of which is typically that of a rather large electric typewriter, see *Figure 61.16*. Electronic typewriters qualify as word processors because of the inclusion in their architecture of electronic controls to store a limited amount of keyed textual data. This storage is restricted in the simplest systems to the provision of a semiconductor buffer memory capable of storing one or two lines of entered text. The operator may backspace through this memory to correct errors in recently keyed text, and the printing mechanism also backspaces in synchronisation. The inclusion of 'lift-off' correction ribbon in several types of printing mechanism enables automatic correction of the typed draft.

In one type of electronic typewriter—the Memory Typewriter* —a non-removable magnetic memory is included. This memory is divided into sections which are addressed by means of a selection control mounted alongside the keyboard. The memory is non-volatile, i.e. retention of the stored data is not dependent on the continuous application of electric power to the unit, and prerecorded textual material may therefore be retained indefinitely, or overwritten with new material according to the operator's requirements.

The recent inclusion of partial-line displays to some new electronic typewriters has enhanced their capability to a point well above that of the simplest units. The display, usually capable of displaying from 16 to 30 characters at a time, is a 'window into memory', providing the operator with a clear view of characters just entered from the keyboard. It is possible in some units to operate in a 'delayed printing' mode, whereby a whole line of text may be entered from the keyboard, viewed in the partial line display, and amended if necessary before it is committed to the printing mechanism.

The falling cost of electronic storage, both semiconductor and bubble memories, is permitting the inclusion of more and more buffer memory within this simplest class of word processing

* IBM registered name.

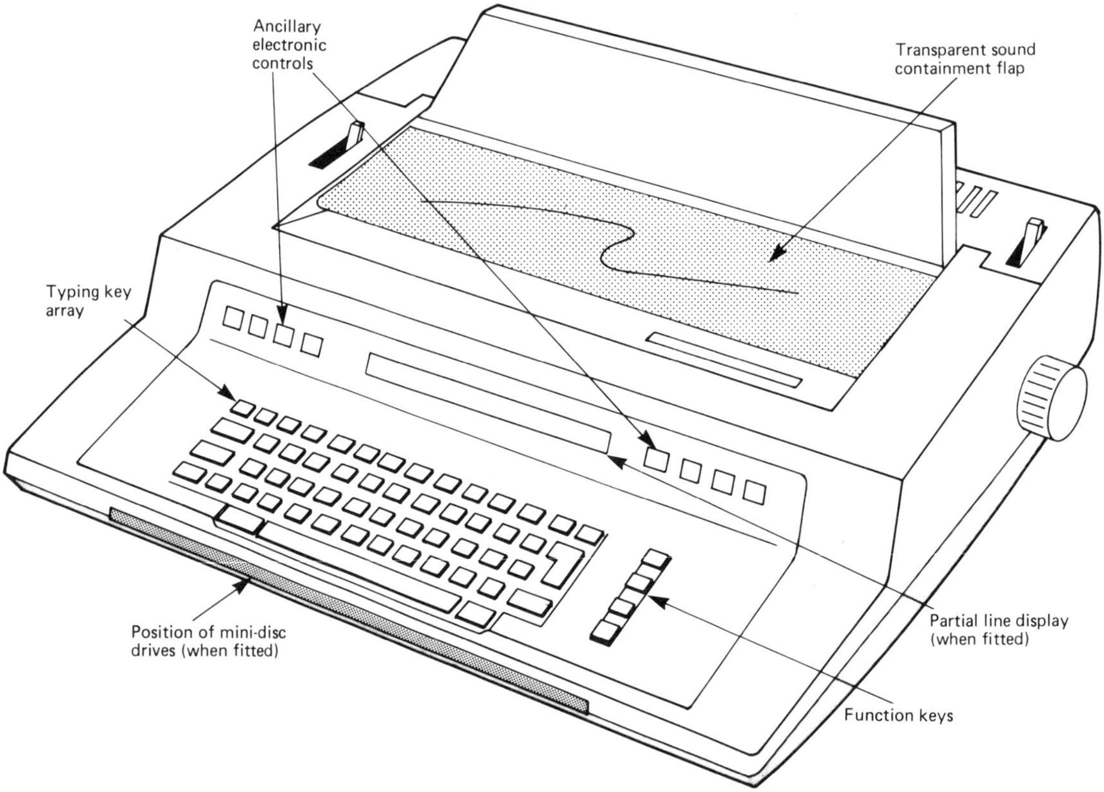

Figure 61.16 Electronic typewriter—general appearance

machines. The result is a trend towards the provision of whole-page buffer storage in products costing relatively little more in real terms than standard electric typewriters of a few years ago.

Some electronic typewriters have taken advantage of the availability of miniature floppy disk drives (the min-diskette). One or more of these may be incorporated within the typewriter, typically underneath the keyboard to provide the useful facility of removable text storage. Machines so equipped can undertake a number of the more complex word processing functions formerly found only on the previous generation of stand-alone display word processors.

61.8.3.2 Stand-alone display word processors

In terms of functional processing power and the ability to handle complex word processing tasks efficiently, these systems occupy a position somewhat above that of electronc typewriters. In view of their general ability to operate on large sections of text, rather than on individual characters and words, they are also referred to as text processors or text editors.

The general appearance of a typical modern stand-alone display word processor is shown in *Figure 61.17*. The major components—control unit, keyboard, display and printer—are interconnected by multi-way cables. The system illustrated has its magnetic disc drives mounted within the control unit, and the keyboard and display are separate units, but different arrangements are used by some machines. For example the disc drives may be mounted above the desk top either within the display unit or in a unit of their own and the display and keyboard may share a common housing.

This class of word processing machine usually employs a much larger control microprogram than simpler machines, and invariably has more than one disc drive. Large buffer memories are used for holding textual data before its transfer to disc storage, or whilst editing and correction are taking place. The repertoire of word processing functins included is extensive, permitting complex operations such as document merge, automatic reformatting to new margin or tabulation settings, automatic searching for specified sequences of characters within previously recorded textual data, and right margin justification. It is also possible to carry out more than one function at once, e.g. output printing of previously recorded text whilst keying in new material, or editing.

Some manufacturers have reduced the total cost of word processor installations, where more than one machine is in prospect, by providing the facility to share certain of the essential component units between more than one operator station (work station).

61.8.3.3 Shared logic systems

These systems use one central controller, or 'central processing unit', together with disc-based data storage facilities, to serve a number of work stations. Each work station consists of some of the elements of a stand-alone display word processor, though local printing may be omitted (early shared logic systems also omit local magnetic media storage facilities). Printing, particularly of final draft text, is carried out by means of a single high-speed printer (100 to 750 lines/min) directly connected to the central processing unit.

Shared logic systems are claimed to be more cost-effective than stand-alone systems for highly structured applications where

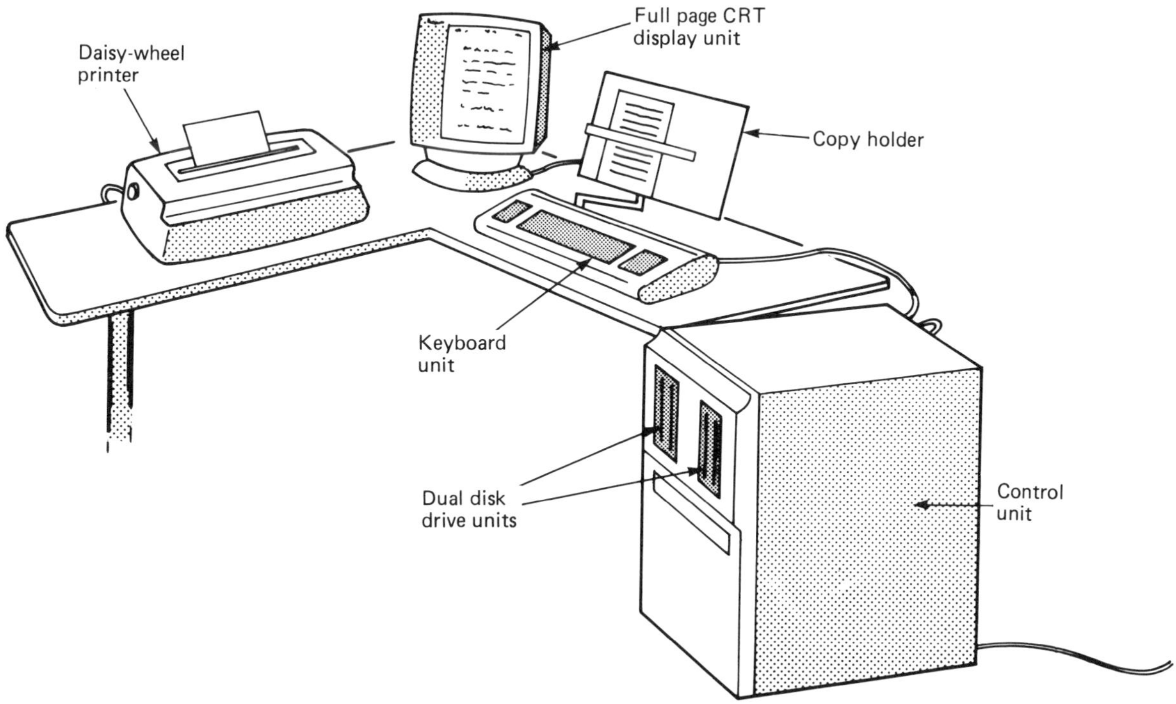

Figure 61.17 Stand-alone display word processor—general arrangement of a heavy-duty unit with separate full page display and keyboard

large volumes of text processing are needed. The central processing unit usually consists of a minicomputer. It employs sophisticated software to provide a wide range of word processing functions and to address the particular application needs of the larger user towards which these relatively complex systems are aimed.

61.8.3.4 Time-sharing systems

This is the word processing analogy of data processing time-sharing systems. Instead of DP terminals, a number of word processing work stations are connected to an in-house or remote mainframe computer via telephone line connections.

A minimum of local text manipulation is possible within each terminal work station, the bulk of the required processing functions taking place at the mainframe in response to data-communicated commands from the terminal.

Archiving of recorded text is carried out centrally on tape or disc files at the mainframe computer. Because of the high-volume, long-document applications for which time-sharing systems are most suitable, printing of the final draft is frequently confined to a high-speed line or page printer directly attached to the central computer.

61.8.3.5 Additional facilities

To extend the application of word processors, a number of optional facilities, accessories and alternative equipment configurations are available from many sources.

Although a majority of word processing equipment is designed within the USA, primarily to meet the needs of the vast domestic market there, a number of manufacturers now offer equipment which will handle without compromise the generally more extensive and varied character sets and typing procedures of the rest of the world (see section 61.8.5 on keyboards).

The growth of data communications in the data processing industry has also begun to be copied, and well proven communication protocols, encoding schemes and data terminal techniques are now available as text communication options on many word processor products. Some of the communications facilities provide a high degree of automation, e.g. unattended receiving. The variety of communications modes available from manufacturers is still somewhat limited. Standard binary synchronous modes (i.e. 'Teletype' mode) predominate, but higher speed and more specialised modes are also offered.

61.8.4 Microprocessor control of word processors

Most word processors introduced since the early 1970s have used microprocessor techniques in their control sections. At first the microprocessor unit was not 'integrated'—rather it was built-up from a number of small and medium scale digital integrated circuits needing considerable space and power within the equipment. These initial applications of microprocessor control have been totally superseded by the integrated microprocessor, and most modern word processors employ several microprocessors to control various sub-systems.

A block diagram of a typical mid-range stand-alone display word processor is shown in *Figure 61.18*. The architecture is characterised by the main data bus to which the various sub-systems, such as keyboard, display, processor and printer, are connected. The example shown uses an 8-bit microprocessor, although there is a trend towards the use of 16-bit units where greater processing speed is sought. This typically enables faster response times to the operator, an increasingly critical factor as the complexity of text manipulation offered has increased.

The size of the control microprogram used by different designs of word processor varies considerably. For mid-range heavy-duty text-editing applications, machines typically deploy 5K to 12K bytes for the basic operating system, and 20K to 40K bytes

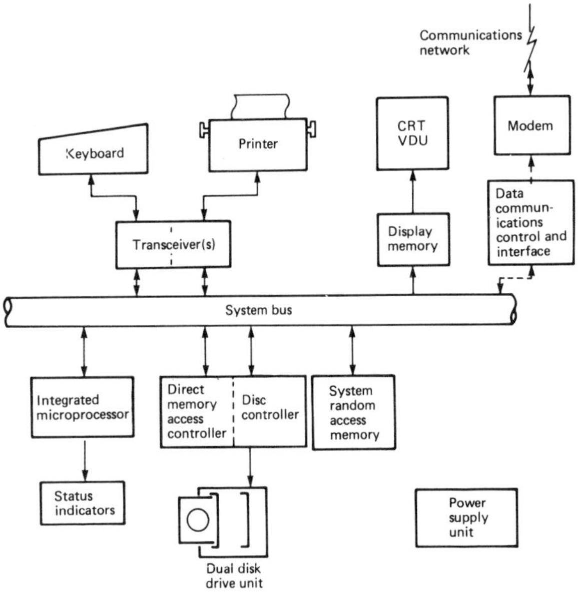

Figure 61.18 Block diagram of typical mid-range stand-alone display word processor

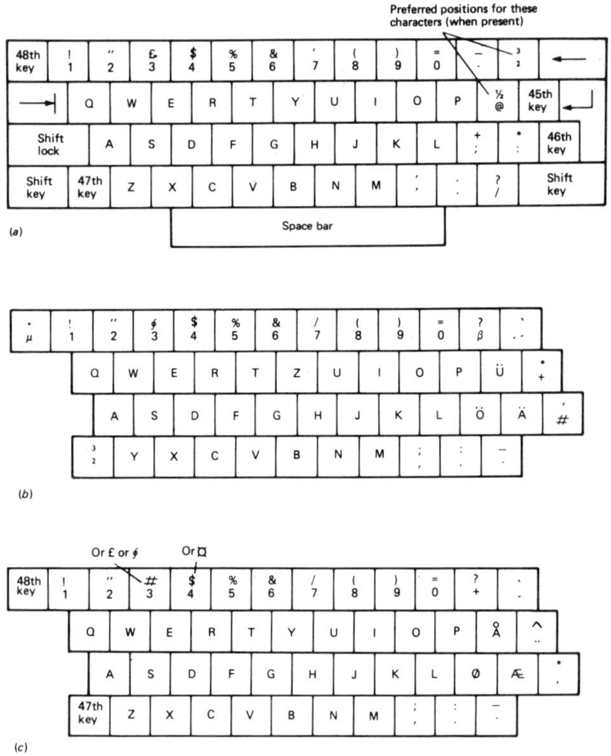

Figure 61.19 Examples of national standard typewriter keyboards. (*a*) British, (*b*) German, (*c*) Norwegian. A 48-key array for the printing characters is shown. The associated non-character keys are shown only in (*a*)

for a general word processing applications package. Such control microprograms are now usually stored as 'system software' on magnetic discs which are loaded into the machine by the operator and automatically 'booted' into RAM. This software resides in RAM for the duration of the work to be performed, but may be replaced by alternative system software for different text processing applications, e.g. mathematical functions, records processing, stock control.

Simpler or earlier word processing machines often utilise a fixed control microprogram, stored in the form of 'firmware' in ROM.

61.8.5 Keyboards

Most word processors have keyboards which emulate those of typewriters, at least so far as the keys for the typing characters are concerned. The familiar 'QWERTY ...' array ('AZERTY ...' in France, 'QWERTZ ...' in Germany, etc) is retained, though equipment featuring modern attempts at more efficient schemes, e.g. Dvorak, is available.

The typing array occupies from 44 to 48 keys, permitting, via the use of a 'shift' key, the input of up to 88 to 96 different characters. Word processing equipment designed primarily for North American markets may use only 44 keys, but the trend in Europe is towards a 48 key, 96 character standard, reflecting the necessary 'extra' characters such as accents and accented letters, and various currency symbols additional to the ubiquitous '$'. *Figure 61.19* illustrates as examples the current British, German and Norwegian national standards for typewriter (and hence word processor) keyboards. It should be noted, however, that these standards are not absolute, many variations being offered by equipment suppliers and used extensively within the countries concerned.

In addition to the keys which constitute the 'typing array', word processers invariably use additional keys for control functions. These keys may be used, for example, for selecting the required unit of stored text to be manipulated (e.g. character, word, line, or paragraph), or for initiating the action to be performed (e.g. delete, merge, skip). Further keys may provide control of data storage and retrieval functions.

Word processor keyboards are now invariably electronic in operation. Each key depression is detected at the key station (i.e. each key button supporting mechanism) by the keyboard control electronics, uniquely encoded in binary electronic form and passed to the system's control microprocessor for interpretation. The use of low force or contactless switches within key stations sometimes causes problems to new operators more familiar with the snap-action and pronounced 'click' of electric or manual typewriters. Such problems are usually short-lived, and the benefits of lower depression force and '**n**-key rollover' are quickly exploited. '*n*-key rollover' is a feature of electronic keyboard systems which enables very rapid and accurate keying-in without regard to the number of keys in the depressed position at one time. Only the order in which keys are depressed is sensed by the system.

6.8..6 Printers

Various printing mechanisms are used in modern word processing equipment, the most widespread of which is the diasy-wheel impact printer. Other impact printing mechanisms include the gold ball and the thimble, which is actually a variatin of the daisy wheel in which the spokes are bent to form a cup shape. Some larger systems use ink jet impactless tedhnology in its most rfined form in order to produce the fully-formed characters demanded by the 'correspondence quality' market for word processors. There is also some use of high-speed 'draft quality' printers, typically employing impact or ink jet matrix tecjniques.

The daisy-wheel printer has raised the typical printing speed achievable with typewriter-like printing mechanisms from the 15–20 characters per second of golfball printers to 30–55 characters per second. It has few moving parts and is capable of operating reliably for long periods with little maintenance. *Figure 61.20* shows the general layout of a daisy-wheel printer and *Figure 61.21* details a typical-daisy wheel. The print wheel, print wheel drive motor, ribbon cartridge, hammer, armature and solenoid are supported in a common carrier which is precisely moved and positioned at each successive horizontal printing position by a servo motor. The print wheel drive motor is commanded by the electronic control system (not shown) to position the relevant character slug on the print wheel at the upper, print position. The solenoid and armature are then energised, causing the hammer to drive the character slug out of the normal plane of the print wheel, thereby impressing the ribbon onto the paper.

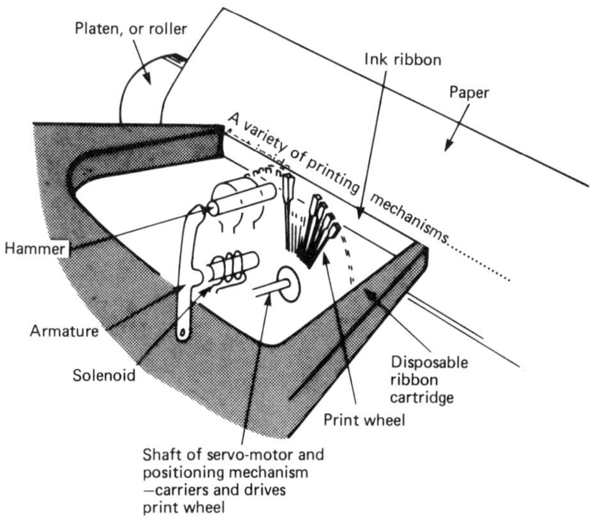

Figure 61.20 General arrangement of characteristic parts of a daisy-wheel printer

61.8.7 Displays

All word processing equipment other than the simplest electronic typewriters now uses some form of electronic display. The purpose of the display is to permit the operator to view the contents of machine memory, and hence to easily carry out complex text preparation and editing tasks and confirm the result prior to committing it to print-out.

Simpler machines may use a 'one-line' display, able to display the character stream forming one line, or part of a line of text. More powerful word processors usually employ multi-line displays, and some extend this capability to displaying the equivalent of a full A4 page of typed text. The necessity to handle alphanumeric characters economically has led to the use of dot matrix techiques to achieve a reasonable electronic rendition of the typescript.

The one-line and partial-line displays usually employ gas discharge techniques, though light-emitting diode matrices, and flattened cathode ray tubes are occasionally encountered. Multi-line and full-page displays are almost all cathode ray tubes and are raster scanned.

All of the character display schemes mentioned abaove are optimised for text character display. Usually the display sub-system of the word processor takes its input from a display memory which can hold a whole display screen's worth of character codes. The display memory is loaded by the control microprocessor. A 'character generator' within the display sub-system converts these character codes sequentially into a stream of pulses which modulates whatever scanning technique is in use in order to create the dots forming the character image on the screen.

Some of the most sophisticated word processing equipment now has the capability to display text in a variety of type styles

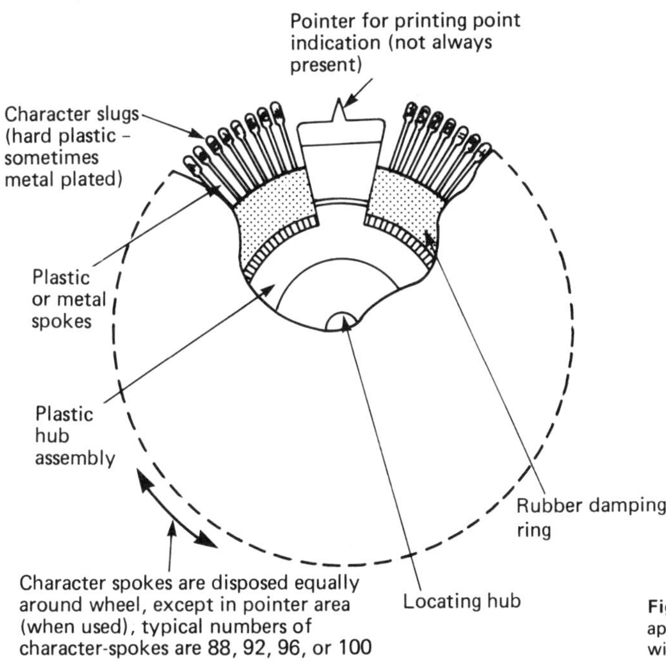

Figure 61.21 Typical daisy-wheel. Such wheels are approximately 75–100 mm in diameter and are available with a wide variety of type styles and fonts (character sets)

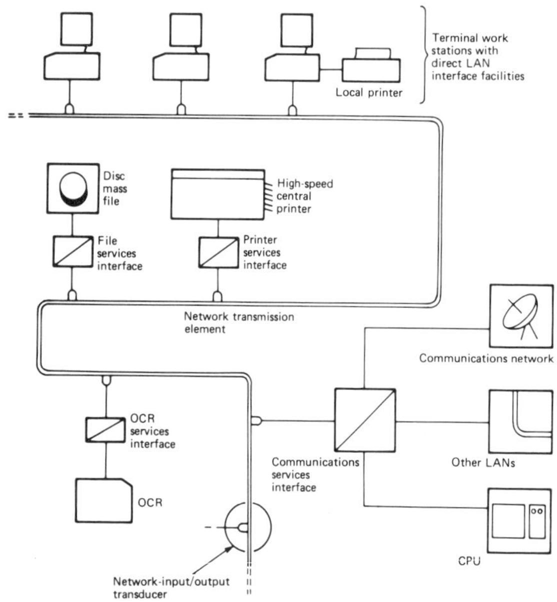

Figure 61.22 A local area network of the Ethernet type

rather than in dot matrix form and, furthermore, to use and to mix different sizes of characters, e.g. Xerox Star series. To present such a variety of textual images a cathode ray display using 'bit-mapped' techniques is used. Here a display memory of far higher capacity is used in order to store the required brightness state of each and every individual picture element ('pixcel') of the display screen. Such a system is also capable of displaying graphical data.

61.9 Local area networks

Reference has been made in section 61.8.1 to the broad concept known as 'the Office of the Future'. The picture briefly presented of interconnected items of office machinery exploiting modern data communications techniques is necessarily simplistic. In the Office of the Future the role of each item of equipment so connected may differ significantly from the familiar role of today. For example, the 'printing engine' of the ubiquitous office copier when teamed with an electronic input device and connected to some sort of 'local area network' (LAN) becomes a raster output scanner (ROS). Any connected device with an input keyboard and output display and/or printing capability can act as a terminal for user communication with any other device on the LAN.

The genesis of LANs, at least for widespread use in the office, has been brought about mainly by two factors: (i) the increasingly rich and varied functions offered in individual items of office machinery (i.e. 'intelligence' made possible by the microprocessor), and (ii) development of low-cost digital data communications technology.

At the present time much attention is being focussed on the 'Ethernet'* system for constructing LANs. Many such systems will be deployed in the near future. The connecting link in the system is an ordinary coaxial cable up to 500 m long, or up to 1500 m with repeaters. There is no controller in the system—attached terminals, etc, simply transmit on the system at any time, relying on the relatively high 10 Mbit s^{-1} transmission rate to afford minimal probability of data collisions. The essential parts of an Ethernet-like LAN are shown in *Figure 61.22.*

* 'Ethernet' is Xerox Corporation's LAN—a system developed in cooperation with Digital Equipment Corporation and Intel.

References

1 COBINE, J. D., *Gaseous Conductors*, Dover Publications (1958)
2 DESSAUER, A. B. and CLARK, *Xerography and Related Processes*, Focal Press (1965)
3 THOURSON, T. L., 'Xerographic Development Processes. A Review', *I.E.E.E. Trans. on Electron Devices*, **ED-19**, 4, 495–511 (1972)
4 SCHEIN, L. B., 'The Electric Field in a Magnetic Brush Developer', *Electrophotography, 2nd Int. Conf.*, SPSE, pp. 65–73 (1974)
5 HENNIKER, J., 'Triboelectricity in Polymers', *Nature*, **196**, p. 474 (1962)
6 PEARSON, D. E., *Transmission and Display of Pictorial Information*, Pentech Press, London (1975)
7 HOLLADAY, T. M., 'An Optimum Algorithm for Half-tone Generation for Displays and hard Copies', *Proc. Soc. Inform. Display*, **21/2**, 185–92 (1980)
8 HUNT, R. W. G., *The Reproduction of Colour*, 3rd Edn, Fountain Press (1975)
9 KUHN, L. and MYERS, R. A., 'Ink Jet Printing', *Scientific American*, **240**, No 4, pp. 120–8, 131–2 (April 1979)
10 KEELING, M. R., 'Ink Jet Printing', *Physics in Technology*, **12**, 196–203 (September 1981)
11 SEYBOLD, J., 'Electronic Printing and the Xerox 9700', *Seybold Report*, **8**, No 5 (16th November 1978)
12 GRAY, R. J., 'Bit Map Architecture Realises Raster Display Potential', *Computer Design*, **19**, part 7, 111–17 (1980)
13 COSTIGDAN, D. M., *Electronic Delivery of Documents and Graphics*, Van Nostrand-Reinhold, New York (1978)
14 COSTIGAN, D. M., 'Facsimile Comes up to Speed', *I.E.E.E. Communications Magazine*, pp. 30—5 (May 1980)
15 JACOBSON, C. L., 'Digital Facsimile Standards', *I.C.C. '78 Conference Record*, pp. 48.2.1–48.2.3 (June 1978)
16 'Telegraph Technique', *Yellow Book*, Vol 7, International Telecommunications Union (1981)
17 'Requirements for Attachments Connected to the Public Switched Telephone Network', *Technical Guide No 30*, British Telecom (1977)
18 'Data Transmission', *Yellow Book*, Vol 8, International Telecommunications Union (1981)
19 FORNEY, G. D. and TAO, W. Y., 'Data Compression Increases Throughput', *Data Communications*, pp. 65–76 (May/June 1976)
20 MUSMANN, H. G. and PREUSS, D., 'Comparison of Redundancy-Reducing Codes for Facsimile Transmission of Documents', *I.E.E.E. Trans. Communications*, **COM-25**, No 11, pp. 1425–33 (November 1977)
21 URBACH J. C., *et al.*, 'Laser Scanning for Electronic Printing', *Proc. I.E.E.E.*, **70**, No. 6, 597 (June 1982)
22 HUNTER, R. and ROBINSON, A. H., 'International Digital Facsimile Coding Standards', *Proc. I.E.E.E.*, **68**, No. 7, 854–867 (July 1980)

Further reading

COLEMAN, V., ERMOLOVICH, T. and VITTERA, J., 'Controlling local area networks', *Electronic Product Design* (October 1982)
HINDIN, H. J., 'Dual-chip sets forge vital link for ethernet local-network scheme', *Electronics* (October 1982)
WANG, F. A., *Office Automation*, Mini-Micro Systems (December 1982)

62 Medical Electronics

D W Hill, MSc, PhD, FInstP, FIEE
North East Thames Regional Health Authority

Contents

62.1 Introduction

Hospitals, in common with many other large organisations are making an increasing use of electronic instrumentation and computing. A good indication of the range of transducers, circuitry and applications is given in the publications.[1,2,3] The purpose of this chapter is to illustrate the broad approach required to obtain a general appreciation of the current role of medical electronics and instrumentation in health care delivery. New developments are covered by abstracting journals[4] and information services.[5] Journals such as 'Medical and Biological Engineering and Computing' cover the field in depth, while books[6,7,8] provide a more detailed account of particular aspects of the subject.

In many cases, but not all, there is a good selection of high-quality apparatus from which to make a choice with a tendency towards 'package' systems and modular arrangements from which a system can be configured to suit an individual application.

An increasing awareness of the high cost of acute beds in hospitals is accelerating health care planners to limit the number of such beds. A consequence of this is the need to provide for better facilities for the care in the community of post-operative patients and also for better diagnostic techniques with a greater availability.

62.2 Diagnosis

Diagnosis is the first task confronting a doctor presented with a patient. At present comparatively little information is used in the general practioner's surgery (office) but in some countries computers are being used to hold patient's records and by decision algorithms and tree branching techniques to assist in arriving at a correct diagnosis.[9,10] It may well happen that a significant contribution to the diagnosis will occur as a result of laboratory analysis of one or more of the patient's body fluids.

An extension of the *ad hoc* diagnostic procedure is the routine bank of tests and questions administered to members of preventive medicine schemes in the form of a multiphasic screening programme which is computer based.[11]

It is evident that the impact of microcomputers on medical instrumentation will be widespread,[12] but it is interesting that the simple mercury-in-glass clinical thermometer still continues to hold its own in spite of the general availability of remote sensing and recording instruments based on thermistors or thermocouples.[13]

Other techniques which may be encountered are as follows.

62.2.1 Radiology[14,15]

One of the first major applications of current electricity in medicine occurred with the generation of x-rays and x-ray sets now make use of sophisticated electronic control systems. In district general and teaching hospitals, the department of diagnostic radiology is one of the most expensively equipped, the others being radiotherapy and the pathology laboratories. Simple mobile x-ray sets are now commonly motor driven from self-contained rechargeable batteries. Electro-optical image intensifiers working in conjunction with television monitors are in widespread use for screening procedures with a contrast medium to outline organs such as the stomach, gut and heart. The contractile action of the heart and peristalsis in the stomach and gut can be recorded on cine film or video tape. Image intensification also substantially reduces the dose of x-radiation received by the patient. Complicated bi-plane image intensifier arrangements are used for heart studies with catheterisation of the chambers of the heart.

In one approach to sharper definition, the conventional silver-halide film has been replaced by a xerographic technique in which the electrostatic image is formed in a fashion similar to that of the well known Xerox photocopier (see Chapter 61). The method gives an enhanced image with soft tissue and is of value in x-rays of the breast mamography.

Undoubtedly the most striking advance seen in radiology in recent years has occurred with the invention of computer-assisted tomography (CT scanning[16]) in which a beam of x-radiation is directed through the patient in a particular direction and falls onto a set of detectors. The first CT scanners operated with a translate–rotate motion. The x-ray tube and the detectors were mounted in a gantry and on opposite sides of the patient. The tube and detectors moved sideways under computer control to scan the patient, the gantry rotated through a known angle and the tube–detector combination moved back to re-scan the patient; the gantry rotates through another similar angle and a further scan is made. Each x-ray image is digitised and stored in the computer's memory. From these images the computer then reconstructs the picture of that particular 'slice' of the patient. The patient is then automatically moved a known distance, typically 1 cm, relative to the gantry and that slice is scanned. In this way a series of pictures of a part of the patient's body is obtained. The first applications were concerned with the head. CT scanning can detect small changes in density and can thus distinguish between a blood-filled haematoma and a fluid-filled cyst. The technique revolutionised the detection of brain tumours and intra-cranial bleeding.

Following its initial success in brain investigations, whole-body CT scanners have been developed and have proved to be of clinical benefit.[16] In order to achieve faster scan times, the x-ray beam is now often fan shaped and falls on a bank of as many as 1200 detectors to produce a scan time of as little as 3 s. However, in practice, 15 s is normally adequate. The slice approach makes it feasible to visualise a tumour in three dimensions and this has led to the development of radiotherapy planning computers which simulate the position of a beam of cobalt unit or linear accelerator radiation on the CT scan picture.[17]

62.2.2 Nuclear medicine techniques

CT scanning is almost unrivalled as an x-ray technique suitable for distinguishing fine details in anatomy, however it cannot follow changes in the function of an organ. CT scanning has superceded radioisotope scanning for localising tumours in the brain, but in other organs CT scans may be made in parallel with isotopic tests of organ function, e.g. the thyroid and kidneys. In the case of isotope imaging, the chosen gamma-emitting radionuclide is administered in a suitable chemical compound which is handled in a desired fashion by the organ concerned. Thus chromium-151 labelled EDTA is used to measure the glomerular filtration rate and I-125 labelled inulin is used to measure the renal plasma flow. A crystal of sodium iodide is placed outside the body and close to the surface above the organ of interest. As the compound passes through the organ, gamma rays emitted pass through a collimator and fall on the crystal where they cause a scintillation of light to be produced which is detected by one or more photomultipliers. The height of the pulse produced by the photomultiplier is proportional to the energy of the gamma ray and a pulse height analyser can be used to distinguish one radionuclide from the other if a double isotope technique is used. Reference 18 gives a clear introduction to the methods of nuclear medicine.

Simple scintilation detectors are of great value in renography for the study of renal function, but the use of a large (28 cm diameter) crystal in the form of a gamma camera is now commonplace for organ imaging, functional studies and the detection of tumour metatastases.[19] The isotope technetium 99 m

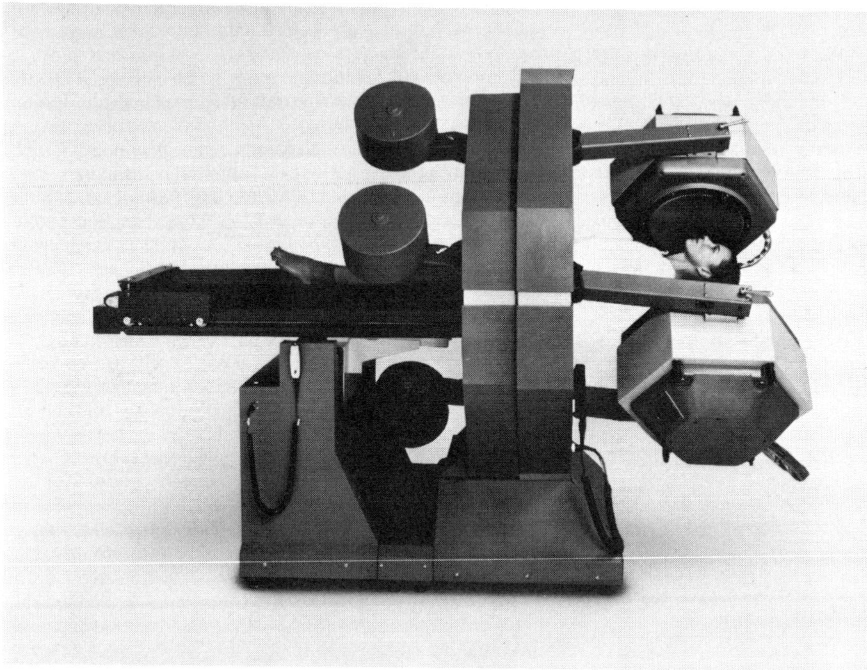

Figure 62.1 Double-head large field of view scanning gamma camera providing facilities for positron emission computed tomographic images (courtesy Siemens Ltd)

with a 6 h half-life and a 140 keV gamma ray energy is widely used for gamma camera studies, but the positron emitting isotope gallium-67 is used to provide tomographic images[20] (*Figure 62.1*).

Another large piece of nuclear medicine equipment is the whole-body counter. As the name implies the equipment is sufficiently large and sensitive that it can detect and analyse the amount of the body burden of radioactive material.[21] Elements such as calcium, phosphorus, sodium, chlorine and nitrogen in the body can be determined with a whole-body counter if they are first made active by subjecting the patient to irradiation with 14 MeV neutrons from a neutron generator.[22]

62.2.3 Thermography

Basically, this is a heat-sensing method of imaging which records the temperature distribution patterns from the surface of the patient's body in the region of interest. A television compatible output format is now adopted, with the line scan accomplished by a rotating polygon mirror and the frame scan by a rocking mirror. A 45° mirror is often placed above the patient and the heat reflected into the scanning system and on to a cooled cadmium mercury telluride detector or a pyroelectric detector.[23,24] Originally, the use of thermography was concentrated on a non-invasive non-ionising radiation method for breast cancer screening. It came clear that it was essential that the subject be thermally in equilibrium with the environment and the interpretation of breast heat patterns is more complicated than was originally envisaged. Progress is being made with the application of computers.[24] Current systems have a resolution of 0.1°C and can produce colour contour displays of the isotherms. Successful applications include the detection of deep vein thrombosis, the effect of drugs on the peripheral circulation and the study of burns.

62.2.4 Ultrasonic imaging

The great advantage of ultrasound over x-rays is that at the power levels required for diagnostic imaging, ultrasound has not been shown to damage tissues. Thus it has to be of inestimable value in the examination of pregnant women since it has not produced deleterious effects on the foetus. A good general account of the subject is given in reference 25. Static B-scanners are still widely used with a hand-held transducer typically operating at 3 MHz and which is moved over the surface above the body region of interest in a raster pattern. Modern systems are capable of producing an image with a grey scale,[26] at least 16 levels are usual. Simple A-scanners with a single transducer serving both as the transmitter and receiver and displaying the echoes received from the various tissue interfaces are used with the echoes displayed on a time base to measure any mid-line shift in patients with a suspected head injury.

As the name implies static scanners cannot be used to observe the real-time motion of body organs, e.g. peristalsis or arterial pulsations or the movements of a foetus. Much recent work in this field has led to the development of four main types of real-time systems: linear array, annular array, rotating crystal and sector scanner.[27] Linear arrays are much used for abdominal and obstetric work, sector scanners for obstetric and cardiac work and sector scanners for cardiac investigations. Reference 28 discusses the methods and terminology encountered in ultrasound imaging systems. Multi-transducer imaging systems for the heart[29] offer the prospect of reducing the number of patients submitted to invasive cardiac catheterisation procedures for the determination of conditions such as incompetent heart valves.

62.3 Electronic instruments

62.3.1 Cardiovascular instruments

The heart is a vital organ of the body and a substantial effort has been devoted to investigating and controlling its pumping action. It may be necessary to record the electrical potentials generated by the beating heart, the pressures developed, the blood flow and volume pumped, the heart and pulse rates and the heart sounds. The rate can be controlled by a suitable form of cardiac pacemaker and if the heart goes into ventricular fibrillation it can usually be restored to a normal sinus rhythm by means of a defibrillator.

62.3.2 Blood flow measurement

Blood flow can be directly detected and measured by two main methods:

(i) Electromagnetic.
(ii) Doppler shift ultrasound techniques.

There are other possibilities such as the use of a laser and the use of thermal flow probes.

Electromagnetic blood flow meters. Electromagnetic blood flow probes have to make a snug fit around the vessel in which the flow is to be measured and thus require a surgical exposure of the vessel. The design of the probe and associated cable and connector has to be such that the assembly can be sterilised and is capable of operating while surrounded with body fluids. The principle of operation depends upon Faraday's law of electromagnetic induction, *Figure 62.2*. The direction of the flow of electrically conducting blood corresponds with the flow of electric current and an electric potential whose magnitude is proportional to the velocity of flow is developed in a third plane mutually perpendicular to the plane of flow and the plane of the applied electromagnetic field. Care is required in the probe design because of the small size of the flow signal consequent upon the limited field strength which can be produced without giving rise to a significant heating of the coil. Early flowmeters used a permanent magnet and generated a DC flow signal. However, artefacts arising from electrode polarisation changes gave rise to baseline shifts and AC systems are now used. The time-dependent magnetic field introduces an unwanted signal which is substantially independent of the blood velocity and is proportional to the rate of change of the field but which is much easier to eliminate than the fluid electrode potentials.[31] This signal arises from both the linkage of flux with the connecting leads to the electrodes and from eddy currents generated in the blood, the vessel wall and the surrounding tissue. It is known as the 'transformer e.m.f.' or 'quadrature voltage'. Both sine wave and square wave excitation of the electromagnet have been employed in practical flow meters. For a 240 Hz sine wave flowmeter, the wanted flow signal was of the order of microvolts and the unwanted quadrature signal was about 10 times larger.[31] A demodulator is employed to extract the flow signal and eliminate the quadrature signal. The electromagnetic flowmeter is able to distinguish forward and reverse flows and to follow the flow pattern during each cardiac cycle. The basic signal is proportional to the blood velocity and thus to the volume flow rate if the vessel's cross sectional area is known. If the volume flow rate signal is integrated for a known time the result is a signal equivalent to a certain volume of blood. Thus an integrating flowmeter can be calibrated by occluding the normal flow through the vessel surrounded by the flow probe and then injecting a known volume of blood through the probe from a syringe. Flow probes which can be chronically implanted have led to detailed studies of drug action on the circulation.[32] An extractable flow probe which can be placed around the thoracic aorta during surgery and post-operatively withdrawn has been used in man.[33,34]

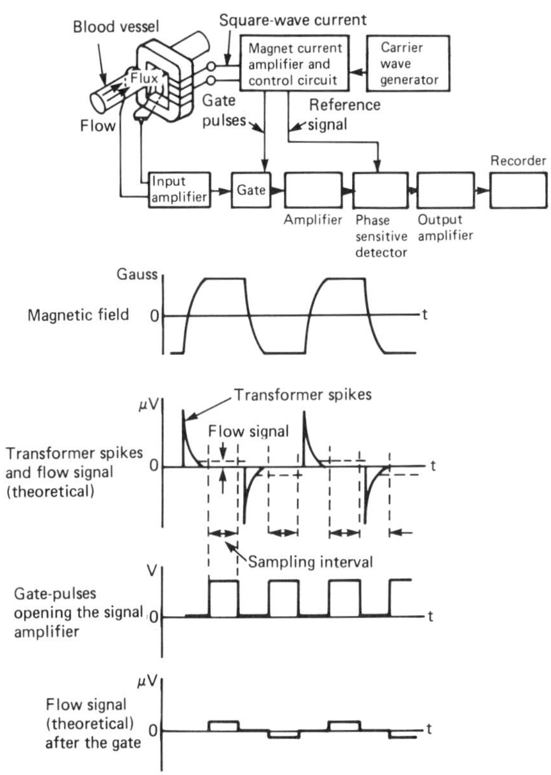

Figure 62.2 Basic principle of operation of a square wave electromagnetic blood flow meter showing the use of gating to eliminate the unwanted transformer spike signals[6]

Doppler shift ultrasound.[35] The apparent shift in the frequency when a relative motion occurs between a source of sound and the observer constitutes the 'Doppler effect'. It has been used in the detection of foetal heart movements and in the study of blood flow.

Two piezoelectric crystals in a suitable housing are placed on the surface of the body over the blood vessel or region of interest. The transmitter crystal generates a continuous beam of ultrasound at about 5 MHz and this is reflected back to the receiver crystal from the moving structure in the body. There is an effective increase in frequency when the reflector moves towards the probe and a decrease when it moves away. The change in frequency, the Doppler shift, is usually in the audio range and after electronic detection can be heard on a loudspeaker. Signal processing of the Doppler shift frequency spectrum can be made in real time to indicate the maximum flow velocity and the blood flow pattern. Dopper flow meters have made an important contribution to the study of occlusive arterial disease.

Lasers are being applied via a microscope for blood velocity studies[36] and a thermal dilution probe has been used for the measurement of venous flows.[37]

Indirect methods. Although indirect methods are not so exact as direct techniques they are most important since they are usually non-invasive and can be applied to people such as out-patients and pregnant women. Plethysmography has proven its worth over the years and is basically a method for measuring the volume changes occurring with time in a particular segment of a limb such as a finger or the calf. In venous occlusion plethysmography, a cuff is applied proximal to the segment and pumped up to a pressure somewhat below the subject's diastolic pressure, i.e. sufficient to occlude the venous outflow from the segment. The arterial inflow continues causing the segment to swell and the rate of volume increase is measured and is proportional to the blood flow into the segmant. A thin mercury filled rubber tube can be placed around the circumference of the segment and the increase in its electrical resistance with the swelling is recorded by means of a bridge circuit.[38] The inflation and deflation of the cuff can readily be automated.[39]

Impedance plethysmography usually operates with a constant current technique with frequencies in the range 25 to 11 kHz. Four disposable band electrodes are placed around the circumference of the segment with the current of about 4 mA r.m.s. passed between the outer 2 electrodes and the resulting voltage change detected at the inner pair.[40] A correlation coefficient of 0.877 has been found in patients[41] between the mercury strain gauge and impedance methods. A block diagram of the commercial Minnesota Impedance Cardiograph is shown in *Figure 62.3*. Impedance plethysmography has been

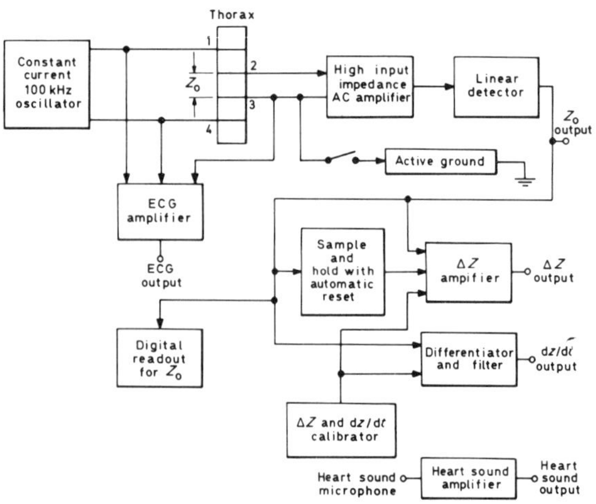

Figure 62.3 Schematic diagram of the IFM/Minnesota impedance cardiograph (courtesy of Instrumentation for medicine Inc.)

successfully used for the detection of deep vein thrombosis.[42]

Photoelectric plethysmography measures the amount of light transmitted or reflected at the ear, finger or vagina. The amount of light received is influenced by the optical density of blood and varies with the cardiac cycle and the amount of blood present in the capillaries.[43]

62.3.3 Blood pressure measurement

Both invasive (direct) and non-invasive (indirect) methods are well established.[44] Direct measurement requires the cannulation of an artery but provides a continuous record of the pressure changes occurring throughout each cardiac cycle and measures the systolic, diastolic and mean pressures. Pressure transducers are employed to convert the pressure into a corresponding electrical signal. They normally operate on capacitance charge, strain gauge bridge or differential transformer principles.[45] In order to obtain a frequency response out to some 20 Hz when connected to a cannula a small volume displacement is necessary for the transducer (0.01 mm^3 per 100 mm Hg). A valuable development has been that of strain gauge microtransducers which can be mounted at the tip of a cardiac catheter and have a frequency response of several kHz which enables them also to record the heart sounds.[46]

Indirect methods make use of a microphone to detect the onset and cessation of the well known Korotkoff sounds corresponding with the systolic and diastolic pressures[47] or the use of a beam of ultrasound to detect the corresponding motion of the arterial wall.[48] Thc operating principle of the ultrasonic arteriosonde monitor is shown in *Figure 62.4*.

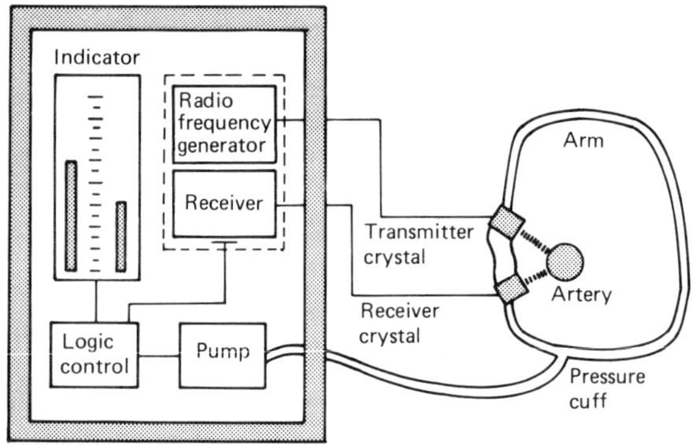

Figure 62.4 Schematic diagram of an ultrasonic indirect blood pressure monitor[6]

62.3.4 Cardiac output

The cardiac output is the volume of blood in litres per minute pumped by the heart. Direct methods for its measurement include the use of an extractable flow probe[34] and dilution methods. The most commonly used are the dye and cool saline dilution methods which require an arterial puncture.[13] The indicator is injected rapidly as a bolus into the right heart via a suitable catheter, it becomes mixed with blood in the heart and pulmonary circulation and is detected in the peripheral arterial circulation by a catheter mounted optical or thermal sensor. The progress of the indicator past the sensor generates the so called dilution curve when the concentration of indicator in the blood is plotted against time. From a knowledge of the amount of indicator injected and the area under the curve the cardiac output can be calculated as an average over several beats. Fibre optic transducers have been developed which will monitor both the dye concentration and the oxygen saturation of the blood.[49]

Iodine-131 labelled human serum albumin can also be used as the indicator, detection being performed by means of an external scintillation counter mounted outside the body over the heart. The method overestimates compared with dye, but only requires a venous injection which makes it suitable for out-patient studies.[50]

The electrical impedance method has been adapted for the monitoring of changes occurring in the stroke volume and cardiac output and contractibility.[51,52] The method is by no means absolute but is useful for following changes on a beat-by-beat basis.

62.3.5 Defibrillators

The life-threatening condition of fibrillation occurs when the contractile action of the fibres (or fibrils) of the heart muscle no longer occurs in a cyclic coordinated fashion. In the majority of incidences the random 'squirming' motion of the heart can be converted back into a regular pumping action by the application of a controlled electric shock to the chest wall, or if the chest is open a smaller shock is applied directly to the heart. In a DC defibrillator a capacitor is charged up to a maximum energy of some 400 J requiring about 7 kV for a 16 μF capacitor. Since irreversible brain damage can occur within a few minutes of the cessation of an effective cerebral circulation, defibrilllators are required to be instantly available and compact battery powered versions are now available.[13] The technique can also be used to reverse atrial fibrillation. It is necessary to ensure that the discharge of the smaller shock used for this application does not coincide with the *T* wave of the patient's ECG, otherwise ventricular fibrillation may be induced. A synchronised defibrillator is required with the discharge fired after a suitable time delay triggered from the preceding *R* wave of the ECG. *Figure 62.5* shows a portable battery powered defibrillator fitted with a cardioscope for observing the ECG.

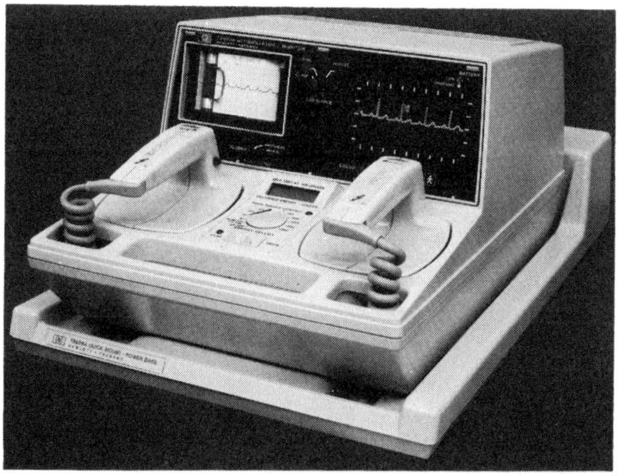

Figure 62.5 Portable battery-powered defibrillator with a built-in cardioscope and ECG recorder (courtesy Hewlett Packard Ltd)

62.3.6 Electrocardiographs

The electrocardiogram (ECG) is the plot against time of the electrical activity of the heart as detected by surface electrodes placed on the limbs of the patient and at defined positions on the thorax. It is capable of giving much information on the history of a 'heart attack' and on the heart rhythm, although it cannot provide information on the pumping capability of the heart except in terms of rhythm. It is a widely used diagnostic technique.

As with all electrical devices connected to the patient and to the mains supply, the question of electrical safety is important and modern ECG recorders are provided with fully floating input circuits which limit the leakage current which can pass through the patient to some 10 μA. Modern circuits have high input impedances in excess of 5 MΩ and common mode rejection ratios better than 60 dB so that interference-free tracings are easily obtainable without mains interference.

Twenty four hour continuous recording of the ECG (ambulatory monitoring) is of value in determining the cause of transient fainting attacks and rhythm disturbances. Miniature FM tape recorders are now available which can record the ECG, EEG, blood pressure and other physiological signals and can be easily worn by the subject.[53]

In intensive care and coronary care units, automatic detection and classification of arrhythmias in the ECG materially assists prompt clinical intervention and the use of medication. Hard-wired systems are now giving place to software systems.[54,55] Another developing application is in the use of computer for the interpretation of ECGs.[56] Until recently, a mini-computers has been used to analyse a standard 12-lead ECG but recent developments make use of a microcomputer mounted in the ECG trolley.

62.3.7 Cardiac pacemakers

With a normal heart, the natural pacemaker is located by the right atrium and supplies regular impulses at the heart rate which travel through the conducting nervous pathways of the heart to cause the ventricles to contract. If the natural pacemaker or the conducting system fails, the ventricles contract spontaneously at a slower rate. This can result in an inadequate cardiac output, particularly under exercise condition as can a partial failure of the conducting system (heart block). The missing stimuli can be provided from a fixed rate (70 pulses per minute) stimulator or from a stimulator which only fires when it detects that a natural stimulus is absent. Such stimulators are known as cardiac pacemakers and they are connected electrically to the heart via a flexible lead. For short-term operation in hospital an external device is used, but for chronic pacing the device is implanted inside the body—usually in the axilla.[57] The circuitry of pacemakers which can respond by inhibiting their stimulus on receipt of an impulse from the heart's own pacemaker is complex and demands great reliability. The circuitry is hermetically sealed inside a metal enclosure covered with epoxy resin and an outer layer of silicone rubber to prevent rejection as a foreign body.

62.3.8 Heart sounds (phonocardiogram)

Multichannel physiological recorders usually make provision for the recording of heart sounds via a crystal microphone placed over the praecordium. The sounds are useful for timing events in the cardiac cycle, e.g. the left ventricular ejection time, for obtaining information on the operation of the heart valves and for calculating the 'systolic time intervals' as an index of cardiac performance.[58] Catheter-tip microphones and pressure transducers capable of handling heart sounds are available for sound recording within the heart. Electronic stethoscopes are not in general use.

62.3.9 Pulse or heart rate meters

A pulse rate meter is actuated by the systolic pressure pulse taken from an arterial pressure transducer or from a photoelectric pick-up placed over a surface artery, e.g. at the wrist. A heart rate meter is usually triggered by each *R* wave of the ECG. It can happen that the heart does not contract each time its natural pacemaker fires due to the presence of a heart block. The difference between the heart and pulse rates is the pulse deficit. Average rate meters are based upon the use of a leaky integrator but a number of designs exist for beat-by-beat meters.[59]

62.3.10 Respiratory instrumentation

Whilst studies of the heart command much attention, it is clearly of equal importance to be able to study respiratory as well as cardiac mechanics and quantify the gas exchange in the lungs and the volumes and flow rates together with the composition of the respired gases. In addition to lung function laboratories run by physicians, anaesthetists are also much interested in respiratory measurements.[6] The main areas are: pulmonary function studies; respiratory gas analysis; blood-gas analysis and ventilators.

Pulmonary function studies. Sophisticated lung function analysers[60] include double spirometers for the measurement of lung volumes and a carbon monoxide analyser for pulmonary diffusing capacity. Simpler spirometers with a timer will measure the peak flow rate, the vital capacity and the forced expiratory volume delivered in one second (FEV_1).

Respiratory gas analysis. A wide range of apparatus is available[6] ranging from simple oxygen analysers, infra-red gas analysers[61] and mass spectrometers[62] capable of monitoring the concentrations of several gases simultaneously on a breath-by-breath basis. A response time of less than 100 ms is required. Recent developments in infrared-emitting diodes and solid-state sensors have produced compact and rugged CO_2 analysers suitable for mounting on a ventilator.

Blood-gas analysis. Modern blood gas analysers can measure the pH, PCO_2 and PO_2 with a blood sample of only 150 μl and can also calculate derived variables such as the base excess and percentage oxygen saturation. The built-in microcomputer checks the correct functioning of all the components and the availability of consumables. Automatic rinsing and flushing occurs after sampling and automatic calibrations make the analyser available on a 24-hour basis.[13] Surveys of the subject are given in references 63, 64. Non-invasive measurements through the skin with heated electrodes are currently receiving much attention.[65,66]

Automatic lung ventilators. Ventilators are encountered in the operating room for pumping anaesthetic gas and vapour mixtures into the lungs of patients during surgery and in the intensive care unit.[67] Those for intensive care use tend to be the more complicated and incorporate alarm systems to draw attention to mains or gas failure, patient disconnection and obstruction. The flow of gas during the inspiratory phase is usually produced by a constant flow generator and a variety of expiratory phases are available including positive, atmospheric and negative end-expiratory pressure. The latest models incorporate electronic gas flow and pressure transducers, infrared CO_2 analysers and a unit for computing pulmonary mechanics.

Figure 62.6 Microprocessor-controlled central station for a patient monitoring system (courtesy Hewlett Packard Ltd)

62.3.11 Patient monitoring systems for intensive care and coronary care units[13]

The aim of patient monitoring systems in the ICU and CCU is to provide a continuous monitoring of one or more important physiological variables such as the ECG and derived values such as the heart rate and to be able to display and record these as required. The ECG and heart rate are virtually always monitored and in post-operative situations one or more blood pressures may be monitored and the systolic, mean and diastolic pressures derived.[68] The respiratory rate and volume and even the body temperature are not so often monitored automatically but by hand as necessary.

Most systems are modular in construction and are configured for the unit concerned. Facilities for interrogating up to 12 bed stations are normal. Non-fade digital storage type visual displays of the original analogue signals from the patients are shown on a visual display screen together with numerical values for the heart and respiratory rates, temperatures and blood pressures. The use of microprocessors has led to sophisticated *R*-wave algorithms, digital filtering and a variety of screen formats. *Figure 62.6* shows a modern central station for a microprocessor-based patient monitoring system.

Radio telemetry systems are employed in some units for the continuous monitoring of signals such as the ECG in coronary care units, to which patients are admitted when they are ambulatory after leaving the CCU, and uterine contractions for patients going into labour.[69]

A number of sophisticated mini-computer based patient monitoring systems have been described,[70] particularly for use with shock patients, but the present trend is to use more limited systems with a built-in microprocessor and good displays.

62.3.12 Instruments for neurological investigations

The electroencephalograph (EEG) is an 8- or 16-channel recorder for recording on a paper chart the electrical activity picked up from the brain via a montage of electrodes carefully placed on the scalp. It has been widely used in the diagnosis of epilepsy[71] and has a part to play in the diagnosis of brain death.[72] A simplified system for monitoring continuously the cerebral activity of patients in the ICU or during surgery is the cerebral function monitor.[73] The monitoring of intra-cranial pressure is important to detect swelling of the brain or an accumulation of cerebrospinal fluid in the skull. Infection problems have been minimised by the development of miniature pressure transducers which can be implanted within the skull and communicate via a radio telemetry link and fibre optic systems are also in use.[74,75]

Whilst the EEG is of help in assessing brain function, neuropathology has benefitted enormously from the advent of the CT head scanner for depicting the extent of cerebral haemorrhage and the presence of lesions.[76]

62.3.13 Instrumentation for obstetric and paediatric investigations

Because it does not involve ionising radiation which can harm the foetus, ultrasound imaging has made a great impact in the investigations of pregnancy problems.[30,77,78] Using a real-time linear array system the beating foetal heart can be detected at 10 weeks, and the position of the foetus can be determined and the foetal head and limb growth monitored throughout gestation.

The cardiotograph is a planar transducer which is worn on the maternal abdomen and picks up the foetal heart rate and maternal uterine pressure contractions. Alternatively, the foetal ECG can be picked up from a foetal scalp electrode and the intrauterine pressure via a catheter and pressure transducer. Both signals can be sent to a central monitoring station via a radio link.[79]

Premature babies can suffer sudden stoppages of their breathing (apnoea) and several types of apnoea monitor have been developed based on the monitoring of transthoracic electrical impedance changes, chest wall movements by the reflection of microwave radiation or the detection of body movements with respiration with the infant laying on a segmented air-filled mattress.[80,81]

62.3.14 Applications in physical medicine and rehabilitation

The electromyograph (EMG) for the recording of the electrical activity developed by muscles is widely used for the assessment of muscle-wasting diseases and in monitoring recovery following trauma, and automatic techniques of analysis are being developed.[82] The impedance cardiograph[40] was developed for the cardiac assessment of post-heart-attack patients by non-invasive means and computer-based systems are available for assessing the recovery of post-stroke patients with respect to their performing manual tasks.[83]

Heat treatment of muscles is available in departments of physical medicine via short wave (27 MHz) diathermies or microwave diathermies.[84] The myoelectric control of prostheses and orthoses has not proved easy, but can offer valuable assistance to some categories of amputees.[85]

62.3.15 Instrumentation for ear, nose and throat investigations

The most commonly encountered instrument in audiology departments is the audiometer for determining the acuity of the ear (equivalent to the frequency response of the ear). The application of computers to audiometry is a recent trend.[86,87]

Uncooperative subjects such as babies can have their response to sound tested by the evoked response method in which signal averaging techniques are applied to the subject's EEG to bring out the response of the brain to an applied audio signal.[88]

Hearing aids are in use by many partially deaf people and the requirements for small, cosmetically acceptable units, have benefitted from the availability of microcircuit amplifiers and miniature batteries.[89] An interesting new development is the attempt to provide a sense of hearing for the totally deaf by stimulating the auditory cortex of the brain.[90]

Electronic techniques are not much used in connection with the nose and its function, but techniques have been developed to measure the response to solfactory stimulation. Quantitative assessments of odour sensitivity have been made.[91]

Patients who have lost their vocal cords following a laryngectomy operation require an artificial larynx in the form of a tone generator applied to the throat in order for them to be able to produce understandable audible sounds.[92]

62.3.16 Electrosurgical instrumentation

Surgical diathermy apparatus (electrosurgery) consists essentially of a radio transmitter operating in the frequency range 0.5–2 MHz and with power outputs of up to a few hundred watts.[6] Many hospitals still use valve and spark-gap equipments, which, although robust, radiate many harmonics and whose output can be 100% modulated at the mains frequency and thus play havoc with physiological recording systems connected to the patient during surgery. Compact solid-state diathermies are now available giving sine-wave outputs at about 500 kHz.[93] Disposable plate electrodes and plate cable continuity monitoring have shown substantial improvements recently.

62.3.17 Psychiatric applications

Electroconvulsive therapy is widely used for the treatment of patients suffering from actue depression and improvements in the instrumentation have been made to provide a closer control of the current levels. The galvanic skin reflex is widely used for monitoring a subject's response to questioning and is based on the measurement of the electrical resistance of the skin.[94] Heart rate meters and biofeedback methods are used in the management of anxiety states and for training in relaxation.[95]

62.3.18 Ophthalmological instrumentation

The movement of the eye generates small electrical potentials which can be recorded on a physiological recording system as the electronystagmogram or the electrooculogram and these can be used to test the visual field of a patient.[96] Eye movement recording with a non-contact method is of importance in ergonomic and time and motion studies of tasks.[97] Television systems have been designed for the continuous recording of pupil diameter (pupilometry) as a variable in behavioural studies.[98]

The pressure developed within the eyeball is used in the diagnosis of glaucoma and a number of types of tonometer have been designed for clinical screening studies.[99] Potentials developed when the eye is illuminated give rise to the electroretinogram.[100]

62.3.19 Instrumentation for gastrology

Radiopills (endoradiosondes) have been employed to track pressure, temperature and pH changes in the gut and to detect the site of bleeding.[101] However, the complexity of the tracking aerial system has not made their use widespread. The availability of electrodes which can be implanted on the surface of the gut has led to detailed studies of the slow wave EMG activity of the smooth muscle of organs such as the colon[102] in investigations of peristalsis.

62.3.20 Urological instrumentation

The measurement of bladder pressure, urine flow rate and volume comprise the investigative technique of cystometry used in the study of urinary incontinence and prostatic disease.[103] In studies on incontinence the bladder and rectal pressures together with their difference and the EMGs of the rectal and urethral sphincters may be recorded.[104] Electrical stimulators have been developed to assist in keeping the urethral sphincter closed in patients with incontinence and in helping patients with a neurogenic bladder to micturate.[105] Bladder stones can be shattered *in situ* with an ultrasonic lithotriptor and then removed per-urethrally.[106]

62.3.21 Pathology laboratory instrumentation

Instrumentation in a hospital pathology department is mainly used in the haematology and biochemistry laboratories. The ever-increasing workload has encouraged the evolution of automatic methods and the advent of microcomputers has made laboratory instruments increasingly 'intelligent' in terms of quality control and ease of operation.

Haematology. Much of the automated work of a haematology laboratory is concerned with blood cell counting and sizing and the measurement of the haematrocrit (percentage of red cells). There are approximately 5 million cells in 1 ml of whole blood. Both red cells (erythrocytes), white cells (leucocytes) and platelets can be counted in samples of diluted blood passed through an orifice containing electrodes. As each cell passes through the orifice the electrical resistance changes momentarily and a count is recorded. The amplitude and shape of the pulse generated by a cell is related to its volume and can be used to produce a size histogram, calculate the mean cell volume and the haematocrit. Platelets can be counted by a dark field optical scatter method and are important in the blood clotting mechanism. Flying-spot microscope differential cell classifiers working with computer pattern recognition algorithms are starting to automate the tedious classification of blood cell types.[107]

The measurement of the blood clotting time is also important, particularly for anti-coagulant therapy. In one approach a reagent is added to the blood in a thermostatted cuvette and the time to form a clot is measured photoelectrically. Other methods use an electro-mechanical approach with a magnetic stainless-steel ball suspended in an optical path of a vertically oscillating cuvette. The formation of a clot moves the ball out of the light beam. Computer analysis of the clotting time curves can yield details of the clotting mechanisms involved.[108]

The haemoglobin content of a patient's blood sample is important since haemoglobin is the constituent of red cells which by combining with oxygen or carbon dioxide effects their transport in the blood. One laboratory method has required the blood sample to be diluted with potassium ferricyanide to form cyanmethaemoglobin and the absorption at a light wavelength of 540 nm of this compound is a measure of the haemoglobin content. Optical methods are increasingly employed, for example in blood-gas analysers to measure the haemoglobin content.

Computer-based systems are now available for automatic blood grouping.[109]

Biochemistry. Analysers capable of automatically analysing a substantial number of specimens fall into two main classes: continuous flow and discrete. In continuous flow analysers, such as the well known Autoanalyser range by Technicon Instruments, up to 20 separate analyses can be performed on one sample such as plasma or urine. The aliquots of the sample fed to each channel are kept separate by interposed air bubbles.

Simple discrete analysers of the 'suck and squirt' type place each sample into a cuvette, add the necessary reagents and then monitor the subsequent chemical reaction, often by measuring a colour change. A new development enables high-speed reactions to be followed in a micro-centrifugal system. The samples and reagents are loaded into a disposable plastic rotor containing perhaps 50 cuvettes. This is spun at a high speed under microprocessor control to mix the sample and reagents and a photometer system measures the colour changes. The arrangement copes well with relatively small batches for different tests rather than the large batches fed through a continuous flow analyser.[110]

Both large continuous flow analysers and centrifugal analysers have their own computers but off-line computer systems are used in the larger laboratories to provide quality control and to generate laboratory worksheets for the technical staff from the patient test requests and to produce reports for the wards and clinics.[111,112]

An important requirement exists for small, dedicated, analysers which can be used out-of-normal hours in the 'emergency' or 'hot' laboratory for urgent samples such as electrolytes and blood-gas values.

62.4 Safety aspects

In intensive and coronary care units and during major procedures such as cardiac surgery, a large amount of electrical equipment can be connected directly to a patient and most of it will be mains powered. Thus it is important that the patient is not exposed to potentially dangerous currents flowing into or out from him to earth or to equipment. Current practice in Europe is to have all the input circuits fully 'floating', i.e. isolated from earth. It is unfortunate that the human heart is most at risk at mains frequencies (50 to 60 Hz) with respect to the initiation of ventricular fibrillation. Direct current should not be permitted at levels in excess of 10 μA on electrodes and transducers since DC is particularly prone to produce burns at electrode–skin interfaces.

The International Electrotechnical Commission has produced its document IEC 601-1 'Safety of Medical Electrical Equipment. Part I: General Requirements' which details three classes of equipment. Class I has protection against electric shock which does not rely only on basic insulation but also has all accessible conductive parts jointed to a protective earth conductor. Class II has double or reinforced insulation and no protective earth. Class III relies for its protection against electric shock on it being supplied only with a safety extra-low voltage (SELV). Having defined the mode of protection, a further classification is possible in terms of application to patients. Type B equipment is class I, II or III equipment with an internal power supply which does not produce more than a specified leakage current and has a reliable protective earth connection. Type BF is type B equipment with an isolated (floating) part for connection to the patient. Type CF is class I or II equipment with an internal power supply and a high degree of protection against leakage currents and a floating part connected to the patient. Type B apparatus is suitable for internal and external application to the patient but not directly to the heart. In that situation type CF should be employed. Type H equipment is class I, II or III equipment with an internal power supply and a degree of protection against electric shock comparable with household electrical equipment. Requirements are also specified for anaesthetic-proof apparatus which is safe to use in explosive anaesthetic atmospheres. For type B equipment the maximum allowable leakage current under normal conditions is 0.1 mA rising to 0.5 mA under single fault conditions. In contrast, for type CF apparatus the corresponding values are reduced to 0.01 and 0.05 mA. Thus equipment connected directly to the heart should not pass leakage currents in excess of 50 μA r.m.s. It is possible to classify patients in hospital in terms of their degree of electrical risk.[13] Details of explosive anaesthetic atmospheres are given in reference 67. Details of levels of 50 Hz current which affect the human heart are given in reference 113.

It is most important that when equipment is purchased it should conform to the national safety requirements and that it should be properly maintained throughout its life. There should also be a proper acceptance test and subsequent quality control measurements.

62.5 Lasers

Small, pulsed, ruby lasers originally aroused much attention when embodied in ophthalmoscopes and used for the 'welding' back in place of detached retinas. However, the shock waves associated with the laser light impact can have deleterious effects and current practice is to make use of a continuous-wave argon laser. The argon light is absorbed by blood vessels and these can be sealed off in diabetic retinopathy, another application concerns the sealing of bleeding gastric ulcers. A fibre optic catheter has been developed for the latter application.[114] Argon lasers are also suitable for the treatment of some patients suffering from port-wine birthmarks. Continuous-wave carbon dioxide lasers with a power output of 60 W into a jointed mirror applicator have been used for surgery on tumours, organs which are prone to bleed such as the liver and for some cutaneous conditions.[115,116] Helium–neon lasers are employed in a number of cell counting and sizing instruments and for anemometry with heart valves.[117] It is most important not to look directly at a laser and to beware of scattered light and burns from stray CO_2 laser radiation in the infrared at 10.6 μm. Safety precautions with lasers are discussed in reference 118. A good introduction to medical applications for various types of laser is reference 119.

62.6 New developments in imaging

Medical imaging represents a growth area for medical instrumentation and this is very much evident with ultrasound where high-resolution real-time systems are edging out the conventional static B-scanner and systems for 'small parts' and cardiac studies are becoming well established. Similarly, the advent of stable, high-resolution wide field of view gamma cameras is opening up the possibility of radioisotope imaging in most large district general hospitals. Computer processing of gamma camera images is widely used and is becoming available for ultrasound images.

The outstanding success with the application of digital techniques to radiography was exemplified in the development of CT scanners. Now that this market is becoming satisfied, major commercial companies are pressing ahead with digital radiography in two major forms as follows.

(*a*) Scanned-beam projection radiography uses a narrow fan beam of x-rays as in CT scanning and a single row of gas-filled or solid-state detectors. The patient lies on a couch which is stepped through the beam. Each line image forms a line of a television display. By exposing each line at two different kilovoltages, typically 85 and 135 kV, it is possible to perform a 'water subtraction' and to remove from the final image the soft tissues leaving only bone and contrast medium. The reverse can also be performed leaving an image of, say, the heart without the rib cage. It is also possible to apply computer image enhancement techniques to sharpen up the outlines of the image once the x-ray signals are stored in digital form.[120,121]

(*b*) The other approach uses a high-quality convential image intensifier system and stores the television signals on video discs. A 'mask' image is first obtained before the contrast medium, is injected and then forms a series of images as the contrast passes through the blood vessels of interest. By subtracting the mask image in turn from the following images, it is possible to subtract out the background image due, for example, to bony structures such as the skull. This has produced excellent images of the cerebral vasculature and these can be obtained with an intravenous rather than an arterial injection, opening up out-patient studies for angiography.[122]

An entirely different growth area in medical imaging and diagnosis occurs with nuclear magnetic resonance techniques which do not employ ionising radiation. The instrumentation is sophisticated requiring a large and very homogeneous electromagnet for whole body imaging, but has the advantage that it can image specific types of tissue and this is being applied to tumour detection.[123] An off-shoot of NMR is TMR, topical magnetic resonance[124] which can follow biochemical changes occurring in living human tissue, for example in the muscles of a forearm following the application of a tourniquet for one hour. Dynamic studies of organ metabolism are feasible and work is proceeding to study kidney preservation and viability for transplant purposes.

Nuclear magnetic imaging depends upon the measurement of the response of the protons within the hydrogen atom to a known radio frequency stimulus. When a patient is placed in a magnetic field, all his protons align with the field and precess around it in a similar fashion to a child's spinning top processing around its vertical gravitational field. When megahertz radio-frequency electromagnetic radiation is directed into the protons, they absorb energy from the beam in a resonance absorption and change their alignment relative to the magnetic field. When the radio frequency is switched off, the nuclei radiate their surplus energy to their surroundings at the same resonant frequency.[125] As the protons fall back (relax) to their original alignment, the resonant frequency emitted can be measured. The length of time required for this relaxation is associated with the environment of the protons since the easier it is for them to pass energy to neighbouring atoms the more rapidly they can return to their original state. The intensity of the emitted radiation falls off exponentially with a time characteristic of the environment—the spin–lattice relaxation time.

To distinguish between different regions of the body having the same resonant frequency a field gradient is applied across the applied electromagnetic field so that the field is greater on one side than the other. The region in the slightly weaker field will resonate at a lower frequency than that in the stronger field and can thus be distinguished from it by the lower frequency emitted at the time of relaxation. The use of the field gradient facilitates the identification of the position of the particular type of tissue from the frequency of the emitted signal, the amount of water (or proton concentration) from the magnitude of the signal and the spin–lattice relaxation time (T_1) from the decay of the signal. Different human tissues have their own values of T_1 and by this means NMR can distinguish primary and secondary malignant tumours and benign lesions.[125] A well illustrated guide to the basic principles of NMR imaging of the human body is provided in reference 126.

The University of Aberdeen NMR imager[125] utilises a four-coil air-cored magnet which can accept the whole human body. It produces a static field of 0.04 T giving a proton NMR frequency of 1.7 MHz for the hydrogen protons of the body. The imager collects data for just over 2 min for the construction of a transverse section image 18 mm thick. Sections are made at 20 mm intervals over the region of interest. Reference 125 reports data from some patients. In one 75 year old man NMR images demonstrated dilated bile ducts and an enlarged gall bladder. The T_1 of the bile in the ducts and gall bladder was normal at about 400 ms. The T_1 of the two kidneys was in the normal range of 300–320 ms. An irregular lesion arising from the anterolateral aspect of the lower pole of the right kidney, however, had a value of T_1 in excess of one second. This value is consistent with a cyst containing clear fluid since the T_1 for both water and urine has been shown to be longer than one second.

References

1 GEDDES, L. A. and BAKER, L. E., *Principles of Applied Biomedical Instrumentation*. 2nd edn. New York, Wiley-Interscience (1975)

2 HILL, D. W. and WATSON, B. W. (eds), *I.E.E. Medical Electronics Monographs*, London, Peter Peregrinus Ltd (1971)

3 COONEY, D. O. (ed), *Advances in Biomedical Engineering*, Parts 1 and 2, Vol. 6, New York, Marcel Dekker (1980)

4 *Excerpta Medica*, Section 27, Biophysics, Bioengineering and Medical Instrumentation, Excerpta Medica, PO Box 1126, 1000 BC, Amsterdam

5 *Topics*, INSPEC Marketing Department, Savoy Place, London WC2

6 HILL, D. W., *Electronic Techniques in Anaesthesia and Surgery*, 2nd edn, London, Butterworths (1973)

7 COBBOLD, R. S. C., *Transducers for Biomedical Measurements —Principles*, New York, Wiley-Intersience (1974)

8 ROLFE, P. (ed), *Non-Invasive Physiological Measurements*, Vol. 1, London, Academic Press (1978)

9 ZIMMERMAN, J. and RECTOR, A., *Computers for the Physician's Office*, Portland, Oregon, Research Studies Press (1978)

10 REINHOFF, O. and ABRAMS, M. E., *The Computer in the Doctor's Office*, Amsterdam: North Holland (1980)

11 COLLEN, M. F., 'Data Processing Techniques for Multitest Screening and Hospital Facilities', *Hospital Information Systems*, eds G. A. Bekey and M. D. Schwartz, New York, Marcel Dekker (1972)

12 PINCIROLI, F. and ANDERSON, J. (eds), *Changes in Health Care Instrumentation due to Microprocessor Technology*, Amsterdam, North Holland (1981)

13 HILL, D. W. and DOLAN, A. M., *Intensive Care Instrumentation*, London, Academic Press (1982)

14 CHRISTENSON, E. E., CURRY, T. S. and DOWDEY, J. E., *An Introduction to the Physics of Diagnostic Radiology*, Lea and Febiger (1978)

15 HILL, D. R. (ed), *Principles of Diagnostic X-Ray Apparatus*, London, Macmillan (1979)

16 EVENS, R. G., ALFIDI, R. J. and HAAGA, J. R., 'Body Computed Tomography: A Clinically Important and Efficacious Procedure', *Radiology*, **123**, 239–240 (1977)

17 SCHERTEL, I., 'Irradiation Planning with Computer Tomography and Treatment Planning Equipment', *Strahlentherapie*, **155**, 757–759 (1979)

18 PARKER, R. P., SMITH, P. H. S. and TAYLOR, D. M., *Basic Science of Nuclear Medicine*, Edinburgh, Churchill Livingstone (1978)

19 LIM, C. B., HOFFER, P. B., ROLLO, F. D. and LICIEN, D. I., 'Performance Evaluation of Recent Wide Field Scintillation Gamma Cameras', *J. Nucl. Med.*, **19**, 942–947 (1978)

20 HAUSER, M. F. and GOTTSCHALK, A., 'Comparison of the Anger Tomographic Scanner and the 15-inch Scintillation Camera Gallium Imaging', *J. Nucl. Med.*, **19**, 1074–1077 (1978)

21 MARISZ, P., JOHNS, E. and SCHOKER, O., 'A Whole Body Counter with an Invariant Response for Whole Body Analysis', *Eur. J. Nucl. Med.*, **3**, 129–135 (1978)

22 WILLIAMS, E. D., BODDY, K., HARVEY, I. and HAYWOOD, J. K., 'Calibration and Evaluation of a System for Total Body *in vivo* Activation Analysis using 14 MeV Neutrons', *Phys. Med. Biol.*, **23**, 405–415 (1978)

23 GARN, L. E. and PENITO, F. C., 'Thermal Imaging with Pyroelectric Vidicons', *I.E.E.E. Trans. Electron. Devices*, **24**, 1221–1228 (1977)

24 NEWMANN, P., DAVISON, M., JACKSON, I. and JAMES, W. B., 'The Analysis of Temperature Frequency Distributions over the Breasts by a Computerised Thermography System', *Acta. Thermographica*, **4**, 3–8 (1979)

25 WELLS, P. N. T., *Biomedical Ultrasonics*, London, Academic Press (1977)

26 MATZUK, T. and SKOLNICK, M. L., 'Ultrasound Grey Scale Display on Storage Oscilloscope System', *Ultrasonics*, **15**, 221–225 (1977)

27 DETER, R. L. and HOBBINS, J. C., 'A Survey of Abdominal Ultrasound Scanners: The Clinicians Point of View', *Proc. I.E.E.E.*, **67**, 664–671 (1979)

28 MAGINNES, M. G., 'Methods and Terminology for Diagnostic Ultrasound Imaging Systems', *Proc. I.E.E.E.*, **67**, 641–653 (1979)

29 PEDERSEN, J. F. and NORTHEVED, A., 'An Ultrasonic Multitransducer Scanner for Real Time Heart Imaging', *J. Clin. Ultrasound*, **5**, 11–15 (1977)

30 WELLS, P. N. T. and ZISKIN, M. L. (eds), *New Techniques and Instrumentation in Ultrasonography*, Edinburgh, Churchill Livingstone (1980)

31 WYATT, D. G., 'Electromagnetic Blood-Flow Measurement', *I.E.E. Medical Electronics Monographs 1–6*, Ed B. W. Watson, London, Peter Peregrinus Ltd (1971)

32 ASTLEY, C. A., HOHIMER, A. R. and STEPHENSON, R. B., 'Effect of implant duration on *in vivo* sensitivity of electromagetic flow transducers', *Am. J. Physiol., Heart Circ. Physiol.*, **5**, H508–H512 (1979)

33 WILLIAMS, B. T., SANCHO-FORRES, S., CLARK, D. B., ABRAMS, T. D. and SCHENK, W. F. Jr, 'Continuous Long-Term Measurement of Cardiac Output in Man', *Ann. Surg.*, **174**, 357–363 (1971)

34 WILLIAMS, B. T., SANCHO-FORRES, S., CLARK, D. B., ABRAMS, I. D., SCHENK, W. G. Jr. and BAREFOOT, C. A., 'The Williams–Barefoot Extractable Blood Flow Probe', *J. Thoracic & Cardiovasc. Surg.*, **63**, 917–921 (1972)

35 WOODCOCK, J. P. and SKIDMORE, R., 'Principles and Applications of Doppler Ultrasound', *New Techniques and Instrumentation for Ultrasonography*, eds P. N. T. Wells and M. C. Ziskin, Edinburgh, Churchill Livingstone (1980)

36 BORN, G. V. R., MELLING, A. and WHITELAW, J. H., 'Laser Doppler Microscope for Blood Velocity Measurements', *Biorheology*, **15**, 363–372 (1978)

37 CLARK, C., 'A Local Thermal Dilution Flowmeter for the Measurement of Venous Flow in Man', *Med. Electron. Biol. Engng.*, **6**, 133 (1968)

38 GREENFIELD, A. D. M., WHITNEY, R. J. and MOWBRAY, J. F., 'Methods for the Investigation of Peripheral Blood Flow', *Brit. Med. Bull.*, **19**, 101 (1963)

39 RUBIN, S. A., QUILTER, R. and BATTAGIN, R., 'An Accurate and Rapid Inflation Device for Pneumatic Cuffs', *Am. J. Physiol., Heart Circ. Physiol.*, **3**, H740–H742 (1978)

40 KUBICEK, W. G., PATTERSON, R. P. and WITSOE, D. A., 'Impedance Cardiography as a Non-Invasive Method of Monitoring Cardiac Function and Other Parameters of the Cardiovascular System', *Ann. N.Y. Acad. Sci.*, **170**, 724–752 (1970)

41 ARENSON, H. M. and MOHAPATRA, S. N., 'Evaluation of Electrical Impedance Plethysmography for the Non-invasive Measurement of Blood Flow', *Brit. J. Anaesth., Proc. Anaes. Res. Soc.* (Sept. 1976)

42 MULLICK, S. C., WHEELER, H. B. and SONGSTER, G. P., 'Diagnosis of Deep Venous Thrombosis by Measurement of Electrical Impedance', *Am. J. Surg.*, **119**, 417 (1970)

43 SARREL, P. M., FODDY, J. and MCKINNON, J. B., 'Investigation of Human Sexual Response Using a Cassette Recorder', *Arch. Sex. Behav.*, **6**, 341–348 (1977)

44 GEDDES, L. A., *The Direct and Indirect Measurement of Blood Pressure*, Chicago, Year Book Press (1970)

45 COBBOLD, R. S. C., *Transducers and Biomedical Measurements: Principles*, New York, Wiley-Interscience (1974)

46 MILLAR, H. D. and BAKER, L. E., 'A Stable Ultra-Miniature Catheter-Tip Pressure Transducer', *Med. Biol. Engng.*, **11**, 86 (1972)

47 GEDDES, L. A. and MOORE, A. G., 'The Efficient Detection of Korotkoff Sounds', *Med. Electron. Biol. Engng.*, **6**, 603 (1968)

48 HOCHBERG, H. M. and SALOMAN, H., 'Accuracy of an Automated Ultrasound Blood Pressure Monitor', *Curr. Ther. Res.*, **13**, 129–138 (1971)

49 COLES, J. S., MARTIN, W. E., CHEUNG, P. W. and JOHNSON, C. C., 'Clinical Studies with a Solid State Fibre Optic Oximeter', *Am. J. Cardiol.*, **29**, 383 (1972)

50 HILL, D. W., THOMPSON, D., VALENTINUZZI, M. E. and PATE, T. D., 'The Use of a Compartmental Hypothesis for the Estimation of Cardiac Output from Dye Dilution Curves and the Analysis of Radiocardiograms', *Med. Biol. Engng*, **11**, 43–54 (1973)

51 KUBICEK, W. G., KARNEGIS, J. N., PATTERSON, R. P., WITSOE, D. A. and MATTSON, R. H., 'Development and Evaluation of an Impedance Cardiac Output System', *Aerospace Med.*, **37**, 377–387 (1966)

52 HILL, D. W. and MOHAPATRA, S. N., 'The Current Status of Electrical Impedance Techniques for the Monitoring of Cardiac Output and Limb Blood Flow', *I.E.E. Medical Electronics Monographs* 23–27, eds D. W. Hill and B. W. Watson, Stevenage, Peter Peregrinus Ltd (1977)

53 WINTERBOTTOM, J. T., 'Equiping an ECG Department', *Brit. J. Clin. Equip.*, **5**, 176–179 (1980)

54 OLIVER, G. C., NOLLE, F. M. and WOLFF, G. A., 'Detection of Premature Ventricular Contractions with a Clinical System for Monitoring Electrocardiographic Rhythm', *Computer*

Biomed. Res., 4, 523–541 (1971)

55 THOMAS, L. J. Jr., CLACK, K. W. and MEAD, C. N., 'Automated Cardiac Dysrhthymia Analysis', *Proc. I.E.E.E.*, **67**, 1322–1337 (1979)

56 MACFARLANE, P. W. and LAWRIE, T. D. V., 'An Introduction to Automated Electrocardiogram Interpretation', London, Butterworths (1974)

57 KENNY, J., 'Cardiac Pacemakers', *I.E.E. Medical Electronics Monographs* 7–12, Stevenage, Peter Peregrinus Ltd (1974)

58 WEISSLER, A. M., HARRIS, W. S. and SCHOENFIELD, C. D., 'Systolic Time Intervals in Heart Failure in Man', *Circulation*, **37**, 149 (1969)

59 MASON, C. A. and SHOUP, J. F., 'Cardiotachometer Designs', *Med. Biol. Engng. Comput.*, **17**, 349 (1979)

60 PRESTON, T. D., 'Automated Lung Function Tests', *Brit. J. Clin. Equip.*, **5**, 194–197 (1980)

61 HILL, D. W. and POWELL, T., 'Non-Dispersive Infra-Red Gas Analysis', London, Adam Hilger (1968)

62 PAYNE, J. P., BUSHMAN, J. B. and HILL, D. W. (eds), 'The Medical and Biological Application of Mass Spectrometry', London, Academic Press (1971)

63 HILL, D. W., 'Electrode Systems for the Measurement of Blood-Gas Tensions and Contents', *Scientific Foundations of Anaesthesia*, eds C. Scurr and S. Feldman, London, Heinemann Medical Books (1974)

64 HAHN, G. E. W., 'Techniques for the Measurement of Partial Pressures of Gases in the Blood', *J. Phys. Elec. Instrum.*, **13**, 470–482 (1980) and **14**, 783–798 (1981)

65 BERAN, A. V., SNIGELAWA, G. Y., YOUNG, H. N. and HUXTABLE, R. F., 'An Improved Sensor and a Method for Transcutaneous CO_2 Monitoring', *Acta. Anaesth., Scand.*, **22**, Suppl. 68, 110–117 (1978)

66 VERSMOLD, H. T., LINDERKAMP, O. and HOLZMANN, M., Limits of tcPO2 Monitoring in Sick Neonates', *Acta. Anaesth. Scand.*, **22**, Suppl. 68, 88–90 (1978)

67 HILL, D. W., *Physics Applied to Anaesthesia*, London, Butterworths (1981)

68 SANDMAN, A. M. and HILL, D. W., 'An Analogue Pre-Processor for the Analysis of Arterial Blood Pressure Waveforms', *Med. Biol. Engng.*, **12**, 360–363 (1976)

69 VERTA, P. S. and KONOPASEK, F., 'The Cardiac Disaster Alarm', *I.E.E.E. Trans. Bio-med. Engng.*, **19**, 248–251 (1972)

70 SHUBIN, H., WEIL, M. H., RALLEY, N. and AFIFI, A. A., 'Monitoring the Critically Ill Patient with the Aid of a Digital Computer', *Computers Biomed. Res.*, **4**, 460–473 (1971)

71 SOREL, I., RONEQUOY PONSAR, M. and HARMANT, J., 'Electroecephalogram and CAT Scan in 393 Cases of Epilepsy', *Acta. Neurol. Belg.*, **78**, 242–252 (1978)

72 SCOTT, D. S., *Understanding EEG*, London, Duckworth (1976)

73 SECHZER, P. H. and OSPINA, I., 'Cerebral Function Monitor Evaluation in Anesthesia/Critical Care', *Curr. Ther. Res. Clin. Exp.*, **22**, 335–347 (1977)

74 LEVIN, A. B., 'The Use of a Fibreoptic Intracranial Pressure Transducer in the Treatment of Head Injuries', *U.S. J. Trauma*, **17**, 767–774 (1977)

75 BEKS, J. W. F., ABARDA, S. and GIELES, A. C. M., 'Extradural Transducer for Monitoring Intracranial Pressure', *Acta. Neurochir.*, **38**, 245–251 (1977)

76 DAY, R. E., THOMSON, J. L. G. and SCHOTT, W. H., 'Computerised Axial Tomography and Acute Neurological Problems of Childhood', *Arch. Dis. Child.*, **53**, 2–11 (1978)

77 WELLS, P. N. T., *Biomedical Ultrasonics*, London, Academic Press (1977)

78 WELLS, P. N. T., 'Ultrasonic Imaging: Pulse-Echo Techniques', *I.E.E. Medical Electronics Monographs* 23–27', eds D. W. Hill and B. W. Watson, Stevenage, Peter Peregrinus Ltd (1977)

79 FLYNN, A. M., KELLY, J., HOLLINS, G. and LYNCH, P. F., 'Ambulation in Labour', *Brit. Med. J.*, *2*, 591–593 (1978)

80 BARETICH, M. F., 'A Tester for Respiration Rate Meters', *J. Clin. Engng.*, **4**, 339–341 (1979)

81 KATONA, P. G., DURAND, D. and STERN, K., 'Microprocessor-Controlled Memory for Cardiopulmonary Monitoring of High Risk Infants', *I.E.E.E. Trans. Bio-med. Engng.*, **24**, 536–538 (1977)

82 BOYD, D. C., BRATTY, P. J. A. and LAWRENCE, P. D., 'A Review of Methods of Automatic Analysis in Clinical Electromyography', *Comput. Biol. Med.*, **6**, 179–190 (1973)

83 LYNN, P. A., PARKER, W. R., REED, G. A. J., BALDWIN, J. F. and PISWORTH, B. W., 'New Approaches to Modelling the Disabled Human Operator', *Med. Biol. Engng. Comput.*, **17**, 344 (1979)

84 GURU, B. S. and CHEN, K. M., 'Hyperthernia in Local EM Heating and Local Conductivity Change', *I.E.E.E. Trans. Bio-med. Engng.*, **24**, 473–477 (1977)

85 SHANNON, G. F., 'A Myoelectrically Controlled Prosthesis with Sensory Feedback', *Med. Biol. Engng. Comput.*, **17**, 73 (1979)

86 MEYER, C. R. and SUTHERLAND, H. C. Jr., 'A Technique for Totally Automated Audiometer', *Japan–Scand. Audiol.*, **7**, 105–109 (1978)

87 SAKABE, N., HIRAI, Y. and ITANI, E., 'Modification and Application of the Computerised Automatic Audiometer', *Japan-Scand. Audiol.*, **7**, 105–109 (1978)

88 BRACKMANN, D. E. and SELTERS, W. A., 'Electrical Response Audiometry: Clinical Applications', *Otolaryngol. Clin. North. Am.*, **11**, 7–18 (1978)

89 WALDEN, B. E. and KASTEN, R. N., 'Threshold Improvement and Acoustic Gain with Hearing Aids', *Audiology*, **15**, 413–420 (1976)

90 DOBELLE, W. H., STENSAAS, S. S., MLADESOVSKY, M. G. and SMITH, J. B., 'A Prosthesis for the Deaf Based on Cortical Stimulation', *Trans. Am. Otol. Soc.*, **61**, 157–175 (1973)

91 VAN DROGELEN, W., HOLLEY, A. and DOVING, K. B., 'Convergence in the Olfactory System: Quantitative Aspects of Odour Sensitivity', *J. Theor. Biol.*, **71**, 39–48 (1978)

92 KNORR, S. G. and ZWITMAN, D. H., 'The Design of a Wireless-Controlled Intra-Oral Electrolaryn'.

93 GEDDES, L. A., SILVA, L. F., DEWITT, D. P. and PEARCE, J. A., 'What's New in Electrosurgical Instrumentation?', *Med. Instrum.*, **11**, 355–359 (1977)

94 HALL, S. H., 'A Bridge for Continuous Resistivity Observations', *J. Phys. E. Sci. Instrum.*, **9**, 728–729 (1976)

95 WATSON, B. W., WOOLEY-HART, A. and TIMMONS, B. H., 'Biofeedback Instruments for the Management of Anxiety and for Realaxation Training', *J. Biomed. Engng.*, **1**, 58–63 (1979)

96 ARMON, H., WEINMAN, J. and PELEG, A., 'Automatic Testing of the Visual Field using Electro-oculographic Potentials', *Docum. Ophthal.*, **43**, 51–63 (1977)

97 HAINES, J. D., 'Non-Contacting Ultrasound Transducer System for Eye Movement Recording', *Ultrasound Med. Biol.*, **3**, 639–645 (1977)

98 USA, S., MURASE, K. and IKEGAYA, K., 'Analysis of the Dynamic Performance of a Television Pupilometer', *Japan J. Med. Electron. Bioo. Engng.*, **16**, 177–183 (1978)

99 THORNBURN, W., 'The Accuracy of Clinical Application Tonometry', *Acta. Ophthalmol.*, **56**, 1–5 (1978)

100 NAKAMURA, Z., 'Human Electroretinogram with Skin Electrode', *Japan. J. Ophthalmol.*, **22**, 101–113 (1978)

101 DELCHAR, I. A. and SMITH, M. J. A., 'An Improved Method for Detecting Passive Pills', *Phys. Med., Biol.*, **21**, 577–583 (1976)

102 SMALLWOOD, R. H., LINKENS, D. A., KWOK, H. L. and STODDARD, C. J., 'Use of Autoregressive Modelling Techniques for the Analysis of Colonic Myoelectric Activity in Man', *Med. Biol. Engng. Comput.*, **18**, 591 (1980)

103 KONDO, A., MITSUYA, H. and TORI, H., 'Computer Analysis of Micturition Parameters and Accuracy of Uroflow Meter', *Japan J. Urol.*, **33**, 337–344

104 DOYLE, P. T., HILL, D. W., PERRY, I. R. and STANTON, S. L., 'Computer Analysis of Electromyographic Signals from the Human Bladder and Urethral and Anal Sphinoters', *Invest. Urol.*, **13**, 205–210 (1975)

105 SEIFERTH, J., HEISSING, J. and KARKAMP, H., 'Experiences and Critical Comments on the Temporary Intravesical Electro-Stimulation of the Neurogenic Bladder in Spina Bifida Children', *GFR Urol. Int.*, **33**, 279–284 (1978)

106 EL FAHIG, S. and WALLACE, D. M., 'Ultrasonic Lithotriptor for Urethral and Bladder Use', *Brit. J. Urol.*, **50**, 255–256 (1978)

107 DURIE, B. G. M., VAUGHT, I. and CHEN, Y. P., 'Discrimination between Human T and B Lymphocytes and Monocytes by Computer Analysis of Digitised Data from Scanning Microphotometry', *Blood*, **51**, 579–589

108 FRANK, H. I. I., DREESEN, V. and HENKER, H. C., 'Computer Evaluation of Performance Curves for the Estimation of Extrinsic Circulation Factors', *Comput. Med. Biol.*, **8**, 65–70 (1978)

109 GOVAERTS, A. and SCHREYER, H., 'Integration of a Groupamatic MG50 in the Routine Work of a Hospital Blood Transfusion Centre', *Rev. Fr. Transfus. Immuno-Hematol.*, **21**, 715–719 (1978)

110 WILLIAMS, D. L., NUNN, R. F. and MARKS, V. (eds), *Scientific Foundations of Clinical Biochemistry*, 2nd edn, London, Heinemann Medical Books (1978)

111 SIMS, G. E., *Automation of a Biochemical Laboratory*, London, Butterworths (1972)

112 CAVILL, I., RICKETTS, C. and JACOBS, A., *Computers in Haematology*, London, Butterworths (1975)

113 RAFTERY, E. B., GREEN, N. L. and GREGORY, I. C., 'Disturbances of Heart Rhythm Produced by Leakage Current in Human Subjects', *Cardiovasc. Res.*, **9**, 256–262

114 KIMURA, W. D., GULACSIK, C. and AUTH, D. C., 'Use of Gas Jet Appositional Pressurization in Endoscopic Laser Photocoagulation', *I.E.E.E. Trans. Bio-med. Engng.*, **25**, 218–224 (1978)

115 HALL, R. R., HILL, D. W. and BEACH, A. D., 'A Carbon Dioxide Surgical Laser', *Ann. R. Coll. Surg.*, **48**, 181–188 (1971)

116 LABANDER, H. and KAPLAN, I., 'Experience with Continuous Laser in the Treatment of Suitable Cutaneous Conditions', *J. Dermatol. Surg. Oncol.*, **3**, 527–530 (1977)

117 YOGANATHAN, A. A. P., REAME, H. H., CORCORAN, W. H. and HARRISON, E. C., 'Laser-Doppler Anemometer to Study Velocity Fields in the Vicinity of Prosthetic Heart Valves', *Med. Biol. Engng. Comput.*, **17**, 38 (1979)

118 ELECCION, M., 'Laser Hazards', *I.E.E.E. Spectrum*, **10**, 32 (1973)

119 KOEBNER, H. K. (ed), *Lasers in Medicine*, Vol. 1, New York, Wiley-Interscience (1980)

120 ALVAREZ, R. E. and MACOVOSKI, A., 'Energy-Selective Reconstructions in X-Ray Computerised Tomography', *Phys. Med. Biol.*, **21**, 733–744 (1978)

121 BRODY, W. R. and MACOVSKI, A., 'Intravenous Angiography Using Scanned Projection Radiography: Preliminary Investigation of a New Method', *Invest. Radiol.*, **1**, 220–223 (1980)

122 GENERAL ELECTRIC COMPANY, *Digital Radiography*, Milwaukee (1980)

123 SMITH, F. W., HUTCHINSON, G. M. S., MALLARD, G. R., JOHNSON, G., REDPATH, G. W., SELBY, R. C., REED, ANN and SMITH, C. C., 'Oesophageal Clinical Use of Nuclear Magnetic Resonance Imaging', *Brit. Med. J.*, **282**, 510–512

124 GORDON, R. E., 'From Molecules to man', *Phys. Bull.*, **32**, 178–180

125 SMITH, F. W., HUTCHINSON, J. M. S., MALLARD, J. R., REID, A., JOHNSON, G. and REDPATH, T. W., 'Renal Cyst or Tumour?', *Diagnostic Imaging*, ed W. Penn, Basel, Karger (1981)

126 GENERAL ELECTRIC COMPANY, *NMR—An Introduction*, Medical Systems Division, Milwaukee (1981)

Index

109 GOVAERTS, A. and SCHREYER, H., 'Integration of a Groupamatic MG50 in the Routine Work of a Hospital Blood Transfusion Centre', *Rev. Fr. Transfus. Immuno-Hematol.*, **21**, 715–719 (1978)

110 WILLIAMS, D. L., NUNN, R. F. and MARKS, V. (eds), *Scientific Foundations of Clinical Biochemistry*, 2nd edn, London, Heinemann Medical Books (1978)

111 SIMS, G. E., *Automation of a Biochemical Laboratory*, London, Butterworths (1972)

112 CAVILL, I., RICKETTS, C. and JACOBS, A., *Computers in Haematology*, London, Butterworths (1975)

113 RAFTERY, E. B., GREEN, N. L. and GREGORY, I. C., 'Disturbances of Heart Rhythm Produced by Leakage Current in Human Subjects', *Cardiovasc. Res.*, **9**, 256–262

114 KIMURA, W. D., GULACSIK, C. and AUTH, D. C., 'Use of Gas Jet Appositional Pressurization in Endoscopic Laser Photocoagulation', *I.E.E.E. Trans. Bio-med. Engng.*, **25**, 218–224 (1978)

115 HALL, R. R., HILL, D. W. and BEACH, A. D., 'A Carbon Dioxide Surgical Laser', *Ann. R. Coll. Surg.*, **48**, 181–188 (1971)

116 LABANDER, H. and KAPLAN, I., 'Experience with Continuous Laser in the Treatment of Suitable Cutaneous Conditions', *J. Dermatol. Surg. Oncol.*, **3**, 527–530 (1977)

117 YOGANATHAN, A. A. P., REAME, H. H., CORCORAN, W. H. and HARRISON, E. C., 'Laser-Doppler Anemometer to Study Velocity Fields in the Vicinity of Prosthetic Heart Valves', *Med. Biol. Engng. Comput.*, **17**, 38 (1979)

118 ELECCION, M., 'Laser Hazards', *I.E.E.E. Spectrum*, **10**, 32 (1973)

119 KOEBNER, H. K. (ed), *Lasers in Medicine*, Vol. 1, New York, Wiley-Interscience (1980)

120 ALVAREZ, R. E. and MACOVOSKI, A., 'Energy-Selective Reconstructions in X-Ray Computerised Tomography', *Phys. Med. Biol.*, **21**, 733–744 (1978)

121 BRODY, W. R. and MACOVSKI, A., 'Intravenous Angiography Using Scanned Projection Radiography: Preliminary Investigation of a New Method', *Invest. Radiol.*, **1**, 220–223 (1980)

122 GENERAL ELECTRIC COMPANY, *Digital Radiography*, Milwaukee (1980)

123 SMITH, F. W., HUTCHINSON, G. M. S., MALLARD, G. R., JOHNSON, G., REDPATH, G. W., SELBY, R. C., REED, ANN and SMITH, C. C., 'Oesophageal Clinical Use of Nuclear Magnetic Resonance Imaging', *Brit. Med. J.*, **282**, 510–512

124 GORDON, R. E., 'From Molecules to man', *Phys. Bull.*, **32**, 178–180

125 SMITH, F. W., HUTCHINSON, J. M. S., MALLARD, J. R., REID, A., JOHNSON, G. and REDPATH, T. W., 'Renal Cyst or Tumour?', *Diagnostic Imaging*, ed W. Penn, Basel, Karger (1981)

126 GENERAL ELECTRIC COMPANY, *NMR—An Introduction*, Medical Systems Division, Milwaukee (1981)

Biomed. Res., **4**, 523–541 (1971)
55 THOMAS, L. J. Jr., CLACK, K. W. and MEAD, C. N., 'Automated Cardiac Dysrhthymia Analysis', *Proc. I.E.E.E.*, **67**, 1322–1337 (1979)
56 MACFARLANE, P. W. and LAWRIE, T. D. V., 'An Introduction to Automated Electrocardiogram Interpretation', London, Butterworths (1974)
57 KENNY, J., 'Cardiac Pacemakers', *I.E.E. Medical Electronics Monographs* 7–12, Stevenage, Peter Peregrinus Ltd (1974)
58 WEISSLER, A. M., HARRIS, W. S. and SCHOENFIELD, C. D., 'Systolic Time Intervals in Heart Failure in Man', *Circulation*, **37**, 149 (1969)
59 MASON, C. A. and SHOUP, J. F., 'Cardiotachometer Designs', *Med. Biol. Engng. Comput.*, **17**, 349 (1979)
60 PRESTON, T. D., 'Automated Lung Function Tests', *Brit. J. Clin. Equip.*, **5**, 194–197 (1980)
61 HILL, D. W. and POWELL, T., 'Non-Dispersive Infra-Red Gas Analysis', London, Adam Hilger (1968)
62 PAYNE, J. P., BUSHMAN, J. B. and HILL, D. W. (eds), 'The Medical and Biological Application of Mass Spectrometry', London, Academic Press (1971)
63 HILL, D. W., 'Electrode Systems for the Measurement of Blood-Gas Tensions and Contents', *Scientific Foundations of Anaesthesia*, eds C. Scurr and S. Feldman, London, Heinemann Medical Books (1974)
64 HAHN, G. E. W., 'Techniques for the Measurement of Partial Pressures of Gases in the Blood', *J. Phys. Elec. Instrum.*, **13**, 470–482 (1980) and **14**, 783–798 (1981)
65 BERAN, A. V., SNIGELAWA, G. Y., YOUNG, H. N. and HUXTABLE, R. F., 'An Improved Sensor and a Method for Transcutaneous CO_2 Monitoring', *Acta. Anaesth., Scand.*, **22**, Suppl. 68, 110–117 (1978)
66 VERSMOLD, H. T., LINDERKAMP, O. and HOLZMANN, M., Limits of tcPO2 Monitoring in Sick Neonates', *Acta. Anaesth. Scand.*, **22**, Suppl. 68, 88–90 (1978)
67 HILL, D. W., *Physics Applied to Anaesthesia*, London, Butterworths (1981)
68 SANDMAN, A. M. and HILL, D. W., 'An Analogue Pre-Processor for the Analysis of Arterial Blood Pressure Waveforms', *Med. Biol. Engng.*, **12**, 360–363 (1976)
69 VERTA, P. S. and KONOPASEK, F., 'The Cardiac Disaster Alarm', *I.E.E.E. Trans. Bio-med. Engng.*, **19**, 248–251 (1972)
70 SHUBIN, H., WEIL, M. H., RALLEY, N. and AFIFI, A. A., 'Monitoring the Critically Ill Patient with the Aid of a Digital Computer', *Computers Biomed. Res.*, **4**, 460–473 (1971)
71 SOREL, I., RONEQUOY PONSAR, M. and HARMANT, J., 'Electroecephalogram and CAT Scan in 393 Cases of Epilepsy', *Acta. Neurol. Belg.*, **78**, 242–252 (1978)
72 SCOTT, D. S., *Understanding EEG*, London, Duckworth (1976)
73 SECHZER, P. H. and OSPINA, I., 'Cerebral Function Monitor Evaluation in Anesthesia/Critical Care', *Curr. Ther. Res. Clin. Exp.*, **22**, 335–347 (1977)
74 LEVIN, A. B., 'The Use of a Fibreoptic Intracranial Pressure Transducer in the Treatment of Head Injuries', *U.S. J. Trauma*, **17**, 767–774 (1977)
75 BEKS, J. W. F., ABARDA, S. and GIELES, A. C. M., 'Extradural Transducer for Monitoring Intracranial Pressure', *Acta. Neurochir.*, **38**, 245–251 (1977)
76 DAY, R. E., THOMSON, J. L. G. and SCHOTT, W. H., 'Computerised Axial Tomography and Acute Neurological Problems of Childhood', *Arch. Dis. Child.*, **53**, 2–11 (1978)
77 WELLS, P. N. T., *Biomedical Ultrasonics*, London, Academic Press (1977)
78 WELLS, P. N. T., 'Ultrasonic Imaging: Pulse-Echo Techniques', *I.E.E. Medical Electronics Monographs* 23–27', eds D. W. Hill and B. W. Watson, Stevenage, Peter Peregrinus Ltd (1977)
79 FLYNN, A. M., KELLY, J., HOLLINS, G. and LYNCH, P. F., 'Ambulation in Labour', *Brit. Med. J.*, *2*, 591–593 (1978)
80 BARETICH, M. F., 'A Tester for Respiration Rate Meters', *J. Clin. Engng.*, **4**, 339–341 (1979)
81 KATONA, P. G., DURAND, D. and STERN, K., 'Microprocessor-Controlled Memory for Cardiopulmonary Monitoring of High Risk Infants', *I.E.E.E. Trans. Bio-med. Engng.*, **24**, 536–538 (1977)
82 BOYD, D. C., BRATTY, P. J. A. and LAWRENCE, P. D., 'A Review of Methods of Automatic Analysis in Clinical Electromyography', *Comput. Biol. Med.*, **6**, 179–190 (1973)
83 LYNN, P. A., PARKER, W. R., REED, G. A. J., BALDWIN, J. F. and PISWORTH, B. W., 'New Approaches to Modelling the Disabled Human Operator', *Med. Biol. Engng. Comput.*, **17**, 344 (1979)
84 GURU, B. S. and CHEN, K. M., 'Hyperthernia in Local EM Heating and Local Conductivity Change', *I.E.E.E. Trans. Biomed. Engng.*, **24**, 473–477 (1977)
85 SHANNON, G. F., 'A Myoelectrically Controlled Prosthesis with Sensory Feedback', *Med. Biol. Engng. Comput.*, **17**, 73 (1979)
86 MEYER, C. R. and SUTHERLAND, H. C. Jr., 'A Technique for Totally Automated Audiometer', *Japan–Scand. Audiol.*, **7**, 105–109 (1978)
87 SAKABE, N., HIRAI, Y. and ITANI, E., 'Modification and Application of the Computerised Automatic Audiometer', *Japan-Scand. Audiol.*, **7**, 105–109 (1978)
88 BRACKMANN, D. E. and SELTERS, W. A., 'Electrical Response Audiometry: Clinical Applications', *Otolaryngol. Clin. North. Am.*, **11**, 7–18 (1978)
89 WALDEN, B. E. and KASTEN, R. N., 'Threshold Improvement and Acoustic Gain with Hearing Aids', *Audiology*, **15**, 413–420 (1976)
90 DOBELLE, W. H., STENSAAS, S. S., MLADESOVSKY, M. G. and SMITH, J. B., 'A Prosthesis for the Deaf Based on Cortical Stimulation', *Trans. Am. Otol. Soc.*, **61**, 157–175 (1973)
91 VAN DROGELEN, W., HOLLEY, A. and DOVING, K. B., 'Convergence in the Olfactory System: Quantitative Aspects of Odour Sensitivity', *J. Theor. Biol.*, **71**, 39–48 (1978)
92 KNORR, S. G. and ZWITMAN, D. H., 'The Design of a Wireless-Controlled Intra-Oral Electrolaryn'.
93 GEDDES, L. A., SILVA, L. F., DEWITT, D. P. and PEARCE, J. A., 'What's New in Electrosurgical Instrumentation?', *Med. Instrum.*, **11**, 355–359 (1977)
94 HALL, S. H., 'A Bridge for Continuous Resistivity Observations', *J. Phys. E. Sci. Instrum.*, **9**, 728–729 (1976)
95 WATSON, B. W., WOOLEY-HART, A. and TIMMONS, B. H., 'Biofeedback Instruments for the Management of Anxiety and for Realaxation Training', *J. Biomed. Engng.*, **1**, 58–63 (1979)
96 ARMON, H., WEINMAN, J. and PELEG, A., 'Automatic Testing of the Visual Field using Electro-oculographic Potentials', *Docum. Ophthal.*, **43**, 51–63 (1977)
97 HAINES, J. D., 'Non-Contacting Ultrasound Transducer System for Eye Movement Recording', *Ultrasound Med. Biol.*, **3**, 639–645 (1977)
98 USA, S., MURASE, K. and IKEGAYA, K., 'Analysis of the Dynamic Performance of a Television Pupilometer', *Japan J. Med. Electron. Bioo. Engng.*, **16**, 177–183 (1978)
99 THORNBURN, W., 'The Accuracy of Clinical Application Tonometry', *Acta. Ophthalmol.*, **56**, 1–5 (1978)
100 NAKAMURA, Z., 'Human Electroretinogram with Skin Electrode', *Japan. J. Ophthalmol.*, **22**, 101–113 (1978)
101 DELCHAR, I. A. and SMITH, M. J. A., 'An Improved Method for Detecting Passive Pills', *Phys. Med., Biol.*, **21**, 577–583 (1976)
102 SMALLWOOD, R. H., LINKENS, D. A., KWOK, H. L. and STODDARD, C. J., 'Use of Autoregressive Modelling Techniques for the Analysis of Colonic Myoelectric Activity in Man', *Med. Biol. Engng. Comput.*, **18**, 591 (1980)
103 KONDO, A., MITSUYA, H. and TORI, H., 'Computer Analysis of Micturition Parameters and Accuracy of Uroflow Meter', *Japan J. Urol.*, **33**, 337–344
104 DOYLE, P. T., HILL, D. W., PERRY, I. R. and STANTON, S. L., 'Computer Analysis of Electromyographic Signals from the Human Bladder and Urethral and Anal Sphinoters', *Invest. Urol.*, **13**, 205–210 (1975)
105 SEIFERTH, J., HEISSING, J. and KARKAMP, H., 'Experiences and Critical Comments on the Temporary Intravesical Electro-Stimulation of the Neurogenic Bladder in Spina Bifida Children', *GFR Urol. Int.*, **33**, 279–284 (1978)
106 EL FAHIG, S. and WALLACE, D. M., 'Ultrasonic Lithotriptor for Urethral and Bladder Use', *Brit. J. Urol.*, **50**, 255–256 (1978)
107 DURIE, B. G. M., VAUGHT, I. and CHEN, Y. P., 'Discrimination between Human T and B Lymphocytes and Monocytes by Computer Analysis of Digitised Data from Scanning Microphotometry', *Blood*, **51**, 579–589
108 FRANK, H. I. I., DREESEN, V. and HENKER, H. C., 'Computer Evaluation of Performance Curves for the Estimation of Extrinsic Circulation Factors', *Comput. Med. Biol.*, **8**, 65–70 (1978)